PHYSICA B

VOLUMES 165 & 166, 1990

PHYSICA B
CONDENSED MATTER

Editors:

F.R. DE BOER
M. DURIEUX
G.H. LANDER

NORTH-HOLLAND

This volume is printed on acid-free paper.

Published monthly

Printed in Great Britain at the Alden Press, Oxford

Proceedings of the

19th International Conference on Low Temperature Physics

LT-19

Part I - Contributed Papers

Brighton, Sussex, UK
16 – 22 August 1990

Editor:
David S. Betts
Division of Physics and Astronomy
University of Sussex
Brighton, UK

Sponsored by:
The International Union of Pure and Applied Physics
The Institute of Physics, London

1990 NORTH-HOLLAND

PREFACE

The 19th International Conference on Low Temperature Physics (LT-19) was held in Brighton, near the campus of the University of Sussex, UK, from 16 to 22 August 1990 inclusive. LT conferences currently take place every three years, the previous two having been held in Karlsruhe (FRG) in 1984 and Kyoto (Japan) in 1987. The Proceedings are organised in the usual pattern for LT conferences, that is, there are three volumes; these are published as special (but consecutively numbered) volumes of Physica B. Parts I and II contain refereed contributed papers corresponding to poster or oral presentations. A small number of papers were accepted for poster presentation but not for inclusion in the Proceedings. Parts I and II, distributed at the Conference, have been prepared directly from camera-ready copies provided in the first instance by the authors. In some cases I have had these retyped or modified so as to improve appearance. Part III, to be issued in February 1991, will contain the rather longer contributions from plenary and invited speakers together with a small number of selected post-deadline contributions which referees regard as of exceptional quality or current interest.

Over a thousand papers were received and all were refereed by one or two of a team of UK low temperature practitioners. This was a substantial task, and it was shared by two overlapping groups under the general supervision of the Publications Committee which included Dr. P.C. Main, Dr. H.M. Rosenberg, Dr. P.V.E. McClintock, Dr. D.J. Sandiford and myself as Chairman. The first of the two groups was the Programme Committee under the chairmanship of Dr. M. Hardiman; the second was a network of referees from the major low temperature departments in the UK. A number of individuals were in both groups.

The Conference was organised as a five-day sequence of sessions of different kinds, with plenary and invited talks, poster and oral presentations. Many of these were in parallel. Parts I and II contain more than 800 2-page contributions. They are not in strict chronological sequence, since we have preferred to group topics in a more logical fashion. There are five broad categories as follows:

Techniques and application
Metals and magnetism
Quantum fluids and solids
Low temperature properties of solids
Superconductivity

The contents pages give a further breakdown of these main headings. The pagination can be related to the sequence code (such as CD-P3) used in the Abstracts Booklet by reference to the paper session index.

Many people helped me enormously in this task; I have named some academic colleagues above and there were numerous others. Also, and crucially, the enterprise would certainly have collapsed without the guidance, organisational support, and sheer hard work put in by Ms. Susan Lippmann of the Institute of Physics and Ms. Mary Carpenter of Elsevier. And I owe a debt of gratitude to several others who helped in various ways when needed at times of high pressure: Ms. Catherine Rogers, Ms. Ann Adams, Mr. Willie Siyanbola, Mr. Mounir Chouder, Ms. Jane Walsh and Ms. Ann Morton-George. My own secretary too, Ms. Nanette Kingan, who was not formally involved at all in LT-19, inevitably had to cope with more work than usual as well as my adverse reactions to stress. Thank you all.

David S. Betts
Editor

ORGANISING COMMITTEE

D.F. Brewer	(Sussex)	Conference Chairman
A.F.G. Wyatt	(Exeter)	Conference Secretary
D.S. Betts	(Sussex)	Chairman, Publications
L.J. Challis	(Nottingham)	
E.R. Dobbs	(London)	Chairman, Finance
H.E. Hall	(Manchester)	
M. Hardiman	(Sussex)	Chairman, Programme
N.W. Kerley	(Oxford Instruments)	
A.L. Thomson	(Sussex)	Chairman, Fund Raising
W.F. Vinen	(Birmingham)	

INTERNATIONAL COMMITTEE (* members of IUPAP Commission C5)

A.F. Andreev *	(USSR)	J.E. Mooij *	(Netherlands)
O. Avenel *	(France)	D.D. Osheroff *	(USA)
G. Chandra	(India)	F. Pobell	(FRG)
F. de la Cruz	(Argentina)	F.B. Rasmussen	(Denmark)
J. Czerwonko	(Poland)	R.C. Richardson	(USA)
G. Deutscher *	(Israel)	C. Rizzuto *	(Italy)
W. Eisenmenger *	(FRG)	I.F. Schegelov	(USSR)
Ø. Fischer *	(Switzerland)	P. Sengupta	(India)
J.P. Franck *	(Canada)	L. Skrbek *	(Czechoslovakia)
G. Frossati	(Netherlands)	T. Shigi	(Japan)
W.-Y. Guan	(China)	T. Tsuneto *	(Japan)
A.J. Leggett	(USA)	W.F. Vinen *	(UK)
E. Lerner	(Brazil)	G.K. White	(Australia)
O.V. Lounasmaa *	(Finland)	J. Winter	(France)

PROGRAMME COMMITTEE (Chairmen)

M. Hardiman	(Sussex)	Chairman
L.J. Challis	(Nottingham)	Co-Chairman (LT properties of solids)
H.E. Hall	(Manchester)	Co-Chairman (Quantum fluids and solids)
B.D. Rainford	(Southampton)	Co-Chairman (Metals and magnetism)
H.M. Rosenberg	(Oxford)	Co-Chairman (LT properties of solids)
J. Saunders	(London)	Co-Chairman (Techniques and applications)
W.F. Vinen	(Birmingham)	Co-Chairman (Superconductivity)

PUBLICATIONS COMMITTEE

D.S. Betts	(Sussex)	Chairman
P.C. Main	(Nottingham)	
P.V.E. McClintock	(Lancaster)	
H.M. Rosenberg	(Oxford)	
D.J. Sandiford	(Manchester)	

SUPPORT

The LT-19 Organising Committee gratefully acknowledges the financial support given by:

Elsevier Science Publishers B.V. (North-Holland)
GEC plc
Oxford Instruments Ltd.
Quantum Design
University of Sussex
US Airforce European Office of Aerospace Research and Development
The Royal Society

CONTENTS

3.8. Flow and vortices

3.9. Nuclear magnetism

4.2. Organic and low dimensional conductors

4.3. Disordered systems

4.4. Low temperature properties of solids

5.3. HTS: microscopic theory – weak correlations

5.9. HTS: photoemission and spectroscopy

PAPER SESSION INDEX (by paper number)

PAPER SESSION INDEX (by page number)

Physica B 165&166 (1990) 1–2
North-Holland

TOWARDS THE REALISATION OF THE FIRST BOLOMETER FOR $\beta\beta$ DECAY RESEARCH

A. Alessandrello, C. Brofferio, D.V. Camin, O. Cremonesi, E. Fiorini, A. Giuliani, G. Pessina

Dipartimento di Fisica, Università di Milano and INFN, Sez. di Milano

Progress in the realisation of the first large mass bolometer for double beta decay research is reported. Results obtained with a 11 g Ge detector (13 keV FWHM resolution at 1 MeV) and a 28 g Te detector are exposed. As temperature sensors for the bolometers, several commercial and NTD germanium thermistors working in the hopping conduction regime have been studied, showing a much more reliability and reproducibility for the NTD type than for the melt-doped ones.

1. INTRODUCTION

We are working on the construction of a thermal detector for double beta decay research. The advantages of the bolometric technique in this kind of physics were already pointed out in several occasions[1]. Let us just recall here that several elements, which have double beta decay candidate isotopes, can in principle work satisfactorily also as bolometers, if they are operated in the form of covalent or ionic crystals with reasonably high Debye temperature. Under these conditions their heat capacity in the mK region is low enough that even the tiny energy delivered in form of heat by an elementary particle produces a detectable enhancement ΔT of the temperature. The thermal signal is converted in a voltage signal by a thermistor working in the hopping conduction regime. Since double beta decay is a rare event the mass of the absorber should be as large as possible, at least tens or hundreds of grams, in weak thermal contact with the heat sink. In our case the sensor is glued on the crystal and electrically and thermally connected to the thermal bath; it is then biased with a voltage V, so that the output signal, neglecting electrical and thermal integrations which can lead to a considerable signal reduction, will be

$$\Delta V = V \cdot A \cdot \frac{\Delta T}{T} \tag{1}$$

where T is the working temperature and $A = -\frac{d\log R}{d\log T}$ depends only on the R vs T characteristics of the thermistor. As thermometers we have studied several commercial melt-doped and some especially made NTD germanium thermistors.

2. STUDIES ON THE THERMISTORS

Fig. 1 shows that for the NTD thermistors the

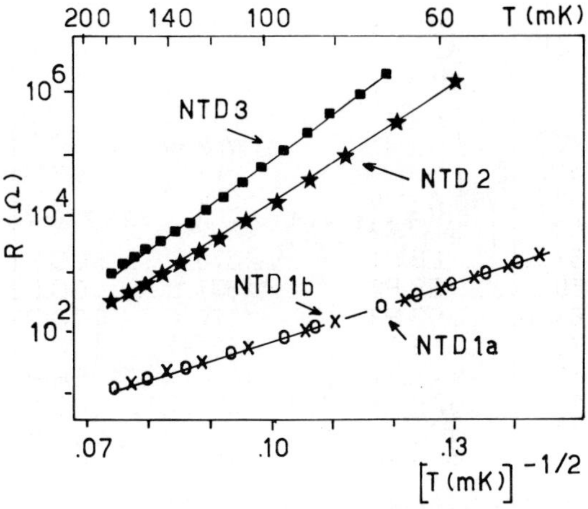

Fig.1. Examples of R vs T characteristics for NTD germanium thermistors. Chips 1a and 1b have the same nominal concentration of dopants

Efros-Shklovskii law

$$R(T) = R_\circ \cdot \exp\left(\frac{T_\circ}{T}\right)^{1/2} \tag{2}$$

is verified very well and that the reproducibility of R vs T characteristic for chips having the same nominal concentration of dopants is quite impressive. We have on the contrary found experimentally that for the melt-doped sensors the discrepancies are considerable. This is probably due to non uniform doping and implies problems in the extrapolation of the R vs T curve to low temperatures, in the reproducibility of the chips and in the reliability of the electrical contacts. As formula (1) shows, in order to maximise the voltage signal it is important to choose a thermistor which can stand a considerable bias with a small increase in the base temperature. Experimentally this seems to depend intrinsically on the thermistor rather than on how it is thermally connected to the bath. This phenomenon can be temptatively ascribed to a thermal decoupling between the phonon and the conduction electron systems[2,3].

3. DETECTORS PERFORMANCES

Unlike other groups working with very small bolometers, we are pushing our efforts at the improvement of the mass of the absorber, still keeping an energy resolution of a few % in the 2-3 MeV region. As a first step we have constructed an 11 g germanium bolometer which worked at 28 mK and gave the ^{60}Co spectrum reported in fig. 2, with a 1 % energy resolution FWHM on the two lines (1173 and 1332 KeV) of the source.

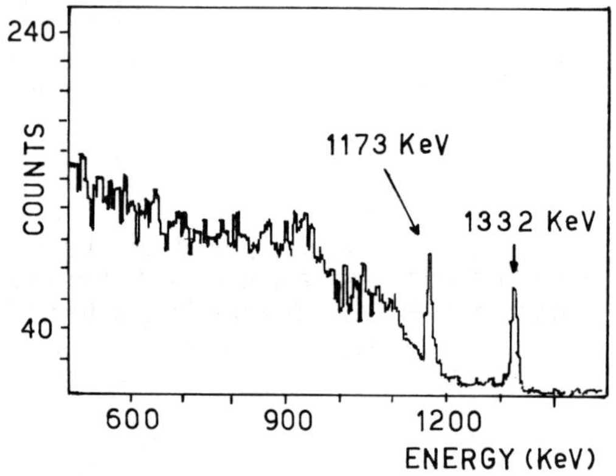

Fig.2. ^{60}Co spectrum obtained with an 11 g Germanium bolometer

We are now working on a 28 g Tellurium bolometer, trying to reach the same energy resolution. As preliminary results we can quote that this detector counts γ-rays with 1 MeV energy threshold: we have exposed the detector to a ^{232}Th source and in fig. 3 a typical impulse for a 2.6 MeV coming from the decay of ^{208}Tl is shown. We hope to improve this resault in the next months reducing microphonic noise and choosing a less resistive sensor which would allow us to decrease the operation temperature.

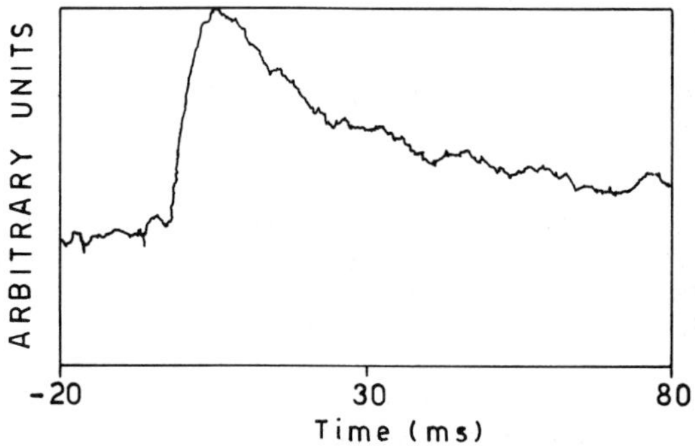

Fig.3. Single impulse of a 2.6 MeV γ-ray in a 28 g Tellurium bolometer

ACKNOWLEDGEMENTS

We are grateful to Prof. E.E. Haller for providing us a few excellent NTD germanium thermistors and for discussions on them.

REFERENCES

[1] E. Fiorini and T.O. Niinikoski, Nucl. Instr. Meth. 224 (1984) 83

[2] N.Wang, F.C.Wellstood, B.Sadoulet, E.E.Haller, J.Beeman, Phys. Rev. B 41 (1990)

[3] A. Alessandrello, C. Brofferio, D.V. Camin, O. Cremonesi, E. Fiorini, A. Giuliani, G. Pessina and E. Previtali, Optimization of large mass low temperature calorimeters, Proceedings of the "III International Workshop on Low Temperature Devices For Neutrinos and Dark Matter", 1989, Gran Sasso, Italy, (in print)

Physica B 165&166 (1990) 3–4
North-Holland

A CRYOGENIC PHONON DETECTOR WITH SIMULTANEOUS MEASUREMENT OF PHONONS AND IONIZATION

N. Wang, T. Shutt, A. Cummings, B. Sadoulet, E. E. Haller, P. Barnes, J. Emes, Y. Giraud-Héraud, A. Lange, J. Rich, R. Ross, and C. Stubbs

Center for Particle Astrophysics and Department of Physics, University of California, and Physics Division, Lawrence Berkeley Laboratory, 1 Cyclotron Rd, Berkeley, Ca 94720

One way to detect dark matter particles from the halo of our galaxy is to detect phonons generated by dark matter particle interaction in a crystal. We report on our efforts to develop a phonon detector operating at 20 mK. We present our recent results on the simultaneous measurement of phonons and ionization, which may provide an effective way to reject background in dark matter search.

1. INTRODUCTION

A dark matter detector requires a low energy threshold (~ 10^3 eV) and good background rejection capability (1). The primary background signals come from radioactive contaminants producing ß rays and Compton scattered photons. These background signals give rise to electron recoils, while dark matter particles are expected to interact with nuclei. Such background signals can be eliminated by simultaneously measuring phonon and ionization signals in a semiconductor. This is because a recoiling nucleus deposits 92% of its energy in phonons and 8% in electron-hole pairs, while a recoiling electron deposits about 67% of its energy in phonons and 33% in ionization (2). As a result, we need to operate a dark matter detector at low temperatures to achieve low threshold and high resolution, and simultaneously detect signals from phonons and ionization .

It is expected that the phonons generated in a crystal by interactions with dark matter particles are mostly high energy phonons (~ 10^{-3} eV) with a lifetime much longer than 1μs. In order to detect these high energy phonons, we have developed a fast and sensitive phonon sensor using neutron transmutation doped Ge (NTD Ge) (3). Although the hot electron effect decreases the sensitivity of a NTD Ge sensor to thermal phonons (4), we have shown that it detects high energy phonons efficiently (5). We are currently investigating possible means of attaching such sensors to large mass crystals, which would act as dark matter detectors.

2. EUTECTIC INTERFACE

Sensors are normally glued to crystals. Since glue usually contains many small grains which absorb and presumably thermalize high energy phonons, it is not a good material for our application. We would like to detect the wavefronts of the high energy phonons which contain position information and perhaps even information on the direction of an incoming particle.

We have developed a eutectic bonding scheme which involves evaporating 1000 Å of Au on polished faces of an NTD Ge sensor and a Ge crystal, pressing the Au-covered faces together and heating in a controlled atmosphere to the eutectic temperature of 356°C. The resulting eutectic bond gives sufficient mechanical strength, and allows for a very thin (~ 2000 Å) and controllable interface.

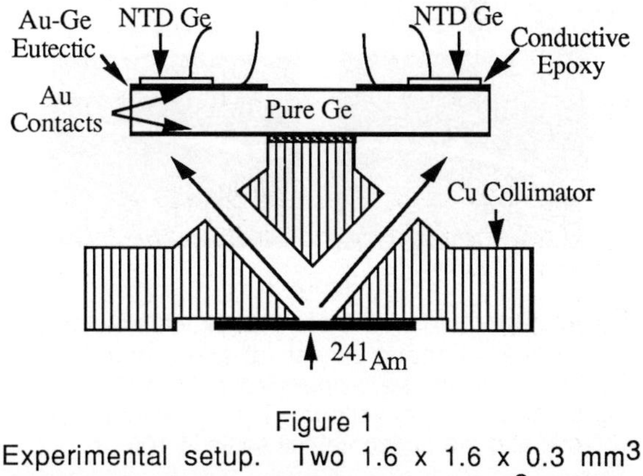

Figure 1
Experimental setup. Two 1.6 x 1.6 x 0.3 mm^3 sensors are connected to a 2 x 8 x 1.2 mm^3 crystal of Ge, one using a Ge- Au -Ge eutectic bond with about 2000 Å of Au, and the other glued with a thin layer of conductive epoxy. The collimation gives irradiated areas of about the same size as the sensor areas.

This work was supported by the Center for Particle Astrophysics, a National Science Foundation Science and Technology Center operated by the University of California at Berkeley under Cooperative Agreement No. ADT-8809616.

2.1 Experimental setup

In order to test the phonon transmission properties of the eutectic bond, we performed experiments on the device shown in Fig 1. Two NTD Ge phonon sensors were attached to a single crystal of n-type Ge with an impurity level less than $10^{11}/cm^3$. At one end, a sensor was eutectically bonded to the crystal. At the other end, a similar sensor was glued to the crystal with conductive epoxy. Electrical connections were made with 0.7 mil Al wires. The Ge crystal was soldered with In to the Cu sample holder, which also served as a collimator. The crystal was irradiated with α particles and photons from ^{241}Am.

2.2 Preliminary results

In Figures 2a and 2b we show pulses observed from 5.5 MeV α particles incident below the eutectically bonded sensor and the glued sensor, respectively. The sample was at about 50 mK and both sensors were biased at a constant voltage. The change in current when energy is deposited in a sensor is amplified with a transconductance amplifier of 1V/μA sensitivity. The sharp peak at the beginning of the trace is the ionization signal, discussed later. What immediately follows the ionization signal is the phonon signal. One sees that the risetime of the phonon signal in Fig. 2a is much shorter than in Fig. 2b, which is a clear indication that a eutectic interface transmits high energy phonons more efficiently than a glued interface.

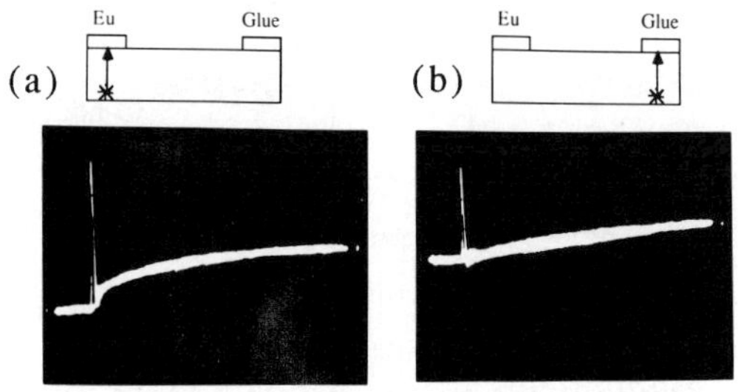

Figure 2

Pulses generated by 5.5 MeV α particles incident on a eutectically bonded sensor (a) and a glued sensor (b). Both sensors were operated at 50 mK and were biased at a constant voltage. The vertical full scale is 800 mV and the horizontal full scale is 100 μs.

3 IONIZATION

When the pulses shown in Fig. 2 were observed, the side of the phonon sensors attached to the Ge crystal were at a positive bias (100 mV). This voltage generates an electrical field of about 1V/cm across the Ge crystal. At 50 mK, the donors and the acceptors are all frozen out, thus very few space charges and thermally activated electron-hole pairs are present. As a result, an electrical field of 1V/cm is sufficient to drift the electron-hole pairs generated by the α particle across the Ge crystal, producing the peak preceding the phonon signal.

We show in Fig. 3 an integral spectrum of 60 keV and 18 keV photons from ^{241}Am. All four contacts of the sensors were shorted together and were connected to the gate of an FET.

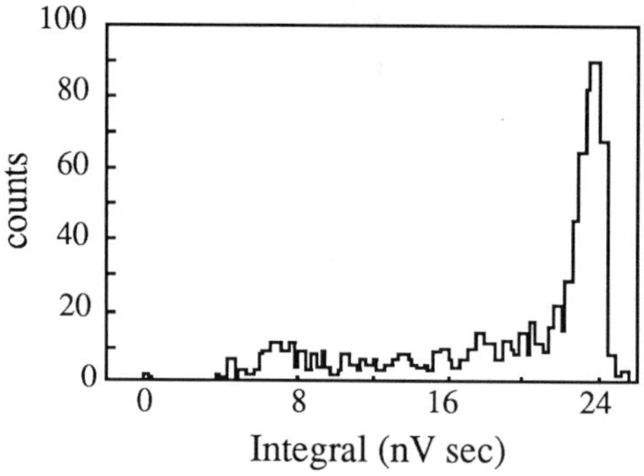

Figure 3

Integral spectrum of ionization signals from photons of ^{241}Am. The detector was operated at about 100 mK and was biased at 0.6V.

The significance of our results (5) is not that we detected ionization, but rather that we observed ionization signals with only 1V/cm of applied electrical field. This means that the increase in phonon energy from the drifting of the electron-hole pairs ($E = n_{charge}\, q_{electron}\, V$) is small, thus the phonon signals won't be severely disturbed.

4 CONCLUSION

We have tested a Ge-Au eutectic interface and have shown that a eutectic bond is faster than a glue bond. We have also simultaneously observed phonon and ionization signals with an electric field of 1V/cm. We are planning more experiments for further understanding of the phonon properties and for a bigger detector.

ACKNOWLEGEMENTS

We thank P. Luke and H. Bauman for their generous help.

REFERENCES

(1) J.R. Primack, D. Seckel, and B. Sadoulet, Ann. Rev.Nucl. Part. Sci. **38** (1988) 571.
(2) G.Gerbier et al., Saclay preprint.
(3) N.Wang et al., IEEE Trans. Nuc. Sci. vol. NS (1989), p.852.
(4) N.Wang, F.C. Wellstood, B. Sadoulet, E.E. Haller, and J. Beeman, Phys. Rev.B **41** (1990) 3761.
(5) T. Shutt et al., in print.

Physica B 165&166 (1990) 5–6
North-Holland

DETECTION OF PARTICLES BY AN AL_2O_3/RUO_2 COMPOSITE BOLOMETER

M. Chapellier[a], G. Chardin[b], M. Fouchier[c], H. Ji, J. Joffrin, J-Y. Prieur and F.B. Rasmussen[d]

Laboratoire de Physique des Solides — Bat. 510, Orsay, France,

We have used a composite bolometer made of a sapphire crystal and ruthenium oxide films to detect individual particles emitted by an ^{241}Am source (5.5 Mev alpha and 60 Kev gamma) The experiment is done in a quick handling dilution fridge at T≈ 80 mK with a cold FET preamplifier. Performances of such detectors lead to consider them as realistic candidates for the detection of Dark Matter.

1. INTRODUCTION

The need to detect recoil events of very low energies in order to detect dark matter (1) by bolometric methods has shed new interest on the few resistive sensors used at low or very low temperatures.Thick film resistors made out of RuO_2.seem interesting as they may be intimately coupled to some crystal by high temperature deposition. The RuO_2 material is commercialized by Dupont de Nemours as a powdered mixture of glass and conductive bismuth ruthenate particles which must be processed at a temperature of ≈800°C to obtain the resistive film. The melting of the two components provides a nice adherence on sapphire and Si crystals, two interesting materials for bolometric systems due to their high Debye temperature.

These thick films have been recognized as reproducible thermometers. They follow Mott's law for variable range hopping : $R = R_0 \exp (T_0/T)^{1/4}$ with a T_0 of the order of a few Kelvins for most of the resistors commercially available. So far, the studies have been concentrated on RuO_2 films deposited on an Al_2O_3 substrate, since this material is interesting for Dark Matter detection due to its non-zero nuclear spin.

On the other hand, Al_2O_3 possesses a large quadrupolar heat capacity (already, at 300 mK, the quadrupolar reservoir heat capacity is equal to that of the phonons; this heat capacity, however, scales as T^{-2} , which should be compared to the T^3 scaling for phonons). But recent experiments (2) have shown that the quadrupolar reservoir is only coupled to the lattice via paramagnetic impurities with such a long time constant that it does not contribute to the transitory heat capacity of the bolometer.

2. RESISTIVE BEHAVIOR OF RuO_2 DEPOSIT

We have compared the behaviour of our homemade film versus a commercial RuO_2 resistor (Dale Electronics) from a calibrated batch. The R(T) law of a typical resistor from this batch is given by:

$$R = 470 \exp (2.6/T)^{1/4}$$

If the homemade film follows the same type of law, an intercalibration in Log/Log scale will be represented by a straight line, with a slope given by the ratio of the two characteristic temperatures.

[a]also DPhG/SRM, IRF, CEN Saclay, Gif/Yvette, France
[b]also DPhPE/SEPh, IRF, CEN Saclay, Gif/Yvette, France
[c]CNET Bagneux, France
[d]H.C. Ørstedt Institute, University of Copenhagen, Denmark

Fig. 1 shows the intercalibration of a "1 kΩ paste" (1 kΩ for a square of thickness ≈15 µm). The first film follows the same type of law as the reference resistance. The 10 kΩ film seems to be very close to the insulating transition. In order to avoid a severe cut-off in the bandwith of the signal (the parasitic capacitance inside the cryostat is

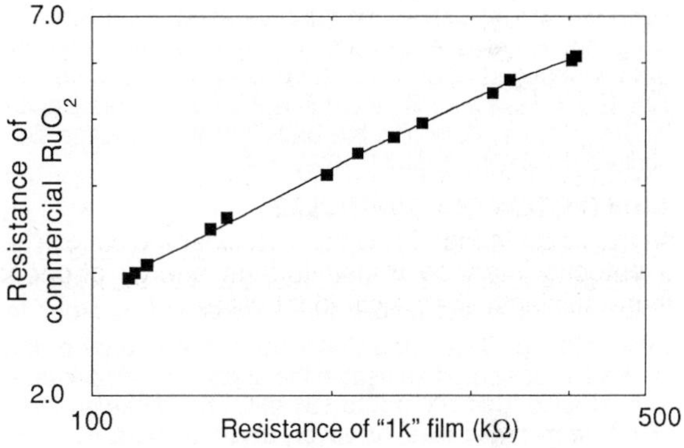

Fig.1 :Variation of the commercial RuO_2 resistance as a function of the "1kΩ" film resistance.

≈300 pF), we have installed a FET to adapt the high impedance of the films.

3. EXPERIMENTAL SETUP

Cryogenics and electronics

As a first step, we are running the experiment using a simple dilution refrigerator, until now limited to an unimpressive 50 mK temperature, but very handy and with a fast cooling. It works without He^4 pot, its shape is that of a cylinder 45 mm in diameter cooled down in a commercial bottle of He^4 with large neck (∅ = 50 mm). This apparatus requires only 4 hours to reach its lowest temperature starting from 300 K.

We are working in a room which is not electromagnetically shielded and with rather noisy surroundings. The acquisition system comprises a Gould digital oscilloscope connected via GPIB to a Macintosh II computer. The FET (CM 860) amplifier working at 100K with low heating (6 mW) is installed inside the vacuum part of cryostat, followed on the top of the cryostat (at room temperature) by a low noise room temperature operational amplifier (Linear Technology 1028). The final bandwith is limited by a PAR 113. amplifier.

0921-4526/90/$03.50

Crystal and thermistor mounting

The thermistor which appears to be best adapted to the bolometric measurements is 0.3 mm wide, 10 mm long with a thickness of ≈10μ. The room temperature resistance is 50 kΩ, increasing to ≈274 kΩ at 77 mK. The thermistor is deposited on a Al_2O_3 crystal 10 mm x 10 mm x 1 mm. Electrical contacts are achieved with Al wires (1% Si) 25 μm in diameter which are pressed on the film by another crystal of Al_2O_3 of size 10 x 5 x 1 mm^3. The thermal contact to the dilution is through 4 copper wires 0.15 mm in diameter and 10 mm long.

This very simple mounting was found to be the most efficient, since the Kapitza resistance between the crystal and copper is large enough to provide isolation with a reasonable time constant at a temperature of ≈100 mK. Copper wires must be considered part of the bath and we believe that this clear-cut system avoids spurious behaviour when an ill-defined intermediate bath couples the crystal to the cold sink.

Measurement of the thermal coupling: We have measured on this film the resistance versus the temperature. By increasing the excitation, and measuring simultaneously the applied power and the resistance of the film, the coupling of the total system (film plus crystal) to the bath.is obtained. It is well described by the standard law $Q/(T_{hot}^4 - T_{cold}^4)$, =1/R where R is the thermal resistance between the system and the bath. For the coupling described previously 1/R=7.5 $10^{-6}WK^{-4}$.

4. DETECTION OF α PARTICLES

A collimated Am^{241} source was used as a source of α particles of energy ≈5.5 MeV. Alpha particles collide the bolometer at a rate limited to ≈1 α/second in order to avoid pile-up. Supposing that most of the energy of the particle is dissipated as heat in the crystal, its temperature should raise from 100 mK to 131 mK. The observed temperature increase was, however, much smaller: ≈1.4mK. We have checked that most of the extra heat capacity comes from the film.

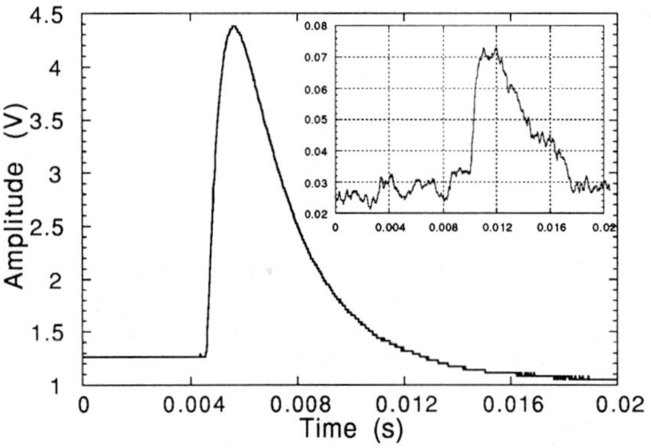

Fig. 2 : Alpha particle recorded on a crystal of 0.8 g. Thermal contacts by four copper wires.Insert shows a signal of gamma ray of 60 Kev of energy

As the final aim of dark matter detection requires honest resolutions in the [0.5, 10] keV energy interval, the signal/noise ratio should be improved by reducing the thickness of RuO_2 film. Accordingly, the heat capacity will be reduced, allowing a higher excitation voltage V for the same dissipation. A reduction of thickness by 2 should increase the signal by 2√2, since the noise is *not* the Johnson noise of the resistor but rather spurious noise such as microphonic or amplifier noise.

Fig. 3 shows the energy distribution obtained for alpha particles using this device. The corresponding resolution defined by ΔE/E, where ΔE is the standard deviation of the distribution, is 1.2 percent.

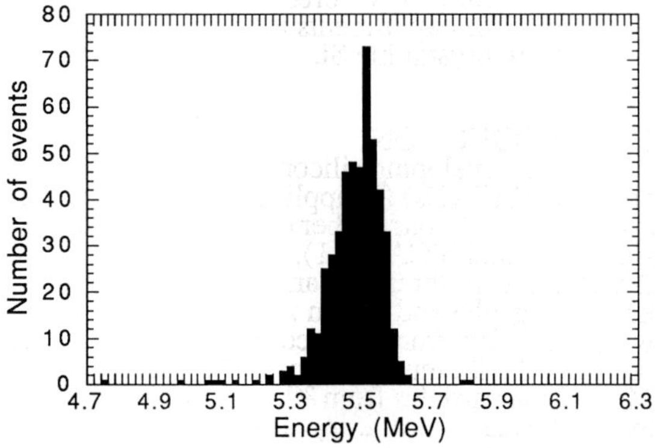

Fig. 3 : Energy distribution of alpha particles from an ^{241}Am source on a RuO_2/Al_2O_3 bolometer.

5. CONCLUSIONS

The effectiveness of RuO_2 films as thermistors in composite bolometers has been demonstrated. Although the feasibility of dark matter detection using bolometers has not yet been demonstrated, the performance improvements achieved suggests that this aim is not unduly unrealistic.

Performances of this type of bolometers will still be greatly improved by reducing the temperature T, the amount of RuO_2 and by increasing the bias factor of the RuO_2 films. We will then be faced with the problem of discrimination of recoil events from the ambient radioactive background, which appears as the main challenge of dark matter detection.

REFERENCES

[1] See E. Fiorini, this Conference.
[2] M. Chapellier, M. Rotter, M. Bassou and M. Bernier, in preparation.

Physica B 165&166 (1990) 7–8
North-Holland

OBSERVATION OF ALPHA-INDUCED BALLISTIC PHONON TIME-OF-FLIGHT USING MULTI-CHANNEL SILICON CRYSTAL ACOUSTIC DETECTORS

Betty A.YOUNG, Blas CABRERA and Adrian T. LEE

Physics Department, Stanford University, Stanford, CA 94305, USA

We have developed a technique for producing double-sided multi-channel silicon crystal acoustic detectors (SiCADs). These detectors, which are operated at ≈ 400 mK and use 40 nm thick films of superconducting Ti as the phonon sensors, have been used in time-coincidence experiments to study the propagation of alpha-induced ballistic phonons through 1 mm of crystalline Si.

1. INTRODUCTION

We are developing silicon crystal acoustic detectors (SiCADs) for applications in particle and astrophysics. In our earlier experiments with single-channel SiCADs (1), we demonstrated a detector energy threshold and resolution of a few keV, using ^{55}Fe and ^{241}Am x-ray sources. Here, we present data from time-coincidence experiments performed with multi-channel SiCADs using ≈ 5.5 MeV alpha particles from ^{241}Am. Phonons generated by the elastic scattering of an alpha in the 1 mm thick Si crystal substrate were simultaneously detected in two independently-instrumented SiCAD channels using an 800 nsec timing gate. The phonon focussing patterns were spatially resolved in one set of experiments (2) and the ballistic time-of-flight confirmed in another, as discussed below.

2. THE DETECTOR

Each channel of the SiCAD consists of a 40 nm thick film of superconducting Ti laid down in a meander pattern on the (100) surface of a 1 mm thick Si wafer substrate. The meander pattern is a series connection of 400 lines (2μm wide, 5 μm pitch) and is aligned with the [110] axes of the wafer. The active area of each pattern is 2 mm × 4 mm, for a total of 16 mm^2 when two immediately adjacent sensors are instrumented. Using variations of standard photolithographic techniques, we deposit and mutually align sensors on both the front and back surfaces of a wafer, for a total of four independent detector channels.

We use cryogenic GaAs MESFET voltage-sensitive preamplifiers with noise $\Delta V_{rms} \approx 1\mathrm{nV}/\sqrt{\mathrm{Hz}}$ at 1 MHz (3), and run the devices in a cryo-pumped ^{3}He refrigerator that can achieve temperatures down to ≈ 250 mK(4). We operate the detectors just below their superconducting transition temperature ($T_c \approx 430$ mK; transition width < 10 mK), and dc current bias the sensors at a level ($I_b \approx 70$ nA) where self-heating effects are small. A self-terminating voltage pulse (≈ 5 μsec long) results when the Ti film is bombarded by phonons generated by a particle event in the silicon crystal substrate. The pulse length is directly related to the thermal relaxation time (≈ 2 μsec) of the Ti sensor on the Si substrate. The pulse risetime is electronics-limited to ≈ 140 nsec. The pulse amplitude is directly proportional to the area of Ti sensor driven normal, except for very large signals which become supressed by the effective input impedence (≈ 300 kΩ) of the preamps.

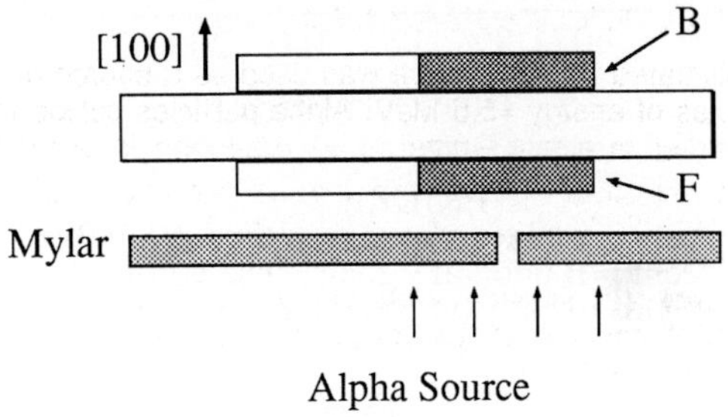

FIGURE 1
Schematic of source/detector geometry for timing experiments with SiCAD operated in F/B mode.

3. TIME COINCIDENCE EXPERIMENTS

The source/detector geometry used in the time-of-flight experiments discussed below is shown in Fig 1. The ^{241}Am alpha source was collimated using a 150 μm diameter hole cut through 250 μm thick mylar, with the hole aligned to the centers of the instrumented frontside (F) and backside (B) sensors.

Over a range of discrete operating temperatures ($T/T_c \approx .90$ - .98), we digitally recorded each signal detected on the frontside of the 1 mm thick detector with the time-coincident signal detected on the backside. In off-line analysis, we determined the time delay between coincident signals at each

0921-4526/90/$03.50 © 1990 – Elsevier Science Publishers B.V. (North-Holland)

operating temperature. The results are presented in Fig 2, where the plotted point size includes both an estimate of the systematic error introduced during the timing analysis and the much smaller statistical errors. The observed time delays between F and B pulses decrease with increasing temperaure, varying from $\Delta t > 300$ nsec at 385 mK (where we first begin to see coincidences) to $\Delta t \approx 166$ nsec at 419 mK (a typical operating temperature). For comparison, the ballistic phonon travel time along the [100] axis of Si (as given by the speed of sound for transverse phonons) is ≈ 168 nsec/mm. For our geometry, ballistic propagation corresponds to a time delay of $\Delta t \approx 166$ nsec between the F/B sensors. This is shown in Fig 2 as a ≈ 4 nsec wide band, which reflects the distribution of alpha interaction depths within our detector.

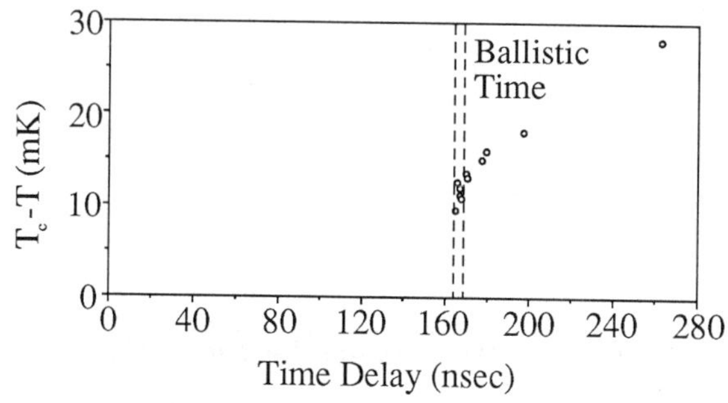

FIGURE 2
Variation with temperature of time delay for coincident F/B phonon signals. The vertical band corresponds to ballistic phonon propagation in the detector.

Fig 3 shows the risetimes of F and B signals as a function of operating temperature. These data are qualitatively consistent with our model of the detector response, and allow us to self-consistently study the effects of the electronics and detector dc resistance while providing additional information on the partitioning of the phonon signal into ballistic and diffusive components.

The F/B data, combined with Monte Carlo calculations and our results obtained using a detector instrumented in the B/B geometry, self-consistently show that ballistic phonons provide ≈1/3 of the signal energy (2). When operating our detector near T_c, the ballistic component alone is sufficient to exceed film threshold. At lower temperatures, the ballistic component lacks the requisite energy density to initiate a transition above T_c, and diffusive phonons which arrive later begin to significantly contribute to the signal. This accounts for the observed shift to longer time delays for the coincident F/B signals at lower temperatures.

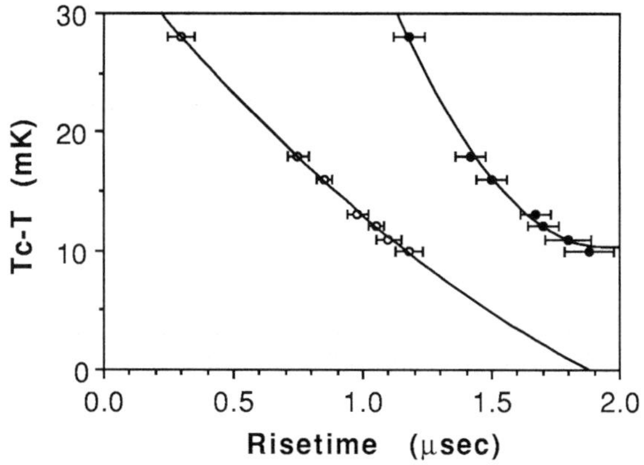

FIGURE 3
Variation with temperature of risetimes for coincident F/B phonon signals. Solid (open) circles correspond to F (B) pulses. The curves are shown to guide the eye.

4. CONCLUSION

In these first F/B coincidence experiments using SiCADs, we have demonstrated a timing resolution of < 2%, and have shown that ≈ 1/3 of the alpha-induced phonon energy absorbed at the backside of the detector arrives ballistically.

ACKNOWLEDGEMENTS

We thank B. Neuhauser for design of the ^{3}He probe and relevent suggestions, and G. Feigl for the high-purity Si used in this work. B.Dougherty and K. Irwin provided many useful comments during the data analysis. This research is funded in part by DOE Grant DE-AM03-76-SF00-326.

REFERENCES

(1) B.A. Young, B. Cabrera and A.T. Lee, presented at III Intl. Workshop on Low Temp. Detectors for Neutrinos and Dark Matter, L'Aquila, Italy, 1989 (in press).
(2) B.A. Young, B. Cabrera and A.T. Lee, submitted to Phys. Rev. Lett..
(3) A.T. Lee, Rev. Sci. Instr. **60** (1989) 3315.
(4) B. Neuhauser, et. al., Stanford Low Temp. Physics Group preprint no. BC80-89, in preparation.

Physica B 165&166 (1990) 9–10
North-Holland

NIOBIUM GRAVITATIONAL RADIATION ANTENNA WITH SUPERCONDUCTING PARAMETRIC TRANSDUCER

N.P. LINTHORNE, P.J. VEITCH, D.G. BLAIR, P.J. TURNER, M.E. TOBAR and A.G. MANN

Department of Physics, The University of Western Australia,
Nedlands, Western Australia, 6009.

Progress and latest results on the development of a niobium gravitational radiation antenna are reported. An improved vibration isolation system has been installed and is currently under test. The previously developed transducer self-calibration procedure is summarized. We show that the power dependent microwave cavity Q limits the maximum achievable antenna-transducer coupling factor, β, and thus the maximum bandwidth of the antenna, B. The self-calibration procedure is combined with new data for the performance of the reentrant cavity transducer to determine the maximum β and B which can be achieved. Results given here show maximum values of $\beta \approx 0.04$ and $B \approx 15$ Hz.

1. INTRODUCTION

The resonant bar gravitational radiation antenna being developed at the University of Western Australia has been described in detail in ref. (1). Briefly, it consists of a 1.5 tonne niobium bar which is supported near its centre by a low loss high isolation mechanical filter. A niobium bending flap, resonant near the antenna's fundamental longitudinal mode, is bonded to the antenna. The relative motion of the antenna and bending flap is monitored by a microwave X-band reentrant cavity parametric transducer, which is also bonded to the antenna. The pump power for the transducer is provided by an ultra low noise oscillator derived from a high Q tunable sapphire loaded superconducting cavity.

The antenna is capable of achieving a noise temperature of less than 1mK, corresponding to a strain sensitivity $h \sim 10^{-18}$, due to the very low acoustic loss of niobium ($Q > 2 \times 10^8$) and the high sensitivity of the superconducting microwave parametric transducer. To achieve this sensitivity requires extremely high vibration isolation from the (noisy) cryogenic environment.

2. ANTENNA MODIFICATIONS AND RESULTS

In initial experiments the vibration isolation was insufficient. Modifications to the vibration isolation have been mostly successful (2). The mode temperature of the antenna's lower normal mode was reduced from 10^4 K to about 8 K, while that of the upper normal mode was still high due to a nearby resonance in the third stage of the antenna suspension.

Low frequency vibrations of the experimental chamber, entering the reentrant cavity via the microwave cable, were found to perturb the cavity resonant frequency. This may produce an additional noise term due to changes in the acoustic Q of the antenna may result. To circumvent this, the cavity frequency is servoed by means of a "force" capacitor mounted between the bending flap and the bar.

Recently, an upper limit to the noise temperature of 58 mK/√Hz has been measured for the microwave transducer. This limit was still set by excess mechanical vibrations internal to the dewar, with the pump oscillator and demodulator not contributing significantly to this result. Greatly improved performance is expected in a run beginning in April 1990.

3. PARAMETRIC TRANSDUCER SELF-CALIBRATION

A method has been developed of calibrating the effective mass of the bending flap and the input power to the transducer. The self-calibration procedure relies on carefully monitoring the modifications to the antenna resonant frequency and acoustic Q as a function of the microwave pump frequency offset in the parametric transducer. The Q variation is due to cold-damping, which reduces the antenna Q without adding to the Brownian motion antenna noise term. In the limit of small pump offset frequencies, the loaded normal mode antenna $\omega_{\pm\ell}$, and acoustic Q's, $Q_{\pm\ell}$, are given by (3)

$$\omega_{\pm\ell} \approx \omega_{\pm} + K_{\pm}\beta\left(\frac{2Q_e\delta\Omega}{\Omega_o}\right) \tag{1}$$

$$Q_{\pm\ell}^{-1} \approx Q_{\pm}^{-1} - \left(1 \mp \frac{\delta_o}{\left(\delta_o{}^2 + \frac{m_e}{M_e}\right)^{\frac{1}{2}}}\right)\left(\frac{4\beta Q_e{}^2\omega_{\pm}\delta\Omega}{\Omega_o{}^2}\right) \tag{2}$$

where $\omega_\pm$ are the unloaded normal mode antenna frequencies, $Q_\pm$ are the unloaded antenna Q's, β is the electromechanical coupling factor, Q_e is the electrical Q of the cavity, M_e is the effective mass of the antenna, m_e is the effective mass of the bending flap, $\delta\Omega$ is the offset of the pump frequency from the cavity resonant frequency Ω_o, δ_o is a measure of the bending flap tuning error, and $K_\pm$ are constants which depend only on ω_+, ω_-, M_e and m_e.

Using Equations 1 and 2, one can uniquely determine m_e, β and Q_e by assuming only M_e; independent of estimates of microwave power level, transmission line losses, microwave couplings and cavity gap spacing. We have previously demonstrated that this gives reasonable agreement with experimental observations (3).

4. ELECTROMECHANICAL COUPLING FACTOR AND ANTENNA BANDWIDTH

The electromechanical coupling of the transducer to the antenna determines the antenna bandwidth, B, through the relation $\beta\omega_a\tau \approx 2$, where $\tau \approx (2\pi B)^{-1}$ is the measurement integration time, and ω_a is the antenna resonant frequency. Clearly a high value of β is necessary to obtain a high bandwidth. This is essential, particularly for reducing the coincidence window with other detectors, and if directional information is to be obtained from the measured propagation time of gravitational wave pulses across the Earth's diameter (propagation time ~ 40 ms).

For our transducer, β is given by (4)

$$\beta \simeq \frac{\frac{1}{2}C_oV_p^2Q_e}{m_e\omega_a^2x_o^2} \qquad (3)$$

with V_p, the voltage across the capacitor, given by

$$V_p^2 = \frac{8Q_e\beta_1P_{cav}}{\Omega_oC_o(1+\beta_1)} \qquad (4)$$

where C_o is the capacitance of the cavity, x_o is the dynamic equilibrium capacitor gap spacing, β_1 is the electrical coupling to the cavity, and P_{cav} is the input power to the cavity. The critical parameters are β_1, Q_e and P_{cav}.

To examine the maximum value of β in more detail we have measured the power dependence of the cavity Q. Figure 1 shows the dependence of Q_e and β ($\propto Q_e^2P_{cav}$) on cavity power . We also find that the cavity Q is dependent on temperature, as shown in Figure 2. The measurement procedures used to obtain these data may have introduced systematic errors up to a factor of two. However, the relative errors are not more than a few percent.

The coupling (and bandwidth) is limited by the Q degradation at high pump power. High cavity fields cannot be used as the concomitant increase in cavity losses is too great. We are testing different cavity geometries and surface preparations to try to understand the mechanism and minimize the cavity losses.

In a recent test we achieved $Q_e = 2.5 \times 10^5$. Combining this result with an antenna acoustic Q, Q_a, of 1×10^7 (limited by losses in the bending flap), we obtain $\beta \approx 0.04$, $\beta Q_a \approx 4 \times 10^5$, and $B \approx 15$ Hz. We hope to confirm this result in a forthcoming run.

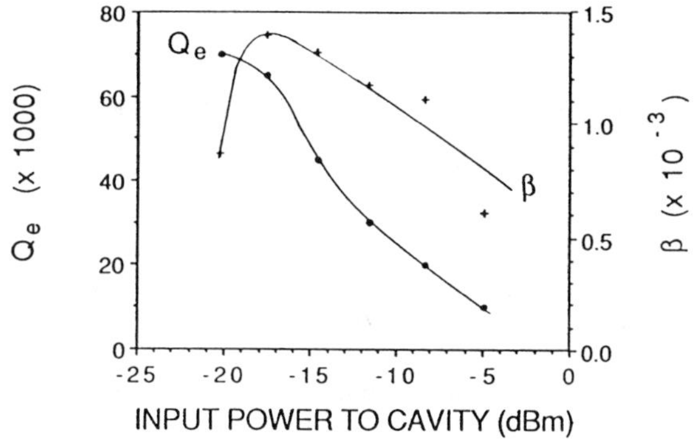

FIGURE 1. Power dependence of cavity Q, Q_e, and antenna-transducer coupling, β ($\propto Q_e^2P_{cav}$).

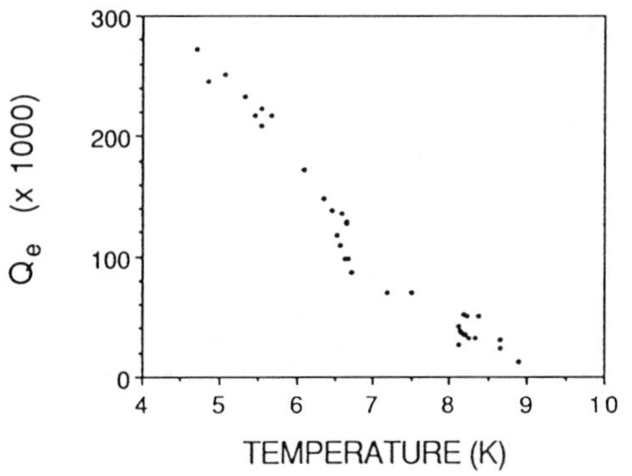

FIGURE 2. Temperature dependence of cavity Q.

ACKNOWLEDGMENTS

This research was supported by the Australian Research Council

REFERENCES

(1) P.J. Veitch, D.G. Blair, N.P. Linthorne, L.D. Mann and D.K. Ramm, Rev. Sci. Instr., 58 (1987) 1910.
(2) P.J. Veitch, D.G. Blair, N.P. Linthorne, M.J. Buckingham, C. Edwards, N.P. Prestage and D.K. Ramm, Proceedings of the Fifth Marcel Grossmann Meeting on General Relativity (World Scientific, Singapore 1989).
(3) N.P. Linthorne, P.J. Veitch and D.G. Blair, J. Phys D., 23 (1990) 1.
(4) D.G Blair, Resonant bar detectors for gravitational waves, in: Gravitational Radiation, eds. N. Deruelle and T. Piran (North-Holland, Amsterdam, 1983) pp 339-385.

Physica B 165&166 (1990) 11–12
North-Holland

FOCUSING OF HYDROGEN ATOMS WITH A CONCAVE He-COATED MIRROR

J.J. Berkhout[+], O.J. Luiten[+], I.D. Setija[*], T.W. Hijmans[*], T. Mizusaki[‡], and J.T.M. Walraven[*]

Universiteit van Amsterdam,
[+]Natuurkundig Laboratorium and [*]Van der Waals Laboratorium,
Valckenierstraat 65/67, 1018 XE Amsterdam, The Netherlands.

We use a concave spherical mirror, coated with liquid He, to focus a highly divergent beam of H-atoms into a small aperture. The temperature dependence of the focussed beam intensity enables us to study the influence of the dynamic surface roughness on the reflection of the H-atoms.

Hydrogen atoms (H) with sufficiently low incident energy, colliding with the surface of liquid ^{4}He, have a probability which approaches unity to undergo purely specular reflection.[1] This is a pure quantum phenomenon related to the large thermal de Broglie wavelength of the atoms at low temperatures and the weak interaction between H and helium. At non-zero temperatures, the small but finite probability for the H-atoms to adsorb onto the He film or to undergo inelastic (non-specular) scattering is governed by processes involving the emission or absorption of ripplons. The theory for scattering of low energy H-atoms from the surface of liquid He is the subject of several papers in the literature.[2,3,4,5]

Indirect evidence for the occurrence of quantum reflection of H from the surface of liquid helium has been obtained experimentally by measurements of the sticking coefficient s[6] defined as the probability for an atom upon collision to enter a surface-bound state. Reflectivities as high as 95% were deduced from these experiments at the lowest temperatures (0.08 K). The reflective properties of He films for H-atoms suggest the feasibility of making near perfect atomic mirrors.

In the present experiment we demonstrate the focusing of a beam of cold ($T < 0.5$ K) hydrogen atoms by the use of such a mirror. In this experiment a buffer volume is filled with H-gas. The atom can escape from this volume through a hole with a diameter of 0.5 mm. A hemi-spherical concave quartz substrate of optical quality with a 9 mm curvature radius is coated with a film of superfluid helium. The mirror is placed in front of the hole and is mounted on a translation stage by means of which its center of curvature can be made to coincide with the exit hole of the buffer volume. The entire cell is linked to the mixing chamber of a dilution refrigerator. When the mirror is in focus particles that scatter from the surface in a small angular range (0.6°) near pure specular reflection will re-enter the buffer volume, thereby reducing the rate at which the density in the volume decays. Hence, by measuring this decay rate as a function of mirror position, the reflectivity of the substrate for normally incident atoms can be obtained. More details of this experiment and its interpretation are given elsewhere.[7]

Fig.1 shows a typical measurement of the decay rate $1/\tau$ versus mirror position, normalized to its value $1/\tau_0$ in the absence of the mirror. The loss factor $\chi \equiv \tau_0/\tau$. The results clearly demonstrate the occurrence of specular reflection of the atoms.

In fig.2 the value χ_{min} of χ when the mirror is in focus, is plotted versus temperature. The triangles represent data for a saturated ^{4}He film of estimated thickness 11.5 nm. Notice that the apparent reflectivity $R_\perp = 1-\chi_{min}$ is decreasing from 73 % at 160 mK to 56 % at 400 mK. The circles represent data in the presence of a ^{3}He monolayer on the film.

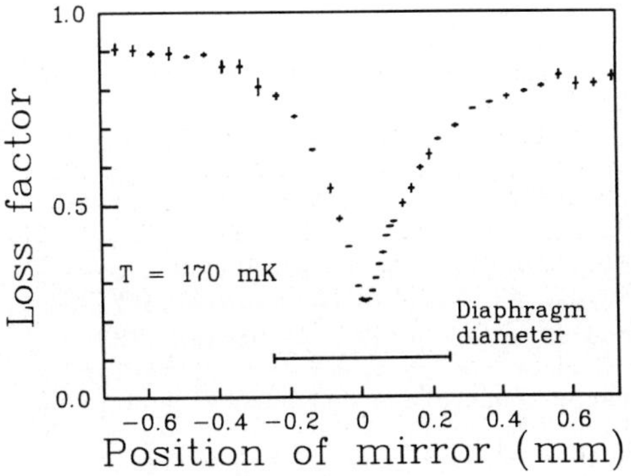

Fig. 1. The loss factor as a function of the vertical mirror position.

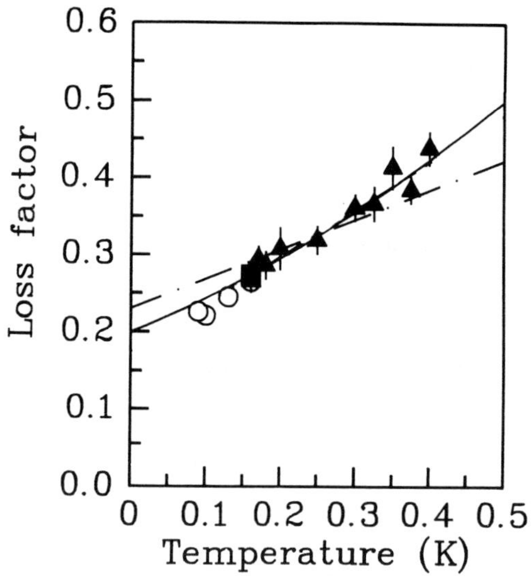

Fig. 2. Measured loss factors as function of temperature. Triangles: results on saturated (115 Å) pure ^4He films; circles: results on saturated ^4He films with full ^3He monolayer coverage. The curves are discussed in the text.

Although the loss factor becomes smaller with decreasing temperature, the extrapolated value for χ at T=0 does not vanish, as was predicted by theory[2] and deduced from measurements of the sticking probability.[6] This is mainly due to errors in the lateral alignment of the mirror. The value 0.2 for χ_{min} at T=0 corresponds to a 40μm off-axis misalignment of an otherwise perfect mirror.

To gain some quantitative insight into the results depicted in fig.2 we assume 3 independent loss mechanisms:

$$\chi_{min} = 1 - \prod_{i=1}^{3} (1 - \chi_i). \tag{1}$$

Here $\chi_1=\gamma_1 T$ with $\gamma_1 = 0.5\,K^{-1}$, is due to sticking and can be obtained[7] from the experimental results of ref.6 χ_3 is a temperature independent loss factor associated with the lateral misalignment of the mirror. As the latter quantity is not precisely known χ_3 is treated as an adjustable parameter. The contribution due to inelastic scattering has a quadratic temperature dependence[5] and is usually assumed to be small. If we set χ_2=0, we obtain the dashed-dotted curve in fig.2. If we include $\chi_2=\gamma_2 T^2$ in the fit we obtain the solid curve in fig.2 with $\gamma_2 = 0.5(1)\,K^{-2}$. This is in reasonable agreement with the calculated value[5] $\gamma_2 = 0.7\,K^{-2}$. Because of the high angular resolution (0.6°) this is the first experiment which reveals the contribution due to inelastic scattering, which strongly emphasizes small angles.

The authors wish to thank O.H. Höpfner for his overall technical support and M. Groeneveld for the construction of the quartz mirror. This work is part of the research program of the "Stichting voor Fundamenteel Onderzoek der Materie (FOM)", which is financially supported by the "Nederlandse Organisatie voor Wetenschappelijk Onderzoek (NWO)".

‡ Department of Physics, Kyoto University, Kyoto, Japan.

References

1. For a review see J.J. Berkhout and J.T.M. Walraven, Spin-Polarized Quantum Systems, S. Stringari, ed. (World Scientific, Singapore, 1989) p. 201.
2. D.S. Zimmerman and A.J. Berlinsky, Can.J.Phys. **61**, 508 (1983).
3. Yu. Kagan and G.V. Shlyapnikov, Phys.Lett. **95A**, 309 (1983).
4. B.W. Statt, Phys.Rev.B **32**, 7160 (1985).
5. T. W. Hijmans and G. V. Shlyapnikov, Phys. Lett. A **142**, 45 (1989).
6. J.J. Berkhout, E.J. Wolters, R. van Roijen, and J.T.M. Walraven, Phys.Rev.Lett. **57**, 2387 (1986).
7. J.J. Berkhout, O.J. Luiten, I.D. Setija, T.W. Hijmans, T. Mizusaki, and J.T.M. Walraven, Phys. Rev. Lett. **63**, 1689 (1989).

Physica B 165&166 (1990) 13–14
North-Holland

The Effect of Thermal Wavelength on the Force experienced by an Atom in an Optical Field

N. P. Bigelow and M. G. Prentiss

A. T. and T. Bell Laboratories, Crawford Corner Rd., Holmdel, NJ 07733-1988 USA

In recent years, there has been a great deal of interest in optical cooling and trapping of atoms. In this paper, we discuss the possibility of a change in the force experienced by an atom in a light field when the atom's thermal wavelength becomes larger than the optical wavelength. We argue that there will be no dramatic changes for an atom at equilibrium with the light field, and that the light force will be effective *even* for an atom whose wavepacket is arbitrarily large.

In general, quantum effects are expected to become important when the thermal deBroglie wavelength, $\lambda_{dB} = (2\pi\hbar^2/mk_BT)^{\frac{1}{2}}$, of a particle becomes comparable to other length scales in a system. Examples of such effects include the onset of superfluidity ($\lambda_{dB} \sim$inter particle spacing), spin waves in spin-polarized atomic hydrogen ($\lambda_{bB} > a_s$ =s-wave scattering length) as well as the transport of electrons in fine wires ($\lambda_{fermi} \sim$wire diameter). With the recent entry of optically cooled and trapped atoms into the field of low temperature physics, a new length scale arises; the optical wavelength λ_{opt}. The logical question is whether the force due to atom-field interaction will change when λ_{dB} becomes comparable to the λ_{opt}. This is a particularly relevant question because it is the atom-field interaction which is ultimately used to cool, trap and study the atoms. Indeed, because some of the lowest temperatures ever achieved in atomic systems, $\sim 2\mu$K (1), are produced using optical cooling and trapping, these questions have attracted increasing interest. In fact in some atom-field interaction experiments, the limit $\lambda_{dB} > \lambda_{opt}$ seems to have already been been exceeded.

In this paper we present an qualitative argument for the case of a two level atom in equilibrium with a plane traveling wave light field. By considering the physical picture both in momentum and configuration space, we conclude that there is, in fact, no reason to expect any dramatic change in the net average spontaneous force experienced by the atom (due to momentum exchange with the field) when $\lambda_{dB} \rightarrow \lambda_{opt}$. The discussion can be generalized to a multi-level atom in an more complex light field (i.e. a standing wave) however this generalization is beyond the scope of this paper.

In general, the interaction of an atom with an electromagnetic field is usually treated in the dipole approximation (2). The atom is modeled as a point like electric dipole $|\vec{d}| \ll \lambda_{opt}$ located at the atom's center of mass. The interaction which gives rise to absorption and emission of photons is the coupling of $\vec{d}$ to the oscillating electric field $\vec{E}(\vec{r},t) = E_o e^{(\vec{k}_{opt}\cdot\vec{r}-\omega t)}$. When λ_{dB} becomes large as compared to λ_{opt}, the point approximation is no longer valid. One might ask, for example, whether the 'large' atom will average over the electric field. If this were the case, it then might be asked whether the atom would decouple from the field, and in fact if the atom's fluorescence would essentially 'turn-off'.

The overall success of laser cooling and trapping seem to provide the ultimate proof that the atom-field interaction does not change significantly when λ_{dB} approaches λ_{opt}. For 'classical' optical molasses (3) the equilibrium temperature for the sodium system is $\sim$ 240μK, making $\lambda_{dB} \sim 234$Å $< \lambda_{opt} = 5890$Å. The more recently reported molasses temperatures from the group at the NIST (4) of 25μK,give $\lambda_{dB} \sim 720$Å $<$ λ_{opt}. In the experiments of Aspect et. al. (1) temperature of $\sim 2\mu$K has been achieved in metastable ^{4}He, for which $\lambda_{dB} \sim 6000$Å whereas $\lambda_{opt} \sim 10830$Å. Each of these comparisons of λ_{dB} to λ_{opt} is sufficiently close that the point dipole approximation may no longer be acceptable, however, the atomic wavepackets are not yet sufficiently large to encompass and optical cycle. From this viewpoint, perhaps the validity of the predictions of the dipole approximation for these experiments is less surprising. In the experiments of Hinds et. al. (5), however, atoms with transverse thermal wavelengths $\sim 1\mu$m are successfully guided along a single node in an light field where $\lambda_{opt} = 5890$Å. The experiments of Moskowitz et. al. (6) which demonstrate the diffraction of sodium atoms by an optical standing wave have also provided clear examples of the efficacy of optical forces *even* when $\lambda_{dB} > \lambda_{opt}$. The experimental evidence, then, is that the force due to the atom-field interaction does persist when $\lambda_{dB} > \lambda_{opt}$.

Consider the interaction of an atom with a plane, monochromatic traveling wave light field $E_o e^{i(\vec{k}\cdot\vec{r}-\omega t)}$. If we model the atom as a free particle plane wave (the limit as T$\rightarrow$0), then its quantum mechanical state is completely specified by its momentum and its internal coordinates (i.e. excited 'e' or ground 'g' state etc.). In particular, in a momentum space representation, the specification of the state of the atom is independent of position (i.e. its state vector may be written $|\vec{k}, i\rangle$ where i=e or g, and $\vec{k}$ is the atomic wave vector). The atom-field interaction hamiltonian involves terms which induce transitions between excited and ground states, change the atomic momentum in units of $\hbar\vec{k}_{opt}$ and create or destroying photons. For a traveling wave, this interaction is also independent of position. Simply put, the photon bath which describes the light field is isotropic in the traveling wave, so that from the point of view of the quantized light field, there is no absolute length scale and the size of the atomic wavepacket should not effect the absorption and emission processes. The simple arguments for the spontaneous molasses force still hold: the atom will preferentially absorb photons and hence gain momentum along $\vec{k}_{opt}$, will reradiate photons isotropically, and will experience a net average force along $\vec{k}_{opt}$.

In configuration space, the atom is depicted as a wavepacket which can extend over many optical wavelengths. In the plane wave limit, the atomic wavefunction ψ has a definite phase at every point in space ($\psi = e^{i(\vec{k}\cdot\vec{r}-\omega t)} = e^{i(phase(\vec{r})-\omega t)}$. We note that the large spatial extent of the atomic wavepacket does not necessarily imply an extended 'physical' dipole moment. If such a physical dipole were to grow as the extent of the wavepacket grew, the radiation length of the dipole would also eventually have to grow. Landau and Lifshitz (8) have remarked on this point: the field (the induced dipole) which is induced in a system (the atom) by an external light field can differ from the external light field only as a result of radiation from the system (which in turn can only occur if the atom is coupling to the field). The only self consistent behavior for the induced dipole is to be in a definite, well behaved phase with respect to the external field, and not to have a 'physical' size which scales with the size of the wavepacket; if the light field induces a dipole moment in the atom it must do so in a manner which depends on the wavelength of the light, and not on the size of the wavepacket.

Consider the two level atom with energies E_1 and E_2, where ψ_1 and ψ_2 are the eigenfunctions for the atom in each of these energy states (7). For an atom in the $\psi = \psi_1$ state, the induced dipole moment along the wavepacket will be in a definite phase with respect to the electric field which is inducing it. For a plane wave light field, the phase between the induced dipole moment $\vec{p}$ and the electric field gradient $\nabla\vec{E}$ will be fixed, and so the spontaneous force experienced by the atom, which is proportional to the gradient of the phase, will be a constant *independent* of the extent of the wave packet. From another viewpoint, the light field induces a phase shift across the wavepacket which is identically $\hbar\vec{k}_{opt}\cdot\vec{r}$, corresponding to the momentum of a quanta of the light field $\hbar k_{opt}$. Again, this picture is independent of λ_{dB}. For an atom which is in a superposition state of ψ_1 and ψ_2, the argument still holds for each amplitude component ψ_1 and ψ_2 independently. When the two levels represent the strong $|s\rangle$ and weak $|w\rangle$ field seekers, the dipole moment of the two components are opposite in sign. The result is a π relative phase shift between the induced dipole moment for each component ψ_s and ψ_w. This difference is related to the 'splitting' of the wavepacket (7,9) in a field gradient (in analogy to a Stern-Gerlach experiment which splits the wavefunction of a spin-$\frac{1}{2}$ atom which is in a superposition state).

In both momentum space and configuration space we thus argue that the net average spontaneous force experienced by an atom in equilibrium with a monochromatic plane wave will not exhibit any drastic change as $\lambda_{dB} \rightarrow \lambda_{opt}$. We note that these arguments can be generalized to more complex fields, and to multi-level atoms. We do not present this discussion as a conclusive derivation, but as a starting point for a more rigorous treatment. An accurate theoretical treatment still remains to be presented. A quantitative analysis must treat, for example, the effect of the dynamics of the internal coordinated of the atom (10), the time scales over which these coordinates can change as well as the allowed interaction time between the atom and the field.

ACKNOWLEDGEMENTS

The authors would like to thank J. Gordon and J. S. Denker for many insightful comments and provocative discussions.

REFERENCES

(1) A. Aspect, E. Arimondo, R. Kaiser, N. Vansteenkiste, C. Cohen-Tannoudji, Phys. Rev. Lett. 61 (1988) 826.
(2) C. Cohen-Tannoudji, B. Diu and F. Laloë, Quantum Mechanics, (Wiley, New York, 1977) p. 1307.
(3) S. Chu, L. Hollberg, J. E. Bjorkholm, A. E. Cable and A. Ashkin, Phys. Rev. Lett. 55 (1985) 48.
(4) P. D. Lett, W. D. Phillips, S. L. Ralston, C. E. Tanner, R. N. Watts, C. I. Westbrook, JOSA B 6 (1989) 2084.
(5) A. Anderson, M. Boshier, S. Haroche, E. Hinds, W. Jhe and D. Meschede, in: Atomic Physics 11, eds. S. Haroche, J. C. Gay and G. Grynberg (World Scientific, New Jersey, 1989), p.626.
(6) P. E. Moskowitz, P. L. Gould, S. R. Atlas and D. E. Pritchard, Phys. Rev. Lett. 51 (1983) 131.
(7) R. J. Cook, Phys. Rev. Lett. 41 (1978) 1788.
(8) L. D. Landau and E. M. Lifshitz, The Classical Theory of Fields, (Permagon, New York, 1975).
(9) V. Letokhov and V. Minogin, Laser Light Pressure on Atoms, (Gordon and Breach, New York, 1987).
(10) see for example C. Cohen-Tannoudji, in: Frontiers in Laser Spectroscopy, eds. R. Balian, S. Harosche and S. Liberman, (North-Holland, Amsterdam, 1977) p. 3-104.

Physica B 165&166 (1990) 15–16
North-Holland

On the Possibility of Observing Quantum Reflection in the Scattering of Atoms from a Nearly Resonant Light Wave

N. P. Bigelow and M. G. Prentiss

A. T. and T. Bell Laboratories, Crawfords Corner Rd., Holmdel, NJ 07733-1988 USA

We describe design considerations for an experiments on the scattering of optically cooled atoms from a standing wave light field. We discuss importance of atom-field interaction times and standing wave intensities in the context of an experiment which seeks to observe quantum reflection of cold atoms from the pseudo potential which describes the atom-field interaction.

Recently, optical methods of cooling atoms have created large densities of ultra-cold atoms. These methods use the atom-field interaction to transfer energy from the atom to the light field, and hence to cool the atomic system. Much of the atom-field interaction which is used to provide the cooling can be described in terms of a force which is experienced by the atom. In fact, there are aspects of the atom-field interaction which can be described by introducing an effective or pseudo-potential to predict the motion of the atom in the light field (1). In many optical cooling experiments, a sample of atoms are produced which are moving so slowly that their thermal deBroglie wavelength λ_{dB} is comparable to the wavelength λ_{opt} of the light with which they interact.

In this paper we discuss the possibility of observing the onset such quantum effects by studying the motion of the atoms in a nearly resonant light field. This discussion is motivated in particular by the possibility of observing a 'quantum reflection' of the atoms from pseudo-potentials created by the light field. By 'quantum reflection' we refer to the reflection of the atom from the light field due to the suddenness of the variation in atom-field pseudo-potential $U(x)$ as compared with λ_{dB} (2). We consider the atom-field interaction potential in a particular limit and discuss interaction time constraints which we are using in planning our experiments.

The effective potential energy $U(x)$ for a two level atom in an optical field due to the pondermotive or dipole force (1) can be written as:

$$U(x) = \frac{1}{2}\hbar\Delta \ln(1 + p(x)) \qquad (1)$$

where Δ is the detuning of the field from the two level resonant frequency ω_o, and $p(x)$ is a saturation factor which depends on Δ and on the intensity profile of the optical field $I(x)$:

$$p(x) = \frac{(I(x)/I_{sat})(\frac{\Gamma}{2})^2}{\Delta^2 + (\frac{\Gamma}{2})^2}. \qquad (2)$$

Here Γ is the natural linewidth of the transition and I_{sat} is its saturation intensity.

For the case of a standing wave light field, where $I(x) = I_o\cos^2(kx)$ detuned below (above) resonance ($\Delta < 0$), the intensity maxima represent the potential minima (maxima).

In the limit where the optical fields are weak ($I_o \ll I_{sat}$), the potential, $U(x)$, is simply proportional to $I(x)$, so $U(x)$varies smoothly as $\cos^2(kx)$.

In contrast, if the optical field is strong ($I_o \gg I_{sat}$), then $U(x)$ will no longer be proportional to I_o. This is the limit where the two level system is saturated, and the dipole moment induced by the light field is effectively decreasing with increasing intensity. In this limit there are a number of important characteristics of the atom-field pseudo potential. For fixed detuning Δ, $U(x)$ will saturate, increasing only very slowly with field intensity (logarithmically). Because of the logarithmic saturation, $U(x)$ can be made to vary rapidly in space, that is, on a scale *much smaller* than the optical wavelength (see FIG. 1). In fact, because $U(x)$ depends on both I_o and Δ, it is possible to maintain a constant potential depth while continuously varying the sharpness of the potential. For the sodium $3^2S_{1/2} \rightarrow 3^2P_{3/2}$transition ($F$=2, $m_F = 2$ to $F', m'_F = 3$), $I_o \approx 6.25\text{mW/cm}^2$, so it is experimentally realistic to sufficiently saturate $U(x)$ that it will resemble a series of delta functions at twice the optical period (where the period is as accurate as the period of the light field). By changing the sign of Δ, it is even possible to invert $U(x)$! In this manner, a variation in the intensity of the light can be used to tune the shape of $U(x)$ from a smooth variation over λ_{opt} to a sharp variation on a length scale much less than λ_{opt}. For atoms whose λ_{dB} is a reasonable fraction (say $\frac{1}{10}^{th}$) of λ_{opt}, it should thus be possible to increase the light field intensity continuously until the interaction potential becomes sufficiently sharp that quantum reflection is observed.

The nature of the atom-field interaction, and hence

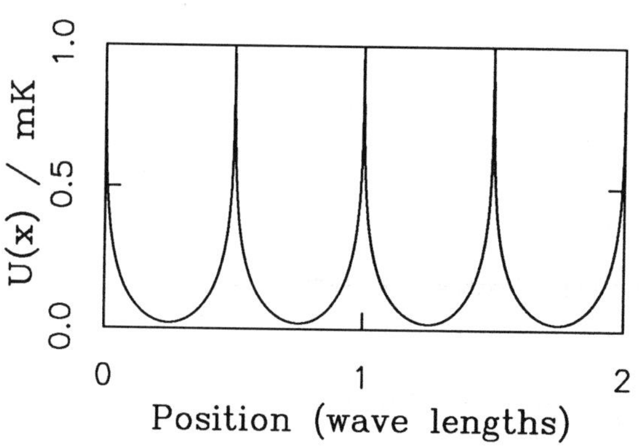

Figure 1. U(x) for sodium $3^2S_{1/2} \rightarrow 3^2P_{3/2}$transition. Here Δ=-10 MHz and I for a 400 mW, 400μm diameter standing wave.

the form of the $U(x)$, depends strongly on the interaction time allowed between the atom and the field. We consider three important time scales for our experiments: the stimulated emission time τ_{st}, the spontaneous emission time τ_{sp}, and the spontaneous heating time τ_{heat}. The stimulated emission time depends on the intensity of the excitation field, and is generally the fastest time scale considered here. The spontaneous time is generally the intermediate time, and for the $3^2S_{1/2} \rightarrow 3^2P_{3/2}$transition $\tau_{sp} \approx$ 16nsec. The spontaneous heating time is essentially the time required for an atom initially at T=0 to come to equilibrium with the stochastic heating processes of spontaneous emission (this is the process which sets the equilibrium temperature and is related to the name 'optical molasses' (3)). For the sodium system this time is on the order of 10 μsec (4). If the two level atom is allowed to interact with the light field for a time on the order the spontaneous heating time, the equilibrium 'temperature' (kinetic energy) for the atom will be comparable to the depth of $U(x)$, confounding scattering experiments off $U(x)$. If the atom is allowed to interact with the light for many stimulated times and many spontaneous times, but less than the spontaneous heating time, then the atom-field interaction will be well described by eq.1. This sets an upper bound on the atom-field interaction time for our scattering experiments which seek to use a strongly saturated potential.

Consider instead a situation where an incident atom, initially in the ground state, is brought into the light field adiabatically, and allowed to interact with the field for a time long compared to the τ_{st}, but shorter than a few τ_{sp}. In this limit, the atoms internal variables are no longer necessarily in equilibrium with the field, and the atom-field interaction is more accurately described in a different picture than used to derive eq. 1; this is the dressed state picture (5). Here the states of the two level atom are replaced by states which include the effects of the field (via stimulated emission processes); the strong ($|s\rangle$) and weak ($|w\rangle$) field seeker states. We can think of the atom-field energy, for say an $|s\rangle$ state, in terms of the potential $U(x) = \hbar[\Delta^2 + \omega_r^2(x)]^{\frac{1}{2}}$. Here $\omega_r(x) = \mu_e E_o/2\hbar$ is the Rabi frequency for the applied field E_o and μ_e is the dipole moment of transition. This potential differs from that of eq. 1 in that it does *not* saturate at high intensities, or for large Δ. The depth of the potential can be made much larger (for a given set of conditions) than that of eq.1, but cannot be made to vary as sharply in space. Again, for an experiment where the saturation of the potential is important, avoiding the dressed state regime places a lower bound of τ_{sp} on the atom-field interaction. In sum then, for an experiment which seeks to scatter atoms via the saturatable potential, eq. 1, the atom-field interaction time must be longer than τ_{sp} but shorter than τ_{heat}. Most importantly, for the sodium system, this window of interaction time is easily achievable.

From the point of view of an experiment to demonstrate quantum reflection of cold atoms from an optical standing wave, the potential described by eq. 1 is an attractive choice. The sharpness of the spatial variation of the potential can be continuously tuned from a regime where it varies smoothly as λ_{opt} to a regime where it is abrupt in comparison to λ_{opt}. For carefully chosen interaction times, and moderately cold atoms ($\lambda_{dB} \sim \lambda_{opt}$) we have shown that it should be possible to observe quantum reflection. Further, unlike many proposed systems for studying quantum reflection, the atom-filed interaction potential is *not* associated with a material surface and is thus free of the subtleties related to the sticking dynamics of cold atoms scattering from a surface (6). We are currently initiating experiments to demonstrate these effects.

REFERENCES

(1) A. Ashkin, Phys. Rev. Lett. 40 (1978) 729, and J. P. Gordon and A. Ashkin, Phys. Rev. A 21 (1980) 1606.

(2) Th. Martin, R. Bruinsma and P. M. Platzman, Phys. Rev. B, 38 (1988) 2257 and Phys. Rev. B Rapid Comm. 39 (1989) 12411

(3) S. Chu, L. Hollberg, J. E. Bjorkholm, A. Cable and A. Ashkin, Phys. Rev. Lett. 55 (1985) 48.

(4) P. D. Lett, W. D. Phillips, S. L. Ralston, C. E. Tanner, R. N. Watts, and C. L. Westbrook, JOSA B 6 (1989) 2084.

(5) J. Dalibard and C. Cohen-Tannoudji, JOSA B 2 (1985) 1707.

(6) J. J. Berkhout, O. J. Luiten, I. D. Setija, T. M. Hijmans, T. Mizusaki and J. T. M. Walraven, Phys. Rev. Lett. 63 (1989) 1689.

Physica B 165&166 (1990) 17–18
North-Holland

THE DEGENERATE-INTERNAL-STATES APPROXIMATION FOR COLD COLLISIONS

A.C. MAAN, E. TIESINGA, H.T.C. STOOF, and B.J. VERHAAR

Department of Physics, Eindhoven University of Technology, Postbus 513, 5600 MB Eindhoven, The Netherlands

The Degenerate-Internal-States approximation as well as its first-order correction are shown to provide a convenient method for calculating elastic and inelastic collision amplitudes for low temperature atomic scattering.

A considerable simplification of (in)elastic collision problems occurs when the energy differences of internal states significantly coupled in a quantum collision are neglected. Usually, this is interpreted classically to correspond to slow internal motion of the collision partners during their interaction time. The first context in which this approximation has been applied was, to the authors' knowledge, the subject of nuclear reactions induced by light projectiles in the fifties (1). In that context it was known as the adiabatic approximation. The same approximation has been applied in connection with heavy ion scattering off nuclei (2) and was then referred to as the sudden approximation. A more recent application occurred under the name (energy-) sudden approximation in atomic scattering (3), in which case the adjective "energy-" was sometimes used to distinguish it from the closely related "centrifugal-sudden" approximation.

In the past our Eindhoven group has contributed to applications of the above approximation (which we used to refer to as adiabatic approximation) in the field of vibrational and rotational excitation of nuclei in scattering (4). We also showed that it could be applied to hyperfine transitions in atom-atom spin-exchange collisions (5), to inelastic scattering of H atoms off the surface of superfluid helium (6) and to scattering of dressed atoms in a microwave trap (7). In the framework of H+H spin-exchange collisions in the H maser (5) we have for the first time avoided the name "adiabatic approximation" in view of the different use of this name in connection with the Born-Oppenheimer method in the same paper. In the present contribution we will adhere to the name "Degenerate-Internal-States" (DIS) approximation for the same reason.

To understand the simplifications of scattering problems brought about by all these identical methods, we write the exact scattering problem in the coupled-channels form

$$\underline{R}''(r) = \mathbf{V}(r)\,\underline{R}(r),$$

where a prime denotes differentiation with respect to the radial distance r, $\underline{R}$ is a vector with components describing the (r-dependent) probability amplitudes for finding the collision partners in each of the possible internal states and **V** is a matrix containing internal energies and coupling potentials. The problem of finding solutions to these equations is simplified considerably if there exist r-independent Hermitian commutators of **V** (constants of the motion): $[\mathbf{C},\mathbf{V}] = 0$. As is well-known, this gives rise to a partial decoupling of the problem. In our situation we consider commutators which **V** does not have by itself, but only after subtraction of a constant diagonal matrix **D**. The elements of **D** are in our case essentially differences of internal energies. The approximation consists of neglecting **D**.

Commutators of the type **C**, considered in previous papers, are the collective nuclear coordinates or the projectile angular momentum along the body-fixed symmetry axis in the case of scattering from deformed nuclei, the surface coordinates in the example of inelastic scattering of H atoms off a surface of superfluid helium and the photon number operator for the scattering of dressed atoms.

To illustrate the simplifications brought about by the above-mentioned methods we consider the example of sub-Kelvin collisions of identical electron-spin 1/2 atoms, for example H, D, and all alkalis. The DIS approximation then amounts to turning off the hyperfine splitting, but retaining its influence on the electron-nucleus spin wave functions. The scattering channels can then be decoupled by transforming to the triplet and singlet channels, leading to the expression

$$S^{\ell,\mathrm{DIS}}_{\{\gamma\delta\},\{\alpha\beta\}} = \left[k_{\{\gamma\delta\}}k_{\{\alpha\beta\}}/k^2\right]^{\ell+1/2} \langle\{\gamma\delta\}|e^{2i\delta^{\ell}_{S}}P_S + e^{2i\delta^{\ell}_{T}}P_T|\{\alpha\beta\}\rangle,$$

the P operators standing for projection operators on singlet and triplet subspaces, and k standing for the wave number associated with an average total internal energy ϵ, but with the same value of the total energy. The problem is thus reduced to the much simpler problem of calculating the phase shifts for (uncoupled) potential scattering.

In view of the considerable pay-offs associated with the DIS approximation, it is worthwhile to extend its range of validity. In 1974 Schulte and Verhaar (4) derived a correction of first order in the non-degeneracy, consisting of a Born type volume integral of the non-degeneracy perturbation, supplemented by a Wronskian surface term. This method was applied to the total three-dimensional scattering wave function. More recently, in connection with frequency shifts of the sub-Kelvin hydrogen maser our group showed this approach to be very helpful when applied to individual partial waves. The first-order perturbation due to the hyperfine splitting $H_{hf}-\epsilon$ turned out to be

$$\Delta S^{\ell}_{\{\gamma\delta\},\{\alpha\beta\}} = \left[k_{\{\gamma\delta\}}k_{\{\alpha\beta\}}/k^2\right]^{\ell+1/2} \Delta^{\ell}$$

$$\langle\{\gamma\delta\}|(P_T-P_S)(H_{hf}-\epsilon)(P_T-P_S)|\{\alpha\beta\}\rangle.$$

For an application of these methods to spin-exchange collisions in the cold deuterium maser we refer to another contribution to this conference (8).

REFERENCES

(1) S.I. Drozdov, Zh. Eksp. Teor. Fiz. 28 (1955) 734, 736 [Sov. Phys. JETP 1 (1955) 588, 591]; D.M. Chase, Phys. Rev. 104 (1956) 838; C.A. Levinson, Nuclear Spectroscopy B (Academic Press, New York, 1960) p. 677.

(2) K. Alder and A. Winther, Mat. Fys. Medd. 32 (1960) No. 8.

(3) P. McGuire and D.J. Kouri, J. Chem. Phys. 60 (1974) 2488.

(4) A.M. Schulte and B.J. Verhaar, Nucl. Phys. A232 (1974) 215; B.J. Verhaar and A.M. Schulte, Phys. Lett. 67B (1977) 381.

(5) B.J. Verhaar, J.M.V.A. Koelman, H.T.C. Stoof, O.J. Luiten, and S.B. Crampton, Phys. Rev. A 35 (1987) 3825; J.M.V.A. Koelman, S.B. Crampton, H.T.C. Stoof, O.J. Luiten, and B.J. Verhaar, Phys. Rev. A 38 (1988) 3535.

(6) E. Tiesinga, H.T.C. Stoof, and B.J. Verhaar, Phys. Rev. B (to appear).

(7) C.C. Agosta, I.F. Silvera, H.T.C. Stoof, and B.J. Verhaar, Phys. Rev. Lett. 62 (1989) 2361.

(8) E. Tiesinga, H.T.C. Stoof, B.J. Verhaar, and S.B. Crampton, Spin-exchange frequency shift of the cryogenic deuterium maser, this volume.

Physica B 165&166 (1990) 19–20
North-Holland

SPIN-EXCHANGE FREQUENCY SHIFT OF THE CRYOGENIC DEUTERIUM MASER

E. TIESINGA, H.T.C. STOOF, and B.J. VERHAAR

Department of Physics, Eindhoven University of Technology, Postbus 513, 5600 MB Eindhoven, The Netherlands

S.B. CRAMPTON

Department of Physics, Williams College, Williamstown, Massachusetts 01267

We derive expressions for the spin-exchange frequency shift and line broadening of the deuterium maser for low temperatures and present the results of a preliminary calculation based on the Degenerate-Internal-States approximation and its first-order correction.

Of the two atomic species expected to remain gaseous down to the absolute zero of temperature (1), hydrogen (H) has been studied much more than deuterium (D). Experimentally, high densities of D have been much harder to achieve (2-5), both because of H contamination and the recently discovered tendency of D to dissolve in liquid helium films (5). Theoretically, H has attracted more interest because of the possibility of achieving Bose condensation of spin-polarized H (1). Electron spin-exchange collisions between H atoms have been studied theoretically because of their important role in determining the decay of trapped H gas (6). In connection with cryogenic hydrogen masers (7-9) such collisions have also been investigated theoretically (10-11) because they produce potentially important frequency shifts limiting the frequency stability. The calculations showed a remarkable sensitivity to non-adiabatic effects at low collision energies (10).

Recently, densities of D up to 10^{15} cm^{-3} have been achieved using liquid helium storage surfaces at temperatures near 1 K (5). In addition, there are possibilities of studying D at higher temperatures using solid neon storage surfaces (12) and at lower temperatures using trapping magnetic fields (13). It therefore seems useful to investigate the influence of hyperfine effects on electron spin-exchange collisions between D-atoms at low temperatures.

We concentrate on the $\beta \leftrightarrow \epsilon$ transition (adopting the usual labels $\alpha\beta\gamma\delta\epsilon\zeta$ of hyperfine levels in order of increasing energy). Its transition frequency goes through a minimum at about 30 mG. Hence, if the maser is operated at this field, the output frequency is quite insensitive to variations in the applied magnetic field. On the basis of the weakness of the field one would expect that a calculation of spin-exchange collisions at zero field would be sufficient. Some care is called for, however, as we will see.

Using the methods and notation of Ref. 10 and 11 we find for the frequency shift $\delta\omega_c$ due to collisions the general expression

$$\delta\omega_c = n\langle v\rangle \Big[\bar{\lambda}_0(\rho_{\epsilon\epsilon}-\rho_{\beta\beta}) + \bar{\lambda}_1(\rho_{\epsilon\epsilon}+\rho_{\beta\beta}) + \bar{\lambda}_2\rho_{\alpha\alpha} + \bar{\lambda}_3\rho_{\gamma\gamma} + \bar{\lambda}_4\rho_{\delta\delta} + \bar{\lambda}_5\rho_{\zeta\zeta}\Big].$$

The collisional line broadening Γ_c is given by a similar expression with λ_i replaced by σ_i. The complex quantities $\sigma_i - i\lambda_i$ are linear combinations of "coherent cross-sections"

$$\sigma_{\beta\epsilon,\upsilon\to\lambda} = \frac{\pi}{k^2}\sum_{\ell}(2\ell+1)\Big[S^{\ell}_{\{\beta\lambda\},\{\beta\upsilon\}}S^{\ell *}_{\{\epsilon\lambda\},\{\epsilon\upsilon\}} - \delta_{\lambda\upsilon}\Big].$$

For deuterium odd (even) ℓ are correlated with (anti-)symmetrized spin states $\{\alpha\beta\}$.

Previous calculations (14,15) of $\delta\omega_c$ and Γ_c for D were based on the high-temperature limit, which allows one to neglect the influence of the hyperfine splitting on the S-matrix, as well as nuclear identity effects. For the lower temperatures, which interest us, this is no longer permitted. We use the Degenerate-Internal-States (DIS) approximation, as well as a first-order correction. An interesting aspect is that for deuterium these approximations have to be extended to inelastic scattering, as pointed out in a separate contribution to this conference (16). The complexity due to nuclear spin 1 prompted us to use computer algebra to derive the relevant expressions.

For B=0 we find the DIS values for λ_1 and λ_4 to be zero, while $\lambda_2 = -1/2\lambda_3 = \lambda_5$, so that

0921-4526/90/$03.50

$$\delta\omega_c^{DIS,B=0} = n\langle v\rangle \left[\bar{\lambda}_0(\rho_{\epsilon\epsilon}-\rho_{\beta\beta}) + \bar{\lambda}_2(\rho_{\alpha\alpha}+\rho_{\zeta\zeta}-2\rho_{\gamma\gamma})\right].$$

All σ_i and λ_0,λ_2 can be expressed in singlet and triplet phase-shifts. In the high-temperature limit the difference of phase-shifts for subsequent ℓ values as well as the difference of initial and final wave numbers for inelastic transitions can be neglected. This yields λ_2=0. The resulting $\delta\omega_c$ differs from that of Ref. (15) by a factor 2. In the same limit the line broadening Γ_c agrees with Ref. (15).

In view of the weakness of the field the B=0 DIS values describe λ_0, λ_2, λ_3, λ_5 and all σ_i approximately also at 30 mG. For λ_1, λ_4 a more subtle treatment is called for. Both the small B≠0 DIS values and the first-order B=0 hyperfine-induced corrections are of interest. In Fig. 1 we present λ_1 and λ_4 (for B = 30 mG) as a function of energy from 0.01 to 10 K. We also show the λ_0 and λ_2 DIS predictions for B=0.

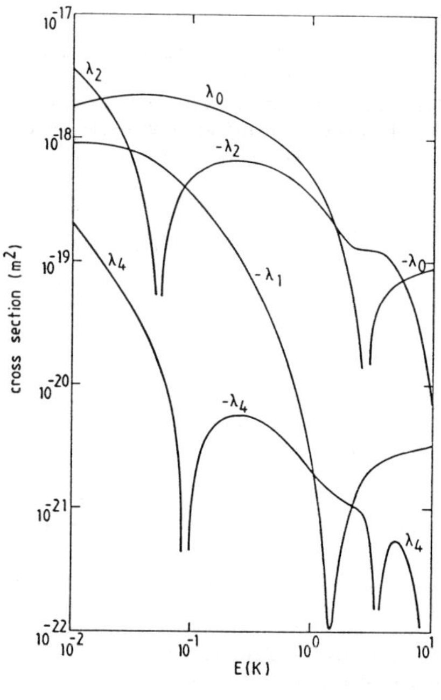

FIGURE 1: Frequency-shift cross-sections

As for atomic H our numerical results show a remarkable sensitivity to non-adiabaticity, this time for ℓ=1. Estimating non-adiabaticity effects by comparing results for the reduced mass equal to half the deuteron mass instead of half the deuterium mass (differing by 0.02%) we find 30% variations of some relevant spin-exchange S-matrix elements in the lower part of the above energy-range. Experimental study of frequency shifts for the cryogenic deuterium maser may therefore be of interest.

REFERENCES

(1) W.C. Stwalley and L.H. Nosanow, Phys. Rev. Lett. 36 (1976) 910.
(2) I.F. Silvera and J.T.M. Walraven, Phys. Rev. Lett. 45 (1980) 1268.
(3) R. Mayer and G. Seidel, Phys. Rev. B 31 (1985) 4199.
(4) I. Shinkoda, M.W. Reynolds, R.W. Cline, and W.N. Hardy, Phys. Rev. Lett. 57 (1986) 1243.
(5) M.W. Reynolds, PhD Thesis, University of British Columbia at Vancouver, February 1989 (unpublished).
(6) A. Lagendijk, I.F. Silvera, and B.J. Verhaar, Phys. Rev. Lett. 33 (1986) 626.
(7) W.N. Hardy, M.D. Hürlimann, and R.W. Cline, Proc. 18th Int. Conf. on Low Temp. Phys., Jap. J. of Appl. Phys. 26 (1987) 2065.
(8) H.F. Hess, G.P. Kochanski, J.M. Doyle, T.J. Greytak, and D. Kleppner, Phys. Rev. A 34 (1986) 1602.
(9) R.L. Walsworth, I.F. Silvera, H.P. Godfried, C.C. Agosta, R.F.C. Vessot, and E.M. Mattison, Phys. Rev. A 34 (1986) 2550.
(10) B.J. Verhaar, J.M.V.A. Koelman, H.T.C. Stoof, O.J. Luiten, and S.B. Crampton, Phys. Rev. A 35 (1987) 3825.
(11) J.M.V.A. Koelman, S.B. Crampton, H.T.C. Stoof, O.J. Luiten, and B.J. Verhaar, Phys. Rev. A 38 (1988) 3535.
(12) S.B. Crampton, Ann. Phys. (Paris) 10 (1985) 893.
(13) J.M.V.A. Koelman, H.T.C. Stoof, B.J. Verhaar, and J.T.M. Walraven, Phys. Rev. Lett. 59 (1987) 676.
(14) S.B. Crampton, H.G. Robinson, D. Kleppner, and N.F. Ramsey, Phys. Rev. 141 (1966) 55
(15) D.J. Wineland and N.F. Ramsey, Phys. Rev. A 5 (1972) 821.
(16) A.C. Maan, E. Tiesinga, H.T.C. Stoof, and B.J. Verhaar, The Degenerate-Internal-States approximation for cold collisions, this volume.

Physica B 165&166 (1990) 21–22
North-Holland

UNSTABLE OSCILLATION OF THE CRYOGENIC H MASER

A.C. MAAN, H.T.C. STOOF, and B.J. VERHAAR

Department of Physics, Eindhoven University of Technology, Postbus 513, 5600 MB Eindhoven, The Netherlands

Paul MANDEL

Université Libre de Bruxelles, Campus Plaine, C.P. 231, 1050 Bruxelles, Belgium

We show that the Maxwell-Bloch equations, describing the dynamics of the sub-Kelvin H maser, predict unstable and chaotic oscillations in a regime which should be easily accessible experimentally.

The room-temperature hydrogen maser (1) is the most stable existing frequency standard for a wide range of measuring times. As such it has been used very successfully for tests of general relativity, for Very Long Baseline Interferometry (VLBI) and for interplanetary navigation (Voyager 2 mission). Since some years work is going on to build a sub-Kelvin H maser for improving the frequency stability by one or more orders of magnitude (2).

A central result governing the operation of the (cryogenic) H maser is the so-called oscillation condition (3)

$$\frac{P}{P_c} = -2q^2\left[\frac{I}{I_{th}}\right]^2 + (1 - Cq)\,\frac{I}{I_{th}} - 1 > 0, \qquad (1)$$

expressing that the total radiated power P is to be positive (see Fig. 1). In Eq. (1) P_c is the critical power, I is the surplus flux of atoms entering the storage bulb in the upper (c) hyperfine level, I_{th} is its threshold value assuming only density-independent relaxation, q is the maser quality factor and $C = 2(T_1{}^0/T_2{}^0)^{1/2} + (T_2{}^0/T_1{}^0)^{1/2}$ with $T_1{}^0$ and $T_2{}^0$ being the density-independent longitudinal and transverse relaxation times.

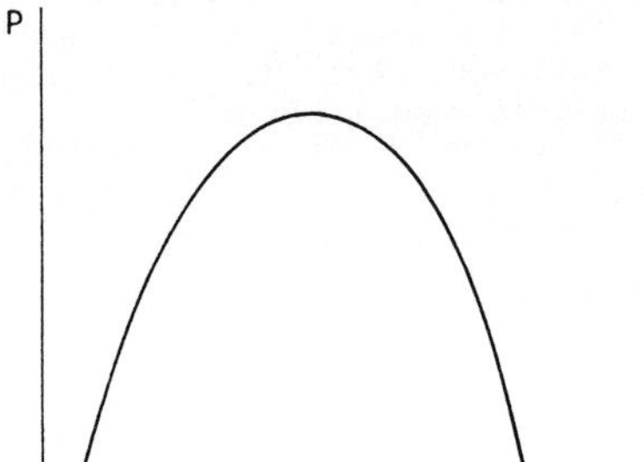

FIGURE 1: Power versus atomic flux

It has not been recognized until now that the condition (1) is only necessary but by no means sufficient. Reformulating the maser dynamics in the form of Maxwell-Bloch equations, it turns out that (1) guarantees only the existence of a steady solution, which is not necessarily stable, however. For stability a second condition $P < P_H$ has to be satisfied, where P_H is a threshold power beyond which unstable behavior begins. Restricting ourselves for definiteness to the case of small detuning $\delta \equiv T_2(\omega_{at} - \omega_m) = T_c(\omega_m - \omega_c)$ this condition has the form

$$\frac{T_1}{T_t}\,\frac{P - P_H}{P_H} = \left[-2q + \frac{T_c}{T_t}\,\frac{1 - 3\delta^2}{(1 + \delta^2)^2}\right]\frac{I}{I_{th}} - \frac{T_2^0}{T_t} < 0, \qquad (2)$$

with $1/T_c = \omega_c/2Q_c$ being the field loss rate and $T_t = (T_1{}^0 T_2{}^0)^{1/2}$. We have calculated the power P according to Eq. (1) as well as P_H, both on resonance, for the parameters of the University of British Columbia cryogenic maser (4). In particular, the cavity quality factor Q_c is equal to 1700. We find that $P < P_H$ for all fluxes satisfying the oscillation condition. Increasing Q_c to the (still modest) value 5×10^5, it is seen that the steady oscillation is unstable except for the very small fluxes (see Fig. 2).

An evaluation of the two conditions has also been carried out for the parameters of the Harvard-Smithsonian cryogenic maser (5). Again, the unstable regime seems easily accessible: Q_c has to be increased by a factor of 20 (see Fig. 3). From a similar comparison it follows that this regime is much more difficult to realize for a room-temperature H maser, mainly due to the much lower maximum atomic densities in the storage bulb allowed by the oscillation condi-

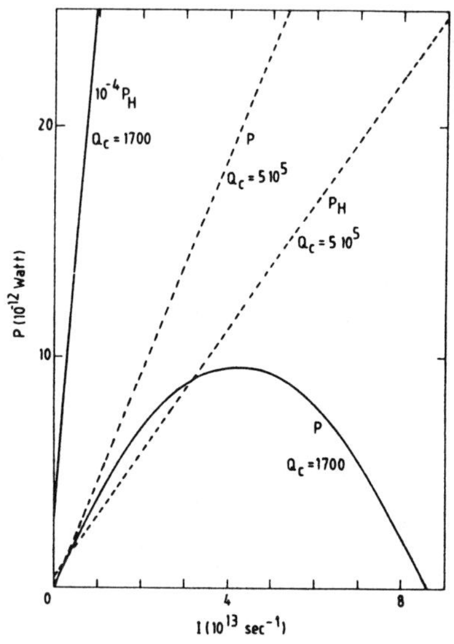

FIGURE 2: Unstable regime for University of British Columbia maser

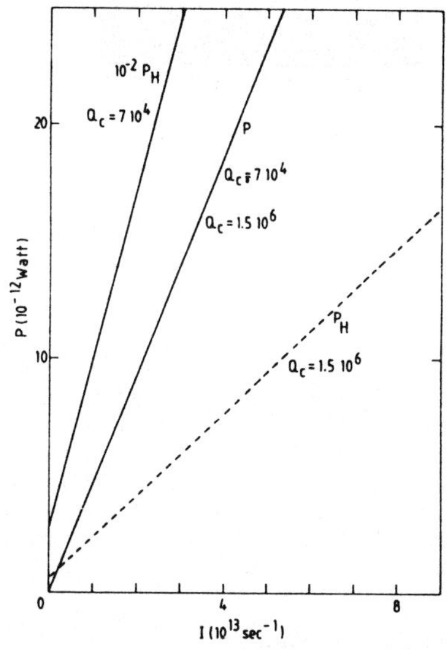

FIGURE 3: Unstable regime for Harvard-Smithsonian maser

tion due to the faster collisional relaxation rates.

The observation of instability may have important applications for obtaining information on the maser which would be very difficult to obtain otherwise. It is a priori to be expected that the nonsteady regime will offer much more information than the quantities frequency and amplitude, obtainable from stationary operation. This is especially welcome in view of the overwhelming number of experimental parameters such as hyperfine populations which determine the maser operation and are notoriously difficult to diagnose.

Apart from the prospect to get insight into the operation of the hydrogen maser, there is an intrinsic interest associated with the possibility to observe deterministic chaos in the cryogenic hydrogen maser. Not only does the derivation of the Maxwell-Bloch equations require far less simplifying assumptions than in the case of a laser, a system well-known to display chaos, but the ratios of the time constants entering the equations are that much different for both systems that the kind of nonsteady behavior to be expected in the H maser will differ largely from that of a laser. Although a detailed study has still to be carried out, it is already clear now at this stage that the operation of the H maser in the nonsteady domain will be characterized by a pulsed output power.

REFERENCES

(1) H.M. Goldenberg, D. Kleppner, and N.F. Ramsey, Phys. Rev. Lett. 5 (1960) 361.

(2) S.B. Crampton, W.D. Phillips, and D. Kleppner, Bull. Am. Phys. Soc. 23 (1978) 86; R.F.C. Vessot, M.W. Levine, and E.M. Mattison, Proc. 9th Annual Precise Time and Time Interval Conference, 1978, p. 549; A.J. Berlinsky and W.N. Hardy, Proc. 13th Annual Precise Time and Time Interval Conference, 1982, p. 547.

(3) D. Kleppner, H.C. Berg, S.B. Crampton, N.F. Ramsey, R.F.C. Vessot, H.E. Peters, and J. Vanier, Phys. Rev. 138 (1965) A972.

(4) W.N. Hardy, M.D. Hürlimann, and R.W. Cline, Proc. 18th Int. Conf. on Low Temp. Phys., Jap. J. of Appl. Phys. 26 (1987) 2065.

(5) R.L. Walsworth, I.F. Silvera, H.P. Godfried, C.C. Agosta, R.F.C. Vessot, and E.M. Mattison, Phys. Rev. A 34 (1986) 2550, and private communication.

Physica B 165&166 (1990) 23–24
North-Holland

A FAST CMN THERMOMETER WITH A WIDE TEMPERATURE RANGE

Dennis S. GREYWALL and Paul A. BUSCH

AT&T Bell Laboratories, 600 Mountain Avenue, Murray Hill, NJ 07974

A compact and portable CMN thermometer is described which has a response time of only 10 sec at 2 mK. The speed of the device derives from the manner in which intimate thermal contact is made to very small particles of the paramagnetic salt.

1. INTRODUCTION

In a recent paper (1) we described a CMN susceptibility thermometer which was designed to have a very fast thermal response and to have an effective temperature range extending from above several hundred mK to below 2 mK. This device, which was used recently in measurements (2,3) of the heat capacity of ^{3}He films adsorbed on graphite, also featured a high sensitivity, a negligible self heating, a small Curie-Weiss constant, a small heat capacity, compactness, and portability. This paper notes several modifications which were made subsequently to further advance its performance.

2. GENERAL DESIGN

The speed of the thermometer derives from the intimate thermal contact made to very small particles of the paramagnetic salt and from an anomalously-small thermal boundary resistance between CMN and silver (1). The good contact was achieved by cold sintering under pressure a mixture (1:1 by volume) of CMN (37 μm diam) and silver (3 μm diam) powders to form thin pads which were simultaneously bonded to strips of thin metal foil, Fig. 1. Many of these elements, separated by paper insulation, were tightly stacked and epoxied (4) into a single mass to form the active portion of the device. Thermal contact to the CMN was therefore made via the foils, which were attached to the base of the thermometer and via the matrix of sintered silver powder surrounding the salt particles, Fig. 2.

A cross sectional drawing of the completed thermometer is shown in Fig. 3. The stack of elements was machined into cylindrical shape to accept the coil needed for the susceptibility measurements. The reference coil required for our self inductance bridge was made as identical as possible to the main coil and was a physical part of the device. Its proximity to the main coil coupled with the counter winding was intended to reduce noise pick up.

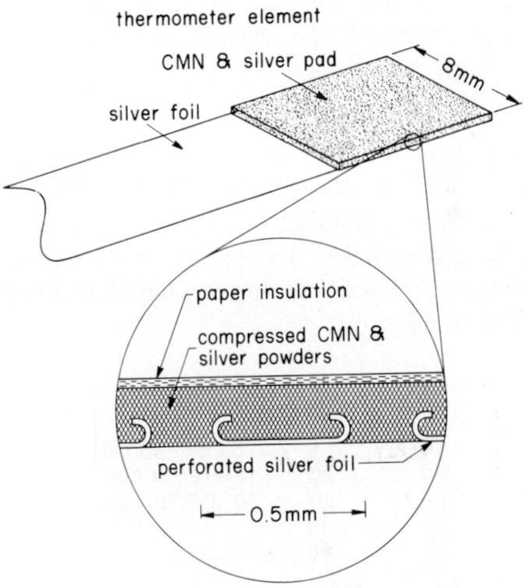

FIGURE 1 Single element of the CMN thermometer.

FIGURE 2 Photograph of a compressed mixture of CMN (large particles) and silver powders.

3. IMPROVEMENTS MADE

In our original design, the CMN-silver pads were bonded to copper foils which had an electroplating of silver. Even though this plating was intentionally made rough, the bond between the pad and the foil was not always completely satisfactory. In our new version we use annealed and etched pure silver foils which have been perforated with 200 0.2 mm diam holes over the area to which the pad is to be bonded. The action of compressing the powder mixture rivots the pad to the foil, Fig. 1, making an extremely reliable attachment.

The jig used to simultaneously punch all of the holes in the foils consisted of a tight packet of headless sewing pins. Punching through the foils into a dense cork backing produced the desired amount of burr surrounding each of the holes.

The free ends of the foils were joined to the thermometer base in our original design using a mechanical clamp. Replacing the clamp on the new device is a more compact crimp joint which is reinforced by a laser weld. The pulsed welding process does not appreciably warm the epoxy and sintered pads at the far end of the foils.

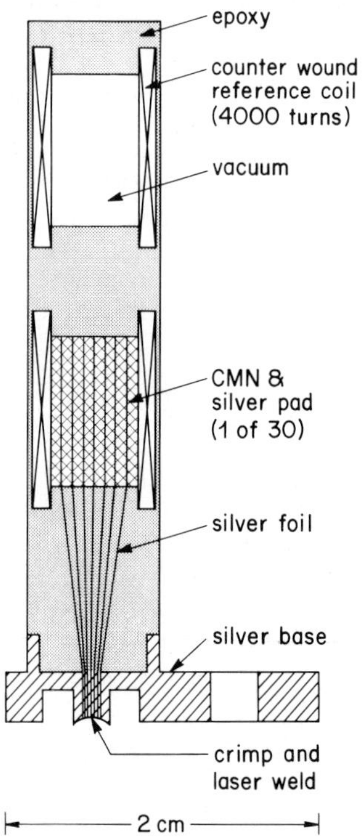

FIGURE 3 Cross-sectional drawing of the completed thermometer.

4. PERFORMANCE

The modifications to the thermometer resulted in a reduction of the response time at 2 mK from more than 30 sec to less than 10 sec. Near 1 mK the time constant is a factor of 2 or 3 larger.

Figure 4 shows a calibration of the thermometer based on the ^{3}He melting-curve temperature scale (5). Here R is the self-inductance bridge ratio which between 7 and 250 mK can be accurately related to the temperature via the simple expression

$$\frac{1}{T-\Delta} = A\left(\frac{R}{1-R}\right) + B$$

with $\Delta = 0.0038$ mK. By including higher-order terms the fit can be extended to near 1 mK. At a bridge frequency of 1000 Hz and at an excitation level sufficient to obtain a temperature resolution of better than 1 part is 1000 at the lowest temperature, the power dissipation is 10 pW and the self heating is comparable to the resolution.

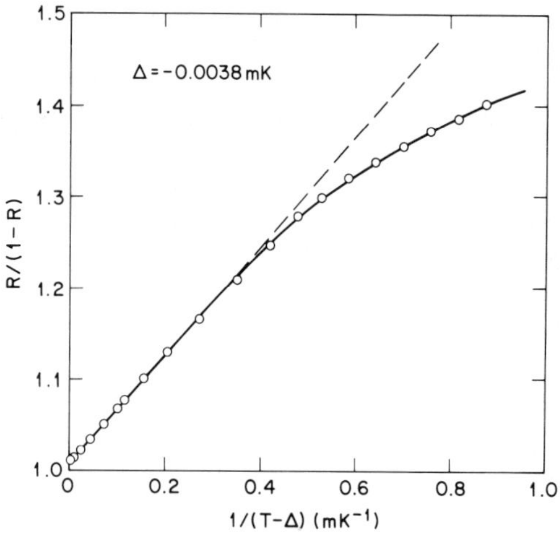

FIGURE 4 Thermometer calibration curve.

REFERENCES

(1) D. S. Greywall and P. A. Busch, Rev. Sci. Instrum. *60* (1989) 471.

(2) D. S. Greywall and P. A. Busch, Phys. Rev. Lett. *62* (1989) 1868.

(3) D. S. Greywall, Phys. Rev. B *41* (1990) 1842.

(4) Emerson and Cuming, MA, type 1266.

(5) D. S. Greywall, Phys. Rev. B *33* (1986) 7520.

Physica B 165&166 (1990) 25–26
North-Holland

SOL - GEL DERIVED GLASS CAPACITANCE SENSORS

Peter STREHLOW

Physikalisch-Technische Bundesanstalt, Institut Berlin, Abbestrasse 2-12, 1000 Berlin 10, Germany

Dielectric measurements are reported on glass capacitance sensors prepared by the sol-gel method. The low temperature dielectric behaviour of glasses can be used as a thermometric property with several advantages over resistance thermometers. The sol-gel process offers the possibility to prepare small glass capacitance sensors of fast response time and sufficient sensitivity down to the µK-range with well-defined chemical composition and morphological structure.

1. INTRODUCTION

New cryogenic technology and low temperature measurement require pratical, reliable and accurate thermometry by means of new sensors and more sophisticated measurement systems. Magnetoresistance and overheating of widely used resistance thermometers, such as carbon resistor, make temperature measurement very difficult in the presence of large magnetic fields and in the temperature range below 20 mK. Glass capacitance thermometers are characterized by a broad operating temperature range, picowatt self-heating, fast response time and insensitivity to magnetic fields (1).

The preparation of glass capacitance sensors of high performance requires a technological procedure such as the sol-gel process allowing to control the chemical composition, the homogenity and the microstructural evolution during the glass formation . Using the sol-gel method it is also possible to prepare high-melting glasses of various shape at relatively low temperatures.

2. EXPERIMENTAL

Monodispersed silica particles were formed from solutions of tetraethylorthosilicate (TEOS), water and ammonia (2). SiO_2 powders synthesized under these conditions had a mean particle size of about 0.5 µm. Thin disks (o.5 mm thick) were successfully prepared and sintered to nearly theoretical density in 2 hrs at 1050 °C. The faces were polished and covered with evaporated gold electrodes.

A second sensor type was prepared by dipping gold sheets into a coating solution consisting of 0.5 mole TEOS, 2 mole H_2O and 0.05 mole HCl in 2 mole ethanol. After coating, the gel-layers were dried and finally densified. The limit to the defect-free thickness for a single dipping is in the range of 0.5 µm. By means of multiple coating, the thickness of the film can be further increased, but a heat treatment is required between dips. The second electrode was prepared by evaporated gold on the glass surface.

Low-loss coaxial cables were soldered to the electrodes of the prepared dielectric sensors. The screened sensors were screwed on the nuclear magnetic copper stage of a $^3He/^4He$-dilution refrigerator . The capacitance and dissipation factor were measured from 700 µK to 4.2 K, using a capacitance bridge in the three-therminal mode. The resolution of the measurement system, consisting of generator, ratio transformer, lock-in amplifier, thermal-controlled reference capacitor C_R, series resistor R and choked coaxial pairs, was determined to be 10^{-8} or 0.1 aF for a 10 pF capacitor.

3. RESULTS

In Figure 1 the capacitance-temperature relation $C(T)/C_R - 1$ and the dissipation-temperature relation $\Delta\tan\delta = \tan\delta(T) - \tan\delta_R$ of the sintered silica sensor are demonstrated, measured during several runs.

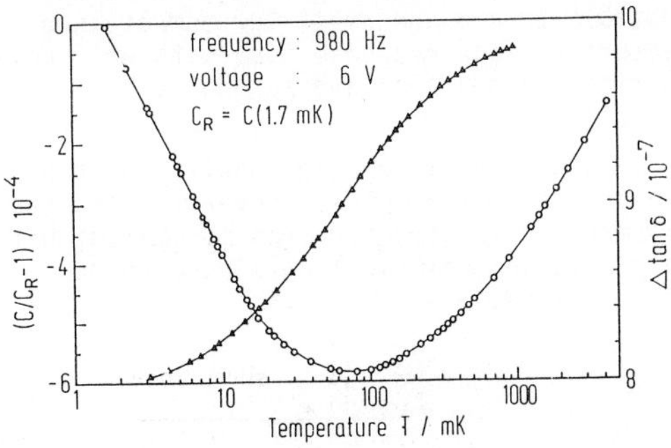

FIGURE 1
The capacitance (circles) and the loss angle (triangle) for the sintered silica sensor as a function of temperature

0921-4526/90/$03.50

A minimum in the capacitance was observed at 75 mK for a frequency of 980 Hz. This minimum depends on the frequency (3). The observed logarithmic temperature dependence of the capacitance below the minimum can be described by assuming the existence of localized two-level excitations (4). The sensitivity at 10 mK, $d((C-C_R/C_R)/dT$, was 3.7 x 10^{-5}/mK. Vitreous silica has a very low dielectric loss at 980 Hz, tan δ being lower than 10^{-6} below 100 mK, which causes self-heating to be negligible with a measuring voltage of 6 V. The response time of the sensor was determined to 10s at a temperature of 10 mK. For temperatures, T < 1 mK, the time constant of the sensor was larger than the retention time of a temperature step of 30 minutes. No relaxation effects were observed from day to day during a given run and also during cycling to room temperature, within a temperature accuracy of about 3 %.

Similar results were obtained for the SiO_2-sensor prepared by dip-coating.The sensitivity at 10 mK was determined to be 3.2 x 10^{-5}/mK. For temperatures below 2 mK, a significant decrease of the sensitivity was observed. Moreover, in contrast to the sintered sensor with each cycling to room temperature, it was necessary to recalibrate the film sensor.

4. DISCUSSION

The most important features of glass capacitance sensors are sufficient temperature sensitivity and negligible self-heating at ultra-low temperatures. The self-heating $\dot{Q}$ can be calculated from the Gibbs equation for a dielectric body and is given by

$\dot{Q} = \pi \times C \times U_0^2 \times f \times \tan\delta$. Using a frequency of f = 980 Hz and a voltage of U_0 = 6 V, the self-heating for the sintered silica sensor at temperatures T < 100 mK is $\dot{Q}$ < 10 pW.

By means of the sol-gel process it was possible to prepare small low-temperature sensors of fast response time with well defined chemical composition and morpholocical structure.

In order to improve the quality of the sensor, the gel to glass conversion and the differences in structure and morphology between polymeric and colloidal gels have to be investigated.

Properties of sintered glasses, such as dielectric permittivity, dielectric loss and heat conductivity, are strongly dependent on the porosity which depends on the particle size. The particle size of alkoxy-derived silica microspheres is determined by the rates of hydrolysis and condensation of the alkoxides and can be calculated by thermodynamic stability criteria for colloidal particles in solution (2).

Differences in structure between gel-derived glasses and melted glasses, if they exist, could affect the localized two-level excitations. Gel-derived glasses are generally expected to have more open structures which are farther from equilibrium than their melted forms.

The hightened sensitivity of the sintered sensor at 10 mK with regard to the film sensor could be founded on a different gel structure (polymeric or colloidal). In both cases the content of OH^- ions was determined to be about 1050 ppm.

The poor reproducibility of the film sensor prepared by dip-coating seems to be caused by thermal stresses. For that reason, the preparation of small sensors by electrophoretic deposition on cylindrical electrodes of different metals will be investigated.

Small low temperature sensors are characterized by a short response time. On the other side, the miniaturization should be restriced to dimensions in the order of the mean free path of phonons. The measured decrease of sensitivity for the film-sensor (thickness 1.5 μm) at temperatures below 2 mK could be a hint on this size effect.

REFERENCES

(1) S.A.J. Wiegers, R. Jochemsen, C.C. Kranenberg, and G. Frossati, Rev. Sci. Instrum. 58 (1987) 2274-2278.

(2) P. Strehlow, J. Non-Cryst. Solids 107 (1988) 55 - 60.

(3) G.Frossati, J.Gilchrist, J. Lasjaunias, and W. Meyer, J. Phys. C: Solid State Phys. 10 (1977) L 515-519

(4) M. von Schickfus, S. Hunklinger, and L. Piché, Phys. Rev. Lett. 35 (1975) 876 - 878.

Physica B 165&166 (1990) 27–28
North-Holland

THE KORRINGA RELATION IN SEMIMETALLIC ARSENIC BELOW 1 K

I. P. GOUDEMOND, J. M. KEARTLAND and M. J. R. HOCH

Department of Physics, University of the Witwatersrand, P O Wits 2050, Johannesburg, South Africa.

Nuclear spin–lattice relaxation in the semimetals has previously been shown to follow the Korringa relation between 4 K and the Debye temperature . We have established procedures for measuring relaxation times at very low temperatures and have made measurements between 150 mK and 1K. The Korringa relation is found to hold within experimental error over the entire range and theoretical calculations show that the dominant contribution to relaxation is from the s–wave carriers. The possibility of using the system in thermometry is briefly mentioned.

1 INTRODUCTION

The carrier density in semimetals is much lower than in metals, and both electrons and holes contribute to conduction. We have previously studied spin–lattice relaxation using NQR methods between 4 K and 520 K. Below the Debye temperature the relaxation time is found to follow a 1/T dependence, consistent with the Korringa relation (1), which applies to metals. In the present work spin lattice relaxation times (T_1) have been measured between 150 mK and 1 K on ^{75}As. This was chosen as the test system because it has spin I = $^3/_2$, and only one naturally occurring isotope, so that only one resonance is observed. Its crystal structure is rhombohedral and the nuclear quadrupole moment couples to the EFG to split the ±m, ±(m–1) states. The purpose of the investigation is to establish whether the Korringa relation holds for semimetals at very low temperatures.

2 EXPERIMENTAL

The annealed powdered As sample (<25μ; 99.9995%) was placed within a Stycast encapsulated coil, which was glued to a copper block for mounting directly into the mixing chamber of an Oxford model 400 dilution refrigerator. Thermal anchoring of the sample was achieved through the use of liquid helium as an exchange fluid and a sintered metal heat exchanger. Temperature was monitored by means of a germanium resistance thermometer mounted on the mixing chamber and by measurement of the maximum echo amplitude $\bar{M}(\infty) \propto 1/T$. Echo amplitudes were measured using a conventional coherent NQR spectrometer. The maximum pulse power available was about 100 W.

In pulsed NQR or NMR experiments at low temperatures, some heating of the sample may occur through the application of rf pulses. The amount of heating is related to the width and amplitude of the applied pulses. NQR echo amplitudes have been calculated for a system of spins I following two pulses of arbitrary width for a powdered sample using the density matrix approach of Das and Saha (2). The details of this calculation will be given elsewhere.

The echo amplitude is found to be given by:

$$\bar{M}(t_w) = \frac{B}{2}\sum_{i=0}^{\infty} A_i[J_{2i+1}(x\omega_0 t_w) - \tfrac{1}{2} J_{2i+1}(x\omega_0 t_w[1-\alpha]) - \tfrac{1}{2} J_{2i+1}(x\omega_0 t_w[1+\alpha])]$$

where ω_0 is the resonance frequency and

$$A_i = \frac{-4}{(2i+3)(4i^2-1)} \quad : x = [(I+m)(I-m+1)]^{\frac{1}{2}}.$$

Expressing the echo amplitude as a function of initial pulse width t_w and pulse width ratio α, enables us to find the optimum combination of pulses to give a measurable echo while minimizing heating effects. Echo amplitudes have been measured as a function of pulse settings at a chosen temperature and the data fitted using the theoretical expression. The theoretical curves for a powder (solid curve) and for a single crystal (dotted curve) as well as the experimental powder data are shown in Fig. 1. Agreement between theory and experiment is excellent for small values of t_w which is the condition used in work at low temperatures. Our calculations and experimental tests show that pulses much shorter than standard pulses may be used in low temperature NQR experiments.

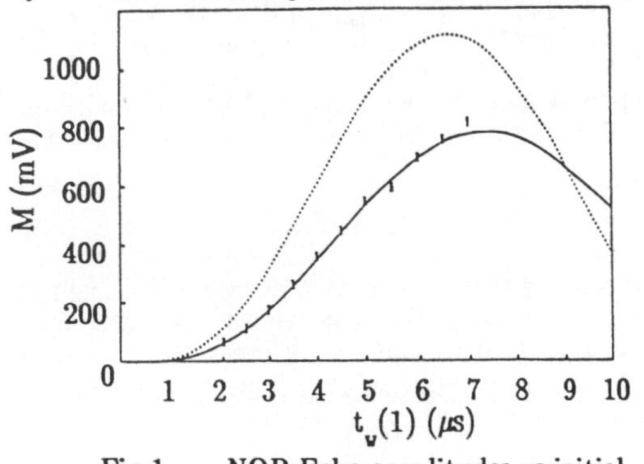

Fig.1. NQR Echo amplitudes vs initial pulse width, $\alpha = 2$ (see text)

3 RESULTS AND DISCUSSION

The Korringa relation states that $T_1T = A$, where A is a constant. T_1 has been found from plots of $-\ln(1-\bar{M}/\bar{M}(\infty))$ versus the delay τ between the saturation pulse and interrogation pulses for temperatures between 150 mK and 1 K.

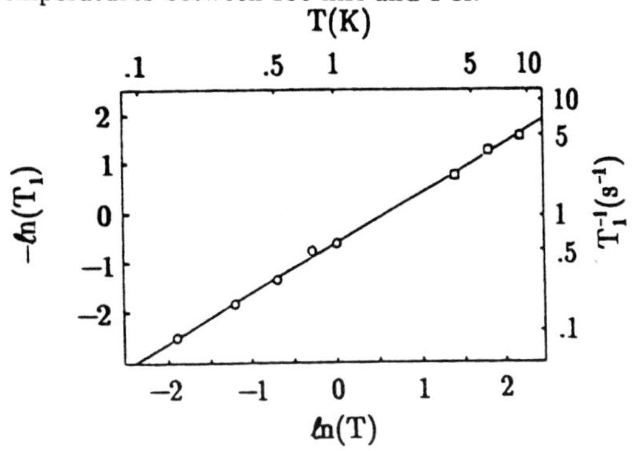

Fig. 2. **Plot of $-\ln(T_1)$ vs $\ln(T)$**

Fig. 2 shows a plot of $-\ln(T_1)$ versus $\ln(T)$. From a linear regression analysis using the expression $T_1 = AT^{-n}$ it was found that $A = 1.81 \pm 0.04$ and $n = 1.0 \pm 0.02$. In both plots the error bars were found to be too small to be shown.

It is clear that the T_1–T data follow the Korringa relation. This indicates that the dominant contributions to spin–lattice relaxation processes arise from interactions with the carriers at the Fermi surface (FS). Mitchell (3) has examined the contributions from contact (s–wave) and non– contact (p–wave) carriers for a spherical FS. We have extended these calculations to the FS of arsenic, which may be considered to consist of three ellipsoidal pockets of electrons and six half–ellipsoidal pockets of holes. The transition rate due to the contact part of the interaction may be written as:

$$W_m = \frac{2\pi}{\hbar}\left[\frac{\mu_0}{4\pi}\frac{8\pi}{3}\mu_B g \mu_N\right]^2 \sum_{kk'} |u_k(0)|^2 |u_{k'}(0)|^2$$

where the sum is over occupied (k) and unoccupied (k′) carrier states. W_m is related to the measured T_1 by the relation:

$$T_1 = \frac{1}{6\,W_m}$$

if we assume that non–contact terms make a negligible contribution (4). We can replace the sum by an integral in k–space and use FS parameters (α_1, α_2, α_3,E_F) quoted by Dresselhaus (5) to obtain

$$W_m = \left[\frac{\mu_0}{4\pi}\frac{8\pi}{3}\mu_B g \mu_N\right]^2 \frac{v_0^2\ m^3\ E_F}{\alpha_1\ \alpha_2\ \alpha_3}\frac{2\pi}{\hbar^7} <|u_k(0)|^2>^2_{E_F}\ kT$$

where v_0 is the atomic volume.

Using the value for A obtained from the T_1–T data we estimate

$$<|u_k(0)|^2>_{E_F} = 6.37 \times 10^{31} m^{-3} \equiv 9.44 \text{ a.u.}$$

We may compare this value with the probability density of the 4 s electrons at the nucleus for the free atom (6)

$$|\psi_{4s}(0)|^2 = 12.4 \text{ a.u.}$$

On this basis we conclude that the dominant contribution to relaxation is due to a contact interaction with the s–wave carriers at the nuclear site.

The system has potential applications in thermometry using the Korringa constant determined above. Advantages include ready availability of high purity arsenic which has a convenient NQR frequency (23 MHz) with no external magnetic field required. A disadvantage in the range below 100 mK is that T_1 becomes inconveniently long (~200 s at 10 mK). It is also necessary to show that the Korringa relation holds at very low temperatures.

4 CONCLUSION

An NQR method for minimizing heating of samples through the use of very short (fractional) rf pulses has been presented. Spin–lattice relaxation times for semimetallic As have been measured between 150 mK and 1 K and analysed together with previous data between 4 and 10 K. It is found that the Korringa relation applies for temperatures above 150 mK and that the dominant relaxation mechanism is due to a constant interaction with the s–wave carriers at the nuclear site. Extension of the present work to lower temperatures is envisaged.

REFERENCES

(1) J. Korringa, Physica 16, 601 (1950).
(2) T.P. Das and A.K. Saha, Phys. Rev. 98 (2), 516 (1955).
(3) A.H. Mitchell, J. Chem. Phys. 26, 1714 (1957).
(4) D.E. McLaughlin, J.D. Williamson and J. Butterworth, Phys. Rev. B 4, 60 (1971).
(5) M.S. Dresselhaus, The Physics of Semimetals and Narrow–Gap Semiconductors, Eds. D. L. Carter and R. T. Bate (Pergamon Press, Oxford, 1971) p.3.
(6) R.E. Watson and A.J. Freeman, Phys. Rev. 124, 1117 (1961).

Physica B 165&166 (1990) 29–30
North-Holland

THERMOMETRY BELOW 1K USING NMR FREQUENCY PULLING

Muneaki FUJII

Department of Physics, Kumamoto University, Kumamoto 860, Japan

At very low temperatures, nuclear magnetic resonance frequency decreases with decreasing temperature in an ordered magnetic medium. This phenomenon (frequency pulling) is caused by a coupled motion of the electronic magnetization and the nuclear magnetization. Application of the NMR frequency pulling to low temperature thermometry is discussed.

1. INTRODUCTION

In an ordered magnetic medium, we usually observe a NMR frequency;

$$\nu_0 = (A/h)\langle S^z \rangle \quad (1)$$

where A is a hyperfine coupling constant between an electron spin S and a nuclear spin I, and <> denote a thermal average. When nuclear spin polarization $\langle I^z \rangle$ is negligibly small at high temperatures, eq.(1) is a good approximation in many cases. However, at very low temperatures where the effective hyperfine field $A\langle I^z \rangle/\gamma_e h$ acting on the electron spin is comparable with the anisotropy field of the magnetic medium, we can find the deviations from eq.(1).[1] On solving the equations of motion for a nuclear spin I and an electron spin S, we find that the nuclear resonance frequency ν is given by

$$\nu \simeq \nu_0 - (A/h)\langle I^z \rangle(\nu_0/\nu_F) \quad (2)$$

where ν_F is the ferromagnetic resonance frequency of electron. Since the shift of ν from ν_0 becomes evident at low temperatures when $\langle I^z \rangle$ becomes large in proportion to T^{-1}, this frequency pulling effect is observed in many magnetic compounds below 1K. However, this effect can be seen above 1K if the hyperfine coupling A is large and the anisotropy field is small.[2,3,4,5,6,7]

In the present study some magnetic compounds were examined by a conventional spin echo method, in order to develop the pulling effect as an NMR thermometer for use below 1K.

2. EXPERIMENTAL

The experimental apparatus is schematically drawn in fig.1. This is the lower part of a dilution refrigerator specially designed and precisely constructed for NMR measurements of magnetic compounds. The mixing chamber(A) and two sample cells(B,C) are all made of STYCAST1266 to prevent an eddy current heating and a reduction in the Q factor of the r.f. circuit(H). The carbon resistors(D,E,F) and a sample(G) are immersed in the dilute phase of the ^3He-^4He mixture in order to achieve a good thermal contact between the thermometers and the sample. Variable capacitor(J) can be adjusted from the top flange with a Cu-Ni connecting rod(I). The carbon resistors (MATSUSHITA 1/4watt) were calibrated against the susceptibility of CMN. For our primary temperature standard, we used the superconductive thermometric fixed point device supplied from N.B.S..

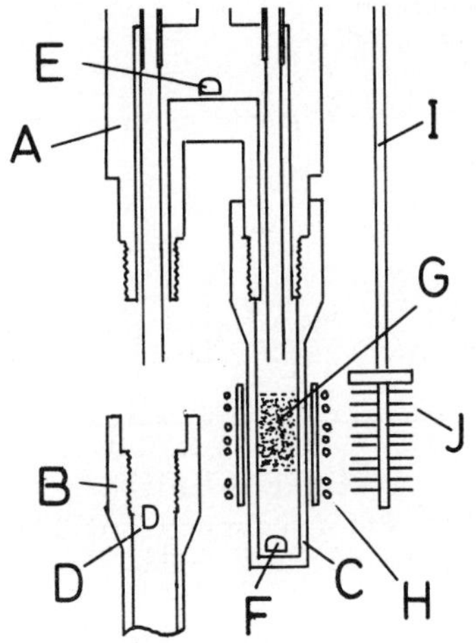

Fig.1
Schematic drawing of the apparatus

0921-4526/90/$03.50

3. RESULTS AND DISCUSSION

3.1. Case(A); Only one NMR mode coupled with ESR mode.

The NMR frequency of ^{153}Eu in EuSe was measured for various temperatures from 1.2K down to 50 mK, which is shown in fig.2. As the temperature decreases, the ^{153}Eu spin echo spectrum becomes very broad making it very difficult to assign sharp resonance frequencies. Thus this substance is not suitable for thermometry.

3.2. Case(B);Two coupled NMR modes

The temperature dependencies of the NMR frequencies of ^{51}V in V_3O_7, measured down to 30 mK, are shown in fig.3. The numerically calculated result based on eq.(2) are shown by the dotted curve and those based on the equations of motion two nuclear spins coupled with an electron spin by the hyperfine coupling constants A_1 and A_2 ($A_1 \simeq A_2$) respectively are shown by the solid curve. When the temperature is lowered, the ^{51}V spin echo spectrum of α-signal becomes sharp whereas that of β-signal becomes broad. This phenomenon has been discussed in detail in the other paper[8]. By measuring the resonance frequencies of α-signal, we can determine the temperatures. Since the half-value width of α-signal is about 0.1 MHz, the sensitivity of this thermometer $\Delta T/T \simeq 0.05$ at 50 mK. Its response time is short because when the temperature of the dilution refrigerator changes ($\dot{T} \simeq$ 1mK/sec) the temperature dependencies of the NMR frequencies agree with fig.3 in the experimental error. The energy input due to measurement of temperature is less than 1 nW.

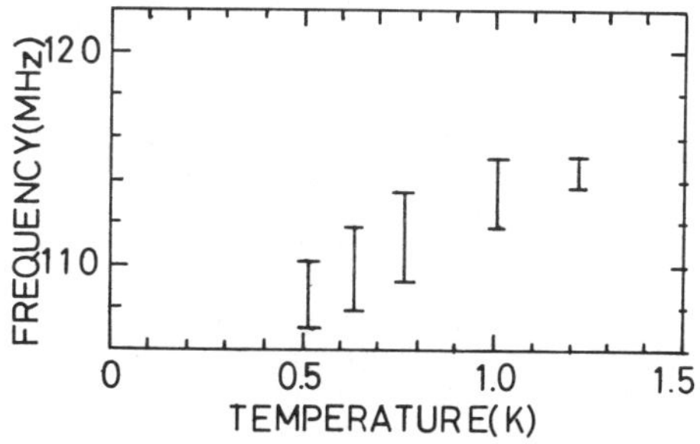

Fig.2 The NMR frequency of ^{153}Eu in EuSe

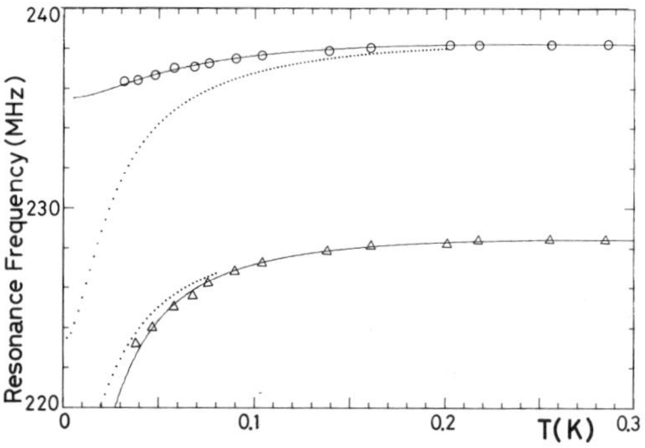

Fig.3 Resonance frequencies of ^{51}V in V_3O_7 as a function of T. Experimental result: o;α-signal, Δ;β-signal.

4. CONCLUSIONS

By measuring the NMR frequencies of ^{51}V in V_3O_7, we can use this magnetic insulator for the thermometer in the millikelvin temperature range. This thermometer has advantages as follows;

(1) Good reproducibility
(2) Short response time
(3) Being equilibrium with the body whose temperature is to be measured
(4) No lead wire which introduce the heat from outside

REFERENCES

(1) P.G. de Gennes, P.A. Pincus, F. Hartmann-Boutron and J.M. Winter, Phys.Rev. 129 (1963) 1105
(2) A.J. Heeger, A.M. Portis and G.L. Witt, J.Appl.Phys. 34 (1963) 1052
(3) G.L. Witt and A.N. Portis, Phys. Rev. 136A (1964) 1316
(4) H. Fink and D. Shaltiel, Phys.Rev. 136A (1946) 218
(5) B.S. Dumesh, JETP Lett. 23 (1976) 14
(6) Yu.M. Bun'kov and B.S. Dumesh, Sov.Phys.JETP 41 (1976) 576
(7) Yu.M. Bun'kov and S.O. Gladkov, Sov.Phys.JETP 46 (1978) 1141
(8) M. Fujii, H. Nishihara, A. Hirai, J.Phys.Soc.Jpn. 52 (1983) 272

Physica B 165&166 (1990) 31–32
North-Holland

CONSISTENCY OF LOW TEMPERATURE SCALES AND THE TEMPERATURE OF THE He-3 MELTING PRESSURE MINIMUM

Gerhard SCHUSTER, Dieter HECHTFISCHER, Wolfgang BUCK and Albrecht HOFFMANN

Physikalisch-Technische Bundesanstalt, Institut Berlin, Abbestr. 2-12, 1000 Berlin 10, Germany

Two independent low temperature scales, based on noise thermometry below 0.6 K and on extrapolation from 1.5 K using magnetic thermometry, have been realized and compared. Both scales coincide at the pressure minimum of the He-3 melting curve within ±0.3 mK and determine its temperature as 315.1 mK.

1. INTRODUCTION

A low temperature scale can be obtained by either extrapolating an existing scale towards lower temperatures or making direct measurements with a primary thermometer. Both methods may fail because of an unknown additional temperature dependence of the extrapolation thermometer or an unexpected and, consequently, uncorrectable systematic error of the primary thermometer. It is, therefore, reasonable to apply both methods to obtain a realistic view of the uncertainties involved. The motive for the investigations described below was a discrepancy of about 1 mK, observed in the range 0.1 K to 0.5 K between temperatures indicated by a noise thermometer and a commercial powder CMN thermometer that had been calibrated at EPT-76 in the interval 0.6 K to 0.9 K. The temperature dependence of the deviation suggested that it was caused by an error introduced during the calibration of the CMN thermometer, but the actual reason could have been a nonlinearity of either the scale or the thermometer itself, because EPT-76 is uncertain within ±1mK at 0.5 K, and the CMN was known to deviate strongly from a Curie law above 1 K.

2. EXPERIMENTS

In order to resolve the discrepancy, experiments began with a check of the internal consistency of EPT-76. The first step was to exclude a scale ambiguity by comparing three versions of EPT-76 (from NPL, KOL and NBS (now NIST))(1) available on calibrated rhodium-iron resistors or calibrated superconductive fixed point devices. All temperatures agreed within ±0.2 mK at 0.5 K and ±0.1 mK above 0.8 K. Several attempts to check the linearity of the scale with powdered CMN were not successful due to large deviations from a Curie law which required use of a single crystal sphere. To prevent potential deviations from the Curie-Weiss law, 1) any superconductive shielding was avoided, 2) the temperature dependence of the empty coil of the mutual inductance bridge was determined which resulted in a correction of 0.5 mK at 1.8 K and negligible below 1.5 K, and 3) two crystals of different materials were used. The first crystal, grown relatively fast from standard chemicals, was cloudy, the second one, made of high purity materials and grown slowly, was clear. The temperature obtained with the less pure material increased only by 0.2 mK above that of the pure material at 0.1 K when both were calibrated above 1 K.

Calibration was made by determination of two constants of a Curie-Weiss law by a fit to EPT-76 between 1.2 K and 1.8 K, whereas the Weiss-constant of 0.23 mK was taken from Hudson and Pfeiffer (2). The uncertainty of the fit to 100 data points, with an rms deviation of 0.04 mK, was negligible compared with the uncertainty of 0.5 mK of EPT-76 in this temperature range. With temperature falling below this range, however, EPT-76 deviates increasingly from the CMN temperatures and the difference reaches a value of +1 mK at 0.5 K (Fig.1), whereas the CMN and noise temperatures agree within (0.1 ± 0.3) mK. These results were obtained with a Josephson junction noise thermometer (3) which runs with a pump frequency of 300 MHz. The carrier frequency of the noise modulation is 70 kHz with an amplifier bandwidth of also 70 kHz, whereas the noise bandwidth - for a gate time of 3 ms - amounts to only 170 Hz, i.e. the noise temperature was not reduced by a bandwidth limitation. This was also confirmed by checks at several temperatures with a gate time of 20 ms corresponding to a noise bandwidth of 25 Hz.

With falling temperatures, however, the CMN and noise scales deviate again reaching a discrepancy of 0.5 mK (Fig.2). This could mean e.g. that an additional - nonthermal - noise increases the apparent noise temperature or that the CMN does not follow the expected Curie-Weiss law. A clue to the problem was found by comparison measurements of the noise thermometer and a platinum NMR-thermometer in the temperature range 10 mK to 100 mK. Although NMR should follow a Curie law, it was only possible to fit a Curie-Weiss law to the noise temperatures with a Weiss constant which corresponds - in magnitude and direction - to the difference between noise and CMN temperatures.

The temperature dependence of the discrepancy is not necessarily caused by the noise

0921-4526/90/$03.50

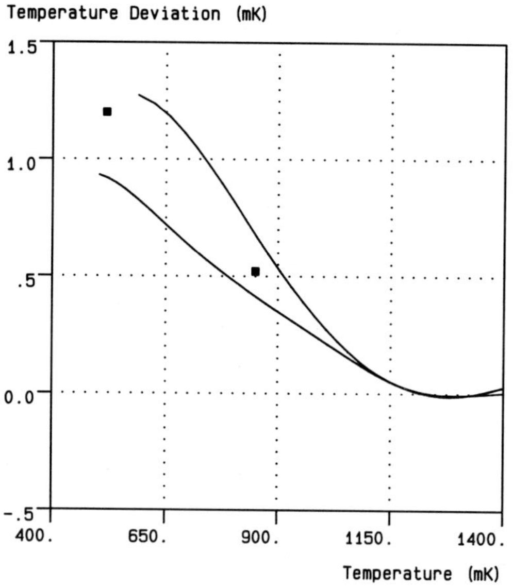

FIGURE 1
Temperature deviation of two rhodium-iron resistance thermometers and two superconductive fixed points, calibrated on EPT-76, from single crystal CMN thermometer calibrated with the same thermometers in the range 1200 mK to 1800 mK.

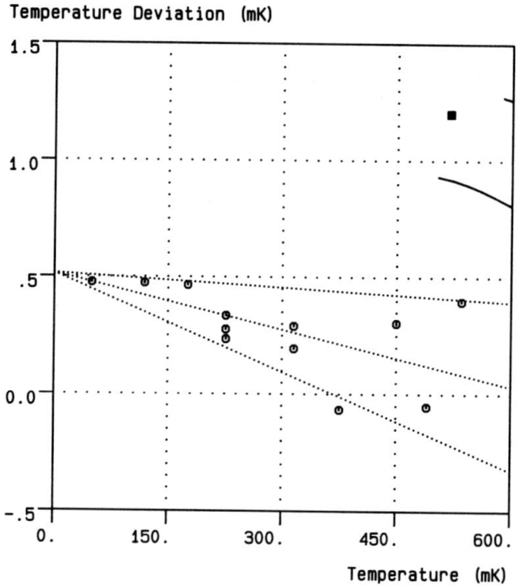

FIGURE 2
Temperature deviation of noise thermometer from single crystal CMN thermometer calibrated on EPT-76 between 1200 mK and 1800 mK. Dotted lines are the mean deviation and ±0.06% limits. Data in upper right corner are those of Fig.1.

thermometer. It could be explained as well by a constant offset of the noise thermometer due to parasitic signals and an offset of +0.6 mK of EPT-76 at 1.5 K compared to thermodynamic temperature. However, since the possible error of the noise thermometer could not be eliminated up to now, it must be regarded as an uncertainty of the combined noise and CMN scale.

3. CONCLUSION

In order to make these results available for comparison with other temperature scales below ETP-76 the pressure minimum of the He-3 melting-curve is a very useful temperature fixed point because it does not require the exchange of calibrated devices between laboratories. Its temperature has been determined 16 times over a time interval of half a year with two physically different melting curve thermometers. The measurement of the relative pressure variations with stabilized temperatures at several 0.5 mK intervals below and above the minimum resulted in an rms deviation of the individual temperatures of 0.060 mK and an average temperature of 315.1 mK on the CMN scale. The random variations are negligible compared with the systematic deviation of the noise thermometer scale which introduces an uncertainty of ±0.3 mK, depending on whether the offset of 0.5 mK determined at low temperatures is due to the noise thermometer and may be subtracted, and whether the deviation of EPT-76 from thermodynamic temperature at 1.5 K is +0.6 mK, a value of the order of its own uncertainty. To exclude gross errors due to a He-4 impurity of the He-3 filling of the melting curve thermometer, the temperature of the mimimum has also been determined after an admixture of 0.1 % He-4. Although difficult to manage experimentally (due to extremely long time constants) the result was the same within the ±0.3 mK error limits.

REFERENCES

(1) The authors gratefully acknowledge that calibrated devices have been made available by R. L. Rusby (National Physical Laboratory, Teddington), M. Durieux (Kamerlingh Onnes Laboratory, Leiden) and J. F. Schooley (National Institute of Standards and Technology, Gaithersburgh).

(2) R. P. Hudson and E. R. Pfeiffer, Temperature, its Measurement and Control in Science and Industry, Vol.4, Instrum. Soc. America, Pittsburgh (1972) p. 1279.

(3) A. Hoffmann and B. Buchholz, J. Phys. E: Sci.Instrum. 17 (1984) 1035.

Physica B 165&166 (1990) 33–34
North-Holland

DEVELOPMENT OF A TEMPERATURE SCALE BELOW 0.5 K

William E. FOGLE, Jack H. COLWELL and R.J. SOULEN, Jr.

National Institute of Standards and Technology, Gaithersburg, MD 20899, USA

We summarize our most recent results on the development of an absolute temperature scale below 0.5 K. It is the most recent of several experiments at our laboratory and represents both our best effort and the one in which all available thermometers were simultaneously and fully operational. Implications of our results for other work will be dlscussed.

1. INTRODUCTION

The thermometers were used:

1) *A Josephson junction noise thermometer,* used to provide the absolute temperature, T_N, from 8mK to 520mK.

2) *A powdered CMN thermometer*, in liquid ^{3}He and measured with a SQUID, used to provide the interpolation temperatures, T_{CMN}, from 8mK to 800 mK.

3) *A ^{3}He melting curve thermometer*, MCT, provided by D. Greywall, used to realize the ^{3}He melting curve pressures, P_m.

4) *Two Rh-Fe resistance thermometers* (RIRTs), calibrated vs EPT-76, used to realize the EPT-76 temperature scale, T_{76}.

5) A *SRM 767a Superconductive Fixed point unit,* used as alternative realization of T_{76}.

6) *A SRM 768 Superconductive fixed point unit*, used to realize the temperature scales below 500mK (NBS CTS-1, and NBS-CTS/83) previously developed at NIST.

2. EXPERIMENTS

We studied the reproducibility of the thermometers at several temperatures. (1) We found that the diaphragm gage used to register the MCT was reproducible to the sub-ppm level throughout the experiment. This was the case even though the MCT was refilled several times during the experiment. (2) We realized the minimum of the melting curve a number of times (~20) with a pressure reproducibility of 0.7ppm. (3) We used a CMN thermometer to achieve precision of 10 ppm from 0.5K-0.7K, and higher sensitivities below 0.5K. (4) We found that the noise thermometer was reproducible to its statistical measurement imprecision (0.1%).

We also studied the inaccuracies of the thermometers. (1).The inaccuracy of the piston gage used to calibrate the capacitance gage was 30 ppm (1mbar). (2) We measured the factors contributing to the inaccuracies of the noise thermometer and conclude that the total systematic error in the noise temperature (T_N) is < 0.1%. (3) The EPT-76 temperature scale, which forms the basis for our RIRT calibrations, is estimated to have a 1mK (0.2%) uncertainty at 0.5K.

TABLE I

T (mK)	T_{CMN} (mK)	%T Difference	
699.847	699.840	+0.0010	'
691.948	691.946	+0.0004	'
683.120	683.116	+0.0006	'
667.169	667.179	-0.0014	EPT-76
657.997	658.006	-0.0014	vs
642.089	642.100	-0.0017	CMN
632.991	633.009	-0.0029	'
616.902	616.912	-0.0017	'
608.028	608.028	+0.0000	'
592.101	592.101	-0.0000	'
583.044	583.069	-0.0042	'
566.975	567.004	-0.0051	'
558.119	558.151	-0.0057	'
540.098	540.134	-0.0067	'
529.967	529.984	-0.0032	'
520.100	519.895	+0.0395	'
204.580	204.613	-0.0162	NOISE
160.340	160.300	+0.0250	vs
99.471	99.508	-0.0374	CMN
44.975	44.967	+0.0169	'
28.329	28.330	-0.0021	'
22.783	22.775	+0.0341	'
18.054	18.059	-0.0273	'

We fitted a Curie-Weiss function to the measured susceptibility, M, of the CMN thermometer from 18mK to 700mK (Table1)

Where

$M = B + C/(T-\Delta)$; $T_{CMN} = \Delta + C/(M-B)$

C=7.949308 ; B=5.932242E-3; Δ= 25.02 μK

Below 18 mK, the CMN susceptibility deviated from the Curie-Weiss law.

We plot $P_m(T)$ in Fig. 1, where the T scale used is T_N for T< 520mK and EPT-76 for T>520mK.

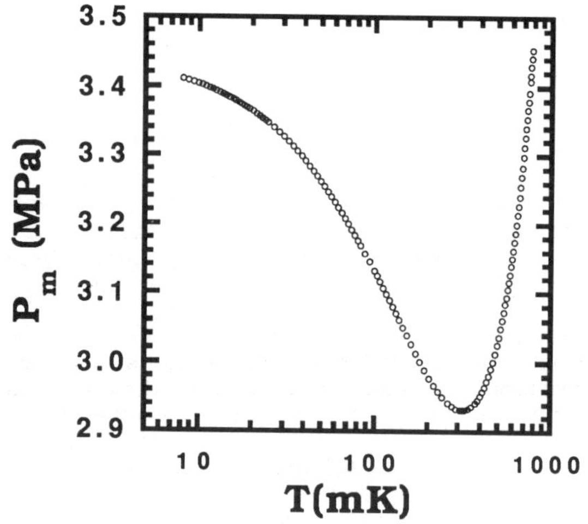

FIGURE 1
This curve consists of 137 separate data points extending from 8mK to 770mK. Not shown are numerous repeat points which checked reproducibility, sensitivity of P_m to filling pressures, etc. Preliminary analysis indicates that the points form a very precise and smooth curve.

To date we have compared the melting P-T relation obtained at NIST as shown above with those published by two other laboratories: AT&T Labs (1) and Cornell (2). The results are displayed in Fig. 2. Relative to the noise thermometer scale, the Halperin scale yields high temperatures, the disparity increasing rapidly at lower temperatures. Between 15mK and 250mK, Greywall's scale agrees well with the NIST scale-a fact not surprising since it is based on temperatures defined by a NIST superconductive fixed point device (the T_c values, ranging from15-204 mK, are shown in the figure).

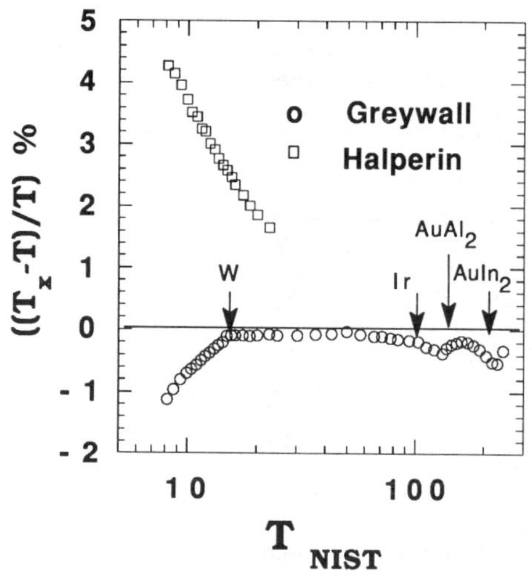

FIGURE 2
Comparison of three MCT scales

4. CONCLUSION

On the basis of the comparison of the ^3He melting curve between NIST and AT&T, we believe that that temperature scales may be successfully transferred between laboratories by combining superconducting fixed points, CMN interpolation and melting curve pressure measurements.

REFERENCES

(1) D. Greywall, Phys. Rev. **33**,7520 (1986)
(2) W.P. Halperin et. al., J. Low Temp. Phys. **31**,617 (1978).

Physica B 165&166 (1990) 35–36
North-Holland

THE INTERNATIONAL TEMPERATURE SCALE OF 1990 AT LOW TEMPERATURES

R L Rusby

Division of Quantum Metrology, National Physical Laboratory
Teddington TW11 OLW, United Kingdom

The basis of the International Temperature Scale of 1990, ITS-90, at low temperatures is outlined. The vapour pressure equations specified for ^{3}He and ^{4}He are given and the prospects for a future extension of the scale to millikelvin temperatures are reviewed.

The International Temperature Scale of 1990, ITS-90 (1), came into effect on 1 January 1990. It replaced the International Practical Temperature Scale of 1968, IPTS-68 (2), and the 1976 Provisional 0.5 K to 30 K Temperature Scale, EPT-76 (3). ITS-90 extends from 0.65 K up to temperatures limited only by the practical application of the Planck law of radiation, well in excess of 2000°C. Values of temperature on ITS-90 have been adopted so as to agree with recent thermodynamic determinations, and many advances in practical techniques developed in the last 20 years have been incorporated.

For the range from the triple point of water, 273.16 K, to the triple point of equilibrium hydrogen, 13.8033 K, ITS-90 is realised using helium-filled capsule-type platinum resistance thermometers of high temperature coefficient. These are calibrated at the triple points water, mercury, argon, oxygen, neon and hydrogen, and at two vapour pressure points of hydrogen near 17 K and 20.3 K. The boiling points of neon, oxygen and water, which were required in IPTS-68, have been replaced by more reproducible triple points. New interpolation formulae are prescribed so as to reduce the 'non-uniqueness' of the scale, that is to say, the differences which arise between different platinum thermometers even when they are calibrated identically. The reference function, which relates resistance ratio to temperature, is based on recent gas thermometry (4,5,6) and total radiation thermometry (7), with some confirmatory data from other sources (8).

Temperatures below the triple point of neon

The ITS-90 is realised within the range 3 K to the triple point of neon (24.5561 K) using an interpolating constant volume gas thermometer calibrated at the triple points of neon and hydrogen, and at a helium vapour pressure point in the range 3 K to 5 K. The overlap with the resistance thermometer range (from 13.8033 K to 24.5561 K) allows the gas thermometer to provide the 17 K and 20.3 K calibration points instead of using hydrogen vapour pressure measurements.

Gas thermometers (constant volume, dielectric constant and acoustic) and magnetic thermometers have been used at various times to develop temperature scales below 24.5 K, but it is only since the advent of germanium and rhodium-iron resistance thermometers that it has been possible to maintain and compare temperature scales in the helium-to-hydrogen range with the precision of the original measurements. The EPT-76 was introduced following an inter-comparison of such scales. The decision to specify the use of the gas thermometer in ITS-90, rather than a magnetic thermometer, reflects greater confidence in its specification and reproducibility. A thermodynamic instrument was preferred over the more 'practical' rhodium-iron resistance thermometer because although it has been shown to be technically

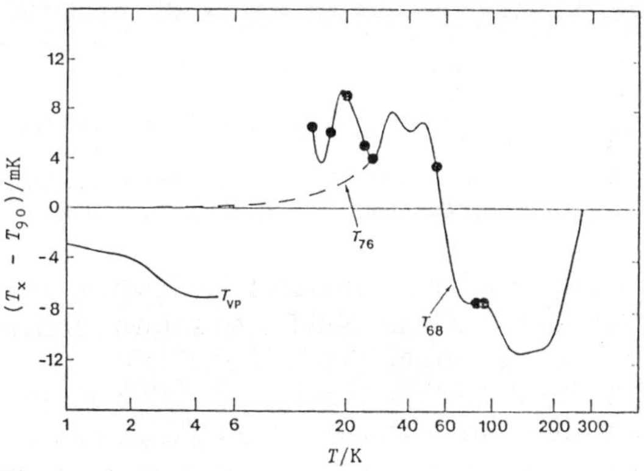

Figure:
Differences between earlier temperatures scales, T_x, and ITS-90, T_{90}, in the range below 273 K. T_x is variously T_{VP}, the helium vapour pressure scales of 1958 (for ^{4}He) and 1962 (for ^{3}He), T_{76}, the EPT-76, and T_{68}, the IPTS-68.

feasible to use the latter for interpolation (9), given a superconductive fixed point for lead (at 7.2 K) or niobium (9.4 K), the limited availability of the thermometers (and of the superconductive devices) was a concern. A definition based on a gas thermometer is both universal and secure, and once realised the scale can be maintained using such secondary thermometers as are available and appropriate.

At lower temperatures ITS-90 specifies vapour pressure equations for ^{3}He (from 0.65 K to 3.2 K) and for ^{4}He (from 1.25 K to the λ-point, 2.1768 K, and from the λ-point to 5 K), with the form

$$T_{90}/\mathrm{K} = \sum_{i=0}^{9} A_i[\{\ln(p/\mathrm{Pa}) - B\}/C]^i$$

These were originally developed for the EPT-76 (10) and the constants, A_i, are given in the Table. The lower limit of 0.65 K for the ^{3}He equation, at which the vapour pressure is 116 Pa, was chosen because of the practical difficulties (chiefly the thermomolecular effect and low sensitivity) of using vapour pressure thermometers at lower temperatures, and in order not to prejudice the future extension of the scale by other means. The equation is in fact valid to 0.5 K, and an earlier equation (in which $\ln(p/\mathrm{Pa})$ was expressed as a function of T) is valid down to 0.2 K (11).

Future developments

With the adoption of ITS-90 a thermodynamically smooth and coherent temperature scale exists down to 0.65 K, which has removed the inconsistencies between earlier scales and should meet the needs of low temperature research and technology for many years to come. Accurate thermodynamic measurements have been made at lower temperatures by noise and nuclear orientation thermometry, and the ^{3}He melting curve has been determined with good precision. (12). As a result of this work a consensus may emerge for specifying the melting curve as the basis for an extension of ITS-90 as far as the ^{3}He-A transition at 2.8 mK, notwithstanding the minimum in the characteristic near 300 mK. How soon this may come about will depend on progress, and the close involvement of those in the field will be required.

References

(1) ITS-90, Metrologia, in print.
(2) IPTS-68, Metrologia 12 (1976) 1-17.
(3) EPT-76, Metrologia 15 (1979) 65-68, and M. Durieux, D.N. Astrov, W.R.G. Kemp and C.A. Swenson, Metrologia 15 (1979) 57-63.
(4) P.P.M. Steur and M. Durieux, Metrologia 23 (1986) 1-18.
(5) R.C. Kemp, W.R.G. Kemp and L.M. Besley, Metrologia 23 (1986) 61-86.
(6) D.N. Astrov, L.B. Beliansky, Y.A. Dedikov, S.P. Polunin and A.A. Zakharov, Metrologia 26 (1989) 151-166.
(7) J.E. Martin, T.J. Quinn and B. Chu, Metrologia 25 (1988) 107-112.
(8) R.P. Hudson, M. Durieux, R.L. Rusby, R.J. Soulen and C.A. Swenson, Document CCT/87-9, Bureau International des Poids et Mesures, 92312 Sèvres, France.
(9) R.L. Rusby, in Temperature, Its Measurement and Control in Science and Industry, Vol. 5, ed J.F. Schooley (AIP, New York, 1982) pp. 829-833.
(10) R.L. Rusby and M. Durieux, Cryogenics 24 (1984) 363-366.
(11) R.L. Rusby and M. Durieux, Proceedings of LT-17, eds U Eckern, A. Schmid, W. Weber and H. Wühl (Elsevier Science Publishers, 1984), pp 399-400, and R.L. Rusby, J Low Temp. Phys. 58 (1985) 203-204.
(12) G. Schuster, A. Hoffmann, L. Wolber, W. Buck and J.-F. March, Proceedings of LT-17 (see Ref. 11), pp403-404.

Table: Constants in the helium vapour pressure equations

Constant	^{3}He 0.65 K to 3.2 K	^{4}He 1.25 K to 2.1768 K	^{4}He 2.1768 K to 5 K
A_0	1.053447	1.392408	3.146631
A_1	0.980106	0.527153	1.357655
A_2	0.676380	0.166756	0.413923
A_3	0.372692	0.050988	0.091159
A_4	0.151656	0.026514	0.016349
A_5	-0.002263	0.001975	0.001826
A_6	0.006596	-0.017976	-0.004325
A_7	0.088966	0.005409	-0.004973
A_8	-0.004770	0.013259	0
A_9	-0.054943	0	0
B	7.3	5.6	10.3
C	4.3	2.9	1.9

Physica B 165&166 (1990) 37–38
North-Holland

Superconducting Transition Thermometry for ^{4}He Lambda Point Experiments

Melora Larson, Guenter Ahlers.

Department of Physics, University of California and CNLS, Santa Barbara, CA 93106, U.S.A.

We report on a high resolution thermometer based on the change in resistance of a metal film near its superconducting transition. This thermometer has an observed absolute temperature resolution of 10^{-8}K. The operating temperature for these thermometers is tuned by an external magnetic field. The temperature resolution is limited by the combination of the thermal noise and the voltage dependence of the resistance in the transition region. Future improvements by an order of magnitude in the resolution seem feasible.

1. Introduction

Thermometers with resolution of a nK or so are needed for the detailed study of the superfluid transition in ^{4}He. For specific heat studies of small samples, for example of ^{4}He in a confined geometry, the devices should have a negligible heat capacity. The toroidal susceptibility thermometers described previously (1) have a relatively large heat capacity since they contain a magnetic salt near its Curie point. Therefore we have developed a Superconducting Transition Thermometer (STT) which is similar to the second-sound detectors (bolometers) previously used in our group. These devices consist of lead-gold thin film resistors with their superconducting transition temperature adjusted by an applied magnetic field (2, 3). The resistance-temperature characteristics of the bolometers suggest that nK resolution should be achievable with the STT's. We report here on the properties of the STT's.

2. Manufacture

We prepared the STT's in pairs in order to have two thermometers with very similar characteristics. The sensing element of the STT's consisted of a lead-gold resistor in a meander pattern of 32 lines, 203μm wide and 203μm apart (Fig. 1). The mask for the resistor pattern was created using standard photolithographic techniques. The resistor made electrical contact with two 200nm thick gold electrodes. It consisted of 50nm of lead evaporated on to 23nm of gold. We used quartz windows 2.54cm in diameter and 0.051cm thick as the substrates for the STT's.

To protect the fragile films of the STT's and to inhibit the oxidation of the lead, we coated each STT with 180nm of MgF_2. This coating made the STT's mechanically robust. The finished slides were then epoxied on to copper posts that could be screwed directly into any experimental platform. The finished thermometers each had resistances of about 25kΩ at room temperature and about 4.35kΩ immediately above their superconducting transition (in zero field, T_c = 2.49 K and 2.51 K for the pair discussed in this report).

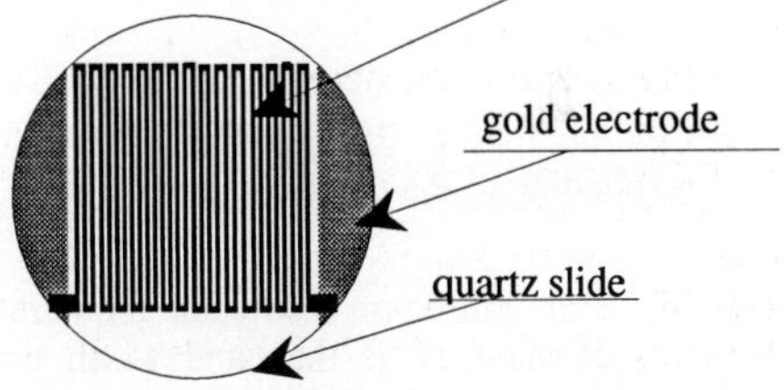

Figure 1: Schematic drawing of STT.

3. Thermometer Characteristics

The STT's were mounted on a thermally isolated platform. Their resistance was calibrated against a previously calibrated germanium thermometer also mounted on the platform. The results are shown in Fig. 2(a). By varying the applied magnetic field from 0 to 650 gauss, we could move the transition temperature of the STT's from 2.5 K to 1.7 K.

For a constant drive voltage, V, the temperature resolution (δT) of the STT's is related to the signal voltage (δV) by

$$\delta V = \alpha_R V \ \delta T , \qquad (1)$$

where $\alpha_R = (1/R)\ dR/dT$ is the figure of merit for the thermometer. Figure 2(b) illustrates how α_R changes as the magnetic field is increased. As the transition is shifted down in temperature, α_R decreases. For example, from 2.17 K (250 gauss) to 1.88 K (467 gauss) α_R at the middle of the transition decreased from 77 K^{-1} to 60 K^{-1}.

The resolution of the STT's is limited by the thermal noise and the figure of merit of the devices. This relation can be expressed as

$$\delta T = \alpha_R^{-1} \left(4\, k_B\, T\, \Delta f / P \right)^{1/2} , \qquad (2)$$

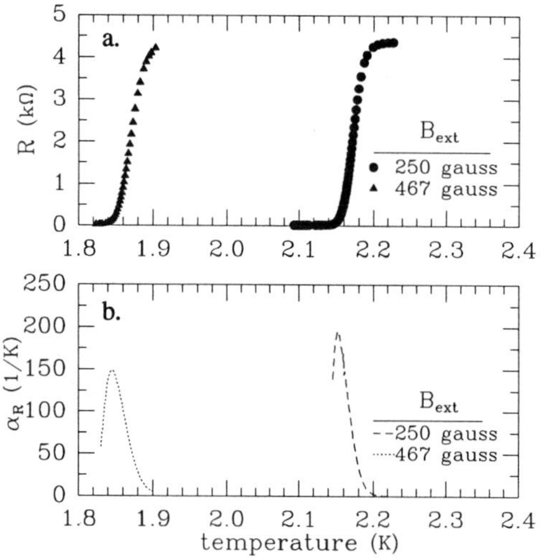

Figure 2: (a) Transition curves for an STT at two external fields. (b) Figure of merit curves corresponding to the transition curves in (a).

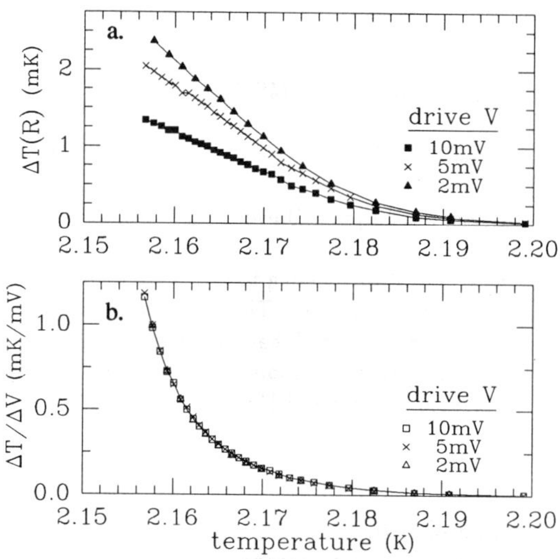

Figure 3:(a) $\Delta T = T(V) - T(20mV)$ at constant R along the transition. (b) $\Delta T / \Delta V$ at constant R through the transition. ΔV is the change in the voltage across the STT from a drive of 20mV.

where δT is the minimum resolvable temperature, k_B is Boltzman's constant, Δf is the band width used for the measurement, and P is the power dissipated in the thermometer (4). The STT's shown in Fig. 2 should be limited to 1nK resolution at the center of their transition when dissipating 10nW.

In order to observe the noise level of the STT's, we kept the power into the platform constant, causing a temperature drift of about 8.5µK/hour. Under these conditions, we observed short term noise in the STT's of 10nK rms for one and 19nK for the other. Due to high lead resistances in our calibration cryostat, about half of the system's Johnson noise was contributed by the leads. Thus, we believe that the noise level observed in the STT's will drop when they are used in an experimental cryostat with lower resistance leads.

Unfortunately, the STT's exhibit a strong dependance on the drive voltage across a thermometer bridge containing the STT and a reference resistor (2.5kΩ). We are interested in the temperature shifts induced by fluctuations of this bridge voltage when we regulate an experimental platform using the STT's, so we are interested in temperature shifts at constant resistance. Figure 3(a) shows the temperature difference at constant R between a drive voltage of 20mV and three lower bridge drive voltages. To see how these shifts depend on the voltage drop across the STT, we divided the temperature shift by the change in this voltage. As can be seen in Fig. 3(b), the three $\Delta T/\Delta V$ curves collapse together. Since the temperature shifts are linear in the voltage across the STT, these shifts are not due to self heating in the thermometers.

These temperature shifts can be analyzed to calculate the dependance of the shift on drifts in the drive voltage. For example, the STT in Fig. 3, centered near 2.17K, showed an apparent shift of 60µK/mV at the center of the transition. So, for 1nK resolution, we must compensate for any drift in the drive voltage greater than 17nV.

4. Conclusion

We believe these STT's can be used to resolve absolute temperatures on the nano-Kelvin level when drifts in the drive voltage are compensated. We have observed as little as 10nK rms noise in our thermometers. The STT's small thermal size and smooth heat capacity, in contrast to the high resolution CAB thermometers (1, 5), make the STT's better suited for high resolution specific heat studies.

This work was supported by National Science Foundation grants DMR 84-14804 and DMR 89-18393.

References

(1) R. V. Duncan, PhD dissertation, University of California, Santa Barbara (1988), unpublished.

(2) R Mehrotra and G. Ahlers, Phys. Rev. B **30** (1984) 5116.

(3) L. Goldner, G. Ahlers, and R. Mehrotra, submitted to Phys. Rev. B.

(4) V. Steinberg and G. Ahlers, J. Low Temperature Phys **53** (1983) 255.

(5) J. A. Lipa, B. C. Leslie, T.C. Wallstrom, Physica Amsterdam **107B** (1981) 331.

Physica B 165&166 (1990) 39–40
North-Holland

GRANULAR NICKEL FILMS AS CRYOGENIC THERMOMETERS

B.M. PATTERSON, J.R. BEAMISH and K.M. UNRUH

Department of Physics, University of Delaware, Newark, Delaware 19716

We have used RF sputtering to prepare a series of granular $Ni_x(SiO_2)_{100-x}$ films with nickel concentrations, x, between 60 and 90 atomic percent. The electrical resistivity of such films depends strongly on composition. We find that films with about 65 percent Ni have a temperature dependence suitable for use as cryogenic temperature sensors down to at least 100 mK. In addition, these films are relatively insensitive to magnetic fields. The largest resistance change we measured was about 1.5% in a field of 6 T at 200 mK, corresponding to a 4% temperature shift.

1. INTRODUCTION

The convenience of resistance sensors has led to their widespread use as secondary thermometers at low temperatures (1). An ideal thermometer would be stable and sensitive over a wide temperature range. For some applications it is important that it also be unaffected by magnetic fields and have a small mass and rapid response time. No commonly used sensor meets all these requirements. Germanium sensors are sensitive to magnetic fields, carbon composition resistors have poor reproducibility, and carbon glass sensors, although stable and relatively insensitive to magnetic fields, have resistances too high for use below 1 K.

Recently there have been reports that ruthenium based "thick film resistors" (2) or zirconium nitride thin films (3) may be useful at low temperatures. The ruthenium devices are made by a proprietary process and are not designed as temperature sensors. The ZrN films may be useful above 2 K, but have substantial magnetoresistance at lower temperatures and may not be sufficiently stable over long periods of time (4). There has also been a report (5) that some granular platinum/alumina films ("cermets") have appropriate temperature characteristics and small magnetoresistances, although such films have not yet been investigated over a range of compositions.

In this paper, we discuss the use of granular nickel/silica films for use as low temperature thermometers. Measurements made between room temperature and 100 mK and in magnetic fields up to 6 T suggest that, by choosing an appropriate composition, useful thin film cryogenic temperature sensors can be made.

2. EXPERIMENTAL METHOD AND RESULTS

For these measurements we deposited films about 1000 Å thick onto room temperature sapphire substrates by RF magnetron sputtering. By using composite targets (mixtures of Ni and SiO_2 powders), $Ni_x(SiO_2)_{100-x}$ films were made with nickel concentrations, x, from 60 to 90 atomic percent. Copper electrodes were deposited and the samples were mounted inside a helium filled copper cell attached to the mixing chamber of a dilution refrigerator. A transverse magnetic field of up to 6 T could be applied to the films. Temperatures were measured using sensors outside the strong magnetic field region. Sample resistances were measured with standard 4-wire AC bridge techniques at power levels where Joule heating was insignificant (below 1 pW at the lowest temperatures).

As the Ni content of the films was decreased from 90 to 60 percent, the room temperature resistivity increased by a factor of about 10^4. In films with less than 70 percent Ni, the resistance increased monotonically as the temperature was lowered and was very sensitive to composition. Films with about 65 percent Ni (room temperature resistivities, ρ, around $6\times10^{-4}\Omega$-m) proved to be the most suitable for cryogenic thermometry. Figure 1 shows the resistance per square (the actual device resistance was about 25 times smaller) of three films with nominal Ni concentrations of 65, 66 and 67 percent. The two lower resistance films have a roughly constant sensitivity d(log R)/d(log T) between 20 and 2 K. However, in contrast to most semiconductor sensors, these films show a crossover to a weaker temperature dependence below 1 K. We see no indication of any upturn in the resistivity, indicating that such films may be useful down to considerably lower temperatures. We have measured the low temperature magnetoresistance of a number of such films in fields up to 6 T. The resistance changes are always small (less than 2%) and negative. Figure 2 shows the effect of magnetic fields on the lowest resistance film in Fig. 1 at temperatures of 0.2, 0.3 and 0.5 K. We plot the relative temperature shift $\Delta T/T$ corresponding to the measured magnetoresistance. The largest shift is about 4% at 0.2 K, decreasing to 2% at 0.5 K. The

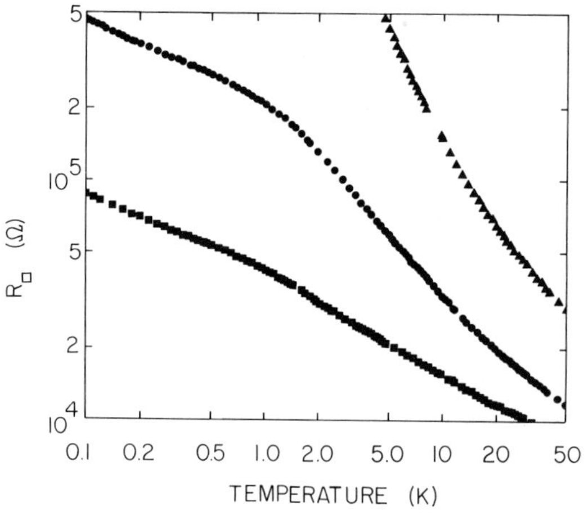

FIGURE 1
The resistance per square of several 1000 Å thick granular $Ni_x(SiO_2)_{100-x}$ films with nickel concentrations, x, close to 65 atomic percent.

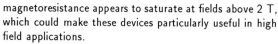

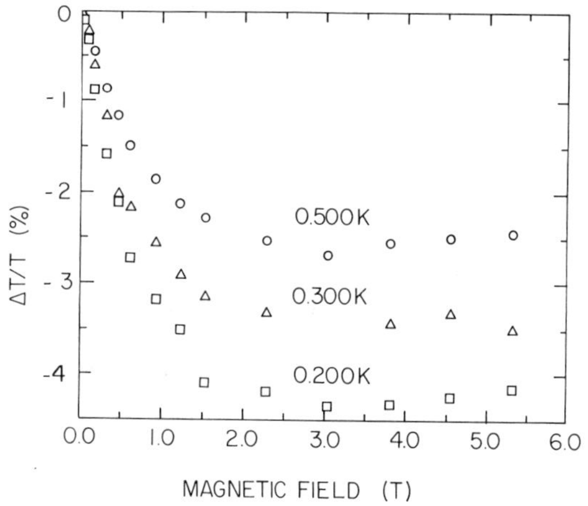

FIGURE 2
The magnetoresistance (converted to relative temperature shift) of the lowest resistance film from Fig. 1.

magnetoresistance appears to saturate at fields above 2 T, which could make these devices particularly useful in high field applications.

We have made some preliminary tests of the stability of these films. Over a period of a month, the sample of Fig. 2 was warmed to room temperature and exposed to air three times without its resistance at low temperatures changing noticeably. We also rapidly cycled this film from room temperature to 4.2 K a number of times. The measured resistances at 4.2 K agreed within 0.004%, corresponding to a temperature uncertainty of 0.2 mK.

Granular metals are expected to (and often do) have a resistivity of the form (6)

$$\rho = \rho_o \exp[(T_o/T)^{\frac{1}{2}}]. \tag{1}$$

Such a temperature dependence describes these films above about 2 K, but the crossover to a weaker temperature dependence below 1 K is not understood. Similar behavior, together with a slightly larger magnetoresistance, has been observed in Pt/alumina cermets (5), suggesting that this behavior is due to the granular structure of the films, rather than to the specific metal/insulator combination. However, our preliminary measurements with a number of other metals indicate that nickel/silica films have advantages in terms of stability and insensitivity to magnetic fields. From our measurements it is clear that RF sputter deposition can be used to produce granular metal films with well-controlled electrical characteristics. They may be most useful in applications such as calorimetry when small mass and insensitivity to magnetic fields are important.

ACKNOWLEDGEMENTS

This work was supported by the Office of Naval Research under contract number N00014-88-K003. We also are grateful to P. Sheng and S.-T. Chui for a number of useful conversations.

REFERENCES

1. L.G. Rubin, B.L. Brandt and H.H. Sample, Cryogenics 22 (1982) 491.
2. Q. Li et al, Cryogenics 26 (1986) 467, and references therein.
3. T. Yotsuya, M. Yoshitake and J. Yamamoto, Appl. Phys. Lett. 51 (1987) 235.
4. K.M. Unruh and J.R. Beamish, unpublished.
5. N.A. Gershenfeld et al, J. Appl. Phys. 64 (1988) 4760.
6. B. Abeles, P. Sheng, M.D. Coutts and Y. Arie, Adv. Phys. 24 (1975) 407.

Physica B 165&166 (1990) 41–42
North-Holland

AN INVESTIGATION ON THE THERMOELECTRIC PROPERTIES OF DILUTE MAGNETIC COPPER-IRON ALLOYS AT LOW TEMPERATURE

Huiling WANG* and Hongchun GAO

Division of Low Temperature, Huazhong University of Science and Technology, Wuhan, China

Based on calibration property experiments on the thermoelectric motive force of dilute magnetic copper-iron alloys at low temperatures (4-273 K) and experiments on dynamic frequency properties of inhomogeneity, an investigation has been made on the influences of different iron impurity concentrations on the low temperature thermoelectric properties and the inhomogeneity properties. It has been found by experiment that, within a fairly wide temperature range, the emf of copper-iron alloy is about 25 percent higher than that of gold-iron. The sensitivity for copper-iron above 16 K is higher than that for gold-iron alloy. The static inhomogeneity of copper-iron alloy is less than 2 μV and the dynamic inhomogeneity calculated statistically is equivalent to that of nickel-chromium alloy.

1. INTRODUCTION

Form a dilute magnetic alloy by using gold and copper as the base, with an insignificantly small amount of ferromagnetic metals dissolved in it. Such an alloy will exhibit extremely high negative emf owing to the interaction between the spinning of the conducting electrons and that of the magnetic ions. But the dilute gold-iron alloy is expensive. Hence an investigation on cheap Kondo alloys such as the dilute copper-iron alloy is of great significance both in theory and application.

2. THE CALIBRATION FOR THE COPPER-IRON

Cover the thermostat rod and the isothermal shade in a cryostat with a glass shade with vacuum lining. When the liquid helium in the glass shade is squeezed out by the evaporated gas, the thermostat rod and the isothermal shade will be in a semi-adiabatic state.

There are two holes of ϕ 5.2 x 50 mm in the center of the thermostat rod made of oxygen-free copper for installing a platinum resistance thermometer. Around them are sic holes of ϕ 3 x 50 mm for installing the thermocouples. The comparative method was adopted. The thermocouples, standard thermo-meters and thermostat were all in good thermal equilibrium.

Fifty-five points have been measured in the temperature range 4.2-273 K. The data were processed with the orthogonal polynomial least square method. The experimental data were fitted seven times.

$$y = \sum_{i=0}^{7} a_i x^i \qquad (*)$$

The mathematical model transformed according to (*) is

$$y = \sum_{i=0}^{n} y_i \frac{W_{n+1}(x)}{(x - x_i)W'_{n+1}(x)}$$

The error introduced into the calculation with the mathematical model is given by the following equation:

$$R_n(x) = \frac{f^{(n-1)}(x)}{(n-1)!} w_{n+1}(x)$$

$$\max_{\leq x_0 \leq x \leq x_n} \left| f^{(n-1)}(x) \right| \left| \frac{W_{n+1}(x)}{(n-1)!} \right|$$

The calculated results show that the maximum error introduced by the mathematical model is 0.02 K. The maximum error in the experimental system using the mean square root equation is smaller than 0.03 K.

The calibration results are made into a calibration table for application. The curves are given in Figs. 1 and 2 in the text.

3. A COMPARISON OF THERMOELECTRIC PROPERTIES BETWEEN COPPER-IRON AND GOLD-IRON

It can be seen from the figures (curves A, B and C) that the emf of copper-iron, is greater than that of the gold-iron (curves D). At temperatures 4-237 K, the emf of NiCr-Cu+0.13 % Fe is 1172 μV higher than that of NiCr-Au+0.07 %Fe. It can be seen from Fig. 2 that the sensitivity of the copper-iron (A, B and C) higher than that of the gold-iron (D). The former is close to the latter only at liquid helium temperature.

* Present address: Physikalisches Institut, Universität Bayreuth, Federal Republic of Germany

0921-4526/90/$03.50

4. THE INFLUENCE OF IRON IMPURITY CONCENTRATION ON emf

The curves representing the relationship of the emf and sensitivity with temperature for copper-iron alloys with three different iron contents (0.05, 0.13 and 0.15 atomic percentage) are shown in Fig. 1 and Fig. 2. From the figures it can be seen that, in general, their emf and sensitivity increase with an increase in the iron content.

5. HOMOGENEITY AND STABILITY

The inhomogeneity of a hot electrode depends on the degree of difference in emf between the homogeneous segment and the nonhomogeneous segment of material and is related to the lengthwise temperature distribution of the thermocouple. In the equation for the parasitic emf that generates inhomogeneity.

$$E_h = \int_0^l H(T, x) \frac{dT}{dx} dx$$

The experimental results are listed in Table. The inhomogeneity measured is smaller than 2 μV.

Experiments on dynamic frequency properties of inhomogeneity have been made with double-material method. The range of velocity for wires is 2-5 mm/s. The experiment reveals that copper-iron alloy exist non-intrinsic factors such as technological influences.

Immerse a thermocouple in liquid nitrogen again and again so that it will undergo repeated cycles of cold and hot tempering from room temperature to 77 K and then measure its emf. We suggested that the evaluation of the stability among the thermocouples should be made in accordance with such a definition:

$$\left.\frac{\Delta E\ (\mu V)}{S\ (\mu V/K)}\right|_{T_i=T_{LN_2}}$$

It can be seen from Table that the stability of the copper-iron alloy is fairly good, the maximum error not exceeding 2 μV. A comparison according to $\Delta E/S$ shows that the copper-iron alloys and gold-iron alloys have identical sensitivity.

Table: Homogeneity and stability

iron impurity	homogeneity (μV)	stability σ_E (μV)	$\Delta E/S$ (K)
Cu + 0.05at Fe	0.5	0.92	0.046
Cu + 0.10at Fe	0.5	0.59	0.031
Cu + 0.13at Fe	1.3	0.28	0.012
Cu + 0.15at Fe	1.4	0.51	0.021
Cu + 0.20at Fe	0.5	0.60	0.026
Cu + 0.25at Fe	0.5	0.82	0.034
Cu + 0.35at Fe	0.5	0.36	0.014
Au + 0.03at Fe	0.7	0.28	0.018

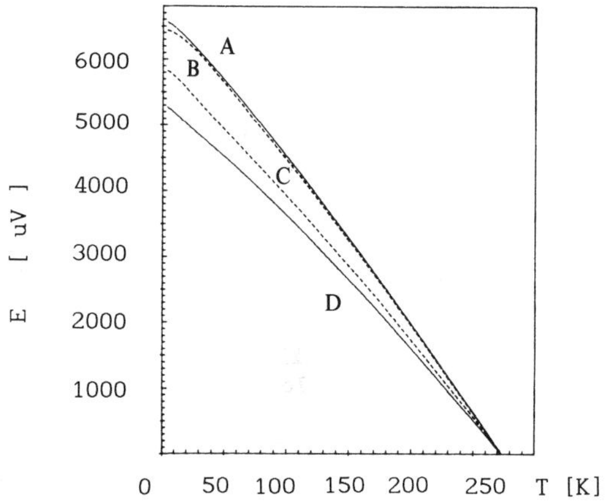

Fig. 1 The relations of the emf with Temperature for different iron contents
A NiCr-Cu+0.15 % Fe, B NiCr-Cu+0.13 % Fe
C NiCr-Cu+0.05 % Fe, D NiCr-Au+0.07 % Fe

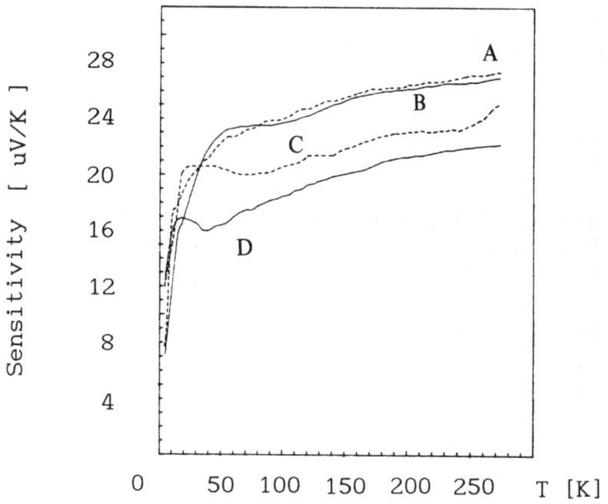

Fig. 2 The curves of the sensitivity for different iron contens

6. CONCLUSION

The calibration table for dilute magnetic alloys with iron concentrations 0.13 and 0.05 at 4-273 K has been obtained by experiment. The maximum value of experimental error is smaller than 0.03 K. The results show that the emf of a copper-iron is 25 % higher than that of a gold-iron. Above 16 K, the sensitivity of a copper-iron is appreciably increased. Experiments show that the static inhomogeneity and stability of copper-iron are both smaller than 2 μV. The dynamic inhomogeneity reveals that, for the copper-iron alloy developed, there exist non-intrinsic factors influences.

REFERENCES

1) Powell, R.L, et al., JR. Res. NBS 76 A (1972)
2) Xian Zhu, Low Temp. phys. No. 4 (1982)
3) Huiling Wang, Low Temp. phys. No. 1 (1985),72

Physica B 165&166 (1990) 43-44
North-Holland

A SMALL PLASTIC DILUTION REFRIGERATOR

Raymond WAGNER and Giorgio FROSSATI

Kamerlingh Onnes Laboratorium, P.O. Box 9506, 2300 RA Leiden, The Netherlands

A small diameter dilution refrigerator especially designed to be used in high magnetic fields has been constructed. It is completely made of plastic, including the heat exchanger, in order to eliminate eddy current heating. Temperatures of 10 mK have been reached in zero field tests.

1. INTRODUCTION

Dilution refrigerators (d.r.) used at high magnetic fields usually have a plastic mixing chamber at the center of the magnet bore, while the rest of the refrigerator is located in the low field region. Viscous heating effects restrict the available cooling power, leading to high final temperatures, given the usually large heat leaks.

By constructing a small diameter d.r. (o.d. 36 mm including the vacuum can) entirely out of plastic, eddy current heating is eliminated, and the entire unit can be placed in the bore of most existing magnets, including Bitter magnets. The refrigerator is designed for circulation rates roughly ten times that of existing d.r.'s of similar size. At this stage only preliminary tests have been conducted and the real capabilities of the refrigerator are not yet known.

2. THE DILUTION REFRIGERATOR

2.1. The still

The still is machined out of Araldite and is 65 mm long. At the top we mounted a film breaker to stop ^{4}He film flow, a precaution that might be important at the lowest temperatures.

2.2. The tubular heat exchanger

Underneath the still there is a tubular heat exchanger, the main purpose of which is to provide thermal insulation from the higher temperature part. The dilute mixture channel is formed by a (6 x 6 mm) spiral groove which is milled in an Araldite rod and has a total length of 0.41 m. A 6 m long, coiled Teflon capillary is placed inside this groove, forming the concentrated mixture channel.

The heat exchanger is closed on the outside by a tight fitting Kapton cylinder, which covers all parts of the refrigerator below the still. We have calculated the temperature at the bottom end of this heat exchanger to be 30 mK at a circulation rate of about 200 μmoles/s.

2.3. The Kapton foil heat exchanger

This heat exchanger consists of a bellows made of Kapton foils which separates

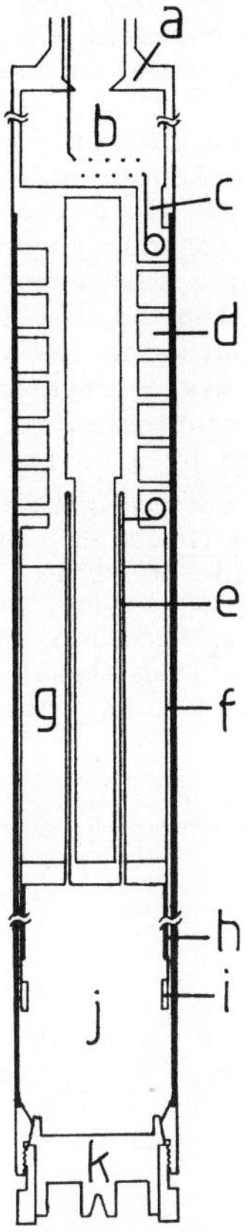

FIGURE 1
Diagram of the plastic dilution refrigerator. The still and mixing chamber are not drawn to scale. a - film breaker, b - still, c - concentrated helium channel of tubular heat exchanger, d - dilute helium channel of tubular heat exchanger, e - concentrated side of foil heat exchanger, f - dilute side of foil heat exchanger, g - Kapton foil area, h - Kapton shield, i - electrical pins, j - mixing chamber, k - conical plug

concentrated mixture (inside) from dilute mixture (outside). To construct the bellows, 600 discs of 12.5 μm Kapton foil were glued together at the edges and machined to a uniform size (i.d. 11, o.d. 26 mm). Most of the central hole was plugged by an Araldite rod fitted with a (0.5 x 5 mm) spiral groove.

We can calculate the mixing chamber temperature T as a function of the circulation rate $\dot{n}$ using the relation (1)

$$T^2 = 7 R \dot{n}/\sigma + 0.012 \dot{Q}/\dot{n}.$$

R denotes the mean Kapitza resistivity (0.008 K^4m^2/W for Kapton foil, (2)), σ the total surface area of the heat exchanger (0.26 m^2), and $\dot{Q}$ the heat leak into the mixing chamber (calculated to be 0.5 μW). Inserting these values yields an expected minimum temperature of 8.5 mK at a circulation rate of about 200 μmoles/s.

2.4. The mixing chamber

A cylindrical Kapton shield is attached to the bottom of the Kapton foil heat exchanger, providing space for the phase boundary. ^{3}He is pumped away along a path external to the shield. A ring is attached to the shield, containing female contacts on which the wires are soldered. The wall of the mixing chamber is formed by the Kapton cylinder mentioned in section 2.2, closed at the bottom end by a conical plug on which an experimental cell can be placed.

3. EXPERIMENTAL PERFORMANCE

In figure 2 we show the mixing chamber temperature as a function of the circulation rate. In an earlier test a stable minimum temperature of 10 mK was obtained at a circulation rate of 270 μmoles/s. In a more recent test instabilities appeared, probably due to a leak in the separating wall of the tubular heat exchanger. Stable temperatures were only obtained at large circulation rates.

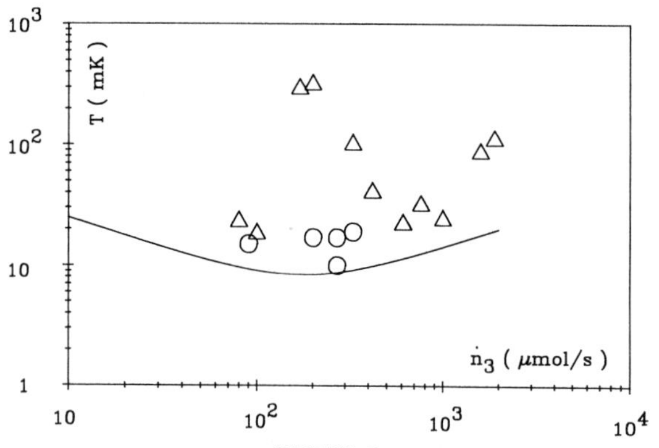

FIGURE 2
The mixing chamber temperature as a function of the circulation rate. The solid line represents the theoretical relation given in section 2.3. The circles represent measurements done with a CuNi capillary in the tubular heat exchanger (instead of Teflon).

Figure 3 shows the mixing chamber temperature as a function of the applied external heat leak at a circulation rate of 110 μmoles/s. The cooling power is fairly high, given the size of the refrigerator, but still lower than what can be expected.

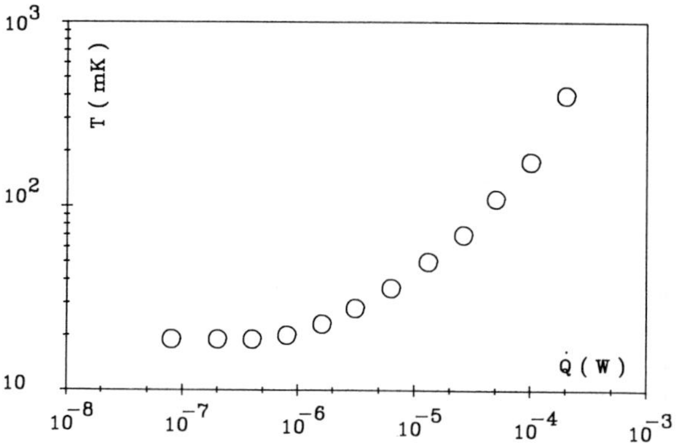

FIGURE 3
The measured mixing chamber temperature as a function of the external heat leak at a circulation rate of 110 μmoles/s.

4. CONCLUDING REMARKS

The use of a new type of plastic heat exchanger has allowed a small diameter dilution refrigerator to reach a temperature of 10 mK. The results are preliminary; once the instability problem has been solved, lower final temperatures and higher cooling powers should be possible. A similar unit with a larger (0.6 m^2) area heat exchanger has been constructed at the University of S. Paulo, Brazil, and is currently being tested.

ACKNOWLEDGEMENTS

We would like to thank J. Flouquet, G. Rémény, E. Wolff, T. Hata and all the technical staff for their help during the early stages of construction at the CRTBT of Grenoble.

We also thank Mr. J. Zijlstra of DuPont de Nemours, Luxemburg, for furnishing the Kapton film free of charge, and acknowledge the skillful technical support of Mr. A.J. Kuyt of our laboratory.

REFERENCES

(1) G. Frossati, J. Physique C 7, (1980), 95.
(2) G. Frossati and D. Thoulouze in: Proc. ICEC 5, ed. K. Mendelssohn (IPC Science and Technology Press, 1974) pp. 229-231.

Physica B 165&166 (1990) 45–46
North-Holland

DILUTION REFRIGERATOR TO OBTAIN POLARIZED LIQUID ^{3}He BELOW 200 mK STARTING FROM OPTICALLY ORIENTED GAS.

Gerard VERMEULEN and Myriam CORNUT

Laboratoire de Spectroscopie Hertzienne(*), Département de Physique de l'ENS, 24 rue Lhomond, 75231 Paris Cedex 05

We describe the design and construction of a dilution refrigerator to obtain polarized liquid ^{3}He in the degenerate region by liquifying and cooling optically pumped gas. This apparatus will extend the temperature region accessible with a ^{3}He cryostat for the study of polarized ^{3}He in our group (1). To establish thermal contact between the mixing chamber and the sample we have designed a non-magnetic heat switch, based on the large thermal conductivity of liquid ^{4}He.

The purpose of the dilution refrigerator, described here, is to extend the temperature region available in our group for the study of polarized liquid ^{3}He down to the degenerate region. The method, currently in use to obtain polarized liquid ^{3}He, is optical pumping of the gas phase with a LNA laser in a magnetic field of about 10 Gauss. Next, the gas is liquefied in a time short compared with the relaxation time by pumping a ^{3}He bath (1, 2). This method, is complementary to the rapid melting of polarized solid, giving access to polarized liquid ^{3}He at higher pressures.

The optical pumping is most efficient at room temperature, T_R, where we are able to obtain polarizations of $\simeq 60$ % . Thus, in our experiments, the ^{3}He is enclosed by a long pyrex container consisting of a pumping cell at T_R, a storage cell at $\simeq 4$ K and an experimental region which can be rapidly cooled as shown in fig. 1. The polarization is transferred to the storage cell from the pumping cell by spin diffusion. To minimize relaxation in the gas phase due to diffusion in magnetic gradients, the field has to be rather homogeneous along the experimental cell. This excludes the use of magnetic materials. The magnetic relaxation time of the liquid is on the order of several minutes, much shorter than that of the gas. Therefore, the gas is polarized before being liquefied in a time short compared with the relaxation time of the liquid.

The cooldown time of a dilution refrigerator is much longer than of a ^{3}He-bath. So, a thermal switch between the mixing chamber and the experimental region is required. The experiment is shown in fig.1.

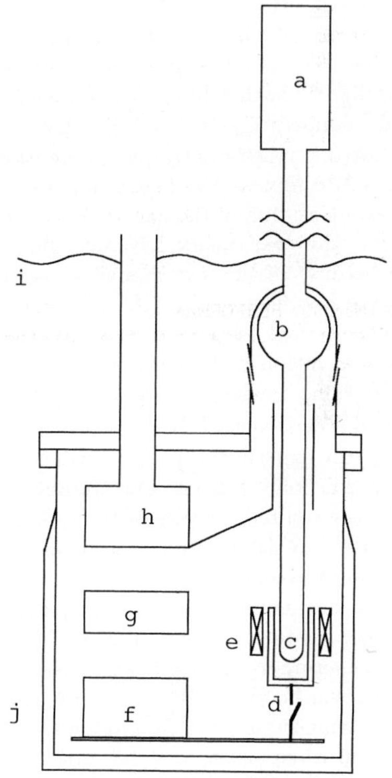

FIGURE 1
Schematic drawing of the experiment: (a) pumping cell, (b) storage cell, (c) experimental region, (d) coil foil and heat switch, (e) NMR coils, (f) mixing chamber, (g) continuous heat exchanger, (h) still, (i) ^{4}He level, (j) double walled vacuum can.

(*) Unité de recherche de l'Ecole Normale Supérieure et de l'Université Pierre et Marie Curie, associé au CNRS (URA 18)

0921-4526/90/$03.50

The dilution refrigerator will be operated continuously, while the sample is being polarized with the heat switch open. The number of ^{3}He atoms is limited by the magnetic relaxation time and the laser power to 2.10^{19} ($\simeq 1\text{mm}^3 \simeq 33\mu$mole of liquid). The enthalpy of the gas beween 1.5 K and 0.5 K is on the order of 600 μJ, the latent heat 1 mJ and the enthalpy between 500 mK and 100 mK 30 μJ. So, the performance of the dilution refrigerator can be relatively modest. A continuous heat exchanger will suffice to cool the mixing chamber to 50 mK with a cooling power of 160 μW at 100 mK and a circulation rate of 200 μmole/s.

Because of the magnetic relaxation in the gas phase the heat switch has to be non-magnetic. We have chosen to use the thermal conductivity of superfluid liquid ^{4}He, which is higher than that of copper above 200 mK. Raising the level of liquid ^{4}He in a stainless steel tube by means of a bellows will establish thermal contact between two silver sinters as shown in fig. 2. Such a heat switch opens only efficiently for temperatures below 1 K. Above 1 K, heat is transferred very efficiently by evaporation of the superfluid ^{4}He film at the high temperature sinter and the condensation of the gas at the low temperature sinter. However, below 1 K this effect becomes neglible because of the low vapour pressure, which inhibits the backflow of the ^{4}He gas. For our purpose this is perfect because the temperature of the experimental region will be above the boiling point of the ^{3}He sample, when the switch is open.

The thermal contact between the heat switch and the pyrex experimental region will be made by a greased coil foil cylinder, slid around the pyrex tube. We use coil foil because of the radiofrequency needed for the NMR to study the polarized liquid. The thermal resistance between the coil foil and the pyrex is the bottleneck in the thermal path from the mixing chamber to the sample region. With proper thermal anchoring to the still the heat leak along the pyrex tube will be about 10 μW. We estimate that the lowest temperature will be limited to about 150 mK. Even at a temperature of 100 mK the thermal equilibrium time of the polarized liquid to the pyrex walls will be about 10 s (2.5 s at 200 mK).

An anticipated difficulty is the production of an H_2-coating on the cold walls of the pyrex cell, to prevent relaxation. The H_2 sticks only well to the walls below a temperature of about 6 K. The temperature of the pyrex cell has to be kept stable at 6-7 K, before rapidly being cooled to below 4 K to freeze the H_2. To have a good temperature control during the initial He transfer, even with exchange gas inside the vacuum, the can is double walled.

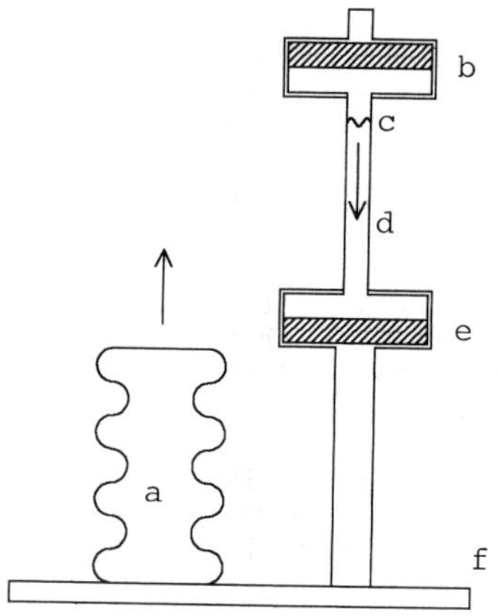

FIGURE 2
The heat switch: (a) bellows, (b) silver sinter connected to the experimental region, (c) ^{4}He level, (d) stainless steel tube, (e) silver sinter at mixing chamber temperature, (f) copper ground plate at the temperature of the mixing chamber. Not shown: a connecting capilary between the ^{4}He in the switch and the ^{4}He in the bellows. The bellows can be stretched mechanically from T_R to lower the ^{4}He level between the two silver sinters.

The dilution refrigerator is still under construction and the performance of the heat switch, which is crucial for this experiment, still has to be tested. It will be used to study the temperature and polarization dependence of the relaxation time of pure liquid ^{3}He. Another project is to study the polarization dependence of the phase separation curve of ^{3}He and ^{4}He, measuring the concentration of ^{3}He by either the dielectric constant, NMR or a vibrating wire viscometer.

REFERENCES

(1) P.J. Nacher, G. Tastevin, L. Wiesenfeld, Rev. Sci. Instrum. 59 (1988), 304.
(2) G. Tastevin, P.J. Nacher, L. Wiesenfeld, M. Leduc and F. Laloë, J. Phys. France 49 (1988), 1.

Physica B 165&166 (1990) 47–48
North-Holland

DEVELOPMENT OF A COMPACT DILUTION REFRIGERATOR WITH A CRYOGENIC ^{3}He J-T CIRCULATION SYSTEM FOR RAPID COOLING DOWN AND CONTINUOUS OPERATION+

Takeshi Igarashi, Masahito Sawano*, Yoshitomo Karaki, and Minoru Kubota,
Institute for Solid State Physics, University of Tokyo, Roppongi 7-22-1, Minato-ku Tokyo 106
Yu Hiresaki, Suzuki Shokan Co., Ltd. Koji-machi 3-1, Chiyoda-ku Tokyo 102
Kiyoshi Akiyama, Rigaku Denki Co., Ltd. Matsubara-cho 3-9-12, Akishima, Tokyo 196
Toyoichiro Shigi Dept. Appl. Phys. Okayama Univ. of Science Ridai-cho 1-1, Okayama 700
JAPAN

A compact dilution refrigerator without pumped ^{4}He bath has been developed to reach as low as 32 mK continuously with a simple counterflow continuous heat exchanger alone below the still. This refrigerator utilizes heat exchange with evaporated ^{3}He from the still and J-T expansion of the incoming ^{3}He in order to condense it. Although similar machines are reported by a few authors, we have investigated other possible paths in the ^{3}He T-H diagram by use of a specially designed film burner heat exchanger. Another dilution cryostat with a cryogenic ^{3}He circulation system is being constructed. Application of such a system for a millikelvin refrigerator without liquid He bath, but with a 4 K refrigerator is also studied.

1. Introduction

A dilution refrigerator is a well established tool to cool samples to millikelvin temperatures and it can be used continuously by circulating ^{3}He. Ordinarily a pumping system at room temperature is used for ^{3}He and also for ^{4}He, the latter is evaporated from a 1K pot to liquify the circulating ^{3}He. A few authors have already proposed[3-5] dilution refrigerators without the 1K pot, but a condensation stage utilizing ^{3}He gas cooling capacity from the still and the J-T expansion of the incomming ^{3}He, which is precooled to 4.2K before entering this stage.

We have studied various paths in the temperature - enthalpy diagramm of ^{3}He in a dilution refrigerator (DR) without 1K pot, using a home refrigerator compressor in the ^{3}He gas handling system.[1] With this room temperature system, we obtain 32 mK in the mixing chamber with simple construction ^{3}He gas heat exchangers including one at a specially designed filmburner. For farther simplication of the system and easier operation, a compact dilution refrigerator with a cryogenic ^{3}He J-T circulation system is being developed, which is applicable either for a system confined in liquid He bath or for a system precooled with a 4K small refrigerator, a Gifford-McMahon cycle plus J-T refrigerator, for example.

2. ^{3}He H-T(P) Diagram and Dilution Refrigerator

The low temperature thermal properties of ^{3}He has been studied in the pioneering work[2] by Wiedemann and his coworkers in Muenchen and the first proposals for dilution refrigerator applications were made by J.Kraus[3] and by A. de Waele, et.al,[4] independently. Much more recently K.Uhlig has reported[5] a very successful dilution machine. Fig.1 shows the enthalpy - temperature diagram from ref.2) on which we added present study's paths and one by K.Uhlig. If the still is kept at 0.6K, the dilute solution in the still has up to 7.7 J/g cooling power per ^{3}He, which can be compensated by a heater power in an ordinary dilution refrigerator. The power can be applied at a film burner whereas in the present study we place a heat exchanger. This film burner is ordinarily kept at about 1 K and offers a good heat exchange position in addition to a ^{3}He gas heat exchanger in the still pumping line.

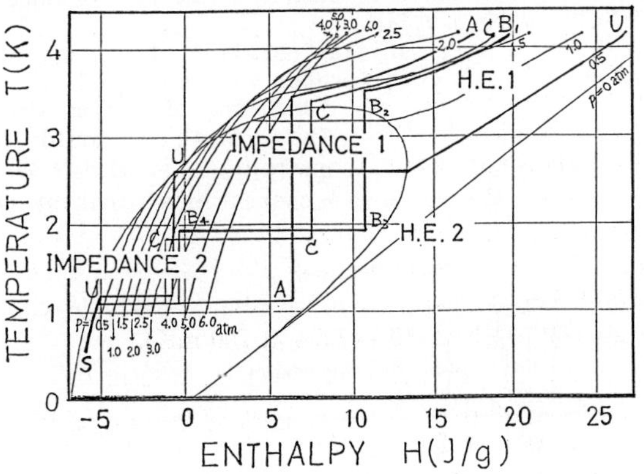

Fig.1 Enthalpy-Temperature diagram of ^{3}He from ref.2. Condensation paths A-S, B-S and C-S are our experiments and U-S is from Uhlig's[5]. The importance of the filmburner heat exchanger (H.E.2) is seen. 7.7 J/g_{3He} is the maximum possible cooling power of the still exchanger.

+ Work supported through Grant in Aid for Developmental Scientific Research, Ministry of Education, Science and Culture, Japan.
* Permanent address: Rigaku Denki Co., Ltd.

0921-4526/90/$03.50

3. Experiments to study ^{3}He condensation Process with room temperature gas handling system (GHS)

Our set up uses a DR vacuum can immersed in a He bath. There is a test section above the still where we have studied heat exchange and pressure and temperature drop relations for more than 10 different arrangements and recorded the corresponding paths in the T-H(P) diagram. (Fig.1) The inlet gas passed through the test section is turned to a mixture of liquid and gas and led to the still heat exchanger where all of ^{3}He is expected to become liquid. This point is S in Fig.1. The liquid ^{3}He is then led to a counterflow heat exchanger and finally to the mixing chamber in just the same manner as in ordinary dilution refrigerators. All of our heat exchangers and flow impedances are made out of CuNi capilaries after simplicity and reliability considerations.

The higher inlet gas pressure at T>3.5K is preferable in the view point of maximum heat exchange possible with the evaporated ^{3}He from the still as see in Fig.1. However when one trys a combination of adiabatic J-T expansion with heat exchange, the high inlet gas pressure needs certain caution, because the path may pass through out side of the ^{3}He inversion curve, That is, by adiabatic expansion the temperature may go up and heat exchange may not be performed properly. This problem can be avoided by the multiple "J-T and heat exchanger" combinations as are tried by us (B,C see Fig.2).

Our heat exchanger at the film burner is expected to function at the lowest temperature namely below 2K before entering to the still exchanger and the actual importance is seen in the results. At the film burner extra heating capability is facilitated by an electric heater. The film burner functions in reasonable manner when the still temperature is below 0.7K. The ratio ^{3}He / ^{4}He was measured for the circulating gas to check the film burner function. We obtain 98% ^{3}He with $\dot{Q}$=0.5mW.

With a still shield installed the cooling power of our refrigerator was measured as a function of temperature. The result could be analyzed as a sum of a constant heat leak of the order of 7 to 10 μW and T^2 dependent cooling power at $\dot{n}_3$~100umol/s.

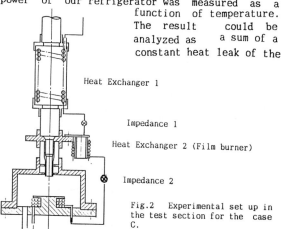

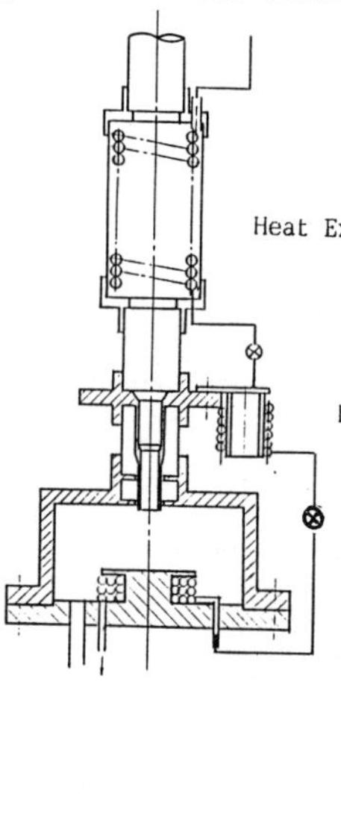

Fig.2 Experimental set up in the test section for the case C.

As to the operation after completion of room temperature setting up it could be precooled with ^{3}He exchange gas to liquid Nitrogen temperature and during the initial liquid He transfer all the mixture gas is precooled in the cryostat to 4.2K. The circulation can be started after 30min pumping of the vacuum can and the initial condensation of mixture gas could be performed quickly with the use of a bypass to the No.1 impedance by a 1/10 impedance. The lowest temperature could be reached within 2.5 hours after the liquid He transfer start.

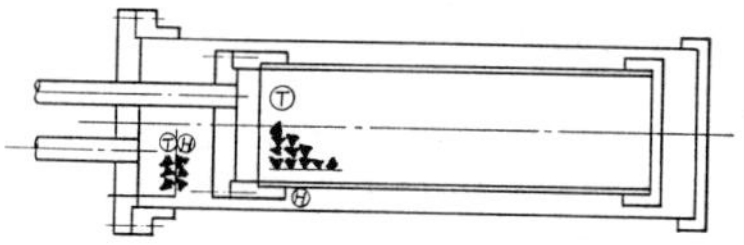

Fig.3 Test adsorption pump. A miniature adsorption pump and a heater controlls the operation.

4. Cryogenic ^{3}He circulation system

Adsorption pumps have been used for cryogenic refirigerators quite extensively.[7][8] Yet ^{3}He circulation system with the regenerated gas at elevated pressure above 0.5 kg/cm^2 has not been reported. Although it may bring obvious advantages, the cyclic adsorption and regeneration process poses a cryoengineering problem. We are studying to achieve short time constant of the system and small heat load to the He bath using a model charcoal pump depicted in fig.3. We have 10 gram of activated charcoal. Each grain of 3mm diameter is made thermal contact to the copper inner chamber wall by means of glueing to copper mesh and then rolling of the Cu mesh to fit tightly to the inner diameter of the chamber. While adsorption is performed reasonablly with moderate heating of 25~40mW, for 100μmol/s circulation, study for regeneration with least possible He consumption is underway.

5. Summary

A ^{3}He condensation stage with J-T and ^{3}He gas heat exchanger functions quite well. Conbination with cryogenic ^{3}He circulation system seems to be promissing yet further study is needed to minimize the heat load to the bath, A 4K refrigerator offers another way of running this dilution refrigerator as being tested by us.

REFERENCE

(1) Preliminary results are given in M.Sawano et.al. Proc.3.Japan-Sino Joint Seminer on Small Refrigerators and Related Topics. Okayama (1989) 27-30. Details will appear in Proc.ICEC13(1990)
(2) J.Kraus, K.Uhlig, W.Wiedemann Cryogenics 14 (1974) 29
(3) J.Kraus. Cryogenics 17 (1977) 173
(4) A.de Waele, et.al. Cryogenics 17 (1977)175
(5) K.Uhlig. Cryogenics 27 (1987) 454
(6) W.P.Kirk, E.D.Adams, Cryogenics 14 (1974) 147
(7) V.A.Mikheev, et.al. Proc. ICEC 10 (1984) 263
(8) R.Salmelin,et.al. J.Low Temp.Phys.76(1989) 83

Physica B 165&166 (1990) 49–50
North-Holland

A NUCLEAR DEMAGNETIZATION CRYOSTAT FOR THERMOMETRY

Wolfgang BUCK, Dieter HECHTFISCHER, and Albrecht HOFFMANN

Physikalisch-Technische Bundesanstalt, Institut Berlin, Abbestrasse 2-12, 1000 Berlin 10, Fed. Rep. of Germany

A single-stage nuclear demagnetization cryostat is reported with 33 moles of copper in a magnetic induction of 8 T developed for the investigation of low temperature thermometry. The lowest temperature measured with a Pt NMR thermometer was 130 μK, and the time dependent heat leak approached 0.3 nW after 80 days.

1. INTRODUCTION

A basic condition for developing thermometers and establishing a temperature scale at very low temperatures is a cooling equipment which meets the requested temperature range and temperature stability. The temperature range has been chosen with respect to intended investigations and according to the range of usefulness of the melting curve of ^{3}He as a temperature standard (1 mK ≤ T ≤ 1 K). In order to stabilize temperatures around 1 mK over a period of several weeks the attainable short-time minimum temperature of the cryostat should approach 100 μK.

To cover the range from 100 μK to 1 K a commercial dilution cryostat was equipped with a single-stage nuclear demagnetization unit (1). Both are connected by a superconducting heat switch made from high-purity aluminum foils. For working temperatures above 500 mK, where the dilution unit is difficult to control, the heat switch can be replaced by a weak thermal conductor.

2. DESIGN OF THE NUCLEAR STAGE

The design of the nuclear demagnetization stage is shown in figure 1. Part (a) is the mixing chamber of the dilution unit. Four Vespel rods ((b), φ 10 mm) bear the thermal link (e) machined from a single massive piece of high-purity copper. Thus, no welded or brazed joints were neccessary. After milling five windows into the thermal link in order to get free experimental space there remained a cross section of 820 mm^2 for the thermal path. To the lower end of the thermal link the nuclear stage (i) is screwed by a M60x2 thread (g). The nuclear stage again is made from one piece of very pure copper (OFHC) and annealed in high vacuum at 800 °C for several days. Its upper part (f) is designed as a mounting post for experimental equipment. The lower part of the stage located within the magnetic field region is longitudinally slit using a circular saw to avoid eddy current heating. The slit pattern is indicated in figure 1 within the outline of the copper stage (i). The effective mass of copper inside the magnetic induction of 8 T is 33 moles (2.1 kg). The total amount of copper located below the heat switch is 118 moles (7.5 kg). No remarkable thermal resistance was observed across the gold plated M60 thread connection during the precooling of the stage.

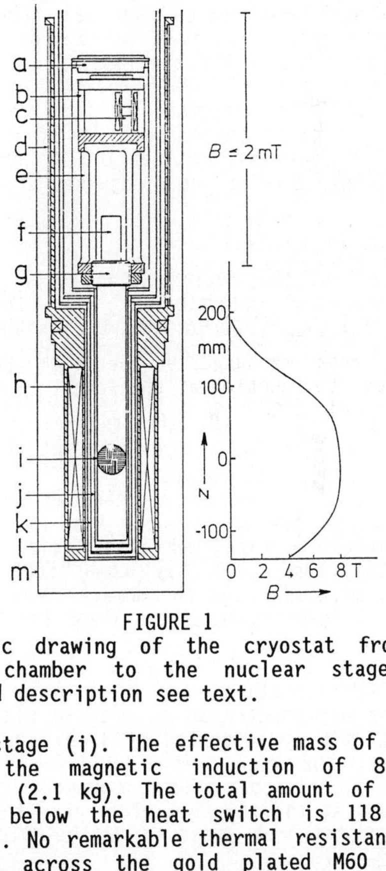

FIGURE 1
Schematic drawing of the cryostat from the mixing chamber to the nuclear stage. For detailed description see text.

The superconducting heat switch (c) is made from 29 aluminum foils (0.1 mm thick) of a purity of 99.999 %. They have been gold plated at their ends in order to improve the thermal contact to the adjacent copper foils, which are the connections to the mixing chamber and the thermal link of the stage, respectively. The switching ratio was determined to be 30 000 at 100 mK being quite comparable to similar designs (2).

Part (h) represents the main solenoid of the superconducting magnet and (d) the compensation coils. Parts (j) and (k) are thermal shields, (l) represents the internal vacuum can and (m) the LHe dewar.

3. THERMOMETRY

During the tests of the cryostat the temperature down to 100 mK was measured with a Ge resistor calibrated against an absolute noise thermometer. Down to 10 mK a thick-film resistor also calibrated against the noise thermometer was used. Both resistance thermometers agree within 1 %. For temperatures below 10 mK a Pt NMR thermometer was developed. Its sensor consists of 100 Pt wires, each 50 μm thick. The electronics are based on a commercially available setup improved by a few homemade components. The Pt NMR thermometer was calibrated against the thick-film resistor between 10 and 40 mK. The agreement of this calibration with the nuclear temperature during demagnetization is again better than 0.5 %, if the nuclear temperature has been adjusted at 40 mK.

4. RESULTS

4.1 Precooling

Precooling the nuclear stage down to 10 mK in a magnetic induction of 8 T takes three days. The temperature of the stage as a function of time behaves proportional to $t^{-1/3}$ as expected. That means that the cooling power of the dilution unit, the conductance of the heat switch, and the thermal conduction through the M60 thread are quite satisfactory.

4.2 Minimum temperature

The lowest temperature measured up to now with the Pt NMR thermometer was 130 μK. From remagnetization experiments it seems, however, that the limitation may come from thermal resistance within the thermometer and not from the final temperature of the stage itself.

4.3 Heat leak

In order to avoid external heat leaks from mechanical vibrations and rf interference the cryostat was mounted on a concrete platform of about 20 000 kg supported by steel springs and damping elements and enclosed into a rf shielded room. The maximum vibration amplitude at the mounting plate near the top of the cryostat was measured to be 0.25 μm at 14 Hz.

The total heat leak to the nuclear stage is plotted as a function of time in figure 2. The origin of the time scale is the date when the temperature fell below 4.2 K first. The heat leak was determined from warming-up curves starting at 10 mK (x) and measured around 1 mK (+) after demagnetization. An external heat leak (constant in time) occurs to be smaller than 0.3 nW. For comparison the time dependence of the heat leaks of the refrigerators of Bayreuth (3) and Jülich (4) is shown in figure 2.

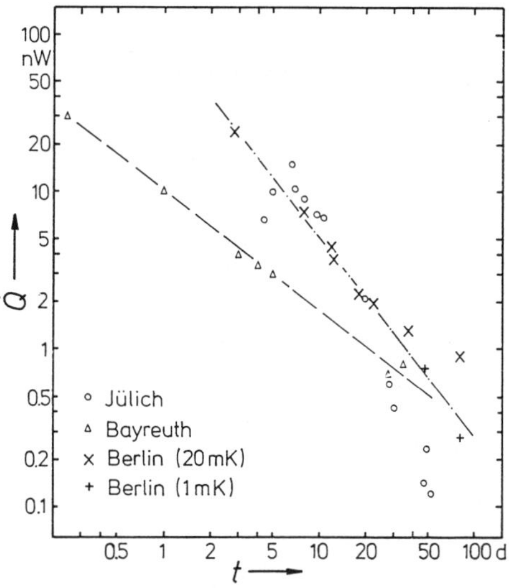

FIGURE 2
Temporal dependence of the heat leak to the nuclear demagnetization stage.

4.4 Thermal conductivity

The thermal conductivity of the nuclear stage was estimated in the temperature range from 2 mK to 9 mK under the assumption of $\lambda = \lambda_0 T$ with $\lambda_0 = 420\ \mathrm{WK^{-2}m^{-1}}$. Since the shape of the nuclear stage is not the best arrangement for thermal conductivity measurements, the agreement with the value of 280 $\mathrm{WK^{-2}m^{-1}}$ calculated from the residual resistance ratio of 180 via the Wiedemann-Franz law ist quite satisfactory. Thus, no contradiction to the Wiedemann-Franz law can be derived from our data in agreement with (5).

5. CONCLUSIONS

At PTB, a cryostat is now available, which is well suited for thermometer development und comparison of temperature scales in the temperature range from about 100 μK up to 1 K. Special precautions have been taken to exclude electrical and radio frequency interferences important for most types of thermometers. Besides the thermometry mentioned above noise and nuclear orientation thermometers will be installed in near future, and a ^{3}He melting curve thermometer.

REFERENCES

(1) W. Buck et al., PTB-Mitteilungen, (F. Vieweg Verlag, Wiesbaden, 1990) in print
(2) R. M. Mueller et al., Rev. Sci. Instrum. 49 (1978) 515
(3) K. Gloos et al., J. Low Temp. Phys. 73 (1988) 101
(4) F. Pobell, Physica 109 & 110B (1982) 1485
(5) K. Gloos et al., Cryogenics 30 (1990) 14

Physica B 165&166 (1990) 51-52
North-Holland

AN ULTRA LOW TEMPERATURE CRYOSTAT BUILT WITH HIGHLY CONDUCTIVE COPPER

T. Arai, T. Igarashi, Y. Karaki, T. Kawae, M. Kubota, S. Ogawa, M. Sawano and K. Shirahama
Inst. for Solid State Phys., Univ. of Tokyo, Roppongi 7-22-1, Minato-ku, Tokyo 106, JAPAN

W. Bergs, J. Hanssen and R. M. Mueller
IFF der KFA Juelich, D-5170 Postfach 1913, Juelich FRG

Features of a cryostat under construction in Tokyo and intended for fast operation at ultra low temperatures will be described. Highly conductive copper is used for both of the following reasons. 1. To achieve the ultimately low temperature and 2. To reduce the entropy loss while precooling of the copper nuclear demagnetization stage by a PrNi5 nuclear demagnetization refrigerator, which achieves continuous cooling by means of two stages which are demagnetized alternately.

1. INTRODUCTION

A compact ultra low temperature cryostat being constructed in ISSP Univ. Tokyo is described. Among number of ULT cryostats this is designed for rapid operation in ULT experiments and for experiments at ultimately low temperatures. The essential importance of the thermal conductivity of the material connecting nuclear demagnetization stage and the sample to attain an ultimate low temperature refrigeration is discussed in ref.1 for example together with the importance of reducing the heat leaks into the system. The high conductivity of the thermal paths is directly connected also to reducing the entropy loss during the entropy pumping between multiple demagnetization stages [2].

In order to realize a rapid operation for a whole experiment we have chosen the following features for the construction of a new cryostat, of which the designing started in 1982 in Juelich, but after many non scientific interferences the site of construction changed repeatedly and finally settled down at ISSP Tokyo owing to many people's helps and encouragements. The construction re-started in 1987.

2. THE CRYOSTAT ENVIRONMENT

The cryostat is built in the basement of the six-storied building of ISSP in the city of Tokyo. An electro-magnetic shielding room of 60 dB rf reduction is a must and the vibration isolation is made by a home made structure. Four of fine sand filled pillars, each of which is sitting on top of a very elastic rubber sheet on the floor, are supporting an aluminium bridge. On this bridge we have three air springs then the cryostat top flange. Fig.1 shows the major appearance of the cryostat. Many of the features inherit that of the one in Juelich which is still alive there.

3. RAPID OPERATION DILUTION REFRIGERATOR

Compact heat exchangers constructed from sintered Cu powders with particle sizes of 80 nm or smaller[3] are one of the main features which realizes rapid cooling to mK temperatures. The operation result to reach below 20 mK within 12 hours from closing up of the vacuum can at room

Fig.1. General appearance of a cryostat being constructed in ISSP, Tokyo. The pumping line size is compared with the moderate scale of the cryostat.

0921-4526/90/$03.50

temperature was once orally reported by MK in 1986[4]. After transportation we have experienced a cold leak which prevented us to obtain mK temperatures till recent time.

Our ^3He circulation system consists of two sealed roots pumps backed up also by a sealed rotary pump. The system capacity allows us to circulate as much as 2 m moles/sec of ^3He. We have experienced blockages between 78K and 4.2K presumably because of similar oil crack products discussed by E. Smith.[5]

4. A COMPACT AND SHIELDED $PRNI_5$ NUCLEAR DEMAGNETIZATION REFRIGERATOR

The importance of the heat conduction path is enphasized in ref.(1)&(2) to reduce the amount of entropy loss. In order to shorten the thermal path in a high magnetic field, a shielded superconducting magnet[6] has been made and tested up to 1.3 T. This magnet is confined in a Nb tube of 51/50 mm diameters and 170 mm length. It has 20 mm diameter bore and should have space for about 0.4 mole of $PrNi_5$.

Highly conductive thermal path makes it possible to perform a continuous refrigeration by two $PrNi_5$ stages which are demagnetized alternately. Fig.2 shows a calculated average cooling power of such a refrigerator together with that of some large scale dilution refrigerators. The importance of the present machine's cooling power will be realized when one notices the entropy reduction of Cu nuclei in a magnetic field of 8 T: 8% at 10 mK, 19% at 6 mK, 36% at 4 mK, 71% at 2 mK, and 95% at 1 mK. The solid line represents an ideal case with a cycling speed of 1 cycle/10 min and the dotted line indicate the effect of finite thermal resistance of a highly conductive copper(rrr=6,000) rod of 10*10*100 mm.

Actual cooling power of the $PrNi_5$ refrigerator depends on the precooling power of the dilution refrigerator through the possible precooling time and also on the time constant of the whole system. They have to be studied experimentally.

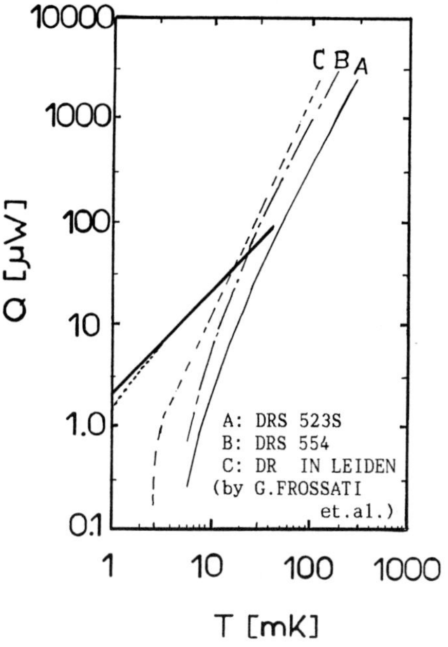

Fig.2. Calculated cooling power of a $PrNi_5$ refrigerator compared with that of some large scale dilution refrigerators. Solid line represents an ideal case whereas dotted line includes the effect of finite thermal conductivity of a high purity Cu rod. Our dilution refrigerator cooling power should resemble curve B.

5. HIGHLY CONDUCTIVE COPPER

We have been constructing ULT thermal paths out of very high purity Cu (nominal 6N7 [7]). In order to estimate the refrigeration properties, some transport characteristics of various samples are being measured. A sample of 3*3*100 mm size was measured to have a residual resistivity ratio of over 8,000 and an electron beam welding junction had no effect to this number after a heat treatment in vacuum.

6. SUMMARY AND ACKNOWLEDGEMENT

A rapid operation ULT cryostat under construction in ISSP Tokyo is described. After many struggles we see some promising features. Experiments presently planed for the new machine are:1] Nuclear magetism under thermal equilibrium. 2] Size effects in Quantum fluids.

Authors wish to thank those who have contributed and helped in various stages of the long term project here and there. Especially the cooperation with K.Akiyama, W.Buenten, J.Lingenbach, B.Radermacher, K.Rogacki, J.Sebek and W.Zinn are heartily aknowledged.

REFRENCES

(1) R.M. Mueller, Hyperfine Interactions 22(1985) 211-220
(2) R.M. Mueller, et.al. Cryogenics 20 (1980) 395
(3) K. Rogacki, et.al. J. Low Temp.Phys. 59 (1985) 397
(4) at "euro ULT seminar" sept.1986 at PTB Berlin
(5) E. Smith, AIP Conference Proc. 194 "Quantum Fluids and Solids-1989",ed.G.Ihas and Y.Takano, p.382-392
(6) R.M. Mueller& M. Kubota, unpublished (1986)
(7) Similar materials are commercially available as "Stressfree-6Nines" audio cables from Nippon Mining Co.,Ltd. Electronic & Specialty Metals Group, Toranomon 2-10-1, Minato-ku, Tokyo 105, Japan

Physica B 165&166 (1990) 53–54
North-Holland

DIFFUSION-WELDED LAMINAR NUCLEAR STAGE

Yu.M.BUNKOV, V.V.DMITRIEV, D.A.SERGATSKOV

Institute for Physical Problems, Kosigyna str.2, 117 334 Moscow, U.S.S.R.

A.FEHER

Department of Experimental Physics, P.J.Šafárik University, 041 54 Košice, Czechoslovakia

E.GAŽO, J.NYÉKI

Institute of Experimental Physics, Slovak Academy of Sciences, Solovjevova str.47,
043 53 Košice, Czechoslovakia

A nuclear cooling stage has been constructed from diffusion-welded copper sheets. We describe details of the technology and the design.

1. INTRODUCTION

Up to the present, mainly three types of copper nuclear cooling stage have been used: the wire bundle (1), the massive machined block (2) and the copper flakes stage (3).

We have developed a nuclear stage using diffusion welding to connect all parts of the nuclear stage, including the heat switch. A similar construction has recently been discussed by Florida group (4). The main advantages of such a design and technique are that all connections are of bulk copper, so that intermetallic alloys are avoided. As a result, the thermal conductivity of the stage and joints should be good, there is a high packing factor, high mechanical strength, absence of unnecessary materials (such as epoxies), and furthermore construction is relatively simple and manufacture is relatively quick.

2. DESIGN

The nuclear stage consists of four parts. The design is shown schematically in the Fig.1.

2.1. The Regrigerant (A)

Sheets for the bundle were cut from commercial oxygen free copper sheets 0.3 mm and 0.6 mm thick. The starting material had a residual resistivity ratio (RRR) of about 35. Strips from the thinner sheet were used to separate the 64 main plates made from the 0.6 mm thick. After diffusion welding and thermal treatment (see below) we have obtained a very rigid structure with a packing factor of 85% and RRR = 300 $\pm$ 30. The total amount of copper is about 6 kg.

2.2. Experimental Flange (B)

Into the bottom of oxygen free copper experimental flange slits were cut by spark erosion,

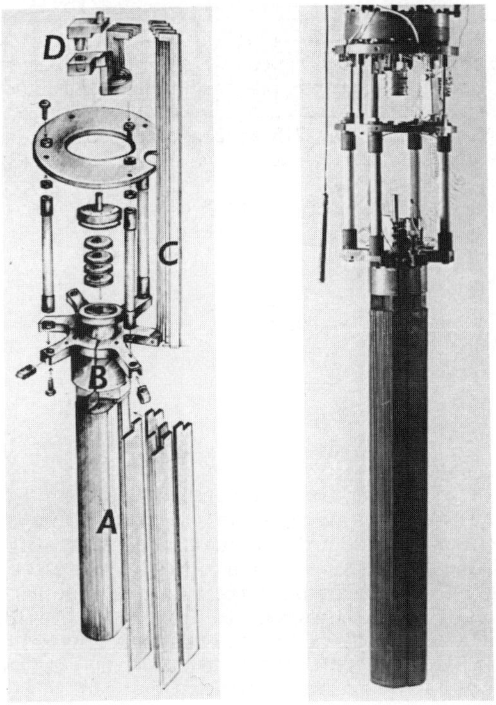

FIGURE 1
The design of the nuclear stage

into which the refrigerant plates were fixed by diffusion welding.

On the upper part, a chamber was machined for the silver heat exchangers of the indium-sealed helium-3 experimental cell. The arrangement of heat exchangers sintered from 100 nm silver powder with surface area of about 60 m^2 is shown in the Figure 2.

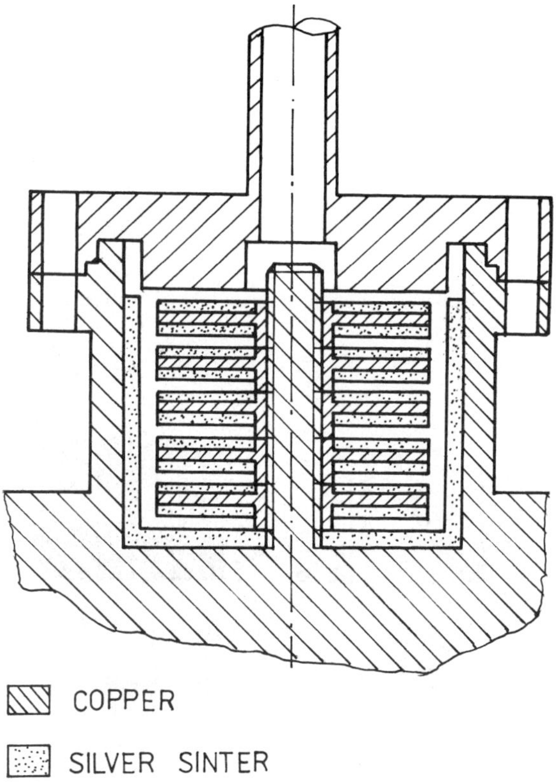

FIGURE 2
The silver heat exchangers arrangement

Ceramic supports were screwed into four of the six arms. A helium-3 melting curve thermometer was placed in a conical hole in one of remaining two arms. The final arm carries the diffusion-welded copper thermal link.

2.3. Thermal Link (C)

The thermal link between the experimental flange and the cold end of the superconducting heat switch consists of five strips, of 5N purity 0.8 mm thick copper sheet. One end of the thermal link was diffusion-welded to the experimental flange and the other to the aluminum heat switch. After heat treatment the RRR of the link was better than 2000.

2.4. Heat Switch (D)

The aluminum heat switch was fused by diffusion welding and its RRR = 1000. The design of the heat switch and details of diffusion welding technology were described in Ref.5.

3. HEAT TREATMENT

All parts of the nuclear stage were joined by diffusion welding under high pressure in a vacuum furnace. The necessary pressure was achieved by differential thermal contractions of the copper jaws and steel supports of a vise-like device. The heat treatment parameters during the stage fabrication were as follows:

Diffusion welding of the copper-copper joints in vacuum at a temperature of 1100 K for 30 minutes. Before welding the contact areas were cleaned by nitric acid and washed in distilled water. In order to prevent unwanted welds between the copper strips molybdenum foil spacers were used;

The copper parts were annealed while the stage was still in the vise after removal of the molybdenum spacers. The procedure of annealing was started in vacuum for 20 min at 1200 K and continued in oxygen at pressure 0.65 Pa for 30 hours at 1120 K. The cooling rate was 45 K/hour;

Diffusion welding of the copper-aluminum joints in vacuum at temperature 790 K for 10 minutes;

Sintering of silver heat exchangers to the walls of the experimental chamber in vacuum at 470 K for 35 min.

4. CONCLUSION

We have manufactured a laminar copper nuclear cooling stage using the relatively simple and reliable method of diffusion welding. We are currently investigating the performances of this stage on the nuclear demagnetization apparatus in Košice.

ACKNOWLEDGEMENTS

The authors would like to express their thanks to academician A.S.Borovik-Romanov for his support of this project and Mr.S.M.Elagin and Yu.M. Kogan for technical help during the manufacturing of the stage at IPP Moscow.

REFERENCES

(1) A.I.Ahonen, M.T.Haikala, M.Krusius and O.V. Lounasmaa, Phys.Rev.Lett.33 (1974) 628.
(2) J.P.Pekola, J.T.Simola and K.K.Nummila: Proc. Int.Cryogenic Engineering Conf. ICEC-10 (Butterworths, Guildford, England, 1984) p.259.
(3) D.I.Bradley, A.M.Guénault, V.Keith, C.J. Kennedy, I.E.Miller, S.G.Mussett, G.R.Pickett and W.P.Pratt,Jr., J.Low Temp.Phys. 57 (1984) 359.
(4) J.D.Kilian, P.B.Chilson, G.G.Ihas and E.D. Adams: Proc.Quantum Fluids and Solids-1989 (AIP New York, 1989) p.393
(5) Yu.M.Bunkov, Cryogenics 29 (1989) 938.

Physica B 165&166 (1990) 55–56
North-Holland

ROTATING CRYOGENIC PLATFORM

M.R. ARDRON, P.G.J. LUCAS, T. ONIONS, M.D.J. TERRETT and M.S. THURLOW

Department of Physics, The University, Manchester, M13 9PL, England.

A description is given of the design, construction and performance of a rotating cryogenic platform utilising continuously operating ^{3}He and ^{4}He refrigerators, with the initial aim of studying the effects of rotation on convection in liquid ^{3}He-^{4}He mixtures.

1. BACKGROUND

A number of cryostats (1,2,3,4) have been constructed in the last decade in which much of the cryogenic plumbing and electronics are mounted on a rotating platform. The motivation in most of these is in studying Coriolis and centrifugal effects on fluid flow states (1) and on the properties of quantised vortices in superfluids (2,3,4). Here we describe one constructed at Manchester aimed initially at studying the effects of rotation on convection in liquid ^{3}He-^{4}He mixtures.

2. DESIGN AND CONSTRUCTION

Many of the experiments that interest us (1,5,6) involve maintaining the temperature of a fluid sample of a few cm^3 of a ^{3}He-^{4}He mixture between 0.6K and 2.5K stable to within a few μK; and require rotation speeds up to 30 r.p.m. Most of the measurements are thermal, utilising resistance thermometry, SQUID-based thermocouple systems (7) and vapour pressure determination, and are often low-level signals susceptible to noise. Thus installation of all low noise electronics and much of the room temperature pipework in the rotating frame is a design feature.

Space requirements for this feature force a large circular platform, in our case a disc of aluminium 2m in diameter and 2.5cm thick. The disc is supported horizontally and centred laterally by two sets of three air-bearings supplied by filtered air at 100 p.s.i. in common with other facilities (2,4). Air-bearing design is described by Powell (8). A novel feature is the use of three AC linear motors to rotate the platform. These act directly on the disc, which has a mild steel ring sandwiched into its perimeter to provide a magnetic return path. Angular speed measurement is made by passing a continuous black-and-white film-strip of alternate 1mm opaque and transparent stripes through an optical encoder at the platform rim providing a frequency of 900Hz at 20 r.p.m.

The cryostat is of conventional design with a ^{4}He reservoir at 4.2K, a ^{4}He refrigerator and a continuously cycled ^{3}He refrigerator. Continuous operation is essential because the measurements often require the fluid sample to remain at the same temperature for months. The ^{3}He and ^{4}He pumps are required off-platform, because of their large size, and it was thus necessary to construct a four-way vacuum seal

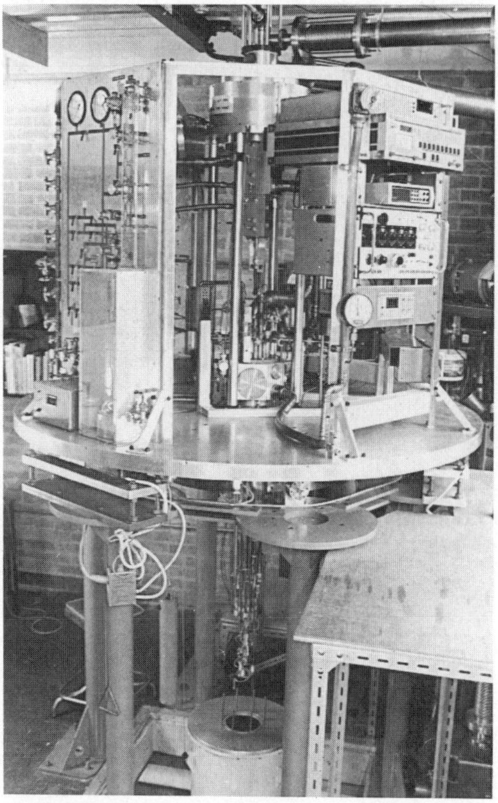

FIGURE 1
View of on-platform facilities.

0921-4526/90/$03.50 © 1990 – Elsevier Science Publishers B.V. (North-Holland)

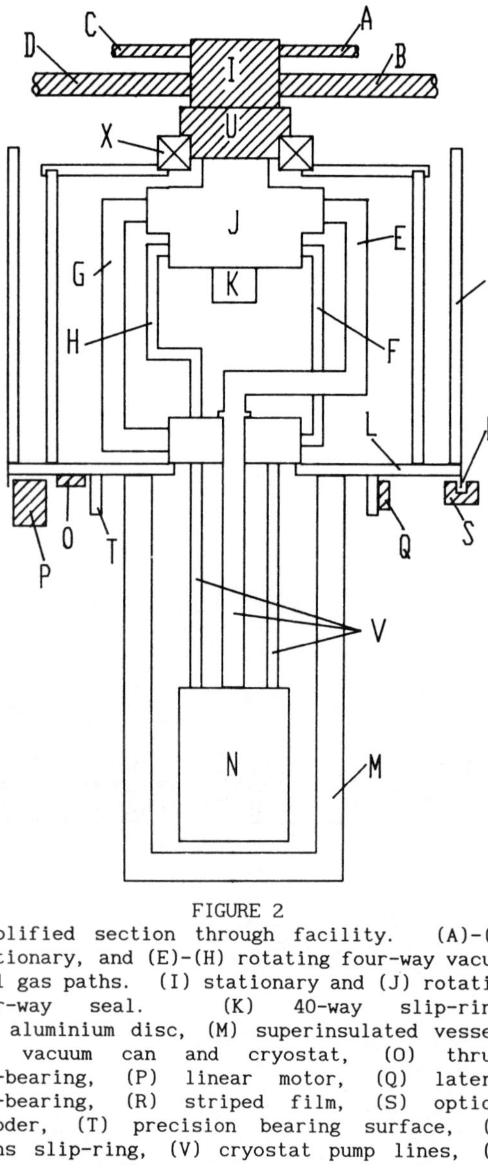

FIGURE 2
Simplified section through facility. (A)-(D) stationary, and (E)-(H) rotating four-way vacuum seal gas paths. (I) stationary and (J) rotating four-way seal. (K) 40-way slip-ring, (L) aluminium disc, (M) superinsulated vessel, (N) vacuum can and cryostat, (O) thrust air-bearing, (P) linear motor, (Q) lateral air-bearing, (R) striped film, (S) optical encoder, (T) precision bearing surface, (U) mains slip-ring, (V) cryostat pump lines, (W) racks for electronics and plumbing, (X) vacuum seal support bearing. Stationary parts are shown shaded.

for (a) pumping ^{3}He and (b) returning it, (c) pumping ^{4}He and (d) recovering ^{4}He boil-off from the main reservoir. This was achieved using coaxial tubes and slightly compressed O-rings as described by Packard and Williams (9). Double oil-filled seals are used. A 240V AC mains slip-ring supplies power to the electronic instruments on the platform which are controlled by passing an IEEE-488 computer bus through a 40-way miniature slip-ring. All analogue measurements and control takes place on-platform, while digital control through the bus is off-platform. Many of the constructional details are shown photographically in Figure 1 and schematically in Figure 2.

3. PERFORMANCE

The cryogenics have been tested as far as the ^{4}He evaporator and have presented few problems. The operation of the IEEE-488 bus through the slip-ring has been straight forward.

Speed control is required since friction from the vacuum seal O-rings varies with platform angular position causing speed variations of ±30% at 4 r.p.m. if the linear motors are driven at fixed power. A thyristor-based linear motor power supply from the manufacturers proved unsuitable as it transmitted severe vibrations to the cryostat. This has been replaced by a set of three power amplifiers delivering up to 3kW of variable amplitude 3-phase supply to the motors. The amplitude is controlled using digital proportional-integral-derivative feed-back from the film-strip encoder. With this system speed variations are ±6% at 4 r.p.m. in the short term with zero long-term drift. An alternative algorithm utilising the feature that the friction variations repeat every revolution may prove more effective still.

ACKNOWLEDGEMENTS

Substantial helpful advice and the loan of equipment from Professor R.E. Packard is gratefully acknowledged. The construction was aided by SERC Grants GRB6973.9 and GRC4548.5

REFERENCES

(1) P.G.J. Lucas, J.M. Pfotenhauer and R.J.Donnelly, J. Fluid Mech. 129 (1983) 251. This cryostat was constructed prior to this reference.
(2) G.A. Williams and R.E. Packard, J. Low Temp. Phys. 39 (1980) 553.
(3) B.C. Crooker, B. Hebral and J.D. Reppy, Physica 108B (1981) 795.
(4) P.J. Hakonen, O.T. Ikkala, S.T. Islander, T.K. Markkula, P.M. Roubeau, K.M.Saloheimo, D.I. Garibashvili and J.S. Tsakadze, Cryogenics 23 (1983) 243.
(5) G.W.T. Lee, P. Lucas and A. Tyler, J. Fluid Mech. 135 (1983) 235.
(6) P.A. Warkentin, H.J. Haucke, P. Lucas and J.C. Wheatley, Proc. Natl. Acad. Sci. 77 (1980) 6983.
(7) Y. Maeno, H. Haucke and J.C. Wheatley, Rev. Sci. Instrum. 54 (1983) 946.
(8) J.W. Powell, Design of Aerostatic Bearings (Machinery Publishing Co. Ltd., 1970).
(9) R.E. Packard and G.A. Williams, Rev. Sci. Instrum. 45 (1974) 1179.

Physica B 165&166 (1990) 57–58
North-Holland

CONTINUOUSLY PUMPED ROTATING MILLIKELVIN CRYOSTAT

J.D. CLOSE, R.J. ZIEVE, J.C. DAVIS and R.E. PACKARD

University of California at Berkeley
California 94720, U.S.A.

We report the construction of a rotating nuclear demagnetization cryostat which can be continuously operated. The refrigerator employs a quadruple concentric rotating vacuum seal. The minimum ^{3}He temperature of this cryostat in the stationary state is 165 μK. In the rotating state there is a relatively large heat leak which depends both on the rotation rate and on the field in the main demagnetization magnet.

1. INTRODUCTION

The study of superfluid ^{3}He under rotation has yielded some of the most exciting results of the last decade (1). This field was pioneered with the construction of the ROTA 1 cryostat (2) at Helsinki, Finland. The dilution refrigerator of ROTA 1 operated in a one shot cryopumped mode. Soon afterwards an existing rotating cryostat at Cornell was modified for operation at millikelvin temperatures (3) which utilized continuous external pumping. These two machines were the only ones of their type until the completion of ROTA 2 in 1988 (4). ROTA 2 can be continuously operated using dual cryopumps. To permit research in this field at Berkeley a rotating millikelvin cryostat has been constructed. This paper gives preliminary results of the operation of this facility. A photograph of the cryostat is shown below in Figure 1.

2. AIR BEARING AND ROTATING VACUUM SEAL

The cryostat is suspended from a single circular steel plate of diameter 1 m. This plate is supported by three air pads pointing in the vertical direction which are 120° apart. Bolted to the bottom of the plate is a steel cylinder about 0.2 m high whose outer surface is positioned by three journal air pads, also 120° apart. These six pads form the air bearing which permits smooth rotation of the cryostat.

The four vacuum lines enter the rotating frame through a quadruple concentric rotating vacuum seal (5). This allows the pot and the dilution refrigerator (as well as the ^{4}He recovery from the bath) to operate continuously during rotation. This is a logical extension of the rotating dilution refrigerator built at Berkeley in 1974 (6).The electrical connections between the laboratory frame and the rotating cryostat are made through a 20 channel slip ring system mounted above the rotating vacuum seal.

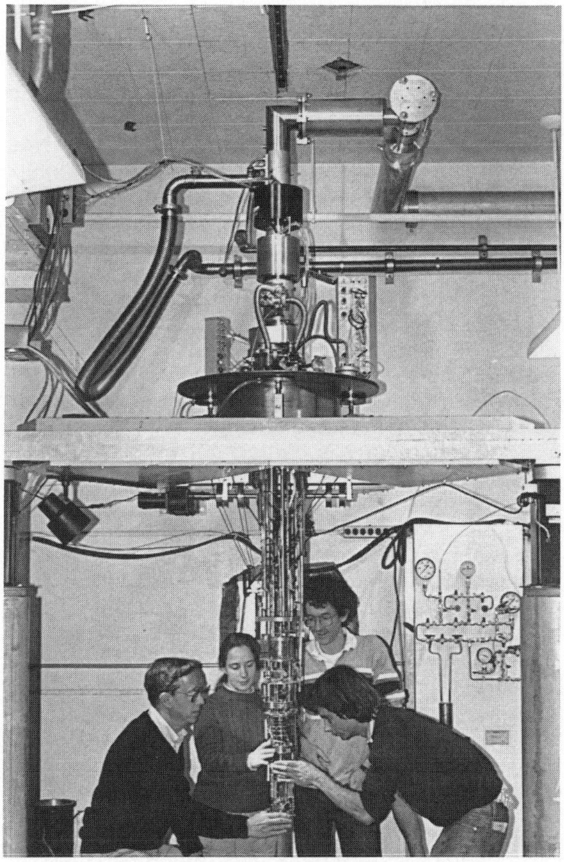

FIGURE 1. Rotating Nuclear Demagnetization Cryostat at Berkeley.

3. VIBRATION ISOLATION

The air bearing is supported on a triangular platform weighing 500 kg which rests on three air springs. Two of these are on top of Aluminum columns filled with sand and the third is bolted to the foundation wall of the building. The pumping system sits on an isolated concrete pad in the adjacent room. The interconnecting tubes first pass through bellows type vibration isolators (7) and then through a 5 tonne concrete block which is cemented to the foundation of the building. The tubes are then bolted to the foundation wall until they enter the rotating vacuum seal through a final set of bellows type isolators. The nuclear stage is connected to the mixing chamber by a 12 cm diameter Stycast 1266 cylinder of wall thickness 2mm which is 16 cm high. This is to increase the resonance frequency of vibration of the nuclear stage.

4. REFRIGERATION AND THERMOMETRY

The dilution refrigerator is a commercial OXFORD-400 model which cools to 6.5 mK with no load. The heat switch is made of aluminum diffusion welded to copper (8), with a compression joint at the bottom of the mixing chamber (9) and a bolted joint at the nuclear stage. This switch allows a precool to 9.5 mK to be carried out in 60 hours at 5.1 T average field. The nuclear stage is of the solid type (10) with 40 moles of copper in the high field region. This stage was annealed in vacuum at 850^0 C. The body of the heat exchanger is a container with a pattern of 2 mm diameter posts, electron discharge milled from a single piece of OFHC copper. This piece is silver plated and packed with silver powder to 50% solid density. To reduce the average distance between the ^{3}He and copper, 2 mm diameter holes are drilled into the sinter between the copper posts. This heat exchanger is electron beam welded to the top of the nuclear stage.

Thermometry is carried out by pulsed NMR on powdered Pt^{195} and utilizes the Greywall (11) temperature scale. The thermometers, as well as the heat switch and experiments have separate persistent superconducting magnets. A cylindrical superconducting shield surrounding each coil. The NMR electronics is completely in the rotating frame. These features simplify the operation of the thermometers during rotation.

5. OPERATION

In the stationary state the minimum ^{3}He temperature (at P=0) is 165 μK with a heat leak 0.7 nW after two weeks of operation. The time dependent heat leak starts at 36 nW and has almost disappeared after 10 days. The refrigeration of ^{3}He is reversible in the magnetic field down to about 0.3 mK.

The operation of the refrigerator under rotation is less satisfactory. At $\Omega = 0.5$ rad s^{-1} and with the main field up to 75 mT the heat leak is below 20 nW. For the same Ω, the heat leak grows rapidly with increasing field to 200 nW at 150 mT and 600 nW at 220 mT. Because of the magnitude of these heat leaks we have not yet addressed the heating due to the Earths field.

During these preliminary tests there is a large torque on the cryostat from the drive belt. We plan to eliminate this force and to improve the support structure for the nuclear stage before we next use the cryostat.

ACKNOWLEDGEMENTS

We would like to thank the following people who worked on the design and construction of this project; T. Pederson for his wonderful machining, L. Garner for the NMR electronics system, T. Kristofferson for electron beam welding and J. Elzey for assistance in assembly. This work was funded by National Science Foundation grants DMR-86-16713 and DMR-88-19110.

REFERENCES

(1) P.J. Hakonen, O. V. Lounasmaa and J. Simola, Physica **B160**,1 (1989).

(2) P.J. Hakonen, O.T. Ikkala, S.T. Islander, T.K. Markkula, P.M. Roubeau, K.M. Saloheimo, D.I. Garibashvili and J.S. Tsakadze, Cryogenics **23**, 243 (1983).

(3) J. Reppy, Cornell University, private communication.

(4) R. H. Salmelin, J. M. Kyynarainen, M. P. Berglund, and J.P. Pekola , J Low Temp. Phys. **76**, 83 (1989).

(5) R. Packard and G.A. Williams, Rev. Sci. Inst. **45**,9 (1974).

(6) G.A. Williams, U. of California at Berkeley Thesis, unpublished 1974.

(7) W. Kirk and M. Twerdochlib, Rev. Sci. Inst. **49**, 765 (1978).

(8) A.S. Borovic-Romanov, Yu. M. Bunkov, V.V. Dmitriev, Yu. M. Mukharsky, Japn. J. Appl. Phys. **26-3**,1719 (1987).

(9) K. Muething, G.G. Ihas, and C. J. Landau, Rev. Sci. Inst. **48**, 906 (1977).

(10) J.P. Pekola , Helsinki U. Of Technology Thesis, unpublished 1984.

(11) D.S. Greywall, Phys. Rev. **B33**, 7520 (1986).

Physica B 165&166 (1990) 59–60
North-Holland

AN IMPROVED DILUTION REFRIGERATOR COLD TRAP

E.N. SMITH

Materials Science Center, Clark Hall, Cornell Unversity, Ithaca, NY 14853 USA

Problems with blockage of the cold trap in a high circulation rate dilution refrigerator have been traced to the production of propene gas by breakdown of the rotary pump oil. A design is presented for an improved cold trap which eliminates the need for frequent cleaning. A brief discussion of some other attempted solutions is included.

1. INTRODUCTION

Traditionally, a trap containing charcoal or zeolite at 77K is placed in the path of the ^{3}He gas returning to a dilution refrigerator in order to remove impurities which could block circulation if they reached the low temperature part of the machine. In more recent, high circulation rate machines using direct drive pumps, it is often found that such a trap of conventional design will develop a partial or complete blockage within several days of use. In one of our machines this blockage has been traced to the production of propene gas, presumably from the cracking of rotary pump oil under high circulation rate conditions. A design is given for a cold trap which has allowed over three months of continuous operation with no sign of blockage, in contrast to the three days with the factory supplied traps, or three weeks with an earlier attempted solution to the problem.

2. DESCRIPTION OF PROBLEM

Over a period of many years, the Cornell low temperature group has operated a number of small dilution refrigerators, sometimes for continuous periods of up to several months or a year, with minimal cold trap blockage problems . Recently, with the acquisition of a pair of new and larger machines (Oxford Instruments Model 600 and Model 2000), we were disturbed to discover that the cold traps, apparently not strikingly different in design to those which we were accustomed to using, were plugging at inconveniently frequent intervals of two or three days. When the cold trap was warmed up to room temperature, the pressure observed in the system would be only 20 or 25 torr. In contrast, it was observed that if one circulated an air-helium mixture through the system it was possible to remove many liter-atmospheres of air, enough to produce a pressure of many atmospheres upon warming the trap, without causing any blockage. In an effort to better understand the nature of the problem, a sample of the trapped material was collected in a suitable tube for injection into a mass spectrograph in the analytical chemistry facility. It was determined that the observed fragments were most consistent with the sample being mainly propene gas, an organic molecule with a liquefaction point (at 1 bar) of 226K and a freezing point of 88K. It was observed that the problem with trap blockage became much worse at higher circulation rates, in a highly non-linear fashion. We had the impression, not well quantified, that the problem was becoming somewhat less severe with time, but with changes of less than a factor of two in several months. The problem was initially confused by a very small air leak in a Swagelock fitting in the refrigerator plumbing beyond all coldtraps.

3. HYPOTHESES

Our first two suspicions were excessive oil mist as a result of the high throughput for the rotary pump, or residual solvents used by the manufacturer in cleaning the gas handling system. Additional oil mist filters had no effect, and no macroscopic evidence of oil was found in the tubing. No solvent residues were found in the analysis of the residual gas in the trap.

Once propene had been specifically identified, it seemed clear that it was condensing on the tube walls, running down as a liquid until slightly above the liquid nitrogen level, then freezing, building up a thicker layer until the tube closed off (the diameter of the Oxford tubes was 10mm for much of the length, but dropped to 6mm at a junction in tubing–much less in one trap which had an overly-exuberant application of silver solder). It seemed plausible that the closing-over was occuring at this restricted point. It remained to determine the source of the propene.

An early concern was that the propene might solidify into small clusters in the gas phase and be blown through the charcoal as a dust. After repair of the small air leak, no blockage of the refrigerator has been

0921-4526/90/$03.50

seen. Thus, this scenario appears not to play a role in this system.

Often the cracking of diffusion pump oil in booster pumps has been suggested as the origin of unwanted materials (particularly hydrogen) in the recirculating gas stream. In fact, this was one of the reasons we had decided to use Roots type pumps rather than vapor booster pumps in our systems. Thus, we were strongly suspicious of the rotary pumps, and the rotary pump oil. The original Balzers sealed, direct-drive pump was replaced by an Alcatel rated at 30% higher capacity, with no apparent effect if the throughput was scaled by the corresponding 30%. We tried replacing the Alcatel 100 oil with a higher grade (and much more expensive) Alcatel 300 oil with only a minimal effect. An initial promising result in a test outside the dilution refrigerator was due to a faulty measurement of the flow rate.

4. INCOMPLETELY SUCCESSFUL SOLUTIONS

Having failed to remove the source of the propene (presumably it is produced by local hot spots in direct drive pumps under heavy loading conditions), we made an initial attempt to cure the blockage problem with a room-temperature adsorption system. A long tube (about 4 meters in length and 12 mm in diameter) was filled with zeolite, which we hoped would adsorb the propene gas. We discovered that this helped for a period of about a week, but then the trap would plug. The adsorption energy at room temperature was not sufficiently high to adequately limit migration of the propene gas, and we had just delayed the arrival of the unwanted material with our chromatographic column.

Several people had suggested to us that we might try to simply put a large diameter open tube at liquid nitrogen temperature as a pre-trap before the gas was allowed to enter the usual trap with its 10mm diameter inlet tube. This was easy to implement, so we tried it. The result was moderately satisfactory, with the time between trap cleanings increased to several weeks rather than three days. However, it was clear that the vapor pressure of the propene was not quite low enough at 77K to completely limit passage through the trap.

5. THE ULTIMATE COLD TRAP

The final implementation which we have tried combined a large (45mm) diameter tube 600mm long with the presence of a significant volume of activated charcoal in the same tube, so that any propene coming through would be likely to contact an exposed surface within a first monolayer of coverage and be likely to be permanently adsorbed before ever reaching a small diameter tube. We filled the tube to a depth of about 300mm with a rather coarse (14 mesh) activated coconut charcoal. Fine screens were placed in the tube at room temperature to limit migration of the charcoal. The trap was connected to the gas handling system with flexible 10mm stainless steel hose. The original equipment trap was simply discarded. This design, illustrated in Figure 1, has proved to be quite satisfactory, and has not yet shown any signs of blockage in over three months of continuous use.

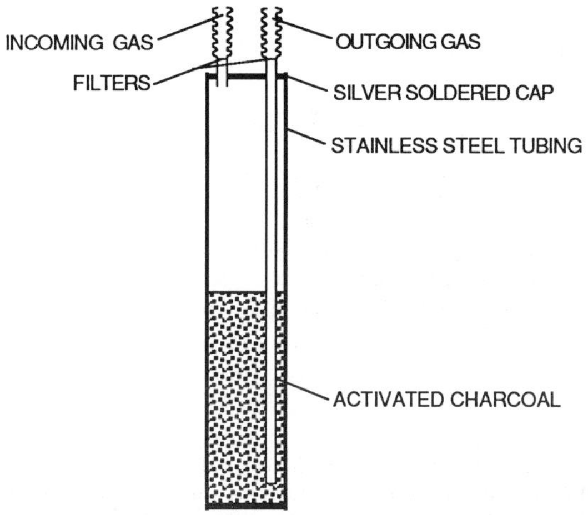

FIGURE 1
Schematic diagram, not drawn to scale, showing the construction of a satisfactorily operating cold trap. The trap is placed in a 25 liter liquid nitrogen dewar which gets refilled approximately once a week.

6. CONCLUSIONS

In our system, a principal source of contamination for the recirculating gas in the dilution refrigerator appears to be propene, produced in the oil of the rotary pump under high throughput conditions. A trap has been constructed which appears to effectively remove this impurity without inconvenient system stoppages.

ACKNOWLEDGMENTS

This work has been supported by the Materials Research Division of the NSF through grant DMR-8818558. The mass spectrographic analysis was performed for us by T. Wachs. Useful discussions with and suggestions from J. Amato, G. Nunes, J. Parpia, L. Pollack, R. Richardson, and E. Varoquaux were greatly appreciated.

Physica B 165&166 (1990) 61–62
North-Holland

FREQUENCY-DETERMINED CURRENT WITH A TURNSTILE DEVICE FOR SINGLE ELECTRONS

V.F. ANDEREGG, L.J. GEERLIGS, J.E. MOOIJ

Department of Applied Physics, Delft University of Technology, P.O. Box 5046, 2600 GA Delft, The Netherlands

H. POTHIER, D. ESTEVE, C. URBINA, M.H. DEVORET

Service de Physique du Solide et de Résonance Magnétique, Centre d'Études Nucléaires de Saclay
91191 Gif-Sur-Yvette, France

We have fabricated a device in which the current is to a high accuracy determined by an external frequency f as I=ef. This device consists of an array of ultrasmall tunnel junctions. An rf voltage is applied to a gate and causes the transfer of a single electron per cycle through the array. The locking of the electron transfer is obtained by using Coulomb blockade of electron tunneling.

1. INTRODUCTION

In recent years the fabrication of submicron tunnel junctions has permitted the observation of a new class of phenomena, charging effects, which are due to the discreteness of electron tunneling. These phenomena occur when the capacitance C of the junctions is small enough to yield a charging energy $E_C = e^2/2C$ larger than the thermal energie k_BT. A review is given by Averin and Likharev (1). We have used the charging effects in a device in which a single electron is transferred per cycle of an externally applied rf voltage. This results in a current equal to the frequency times the electron charge. The effect is qualitatively similar to the Shapiro steps observed in "classical" (large capacitance) Josephson junctions, and seems a candidate for a standard of dc current. Together with the quantum Hall effect ($V/I=R_K=h/e^2$) and the Josephson effect ($V=hf/2e$) , this device with I = ef provides a metrological triangle to check these relations for inconsistencies.

2. CRITICAL CHARGE FOR ELECTRON TUNNELING

In a voltage biased linear array of ultrasmall tunnel junctions the current-voltage characteristic exhibits zero conduction below a certain voltage, at zero temperature. This so-called Coulomb gap arises because of the discreteness of charge transfer in tunnel junctions. It can be conveniently described by introducing a critical charge Q_c. In a junction with charge Q, no tunneling can occur for $|Q|<Q_c$. Q_c depends on the junction capacitance C and on the equivalent capacitance C_e of the circuit in parallel with the junction (2) :

$$Q_c = e/[2(1+C_e/C)] .$$

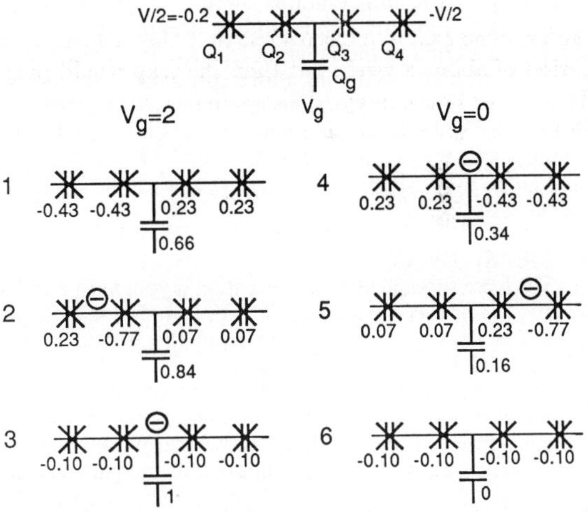

FIGURE 1
Principle of controlled single electron transfer through a linear array of small tunnel junctions. Junctions, with capacitance C, are denoted by crossed capacitor symbols, the gate voltage V_g is applied via a true (non-tunneling) capacitance C_g. If $C_g=C/2$, tunneling across any junction can only occur if for that junction $|Q|>Q_c$, with $Q_c=e/3$. The voltages and charges are indicated in units of e/C and e. 1-6 indicate consecutive times in the cycle. Left: First half of the cycle, $V_g=2$. An elementary charge (- in a circle) ends up trapped on the central electrode. Right: Second half of the cycle, $V_g=0$. The charge can only leave on the right hand side. No further tunneling can occur in the emptied array .

0921-4526/90/$03.50

3. TURNSTILE DEVICE

Our device is an array of 4 tunnel junctions of capacitance C. The gate voltage is capacitively coupled to the central island between junctions 2 and 3. If the gate capacitor C_g is chosen equal to C/2 all critical charges have the same value: $Q_c = e/3$.

The principle of our experiment is illustrated in Fig. 1. The first half of the rf cycle a bias voltage V and gate voltage V_g are applied such that the critical charge is only exceeded in the left arm of the array. An elementary charge will then tunnel across junctions 1 and 2. Once on the central island part of it will polarize the gate capacitor and the charges on the junctions will all be lower than Q_c. The charge is therefore trapped on the middle island until the bias conditions are changed. We emphasize that no other tunneling event can occur after the electron has been trapped. In the next half of the cycle the gate voltage is decreased, resulting in an increase of the charges on the junctions. Because of the asymmetry caused by the bias voltage the elementary charge will tunnel through junctions 3 and 4. Cyclically changing the bias conditions by applying an alternating voltage to the gate moves one electron per cycle through the array.

Electron tunneling is a stochastic process: if the tunneling is energetically favorable it will happen within a typical time RC where R is the tunneling resistance of the junction.This poses a restriction on the range of frequencies we can use. To avoid losing cycles f must be much smaller than $(RC)^{-1}$. At finite temperatures there is a probability of thermal activation out of the middle island. This gives a second restriction to the range of frequencies: a thermally assisted transfer will be more probable for lower frequencies. These restrictions are treated quantitatively elsewhere (3). Here we only quote the result that with the smallest present-day junctions ($C \approx 10^{-16}$ F), for $T \leq 75$ mK and $f \leq 30$ MHz the error in the relation $I = ef$ is expected to be less than 10^{-8}.

4. MEASUREMENTS

We have fabricated a device with a layout very similar to Fig. 1. For this device C = 0.5 fF, R = 340 kOhm and C_g = 0.3 fF. C_g is determined from the period of the current modulation by a gate voltage (2).

In Fig. 2 we show three I-V curves of our device. The dotted curve shows the measurement without ac gate voltage applied. The solid curves show that when a voltage of frequency 5 and 10 MHz is applied to the gate, a wide current plateau develops in the Coulomb gap at a value of I=ef. In this figure the amplitudes of the ac voltages were adjusted to obtain the widest plateaus. The height of the plateau is independent of the amplitude. The inset of Fig. 2 shows that the agreement with the relation I = ef is excellent up to frequencies of 20 MHz. The markers give the measured current plateau versus frequency and the solid line is I = ef with e the value of the elementary charge. The measured current coincides with ef within experimental accuracy, which is around 0.3 %. We simulated the dependence of the I-V curves on ac amplitude and found very good agreement with our measurements (3).

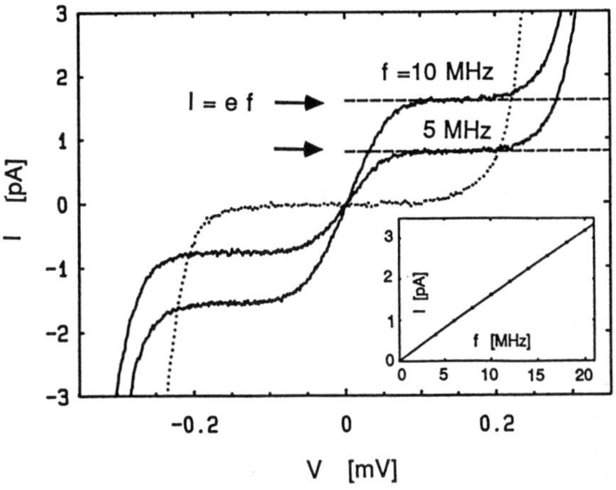

FIGURE 2
Current-voltage characteristics without ac voltage (dotted) and with applied ac gate voltage at frequencies f equal to 5 and 10 MHz. Current plateaus are seen at I = ef. The inset shows measured current versus frequency (markers) and the line I = ef with e the elementary charge.

5. CONCLUSIONS

We have fabricated a device in which the transfer of single electrons is controlled by an externally applied ac voltage. Accuracy considerations predict that it is a promising device for a new standard of dc current.

ACKNOWLEDGEMENTS

This work was supported by the Dutch Foundation for Fundamental Research on Matter (FOM) and the Commissariat à l'Energie Atomique (CEA). We thank the Center for Submicron Technology (CST) in Delft for the use of their nanolithographic facilities and P.F. Orfila for technical assistance.

REFERENCES

(1) D.V. Averin and K.K. Likharev, Single electronics, in: Quantum Effects in Small Disordered Systems, eds B.L. Al'tshuler, P.A. Lee, and R.A. Webb (Elsevier, Amsterdam) to be published.
(2) L.J. Geerligs et al., Tunneling time and offset charging in small tunnel junctions, this volume.
(3) L.J. Geerligs et al., Frequency-locked turnstile device for single electrons, submitted to Phys. Rev. Lett.

Physica B 165&166 (1990) 63–64
North-Holland

SINGLE-ELECTRON TUNNELING IN POINT-CONTACT TUNNEL JUNCTIONS

R.T.M. Smokers, P.J.M. van Bentum and H. van Kempen.

Research Institute for Materials, University of Nijmegen, Toernooiveld, NL-6525 ED Nijmegen, the Netherlands

SET effects have been observed in point-contact tunnel junctions on surface doped Si. Changing the electromagnetic environment does not appreciably affect the junction capacitance. Furthermore the influence of a Coulomb gap on STM-spectroscopy on superconductors is discussed.

1. INTRODUCTION

In small tunnel junctions the condition can be reached that the charging energy $e^2/2C$, with C the capacitance of the junction, is comparable to, or larger than the thermal energy k_BT. For low voltages across the junction Coulomb interactions strongly suppress the tunneling probability (Coulomb blockade). For the ideal case of a single current biased normal metal tunnel junction a full quantum-mechanical calculation [1] predicts an $I(V)$ characteristic, which is parabolic near the origin, and for higher voltages approaches linear asymptotes, which are displaced from the origin by $\Delta V = e/2C$. For small currents tunneling events are correlated in time, leading to oscillations of the voltage across the junction (SET-oscillations), with frequency $f_{\rm SET} = I/e$ and amplitude $e/2C$. A semi-classical approach [2,3] , based on a Fermi golden rule calculation of the tunneling probability and Monte Carlo simulation of the tunneling events, gives essentially the same results.

In practice stray capacitances of the leads tend to make the junction effectively voltage biased, thus suppressing the SET oscillations. Decoupling from these stray capacitances can be achieved in nanofabricated planar tunnel-junction arrays [4] and in point contact tunneling through small isolated particles [5].

Solitary point-contact tunnel junctions on various materials such as Al, Sn, stainless steel, and even high-T_c superconductors such as $YBa_2Cu_3O_{7-\delta}$ show clear signs of a Coulomb blockade, indicating capacitances of order 10^{-17}F [6]. This was understood by treating the junction environment as a transmission line, assuming that only interactions in a space region $c\tau_{\rm T}$ accessible within the tunneling time $\tau_{\rm T}$ can contribute to the Coulomb blockade.

A subsequent analysis [7] showed that the stray capacitance effectively interfering with the tunneling process is determined by the space region $v\tau$ that can be probed by the electromagnetic field (propagation velocity v) produced by virtual tunneling events in a time span $\tau = \max(\tau_{\rm T}, \Delta t)$ before actual tunneling. Δt stems from the uncertainty relation $\Delta E \cdot \Delta t = h$ with ΔE the energy gain of the electron after tunneling. For clean junctions $\tau_{\rm T} \approx 10^{-15}$s, while $\Delta t \approx 10^{-12}$s. For single junctions with tunnel resistance $R_{\rm T} > R_{\rm Q} = h/2e^2$ and lead impedance $S(\omega) \ll R_{\rm T}$, $I(V)$ is predicted to be almost linear with only a small trace of Coulomb interactions in a small conductance dip around zero bias. Measurements on a solitary planar junction [4] are in good agreement with the predictions of ref.[7].

In this contribution we analyse the effect of the electromagnetic environment by tunneling into surface doped Si, at temperatures close to the metal-insulator transition. Also we investigate the influence of a Coulomb blockade in STM-spectroscopy on classic and high-T_c superconductors.

2. TUNNELING ON SURFACE DOPED Si

The samples in this experiment are made on a single crystalline Si wafer by standard integrated circuit techniques. By implantation of As-ions and succesive diffusion at 1000 °C a constant dopant density of $\approx 8 \cdot 10^{18}\mathrm{cm}^{-3}$ is obtained down to 0.1μm below the surface. At $T = 4.2$K $R_\square \approx 50$kΩ, while at $T = 1.2$K $R_\square \approx 30$MΩ. Using differently patterned structures in combination with temperature variation provides a wide range of series resistances. By exposure to air an oxide layer of $\approx$ 10Å is formed. Using a low temperature STM-like set-up a tungsten tip is allowed to make mechanical contact with this oxide layer, thus forming a stable point-contact tunnel junction.

Fig.1a shows an example of the $I(V)$ characteristics of a point contact at $T = 4.2$K. The tunnel resistance is $R_{\rm T} =$ 7.2MΩ. Although at this temperature the series resistance is still much smaller than $R_{\rm T}$, we find an obvious Coulomb gap. The asymptotes are offset by $e/2C = 1.4$mV, yielding $C = 5.7 \cdot 10^{-17}$F. At lower voltages the characteristic is clearly parabolic, as can be seen from the linear parts in the $dI/dV(V)$ curve (fig. 1b). The rounding of the $dI/dV(V)$ curve near the origin is slightly larger than expected from temperature broadening at 4.2 K, which might be due to

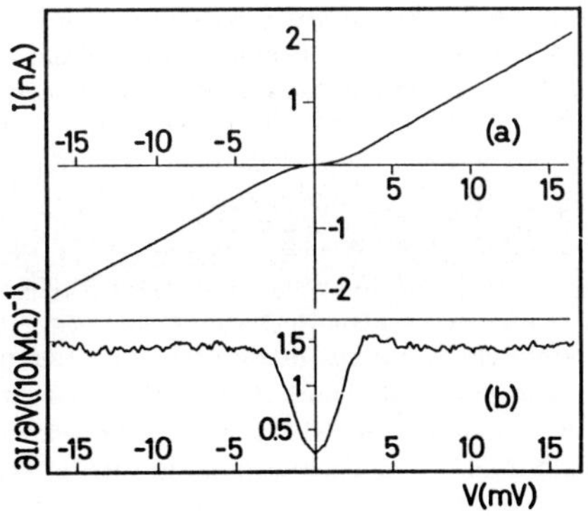

Fig.1 $I(V)$ (a) and $dI/dV(V)$ (b) curves of a point contact on a surface doped Si sample measured at $T = 4.2$K.

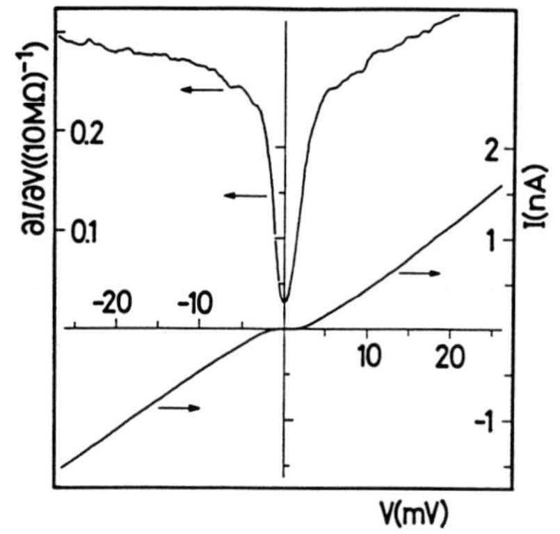

Fig.2 $I(V)$ and $dI/dV(V)$ curves of a point contact on a surface doped Si sample measured at $T = 1.5$K.

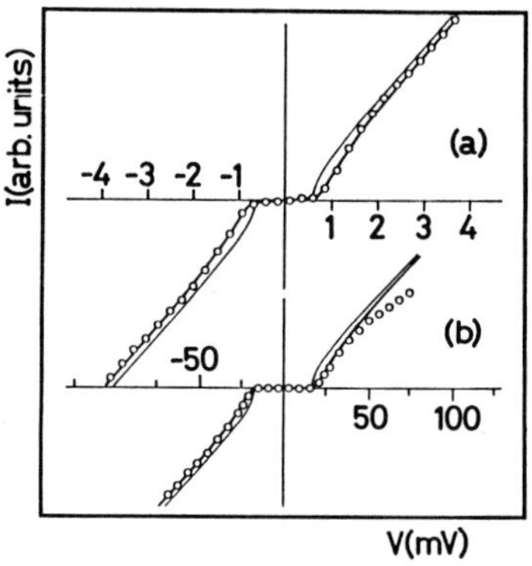

Fig.3 $I(V)$ characteristics (circles) of point contacts on (a) a Sn film at $T = 1.2$K, and (b) ceramic $YBa_2Cu_3O_{7-\delta}$ at $T = 4.2$K, compared with B.C.S. curves (thin lines) and fits including SET-effects (thick lines).

modulation by stray rf signals. Extrapolations of the linear parts however do intersect the origin. C and R^{-1} could be increased by pushing the tip further into the oxide layer.

Fig.2 shows the $I(V)$ and $dI/dV(V)$ curves of another junction measured at $T = 1.5$K. We find $R_T = 36$MΩ and $C = 2.7 \cdot 10^{-17}$F. Cooling this contact down from $T = 4.2$K to $T = 1.5$K did not alter the $I(V)$ characteristics, although the series resistance increased up to $\approx 2 \cdot R_T$. The resistance and apparent capacitance of the junction stayed roughly the same.

This suggests that it is not the high series resistance that provides the decoupling from stray capacitances, but possibly the relatively slow hopping into localised dopant states. In point-contact tunneling on other materials impurity states in the oxide or at the surface of the material probably have the same effect.

3. TUNNELING SPECTROSCOPY

In STM spectroscopy one wants to distinguish between intrinsic properties of the materials under investigation, and effects associated with the nature of the tunneling process in these low capacitance point-contact tunnel junctions. In general one expects that spectroscopic features will be displaced in energy $E' \approx E_0 + e/2C$ and, because of the voltage oscillations across the junction, the results are averaged over a bandwidth of order e/C [8].

Fig. 3a shows (open circles) the $I(V)$ characteristics of a point-contact tunnel junction between a tungsten tip and a 3000 Å Sn film, measured at $T = 1.2$K. The tip is in mechanical contact with the oxide layer of the Sn film. The Sn film was part of a planar tunnel junction with a 150 Å Al film. From the tunneling characteristics of this planar junction the energy gaps of both films could be inferred.

The thin line in fig.3a is the B.C.S. tunneling curve with $\Delta = 0.59$meV from the measurement on the planar junction. The thick line is a Monte Carlo calculation of the $I(V)$ curve including SET effects with the capacitance C as a single fitting parameter. We find $C = 3.2 \cdot 10^{-16}$F.

In fig.3b the $I(V)$ characteristic is depicted of a tunnel junction between a tungsten tip and the ceramic high-T_c superconductor $YBa_2Cu_3O_{7-\delta}$, measured at $T = 4.2$K. The thick line is again a Monte Carlo fit, this time with both Δ and C as fitting parameters. We find $C = 1.6 \cdot 10^{-17}$F and $\Delta = 16$meV. The thin line represents the B.C.S. ($T = 0$) curve for this Δ.

Material intrinsic effects cannot explain the measurement on Sn, as they would also show up in the planar junction. Other effects such as rf pick-up or current induced nonequilibrium can explain a broadening of the gap singularity, but not the shift to higher energy.

4. CONCLUSIONS

Clear manifestations of a Coulomb blockade are observed in STM measurements on semiconductors, metals and superconductors. We find no appreciable effect of the long range electromagnetic environment when changing the series resistance from 50kΩ to 80MΩ. Decoupling from stray capacitances may be due to tunneling through impurity states.

Part of this work was financed by the Stichting voor Fundamenteel Onderzoek der Materie (FOM).

References

[1] D.V. Averin and K.K. Likharev, J. Low Temp. Phys. **62** (1980) 345.

[2] P.J.M. van Bentum, L.E.C. van de Leemput, R.T.M. Smokers and H. van Kempen, Phys. Scripta **T 25** (1989) 122.

[3] K.Mullen, E. Ben-Jacob, R.C. Jacklevic and Z. Schuss, Phys. Rev. **B 36** (1988) 98.

[4] P. Delsing, K.K. Likharev, L.S. Kuzmin and T. Claeson, Phys. Rev. Lett. **63** (1989) 1180.

[5] P.J.M. van Bentum, R.T.M. Smokers, and H. van Kempen, Phys. Rev. Lett. **60** (1988) 2543.

[6] P.J.M. van Bentum, H. van Kempen, L.E.C. van de Leemput, and P.A.A. Teunissen, Phys. Rev. Lett. **60** (1988) 369.

[7] Yu. V. Nazarov, Sov. Phys. JETP **68** (1989) 561.

[8] P.J.M. van Bentum, H.F.C. Hoevers, L.E.C. van de Leemput and H. van Kempen, J. Magn. Magn. Mat. **76-77** (1988) 561.

Physica B 165&166 (1990) 65–66
North-Holland

MAGNETIC VISCOSITY IN r.f.SQUIDs COUPLED TO FERROMAGNETIC CORES

M.Cerdonio,[a,b] P.Falferi,[b] G.Durin,[c] A.Maraner,[c] G.A.Prodi,[a,b] R.Tommasini,[a] and S.Vitale[a]

[a] Dipartimento di Fisica, Università di Trento, and Gruppo Collegato di Trento, Istituto Nazionale di Fisica Nucleare, 38050 Povo, Trento, Italy
[b] Laboratori Nazionali di Legnaro, Istituto Nazionale di Fisica Nucleare, 35020 Legnaro, Padova, Italy
[c] Dipartimento di Fisica, Politecnico di Torino, 10139 Torino, Italy

We are investigating on r.f.SQUIDs coupled to soft ferromagnetic cores operating in the linear reversible regime at liquid helium temperatures. The excess noise found over the empty SQUID background is generated by the thermal fluctuations of magnetization of the core and obeys the fluctuation-dissipation theorem. In particular, the experiments demonstrates the equivalence between 1/f magnetization noise and magnetic viscosity at low fields. Applications of tested ferromagnets to gyromagnetic gyroscopes and impedance matching are briefly discussed.

1. INTRODUCTION

Applications of SQUID magnetometry as for gyromagnetic gyroscopes (1), strain transducers (2) and impedance matching (3) involve the coupling to ferromagnetic cores. In connection with these issues, we recently presented and tested experimentally a detailed model for the operation of radiofrequency SQUIDs coupled to magnetic cores (4). Under conditions of thermal equilibrium and of linear reversible regime, the coupling to the SQUID can be described in terms of a frequency-dependent complex inductance. The imaginary part of this inductance describes the core losses and therefore determines the spontaneous fluctuations of magnetization of the core as given by the fluctuation-dissipation theorem. This thermal noise simply adds to the SQUID background noise under usual operating conditions (4).

2. RESULTS

The experimental apparatus and measurement techniques have been described in ref. (4). In brief, the r.f.SQUID is coupled to the core by a superconducting transformer which provides for in-situ measurements of its inductance. As discussed in ref. (4), the excess flux noise at the SQUID input has a power spectrum given by

$$S_\phi(\omega) = -4k_B T M_{SQ}^2 \frac{Im\{L_T(\omega)\}}{\omega \mid L_T(\omega) \mid^2} \qquad (1)$$

where T is the core temperature, k_B is the Boltzmann constant, M_{SQ} is the SQUID-transformer mutual inductance, L_T is the complex inductance of the superconducting transformer and is linearly related to the initial permeability of the core $\mu(\omega)$. This noise prediction is in detailed agreement with the noise measurements at equilibrium.

Two contributions are mainly present in the excess noise. The first one is due to the roll-off behaviour at some frequency ω_c of the initial permeability $\mu(\omega)$. In the range $\omega \ll \omega_c$ this corresponds to a $Im\{\mu(\omega)\}$ proportional to ω and to a constant $\mid \mu(\omega) \mid = \mu_{lf}$ typically ~ 1000 for the samples tested at liquid helium temperatures. The related Nyquist noise is therefore flat in the same frequency range. A detailed agreement with experiments has been reported in ref. (4) in the temperature range $1.4 \div 4.2K$.The second contribution to magnetization noise comes from the so called magnetic viscosity of the material. It is described by a constant non-zero $Im\{\mu(\omega)\}$ below some frequency and corresponds to a 1/frequency Nyquist noise, as follows from equation 1 if $Im\{\mu(\omega)\} \ll \mu_{lf}$. In ref (5) we reported measurements in the frequency domain, $0.1 \div 1000Hz$, which entirely support the thermodinamic origin of the observed 1/f noise at liquid helium temperatures. The samples tested are toroidal ribbon wound cores of soft ferromagnetic amorphous and polycristalline permalloy materials (6) with negligible magnetic flat noise contribution. A useful phenomenological parameter to describe the 1/f noise tail is α, defined as the low frequency value of the $-arg\{\mu(\omega)\}$. We found α material dependent but sample independent within each composition; however, data are not yet sufficient for making correlations with materials properties. Attempts to test materials with very different structures such as ferrites, both ferroxcube and ferroxplana (7), and spinels (8) resulted unsuccesful because their permeability decays to ~ 1 at liquid helium temperature. Up to now the lowest dissipation found corresponds to an $\alpha \simeq 1 \times 10^{-3}$.

For a linear system, a constant $Im\{\mu(\omega)\}$ over a wide frequency range implies a time-domain response to a step excitation proportional to log(t), for delays t within the corresponding time range. Figure 1 shows such a typical logarithmic relaxation of magnetization. The speed of the logarithmic relaxations depends linearly on the amplitude of the applied step field, as it is shown in figure 2, even at these small variations of $\sim 1\mu A/m$. These results experimentally demonstrate also the equivalence between magnetic viscosity and 1/f thermal noise at fields and temperatures well below those usually involved in magnetic viscosity experiments. In the framework of the widely accepted theoretical explanation, based on thermal activated process with a flat energy barrier distribution, our experiments

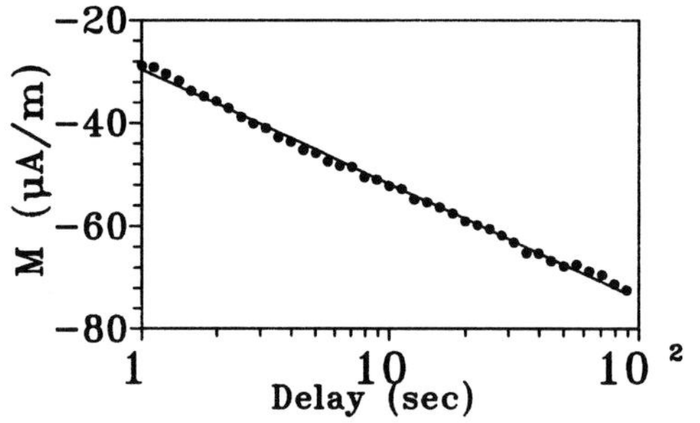

Fig.1: a typical relaxation of the magnetization M of an amorphous toroidal core at $4.2K$ after a step change in the applied field of $\sim 1\mu A/m$. Circles are data points and the straight line is shown to demonstrate the logarithmic behaviour in the time delay from the step. The $Im\{\mu(\omega)\}$ is nearly constant in the corresponding frequency range $0.01 \div 1Hz$.

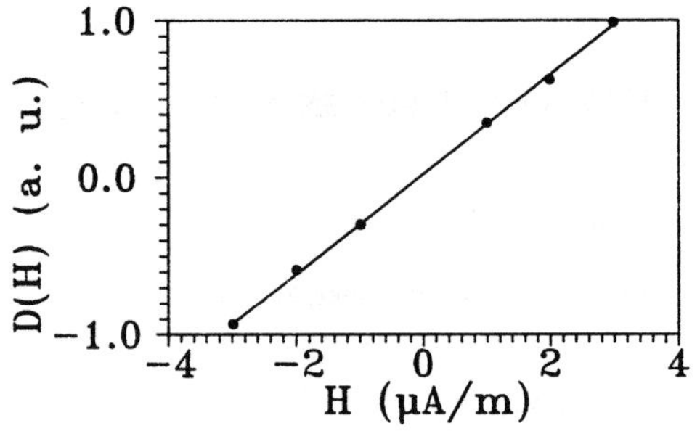

Fig.2: the speed D of the logarithmic relaxations, defined as $M(t) = D \times \log(t)$ as a function of the amplitude of the step excitations in applied field H. Circles are data points for the same sample of fig.1 and are fitted by the straight line.

scan much lower activation energies. Further investigations at lower temperatures are in progress.

3. DISCUSSION

The value of the phenomenological parameter α gives a picture of the 1/f noise contribution to the total thermal noise in a r.f.SQUID coupled to a ferromagnetic core. In fact,the product $\omega_c\alpha$ discriminates between frequency regions where 1/f or flat noise dominates. For the tested cores with the lowest losses this critical frequency is of the order of $1kHz$.

For what concerns gyromagnetic gyroscopes, the operation frequency falls well inside the 1/f tail. By describing the thermal fluctuations of magnetization in terms of an equivalent applied field, the noise power spectrum is given by the fluctuation-dissipation theorem as

$$S_B(\omega) = \frac{4k_BT\mu_o}{V}\frac{\alpha}{\omega\mu_{lf}} \qquad (2)$$

where V is the total volume of the ferromagnetic maerial. If the empty SQUID noise is negligible in respect to the magnetization noise, a condition which is met if μ_{lf} is high enough, then the gyrosope is optimized and its limiting sensitivity in angular velocity follows from equation 2. In fact, $S_\Omega = (e/m)^2 S_B$, where e/m is the spin gyromagnetic factor for electrons. For the prototype under construction $V \sim 0.1l$, $\alpha \sim 10^{-3}$ and $\mu_{lf} \sim 10^3$, and the minimum detectable angular velocity at $1K$ is calculated to be

$$\Omega_{min} \simeq 6 \times 10^{-5}\frac{Hz}{\nu}\frac{rad}{s\sqrt{Hz}}$$

where ν is the modulating frequency of the gyroscope.For what concerns impedance matching, the advantage of the use of low losses ferromagnetic cores in coupling transformers rests mainly in the realization of larger inductances in smaller volumes. For instance a capacitive resonant transducer of a resonant gravitational antenna can be read by a SQUID through an impedance matching stage. However, the best ferromagnets tested up to now are still a hundred times more dissipative at the operating frequency $\sim 1kHz$ than the empty superconducting transformer employed (9). This fact prevents actual applications of ferromagnetic cores where high quality factors are required.

ACKNOWLEDGMENTS

We thank Proff. P.Mazzetti, H.Warlimont and H.R.Hilzinger for helpful discussion and advice. We are indebted also to Vacuumschmelze, Philips and Siemens for providing us the samples and to R.Dallapiccola and F.Gottardi for technical help.

REFERENCES

(1) S. Vitale, M. Bonaldi, P. Falferi, G.A. Prodi and M. Cerdonio, Phys.Rev. B 39, (1989), 11993
(2) L. Adami, M. Cerdonio, F.F. Ricci and G.L. Romani, Appl. Phys. Lett. 30, (1977), 240
(3) M. Cerdonio, F.F. Ricci and G.L. Romani, J. Appl. Phys. 48, (1977), 4799. S. Barbanera, P. Carelli, I. Modena and G.L. Romani, J. Appl. Phys. 49, (1978), 905. S.Q. Xue, P. Gutmann and V. Kose, Rev. Sci. Instrum. 52, (1981), 1901.
(4) S. Vitale, G.A. Prodi and M. Cerdonio, J. Appl. Phys. 65, (1989), 2130.
(5) G.A. Prodi, S. Vitale, M. Cerdonio and P. Falferi, J. Appl. Phys. 66, (1989), 5984.; G.A.Prodi, S. Vitale, M. Cerdonio and P. Falferi, J.M.M.M., in print.
(6) Courtesy of Vacuumschmelze GmbH, Hanau, BRD.
(7) Courtesy of Philips, Eindhoven, NL
(8) Courtesy of Siemens Matsushita GmbH, Muenchen, BRD
(9) P. Carelli, M.G. Castellano, C. Cosmelli, V.Foglietti, I. Modena, Phys. Rev.A 32, (1985),3258.

Physica B 165&166 (1990) 67–68
North-Holland

NOISE AND CHAOS IN THE rf SQUID

A. R. BULSARA, and E. W. JACOBS

Naval Ocean Systems Center, San Diego, CA 92152, USA.

W. C. SCHIEVE

Physics Dept. Univ. of Texas, Austin, TX 78712, USA.

The response of a chaotic nonlinear dynamic system to weak external noise has recently been the subject of several investigations.[1–6] In this work we consider the rf SQUID, driven by a combination of deterministic (periodic) and random forcing terms. In its simplest form, the rf SQUID consists of a single Josephson junction shorted by a superconducting loop. An external magnetic field produces a geometrical magnetic flux across the loop together with a circulating supercurrent in the loop. Defining by $x(t)$ the net flux (in units of the flux quantum) sensed by the SQUID and assuming the external flux (in the same units) to be made up of a deterministic and fluctuating component, one may write down an equation for $x(t)$:[3]

$$\ddot{x} + k\,\dot{x} + \omega_0^2 x + \beta \sin 2\pi x = q \sin(\omega t) + F(t),$$

where the damping k, natural frequency ω_0 and the nonlinearity parameter β may be defined in terms of the inductance, capacitance and normal state resistance of the SQUID loop and we assume $F(t)$ to be Gaussian delta-correlated noise with zero mean and variance σ^2. In this work we consider the effects of weak Langevin noise on the dynamics of the system *above* the homoclinic threshold i.e. in the chaotic regime.

In figure 1 we show the maximal Liapounov exponent λ as a function of the periodic driving amplitude q with $(\omega_0, \beta, k, \omega) \equiv (1.0, 2.0, 1.0, 2.25)$. The homoclinic threshold for this parameter configuration occurs at $q \approx 1.23$. One observes positive and negative values of λ corresponding respectively to chaotic and periodic solutions. The effects of a small amount of Langevin noise are shown via the dotted curve. One observes a "smoothing" of the Liapounov exponent spectrum. Further, it is evident, in the chaotic regime, that for large enough values of the noise the Liapounov exponent λ can always be made positive. Important differences exist, however, between the well-known phenomenon of deterministic chaos (which occurs in the absence of noise) and the "noisy chaos" corresponding to the change in sign of the maximal Liapounov exponent, induced by noise.

The Poincare plot of figure 2 shows a section of a deterministic chaotic attractor of the system; the attractor possesses the well-known self-replication property and its power spectral density resembles that of broadband noise on which peaks corresponding to the driving frequency and its harmonics are superimposed. In the presence of small

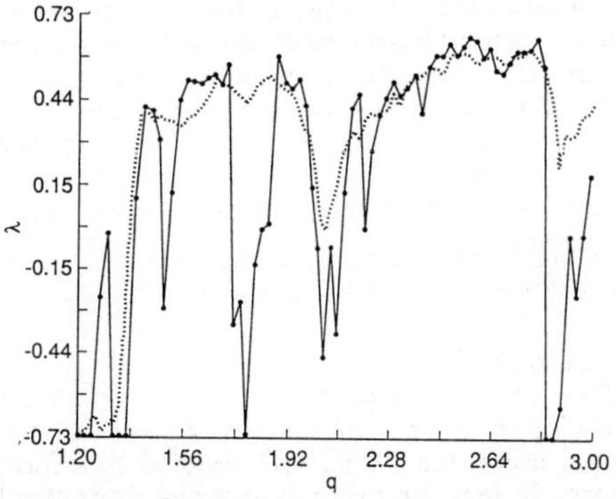

Fig 1: Maximal Liapounov exponent in the absence of noise (solid line) and for noise variance $\sigma^2 = .0025$ (dotted line). $(\beta, \omega_0, k, \omega) \equiv (2,1,1,2.25)$.

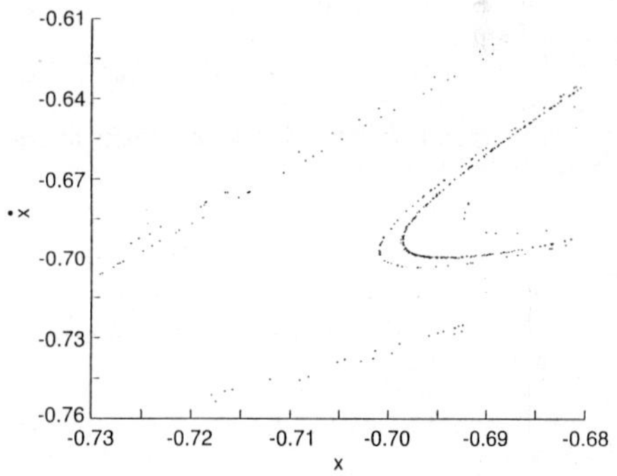

Fig 2: Section of deterministic attractor corresponding to (1), showing self- replication property. $q = 1.43$; other parameters same as in fig 1.

amounts of noise, a smearing occurs in the deterministic attractor, an increased area of phase space is now accessible to the system and the self-replicating property is destroyed (figure 3). One may numerically compute the probability density function $P(x)$ corresponding to the deterministic and noisy attractors. For the deterministic case, the probability density displays multiple maxima reminiscent of the non-differentiable potentials suggested by Graham and Tel[7]. In the presence of noise, the peaks in the its harmonics are superimposed. In the presence of small amounts of noise, a smearing occurs in the deterministic attractor, an increased area of phase space is now accessible to the system and the self-replicating property is destroyed (figure 3). One may numerically compute the probability density function $P(x)$ corresponding to the deterministic and noisy attractors. For the deterministic case, the probability density displays multiple maxima reminiscent of the non-differentiable potentials suggested by Graham and Tel[7]. In the presence of noise, the peaks in the probability density function are seen to have a greater width but smaller height; the noise tends to "smooth" the probability density function, coarse-graining the deterministic randomness of the attractor itself. This feature is a direct consequence of the greater region of phase space that is made available to the system in the presence of noise. Ultimately, if the noise is increased even further, the probability density function collapses into a set of broad well-defined peaks corresponding to the wells in the potential function describing the deterministic dynamics of the SQUID. The maximal Liapounov exponent λ may be examined as a function of increasing noise variance for fixed deterministic driving. As σ^2 is increased, λ decreases, eventually approaching zero; this limit corresponds to the pure noise-driven case.

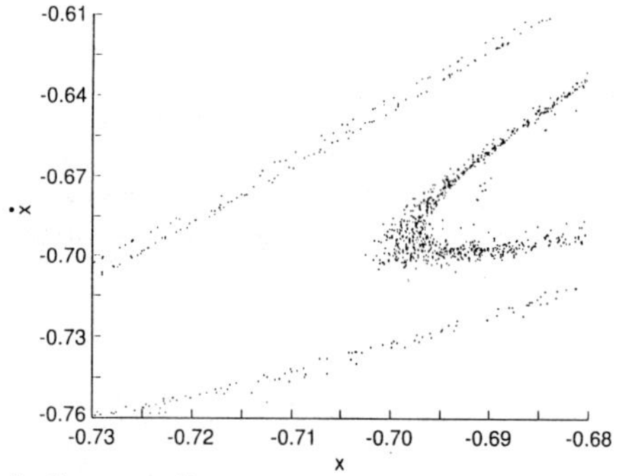

Fig 3: Same as fig 2 with Langevin noise of variance $\sigma^2=.0025$; no self-replication.

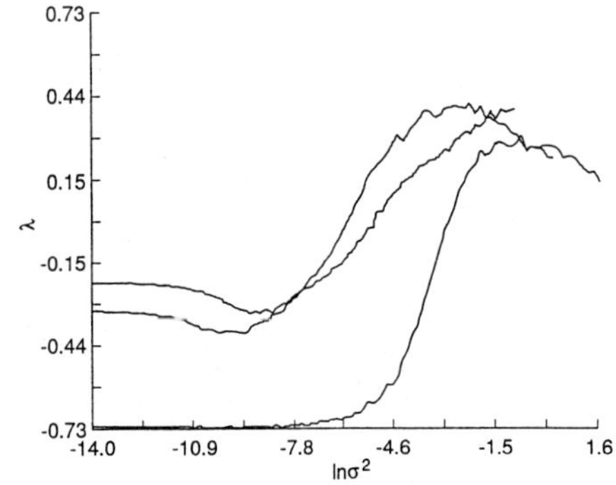

Fig 4. Liapounov exponent as function of noise variance for q=5.35 (bottom curve), 4.001 (middle curve) and 2.05 (top curve). Other parameters same as in fig 1.

We now consider the effect of noise on a periodic solution of (i.e. a value of the modulation amplitude q for which the noise-free Liapounov exponent λ in figure 1 is negative). Figure 4 shows the maximal Liapounov exponent λ as a function of the noise variance σ^2 for three different values of the modulation amplitude q (all other system parameters being held constant). The cases considered correspond to motions of period one ($q=5.35$), two ($q=4.001$) and three ($q=2.05$) in the absence of noise. One observes that λ crosses zero at some critical value σ_c^2 of the noise strength. The critical noise strength depends on the system and driving parameters as well as on the value λ_0 of the Liapounov exponent in the absence of noise. This relationship will be further quantified in an upcoming publication. Similar behavior is observed for all the periodic points characterized by a negative exponent in figure 1. It is instructive to consider the Poincare dynamics of the system for different values of the noise strength. In each case the noise-free Poincare plot consists of discrete points (the number corresponds to the periodicity of the motion) with each point having associated with it a delta-function peak on the probability density plot. As the noise strength is gradually increased the Poincare plots display the transition from noisy periodic motion (characterized by fuzzy dots and a corresponding broadening of the delta-function peaks in the probability density function) to purely noise-driven transitions between the potential wells. The power spectral density consists (in the absence of noise) of a series of peaks at multiples of ω/p (p being the periodicity of the noise-free motion). As the noise level increases these peaks are superimposed on the broadband noise background; ultimately, for very large values of the noise strength the noise floor will rise up to obscure the peaks. In all these cases there exists an intermediate regime ($\lambda \geq 0$) in which the motion is seemingly random but not purely noise-driven. The probability density function $P(x)$ may have multiple peaks showing that a much greater area of phase space is being traversed by the system than in the periodic or noisy-periodic cases. This case which must be contrasted with the case of pure noise-induced motion (occurring at very high σ^2 values and characterized by a broad probability density function with as many peaks as there are wells in the potential) may be considered "noise-induced chaos" and is characterized by a positive Liapounov exponent as well as a broadband power spectral density that includes peaks at multiples of ω/p, p being the periodicity of the motion in the absence of noise. Although the Poincare plot $(\dot{x},x)$ in the regime of noisy chaos may often resemble the attractors that characterize deterministic chaos, it lacks the self-replication property; the probability density function $P(x)$ for the noisy chaotic case is also broader than its deterministic chaotic counterpart, reflecting the "smoothing" effects of noise referred to earlier.

References

1. J. Crutchfield, J. Farmer and B. Huberman; Phys. Repts. 82, 45 (1982).
2. T. Kapitaniak; "Chaos in Systems with Noise", (World Publishing, N.Y. 1989).
3. G. Mayer-Kress and H. Haken; J. Stat. Phys. 26, 149 (1981).
4. H. Herzel, W. Ebeling and Th. Schulmeister; Z. Naturforsch. 42a, 136 (1987).
5. A. Bulsara, W. Schieve and E. Jacobs; Phys. Rev. A41, 668 (1990).
6. W. Schieve and A. Bulsara; Phys. Rev. A41, 1172 (1990).
7. R. Graham and T. Tel; J. Stat. Phys. 35, 729 (1984).

Physica B 165&166 (1990) 69–70
North-Holland

HIGH T_C SUPERCONDUCTING DIFFUSION TYPE WEAK LINKS

Z. IVANOV*+ and T. CLAESON +

+ Department of Physics, Chalmers University of Technology, 412 96 Göteborg, Sweden
* Institute of Electronics BAS, 1784 Sofia, Bulgaria

The properties of weak links with artificial grain boundary barriers formed by a diffusion of material from narrow lines into high j_c microbridges have been studied. They showed Josephson properties as measured by dc and microwave response. I_cR products of up to 3 mV were measured at 4.2K for a Pb-BSCO film (T_c=64K) and 1 mV at 77K for an YBCO film (T_c=87K).

1. INTRODUCTION

To develop high T_c superconducting (HTS) Josephson junctions and to assure effective Josephson current transport through an artificial barrier, a smooth and sharp interface on an atomic scale is required. The elevated temperatures needed to obtain HTS films with high values of T_c and j_c tend to give oxygen deficit interfaces. There are also problems with the coherence length and the current transport along the c-axis of a film. Merely low I_cR-products have been achieved (R being the junction normal state resistance, I_c the critical current) (1). Better values have been reported for native grain boundaries, up to 3 mV (2), but it is difficult to have a sufficiently high j_c in the electrodes that are needed for many applications.

We have studied the dc and rf properties of weak links with artificial grain boundary barriers formed by diffusion of material from narrow lines into high j_c microbridges.

2. RESULTS AND DISCUSSION

Post annealed Pb-BSCO and in-situ YBCO thin films (300 nm) were deposited by laser ablation on MgO (100) substrates (3). The films were highly c-axis oriented with T_c=83K and $j_{c77}=1\cdot10^4$ A/cm^2 for Pb-BCCO and T_c=92K and $j_{c77}=7\cdot10^5$ A/cm^2 for YBCO. Three 20 μm long and 4-6 μm wide bridges and corresponding contact pads were patterned on each substrate by e-beam lithography and Ar ion etching. The process was optimized to minimize the degradation of T_c. A narrow (70-130 nm) line was formed across the bridge using an e-beam exposed double layer resist, thermal evaporation of Al or Cu (100 nm) and lift-off (Fig. 1a). A weak link in the bridge was formed by diffusion of an Al(Cu) line into the HTS material at T_a=300-450 ^{0}C for t_a=10-20 min. (Fig.1b). As the width ,W, of the Al(Cu) line is less than the film grain size, d (which is about 500 nm for Pb-BSCO and 200 nm for YBCO,cf. Fig. 1c), we assume that diffusion occurs mainly along one grain boundary plane (Fig. 1b) and that the length , L, of the weak link mainly depends on T_a and t_a and not on W as long as W<<d .

I-V, dV/dI vs. V, I_c(T) and the microwave response were measured for the weak links.

Three Pb-BSCO weak links were tested. T_c for the bridge was depressed during the processing. An anneal at 450 ^{0}C for 15 min. gave a full depression of the superconductivity of the weak link

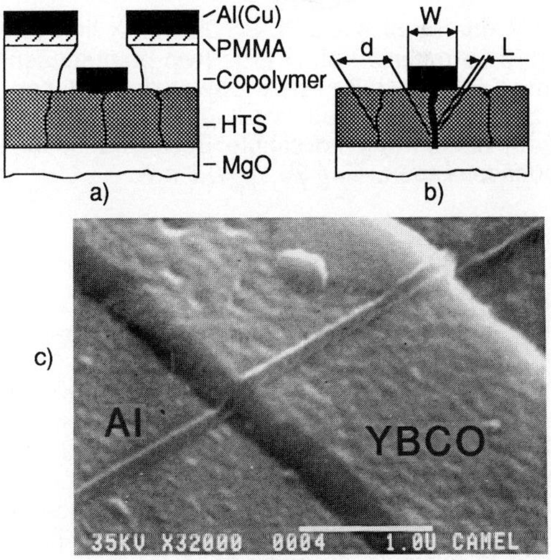

FIGURE 1
(a) Fabrication of 70-130 nm narrow Al or Cu lines across HTS bridges; (b) grain boundary diffusion type weak link; (c) SEM photograph of the weak link (before annealing).

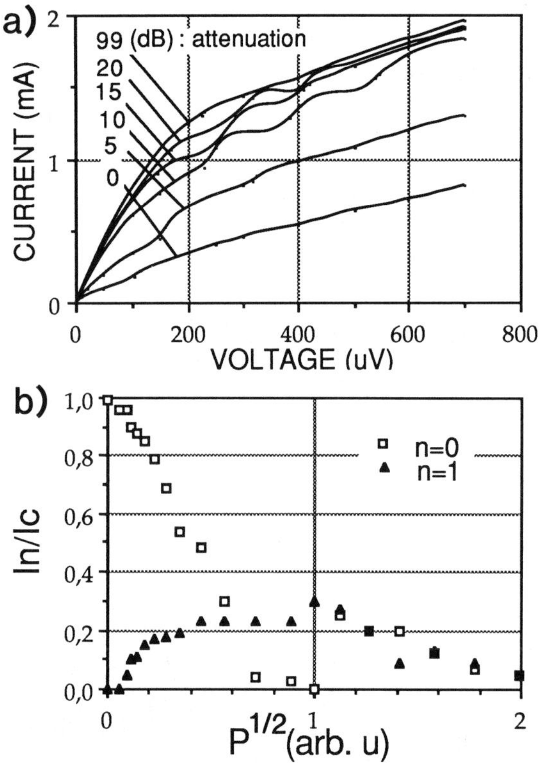

FIGURE 2
(a) I-V curves of a Cu/Pb-BSCO weak link under microwave radiation of 37 GHz (two point measurement). (b) The microwave power dependence of the zeroth and first step amplitudes. The junction is current biased and no simple Bessel function dependence is seen. T=4.2K. I_c=0.6 mA.

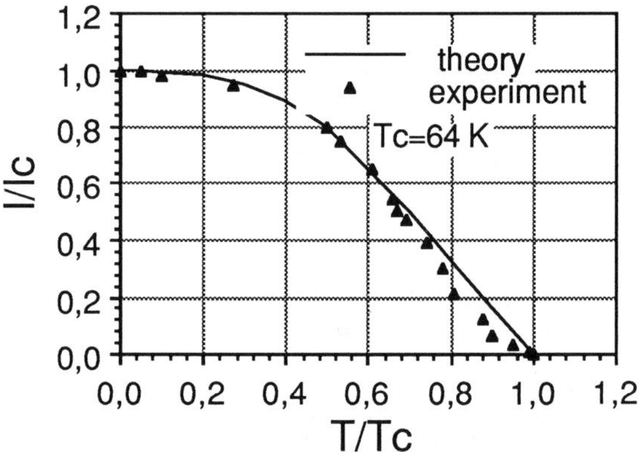

FIGURE 3
Temperature dependence of the critical current of the Cu/Pb-BSCO weak link compared with theoretical data for an SNS weak link (4).

with an Al line. Decreasing T_a to 300 ^{0}C for 10 min. gave a supercurrent with I_cR=80 μV at 4.2K. The same annealing procedure gave I_cR=3 mV, R=3 Ohm and good microwave response for a Cu line. Fig.2a shows microwave induced steps for the latter junction in a two point measurement. The step heights vary continously with microwave power going through maxima and minima (Fig. 2a). As the Josephson junction had a low resistance it is current rather than voltage biased. Thus the step height to microwave amplitude does not display a simple Bessel function dependence. A comparison of the temperature variation of the critical current of this weak link, $I_c(T)$, with a theoretical curve for an SNS bridge (4) shows a very good qualitative agreement, see Fig. 3. Hence, we conclude that Cu(Al) diffuses along one of the grain boundaries and forms a weak link. The Cu is preferable to Al, which tends to form too weak a link in Pb-BSCO.

Al is the preferable overlay for YBCO. Cu gave too low R. The YBCO films were generally of higher quality than the Pb-BSCO ones. By capping the YBCO film before covering it with resist and processing, it was possible to avoid severe degradetion of the superconductivity in the bridge. Six weak links were fabricated. I_cR values at 77K ranged between 0,15 and 1 mV. The microwave response was as expected for a Josephson junction.

3. CONCLUSIONS

Artificial grain boundary weak links were formed by diffusion of 70-130 nm narrow lines of Al or Cu into HTS bridges. They showed Josephson properties with I_cR products of up to 3 mV at 4.2K for Pb-BSCO (T_c=64K) and 1 mV at 77K for YBCO (T_c=87K).

ACKNOWLEDGEMENT

The support from the Swedish Board of Technical Development and the Bulgarian Academy of Sciences is acknowledged.

REFERENCES

(1) M. Matsuda, A.Matachi, N. Hashimoto, and S. Kuriki, Extended Abstracts of 1989 International Superconductivity Electronics Conf. (ISEC'89) 497.
(2) I.S. Gergis, J.A. Titus, P.H. Kobrin, and A.B. Harker, Appl. Phys. Lett. 53 (1988) 2226.
(3) Z. Ivanov and G. Brorsson, Appl. Phys. Lett. 55 (1989) 2123.
(4) K.K. Likharev, Rev. Mod. Phys. 51 (1979) 101.

Physica B 165&166 (1990) 71–72
North-Holland

FABRICATION AND PROPERTIES OF YBCO/Ag/YBCO PLANAR SNS JUNCTIONS

David COHEN and Emil Polturak

Physics Department, Technion-Israel Institute of Technology, Haifa 32000, Israel.

We fabricated planar YBCO/Ag/YBCO junctions having lateral dimensions of 1mm × 1mm and a nominal thickness of 300Å. The junctions were prepared in situ using the laser ablation technique and contact masks to define the geometry. The critical current of the junctions in a magnetic field and in a microwave radiation shows response characteristic of an SNS Josephson Junction. In addition, the junctions show a novel nonlinear effect which to the best of our knowledge was not observed with conventional superconductors.

The fabrication of HTSC thin film Josephson devices is difficult because of the short coherence length. This problem can be circumvented to some extent by using a normal metal for the barrier. The coherence length of the normal metal is dictated by the mean free path, and is significantly larger than that of the HTSC, allowing for a fabrication of junctions having thickness of a few hundred Angstrom[1]. We have fabricated planar junctions of YBCO/Ag/YBCO using laser ablation[2]. By interchanging targets for the laser and stainless steel contact masks we defined a cross geometry for the junction having lateral dimensions of 1mm × 1mm and a nominal thickness of 300Å. All of this was done in situ at a substrate temperature of 650C, using external manipulators. In this way, we're protecting the various interfaces against degradation caused by the exposure to the atmosphere. Due to the high temperature at which the junctions were prepared, we found a considerable interdiffusion of Y, Ba and Cu into the junction region. The resistance of the junctions showed a metallic behaviour, with a value of 0.1-1 mohm at low temperatures. In some junctions, we used a mixture of Zr and Ag for the normal part, in order to test the behaviour of the junctions made of an alloy. All of the work reported was done with films grown on sapphite substrates.

The T_c for the junctions was typically 77K. We measured the dependence of I_c on magnetic field directed in the plane of the junction. This is shown in Fig. 1. One can interpret this dependence as that of a Josephson junction through which the current density is not uniform, having a maximal value at the center of the junction and falling off towards the edges. In addition, taking the penetration depth as 1000Å, we estimate the effective width of the junction as 0.2 mm. This is consistent with a picture where the interdiffusion produces some alloy which has a larger concentration at the edges and does not superconduct.

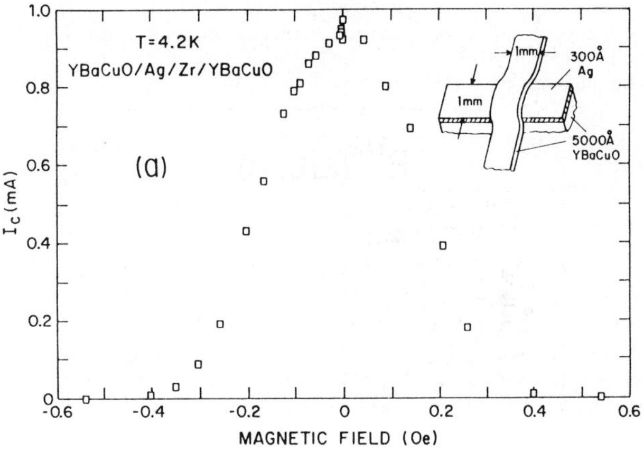

FIGURE 1
Magnetic field dependence of the critical current. Inset shows the geometry of the junction.

We also measured the dependence of I_c on the amplitude of 10.3 GHz radiation incident at the junction. This dependence is in accord with the prediction of the RSJ model[3], as shown in Fig. 2. It seems, therefore, that the junctions behave in several important respects as Josephson junctions. What we have not observed were Shapiro steps which should appear on the I-V characteristic. Instead, we observed the appearance of a region where the voltage across the junction has an opposite sign to the current. The magnitude of the opposite sign voltage incrased with the microwave power. This novel

0921-4526/90/$03.50

effect, shown in Fig. 3, was reproduced in several junctions having either cross geometry or edge geometry. We found that a modification of the RSJ model to include the possibility of a nonlinear interaction between the flux threading the junction and the current flowing through it can reproduce I-V characteristics in a qualitative agreement with experiment.[4]

REFERENCES

(1) G. Deutscher and P. de Gennes, in Superconductivity, Ed. Parks, Vol. 2, Marcel Dekker, 1969, Ch. 17.
(2) G. Koren, E. Polturak, B. Fisher, D. Cohen and G. Kimel, App. Phys. Lett. 53, 695 (1988).
(3) P. Russer, J. App. Phys. 43, 2008 (1972).
(4) D. Cohen and E. Polturak, to be published.

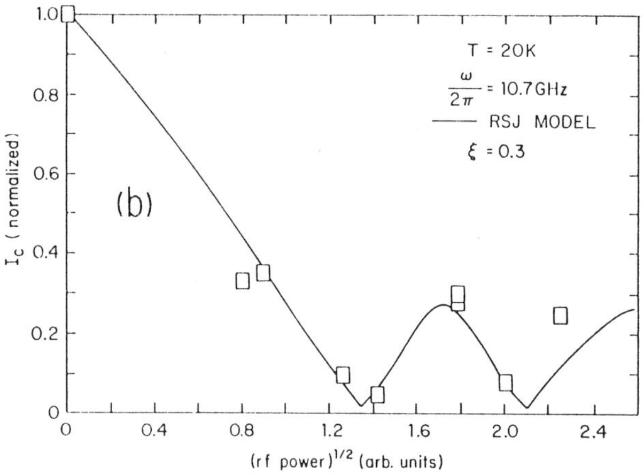

FIGURE 2
Dependence of the critical current on the amplitude of the microwaves. Solid line is a fit to the RSJ model with $h\omega/2eI_cR = 0.3$.

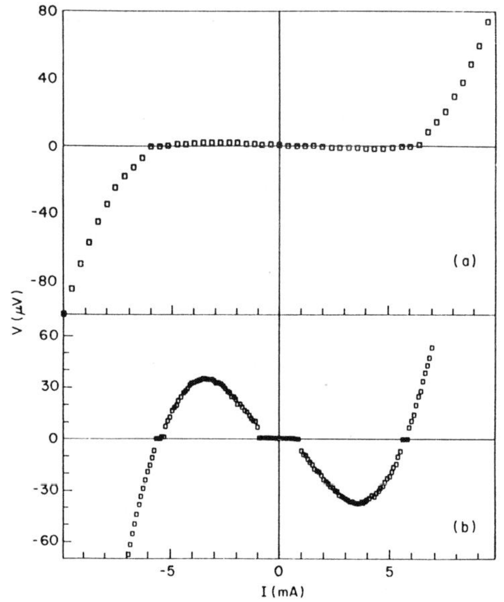

FIGURE 3
I-V curves at 50K with microwave radiation. In (a) the microwaves are attenuated by 18 dB with respect to (b).

Physica B 165&166 (1990) 73–74
North-Holland

MICROWAVE RESPONSE OF YBaCuO/MgO/Pb PLANAR JOSEPHSON TUNNEL JUNCTIONS

I. IGUCHI, Z. WEN and M. FURUYAMA

Institute of Materials Science, University of Tsukuba, Tsukuba, Ibaraki 305, Japan

Microwave response of YBaCuO (YBCO)/MgO/Pb planar Josephson tunnel junctions fabricated by an in situ deposition technique is presented. The observed Josephson critical current is 20 - 400μA. The current-voltage characterisitcs under microwave irradiation of 9GHz yield the Shapiro steps at integer multiples of hf/2e where f is the microwave frequency. Dependence of the Shapiro step height on the microwave voltage amplitude is consistent with that of the conventional theory. A series of subharmonic steps at half integer multiples of hf/2e are also observable, although their structures are weak.

1. INTRODUCTION

The thin film Josephson tunnel junctions utilizing high T_c oxide superconductors are quite important for both basic studies and device applications. The fabrication of high T_c tunnel junctions is, however, very difficult due to the extremely short coherence length and the surface degradation effect of these materials. Quite recently, we have reported the reproducible tunnel junctions by an in situ fabrication technique utilizing an electron-beam coevaporation system [1]. The observation of Josephson effect in thin film YBCO tunnel junctions has been also reported very recently [2,3].

In this paper, we present the measurements of high frequency properties of YBCO/MgO/Pb planar Josephson tunnel junctions with an MgO insulating barrier.

2. JUNCTION FABRICATION

The junctions were fabricated by the in situ deposition of three layers of films utilizing an electron-beam coevaporation chamber equipped with a mask changing system. First, a YBCO film of 0.2mm wide and 150-200nm thick was deposited onto MgO(100) and $SrTiO_3$(100) substrates at 600-700°C, then changing masks, a MgO barrier film of < 6nm thick and subsequently a Pb film of 0.2mm wide and 300nm thick were deposited. The fabricated junctions were cross-type and quite reproducible. The tunnel resistance strongly depended on the MgO barrier thickness. Josephson current was observed for the samples with the barrier thickness less than 4nm. From the X-ray diffraction analysis, YBCO films were found to be highly oriented along the c-axis direction.

3. RESULTS

Figure 1 shows the observed Josephson current-voltage characteristics under microwave irradiation of 9GHz. The appearance of Shapiro steps are evident. The tunnel resistance of Josephson junctions was very small and 0.01 - 0.3Ω. The observed critical current was 20-400 μA. A hysteretic behavior of the current-voltage characteristic was also occasionally seen.

The RI_c product of the junction was very small (of order of μV), whose beha-

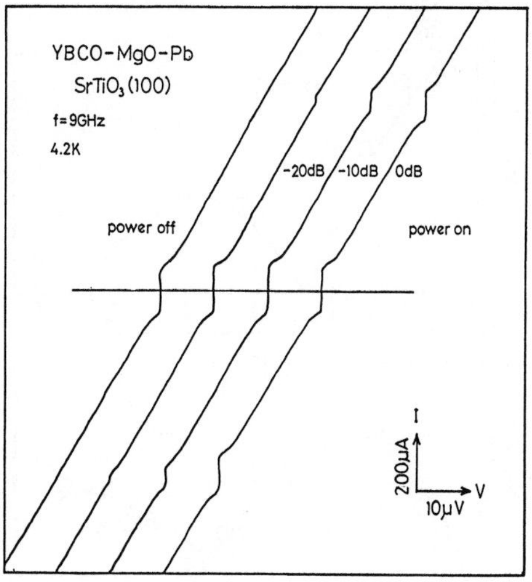

Fig. 1. Current-voltage characteristics of Josepshon tunnel junction under microwave irradiation of 9GHz.

0921-4526/90/$03.50

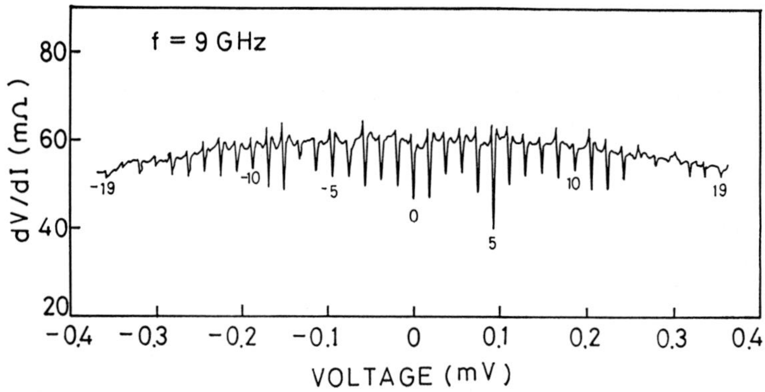

Fig. 2. Differential tunnel resistance as a function of voltage under microwave irradiation of 9GHz.

vior was rather strange since a tunnel gap of about 20meV was clearly observed for a 1Ω junction. In the I-V curves of a Josephson junction, a Pb gap structure was observable about at 1mV, but a YBCO gap structure was not because the critical current of a YBCO film was attained before it became observable.

For the sample shown in Fig. 1, the Shapiro steps were observed up to the fourth one. For the other sample, they were observed up to at least the 19th step. Figure 2 shows such an example. The trace shows the differential tunnel resistance dV/dI vs V curve. The number of observable Shapiro steps increased as the microwave power was increased.

Figure 3 shows the dependence of the height of the n-th Shapiro step (n=0,1)

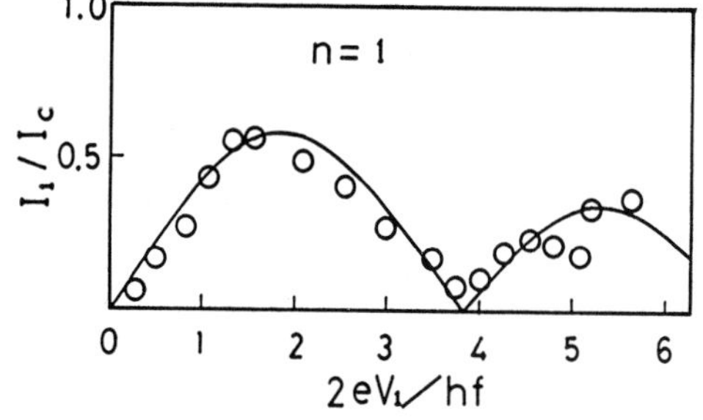

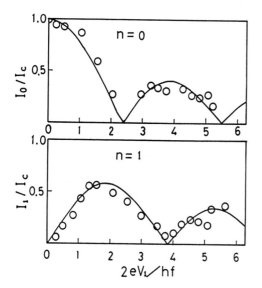

Fig. 3. Dependence of the n-th Shapiro step height on the microwave voltage.

on the microwave voltage V_1. The solid lines correspond to the theoretical amplitude of the n-th order Bessel function. The agreement between the experiment and the theory is reasonable.

It is remarkable to point out here that the Shapiro steps were also observed at half integer multiples of hf/2e although their structures were weak as shown in Fig. 4. At the present time, the appearance of these subharmonic step structures is not clear, but we expect that it may be related to some physical properties of high T_c superconductors.

REFERENCES

(1) M. Furuyama, I. Iguchi, K. Shirai, T. Kusumori, H. Ohtake, S. Tomura and M. Nasu, Jpn. J. Appl. Phys. 29 (1990) No. 3 in press.

(2) J. Kwo, T. A. Fulton, M. Hong and P. L. Gammel, Appl. Phys. Lett. to be published.

(3) I. Iguchi, M. Furuyama, T. Kusumori, K. Shirai, S. Tomura, M. Nasu and H. Ohtake, Jpn. J. Appl. Phys. 29 (1990) No. 4 in press.

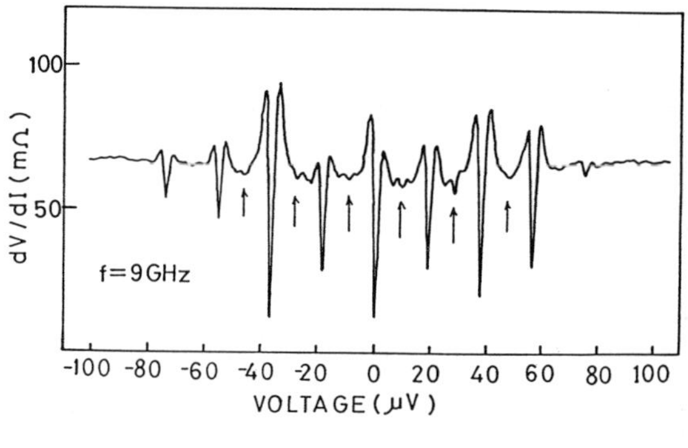

Fig. 4. Differential tunnel resistance as a function of voltage under microwave irradiation of 9GHz.

Physica B 165&166 (1990) 75–76
North-Holland

PERFORMANCE OF ALL-NbN SUPERCONDUCTIVE TUNNEL JUNCTIONS AS MIXERS AT 205 GHz

W.R. McGrath[1], J.A. Stern[2], H.G. LeDuc[1]

1. Jet Propulsion Laboratory, California Institute of Technology 4800 Oak Grove Dr. Pasadena, CA 91109
2. California Institute of Technology, Pasadena, CA

We have fabricated small area ($1x1\mu^2$), high current density NbN-MgO-NbN tunnel junctions with I-V characteristics suitable for high frequency mixers. These junctions are integrated with superconducting microstrip lines designed to resonate out the large junction capacitance. We have studied the mixer gain and noise performance near 205GHz as a function of the inductance provided by the microstrip. This has yielded values of junction capacitance of $85fF/\mu^2$ and magnetic penetration depth of 3800 angstroms. Mixer noise as low as 133K has been obtained for properly tuned junctions. This is the best noise performance ever reported for an NbN SIS mixer.

1. INTRODUCTION

Heterodyne receivers employing Nb-based or Pb-alloy SIS tunnel junctions as the nonlinear mixing element have been developed in recent years for frequencies up to a few hundred GHz(1). At higher frequencies, in the submillimeter wave band (300-3000GHz) NbN tunnel junctions are the best choice due to the high energy gap of about 5 mV which implies an energy gap frequency of 1200 GHz. We have fabricated small area, high current density NbN junctions with I-V characteristics suitable for high frequency mixers. These junctions are integrated with superconducting microstrip lines designed to resonate with the large junction capacitance. We have studied the mixer gain and noise performance at 205 GHz as a function the inductance provided by the microstrip. This has yielded values for junction capacitance and the magnetic penetration depth. Mixer noise temperature as low as 134K has been obtained for properly tuned junctions. This is the best noise performance ever reported for an NbN mixer.

2. EXPERIMENTAL DESIGN

High quality NbN-MgO-NbN tunnel junctions have been fabricated using a recently developed trilayer process which is fully described elsewhere(2,3). The junctions have an area 1x1 μ^2, a current density Jc = 5000 - 10,000 A/cm2, and a normal state resistance R_n = 60-80 Ω. The gap voltage is V_g = 4.8 mV at 4.2 K, and the width ΔV_g of the quasiparticle current rise at V_g is typically about 1 mV.

The junctions are tested at several frequencies between 200 and 210GHz in a full height waveguide mount. Specially designed variable temperature noise sources(4) allow mixer gain and noise to be measured to better than ±10%. Measurements on a 48X scale model of the mixer block indicate that it can properly impedance match junctions with values of the relaxation parameter $\omega R_n C$ up to about 3 (where $\omega = 2\pi f$, f=205 GHz, and C is the junction capacitance). Previous reports (5) give a junction capacitance of 80 fF/μ^2 for these junctions. This yields a value $\omega R_n C$ = 8 for $R_n=50\Omega$, which implies that most of the rf current will be shunted by the junction capacitance.

To properly tune this large capacitance, we have integrated a superconducting open-circuited microstrip line (stub) with the NbN junctions. The stub length is designed to provide a inductive susceptance to resonate out the junction capacitance. The propagation velocity υ on the line determines the length and is calculated using the expressions given in ref. (6). Stubs 75μ, 80μ, and 86μ long by 4.5μ wide, and 75μ by 7μ were tested. The total capacitance each stub would resonate with is calculated using $C = 1/(\omega^2 L)$ where L is the stub inductance. For our stubs, the magnetic penetration depth λ (3000-4000Å), film thickness (3000-4000Å), and dielectric thickness (1200Å) are comparable. In this case, υ is sensitive to λ, and hence it must be known accurately to properly design the resonant circuit.

3. RESULTS AND DISCUSSION

Figure 1 shows the mixer noise temperature versus the capacitance tuned by the stub. Without the stub, values above 1000K were obtained. However, T_m below 200K was obtained for stubs which tuned a capacitance near 85fF which agrees well with previously reported data(5) using a different technique. Mixer gain was also found to improve from -19dB without a stub to -11dB with one.

The propagation velocity υ on the stub was determined by a method (6) using the ac Josephson Effect in the junctions as a voltage controlled oscillator to sample the resonances of the stub. The frequency spacing of these resonances is given by $\upsilon/2l$ where l is the length of the stub. This method gave λ = 3800Å for our films, which is significantly larger than the previously reported (5) value of 2800Å. This led to stub lengths of 75-80μ for best mixer performance.

Figure 1 also shows the mixer noise temperature when the junction temperature was reduced to 1.5K by pumping on the LHe bath. Mixer noise improved by about 50K for the best results near 200K. The lowest value obtained was 134K at 202.5GHz. The mixer gain also improved by 2-3dB. Mixer performance is expected to change with changes in the shape of the I-V curve. However, we see very little change in the I-V curve. Another possibility is ac losses in the superconductors. Kautz (7) has published calculations for losses in superconducting microstrip and shown that high losses can exist at high frequencies and these losses can be strongly temperature dependent even below half the transition temperature. We are currently investigating the losses in our microstrip lines.

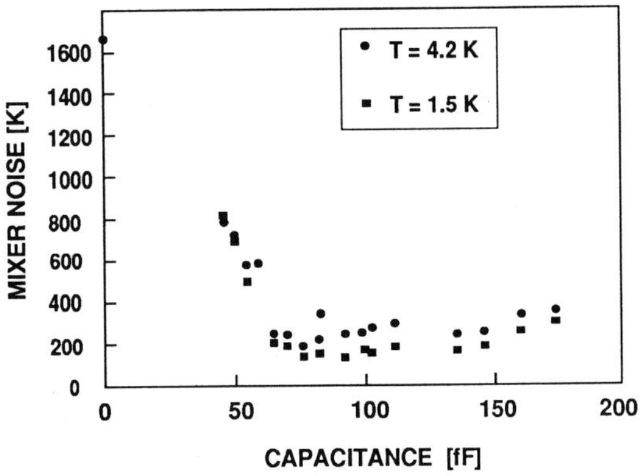

Fig. 1 Mixer noise versus capacitance tuned by microstrip stub.

4. SUMMARY

We have made a systematic study of the gain and noise of SIS mixers at 205 GHz using NbN-MgO-NbN junctions shunted with microstrip lines as rf tuning elements. This has given values of junction capacitance of 85fF/μ^2 and magnetic penetration depth of 3800Å. Mixer noise temperature as low as 134K has been achieved which is the best reported to date.

ACKNOWLEDGEMENTS

This work was supported in part by the Jet Propulsion Laboratory, California Institute of Technology under contract with the National Aeronautics and Space Administration; and the Strategic Defense Initiative Organization.

REFERENCES

(1) J.R. Tucker and M.J. Feldman, Rev. Mod. Phys. **57** (1985) 1055.

(2) H.G. LeDuc, J. Stern, S. Thakoor, and S. Khanna, IEEE Trans. Magn. **MAG-23** (1987).

(3) J. Stern, B. Hunt, H. Leduc, A. Judas,W. McGrath, S. Cypher, and S. Khanna, IEEE Trans. Magn. **MAG-24** (1989).

(4) W.R. McGrath, A.V. Raisanen, and P.L. Richards, Int. J. IR and Millimeter Waves **7** (1986) 543.

(5) A. Shoji, M. Aoyagi, S. Kosaka, F. Shinoki, and H. Hayakawa, Appl. Phys. Lett. **46** (1985) 1098.

(6) A. Raisanen, W. McGrath, P. Richards, and F. Lloyd, IEEE Trans. Microwave Theory Tech. **MTT-33** (1985) 1495.

(7) R. Kautz, J. Appl. Phys. **49** (1978) 308.

Physica B 165&166 (1990) 77–78
North-Holland

EXCESS LOW-FREQUENCY FLUX NOISE IN dc SQUIDS INCORPORATING Nb/Al-OXIDE/Nb JOSEPHSON JUNCTIONS*

Martin E. HUBER and Michael W. CROMAR

National Institute of Standards and Technology, 325 Broadway, Boulder, CO 80303-3328 U. S. A.

We have fabricated thin-film dc SQUIDs (Superconducting QUantum Interference Devices) incorporating Nb/Al-Oxide/Nb tunnel junctions. The spectral density of the voltage noise, S_V, of stripline SQUIDs is characterized between 1 Hz and 2000 Hz. In this frequency range, S_V is proportional to the square of the responsivity $(\partial V/\partial\Phi)$ over a significant range of bias conditions with an unusual frequency dependence. In a 7 pH SQUID, the spectral density of the flux noise, S_Φ, at 1 Hz is less than $10^{-11}\ \Phi_0^2/\mathrm{Hz}$, where $\Phi_0 \equiv h/2e$. The observed noise does not appear to be environmental; also, it is independent of the value of the junction shunt resistance and whether the stripline material is a PbIn alloy or Nb. Subject to the constraint of a constant product of the junction area, critical current density, and SQUID self-inductance in the SQUIDs studied, S_Φ is inversely proportional to the junction area.

1. INTRODUCTION

The intrinsic noise level of dc SQUIDs (Superconducting QUantum Interference Devices) can, in theory, be very low (1), making them ideal sensors for electric currents and magnetic fields (2). In practice, excess noise of one form or another is often observed. We are studying the intrinsic noise of dc SQUIDs (3) incorporating low-leakage Nb/Al-Oxide/Nb Josephson tunnel junctions (4).

One type of excess noise which has been observed before (5,6) is flux-like. If biased with a constant current, the voltage V across the dc SQUID varies with applied magnetic flux Φ with a small-signal responsivity $\partial V/\partial\Phi$, which is a function of Φ. According to models of dc SQUID dynamics (1), the spectral density of the voltage noise S_V is not simply related to $(\partial V/\partial\Phi)^2$. Proportionality between the two thus indicates the presence of a flux noise with spectral density $S_\Phi = S_V/(\partial V/\partial\Phi)^2$. There are usually no obvious sources of flux noise present, however, which is why the noise is called "flux-*like*".

We, too, observe a flux-like noise, for which S_Φ is independent of the operating point and has an unusual frequency dependence (neither frequency independent nor a power law function). We report on the dependence of the value of S_Φ on physical parameters such as the shunt resistance, SQUID geometry, and SQUID-loop materials.

2. SQUID PREPARATION AND MEASUREMENT

The SQUIDs are fabricated on Si in a thin-film process (3) incorporating Nb/Al-Oxide/Nb Josephson junctions. Resistors are $AuIn_2$ and the insulation is SiO. The upper wiring layer is either a PbIn alloy or Nb. All SQUIDs studied are of a low-inductance, self-shielded (stripline) geometry (3). The two junctions are located at opposite ends of a long, narrow stripline (the SQUID "body") and connect the upper and lower conductors of the stripline. The current is injected midway between junctions in one layer and is collected from a symmetric point on the other layer. The nominal design parameters are listed in Table 1. The insulation thickness is 300 nm and the critical current density J_c is about 700 A/cm^2. We measure S_V in the range 1 Hz to 2 kHz with the SQUID biased at a constant current. The SQUID is connected to a resistor in series with a transformer for current gain; the secondary winding is connected to a commercial dc SQUID operating in a flux-locked mode.

3. RESULTS AND DISCUSSION

To determine whether the noise was in fact flux-like, SQUIDs were first biased at three different drive currents. At each drive current, S_V and $\partial V/\partial\Phi$ were measured at various values of the applied flux. We observed proportionality between S_V and $(\partial V/\partial\Phi)^2$, where the value of the slope is S_Φ (Fig. 1).

Since such noise could result from external sources, it was necessary to rule out that possibility before considering intrinsic sources. The standard SQUID mount is a circuit board with BeCu contacts,

TABLE 1
Nominal parameters for the stripline SQUIDs tested. R is the shunt resistance for a single junction and L is the SQUID self-inductance.

ID	Junction Area (μm^2)	Junction Separation (μm)	SQUID Width (μm)	R (Ω)	L (pH)
A	4.0	500	10	5.0	26.5
B	7.8	350	15	2.4	12.4
C	16.0	250	20	1.2	6.7
D	32.5	175	30	0.6	3.2
E	16.0	500	40	1.2	6.8

* Contribution of the U.S. Government, not subject to copyright in the United States.

1990 – Elsevier Science Publishers B.V. (North-Holland)

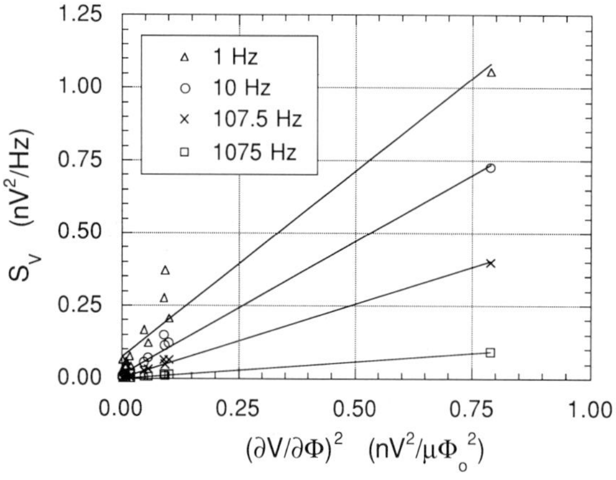

FIGURE 1
S_V vs. $(\partial V/\partial \Phi)^2$ for a type C SQUID at various bias points and frequencies ($\Phi_o \equiv h/2e$). Although the data are clustered near the origin, a linear relationship also exists on an expanded scale.

sandwiched between solid Nb disks. The SQUID mount is enclosed in a cylinder of high permeability alloy specially designed for use at low temperatures, and the cylinder is lined with Pb foil. There are also three layers of room temperature high-permeability shielding and the Dewar is non-magnetic metal. We reduced the low-frequency shielding in stages, including replacing the mount with an unshielded circuit board incorporating In contacts. No increase in S_Φ was observed until the SQUID was operated with neither superconducting nor normal-metal magnetic shielding. SQUIDs with half the normal insulation thickness (and hence less area to couple

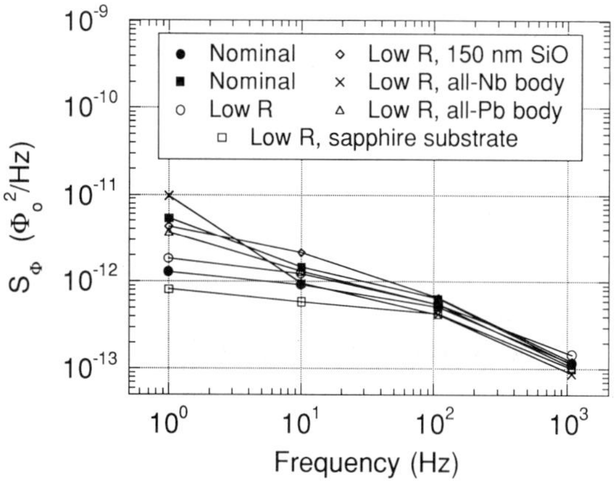

FIGURE 2
S_Φ vs. frequency for type C SQUIDs at selected frequencies for all fabrication runs. The bias current is about 0.8 times the maximum zero-voltage current.

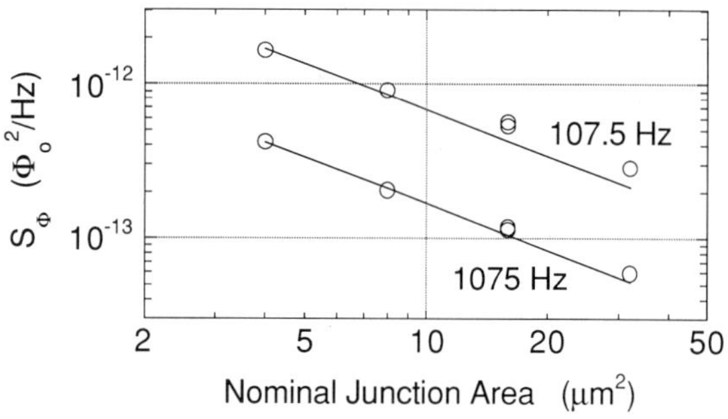

FIGURE 3
S_Φ vs. nominal junction area "a" for SQUID types A through E. The line is a fit to $S_\Phi \propto 1/a$.

to external fields) exhibited no reduction in S_Φ (Fig. 2). In addition, SQUIDs fabricated on sapphire rather than Si exhibited no variation in S_Φ. As a whole, these results implicate non-environmental sources as the origin of the excess noise.

Figure 2 summarizes the results of all fabrication runs, including those in which the junction shunt resistance and stripline materials were changed. Notice that although S_Φ is process dependent below 100 Hz, it is not correlated to the features studied; variation is observed even in SQUIDs processed under identical conditions on different wafers (the "nominal" points). Above 100 Hz, S_Φ is the same for all fabrication runs. The shunt resistances of the "low-R" SQUIDs were 0.4 times the values listed in Table 1; we conclude that S_Φ is not a function of the shunt resistance. Also, although Wellstood et al. (6) have observed less noise in SQUIDs with Pb loops than those with Nb loops, there is no such dependence in these SQUIDs.

Finally, at 107.5 and 1075 Hz, we average the value of S_Φ for all SQUIDs of a given type and plot S_Φ vs. the nominal junction area (Fig. 3). Subject to the constraint $a \cdot J_c \cdot L$ = constant for the SQUIDs studied, we observe $S_\Phi \propto 1/a$ ("a" is the junction area). The specific parameter dependence of the noise is under investigation.

REFERENCES

(1) C. D. Tesche and J. Clarke, J. Low Temp. Phys. 29 (1977) 301.

(2) T. Ryhänen, H. Seppä, R. Ilmoniemi, and J. Knuutila, J. Low Temp. Phys. 76 (1989) 287.

(3) M. W. Cromar, J. A. Beall, D. Go, K. A. Masarie, R. H. Ono, and R. W. Simon, IEEE Trans. Mag. 25 (1989) 1005.

(4) H. A. Huggins and M. Gurvitch, J. Appl. Phys. 57 (1985) 2103..

(5) R. H. Koch, J. Clarke, W. M. Goubau, J. M. Martinis, C. M. Pegrum, and D. J. Van Harlingen, J. Low Temp. Phys. 51 (1983) 207.

(6) F. C. Wellstood, C. Urbina, and J. Clarke, IEEE Trans. Mag. 23 (1987) 1662.

Physica B 165&166 (1990) 79-80
North-Holland

DC SQUID WITH VARIABLE THICKNESS BRIDGE

H.FURUKAWA, S.FUJITA, A.MATANI, M.YOSHIDA, and K.SHIRAE

Faculty of Engineering Science, Osaka University, Toyonaka, Osaka, 560, JAPAN+

We tried to fabricate dc SQUID with variable thickness bridge (VTB), instead of the tunnel junction type dc SQUIDs. Bridge was realized on a clean surface of substrate by double oblique depositions of Nb in a simple process. VTB type dc SQUIDs work well in our multichannel SQUID system.

1. INTRODUCTION

The use of dc SQUID magnetometer is rapidly spreading in Biomagnetism, and the multichannel dc SQUID system, such as 100 channels or more, is needed for one time measurement of brain or heart magnetic fields. For this purpose, we have developed a simple fabrication method of dc SQUID with variable thickness bridges (VTB) instead of tunnel junctions.

2. FABRICATION METHOD

Figure 1 shows the pattern of our VTB type dc SQUID. Outward is similar to the washer type(1). VTBs are formed near the edge of a Nb step. Electron beam evaporation, photo-lithography and chemical etching were used.

Figure 2 shows the fabrication steps of VTB; (1) 400 nm thick Nb step is shaped on a quartz substrate, (2) 15 nm thick Nb film is deposited, (3) 200nm thick Nb film is obliquely deposited with an angle of 60° from the right hand, (4) finally, 200 nm thick Nb film is deposited obliquely with an angle of θ from the left hand. The shadow region left behind the final oblique deposition acts as a short bridge. Bridge width was determined as 5 μm for the reasons of photo-etching technique.

3. RESULTS AND DISCUSSIONS

SEM photograph of a VTB is shown in Figure 3. Bridge channel is clearly observed, and the bridge length is read as about 400 nm. Irregular notches are seen everywhere, however, we can say VTB has been successfully fabricated.

Typical voltage vs. current and voltage vs. magnetic flux curves of the VTB type dc SQUID are shown in Figure 4.

Although the critical currents are dependent on various parameters such as

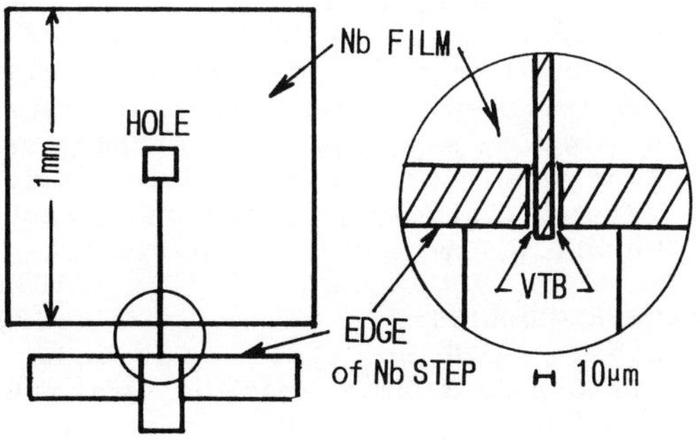

FIGURE 1
Patterns of dc SQUID with variable thickness bridges.

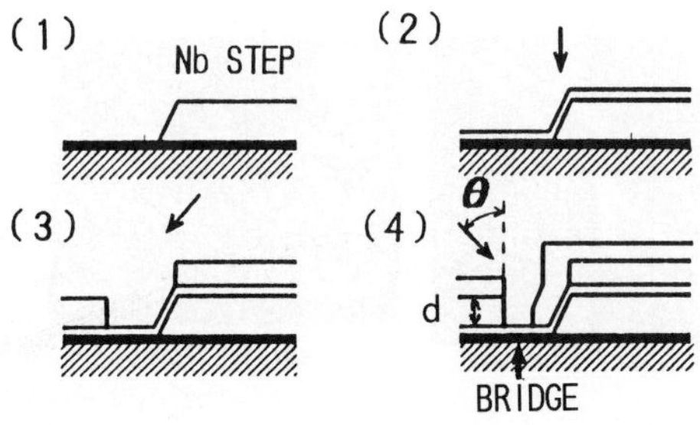

FIGURE 2
Fabrication steps of variable thickness bridge using oblique depoisitions of Nb from right and left side.

+ This work is partly supported by SUZUKI MEDICAL SCINECE RESEARCH FOUNDATION.

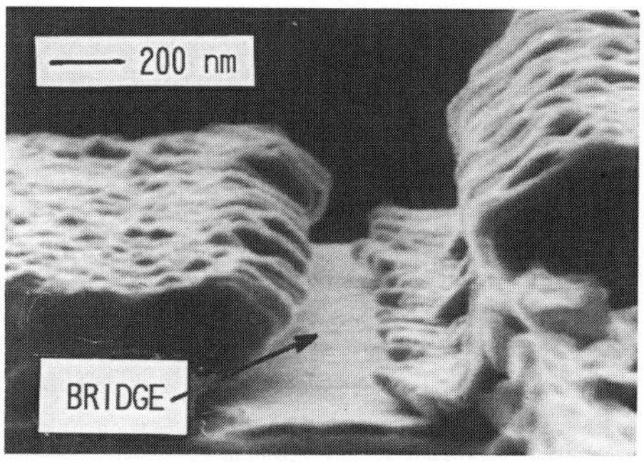

FIGURE 3
A SEM photograph of a variable thickness bridge.

bridge thickness, width, substrate temperature etc., the dominant parameter is the incident angle θ . Figure 5 shows the critical current dependencies on the incident angle θ . With the increasing θ , critical current Ic increases almost exponentially. Full rectangles represent samples, which did not response to the magnetic field because of large Ic, and blank rectangles represent samples responding to the magnetic field. Although fluctuations of Ic are large at θ =20°, all the samples fabricated at θ =20° worked satisfactorily. At θ =10°, the critical currents dropped below 1 μA.

Critical temperatures of the bridges were 7±1 K regardless of the incident angle.

Bridge length is given by d·tanθ in the model of Figure 2, where d is the

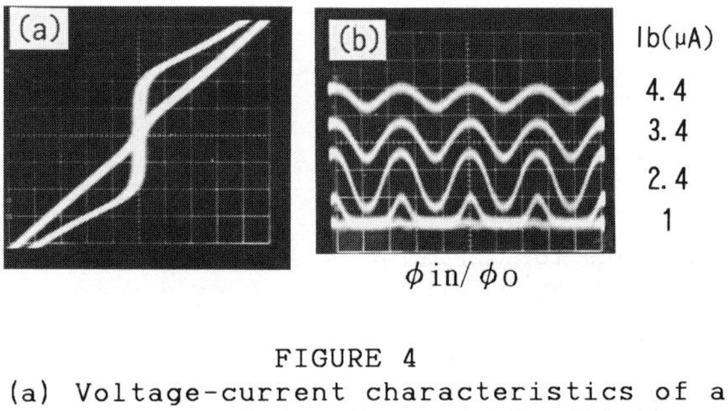

FIGURE 4
(a) Voltage-current characteristics of a typical dc SQUID with variable thickness bridges; vertical scale, 1μ A/div., horizontal scale, 10μ V/div., (b) voltage modulation characteristics by the input magnetic flux at various bias current Ib, vertical scale same as in (a).

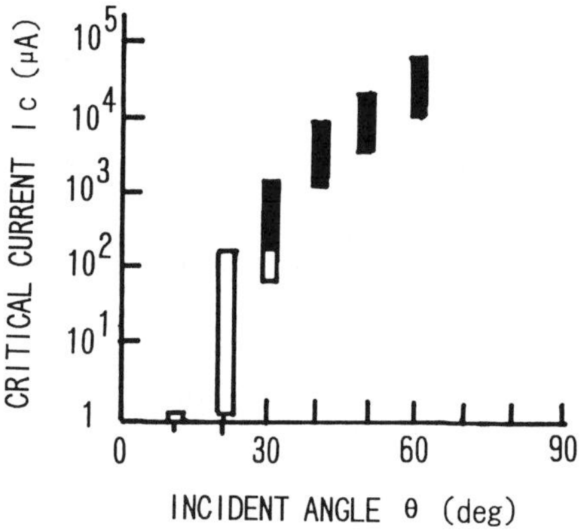

FIGURE 5
Critical current dependence on the incident angle θ of the oblique deposition from left side.

thickness of the first obliquely deposited film. As d is set to 200 nm, the bridge length is 70 nm for θ =20°. In Figure 3, a sample with longer bridge was chosen on purpose of clear presentation of bridge.

Noise power spectrum at 1 Hz (in FLL state) is about 9000h, 7 times higher than the low noise tunnel junction dc SQUID(2). We are expecting, however, the noise reduction is possible because of the pre-amplifier noise is dominant at present experiment.

4. CONCLUSION

DC SQUID with variable thickness bridge has been fabricated using double oblique deposition of Nb. Bridge length and critical current can be controlled by the incident angle. Precise fabrication means such as dry etching and electron beam lithography would improve the reproducibility and the performance of the SQUID.

ACKNOWLEDGEMENTS
Authors express their gratitude to Mr.I.Tamura and Mr.M.Wada of Osaka Gas Company for the fine SEM photographs.

REFERENCES

(1) M.B.Ketchen, IEEE Trans. on Magnetics, MAG-23 (1987) 1650.
(2) P.Carelli, IEEE Trans. on Magnetics, MAG-25 (1989) 1026.

Physica B 165&166 (1990) 81–82
North-Holland

COMPARISON OF SHUNTED DC-SQUIDs WITH LARGE β

E.P.HOUWMAN, D.VELDHUIS, J.FLOKSTRA, and H.ROGALLA

University of Twente, Faculty of Applied Physics, Low Temperature Group,
P.O.Box 217, 7500 AE Enschede, The Netherlands.

The performance of DC-SQUIDs with inductively and resistively shunted inductances is studied theoretically and experimentally and compared to the performance of a standard (unshunted) SQUID. By shunting the inductance the voltage modulation depth remains unaffected for large β. The consequences for the flux-voltage transfer and the noise performance are discussed.

1. INTRODUCTION

Since the description of the standard DC-SQUID by Tesche and Clarke several modifications on the SQUID design were proposed to overcome the fairly strict constraint for the screening parameter $\beta = 2LI_0/\Phi_0 \approx 1$. L is the SQUID inductance, I_0 the critical current of a junction and Φ_0 the flux quantum. Thermal noise restricts I_0 to values larger than about 4 μA, for practical devices operated at LHe temperature. This limits L to about 250 pH at the most.

We are developing a fabrication process for DC-SQUIDs for the application to neuromagnetometry (1). In that case the SQUID is coupled to an inductance of about 1 μH. In the Ketchen-Jaycox washer design this can be achieved by a planar, tightly coupled coil with many (n_i = 30-60) turns. Apart from fabrication problems, the capacitive coupling between SQUID and input circuit may cause resonances in the system, that can give rise to excess noise.

The SQUID inductance can be increased, thus n_i decreased, by the introduction of a shunt across L, on the expense of some loss of sensitivity, that may be acceptable for many neuromagnetic experiments. On the other hand excess noise due to resonances may be avoided.

We are studying the properties of the standard, the resistively and the inductively shunted SQUID experimentally.

2. THEORETICAL CHARACTERISATION

The voltage modulation depth of a standard SQUID in the optimum bias point depends on β as $\Delta V_m \approx I_0R/(1+\beta)$ (2), where R is the shunt resistance of the junction. A simple estimate of the noise leads to an energy sensitivity $\varepsilon/(2k_BT\Phi_0/I_0R) \approx (\beta+1)^2/\pi^2\beta + \beta/4$. This expression underestimates the noise, but describes the functional behaviour fairly well.

For the resistively shunted SQUID it was shown that $\Delta V_m \approx I_0R/2$ is nearly independent of β, if $\gamma = R/R_{sh}$ is about 1 or larger, where R_{sh} is the shunt resistance over L (3). The energy resolution is analytically obtained as $\varepsilon/(2k_BT\Phi_0/I_0R) \approx 4/\pi^2\beta + \beta(1+2\gamma)/4$. For low β the estimate is too low, but it becomes rather good for larger β and $\gamma > 0.5$. For large β optimum performance is obtained for $\gamma \approx 1$.

For the inductively shunted SQUID the flux quantisation condition can be written as (4) $\delta_1-\delta_2 = \pi[\beta_{sh}-\beta_{sh}^2/(\beta_p+\beta_{sh})]j + 2\pi[\Phi_p\beta_{sh}/(\beta_p+\beta_{sh})]$ where δ_i is the phase difference over junction i, j the circulating current through the junctions and Φ_p the flux in the large pick-up loop, formed by the loop inductance L_p and the shunt inductance L_{sh}. It is assumed that no flux is applied to the small loop, made up of L_{sh} and the junctions. For $\beta_p \gg \beta_{sh}$ and $\beta_{sh} \gg 1$ this condition is that of a standard SQUID, to which an effective flux $\Phi_{eff} \approx \Phi_p/(\beta_p+1)$ is applied, but with effective screening parameter $\beta_{sh} \approx 1$. Thus $\Delta V_m \approx I_0R/2$, but the transfer becomes $\partial V/\partial\Phi_p \approx (\partial V/\partial\Phi_{eff})/(\beta_p+1) \approx (\pi I_0R/\Phi_0)/(\beta_p+1)$. The energy resolution is obtained, neglecting the mixing due to the junctions and assuming that the noise current is shunted by L_{sh}, as $\varepsilon/(2k_BT\Phi_0/I_0R) \approx 0.35(\beta_p+1)^2/\beta_p$

Figure 1 shows the voltage modulation depth as function of β, respectively β_p, for the standard and the shunted SQUIDs, calculated from an analytical model based on the work of Enpuku *et al.* Theoretically the effect of the shunting is very pronounced in the used approximations.

Figure 2 shows the energy resolution according to the above approximate formulas. In all three cases ε shows a clear minimum for $\beta \approx 1$ and a linear increase with β for larger β. The price for larger β is some increase of ε, that is comparable for the three designs.

3 FABRICATION AND EXPERIMENTAL RESULTS

10 SQUIDs of the three designs, are fabricated in a single run. The SQUID washer is formed by a Nb/Al,AlO_x/Nb tri-layer process, using DC magnetron sputtering and thermal oxidation. The junction areas (5x5 μm^2) are defined by a standard SNAP process, by which also an insulation layer is formed on the washer. The employed

0921-4526/90/$03.50 © 1990 – Elsevier Science Publishers B.V. (North-Holland)

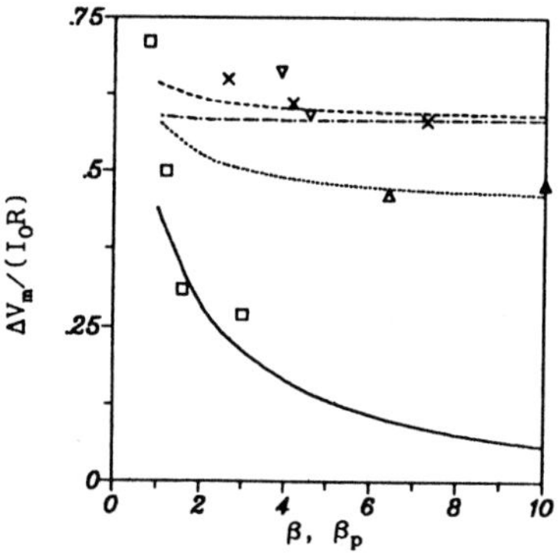

Fig. 1 : Voltage modulation depth as function of β for noiseless SQUIDs with β_c=0, at I_B=2.1 I_0. Lines : theory, symbols : experimental data. ——,□ standard SQUID; - · - · -,x resist.shunted SQUID (γ=1); ·········,Δ induct.shunted SQUID (β_{sh}=1), ----,∇ (β_{sh}=o.5).

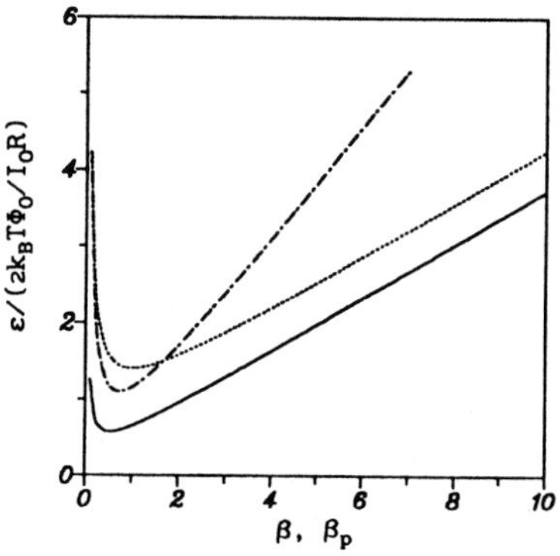

Fig. 2 : Energy sensitivity as function of β. —— standard SQUID; - · - · - resist.shunted SQUID (γ=1); ········· induct.shunted SQUID (β_{sh}=1).

junctions are fabricated with a double oxide barrier. The junctions are of high quality with V_m up to 80 mV at 4.2 K and 600-1100 mV at 1.6 K. The fabrication and characterisation of the junctions is described elsewhere (5). A second insulation and planarisation layer of RF sputtered SiO_2 is deposited. The final Nb layer is patterned to form the contacts to the counter electrode of the junctions and the modulation and signal coils.

The resistively shunted SQUIDs have inductances 240, 470, 1200 and 2400 pH. We chose $\gamma \approx 1$ and R, such that $\beta_c = 2\pi I_0 R^2 C/\Phi_0 \leq 0.3$, throughout.

The two standard SQUIDs have inductances 110 and 240 pH. The junctions are placed at the inside of the washer to avoid the parasitic slit inductance.

The inductively shunted SQUIDs consist of a small washer with a hole inductance of 110 pH, on which the modulation coil is placed. This washer is attached to a large washer, with an inductance of 940 or 2400 pH, on which the input coil is placed.

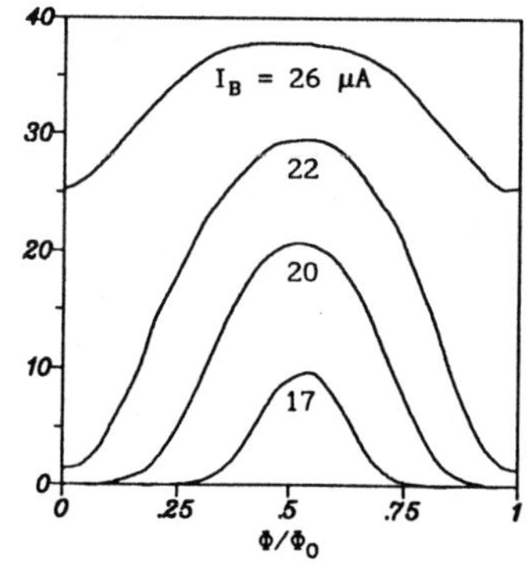

Fig. 3 : Experimental flux-voltage transfer of inductively shunted SQUID.

In figure 1 some first results on the experimental ΔV_m (at $I_B \approx 2.1\ I_0$), obtained for the three SQUID designs, are shown. The strong dependence of ΔV_m on β for the standard SQUID is clearly seen, in contrast with the relative independence of ΔV_m on β for the shunted SQUIDs.

Figure 3 shows the flux-voltage transfer of a typical inductively shunted SQUID, with $\beta_p \approx 10$, $\beta_{sh} \approx 1.2$, $I_0 \approx 11\ \mu A$ and $R \approx 4.3\ \Omega$. The large voltage swing and the high symmetry of this curve, make this SQUID very suitable for application in flux-locked loop systems.

The sensitivity is still to be determined and will be reported elsewhere.

4. CONCLUSIONS

The dependence of ΔV_m and ε of the standard, resistively and inductively shunted SQUIDs on β is calculated. For the shunted SQUIDs ΔV_m is nearly independent of β, in contrast with the strong dependence for the standard SQUID. The dependence of ε on β is comparable for the three designs.

A fabrication process for high-quality Nb/Al,AlO_x/Nb tunnel junctions (V_m up to 80 mV at 4.2 K and over 600 mV at 1.6 K) was developed and applied to standard and shunted DC-SQUIDs.

The ΔV_m of the SQUIDs was determined as function of β and is found to be in accordance with the theoretical expectations.

REFERENCES

(1) H.J.M.ter Brake *et al.*, Design and Construction of a 19-Channel DC-SQUID Neuromagnetometer, this Volume.

(2) R.L.Peterson and C.A.Hamilton, J.Appl. Phys.**50**, (1979) 8135.

(3) K.Enpuku *et al.*, J.Appl.Phys.**57**,(1985) 1691.

(4) H.Koch, ICEC **10**, (1984) 834.

(5) E.P.Houwman *et al.*, accepted for publ. in J.Appl.Phys. (1990).
D.J.Adelerhof *et al.*, Conductance Studies on Different Types of Nb/Al,AlO_x(/Al)/Nb Josephson Tunnel Junctions, this Volume.

Physica B 165&166 (1990) 83–84
North-Holland

A NEW TYPE OF SQUID

G.S.KRIVOY and V.A.KOMASHKO*

Omsk Polytechnical Institute, 644050 Omsk, Mir prospect, 11, USSR

We elaborated and investigated the rf SQUID in which a dc SQUID is used as a weak link. The magnetic flux being measured is applied to the dc SQUID loop while the output signal is taken by means of the typical rf SQUID circuitry. The considerable increase of the output signal is achieved in comparison with the conventional dc and rf SQUIDs.

1. INTRODUCTION

At present rf SQUIDs as well as dc SQUIDs are widely used for magnetic and other measurements [1]. rf SQUIDs contain one weak link while dc SQUIDs contain two weak links in a superconducting ring.

The signal output from the rf SQUID is realized by means of a circuit fed from the oscillator at the resonant frequency ω_0. The circuits for the signal output from dc SQUID are characterized by some diversity [2]. However the typical signal level from both rf and dc SQUIDs possesses the order of several tens microvolt and requires sufficiently complex low noise preamplifiers.

The present paper deals with a new type of the SQUID combining the dc SQUID and the rf SQUID and possessing a larger value of the output signal than conventional SQUIDs do.

2. CONFIGURATION OF THE SQUID

The SQUID circuit called below "a double SQUID" or, for short, dSQUID, is shown in Fig.1.

Its difference from the traditional circuits lies in the fact that it is dc SQUID that is used in the rf SQUID as a weak link. The external magnetic flux being measured is applied to the loop of the dc SQUID while the coupling with the tank circuit is provided by the rf SQUID loop inductance.

The present device will be characterized by the following properties:

1) the modulation of the rf voltage on the tank circuit is provided by changes of the plateau height of the IV-characteristics of the rf SQUID - tank circuit system owing to the critical current changes of the dc SQUID influenced by the external magnetic flux Φ_X being measured.

2) dc SQUID in contrast to the traditional circuitry [1,2] is in a superconducting state converting into resistive state only for a short time (the order of several picoseconds) when a jump into another quantum state occurs in the rf SQUID.

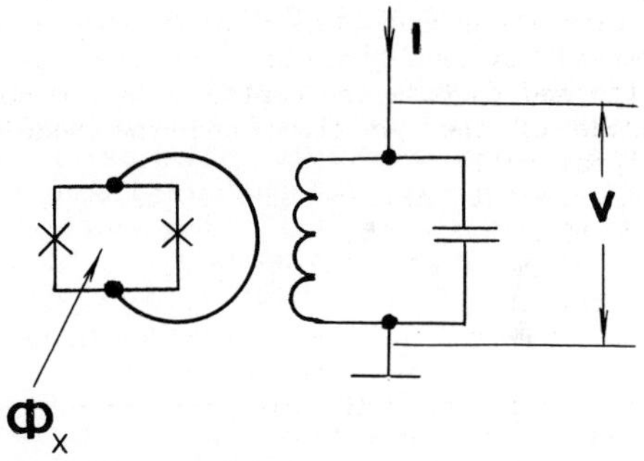

FIGURE 1
dSQUID circuit and its circuitry. I is the pumping current, V is the output signal, Φ_X is the external magnetic flux being measured.

From the aforesaid it follows that it is possible to obtain a signal exceeding in amplitude the ordinary signal levels from the rf SQUIDs and dc SQUIDs (owing to increasing the modulation depth) and to get lesser intrinsic noise in comparison with a traditional one. It is made possible because it is rf SQUID that is a signal amplifier from dc SQUID operating most of its time in a superconducting state, and the influence of the tank circuit with the subsequent amplifier upon dc

* Permanent address: Institute of Cybernetics, Ukrain. SSR Academy of Sciences, Kiev, USSR

SQUID noise will be decreased.

The desired signal ΔV from dSQUID equals to

$$\Delta V = \Delta I_c \omega_o L_t L/M, \qquad (1)$$

where ΔI_c is the change of the dc SQUID critical current under the influence of the external magnetic flux, L_t,L are the dSQUID and the tank circuit coil inductances, respectively, M is the mutual inductance between L_t and L. The desired signal ΔV_{rf} of the rf SQUID equals to [1]

$$\Delta V_{rf} = (\Phi_o/2)\omega_o L_t/M, \qquad (2)$$

where Φ_o is a magnetic flux quantum. From the comparison (1) and (2) it is clear that the dSQUID signal is $2\Delta I_c L/\Phi_o$ times greater than that of the ordinary rf SQUID. For obtaining $\Delta V > \Delta V_{rf}$ under the conditions of the typical inductance L = 0.5 nH it is required $\Delta I_c > 2$ μA. It is quite a real range of the dc SQUID critical current change.

3. EXPERIMENTAL PROCEDURE AND RESULTS

For experimental investigations of the dSQUID a thin-film dc SQUID with a quantization loop up to 40x40 μm^2 was used which was placed instead of a weak link into the bulk design unit from niobium analogous to the traditional rf SQUID. The external magnetic flux was produced by a solenoid located above the dSQUID. This magnetic flux was taken simultaneously by the dc SQUID and the dSQUID loop. It was realized in such a way in order to show the ratio of ΔV and ΔV_{rf} signals on one and the same oscillogram. The reason is that the dSQUID response to the magnetic flux in its loop is equivalent to the rf SQUID response (ΔV_{rf} signal) of the same topology.

The periodicity of the signal ΔV_{rf} in magnetic flux was considerably less than that of the signal ΔV. It is due to a more tight coupling of the solenoid and dSQUID loop than that of the dc SQUID.

The IV-characteristics of the tank circuit coupled with the dSQUID are shown in Fig.2. This figure shows two extreme IV-curve positions at various values of the external magnetic field corresponding to the integer and halfinteger value of the magnetic flux in a dc SQUID being part of a dSQUID. These IV-curve positions cause the output signal swing ΔV. The tank circuit voltage changes within the range of any single plateau represent a conventional signal ΔV of the rf SQUID.

The two rf pumping modes are available: external and autodyne. The output signal ΔV in the autodyne mode will be larger than that in the external mode because in the autodyne mode the dSQUID operates all time on the first plateau where a change of the voltage on the tank circuit is the largest. The signal levels ΔV = 690 μV and ΔV_{rf} = 43 μV were determined in the autodyne mode.

The signal level ΔV corresponds to the hysteretic parameter $l = 2\pi L I_c/\Phi_o$ change of the dSQUID in the range of 14 to 63. The results of the experiments presented give the possibility to assume that the

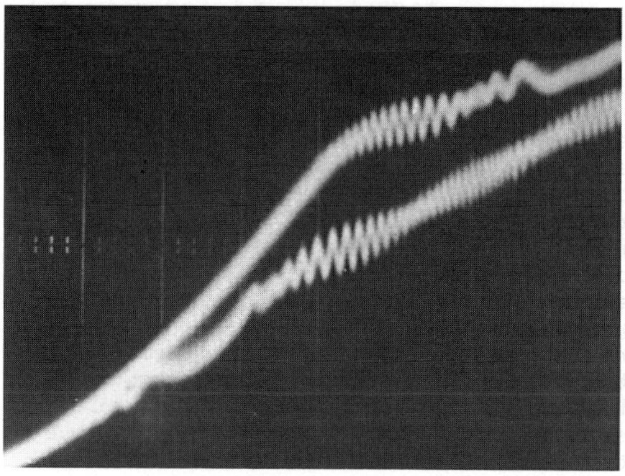

FIGURE 2
The oscillogram of the dSQUID IV-characteristics coupled to the tank circuit. The current I is on the horizontal axis, the voltage V is on the vertical axis.

SQUID configuration under consideration may find application in magnetometers. The design of special constructions of such dSQUIDs results in increasing conversion transconductance in addition to increasing of the output signal.

In case of employing this configuration with the resistive dc SQUID as a weak link it is quite possible to create rather low noise sensors of the noise thermometer.

ACKNOWLEDGEMENTS

We would like to thank N.A. Gopkalo for help in preparing this paper.

REFERENCES

[1] A. Barone and G. Paterno, Physics and Applications of the Josephson Effect (J. Wiley & Sons, New York, 1982).

[2] V.V. Danilov, K.K. Likharev, O.V. Snigirev and E.S. Soldatov, Radiotekhnika i elektronika 22 (1977) 2383.

Physica B 165&166 (1990) 85–86
North-Holland

HIGH TEMPERATURE SQUID SUSCEPTOMETER AND ITS APPLICATIONS

Luwei ZHOU, Jinwu QIU, Xianfeng ZHANG, Zhimin TANG, Yimin CAI, and Yongjia QIAN

Department of Physics, Fudan University, Shanghai 200433, China

A SQUID susceptometer using high temperature rf SQUID and normal metal pick-up coil is be employed in testing weak magnetization of sample. The magnetic moment resolution of the device is $1X10^{-6}$emu, and that of the susceptibility is $5X10^{-6}$emu/cm^3.

1. INTRODUCTION

Since Colcough *et al.*(1) first revealed the existence of intrinsic rf SQUID effect in high T_c superconductor Y-Ba-Cu-O, Zimmerman *et al.*(2) are the first who deviate the applicable high temperature rf SQUID. After that, many laboratories, including ours, have made rf SQUIDs with their own features.(3-8) Before a liquid nitrogen superconducting flux transformer can be made, a pick-up coil made of normal metal is used in this device. The intention of this work is to substitute the low temperature liquid helium device with the high temperature SQUID while keeping the extraordinary features of a SQUID variable temperature susceptometer VTS.

The temperature range of the VTS developed in this laboratory is from 77K to 300K, and the magnetic field range 0-0.1T. The device has been used to characterize superconducting thin films.

2. EXPERIMENTAL SETUP AND OPERATIONS

The VTS includes the SQUID system, sample chamber, magnet, dewar, and a computerized data acquisition system. The SQUID system consists of SQUID sensor, signal pick-up coil and the correspondent electronics. The SQUID used here is a double-hole rf SQUID made of a YBCO bulk material working at 19 MHz. The flux resolution at 77K is $5X10^{-4}$ Φ_0 /$\sqrt{}$Hz at the frequency range of 20-200 Hz, and $2X10^{-4}$ Φ_0 /$\sqrt{}$Hz near dc end.

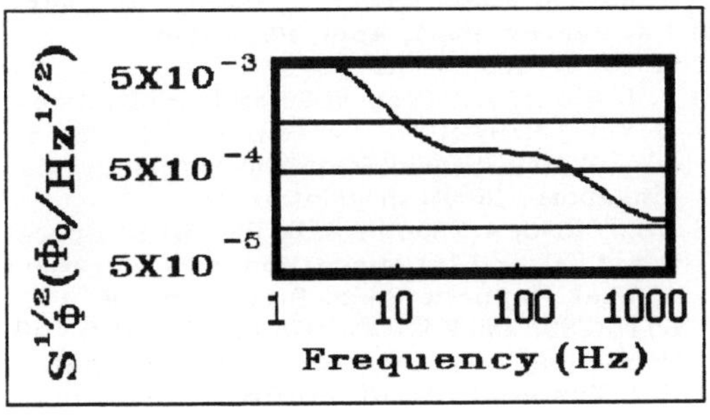

Fig.1

Fig. 1 gives the noise spectrum of the device at 77K. The inset is its noise output characteristics near dc end. To reduce the external interferences, the SQUID is placed in a YBCO superconducting screening hollow cylinder. In this experiment, a pick-up coil made of normal metal is used to substitute the superconducting coil before a satisfactory superconducting flux transformer working in liquid nitrogen can be made. The inductances of the pick-up coil and the input coil do not have to be matched in the 77K SQUID. When a signal is picked up with a normal metal coil, the amount of the signal depends not only on the moving speed of a sample, but also on its relative position. If a pick-up signal is calibrated with a standard sample in the same way that a sample is measured, then a normal metal pick-up coil can also generate a signal proportional to magnetic moment. This idea is the main feature of this apparatus, which bursts the idea that a superconducting flux transformer must be used in the application of a SQUID. Our recent test of a superconducting flux closed circuit in liquid nitrogen temperature using the SQUID shows that a closed flux transformer can be made(9), although it is not yet mature enough to be used in a susceptometer. The normal metal coils used in this experiment are a pair of counter-wound coils 10mm apart from each other and with a same number (N=60) of rings on each coil. The electronics employed here are those used in a usual liquid helium rf SQUID similar to S.H.E. 330 SQUID system. Sample chamber system consists of a thermostat, temperature measuring and controlling part, and sample transmission mechanism. The temperature deference between the sample space and the sample itself is corrected by the calibration.

A normal metal solenoid is used to supply the magnetic field for the system before a satisfied high T_c superconducting magnet can be made. The normal magnet used here supplies a center field as high as 0.1 T when the current is 2 A.

Vacuum adiabatic method is used to keep constant temperature. The temperature decreasing is obtained with a phosphorus-copper ring mounted between the sample space and the liquid nitrogen

* This work is supported by China National Center for R & D on superconductivity.

0921-4526/90/$03.50

chamber. The sample which is suspended on a string enters the sample space from the top of the dewar. When a heater is not turned on, the typical liquid nitrogen consumption is about 2 liters/day.

The procedures of sample measuring is similar to those for liquid helium SQUID susceptometer. The differences of the measurements with a superconducting flux transformer from those without is that in the former case the amount of the signal has to be calibrated using a standard sample while keeping exactly same way as a sample moves. The standard samples are annealed high purity aluminum and gallium cylinders.

3. EXPERIMENTAL RESULTS

The experimental results demonstrates, that the design of the normal metal coils does work, except the absolute value of the susceptibility has to be calibrated.

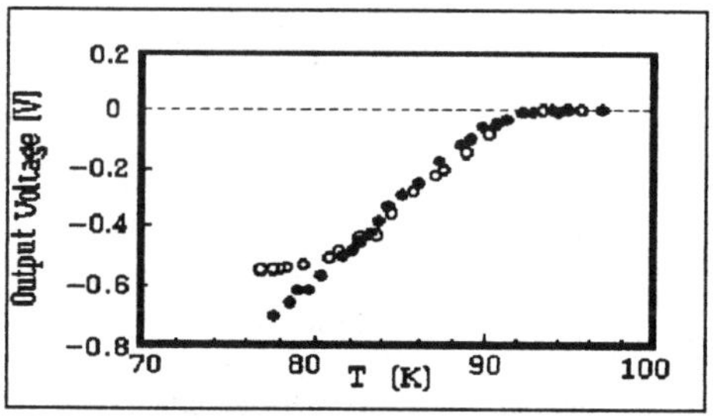

Fig.2

Fig.2 displays the Meissner and magnetic screening effects of a small piece of Y-Ba-Cu-O sample (3 mg) detected by this device when the external field is 100 Oe. In measuring the magnetic screening effect, the magnetic field is lowed to zero first, the sample temperature is then dropped to 77K from room temperature. After a certain magnetic field is applied, magnetic moment signal is measured at increasing temperatures. In measuring Meissner effect, the magnetic moment signals are measured at a given field while the sample temperature is lowering from a temperature well above then T_c. The results in Fig. 2 shows that the Meissner signal is about 60% of that of screening effect, which gives useful information for improving the sample making technology.

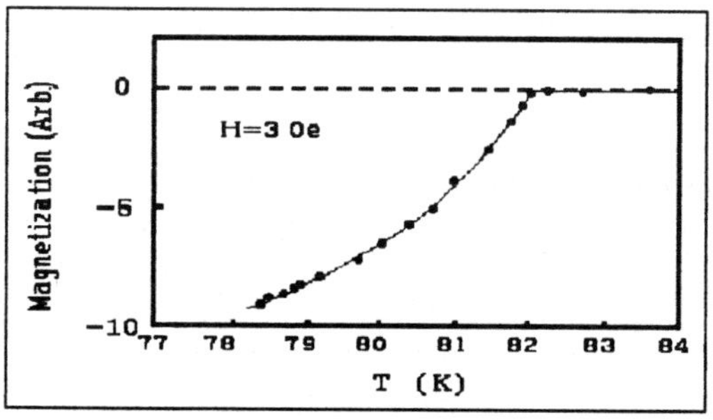

Fig.3

Fig.3 gives a magnetic transition curve of a YBCO superconducting thin film detected by this apparatus. The film thickness is about 3000 - 5000 A, and the applied field is 3 Oe. The transition temperature determined by this VTS is 82.3 K which is very close to the value T_{c0}, zero resistance temperature found by four-probe method. In the case when SQUID is functioning as a sensor, the measured field is in principle a relative change, and a calibration has to be done to determine the absolute value of moment and susceptibility from a signal even in a liquid helium SQUID susceptometer. It is more important when normal metal pick-up coils are used in the liquid nitrogen device. A special moment calibrating coil is employed to supply a certain value of a magnetic moment. The moment m of the coils calculated as m=NSI, where N is the number of the circles of the coil, S the average area of the coil, and I the current through the coil. In this experiment N=30, the average diameter of the coil is about 0.3cm, and the moment resolution of the device is about $1X10^{-6}$emu due to the signal size and the device resolution. Some metallic diamagnetic materials with high purity, such as aluminum and gallium are used to calibrate the susceptibility values. The calibration shows that the susceptibility resolution of the device is consistent to that estimated from the measurement on a small sample of high T_c superconductor, namely about $5X10^{-6}$emu/cm^3.

In summary, the SQUID susceptibility using high temperature rf SQUID and normal metal pick-up coil can be employed in testing weak magnetization of sample. The susceptibility transition curve of superconducting thin film and the calibrations of metallic diamagnetic samples show the feasibility of usage of the non superconducting transformer on SQUID. The magnetic moment resolution of the device is $1X10^{-6}$emu, and that of the susceptibility is $5X10^{-6}$emu/cm^3.

REFERENCES

(1) M.S. Colcough *et al.*, Nature 328 (1987) 47.
(2) J.E. Zimmerman *et al.*, Appl. Phys. Lett. 1 (1987) 617.
(3) C.M. Pegrum *et al.*, Appl. Phys. Lett. 51 (1987) 1364.
(4) S. Harrop *et al.*, Physica C 153-5 (1988) 1411.
(5) I.K. Harvey *et al.*, Appl. Phys. Lett. 52 (1988) 1634.
(6) J.W. Qiu *et al.*, Properties and Applications of YBCO RF-SQUID, Proceeding of BHTSC'89, ed. Z.Z. Gan, (World Scientific Publishing Co., Singapore, 1990), in print.
(7) W.C. Chiao, A YBaCuO Bulk RF-SQUID Operated at 77K and Its Application in Measurement of Weak Magnetic Field, Proceeding of BHTSC'89, ed. Z.Z. Gan, (World Scientific Publishing Co., Singapore, 1990), in print.
(8) S.Q. Xue *et al.*, An RF-SQUID Operated in Liquid Nitrogen up to 85K Made of High T_c YBCO Bulk, Proceeding of BHTSC'89, ed. Z.Z. Gan, (World Scientific Publishing Co., Singapore, 1990), in print.
(9) To be reported elsewhere.

Physica B 165&166 (1990) 87–88
North-Holland

GAS FLOATING TECHNIQUE FOR DETECTION OF TARPPED FLUX QUANTA

Q.GENG*, H.MINAMI*, K.CHIHARA*, J.YUYAMA* AND E.GOTO *,§,+

* Quantum Magneto Flux Logic Project, Research Development Corporation of Japan.
§ Faculty of Science,Tokyo Univ.. + Information Science Lab,The Institute of Physical and Chemical Research.

A gas flow-controlled SQUID pick-up coil system for detection of trapped flux quanta in superconducting film is described. Preliminary results, mainly the gap measurement, are given.

1. INTRODUCTION

Information about the distribution of magnetic flux quanta trapped in superconducting films is of importance for the development of Josephson electronics(1) and magnetic shielding(2). Resent advances by using laser beams(3), electron beams(4) and the Fraunhofer diffraction pattern(5) made it possible to image the trapped flux quanta(3,4,5) in Josephson junctions. In addition to these methods, we presently have been involved in developing a gas flow-controlled technique which utilizes a SQUID to specify the location of trapped fluxon directly. Since this technique requires no special environment and set-up is simple, it is a potentially useful procedure for detecting the presence of flux in a superconducting circuit lain on a flat substrate or in a superconductor plane, where it is impractical to use other methods.

2. DESCRIPTION OF THE FLOATING SYSTEM

The head used in the experiments is shown in Fig.1. It was made of FRP and had a nozzle of 4 mm diameter. The overall length of the head was 28 mm and the outside diameter was 20 mm. A differential SQUID pick-up coil, with a separation of 3 mm, was wound with Nb-Ti wire around a 1 mm diameter quartz which was located at the center of the head. Mechanical support of the head was obtained through a 8 mm long 2 mm inner diameter bellows that also served as a connection with gas supply system. With this arrangement the floating head is loosely suspended, thus no external force will be acted on the head along the normal direction except hydrodynamical one, which is given(2) by

$$F = \int_{r_1}^{r_2} 2\pi r[p(r,\dot{m},h)-p_a]dr + \pi r_1^2[p_s(\dot{m})-p_a] \qquad (1)$$

where h is the gap height, $\dot{m}$ the mass flow rate, p_a the surrounding pressure, and p_s the supplying pressure at the nozzle. The first term of Eq.(1) denotes the negative contribution due to Bernoulli theorem (2,6), which comes from the flow through the gap between the head and the superconductor, while the second term has a positive contribution.

The boundary layers affect the velocity profile in the gap region through the ratio of their thickness to h. By using a simplified model(2) derived from analogy to the parallel flow along a flat plate(6), we obtain the displacement thickness

$$\delta(r) = 1.721\alpha(\nu(r-r_1)/U_m)^{1/2} \qquad (2)$$

where ν is the kinematic viscosity, α is an experimental parameter describing the axial symmetry and velocity distribution in the gap, and U_m is the velocity at the middle point of the gap region, which is related to the mass flow rate through

$$\dot{m} = \rho_m A_m U_m = \rho_1 A_1 U_1 = \rho A U \qquad (3)$$

where ρ is the density of gas, A is the cross area of flow path, and subscripts m and 1 designate the middle and the initial position in the gap region respectively. The velocity U_1 at the nozzle depends on effusion model and flow coefficient C_0 arising from stream line bending at the inlet of gap.

By using the adiabatic effusion model for U_1 and combining Eqs.(1), (2) and (3), we predict it is possible to control the gap between the floating head and the superconducting plane by changing the mass

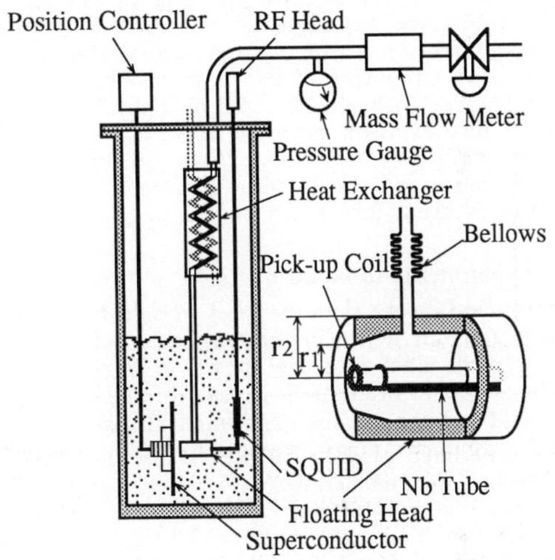

FIGURE 1
Schematic diagram of gas flow-controlled pick-up coil system.

flow rate at room temperature. The calculation results for nitrogen gas are shown in Fig.2 by solid and dashed line for 77K and 288K, respectively.

3. EXPERIMENTS AT 288K AND 77K

3.1 Experiments at room temperature

In order to test our idea and get acquainted with the floating head, experiments were first performed at room temperature. In the experiments, both air and nitrogen gas were used. The gap was measured by a gap sensor(Sentec) and a capacitance bridge (AH), which was calibrated against the thickness of teflon sheets at both room temperature and 77K. The results were consistent with each other and reproducible at high ($\geq 12g$/min) mass flow rates. One of those results is shown in Fig.2. The main conclusion of these experiments is that it is possible to get an attraction in such system and the behavior of the gap as a function of flow rate can be qualitatively explained by a model based on boundary layer theory. The deviation between theory and experiments at small flow rate depends on the initial gap setting, because the effects of the stiffness of bellows and other external forces can not be neglected completely.

3.2 Experiments in liquid nitrogen

Experiments were then made at 77K in the cryostat later used for liquid helium experiments. Results of the gap vs. flow rate are shown in Fig.2, where solid line is the prediction of the boundary layer theory and squares are the experimental results.

The experimental data, measured by the capacitance bridge, can be divided into three regions. At small mass flow rate(region I), the gap is increasing as $\dot{m}$ increases. The magnitude of the change in this region depends on the value of initial gap. In region II, the gap is almost constant. This is clearly contrary to the theory(solid line). A possible explanation for the difference could be the existence of liquid phase in the experiments while in theory only gas phase is considered. In region III, the gap has shown the tendency predicted by the theory. However in the most part of this region, the flow speed in the gap may excess sound velocity, so our simplified model can not be applied. At present we do not understand the reason for a big jump occurred at the value about $24g$/min mass flow rate, further study of this phenomenon is under way in our lab.

4. OUTLINE EXPERIMENTS AT 4.2K

Experiments at 4.2K have been just started with a slightly different head supporting system. Instead of the bellows, three 0.4 mm inner diameter and 0.5 mm outer diameter stainless steel tubes were used to connect the head through a hollow ring frame to the gas supply system. In such way, the pick-up coil can be fixed tightly on the parallel direction between the head and the superconductor, while having reasonable flexibility along the normal direction. Furthermore, to insure the temperature of incoming gas below 6K, an extra heat exchanger immersed in liquid He was employed.

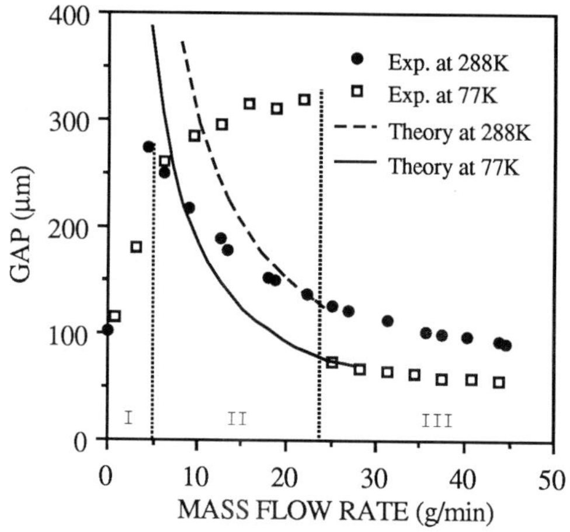

FIGURE 2
Gap versus mass flow rate for nitrogen gas.

Preliminary results have shown that up to $18g$/min He mass flow rate ($\approx 100l$/min), the gap is still in region II of Fig.2. For instance, at $40l$/min ($\approx 7.14g$/min) He flow rate, the distance between the pick-up coil and the superconductor is 130 ± 4 μm, which is slightly larger than the value of initial gap and much larger than 20μm predicted by the theory(2). Fortunately, the variation of the gap upon lifting and rotating the sample at those flow rates is smaller than 3%, which is within the calibration accuracy of the gap measurements. And the noise level on the SQUID measured by a spectra analyzer(Advantest) is 2.2×10^{-4} $\phi_0/\sqrt{\text{Hz}}$.

Using this system, we are able to map out some magnetic field distribution on several superconducting films by lifting and rotating the sample at different flow rates. The results are still being analyzed and the experiments are being continual.

ACKNOWLEDGEMENTS

We thank Dr. F. Naruse for his fruitful discussion on the boundary layer model.

REFERENCES

(1) R.P. Huebener, Magnetic Flux Structures in Superconductor (Springer, Berlin, 1979).

(2) H. Minami, J. Yuyama, E. Goto and F. Naruse, Proceedings of the 6th Symposium on Josephson Electronics (1989) 89 (in Japanese).

(3) M. Scheuermann, J.R. Lhota, P.K. Kuo and J.T. Chen, Phys. Rev. Lett. 50 (1983) 74.

(4) J. Bosch, R. Gross, M. Koyanagi and R.P. Huebener, Phys. Rev. Lett. 54 (1985) 1448.

(5) O.B. Hyun, J.R. Clem, L.A. Schwartzkopf and D.K. Finnemore, IEEE Trans. Magnetics Mag-23 (1987) 1176.

(6) H. Schlichting, Boundary-Layer Theory 7th edn. (McGraw-Hill, New York, 1979).

Physica B 165&166 (1990) 89–90
North-Holland

PRINCIPLES OF SUPERCONDUCTOR IMAGING SURFACE GRADIOMETRY*

D B van Hulsteyn, W. C. Overton jr, E R Flynn

Los Alamos National Laboratory, Los Alamos, New Mexico 87545, USA

Magnetic sources in the vicinity of superconducting surfaces produce images in such a way that a single pick-up coil parallel to the surface acts as a first order gradiometer. The principles will be described, together with specific discussions of the imaging by planar and spherical surfaces. The fact that these surfaces also deflect noise due to remote sources make this concept particularly appealing as a method for detecting extremely weak sources in a hostile environment. Possible applications to neuromagnetometry, corrosion detection and non-destructive evaluation will be discussed.

1. INTRODUCTION

In electrostatic applications, if two or more sets of charge distributions produce an equipotential surface on which $\phi = 0$, the principle of imaging can often be applied. For example, two equal but opposite charges generate a planar surface at the perpendicular bisector that acts as a grounded plane. One speaks, then, of a charge in front of such a surface as having an image plane of equal but opposite charge on the opposite side and obtains the solution by treating the two problems as being equivalent. The principles of electrostatic imaging have been applied successfully in a number of practical situations, a notable example being the propagation of radio waves over a conducting earth.

We have extended the imaging analysis to incorporate currents rather than electric charges, taking advantage of the fact that a superconducting surface is an excellent approximation to a perfect conductor. In our analysis, if a set of current carrying elements give rise to an isomagnetic contour on which the normal component of the magnetic field is zero, the situation is wholly analogous to the electrostatic case. The simplest example is of two parallel wires in which the currents flow in opposite directions. The magnetic field normal to the perpendicular plane half-way between is zero everywhere, so that one can treat the case of a current parallel to a superconducting surface as being equivalent to the two wire problem. Indeed, a wire in the vicinity of the plane will be repelled by its image wire; this behavior is consistent with the Meissner efect.

---

* This work was conducted under the auspices of the U. S. Department of Energy

2. PLANAR MODELING

In this presentation we shall discuss the simpler problems of current and loop dipoles because of their potential applications to non-destructive evaluation, corrosion studies and magnetoencephalography (MEG), to name a few. The loop dipoles, for which the magnetic moment M is equal to NIA (amp-m2) where N is the number of turns, is well known. Less familiar is the so-called current dipole which is modeled as a short (typically a few millimeters) current-bearing filament. Although this type of dipole is physically perplexing, it does appear to be a good representation of neural currents generated in the brain and the heart. Because of our experience and interest in neuromagnetism, the sources we will discuss will be primarily current dipoles, although it will be convenient to approximate the sensing coils by loop dipoles.

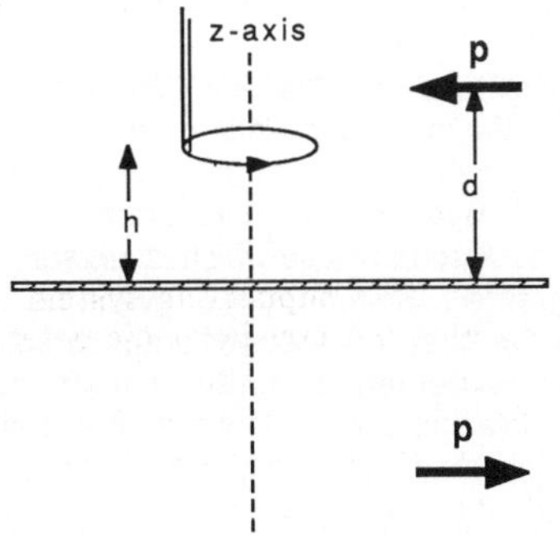

FIGURE 1

The dipole p and the sensing loop are above the superconducting plane; the image dipole is below.

Consider, as a first example, a current dipole oriented parallel to a superconducting plane d cm away (Fig. 1). The image dipole that allows the boundary condition to be satisfied has the same moment 'p', is equidistant on the opposite side and is oriented in the opposite direction (in this example, along the -y axis) from the source. To understand the import of this statement, consider a pick-up coil parallel to the suraface at (0,0, h). It senses both the source and its image, so that the average flux density it detects is

$$B_{no}=(\mu/4p)px\{1/[x^2-(d-h)^2]^{1.5}-1/[x^2+(d+h)^2]^{1.5}\}$$

Consider, now a different scenario in which the source and the loop are configured the same way but in which the superconducting plane and the image have been removed and a second pick-up coil parallel to the first is placed at (0,0,-h). If the two coils are wound in series opposition as shown in Fig. 2, the net average flux density detected is precisely the same as the result presented in the preceding equation. In other words, for this simple planar case, the single loop in the vicinity of the superconducting plane observes the current dipole in a way that makes the loop magnetometer act like a first order gradiometer.

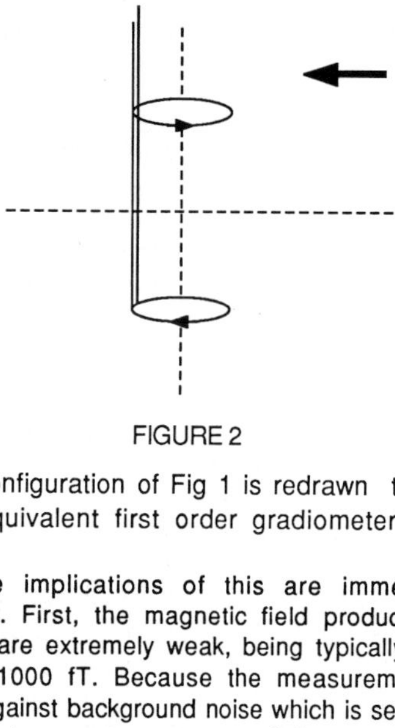

FIGURE 2

The configuration of Fig 1 is redrawn to show how the equivalent first order gradiometer is achieved.

The implications of this are immense for two reasons. First, the magnetic field produced by neural activity are extremely weak, being typically in the range of 10 -1000 fT. Because the measurements must be made against background noise which is several orders of magnitude greater, commercially available sensing systems rely on first or second order gradiometers which require extremely accurate construction and balancing. The imaging concept indicates that a single coil and planar superconductor will produce the equivalent result. Second, the presence of the superconducting surface in itself provides further magnetic noise reduction by steering or deflecting the undesired noise away from the sensing loop. A numerical example of this will be presented.

The issue of mutual inductance has, so far, not been introduced because, in deriving the equation for the average flux, the current induced in the sensing coil was ignored. In reality such a current must be included and in so doing, the coil itself must be imaged. For purposes of this discussion, the coil and its image will be assumed to be simple loop dipoles. Note that their orientations must be in opposite directions in order to satisfy the boundary condition. Because of the additional contibution from the image loop, the flux density sensed is modified to

$$B_n = B_{no} - (N/4)(2\pi a\mu)(M/a^2)(a/2h)^3$$

where a is the radius of the coil.The reduction, which will be of order 10% or less for cases of practical interest, is a result of the mutual inductance between the loop and the superconducting plane.

3. SPHERICAL GEOMETRY

A final point, which is really the essence of the discussion, is that we are pursuing more complicated geometries than the simple plane discussed so far. It is well-known in electrostatics, for instance, that a charge in the vicinity of a conducting sphere has a relatively simple image. It is not surprising that a tangentially oriented current dipole located at a point (0,0,d) on the Z-axis inside a superconducting sphere of radius a produces an image $\mathbf{p'} = -(a/d)\mathbf{p}$ at $(0,0,a^2/d)$. What is unexpected is that a loop dipole **M** at the same position produces an image $\mathbf{M'} = -(a/d)^3\mathbf{M}$, also at $(0,0,a/d^2)$. The significance of these effects and their implication to gradiometry will be discussed.

The reason for pursuing this course is two-fold. First, a spherical (more realistically a hemispherical) configuration, if practicable, would be ideal for studies of brain neural activity. It is straightforward to envision an array of loops that would permit researchers to obtain all of their data on a single set of measurements, obviating the need to move the dewar from position to position as is now done. Second the noise reduction from distant sources would be far greater in this geometry than in the planar case. In a companion paper, some of the experimental efforts to evaluate the spherical case will be described.

REFERENCES

(1) W. C. Overton jr, D. B. van Hulsteyn and E. R. Flynn, Use of Superconducting Plates and Shell to Deflect Magnetic Noise Fields: Application to MEG, in: 11th Annual International Conference of the IEEE Engineering in Medicine & Biology Society, Seattle, Washington, November 1989.

(2) J. H. Tripp, Biomagnetism: An Interdisciplinary Approach (NATO ASI Series A:Life Sciences, Vol 66, Plenum Press, New York and London, 1982).

Physica B 165&166 (1990) 91–92
North-Holland

EXPERIMENTS ON SUPERCONDUCTOR IMAGING SURFACE MAGNETOMETRY *

W. C. OVERTON, jr., D. B. van HULSTEYN and E. R. FLYNN
Neuromagnetometry Laboratory, Los Alamos National Laboratory, MS M715
Los Alamos, New Mexico 87545, USA

The recent development of a new concept in magnetometry has led to experiments to evaluate the performance of such systems. The superconductor imaging surface magnetometer employs superconducting plates and shells to deflect magnetic noise and to image nearby dipole sources while the SQUID pick-up coils are located in areas where the noise has been greatly reduced. The experiments herein reported confirm the principles of imaging magnetometry for detecting nearby dipole sources.

1. INTRODUCTION

London's analysis (1) of the superconducting sphere in a uniform magnetic field relates the current density, penetration depth and internal field to the external field and leads to an understanding of the Meissner effect. London's procedure can be extended to the analysis of the field of superconductors of various shapes such as cylindrical rods, plates with circular edges, and curved shells. Since the magnetic noise fields usually encountered in SQUID magnetometry are analogous to London's uniform field, one can adapt the principles involved in the deflection of bodies into a new concept for SQUID magnetometry. A first implication is that one should place the magnetometer pick-up coil at a location near the superconducting body where the noise field is a minimum. In some configurations, the noise may be reduced by as much as 1,000.

We show in Fig 1 a schematic of a magnetometer system in which a flat superconducting plate is in a noise field coming from 45° with respect to the plate normal (the z-axis). A weak magnetic field due to a nearby dipole source, such as *m* in Fig 1, may have an amplitude of only 10 fT to 1,000 fT, which is typical for brain waves and fields caused by flaws in materials. Meanwhile, the noise fields may have amplitudes of a million to ten million fT.

Since the superconducting plate of Fig 1 has essentially infinite conductivity, the target dipole *m* will exhibit an image *m'* behind the front face. Thus, the field at the pick-up coil A-A' will be the sum of fields due to the source and its image. We have named the system the Superconductor Imaging Surface Magnetometer (2).

The purpose of this paper is to report on experiments done to test the performance of this novel magnetometer system in the uniform field of a Helmholtz coil and in the dipole field of a wire loop dipole that can be moved to arbitrary coordinate positions.

---

* Work performed under the auspices of the U.S. Department of Energy

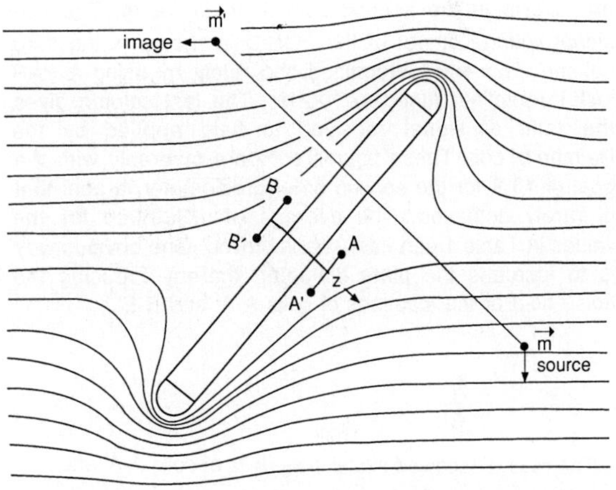

FIGURE 1
Schematic of one of many configurations of the superconductor imaging surface magnetometer. A-A' denotes the location of signal pick-up coil in front of the front face. B-B' is the noise cancellation coil the same distance behind the back face. Line of flux due to a distant source are deflected by superconducting plate as shown. For an ordinary wire loop dipole, the image *m'* is the mirror image of *m*.

2. EXPERIMENTAL DISCUSSION

The superconducting plate illustrated in Fig 1 was actually a 7.8 cm diameter by 0.95 cm thick plate of lead with edge rounding. The pick-up coil A-A' was a 7-turn coil of NbTi wound on a 1.27 cm diameter form and had an inductance of 1.7 µHy. The identical coil at B-B' was connected in series opposition to the one at A-A'. The shielded pair of connecting lines was then attached to a B.T.I. hybrid rf SQUID located approximately 50 cm higher in the LHe dewar. The total circuit inductance was therefore

$$L(A\text{-}A') + L(B\text{-}B') + L(\text{leads}) + L(\text{SQUID}),$$

which was equal to about 5.5 µHy. The sensitivity was therefore reduced from the typical 17 fT/√Hz for an rf SQUID magnetometer to approximately 49 fT/√Hz for our system.

In an expeiiment to test the noise rejection capability of the system, we used a Helmholtz coil pair with a field factor of 31.5 µT/A to produce a 100 Hz uniform applied field. This was driven at 2 to 6 mA peak-to-peak. The resulting B_z values are shown in the first column of Table 1. The corresppondíng flux ratios are shown in the second column in ϕ_0 units. The rms signal voltage output of the SQUID is shown in the third column. This was determined accurately by using a PAR lock-in amplifier tuned to 100 Hz. The last column gives the ratio of signal received to field applied by the Helmholtz coil. These figures compare favorably with the goal of 10^{-6} for the second order gradiometer, a goal that is rarely achieved. The average of 1/330,000 for the values in Table 1 can easily be improved. One obvious way is to increase the plate diameter, thereby reducing the noise field at the locations of coils A-A' and B-B'.

Table 1
The ratio of received signal in ϕ_0 units of coils A-A' and B-B' to the amplitude produced by the Helmholtz coil.

Bz (p-p) (units of fT x 10^7)	ϕ/ϕ_0 rms (x 10^3)	Signal ϕ_0 units	Signal/ drive
2.42	3.76	0.010	1/376,000
2.42	3.76	0.016	1/235,000
1.21	1.88	0.0055	1/342,000
2.42	3.76	0.011	1/342,000
3.64	5.66	0.016	1/354,000

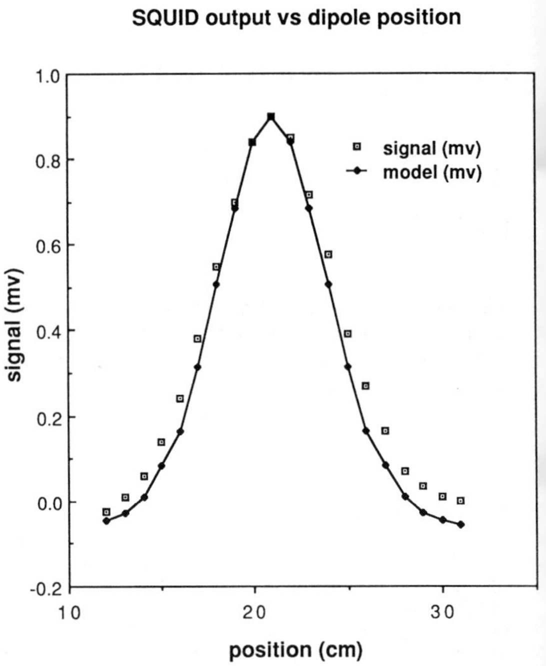

FIGURE 2

Response of the image magnetometer to a loop dipole driven at 100 Hz and located 9.7 cm below coil A-A'. The x-position is varied from 12 to 31 cm, with the center position at x = 21 cm.

We show in Fig 2 the response of the magnetometer to a wire-loop dipole driven ar 100 Hz. The theoretical solid curve is the same for the image magnetometer and an ordinary first-order gradiometer with baseline 2d, where d is the distance of the coil A-A' from the front surface. The experimental points (squares) are in good agreement with the theory.

3. SUMMARY

The Superconductor Imaging Surface Magnetometer is shown to be suitable for detecting weak near-source magnetic signals in the presence of overwhelming magnetic noise without requiring a magnetically shielded room.

REFERENCES

(1) F. London, Superfluids, Vol. I, Superconductivity, J. Wiley & Sons, New York (1950).
(2) W. C. Overton, D. B. van Hulsteyn and E. R. Flynn, Use of Superconducting Imaging Surfaces to Enhance Detection of Weak Magnetic Sources by SQUID Systems, Los Alamos National Lab. Report LA-MS 11473, Jan. 1989.

Physica B 165&166 (1990) 93–94
North-Holland

LABORATORY TESTS OF A MOBILE SUPERCONDUCTING GRAVITY GRADIOMETER

F.J. van KANN, M.J. BUCKINGHAM, M.H. DRANSFIELD, C. EDWARDS, A.G. MANN, R. MATTHEWS, R.D. PENNY and P.J. TURNER

Physics Department, The University of Western Australia, Nedlands, Western Australia, 6009.

We describe a gravity gradiometer designed to measure off-diagonal components of the earth's gravity gradient tensor, and intended for airborne geophysical exploration. The instrument consists of an orthogonal pair of mass quadrupoles pivoted about a common vertical axis. Each quadrupole is a rectangular bar of niobium having square cross section and with integrally machined flexural micro-pivots. Superconducting niobium pancake coils form the basis of the SQUID based transducers and provide supplementary magnetic springs to enable matching of mechanical parameters.

1. INTRODUCTION

Transducer technology deriving from the second generation of gravitational radiation detectors lends itself to the construction of sensitive gravity gradiometers. For example, Paik's (1) transducer consisted of a flat pancake coil wound from superconducting wire and coupled into a SQUID magnetometer. The inductance of the coil depends on the proximity of a nearby superconducting ground plane and relative motion of the plane and coil therefore appears as a current modulation in the flux conserving loop - providing the basis of a sensitive superconducting accelerometer.

Gradiometers may be classified by the nature of the nominal acceleration environment in which they must operate i.e. g_E (earthbound) or zero g_E (orbital); both the level and spectral range of the signal are then effectively determined. For example a proposed orbital instrument (2), intended for geodesy, requires, and can in principle attain, a sensitivity of 10^{-4}Eö (1 Eötvös unit = 10^{-9} s^{-2}) at wavelengths above about 10km. Our own work (3) however, has focussed on instruments for wavelengths in the 50m-50km range - the interval of interest to exploration geophysics, and covered by an instrument of bandwidth 1 to 10^{-3} Hz used in a low flying aircraft

2. THE GRADIOMETER

The instrument we describe here has a target sensitivity of 1Eö/√Hz in a frequency band below 1Hz, and is the laboratory prototype of an instrument intended for operation from an aircraft. The **OQR** instrument, which measures an off-diagonal [xy] component of the gradient tensor, consists of two perpendicular mass-quadrupole sensors (rectangular bars) with common pivot axes aligned along the vertical z-axis. This configuration, the **O**rthogonal **Q**uadrupole **R**esponder [**OQR**], simplifies the problem of dealing with angular accelerations about the pivot axis.

An extended rigid body, at rest in a static non-uniform gravitational field and supported by an ideal pivot, experiences a torque due to both its *dipole* and *quadrupole* moments. The gravity vector **g** couples to the dipole, and the gravity gradient tensor Γ couples via the quadrupole moment. If the responder, in the present case a rectangular bar of niobium 90mm long and 30x30mm square cross section, is pivoted about its centre of mass the torque due to the dipolar term vanishes. In practice, precision electron discharge machining [EDM] of the microscopic scale pivot, which is integral to the bar, ensures that the dipole moment is indeed small; each bar can be mechanically balanced to better than 1 part in 10^6, and in the **OQR** the net effect of the *residual* dipole can be reduced electronically by another three orders as discussed below.

In a moving frame, the vehicle acceleration **a** must be added to **g**, and the dipolar interaction is dominated by the *induced*, rather than the residual dipole. The acceleration **a** deforms the responder, resulting in a dipole proportional to |**a**| and thus a torque quadratic in the acceleration amplitude. This becomes a serious design constraint in too harsh acceleration environments. For example, in the prototype **OQR**, the differential induced dipolar term corresponds to a 10Eö gradient at |**a**| ~ $10^{-2}g_E$. Good vibration isolation and accurate measurement of **a** (for post processing the data) are then essential.

Angular motions are sensed by pairs of superconducting pancake coils arranged in current differencing configurations (2). In principle, the main readout would consist of just the two angle sensing loops coupled differentially to a SQUID. However, apart from the desired gradient information, the signal from the primary coils contains small residual terms resulting from the common mode accelerations perpendicular to the pivot axis. To account for these, the main readout includes two further loops which allow the passive elimination of the effects of the common mode acceleration vector. The scale factor for each of these secondary loops depends on a persistent current which is adjusted empirically to make the response of the main readout independent of transverse acceleration.

A fifth loop is included to trim out residual thermal effects in a similar way.

The effect of vehicle rotation is twofold. The rotation tensor **R** may be decomposed into its symmetric part $\mathbf{R}^s$ (centrifugal and Coriolis terms) and antisymmetric, $\mathbf{R}^a$ (angular accelerations) part. The former is indistinguishable from Γ which is also symmetric, so the "gradient" which is measured in the non-inertial frame is actually $\mathbf{G} = \Gamma + \mathbf{R}^s$. In the present case $G_{xy} = \Gamma_{xy} - \omega_x\omega_y$, where ω is the angular velocity.

Rotation about the x- and y-axes, ω_x and ω_y is measured optically and controlled by servos referenced to a room temperature inertial system. The latter is a gimballed platform stabilised by a pair of phase modulated fibre optic gyros to about 2.10^{-5} (rad/sec)/$\sqrt{\text{Hz}}$. The cold gradiometer package is also mounted on gimbals, in this case driven by diamagnetic actuators.

The effect of the antisymmetric part of **R**, corresponding to angular acceleration $\dot{\omega}_z$ about the pivot (z- axis), is minimised in the **OQR** configuration if the two responders are closely matched, . The fundamental torsional mode of each responder is about 3Hz but limitations in the EDM machining of the micro-pivots are such that the natural frequencies are matched to only a few percent. The pair must therefore be further trimmed to within a few parts in 10^5 using ancillary superconducting loops which act as magnetic springs, adjusted by selecting the appropriate persistent currents. These loops have an important secondary role; each is coupled to a SQUID which then provides a signal proportional to the $\dot{\omega}_z$. This is used as input to a rotational stabilisation servo system which reduces angular acceleration of the gradiometer assembly by 80dB around 3Hz. Together with the matching factor, this yields rejection of $\dot{\omega}_z$ by some nine orders of magnitude.

The thermal environment in which the package is mounted is carefully controlled and maintains an operating temperature constant to within a few μK at about 5K .

3. PERFORMANCE

The gradiometer package has been extensively studied under controlled acceleration conditions i.e. with the dewar mounted on a shakeable platform on the laboratory floor. A typical gradient signal, arising from a 300kg mass, is shown below.

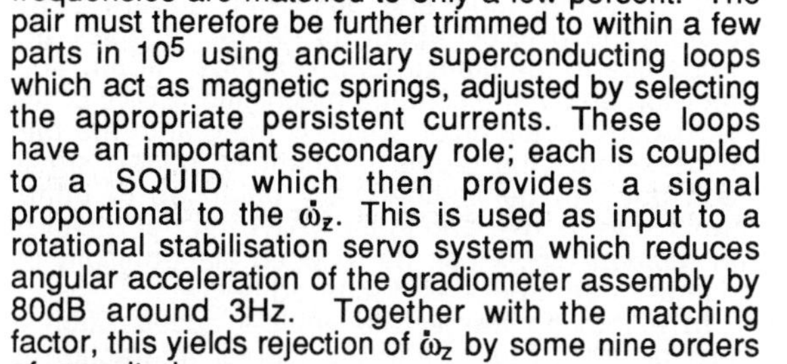

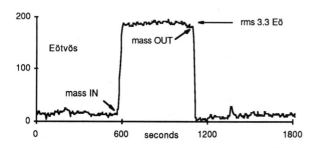

The rms noise shown here is an order of magnitude larger than that predicted from the SQUID and coupling parameters and is largely due to unwanted structural resonances in the dewar probe and attitude control systems. For example, the drive stirrup of each cryogenic actuator has a high Q mode at ~350Hz which, if deliberately excited, shows clear evidence of down-conversion (4) of the Coriolis $(\omega_x\omega_y)$ term. Several other bright lines in the structural spectrum are in evidence so that under severe operational conditions the control servos could be responsible for a considerable amount of down conversion noise.

Structural resonances also contribute terms which arise from the frictional conversion of mechanical energy to heat. Since neither the heat capacity of the gradiometer assembly, nor its internal thermal response time, are at present properly optimised to minimise the effect of these tiny heat currents, the low frequency noise i.e. <10mHz is dominated, through the well-known temperature dependence of the penetration depth, by these terms rather than 1/f SQUID noise.

Finally, although under "quiet" conditions the z-axis servo and both x- and y-axis servos maintain lock, their performance, and thus the level of platform acceleration/rotation which can be tolerated, is limited by the same unwanted structural resonances. We are at present engaged in a detailed search and elimination programme prior to transferring into a mobile laboratory in readiness for field trials.

4. CONCLUSIONS

We have successfully demonstrated the operation of a superconducting gradiometer with a sensitivity of a few Eö/$\sqrt{\text{Hz}}$. We have identified the major noise sources which limit low frequency performance; many of these are associated with the attitude control system and we are confident that they can be brought under control.

ACKNOWLEDGEMENTS

This work was initiated under the NERDDP programme and supported by BP Minerals Australia.

REFERENCES

(1) H.J. Paik, J. Appl. Phys., **47**, 1168 (1976)

(2) S.H. Morgan and H.J. Paik, NASA Tech. Memo. **TM-4091**, (1989). See also H.J. Paik and J-P. Richard, NASA Report **4011**, (1986).

(3) F J. van Kann, C. Edwards, M.J. Buckingham and R.D. Penny, IEEE Trans. Mag. **MAG-21**, 610 (1985).

(4) F.J. van Kann, M.J. Buckingham, M.H. Dransfield, C. Edwards, A.G. Mann, R. Matthews, R.D. Penny and P.J. Turner, Proc. 15th Gravity Gradiometry Conference. USAF Academy, Colorado Springs, (1987). See also same authors in Proc. 17th Gravity Gradiometry Conference, USAF Geophys. Labs. , Hanscom, Mass. (1989).

Physica B 165&166 (1990) 95-96
North-Holland

DESIGN AND CONSTRUCTION OF A 19-CHANNEL DC-SQUID NEUROMAGNETOMETER

H.J.M. TER BRAKE, J. FLOKSTRA, E.P. HOUWMAN, D. VELDHUIS, W. JASZCZUK, A. MARTINEZ* and H. ROGALLA

University of Twente, Faculty of Applied Physics, Low Temperature Group, P.O.Box 217, 7500 AE Enschede, The Netherlands.
*University of Zaragoza, Dept. of Electrical Engineering and Informatics, Maria Zambrano 50, 50015 Zaragoza, Spain.

The 19-channel DC-SQUID neuromagnetometer which is under construction at the University of Twente is described. Several aspects of this development are considered: the DC-SQUID sensor, the design of the SQUID module, the arrangement of the 19 gradiometers, and the electronics, including the output transformer, the preamplifier and the control and detection section. The completed system will be installed in our magnetically shielded room.

1. INTRODUCTION

During the last few years biomagnetic instrumentation has changed from single-channel SQUID systems to multichannel magnetometer units (1). In this way the magnetic field distribution around a subject's body can be measured much faster and more reliable. Furthermore, spontaneous activity in the body can now be studied. Multichannel systems of about 20 to 30 channels are developed by or are under construction in university groups as for instance in Helsinki and Rome and in industries like BTi, Siemens and Philips.

At the University of Twente a 19 channel DC-SQUID magnetometer for brain research is under construction. In this paper several aspects of the system are shortly described. These concern the sensing coils, which have been optimized with respect to the signal-to-noise ratio of the system, the cross-talk elimination by means of external feedback, electronic noise cancellation, the DC-SQUID sensor and the design of the SQUID module, and finally the electronics including the output transformer, the preamplifier and the control and detection section.

2. SENSOR ASSEMBLY

The gradiometers are optimized with respect to the signal-to-noise ratio according to the method presented in ref. 2. The resulting design, based on a system-noise level of $(2\text{-}6)\cdot 10^{-6}$ $\phi_0/\sqrt{Hz}$ and an input inductance of (0.2-0.4) μH is: diameter 20 mm, baseline 40 mm, 2 times 3 turns separated by 0.5 mm and 6 mm for the proximal and the compensating coil section, respectively. The gradiometers are distributed in a hexagonal configuration over the bottom of a CTF-SST 140 cryostat having a curvature with a radius of 125 mm. The sensing-coil holders and the support are presented in fig. 1. The material used is tufnol, a textile/epoxy composite. Because the calculated cross-talk between two neighbouring channels is about 5 % we apply external feedback. The output of each SQUID system is in this case fed back to the corresponding flux-transformer circuit (3). Our shielded room (having only one μ-metal wall) suppresses low-frequency fields by a factor of about 10. As a further reduction, electronic noise cancellation can be applied (4). For this purpose three orthogonal 1-cm loops, connected to SQUID systems, are installed as references for the ambient magnetic noise. A first-order gradiometer may be incorporated as an additional noise reference. All in all the magnetometer therefore comprises 22 to 23 SQUID systems.

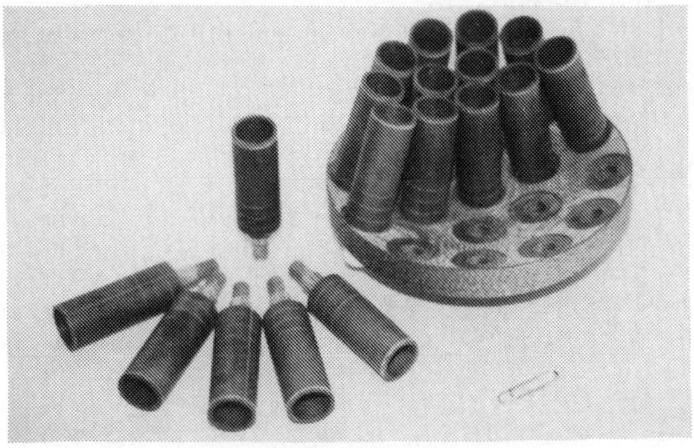

fig. 1: The sensing-coil holders and the support, partly assembled.

The DC-SQUIDs are fabricated based on the Nb/Al,AlO_x/Nb technology. The junctions are of high quality with a gap voltage of 2.9 mV, an I_cR_N product that is 80 % of the theoretical value and a high subgap resistance leading to V_m=70 mV at 4.2 K. The junctions are shunted by

0921-4526/90/$03.50

a Pd resistor of about 3 Ω and are part of a standard washer type SQUID configuration. DC-SQUIDs with different inductances, varying from 110 to 240 pH, have been fabricated. The planar input coil consists of 30 to 42 turns corresponding to an input inductance of 0.2-0.4 μH. Apart from this standard type SQUID, also resistively and inductively shunted configurations were made,which allow for a larger SQUID inductance.Thus the number of turns of the input coil may be reduced. These different approaches are compared to come to the best SQUID design for our application (5).

Each SQUID is placed on a solid Nb body which fits into a cylindrical superconducting shield. As depicted in fig.2, this module is divided into compartments to maximize the shielding of the different sections.

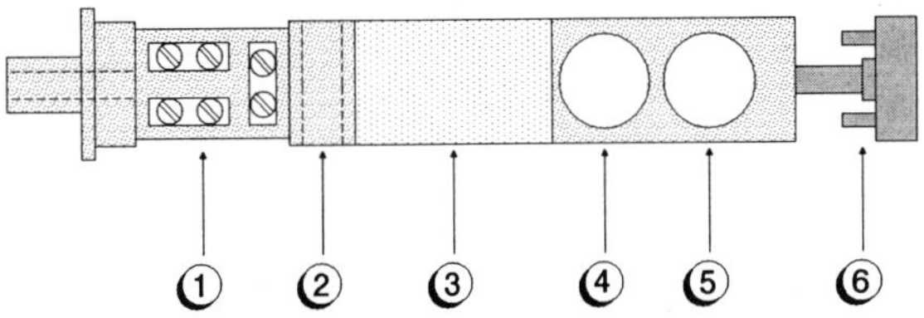

fig. 2: Compartments of the SQUID module

1: screw connections for sensing coil and feedback inductor
2: hole for feedback inductor
3: plane for SQUID chip
4: hole for output transformer
5: hole for capacitor (tuning the transformer)
6: connector

3. ELECTRONICS

The preamplifier section is placed directly on top of the cryostat in an aluminium enclosure. It consists of U311 FETs in a cascode configuration, loaded with an RCL-network resonating at the modulation frequency of 100 kHz. The quality factor is about 5 and the preamplifier gain about 50. In order to reduce the effect of the preamplifier noise (1 nV/√Hz), a small transformer is used between the SQUID and the preamplifier. It is tuned to 100 kHz via a capacitive shunt. Without the transformer, the preamplifier would give an equivalent noise in the SQUID of about 2.10^{-5} $\phi_o/\sqrt{Hz}$, the forward transfer of the SQUID being roughly 70 $\mu V/\phi_o$ (5). In order to limit the noise contribution to 10^{-6} $\phi_o/\sqrt{Hz}$, the tuned transformer should thus have a gain of at least 20. A moderate Q-factor of 10 was chosen to give a bandwidth of 10 kHz. To realize an overall gain of more than 20, the transformer voltage-gain should therefore be larger than 2. A prototype transformer has been made of 0.1 mm copper wire. It has a length of 5 mm and an outer diameter of 5.3 mm, thus fitting in the 7-mm hole of the SQUID module. The prototype consisted of 94 primary and 371 secondary turns. The measured inductances are respectively 14.9 μH (20.6 μH) and 204 μH (297 μH) whereas the coupling factor appeared to be as high as 0.94 (0.97) (transformer voltage-gain 3.5 (3.7)). The values between brackets were measured outside the superconducting enclosure.

The control and detection electronics is realized in a miniaturized version on a single Eurocard for each channel. Because of the miniaturization and the fact that the system is not designed to be a general-purpose magnetometer, several parameters have been fixed among which the sensitivity and the frequency bandwidth (5 kHz). The sensitivity is 1 V at the output for 1 ϕ_o in the SQUID. The dynamic range of 10 V corresponds to an effective field of about 3 nT in the gradiometer. The power supplies and the bias-current supplies are independent for each channel, whereas the 100 kHz modulation for all channels is obtained from an external oscillator. The electronics have been developed independently from the DC-SQUID devices. System parameters as dynamic behaviour, the gain of the feedback path, and the open-loop gain have been adjusted by means of a DC-SQUID simulator (6).

Before A/D conversion some analogue filtering is necessary. For this purpose we developed a 50 Hz adaptive notch filter that suppresses the mains frequency by more than 60 dB with a bandwidth of only 1 Hz. Further, high-pass and low-pass filters will be applied, the latter also functioning as anti-alias filters for the 1 kHz sampling frequency. Thus signals can be measured without significant phase-shift up to a frequency of 100 Hz, which is adequate for most biomagnetic experiments.

4. CONCLUSION

We presented the design of our 19-channel DC-SQUID neuromagnetometer. Various aspects concerning the sensor assembly and the electronics are shortly discussed. After the first test experiments of the completed system, an extensive description will be given elsewhere. The neuromagnetometer will be installed in the magnetically shielded room of the Biomagnetic Center Twente. It will be used among other applications for studies on epilepsy and cognitive processes.

REFERENCES

(1) see e.g. G.L.Romani, Physica **126 B** (1984) 70.
(2) H.J.M.ter Brake, Sensing-coil optimization based on signal-to-noise ratio, to be published in Proceedings Biomagnetism VII, New York 1989.
(3) H.J.M. ter Brake et al., Cryogenics **26** (1986) 667.
(4) H.J.M. ter Brake et al., J. Phys. E: Sci. Instrum. **22** (1989) 560.
(5) E.P.Houwman et al., Comparison of shunted DC-SQUIDs with large β, this volume.
(6) A. Martinez et al., A low-noise multi-step SQUID simulator, to be published in Cryogenics.

Physica B 165&166 (1990) 97–98
North-Holland

A 24-SQUID GRADIOMETER FOR MAGNETOENCEPHALOGRAPHY

S. AHLFORS, A. AHONEN, G. EHNHOLM*, M. HÄMÄLÄINEN, R. ILMONIEMI, M. KAJOLA, M. KIVIRANTA, J. KNUUTILA, O. LOUNASMAA, J. SIMOLA, C. TESCHE**, and V. VILKMAN

Low Temperature Laboratory, Helsinki University of Technology, 02150 Espoo, Finland
*Instrumentarium Imaging Ltd, Teollisuuskatu 27, 00510 Helsinki, Finland
**IBM Thomas J. Watson Research Center, Yorktown Heights, New York 10598, USA

In this contribution we briefly describe our newest 24-channel neurogradiometer, employing dc SQUIDs as magnetic flux sensors, and present an example of its performance. The instrument is used in a magnetically shielded room. This state-of-the-art apparatus is able to locate the stimulus-activated area in the brain by just one measurement, without moving the dewar.

1. INTRODUCTION

The human brain is, by far, the most complex machine in existence, and it is also the most important. There are 10^{11} neurons in the cerebral cortex. These nerve cells are the active units in a vast network, consisting of 10^{14} interconnections or synapses. Information transfer within the neuronal system takes place by means of electric current pulses which, in turn, produce weak magnetic fields. Measuring these is what magnetoencephalography (MEG) is about: the aim is usually to locate the area in the cortex that has been activated by the stimulus. The method has a spatial resolution of 1-2 mm, a temporal resolution of better than 1 msec, and it is completely harmless to the subject. No other technique offers these advantages at present .

Typically, for source localisation, the magnetic field must be measured at 20-50 positions and at each site the stimulus must be repeated 30-300 times in order to obtain a satisfactory signal-to-noise ratio. An instrument which produces the whole magnetic field pattern at once, without moving the sensors, has been awaited urgently.

2. MECHANICAL STRUCTURE

Our gradiometer is kept at 4.2 K in a fiberglass dewar, having a slanted spherical cap (see Fig. 1). The dewar was specially designed for this application and manufactured by CTF Inc., Canada. The electronics has been divided into three identical groups of 8 channels. The preamplifiers, located in three detachable aluminium boxes on top of the dewar, are connected to the SQUIDs via twisted pairs of superconducting wires inside shielded tubes.

3. COIL CONFIGURATION

The pickup coil array consists of 12 sensor units, each measuring the two orthogonal tangential derivatives $\partial B_z/\partial x$ and $\partial B_z/\partial y$ of the radial field component B_z. Wire-wound coils were chosen for the gradiometers

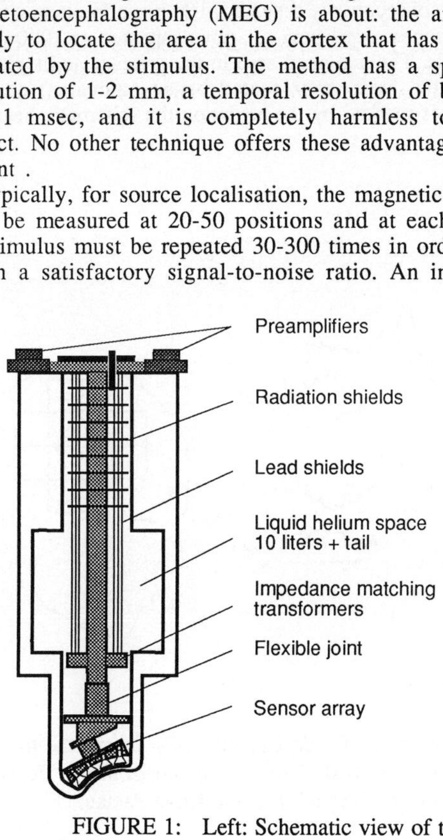

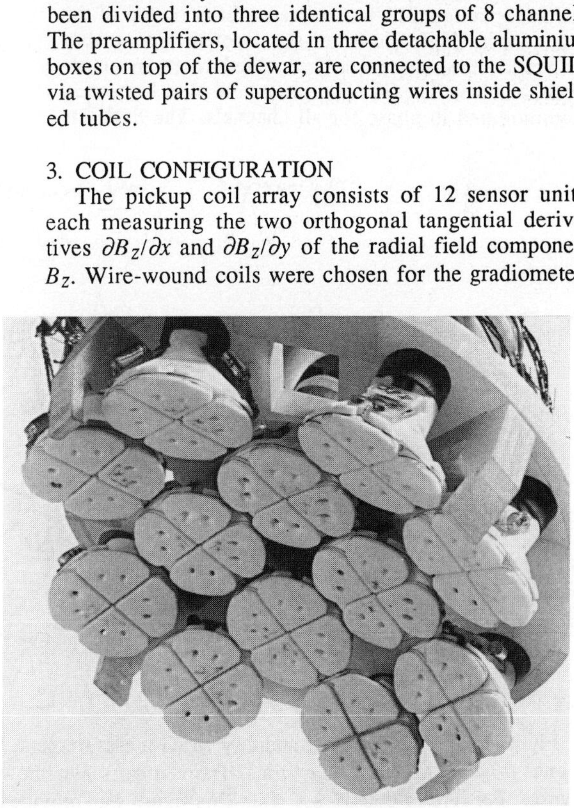

FIGURE 1: Left: Schematic view of the dewar and insert. Right: Photograph of the sensor array.

because of technical difficulties encountered in fabricating reliable thin film structures in large quantities. Our instrument employs dc SQUIDs provided by IBM.

The advantage of a planar coil configuration is its compact size with unimpaired source positioning accuracy. The locating capability of our device and the amount of information gained in measurements have been compared with several axial gradiometer coil configurations, and found entirely satisfactory for practical work.

The 12 two-SQUID units are placed on the spherical bottom inside the dewar; the radius of curvature is 125 mm and the coils cover an area of 125 mm in diameter. Each unit consists of two Nb wire-wound figure-of-eight gradiometer coils and the SQUIDs they are connected to. The area of both loops is 3.7 cm^2 and the effective base length is 1.3 cm. The design of the units is illustrated in Fig. 2. The gradient noise level is 3-5 $fT/(cm\sqrt{Hz})$. The balance against homogeneous fields is, on the average, about 1 %. The coupling between the orthogonal coils in a single unit is about 2% and between adjacent units less than 1%.

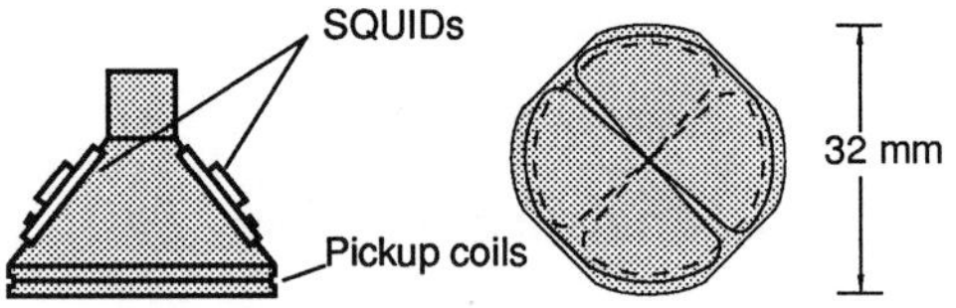

FIGURE 2: One of the two-channel sensor units, viewed from the side and from below.

4. ELECTRONICS AND SIGNAL MONITORING

The design of the electronics is a slightly modified version of that used in our earlier seven-channel system. For each SQUID, there is an impedance matching transformer in liquid helium, a preamplifier on top of the dewar, and a single Eurocard-sized detector-controller unit outside the magnetically shielded room in an rf-shielded cabinet. The 90-kHz flux modulation frequency is common and in phase for all channels. The SQUIDs can be operated either in an open loop or in a flux-locked loop mode, and the flux-voltage characteristics can be monitored for adjustment of the bias current. The operating mode, the output gain, and resetting of the feedback integrator are controlled remotely by a computer. The main difference to our earlier system is the balancing of all lines between the SQUIDs and preamplifiers and between the preamplifiers and controllers. The structure of the impedance matching transformers is simplified. The real-time signal on 8 selectable channels can be monitored and the on-line averaged responses of all 24 channels can be displayed during the measurements.

5. AN EXAMPLE OF DATA

In Fig. 3 we present an auditory measurement over the left side of the human head. The stimulus consisted of 120 consecutively repeated, 100 msec frequency modulated tones, pew, pew, ..., pew, repeated at 1000 msec intervals. The activated area in the primary auditory cortex of the brain, represented by an equivalent current dipole, is marked by the arrow on the schematic head.

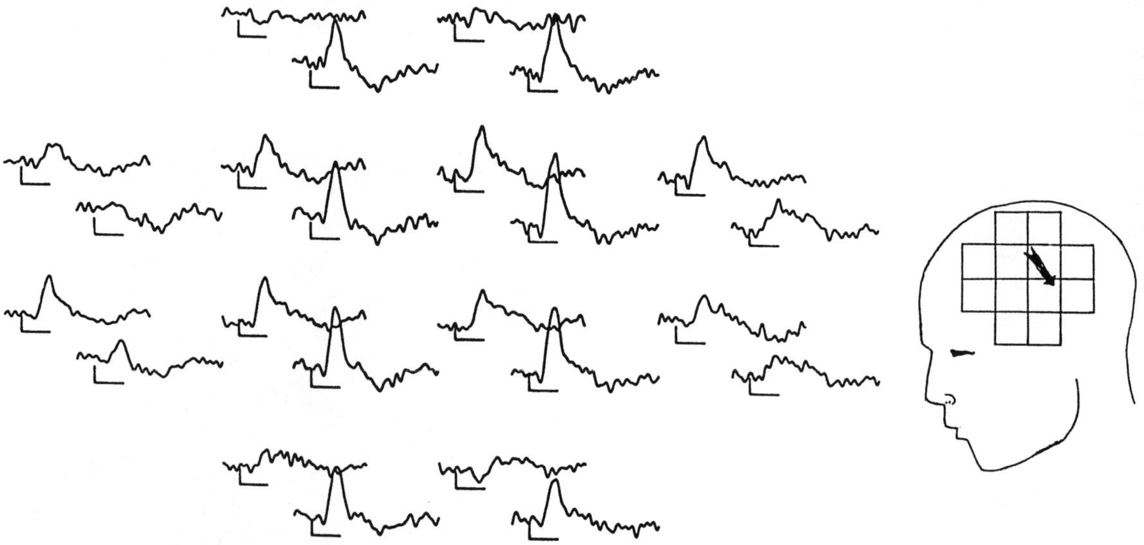

FIGURE 3: A 24-channel auditory MEG mesasurement. The upper traces of each pair give $\partial B_z/\partial x$ and the lower ones $\partial B_z/\partial y$ as a function of time. The vertical scale, shown by the L´s, is 1000 fT/m, and the horizontal scale 100 msec. For further details, see text. This figure was provided by Riitta Hari, Sing-teh Lu, and Ritva Paetau.

Physica B 165&166 (1990) 99–100
North-Holland

MULTICHANNEL DC SQUID SYSTEM WITH TWO STAGE CONFIGURATION

H.FURUKAWA, S.FUJITA, A.MATANI M.YOSHIDA and K.SHIRAE

Faculty of Engineering Science, Osaka University, Toyonaka, Osaka 560, Japan*

For experimental evaluation, two channel dc SQUID system of the parallel type has been constructed. We have confirmed the following performances; voltage bias operation can accept any dc SQUID with different critical current, crosstalk between channels is negligibly small, and noise reduction can achieve owing to the flux gain.

1. INTRODUCTION

SQUID magnetometer has been used as an essential instrument for the measurement of weak magnetic field such as biomagnetic field. Recently, the development of the multichannel SQUID has become an important subject, especially in the biomagnetic applications, because the multichannel SQUID system is expected to obtain more precise informations compared with a single channel one. We have proposed various SQUID systems. They are multichannel rf SQUID system(1), multichannel dc SQUID system of serial type(2), and multichannel dc SQUID system of parallel type(3). Here, we report the experimental evaluation of our multichannel dc SQUID system.

2. PRINCIPLE OF OPERATION

The basic idea of the multichannel dc SQUID system of parallel type has reported previously. In order to evaluate the operation experimentally, two channel dc SQUID system has been developed as shown in figure 1. It consists of two stages of the dc SQUIDs. In the 1st-stage, two dc SQUIDs(S_1, S_2) are connected in parallel. The dc current I_d produces the voltage drop V_b at a small resistance r. The 1st-stage dc SQUIDs are biased by V_b. Therefor, the output of the 1st-stage dc SQUIDs is not a voltage but a current. The output current flows into the inductance L_g and is applied to a 2nd-stage dc SQUID. The ac fluxes with separate frequency(ω_1,ω_2) and a constant amplitude($\simeq \phi_0/4$;ϕ_0 is the flux quantum) are applied to S_1 and S_2 to mark the input signals(ϕ_1,ϕ_2). The output voltage of the 2nd-stage can be expressed as(3)

$$\Delta V = K_0 M \left[\frac{K_1\phi_1 \sin\omega_1 t}{1+\dfrac{2j\omega_1 L_g}{R}} + \frac{K_2\phi_2 \sin\omega_2 t}{1+\dfrac{2j\omega_2 L_g}{R}} \right] (\sin\omega_0 t)$$

where K_1, K_2 are the constant of the 1st-stage dc SQUIDs, and K_0 is the constant of the 2nd-stage dc SQUID, R is the normal resistance of the dc SQUID, and M is the mutual inductance between L_g and L_s(the 2nd-stage dc SQUID inductance). The 2nd-stage carrier frequency ω_0 is set much higher than ω_1 and ω_2. Above equation shows that the output voltage of the 2nd-stage is AM-AM wave. This output voltage, after amplification, is demodulated by PSD_0(phase sensitive detector). The output of PSD_0 is separated by PSD_1 and PSD_2 into the original input signals.

3. EXPERIMENTAL RESULTS

The dc SQUID of a Nb thin film square washer type(4) with the variable thickness bridge are used (to be reported in this conference).

Figure 2 shows the V-I characteristics of the 1st-stage dc SQUIDs . As shown, the critical currents are different

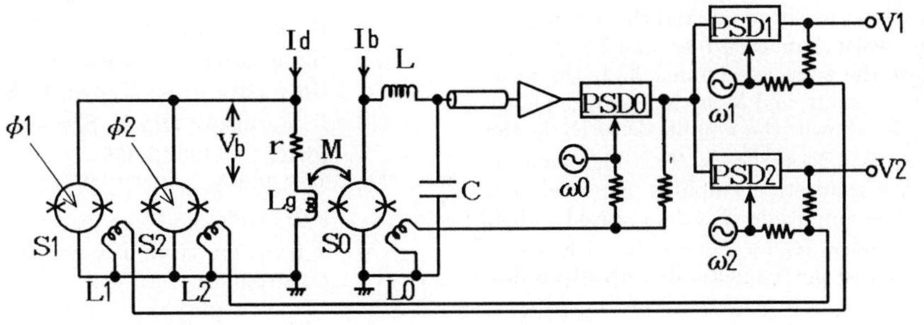

FIGURE 1 Configuration of two channel dc SQUID system

0921-4526/90/$03.50

with each other. In case of the current bias operation, different current I_1 and I_2 are required, however in case of the voltage bias operation, two dc SQUIDs can be

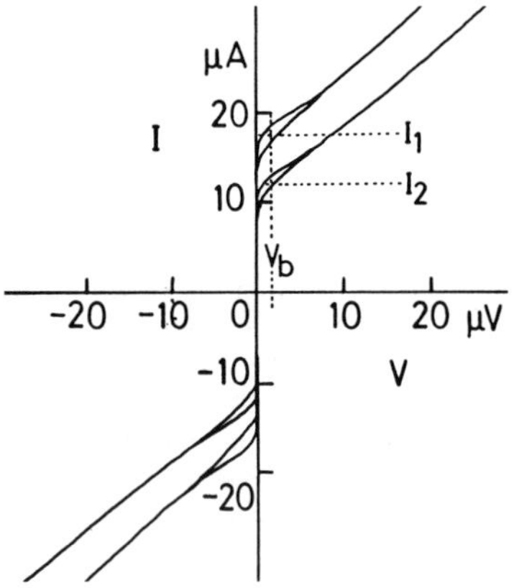

FIGURE 2 V-I characteristics of 1st-stage dc SQUIDs

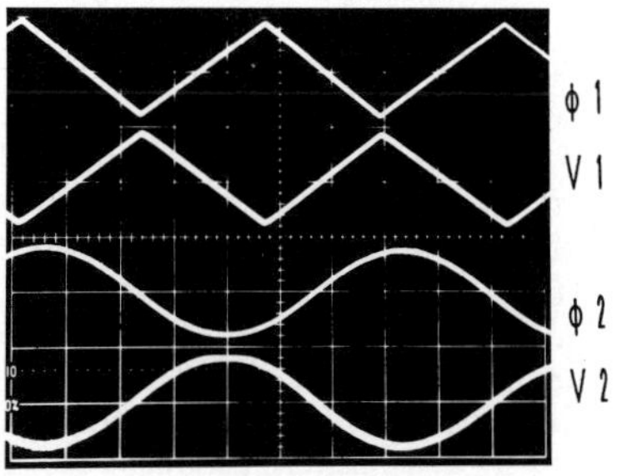

FIGURE 3 Wave forms of applied signals(ϕ_1, ϕ_2) and the outputs(V_1,V_2). H:2msec/div

operated by a single bias voltage $V_b(\simeq 2\mu V)$. The inductance L_g prepared by thin film on separate substrate is coupled to the 2nd-stage dc SQUID, and M is 0.8 nH. The feed back coils L_0, L_1 and L_2 are winding type. The carrier frequency in the 1st-stage are 30.6 KHz and 50.9 KHz, and the 2nd-stage frequency is 21.2 MHz.

Figure 3 shows the wave forms of the applied input signals(ϕ_1,ϕ_2) with the amplitude ϕ_0 and the outputs(V_1, V_2). Crosstalk between channels is less than $7 \times 10^{-5}\phi_0$.

Figure 4 shows the noise spectrums; N_0 is the measured noise at the S_0 input, and N_1 is the measured noise at the S_1 input. As shown, the amplitude of N_0 in the white noise region is $6 \times 10^{-5}\phi_0/\sqrt{Hz}$. This value is nearly equal to the solid state amplifier noise referred to the S_0 input. The amplitude of N_1 is reduced to half of N_0 in the white noise region. The reduced factor is approximately equal to the magnetic flux gain from the

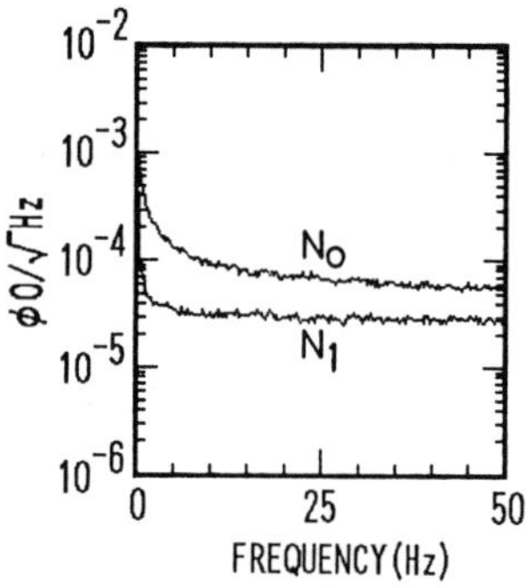

FIGURE 4 Noise spectrums
N_0:measured noise at S_0 input
N_1:measured noise at S_1 input

1st-stage to the 2nd-stage. The low frequency fluctuation of I_d does not produce additional noise because the 1st-stage outputs are AM wave with high frequency.

4. DISCUSSION

The experimental results show that the two stage configuration of dc SQUIDs reduces the noise by the flux gain, however the reduced noise is much larger than the noise of an ordinary single dc SQUID. To improve the noise characteristics of our system, it is most important to reduce the preamplifier noise.

We are facing to a problem; when the current I_d is cut off, the noise at the 2nd-stage increase to six times in the white noise region. This phenomenon also occur when the coil L_g is shunted with very low resistance. This problem similar to the experiment reported by Ryhanen et al(5).

5. CONCLUSIONS

The multichannel method using the parallel configuration of the dc SQUIDs was verified experimentally. The common voltage bias operation for the dc SQUIDs could accept any dc SQUID regardless of the amplitude of critical current. The system noise was reduced owing to the two stage configuration of the dc SQUIDs.

REFERENCES

(1) K. Shirae et al, Cryogenics 21 (1981)707
(2) K. Shirae et al, Proc. 6th Int. Conf. Biomagnetism, Tokyo, 1987 (Tokyo Denki University Press, Tokyo, 1988)470
(3) H. Furukawa and K. Shirae, Japanese. J. Appl. Phys. 28 (1989)L456
(4) M.B.Ketchen, IEEE Trans. Magn., MAG-23 (1987)1650
(5) T.Ryhanen et al, J. Low. Temp. Phys., 76 (1989)287

* This work is partly supported by SUZUKI MEDICAL SCIENCE RESEARCH FOUNDATION.

Physica B 165&166 (1990) 101–102
North-Holland

Thermal Effects on the Josephson Series-Array Voltage Standard

Robert V. Duncan

Sandia National Laboratories, Albuquerque, New Mexico 87185, U.S.A.

A series-array voltage standard containing 2,076 Josephson junctions has been operated in a liquid helium bath maintained within a refrigerated Dewar over the temperature range 1.56K ≤ T ≤ 4.54K. No systematic variation in the array voltage near 1.018V with temperature was detected over this entire range, indicating that any temperature coefficient of the array voltage must be less than 2×10^{-8} V/K. The critical current and gap energy of the weakest junction within the array in the absence of millimeter wave power were measured as a function of temperature over this same range. The noise sensitivity of the array was observed to change abruptly but by a small amount as the helium bath was pumped superfluid. Voltage calibrations were unaffected by the bath's superfluid transition.

1. INTRODUCTION

Microfabricated series-arrays of Josephson junctions have been developed which are capable of producing quantized voltage levels over a wide voltage range (1). These arrays have been used in a calibration system since February 10, 1987 to maintain the U.S. Legal Volt (2) at the National Institute of Standards and Technology (NIST, formerly NBS). A similar system within the Primary Standards Laboratory (which is operated for the Department of Energy, Albuquerque Operations Office by Sandia National Laboratories (SNL)) has been in operation since July, 1989. Measurements of the temperature dependence of the array's quantized voltage states and DC characteristics are reported here.

2. EXPERIMENTAL DESIGN AND CALIBRATIONS

The array used in this work contains 2,076 Josephson junctions connected in series (3). It was manufactured at NIST in Boulder, CO, USA on February 24, 1987 and was numbered 211-03. The SNL array voltage calibration system is quite similar to the NIST system (2). The differences between these systems have been described previously (4). In the SNL system a refrigerated Dewar which has been described previously (4,5) was used to maintain the array voltage standard at 4K. A diode thermometer was located 8.9 cm above the bottom of the tail section of the Dewar. This thermometer was placed such that it would be at the same height as the array voltage standard during data taking. It was calibrated against the saturated vapor pressure of the liquid helium within the Dewar space using the T_{58} temperature scale (6). The three closed cycle refrigeration stages built into the Dewar eliminated most of the radiation and conduction heating to the helium bath. This made it possible to pump the bath to 1.54K with only a small roughing pump and a valved pumping orifice of 0.44 cm diameter. The indicated saturated vapor pressure at the superfluid transition temperature agreed with the known value (6) to within the resolution of our pressure gauge. Careful calibration of the tail section thermometer resulted in a +/- 0.01K array temperature uncertainty throughout the full three Kelvin temperature range reported here.

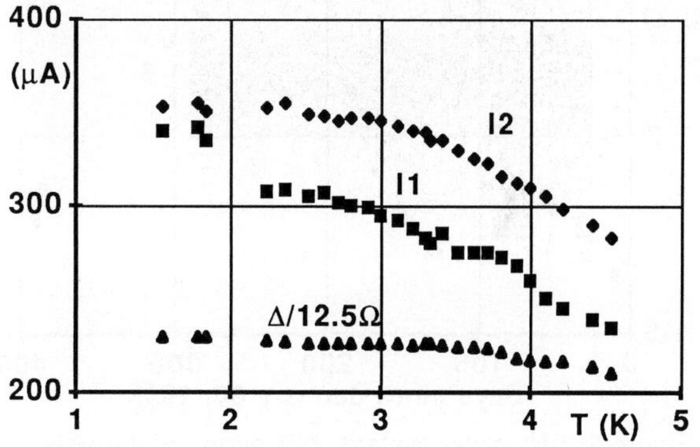

Figure 1: The gap voltage Δ divided by 12.5Ω and the critical currents I_1 and I_2 (see text) vs. temperature.

3. RESULTS AND DISCUSSION

The array bias current was supplied by a JBS-200 source manufactured by NIST in Boulder, CO (7). This source provided an adjustable DC current upon which an oscillating current of variable frequency and amplitude was superimposed. With no millimeter wave bias, and no DC current, the magnitude of a 50Hz oscillating current was increased slowly from zero up to the critical current of the weakest junction in the array. As the current magnitude was increased beyond this critical current the weakest junction switched normal, creating a voltage across the array which equalled the junction's gap voltage Δ ≈ 2.7 mV (8). The repeatability of the array current

0921-4526/90/$03.50

and voltage measurements were +/- 4μA and +/- 30μV, respectively. The total current through the array was the sum of the applied current and random currents created from various noise sources. Fig. 1 displays the variation of Δ and the critical current with bath temperature over a three Kelvin range. In Fig. 1 the critical current I_1 is the peak-to-peak current amplitude which may be driven through the array without causing the weakest of the 2,076 junctions to switch normal. As the drive current was increased from I_1 to I_2 the noise current would occasionally cause this junction to switch. The current I_2 is the critical current (in the absence of noise) of the weakest junction in the array. The value of I_1 must be determined by approach from lower currents so that power dissipation in the normal junction (≈ 1μW) will not affect the measurement. Notice that I_1/I_2 is typically 0.85 when the liquid helium bath is normal, and that it becomes about 0.95 when the bath is superfluid. This is presumably due to the improved array cooling provided by the superfluid bath.

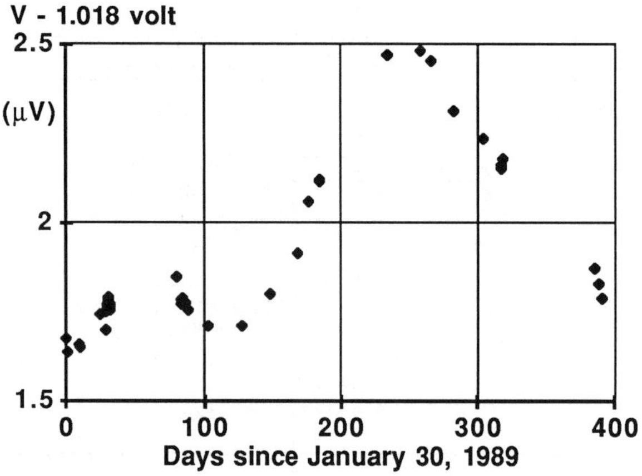

Figure 2: SNL array system calibrations of the voltage reference vs. time. The data were taken at NIST through day 32, at SNL through day 304, at RMC through day 318, and at UNM through day 391.

Throughout the last year the entire SNL array system has been transported from NIST in Gaithersburg, MD to SNL in Albuquerque, NM, then to the Research and Manufacturing Company, Inc. (RMC) in Tucson, AZ, and finally to the University of New Mexico (UNM) in Albuquerque. The SNL system was operated with the refrigerated Dewar only at RMC and at UNM. A conventional evaporating Dewar was used at NIST and SNL. The results of using the SNL system to calibrate a single DC voltage reference over the last year are displayed in Fig. 2. Both the SNL system and the DC reference appear to transport nicely, as evidenced by the good repeatability between the different locations.

The refrigerated Dewar was used with the SNL system at RMC and UNM to check for any dependence of the quantized array output voltage on temperature. The DC reference of Fig. 2 was calibrated five times at RMC and three times at UNM with the array at different temperatures. The DC reference displayed a drift in its nominal 1.018V output of -0.00465 μV per day during the time it was used with the refrigerated Dewar. The results of these eight calibrations (after being corrected for the DC reference's drift) show no systematic variation of the array voltage with temperature. These measurements were taken as the array temperature was raised, lowered, and raised again. This indicates that any average temperature coefficient of the array voltage must be less than 0.02 μV/K over the range $1.71K \leq T \leq 4.48K$. See also Ref. 4. Two measurements were made at 3.66K, one each at 1.0 bar and 1.7 bar, and no systematic variation of the array voltage with pressure was detected.

Similar array voltage standards are in use in laboratories at altitudes ranging from sea level to about 2,000 meters. The corresponding temperature of the liquid helium which bathes these arrays during calibration varies from 4.21K at sea level to 3.94K at 2000 meters (6). (Here standard pressure conditions and equilibrium of the liquid helium with its saturated vapor are assumed). The results presented here show that no more than a 0.006μV systematic variation between these laboratories may exist in ≈1V calibrations due to array temperature effects. Since the noise level of these measurements is about 0.01 ppm such a tiny systematic error with altitude would be of little consequence.

ACKNOWLEDGEMENTS

The assistance of the NIST Electricity Division, of RMC Cryosystems, and of the UNM Department of Physics and Astronomy is gratefully acknowledged.

REFERENCES

(1) F.L. Lloyd et al., IEEE Elect. Dev. Let. EDL-8 (1987) 449.
(2) R.L. Steiner and B.F. Field, IEEE Instru. and Meas. IM-38 (1989) 296.
(3) R.L. Kautz, C.A. Hamilton, and F.L. Lloyd, IEEE Magnetics MAG-23 (1987) 883.
(4) R.V. Duncan, submitted for presentation at CPEM 1990.
(5) R. Duncan and E.T. Swartz, Bul. Am. Phys. Soc. 35 (1990) 801.
(6) F.G. Brickwedde et al., J. Res. Natl. Bur. Stand. 64A (1960) 1.
(7) C.A. Hamilton, 'Instruction Manual for JBS-200 Josephson Bias Source' available through NIST, Boulder, CO.
(8) C.A. Hamilton, 'Series-Array Voltage Standard: Theory and Operating Instructions' available through NIST, Boulder, CO.

Physica B 165&166 (1990) 103–104
North-Holland

JOSEPHSON TUNNEL JUNCTION COUPLED TO SUPERCONDUCTING THIN-FILM RESONATOR

Anders LARSEN

Physikalisch-Technische Bundesanstalt, Abt. 2.24, D-3300 Braunschweig, Federal Republic of Germany

Hans DALSGAARD JENSEN

Danish Institute of Fundamental Metrology, DK-2800 Lyngby, Denmark

Jesper MYGIND

Technical University of Denmark, Physics Laboratory I, DK-2800 Lyngby, Denmark

Self-pumped parametric excitations in small Nb-Al_2O_3-Nb Josephson tunnel junctions strongly coupled to superconducting thin-film resonators have been studied. The large and nearly constant-voltage steps observed in the dc-IV characteristic at voltages corresponding to harmonics of the fundamental resonator-induced mode demonstrate sub-harmonic pumping of this mode.

1. INTRODUCTION

Most of the self-induced steps observed in dc I-V characteristics of Josephson junctions are theoretically well understood (1). We report on parametrically generated steps in small Nb-Al_2O_3-Nb tunnel junctions strongly coupled to a superconducting stripline resonator. The steps occur at voltages where the Josephson frequency equals harmonics of the fundamental resonance frequency of the resonator loaded by the junction. The steps found in earlier measurements (2,3,4) corresponded to frequencies of the higher resonator modes. This we believe is due to a smaller value of the coupling parameter. To our knowledge self-pumped sub-harmonic parametric generation has previously been observed only in inductively compensated junctions (5,6).

2. SYSTEM AND EXPERIMENTAL SET-UP

Figure 1 shows the experimental system. The dimensions of the junction and the microstrip resonator designed to have its fundamental resonance frequency around 8 GHz are L*W = 30*20 μm^2 and l_r*w_r = 5500*200 μm^2, respectively (sample H5-2A). The high inductance of the dc and rf connections also shown in the figure effectively decouples them from the energy of the resonator. An external dc magnetic field can be applied in the plane of the structure perpendicular to L.

The samples are fabricated from a Nb-Al_2O_3-Nb sandwich with typical current density 100 A/cm^2 and specific capacitance 0.047 pF/μm^2. The niobium films are rather thin and the effective London penetration depth is 105 nm at 4.2 K. The bottom niobium film is used as ground-plan in the microstrip resonator. The thick (600 nm) Pb-In(9.5%)-Au(4.8%) top layer forming the strip

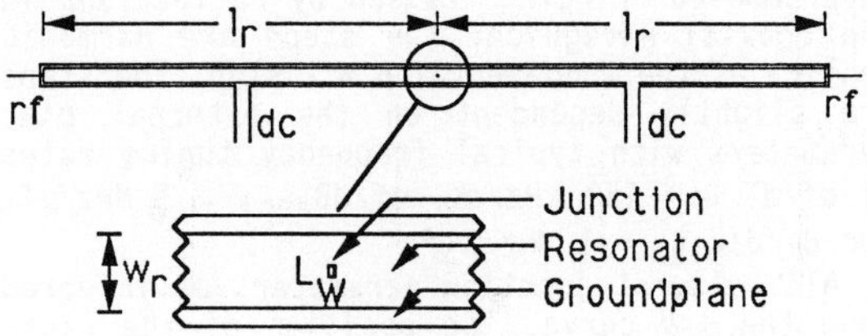

FIGURE 1
Sketch (not to scale) of sample geometry. The small tunnel junction (L*W) is placed in the middle of an open-ended superconducting microstrip resonator (2 l_r*w_r).

also provides the galvanic contact to the junction sandwich through a 10*10 μm^2 window in the h = 400 nm thick SiO layer separating the two electrodes.

The experimental set-up is installed in a shielded room and large efforts have been made to reduce disturbances from external noise sources. The sampleholder is mounted in a superinsulated aluminium cryostat with double magnetic shield. The dc bias module including pre-amplifiers are battery powered and all wires connecting to the junction are carefully twisted and filtered. A coaxial waveguide allows for application of external microwave signals and detection of signals generated by the system.

3. EXPERIMENTAL RESULTS AND DISCUSSION

The characteristic impedance Z_0 of the superconducting transmissionline is considerably smaller than the impedance of the small junction, and hence, the modes where the open-ended res-

0921-4526/90/$03.50

onator contains an even number of half-wavelengths couple better to the junction. The two line sections provide the inductive load which compensate the junction capacitance C_J at a resonance frequency $f_{r,n}$ being slightly lower than the resonance frequency of the unloaded resonator ($2\,l_r \sim n\,\lambda$). In the low-loss limit the resonance condition is

$$2\pi f_{r,n} C_J + (2/Z_0)\tan(2\pi f_{r,n} l_r / v) = 0$$

where v is the wave velocity $c/(\epsilon_r)^{1/2}$, ϵ_r being the relative dielectric constant of the SiO insulator.

The dc I-V characteristic contains different types of current steps, the size of which depends strongly on temperature and applied magnetic field. Most prominent are the Fiske steps and the resonator-induced steps. A typical set for one of our best coupled samples (H5-2A, B_{appl}= 260 μT, T = 4.1 K) is shown in figure 2. Narrow linewidth microwave emission at $f_{r,1}$ = 8.3 GHz and $f_{r,2}$ =16.6 GHz have been observed from all the steps. Within experimental uncertainty mainly caused by the slope of the steps and a small superimposed "ringing" caused by reflections in the coaxial waveguide, the steps are harmonic replica of the fundamental n = 1 step. The steps are slightly dependent on the external bias parameters with typical frequency tuning rates of df/dT = - 150 kHz/mK, df/dB_{appl} = 1 MHz/μT, and df/dI_{bias} = 1 MHz/μA.

All relevant junction parameters as inferred from the I-V curve, the position of the Fiske steps, magnetic field dependence of the critical current I_c, and escape rate from the V = 0 state, are internally consistent. For sample H5-2A we find at 4.2 K the magnetic thickness d = 225 nm, the normalized length L/λ_J = 1.0, and the normalized Swihart velocity v_S = 0.029. In this sample we also observe current steps at voltages $V = V_{FS1} \pm p\,V_{r,1}$ and $V = V_{FS1} + q\,V_T$, (p = 1, 2 and q = 1, 2, ..., 5) indicating higher order mixing of the internal Fiske resonance with both the fundamental longitudinal and the transversal (V_T = 225 μV) modes of the loaded resonator. For B_{appl} = 0 we clearly resolve the first Zero-Field Step at 2 V_{FS1}.

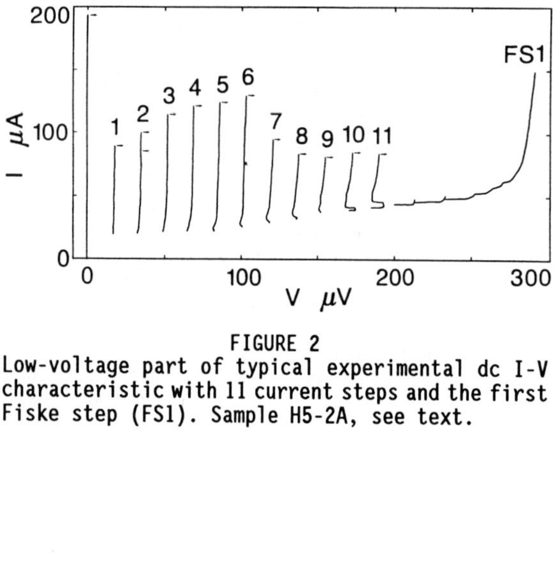

FIGURE 2
Low-voltage part of typical experimental dc I-V characteristic with 11 current steps and the first Fiske step (FS1). Sample H5-2A, see text.

The microwave properties of the loaded passive resonator are determined *in situ* with the junction used as detector biased near the gap voltage. Also the quantities in the resonance condition as determined from the resonance frequencies, the junction capacitance, the magnetic thickness d, and the accepted value of ϵ_r = 5.7 form a closed set. For sample H5-2A the normalized stripline wave velocity is 0.30, Z_0 = 320 mΩ, d = 225 nm, and C_J = 29 pF.

In samples with smaller junctions and lower values of Z_0 we measure quality factors Q > 500 at 4.2 K (> 1200 at 2.1 K) for modes with an odd number of half-wavelengths in the resonator. For full-wavelength modes n = 1, 2, 3 the coupling is much stronger and the Q is as low as 150. In sample H5-2A we see only resonances with n = 1, 2, 3. At 4.2 K we find $Q_{r,n}$ = 32.4, 76, and 300, respectively. The losses in the resonator can be ignored. We have not been able to associate a relevant junction resistance to the derived value of the load resistance 20 mΩ, however, using the measured $Q_{r,1}$, $V_{r,1}$, I_c, and the found Z_0 we can estimate the coupling parameter z , (ref. 1, p. 382) to 460. This corroborates our conclusion, that the observed current steps are due to parametric sub-harmonic pumping of the fundamental loaded resonator mode.

ACKNOWLEDGEMENTS
We thank R. Fromknecht, J. Niemeyer, R. Pöpel, and M.R. Samuelsen for stimulating discussions, A. Grunnet-Jepsen and A. Morgenstjerne for contributing to the measurements, and several members of the PTB Josephson group for assistance with the sample preparation. The work was partly funded by the Commission of the European Communities, the Danish Research Academy, and the Danish National Council of Metrology.

REFERENCES
(1) K.K. Likharev, Dynamics of Josephson Junctions and Circuits (Gordon and Breach Science Publ., New York, 1986).
(2) A.D. Smith, B.J. Dalrymple, A.H. Silver, R.W. Simon, and J.F. Burch, IEEE Trans. Magn. MAG-23 (1987) 796.
(3) G. Paterno, A.M. Cucolo, and G. Modestino, in LT-17: Contributed Papers, eds. U. Eckern et al. (Elsevier Science Publ. B.V., Amsterdam, 1984) pp. 215-216.
(4) V.Yu. Kistenev, L.S. Kuzmin, A.G. Odintsov, E.A. Polunin, and I.Yu. Syromyatnikov, in SQUID-85, eds. H.-D. Hahlbohm and H. Lübbig (W. de Gruyter, Berlin, 1985) pp. 113-118.
(5) L. Kuzmin, H.K. Olsson, and T. Claeson, *ibid.* pp. 1017-1022.
(6) H.K. Olsson and T. Claeson, Jap. Journ. Appl. Phys. 28 (1987) 1577.

Physica B 165&166 (1990) 105-106
North-Holland

SYNCHRONIZATION OF RADIATING INTRINSIC HIGH-Tc JOSEPHSON JUNCTIONS

Grzegorz JUNG*, Janusz KONOPKA* and Stefano VITALE

Dipartimento di Fisica, Universita' di Trento, 38050 Povo (TN), ITALY

Superconducting high Tc granular films radiate a pronounced electromagnetic energy up to millimeter wavelength range. Emission is very sensitive to the bias current, temperature and applied magnetic field. From the periodicity and shape of the emission dependence on external magnetic field the structure of radiating clusters was revealed. The most powerful emissions were ascribed to multi-link quantum interferometer clusters. Spectral properties of those emissions indicate that multijunction clusters get synchronized and radiate coherently.

1. INTRODUCTION

Electromagnetic properties of granular high-Tc superconductors are determined by the properties of inter-granular Josephson junctions network that links superconducting grains [1]. The existence of intrinsic Josephson junctions was confirmed in various experiments e.g. by observations of low field microwave absorption, reduced critical currents, relaxation of magnetization, and quantum interference effects in the intrinsic rf and dc SQUIDs. Recently some of us have provided a direct evidence for the Josephson microwave emission from current biased high-Tc granular films [2,3]. In this study we report on spectral properties of electromagnetic emissions from high-Tc superconductor that demonstrates the synchronization of radiating intrinsic junction clusters.

2. EXPERIMENTAL

High-Tc YBaCuO granular superconductor samples in the form of 4 mm long and 0.1 to 0.5 mm wide thin film strips were deposited onto unheated zirconia oxide substrates by dc magnetron sputtering or alternatively, by chemical pyrolitic technique. The details of sample fabrication were reported elsewhere [2]. In the search for microwave emission samples were dc current biased in a waveguide holder capacitively coupled to the fundamental TE01 mode of a standard X-band R-100 waveguide. Radiation was detected by a low noise superheterodyne receiver of the 12.6 GHz center frequency and 250 MHz 3 db bandwidth, followed by a wide band bolometric head as a video detector. Frequency spectra of the emitted radiation were measured in a microstrip arrangement constituted of the cooper ground plane on which the substrate was fixed and the superconducting strip with one end grounded to the ground plate and the opposite one directly connected to the semirigid coaxial line and to the dc current source. The detected radiation was amplified in a low noise two stage wide band amplifier and processed further by the HP spectrum analyzer.

3. RESULTS AND DISCUSSION

Typical microwave emission curves recorded in function of the dc current flowing in the YBCO strip at 4.2K is shown in Fig.1 (bottom curve).

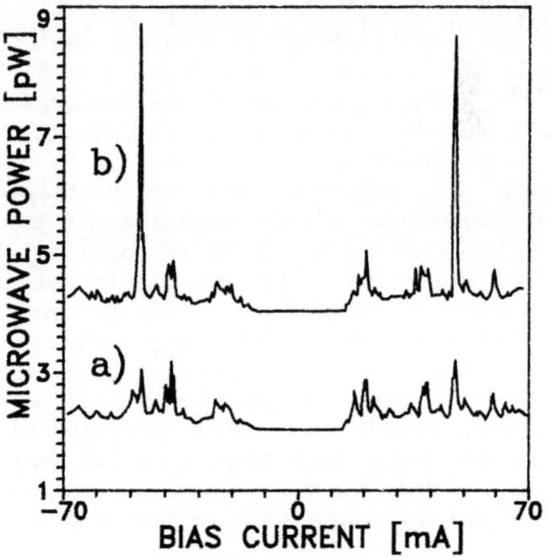

FIGURE 1
X-band microwave emission versus bias current for YBaCuO on ZrO_2 film, recorded at 4.2 K. a)-recording performed at the terrestial field. b)-recording performed at the field adjusted to maximize the amplitude of the lines at +/- 47mA. The curve b is vertically shifted by 2 pW upwards with respect to the curve a.

---

*Permanent address: Instytut Fizyki PAN, Lotnikow 32, 02-668 Warszawa, POLAND

By careful examination of the temperature dependence of radiation characteristics two different groups of excess emission lines, voltage stable lines (VSL) and current stable lines (CSL), can be distinguished. VSLs appear at the fixed voltage drops across the strip while the CSLs show out always at the same bias currents for relatively large temperature range. Amplitudes of both types of lines are very sensitive to the applied magnetic field. In particular, CSL amplitude can be magnetically tuned by more than two orders of magnitude.

Typically, a VSL amplitude vs. magnetic field demonstrates a single broad maximum resembling the Ic(B) pattern of a large Josephson junction with strong structural fluctuations in the barrier. CSL power pattern is periodic in the applied magnetic field and usually exhibits two periods; large period of narrow and high spikes and small one of low periodic ripples between the spikes. To explain the observed patterns we have proposed a simple model which assumes that the radiating cluster is composed of a parallel connection of coupled identical two-junction quantum interferometers [3].

Knowing the periodicity of the power vs. magnetic field pattern, and assuming that the radiated energy is proportional to the square of the effective critical current, one can estimate the structure and the size of the radiating Josephson cluster. It turned out that various patterns have nearly the same periodicity of large spikes corresponding to a single interferometer area. Therefore, surprisingly, all basic interferometer have almost the same average loop area. Single loop radius was estimated to be of the order of several μm, i.e. of the order of the average grain size of the investigated film. The ripple period, determining the area of the entire radiating structure varies from line to line, but generally it is relatively small. In most cases cluster was composed of only several active junctions. In order to get the best fit of our experimental data to the above model (3), we had to assume that the cluster is composed of two groups of parallel junctons separated by a large inductance, e.g. a kind of a long superconducting grain. The larger group is typically composed of 5 to 20 junctions, while the smaller one, which oscillates in a rather opposite phase with respect to the larger one, consists of only 2 to 5 junctions. The proposed structure of the radiating cluster is shown in Fig.2.

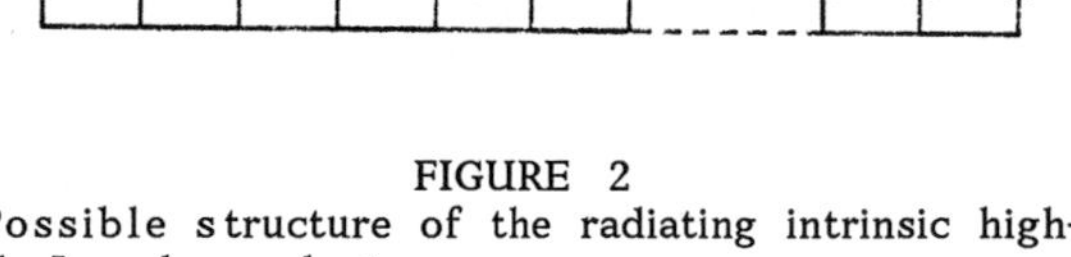

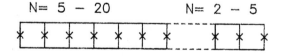

FIGURE 2
Possible structure of the radiating intrinsic high-Tc Josephson clusters.

The obvious consequence of the proposed model is an assumption of synchronization of the radiating junctions. Synchronization of the radiating junctions should, however, be reflected in narrow spectral lines of the emitted radiation. Figure 3 demonstrates the examples of frequency spectra of the UHF emissions recorded at different bias and magnetic field conditions.

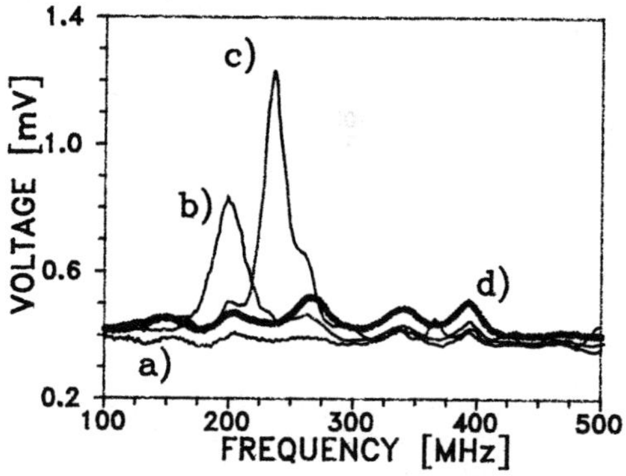

FIGURE 3
UHF frequency spectrum of the Josephson emission at 4.2 K. a)-zero bias current; b)- bias current close to the optimum corresponding to the CSL position; c)-optimum bias current at CSL position; d)- the same bias current as in c, but with external magnetic field increased by 20 mG (bold curve).

At zero bias the spectrum is flat. When bias current gets close to the CSL position a spectral line emerges from the white noise. With further current tuning the frequency and the height of the line changes. At optimum current, corresponding to the maximum emission, the linewidth in Fig.2 curve c is only 21.5 MHz. Observe that the minimum linewidth of a classical Josephson radiation is 170 MHz per each ohm of junction normal resistance at 4.2 K. Spectral line completely disappears if the magnetic field is changed by more than several mGauss (bold curve d in Fig.3). Such spectral features and a sharp current dependence of line amplitudes (see Fig.1) indicate that intrinsic junctions radiate coherently. Synchronization conditions depend on the bias and external magnetic field. Locking mechanism may be just a uniform vortex flow across entire cluster. Available vortices are released from pinning sites by a Lorentz force of a bias current. Therefore, significant coherent emissions shows out at some fixed current values. Each CSL line in that picture corresponds to a certain strength of a particular group of the pinning centers.

4.REFERENCES

1. J.Clark and R.Koch, Science 242, 217, 1988
2. J.Konopka and G.Jung, Europhys. Lett. 8, 549, 1989
3. G.Jung and J.Konopka, Europhys. Lett. 10, 183, 1989

Physica B 165&166 (1990) 107–108
North-Holland

MILLIMETER WAVE DETECTORS BASED ON FUTURE THIN FILM HIGH Tc JOSEPHSON JUNCTIONS.

N.F. Pedersen

Physics laboratory I, The technical University of Denmark, DK-2800 Lyngby, Denmark.

The properties of some millimeter wave detectors are evaluated by assuming the existence of (hypothetical) ideal BCS like thin film Josephson tunnel junctions made of high Tc superconductors. These extrapolations from the behaviour of low Tc junctions may or may not be justified depending on future progress in junction fabrication.

1. INTRODUCTION.

Although the progress in making high quality thin film Josephson tunnel junctions in high Tc superconductors has been slow, it is interesting to speculate about the potential gains in millimeter wave devices should such junctions become available. Thus we will assume the existence of thin film Josephson junctions with energy gaps of the order 10 - 20 meV, i.e. value 10 times larger than for a typical low Tc superconductor. We will consider parametric amplifiers, SIS mixers, and Josephson mixers. Implicitly we think of using the junctions at helium temperatures and take advantage of the increased frequency range.

2. RELEVANT INTRINSIC FREQUENCIES FOR THE JOSEPHSON JUNCTION.

The dimensioning and design of several types of Josephson junction rf devices utilising low Tc thin film Josephson junctions has been described by several authors (1-3). For our purpose the discussion in (1) is particularly useful since it relates both parametric amplifiers, SIS mixers and Josephson mixers directly to junction properties such as the current density and the energy gap.

The usual Josephson junction equivalent circuit diagram for a (low Tc) tunnel junction contain three circuit elements: A capacitance C, a tunnel resistance R, and a supercurrent $I_o \sin\phi$ where ϕ is the phase difference. For an ideal junction all the relevant high frequency properties are described by the three system frequencies - the plasma frequency ω_o, the RC cutoff frequency ω_{RC}, and the characteristic frequency ω_C defined through the circuit elements by

$$\omega_o = \sqrt{2eIo/\hbar C}$$

$$\omega_{RC} = 1/RC \text{ , and} \qquad (1)$$

$$\omega_C = 2eRI_o/\hbar$$

Further, the McCumber loss parameter βc is

$$\beta c = 2eR^2 I_o C/\hbar = (\omega_o/\omega_{RC})^2$$

Figure 1 shows how those three system frequencies vary with the current density J. From Eq. (1) we note that ω_o increases as $\sqrt{J}$ and ω_{RC} as J whereas ω_C is independent.

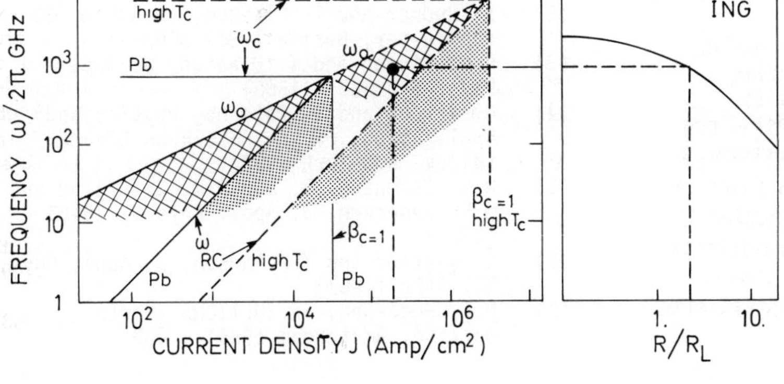

Fig. 1. (a) The three system frequencies ($\omega/2\pi$) as a function of the current density. Full lines: Pb-Pboxide-Pb junction. Dashed lines: High Tc junction. (b) The ING-curve.

0921-4526/90/$03.50 © 1990 – Elsevier Science Publishers B.V. (North-Holland)

3. PARAMETRIC AMPLIFIERS.

In a Josephson junction parametric amplifier the nonlinear inductance connected with the supercurrent is modulated by the pump and tuned to resonance at the signal frequency by the dc bias current i or an external magnetic field. This implies that $\omega < \omega_o$ which is shown as the crosshatched area in Fig. 1. Let us now assume a high Tc superconductor with an energy gap ten times that of lead, and a barrier with the same properties as those of lead oxide. From Eq. (1) and Fig. 1 we se that for a given current density ω_C increases a factor of ten, ω_{RC} decreases a factor of ten, and the plasma frequency is unchanged. The simple extrapolations are shown as the dashed lines. For the low Tc junction the intersection point of all three frequencies represents an upper limit for the use of almost any kind of device. This limit, in which also $\beta c = 1$, and non-hysteretic junctions occur, may be reached by increasing the junction current density (to the order of $5{*}10^4 A/cm^2$ for Pb). Fig. 1 shows that to reach the corresponding limit for the high Tc junction (increasing the frequencies by a factor of ten) will require a factor of a hundred in the current density, thus reaching the level of $5{*}10^6 A/cm^2$. At the moment this tunneling current density appears somewhat unrealistic, since this level is only obtained for the best superconducting high Tc films.

For the 3 photon parametric amplifier the gain G may be expressed (ref.5)

$$G = N/D^2 \qquad (2)$$

$$D = ((\omega/\omega_C)(1+r))^2 + ((\omega/\omega_o)^2 - \cos\Phi_o)^2 - (J_1 \sin\Phi_o)^2$$

where the nominator N is non-resonant. The phase Φ_o is determined from the normalized dc bias current, i, by $\Phi_o = \arcsin i$ and $r = R/R_L$, where R_L is the load resistance of the external circuit. The argument of the Bessel function J_1 is determined by the pump amplitude. The curve for infinite gain is obtained by setting D = 0. This socalled ING curve (1) is shown in Fig. 2b for our hypothetical high Tc Josephson junction.

With the help of the ING curve we can now do the dimensioning of a 1000 GHz parametric amplifier (note from Fig. 1 that this is impossible with low Tc superconductors). From Fig. 1 at 1000 GHz and $\omega < \omega_o$ we may choose a current density of $2{*}10^5 A/cm^2$. With junction dimensions of $1\mu m{*}1\mu m$ this gives a junction resistance of about 10 Ω and a load resistance of about 3 Ω to match the ING curve. These parameters seems not to be a priori unrealistic in the future.

4. SIS-MIXERS.

SIS mixers are described for example in (1,2). The SIS mixer is a quantum mechanical device, which uses the sharp current onset at the gap voltage to produce an down converted, amplified signal with low noise. It has been shown (6) that SIS mixing may occur up to frequences near the gap frequency, i.e. potentially 10 THz in high Tc junctions. An important design rule for SIS mixers is to avoid the junction capacitance to shunt the signal; this is done by having $\omega < \omega_{RC}$, which corresponds to the dotted areas in Fig. 1. We note that the interesting region above 1 THz requires very high (unrealistic?) current densities. Finally tunneling experiments so far only give little hope for the sharp current onset at the gap voltage, which is absolutely necessesary for the SIS mixer.

5. THE JOSEPHSON MIXER.

The Josephson mixer (1,2,4) is known from low Tc devices to extend to very high frequencies. Since the effect has been demonstrated in high Tc point contacts, it leaves us with hope for future use. It requires a non-hysteretic IV-curve, which for our model exists in the area to the right of the vertical line in Fig. 1. The current density there is in excess of $5{*}10^6 A/cm^2$ for our hypothetical high Tc tunnel junction. By comparing to the presently obtainable thin film current densities this seems somewhat unrealistic.

6. CONCLUSION.

Under the assumption that in the future it will be possible to make high quality thin film Josephson junctions with energy gaps ten times higher than that of lead, we conclude that there may be very interesting new possibilities for high frequency devices. Particularly parametric amplifiers look promising.

REFERENCES.

(1) N.F. Pedersen, in SQUID '80 (deGruyter, Berlin 1980) p907
(2) S. Rudner and T. Claeson, in SQUID '85 (deGruyter, Berlin 1986) p963
(3) H.K. Olsson and T. Claeson, J. Appl. Phys. 64, 5234 (1988)
(4) A. Barone and G. Paterno, Physics and Applications of the Josephson Effect (Wiley, New York 1982)
(5) O.H. Sorensen, B. Dueholm, J. Mygind and N.F. Pedersen, J. Appl. Phys.51, 5483 (1980)
(6) D. Winkler and T. Claeson, J. Appl. Phys. 62, 4482 (1987)
(7) N.F. Pedersen, in SOLITONS (North Holland, Amsterdam 1986) p 469

Physica B 165&166 (1990) 109–110
North-Holland

SIGNAL PROCESSING FOR SQUID-MAGNETOMETRY OF HIGH-T_C SINGLE CRYSTALS*

W. KRITSCHA, F.M. SAUERZOPF and H.W. WEBER

Atominstitut der Österreichischen Universitäten, A-1020 Wien, Austria

In the present contribution we show that reliable results on the magnetic moments of very small samples can be achieved by digitally processing, storing and subsequently analyzing the complete signal cycles of the SQUID sensor. Compared to conventional peak-to-peak detection, influences of residual magnetic moments of the sample holder, of drift effects in high magnetic fields and of occasional flux jumps in the sensor can be accounted for. In this way, the nominal sensitivity of the system can be reached even under adverse experimental conditions.

1. INTRODUCTION

SQUID-magnetometry has received increased attention during the past few years because of its unique capability to investigate the magnetic properties of high-T_C superconductors. Especially single crystals of high-quality $YBa_2Cu_3O_{7-\delta}$ are generally so small (20 - 1000 µg), that their magnetic moments fall below the resolution of most other techniques. However, even with a SQUID magnetometer, magnetization measurements on a 20 µg-single crystal are not completely straightforward, in particular at higher temperatures (> 50 K, small diamagnetic signals) and at higher magnetic fields (> 2 T, magnetic signals of the sample holder, drift of the system). Extensive measurements (1) employing a commercial SQUID magnetometer (2) equipped with a specially designed 8 T superconducting magnet have shown that the commonly employed peak-to-peak detection of the signals, which result from a continuous up-and-down movement of the sample through a pair of compensated pick-up coils, fails because the magnetic moments of the superconductor are masked by contributions of the sample holder (which cannot be compensated in many systems because of too short sample tube lengths) and by drift or flux jumping events in the SQUID sensor itself. Procedures devised to eliminate these problems will be discussed in the following.

2. EXPERIMENTAL

The sample is mounted on a sample holder of the smallest possible dimensions (quartz tube, etc.), which is connected to a motor-driven rod with a thread. After establishing a certain temperature, the magnetic field is ramped to the desired value and latched after a waiting time of 4 minutes by a superconducting switch. Following a further waiting time of 8 minutes, the motor is started and the sample driven through the pick-up coils at an exactly constant rate for three complete cycles. The time is recorded at the beginning and at the end of the mesurement. About 2400 data-points referring to the SQUID-output-voltage and a voltage assigned to the sample position within the measuring system, are recorded and stored in a computer. A typical set of data obtained under rather adverse experimental conditions (high temperature, high fields, magnetization with H parallel to the basal plane of the superconductor) and plotted versus time is shown in Fig. 1. In addition to the strong drift, the contribution of the sample holder (arrows) can be observed. Clearly, a peak-to peak evaluation, i.e. detecting the individual maxima and minima of the curve and averaging them, would lead to completely wrong magnetic moments.

3. EVALUATION PROCEDURES AND RESULTS

Having completed the entire measuring program, the data are transferred to another computer for evaluation. The first correction refers to the signals contributed by the sample holder. These are measured for each sample holder at $T > T_C$ (T = 95 K for YBCO single crystals) in the entire field range from 0 to 8 T. Having established in a separate experiment that the magnetic moments of the sample holder are only marginally temperature dependent, it is assumed furtheron that the field-dependent 95 K signals can be subtracted from the experimental data at all temperatures in the same way. This results in the data shown in Fig. 2.

* Work supported in part by Fonds zur Förderung der Wissenschaftlichen Forschung, Wien, under contract # 6837.

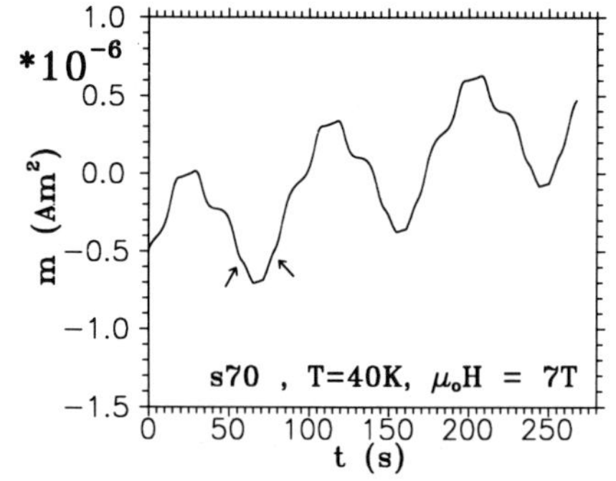

FIGURE 1
Uncorrected SQUID signals as a function of time. YBCO single crystal (112 μg), 40 K, 7 T, H parallel a,b.

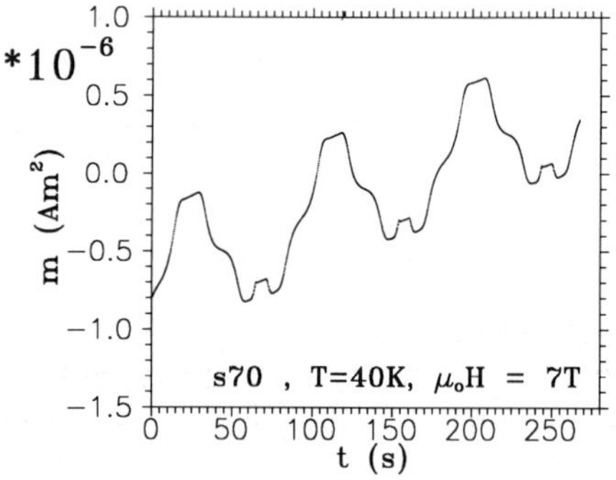

FIGURE 2
Data of Fig. 1 corrected for sample holder moments.

As the next step, the "best" set of experimental data (in general one obtained at low temperatures and in zero field following a complete magnetization cycle, where the signals are large and the sample holder contribution is minimal because of H = 0) is selected to establish the "ideal" dependence of the SQUID output voltage on the position within the pick-up coils. This is achieved in the following way (Fig. 3). SQUID signal voltages of one cycle (arrows) are selected and analyzed as a function of potentiometer voltages referring to the sample postion. This set of data is interpolated using about 400 data points. The maximum SQUID voltage is normalized to 1, the minimum to -1 (Fig. 3b).

Finally, the normalized signal is used as an input function for fitting the following equation to a complete set of data (Fig. 2):

$$m\,(t,\, x + x_{off}) = m_0\,(t = 0,\, x)\,.$$
$$\left[1 - \ln\left(1 + \frac{t}{\tau}\right)\right] + dl.t + dq.t^2 + m_{off}$$

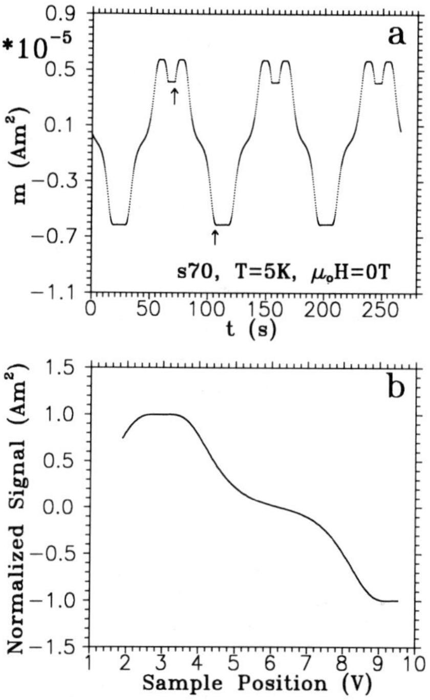

FIGURE 3
SQUID signals (remanent magnetization) as a function of time (a) and normalized signals of one cycle (between arrows) as a function of position (b).

where m_0 denotes the magnetic moment (Am^2), τ the time constant (s) of magnetic relaxation, m_{off} the offset of the SQUID sensor (Am^2), x_{off} the offset (measured on a potentiometer) in the direction of rod propagation (V), and dl (s^{-1}) and dq (s^{-2}) drift effects developing proportional to the time t or to t^2, respectively. This allows us to correct for drift effects (dl, dq) and to include possible effects of magnetic reaxation (τ), although they are generally small because of the long time span (12 minutes) which has elapsed between the end of the field change and the start of the measurement.

In summary, we have shown that unfavorable experimental conditions resulting from generally very small moments in high-T_c single crystals, from non-ideal arrangements of pick-up coils leading to signifcant sample holder contributions, as well as from drift effects in high magnetic fields can be accounted for by employing computer controlled data acquisition and processing procedures.

REFERENCES

(1) F.M. Sauerzopf et al., to be published
(2) S.H.E. Corporation, Model VTS 800

Physica B 165&166 (1990) 111–112
North-Holland

SYSTEMATIC STUDIES OF THE EFFECT OF A POST-DETECTION FILTER ON A JOSEPHSON-JUNCTION NOISE THERMOMETER

Robert J. SOULEN, Jr., William E. FOGLE, Jack H. COLWELL

National Institute of Standards and Technology, Gaithersburg, MD 20899, USA.

We present the results of an extensive study of the effect of a filter upon the performance of a resistive SQUID noise thermometer used to define an absolute temperature scale below 1 K. Agreement between the theory and the experimental results indicates that the scale defined by this noise thermometer is accurate to 0.1%.

1. INTRODUCTION

At the NIST we have been developing a temperature scale below 1 K. Absolute temperature is defined by a Josephson-junction noise thermometer comprised of a resistor R at temperature T connected in parallel with a Josephson junction (1). Voltage fluctuations due to the Johnson noise in R are converted by the Josephson junction into frequency fluctuations which are repeatedly measured by a frequency counter, and which are used to calculate the variance, σ^2(2). If the bandwidth of the measurement circuit is limited by a Butterworth filter of the form

$$A(\omega) = 1/(1+\{(\omega-\omega_o)/\omega_1\}^2) \qquad (1)$$

then the variance is given by (2)

$$\sigma^2 = \sum_{n=1}^{N}(f_i - \hat{f})^2 = \frac{\sigma_T^2}{\tau}\left[1 - \frac{1}{\tau}\left(\frac{1}{\omega_1} - C\omega_1\right)\left(1 - e^{-\omega_1 \tau}\right)\right]$$

where

$$\sigma_T^2 = \frac{2k_B R T}{\phi_o^2} \quad \text{and} \quad C = \frac{\phi_o^2}{16\pi^2\omega_1 R T}\frac{<V_a^2>}{V_s^2} \qquad (3)$$

and where f_i is an individual frequency reading, <f> is the average frequency for N readings, k_B is Boltzmann's constant, τ is the measurement time of the frequency counter, and ϕ_o is the flux quantum. We have made several improvements over our previous studies (2) : (a) C is now given in Eqn. 3 in terms of the measured signal-to-noise $V_s^2/<V_A^2>$ whereas previously it was a fitted parameter; (b) we varied T over two orders of magnitude; (c) and we reduced the statistical uncertainty of the variance from 1% to 0.1% to better test this model.

2. EXPERIMENTS

Since the variance depends critically on the character of the filter, we carefully measured its frequency response and fitted the data to Eqn. 1. A comparison of the data and the fit is shown in Fig.1. We see that the agreement is very good when the frequency lies within ± 40 kHz of f_o, while it deteriorates progressively as the frequency is increased outside this range. Inspection of the model (2) used to derive Eqn. 2 leads us to conclude that this deviation in $A(\omega)$ will cause a deviation in the the variance from Eqn. 2 for gate times shorter than the inverse of the frequency limit where $A(\omega)$ deviates from Eqn. 1, i.e., for τ<25μs (i.e., τ= $(40\text{kHz})^{-1}$).

The variance was measured as a function of τ at several values of T using this filter. An example is shown in Fig. 2. We see that the fit to Eqn. 2 is within the statistical imprecision of 0.1% for all values of τ, except for the value at 12μs, where the deviation is quite large. We consider this curve a manifestation of the deviation of $A(\omega)$ from Eqn. 1 as discussed above.

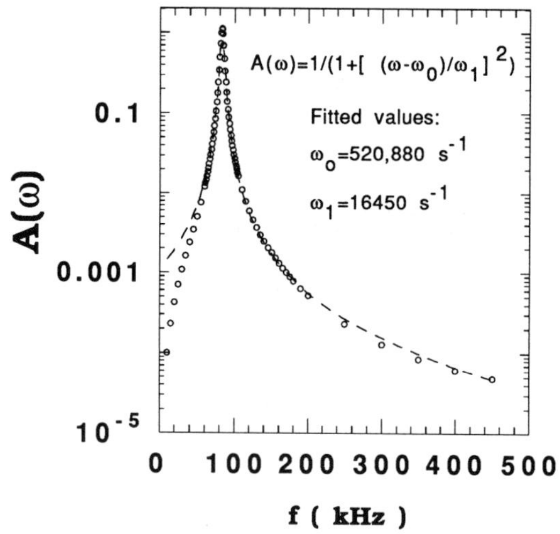

FIGURE 1
The frequency response of the audio frequency filter determining the bandwidth of the noise measurements. Experimental points (o) and fit to Eqn.1(--). Fitted parameters shown in insert.

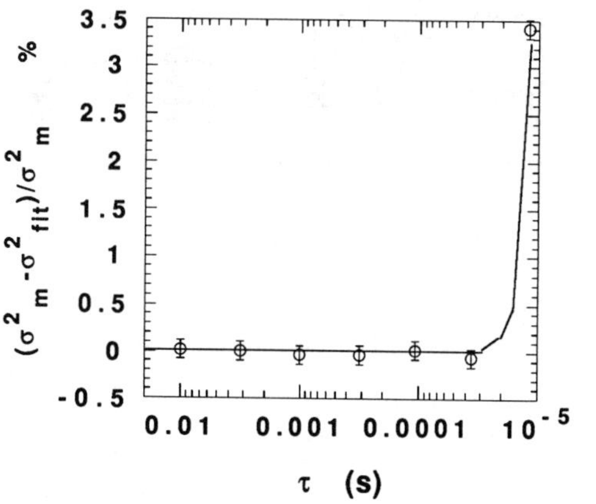

FIGURE 2
The difference between the measured variance and the fit to Eqn. 2 as a function of τ for data set 3543. The statistical imprecision (0.1%) is shown. Fitted values: $\sigma_T{}^2 = 1.5062\ s^{-1}$, $C = 1.72 \times 10^{-10}\ s^{-2}$, and $\omega_1 = 16{,}824\ s^{-1}$. From the value of $\sigma_T{}^2$ we calculate a temperature of 18.059 mK which agrees to within 0.1% of other thermometers used in the experiments described elsewhere in this conference (3).

In all, 84 data sets like the one described above were taken, which spanned the range 8mK to 520mK. They showed the same behavior with respect to τ; the fitted values of ω_1 scattered about a value of 17,207 s^{-1} with a standard deviation of 290; C scaled as $V_S{}^2/\langle V_A{}^2\rangle$ over two decades and as T^{-1} over three.

3. CONCLUSION

The measured variance conforms to Eqn. 2 to within its statistical precision for $\tau > 12\mu s$, supporting the argument that the effect of the filter on a R-SQUID noise thermometer contributes no more than a 0.1% systematic error to the temperature scale obtained from it.

4. REFERENCES

(1) R. A. Kamper and J. E. Zimmerman, J. Appl. Phys. 42 (171) 132.

(2) R. J. Soulen, Jr., Deborah Van Vechten, and H. Seppa, Rev. Sci. Instr. 53 (1982) 1355.

(3) W. E. Fogle, J. H. Colwell, and R. J. Soulen, Jr, this conference.

Physica B 165&166 (1990) 113–114
North-Holland

CORRELATION OF FLUX STATES GENERATED BY OPTICAL SWITCHING OF A SUPERCONDUCTING CIRCUIT*

Charles E. CUNNINGHAM, George S. PARK, Blas CABRERA, and Martin E. HUBER §

Physics Department, Stanford University, Stanford, California 94305, USA; § National Institute of Standards and Technology, Electromagnetic Technology Division, Boulder, Colorado 80303, USA

We pulse a superconducting microbridge with light, changing the quantum flux state of a superconducting circuit. Long sequences of pulses are used to measure the degree of correlation between successive flux states. In a series of runs, the pulse length was changed over six decades from 6 ns to 10 ms. The correlations fit a simple Fokker-Planck conditional probability model.

1. INTRODUCTION

We have created a fast, low noise switch using light to drive a superconducting microbridge normal.(1) We plan to use a network of such switches to avoid the 1/f noise in SQUID magnetometers by chopping the input circuit.(2) A switch was connected across the terminals of an rf SQUID for characterization. With the switch open, the magnetic flux in the circuit is arbitrary. With the switch closed, the flux falls into the nearest quantum state, $\phi = n\,\phi_0$, where $\phi_0 = h/2e$ and n is an integer. We noted previously that the correlation between successive flux states was strong with 100 ns pulses and absent with 30 ms pulses.(1) In this experiment, the pulse length was varied over the intermediate range to test a proposed model.

2. EXPERIMENT

The switch (Fig. 1a) is a Nb line 1.5 mm long by 2.2 µm wide by 24 nm thick connecting two Nb contact pads. It is deposited on a sapphire chip (6.3 mm x 6.3 mm x 350 µm). These chips are fabricated at NIST, Boulder using dc sputtering and plasma etching. We make superconducting contact with the chip by compressing dimpled Nb foils against the pads. The foils are spot welded to Nb wires that are connected to an rf SQUID. An optical fiber with a 50 µm core is held directly over the line by a ruby ferrule which is bonded to the chip with epoxy.

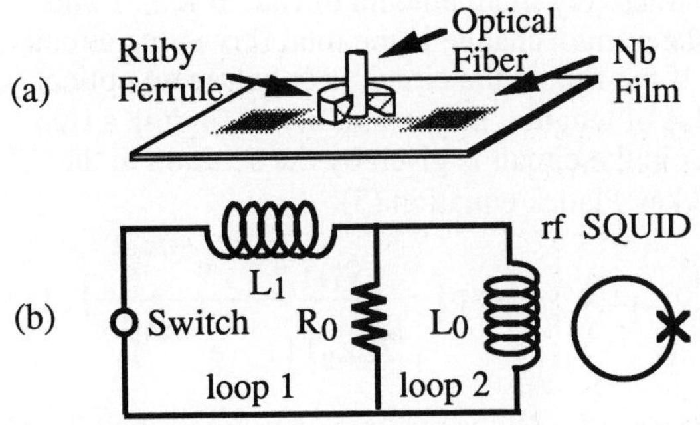

Fig. 1 (a) Laser driven switch. (b) Test circuit for switch. $L_0 = 2.24\ \mu H$, $R_0 \approx 2\ \Omega$, $L_1 \approx 0.4\ \mu H$, $R_{switch} \approx 250\ \Omega$.

The switch is opened by sending 850 nm light from a 2 mW cw laser diode down the fiber. A threshold intensity of about 0.25 µW/µm^2 is necessary to drive the illuminated section of the line normal, opening the switch. Flux can be trapped in metastable pinning sites in the switch, resulting in a nonquantized value of the flux in the circuit. When the switch is superconducting, the laser intensity is biased just below threshold to provide thermal energy which allows flux to detrap. Unfortunately, 50 µW peak-to-peak noise in the laser intensity moves the bias level into and out of the detrapping range during a run.

* Contribution of the U. S. Government, not subject to copyright in the United States.

Using ^4He exchange gas (62 Pa), the chip was heat sunk to a liquid ^4He bath. The circuit is surrounded by a Nb shield within a low temperature and high permeability alloy cylinder that attenuates the ambient magnetic field to less than 0.1 μT. The SQUID output is recorded on a digital oscilloscope, and the data is sent to a computer for averaging over 16.7 ms intervals (to remove 60 Hz noise) and for storage.

3. THEORY AND DATA ANALYSIS

The circuit shown in Fig. 1b has two paths for circulating current. Loop 1 has a 1.5 ns time constant (switch normal), and loop 2 has a 1.3 μs time constant. When the switch is normal, the Johnson noise from R_0 can change the flux in either loop no faster than its time constant. Since $L_0 > L_1$, the majority of the magnetic flux is in loop 2. Thus, optical pulses between 1.5 ns and 1.3 μs affect only the relatively small amount of flux in loop 1 and make a small change in the total flux in the circuit.

If the flux in the circuit is ϕ_i before an optical pulse of length t, the probability of finding a flux ϕ_{i+1} in the circuit is given by the solution to the Fokker-Planck equation (3):

$$P(\phi_{i+1},t|\phi_i) \propto \exp\left[-\frac{\{\phi_{i+1}-\phi_i e^{-t/\tau}\}^2}{2Lk_BT\{1-e^{-2t/\tau}\}^2}\right], \quad (1)$$

where $\tau = L_0 / R_0$ is the time constant of loop 2 and T is the temperature. For a given set of flux data $\{\phi_i\}$ generated by pulses of length t, Eqn. 1 predicts a point-to-point correlation function:

$$R \equiv \left(\sum_i \phi_i \phi_{i+1}\right)\left(\sum_i \phi_i^2\right)^{-1} = e^{-t/\tau}. \quad (2)$$

We took two data sets of 20 000 flux averages at pulse lengths of 6 ns, 10 ns, 30 ns, 100 ns, 300 ns, 1 μs, 3 μs, 10 μs, 100 μs, 1 ms, and 10 ms. Fig. 2b shows ϕ modulo ϕ_0 plotted over two periods for a typical data set; the SQUID noise is responsible for the finite width to the flux states. For each data set, the flux data were binned as ϕ modulo ϕ_0 and were fit to a Gaussian. In an iterative process known as Chauvenet's criterion (4), the points lying outside 2.5 σ were eliminated and the remaining data were refit until the change in σ was less than 0.1%. We calculated R for each data set and determined the best fit to Eqn. 2 (Fig. 2a).

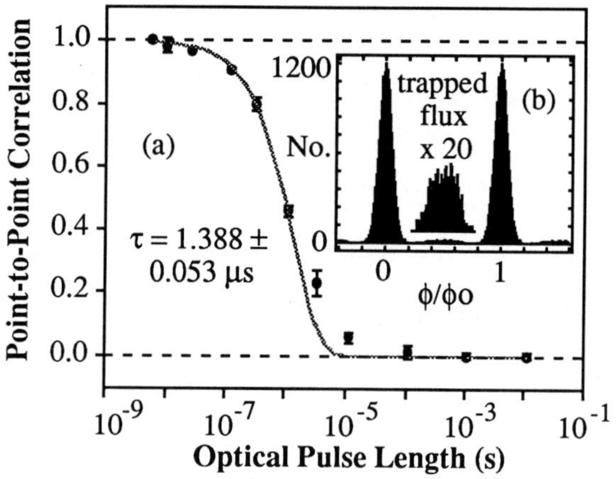

Fig. 2 (a) Point-to-point correlation of flux states with best exponential fit to Eqn. 2. (b) Histogram of 20 000 points of flux data plotted modulo ϕ_0. Inset: trapped flux points magnified x 20.

4. CONCLUSION

The proposed model fits the data well. Theory predicts a time constant of 1.3 ± 0.1 μs, with accuracy limited by the measurement of the shunting resistance R_0. The fit to Eqn. 2 indicates a time constant of 1.388 ± 0.053 μs. The flux trapping appears to be due to the laser diode, as previous operation with a Nd:YAG laser (1,5) did not trap flux. We are now working on the laser stability and, once stabilized, will use the switch to study the intriguing dynamics of the flux trapping process by varying the laser parameters.

ACKNOWLEDGEMENTS

This work was funded in part by the Office of Naval Research, Contract #N00014-87-K-0135.

REFERENCES

(1) Charles E. Cunningham *et al.*, IEEE Trans. Mag. **25** (1989) 1022.
(2) J. T. Anderson *et al.*, Rev. Sci. Instrum. **60** (1989) 202 and 209.
(3) F. Reif: *Fundamentals of Statistical and Thermal Physics* (McGraw-Hill, New York, 1965), pp. 560-582.
(4) H. D. Young: *Statistical Treatment of Experimental Data* (McGraw-Hill, New York, 1962), pp. 78-80.
(5) B. Cabrera, C. E. Cunningham, and D. Saroff, Phys. Rev. Letters **62** (1989) 2040.

Physica B 165&166 (1990) 115–116
North-Holland

SPUTTERED YBCO FILM BOLOMETERS

S.J. ROGERS and A.S.S. AL-AMEER

Physics Laboratory, University of Kent at Canterbury, Kent, CT2 7NR, UK

J.F.LANCHBERRY

E.R.A. Leatherhead, Surrey, UK.

Studies of disordered sputtered films of YBCO, show a multiphase transition to the super-conducting state that can extend from the temperature of liquid helium to that of liquid oxygen. We have tested the use of such thin films as bolometers in the temperature range 4K to 15K as detectors for thermally generated acoustic waves in ^{4}He gas. In the experimental geometry, the response time of the device is typically 2μs. Temperatures determined from the acoustic velocities are compared to those from resistance thermometry. We have examined the stability of the films under thermal cycling and find some aging effects.

1. INTRODUCTION

Superconducting thin films when used as thermal energy flux detectors, combine the advantage of high sensitivity with that of a rapid response, but until the advent of high T_c superconductors their use was limited to relatively low temperatures. Although it is impracticable, in the case of the high T_c superconductors, to provide the fields needed to move T_c significantly, early observations (1) of the resistivity of YBCO sputtered films suggested that, for their use as bolometers, this might not be necessary. Such films are often multi-phase, and the transition to the superconducting state can extend all the way from the boiling point of helium up to that of oxygen. We have tested the use of such films in the detection of thermally generated acoustic waves in ^{4}He gas in the temperature range from 4K to 15K.

2. FILM PREPARATION

The samples used were prepared at the E.R.A. Laboratory by radio-frequency diode sputtering from pressed powder targets of YBCO prepared at Kent by Dr.A.V.Chadwick. A radio frequency of 13.6 MHz was used with a power density at the target of 2.2 W cm^{-2}. During sputtering, the oxygen pressure was 10^{-2} torr and the substrate was held at a temperature of about 20°C.
Films on both alumina and sapphire substrates were studied, but, for the bolometers, 0.6mm alumina was used. The film samples were typically 1 × 0.5 cm^2, with thicknesses in the range 0.1 to 3μm

3. RESISTIVITY DATA

4-terminal resistivity measurements for a typical YBCO film on alumina show a maximum at about 85K. At higher temperatures the fall in resistance is semiconductor-like. Below the maximum, there are two steps in the transition to zero resistance which show that the film is a mixture of two phases; one phase has a transition centred on 80K, and a second transition is centred on 25K. The temperature at which zero resistance is achieved varies with sample and is moved to slightly higher temperatures by annealing in oxygen at 850°C.

When used as bolometers, the films were connected in a 2-terminal configuration which included surface and lead resistances. Fig.1 shows the effect of thermal cycling and aging on the resistance of a bolometer film measured in this way. The eleven curves from n = 0 to n = 10 cover a period of more than two months, and between each set of measurements the film was cycled to room temperature. Throughout this period, the film was kept either in helium gas at various pressures or under vacuum. For clarity, successive curves are displaced vertically by adding n × 50 ohms. The n = 10 curve is thus displaced by 500 ohms. It will be seen that the average difference between the n = 0 and n = 10 curves is about 1000 ohms; some 500 ohms of this is due to thermal cycling and aging. In addition to an overall upward shift in resistance - probably associated with surface and contact resistance - cycling and aging increased the difference between the low temperature and normal state resistance of the film itself, mainly by increasing the slope of the R-T characteristic at the lowest temperatures. This change is advantageous in the present application. The data in Fig.1 are for small sensing currents - typically 10 μA.

4. ACOUSTIC WAVE DETECTION IN ^{4}He GAS

The resistivity data suggests that such disordered YBCO films may serve as useful bolometers from 4K to 80K.. The sensitivity (1/R)(dR/dT) is typically $0.015 K^{-1}$.

In the tests, the bolometer is set facing a thin film heater. A short heating pulse (~ 1 μs) serves to generate an acoustic pulse in the helium gas by the process of the thermally activated desorption of helium atoms from the heater surface. The heating effect of the adiabatic wave is seen as a resistance

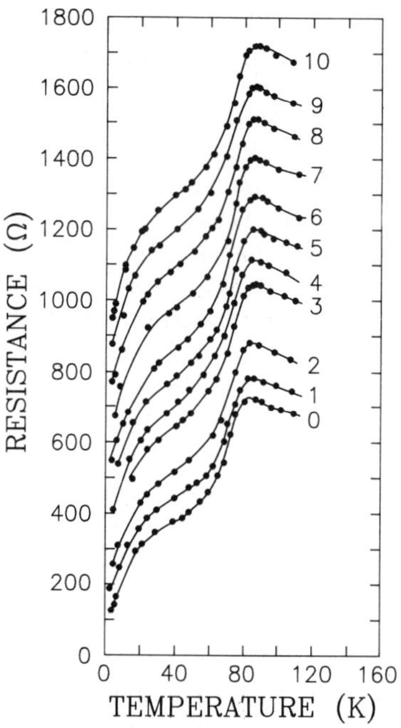

FIGURE 1: Aging effects for a 2.7 μm film. The values R+50n are plotted. Counting n = 0 as day zero, successive runs (n=1 to 10) were on days 2, 16, 21, 27, 35, 44, 50, 56, 59, & 63.

increase in the bolometer film. The energy transfer mechanism from the heater to the gas is efficient in the liquid helium temperature range, but less so at higher temperatures where we needed to use a microcomputer-based system to average many signals.

With a distance of 2.44mm between heater and detector, for the direct acoustic pulse the rise time for the detected signal was 2μs. This time, which is not limited by the time resolution of the 20MHz transient recorder, is much longer than the calculated time for the relaxation of temperature differences in the thickness of the bolometer film. For the YBCO material, the specific heat data of Ayache et al (2) and the thermal conductivity data of Jezowski et al (3) give for the diffusivity,$\alpha = 2 \times 10^{-3}\, m^2 s^{-1}$. For a film of thickness 2μm, the expression $(2\, \mu m)^2/\alpha$ yields 2×10^{-9}s as a measure of the relaxation time. We believe the time resolution is limited by geometry.

Even so, it is sufficient for quite precise measurements of the velocity of sound in the helium gas, as determined from the time between echoes. This is seen in Table 1, which presents a comparison of the temperature, T_o, determined from v_o, our velocity at limitingly low pressures, with T_{Ge}, the temperature provided by a precalibrated precision germanium resistance thermometer.

Table 1. Sound Velocities and Temperatures

v_o (ms^{-1})	T_o (K)	T_{Ge} (K)
143.3	5.93	5.95
157.9	7.20	7.13
179.0	9.26	9.12
199.1	11.45	11.21

REFERENCES

(1) J.F.Lanchberry, J.Phys.D:Appl.Phys., 21 (1988) 538

(2) C.Ayache, B.Barbara, E.Bonjour,P.Burlet, R.Calemczuk,M.Couach, M.J.G.M. Jurgens,J.Y.Henry and J.Rossat-Mignod, Physica, 148B (1987) 305.

(3) A.Jezowski, J.Mucha, K.Rogacki, R.Horyn, Z.Bukowski, M.Horobiowske, J.Rafalowicz, J.Stepien-Damm, C.Sulkowski, E.Trojnar, A.J.Zaleski and J.Klamut, Phys.Lett., 122 (1987) 431.

Physica B 165&166 (1990) 117–118
North-Holland

CHAOS IN A SUPERCONDUCTING SAUCER

G DAVIES, C J LAMBERT, N S LAWSON, R A M LEE, R MANNELLA, P V E McCLINTOCK and N G STOCKS.

Department of Physics, Lancaster University, Lancaster, LA1 4YB.

The dynamical behaviour of a very small permanent magnet levitating in a vacuum above a superconducting saucer at 4.2K has been investigated. The arrangement, which constitutes a near-Hamiltonian system, able to undergo both rotational and orbital motions, is found to exhibit a rich variety of behaviours, including chaos. Its relationship to chaotic motion in the solar system is discussed.

1. INTRODUCTION

The motions of some members of the Solar System - notably the tumbling of Hyperion (1) and, arguably, the orbital dynamics of the inner planets (2) - are believed to be chaotic. Such behaviour is difficult to investigate directly because of the extremely long time-scales involved. In this paper we describe an experiment in which some aspects of the motion can, however, be modelled conveniently in the laboratory.

2. EXPERIMENTAL DETAILS

Following a suggestion by A B Pippard (3), the investigation is based on measurements of the dynamical behaviour of a very small magnet levitating in a vacuum above a superconducting lead saucer, of 50 mm radius curvature, immersed in liquid helium at 4.2 K as sketched in Fig 1. Such an arrangement constitutes a near-Hamiltonian system, able to display both rotational and orbital motions, and with a finite spin-orbit coupling.

In practice, the motion of the magnet is adjusted initially with the aid of a periodic drive applied to a set of four coils mounted slightly above the plane of the lead saucer. With appropriate choices of frequency, and of the phase shifts between the voltages applied to the coils, the magnet could be made to move in an elliptical orbit or (with some difficulty) to spin rapidly on its axis in the centre of the bowl. In the most general case, the magnet would be spinning as well as orbiting the inside of the bowl.

Once any particular desired motion has been achieved, the drive was switched off. An opposed pair of the same coils was then used, instead, to pick up the induced emf from the moving magnet. The signal was taken to a high input impedance (to reduce damping of the motion) pre-amplifier and thence to a Nicolet 1180 data processor, able to record 512 K points of a discrete time series at uniform (typically 1 ms) intervals over an interval of several minutes, storing the data continuously on disk.

Results were recorded for several different neodymium magnets of different aspect ratio, all of which were of square cross-section and symmetric about their magnetic axes. The ratio of the lengths along their magnetic axes to either of the other two (equal) dimensions varied from 0.34 to 2.5: their shapes therefore varied between those approximating a "flake" and a "rod" respectively.

3. RESULTS

Some six traces of typical data obtained from a ~ 1 mm cubical magnet are shown in Fig 2; the full time series consists of 94 such traces. It must be emphasized that the measurements are made in the absence of any externally applied driving force, with the magnet orbiting freely in the vacuum, almost without dissipation. Thus the various strikingly sudden changes in behaviour (intermittency) observable in Fig 2 are all occurring quite spontaneously and not as the result of outside intervention.

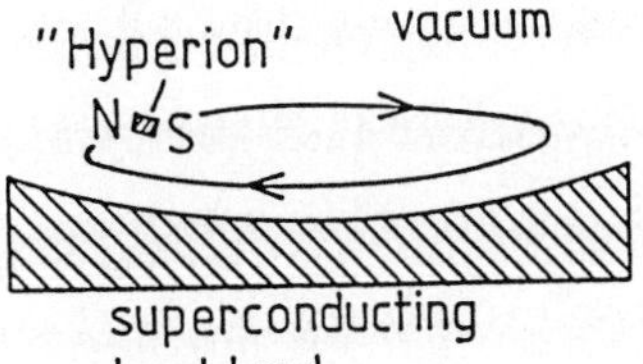

FIGURE 1

Sketch of the experimental arrangement. The magnet labelled "Hyperion" orbits in an approximately horizontal plane; it also tumbles about its own axes.

Damping of the motion though small, was not zero: the rate of decay of amplitude was found to depend on the type of motion, but its time constant was typically ~ 15 - 30 minutes. In practice, unless the coils (of resistance ~ 4Ω) were deliberately short-circuited, so as to damp the motion and bring the system quickly to rest, the magnet was always found to be moving. It is believed that the observed damping arose mainly as the result of vertical movements, inducing dissipative eddy currents within the body of the magnet itself: this effect was naturally more pronounced in some types of orbit than in others.

4. DISCUSSION

To try to obtain some quantitative measure of the extent to which the motion was chaotic, the largest Lyapunov exponent was calculated for each time series by use of the algorithm proposed by Wolf, Swift, Swinney and Vastano (4). Essentially, an embedding space was constructed by extracting multidimensional "points" from the time series. This was done by taking as a "point" an array of values separated by a constant delay; the multidimensional point was then followed for a certain time and its rate of divergence was calculated.

The dominant Lyapunov exponent, calculated in this way for each of the 94 traces of the (single, continuous) time series of which part is shown in Fig 2, is plotted against trace number (time) in Fig 3. It can be seen: (a) that the exponent is always positive, immediately indicating chaotic behaviour; and (b) that it tended to decrease with the passage of time. The latter effect probably arose because the more extravagant vertical tumbling motions suffer greater damping (see above) and consequently tend to die away relatively fast.

The particular example illustrated (Fig 2) was selected because it looked especially chaotic to the eye.

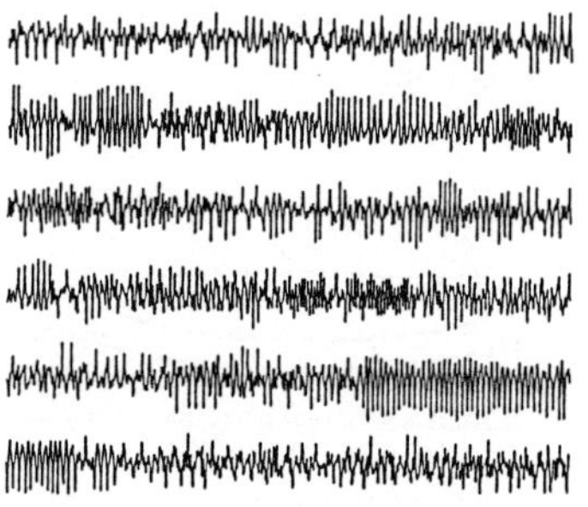

FIGURE 2
The emf induced in a pair of fixed coils by the moving magnet, plotted against time. For convenience of plotting, the complete time series has been broken into 94 horizontal traces of which 6 are reproduced here.

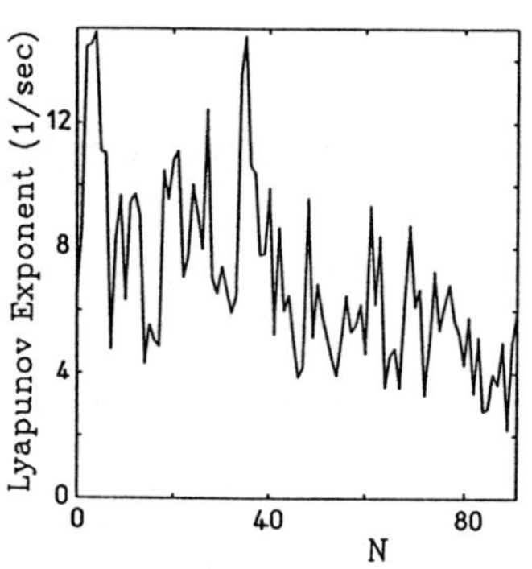

FIGURE 3
The dominant Lyapunov exponent calculated for individual traces of the time-series shown in part in Fig 2, plotted against the trace number N. Each trace represents an interval of 2.8 s.

Other, more regular looking, motions were also analysed. They yielded smaller Lyapunov exponents that were still positive but which did not significantly decrease during the acquisition of a time series.

The Hamiltonians of Hyperion and of the magnet, which are of a similar structure, are not of course, the same, even in the limits of zero damping and of dipole coupling between the magnet and its image in the saucer. One must be careful not to push the analogy too far. Nontheless, the experiments have revealed the characteristic chaotic behaviour of a body tumbling in its orbit, with a maximum of 12 degrees of freedom, under almost dissipationless conditions, in the presence of finite spin-orbit coupling.

5. CONCLUSION

We have demonstrated dynamical chaos in the laboratory in a near-Hamiltonian system. The observed phenomena may be of relevance to comparable behaviour observed in celestial mechanics.

ACKNOWLEDGEMENTS

It is a pleasure to acknowledge stimulating discussions or correspondence with N B Abraham, G P King and A B Pippard.

REFERENCES

1. J. Wisdom, S. F. Peale and F. Mignard, Icarus 58 (1984) 137.
2. J. Laskar, Nature 338 (1989) 237.
3. Suggestions made at the Royal Society Discussion Meeting on Dynamical Chaos, 5 February 1987.
4. A. Wolf, J. B. Swift, H. L. Swinney and J. A. Vastano, Physica 16D (1985) 285.

Physica B 165&166 (1990) 119–120
North-Holland

VISCOMETER UTILIZING A FLOATING CHARGED MAGNETIC PARTICLE

K. GLOOS, J.H. KOIVUNIEMI, W. SCHOEPE[#], J.T. SIMOLA, and J.T. TUORINIEMI

Low Temperature Laboratory, Helsinki University of Technology, 02150 Espoo, Finland[*]

A new type of high sensitivity viscometer is presented, reaching Q-values of up to one million at 0.3 kHz resonance frequency. Possible applications to thermometry and for the creation and detection of vortex rings in superfluid helium are discussed.

1. INTRODUCTION

Typical low temperature viscometers, like vibrating wires and torsional oscillators, have a sensor which is mechanically connected to the surroundings. This introduces some parasitic damping. By using a floating magnetic particle, levitated by the Meissner repulsion between two superconducting plates, this problem can be largely overcome. Because the particle has a small but macroscopic size, it should be possible to create and study vortex rings in superfluid helium with a device of this type.

2. EXPERIMENTAL

Our experimental cell (Fig. 1a) is made of niobium. Top and bottom plates act as a capacitor with a gap of about 1 mm. The lower plate contains an electrically insulated 1 mm diameter Nb wire as a collector.

The magnetic particle, cut from a $SmCo_5$ permanent magnet, was ground to about 0.1 mm diameter. The particles that we have investigated are not ideally spherical but have a surface roughness of about 0.01 mm. Above T_c of Nb, the capacitor is charged to about 500 V. By this the particle, resting on the bottom plate, is also charged and starts to move between the two capacitor plates as discussed in Ref. 1. From the integrated current sensed by the collector when the particle hits it, one obtains an estimate of the charge on the particle, typically 1 pC. When cooling through T_c the signal vanishes since, owing to the Meissner effect, the charged particle is now floating between the superconducting capacitor plates. Because of the finite charge left on the particle, its motion can still be detected via the ac-current signal, induced on the collector.

The resonance frequency for the vertical motion of the charged particle can be found, e.g., by gently knocking on the device and by Fourier transforming the signal. Over a very wide range the frequency depends linearly on the dc bias voltage, which determines the particle's vertical equilibrium position inside the gap. The bias-dependent resonance frequency, with a typical sensitivity of about 50 mHz/V, offers the possibility to drive the particle at constant excitation frequency in a feedback loop, using the cell itself as a voltage controlled oscillator (Fig. 1b). In this mode parasitic frequency-dependent pickup signals can be handled easily.

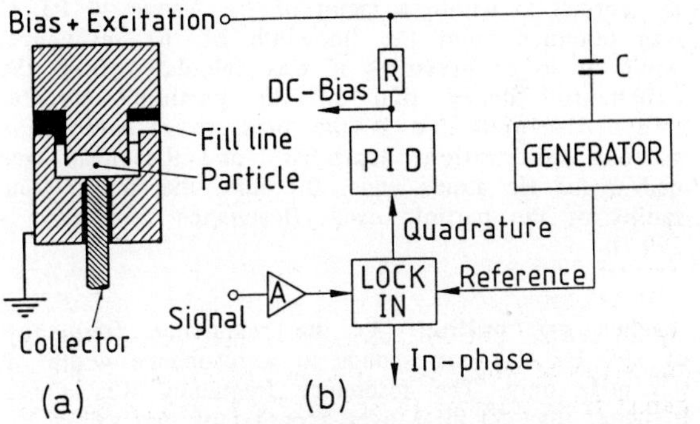

FIGURE 1
(a) The experimental cell, made of three Nb parts which are insulated from each other by Stycast 1266 (black area). (b) The constant-frequency phase-locked loop.

3. RESULTS

To demonstrate the feasibility of our viscometer, the pressure-dependent damping of the particle's movement in ^4He gas at 4.2 K was investigated. Results are shown in Fig. 2. As expected for an ideal gas at high pressures, the quality factor $Q \propto 1/\sqrt{P}$ as long as the viscous penetration depth is smaller than the particle size; at lower pressures, when the mean free path of the He atoms is larger than the particle size, $Q \propto 1/P$. At 0.1 Pa the measured Q-value

* Work supported by the Academy of Finland and by the Körber-Stiftung (Hamburg). # Permanent address: Institut fur Angewandte Physik, Universität, 8400 Regensburg, West Germany.

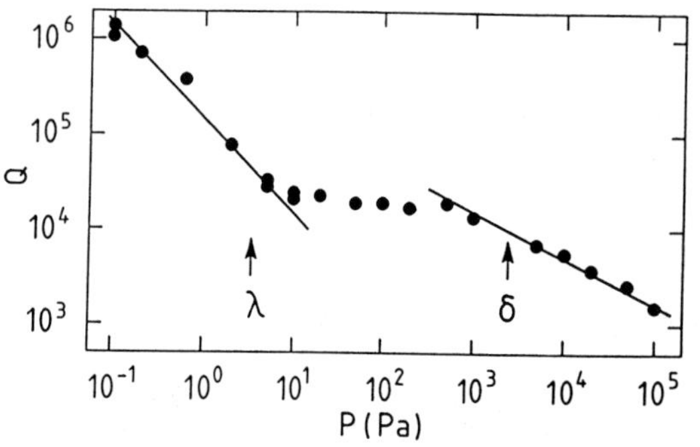

FIGURE 2
Measured Q-value in ^{4}He-gas at 4.2 K. The pressure is correct to within a factor of 2. Above 10 Pa, Q was obtained from the linewidth of the resonance, while at lower pressures it was calculated from the exponential decay time of the particle's oscillations. Also indicated is the pressure at which the viscous penetration depth δ and the mean free path λ of He atoms equal 0.1 mm, the approximate radius of the particle used. Resonance frequency is 299 Hz.

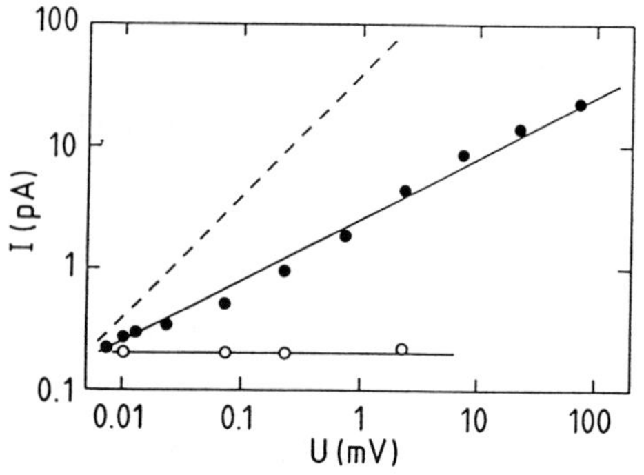

FIGURE 3
Measured RMS current induced by the moving particle, as a function of the excitation voltage at 299 Hz in ^{4}He gas at 4.2 K and under a pressure of about 0.1 Pa. 1 pA corresponds to about 1 μm displacement. Solid and open circles refer to translational and rotational modes, respectively. The dashed line is the signal expected for an ideal harmonic oscillator at Q = 10^6, extrapolated from the low-Q response.

reaches one million. At the resonance frequency of 299 Hz this corresponds to a resonance width of 0.3 mHz only. The resonance frequency was stable to better than 50 mHz over a period of two weeks.

In this regime, the response of the particle no longer depends linearly on the excitation (Fig. 3). We suspect that this effect is caused by the anharmonicity of the potential well in which the particle moves, also responsible for the bias-dependent resonance frequency.

Another interesting feature is the appearance of a second mode, the signal size of which is independent of the excitation amplitude. This is very likely caused by rotation of the particle around its magnetic axis. The electrical dipole moment, estimated from the signal size, is about 10^{-18} Cm, and is due to the particle's deviation from spherical symmetry.

4. POSSIBLE FURTHER APPLICATIONS

Because of its very low residual damping, the device is ideally suited for studies of low dissipation processes.

We have developed this measuring technique to investigate ^{4}He down to mK-temperatures. Since the phonon density in ^{4}He below 0.5 K varies strongly with temperature (2), one has a very sensitive thermometer. The same would apply to ^{3}He at μK-temperatures (3).

Because the particle has a small mass, the oscillator can resolve very small energy changes, e.g., those due to vortex ring production (4,5).

Another large area of investigation would be provided by the nonresonant rotational mode. For pure rotation the flow is stationary and the velocity can be chosen at will.

An alternative would be to employ a superconducting particle and replace the superconducting suspension by an active magnet system. The essential advantage of this method would be the possibility to adjust the particle's resonance frequency by simply changing the magnetic field gradient.

ACKNOWLEDGEMENTS
We gratefully acknowledge discussions with Prof. M. Krusius.

REFERENCES

(1) J.T. Simola and J.T. Tuoriniemi, this conference
(2) M. Morishita, T. Kuroda, A. Sawada, and T.Satoh, J. Low Temp. Phys. **76** (1989) 387
(3) J.P. Carney, K.F. Coates, A.M. Guenault, G.R. Pickett, and G.F. Spencer, J. Low Temp. Phys. **76** (1989) 417
(4) A.L. Fetter, Vortices and Ions in Helium, in: The Physics of Liquid and Solid Helium, Part I, eds. K.H. Bennemann and J.B. Ketterson, (J. Wiley and Sons, New York, 1976) pp. 207
(5) P.C. Hendry, N.S. Lawson, P.V.E. McClintock, C.D.H. Williams, and R.M. Bowley, Phys. Rev. Lett. **60** (1988) 604

Physica B 165&166 (1990) 121–122
North-Holland

TIME-OF-FLIGHT SPECTROMETER FOR MEASURING THE VISCOSITY OF FLUIDS BY USING CHARGED MICROPARTICLES

J.T. Simola and J.T. Tuoriniemi

Low Temperature Laboratory, Helsinki University of Technology, 02150 Espoo, Finland*

Our method for measuring viscosities of nonconductive fluids in the range $\eta \geq 10^{-4}$ Pa·s is characterized by a well defined flow geometry, a very small dissipative heating of the fluid, $P \sim 10^{-12}/\eta$ Watts , and a small sample volume, $V \sim 10\ mm^3$, required. The method is applicable in high viscosity cryogenic fluids, like ^3He and ^3He-^4He mixtures, where the vibrating viscometers are strongly damped.

1. INTRODUCTION

We have developed a viscosimetric technique of Stokes type where a small charged sphere is driven by a uniform dc-electric field through the medium studied. The metallic microsphere is placed between two parallel capacitor plates, immersed in the fluid. (Fig. 1). When the particle is resting on the lower plate and the capacitor is biased to field E the particle becomes charged and, beyond a critical electric field determined by gravitation and adhesion, it starts to move up and down between the plates.The sphere is charged up to $Q = (2\pi^3/3)\varepsilon E R^2$ where R is the radius of the sphere and ε is the dielectric constant of the medium (1). Our results verify this relation in the range R = 2 ... 100 μm. The motion of the particle is followed by detecting the current induced on the collector by the moving charge.

2. TECHNIQUE

For a particle approaching a circular collector axially at velocity v the current induced on the collector is

$$I(t) = \frac{Qv}{r}\left[1+\left(\frac{z(t)}{r}\right)^2\right]^{-3/2} \quad (1)$$

where z(t) is the time dependent distance of the particle from the collector of radius r. The technically appealing feature here is that, by decreasing r, we can increase the height (decrease the width) of the current spike I(t), marking the instant when the particle hits the collector. In our studies we used collectors with r = 0.1 ... 1 mm.

An oscilloscope record of I(t), when the particle is moving under constant bias conditions through viscous mineral oil, is shown in Fig. 2. From data like this we get the time of flight τ of the particle and the charge $Q \approx \int I(t)dt$. This information on the velocity $v = d/\tau$ and on the force $F = EQ \pm Mg$ acting on the particle (d=gap of the capacitor, M=mass of the particle, g=acceleration of gravity) can be used to obtain the viscosity η from the Stokes law $F = 6\pi R\eta v$, valid at low velocitites. The effect of the walls, i.e. the deviation from the ideal "sphere-moving-in-infinite-fluid" geometry, is properly taken into account (1) by dividing the observed η by (1+10R/d).

* Work supported by the Academy of Finland and by the Körber-Stiftung (Hamburg)

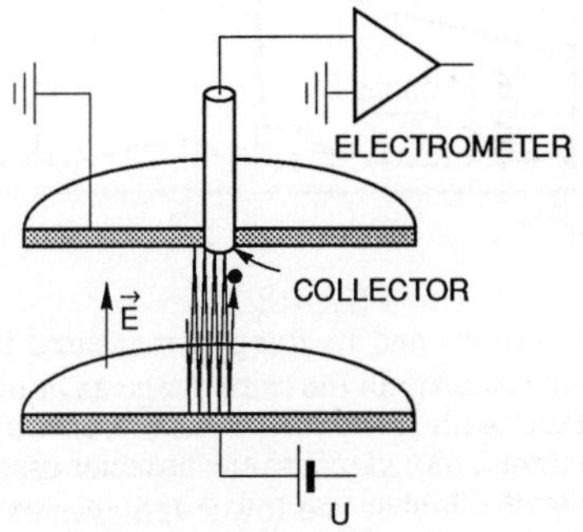

FIGURE 1
Time-of-flight viscosimeter.

The collisions of the particle with the capacitor plates are inelastic and in each event the velocity drops below the viscous limiting value $v = F/(6\pi R\eta)$. The velocity drop observed in our experiments with copper and nickel spheres varied from typically 40% at room temperature to 1 ... 5% in liquid helium. Because of this competing dissipation mechanism the resolution of our method is limited to viscosities higher than

$$\eta_c \approx \frac{1}{6\pi}\left(\frac{FM}{dR^2}\right)^{1/2} \quad (2)$$

When $\eta = \eta_c$ the viscous relaxation time is equal to the time-of-flight across the gap and the energy lost in the collisions is roughly equal to the viscous energy loss.

By plotting the observed time-of-flight vs. the driving force one can determine whether $\eta < \eta_c$ or $\eta > \eta_c$: in the former case the particle experiences free acceleration and $\tau \propto 1/\sqrt{F}$ whereas in the latter case $\tau \propto 1/F$ by the Stokes law (see Fig. 3). Our experiments in liquid helium always gave $\tau \propto 1/\sqrt{F}$ because for helium $\eta \approx 10^{-6}$ Pa·s<< η_c. According to Eq. (2) one can increase the sensitivity, i.e. measure lower viscosities, by decreasing the driving force. This, however, has a lower limit obtained as follows. The total force on the sphere resting on the lower plate is

$$F = EQ - Mg - F_a = \frac{2\pi^3}{3}\varepsilon E^2R^2 - \frac{4\pi}{3}\rho g R^3 - 2\pi R\sigma \quad (3)$$

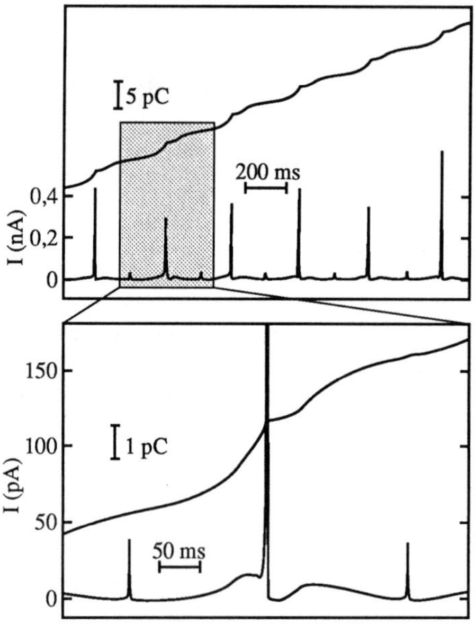

FIGURE 2
Electrical current and its integral measured in pump oil which has a viscosity in the same range as liquid ^3He. The nickel sphere, with R = 62 μm, was driven with E = 1100 V/mm across a 2 mm gap. The electrometer used had a gain of 10^9 V/A. The higher and lower spikes mark the charge exchange at the upper and lower electrodes, respectively

This includes, in addition to the electrical and gravitational forces, the adhesion F_a; σ is the surface tension at the fluid-metal interface. Consequently the threshold field for the particle to start moving is

$$E_c = \frac{1}{\pi}\left[3\frac{\sigma}{\varepsilon R} + 2\frac{\rho g R}{\varepsilon}\right]^{1/2} \quad (4)$$

This critical electric field has the minimum strength $[24\,\rho g\sigma/(\pi^4\varepsilon^2)]^{1/4} \approx 560$ V/mm at the optimal radius $[3\sigma/(2\rho g)]^{1/2} \approx 90$ μm, if σ and ε of liquid ^4He and a typical metallic density $\rho = 8000$ kg/m^3 are assumed. When the particle starts moving at $E = E_c$, the adhesion is "turned off" and the net driving force $F = E_cQ - Mg = F_a$. This leads, under constant bias conditions, to the minimum resolvable viscosity

$$\min(\eta_c) = [2\sigma\rho/(27d)]^{1/2}R \quad (5)$$

Taking into account that when the particle is flying down $F = F_a + 2Mg$, we obtain for the viscous power dissipation at the threshold $E=E_c$

$$P = (2\pi/3)(\sigma^2 R/\eta)[1+4\rho g R^2/(3\sigma)] \quad (6)$$

According to Eq. (4) a metallic particle with R = 10 μm requires a lifting field $E > E_c \approx 750$ V/mm. The viscosity resolution in 1 mm gap would extend down to $\eta_c \approx 10^{-4}$ Pa·s, see Eq (5). At millikelvin temperatures, however, the heating power $P = 10^{-12}/\eta$ Watts, from Eq(6), would pose a lower limit to the measurable viscosities.

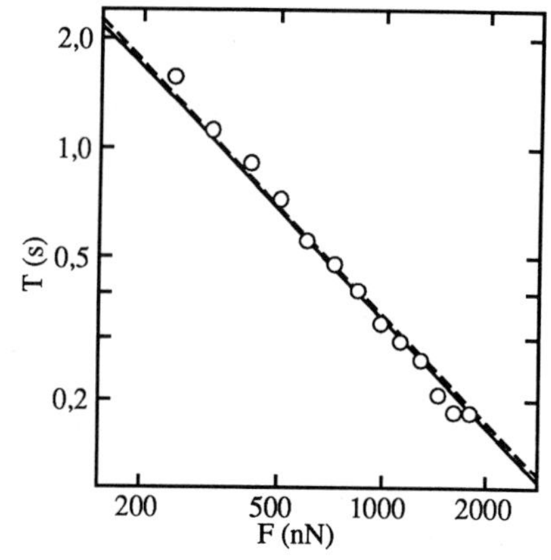

FIGURE 3
Time-of-flight vs. driving force in the experiment of Fig.2 obtained by varying the driving electric field. We find $\tau \propto 1/F$ as expected when the viscous drag dominates the dissipation. Solid line corresponds to η = 0.1 Pa·s.

3. CONCLUSION

The essential differencies between our time-of-flight viscometer and the vibrating wire method used widely in viscosity measuments at low temperatures (2) are that our method 1) works at high viscosities when the damping of the relatively large wire would cause too much heating 2) is independent of magnetic field, and 3) has a good signal to noise ratio, nearly independent of η. The method also represents a well defined flow geometry with a single characteristic velocity v. These properties make our system especially suitable for measurements in the high viscosity regime of ^3He and ^3He-^4He mixtures. A "vibrating particle" method for measurement of low viscosities is described in a companion paper (3).

ACKNOWLEDGEMENTS
We thank W. Schoepe for ideas and encouragement, and K. Gloos for stimulating discussions.

REFERENCES

(1) J.T. Tuoriniemi, Diploma Thesis, Helsinki University of Technology (1989)
(2) A.M.Guenault,V.Keith,C.J.Kennedy,S.G.Mussett, G.R. Pickett , J. Low Temp. Phys. 62 (1986) 511
(3) K. Gloos, J.H. Koivuniemi, W. Schoepe, J.T. Simola, and J.T. Tuoriniemi, this conference

Physica B 165&166 (1990) 123–124
North-Holland

BONDING Si WAFERS AT UNIFORM SEPARATION *

Ilsu Rhee, Athos. Petrou, David J. Bishop[a] and Francis M. Gasparini

Department of Physics and Astronomy, State University of New York at Buffalo, Buffalo, N.Y. 14260, U.S.A.
[a]AT & T Bell Laboratory, Murray Hill, N.J. 07974, U.S.A.

We have developed a bonding process for Si wafers which can be used to space wafers uniformly at distances $\geq 0.1\mu m$ over an area as large as $20 - 40cm^2$. Wafers bonded this way have been used at cryogenic temperatures and have been recycled successfully. The bonding process is one whereby a wafer with an SiO_2 pattern is brought into intimate contact with another wafer by using hydrostatic pressure. Bonding takes place at a temperature of 1150^0C and involves a chemical reaction between Si and SiO_2. The resulting structure can be evaluated by using the infrared optical techniques and by filling the space between the wafers with superfluid helium.

For work on finite-size effects at the superfluid transition of ^{4}He [1] we have had to develop means to confine liquid helium in a well characterized and uniform geometry. For this we developed an experimental cell consisting of two Si wafers bonded at a separation dictated by a SiO_2 pattern. Bonding of Si to SiO_2 or SiO_2 to SiO_2 is not a new process[2]; however, for our purposes we desired homogeneity in spacing at a level not achieved before: a few 100Å over an area of $20 - 40cm^2$. To achieve this we started with wafers of exceptional free-state flatness which were either preselected from a large batch or cut from a larger wafer to retain only the central flatter region. An example of such a wafer(2″) is shown in Fig 1.

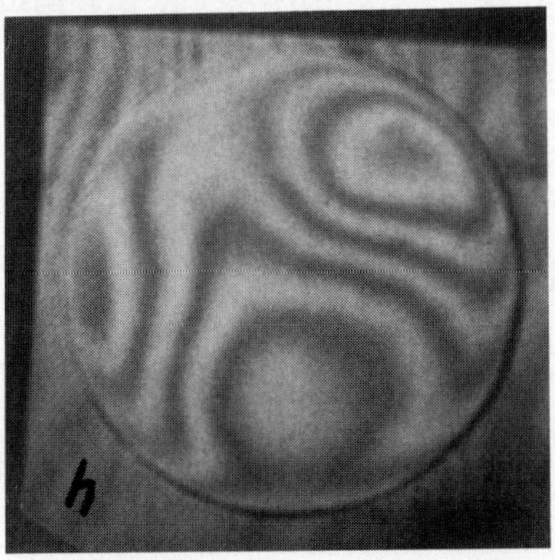

Fig 1. Interference fringes between a free standing wafer and an optical flat. Three fringes correpond to $\sim 1\mu m$.

* Work supported by NSF, Grant Numbers DMR 8601848 and DMR 8905771

To make a cell we oxidize one wafer then pattern it using photolithography. We then bond this to a second wafer in which a hole is spark-cut in the center. This arrangement is shown in Fig 2. The oxide pattern consists of a 4 mm border and $0.8 \times 0.8mm^2$ rectangles spaced about 2 mm apart. This distance was chosen after examining the typical variations from flatness of a number of wafers[3]. The thickness of the oxide determines the wafers' separation after bonding. This is done as follows. The wafers are first cleaned and staged together in a clean room. Then the space between wafers is evacuated through the hole using a quartz vacuum fixture. The wafers tend to self seal under the influence of the hydrostatic pressure, so that there is very little leak rate through the side.

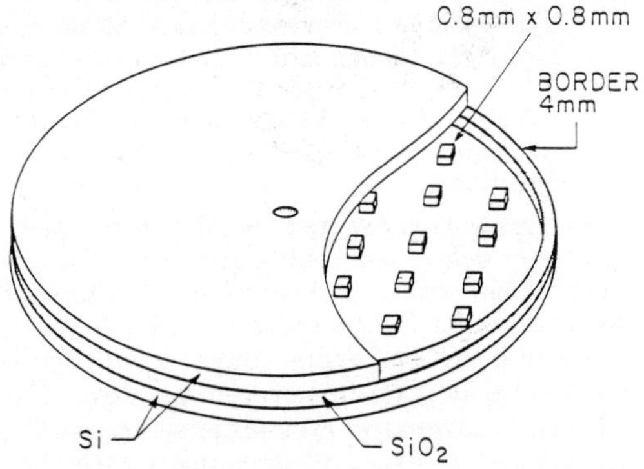

Fig 2. Schematic picture of two bonded wafers. The shaded areas are patterned SiO_2. The height of the oxide is not to scale.

This pressure also forces the wafers to contour to each other. They are then placed in a furnace while still under vacuum and brought to $\sim 1150^0C$.

This temperature is held for a period of 6 – 8 hours sometimes at an overpressure of about 8 – 10 psi. The mechanism of bonding is believed to be a reaction involving water and OH^- ions which are present in the SiO_2 from the wet oxidation process[(4)]. The resulting spacing between wafers, d, can be determined by measuring interferometer fringes at wavelengths greater than $1\mu m$, the band edge of Si. The two wafers acts as a Fabry-Perot interferometer giving intensity maxima in the transmitted light at wavelengths satisfying the condition $2ndcos\theta = N\lambda$[(5)]. Here n is the index of refraction of the medium between the wafers, θ is the angle of incidence and N is the order of interference. In Fig 3 we show results of a series of measurements for a 3″ diameter cell with expected spacing of $3.85 \pm 0.02 \mu m$ based on measured oxide thickness before bonding.

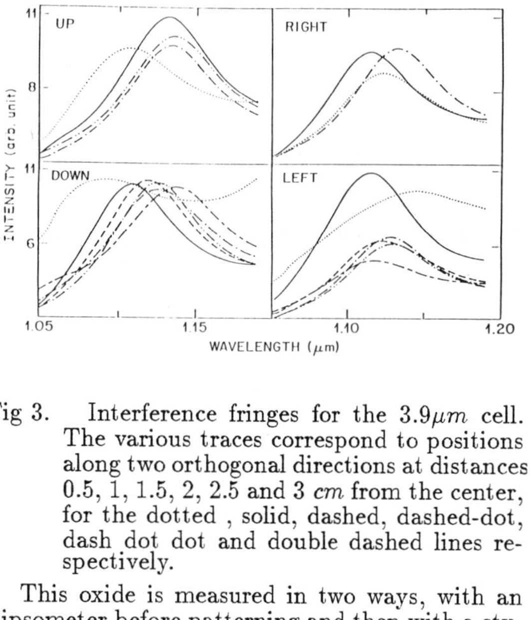

Fig 3. Interference fringes for the $3.9\mu m$ cell. The various traces correspond to positions along two orthogonal directions at distances 0.5, 1, 1.5, 2, 2.5 and 3 *cm* from the center, for the dotted , solid, dashed, dashed-dot, dash dot dot and double dashed lines respectively.

This oxide is measured in two ways, with an ellipsometer before patterning and then with a stylus instrument after lithography and etching. The various traces on Fig. 3 represent 19 different positions along two orthogonal directions of the cell. These yield a *local spacing* over an area of $\sim 1mm^2$ which can be averaged over all positions and give $3.93 \pm 0.04\mu m$ in excellent agreement with the expected value and with very little scatter.

We note that in Fig 3 the worse deviation from the average value comes from the dotted traces which represent positions $0.5cm$ away from the center hole. One might expect that this would be so, certainly due to the spark-cutting of the hole, and perhaps the stress associated with the quartz vacuum chuck in the bonding process.

A further diagnostic of the quality of the cells and the bonding can be done by filling the cell with helium and determining the total mass required. This is done by incorporating the cell as part of a high-Q torsional oscillator, and determining the mass loading as one scans the temperature from above the superfluid transition to $\sim 1.2K$[(1)]. For the cell discussed above this yields an *average* spacing of $3.68 \pm 0.15\mu m$, in good agreement with the optical measurements and the designed oxide thickness.

The loading of the cells with liquid helium also allows us to evaluate the quality of the bonding. If this were not complete or strong enough, then, as temperature changes and the pressure built up in the cell, it would expand and admit more helium. This would cause a shift of the resonant frequency of the torsional oscillator at temperatures above the superfluid transition. Only one of the four cells we used, $2.8\mu m$ spacing, showed such symptons[(3)]. In none of our experiments did we exceed 200 torr differential pressure in the cells. This was more of a precaution than a tested upper bound. We have made and tested a number of cells following the above procedure[(6)], of these, four yielded useful data. These were for separation of 0.106, 0.519, 2.8 and 3.9 μm. In a number of cases the cells were found not to be useful for our purposes because of very high Q fourth sound resonances.

(1) I. Rhee, F.M. Gasparini and D.J. Bishop, Phys. Rev. Lett. **63**, 410(1989)

(2) R.C. Frye, J.E. Grifith and Y.H. Wong, J. Electrochem. Soc. **133**, 1673(1986)

(3) I. Rhee, A. Petrou, D.J. Bishop and F.M. Gasparini, to appear in Rev. Sci. Inst. (1990)

(4) R.K. Iler, *The Chemistry of Silica* (Wiley, N.Y., 1979)

(5) See, for example, G.F. Fowles, *Introduction to Modern Optics* (2nd ed., Holt, Rinehart and Winston Inc., 1975)

(6) I. Rhee, PhD thesis, SUNY at Buffalo (1989), unpublished

Physica B 165&166 (1990) 125–126
North-Holland

ANOMALOUS LOW TEMPERATURE MECHANICAL PROPERTIES OF SINGLE-CRYSTAL SILICON

R.E. MIHAILOVICH, J.M. PARPIA

Laboratory of Atomic and Solid State Physics, Cornell University, Ithaca, New York 14853-2501, USA

We have measured the low temperature mechanical properties of a high Q oscillator fabricated from single crystal silicon. Operated at strains of 10^{-7}, the oscillator exhibits unusual behavior in both the dissipation and resonant frequency. In addition, we have performed systematic frequency sweeps through resonance at varying strain amplitudes. We observe the dramatic onset of non-linear behavior at temperatures below a few hundred millikelvin.

1. INTRODUCTION

Anomalies in the low temperature mechanical properties of metal oscillators have been observed in many experiments. These oscillators were typically machined from alloys of BeCu or AgCu and then heat treated to enhance the Q. Similar anomalous results were obtained with a silicon oscillator by Kleiman et al.(1) We have undertaken a further study of these unusual mechanical properties in order to better understand the behavior of these devices. Silicon is ideal for such studies, as it is available in high-purity-single-crystal wafers which do not require treatment to increase the Q. The absence of conduction electrons also simplifies the system. Appropriate design allows us to operate at high frequencies, an advantage in high Q oscillators as it shortens the response time.

2. APPARATUS

The oscillator is fabricated from a double-side polished, 350μm thick < 100 > wafer, doped to a resistivity of 4-8 ohm-cm. We used photolithographic patterning and anisotropic etching techniques to fabricate the oscillator, which is of a double-paddle design (see Figure 1) with two inertial elements (the head and paddles) and two torsional elements (the neck and tail).

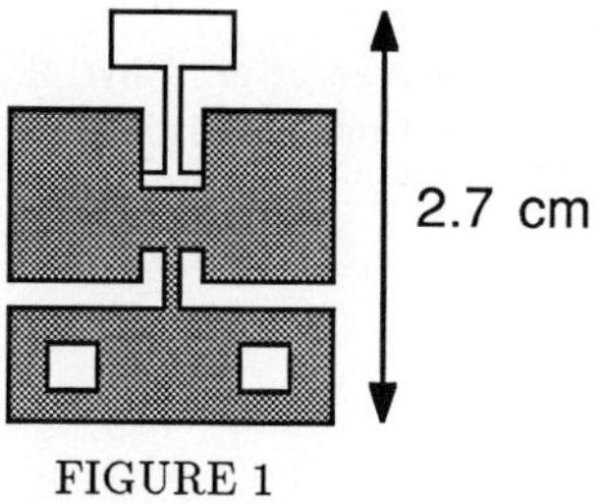

FIGURE 1
Double paddle silicon oscillator with evaporated gold (shaded region) covering paddles, tail and holed base.

Anisotropic etching of the < 100 > wafer produces bevelled edges, which when properly aligned during etching, produces neat corners at the junction of torsion rods and inertial elements. A gold film is evaporated on the paddles, tail and base of the oscillator, to provide a ground plane for capacitive drive and detection as well as a thermal conduction path. The oscillator is clamped to a copper base with screws after the base is coated with a thin layer of vacuum grease.

3. RESULTS

The double paddle oscillator has several resonances. The experiments reported here were obtained by operating in the antisymmetric torsional mode, where the strain is confined to the upper torsion rod, thus minimizing potential clamping losses. At low temperatures, the typical Q is $> 2 \times 10^6$. We plot the dissipation and resonant frequency of the oscillator measured at a constant strain amplitude of 10^{-7} in Figure 2.

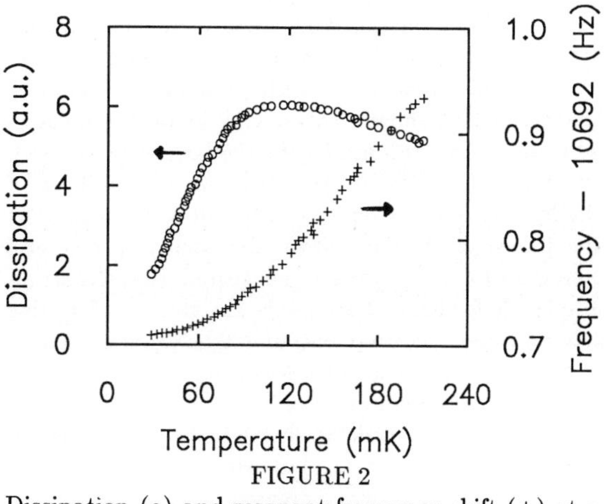

FIGURE 2
Dissipation (o) and resonant frequency shift (+) at a constant strain amplitude of 10^{-7}.

0921-4526/90/$03.50

Both the measured quantities exhibit the anomalous behavior in the temperature range below 200 mK. The dissipation shows a gentle peak, and then decreases linearly with temperature below 100 mK. The shift of the resonant frequency is also proportional to the temperature. Both of these results are qualitatively similar to those reported by Kleiman et al.(1)

To better characterize this anomalous behavior, we used a frequency synthesizer to continuously sweep through the resonance. These sweeps were done at constant drive levels, producing peak strains as large as 2×10^{-6}. Figure 3 shows results obtained for scans conducted at 800 mK. The response is characteristic of a nearly linear oscillator, with a symmetric amplitude and little reduction in the Q with higher strain.

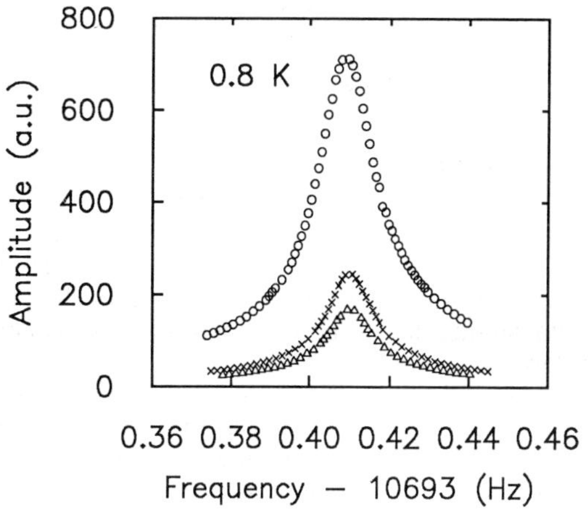

FIGURE 3
Frequency sweeps at T=0.8K for drives of 780 mV (o) 210 mV (×) and 156 mV (△).

Figure 4 shows the results of scans taken at a temperature of 15 mK at different drive levels. The response is typical of an oscillator with pronounced nonlinear characteristics, showing a skew to higher frequencies with increasing amplitude. At the largest drive level, the response is clearly hysteretic. An examination of the hysteretic trace serves to rule out self-heating. At the lowest excitation shown, the oscillator response continues to be asymmetric, suggesting that the strain amplitude for linear response is less than 10^{-8} at this temperature. Measurements at 200 mK, not presented in this work, show a much less pronounced skewness and only at the highest drive levels. A frequency shift of the same magnitude as shown in Figure 4 was observed by Kleiman et al., but with a decrease in frequency as the drive was increased.

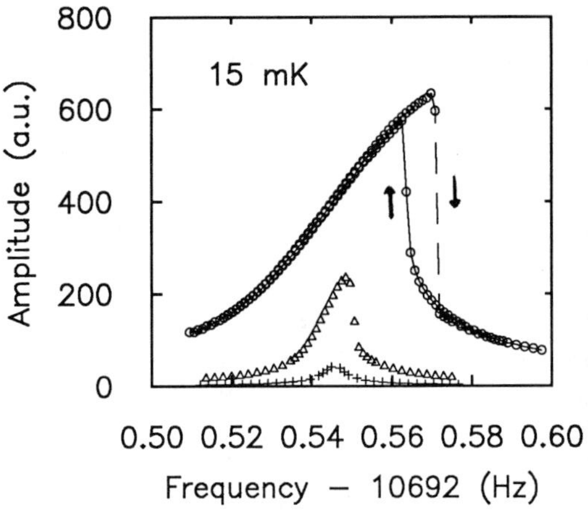

FIGURE 4
Frequency sweeps at T=15 mK for drives of 780 mV (o) 110 mV (△) and 20 mV (+). At the 780 mV drive level, the hysteretic response is shown as the dashed line (sweeping up in frequency) and as the solid line (sweeping downward).

4. CONCLUSION

It is our contention that the anomalous mechanical behavior is accompanied by a remarkable nonlinear response of the oscillator. We base this on the absence of the skewed behavior at high temperatures where the response is "normal" and the onset of the nonlinear response in the regime where the frequency and dissipation behave anomalously. We are continuing the investigation to gather more data for comparison to theoretical models and to previous experimental work.

ACKNOWLEDGEMENTS

This research is supported through a grant from the NSF through the Cornell Materials Science Center DMR-8818558 and the NSF through DMR-8820170. We would like to acknowledge the support and collaboration of L. Pollack, E. Smith and R.C. Richardson.

REFERENCES

(1) R. N. Kleiman et al., Phys. Rev. Lett., 59, 2079, (1987) and references therein.

Physica B 165&166 (1990) 127–128
North-Holland

High-Resolution Acoustic Measurements by Path Length Modulation

P.J.Hamot, H.H.Hensley, Y. Lee, and W.P. Halperin

Department of Physics and Astronomy, Northwestern University, Evanston, IL, 60208, USA

An acoustic technique is described which permits precise measurement of phase velocity in liquids or gases with high attenuation (α >10 cm^{-1}) utilizing a thin, variable path length acoustic cavity formed by a quartz transducer and a piezoelectric bimorph. The path length can be changed manually or in feedback where the bimorph motion is driven by the output of a cw spectrometer. Measurement of phase velocity is then simply measurement of cavity size, which is done capacitively to high precision. The path length modulation method can allow absolute measurement of attenuation as well as resolution of ~10^{-5} in phase velocity with a path length as short as 50μ. It is anticipated that this approach can be used successfully to explore regimes of collective modes and pair breaking in superfluid ^{3}He, as well as textural effects in restricted geometries.

1. INTRODUCTION

Ultrasound is a useful probe of the microscopic features of a quantum fluid such as ^{3}He where anomalies in attenuation (α) and phase velocity (v_ϕ) have been associated with collective excitations of the superfluid order parameter. Unfortunately, the high attenuation associated with these features necessitates the use of short sample thicknesses to recover a transmitted signal, leading to difficulties in the measurement of sound velocity with conventional pulsed or cw techniques. We have developed a new cw approach which allows a very short (down to 50 μ) path length to explore highly attenuating (α~1000cm^{-1}) media while preserving the high resolution in sound velocity of a long path length interferometric experiment. In addition, the technique establishes a geometry in the sample region which can be useful in the study of textural related phenomena.

2.EXPERIMENTAL

2.1. Spectrometer

The central element of the experimental apparatus is a cw acoustic impedance spectrometer described elsewhere (1). Changes in phase velocity of the sample manifest themselves as oscillations at the spectrometer output resulting from interference between outgoing and incoming waves in the cavity. The period of the oscillations relates to the phase velocity while the amplitude relates to attenuation. Quantitative measurement of phase velocity requires precise knowledge of the path length, while attenuation is only measured in a relative fashion.

2.2. Path Length Modulation (PLM)

The *path length modulation* method (2) employs the same spectrometer as above, but allows for a variable path length creating the same oscillatory output. This undergoes one oscillation when the number of half-wavelengths in the cavity changes by one. The cavity size is varied by the use of a piezoelectric bimorph as one wall of the cavity, which flexes when charged displacing its center relative to the transducer. A capacitor plate is attached to the bimorph, and thus precise measurement of the position of the bimorph with a capacitance bridge is a measure of phase velocity. The voltage applied to the electrodes of the bimorph can be ramped to create oscillatory output which can be analyzed in a similar way to the method described in 2.1. However, this mode also allows for absolute measurement of attenuation since the amplitude of the received signal is known as a function of change in path length. The spectrometer output can also be amplified through a voltage gain g and used to drive the bimorph. This effectively locks a constant number of half-wavelengths in the cavity, and so the size of the cavity, z , is proportional to the

Work supported by the Low Temperature Physics Program of the NSF under contract DMR-8620275.

wavelength of sound. This mode is particularly sensitive to detection of *changes* in phase velocity, which are characteristic signatures of a collective mode. For small changes, the resolution is

$$\frac{\Delta\lambda}{\lambda} = \frac{\Delta z}{z} + \text{offset}$$

where the feedback offset term is a function of the wave vector , z, and is inversely proportional to g. Since the change in bimorph position is the change in capacitor spacing, the resolution can be made to be on the order of the resolution in capacitance, which is as good as 10^{-5} in feedback.

3. DATA AND RESULTS

3.1. Sound Velocity in ^{4}He

To prove the feasiblity of the PLM method at cryogenic temperatures, we show results obtained for sound velocity as a function of pressure in ^{4}He at 4.2K. Data were taken in the feedback mode presented in Figure 1 where each line represents a pressure sweep with different numbers of half-wavelengths locked in the cavity. The wavelength of sound at a particular pressure can be determined from the separation of the sweep lines. The change in sound velocity over the range shown is 1.5 x 10^{-2} and the resolution is roughly two orders of magnitude better for this nominal z = 450 μ. The absolute values of sound velocity obtained (Figure 2) are in excellent agreement with the existing data (3).

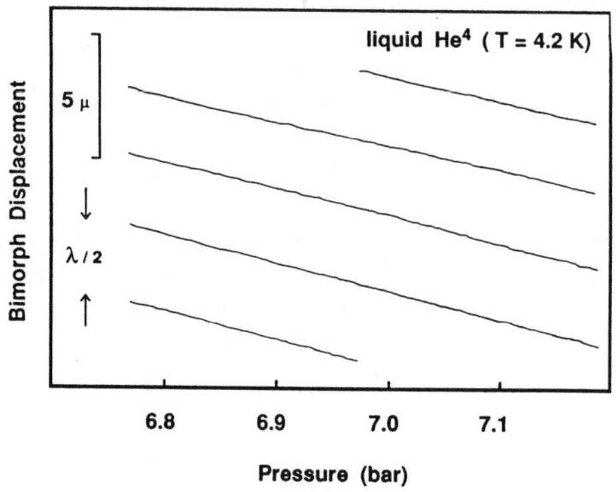

Figure 1. A data specimen of sound velocity in ^{4}He at 44.4 MHz taken in feedback mode.

3.2. Attenuation Measurements

We have also demonstrated the ability to measure absolute attenuation using N_2 gas at room temperature, a medium selected owing to its relatively high attenuation. This data is taken in the manual mode and is of the form shown in Figure 3, where the amplitude of the signal decreases with increasing path length. While these results are encouraging, it is too preliminary to make quantitative statements.

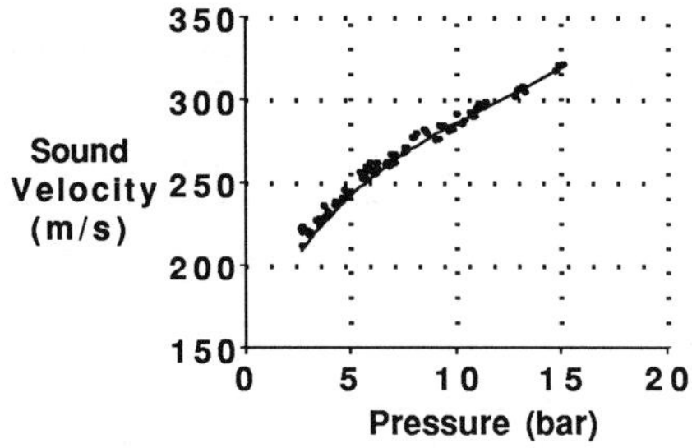

Figure 2. Sound velocity data for ^{4}He at 4.2K and 26.6MHz. The data points are from this work, while the line is a fit to the data of Atkins and Stasior (3).

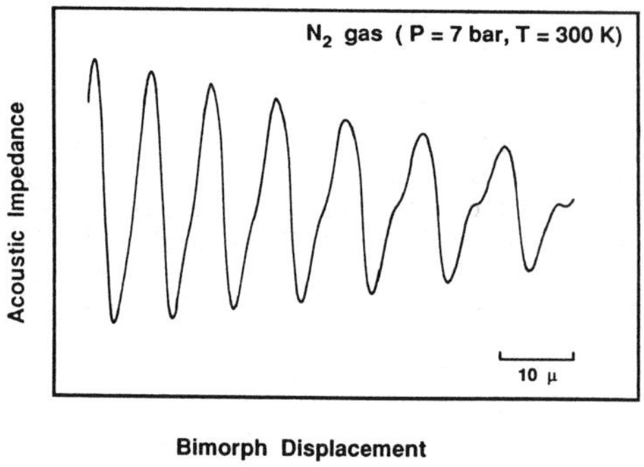

Figure 3. Sample data for attenuation measurements in manual mode.

4. CONCLUSIONS

We have demonstrated the PLM technique allows accurate sound velocity measurements in short path lengths as well as shows promise in measuring absolute attenuation. We are currently designing experiments for superfluid ^{3}He where textures in the B phase would be fixed by the cavity geometry. The technique also has potential for use in other areas of ultrasonics.

REFERENCES

1)B.S.Shivaram, M.W.Meisel, B.K.Sarma, W.P.Halperin, and J.B. Ketterson, Journal of Low Temperature Physics, 63 (1986) 57.
2)P.J.Hamot, H.H.Hensley, W.P.Halperin, Journal of Low Temperature Physics, 77 (1989) 429.
3)K.R. Atkins and R.A.Stasior, Canadian Journal of Physics, 31 (1953) 1156.

Physica B 165&166 (1990) 129–130
North-Holland

OSCILLATORY HE II MASS FLOW BEHAVIOUR OF FEPS IN SELF SUTAINED MODE

S. KASTHURIRENGAN[$], S. JACOB[$], R. KARUNANITHI[$] AND A. HOFMANN[*]

Department of Physics, Indian Institute of Science, Bangalore 560 012, India[$]
Institut für Technische Physik, KFK, D7500 Karlsruhe 1, West Germany[*]

He II flow driven by Fountain effect pumps(FEPs) operating in self sustained mode in test loops of different impedances have shown oscillatory behaviour for test section heat loads upto 1.2W. The period of oscillation depends on the applied heat load, bath temperature of He II (T_{bath}) and the impedance of the test loop. The occurrence of oscillations is qualitatively understood due to the existence of an "*onset temperature*" to generate convection in the loop. The boundary between the stable and oscillatory states have been mapped. A combined normal and self sustained mode of operation of FEP eliminates these oscillations for practical applications.

1. INTRODUCTION

Due to the outstanding heat transfer characteristics forced flow HeII has been considered as an excellent coolant for superconducting magnets. FEPs have been examined (1) as an alternative to mechanical pumps to generate such a flow. They are based on thermomechanical effect of He II in narrow pored filters. A heat load in front of the filter develops a fountain pressure which drives He II through the filter.

Srinivasan et al(2) have shown FEPs can be operated either in the normal mode (direct heating in front of the filter) or in self sustained mode (bringing the heat applied at test section in front of the filter using a heat exchanger). The present study relates to the latter mode of operation, wherein the He II mass flow shows oscillatory behaviour for heat loads below 1.2W. The paper discusses these results along with a qualitative explanation of these oscillations.

2. EXPERIMENTAL SET UP

The schematic of the experimental arrangement (figure 1) is similar to those used in earlier studies (2-4). A 1μ grain size Al_2O_3 filter (31 mm dia. and 100mm length) is used as FEP. Mass flow generated by FEP flows through an acoustic mass flow meter FM in between the two heat exchangers HX1 and HX2, impedance loop FI1, test section TS, impedance loop FI2 and the warm end heat exchanger WHX. The test section heater HTSM operates the FEP in the self driven mode, while HFEP, heater in front of filter operates FEP in normal mode. This configuration is the High Impedance Loop (HIL) and the system without FI1 and FI2 is the Low Impedance Loop (LIL). The entire loop can be isolated from the main He II pool by a cold valve so that it can be operated at any pressure below 7 bar.

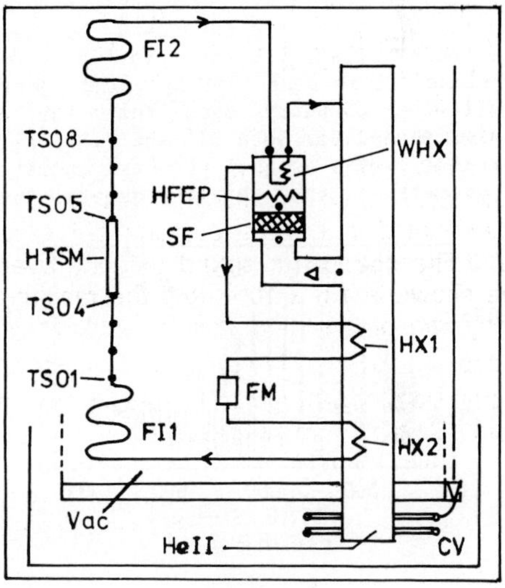

FIGURE 1

3. OSCILLATORY MASS FLOW BEHAVIOUR

For both loop impedances, on increasing the power Q_{FEP}, the mass flow increases and reaches a saturation value ($\cong$ 2.8g/s for LIL and $\cong$ 1.2g/s for HIL). On increasing test section heater power Q_{TSM}, similar behaviour is observed but with slightly lesser mass flow values for both loops, except below 1.2W, where the mass flow becomes *oscillatory*.

3.1 Effect of Q_{TSM} on oscillations

Figure 2 shows a typical mass flow oscillation of HIL at 1.9K when Q_{TSM} is varied from .2W to 1.2W. It is seen that the period of oscillation decreases with increasing Q_{TSM}, until at a particular power $Q_{TSM(stable)}$ the oscillations totally disappear. The same behaviour is observed for both loops.

3.2 Effect of loop impedance and T_{bath}

At T_{bath} = 2K and $Q_{TSM} \cong$ .4W, the oscillation period is $\simeq$ 50s for LIL and $\simeq$ 150s for HIL. At

0921-4526/90/$03.50

all combinations of Q_{TSM} and He II bath temperatures (T_{bath}) where the oscillations occur, one notes that the oscillation period is longer for HIL than for LIL. With a constant Q_{TSM}, both the amplitude and the period of oscillations decrease with decreasing T_{bath}, for both the loops until at a particular $T_{bath(stable)}$ the oscillations totally disappear to reach stable state. The LIL does not become oscillatory as long as T_{bath} is below 1.8K. This limit is ≅ 1.6K for HIL.

3.3 Effect of HFEP power on oscillations

It has been observed that application of a small heat load Q_{FEP} decreases the amplitude of oscillations considerably. When Q_{FEP} is in the range .4 to .6W, the oscillations is nearly prevented in all operational regimes in the loops. This power is quite small when compared to the peak power of Q_{TSM} ≅ 6W which can be removed by self sustained mode. So for practical applications the oscillations can be avoided by combining the normal and self sustained modes.

Figure 3 shows the boundary between the stable and oscillatory states plotted in Q_{TSM} - T_{bath} plane. It is seen that by using Q_{FEP} one can shift the boundary such that the stable state is reached for both HIL and LIL. Although the present curves cannot be extrapolated to other geometries, they show the general trend.

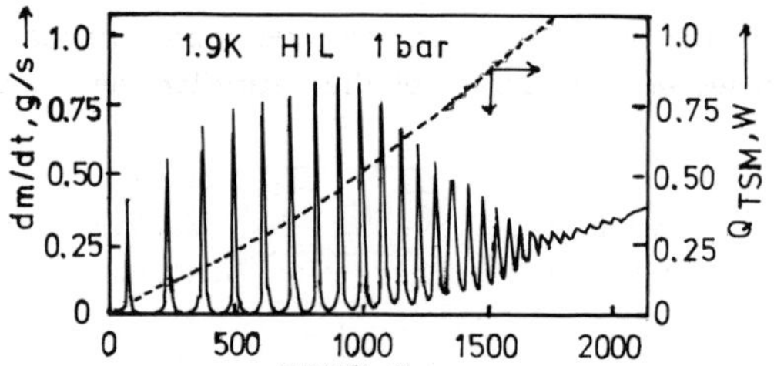

FIGURE 2

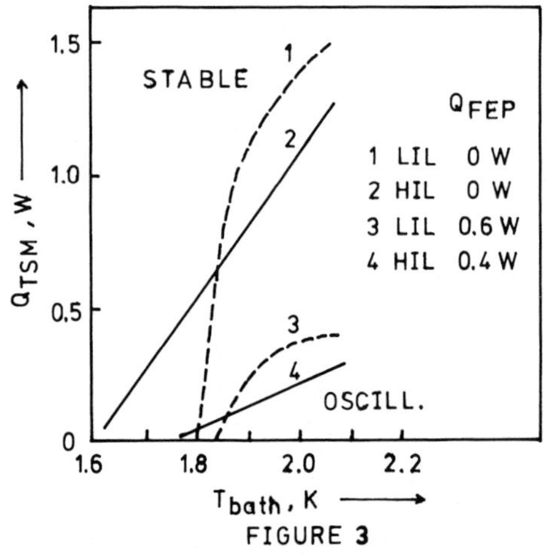

FIGURE 3

4. QUALITATIVE UNDERSTANDING OF OSCILLATIONS

The oscillations in FEP loops arise due to the "*onset temperature*", T_{ON}, to generate convection in the loop. This results by solving the energy balance equations at WHX. It is found that no convection is possible as long as fluid temperature at its inlet (TFPX1) is below T_{ON}. Reference (5) gives the calculation details and plots T_{ON} as a function of T_{bath} at different pressures.

Assume that at T_{bath} = 2K, Q_{TSM} is applied. Due to the heat transfer by Gorter Mellink (GM) conduction, TFPX1 rises above T_{ON}, causing a flow. Now by the flow, the heated zone in front of the filter moves away from HTSM and a cold front takes its place. This causes decrease in TFPX1 and when this falls below T_{ON}, the flow stops. This cycle repeats causing the oscillations in the loop. As Q_{TSM} is increased, TFPX1 rises and increases the mass flow rate (oscillation period decreases). At a given Q_{TSM}, TFPX1 is higher than T_{ON} continuously, and this establishes the steady state.

It is clear that when T_{bath} is decreased, T_{ON} also falls, and a situation favourable to steady state is reached. In the HIL case, since less heat is extracted by GM conduction, oscillation periods are longer due to the decreased mass flow and hence lower T_{bath}s are needed for stabe state. When Q_{FEP} is applied TFPX1 rises by direct heating and causes stable state.

5. CONCLUSION

Forced He II flows in test loops of different impedances driven by self sustained FEPs show oscillations below 1.2W of heater power. The dependence of oscillation period on Q_{TSM} and T_{bath} and pressure has been studied and results have been qualitatively understood due to the existence of an onset temperature for convection in the test loops. The oscillations are controllable for practical applications.

REFERENCES

(1) A. Hofmann, Proc. ICEC 11 (Butterworths, Guildford, U.K.,1986) 306.
(2) R. Srinivasan, Cryogenics 21 (1985) 641.
(3) A. Hofmann,"A study of nuclear heat loads tolerable for NET/TF coils cooled by forced flow of He II" KFK Report 4365, Karlsruhe, 1988.
(4) A. Hofmann, A. Khalil and H.P. Kraemer, Adv. in Cryog. Engg.33 (1988) 471.
(5) A. Hofmann and S. Kasthurirengan, CEC/ICMC, 1989, paper HA01.

Physica B 165&166 (1990) 131–132
North-Holland

OBSERVATION OF NEW NONLINEAR PHENOMENA IN FOCUSED ACOUSTIC BEAM IN PRESSURIZED SUPERFLUID HELIUM-4

Koichi KARAKI, Takehiko SAITO[+], Koichi MATSUMOTO[+], Yuichi OKUDA[+]

OLYMPUS Opt. Co. Ltd., Kuboyamacho, Hachioji, Tokyo 192, JAPAN
[+]Department of Applied Physics, Tokyo Institute of Technology, O-okayama, Meguro-ku, Tokyo 152, JAPAN

We have studied the propagation properties of focused acoustic beam in superfluid ^{4}He at 300mK under several pressures. We observed new nonlinear phenomena in the reflected and transmitted signals. Under low pressures (anomalous dispersion region), the received power is saturated by harmonic generations at some input power and is depleted by parametric amplification of decay products for higher input power. Under higher pressures (normal dispersion region), it is depleted more rapidly, and moreover, exhibits several sharp dips as the input power increased further. It cannot be explained by a simple 4-phonon process proposed so far.

As the density and the sound velocity of ^{4}He are both very low, the sound propagation in superfluid ^{4}He can easily become nonlinear at a moderate sound energy density. Moreover, the nonlinear effect is extremely enhanced at the focusing point in the converging acoustic beam. It is very interesting to investigate what is happening just at the focal point. In addition, by changing the pressure of superfluid ^{4}He, we can change the dispersion curve of phonon in ^{4}He from the anomalous one to the normal one. That is, we can control the phonon scattering mechanism by the pressure. To clarify the relation between the nonlinear effect in the high power density acoustic field and the dispersion character is the main motive force of the present study.

We have measured the propagation properties of focused acoustic pulse in superfluid ^{4}He at about 300mK under several pressures using both the reflection and the transmission modes of an acoustic microscope.

An acoustic lens is mounted on the top of the focusing mechanism and a flat detector is fixed in a closed cell. The cell is bolted on the bottom of a dilution refrigerator. The structure of the focusing mechanism and the acoustic lens have been described elsewhere[(1)]. The flat detector is a columned sapphire rod (5mm long, 6.1mm diameter) and its both ends are finished flat. The W-SiO_2 double quarter-wavelength antireflection coating is affixed on one end. The ZnO transducer is affixed on the other end.

In the experiment, a 386MHz electromagnetic pulse (pulse duration 100ns) is applied to ZnO transducer of the lens. This electromagnetic pulse is truncated by pin-diode gate from cw wave generated from the oscillator. In the reflection mode, we focused on the surface of the antireflection coating of the flat detector.

When the input power is high in the reflection mode, internal reflections from the spherical surface of the lens become large and interfere with a reflected signal from the flat detector. Therefore, the transmission mode is necessary to do an accurate measurement.

Typical experimental results (300mK ; 11atm and 24 atm) of reflection mode are shown in Fig.1. The input power (horizontal axis in the figure) means the output power of cw oscillator.

The received power increased with the input power higher than +15dBm for both 11atm and 24atm. These are the reflected signals from the internal surface of the lens, which are invisible in lower input power region. They are not seen in the transmission mode. In the other regions, the results of the reflection mode and the transmission mode are almost the same, except the constant loss about 5dB (the efficiencies of ZnO transducers and antireflection coatings etc.).

Under low pressures, simple saturation and depletion were observed. For the input power below -20dBm, the received power increased in proportion

0921-4526/90/$03.50

to the input power. At about -20dBm, it was saturated. For higher input power, it decreased monotonously with the input power. These results agree well with those of Wright et al.(2) and Foster et al.(3) Since the frequency used in our experiment is lower, we think that the parametric amplification of the decay products by 3-phonon process is weak, and the saturation by the harmonic generation is dominant.

Under the pressure higher than about 18atm, the behavior of received signals changed dramatically. In the low input power region below -15dBm, the received power increased in proportion to the input power as well as the case for low pressure. As seen in Fig.1, the strong depletion with complex resonance-like structure appeared at the input power higher than -10dBm. Three very sharp dips (-5, +5, +10dBm) overlapped on the simple decreasing curve. In other words, the received signal decreased rapidly at -5dBm, and recovered ; it decreased again at +5dBm and +10dBm.

Here, the first question is whether this behavior originates from the propagation of sound in superfluid ^{4}He or not. As this phenomenon depends on the pressure, it should originate from the property of superfluid ^{4}He. However, it may be due to the interference of the reflected signals and internal reflections of lens, the effect of the antireflection coating of reflector, or the property of

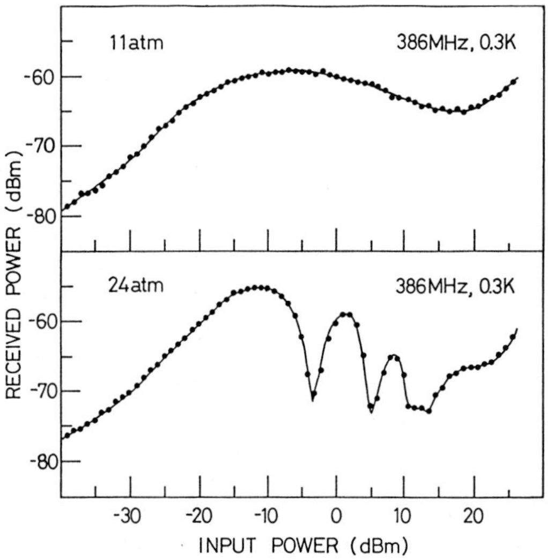

Fig.1. The received power in the reflection moae in superfluid ^{4}He as a function of input power of cw oscillator.

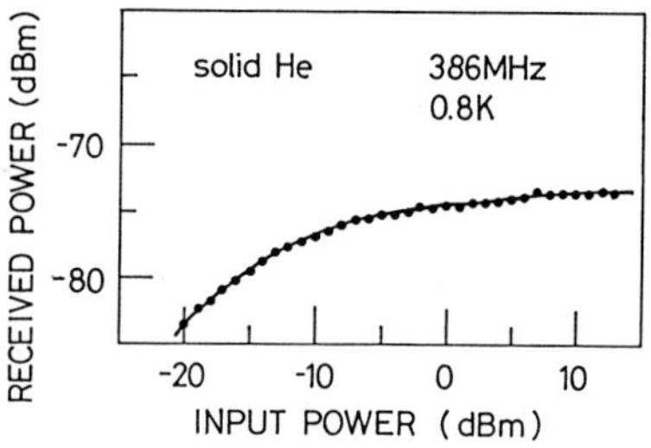

Fig.2. The received power in the reflection mode in solid ^{4}He.

ZnO at low temperature under pressure. The first possibility is ruled out by the agreement between the reflection and the transmission modes. The second is not consistent with the fact that we obtained quite the same results using a flat bare sapphire without W-SiO_2 coating as a reflector. The third possibility is excepted by the following two facts. The behavior of the internal reflection of lens was normal when we measured it by the same manner. Moreover, we obtained the result as shown in Fig.2. in solid ^{4}He at about 800mK, 26atm. It does not contain any resonance-like structures.

As described above, we believe that these resonance-like phenomena originate from the property of superfluid ^{4}He. The problem is whether they are medium-altering effects (cavitation, streaming, vortex formation and rapid melting and freezing) or not. About this problem, we think that they can be ruled out by the same reasons described in ref.(2).

Then what nonlinear phenomenon does happen in the pressurized superfluid ^{4}He ? The simple 4-phonon process cannot explain that the received signal decreases and recovers and decreases again. There must be some different nonlinear effects mixing or overlapping in this phenomenon.

This work was supported by a Grant-in-Aid of from the Ministry of Education, Science and Culture of Japan.

REFERENCES

(1) K. Karaki, M. Suzuki, Y. Okuda : J. Appl. Phys. 67 (1990) 1680
(2) D.R. Wright, B. Hadimioglu, L.J. La Comb, Jr., C.F. Quate and J.S. Foster : Jpn. J. Appl. Phys. suppl. 26-3 (1987) 9 (Proc. LT18)
(3) J.S. Foster and S. Putterman : Phys. Rev. Lett. 54 (1985) 1810

Physica B 165&166 (1990) 133–134
North-Holland

HIGH RESOLUTION CALORIMETRY IN LIQUID HELIUM

Gane K. S. WONG, Paul A. CROWELL, H. Adrian CHO, and John D. REPPY

Laboratory of Atomic and Solid State Physics, and the Material Sciences Center, Clark Hall, Cornell University, Ithaca, New York, 14853-2501, U.S.A.

We report advances in the adiabatic calorimetry technique that permit a determination of heat capacity to at least 100ppm. The resolution is ultimately limited not only by the thermometry, but also by how well we can control the heat input to the cell.

In the adiabatic calorimetry technique, one applies a known heat pulse to an otherwise thermally isolated cell and then measures the resultant temperature rise. The temperature rise must be large enough to be measurable, but small enough to avoid smearing out any temperature dependent heat capacity features. Experiments using resistance thermometry might typically resolve temperature to 6 digits. A reasonable compromise would allow for 3 digits of resolution along each of the heat capacity and the averaged temperature axes. Our work on superfluid helium in porous media, particularly aerogel glass, has kindled our interest in much better resolution along both axes.

As a first step, we developed a SQUID based magnetic thermometer in the style of Lipa and coworkers[1]. We measure the temperature dependent magnetization of a $0.016cm^3$ sample of $Cu(NH_4)_2Br_4.2H_2O$ salt, which is grown from solution in a *glass* beaker because the solution will react with most metals. The salt is crushed into a 3mm length of a 4-40 tapped hole and subjected to the 375G magnetic field that we trap inside a long superconducting niobium tube, 15D×64H mm. Our sensor is a superconducting flux transformer, featuring a 2μH astatic coil pair at the salt pill and a matching 2μH coil at the rf SQUID. The rf SQUID and all of the associated electronics, the rf head, the flux lock loop, and the flux counter, are purchased from BTI[2]. There is a compromise between signal to noise ratio and slew rate that is determined by the output filter setting on the flux lock loop. At 10Hz, we can resolve $2\times10^{-3}\phi_0$(rms) and track up to approximately $50\phi_0$/sec.

Screening currents in the superconducting flux transformer couple $0.74\phi_0/\mu$K into the rf SQUID at 2.17K. The thermometer works even below the 1.8K ferromagnetic ordering transition in the CAB salt[3], which would not have been the case if we had measured the zero-field AC susceptibility instead. We have at least 3nK(rms) resolution from 1.4 to 2.2K. Sweeping this range causes the flux counter to overflow and wrap around 11 times, for a total count of over $11 \times 10^5\phi_0$. Figure 1 shows that the ferromagnetic transition is smeared out as the external field is increased. Increasing the field improves the average sensitivity above the CAB ordering temperature, which is where we operate, but only at the expense of the peak sensitivity. There is also a very small heat capacity anomaly associated with the ferromagnetic transition, with an an integrated area of order 0.1mJ, prominent over a 0.5K regime. This anomaly, like the magnetization, is also smeared out as the external field is increased.

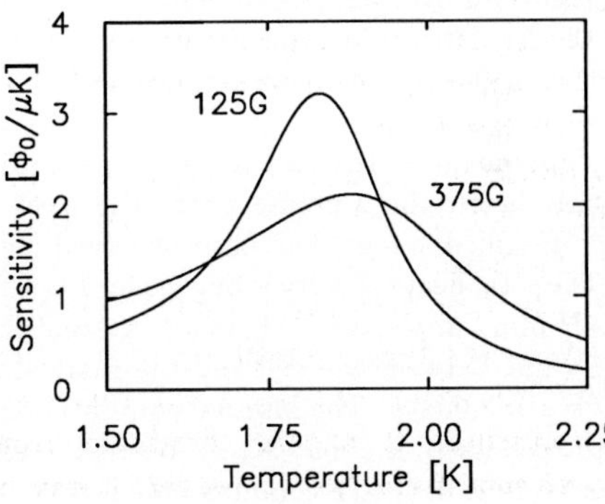

Figure 1: The thermometer sensitivity is depicted at two values of the external field, which is normally 375G.

Temperature stability requires that the cell be well isolated from the outside world. Incorporated into our cell is a bellows-actuated low temperature valve with an indium-coated 60° needle and 120° seat arrangement that can be made to stick by applying 250N of force. The needle is not attached to the actuator mechanism, so it is not pulled back when the bellows are pumped out. It pops off quite safely at the end of the run when the cell is warmed up. We have not systematically searched for the optimal force or angles, nor have we tried to reuse the needle and seat. The niobium capillary shielding

0921-4526/90/$03.50 © 1990 – Elsevier Science Publishers B.V. (North-Holland)

the flux transformer wires should be broken in two to improve the thermal isolation. Shield integrity can be protected by an overlapping NbTi sleeve. The strongest thermal link between our cell and the main stage is now due to 3 hollow Vespel support rods.

Scatter in a heat capacity measurement is also limited by how well one can measure the heat pulses and how well one can control the fluctuations in the heat leak to the cell. We use rectangular heat pulses and measure the voltage across a room temperature 4ppm/K current sense resistor. Every wire or fill line used is extensively heat sunk at the main stage, glued or soldered over a 20-50cm length, before being coiled up and attached to the cell. The cell is enclosed by a radiation shield that is thermally anchored at this main stage. The main stage is regulated by a germanium resistance thermometer and the regulator setpoint is iteratively adjusted until the drift in the cell temperature, generally less than 0.1nK/sec, is minimized.

Our experiment is designed to work a few mK to a few 100mK below the λ-point, where the cusp singularity in ^{4}He-filled aerogel has been observed (see Figure 2). The total heat capacity is huge, over 1J/K for our 19D$\times$2.6H mm sample, resulting in an external thermal isolation time, between the cell and the main stage, of 2.6×10^5sec or 3.0days. The internal equilibration time, within the cell itself, is about 30sec. We fit the temperature versus time data to a common linear drift on both sides of the pulse, plus a quasi-exponential decay after the pulse. Deviations from exponential decay appear in the 5th digit because the external isolation time is temperature dependent. We avoid a recursion problem by modeling the isolation time as a linearly time dependent quantity, which is reasonable because the temperature change is very nearly linear.

We acknowledge the support of the National Science Foundation through grant NSF-DMR84-18605 and the Cornell Material Sciences Center through grant NSF-DMR85-16616-A02. G.K.S.W. acknowledges the support of an A.D. White Fellowship from Cornell University and a 1967 Postgraduate Fellowship from the Natural Sciences and Engineering Research Council of Canada. P.A.C. acknowledges the support of an AT&T Bell Laboratories Ph.D. Scholarship.

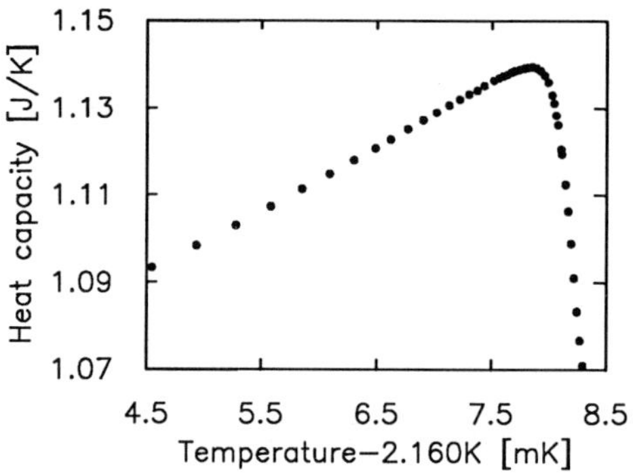

Figure 2: The cusp singularity in a porous aerogel that is completely filled with superfluid helium can be resolved to 100ppm using temperature rises of only 30μK.

REFERENCES

[1] J. A. Lipa, B. C. Leslie, and T. C. Wallstrom, Physica **107B**, 331 (1981); T. C. P. Chui and J. A. Lipa, in *Proceedings of LT-17*, ed. by V. Eckern, A. Schmid, W. Weber, and H. Wühl (North-Holland, Amsterdam, 1984), p. 931; D. Marek, Jap. J. Appl. Phys. **26**, 1683 (1987)

[2] Biomagnetic Technologies Inc. of San Diego, California

[3] A. R. Miedema, R. F. Wielinga, and W. J. Huiskamp, Physica **31**, 1585 (1965); R. F. Wielinga and W. J. Huiskamp, Physica **40**, 602 (1969); L. J. de Jongh, A. R. Miedema, and W. J. Huiskamp, Physica **46**, 44 (1970)

Physica B 165&166 (1990) 135–136
North-Holland

CAPTURE OF FOREIGN ATOMS BY HELIUM CLUSTERS

A. SCHEIDEMANN, B. SCHILLING AND J. P. TOENNIES

Max Planck Institut für Strömungsforschung, Göttingen, FRG

J. A. NORTHBY

Physics Department, University of Rhode Island, Kingston, RI 02881 USA

Foreign atoms can be easily introduced into free helium clusters by several methods. The process depends strongly on the conditions under which the clusters were formed, however. Expansions through the critical region produce the largest capture effect.

The study of liquid helium in restricted geometries and in films has provided much insight into the nature of this unique fluid, and in particular into the nature of the superfluid phase transition. A natural and obvious extension of this effort is the study of free helium clusters. They provide in many ways an ideal finite system. Bound state clusters in a vacuum can be easily formed. They are free of substrate interactions, and the energy and mass are constants of the motion. The interparticle interaction is well known, and the theoretical apparatus for calculating their properties is available. Indeed, much progress has been made in this direction recently by several groups (1-3).

Unfortunately, from an experimental point of view, this system is not quite so ideal. While generating a helium cluster beam by nozzle expansion methods is simple, measuring the propereties of those clusters is a very difficult problem. The interaction of microscopic neutral helium clusters with external fields is too small to be useful, and it will undoubtedly be necessary to utilize more strongly interacting probes in order to extract useful information about the state of the system. Such probes, for example, might consist of charged or metastable electronic states of the helium atom, or of one or more foreign atoms attached to the cluster. It is true that such probe particles will have a significant perturbing effect on the cluster for small clusters, but that is a problem which appears to be unavoidable. It means that theoretical models will necessarily have to include the probe if they are to make experimentally verifiable predictions. On the experimental side, it will be helpful to utilize many different probes in order to identify which properties are characteristic of the system itself. It is likely that neutral impurity particles will provide the least perturbing probes, however.

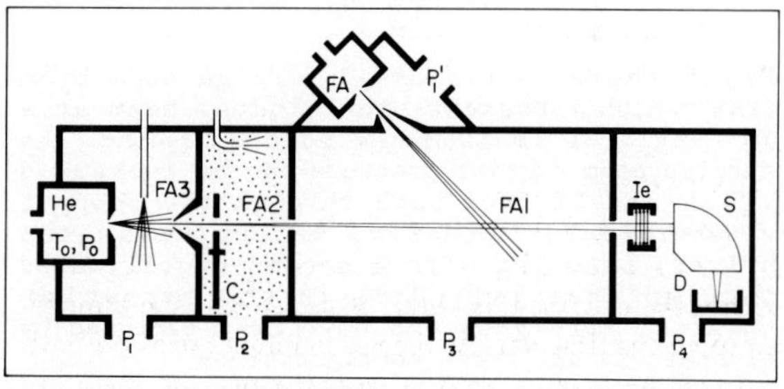

Figure 1. Schematic of apparatus showing interaction configurations FA1, 2 and 3.

We are presently attempting to develop a range of experimental methods for generating and studying helium clusters. We have previously described our studies of the production of neutral cluster beams (4), and briefly reported on studies of the production by electron impact of ionic and metastably excited clusters (5). It is the purpose of the present note to summarize our more recent efforts to introduce other neutral particles into helium clusters. Development of such techniques is a necessary precursor step to utilizing foreign atoms as probes of cluster properties.

A schematic of the relevant features of our apparatus is given in Fig. 1. The primary helium cluster beam is produced by expansion of pure helium gas from a high pressure and low temperature stagnation volume through a 5μ sonic nozzle into a differentially pumped vacuum. The beam is chopped and after a 1.4m flight path the clusters are ionized by electron impact (I_e), and the lighter charged fragments are mass-analyzed in a magnetic spectrometer (B) and detected by an open multiplier (D). Provision is made for interaction with foreign atoms in three places.

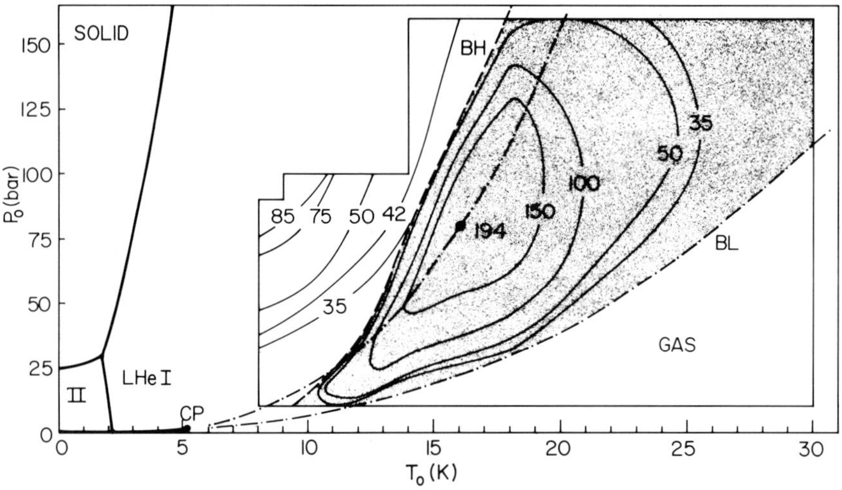

Figure 2. Contour plot of $^{22}Ne^+$ counting rate as a function of cluster source conditions (P_0, T_0).

First there is a crossed foreign atom beam (FA1) which intersects the cluster beam at a 40° angle 61 cm from the nozzle. Second, the foreign atom partial pressure can be increased along the 12.5 cm path through the chopper chamber (FA2). Finally, we can cross the primary beam jet with a second uncollimated foreign atom jet (FA3) in the expansion chamber. Foreign atoms have been captured by clusters in all three configurations.

The presence of captured foreign atoms is detected by an increased signal at the mass of the foreign atomic ion in the spectrometer. Confirmation that they are captured and not just scattered into the detector is found in the observation that the foreign atom signal is coincident with the chopped helium cluster signal. Further confirmation is given by our observation of both foreign atom cluster ions $(FA)_n^+$ and mixed cluster ions $(FA)_nHe_m^+$ as well. These could not form unless many foreign atoms are already present in the helium cluster as a result of multiple capture prior to ionization. In the case of some species we have observed very large foreign atom cluster ions (e.g., $(H_2O)_{18}^+$) indicating a very high degree of multiple capture prior to ionization. To date we have successfully captured Ne, Ar, Kr, H_2, O_2, H_2O, CH_4 and SF_6 in helium clusters. Ne has been captured in all three experimental configurations described above. The others have been captured in only one or two configurations. However, every combination we have tried has worked.

Perhaps the most interesting observation we have made (6) is that the capture process depends very sensitively on the helium cluster beam stagnation conditions. In particular, we find that clusters formed in expansions which pass through the critical region are much better able to capture foreign atoms and deliver them to the detector than those formed under other conditions. This behavior is illustrated for the case of ^{22}Ne capture from the secondary beam FA1 in Fig. 2. Here we give a contour plot of the $^{22}Ne^+$ counting rate as a function of cluster beam stagnation conditions, for a constant Ne atomic beam. It shows a sharply defined region of strong capture with a local maximum at P_0=80 bar and T_0=16 K. The dashed curve passing through this point is the S=6.5 KJ/Kg-K isentrope, which is close to the critical point entropy (5.6±.5 KJ/Kg-K). The line BL is the 9.2 KJ/Kg-K isentrope. Below this line we see no capture even though there are clusters present in the beam. The upper boundary of this region, BH, is sharply defined but isn't exactly an isentrope. Whether these "critical clusters" are simply much larger, or differ in more complex ways from those produced under other conditions is at the moment an open question.

In summary, we have found that for certain expansion conditions it is very simple to introduce foreign atoms into helium clusters. In fact, it is so easy that the greater problem may well be to prepare pure He clusters under these conditions.

One of us (J.N.) acknowledges support from the NSF (INT8822637 and DMR8816482).

(1) V.R. Pandharipande, S.C. Pieper and R.B. Wiringa, Phys. Rev. **B34**, 4571 (1986).
(2) S. Stringari and J. Treiner, J. Chem. Phys. **87**, 5021 (1987).
(3) P. Sindzingre, M.L. Klein and D.M. Ceperly, Phys. Rev. Lett. **63**, 1601 (1989).
(4) H. Buchenau, E.L. Knuth, J.A. Northby, J.P. Toennies, and C. Winkler, J. Chem. Phys., to be published.
(5) H. Buchenau, J.A. Northby, C. Winkler and J.P. Toennies, Jap. Jour. Appl. Phys. **26** (Sup. 26-3), 11 (1987).
(6) A. Scheidemann, J.P. Toennies and J.A. Northby, to be published.

Physica B 165&166 (1990) 137–138
North-Holland

OPTICAL SPECTROSCOPY OF ALKALI AND ALKALI-LIKE IONS IN SUPERFLUID ^{4}HE

H. Bauer, M. Beau, J. Fischer, H. J. Reyher*, J. Rosenkranz, K. Venter

Physikalisches Institut der Univ. Heidelberg, Philosophenweg 12, 6900 Heidelberg, F.R.G.
*Fachbereich Physik der Universität Osnabrück, Barbarastraße 7, 4500 Osnabrück, F.R.G

Recent experimental work by the authors on optical microprobes in superfluid ^{4}He is extended to the recombination of snowball structured ions (alkali ions) and electron-bubbles. First results for Li, Na and K are indicating radiationless transitions for these ions, while in the case of the transition elements Cu, Ag and Au with similar valence electron configuration, several recombination bands have been detected optically.

INTRODUCTION:

In recent papers /1, 2/ we reported on the laserspectroscopy of alkaline earth atoms (Ca, Sr, Ba) in liquid Helium. On the one hand, these optical microprobes give information of the microscopic structure of the defects in liquid helium and on the other hand, they can be new tools for the investigation of experiments like trapping of ions or atoms in vortex lines or on ^{4}He films /3/. The atoms used are produced directly in the liquid via recombination of the corresponding ions and electrons (see fig. 1). The ions are produced in a specially developed ion-source working a few centimeters above the liquid level. The electrons are provided by means of tip emission in the liquid /4/.

The investigated alkaline-earth (Ca, Sr, Ba) /2/ atoms should form bubble states in the liquid as can be deduced from the small optical line shifts of the excitation and emission bands. An extension of the bubble model calculation of /5/ give additional confirmation of this assumption. In former studies /6/ the optical data of the Ba^+ in liquid helium had been interpreted also in the frame of a bubble model calculation. Therefore in the case of the alkaline earth atoms the recombination takes place between two bubble type ions. As has been reported by /7/ and /8/ the alkali ions should form snowball states such as He^+ in liquid helium, because of the stronger localisation of the valence electron. In this case the recombination between snowball-type (alkaline) ions and bubble-type ions (electrons) can be studied in liquid helium.

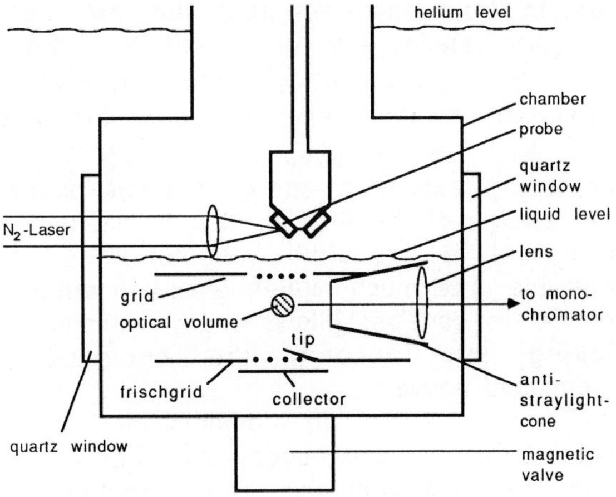

Fig. 1: Experimental chamber

RESULTS:

Preliminary results for Li, Na, and K indicate, that these ions recombine radiationless because no fluorescence has been detected in the range of 250 - 750 mm. The corresponding ion currents have been measured in a drift cell providing an independent detection method for the presence of the defect ions in the liquid. The transition elements Cu, Ag, and Au with the same outer electron configuration but filled d-orbitals, in contrast, show many optical transitions, listed in tab. 1, which can be unambiguously attributed to the recombination lines of these defects in liquid helium. The defect structure of these ions is not yet clear. A possible explanation for the different behavior of alkali and Cu, Ag, Au ions may come from the energy loss of the metal ion (m^+) during relaxation after penetrating the liquid. While this energy loss is about 3.5 eV for Li^+ and about 2 eV for Na^+ and Cu^+, the remaining excitation energy of the m^+ - e^- complex is about 1.5 eV for Li, 3 eV for Na but about 5.5 eV for Cu. Assuming that the recombination takes place via tunneling, an estimate of the tunnel probability as a function of the m^+ - e^- distance predicts an additional energy loss of about 0.7 eV in the mutual coulomb field before recombination.

An additional energy loss of about 0.7 eV results from the fact that the m^+ - e^- complex relaxes into an atomic-like state, from which the optical emission starts. Therefore the total energy loss of Li^+ - e^- and Na^+ - e^- recombination results in a remaining excitation energy located below the energy of the first excited atomic level, therefore no radative recombination is possible. The much higher ionisation energy of Cu in contrast allows many transitions taking into account the corrections mentioned above.

The most promising elements for optical recombination spectroscopy of alkali ions, are the ones with the highest remaining excitation energies like Rb and Cs. The work on these ions is in progress in our laboratory.

Tab. 1.: Data of the recombination lines of Cu in LHe II

free atom/9/		atom in LHeII	
transition	λ (nm)	shift (nm)	FWHM (nm)
5s $^2S \to 4p^2P$	809.26	+0.29(16)	2.40(2)
$4p'^4P \to 4s^2S$	249.21	-0.14(4)	0.44(4)
$4p'^4D \to 4s'^2D$	309.39	-0.36(5)	0.35(5)
$5s'^4D \to 4p'^2P$	427.51	+0.11(7)	0.47(2)
$\to 4p'^2F$	465.11	+0.09(8)	0.59(2)
$\to 4p'^2D$	529.25	+0.12(9)	0.79(2)
$4d'^4D \to 4p'^4D$	360.20	-0.98(5)	0.43(5)
$4d'^4F \to 4p'^4D$	359.91	-1.27(5)	0.43(5)

(1) H. Bauer, M. Beau, A. Bernhardt, B. Friedl and H. J. Reyher. Phys. Lett. A 137 (1989) 217.

(2) H. Bauer, M. Beau, B. Friedl, C. Marchand, K. Miltner and H. J. Reyher, submitted to Phys. Lett.

(3) H. Etz, W. Gombert, P. Leiderer; Proceed of the LT 17 (North Holl; Amsterdam, 1984) Part 1, S. 295

(4) P.V.E. McClintrock, J. Low. Temp. Phys. 11 (1973) 15.

(5) A. P. Hickmann, W. Steets and N. F. Lane. Phys. Rev. B 12 (1975) 3705

(6) H. J. Reyher, H. Bauer, C. Huber, R. Mayer, A. Schäfer and A Winnacker. Phys. Lett. A. 115 (1986) 238

(7) W. I. Glaberson and W. W. Johnson. J. of Low. Temp. Phys. 20 (1975) 313.

(8) M. W. Cole and R. A. Bachmann. Phys. Rev. B 15 (1977) 1388.

(9) W. L. Wiese and G. A. Martin, in: CRC Handbook of Chemistry and Physics, 62nd ed. (CRC Press, Boca Raton, 1981) p. E-335; J. Reader and C. H. Corliss, ibid., p. E-205

Physica B 165&166 (1990) 139–140
North-Holland

A HEAT SWITCH AT VERY LOW TEMPERATURE AND HIGH MAGNETIC FIELD

S.A.J. WIEGERS, P.E. WOLF, L. PUECH

Centre de Recherches sur les Très Basses Températures, C.N.R.S., BP 166 X, 38042 Grenoble-Cédex, France

We describe a heat switch operating between 3 mK and 100 mK in magnetic fields up to 10 T. The lowest temperature achieved in the closed state is 5 mK. The heat leak from the sample at 100 mK to the cold source at 3 mK is less than 25 nW.

1. INTRODUCTION

Our present experiment on transiently spin-polarized liquid ^3He [1] requires the use of a heat switch with the following constraints : (i) Operation in a field up to 10 T, (ii) minimal temperature reached in the closed state below 5 mK, (iii) heat leak in the open state smaller than 100 nW for the two sides of the switch at respective temperatures 3 mK (cold source) and 100 mK (sample). Points (i) and (ii) are necessary to obtain large (~ 70 %) nuclear polarizations of solid ^3He. Point (iii) is needed to measure the heat released during spin relaxation of polarized liquid ^3He at 100 mK [2], obtained by fast melting of the former solid. This measurement will give access to the magnetic susceptibility of the highly polarized liquid.

Since we could not use a superconducting heat switch due to the high field and geometrical constraints, we developped a switch of principle similar to that first used by Reinstein and Gerber [3]. Here, thermal contact with the cold source is provided by liquid ^3He, which can flow in (out) the switch from (to) a lower reservoir. We found that, provided several essential modifications with respect to the original solution were made, such a switch could met the above requirements . We here describe design and performances of our switch. We believe that its principle can be adapted to many experiments requiring good thermal insulation below 100 mK in a high magnetic field.

2. SWITCH DESIGN

Figure 1 shows the final version of our switch. The switch itself consists of two silver pieces sintered with 700 Å silver powder and connected by a thin (0.2 mm) Kapton tube 28 mm in diameter and 2 cm long. The upper piece is connected to the mixing chamber. The lower one is thermally connected to the sample (in our experiment, the polarized ^3He cell). Closing the switch is obtained by filling it with liquid ^3He from the lower reservoir. The CuNi capillary connecting this reservoir to the switch is level with the top of the lower sinter. Thus, in the open state, the lower part of the switch remains filled with ^3He, which acts as a heat tank in our experiment. The gap between the two sinters is 2.5 mm. This is large enough compared to the

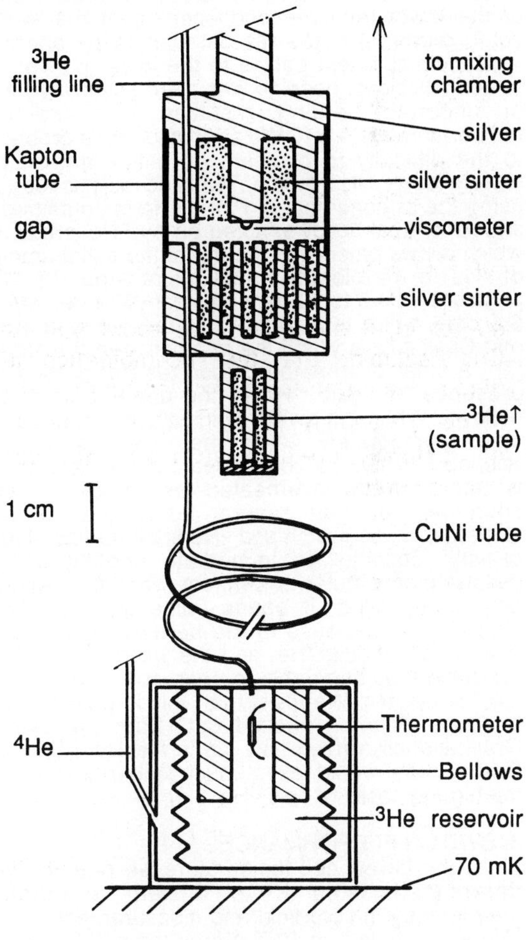

FIGURE 1
Switch design (simplified)

capillary length of liquid ^3He ($\simeq$ 0.6 mm [4]) to avoid any short circuit through the open switch, due to a drop of ^3He hanging from the top sinter.

Thermometers and heaters are glued at both ends of the switch. A ^{3}He viscometer (a CuNi 70 µm diameter wire bent into a small semi circle 1.7 mm in diameter) is located just below the upper sinter. It is used as a field-independant thermometer [5], when the switch is closed and as a level gauge in the course of switch operation (opening or closing). Its resonance freqency and quality factor in vacuum are respectively 15 Khz and 2500.

The lower reservoir, located 14 cm below, is connected to a 70 mK plate. In a first series of experiments, we used a reservoir only slightly thermally coupled to the plate so that the switch could be closed (as in ref.[3]) by heating the reservoir up to .65K (corresponding to a saturated vapour pressure of 0.9 Torr or 15 cm of liquid ^{3}He). However, opening appeared to be a problem under these conditions. The time elapsed between cooling of the lower reservoir and opening of the switch could reach 24 hours (as determined by applying repeatly small heat pulses to the lower part of the switch). Besides, opening was often associated with an important heating, resulting in a final cell temperature of 20 mK. We attributed these problems to the difficulty to nucleate an initial void in the initially totally filled switch. We checked this idea by using the arrangement of Fig.1. ^{3}He is contained in stainless steel bellows, acted on by ^{4}He pressure, which allows precise [6] control of the initial amount of ^{3}He in the closed switch (increasing the ^{4}He pressure from 0 to 10 bars squeezes the bellows by 5.5 cm^3, which is enough to fill the 1m long 1.5x2 mm CuNi capillary and the gap between the sinters [7]). A small overheated carbon resistor probes the presence of liquid ^{3}He in the upper part of the bellows. When the switch is initially full, it does not open immediately when the ^{4}He pressure is released. Instead, a void appears in the bellows, as is clear from the overheated resistor signal. This situation may last several hours until a void nucleates in the switch and ^{3}He starts to flow due to gravity, causing an immediate cooling of the bellows thermometer. In contrast, when the switch is only partly filled, it opens immediately. This is shown by a decrease of the bellows temperature, due to flow of cold ^{3}He, and the viscometer signal. We have thus been able to open our switch in less than 5 minutes. We insist that this is possible only because ^{3}He is "sucked out" from the switch. Gravitationally driven flow, as in the Reinstein and Gerber case, would take a much longer time, due to the high viscosity of cold ^{3}He.

3. SWITCH PERFORMANCES

i) the lowest cell temperature we reached was determined using ^{51}Co nuclear orientation thermometry in zero field and measurements of the ^{3}He viscosity in 1 to 10 T. We found a value 5± 0.5 mK,independant on the field. This was obtained for a mixing chamber temperature of 3 mK and silver sinter exchangers of large area (30 m^2 for the upper sinter and 100 m^2 for the mixing chamber).

ii) The heat leak from the lower reservoir to the switch (T<20 mK) was determined from the warming rate of the open switch [8], together with the measured specific heat of the switch. It is less than 5 nW up to 10T. This shows that Foucault heating through vibrations of our silver cell is negligible.

iii) We have determined the insulating ability of the open switch. The Kapton tube conductivity was measured in zero field by heating the lower part of the empty switch and recording its temperature T. We find $(T/37\ mK)^3 = (Q+2\ nW)$ for an applied power Q ranging from 1 to 1000 nW and an upper switch temperature below 10 mK. This T^3 dependance is expected for a plastic. The figure of 2 nW is consistent with the estimated parasitic heat input through the CuNi capillary. When the switch is open, but its lower part filled with liquid ^{3}He, which is the case in our experiment, there is parallel conduction by the ^{3}He gas. Due to the very rapid increase of the saturated vapour pressure with temperature, this process sets up quite sharply. At 200 mK, its contribution to the heat flux is 4 µw but is <<1 nW at 100 mK, the temperature of interest to us. This gives a total heat leak at 100 mK less than 25 nW, allowing precise energy production measurements

4.CONCLUSION

The switch here described allows to achieve good thermal insulation below 100 mK (heat leak < 25 nW) under strong magnetic field. The bellows arrangement appears essential in order to obtain a short opening time and an initial low temperature (5 mK) for the open switch. Due to its limited size, this switch may have many applications, e.g in nuclear demagnetization.

REFERENCES

[1] L.Puech, S.A.J. Wiegers and P.E. Wolf, to be published.

[2] L.Puech, G. Bonfait, B. Castaing, C. Paulsen, J.J Prejean and P.E. Wolf, in Spin Polarized Quantum Systems eds ISI-S. Stringari (World Scientific, Singapore, 1989) pp. 27-34

[3] L.E. Reinstein and J.A. Gerber, LT13 Proceedings (Plenum Press, 1973) pp 622-625

[4] With a surface tension of 0.156 ergs/cm^2, as measured by A.J. Ikushima, M.Iino and M.Suzuki, Can. J.Phys.65 (1987) 1505

[5] The change of the ^{3}He viscosity under 10 T (0.3%) is negligible compared to its variation with temperature : see G.A. Vermeulen et al, Phys. Rev. Lett.60 (1989) 2315

[6] This precise control is difficult in the thermally driven switch, which goes from empty to full for a lower reservoir temperature variation of less than 0.01K.

[7] The upper sinter remains filled in the open state, due to capillarity.

[8] In this temperature range, conduction towards the mixing chamber is negligible.

Physica B 165&166 (1990) 141–142
North-Holland

DIGITAL BRIDGE FOR ULTRALOW TEMPERATURE THERMOMETRY*

Inseob HAHN and C.M. GOULD

Department of Physics, University of Southern California, Los Angeles, CA 90089-0484 USA

A transparent replacement for the usual set of precision ratio transformers within ultralow temperature thermometry systems is described. This new device uses digital electronics to both lower costs and to provide local intelligence for greater flexibility. The device has been used in our experiments on superfluid ^{3}He without the difficulties usually attendant with interfacing computers to ultralow temperature apparatuses.

1. INTRODUCTION

The standard method of precision ultralow temperature thermometry involves a measurement of the magnetic susceptibility of a weakly paramagnetic salt by means of a mutual inductance bridge operated in a null configuration. The bridge's arms are traditionally driven by a set of precision ratio transformers, which are themselves driven from a common source. There are two problems associated with this technique. The first is that in practice, humans must be present to continually rebalance the bridge near the null condition, and the second is that the cost of this device is a significant consideration in limited laboratory budgets. The first problem has been overcome in certain laboratories by introducing computer controlled ratio transformers, but this aggravates the second problem of cost.

We have developed a new device, which we term a digital bridge, which solves both of these problems. We replace the ratio transformers in the traditional system with a purely digital implementation, which substitutes without requiring any other changes in laboratory instrumentation. One of our design criteria was that the device should also perform at a level superior to the traditional solution. The key to the success of our device is the ready availability of inexpensive fast high-precision D/A converters which are a byproduct of the current boom in the compact disc market.

2. STRATEGY

The conventional thermometry bridge employs three ratio transformers to generate three independent sine waves of rigorously controlled relative phase. Our strategy is to replace each of these independent sine wave sources with an D/A converter. The input of each converter is computed, in real time, to be the digital representation of a sine wave of a selectable amplitude. The amplitudes are controlled either manually, as in the traditional system, or automatically by having the digital bridge sense the off-balance condition and adjust itself accordingly. With the exception of the front panel buttons used in a manual control mode, this system has no moving parts.

3. IMPLEMENTATION

There are three important sections of the digital bridge: a single-board computer which provides the local intelligence with which to monitor the state of the thermometry bridge and issues commands to keep it near balance; a home-built dedicated multiplier board to compute the digital representation of the sine waves; and the analog section consisting of a set of three D/A converters and associated filters. A schematic diagram of the digital bridge is shown in Figure 1. These three sections will now be discussed in reverse order.

3.3. Analog Section

This section generates each output sine wave as a series of 1024 voltage steps by a high-speed 18-bit D/A converter (Analog Devices DAC 1146). We use only the most significant 16 bits since this is sufficient to satisfy the design criterion that we improve upon the conventional technique, and it greatly simplifies the digital implementation. This converter is specified to settle to better than 17-bit accuracy in 2 μs, which is adequate even at our highest generated frequency of 224 Hz. Inspection of the D/A output shows this settling time to be a conservative guarantee. We use three independent D/A converters to produce the three required sine waves, and reference them all to a common voltage source (analogously to feeding the usual ratio transformers from a common source).

We feed the D/A outputs to carefully matched third order Butterworth low-pass filters (f_c = 1 kHz) in order to block high-frequency digital noise which seriously degrades the SQUID detector's performance. The filter's output impedance is 100 kΩ. The output of the entire box goes through a connector identical to the BTI RBU, which facilitates easy interchanging to test the device.

* Supported by the National Science Foundation through grant DMR89-01701.

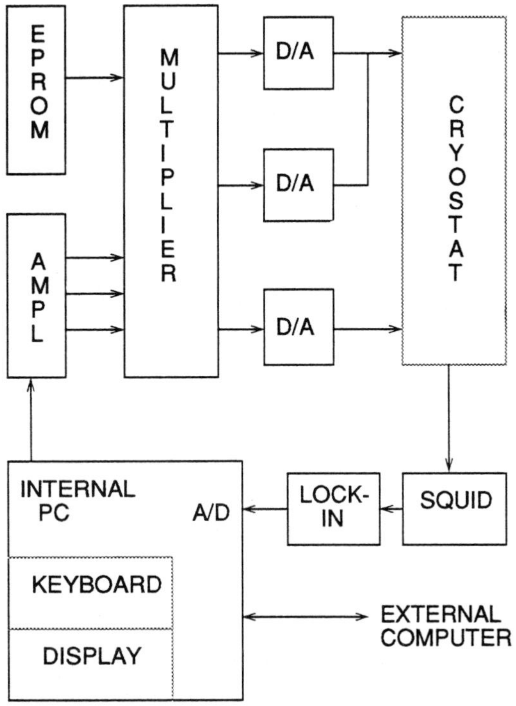

Figure 1 Block diagram of the digital bridge as used in our thermometer showing the flow of signals and control.

3.2. Multiplier Section

The numbers fed to the D/A converters are the 16 high order bits of the 32 bit product taken from a multiplier chip (Analog Devices model ADSP-1016A) whose two inputs are the 16 bit representations of (a) a full amplitude sine wave, and (b) a fractional amplitude at which we desire the sine wave to be produced. The full amplitude sine wave is taken from two EPROM chips, while the fractional amplitude is acquired from the internal computer. In order to generate the sine wave, a simple counter steps through the EPROM. This simple addressing scheme facilitates the subterfuge of introducing a π phase shift by inverting one bit. This extends the resolution of two of the sine waves effectively to 17 bits because their amplitudes may thus be made negative. The most significant bit of the address counter is also converted into a reference signal for use by our lock-in.

The three separate sine waves are generated by time-multiplexing the EPROM and multiplier chip, and then redirecting the output to the D/A converters. We offset the phase between adjacent sine waves stored in the EPROM by $2\pi/4096$ so that the phases of the outputs of the D/A converters are rigorously equal.

3.1. Internal Computer

Primarily for ease in developing the software, the internal computer is a single-board PC which fits in a standard PC backplane. (We use the BUS-PC from Faraday Electronics, but any PC would work as well. We used the cheapest PC available because speed of the processor is of minimal significance in this application.) The purpose of the computer is to load the appropriate fractional amplitude registers on the multiplier board when instructed by the user from the front panel or by the external computer, or when dictated by the bridge balance, and to communicate the state of the thermometer to the external computer and to the user through a simple display. The control software can be run from an external computer (when testing) or loaded into an on-board EPROM.

4. RESULTS

The digital bridge has been in continuous use in our laboratory for several months during our current study of superfluid ^{3}He. The impact of noise from the digital electronics is minimal owing to the extent to which we shield the analog outputs from the rest of the system. We use separate power supplies to feed the digital and analog sections, and heavily rf block each of the power supply lines. The filters are purely passive and reside in their own shielded box. The A/D converters reside in their own box, with the 51 signal lines (3×16 data lines plus 3 control lines) each optically isolated. Even the connections to the front panel keyboard and liquid crystal display are shielded to prevent interference.

The linearity of the device is excellent. Computer simulations of expected nonlinearities in the D/A converters suggested that when detecting the amplitude of a multi-step sine wave, both differential and integral nonlinearities would be less than the smallest step. When we originally tested the device we discovered nonlinearities on the percent level which was traced to the 50 Ω input impedance of the filters. After buffering with a transconductance amplifier (and concomitantly modifying the filters) any remaining nonlinearity is smaller than present in the best 5-digit ratio transformers in our laboratory.

In short, the digital bridge is a useful addition to our laboratory because of its superior performance, its easy communication of temperature information with other laboratory computers, its usefulness in self-balancing, and certainly its rather modest impact (after the initial investment in development) on our laboratory budgets.

Physica B 165&166 (1990) 143–144
North-Holland

HIGH RESOLUTION DC SIZE EFFECT MEASUREMENTS OF RESISTIVITY IN ALUMINIUM UP TO 30K USING RF SQUID AND HTSC WIRING.

José ROMERO*, Thomas FLEISCHER and Robert HUGUENIN

Institut de Physique Expérimentale, BSP, 1015 Lausanne, Switzerland

An application of high temperature superconductor (HTSC) is presented in an electrical circuit in which no high density current flows. The use of HTSC as part of the wiring circuit in a high resolution bridge with a direct current comparator and a RF SQUID allowed to make high resolution measurements of the electrical resistivity of thin samples up to 30 K. The results exhibit a clear difference of the T-dependence of the resistivity for polished and rough samples

1. INTRODUCTION

The influence of surface scattering on the electrical resistivity of thin metallic samples is a long standing problem but a number of questions have not yet been clearly answered, like the deviations from Matthiessen's rule(1) due to the surface or the determination of the so-called specularity parameter p which describes the reflection of the electrons hitting the surface. In the present report we will propose a way to extend the temperature range of high resolution DC size effect measurements towards high temperatures (30K) and apply it to test the usual assumption of diffuse scattering (p=0) at the surface of thin Al samples (resistance ranging from 0.01 to 50 μΩ). In such measurements RF SQUIDs are used as null detectors together with room temperature current comparators (DCCC). The resolution of modern bridges is of the order of 10^{-6} or even 10^{-8} but the temperature range for which this resolution is available is limited by the use of traditionnal superconducting wires in the input circuit to the SQUID, that is $T_C = 9K$.

2. RESISTANCE BRIDGE AND THE SQUID

Figure 1 shows schematically the part of our circuit which includes the sample. The DCCC Guildline Model 9575 delivers the two currents I_X and I_S, respectively in the unknown R_X and comparison R_S resistances, maintaining a stability of 10^{-7} in the ratio I_X/I_S. R_X is obtained from the balance condition of the bridge:$R_S I_S = R_X I_X$.This equipement allows to measure electrical resistivities between 1 and 9 K with a resolution of 10^{-6}. The null detector in the bridge is a RF SQUID 330X from SHE. The wires between AB and DC are superconducting Niomax S (T_C= 9 K). When using the bridge at temperatures higher than 9K, these conventionnal superconducting wires become resistant and all resolution is lost.
Using typical numerical values appropriate to the present system(2) (R_X=10 nΩ, R_S= 1 mΩ, I_X= 100 mA,

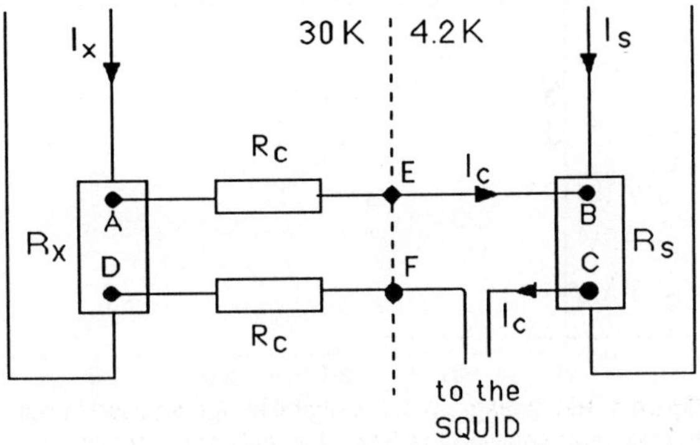

Figure 1. Part of the circuit with the sample and the comparison resistance. R_C represents the resistance of the contacts. In the improved circuit it also includes the resistance of the Ag wires (see text).

I_S=1mA, I_C=0.25nA), the maximum resistance ($2R_C$) (see fig. 1) allowed is 3μΩ for a resolution of 10^{-6}, and 300 μΩ for a resolution of 10^{-4}. Attempts to control the electrical resistance and thermal gradient when R_X was at 30 K with different wirings of normal metals revealed unsuccessful and we decided to use HTSC between AB and DC.

3. HTSC CONTACTS

The HTSC used in the experiment is a parallelipiped 20x5x3 mm of YBaCuO. Between the HTSC and the sample the contact was made with a pure silver wire 1 cm in length along AE and DF in Figure 1. This wire was fixed to the HTSC with silver paint before heat treatment of the ceramics and then spot-welded to the sample. The total resistance was 65 μΩ at 4.2K, due essentially to the silver wire. With this arrangement and R_X=10 nΩ, a resolution of 5×10^{-6} was obtained at 4.2 K, which drops to 5×10^{-4} at 30 K, due to the increase of resistivity of the silver wires and the reduced stability of the sample temperature at 30 K.

* Present address : World Radiation Center / PMOD CH-7260 Davos-Dorf, Switzerland

0921-4526/90/$03.50

4. MEASUREMENTS

To illustrate the use of the improved measuring system at high temperature a plot is shown in Figure 2 of the difference as a function of T in the resistivity of the same sample, having approximately the same thickness but whose surfaces have been treated differently. One of the surface has been electro-chemically polished and the other one etched in $FeCl_3$. It is seen that the data are significant at the level of 5×10^{-4} over the temperature range 20-30 K. One should remark that the maximum temperature of 30 K is not a limiting value due to the circuitry : it is here the maximum temperature at which measurements were taken because of the long time needed to reach a good thermal stability in the sample.

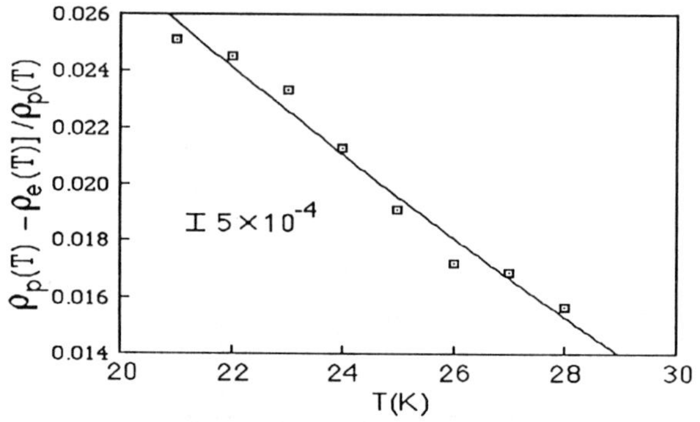

Figure 2. Difference in the resistivity of two samples with polished (p) and rough (e) surface divided by the resistivity of the polished sample : $(\rho_p(T)-\rho_e(T))/\rho_p(T)$ as a function of temperature; thickness d≈1 mm. The line is to guide the eye.

Figure 3 shows the drastic changes in the temperature dependence of the surface resistivity due to a change in the surface condition. It appears that the surface resistivity of etched samples increases with temperature, the variation being more pronounced in thinner samples. On the contrary the surface resistivity of samples with polished surfaces remains approximately constant in thick samples (curve □ in fig.3) while it decreases with increasing temperature when the thickness becomes smaller. One should remark that in all our samples the ratio d/l of thickness to mean free path is always larger than 1. The observed behaviour is not completely explained by the standard size effect theories(1) and also contrasts with the observation made in cadmium where the surface resistivity shows a drastic decrease in the temperature dependence when a polished sample is etched (3). This was taken as a proof that the specularity parameter should depend on the angle of incidence of the electrons hitting the surface. The present investigation shows that the behaviour observed in cadmium is not a general rule. This may be connected with the fact that the usual parameter

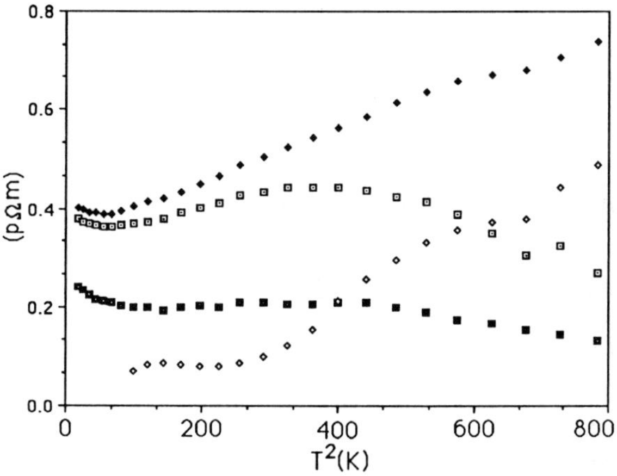

FIGURE 3: Surface resistivity $\rho - \rho_\infty$ vs T^2 for samples with polished (▣,d=0.6mm ;□ ,d=0.796mm) and rough (♦,d=0.725 ;◇,d=1.008)

d/l of the current theories is not sufficient to describe the details of the experimental observations. This is also suggested by the difference in the enhancement of the resistivity due to the surface by either sample size or temperature (4).

5. CONCLUSIONS

By inserting pieces of HTSC in the circuit of a high resolution resistance bridge, using a current comparator and a RF SQUID as a null detector, it is possible to work from very low temperatures up to 30 K (not limiting) . These materials allow electrical contact where low density currents flow and work as a thermal insulator, in this case with a gradient 20 K cm-1. The resolution of the bridge drops from 5×10^{-6} at 4.2 K to 5×10^{-4} at 30 K. A method is indicated for making the contacts between the sample and the HTSC. This device has been used to investigate DC size effect in the electrical resistivity of Al samples having received various surface treatments. In contrast to the usual assumption of diffuse scattering at the surface of Al because of the presence of an oxyde layer, different results are obtained for the temperature dependence of the surface resistivity for rough and polished samples.

REFERENCES

(1) J.R. Sambles and K.C Elsom J. Phys. F : Met. Phys. 15 (1985) 161.
(2) J. Romero, T. Fleischer and R. Huguenin Cryogenics in print
(3) J. Romero and R. Huguenin Helv. Phys. Acta 62 (1989) 758
(4) J. van der Maas and R.Huguenin Jap. J. of Appl. Phys. 26 (1987) 653

Physica B 165&166 (1990) 145–146
North-Holland

A VERY HIGH STABILITY SAPPHIRE LOADED SUPERCONDUCTING CAVITY OSCILLATOR

A.J. GILES, A.G. MANN, S.K. JONES, D.G. BLAIR, AND M.J. BUCKINGHAM,

Department of Physics, The University of Western Australia, Nedlands, Western Australia, 6009.

We have implemented a sapphire loaded superconducting cavity resonator in a novel phase stabilised loop oscillator at 10 Ghz and achieved a fractional frequency stability of 10^{-14} for 3 to 300 seconds integration time and a resonator Q of 3×10^8. We present here an overview of the oscillator.

1. INTRODUCTION

The ultra-high Q and low temperature coefficient obtainable in superconducting cavities at low temperatures has enabled the construction of microwave oscillators of exceptional short term frequency stability. Using a niobium cavity with a loaded Q of 10^{10} at 1.38K, Stein and Turneaure (1) achieved a short term square-root Allan Variance (SRAV) of $\sigma_y(2,\tau) = 6\times10^{-16}$ in their superconducting cavity stabilised oscillator (SCSO). Stability limitations set by mechanical deformation of the cavity due to vibration and tilt (2) and the relative difficulty in obtaining ultra-high Q cavities have restricted the practical development of this oscillator. Braginsky et al (3) has shown that high purity sapphire is an excellent dielectric for high stability resonator applications. We have developed a sapphire loaded superconducting cavity (SLOSC) resonator exploiting the high energy confinement factor of sapphire (~ 95%) and turning points in the frequency versus temperature (and power) curves to produce a very high stability oscillator.

2. SLOSC Resonator

Design considerations and stability estimates for the SLOSC resonator (Figure 1) have been previously communicated (4),(5). Most of the research has concentrated on the properties of a mode at 9.73 GHz which displays a Q in excess of 10^9 at 4K (5) that remains relatively high until near the superconductor's transition temperature (see Figure 2a). The temperature dependence of the 9.73GHz mode frequency (Figure 2b) reveals a turning point at about 6K. This has been explained (6) by the competing effects of the temperature-dependent superconducting penetration depth and the Curie law of paramagnetic impurities (Cr^{3+}) in the sapphire. The turning point considerably relaxes temperature control requirements: only 0.1 mK regulation is needed for fractional frequency stability ~ 3×10^{-16}, at an offset of 1mK from the peak.

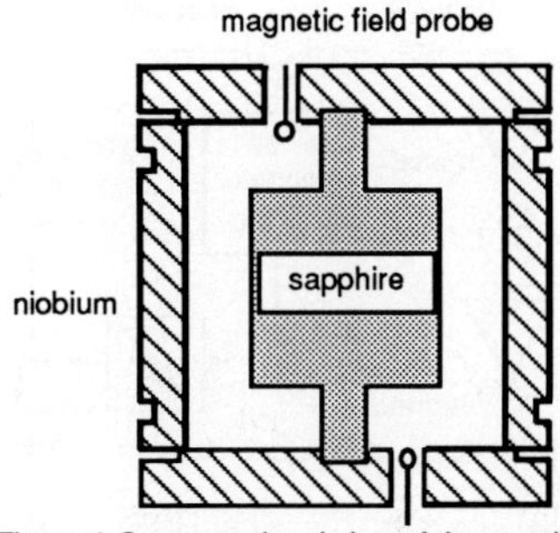

Figure 1 Cross-sectional view of the sapphire loaded superconducting cavity resonator (SLOSC).

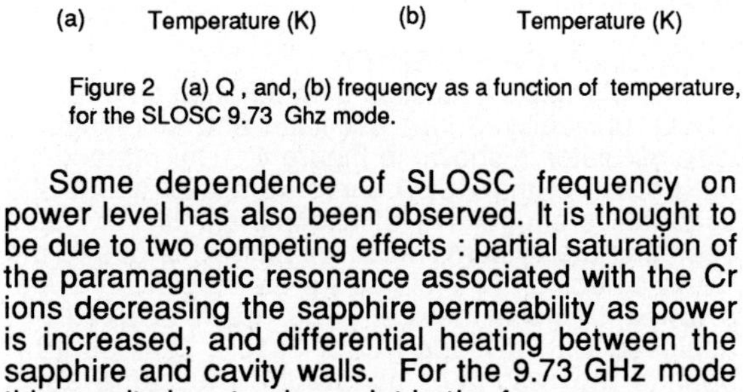

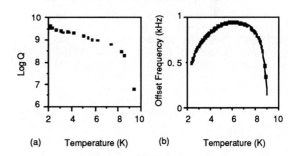

Figure 2 (a) Q , and, (b) frequency as a function of temperature, for the SLOSC 9.73 Ghz mode.

Some dependence of SLOSC frequency on power level has also been observed. It is thought to be due to two competing effects : partial saturation of the paramagnetic resonance associated with the Cr ions decreasing the sapphire permeability as power is increased, and differential heating between the sapphire and cavity walls. For the 9.73 GHz mode this results in a turning point in the frequency versus power curve at about 2 mW. Further details will be published elsewhere.

0921-4526/90/$03.50

3. PHASE STABILISED LOOP OSCILLATOR

The initial configuration was an all cryogenic loop oscillator using GaAsFET amplifiers at 6K. However the best stability achieved was only $\sigma_y(2,\tau) = 6\times10^{-14}$ at 10 seconds integration time and probably limited by amplifier flicker noise since the corresponding flicker phase noise, $S_\phi(f_m) = -87\text{dBc}/f_m$ rad/Hz, is similar to that measured in an 8.5 GHz GaAsFET amplifier (7).

The desire for greater stability led to the concept of a phase stabilised loop oscillator (PSLO) circuit, shown schematically in Figure 3. The resonator is both the frequency determining element of the loop oscillator and the dispersive element in a Pound type discriminator. The circuit is similar to that of Galani et al (8), which actively degenerates the phase noise of a loop oscillator, but has intrinsically better dc performance owing to the use of RF modulation.

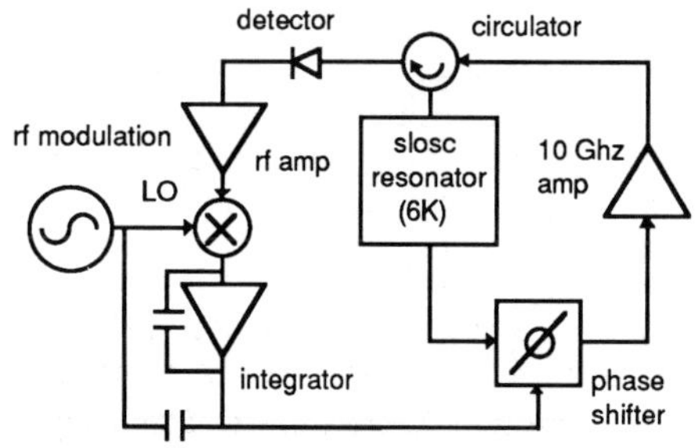

Figure 3 Block diagram of the phase stabilised loop oscillator circuit.

All of the loop oscillator components are at room temperature except for the circulator and other isolators that are placed near the SLOSC to minimise frequency pulling effects resulting from changes in cable VSWR. Stainless steel coaxial cables carry the signals to and from the SLOSC. The microwave power dissipated in the SLOSC is typically 2mW. The transfer function of the stabilization loop is essentially that of a single integration and has a unity gain crossover point at about 3 kHz.

4. OSCILLATOR STABILITY

The SRAV as a function of integration time for the PSLO, unstabilised loop oscillator and all cryogenic loop oscillator is shown in Figure 4. The remarkably flat portion of the PSLO curve from 3 to about 300 seconds is suggestive of a dominant flicker noise source in the system, which we believe is due to the GaAs detector. To test this we are replacing it with a lower 1/f noise device. The result of 9×10^{-15}, achieved with a loaded Q of 3×10^8, represents 3 ppm of the resonator bandwidth, the same as in Stein and Turneaure's SCSO.

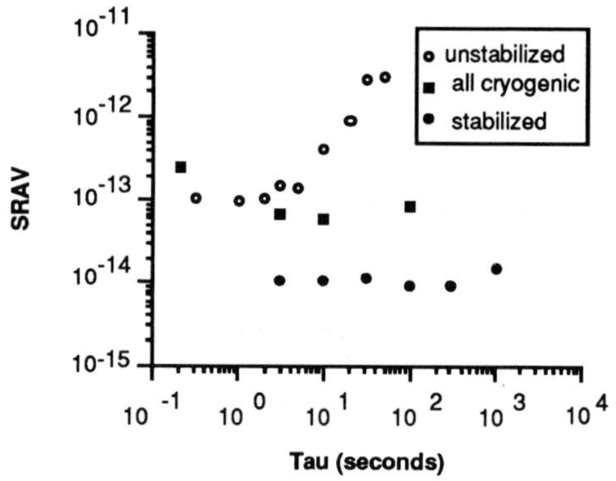

Figure 4 Measured square root Allan variances for various oscillators.

For integration times longer than about 300 seconds, the performance begins to degrade. We have identified a strong correlation with the liquid helium level (residual frequency pulling) and diurnal variations of room temperature. There is preliminary evidence of a large acceleration sensitivity, of the order of 10^{-9} per g , but further tests are required to determine its cause.

5. CONCLUSIONS

We have demonstrated the very high stability of a PSLO based on a SLOSC and have identified sources of frequency instability which need further investigation.

ACKNOWLEDGEMENTS

We are indebted to Dr Alan Young of the CSIRO Division of Radiophysics for the design of the frequency control electronics. This research was funded by the Australian Research Council.

REFERENCES

(1) S.R.Stein and J.P.Turneaure, Proc. IEEE, **63** (1975), pp. 1249-50.

(2) B.Komiyama, IEEE Trans., **IM-36,** No.1, (1987), pp.2-8.

(3) V.B.Braginsky, V.P. Mitrafanov, and V.I. Panov, "Systems with Small Dissipation", (University of Chicago, Chicago, 1985)

(4) D.G.Blair and S.K.Jones, IEEE Trans. Mag., **MAG-21** (1985) , pp.142-145,

(5) S.K.Jones and D.G.Blair, Electron. Lett., **23** (1987), pp. 817-818.

(6) S.K.Jones, D.G.Blair and M.J.Buckingham, Electron Lett., **24** (1987), pp. 346-347.

(7) C.P.Lusher and W.N.Hardy,IEEE Trans. Microwave Theory Tech., **MTT-37** (1989), pp. 643-646.

(8) Z.Galani, M.J.Bianchini, R.C.Waterman Jr., R.W.Laton and J.B.Cole, IEEE Trans. Microwave Theory Tech., **MTT-32**, (1984), pp. 1556-1565.

Physica B 165&166 (1990) 147–148
North-Holland

MAGNETIC SUSCEPTIBILITY OF INSTRUMENT MATERIALS BELOW 10K

J.M. Lockhart[§#], R.L. Fagaly[+], L.W. Lombardo[*], and B. Muhlfelder[#]

[§]Physics & Astronomy Dept., San Francisco State Univ., 1600 Holloway Ave., San Francisco, CA 94132 U.S.A.
[#]GP-B Program, Hansen Laboratories of Physics, Stanford University, Stanford, CA 94305 U.S.A.
[+]Fusion Division, General Atomics, P.O. Box 85608, San Diego, CA 92138-5608 U.S.A.
[*]Department of Applied Physics, Stanford University, Stanford, CA 94305-4090 U.S.A.

Measurements were made of the magnetic susceptibility in the temperature range 2K - 10K of twenty-nine samples of materials and components used in the construction of cryogenic instruments. The use of SQUID-based superconducting susceptometers allowed high resolution. The data was fitted to a Curie law form and the values of the coefficients were determined. The remanent magnetization at 2K of several of the samples was also measured.

1. INTRODUCTION

In the course of the construction of cryogenic apparatus for sensitive magnetic measurements we have found it necessary to determine the magnetic properties of several materials and components in the temperature range 2K - 10K. In general, the results given here complement the cryogenic susceptibility measurements reported by Salinger and Wheatley (1), Commander and Finn (2), and Simmonds (3), the low-temperature remanent magnetization measurements described by Cabrera (4), and the room temperature susceptibility measurements of Keyser and Jefferts (5). The materials which we tested are normally considered to be non-magnetic. However, many of the samples show susceptibility coefficients and remanent magnetization values which would make them unsuitable for use in any significant quantity in applications such as high resolution NMR, CMN thermometry, and precision SQUID measurement systems.

2. APPARATUS AND METHOD

The measurements were performed using either a S.H.E. Model VTS-905 SQUID susceptometer (6) or a Quantum Design Model MPMS SQUID susceptometer (7). In each case, samples were given a light acid etch (using HCL or HNO_3) to remove surface contamination. Samples were subjected to several cycles of vacuum pumping and flushing with He gas prior to measurement to minimize oxygen contamination. The sample sizes, shapes, and susceptibilities allowed demagnetizing effects to be ne-glected. The S.H.E. susceptometer and techniques for its use are described in the excellent paper by Day et al. (8)

The magnetic moments of the samples were measured at several temperatures in the range 2K - 10K using applied fields of either 100 G, 1.0 kG, or 10.0 kG whose values were known to 0.5%. The susceptometers are capable of temperature resolutions better than 0.05K, but shorter thermal equilibration times used here gave a temperature uncertainty of 0.1K. For the remanent magnetization measurements, the samples were degaussed with an ac degaussing coil having a peak field of 400 G. The samples were then exposed to a 100 G dc magnetizing field which was removed prior to the moment measurements. Each susceptometer was calibrated using samples of pure lead and/or pure niobium.

3. RESULTS AND DISCUSSION

Let χ be the magnetic susceptibility defined by $\chi = M/H$ where M is the magnetic moment per unit volume and H is the magnetic intensity, V be the volume of the sample, and ρ be its density. We generally work with values of χ/ρ, which are obtained by dividing χV by the mass of the sample. (In a few cases the value of χV per piece was used.) We fit the χ/ρ data at each temperature to the Curie law form

$$\chi/\rho = B + (C/T)$$

and extract the parameters B and C and their errors. For a few samples susceptibility data was taken at only two temperatures; in these cases the error estimates for B and C were determined from the scatter in the χ/ρ data rather than from the fit. We also give the remanent moment M *per gram* at a temperature of 2K for some samples.

The principal intent in this work was to qualify materials for use in apparatus construction and no special efforts were made to obtain minimal uncertainties. Results are *representative.*Repeated measurements were made on each sample, but multiple samples of a given material were not checked. The data in Table 1 on materials measured by others is generally consistent with the earlier work. For example, our data on quartz is consistent with the results of Simmonds and consistent with the coefficient C reported by Salinger and Wheatley, but inconsistent with the B coefficient of the latter paper. Our values for GE 7031 varnish are consistent with Simmonds' results, while our results for G-10 fiberglass/ epoxy are somewhat smaller than his.

Table 1. Magnetic susceptibility and remanence in the range 2K - 10K of materials useful in cryogenic apparatus construction.

Material	Supplier	C 10^{-6} cm^3 K/g	B 10^{-6} cm^3/g	M(2K) 10^{-6} emu/g	Notes
Dielectric Structural Materials					
Fused Quartz, Supersil	Amersil-Heraus	0.046±0.02	-0.383±0.004		a
MACOR (machinable glass ceramic)	Corning Glass	24.5±0.6	2.87±0.18		
Stycast 2850	Emerson & Cuming	17.1±0.9	2.97±0.15		b
Stycast 1266	Emerson & Cuming	-0.63±0.03	0.176±0.009		
Varnish, Type 7031, Cured	General Electric	2.17±0.11	-0.54±0.03		
Fiberglass/Epoxy, Type G-10	Synthane-Taylor Corp.	52.±2.6	1.1±0.06		c
Epoxy, Epibond 1210A/9615-10	Furane Plastics	80.0±4	7.00±0.35	50.±2.5	d
Epoxy, Silver-Filled, Eccobond 83C	Emerson & Cuming	0.65±0.03	0.38±0.015	13.±0.65	
Epoxy, Armstrong A-12	Armstrong	3.75±0.2	8.13±0.4	5300.±265	d, e
Metals					
Beryllium-Copper, Type 25 (Cu-2Be)	Unknown	1.25±0.06	9.38±0.5	6.5±0.3	d
Beryllium-Copper Foil, .0005"	Unknown	3.54±0.18	0.234±0.012		
Coil Foil	See note	0.343±0.017	-0.026±0.001		i
Ti/6Al/4V	Robin Materials	N/A	N/A	12,000.±1200	f
Phosphor-Bronze Wire, 1/2 hard	MWS Wire Industries	0.225±0.027	1.89±0.09	9.7±0.5	
Phosphor-Bronze Wire, .005" Formvar	Lakeshore Cryotronics	0.22±0.013	-0.035±0.012	2.9±0.3	
Phosphor-Bronze C Tube, .060" dia.	Unknown	0.28±0.03	0.09±0.008	3.2±0.2	
Manganin Wire, Low Ni	Sigmund Cohn	N/A	(318.)	92.±5	j
Platinum-8 Tungsten Wire, Alloy 479	Sigmund Cohn	0.43±0.018	0.24±0.01	10.7±0.5	
Platinum- 8 Tungsten Wire	California Fine Wire	1.0±0.05	0.9±0.05	540.±27	d
Tapes and Strings					
Mylar Tape, .002"	Unknown	0.341±0.017	-0.431±0.022		
Teflon Tape (PTFE)	Norton Chemplasr	0.063±0.003	-0.33±0.02		
Cotton Thread, #30	Coates & Clark	0.00071±0.000	-0.0016±0.000		g
Dental Floss, Unwaxed	Johnson & Johnson	0.037±0.002	0.0088±0.000		g
Dental Floss, Waxed	Johnson & Johnson	-0.0294±0.0015	0.0449±0.002		g
Special					
Ge Thermometer, GR200B-1000	Lakeshore Cryotronics	0.164±0.007	0.949±0.050	8.0±0.4	h
Carbon Resistor, 470 ohm, 1/2 W	Spear Carbon	7.81±0.67	18.4±0.9	620.±31	h
Metal Film Resistor, CGW 1000, 1%	Unknown	N/A	(1301.)		h, j
Titanium Screw, 2-56, Grade 1 (pure)	Olander Corp.	0.051±0.003	0.442±0.02	1.0±0.2	h
Aluminum Screw, 0-80	Olander Corp.	N/A	(0.024)	1.6.±0.3	h, j

[a]An HF etch of up to 3 days may be needed for complete removal of surface contamination (E.P. Day and J. Peterson, personal communication).

[b]As mentioned by S&W, the thermal expansion coefficient of 2850GT closely matches that of copper.

[c]The exact composition of G-10 varies considerably. Data presented is the average of three samples.

[d]Susceptibility data were taken at only two different temperatures for this sample.

[e]Armstrong A-12 epoxy is frequently used in the manufacture of liquid helium dewars.

[f]This sample showed partial superconducting behavior over part of the temperature range used, likely due to pockets of V.

[g]Susceptibility data per unit length; units of B, C and M are 10^{-6} emu K/cm, 10^{-6} emu/cm, and 10^{-6} emu/cm respectively.

[h]Susceptibility data is per piece; units of B, C, and M are 10^{-6} cm^3 K, 10^{-6} cm^3, and 10^{-6} emu respectively.

[i]For details, see A.C. Anderson, G.W. Salinger, and J.C. Wheatley, Rev. Sci. Instrum. 32 (1961) 1110.

[j]Susceptibility does not fit the Curie form. Value given in parentheses is the mass susceptibility at 2K.

REFERENCES

(1) G.L. Salinger and J.C. Wheatley, Rev. Sci. Instrum. 32 (1961) 872.
(2) R.J. Commander and C.B.P. Finn, J. Phys. E 3 (1970) 78.
(3) M.B. Simmonds, Proceedings of SQUID Magnetometry Workshop, eds. D.U. Gubser and S.A. Wolf (U.S. Naval Research Laboratory, 1984) p. 89.
(4) P.T. Keyser and S.R. Jefferts, Rev. Sci. Instrum. 60 (1989) 2711.
(5) B. Cabrera, Ph.D. Thesis, Stanford University, 1975.
(6) Manufactured by Biomagnetic Technologies, Inc., 9727 Pacific View Dr., San Diego, CA 92121
(7) Manufactured by Quantum Design, Inc., 11578 Sorrento Valley Rd., Suite 30, San Diego, CA 92121
(8) E.P. Day, et al., Biophys. J. 52 (1987) 837.

Physica B 165&166 (1990) 149–150
North-Holland

LOW FREQUENCY AC SHIELDING EFFECTIVENESS OF A BULK NORMAL METAL / HIGH T_C SUPERCONDUCTING COMPOSITE.

J. J. CALABRESE and J. C. GARLAND

Department of Physics, Ohio State University, Columbus, Ohio 43210 U. S. A.*

We have investigated the electromagnetic shielding effectiveness of a random bulk Ag/$YBa_2Cu_3O_{7-\delta}$ composite material containing 40% Ag. A cylindrical enclosure was fabricated out of the composite material, and the shielding effectiveness of the material was determined by measuring the frequency dependence of the r.f. coupling between two coils separated by the enclosure walls. At room temperature, the enclosure was essentially transparent to r.f. at frequencies below 1kHz, while at higher frequencies, eddy currents in the connected Ag matrix of the composite attenuated the coupled r.f. in accord with the classical $f^{1/2}$ skin depth expression. At 77K, however, the superconducting phase introduced a minimum of 70 dB additional attenuation below 1kHz, with a smooth crossover into the high-frequency skin depth regime.

1. INTRODUCTION

Electromagnetic shielding is a potentially useful application of high temperature superconductors that does not rely on the material having high critical current density. Although it is well known that pure sintered oxide superconductors, e.g., $YBa_2Cu_3O_{7-\delta}$, have undesirable mechanical properties, we have developed a normal metal-superconducting bulk composite that retains the desirable electrical and magnetic properties of high T_c superconductors, but whose mechanical properties are modified to reduce the poor strength, brittleness, lack of machinability, and poor environmental stability of the pure oxide. The electromagnetic shielding effectiveness of this material has been studied by fabricating a closed cylindrical enclosure of the composite and by measuring the frequency dependence of the r.f. coupling between two coils separated by the enclosure walls. The attenuation of an a.c. signal transmitted by a coil in the enclosure was measured as a function of frequency from 10 Hz to 200 kHz. Our results indicate that this composite is an extremely effective shielding material over a wide frequency bandwidth.

2. SAMPLE PREPARATION

The composite was made by mixing 60% by volume of fully reacted 10 micron average diameter $YBa_2Cu_3O_{7-\delta}$ powder (1) and 40% 5-8 micron silver powder (2) in a jar mill for four hours. A paraffin plug was placed in the die and the mixed powder was poured into the die around the plug to form the rough shape of the enclosure. The paraffin plug enabled the can to be compressed under quasi-isostatic conditions (3). The mixed powder was then compressed in a steel die at a pressure of 25,000 psi for 10 minutes. After removal from the die, the paraffin was melted and removed from the compaction. The lid of the can was fabricated separately as a disk. Both parts of the enclosure were then sintered at 920°C in flowing O_2 for 12 hours, and then annealed at 500°C in flowing O_2 for an additional 12 hours. The furnace was then turned off and the composite pieces allowed to cool to room temperature over several hours.

Figure 1 Photograph of Ag/$YBa_2Cu_3O_{7-\delta}$ composite enclosure.

After cooling, the enclosure and the lid were machined on a lathe into final form, with a pressure fit between the two pieces. A 3-turn copper transmission coil was placed in the enclosure, with connection to the outside environment through a minature coaxial connector. A 20 turn copper detector coil was mounted on a nylon fixture that was attached to the outside bottom of the enclosure. The enclosure could be removed from the fixture in order to measure the direct (unscreened) mutual inductance between the two coils. The inductive reactance of each coil was negligible at the frequency range of interest. Figure 1 is a photograph of the completed enclosure, whose diameter

* supported in part by Superconductive Components, Inc.

was 25 mm and length 15.9 mm. The enclosure's internal cavity was 13.4 mm deep, with wall thickness of 2.6 mm.

3. EXPERIMENTAL DETAILS

An AC current was applied to the transmitter coil (inside the enclosure) with an HP 3335A frequency synthesizer. The voltage across this coil was measured by a PAR 126 lock-in amplifier, while the voltage induced in the detector coil was measured by a PAR 124 lock-in. The voltage amplitude and phases across each coil were recorded as a function of frequency from 10 Hz to 200kHz. To calibrate the coil configuration, the "bare" mutual inductance (with no enclosure in place) was measured both at room temperature and 77K.

4. RESULTS

The ratio of the voltages measured across the transmitter coil and detector coil is plotted as a function of frequency

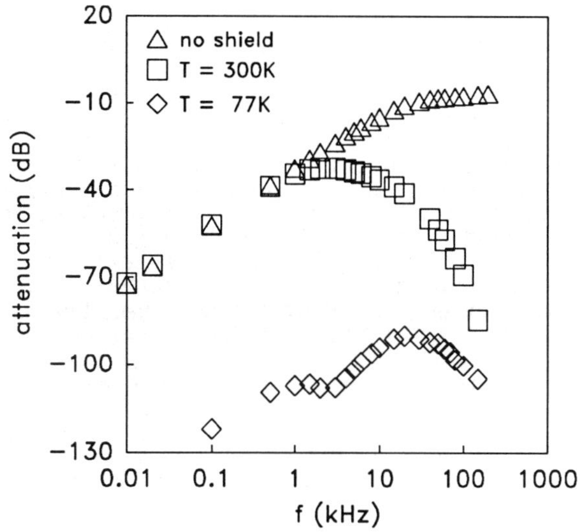

Figure 2 Ratio of detector voltage to transmitter voltage as a function of frequency at 300k and 77K. The triangles are for bare (unscreened) coupling.

in Figure 2. With no shield in place, the voltage ratio reflects the linear frequency dependence and temperature independence of direct magnetic coupling. With the enclosure present, the room temperature data shows additional attenuation at frequencies above 1 kHz that results from eddy currents induced in the normal metal randomly connected matrix in the composite. This additional attenuation has roughly the $f^{1/2}$ dependence expected from the classical skin depth (4). Figure 3 shows the additional attenuation introduced by the shield; in this figure, the direct (unscreened) coupling has been subtracted, demonstrating more clearly the $f^{1/2}$ dependence of the room temperature data. A fit to the room temperature data of the skin depth expression yields a normal-state average resistivity of 5 μΩ-cm, about a factor of two lower than the value of 12 μΩ-cm expected for this composition. The 77K data shows an additional attenuation of at least 70 dB below 1kHz resulting from the presence of the superconducting phase in the composite. Because of leakage around the enclosure and its lid and through the normal metal coaxial cable connector, attenuation of greater than 70db could not be measured. The small structure in the 77K data in Figure 3 is not understood but is most likely a spurious resonance.

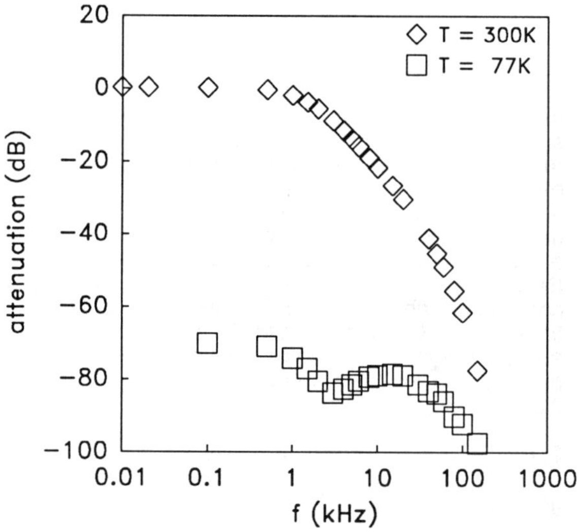

Figure 3 Attenuation of superconducting-normal metal composite as a function of frequency at 300K and 77K.

In conclusion, we find that the bulk metal particle - superconducting composite displays excellent electromagnetic shielding effectiveness over a broad frequency range extending down to 0 Hz.

REFERENCES

(1) supplied by Superconductive Components Inc.
(2) Alfa Products, Inc.
(3) R. L. Sands and C. R. Shakespeare, Powder Metallurgy (CRC Press, Cleveland,Ohio, 1966)
(4) J. D. Jackson, Classical Electrodynamics (John Wiley & Sons, New York, 1975)

Physica B 165&166 (1990) 151–152
North-Holland

LOW FIELD FLUX CREEP IN NIOBIUM SUPERCONDUCTING WIRES

A.G. MANN, N.J. McDONALD, and F.J. van KANN

Physics Department, The University of Western Australia, Nedlands, Western Australia 6009.

Using a SQUID magnetometer we have measured flux creep in superconducting persistent current loops made of niobium and niobium-titanium wire. The observed creep is 4 orders of magnitude less than that reported in niobium alloys above H_{c1}, but 2 to 4 orders of magnitude larger than reported for trapped flux in niobium tubes well below H_{c1}. A novel ' soak ' technique, which eliminates the need for a persistent current and precise temperature regulation, is described.

1. INTRODUCTION

Flux creep in persistent current circuits is a potential limitation to the sensitivity of superconducting gravity gradiometers (1), which require current stability to better than 1 part in 10^9 over hours. Flux creep is readily observed in hard niobium alloys above H_{c1} (1),(2),(3),(4) where the trapped field or current decays logarithmically with a decay constant , - d lnB /d lnt , of order 10^{-4}. Below H_{c1}, in single core NbTi wire (5) the decay constant is about 2×10^{-7}, while in pure niobium tube it may be smaller than 10^{-10} for small fields and of order 10^{-8} for fields above 0.2 H_{c1} (6). To our knowledge flux creep has not been reported in niobium wires up to now.

2. EXPERIMENTAL APPARATUS

Flux creep was initially investigated in the gradiometer using the (modified) circuit of Figure 1.

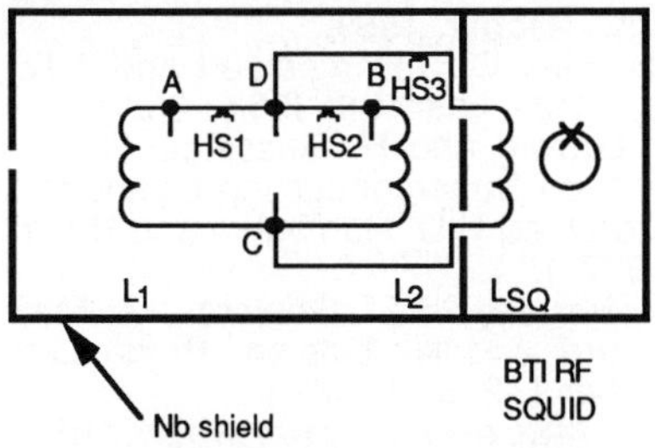

Figure 1 Circuit for measuring flux creep.

This employs at least one spiral-wound "pancake" coil, L_1, positioned rigidly 0.5 mm from a niobium ground plane . L_2 is either a spiral coil similar to L_1 or a toroidal ballast coil. The wire used was 0.13 mm niobium wire (annealed) from California Fine Wire Co. The entire circuit is enclosed by a niobium shield. Omitted for clarity are a 0.01 ohm shunt across the RF SQUID input coil (L_{SQ}) and low-pass filtering on the leads to prevent RFI unlocking the SQUID. L_1 and L_2 may be measured in situ by applying a very low frequency ac current (~ μA) to the appropriate pair of power leads in conjunction with operation of a heat switch.

Due to the temperature dependence of the penetration depth in wire and groundplane surfaces (of order 100 ppm/K) any flux creep after setting up a persistent current (in the loop formed by L_1 and L_2) will be hard to see unless temperature regulation is very precise. The gradiometer temperature however is normally regulated to a precision of a few μK, which is just adequate for observe flux creep out to 10^5 seconds. If, after sufficient time, the current is destroyed, the flux which has "soaked" into the superconducting surfaces will reemerge, giving rise to an observed creep having the same time dependence (and magnitude) as initially observed but with the opposite sign. This latter creep can be measured with far greater precision than that with the persistent current present, even with poor thermal control, since the residual trapped current can be very small, limited only by thermal emfs generated during operation of the heat switches. When a large toroidal coil was soaked the decay constant was essentially the same as that observed in a spiral coil, implying that most of the creep was due to the wire.

It is not even necessary to wait to decay the persistent current. Measurable flux will soak into the superconductor in just a few minutes if a constant current (typically 1A) is supplied to the coil under test (e.g. via leads A and C for L_1) while rest of the circuit (and SQUID input) is isolated by the appropriate heat switch(es). This observation led us to a simplified circuit , with only a single test coil L_1 and heat switch in series across L_{SQ}, mounted in a small liquid helium dip probe, for rapid comparison of wire samples by the "soak" method. The Nb shield for the test coil, heat switch and superconducting joints was the same tube system used in the BTI RF SQUID. The test coil was either a

0921-4526/90/$03.50 © 1990 – Elsevier Science Publishers B.V. (North-Holland)

6mm diameter solenoid up to 30mm long, or a small toroid in a Nb shield. Initially, with solenoidal coils, anomalously high creep rates were observed, due, we suspect, to ferromagnetic impurities in a 1mm thick brass mounting cradle. Substitution of a perspex one dropped the drift rates in solenoidal coils by more than an order of magnitude, to levels close to those of the shielded toroidal ones. The 5 small (1.6 mm) brass bolts and nuts used in the superconducting joints did not contribute significantly to the observed drift, as was confirmed in one experiment with spot-welded joints. The superconducting leads (from the coil, heat switch and SQUID) were secured to the half-cylindrical mounting cradle by means of a "magnetically transparent" material such as wax or a re-usable putty (8)

Due to residual trapped current (observed by overpressurizing the LHe storage dewar) and mechanical disturbances, the flux creep in this apparatus could not be followed for times much longer than 10^3 to 10^4 seconds.

3. RESULTS

In the gradiometer and dip probe circuits we observed :

1. dB/dt is linear in the persistent or soak current from 0.1 to 4A (magnetic fields of 10 to 400 Gauss).
2. dB/dt varies approximately as the logarithm of the soak time, from 15 to 600 s: a soak of 5 minutes is equivalent to waiting 10 hours before decaying a persistent current half the magnitude of the soak current.

Figure 2 shows the nearly logarithmic creep observed in a gravity gradiometer with Nb coils (8) after a persistent current corresponding to 0.2 H_{c1} was suddenly destroyed. The decay constant is about 8 x 10^{-8} .

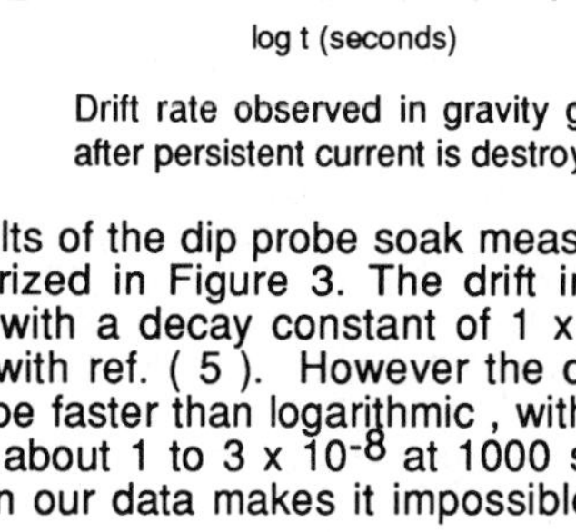

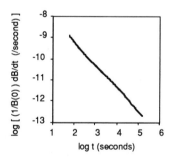

Figure 2 Drift rate observed in gravity gradiometer after persistent current is destroyed.

The results of the dip probe soak measurements are summarized in Figure 3. The drift in NbTi is logarithmic with a decay constant of 1 x 10^{-7} , in agreement with ref. (5). However the drift in Nb appears to be faster than logarithmic , with a decay constant of about 1 to 3 x 10^{-8} at 1000 sec. The dispersion in our data makes it impossible to tell if there is any significant difference between the two brands of Nb wire tested, California Fine Wire and Supercon wire.

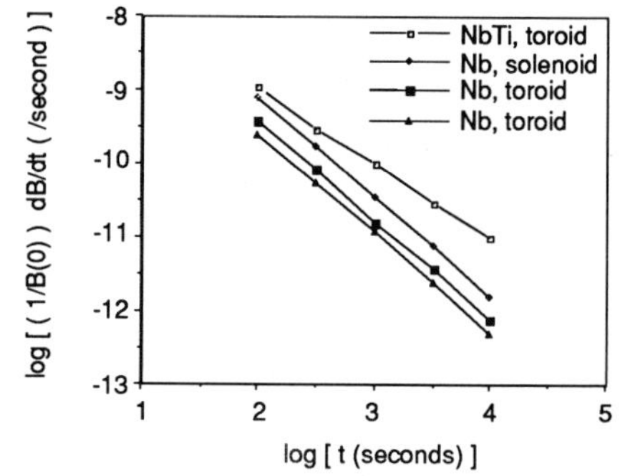

Figure 3 Drift rate as a function of elapsed time after soaking test coil with 1A for 300s. Comparison of Nb wires with NbTi.

4. CONCLUSIONS

In summary, we observe flux creep in niobium wires up to an order of magnitude less than in NbTi wires. Further work needs to be done on wire coil samples of different diameter, purity and surface treatment.

ACKNOWLEDGEMENTS

This work was initiated under the NERDDP program and supported by BP Minerals Australia.

REFERENCES

(1) C. Edwards, M. J. Buckingham, M.H. Dransfield, F. J van Kann, A. G. Mann, R. Matthews and P.J. Turner, "Development of a Mobile Gravity Gradiometer for Geophysical Exploration", Proceedings of the seventeenth Gravity Gradiometry Conference. U. S. Air Force Academy, Hanscom, Massachusetts, USA, October, 1989.

(2) Y.B. Kim, C.F. Hempstead and A.R. Strnad, Phys. Rev. Letters **9**, 306 (1962).

(3) W. Ludwig and H. Gessinger, Proc. 1st Int. Conf. on Superconducting Quantum Devices, SQUID, ed H.D. Hahlbohm and H.Lubbig, 239 (1977).

(4) J. Chaussy, J.L. Genicon, J. Mazuer, M. Renard and A. Sulpice, Phys. Lett., **87A**, number 1, 2, 61 (1981).

(5) E. Mapoles, Development of a Superconducting Gravity Gradiometer for a Test of the Inverse Square Law, Ph.D. thesis, Stanford University, 1981 (unpublished).

(6) D.Marek, Proc. LT 18 , Jap. J. Appl. Phys., **26**, 1683 (1987).

(7) 'Blu-Tack', Bostik Division, Emhart Australia Pty Ltd.

(8) N.J. McDonald, Honours Thesis, University of W.A. (1989) (Unpublished).

Physica B 165&166 (1990) 153–154
North-Holland

A MECHANISM FOR PRECISE ROTATION OF NUCLEAR TARGETS AT LOW TEMPERATURES

J. E. KOSTER*, C. R. GOULD*, D. G. HAASE*, N. R. ROBERSON+

*Physics Department, North Carolina State University, Raleigh, N. C. 27695-8202 and Triangle Universities Nuclear Laboratory, Durham, N. C. 27706

+Physics Department, Duke University, Durham, N. C. 27706 and Triangle Universities Nuclear Laboratory, Durham, N. C. 27706.

We have built an aligned ^{165}Ho nuclear target for studies of fundamental symmetries in neutron scattering. A feature of the target cryostat is a mechanism to rotate the target reproducibly to less than 1/2° under computer control while the target is maintained below 0.3 K. We discuss the considerations of torsion and heat flow necessary to the design of the mechanism.

1. INTRODUCTION

We have constructed a rotating aligned ^{165}Ho single crystal target for neutron transmission experiments. The main design criteria was that the target could be reproducibly and precisely rotated under remote control. Good thermal contact must be maintained to the mixing chamber of our dilution refrigerator to remove the large ^{165}Ho nuclear enthalpy. The sample is maintained in vacuum, for convenience, as well as to avoid spurious scattering effects.

The target is a single crystal of holmium cut and ground to a 101.6 gm. cylinder 2.8 cm high and 2.3 cm diam. The nuclei are aligned along the hexagonal c axis through the hyperfine interaction. The nuclear heat capacity of the target is 4.25 J/K at 0.3 K; it is necessary to remove 2.3 J of heat to cool it from 1 K to 0.3 K.[1] The target is being used for tests for violations of time reversal invariance (TRI) in neutron transmission, using the "five-fold correlation".[2] The magnitude of the TRI violating scattering amplitude depends on the angle between the target nuclear alignment axis and the neutron beam momentum. To distinguish TRI mimicking effects it is important to vary the alignment axis to beam direction often and precisely during the transmission measurement. The experiment also requires the aligned target have zero nuclear polarization and no external magnetic field to cause precession of the neutron spins.

2. ROTATION SYSTEM

The target is cooled by a dilution refrigerator having two heat exchangers connected in series with the mixing chamber. Each component of the refrigerator has a central axis hole to allow a straight shaft to pass from room temperature to the target. For rigidity, all of the support rods of the refrigerator are 1/8" to 3/8" x 0.010" OD wall stainless steel tubes. In tests without the holmium target the refrigerator cooled to 42 mK and achieved a cooling power of 85 μW at 110 mK and a flow rate of 120 μmol/sec.

The end of the holmium cylinder is soldered[3] to a copper holder, which is supported by a stainless steel shaft held in a pair of precision bearings[4] on the mixing chamber. The shaft is in turn connected to a room temperature stepping motor through two stainless steel tubes.

The upper end of the holmium target was machined to align the holder and target to the center of the rotation shaft to within ± 0.0005". Thermal connection to the sample is made through three 10 cm. long flexible copper braids (each braid 110 x 0.006" OD wires) screwed to the mixing chamber and the copper cold finger. The target can be rotated through a 270 degree arc.

The shaft is rotated by a 14" long x 3/16" OD x 0.010" wall stainless steel tube from the mixing chamber to the coldplate, connected through a bellows coupler to a 45" long x 3/8" x 0.030" wall stainless steel tube through a 1/2" vacuum tube to room temperature. At several points the 3/8" axle is interrupted by copper collars which rub against the 1/2" tube to provide thermal contact to 4.2K and to shield infrared radiation from room temperature. Inside the vacuum can of the refrigerator the axle is connected by copper braids to thermal sinks at 4.2 K and on the coldplate. The temperature of the axle at the connection to the coldplate was found to be about 1.5 K. The axle did not produce any noticeable effect on the liquid helium boiloff rate or the operating temperature of the coldplate. The heat leak down the 3/16" tube from 1.5 K to the mixing chamber is calculated to be 0.35 μW.

At room temperature the 3/8" tube is connected to a flexible coupler and thence a Ferrofluidics rotation seal.[5] The end of the shaft is rotated by a computer controlled stepping motor[6] through a 25 : 1 anti-backlash step down gearbox.[7] The shaft angle is monitored by a 13-bit absolute angle encoder connected to the computer system.[8] Through a feedback arrangement the orientation of the upper end of the shaft can be selected and controlled to ±. 0.04°.

The limitation in cooling the target is the thermal conductance κ of the braids. If we define T_s and T_{mc} as the temperatures of the sample and mixing chamber respectively then the heat flow to the mixing chamber is

$$\dot{Q} = \kappa (T_s - T_{mc}).$$

The conductance κ is a linear function of temperature $\kappa = (\lambda A/l)\,T$ where $\lambda \sim 100$ W/mK^2 for copper. Therefore

$$\dot{Q} = (\lambda A/2l)(T_s + T_{mc})(T_s - T_{mc})$$

This heat must be removed by the ^3He circulation cooling power $\dot{Q}_{mc} = RT_{mc}^2$ where $R \sim 6 \times 10^{-3}$ W/K^2 for our refrigerator. Since $\dot{Q} = \dot{Q}_{mc}$ we can solve for T_{mc}^2 and then calculate $\dot{Q}$ in terms of the sample temperature, T_s^2:

$$\dot{Q} = R/\{1 + R/(\lambda A/2l)\}\, T_s^2.$$

Therefore the effective cooling coefficient of the dilution refrigerator has been reduced to $R_{eff} = R/\{1 + R/(\lambda A/2l)\}$. For our copper braids $(\lambda A/2l)$ was about 3×10^{-3} W/K and therefore $R_{eff} = 2 \times 10^{-3}$ W/K^2.

Several tests of the rotation system were made at room temperature. It was found that the target returned to its original orientation to within ± 1/2° after repeated rotations. The radial play of the target during rotation was about 0.024" and average radial vibration about 0.0001". The torque required to turn the target was measured to be about 5×10^{-2} N-m without the braids in place. We calculate this torque would produce at worst a 0.5° twist in the tubes connecting the target to room temperature and generate 40 mJ of heat while rotating the target 45°.

3. ROTATION OF ALIGNED TARGET

To test of the ability of the cryostat to align and precisely rotate a target, we measured the nuclear deformation cross section σ_{def} of ^{165}Ho by the transmission of unpolarized neutrons. Because of the shape of the ^{165}Ho nucleus the neutron transmission varies as the $\cos^2$ of the angle between the beam momentum and the nuclear alignment direction. This behavior is shown in Figure 1. The target cooled to about 0.25 K in about 12 hours. The target alignment of 67% was maintained for over 15 hours while the target was rotated in 45° steps every 15 minutes. Our measurements gave values of σ_{def} consistent with the literature and indicated reproducible control of the alignment angle.

ACKNOWLEDGEMENT

This work was supported by the U. S. Department of Energy, Office of High Energy and Nuclear Physics, under Contracts No. DE-AC05-76ER01067 and DE-FG05-88ER40441.

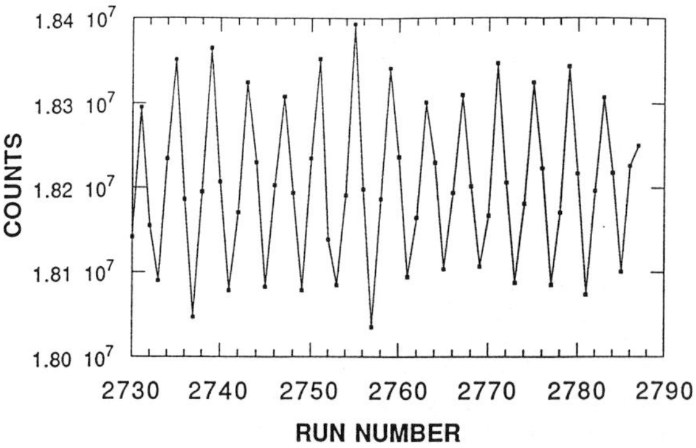

FIGURE 1
Transmission of unpolarized 1.9 MeV neutrons through aligned holmium target, showing the nuclear deformation effect. Each data point was measured at one angle. The target was rotated in sequence -90°, -45°, 0°, 45°, 90°, 45°, 0°, -45°, etc.

REFERENCES

1. M. Krusius, A. C. Anderson, and B. Holmstrom, Physical Review 177, 910(1969).
2. C. R. Gould, et al, in "Tests of Time Reversal in Neutron Physics," N. R. Roberson, C. R. Gould and J. D. Bowman, eds. (World Scientific, 1987) p. 130.
3. 33% Cd-67% Bi eutectic solder, see William A. Steyert, Jr. Rev. Sci Instrum 38, 964(1967).
4. PIC E1-9, Precision Industrial Components, Middlebury, Ct. 06762.
5. Ferrofluidics Seal MB-188-L-N-090 Instrumentation Feedthrough. Ferrofluidics Corp., Nashua, New Hampshire, 03061.
6. Vextra 2 phase stepping motor, PH266M-E1.2
7. Berg 25:1 servogear box, SX-B4-14.
8. Model M25 BEI Motion Systems Co., Position Controls Division, Santa Barbara CA 93103.

Physica B 165&166 (1990) 155–156
North-Holland

AN ULTRA HIGH VACUUM LOW TEMPERATURE GYROSCOPE CLOCK

T. Walter, J.P. Turneaure, S. Buchman, C.W.F. Everitt and G.M. Keiser

GP-B Program, Hansen Laboratories of Physics, Stanford University, Stanford, CA 94305 U.S.A.

We propose to perform a null-gravitational redshift experiment by comparing a mechanical gyroscope clock with atomic clocks. The Gravity-Probe-B Relativity Gyroscope Experiment provides the opportunity for this co-experiment. The goal is to measure the effect to an accuracy of 0.01% of the gravitational redshift due to the eccentricity of the orbit of the earth about the sun. This corresponds to an integrated frequency measurement over one year of $\Delta\nu/\nu=3*10^{-14}$. A major disturbance torque on the gyroscope is due to fluctuations in the molecular drag of the residual gas caused by temperature variations. We propose to use a low temperature bake-out technique in order to achieve the required vacuum of 10^{-17}torr.

1. BACKGROUND

The GP-B Relativity Gyroscope Experiment, whose goal is to measure the geodetic and frame-dragging precessions of gyroscopes in earth orbit, provides the opportunity for a proposed co-experiment to perform a null gravitational redshift measurement. In this experiment the spin frequencies of the orbiting gyroscopes are compared with earth based atomic clocks. A null result means equal redshift for the atomic and gyroscope clocks, therefore verifying the equivalence principle. These measurements will extend gravitational redshift observations to clocks based on the spin speed of a rotating mass. This experiment can also be interpreted as a check of the invariance of the fine structure constant α; atomic clock frequencies depend on α^4, while the mechanical gyroscope frequency varies as α^2.

The principal change in gravitational potential experienced by the gyroscopes and the atomic clocks over a year is due to the eccentricity of the orbit of the earth about the sun. This leads to a peak-to-peak variation in the gravitational redshift of $\Delta\nu/\nu=3*10^{-10}$ over the year. The goal of the null-gravitation redshift experiment is to measure the gravitational redshift of the gyroscope clock relative to that of atomic clocks with an accuracy of 0.01% of the value of the gravitational redshift. A previous null-gravitational redshift experiment compared the shifts of a hydrogen maser and a superconducting cavity stabilized oscillator to an accuracy of 2% (1).

2. EXPERIMENTAL APPARATUS

The gyroscope is a very homogeneous ($\Delta\rho/\rho<1*10^{-6}$), very uniform ($\Delta r/r<1*10^{-6}$) 38mm diameter sphere of fused quartz (or single crystal silicon) coated with a 250µm layer of niobium of better than 2% uniformity. It is electrostatically suspended and then spun with He gas to 170Hz. The operational temperature is 2K. Spin speed is measured by observing the rotating magnetic flux trapped in the superconducting rotor coating. The measuring pick-up loop is located on the gyroscope housing and referenced to the fixed stars. Figure 1 is a schematic diagram of the gyroscope clock and its instrumentation and data acquisition system.

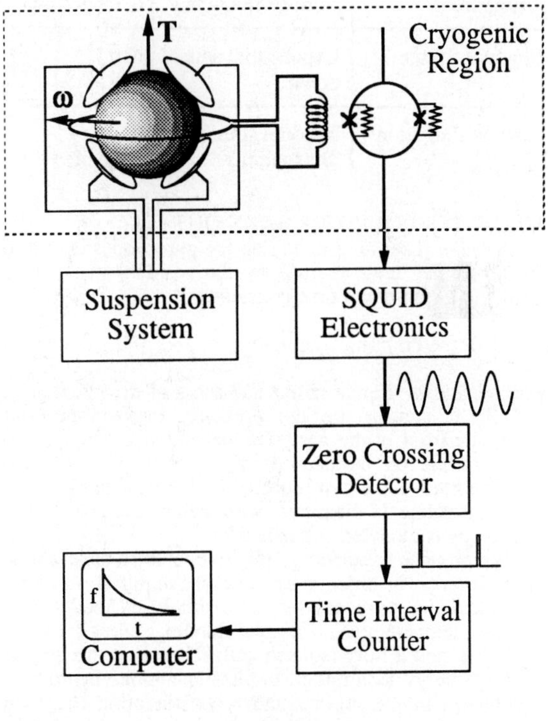

FIGURE 1
Schematic diagram of the gyroscope clock.

The GP-B experiment already incorporates most of the essential characteristics needed for a gyroscope clock. These are: *a*) a drag free (10^{-11}g) environment which reduces the external torques on the gyroscope, *b*) a gyroscope read-out which can be used to measure spin phase relative to the fixed stars, *c*) a Global Positioning System receiver which allows time transfer from earth based atomic clocks with an accuracy of about 10ns, and *d*) a stable low temperature environment ($\Delta T<1mK$ over one year). Table 1 gives a summary of an analysis of the principal gyroscope disturbing torques, their effect, and the corresponding uncertainty in spin speed.

TABLE 1.
Summary of the principal gyroscope disturbing torques.

Torque	Effect	($\Delta\omega/\omega$)
Molecular Drag	Exponential spin-down with a time constant of $\tau = (3x10^{-7}/P)$ years	1×10^{-14} for 10^{-17} torr
Mass Unbalance & Rotor Asphericity	Spin variations modulated at the polhode frequency	$<10^{-14}$ with filter & modelling
Readout System	Exponential spin-down	$<10^{-14}$
Cosmic Radiation	Random spin fluctuations	3×10^{-14} modelled

3. LOW TEMPERATURE BAKE-OUT

A major disturbing torque on the gyroscopes is due to the molecular drag exerted by the residual gas. This results in a spin-down time constant:

$$\tau = 3/10*(M/Pr^2)*(k_BT/2\pi m)^{1/2} \quad (1)$$

where r and M are the radius and mass of the gyroscope, and P, T, and m are the pressure, temperature and molecular mass of the gas. The dependence of the spin-down torque on temperature fluctuations sets the He pressure limit in the gyroscope to 10^{-17}torr, assuming that the temperature is measured with an accuracy of 10μK and torque is modeled at this level.

We propose to perform a low temperature bake-out of the gyroscope in order to achieve the required pressure. For calculation purposes we use a simplified model with uniform temperatures, a unique binding energy to the walls E_B, and a fully isolated cell. Note that the unique binding energy assumption implies sub mono-layer wall coverage. In the regime under consideration the total number of atoms in the cell is very well approximated by the number of atoms on the wall. The condition for this approximation is:

$$V/A << \lambda(T)*\exp(E_B/k_BT) \quad (2)$$

$$\lambda(T) = (h^2/2\pi mk_BT)^{1/2} \quad (2')$$

where V/A is the volume to area ratio of the cell and $\lambda(T)$ is the thermal wavelength of the gas. A cell pumped out to P_1 at T_1 will have a pressure P_2 at T_2 equal to:

$$P_2 = P_1*(T_2/T_1)^{3/2}*\exp[(E_B/k_B)*(1/T_1-1/T_2)] \quad (3)$$

We use E_B = 150K, the average measured value for the binding energy of He on copper (2). Under these conditions a cell with sub mono-layer wall coverage at 6K will experience a drop in pressure of over twenty orders of magnitude when cooled to 2K. The surface coverage condition is satisfied at 6K for He pressures below about 10^{-7}torr. The bake-out temperature of 6K is limited by the requirement to keep the gyroscope coating below its superconducting transition temperature.

Next we address the question of the pump-out time during the bake-out at 6K. A Monte Carlo simulation indicates that the fraction β of wall collisions inside the gyroscope which result in a He atom escaping into the pumping line is of the order of 10^{-4}. The time dependence of the surface density σ while pumping on the atoms in the bulk is given by:

$$\sigma = \sigma_0\exp(-t\beta/\tau_S) \quad (4)$$

$$\tau_S = (4\lambda(T)/vs)*\exp(E_B/k_BT) \quad (4')$$

where τ_S is the atom residency time on the surface, v is the thermal velocity, and s the sticking coefficient ($s \approx 1$). We make the assumption that the pumping speed is limited by β. The resulting pump-out time for the gyroscope is of the order of 10h, consistent with preliminary experimental results (2).

In conclusion we find that the opportunity and the experimental techniques exist for the performance of the null-gravitational redshift experiment using a gyroscope clock. The low temperature bake-out at 6K with subsequent cooling to 2K is a suitable method for achieving the vacuum level needed for the operation of this clock.

REFERENCES

(1) J.P. Turneaure, C.M. Will, B.F Farrel, E.M. Mattison, and R.F.C. Vessot, Phys. Rev. D 27 (1983) 1705.
(2) J.P. Turneaure, E.A. Cornell, P.D. Levine and J.A. Lipa, Ultrahigh Vacuum Techniques for the Experiment, in: Near Zero, eds. J.D. Fairbank, B.S. Deaver Jr, C.W.F. Everitt and P.F. Michelson (W.H. Freeman and Co., New York) pp 671-678.

Physica B 165&166 (1990) 157–158
North-Holland

TEST OF THE EQUIVALENCE PRINCIPLE FOR NUCLEAR–POLARIZED BODIES AT LOW TEMPERATURE

WEI–TOU NI

Department of Physics, National Tsing Hua University, Hsinchu, Taiwan, Republic of China

Equivalence principle is an important cornerstone of general relativity and metric theories of gravity. To test the equivalence for nuclear–polarized bodies, we propose to use intermetallic compounds of Pr and Cu suitably polarized and magnetically shielded as our polarized bodies. We present the scheme of our experiment, and aim at a precision of 10^{-5}–10^{-8} on the equivalence of polarized nuclei.

1. INTRODUCTION

Equivalence and universality is the characteristic that distinguishes gravity and other interactions so far. The vast range of validity of equivalence principles poses challenges for both experimentalist and theorist. For experimentalist, the challenge is to push the precision higher and find new ways to test them. For theorist, the challenge is to find the microscopic origin of gravity with this vast range of validity of equivalence principles. Actually this precision provides both a guide and a challenge to push our understanding to smaller and smaller distances — and hopefully to Planck distance.

How far will this precision of validity go? This is the question we want to address. According to our present understanding of physics, particles and fields transform appropriately under inhomogeneous Lorentz transformations. These inhomogeneous Lorentz transformations form the Poincaré group. The only invariants characterizing irreducible representations of the Poincaré group are mass and spin (or helicity in the case of zero mass). Both electroweak and strong interactions are strongly polarization–dependent. The question comes whether the gravitational interaction is polarization–dependent.

To test the equivalence principle for electron–polarized bodies, we have made magnetically shielded spin–polarized bodies of Dy_6Fe_{23}. We use both a beam balance(1,2,3) and a torsion balance(4) to test the equivalence principle.

Since the nuclear–ordering temperature is much lower than the electron–spin ordering temperature, low temperature facility is required for the equivalence experiments for nuclear–polarized bodies. In the next section, we address to the problem of making nuclear–polarized bodies. In the third section, we present the design of our dilution refrigerator and the experiment.

2. NUCLEAR–POLARIZED BODIES

There are two general kinds of methods in making polarized targets in high–energy and nuclear experiments — the thermal–equilibrium methods (brute–force methods) and the nonequilibrium methods. In the thermal equilibrium methods, usually one applies a high magnetic field to reduce the low–temperature requirement. The most common nonequilibrium methods use dynamical nuclear polarization (DNP). To reduce the effects of magnetic field on particle scattering, the frozen spin targets are used by various groups(5,6,7). In these experiments, butanol, propanediol and pentanol are cooled by dilution refrigerator in a high magnetic filed, dynamically polarized by microwaves and moved to a low field area (typecally ~0.3T). In the low holding magnetic field, the relaxation of the spin polarization is typically about 20 days at about 50 mK.

In the equivalence–principle experiment, we need magnetically shielded polarized bodies. The frozen–spin method is OK if we can magnetically shield the target plus the holding magnet. For this purpose we can use either permanent magnets or superconducting magnets and shield them with mangnetically soft materials plus superconducting coating. However the mechanical difficulties in manipulating the shields inside the dilution refrigerator would be rather sophisticated.

We now turn to the thermal equilibrium methods. To obtain significant polarization, we need either a high B/T ratio or a nuclear–ordered state. High B/T ratio is impractical for the equivalence–principle experiment. However, there are systems which have higher than usual nuclear–ordering temperatures. One such system is solid ^{3}He. At 40 atm, ^{3}He becomes solid below 1K. The spin g–factor for ^{3}He is –4.4. At 1mK, above 0.5T, ^{3}He is in a high field phase with a large magnetization, 0.6 M_{sat}. In this state, there are about 20% nucleons polarized. In reference 8 we propose to use this material for doing equivalence–principle experiments. Although a high polarization per nucleon can be achieved in principle, the experimental procedure to obtain a large fraction of solid ^{3}He is not easy.

In some other systems, there are enhanced hyperfine and exchange interactions. In these systems, the nuclear–ordering can take place in the mK range. One such system consists of the intermetallic compounds of Pr and Cu. If magnetic ordering occurs above 1K in singlet ground–state systems, it is probably

0921-4526/90/$03.50

to a pure electronic state; if the critical temperature lies between 10 mK and 1K, the state is probably a mixed electronic–nuclear state; if the critical temperature is below 10 mK, it is probably a nuclear state(9).

Praseodimium has only one stable isotope ^{141}Pr with I=5/2 and $\mu_I=4.3\mu_N$. $PrCu_5$ transits to a mixed electronic–nuclear ferromagnetic state at 24 mK (10,11). $PrCu_6$ transits to a nuclear ferromagnetic state at 2.5 mK(9). To make intermetallic compound of Pr and Cu, arc melting in Ar and proper annealing will do. The body of $PrCu_5$ (or $PrCu_6$) will be polarized with two permanent magnets. A magnetic field of tens of mT is enough to polarize it. This field is shielded by two high–μ shielding cups. A thin layer of superconducting coating will then shield the magnetism well.

3. LOW–TEMPERATURE FACILITY AND THE EXPERIMENT

Our $^3He/^4He$ Dilution Refrigerator from CryoVac is scheduled to arrive in August, 1990. The cooling power at 100 mK is greater than 300 μW and the minium temperature is 4.5 mK. A space is reserved for the nuclear demagnetization stage. The experimental space (in the vacuum can) is 400 mm x 65 mm ID. To cool $PrCu_6$ below 24 mK, the dilution refrigrator is enough. To cool $PrCu_6$ below 2.5 mK, a nuclear demagnetization stage is needed in addition. We will first do experiment with $PrCu_5$. To perform the equivalence–principle experiment, we insert the polarized body of $PrCu_5$ or $PrCu_6$ inside the cylindrical hole of nonmagnetic cylinder as in Fig. 1. The size and density are properly matched to minimize the local gravity gradient effects. The filled cylinder is hung from a fibre. The angle position of the fibre is sensed by an optical lever, capacitance plates or some other mechanism. In solar gravitational field, the nonmagnetic cylinder is served as the unpolarized body to be compared with. Any deviation from equivalence will give a periodic signal in the fibre angle position. Planned accuracy of the first experiments are in the range 10^{-7}–10^{-10} in the Eötvös parameter.

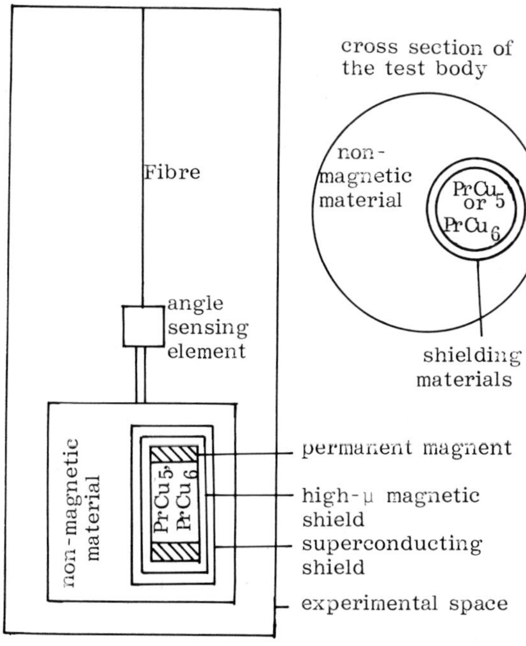

Figure 1. Schematic for the polarized $PrCu_5$ (or $PrCu_6$) equivalence-principle experiment.

ACKNOWLEDGEMENTS

I would like to thank James M. Daniels and Jow–Tsong Shy for helpful discussions. This work is supported in part by the National Science Council of the Republic of China NSC–79–0208–M007–110.

REFERENCES

(1) W.–T. Ni, P.–Y. Jen, C.–H. Hsieh, K.–L. Ko, S.–C. Chen, S.–s. Pan and M.–H. Tu, Test of the Equivalence Principle for Spin–Polarized Bodies in: Proceedings of the 2nd ROK–ROC Metrology Symposium, eds. J. W. Won and Y. K. Park, published by Korea Standards Research Institute (1988) pp. VII–2–1 to VII–2–5.

(2) C.–H. Hsieh, P.–Y. Jen, K.–L. Ko, K.–Y. Li, W.–T. Ni, S.–s. Pan, Y.–H. Shih and R.–J. Tyan, Mod. Phys. Lett., 4, (1989) 1597.

(3) W.–T. Ni, Y. Chou, S.–s. Pan, C.–H. Lin, T.–Y. Hwong, K.–L. Ko and K.–Y. Li, An Improvement of the Equivalence Principle Test for Spin–Polarized Bodies and the Mass Loss, in: Proceedings of the 3rd ROC–ROK Metrology Symposium, Hsinchu, May 22–24, 1990, published by the Industrial Technology Research Institute, Hsinchu, Taiwan, Republic of China.

(4) Y. Chou, W.–T. Ni and S.–L. Wang, Torsion Balance Equivalence–Principle Experiment for the Spin–Polarized Dy_6Fe_{23}, National Tsing Hua University Preprint GP–003, March, 1990.

(5) T. O. Niinikoski and F. Udo, Nucl. Instr. Meth., 134 (1976) 219.

(6) P. P. J. Delheij, D. C. Healey and G. Wait, J. Phys. Soc. Jpn., 55, (1986) Suppl. p. 1090.

(7) P. Chaumette, H. Desportes, J. Derégel, G. Durand, J. Fabre, L. van Rossum, Dilution Refrigerator and Solenoid for the Fermilab Spin Physics Facility, in: High–Energy Spin Physics, eds. K. J. Heller (Amer. Inst. Phys. 1989) pp. 1331–1333.

(8) W.–T. Ni, Equivalence Principles and Polarized Experiments, in: Third Asia Pacific Physics Conference, eds. Y. W. Chan, A. F. Leung, C. N. Yang and K. Young (World Scientific, 1988), pp. 315–326.

(9) J. Babcock, J. Kiely, T. Manley, and W. Weyhmann, Phys. Rev. Lett., 43 (1979) 380; and references therein.

(10) K. Andres, E. Bucher, P. H. Schmidt, J. P. Maita, and S. Darack, Phys. Rev. B11 (1975) 4364.

(11) J. L. Genicon, J. L. Tholence, and R. Tournier, J. Phys. (Paris), Colloq. 40 (1978) C6–798.

Physica B 165&166 (1990) 159-160
North-Holland

MULTIPURPOSE CLOSED-CYCLE CRYOCOOLER FOR LIQUEFYING HYDROGEN, HELIUM-4 OR HELIUM-3

Calvin WINTER

QUANTUM TECHNOLOGY CORP., 6237 - 148 St., Surrey, B.C., CANADA V3S 3C3

A cryogenic refrigerator utilizing helium-4 gas in closed-cycle Gifford-McMahon and Joule-Thomson cooling loops was built and achieves continuous operating temperatures of 2.8K. The object cooled is a thin walled (0.1mm) seamless electroformed nickel target sample cell with a volume of 160ml. Room temperature hydrogen, helium-4 or helium-3 gas, supplied at a pressure slightly above atmospheric, is liquefied by the cryocooler and fills the cell. Unusual features include: horizontal operation; a long narrow extension on the vacuum shroud (900mm long, 76mm diameter) and special valves to select an operating temperature appropriate to the sample gas and maximize the cooling power available at that temperature.

1. INTRODUCTION

It is common practice to utilize liquid helium supplied from an external source as a coolant. This becomes inconvenient and expensive when long run times are required or when stable operating temperatures above or below 4.2K are required. An additional disadvantage is the large liquid helium reservoir required which must be kept vertical.

Reliable modern closed-cycle cryocoolers allow operation below liquid helium-3 temperature and allow greater flexibility to suit the experimental requirements (1-3).

2. DESIGN REQUIREMENTS

The nuclear physics experiment required a liquid gas target located in a horizontal photon beam. Protons and neutrons produced by nuclear reactions within the liquid travel through: the liquid; the sample cell wall; vacuum; the heat shield; vacuum; the vacuum shroud; air and are then counted in detectors.

Due to the high cost of helium-3, the shape of the sample cell was optimized to maximize the interaction length with a minimum total volume. The resulting shape was a cylinder 50mm in diameter and 50mm long capped with two semi-spherical end caps. The experimental requirement of predictable energy loss by the secondary particles dictated a uniform wall thickness. The cell complete with a fill tube 100mm long 12.7mm diameter is seamless, Figure 1. It was made by electroforming nickel with a wall thickness of only 0.1mm.

The experimental requirement of being able to measure secondary particles emerging from the target in any direction meant that the sample cell had to be some distance from the cryocooler. The sample cell is located in a horizontal beam pipe 76mm in diameter at a distance of 650mm from the body of the cryocooler.

3. SAFETY CONSIDERATIONS

Hydrogen boils at 20K and freezes at 14K. This cryocooler could easily fill the sample cell with liquid hydrogen, then freeze hydrogen in the fill line. On warming the confined sample excessive pressures would result in cell rupture. Safety features to prevent this, implemented when using hydrogen, include: a temperature controller heating the second stage to 19K; an independent thermometer with a safety interlock to switch off the cryocooler if the second stage temperature goes below 18K; a safety vent line separate from the normal fill/vent line attached to the sample cell. All vent lines, and a relief valve on the vacuum space are vented outside of the laboratory.

4. CLOSED-CYCLE CRYOCOOLER

The 2.8K closed-cycle cryocooler, Figure 2, consists of a Quantum Technology Corp. QUANTUMCOOLER Model LQ3-E6-J10 Joule-Thomson (J-T) loop. Pre-cooling of the J-T loop is accomplished with a Cryomech Model GB04 Gifford-McMahon (G-M) cryocooler. The G-M cold head has two independent free (gas driven) displacers and a stationary second stage regenerator to provide very low second stage temperatures.

The J-T, G-M and sample gas loops are thermally interconnected at heat stations #1, #2 and #3 but are otherwise totally separate.

Due to experimental design constraints, heat station #3 is 13mm in diameter, 600mm long and serves to support the sample cell. It consists of three concentric tubes with the sample liquid in the space between the two outer tubes and the J-T loop helium-4 liquid/gas cooling mix flowing inside the two inner tubes. Cooling of the sample to well below its boiling point (to eliminate bubbles) throughout the sample cell is achieved solely by natural convection from this heat exchanger. (Thermal conduction along the cell wall is negligible).

5. OPERATING PROCEDURES

Liquefying hydrogen is done with the bypass valve open and the J-T valve closed. This allows nearly all of the second stage cooling capacity (7W) to be utilized on the sample.

To liquefy helium-4 the J-T valve is open, after cooling to 20K the bypass valve is closed.

To liquefy helium-3, after reaching 4K the J-T valve is partially closed. This reduces the cooling power, the J-T loop flow and the J-T return pressure (which is limited by the finite displacement of the external J-T compressor and by the back pressure in the heat exchangers). The ultimate temperature reached is equal to the boiling temperature of liquid helium-4 at the pressure in heat station #3.

6. EXPERIMENTAL RESULTS

Two difficulties were encountered in initial testing. First, the system vibrated due to the movement of the heavy first stage G-M displacer. This was dramatically reduced by reducing the stroke of the first stage displacer and by rigid mounting of the cryocooler.

Second, a Taconis oscillation in the nearly horizontal sample fill tube between heat stations #2 and #3 occurred as soon as the first few drops of helium were liquefied. Heat loading from the oscillation prevented further liquefaction. Oscillations were eliminated by lengthening the fill tube with "U" bends and putting wool in the tube for damping.

Initial cool down takes 9 hours, this is partly due to the thermal diffusion time constant of the long (1.3m) aluminum heat shield. Typical heat shield tail, first stage and second stage temperatures are 90K, 50K and 10K respectively. Third stage cooling powers of 100mW at 2.8K and 600mW at 3.6K were measured. These correspond to heat station #3 pressures of 0.17 and 0.5 (x100kPa) respectively. The J-T supply pressure was 4 (x100kPa). Increased cooling power can be achieved by increasing the J-T flow or J-T supply pressure or both.

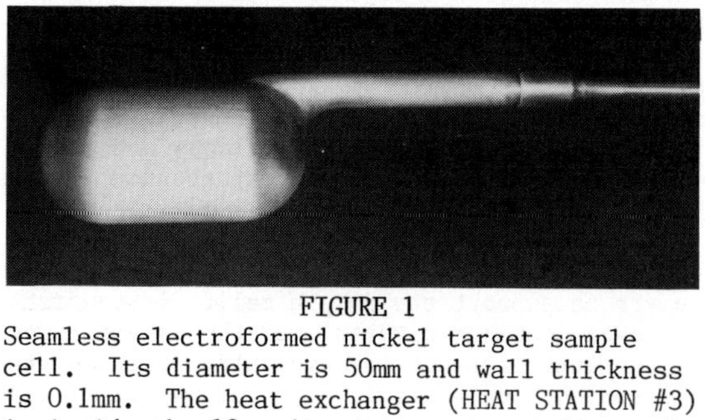

FIGURE 1
Seamless electroformed nickel target sample cell. Its diameter is 50mm and wall thickness is 0.1mm. The heat exchanger (HEAT STATION #3) is inside the 13mm diameter support tube on the right. The incident photon beam is horizontal and is 20mm in diameter and so passes under the heat exchanger. The entire assembly, covered with an aluminum heat shield fits inside a 76mm diameter vacuum shroud extension 900mm long.

ACKNOWLEDGEMENTS

The assistance of A. Swiecicki, L. Robinson and others in this work is gratefully acknowledged. Financial support was received from the Science Council of British Columbia, Revenue Canada-Taxation (SR&ED-ITC) and the Instituto Nazionale di Fisica Nucleare.

REFERENCES

(1) W.H. Higa and E. Wiebe, One Million Hours at 4.5K, in: NBS Special Publication 508 (1978) page 99.
(2) D.K. Milne, B. Wilcockson and J. Griffiths, The Radiophysics MK1 Helium Liquefier, Internal report RPP2474, Division of Radiophysics, CSIRO, Sydney (1981).
(3) Troy Henderson, 3K Helium Refrigerator, Telescope Services Division Report No. 12, National Radio Astronomy Observatory, Green Bank, West Virginia (1985).

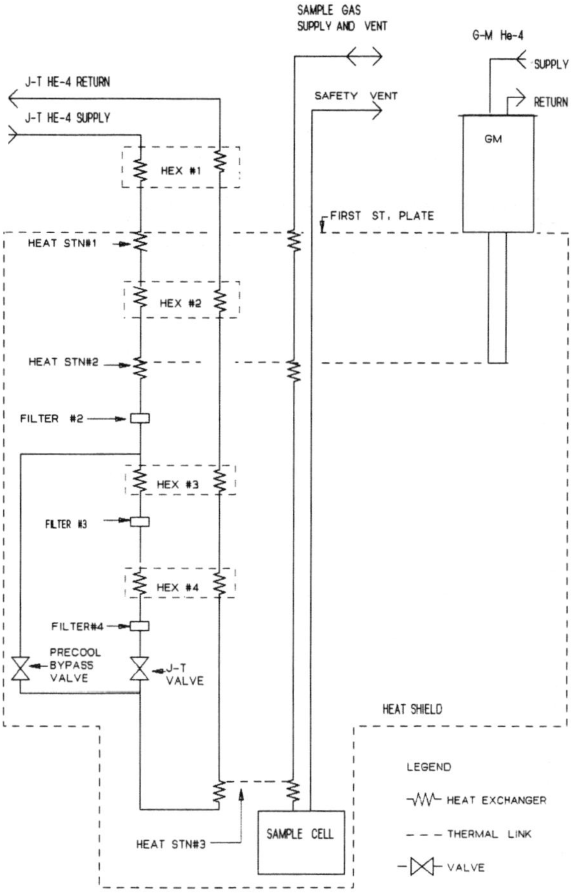

FIGURE 2
Cryocooler plumbing internal to the vacuum dewar.

Physica B 165&166 (1990) 161-162
North-Holland

CLEANER CATALYST FOR THE PARA TO ORTHO CONVERSION IN HIGH PURITY DEUTERIUM AT LOW TEMPERATURES

D.I. HEAD and R.L. RUSBY

Division of Quantum Metrology, National Physical Laboratory, Teddington, Middlesex, TW11 OLW, United Kingdom

The triple-point of high purity deuterium (99.96% D_2), sealed in high pressure cells has been measured. The traditional spin catalyst of hydrous ferric oxide has been found unsuitable due to hydrogen contamination. Results using Gd_2O_3 are reported, which suggest it may be a satisfactory alternative catalyst.

1. INTRODUCTION

In order to improve the thermodynamic accuracy of temperature measurement a new International Temperature Scale, ITS-90, was introduced on 1 January 1990 (1). In the cryogenic range of the scale, the fixed points are almost all triple points of elemental gases. However two hydrogen boiling points (near 17 and 20 K) are included. It has been suggested (2) that one or both might be replaced by the deuterium triple point at 18.7 K, but this idea has not been fully proven because of the problems in accurately realising this triple point.

To obtain accurate realisations of triple points in sealed cells, gas of the highest purity is required at a high pressure. Pavese and McConville (4) and now Ancsin (5) and Khnykov et al (6) claim to have sources of high purity deuterium. Pavese and McConville (4) have also prepared a container which does not contaminate the deuterium over time. Deuterium, like hydrogen (3), requires a catalyst to speed up the spin conversion to the equilibrium ortho/para ratio. Failure to achieve this will produce gross errors.

Pavese's value of $18.688_8 \pm 0.001$ K is the present best estimate of the thermodynamic temperature for equilibrium deuterium and he obtained $18.723_2 \pm 0.001$ K for the triple point of normal deuterium. After conversion to thermodynamic temperature, Khnykov et al have reported a value for equilibrium deuterium which agrees with the result of Pavese. Similarly the initial measurements of Ancsin gave a comparable value, when using his purest gas, but his cell slowly became contaminated with time.

2. THE CATALYST

Over the long term the problem of hydrogen contamination by the catalyst has not yet been solved. As these triple point cells will probably remain sealed for many years the traditional catalyst, hydrous ferric oxide, has been found unsuitable, as has the commercially available nickel silicate (4).

Khnykov et al (6) have reported that $NiO.Cr_2O_3$ is suitable as it contains no water of hydration, but have so far produced no long term measurements.

In an attempt to provide a non-contaminating catalyst we selected gadolinium oxide (Gd_2O_3) because of its magnetic properties. It has previously been investigated by Weitzel et al (7) with hydrogen.

Three deuterium cells were initially measured, one with no catalyst. The others contained one gram of de-hydrated ferric oxide and one gram of gadolinium oxide respectively. Both the catalysts and the cells were vacuum baked at 400°C prior to filling with deuterium of purity 99.96% D_2 (ie 0.04% HD) (8). The details of the triple point cells and the cryogenic measurement system will be published elsewhere.

3. RESULTS

Based on the manufacturer's measurement of the HD impurity (8) we expected a depression in all three cells of about 1 mK. The uncatalysed (normal) NPL cell gave a value 2.4 mK below Pavese's best estimate. Our lower value may suggest further hydrogen contamination during storage or in the filling process, though some spin conversion may have occurred in the cooldown. Later measurements showed a conversion rate of less than 0.3 mK per day.

Repeated measurements of the cell containing de-hydrated ferric oxide over a week gave the same result as the initial measurement. This indicated that the catalyst operated efficiently even though it had been vacuum baked. Meanwhile the results from the cell containing Gd_2O_3 took a few days to equilibrate, indicating that gadolinium oxide acts as a slow catalyst.

The triple point of the cells containing de-hydrated ferric oxide (Fe cell) and the gadolinium oxide (Gd cell) were lower than the best thermodynamic value of Pavese by 4.0 mK and 1.6 mK respectively.

A repeat measurement was made of the two cells containing one gram of catalyst one year after the first experiments. The triple point value of the Fe cell fell by another 0.5 mK while that for the Gd cell was a further 0.6 mK lower. These depressions presumably arise from gradual in situ contamination with hydrogen.

A third set of measurements was made on the catalysed cells after an interval of another year. The normal cell was also re-measured. The initial measurement of the normal cell was lower than that measured two years previously by 1.2 mK. As it contains no catalyst, we assume that it is being contaminated by the cell body. (The cell is made from steel and copper with a small indium seal. As the indium cannot be baked at 400°C it is a possible source of contamination).

Once equilibrated the Gd cell gave a value a further 0.4 mK lower, making a total two year drop of 1.0 mK. There is a similarity in the depressions of the normal and Gd cells. The slightly larger value for the normal cell is not significant in view of the occurrence of conversion.

The triple point of the Fe cell showed another fall of 0.3 mK, bringing a fall over two years of 0.8 mK. Again this is of the same magnitude as the other cells.

4. CONCLUSIONS

The traditional hydrated ferric oxide catalyst is not suitable for high purity deuterium even after de-hydration as it causes too much contamination.

The gadolinium oxide works as a catalyst with deuterium, though its action is comparatively slow. As the increase in contamination, over the two years in both the Gd and normal cells were similar we assume these were due to hydrogen coming from the identical cell bodies. If this is correct it suggests that Gd_2O_3 does not contaminate the gas and could be an acceptable catalyst for high purity deuterium.

It is proposed to make some measurements using the Gd_2O_3 catalyst and specially prepared deuterium gas of higher purity (4) in collaboration with Dr Pavese. These cells will be pinch welded instead of being sealed with indium, which would remove this source of contamination.

ACKNOWLEDGEMENTS

We would like to thank Dr K Nara of NRLM, Japan for his assistance in the first set of measurements and Alison Jones who prepared the hydrated ferric oxide catalyst.

REFERENCES

(1) H. Preston Thomas, Metrologia 27 (1990) 3.
(2) F. Pavese and G.T. McConville, Temperature Measurement (Proceedings of the International Symposium on Temperature Measurement in Industry and Science), China Academic Publishers, Beijing, China, (1986) p. 140.
(3) J. Ancsin, Metrologia 13 (1977) 79.
(4) F. Pavese and G.T. McConville, Metrologia 24 (1987) 107.
(5) J. Ancsin, Metrologia 25 (1988) 155.
(6) V.M. Kynykov, M.I Loev, V.S. Parbusin, L.I. Rabuch and D.N. Astrov, BIPM Doc. CCT/89-9, Bureau International des Poids et Mesures, Sèvres, France.
(7) D.H. Weitzel, J.W. Draper, O.E. Park, K.D. Timmerhaus and C.C. Van. Valin, Advances in Cryogenic Engineering 2 (1957) 12.
(8) Cambridge Isotope Laboratories, 20 Commerce Way, Woburn, Massachusetts 01801.

Physica B 165&166 (1990) 163–164
North-Holland

THE EFFECTS OF EXCHANGE GAS TEMPERATURE AND PRESSURE ON THE BETA-LAYERING PROCESS IN SOLID DEUTERIUM-TRITIUM FUSION FUEL

James K. HOFFER and Larry R. FOREMAN

Los Alamos National Laboratory, Los Alamos, New Mexico 87545, USA

John D. SIMPSON and Ted R. PATTINSON

KMS Fusion, Inc., Ann Arbor, Michigan 48106, USA

It has recently been shown that when solid tritium is confined in an isothermal enclosure, self-heating due to beta decay drives a net sublimation of material from thick, warmer layers to thin, cooler ones, ultimately resulting in layer thickness uniformity. We have observed this process of "beta-layering" in a 50-50 D-T mixture in both cylindrical and spherical enclosures at temperatures from 19.6 K, down to 11.6 K. The measured time constants are found to depend on the ^{3}He content as suggested by recent theoretical predictions. When using an enclosure having low thermal conductivity, the ultimate layer uniformity is found to be a strong function of the exchange gas pressure. This is due to the presence of thermal convection in the exchange gas and consequent temperature anisotropy at the solid layer surface.

1. INTRODUCTION

The concept of beta-layering was first introduced by Miller(1) in 1975. A one-dimensional theory was formalized by Martin et al.(2) in 1985. It proposed that in a chamber partially filled with deuterium and tritium (DT) fuel and cooled to below the triple point, an automatic redistribution of the solid occurs. Because of self-heating due to the absorption of beta particles, thick layers of solid DT tend to have warmer interior surfaces than do thin ones. A net sublimation of material from thick sections to thin ones takes place, expressed as

$$\delta(t) = \delta_0 \cdot e^{-t/\tau} , \qquad (1)$$

where δ_0 is the initial anisotropy in the layer thickness and the rate constant $\tau \equiv H_s/q$, where H_s is the heat of sublimation and q is the self-heating rate. Hoffer and Foreman(3) first observed the effect in pure tritium, confirming eq. 1. Those experiments, and subsequent ones in a 50-50 mixture of DT(4), showed that the equilibration rate was also a function of the age of the material, due to the accumulation of ^{3}He in the vapor space. Recently, Bernat(5) has extended Martin's formalism to account for the diffusion of DT vapor across the interior void space in the presence of stagnant ^{3}He gas. He also arrives at Eq. 1, but with $\tau = H_s/q + \tau_3$, where

$$\tau_3 \equiv \frac{(r-d)\cdot k_s \cdot R \cdot T \cdot P_3}{d \cdot q \cdot D \cdot P \cdot (dP_{DT}/dT)} . \qquad (2)$$

In this equation, r is the inner radius of the enclosure, d is the equilibrium solid layer thickness, k_s is the thermal conductivity of the solid, R is the gas constant, T is the temperature and and P_3 the partial pressure of ^{3}He in the vapor space. D is the diffusivity of the DT-^{3}He binary gas system. P_{DT} is the vapor pressure of DT, and $P \equiv P_{DT} + P_3$ is the total pressure. If we assume that all the ^{3}He produced by tritium decay is in the vapor space, then P_3 will increase every day by

$$dP_3/dt \approx 1.537 \cdot 10^{-4} \cdot \sigma_s \cdot R \cdot T \cdot f_s/(1-f_s), \qquad (3)$$

where, σ_s is the solid density, and f_s is the fraction of volume occupied by the solid. Using properties for DT compiled by Souers(6), we have estimated τ_3 for both cylindrical and spherical geometries, for various values of f_s as shown in Table 1. The value f_s = 0.88 corresponds to the case where an enclosure is filled with liquid DT and then frozen. Note that the effects of ^{3}He increase dramatically below ~16 K.

Table 1. Time Constants for Beta-Layering

T (K)	H_s/q (min)	τ_3(cyl.)	τ_3(sph.) (min/day)	τ_3(sph.)
		f_s=.88	f_s=.88	f_s=.2
10.0	24.74	51300	40800	18600
12.0	25.18	2740	2180	992
14.0	25.62	343	273	124
16.0	26.07	73.2	58.3	26.5
18.0	26.51	22.2	17.7	8.05
19.0	26.73	13.5	10.7	4.89
19.79	26.91	9.45	7.52	3.42

0921-4526/90/$03.50

2. TEMPERATURE EFFECTS - EXPERIMENTAL PROCEDURES

We made use of the previous apparatus and optical technique(3,4), i.e., a copper cylinder sealed with sapphire windows, to insure isothermal boundaries and good optical access. To minimize the effects of ^3He accumulation, the DT mixture was purified in a Pd bed prior to each series of experiments. This removed the ^3He so well that we could fill the cylinder completely with liquid DT prior to freezing. However, it was impractical to purify the DT mixture following beta-layering at each different temperature, and thus slight accumulations of ^3He gas could not be prevented.

3. TEMPERATURE EFFECTS - EXPERIMENTAL RESULTS

We initially interpreted the results shown in Fig. 1 as a temperature effect in "fresh" DT (i.e., DT having no ^3He impurity) and we fitted the data with an Arrhenius expression. However, the Bernat model outlined above suggests a better interpretation, namely that there is virtually no effect of temperature on the rate of beta-layering in pure DT, but that the effect of ^3He accumulating in the vapor space is so strongly temperature dependent that even miniscule amounts of ^3He will dominate the rate of beta-layering below ~14 K. Figure 1 shows values of τ calculated from Eq. 2 at times corresponding to the beginning and the end of each experiment. The close agreement with our measured values shows that the model describes the physics of beta-layering in isothermal enclosures remarkably well.

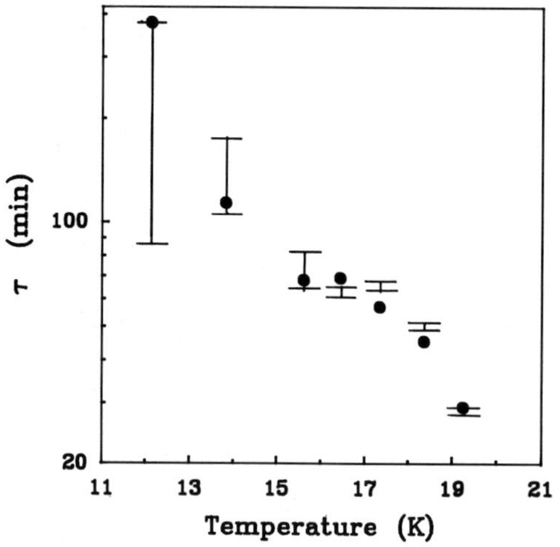

FIGURE 1
Equilibration rate constants in "fresh" DT solid at various temperatures. The horizontal bars represent values of τ predicted by Eq. 2 at the beginning and at the end of each experiment. The data are measured at different times and thus do not correlate directly with temperature.

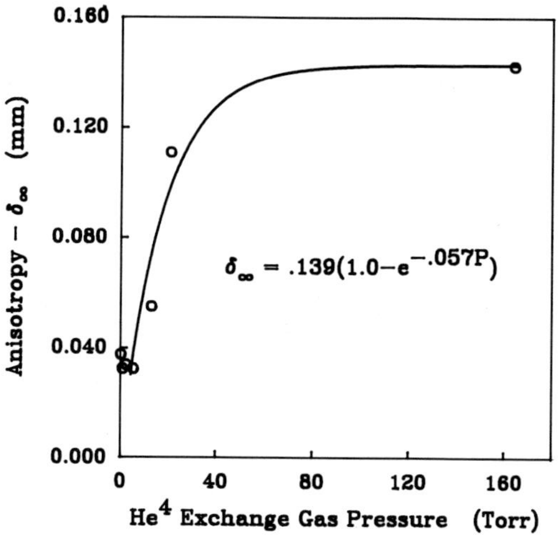

FIGURE 2
The dependence of the exchange gas pressure on the ultimate symmetry observed following beta-layering of solid DT in a spherical lexan shell.

4. PRESSURE EFFECTS - PROCEDURES AND RESULTS

In the second set of experiments, the copper cylinder was replaced by a hollow lexan sphere having the same volume. This is characteristic of prototype inertial confinement fusion (ICF) reactor targets. In both sets of experiments, the DT containers were thermally linked to the cryostat by ^4He gas at a pressure of ~600 torr. But, with the poorly conducting lexan sphere, a large asymmetry in the equilibrium layer thickness, δ_∞, was observed. Suspecting that this was an effect of convection in the ^4He exchange gas, we conducted a series of experiments at decreasing pressures until relatively symmetric layers were achieved, as shown in Fig. 2. The slight anisotropy remaining below ~5 torr is probably due to the metal fill tube and support tube attached to the poles of the sphere.

REFERENCES

(1) J. R. Miller, "Progress Report on the Laser Fusion Program at LASL," LA-6245-PR, Los Alamos National Laboratory, Los Alamos NM, 87545, USA (1976) p. 87.
(2) A. J. Martin, R. J. Simms, and R. B. Jacobs, J. Vac. Sci. Technol. A6 (1988) 1885.
(3) J. K. Hoffer and L. R. Foreman, Phys. Rev. Letters 60 (1988) 1310.
(4) J. K. Hoffer and L. R. Foreman, J. Vac. Sci. Technol. A7 (1989) 1161.
(5) T. P. Bernat, private communication. E. R. Mapoles, J. J. Sanchez, and T. P. Bernat, to be published.
(6) P. C. Souers, Hydrogen Properties for Fusion Energy (University of California Press, Berkeley CA, 1986).

Physica B 165&166 (1990) 165–166
North-Holland

SURFACE TENSION OF LIQUID N_2 NEAR THE LIQUID-VAPOR CRITICAL POINT

Akiko SHIMOYANAGITA, Nobutaka ITAGAKI, Akira SATO, and Masaru SUZUKI

The University of Electro-Communications, 1-5-1, Chofugaoka, Chofu-shi, Tokyo 182, Japan.

The surface tension of liquid N_2 was measured in the temperature range from 80 K up to the liquid-vapor critical point by the surface wave resonance method. This new method has the advantage of the precise measurement near the critical point at low temperatures. The present data are in agreement with the previous data obtained by the capillary rise method within the experimental errors. The critical exponent for the surface tension was found to be $\mu = 1.23 \pm 0.04$. The value is consistent with the recent experimental values of other liquids.

1. INTRODUCTION

Although the surface tension is very well known physical quantity for a long time, its behavior near the liquid-vapor critical point Tc has been attracting interest. Because the critical exponent for the surface tension is believed to be universal and be related with other critical exponents of thermodynamic properties in bulk. The phenomenological theory of Fisk and Widom (3) leads the scaling relation of the critical exponents,

$$\mu = \gamma' + 2\beta - \nu',$$

where, μ, γ', β and ν' are the critical exponents below Tc for the surface tension, the isothermal compressibility, the difference between the coexistent liquid and vapor densities and the correlation length, respectively. The theoretical value of calculated by the renomalization group is 1.26.

The recent experimental values are in the range $\mu = 1.28 \pm 0.06$. However, there are hardly any precise measurements of simple liquids and it is necessary to measure the surface tension more precisely and systematically to study the liquid surface. So we have measured the surface tension of liquid N_2 at first.

2. EXPERIMENTAL

The capillary rise method was usually used to measure the critical behavior of the surface tension. It can have very high resolution, if a small capillary is used. However, it is difficult to ensure that a fluid in a small capillary has the same temperature and density as the fluid in bulk and to estimate the effects of wetting layers. These problems become more serious near Tc. Furthermore, in this method it is not the exponent for surface tension μ but μ-β that is measured directly. So some of variability in μ might come from the assumed β.

These problems are avoided by using the surface waves (4). In this experiment the surface tension was obtained by measuring the surface wave resonance on liquid N_2 contained in a rectangular cavity. The surface tension σ was calculated from the resonant frequency through the dispersion relation,

$$\omega^2 = \{(\rho_L - \rho_V)gk + \sigma k^3\}/(\rho_L + \rho_V),$$

where, ρ_L and ρ_V are the densities of coexistent liquid and vapor, respectively, g is the acceleration due to gravity, and k is the wave number. The wave number is determined by

$$k = n\pi/(d+\Delta d),$$

where n is the mode number of resonance, d is

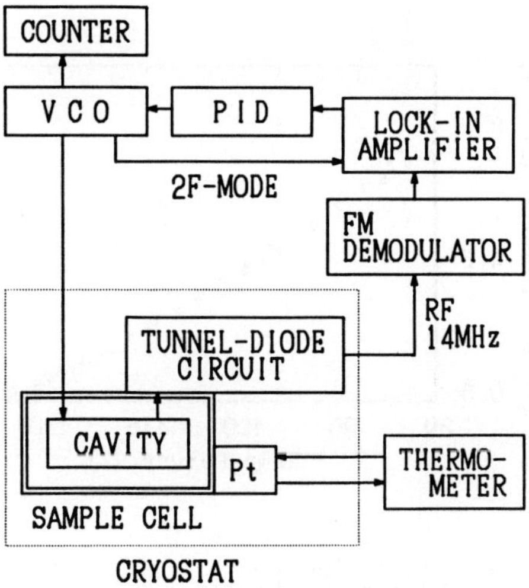

Figure 1
Block diagram of the measuring system.

the distance of cavity, and Δd is the correction of meniscus. To estimate the correction of meniscus, two sets of paralleled cavities measuring 22×15×5 mm and 22×15×10 mm were used.

For the excitation and the detection of the surface waves, sets of 500 small capacitors on glass plates were used in the interleaved structure. The interelectodes made of photoetched aluminum had a distance of 10 μm. The surface waves were excited by applying an alternating voltage of a voltage-controlled oscillator (VCO) to the exciter. The standing wave in the cavity was detected as a periodic capacitance change of the detector, which has twice the frequency of the excitation.

A block diagram is shown in Fig.1. The detected capacitance change was converted to a frequency-modulated RF signal by a tunnel-diode circuit in a cryostat. The inphase output of the lock-in amplifier was returned to the VCO through the proportional-integral-differential (PID) control circuit so that the output of the lock-in amplifier becomes zero. That is, the total system, working as a phase-locked-loop (PLL) circuit, locked the VCO at the resonant frequency of the surface wave.

3. RESULTS AND DISCUSSION

It is shown in Fig.2 that the temperature dependence of the surface tension of liquid N_2 over a wide temperature range. Circles in the figure denote the present data. The error in this experiment is considered to be mainly due to ambiguities in the resonant wave numbers. Triangles and crosses denote the data of Stansfield (5) and Gielen et al. (6) using the capillary rise method. We have found that the two different methods yield consistent results

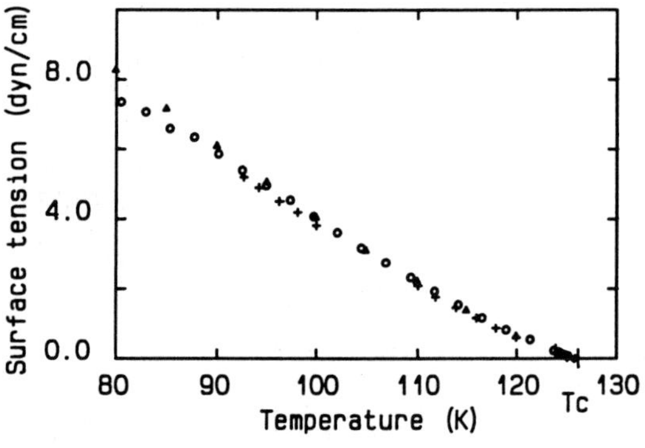

Figure 2
Comparison between the present data ○ and the data obtained by the capillary rise method Δ,+.

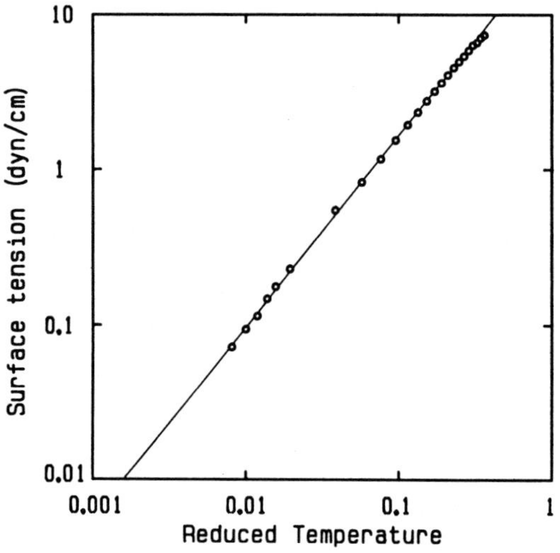

Figure 3
Critical log-log plots of the present data. The straight line represents $\varepsilon^{1.23}$.

within the experimental errors, although they are subject to different kinds of systematic errors.

Figure 3 shows plots of the surface tension against the reduced temperature $\varepsilon = (Tc-T)/Tc$ on a log-log scale. A linear relationship on log-log scale is obtainable by the present data. We have analysed them assuming the single power-law formula. The critical exponent for the surface tension was found to be $\mu = 1.23 \pm 0.04$. The value is consistent with the recent experimental values for other liquids.

ACKNOWLEDGEMENTS

This work was partly supported by a Grant-in-Aid for Scientific Research of the Ministry of Education, Science and Culture of Japan.

REFERENCES

(1) B. Widom, Phase Transition and Critical Phenomena, C. Domb and M.S. Green, eds. (Academic Press, New York, 1972). Vol 2.
(2) M.R. Moldover, Phys. Rev. **A31** (1985) 1022.
(3) S. Fisk and B. Widom, J. Chem. Phys. **50** (1969) 3219.
(4) M. Iino, M. Suzuki and A.J. Ikushima, J. Low Temp. Phys. **63** (1986) 495.
(5) D. Stansfield, Proc. Phys. Soc. **72** (1958) 845.
(6) H.L. Gielen, O.B. Verbeke and J. Thoen, J. Chem. Phys. **81** (1984) 6154.

Physica B 165&166 (1990) 167–168
North-Holland

IMAGING OF SPATIAL STRUCTURES OF FISKE-, SHAPIRO-, AND PHOTON-ASSISTED TUNNELLING STEP STATES IN JOSEPHSON TUNNEL JUNCTIONS

T. DODERER, B. MAYER, C. KRUELLE, R. P. HUEBENER, D. QUENTER, J. NIEMEYER[+], R. FROMKNECHT[+], R. POEPEL[+], A. V. USTINOV[*], U. KLEIN[#], J. H. HINKEN[$]

Physikalisches Institut, Lehrstuhl Experimentalphysik II, Universität Tübingen, FRG.

Using Low Temperature Scanning Electron Microscopy (LTSEM) the Fiske-Steps generated in a Josephson tunnel junction in the presence of a magnetic field have been studied. Spatial structures of the beam-induced change of the maximum step currents and also of the voltage signals of current biased step states have been observed with a spatial resolution of a few µm. In combination with a microwave source LTSEM offers the possibility to study the spatial distribution of rf-induced constant-voltage steps (Shapiro-steps) and of photon-assisted tunnelling steps. In the former case the spatial distribution of the maximum step current of various Shapiro-steps has been imaged. In the latter case we obtained evidence for the nucleation of an inhomogeneous energy gap state in the junction electrodes due to microwave irradiation.

1. INTRODUCTION

Josephson tunnel junctions show interesting behaviour in a magnetic field or if they are coupled to a rf-source (1). In both cases one observes not only a change of the dc-Josephson current but also structures at finite voltages in the current-voltage-characteristics (IVC). With a magnetic field applied parallel to the junction barrier Fiske-steps are generated at discrete voltages. As resonant dynamic modes of the junction, the corresponding voltages are only determined by the dimensions and the material of the junction. On the other hand, by applying microwave irradiation Jospehson tunnel junctions show at least two distinct peculiarities. First, due to phase locking of the Josephson oscillator to the external oscillator rf-induced constant-voltage steps (Shapiro-steps) develop. These steps are the basis of voltage standards (2). Second, near the energy gap region photon-assisted tunnelling steps in the IVC of the quasiparticles can be observed. The IVC contains only spatially averaged information of the static or dynamic processes inside the junction. By applying LTSEM for investigating these various states it is possible to get a deeper insight into the corresponding junction behaviour. Here we present the results of the first spatially resolved measurements dealing with Josephson tunnel junctions in Fiske-, Shapiro-, or photon-assisted tunnelling step states with a spatial resolution of a few µm.

2. EXPERIMENTAL PROCEDURES, RESULTS, AND DISCUSSION

2.1. Fiske-steps

The sample used in our experiments was a 50x50 μm^2 Nb/AlOx/Nb-tunnel junction. The relative length ℓ and width w of the junction was $\ell/\lambda_J = w/\lambda_J = 2.0$. We have imaged the spatial distribution of the maximum Josephson current I_c by scanning the junction with the electron beam and by recording the beam induced change of I_c as a function of the position of the beam focus (3). We have observed only a slight deviation from a homogeneous current distribution due to the self field of the Josephson current. The same imaging procedure was applied to the maximum Fiske-step current for various steps. For current biased step states we observed a change of the corresponding step voltage due to electron beam irradiation. Fig.1 shows these changes of the maximum step current (part a) and of the step voltage (part b) for two different modes (4). The change of the maximum step current for steps up to the 7th Fiske-step clearly indicates the cavity-mode excitation (5). If the magnetic field is parallel to the edge of the junction, the spatial variation of the beam-induced current and voltage signal appears only perpendicular to the direction of the field (Fig. 1.a). On the other hand, we observed two-dimensional cavity-mode excitations for nearly diagonal direction of the magnetic field (Fig. 1.b).

[+]Physikalisch-Technische Bundesanstalt, Braunschweig, FRG. [*]Permanent address: Institute of Solid State Physics, USSR Academy of Sciences, Chernogolovka, Moscow district, USSR. [#]Technische Universität Braunschweig, Institut für Hochfrequenztechnik, FRG. [$]FUBA-Forschungszentrum, Bad Salzdetfurth,FRG.

0921-4526/90/$03.50 © 1990 – Elsevier Science Publishers B.V. (North-Holland)

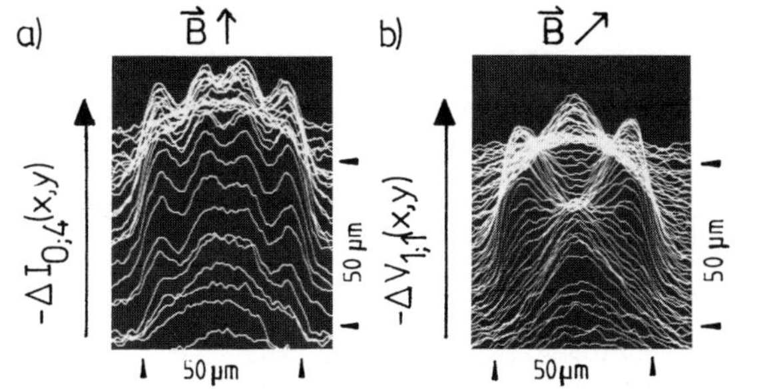

FIGURE 1
(a) Spatial distribution of the electron beam induced change of the maximum Fiske-step current of the 4th step $I_{0;4}$. (b) Spatial distribution of the electron beam induced change of the Fiske-step voltage of the first mixed mode step $V_{1;1}$.

2.2. Shapiro-steps

The samples consisted of Pb-alloy- and Nb/AlOx/Nb-tunnel junctions embedded in a microstripline well coupled to a receiving antenna and terminated by a resistive load. The length of the junctions in the direction of the microwave propagation was 20 µm, and the width was 40 µm. These dimensions are nearly the same for tunnel junctions used as voltage standards (2). For details of the experimental set-up we refer to (6). The samples with homogeneous rf input across the whole width showed the same distribution of the maximum step current densities as the Josephson current density up to the 4th Shapiro-step. In small Josephson junctions the distribution of the Josephson current reflects the distribution of the conductivity across the junction (3). One sample with inhomogeneous coupling to the stripline (only near one corner) showed an inhomogeneously distributed current density of the 2nd step whereas the 1st step and the Josephson current were distributed quite homogeneously. By imaging the distribution of the Josephson current density with a trapped flux quantum inside the junction, we obtained images such as in ref. (3). In contrast, the distribution of the Josephson current, reduced to about the same value due to the rf-irradiation, was quite homogeneous.

2.3. Photon-assisted tunnelling steps

We used 50x50 µm^2 and 100x100 µm^2 Nb/AlOx/Nb-junctions. The single junction was irradiated directly with 70 GHz microwaves with the help of a horn antenna. The well-known photon-assisted tunnelling steps were generated in the IVC of the quasiparticles. By scanning the electron beam across the current-biased junction and recording the electron-beam induced voltage signals we obtained the energy gap distribution of the junction electrodes (3). Without microwave irradiation the energy gap value was rather homogeneously distributed across the junction. However, biasing the junction at photon-assisted tunnelling steps we observed inhomogeneous energy gap states due to microwave irradiation (Fig. 2). From the 2nd step below up to the 2nd step above the unperturbed energy gap different states along the IVC were investigated. Biasing the smaller junction on a single step, the energy gap distribution did not change qualitatively for different current values. However, it changes significantly from one step to another (Fig.2). The larger sample showed a more complicated behaviour. Using the samples from section 2.2. we did not observe any inhomogeneous distribution of the energy gap. This probably results from the smaller dimensions of those junctions (7,3). The way of rf-coupling the junction also may play a role.

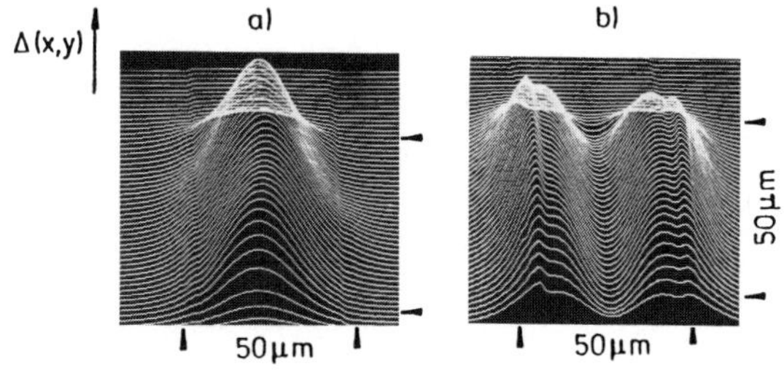

FIGURE 2
Measured spatial distribution of the energy gap of the tunnel junction electrodes of the smaller junction. In (a) the junction was current-biased at the centre of the central step. In (b) the same junction was current-biased at the centre of the 1st step, i.e., the corresponding voltage was about 290 µV higher. For both images the same microwave power was used. The signal-peaks at the left- and right-hand edge of the junction are due to thickness-variations of the sample.

ACKNOWLEDGEMENTS

We would like to thank M. Koyanagi for manufacturing and sending the quadratic junctions of high quality. Financial support of this work from the Deutsche Forschungsgemeinschaft is gratefully acknowledged. One of the authors (A.V.U.) is pleased to acknowledge support of the Alexander von Humboldt-Stiftung.

REFERENCES

(1) A.Barone, G.Paternò, Physics and Applications of the Josephson Effect (Wiley&Sons, New York, 1982)
(2) J.Niemeyer, in: Superconducting Quantum Electronics, ed. V.Kose (Springer, Berlin, 1989) pp. 228-254
(3) R.P.Huebener, this volume, pp. 205-225
(4) H.Svensmark, in: SQUID '85, eds. H.D.Hahlbohm, H.Lübbig (de Gruyter, Berlin, 1985) pp.471-475
(5) S.Meepagala, et al. Phys.Rev.B36 (1987) 809
(6) T.Doderer, et al. Cryogenics 30 (1990) 65
(7) H.Akoh, K. Kajimura, Phys. Rev. B25 (1982) 4467

Physica B 165&166 (1990) 169–170
North-Holland

THE ANGULAR DEPENDENCE OF THE MAGNETIC PHASE DIAGRAM OF $CsMnBr_3$

Alain CAILLE, Martin L. PLUMER, Mario POIRIER and Bruce D. GAULIN*

Centre de Recherche en Physique du Solide, Département de Physique, Université de Sherbrooke, Sherbrooke, Québec J1K 2R1 Canada.

The angular dependence of the magnetic phase diagram of $CsMnBr_3$ for H in the ac plane is determined using an ultrasonic velocity technique. These results are well reproduced using a previously developed non local Landau theory.

1. INTRODUCTION

$CsMnBr_3$ is an hexagonal quasi-one-dimensional magnetic insulator whose crystal structure is viewed as a layered triangular lattice with antiferromagnetic exchange interactions. In this material, strong in-plane magnetic anisotropy forces the magnetic moments to lie in the ab plane at low temperatures. The inherent magnetic frustration in this triangular lattice imposes a 120° magnetic structure in the ground state with the $\sqrt{3}\times\sqrt{3}$ lattice periodicity. The order parameter space has $Z_2 x S_1$ symmetry making this material a member of the new chiral universality class exposed by Kawamura (1). A magnetic field in the ab plane reveals a new type of multicritical point at $T=T_N$, H=0. The field acts to split the transition at T_N into two second order transitions which, with decreasing temperature, belong respectively to the XY and Ising universality classes (2). The magnetic phase diagram with a field in the ab plane was first determined using neutron scattering by Gaulin et al (3) and confirmed by the ultrasonic study of Poirier et al (4). A non local Landau theory has been used successfully to understand these experimental results (5). In this paper we present the experimentally determined phase diagram for a magnetic field sweeping the ac plane and show these data to be well reproduced by the non local Landau theory.

2. EXPERIMENTAL RESULTS

The ultrasonic velocity measurement technique and the crystal used in these experiments are described in Ref.(4). Figure 1 shows the single phase boundary between the paramagnetic phase (phase 1) and the helically polarized magnetic phase (phase 4) for a magnetic field in the c-direction. Figure 2 presents the angular dependence of the two critical temperatures for a field sweeping the ac plane at H = 4.245 T. The limitations and the general conditions for the experiments may be found in Ref.(4).

3. NON LOCAL LANDAU THEORY

The spin density $\vec{\rho}(\vec{r})$ is expressed by

$$[1] \quad \vec{\rho}(\vec{r}) = \vec{m} + \vec{S}e^{i\vec{Q}\cdot\vec{r}} + \vec{S}^{*}e^{-i\vec{Q}\cdot\vec{r}}$$

where m is the field induced uniform magnetization which is assumed to be parallel to H. S and Q characterize the long range magnetic order. Assuming that the magnetic anisotropy energy keeps the polarization S in the ab plane, the non local Landau free energy expansion to fourth order becomes

$$[2] \quad F = A_Q S^2 + \frac{1}{2}A'_0 m^2 - \frac{1}{2}A_{z0} m^2 \sin^2\theta + B_1 S^4 + \frac{1}{2}B_2|\vec{S}\cdot\vec{S}|^2 + \frac{1}{4}B_3 m^4 + 2B_4|\vec{m}\cdot\vec{S}|^2 + B_5 m^2 S^2 - mH$$

where $S^2 = \vec{S}\cdot\vec{S}^{*}$ and $\vec{H}$ is at an angle θ with the ab plane. A_{z0} is the only term which is not due to an isotropic exchange; anisotropy terms fourth-order in $\vec{\rho}$ have been omitted for simplicity (6). The term A_{z0}, which should be large and negative, is the only feature added to the free energy of Ref.5 reflecting the angular dependence of H in the ac plane. Proceeding as in Ref.(5), the phases boundaries between phase 1 and the linear phase 5 and between 5 and the elliptical phase 7 are easily determined and are given by

$$[3] \quad H_{15}^2 = \left(-\frac{A_Q}{B_5}\right)\left[\Delta(\theta) + \left(1 - \frac{B_3}{B_5}\right)A_Q\right]^2$$

$$[4] \quad H_{57}^2 = \left(-\frac{A_Q}{B_{12}(\theta)}\right)\left[\Delta(\theta) + \left(1 - \frac{B_{13}(\theta)}{B_{12}(\theta)}\right)A_Q\right]^2$$

*Department of Physics, McMaster University, Hamilton, Ontario, Canada L8S 4M1.

0921-4526/90/$03.50 © 1990 – Elsevier Science Publishers B.V. (North-Holland)

The different parameters entering [3] and [4] have the following definition.

[5] $A_Q = a(T-T_Q)$; $A_0' = a(T-T_0')$; $T_Q = T_N$

[6] $\Delta(\theta) = A_0' - A_{Z0}\sin^2\theta - A_Q$

[7] $B_{12}(\theta) = B_5 + \dfrac{BB_4(\theta)}{B_2}$

$B_{13}(\theta) = B_3 + \dfrac{2B_4(\theta)B_5}{B_2}$

[8] $B = 2B_1 + B_2$

[9] $B_4(\theta) = B_4\cos^2\theta$

A single line of transition points is obtained between phases 1 and 4 if $\theta = \pi/2$, $\vec{H}//\vec{c}$

[10] $H_{15} = H_{57} = H_{14}$; $B_4(\pi/2) = 0$

[11] $H_{14}^2 = \left(-\dfrac{A_Q}{B_5}\right)\left[\Delta + \left(1 - \dfrac{B_3}{B_5}\right)A_Q\right]^2$

$\Delta = \Delta(\theta = \pi/2) = A_0' - A_Q - A_{Z0}$

All the parameters appearing in [11], except for A_{Z0}, were fitted to the data for H in the ab plane and may be found in Ref.(5). The experimental data for H // c were then fitted to [11] with the single parameter A_{Z0}. The result is shown as the full drawn curve in figure 1. The best fit is obtained with $A_{Z0} = -2.35 \times 10^{+4}$ in cgs units.

This value is large ($\Delta(0) \cong 2.1 \times 10^{+4}$), reflecting the strong planar anisotropy. In figure 2, we have drawn the predicted phase boundaries H_{15} and H_{57} using [3] and [4] respectively as a function of the angle Θ for H = 4.245 T.

4. CONCLUSION

As seen above (and in Ref.5), the non local Landau free energy expansion to fourth order is very successful in explaining the phase diagram of $CsMnBr_3$. Not only is it possible to obtain a good fit to the experimental data for H(T) with a reasonable choice of the expansion coefficients but the angular dependence, for H in the ac plane (which is a no free parameter prediction), shows the same general features as the experimental data. Better agreement would result from including the fourth-order anisotropy terms omitted in the present model, but at the cost of having to fit more unknown coefficients.

ACKNOWLEDGEMENTS

We acknowledge the financial support of CRSNG and FCAR and the technical aid of M. Castonguay.

REFERENCES

(1) H. Kawamura, J. Appl. Phys. 63, 3086 (1988); Phys. Rev. B. 38, 4916 (1988).
(2) H. Kawamura, A. Caillé and M. L. Plumer, Phys. Rev. B. (to be published).
(3) B. D. Gaulin, T.E. Mason, M.F. Collins, J.Z. Larese, Phys. Rev. Lett. 62, 1380 (1989).
(4) M. Poirier, M. Castonguay, A.Caillé, M.L. Plumer and B.D. Gaulin, (these proceedings).
(5) M. L. Plumer and A. Caillé, Phys. Rev. B. 41, 2543 (1990).
(6) M.L. Plumer, A. Caillé and K. Hood, Phys. Rev. B. 39, 4489 (1989).

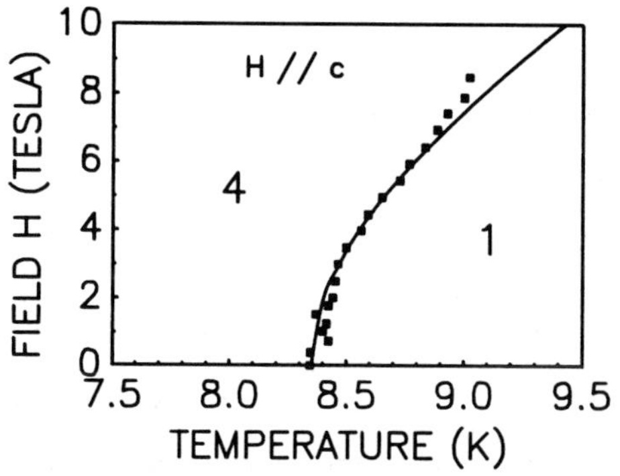

FIGURE 1
Magnetic phase diagram for H // c. The full line is the fitted theory; phases are labelled as in the text

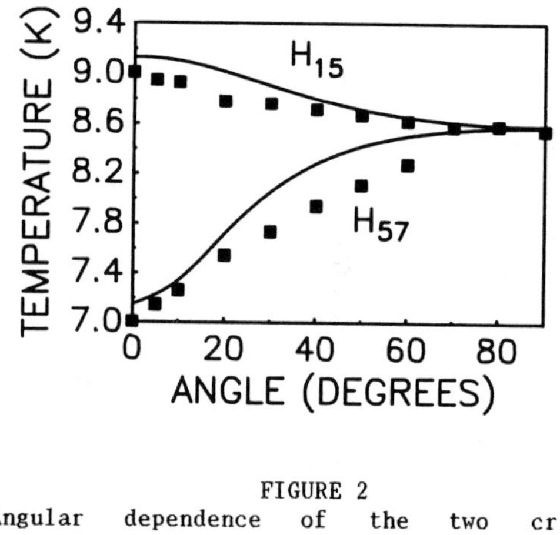

FIGURE 2
Angular dependence of the two critical temperatures for H in the ac plane at 4.245 T. The full lines are the no-free-parameter predictions.

Physica B 165&166 (1990) 171–172
North-Holland

ULTRASONIC STUDY OF QUASI-ONE-DIMENSIONAL MAGNETIC SYSTEM $CsMnBr_3$

Mario POIRIER, Mario CASTONGUAY, Alain CAILLE, Martin L. PLUMER and Bruce D. GAULIN*

Centre de Recherche en Physique du Solide, Département de Physique, Université de Sherbrooke, Sherbrooke, Québec J1K 2R1 Canada.

We report the elastic modulus investigation of the quasi-one-dimensional magnetic system $CsMnBr_3$ at low temperatures and moderate magnetic fields with an ultrasonic velocity measurement technique. The magnetic phase diagram with H in the basal plane is obtained.

1. INTRODUCTION

$CsMnBr_3$ is an insulator having hexagonal lattice structure, space group $P6_3/mmc$ (1), and exhibiting quasi-one-dimensional magnetic behavior above 15 K as a result of a stronger directional dependence of the exchange interaction. As the temperature is decreased, short range antiferromagnetic order develops within the chain of the spin-5/2 Mn^{2+} magnetic moments; below 20 K the spins are restricted to the ab plane by a strong in-plane magnetic anisotropy. In zero magnetic field, three-dimensional spin ordering occurs at T_N = 8.3 K. The application of a magnetic field in the basal plane breaks the transition into two lines of transitions. The magnetic phase diagram for H in the basal plane has recently been measured by Gaulin et al (2) and theoretically explained by Plumer et al (3). In this paper we present the results of magneto-elastic coupling experiments on $CsMnBr_3$ using a phase comparison method to measure the ultrasonic velocity. An anomaly is observed in the temperature dependence of the elastic modulus C_{44} around T_N. The splitting of the anomaly by a magnetic field aligned in the basal plane is used to obtain the magnetic phase diagram.

2. EXPERIMENT

The crystal used in the ultrasonic experiment was grown by the standard Bridgeman technique. Its tendency to cleave along the {1120} plane facilitated its orientation for acoustic propagation along the c-axis [0001]. Parallel faces, approximately 10 mm apart with normals parallel to the c-axis, were polished to receive the transducers. The longitudinal and transverse acoustic pulses were generated, at 30 MHz and odd overtones, by Y- and X-cut co-axially plated $LiNbO_3$ transducers bonded to the sample with Silicone Sealant. Acoustic velocity measurements were made at 90 MHz by a phase comparison method: a precision of 1 ppm may be obtained with this ultrasonic interferometer. Magnetic field amplitudes between 0 and 9 Tesla were applied perpendicular to the c-axis using a superconducting coil. The temperature was monitored with a Si diode and a $SrTiO_3$ capacitance sensors with a precision of 10 mK. Thermal expansion effects were neglected as they are at least one order of magnitude smaller than the measured velocity variations.

3. RESULTS AND DISCUSSION

For propagation along the c-axis, the longitudinal and transverse velocities yield respectively the elastic moduli C_{33} and C_{44}. At room temperature we obtain C_{33} = 3.93(.05) and C_{44} = 0.29(.01) in units of 10^{11} dyne/cm^2. The small value for C_{44} is consistent with the values generally obtained for similar ABX_3 quasi-one-dimensional systems having antiferromagnetic intrachain coupling. Both moduli decrease monotonically with temperature from 300 K down to 4 K with no clear structure associated with the onset of short-range order along the chains as in $CsNiCl_3$ (4). This may be an indication of a weak magneto-elastic coupling for this compound.

Figure 1 shows the relative variation of C_{44} as a function of temperature between 4 and 10 K. An anomaly is clearly seen at T_N = 8.34 K where three-dimensional antiferromagnetic order develops; the gradual slope variation at T_N is consistent with the second-order nature of the transition. The application of a magnetic field in the basal plane splits this zero-field transition; this can be clearly seen in figure 2 where field scans have been performed for two sample's temperature 4.3 and 10.3 K. There are clear indications of two different phase transitions. Various temperature (at fixed field) and field (at fixed temperature) scans have been performed in order to map out the phase diagram of $CsMnBr_3$ for a field applied in the basal plane. The results are shown in figure 3.

The phase diagram deduced from ultrasonic propagation is identical to that obtained by

*Department of Physics, McMaster University, Hamilton, Ontario L8S 4M1, Canada

0921-4526/90/$03.50

Gaulin et al (2) with elastic neutron scattering. Three phases numbered 1, 5 and 7 by Plumer and Caillé (3) are separated by two transition lines emanating from a novel kind of multicritical point (5) at H = 0, T_N = 8.34K. Phase 1 is paramagnetic and the ordered phases 5 and 7 have respectively linear and elliptical polarizations. Transitions between the phases are all second order; this seems to be confirmed by the continuous slope variation at the transition observed in figures 1 and 2. As the field is increased up to 9 Tesla, the modulus softens progressively for all temperatures (figure 2); this is a manifestation of magneto-elastic coupling and field effects should then be confirmed at higher temperatures by further experiments. These acoustic velocity data demonstrate the versality and sensitivity of the technique to map out the magnetic phase diagram of quasi-one-dimensional magnetic systems. This has also been shown previously by us with $CsNiCl_3$ (6).

4. CONCLUSION

The ultrasonic velocity data presented in this paper constitute the first study of the elastic behavior of the quasi-one-dimensional antiferromagnet $CsMnBr_3$. Although no anomaly is observed in the temperature dependences of C_{33} and C_{44} corresponding to the onset of short-range order along the chains, a small one is seen at T_N = 8.34 K in zero field when three-dimensional ordering develops. The field behavior of this anomaly has been studied up to 9 Tesla and the magnetic phase diagram of $CsMnBr_3$ has been mapped out (7). Comparison with elastic neutron scattering data is quite remarkable.

ACKNOWLEDGEMENTS

We acknowledge the financial support of NSERC and FCAR.

REFERENCES

(1) J.Goodyear and D.J. Kennedy, Acta Cryst. Sect.B, 28,1640 (1974).
(2) B.D. Gaulin, T.E. Mason, M.F. Collins and J.Z. Larese, Phys. Rev. Lett. 62, 1380 (1989)
(3) M.L. Plumer and A. Caillé, Phys. Rev. B41, 2543 (1990).
(4) K.R. Mountfield and J.A. Rayne, J. de Physique. 42, 468 (1981).
(5) H. Kawamura, A. Caillé and M.L. Plumer, Phys. Rev. B (to be published).
(6) M. Poirier, A. Caillé and M.L. Plumer, Phys. Rev.B. (to be published april 1990)
(7) A. Caillé, M.L. Plumer, M. Poirier and B.D. Gaulin, (these proceedings)

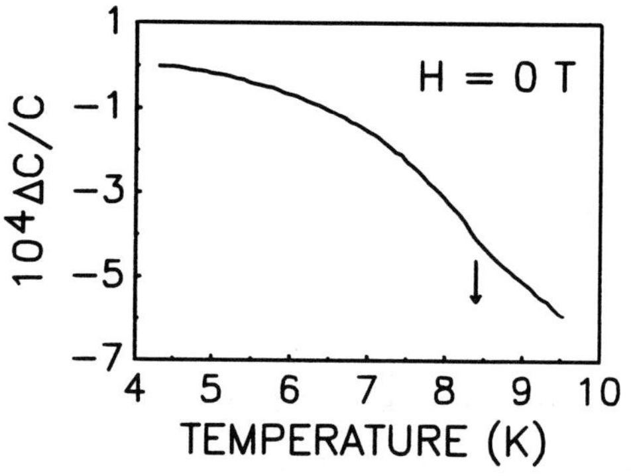

FIGURE 1
Variation of C_{44} as a funtion of temperature for zero magnetic field. The arrow indicates T_N.

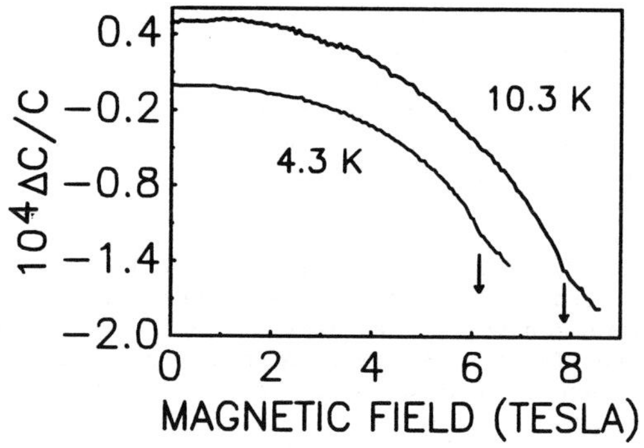

FIGURE 2
Variation of C_{44} as a function of magnetic field for two temperatures. The arrows indicate the transition temperatures.

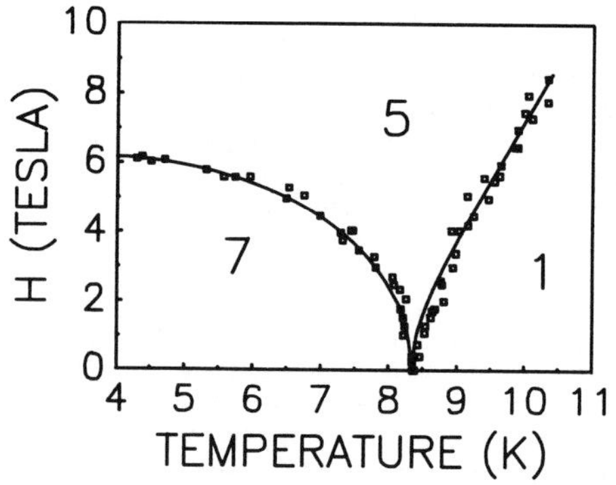

FIGURE 3
Magnetic phase diagram of $CsMnBr_3$ for the magnetic field H oriented in the basal plane.The full lines are a non local Landau theory fit (Ref.3).

Physica B 165&166 (1990) 173–174
North-Holland

EFFECT OF FIELD-INDUCED ORDER ON THE GROUND STATE OF A SINGLET GROUND STATE SYSTEM

J. R. Fletcher, S. S. U. Kazmi, K. J. Maxwell and J. R. Owers-Bradley

Department of Physics, University of Nottingham, University Park, Nottingham NG7 2RD, U.K.

Susceptibility measurements on the singlet ground state dimer system $Cs_3Cr_2Br_9$ at low temperatures are used to predict the ground state behaviour in the region of a field-induced phase transition as $T \rightarrow 0$. Results are compared with optical data. High field magnetic effects, possibly associated with the population of higher spin states, are also reported.

1. INTRODUCTION

The compound $Cs_3Cr_2Br_9$ belongs to a class of materials where trivalent transition metal ions couple strongly in pairs by exchange interactions to form magnetic dimers. The Cr-Cr separation is 3.3 Å within the dimers and 7.4 Å (same sublattice), 7.5 Å (different sublattices) between nearest neighbour dimers. Antiferromagnetic coupling between ground state Cr^{3+} ions leads to a ground dimer manifold consisting of singlet, triplet, quintet and septet levels at 0, 6.3, 17.5 and 31.0 cm^{-1} respectively as measured by hot-band optical transitions (1,2). The nature of the coupling within the dimers is relatively well understood from optical studies (2), and most absorption and Zeeman data at temperatures of 4 K and above may be explained using an isolated dimer model. Evidence for weak inter-dimer interactions is provided by neutron scattering studies (3,4) which show that the singlet-triplet excitations involving the two lowest dimer levels exhibit significant dispersion.

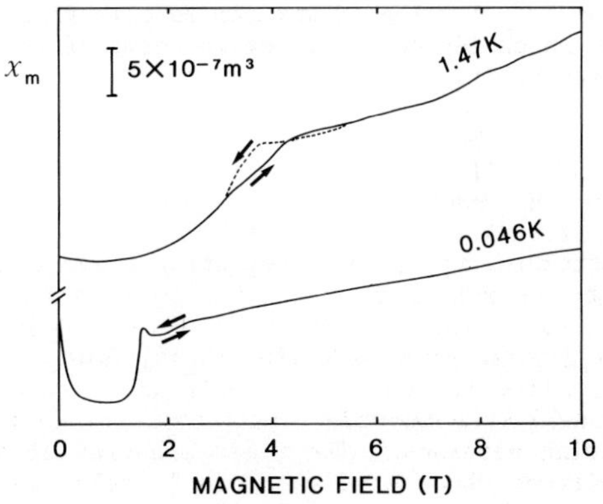

FIGURE 1
Mass susceptibility of polycrystalline $Cs_3Cr_2Br_9$ as a function of applied magnetic field for several temperatures. Arrows denote the direction of field sweep.

2. RESULTS AND DISCUSSION

Low temperature susceptibility results (5) show that, in an applied magnetic field, the inter-dimer coupling leads to an ordered state, as shown by the distinctive cusp in χ as a function of field shown in Figure 1 for several temperatures. This cusp, which moves to higher fields with increasing temperature, may be followed to temperatures of about 2 K. Various models have been suggested to account for the transition to order, including a soft-mode model in the random phase approximation (4), and a static molecular field model (6). In each case, the system is treated as consisting of just two levels, a ground state singlet and an excited triplet. The soft-mode picture is one in which the minimum of the singlet-triplet excitation in k-space near the K-symmetry point is driven to zero energy in applied field. The static model gives rise to an ordering of the spin component perpendicular to the field in the vicinity of the point of level crossing between the singlet and lowest component of the triplet at sufficiently low temperatures. In either instance, below a critical temperature, the system is predicted to order at a critical field (identified by the susceptibility cusp), but to return to a disordered state at some higher field. Both models suffer from neglect of the quintet and septet levels which make no contribution at temperatures in the liquid helium range at zero field. However, components of these levels move rapidly to lower energy in applied field and become populated at high fields. The higher spin states are not expected to produce significant effects at moderate fields and for temperatures of order 1 K or less. From our measurements down to temperatures below 50 mK, it is found that the transition to order occurs at much lower fields than predicted as $T \rightarrow 0$, but at higher fields than expected for temperatures in the range 1-2 K. This seems to be associated with the band nature of the triplet state with susceptibility data suggesting that the bottom

0921-4526/90/$03.50

of the band has an energy of 1.4 cm^{-1} at T = 0, much lower than the prediction of the RPA calculations from neutron studies at 1.6 K.

Here we show that the susceptibility may be used to deduce the nature of the ground state as a function of applied field as $T \to 0$. There is little detectable change in χ for temperatures below 0.15 K. The initial decrease observed at low temperatures and low fields is attributed to trace amounts of paramagnetic impurity. Thus to a good approximation, the mass susceptibility as $T \to 0$ may be represented by the function

$$\chi_m = a + b(H - H_c) \text{ for } H \geq H_c$$

where $a = 6.18 \times 10^{-7}$, $b = 7.3 \times 10^{-8}\ \mu_o$ and $H_c = 1.5\ \mu_o^{-1}$. The energy per mole is given by $-\int MdH$ which may be rewritten in terms of the susceptibility

$$-\mu_o \int_{H_c}^{H} \{\int_{H_c}^{H} \chi\, dH\}\, dH_1$$

Substituting for χ and integrating leads to an energy per mole

$$-\mu_o [a/2(H - H_c)^2 + b/6(H - H_c)^3] \text{ Joules}$$

The energy per dimer is

$$-2.52 \times 10^{-2}\ (B - 1.5)^2 - 9.9 \times 10^{-4}\ (B-1.5)^3 \text{cm}^{-1}$$

where B is in Tesla, giving the ground state behaviour as $T \to 0$ shown in Figure 2. Optical absorption studies in the ordered region reveal an overall shift to higher energy of the spectrum as field increases, consistent in magnitude with the depression of the ground state energy given by this calculation. On a simple model, the depression of the coupled ground state in field can be treated in terms of the interaction of singlet and triplet states on neighbouring sites. A more complete treatment must involve the dispersive aspects of the excited states. This will be the subject of a future publication in which details of optical studies in the ordered region will also be reported. Evidence for the effect of higher spin states at very high fields has not been presented here. However, it is worth noting two experimental effects which may be associated with the large magnetic moments induced when these states are populated. Firstly, repeated cycles to very high fields of a polycrystalline boule resulted in a breakup of the sample into several pieces, probably due to magnetostrictive forces on individual crystallites. Secondly, results taken on finely-ground powder samples were found to be qualitatively different from the polycrystalline results. Evidence for magnetic

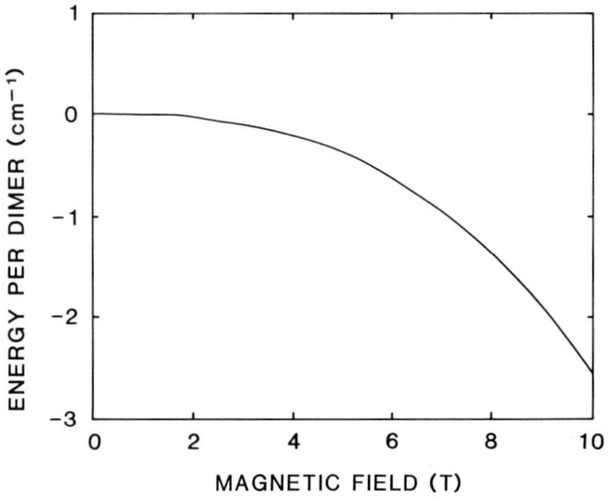

FIGURE 2
The ground state behaviour as deduced for the susceptibility data.

order associated with the triplet states was still observed at the same field and temperature as before, but up and down field sweeps gave quite different values of χ in the high field region when the maximum field exceeded 9.3 T, the field at which appreciable population of a component of the quintet state would be expected for temperatures of order 1 K. Once the powder was subjected to fields greater than this, recovery of the up-sweep signal was achieved only upon reduction of the field to about 5 T, suggesting alignment of powder grains at high field due to population of states with large magnetic moments. Such effects may well be responsible for disruption in the development of the ordered state back to disorder as predicted by two-level models.

ACKNOWLEDGEMENTS
One of us (S.S.U.K.) would like to thank the Government of Pakistan for a C.O.T. Scholarship.

REFERENCES
(1) N. J. Dean and K. J. Maxwell, J. Magn. Magn. Mater. 31-34 (1983) 563.
(2) N. J. Dean, K. J. Maxwell, K. W. H. Stevens and R. J. Turner, J. Phys. C18 (1985) 4505.
(3) B. Leuenberger, A. Stebler, H. U. Gudel, A. Furrer, R. Feile and J. K. Kjems, Phys. Rev. B30 (1984) 6300.
(4) B. Leuenberger, H. U. Gudel, R. Feile and J. K. Kjens, Phys. Rev. B31 (1985) 597.
(5) S. S. U. Kazmi, K. J. Maxwell and J. R. Owers-Bradley, J. de Physique Colloque C8 (1988) 1509.
(6) B. Leuenberger, H. U. Gudel, R. Horne and A. J. van Duyneveldt, J. Magn. Magn. Mater. 49 (1985) 131.

Physica B 165&166 (1990) 175–176
North-Holland

THE ACCURATE EXPRESSION OF THE GROUND STATE FOR SQUARE LATTICE S=1/2 HEISENBERG ANTIFERROMAGNET BY RVB-STRINGS SPREADING OVER ANY N-TH N.N.

Y.NATSUME, T.HAMADA, S.NAKAGAWA, J.KANE and M.KATAGIRI

Department of Physics, Faculty of Science, Chiba University, Yayoi-cho, Chiba 260, Japan

The method of constructing the ground state of square lattice quantum Heisenberg antiferromagnet is proposed on the basis of the covering by RVB-strings, where RVB with larger distance than nearest neighbor is considered. By introducing only a few geometrical parameters in the variational procedure, we obtain the quite accurate expression.

1. INTRODUCTION

Recent extensive investigations (1,2) for properties on CuO_2 plane in high-Tc superconducting oxide compounds suggests that the accurate expression of the ground state for square lattice S=1/2 Heisenberg antiferromagnet (SQHAF) is essentially required to understand the nature of the superconductivity, as well as magnetic properties. In fact, much theoretical work (3) has been devoted to the problem of the ground state Ψ_g in SQHAF, whose Hamiltonian is

$$\mathcal{H} = 2J \sum_{i,j} S_i S_j \qquad (J>0).$$

Here, sites i and j are nearest neighbors (NN). In short, if we concentrate our attention on numerical work, we can divide it into Monte Carlo study (4) and exact-diagonalization study (5,6). Though the former has provided many fruitful results, there remains room for the investigation of Ψ_g itself in consideration of the inherent restriction of its method. On the other hand, we can obtain the exact solution Ψ_g for finite spin number N by the latter as a list of numerical values for an eigenvector. Thus, we are faced with the problem of how the physical picture is constructed in such a list.

In the present paper, we propose (3,7) an accurate expression Φ for Ψ_g on the basis of the covering of RVB-strings, where we take into account the singlet pairs with far distance as well as NN. Introducing only a few geometrical parameters, we can obtain the quite well

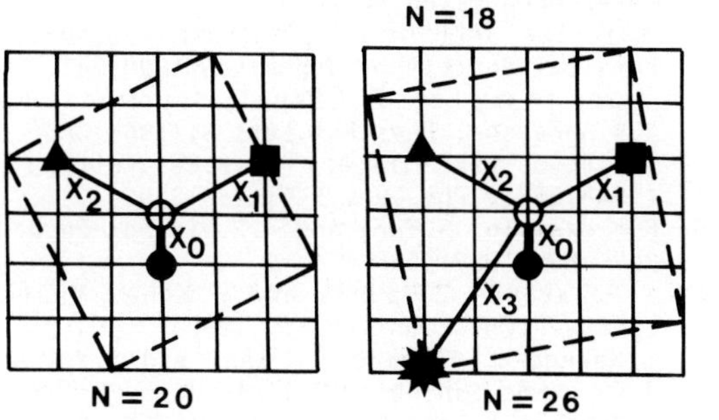

Fig.1 Classification of weight factors for RVB strings by geometrical characters for systems N=16,18,20,26.

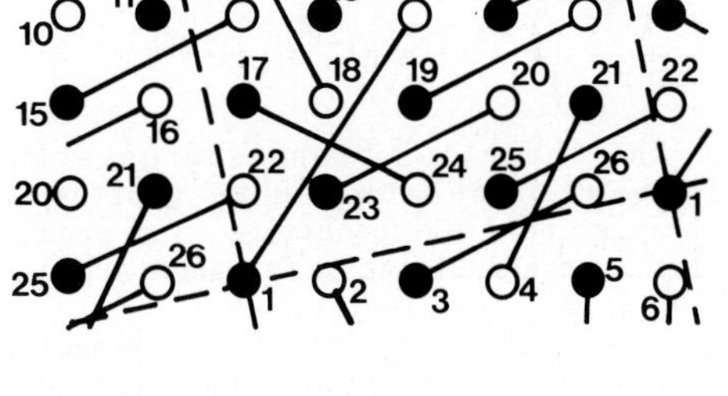

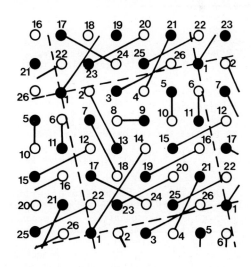

Fig.2 An example in N=26 for the covering of RVB-strings with any n-th nearest neighbor. This weight factor is $X_0^3, X_1^7, X_2^2, X_3^1$, according to geometrical parameters in Fig.1.

Table I . Obtained values of f and ratios for X_1/X_0, X_2/X_0 and X_3/X_0 in the procedure of f→1. Values for NNRVB represent f for the trial function including only NNRVB.

N	f	X_1/X_0	X_2/X_0	X_3/X_0	NNRVB
16	0.999819	0.276			0.85671
18	0.999928	0.268	0.246		0.85588
20	0.999839	0.255	0.230		0.81524
26	0.999528	0.201	0.185	0.135	0.71578

Table II . The ground-state energy E_g per N obtained by each function. The unit is J. Values in parentheses are ratios to exact ones.

N	exact	Φ^N	NNRVB
16	-1.40356	-1.40342(0.999897)	-1.3373(0.9528)
18	-1.38797	-1.38792(0.999963)	-1.3235(0.9536)
20	-1.38162	-1.38151(0.999923)	-1.3103(0.9484)
26	-1.36890	-1.36867(0.999827)	-1.2844(0.9383)

Table III. The staggered magnetization obtained by each function. Values in parentheses are ratios to those of exact solution.

N	exact	Φ^N	NNRVB
16	0.092176	0.092151(0.999736)	0.071389(0.7745)
18	0.089534	0.089507(0.999689)	0.070505(0.7875)
20	0.085910	0.085886(0.999716)	0.064226(0.7476)
26	0.077966	0.078002(1.000458)	0.051925(0.6660)

agreement between Φ and Ψ_g, i.e. $f = |\langle\Phi|\Psi_g\rangle|^2$ gets close to 1.

2. OUR EXPRESSION

In consideration of Marshall's condition (8), we propose the following expression;

$$\Phi = \sum_{m,n} \prod_{p}^{N/2} W_p(\vec{r}_{m(p)}-\vec{r}_{n(p)})[m(p),n(p)].$$

Here, the summation means the covering of singlet pairs [m,n] between m and n sites, where m(n) belongs to A(B) sublattice. The product is made for each RVB-string p for such a m-n pair, where W_p is a weight factor.

Here, we classify W_p into a few parameters X_j considering the geometrical character of each string as shown in Fig.1. An example is shown in Fig.2. This classification of W_p is naturally induced especially from one of Marshall's condition (8); Ψ_g should be totally-symmetric. In numerical calculation, variational parameters are ratios of X_1/X_0, X_2/X_0 and X_3/X_0.

3. RESULTS OF CALCULATION

We make the variational calculation for f→1 using the exact solution in systems of N=10~26 (6). The obtained parameters and f are listed in Table I . The value for NNRVB is given in the covering only by RVB for NN. While f for NNRVB becomes distant from 1 with increased N, our proposed expression Φ can stay close to 1. We would like to remark that RVB with distance than NNRVB contributes more substantially as N increases. Though Anderson et al. have pointed out (9) the possibility of the contribution of RVB with distance more than NN, much work has been devoted to NNRVB models. Any expression by such models fail to reflect Ψ_g itself.

The ground-state energy E_g per N is shown in Table II . The extrapolated value E_g for N →∞ is -(1.358±0.010)J. Further, Table III shows the staggered magnetization

$$\kappa_{stag} = (1/N^2) \ (\sum_i \eta_i S^z)^2 \quad ,$$

where η = +1(-1) for A(B) sublattice. It becomes 0.25 for classical Néel state. Because Φ can express accurately Ψ_g itself, the agreement as to κ_{stag} is also quite well. We obtain the extrapolated value for N →∞ as 0.063. For NNRVB, it vanishes. The ground state Ψ_g in SQHAF is neither the classical Néel state nor the spin-liquid state expressed by NNRVB. It is very the quantum state expressed quite well by the proposed covering of RVB strings spreading any n-th nearest neighbor.

REFERENCES

(1) One of the most noteworthy experimental work for the relationship between magnetic properties and the superconductivity is G.Shirane et al; Phys.Rev.Lett.63(1989)330. and references cited therein.

(2) Several magnetic mechanisms of the superconductivity have been proposed. For example, A.Ohta, M.Nojima, T.Suzuki and Y.Natsume; J.Phys.Soc.Jpn.58(1989)4147. and references cited therein.

(3) A review article by Y.Natsume; Butsuri(The Physical Society of Japan) 45(1990)241.

(4) For example, S.Liang, B.Doucot and P.W.Anderson; Phys.Rev.Lett.61(1988)365. D.A.Huse and V.Elser; Phys.Rev.Lett.60 (1988)2531, Phys.Rev.B37(1988)2380.

(5) For example, S.Tang and J.E.Hirsch; Phys. Rev. B39(1989)4548.

(6) S.Nakagawa, T.Hamada and J.Kane; Master Thesises(Chiba University,1989).

(7) S.Nakagawa, T.Hamada, J.Kane and Y.Natsume; J.Phys.Soc.Jpn.59(1990)1131.

(8) W.Marshall; Proc.Roy.Soc.A232(1955)48.

(9) P.W.Anderson; Science 235(1987)1196. G.Baskaran, Z.Zou and P.W.Anderson; Solid State Commun.63(1987)973.

Physica B 165&166 (1990) 177–178
North-Holland

Pulsed NMRON Relaxation Measurements and Thermometric NMR in the Quasi-2 Dimensional Ferromagnet: $Mn(COOCH_3)_2 \cdot 4H_2O$

M. Le Gros, A. Kotlicki* and B.G. Turrell

Department of Physics, University of British Columbia, Vancouver, B.C., Canada V6T 2A6

The measurement of the field dependence of the nuclear spin-lattice relaxation time of ^{54}Mn in the two manganese sites in the quasi-2 dimensional ferromagnet $Mn(COOCH_3)_2 \cdot 4H_2O$ obtained by the pulsed NMRON technique is reported. This technique allows the observation in low fields of the higher frequency resonance which previously could not be measured by CW methods. The anomaly in the ^{54}Mn relaxation time observed in the ^{55}Mn level crossing regime is discussed, and the thermometric observation of the field dependence and line width of the resonance lines from the abundant ^{55}Mn spin systems is reported and related to the ^{54}Mn spin-lattice relaxation behavior.

1. INTRODUCTION

Manganese acetate tetrahydrate has a crystallographic layered structure (1). There are two manganese sites Mn1 and Mn2 in a planar structure with Mn2-Mn1-Mn2 triplet groups strongly bound by antiferromagnetic exchange (2). These triplet groups each of effective spin 5/2 are coupled ferromagnetically in the plane, the ordering temperature being $T_c = 3.19$ K. There is a much weaker interlayer interaction so that the system can be considered as a quasi-2 dimensional ferromagnet. The interlayer interaction is antiferromagnetic, but a very low field of $B_0 = 6 \cdot 10^{-4}$T drives the antiferromagnetic planar array into a ferromagnetic structure.

We previously reported (3) two, magnetic pseudo-quadrupole split, hyperfine multiplets corresponding to the Mn1 (lower frequency) and Mn2 (higher frequency) sites observed in an applied magnetic field by a cw NMRON technique. A large difference between the values of the spin lattice relaxation time T_1 for the two sites was measured. For $B_0 = 0$ only the lower frequency Mn1 resonance could be observed. It was suggested that the pulsed NMRON method could be utilized to observe the higher frequency Mn2 resonance and to investigate in detail the field dependence of the spin-lattice relaxation times.

The results of these measurements are reported in this paper. Since the hyperfine fields are negative an applied magnetic field reduces the separation of the two sextuplets of resonance lines leading in sufficiently high fields to level crossing. The field dependence of the ^{54}Mn spin lattice relaxation was studied in the level crossing regime. The thermometric NMR technique (4) was used to investigate the ^{55}Mn nuclear resonance lines and measure the magnetic field at the level crossing of the ^{55}Mn1 and ^{55}Mn2 lines.

2. EXPERIMENTAL

The sample was a ^{54}Mn-$^{55}Mn(COOCH_3)_2 \cdot 4H_2O$ crystal with 4 μCi of ^{54}Mn activity. It was mounted on a copper cold finger connected to the mixing chamber of a dilution refrigerator so that a magnetic field could be applied along the a-axis. The apparatus used for pulsed NMRON has been described previously (5). In order to measure T_1 a rotation pattern was first obtained for each field. This determined the length of a 180° pulse which was then applied to the crystal, causing a well-defined change in the gamma anisotropy. The relaxation of the anisotropy to the initial condition was monitored by the standard multiscaling method. The r.f. pulse from the synthesizer controlled by the programmable pulser was synchronized with the trigger of the multiscaler in such a way that the pulse was always applied in a preset channel. This procedure allowed the use of signal averaging to obtain good counting statistics in very narrow (few milliseconds) channels. The apparatus was fully controlled by a computer, making the data acquisition completely automatic. The multiexponential decay curve was then fitted to the experimental points using T_1 as an adjustable parameter.

The thermometric NMR measurements were performed in a way similar to that described in our work on $MnCl_2 \cdot 4H_2O$ (4) except that in the case of manganese acetate T_1 is sufficiently short that thermometric NMR can

* Permanent address: Institute of Experimental Physics, Warsaw University, Hoza 69, 00–689, Warsaw, Poland.

0921-4526/90/$03.50

be observed directly, without the necessity of field-induced thermal contact between ^{54}Mn and ^{55}Mn spins.

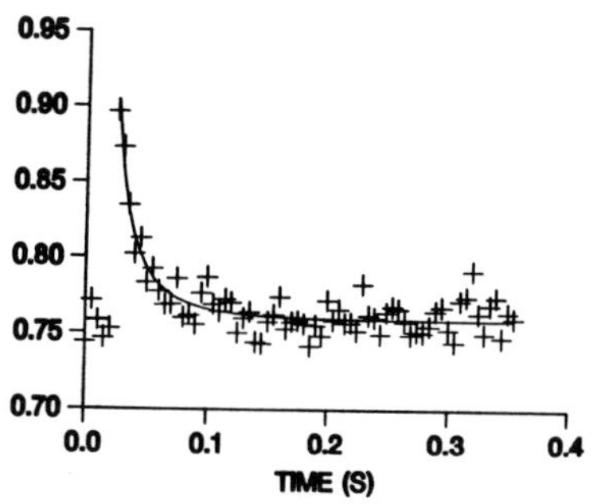

Figure 1

A spin-lattice relaxation curve for the ^{54}Mn M=–3→–2 transition at B_0=0 for the Mn2 site.

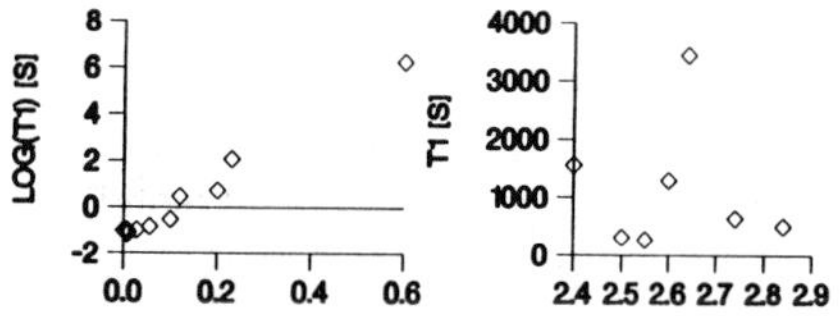

Figure 2

Field dependence of T_1 for ^{54}Mn at the Mn2: (a) in the range B_0= 0–0.6T; (b) in the level crossing regime.

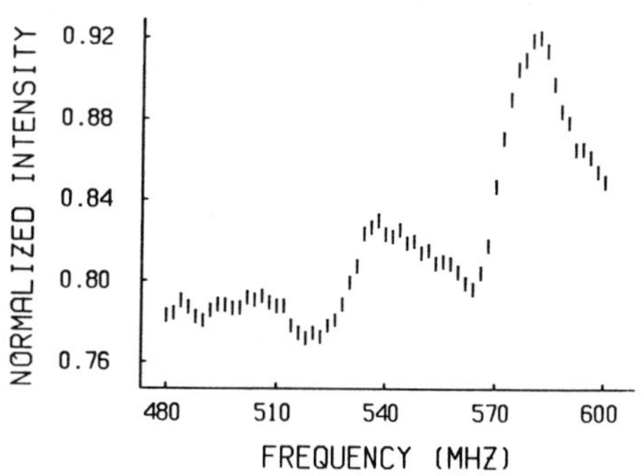

Figure 3

Thermometric NMR lines for ^{55}Mn in B_0 = 0.004T.

3. RESULTS AND DISCUSSION

A typical relaxation curve obtained for the higher frequency M = –3 → –2 transition at 476 MHz in B_0 = 0 is shown in Fig. 1. The channel width was 5ms and a relaxation time T_1 = 100±10ms was obtained. The field dependence of the T_1 is shown in Fig. 2. Note that the relaxation time changes by 7 orders of magnitude between 0T and 0.6T.

A typical thermometric measurement is shown in Fig. 3. The level crossing field was found to be 2.64±.05 T. The ^{54}Mn spin lattice relaxation times were then measured in the 2.4T - 2.8T field region (Fig. 2b).

Highly polarized nuclear spins with similar resonance frequencies are strongly coupled by the Suhl-Nakamura (SN) interaction (6,7). In manganese acetate this indirect interaction has a long range and the magnitude of the coupling depends strongly on the nuclear spin concentration and on the ratio of the hyperfine coupling to the electronic magnon energy gap. Nuclear spins on the Mn1 and Mn2 sites form two interpenetrating nuclear ferromagnets with nuclear magnon excitations, the lifetime of which is strongly dependent on the strength of the SN interaction.

We suggest that the ^{54}Mn spin-lattice relaxation is caused by high order, nuclear magnon scattering processes (9). Above the paramagnetic saturation field of 0.014T the SN interaction is considerably weakened by the increased gap in the electronic magnon spectrum leading to an increase in T_1. In the level crossing regime the magnon spectra from the two nuclear ferromagnets begin to overlap causing a T_1 minimum. Exactly at the level crossing field the antiparallel nuclear ferromagnets coalesce to form a nuclear ferrimagnet with increased magnon lifetimes as indicated by the decreased width of the thermometric resonance lines. The corresponding increase in magnon lifetime would cause the sharp increase in T_1. This effect is similar to that seen in the resonance behavior of other exchange narrowed systems (10).

REFERENCES

1. P. Burlet, P. Burlet and E.F. Bertaut, Solid State Commun. 14 (1974) 665.
2. Y. Okuda, M. Matsuura and T. Haseda, J.Phys.Jap. 44 (1978) 371.
3. M. Le Gros, A. Kotlicki and B.G. Turrell, Hyp.Int. 43 (1988) 371.
4. A. Kotlicki and B.G. Turrell, Phys.Rev.Lett. 56 (1986) 773.
5. M. Le Gros, A. Kotlicki and B.G. Turrell, Hyp.Int. 36 (1987) 161.
6. H. Suhl, Phys. Rev. 109 (1958) 606.
7. T. Nakamura, Progr. Theoret. Phys. (Kyoto) 20 (1958) 542.
8. P.G. De Gennes, P.A. Pincus. F. Hartmann-Boutron and J.M. Winter, Phys.Rev.129, 3, (1963) 1105.
9. P.G. De Gennes, P.A.Pincus, F.Hartmann-Boutron and J.M. Winter, Phys.Rev.129, 3, (1963) 1105.
10. P. Pincus, Phys.Rev.131, 4 (1963) 1530.

Physica B 165&166 (1990) 179–180
North-Holland

NET SPONTANEOUS MAGNETISATION IN THE DILUTE ISING ANTIFERROMAGNET $Fe_{.46}Zn_{.54}F_2$

M.Lederman, J.Hammann*, R.Orbach

Physics Department and Solid State Science Center, UCLA, Los Angeles 90024,USA
* On Leave from DPhG/SPSRM, CEN Saclay, 91191 Gif-sur-Yvette, France

Low field dc magnetisation measurements reveal the existence of a net spontaneous magnetisation in the antiferromagnetic ground state. Its temperature dependence is that of the order parameter. It is ascribed to the onset of strains which induce inequivalent sublattice moments.

The dilute antiferromagnetic system $Fe_xZn_{1-x}F_2$ has been very extensively studied as a model system for the random exchange (RE) and random field (RF) Ising problems (1). Neutron diffraction measurements showed that the antiferromagnetic long range order (LRO) was not destroyed by the random field (RF) as long as the field is applied at low temperatures. If however the system is cooled down with the field on, a metastable domain state appears.

These procedure dependent properties are well detected in dc magnetisation measurements(2). Cooling in zero field and applying the field at low temperature (ZFC) does not give the same response as the field cooled (FC) procedure.

These irreversible effects due to RF should decrease with the amplitude of the field. We have performed magnetisation measurements parallel to the cristallographic c axis down to very low fields (H<50 Oe) in $Fe_{.46}Zn_{.54}F_2$, using a SQUID magnetometer. We still found large irreversible effects. However in this low field range we ascribe them to the existence of a net spontaneous magnetisation in the zero field antiferromagnetic state.

In the inset of fig.1 we show the temperature dependence of dM/dH (dH=5 Oe) for our particular sample. The sample was a single crystal from the same batch as the cristals of ref(3). The concentration gradient was smaller than 0.5%/cm as checked by birefringence measurements. The antiferromagnetic peak in dM/dH is indeed very narrow indicating a well defined transition temperature.

In Fig.1 we have plotted the ZFC and FC dc magnetisations (in arbitrary units) as a function of temperature. The sample was cooled down in the residual field (h=-0.6 Oe), an extra field of 5 Oe was then applied and the magnetisation recorded as the temperature is first increased up to $T>T_c$ (ZFC curve) and then decreased again to below T_c (FC curve). The resulting zero field cooled curve is characteristic of an antiferromagnetic response in a field parallel to the sublattice magnetisation.

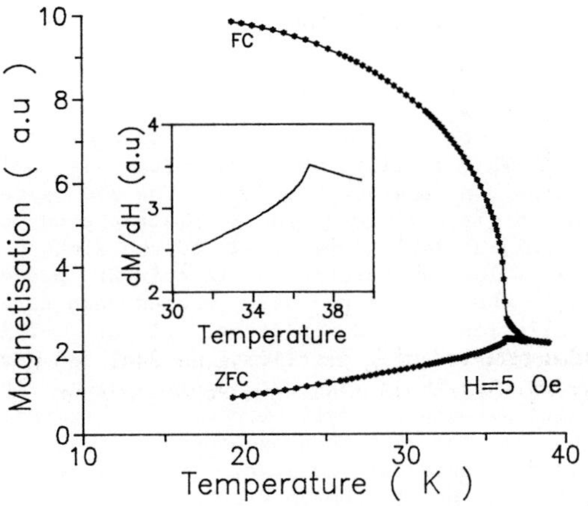

Figure 1

The shape of the field cooled curve is quite different. At T_c the magnetisation increases very sharply and saturates at low temperatures. The excess magnetisation is time independent. Increasing the temperature back up to T_c yields the same magnetisation curve.The large response in the FC procedure suggests the existence of a spontaneous magnetisation.

This is further supported by the results shown in Fig.2. In this figure, the S-shaped curve corresponds to the values of the FC magnetisation as a function of field at T=20K (The residual field has been taken into account). The magnetisation increases linearly with field up to H=5 Oe and then saturates. It is reversible and symmetric in field. A rough estimate of the value of the slope yields M/H=0.06emu. This is more than 10 times larger than the paramagnetic susceptibility at T=39K, and is only 4 times smaller than the inverse of the demagnetising factor which for the cubic shape used should be of order $3/4\pi$ emu. The observed behavior is that of a weak ferromagnet with a nearly equilibrium distribution of up

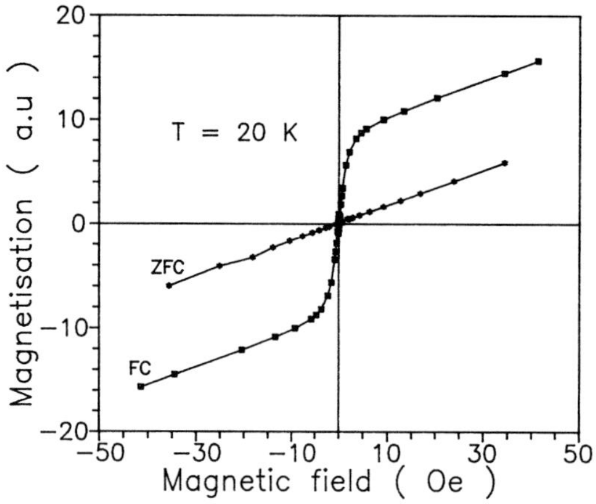

Figure 2

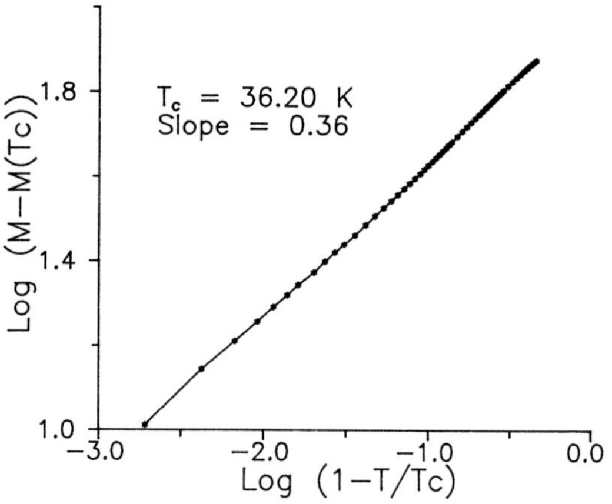

Figure 3

and down domains minimising the demagnetising energy. This equilibrium distribution is only observed in the FC procedure. The ZFC values shown in fig.2 yield a linear field dependence with only a small slope equal to the slope of the FC values at saturation. In this ZFC procedure, the sample has been cooled down after cancellation of the residual field. This results in an equal distribution of up and down domains thus in a zero net magnetisation. The domains remain strongly pinned at low temperature, and the only response measured, as the field is changed at constant temperature, is the regular antiferromagnetic response.

The approximate amount of the spontaneous net moment along the c direction corresponds to a moment per Fe ion of 10^{-3} times its saturation value.

Fig.3 is a log-log plot of the excess FC magnetisation as a function of the reduced temperature $t=1-T/T_c$ in the residual field $h=-0.6$ Oe. The observed linear behavior yields a slope of 0.36. This compares well with the critical exponent of the RE antiferromagnetic order parameter as observed by Mossbauer experiments (4). The net spontaneous moment is thus related to the sublattice magnetisation.

The picture which emerges is that of a domain state in which the domains are essentially antiferromagnetic but have a small net magnetic moment. The distribution of up and down domains is defined by the field in which the system has been cooled down to below T_c. For $T<T_c$ this distribution remains unaffected by temperature or field variations. The existence of a domain structure is favored by the presence of the Zn ions as the wall energy is minimised when they are located on these non magnetic ions.

The observed low field behavior is probably not peculiar to $Fe_{.46}Zn_{.54}F_2$. In $Mn_{.45}Zn_{.55}F_2$, the temperature dependence of the ZFC and FC magnetisations would also be consistent with the existence of a domain structure and a net spontaneous magnetic moment. In this compound a time dependence has been reported for the ZFC magnetisation (5). This would be due to a smaller pinning of the domain walls as expected from a system with a smaller anisotropy.

The existence of a net spontaneous magnetisation is probably related to the observed symmetry breaking of the cristal at T_c (6). A deformation perpendicular to the c axis appears below T_c due to the magneto-elastic coupling. It has the same temperature dependence as the staggered magnetisation. Such a deformation can break the equivalence of the antiferromagnetic sublattices and thus give rise to a net magnetisation (7).

ACKNOWLEDGMENTS: We would like to thank G.Kenning and R.Hoogerbeets for their help in the SQUID experiments, V.Jaccarino, A.King I.Ferreira, N.Nighman for fruitful discussions and for providing the samples. The research was supported in part by NSF grants DMR-81-18968 and DMR-86-12022.

REFERENCES

(1) V.Jaccarino, A.R.King, Statphys 17 (1989)
(2) P.Pollak, W.Kleeman, D.P.Belanger, Phys.Rev.B, 38 (1988) 4773
(3) A.R.King, I.B.Ferreira, V.Jaccarino, D.P.Belanger, Phys.Rev.B, 37 (1988) 219
(4) N.Rosov, A.Kleinhammes, P.Lidbjork, C.Hohenemser, M.Eibschutz, Phys.rev.B, 37 (1988) 3265
(5) H.Ikeda, K.Kikuta, J.Phys.C, 17,(1984) 1221
(6) C.A.Ramos, A.R.King, V.Jaccarino, S.M.Rezende J.de Physique C8 49 (1988) 1241
(7) T.Moriya, J.Phys.Chem.Solids, 11 (1959) 73
J.Nasser, J.de Phys. 40, (1979) 51

Physica B 165&166 (1990) 181–182
North-Holland

MAGNETIC HYSTERESIS BEHAVIOUR OF ANISOTROPIC SPIN GLASS $FeMnTiO_5$

J.K. SRIVASTAVA*, K. ASAI and K. KATSUMATA

The Institute of Physical and Chemical Research (RIKEN), Wako-shi, Saitama, 351-01, Japan

Magnetic hysteresis measurements ($4.2 \leqq T \leqq 300K$, $-20 \leqq H \leqq 20$ kOe) support the anisotropic spin glass nature of $FeMnTiO_5$ and reveal the hysteresis loop regions over which the spin glass behaviour remains partially destroyed at 4.2K.

1. INTRODUCTION

Fe_2TiO_5 (orthorhombic, cations distributed on tetrahedral 8f- and octahedral 4c- sites) is a well known anisotropic spin glass (ASG) in which the Fe^{3+}- spin component parallel to c-axis freezes at $T_\ell \sim 50K$ (longitudinal freezing) and the perpendicular spin components freeze at $T_t \sim 8K$ (transverse freezing (TF)) (1-4). $FeMnTiO_5$, obtained by replacing one Fe^{3+} by Mn^{3+} in Fe_2TiO_5, has more orthorhombic distortion than Fe_2TiO_5 (lattice parameters (± 0.02 A°) for $FeMnTiO_5$ are: a = 9.59 A°, b = 9.96 A° and c = 3.68 A°, and for Fe_2TiO_5 they are: a = 9.80 A°, b = 9.97 A° and c = 3.73 A°) (5). Our earlier work (5), using Mössbauer effect, d.c. magnetisation (M) and a.c. susceptibility techniques, showed $FeMnTiO_5$ to have ASG properties much different from those of Fe_2TiO_5. In the present paper, we report its magnetic hysteresis behaviour and discuss the results in the light of the earlier work (5).

2. EXPERIMENTAL

The sample (powder) which was used in our earlier study (5), has been used for the present vibrating sample magnetometer measurements also.

3. RESULTS AND DISCUSSION

The earlier study (5) showed no TF in $FeMnTiO_5$ down to 4.2K. The spin components along the a,b- axes remain paramagnetic at least upto 4.2K on cooling ($T_t < 4.2K$) and all the transitions occur for the c-axis (longitudinal) spin component only. The system is found to have the cation distribution as $(Fe^{3+}Mn^{3+}_{0.3}Ti^{4+}_{0.7})_f\,[Mn^{3+}_{0.7}\,Ti^{4+}_{0.3}]_cO_5$. The Mn^{3+} and 40% Fe^{3+} f-site ions couple with the c-site ions via the f-O-c superexchange path and show the following transitions: paramagnet (P) $\rightleftarrows$ ($T_C \sim 400K$) $\rightleftarrows$ ferrimagnet (F) $\rightleftarrows$ ($T_\ell \sim 72K$) $\rightleftarrows$ $(ASG)_1$. The remaining 60% f-site Fe^{3+} ions, which were initially uncoupled, show the transition: P $\rightleftarrows$ ($T'_\ell \sim 42K$) $\rightleftarrows$ $(ASG)_2$. Mössbauer spectra show that for the ASG state $(ASG)_2$, the average size of the randomly frozen clusters is much bigger (5).

From above, it is clear that the $FeMnTiO_5$ hysteresis loop (HL) saturation will be most difficult at 4.2K when all c-axis spins have the spin glass (SG) freezing. Fig.1 shows this

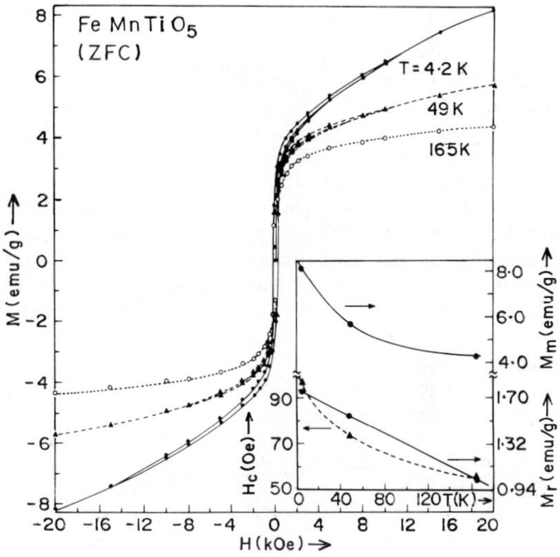

FIGURE 1

where typical loops for T = 4.2K ($T < T'_\ell$), 49K ($T'_\ell < T < T_\ell$) and 165K ($T_\ell < T < T_C$) are displayed; ZFC = zero field cooled, H = external magnetic field. Fig.1 inset shows an increase in M_r, H_c and M_m with decreasing T; M_r = remanent magnetisation, H_c = coercive field, M_m = M value at maximum H (20 kOe). When the SG freezing occurs, a local anisotropy energy develops at the spin (cluster) site, owing to frustration, which hinders spin (cluster) rotation making its alignment along $\vec{H}$ difficult (6). This increases M_r, H_c and makes the HL saturation difficult (7). M_m increase with decreasing T is due to the increase in transverse spin components' contribution (paramagnetic) to M (3-5). This contribution will start saturating only below ~ 10K (Fe^{3+}, Mn^{3+} in orthorhombic crystal field). However M_r/M_m

* Present address: Tata Institute of Fundamental Research, Bombay-400005, India.

0921-4526/90/$03.50 © 1990 – Elsevier Science Publishers B.V. (North-Holland)

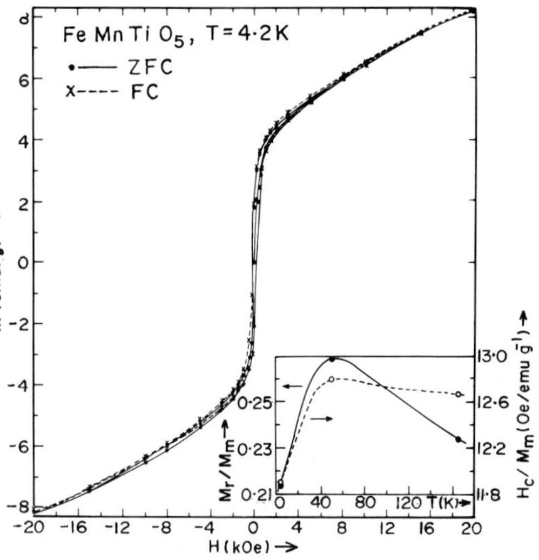

FIGURE 2

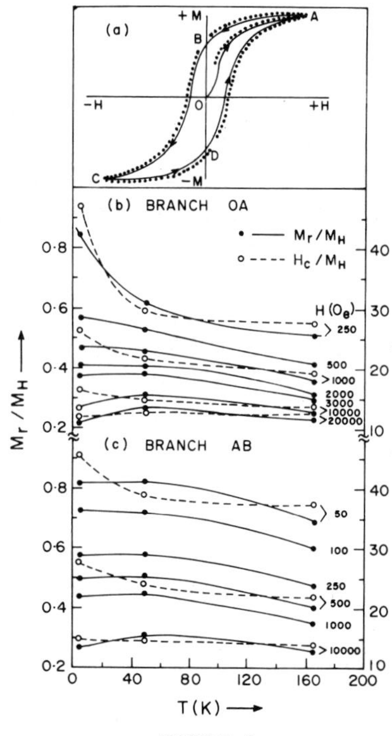

FIGURE 3

and H_c/M_m (Fig.2 inset) show a decrease at 4.2K indicating a relatively faster M_m increase at 4.2K. This indicates a partial destruction of the SG state, actually of $(ASG)_2$ (due to probably its bigger clusters), at 4.2K by 20 kOe field resulting in a larger alignment of spins (clusters) along $\vec{H}$. M_r/M_H and H_c/M_H vs. T plots (Figs. 3,4) corresponding to different loop branches, for various H, confirm this conclusion and yield a critical H for each loop branch

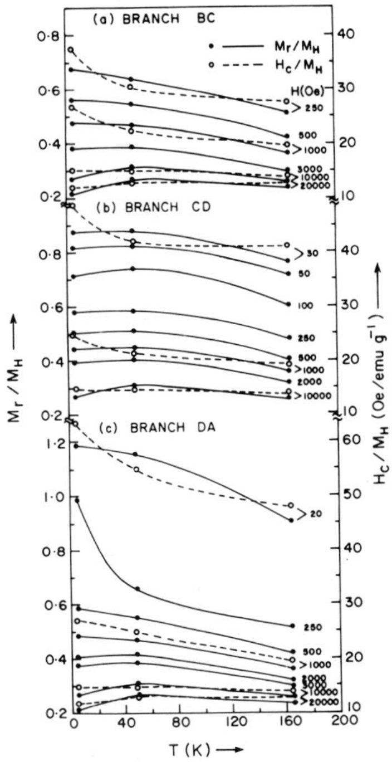

FIGURE 4

above which $(ASG)_2$ starts getting destroyed; M_H = M value for a particular H. The dotted portions of Fig.3a schematically represent the HL regions over which $(ASG)_2$ remains destroyed.

The 4.2K - field cooled (FC) HL (Fig.2 dashed line, FC from paramagnetic state in +20 kOe) is similar to the ZFC one (Fig.2 full line) but shows a centre shift (7) of -35 Oe.

REFERENCES

(1) U. Atzmony, E. Gurewitz, M. Melamud, H. Pinto, H. Shaked, G. Gorodetsky, E. Hermon, R.M. Hornreich, S. Shtrikman and B. Wanklyn, Phys. Rev. Letters 43 (1979) 782.

(2) Y. Yeshurun and H. Sompolinsky, Phys. Rev. B 31 (1985) 3191.

(3) J.K. Srivastava and J. Hammann, Japan. J. Appl. Phys. 26 (supplement 26-3) (1987) 787.

(4) J.K. Srivastava and J. Hammann, Japan. J. Appl. Phys. 26 (supplement 26-3) (1987) 789.

(5) J.K. Srivastava, S. Ramakrishnan, G. Rajaram, G. Chandra, V.R. Marathe, R. Vijayaraghavan, V. Srinivas, J.A. Kulkarni and V.S. Darshane, Hyperfine Int. 34 (1987) 537.

(6) J.K. Srivastava, G. Jéhanno and J.P. Sanchez, Phys. Letters A 121 (1987) 322.

(7) J. Hubsch and G. Gavoille, Phys. Rev. B 26 (1982) 3815.

Physica B 165&166 (1990) 183–184
North-Holland

Dynamic Scaling in the Ising Reentrant Spin Glass $Fe_{0.62}Mn_{0.38}TiO_3$

J.-O. Andersson, K. Gunnarsson, P. Svedlindh, P. Nordblad , L. Lundgren , H. Aruga* and A. Ito*
Department of Solid State Physics, Institute of Technology, Uppsala University, S-751 21 Uppsala, Sweden
*Department of Physics, Faculty of Science, Ochanomizu University, Bunkyo-ku, Tokyo 112, Japan

The dynamic susceptibility of the Ising reentrant spin glass system $Fe_{0.62}Mn_{0.38}TiO_3$ has been investigated using a SQUID magnetometer. The ac susceptibility was measured in the frequency interval 5 x 10^{-3} - 2 x 10^3 Hz. The data have been analyzed assuming critical slowing down, yielding a transition temperature T_{RSG} = 26.8 K , a dynamic exponent $z\nu$ = 8.2 and an order parameter exponent $\beta = 0.55$. Also, the wait time dependence of the zero field cooled magnetization was studied.

The field of disordered magnetism provides a number of challenging problems. One phenomenon of great interest is the reentrant spin glass (RSG) transition, where the system undergoes a transition from an ordered state to a spin glass state. Most RSG reported hitherto are of the ferromagnetic to spin glass order type. However, some RSG systems with antiferromagnetic (AF) order also exist, e.g. $Fe_{1-x}Mg_xCl_2$ (1) and $Fe_xMn_{1-x}TiO_3$. (2)

In this paper we report ac susceptibility measurements of the short range Ising RSG $Fe_{0.62}Mn_{0.38}TiO_3$ using a SQUID magnetometer. Also the time dependence of the zero field cooled (ZFC) magnetization has been investigated, in particular the wait time dependence of the relaxation, i.e. the ageing behaviour (3). Our results are in agreement with neutron scattering measurements by Yoshizawa et al (2), indicating a transition from pure AF to mixed AF and RSG state at a temperature, T_{RSG} .

The sample, a single crystal of $Fe_{0.62}Mn_{0.38}TiO_3$, was in the shape of a rectangular parallelepiped, 4x2x2 mm^3. The magnetic structure is well described by hexagonal layers, which contain the magnetic ions, having their moments parallel to the hexagonal c-axis (2). The origin of the RSG behaviour is the competition between antiferro and ferromagnetic exchange interaction within the layers. The ac susceptibility, $\chi(\omega) = \chi'(\omega) + i\chi''(\omega)$ was investigated in a wide frequency interval, viz. 5 x 10^{-3} Hz $\leq \omega/2\pi \leq$ 2 x 10^3 Hz. The ac susceptibility was measured by stepwise cooling the sample in a small ac field (h $\approx$ 0.1 G, ω constant). After stabilizing the temperature, $\chi'(\omega)$ and $\chi''(\omega)$ were detected by a lock in amplifier. The temperature was varied within the range 19 K $\leq$ T $\leq$ 35 K. For some frequencies, the interval was extended down to 7.3 K. The ZFC magnetization was measured by cooling the sample in zero magnetic field . After a certain wait time, t_w, at constant temperature, a static field, H=1 G, was applied and the time dependence of magnetization was recorded in the interval 3 x 10^{-1} s $\leq$ t $\leq$ 3 x 10^3 s .

Figure 1 shows the real and imaginary parts of the ac susceptibility versus temperature for different frequencies of the ac field. The real part χ' peaks at $T_N \approx$ 34 K and becomes frequency dependent below ~ 30 K. Unlike the results obtained for pure spin glasses , e.g. $Fe_{0.5}Mn_{0.5}TiO_3$ (4) , we get two maxima in χ'' - one near the AF-RSG transition and one broad peak at ~ 10 K (see inset of fig 1(b)). According to results from neutron scattering (2) some spins are removed from the antiferromagnetic long range order to the spin glass ordering with decreasing temperature below T_{RSG}. Our results are consistent with such an interpretation.

If we assume conventional critical slowing down on approaching T_{RSG} from above, the temperature dependence of the relaxation times is described by (5)

$$\tau_{max}=\tau_0\left(\frac{T_f-T_g}{T_g}\right)^{-z\nu} \qquad [1]$$

,where $z\nu$ is the dynamic exponent and τ_0 is a

microscopic relaxation time. τ_{max} corresponds to the maximum relaxation time of the spin system at a temperature T_f. Experimentally T_f can be defined by use of the inflection point of the imaginary part of the ac susceptibility , χ''. τ_{max} is then set equal to ω^{-1}.Fitting to eq. [1] yields T_{RSG} = 26.8±0.1 K , log τ_0=11.6±1.0 and zν = 8.2±1.0. According to a new approach to dynamic scaling proposed by Geschwind et al (6) $\chi''(\omega)$ may be scaled as follows

$$\chi''T = \omega^{\beta/z\nu}\,\widetilde{F}(t/\omega^{1/z\nu}) \qquad [2]$$

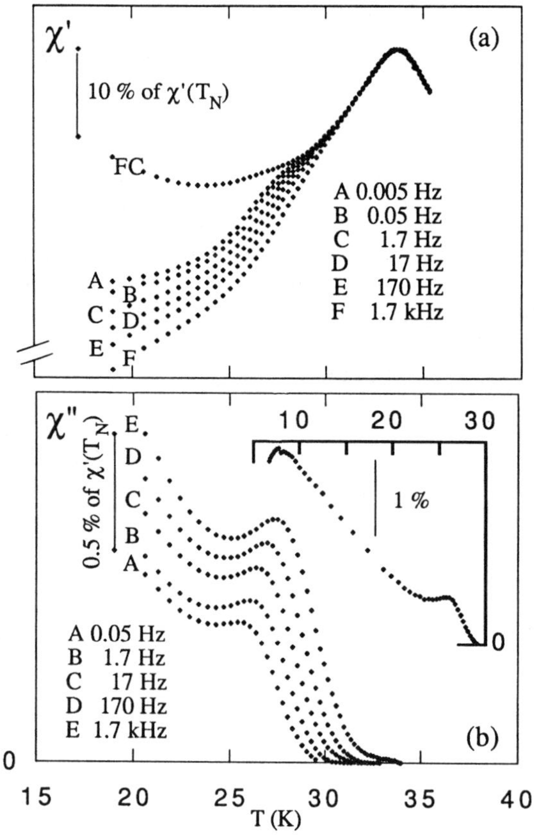

Fig 1. The real part , χ' , and the imaginary part , χ'' , of the c-axis susceptibility and the field cooled (FC) susceptibility vs temperature. The different curves correspond to different frequencies of the applied ac field. (a)χ' vs temperature. 10 % of $\chi'(T_N)$ is indicated. (b) χ'' vs temperature. 0.5 % of $\chi'(T_N)$ is indicated. inset: χ'' vs temperature for the frequency 0.05 Hz for an extended temperature range. 1 % of $\chi'(T_N)$ is indicated.

Thus the (first) peak of $\chi''(\omega)$ scales as

$$\chi''_{peak}T_{peak} \sim \omega^{\beta/z\nu} \qquad [3]$$

Fitting to eq. [3] yields β/zν = 0.067±0.005 which gives β = 0.55±0.10 .As a comparison zν = 10 and β = 0.60±0.10 for $Fe_{0.5}Mn_{0.5}TiO_3$ (4).
Fig.2 shows the relaxation rate S(t)=∂M(t)/∂logt of the ZFC magnetization M(T) at 19.0 K , i.e. well below T_{RSG} , for four different wait times. As can be seen the relaxation rate curves exhibit maxima at $t \approx t_w$, which shows that aging behaviour (3) exists in the reentrant spin glass phase.

Financial support from the Swedish Natural Science Research Council is gratefully acknowledged.

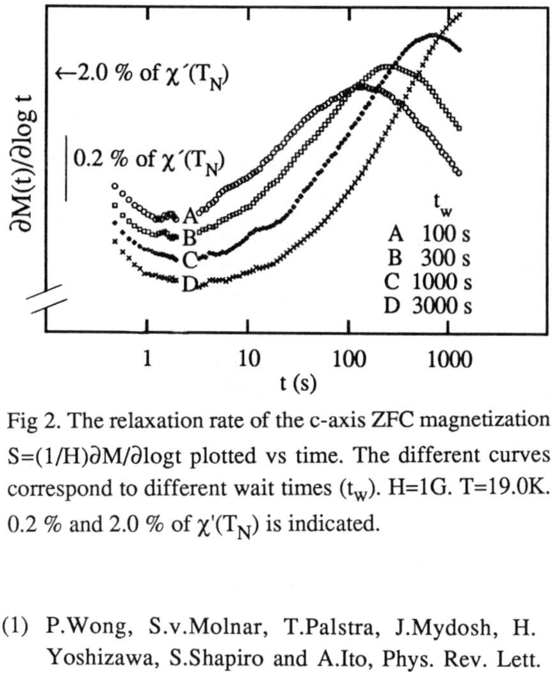

Fig 2. The relaxation rate of the c-axis ZFC magnetization S=(1/H)∂M/∂logt plotted vs time. The different curves correspond to different wait times (t_w). H=1G. T=19.0K. 0.2 % and 2.0 % of $\chi'(T_N)$ is indicated.

(1) P.Wong, S.v.Molnar, T.Palstra, J.Mydosh, H. Yoshizawa, S.Shapiro and A.Ito, Phys. Rev. Lett. **55**, 2043 (1985).

(2) H. Yoshizawa, S. Mitsuda, H. Aruga and A. Ito, J. Phys. Soc. Jpn. **58**, 1416 (1989).

(3) L.Lundgren , P.Svedlindh , P.Nordblad and O.Beckman , Phys Rev Lett. **51**, 911 (1983)

(4) K.Gunnarsson , P.Svedlindh , P.Nordblad , L.Lundgren , H.Aruga and A.Ito , Phys. Rev. Lett. **61**, 754 (1988)

(5) P.C.Hohenberg and B.I.Halperin , Rev.Mod.Phys. **49**, 435 (1977)

(6) S.Geschwind , D.A.Huse and G.E.Devlin ,preprint

Physica B 165&166 (1990) 185–186
North-Holland

TIME RESOLVED SPECIFIC HEAT OF THE INSULATING SPIN-GLASS $Eu_{.4}Sr_{.6}S$

C. Pappas[1], M. Meissner[1,2], G. Schmidt[1] and N. Bontemps[3]

[1] Hahn-Meitner-Institut Berlin, Neutronenstreuung II, D-1000 Berlin 39, FRG
[2] Institut für Festkörperphysik, TU Berlin, D-1000 Berlin 12, FRG
[3] E.S.P.C.I., 10 rue Vauquelin, F-75231 Paris, France

We present time-resolved specific heat measurements of $Eu_{.4}Sr_{.6}S$ below T_g (50mK $\leq$ T $\leq$ 1000mK) on a time scale of 10μs < t <10^2s. For T$\leq$ 500mK and short times (t < 100ms), we found complex thermalization profiles which suggest a time-dependent spin-glass specific heat.

1. INTRODUCTION

The dynamics of spin-glasses below T_g are characterized by aging and a broad distribution of relaxation times. These two features are closely related and have been mainly investigated through the decay of the remanent magnetization (M_R), the imaginary part of the susceptibility and the spectrum of magnetic fluctuations (noise) (1).

It has already been shown that the decay of M_R leads to a release of magnetic energy, which can be seen in specific heat measurements (2). Aging should also affect the specific heat but on a weaker level because it involves only a small amount of the total remanent magnetization (typically $10^{-3}M_R$). However, up to now, the specific heat of spin-glasses has shown no age dependence (2,3).

In order to optimize the experimental conditions under which we searched for an age(time)-dependent specific heat, we chose to extend the time window of our measurements to much shorter times (t > 10μs) than those used previously (t > 2sec, (3,4)). We also focused on the insulating system $Eu_{.4}Sr_{.6}S$ ($T_g \sim$ 1.5K), which shows aging on a much faster time scale than is usual in spin-glasses (4). Therefore, we assumed that it would be easier to detect an energy release due to aging when it takes place on the short time scale, as in $Eu_{.4}Sr_{.6}S$, than when it is spread over several days, as usual in spin-glasses. Here we present the first specific heat results using time-resolved thermometry (5) for 50mK $\leq$ T $\leq$ 1000mK and 10μs < t <10^2s.

2. EXPERIMENTAL METHODS

Our sample is the single crystal previously used for Faraday-rotation experiments (1,4). It is a platelet of thickness ~0.5mm and mass 22.6mg. The sample was heated by a short voltage pulse of duration ~50μs applied to the gold film heater mounted on one surface of the sample. The subsequent temperature vs. time profiles were determined with a carbon film thermometer on the opposite side. The sample was clamped onto two copper posts (∅ 1mm). The thermal resistance R_{SB} of the contact area between these posts and the sample imposes the sample-to-bath time constant (t_{SB}).

After every temperature cycle the carbon film thermometer was calibrated against a Ge-resistance. The temperature vs. time profiles were measured either with a resistance bridge (time constant ~0.2sec) or with a wide-band amplifier (DC-1MHz). In the latter case we used electronic filters adapted to the time windows investigated, and the signal was averaged on a multichannel analyzer over several hours. In order to minimize drifts due to non-equilibrium effects of thermometer and sample, the system was kept at 50mK for several days before the measurements, then warmed up to the working temperatures.

3. EXPERIMENTAL RESULTS

3.1 Long Time (t > 1s) Specific Heat

For 1000mK > T > 60mK, all temperature vs. time profiles measured with the 'slow thermometry' follow the exponential:

$$\Delta T(t) = \Delta T_C * exp(-t/t_{SB}) \quad (1)$$

which implies a time-independent specific heat, C_P, at the 'long-time' (t > 1s) limit. The resulting values of C_P are given in fig.1. They are in agreement with previous measurements made for T > 300mK (6). The specific heat shown in fig.1, decreases continuously with decreasing temperature, down to ~ 130mK, where the nuclear hyperfine contribution becomes comparable to the

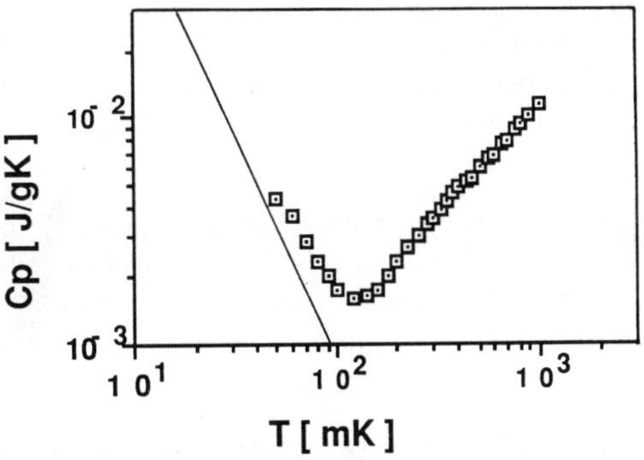

FIGURE.1 : Specific heat as a function of temperature, on a log-log scale. The solid line represents the estimated nuclear specific heat (see text).

magnetic one and C_P is at a minimum. At $T \leq 100mK$, C_P varies as T^{-2} in agreement with C_N=8.71*$10^{-6}T^{-2}$ J/gK (continuous line on fig.1), the nuclear specific heat calculated for an estimated hyperfine splitting of ~30T (7).

Fitting the data to eq.1 also gives the characteristic time: $t_{SB} = C_T * R_{SB}$ where C_T is the total specific heat of the sample (in J/K) and R_{SB} is given in K/W. We measured R_{SB} separately, with a stationary heat flow method, and found $R_{SB} \sim T^{-3}$, as expected for a phonon mediated boundary resistance. Thus t_{SB} increases continuously with decreasing temperature. Typically: t_{SB}=6s at 1000mK, 90s at 130mK and 1200s at 60mK.

3.2 Short Time (1s > t > 10 μs) Specific Heat

Fig.2 shows, on a log-log scale, temperature profiles for 500mK > T > 80mK. For the sake of comparison, the data are normalized as suggested by eq.1: $\Delta T/\Delta T_C$ is plotted versus t/t_{SB}. This scaling accounts for the temperature dependence of C_P, therefore all long-time data on fig.2 fall onto a single exponential (eq.1). However, this is not the case for the short-time data.

If the short time specific heat were time-independent we would expect, after the heat pulse, a heat diffusion thermalization of the sample given by the exponential:

$$\Delta T(t) = \Delta T_C * exp(-t_D/t) \quad (1)$$

with t_D the characteristic diffusion time. Although eq. 2 does not describe our data at any temperature, the deviations from the exponential are more dramatic at low temperatures than at 500mK. At 500mK we observe a steady, however non-exponential, rise of ΔT towards ΔT_C. At 200 and 300mK, we found well defined minima of the ΔT vs t curves. At T≤200mK, we first see a sharp increase of the temperature up to $\sim 10 * \Delta T_C$ followed by a subsequent slow decrease towards ΔT_C. Therefore we deduce that the specific heat of $Eu_{.4}Sr_{.6}S$ immediately after the heat pulse is strongly time-dependent. This effect becomes more visible at low temperatures, where the relaxation times are longer and the short-time specific heat is at least one order of magnitude smaller than its long-time limit C_P.

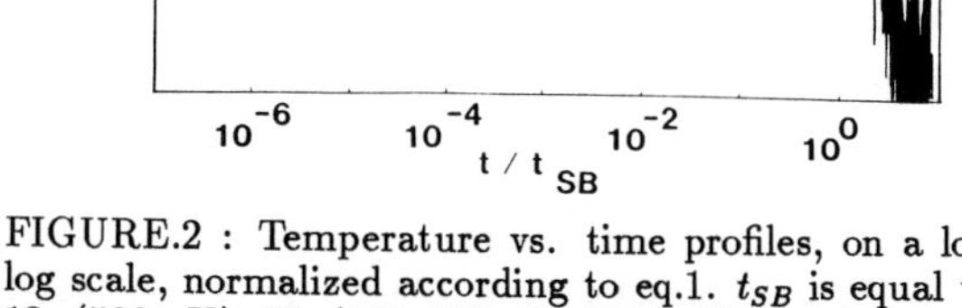

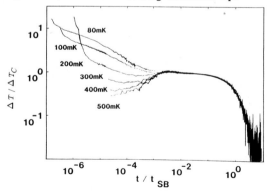

FIGURE.2 : Temperature vs. time profiles, on a log-log scale, normalized according to eq.1. t_{SB} is equal to: 12s (500mK), 18s (400mK), 25s (300mK), 42s (200mK), 180s (100mK) and 385s (80mK).

4. DISCUSSION

To our knowledge this is the first time that profiles like those seen in fig.2 have been observed. They are different from those seen in the systems we have investigated with time-resolved thermometry (glasses, crystalline materials) (5). The peculiar features of the ΔT vs. T profiles shown in fig.2 cannot be simply explained in the framework of a single (electronic, nuclear) spin reservoir coupled to the lattice. We believe that fig.2 reveals the response of the spin-glass phase to a temperature pulse. This assumption is also supported by the fact that we have obtained similar temperature profiles with the spin-glass system $CsNiFeF_6$ (8).

More specifically, the measurements at 200mK and 300mK show that, at short times, only a small fraction of the system (or of the specific heat) responds to the heat pulse whereas the rest follows on a much longer scale. Therefore we deduce that the specific heat of $Eu_{.4}Sr_{.6}S$ immediately after the heat pulse is strongly time dependent. This is seen at a time-scale which shifts to longer times with decreasing temperature. However, the relation to aging and to the large distribution of relaxation times recalled in the introduction is not yet clear. It could be, as previously shown in spin-glasses that a temperature step tends to sweep out the memory of the age of the system (9), which then recovers on a time scale which gets longer as the temperature decreases. Further experiments, varying the age of the sample and the elevation of temperature after the heat pulse, are needed to compare our results with those obtained with magnetic measurements. Nevertheless, these results suggest that the specific heat of the zero-field cooled spin-glass state is influenced by the spin-glass dynamics.

ACKNOWLEDGEMENTS

We wish to thank F. Mezei, P. Monod and M. Ocio for helpful discussions.

REFERENCES

(1) For a review see : J. Ferré and N. Bontemps, "Relaxation and critical dynamics in spin-glasses", in Trans. Tech. Publications Material Sciences and Technology (1989), and references therein.
(2) A. Berton, J. Chaussy, J. Odin. J. Peynard, J.J. Préjean and J. Souletie, Solid State Commun. 37, (1981) 241
G.J. Nieuwenhuys, J.A. Mydosh, Physica 86-88 B, (1977) 880
(3) P. Nordblad, L. Sandlund, P.Granberg and L. Lund gren, Journal de Phys. 49 Coll. C8, (1988) 1041
(4) W. Luo, M. Lederman, R. Orbach, N. Bontemps and R. Nahoum, Phys. Rev. B, in print
(5) J.J. DeYoreo, W. Knaak, M. Meissner, R.O.Pohl, Phys. Rev. B 34, (1986) 8828
(6) D. Meschede, F. Steglich, W. Felsch, H. Maletta and W. Zinn, Phys. Rev. Lett. 44, (1980) 102
(7) G. Eiselt, J. Kötzler, H. Maletta, D. Stauffer and K. Binder, Phys. Rev. B19, (1979) 2664
(8) C. Pappas, M. Meissner, B. Wanklyn, unpublished
(9) P. Refregier, E. Vincent, J. Hammann, and M. Ocio, J. de Phys. 48, (1987) 1533

Physica B 165&166 (1990) 187–188
North-Holland

ON THE MAGNETIC ORDERING IN AMORPHOUS AND QUASI-CRYSTALLINE Al-Mn PHASES

J.C. LASJAUNIAS+, M. GODINHO+, and C. BERGER◊

+Centre de Recherches sur les Très Basses Températures, ◊Laboratoires d'Etudes des Propriétés Electroniques des Solides, C.N.R.S., BP 166 X, 38042 Grenoble-Cédex, France

Both the nuclear hyperfine specific heat at very-low temperature and the freezing temperature investigated in the whole available concentration range of Mn ($15 \leq c \leq 22$ at.%) of Al-Mn alloys indicate similar spin-glass properties for amorphous and quasi-crystalline phases, that hints for similar structural local environments of the Mn atoms which are at the origin of the magnetic moment.

One of the typical properties of the quasi-crystalline (Q.C.) (icosahedral and decagonal-T) phases of AlMn(Si) alloys in comparison to their crystalline counterpart is the presence of localized magnetic moments with a spin glass (S.G.)-like ordering at a few Kelvin for $c \simeq 20$ at.% Mn [1-6]. However the role of the quasi-crystalline structure for the appearance of magnetic Mn's is still not cleared up ; moreover, a similar origin seems very probable in the case of amorphous (sputtered) phases in view of the similarity of many of their magnetic properties investigated up to now.

Below the freezing temperature T_f, measured here by the maximum of the a.c. susceptibility at frequency $\simeq$ 110 Hz [4], the electronic spins are stabilized in permanent moments and give rise at very low temperature (Fig. 1) to a nuclear hyperfine contribution to the specific heat, even in the absence of an applied magnetic field. In the case of 3d transition impurities in an Al host, this interaction is mainly due to a core polarization effect between the effective magnetic field $\mathbf{H}_{eff}$, of amplitude proportional to the electronic magnetization (spin $\mathbf{S}$), and the nuclear magnetic spin $\mathbf{I}$ of the *same* atom. At temperatures high in comparison to the Zeeman levels spacing, the nuclear specific heat is :

$$C_p = x^* N k_B \frac{I+1}{3I} \left(\frac{\mu_n H_{eff}}{k_B T} \right)^2$$

$$= C_N T^{-2} \text{ (per mole of alloy) ,} \qquad (1)$$

where I and μ_n are the spin and the nuclear magnetic moment of ^{55}Mn nuclei, x^* the molar concentration of magnetic Mn *atoms*, each bearing a magnetic moment μ. Hence C_N is directly proportional to the number of individual Mn atoms with a moment close to their saturation :

$$C_N \sim x^* |H_{eff}|^2 \sim x^* \langle S \rangle^2 \qquad (2)$$

We assume that origins other than nuclear magnetic from the magnetic Mn atoms are negligible or unlikely [4].

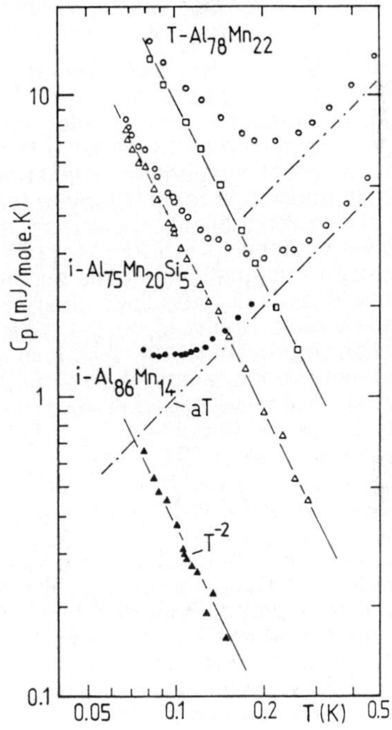

FIGURE 1
Analysis of the very-low temperature specific heat into a nuclear hyperfine ($\sim T^{-2}$) term and an electronic or magnetic ($\sim aT$) contribution, for three quasi-crystalline alloys : i-$Al_{86}Mn_{14}$, i-$Al_{75}Mn_{20}Si_5$ and T-$Al_{78}Mn_{22}$.

We have measured this contribution by a transient heat-pulse technique, in a series of amorphous and quasi-crystalline (icosahedral and decagonal-T) alloys, prepared by sputtering and melt-spinning respectively [4,7,8] ; in the case of amorphous and decagonal-T, they are single phase. The time span for this measurement at the lowest temperatures ($T \leq 100$ mK) is usually longer than 10 mn and can reach about 1 hr for large values of C_N. In Fig. 1, one can see the very rapid increase of C_N between the i-$Al_{86}Mn_{14}$ alloy (due to about 70 vol. % of icosahedral grains of composition close to $Al_{72}Mn_{18}$ [9]) and the decagonal T-$Al_{78}Mn_{22}$, that corresponds to an increase of only 4 at.% for the actual Mn concentration in the QC phase. From the magnetic susceptibility which obeys a Curie law in the low temperature range not too far from T_f, we can obtain a molar concentration x^* of magnetic Mn atoms by assuming an electronic spin S of about 2 which corresponds to $\mu_{eff} \simeq 5\ \mu_B$, like in canonical spin-glasses Ag-Mn or Cu-Mn. One finds $x^* = (8.5 \pm 1.5)10^{-3}$, corresponding to only 4 % of total Mn atoms, and from relation (1) an effective field $|H_{eff}| = 270 \pm 25$ kOe, in excellent agreement with the values for dilute ($c < 1$ at.%) Ag-Mn or Cu-Mn [4].

Indeed, all magnetic data investigated for this concentration of about 20 at.% such as high field magnetization, low temperature magnetic hysteresis [5], frequency dependence of the freezing temperature [4] indicate great similarities with canonical metallic S.G. ; in addition, a recent study of the non-linear terms of d.c. magnetization proves the existence of a phase transition in i-$Al_{73}Mn_{21}Si_6$ [10]. However, instead of magnetic moments occuring on isolated impurities as in diluted alloys, their origin in clusters of a few Mn's seems to be the most plausible explanation at present [7,10]. This is in agreement with the high Kondo temperature (~ 600 K) of the Al-Mn system, since we expect a lowering of T_K by clustering of the Mn impurities, as it generally occurs in other Kondo alloys [11].

Moreover, when we plot the variation of T_f versus the amplitude of the nuclear specific heat C_N (Fig. 2) in the whole available concentration range, it roughly obeys a linear variation, more exactly in $C_N^{0.8}$, as well for QC as for amorphous alloys. Since C_N is directly proportional to the concentration x^* of atomic moments, with the hypothesis of a value of H_{eff} little dependent on the concentration as in canonical S.G., this scaling relation is very similar to $T_f \sim x^{0.7}$ observed in Cu-Mn or Au-Mn S.G. for $x \gtrsim 1$ at.% [12], and which results from the $1/r^3$ long range magnetic interactions of the R.K.K.Y. model. We note that x^* is also close to 1 at.% for a nominal Mn concentration of 20 %. Hence there is an overall similarity of the magnetic ordering in both amorphous and QC phases, which suggests that similar local environments of the Mn atoms are at the origin of the magnetic moment. This is also supported by the similarity of the electronic properties of both phases up to 20 at.% Mn [9].

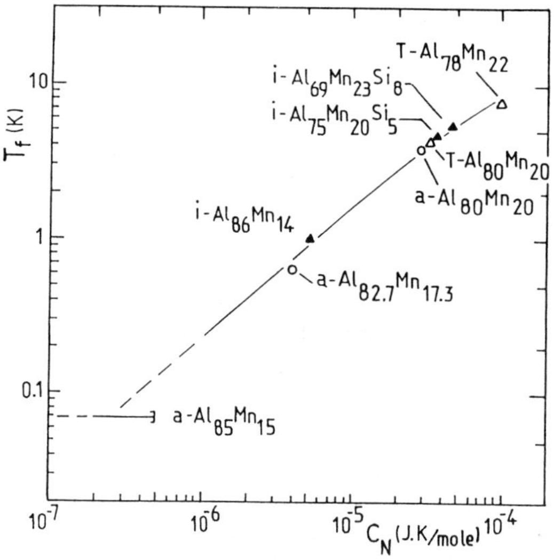

FIGURE 2
Freezing temperature T_f as measured by a.c. susceptibility at 110 Hz versus the amplitude of the nuclear hyperfine term for all the amorphous and quasi-crystalline (icosahedral and decagonal-T) samples.

REFERENCES

[1] J.J. Hauser, H.S. Chen, and J.V. Waszczak, Phys. Rev. B **33** (1986) 3577.

[2] K. Fukamichi, T. Goto, T. Masumoto, T. Sakakibara, M. Oguchi, and S. Todo, J. Phys. F **17** (1987) 743.

[3] C. Berger, and references quoted therein, "*Fundamental and Applicative Aspects of Disordered Magnetism*", P. Allia *et al.* ed., World Scientific (1989).

[4] J.C. Lasjaunias, J.L. Tholence, C. Berger, and D. Pavuna, Solid State Commun. **64** (1987) 425.

[5] R. Bellissent, F. Hippert, P. Monod, and F. Vigneron, Phys. Rev. B **36** (1987) 5540.

[6] K. Wang, P. Garoche and Y. Calvayrac, J. Physique (Paris) C8 **49** (1988) 237.

[7] C. Berger, K. Hasselbach, J.C. Lasjaunias, C. Paulsen, and P. Germi, J. Less-Common Metals **145** (1988) 565.

[8] A. Gozlan, C. Berger, G. Fourcaudot, J.C. Grieco, F. Cyrot-Lackmann and P. Germi, Solid State Commun. **73** (1990) 417.

[9] M. Godinho, C. Berger, J.C. Lasjaunias, K. Hasselbach and O. Béthoux, 7th Int. Conf. on Liquid and Amorphous Metals (Kyoto, 1989), to be published in J. Non Crystal. Solids.

[10] C. Berger and J.J. Préjean, submitted to Phys. Rev. Lett. (1990).

[11] M. Miljak and J.R. Cooper, Physica **86-88B** (1977) 476.

[12] See, e.g., V. Cannella, in *Amorphous Magnetism-I*, H.O. Hooper and M. de Graaf, ed., Plenum Press (1973), 195.

Physica B 165&166 (1990) 189–190
North-Holland

A POSSIBLE FERMI LIQUID REGIME IN CuMn AND SIMILAR SPIN GLASS ALLOYS

Selman HERSHFIELD

Joseph Henry Laboratories of Physics, Jadwin Hall, Princeton University, Princeton, NJ 08544, USA

It is proposed that when the spins are sufficiently frozen out in metallic spin glass alloys like CuMn, there is a Fermi liquid regime.

1. INTRODUCTION

Systems of randomly distributed magnetic impurities in metals like the more concentrated alloys of CuMn and AuFe undergo a spin glass transition. In spite of many experimental and theoretical studies the nature of the very low temperature excitations about the ground state or a metastable state is still unknown. Low temperature here means far below the transition temperature, T_c. At these temperatures the decay from a given local energy minimum becomes very slow on the experimental time scale, and it is possible to probe just the excitations about a given energy minima.

From studies of other frozen magnetic systems there are a number of reasonable possibilities for these low temperature excitations. Since many of the alloys, including CuMn, are Heisenberg systems with very weak anisotropy, spin waves may be important at low temperatures. Because these are random systems, there will be some spins which are more weakly connected to the rest of the system than a typical spin. At low temperatures these few spins will be more active and may dominate the temperature dependence of various experimental quantities. Yet a third kind of excitation which may be important are large clusters of spins, called droplets, which flip together. Such droplet excitations are believed to be important in Ising short range spin glasses.

All three of the above are purely spin excitations. In addition to the spins there are the conduction electrons, which have a constant density of states for excitations down to zero energy. In other systems of magnetic impurities in metals at low temperatures when the impurity degree of freedom is frozen out, the spins primary role is to renormalize the electronic properties and induce interactions between the electrons(1). The region in temperature where this happens is called a Fermi liquid regime.

The hypothesis presented here is that at low temperatures in CuMn and similar spin glass alloys there is such a Fermi liquid regime. Since these systems are very complicated, rather than show this by some purely theoretical argument, we examine three experimental properties, the remanent magnetization, the resistivity, and the specific heat, and show that a Fermi liquid regime is consistent with all the qualitative features of the data. Each of the three other experimental properties mentioned above are either inconsistent with the data or have problems explaining it.

2. COMPARISON TO EXPERIMENT

2.1 Remanent Magnetization

Alloul et. al. (2,3) have done a series of measurements on the $T \ll T_c$ temperature dependence of the remanent magnetization. The remanent magnetization is created by cooling the spin glass in a magnetic field to a temperature below the transition temperature and then turning off the magnetic field. For T less than T_c the remanent magnetization will decay logarithmically in time. Alloul et. al. (2) find that for $T < 0.1\ T_c$ the temperature dependence of the remanent magnetization, M(T), is reversible on experimental time scales, indicating that at these temperatures one is indeed probing the "in well" excitations. In this low temperature regime the magnetization reduction, M(T)-M(0), is proportional to $-T^2$, as required for a Fermi liquid regime in these alloys. A Pauli susceptibility is not required in a Fermi liquid regime for these alloys because the spins can rotate by small amounts to a changing magnetic field.

By introducing a nonmagnetic impurity (Pt), Alloul et. al. were able to increase the anisotropy until the characteristic temperature of the anisotropy became much greater than the temperature. No change in M(T) - M(0) was seen. This rules out conventional spin waves which should show a gap opening up for such large anisotropies. Since Fermi liquid effects involve virtual processes, anisotropies should not suppress them.

The spin reduction was also measured via NMR (3) which measures the expectation value of each spin individually because the resonant frequency is proportional to the expectation value of the spin. In the NMR experiments the same behavior was seen as in the direct magnetization experiments. This means that the spin reduction is occuring fast on the time scale of the NMR measurement (10^{-9} sec.), which makes it difficult to account for the magnetization reduction via large droplet excitations which fluctuate on long time scales. The spin reduction due to Fermi liquid effects occur on electronic times, which are short.

The whole frequency spectrum of the NMR signal shifts to higher frequencies as one decreases the temperature, indicating that the spin reduction takes place fairly uniformly throughout the sample. In particular Alloul and Mendels (3) rule out that only twenty percent of the spins cause the spin reduction. Again this is consistent with Fermi liquid effects because each spin participates in the virtual processes which renormalize the electronic properties.

2.2 Resistivity

The resistivity at the lowest temperatures has been reported to be quadratic in the temperature by a number of workers (4,5). More data is needed to show that this behavior is generic in metallic spin glasses. Within a Fermi liquid regime a T^2 resistivity is required. It is due to an electron-electron interaction mediated by the spins.

2.3 Specific heat

The magnetic contribution to the specific heat is defined as the specific heat of the alloy minus the specific heat of the pure metal and the nuclear specific heat. It is roughly linear starting above the transition temperature and continuing down far below T_c. This linear specific heat, however, extrapolates to zero at a finite temperature. At approximately $0.1T_c$ in CuMn there is evidence for a second linear regime which does extrapolate to zero at zero temperature (6). In the more dilute alloys a Schottky anomaly in the nuclear specific heat prevents one from getting far into this regime; however, by going to more concentrated alloys one will be able to get to lower reduced temperatures, T/T_c, before the Schottky anomaly becomes important. A linear specific heat is required within a Fermi liquid scheme. It should be viewed as an enhancement to the electronic specific heat due to the interaction with the spins. We also note that a linear specific heat is not possible for spin waves because the spin wave ground state is unstable for a nonzero density of states for excitations at zero energy.

3. CONCLUSION

We have seen that a Fermi liquid regime is consistent with all the qualitative features of the data, while three other possible excitations are not. Further experimental and theoretical work is needed to determine conclusively if there is a Fermi liquid regime in these alloys.

ACKNOWLEDGEMENTS

I would like to thank P. DeVegvar, D. S. Fisher, C. Henley, H. R. Krishnamurthy, and L. P. Levy for useful discussions. This work was supported by NSF grant DMR-8719523.

REFERENCES

(1) P. Nozieres, J. Low Temp. Phys. 17 (1974) 31.
(2) H. Alloul et. al. , Europhys. Lett. 1 (1986) 595.
(3) H. Alloul and P. Mendels, Phys. Rev. Lett. 54 (1985) 1313.
(4) O. Laborde and P. Radhakrishna, J. Phys. F 3 (1973) 847.
(5) H. Albrecht et. al., Z. Phys. B 49 (1983) 213.
(6) D. L. Martin, Phys. Rev. B 20 (1979) 368; Phys. Rev. B 21 (1980) 21.

Physica B 165&166 (1990) 191–192
North-Holland

CRITICAL DYNAMICS OF AuFe REENTRANT FERROMAGNETS CLOSE TO THE PERCOLATION THRESHOLD

C. Pappas[1], C. Lartigue[1], M. Alba[2] and F. Mezei[1]

[1] Hahn-Meitner-Institut, Neutronenstreuung II, D-1000 Berlin 39, FRG
[2] Institut Laue Langevin, Avenue des Mertyrs, 156X - 38042, Grenoble Cedex, France

We present inelastic neutron scattering data on critical dynamics of AuFe 18% and 16% disordered ferromagnets. The concentrations of samples were close to the critical value ($x_c \sim 15.5\%$) where ferromagnetism disappears and spin-glass sets in. The value of the dynamic critical exponent z for AuFe 18% ($z \sim 2.3$) is close to that of a 3-D Heisenberg ferromagnet, i.e. is not affected by disorder. On the other hand, we found $z \sim 1.8$ for AuFe 16%. This drastic decrease of the dynamic critical exponent, for a small change of the Fe concentration, suggests an influence of disorder on critical dynamics.

1. INTRODUCTION

One important question in disordered magnetism is the influence of disorder on the critical behaviour and critical exponents. The Harris criterion (1), which is valid only in the limit of weak disorder, states that disorder changes the values of critical exponents only when $\alpha \geq 0$, where α is the exponent of the specific heat of the corresponding non-disordered system. However, these predictions apply only in the case of weak disorder, and the question of strong disorder is still open.

Several measurements have been made with disordered ferromagnets (2). Major result of these studies is that critical exponents tend towards the values of the spherical model ($\gamma = 2$, $\beta = 0.5$ and $\nu = 1$) when approaching the percolation threshold x_c. Recent specific heat experiments also confirm this conclusion (3). However, all these results concern only the static critical behaviour. Experimental as well as theoretical information on the critical slowing down is lacking.

We focussed on AuFe because it is an extensively studied system which shows, close to the percolation threshold x_c ($\sim$ 15.5 Fe at %), either ferromagnetism ($x > x_c$) or a spin glass phase ($x < x_c$) (4). In the vicinity of x_c, the ferromagnetic phase is strongly disordered, and a reentrant spin glass behavior is seen at low temperatures (4,5). Neutron scattering has already been used to investigate several properties of this system except critical dynamics (4). Here we present inelastic neutron scattering results showing critical slowing down of the dynamics close to the ferromagnetic critical temperature T_c.

2. EXPERIMENTAL DETAILS

We studied two reentrant ferromagnetic samples with concentrations close to x_c: x=18 % and 16 %. The samples were polycrystalline disks prepared by arc melting of the constituents (0.5mm thick, diameter 37mm). They were annealed over 12 hours at 900^0 and then for 1 hour at 550^0. They were then quenched and kept in liquid Nitrogen between the measurements. The homogeneity of the samples was tested with susceptibility measurements.

The measurements were done at the cold-neutron three-axes spectrometer IN12 of the ILL, Grenoble. We used the collimation 20'-20'-20'. and covered a q range between $0.030Å^{-1}$ and $0.100Å^{-1}$. In order to maximize the counting rate, we used two different configurations: (1) $0.030Å^{-1} \leq q \leq 0.063$ $Å^{-1}$, k_i=$1.1Å^{-1}$ $\Gamma_{res} = 24$ μev FWHM, and (2) $0.085Å^{-1} \leq q \leq 0.100$ $Å^{-1}$, k_i=$1.4Å^{-1}$ $\Gamma_{res} = 60\mu$ev FWHM. The values of Γ_{res} were determined from independent measurements on a Vanadium disk.

The background contribution was corrected using either high temperature spectra (at $\sim 2T_c$) or separate empty can measurements. Within the experimental accuracy, both procedures lead to the same results.

3. EXPERIMENTAL RESULTS

Our investigation was focussed on the critical scattering at T_c. The ferromagnetic ordering temperatures (T_c = 140.5K and 92.5K for AuFe 18% and 16% respectively) were determined from the peaks of the critical scattering, as a function of temperature. Assuming the Ornstein-Zernike form, $S(q) = \chi/(1+q^2/\kappa^2)$, where χ is the bulk susceptibility and κ the inverse correlation length, we deduced $\nu \sim 1$, in agreement with previous SANS experiments on AuFe 17% (5) and with results on other disordered systems (2).

Critical fluctuations close to T_c are related to the Fourier transformation of the spin-spin correlation function:

$$S(q,\omega) = S(q)\frac{\Gamma(q)}{\Gamma(q)^2+\omega^2} \qquad (1)$$

The experimental determination of this function implies accurate measurement of the quasi-elastic signal down to very low values of the scattering vector q. After correction, all $S(q,\omega)$ vs. ω spectra can be fitted to a single Lorentzian, in agreement with eq.1. Fig. 1 shows a typical spectrum of AuFe 18% at T_c, after subtraction of the background, as well as the best fit to eq.1 convoluted with the resolution. Within the investigated temperature range ($T/T_c \leq 1.06$), Γ does not depend on the temperature. This differs from pure Fe, for which Γ slightly decreases with increasing T, in both the same q and T/T_c range [6]

In the following we focus on the q-dependence of Γ at T_c. According to dynamic scaling we expect:

$$\Gamma(q) = A\,q^z \quad (2)$$

with z=2.5 for Heisenberg ferromagnets (7).
In fig.2, Γ is plotted versus q on a log-log scale. For the sake of comparison the curve of Fe at Tc is also shown. The behavior for AuFe 18% is similar to that of Fe and the resulting value of $z = 2.3 \pm 0.1$ is very close to the one expected for a pure Heisenberg ferromagnet. On the other side in AuFe 16%, z decreases down to $z = 1.8 \pm 0.1$, which is much lower than the value found in both pure Fe and AuFe 18%.

4. DISCUSSION

According to dynamic scaling (7), the values of z depend on the balance between exchange and spin non-conserving interactions (e.g dipolar with z=2). However, as far as the interactions are concerned, AuFe 18% and 16% are very similar and show comparable Curie temperatures. For the sake of comparison we note that, for pure Fe, $T_c \sim 1043$K. Therefore, we deduce that the concentration dependence of z, seen in fig.2 can be attributed to disorder. AuFe 16% is very close to x_c, which is the limit of strong disorder for the ferromagnetic phase.
Disorder not only affects the value of z , but also the temperature dependence of Γ, which differs from that of pure Fe for both samples. This implies a modification of the scaling relation $\Gamma = Aq^z F(\kappa/q)$ in our

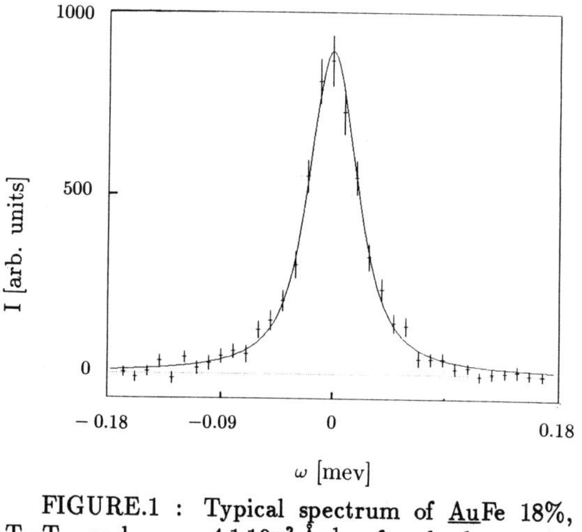

FIGURE.1 : Typical spectrum of AuFe 18%, at $T=T_C$ and $q = 4.1\,10^{-2}$ Å^{-1}, after background correction. The continuous line shows the best fit to eq.1 with Γ=36.5 μev (FWHM).

disordered AuFe system with respect to pure Fe. Further experiments are planned in order to extend the q-range covered by our data to smaller values of q and accurately determine κ above T_c. This will allow verification of scaling between Γ and κ in the limit of strong disorder.

ACKNOWLEDGMENTS
We wish to express our thanks to C. Förster and W. Rönnfeldt for helping in preparing the samples and to U. Zinngrebe and M. Meissner for testing the samples with susceptibility measurements.

REFERENCES

(1) A.B. Harris, J. of Phys. C7, (1974) 1671
(2) for a review see : K. Westerhold and G. Sobota, J. of Phys. F 13, (1983) 2371
(3) J. Wosnitza and H. von Löneysen, J. de Phys. C8, (1989) 1203
(4) B.V.B. Sarkissian, J. Phys. F 11, 2191 (1981)
(5) A.P. Murani, Materials Science Forum, 27/28,(1988) 195
(6) F. Mezei, Phys. Rev. Lett., 49, (1982) 1096
(7) P.C. Hohenberg, B.I. Halperin, Review of Modern Phys. 49, (1977) 467

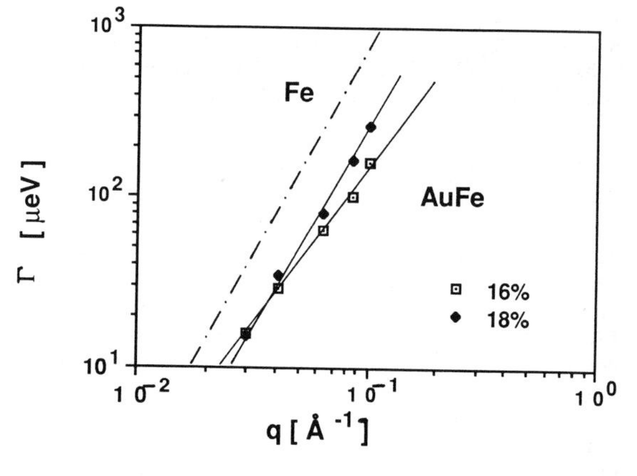

FIGURE.2 : Log-log plot of Γ versus q at T_c.

Physica B 165&166 (1990) 193–194
North-Holland

MAGNETIC AND ELECTRON SPIN RESONANCE STUDIES OF $(Ce_{1-c}Gd_c)Rh_2$

A. TARI

Department of Physics, King Fahd University of Petroleum & Minerals, Dhahran 31261
Saudi Arabia

Magnetic and ESR measurements have been performed on several samples of the $(Ce_{1-c}Gd_c)Rh_2$ series in the interval $0 \leqq c \leqq 1$. It is found that only those Ce atoms with at least 3nn Gd atoms are magnetic. The cerium sublattice appears to reach the magnetic percolation limit at about c=0.3. Using the lattice parameters and the RKKY model a value of 0.19eV is computed for the effective exchange integral J_o at the $GdRh_2$ end of the series.

1. INTRODUCTION

This work is a part of an ongoing investigation regarding the valence state of Ce in $(Ce_{1-c}Gd_c)Rh_2$ and its effect on the magnetic, thermal and resonance properties of these compounds. The present paper deals with the magnetic properties of these compounds and the electron spin resonance (ESR) of Gd in this series.

Cerium is in an intermediate valence state in a number of intermetallic compounds(1). Recent L_{III} experiments carried out by Wohlleben and Rohler(2) and Mihalisin et al(3) put the valency of Ce in $CeRh_2$ at between 3.13 and 3.21. From a high temperature susceptibility study Weidner et al(1) put an upper limit of 3.4 at 1200K for the valency of Ce in this compound.

$(Ce_{1-c}Gd_c)Rh_2$ is, perhaps, the simplest system in which to investigate the valence state of Ce. Gd is an S-state ion and is therefore free from the complications of the crystal field (CF) effects. Hence its magnetic moment remains constant when its concentration is varied.

The strong exchange field produced by Gd is expected to drive the Ce ions towards the magnectic Ce^{3+} state. In this series, therefore, the effect of the local environment is expected to be important.

2. SAMPLES AND EXPERIMENTAL TECHNIQUES

The compounds were made by melting together stoichiometric amounts of 99.9% pure Rh and 99.99% pure rare earths. The resulting buttons were melted a number of times for good homogeneity under a high purity argon atmosphere and were subsequently heat treated under an argon pressure of $p \leqq 10^{-7}$ torr for three days. The cubic Laves structure of all the compounds was confirmed by x-ray diffraction.

Magnetisation measurements were made at temperatures between 2K and 300K and in fields of up to 50kOe and the ESR measurements were performed at the X-ban frequency at temperatures between 100K and 300K on powder samples.

3. RESULTS

Figure-1 shows the variation of the saturation magnetisation per formula unit μ with c. The straight line in this figure represents the variation of the gadolinium magnetisation alone i.e. assuming that cerium atoms are not magnetic in this series. As the figure shows the experimental points start to deviate from this line at about c=0.3. The deviation is largest in the interval $0.3 \leqq c \leqq 0.8$. This suggests that in compounds with c > 0.3 the cerium atoms are magnetic and couple antiparallel to the gadolinium sublattice. However, as will be shown presently not all the cerium atoms become magnetic at this concentration but a magnetic percolation limit is reached in the cerium sublattice whereby a continuous chain of magnetic atoms is established throughout the compounds.

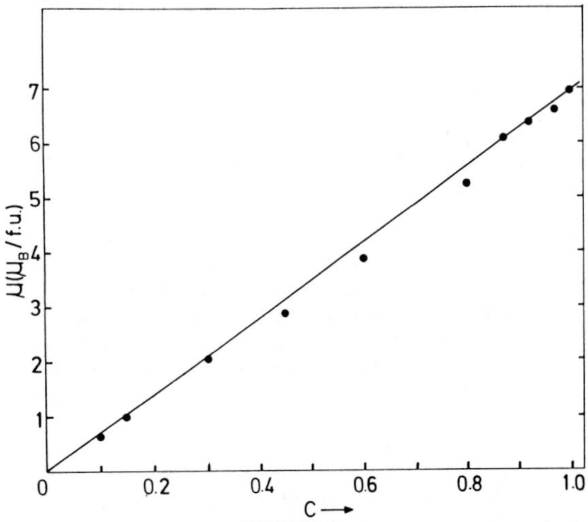

FIGURE 1
Change of saturation magnetisation with c

0921-4526/90/$03.50 © 1990 – Elsevier Science Publishers B.V. (North-Holland)

Although our ESR investigation is not yet complete, the available data confirms the conclusion reached above. This is shown in Figure-2 where the variation of the g-value and the temperature gradient of the linewidth, $D=d(\Delta H)/dT$, with c is displayed. As the figure shows D increases with the increasing gadolinium content, peaks at about c=0.3 and then decreases towards D=2.6(G/K). We shall interprete this variation by starting from the $GdRh_2$ end where the system is bottlenecked.

Cerium is magnetic at this end of the series and hence acts as a strong scattering center and thus help to open the bottleneck. With increasing Ce concentration the bottleneck opens further. The decrease in the gadolinium concentration also contributes to this. At c ≅ 0.3 where D has its maximum value a large number of Ce atoms have spin fluctuation temperatures in or below the temperature range of our ESR measurements. These cerium atoms and those already magnetic act as scattering centers thereby open the ESR bottleneck considerably leading to the observed peak.

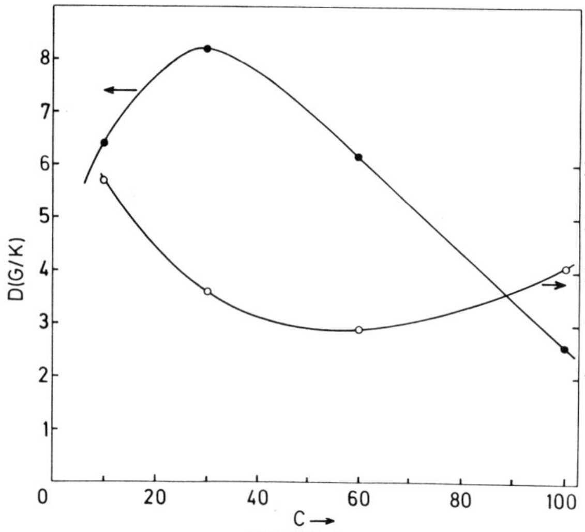

FIGURE 2
Change of the g-value (open circles) and the temperature gradient of the linewidth D (solid circles) with gadolinium content.

In the concentration range c < 0.3 the number of cerium atoms acting as strong scattering centers becomes fewer and fewer while the gadolinium concentration is still high. This again increases the effect of the bottleneck. This trend continues until the gadolinium concentration is sufficiently dilute. For c ≦ 0.01 Barberis et al(4) obtained D ≅ (18 ± 2)G/K which is the non-bottleneck limit.

Shown in Figure-3 is the variation of the mean reduced moment for Ce, $\bar{\mu} = \mu_{Ce}/\mu_o (\mu_o = 2.14\mu_B$ for Ce^{3+}) and P(n,c) with c. P(n,c) is the probability of finding n nearest neighbour (nn) Gd atoms to a given Ce site. μ_{Ce} is computed by assuming a saturation moment of $7\mu_B$ for Gd in this series. As the figure shows n=3 gives the best fit. Thus only those Ce atoms with n ≧ 3 nn Gd are magnetic. It should be added however, that the transition from a non-magnetic to a magnetic state is not continuous but a gradual process and that the magnetic Ce atoms are also expected to take part in this process of driving the remaining non magnetic Ce atoms towards the magnetic Ce^{3+} state.

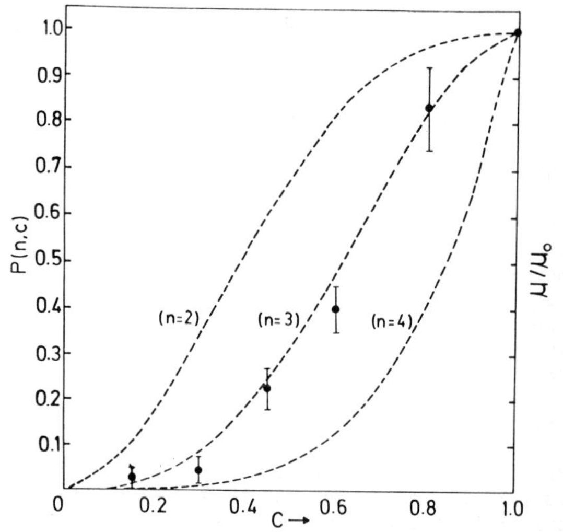

FIGURE 3
Variation of the $\bar{\mu}$ and P(n,c) with c (for the meaning of $\bar{\mu}$ and P(n,c) see the text)

Using the lattice spacing data and the RKKY model and assuming 3 conduction electrons per formula unit we have calculated the effective exchange integral J_o for Gd in $GdRh_2$ and found a value of 0.19eV which is reasonable.

ACKNOWLEDGEMENT

The author acknowledges gratefully the support of KFUPM for this work.

REFERENCES

(1)P. Weidner, B. Wittershagen, B. Roden and D. Wohlleben, Solid State Commun. 48 (1983) 915

(2)D. Wohlleben and J. Rohler, J. Appl. Phys. 55 (1984) 1904

(3)T. MUhalisin, A. Harrus, S. Raaen and R.D. Parks, J. Appl. Phys. 55 (1984) 1966

(4)G.E. Barberis, D. Davidov, J.P. Donosa, C. Rettori, J.F. Suassuna and H.D. Dokter, Phys. Rev. B19 (1979 5495

Physica B 165&166 (1990) 195–196
North-Holland

THE EFFECTS OF HYDROGEN ON THE LOW TEMPERATURE MAGNETIC PROPERTIES OF PdNi

Z. WANG, H.P. KUNKEL and Gwyn WILLIAMS

Department of Physics, University of Manitoba, Winnipeg R3T 2N2, Canada

We report the effects of hydrogen on PdNi through measurements of the ac susceptibility, the dilation at room temperature and the coercive field at 4.2 K.

In a series of recent papers (1-3), the onset of magnetic ordering in dilute PdNi has been investigated using a combination of spontaneous resistive anisotropy (SRA) (basically the difference between the longitudinal and transverse magnetoresistivities extrapolated to zero induction B (4)) and ac susceptibility measurements. The occurrence of a non-zero SRA requires two principal ingredients, a substantial polarizing field (here the exchange field from inter-impurity interactions) and spin-orbit coupling at Ni sites. While an SRA will not appear without the latter coupling, its presence complicates considerably the analysis of susceptibility data; it causes a significant "regular" contribution to the magnetic response to persist in an applied field masking the critical contribution (3). Thus although Curie temperatures T_c can be evaluated, critical exponents are much more difficult to determine reliably (3). The influence of hydrogen (H) on the magnetic properties of Pd containing Fe, Cr and Mn has been studied for some time eg (5). Specifically in the case of PdFe the introduction of H causes the well known giant moment ferromagnetism to be replaced by spin-glass like freezing (5), an effect which has been attributed to a suppression of the host's exchange enhanced susceptibility (with its attendant long range spin polarization) due to a hydrogen induced filling of the 4d band. Here we present a summary of the influence of H on the low temperature magnetic properties of PdNi which appears somewhat different from that reported for other transition metal impurities.

PdNi samples were prepared by arc melting using high purity starting materials as described previously (3). Hydrogen was introduced by a cathode polarization technique using a $NaCO_3$ electrolyte and a constant current density of approximately 100 mA/cm^2 for various exposure times. Continuous measurements of the ac susceptibility were made using a previously described phase-locked susceptometer (3).

Figure 1 indicates the substantial hydrogen induced elongation of a Pd+5at.% Ni sample as a function of charging time; these data can be used to estimate the H-to-metal ratio if this ratio is taken to be about 0.7 in the plateau

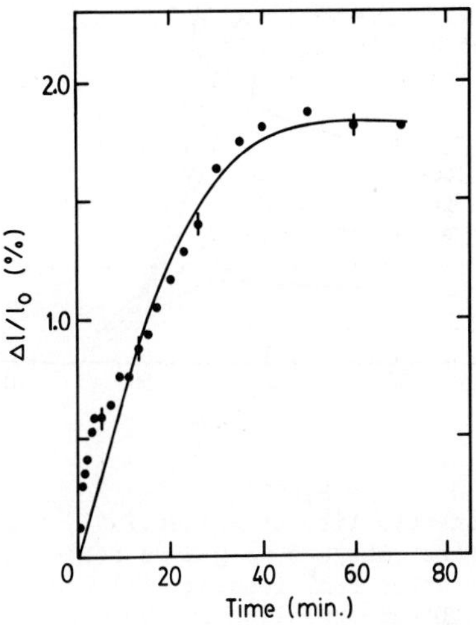

FIGURE 1
The hydrogen induced elongation of a Pd+5at.% Ni sample as a function of charging time.

region (5). Figure 2 shows the ac susceptibility of this same sample (measured in zero static biasing field) as a function of temperature in a hydrogen free state (insert) and as a function of hydrogen loading (main figure). The hydrogen free sample displays a susceptibility which increases rapidly with decreasing temperature below 65 K, with a Curie temperature T_c estimated at 62.7 ± 1 K (close to the infexion point on this curve) and with asymptotic critical exponents ($\gamma \approx 1.39$, $\beta \approx 0.4$, $\delta \approx 4.1$) that confirm the ferromagnetic nature of the order. The susceptibility continues to climb below T_c, passing through the Hopkinson peak at about 62 K. The measured peak susceptibility represents only 1% of the limit set by demagnetising constraints and furthermore is up to a factor of 50 smaller than that measured in soft ferromagnets (i.e. PdMn (6)) with comparable dimensions, and this offers indirect confirmation of the role played by (spin-orbit induced) anisotropy in the present system.

The main body of figure 2 reproduces the

0921-4526/90/$03.50

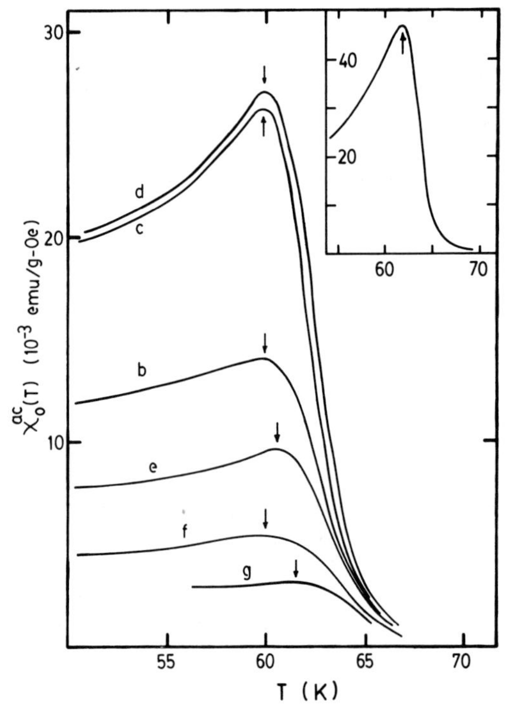

FIGURE 2
The ac susceptibility of Pd+5at.% Ni, insert-hydrogen free; steps b to g were followed sequentially after full loading. (b) partial discharge by vacuum annealing at 500°C for 3 hrs. (c) further discharge by reverse electrolysis for 2 min. (d) reverse electrolysis for an additional 2 min. Partial reloading by direct electrolysis for (e) 3 min. (f) 6 min. and (g) 9 min. The hydrogen content increases from top to bottom.

effects of hydrogenation and, while the loading sequence was slightly complicated, with increasing hydrogen content, the peak susceptibility is progressively suppressed, but the peak temperature remains essentially unchanged (at 61 ± 1 K). While this much suppressed magnetic response precludes a reliable analysis of these data for both critical temperatures and exponents (as discussed below), the behaviour of the Hopkinson peak temperature provides a qualitative indication to the behaviour of the ordering temperature. Consequently we claim that the magnetic ordering temperature is not being significantly depressed by the addition of hydrogen. What is being affected is the anisotropy/coercivity of the system, as is shown in figure 3. Here the ac susceptibility at 4.2 K is shown as a function of static biasing field. The coercive field H_c is essentially half the separation in field between the maximum slopes of the hysteresis loop; in the

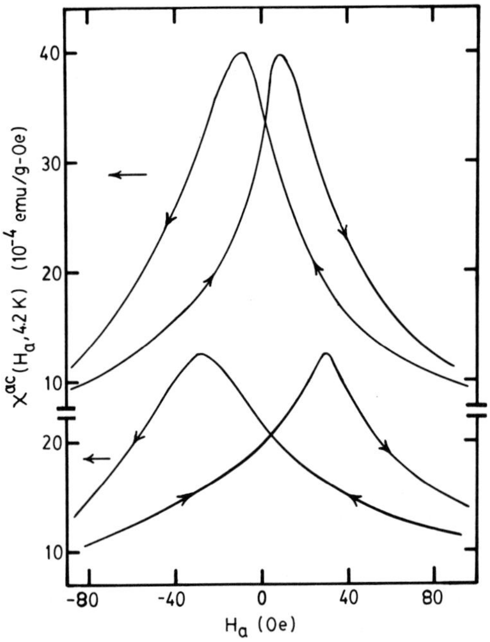

FIGURE 3
The ac susceptibility as a function of static biasing field (hysteresis loop) at 4.2 K for hydrogen-free (top portion) and hydrogenated (5 min., lower portion) Pd+5at.% Ni.

hydrogen free sample $H_c \approx 18$ Oe (the system is not magnetically soft), climbing by a factor of 3 to ≈56 Oe after just 5 mins. exposure. This increasing anisotropy/ coercivity means that the regular component in the susceptibility (arising from coherent rotation, domain wall motion, etc.) becomes progressively more difficult to saturate, further obscuring emerging critical effects. This is the cause of our inability to extract estimates for critical exponents in hydrogen containing samples, which means that while these data suggest the magnetic ordering temperature is not changing substantially with hydrogen loading, any changes in the nature of this ordering cannot be ascertained. Such questions might be answered by microscopic measurements on this system.

REFERENCES

(1) H.P. Kunkel et al., J. Phys. F: Met. Phys. 17 (1987) L157.
(2) H.P. Kunkel et al., J. Phys.: Condens. Mat. 1 (1989) 3381.
(3) Z. Wang et al., J. Phys.: Condens. Mat., in print.
(4) J.W.F. Dorleijn, Philips Res. Rep. 31 (1976) 287.
(5) J. Mydosh, Phys. Rev. Let. 33 (1974) 1562.
(6) S.C. Ho et al., J. Phys. F: Met. Phys. 11 (1981) 699.

Physica B 165&166 (1990) 197–198
North-Holland

MOMENT-VOLUME INSTABILITIES IN FCC FE-BASED ALLOYS*

M.M. ABD-ELMEGUID**, A. BLOCK, and H. MICKLITZ

Experimentalphysik IV, Ruhr-Universität, D-4630 Bochum, FRG

High pressure ^{57}Fe Mössbauer experiments at low temperature (4.2 K) offer the possibility to study pressure-induced moment instabilities, i.e. transitions from a "high-moment"- to a "low-moment"- state, in fcc Fe-based alloys. These results are compared with recent theoretical predictions of the magnetic ground state in these systems.

1. INTRODUCTION

The study of pressure effects on the magnetic properties of 3d transition metals and their alloys is a fundamental issue in understanding of itinerant electron magnetism. This is related to the fact that an external pressure varies the 3d correlation between the magnetic neighbouring atoms which results in a modification of the magnetic state of the system. A strong mutual relationship between magnetism and volume, therefore, has been found in a variety of fcc Fe-based alloys such as fcc $Fe_{1-x}Ni_x$ (1) and fcc Fe_3Pt (2). High-pressure ^{57}Fe Mössbauer studies performed at low temperature (4.2 K) offer the possibility to study in more detail the Fe moment instabilities of the magnetic ground state in such systems.

2. EXPERIMENTAL

High-pressure ^{57}Fe Mössbauer effect (ME) experiments were performed with a Chester-Jones type high-pressure set-up with specially designed B_4C anvils which allow ME transition experiments at liquid helium temperature for pressures up to $\simeq$ 8 GPa (3). We have studied the following systems: $Fe_{0.65}Ni_{0.35}$, $Fe_{0.685}Ni_{0.315}$, $Fe_{0.72}Pt_{0.28}$ (disordered and ordered), $Fe_{0.28}Cu_{0.72}$ and fcc-Fe precipitates in fcc-Cu. All ME spectra were fitted by using a modified histogram method (4). The analysis of the data reveals the pressure dependence of the Fe magnetic moment μ (p) [$\propto \overline{B}_{eff}$, the effective magnetic hyperfine field at the Fe nucleus]. Thermal scanning experiments additionally allow the determination of the pressure dependenc of the magnetic ordering temperature, T_C (p).

* This work was supported by the Deutsche Forschungsgemeinschaft (SFB 166).

** present address: IFF, KFA-Jülich, D-5170 Jülich, FRG

3. RESULTS

While *no* pressure-induced magnetic phase transition is observed in $Fe_{0.65}Ni_{0.35}$ for pressures up to $\simeq$ 7 GPa, such a transition is found in the more Fe-rich alloy $Fe_{0.685}Ni_{0.315}$ at a cricital pressure $p_c \simeq 5.8$ GPa (5). This transition is interpreted as a pressure-induced phase transition from a "high-moment" (HM)- to a "low-moment" (LM)-state.

Such a pressure-induced HM $\rightarrow$ LM transition also is observed in disordered and ordered $Fe_{0.72}Pt_{0.28}$ at $p_c \simeq 4.2$ GPa and 6.0 GPa, respectively (6). The magnetic hyperfine field distribution, obtained from the ME spectra analysis, clearly shows the existence of two *distinct* magnetic states (HM and LM) with a depopulation of the HM-state and population of the LM-state with increasing pressure.

The metastable fcc-$Fe_{0.28}Cu_{0.72}$ alloy (7), which has a relatively low Fe-content, shows only a weak pressure dependence of both $\overline{B}_{eff}$(p) and T_C (p) [dln $\overline{B}_{eff}$(p)/dp $\simeq -1.7 \cdot 10^{-2}$ GPa^{-1}, dlnT_C(p)/dp $\simeq -3.6 \cdot 10^{-2}$ GPa^{-1}] (8).

Antiferromagnetically ordered fcc-Fe precipitates (size $\simeq$ 22 nm) in fcc Cu (9), which are in the LM-state at ambient pressure, seem to undergo a transition to a non-magnetic (NM) state at a pressure of $\simeq$ 1.0 GPa (8).

4. DISCUSSION

The quite different behavior of weak ($Fe_{0.65}Ni_{0.35}$) and strong ($Fe_{0.72}Pt_{0.28}$) itinerant ferromagnetism at low pressures (p $\lesssim$ 2 GPa) (10) disappears at high pressure: both systems reveal an instability of the Fe magnetic moment at high pressure (small volume) and the occurence of a pressure-induced HM $\rightarrow$ LM transition at a critical pressure p_c. This finding is consistent with the results of recent band calculations of the magnetic ground state in ordered Fe_3Ni (11) and Fe_3Pt (12) using the so-called "local spin den-

0921-4526/90/$03.50

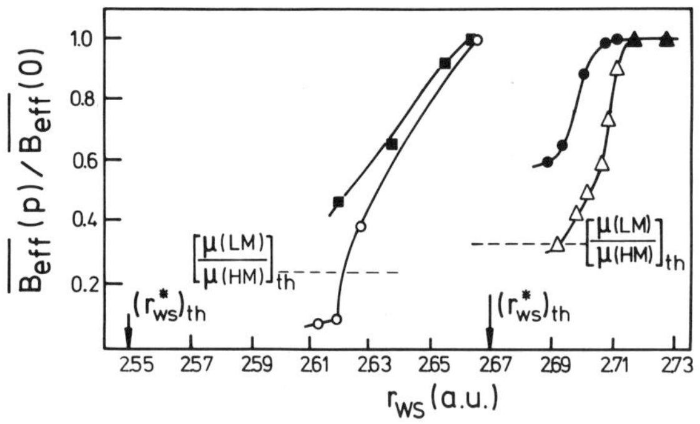

FIGURE 1

Measured values of $\overline{B}_{eff}(p)/\overline{B}_{eff}(0)$ as a function of Wigner-Seitz radius r_{ws} of Fe WS sphere for $Fe_{0.65}Ni_{0.35}$ (■), $Fe_{0.685}Ni_{0.315}$ (○), ordered (●) and disordered (Δ) $Fe_{0.72}Pt_{0.28}$. Theoretical predictions of r^*_{ws} and $\mu(LM)/\mu(HM)$ for ordered Fe_3Ni (Ref. 11) and Fe_3Pt (Ref. 12) are given for comparison.

sity approach" (LSDA). We have plotted in fig. 1 our high-pressure results together with these theoretical predictions. The Wigner-Seitz (WS) radius r_{ws} of the Fe WS sphere was estimated from the lattice constants at ambient pressure and the compressibility data. The theoretical value of r^*_{ws} where the HM → LM transition should occur in ordered Fe_3Ni and Fe_3Pt and the theoretical values of the ratio μ (LM)/μ(HM) are also given in fig. 1. Our experimental resutls essentially agree with respect to the ratio μ (LM)/ μ (HM) and the prediction that r^*_{ws} for Fe_3Ni is smaller than that for Fe_3Pt. The fact that $(r^*_{ws})_{exp} > (r^*_{ws})_{th}$ probably is due to an intrinsic problem of the LSDA calculations: they always give too small lattice constants (13).

The pressure-induced instability of the HM state is found to be enhanced upon increasing the number n of Fe nearest neighbor atoms, i.e. the Fe concentration: $Fe_{0.65}Ni_{0.35}$ ($\overline{n} \simeq 7.8$) is more stable than $Fe_{0.685}Ni_{0.315}$ ($\overline{n} \simeq 8.2$), ordered $Fe_{0.72}Pt_{0.28}$ ($n = 8.0$) is more stable than disordered $Fe_{0.72}Pt_{0.28}$ ($\overline{n} \simeq 8.4$), and $Fe_{0.28}Cu_{0.72}$ ($\overline{n} \simeq 3.4$) does not show any instability of the HM state at all.

The pressure-induced transition from a LM- to a NM-state, as indicated from our high-pressure experiments on fcc-Fe precipitates, also is predicted by theory (14). However, more experimental data are needed to confirm these first results.

REFERENCES

(1) R.C. Wayne and L.C. Bartel, Phys. Lett. A28 (1968) 196
(2) Y. Nakamura, K. Sumiyama and M. Shiga, J. Mag. Magn. Mat. 12 (1979) 127
(3) A. Eichler and J. Wittig, Z. Angew. Phys. 25 (1968) 319
(4) R.A. Brand, J. Lauer, and D.M. Herland, J. Phys. F13 (1983) 675
(5) M.M. Abd-Elmeguid, B. Schleede, and H. Micklitz, J. Mag. Magn. Mat 72 (1988) 253
(6) M.M. Abd-Elmeguid and H. Micklitz, Phys.Rev. B40 (1989) 7395
(7) K. Sumiyama, T. Yoshitake, and Y. Nakamura, J. Phys. Soc. Jpn. 53 (1984) 3160
(8) A. Block, Diplom Thesis, Ruhr-Universität Bochum, unpublished
(9) T. Ezawa, W.A.A. Macedo, U. Glos, W. Keune, K.P. Schletz, and U. Kirschbaum, Physica B161 (1989) 269
(10) K. Hayashi and N. Mori, Solid State Commun. 38 (1981) 1057
(11) V.L. Moruzzi, Physica B161 (1989) 99
(12) P. Entel and M. Schröter, Physica B161 (1989) 160
(13) M. Podgorny, Physica B161 (1989) 105
(14) V.L. Moruzzi, P.M. Marcus, K. Schwarz and P. Mohn, Phys. Rev. B34 (1986) 1784

Physica B 165&166 (1990) 199–200
North-Holland

THERMAL SPONTANEOUS FERROMAGNETISM OF $TiBe_{2-x}Cu_x$

David M. IOSHPE
Racah Institute of Physics, Hebrew University, Jerusalem, 91904 Israel and
Gilo, Block 63/12, Shabtay Negbee St. 15, Jerusalem, 93384, Israel, (present address).

We have proposed the thermal spontaneous ferromagnetism (TSF) of $TiBe_2$ with two first order transitions at temperatures T_c=1.8K and T_s=1.55K on a basis of our CESR measurements and other results. We propose now the TSF of $TiBe_{2-x}Cu_x$ and show that the cooper addition x increases smoothly the TSF temperature area and it requires a new definition of the critical concentration x_{cr} which we give here.

We found a narrowing of the cubic Laves phase $TiBe_2$ CESR line between the temperatures T_s=1.55K and T_c=1.8K, (the temperature area 0.25K)[1]. The line width decreased from 12G at T_c and T_s to 8G at T_{min}=1.7K (Fig.1). The nature of the revealed narrowing is the exchange one at most and we have concluded that a weak itinerant ferromagnetic order exists in $TiBe_2$ at T=1.55+1.8K. Because it is a strongly exchange enhanced paramagnet out of this temperature area, we have proposed the thermal spontaneous ferromagnetism (TSF) in $TiBe_2$ at this area [1,2]. The occurance of *two* phase transitions from para- to ferromagnetism at T_c and vice versa at T_s is in good agreement with *two* kinks at the same temperatures in the temperature dependence of the $TiBe_2$ resistivity [3]. Both phase transitions are of the first order, because the spontaneous magnetic moment reaches its maximum value at T_{min}=1.7K, ie. at the reduced temperature T_{rc}=0.056 and T_{rs}=0.097. The magnetic measurements must be done in the weak fields because of a very low $TiBe_2$ spin fluctuation temperature T_{sf}=20K [4]; our CESR measurement was done in field 220G.

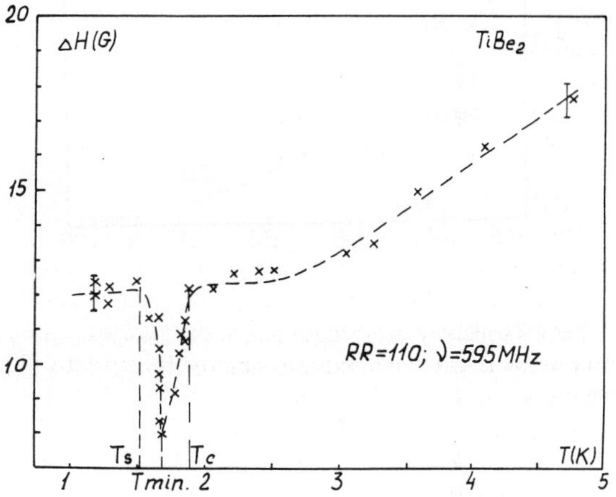

Fig.1 The temperature dependence of the $TiBe_2$ line width in the field 220G [1].

We propose now, that the cooper addition x to $TiBe_2$, ($TiBe_{2-x}Cu_x$), only increases smoothly its TSF area with T_c=1.8K (x=0) to the area with T_c=21K (x=0.5) [5], and this compound is TSF too. This explains well, the coincidence of the calculated value for the $TiBe_2$ magnetic moment 0.22, of Bhor magneton [6], with experimental value 0.23 of Bhor mgneton for $TiBe_{2-x}Cu_x$ which does not change with a change of x from 0.15 to 0.5 [7].

Fig.2, (branches 1 and 2), presents our fit of the exponential function to the experimental points [5] of the Curie temperature from Cu concentration dependence:
$T_c(x)=T_c(0)e^{n_c x}=1.8e^{5.25x}$. It can be seen that this fit may be an acceptable first approximation for $T_c(x)$. We can now evaluate the increase of the of the TSF temperature area with increase of x, if, suppose, the temperature of the second transition $T_s(x)=T_s(0)e^{n_s x}=1.55e^{-5.25x}$, (Fig.2, branches 1a and 2a), ie. it is an exponential function too and it decreases at the same rate that $T_c(x)$ increases: n_s=-n_c. In this case, $T_s(0.5)$=0.1K and the TSF temperature area is 21K.

The critcal concentration x_{cr} is defined by the relation $T_c(x_{cr})=0$ and it is accepted now, that x_{cr}=0.03+0.05, received by the linear extrapolation of $T_c(x)$ to T_c=0 [7]. Because we saw that $T_c(0)=1.8K\neq 0$, it is not any x_{cr} for $TiBe_{2-x}Cu_x$, if the old definition will be accepted.

But because we propose here, the TSF of this compound, we must define that $x=x_{cr}$ if the second transition temperature $T_s(x_{cr})=0$. Therefore, if it is valid, our consideration above exponential fit for $T_s(x)$, it is not any critical concentration for $TiBe_{2-x}Cu_x$ by new definition too.

Fig.2, branch 3, shows that the linear fit to the experimental points of [5] for x=0.15+0.5 is better than the exponential one [7]. If we fit $T_c(x)$ by branches 1 and 3 (Fig.2), we can suppose that $T_s(x)$ is represented by branches 1a and 3a with the slope of 3a branch equal in magnitude but opposite in sign to the slope of branch 3. In this case $T_s(x_{cr})=0$ is x_{cr}=0.17, ie. it is x_{cr} on the basis of the results of [5] by a new definition. A consideration of the other experimental results reveal small deviation from this value [8].

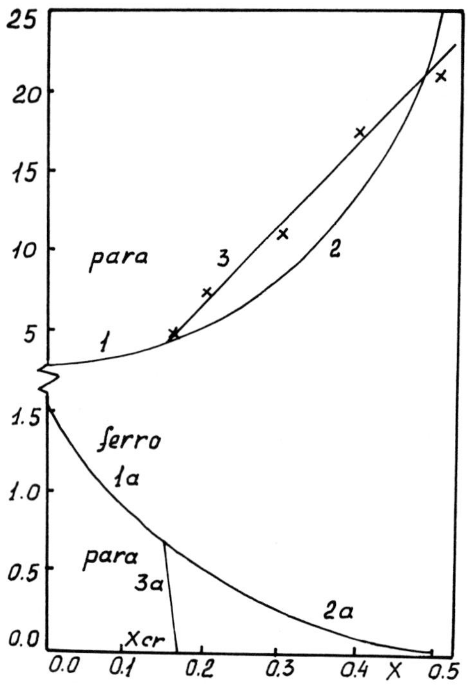

Fig.2 The $TiBe_{2-x}Cu_x$ $T_c(x)$ and $T_s(x)$ dependences for two cases of the fit to the experimental results (+) from [5]. (see text).

The measurements of $T_c(x)$ for x=0+0.15 in the weak fields and $T_s(x)$ for x=0+0.5, have never been done and it is necessary to do them to check our results.

The author perfomed the experimental part of this research during his time at the Hebrew University on a grant from the Ministry of Absorbtion in Israel and he would like to thank Prof. D Shaltiel, Mr A. Grayevsky and Prof. V. Zevin for discussions; he is indebted to Dr J.L. Smith for the $TiBe_2$ sample.

REFERENCES

[1] D.M. Ioshpe. Euophys. Conf. Abstr. 13A (1989) 81 and Phys. Lett. to be published.
[2] M. Shimizu, J. Physique 43 (1982) 681.
[3] F. Acker et al. J. Magn. Magn. Mater. 22 (1981) 250.
[4] H. Alloul et al. J. Magn. Magn. Mater. 54-57 (1986) 1017; M.T. Beal-Monod. Phys. Rev. B24 (1981) 261.
[5] J.L. Smith. Physica 107B (1981) 251.
[6] T. Jarlborg et al. J. Magn. Magn. Mater. 23 (1981) 291.
[7] D. McK. Paul et al. J. Magn. Magn. Mater. 42 (1984) 201.
[8] D.M. Ioshpe, to be published.

Physica B 165&166 (1990) 201–202
North-Holland

THE LOW TEMPERATURE MAGNETIZATION OF $ZrZn_2$

S. M. Hayden† and C. C. Paulsen§

†Institut Laue-Langevin, B.P. 156X, 38042 Grenoble Cedex, France.
§Centre de Recherches sur les très Basses Températures, CNRS, B.P.166X Grenoble, France.

We report high sensitivity measurements of the initial temperature variation of the magnetization of $ZrZn_2$ for temperatures T and applied fields B in the ranges $0.8 \le T \le 5$K and $0 \le B \le 8$ Tesla respectively. Our measurements are inconsistent with the initial variation being governed solely by single particle excitations.

1. INTRODUCTION

The initial variation of the magnetization with temperature in a ferromagnet can be a powerful probe of the nature of the magnetic excitations from the ground state. In ferromagnetic metals close to the magnetic instability at low temperature the magnetization falls initially as T^2. This behaviour can arise because of the presence of strong long-wavelength low energy fluctuations [1] or else in systems where the band structure is such that single-particle excitations are important [2]. We describe here a investigation of the initial decrease of the magnetization with temperature as a function of magnetic field in $ZrZn_2$. This system is perhaps the archetypical weak ferromagnet, however, the nature of the low-energy magnetic excitations remains unclear. Bernhoeft *et al.* [3] have shown that well defined propagating modes are present at low wavevectors q but the situation at larger q, where single-particle excitations would exist is still uncertain.

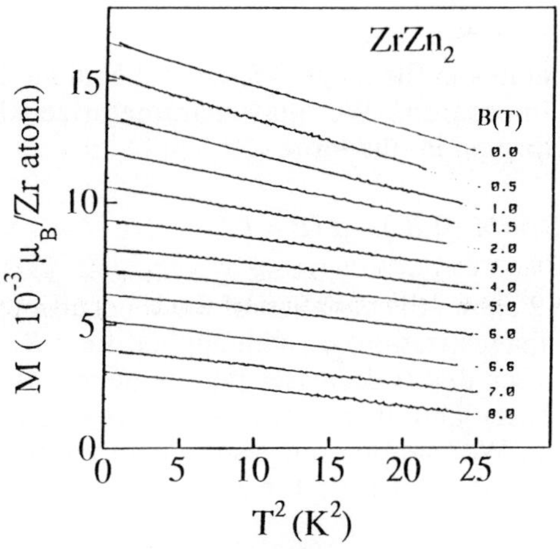

FIGURE 1.
The variation of magnetic moment with temperature of $ZrZn_2$ for applied fields B =0, 0.5, 1.0, 1.5, 2, 3, 4, 5, 6, 6.6, 7, 8 Tesla. The curves have been displaced for clarity.

2. EXPERIMENTAL

Our samples were prepared by melting high purity zone refined zirconium and zinc contained by a Y_2O_3 crucible inside a tantalum bomb at 1200°C. The mixture was cooled slowly through the melting point and then annealed at various temperatures above 500°C over a period of 5 days. The resulting polycrystalline ingot was found to have resistance ratios [ρ(293K)/ρ(4.2K)] in the range 30-50. The Curie temperature T_c and magnetic moment at zero field $M(0,0)$ determined from magnetization data in the field range $2 < B < 4$ Tesla were 26.8(2) K and 0.157(3) μ_B (Zr atom)$^{-1}$ respectively.

3. RESULTS

Figure 1 shows the temperature dependence of the magnetization measured by an a.c. SQUID technique for various applied fields. The magnetization exhibits a T^2 behaviour for

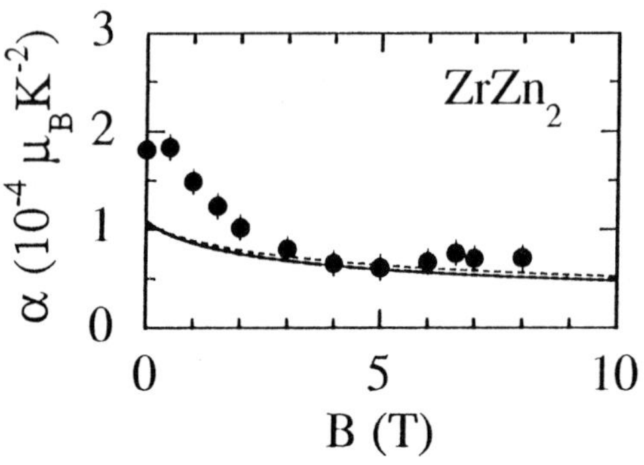

FIGURE 2.
The T^2 coefficient of the magnetization, α, as a function of magnetic field for $ZrZn_2$.

temperatures in the range $0.8 < T < 5$ K at all the fields measured. We may parameterize the magnetization in the form:

$$M(T,B) = M(0,B) - \alpha(B)\, T^2. \quad (1)$$

Figure 2 shows the variation of the T^2 coefficient of the magnetization, α, with applied field B. It can be seen that α decreases rapidly up to a field of 3 Tesla and then becomes approximately constant. We interpret our results with reference to the equation of state for a weak ferromagnet [1]:

$$B = A(T)\, M(T,B) + b\, M^3(T,B)\,. \quad (2)$$

In a conventional band theory where the magnetic excitations are uncorrelated electron-hole pairs or 'single-particle' excitations, $A(T)$ and b are expressed in terms of the one-particle density of states at the Fermi energy and its derivatives. The solid line in figure 2 represents the prediction of such a model, where $A(T)$ and b are assumed to be field independent. We have used values for $A(T)$ and b corresponding to B =4 Tesla, these were determined from fits of equation (2) to magnetization isotherms with $5 < T < 30$K. Our procedure describes well the variation of α for larger fields.

The observed deviation at low fields can be explained by considering the field-dependent. corrections to equation (2) due to long-wavelength low-energy spin fluctuations. Following Lonzarich and Taillefer [1,2]:

$$A = \left[\, a(T) + 3b\langle m_{\parallel}^2\rangle + 2b\langle m_{\perp}^2\rangle \right], \quad (3)$$

where $\langle m_{\parallel}^2\rangle$ and $2\langle m_{\perp}^2\rangle$ are the thermal variances of the local magnetisation parallel and perpendicular to the average magnetisation **M**, respectively. For strong magnetic fields the contribution from the transverse fluctuations $\langle m_{\perp}^2\rangle$ at small wavevectors would be suppressed due to the introduction of a gap into the spin-wave excitation spectrum [2]

$$\hbar\omega(q) = g\mu_B B + Dq^2. \quad (4)$$

Thus α is reduced in the field and its value in high magnetic fields is determined by single-particle excitations and enhanced longitudinal magnetic fluctuations: the first two terms in equation (3). The solid line in figure 2 would then not include the effects of the last term in equation (3) since the values for $A(T)$ and b were determined for B=4Tesla.

To summarize, our data are consistent with the magnetic excitation spectrum of $ZrZn_2$ not being completely dominated by single-particle excitations at low temperature and zero field. We observe a rapid variation of the T^2 coefficient of magnetization with an applied field which we attribute to the suppression of the spin wave contribution.

ACKNOWLEDGEMENTS
We would like to thank N. R. Bernhoeft, J.-L. Tholence and M. Wulff for their assistance.

REFERENCES

(1) G. G. Lonzarich and L. Taillefer, J. Phys. C: Solid State Phys. **18**, 4339 (1985).
(2) D. M. Edwards and E. P. Wohlfarth, Proc. Roy. Soc. A**303**, 127 (1968).
(3) N. R. Bernhoeft, S. A. Law, G. G. Lonzarich and D. McK. Paul, Physica Scripta **38**, 191 (1988).

Physica B 165&166 (1990) 203–204
North-Holland

QUENCHING OF SPIN FLUCTUATIONS BY HIGH MAGNETIC FIELDS IN HEAT CAPACITY OF $Zr_{1-x}Hf_xZn_2$

Kôki IKEDA and Masahito YOSHIZAWA
Department of Metallurgy, Faculty of Engineering, Iwate University, Morioka 020, Japan

Kenzo KAI
Institute for Materials Research, Tohoku University, Sendai 980, Japan

Toshio NOMOTO
Miyagi Technical College, Natori 981-12, Japan

The low-temperature (1.5-40 K) heat capacity of the cubic Laves phase $Zr_{1-x}Hf_xZn_2$ (x=0-1) compounds was measured in magnetic fields of 0, 6, 10, and 15 T. The pronounced decrease of the electronic specific-heat constant with increasing field (by 27-33 % at 15 T) was found in the weak itinerant-electron ferromagnetic compounds with x=0-0.16. The results were discussed in terms of the quenching of spin fluctuations under high magnetic fields.

The quenching of spin fluctuations by high magnetic fields has been observed in the low-temperature heat-capacity measurements for highly enhanced Pauli-paramagnets, RCo_2 (R=Sc, Y and Lu) (1), $CeSn_3$ (2), $TiBe_2$ (3), and UAl_2 (4) and for weak itinerant-electron ferromagnets Sc_3In (5) and Ni_3Al (6). The pseudobinary $Zr_{1-x}Hf_xZn_2$ (x=0-1) compounds are the prototypical materials for this investigation. The cubic Laves phase ($MgCu_2$ type) compound $ZrZn_2$ is the weak ferromagnet with the Curie temperature T_c =21 K and the spontaneous magnetization extrapolated to 0 K and zero field is 0.12 μ_B/(Zr atom) (7). On the other hand, the isostructual $HfZn_2$ compound is the strong Pauli-paramagnet with the Stoner-enhancement factor S=3.6 (8). According to Ogawa (7), the Curie temperature and magnetization of $Zr_{1-x}Hf_xZn_2$ compounds both decrease with increasing Hf content (x) and become zero around x=0.16. In the present work, the results of low-temperature heat-capacity measurements of the $Zr_{1-x}Hf_xZn_2$ compounds with x=0, 0.10, 0.13, 0.16, and 1 in the magnetic fields of 0, 6, 10, and 15 T are reported and discussed in terms of the quenching of spin fluctuations under high magnetic fields.

The $Zr_{1-x}Hf_xZn_2$ samples were prepared by the solid-state diffusion (sintering) method. The measurements were carried out by using a calorimeter of the isolation heat-pulse type at the High Field Laboratory for Superconducting Materials, IMR, Tohoku University. Fig. 1 is the results of the heat capacity measurements for $ZrZn_2$ and $HfZn_2$ from 1.5 to 6.3 K at four magnetic fields of 0, 6, 10, and 15 T. As can be seen, all C/T vs. T^2 curves are linear and there is no evidence for a $T^3\ln T$ term. For $ZrZn_2$, one can distinctly see that the curves are parallel to one another and that they fall with increasing magnetic field. However, for $HfZn_2$ there is no difference between the zero-

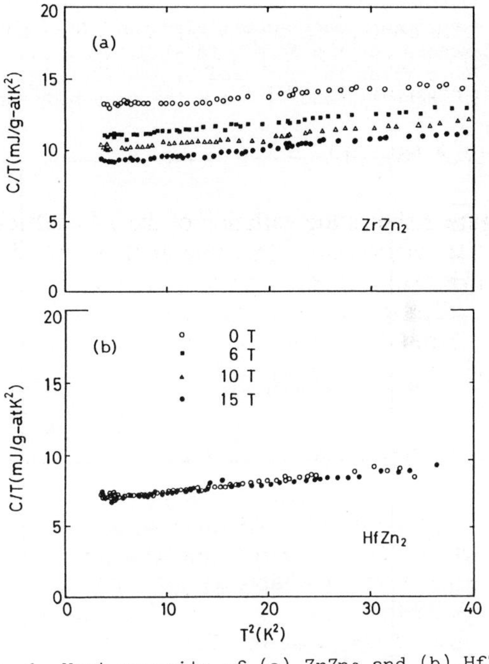

Fig. 1. Heat capacity of (a) $ZrZn_2$ and (b) $HfZn_2$.

field data and the 15 T-data. In order to examine the influence of the magnetic field on the heat capacity of $Zr_{1-x}Hf_xZn_2$ compounds, the electronic specific-heat constant, γ , and the Debye temperature at 0 K, Θ_D, were calculated from a least-squares fitting of the data for each field to the equation: $C/T=\gamma+\beta T^2$, where $\beta=1944/\Theta_D{}^3$ (J/g-at.K^4).

Fig. 2 shows the electronic and lattice specific-heat constants, γ and β, respectively, which were obtained from the least-squares fitting of our zero-field data of $Zr_{1-x}Hf_xZn_2$

0921-4526/90/$03.50

to $C/T=\gamma+\beta T^2$, as a function of x. With increasing x, the γ value increases and reaches a maximum at x=0.10. This $Zr_{0.90}Hf_{0.10}Zn_2$ compound is ferromagnetic at 4.2 K according to our magnetization measurements and the critical composition between ferromagnetism and paramagnetism is x≈0.16 (7). On the other hand, the Debye temperature, Θ_D, monotoneously decreases from Θ_D=327 K for $ZrZn_2$ to 289 K for $HfZn_2$ with increasing x, as can be seen from the composition dependence of β in fig. 2.

The γ value of $ZrZn_2$ (12.7 mJ/g-at.K^2) is about twice larger than that of $HfZn_2$ (6.6 mJ/g-at.K^2). This may be reasonable because the $ZrZn_2$ compound has a large mass enhancement. In the pulsed field de Haas-van Alphen measurements up to 35 T for a $ZrZn_2$ single crystal, Ruitenbeek et al. (11) have found that the spin-split Fermi surface is field dependent and the mass enhancement factor is 2.9. If the $HfZn_2$ compound has the same value of the electronic specific-heat constant determined from the band-structure density of states, γ_0, as that in the $ZrZn_2$ compound, which has isostructure and the same number of outer-shell electron, the mass enhancement of the former is ~1.0.

The γ value is expressed by $\gamma = \gamma_0(1+\lambda_{sf}+\lambda_{sf})$. Here λ_{ph} and λ_{sf} are the many-body mass

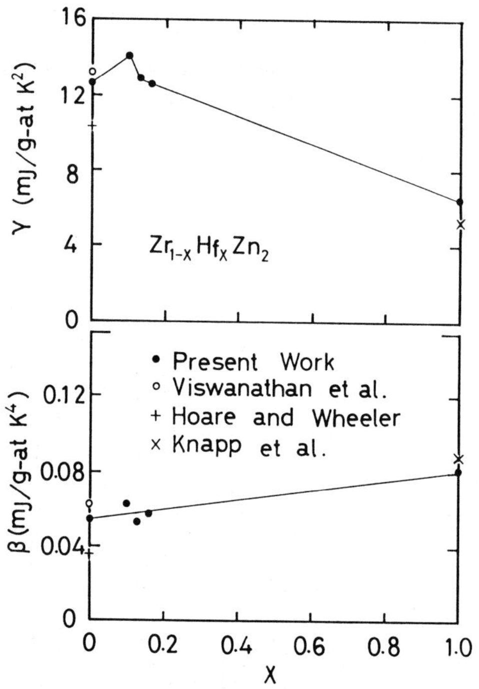

Fig. 2. Electronic and lattice specific-heat constants of the $Zr_{1-x}Hf_xZn_2$ compounds obtained in the present work and those reported previously (8-10).

enhancement factors which are respectively caused by the electron-phonon interaction and the spin fluctuations. If the magnetic field is sufficiently large so that the Zeeman splitting energy of opposite spin states is comparable to or larger than the characteristic spin-fluctuation energy, k_BT_{sf}, the spin fluctuations no longer have sufficient energy to flip spins, and therefore, the inelastic spin-flip scattering is quenched. Thus the electronic specific-heat enhancement due to spin fluctuations, λ_{sf}, decreases with increasing magnetic field.

The decrease of γ with increasing field in $ZrZn_2$, where the decrement ratio is 16, 23 and 31 % at 6, 10 and 16 T, respectively, is probably due to the suppression and quenching of spin fluctuations. However, this phenomenon in $ZrZn_2$ disappears at higher temperatures than ~20 K. The γ decrease under magnetic fields was found in the $Zr_{1-x}Hf_xZn_2$ compounds with $0\leq x\leq 0.16$, and most pronouncedly in $Zr_{0.90}Hf_{0.10}Zn_2$, which has a maximum γ value. In this compound, the decrement ratio of γ reaches 33 % at 15 T and the C/T value shows a pronouncedly decrease over all temperature range up to high temperature limit of measurements (40 K). The γ decrease under magnetic fields in $Zr_{1-x}Hf_xZn_2$ compounds is semiquantatively explained by the numerical caluculation on the quenching of spin fluctuations by Hertel et al. (12), if T_{sf} is assumed to be 53-65 K for these compounds.

REFERENCES

(1) K. Ikeda and K.A. Gschneidner, Jr., Phys. Rev. Lett. 45 (1980) 1341.
(2) K. Ikeda and K.A. Gschneidner, Jr., Phys. Rev. B 25 (1982) 4623.
(3) G.R. Stewart, J.L. Smith and B.L. Brandt, Phys. Rev. B 26 (1982) 3783.
(4) G.R. Stewart, A.L. Giorgi, B.L. Brandt, S. Foner, and A.J. Arko, Phys. Rev. B 28 (1983) 1524.
(5) K. Ikeda and K.A. Gschneidner, Jr., J. Magn. Magn. Mat. 22 (1981) 207, 30 (1983) 273.
(6) S.K. Dhar and K.A. Gschneidner, Jr., Phys. Rev. B 39 (1989) 7453.
(7) S. Ogawa, Researches of the Electrotechnical Laboratory No. 735 (Tokyo, 1972).
(8) G.S. Knapp, B.W. Veal and H.V. Culbert, Int. J. Magnetism 1 (1971) 93.
(9) F.E. Hoare, and J.C.G. Wheeler, Phys. Lett. 23 (1966) 402.
(10) R. Viswanathan, H. Luo and D.O. Massetti, AIP Conf. Proc. 5 (1972) Pt.2, 1290.
(11) J.M. van Ruitenbeek, W.A. Verhoef, P.G. Mattocks, A.E. Dixon, A.P.J. Deursen, and A.R. de Vroomen, J. Phy. F 12 (1982) 2919.
(12) P. Hertel, J. Appel and D. Fay, Phys. Rev. B 22 (1980) 534.

Physica B 165&166 (1990) 205–206
North-Holland

ANOMALOUS DIAMAGNETISM AT THE TWINNING PLANE

Serguei N. BURMISTROV and Leonid B. DUBOVSKII

I.V. Kurchatov Institute of Atomic Energy, Moscow 123182, USSR

The twin in magnetic field parallel to the twinning plane is considered In weak fields the surface diamagnetic susceptibility diverges as inverse squared magnetic field. Such behavior of susceptibility leads instability and results in formation of diamagnetic domain at the twinning plane.

At present a number of anomalous physical properties in twins of anisotropic metals are known to exist (1-3). One possible origin of the anomalies lies in a variation of the electron spectrum occuring near the twinning plane. For the isotropic electronic spectrum no anomalies appear during the passage of electrons through the twinning plane. The situation changes drastically in the presence of an anisotropy of the electron spectrum. This is the reason why in this case a part of the electrons is subjected to reflection from the twinning plane. The reflection is caused by the restrictions associated with the fulfillment of the conservation laws for the energy and the quasimomentum component parallel to the plane. This results (4,5) in the kinematic constraints for an electron moving from one monobloc of the twin to another one and in the appearence of an effective potential barrier at the boundary for the motion of an electron. The height of the potential barrier depends on the quasimomentum component lying in the plane parallel to the boundary.

For the study of the electron spectrum we assume the quadratic anisotropic dispersion law:

$$\hat{H}=-1/2\partial/\partial x(m_{11}^{-1}\partial/\partial x)-1/2[\partial/\partial x(m_{12}^{-1})+$$

$$+m_{12}^{-1}\partial/\partial x]\partial/\partial y-1/2m_{22}^{-1}\partial^2/\partial y^2-1/2m_3^{-1}\partial^2/\partial z^2,$$

where x is coordinate running normal to the twinning plane. The mass tensor elements are expressed in terms of masses m_1, m_2, m_3 along the principal crystallographic axes as follows:

$$m_{11}^{-1}=m_1^{-1}\cos^2\phi+m_2^{-1}\sin^2\phi,$$

$$m_{22}^{-1}=m_1^{-1}\sin^2\phi+m_2^{-1}\cos^2\phi,$$

$$m_{12}^{-1}=(m_1^{-1}-m_2^{-1})\sin\phi\,\cos\phi,$$

where $\phi=\phi(x)$ is the angle through which the principal crystallografic axes 1 and 2 are rotated with respect to the spatial axes x and y. The angle $\phi(x)$ is the continuous function which varies from the value of $-\alpha$ in the left half-space to the value α in the right one over a distance b which is characteristic of the change in the structure of the crystallographic axes near the twinning plane from one single-crystal block to the other (b»a, where a is the interatomic distance).

To study the energy spectrum of the Hamiltonian, we transform to Fourier components along the coordinates y and z along the twinning plane and then perform a gauge transformation of the phase of the wave-function. The result becomes

$$\hat{H}=-1/2d/dx(m_{11}^{-1}(x)d/dx)+\frac{m_{11}(x)}{2\,m_1m_2}k_y^2+k_z^2/2m_3,$$

where k_y and k_z are the conservative components of the momentum along the twinning plane. The kinetic energy $K(x)=1/2k_y^2[m_{11}(x)/m_1m_2]$ with the variable mass is an effective potential energy in which an electron moves. As a result, either potential barrier or a potential well arise, depending on the sign of the mass difference m_1-m_2.

The including in the consideration of a magnetic field parallel to the twinning plane gives rise to new characteristic features of the behavior of the system. The effective potential causes some of the electron moving near the internal boundary not to follow the circular path as in the bulk of the system but skipping paths are involved by reflection from the boundary. The skipping paths induce magnetic surface levels (6), exactly as in the well-known case of a mirror surface of a metal (7).

Here we shall analyze the anomalous behavior of the diamagnetic susceptibility at the twinning plane. The description of the motion of electrons in a magnetic field parallel to the twinning plane is as follows:

$$\hat{H}=-1/2d/dx(m_{11}^{-1}(x)d/dx)+\frac{m_{11}(x)}{2\,m_1m_2}(k_y-eA_y/c)^2+$$

0921-4526/90/$03.50

$+k_z^2/2m_3$

The vector potential $A_y(x)$ can be expressed in terms of the z component of the magnetic field:

$$A_y(x)=\int dx' B_z(x'),$$

which gives rise to the microscopic density of the current

$$j_y(x)=2e/m_y(2\pi)^{-2}\int dk_y dk_z(k_y-eA_y/c)\sum_n \psi_n^2(x)*$$
$$*[1+\exp(E_n(k_y k_z)-\mu)/T]^{-1}; \quad m_y=m_1 m_2/m_{11}(x).$$

The calculation of the averaged over atomic distances $j_y(x)$ can be performed semiclassically. The energy spectrum is :

$E_n(k_y,k_z)=\varepsilon_n(k_y)+k_z^2/2m$,

where $\varepsilon_n=\hbar\Omega(n+1/2)$, $\Omega=eB/mc$, $m=(m_1 m_2)^{1/2}$. This equation is valid for an electron, which is not reflected by the twinning plane, what occurs if $k_y^2/2m_1>\varepsilon_n$ or $k_y^2/2m_2<\varepsilon_n$, $k_y>0$ $(m_2>m_1)$. For the electrons, which are reflected from the plane $(k_y^2/2m_2>\varepsilon_n)$ the energy spectrum changes:

$\varepsilon_n=\pi\hbar\Omega(n+3/4)/\sigma(p)$; $\sigma=\pi/2+\arcsin p-p\sqrt{1-p^2}$;

$p=k_y/\sqrt{2m\varepsilon_n}$. Then the averaged over atomic distances $j_y(x)$ is as follows:

$$j_y(x)=e\int_0^\infty dE J(E,x)/(1+\exp(E-\mu)/T)$$

$$J(E,x)=-\sqrt{m_1 m_3}/(\pi\hbar)^3\int_0^E \sqrt{\varepsilon/(E-\varepsilon)}\,d\varepsilon\,\theta(X)\theta(2-X)$$

$$*\int_{-1}^{1} d\xi\,\theta(\xi^2 m_2/m_1-1)\xi(1-\xi^2)^{1/2}(\xi-X)/Q(\xi,X);$$

$$Q(\xi,X)=(\pi/2+\arcsin\xi)(1-(\xi-X)^2)^{1/2},$$

$$X=x/a,\quad a=c\sqrt{2m_2\varepsilon}/eB.$$

The obtained expression for the current density leads to anomalous susceptibility near the twinning plane: $\chi\simeq\partial M/\partial B=c^{-1}j/dB/dx$. As we have $dB/dx\simeq B/r_L$ $(r_L=cp_F/eB)$, then $\chi\simeq 1/B^2$. If $B\Rightarrow 0$, we have $\chi\Rightarrow\infty$. This behavior of χ causes a domain diamagnetic structure, exactly as in the known case of diamagnetic domains in the bulk (8), when $\chi\simeq B^{-3/2}$. However, there is an important difference between the diamagnetic structure in the bulk of a metal and at the twinning plane. In the bulk of the metal the domain structure exists only at low enough temperatures $(2\pi^2 T<\Omega)$. The existence of the domain at the twinning plane is restricted only by the condition $T<\mu$. The magnitude of the magnetic field in the domain is of the order $B\simeq m\mu v_F/e$ and its dimension is of the order $r_L\simeq c/\mu$. For the amplitude of the magnetic field and the dimension of domain the estimate gives the order of 10^6-10^7Oe, r_L 10nm for common metals and 10^4-10^5Oe, $r_L=10^2$-10^3nm for semimetals. The diamagnetic domain is not restricted by low temperatures. As temperature is decreasing, the form of domain is changed due to the fulfilling of the conditions for the de Haas-van Alfen effect in the bulk at low enough temperatures $T\simeq(2\pi^2)^{-1}\mu v_F/c$. In the region near the twinning boundary the domain induces the magnetic field distribution varying with temperature. This temperature may reach the order of 10K because of the large amplitude B.

The observation of the considered phenomenon may be possible with the help of measuring Knight shift, μSR-method, the neutron scattering or by decoration used successfully in the investigation of the Abrikosov vortex lattice in high temperature superconductors. The appearance of induced magnetic field in the system leads to the variation of properties from the macro scopic point of view. It occurs the break of symmetry with respect to time-reverse. This results in various effects analogous those in systems without the center of inversion. The preliminary report will appear elsewhere (9).

(1) I.N.Khlyustikov and A.I.Buzdin, Adv. Phys. 36 (1987) 271.
(2) S.N.Burmistrov and L.B.Dubovskii, Phys. Lett. 98A (1983) 129.
(3) V.S.Bobrov and S.N.Zorin, JETP Lett. 40 (1984) 1147.
(4) S.N.Burmistrov and L.B.Dubovskii, JETP Lett. 45 (1987) 547.
(5) M.I.Kaganov and V.B.Fiks, Sov. Phys. JETP 46 (1977) 393.
(6) S.N.Burmistrov and L.B.Dubovskii, Phys. Lett. 127A (1988) 120; Sov. Phys. JETP 67 (1988) 1831.
(7) M.S.Khaikin, Sov. Phys. Usp. 11 (1969) 785.
(8) A.A.Abrikosov, Fundamentals of the Theory of Metals (North-Holland, Amsterdam, 1988).
(9) S.N.Burmistrov and L.B.Dubovskii, Pis'ma Zh. Eksp. Teor. Fiz. 51 (1990) 'in print'.

Physica B 165&166 (1990) 207–208
North-Holland

LOW-TEMPERATURE LATTICE PROPERTIES OF TRANSITION METALS : PSEUDOPOTENTIAL MODEL COUPLED WITH BAND STRUCTURE CALCULATION

V.Yu. MILMAN, V.N.ANTONOV, and V.V.NEMOSHKALENKO

Institute of Metal Physics, Academy of Sciences Ukr. SSR, KIEV 252142, USSR

Recently we have shown that pseudopotential (PP) model is applicable in the study of lattice statics, dynamics, and thermodynamics of transition metals. Here this model is summarized and new results are given obtained e.g. for metastable FCC phases of *Co* and *Fe*, for high pressure equation of state (EOS) of *Cu*. For the first time our model is used in the study of disordered alloys.

1. INTRODUCTION

Though total energy calculations became at present a powerful means of metal lattice properties study nevertheless *ab initio* study of thermodynamics still presents a challenge for scientists (1). Hence different models are still useful in e.g. Debye temperature $\Theta(T)$, lattice Grüneisen parameter $\gamma(T)$ etc. investigation.

Model PP approach was used previously by us in study of phonon spectra (2-4), electron-phonon interaction (4), high pressure EOS (2-5) for transition metals. Full review of our results together with model parameters values will be published elsewhere (4). Similar model was also suggested recently (6,7). Thus only summary of our approach and some new results will be given here.

2. PSEUDOPOTENTIAL MODEL

Transition metal electron density is formed by superposition of *s*-like and *d*-like electrons. Contribution from *s*- electrons to total energy is taken in the second order perturbation theory with dielectric function of Taylor and local PP of Heine-Abarenkov. Interaction of *d*- electrons was simulated by repulsive pair potential in the Born-Mayer form.

Four model parameters were fitted to elastic constants under lattice equilibrium condition. main difficulty concerning with the choice of effective valence *Z*. We took *Z* as the sum of *s* and *p* occupancies from our *ab initio* RAPW or LAPW data resulting *Z* being in the range 1.4-2.0. PP for 3*d* elements had usual form whereas PP for *Ir* ,*Pt*, *Pd*, *Rh* was unexpectedly deep and short-ranged. Mentioned above approach (6,7) is conceptually very close to our method but dielectric function was taken in a rationalized form, Kohn anomalies being intrinsically absent.

3. RESULTS

Here we present results not included in the review of our PP model (4).

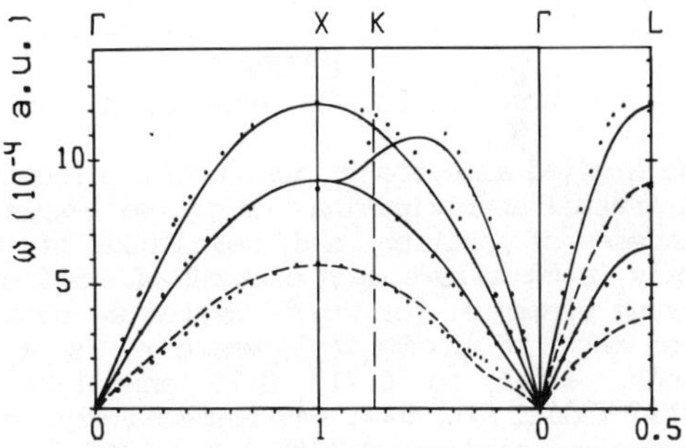

FIGURE 1
Phonon dispersion for FCC *Co* (solid) and *Pt* (dashed), experiment from (8) and (4) respectively. Only those branches for *Pt* are shown which do not intersect curves for *Co*.

Phonon dispersion curves for FCC *Co* (Fig. 1) and *Fe* (Fig. 2) are typical for 3*d* FCC metals. Another kind of $\omega(q)$ curve is illustrated by Fig. 1 where some branches for *Pt* are given. The most prominent feature there is Kohn anomaly in T_1 [110] branch. Similar anomaly was observed both experimentally and theoretically in *Pd* and it was also predicted in *Rh* (4).

High pressure lattice response is also well described in PP approach (5). Zero isotherms shown in Fig. 3 agree with measured data up to the highest presently attainable pressures (9).

From thermodynamic study we inferred that Kohn anomaly caused sharp increase of γ in *Pd*, *Pt*, *Rh*, *Ir* as *T* rises in the range 0-50 K while in metals with normal dispersion (*Cu*, *Ni*) $\gamma(T)$ appeared to be nearly constant (4). Anomalous negative dispersion near Γ point in *Rh* and *Ir* produces abrupt change of Θ at low temperatures resulting in $|\Theta(0)-\Theta(300)|$ values 195 and 155 K for *Rh* and *Ir* respectively (170 and 140 K in experiment). For *Pd* and *Pt* this value is nearly zero, and for *Ni* it equals to 60 K (4).

Our model gives likely unique possibility for qualitative study of mentioned above anomalies comparing e.g. with (6). Note that correct description of anomaly position is due here merely to the choice of Z value.

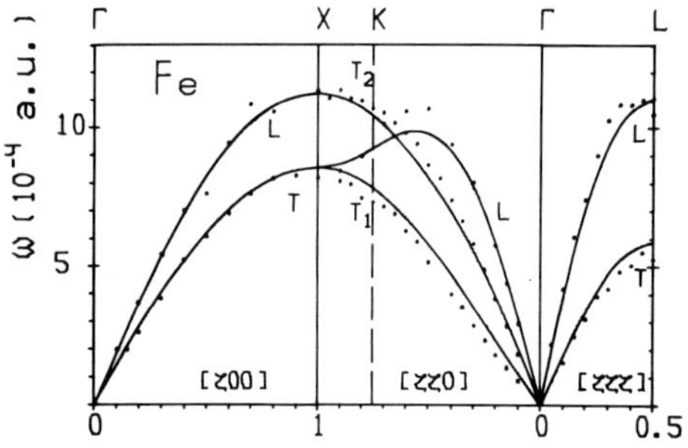

FIGURE 2
Phonon spectrum for FCC *Fe*, experiment (10).

We studied also transition metal alloys in mean crystal approximation, e.g. we explained dependence of position and magnitude of Kohn anomaly in *Pd* alloys upon content of different alloying elements. For *Cu-Ni* system we obtained values $X_{ik}=(1/B_{ik})(dB_{ik}/dx)$, where x is a *Ni* content, equal to 0.71, 0.75 and 1.32 for elastic moduli B_{11}, B_{44}, B_{33} respectively, close to experimental data (0.36±0.1, 0.62±0.2, and 1.07±0.2).

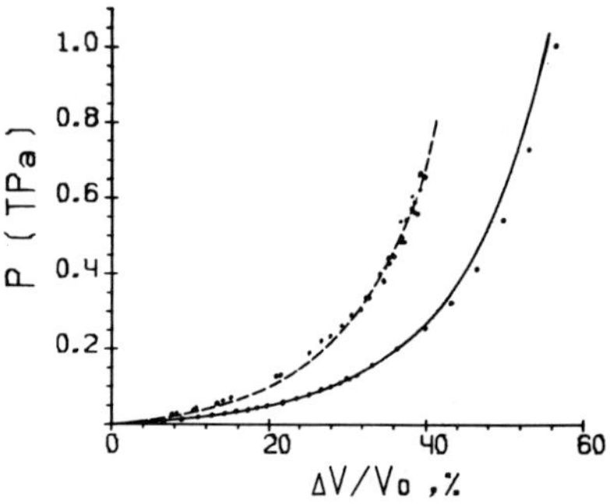

FIGURE 3
EOS for *Cu* (solid) and *Pt* (dashed), experiment from (9) and (5), respectively.

In the same system recently phonon group velocities were measured and we managed even to describe these curves (Fig. 4) though without slight anomalies detected previously at the limit of the accuracy of experiment (11).

4. CONCLUSIONS

We presented main concepts of simple PP transition metal model. Its possible application for calculation of binding energy at different atomic displacements was briefly illustrated for a number of transition metals and their alloys. We suggest that the model correctly describes main features of interatomic forces and electron density distribution in transition metals.

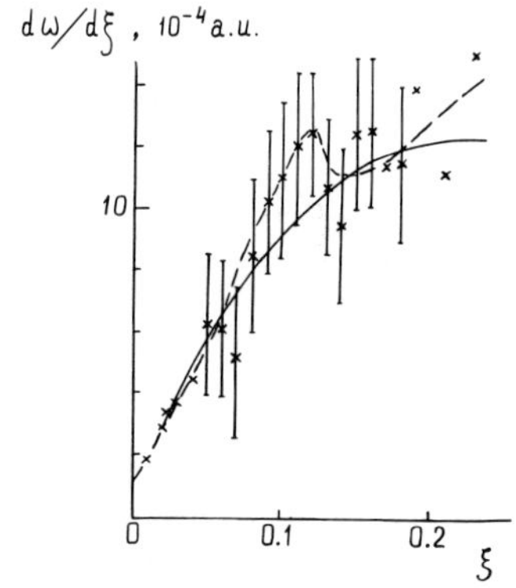

FIGURE 4
Dispersion of group velocity for transverse phonons in $Cu_{0.9}$ $Ni_{0.1}$ alloy in the (0.45+ξ/2, 0.45+ξ/2, 0.45+ξ) direction, crosses and dashed line from (11).

ACKNOWLEDGEMENTS

We would like to thank Prof. V.G.Vaks, Prof. E.V.Zarochentsev , Prof. V.Heine and Prof. W.A.Harrison for helpful discussions. Great help of Dr. A.V.Zhalko-Titarenko at the beginning of this work is highly appreciated.

REFERENCES

(1) V.L. Moruzzi et al. Phys. Rev. B37 (1988) 790.
(2) V.V. Nemoshkalenko et al. Ukrain. Phys. J. 30 (1985) 1372.
(3) V.V. Nemoshkalenko et al. JETP Lett 47 (1988) 295.
(4) V.N. Antonov et al. Z. Phys., in print.
(5) V.Yu. Milman et al. Equation of state of transition FCC metals up to 1 TPa, in: Shock Compression of Condensed Matter, eds. S.C.Schmidt, J.N.Johnson and L.W.Dawison (Elsevier, Amsterdam) in print.
(6) N. Singh et al., Phys. Rev. B38 (1988) 7415.
(7) N. Singh, Phys. Stat. Sol. 156B (1989) K33.
(8) E.C. Svensson et al. Can. J. Phys. 57 (1979) 253.
(9) W.J. Nellis et al. Phys. Rev. Lett. 60 (1988) 1414.
(10) J. Zaretsky et al. Phys. Rev. B35 (1987) 4500.
(11) A.Yu. Rumyantsev et al. Phys. Stat. Sol. 143B (1987) 625.

Physica B 165&166 (1990) 209–210
North-Holland

THE SHERRINGTON-KIRKPATRICK ISING SPIN GLASS IN A TRANSVERSE FIELD: STABILITY ANALYSIS AND BREAKING OF THE REPLICA SYMMETRY

G. BÜTTNER and K.D. USADEL

Theoretische Physik and SFB 166, Universität Duisburg, Lotharstraße 1, 4100 Duisburg 1 / FRG

We investigate the stability of the replica-symmetric solution of the Sherrington-Kirkpatrick(SK)-Ising spin glass in a transverse field. It is shown that the replica-symmetric solution is unstable in the whole low-temperature phase. The low temperature behaviour is investigated by applying the first stage of replica-symmetry breaking to the present model.

Recently the theory of quantum spin glasses has received a growing interest. Especially the SK-model in a transverse field, a model describing proton-glasses, became the source of numerous investigations. This model is particular suitable for dealing with the problem of quantum fluctuations, because with increasing transverse field the quantum nature of the dynamical variables becomes more important.

The exact phase boundary was obtained by Yamamoto and Ishii(1) using a perturbation expansion method, for a short-range model studied in mean-field theory by Dobrosavljević and Stratt(2) using a path-integral approach and by Usadel and Schmitz(3) using the Trotter-Suzuki formalism. The critical value $\Delta_c(T=0)$ is best obtained as $\Delta_c(T=0)=1.506\,J$ by Yamamoto and Ishii.

The interesting question of replica symmetry breaking in the low temperature phase has been addressed in some recent investigations. Using the static approximation Thirumalai et al.(4) found stable replica symmetric solutions in a small region close to the spin glass freezing temperature while Ray et al.(5) using the approximation of Yokota(6) found stable replica symmetric solutions in the whole low temperature phase.

Contrary we find that replica-symmetry is broken throughout the low temperature phase. In the present note we outline the corresponding calculations and we construct the first stage of replica symmetry breaking.

The Hamiltonian of the present model of N interacting spins in a transverse field Δ reads

$$\mathcal{H} = -\sum_{i,j}^{N} J_{ij}\sigma_{iz}\sigma_{jz} - \Delta\sum_{i}^{N}\sigma_{ix} \quad ,$$

where σ_j are the Pauli matrices for the jth spin. The exchange interactions J_{ij} are Gaussian distributed with zero mean and a variance of J^2/N.

Following Usadel and Schmitz(3) by using the replica method and the Trotter-Suzuki formula one obtaines the following expression for the free energy functional

$$F[\mathbf{Q}] = \frac{1}{4}\left(\frac{\beta J}{M}\right)^2 \sum_{k\alpha k'\alpha'} Q^2_{k\alpha k'\alpha'} - \ln \mathrm{Tr}\,\exp\left(H_{eff}[\mathbf{Q}]\right)$$

with an effective Hamiltonian

$$H_{eff}[\mathbf{Q}] = \frac{1}{2}\left(\frac{\beta J}{M}\right)^2 \sum_{k\alpha k'\alpha'} Q_{k\alpha k'\alpha'}\,\sigma_k^\alpha\,\sigma_{k'}^{\alpha'} + B\sum_{k\alpha}\sigma_k^\alpha\,\sigma_{k+1}^\alpha$$

and

$$B = -\frac{1}{2}\ln\tanh\frac{\beta\Delta}{M} \quad ,$$

where $\mathbf{Q}$ are the order parameters, α, α' the replica-indices and k, k' the locations in Trotter-direction. Stationarity of F with respect to $\mathbf{Q}$ gives the selfconsistency equations for these quantities. There are two types of these variables, the order parameters for $\alpha \neq \alpha'$, denoted by $Q_{\alpha\alpha'}$, which are independent of k and k', and the spin self interactions $R_{\Delta k} = Q_{k\alpha,k+\Delta k,\alpha}$. The self interactions depend on the difference $\Delta k = |\,k - k'\,|$ only because of translational invariance in Trotter space.

For the replica-symmetric solution the order parameters $Q_{\alpha\alpha'}$ are independent of α and denoted by q. After some tedious algebra one obtains as a new effective Hamiltonian

$$H_{rs} \;=\; \frac{\beta J}{M}\sqrt{q}\,z\sum_{j=1}^{M}\tau_j + \frac{1}{2}\left(\frac{\beta J}{M}\right)^2\sum_{i,j=1}^{M}(R_{i-j}-q)\tau_i\tau_j + B\sum_{i=1}^{M}\tau_i\tau_{i+1}$$

The selfconsistency equations for q and $R_{\Delta k}$ follow from the stationarity of the corresponding free energy. The pseudo spins τ_j occur by mathematical reasons and have no direct relation to the original spins σ_k^α.

The derivation of the stability condition for the replica-symmetric solution follows closely de Almeida and Thouless(AT)(7). The resulting matrix of the second derivatives of the free energy has the same structure as the one obtained by AT. In agreement with the calculations of AT we find the eigenvalue $\lambda = P - 2Q + R$ to be the smallest one near the phase transition. To study the behaviour of λ near T_f, λ and q have to be expanded for small q, giving to the leading order

$$\lambda = (\beta_f J)^2\left(-\frac{1}{3}V^2 + 2V - 3\right)q^2 .$$

0921-4526/90/$03.50

The problem to show λ to be negative reduces to the question whether V at T_f is equal to three or not, where the quantity V is a mean of four-spin correlation functions,

$$V = (\beta_f J)^2 \frac{1}{M^4} \sum_{ijk\ell} \langle \tau_i \tau_j \tau_k \tau_\ell \rangle.$$

Note that $\beta_f = 1/k_B T_f$ and the thermal average with H_{rs} is taken at $T = T_f$. The behaviour of V as a function of the transverse field Δ is shown in Fig.1. Only in the limit $\Delta = \Delta_c$ V approaches the value of three.

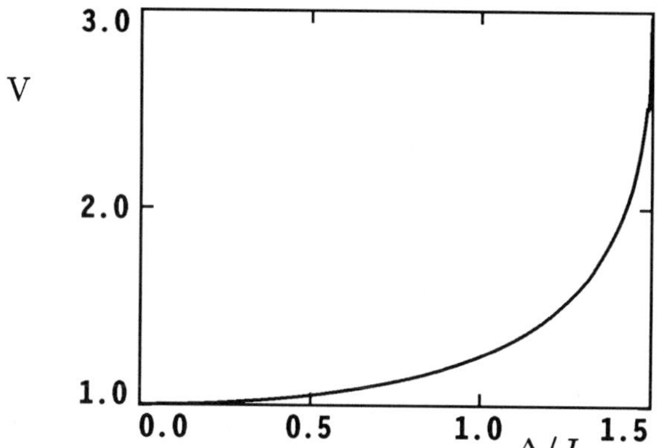

FIGURE 1

Therefore we can conclude that for the transverse SK-Ising spin glass replica symmetry is broken within the whole low temperature phase.

Since replica symmetry is broken in the whole spin glass phase it appears to be straightforward to construct a replica-symmetry broken solution following the work of Parisi(8). But because of the complexity of the problem caused by the Trotter-Suzuki formalism we have to restrict ourselves to the first stage of replica-symmetry breaking. Within this theory the order parameters $Q_{\alpha\alpha'}$ are substituted by only two different values q_0 and q_1, respectively. In the limit $n \to 0$, where n is the number of replicas, a distance between 0 and 1 of different replicas in the ultrametric space is defined and a quantity m_1 in this picture denotes the part of order parameters being q_0.

In analogy to the replica-symmetric solution we obtain a symmetry broken effective Hamiltonian

$$\begin{aligned} H_{sb}(\tau) &= \frac{\beta J}{M}\left(\sqrt{q_1}\, z + \sqrt{q_0 - q_1}\, y\right) \sum_{j=1}^{M} \tau_j \\ &+ \frac{1}{2}\left(\frac{\beta J}{M}\right)^2 \sum_{i,j=1}^{M} (R_{i-j} - q_0)\tau_i\tau_j + B \sum_{i=1}^{M} \tau_i \tau_{i+1}. \end{aligned}$$

and a set of selfconsistency equations for q_0,q_1,m_1 and $R_{\Delta k}$. From the solutions of the selfconsistency equations thermodynamic quantities are obtained readily.

We have calculated entropy, internal energy and susceptibility and the corresponding replica-symmetric quantities as well. The entropy for $\Delta = 0.1$ (left) and $\Delta = 0.5$ (right) is shown in Fig.2.

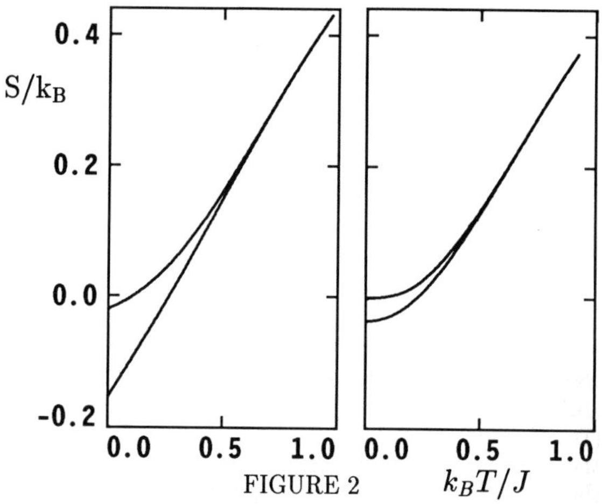

FIGURE 2

We find that for small transverse fields Δ the qualitative behaviour of all of these quantities is nearly the same as for the SK-model. With increasing Δ, however, there appears a distinct change in the qualitative behaviour, i. e. the replica symmetric result becomes much better especially what the entropy and presumably also the internal energy is concerned. Both the replica symmetric and the replica symmetry broken zero temperature entropies increase very quickly with increasing Δ, but they remain negative, so the two corresponding entropies converge. This convergence is even faster than it might be expected from the calculation of the instability eigenvalue. The same behaviour is found for the internal energies, but not for the susceptibilities, where both quantities increase moderately towards one.

REFERENCES

(1) T. Yamamoto and H. Ishii, J. Phys. C20 (1987) 6053.

(2) V. Dobrosavljević and R. Stratt, Phys. Rev. B36 (1987) 8484.

(3) K. D. Usadel and B. Schmitz, Sol. St. Commun. 64 (1987) 975.

(4) D. Thirumalai, Q. Li, and T. R. Kirkpatrick, J. Phys. A22 (1989) 3339.

(5) P. Ray, B. K. Chakrabarti, and A. Chakrabarti, Phys. Rev. B39 (1989) 11828.

(6) T. Yokota, Phys. Lett. A125 (1987) 482.

(7) J. R. L. de Almeida and D. J. Thouless, J. Phys. A11 (1978) 983.

(8) G. Parisi, Phys. Lett. A73 (1979) 203.

Physica B 165&166 (1990) 211–212
North-Holland

SLOW RELAXATION OF DILUTED ANTIFERROMAGNETS

U. NOWAK and K.D. USADEL

Theoretische Physik and SFB 166, Universität Duisburg, Lotharstraße 1, 4100 Duisburg 1 / FRG

Threedimensional diluted Ising-antiferromagnets cooled in a uniform external magnetic field develop a frozen domain state. The relaxation of the domain state after switching off the external field is investigated in detail by a Monte Carlo simulation. A new interpretation of the results leads to a possible connection to the theory of self-organized criticallity proposed by Bak et.al.

In recent years, there have been several theoretical studies on the behavior of random-field systems. Experiments are often performed on diluted Ising-type antiferromagnets in a uniform external magnetic field (DAFF) which belong to the same universality class as the random-field Ising model (RFIM) (1). Like the RFIM the DAFF develops a metastable domain state if the system is cooled in a field from the paramagnetic phase (2). The domain state relaxes towards long-range order after switching off the field. This relaxation is non-exponential due to pinning of the domain walls at vacancies. Since the domain state carries surplus magnetization and internal energy a slow decay of these remanent quantities can be observed. The remanent magnetization of an experimental realization of the DAFF was found by Kleemann et. al. (3). In an earlier publication we investigated the mechanisms for this relaxation by a monte carlo simulation (4) which showed that processes which are relevant for the relaxation primarily take place in the domain walls confirming ideas published earlier by Nattermann and Vilfan (5). From the results of our simulations we could not distinguish whether the decay of the magnetization follows a power-law or a logarithmic law. Therefore, we increased the time of a Monte Carlo run to up to 100000 Monte Carlo steps per spin (MCS). In analysing our data carefully we observed a structure in the fluctuations of the data leading to a new kind of interpretation of the relaxation which will be reported in this letter.

As in our earlier publication we have performed monte carlo simulations on a threedimensional simple cubic lattice with a size of $61 * 61 * 60$ and a site-dilution of 50%. Only nearest neighbour interaction is considered. We used helical boundary conditions and the heat-bath algorithm. As in our earlier work we prepared the domain state choosing a fixed temperature T and field B and performing 1000 MCS starting from a completely random spin configuration. Within this time the random spin configuration relaxes to a practically frozen domain state serving as the initial configuration for the computation of the remanent magnetization and energy.

Analysing our data we observed that fluctuations of the remanent quantities which made the data fitting ambigous in the earlier simulations occur in the longer simulations as well but now also on longer time scales. The decaying remanent quantities fluctuate around their mean value in a very broad range of time scales. This is demonstrated in fig. 1 which shows a log-log plot of the remanent magnetization versus time.

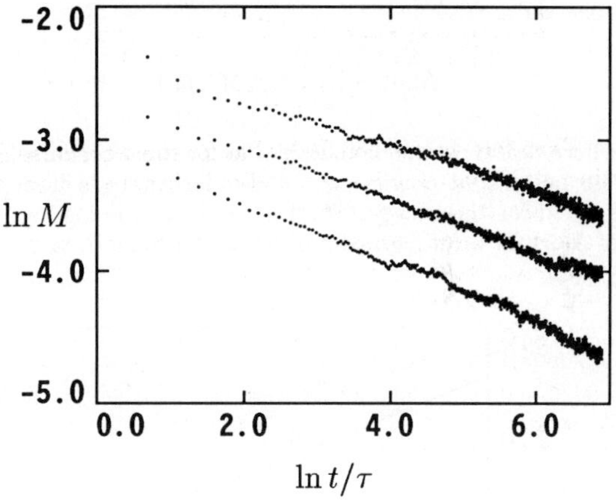

FIGURE 1

The upper curve shows the magnetization of the first 1000 MCS after switching off an external field of $B_i = 2.0$ at a temperature $T = 0.8$. The intermediate curve shows the magnetization of the first 10000 MCS of the same run rescaled in time by a factor of 10, i. e. every point is an average of 10 consecutive MCS. The lower curve is again rescaled by a factor of 10 and every point is an average of 100 consecutive MCS. Similar behaviour was found for the energy and for different temperatures and fields. From these results we are led to conclude that there are relaxation processes on all time scales leading to a fluctuating magnetization $M(t)$ which has a self similar structure, i. e. it shows the same qualitative behavior on all time scales τ. This lack of a typical time scale in our opinion can be explained by a lack of a typical length scale in the system. Above the percolation threshold p_c for the vacancies (31%vacancies) there exists a percolating vacancy cluster which is a fractal below a length scale $\xi \sim (p - p_c)^{-\nu}$. Apart from this the domain wall by itself is a fractal as has been shown by Cambier and Nauenberg (7) for the RFIM. Since the relaxation of the domain state takes place in the domain wall or is caused by movements of the domain walls the fractality of the domain wall as well as the fractality of the vacancy clusters result in a lack of typical length scales in the system leading to a wide range of energy barriers with long range temporal correlations. This lack of typical time scales during the relaxation naturally involves a power-law decay for the remanent quantities. Indeed we could confirm the power-law as best fitted by averaging over several dilution configurations. We found the exponent of the power-law to be proportional to the temperature only slightly depending on the initial field but different for magnetization and energy.

Within a recent theory of self organized criticallity (SOC) from Bak et. al.(6) the lack of typical time scales leads to a $\omega^{-\alpha}$- law for the power noise spectrum with $\alpha \approx 1$. In order to analyse our simulation data along these lines we have calculated the power spectrum of the noise data $S(\omega)$:

$$S(\omega) = | \int e^{i\omega t} \Delta M(t) dt |^2 \qquad (1)$$

where $\Delta M(t)$ is the deviation of magnetization data from the best-fitted power-law. A similar analysis was performed for the energy ΔE. Fig. 2 shows the log-log plot of $S(\omega)$ for a temperature of 0.6 and a field B_i of 2.0.

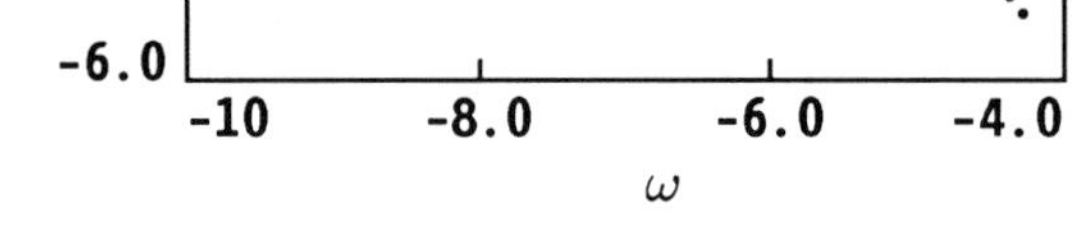

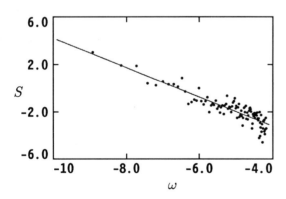

FIGURE 2

The possible values for ω varied from $\omega_{min} = \frac{2\pi}{T}$ to $\omega_{max} = \frac{\pi}{\Delta t}$ where T is the simulation time (100000 MCS for the long time simulation) and Δt is the smallest computed time step (100 MCS). Even though we have averaged the data of $S(\omega)$ over slightly different ω the data are quite unexact and therefore the resulting exponent is only roughly determined.

In order to get a higher accuracy we have analysed the data for different temperatures and initial fields B_i The values of α fluctuate arround $\alpha \approx 1$ in agreement with predictions by SOC theory. No systematical temperature dependence of α is observed and the results for magnetization and energy are practically the same as far as the exponent α is concerned.

The analogy between the theory of SOC from Bak et. al. and our simulations now becomes apparent. A crucial ingredient of the SOC model is an underlying dissipativ transport. This dissipativ transport is realized in our relaxation process by the disappearence of magnetization and internal energy stored in the unsatisfied bonds of the domain wall. However, there is a difference between the dynamics in our situation compared to the usual setup investigated by Bak and others: our system is not an open system in a steady state. However, our system is out of equilibrium and far away from reaching equilibrium for the times investigated and it is open in the sence that there is a dissipativ transport of energy into the heat-bath. Therefore in our opinion the analogy is strong enough to consider the DAFF to be in a SOC state during the relaxation at least for the time scales investigated.

REFERENCES

(1) S. Fishman and A. Aharony, J.Phys. **C12**, (1979), L729

(2) G. S. Grest, C. M. Soukoulis, and K. Levin, Phys.Rev.B **33**, (1986), 7659

(3) U. A. Leitão, W. Kleemann, and I. B. Ferreira, Phys.Rev.B **38**, (1988), 4765

P. Pollak, W. Kleemann, D. P. Belanger, Phys.Rev.B **38**, (1988), 4773

(4) U. Nowak and K. D. Usadel, Phys.Rev.B **39**, (1989), 2516

(5) T. Nattermann and I. Vilfan, Phys.Rev.Lett. **61**, (1988), 223

(6) J. A. Cambier and M. Nauenberg, Phys.Rev.B **34**, (1986), 7998

(7) P. Bak, C. Tang, and K. Wiesenfeld, Phys.Rev.Lett. **59**, (1987), 381

C. Tang and P. Bak, Phys.Rev.Lett. **60**, (1988), 2347

T. Hwa and M. Kardar, Phys.Rev.Lett. **62**, (1989), 1813

D. Dahr and R. Ramaswamy, Phys.Rev.Lett. **63**, (1989), 1659 and references therein

Physica B 165&166 (1990) 213–214
North-Holland

DIMERISATION IN FRUSTRATED ANTIFERROMAGNETS

N JOAD† and G A GEHRING‡

†Department of Theoretical Physics, Oxford University, Oxford OX1 3NP, UK

‡Department of Physics, Sheffield University, Sheffield S3 7RH, UK

The breakdown of antiferromagnetism in the pseudospin formulation of the t-J model into a novel dimerised state is considered. The method used is a simplified Wilson renormalisation scheme. The state has a lower energy than Hartree Fock for some parameter values in both one and two dimensions.

1. Introduction

The cuprate superconductors have an antiferromagnetic phase which is rapidly destroyed by doping with mobile vacancies. In this paper we investigate a model which shows the breakdown of antiferromagnetic order into a novel dimerised state. In the usual theories of dimerisation the dimer state is non-magnetic. In this case we map the dimers onto a spin that interacts again with other dimers so that the whole sample becomes one highly correlated dimer.

We consider an S=1 model with Heisenberg and biquadratic coupling,this is the most general form consistant with full rotational symmetry.

$$H = J\sum_{ij}[\cos\gamma(\mathbf{S}_i.\mathbf{S}_j) + \sin\gamma(\mathbf{S}_i.\mathbf{S}_j)^2] \qquad (1)$$

The states of $S = 1$ can be used to describe an antiferromagnet with mobile vacancies[1,2]. $S^z = 0$ is a vacancy while $S^z = \pm 1$ are the spins. We shall be interested in the region $\frac{\pi}{4} < \gamma < \frac{\pi}{2}$ in which both terms are positive and the biquadratic term is large enough to frustrate the antiferromagnetism. In one dimension there is an exact solution at $\gamma = \frac{\pi}{4}$[1].Papanicolaou [3] and Chubukov [4] have developed Hartree Fock theories in this region,in which the ground state is written as a product of single site terms.

$$|\psi_i> = \alpha_i|1> + \beta_i|0> + \gamma_i|-1>$$

The ground state for $\frac{\pi}{4} < \gamma < \frac{\pi}{2}$ is given by $\beta_i = 1$ at even sites and $\beta_i = 0$ at odd sites. There is a whole branch of spin wave excitations that have zero energy [4]. Thus the Hartree Fock state is not an easy starting point from which to construct a better theory.

McCoy and Itoyama [5] showed that this Hamiltonian could have a parity violating ground state and they suggested some sort of spiral. The classical ground state does have spins arranged with $\cos\theta = -\frac{1}{2}\cot\gamma$ but these could be puckered or in a spiral. As a result of this degeneracy there are no well behaved spin wave excitations away from the classical ground state [4].

In this paper we investigate an approximate non-degenerate ground state of this model built using a renormalising valence bond picture. It has a lower energy than the Hartree Fock [3] theory for $\frac{1}{2} < \tan\gamma < 4$ in one dimension and $\frac{25}{26} < \tan\gamma < \frac{6}{5}$ in two dimensions. The method involves solving a finite problem and keeping only the lowest states before going onto the next level. Unlike the dimerisation schemes proposed by Anderson [7] and Affleck [8] the dimer spin S_D is $S_D = 1$ rather than $S_D = 0$.

2. Decimation of a chain

Consider the Hamiltonian (1) for a pair.The energy of the states of definite angular momentum and the regions where they are the ground states are given by

$$S_T = 1 \quad E_p = 4\sin\gamma - 2\cos\gamma \quad \tfrac{3\pi}{2} < \gamma < \arctan(\tfrac{1}{3})$$
$$S_T = 1 \quad E_p = \sin\gamma - \cos\gamma \quad \arctan(\tfrac{1}{3}) < \gamma < \tfrac{\pi}{2}$$
$$S_T = 2 \quad E_p = \sin\gamma + \cos\gamma \quad \tfrac{\pi}{2} < \gamma < \tfrac{5\pi}{2}$$

The state with S_T overlaps with the ferromagnetic state. The $S_T = 1$ state overlaps with both the Neel state and the partially ordered state. The state $S_T = 0$ overlaps with the quadrupolar and the Neel states. In the region of interest the dimer ground state has $S_T = 1$. We consider the projection of (1) on to the subspace of interacting dimers. Since we have preserved the rotational symmetry the new Hamiltonian written in terms of the dimer spins will have the same form as (1) with the addition of a constant term.

$$H = E_p + J' \sum_{dimers} [\cos\gamma'(\mathbf{S}'_i.\mathbf{S}'_j + \sin\gamma'(\mathbf{S}'_i.\mathbf{S}'_j)^2] + \epsilon_0 \quad (2)$$

where E_p is the sum of the non interacting dimer energies. The coefficents J', γ', ϵ_0 are evaluated by taking the matrix elements of (1) between the dimers: $J' = J/4 \quad \gamma' = \gamma \quad \epsilon_0 = 2N\sin\gamma$.

The decimation is sketched below.

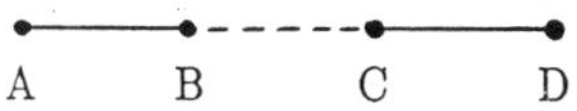

We note that in the original chain each site has two bonds. One bond is optimised in a dimer. The other bond is higher in energy since it has been evaluated in a restricted basis set (this bond energy would be zero for two $S = \frac{1}{2}$ coupled to $S_T = 0$). We also note that the number of bonds has been reduced by a factor of two. We now have a renormalisation procedure which can be iterated infinitely. At each stage the effective interaction is reduced by $\frac{1}{8}$ ($J' = J/4$ and there are only half as many bonds). This is equivalent to a simplified Wilson R.G.[6]procedure in which the Hamiltonian is diagonalised for each dimer and arestricted basis set, $S_T = 1$,is carried forward and the higher statesdiscarded. After many iterations the system has $J \rightarrow 0$ and the ground state energy is found to be:

$$E_D = \frac{4N}{7}(2\sin\gamma - \cos\gamma) \quad (3)$$

This is lower than the Hartree Fock partially ordered state for $\tan\gamma < 4$ and lower than the Neel state for $\tan\gamma > \frac{1}{2}$. This range includes the point $\tan\gamma = 1$ which is exactly soluble (Permutation Model) [1]. The energy of the fully dimerised state at this point is $\frac{4}{7}$N (cf. the partially ordered state N) while the exact result is $(2 - \frac{\pi}{3\sqrt{3}} - \ln 3)N$. The dimerised state is thus considerably lower in energy than the Hartree Fock state. The partially ordered state is $2^{N/2}$ fold degenerate.The fully dimerised state has a degeneracy of N/2 because of the choice of N spins from which to start the dimerisation from. It should be possible to construct a *resonating* dimer state [7] from a linear combination of these fully dimerised states. The fully dimerised state has all the N spins coupled to $S_T = 1$ so that in the thermodynamic limit it is non-magnetic. It is also formed from pairs of sites in an $S = 1$ state so that it contains a coherent mixture of holes and antiferromagnetically coupled spins without long range order.

3. Decimation of a square lattice

In the case of a square lattice a composite dimerisation in two perpendicular directions is carried out.The effective Hamiltonian is then:

$$H = E_1 + 1/8 \sum_{i''j''} \cos\gamma(\mathbf{S}''_i.\mathbf{S}''_j) + \sin\gamma(\mathbf{S}''_i.\mathbf{S}''_j)^2$$

where $E_1 = N(\frac{19}{8}\sin\gamma - \frac{3}{8}\cos\gamma)$

After many iterations the energy becomes

$$E = \frac{72}{31}\sin\gamma - \frac{12}{31}\cos\gamma \quad (4)$$

this is lower in energy than the Hartree Fock result for $25/26 < \tan\gamma < 6/5$. In general the dimer state should be less favourable with increasing coordination number. If the same procedure is carried out in 3 dimensions the fully dimerised state is always higher in energy than the partially ordered state.

4. Conclusions

We have shown the existence of a mixed defect antiferromagnetic phase in which mobile vacancies are coupled coherently to the spin correlations. It is more stable than Hartree Fock in one dimension and for a small parameter range in two dimensions.

Acknowledgements

We would like to thank N Papanicolaou for introducing us to this problem and J B Parkinson for very helpful discusions on the 1-d problem. N J thanks SERC for financial support.This work was financed in part by ESPRIT P3041-MESH.

References

[1] Bill Sutherland Phys Rev B12 3795 (1975)
[2] Mader, Psaltakis, and Papanicolau Preprint
[3] N Papanicolau Nucl Phys B305 367 (1988)
[4] A V Chubukov J Phys Condens Matt 2 1593 (1990)
[5] B McCoy and Itoyama Preprint
[6] K G Wilson Rev Mod Phys 47 773 (1975)
[7] P W Anderson Proc of the Enrico Fermi International School of Physics, Varenna (July 1987)
[8] Ian Affleck, Tim Kennedy, Elliot H Lieb and Hal Tasaki PRL 59 799 (1987) Commun Math Phys 115 437 (1988)

Physica B 165&166 (1990) 215-216
North-Holland

MONTE CARLO STUDIES OF CRITICAL PHENOMENA NEAR A LIFSHITZ POINT IN AN ISING FERROMAGNET WITH DIPOLAR INTERACTION

Seizo WATARAI

Department of Mathematics and Physics, Setsunan University, Neyagawa, Osaka 572, Japan

Monte Carlo method is applied to a simple cubic Ising model with the n.n. exchange and dipolar interactions. In addition to the anisotropic long range nature of the interaction, this model possesses Lifshitz points leading to the multicritical behaviors. Simulations are performed in the region of the ferromagnetic (or ferroelectric) phase transitions, and the standard finite size scaling method is used to obtain the critical indices. The results seem to show the existence of much complicated crossover behaviors and Tc tending to zero near the Lifshitz point.

1. INTRODUCTION

We take a simple cubic Ising lattice possesing the n.n. exchange (energy of J) and the dipolar interactions (energy of μ^2/d^3, μ=dipole strength and d=lattice spacing), i.e.

$$H=-\frac{1}{2}J\sum_{\langle ij\rangle}\sigma_i\sigma_j-\frac{1}{2}\mu^2\sum_{i\neq j}(3\cos^2\theta_{ij}-1)\sigma_i\sigma_j/r_{ij}^3. \quad (1)$$

Here, the Ising spin $\sigma_i=\pm1$ stands for i-th dipole orientation confined only to +z or -z axis, r_{ij} is the distance between dipoles i and j, and θ_{ij} the angle between the z axis and r_{ij}.

Using the Fourier transform of $\sigma_q=\sqrt{N}^{-1}\sum_i\sigma_i\times\exp(-iqr_{ij}/d)$, H can be expressed as $H=-\frac{1}{2}\sum_q K(q)\times|\sigma_q|^2\mu^2/d^3$ with

$$K(q)=r(6-q^2)+\frac{4}{3}\pi+0.164q^2+4\pi\frac{q_z^2}{q^2}+0.491q_z^2+O(q^4), \quad (2)$$

where ratio $r=J/(\mu^2/d^3)$ plays an important role in this system.

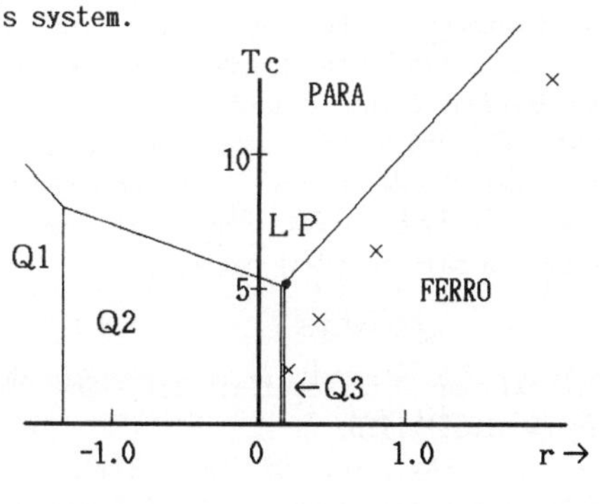

FIG.1 Phase diagram obtained from MFA. Q1,Q2 and Q3 are antiferromagnetic states with wave vectors (1,1,1),(1,1,0) and (1,0,0),respectively. LP is a biaxial Lifshitz point at r=0.164. Tc's obtained in present simulations are shown by (×), see §4.

Larkin-Khemel'nitskii[1)] derived the Landau-like critical exponents (but with log-correction terms) for the phase transition in the similar model to EQ.(1). Applying the renormalization group method to the model of EQ.(1), Aharony[2)] obtained the same results as Larkin-Khemel'nitskii and also predicted a crossover between the 3-dimensional Ising (3DI) and the dipolar critical (DP) regimes.

As for the Monte Carlo studies in the system of EQ.(1), Vorontsov•Vel'yaminov-Favorskii[3)] reported the data for the case of J=0, where the antiferromagnetic transition takes place. They emphasized the strong 1-dimensional natures appeared, due to the anisotropic behavior of dipolar interaction.

Kretschmer-Binder[4)] applied the mean field approximation(MFA) to EQ.(1),and also performed the Monte Carlo simulations. They derived the phase diagram shown in FIG.1, and pointed out the possibility of the existence of Lifshitz points (one of them is noted by LP). They also suggested that Tc→0 as r→0.164 (i.e. near LP, see EQ.(2)) and their Monte Carlo data seemed to support this. But, their data were insufficient for analyzing critical indices accurately, since the computers in those days needed vast computing-time for the long range interactions as in EQ.(1).

Now, by using high speed vector-processors, we have re-examined the Monte Carlo approach to the above system in order to study the critical and multicritical natures in the phase transitions. The preliminary results are reported in REF.5.

2. MONTE CARLO SIMULATION

Simulations have been done for r=0.2, 0.4, 0.8, 2.0 and 8.0 in the model of EQ.(1). We have taken the linear dimension L(system size $N=L^3$) as L=6~16, and the dipolar interactions were treated by the Ewald method, which inevitably introduced the periodic boundary condition.

After discarding 5000~10000 Monte Carlo steps per spin(MCS) from the initial ferromagnetic state, we used more than 10^5 MCS to calculate equilibrium quantities such as $\langle|\sigma|\rangle$, $\langle\sigma^2\rangle$, $\langle\sigma^4\rangle$, $\langle H\rangle$ and $\langle H^2\rangle$, where $\sigma=N^{-1}\Sigma_i\sigma_i$ and $\langle\cdots\rangle$ means the statistical avarage. From these avarages, we can obtain the order parameter $M_L=\langle|\sigma|\rangle$, susceptibility $\chi_L=(\langle\sigma^2\rangle-M_L^2)L^3/T$, specific heat $C_L=(\langle H^2\rangle-\langle H\rangle^2)L^3/T^2$ and 4-th order cumulant $g_L=(\langle\sigma^4\rangle-3\langle\sigma^2\rangle^2)/\langle\sigma^2\rangle^2$, where the temperature T is normalized by μ^2/d^3.

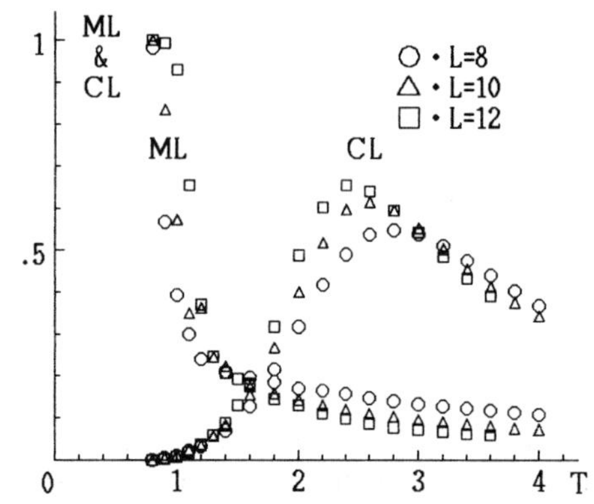

FIG.2 M_L and C_L plotts for r=0.2.

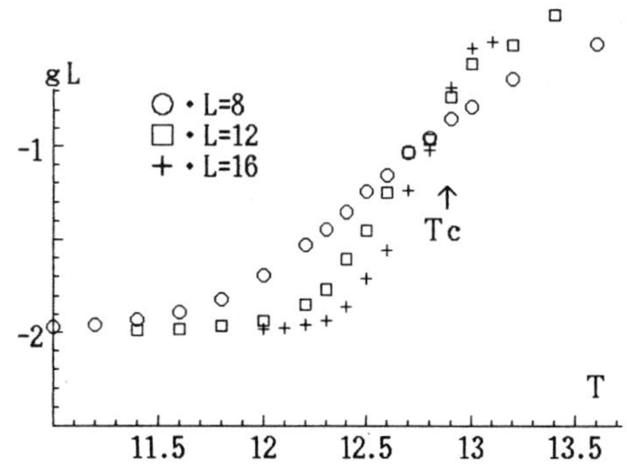

FIG.3 g_L plotts for r=2.0.

We applied the finite size scaling theory to obtain Tc and critical exponents ν, β and γ:

$$g_L \sim Q_g(tL^{1/\nu}),\ M_L \sim L^{-\beta/\nu} Q_M(tL^{1/\nu}) \text{ and } \chi_L \sim L^{\gamma/\nu} Q_\chi(tL^{1/\nu}), \tag{3}$$

where t=(T-Tc)/Tc is reduced temperature, Q_g and so on are the universal functions.

FIG.2 shows the data of M_L and C_L for r=0.2, which is very close to the case of r=0.164, where the multicritical point may appear. The peak positions in C_L might not correspond to Tc, rather to the Schottky type anomalies concerning to the destruction of the chain ordering along the z axis.

3. FINITE SIZE SCALING

From the cross points of g_L, Tc can be obtained and $1/\nu$ from the slopes near Tc(see EQ.(3)). In FIG.3, an example case of r=2.0 is shown. However, as r→0.164, the system becomes to show so strong 1-dimensionality that the data g_L, in the case of r=0.2 for example, cannot be used to determine ν. After setting Tc and ν, β and γ can be obtained from M_L and χ_L in EQ.(3), respectively. In TABLE 1, these values are summarized, which satisfy the scaling relation $(2\beta+\gamma)/\nu=3$ very well.

r	Tc	$1/\nu$	β/ν	γ/ν
0.2	2.012±0.55	–	–	–
0.4	3.867±0.017	2.04±0.21	0.49±0.01	2.03±0.02
0.8	6.429±0.028	1.45±0.16	0.49±0.02	2.01±0.11
2.0	12.75±0.07	1.56±0.03	0.48±0.04	2.00±0.07
8.0	41.36±0.14	1.65±0.14	0.48±0.05	2.01±0.08

Table 1. Critical indices obtained.

4. SUMMARY

From Monte Carlo simulations, we have obtained the critical indices in the ferromagnetic phase transition in the model of EQ.(1). Tc's are shown in FIG.1 by cross, which seem to support the conjecture made by Kretschmer-Binder[4], namely, Tc→0 as r→0.164. The values of $1/\nu$, β/ν and γ/ν in the case of r=2.0 and 8.0 (corresponding to weaker dipolar interaction) are quite reasonable, because they are approximately equal to the 3DI exponents ($1/\nu=1.6$, $\beta/\nu=0.5$ and $\gamma/\nu=2.0$). In the case of lower values of r, on the other hand, they show complicated manners. For r=0.4, the values are very close to those of the Landau-like exponents of the tricritical point ($1/\nu=2.0$, $\beta/\nu=0.5$ and $\gamma/\nu=2.0$). For r=0.8, they seem to be of the 3DI within the statistical errors, but the crossover effect might not be excluded, due to closeness by both the DP and multicritical regions.

REFERENCES

1) A.I.Larkin and D.E.Khemel'nitskii : Sov. Phys. JETP 29 (1969) 1123.
2) A.Aharony : Phys. Rev. B8 (1973) 3349.
3) P.N.Vorontsov-Vel'yaminov and I.A.Favorskii : Sov. Phys. Solid State 15 (1974) 1973.
4) R.Kretschmer and K.Binder : Z. Physik B34 (1979) 375.
5) S.Watarai and A.Nakanishi : J. Phys. Soc. Jpn 55 (1988) 1508.

Physica B 165&166 (1990) 217–218
North-Holland

ELECTRICAL PROPERTIES OF THULIUM THIN FILMS AT LOW TEMPERATURES

J. DUDÁŠ, D. KNEŽO

Technical University, 043 89 Košice, Czechoslovakia

A. FEHER

Šafárik University, 041 54 Košice, Czechoslovakia

H. RATAJCZAK

Institute of Molecular Physics, PAS, 60-179 Poznaň, Poland

A temperature dependence of the electrical resistance of thulium thin films has been studied in the temperature range from 4.2 K to 300 K. Thin films prepared in UHV in an thickness interval from 16 nm to 370 nm exhibit the increase of spin disorder resistivity and the decrease of Néel temperature with decreasing film thickness.

1. INTRODUCTION

Thin films (TF) containing RE metals are studied as promising for industrial applications, e.g. RE-TM films as materials for data storage or HTS films for electronics - as Josepson junction devices and chip interconnectors, the GdBaCuO films on Si substrates without buffer layer being very promising.

Electrical and magnetic properties of RE metal films were studied at low temperatures mostly in Dy and Sm. Our first attempts to prepare pure Tm films proved difficulties (1) because of hydrogen in them, which causes anomalies overlapping the effect of magnetic structure on the electrical resistance (2).

The aim of this paper is to refer on preparing pure Tm films, on study of the influence of temperature on electrical resistance and on influence of thickness on spin disorder resistivity and on paramagnetic-antiferromagnetic transition.

2. EXPERIMENTAL

Thulium thin films were evaporated in UHV $\sim 10^{-7}$ Pa from pure bulk Tm sample (residual resistance ratio RRR = 143) onto glass substrates heated up to $\sim 250^{o}$C. Protective SiO layers were deposited on them against contamination with air. Current leads and potential probes were cemented at appropriate positions using silver paint. Conventional dc arrangement was used to measure the electrical resistance in the temperature range from 4.2 K to 300 K using digital Keithley programmable current source 224 and nanovoltmeter 181. The temperature of the films was measured using calibrated Ge and Pt thermometers. Optical Tolansky method was used to measure film thickness. The crystal structure was determined using X-ray diffraction techniques.

3. RESULTS AND DISCUSSION

We have studied thulium TF in a thickness range from 16 nm to 370 nm. Size effect of the electrical resistivity of these films doesn´t exhibit anomalous behaviour and is in accordance with Fuchs-Sondheimer theory. Typical resistance (R) vs. temperature (T) dependences of four of these films are illustrated in figure 1. "Hump-backed" anomaly for 364 nm thick film and a change of the slope for 16,89 and 189 nm thin films, corresponding to the paramagnetic-antiferromagnetic transition, are clearly seen in this figure. The values of Néel temperature (T_N) of our film and bulk samples were estimated using numerical analysis and are illustrated in figure 2. We can see in this figure, T_N values of all films are lower thant that of bulk and, T_N value decreases with decreasing film thickness (t).

We assume the observed T_N vs. t dependence is caused by one of the following reasons or by their combinations:

1) decrease of T_N value is caused by increasing internal stress with decreasing film thickness
2) decrease of T_N value is caused by increasing relative contamination with decreasing film thickness (caused probably by hydrogen)
3) decrease of T_N as predicted by theory (3).

Further experiments are needed to resolve the prevailing mechanism, although in the case of Dy and Sm thin films, prepared in high vacuum, the RRR vs. t dependences preffered the second mechanism (4).

Spin disorder resistivity ρ_m of Tm bulk and

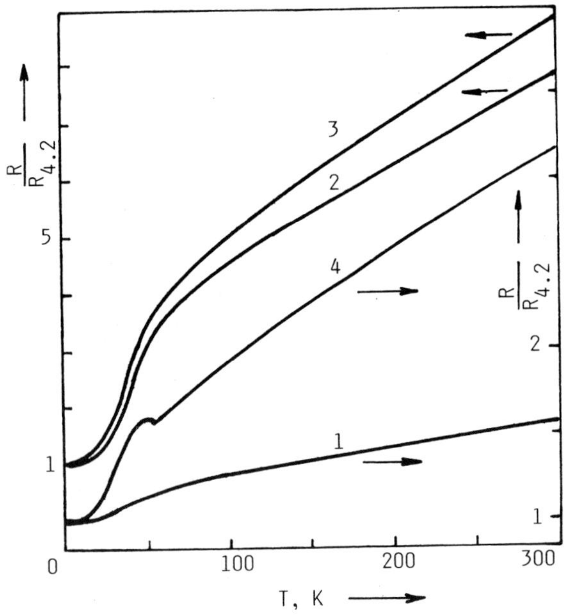

FIGURE 1
Resistance ratio $R/R_{4.2}$ as a function of temperature of four thulium films: 1 - 16 nm, 2 - 89 nm, 3 - 189 nm, 4 - 364 nm.

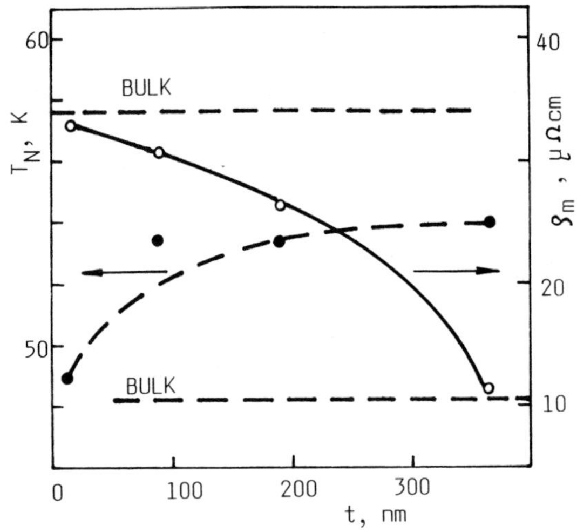

FIGURE 2
Thickness dependence of the Néel temperature T_N and spin disorder resistivity ϱ_m of thulium thin films.

thin films samples was determined by the method of ref.(5) and its thickness dependence is illustrated in fig.2. The ϱ_m value of all films is larger than that of our bulk sample and this value increases with decreases film thickness.

Temperature dependences of resistance for thinner films resemble that found for the basal plane of Tm single crystal, whereas "hump-backed" anomaly found in thicker films is similar to that found for the single crystal c-axis. This suggests the preferential orientation of our Tm films. The X-ray diffraction study revealed the various crystal orientation in thinner and thicker films - the majority of crystallites in thinner films have their basal plane parallel to substrate plane, just opposite to thicker films. As the ϱ_m value for basal plane of Tm single crystal was found to be more than three times larger than that found for c-axis (5), we assume, the increase of ϱ_m value with decreasing film thickness found in our Tm films was caused by various preferential crystal orientation in thinner and thicker films.

Finally, such a thickness behaviour of the spin disorder resistivity was experimentally observed in thin Dy films (6).

ACKNOWLEDGEMENT
We are grateful to Prof.Š.Jánoš for helpful discussions.

REFERENCES
(1) J.Dudáš, A. Feher, H. Ratajczak, Int.Conf. Low Temp.Physics, Budapest 1987, Digest F, 47.
(2) J.N. Daon, P. Vajda, A. Lucasson, Sol.St. Commun. 34 (1980), 959.
(3) M.I. Klein, R.S. Smith, Phys.Rev. 81 (1951), 378.
(4) J. Dudáš, A. Feher, Š. Jánoš, Acta Phys. Hung. 62 (1987), 77.
(5) K.N.R. Taylor, Contemp.Physics 11 (1970), 423.
(6) J. Dudáš, A.Feher, V. Kavečanský, Phys. Lett. 109A (1985), 113.

Physica B 165&166 (1990) 219–220
North-Holland

ABSENCE OF RESONANT SCATTERING IN MIXED VALENT $Hg_{1-x}Fe_xSe$.

A. Lenard, M. Arciszewska, T. Dietl, M. Sawicki, W. Plesiewicz, T. Skośkiewicz, W. Dobrowolski, and K. Dybko

Institute of Physics, Polish Academy of Sciences, Al. Lotników 32/46, PL–02668 Warszawa, Poland

Resistivity and magnetic susceptibility of $Hg_{1-x}Fe_xSe$ have been measured down to 30 mK, in the magnetic field range of 50 mOe – 3 kOe. The data substantiate the classification of $Hg_{1-x}Fe_xSe$ as an inhomogeneous mixed–valence system in which inter–site Coulomb interactions among Fe resonant donors charges impose their spatial correlations, thereby reducing the efficiency of electron scattering. Resistive and inductive effects indicative of superconductivity have been observed in some samples. The phenomenon is assigned to the presence of Hg–related inclusions and the associated proximity effect.

1. INTRODUCTION

It has been demonstrated that substitutional Fe impurities in HgSe give rise to resonant donor levels located at $\approx$ 0.2 eV above the bottom of the conduction band (1). Therefore, for sufficiently high Fe concentration, $N_{Fe} \gtrsim 4 \times 10^{18}cm^{-3}$, the Fermi level is pinned to the Fe^{2+} level, and two charge states of Fe (Fe^{2+} and Fe^{3+}) coexist. One may expect a strong suppression of the low temperature electron mobility in this mixed–valence region of Fe donors, caused by Friedel's resonant scattering and the Kondo effect. The experimental data (2–5) contradict this expectation: the electron mobility increases when the temperature is lowered (4,5) attaining at 4.2 K values which can be as much as 4 times greater than those of HgSe:Ga with similar electron concentrations.

These findings can be understood in terms of the model introduced by J. Mycielski (6). This model assumes the Fe resonant states to be long living, which means that effects of **k**–d hybridization can be disregarded. If it is the case, the inter–impurity Coulomb interactions become the dominant interactions in the system. These interactions tend to keep ionized donors apart, leading, in an extreme case, to the formation of a kind of pinned Wigner crystal. It is the correlation in the position of donor charges which is thought (6) to lead to the reduction of the ionized–impurity scattering rate in the mixed–valence regime. Furthermore, it has been pointed out (7,8) that the inter–site interactions lead to the appearance of a Coulomb gap in the one–electron density of impurity states, thereby diminishing the role of **k**–d hybridization. Additional reduction of the mixing between **k** and d states may arise from the coupling between the impurity and the lattice. In this paper we present low–temperature conductivity and magnetic susceptibility measurements of $Hg_{1-x}Fe_xSe$ (9). Our data provide new evidences for the weak **k**–d hybridization.

2. CONDUCTIVITY

Electron concentration n and mobility μ of our samples, deduced from the Hall effect measurements at 4.2 K, are depicted in Fig. 1, together with the experimental results of other authors. It is seen that for $N_{Fe} \gtrsim 4 \times 10^{18}cm^{-3}$ ($x \gtrsim 3 \cdot 10^{-4}$), n becomes independent of N_{Fe} indicating pinning of the Fermi level by Fe states.

As already mentioned, the large mobilities of electrons in HgSe:Fe in the mixed–valence region are thought (6) to reflect spatial correlations of resonant–donor charges, driven by inter–site Coulomb interactions. Results of recent calculations (8), shown by solid line in Fig. 1, support this conjecture.

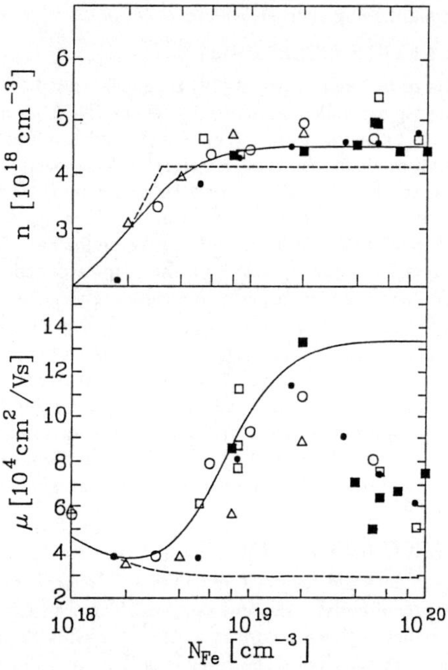

Figure 1. Electron concentration and mobility in HgSe:Fe as a function of Fe concentration. Experimental points at 4.2 K are from: open circles (2), squares (3), triangles (5); solid circles (4) and solid squares – our data. Broken lines were calculated without taking into account the inter–donor interactions, and the solid lines include these interactions in the short–range correlation model (8).

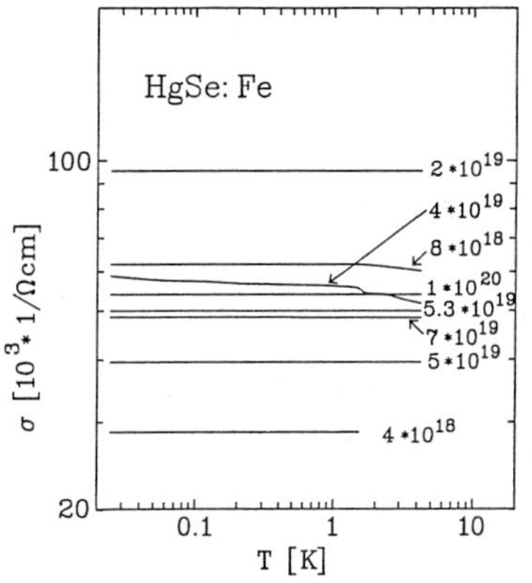

Figure 2. Temperature dependence of conductivity for various values of iron concentrations shown in cm^{-3}.

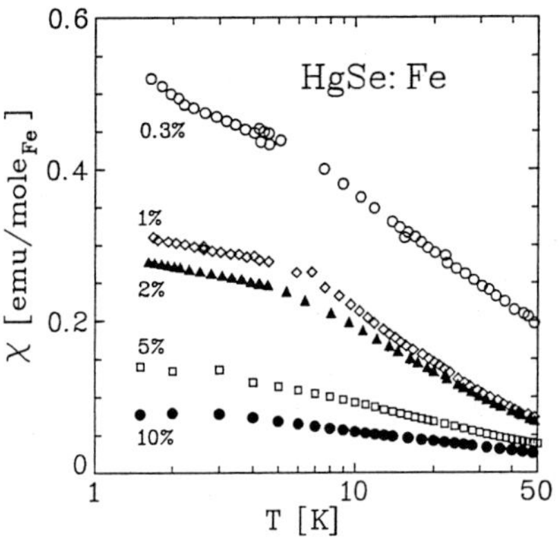

Figure 3. Mole magnetic susceptibility as a function of temperature in $Hg_{1-x}Fe_xSe$. Data for x=5 % and 10 % are taken from (10).

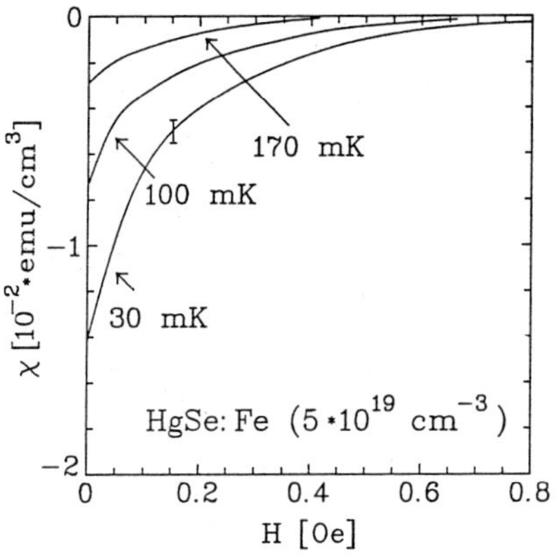

Figure 4. Apparent magnetic susceptibility at various temperatures of a HgSe:Fe sample which shows the superconductivity.

Results of our conductivity measurements, carried out between 4.2 K and 30 mK are shown in Fig. 2. It can be seen that σ changes by less than 5 % in this temperature range (with an exception of the sample with $N_{Fe} \approx 4 \times 10^{19} cm^{-3}$). This weak temperature dependence points to the absence of resonant scattering and the Kondo effect, even for quasiparticles having energies as low as $\sim 3\mu eV$.

3. MAGNETIC SUSCEPTIBILITY

Results of our and previous (10) magnetic susceptibility measurements are collected in Fig. 3, where the mole magnetic susceptibility χ_m of $Hg_{1-x}Fe_xSe$ is plotted as a function of temperature. It is seen that, contrary to Mn-based semimagnetics, the increase of χ_m with decreasig temperature, becomes less steep at lower temperatures, where χ_m tends to level of. Such behavior is, of course, specific for Van Vleck ions, such as Fe in the 2+ charge state. However, the absence of full leveling of, especially in samples with the lowest x-values, is worth noting. We assign the visible increase of susceptibility at the lowest temperatures to the presence of Fe^{3+} ions. A lower limit of their concentration can be estimated neglecting spin-spin interactions, i.e., assuming that their susceptibility obeys the Curie law. Such an assumption yields $N_{Fe^{3+}} \simeq 3 \times 10^{18} cm^{-3}$, a value close to the number of states in the conduction band below the iron level, i. e., $4 \times 10^{18} cm^{-3}$.

4. SUPERCONDUCTIVITY

We have detected resistive and inductive effects indicative of superconductivity in some samples. The transition to a superconductive state occurs in a relatively wide temperature range. Furthermore, the resistance drop and diamagnetic signal are quenched by rather small electric current and magnetic fields (see Fig. 4).

In order to elucidate the origin of this phenomenon the experiments were carried out for samples prepared in various ways. It was found that the low-temperature drop of the resistance was suppressed in samples grown from stoichiometric melts, in which Fe impurities were compensated by an appropriate amount of Se. On the other hand, the low-temperature anomaly appeared in the HgSe sample that contained an excess of Hg. The anomaly did not show-up, however, in those HgSe:Fe samples which were annealed in saturated Se vapor, the procedure known to release the excess of Hg from the material.

The above series of experiments suggest strongly that the phenomenon is to be related to the presence of superconductive Hg inclusions, which by means of the proximity effect lead to the apparent macroscopic superconductivity of the bulk. We estimate that the penetration radius of the Cooper pairs, because of the extremely long mean free path of electrons in HgSe:Fe, can be as large as 40 μm at the lowest temperature and magnetic field accessible to us (30 mK and 50 mOe). For this radius, 10^8 Hg-related inclusions per ccm may give rise to the development of a superconducting percolation cluster.

REFERENCES

(1) A. Mycielski et al., J. Phys. C **19** (1986) 3605.
(2) W. Dobrowolski et al., *Proc. 18th Int. Conf. Physics of Semiconductors*, Stockholm 1986, ed. O. Engström (World Scientific, Singapore 1987) p. 1743.
(3) N. G. Gluzman et al., Fiz. Tekh. Poluprov. **20** (1986) 94; ibid. **20** (1986) 1996.
(4) F. Pool, J. Kossut, U. Debska, and R. Reifenberger, Phys. Rev. **B35** (1987) 3900.
(5) C. Skierbiszewski et al., Semicon. Sci. Technol. **4** (1989) 293.
(6) J. Mycielski, Solid State Commun. **60** (1986) 165.
(7) T. Dietl, Japan. J. Appl. Phys. **26** (1987) 1907, Suppl. 26-3.
(8) Z. Wilamowski, K. Świątek, T. Dietl, and J. Kossut, Solid State Commun., in print.
(9) Some of the results presented here are discussed in more detail in: A. Lenard et al., Acta Phys. Polon. **A75** (1989) 249; J. Low Temp. Phys., submitted; M. Arciszewska et al., Acta Phys. Polon. **A77** (1990) 155.
(10) A. Lewicki, J. Spałek, and A. Mycielski, J. Phys. C. **20** (1987) 2005.

Physica B 165&166 (1990) 221–222
North-Holland

LOW TEMPERATURE MAGNETIC AND THERMAL PROPERTIES OF $Ce(CF_3SO_3)_3{\cdot}9H_2O$

G. H. BELLESIS, Satoru SIMIZU, and S. A. FRIEDBERG

Physics Department, Carnegie-Mellon University, Pittsburgh, PA 15213 U.S.A.

$Ce(CF_3SO_3)_3{\cdot}9H_2O$ or CeTFMS forms hexagonal crystals structurally similar to those of $Ce(C_2H_5SO_4)_3{\cdot}9H_2O$ or CeES. We have measured the magnetic susceptibilities of CeTFMS between ~0.09 K to 20 K and the heat capacity C_P from ~0.9 K to 70 K. We find that the Ce^{3+}ground state is a Kramers doublet $|J = 5/2; J_z = \pm 5/2\rangle$ with $g_{\parallel} = 3.78$ and $g_{\perp} \approx 0$ separated from a second doublet by an energy $\Delta/k = 20.6$ K. The spin-spin interaction is primarily magnetic dipolar. The effects of spin-lattice coupling, so important in CeES with $\Delta/k = 7$ K, are reduced by the larger Δ in CeTFMS.

We have recently studied (1) the low temperature magnetic and thermal behavior of members of the series of rare earth trifluoromethanesulfonate nonahydrates, $R(CF_3SO_3)_3{\cdot}9H_2O$, or *R*TFMS. These compounds form hexagonal crystals very similar in structure (2) to those of the familiar rare earth ethylsulfate nonahydrates, $R(C_2H_5SO_4)_3{\cdot}9H_2O$, or *R*ES. They belong to the same space group, $P6_3/m$, with Z = 2. EPR measurements (3) on several R^{3+} ions substituted in YTFMS crystals show the R^{3+} site symmetry to be C_{3h}, as in the *R*ES salts. The shape and dimensions of the $[R(H_2O)_9]^{3+}$ coordination polyhedra are nearly identical in the two series and the molar volumes differ by only about 3%. However, the unit cell dimensions are significantly different. In CeTFMS, the subject of this paper, $a = 13.928$ Å and $c = 7.449$ Å as compared with $a = 14.0706$ Å and $c = 7.1346$ Å for CeES. Thus, typically, the c/a ratio for *R*TFMS is ~5-10% larger than that of the corresponding *R*ES.

In a crystalline electric field of C_{3h} symmetry the $^2F_{5/2}$ ground state of the free Ce^{3+} ion splits into three Kramers doublets described approximately as $|J_z = \pm 5/2\rangle$, $|J_z = \pm 3/2\rangle$, and $|J_z = \pm 1/2\rangle$. In concentrated CeES (4), the $|\pm 5/2\rangle$ doublet is lowest, separated at low temperature by ~7 K from the first excited doublet, $|\pm 1/2\rangle$, and by ~200 K from the third doublet, $|\pm 3/2\rangle$. CeES exhibits a number of unusual properties. For example, the heat capacity maximum (5) near 3 K associated with thermal excitation from the lowest to the first excited Kramers doublet is much larger than that of a simple two-doublet Schottky anomaly. This was explained (6) in terms of the coupling of the Ce^{3+} ions to static lattice strain and a resultant temperature dependence of the splitting Δ of the two lowest doublets. This theory has been reformulated (7) as a cooperative Jahn-Teller effect. Equally interesting and closely related is the mechanism of magnetic interactions between Ce^{3+} ions in CeES. Unlike the magnetic coupling in many of the ES salts of the heavy rare earths which is almost purely dipolar, the weak interionic interaction in CeES has been shown to contain significant non-dipolar contributions. These are now believed to occur via the phonon field through a process of virtual phonon exchange (8). We have been curious to see whether these or similar effects are observable in CeTFMS.

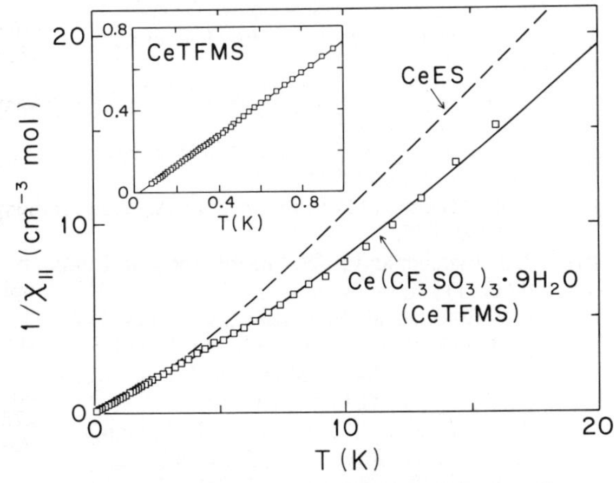

FIGURE 1 Magnetic susceptibility of CeTFMS.

The magnetic susceptibilities of single crystal CeTFMS were measured parallel and perpendicular to the c-axis down to ~0.089 K by a mutual inductance technique at a frequency of 80 Hz. Details of specimen preparation and experimental technique have been previously reported (9). The $\chi_{\parallel}$ data shown in Fig. 1 refer to a specimen in the form of an infinite needle. Data taken between 0.089 K and 1 K are plotted as $1/\chi_{\parallel}$ vs. T in the inset in Fig. 1 together with a straight line representing a Curie-Weiss law $\chi_{\parallel} = C_{\parallel}/(T - \theta_{\parallel})$. Least squares fits to data over three ranges, 0.08 K - 1 K, 0.5 K - 4 K, and 0.08 K - 4 K yield essentially the same parameters. The Weiss constant, $\theta_{\parallel} = 0.022$ K, is of ferromagnetic sign. Assuming an effective spin $S' = 1/2$, the Curie constant $C_{\parallel}$ yields $g_{\parallel} = 3.78$. $\chi_{\perp}$ below 4 K is very small and nearly temperature-independent corresponding to $g_{\perp} \approx 0$. Comparison with the splitting factors for concentrated CeES (4), namely $g_{\parallel} = 3.80$ and $g_{\perp} \approx 0$, indicates that the ground doublet of Ce^{3+} in CeTFMS is also $|\pm 5/2\rangle$. The

0921-4526/90/$03.50 © 1990 – Elsevier Science Publishers B.V. (North-Holland)

measured $\theta_\parallel$ may be compared with a theoretical Weiss constant $\theta_\parallel^{dip}$ calculated (10) in a mean field approximation assuming only magnetic dipolar interaction. One finds $\theta_\parallel^{dip} = (N_0 g_\parallel^2 \mu_B^2/4kV_{mol})(p + 4\pi/3)$ where p is a constant for a given structure obtained by dipolar summation. In CeTFMS $p = 3.47$ and $\theta_\parallel^{dip} = 0.027$ K which is close to the observed value. This suggests that interionic interactions in the salt are predominantly of magnetic dipolar character. It should be noted that in CeES where $p = 4.22$ and one calculates $\theta_\parallel^{dip} = 0.031$ K, the measured value (11) is $\theta_\parallel = -0.048$ K. The antiferromagnetic sign and the discrepancy in the calculated and measured magnitudes of $\theta_\parallel$ are consistent with the well-established presence in CeES of important non-dipolar interactions.

Fig. 1 also shows a $1/\chi_\parallel$ vs T plot of data for CeTFMS over a wide range, 0.08 K to 16 K. The breakdown of Curie-Weiss behavior above ~5 K is evident as are significant differences with $\chi_\parallel$ for CeES (dashed curve). The solid curve has been fitted to the CeTFMS data using a model in which interaction among Ce^{3+} ions is treated in a mean-field approximation. Thus we take $1/\chi_\parallel = 1/\chi_\parallel^0 - n_{mol}$ where n_{mol} is a mean field coupling constant and $\chi_\parallel^0$ is the interaction-free susceptibility. Using Van Vleck's formula and taking into account the ground doublet with g-factor $g_1^\parallel$ as well as the first-excited doublet, $|\pm 1/2\rangle$, with energy Δ and g-factor $g_2^\parallel$ one obtains (4)

$$\chi_\parallel^0 V_{mol} = \frac{N_0\mu_B^2}{4kT}\,\frac{(g_1^\parallel)^2 + (g_2^\parallel)^2 e^{-\Delta/kT}}{1 + e^{-\Delta/kT}}.$$

The fit proved not to be very sensitive to the choice of n_{mol} or $g_2^\parallel$. Thus n_{mol} was fixed at the pure dipolar value $n_{mol} = (p + 4\pi/3)/V_{mol} = 0.0203\ \mathrm{cm}^{-3}$ and $g_2^\parallel$ was set equal to 1.0, its value in CeES. The best values of the two remaining parameters are $g_1^\parallel = 3.81$ and $\Delta/k = 20.6$ K. Thus in CeTFMS Δ is nearly three times its value in CeES, namely, $\Delta/k \approx 7$ K.

The heat capacities of CeTFMS and diamagnetic LaTFMS have been measured between 1 K and 70 K in a vacuum calorimeter by an automated heat pulse technique. C_P data on LaTFMS, slightly adjusted by a corresponding states argument, were used to correct the CeTFMS results for the contibution of lattice vibrations. The resulting values of C_P(mag) are plotted vs. T in Fig. 2. A dashed curve representing the analogous data (5) for CeES is also shown. Solid curves representing the best fit of a two-level Schottky anomaly to each set of data are also shown. In the case of CeES, the discrepancy near the maximum is quite pronounced. However, for CeTFMS the scatter of the data above the C_P maximum does not allow us to say with certainty that the Schottky value is significantly exceeded. The fitted value of Δ/k is 21.5 K. Two theories (5,6) based upon spin-lattice interaction provide quantitative fits of the CeES data. In both of them the doublet separation Δ becomes T-dependent and is calculated self-consistently in a mean-field approximation as $\Delta(T) = \Delta(\infty) + \lambda \tanh(\Delta(T)/2kT)$ where λ is a spin-lattice coupling constant. The magnetic heat capacity is then

$$C_P(\mathrm{mag}) = Ry^2\mathrm{sech}^2 y/\{1 - \lambda y\,\mathrm{sech}^2 y/(\Delta(T))\}$$

where $y = \Delta(T)/2kT$. For $\lambda = 0$, this is just the Schottky curve shown in Fig. 2. A fit of the CeTFMS data up to 18.5 K with this formula yields $\lambda/k = 1.32$ K and $\Delta(\infty)/k = 20.2$ K and is represented by a dashed curve. A non-zero value of λ is probably allowed by the data and possibly even one slightly larger than that quoted here. Whether the necessity of $\lambda \neq 0$ has been established is less clear. One can say that the measurements place an upper limit on the coupling constant i.e. $\lambda/k \lesssim 1.5$ K. The parameters required to fit the C_P(mag) data for CeES with the same theory (6) are $\Delta(\infty)/k = 4.2$ K and $\lambda/k = 2.9$ K. Thus, spin-lattice coupling in CeTFMS is at most about half as great as that in CeES. Presumably the larger doublet splitting Δ in CeTFMS suppresses or reduces the effectiveness of this coupling. This would account also for the apparently reduced role of the phonon mediated spin-spin interactions in CeTFMS.

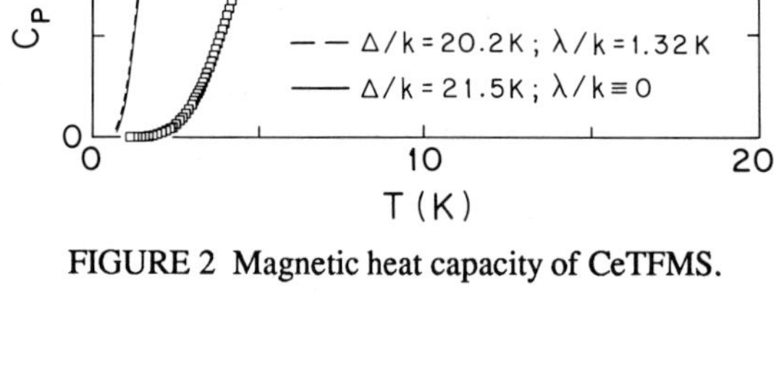

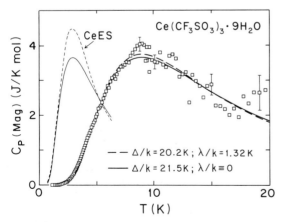

FIGURE 2 Magnetic heat capacity of CeTFMS.

REFERENCES

(1) G. H. Bellesis, Satoru Simizu, and S. A. Friedberg, Jpn. J. Appl. Phys. 26, Suppl. 26-3 (1987) 811.
(2) J. M. Harrowfield, et al., Aust. J. Chem. 36 (1983) 483.
(3) G. H. Bellesis, Satoru Simizu, and S. A. Friedberg, J. Appl. Phys. 61 (1987) 3286.
(4) G. S. Bogle, A. H. Cooke, and S. Whitley, Proc. Phys. Soc. A 64 (1951) 931.
(5) Horst Meyer and P. L. Smith, J. Phys. Chem. Solids 9 (1959) 285.
(6) J. R. Fletcher and F. W. Sheard, Solid St. Commun. 9 (1971) 1403.
(7) R. J. Elliott, et al., Proc. Roy. Soc. A 328 (1972) 217.
(8) J. M. Baker, Rep. Prog. Phys. 34 (1971) 109.
(9) G. H. Bellesis, Satoru Simizu, and S. A. Friedberg, J. Appl. Phys. 61 (1987) 3283.
(10) A. H. Cooke, D. T. Edmonds, C. B. P. Finn, and W. P. Wolf, Proc. Roy. Soc. A 306 (1968) 313.
(11) H. W. J. Blöte, Physica 61 (1972) 361.

Physica B 165&166 (1990) 223–224
North-Holland

QUANTUM XY AND HEISENBERG MODELS ON THE PLANAR LATTICES - THE RENORMALIZATION GROUP APPROACH

Andrzej DRZEWIŃSKI and Józef SZNAJD

Institute of Low Temperature and Structure Research, Polish Academy of Sciences, P.O. Box 937, 50-950 Wrocław 2, Poland

Within a real-space renormalization group framework, the critical behaviour of the spin-½ quantum XY and Heisenberg models on the three planar lattices has been studied. We have restricted ourselves to ferromagnetic-like case. It has been shown that there is a nontrivial fixed point and corresponding critical temperature for the nearest-neighbour XY model for each 2d lattice. Concerning the Heisenberg model, to first order in cumulant expansion there is no nontrivial fixed point, however, it appears in the second and third order calculations. This suggests that 2d Heisenberg model also exhibits some kind of a critical behaviour.

1. INTRODUCTION

During the last years a variety of the real space renormalization group techniques have been applied with considerable success to study spin systems. The situation appears to be less clear in the case of quantum approach where the character of transitions is more complicated. Our paper is concerned with generalization of the Niemeijer and Van Leeuwen method and cell-type transformation to the 2d quantum XY and Heisenberg models. The main question is whether in these systems the phase transitions appear. The usage of three planar lattices (triangular, square and hexagonal) enables us to get more reliable answer.

2. TRANSFORMATION

We have studied the Hamiltonian in the following form:

$$H = K_1^{xy} \Sigma (S_i^x S_j^x + S_i^y S_j^y) + K_1^z \Sigma S_i^z S_j^z \qquad (1)$$

where the sum is over the nearest-neighbours and the Pauli-operators S^α ($\alpha = x,y,z$) are normalized to unity. When K_1^z equals zero we obtain the XY model, while for K_1^{xy} equals K_1^z we obtain the isotropic Heisenberg model. As usual, the renormalization procedure will produce a whole set of new interactions between the second, third, etc., neighbouring spins described by XY-like and Ising-like parameters K_2^α, K_3^α ...

Formally the RSRG transformation is represented by the equation:

$$\exp[H'(\bar{\sigma})] = \mathrm{Tr}_{\bar{S}}\, P(\bar{\sigma},\bar{S}) \exp[H'(\bar{S})] \qquad (2)$$

where H and H' are two succeeding Hamiltonians, and the weight operator P relates the cell spins σ and site spins S. The choice of the weight operator is to a considerable extend arbitrary (1) but it should quarantee that all symmetries of the original problem are preserved after transformation (2). For an arbitrary odd number (N) of spins in a cell our weight operator can be written in the form (2):

$$P(\bar{\sigma},\bar{S}) = \frac{1}{2} \prod_\beta [1 + (\sigma_\beta M_\beta)/2] \qquad (3)$$

where

$$\bar{M}_\beta = \sum_{n=0}^{(N-1)/2} A_{2n+1}/(2n+1)!!\, \Sigma\, \bar{S}_i (\bar{S}_j \bar{S}_k)(\bar{S}_l \bar{S}_m)\ldots \qquad (4)$$

with all possible permutations of the indices i, j,k ... and $(\bar{S}_j \bar{S}_k)$ denotes the scalar product of the spin operators. The coefficients A_{2n+1} for cells can be find in our previous paper (2).

Using a standard cumulant expansion we divide the Hamiltonian into two parts H_o and V, containing all the intracell and intercell interactions, and obtain the renormalized Hamiltonian

$$H'(\sigma) = \ln \mathrm{Tr}_{\bar{S}}\, P(\bar{\sigma},\bar{S}) \exp(H_o) + \langle W\rangle + \ldots \qquad (5)$$

where W is given by the Baker-Cambell-Hausdorf formula. The angular brackets denote a partial expectation value.

3. FIXED POINTS

3.1. Triangular lattice

The size and form of the cells are connected with approximation, the larger cells should give better results. Since the size of the cell is limited by the complexity of calculation we decide to use the seven spin cell. Additionally, this choice preserves the symmetry of triangular lattice.

In the first order calculation of such an approximation we find only one XY-like point: $K_1^{xy} = 0.59$. In the second order there are two "ferromagnetic" fixed points.

	$(K_1^{xy})^*$	$(K_1^z)^*$	$(K_2^{xy})^*$	$(K_2^z)^*$	$(K_3^{xy})^*$	$(K_3^z)^*$
XY	0.33	0.05	0.03	≃0	≃0	≃0
H	0.46	0.46	0.07	0.07	0.03	0.03

The inverse critical temperature of the XY-like transition $K_c = 0.36$ is in a good agreement with the exact (HTSE) result: $K_c = 0.38$ (3). Then, the existence of the Heisenberg-like point is rather controversial and as it was suggested (2) this point may be spurious effect of the second order cumulant expansion. For this reason we performed the calculation up to the third order. In such approximation this point does not disappear and is located at $K_1^* = 0.34$, $K_2^* = 0.12$ and $K_3^* = 0.03$. Moreover, the comparison of the contributions to K_1 around the fixed point H coming from several orders of the calculations shows that the contribution from the first order is, in fact equal to that found in the second one. At the same time the contribution from the third order is essentially smaller. It could explain the fact why this fixed point does not appear in the first order calculation.

3.2. Square lattice

For this lattice we use the five spin cells (4,5). In the first order calculation we have $(K_1^{xy})^* = \pm 1.2$. In the second order calculation we obtain the Heisenberg-like points again.

	$(K_1^x)^*$	$(K_1^z)^*$	$(K_2^{xy})^*$	$(K_2^z)^*$	$(K_3^{xy})^*$	$(K_3^z)^*$
XY	0.33	0.09	0.2	0	≃0	0
XYAF	-0.46	0.21	0.22	0	-0.03	0
H	0.73	0.73	0.75	0.75	0.11	0.11
HAF	-0.51	-0.51	0.16	0.16	-0.05	0.05

The ferro- and antiferro- points are not located symmetrically with respect to $K_i = 0$ because, unfortunately, our transformation does not preserve the symmetry under the transformation $K_i \to -K_i$. The deviation from the symmetry is caused by the choice of the cells with non-equivalent spins. In our case the central spin of the five-spin cell is discriminated from the others and partial average over this spin is not invariant under the transformation $K_i \to -K_i$.

3.3. Hexagonal lattice

The smallest cell having all symmetries of the hexagonal lattice is a thirteen-spin cell which is too complicated to our numerical calculations. In this situation the four-spin cell which state is defined by the three outside spins seems to be the only possibility, though the break of the spin equivalence can influence certain results.

In the first order calculation we find in the standard way two nontrivial XY-like points: $(K_1^{xy})^* = \pm 2.77$. In the second order, in contrary to the previous lattices, the Heisenberg-like point does not appear. Our transformation exhibits only XY-like nontrivial fixed points

	$(K_1^{xy})^*$	$(K_1^z)^*$	$(K_2^{xy})^*$
XY	0.72	0.37	0.24
XYAF	-0.72	0.37	0.24

The inverse critical temperature $K_c = \pm 1.25$ seems to be too high to justify the truncation procedure, but unfortunately, there are no exact results (HTSE) for comparison. We ought to emphaseze that the symmetry of the location of the fixed point under transformation $K_i \to -K_i$ is unsatisfactory for the Heisenberg model. It suggests that our choise of the cells and weight operator is unsuitable in this case and it may be a reason of absence of the Heisenberg point.

4. CONCLUSIONS

This paper was intended to check the existence of the phase transitions on the planar quantum models. Unfortunately, within the NVL majority rule spin assignment, there are no such divisions, except for some very special cases, for which all symmetries of the original problem would be preserved in the process of the RSRG procedure. However we can draw a few conclusions.

The XY model on each 2d lattice exhibits a nontrivial fixed point. The corresponding values of the inverse critical temperature agree with results of HTSE very well.

The situation is even less clear for the Heisenberg model. In the first order cumulant expansion there is no Heisenberg-like point for any lattice. On the other hand such a fixed point appears in the second order, apart from the case of the hexagonal lattice. However, the approximation on this lattice seems to be rather poor from the symmetry point of view as well as truncation procedure. It suggests that the 2d Heisenberg model also undergoes same kind of a phase transition, although further investigations are necessary to confirm this result.

REFERENCES

(1) J.Th. Niemeijer and J.M.J. Van Leeuwen, Physica 71 (1974) 17.
(2) J. Sznajd, Z. Phys. B 62 (1986) 349.
(3) J. Rogiers, E.W. Grundke and D.D. Betts, Can. J. Phys. 57 (1979) 1719.
(4) D.D. Betts and M. Plischke, Can. J. Phys. 15 (1976) 1553.
(5) A. Drzewiński, J. Sznajd, Phys. Lett. 138 (1989) 143.

Physica B 165&166 (1990) 225–226
North-Holland

NEUTRON DIFFRACTION STUDY OF THE FERROMATNETIC SPINEL $LiZn_{0.5}Mn_{1.5}O_4$

R. PLUMIER, M. SOUGI

Service de Physique du Solide et de Résonance Magnétique, CEN-Saclay, 91191 Gif-sur-Yvette Cedex, France.

Elastic neutron diffraction experiments show that the cubic spinel $LiZn_{0.5}Mn_{1.5}O_4$ which has an O^7 space group is a ferromagnet up to $T_c = 22K$ where the moment drops from $1.1\mu_B$ to zero. Another first order magnetic transition takes place at $T_1 = 15K$ with 20% moment reduction. Comparison is made with magnetization measurements and the unusual ferromagnetism of this spinel is discussed.

Elastic neutron diffraction (n.d.) experiments have recently been performed at ILL high flux reactor on a powdered sample of the cubic ($a = 8.19\AA$) spinel $LiZn_{0.5}Mn_{1.5}O_4$ reported to be a ferromagnet with $T_c = 22K$(1). The n.d. results are obtained at zero magnetic field and various temperatures on both sides of T_c where the magnetic contributions to the Bragg reflections disappear (Fig.1,2).

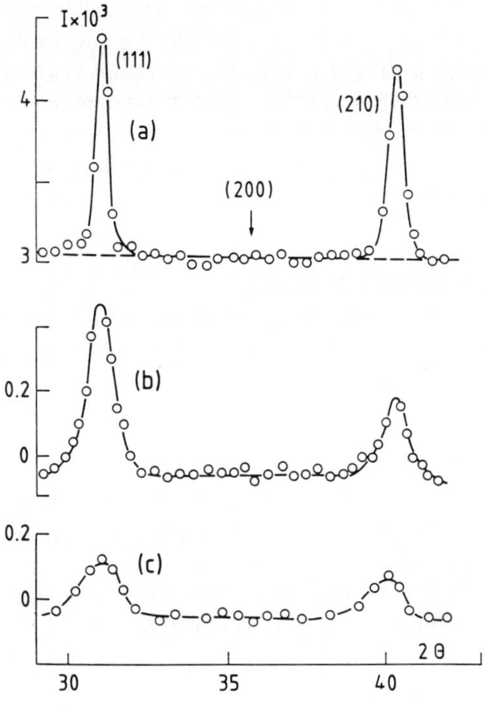

FIGURE 1

Part of the neutron diffraction spectrum obtained at $\lambda = 2.52\AA$ (a) at $T = 50K$; (b) difference spectrum $I(1.5K) - I(50K)$; (c) $I(18.5K) - I(50K)$.

The negative scattering lengths of both Mn and Li versus the positive one of Zn ($b_{Li} = -0.214$, $b_{Mn} = -0.39$, $b_{Zn} = 0.57$ in 10^{-12}cm unit) makes elastic neutron scattering a much better tool than X rays to study the distribution of the various cations among the A and B sites of the spinel structure. From the intensities of the nuclear reflections observed at $T > T_c$, the best reliability factor ($R = 6\%$) is reached for the distribution $Li(Zn_{0.5}Mn_{1.5})O_4$ ruling out the previously reported $Li_{0.5}Zn_{0.5}(Li_{0.5}Mn_{1.5})O_4$. The overall space group is O^7 although 1:3 order is not completely achieved on the B sites since the best R factor corresponds to $(1-x)Zn$ and x $Mn(x = 0.05)$ on one site with the remainder Zn and Mn statistically distributed amongst the other three B sites.

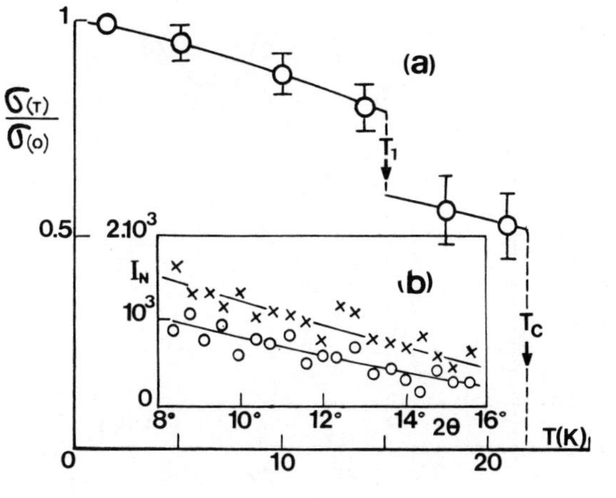

FIGURE 2

(a) Ratios to the value at $T = 1.5K$ of the moment length obtained in neutron diffraction using the (111) magnetic reflection; (b) low angle difference spectrum: $(\circ\circ\circ) I(18.5K) - I(14K)$; $(\times\times\times) I(23K) - I(18.5K)$.

From the intensities of the magnetic reflections at $T = 1.5K$, a Mn^{4+} moment length $\sigma(0) = 1.98\mu_B$ is obtained to compare with the expected value $\sigma_0 = 3\mu_B$ for a spin only $(3d^3)$ ion. At increasing temperatures, a regular moment decrease is observed up to $T_1 = 15K$ where an abrupt 20% moment reduction accompanied by an increase of the low angle neutron scattering takes place (Fig.2).
In the range $T_1 < T < T_c$, a regular moment decrease is again observed whereas at $T_c = 22K$, the moment drop from $\sigma = 1.06\mu_B$ to zero with a further increase of the low angle neutron scattering (Fig.2). On the other hand, we observe that all the magnetic reflections display a larger line width than the nuclear ones (Fig.1) in the temperature range $1.5K < T_i < 15K$ and a still larger one in the range $15K < T_j < 22K$. Assuming Gaussian profiles for both the experimental and natural contributions to the total line width, the average sizes $d_i = 170Å$ and $d_j = 130Å$ of the ferromagnetic domains are obtained. The reduction of the Mn^{4+} observed in n.d. and magnetization experiments (2) may be attributed to the finite size of the ferromagnetic domains, the slightly larger value obtained from magnetic measurements being explained by an increase of the domain size under an applied magnetic field. The

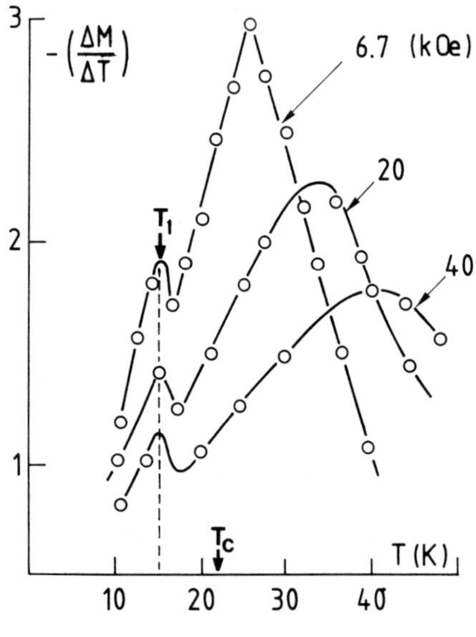

FIGURE 3
Derivatives $-(\Delta M/\Delta T)_{H_i}$ in emu.$g^{-1}K^{-1}$ unit plotted as a function of temperature.

derivatives $(\Delta M/\Delta T)_{H_i}$, readily obtained from isofield magnetization curves determined in our laboratory up to $H = 10.5$ Tesla on the same powdered specimen, clearly show (Fig.3) the first order character of the transition taking place at T_1.
On the other hand the maximum occurring (Fig.3) at $T \geq T_c$ corresponds to (H_i, T_i) set of values where a kink is observed on the isofield magnetization curves. It reveals (2) the reconstruction of small magnetic domains under the application of a magnetic field which takes place in the range $T_c < T < 38.5K(2)$.

Although reported (1) in a few isomorphous compounds ($CuMg_{0.5}Mn_{1.5}O_4$, $LiMg_{0.5}Mn_{1.5}O_4$) which also have $Mn^{4+} - Mn^{4+}$ magnetic interactions on B sites only, ferromagnetism is unusual in oxygen spinels. It is related (3) to the orthogonality of the d_{xy} orbitals which leads to a 90° positive interaction between Mn^{4+} first nearest neighbours larger than the antiferromagnetic interaction due to the direct overlap of these orbitals. On the other hand, the partial x inversion observed on B sites may explain why small ferromagnetic domains are observed. Such sites are normally unoccupied by magnetic ions in the case of perfect $\underline{1} : \underline{3}$ order. In this case, sites $\underline{1}$ are under ferromagnetic interactions with six first nearest neighbours whereas sites $\underline{3}$ have only four interactions of this type. It is around these "magnetic impurities", on sites $\underline{1}$, acting as nuclei, that ferromagnetic domains are more likely to grow (4).

ACKNOWLEDGEMENTS
Thanks are due to Dr. A. Hewat and J.L. Soubeyroux for help in collecting the neutron diffraction data at ILL.

REFERENCES
(1) G. Blasse, J. Phys. Chem. Solids 27 (1966) 383.
(2) R. Plumier, M. Sougi, J. Appl. Phys. 60 (1990) May 1 issue.
(3) J.B. Goodenough, Phys. Rev. 117 (1960) 1442.
(4) J. Villain, Z. Phys. B 33 (1979) 31.

Physica B 165&166 (1990) 227–228
North-Holland

^{151}Eu AND ^{119}Sn MÖSSBAUER STUDIES IN $SnEu_3Rh_4Sn_{12}$

G. BALAKRISHNAN, R. NAGARAJAN, S.K. PAGHDAR, L.C. GUPTA and R. VIJAYARAGHAVAN.

Tata Institute of Fundamental Research, Bombay 400 005, India

Single crystals of the compound $SnEu_3Rh_4Sn_{12}$ have been grown. The magnetic properties of this compound have been investigated by magnetic susceptibility and Mössbauer studies.

1. INTRODUCTION

$SnEu_3Rh_4Sn_{12}$ belongs to a class of rare earth(RE) intermetallic stannides, $RE_xRh_ySn_z$, which exhibit remarkable superconducting and magnetic properties (1). A variety of ground state properties are obtained by varying the RE constituent as well as the relative occupancy of RE/Sn of one of the sublattices (2). For example, $(Sn_{1-x}RE_x)RE_4Rh_6Sn_{18}$ exhibits superconductivity, magnetic order or reentrant behaviour depending on x (2). The compounds adopt different crystallographic phases for differing RE. Phase I exists for larger RE, being superconducting when RE = La to Gd, Yb (and also when RE is substituted by Ca, Sr, Th) and magnetic for RE = Eu and Gd. Of this series, the magnetic compounds have not been as well studied as the superconducting members. We have undertaken the study of the properties of the compound $SnEu_3Rh_4Sn_{12}$ through magnetic susceptibility, dc resistivity, ^{119}Sn and ^{151}Eu Mössbauer techniques. The results are presented in this paper.

2. EXPERIMENT

Single crystals of the compound were grown by the "tin flux" method starting with the constituent elements in the molar proportion Eu:Rh:Sn::3.97:5.29:90.74%. Several large crystals were obtained, and X-Ray powder diffraction pattern showed that the compound crystallised in the "phase 1" (cubic) structure as reported (3).

Magnetic susceptibility measurements were made on two small crystals using a Faraday Balance from 4.2 K to 300 K. Resistivity measurements were made on one of the crystals (4.5 * 3 * 2 mm) from 4.2 K to 300 K.

^{151}Eu and ^{119}Sn Mössbauer spectra were recorded over the temperature range 4.2 K to 300 K using a constant acceleration spectrometer with a multiscalar analyser developed in our laboratory (4).

3. RESULTS AND DISCUSSION

The magnetic susceptibility measured above the ordering temperature showed a Curie Weiss behaviour, giving an effective moment of 8.15 μ_B, consistent with Eu being divalent. A clear indication of the antiferromagnetic ordering is seen at 12 K. The ordering temperature is the same as that obtained in a previous neutron diffraction study on this compound (5).

The resistivity results show no anomalous behaviour except for the drop near 12 K due to the magnetic order.

The ^{151}Eu Mössbauer spectrum at room temperature shows the presence of a quadrupolar interaction. The isomer shift (IS) w.r.t. EuF_3, is -10.8 mm/s and is temperature independent. The value of IS is characteristic of divalent Eu, and is similar to that reported in these types of compounds (6). The spectra suggest that the magnetic ordering occurs below 15 K and a hyperfine split spectrum due to magnetic order is observed at 4.2 K (Fig. 1). The internal field sensed by the Eu nucleus is estimated to be about 300kG.

^{119}Sn Mössbauer spectra show a temperature independent quadrupole split assymetric pattern. Figure 2 shows the spectrum obtained at room temperature. In this system of $Sn(1)Eu_3Rh_4Sn(2)_{12}$, Sn(1) and Sn(2) are two distinct chemically inequivalent sites. Sn(1) occupies a cubic site in the A15 like sublattice of $SnEu_3$, while

0921-4526/90/$03.50

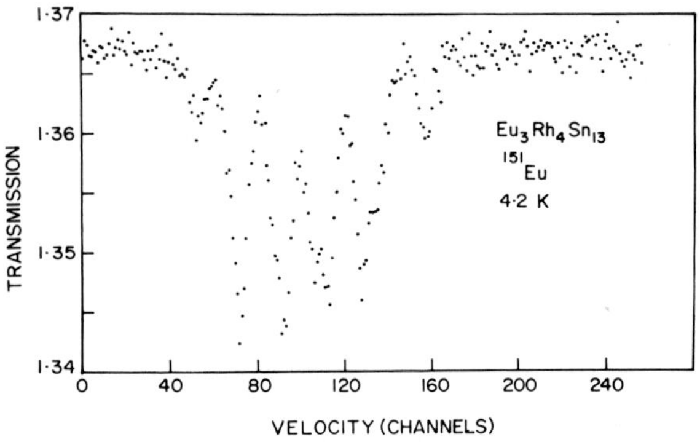

FIGURE 1

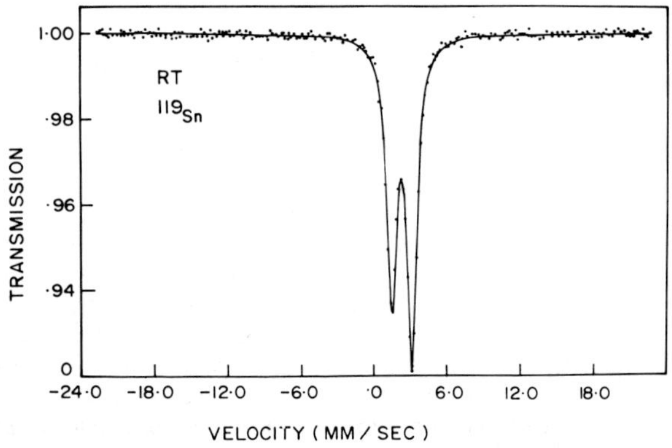

FIGURE 2

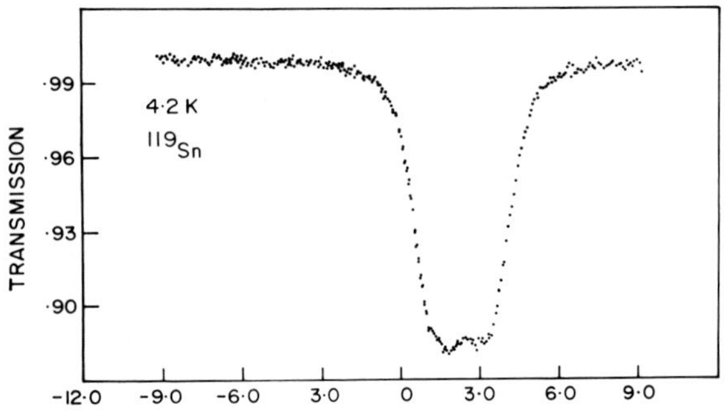

FIGURE 3

Sn(2) has a non-cubic site symmetry. The Mössbauer pattern obtained was hence fitted to a superposition of a single line and a doublet. The fit results in an IS value (w.r.t. $CaSnO_3$) of 3.4 mm/s and 2.2 mm/s for Sn(1) and Sn(2) respectively, with an EFG of 1.68 mm/s at the Sn(2) site. The intensity ratio obtained from the fit was 1:11 for the two Sn sites which compares favourably with the actual occupancy ratio of 1:12. The ^{119}Sn Mössbauer spectrum at 4.2 K, (Fig. 3) in the ordered state shows a transfered field of about 28 kG.

4. CONCLUSIONS

The magnetic susceptibility studies on single crystals of $SnEu_3Rh_4Sn_{12}$ show that the compound orders antiferromagnetically at 12 K. The Mössbauer spectrum of ^{119}Sn reveals the presence of two distinct sites for Sn. The ^{151}Eu Mössbauer study indicates magnetic ordering of the divalent Eu.

REFERENCES

(1) J.P. Remeika, G.P. Espinosa et al. Solid State Commun. 34 (1980) 923.
(2) S. Miraglia, J.L. Hodeau et al., Solid State Commun. 52 (1984) 135.
(3) A.S. Cooper, Mater. Res. Bull. 15 (1980) 799.
(4) R. Nagarajan and S.K. Paghdar, Proc. S.S.P. Symp. (India) 27C (1984) 322.
(5) P. Bordet et al., Physica B+C 136 (1986) 432.
(6) G.K. Shenoy et al., Solid State Commun. 37 (1980) 53.

Physica B 165&166 (1990) 229–230
North-Holland

ON THE COEXISTENCE OF LOW- AND HIGH-SPIN STATES IN TRANSITION METAL SYSTEMS

M. SCHRÖTER and P. ENTEL

Theoretische Physik and SFB 166, Universität-GH-Duisburg, Postfach 10 15 03, 4100 Duisburg 1, FRG

We discuss the unusual magnetic behaviour of itinerant magnetic systems close to magnetovolume instabilities. It is shown that the unusual properties result from transitions between, and coexistence of, high-spin (HS) and low spin (LS) states. We argue that the Invar effect is related to the coexistence of LS and HS states over a substantial temperature range. For the theoretical description an expansion of the free energy in terms of the magnetization and volume is used, which at zero temperature reproduces one-electron band structure results obtained with the fixed-spin-moment (FSM) method (1).

Ground states with different spin multiplicities are known to occur in many d^4-d^7 transition metal complexes as function of the ligand field strength. Compounds like cis-dithiocyanatobis (1,10-phenanthroline) iron(II), $[Fe(phen)_2(NCS)_2]$ are nice examples of reversible conversions between HS and LS states at a *critical* temperature with a pronounced peak in the molar heat capacity (2). At low T the LS $^1A_{1g}$ (t_{2g}^6) state is stable in the presence of a strong ligand field, while at high T the HS $^5T_{2g}$ ($t_{2g}^4 e_g^2$) becomes favourable because of a frozen in axial distortion. This observation has lead to a microscopic theory for such complexes, in which the coupling of electronic and lattice degrees of freedom is thought to arise from the Jahn-Teller coupling of the HS threefold degenerate $^5T_{2g}$ state and the local molecular distortion (3) (here, the spin-orbit coupling is important). Cooperative molecular and lattice distortions then cause that many Fe ions undergo simultaneouly this transition. For brevity we only note that very early also phenomenological models like extended versions of the compressible Ising models were used to describe HS $\rightleftharpoons$ LS transitions in similar insulating complexes.

In the past years it has become evident that HS $\rightleftharpoons$ LS transitons can also occur in many metallic 3d compounds, including elemental systems (1,4). For a recent review of the properties of such systems and further references see (5). In spite of the differences between metallic 3d systems and insulating 3d complexes, there are similarities concerning the HS $\rightleftharpoons$ LS transition. At the LS→HS transition in $[Fe(phen)_2(NCS)_2]$ crystals electronic levels cross according to $E(^5T_{2g}) > E(^1A_{1g}) \rightarrow E(^1A_{1g}) > E(^5T_{2g}, ^5B_{2g})$, while in metallic systems the LS→HS transition is accompanied by a crossing of the gravity centers of two narrow bands, a spin-down polarized band of mainly e_g symmetry and a spin-up polarized band of mainly t_{2g} symmetry, which in the HS state lies slightly on top of the e_g band. It seems that right after the transition to the HS state these two spin-polarized bands are pinned just below and above the Fermi level. For details see (6).

The ability of metallic 3d compounds to undergo HS $\rightleftharpoons$ LS transitions is connected with the fact that also in itinerant electron systems the spin state of an individual 3d ion depends on the coupling of electronic to lattice degrees of freedom. This can lead to a variety of anomalous properties like negative or vanishing thermal expansion (Invar effect), negative pressure dependence of T_c, etc. (5). In this communication we discuss a simple model, which allows to describe stable, metastable and (nearly) unstable (systems with HS $\rightleftharpoons$ LS transitons) metallic 3d compounds. Since there is no realistic miscroscopic theory, valid at all temperatures, we make use of a Landau expansion in terms of magnetization and volume (similar expansions have often been used and for many of the essential references we refer to (6-8)). The new feature in our work is that band structure data serve as input leading to a quasi-exact theory at $T=0$ K, while finite temperature results are obtained with the help of functional integration. We start with the free energy functional,

$$\mathcal{F} = \frac{1}{\beta} \ln \int \mathcal{D}m\, d\omega e^{-\beta\mathcal{H}[\mathbf{m},\omega]} \qquad (1)$$

with a hamiltonian of the form

$$\begin{aligned} \mathcal{H} \;=\; & \frac{1}{V_0(T)} \int_{V_0(T)} d^3r \left\{ g_\omega (\nabla\omega)^2 + \frac{\kappa}{2}\omega^2 + \gamma\omega^3 \right. \\ & + g_m \sum_{\alpha\beta} (\nabla_\alpha m_\beta)^2 + a(\omega_c - \omega)\mathbf{m}^2 \\ & \left. + b\mathbf{m}^4 + c\mathbf{m}^6 + \ldots \right\}. \qquad (2) \end{aligned}$$

Volume and magnetization are split into static and fluctuating parts, $\omega(\mathbf{r},T) = \omega_0(T) + \omega'(\mathbf{r},T)$ ($\omega_0 = (V_0(T) - \Omega_0)/\Omega_0$, V_0 is the actual volume and Ω_0 a reference volume) $\mathbf{m}(\mathbf{r},T) = \mathbf{m}_0(T) + \mathbf{m}'(\mathbf{r},T)$. Homogeneous magnetization, relative volume and mean square amplitudes of volume and longitudinal and transverse spin fluctuations are determined self-consistently in gaussian approximation. Parameters in $\mathcal{H}$ can be obtained from $T=0$ binding surfaces calculated with the FSM method (1). This unequivocally fixes each itinerant magnetic system (6). In the following we briefly discuss the main results.

The classical gas of spin and volume fluctuations can undergo magnetic phase transition of first and second order. The essential parameters are b and ω_c. ω_c is given by

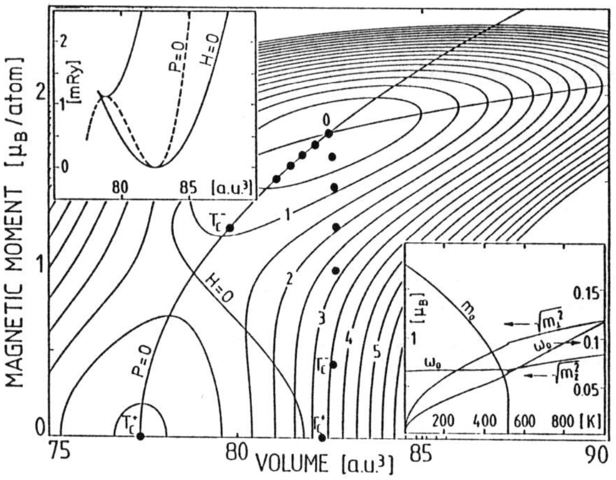

FIGURE 1

Total energy curves $E(m_0, V_0)$ for Fe_3Pt at 0.5 mRy intervals (ground state is reference state). Along the zero pressure line one finds two local minima, the ground state with $m_0 \neq 0$ and crossing of $p, H = 0$ lines and the metastable state with $m_0 = 0$. These two local minima are seperated by a saddle point where again the $p, H = 0$ lines cross. Note that in stable magnetic systems like bcc Fe the second local minimum as well as the region to the right of this minimum (where the system can sontaneously polarize without change in energy) are absent. Black dots: temperature evolution of the system from $T = 0$ to $T = T_c$, left: without anharmonic, right: with anharmonic lattice contributions. Upper inset: $E(m_0, V_0)$ along the $p, H = 0$ lines. Lower inset: temperature dependence of spontaneous magnetization, relative volume change and mean square amplitudes of longitudinal and transverse spin fluctuations.

by $H = \partial\mathcal{H}/\partial m = 0$ for $m \to 0$ (for example, $(\omega_c - \omega_0)$ is negative for stable bcc Fe and positive for unstable fcc Fe and is close to become positive for all nearly unstable Invar systems. Stable magnetic systems possess only one local minimum on the binding surface (i.e. $E(m_0, V_0)$), while all systems being on the verge of a magnetic volume collapse have (at least) two close in energy lying local minima. Two minima always exist if $\Omega_0 < \Omega_c \approx V_0(0)$, $b < 0$, where Ω_0 is now the hypothetical nonmagnetic reference volume. Obviously the binding surface is a useful tool to characterize magnetic systems. Especially the tempeature evolution of the binding surface is of interest. Here, we present for the first time the temperature dependence of the binding surface of Fe_3Pt. Fe_3Pt is on the verge of a magnetic volume collapse. A reminiscence of this is found up to high temperatures in the paramagnetic state, in which the system can change its volume or get polarized without much cost in elastic energy (see Fig. 2 for details). Hence the paramagnetic state of systems close to a magnetic volume collapse possesses many HS states which quasi coexist with the LS state. Note also that at very high T the binding surfaces of Fe_3Pt and fcc Fe are nearly identical if the axis are appropriatly scaled.

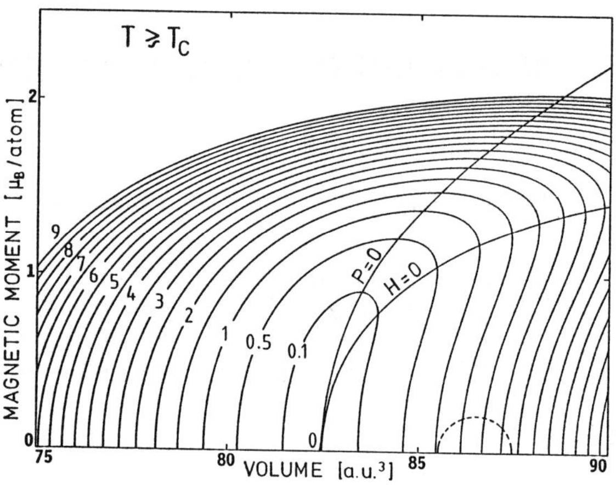

FIGURE 2

Constant energy curves $E(m_0, V_0)$ for Fe_3Pt in the paramagnetic state ($T \geq T_c$) at 0.5 mRy intervals (equilibrium state is reference state). Note that at T_c the saddle point and the stable local minimum meet on the $m_0 = 0$ line, the second local minimum from the $T = 0$ binding surface starts to disappear at $T_c/2$. Note further that the broad shallow ground state minimum, $E(m_0 = const., V_0)$, survives in the paramagnetic phase and corresponds now to the $E(m_0 = 0, V_0)$ curve. Of particular interest is the fact that the system can polarize above T_c without substantial cost in energy. An increase of the equilibrium lattice constant by 0.5% costs 0.1 mRy ($\approx$ 15.7 K) and leads along the $p = 0$ line to a polarization of 1 μ_B. Thus any experiment which sufficiently disturbs the homogeneous paramagnetic state can probe many close in energy lying HS states. With further increase in T this property is lost, for comparison we show the 0.1 mRy contour line at 1000 K (dashed curve).

REFERENCES

(1) V.L. Moruzzi, P.M. Marcus, K. Schwarz and P. Mohn, Phys. Rev. B 34 (1986) 1784.

(2) M. Sorai and S. Seki, J. Phys. Chem. Solids 35 (1974) 555.

(3) T. Kambara, J. Chem. Phys. 74 (1081) 4557.

(4) G. Fuster, N.E. Brener, J. Callaway, J.L. Fry, Y.Z. Zhao and D.A. Papaconstantopoulos, Phys. Rev. B 38 (1988) 423.

(5) E.F. Wassermann, in *Ferromagnetic Materials*, Vol. 5 (North-Holland, Amsterdam, 1989).

(6) P. Entel and M. Schröter, Physica B 161 (1989) 160;
M. Schröter, P. Entel and S.G. Mishra, J. Magn. Magn. Mat. (1990).

(7) D. Wagner, J. Phys.: Condens. Matter 1 (1989) 4635.

(8) P. Mohn, K. Schwarz and D. Wagner, Physica B 161 (1989) 153.

Physica B 165&166 (1990) 231–232
North-Holland

CROSS-RELAXATION IN PARAMAGNETIC CRYSTALS AT LOW TEMPERATURES

Dmitrii A. TAYURSKII

Physics Department, Kazan State University, 420 008 Kazan, USSR

The well- known phenomena of cross-relaxation between two spin-systems in paramagnetic crystals is theoretically investigated at low temperatures when the Zeeman energy of spins exceeds their heat energy, but there is no order due to spin-spin interactions in spin system.

1. INTRODUCTION

The cross-relaxation (CR) takes place in paramagnetic crystals containing two kinds of spins (S spins and I spins) with close resonant frequencies ω_s and ω_i. This phenomena consists in an energy exchange between the Zeeman subsystems and the reservoir of spin-spin interactions (SSI). The kinetics of spin system is determined by its thermodynamic properties only.

In the high temperature (HT) approximation ($h\omega_m << kT_o$, where m=s,i, T_o is a lattice temperature) the theory of CR was developed in (1). Accordingly to this theory the Zeeman subsystems and the reservoir of SSI are statistically independent. A coupling between subsystems appears in the result of CR.

At low temperatures (LT) ($h\omega_m > kT_o$) the subsystems become thermodynamically dependent (2,3). The strong thermodynamic coupling causes that any change of one subsystem state must be accompanied by changes of states of others. So we can expect the essential increase of CR efficiency.

2. THERMODYNAMICS OF SPIN SYSTEM AT LT

Let us consider a concentrated paramagnetic crystal containing S and I spins (S=I=1/2).The Hamiltonian of problem is:

$$H=H_s+H_i+H_{ss},\ H_s=\omega_s\sum_j S_j^z,\ H_i=\omega_i\sum_\alpha I_\alpha^z$$

$$H_{ss}=\tfrac{1}{2}\sum_{ij}(A_{ij}S_i^zS_j^z+B_{ij}S_i^+S_j^-)+$$

$$+\tfrac{1}{2}\sum_{\alpha\beta}(A_{\alpha\beta}I_\alpha^zI_\beta^z+B_{\alpha\beta}I_\alpha^+I_\beta^-)+\sum_{i\alpha}A_{i\alpha}S_i^zI_\alpha^z$$

The SSI constants include the exchange and the dipole-dipole interactions between spins.

If a lattice temperature is equal to zero then the magnetizations of spin sublattices

$$p_s=-2\langle S_j^z\rangle \text{ and } p_i=-2\langle I_\alpha^z\rangle$$

are full. At low temperatures magnetizations are about full and number of spins of every kind turned over a direction of magnetic field is small. That is why we can speak about elementary excitations, which can be named by spin excitons(4). It is necessary to note that such excitations result from the Zeeman interaction with magnetic field by contrast to magnons existing due to SSI. Thus we have two non-interacting types of spin excitons.

Any non-equilibrium state of spin system can be considered as the state with some number of non-equilibrium excitons of every type. In the quasi-equilibrium density operator we can introduce the chemical potentials μ_s and μ_i (2,3) describing a number of flipped spins:

$$\rho_q = Q_q^{-1}\exp(-\beta(H-\mu_s\sum_j S_j^z-\mu_i\sum_\alpha I_\alpha^z))$$

where β is an inverse spin temperature. Such form of quasi-equilibrium distribution takes place due to the fact of the availibility of three first integrals. A physical mechanism establishing this quasi-equilibrium is a spin rotation in local fields like the HT theory. The only characteristic feature concludes in an appearence of the molecular field at LT, which causes the ordering of local fields.

In the regular paramagnetic crystal a wave vector is considered to be a good quantum number and S-excitons and I-excitons are characterized by wave vectors $\mathbf{k}_s$ and $\mathbf{k}_i$ respectively. The energy spectrum of each type can be found in the molecular field approximation. The obtained expressions describe two exciton zones. For each zone we can find out the average energy. As we not interested in the effects connected with phase transition into magnetic ordered state then we can assume that $\beta\langle H_{ss}\rangle \ll 1$. In this approximation we have the following expressions for the average energies:

$$E_s=\omega_s-\tfrac{1}{2}p_s\Sigma_j A_{ij}-\tfrac{1}{2}p_i\Sigma_\alpha A_{i\alpha}$$

$$E_i=\omega_i-\tfrac{1}{2}p_i\Sigma_\alpha A_{\alpha\beta}-\tfrac{1}{2}p_s\Sigma_i A_{i\alpha}$$

Due to the molecular field appearence the average energy of one type of excitons(and, consequently, their number and the Zeeman energy or magnetization of corresponding subsystem) depends on a value of the molecular field which is created by spins of this kind as well as by spins of other kind. So we can conclude that the Zeeman energy of S spins depends on the Zeeman energy of I spins and the energy of SSI (the last in the disordered state ts determined by the energy of spins in the molecular field) and vice versa. Then the Zeeman subsystems and the reservoir of SSI are dependent at LT.

3. CROSS-RELAXATION AT LOW TEMPERATURES

Let us consider the evolution under the action of the perturbation

$$H_{cr}=\tfrac{1}{4}\Sigma_{i\alpha}B_{i\alpha}(S_i^+I_\alpha^-+S_i^-I_\alpha^+)$$

which causes flip-flops of spins of different kinds. We can obtain kinetic equations by the method of the non-equilibrium statistical operator. They are

$$dp_s/dt = -\tfrac{1}{8}N_s\{1-\exp\beta(\mu_s-\mu_i)\}W(\mu_s-\mu_i)$$

$$dp_i/dt = -N_iN_s^{-1}dp_s/dt$$

$$d\langle H_{ss}\rangle/dt = \tfrac{1}{2}N_s^{-1}(\omega_s-\omega_i)\ dp_s/dt$$

Here $W(\omega)$ is the probability of flip-flop process induced by H_{cr}. An explicit form of $W(\omega)$ is very complicated. Now it is important only that this probability has a sharp maximum at the frequency $\omega'=\omega_i-\mu_i-\omega_s+\mu_s+\delta$ where δ is the first moment of the corresponding function of CR form. The location of this maximum and the difference of average energies of excitons don't coincide in a frequency space.

It follows from the kinetic equations that CR leads to a complicated kinetics of the Zeeman subsystems and the reservoir of SSI at LT. A flip-flop process changes the magnetizations of sublattices. Using the terms of elementary excitations we can say that one flip-flop is the change of different type excitons numbers on the values +1 and -1 and vice versa. The appearing disbalance of energies of excitons E_s-E_i is compensated by the reservoir of SSI. So at LT the transformation of the Zeeman energy into the energy of SSI is carried out by flip-flops of spins of different kinds like HT theory. At the same time we have an essential distinction. At HT for the each pair of spins S and I giving some energy to the reservoir of SSI in the result of the mutual flips there is a pair of spins which takes this energy away. Thus at HT the reservoir of SSI compensates the difference between the Zeeman energy only. At LT the reservoir of SSI also compensates the difference between the average energies of excitons Besides at LT the reservoir of SSI has to compensate the difference between the disbalance of average energies of excitons and the disbalance of energies of excitons which is created and is annihilated with the maximum probability.

It follows from stationary solution of the kinetic equations that CR leads to the equality of chemical potentials at any temperature. It is understood quite easily. μ_s and μ_i have the meanings of the energies of non-equilibrium excitons and we have two thermodynamic subsystems with equal energies at the same temperatures. Apparently such subsystems must be in the state of dynamic equilibrium.

In the case of significant deviations from an initial state when spins S are saturated while the magnetizations of I spins and the inverse spin temperature are equal to equilibrium values the kinetic equations yield:

$$\beta-\beta_0=-\Delta\exp\beta_0(\omega_s-\omega_i)$$

$$\Delta=\omega_s-\omega_i+\tfrac{1}{2}p_i^0\Sigma_\alpha A_{i\alpha}-\tfrac{1}{2}p_s^0\Sigma_i B_{i\alpha}$$

At the essential difference of resonant frequencies we shall have an exponential increase or decrease of the spin temperature's deviation at LT. Such behavior of spin system is the LT effect only.

REFERENCES

(1) B.N. Provotorov, JETP Sov.Phys. 42 (1962) 882.

(2) J. Philippot, Phys. Rev. 133A (1964) 471.

(3) D.A. Tayurskii, phys.stat.sol.(b) 152 (1989) 645.

(4) A.I. Akhiezer and I.Y. Pomeranchuk, J. Phys. USSR 8 (1944) 206.

Physica B 165&166 (1990) 233–234
North-Holland

LINEAR UPTURN IN THE LOW-TEMPERATURE RESISTIVITY OF A NON-REENTRANT AMORPHOUS Ni-Fe-Si-B SPIN GLASS

G. Thummes*, R. Carloff, and J. Kötzler

Institut für Angewandte Physik, Universität Hamburg, Jungiusstr.11, D-2000 Hamburg 36, F.R.Germany

The distinct linear increase of the electrical resistivity $\rho(T)$, which has been measured below the paramagnetic to spin glass transition temperature $T_f = 6.4$ K of the $Ni_{80-x}Fe_xSi_8B_{12}$ metallic glass with x = 2.4, is related to elastic scattering from the spin glass order parameter. This behaviour contrasts to the much weaker $-\sqrt{T}$ rise of $\rho(T)$ in the reentrant spin glass phases (x > 4), which is due to modification of the electron-electron interaction in the presence of structural disorder.

1. INTRODUCTION

The origin of the minimum in the electrical resistivity $\rho(T)$ observed at low temperatures in amorphous metallic alloys with magnetic constituents has been the subject of many investigations. However, the question whether the resistivity rise $\rho(T \to 0)$ is related to the magnetic state or arises from structural disorder has not yet been fully answered.

In preceding work on $Ni_{80-x}Fe_xSi_8B_{12}$ metallic glasses (1,2) we have investigated the influence of the paramagnetic (PM), non-reentrant spin-glass (SG), ferromagnetic (FM), and reentrant spin-glass (RSG) phases for Fe contents between x = 2.4 and 16 on $\rho(T)$. Our results indicated different mechanisms behind the low-temperature upturns of $\rho(T \to 0)$ in the SG-state of the x = 2.4 alloy and in the FM- and RSG-states both of which occur in alloys with x > 4 (3). In the two latter cases, a quasi-universal $-\sqrt{T}$ increase of $\rho(T \to 0)$, unaffected by the FM - RSG transition, was quantitatively attributed to quantum contributions (see (4)) due to the modification of electron-electron interaction in the presence of structural disorder (2).

In contrast, in the SG-state ($T < T_f = 6.4$ K) of the x = 2.4 alloy we have observed a distinct linear rise of $\rho(T)$, as shown in fig. 1. By applying a magnetic field of B = 5 T, the quasi-universal $\sqrt{T}$-law, indicated by the dashed line in fig. 1, is recovered. These novel effects were tentatively associated with weak localisation from quantum interference of conduction electrons limited by an inelastic scattering rate $\tau_i^{-1} \sim T^2$ (2). This supposition was supported by our finding that the evaluation of the linear slope in terms of quantum interference, $\Delta\rho(T) \sim (D\tau_i)^{-1/2}$ (D : diffusion constant, see (4)), yielded a τ_i^{-1} which is in reasonable agreement with electron-phonon scattering rates reported for other amorphous metals. However, recent magnetoresistance and magnetisation measurements in the PM- and SG-state of the x = 2.4 alloy are at variance with a weak localisation mechanism (5). This led us to an alternative interpretation of the linear T-law, to be presented below.

*Present address: Institut für Angewandte Physik, Universität Giessen, D-6300 Giessen, FRG.

2. LINEAR T-LAW IN $Ni_{77.6}Fe_{2.4}Si_8B_{12}$

The fact that the linear variation of $\rho(T)$ has been observed only in an alloy with x < 4 suggests that it is intimately related to the magnetic structure of the SG-phase. We recall here the much weaker $\sqrt{T}$-behaviour in the RSG-states occuring in the ferromagnetically ordering alloys (x > 4).

To discuss our central result, we start with a theoretical prediction by Fischer (6), who has calculated $\rho(T)$ of spin glasses due to elastic and inelastic scattering of conduction electrons from diffusive low-temperature excitations in the frustrated spin system. In a nearly free electron model the elastic contribution to the temperature-dependent part $\Delta\rho(T) = \rho(T)-\rho(0)$ of the resistivity can be written as

$$\Delta\rho_e(T) = -\alpha \frac{3\pi}{8} \frac{m^* v_0}{e^2\hbar} \frac{(JS)^2}{E_f} \frac{c}{T_f} T \quad , \qquad (1)$$

where m^* is the effective mass, E_f the Fermi energy, v_0 the atomic volume, c and S the concentration and spin of the magnetic impurities, J the exchange constant between conduction electrons and the spins, and $\alpha \simeq 1$. In eq. (1) the proportionality to T originates from the SG order parameter $q(T) = S^2(1-\alpha T/T_f)$. For the inelastic scattering Fischer obtained $\Delta\rho_i = A_1T^2 - A_2T^{5/2}$, the magnitude of which is difficult to estimate since A_1 and A_2 depend on the unknown spin diffusion constant.

Eq. (1) is valid for not too large spin concentrations, where the moments can still be considered to scatter independently, and for temperatures where the Kondo effect is negligible. The inset to fig. 1 illustrates the absence of any logarithmic T-dependence so that we can rule out dominant contributions of the Kondo type.

In order to discuss the magnitude of the experimental slope, $\Delta\rho/T = -0.0464\ \mu\Omega cmK^{-1}$ from fig. 1, we use the following parameters: $S \simeq 1$, as reported for Fe spins in various Ni-rich amorphous alloys (e.g. (7)); $v_0 = 1.054 \times 10^{-29}\ m^3$, c = 2.4/100 and $T_f = 6.4$ K for $Ni_{77.6}Fe_{2.4}Si_8B_{12}$. For the Fermi energy we assume $E_f \simeq 8.4$ eV, as calculated for liquid Ni (8). Inserting these values and the measured slope into eq. (1) yields $(m^*/m)^{1/2}J \simeq 1.25$ eV, where m is the

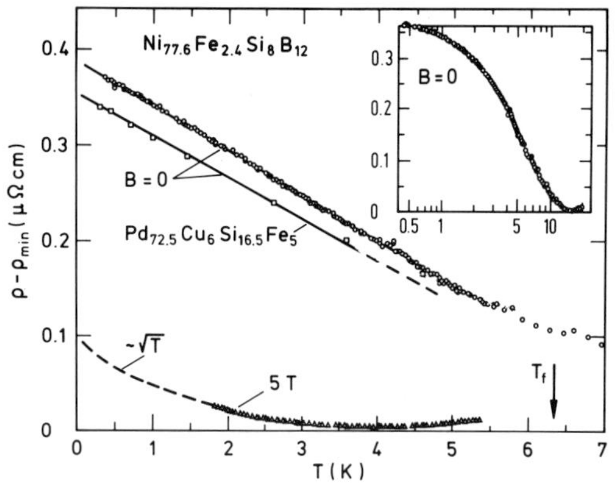

FIGURE 1

$\rho(T)$-ρ_{min} vs T for the spin glass $Ni_{77.6}Fe_{2.4}Si_8B_{12}$ at B = 0 (circles) and 5 T (triangles),($\rho_{min} \simeq 105$ $\mu\Omega$cm, from Ref. (2)) and for the spin glass $Pd_{72.5}Cu_6Si_{16.5}Fe_5$ (squares), ($\rho_{min} \simeq 113$ $\mu\Omega$cm, from Ref. (11)). Straight lines represent linear fits and dashed line -$\sqrt{T}$ variation. Inset: $\rho(T)$-ρ_{min} vs logT for $Ni_{77.6}Fe_{2.4}Si_8B_{12}$

free electron mass. Our previous evaluation (2) of the $\sqrt{T}$ - behaviour of $\rho(T)$ in terms of the interaction theory (4) revealed a high density of states of conduction electrons $N(E_f) \simeq 4/(\text{atom eV})$ for all Fe concentrations, which implies $m^*/m \approx 10$. Hence, we obtain $J \approx 0.4$ eV, which is in reasonable agreement with $J \simeq (0.6 \pm 0.4)$ eV for Fe impurities in various host metals (see e.g. (9)). Thus eq. (1) also accounts quantitatively for the magnitude of the linear T-law in the SG phase of the x =2.4 alloy.3.

3. DISCUSSION AND CONCLUSIONS

The different low-temperature variations of $\rho(T)$ in the RSG-and SG-phases of $Ni_{80-x}Fe_xSi_8B_{12}$ reflect the different magnetic structures on a length scale of the electron mean free path $l_e \approx 8$ Å (2).

In the RSG state the weak -$\sqrt{T}$ variation of $\rho(T\rightarrow0)$ is a result of the structural disorder within l_e, giving rise to a modified electron-electron interaction. The -$\sqrt{T}$ behaviour is unaffected by the FM-RSG transition and is also insensitive to applied magnetic fields (2). This indicates the presence of considerable FM order still prevailing in the RSG phase on length scales larger than l_e. Further support for this comes from low-field longitudinal magnetoresistance measurements in the FM-and RSG-phases, showing a positive contribution, characteristic for ferromagnetic alloys, which is due to spin-orbit scattering (5).

In contrast, in $Ni_{77.6}Fe_{2.4}Si_8B_{12}$ at the PM-SG transition neither long- nor short-range order occurs. In the SG-phase the scattering by the spin system, which is disordered within l_e, determines the T-dependence of $\rho(T)$, as predicted by Fischer (6). Obviously, for the x = 2.4 alloy the elastic scattering from diffusive spin excitations, eq. (1), dominates the inelastic contribution, which might be an indication of a rather high spin diffusion constant (6).

We note that in the SG-phase of $Ni_{77.6}Fe_{2.4}Si_8B_{12}$ the negative low-field magnetoresistance can also be well explained by a spin-flip scattering mechanism (5). Any contributions from weak localisation are completely masked. This conflicts with our earlier conjecture about weak localisation of electrons in the frozen spin system being the origin of the linear T-law. Also recent calculations (10) show that below T_f, at least in an isotropic SG, dephasing magnetic scattering is reduced only by a factor S/(S+1).

There is evidence that eq. (1) can also explain $\rho(T\rightarrow0)$ in other transition-metal-based spin glasses. In fig. 1 we have included $\rho(T)$ data for an amorphous $Pd_{72.5}Cu_6Si_{16.5}Fe_5$ SG with T_f = 15 K taken from fig. 2 in Ref. (11). Clearly, a linear T-law holds below 4 K with a slope $\Delta\rho/T = -0.044$ $\mu\Omega$cmK^{-1} which is 6% smaller than that of $Ni_{77.6}Fe_{2.4}Si_8B_{12}$. Since the ratio between the slopes corresponds approximately to the ratio between the c/T_f-values of 0.37 and 0.33 for $Ni_{77.6}Fe_{2.4}Si_8B_{12}$ and $Pd_{72.5}Cu_6Si_{16.5}Fe_5$, respectively, eq. (1) implies that the prefactor is nearly the same for both materials

In conclusion, we have shown that the linear rise of $\rho(T\rightarrow0)$ in the SG-phase of amorphous $Ni_{77.6}Fe_{2.4}Si_8B_{12}$ can be quantitatively related to elastic scattering from the spin glass order parameter. We have given evidence that the same mechanism may also account for $\rho(T\rightarrow0)$ in other non-reentrant spin glasses.

ACKNOWLEDGMENT

We thank K.H. Fischer, Jülich, FRG, for drawing our attention to Reference (6).

REFERENCES

(1) G. Thummes,R. Ranganathan. and J.Kötzler, Z. Phys. Chem., New Series **157** (1988) 699

(2) G. Thummes, J. Kötzler, R. Ranganathan, and R. Krishnan, Z. Phys. B **69** (1988) 489

(3) R. Ranganathan, J.L. Tholence, R. Krishnan, and M. Dancygier, J. Phys. C **18** (1985) L1057

(4) P.A. Lee and T.V. Ramakrishnan, Rev. Mod. Phys. **57** (1985) 278

(5) R. Carloff, diploma thesis, Hamburg (1989), and R. Carloff et al., to appear

(6) K.H. Fischer, Z. Phys. B **34** (1979) 45

(7) M. Goto, T. Tokunaga, H. Tange, and T. Hamatake, Japan. J. Appl. Phys. **19** (1980) 51

(8) O. Dreirach, R. Evans, H.-J. Güntherodt, and H.-U. Künzi, J.Phys. F **2** (1972) 709

(9) A.K. Nigam and A.K. Majumdar, Phys. Rev. B **27** (1983) 495

(10) W. Wei, G. Bergmann, and R.-P. Peters, Phys. Rev. B **38** (1988) 11751

(11) J. Kästner and D.M. Herlach, J. Magn. Magn. Mater. 51 (1985) 305

Physica B 165&166 (1990) 235–236
North-Holland

INELASTIC-LIGHT-SCATTERING IN $Cd_{0.95}Mn_{0.05}Se$:In NEAR THE METAL-TO-INSULATOR TRANSITION

E.D. ISAACS
AT&T Bell Laboratories, Murray Hill, NJ 07974 USA

T. DIETL and M. SAWICKI
Institute of Physics, Polish Academy of Sciences, Warsaw, Poland

D. HEIMAN and M. DAHL
Francis Bitter National Magnet Laboratory, MIT, Cambridge, MA 02139 USA

M.J. GRAF
Department of Physics, Boston College, Chestnut Hill, MA 02167 USA

S.I. GUBAREV and D.L. ALOV
Institute of Solid State Physics, Academy of Sciences of the USSR, Chernogolovka, 142432 USSR

Inelastic-light-scattering measurements on $Cd_{1-x}Mn_xSe$:In, x=0.05, near the metal-to-insulator transition are made in a dilution refrigerator down to T~0.25 K. The electron spin splitting increases dramatically for decreasing temperature, and at low fields can be expressed as an effective g-factor increasing to a value of g~500. Spectral linewidths in back scattering (large q-vector) are broader than for forward scattering, demonstrating the diffusive character of the delocalized electrons. Comparison of these linewidths give values for the spin-relaxation time T_2 and spin-diffusion coefficients D_s. We discuss the temperature and field dependencies of these quantities on the metallic side of the metal-to-insulator transition.

Raman scattering measurements of the spin dynamics of electrons in n-$Cd_{0.95}Mn_{0.05}Se$ are probed near the metal-to-insulator transition in magnetic fields up to 8 T and temperatures down to 0.25 K (1). The observed spin-flip scattering line is dominated by itinerant (diffusive) electrons. The spin-split energies (Stokes shifts), being determined by the s-d exchange interaction, increase strongly with lowering temperature, and at 0.25 K attain values which correspond to a large effective Landé g-factor.(2) Variations of the spin-split energy with the magnetic field follow a modified Brillouin function suitable for this diluted magnetic semiconductor in a paramagnetic phase. Additional magnetization observed here, which depends only weakly on the external magnetic field, could be induced by bound magnetic polarons and, therefore, its appearence points indirectly to the presence of local s-spins on the metallic side of the metal-to-insulator transition. (3)

Studies of the spin-flip linewidth (HWHM) $\Gamma = T_2^{-1}+D_sq^2$ in forward- and back-scattering geometry provide information on spin-relaxation time T_2 and spin-diffusion coefficient D_s of itinerant electrons. (4) Figure 1 shows examples of the Stokes shift spectra, which we have attributed to the electron spin-flip transitions. Since the excitation power had to be reduced when the temperature was lowered, the intensity of scattered radiation diminished accordingly. It is seen that the Stokes shift maximum at a given magnetic field increases with decreasing temperature, and at a given temperature increases with the magnetic field. The Stokes shift of ~3 meV at 0.25 K and 0.1 T points to an effective Landé factor $g \approx 500$. This directly confirms a dramatic influence of the s-d exchange interaction upon the spin-splitting of electronic states. Comparing results displayed in Fig. 1 with those for bound electrons demonstrates that the scattering arises from itinerant electrons. First, the linewidth is larger in backscattering geometry, indicating a diffusive character of the spin states involved. Second, the lineshape is Lorenzian, in contrast to the Gaussian lineshape observed for bound electrons.

We expect that T_2 is primarily determined by spin-flip scattering of electrons by magnetic ions and motionally narrowed inhomogeneous broadening caused by compositional fluctuations of magnetization. The electron-electron interactions are most apparent in the temperature dependence of the electron diffusion process, where D_s decreases below T on the order of 1 K. It can be shown that, in agreement with theoretical expectations, both T_2

Work supported by NSF DMR-8813164 and DMR-8807419.

and D_s are strongly affected by electron-electron interaction and localization corrections.

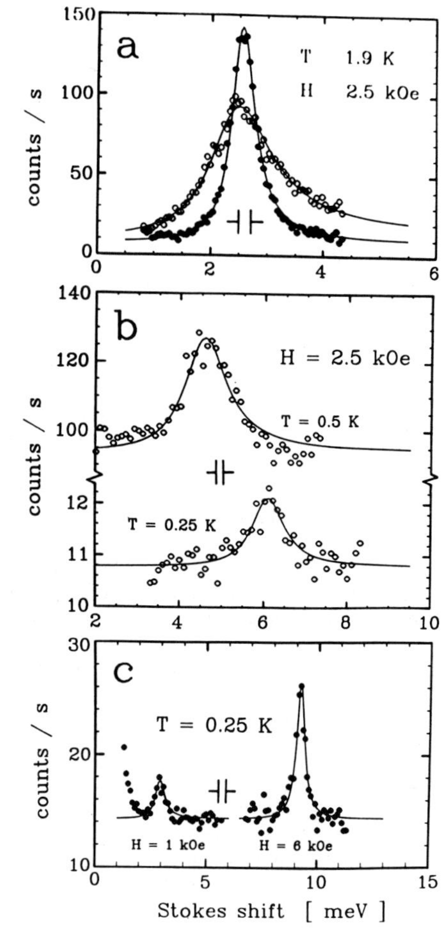

Fig. 1. Spectra from spin-flip Raman scattering in $Cd_{1-x}Mn_xSe$:In, x=0.05, doped to just above the metal-to-insulator transition with n = 7×10^{17} cm^{-3}. The narrow feature in (a) is from forward scattering, while the broader feature in back scattering shows the addition of diffusion. The temperature dependence is shown in (b), and the magnetic field dependence is shown in (c).

REFERENCES

(1) D. Heiman, X.-L. Zheng, S. Sprunt, B.B. Goldberg, and E.D. Isaacs, SPIE **1055**, 96 (1989).

(2) E.D. Isaacs, D. Heiman, R. Kershaw, D. Ridgley, K. Dwight, A. Wold, and J.K. Furdyna, Phys. Rev. B**37**, 7018 (1988).

(3) M. Sawicki, T. Dietl, J. Kossut, J. Igalson, T. Wojtowicz, and W. Plesiewicz, Phys. Rev. Lett. **56**, 508 and 2419 (1986).

(4) S. Geschwind and R. Romestain, in *Light Scattering in Solids IV*, ed. by M. Cardona and G. Guntherdot (Springer-Verlag, New York), p.164.

Physica B 165&166 (1990) 237-238
North-Holland

HIGH FIELD MAGNETIZATION OF ONE-DIMENSIONAL ISING ANTIFERROMAGNETS $CsCoCl_3$, $CsCoBr_3$ AND $RbCoCl_3$

Hidenobu HORI, Hideto MIKAMI, Muneyuki DATE and Kiichi AMAYA*

Faculty of Science and Faculty of Engineering Science*, Osaka University, Toyonaka, Osaka 560, Japan

The magnetization processes of quasi one-dimensional Ising antiferromagnets $CsCoCl_3$, $CsCoBr_3$ and $RbCoCl_3$ are studied in non destructive pulsed high magnetic fields up to 62T. The characteristic metamagnetic transitions are commonly observed at 4.2 K well below the Neel temperatures. The magnetization processes are analysed by the molecular field approximation taking the inter-chain interaction into account and also by Yang and Yang's rigorous theory of a purely one dimensional antiferromagnet with an anisotropic intra-chain exchange interaction.

$CsCoCl_3$ is well known to be a quasi one-dimensional Ising antiferromagnet along the crystallographic c-axis (1) and also to show the successive phase transitions from a disordered to a partially disordered state (2) at 21 K followed by the lowest ferromagnetic state below 9 K.

One of our interests on the high field magnetization measurement is to observe both the metamagnetic transition expected at the critical field of $H_c = 2J_0/g_{/\!/}\mu_B$ and the saturation magnetization $M_s = g_{/\!/}\mu_B S$, where the J_0 and $g_{/\!/}$ are the Ising antiferromagnetic intra-chain exchange interaction and the g value of the lowest doublet (S=1/2) along the c-axis. Observation of the saturation magnetization and the critical field enables us to determine the parameters J_0 and $g_{/\!/}$ most directly.

The experimental results (3)-(7), however, do not show an ideal stepwise transition as expected and instead, the data show a characteristic broad transition width bounded by a lower- and an upper-critical field of H_{c1} and H_{c2}, respectively.

We pointed out in the former reports (6), (7) that there are two theoretical approaches to explain the experimental results. One is the rigorous theory (8) of purely one-dimensional system with an anisotropic intra-chain exchange interaction. The other is the molecular field approximation supposing the isotropic antiferromagnetic intra-chain exchange interaction J_0 and taking the antiferromagnetic inter-chain exchange interaction J_1 into account for the triangular configuration of Co^{2+} in the plane perpendicular to the linear chain c-axis. J_1 is very important, even if it is smaller than J_0 by an order of magnitude, at around the critical field where the main exchange field is just canceled by the external field. In fact, it is found that an intermediate state with ferrimagnetic configurations should exist and the lowest state would show the magnetization of $M_s/3$ with the field span of $6J_1/g_{/\!/}\mu_B$, where M_s is the saturation magnetization at the ferromagnetic state.

In this paper, we performed the high field magnetization measurements of $CsCoCl_3$, $CsCoBr_3$ and $RbCoCl_3$, to compare the above two approaches.

The Figure 1 shows the magnetizations measured at 4.2 K in pulsed magnetic fields applied along the c-axis. In all cases, characteristic metamagnetic magnetization processes are observed in the field region bounded by H_{c1} and H_{c2} as shown in the case of $CsCoCl_3$. The linear magnetizations with respect to the magnetic field observed at $H<H_{c1}$ are attributed to be due to the Van-Vleck temperature independent susceptibility χ_{vv}. dM/dH curves are also measured in both increasing and decreasing mode of the pulsed magnetic field and we found a similar anomalous structure (↓) and the hysteresis with the case of $CsCoCl_3$ as shown typically in the case of $CsCoBr_3$ in Fig. 2.

The critical field values H_{c1} and H_{c2}, the field span $\Delta H = H_{c2} - H_{c1}$, and the $g_{/\!/}$ values along c-axis are determined from the observation of the metamagnetic transition and the saturation magnetization M_s, respectively.

According to our molecular field approximation, H_{c1} and H_{c2} are related to the intra- and inter-chain antiferromagnetic exchange interaction J_0 and J_1 by $H_{c1} = 2J_0/g_{/\!/}\mu_B$ and $H_{c2} = H_{c1} + 6J_1/g_{/\!/}\mu_B$.

On the other hand, Yang and Yang's theory gives an anisotropy factor of the intra-chain exchange interaction $J\perp/J_{/\!/} = \gamma$ defined in the following Hamiltonian

0921-4526/90/$03.50

Table 1. Parameters determined by experiments. χ_{vv}'s are given in the unit of $10^{-2}\mu_B$/Tesla$\cdot Co^{2+}$.

	H_{c1}(T)	H_{c2}(T)	ΔH(T)	$g_{/\!/}$	J_0/k(K)	J_1/J_0	$J_{/\!/}/k$(K)	γ	χ_{vv}
$CsCoCl_3$	33.0	44.6	11.6	5.9	68.7	0.12	80.6	0.097	1.04
$RbCoCl_3$	37.8	50.2	12.4	6.0	76.2	0.11	92.7	0.091	1.08
$CsCoBr_3$	40.6	56.6	16.0	4.5	62.2	0.13	78.4	0.106	1.21

$$\mathcal{H} = \sum_i 2J_{/\!/}[S_i^z S_{i+1}^z + \gamma(S_i^x S_{i+1}^x + S_i^y S_{i+1}^y)]$$

In the present analysis, the Van-Vleck temperature independent magnetization at low fields given by $M = \chi_{vv} H$ is extrapolated up to the saturation field H_{c2} and subtracted from the total magnetization and thus obtained parameters are listed in Table 1.

In comparison with molecular field theory, it is noted that J_1/k are about 8 K and J_1/J_0 are about 0.1. This is in contradiction with the reported values of J_1/k = 0.42 K and 1.0 K for $CsCoCl_3$ and $CsCoBr_3$, respectively. However, the observed magnetizations can be interpreted to show an expected double step process, even if it is not so clear possibly due to transient effects.

On the other hand, the magnetization gives systematic deviation (7) against the theoretical prediction of a linear chain magnet. The deviation is expected to come most likely from the finite inter-chain interaction. Systematic difference between J_0 and $J_{/\!/}$ comes from γ. $J_{/\!/}$ and γ are close to the reported value. This fact supports the linear chain model.

Origin of the anomaly in the dM/dH signal is not clear yet. There are no available theoretical estimation of χ_{vv} at present.

Concludingly, the remained discrepancy between theories and experiments may be resolved by taking both the strong 1-dimensionality with an anisotropic intra-chain exchange interaction and relatively weak inter-chain interactions characterized by the triangular configuration into account.

REFERENCES

(1) N. Achiwa, J. Phys. Soc. Jpn. 27 (1969) 561.
(2) M. Mekata, J. Phys. Soc. Jpn. 42 (1977) 76. H. Shiba, Prog. Theor. Phys. 64 (1980) 466.
(3) K. Amaya, S. Takeyama, T. nakagawa, M. Ishizuka, K. Nakao, T. Sakakibara, T. Goto N. Miura, Y. Ajiro and H. Kikuchi, Physica B155 (1989) 396.
(4) S. Takeyama, K. Amaya, T. Nakagawa, M. Ishizuka, K. Nakao, T. Sakakibara, T. Goto, N. Miura, Y. Ajiro and H. Kikuchi, J. Phys. E: Sci. Instrum. 21 (1988) 1025.
(5) H. Hori, K. Amaya, J. Nakahara, I. Shiozaki, M. Ishizuka Y. Ajiro, T. Sakakibara and M. Date, J. de Phys. C8 (1988) 1455.
(6) H. Hori, K. Amaya, H. Mikami and M. Date, to be published.
(7) K. Amaya, H. Hori, I. Shiozaki, M. Date, M. Ishizuka, T. Sakakibara, T. Goto, N. Miura, H. Kikuchi and Y. Ajiro, to be published in J. Phys. Soc. Jpn.
(8) C.N. Yang amd C.P. Yang. Phys. Rev. 151 (1966) 258.

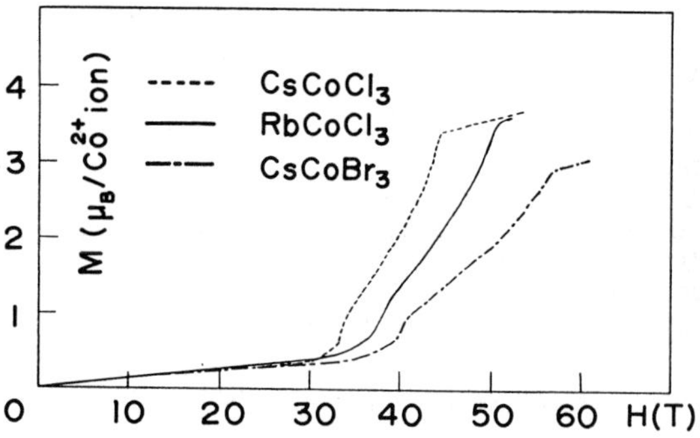

Fig. 1 Magnetization curves of $CsCoCl_3$ and $CsCoBr_3$ at T=4.2K in the pulsed magnetic field H(//c).

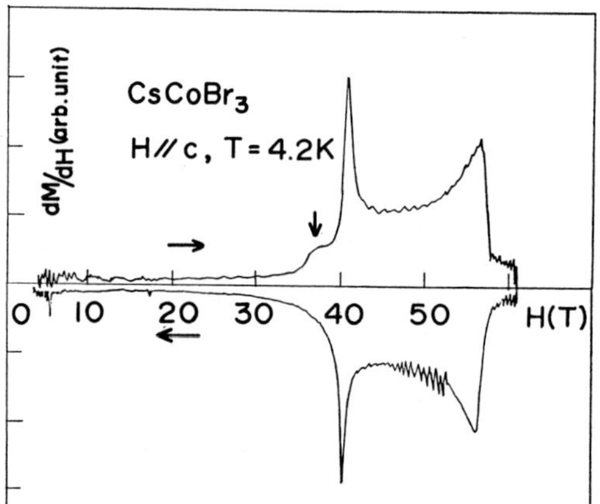

Fig. 2 dM/dH curves of $CsCoBr_3$ at T=4.2 K in increasing (→) and decreasing (←) mode of the pulsed magnetic field H(//c).

Physica B 165&166 (1990) 239–240
North-Holland

THE THEORY OF SPIN FLUCTUATIONS IN MEDIUM AND HIGHLY STONER-ENHANCED PARAMAGNETIC METALS

H. WINTER, Z. SZOTEK* AND W.M. TEMMERMAN*

Kernforschungszentrum Karlsruhe, Inst. f. Nukleare Festkörperphysik, P.O.B. 3640, D-7500 Karlsruhe, FRG; *SERC Daresbury Laboratory, Warrington WA4 4AD, UK

We discuss the successful application of the LDA-RPA approach to the evaluation of the wave vector- and frequency-dependent spin fluctuation spectrum and related quantities of systems representing a wide class of materials.

The spin density fluctuation spectrum, χ^s, can be probed by magnetic neutron scattering provided the Stoner-enhancement, S, of the system in question is high enough (1,2). For materials with lower values of S χ^s is also of interest due to its influence on various quantities. From the theoretical point of view the LDA-RPA approach gives a reasonable description of χ^s for various substances characterized by quite different fluctuation strengths (3). It leads to the following integral equation for the lattice transform, $\chi^s_{\mathbf{q}}$:

$$\chi^s_{\mathbf{q}}(\rho\tau,\rho'\tau';\omega)=\chi^P_{\mathbf{q}}(\rho\tau,\rho'\tau';\omega)+\sum_{\bar\tau}\int d\bar\rho\;\chi^P_{\mathbf{q}}(\rho\tau,\bar\rho\bar\tau;\omega)\,K^s_{xc}(\bar\rho,\bar\tau)\,\chi^s_{\mathbf{q}}(\bar\rho\bar\tau,\rho'\tau';\omega)$$

Here, ρ, ρ', $\bar\rho$ are local coordinates within the WS-cells of the sites τ, τ', $\bar\tau$ in the unit cell, χ^P is the non-interacting susceptibility and $K^s_{xc}(\bar\rho,\bar\tau)$ is the derivative of the LDA exchange correlation potential at site $\bar\tau$ with respect to the magnetization density.
As an example for a medium Stoner-enhanced material we treated Vanadium ($S\approx2.75$). The low frequency part of its fluctuation spectrum for some wave vectors **q** in (1,0,0)-direction is exhibited in Fig. 1. Whereas the features at small $|\mathbf{q}|$ are mainly due to intraband transitions in the neighbourhood of the Fermi surface, interband transitions within the six lowest conduction bands dominate at larger wave vectors.

Coupling the electrons to χ^s via the potential resulting from the LDA we calculated the spin-fluctuation contribution, δm^{sp}, to the elctronic mass. We obtain $\delta m^{sp}/m_0 = 0.11$, a value which is compatible with experiment. To investigate the role of spin fluctuations in depressing the superconducting transition temperature, T_c, of V through an approach superior to the jellium-model treatment of (4) we evaluated the spin fluctuation kernel of the gap equation in addition to its

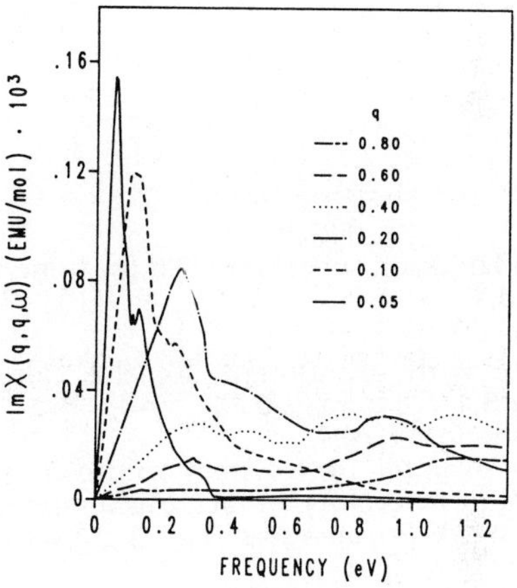

FIGURE 1
The fluctuation spectrum of V for some **q**-vectors in (1,0,0)-direction in the energy range of ≈ 1 eV.

phonon part. The resulting T_c as a function of the frequency-cutoff ε_{max} is shown in table 1..

ε_{max} [eV]	T_c ignoring spin fluctuations [K]	T_c including spin fluctuations
0.19 ≈ 5 ω_D	19.43	8.43
0.95 ≈ 25 ω_D	19.48	9.23
9.1 ≈ 240 ω_D	19.46	9.46
12 ≈ 320 ω_D	19.44	9.48

TABLE 1
Results of T_c-calculations for Vanadium

The remaining difference to the measured T_c may be caused by an underestimation of the spin independent part of the Coulomb interaction which has been described by the McMillan parameter $\mu^* = .13$.

The appropriateness of the LDA--RPA for highly Stoner-enhanced materials is demonstrated by our results for Pd ($S \approx 9$). Fig. 2 shows as an example the fluctuation spectrum for some small **q**-vectors.

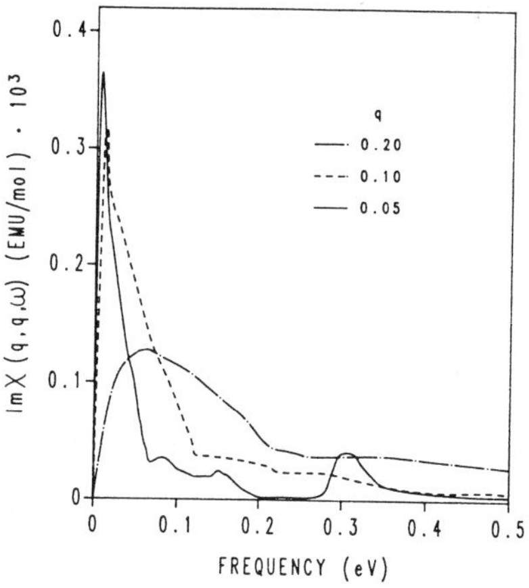

FIGURE 2
The spin fluctuation spectrum of Pd for some small wave vectors.

Compatible with experiment are the calculated data: $S = 8.4$, $\delta m_s/m_o = 0.16$ and the static coherence-length $\xi \approx 4$ Å.

Ni3Ga is an example for a system on the verge of a ferromagnetic instability. In fact, the similar system Ni3Al is a weak ferromagnet. Fig. 3 shows results for Ni3Ga at small **q**-vectors.

For the **q**-dependent Stoner-enhancement we obtain $S(\mathbf{q}) = 31.1$ at $\mathbf{q} = 0.02*2\pi/a$. This number together with the values of $S(\mathbf{q})$ at larger $|\mathbf{q}|$ might well extrapolate to ≈ 100, a value suggested by experiment for the limit $|\mathbf{q}| \to 0$. It should be pointed out that these kinds of calculations are sensitive tests onthe accuracy of the bandstructure methods employed to gain the one-particle quantities. In the case of Ni3Ga we used the LMTO with s,p,d-partial waves at the Ni- and s- and p-components at the Ga-sites. The inclusion of Ga d-basis states would have risen the DOS at the Fermi energy to ≈ 103 states/ryd unit cell, implying that Ni3Ga is predicted to be in the ferromagnetic state. Both the spin-polarized and the non-spin-polarized bandstructure calculations of (5) for Ni3Al suggest that the LAPW is a more reliable method to treat the magnetic properties of such systems. However, as our way of calculating susceptibilities does not depend on the use of a particular bandstructure method, this drawback of the LMTO is no argument against the ability of the LDA-RPA approach to give a realistic description of magnetic fluctuations in a wide class of materials.

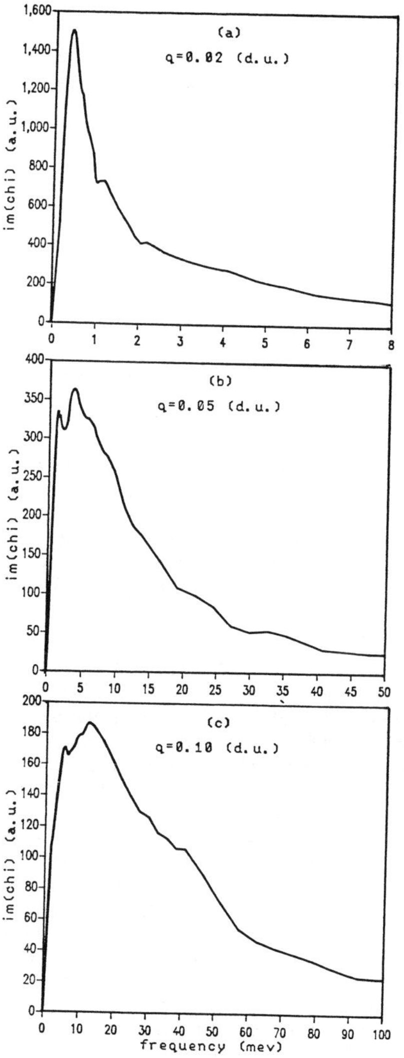

FIGURE 3
The spin fluctuation spectrum of Ni3Ga for small wave vectors. **q** is given in units of $2\pi/a$.

REFERENCES

1. J.W. Cable and E.O. Wollan, Phys. Rev. **140** (1965) 2003
2. N.R. Bernhoeft, S.M. Hayden, G.G. Lonzarich, P. Mck. Paul, and E.J. Lindley, Phys. Rev. Lett. **6** (1989) 657
3. E. Stenzel and H. Winter, Z. Phys. F: Met. Phys. **15** (1985) 1571
4. H. Rietschel and H. Winter, Phys. Rev. Lett. **43** (1979) 1256
5. B.I. Min, A.J. Freeman, and H.J.F. Jansen, Phys. Rev. B **37** (1988) 12.

Physica B 165&166 (1990) 241–242
North-Holland

THE POSSIBILITY OF SURFACE SPINS' QUANTIZATION AXES RESOLUTION BY MEANS OF SCANNING TUNNELING MICROSCOPE WITH MAGNETIC TIP.

A. A. MINAKOV, I. V. SHVETS

General Physics Institute of the Academy of Sciences of the USSR.
USSR.117942, Moscow, Vavilov street 38.

It is shown that scanning tunneling microscope with magnetic tip gives us a principle possibility to determine the direction of spins of surface magnetic ions with an atomic resolution. The tip should not necessarily be a ferromagnet. For instance, the tip can as well be an antiferromagnet with a magnetic structure of spin density wave type such as observed in chromium. The advantage of the antiferromagnetic tip is that it doesn't influence upon the studied surface via the magnetostatic field. The analogous consideration based on exchange interaction is possible to carry out for atomic force microscope with magnetic tip.

1. INTRODUCTION

Up to now only few methods of the investigation of magnetic properties of magnets' surfaces are developed. Practically all of them are based on the measurements of the polarization of electrons emitting from the surface. Polarization of the emitting electrons is usually measured by means of the Mott detector. These methods allow one to determine the average magnetization of a surface layer but not its magnetic structure. The present paper shows that the Scanning Tunneling Microscope (STM) with magnetic tip gives us a principle possibility to determine the directions of the spins of separate magnetic ions of a magnetic surface. In fact, the magnetic tip operates as spin analyzer of electrons tunneling from the studied surface.

2. MAIN RESULTS

One-dimensional electron tunneling from ferromagnet to the antiferromagnetic tip over rectangular barrier is considered similarly to (1) (Fig.1). The ferromagnet's gamiltonian is $H=-(\hbar^2/2m)(d/_{d\xi})^2+eU_1-\vec{\hbar}\vec{\sigma}$, here $\vec{\hbar}\vec{\sigma}$ is the exchange term, σ is Pauli spin operator. The effective field $\vec{\hbar}$ defines the spin quantization axes Z'. Consider the wave $\chi_\uparrow \exp(ik_1\xi)$ with the energy E coming to the barrier. Here $\chi_\uparrow$ is the

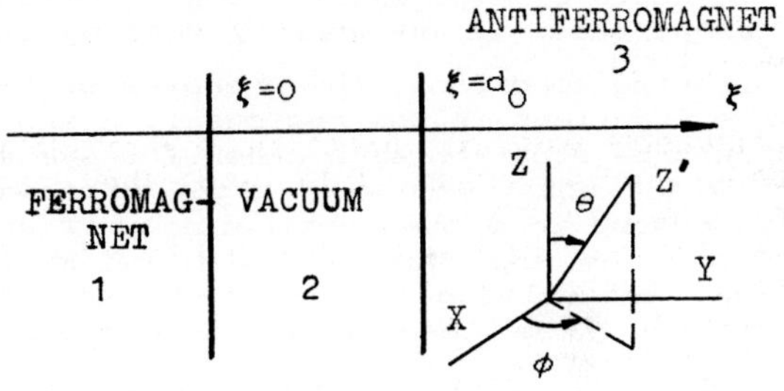

FIGURE 1
Plane ferromagnet and SDW antiferromagnet separated by an insulating rectangular barrier.

coming to the barrier. Here $\chi_\uparrow$ is the
spin amplitude of the state with the spin projection on Z' axes equal to $\hbar/2$. The wave reflected from the boundary $\xi=0$ has the nonzero amplitudes of spin states $\chi_\uparrow$ and $\chi_\downarrow$. The wave with $\chi_\downarrow$ is supposed to be a vanishing one as the exchange field $\vec{\hbar}$ is rather high ($\vec{\hbar}>E-e\cdot U_1$), i.e. the ferromagnet is the one-band one. So, the wave function at $\xi<0$ is $\Psi_{1\uparrow}=\chi_\uparrow\cdot[\exp(ik_1\xi)+R_\uparrow\exp(-ik_1\xi)]$,

$$\Psi_{1\downarrow}=\chi_\downarrow R_\downarrow \exp(-k'_1\xi).$$

The hamiltonian in the region 2 is $H=-(\hbar^2/2m)(d/_{d\xi})^2+eU_0$ and the corresponding wave functions

$$\Psi_{2\uparrow}=A_\uparrow\exp(k_2\xi)\left|{1 \atop 0}\right|+B_\uparrow\exp(-k_2\xi)\left|{1 \atop 0}\right|,$$

$$\Psi_{2\downarrow}=A_{\downarrow}\exp(k_2\xi)\begin{vmatrix}0\\1\end{vmatrix}+B_{\downarrow}\exp(-k_2\xi)\begin{vmatrix}0\\1\end{vmatrix}.$$

Here $\begin{vmatrix}1\\0\end{vmatrix}$ and $\begin{vmatrix}0\\1\end{vmatrix}$ are the spin states amplitudes in the XYZ coordinates. The wave function in the antiferromagnet is supposed to be

$$\Psi_3=C\cdot\exp(i\alpha(\xi-d_0))[\begin{vmatrix}1\\0\end{vmatrix}\sin(\lambda(\xi-d_0)+\delta)+$$
$$+\begin{vmatrix}0\\1\end{vmatrix}\cos(\lambda(\xi-d_0)+\delta)]$$

and its energy is E_{SDW}. This function describes the circular polarized spin density wave (SDW) with the period π/λ and phase δ. One can convince that such a function satisfies the constancy of the particle flow $I\sim\sum_{\sigma}(\Psi^{*}\cdot d\Psi/_{d\xi}-\Psi\cdot d\Psi^{*}/_{d\xi})$, here σ is the spin component. The change of the coordinate system at $\xi=0$ requires the spinor transformation:

$$\chi_{\uparrow}=\begin{vmatrix}\cos(\theta/2)\\ \sin(\theta/2)\cdot\exp(i\phi)\end{vmatrix},$$

$$\chi_{\downarrow}=\begin{vmatrix}\sin(\theta/2)\\ -\cos(\theta/2)\cdot\exp(i\phi)\end{vmatrix}.$$

The values k_1, k'_1, k_2 can be deduced by meeting the following condition: the energies of the wave functions in the regions 1, 2 and 3 should be equal to each other. The system of eight equations with eight unknowns is deduced from the continuity of Ψ and $d\Psi/d\xi$. The amplitude C vs. θ and ϕ is found by solving the system. As was expected from the symmetry considerations, $|C|^2$ doesn't depend upon θ if $\phi=0$. The less the ferromagnet's spin projection on the SDW Polarization Plane (WPP), the less tunneling probability ω: $\omega\sim|C|^2$ (Fig.2). The tunneling current also depends essentially upon the magnitude of ferromagnet's spin projection on the plane OYZ. It can be immediatelly calculated from the dependences $|C(\theta,\phi)|^2$ and $\delta(\theta,\phi)$.

3. DISCUSSION

These calculations illustrate the following idea. Scanning tunneling microscope (2) with magnetic tip enables one to determine the magnetic structure of a surface. Besides, the tip shouldn't necessarily be a ferromagnet. For instance, the tip can be an antiferromagnet with the SDW type magnetic structure. Such a structure is observed in chromium in which the SDW is formed by the Fermi-surface electrons. Let us show, how the magnetization direction of the ferromagnetic surface of oxide can be determined in principle. Unlike the magnetic cation, the tunneling current

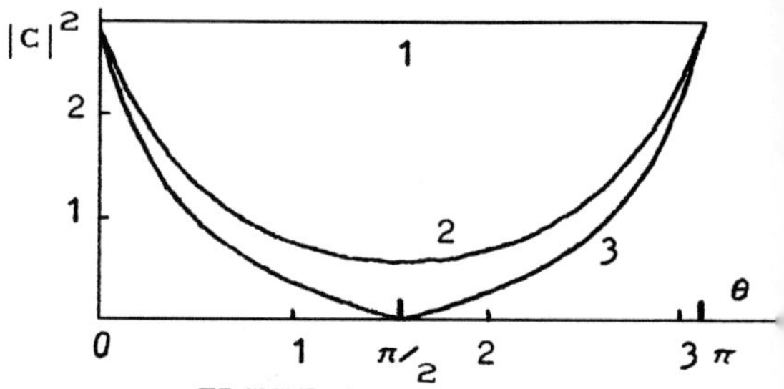

FIGURE 2

Magnitude of $|C|^2$ vs. the angle θ at different values of the angle ϕ: curve 1-$\phi=0$, 2-$\phi=\pi/_4$, 3-$\phi=\pi/_2$, and at the following parameters: $\alpha=2\cdot10^8\,cm^{-1}$, $d_0=5\cdot10^{-8}\,cm$, $\lambda=-2\cdot10^8\,cm^{-1}$, $E_{SDW}=4.2eV$, $U_2=7.1V$, $U_1=4.15V$, $|\vec{h}|=0.15eV$.

from the nonmagnetic anion doesn't depend upon direction of magnetization in the tip (or WPP direction). So, the corrugation measured by STM with the magnetic tip, will depend upon the angle β_{ts} between magnetization (or WPP) direction in the tip and the direction of cations' spins. If the direction of magnetization (or WPP) in the tip is known, one can determine the direction of surface spins by measuring the corrugation vs. β_{ts}. If the surface of oxide is noncollinear antiferromagnet, the corrugations between anion and magnetic ions with different quantization axes measured by such a STM, will differ from each other. Besides, the corrugation will depend upon the angle between the direction of antiferromagnetism vector and the SDW polarization plane. The advantage of the antiferromagnetic tip over the ferromagnetic one is that it doesn't influence upon the studied surface via the magnetostatic field.

These considerations can be also applied to atomic force microscope with magnetic tip in which the exchange interaction of the tip with the surface enables one to determine the surface magnetic structure of a dielectric. As well as in the spin-resolving scanning tunneling microscope, in the atomic exchange-force microscope, the antiferromagnetic tip is supposed to be more preferable than the ferromagnetic one. The reason is that, the magnetostatic interaction force is expelled from STM experiment with antiferromagnetic tip.

REFERENCES

(1) J.C.Slonczewski, Phys.Rev.B 39 (1988) 6995.
(2) G.Binnig, H.Rohrer, Ch.Gerber, E.Weibel, Phys. Rev.Lett. 49 (1982) 57.

Physica B 165&166 (1990) 243–244
North-Holland

LOW TEMPERATURE ANTIFERROMAGNETIC DOMAINS DYNAMICS IN HELICAL ANTIFERROMAGNETS.

A. A. MINAKOV, I. V. SHVETS, V. G. VESELAGO

General Physics Institute of the Academy of Sciences of the USSR.
117942, Moscow, Vavilov street 38.

Magnetostriction of helical antiferromagnets of the system $Zn_xCd_{1-x}Cr_2Se_4$ is measured by means of film strain gauge spattered at the studied sample. It is shown that magnetostriction measurements are a convenient method of antiferromagnetic domains dynamics investigations. The domains reorientation in helical antiferromagnets is shown to occur by means of thermofluctuative creation of reversal centres. It is shown that an interaction between different regions of domain structure exists. The interaction is caused by the spontaneous magnetostriction distortions. The surface density of domain wall energy and crytical size of reversal centre are calculated from the experimental results.

1. INTRODUCTION

Most of physical properties of an antiferromagnet depend on its domain structure. So, antiferromagnetic domain structure (DSA) investigations are of great importance for understanding the properties of antiferromagnets (AFM). Domain structures of antiferromagnets have not been studied well enough by now, due to experimental problems arising in the investigations (1). Antiferromagnetic domains dynamics has not been studied at all.

The crystals of the system $Zn_xCd_{1-x}Cr_2Se_4$ are helical cubic AFM if $x \geq 0.5$. As far as one can gradually change parameters of the AFM by varying the concentration x, this system is a convenient object for the studies aimed at understanding the general features of the antiferromagnetic domains (2).

We have found that magnetostriction measurements provide a convenient method of DSA dynamics studies. As shown in (3), the cubic helical AFM $ZnCr_2Se_4$ (T_N=21K) consists of domains, the helix vector $\vec{q}$ of each domain is parallel to one of [100] type axes. The lattice is compressed along $\vec{q}$ at $\Delta l/l \simeq 5\cdot10^{-4}$. When sufficiently strong magnetic field $\vec{H}$ is applied along [100] direction, the $\vec{q}$ vectors aligne along $\vec{H}$ as far as the susceptibility $\chi_{\parallel}$ along $\vec{q}$ is an order of magnitude higher then the susceptibility $\chi_{\perp}$ in the plane perpendicular to $\vec{q}$. This DSA reorientation is accompanied by the striction along [100] direction.

2. EXPERIMENTAL

As DSA is very sensitive to external stress applied to sample, we have to develop a special technique of magnetostriction measurements based on resistance film strain gauge spattered at the studied sample. The gauge doesn't deform the sample under study.

3. RESULTS AND DISCUSSION

The results of the following experiment are presented in Fig.1. Magnetostriction of the single crystal $Zn_{0.5}Cd_{0.5}Cr_2Se_4$ (T_N=14K) is measured along [100] axis at 12.9K (solid line-tetragonal striction) and at 14.5K (dotted line-isotropic striction). The only magnetostricion observed at $T>T_N$ is isotropic. Magnetic field is applied along [010] (curve OCB) or along [100] (curve OFB) directions. The rate of the field scanning is about 1kOe/s. The vertical parts of the curve correspond to 3s pauses in the field scanning. The

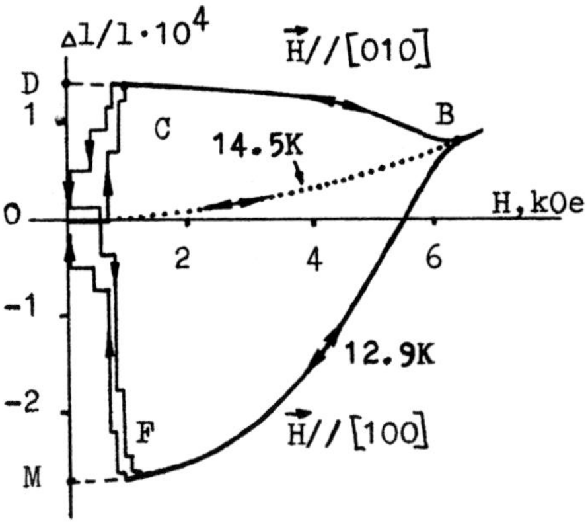

FIGURE 1
Magnetostriction of single crystal $Zn_{0.5}Cd_{0.5}Cr_2Se_4$ along direction [100].

slow DSA reorientations are observed at constant field during the pauses. The reorientation rate drastically depends upon the temperature. Magnetostriction corresponding to DSA reorientation is tetragonal one (MD/OD=3, MD $\stackrel{def}{=}$ 3δ). The paraprocess magnetostriction corresponding to distortions of the helical magnetic structure is observed at this temperature at field greater than 1kOe (curves FB and CB). This is the isotropic volume expansion. Point B corresponds to boundary field at 12.9K on H-T phase diagram of the helical antiferromagnet. So, H-T phase diagram of helical antiferromagnet can be deduced from the magnetostriction measurements.

The analogous results have been obtained for $ZnCr_2Se_4$ and for single crystals with x=0.5-1.

The DSA dynamics accompanied by crystal tetragonal deformation has been studied in the temperature range 4.2K-T_N. The processes of spontaneous DSA relaxation in zero field and forced reorientation in external magnetic field were investigated by the dependences of magnetostriction vs. time. The domains reorientation is shown to occur by means of thermofluctuative creation of reversal centres. Really, as the temperature increases from 4.2K to 8K, the DSA reorientation rate increases by some orders of magnitude. At the same time, the parameters of magnetic structure of the system remain practically constant. In spite of expectations, the DSA dynamics doesn't obey the exponential but logarithm-like law. This means that an interaction between different regions of domain structure exists. DSA dynamics is

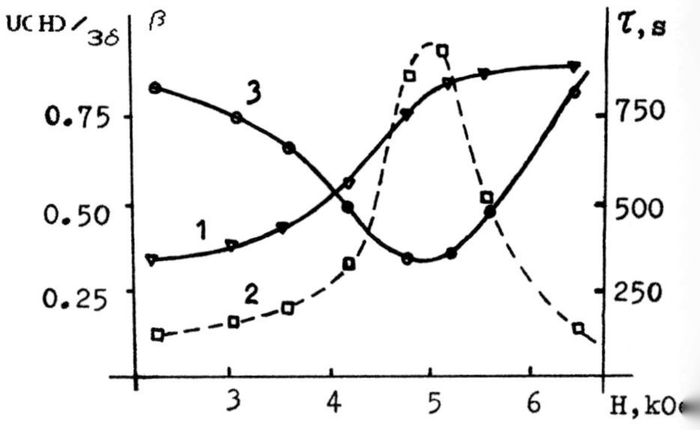

FIGURE 2
Parameters U-1, τ-2, β-3 of stretched exponent approximating forced reorientation of DSA at $ZnCr_2Se_4$ at T=4.2K.

well approximated by "stretched" exponent $\Delta l(t)/l = U(H)\cdot\exp(-(t/\tau)^{\beta})$, $0<\beta<1$. From the dependences of U, τ, β vs. field (Fig.2), temperature and from the magnitudes of spontaneous magnetostriction elastic energy, the following conclusion is made. DSA dynamics in helical AFM can be explained by hierarchically constrained dynamics model (4), which was developed to explain magnetization relaxation processes at spin glasses. The interaction between different regions of DSA and, hence, hierarchy of dynamics are caused by the spontaneous magnetostriction distortions. The density of the domain wall energy ($\sigma \simeq 5\cdot10^{-3}$ erg·cm^{-2}) and critical size of the reversal centre (≃20 Å) are calculated from the experimental results. It is shown that if the magnitudes of spontaneous magnetostriction and domain wall energy of an antiferromagnet are given, one can determine whether the single domain state of a real (with dislocations) antiferromagnet would be preferable or not.

REFERENCES

(1) B.K.Tanner, Comtemp. Phys. 20 (1979) 187.

(2) A.A.Minakov, Proc. of Lebedev Physical Institute 139 (1982) 97.

(3) R.Plumier, J.de Phys. 27 (1966) 213.

(4) R.G.Palmer, D.L.Stein, E.Abrahams and P.W.Anderson, Phys. Rev. Lett. 53 (1984) 958.

Physica B 165&166 (1990) 245–246
North-Holland

COMMENSURABILITY EFFECTS AND ELECTRON STRUCTURES OF HEAVY RARE EARTH METALS

Leonid A. BOYARSKY
Institute of Inorganic Chemistry, 630090, Novosibirsk, USSR

Anomalies of hyperfine field, electric and magnetic properties in heavy REM are considered in the points of commensurability of the periods of magnetic and crystal structures. It is shown that the anomalies are associated with the transformation of the electronic spectrum on the Fermi surface.

The period lengths of magnetic structures of the heavy REM are incommensurate in the general case with the periods of the crystal lattice. With temperature variation, the period lengths change and the lock-in transition into the commensurate phase may become energetically advantageous, the latter phase remaining stable in some temperature interval. There are also isolated temperature points in which capture into the commensurate phase does not take place, at least within the accuracy of experiment. These points were called the cross-points [1,2].

Studying temperature dependence of magnetic structure wave vector q(T) with the help of synchrotron radiation [3,4], led to the conclusion that in most cases of commensurability, the lock-in transition takes place, and the weak ferromagnetism appears which is caused by the spin-slip. The Mössbauer method revealed a vast domain of magnetic hysteresis in Holmium and Erbium metals which contradicts the spin-slip model. More consistent is the model of additional domain structure appearing near the commensurability points because of competition of exchange interaction with crystal anisotropy.

Thermodynamic analysis [6] shows that the anisotropic terms of the Hamiltonian leads to the strong nonlinearity of the order parameter – the mean magnetization S(r). This results in the growth of the higher harmonics of q and formation of the domain structure.

Conduction electrons experience diffraction on the magnetic moment density wave and change their energy spectrum. The change consists in appearance of a system of microgaps on the Bragg diffraction planes in k-space at

$$k = (nG + mq)/2$$

where G is the vector of the crystal reciprocal lattice, m and n are the integers.

Most essential is the topology change of the Fermi surface of heavy REM near the horizontal parts on the 'sleeves' of the quasi-cylindrical form which are responsible for the electron susceptibility peak and the arising of the first harmonic of the magnetic structure.

Near the commensurability point, the microgap merges with the crystal main gap, peculiarities in q(T) behavior and electrical resistance R(T) are mutually proportional, and the increase of the effective area of the parallel parts on the Fermi surface causes growth of the magnetic susceptibility and anomaly of the hyperfine field in the nucleus [7].

The cross-point was observed in Erbium metal at 33K [1,2], it corresponds to a magnetic to lattice period ratio m:n = 4:15. Electroconductivity along the hexagonal axis has an derivative minimum at this point. Wave vector q(T) behaves in the same way. In [7] these peculiarities were shown to be proportional.

In the commensurability point, disappearance of the microgap of the low order near the main gap leads to the increase of the effective area of contribution of the parallel parts of the Fermi surface into the electron susceptibility and to the display of the logarithmic peculiarity of the quasi-one-dimensional electron spectrum. Additional electron susceptibility peak causes the increase of the indirect exchange integral, magnetic susceptibility, and the sample magnetization.

Exchange polarization on conduction electrons by magnetic ions leads to the anomaly of the hyperfine field H_{hf} on the admixture. The abnormal behavior of H_{hf} in the Holmium metal cross-point was observed in the Mössbauer experiments [5].

Thus the reconstruction of the electron spectrum in the commensurability points of the magnetic and crystal structures can manifest itself in many characteristics of the sample determined by the behavior of the conduction electrons on the Fermi surface.

REFERENCES

[1] A.G.Blinov, L.A. Boyarsky, E.M. Savitsky et al., Fiz. Tverd. Tela 25 (1983) 980.
[2] A.G. Blinov, L.A. Boyarsky, N.B. Koltchugina et al., Neodnorodnye electronnye sostojania, Novosibirsk, (1984) 50.
[3] D. Gibbs, J. Bohr, J.D. Axe et al., Phys. Rev. B34 (1986) 8182.
[4] D. Gibbs, D.E. Moncton, K.L.D'Amico et al., Phys. Rev. Lett 55 (1985) 234.
[5] S.K. Godovicov, Fiz. Tverd. Tela 27 (1985) 1291.
[6] Yu. A. Izumov, Uspekhi Fiz. Nauk 144 (1984) 439.

[7] L.A. Boyarsky, Yu. G. Pejsakhovich, Materialy Vtoroy Vsesojuznoy Conferencii po Yaderno-Spectroscop. Issledovaniam STV, Grozny 7-12 Sept.1987. (Moscow State Univ. Edition 1988) 49.

Physica B 165&166 (1990) 247–248
North-Holland

VERY LOW TEMPERATURE THERMOELECTRIC RATIO OF K̲Rb, K̲Na, AND L̲iMg ALLOYS

J. ZHAO,* Z.-Z. YU,** H. SATO,+ W.P. PRATT Jr., P.A. SCHROEDER, and J. BASS

Department of Physics and Astronomy, Michigan State University, East Lansing, MI, 48824, USA

Measurements of the thermoelectric ratio, G, for K̲Rb, K̲Na, and L̲iMg alloys at temperatures below 1K reveal no evidence of anomalies. The characteristic thermoelectric ratio for impurity scattering in L̲iMg, G_o^a(L̲iMg) = - 0.2 ± 0.2 V^{-1}, does not require many-body effects for its explanation.

Recent measurements [1] of the electrical resistivities of K̲Rb, K̲Na, and L̲iMg alloys at temperatures below T = 1K uncovered an unexpected temperature-dependent anomaly. This led us to examine the thermoelectric ratio G of these alloys for anomalous behavior. G is related to the thermopower S by

$$G = S/LT, \qquad (1)$$

Below 1K, the Lorenz ratio L in the alloys is nearly constant at the Sommerfeld value L_o [2], so that G ∝ S/T.

For the alkali metals, with their nearly free electron Fermi surfaces completely enclosed within the first Brillouin zone, G should vary at low temperatures as [3]

$$G = (L_o/L)[G_o + bT^2 + c(1/T)\exp(-\theta*/T)], \quad (2)$$

where G_o arises from scattering of electrons by the impurities in the sample, bT^2 is due to Normal phonon-drag, and $c(1/T)\exp(-\theta*/T)$ is the form for Umklapp phonon-drag in a metal where a minimum phonon-wavevector is needed for Umklapp scattering at very low temperatures. We expect b to be negative and c positive.

Figs. 1, 2, and 3 show G as a function of T for K̲Rb, K̲Na, and L̲iMg alloys, respectively. Typical values for the high purity host metals are included for comparison. The expected T^2 variation below 1K is found for pure K and dilute (≤ 1.3%) K̲Rb alloys, as shown elsewhere [2]. The data are compatible with the form of Eqn. 2: at the lowest temperatures they become constant, and with increasing T the K̲Rb and K̲Na data first become more negative and then turn over in the vicinity of 3K and become more

Present Addresses:* EMCORE Corp., Somerset, NJ, USA;** Xerox Corp., Webster, NY, USA;+ Physics Department, Tokyo Metropolitan University, Tokyo, Japan.

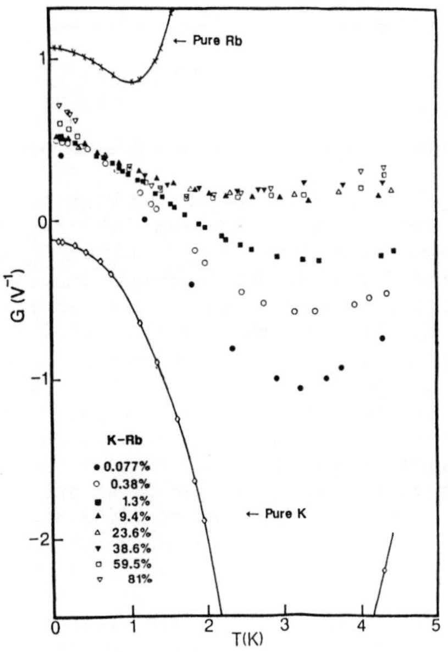

Figure 1. G vs T for K, Rb, and K̲Rb alloys.

positive. The decreases in magnitude of the resulting negative "peaks" with increasing impurity content are due to quenching of phonon-drag--especially the Normal component--by the impurities [3]. The absence of a phonon-drag component in the L̲iMg alloys is due to Li's high Debye temperature.

At low temperatures, the Gorter-Nordheim relation [3] predicts for G_o of an alloy:

$$G_o = (\rho_a G_o^a + \rho_p G_o^p)/\rho_o, \qquad (3)$$

where the sub- and superscripts represent the known alloy impurity "a" and the unknown

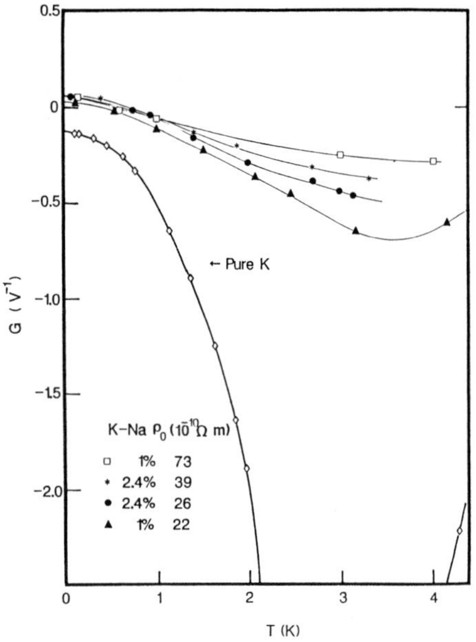

Figure 2. G vs T for K and K̲Na alloys.

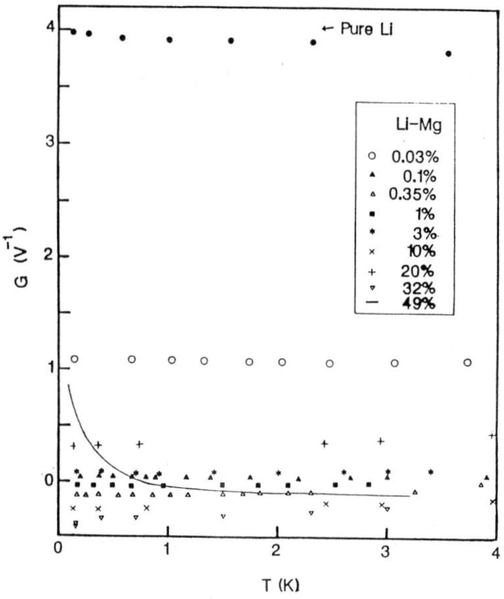

Figure 3. G vs T for Li and L̲iMg alloys.

residual impurities "p" in the high purity host metal, respectively. The residual resistivity is given by $\rho_o = \rho_a + \rho_p$. Since G_o^p will be different for different (unknown) impurities in the high purity host, we are interested only in G_o^a. Eqn. (3) shows that if the alloy impurity content is large enough so that $\rho_a \gg \rho_p$, then $G_o \to G_o^a$; i.e., G_o "saturates" with increasing impurity content, until the impurity content becomes so large that the Fermi surface of the alloy greatly deviates from that of the host.

From Figs. 1-3: G_o in K̲Rb first saturates at $+ 0.50 \pm 0.05$ V^{-1} for up to 38.6% Mg, but then rises toward the G_o for pure Rb at higher concentrations; G_o in K̲Na saturates at $+0.07 \pm 0.05$ V^{-1}, and G_o in L̲iMg saturates at -0.2 ± 0.2 V^{-1} up to 32% Mg, but may rise again for 49% Mg.

These values of G_o^a for K̲Rb and K̲Na are slightly more positive than those estimated by Guenault and MacDonald [4] from thermopower measurements. The value of G_o^a for L̲iMg is the first published.

For a free electron metal with a spherical Fermi surface, we expect [3,5]:

$$G_o = (e/\epsilon_f)[1 + d\ln\ell(\epsilon)/d\ln\epsilon + \Delta\xi]_{\epsilon_f}, \qquad (4)$$

where e is the electronic charge, ϵ_f is the alloy Fermi energy, $\ell(\epsilon)$ is the energy dependent electron mean-free-path, the logarithmic derivative is to be evaluated at the Fermi energy, and many-body effects predicted by Nielsen and Taylor (NT) [5] are separated off into $\Delta\xi$. For simple elastic electron-

impurity scattering, ℓ should be independent of ϵ [3], and without the NT effects we expect:

$$G_o = (e/\epsilon_f). \qquad (5)$$

With the known values of ϵ_f for K and Li, and assuming that ϵ_f does not change upon alloying, Eqn. 5 predicts G_o^a(K̲Rb) $= G_o^a$(K̲Na) ≈ -0.5 V^{-1} and G_o^a(L̲iMg) ≈ -0.2 V^{-1}. For K̲Rb and K̲Na, the more positive values of the experimental data may plausibly be attributed [5] to the NT many-body terms. No such terms are needed for L̲iMg.

In summary, we find no evidence of anomalous temperature dependences in G for K̲Rb, K̲Na and L̲iMg, and no need for many-body effects in G_o^a(L̲iMg).

ACKNOWLEDGMENTS
This work was supported in part by the US NSF under Low Temperature Physics grant 87-00900.

REFERENCES
(1) J. Zhao, Y.J. Qian, Z.-Z. Yu, M.L. Haerle, S. Yin, H. Sato, W.P. Pratt Jr., P.A. Schroeder and J. Bass, Phys.Rev.B 40 (1989) 10,309.
(2) J. Zhao, Ph.D. Thesis, Michigan State University, (1988) (Unpublished).
(3) F.J. Blatt, P.A. Schroeder, C.L. Foiles and D. Greig, Thermoelectric Power of Metals, Plenum Press, NY, 1976.
(4) A.M. Guenault and D.K.C. MacDonald, Proc. Roy. Soc. (London) A264 (1961) 41.
(5) P.E. Nielsen and P.L. Taylor, Phys.Rev.B 10 (1974) 4061.

Physica B 165&166 (1990) 249–250
North-Holland

EXPLANATION FOR THE ANOMALOUS ELECTRICAL RESISTIVITY DATA FOR THIN WIRES OF POTASSIUM BELOW 1 K

David MOVSHOVITZ* and Nathan WISER#

*Department of Physics, Bar-Ilan University, Ramat-Gan, Israel
#Cavendish Laboratory, University of Cambridge, Cambridge, England

An explanation is presented for the negative temperature derivative $d\rho(T)/dT$ recently measured for the electrical resistivity of thin wires of potassium below 1 K. The explanation is based on the contributions to $\rho(T)$ due to normal electron-electron scattering and normal electron-phonon scattering. These two scattering processes are non-resistive for thick wires, but each makes a negative contribution to $\rho(T)$ for thin wires at low temperatures. The calculated values of $d\rho(T)/dT$ are in excellent agreement with the data.

1. RESISTIVITY DATA FOR THIN WIRES

Below about 1 K, the temperature-dependent part of the electrical resistivity $\rho(T)$ of a pure thick wire of a non-transition metal is expected to exhibit a quadratic temperature dependence,

$$\rho(T) = AT^2 \quad (1)$$

At these low temperatures, the electron-phonon scattering term is negligible, leaving only the umklapp electron-electron scattering term given by eq. (1). This is illustrated in Fig. 1, where the empty diamonds correspond to the data (1,2) for a 'thick' wire of potassium, having diameter d = 1.5 mm. The dashed straight line through the diamonds shows that one indeed obtains the predicted straight-line behavior for $d\rho/dT$.

However, when one measures $d\rho/dT$ for thin wires, significant deviations from the thick-wire results are observed (1,2). This is shown by the full symbols in Fig. 1. It is seen that for thinner wires, $d\rho/dT$ becomes progressively smaller in magnitude. Moreover, the temperature dependence of the data also changes; simple straight-line behavior is no longer observed. Most surprising is the fact that even for wires as thick as d = 0.9 mm, clear deviations are seen from the temperature dependence predicted by eq. (1). Our goal here is to account for these data quantitatively - both their magnitude as well as their temperature dependence.

2. NON-RESISTIVE ELECTRON SCATTERING

For a thick wire of a non-transition metal, the contribution to $\rho(T)$ due to normal electron-electron scattering (NEES) vanishes because the total electron momentum is conserved at each NEES collision. For a thick wire of an alkali metal at low temperatures, the contribution to $\rho(T)$ due to normal electron-phonon scattering (NEPS) vanishes because of phonon drag (3). However, for a thin wire, both NEES and NEPS do contribute to $\rho(T)$ by altering the direction of the electron trajectory.

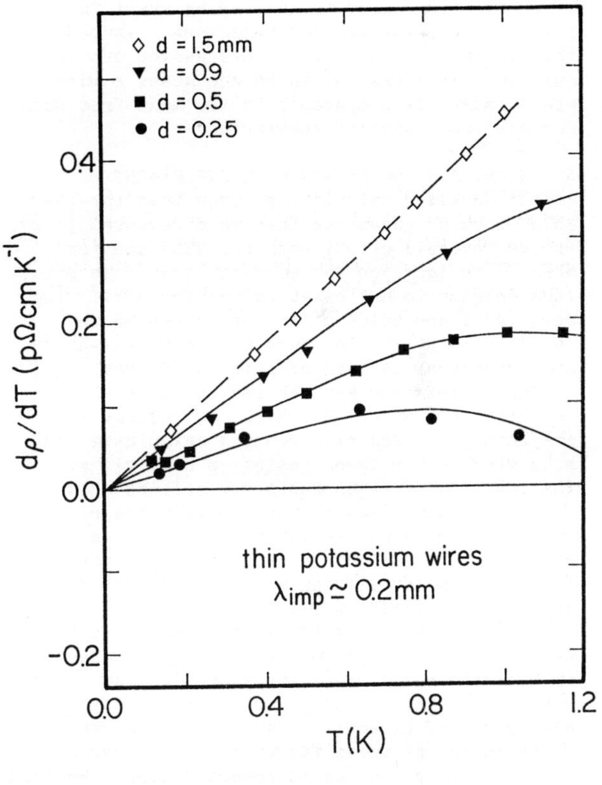

FIGURE 1
Temperature dependence of $d\rho/dT$ for thin wires of potassium. The data points were taken directly from Fig. 1a of Ref. 2. All the wires have the same electron-impurity scattering mean free path λ_{imp}. The curves give the calculated values.

0921-4526/90/$03.50

Figure 2 illustrates the situation. An electron undergoes a non-resistive collision at point P. This collision alters the direction of the trajectory, causing the electron to strike the surface of the wire at the more distant point B, rather than at point A. The electron mean free path is thereby lengthened (PB > PA) and hence $\rho(T)$ is <u>decreased</u>. Alternatively, the point B could be nearer to point P than is point A (PB < PA). In the latter case, the electron mean free path is shortened and $\rho(T)$ is <u>increased</u>. The net effect of non-resistive NEES and NEPS on $\rho(T)$ is, of course, determined by which of these events dominates when one sums over all possible electron trajectories.

The role of impurity scattering is very important, because electron-impurity scattering causes the electron to lose all memory of its previous trajectory. Therefore, the effect of non-resistive electron scattering described above (and illustrated in Fig. 2) does not occur if an electron-impurity scattering event takes place anywhere along the trajectory. Indeed, this is the reason that the discussion applies only to thin wires, by which one means a wire whose radius is comparable to the mean free path for electron-impurity scattering.

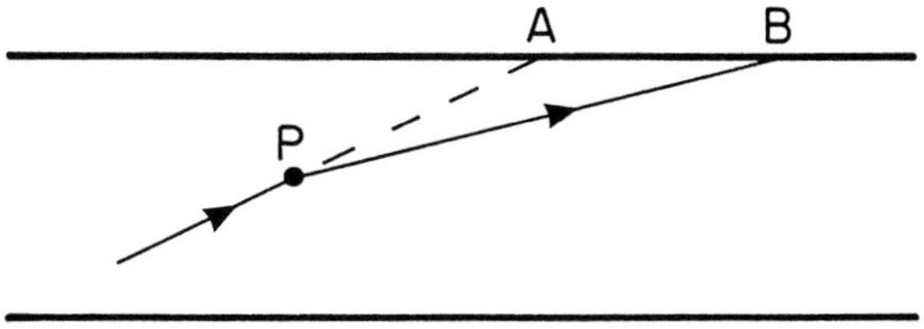

FIGURE 2
The electron undergoes a non-resistive collision at point P, which causes the electron to strike the surface of the thin wire at point B, rather than at point A.

3. THEORY AND COMPARISON WITH EXPERIMENT

Our detailed calculations show that for the thin wi es of potassium that were measured (1,2), the ne overall effect both for NEES and for NEPS i to <u>increase</u> the electron mean free path, thus leading to decreased values for $\rho(T)$. The results of our calculations are given by the curves in Fig. 1. The agreement between theory and experiment is evident from the Figure.

The calculation was carried out using 'electron dynamics', in which one follows the trajectory of each electron and calculates its mean free path between <u>resistive</u> collisions. The function of a <u>non-resistive</u> collision is to change the direction of the electron trajectory, as it continues on its way to the next resistive collision. A suitable average over all possible trajectories for all the electrons on the Fermi surface then yields $\rho(T)$.

An important parameter in the theory is the surface-roughness parameter, usually denoted by p, where p = 1 means specular electron-surface scattering and p = 0 means diffuse electron-surface scattering. It has become clear in recent years (4) that it is inadequate to just take p = 0, as is commonly done. We find that it is important to use Soffer's approach of incorporating the fact that p depends on the angle at which the electron strikes the surface of the wire. The fraction of electrons that are reflected specularly becomes quite appreciable at grazing incidence. In practice, we used the following recommended expression (4)

$$p(\theta) = \exp\left[-(4\pi\alpha\cos\theta)^2\right] \qquad (2)$$

where θ is the angle between the direction of the electron trajectory and the normal to the surface of the wire, and α is the ratio of the electron wavelength to the root-mean-square surface roughness. One expects α to be of order unity; we used α = 2 for each wire.

4. PREVIOUS WORK AND SUMMARY

The idea that non-resistive NEPS contributes to $\rho(T)$ for thin wires of the alkali metals at low temperatures was first proposed by Kaveh and Wiser (5). The idea that non-resistive NEES makes a negative contribution to $\rho(T)$ because of the presence of impurities in the wire was first proposed by De Gennaro and Rettori (6). We subsequently criticized the De Gennaro-Rettori paper on the basis of a Monte-Carlo calculation of NEES (7), but our present work shows that this criticism of Ref. 6 is incorrect.

In summary, we have calculated the contributions to $\rho(T)$ of <u>both</u> NEES and NEPS within a unified theoretical framework. Moreover, we have taken explicit account of the important angular dependence of the surface-roughness parameter $p(\theta)$. Our calculated curves presented in Fig. 1 show that the resistivity data have now been accounted for quantitatively, both their magnitude and their temperature dependence.

REFERENCES

(1) Z.-Z. Yu, M. Hearle, J.W. Zwart, J. Bass, W.P. Pratt, Jr., and P.A. Schroederm Phys. Rev. Lett. 52 (1984) 368.
(2) J. Zhao, W.P. Pratt, Jr., H. Sato, P.A. Schroeder, and J. Bass, Phys. Rev. B 37 (1988) 8738.
(3) M. Kaveh and N. Wiser, Phys. Rev. Lett. 29 (1972) 1374.
(4) See, for example, J.R. Sambles and T.W. Priest, J. Phys. F 12 (1982) 1971.
(5) M. Kaveh and N. Wiser, J. Phys. F 15 (1985) L195.
(6) S. De Gennaro and A. Rettori, J. Phys. F 14 (1984) L237.
(7) D. Movshovitz and N. Wiser, J. Phys. F 17 (1987) 985.

Physica B 165&166 (1990) 251–252
North-Holland

DISLOCATION-ENHANCED ELECTRON-ELECTRON SCATTERING IN THE RESISTIVITY OF COPPER

J SPRENGEL and G THUMMES*

Institut für Angewandte Physik, Universität Hamburg, Jungiusstr.11, D-2000 Hamburg 36, F.R.Germany

From our measurements of the electrical resistivity of gradually strained Cu-whiskers between T=0.4 and 2 K the variation of the T^2-coefficient A of the electron-electron scattering term with dislocation density is determined. Furthermore, we calculate the enhancement of A by dislocation-induced anisotropy using a "two band" model for the electron-dislocation relaxation time. For dislocation densities less than 10^{10} cm^{-2} the measured increase of A upon straining is in quantitative agreement with our model. At higher densities we observe a decrease of A which may be attributed to a cross-over from small- to large-angle scattering.

1. INTRODUCTION

One of the puzzling features of the low-temperature electrical resistivity $\rho(T)$ of the simple metals is the observed sample dependence of the electron-electron scattering contribution, $\rho_{ee} = AT^2$, in particular in K and Cu as recently reviewed (1,2). According to Kaveh and Wiser (KW) (3) an increase in the magnitude of A may be related to an anisotropic relaxation time $\tau_0(\mathbf{k})$ on the Fermi surface arising from dislocations. Such an enhancement of A has been observed in strained samples of Cu and CuAg alloys (4-6) and for Cu also in a high magnetic field (6). The latter effect can be explained by a magnetic-field-induced anisotropy (7). Although the effect of anisotropy on A is now beyond doubt, the maximum possible value of A in the presence of dominant dislocation scattering is still disputed. In order to understand the dislocation effect in Cu, we present here $\rho_{ee}(T)$ measurements on pure single-crystalline copper whiskers and calculations of the enhancement of A for a "two-band" model of anisotropic dislocation scattering.

2. MODEL CALCULATION FOR Cu

In the noble metals, small-angle scattering by the long-range strain fields of dislocations causes a relaxation time $\tau_{dis}(\mathbf{k})$, which is much lower in the neck regions on the Fermi surface than in the nearly spherical belly regions (8). Within a "two-band" relaxation-time model (8) we approximate $\tau_{dis}(\mathbf{k})$ by a step function,

$$\tau_{dis}(\mathbf{k}) = \eta \langle\tau_{dis}\rangle_{tr} \begin{cases} 1 & \text{in belly regions} \\ \tau_n/\tau_b & \text{in neck regions} \end{cases} \quad (1)$$

Here τ_n and τ_b are the relaxation times for dislocation scattering in the neck and belly regions,respectively, $\langle..\rangle_{tr}$ denotes the transport average (7), and $\eta = \tau_b/\langle\tau_{dis}\rangle_{tr}$.

*Present address: Institut für Angewandte Physik, Universität Giessen, D-6300 Giessen, FRG.

For a metal sample containing impurities and dislocations, the inverse of the total relaxation time for scattering from static lattice defects can be written as $1/\tau_0(\mathbf{k}) = (1/\tau_{dis}(\mathbf{k})) + 1/\tau_{imp}$, where τ_{imp} is due to isotropic impurity scattering. The residual resistivity is given approximately by $\rho_0 = \rho_{imp} + \rho_{dis}$. The ratio $\rho_{imp}/\rho_{dis} = \langle\tau_{dis}\rangle_{tr}/\tau_{imp}$ is a measure of the anisotropy (3,7). From this, on the basis of our recent theoretical results (eqs. (4.1) and (4.2) in Ref. (7)), it follows immediately from the form of eq. (1) that for Cu the variation of A with dislocation-induced anisotropy is given exactly by

$$A = A_i \{1 + S/(1 + \Gamma\, \rho_{imp}/\rho_{dis})^2\}\,. \quad (2)$$

Here $\Gamma \equiv (\tau_n/\tau_b)\eta^2$ and $S \equiv (A_a - A_i)/A_i$ is the sample dependence (3,7). The coefficients A_a and A_i are the values of A in the anisotropic and isotropic limits, respectively. From a numerical integration for a spherical Fermi surface we find that η agrees within < 4% with the simple average

$$\eta \simeq \{1 - 4\,(1 - \cos D)(1 - \tau_n/\tau_b)\}^{-1}, \quad (3)$$

where D is the half angular width of the neck regions. Eq. (3) implies that the effect of dislocation scattering on A is nearly independent of crystallographic orientation, in contrast to a magnetic-field-induced anisotropy (7).

For the calculation of Γ we take $\tau_n/\tau_b \simeq 0.24$ (9) and $D \simeq 20^0$ (10) for Cu, which yields $\Gamma \simeq 0.4$. For $\Gamma = 1$ the functional dependence of A in eq. (2) is identical with that in the expression which KW (3) originally derived for the alkali metals and later also used to explain the effect of dislocations on A in Cu.

3. RESULTS AND DISCUSSION

High resolution (1-10 ppm) resistivity measurements have been made on single-crystalline Cu-whiskers in a ^{3}He-^{4}He dilution refrigerator, as described elsewhere (6,11). Here we report results for three samples with diameters $d \simeq 10$-20 μm which, in order to test eq. (2), were gradually strained in situ at $T < 4$ K by means of the Lorentz force resulting from a high dc current (± 100 mA) supplied to the whiskers

0921-4526/90/$03.50

in a transverse magnetic field. The amount of straining is monitored by the resistivity increase. The induced dislocation densities N_{dis} are estimated using $\rho_{dis}/N_{dis} \simeq 2 \times 10^{19}$ Ωcm^3 in Cu (e.g.(9))

As with our previous measurements (6), for all samples at T < 2 K the T-dependent part of the resistivities obeys a T^2 - law, $\Delta\rho_d(T) = A_d T^2$. The coefficients A_d depend on diameter owing to appreciable surface scattering in the thin samples (d << bulk electron mean free path l_e). The evaluation of the coefficients A for bulk Cu is based on surface enhancement factors $G = A_d/A$, where G depends on d/l_e and the surface roughness (11,12). For the unstrained samples, G can be determined independently from measurements of $\Delta\rho_d(T)$ between 2 and 4 K, as we have shown in a previous study on a large number of Cu-whiskers (11). After straining, the change in G is calculated from that in $\rho_d(0)$ (i.e. in d/l_e) using a recent size-effect theory (12). The ratio ρ_{imp}/ρ_{dis} has been obtained as follows: for the undeformed whiskers, with $N_{dis} < 10^6$ cm^{-2}, the bulk residual resistivity $\rho(0) = 0.272$ nΩcm (11) is mainly due to impurities, $\rho(0) \simeq \rho_{imp} \geq 10^2\rho_{dis}$. Then, after straining, ρ_{dis} can be evaluated from the increase in $\rho(0)$.

Fig. 1 clearly shows that for $\rho_{imp}/\rho_{dis} \geq 0.4$ the experimental bulk coefficients A increase upon straining. However, for $\rho_{imp}/\rho_{dis} < 0.3$ (corresponding to $N_{dis} > 10^{10}$ cm^{-2}) A again decreases. We first discuss the value of A for the unstrained whiskers. From all three samples (with d = 10.3, 13.2, 19.1 μm and G = 4.8, 2.6, 1.9, respectively) we obtain an average value of A = (27±1) fΩcmK^{-2}. This number is in excellent agreement with our previous reported value for two whiskers (6) and also with A = (26.8±0.5) fΩcmK^{-2} measured on an annealed bulk sample of Cu (4).

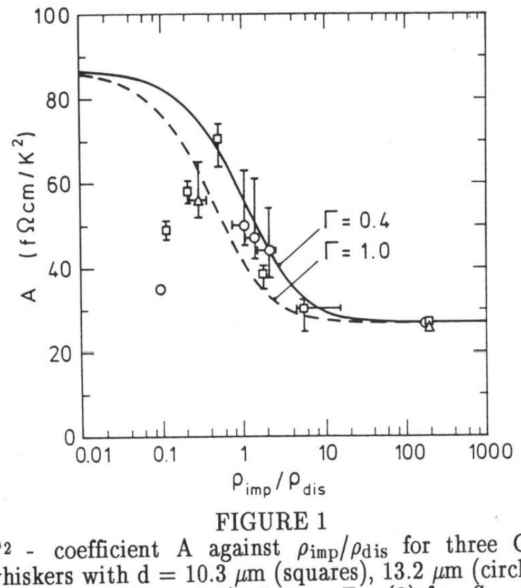

FIGURE 1
T^2 - coefficient A against ρ_{imp}/ρ_{dis} for three Cu-whiskers with d = 10.3 μm (squares), 13.2 μm (circles) and 19.1 μm (triangles). Curves: Eq. (2) for S = 2.2 and A_i = 27 fΩcmK^{-2}.The value $\Gamma = 0.4$ (solid curve) is appropriate to Cu (see text).

Taking A_i = 27 fΩcmK^{-2} for the electron-electron scattering coefficient in the isotropic limit and $\Gamma = 0.4$ for Cu, we obtain from a fit of the data to eq. (2) a sample dependence of S = 2.2 (solid line in fig. 1). S can be estimated independently as follows: as dicussed in (7), S is related to the "umklapp fraction" Δ of electron-electron scattering. Using $\Delta = 0.05$, which follows from our analysis of the magnetic-field effect on A, we obtain nearly the same value S $\simeq$ 2.2 ± 0.2, where the error is due to the uncertainty in the relaxation time approximation (see table 2 in (7)).

Therefore, we conclude that, for not too high N_{dis}, the dislocation-effect on A is in satisfactory agreement with our model calculation for Cu, but that the KW expression (3) (dashed line in fig. 1) provides only a qualitative description of $A(\rho_{imp}/\rho_{dis})$. We note that the data for CuAg alloys (5) follow qualitatively eq. (2), but exhibit in the isotropic limit a large scatter of A that makes a quantitative comparison difficult.

Finally, we suggest a mechanism for the decrease of A at $N_{dis} > 10^{10}$ cm^{-2}. There is strong experimental evidence (e.g. from de Haas-van Alphen amplitudes in Cu crystals (13)) that at high dislocation densities a compensation of the long-range strain-fields occurs. In this case only large-angle scattering from the dislocation cores prevails and the model of anisotropic dislocation scattering, eq. (1), is no longer valid. This might be the reason for the decrease of A at $\rho_{imp}/\rho_{dis} < 0.3$ in fig. 1.

ACKNOWLEDGMENT
Financial support from the Stiftung Volkswagenwerk is gratefully acknowledged.

REFERENCES

(1) M.Kaveh and N.Wiser, Adv.Phys.33 (1984) 257
(2) R.J.M. van Vucht, H. van Kempen, and P.Wyder, Rep.Prog.Phys.48 (1985) 853
(3) M.Kaveh and N.Wiser, J.Phys.F 12 (1982) 935 M.Kaveh and N.Wiser,J.Phys.F 13 (1983) 1207
(4) S.D.Steenwyck,J.A.Rowlands, and P.A.Schroeder, J.Phys.F 11 (1981) 1623
(5) J.Zwart, W.P.Pratt Jr., P.A.Schroeder, and A.D.Caplin, J.Phys.F 13 (1983) 2595
(6) G.Thummes and J.Kötzler, Phys.Rev.B 31 (1985) 2535
(7) J.Sprengel, G.Thummes, and J.Appel, J.Phys.: Condens. Matter 1 (1989) 3621
(8) J.S.Dugdale and Z.S.Basinski, Phys.Rev.157 (1967) 552
(9) A.Bergmann, M.Kaveh, and N.Wiser, J. Phys. F 12 (1982) 2985 and 3009
(10) J.S.Dugdale and L.D.Firth, J.Phys.C 2 (1969) 1272
(11) G.Thummes, V.Kuckhermann, and H.H.Mende J.Phys.F 15 (1985) L65
(12) J.R. Sambles and T.W. Preist, J.Phys.F 12 (1982) 1971
(13) Z.S. Basinski et al., J.Phys.F 13 (1983) L233

Physica B 165&166 (1990) 253–254
North-Holland

Scattering Rates in Cadmium from Radio-Frequency-Size-Effect Measurements below 1K

Alain Jaquier, Luc Patthey, Robert Huguenin and Pierre-Alain Probst *

Institut de Physique Expérimentale de l'Université de Lausannne, CH-1015 Lausanne

High precision measurements of the temperature dependent scattering rates have been made in Cd down to 200 mK. The data show the presence of contributions from electron-phonon scattering superimposed to a term linear in T^2. The coefficient of the latter is of the right order of magnitude to be attributed to electron-electron scattering for electrons on the lens of the Fermi surface, but is a factor 10 larger for some orbits on the first and second bands.

1. INTRODUCTION

Previous measurements (1,2) of the scattering rates ν(T) in Cadmium have been made with the precision of a few percent, typical of standard Radio-Frequency-Size-Effect (RFSE) (3) equipments. They have shown a wealth of different variations with temperature depending on the electron orbit considered. It was possible in particular to attribute the high temperature behaviour of ν(T) for lens electrons ($T \geq 3K$) to intersheet scattering (1) from the third band of the Fermi surface (FS) to the first and second bands and not to ineffective electron-phonon (e-p) scattering when the phonon wave vectors become small (2). A contribution from electron-electron (e-e) scattering could only be demonstrated with some degree of certainty for one orbit in the vicinity of the cap of the lens. A better determination of this term needs measurements at lower temperatures and higher precision. On the other hand very large T^2 contribution were observed on orbits of the first and second bands whose origin is not clear. In a recent calculation (4) it was proposed that these terms are due to e-p scattering together with the particular shape of the FS of Cd.

We present here new measurements extending down to 200 mK to obtain more informations on e-e scattering and its expected isotropy and possibly on the origin of the large T^2 terms which should drop with decreasing temperature if they are indeed due to e-p scattering.

2. EXPERIMENT

The change in the derivative of the real part of the surface impedance is measured as a function of a dc magnetic field using a marginal oscillator (5) and modulation techniques. The resolution, with a digital lock-in amplifier (6) is 10^{-4} under good conditions. Low temperatures are obtained with a dilution refrigerator. The temperature is measured with GE thermometers located on both sides of the sample and is controlled digitally (7) with a precision ≤ 1mK. A computer controls the digital lock-in, the temperature, the magnetic fields and performs averages over a large number of amplitudes of the RFSE-signal. The two samples used are single crystals cut from 6N rods, 537 μm thick with a normal n parallel to $\langle 11\bar{2}0\rangle$ and 240 μm with n parallel to $\langle 10\bar{1}0\rangle$. The orientations are chosen so that the electron orbits are symmetrical.

3. RESULTS AND DISCUSSION

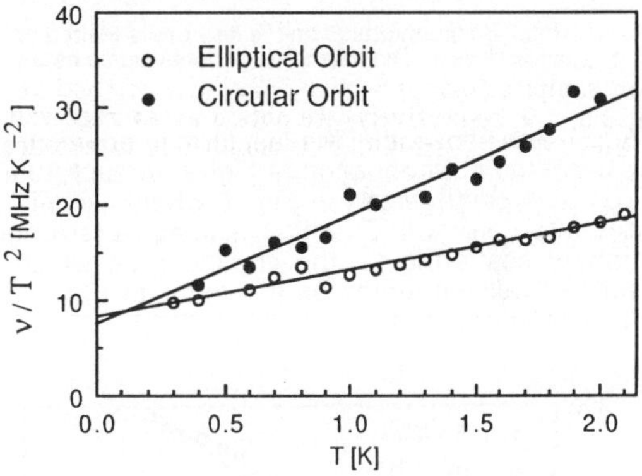

Fig.1 Scattering rates on two typical orbits on the lens (third band) plotted as $\nu(T)/T^2$ vs T.
(o) Elliptical orbit : $\alpha = 8.1 \pm 0.2$ MHz/K^{-2}.
(•) Circular orbit : $\alpha = 7.3 \pm 0.4$ MHz/K^{-2}.

The scattering rate ν(T) averaged over an orbit is obtained from the amplitude A(T) of the RFSE resonance signal from the expression : ν(T) = 1/t {ln A(0) -ln A(T)}, where t is the time for the electrons to cross the sample. In general ν(T) is a complicated function of temperature which can be written as :

$$\nu(T) = \alpha T^2 + \beta T^3 + \gamma f(T, T_t) \qquad [1]$$

The last term is observed for the electrons on the lens of the FS and represents the onset of intersheet scattering with a characteristic temperature, T_t, determined by a threshold phonon wave vector.

* Present Address : H.H. Wills Physics Laboratory, University of Bristol, Tyndall Avenue BRISTOL BS8 1TL

The low temperature data below 2K for two electron orbits on the lens of the third band of the FS are presented in Fig.1, where $\nu(T)/T^2$ is plotted as a function of T. The intercepts for T = 0 K have nearly the same value, α being 7.3 10^{-6} and 8.1 10^{-6} $s^{-1}K^{-2}$ for the circular and elliptical orbits respectively. This is of the right order of magnitude to be attributed to e-e scattering. The common value for the two orbits is also in agreement with the expected isotropic e-e scattering.

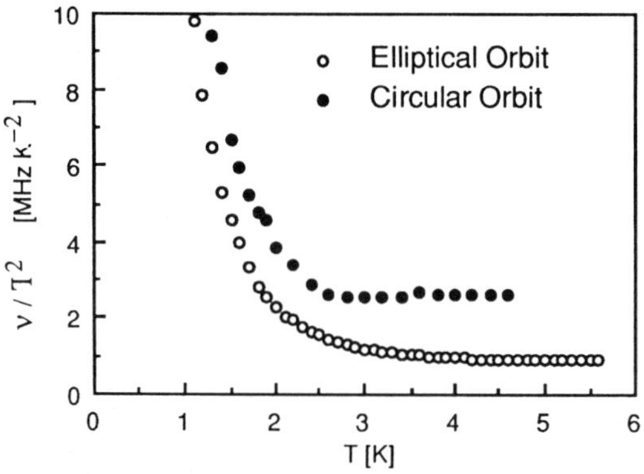

Fig.2 Scattering rates on the lens orbits plotted as $\nu(T)/T^5$ vs T. The asymptotical behaviour is nearly T^5.

Fig. 1 also shows that the last term in equation [1] only becomes of importance at higher temperatures. This is also clearly seen in Fig. 2 where $\nu(T)/T^2$ is represented versus T which shows a stronger intersheet scattering for the circular orbit which is closer to available states on the monster of the FS than the elliptical orbit.

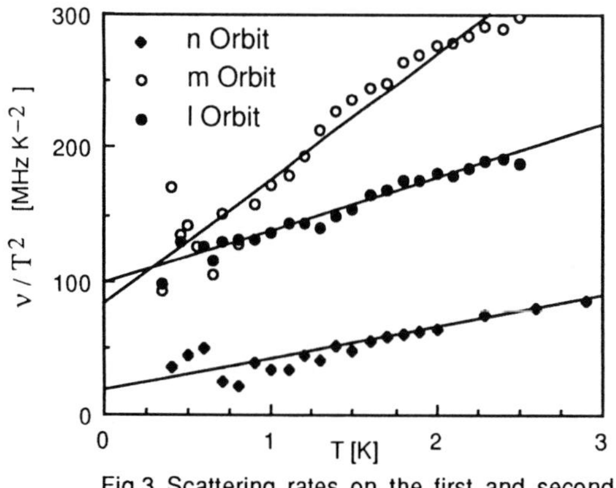

Fig.3 Scattering rates on the first and second bands orbits plotted as $\nu(T)/T^2$ vs T with B // ⟨1000⟩ + 10°.
(◆) n orbit: $\alpha = 17 \pm 5$ MHz/K^{-2}.
(o) m orbit. : $\alpha = 82 \pm 10$ MHz/K^{-2}.
(●) l orbit : $\alpha = 97 \pm 5$ MHz/K^{-2}.

The experimental data for the first (l orbit) and second (m,n orbits) bands of the FS are shown in Fig. 3. Here the resonance signal disappears above about 3 K suggesting a strong e-p scattering. The data can be analysed with only the first two terms in equation [1]. The origin of the contribution αT^2 is however not clear in this case. The coefficient α is much larger than for the lens electrons, about 2 times for the n orbit and 15 times for the l and m orbits. This makes it very unlikely to be due to e-e scattering and confirms the data of reference (1) which were however only significant above about 2 K.

A large T^2 term has been predicted recently (4) from e-p scattering but one would expect that it would drop somewhere below 1 K.

4. CONCLUSION

We have observed that the temperature dependent scattering rates in Cd can be analysed as the superposition of two terms $\nu(T) = \alpha T^2 + \beta T^3$, in the temperature interval 0.2 to about 3 K. The T^3 terms can be attributed to e-p scattering for all electron orbits considered while the origin of αT^2 is not completely clear yet. For the electrons on the lens of the FS the isotropy of the coefficient α and the values obtained for α allows to attribute αT^2 to e-e scattering. α has the order of magnitude of the theoretical calculations by Lawrence (8) and MacDonald (9). For the orbits measured on the first and second bands of the FS (orbit n,m,l) α is however between twice and 13 times larger than for the lens orbit and is very likely due to an other mechanism.

ACKNOWLEDGEMENTS

The financial support of the Swiss National Science Foundation is gratefully acknowledged.

REFERENCES

(1) P A Probst, W M MacInnes and R Huguenin, *J. Low Temp. Phys.* **41**, 115 (1980)
(2) A Myers, R Thompson and Z Ali, *J. Phys. F: Met. Phys.* **4**, 1707 (1974)
(3) D K Wagner and R Bowers, *Adv. in Phys.* **27**, 651 (1978).
(4) W.E. Lawrence, Wei Chen and James c. Swihart, *J. Phys. F: Metal Phys.* **16** L49-L54 (1986)
(5) P A Probst, B Collet and W M MacInnes, *Rev. Sci. Instr.* **47**, 1522 (1977).
(6) P A Probst and B Collet, *Rev. Sci. Instr.* **56**, 466 (1985).
(7) P A Probst and J Rittener, *HPA* **62**, 298-301 (1989).(3)
(8) W E Lawrence, J W Wilkins, *Phys. Rev. B.* **7**, 2317 (1973)
(9) A H Mac Donald and D J W Geldart, *J. Phys. F : Met. Phys.* **10**, 677 (1980)

Physica B 165&166 (1990) 255–256
North-Holland

NONLINEAR DOPPLER-SHIFTED CYCLOTRON RESONANCE IN METALS

I.F.Voloshin, N.A.Podlevskikh, V.G.Skobov, L.M.Fisher, A.S.Chernov

Lenin All-Union Electrical Engeneering Institute, Krasnokazarmennaja st. 12, Moscow, 11250, USSR

Nonlinear reduction of collisionless cyclotron absorption reduced the permittivity of an electron-hole plasma in cadmium so much that a new Doppler wave was observed. The new doppleron was due to a Doppler-chifted cyclotron resonance of "lens" electrons, but its circular polarization was opposite to that of the familiar doppleron. A linear plasma bleaching effect was observed also in the samples of tungsten.

The properties of doppleron (electromagnetic wave capable of propagating in metals in magnetic field) have been studied in detail (1). Nonlinear behaviour of the rf impedance of cadmium plate in a static magnetic field H normal to its plane (H ∦ [0001]) was investigated earlier (2). In this paper we inform about investigation of the surface resistance of a cadmium plate subjected to a static magnetic field H normal to its surfase (H ∦ [0001]) in a wide region of ac field amplitude. The plate with thickness d=1.7 mm was cut from a single crystal characterized by the resistance ratio $0.5\text{x}10^5$. The surface resistance was measured in circular polarized exciting ac field at the temperature 1.4 K. The surface resistance R was determined as a function of H. The results obtained for the negative circular polarization are shown in Fig. 1a, whereas those obtained for the positive polarization are given in Fig. 1b. Oscillations of curve 1 are associated with the excitation of an electron doppleron in the plate and this was due to a Doppler-shifted cyclotron resonance (DSCR) of electrons at a limiting point of the Fermi-surface lens. In accordance with theory this doppleron exists in negative polarization only. Curve 3 shows no oscillations in opposite polarization. There are Gantmakher-Kaner oscillations in both circular polarizations of rf field also but its magnitude is negligible in comparison with doppleron. The amplitude of the hole doppleron oscillatios, due to the DSCR of holes from the Fermi-surface monster observed in the positive polarization, was also small and had another period. Curve 4 was obtained when the amplitude of the ac magnetic field was 40 Oe, and curves 2 and 5 were recorded at amplitude 75 Oe. Curve 2 does not differ signficantly from curve 1. The changes between curves 3 and 5 were more striking. In the case of curve 4 in the range

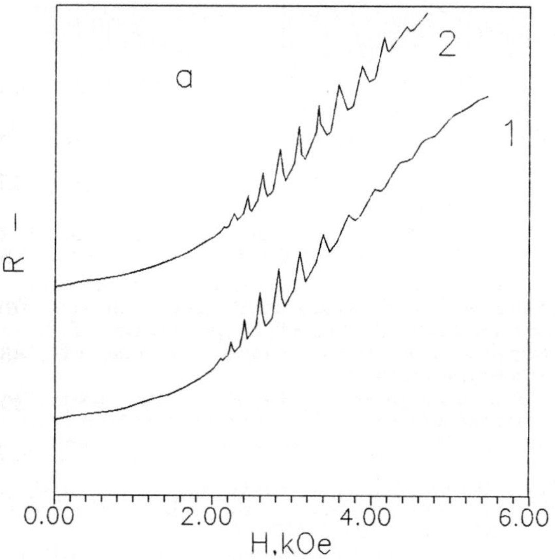

FIGURE 1a

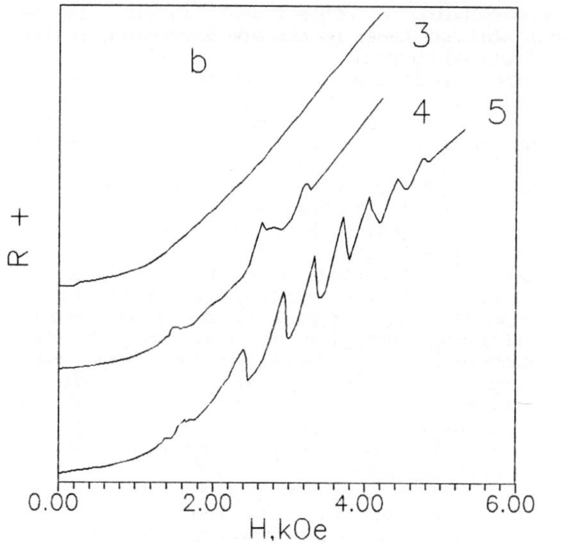

FIGURE 1b

of field where the electron doppleron oscillations were observed in the negative polarization, we see two irregular maxima. Under strongly nonlinear conditions (curve 5) the singularities in the range of existence of the electron doppleron become regular. The oscillations begin above the hole doppleron threshold and, as in the negative polarization case, extend to values of H which increases on increase in the amplitude of the exciting field. The period of these oscillations decreases monotonically whith increase in H and reaches a limiting value very close to the limit of the doppleron oscillation period in the negative polarization (compared whith curve 2).Therefore, the properties of the oscillations of curve 5 indicate that a new doppleron appears in the nonlinear case when the polarization is positive. The proximity of the limiting values of oscillations period of curve 2 and 5 indicates that the new doppleron is due to a DSCR of electrons which have the shift beyond the cyclotron period very close to the shift of the electrons at the limiting point.

The appearance of a new doppleron in positive polarisation with wave length approximately the same as for electron dopleron is very extraordinariry because in this polarisation the dispersion equation has no corresponding solution in linear regime. This wave appears because the majority of electrons on the Fermi surface were trapped by the wave field and ceased to contribute the collisionless absorption process. The only electrons which were not trapped are those in the vicinity of a certain limiting point of the Fermi-surface lens. Consequently, the region of collisionless absorption is limited on the short-wavelegth side, which gives rise to a new doppleron. Such approach makes it possible to calculate the spectrum of a new doppleron. Curves 1 and 2 in fig 2 represent the calculated dependences of these periods on the field H. The points in fig 2 represent the experimental results corresponding to curves 2 and 5 in Fig. 1. We can see that these points are very close to the calculated curves.

When the frequency is reduced, the range of existence of doppleron shift toward weaker fields and becomes narrower. Narrowing of this range reduces the number of oscillations in the experimental curves. However, a shifts toward weaker fields is accompanied by enhancement of the influence of the nonlinearity on the oscillation profile. Fig.3 shows the magnetic field dependences of surface resistance in negative polarization obtained for different amplitudes of the exiting rf field at a frequency 8 kHz. Curves 1, 2, and 3 correspond an rf field intencity 4, 126, and 200 Oe, respectively. This curves manifest even more strongly the nonlinear changes for negative polarization. One can see that the oscillations become sharper and asymmetric. The slopes of the maxima exhibit abrupt jumps indicating resonance "flipping". The same effect takes place in the case of new nonlinear oscillations in plus polarization. In the region of such a jump the suface resistance becomes a multivalued function of H and behaves differently when the magnetic field is reduced or increased. This hysteresis is demonstrated in the insert in Fig.3 which shows, on a scale magnified by a factor of 10 in respect of H, a fragment of a record obtained for one of the extrema of curve 3.

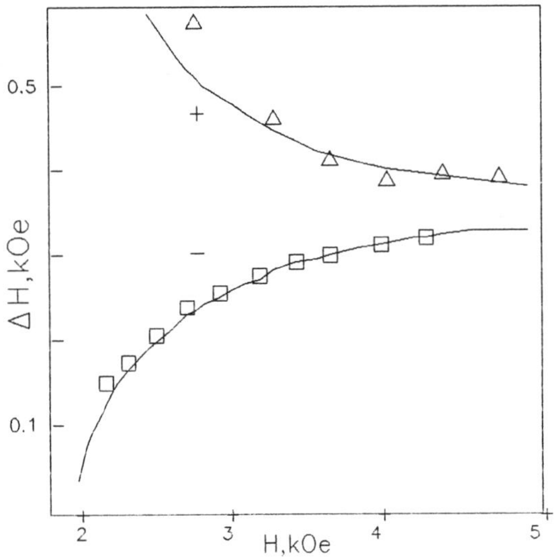

FIGURE 2

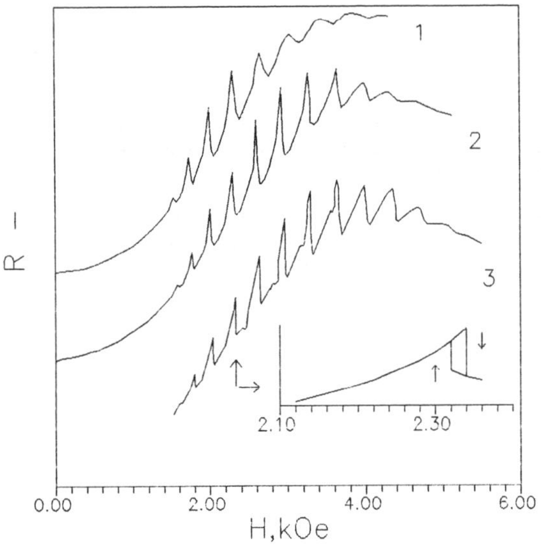

FIGURE 3

In conclusion let us notice that the bleaching effect in the nonlinear regime exists not only for cadmium but also for tungsten.

(1) L.M.Fisher, V.V. Lavrova, V.A. Yudin, O.V.Konstantinov, V.G. Scobov, Sov.Phys. JETP 33 (1971) 410.
(2) I.F. Voloshin, G.A. Vugal'ter, V.Ya.Demikhovskii, L.M.Fisher, V.A.Yudin, Sov.Phys. JETP 46 (1977) 790.

Physica B 165&166 (1990) 257–258
North-Holland

THE GALVANOMAGNETIC PROPERTIES OF COMPENSATED METALS UNDER CONDITIONS OF A STRONG STATIC SKIN EFFECT

A.N.CHEREPANOV,* V.V.MARCHENKOV,* V.E.STARTSEV+

International Laboratory of High Magnetic Fields and Low Temperatures, Wroclaw, Poland
+Institute of Metal Physics, Ural Department, USSR Academy of Sciences, Sverdlovsk, USSR

The field and angular dependences of the transverse magnetoresistance and the Hall coefficient of tungsten and rhenium single crystals with the resistance ratio up to 100000 were measured at T=4.2K in the magnetic fields up to 150 kOe. The role of the static skin effect in the galvanomagnetic properties of the compensated metals is evaluated through the experiments with tungsten and rhenium.

1. INTRODUCTION

The scattering of conduction electrons by the surface of a compensated metal under the condition $r \ll d < l$ (r is the Larmor radius, d is the transverse dimension of the sample, and l is the electron mean free path) may lead to a displacement of a direct electric current into a subsurface layer with thickness of the order of r , i.e. to the static skin effect (SSE) (1,2). The role of electron-surface scattering under SSE was experimentally investigated only for transverse magnetoresistance (MR), mainly for metals with the closed Fermi surface (FS) (3,4). Therefore the study of possible existence of SSE in metals with open FS and the role of electron-surface scattering in other kinetic coefficients, in particular in the Hall coefficient (HC), is of great interest. Our paper is devoted to these problems.

2. SAMPLES

The MR and the HC of tungsten and rhenium single crystals were investigated. 'Pure' and 'dirty' tungsten single crystals with residual resistance ratio RRR=100000 and RRR=1000 respectively and rhenium crystals with RRR=30000 were used. In this case at T=4.2K the electron mean free path l=4mm for 'pure' tungsten, l=0.04mm for 'dirty' tungsten, and l=1mm for rhenium. The samples were the bars and the plates with d=(0.3-3.0)mm in section. In 10-150 kOe magnetic fields the value of r was changed within the range $(0.5\text{-}8.0)\cdot 10^{-3}$ mm, i.e. for 'pure' tungsten and rhenium crystals the SSE conditions were fulfilled: $r \ll d < l$.

3. RESULTS

3.1. To prove the SSE existance in tungsten, i.e. in the metal with the closed FS, the following experiment was carried out. The MR of 'pure' and 'dirty' samples in the form of the bars was measured. Then some part of volume of the both samples cores was extracted by spark-cutting a hole along the sample axis. The exterior surface in this process was not broken. The MR R_{xx} of these hollow samples was then measured again. Figure 1 shows that for 'dirty' tungsten

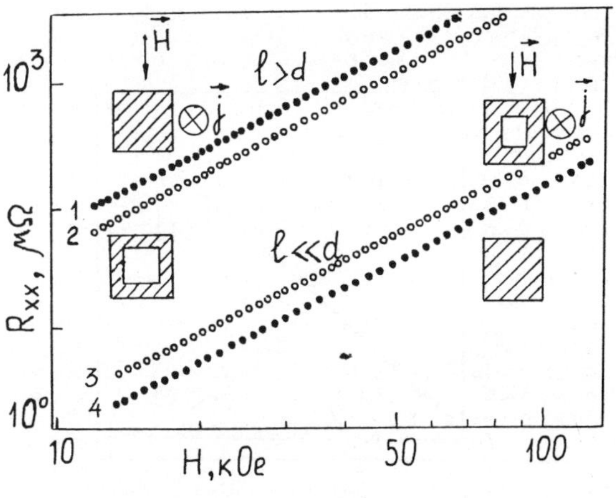

Fig.1. The field dependences of the MR of the 'pure' [1,2] and the 'dirty' [3,4] tungsten crystals.

---

* Permanent addres: Institute of Metal Physics, USSR Acad. Sci. Ural Dept., Sverdlovsk, USSR

the transition from a continuous crystal to a hollow one leads to MR rise as it should be in case of the uniform current distribution in a conductor volume. For 'pure' tungsten crystals (l>d) the transition from a continuous crystal to a hollow one R_{xx} value is seen not to grow, as it would be with the uniform current distribution in a conductor, but reduces due to the appearance of the additional interior surface. The estimations in the field H=150 kOe show that the current density near the sample surface in the layer of thickness of the order of r is 10^3 -10^4 times higher than that in the crystal volume. This proves the existance of SSE under these conditions.

3.2. Figure 2 shows the field dependences of the tungsten HC. Under SSE the HC indicates linear increase with the magnetic field growth. In the absence of SSE, when l<<d, the HC is independent of the magnetic field as it is to be expected according to (5).

3.3. In rhenium, i.e. in the metal with the open FS it can create several types of electron trajectories: A-closed, B - two-dimentional net of open trajectories, C - thick layer of open trajectories, and D - space-net of open trajectories. That's why for the rhenium samples the anisotropy $R_{xx}(\varphi)$ and the field dependences $R_{xx}(H)$ of MR and the field dependences of the HC for the above mentioned types of trajectories were measured.

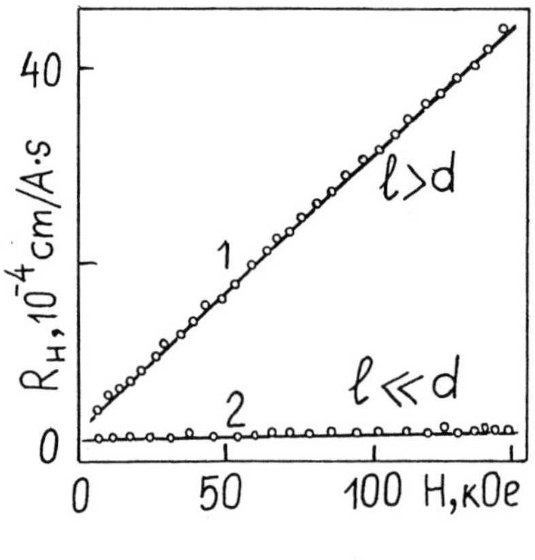

Fig.2. The field dependences of the HC of the 'pure' [1] and the 'dirty' [2] tungsten crystals.

Type A. One of the way of the SSE manifestation in the compensated metals is the sample shape effect (SE), i.e. dependence of the magnitude and the anisotropy of the MR on the shape and the dimensions of the sample. For $R_{xx}(\varphi)$ and $R_{xx}(H)$ the SE is observed, and the SSE takes place. The HC shows linear increase with the magnetic field growth.
Type B. The SE is not observed neither for $R_{xx}(\varphi)$ nor for $R_{xx}(H)$.The SSE is absent. The HC is independent of the field.
Type C. There are two cases.
1. The open direction is perpendicular to the electric current in p-space. The SE is not observed neither for $R_{xx}(\varphi)$ nor $R_{xx}(H)$. The SSE is absent. The HC is independent of the field.
2. The open direction is parallel to the electric current in p-space. Within the error of measurements the SE is not observed neither $R_{xx}(\varphi)$ nor $R_{xx}(H)$.The SSE is absent, that can be explained by the diffuse surface scattering of electrons (see ref.(2)). The HC shows linear increase with the magnetic field growth.
Type D. For $R_{xx}(\varphi)$ and $R_{xx}(H)$ the SE is observed. The SSE takes place.

4. CONCLUSIONS

The investigations of the MR and the HC of tungsten and rhenium single crystals under SSE and the comparison of the results with theory (1,2,6) lead to the following conclusions.

1. The MR of a hollow crystal is less than that of a continuous one with the same exterior sample dimensions.
2. The SSE is observed in the metals with the open FS and its manifestation depends on the electron trajectory type and the character of the conduction electron interaction with the sample surface.
3. In compensated metals the anomalous increase of the HC with the magnetic field growth has been discovered. This effect is caused by a strong surface scattering of the conduction electron.

REFERENCES

(1) M.Ya. Azbel', Zh. Eksp. Teor. Fiz. (1963) 983.
(2) O.V.Kirichenko, V.G.Peschanskii et. al., Zh.Eksp. Teor. Fiz. (1979) 2045.
(3) O.A.Panchenko, P.P.Lutsishin, Zh. Eksp. Teor. Fiz. (1969) 1555.
(4) M.Suzuki and S.Tanuma, J. Phys. Soc. Jap. (1978) 1539.
(5) L.M.Lifshitz, M.Ya.Azbel', and M.I. Kaganov, Electronic Theory of Metals (Nauka, Moskow, 1971).
(6) A.I.Kopeliovich, Zh. Eksp. Teor. Fiz. (1980) 987.

Physica B 165&166 (1990) 259–260
North-Holland

AN APPLICATION OF NEW TRANSPORT EQUATION TO DILUTE ALLOYS

Yi Lin and Yao Kai-Lun

Center of theoretical physics, CCAST (world Lab), Beijing, China.
Department of physics, Huazhong University of Science and Technology, Wuhan, China.

We present a method of calculating the low-temperature resistibity by means of the new transport equation for many particle systems. With only very simple calculations, the conductivity can be easily found. Our results show that the resistivity increases with decreasing temperature and does not exhibit logarithmically divergent behavior at $T=0$. Compared with the traditional theory, the kondo temperature T_k and the density of state are also obtained.

1. INTRODUCTION

Based on the s-d interaction model for dilute magnetic alloys, kondo showed that for $J<0$ the low-temperature resistivity of dilute alloys appears a term proportional to lnT. The logarithmic term explains the phenomenon of the resistance minimum in dilute alloys, bot it diverges logarithmically as the temperature approaches zero. The logarithmically divergent behavior has attracted a lot of authors to investigate how to remove the divergence at $T=0$. There have been further developments in the theory of the Kondo problem. Especially, the new progress is noticeable in the 'exact'solution of the s-d Hamiltonian.

Hamann studied the low-temperature resistivity by using equations of motion for the double-time Green's function and reducing the coupled equations to a single nonlinear integral equation for the conduction-electron t matrix. After a series of calculations, he found that the resistivity has a finite value at $T=0$.

Usually, the transport equantities, such as conductivity, can also be given by transport equation. Kadanoff and Baym and other authors developed a transport equation for many particle systems in the absance of the external field. Recently, Mahan and Hansch investigated the Kadanoff-Baym equation and derived the new transport equation in the sense of linear response to electric fields, which has been solved in several cases for experimental interest.

In this paper, a method of calculating low-temperature resistivity for $J<0$ is presented by using the new transport equation. In treating the spin-flip scattering, we considered the case of electrons scattering near the Fermi energy. In this case, the scattering center has resonance because electrons for energies near the Fermi level are scattered a lot. This makes the resistivity logarithmically temperature-dependent. By simple calculations, we found that the resistivity has a form similar but not identical to that of Hamann's. The resistivity increases with decreasing temperature and does not have logarithmically divergent behavior at $T=0$. The Kondo temperature T_k and the density of state are also obtained.

2. MAHAN-HANSCH EQUATION APPROXIMATION TO DILUTE ALLOYS

We consider a system of electrons interacting with randomly located localized impurity spins via a spin-dependent interaction. We assume that, in the limit of low impurity concentration, interference effects for interactions between two or more impurity electrons can be neglected. The so-called s-d exchange interaction acts between the conduction electrons and localized moment. The Hamiltonian of the system is given by

$$H=\sum_{\vec{k}\sigma}\varepsilon_{\vec{k}}C^{+}_{\vec{k}\sigma}C_{\vec{k}\sigma}+\sum\varepsilon_{L\sigma}C^{+}_{L\sigma}C_{L\sigma}+H_{sd}\ ,$$

$$H_{sd}=-\sum_{\vec{k}\vec{k}'}\left(\frac{J_{\vec{k}\vec{k}'}}{N_0}\right)\left[S_L^{(z)}\left(C^{+}_{\vec{k}\uparrow}C_{\vec{k}'\uparrow}-C^{+}_{\vec{k}\downarrow}C_{\vec{k}'\downarrow}\right)+S_L^{(+)}C^{+}_{\vec{k}\downarrow}C_{\vec{k}'\uparrow}+S_L^{(-)}C^{+}_{\vec{k}\uparrow}C_{\vec{k}'\downarrow}\right], \tag{2.1}$$

where $S_L^{(z)}$ is the z component of the spin operator associated with the impurity, $S_L^{(+)}=C^{+}_{L\uparrow}C_{L\downarrow}$ and $S_L^{(-)}=C^{+}_{L\downarrow}C_{L\uparrow}$ are the raising and lowering operators for the localized spin, and $J_{\vec{k}\vec{k}'}$ is the strength of the exchange interaction which is assumed to be independent of $\vec{k}$ and $\vec{k}'$, N_0 denotes the atomic number of the system.

The rate of spin scattering can be expressed in terms of self-energies $\Sigma^{<}(\vec{Q},\omega)$ and $\Sigma^{>}(\vec{Q},\omega)$.

* Work supported by China National Natural Science Foundation

$\Sigma^{<}(\vec{Q},\omega)$ may be regarded as the scattering rate of a conduction electron which is scattered into wave vector $\vec{Q} = \vec{k} + e\vec{E}_0 T$ and energy ω, assuming that the state is initially unoccupied, and $\Sigma^{>}(\vec{Q},\omega)$ is the collision rate of a conduction electron with momentum-energy configuration $\vec{Q},\omega$. Consider the case that the scattering center has resonance for scattering, so that electrons for energies in the vicinity of the resonance are scattered a lot, while other electrons are scattered a little. To the second approximation, $\Sigma^{<}$ and $\Sigma^{>}$ are given by

$$\Sigma^{\gtrless}(Q,w) = \mp i n_i (J/N) S_L^{(Z)} + n_i (J/N_0)^2 \sum_{\vec{q}} \int \frac{dw'}{w-w'} \times [S_L^{(\mp)} S_L^{(\pm)} + S_L^{(Z)}] G^{\gtrless}(\vec{Q}',w') \qquad (2.2)$$

where Green's function $G^{<}(\vec{Q},\omega)$ can be identified as the density of particles in the system with momentum $\vec{Q}$ and energy ω, $G^{>}(\vec{Q},\omega)$ is the density of states available for the addition of an extra particle with $\vec{Q}$ and ω, and n_i is the density of impurities.

The problem at hand is to solve $G^{<}(\vec{Q},\omega)$, in order to derive the Kondo resistivity from

$$\vec{J}_c(\vec{R},T) = -\frac{ie}{m(2\pi)^4} \int d\omega \int d^3Q \, \vec{Q} G^{<}(\vec{Q},\omega;\vec{R},T), \qquad (2.3)$$

we follow the steps of Mahan and Hansch and write the transport equation for homogeneous steady state systems as

$$\frac{e}{m}\vec{E}_0 \cdot \vec{Q} \left[-\frac{\partial n(\omega)}{\partial \omega}\right] \Gamma(\omega) [A(\vec{Q},\omega)]^2 = \Sigma^{>} G^{<} - \Sigma^{<} G^{>}, \qquad (2.4)$$

where $\Gamma(\omega) = -\mathrm{Im}\Sigma^{r}(\vec{Q},\omega)$. In order to solve the transport equation, we write the Green's function $G^{<}$ as its equilibrium term plus another term linear in the electric field

$$G^{<}(\vec{Q},\omega) = iA(\vec{Q},\omega) \left[n(\omega) - \frac{\partial n(\omega)}{\partial \omega} \frac{e}{m} \vec{E}_0 \cdot \vec{Q} \Lambda(\vec{Q},\omega) \right], \qquad (2.5)$$

where $\Lambda(\vec{Q},\omega)$ is an unknown function, which will be determined selfconsistently. Once we solve the unknown function Λ, we can get the conductivity tensor from it.

In solving the unknown funciton $\Lambda(\vec{Q},\omega)$, inserting the expressions for $G^{<}$ and $G^{>} = G^{<} - iA$ into the transport equation and using the self-energy contributions, we have the following fundamental integral equation for $\Lambda(\vec{Q},\omega)$:

$$\Lambda(\vec{Q},\omega) = \frac{1}{2} A(\vec{Q},\omega) - \frac{n_i (J/N_0)^2 [2nS_L^{(Z)} + S(S+1) + S_L^{(Z)}]}{2(2\pi)^3 V_F \Gamma(\omega)} \times \int dq \, q \left(1 - \frac{q^2}{2k_F^2}\right) \int d\omega' \frac{A(\vec{Q}',\omega')\Lambda(\vec{Q}',\omega')}{\omega - \omega'} \qquad (2.6)$$

The equation of this form is known as an integral equation of the second kind which can be solved by using the iterative method. After doing the integrals, we have

$$\Lambda(\vec{Q},\omega) = \frac{1}{2}A(\vec{Q},\omega) + \frac{n_i (J/N_0)^2 [2nS_L^{(Z)} + S(S+1) + S_L^{(Z)}]}{2(2\pi)^3 V_F \Gamma(\omega)} \, In\frac{\omega-\omega_F}{D} \left(1 - \frac{q^2}{2k_F^2}\right) \frac{1}{2} [A(\vec{Q}',\omega_F)]^2 q dq \Bigg/ \left[1 - \frac{n_i (J/N)^2 [2nS_L^{(Z)} + S(S+1) + S_L^{(Z)}]}{2(2\pi)^3 V_F \Gamma(\omega)} \, In\frac{\omega-\omega_F}{D} \int A(\vec{Q}',\omega) \left(1 - \frac{q^2}{2k_F^2}\right) q dq\right]. \qquad (2.7)$$

The next step is to evaluate the T-dependent conductivity of the sytem:

$$\sigma = \frac{e^2 n_0}{m} \int \frac{d\omega}{2\pi} - \frac{\partial n(\omega)}{\partial \omega} \int d\omega' A(\vec{Q}',\omega) \Lambda(\vec{Q}',\omega). \qquad (2.8)$$

The integral over kinetic energy can be readily carried out by considering the behavior of the spectral functions

$$\int d\omega' A(\vec{Q}',\omega) = 2\pi,$$

$$\int d\omega' A(\vec{Q}',\omega)^2 = \frac{2\pi}{\Gamma(\omega)}.$$

After doning these integrals, we obtain the conductivity in the form

$$\sigma = C + X/\left(1 - Z \ln\left(\frac{KT}{D}\right)\right) \qquad (2.9)$$

where

$$C = \frac{e^2 n_0}{m} \int d\xi \left[-\frac{\partial n(\xi)}{\partial \xi}\right] \frac{1}{\Gamma(\omega)} (1 - M),$$

$$X = \frac{e^2 n_0}{m} \int d\xi \left[-\frac{\partial n(\xi)}{\partial \xi}\right] \frac{1}{\Gamma(\omega)} \frac{M}{1 - N \ln\xi}$$

$$Z = \frac{e^2 n_0}{Xm} \int d\xi \left[-\frac{\partial n(\xi)}{\partial \xi}\right] \frac{1}{\Gamma(\xi)} \frac{MN}{(1 - N \ln\xi)^2}$$

$$M = \frac{1}{2\Gamma(\omega)} \int \left(1 - \frac{q^2}{2k_F^2}\right) [A(\vec{Q}',\omega)]^2 q dq \Big/ \int \left(1 - \frac{q^2}{2k_F^2}\right) A(\vec{Q}',\omega_F) q dq,$$

$$N = \frac{n_i (J/N_0)^2 [2nS_L^{(Z)} + S(S+1) + S_L^{(Z)}]}{2(2\pi)^3 V_F \Gamma(\omega)} \int \left(1 - \frac{q^2}{2k_F^2}\right) A(\vec{Q}',\omega_F) q dq.$$

Thus, the conductivity of the system we evaluated has a term which is inversely prooprtional to $1 - Z\ln(kT/D)$. It is obvious that the Kondo temperature is $T_K = \frac{D}{k}\exp(1/Z)$ which has the form similar to other authors'. Compared with the traditional Kondo theory, one can get the density of state near the Fermi surface

$$\rho(\omega_F) = -Z/4J \qquad (2.10)$$

where $\rho(\omega_F)$ denotes the density of state.

REFERENCES

(1) J.Kondo, Prog. Theor, phys. (Kyoto), 32 (1964) 37.

Physica B 165&166 (1990) 261–262
North-Holland

LOW TEMPERATURE HALL CONDUCTIVITY OF DILUTE ALLOYS

Yao Kai-Lun and Yi Lin

Center of theoretical physics, CCAST (world Lab), Beijing, China.
Department of physics, Huazhong University of Science and Technology, Wuhan, China.*

In this paper, a method of calculating low-temperature Hall conductivity in a weak magnetic field by means of Mahan-Hansch transport equation including the external magnetic field is presented. Hall conductivity and Hall coefficient based on the s-d exchange interaction model are obtained. Compared with other calculations, our method overcomes the defect of the divergence of Hall conductivity at T = ok and kondo temperature T_K is obtained. Our result is in good agreement with experiments.

1. INTRODUCTION

Kondo's effect in dilute alloys has been explained without the external field. However, it is more complicated for dilute megnetic alloys in the magnetic field. In general, assumed that the discrete structure of the Landau levels is not important for the transport quantities, most of authors, obtained Hall conductivity based on the s-d exchange interaction model using a kubo-type formula. Their results in part exhibit the relation between Hall conductivity and the external field, but the anomalous temperature dependence for isotropic scattering in kondo alloys at the low temperature is not well understood.

In this paper, a method of calculating low temperature Hall conductivity tensor for dilute magnetic alloys in a weak magnetic field is presented by means of Mahan-Hansch equation in external magnetic field. Compared with other calculations, Our method overcomes the defect of the divergence of Hall conductivity at T = ok and Kondo temperature T_K is easily obtained. Our results are in good agreement with the experiments

2. MAHAN-HANSCH EQUATION APPROXIMATION

Mahan and Hansch derived the new transport equation including the external magnetic field in the sense of linear response to electric fields, by means of Kadanoff and Baym or equivalent keldysh procedure. We follow the steps of Mahan and Hansch to write the transport equation for steady state system. In order to make the derivation simple, we assumed that the system to deal with has a spherical Fermi surface. Also we make the standard approximation that the retarded self-energy has a negligible dependence upon wave vector $\vec{q}$, but it depends significantly upon energy, thus we neglect terms in derivation with respect to $\vec{q}$, and use the shorthand notation $\Gamma(\omega) = -\mathrm{Im}\Sigma^r$. One find

$$G^>\Sigma^< - \Sigma^> G^< = i\frac{e}{c}\vec{v}_q\cdot(\vec{H}\times\vec{\nabla}_q G^<) - e\vec{E}\cdot\vec{v}_q\Gamma(\omega)\frac{\partial n}{\partial\omega}A^2 \quad (2.1)$$

In Eq. (2.1), G and Σ are Green's function and self-energy, the superscripts r denotes respectivity the retarded Green's function, n is distribution function, $\vec{v}_q$ is given by $\vec{\nabla}_q\varepsilon$, for mor more details we refer to the work by Mahan. We shall solve equation (2.1) in sec. 3 and adopt self-consistnetly with an ansatz

$$G^< = in\,A - i\,\frac{\partial n}{\partial\omega}\,\gamma \quad (2.2)$$

the first term in (2.2) is the equilibrium expression and γ is an unknown function (vertex) which is linear in a electric field. Once we know this function, we can calculate the current density $\vec{J}_c$.

$$\vec{J}_c = e^2\int\frac{d^3q}{(2\pi)^3}\int\frac{d\omega}{2\pi}\left[-\frac{\partial n}{\partial\omega}\right]\vec{v}_q\gamma \quad (2.3)$$

Inserting (2.2) into (2.1), we obtain an eq equation for the function γ

$$G^>\Sigma^< - \Sigma^> G^< = -\frac{\partial n}{\partial\omega}[A^2\Gamma e\vec{E}\cdot\vec{v}_q + \frac{e}{c}\vec{v}_q\cdot(\vec{H}\times\vec{\nabla}_q\gamma)] \quad (2.4)$$

3. HALL CONDUCTIVITY IN A WEAK MAGNETIC FIELD

We consider the so-called s-d exchange interaction in dilute magnetic alloys acting between

* Work supported by China National Natural Science Foundation.

conduction electrons and Localized moment. To the second scattering approximation, self-energies can be written as Ref. (3). At last, we can deduce a simplified Version of Eq. (2.4).

$$\gamma=\frac{1}{2}A^2 e\vec{E}\cdot\vec{v}_q\vec{v}_q(\vec{h}\times\vec{\nabla}_q\gamma)-\frac{n_i[2nS_L^{(Z)}+S(S+1)+S_L^{(Z)}]A}{2\Gamma(\omega)}$$

$$\times\sum_q\int\frac{d\omega'}{\omega-\omega'}(J/N_0)^2\gamma\ , \quad \text{and } \vec{h}=\frac{\vec{H}e}{2\Gamma c}\ . \qquad (3.1)$$

Setting the magnetic field $\vec{H}$ in the Z direction of our coordinate system. In the limit of a weak magnetic field $h \ll 1$, it can be used as a small expansion parameter, and $\vec{v}_q=\vec{Q}/m$ is reasonable.

The "exact" solution of Eq. (3.1) is represented by an expansion of the type

$$\gamma=\sum_{n=0}^{\infty} h^n\,\gamma^{(n)} \qquad (3.2)$$

Inserting (3.2) into (3.1) and collecting contrbutions of equal order, we generate a sequence of equations. In principle, we can solve all of the function $\gamma^{(n)}$, and in fact in weak magnetic fields, we are interested in the terms of the zero order $\gamma^{(1)}$ and the first order $\gamma^{(0)}$. Obviously, $\gamma^{(0)}(\vec{Q},\omega)$ is a vertex function in absence of the external field, and $\gamma^{(1)}(\vec{Q},\omega)$ is a vertex function including the external magnetic field within the first order h. We let

$$\gamma^{(0)}(\vec{Q},\omega)=Ae\vec{E}\cdot\vec{Q}\Lambda(\vec{Q},\omega)/m, \qquad (3.3)$$

$$h\gamma^{(1)}(\vec{Q},\omega)=\frac{A(\vec{Q},\omega)}{m^2}\ e\vec{E}\cdot(\vec{h}\times\vec{Q})\Lambda(\vec{Q},\omega)\ . \qquad (3.4)$$

where $\Lambda(\vec{Q},\omega)$ has been obtained in Ref (3).

As calculated above, we get the vertex function in the first order approximation

$$\gamma(\vec{Q},\omega)=\gamma^{(0)}+h\gamma^{(1)}\ . \qquad (3.5)$$

In (3.5), the first term $\gamma^{(0)}(\vec{Q},\omega)$ stands for the vertex function without the magnetic field, the second term $\gamma^{(1)}(\vec{Q},\omega)$ is the modification of the magnetic field to $\gamma(\vec{Q},\omega)$. In what follows, we calculate Hall conductivity tensor. Substituting (3.5) into (2.4), with some treatment, one has

$$\sigma_{xx}=\frac{1}{3}[C+X/(1-Z\ \ln(\frac{KT}{D}))],$$
$$\sigma_{xy}=\frac{1}{3}[C'+X'/(1-Z\ \ln(\frac{KT}{D}))]. \qquad (3.6)$$

where

$$C'=\frac{e^2n_0}{m}[\frac{He}{2C}]\int d\xi[-\frac{\partial n(\xi)}{\partial\xi}]\frac{1}{\Gamma(\xi)^2}(1-M),$$

$$X'=\frac{e^2n_0}{m}[\frac{He}{2C}]\int d\xi[-\frac{\partial n(\xi)}{\partial\xi}]\frac{1}{\Gamma(\xi)^2}\frac{M}{1-N\ \ln\xi}\ .$$

It is obvious that the kondo temperature is $T_K=\frac{D}{K}\exp(\frac{1}{Z})$ from (3.6), which has the form similar to other authors'. Our results do not diverge at T = ok. Hall coefficient R_H is related to the conductivity components σ_{xx} and σ_{xy} through

$$R_H=\frac{1}{H}\ \frac{\sigma_{xy}}{\sigma_{xy}^2+\sigma_{xx}^2}\ . \qquad (3.7)$$

Substituting (3.6) into (3.7), R_H can be written as

$$R_H=R_0(1-H^2R_0^2\sigma_{xx}^2),$$
$$R_0=\sigma_{xy}/H\ \sigma_{xx}^2\ . \qquad (3.8)$$

4. CONCLUSIONS

In this paper, a method of calculating low temperature Hall conductivity in a weak magnetic field by using Mahan-Hansch transport equation in the external magnetic field for many particle system is presented. In our calculations, we assumed that the scattering center has a resonance for scattering near the Fermi energy, hence, the electron for their energies near the Fermi surface are scattered a lot, while the other electrons are scattered a little. This is the reason why the Hall conductivity is greatly temperature-dependent. In the limit of a weak fie field, $h=\frac{He}{2\Gamma C}$ works as an expansion parameter to solve the vertex function $\gamma(\vec{Q},\omega)$. Finally, Some valuable results for low temperature Hall conductivity and Hall coefficient are obtained. The conductivity does not suffer from divergence at zear temperature. Our result is in good agreement with the experiments.

REFERENCES

(1) W. Hansch and G. D. Mahan, phys., Rev. B 28 (1983) 1986.

(2) A. Fert, A. Friederich, and A. Hamzic J. Magn. 24 (1981) 231.

(3) L. Yi and K-L. Yao, An application of new transport equation to dilute alloys, this Volume.

Physica B 165&166 (1990) 263–264
North-Holland

"LAMINAR" AND "TURBULENT" DIRECT CURRENT FLOW IN THIN METAL PLATES

L.M.FISHER, S.V.KRAVCHENKO (*), N.A.PODLEVSKIKH, I.F.VOLOSHIN and S.I.ZAKHARCHENKO

All Union Electric Engineering Institute, Krasnokazarmennaja st., 12, Moscow, 111250, USSR

Nonlinear and nonstationary effects in the dc resistance of thin metal plates are studied experimentally. The bending of the electron trajectories in the self magnetic field of the current causes the resistance R to depend on the current. Current instability, in the form of periodic or random (parameter-dependend) self-oscillations of the voltage, was observed when current exceeds a certain value depending on the temperature and the width of the sample. The transition from monochromatic to random oscillations usually proceeds in accordance with the model proposed for arbitrary non-linear systems by Ruelle and Takens. The possible physical origin of the observed effect is the inhomogeneity of the current distribution over the sample cross section.

Some years ago Kaner et al. (1) demonstrated theoretically that pronounced magnetodynamic nonlinearity occurs in thin metal plates which carry a direct current. Nonlinearity taking place in the case $d \ll (rd)^{1/2} \ll l$ is due to the influence of self magnetic field, $\mathscr{H}$, of the current on the electron trajectories (here $2d$ is the thickness of the plate, r is the radius of curvature of electron trajectory in the magnetic field of the current, and l is the electron mean free path). Influence of $\mathscr{H}$ leads to decrease of electrons surface scattering and to creation of effective group of electrons realizing their full free path. As a result, the conductivity of plate increases. This effect had been observed in gallium previously (2).

We measured the differential resistance R of tungsten plate by four-probe method used modulation technique (3,4). The plates were cut from high-quality single crystal. The samples were immersed directly in liquid helium.

The dependence $R(I)$ is shown in figure 1 by upper curve. The resistance depends nonmonotonically on the current: it decreases at small I to a minimum at $I = I_{min}$ and then increases. The values of I_{min} are the same for the samples with different dimensions.

The qualitative behaviour of the nonlinear resistance for $I < I_{min}$ is as predicted by the theory (1). We consider that nonmonotonic behaviour of $R(I)$ is connected with the magnetic field component $\mathscr{H}_z$ perpendicular to the plate. The influence of $\mathscr{H}_z$ existing for all plates with finite dimensions is especially pronounced when the characteristic radius of curvature of the electron trajectory in the xy-plane became less then the half-width D of plate. This occurs for current value $I_D = k\,c^2\,p_F\,/\,e \simeq (6 - 8)$ A which is independent on D, d and temperature, and is in accordance with experimental value I_{min} (here c is the speed of light, p_F is the Fermi momentum, e is the electron charge, and the k is dimensionless constant of order unity). At higher current, $I > I_D$, the region of effective electrons inside the sample is diminished, and resistance increases. In this condition the distribution of current in plate becomes inhomogeneous like in pinch effect.

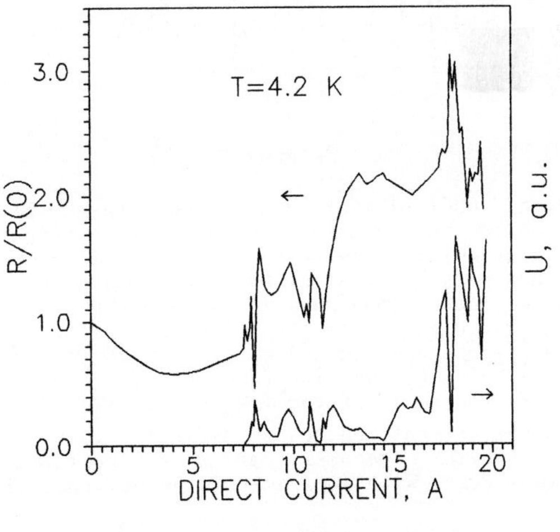

FIGURE 1

Three regions can be distinguished on $R(I)$ curve: a region of smooth variation ($I < I_1 = 7.5$ A), region of reproducible abrupt peaks ($I_1 < I < I_2 = 19$ A), and a region $I > I_2$ in

(*) Permanent address: Institute of Metrological Service, Andreevskaya nab. 2, Moscow 117334, USSR which the resistance varies in time with period of several seconds. These random oscillations suggested to us that the sharp $R(I)$ peaks observed at the smaller currents are also due to nonstationary processes. To check on this assumption, we measured the *ac* component U of the voltage drop on the sample in the absence of current modulation. The dependence of this signal on the *dc* current is shown by lower curve in figure 1. Comparison of upper and lower curves shows that the most abrupt singularities on $R(I)$ and $U(I)$ correlate with one another.

To track the instability development we have investigated the self-oscillation spectra of the voltage on the sample. Some results are shown in figure 2. It was found that the

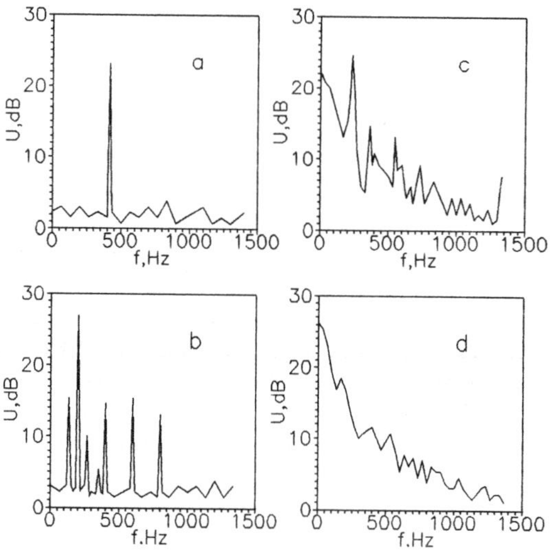

FIGURE 2

voltage oscillations near the instability threshold are almost monochromatic (figure 2,a). At a larger current, a second independent frequency f_2 appears in the spectrum and is in general not commensurable with the first (figure 2,b). Combination frequencies of the form mf_1+nf_2 where m and n are integers, can also appear. Further increase of the current changes the ratio of the frequencies f_1 and f_2. When the current reaches a definite value I_2 the spectrum becomes abruptly continuous (figure 2,c). With increase of the current, the noise amplitude increases noticeably (figure 2,d) and its dependence on the frequency is approximately f^{-1}. Following the "chaos" region, the oscillation spectrum again becomes discrete in a number of cases. We remark that it was usually possible to note a correspondence between the abrupt restructuring of the spectrum and the peaks of the differential resistance.

The sequence of spectra shown in figure 2 recalls the transition to turbulent liquid flow between two rotating cylinders (Couette flow). A similar picture is explained within the framework of the model proposed by Landau and Lifshitz for the transition to turbulence. According to this model, a sequence of Hopf bifurcations takes in many nonlinear dissipative systems upon variation of any parameter that influences the system; these bifurcations are accompanied by creation of new independent oscillation frequencies. According to Ruelle and Takens the second Hopf bifurcations should be followed by a strange attractor, and the oscillation spectrum should turn into a continuum. This is precisely the system behavior observed in most of our experiments.

Variation of temperature and of the external magnetic field effects strongly on the instability, changing not only the parameters of spectrum but the whole picture of instability development. In particular, Feigenbaum model had been realized. All these situations correspond to the different models of development of stochastization in nonlinear systems (5).

Since the conductivity is essentially nonlocal in the situation considered, no satisfactory explanation can be obtained for the observed self-octilations by using the known mechanisms that bring about instability in a conducting medium. The physical nature of the effect may be connected with stability loss of current "pinch" (4).

(1) E.A.Kaner, N.M.Makarov, I.B.Snapiro and V.A.Yampol'skii, Sov. Phys. JETP 60 (1984) 1252

(2) M.Yaqub and I.F.Cochran, Phys. Rev. Lett. 10 (1963) 390

(3) I.F.Voloshin, S.V.Kravchenko, N.A.Podlevskikh and L.M.Fisher, Sov. Phys. JETP 62 (1985) 132

(4) S.I.Zakharchenko, S.V.Kravchenko and L.M. Fisher, Sov. Phys. JETP 91 (1986) 660

(5) J.P.Golub, S.V.Benson and I.Steinman, in: Nonlinear dynamics, Ann. N.Y. Acad. Sci., Vol. 22, (1980) 357.

Physica B 165&166 (1990) 265–266
North-Holland

LOW-TEMPERATURE CONDUCTIVITY OF THIN METALLIC SLABS AT ARBITRARY FUCHS BOUNDARY CONDITIONS

Jerzy CZERWONKO

Laboratory of Theoretical Physics, Joint Institute for Nuclear Research, Head P.O.Box 79, 101000 Moscow, USSR *

The conductivity of three- and two-dimensional slabs, thin with respect to the mean free path, is calculated in the isotropic metal model at arbitrary disordered impurity scattering and an arbitrary specularity degree of the border. Three first asymptotic terms are calculated. It is shown that the first two of them are not influenced by the go-in part of the scattering integral.

The formulated problem has been solved by Fuchs (1) more than half a century ago but in the relaxation time approximation leading to the differential kinetic equation, at dimension d=3. We consider the integrodifferential kinetic equation because of the "go-in" term in its scattering integral describing impurity scattering but only at diffusive boundary conditions (2). We succeeded in solving the problem of slabs thin with respect to the mean free path. Now we present the generalization for arbitrary border scattering in the Fuchs sense (1). Similarly as in (2), we will confirm his result for average conductivity in our first two asymptotic terms though there is some surprise with Fuchs asymptotic expansion. We are unable to apply the Falkovsky integral boundary conditions (3) in their general form, (2), though introduction of angle-dependent specularity coefficient, (4), is possible. In (2), we are extended electric fields to time-dependent ones, at the normal skin effect. To do this, some Fermi liquid techniques were applied, cf. e.g. (5).

Let us define the function $\Psi(\mathbf{n},z)$ through the time Fourier transform of the deviation of phase space electron distribution function from its local equilibrium value, δf, (5). We have

$$\delta f = ev\tau(-\partial f/\partial\varepsilon)\Psi(\mathbf{n},z)E_\omega .$$

Here e is the electron charge, v and f are the Fermi velocity and distribution function, $\mathbf{n}$ is the unit vector directed along electron momentum $\mathbf{p}$ and E_ω is the time Fourier transform of the homogeneous electric field directed along the border of the slab. The Cartesian coordinate z is perpendicular to it; z=0 in the middle of the slab. The quantity $\tau=1/\langle \mathbf{W}(\mathbf{nn}')\rangle_{\mathbf{n}}$, where $\mathbf{W}$ describes the probability density of impurity scattering per unit time and spherical angle (d=3) or angle (d=2). The symbol $\langle\ldots\rangle_{\mathbf{n}}$ denotes average over spherical angles or angles at d=3 or d=2.

The kinetic equation can be written as follows:

$$a\Psi(\mathbf{n},z)+\Psi'(\mathbf{n},z)\cos\theta=\sin\theta\sin\varphi+\langle\mathbb{G}_\omega(\mathbf{nn}')\Psi(\mathbf{n}',z)\rangle_{\mathbf{n}'},$$

cf. (5) and (2). Here $\cos\theta$ is the z-th component of the vector $\mathbf{n}$ and φ is its azimuthal angle; at d=2, $\varphi=\pi/2$. The prime denotes the partial derivative with respect to dimensionless z taken in units of a mean free path $l=v\tau$, $a=1+i\omega\tau$ with ω being the frequency. The function $\mathbb{G}_\omega(x)$ is defined as $\mathbb{G}(x)+i\omega\tau\mathbb{A}^s(x)$ with $\mathbb{G}(x)\equiv\tau\mathbf{W}(x)-1$ and $\mathbb{A}^s(x)$ being the spin-symmetric dimensionless forward scattering amplitude of quasiparticles (5). We will work with the series expansion of the function $\mathbb{G}_\omega(\mathbf{nn}')$, on Legendre polynomials at d=3 and on the cosine Fourier series at d=2. The expansion coefficients will be denoted as $\mathbb{G}_{\omega l}$ in both cases. We can put $\mathbb{G}_{\omega 0}=0$, (2). In what follows the shorthand notation $\cos\theta = c$, $\sin\theta = s$, $\cos\theta' = c'$, etc. will be introduced.

Let us look for the solution in the form

$$\Psi(\mathbf{n},z)=\sin\varphi\; s[\mathbb{A}(c,z)\exp(-za/c)+1]/(a-\mathbb{G}_{\omega 1}/d),$$

with $\varphi=\pi/2$ at d=2. After substitution one finds

$$c\,\mathbb{A}'(c,z)\exp(-za/c)=\sum_{l=1}\tilde{\mathbb{G}}_{\omega l}R_l(c)\,\langle R_l(c')\,\mathbb{A}(c',z)\exp(-za/c')\rangle_{c'},$$

Here $\langle f(c)\rangle_c$ is an average of $(1-c^2)f(c)$, at d=3 over c on the interval (-1,1) and, at d=2 over θ on the interval $(0,2\pi)$. The quantity $R_l(c)=P_l'(c)$ at d=3 and $U_{l-1}(c)$ at d=2 with P_l and U_l being respectively the l-th Legendre and Tchebyshev polynomial of the second kind, cf. (6). The quantity in the last line of the previous formula will be denoted as $B_l(z)$. Moreover, $\tilde{\mathbb{G}}_{\omega l}=\mathbb{G}_{\omega l}$ at d=2 and $\tilde{\mathbb{G}}_{\omega l}=\mathbb{G}_{\omega l}/l(l+1)$ at d=3.

The Fuchs boundary conditions have the form

*Permanent address: Institute of Physics, Technical University of Wrocław, Wybrzeże Wyspiańskiego 27, 50-370 Wrocław, Poland

0921-4526/90/$03.50

$\varepsilon\Psi(\mathbf{n},b)=\Psi(\mathbf{n}^r,-b)$, $\varepsilon\Psi(\mathbf{n}^r,-b)=\Psi(\mathbf{n},-b)$, $n_z\equiv c>0$, (1).
Here $\mathbf{n}^r$ is the vector $\mathbf{n}$ reflected in the xy plane, $z = \pm b$ determines the border of the slab and ε is the specularity coefficient. The solution of the equation for $\mathbb{A}$, satisfying our boundary conditions takes the form

$$\mathbb{A}(c,z)=\{1-\varepsilon-\varepsilon\sum_{l=1}\tilde{\mathbb{G}}_{\omega l}R_l(|c|)|c|^{-1}\int_{-b}^{b}dy\, B_l(y)\exp[-(b-y)/a|c|]\}/[\exp(ab/|c|)-\varepsilon\exp(-ab/|c|)]+\sum_{l=1}\tilde{\mathbb{G}}_{\omega l}R_l(c)c^{-1}\int_{-bc/|c|}^{z}dy\, B_l(y)\exp(ya/c).$$

Substituting this function into the definitions of the functions $B_l(z)$ one finds

$$B_k(z) = -(1-\varepsilon)\sum_{\alpha=\pm1}\alpha^{k-1}\mathbf{V}_k(\varepsilon,a(b+\alpha z),q) + \sum_{l=1}\tilde{\mathbb{G}}_{\omega l}\int_{-b}^{b}dy\, B_l(y)[\mathrm{sign}(z-y)^{1+k}\mathbf{W}_{lk}(a|z-y|) + \varepsilon\sum_{\alpha=\pm1}\alpha^{k-1}\mathbf{W}_{lk}(\varepsilon,a(q+\alpha z-y),q)],\quad q\equiv2b,$$

$\mathbf{V}_k(\varepsilon,u,q)\equiv\langle R_k(c)\exp(-u/c)/[1-\varepsilon\exp(-aq/c)]\rangle$ and in the definition $\mathbf{W}_{lk}(\varepsilon,u,q)$ instead of $R_k(c)$ is $R_k(c)R_l(c)/c$. Moreover, $\mathbf{W}_{lk}(u)\equiv\mathbf{W}_{lk}(0,u,q)$ and the average $\langle\mathbf{X}\rangle$ is now defined by

$$\langle\mathbf{X}\rangle=\left\{\frac{1}{2}\int_0^1 dc(1-c^2)\mathbf{X},\ d=3;\ \frac{1}{\Pi}\int_0^1 dc(1-c^2)^{1/2}\mathbf{X},\ d=2\right\}.$$

The ω, z-dependent conductivity is given by

$$\sigma_\omega(z)=[e^2 n\tau/m^*(a-\mathbb{G}_{\omega1}/d)][1+B_1(z)d/(d-1)]\equiv(e^2 n\tau/m^*)\,D(z),$$

where $n = p_F^d/d\,\pi^{d-1}\hbar^d$ is the density of carriers and $m^*=p_F/v$ their effective mass.

Asymptotic properties of the functions $\mathbf{V}_l$ and kernels $\mathbf{W}_{lk}$ at $q \ll 1$ can be obtained only via their series expansion in powers of ε and application the Sommerfeld-Watson transformation to the series terms (2). The quantity $(1-\varepsilon)\mathbf{V}_l(\varepsilon,u,q)$ tends to zero at $\varepsilon=1$ but it does not appear for the series expansion in the asymptotic form at $q \ll 1$ because of its unit convergence radius. Asymptotic values of functions $\mathbf{V}_l$ and $\mathbf{W}_{lk}$ cause that the system of integral equations can be restricted to the functions B_{2m+1} with the accuracy $O(q^2\mathrm{Lnq})$ and, moreover, it can be solved by two iterations with this accuracy (2). Taking into account that the free term of the equation at $k=2m+1$ is $-(1-1/d)\delta_{m0}+O(q\mathrm{Lnq})$ and performing iterations and elementary integrations one finds, at once for "dimensionless conductivity"

$$D(z)=d\{Q_0(z)+\sum_{n=0}\tilde{\mathbb{G}}_{\omega,2n+1}\int_{-b}^{b}dyQ_n(y)[\mathbf{W}_{2n+1,1}(a|z-y|)+\varepsilon\sum_{\alpha=\pm1}\mathbf{W}_{2n+1,1}(\varepsilon,a(q+\alpha z-y),q)\}/(d-1)(a-\mathbb{G}_{\omega1}/d),$$

$$Q_n(z)\equiv\delta_{n0}(1-1/d)-\sum_{\alpha=\pm1}\{(1-\varepsilon)\mathbf{V}_{2n+1}(\varepsilon,a(b+\alpha z),q) - (\mathbb{G}_{\omega,1}/da)[\mathbf{V}_{2n+1}(a(b+\alpha z))-\varepsilon\mathbf{V}_{2n+1}(\varepsilon,a(b+\alpha z),q)+\varepsilon\mathbf{V}_{2n+1}(\varepsilon,a(3b+\alpha z),q)]\}+O(q^2),\quad \mathbf{V}_k(u)\equiv\mathbf{V}_k(0,u,q).$$

Disregarding the terms $O(q^2\mathrm{Lnq})$ and integrating the asymptotic expansions, (2), one finds

$$D(z)=dQ_0(z)/(d-1)(a-\mathbb{G}_{\omega1}/d)+\Delta D_d,$$

where $\Delta D_d = \Phi_d(1+2\varepsilon)(1+\varepsilon)[q\mathrm{Lnq}/(1-\varepsilon)]^2$ and $\Phi_3 = 3\sum_{n=0}[\,\mathbb{G}_{\omega,2n+1}/16(2n+1)(n+1)\,][\,(2n+1)!!/2n!!\,]^2$, $\Phi_2=[\mathbb{G}_\omega(1)-\mathbb{G}_\omega(-1)]/\pi^2$.

Averaging $Q_0(z)$ one can write

$$Q_0(z)=(a-\mathbb{G}_{\omega1}/d)\{1-1/d-2(1-\varepsilon)\langle c\rangle/aq + 2(1-\varepsilon)^2\langle c\exp(-aq/c)/[1-\varepsilon\exp(-aq/c)]\rangle\}/a,$$

in agreement with formula (21) of (1), at d=3. The function Q_0 describes also the same distribution of current, but at $\mathbb{G}_{\omega1} = 0$. The " go-in " term influences, via $\mathbb{G}_{\omega1}$, the last quantity but not the average conductivity. Asymptotic expansion of the integral in $\bar{Q}_0$, (2), gives

$$\overline{D(z)}=-E_dq[(\mathrm{Lnaq}+\mathbb{C}_d)(1+\varepsilon)/(1-\varepsilon)+(1-\varepsilon)^2\xi(\varepsilon)]+\Delta D_d,$$

where $E_3=3/4$, $\mathbb{C}_3=C-1$; $E_2=2/\pi$, $\mathbb{C}_2=C-\mathrm{Ln}2-1/2$ with C being Euler's constant and $\xi(\varepsilon)$ defined as the series expansion of the terms $\varepsilon^n(n+1)^2\mathrm{Ln}(n+1)$, n=1,2,.... Our small parameter is $-q\mathrm{Lnq}/(1-\varepsilon)$, as we see. The asymptotic formula of (1), disregarding O(q) terms is $\bar{D}=-E_3(1-\varepsilon)q\mathrm{Lnq}$, a=1, and is a physical nonsense because the conductivity can not diminish at growing specularity and become zero instead of σ_{bulk} at $\varepsilon=1$. On the other hand, the result of (1) about zero conductivity at q=0 is true and of simple physical meaning. In fact, if the probability to be caught at the border is nonzero and the way to go in the x,y - direction is infinite, then the electron moving not parallelly to the border will surely be caught after multiple border scattering.

REFERENCES

(1) K. Fuchs, Proc. Camb. Phil. Soc. 34 (1934) 100.
(2) J. Czerwonko, Z. Phys. B- Condensed Matter, in print.
(3) L.A. Falkovsky, Zh. Eksp. Teor. Fiz. 58 (1970) 1830 [Sov.Phys. JETP 31 (1970) 981].
(4) L.A. Falkovsky J. Low Temp. Phys. 36 (1979) 713.
(5) D. Pines and P. Nozières, Theory of Quantum Liquids (Benjamin, New York, 1966).
(6) I.S. Gradstein and I.M. Ryzhik, Tables of Integrals, Sums, Series and Products (Nauka, Moscow, 1971).

Physica B 165&166 (1990) 267–268
North-Holland

EFFECT OF ELECTRON ENTRAINMENT AT LOW TEMPERATURE DEFORMATION OF METALS: KINETICS AND STATISTICS OF DYNAMICAL PROCESSES

M A LEBYODKIN, V Ya KRAVCHENKO, V S BOBROV

Institute of Solid State Physics of the USSR Academy of Sciences, 142432, Chernogolovka, USSR

The results of the investigations of the effect of electron entrainment at low temperature twinning of Nb and "catastrophic" dislocation glide of Al are reported. Electric pulses, related to this effect, are employed for an analysis of kinetics and statistics of dynamic deformation proceses.

1. INTRODUCTION

In (1) the effect of conduction-electron entrainment by the dislocation flux in crystals was predicted. It was supposed that dislocation dynamics may be studied from the electric effects. To observe the effect of electron entrainment one has to achieve high rates of the plastic flow $\dot{\varepsilon}$ in conditions of directed motion of defects. In (2-4) the electric effects were studied upon an abrupt deformation of metals (Nb, Al, Mo, Re, W). This work reports the results of the investigations of the effect of electron entrainment at low temperature twinning of Nb and "catastrophic" dislocation glide of Al.

2. RESULTS AND DISCUSSION

At low temperature twinning of Nb at the moments of loading jumps one can register on the faces of the deformed samples trains of short (~ 2-3 μs) electric pulses of unlike sign, connected with the motion of twins across the system of intersecting twinning planes (e.g. fig. 1a). With the electric contacts normal to the twin motion direction only individual pulses of noticeably lower amplitude (fig.1b) were observed which indicates the entrainment effect anisotropy connected with that of the twinning crystallography. When the conditions were offered for the motion of parallel twin packets from the region of one of the contacts, pulse trains of the same sign were registered. Their polarity corresponded to the electron entrainment in the twin motion direction.

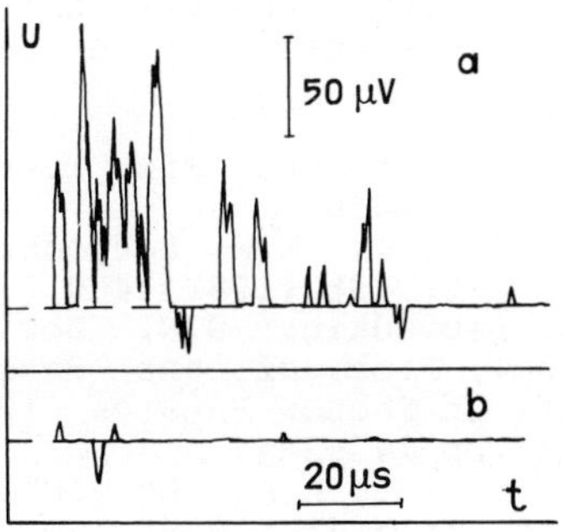

FIGURE 1
Pulse trains at the moments of loading jumps at twinning of Nb: a - parallel, b - perpendicular arrangement of electric contacts to the twin motion direction.

The electric effects are observable not only at deformation twinning but, also, in conditions of "catastrophic" dislocation glide. Examples of pulse trains, registered at abrupt deformation of Al, are shown in fig.2. The pulses were recorded at the background of thermo-EMF signals, their amplitude was much smaller against that of Nb.

As contrasted from Nb, at deformation of Al we were able to observe individual

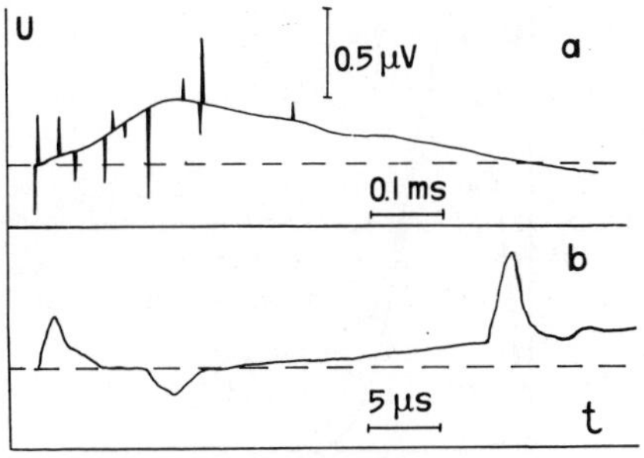

FIGURE 2
Pulse trains at "catastrophic" dislocation glide of Al.

pulses too (e.g. fig.3) recorded at low loading jumps. These are associated with the motion of individual dislocation aggregates. The correlation was established between these pulses amplitude and the deformation increment rate at the moments of loading jumps. Along with the observation of anisotropy and polarity of pulses upon twinning, these results qualitatively agree with the theoretical predictions (1). The absence of the temperature dependence of the amplitude and the pulse duration in Nb and Al also agrees with the theoretical results (4).

The data analysis and comparison with theory (1) suggest the conclusion concerning the observation of the electron entrainment effect at the motion of twins and dislocation aggregates. There is also a qualitative and quantitative agreement with the theoretical results.

Investigations of electrical pulses enable one not only to estimate the theoretical parameters but, also, to judge about the character of evolution of high velocity deformation processes (e.g. (4)). The shape of individual pulses and the duration of pulse trains may reflect the kinetics of formation and motion of deformation carriers at the moment of loading jumps. Fig.3 presents, as an example, the results of calculation of an individual pulse shape in Al using the theory (1) and relationships (5), describing the glide band evolution.

The data on registration of electric pulse trains were also employed for a statistic analysis of dynamical deformation processes (6). A qualitative similarity of the pulse distribution function density with respect to the D(U) amplitude was found at an abrupt deformation of Nb and Al. Normalized functions $D^*(U/\langle U\rangle) = D(U)\langle U\rangle$, $\langle U\rangle$ being a mean value of the amplitude of statistically selected pulses, were used. The results of comparison (fig.4) may indicate the universality of the regularities, determining the statistics of dynamical processes at deformation twinning and "catastrophic" dislocation glide.

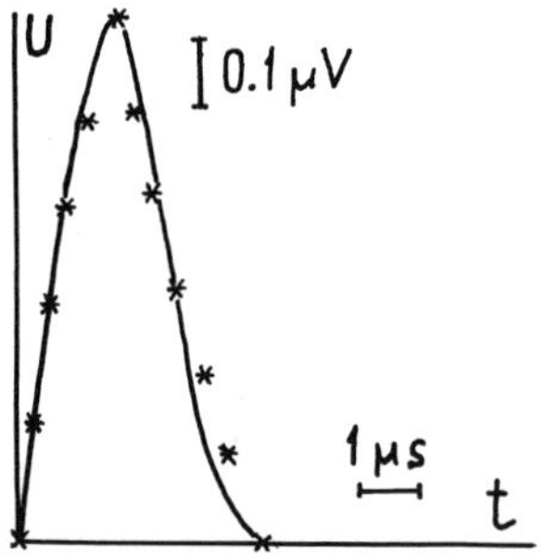

FIGURE 3
Individual pulse at a low loading jump of Al (* stands for the results of the pulse shape calculation).

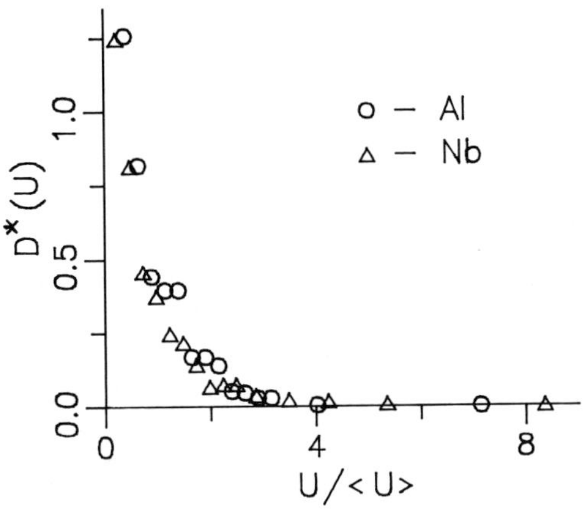

FIGURE 4
Normalized distribution function densities $D^*(U/\langle U\rangle)$ of the pulses with respect to the amplitudes for Nb and Al

3. CONCLUSIONS

The investigations of the electron entrainment effect not only confirm the results of the theory of electron-dislocation interaction but, also, may be a highly informative and quick-acting technique to study deformation processes.

ACKNOWLEDGEMENTS

The work is carried out under the USSR Ac Sci grant "Dislocation". The authors are grateful to D.E. Khmel'nitskii, S.P. Obukhov and S.I. Zaitsev for the helpful discussion of the results.

REFERENCES

(1) V.Ya. Kravchenko, Fiz. Tverd. Tela 9 (1967) 1050.
(2) V.S. Bobrov, M.A. Lebyodkin, JETP JETP Lett. 38 (1983) 400.
(3) M.A. Lebyodkin, S.N. Zorin, V.S. Bobrov, Proc. of the Second Int. Conf. on Phonon Physics (Budapest, 1985) pp.242-244.
(4) V.S. Bobrov, M.A. Lebyodkin, Fiz. Tverd. Tela 31 (1989) 120.
(5) B.I. Smirnov, Dislocation Structure and Hardening of Crystals (Nauka, Leningrad, 1982).
(6) V.S. Bobrov, S.I. Zaitsev, M.A. Lebyodkin, Zh. Eksp. Teor. Fiz. (1990), in print.

Physica B 165&166 (1990) 269–270
North-Holland

ELECTRON-HOLE CORRELATION IN GIANT QUANTUM ATTENUATION OF SOUND WAVES IN BISMUTH

Masatoshi MORI, Masao KOGA† and Nobuyuki GOTO†

Kinki University, Department of Management Engineering, Iizuka 820, Japan
†Nagasaki University, Department of Physics, Nagasaki 852, Japan

In a strong magnetic field, the deformation potential strongly depends on the density of states. Hence, there appears a strong correlation between the peaks as the peaks due to the electrons and holes come closer to each other. We made numerical calculations taking the density-of-states dependence of deformation potentials into account. The results could quantitatively explain the correlation shown by the experiments.

1. INTRODUCTION

At low temperatures, the carriers at the Fermi level alone contribute to the physical quantities such as electric resistance, specific heat, and so on. Therefore, these physical quantities undergo quantum oscillations reflecting the density of states at the Fermi level in strong magnetic fields. Compared with this, the quantum oscillations in the attenuation coefficient of the sound waves are characteristic for the following two points and show unique behavior not seen in the quantum oscillations of other physical quantities. The first point is proposed by Gurevich *et al.*(1) :The carriers which contribute to the attenuation of sound waves are limited to the carriers at the bottom of the Landau level due to the conservation laws of energy and momentum. Hence, the quantum oscillations of attenuation coefficient show spike-like peaks at the magnetic field where the bottom of the Landau level and the Fermi level coincide, so called Giant Quantum Oscillations. The second point is proposed by Goto(2) :The deformation potential which determines the amplitude of attenuation peaks itself depends on the density of states due to the charge neutrality condition. Therefore, giant quantum oscillations are amplitude-modulated with de Haas type quantum oscillations(2,3). Then various effects are given on the attenuation curves(4,5), and effects are especially large when the electron peak and the hole peak come closer to each other.

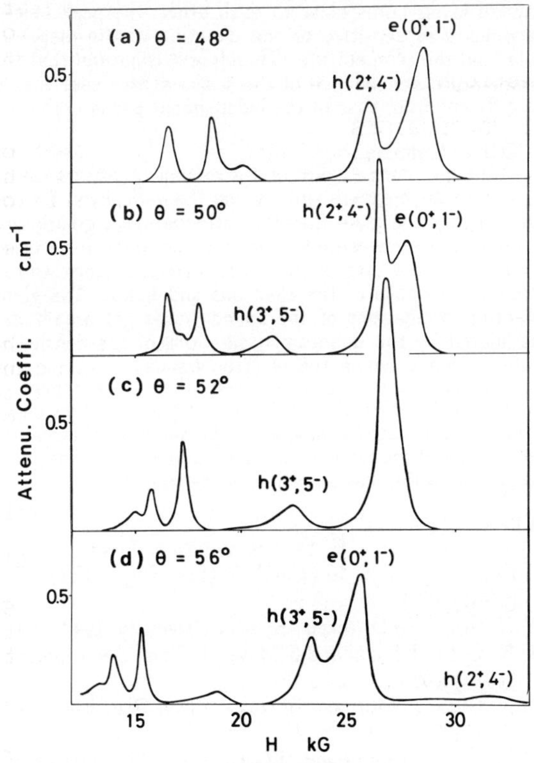

FIGURE 1
Experimental results by Goto(2) in the case of *q//z*-axis, *H//yz*-plane, T=1.1 K and the frequency of the sound is 60 MHz.

Figure 1 shows the attenuation curves of the sound waves in the bismuth single crystal obtained by Goto(2). Sound waves are propagated along the trigonal axis of the crystal and the direction of the magnetic field is changed in the mirror plane. The electron peak e(0+,1-) and the hole peak h(2+,4-) come closer to each other as the angle of magnetic field direction from the trigonal axis θ increases. The amplitude of the hole peak h(2+,4-) abruptly changes as it comes closer to the electron peak e(0+,1-) from the lower magnetic filed side.

Taking the density-of-states dependence of deformation potentials into account in the theory by Gurevich *et al.*, we made numerical calculations of attenuation curves. The result could explain the correlation of electron peaks and hole peaks quantitatively.

2. CALCULATION

2.1 Theory

The expression given by Gurevich *et al.* is expressed for the *j*-th Fermi ellipsoid as follows:

$$\alpha_j \sim |D_j|^2 \omega \frac{H}{T} \sum_{n,s} \int_{-\infty}^{\infty} \frac{dy}{\pi} \frac{B_j/\cos\theta}{1+B_j^2 y^2} \cosh^{-2}[\frac{1}{2}(y^2 - \frac{A_{n,s}^j}{k_B T})], \quad (1)$$

where θ is the angle between the wave vector q and the magnetic field H, and the definitions of quantities such as $A^j_{n,s}$ and B_j are the same as in ref.4.

According to the band model for bismuth, there are three electron ellipsoids; a, b, c at **L** points and a hole ellipsoid; h at **T** point in the first Brillouin zone. Then we must sum up α_j over j to get the total attenuation.

The deformation potential of an ellipsoid depends not only on its own density of states but also on those of all other ellipsoids because of the charge neutrality condition. When the sound waves are longitudinal waves and propagate in parallel with the trigonal axis, the deformation potentials of electrons and holes are written as $D_e \sim n_h/(n_e+n_h)$, $D_h \sim n_e/(n_e+n_h)$. Since the deformation potential of the holes strongly depends on the density of states of the electrons; n_e, the height of the hole peak will be modulated by the density of states of electrons as the electron level gets close.

2.2 Calculation

Figure 2 is the result of the calculation under the same conditions as in the experiment Fig.1. There is a small discrepancy in angles between Fig.1 and Fig.2., but this is not a serious problem in the present argument because the discrepancy is due to the choice of mass parameters.

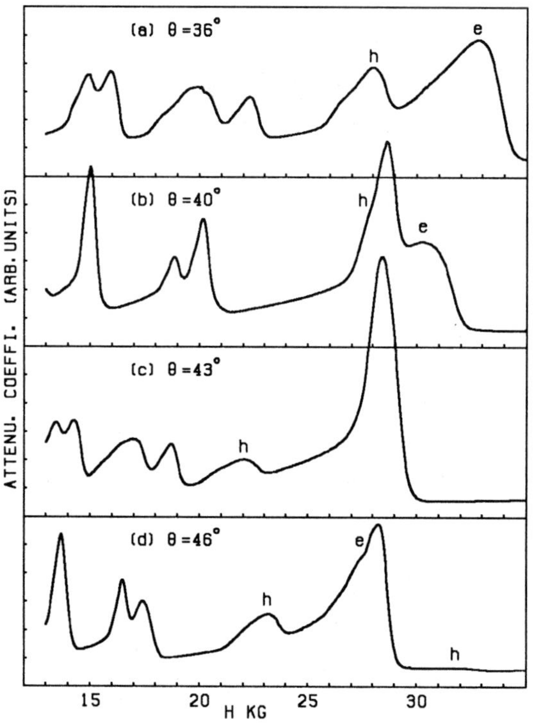

FIGURE 2
Calculated curves of the attenuation coefficient under the condition; $q//z$-axis, T=1.1 K, the frequency of the sound is 60 MHz and $\tau=1.0 \times 10^{-11}$sec.

3. DISCUSSION

The numerical calculation Fig.2 could reproduce the correlation between the peaks shown in the experimental result Fig.1 very well. The reason why the hole peak (2+,4-) becomes large as it goes from Fig.1 (a) to (b) is because the density of states of the electron level (0+,1-) becomes large. On the other hand, the reason why the electron peak(0+,1-) becomes small is not because the hole level (2+,4-) comes closer from the lower magnetic field side but because the density of states of the holes decreases as the hole level(1+,3-) with a lower quantum number goes remote. As in (c) where the two peaks completely coincide, the height becomes abruptly large since both of the densities of the states take maxima. In (d), when h(2+,4-) goes to the higher magnetic field side of e(0+,1-) the hole peak h(2+,4-) becomes extremely small because the density of states of e(0+,1-) is small and there is no electron level in the higher magnetic field side.

Mase *et al.*(6) measured the temperature dependence of the peak height in the condition where the electron peak and the hole peak come extremely close to each other. They observed that the temperature dependence of the peaks which come close to each other is different from that of the independent peaks and stressed that the electrons and the holes underwent the exitonic phase transition at temperatures lower than a certain temperature. When two Landau levels come close to each other, the peak height becomes very sensitive to the direction of the magnetic field, and the temperature. Therefore it is natural that the temperature dependence of the peaks of this case should be different from that of the independent peaks.

4. CONCLUSIONS

Although many studies on the exitonic phase transition have been done experimentally and theoretically so far, we can say that the giant quantum attenuation is greatly unsuitable for the research for the exitonic phase, for even in the normal state there appears the correlation between the attenuation peaks of the electrons and holes. The giant quantum oscillations of the sound waves get amplitude-modulated by the quantum oscillations of the density of states. Hence, when the electron Landau level and the hole Landau level are close to each other, there appears a correlation between the peaks and the attenuation curve exhibits complicated behaviors as the direction of the magnetic field and the temperature are changed. This has given many people the illusion of the phase transition.

REFERENCES

(1) V.L. Gurevich, V.G. Skobov and Yu.A. Firsov: Zh. Eksp. Teor. Fiz. 40 (1961) 78 (Sov. Phys. JETP 13 (1961) 552).
(2) N. Goto: Ph.D. thesis, Kyushu University, 1974.
(3) N. Goto, T.Sakai and S.Mase: J. Phys. Soc. Jpn. 38 (1975) 1653.
(4) N. Goto, M.Mori and M.Koga: Phys. Rev. B36 (1987) 1262.
(5) M. Mori, M.Koga and N.Goto: Proc. 18th Int. Conf. on Low Temperature Physics, Kyoto, 1987 E005. (J.J.A.P. 26 (1987) Suppl.26-3 659).
(6) S. Mase and T. Sakai: J. Phys. Soc. Jpn. 31 (1971) 730.

Physica B 165&166 (1990) 271–272
North-Holland

DE HAAS-VAN ALPHEN EFFECT IN Pd_2Si

E.G. HAANAPPEL and W. JOSS

Max-Planck-Institut für Festkörperforschung, Hochfeld-Magnetlabor,
B.P. 166X, F-38042 Grenoble Cédex, France

R. MADAR and A. ROUAULT

Institut National Polytechnique de Grenoble, E.N.S.I.E.G., U.R.A. 1109, C.N.R.S.,
B.P. 46, F-38402 Saint-Martin-d'Hères Cédex, France

A de Haas-van Alphen study of the disilicide Pd_2Si has been performed. The sample is rotated in two perpendicular symmetry planes. In the basal plane five frequency branches are observed, in the other plane only two. A Fermi surface geometry is deduced which is characterized by one large open sheet and several small pockets. This geometry is in reasonable agreement with a band structure calculation if in the latter the Fermi energy is slightly decreased.

1. INTRODUCTION

It is well known that the reaction of transition metals with silicon can lead to the formation of a class of stable compounds: the silicides. Different phases are formed depending on temperature. These compounds are not only of physical interest, but also of considerable technological interest due to device oriented applications in VLSI industry. In this article we will discuss one of these compounds: Pd_2Si.

A few theoretical studies have been performed on this material. Bisi and Calandra (1) have calculated partial and total densities of state for the silicides using the extended iterative Huckel method. Bisi *et al.* (2) have done a detailed band structure calculation of Pd_2Si using the linear muffin-tin orbitals method in the atomic-sphere approximation. From the experimental side mainly spectroscopic studies have been performed (3). Also specific heat and resistivity data have been reported (4). In this article we report the results of a de Haas-van Alphen (dHvA) study, which directly tackles the Fermi surface.

Pd_2Si crystallizes in a hexagonal Fe_2P structure with lattice constants a = 6.49 Å and c = 3.43 Å. This structure is characterized by two nonequivalent hexagonal planes and contains three molecules per unit cell. The Brillouin zone of this structure and the main symmetry points are shown in Fig. 1. The x- and z-directions coincide with respectively the a- and the c-axis; the y-direction is perpendicular to both.

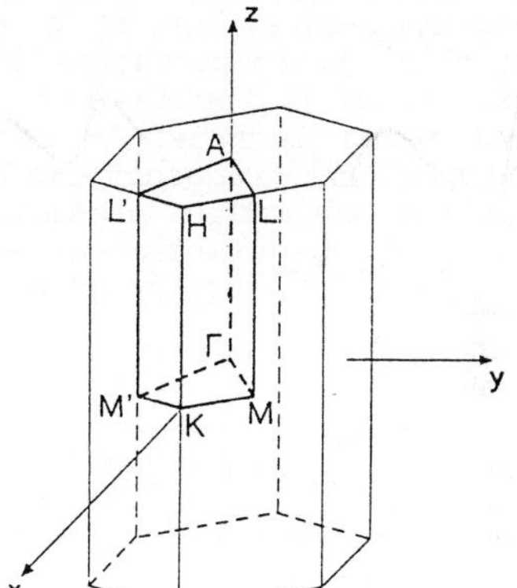

FIGURE 1
The Brillouin zone of Pd_2Si and the main symmetry points.

2. EXPERIMENTAL CONSIDERATIONS

The dHvA measurements were performed on two crystals, both with a size of approximately 3 x 1 x 1 mm. The samples were mounted inside a tightly fitting coaxial set of pick-up coils. The assembly of sample and pick-up coils can be rotated in one plane. The low frequency high amplitude field modulation technique was used with modulation amplitude proportional to B^2 and with detection on the second harmonic. The temperature was 4.2 K and fields were used up to 10 T in a superconducting coil.

3. RESULTS AND DISCUSSION

The main features of the observed dHvA frequencies are shown in Figures 2 and 3. In Fig. 2 can be seen that for rotation of the sample in the (zx)-plane five branches of frequencies are observed. A high frequency branch is observed, which is not split. Furthermore four branches of lower frequencies are found, which are split in two or three. Not always is this splitting resolved. It is not clear whether this splitting is due to details of the Fermi surface or, e.g., to the effect of a slight sample misalignment to degenerate frequencies. In Fig. 3 two branches of rather low frequencies are observed for rotation of the sample in the (yx)-plane.

The main feature in the (zx)-plane is the high frequency branch at about 3800 T. This branch is nearly isotropic and shows hardly any dispersion. In the (yx)-plane, however, this branch is absent. This suggests a open sheet of the Fermi surface which has a nearly spherical part, presumably centered at Γ, and which extends more or less cylindrically along the z-axis. Such a sheet would also gives rise to a frequency branch with upward curvature as is found at 1700 T with the field along the z-direction. It is easily verified that this upward curvature is stronger than $1/(\cos \phi)$ as is the case for a truely cylindrical Fermi surface. In addition the value 3800 T of this frequency can be brought into agreement with that predicted by the band structure calculation in Ref. 2 by lowering the Fermi energy by about 20 meV. A slightly smaller correction was suggested by Laborde *et al.* (4).

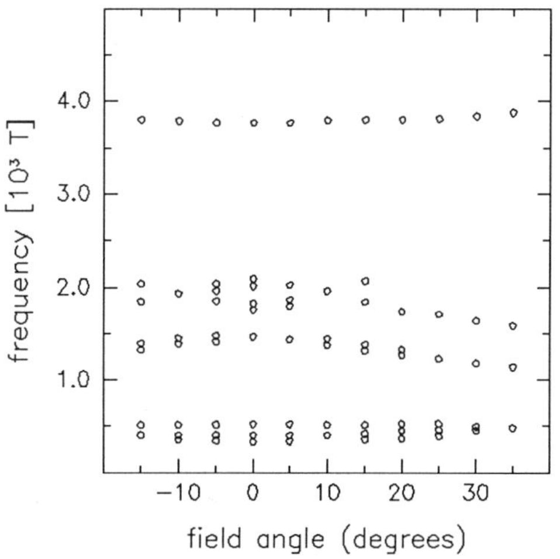

FIGURE 2
dHvA frequencies for rotation in the (zx)-plane. The field direction is defined with respect to the z-direction.

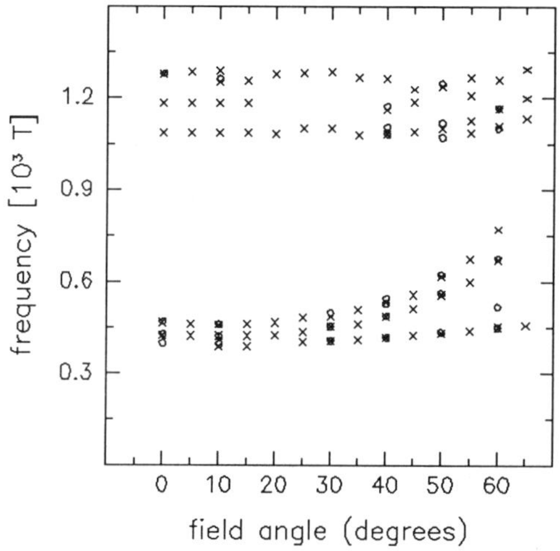

FIGURE 3
dHvA frequencies for rotation in the (yx)-plane. The field direction is defined with respect to the y-direction. Open circles represent high resolution measurements, crosses lower resolution ones.

The other three frequency branches in the (zx)-plane are somewhat harder to interpret in view of the limited angular range of the data. An extrapolation from the plot suggests that these frequencies will be connected to the lowest three frequencies in the (yx)-plane, but an extension of the angular range is necessary to prove this. These frequencies can be interpreted according to the band structure calculation as given by pieces of Fermi surface centered along the ΓA-axis.

The three frequencies around 1200 T which are found for rotation in the (yx)-plane are difficult to relate to sheets of the Fermi surface. An extension of the present study should shed light on this problem.

In conclusion a preliminary study of the Fermi surface of Pd_2Si is reported. Several features of the observed frequencies can be brought into agreement with a band structure calculation by a small decrease in the Fermi energy.

REFERENCES

(1) O. Bisi and C. Calandra, J. Phys. C: Solid State Phys. 14 (1981), 5479
(2) O. Bisi, O. Jepsen and O.K. Andersen, Phys. Rev. B 36 (1987), 9439
(3) For discussion and references of these experiments refer to (1) and (2)
(4) O. Laborde, J.C. Lasjaunias, R. Marani, A. Rouault and R. Madar, Phys. Rev. B (in print)

Physica B 165&166 (1990) 273–274
North-Holland

SKEW SCATTERING EFFECT ON THE HALL CONDUCTANCE FLUCTUATION IN THE HIGH Tc SUPERCONDUCTORS

A.G.ARONOV* and S.HIKAMI

Department of Pure and Applied Sciences, University of Tokyo, Komaba, Meguro-ku 3-8-1, Tokyo 153, Japan
*Leningrad Nuclear Physics Institute, Gatchina, Leningrad, U.S.S.R.

Near the transition point of the high temperature superconductor, the fluctuation effect of the Hall conductance is observed. We newly considered a skew scattering effect due to the spin-orbit interaction to explain this Hall conductance fluctuation. The magnetic field dependence of this effect is investigated.

1. INTRODUCTION

The temperature dependence of the Hall constant of the high temperature superconductors seems to be anomalous. This is considered as the effect of the anomalous skew scattering. Recently, near the transition temperature, the Hall conductance fluctuation has been observed.[1)] Aslamazov-Larkin fluctuation term usually does not provide Hall conductance fluctuation part. Therefore, it may be reasonable to speculate the skew scattering due to the spin-orbit exists and it provides the Hall conductance fluctuation near the transition temperature. We have newly investigated the Hall conductance fluctuation beyond the Aslamazov-Larkin term due to the skew scattering and its effect on the Maki-Thompson term[2)] . We also show that higher order of Aslamazov-Larkin term without the skew scattering does not give any Hall conductance fluctuation by the symmetrical reasons. For the analysis of the fluctuation part of the Hall conductance, it is difficult to know the normal value of the Hall conductance which should be subtracted to obtain the fluctuation part. The analysis of the magnetic field dependence can provide the fluctuation part as same as a diagonal magnetoresistance of the high temperature superconductors are able to provide the correct fluctuation part.[3)] For this reason, we evaluate the magnetic field dependence of Hall conductivity fluctuation.

2. SKEW SCATTERING DUE TO SPIN-ORBIT INTERACTION

We consider a skew scattering due to the spin-orbit interaction in the presence of the impurity. The scattering amplitude is given by the following momentum dependent form,

$$f_{pp'} = a_{pp'} + ib_{pp'}\vec{\sigma}\cdot(\vec{p}\times\vec{p}'), \tag{1}$$

where p and p' are coming and outgoing momentum and $\vec{\sigma}$ is a spin vector. The skew scattering appears after the average of the impurity, and its lifetime denoted by τ_{sk} is given by

$$1/\tau_{sk}{}^{\mu\nu} = i\epsilon_{\mu\nu\gamma}\langle(a_{pp'}b_{p'p} - a_{p'p}b_{pp'})p_\mu{}^2 p_\nu{}^2/m^2\rangle_{imp} \times \frac{2\pi\nu_1 n_{imp}}{v_F^2/3}\langle\sigma_\gamma\rangle , \tag{2}$$

where $\epsilon_{\mu\nu\gamma}$ is antisymmetric tensor, μ,ν are space indices, ν_1 is a density of state for a single spin, and $\langle\sigma_\gamma\rangle$ is the γ component of the average of the spin which is proportional to the external magnetic field. It should be noted that a and b must have imaginary parts, and if they are real, the skew scattering is absent.

We find that the usual Aslamazov-Larkin term with a skew scattering does not provide a fluctuation part, hence we investigated the next term with a skew scattering vertex correction of Aslamazov- Larkin process (Fig.1).

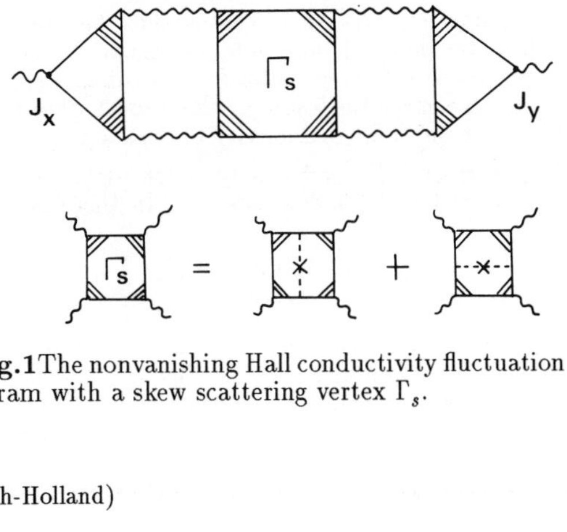

Fig.1 The nonvanishing Hall conductivity fluctuation diagram with a skew scattering vertex Γ_s.

We denote this term by $\Delta\sigma^{sk}_{xy}$. This quantity is calculated[2] and it is related to the diagonal conductivity fluctuation of Aslamazov-Larkin term as

$$\Delta\sigma^{sk}_{xy} = -\Delta\sigma^{AL}_{xx}\tau B^{yy}/\sigma_0\tau^{xy}_{sk}, \qquad (3)$$

where σ_0 is the normal conductivity and τ is the impurity scattering lifetime. In two dimensions, the Hall conductance fluctuation Δg^{sk}_{xy} becomes

$$\Delta g^{sk}_{xy} = -\frac{2e^2}{\pi^2\hbar}\frac{\tau}{\tau_{sk}}(\frac{\Delta\sigma^{AL}_{xx}}{\sigma_0})\log\frac{\pi^2 e}{4\epsilon}, \qquad (4)$$

where $\Delta\sigma^{AL}_{xx} = e^2/16\hbar d\epsilon$, d is the thickness of the two dimensional film. In three dimensions, we have, $B^{yy} = 5e^2(2\pi T/D)^{1/2}/12\pi^2$, D is a diffusion constant. Above formula is applied also in the presence of the magnetic field and the field dependence of ${\sigma_{xx}}^{AL}$ was discussed before.[3]

From these result, we find that the singularity of the temperature dependence of the Hall conductance fluctuation is same as the diagonal conductivity in three dimensions and only logarithmic correction appears in two dimensions. The sign of the fluctuation part is determined by the above formula and it has opposite sign to the normal part or to the Aslamazov-Larkin diagonal fluctuation. This is due to the fact that we consider the next order of the Ginzburg-Landau Hamiltonian and the opposite sign does not depend on the sign of carriers, hole or electron.

We also investigated the skew scattering effect on Hall conductivity fluctuation due to the Maki-Thompson term. We denote this contribution by $\Delta\sigma^{MT}_{xy}$ and we obtain from the diagrams of Fig.2 as

$$\Delta\sigma^{MT}_{xy} = (2\omega_c\tau + \tau/\tau_{sk})\Delta\sigma^{MT}_{xx},$$

where ω_c is a cyclotron frequency. We notice that the sign is same as the diagonal case for the Maki- Thompson process. Therefore, our skew scattering effect has a different sign for the Aslamazov-Larkin term and for the Maki-Thompson term. We also notice that the crossover point from the Maki-Thompson to Aslamazov-Larkin region appears at a different point from the diagonal conductivity case, since the leading Aslamazov-Larkin term is absent for the Hall conductivity fluctuation.

The normal Hall conductivity is considered also to be influenced by the skew scattering. In this case, we have

$$\sigma_{xy} = (\omega_c\tau + \tau/\tau_{sk})\sigma_{xx}.$$

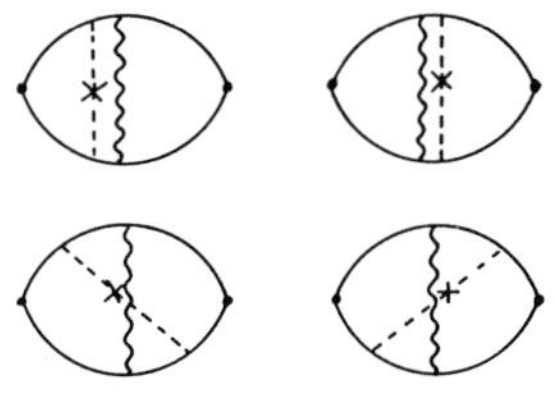

Fig.2 Skew scattering effect on Maki-Thompson fluctuation

3.DISCUSSION

We consider that it may be possible to estimate the value of the ratio τ/τ_{sk} by the comparison of the formula presented here to the experimental value of the fluctuation. The skew scattering of high temperature superconductors may be different from the conventional anomalous Hall effect in which the saturation effect is observed for a strong magnetic field. We consider the possibility of the spin-orbit interaction where the spin belongs to the conduction carrier and the impurity has no spin. In this case, the saturation for the strong magnetic field may be not observed. The imaginary character of a and b for the scattering amplitude mean the resonance and it may be of interest to see the difference from the other similar cases.[5,6] We emphasize that the existence of the skew scattering effect is important as the basic transport property of high temperature superconductors.

ACKNOWLEDGEMENTS

We acknowledge the support by Yamada Science Foundation and also by Mitsubishi Foundation.

REFERENCES

1) Y. Matsuda, a private communication.

2) A. G. Aronov and S. Hikami, Phys. Rev. **B41** in press.

3) S. Hikami and A. I. Larkin, Mod. Phys. Lett. **B2** (1988), 693.

4) A. G. Aronov, S. Hikami and A. I. Larkin, Phys. Rev. Lett. **62**(1989), 965.

5) P. Coleman, P. W. Anderson and T. V. Ramakrishnan, Phys. Rev. Lett. **55** (1985), 414.

6) A. Fert and O. Jaoul, Phys. Rev. Lett. **28** (1972), 303.

Physica B 165&166 (1990) 275–276
North-Holland

On the Self-Interaction Correction of Localized Bands: Application to the 4p semi-core states in Y.

Z. SZOTEK, W.M. TEMMERMAN and H.WINTER*

SERC Daresbury Laboratory, Warrington WA4 4AD, UK
*Kernforschungszentrum Karlsruhe, INFP, P.O.Box 3640, D-7500 Karlsruhe, FRG

We apply the self-interaction correction to the Local Density Approximation, implemented within the Linear-Muffin-Tin-Orbital Atomic-Sphere-Approximation band structure method, to the 4p semi-core states in Yttrium. We show that this correction improves the theoretical lattice constant indicating the importance of the self-interaction correction for the bonding properties.

1. INTRODUCTION

In the Density Functional (DF) theory each electron interacts with itself through the Coulomb electrostatic energy. In the exact formalism this unphysical interaction would be canceled by the exchange and correlation energy contribution (1). However, in the commonly practised Local Density Approximation (LDA) to the DF theory this cancelation is not exact due to the use of approximate functionals. One can improve on the LDA by considering the self-interaction corrected (SIC) functionals (2).

Contrary to the application of the SIC to atoms (3) the SIC in solids is less well founded. The magnitude of the correction depends on the degree of localization of the one-electron wave functions; eg a Wannier function giving rise to a larger SIC than a Bloch function. In principle one should minimize the total energy with respect to all possible one-electron wave functions, however, in practice one uses the Wannier functions.

The bonding properties of the elemental metals are well understood within the LDA (4,5). However, there is a tendency for overbonding at the beginning of the transition metal series. For example the theoretical lattice constant of Yttrium is underestimated by more than 2%. In a recent publication (6) we have demonstrated that a major contribution to this overestimate in the bonding comes from the 4p semi-core states. Here we apply the self-interaction correction to these 4p states and show that it can improve the bonding properties of Y metal.

From numerous studies it has been established that the SIC will further localize localized states and is zero for fully delocalized states (1). Recently, we have implemented the SIC in the LMTO-ASA bandstructure method, and have applied it to the rare gas solids Ne, Ar, Kr and Xe (7). In particular, we have been able to demonstrate that the SIC reduced the LDA bonding in these systems by about 50%. With respect to Y metal the scenario is that upon the application of the SIC to the 4 p semi-core states, these states will further localize and the lattice constant will expand.

2. FORMALISM

We derived in a previous paper (7) the following LDA-SIC LMTO-ASA Hamiltonian:

$$
\begin{aligned}
< \mathrm{X}^{\mathbf{k}}_{\mathrm{L}}(\mathbf{r}) \mid \mathrm{H}_{\mathrm{u}} \mid \mathrm{X}^{\mathbf{k}}_{\mathrm{L'}}(\mathbf{r}) > = & < \mathrm{X}^{\mathbf{k}}_{\mathrm{L}}(\mathbf{r}) \mid \mathrm{H}_0 \mid \mathrm{X}^{\mathbf{k}}_{\mathrm{L'}}(\mathbf{r}) > - \\
& \sum_{\mathrm{i,j(i.ne.j)}}^{\text{occupied}} \mathrm{M}^{\mathbf{k}}_{\mathrm{Li}} < \Psi_{\mathbf{k},\mathrm{i}} \mid \mathrm{H}_0 \mid \Psi_{\mathbf{k},\mathrm{j}} > \mathrm{M}^{\mathbf{k}}_{\mathrm{jL'}} + \\
& \sum_{\mathrm{i}}^{\text{occupied}} \mathrm{M}^{\mathbf{k}}_{\mathrm{Li}} < \Psi_{\mathbf{k},\mathrm{i}} \mid \widehat{\mathrm{V}}^{\mathrm{sic}} \mid \Psi_{\mathbf{k},\mathrm{i}} > \mathrm{M}^{\mathbf{k}}_{\mathrm{iL'}} + \\
& \sum_{\mathrm{i}}^{\text{occupied}} \sum_{\mathrm{j}}^{\text{unoccupied}} \mathrm{M}^{\mathbf{k}}_{\mathrm{Lj}} < \Psi_{\mathbf{k},\mathrm{j}} \mid \widehat{\mathrm{V}}^{\mathrm{sic}} \mid \Psi_{\mathbf{k},\mathrm{i}} > \mathrm{M}^{\mathbf{k}}_{\mathrm{iL'}} + \\
& \sum_{\mathrm{i}}^{\text{occupied}} \sum_{\mathrm{j}}^{\text{unoccupied}} \mathrm{M}^{\mathbf{k}}_{\mathrm{Li}} < \Psi_{\mathbf{k},\mathrm{i}} \mid \widehat{\mathrm{V}}^{\mathrm{sic}} \mid \Psi_{\mathbf{k},\mathrm{j}} > \mathrm{M}^{\mathbf{k}}_{\mathrm{jL'}}
\end{aligned}
$$

where $\mathrm{M}^{\mathbf{k}}_{\mathrm{Li}} = < \mathrm{X}^{\mathbf{k}}_{\mathrm{L}}(\mathbf{r}) \mid \Psi_{\mathbf{k},\mathrm{i}}(\mathbf{r}) >$,

$$
\begin{aligned}
< \Psi_{\mathbf{k},\mathrm{j}}(\mathbf{r}) \mid \widehat{\mathrm{V}}^{\mathrm{sic}} \mid \Psi_{\mathbf{k},\mathrm{i}}(\mathbf{r}) > = & \sum_{m,n} < \Psi_{\mathbf{k},\mathrm{j}}(\mathbf{r}) \mid \Psi_{\Gamma,m}(\mathbf{r}) > \\
& < \mathrm{w}_m \mid \mathrm{V}^{\mathrm{sic}}_n \mid \mathrm{w}_n > < \Psi_{\Gamma,n}(\mathbf{r}) \mid \Psi_{\mathbf{k},\mathrm{i}}(\mathbf{r}) >
\end{aligned}
$$

and $< \mathrm{X}^{\mathbf{k}}_{\mathrm{L}}(\mathbf{r}) \mid \mathrm{H}_0 \mid \mathrm{X}^{\mathbf{k}}_{\mathrm{L'}}(\mathbf{r}) >$ is the standard LDA LMTO-ASA Hamiltonian (8). In this context 'occupied' means all SIC corrected bands, and 'unoccupied' refers to the bands which are not SIC corrected. The matrix elements can readily be evaluated and we found most useful to write $\mathrm{X}^{\mathbf{k}}_{\mathrm{L}}(\mathbf{r})$ as

$$
\mathrm{X}^{\mathbf{k}}_{\mathrm{L'}}(\mathbf{r}) = \sum_{\mathrm{L}} \left(\phi_{\nu\mathrm{L}}(\mathbf{r}) \Pi^{\mathbf{k}}_{\mathrm{LL'}} + \dot{\phi}_{\nu\mathrm{L}}(\mathbf{r}) \Omega^{\mathbf{k}}_{\mathrm{LL'}} \right)
$$

with Π, Ω and ϕ_ν described in (8) , and $\dot{\phi}_\nu$ being an energy derivative of ϕ_ν. The Wannier states are obtained from symmetrized Bloch states:

$$
\Psi^{s}_{\mathbf{k},n}(\mathbf{r}) = \sum_{m=1}^{N} \Psi_{\mathbf{k},m}(\mathbf{r}) < \Psi_{\mathbf{k},m}(\mathbf{r}) \mid \Psi_{\Gamma,n}(\mathbf{r}) >
$$

by performing the following Brillouin zone integral

$$
\mathrm{w}_n(\mathbf{r}) = \int \mathrm{d}\mathbf{k}\, \Psi^{s}_{\mathbf{k},n}(\mathbf{r})
$$

Here $\Psi_{\mathbf{k},n}(\mathbf{r})$ are the unsymmetrized Bloch functions with band index n and at a **k** point in the Brillouin zone, and $\Psi_{\Gamma,n}(\mathbf{r})$ is the Bloch function at the Brillouin zone center (Γ point). The charge per band n is then obtained as

$$\rho_n(\mathbf{r}) = w_n^*(\mathbf{r}) \, . \, w_n(\mathbf{r})$$

and the SIC potential is defined as

$$V_n^{sic}(\mathbf{r}) = V_H[\rho_n(\mathbf{r})] + V_{xc}[\rho_n(\mathbf{r})]$$

consisting of a Hartree (H) and exchange and correlation (xc) term (3).

3. RESULTS

In the table below we present the results of our calculations for the percentage deviation of the theoretical lattice constant from the experimental value and the equilibrium volume. These have been performed using s,p and d muffin-tin orbitals. For simplicity we have performed the calculation for Y metal in the fcc structure, which should not affect our conclusions regarding the bonding properties (4). From column (a) we see that a frozen core LDA calculation results in a lattice constant which is only 0.55% smaller than the experimental value. Relaxing the core (column b) reduces the lattice constant by 0.85%. This relaxation consists of an SCF-LDA of all the core states from the 1s to the 4s with a relativistic atomic/ code, and all the higher states are treated as bands through the use of panels. The lower panel consists of the 4p bands, whilst the higher panel is made up of 5s, 5p and 4d bands. Applying the SIC to the core states and the lower panel (column c) expands the lattice by 0.58% and a result similar to the LDA frozen core calculation is obtained. Freezing all the core levels up to the 4s inclusive, and applying the SIC to the lower panel only (coulumn d), slightly affects the results of the LDA-SIC (similar observation is true in the case of the LDA), demonstrating that the semi-core 4p bands, together with the 5s, 5p and 4d bands, determine the bonding properties in Y metal. Even though the SIC increases the deeper the core states are, these states do not partake in the bonding and therefore the bonding properties of these states are essentially unaffected by the SIC. For the 4p states however the SIC is sufficiently big (ΔE=5.5 eV), and these 4p states are just shallow enough to participate in the bonding. Therefore SIC correcting the 4p states in Y metal affects the bonding properties: it corrects for overbonding due to the 4p states in the LDA.

Property	LDA (a)	LDA (b)	LDA-SIC (c)	LDA-SIC (d)
% deviation from exp. lattice constant	-0.55	-1.4	-0.82	-0.66
equilibrium volume (a.u.)	219.2	213.7	217.4	218.4

4. CONCLUSION AND OUTLOOK

We have demonstrated that the SIC plays a role in the bonding properties of Y metal, improving the theoretical lattice constant by 40% and we are presently doing a systematic investigation of the influence of the SIC on the bonding properties of the early transition metals.

The SIC could also provide an explanation of the success of the calculations which put the f states in the core such as for some rare earths, Pr (9) and rare earth metal compounds, $CeAl_2$ (10). In a SIC corrected band structure the occupied and localized f states are pulled below the bottom of the conduction band. Consequently, these f states do not interact directly with the conduction bands. Freezing these f states in the core in the LDA calculation mimics this scenario.

REFERENCES

(1) R.O. Jones and O. Gunnarsson, Rev. of Mod. Phys. 61 (1989) 689.
(2) R.A. Heaton,, J.G. Harrison, and C.C. Lin, Phys.Rev. B28 (1983) 5992.
(3) J.P. Perdew and A. Zunger, Phys.Rev. B23 (1981) 5048.
(4) V.L. Moruzzi, J.F. Janak, and A.R. Williams, Calculated Electronic Properties of Metals (Pergamon, New York, 1978).
(5) O.K. Andersen, O. Jepsen and D. Glotzel, Highlights of Condensed Matter Theory, eds. F. Bassani, F. Fumi and M.P. Tosi (North-Holland, Amsterdam, 1985).
(6) W.M. Temmerman and P.A. Sterne, in print.
(7) Z. Szotek, W.M. Temmerman and H. Winter, to be published.
(8) O.K. Andersen, Phys.Rev. B12 (1975) 3060.
(9) M. Wulff, Ph.D. Dissertation (University of Cambridge, 1985).
(10) T. Jarlborg, A.J. Freeman and D.D. Koelling, JMMM 60 (1986) 291.

Physica B 165&166 (1990) 277–278
North-Holland

STATIC SKIN EFFECT AT HIGH CURRENT DENSITIES.

K.Oyamada, V.G.Peschansky, D.I.Stepanenko.

Institute for Low Temperature Physics & Engineering, UkrSSR Academy of Sciences, Kharkov, USSR.
Kharkov State University, USSR.

The nonlinear effects in the conductivity of metals and intrinsic semiconductors placed into homogeneous magnetic field Ho in the conditions of static skin effect are investigated. The electric current distribution, Hall field, and intrinsic magnetic field of current H_j in the sample are obtained by the asymptotic solution of the system of Maxwell equations and kinetic equation in the case of slowly changing of total magnetic field at a distance of the order of electron Larmor radius. The competition between the static skin effect and the pinch effect is analyzed. It is shown that by varying Ho one can study separately the dynamic nonlinear effect associated with the H_j influence on the ballistics of conduction electrons, and the deviation from the Ohm's law due to charge carriers heating by electric current.

In conductors whose resistance increase infinitely with increasing external magnetic field $\mathbf{H_o}$ (metals with open Fermi surfaces, compensated metals and intrinsic semiconductors with equal numbers of electrons N_1 and holes N_2) the electric current is displaced to the sample surface. The static skin effect (1-3) is related to a higher mobility in strong magnetic field (the electron trajectory radius r is much smaller than its mean free path l), of those charge carriers that interact with the conductor surface in contrast to bulk electron which do not collide with the sample boundary. For a magnetic field parallel to the plate faces $x_s = \pm d/2$, the conduction electrons reflected specularly from these faces can drift in the plane normal to $\mathbf{H_o}$ until they are scattered in the bulk even with no open section available. The electrical conductivity of the subsurface layer, roughly $2r \sim H_o^{-1}$ thick, turns to be of the order the bulk electrical conductivity σ_o (d>>l) with $H_o=0$ while the electric current inside the sample is strongly damped by the magnetic field for r<<l,d if the angle θ between the current density $\mathbf{j}$ and the field $\mathbf{H_o}$ is much larger then r/l. For $r \ll l^2\sin^2\theta/d$ the electric current is almost completely concentrated near the sample surface and the resistivity, being inversely proportional to the current-carrying layer thickness 2r, increases linearly with magnetic field H_o.

Presence of surface defects such as roughness or adatoms results in a decrease in the transport free path of conduction electrons colliding with the sample surface and hence in a decrease in the skin-layer electrical conductivity:

$$\sigma_{skin} = \sigma_o \frac{r}{r+wl\sin^2\theta} \quad (1)$$

where w is the effective span of the charge scattering indicatrix by sample boundary, which accounts for deviations from specularity. Here and below we omit insignificant numerical factors of the order of unity which depend of specific form of the dispersion relation of conduction electrons.

For high current densities, effects of the intrinsic magnetic field $\mathbf{H_J}$ on electron ballistics can cause essential changes in the current distribution over the sample. In inhomogeneous magnetic field $\mathbf{H_J}$ the charge carriers start to drift normal both to magnetic field $\mathbf{H}=\mathbf{H_o}+\mathbf{H_J}$ and to its gradient, thereby increasing the electrical conductivity in the bulk and brining about a competition between the skin and pinch effects (4). Electric current distribution, magnetic field $\mathbf{H_J}$ and electric field $\mathbf{E}$ for $(r/H)dH/dx \ll 1$ can be then found by solving self-consistently the Maxwell equations and kinetic equation with a boundary condition involving the charge carriers scattering by the sample boundary.

Another mechanism of deviation from the Ohm law, connected with the current heating of charge carriers, produces essentially no effects on the current distribution profile. For semiconductors and semimetals this nonlinear mechanism is of basic importance. If the drift velocity of charge carriers $\mathbf{u}=c[\mathbf{EH}]/H^2$ becomes comparable sound velocity s, a kink in I-V curve of the sample can be observed (the Esaki effect). Several kinks will be observed under the static-skin conditions ($wd/l<1$, $\theta>>r/l$). As j is increased, an "avalanche" emission of phonons at $u=s$, occurs first in the skin layer for

$$j = j_1 = \frac{Nes}{\sin^2\theta}\frac{r}{d}\frac{r+wl}{r+wl\sin^2\theta}$$

and then in the sample bulk for

$$j = j_2 = \frac{Nes}{\sin^2\theta}\frac{r}{d}\frac{l}{r+wl\sin^2\theta}$$

(e being the electron charge, c the velocity of light, $N=N_1+N_2$).

A weak dependence of j_1 on w with $r<<d$ is due to the fact that the current is almost completely concentrated within the skin layer and amount of its small part outside the skin layer depends strongly on the surface condition: the higher is the degree of specularity, the lesser is the part let to the sample bulk. The surface condition can be judged upon from position of the kinks in the I-V curve:

$$\frac{j_1}{j_2} = w + \frac{r}{l} \tag{2}$$

For metals with charge densities of order of one per atom, the conduction electron drift caused by the inhomogeneous magnetic field is more appreciable. In wide range of external magnetic field H_0 the weak heating of conduction electrons can be neglected as compared to the effects of the intrinsic magnetic field as far as electrical conductivity is concerned. In this case the nonlinear electrical conductivity is determined mainly by the behavior of conduction electrons at the Fermi surface in inhomogeneous magnetic field whereas the nonlinear effect becomes less pronounced with increasing H_0.

For small current densities when $H_J<<H_0$ the electrical conductivity of the sample bulk σ_{vol} is of the form:

$$\sigma_{vol} = \sigma_0 \frac{(r/l)^2}{(r/l)^2+\sin^2\theta} +$$

$$+ \sigma_0\left(\frac{j}{Nes}\right)^2\left(\frac{\omega_0}{\Omega cl\sin\theta}\right)^4 (svd)^2\,(r+wl\sin^2\theta)^2 \tag{3}$$

where ω_0 is plasma oscillations frequency of the electron gas, Ω is the gyrofrequency in magnetic field H_0.

The intrinsic magnetic field of current increases considerably when approaching the sample surface because the current density in the skin layer is much higher than that far from the conductors boundaries. Allowance for the nonlinear effects, however, does not affect the σ_{skin} value.

For

$$j \geq Nes\,\frac{c^2vl\sin\theta}{s\omega_0^2rd(r+wl\sin^2\theta)}$$

account of the nonlinear effects not only in the skin layer but also inside the sample is of essential importance. If $H_J>H_0\sin\theta$ the current is pinched. The shape and width Δ of the pinch differ greatly for different angles θ. The pinch becomes narrower with increasing j, however with $\theta<r/l$, Δ is still of the order of d if $H_J\simeq H_0$. The current profile across the sample for $\theta<r/l$ is of the form:

$$j(x) = \begin{cases} \dfrac{\sigma_0 E}{1+\mu\,(\alpha x)^2} & |\alpha x\mu^{1/2}| < 1; \\[2ex] \left(\dfrac{\sigma_0 E}{\mu}\right)^{1/3}\left(\dfrac{cH_0}{x}\right)^{2/3} & |\alpha x\mu^{1/2}| >> 1; \end{cases} \tag{4}$$

$$\alpha = 4\pi\sigma_0 E/H_0, \quad \mu = \frac{\Omega^2\tau^2}{1+\Omega^2\tau^2}, \quad \tau = 1/v.$$

Under the static skin conditions ($wd/l<1$, $\theta>>1$), the region of maximum current in sample is fairly narrow, so that for almost any attainable current densities $r\sigma_{skin}/d$ is much larger than σ_{vol}.

In high enough magnetic fields when $\Omega > (\omega_0/c)(svd/l)^{1/2}$, which for $d=l=0.01$ cm corresponds to $H_0>10^5$ Oe, the dynamic nonlinear effect associated with effects of the intrinsic magnetic field on the ballistics of conduction electrons is suppressed and one may expect the appearance of the kink in the I-V curves to appear due to the heating of the charge carriers.

Thus, the dynamic nonlinear effect and deviations from the Ohm law related to current heating of charge carriers in metals can be studied separately by varying external magnetic field H_0.

REFERENCES.

(1) M.Ya.Azbel, and V.G.Peschansky, JETP, 49 (1965) 579; 52 (1967) 1003; 55 (1968) 1980.
(2) O.V.Kirichenko, V.G.Peschansky, and S.N.Savelieva, JETP, 77 (1979) 2045.
(3) V.G.Peschansky, J.Stat.Phys., 38 (1985) 253.
(4) V.G.Peschansky, K.Oyamada, V.V.Polevich, JETP, 67 (1974) 1989.

Physica B 165&166 (1990) 279–280
North-Holland

NONLINEAR RESPONSE OF A MESOSCOPIC SYSTEM

J. Liu and N. Giordano

Department of Physics, Purdue University, West Lafayette, IN 47907, USA

The response of a mesoscopic Sb film to microwave radiation has been studied. We have found that the system acts as a rectifier; i.e., a dc current, I_{dc}, is produced when the sample is irradiated with a microwave field. Results for the dependence of I_{dc} on microwave power, temperature, and magnetic field are presented, and shown to be in good agreement with the theory of Fal'ko and Khmel'nitskii.

1. INTRODUCTION

It has recently been pointed out by Fal'ko and Khmel'nitskii (1,2) that, due to the random nature of the impurity distribution, a disordered mesoscopic system will not exhibit a center of inversion. Strictly speaking, this is true of all disordered systems, but in macroscopic samples it is appropriate to average over the impurity distribution, which yields, to what is usually a very good approximation, a symmetric system. However, such averaging is not appropriate for a mesoscopic system. It is well known that a system which does not have inversion symmetry will exhibit a dc current (i.e., rectification) when an ac field is applied. Such nonlinear effects have been discussed theoretically (1-4), and the associated harmonic generation has been observed experimentally (5-7). The previous experiments have all been conducted at fairly low frequencies (typically $\lesssim$100 Hz). In the present work we study the behavior at frequencies such that $\omega \gtrsim \tau_\phi^{-1}$, where τ_ϕ is the electron phase coherence time. At these frequencies the nonlinearities are due to electrons which are excited by the ac field, and the effect is predicted to be a strong function of ω. This is in contrast to the low frequency limit, in which the effect is independent of frequency. Here we report the observation of this rectification effect in mesoscopic Sb films in the presence of 8 GHz radiation.

2. RESULTS

The samples were made from evaporated Sb films, and had sheet resistances of ~100 Ω. The particular sample considered below was 2 μm×2 μm×100 Å, and was connected to much larger regions of the same film on two ends; the measurements were thus performed in a two lead geometry. Previous studies in our laboratory (8) have found electron phase coherence lengths $L_\phi \sim 3000-8000$ Å for macroscopic Sb films in the temperature range studied here. Hence, the sample size was slightly larger than L_ϕ. The apparatus was similar to that described in (8), and the absolute magnitude of the microwave field in the sample was accurately known from our previous work.

The inset in Fig. 1 shows results for I_{dc} as a function of E_{ac}^2 at 4.2 K, where E_{ac} is the amplitude of the ac electric field. It is seen that I_{dc} is proportional to E_{ac}^2, in good agreement with the theory, which in two dimensions predicts (1,2)

$$I_{dc} = 4\pi\sqrt{\zeta(3)}\, e \left(\frac{\omega}{\tau_\phi}\right)^{1/2} \left(\frac{eE_{ac}\, L}{\pi\hbar\omega}\right)^2 , \quad (1)$$

where L is the length of the sample. The magnitude of I_{dc} at the largest fields in Fig. 1 is ~0.3 nA. Using our best estimates (8) for the parameters which enter Eq. 1, we calculate I_{dc} =0.3 nA, in good agreement with the experimental value. However, the theory, Eq. 1, assumes $L \lesssim L_\phi$, whereas for our sample $L/L_\phi \sim 4$ at 4.2 K. This leads to a factor of 4 reduction in the expected value of I_{dc} (9), but the agreement is still quite acceptable, especially since

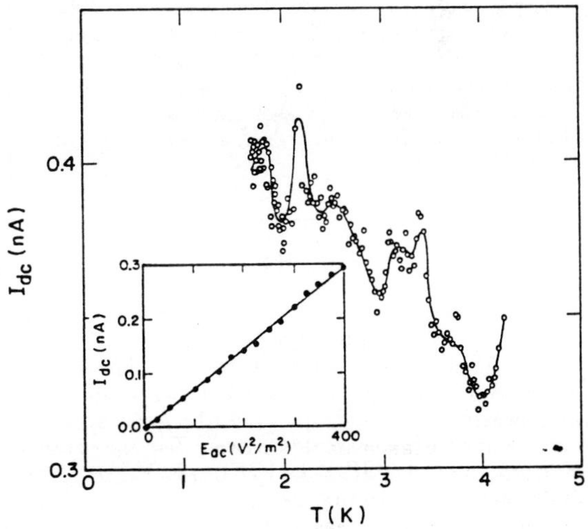

FIGURE 1
I_{dc} as a function of temperature. The microwave frequency was 8.4 GHz, with E_{ac}=20 V/m. The inset shows I_{dc} as a function of E_{ac}^2, at 4.2 K. The solid lines are guides to the eye.

the theory predicts only the *rms* value of I_{dc}. The main part of Fig. 1 shows the variation of I_{dc} with T. This variation is seen to be nonmonotonic, and quite irregular. This is expected, since the actual magnitude of I_{dc} will depend on the particular impurity distribution which is present. If the temperature is changed by a sufficiently large amount, one effectively samples a different distribution (3). For changes $\Delta T \sim E_c / k_B$, where the correlation energy is given by $E_c \sim \pi^2 \hbar \tau_\phi^{-1}$, the value of I_{dc} is predicted to vary considerably. For the sample considered in Fig. 1, $\Delta T \sim 0.3$ K at $T = 1.5$ K, which is consistent with our results.

Figure 2 shows results for I_{dc} as a function of a magnetic field, H, applied perpendicular to the plane of the sample. Again one sees large fluctuations in I_{dc}. These fluctuations are *not* due to random time-dependent noise. They are completely reproducible, and are quite reminiscent of the aperiodic fluctuations observed in the conductance of mesoscopic systems. Their reproducibility is demonstrated by comparing the results for two separate measurements; in one case the field was swept up, while in the other it was swept down. The fluctuations in the two cases are seen from Fig. 2 to be very similar. The theory predicts that changes in the magnetic field such that

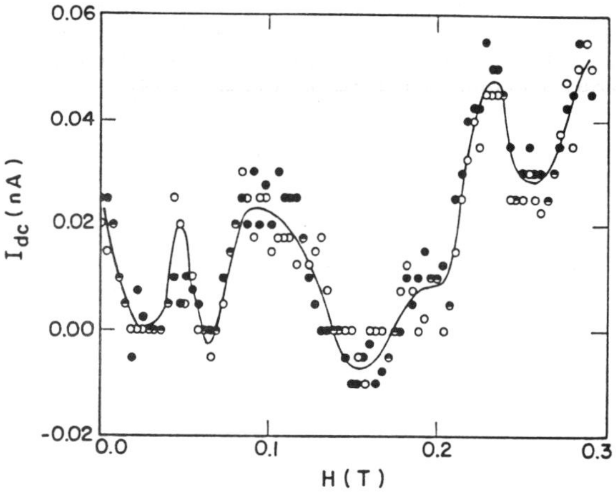

FIGURE 2
I_{dc} as a function of H, at 4.2 K for the sample considered in Fig. 1. The solid symbols were obtained while sweeping the field up, while the open symbols were obtained while sweeping down. The microwave frequency was 8.4 GHz, and $E_{ac} = 20$ V/m. The solid line is a guide to the eye.

$(\Delta H) L_\phi^2 \sim \phi_0$ (where $\phi_0 = hc/e$ is the flux quantum) will produce a large change in I_{dc}. For the case shown in Fig. 2, we calculate $\Delta H \sim 0.04$ T, which is quite consistent with our results. The fluctuations as a function of H were reproducible, so long as the sample was treated with "care." However, if it was exposed to electrical transients, or the temperature was cycled from 4 K to 77 K and back, the fluctuation pattern was completely different (although it was qualitatively similar). This was probably due to impurity motion within the sample, which is equivalent to producing a "new" sample (3).

3. CONCLUSIONS

We have studied the nonlinear response of a mesoscopic Sb film through measurements of the dc current induced by an applied microwave field. The magnitude of I_{dc} is in good agreement with the theory, and the sample specific fluctuations as a function of T and H are in accord with our general expectations for a mesoscopic system.

ACKNOWLEDGMENTS

We thank V. I. Fal'ko for helpful correspondence. This work was supported in part by the National Science Foundation through grants DMR-8614862 and DMR-8915574.

REFERENCES

(1) V. I. Fal'ko and D. E. Khmel'nitskii, Zh. Eksp. Teor. Fiz. **95** (1989) 328 [Sov. Phys. JETP **68** (1989) 186].

(2) V. I. Fal'ko, Europhys. Lett. **8** (1989) 785.

(3) B. L. Al'tshuler and D. E. Khmel'nitskii, Pis'ma Zh. Eksp. Teor. Fiz. **42** (1985) 291 [JETP Lett. **42** (1985) 359].

(4) A. I. Larkin and D. E. Khmel'nitskii, Zh. Eksp. Teor. Fiz. **91** (1986) 1815 [Sov. Phys. JETP **64** (1986) 1075].

(5) P. G. N. de Vegvar, G. Timp, P. M. Mankiewich, J. E. Cunningham, R. Behringer, and R. E. Howard, Phys. Rev. B **38** (1988) 4326.

(6) R. A. Webb, S. Washburn, and C. P. Umbach, Phys. Rev. B **37** (1988) 8455.

(7) S. B. Kaplan, Surf. Sci. **196** (1988) 93. This measurement actually did not involve an ac field, but was sensitive to essentially the same nonlinear effects observed in (5) and (6).

(8) J. Liu and N. Giordano, Phys. Rev. B **39** (1989) 9894.

(9) For a sample with $L > L_\phi$, the dc current of a region of size $L_\phi \times L_\phi$ will be given by Eq. 1, but the contributions of different regions will add incoherently. This will reduce the measured I_{dc} by a factor L / L_ϕ, which is ~4 for our sample at 4.2 K.

Physica B 165&166 (1990) 281–282
North-Holland

MODIFICATION OF WEAK LOCALIZATION IN THIN ANNEALED PD FILMS

Nicos PAPANDREOU and Pierre NEDELLEC

Centre de Spectrométrie Nucléaire et de Spectrométrie de Masse, Bâtiment 108, 91405 Campus ORSAY, FRANCE

We have measured the temperature and magnetic field dependance of the conductance of thin annealed Pd films. Our data show modification to the behavior of corresponding homogeneous films. We analyze the results in term of interplays between localization and percolation.

1. INTRODUCTION

Unusual electrical transport behaviour appears in metallic films when two characteristic scales of disorder are present (1) : an intrinsic atomic disorder responsible for electron localization (2) at low temperature and a macroscopic disorder at larger scale related, for instance, to the presence of holes in our metallic samples. An electric current flows through the film as long as the metallic surface coverage p exceeds the percolation threshold p_c (3). Near p_c, the conductance goes to zero :

$$C \sim (p - p_c)^{\mu} \qquad 1$$

and μ is typically 1.3 for 2D system.

When treating the electronic localization in such an inhomogeneous metal, different lengths must be considered : i) the temperature T and the magnetic field H induced scales respectively the inelastic collision length which varies as $L_\varphi \sim T^{-\alpha/2}$ (2) and the magnetic length L_H related to H by (4) :

$$L_H^2 = \frac{\hbar}{4eH} \qquad 2$$

Note that L_H is relevant only if $L_H < L_\varphi$. ii) the percolation length ξ_p, the scale above which the system appears to be homogeneous.

At low temperature L_φ can be of the same order than ξ_p and the electronic properties are sensitive to both percolation and localization. When the atomic disorder is sufficiently diluted, the theory of weak localization (2,4) describes the modifications of the conductance and the magnetoconductance (MC). Two behaviours are distinguished (5,6). In the homogeneous regime (L_H, $L_\varphi > \xi_p$), the inhomogeneities are averaged and the conductance is that of a homogeneous 2D metal corrected by a geometrical form factor :

$$C(T) \sim (p-p_c)^{\mu} \log(T) \qquad 3$$

The MC $\Delta C(H) = C(H) - C(H{=}0)$ is given by :

$$\Delta C = -(p-p_c)^{\mu} \frac{e^2}{4\pi^2\hbar}\left[\Psi\left(\frac{1}{2} + \frac{H_\varphi}{H}\right) - \log\left(\frac{H}{H_\varphi}\right)\right] \qquad 4$$

where H_φ is related to the inelastic length L_φ (equ. 2). The MC of Pd is negative because of the strong spin-orbit interaction (4). The second regime is the inhomogeneous regime (L_H, $L_\varphi < \xi_p$). The conducting cluster is seen by the electrons as a geometrical environment with a fractal dimension. The conductance is expected to vary as (5) :

$$C(T) \sim (p-p_c)^{\mu}\, T^{-0.34\,\alpha} \qquad 5$$

No theoretical predictions are available on the MC variation in the inhomogeneous regime.

2. FILM PREPARATION

In order to obtain a model system suitable for the study of the percolation-localization crossover, we have used annealed Pd films. Polycrystalline homogeneous pure Pd films are electron gun evaporated onto a quartz (transport measurement) and NaCl substrate (TEM analysis) at room temperature in high vacuum ($<10^{-8}$ torr). On the same substrate different films are obtained with thicknesses from 45 to 75 Å. The samples are then annealed in an Ar atmosphere, the temperature rising step by step from 150 °C to 300 °C.

The film topology is drastically altered by this procedure : heating causes an increase in the size and thickness of the crystallites resulting in the appearance of voids between the metallic grains. We obtain a distribution of the metallic surface coverage leading to a percolative transition; the thinner the initial film, the closer it is to the percolation threshold.

TEM analysis has been performad on films of different thicknesses but annealed during the same run. It shows the critical surface coverage is between 0.4 and 0.5. Thick films (> 60 Å), above the threshold, exhibit a well-connected metallic cluster with typically 500 Å wide channels. The thinner films present smaller grains. A first attempt to determine the fractal dimensionnality gives 1.7 ± 0.1.

3. LOW TEMPERATURE MEASUREMENTS

We have measured the low-temperature conductance and MC of four annealed samples lying at different distances of the percolation threshold.

The MC can be accounted for by equ. 4 and we deduce the geometrical form factor independant of T (the values are normalized assuming a factor 1 for the less resistive film), H_φ, thus the inelastic length L_φ. Table 1 summarizes our results at 4.2K.

$R_\square$ (4.2K) Ω	$L\varphi$(Å)	Geometrical factor ΔC(H)	C(T)
250	620	1	1
380	560	0.94	0.79
1250	270	0.56	0.59
6500	300	0.43	0.47

Table 1

$L\varphi$ follows the approximate law $L\varphi \sim T^{-0.75}$ which leads to $\alpha \sim 1.5$. This value deduced on a small temperature range (1K - 5K) can be interpreted as the sum of a linear ($\alpha = 1$) part due to electron-electron collisions and a higher power term relative to electron-phonon interaction as reported for homogeneous Pd films (7).

The variation ΔC(T) = C(T) - C(1.3 K) of the conductance for three annealed samples is presented in Figure 1. The slope of ΔC(T) is not universal as in the homogeneous case but decreases as the film approaches the percolation threshold. The geometrical factor values are in good

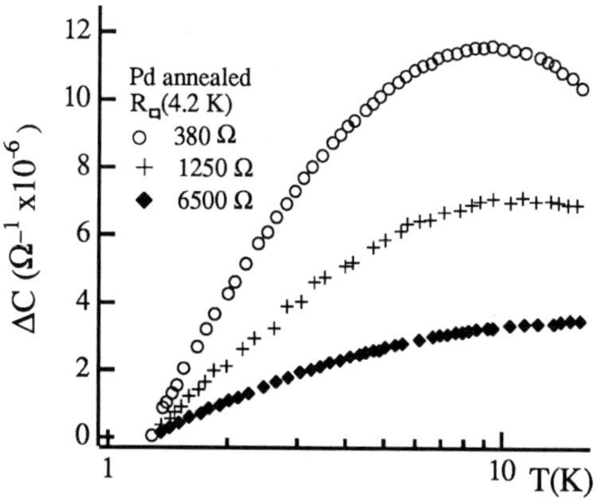

FIGURE 1

agreement with those deduced independently from the MC (Table 1). In Figure 2 a negative power law

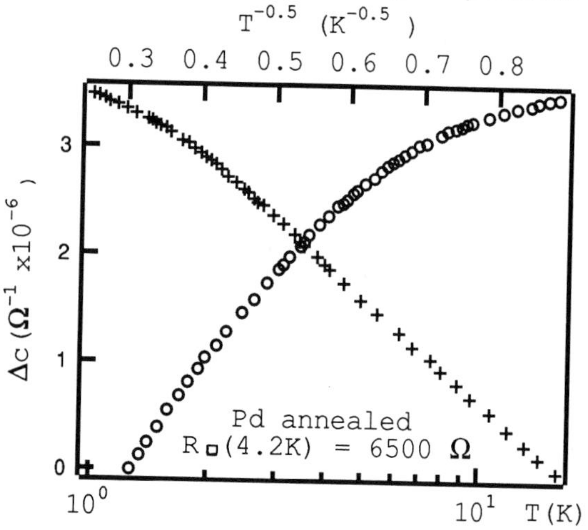

FIGURE 2

describes the experimental variation of ΔC. The same exponent is obtained with $\alpha = 1.5$. in equ. 5.

In the investigated temperature range, L_φ is smaller than 1000 Å and ξ_p is estimated above 2000 Å in a preliminary analysis of the TEM pictures. It seems reasonable to attribute the new behaviour of the conductance to a percolation-localization crossover.

To summarize, we have given evidence for a change of the low temperature electrical transport in inhomogeneous thin metallic films where a percolation-localization crossover occurs. These results are coherent with another similar study performed on perforated Pd films prepared by heavy-ion irradiation (8).

REFERENCES

(1) A. Palevski and G. Deutscher, Phys. Rev. B34 (1986) 431
(2) P.A. Lee and T.V. Ramakrishnan, Rev. of Mod. Phys. 57 (1985) 287
(3) D. Stauffer, Introduction to Percolation Theory (Taylor and Francis, London, 1985)
(4) G. Bergmann, Phys. Reports 107 (1984) 1
(5) Y. Gefen, D.J. Thouless and Y. Imry, Phys. Rev. B28 (1983) 6677
(6) D.E. Khmelnitskii, JETP Lett. 32 (1980) 229
(7) H. Raffy, P. Nédellec, L. Dumoulin, D.S. McLachlan and J.P. Burger, J. Physique 46 (1985) 627
(8) N. Papandreou and P. Nédellec, to be published

Physica B 165&166 (1990) 283–284
North-Holland

ANOMALOUS TRANSPORT AND MAGNETIC PROPERTIES OF $Al_{63}Cu_{25}Fe_{12}$ ICOSAHEDRAL SINGLE GRAINS

T. KLEIN, A. GOZLAN, C.BERGER, F.CYROT-LACKMANN,Y. CALVAYRAC*, A. QUIVY*, G. FILLION+

Laboratoire d'Etudes des Propriétés Electroniques des Solides, CNRS, B.P.166X - 38042 GRENOBLE Cedex, France
*Centre d'Etudes et de Chimie Métallurgique, CNRS, 15 rue Urbain -94407 VITRY Cedex, France
+Laboratoire L. Néel, CNRS, B.P.166X -38042 GRENOBLE Cedex, France

We report the first measurements of transport and magnetic properties on a monoquasicrystalline grain and on single icosahedral melt-spun phases of $Al_{63}Cu_{25}Fe_{12}$. The most salient feature is the very high resistivities measured for all samples : $\rho_{4K} \sim 4000\mu\Omega cm$. The anomalous $\rho(T,H)$ and $\chi(T)$ dependences are also discussed.

The new stable $Al_{63}Cu_{25}Fe_{12}$ quasicrystalline phases allow, for the first time, measurements on single-phase of very high structural quality (1,2), in contrast to other quasicrystalline materials such as AlMn (3). We present here measurements of $\rho(B,T)$ and $\chi(T)$ on single grains and i-melt-spun phases.

We show that the i-AlCuFe phase has a very high resistivity (ρ 300mK ~ 4000 $\mu\Omega$cm), and we present its temperature and magnetic field dependence.

Three kinds of samples were measured :
- a monograin of millimeter size resulting of a coarsening of small icosahedral grains during long annealing treatment (several days between 1133 and 1143K) (1) ;
- a polycrystalline sample obtained by conventional casting;
- rapidly solidified and annealed samples. Pure icosahedral phases of very high structural quality, are produced after annealing (800°C-1H) whereas the as-quenched samples are two-phased (i-phase + cubic AlFe-type phase) (1).

The resistivity of all samples was measured down to 300 mK and under magnetic fields up to 8 Tesla, using the four probes method. The electrical contacts were insured by silver glue or ultrasonic soldering. The most salient feature is the very high resistivity of all of the annealed samples (~ 4000 $\mu\Omega$cm) which is about one order of magnitude higher than in other quasicrystals (for instance $\rho \sim$ 200-400$\mu\Omega$cm in melt-spun AlMn quasicrystals (3), $\rho \sim$ 80$\mu\Omega$cm in the AlMgZn type i-phases (4), $\rho \sim$ 800$\mu\Omega$cm in the stable i-AlLiCu (5)). Our resistivities (measured over a set of a dozen of samples of each type) are also in accordance with the values published on an other composition ($Al_{65}Cu_{20}Fe_{15}$) (6-10). Furthermore a non expected feature is that the residual resistivity is increased by a factor of at least 3 upon annealing ($\rho_{4K} \sim 1300\mu\Omega cm$ for the as-quenched samples). This behavior is in striking contrast with that of crystals, for which the improvement of the crystalline state (elimination of defects) reduces the resistivity. Therefore, are those high resistivities an intrinsic property of quasicrystal that would be recovered in defect-free materials ?

The temperature dependence of resistivity (see figure 1) is strong and negative in the whole temperature range (0.3-300K) for all the high quality samples. Note that the temperature coefficient $\alpha = (1/\rho)\, \Delta\rho/\Delta T \sim 2.10^{-3}$ is much higher than that of disordered materials.

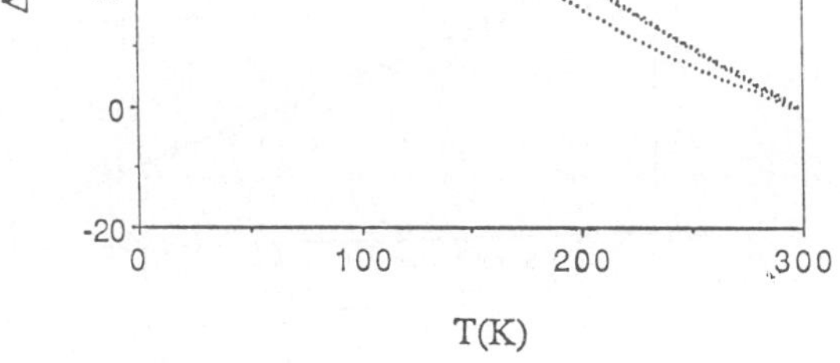

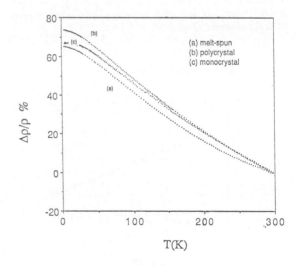

FIGURE 1
Temperature dependence of resistivity in i-$Al_{63}Cu_{25}Fe_{12}$.

The low temperature longitudinal magnetoresistance of i-$Al_{63}Cu_{25}Fe_{12}$ is positive down to the lowest temperature of measurement, and decreases with temperature, as seen in figure 2. This behavior and the $\sqrt{B}$ dependence at high field are reminiscent of weak localization effects, although those effects are generally observed in metallic phases of much lower resistivity.

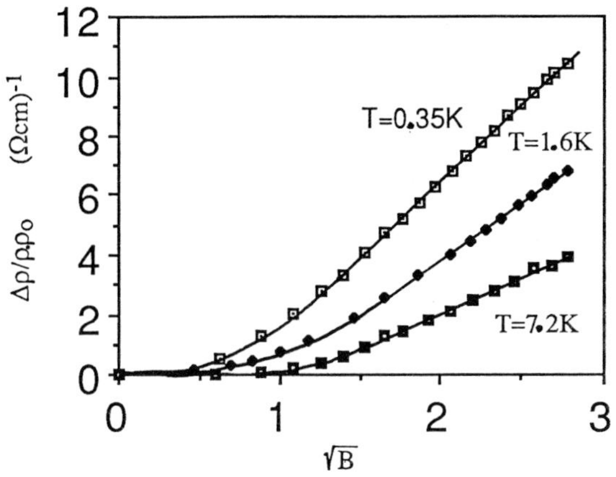

FIGURE 2
Magnetoresistance in i-$Al_{63}Cu_{25}Fe_{12}$.

The magnetoresistance does not show the negative contributions that was observed in AlMn quasicrystals and attributed to scattering of conduction electrons on magnetic sites (7). The magnetic susceptibility of i-$Al_{63}Cu_{25}Fe_{12}$ was thus measured at low temperature (2-300 K) on a long annealed polycrystalline sample. A diamagnetic behavior (χ=-4.10^{-7}emu/g), and a $\sqrt{T}$ dependence between 2 and 40 K are observed (Figure 3).

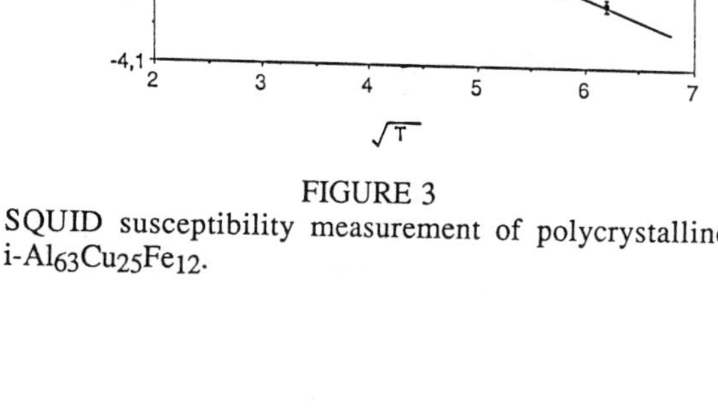

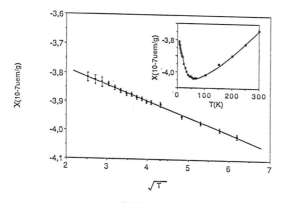

FIGURE 3
SQUID susceptibility measurement of polycrystalline i-$Al_{63}Cu_{25}Fe_{12}$.

We note that contradictory results are found in litterature (with $\chi > 0$ (10) or $\chi < 0$ (8-9)) in the biphased i-$Al_{65}Cu_{20}Fe_{15}$, but we emphasize here that a diamagnetic behavior is observed in our pure high quality sample.

CONCLUSION

The most striking result is the very high resistivity measured in a system containing only metallic elements. We note that, the $\sqrt{B}$ dependence of the magnetoresistance and the $\sqrt{T}$ dependence of the susceptibility (and of the resistivity at very low temperature 0.3-3K) are reminiscent of weak localisation (including spin-orbit scattering) and enhanced electron-electron interaction effects. However, are those effects compatible with such high values of resistivity and with a scattering by a quasiperiodic potential? Finally, can we consider these high values of resistivity observed in an almost perfect i-phase, as the first signature of quasiperiodic order ?

ACKNOWLEDGEMENTS :

One of us (T.Klein) acknowledges the CEA for financial support.

REFERENCES :

(1) Y. Calvayrac, A.. Quivy, M. Bessiere, S. Lefebvre, M. Cornier-Quiquandon, D. Gratias, J.Phys.51, (1990), 417.
(2) P.A. Bancel, Phys. Rev. Lett. 63, (1989), 2741.
(3) C. Berger, G. Fourcaudot, A. Gozlan, F. Cyrot-Lackmann, J.C. Lasjaunias, 10th Gal Conf. Condensed Matt. Div. E.P.S. (Lisbon, april 90).
(4) J.E. Graebner, H.S. Chen, Phys.Rev.Lett. 58,(1987), 1945.
(5) K. Kimura, H. Iwahashi, T. Hashimoto, S. Takeuchi, U. Mizutani, S. Ohashi, G.Itoh, J. Phys. Soc. Jpn 58, (1989), 2472.
(6) A.P. Tsai, A. Inoue, T. Masumoto, J.Mat. Sci.Lett. 7, (1988), 322.
(7) C. Berger, J.C. Lasjaunias, J.C. Tholence, D. Pavuna, P. Germi , Phys. Rev. B37, (1988), 6525.
(8) S. Matsuo, H. Nakano, T. Ishimasa, Y. Fukano, J.Phys. Cond. Matter 1,(1989) 6893.
(9) Z.M. Stadn ik, G. Stroink, H.Ma, G. Williams, Phys.Rev. B 39, (1989), 9797.
(10) J.L. Wagner, K.M. Wong, S.J.Poon, Phys.Rev.39, (1989), 8091.

Physica B 165&166 (1990) 285–286
North-Holland

MAGNETIC FIELD DEPENDENCE OF THE SPECIFIC HEAT OF Si:P NEAR THE METAL-INSULATOR TRANSITION

H. v. LÖHNEYSEN and M. LAKNER

Physikalisches Institut der Universität Karlsruhe, D-7500 Karlsruhe, FRG

The specific heat of uncompensated Si:P with P concentration 0.79 and $4.5 \cdot 10^{18}$ cm^{-3}, i.e. in the vicinity of the critical concentration $N_c = 3.2 \cdot 10^{18}$ cm^{-3}, has been measured between 0.05 and 3 K and in fields up to 5.7 T. A magnetic-field induced increase of the number of localized electrons exhibiting a Schottky anomaly is observed together with a concomitant decrease of the contribution from delocalized electrons.

The metal-insulator (MI) transition in Si:P occurring as a function of P concentration N has been studied for more than a decade with various methods. Very recently, we reported very accurate measurements of the low-temperature specific heat of uncompensated Si:P [1] extending earlier work [2-5]. In particular, a quantitative analysis in terms of localized and delocalized electrons based on a phenomenological two-component model [2] was given. The localized electrons give rise to an anomalous contribution $\Delta C \sim T^{\alpha}$ to the specific heat C between ~0.05 and ~0.5 K. This contribution arises from excitation of exchange-coupled clusters of localized spins. It develops towards a Schottky anomaly in a magnetic field. The delocalized electrons yield a contribution γT. As observed before [4,5], γ varies smoothly through the MI transition which occurs at $N_c = 3.2 \cdot 10^{18}$ cm^{-3} [6]. There appears to be a finite γ at concentrations as low as $N_c/10$. In the present work, we report on a detailed study of the magnetic field dependence of C for two uncompensated samples, one metallic and one insulating, with N = 4.5 and $0.79 \cdot 10^{18}$ cm^{-3}, in order to elucidate the relative contributions of localized and delocalized electrons in an external magnetic field B.

Fig. 1 shows C vs. T for the insulating sample in various magnetic fields. The anomalous contribution ΔC (with $\alpha = -0.04$ for this sample) is suppressed in a magnetic field, and a Schottky anomaly develops. The solid lines are fits to the data (except for B = 0) of the form

$$C(B,T) = \gamma(B)T + \beta T^3 + C_{Sch}(B,T) \qquad (1)$$

βT^3 represents the phonon contribution which will not be considered further. $\gamma(B)T$ represents the contribution due to delocalized electrons in the impurity band and $C_{Sch}(B,T)$ is a Schottky anomaly for two non-degenerate levels of energy $E = g\mu_B B_{eff}$ arising from the Zeeman splitting for localized electrons. Here $B_{eff} = B + B_i$ is a fit parameter with $B_i > 0$ (<0) corresponding to predominantly ferromagnetic (antiferromagnetic) interactions between localized electrons. The upturn of the data at very low temperatures (T < 0.2 K) is attributed to the Zeeman splitting of ^{31}P nuclei [1]. To illustrate the reliability of our data

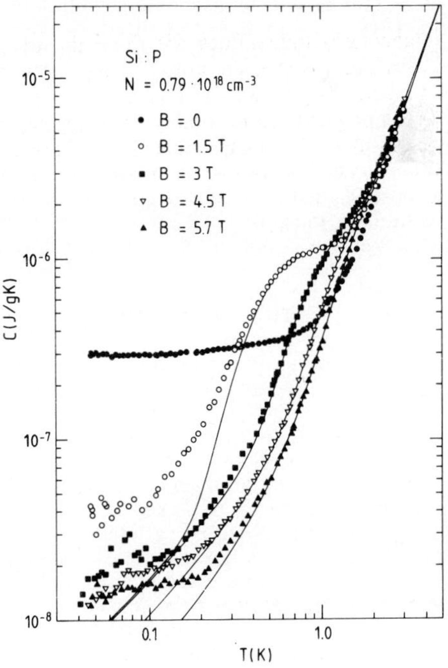

FIGURE 1
Specific heat C vs. temperature T for Si:P with $N = 0.79 \cdot 10^{18}$ cm^{-3}. Solid lines indicate fits of Equ. (1) to the data.

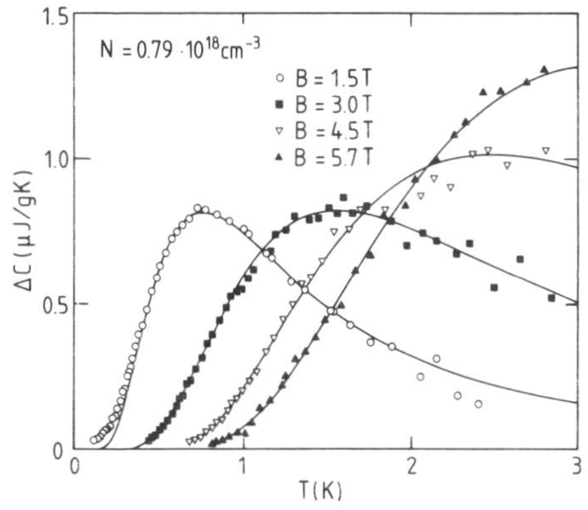

FIGURE 2
Excess specific heat ΔC vs. temperature T for the same sample as Fig. 1. Solid lines indicate fits of two-level Schottky anomalies with B_{eff} = 1.38, 2.8, 4.45 and 5.8 T.

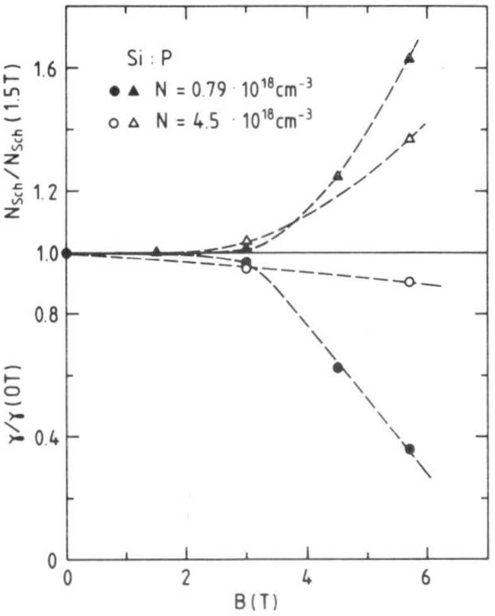

FIGURE 3
Magnetic-field dependence of $\gamma/\gamma(0)$ (circles) and N_{Sch}/N_{Sch} (1.5 T) (triangles) for two Si:P samples.

evaluation, we show in Fig. 2 $\Delta C = C - \gamma T - \beta T^3$. For $B \geq 3T$ ΔC resembles quite closely a Schottky anomaly with $B_{eff} \approx B$. For $T << g\mu_B B$, ΔC is considerably wider than C_{Sch} in 1.5 T, even taking into account the nuclear contribution (see also Fig. 1). The position and width of the anomaly prohibits a determination of γ for this field. These data are therefore fitted with the zero-field value $\gamma(0)$. Comparison with $\gamma(3T)$ justifies this analysis (see Fig. 3). The larger width of ΔC for $B \leq 1.5$ T is simply due to the fact that a small field cannot completely suppress the thermal excitation of exchange-coupled localized electrons (e.g. singlet-triplet transitions for antiferromagnetic pairs). The number of localized electrons N_{Sch} as determined from the fit C_{Sch} to ΔC depends on a magnetic field in high fields which was not observed in previous work [5]. A similar quality of fit was obtained for the metallic sample, although the scatter in ΔC is larger because of the much larger γT contribution to ΔC. We note in passing that only for low-N samples $B_{eff} \approx B$, while for high N, $B_{eff} < B$, indicating the importance of antiferromagnetic interactions among localized electrons.

Fig. 3 shows the dependence of γ and N_{Sch} on a magnetic field. The data are normalized to $\gamma(0)$ which is obtained from a linear fit of C/T vs. T^2 to the zero-field data above 1.5 K. $N_{Sch}(B)$ varies in the opposite direction as $\gamma(B)$, i.e. a large magnetic field causes a shift from delocalized to localized electrons. This shift is much more pronounced for the insulating than for the metallic sample. Although an explanation for the γT term on the insulating side is lacking, the data clearly point to a magnetic-field induced MI transition observed here in a thermodynamic quantity. The density of delocalized electrons N_γ calculated from γ and of localized electrons obtained from N_{Sch} do not simply add up to yield the total density N of donor electrons. Hence the phenomenological two-component model [2] appears to be incomplete. An extended review of recent work on the MI transition in Si:P will appear elsewhere [7].

We thank Dr. W. Zulehner, Wacker Chemitronic for kindly providing the samples. This work was supported by Deutsche Forschungsgemeinschaft.

REFERENCES

[1] M. Lakner and H. v. Löhneysen, Phys. Rev. Lett. 63, 648 (1989)

[2] M.A. Paalanen, J.E. Graebner, R.N. Bhatt and S. Sachdev, Phys. Rev. Lett. 61, 597 (1988)

[3] J. R. Marko, J.P. Harrison and J.D. Quirt, Phys. Rev. B10, 2448 (1974)

[4] N. Kobayashi, S. Ikehata, S. Kobayashi and W. Sasaki, Solid State Commun. 24, 67 (1977)

[5] N. Kobayashi, S. Ikehata, S. Kobayashi and W. Sasaki, Solid State Commun. 32, 1147 (1979)

[6] Here we use the Thurber scale, see footnote in Ref. [1]. See, however, also U. Thomanschefsky, J. Appl. Phys. (to be submitted)

[7] H. v. Löhneysen, in Festkörperprobleme/Advances in Solid State Physics Vol. 30 (1990), forthcoming

Physica B 165&166 (1990) 287–288
North-Holland

MAGNETORESISTANCE AND WEAK LOCALISATION OF AMORPHOUS $Cu_{65}Ti_{35}$

Peter LINDQVIST

Physikalisches Institut, BauV, Universität der Bundeswehr München, 8014 Neubiberg, FRG

The weak localisation theory is used to determine the inelastic scattering time τ_{ie} and the spin–orbit scattering time τ_{so} from the magnetoresistance of amorphous $Cu_{65}Ti_{35}$. Problems appearing in the analysis of comparable alloys are avoided by studying $\Delta\rho/\rho$ in the limit $\Delta\rho/\rho \sim B^2$. The temperature dependence of τ_{ie} is found to be similar to that found in simpler systems (only s– and p–electrons at the Fermi level), i.e. weak for $T < 5$ K (here: $\sim T^{-1.15}$) and much stronger for $T > 20$ K (here: $\sim T^{-3.7}$).

1. INTRODUCTION

The description of the magnetoresistance $\Delta\rho/\rho$ of 3D amorphous metals in terms of weak localisation WL [1] has been established for some years. Using this model the inelastic scattering time τ_{ie} and the spin–orbit scattering time τ_{so} can be extracted. $\Delta\rho/\rho$ is normaly of the order $10^{-4} - 10^{-3}$ in a magnetic field of a few Tesla and at temperatures around and below the temperature of liquid He. Because it is very easy to measure such a magnitude of $\Delta\rho/\rho$, most studies are performed in this temperature range. $\Delta\rho/\rho$ is often $\sim \sqrt{B}$ in this case.

Systems with only s– and p–electrons at the Fermi level are generally well described at a qualitative as well as quantitative level. For these systems the temperature dependence of τ_{ie} is weak at low temperatures ($T < 5$ K), which indicates a possible saturation of τ_{ie} when $T \to 0$ K. At higher temperatures a power law temperature dependence $\tau_{ie} \sim T^{-p}$ with $2 \lesssim p \lesssim 4$ is found, in accordance with theory. Systems with additional d–electrons at the Fermi level are not as easy to describe within theory on a quantitative level in the low temperature range. The presence of electron–electron interactions EEI effects in $\Delta\rho/\rho$ which cannot be described correctly by theory may be a reason for the difficulty, but a break down of the WL theory is also possible. In these cases it can be of interest to study $\Delta\rho/\rho$ in the low magnetic field limit. In this range it can be assumed that the theory describes data better. The analysis is also easier since $\Delta\rho/\rho = \alpha B^2$ with $\alpha = \alpha(\tau_{ie}, \tau_{so})$. This demands a precision better by about 2 orders of magnitude than usually used in resistance measurements. A study of amorphous $Cu_{65}Ti_{35}$ along these lines is presented in this paper. The alloy has previously been investigated in the high field range [2]. The magnetoresistance $\Delta\rho/\rho$ show features normal for a amorphous metal with d–electrons at the Fermi surface.

2. EXPERIMENTAL METHOD AND ANALYSIS

The magnetoresistance of amorphous $Cu_{65}Ti_{35}$ was measured in a wide temperature range from 1.5 K to 46.5 K in the low magnetic field limit. The maximum field used varied from 26 mT at the lowest to 9.3 T at the highest temperature. The resistance measurements were done with a Guildline 9970 resistance bridge with a resolution better than 0.5 nV. $\Delta\rho/\rho$ was of the order $10^{-6} - 10^{-5}$ with a r.m.s. of about $5 \cdot 10^{-8}$ (0.25 nV) for fitted B^2 lines. Temperature was controlled by carbon resistors (Allen–Bradley). The magnetoresistance of the carbon resistors introduce a temperature error. At the lowest B, this is negligible, but at higher B it is necessary to make some corrections. The sample has, however, a weak temperature dependence between 3 K and 13 K, which makes corrections unnecessary at these temperatures. At still higher T the temperature error was estimated to be < 30 mK, by measuring the magnetoresistance of the carbon resistor. This corresponds to additive corrections rising from 5 to 30% of $\Delta\rho/\rho$ in the range 20 to 45 K. Electron–electron interaction effects [3,4] may also give some contribution to $\Delta\rho/\rho$. Calculations showed that EEI in the particle–particle channel could be neglected at all temperatures as well as magnetic fields studied here, but that EEI in the particle–hole channel should give some contribution to $\Delta\rho/\rho$ above 20 K. The values of the corrections depend on the value of the electron–screening constant F. From $\rho(T)$ below 1 K, F was estimated to be below 0.45 [5]. At 46.5 K this contribution to $\Delta\rho/\rho$ may dominate $\Delta\rho/\rho$. Although large, the exact values of both these corrections are of minor importance in the analysis of a power law dependence.

The expression for weak localisation including spin–splitting [6] in the low B limit is:

$$\Delta\rho/\rho = \rho \frac{e^4}{2\pi^2\hbar^3} (D\,\tau_{so})^{3/2} f(t)\, B^2$$

$$f(t) = \frac{1}{96}\left[t^{-3/2} - 3(t+1)^{-3/2}\right] + \frac{\gamma}{8}\left[t^{-1/2} + (t+1)^{-1/2}\right] + \frac{\gamma}{2}\left[t^{1/2} - (t+1)^{1/2}\right]$$

$$\gamma = \left(\frac{g^* \mu_B}{2D\hbar}\right)^2$$

$$t = \frac{\tau_{so}}{4\tau_{ie}}$$

g^* is a effective g–factor and D the electronic diffusion constant.

0921-4526/90/$03.50

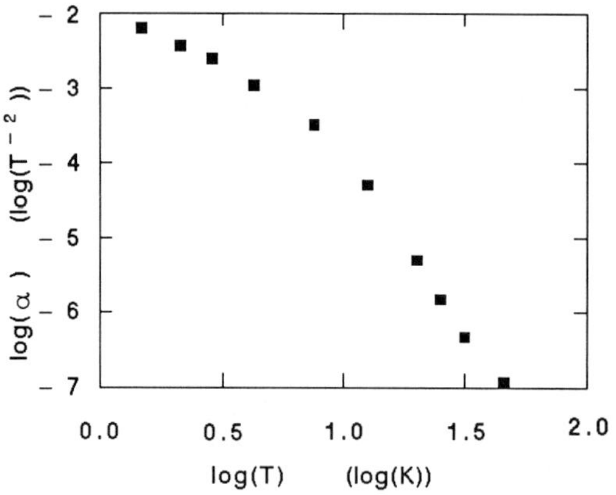

FIGURE 1
Plot of log α vs. log T, using fits of experimental data to $\Delta\rho/\rho = \alpha B^2$. Corrections for magnetoresistance of the thermometer are made.

For the evaluation of data, at least one relation between temperature and t has to be known. For this purpose $\rho(T)$ was used. $\Delta\rho(T)_{wl}$ has a maximum for t = 1/8. A maximum in the experimental $\rho(T)$ is found at about 10 K. However, EEI are present in $\rho(T)$ as well, which produce a maximum in $\Delta\rho(T)_{wl}$ somewhere above 10 K. Using $F \lesssim 0.45$ the maximum temperature T_{max} is somewhere above 18 K. It turns out that the temperature dependence of τ_{ie} is only weakly dependent on T_{max} at low T. Assuming $\tau_{ie} \sim T^{-p}$; we get $p = 1.15 \pm 0.1$ ($T \lesssim 5$ K). At high temperatures, the situation is slightly more complicated. Depending on the assumptions for T_{max} and F , the range for p is larger with $p = 3.7 \pm 0.5$ for $T > 20$ K. τ_{so} was evaluated with t = 1/8 and found to be in the range 0.2 ps to 2.5 ps.

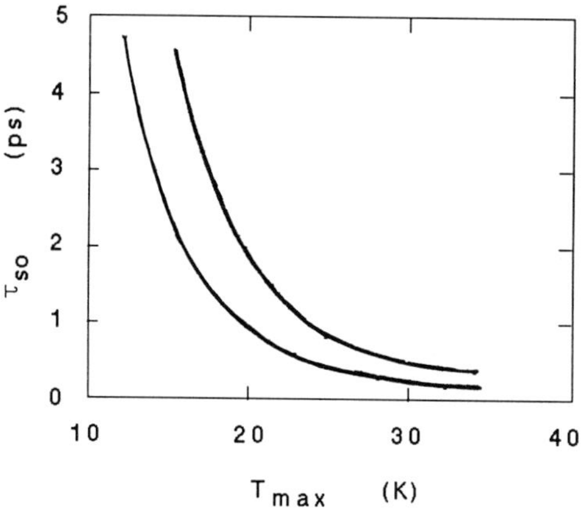

FIGURE 2
Spin orbit time τ_{so} vs. T_{max}. τ_{so} is calculated from theory with $\rho = 182\ \mu\Omega$cm and $D = 0.315$ cm²/s [2]. The upper curve is for $g^* = 0$ (no spin splitting) and the lower curve is for $g^* = 2$ (free electron value).

3. DISCUSSION AND CONCLUSIONS

The temperature dependence of τ_{ie} is similar to what is found for alloys with only s– and p–electrons at the Fermi surface. It is much weaker at low T than at high T, which may suggest that τ_{ie} saturates when $T \to 0$ K, either caused by magnetic impurities [1] or by zero point motion [7]. The data can also be interpreted with the help of a temperature dependent exponent p. For $Cu_{50}Y_{50}$ Bieri et al [8] found that $\tau_{ie} \sim T^{-0.5}$ below 4 K. However, such a temperature dependence has no straight forward theoretical explanation. At high T, $\tau_{ie} \sim T^{-3.7+-0.5}$ indicating that electron–phonon scattering is responsible for the phase breaking of the electronic wave function. It can be of interest to compare this result with the ones from Schulte and Fritsch [9] on $Cu_{44}Ti_{56}$. They studied this alloy from 6 K to 20 K at low magnetic fields and found $2.20 < p < 2.75$. This range of p agrees well with the present data if only the overlapping temperature range is considered. This suggests that p is underestimated in investigations where $\Delta\rho/\rho \approx 10^{-4} - 10^{-3}$, since these can only be performed for temperatures below 20 K. Thus, higher values of p, $3 \lesssim p \lesssim 4$, are favoured rather than values close to two for temperatures above 20 K. Only the order of magnitude could be obtained for τ_{so} from such an analysis. However, this range agrees well with other investigations in the CuTi system [2].

ACKNOWLEDGEMENTS

I would like to thank M. Andersson for the magnetoresistance calibration of the thermometer. Part of this work was done at the Dept. of Solid State Physics, The Royal Institute of Technology, Stockholm.

REFERENCES

(1) For review see: B.L. Al'tshuler and A.G. Aronov, Electron–Electron Interactions in Disordered Systems (North–Holland, Amsterdam, 1984);
P.A. Lee and T.V. Ramakrishnan, Rev. Mod. Phys. 57 (1985) 287.
(2) P. Lindqvist and G. Fritsch, Phys. Rev. B, 40 (1989) 5792.
(3) B.L. Al'tshuler, A.G. Aronov, A.I. Larkin and D.E. Khmel'nitzkii, Sov. Phys.–JETP 54 (1981) 411.
(4) P.A. Lee and T.V. Ramakrishnan, Phys. Rev. B 26 (1982) 4009.
(5) P. Lindqvist and Ö. Rapp, Proc. Symp. on Anderson Localisation, Tokyo, 1984.
(6) H. Fukuyama and K. Hoshino, Phys. Soc. Japan, 50 (1981) 2131.
(7) N. Kumar, D.V. Baxter, R. Richter and J.O. Strom–Olsen, Phys. Rev. Lett. 59 (1987) 1853.
(8) J.B. Bieri, A. Fert, G. Creuzet and A. Schuhl, J. Phys. F 16 (1986) 2099.
(9) A. Schulte and G. Fritsch, J. Phys. F 16 (1986) L 55

Physica B 165&166 (1990) 289–290
North-Holland

CONTRIBUTIONS TO THE DIFFUSION THERMOPOWER OF AMORPHOUS Cu_xSn_{100-x} AND $Mg_{70}Zn_{30}$ FILMS

C. LAUINGER, J. FELD, E. COMPANS, P. HÄUSSLER and F. BAUMANN

Physikalisches Institut der Universität Karlsruhe, D-7500 Karlsruhe FRG

In situ thermopower $S(T)$ measurements of quench–condensed amorphous Cu_xSn_{100-x} ($25 \leq x \leq 70$) and $Mg_{70}Zn_{30}$ films are reported. $S(T)$ shows a knee at about 30 K. We explained this structure as being caused by electron–phonon interaction and so–called roton–like collective density excitations. In our analysis the sign as well as the absolute value of $S(T)$ at low temperatures agree for the Sn rich samples with the free electron model.

1. INTRODUCTION

According to Mott [1] the diffusion thermopower $S_D(T)$ of metals is

$$S_D(T) = -\frac{\pi^2 \cdot k_B^2}{3|e|E_F} \cdot T \cdot \xi_D \text{ with } \xi_D = \left.\frac{\partial \ln \sigma(E)}{\partial \ln E}\right|_{E_F} \quad (1)$$

where ξ_D includes all informations about the scattering processes. Using the conductivity $\sigma(E)$ from Faber–Ziman theory [2] for disordered metals $\xi_D = 3 - 2q - \frac{1}{2}r$ with notations as in [3]. The q term depends on the structure function $a(2k_F)$ (where $2k_F$ is the diameter of the Fermi sphere) and the pseudopotential $v(2k_F)$. The r term depends additionally on the derivative of $v(K)$. In the free electron model (FEM) ξ_D=3 and $S_D(T)$ is negative and proportional to T.

$S(T)$ of most amorphous (a–)metals, however, shows a characteristic knee at low temperature. This is usually interpreted as an enhancement of the bare diffusion thermopower $S_D(T)$ due to electron–phonon interaction. $S(T)$ is [4]

$$S(T) = S_D(T) + \left[\frac{S_D(T)}{T} + 2\alpha + \gamma\right] \cdot T \cdot \lambda(T) \quad (2)$$

where the last term, which we denote $S_{add}(T)$, summarizes these additional contributions to $S_D(T)$. $\lambda(T)$ describes the electron–phonon interaction, 2α accounts for velocity and relaxation time renormalization and γ for higher order contributions. S_{add}, can be either positive [5] or negative [6].

While most authors assume ξ_D to be temperature independent, others [7,8] have pointed out, that this may not be valid causing $S_D(T)$ to be non-linear in T. In particular, roton–like excitations (which are a typical feature of a–metals) are expected to cause a pronounced low temperature structure in $S_D(T)$ at a characteristic temperature T_{rot} [8].

We have studied the influence of both electron–phonon interaction and roton–like excitations on the thermopower of a–metals. For this purpose we performed measurements of $S(T)$ of a–Cu_xSn_{100-x} and a–$Mg_{70}Zn_{30}$ since for the latter roton–like excitations at low energy and large wave vectors are well established by inelastic neutron scattering [9]. Experimental details are given elsewhere [10].

2. RESULTS AND DISCUSSION

Fig. 1 shows $S(T)$ of a–Cu_xSn_{100-x} films. For $x \leq 65$ $S(T)$ shows a pronounced knee at about 30 K.

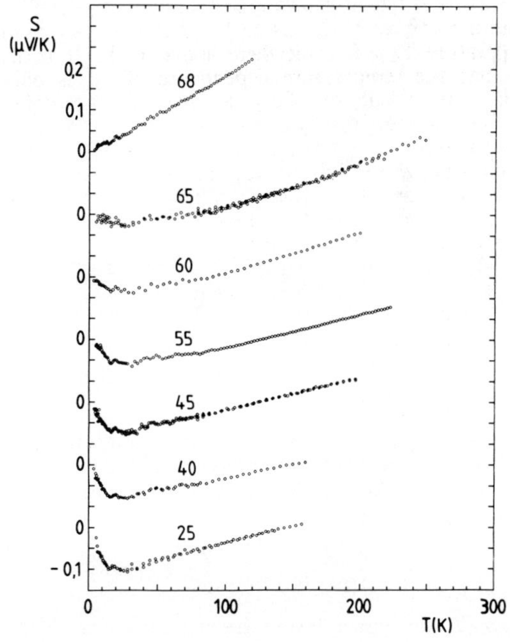

Figure 1: $S(T)$ of a–CuSn after annealing. The numbers give the Cu content x in at%.

First we compare the data with eq.(2) under the usual assumption ξ_D = const. The bracket in eq.(2) then is constant and the temperature dependence of S_{add} is $T \cdot \lambda(T)$. From our measurements we determined

$\lambda(T)/\lambda(0)$ according to eq.(2). Fig. 2 shows these data versus the reduced temperature T/Θ_D. The Debye temperature Θ_D is taken from specific heat measurements [11]. The solid line is calculated according to [4]. The experimental data agree qualitatively with the theoretical calculation. At high temperatures, however, the experimental data lie clearly above the theoretical curve. This discrepancy does not depend on Cu concentration and is observed for many a–metals. We interpret this discrepancy as a hint, that the assumption of a constant ξ_D is not justified. In addition, the above analysis yields a positive sign of $S_D(T)$, while the FEM predicts a negative sign (the Hall coefficient of Sn rich a–CuSn is also negative [8], as predicted in the FEM).

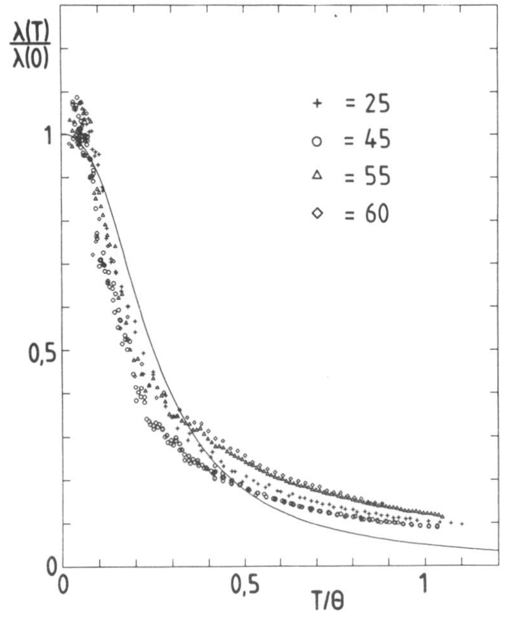

Figure 2: $\lambda(T)/\lambda(0)$ versus T/Θ_D of a–CuSn.

Extending our discussion, we first concentrate on the low temperature thermopower $S_{low}(T)$, below the knee. From eq.(1) and (2) follows (neglecting α, γ, r and taking $\lambda(T) \simeq \lambda(0)$)

$$1 - \frac{S_{low}(T)}{[1+\lambda(0)]\cdot S_o(T)} = \frac{2}{3}q \propto a(2k_F) \qquad (3)$$

with $S_o(T)$ the bare diffusion thermopower of the FEM. Fig.3 shows the quantity on the left side of eq.(3) as taken from our measurements versus $a(2k_F)$ as taken from [12] (each film is represented by one data point). The predicted proportionality is nicely confirmed. Furthermore this interpretation for the Sn rich samples yields negative $S_D(T)$ as expected. At temperatures above T_{min}, $S(T)$ increases strongly. We interpret this increase as being caused by inelastic scattering of the

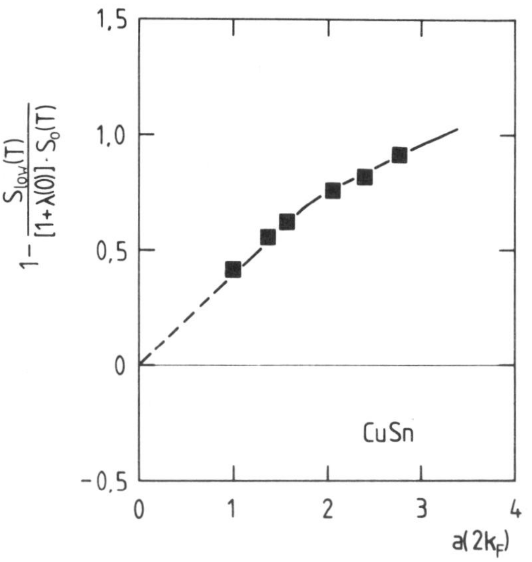

Figure 3: Left hand side of eq.(3) vs. $a(2k_F)$ with $\lambda(0)$ calculated using McMillan's formula [13].

electrons by roton–like excitations together with the influence of electron–phonon interactions. This interpretation is supported by measurements of $S(T)$ of a–$Mg_{70}Zn_{30}$ which is very similar to a–Cu_xSn_{100-x} with $x \approx 55$. We find qualitative agreement between T_{min} and T_{rot} [9].

Precise measurements of $R(T)$ of a–$Mg_{70}Zn_{30}$ films show at low temperatures a minimum followed by a maximum. For these films T_{min} determined from $S(T)$ agrees well with the temperature of the inflection point in $R(T)$.

REFERENCES

[1] N. F. Mott, H. Jones, The Theory of the Properties of Metals and Alloys (Clarendon, Oxford 1936)
[2] T. E. Faber, J. M. Ziman, Phil. Mag. 11 (1965) 153
[3] R. D. Barnard, Thermoelectricity in Metals and Alloys (Taylor & Francis, London 1972)
[4] A. B. Kaiser, G. E. Stedman, Sol. Stat. Comm. 54 (1984) 91
[5] B. L. Gallagher, J. Phys. F 11 (1981) L207
[6] E. Compans, F. Baumann, Jpn. J. Appl. Phys. 26 (1987) 805
[7] R. Harris, G. Mulimani, Phys. Rev. B 27 (1983) 1382
[8] P. Häussler, submitted to "Glassy Metals III", eds. H. Beck and H.–J. Güntherodt, in print
[9] J.-B. Suck, H. Rudin, H.-J. Güntherodt, H. Beck, Phys. Rev. Lett. 50 (1983) 49
[10] E. Compans, Rev. Sci. Instrum. 60 (1989) 2715
[11] M. Hofacker, Thesis Universität Aachen (1986)
[12] H. Leitz, Z. Phys. B 40 (1980) 65
[13] W. L. McMillan, Phys. Rev. 167 (1968) 331

Physica B 165&166 (1990) 291–292
North-Holland

LOW TEMPERATURE RESISTIVITY ANOMALY IN TiNb AND TiV ALLOYS

Takako SASAKI and Yoshio MUTO

Institute for Materials Research, Tohoku University, Sendai 980, Japan

High precision measurements of the electrical resistivity (ρ) of TiNb and TiV alloys in magnetic fields up to 23T revealed a slight enhancement of the resistivity (Δρ) below T_0 = 5-32 K. The temperature dependence and the magnetic field dependence of Δρ depend on the magnitude of the resistivity ρ_0 at T_0. The lnT dependence of Δρ is observed in samples with ρ_0 lower than about 110 μΩcm, while the √T dependence is dominant in samples with ρ_0 higher than it, where the negative magnetoresistance proportional to √H appears. These characteristics are roughly explained considering both the two-level-system and the weak localisation effect.

We(1,2) reported that the low temperature electrical resistivity ρ of the superconducting TiNb and TiV alloys slightly increased with decreasing temperature and indicated that this anomalous resistivity was closely related to the hexagonal ω-phase precipitation in the bcc β-phase matrix. In order to clarify the mechanism of the anomalous electron scattering, the temperature dependence and the magnetic field dependence of the anomalous resistivity have been studied for both quenched and annealed TiV alloys and quenched TiNb alloys.

The quenched samples were prepared by quenching into ice water from 1000°C after the solution treatment at 1300°C for 10 hours. We have also annealed the quenched samples with 17-27 at% V at 300°C for 1 hour. The change of a microstructure caused by the anneal was observed by an electron diffraction method. The temperature dependence of ρ was measured in a magnetic field of 15 or 20.6 T which is high enough to suppress the superconducting transition. The magnetic field dependence of ρ was measured at 1.8 K up to 23 T.

The temperature dependence of ρ/ρ_0 for quenched TiV alloys are shown in Fig.1 as a function of lnT, where ρ_0 is a resistivity at T_0 below which the enhancement of ρ is observed. The value of T_0 is between 5 and 32 K in each sample. The logarithmic temperature dependence of ρ is observed in samples with 70-36 at% V. In samples with V concentration less than about 30 at%, ρ varies as a function of √T rather than lnT below about 15 K as shown in Fig.2. On the other hand in quenched TiNb alloys the temperature dependence of ρ is expressed by the formula $\rho/\rho_0 \propto -A\ln T$ independently of Nb concentration (1). In order to compare the magnitude of the enhancement of ρ with ρ_0 and the metastable ω-phase precipitation in β-phase matrix, We show the concentration dependence of a quantity Δρ which is a increase of ρ between 5 K together with ρ_0 and the phase diagram of the microstructure in

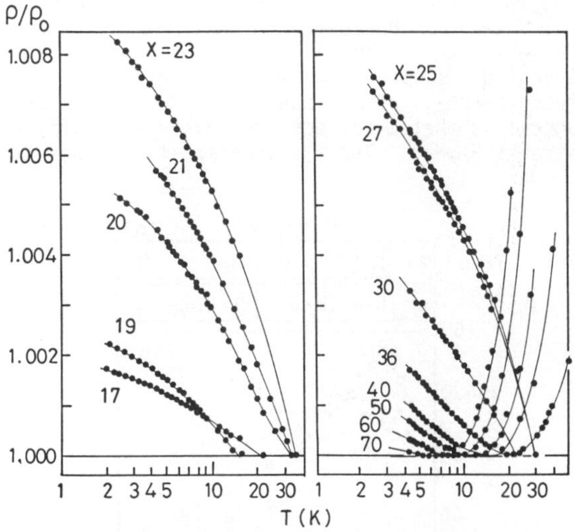

Fig.1. Resistivity ρ/ρ_0 vs lnT for quenched $Ti_{100-X}V_X$.

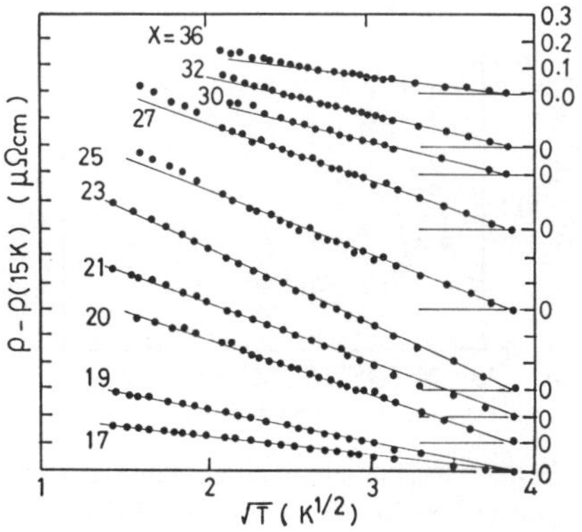

Fig.2. Resistivity ρ-ρ(15K) vs √T for Quenched $Ti_{100-X}V_X$.

Fig. 3. It can be seen that $\Delta\rho$ increases with increasing ρ_0 and the $\sqrt{T}$ dependence is observed for the ρ_0 above 110 $\mu\Omega$cm. As pointed out in the previous paper (1), the enhancement of ρ with lnT dependence becomes remarkable with the increase of the intensity of diffuse streaks characteristic of the precursory lattice distortion for β-ω transformation (which shown by ($\beta\rightarrow\omega$) in Fig.3.a).

These two kinds of anomalous resistivity show a contrast in the magnetic field dependence. The temperature coefficient A in the formula $\rho/\rho_0 \propto -A\ln T$ dose not depend on a magnetic field, while another anomalous resistivity with $\sqrt{T}$ dependence exhibits the negative magnetoresistance (MR) proportional to $\sqrt{H}$.

In Fig.4 the temperature dependence of ρ for annealed TiV alloys is shown. After the anneal, ρ_0 becomes lower than about 110 $\mu\Omega$cm and the temperature dependence of ρ changes to lnT from $\sqrt{T}$. Moreover the negative MR cannot be observed. The MR is almost zero for annealed samples within an accuracy of measurements. The structural change by the anneal is that ω reflections weaken and diffuse streaks appear. These characteristics observed

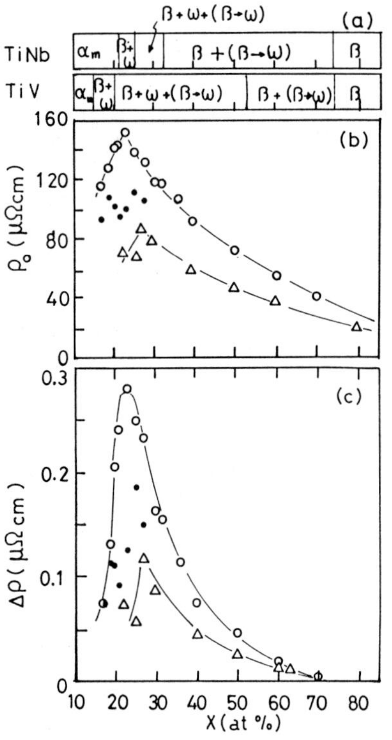

Fig.3. (a)Microstructures. β:β-phase matrix, ω:ω reflections, ($\beta\rightarrow\omega$):diffuse streaks. (b),(c)concentration dependence of ρ_0 and $\Delta\rho$ for quenched $Ti_{100-X}Nb$(Δ), $Ti_{100-X}V_X$(o) and annealed $Ti_{100-X}V_X$(o).

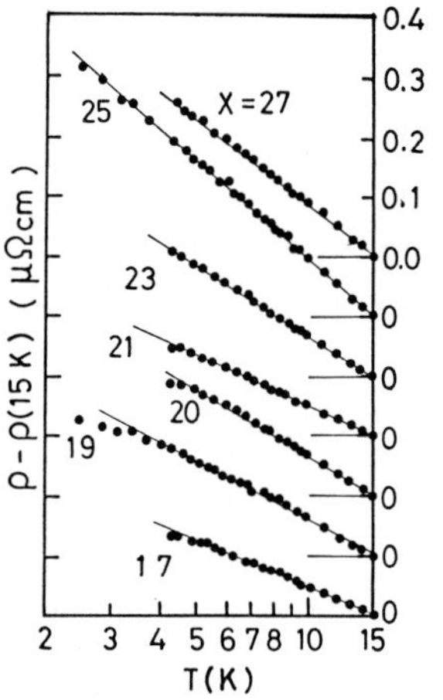

Fig.4. Resistivity ρ-ρ(15K) vs lnT for annealed $Ti_{100-X}V_X$.

in annealed samples is the same as in quenched samples with ρ_0 lower than 110 $\mu\Omega$cm. These results suggest that the anomalous resistivity consists of two kinds of component. One is characterized by the lnT and non magnetic field dependence and closely related with the β-ω structural transformation, which is considered to be caused by the electron scattering by the atomic tunneling between ω and β configurations (two-level- system (TLS)). The ω-phase results from simultaneous displacement of two atoms in a β-phase unit cell by only about 0.5 Å along the [111] direction, therefor at low temperatures atomic tunneling can conceivably occur. Another is characterized by the $\sqrt{T}$ and $\sqrt{H}$ dependence and dominant in high resistive samples. Moreover the magnitude of the magnetoconductivity $\sigma(H)$, $d\sigma(H)/d\sqrt{H}$, estimated from a experimental result for $Ti_{83}V_{17}$ is 1.4 which is the same order as expected in the weak localisation (WL) theory(3). Therefor this component can be attributed to the WL effect.

In conclusion, the anomalous enhancement of ρ between the temperature T_1 and T_2 ($T_1<T_2$) is expressed as follows,

$$\Delta\rho \equiv \rho(T_1)-\rho(T_2)=A\ln(T_2/T_1)+\rho_0^2B(\sqrt{T_2}-\sqrt{T_1})$$

The first term is a contribution of the scattering by TLS, and the second term is that of the WL effect. Because the second term is proportional to ρ_0^2, the WL effect appears remarkably in high resistive TiV alloys. In TiNb and TiV alloys with low resistivity, this effect is negligible, therefor the lnT dependence is mainly observed.

REFERENCES

(1) T.sasaki, S Hanada and Y.Muto, LT-18, part 2(1987)923
(2) T.Sasaki,S.Hanada and Y.Muto, Physica 148B(1987)513.
(3) A.Kawabata, Solid State Commun.34(1980)431.

Physica B 165&166 (1990) 293-294
North-Holland

Temperature and magnetic field dependence of resistance in Zr-N sputtered films

Masahito YOSHIZAWA, Kôki IKEDA, Naoki TOYOTA[a], Tsutomu YOTSUYA[b], Thomas Müller[c] and Peter Wyder[c]

Department of Metallurgy, Faculty of Engineering, Iwate University, Morioka 020, Japan
Osaka Prefectural Industorial Technology Research Institute, Osaka,550, Japan[a]
Institute for Material Research, Tohoku University, Sendai 980, Japan[b]
Hochfeld-Magnetlabor, Max-Planck-Institut für Festkörperforschung, F38042 Grenoble, France[c]

The Zr-N sputtered films fabricated with various substrate temperatures and nitrogen partial pressures were devided into two groups by the ratio of the resistance at helium and room temperatures. We found a clear boundary and remarkable differences both in the temperature dependence of the resistance and in the magnetoresistance between the two groups. The characteristics of the Zr-N films are considered to be due to Anderson localization, though the oscillatory behaviour in the magnetoresistance is not fully understood.

1. INTRODUCTION

Recently, Zr-N sputtered films (1) has drawn attention as thermometer in the cryogenic temperature region with a magnetic field. These films have sufficiently large sensitivity as resistance thermometer (2) and small magneto-resistance (MR) for the use in the magnetic field. The MR causes only 4.5mK temperature change in the magnetic field up to 6T (2). In spite of their interesting characteristics, they have not been precisely studied so far. We present the electrical resistance measurements of Zr-N films, which were deposited under various conditions, in the magnetic fields up to 30T.

2. SAMPLE PREPARATION

The Zr-N films have been deposited by rf magnetron sputtering method in the argon and nitrogen mixed gas atmosphere (1). The thickness of the films is about 200nm. The characteristics of the sputtered films are governed by the substrate temperature (T_S), the nitrogen partial pressure (P_{N2}) and the plane of the substrate. The samples reported here were deposited on the R-plane,(1102), Sapphire in order to avoid the plane dependence of the substrate. The fabrication conditions of the films are shown in Fig.1.

3. RESISTANCE

The value of the resistance depends on T_S and P_{N2}. The samples shown in Fig.1 can be devided into two groups by the resistance ratio R.R. of R_{HeT} to R_{RT}. Here R_{HeT} and R_{RT} are the resistance at helium and room temperatures. The samples with the value of R.R.>9 and R.R.<3 are hereafter expressed as the groups I and II, respectively. The difference of the R.R. between the two groups is reflected on the temperature dependence of the resistance, which is shown in Fig.2 in logarithmic scales. Two types of the temperature dependence were found: logT

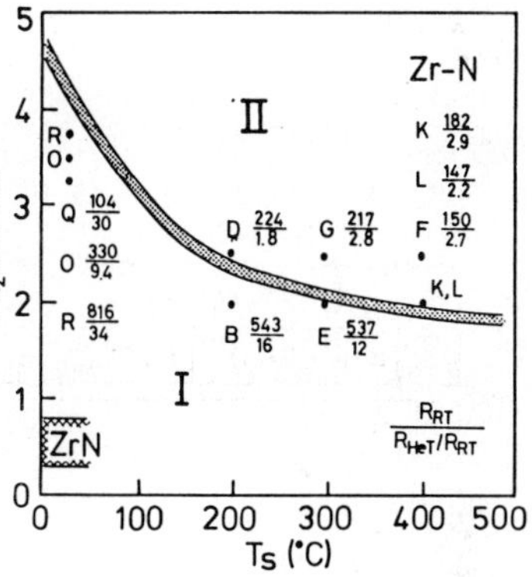

Fig.1. Characteristics of Zr-N films fabricated under various substrate temperatures and nitrogen partial pressures. The dotted area indicates a schematically illustrated boundary between the groups I and II.

dependence for the sample L of the group II (see inset of Fig.2), and $\sqrt{T}$ dependence in the conductance of the group I samples B and O. We suppose that the relation between R.R. and the temperature dependence holds for all the samples, because between 300K and 4.2K the logT dependence gives 1.85 in R.R. and the $T^{-1/2}$ 8.45. It is remarkable that there is a clear boundary between the two groups, as schematically illustrated by the dotted area.

4. MAGNETORESISTANCE

Magnetoresistance (MR) of the Zr-N films at low temperature is shown in Fig.3. The samples B,L and R were measured at 1.7K, D,G and K at 1.76K and O at 1.3K. we found a few general features on the MR. [1] The samples of the group I show larger negative MR than the group II. [2]MR of the group I shows $\sqrt{H}$ dependence in the high magnetic field. [3]Except a few samples, an

0921-4526/90/$03.50 © 1990 – Elsevier Science Publishers B.V. (North-Holland)

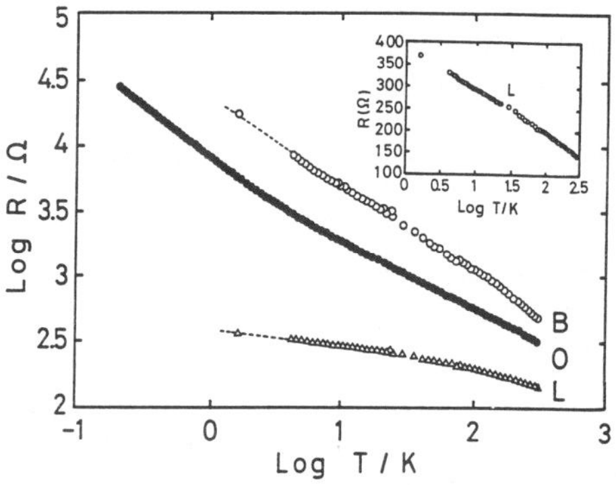

Fig.2. Temperature dependence of resistance of Zr-N films in logarithmic scales.

oscillatory MR is observed in the weak field region as shown in the inset of Fig.3.

5. CHARACTERISTICS AS THERMOMETER

Zr-N films have already been successfully used as thermometer. Although the group I has larger MR than II, the group I is more suitable for the thermometer because of its larger sensitivity d(lnR)/dT. The sensitivity of the sample O is -1.25×10^5ppm/K at 4.2K and -3000ppm/K at 300K. Equivalent temperature change of MR is -5mK at 10T, 78mK at 20T and 187mK at 30T at 1.7K, and -81mK at 10T, 51mK at 20T and 352mK at 30T at 4.2K for the sample B.

6. DISCUSSIONS

On the Group I, the $T^{1/2}$ in conductance and $H^{1/2}$ in MR are considered to be due to 3-dimensional (3-d) localization. The dimension of the film system is determined by comparing the film thickness d with the characteristic length $\bar{d}=\sqrt{D\tau_\varepsilon}$, where D and τ_ε are the diffusion constant and the inelastic scattering time, respectively. The system can be regarded as 2-dimensional in the case of $d<<\bar{d}$ and 3-d for $d>\bar{d}$. We estimated $\bar{d}$=45nm and $\tau_{so}=2\tau_\varepsilon$ at 4.2K for the sample O. These values confirm that is is a 3-d system.

MR is caused from quantum interference, spin-orbit and Coulomb interactions for the localization systems. According to the localization theory for the weakly localized regime (3), MR shows negative in the high magnetic field for the both cases $\tau_\varepsilon<<\tau_{so}$ and $\tau_\varepsilon>>\tau_{so}$ for the 3-d system. Here τ_{so} is the spin-orbit scattering time. Positive MR is expected in the case of $\tau_\varepsilon\simeq\tau_{so}$, which corresponds to our cases. Therefore the positive MR in the middle field and the negative one in the high field region are considered to be due to the localization. The negative MR in the weak field is not explained by either the present localization theory or Coulomb interaction. The Coulomb interaction causes a positive MR except exceptional cases with unrealistic set of parameters.

From the analogous discussion made on the group I, the logarithmic temperature dependence of the resistance of the sample L may indicate a sign of the 2-dimensionality of the system. It is supported by the facts that smaller R_{RT} and R_{HeT} in the group II cause a longer electronic mean free path and a larger $\bar{d}$, and that there is a difference between in the parallel and perpendicular fields in the MR. It is the problem what the origin of the large difference in the resistance between the group I and II is. The Zr-N films have stable ingredients of 40%-Zirconium and 60%-Nitrogen almost independent of P_{N2} (4), and have FCC structure the same as the stoichiometric ZrN [1]. It suggests that the films are constructed mainly from the stoichiometric ZrN cluster with excess N_2 and vacancies. Since the high T_s and the high P_{N2} prevent the substitution of the vacancies and promote recrystallization, the low resistance in the group II may result from better quality of the films, some kind of phase chage and/or surface effects.

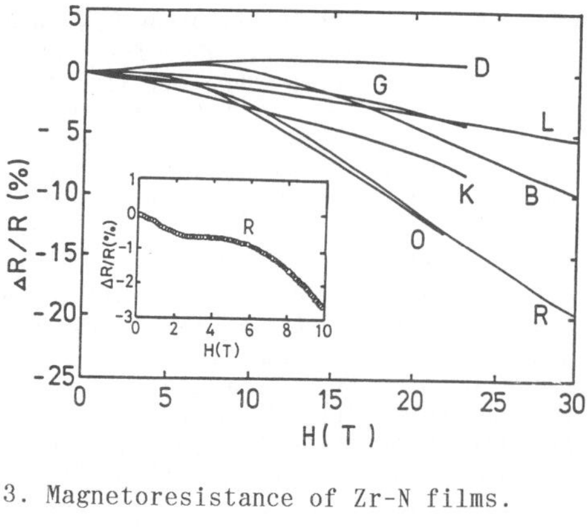

Fig.3. Magnetoresistance of Zr-N films.

ACKNOWLEDGMENTS

One of us(M.Y.) thanks Max-PLanck-Geselschaft for the financial support. We are grateful to Dr.T.Saso and Prof.S.Morita for the helpful discussions. The help of Mr.H.Kawazoe and Mr.T.Kawamura would be appreciated.

REFERENCES

(1) T.Yotsuya, M.Yoshitake, Y.Suzuki, S.Ogawa and J.Yamamoto : Proc.5th Sensor Symposium (1985)9.
(2) T.Yotsuya, M.Yoshitake and J.Yamamoto : Appl.Phys.Lett. *51*(1987)235.
(3) see for example, S.Kobayashi F.Komori : Prog.Theor.Phys. Suppl. *84*(1985)224.
(4) T.Yotsuya, M.Yoshitake, S.Ogawa, Y.Suzuki and J.Yamamoto : Jap.J.Appl.Phys. *26*(1987) Suppl.26-3,1743.

Physica B 165&166 (1990) 295–296
North-Holland

STRUCTURAL DISORDER AND INCIPIENT LOCALIZATION EFFECTS IN NbC THIN FILMS

Yoshinori KUWASAWA, Shoujun HINO*, Shigeru NAKANO, Atsushi KOYAMA** and Tadashi MATSUSHITA**

Department of Physics, Faculty of Science,
*Department of Image Science Technology, Faculty of Engineering,
Chiba University, Yayoicho 1-33, Chiba, Chiba 280, Japan
**Photon Factory, National Laboratory for High Energy Physics, Ohomachi, Tsukuba, Ibaragi 305, Japan

Films used in this study were classified into three groups, metallic, weak nonmetallic and non-metallic from the measured results of the temperature dependence of the resistivity and superconducting transition temperature. The results of X-ray diffraction and photoelectron spectroscopy showed that films were NbC of single phase NaCl type structure with vacancies and oxygen atoms at carbon sites. For the nonmetallic type samples, three different Nb-Nb atomic distances were observed in the fluorescence EXAFS mesurements, one of which was much longer than that of bulk NbC. It can be said that such large distortions in lattice and Nb-chain are responsible for the incipient locallization effect.

1. INTRODUCTION

Negative temperature coefficient of resistivity has been found in the crystalline or amorphous metals, in which disorder plays an important role. R.C.Dynes et al. (1) found that the temperature coefficient of resistivity (TCR) of $LuRh_4B_4$ film, when irradiated by α particles decreases with increasing dose and finally changes its sign. Clearly it is due to defects formed by irradiation.

In this study, the similar behavior was observed for NbC films, where vacancies and O-atoms at C sites distorted the lattice.

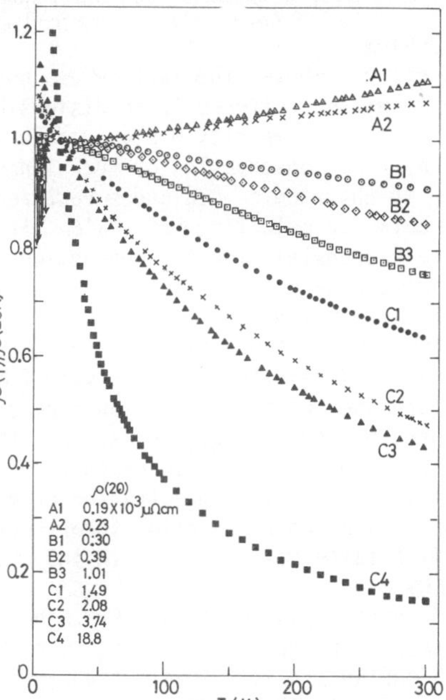

Fig.1 The temperature dependence of the resistance of the samples. The resitance are normalized at 20K. Note that A's and B's become superconducting at low temperature.

Table 1 Characteristics of the samples

	T_s	T_c	ρ(20)	ρ(300)	a	C_{ox}
A1	650	8.2	0.19	2.21	4.462	10.2
A2	650	1.7	0.23	1.74	4.470	
B1	600	5.8	0.30	3.86	4.422	18.2
B2	510	5.7	0.39	4.52	4.360	
B3	650		1.01	3.32	4.348	
C1	550	--	1.49	9.56	4.330	
C2	575	--	2.08	10.54	4.332	30.7
C3	300	--	3.74	16.26	4.350	
C4	500	--	18.86	28.09	4.303	37.6

T_s: Substrate temerature during deposition(℃)
T_c: Superconducting transition Temperature(K)
ρ(20), ρ(300): Resistivity at 20 and 300 K resp. ($\times 10^3$ $\mu\Omega$cm)
a: Lattice constant(Å)
C_{ox}: Oxygen content(%)

2. SAMPLE PREPARATIONS

The films used in this study were prepared by ion beam sputtering. A Nb target was bombarded by Ar ion beam of 8 kev in the gas of carbon monoxide, of which the partial pressure was 3-8 x 10^{-6} Torr. Films were grown on the random oriented single crystal sapphire sabstrates at a deposition rate of 20-80 Å/h. The temperatures of substrates during deposition were 300-650° C. The size of samples was 12x10 mm^2 and the thickness of the films were 500-4000 A. As mentioned later,the temerature of substrate were determinative factor of the properties of the films.

3. ELECTRICAL RESISTANCE AND T_c's

Standared 4 probe technique were used in the measurement of resitivities and T_c's of the samples. The results were given in Fig.1 and in Table 1. As seen from Fig.1, A_1 and A_2 have the positive temperature coefficient of resistivity (TCR) and become superconducting at low temperature(metallic behavior),whereas C_1 to C_4 have negative TCR and are not superconducting till 1.5K(nonmetallic behavior). B1 to B3 shows the negative TCR but are superconducting at low temperature(weak nonmetallic behavior).

4. X-RAY DIFFRACTION AND FLOULECENCE EXAFS

The structure of the films were obtained from Xray diffraction analysis. Films deposited at $T_s > 300°$ C showed single phase NaCl crystalline structure,but films deposited at $T_s < 300°$ C were amorphous. The lattice constant are given in Table 1.Floulecence EXAFS mesurement for NbK absorption edge were made at Photon Factory, Tsukuba. Special care was taken to avoid the effect of Bragg reflection from the substrate. As shown in Fig.2, only one peak of Nb is observed for A_1,though the peak is broad. For C_1, three peaks and one peak are observed corresponding Nb-Nb and Nb-O atomic distances respectively. The results are given in Table 2.

5. PHOTOELECTRON SPECTROSCOPY

Mearsurment of XPS and UPS were made for these samples. Mg$K\alpha$ and Al$K\alpha$ were used as

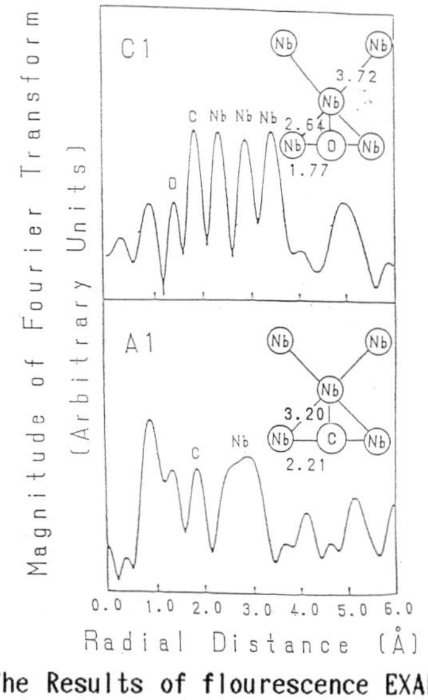

Fig.2 The Results of flourescence EXAFS

Table 2 Atomic Distances

	Nb-O	Nb-C	Nb-Nb		
bulk		2.24	3.16		
A1		2.21	3.20		
C1	1.77	2.20	2.64	3.19	3.72

light sources. The chemical shift of C1s and O1s were 2.3 eV and 2.2 eV, respectively, which are due to Nb-C and Nb-O bond. Therefore,it can be said that O-atom substitute C-atom of NbC.

The energy shift of spectral lines, $Nb_{3d5/2}$ and $Nb_{3d3/2}$, from those of the bulk Nb for A's are larger than that for C's. This suggest that A's have less vacancies at C sites than C's. The relative intensity of the spectral line C_{1s} to that of $Nb_{3d5/2}$ can be approximately expressed as the product of x $(=n_C/n_{Nb})$ and $\sigma(C_{1s})/\sigma(Nb_{3d5/2})$, where n's and σ's are number of atoms per unit volume and the total cross section for photoelecric absorp-tion respectively. In this way,x can be calculated from the experimental data and $\sigma(Nb_{3d5/2})/\sigma(C_{1s})=5.01$, a value calculated by Scofield(2).Similarly, using the value $\sigma(Nb_{3d5/2})/\sigma(O_{1s})=1.76$ (2), $y=n_O/n_{Nb}$ are also calucualted and listed in Table 1. The estimated value of x+y are about 0.9 for A's and 0.7 for C's, that suggest many vacant C-sites in the lattice of C-group sample.

6. DISCUSSIONS

As mentioned above, the lattice of nonmetallic sample may be suffered large distortions due to vacancies and impurity O-atom. It was confirmed by the observation of much longer and shorter Nb-Nb atomic distances compared with that of bulk or metallic NbC film. If Nb-Nb bonds are responsible for the conduction in NbC, the abrupt change of bond length will be the cause of locallization.

For the sample of nonmetallic group(C's), the electrical conductivity increases with T at high temperature,and with $T^{1/2}$ at low temperature. This dependence on temperature agrees with that derived from locallization theory(3).

A limit of disorder concentration of metallic behavior based on percolation theory will be about 30%. Detaile will be given in the Session.

REFERENCES

(1) R.C. Dynes, J.M.Rowell and P.H.Schmidt,in: Tenary Superconductors, eds. G.K. Shnoy, B.D.Dunlap and F.Y. Fradin (North-Holland, Amsterdam 1981) pp.169

(2) H.Scofield, J. Electron Spectroscopy 8 (1976) 129.

(3) P.A.Lee and Ramakrishnan, Rev. Mod.Phys. 57 (1985),322

Physica B 165&166 (1990) 297–298
North-Holland

MAGNETORESISTANCE IN THE INCIPIENT ANDERSON LOCALIZATION REGION FOR GRANULAR Pd_xC_{1-x} FILMS

A. CARL, G. DUMPICH and E. F. WASSERMANN

Tieftemperaturphysik, Universität Duisburg, 4100 Duisburg 1, FRG

We report on magnetoresistance (MR) measurements for thin granular Pd_xC_{1-x} films with $0.3 < x < 0.34$ on the metallic side near the metal-insulator-transition ($x_c = 0.3$). The resistance versus temperature behavior for these films can be well described as resulting from incipient Anderson localization. MR-data for the films are analysed in the framework of localization and electron-electron interaction, within the limit of weak disorder ($k_Fl_e >> 1$). This leads to considerable agreement between experimental data and theory, from which we obtain reliable values for the phase coherence length L_ϕ and its temperature dependence, although the films are in the strong scattering limit ($k_Fl_e \simeq 1$).

1. INTRODUCTION

The metal-insulator-transition (MIT) in 3 dimensional (3dim) disordered systems has attracted considerable interest in recent years, both experimentally and theoretically (1). The MIT in microscopically disordered systems like random mixtures or granular systems is often found to be related to Anderson localization (AL) (2) and/or electron-electron interactions (EEI) (3). For both cases, with increasing disorder, the zero temperature conductivity $\sigma(0)$ continuously scales to zero at a critical concentration x_c with $\sigma(0)\propto((x/x_c)-1)^\nu$ (1). $\sigma(0)=Ae^2/\hbar\xi$ is determined by the localization correlation length ξ, that diverges at the MIT like $\xi\propto((x/x_c)-1)^{-\nu}$. The critical exponent ν should be $\nu=1$ for pure AL, for pure EEI it should be $\nu=1$ or $\nu=1/2$. Thus, the question arises, what indeed is the relevant process that determines the nature of the MIT. A possible answer may result from the extension of the zero temperature scaling theory (2) to finite temperatures (4), where ξ in $\sigma(0)=Ae^2/\hbar\xi$ has to be replaced by the relevant length scale L(T) for T>0, with $\sigma(T)=Be^2/\hbar L(T)$ L(T) is usually given by inelastic electron scattering processes. L(T) is replaced by the phase coherence length $L_\phi=(D\tau_\phi)^{1/2}$ for localization, where D is the electron diffusion constant and $\tau_\phi\propto T^{-p}$ (p=1,2,3) the phase coherence time, or by $L_T=(D\hbar/k_BT)^{1/2}$ for EEI. A theory for the MIT that includes both AL and EEI, is still missing.

On the other hand, $L_\phi(T)$ can be obtained by fitting magnetoresistance (MR) measurements at various temperatures to localization theory (5). However, since the theory is valid only for homogeneous samples ($k_Fl_e>>1$), one has to check, whether the theory holds also for disordered samples ($k_Fl_e\simeq1$), where l_e is the elastic electron mean free path. Therefore we have analysed MR-data, obtained at different temperatures between 1.5K<T<20K for granular Pd_xC_{1-x} films with $k_Fl_e\simeq1$ in the direct vicinity of the MIT.

2. EXPERIMENTAL

Granular Pd_xC_{1-x} films with 0.1<x<1 and thicknesses 10nm<t<70nm have been prepared by co-deposition of pure palladium and high purity carbon at T=300K in an UHV-system. The composition of the films is maintained by different evaporation rates for Pd and C, where the deposition of Pd and C is independently monitored by two separate quartz crystal oscillators. The temperature dependence of the film resistance is measured with four-terminal DC-technique ($\delta R/R\simeq10^{-5}$) within 1.6K<T<300K in situ as well as in a separate ^{4}He-cryostat, where magnetic fields up to 5T can be applied perpendicular to the film plane.

3. RESULTS

As a typical example for Pd_xC_{1-x} films near the MIT, showing incipient Anderson localization effects (6), Fig.1 shows $\Delta\sigma(B)=-(R_\square(B)-R_\square(0))/R_\square(0)^2$

0921-4526/90/$03.50

versus $\sqrt{B}$ for a film with x=0.33, t=27nm and $k_F l_e$=0.61 between 0.1T<B<5T. As one can see from Fig. 1, the MR is positive with $\Delta\sigma(B) \propto B^2$ at low fields and $\Delta\sigma(B) \propto B^{1/2}$ at high fields for temperatures between 1.5K<T<20K, where absolute values for $\Delta\sigma$ decrease with increasing temperature.

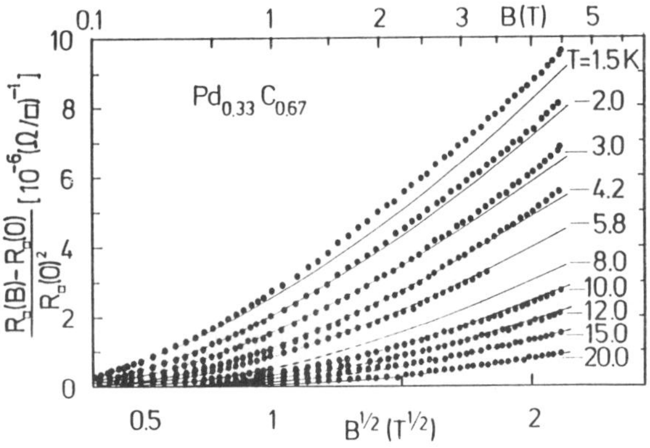

FIGURE 1
$\Delta\sigma(B)$ vs $\sqrt{B}$ at different temperatures for a Pd_xC_{1-x} film with x=0.33. Solid lines are calculated by theory (5)

For the analysis of MR-data it is essential to check, whether the film behaves 2dim or 3dim with respect to localization and/or EEI in the presence of a magnetic field. This can be done in comparing the relevant length $L_B=(\hbar/(4eB))^{1/2}$, resulting from the magnetic field B, with t. For L_B<t the film behaves 3dim, and for L_B>t 2dim with respect to localization effects. The film, as shown in Fig. 1, behaves therefore 3dim for magnetic fields $B>\hbar/(4et^2)\simeq 0.23T$. Thus, we analysed the MR-data, using 3dim localization (5) as well as 3dim EEI theory (3) (solid lines in Fig. 1) (7). As one can see from Fig. 1 we find good agreement between experimental data (low field) and theory within 1.5K<T<20K.

4. DISCUSSION

Fig. 2 shows the phase coherence length $L_\phi(T)$ versus temperature in a log-log plot between 1.5K<T<20K, as obtained from MR-measurements for the Pd_xC_{1-x} film with x=0.33 from Fig. 1 (full dots). Additionally, $L_\phi(T)$ for a pure Pd film with t=20nm and $k_F l_e$=12.3 is plotted in Fig.2 (open dots).
As one can see from Fig.2, for the pure Pd film, $L_\phi(T)$ is found to vary like $L_\phi(T) \propto T^{-0.75}$ for T < 8K and like $L_\phi(T) \propto T^{-1}$ for T > 8K as being typical

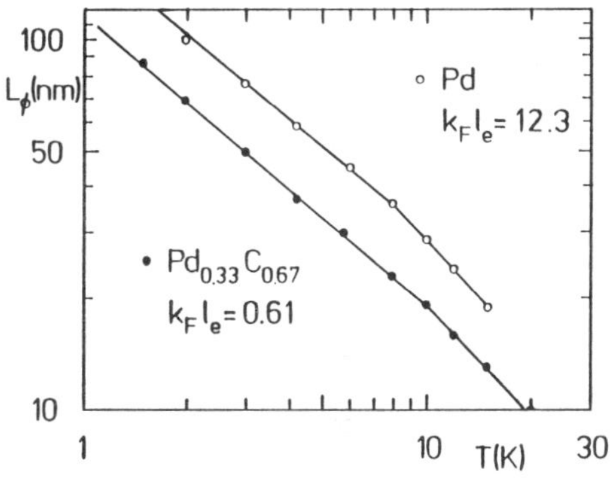

FIGURE 2

log-log plot of L_ϕ vs. temperature for a Pd film and Pd_xC_{1-x} film with x=0.33

for 3dim thin films (1). However, not only the same temperature dependence, but also reliable absolute values are found for the more disordered film with x=0.33, although $k_F l_e \simeq 1$. From this we conclude, that localization as well as EEI theory remain valid also for films with $k_F l_e \simeq 1$, justifying the present analysis. A more detailed analysis will be given elsewhere (7).

ACKNOWLEDGEMENT
This work was supported by the DFG (Deutsche Forschungsgemeinschaft)

REFERENCES

(1) T. Ando and H. Fukuyama, Anderson Localization,(Springer,Berlin,1988) and refs. cited therein

(2) E. Abrahams, P.W. Anderson, D.C. Licciardello and T.V. Ramakrishnan, Phys. Rev. Lett **42** (1979) 673

(3) B.L. Altshuler, A.G. Aronov and P.A. Lee Phys. Rev. Lett. **44** (1980) 1288

(4) Y. Imry J. Appl. Phys. **52** (1981) 1817

(5) A. Kawabata Solid State Commun **34** (1980) 432

(6) A. Carl, G. Dumpich and E.F. Wassermann to be published in Vacuum

(7) A. Carl, G. Dumpich and E.F. Wassermann unpublished

Physica B 165&166 (1990) 299–300
North-Holland

CONDUCTION MECHANISM IN GRANULAR RuO_2-BASED THICK-FILM RESISTORS

Wilfried SCHOEPE

Institut für Angewandte Physik, Universität Regensburg, 8400 Regensburg, W.Germany

The conductivity of a commercial thick-film resistor is measured between 4 K and 15 mK and in magnetic fields up to 7 Tesla. The data can be described by the variable-range hopping mechanism with a Coulomb gap in the density of states. The negative magnetoresistance may be attributed to quantum-interference effects in the strongly localized regime.

Commercial RuO_2-based thick-film resistors are useful low-temperature thermometers because of their excellent stability and small magnetoresistance (1-4). An understanding of the temperature and field dependences of these devices, however, is lacking. Conduction in granular systems may be affected by manufacturing processes. Furthermore, the applicability of theoretical models originally developped for doped semiconductors is still under discussion. The data obtained in this work actually can be described in terms of a hopping conduction which is affected by temperature and magnetic field as expected for a strongly localized semiconductor.

The sample was taken from a batch of commercial 10 kΩ resistors of Siegert GmbH (5). It was mounted inside the mixing chamber of a home-made dilution refrigerator. A magnetic field up to 7 Tesla could be applied either perpendicular or parallel to the conductive layer. Fig. 1 shows that the temperature dependence in zero field follows a $\exp(T_o/T)^{\frac{1}{2}}$ law below 4 K with $T_o =$ 0.48 K. (The deviation below 20 mK is due to heating effects even though the measuring current was reduced to 15 pA.) This behaviour is observed rather often in granular systems and may be attributed to variable-range hopping with Coulomb interaction (6). Strictly, the model is valid only for $T \ll T_o$ but no deviation is visible up to 4 K. From $T_o \sim e^2/4\pi\epsilon_o k\epsilon a$ (where ϵ is the dielectric constant and a the radius of localisation) one finds $\epsilon a \sim 3 \cdot 10^{-5}$ m. Assuming $\epsilon \sim 10$ gives $a \sim 3 \cdot 10^{-6}$ m, which corresponds to the size of the metallic grains.

The magnetoresistance is negative at all temperatures and fields, see Fig. 2. This is in striking contrast to the usual case of a large positive magnetoresistance in the strongly localized regime caused by orbital shrinking of the wavefunction in a magnetic field. The absence this effect indicates, that the long-range behaviour of the wavefunction is not relevant for transport in this case.

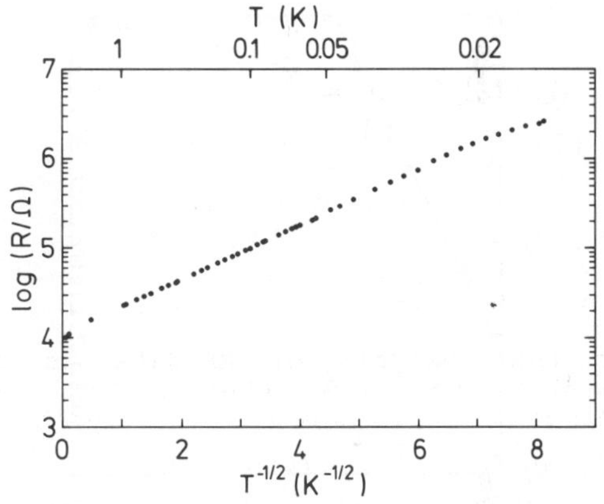

FIGURE 1
Temperature dependence of the resistance in zero field. The straight line behaviour indicates a $\exp(T_o/T)^{\frac{1}{2}}$ law with $T_o = 0.48$ K. The deviation below 20 mK is due to Joule heating by the measuring current of 15 pA.

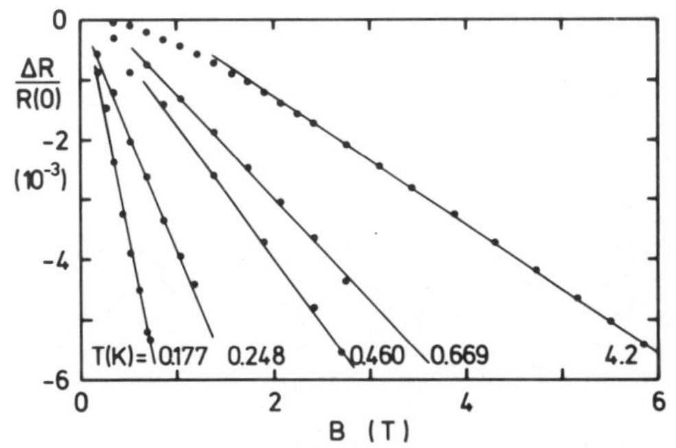

FIGURE 2
Magnetoresistance at various constant temperatures. Except for small fields the magnetoresistance varies linearly as B/B_c.

A negative magnetoresistance of the hopping mechanism has recently been calculated on the basis of quantum interference effects of the tunneling electron (7-9). The magnetic field dependence is predicted to be linear except for very small fields where it should increase quadratically. This is in agreement with the data of Fig. 2. The temperature dependence of the linear part should scale as $r^{\frac{7}{2}}$, where $r \propto (T_o/T)^{\frac{1}{2}}$ is the temperature dependent hopping length (9). Therefore, the inverse of the slopes of the straight lines should vary as $T^{\frac{7}{4}}$. The experimental results, see Fig. 3, are approximately described by a T^2 law for $T \lesssim T_o$ and tend to saturate above 1 K. This is consistent with the above mechanism. A more detailed analysis requires a calculation of the prefactor which is not yet available.

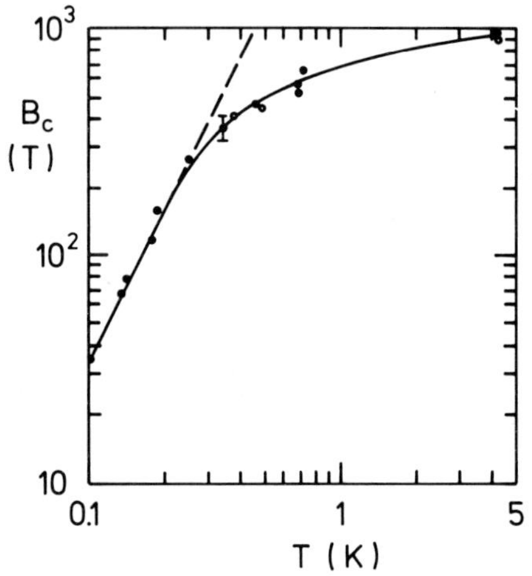

FIGURE 3
The temperature dependence of B_c (i.e. of the inverse of the slopes in Fig. 2).

ACKNOWLEDGEMENTS

It is a pleasure to thank W. Pfab of Siegert GmbH for supplying the sample. I have had very helpful discussions on the theory with W. Schirmacher and B. Shklovskii. Similar experiments were performed simultaneously and independently by K. Neumaier and I am grateful to him for cummunicating his results prior to publication.

REFERENCES

(1) H. Doi, Y. Narahara, Y. Oda and H. Nagano, Proceedings LT 17, eds. U. Eckern et al. (North Holland, Amsterdam, 1984) p. 405.
(2) W.A. Bosch, F. Mathu, H.C. Meijer and R.W. Willekers, Cryogenics **26** (1986) 3.
(3) Q. Li, C.H. Watson, R.G. Goodrich, D.G. Haase and H. Lukefahr, Cryogenics **26** (1986) 467.
(4) M.W. Meisel, G.R. Stewart and E.D. Adams, Cryogenics **29** (1989) 1168.
(5) Siegert GmbH, D-8501 Cadolzburg, W.Germany.
(6) B.I. Shklovskii and A.L. Efros, Electronic properties of doped semiconductors (Springer-Verlag, Berlin, 1984).
(7) V.L. Nguen, B.Z. Spivak and B.I. Shklovskii, Zh. Eksp. Teor. Fiz **89** (1985) 1770 (English translation Sov. Phys. JETP **62** (1985) 1021).
(8) U. Sivan, O. Entin-Wohlman and Y. Imry, Phys. Rev. Letters **60** (1988) 1566.
(9) W. Schirmacher, Phys. Rev. B., to be published.

Physica B 165&166 (1990) 301–302
North-Holland

MAGNETOCONDUCTIVITY OF METALLIC InSb IN THE EXTREME QUANTUM LIMIT

S.S. Murzin*, A.G.M. Jansen and U. Zeitler.

Max-Planck-Institut für Festkörperforschung, Hochfeld-Magnetlabor, 166X, F-38042 Grenoble Cedex, France.

In the extreme quantum limit of the applied magnetic field, the resistivity tensor of n-type InSb has been measured in magnetic fields up to 20 T and temperatures down to 50 mK. For decreasing temperatures below 1 K, the transverse resistivity shows an anomalous decrease with a logarithmic temperature dependence, but the Hall effect is temperature independent. Because of the constant number of electrons, the observed phenomenon is totally different from the usually observed metal-insulator transition induced by a magnetic field.

For metallically doped semiconductors in strong magnetic fields, the extension of the electronic wavefunction at the impurity shrinks and a metal-insulator transition can be induced by the magnetic field (1). Experimentally this magnetic freeze-out of the electrons is observed in an applied magnetic field with a strong increase in all the elements of the resistivity tensor. This increase is more pronounced at lower temperatures. Ignoring the disorder which is inherently present in a doped semiconductor, correlated phenomena for the interacting electron system in a magnetic field (spin-density-wave, charge-density-wave, Wigner crystal) have been predicted (2) with similar effects on the transport properties. Well below the metal-insulator transition, we investigated the magnetoconductivity of doped InSb in magnetic fields in the extreme quantum limit with only one spin-splitted Landau level occupied by the electrons, and observed an anomalous behaviour in the magnetotransport data.

The studied InSb samples were single crystalline with electron concentrations n above 10^{16} cm^{-3} (InSb 1: 1.4 10^{16}; InSb 2: 5.6 10^{16}; InSb 3: 5.4 10^{16}). The dimensions of the bulk samples were of the order of 1 mm^2 in cross section and 10 mm in length. The various components of the resistivity tensor were measured in a ^{3}He/^{4}He dilution refrigerator in fields up to 20 T using phase-sensitive detection techniques.

In Figure 1 we have plotted the transverse (ρ_{xx}) and Hall (ρ_{xy}) resistivities for InSb 3 as a function of the magnetic field (parallel to the z-axis) at $\simeq$ 1 K and 100 mK. The Shubnikov-de Haas oscillations in ρ_{xx} end up in a strong increase of the resistivity above 8 T as a manifestation of the extreme quantum limit. In the extreme quantum limit the measured transverse resistivity decreases with decreasing temperature. The amount of

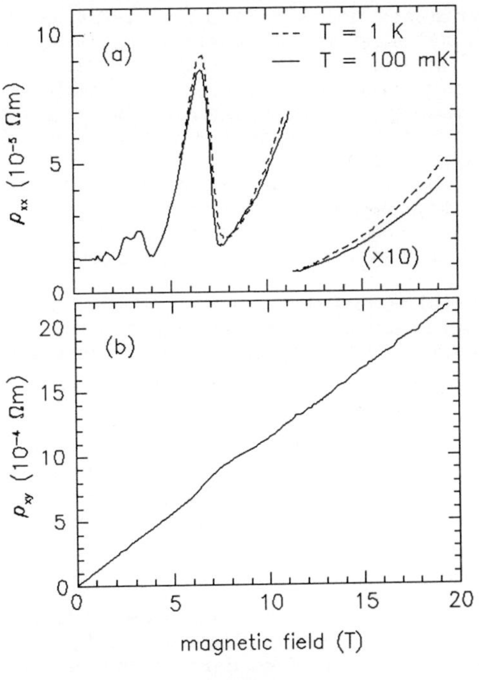

FIGURE 1
Magnetic field dependence of the transverse resistivity ρ_{xx} (a) and Hall resistivity ρ_{xy} (b) for the n-InSb 3 sample at $\simeq$ 1 K and $\simeq$ 100 mK. The Hall resistivity was independent of temperature.

* permanent address: Institute of Solid State Physics, Chernogolovka, 142432 USSR.

0921-4526/90/$03.50

temperature dependence in ρ_{xx} increases with increasing magnetic field. The Hall resistivity ρ_{xy} varies approximately linearly with the magnetic field. The structure in ρ_{xy} around 8 T can be ascribed to magnetic quantum oscillations in the Hall effect. The Hall coefficient $R = \rho_{xy}/H$ reveals an oscillatory behaviour similar to the Shubnikov-de Haas effect in the transverse resistivity.

The observed temperature dependence of ρ_{xx} and ρ_{xy} in the extreme quantum limit differs from related work on the metal-insulator transition, where both ρ_{xx} and ρ_{xy} increase for decreasing temperatures (3). The constant Hall coefficient can be interpreted as a constant number of electrons without any influence of magnetic freeze-out. The field for magnetic freeze-out to occur can be found from the Mott criterion by comparing the size of the electron wavefunction with the electron density, $na_{//}a_{\perp}^2 = \delta^3$ with $\delta = 0.3$, $a_{//} = a_B/\ln(a_B/\Lambda)^2$ and $a_{\perp} = 2\Lambda$ (a_B is the effective Bohr radius and $\Lambda = (\hbar/eB)^{1/2}$ the magnetic length in a magnetic field B) (4). The applied magnetic fields are not sufficiently high to reach the Mott criterion for the investigeted samples. In the presented experiments holds $\rho_{xy} > \rho_{xx}$, pointing to a good metallic system where $\rho_{xy}/\rho_{xx} = \omega_c\tau$ ($\omega_c = eB/m^*$ is the cyclotron frequency with m^* the effective mass). However, this condition is not observed in experimental data near the known metal-insulator transition induced by a magnetic-field.

In Figure 2 we have plotted the temperature dependence of the transverse resistivities for the investigated samples at constant magnetic fields in the extreme quantum limit. The resistivity decreases with a logarithmic temperature dependence for temperatures below 2 K. At the magnetic fields indicated in Figure 2, the variation in the Hall resistivity ρ_{xy} was much smaller ($\lesssim$ 1 %). Because $\rho_{xy} >> \rho_{xx} \geq \rho_{zz}$ in the applied magnetic fields, the conductivity components can be written as $\sigma_{xx} \simeq \rho_{xx}/\rho_{xy}^2$, $\sigma_{xy} \simeq 1/\rho_{xy}$ and $\sigma_{zz} = 1/\rho_{zz}$ with $\sigma_{zz} >> \sigma_{xy} >> \sigma_{xx}$. Hence, the conductivity σ_{xx} decreases also with a logarithmic temperature dependence upon cooling down. In a measurement of σ_{zz}, a similar decrease with logarithmic temperature dependence has been observed.

The observed log-T dependence is difficult to explain with the dimensionality of our bulk samples. In disordered systems a log-T dependence is expected for two-dimensional systems in view of diffusion related quantum phenomena in the transport properties (electron-electron interaction and weak localization). In a three-dimensional system, diffusional transport yields rather a $T^{1/2}$-dependence.

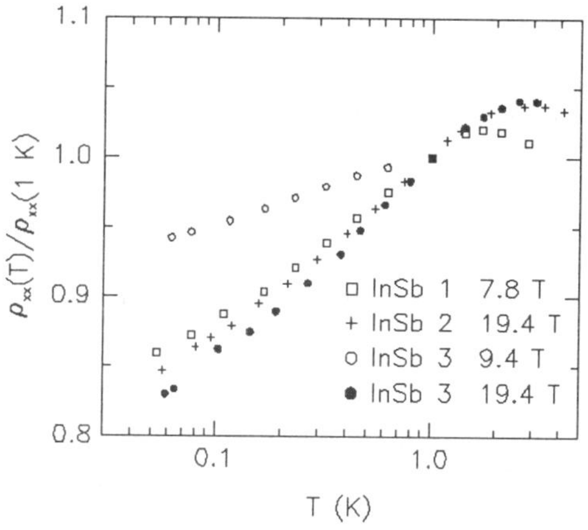

FIGURE 2
The transverse resistivity ρ_{xx}(T) normalized at T = 1 K as a function of the temperature on a logarithmic scale for the different samples in the indicated magnetic fields.

The observed log-T dependence in ρ_{xx} yields $\sigma_{xx} \rightarrow 0$ in the limit $T \rightarrow 0$. This situation describes a Hall insulator in three dimensions, with a conductivity tensor analogously to that of the quantum Hall effect (zeros on the diagonal and constant σ_{xy}). For the possible occurence of such a novel state, the importance of the electron-electron interaction in a disordered system (5) in the extreme quantum limit of the magnetic field has been put forward (6). Especially in the extreme quantum limit, electron-electron interaction effects could give a large correction to the conductivity.

REFERENCES

(1) B.I. Shklovskii and A.L. Efros, Electronic Properties of Doped Semiconductors (Springer Verlag, Berlin, 1984).
(2) A.H. MacDonald and G.W. Bryant, Phys. Rev. Lett. 58 (1987) 515 and references therein.
(3) S.B. Field, D.H. Reich, T.F. Rosenbaum, P.B. Littlewood and D.A. Nelson, Phys. Rev. B 38 (1988) 1856; M. Shayegan, V.J. Goldman, H.D. Drew, N.A. Fortune and J.S. Brooks, Solid State Commun. 60 (1986) 817.
(4) S. Ishida and E. Otsuka, J. Phys. Soc. Japan 43 (1977) 124.
(5) B.L. Al'tshuler and A.G. Aronov, Zh. Eksp. Teor. Fiz. 77 (1979) 2028 [Sov. Phys. JETP 50 (1979) 968].
(6) S.S. Murzin, Pis'ma Zh. Eksp. Teor. Fiz. 44 (1986) 45 [JETP Lett. 44 (1986) 56].

Physica B 165&166 (1990) 303–304
North-Holland

ELECTRICAL TRANSPORT IN As ION–IMPLANTED Si IN THE METAL–INSULATOR TRANSITION RANGE.

CHEN GANG, R.W. VAN DER HEIJDEN, A.T.A.M. DE WAELE, H.M. GIJSMAN AND F.P.B. TIELEN*.

Department of Physics, Eindhoven University of Technology, P.O. Box 513, 5600 MB Eindhoven, The Netherlands. *Department of Electrical Engineering, EFFIC.

Resistance measurements are reported for silicon, ion–implanted with As, for various concentrations within a few percent of the metal–insulator transition at temperatures down to 0.1 K. At the lowest temperatures, the temperature dependence of the resistance is dominated by inhomogeneities in the dopant distribution. The observed current dependence of the resistance is discussed.

Recently, the metal–insulator transition (MIT) in bulk As–doped Si has been investigated in detail by electrical transport measurements (1–4). Very close to the MIT, the characteristic length, the localization length, becomes large. It is of interest therefore to investigate the MIT in systems of small dimensions. From experiments on the related system of P–doped Si, it was suggested that length scales as large as 20 μm play a role (5).

In this work, electrical transport properties are reported on a series of samples of As–ion–implanted Si with concentrations very close to the MIT. The implanted layer thickness is less than 0.1 μm, while in the plane the typical dimension has been chosen as small as 5 μm. Such devices may have applications as low–temperature thermistors (6–8) for particle detection purposes (9).

The sample parameters are listed in Table 1. The samples are not intentionally compensated. The electrode structure is shown in the inset in Fig. 1. The interdigit structure effectively corresponds to electrodes 5 μm apart and 7 mm wide. The digits were not metallized for practical reasons and are so responsible for a series resistance of about 10 Ω. Therefore only conductivities smaller than about 1 $(\Omega cm)^{-1}$ can be measured, leaving an interesting range near the "minimum metallic conductivity" ($\sim$ 25 $(\Omega cm)^{-1}$) unfortunately out of range. Data were taken using a commercial resistance bridge operating at 25 Hz.

A typical set of data is shown in Fig. 1. These

No.	impl. 1 160 keV (10^{13} cm^{-2})	impl. 2 60 keV (10^{13} cm^{-2})	Diff. 1000 °C (min)	Peak dens. (10^{18} cm^{-3})
1	8.2	3.5	240	6.4
2	8.8	3.9	240	6.9
3	8.2	3.5	100	7.6
4	8.8	3.9	100	8.1
5	11.4	4.3	240	8.4
6	11.4	4.3	160	8.9
7	11.4	4.3	140	9.1
8	11.4	4.3	120	9.3
9	11.4	4.3	100	9.7

Table 1. *Process parameters and final peak concentrations. Column 1 designates the wafer number, columns 2 and 3 the doses of the two subsequent implants for each wafer. Column 3 gives the diffusion time at 1000⁰ C in N_2 atm. for each wafer. Column 4 gives the nominal peak density. For wafers 1–5 and 9, the densities were obtained from computer–simulations of the process, for wafers 6–8 from suitable interpolations between the simulation results for wafers 5 and 9.*

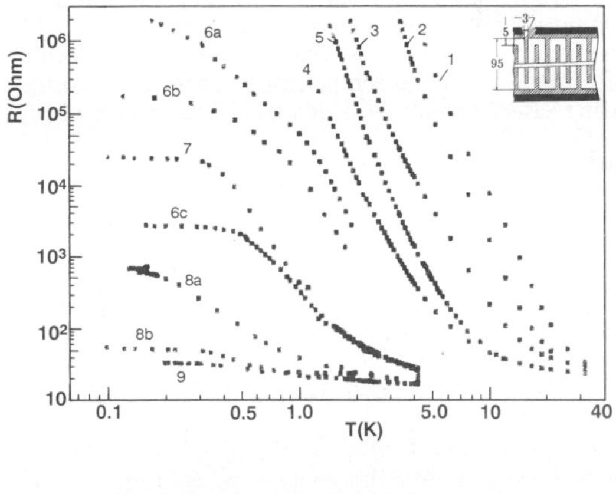

Fig. 1. *Resistance–temperature characteristics for several samples on logarithmic scales. The numbers correspond to the wafer numbers in Table 1. Samples 6a–6c are three different samples from the same wafer 6. Samples 8a and 8b are two samples from two different but identically processed, wafers 8. The inset shows the structure: the white area contains the relevant dope, dashed are strongly metallic doped regions and black are the Aluminum contact pads. Note that the Aluminum does not extend over the digits.*

data were taken at low enough current so that they did not depend on current. Curves 1–5 and curve 9 are the same as those reported earlier (10). Samples 1–5 are clearly on the insulating side of the MIT and are well described by a modified variable range hopping law (6). Sample 9 is very close to, or across, the transition. Samples 6–8 were intentionally prepared to lie in the transition range between 5 and 9. They are all three identically processed as sample 5, except for different diffusion times to control the final peak density (see Table 1). Samples 6–8 were made independently of the other 6 samples in a different process run. From the positions of the curves in Fig. 1 (and those from five other wafers not included in Fig. 1), it is concluded that the reproducibility and homogeneity of the peak density of the samples is at least as good as a few percent.

The strong dependence on temperature T of samples 6–8 indicates that they must be on the insulating side. The absolute value of the densities is therefore lower than the nominal values given in Table 1, as the MIT occurs at $8.5 \cdot 10^{18}$ cm^{-3} (1). A surprising feature is the flattening of the curves below a certain temperature. A similar unexplained flattening has been observed in some bulk samples (4). For the present case, however, it is plausible that they are caused by metallic "shorts". Based on bulk data (1), it can be estimated that areas of the order of 100 μm^2, that received a slightly higher dose (of the order of only a percent or so) can explain the observed saturated resistances. Because of the small electrode separation, the present devices are sensitive to inhomogeneities on this scale. It is then also evident that the magnitudes of the saturating low–temperature resistance varies by several orders of magnitudes for different samples, even from the same wafer.

Fig. 2 shows for sample 6a that the measured resistance strongly depends on measuring current. The dependence is much less strong as compared to data on bulk Si:P in the same temperature range (5). A simple heating effect cannot be ruled out. The rather abrupt

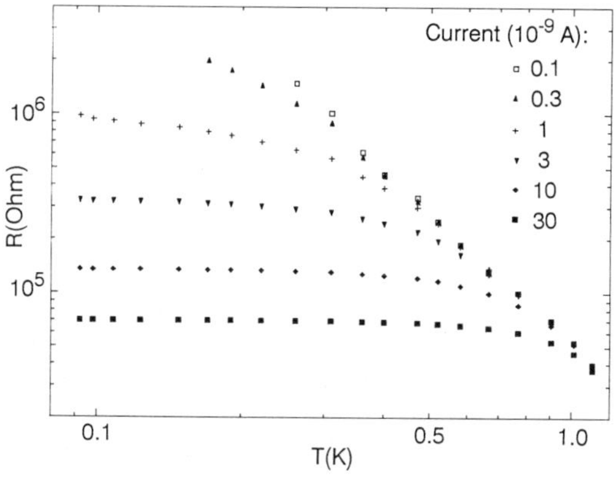

Fig. 2. *Resistance–temperature characteristics of sample 6a for six different measuring currents.*

change from the temperature dependent, current independent, curve to a nearly temperature independent behaviour points to an intrinsic effect due to the electric field E. Activationless hopping at large E was recently reported for compensated GaAs, doped near the MIT (11). A characteristic electric field E_c, where the resistance starts to depend on field, may be estimated from the data in Fig. 2. E_c can be converted to a characteristic length L through $eE_cL \sim kT$ (e electron charge, k Boltzmann constant), where theoretically L should be the localization length (12). Values obtained for L in this way are a few hundred nm, about two orders of magnitude smaller than for bulk Si:P (5). The L could be associated with an enlarged localization length near the MIT, but it is also of the order of the sample dimensions.

In conclusion, the MIT in ion–implanted Si:As is sharp within a few percent of the impurity concentration, consistent with the bulk behaviour. Because of the small lateral dimensions used in the present work, it could be shown that the transition occurs over a range narrower than the typical inhomogeneity range of an As–implanted layer on a 10 μm scale. From the field dependence of the resistance, a characteristic length of the order of the implantation depth is found.

We would like to thank P.A.M. Nouwens for help with the sample preparation.

REFERENCES.

1. P.F. Newman and D.F. Holcomb, Phys. Rev. B **28** (1983) 638
2. W.N. Shafarman and T.G. Castner, Phys. Rev. B **33** (1986) 3570
3. W.N. Shafarman, T.G. Castner, J.S. Brooks, K.P. Martin and M.J. Naughton, Phys. Rev. Lett. **56** (1986) 980
4. W.N. Shafarman, D.W. Koon and T.G. Castner, Phys. Rev. B **40** (1989) 1216
5. T.F. Rosenbaum, K. Andres and G.A. Thomas, Solid State Commun. **35** (1980) 663
6. A. Alessandrello, D.V. Camin, G.F. Cerofolini, E. Fiorini, A. Giuliani, C. Liguori, L. Meda, T.O. Niinikovski and A. Rijllart, Nucl. Instr. and Meth. A **263** (1988) 233
7. C.C. Zammit, A.D. Caplin, R.A. Stradling, T.J. Sumner, J.J. Quenby, N.J.C. Spooner, S.F.J. Read, G.J. Homer, J.D. Lewin, P.F. Smith, J. Saunders and M.J. Lea in: *Superconducting and Low Temperature Particle Detectors*, Eds. G. Waysand and G. Chardin (Elsevier, New York, 1989)
8. M.I. Buraschi and G.U. Pignatel in: same as Ref. 7.
9. D. McCammon, M. Juda, J. Zhang, S.S. Holt, R.L. Kelley, S.H. Moseley and A.E. Szymkowiak, Japan. J. Appl. Phys. Suppl. **26–3** (1987) 2084
10. Chen Gang, H.D. Koppen, R.W. van der Heijden, A.T.A.M. de Waele, H.M. Gijsman and F.P.B. Tielen, Solid State Commun. **72** (1989) 173
11. F. Tremblay, M. Pepper, R. Newbury, D. Ritchie, D.C. Peacock, J.E.F. Frost, G.A.C. Jones and G. Hill, Phys. Rev. B **40** (1989) 3387
12. B.I. Shklovskii, Sov. Phys. Semicond. **6** (1973) 1964

Physica B 165&166 (1990) 305–306
North-Holland

THE ^{29}Si KNIGHT SHIFT IN Si:P AND Si:P,B IN THE VICINITY OF THE METAL–INSULATOR TRANSITION

M. J. R. HOCH*, U. THOMANSCHEFSKY AND D.F. HOLCOMB

Laboratory of Atomic and Solid State Physics, Cornell University, Ithaca, N Y, 14853, U S A.

^{29}Si Knight shift measurements have been carried out on samples of Si:P and Si:P,B in the vicinity of the metal–insulator transition at temperatures of 4.3 K and 1.5 K. The Knight shift <K> appears to change in a continuous way through n_c. Comparison of the results for the uncompensated and compensated samples reveals that, for a given n/n_c value, <K> is significantly lower in Si:P,B than in Si:P. The difference in <K> values increases as the transition is approached from the metallic side. Possible explanations for this difference in behaviour are briefly discussed.

1 INTRODUCTION

Low temperature electrical conductivity measurements on Si:P and Si:P,B (1) have shown that these materials have different zero temperature conductivity exponents in the vicinity of the metal–insulator (MI) transition. This could mean that the compensated and uncompensated materials belong to different universality classes (2).

Knight shift measurements offer an alternative means to the conductivity for probing the properties of the delocalised electrons. Alloul and Dellouve have recently (3) measured the ^{31}P Knight shifts in metallic Si:P close to the transition and have interpreted their results in terms of an inhomogeneous two–fluid magnetic model. They suggest that only those ^{31}P nuclei which are somewhat distant from regions of spin localisation are observed in the NMR signals.

We have measured ^{29}Si Knight shifts for Si:P, for which previous data is available, and also for Si:P,B, for which no previous results appear to have been published. The objective has been to obtain further information on the effects of disorder on the properties of the delocalised electrons.

2 EXPERIMENTAL DETAILS

Measurements were made on crushed (<150 μ) Si:P and Si:P,B wafers whose properties had been characterized by conductivity and neutron activation analysis. The samples were placed in small teflon cylinders, which were inserted into an NMR probe mounted on a Janis variable temperature helium cryostat. The probe permitted a train of three samples immersed in liquid helium to be moved through a fixed coil embedded in a Stycast epoxy cylinder by means of a stainless steel rod brought out through the top of the cryostat. For all experiments the central sample (insulating Si:As) was used as a reference. Fourier transform pulsed NMR methods at 8.5 MHz were used for obtaining the line profiles. Measurements could be made rapidly on all three samples. This minimized any effects due to possible magnetic field drift of the electromagnet.

3 RESULTS AND DISCUSSION

The measured Knight shifts are small, $(10^{-4} - 10^{-5})$, and have a Poisson distribution of values for the inhomogeneous system. Fig. 1 shows representative spectra for Si:P and Si:P,B at 4.3 K. Similar results were found at 1.5 K. It is convenient to discuss the average Knight shift <K> for the distribution and Fig. 2 shows a plot of <K> versus n for the two sets of new samples together with previous data for Si:P obtained by other workers. For comparison the measured magnetic susceptibility, χ, for Si:P is shown as a function of n.

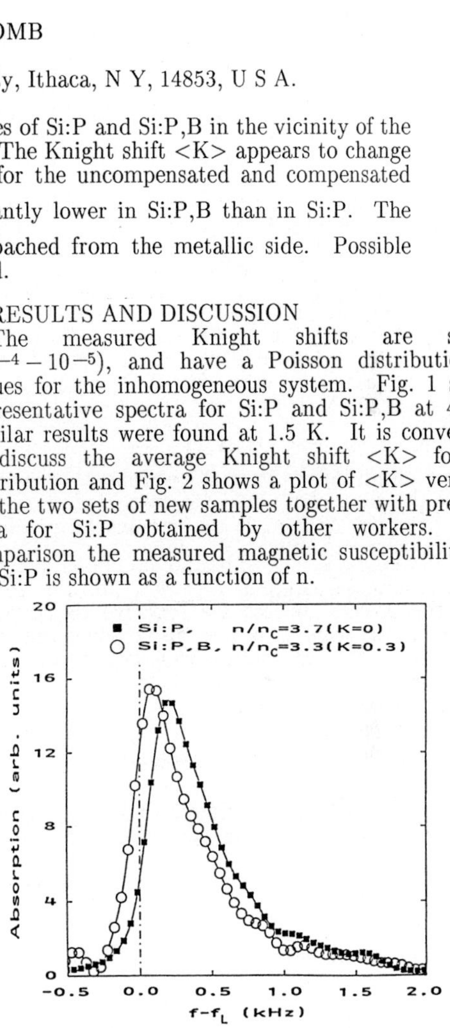

Figure 1. Absorption lineshapes for Si:P and Si:P,B (compensation ratio K = 0.3) in a field of 1 T at 4.3 K.

A number of observations can be made about the data of Fig. 2. At fairly high concentrations ($n > 10^{19}cm^{-3}$) the <K> values for the uncompensated and compensated samples converge. For $n < 10^{19}cm^{-3}$ the <K> values for Si:P,B are somewhat lower than

*Permanent address: Department of Physics, University of the Witwatersrand, P O Wits, 2050, Johannesburg, South Africa.

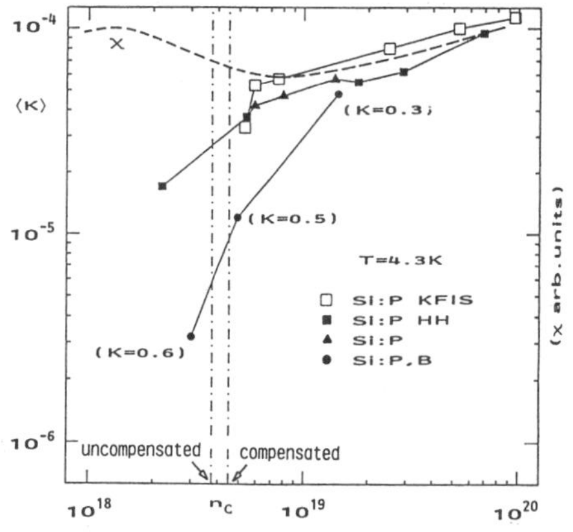

Figure 2. Mean Knight shift $\langle K \rangle$ as a function of electron concentration n at 4.3 K. The behaviour of the measured susceptibility, χ, is shown for comparison. Previous data for Si:P obtained by Kobayashi et al (KFIS) (4) and Hirsch and Holcomb (HH) (5).

for Si:P. Both curves suggest continuous behaviour through the transition. This is consistent with the two–fluid model applying in the just–insulating phase as well as on the metallic side.

For homogeneous simple metals the Knight shift may be expressed as follows:

$$\langle K \rangle = \frac{8\pi}{3} \langle |u(o)|^2 \rangle_{E_F} \chi_P$$

where χ_P is the Pauli susceptibility and $\langle |u(o)|^2 \rangle_{E_F}$ is the electron probability density at the nucleus averaged over the Fermi surface. For a disordered metal the above expression must be modified by replacing χ_P by the average local susceptibility $\langle \chi_i \rangle$ for spins in region i. It is also possible that the conduction electron density at the nuclear sites has a distribution of values $\langle |u(o,i)|^2 \rangle_{E_F}$. The Pauli susceptibility may be expressed in the usual way in terms of the density of states at the Fermi surface $N(E_F)$.

We suggest the following explanation for the data of Fig. 2. At high concentrations ($n >> 10^{19} cm^{-3}$) both compensated and uncompensated systems approximate dirty metals and the Knight shifts follow the Pauli susceptibility behaviour ($\chi_P \propto n^{1/3}$). At lower concentrations on the metallic side of the transition the measured susceptibility is enhanced by local moments. The Knight shifts continue to decrease in contrast to χ. This is because the $\langle K \rangle$ values are sensitive to the delocalised electron–nucleus interactions. For Si:P,B the decrease in $\langle K \rangle$ with reducing n is more rapid than for Si:P. If it is assumed that $\langle |u(o,i)|^2 \rangle_{E_F}$ is roughly the same for both compensated and uncompensated samples then this implies that, for the delocalised electrons, $\langle \chi_i^{comp} \rangle < \langle \chi_i^{uncomp} \rangle$. This in turn suggests a lower density of states for the delocalised electron fluid in the compensated system than in the uncompensated system.

It is interesting to note that the results of Hirsch and Paalanen (6) show that, for metallic samples with similar free electron densities, $\chi^{comp} > \chi^{uncomp}$. The difference is strongly temperature dependent. The decrease in the delocalised electron fluid susceptibility with compensation, as inferred from the present Knight shift results, is therefore to be contrasted with the corresponding increase in the total susceptibility, which, in turn, is dominated by the localised fluid component.

Compensation increases the degree of disorder and strengthens the Anderson character of the transition. The present results suggest that this is associated with a lower density of states at the Fermi level, which may be linked with the enhanced fluctuations in the lattice potential produced by compensation.

4 CONCLUSION

Knight shift measurements on Si:P and Si:P,B samples near the MI transition show that on the just–metallic side of the transition the Knight shift is significantly smaller in the compensated samples than in the uncompensated samples. The difference tends to decrease as n/n_c increases. For both cases the evidence suggests that the Knight shift changes in a continuous way through the transition.

The compensated samples have enhanced disorder and therefore increased Anderson character when compared with the uncompensated samples. This change appears to be associated with a decrease in the density of states at the Fermi level in the compensated material for $n \sim n_c$.

REFERENCES

(1) M.J. Hirsch, U. Thomanschefsky and D.F. Holcomb, Phys. Rev. B 37, 82571 (1988)
(2) G.A. Thomas, Phil. Mag. B 52, 479 (1985).
(3) H. Alloul and P. Dellouve, Phys. Rev. Lett. 59, 578 (1987).
(4) S. Kobayashi, Y. Fukagawa, S. Ikehata and W. Sasaki, J. Phys. Soc. Japan 45, 1276 (1978).
(5) M.J. Hirsch and D.F. Holcomb, Phys. Rev. B 33, 2520 (1986).
(6) M.J. Hirsch and M.A. Paalanen, Bull. Am. Phys. Soc. 33, 385 (1988).

Physica B 165&166 (1990) 307–308
North-Holland

ANOMALOUS BEHAVIOR OF DISCRETE RESISTANCE FLUCTUATION LIFETIMES IN NOVEL METAL INSULATOR-METAL TUNNEL JUNCTIONS

Xiuguang JIANG and J.C. GARLAND

Department of Physics, Ohio State University, Columbus, Ohio 43210 U.S.A.*

Measurements are reported of the switching lifetimes of the discrete resistance fluctuations in Au-Al_2O_3-Au tunnel junctions which have small Al particles randomly embedded within the insulating barrier. The tunnel junctions are 50x50 μm^2 and the Al particles are nominally disk-shaped with thickness about 30Å and radius 500Å. Tunnel M-I-M junctions which do not contain embedded metal particles typically have switching lifetimes that decrease exponentially with applied d.c. bias voltage. However, we have found that the presence of embedded particles leads to anomalous behavior, in which the switching lifetimes increase with applied bias voltage. We ascribe this behavior to the interaction of defects with the charge state of the embedded Al particles.

1. INTRODUCTION

Discrete resistance fluctuations, or random telegraph signals (RTS), are frequently observed in metal-insulator-metal (MIM) tunnel junctions and are generally attributed to the trapping of single electrons by ionic defects within the insulator. Although the detailed mechanism governing the trapping process is not well understood, the dependence of the switching lifetimes on both temperature and applied d.c. bias voltage are generally consistent among a large variety of MIM junctions. (1,2,3) Typically, both the average high and low conductance lifetimes τ_H and τ_L decrease exponentially with increasing d.c. bias voltage. Ordinarily a bias voltage increase of 150-200mV shortens the lifetime by about an order of magnitude.

Previous work (1-3) on switching RTS has centered on tunnel junctions with homogeneous insulating barriers. In this paper, however, we consider switching RTS in MIM tunnel junctions with small metal particles randomly embedded inside the insulating barrier. We have observed that, unlike homogenous junctions, the switching lifetimes of our samples *increase* exponentially with d.c. bias voltage, apparently as a result of charge state of the embedded Al particles. In addition, we find that the charge state induces structure into the plots of lifetime as a function of bias voltage.

2. EXPERIMENTAL DETAILS

As shown in Figure 1, our samples were large 50x50 μm^2 Au-Al_2O_3-Au tunnel junctions, with the thickness of the Au films nominally 1000Å and the thickness of the Al_2O_3 insulating barrier typically 30-60Å. The Al_2O_3 barriers were fabricated by first depositing 50-70Å of Al onto a 50x50 μm^2 area of an Au electrode. This Al film was then oxidized in oxygen at a pressure of 10^{-2} Torr for 5 minutes. Then another 40-50Å of Al was deposited over the previous layer of Al and allowed to oxidize in air for 2 hours. Finally, the top Au electrode was deposited to complete the junction.

The second oxidation step does not completely oxidize the second Al film, resulting in clusters of disconnected Al

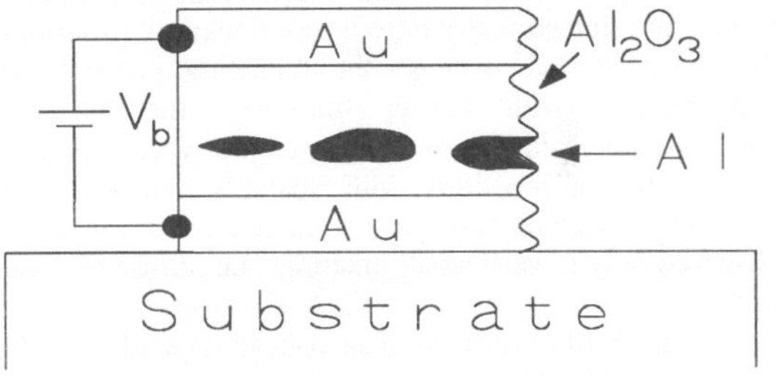

Figure 1. Schematic diagram of tunnel junctions illustrating embedded Al particles.

disk-shaped particles. The tunneling current therefore is governed by two channels: a direct channel between the two electrodes, and an indirect channel, via an Al particle as an intermediate step. For the indirect mode, the junction acts essentially as two junctions in series, with the Al particle forming a common electrode. For our samples, one of these series junctions (the "bottom" junction) has a slightly thinner barrier thickness than the "top" junction. Moreover, the indirect channel always dominates the overall tunneling process, so that as a practical matter only discrete resistance fluctuations arising from defects located within the junction areas separated by Al particles can be observed.(4) In determining average lifetimes of the high and low

* Supported by the National Science Foundation grant # DMR-8821167.

0921-4526/90/$03.50

conductance states, typically $1.5x10^5$ measurements of the sample current are recorded, which correspond to about 2000 conductance switching transitions.

3. RESULTS

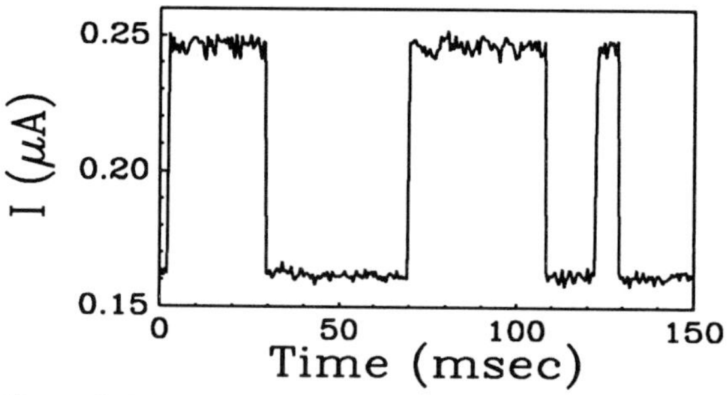

Figure 2. Plot of time dependence of the sample current. The sample temperature and bias voltage were 44.7 K and 178 mV, respectively.

Figure 2 shows the time dependence of the current for a discrete RTS of a sample biased at a constant d.c. voltage of 178mV. For this sample, the current fluctuated about 50% between the two conductance states, a typical amount for these large samples. As discussed elsewhere in detail, (4) the large amplitude of the fluctuations is believed to result from (intentionally introduced) thickness variations of the oxide barrier; because the tunneling current varies exponentially with barrier thickness, the current is concentrated at the region of the sample where the barrier is thinnest. In junctions with spatially uniform barrier thickness, discrete resistance fluctuations are ordinarily observed only in very small junctions, i.e., those of area <1 μm^2.

Figure 3 illustrates the bias voltage dependence of the lifetimes of the high and low conductance states for the sample whose data were shown in Figure 2. Figures 3(a) and 3(b) are the results from two sets of measurements, taken at a temperature of 44.7K, and illustrate two distinct behaviors. First, the lifetime of the high conductance state decreases exponentially with bias voltage, as is consistent with previous results from ordinary MIM tunnel junctions. (1,2,4) Second, the lifetime of the low conductance state shows the reverse behavior, increasing approximately exponentially with increasing voltage. This latter, unexpected behavior has been observed for RTS associated with four different defects, and is accompanied by an unusually rapid decrease of the high conductance lifetime with voltage (about an order of magnitude decrease a for 30mV increase in voltage). Note further, that in contrast to the high-conductance lifetimes the low conductance lifetimes showed an unusual structure that did not reproduce between the two data runs. This structure in the plots is far

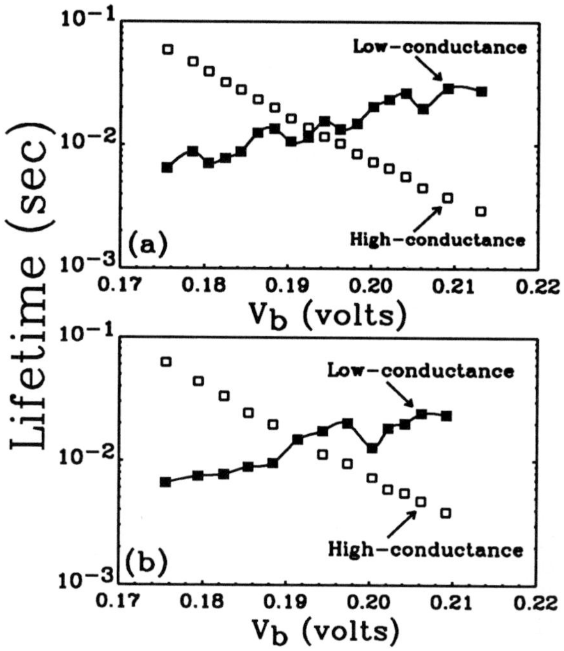

Figure 3. Plots of the lifetimes as a function of bias voltage. Both (a) and (b) represent results of different measurements of the same RTS with T=44.7K.

outside experimental errors, which are only about 3%. In another paper, we will present a model that accounts for this structure and the anomalous lifetime dependence in terms of the interaction between the switching defect and the charge state of the Al particles. According to our model, the charges on the Al particles alter the potential energy distribution surrounding the switching defect, thus modulating the lifetimes of the defect states.

ACKNOWLEDGEMENTS

The authors wish to thank T.K. Xia for useful conversations.

REFERENCES

(1). C.T. Rogers and R.A. Buhrman, Phys. Rev. Lett. 55, 859 (1985).
(2). R.T. Wakai and D.J. Van Harlingen, Phys. Rev. Lett. 58, 1687 (1987).
(3). M.J. Kirton and M.J. Uren, Advances in Phys. 38, 367 (1989), and references therein.
(4). X. Jiang, M.A. Dubson, and J.C. Garland, Observations of Giant Discrete Resistance Fluctuations in Normal Metal Tunnel Junctions, to be published.

Physica B 165&166 (1990) 309–310
North-Holland

ON THE SCALING THEORY OF THE MOTT TRANSITION

Mucio A. CONTINENTINO*

Instituto de Física, Universidade Federal Fluminense, Outeiro de São João Batista s/n, Niterói, 24020, Rio de Janeiro, Brasil

A recently proposed scaling theory of the metal insulator transition due to correlations is extended to take into account the multicritical character of this instability at zero temperature as required for the vanishing of the compressibility.

Recently we have proposed a scaling theory of the metal-insulator transition due to correlations, the so called Mott transition (1). The theory is based on a renormalization group treatment of the Hubbard model which shows that this transition at zero temperature is continuous and associated in this approach with an unstable fixed point (2) at $(U/W)_c = \tilde{U}_c$ and $T = 0$. Here U is the Hubbard on site Coulomb repulsion and W the bandwidth. We considered an expansion of the renormalization group equations close to the fixed point and introduced temperature as a relevant "field" scaling as $(T/W)' = b^y(T/W)$ where b is a scaling factor and $y > 0$, a critical exponent (1). Using general properties of the zero temperature fixed point (3), we obtained several results. Firstly we have shown that due to dimensional shift the effective dimensionality of this problem $d_{eff} = d + y$, where d is the space dimension (4,3). Also the quantum character of the critical fluctuations close to $\tilde{U}_c$ $(T = 0)$, imply $y = z$ where z is the dynamic exponent (4). The scaling form of the correlation length was found to be: $\xi \propto (U - U_c)^{-\nu} f(T/T_c)$. The crossover temperature $T_c = A(U - U_c)^{\phi_t}$ with $\phi_t = \nu y$ where ν is the correlation length exponent (1). In our scaling approach, *where the metal insulator transition is accompanied by a transition to an antiferromagnetic state*, ξ can be interpreted as the typical size of magnetically correlated regions. The crossover line in the non-critical part of the phase diagram, for $U < U_c$, marks the onset, with decreasing temperature $(T < T_c)$, of the Fermi liquid regime which we identified with the so called (5) *highly correlated eletron gas*. The enhancement of the effective mass and susceptibility of this electron gas due to the incipient instability were obtained (1).

In this report we explore the consequences of the scaling theory when the fixed point at $\tilde{U}_c$, $T = 0$, has tricritical character. We assume this is the case for the situation considered here where the number of electrons is fixed in one electron per atom. The results derived in Refs. 1 and 3 for the critical behavior associated with the zero temperature fixed point also hold in this case. However the upper critical dimensionality above which the exponents attain their classical values is now $d_c > 3$. Furthermore these exponents assume for $d_{eff} = d + z \geq 3$ their classical *tricritical* values (6). In particular $\alpha = 1/2$ different from the classical value $\alpha = 0$ associated with a normal critical point (1). This has important implications for the behavior of the compressibility K close to the transition as we show below. The singular part of the ground state energy (1) $E \propto (\tilde{U}-\tilde{U}_c)^{2-\alpha}$ where $\tilde{U}$ is a smooth function of volume V. The compressibility is given by $1/K = V(\partial^2 E/\partial V^2) \propto a(U - U_c)^{-a} + g(U - U_c)^{1-a}$ where a and g are constants. When $a=0$ as in the case the fixed point at U_c, T=0 is a normal critical point, the compressibility has a logarithmic singularity but remains finite at the transition. This situation is appropriate to describe magnetic phase transitions where the metallic character of the system does not change as for example in the Stoner theory of the paramagnetic to ferromagnetic transition (4) or in Kondo lattices (7). In the tricritical case where $a = 1/2$, we get $K \propto (U - U_c)^{1/2}$ and consequently vanishes at U_c, implying a divergence of the bulk modulus, at the T=0 transition, as is expected to occur (8) for a metal-insulator transition.

The enhancement of the effective mass and susceptibility of the Fermi liquid for $T \leq T_c$ is given by: $m^* \propto (U - U_c)^{3/2-y}$ and $\chi \propto (U - U_c)^{-1}$

*CNPq Research Fellow

respectively where $y = z$. For $z = 5/2$ the ratio m^*/χ becomes independent of pressure. This value of z is intermediate between the values expected for antiferromagnetic spin fluctuations ($z = 2$) and that for paramagnons ($z = 3$) and implies a weak wavevector dependence of the lifetime of the spin fluctuations.

We have shown that the scaling theory distinguishes between two situations in which the bulk modulus either remains finite or diverges at a zero temperature transition. The latter case is the one appropriate to describe the metal insulator transition due to correlations and is associated with the tricritical character of the zero temperature fixed point.

REFERENCES

(1) M.A.Continentino, Europhysics Lett. 9, 77 (1989).
(2) J.Hirsch, Phys.Rev.B32, 5259 (1980).
(3) A.J.Bray and M.A.Moore, J.Phys.C 18, L927 (1985).
(4) J.A.Hertz, Phys.Rev.B 14, 1165 (1976).
(5) N.F.Mott, *Metal-Insulator Transitions* (Taylor and Francis, London, 1974).
(6) G.Toulouse and P.Pfeuty, *Introduction au Groupe de Renormalisation et a ses applications* (Presses Universitaires de Grenoble, Grenobles, 1975).
(7) M.A.Continentino, G.Japiassu and A.Troper, Phys.Rev.B 39, 9734 (1989).
(8) P.W.Anderson in Lecture Notes given at Varenna summer school. "Frontiers and Bordelines in Many Particle Physics", july 1987.

Physica B 165&166 (1990) 311–312
North-Holland

RESISTANCE SCALING IN INHOMOGENEOUS MEDIA IN THE VICINITY OF A METAL-INSULATOR TRANSITION

O.I.Barkalov, I.T.Belash, S.E.Esipov, V.F.Gantmakher, E.G.Ponyatovskii and V.M.Teplinskii

Institute of Solid State Physics, Academy of Sciences of the USSR, Chernogolovka, 142432, USSR

Scaling relation between residual R_0 and temperature-dependent R_1 parts of the sample electrical resistance in the vicinity of a metal-insulator transition can be a guide in examining the structure of the material. In the relation $R_1 \propto (R_0)^{\nu}$ the exponent value ν=1 is typical for a random mixture of metallic and insulating domains, ν=0 for granular metals. A special case ν=0.75 has been observed recently. It corresponds to a fractal structure of the insulating phase with the classical size effect governing the conductivity of the metallic channels. Experimental data on Zn-Sb and Al-Ge alloys are presented.

1. INTRODUCTION

Transport phenomena in inhomogeneous media are usually considered by applying the percolation theory (1). However, the material science has at its disposal a lot of geometrical patterns which are not of a simple random nature. For instance, granular metals contain in essence a correlated system of rather regularly alternating metallic and insulating regions. Much more complicated structures may occur when the solid mixture arises as a result of a phase transition in some parts of the sample. If these two phases have different conductivities, for instance, when a metal-insulator transition takes place, one should describe conducting network and the total conductance of the sample.

In this content the term "metal-insulator transition" has two meanings: "point transition" when a small volume becomes dielectric, and "sample transition" when the conductance of the whole sample drops to zero. We shall concern ourselves only with studying the metallic side of the sample transition. The main idea is that while analyzing scaling relations for the sample resistance one can judge what type of the structure is realized in the sample and distinguish, for example, a fractal pattern from a random one.

2. RESISTANCE SCALING

Let's represent the resistance R as a product of a resistivity ρ by a geometrical factor φ

$$R = \rho(T)\varphi = [\rho_0+\rho_1(T)]\varphi = \frac{p_F}{ne^2}\left(\frac{1}{l_0}+\frac{1}{l_T}\right)\varphi \qquad /1/$$

(p_F is the Fermi momentum, mean free path $l^{-1}=l_0^{-1}+l_1^{-1}$). When the changes in the resistivity are controlled by the carrier density n as well as when the resistance changes are due only to the alternation in the shape of the conductive channels, i.e. of the factor φ, then relation

$$R_1 \propto R_0 \equiv [R_0]^1 \qquad /2/$$

holds. The latter case apparently takes place during approaching the transition in a random mixture of two phases. Indeed, when approaching the threshold, what alters is the size of the cells of the conductive backbone in the infinite cluster, i.e. the factor φ.

Insulating films in an ideal granular metal are connected in series with metal volumes. As the tunnel resistance of the films does not depend on T one gets

$$R_1 = const \propto [R_0]^0. \qquad /3/$$

3. FRACTAL PATTERN OF THE Zn-Sb ALLOY STRUCTURE

The exponent values ν=1 from Eq./2/ and ν=0 from Eq./3/ are not the only possible ones. Below follows an example with ν=0.75: an inhomogeneous structure created in the course of a special phase transition, namely, the amorphization of the metastable phase of alloy Zn-Sb quenched under pressure (2,3). The initial state is crystalline and metallic while the final one is amorphous and insulating. The amorphization process could be led slowly by the low temperature annealing. It could be repeatedly interrupted by returning to nitrogen temperatures.

This transition to the amorphous state has the following important features:

1. the specific volume increases significantly during the transition ;

2. the sample resistance R increases by orders of magnitude while the sample remains metallic at low temperatures, i.e. while it retains $\partial R/\partial T>0$;

3. the large increase of the resistance is accompanied by a very small decrease of the temperature T_c of the superconducting transition. This point contains the evidence that the sample is macroscopically inhomogeneous in the intermediate states of the transition (2) .

A fractal-like model of the intrinsic structure of the sample was proposed in (3). It supposes the growing insulating amorphous inclusions to be cactus-like with leaves branching many times. These leaves, or sheets, are supposed neither to intersect each other nor to merge. This maintains the existence of current paths, at the far-gone stages of the transition, in contrast with the percolation model.

The development of the "cactus" structures can be described by a parameter d, mean distance between the leaves. According to (3), the current paths are located at the surface with fractal dimension three which separates two "cactuses"; from the mathematical point of view the current paths are similar to trajectories of brownian particles. The fractal dimension of such a trajectory is two: its length is proportional to squared radius K of the domain it occupies.

So, as far as electrical resistance is concerned, the conductive channels are conductive brownian trajectories. Now d becomes the step length of the random walk and, with the distance between the contacts L being constant, we have the channel length $\lambda \propto d^{-1}$. Supposing in addition that the cross-section of the channel is d^2 we obtain $\varphi \propto d^{-3}$.

To get an exponent in Eq./2/ different from $\nu=1$ we need some dependence of φ on d. Such dependence can exist due to the dc size effect. For a wire with diameter $d \ll l_0$ the mean free path l_0 can be, within a rather good accuracy, replaced in Eq./1/ by d (5). Then, with whatever relation between l_T and d, one has

$$\rho_0(d) \propto d^{-1}, \quad \rho_1(d) \propto d^0. \qquad /4/$$

Combining this with $\varphi \propto d^{-3}$ we finally get

$$R_0 \propto d^{-4}, \quad R_1 \propto d^{-3}, \quad R_1 \propto [R_0]^{0.75}. \qquad /5/$$

That is just what follows from the experiment (Fig.1). Note, in passing, that in the percolative system the links between the nodes of its backbone always contain so called red bonds, regions with minimal possible cross-sections (1). This means that the dc size effect cannot influence the ν value.

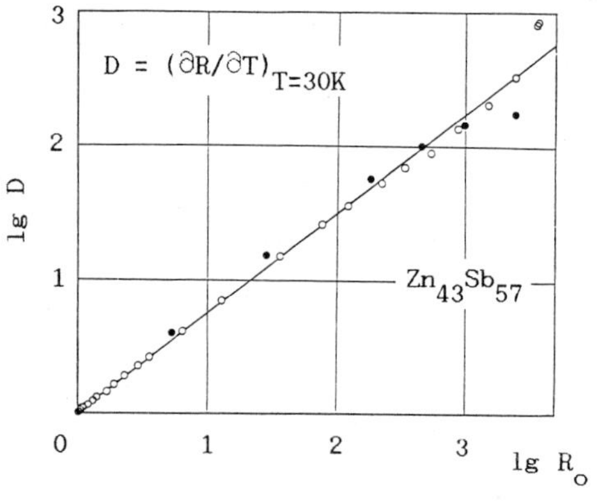

FIGURE 1

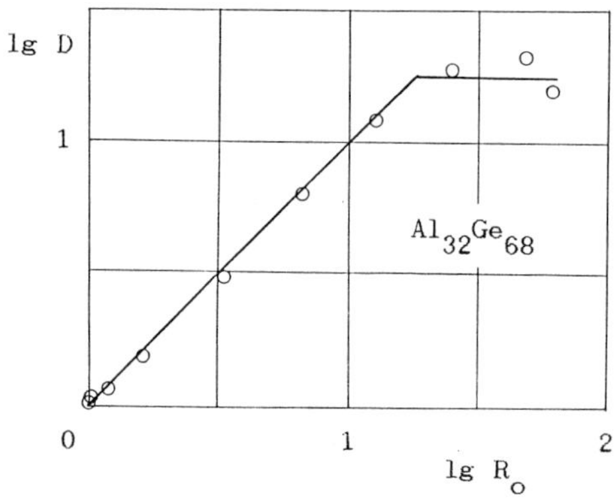

FIGURE 2

4. Al-Ge DATA

The experiment with the Al-Ge alloy was performed similarly to those with Zn-Sb. The differences were only quantitative. However, the results are quite different, as it can be seen from the comparison of Figs.1 and 2. Note that Fig.2 demonstrates the exponent $\nu=1$ instead of $\nu=0.75$. Similarity of the processes in the both alloys gives reason to suggest that the structures in both are the same and that the relation between l and d is the main source of the difference.

Exponent 0.75 is the result of the combination of a fractal structure of the insulating phase and a rather long mean free path l. If, instead, $l \ll d$ holds one will find $\nu=1$.

CONCLUSIONS

In conclusion, the scaling relations between the parts of the resistance in the vicinity of the metal-insulator transition contain information about the macrostructure of the sample. Eqs /2/ and /3/ express the two limiting cases which correspond to a random mixture of phases and to a granular metal. The exponent ν can be changed by the dc size effect. The Zn-Sb experimental data give such an example.

REFERENCES

(1) D.Stauffer. Introduction to percolation theory (Taylor and Fransis, London, 1985).
(2) O.I.Barkalov, I.T.Belash, V.F.Gantmakher, E.G.Ponyatovskii and V.M.Teplinskii, Pis'ma JETF 48 (1988) 561 [JETP Letters 48 (1988) 609].
(3) V.F.Gantmakher, S.E.Esipov and V.M.Teplinskii. Zh. Eksp. i Theor. Fiz. 97 (1990) 373 [Sov. Phys. JETP 70, No 1 (1990)].
(4) J.Feder. Fractals (Plenum Press, New York, 1988).
(5) R.G.Chambers in : The Physics of Metals. 1.Electrons, ed. J.M.Ziman, (Cambr. Univ. Press, 1969).

Physica B 165&166 (1990) 313–314
North-Holland

THE RANGE OF THE DYNAMICAL COULOMB INTERACTION IN JUNCTIONS OF THIN DISORDERED FILMS

GERD BERGMANN and WEI WEI

University of Southern California, Los Angeles, California 90089-0484

We have experimentally searched for the long range of the (dynamical) Coulomb interaction in sandwiches of two disordered metal films separated by an insulating layer. We expect the two films to assist each other in screening but our measurements do not support the long range character of the dynamical Coulomb potential and challenge the theoretical description of the latter.

Disordered metals have been studied very intensively in recent years. One of their fascinating properties is the fact that the dynamic Coulomb interaction is dramatically different from that in periodic metals (see for example (1). One obtains for the screened Coulomb energy between two electrons in a thin film for small wave number q (large distances in real space):

$$U(q,\omega) = \frac{1}{Nd}\ \frac{Dq^2 - i\omega}{Dq^2} \quad (1)$$

where N is the density of states, D the diffusion constant, d the thickness of the film and q and ω the wave number and frequency. Eq. (1) shows that for finite frequency the potential has a divergent part $i\omega/NdDq^2$ which increases with decreasing film thickness beeing equivalent to a long range in real space. By increasing the film thickness the effective or screened Coulomb potential is reduced.

In thin disordered films this imperfectly screened Coulomb potential yields anomalies in the temperature dependence of the resistance and the Hall effect. For example, the temperature dependence of the resistance (per square) of such a film is modified by

$$\frac{\Delta R}{R^2} = -\ [1-F']\ \frac{e^2}{2\pi^2\hbar}\ \ln\left(\frac{k_BT\tau_o}{\hbar}\right) \quad (2)$$

Here F' is a parameter which is of the order of .2.

We decided to test the long range character of the Coulomb interaction in a direct experiment. For this purpose we investigated a sandwich consisting of two metal films which are separated by an insulating layer. (The latter serves as a tunneling barrier.) With respect to the time scale of the Coulomb anomaly we found that it is very easy to decouple the two films. As soon as the tunneling time τ_t is longer than the thermal coherence time $\tau_T=\hbar/k_BT$ the electrons in the two films are uncoupled with respect to the Coulomb anomaly. On the other hand the two films have to assist each other in screening because the Coulomb interaction is supposed to be long ranged.

Now we consider the case of two films which are separated only by a very small distance less than the individual film thickness. For q much less than the inverse total sandwich thickness the electrons in both films experience the same potential. If the two films have the same diffusion constant then the screened potential in eq. (1) is reduced by the factor R/R_1 where R_1 is the resistance of the film with the charge wave, R_2 is the resistance of the other film and $R=[1/R_1+1/R_2]^{-1}$ is the total resistance of the sandwich.

Such a sandwich of two films separated by a tunneling barrier therefore has the interesting property that the electrons are confined to the individual films during the relevant time scale of the Coulomb anomaly. On the other hand the effective Coulomb interaction experiences the screening of the other film and its magnitude is reduced. Therefore we expect to measure for the Coulomb anomaly the added contribution of the two films. However, each contribution is proportional to its individual Coulomb matrix element and is, therefore, reduced due to the presence of the other film by R/R_i (i=1,2). One easily recognizes that the total contribution remains the same as for a single film. There is a perfect compensation of the two effects.

0921-4526/90/$03.50

The tunneling rate can be determined from the magneto-resistance curves using the theory of weak localization (for two films separated by a tunneling barrier (2). Experimentally, this is achieved in the system of two then layers of Mg separated by an insulating Sb layer. The samples are made in situ by quench condensing thin metal films onto a crystalline quartz substrate held at helium temperature in a vacuum better than 10^{-11} torr. The films made this way are thin, homogeneous and disordered and are very suitable for the present investigation. Magneto-resistance measurements are made after various steps of evaporations.

In a series of experiments we have determined the tunneling time for a Mg/Sb/Mb sandwich as a function of the thickness of the Sb layer. The tunneling rate depends essentially exponentially on the thickness of the Sb barrier. We investigated about five sandwiches of Mg/Sb/Mg/Sb/Mg.. with repeated condensation of Sb/Mg. In Table I we have collected the data for one of these experiments. Column one gives the composition of the sandwich in several stages of the experiment. The third column gives the thickness of the upper element of the sandwich.

Since both weak localization and Coulomb anomaly contribute to the temperature dependence of the film resistance at low temperatures, the Coulomb anomaly can be investigated only when weak localization is suppressed. This is achieved by a strong magnetic field which destroys weak localization, while the Coulomb anomaly is essentially field independent. Experimentally, we measured the temperature dependence of the resistance from 4.2 K up to 25 K in the magnetic field of 7 T after each step of evaporation and annealing in all the experiments. For T > 10 K, the phonon contribution to the resistance increases quickly. Below 10 K, the temperature dependence of the conductance satisfies the logarithmic law of eq. (2). We denote the slope of the logarithmic temperature dependence of the film by A_C. For a single film one has $A_C = (1-F') \simeq .8$. The second column in Table I shows the values of A_C for the repeated sandwich at various steps of its construction. The important result is that A_C increases roughly by .8 whenever a Mg/Sb/Mg junction is prepared consisting of a Sb layer thicker than about 4 atola. Only for a Sb thickness of less than about 4 atola, does the sandwich of Mg/Sb/Mg behave as a single film with an A_C value of the order of 0.8. It is also seen that the prefactor A_C does not depend on the thicknesses of the individual Mg layers as long as each of them is thick enough to be conducting.

sandwich	A_C	d atola	R Ω
Mg	0.78	37.4	102.6
Mg/Sb	0.81	1.9	106.8
Mg/Sb/Mg	0.85	3.6	103.7
Mg/Sb/Mg	0.83	23.0	55.9
Mg/Sb/Mg/Sb	0.86	4.2	57.5
Mg/Sb/Mg/Sb/Mg	1.64	3.9	53.6
Mg/Sb/Mg/Sb/Mg/Sb	1.69	6.2	54.9
Mg/Sb/Mg/Sb/Mg/Sb/Mg	2.67	5.6	54.6

Table I: The parameters for the different sandwiches in one set of experiments.

The thermal coherence time $\hbar/k_BT$ is 1.5 ps at 5 K. The tunneling time is at least twice the thermal coherence time at 5 K, if the thickness of the Sb layer is bigger than 4 atola. For the Sb layer thinner than about 4 atola, the tunneling time is shorter than or at most similar in magnitude to the thermal coherence time. Then the two Mg films separated by this thin layer of Sb behave as one single film since the value of A_C remains unchanged. With the fact that the value of A_C increases in steps of about .8 when the two Mg layers are separated by a Sb layer thicker than about 4 atola, it is easily concluded that the Coulomb anomaly is decoupled when the tunneling time is much longer than the thermal coherence time. In this case one obtains for the Coulomb anomaly the sum of two independent Mg films. There is no reduction due to mutual screening. We are quite puzzled about these experimental results and consider them to be a disturbing challenge of our present understanding of the Coulomb interaction at finite frequencies.

REFERENCES

(1) B.L.Altshuler and A.G.Aronov, in Modern Problems in Condensed Sciences, edited by A.L.Efros and M.Pollak North-Holland, Amsterdam, 1985), p.1
(2) G.Bergmann, Phys.Rev. B39, 11280 (1989)

Physica B 165&166 (1990) 315–316
North-Holland

DISORDER-AVERAGING AND THE DYNAMIC NONLINEAR σ-MODEL OF LOCALIZATION THEORY

Marc Horbach and Gerd Schön
Department of Applied Physics, Delft University of Technology, Lorentzweg 1, 2628CJ Delft, The Netherlands

The dynamic nonlinear σ-model of localization theory is an alternative for the formulations in terms of replica's or superfields. In this paper the similarities and differences between these methods are pointed out.

1. THE DYNAMIC NONLINEAR σ-MODEL.

In order to calculate disorder-averaged quantities essentially two different averaging procedures are possible: 'afterward'- and 'initial-stage'-averaging. The advantage of afterward averaging is that it is conceptually simpler. The disadvantage, however, is that it does not lead to an effective Hamiltonian from which directly disorder-averaged quantities can be calculated. Such an effective Hamiltonian has proved to be powerful in localization theory, since by using saddle-point techniques a theory describing the critical diffusion modes can be obtained. This theory, a nonlinear σ-model for a matrix field (2,3,4,5) can then be studied by renormalization group techniques. In this way the Anderson transition for the mean conductance and, with the so-called extended nonlinear σ-model, a crossover from a Gaussian distribution of conductances over the ensemble in the metallic to a lognormal distribution on approaching the Anderson transition have been described (6).

Technically initial-stage averaging methods in localization theory involve some important nontrivial aspects. The averaging is most commonly performed by assigning either replica- (2,3,4) or superfield-indices (5) to the fields in order to solve the denominator problem. These fields are integration variables in a functional integral representation of the quantity one wants to calculate. The critical quantities are those which involve products of retarded and advanced Green functions at slightly different energies. In order to generate such quantities one assigns an extra index to the fields, namely the label retarded or advanced. The two different energies are implemented as parameters, their mean value being the Fermi energy, and their difference the external frequency.

The disorder-averaging has as a consequence that originally uncoupled fields couple: fields in different replica's or fermions and bosons in the superfield method couple.

It was shown in reference (1) that the unphysical replica's or superfields can be avoided by using the time-path formalism, which is a finite temperature real-time quantum field theory (7). In this method the generating functional involves a time integration over a path C in the complex time plane (figure 1). In calculating real-time quantities the vertical parts of the contour do not contibute since the generating functional can be shown to factorize into two parts, one part involving the horizontal and one part involving the vertical branches. Since the generating functional involving only the horizontal branches, $Z_H[j]$ has the property that for vanishing source fields it is normalized to one, $Z_H[0]=1$, hence the denominator problem is absent.

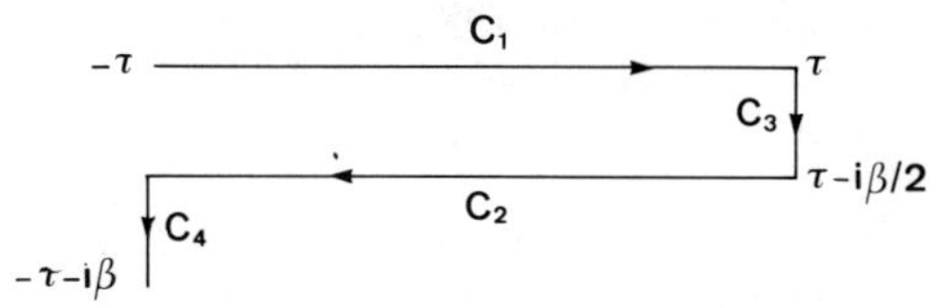

FIGURE 1
The contour C in the complex time plane

The fields $\varphi(t)$ involved in Z_H have their time arguments at branch 1 or branch 2, distinguished by an index: $\varphi_1(t)$ or $\varphi_2(t)$. Here t denotes the real part of the time argument. One can consider these indices as spanning a space, so called thermal-space. A Green function matrix is defined by $G_{\alpha\beta}(t,t') \equiv \langle T_C \varphi_\alpha(t)\, \varphi_\beta(t') \rangle$ with reference time $\tau \to \infty$, where T_C denotes the ordering-operator along the contour C. Via a temperature dependent rotation in the thermal space it is possible to arrive at the 'thermo-fields' $\psi_1(t)$ or $\psi_2(t)$, which have the property that the free electron Green function matrix acquires the form

$$G_{0;\alpha\beta}(t,t') = \begin{pmatrix} \frac{1}{E-\varepsilon(k)+i\eta} & 0 \\ 0 & \frac{1}{E-\varepsilon(k)-i\eta} \end{pmatrix} \qquad [1]$$

The elements on the diagonal are the free retarded and advanced Green functions, which thus arise naturally in the time-path formalism. To summarize, this formalism solves the denominator problem, contains the concept of retarded and advanced fields $\psi_1(t)$ and $\psi_2(t)$, and contains the energy as a variable rather than a parameter, therefore offering a natural starting point for studying interacting disordered systems and time-dependent phenomena.

Initial-stage averaging over a Gaussian distribution of random potentials, which are taken to be δ-correlated in space, leads to a quartic interaction with different energies and different thermal indices coupled. This is analogous to the coupling of different replica's or superfield components.

Decoupling of the interaction term with help of a matrix field $Q_{\alpha\beta EE'}(r)$, where the energy variables are written as matrix indices, eventually leads (after a saddle-point approximation and small gradient, small frequency expansion) to the effective free energy functional (1)

$$F[Q] = \int dr \; Tr\{\hat{D}_0 \nabla_r Q(r) \nabla_r Q(r) - \hat{\omega}_0 Q(r)\} \qquad [2]$$

Here Tr stands for the trace over the thermal indices and over the energy indices , $\hat{D}_0 = 2\pi D_0 N\tau^2$ and $\hat{\omega}_0 = 4\pi N\tau(i\omega_0+\eta)$ where D_0, N and τ are the classical diffusion constant, the density of states and the lifetime in an eigenstate of momentum, all calculated at the Fermi energy. The frequency ω_0 is the difference of the energy arguments of the matrix field and satisfies $\omega_0 << 1/\tau$.
The matrix field Q satisfies the geometric constraints

$$Q^2(r) = (1/\tau^2(E))\, \delta_{\alpha\beta EE'} \quad \text{and} \quad Tr\, Q(r) = 0. \qquad [3]$$

The expression [2] and [3] constitute the 'dynamic' nonlinear σ-model. The natural way to parametrize Q is

$$Q_{\alpha\beta EE'}(r)=[\exp(-W/2)]_{\alpha\gamma EE''}Q_{0;\gamma\gamma E''E''}[\exp(W/2)]_{\gamma\beta E''E'} \qquad [4]$$

where the matrix W is off-diagonal in the thermal space and satisfies $W^+ = -W$ (1). The free W-propagator then is

$$\langle W_{\alpha\beta EE'}(q) W_{\beta\alpha E'E}(-q)\rangle = \frac{1}{2\pi N\tau}\frac{1}{D_0 q^2 - i(E-E'+i\eta)\sigma_{z;\alpha\alpha}} \qquad [5]$$

The W-propagator thus also has a matrix structure. The σ_z distinguishes between a retarded and an advanced diffusion mode. Their role in obtaining the correct results in the calculation of physical quantities will be shown in the next section.

2. RENORMALIZATION AND THE VANISHING OF ELECTRON-LOOP CONTRIBUTIONS.

Physical quantities should be calculated from the renormalized theory . The theory can be renormalized with two renormalization constants, Z_1 and Z_2, which are defined in 2+ε dimensions by (8,9)

$$\frac{1}{D_0} = \mu^{-\varepsilon} Z_1 \frac{1}{D} \quad \text{and} \quad \omega_0 = \mu\omega\sqrt{Z_2} \qquad [6]$$

Here D and ω are the renormalized parameters and μ is a momentum scale. The renormalization constants on the one-loop level follow from the two-point W-vertex function $\Gamma^{(2)}(q,E,E')$ which has two contributions, shown in figure 2.

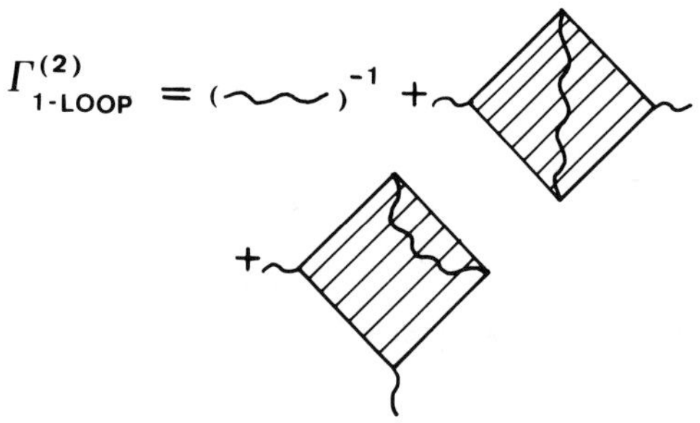

FIGURE 2
The 1-loop 2-point W-vertex function.

The last contribution in figure 2 contains an unspecified energy E'' over which an integration is performed. This signals that one of the electron Green functions involved in the diffusion mode belongs to an electron loop. The energy integration is analogous to the summation over replica's or over superfield components. Since it furthermore contains a summation over the thermal indices this contribution vanishes

$$\int dE \sum_{\alpha} \frac{1}{2\pi N\tau}\frac{1}{D_0 q^2 - i(E-E'+i\eta)\sigma_{z;\alpha\alpha}} = 0. \qquad [7]$$

This shows that the matrix structure of the diffusion mode and the energy integration are responsible for the vanishing of the electron-loop contribution. The first two contributions in figure 2 lead to (8,9)

$$Z_1 = 1 - 2\frac{1}{D\varepsilon} \quad \text{and} \quad Z_2 = 1 \qquad [8]$$

for the orthogonal case, leading to the well-known one-loop β-function.

REFERENCES

(1) Horbach M., Schön G., Physica A, in print.
(2) Wegner F., Z.Phys.B**35** (1979), 207.
(3) Efetov K.B., Larkin A.I., Khmelnitskii D.E., Sov.Phys.JETP**52** (1980), 568.
(4) Pruisken A.A.M., Schäfer L., Nucl.Phys.B**200** (1982), 20.
(5) Efetov K.B., Adv.Phys.**32** (1983), 53.
(6) Kravtsov V.E., Lerner I.V., Yudson V.I., Sov.Phys.JETP**67** (1988), 1441.
(7) Niemi A.J., Semenoff G.W., Ann.Phys.**152** (1984), 105.
(8) Hikami S., Phys.Rev.B**24** (1981), 2671.
(9) Pruisken A.M.M., Z. Wang, Columbia preprint (1989).

Physica B 165&166 (1990) 317–318
North-Holland

VARIABLE RANGE HOPPING CONDUCTIVITY IN Si:As BOLOMETERS

C. C. ZAMMIT, A. D. CAPLIN

Department of Physics, Imperial College of Science, Technology and Medicine, London SW7 2BZ, UK.

M. J. LEA, P. FOZOONI, J. KENNEFICK and J. SAUNDERS.

Department of Physics, Royal Holloway and Bedford New College, University of London, Egham, Surrey, TW20 0EX, UK.

The temperature dependent resistance $R(T)$ of ion-implanted Si:As bolometers has been measured in the variable range hopping regime, with $R(T) = CT^{2m}exp(T_m/T)^m$, for $1.6 > T > 0.2$K. The exponent m changed from 0.5 to 0.28 as the dopant concentration approached the metal-insulator transition. These sensitive bolometers, with an implant depth of 0.1 microns, are limited by the thermal contact between electrons and phonons below 0.1K.

Ion-implanted Si bolometers are being developed for use as infrared, X-ray and Dark Matter detectors (1). We present measurements of the temperature dependent resistance $R(T)$ of Si:As bolometers in the variable range hopping (VRH) regime. The bolometers were fabricated by the Microelectronics Centre of the University of Southampton, using a two step implant process, followed by annealing and diffusion stages, to produce an almost rectangular concentration profile with a nominal depth of 0.1 microns. The doping density n and profile were determined by Secondary Ion Mass Spectroscopy (SIMS). A sequence of doping levels was investigated.

Measurements of the resistance $R(T)$ of seven bolometers (typically 0.6 × 0.6 mm) were made from 1.6 K to 30 mK using DC currents I from 0.2 to 50 nA as shown

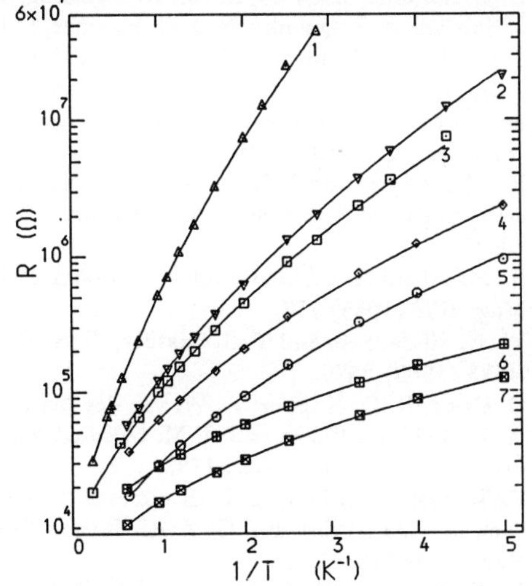

Figure 1 - The resistance $R(T)$ for 7 Si:As bolometers. The solid lines show the fit to equation (1).

in figure 1. As expected, $R(T)$ decreases rapidly as n increases towards n_c, the critical concentration for the metal-insulator (M-I) transition.

For hopping conductivity, Shklovskii and Efros (2) showed that

$$R = CT^s exp(T_m/T)^m \quad (1)$$

where T_m is a scale temperature. The exponent m depends on the electron density of states $g(\epsilon)$ at an energy ϵ relative to the Fermi energy which can be written as:

$$g(\epsilon) = N_p|\epsilon|^p \quad (2)$$

For a constant density of states, $p = 0$, $m = 1/4$, $s = 1/2$ (Mott VRH) while for a Coulomb gap, $p = 2$, $m = 1/2$, $s = 1$. In general, $m = (p+1)/(p+4)$ and $s = 2m$ (see Mansfield et al (3)). Many authors neglect the T-dependent prefactor in equation (1) but it can be significant, unless $T_m \gg T$. We fit our data to equation (1), with C, T_m, and m as adjustable parameters and with $s = 2m$, as shown in Table 1. The most striking result is the decrease in m from 0.5 to 0.28 as the M-I transition is approached as shown in figure 2.

Table 1 Si:As bolometers

Sample	n 10^{18}cm^{-3}	$T_m^{(1)}$ K	m	$T_{0.25}^{(2)}$ K	$T_{0.5}^{(2)}$ K
1	7.0	74	0.490 ±0.017	18700	36.2
2	8.5	48	0.466 ±0.021	8120	17.0
3	8.5	72	0.428 ±0.019	3610	13.5
4	9.0	101	0.380 ±0.030	2260	8.68
5	9.0	95	0.380 ±0.031	1990	8.14
6	9.5	304	0.285 ±0.018	257	2.82
7	9.5	321	0.285 ±0.027	281	2.96

(1)Assuming $s = 2m$, (2)Assuming $s = 0$

The errors given are the 90% confidence limits using a χ^2 analysis. In bulk Si:As Shafarman and Castner (4) found $m \simeq 0.25$ (with $s = 0$) for $0.85 < n/n_c < 0.99$ while $T_{0.25}$ changed by a factor of 10^5 over the same region. An analysis of our data using $m = 0.25$, $s = 0$ gives values of $T_{0.25}$ shown in Table 1 which are larger (and hence further from the M-I transistion) than all but two of their bulk samples. Gang et al (5) have also measured m in ion-implanted Si:As bolometers. Neglecting the prefactor in equation (1) they found $m \simeq 0.5$ with $T_{0.5} > 250$K for their dopant levels. An analysis of our data with $m = 0.5$, $s = 0$ gives values of $T_{0.5} < 40$K, shown in Table 1. Hence, relative to the M-I transition, the carrier densities in our samples lie between these two sets of measurements and show the crossover from $m = 0.5$ towards $m = 0.25$. Samples with peak doping densities slightly above the bulk n_c in Si:As of 8.6×10^{18} cm^{-3} (4) are still below the M-I transition, as observed also by Gang et al (5). Systematic errors in the SIMS measurements, the need to consider an average density, and a true low dimensional effect are all possible explanations.

A similar decrease in m was reported by Trembley et al (6) in GaAs near the M-I transition. This was interpreted as the narrowing of the Coulomb gap near the M-I transition. We also fitted our data to equation (1) by adjusting s and m independently. The best fits were obtained for $s \simeq 2m$ though the error bars on s were then substantial. We therefore used the theoretical value $s = 2m$ for VRH hopping in our analysis. Further experiments and theory are needed to elucidate the effect of the T-dependent prefactor. The exponent s may scale to zero close to the M-I transition (4).

Below 0.2K significant self-heating was observed. For a power-law dependence of the thermal contact between hot electrons at T_e and the thermal bath at T, the power dissipation P is given by:

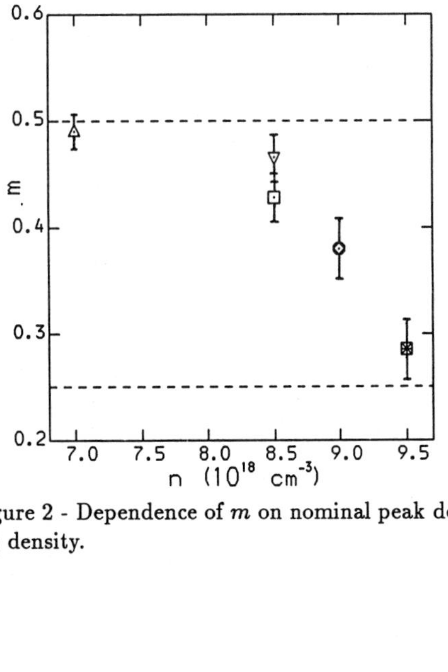

Figure 2 - Dependence of m on nominal peak doping density.

$$P = I^2 R = K(T_e^x - T^x) \quad (3)$$

A plot of $\log P$ versus T_e for $T_e >> T$ is shown in figure 3. Above 0.2K the exponent x is close to 5 which would give a differential thermal resistance $\propto T^{-4}$. This exponent is also observed in planar Ge thermometers (7) due to the coupling between the hot electrons and the thermal phonons. Despite the relatively poor thermal contact, good bolometric sensitivity can be achieved by minimising the bias power below 0.1 K. The estimated best energy detection threshold for a composite bolometer consisting of a 1 gram Si target and sensor (sample 4) is in the region of 10 keV with a bias current < 1 nA.

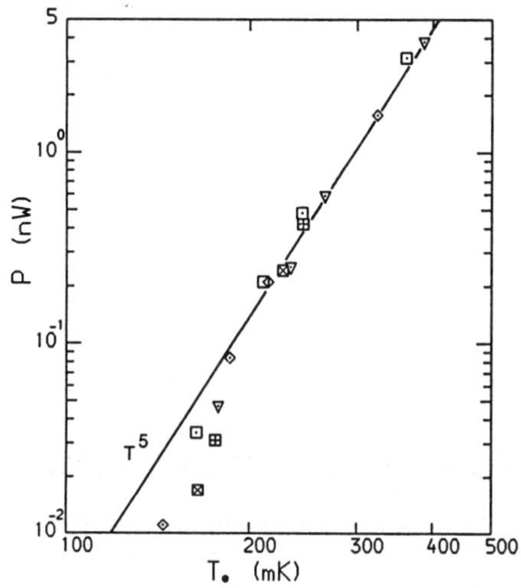

Figure 3 - Electron self heating in Si:As bolometers.

REFERENCES

(1) C. C. Zammit, A.D. Caplin, R. A. Stradling, T. J. Sumner, J. J. Quenby, N. J. C. Spooner, S. F. J. Read, G. J. Homer, J. D. Lewin, P. F. Smith, J. Saunders and M. J. Lea, in "Superconducting and Low Temperature Particle Detectors" (Elsevier, Amsterdam, 1989) 85.

(2) B. I. Shklovskii and A. L. Efros "Electronic Properties of Doped Semiconductors" (Springer-Verlag, 1984).

(3) R. Mansfield, S. Abboudy and P. Fozooni, Phil. Mag. B57 (1988) 777.

(4) W. N. Shafarman and T. S. Castner, Phys. Rev. B. 33 (1986) 3570.

(5) C. Gang, H. D. Kopper, R. W. van der Heijden, A. T. A. M. de Waele and H. M. Gijsman, Solid State Commun. 72 (1989) 173.

(6) F. Trembley, M. Pepper, D. Ritchie, D. C. Peacock, J. E. F. Frost and G. A. C. Jones, Phys. Rev. B. 39 (1989) 8059.

(7) G. X. Mack, A. C. Anderson and P. R. Swinehart, Rev. Sci. Instrum. 54 (1983) 949.

Physica B 165&166 (1990) 319–320
North-Holland

STRONG COUPLING D-WAVE SUPERCONDUCTIVITY STABILIZED BY ANTIFERROMAGNETIC SPIN FLUCTUATIONS

E. SCHACHINGER† and J.P. CARBOTTE

Physics Department, McMaster University, Hamilton, Ontario, Canada L8S 4M1

We present numerical results for the specific heat of a D-wave superconductor which fully include the anisotropy and strong coupling. This represents a generalization of previous more simplified models for D-wave superconductivity stabilized by antiferromagnetic spin fluctuations, in which the anisotropy is treated only approximately.

In previous studies, Millis *et al.* (1) and Williams and Carbotte(2) have calculated some of the properties of a D-wave superconductor stabilized by antiferromagnetic spin fluctuations(3-5) which could be applicable to the heavy fermion systems(6,7) but probably not to the oxides(8). In this model, the anisotropy is kept only in the numerator of the Eliashberg equations. This approximation simplifies the mathematics enormously. Here we retain the anisotropy in the square root denominator and find large corrections to the previous results for the specific heat near T_c, as well as, at lower temperatures.

The interaction between conduction electrons mediated by the spin fluctuation is modelled by

$$I^2\chi(\vec{k}_F-\vec{k}'_F,\omega_n-\omega_m) \equiv I^2\chi_0(\vec{k}_F-\vec{k}'_F)\Phi(\omega_n-\omega_m) \quad (1)$$

where I is an interaction strength, χ_0 is the static magnetic susceptibility and Φ its dynamic part with spectral representation:

$$\Phi(\omega_n-\omega_m) = \frac{2}{I^2N(0)\pi}\int_0^\infty d\omega \frac{\omega A_{SF}(\omega)}{\omega^2+(\omega_n-\omega_m)^2} \quad (2)$$

with spectral density $A_{SF}(\omega)$ equal to $A\delta(\omega-\omega_E)$ where A is a parameter that scales T_c and ω_E is the spin fluctuation frequency. For the susceptibility we use(3-5)

$$I^2\chi_0(\vec{k}_F-\vec{k}'_F) = J_0 - J_1\eta_i(\hat{k})\eta_i(\hat{k}') \quad (3)$$

where the J's are the exchange constants, $\vec{k}_F$ ($\hat{k}$) is momentum (angles) on the Fermi surface and the ratio J_1/J_0 is denoted by g. Two possible simplified models for $\eta_i(\hat{k})$ are

$$\eta_1(\hat{k}) = \frac{\sqrt{15}}{2}(\hat{k}_x^2-\hat{k}_y^2)$$

and

$$\eta_2(\hat{k}) = \frac{\sqrt{5}}{2}(\hat{k}_x^2+\hat{k}_y^2-2\hat{k}_z^2) \quad (4)$$

Within the above simple model, the Eliashberg equations and corresponding free energy formula are treated without approximations. Due to lack of space the equations cannot be written down here. Only results can be given. In Fig. 1 we show results for the normalized specific heat jump at T_c, $\Delta C(T_c)/\gamma T_c$, where $\Delta C(T)$ is the specific heat difference between superconducting and normal state and γ is the Sommerfeld constant. The horizontal axis is the strong coupling parameter, T_c/ω_E and BCS theory corresponds to the limit, $T_c/\omega_E \to 0$. The curves are for the η_1 model. The results for η_2 are qualitatively the same. Various values of g are considered, namely, 1.2, 1.1, 1.0, 0.9, 0.8, 0.7. We note first that g has an influence on $\Delta C(T_c)/\gamma T_c$ only when the coupling is strong and that it varies from the BCS ($T_c/\omega_E \to 0$) value of 0.67 (to be compared with the isotropic BCS value of 1.43) to slightly above 1.4 at $T_c/\omega_E = 0.2$ in the η_1 model and about 1.3 in the η_2 model. Strong coupling corrections can therefore be large and of the order of a factor of 2. The normalized specific heat slope at T_c given by $T_c(d\Delta C(T_c)/dT)/\Delta C(T_c)$ is presented as a function of coupling strength T_c/ω_E for $g = 1.2$, 1.1, 1.0, 0.9, 0.8 and 0.7 in Fig. 2. It starts at 2.33 and 2.04 for the η_1 and η_2 models, respectively, and reaches approximately 3.8 and 3.33 (for $g = 0.7$) for

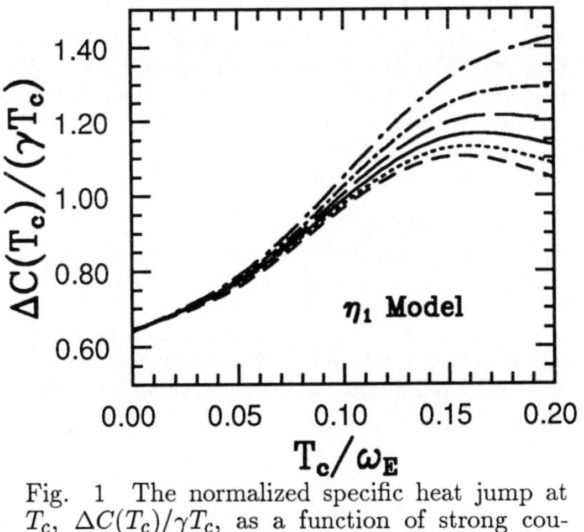

Fig. 1 The normalized specific heat jump at T_c, $\Delta C(T_c)/\gamma T_c$, as a function of strong coupling parameter T_c/ω_E for $g = 1.2$ (intermediate dashed), 1.1 (short dashed), 1.0 (solid), 0.9 (long dashed), 0.8 (intermediate short dashed) and 0.7 (long short dashed).

† Institut für Theoretische Physik, Technische Universität Graz, A-8010 Graz, Austria.

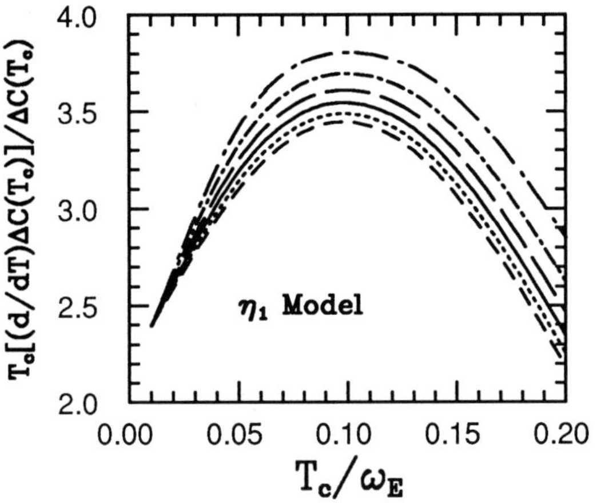

Fig. 2 Normalized specific heat slope at T_c, $T_c(d\Delta C(T_c)/dT)/\Delta C(T_c)$, as a function of coupling parameter T_c/ω_E. For further identification of curves, see Fig. 1.

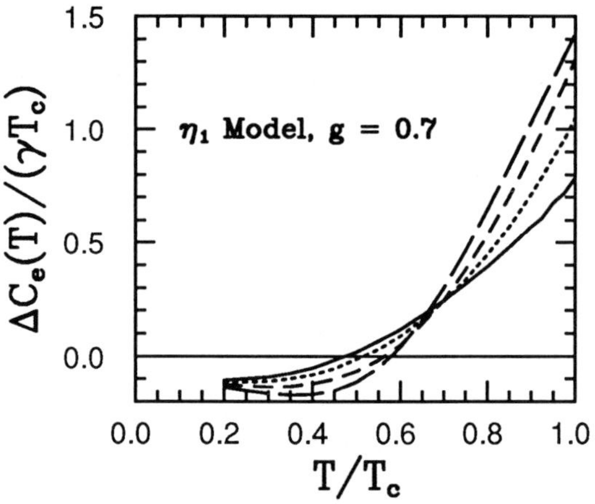

Fig. 3 The temperature dependence of the normalized specific heat difference $\Delta C(T)/\gamma T_c$ for $g = 0.7$ and different coupling strengths: $T_c/\omega_E = 0.05$ (solid), 0.1 (short dashed), 0.15 (intermediate dashed) and 0.2 (long dashed).

$T_c/\omega_E \simeq 0.1$ and 0.08, repectively. We conclude that strong coupling effects do not change the normalized slope as effectively as they do the normalized jump of Fig. 1. Finally, in Fig. 3, we show the temperature variation of the specific heat difference $\Delta C(T)/\gamma T_c$ for various values of T_c/ω_E, namely, 0.05, 0.1, 0.15 and 0.2. We note that as the coupling is increased the curves start at a higher value for $T = T_c$ and start falling more rapidly. For $T/T_c = t$ between 0.6 and 0.7 the curves cross and the higher coupling curves dip negatively more deeply than the lower coupling curves before going to zero at $t = 0$. This is expected by the entropy sum rule.

Strong coupling corrections can have a profound effect on the value of the normalized specific heat at T_c (a factor of ~ 2), in the normalized slope (a factor of ~ 1.5) and in the temperature variation of the specific heat. Such effects need to be included in any comparison of D-wave results with experimental data.

ACKNOWLEDGEMENT

This research was supported in part by the Natural Sciences and Engineering Research Council of Canada (NSERC), by the Canadian Institute for Advanced Research (CIAR), and by the Ontario Center for Materials Research (OCMR). E.S. acknowledges support by the Wissenschaftsfonds of the Styrian Government, Austria.

REFERENCES

(1) A.J. Millis, S. Sachdev, and C.M. Varma, *Phys. Rev. B* **3**, 4975 (1988).
(2) P.J. Williams and J.P. Carbotte, *Phys. Rev. B* **3**, 2180 (1989).
(3) K. Miyake, *Jour. of Magn. and Magn. Mat.* **63** and **64**, 411, (1987)
(4) K. Miyake, S. Schmitt-Rink, and C.M. Varma, *Phys. Rev. B* **34**, 6554 (1986).
(5) K. Miyake, T. Matsuura, H. Jichu, and Y. Nagaoka, *Progr. Theor. Phys.* **72**, 1063 (1984).
(6) F. Steglich, in Theory of Heavy Fermions and Valence Fluctuations, edited by P. Fulde (Springer Series in Solid State Sciences, Springer New York, 1985), p.23.
(7) U. Rauchschwalde, U. Ahlheim, C.D. Bredl. H.M. Mayer, and F. Steglich, *Jour. of Magn. and Magn. Mat.* **63** and **64**, 447, (1987)
(8) C.M. Varma in High-T_c Superconductors, edited by H.W. Weber (Plenum Press, New York, 1988), p.13.

Physica B 165&166 (1990) 321–322
North-Holland

OBSERVATION OF MAGNETIC ORDER IN THE HEAVY FERMION SUPERCONDUCTOR UBe_{13}

R. N. KLEIMAN,* D. J. BISHOP

AT&T Bell Laboratories, Murray Hill, New Jersey 07974

H. R. OTT

ETH Zurich

Z. FISK, J. L. SMITH

Los Alamos National Laboratory

We have measured the magnetostriction, L(H), of a single crystal of the Heavy Fermion superconductor, UBe_{13}, using an all silicon high precision capacitance dilatometer. We find clear evidence for a transition to an antiferromagnetic state at $T_N \sim 8.8K$, which is suppressed in a field by $dT_N/dH \sim 0.36K/T$. At low temperatures we observe pronounced magnetostrictive oscillations, which we believe are de Haas-van Alphen oscillations due to an unusual aspect of the Fermi surface.

1. INTRODUCTION

Recent work has shown the strong relationship between magnetism and superconductivity in the heavy fermion superconductors. In UPt_3, with $T_c \sim 0.55K$, an antiferromagnetic transition is observed (1) with $T_N \sim 5.0K$. In URu_2Si_2, with $T_c \sim 1.2K$, there is a transition (2) to an antiferromagnetic state at $T_N \sim 17.5K$. In UBe_{13}, with $T_c \sim 0.8K$ a magnetic transition above T_c has not been reported previously (3). We find clear evidence for an antiferromagnetic transition at $T_N \sim 8.8K$. This strongly suggests that the presence of a magnetic transition with $T_N \sim 10\ T_c$ is a general feature of superconductivity in this class of materials. At low temperatures we see unusual magnetostrictive oscillations, not previously observed in any system, which we believe are related to the de Haas-van Alphen effect and to the magnetism we observe.

2. TECHNIQUE AND MATERIALS

Dilatometry measurements are valuable for the study of weakly magnetic systems due to their high sensitivity. This technique is especially justified for the heavy fermion systems, since they not only have strongly enhanced electronic properties due to the mass renormalization, but also have strongly enhanced Gruneisen parameters (4), typically $\gamma_e \equiv \frac{d\ell n\gamma}{d\ell nV} \sim 100$, where γ is the linear specific heat coefficient. Assuming that $\vec{M}$ points along $\vec{H}$, $m = \chi H$, and that the dilatations are isotropic, it follows that (5) $\frac{1}{L}\frac{dL}{dH} = \frac{1}{3}\kappa\gamma_m m$ where $m = M/V$ is the magnetization/unit volume, κ is the compressibility, and γ_m the magnetic Gruneisen parameter $\gamma_m \equiv \frac{d\ell n\chi}{d\ell nV}$.

The dilatometer itself is constructed *only* from single crystal silicon to achieve high stability for magnetostrictive measurements, and is ideal for thermal expansion measurements as well, since its length changes at low temperatures are negligible. With this cell design we achieved a sensitivity of $\Delta L/L \sim 3\times10^{-9}$ for measurements either as a function of H or T.

The sample was grown from an Al melt using standard Czochralski techniques having dimensions $\sim5\times5\times5\ mm^3$. The magnetic field was applied along a [100] direction and the dilatation measured along a perpendicular [100] direction. The capacitance ($C_o \sim 5$ pF) was measured using a 3-terminal capacitance bridge.

The thermal expansion of UBe_{13} has been measured by Ott, (6) et al. at $H = 0$ for $0.3 < T < 10K$. Our study of the thermal expansion in the superconducting state and up to 20K is the subject of a later work, (7) but is generally consistent with their results.

3. RESULTS

In order to directly measure the quantity of interest, dL/dH, we added a very low frequency ($f \sim 10^{-3}$ Hz) field modulation of ~1 kG to the applied field, while recording the consequent modulations of the length of the sample. The amplitude was measured at a number of temperatures, tracing out a curve equivalent to M(T) as shown in Fig. 1 for $H_o = 3T$ and $H_o = 7T$. The curves have the characteristic shape of an order parameter, being constant at low temperatures and vanishing at 7.65 and 6.33K respectively. We believe that the shape of the dL/dH curves is clear evidence for magnetic ordering and the suppression of the transition temperature ($dT_N/dH \sim 0.36T/K$) in a field indicates that it is antiferromagnetic in nature. The transition temperatures extrapolate to $T_N = 8.8 \pm 0.2K$ for H=0.

We have measured dL/dH vs. T for a number of fields between 5 and 78.5 kG. The low temperature values of 1/L(dL/dH) are plotted vs. H in the inset to Fig. 1 showing a linear variation with field. From the slope of this line, and using $\kappa = 8\times10^{-13} cm^3/erg$, we find that $\gamma_m \chi \sim 2.9\times10^{-3} emu/cm^3$. which is approximately 16 times higher than the measured low temperature value for the Pauli susceptibility, χ_o.

While no hysteresis is observed in the curves of dL/dH as a function of temperature, there is significant hysteresis as a function of magnetic field for $T < T_N$. In Fig. 2 we show plots of L(H), taken at constant temperature, and sweep rates dH/dt~0.3mT/s. Fig. 2(a) is for $T > T_N$, (T=10.0K) and shows no magnetic hysteresis. Fig. 2(b) is for $T < T_N$ (T=4.0K) and shows significant magnetic hysteresis. Fig. 2(c) is for T = 1.25K and shows sharp jumps in L(H) at specific fields, which fall within an envelope like that of Fig. 2(b). Upon repetition such a trace appears similar, but the jumps do not occur at precisely the same field. Fig. 2(d) is for T = 0.825K, and shows that the frequency of the jumps has increased. They are now reduced in amplitude, but appear more convincingly to be oscillatory, perhaps periodic in 1/H. With Fig. 2(e) at T = 0.600K this trend continues. The anomaly below 2.6T is associated with H_{c2} (0.600K) and is discussed elsewhere. (7)

*Also at Cornell University, Ithaca, NY 14853

0921-4526/90/$03.50

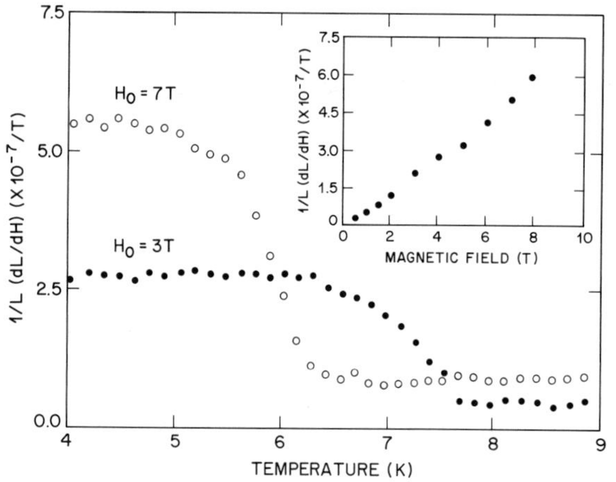

FIGURE 1

4. DISCUSSION

The oscillatory magnetostriction shown in Fig. 2 is reminiscent of the de Haas-van Alphen effect. However, our observations are not consistent with conventional expectations for it. The oscillation frequency is given by $\omega_{1/B} \approx \phi_0 S$ where $\phi_0 = hc/e$, and S is an extremal area of the Fermi surface. The Fermi surface topology determines S and rarely deviates by more than 10^{-4} . In our measurements $\omega_{1/B}$ changes drastically, with $\omega_{1/B} \propto 1/T^3$ for $0.6 < T < 1.25K$. The oscillation amplitude is predicted to have a T and H dependence which we do not observe. Finally, de Haas-van Alphen oscillations are sinusoidal and non-hysteretic, whereas here the oscillations appear as first order jumps.

We believe that modifications appropriate to UBe_{13} can help to explain our results. The amplitude of the magnetostrictive oscillations can be directly related (6) to the magnetization oscillations by replacing γ_m with γ_S, where $\gamma_S = \frac{d\ell nS}{d\ell nV}$ is a Gruneisen parameter associated with the extremal orbit area S. This can result in a tremendous enhancement over the usual oscillation amplitude. The first order nature of the jumps can be understood in terms of magnetic interaction (8). The oscillations are governed not by the applied field, H, but by the internal field $B = H + 4\pi M$. If M becomes comparable to the oscillation period the system is in a metastable state during part of the cycle and undergoes a transition to a lower energy state only after a suitable nucleation process. The strong temperature dependence to the oscillation frequency may be due to the formation and opening up of a neck between parts of the Fermi surface. In this temperature range the effective mass renormalization is rapidly increasing, which may be reflected in the Fermi surface topology. Finally, the change in oscillation rate may only be an apparent one, due to a change in nucleation rate with temperature.

5. CONCLUSION

We have found that UBe_{13} orders into an antiferromagnetic state with $T_N \sim 8.8K$, which is suppressed in a field with $dT_N/dH \sim 0.36$ K/T. Below ~3K we observe very unusual magnetostrictive oscillations, which we believe can be understood in the context of the de Haas-van Alphen effect.

ACKNOWLEDGEMENTS

We gratefully acknowledge helpful discussions with B. Batlogg, D. M. Lee, L. Levy, A. J. Millis, and C. M. Varma.

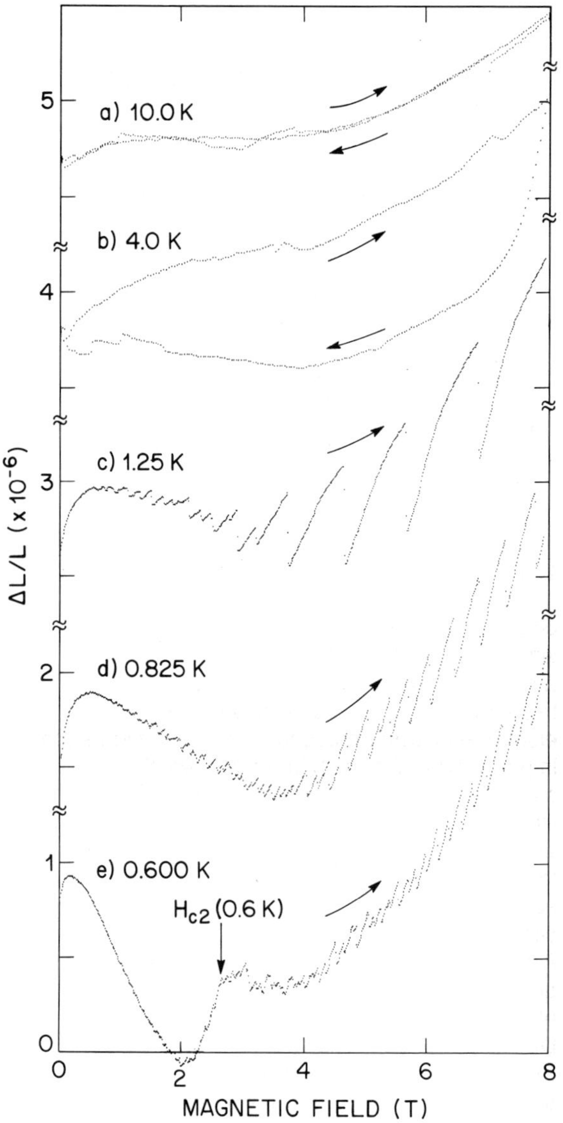

FIGURE 2

REFERENCES

(1) G. Aeppli, et al. Phys. Rev. Lett. 60, 615 (1988).

(2) T. T. M. Palstra, et al. Phys. Rev. Lett. 55, 2727 (1985); C. Broholm, et al. Phys. Rev. Lett. 58, 1467 (1987).

(3) It has been suggested however that pure UBe_{13} becomes magnetically ordered for $T < T_c$ as in $U_{1-x}Th_xBe_{13}$.

(4) B. S. Chandresekhar and E. Fawcett, Adv. Phys. 20, 775 (1971).

(5) A. de Visser, et al. Physica 147 B, 81 (1987).

(6) H. R. Ott, Physica B 126, 100 (1984).

(7) R. N. Kleiman, et al. (1989) to be published.

(8) A. B. Pippard, Proc. Roy. Soc. A 272, 192 (1963); Electrons at the Fermi Surface, ed. M. Springford (Cambridge University Press, Cambridge, 1980) Chap. 4.

Physica B 165&166 (1990) 323–324
North-Holland

MAGNETIC OSCILLATION PHENOMENA IN THE HEAVY FERMION SUPERCONDUCTOR $CeCu_2Si_2$

M.HUNT, P.MEESON, P-A.PROBST, P.REINDERS§ AND M.SPRINGFORD.

H.H. Wills Physics Laboratory, Royal Fort, Tyndall Ave., Bristol BS8 1TL. UK.

W. ASSMUS AND W.SUN

Physikalisches Institut, J.W. Goethe-Universitat, Robert-Mayer-Str, 2-4, 6000 Frankfurt am Main, Fed. Rep. of Germany.

Measurements of the magnetisation of single crystals of $CeCu_2Si_2$ at ~20mK and ~13T clearly demonstrate oscillations periodic in 1/B (the de Haas van Alphen effect). This is direct evidence for the existence of long lived charged quasiparticles in a Fermi Liquid like environment. We provide details of quasiparticle masses and the variation of the dHvA frequencies as a function of crystal orientation.

1. INTRODUCTION

The so called Heavy Fermion alloys are characterised by a low temperature specific heat which is typically 1000 times that of ordinary metals, and is roughly proportional to temperature. From this and other evidence (1) it is inferred that the electron-electron interactions in these metals are important and that quasiparticles, which are surprisingly heavy, form as a result of Fermi Liquid like effects. Some alloys are also superconducting, and it is presumably the heavy quasiparticles which form the superconducting state since the jump in the specific heat at the transition temperature, T_c, is comparable to the specific heat of the normal metal. A consequence of the heaviness of the quasiparticles is that the Fermi temperature of the electron system is low, approximately 1/20 of the Debye temperature, hence Heavy Fermion Superconductors cannot be described by straightforward BCS theory, and are therefore possible "novel" superconductors (2). We present new results on the representative alloy, $CeCu_2Si_2$.

2. SAMPLE CHARACTERISATION

Our sample is a single crystal of $CeCu_2Si_2$ grown in West Germany using the cold boat technique (3). The normal state transverse magnetoresistance was measured and closely follows a T^2 law in the temperature region of interest, for further details and a description of the complex phase diagram we refer to a companion paper(4). Although the properties of $CeCu_2Si_2$ are known to be sensitive to the precise stoichiometry, the behaviour of our sample, (for instance T_c, H_{c2}), is typical of previously available material (4,5,6). The crystal structure is tetragonal.

3. MAGNETIC OSCILLATIONS

Oscillations in the magnetisation of the sample as a function of applied magnetic field were observed and found to be periodic in 1/B (the de Haas-van Alphen (dHvA) effect). Oscillations were observed throughout the a-c plane using an in situ crystal rotation mechanism and a standard second derivative technique. The dHvA frequencies lie between 130-550T, corresponding to an area of approximately 5% of the Brillouin zone, and are plotted as a function of crystal orientation in Fig.1.

4. QUASIPARTICLE MASSES

The masses of the quasiparticles were determined in the standard way from the temperature dependence of the dHvA oscillations and are displayed in Table 1. Throughout most of the crystal plane investigated it is possible to observe dHvA oscillations both above and below a metamagnetic transition. In the a-axis orientation conditions are such that quasiparticle masses could be determined both above and below the transition region and, as can be seen from Table 1, a small change in both the observed frequency and the mass of the quasiparticles occurs. Given that the small regions of Fermi surface we are investigating are likely to be sensitive to a restructuring of the whole Fermi surface, we infer that only a small change in the bandstructure is taking place.

Although the mass values as tabulated are not

§ Present address: Institut für Festkörperphysik, Technische Hochschule Darmstadt, Fachbereich 5, Hochschulestrasse 8, 6100 Darmstadt, Fed. Rep. of Germany.

strikingly large by Heavy Fermion standards it must be remembered that they refer to only small portions of the Fermi Surface. The values are reminiscent of our earlier work in $CeCu_6$, in which mass enhancements by factors of order 200 are seen (8), and we believe that here too the observed quasiparticles are also strongly renormalised..

5. COMPARISON WITH THEORY

We may compare our results with the quasiparticle band structure calculated by Sticht et. al (7). In this model the bandstructure is determined by regarding the Ce 4f electrons as part of a Kondo Lattice, which, when added to an LDA calculation for the remaining electrons, leads to a Fermi surface consisting of sheets of "light" electrons coexisting with sheets of strongly renormalised "heavy" quasiparticles. In the present experiment the presence of sheets of light electrons would be expected to have a dominant signal amplitude, compared to the heavy quasiparticles. From the absence of such 'light' electrons in the present experiment we conclude that the band structure calculation of Sticht et. al. is wrong.

6. CONCLUSIONS

We believe that the present investigations have revealed the existence of strongly renormalised quasiparticles in a Fermi liquid like environment. The observed T^2 resistivity dependence in the same sample demonstrates that the superconducting state is formed out the "coherent" Fermi liquid. This point is not clear in the existing literature (1).

Although we have not fully investigated all crystal orientations, we believe that the present results are sufficient to exclude the possible presence of "light" electrons, and that current band structure calculations do not agree with experiment.

Our measured quasiparticle masses are not sufficient to explain the large heat capacities seen elsewhere in this material, and we conclude that much of the Fermi surface is at present undetected. This is due in part to imperfect samples, and in part to the presumed even higher quasiparticle masses.

TABLE 1. Measured quasiparticle masses and dHvA frequencies in $CeCu_2Si_2$. The table demonstrates a small change in quasiparticle behaviour through a magnetic phase transition region in the a-axis.

Orientation	Field (Tesla)	Frequency (Tesla)	Mass (m^*/m_e)
a-axis	4.8-5.8	171±2	4.62±0.07
a-axis	7.5-11.5	162±2	5.15±0.10
c-axis	9.5-11.5	295±5	5.81±0.05

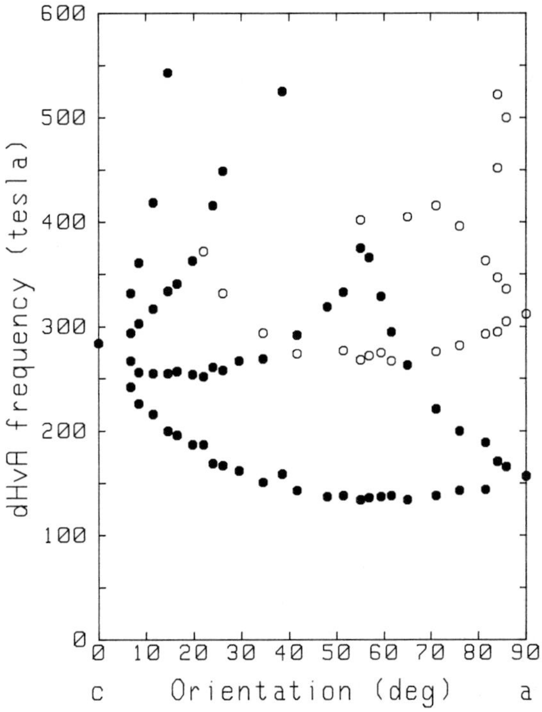

FIGURE 1. dHvA Frequency as a Function of Crystal Orientation in the c-a Plane of $CeCu_2Si_2$. The open circles are considered to be harmonics.

ACKNOWLEDGEMENTS
This work is supported financially by the SERC, and one of us (P-A. P) acknowledges partial support from Swiss National Funds.

REFERENCES

(1) See for example H.R.Ott, Characteristic Features of Heavy Electron Materials, in Progress in Low Temperature Physics, Vol. XI, ed D.F.Brewer (Elsevier Science Publishers 1987)
(2) F.Steglich, J.Aarts, C.D.Bredl, W.Lieke, D.Meschede, W.Franz and H.Schaffer, Phys. Rev. Lett. 43 1892 (1979).
(3). W.Sun, M.Brand, G.Bruls, and W.Assmus, to be published in Z. fur Physik.
(4).M.Hunt, P.Meeson, P-A.Probst, P. Reinders, M.Springford, W.Sun and W.Assmus, Phase Diagram of $CeCu_2Si_2$ in the B-T Plane, this volume.
(5).U.Rauchswalbe, Ph.D. Thesis, Technische Hochschule, Darmstadt, (1986) Unpublished.
(6) Y.Onuki, T.Hirai, T.Kumazawa, T.Komatsubara and Y.Oda, J. Phys. Soc. Jap. 56, 1454, (1987).
(7). J.Sticht, N.d'Ambrumenil, and J.Kubler, Z. Phys. B.- Condens.Matt. 65, 149, (1986).
(8). S.Chapman, M.Hunt, P.Meeson, P.Reinders and M.Springford, to be published.

Physica B 165&166 (1990) 325–326
North-Holland

FEATURES IN THE UPPER CRITICAL FIELD OF UBe_{13}: EVIDENCE FOR TWO NEW PHASE TRANSITIONS

G.M. SCHMIEDESHOFF[1], R.T. TISDALE[1], J.A. POULIN[1], S. HASHMI[1], J.L. SMITH[2]

[1] Department of Physics, Bowdoin College, Brunswick, Maine 04011, USA.†
[2] Los Alamos National Laboratory, Los Alamos, New Mexico 87545, USA.‡

High precision measurements of the upper critical magnetic field of polycrystalline UBe_{13} are deduced from the dc magnetization. Two sharp features consistent with the presence of phase transitions are observed. Anomalous behavior of the normal–state magnetization leads us to suggest that the warmer feature is associated with a field–induced magnetic state. The absence of additional anomalous behavior of the magnetization and the presence of a qualitatively similar feature in the critical field of a 3% thoriated UBe_{13} sample suggest that the colder transition may be to a second superconducting state.

The upper critical magnetic field, H_{c2}, of polycrystalline UBe_{13} has been investigated by several groups (1,2). The temperature dependence of H_{c2} is unusual and has been difficult to describe theoretically. An enormous initial slope is followed by strong negative curvature which precedes a region of positive curvature as the temperature is lowered. At $T = 0$, H_{c2} is surprisingly large, reflecting a small superconducting (SC) coherence length. In this paper, we present high–precision measurements of the dc magnetization, M, from which we deduce $H_{c2}(T)$. Positive curvature in $H_{c2}(T)$ is not observed. Two sharp breaks in the slope of $H_{c2}(T)$, consistent with the existence of phase transitions, are resolved.

The dc magnetization is measured using an *in situ* capacitive magnetometer (3). We calibrate the magnetometer by assuming a one– to–one correspondence between the magnetic force on the sample, $F = M \cdot \nabla H$, and the change in capacitance ΔC. Vibrating sample magnetometer (VSM) measurements show that M(H) is nearly linear at 4.2K to about 20 T (4). A measurement of $\Delta C(H)$ at 4.2 K defines the calibration function.

We define H_{c2} to be the point at which magnetic hysteresis, resulting from flux pinning effects, disappears in a transition to the normal state as described previously (2). Our thermometer is a carbon resistor mounted near magnetic center (5). Our upper critical field measurements are shown in Fig. 1 (small dots). The initial slope near T_c, about –38 T/K, and the T = 0 intercept, $H_{c2}(0) \simeq 12$ T, allows an estimation of the SC coherence–length, $\xi_o \simeq 160$Å, in the dirty–limit (using a typical resistivity at T_c of 150 $\mu\Omega$–cm). These results are consistent with earlier evaluations (1).

Two sharp features are observed in $H_{c2}(T)$ at 0.47 K, 6.2 T and 0.32 K, 8.2 T. To accentuate these features, we make use of the general low–temperature linearity

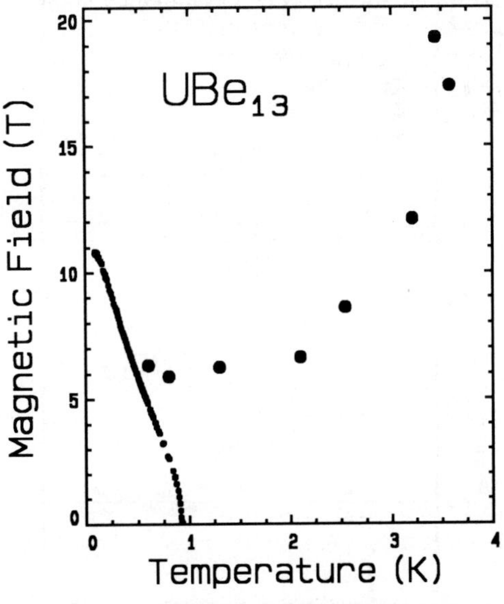

Figure 1. The low temperature phase diagram of pure UBe_{13}. Small solid circles represent $H_{c2}(T)$. Large solid circles represent the onset of anomalous behavior in M(T,H) (see text).

† Supported by a Cottrell College Science Grant from the Research Corporation.
‡ Supported by the USDOE, Office of Basic Energy Sciences, Division of Materials Sciences.

of $H_{c2}(T)$. We define ΔH_{c2} as the difference between all the H_{c2} data in the vicinity of a particular feature and a linear fit to the data just above the transition. ΔH_{c2} data for each feature are shown in Figs. 2a and 2b. The sharpness of these features is consistent with the existence of two previously unobserved phase transitions. Note the absence of any positive curvature.

Sharp features in $H_{c2}(T)$ can result from phase transitions in either the SC or the normal state. To elucidate the origin of these features, we studied the behavior of the dc–magnetization in both states. We do not find any structure in the SC state M(T,H) associated with the features in $H_{c2}(T)$. In this state, the magnetization is dominated by flux pinning which is known to "smear out" the magnetic signature of phase transitions (such as the lower critical field (6)).

M(T,H) measurements in the normal state, described in detail elsewhere (7), show anomalous behavior above a threshold field of about 6 T. This threshold field is in excellent agreement with the position of the warmer feature. This anomalous behavior of M(T,H) has also been observed in a high purity sample of UBe_{13} fabricated elsewhere. The design of our magnetometer is such that we cannot reliably identify the origin of the anomalous behavior. However, its onset at nearly the same field as the warmer feature in $H_{c2}(T)$ leads us to believe that a field–induced phase transition with a magnetic character does indeed exist in UBe_{13}. This new state appears to co–exist with the SC state in fields above about 6 T. An approximate phase diagram is also shown in Fig. 1 where the large dots represent the onset of anomalous behavior in M(T,H).

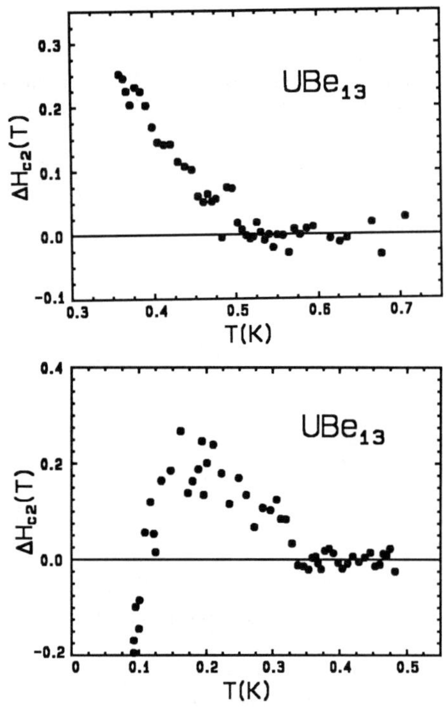

Figure 2. $\Delta H_{c2}(T)$ vs. temperature (see text) of (a) the warmer feature in $H_{c2}(T)$ and (b) the colder feature in $H_{c2}(T)$.

We have not found any feature in our magnetization data associated with the colder feature in $H_{c2}(T)$ shown in Fig. 2b. The qualitative shape of this feature is similar to that of a feature previously observed in $H_{c2}(T)$ of 3% thoriated UBe_{13} (2). These observations suggest that the colder feature is the result of a phase transition within the SC state itself, a state similar to that previously observed in the thoriated system (8).

We would like to thank J. S. Brooks and G. R. Stewart for many useful discussions (the latter also for providing a high-quality sample of UBe_{13}), and S. Foner, E. J. McNiff, and B. Andraka for their respective magnetization measurements. We also thank B. S. Held, P. W. C. Emery and R. L. Stevens for their assistance.

REFERENCES

(1) J. W. Chen *et al.*, J. Appl. Phys. **57**, 3076 (1985). U. Rauchschwalbe *et al.* Z. Phys. **B60**, 379 (1985). J. P. Brison *et al.*, J. Low Temp. Phys. **76**, 453 (1989).

(2) G. M. Schmiedeshoff *et al.*, Phys. Rev. **B38**, 2934 (1988).

(3) J. S. Brooks *et al.*, Rev. Sci. Instrum. **58**, 117 (1987).

(4) All data presented in this paper are derived from measurements on sample #3317 which was prepared at Los Alamos National Laboratory.

(5) For temperatures below 1 K we used a Speer 220 resistor. See M. J. Naughton *et al.*, Rev. Sci. Instrum **54**, 1529 (1983). For temperatures above 1 K we use a carbon glass resistor with the manufacturers calibration.

(6) A. M. Campbell and J. E. Evetts, Adv. Phys. **21**, 199 (1972).

(7) G. M. Schmiedeshoff *et al.*, to be published.

(8) H. R. Ott, H. Rudigier, Z. Fisk, and J. L. Smith, Phys. Rev. **B31**, 1651 (1985).

Physica B 165&166 (1990) 327–328
North-Holland

FERMI SURFACE OF $CeSn_3$

Akira HASEGAWA and Hiroshi YAMAGAMI†

College of General Education, Niigata University, Niigata 950-21, Japan
† Graduate School of Science and Technology, Niigata University, Niigata 950-21, Japan

Based on an itinerant model for the 4f electrons, the energy band structure is calculated for $CeSn_3$, known to be the heavy-electron system having the Kondo temperature of about 200 K, by a self-consistent relativistic APW method with the exchange and correlation potential in the local-density approximation. Large hole and electron sheets of the Fermi surface clarify origins of all of the major frequency branches of the order 10^7 Oe of the de Haas-van Alphen effect for the field directions on both {100} and {110} planes.

1. INTRODUCTION

The cerium intermetallic compound $CeSn_3$ forms the heavy-electron system having the Kondo temperature of about 200 K. In 1981, Johanson et al. (1) succeeded in measuring the de Haas-van Alphen (dHvA) effect in $CeSn_3$. Based on an itinerant model for the 4f electrons, the band structure calculations were carried out, and it was shown that the Fermi surface consists of a large hole sheet centered at the R point and a large, spherical but distorted electron sheet centered at the Γ point (2-4).

Recently, Umehara et al. (5) have measured the dHvA effect for the field directions on both {100} and {110} planes, and revealed by measuring the magnetoresistance that $CeSn_3$ is a compensated metal. Therefore, all of the major frequency branches of the order 10^7 Oe should be explained by the large hole and the large electron sheets of the Fermi surface. Though one of the branches may be explained by the large hole sheet, origins of other branches cannot be explained by any electron sheets proposed previously. Here, we investigate the shape of the large electron sheet more carefully than done by any previous calculations, and try to clarify origins of the major frequency branches.

2. THE FERMI SURFACE

The energy band structure is calculated for $CeSn_3$ self-consistently by a symmetrized, relativistic APW method with the exchange and correlation potential in the local-density approximation (6).

The large hole sheet, which belongs to the eighth band, is centered at R and bulges towards <100> directions. It contains about 0.44 holes /cell and has the cyclotron effective mass, the absolute magnitude of which is quite anisotropic, i.e. $2.1m$, $1.7m$ and $0.9m$ in <100>, <110> and <111> directions, respectively, where m is the free-electron mass. The large hole sheet is similar to those predicted by previous calculations (2-4).

The large electron sheet centered at Γ in the ninth band is essentially spherical but deeply concave in <111> directions. It has a small hollow at the center. Eight convavities are connected at the hollow, because there is no occupied states along the Λ axis in the ninth band. The shape of the large electron sheet we predict here is different from those of any previous calculations. The cyclotron mass on the large electron sheet ranges from $1.4m$ to $5.2m$. Figure 1 shows the cross sections of the large hole and electron sheets.

In addition, there are several small hole and electron sheets in the seventh, the eighth, the ninth and the tenth bands. Description of these small sheets will be published elsewhere.

3. COMPARISON WITH THE DE HAAS-VAN ALPHEN EFFECT

The large hole sheet in the eighth band and the large electron sheet in the ninth band produce various frequency branches of the dHvA effect. The theoretical branches of the order 10^7 Oe are compared with the experimental results in Fig. 2. Here is a description of the theoret-

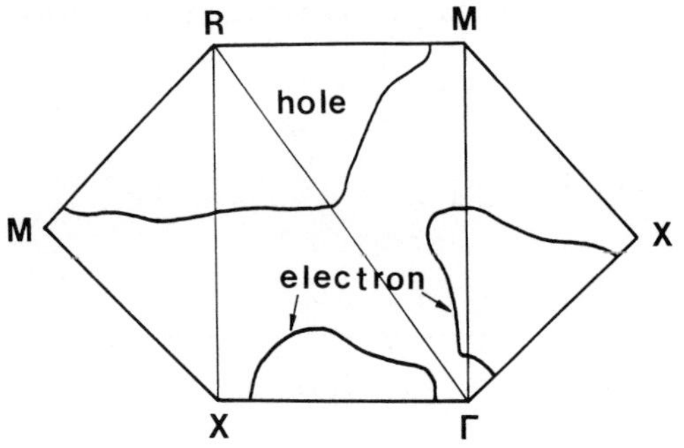

FIGURE 1
The cross sections of the eighth-band large hole sheet and the ninth-band large electron sheet.

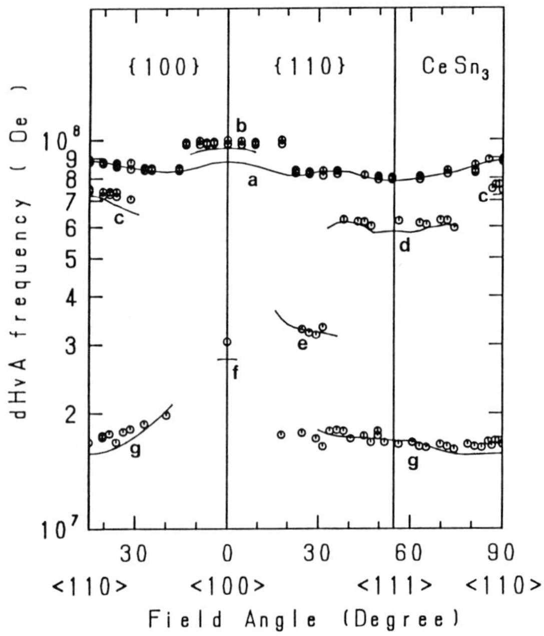

FIGURE 2
Theoretical and experimental frequencies of the de Haas-van Alphen effect in $CeSn_3$. The branches *a*, *b*, *c*, *d*, *e*, *f* and *g* show theoretical results, and small circles show the experimental results of ref. (5).

ical branches together with their origins.

a: an orbit on the large hole sheet, existing in the whole range of angles.

The following branches, *b*, *c*, *d*, *e*, *f* and *g*, originate from the large electron sheet.

b: an outermost orbit centered at Γ, existing in a narrow range of angles around <100>,

c: a rhombus-like orbit which is centered at about $(0.1,0.1,0)(2\pi/a_0)$, where a_0 is the lattice constant, and exists in a narrow range around <110>,

d: a nearly triangular orbit which centered at about $(0.13,0.13,0.13)(2\pi/a_0)$ and exists in a limited range around <111>,

e: an orbit like dog's bone which exists at angles between 15° and 35° on <110> plane,

f: an orbit like a cross which is centered at about $(0.3,0,0)(2\pi/a_0)$ and exists in an extremely narrow range of angles around <100>,

g: an orbit which runs near X and Γ through the hollow and exists in a wide range of angles except around <100>. (An example is seen on the ΓXR plane in Fig. 1.)

Both the magnitude and the angle range of the theoretical branches agree reasonably well with that of the experimental frequency branches, supporting the validity of both the large hole and the large electron sheets of the Fermi surface predicted by this calculation. The theoretical results for the cyclotron effective mass for these branches are, however, far smaller than the experimental results. Disappearance of the branch *a* in the vicinity of <100> may be ascribed to large enhancement of the effective mass due to the many-body effect.

ACKNOWLEDGEMENTS

We are grateful to Professor Y. Onuki for a permission of use of the experimental data for the de Haas-van Alphen effect in $CeSn_3$ prior to publication.

REFERENCES

(1) W. R. Johanson, G. W. Crabtree, A. S. Edelstein and O. D. McMasters, Phys. Rev. Letters 46 (1981) 504.

(2) D. D. Koelling, Solid State Commun. 43 (1982) 247.

(3) A. Yanase, J. Magn. & Magn. Mater. 31-34 (1983) 453.

(4) P. Strange and D. M. Newns, J. Phys. F: Metal Phys. 16 (1986) 335.

(5) I. Umehara, Y. Kurosawa, N. Nagai, M. Kikuchi, K. Satoh and Y. Onuki, J. Phys. Soc. Jpn. (1990).

(6) H. Yamagami and A. Hasegawa, J. Phys. Soc. Jpn. (1990).

Physica B 165&166 (1990) 329–330
North-Holland

DE HAAS–VAN ALPHEN EFFECT IN $CeCu_2$

Kazuhiko SATOH, Izuru UMEHARA, Yoshiko KUROSAWA and Yoshichika ŌNUKI

Institute of Materials Science, University of Tsukuba, Tsukuba, Ibaraki 305, Japan

We have observed the de Haas-van Alphen oscillation in the antiferromagnetic Kondo lattice compound $CeCu_2$. Fermi surface of $CeCu_2$ is found to be different from that of YCu_2. Cyclotron effective masses of $CeCu_2$ range from 0.5 to 5.3 m_0, which are larger than 0.1-0.7m_0 in YCu_2.

Rare earth intermetallic compounds RCu_2 form an interesting series of compounds whose magnetic properties variously change with the rare earth element R. Most of these compounds crystallize in the $CeCu_2$-type orthorhombic structure except $LaCu_2$ which possesses the AlB_2-type hexagonal structure. $CeCu_2$ can be classified into the antiferromagnetic Kondo lattice substance with the Néel temperature T_N, 3.4 K (1,2). Below T_N, magnetization of $CeCu_2$ shows the metamagnetic behavior around 17 kOe when the magnetic field is applied along the a-axis (3). The magnetic susceptibility and magnetization show a large anisotropy at low temperatures, reflecting the orthorhombic structure (1,4). To clarify the Fermi surface properties, we have done the experiment of the de Haas-van Alphen (dHvA) effect, using ^{3}He cryostat and 160 kOe superconducting magnet.

We show in Fig. 1 (a) a typical dHvA oscillation of $CeCu_2$ along the a-axis and (b) a corresponding fast Fourier transformation (FFT) spectrum between 60 and 90 kOe. Two signals, denoted δ and η, are dominant and are observed above 25 kOe.

Figure 2 shows the angular dependence of the dHvA frequencies for $CeCu_2$. As we have done the dHvA experiment above 60 kOe, this substance is in the paramagnetic (or field-induced ferromagnetic) state around the a-axis. On the other hand, it is antiferromagnetic around the b- and c-axes and in the a-plane. We note that about ten kinds of dHvA branches are observed around the a-axis, whereas only one branch is detected around the c-axis.

We briefly describe the characteristic natures for main branches:

α :observed only around the a-axis, possessing the heavy mass of 5.26 m_0

β_i:observed in a wide field range except around the c-axis

γ_i:observed only around the a-axis

δ :observed in a wide field range and its intensity is strong around the a-axis and weak around the b-axis.

ε :observed in a wide field range and its intensity becomes weak around the symmetrical axes

ζ :only closed orbit

η :nearly cylindrical Fermi surface.

Here, several similar branches such as β_i and γ_i are observed around the a-axis, probably due

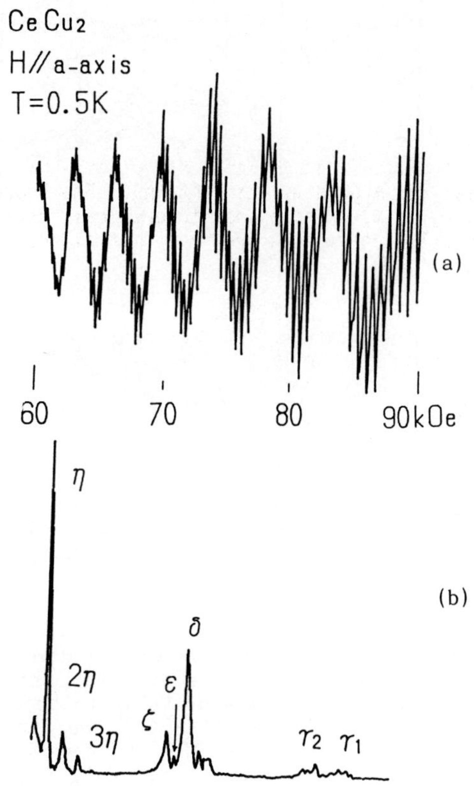

FIGURE 1
(a) A typical dHvA oscillation of $CeCu_2$ at 0.5 K and (b) the corresponding FFT spectrum between 60 and 90 kOe. The magnetic field is applied along the a-axis.

0921-4526/90/$03.50

TABLE I
Cyclotron effective masses and corresponding dHvA frequencies for $CeCu_2$.

H//a-axis		H//b-axis		H//c-axis	
F_i ($\times10^7$Oe)	m_c^*/m_0	F_i ($\times10^7$Oe)	m_c^*/m_0	F_i ($\times10^7$Oe)	m_c^*/m_0
α 8.25	5.26				
β_2 5.26	3.52	β_2 2.54	4.35		
γ_1 2.38	2.25				
γ_2 2.18	2.19				
1.34	1.68				
ε 1.19	1.52				
ζ 1.04	1.71	ζ 0.864	1.19	ζ 0.368	0.50
η 0.126	0.76				

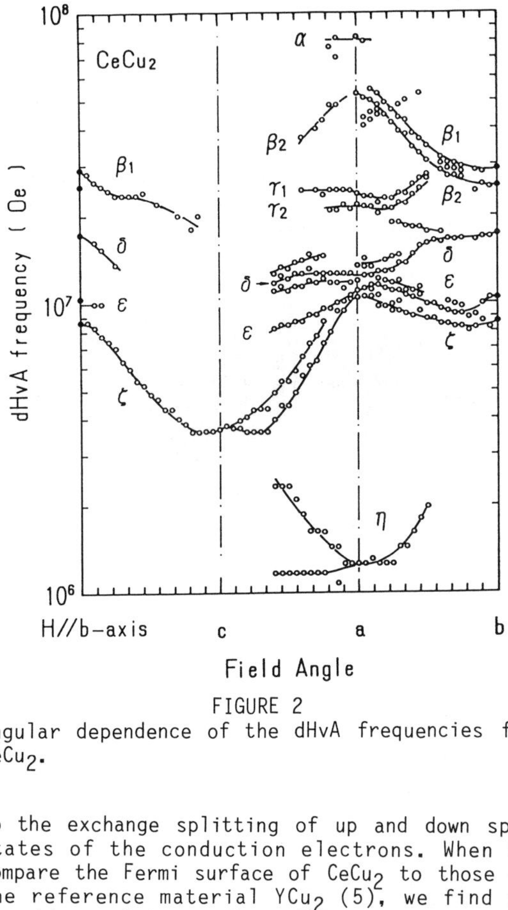

FIGURE 2
Angular dependence of the dHvA frequencies for $CeCu_2$.

to the exchange splitting of up and down spin states of the conduction electrons. When we compare the Fermi surface of $CeCu_2$ to those of the reference material YCu_2 (5), we find no resemblance between them. This is contrast to those of the similar materials CeB_6 (6) and $CeAl_2$ (7) whose Fermi surfaces are similar to those of La-compounds. Three reasons are considered to explain the discrepancy of the Fermi surfaces between $CeCu_2$ and YCu_2: difference of the ionic potentials between Ce^{3+} and Y^{3+}, hybridization of 4f electrons to other s, p, d conduction electrons and the antiferromagnetic ordering in $CeCu_2$.

The heavy-fermion state in $CeCu_2$ is reflected in the cyclotron masses even if the antiferromagnetic ordering overcomes the Kondo state. In Table I, we list the dHvA frequencies for $CeCu_2$ along the symmetrical axes and the corresponding cyclotron effective masses, m_c^*, which are determined from the temperature dependence of the amplitudes of the dHvA oscillations. The cyclotron effective masses in $CeCu_2$ range from 0.5 to 5.3m_0, which are larger than those of YCu_2, 0.1-0.7m_0. This mass enhancement is roughly consistent with that of the low-temperature specific heat ratio, namely 50 mJ/K^2 mol under 80 kOe in $CeCu_2$ (8) and 6.7 mJ/K^2 mol in YCu_2.

In summary, we have observed the dHvA oscillation of $CeCu_2$. Fermi surface of $CeCu_2$ is different from that of YCu_2. The cyclotron effective masses of $CeCu_2$ are roughly eight times larger than those of YCu_2, indicating the heavy-fermion state.

We are grateful to Dr. H. Harima, Prof. A. Yanase and Prof. T. Goto for their fruitful discussions. We have utilized the 160 kOe superconducting magnet at Cryogenic Center, University of Tsukuba. We are very grateful to Prof. R. Yoshizaki for his kind support. This work was supported by Grant-in-Aid for the Scientific Research from the Ministry of Education, Science and Culture, and also by University of Tsukuba Project Research.

REFERENCES

(1) Y. Onuki et al. J. Phys. Soc. Jpn. **54** (1985) 3562.
(2) E. Gratz et al. J. Phys. F **15** (1985) 1975.
(3) Y. Onuki et al. J. Mag. Mag. Mat. **76&77** (1989) 191.
(4) Y. Onuki et al. printed in Physica B.
(5) Y. Onuki et al. J. Phys. Soc. Jpn. **58** (1989) 4552.
(6) Y. Onuki et al. J. Phys. Soc. Jpn. **58** (1989) 3698.
(7) P.H.P. Reinders and M. Springford, J. Mag. Mag. Mat. **79** (1989) 295.
(8) C.D. Bredl, J. Magn. Magn. Mat. **63&64** (1987) 355.

Physica B 165&166 (1990) 331–332
North-Holland

MAGNETORESISTANCE AND DE HAAS–VAN ALPHEN EFFECT IN $CeIn_3$

Izuru UMEHARA, Yoshiko KUROSAWA, Makoto KIKUCHI, Nobuyuki NAGAI, Kazuhiko SATOH and Yoshichika ŌNUKI

Institute of Materials Science, University of Tsukuba, Tsukuba, Ibaraki 305, Japan

We have measured the transverse magnetoresistance and de Haas-van Alphen (dHvA) effect in the antiferromagnetic Kondo lattice substance $CeIn_3$. The magnetoresistance increases in the wide field region, suggesting a compensated metal. Open orbits are found along the <110> and <112> directions. We have detected one dHvA branch which corresponds to the electron sheet of sixth band in $LaIn_3$, centered at the Γ point.

$CeIn_3$ which crystallizes in the $AuCu_3$ type structure is a well known Kondo lattice compound with an antiferromagnetic order at 10 K (1). The ordered moment 0.65 μ_B per cerium atom is comparable to the value 0.71 μ_B expected from the Γ_7 ground state. Nevertheless, $CeIn_3$ possesses the large electronic specific heat coefficient of 130 mJ/K^2 mol at low temperature, indicating a heavy electron system (2).

In this paper we report the magnetoresistance and de Haas-van Alphen (dHvA) effect in the field up to 150 kOe and at temperature down to 0.45 K to clarify the Fermi surface topology and the cyclotron effective mass of the heavy electron.

We have measured the transverse magnetoresistance at 0.5 K. Figure 1 shows the angular dependence of the magnetoresistance $\Delta\rho/\rho = \{\rho(H)-\rho(0)\}/\rho(0)$ under the constant field H, 150 kOe in the {100} and {110} planes. The electrical resistivity at 0.5K is 0.45 μΩ•cm and the residual resistivity ratio $\rho_{RT}/\rho_{0.5K}$ is 90, indicating a sample with high quality. The magnetoresistance increases in a wide field region, for example indicating the $H^{1.45}$ dependence along the <100> direction. This result suggests that this material is a compensated metal with an equal carrier concentration of electrons and holes. This is simply expected from the total number of valence electrons, 12 per unit cell if the Ce ion is trivalent. Here we remark the magnetoresistances along the <110> and <111> directions which saturate at high fields. Open orbits are present along the <110> and <112> directions, respectively.

Next we have measured the dHvA effect. We show in Fig. 2 (a) the typical dHvA oscillation at 9° from the <100> direction in the {110} plane and (b) the corresponding fast Fourier transformation (FFT) spectrum. One signal with the dHvA frequency of 2.93×10^7 Oe, namely the extremal cross-sectional area $S\{(2\pi/a)^2\}$ of 0.156 is observed. Here, the dHvA frequency F(Oe) is related to the extremal cross-sectional area of the Fermi surface $S\{(2\pi/a)^2\}$ as $F = 1.882 \times 10^8 S$, where the lattice constant a is 8.859 a.u. or 4.688 A. The other signals of 5.89×10^7 Oe and 8.79×10^7 Oe are second and third harmonics, respectively.

Figure 3 shows the angular dependence of dHvA frequency. The dHvA branch is observed in the

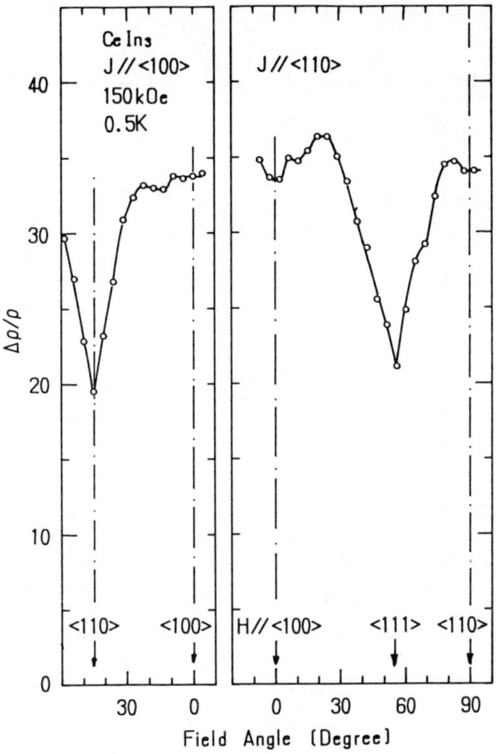

FIGURE 1
Angular dependence of the transverse magnetoresistance in $CeIn_3$.

0921-4526/90/$03.50 © 1990 – Elsevier Science Publishers B.V. (North-Holland)

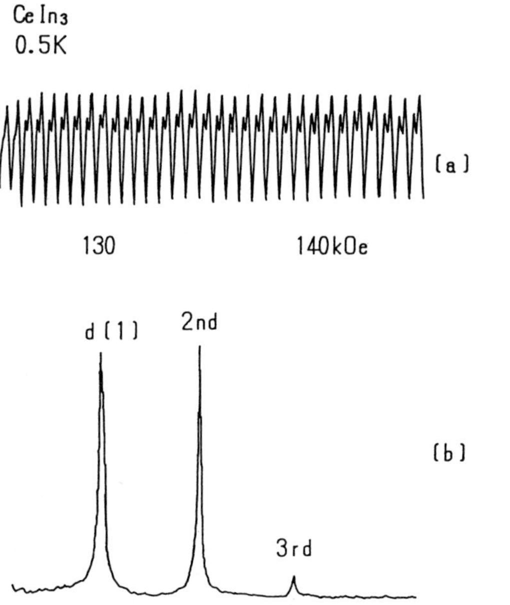

FIGURE 2
Typical dHvA oscillation and the corresponding FFT spectrum in $CeIn_3$.

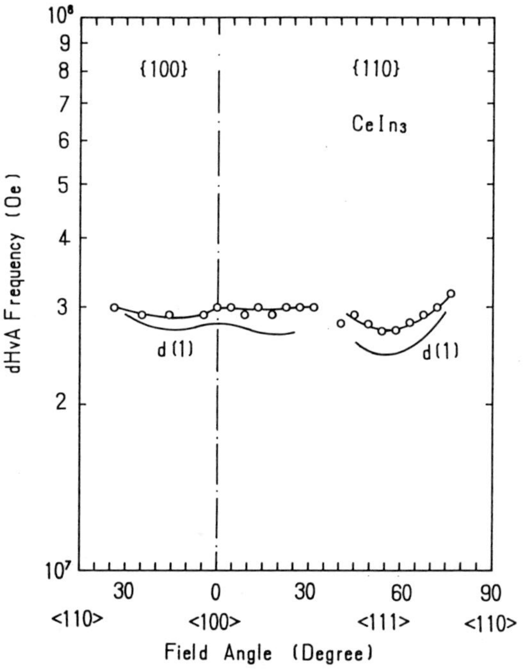

FIGURE 3
Angular dependence of dHvA frequencies in $CeIn_3$.

region centered at the <100> and <111> directions. Its cyclotron mass has been determined as $2.02m_0$ at <100> and $2.88m_0$ at <111> from the temperature dependence of the dHvA amplitude.

We compare our data to the Fermi surface which is constructed from the APW energy band structure for $LaIn_3$ (3-5). In Fig. 3 we show the theoretical branch labeled as d(1) in $LaIn_3$, where the number in parentheses shows the degeneracy of the branch. This frequency branch originates from the electron sheet in the sixth band which is essentially a sphere centered at the Γ point, bulges out toward the <100> directions and connects with another sheet centered at the R point by arms along the <111> directions. From the absolute magnitude and angular dependence of the dHvA frequencies in $CeIn_3$, we conclude that the observed branch corresponds to this electron Fermi surface of $LaIn_3$.

The Kondo effect is highly reflected in the cyclotron mass at low temperature even if the magnetic order overcomes the Kondo effect. The cyclotron mass for the d(1) orbit is not yet determined in $LaIn_3$ or $PrIn_3$, but is calculated as $0.53m_0$ and $0.39m_0$ in the <100> and <111> directions, respectively for $LaIn_3$. The mass of $2.02m_0$ or $2.88m_0$ in $CeIn_3$ is thus four to seven times heavier than those of $LaIn_3$.

In conclusion, the Kondo effect has minor influence on the geometry of the Fermi surface. An important point is that the energy band becomes rather flat at the Fermi energy, which brings about the heavy mass.

We are grateful to Professors A. Hasegawa and K. Ueda for their fruitful discussions. We have utilized the 16 T-superconducting magnet at Cryogenic Center, University of Tsukuba. We are grateful to Professor R. Yoshizaki for his kind support. This work was supported by Grant-in-Aid for the Scientific Research from the Ministry of Education, Science and Culture, and also by University of Tsukuba Project Research.

REFERENCES

(1) J.M. Lawrence and S.M. Shapiro, Phys. Rev. **B22** (1980) 4379.
(2) S. Nasu, A.M. Van Diepen, H.H. Newmann and R.S. Craig, J. Phys. Chem. Solids **32**(1971)2773.
(3) A. Hasegawa, Crystalline Electric Field Effects in f-electron Magnetism ed. R.P. Guertin, W. Suski and Z. Zolnierek (Plenum, New York, 1982) p. 201.
(4) Z. Kletowski, M. Glinski and A. Hasegawa, J. Phys. F . Metal Phys. **17**(1987)993.
(5) H. Kitazawa, Q.Z. Gao, H. Ishida, T. Suzuki, A. Hasegawa and T. Kasuya, J. Mag. Mag. Mater. **52**(1985)286.

Physica B 165&166 (1990) 333–334
North-Holland

FERMI SURFACE IN $CeSn_3$

Yoshichika ŌNUKI, Izuru UMEHARA, Yoshiko KUROSAWA, Nobuyuki NAGAI, Makoto KIKUCHI and Kazuhiko SATOH

Institute of Materials Science, University of Tsukuba, Tsukuba, Ibaraki 305, Japan

We have measured the magnetoresistance and the de Haas-van Alphen (dHvA) effect in $CeSn_3$. The magnetoresistance increases in all field directions. This result claims that $CeSn_3$ is a compensated metal. Compared to an uncompensated metal $LaSn_3$, the 4f electrons in $CeSn_3$ become itinerant electrons. This is consistent with the dHvA data which are almost explained by the modified Fermi surface model based on the result of band calculation where the 4f electrons are treated the same as the usual s, p, d conduction electrons.

$CeSn_3$, which crystallizes in the cubic $AuCu_3$ type structure is the first heavy electron compound in which the dHvA effect was observed(1,2). To clarify the Fermi surface topology of $CeSn_3$, we have done the experiments of dHvA and magnetoresistance for $CeSn_3$ in the {100} and {110} planes in the field up to 150 kOe and at temperature down to 0.45 K.

First we show in Fig. 1 the angular dependence of magnetoresistance under the constant field H, 150 kOe. The magnetoresistance at 150 kOe is large in magnitude, 30-45, and increases in all field directions, showing a $H^{1.3-1.4}$ dependence. This result claims that $CeSn_3$ is a compensated metal with an equal carrier concentration of electrons and holes, and possesses no open orbits.

If the Ce ion is trivalent, $CeSn_3$ should be the uncompensated metal as well as $LaSn_3$ because the total number of valence electrons per unit cell is 15. The present result of magnetoresistance claims that the 4f electron of the Ce atom becomes an itinerant electron so as to keep the carrier concentration of electrons and holes equal.

Figure 2 shows the angular dependence of the dHvA frequencies in the {100} and {110} planes. Here, the frequency F(Oe) is related to the cross-sectional area of Fermi surface S as $3.742\times10^8 S(a.u.)\{= 1.856\times10^8 S(2\pi/a)^2\}$, where the lattice constant a is 8.921 a.u. or 4.721 A. About twenty branches are observed in a wide frequency range of 10^5 Oe to 10^8 Oe.

We discuss the topology of the Fermi surface on the basis of the band calculation (3,4). The branch α corresponds to a nearly spherical 8th band-hole Fermi surface, bulges along the <100> direction. The orbit along the <100> field direction circulates around four bulges, while only along <111> it circulates around no bulge This is the reason why the dHvA oscillation can not be observed around <100>.

The β, γ, δ and ε branches are probably due to the complicated electron Fermi surface in the 9th band.

We propose a simple Fermi surface model to explain these dHvA branches, as shown in Fig. 3. We postulate simply that the γ branch is due to the spherical Fermi surfaces which are located in the corners of the octahedron. These spheres are connected by the arms. The location of the arm is determined by the observed field angles

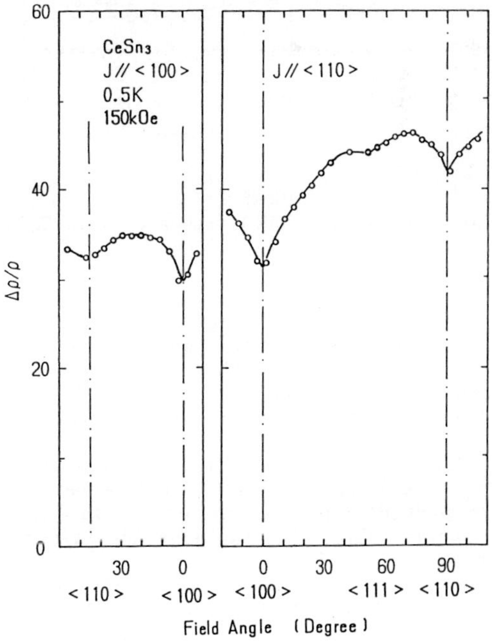

FIGURE 1
Angular dependence of transverse magnetoresistance in $CeSn_3$.

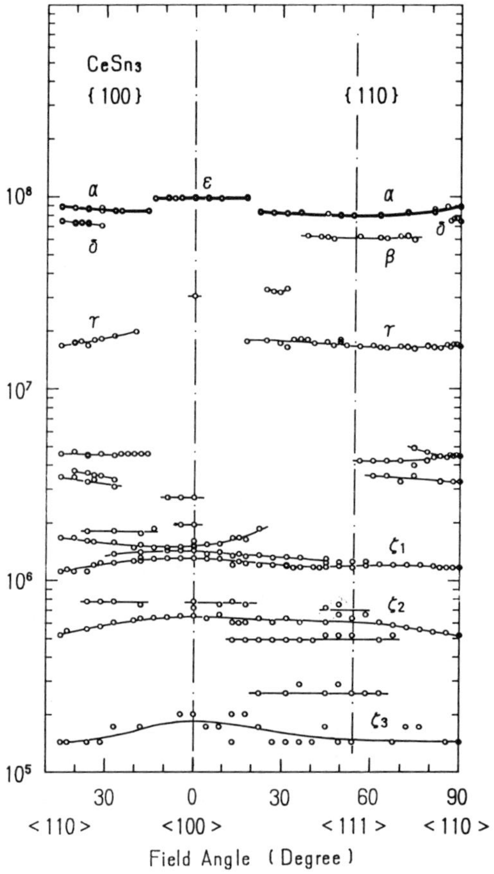

FIGURE 2
Angular dependence of dHvA frequencies in $CeSn_3$.

of the γ branch. The length of the arm is determined by the ε branch which circulates along the outside of the Fermi surface, containing four spheres and four arms. From this Fermi surface, we can roughly calculate the β and δ branches shown in Fig. 3 as 5×10^7 Oe and 7×10^7 Oe, respectively, which are almost the same as 6.2×10^7 Oe and 7.4×10^7 Oe, respectively in the present dHvA experiment. This simple Fermi surface model thus explains the main dHvA branches.

Here we note that the volume of the hole is $0.23(2\pi/a)^3$, while the electron is $0.15(2\pi/a)^3$. To keep the carrier concentration of electrons and holes equal, the present simple arm is necessary to be modified. Moreover, the cylindrical arm is not detected in the experiment. We suppose that the arm is big, bulging towards the inside.

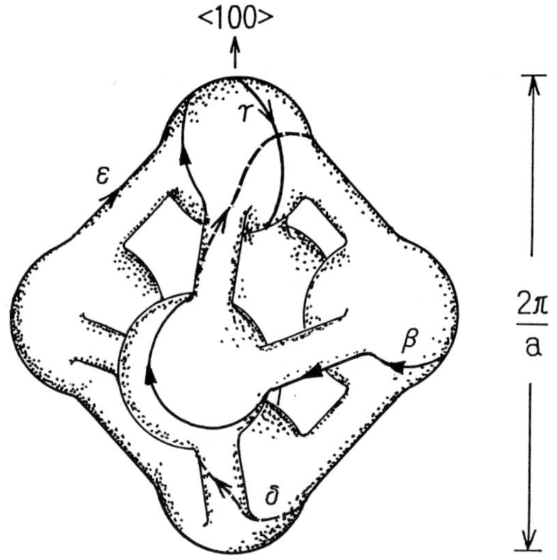

FIGURE 3
Simplified electron Fermi surface model in the 9th band of $CeSn_3$.

The cyclotron mass which has been determined from the temperature dependence of dHvA amplitude is in the range of 0.4 to $6.3m_0$. For example, the branches of α, β, γ, δ, ε, ρ_1 and ρ_2 are 3.83, 6.25, 4.07, 6.30, 2.72, 0.37 and $0.65m_0$, respectively, for the symmetry directions. The hybridization of 4f electron to the s, p, d bands spreads widely over every bands.

We are grateful to Professor A. Hasegawa for his critical reading of a manuscript and helpful discussion. We have utilized the 16 T-superconducting magnet at Cryogenic Center, University of Tsukuba. We are grateful to Professor R. Yoshizaki for his kind support. This work was supported by Grant-in-Aid for the Scientific Research from the Ministry of Education, Science and Culture, and also by University of Tsukuba Project Research.

REFERENCES

(1) W.R. Johanson, G.W. Crabtree, A.S. Edelstein and O.D. MacMasters, Phys. Rev. Lett. **45** (1981) 504.
(2) W.R. Johanson, G.W. Crabtree, D.D. Keolling, A.S. Edelstein and O.D. McMasters, J. Appl. Phys. **52** (1981) 2134.
(3) D.D. Keoling, Solid State Commun. **43** (1982) 247.
(4) A. Hasegawa, private commun.

Physica B 165&166 (1990) 335–336
North-Holland

ON THE FERMI SURFACE OF $CeAl_2$

G. Y. Guo

SERC Daresbury Laboratory, Warrington WA4 4AD, UK

The electronic band structure of $CeAl_2$ in the cubic Laves structure has been calculated. The Ce f–electrons were explicitly treated as core electrons. The calculated Fermi surface is found to be consistent with recent de Haas–van Alphen effect results, thereby corroborating that in $CeAl_2$, the f–electrons are essentially localised and contribute little to the formation of the Fermi surface.

The intermetallic cerium compound $CeAl_2$ shows a variety of unusual and interesting behaviours. For example, it undergoes a phase transition to a modulated antiferromagnetic structure at a low temperature of about 3.8K [1]. It also exhibits a Kondo effect (T_K = 7K) [2]. Like elemental Ce metal, $CeAl_2$ transforms to a volume–collapsed state under high pressures (P > 65kbar) [3]. Moreover, $CeAl_2$ has a large electronic specific heat coefficient (γ = 135 mJ/mol–K/K) [4]. Therefore, $CeAl_2$ is often refered to as a Kondo lattice or a 'mixed–valence' compound or even a 'heavy' fermion system.

Of crucial importance to the understanding of the fascinating properties of $CeAl_2$ is its electronic band structure (BS) in the vicinity of the Fermi level (E_f) and the role of the Ce 4f electrons. The BS of $CeAl_2$ has been calculated by Pickett and Klein [5] and by Jarlborg et al [6]. These authors found that in $CeAl_2$, the narrow f–bands are situated mainly above but crossing E_f. Consequently, the Fermi surface (FS) bears little resemblance to that of the isostructural $LaAl_2$ which contains no f–electrons. Very recently, de Haas–van Alphen (dHvA) effect, an ideal probe for the BS near E_f, has been measured by Lonzarich and co–workers [7] and also by Springford and Reinders [8]. However, both Lonzarich [7], and Springford and Reinders [8] found that main FS features are similar to those observed in $LaAl_2$. Lonzarich [7] and Springford and Reinders [8] therefore concluded that in $CeAl_2$, the f–electrons are sufficiently localised that they contribute negligibly to the FS formation. Motivated by these dHvA experiments [7,8], we reinvestigated the BS of $CeAl_2$ within the density functional theory with the standard local density approximation. However, unlike previous theoretical calculations [5,6], we explicitly treated the Ce 4f electrons as 'localised' states. This is implemented in the linear muffin–tin orbital (LMTO) method [9] by treating the 4f electron as a core state. We found, among other things, that most of the high frequency dHvA oscillations observed in $CeAl_2$ can be quantitatively interpreted in terms of our calculated ferromagnetic FS. In this short paper, we report salient results of this work.

In the present calculations, we used the LMTO method. The basis functions were s, p, d, 5f LMTOs for Ce sites and s, p, d for Al sites. The core charge densities were recalculated in every self–consistent (SC) iteration. The Ce 5p electrons were also treated as band states through the use of a second energy window. As mentioned before, we assumed Ce ions each having one localised 4f electron treated as a core state. The lattice constant used is 8.06A and the atomic radii chosen are 1.97A for Ce and 1.58A for Al. The tetrahedron technique [9] was used to calculate the density of states (DOS), dHvA frequency and cyclotron mass. In the DOS (dHvA frequency and cyclotron mass) calculations, 95 (505) k–points in the irreducible wedge of the Brillouin zone (BZ) were included in the BZ integration.

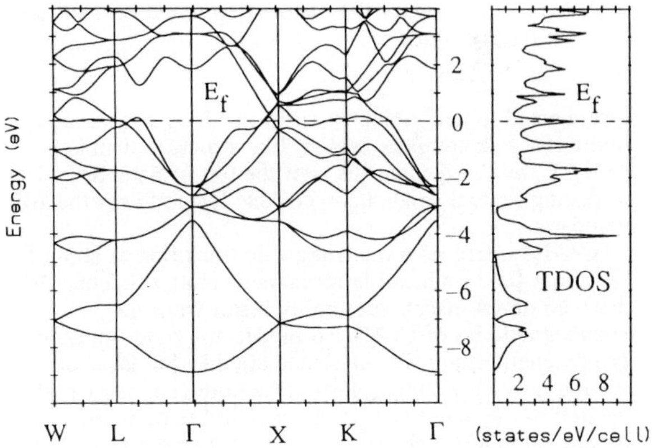

Figure 1 Band structure and total density of states (TDOS) for $CeAl_2$ in the cubic Laves structure.

In fig. 1, we show the BS and total DOS (TDOS) for $CeAl_2$ in the paramagnetic state. We note that the general features in the energy below 2 eV are very similar to those of $LaAl_2$ [6,10]. The calculated TDOS at E_f is 2.95 states/eV/Ce. The dominant contributions are from Ce 5d (43%) and Al 3p (30%). The corresponding electronic specific heat coefficient, γ, is 6.94 mJ/mol–K/K, which is about 20 times smaller than the experimental value [4]. There are three bands crossing E_f (bands 8, 9 and 10) in $CeAl_2$. Three dimensional (3–d) plots (not included in this paper) show that, as in $LaAl_2$, the FS of band 10 is a sphere–like surface centred at the BZ centre (Γ), with an additional tiny pocket located near L–point. Band 9 forms a large star–like open surface also centred at Γ. The surface

0921-4526/90/$03.50 © 1990 – Elsevier Science Publishers B.V. (North-Holland)

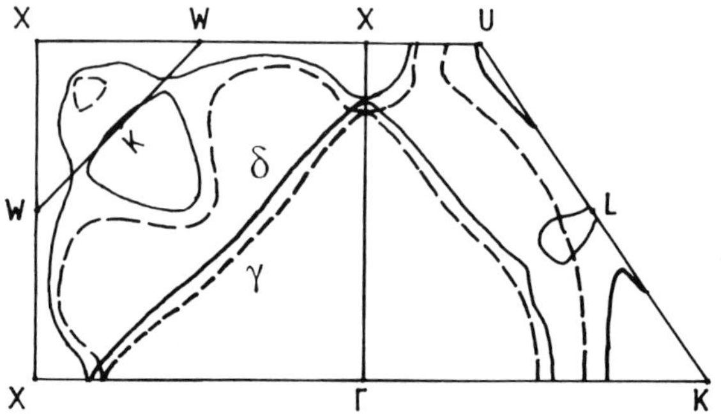

Figure 2 Cross sections of the band 9 and 10 Fermi surface of $CeAl_2$ (solid lines for the majority spin and dashed lines for the minority spin).

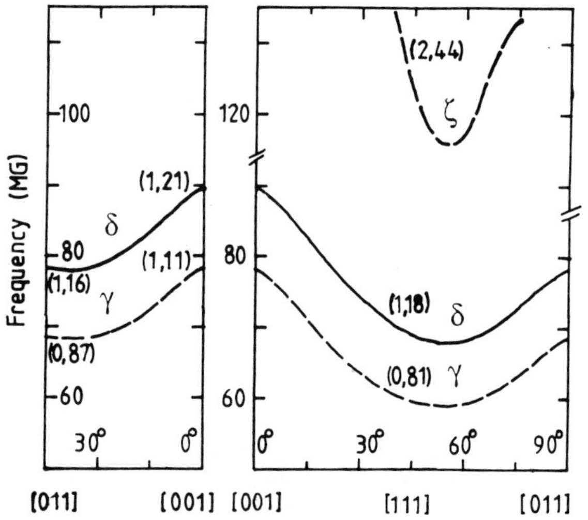

Figure 3 Calculated angular dependence of dHvA frequencies in $CeAl_2$ (see text). The associated cyclotron masses for [011], [001] and [111] directions are also given (in round brackets) (in units of the free–electron mass).

is connected to near neighbor surfaces along x, y, z axes, giving rise to a complex zeolite–like structure. Unlike $LaAl_2$, E_f in $CeAl_2$ also cuts near the top of band 8, producing several small 'hole' pockets located near the BZ boundary.

$CeAl_2$ undergoes a metamagnetic transition at about 5 Tesla to a field–induced ferromagnetic state [8]. Thus the observed dHvA effects presumbly result from the ferromagnetic FS of $CeAl_2$. To model this field–induced ferromagnetic state, we performed another SC BS calculation with a spin–polarised Ce core (i.e. one Ce 4f electron in the majority spin channel and none in the minority spin channel). We found BSs for both spins to be very similar to the paramagnetic one (fig. 1). The spin–splittings for bands near E_f varies from a few mRyds up to 10 mRyds. The TDOS at E_f is slightly reduced (now 2.43 states/eV/Ce). However, there are now six sets of FS. The FS cross sections for bands 9 and 10 on high symmetry planes are shown in fig. 2. The topology of the ferromagnetic FS for band 8, for band 9 of minority spin and for band 10 of majority spin remains the same as that in the paramagnetic state. The small pocket near L–point of band 10 in the paramagnetic state now disappears from the FS of band 10 of minority spin (fig. 2). The FS topology of band 9 of majority spin changes dramatically (fig. 2). It is now similar to that of band 9 in $LaAl_2$ [10]. The observed high frequency dHvA branches labelled as δ and γ in ref. 7 are clearly due to orbitals on two large co–centre surfaces of the spin–splitted band 10, respectively. Examining 3–d FS plots (not shown here) suggests that the ζ branch can be assigned to orbitals on the minority spin FS of band 9 with surface normal in the vicinity of [111] direction. The calculated dHvA frequencies for these three branches are shown in fig. 3. Both the shape and value of these frequency–angle curves (fig. 3) are in good agreement with those in fig. 6 of ref. 7. However, other dHvA oscillations, low frequency branches in particular, are not obvious. Detailed analyses are being carried out and the results will be reported in a future publication.

The calculated cyclotron masses for the ζ, δ and γ branches in [001], [011] and [111] directions are also given in fig. 3 (in round brackets). It is interesting to note that the ratio of the observed mass [8] over the calculated mass, varying from about 12 to 17, is comparable to the specific heat enhancement factor (about 20). Therefore, our results also resolve the apparent large discrapancy (a factor of about 5) between the mass enhancement determined in a dHvA experiment and that observed in the specific heat [8]. In summary, we found that the standard local density functional band model in which the f–electrons are regarded as core states, is essentially consistent with the experimental FS of $CeAl_2$.

REFERENCES

[1] B. Barbara et al., Solid State Commun. 24 (1977) 481.
[2] M. Nicolas–Francillon et al, Solid State Commun. 11 (1972) 845.
[3] B. Barbara et al., Physica B 86–88 (1977) 177.
[4] C. D. Bredl, F. Steglich and K. D. Schotte, Z. Phys. B 29 (1978) 327.
[5] W. E. Pickett and B. M. Klein, J. Less–Common Metals 93 (1983) 219.
[6] T. Jarlborg, A. J. Freeman and D. D. Koelling, J. Magn. Magn. Mat. 60 (1986) 291.
[7] G. G. Lonzarich, J. Magn. Magn. Mat. 76–77 (1988) 1.
[8] M. Springford and P. H. P. Reinders, J. Magn. Magn. Mat. 76–77 (1988) 11; P. H. P. Reinders, J. Magn. Magn. Mat. 76–77 (1989) 122.
[9] H. L. Skriver, The LMTO method (Springer, Berlin, 1984).
[10] A. Hasegawa and A. Yanase, J. Phys. F 10 (1980) 847.

Physica B 165&166 (1990) 337–338
North-Holland

PHASE DIAGRAM of $CeCu_2Si_2$ in the B-T plane

M. HUNT, P. MEESON, P.-A. PROBST, P. REINDERS* and M. SPRINGFORD

H.H. Wills Physics Laboratory, University of Bristol, Tyndall Avenue, Bristol, BS8 1TL, U.K.

W. ASSMUS and W. SUN

Physikalisches Institut, J. W. Goethe-Universitat, Robert-Mayer-Str, 2-4, Frankfurt am Main, F.R.G.

A combination dHvA and resistivity measurements investigations on a single crystal of the heavy fermion superconductor $CeCu_2Si_2$ has revealed a hitherto hidden complexity in the B-T phase diagram. As well as the superconducting-normal phase boundary, with T_c=0.72K and B_{c2}=1.5T, a double magnetic transition is observed which, depending on the orientation, occurs between 7.2T and 2.8T at 20mK, and which near the c axis accompanies a possible third transition. In the c-direction at least, the two transitions merge and then vanish as the temperature rises to 470mK so revealing an occlusion in the B-T phase plane. The nature of the phase transitions is not revealed by these experiments but a minor reconstruction of the Fermi surface and a change in quasiparticle mass is seen to occur across the occlusion.

1. INTRODUCTION

Measurements of the angle resolved de Haas van Alphen (dHvA) effect and of transverse resistivity on a single crystal of the heavy fermion superconductor $CeCu_2Si_2$, grown using the cold boat technique (1), have yielded information on the B-T phase diagram, the Fermi surface topology and quasiparticle masses. In this report we focus on the complexities of the phase diagram while the other results are discussed more fully in an accompanying paper (2).

$CeCu_2Si_2$ has the body centred tetragonal structure of $ThCr_2Si_2$ with the lattice constants a = 4.094 Å and c = 9.930 Å at room temperature (3). The transverse resistivity of a single crystal was measured as a function of applied magnetic field in the temperature range 20 - 720 mK with the current in the c-direction and the field in the a-direction. A standard ac, 4 wire technique was used, with pressure contacts to the sample.

2. RESULTS

Representative results are displayed in Figure 1. Notable features include the superconducting-normal phase transition, the as yet unexplained bump above B_{c2} and, at the lowest temperatures, the two distinct regions of positive, almost linear magnetoresistance. By extrapolation, the zero field residual resistivity was found to be described to within 1% by $\rho = \rho_0 + AT^2$ with $\rho_0 = 4.5 \pm 0.5$ μΩcm and A = 2.5 ± 0.2 $\mu\Omega cmK^{-2}$.The error in A arises principally from uncertainties in the sample geometry. This T^2 dependence with such a large value of A, together with the observation of quantum oscillations due to heavy long lived charged quasiparticles shows that the superconducting state is formed from a coherent Fermi liquid of such heavy quasiparticles, contrary to the assertion of Ott (4).

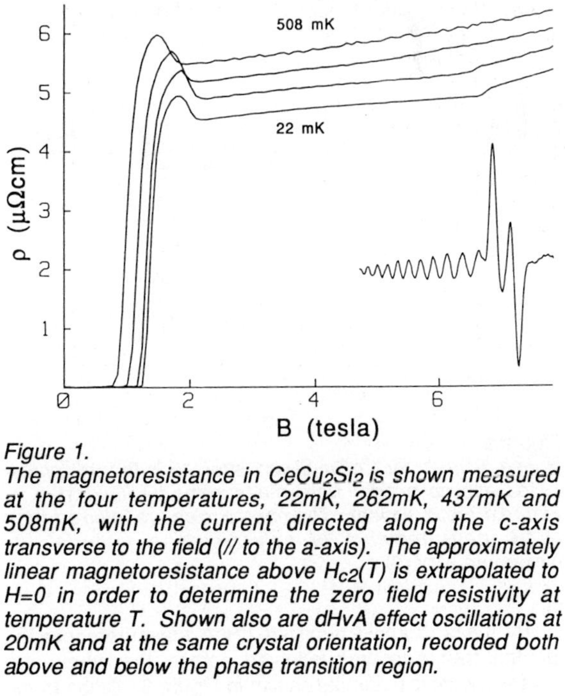

Figure 1.
The magnetoresistance in $CeCu_2Si_2$ is shown measured at the four temperatures, 22mK, 262mK, 437mK and 508mK, with the current directed along the c-axis transverse to the field (// to the a-axis). The approximately linear magnetoresistance above $H_{c2}(T)$ is extrapolated to H=0 in order to determine the zero field resistivity at temperature T. Shown also are dHvA effect oscillations at 20mK and at the same crystal orientation, recorded both above and below the phase transition region.

* Present address: Institut für Festkörperphysik, Technische Hochschule Darmstadt, Fachbereich 5, Hochschulstrasse 8, 6100 Darmstadt W. Germany

Included as an inset in Figure 1 is an example of dHvA effect oscillations observed in the same sample at 20mK using a second derivative technique at the same crystallographic orientation. The large double feature that aligns with the abrupt change in magnetoresistance behaviour indicates that a double magnetic transition is occurring here. With increase in temperature the double transition was seen to merge to one and then to disappear above 470mK, which suggests that a separate phase exists as a small occlusion in the B-T phase diagram for this orientation. The low temperature kink in the magnetoresistance showed a similar temperature dependence. From the resistivity measurements we were also able to delineate the superconducting-normal transition, with $T_c = 0.72 \pm 0.02$ K, $B_{c2} = 1.5$ T and $(\partial B_{c2}/\partial T)_{T_c} = -4.7$ T/K, but no evidence was found for the higher temperature phase boundary inferred from NMR measurements on a polycrystalline sample by Nakamura et. al. (5). This may reflect the anisotropy of the material. However they further inferred from the presence of a large distribution of hyperfine fields at the Cu sites that a single magnetic ordered state exists above T_c, of SDW type. Our measurements, which we present in Figure 2, show that the B-T phase diagram is more complex.

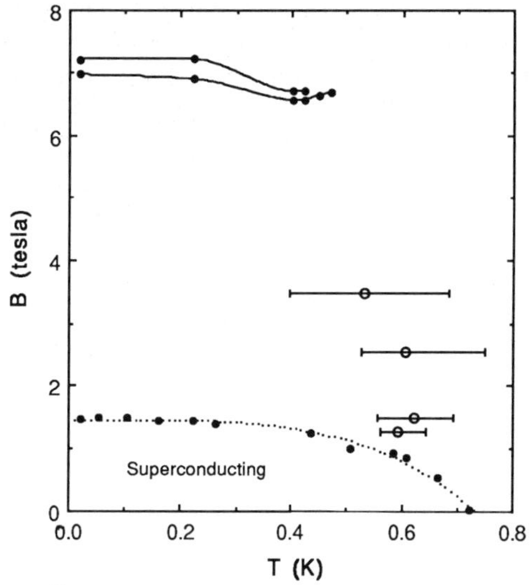

Figure 2.
Phase diagram for $CeCu_2Si_2$ constructed in the B-T plane (B parallel to a-axis) from measurements illustrated in Figure 1 (solid circles). The open circles are inferred from NMR measurements in polycrystalline material by Nakamura et al (5).

By mass of angle resolved dHvA measurements the double transition was followed across the whole c-a crystallographic plane, as shown in Figure 3. Close to the c-axis a third transition is also seen, and the anisotropy of the material is apparent. Analysis of the angle resolved dHvA oscillations (2) indicates a symmetry appropriate to a single crystal so that we believe these transitions to be an intrinsic property of $CeCu_2Si_2$.

Whilst we cannot determine from our measurements the nature of whatever magnetic or other ordering is occuring, analysis of the dHvA oscillations shows that a minor distortion of the Fermi surface and a change in quasiparticle mass occurs across the occlusion in Figure 2. These experiments will be more fully described elsewhere (6).

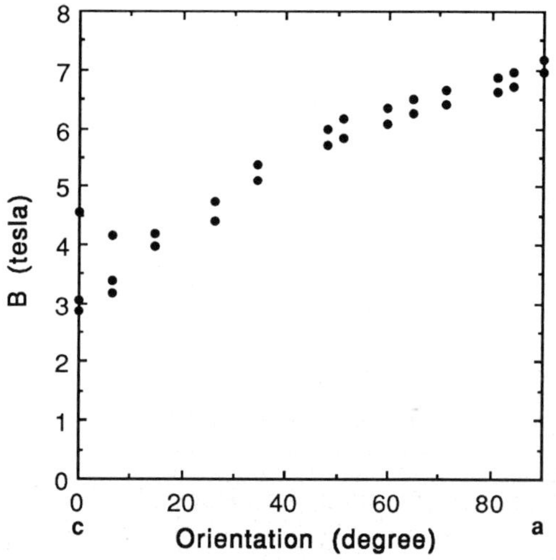

Figure 3.
Anisotropy of part of the phase diagram of $CeCu_2Si_2$ measured for magnetic field directions in the c-a plane.

ACKNOWLEDGEMENTS

This work is supported financially by the SERC, and one of us (P.-A. P) acknowledges partial support from Swiss National Funds.

REFERENCES

(1) W.Sun ,M.Brand,G.Bruls and W.Assmus , to be publlished in Zeitschrift fur Physik.
(2) M.Hunt , P.Meeson, P-A.Probst, P.Reinders, M.Springford, W.Assmus and W.Sun, this volume.
(3) T Jarlborg, H F Braun, and M Peter, Z. Phys. B **52**, 295-301 (1983)
(4) H.R.Ott,Characteristic features of heavy-electron materials, in: Progress in Low Temperature Physics, Vol **XI**, ed. D.F.Brewer (Elsevier Science Publishers 1987) pp 215-290.
(5) H.Nakamura, Y.Kitaoka, H.Yamada and K.Asayama, J.M.M.M. **76 & 77** (1988) 517.
(6) M.Hunt, P.Meeson, P-A.Probst, P.Reinders, M.Springford, W.Assmus and W.Sun, Submitted to Phys. Rev. Lett.

Physica B 165&166 (1990) 339–340
North-Holland

DE HAAS-VAN ALPHEN EFFECT IN $LaGa_2$ AND $SmGa_2$

I. Sakamoto, T. Miura, K. Miyoshi and H. Sato

Department of Physics, Tokyo Metropolitan University
Fukazawa, Setagaya-ku, Tokyo 158, Japan

We have measured de Haas-van Alphen (dHvA) effect in $LaGa_2$ and $SmGa_2$ in magnetic field up to 80kG. Angular dependence of the dHvA frequencies were obtained in the three principal crystallographic planes in $LaGa_2$ and in the [0001] plane for $SmGa_2$. Observed frequencies range from 10 MG to 40 MG for both compounds. The cyclotron masses range from $0.3m_o$ to $0.7m_o$ for $LaGa_2$ and from $0.5m_o$ to$1.5m_o$ for $SmGa_2$. The mass enhancement in $SmGa_2$ indicates that 4f electrons hybridize with conduction electrons appreciably.

1. INTRODUCTION

The RGa_2 compounds (R=rare earth element) crystallize in the simple hexagonal AlB_2 type structure (P6/mmm space group). RGa_2 (R=Ce,Pr, Nd, Gd, Tb, Ho and Er) have been known as antiferromagnets with T_N ranging from 7K for $PrGa_2$ to 15K for $TbGa_2$ (1-3). The antiferromagnetic order has been ascribed to the RKKY interaction between well-defined 4f moments. Recently the existence of hybridization of 4f electrons with conduction electrons has been suggested from the sine modulated magnetic structure in $CeGa_2$ (4,5).

The aim of the present work is to find out the evidence of 4f electron hybridization and the Fermi surface (FS) properties in RGa_2 compounds using dHvA effect. This paper reports the results of $LaGa_2$ and $SmGa_2$. $LaGa_2$ is the reference substance of RGa_2 compounds. For $SmGa_2$ we have very little knowledge of magnetic and electrical properties.

2. EXPERIMENTAL

Single crystals of $LaGa_2$ and $SmGa_2$ were prepared using tri-arc Czochralski technique. Crystals were annealed at 1000°C for 50 hr under a vacuum of 10^{-7} torr to improve the residual resistance ratio RRR. During annealing, each sample was wrapped successively with Ta, Zr and Ta foils. The RRR values are about 200 for LaGa2 and 400 for SmGa2.

Usual four probes ac method was used for resistivity measurement. Magnetization measurements were made by SQUID magnetometer of QD Inc. The dHvA effect was measured in magnetic field up to 80kG using a field modulation technique in a ^4He cryostat down to 1.3K.

3. RESULTS AND DISCUSSION

Figure 1 (a) shows the temperature (T) dependence of resistivity ρ in $LaGa_2$ and $SmGa_2$ single crystals, where the current flows along a-axis. ρ(T) for $SmGa_2$ shows a

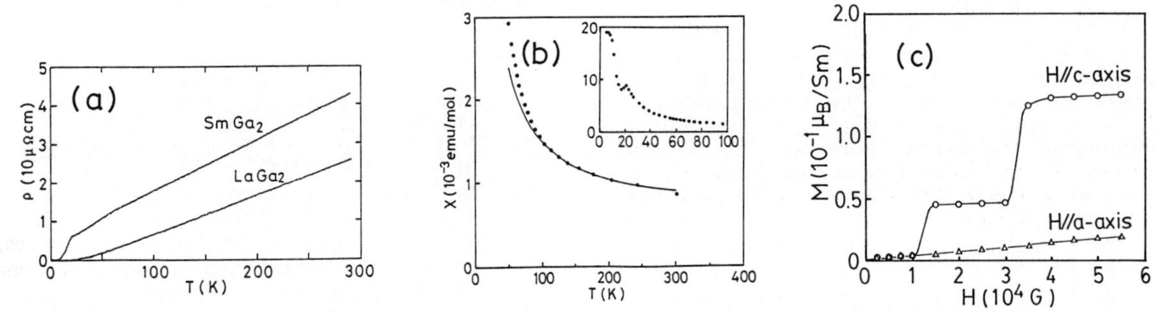

Figure 1. (a)Temperature dependence of resistivity ρ of $LaGa_2$ and $SmGa_2$ single crystals. The current flows parallel to the a-axis. (b) Temperature dependence of susceptibility χ of $SmGa_2$ measured at 5kG. The field is along the c-axis. The inset shows the low temperature part. (c) Field dependence of magnetization of $SmGa_2$ measured at 1.8K.

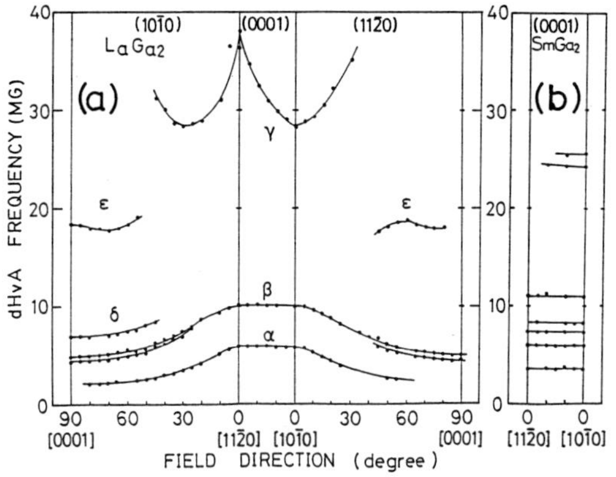

Figure 2. Angular dependence of the dHvA frequencies of $LaGa_2$ (a) and $SmGa_2$ (b).

steep decrease below 20K, where the antiferromagnetic order sets in, as seen in a later. Figure 1(b) shows the temperature dependence of the magnetic susceptibility χ in magnetic field (H) of 5 kG parallel to the c axis. The curve has a maximum at T=20K, where the resistivity shows a sharp decreases. This shows that $SmGa_2$ is an antiferromagnet with T_N=20K. The solid line in Fig. 1(b) shows the calculated magnetic susceptibility of Sm^{3+} free ion using a formula of Van-Vleck with the spin-orbit coupling constant of 520K. The curve well fits the experimental results above 100K, which shows that Sm is in the tri-valent state.

Figure 1(c) shows the magnetization M of $SmGa_2$ measured at 1.8K with H//a and c axes. The M(H) result for H//a-axis varies linearly up to 50kG, while M(T) for H// c-axis shows a metamagnetic behavior: spin flip occurs at about 15kG and 35kG. The magnetic moment tends to 0.13μB/ion at 55kG. This value is about 15% of that expected from J=5/2 for tri-valent Sm ion.

Figure 2 shows the angular dependence of dHvA frequencies measured in field up to 80kG for $LaGa_2$ and $SmGa_2$. For $LaGa_2$ measurements were made in the three principal crystallographic planes. We found the five fundamental frequencies denoted α, β, γ, ε and δ in Fig. 2(a). The maximum frequency observed is about 40MG, which is 20 % of the Brillouin zone area.

The α frequency is assigned to a cigar-shaped FS, the principal axis of which is parallel to the [0001] direction. We note the β frequency splits into two branches at the field angle > 30° in (10$\bar{1}$0) and (11$\bar{2}$0) planes. This implies that FS of the β frequency has a dumb-bell shape with the principal axis parallel to the [0001] direction. The γ frequency disappears for the field angle >30° in the (11$\bar{2}$0) plane and >50° in the (10$\bar{1}$0) plane. This indicates that FS of the γ frequency is a open surface. A hexagonal network of tubes is one of the model which qualitatively explains the angular dependence of the γ frequency.

Figure 2(b) shows the dHvA frequency of $SmGa_2$ in the (0001) plane. The magnitudes of the frequencies are very similar to those of $LaGa_2$ in that plane. This suggests that there is some resemblances between their FS pieces.

Table I De Hass-van Alphen frequencies F and the cyclotron masses m_c^* of $LaGa_2$ and $SmGa_2$ for fields along the [10$\bar{1}$0] direction.

LaGa2		SmGa2	
F(MG)	m_c^*/m_o	F(MG)	m_c^*/m_o
6.0	0.3	8.9	0.5
10.1	0.51	11.1	1.1
28.9	0.67	25.7	1.5

m_o is the free electron mass.

Table I shows the cyclotron effective masses of some frequencies for the both compounds in H// [10$\bar{1}$0]. The masses of $LaGa_2$ are light and are in the range (0.3-0.7)m_o, whereas the $SmGa_2$ masses are 2-3 times larger . This indicates the hybridization of 4f electrons with conduction electrons to occur in $SmGa_2$. If 4f electrons are completely localized, such a mass enhancement can not be expected.

We are now planning to make the measurement of dHvA in $LaGa_2$, $SmGa_2$ and other RGa_2 compounds in fields up to 150 kG to study completely the Fermi surface properties.

We ackowledge helpfull discussions with I.Shiozaki, T. Fukuhara and Y. Onuki.

REFERENCES

(1) T.H. Tsai, J.A. Gerber, J.M. Weymouth, and D.J. Sellmyer, J. Appl. Phys. **49** (1978) 1507.

(2) T.H. Tsai, and D.J. Sellmyer, Phys. Rev. **20**B (1979) 4577.

(3) M. Takahasi, H. Tanaka, T. Satoh, M. Kohgi, Y. Ishikawa, T. Miura and H. Takei, J. Phys. Soc. Jpn. **57** (1988) 1377.

(4) P. Burlet, M.A. Frémy, D. Gignoux. G. Lapertot, S. Quezel, L.P. Regnault, and E. Roudaut, J. Mag. Mag. Mat. **63&64** (1987) 34.

(5) M. Jerjini, M. Bonnet, P. Burlet, G. Lapertot, J. Rossat-Mignod, J.Y. Henry and D. Gignoux, J. Mag. Mag. Mat. **76&77** (1988) 405.

Physica B 165&166 (1990) 341–342
North-Holland

Energy Band Calculation of $LaCu_6$

Hisatomo HARIMA, Akira YANASE and Akira HASEGAWA*

College of Integrated Arts and Science, University of Osaka Prefecture, Sakai Osaka 591, Japan
**College of General Education, Niigata University, Niigata 950-21, Japan*

We have carried out a band calculation of $LaCu_6$, which is a reference material of a heavy fermion compound $CeCu_6$, by an LAPW method and the local density approximation. We have also calculated Fermi surfaces and extremal areas in order to compare with the de Haas-van Alphen effects. There are many Fermi surfaces in $LaCu_6$, and the calculated results explain almost of the measured frequency.

For many RCu_6 compounds, where R is La, Ce, Pr, Nd and Sm, de Haas-van Alphen (dHvA) effects are measured [1-4]. Among them, $CeCu_6$is widely regarded as an example of a heavy fermion metal and has heavy cyclotron masses in the range of 6 to $80m_0$ [1]. However, no band structure calculation were published because they have a complicated crystal structure. We have calculated a band structure and extremal areas of Fermi surfaces (FS) of $LaCu_6$, which is a reference material, and compare with the measured de Haas-van Alphen frequencies.

$LaCu_6$ possesses a monoclinic structure, which is a slight modification of the orthorhombic $CeCu_6$ structure [5]. As the crystal distortion is very small, the orthorhombic notation are used in the dHvA measurements. Therefore we use the orthorhombic structure in a band structure calculation for $LaCu_6$. The unit cell contains four formula units.

We have carried out a energy band structure calculation of $LaCu_6$ using an LAPW method with an exchange and correlation potential in the local density approximation. Spin orbit interactions are included in the sense of a perturbation for usual one-electron eigenstates including the other relativistic effects. The final band structure in fig. 1 is given from eigenstates at 225 points in the irreducible Brillouin Zone. About 1200 basis functions for each point are needed especially for 120 d-bands on 24 Cu. The specific heat coefficient γ calculated from a density of states in fig. 2 is $6.2mJ/mol \cdot K^2$ per one formula unit while experimental value is $8mJ/mol \cdot K^2$ [2].

As shown in fig. 2, 120 d-bands on Cu in the range of 0.1 to 0.3 Ry and 28 f-bands around 0.65 Ry are well localized and are separated from the Fermi level.

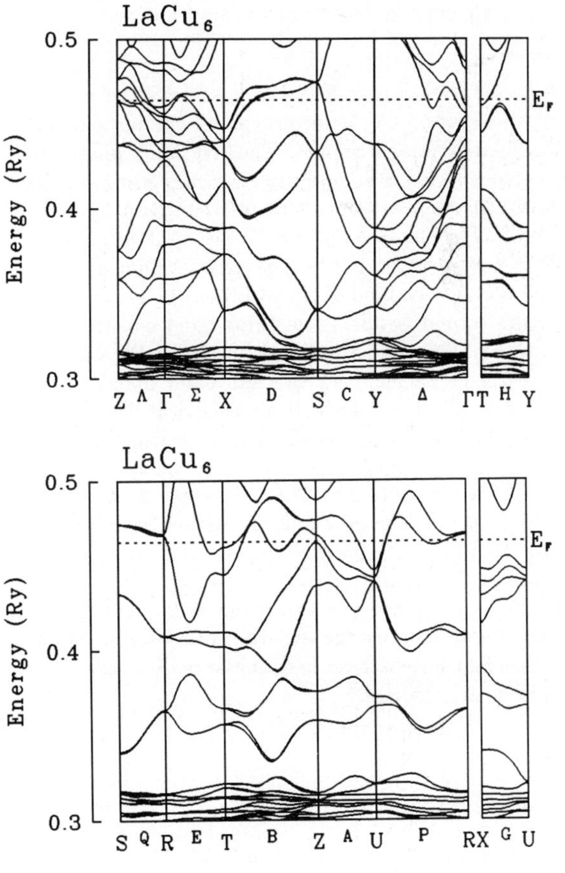

Figure 1. Energy-band structure of $LaCu_6$ in the vicinity of the Fermi energy denoted by E_F.

0921-4526/90/$03.50

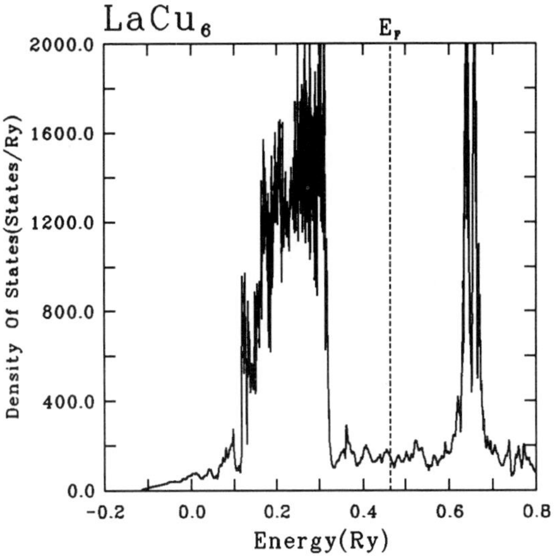

Figure 2. Calculated density of states of $LaCu_6$

The Fermi level crosses four bands which have mainly 4s components on Cu, however, consist complicated FS reflecting its crystal structure. The first band constructs three kinds of closed FS and a cylindrical FS around the Q-axis. The angular dependence of the dHvA frequencies on $LaCu_6$ shows no closed FS [3], so it seems that all extremal areas can not be measured. A multiply connected FS and one kind of closed FS on the P-axis stem from the second bands. The third band constructs a cylindrical FS around the H-axis, a closed FS and the largest FS in $LaCu_6$, and the last band makes three kinds of small closed FS and a cylindrical FS around H-axis, which is similar to one from the third. Cylindrical FS on the Q- and H-axes have extremal areas in the range of 5×10^5 Oe to 2×10^6 Oe corresponding experimental branches near the c-axis. A calculated frequency 2×10^7 Oe from the largest FS explains the branch near the b-axis. There are too many extremal areas (more than 20 on the c-axis) in the range of 2×10^5 Oe to 3×10^7 Oe to identify the experimental results. Many branches may have not yet been measured because their intensities are too weak.

We are grateful to professor Y. Onuki for a useful discussion, and to Dr. K. Takegahara for kindly support. Numerical computation was partly performed at the Computer Center of the Institute of Molecular Science, Okazaki National Research Institute. This work was partly supported by the project for the physics on actinide compounds, a Grand-in-Aid for Scientific Research on Priority Area from The Ministry of Education, Science and Culture.

References

(1) P.H.P. Reinders, M. Springford, P.T. Coleridge, R. Boulet and D. Ravot, Phys.Rev.Lett. 57 (1986) 1631
P.H.P. Reinders, M. Springford, R. Boulet and P.T. Coleridge, Jpn.J.Appl.Phys. 26 (1987) Suppl.26-3 505
P.H.P. Reinders, M. Springford, P.T. Coleridge, R. Boulet and D. Ravot, J.Magn.Magn.Mater. 63&64 (1987) 297
M. Springford and P.H.P. Reinders, J.Magn.Magn.Mater. 76&77 (1988) 11
(2) Y. Onuki, M. Nishihara, Y. Fujimura and T. Komatubara, J.Phys.Soc.Jpn. 55 (1986) 21
(3) Y. Onuki, A. Umezawa, W.K. Kwok, G.W. Crabtree, M. Nishihara, T. Komatubara, K. Maezawa and S. Wakabayashi, Jpn.J.Appl.Phys. 26 (1987) Suppl.26-3 509
(4) Y. Onuki, M. Nishihara, Y. Fujimura, T. Yamazakiand T. Komatubara, J.Magn.Magn.Mater. 63&64 (1987) 317
D. Endoh, T. Goto, T, Suzuki, T. Fujimura, Y. Onuki and T. Komatubara, J.Phys.Soc.Jpn. 56 (1987) 4489
Y. Onuki, Y. Kurosawa, T. Omi, T. Komatubara, R. Yoshizaki, H. Ikeda, K. Maezawa, S. Wakabayashi, A. Umezawa, W.K. Kwok and G.W. Crabtree, J.Magn.Magn.Mater. 76&77 (1988) 37
(5) H. Asano, M. Umino, Y. Hataoka, Y. Shimizu, Y. Onuki, T. Komatubara and F. Izumi, J.Phys.Soc.Jpn. 54 (1985) 3358

Physica B 165&166 (1990) 343–344
North-Holland

Fermi Surfaces of UB_{12}

Hisatomo HARIMA, Akira YANASE, Yoshichika ONUKI*, Izuru UMEHARA*, Yoshiko KUROSAWA*, Nobuyuki NAGAI*, Kazuhiko SATOH*, Mitsuo KASAYA† and Fumitoshi IGA‡

College of Integrated Arts and Science, University of Osaka Prefecture, Sakai Osaka 591, Japan
**Institute of Material Science, University of Tsukuba, Tsukuba Ibaraki 305, Japan*
†*Department of Physics, Tohoku University, Sendai 980, Japan*
‡*Electrotechnical Laboratory, Tsukuba Ibaraki 305, Japan*

We have measured the de Haas-van Alphen(dHvA) oscillation, and calculated the Fermi surfaces from the band structure of UB_{12}. The angular dependence of the dHvA frequencies are consistent to the calculated extremal areas. Fermi surfaces consist of electron and hole ones, being a compensated metal. The measured cyclotron masses in the range of $0.6m_0$ to $6m_0$ are also agreement with the calculations.

UB_{12} is one of the cubic RB_{12} type compounds [1], where R is a heavy rare earth element, U, Pu, Np, Y and Zr. This material possesses a high melting point of 2235°C and is a congruently melting material. The previous magnetic susceptibility data show a Pauli paramagnetic nature, although the distance between the U atoms, 5.28Å exceeds the Hill limit value of 3.4Å. The electrical property is, however, small in knowledge. To clarify the Fermi surface (FS) property we have calculated the band structure and have measured the de Haas-van Alphen (dHvA) effect.

We have carried out a energy band structure calculations of UB_{12} [2] using an LAPW method with an exchange and correlation potential in the local density approximation. Spin-orbit interactions are included in the sense of a perturbation for usual one-electron eigenstates including the other relativistic effects. The self-consistent potential and the final band structure, as shown in fig.1, are given from eigenstates at 85 points in the irreducible Brillouin Zone, which are calculated using about 700 basis functions. 5f bands on U of UB_{12} have large band-width of about 0.2Ry, which is comparable to its spin-orbit interaction, because of its hybridization with 2p-electron on B. This itinerant character of 5f-electron on UB_{12} is consistent with its magnetic behavior. The Fermi level crosses two bands among these wide 5f bands, which have mainly $j = 5/2$ components. These two bands construct multiply connected FS as shown in fig.2. UB_{12} is a compensated metal and a number of the carrier is calculated as 0.317 for each hole and electron bands. The calculated specific heat coefficient γ is $12.1 mJ/mol \cdot K^2$ [3], while the experimental γ is $20 mJ/mol \cdot K^2$, so a mass enhancement factor is concluded about 1.65 in zero magnetic field.

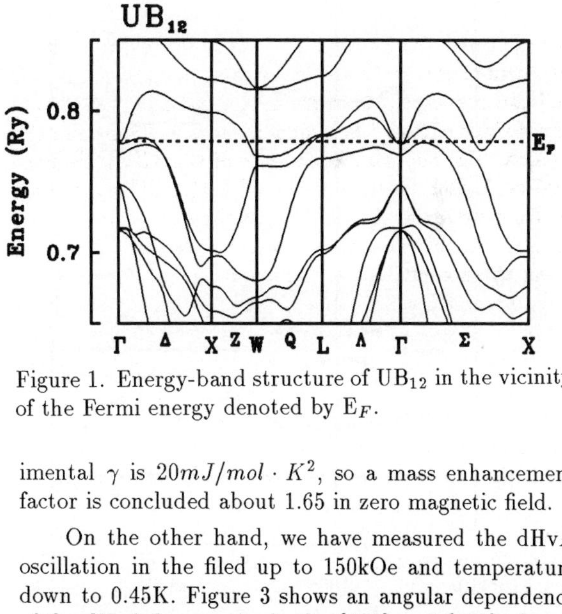

Figure 1. Energy-band structure of UB_{12} in the vicinity of the Fermi energy denoted by E_F.

On the other hand, we have measured the dHvA oscillation in the filed up to 150kOe and temperature down to 0.45K. Figure 3 shows an angular dependence of the dHvA frequencies in the {100} and {110} planes. About fifteen branches are observed in a wide frequency range of 10^6 Oe to 10^8 Oe.

Calculated FS mentioned above explains the dHvA signals as follows. the α branch stems from the outside orbits on the empty tunnel centered at the X points in fig.2b. There are two candidates for the β branch, one is the inside orbits in the empty tunnel as mentioned above with the frequency of 3.90×10^7 Oe, and the other is the orbits with the frequency of 4.86×10^7 Oe on the center of the cube around Γ point in fig.2a. As the β branch

0921-4526/90/$03.50

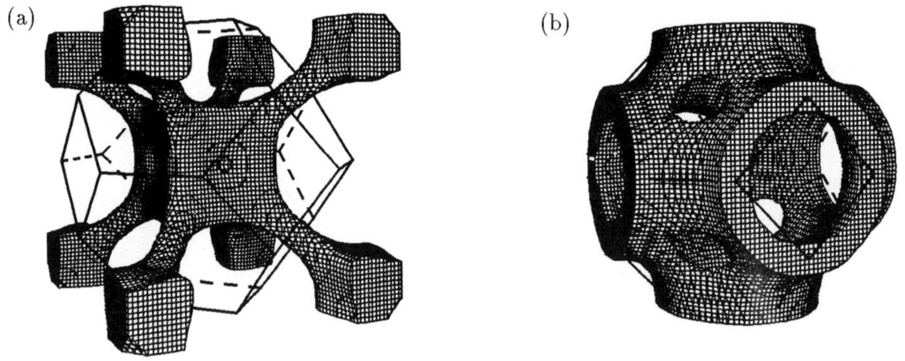

Figure 2. Calculated hole sheet (a) and electron sheet (b) of the Fermi surface of UB_{12}.

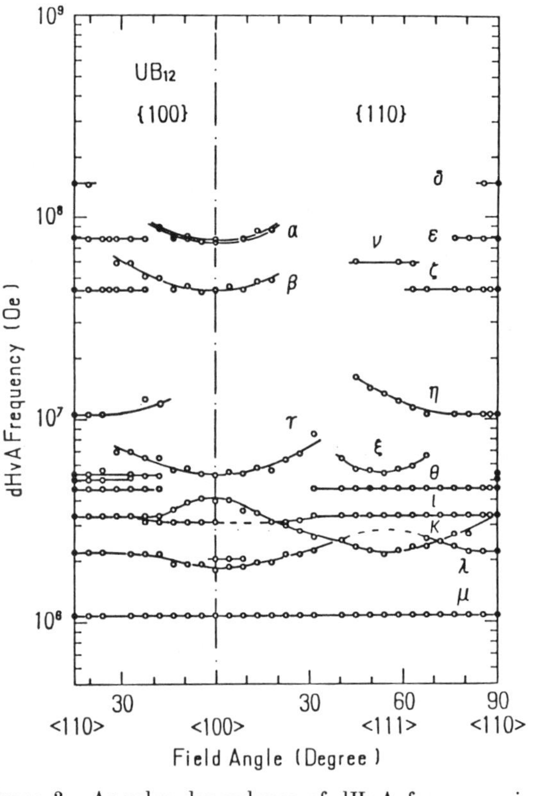

Figure 3. Angular dependence of dHvA frequency in UB_{12}

consists of two signals of 4.38×10^7 Oe and 4.29×10^7 Oe, there is a possibility that two orbits mentioned above possesses the almost the same extremal areas. The ν branch is accepted to the orbits on the cube in <111> direction. In fig.2a, the slender arms which is a little thick around L points, may explain γ, ι and κ branches. Small two closed FS around Γ point correspond to λ and μ branches. δ branch has a large frequency, so it may be due to the orbits between 4 cubes through the slender arms. An opening around L point of electron FS causes an orbit of ξ branch. The multiply connected electron FS has some other extremal areas, in which a orbit from L point to L point through Σ points may assigned to η branch.

The measured cyclotron masses are in the range of 0.6 to $6m_0$. For example, the masses for the α, β, η, λ, are 1.75(and 2.01), 1.93, 2.33, 1.00, respectively. On the other hand, they are roughly agreement with the calculated ones of 1.67, 2.8(and 2.6), 1.26, 1.11, respectively.

We are grateful to professor T. Goto for a helpful discussion and providing us with the experimental data prior to publication. Numerical computation was performed at the Osaka University Computer Center and at the Computer Center of University of Osaka Prefecture. We have utilized a 16T-superconducting magnet at Cryogenic Center, University of Tsukuba. This work was supported by the project for the physics on actinide compounds, a Grand-in-Aid for Scientific Research on Priority Area from The Ministry of Education, Science and Culture, and by University of Tsukuba Project Research.

References

1) P. Blum and F. Bertaut Acta. Crysta. 7 (1954) 81
2) H. Harima, S. Miyahara and A. Yanase, Physica in press as the proceedings of the International Conference on the Physics of the Highly Correlated Electron System, Santa Fe, 1989
3) $15mJ/mol \cdot K^2$ reported as γ value in 2) was roughly estimated.

Physica B 165&166 (1990) 345–346
North-Holland

Superconducting Lower Critical Fields in UPt_3 *

Zuyu Zhao*, F. Behroozi+, J.B. Ketterson*, Yongmin Guan*, Bimal K. Sarma† and D.G. Hinks•

*Department of Physics and Astronomy, Northwestern University, Evanston, IL 60208 USA
+Department of Physics, University of Wisconsin–Parkside, Kenosha, WI 53141 USA
†Department of Physics, University of Wisconsin–Milwaukee, Milwaukee, WI 53201 USA
•Material Science Division, Argonne National Laboratory, Argonne, IL 60439 USA

DC magnetization curves of a single crystal of UPt_3 (T_c = 542 mK) are obtained with the external field parallel and perpendicular to the hexagonal c-axis. The magnetization curves are analyzed to yield the lower critical field H_{c1} as a function of temperature and orientation. For both the c-axis and basal plane orientations, H_{c1} shows an apparent kink in slope about 150 mK below T_c. This behavior appears to confirm the recently proposed model for the superconducting states in UPt_3 in which a two dimensional order parameter leads to the formation of two distinct superconducting phases.

Several cerium and uranium compounds such as $CeCu_2Si_2$, UBe_{13}, and UPt_3 exhibit very large electronic specific heats (two to three orders of magnitude larger than copper). Strong on site repulsion causes the f electrons to behave essentially as uncorrelated moments at high temperature. However the combined effect of falling temperature and hybridization results in the formation of a highly correlated state at low temperatures, a Fermi liquid, with a Fermi surface well re-presented by the local density band model(1); this Fermi liquid has a characteristic temperature of order 10K. Further, several of these so called heavy fermion systems become superconducting at about T ≈ 1K.

There is considerable speculation on the nature of superconductivity in these systems(2). It has been suggested that the nearly magnetic character of these systems may lead to pairing via magnetic excitations.

Much recent activity has focused on the superconducting behavior of the heavy fermion system UPt_3 due to its many intriguing properties. Recent heat capacity data(3) show the signature of two phase boundaries near T_c which converge to a critical point at a field H ≈ 0.5 T and a temperature T ≈ 0.4 K. Neutron scattering(4) and ultrasonic attenuation(5) data also indicate several unusual features which may be attributed to the interaction of the magnetic and superconducting order parameters, further adding to the evidence for the unconventional nature of superconductivity in UPt_3.

Recently Hess, Tokuyasu, and Sauls (HTS) proposed(6) a model for the superconducting states of UPt_3 in which a two dimensional order parameter gives rise to two superconducting phases of different symmetry which exist in adjacent temperature domains. Hence this model accounts for the existence of two jumps in the specific heat. Further, the model predicts an abrupt change in the slope of the upper critical field phase line when the field is in the basal plane, signaling the transition between the two superconducting phases at a finite field. The lower critical field phase line is also expected to display a kink for all field orientation at a temperature very close to the zero-field transition.

With an rf resonance technique which only probes the surface (either the skin or London depths) of the sample, Shivaram et al. recently reported a kink in H_{c1} with the field in the basal plane; however the ratio of the slopes is much larger than that predicted by the theory. No kink was confirmed with the field along the c-axis.

Here we present new lower critical field data which support the predictions of the HTS model for all field orientations. Our data consist of low field magnetization curves of a single crystal spherical sample of UPt_3. The sample was produced by spark cutting techniques from a high quality single crystal ingot. The UPt_3 sphere, (4.03 ± 0.01)mm in diameter, was etched chemically to remove any surface damage and annealed before use.

We emphasize that in order to obtain the lower critical fields reliably, it is essential to perform the magnetization measurements on a single crystal ellipsodal sample with a well known demagnetizing factor. A spherical sample

*Work supported by the Low Temperature Physics Branch of the National Science Foundation under grant DMR89-07396.

is perhaps the most convenient with a demagnetizing factor of 1/3 for all crystallographic orientations.

A top loading Oxford dilution refrigerator with a 12T magnet was used for the dc susceptibility measurements. The dc technique uses two balanced opposing coils, one of which contains the sample. When the external field is ramped, the net emf from the coils is proportional to the dc susceptibility of the sample(8). Direct integration of this signal gives the magnetization. Since the demagnetization factor of our spherical sample is accurately known, the Meissner slope can be used to calibrate the magnetization scale precisely.

Typically the low field magnetization data were taken by warming the sample above T_c in zero field to drive off any trapped flux, then cooling to a fixed temperature. The field was then ramped at about 1 G/s to obtain the low field magnetization data.

The lower critical field of the sphere, referred to as $(H_{c1})_{sphere}$, was taken as the field at which the initial slope of the magnetization curve just began to deviate from the Meissner value; then

$$H_{c1} = 3/2(H_{c1})_{sphere} \tag{1}$$

Figure 1 shows the lower critical fields as a function of temperature for the field parallel to the c-axis, while Figure 2 shows the data for the field parallel to the a and b-axes. In all cases a kink near 0.4 K is apparent.

The straight lines for the combined a- and b-axes data were fitted by the following procedure. The data were partitioned into low and intermediate field groups and separately fitted to straight lines. The total r.m.s. error was then minimized as a function of the partitioning point. The fitting gives T_c^*, $T_{c(+)}$ and $T_{c(-)}$ respectively as 395.0 mK, 541 mK and 517 mK. Points lying in the dashed region of the lines were not included in the fit. The slope ratios of the H_{c1} vs. T curve above and below T_c^* for the external field in the basal plane is estimated to be 1.19 and is in good agreement with the HTS model. The c-axis data appears to exhibit some curvature; however for comparison with the existing theory (which predicts straight line behavior) the same two-straight line procedure was also adopted for the c-axis data. The intersection occurs at $T_c^* = 395.5$ mK. Also the ratio of the slope of the two fitted lines (Fig. 1) is estimated to be 1.36. Although the existence of the kink at T_c^* is in general agreement with the HTS theory, the details need further exploration.

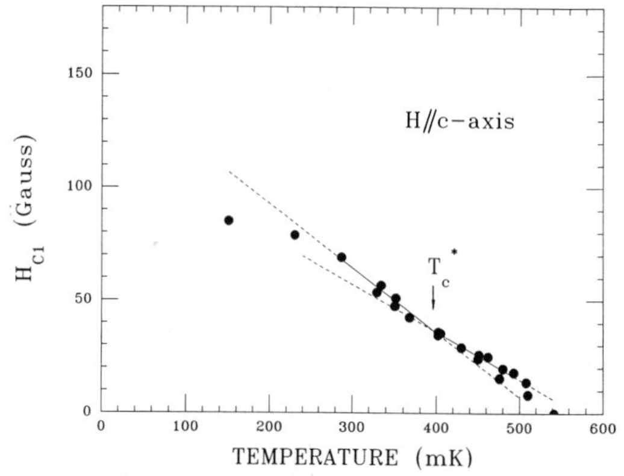

FIGURE 1
The lower critical fields vs. temperature with $\vec{H} \parallel$ c-axis. The filled circles and the straight lines are respectively the experimental data and the results of a linear least squares fit (see text).

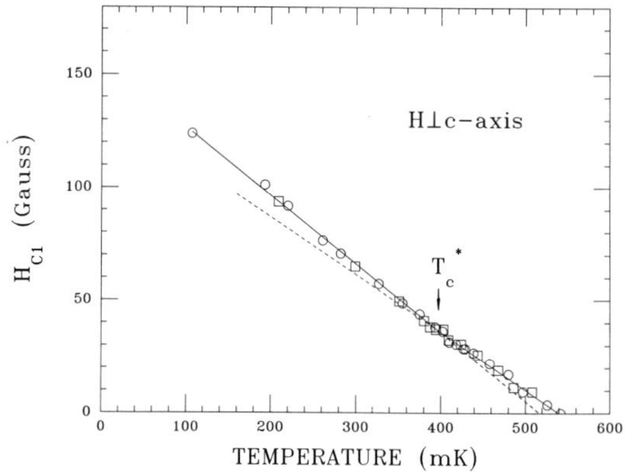

FIGURE 2
The lower critical fields vs. temperature for $\vec{H} \perp$c. The squares (circles) are the experimental data taken with $\vec{H} \parallel$ to the a-(b-) axis. The straight lines are a linear least squares fit (see text).

REFERENCES

(1) T. Oguchi and A.J. Freeman, J. Mag. and Mag. Mater. 52, 174 (1985). M.R. Norman, R.C. Albers, A.M. Boring and N.E. Christensen, Solid State Commun. 68, 245 (1988).
(2) Z. Fisk, D.W. Hess, et al. Science 239, 33 (1988).
(3) K. Hasselbach, L. Taillefer, and J. Flouquet, Phys. Rev. Lett. 63, 93 (1989). R.A. Fisher, S. Kim, et al. Phys. Rev. Lett. 62, 1411 (1989).
(4) G. Aeppli, D. Bishop, et al. Phys. Rev. Lett. 63, 676 (1989).
(5) A. Schenstrom, M-F. Xu, et al. Phys. Rev. Lett. 62 332 (1989).
(6) D.W. Hess, T.A. Tokuyasu and J.A. Sauls, J. Phys. Condens. Matter 1, 8135 (1989).
(7) B.S. Shivaram, J.J. Gannon, Jr. and D.G. Hinks, Phys. Rev. Lett. 63. 1723 (1989).
(8) F. Behroozi, Am. J. Phys. 51. 28 (1983).

Physica B 165&166 (1990) 347–348
North-Holland

STABILITY OF DOUBLY QUANTIZED VORTICES IN UNCONVENTIONAL SUPERCONDUCTORS*

T.A. TOKUYASU and J.A. SAULS

Science and Technology Center for Superconductivity and Department of Physics and Astronomy, Northwestern University, 2145 Sheridan Road, Evanston, IL 60208, USA

The mixed phase of a conventional, isotropic superconductor is well understood in terms of the Abrikosov state — a two-dimensional hexagonal lattice of vortices each supporting a single flux quantum $\Phi_o = hc/2e$. Much less is known about the vortex state of unconventional superconductors (we use the term 'unconventional' to refer to superconductors which break a discrete symmetry in addition to gauge symmetry). A strong candidate for unconventional superconductivity is the uniaxial heavy fermion system, UPt_3, which exhibits a complex phase diagram in the H *vs.* T plane [1,2]. The multiplicity of superconducting states in UPt_3 require an unconventional order parameter belonging to a higher dimensional representation (or more than one representation). For UPt_3 there are but a small number of possible unconventional states. A model that accounts for the thermodynamic properties of superconducting UPt_3 in low fields near T_c is defined by a two-dimensional 'vector' order parameter, $\vec{\eta} = (\eta_x, \eta_y)$, lying in the basal plane of the uniaxial crystal [3]. We present single vortex and vortex lattice solutions of the Ginzburg-Landau equations for a uniaxial superconductor belonging to the E_1 (or E_2) class (*i.e.* the vector representation). See Ref. [4] for details of the relevant GL equations. One of the more striking results is that the mixed state of such an unconventional superconductor is defined by an array of *doubly* quantized vortices for a large range of magnetic fields and gradient coefficients defining the GL free energy functional. However, near the upper critical field the vortex lattice is always described by a lattice of singly quantized vortices.

We have considered rectilinear vortices along the c-axis in the ground states with broken time-reversal symmetry, $\vec{\eta}_\pm \sim (\hat{x} \pm i\hat{y})e^{i\phi}$. An important class of singly quantized vortices breaks the rotational symmetry of the basal plane, giving rise to a density and current density with triangular symmetry [4].

The vortex lattice formed by these triangular vortices at low field ($H \simeq 0.06H_{c2}$) is shown in Fig. 1. Note that the broken rotational symmetry of these vortices gives rise to a short-range anisotropic vortex-vortex interaction that competes with the isotropic long-range interaction and leads to *orientational frustration* of the triangular vortices and distortion of the hexagonal lattice.

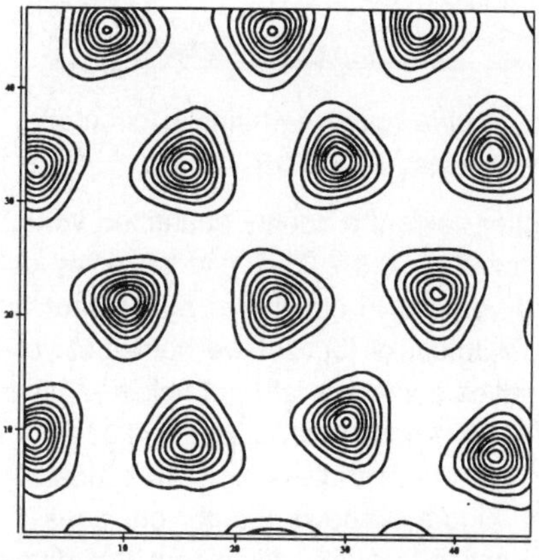

Figure 1. $|\vec{\eta}(x,y)|$ for a lattice of single quantum vortices.

While the frustrated lattice of triangular vortices is stable for a range of gradient coefficients of the GL free energy, other novel solutions exist in a large portion of the GL phase diagram. In particular, we find that a lattice of singly quantized vortices is unstable to the formation of an array of doubly quantized vortices for fields $H < 0.3H_{c2}$. Figure 2 shows a vortex array at a field of $H \simeq 0.3H_{c2}$ with $16\Phi_o$, generated by relaxing the GL equations from a starting solution of singly quantized vortices (the starting solution was the stable solution in the other region of the GL phase diagram). The large oval regions in the contour plot of $|\vec{\eta}(x,y)|$ contain two quanta of flux, while the smaller circular defect is a singly quantized vortex. The appearance of

* Supported by NSF (DMR 88–09854) through the Science and Technology Center for Superconductivity.

a singly quantized vortex among the doubly quantized vortices results from the choice of starting solution — occasionally a singly quantized vortex is left unpaired. The double quantum vortices are oval shaped only for $H \simeq 0.3H_{c2}$.

Figure 2. Double (oval) and single (circular) quantum vortices; $H = 0.3H_{c2}$.

At higher fields the doubly quantized vortices dissociate; for $H > 0.3H_{c2}$ we find an array of singly quantized vortices in qualitative agreement with the result of Zhitomirskii [5], but we have not yet identified the lattice symmetry at high fields. At low fields $H \ll 0.3H_{c2}$ the doubly quantized vortices are axially symmetric, but nevertheless have an intricate internal structure. Figure 3 shows the current density of the doubly quantized vortex. Note that the *direction* of current flow inside and outside of the vortex core (of order 10ξ in diameter) is reversed.

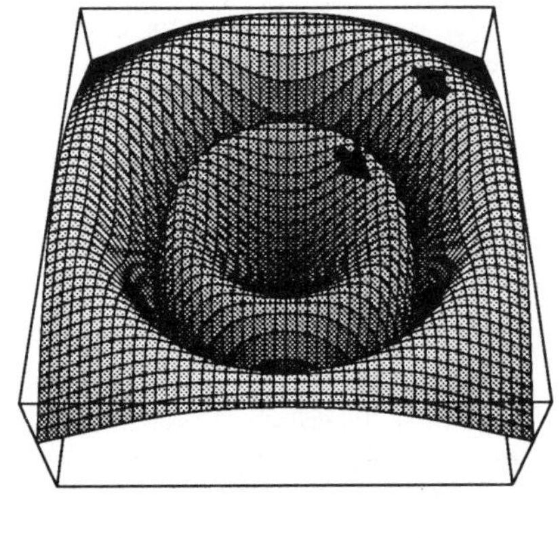

Figure 3. Current density, $|\vec{j}(x,y)|$.

This reversal of current flow is related to the presence of an order parameter in the core with *internal* orbital momentum opposite to that in the background order parameter far from the core, and gives rise to a magnetic field distribution with a *minimum* at the center of the vortex, shown in Fig. 4.

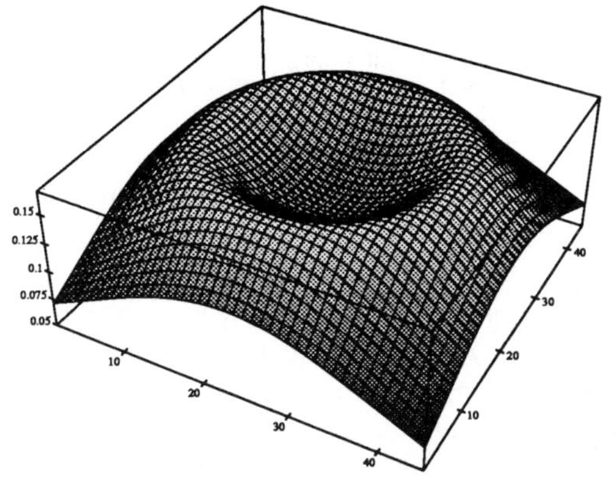

Figure 4. Field distribution, $h(x,y)$ for a two quantum vortex.

We suggest that the transition observed in high fields in UPt_3 [6,7] is a transition between a lattice of singly and doubly quantized vortices described by this two-dimensional superconducting order parameter. A Bitter pattern experiment, an STM image, or perhaps NMR linewidth should be an unmistakable test for a lattice of doubly quantized vortices.

[1] A. Schenstrom, M-F Xu, Y. Hong, D. Bein, M. Levy, B. Sarma, S. Adenwalla, Z. Zhao, T. Tokuyasu, D. Hess, J. Ketterson, J. Sauls, and D. Hinks, *Phys. Rev. Lett.*, **62**, 332 (1989).

[2] K. Hasselbach, L. Taillefer, J. Flouquet, *Phys. Rev. Lett.*, **63**, 93 (1989).

[3] D.W. Hess, T.A. Tokuyasu, and J.A. Sauls, *J. Phys. Cond. Matt.*, **1**, 8135 (1989).

[4] T.A. Tokuyasu, D.W. Hess, and J.A. Sauls, *Phys. Rev.*, **B41** (1990).

[5] M. Zhitomirskii, *Pis'ma Zh. Eskp. Teor. Fiz.*, **49**, 5144 (1989).

[6] Y. Qian, M-F Xu, A. Schenstrom, H-P Baum, J.B. Ketterson, D. Hinks, M. Levy, and B. Sarma, *Sol. State Comm.*, **63**, 599 (1987).

[7] V. Müller, C. Roth, D. Maurer, E. Schiedt, K. Luders, E. Bucher, and H. Bommel, *Phys. Rev. Lett.*, **58**, 1224 (1987).

Physica B 165&166 (1990) 349–350
North-Holland

MAGNETIZATION AND SOUND VELOCITY MEASUREMENTS NEAR THE ANTIFERROMAGNETIC TRANSITION IN THE HEAVY FERMION SYSTEM UPt_3

S. Adenwalla[1], S. W. Lin[2], Q. Z. Ran[2], Jia-Qi Zheng[1], M-F. Xu[2], K. J. Sun[3], D. G. Hinks[4], J. B. Ketterson[1], M. Levy[2] and Bimal K. Sarma[2],

[1] Physics Department, Northwestern University, Evanston, IL 60208, USA,
[2] Physics Department, University of Wisconsin-Milwaukee, Milwaukee, WI 53201, USA,
[3] NASA-Langley Research Center, MS 231, Hampton, VA 23602,
[4] Materials Science Division, Argonne National Laboratory, Argonne, IL 60439, USA.

The heavy fermion superconductor UPt_3 has a superconducting transition at 0.5 K and an antiferromagnetic transition at 5 K. We present measurements in the vicinity of the antiferromagnetic transition of both the sound velocity and magnetisation.

1. INTRODUCTION

Upt_3 is a heavy fermion superconductor in which magnetism and superconductivity coexist. Neutron diffraction measurements (1) reveal an antiferromagnetic (AFM) ordering with a Neel temperature of 5 K, below which a weak AFM moment of 0.02 μ_B develops. The experiments reveal the existence of AFM spin fluctuations with energies comparable to the superconducting transition temperature, Tc ~0.5 K. Various measurements suggest an exotic and rich phase diagram (see ref. 2) in the H-T plane. The pairing is thought to be d-wave (3,4) with a vector order parameter. It is thought that the coupling between the AFM and the superconducting order parameters gives rise to the richness of the superconducting states in UPt_3. As evidenced from neutron diffraction measurements, the antiferromagnetic ordering in UPt_3 is in the basal plane. In order to study this antiferromagnetic ordering we have performed dc magnetization and ultrasonic velocity measurements near the Neel transition. We see a weak signature in the magnetization but no resolvable feature in the sound velocity.

2. EXPERIMENTAL

Dc magnetization measurements were performed on a QUANTUM DESIGN squid magnetometer, in the temperature range 2 K to 25 K. Fields ranging from 50 gauss to 50,000 gauss were applied. The velocity was measured by a phase sensitive method, using a computer controlled MATEC MBS 8000 unit, which has a velocity resolution of the order of ~1-2 ppm. Magnetization and longitudinal and shear velocity wave measurements were performed along both the a-axis and the c-axis. The dc magnetization measurements were performed on various UPt_3 samples, all of which were known to be superconducting from earlier measurements. The measurements reported in this paper are on our first UPt_3 sample on which attenuation measurements revealed the existence of multiple superconducting phases (5).

3. RESULTS AND DISCUSSION

The magnetization measurements, reveal a very weak signature at 5 K only for fields along the c-axis. There is a slight increase in slope of the M vs T curve at 5 K and as the field is increased (200 gauss) the signature disappears. There is also a change of slope at 3.5 K. Measurements were performed on various UPt_3 samples, and the strength of the magnetization anamoly varies from sample to sample. For fields along the basal plane, the magnetization decreases with temperature upto a temperature of about 3 K where it reveals an upturn.

We cannot resolve any signature in the velocity measurements for longitudinal sound propagated along both the b and c axes at the Neel temperature, and for transverse sound along the c-axis. There is a gradual change in the velocity over this temperature range, but no abrupt change results from the AFM ordering, implying that the coupling between the AFM ordering and the lattice distortion is extremely small.

In Fig. 1 we plot the magnetization as a function of temperature. The straight line is a linear fit to the data above 5 K. There is a small increase in the magnetization. We also plot the magnetization below T_N after subtracting out the background obtained by extrapolating the straight line fit to the higher temperature data. The ratio $(\Delta M)_{max}/M$ is ~1%. On the application of a magnetic field (~200 gauss) this feature disappears (probably becomes too broad for us to resolve).

In Fig. 2 we show the magnetization data on

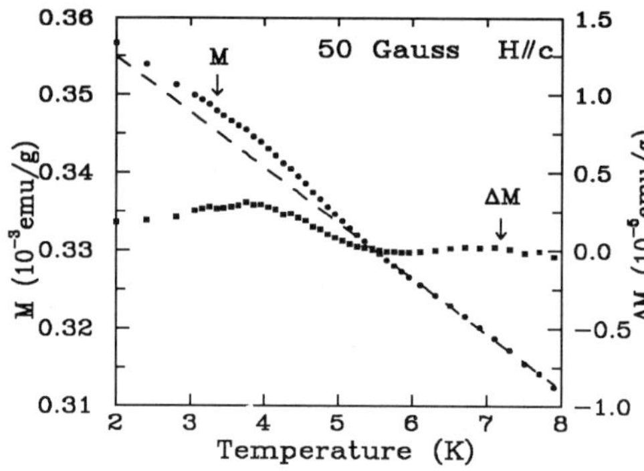

Fig. 1 dc magnetization for H//c

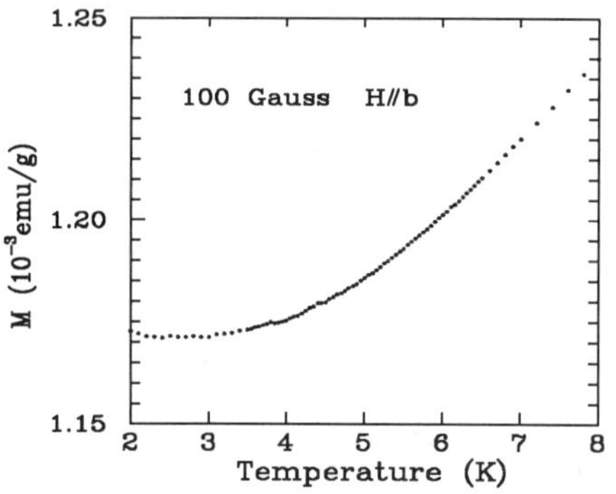

Fig. 2 dc magnetization for H⊥c

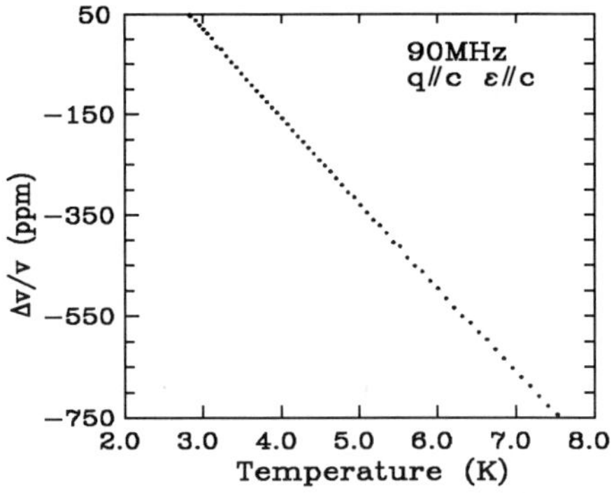

Fig. 3 longitudinal sound velocity along the c-axis

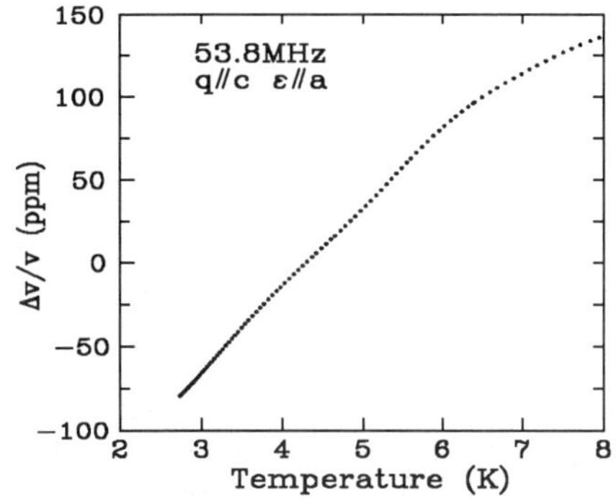

Fig. 4 transverse sound velocity along the c-axis

the same sample measured along the b-axis. Again around 5 K the magnetization seems to deviate from a straight line showing an upturn. However, one should be careful in interpreting this data, since the basal plane magnetization goes through a broad peak at ~17 K (6), which is due to a Kondo-like effect. At 5 K we are at the tail end of this peak, and the change in magnetization around 5 K may be just the effect of this peak.

In Fig. 3 we plot the longitudinal sound velocity along the c-axis. Even though the relative velocity measurements have a sensitivity of ~1-2 ppm, there is a large background velocity change, which makes it difficult to resolve unambiguously any velocity anamoly in the longitudinal sound velocity at around 5 K. In Fig. 4 we plot the shear wave velocity for sound propagating along the c-axis. We note here that the background velocity change is very small.

From these measurements it is clear that the effect of the antiferromagnetic ordering on both the magnetization and sound velocity is very small. This is to be compared to that observed near the AFM ordering in URu_2Si_2, where the ordering moment is only about twice that in UPt_3, but the magnetization and velocity anomalies are much larger (7).

ACKNOWLEDGEMENTS:

This work at was supported in parts by the Office of Naval Research (UWM), by the National Science Foundation under grant NO. DMR 89-07396 (Northwerstern) and by the Department of Energy, Office of Basic Energy Sciences-Materials Sciences Grant No. W-31-109-ENG-38.

REFERENCES

(1) G. Aeppli, et al., Phys. Rev. Lett. 58, 808 (1987). G. Aeppli, et al., Phys. Rev. Lett., 63, 676 (1989).
(2) S. Adenwalla, Z. Zhao, J. B. Ketterson, M. Levy and B. K. Sarma, "Phase Diagram of Superconducting UPt_3", this volume.
(3) R. Joynt, Supercond. Sci. Technol., 1, 210 (1988).
(4) D. W. Hess, T. A. Tokuyasu, and J. A. Sauls, J. Phys. Cond. Matt. 1, 8135 (1989).
(5) Y.J. Qian, M-F. Xu, A. Schenstrom, H-P. Baum, J.B. Ketterson, D.G. Hinks, M. Levy and Bimal K. Sarma, Solid State Commun. 63, 599 (1987).
(6) A. de Visser, A. Menovsky, J. J. M. Franse, Physica 147 B, 81 (1987).
(7) S. W. Lin, et al., "Sound Velocity and Magnetization Measurements ..on URu_2Si_2", this volume.

Physica B 165&166 (1990) 351–352
North-Holland

PHASE DIAGRAM OF SUPERCONDUCTING UPt_3

S. Adenwalla[1], Z. Zhao[1], J. B. Ketterson[1], M. Levy[2] and Bimal K. Sarma[2]

[1] Physics Department, Northwestern University, Evanston IL 60208, USA,
[2] Physics Department, University of Wisconsin-Milwaukee, Milwaukee, WI 53201, USA.

The heavy fermion superconductor UPt_3 shows evidence of a complex phase diagram in the H-T plane. A variety of experiments reveal the existence of at least two and maybe more superconducting phases. The phase transition line(s) obtained from the different experiments is (are) in qualitative agreement.

There is increasing experimental evidence showing that UPt_3 is an unconventional superconductor. Heat capacity and ultrasonic attenaution measurements show power-law temperature dependences. In addition, a variety of experiments reveal the existence of more than one superconducting phase; these include ultrasonic attenuation, torsional oscillator, neutron diffraction and heat capacity measurements.

Ultrasonic attenuation measurements reveal at least two and possibly three separate superconducting phases (1,2,3). An attenuation peak is seen in a magnetic field sweep at low temperatures at H_{FL}~0.6 H_{c2}. This is seen for all orientations of the magnetic field with respect to the hexagonal c-axis. The position of this H_{FL} peak is frequency independent, and divides the H-T plane into two regions indicating the existence of a low field and a high field phase. In addition in a temperature sweep a λ-shaped peak (2,3) is seen in certain cases just below Tc. The λ-peak is observed 20-30 mK below Tc. Both the λ-peak and the H_{FL} peak are seen only with longitudinal sound and not with transverse sound. Since longitudinal sound involves compressional waves, this phase transition must be associated with a density change. The H_{FL} phase transition line is determined by the position of a peak in the ultrasonic attenuation in a field sweep and the line of λ-peaks are detected only during temperature sweeps.

Recent heat capacity measurements (4) in a magnetic field show further thermodynamic evidence of these phases. A distinct double heat capacity jump is seen in a low field temperature sweep, showing the presence of two low field stable superconducting phases. On the application of a magnetic field, both heat capacity jumps shift to lower temperatures, and the seperation between the two decreases. In the case of the magnetic field applied along the c-axis, the two lines meet at a field of 0.5 T, whereas in the case of H//c-axis they do not meet for fields up to 0.8 T. The observation of a double heat capacity jump has been explained by Hess et al. (5) as a split superconducting transition in UPt_3. This could be caused by a symmetry breaking field, a likely candidate being the strain in the lattice arising from the antiferromagnetic ordering at 5 K. However recent heat capacity measurements by the Chicago group (6) do not show the double heat capacity jump in a temperature sweep. These measurements were done on a sample, that is a piece of one of the crystals used in the neutron scattering work (7), which showed AFM ordering.

Other experiments involving neutron diffraction (7) and torsional oscillator measurements (8) also show evidence of multiple phases. Even though there is qualitative agreement for the multiple phase diagram, there seems to be quantitative disagreemnt between the various measurements.

In Fig. 1 we plot the various sets of data that observe a signature of a phase transition, with the magnetic field H applied along the c-axis. These include ultrasonic attenuation measurements, H_{FL} (closed squares) and H_{c2} (closed circles) from ref. 1, heat capacity measurements, C_p, of ref. 4 (low field) (closed triangles, and those from ref. 6 (high field) (open triangles), and torsional oscillator measurements, TO,(diamonds), from ref 8. These measurements were made on different samples, where the transition temperatures range from 485mK to 540 mK. To account for the differences in Tc, and thermometry, the data are plotted as a function of T/Tc.

In Fig. 2 we plot the heat capacity data and the ultrasonic H_{FL} and H_{c2} data as a function of T/Tc, for the magnetic field in the basal plane. The symbols are the same as in Fig. 1. We also plot the the λ-peak features as seen in ultrasonic attenuation (open squares). It is questionable whether the λ-peak feature (2,3) and the second heat capacity jump (4) are signatures of the same phase transition.

Even though there is quantitative disagreement, it is obvious from these various

0921-4526/90/$03.50

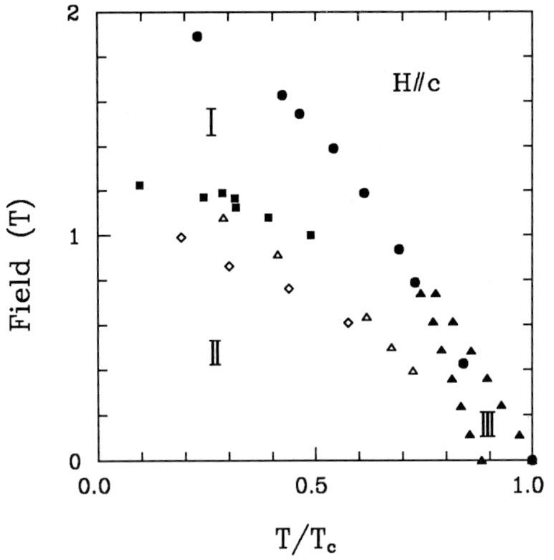

Fig. 1 UPt_3 phase diagram for H $\parallel$ c.

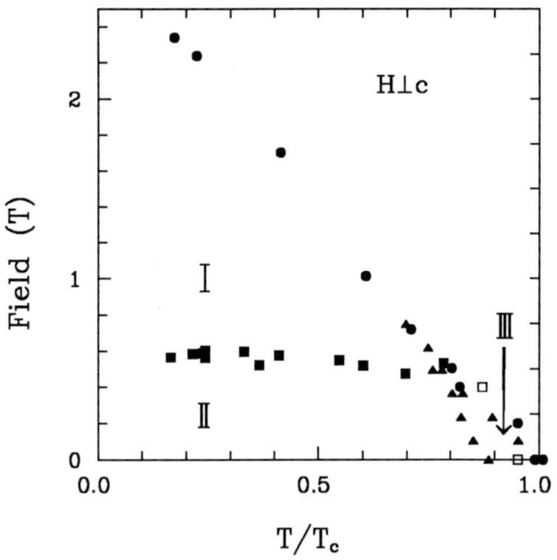

Fig. 2 UPt_3 phase diagram for H $\perp$ c.

measurements, that there are distinct superconducting phases. These are labelled I, II and III. At low temperatures there is a high field phase (I) and a low field phase (II). According to Tokuyasu and Sauls (9) the low field phase corresponds to a mixed state, with doubly quantized vortices. The high field phase has the singly quantized vortices, and the transition is a disassociation of these doubly quantized vortices.

Clearly, to obtain details of the phase diagram one must perform the various measurements on the same sample. And to ascertain the nature of the phases, a Scanning Tunneling Microscope (STM) experiment must be done. A STM experiment would in principle be able to image the structure of a single vortex core as well as the flux lattice. Neutron diffraction study to obtain details of the vortex lattice is possibly ruled out. Neutrons probe only the field distribution in the vortices, and since the London penetration depth in UPt_3 is exceptionally large (ξ~330A and λ_L~3000A), and the lattice spacing of the vortices at the H_{FL} transition is ~ξ, the experiment will not be sensitive enough.

ACKNOWLEDGEMENTS

We thank S. W. Lin for help in the preparation of this manuscript. The work was supported in parts by the Office of Naval Research (at UWM) and the National Science Foundation under grant No. DMR-8907396 (at Northwestern).

REFERENCES

(1) Y.J. Qian, M-F. Xu, A. Schenstrom, H-P. Baum, J.B. Ketterson, D.G. Hinks, M. Levy and Bimal K. Sarma, Solid State Commun. 63, 599 (1987).

(2) V. Muller, D. Maurer, E.W. Scheidt, K. Luders, E. Bucher and H.E. Bommel, Phys. Rev. Lett. 58, 1224 (1987).

(3) A. Schenstrom, M-F. Xu, Y. Hong, D. Bein, M. Levy, B.K. Sarma, S. Adenwalla, Z. Zhao, T.A. Tokuyasu, D.W. Hess, J.B. Ketterson, J.A. Sauls and D. Hinks, Phys. Rev. Lett. 62, 332 (1989).

(4) K. Hasselbach, L. Taillefer and J. Flouquet, Phys. Rev. Lett. 63, 93 (1989); L. Taillefer, preprint.

(5) D. W. Hess, T. A. Tokuyasu and J. A. Sauls, J. of Phys. Condens. Matter. 1, 8135 (1989).

(6) B. Ellman, J. Yang, T.F. Rosenbaum and E. Bucher, preprint.

(7) G. Aeppli, D. Bishop, C. Broholm, E. Bucher, K. Siemensmeyer, M. Steiner and N. Stusser, Phys. Rev. Lett. 63, 676 (1989).

(8) R.N. Kleiman, P.L. Gammel, E. Bucher and D.J. Bishop, Phys. Rev. Lett. 62, 328 (1989).

(9) T.A. Tokuyasu and J.A. Sauls, this volume.

Physica B 165&166 (1990) 353–354
North-Holland

SPECIFIC HEAT OF NON-STOICHIOMETRIC, NON-SUPERCONDUCTING $CeCu_2Si_2$ UNDER PRESSURE

Andreas EICHLER, Kay-Rüdiger HARMS, and Friedrich-Wilhelm SCHAPER

Inst. Tech. Physik, TU Braunschweig, D-3300 Braunschweig, Fed. Rep. Germany

Polycrystalline samples of $CeCu_2Si_2$ with a Cu deficiency of about 5 %, which are not superconducting at normal pressure, have been investigated by specific heat measurements under pressure up to 9.5 kbar. The pressure dependence of the electronic specific heat and of the Kondo temperature is markedly reduced from the high values found in a former investigation of superconducting material not deficient in Cu. No traces of pressure induced superconductivity could be detected, not even in ac-susceptibility measurements up to 20 kbar and temperatures down to 120 mK.

1. INTRODUCTION

The heavy fermion (HF) compound $CeCu_2Si_2$ is especially interesting because of becoming superconducting (sc). There is no doubt that the heavy quasiparticles take part in the superconductivity in this material.(1) The mechanism, however, leading to superconductivity is by no means evident. Experimental search for correlations between sc parameters and such of the HF state might help to find the right theory; it has shown a strong dependence of both normal and sc properties on deviations from stoichiometry, purity, and lattice order.(3,4)

High pressure experiments are believed to be a valuable tool in the investigation of HF materials mainly for two reasons: 1) the purity or chemical composition of a specific sample is unaffected by pressure (p), and 2) the characteristic energies of the HF state vary considerably with p. Thus, it has been found that the Kondo temperature T_K of sc $CeCu_2Si_2$ rises rapidly with p, corresponding to a Grüneisen-parameter $\Gamma_K = -d\ln T_K/d\ln V = 60 - 70$.(5,6) It has been shown by Razafimandimby et al.(RFK, 7) that this large Γ_K determines the electron-phonon coupling parameter λ, and, therefore, is finally responsible for the superconductivity. Their estimates lead to the right order of magnitude for T_c.

Bleckwedel (8) could show that the trend of $T_c(p)$ can be correlated to $T_K(p)$ within the model of RFK if the pressure dependence of Γ_K is considered, too. The sign of dT_c/dp turned out to be very sensitive on the balance between the first and the second derivative of $T_K(p)$.

The present investigation on $CeCu_{1.9}Si_2$ was performed with mainly two objectives: 1) we wished to study the HF properties of this material under pressure, compare them to our former results on sc $CeCu_2Si_2$ (6), and look for any peculiarities in correlation with the absence of superconductivity; 2) we wanted to search for p-induced superconductivity like in the single crystal material of ref.(5), which was supposed to have grown with a slight deficiency in Cu from a stoichiometric melt.(4)Such a finding of p-induced superconductivity would support the idea of the existence of a critical volume.(4)

2. EXPERIMENTAL DETAILS AND RESULTS

The sample of composition $CeCu_{1.9}Si_2$ was made available by the Darmstadt group and had comparable properties to that of the sample in ref. (3). Specific heat was measured under pressure in the same way as with the sc material (6), using an ac-method described earlier.(9) These experiments were supplemented by measurements of the ac-susceptibility on a larger piece of material of the same batch extending to 20 kbar. With a small dilution refrigerator calorimetry was possible from ≃0.2 K to ≃2 K.

Original data of c(T,p) like those shown in fig. 1 have been measured at seven different pressures. The data at the lowest p compare very well to those of ref.(3) at p = 0, both in absolute value and in relative trend with T. The overall shape of the curves is conserved under pressure, while the absolute value decreases with p. This qualitative trend was expected and is similar to that found in the sc material. The quantitative effect, however, is much smaller.

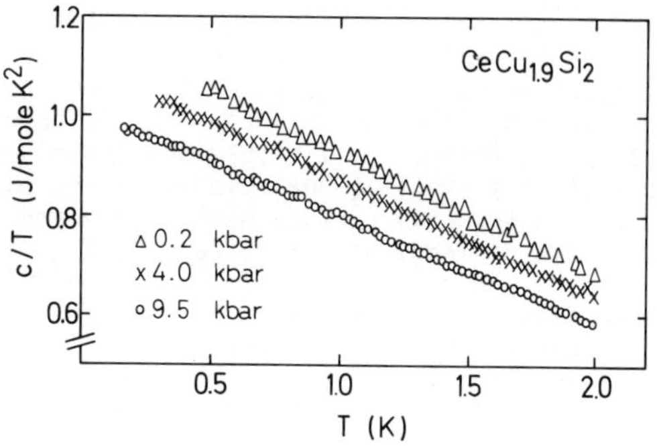

Fig. 1. Specific heat of $CeCu_{1.9}Si_2$at three different pressures.

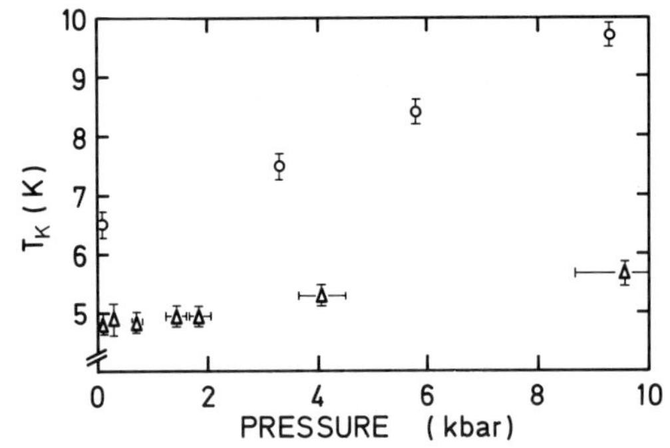

Fig. 2. Pressure dependence of the Kondo temperature of $CeCu_2Si_2$ as derived from the electronic specific heat coefficient.
o: sc $CeCu_2Si_2$ (6), Δ: non-sc $CeCu_{1.9}Si_2$.

Extrapolating the c/T values to T=0 and converting to Kondo temperatures using the single ion model (10) leads to the data shown in fig. 2. $T_K(p)$ can be well approximated linearly with $\Gamma_K = -\, d\ln T_K/d\ln V = 17$, which is ≃4 times smaller than that of superconducting $CeCu_2Si_2$.

c/T vs. T does not show any significant structures (fig. 1). $CeCu_{1.9}Si_2$ is no bulk superconductor at p=0. Our results prove that p does not induce superconductivity for T > 0.2 K; this is supported by ac-susceptibility data, sensitive to small traces of sc material, up to 20 kbar.

3. DISCUSSION

The most surprising result is the so much smaller dependence of c on p as compared to the sc material. This shows once more the sensitivity of HF properties on small changes in composition; as an open question remains, however, why Γ_K is even smaller than in many diluted Kondo systems. We are not able to give an explanation, but we think it might have to do with the absence of coherence effects, which should show up in a maximum in c/T. As known (3,4), coherence is not fully developped in $CeCu_{1.9}Si_2$. In $CeAl_3$ the coherence structure is extremely sensitive on p, being already washed out below 1 kbar (11) which corresponds to a closing of the pseudogap in the quasiparticle band structure. This band structure, we think, still determines the p dependence of the specific heat even when the coherence structure has already been eliminated, and may lead to large values of Γ_K. In sc $CeCu_2Si_2$ the coherence structure is buried in the transition to the sc state, but at the lowest pressure in our former paper (6) vestiges of a structure were seen. Furthermore, the observed variation of $\Delta c(T_c)$ with p was interpreted to reflect the closing of the pseudogap. It is plausible, therefore, to assume a contribution of the coherence BS to the large Γ_K in sc $CeCu_2Si_2$. On the other hand, the lack of coherence may be one reason for the relative smallness of Γ_K in $CeCu_{1.9}Si_2$.

The absence of superconductivity may be a direct result of the lack of coherence. Up to now, there is no theoretical proof, however, that coherence is necessary for superconductivity.

The idea, that superconductivity is observed in $CeCu_2Si_2$ only if the specific volume is smaller than a critical value, cannot be ruled out by our results, since 20 kbar may not be enough to reach this estimated critical volume (4). Based on our data of $T_K(p)$, however, we think it is rather unlikely that superconductivity might be induced by pressure. From the model of RFK (7), and allowing for a p dependence of the parameters (8), we find above all that dT_K/dp is too small to lead to measurable values of T_c. Moreover, from the approximate linearity of $T_K(p)$ we derive a decrease of T_c with pressure using the empirical relationship

$$d\ln T_c/dp = \{1 - 1/\lambda(p)\}\cdot d\ln T_K/dp + \{2/\lambda(p)\}\cdot T_K''(p)/T_K'(p)$$

with $T_K'(p) = dT_K/dp$ and $T_K''(p) = d^2T_K/dp^2$. Only a very rapid upturn of $T_K(p)$ at p > 20 kbar might bring up T_c to measurable values.

ACKNOWLEDGEMENTS

We are very grateful to Prof. Steglich and his coworkers for leaving the sample material and helpful discussions. Thanks are also due to Prof. Gey for making the dilution refrigerator available to us. Support by the Deutsche Forschungsgemeinschaft is gratefully acknowledged.

REFERENCES

(1) For a review on $CeCu_2Si_2$ and HF see (2-4).
(2) F. Steglich, J.Phys.Chem.Solids 50 (1989) 225.
(3) C.D.Bredl, W.Lieke, R.Schefzyk, M.Lang, U. Rauchschwalbe, F.Steglich, S.Riegel, R.Felten, G.Weber, J.Klaase, J.Aarts, and F.R. de Boer, J.Magn.Magn.Mat. 47&48 (1985) 30.
(4) U.Rauchschwalbe, Physica 147 B+C (1987) 1 and references herein.
(5) F.G.Aliev, N.B.Brandt, V.V.Moshchalkov, and S.M.Chudinov, Solid State Commun. 45 (1983) 215.
(6) A.Bleckwedel and A.Eichler, Solid State Commun. 56 (1985) 693.
(7) H.Razafimandimby, P.Fulde, and J.Keller, Z. Phys. B 54 (1984) 111.
(8) A.Bleckwedel, Thesis, TU Braunschweig 1986.
(9) A.Bleckwedel and A.Eichler, in: Physics of Solids under High Pressure, eds. J.S.Schilling and R.N.Shelton (North-Holland, Amsterdam, 1981) p. 323.
A.Eichler and W.Gey, Rev. Sci. Instrum. 50 (1979) 1445.
(10) V.T.Rajan, Phys. Rev. Lett. 51 (1983) 308.
(11) G.E.Brodale, R.A.Fisher, N.E.Phillips, and J.Flouquet, Phys. Rev. Lett. 56 (1985) 390.

Physica B 165&166 (1990) 355–356
North-Holland

Anisotropic field dependence of thermal conductivity in superconducting UPt3

K. BEHNIA, L.TAILLEFER , J. FLOUQUET

Centre de Recherches sur les Très Basses Températures, CNRS - BP 166 X F- 38042 Grenoble cedex, France

D. JACCARD

Départemant de Physique de matière condensée, Université de Genève, 1211 Genève 4, Switzerland

and Z. FISK

Los Alamos National laboratoaries, Los Alamos, New Mexico 87545, USA

The thermal conductivity of the heavy fermion superconductor UPt_3 has been measured as a function of magnetic field in a single crystal for a heat current along the c-axis. It is found to depend strongly on the relative orientation of current and field. Moreover for a field in the basal plane, anomalies are observed along a line in (H-T) plane where anomalies in ultrasonic attenuation have been reported.

Several recent experiments have revealed the existence of anomalies in the temperature and magnetic field dependence of various physical properties of superconducting UPt_3 (1,2,3,4) which have been attributed to a multiplicity of superconducting phases (see for example ref. 5). Such a complex phase diagram has come to be the strongest argument in favor of an unconventional superconductivity in this compound. Howevever, the right symmetry class of what seems to be a multi-component superconducting order para–meter has not been determined. In this respect, thermal conductivity measurements on single crystals may prove particularly useful as, for example, they can lead to a precise identification of the superconducting order parameter in the case of a resonant impurity scattering (6,7). Previous measurements of the thermal conductivity (8), performed on a polycrystal, have revealed an unusual T^2 dependence at low temperatures pointing to an unconventional gap function in UPt_3. In this paper we present the first part of a detailed study of the thermal conductivity on single crystals, where the different configurations for the relative orientation of heat current, crystalline axes and magnetic field are investigated.This first part deals with the field dependence of the thermal conductivity (κ) when the heat current is applied along the c-axis.

The sample used here was a monocrystalline whisker, 4 mm in length with a cross-sectional area of 0. 2 x 0.2 mm^2. Two RuO_2 thermometers were used to measure the temperature difference between two points 2.5 mm apart along the whisker .To investigate the field dependence of κ, the temperature of the cold finger of a dilution refrigerator fixed at one end of the sample was kept constant and the thermal gradient arising from an invariant heat source installed on the other end was measured as a function of magnetic field. The temperature difference along the sample was always between 5 to 12 mK Thus, the change in average sample temperature is small and does not yield any significant alteration of the general behaviour of the κ(H) curves. The experimental set-up was realised in a way to permit a resistivity measurement of the sample in the same conditions.

At zero field, the residual resistivity (ρ_0) was found to be 0.8 $\mu\ \Omega$ cm (with $\rho=\rho_0+AT^2$) of and the resistive critical temperature 560 mK. The Lorentz number L (defined as $\kappa\rho$ /T) was measured to be equal to 2.5 X 10^{-8} $\Omega W/K^2$ at H= 30KOe and T= 100 mK very close to the sommerfeld value (L_0) . At zero field , just above the transition a L/L_0 ratio of 0.8 was found which tends to confirm that at the onset of the superconductivity the phonon contribution to the thermal conductivity can be neglected when compared to the electronic term.

Figure 1 shows the field dependence of κ for two orientations of the magnetic field at 210 mk. There are two salient features in this data The first one is the apparent anisotropy of the thermal conductivity in the superconducting phase compared to a fairly isotropic behaviour in the normal state. What is rather surprising is the sign of this anisotropy: The conductivity is enhanced when the field , **H**, (that is, the vortices) and the heat current, $\mathbf{J_Q}$, are perpendicular. According to a simplified picture, one could expect a short-circuiting of the superconducting contribution when the vortex cores are parallel to the heat current , leading to a better heat conduction in the latter case. However the opposite is observed experimentally, which implies either an anisotropy of the electron scattering, which is sensible to the field direction, or the existence of an additional heat carrier only present in the $\mathbf{H \perp J_Q}$ geometry. In the latter case, one such possiblity is a heat transport mechanism associated with the displacement of vortices under the influence of a thermal force $\mathbf{F}=-S_D \nabla T$; where S_D is the entropy of a single moving flux line. The flux flow is then accompanied by a transport of entropy proportional to the velocity and density of moving vortices. To be taken seriously, such a process would have to be reconciled with the strong pinning of vortices one is naturally led to infer from the virtual absence of a meissner effect (5).

The second point of interest in the data of figure 1 is the sudden break in slope for the transverse configuration ($\mathbf{H \perp J_Q}$, $\mathbf{H \perp c}$) occuring at a field H* which varies only slightly with temperature. Indeed, H*= 6.1, 5.6, 5.3 at T= 110, 210, 310 mk, respectively. The anomalies seem to coincide with peaks in the field dependence of the ultrasonic attenuation for the same configuration reported by Schenstrom and his collaborators(1). The quasi-horizontal line in the (H-T) diagram on which these two types of anomalies lie is thought to correspond to a field-induced phase transition, evidence for which have been also found in torsional oscillator experiments (2) for H||c configuration. Before further experiments with the heat current in the basal plane, it is too early to say whether this transition involves a change in vortex lattice rather than a change in the symmetry of the order parameter.

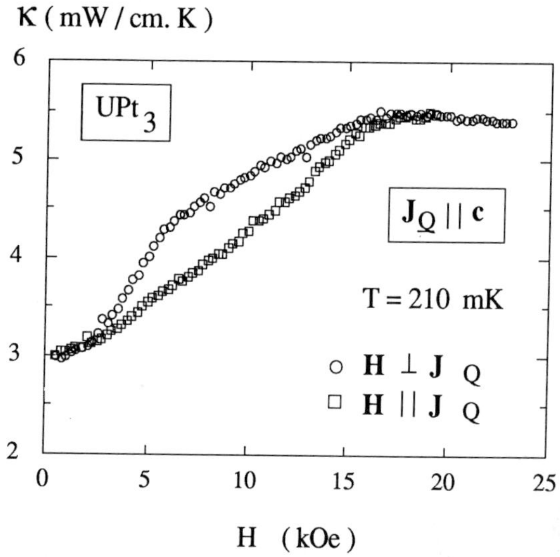

Figure 1- The field dependence of the thermal conductivity of UPt_3 for a field parallel (●) and normal (○) to the heat current .Note the anomaly at H* for the $J_Q \perp H$ curve.

ACKNOWLEDGEMENTS

K.B. acknowledges helpful and stimulating discussions with A. Buzdin and H. Jensen.

References:

(1) A.S. Schenstrom, M-F. Xu,Y. Hong, D. Bein, M. Levy, B.K. Sarma, S.Adenwalla, Z. Zhao, T. Tokuyasu, D.W. Hess, J.B. Ketterson, J.A. Sauls and D.J. Hinks, Phys. Rev. Lett. **62**, 332 (1989).
(2) R.N.Kleiman, P.L.Gammel, E.Bucher, and D.J.Bishop, Phys. Rev. Lett. **62**,328 (1989)
(3) R.A. Fisher, S. Kim, B.F. Woodfield, N.E. Phillips, L. Taillefer, K. Hasselbach, J. Flouquet, A.L. Giorgi and J.L. Smith, Phys.Rev.Lett. **62**, 1411 (1989).
(4) K. Hasselbach, L. Taillefer, J. Flouquet, Phys. Rev. Lett. **63**, 93 (1989).
(5) L.Taillefer,Proc. ICPHCES (Santa Fe, Sept . 1989) To appear in Physica B.
(6) P.J.Hirschfield, P.Wölfle, and D. Einzel, Phys. Rev. B **37,** 83 (1988).
(7) B. Arfi, H.Bahlouli and C.J. Pethic, Phys. Rev. B **39** , 8959 (1989).
(8) A.Sulpice, P.Gandit, J.Chaussy, J.Flouquet, D.Jaccard, P.lejay and J.L.Tholence ,J. Low Temp. Phys. 62, 39 (1986)
(9) R.Joynt, Supercond. Sci. Technol. (1988) 210

Physica B 165&166 (1990) 357–358
North-Holland

THE PHASE DIAGRAM OF SUPERCONDUCTING UPt_3 FOR B // c

K. Hasselbach, L. Taillefer and J. Flouquet

Centre de Recherches sur les Très Basses Températures, CNRS, BP 166X, 38042 Grenoble-Cedex, France

We have measured the specific heat (0.3 < T < 0.6 K) of superconducting UPt_3 in an external magnetic field (B< 1T) applied along the c-axis of the hexagonal crystal. We studied the field dependence of the two superconducting phases reported before in zero field (1). The phase diagram for B//c has a critical point at 0.75 tesla and 0.375 K.

1. INTRODUCTION

The coexistence of superconductivity and highly correlated electrons in the heavy fermion compounds has intrigued many researchers in the field of low temperature physics. The hexagonal UPt_3 belongs to this class of compounds. Its anisotropic structure is reflected in its physical properties. Recently specific heat measurements have proven the existence of two superconducting transitions at T_{c1} and T_{c2}, about 60 mK apart (1). The splitting of the superconducting transition has been ascribed to the lifting of the order parameter's degeneracy by a small symmetrie breaking field (6,8). In a previous paper (4) we have investigated the field dependence of the double transition for a field in the basal plane (B⊥c). Our specific heat data yield a kink in $H_{c2}(T)$ at $(T^*, B^*_\perp)$=(0.37K ,0.5 tesla). The upper transition at T_{c2} in the specific heat disappears at $(T^*, B^*_\perp)$. A kink in H_{c2} for B⊥c has previously been observed by resistivity measurements (5). Further anomalies in the field dependence of the superconducting properties of UPt_3 have been established by ultrasound measurements (2,3). A peak of absorption has been observed within the superconducting phase. For B⊥c the peak appears at 0.5 tesla and is nearly temperature independent. Combining the information from specific heat, resistivity and ultrasound we presented a superconducting phase diagram of UPt_3 for B⊥c (4). It contains three superconducting phases and a tetracritical point. From theoretical work (7) it has been inferred that the small symmetrie breaking field could be the antiferromagnetic moment (~0.02 μ_B) in the plane coupling to the vectorial order parameter. In this paper we investigate by means of specific heat measurements the multiplicity of the superconducting phase for a field along the hexagonal axis.

2. EXPERIMENTAL

The experiments have been performed on the same high purity UPt_3 sample which was studied in (4). The monocrystalline sample (270 mg) is shaped in form of a disc. The c-axis is tilted 25 degrees out of the disc plane. The sample was mounted with grease on a silicon support, which carried also a Si P-doped thermometer. The heat capacity was measured by a heat puls relaxation technique. The heat capacity of the sample holder is estimated to be inferior to 3 percent of the total heat capacity at 0.5 K and is neglected in the present experiment.

3. RESULTS

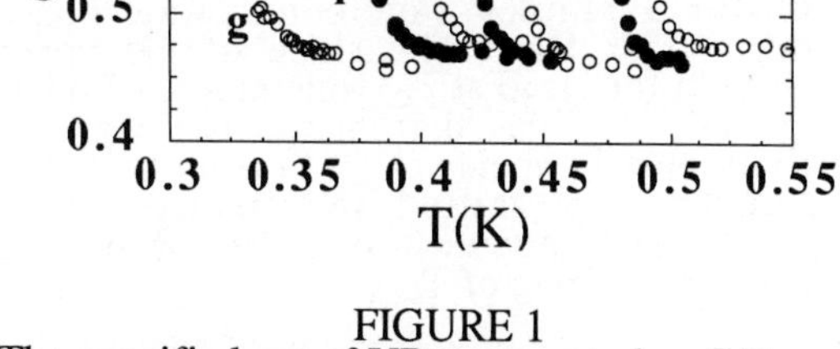

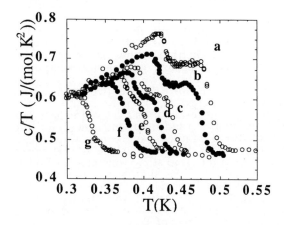

FIGURE 1
The specific heat of UPt_3 presented as C/T versus T for various applied fields B along the hexagonal axis: a->0, b->0.125, c->0.375, d->0.5, e->0.625, f->0.75, g->1 tesla

In Fig. 1 we show the specific heat of UPt_3 for $0.3<T<0.6$K and for magnetic fields ($B\leq 1$ T) along the c-axis. The sharp double structure at B=0 shifts towards lower temperatures with increasing field and is clearly distinguishable up to 0.625 T. At 0.75 T an inflection point is still discernable. For higher fields the upper transition at T_{c2} has either vanished or became too small as we infer from the continuous decrease of the overall jump height.

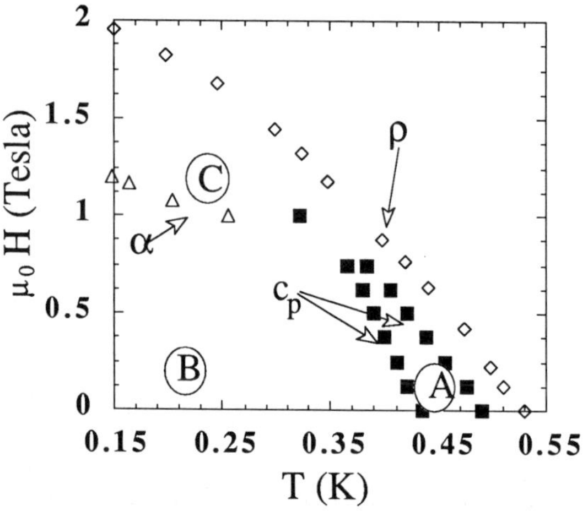

FIGURE 2
Full squares represent the midpoint of every jump in the specific heat, triangles indicate the line of anomalies detected by ultrasound attenuation α (2), diamonds indicate the H_{c2} line established through resistivity measurements on high quality single crystals (5), A, B, C indicate the phases according to (4).

Note in particular that the discontinuity in the specific heat at T_{c2} , the high temperature superconducting transition, shows a steep initial decrease and an apparent quasiconstant value up to its collapse.In Fig. 2 we show T_{c1} and T_{c2} as function of field for B//c. The two specific heat anomalies merge at 0.35 K , and 0.75T. Combining our results from the specific heat data with resistivity (5) and ultrasound data (2) for B//c we obtain a similar superconducting phase diagram as for B⊥c. The slope of the upper critical field for $H\rightarrow 0$ at T_{c2} amounts to -7.5T/K and -7.7T/K as deduced from specific heat and resistivity measurements. $H_{c2}(T)$ from the latter measurements is shifted to slightly higher temperatures because of the different experimental probing of T_c.

4. DISCUSSION

The collection of Data in Fig. 2 suggests that also for B//c a tetracritical point is found in the phase diagram. The critical point is shifted to higher fields and lower temperatures ($T^*,B^*_{//}$)= (0.35K,0.75T), where as ($T^*,B^*_{\perp}$) = (0.37K,0.5T). This would imply that H_{c2} for B//c also exhibits an anomaly at ($T^*,B^*_{//}$). Resistivity measurements did not show up to now such an anomaly, although a faint indication of it may be visible in the data of (5). Magnetoresistance measurements are in progress. The similarity of the phase diagrams for B//c and B⊥c suggests that the same phases are present for both field directions, insinuating a much weaker field orientation effect of the phases than suggested in (7). This puts some further question marks as to the applicability of the weak antiferromagnetic order as the symmetrie breaking field. In order to shed more light on this it is important to investigate the field orientation dependence of the symmetrie breaking field.

ACKNOWLEDGEMENTS
K. H. thanks the European Community for its support.The authors thank A. de Visser for discussion.

REFERENCES

(1) R.A. Fisher, S.Kim, B.F. Woodfield, N.E.Phillips, L. Taillefer, K. Hasselbach, J. Flouquet, A. L. Giorgi and J.L. Smith, Phys. Rev. Lett 62 (1989) 1411.
(2) A. Schenstrom, M-F. Xu, Y. Hong, D. Bein, M. Levy, Bimal K. Sarma, S. Adenwalla, Z. Zhao, T. Tokuyasu, D. W. Hess, J. B. Ketterson and J. A. Sauls, Phys. Rev. Lett. 62 (1989) 332.
(3) V. Müller, Ch. Roth, D. Maurer, E.W. Scheidt,K. Lüders, H. E. Bömmel, Phys. Rev. Lett. 58 (1987) 1224.
(4) K. Hasselbach, L. Taillefer, J. Flouquet, Phys. Rev. Lett.63 (1989) 93.
(5) B.S. Shivaram, T. F. Rosenbaum, D. G. Hinks, Phys Rev. Lett 57 (1986) 1259.
(6) R. Joynt, V.P. Mineev, G.E. Volovik, M.E. Zhitomirsky submitted to Phys. Rev. B.
(7) E. I. Blount, C. M. Varma and G. Aeppli, Santa Fe Conference, to appear in Physica B.
(8) D. W. Hess, T. A. Tokuyasu and J.A. Sauls, J. Phys.: Condens. Matter 1 (1989) 8135.

Physica B 165&166 (1990) 359–360
North-Holland

MAGNETORESISTANCE OF THORIATED UBe_{13}

B. A. ALLOR and M. J. GRAF

Dept. of Physics, Boston College, Chestnut Hill, MA 02167 USA

J. L. SMITH and Z. FISK

Center for Materials Science, Los Alamos National Laboratory, Los Alamos, NM 87545, USA

We have measured the magnetoresistance for the heavy-fermion system $U_{1-x}Th_xBe_{13}$ for x=0, 0.006, 0.015, and 0.033 at temperatures between 1.2 K and 6K. The samples with x=0.006 and 0.015 develop a positive magnetoresistance as the temperature is *increased* above approximately 2 K. We observe the large negative magnetoresistance for the pure UBe_{13} and the sample with x=0.033, in agreement with the results of previous studies.

1. INTRODUCTION

Studies of the magnetoresistance (MR) have been useful in describing the properties of heavy-fermion systems. For many of the Ce-based compounds, such as $CeCu_{2.2}Si_2$ and $CeAl_3$, the high-temperature ($T > 1K$) negative MR characteristic of incoherent Kondo scattering of the conduction electrons gives way at lower temperatures to a low-field positive MR This has been interpreted as the onset of coherent Kondo scattering from the magnetic lattice sites. This explanation agrees with measurements of the specific heat.(2) The situation is much different for UBe_{13}, however, where the strong negative MR has been observed down to the lowest temperatures.(3,4) Samples studied with 3.3 % thorium substituted for uranium show similar behavior.(4) This has been attributed to the fact that the coherent Kondo state is not fully formed at temperatures above T_c (about 1 K), an interpretation that has been supported by studies of MR above 1 K and at high pressures,(5) and possibly by high-field, low-temperature specific heat measurements(6).

In this work we present measurements of the magnetoresitance of $U_{1-x}Th_xBe_{13}$ and find a *positive* MR for samples with x = 0.006 and 0.015 at temperatures above about 2-3 K, while the results for x= 0 and 0.033 show the typical strong negative MR at all temperatures.

2. EXPERIMENT

2.1 Samples

The four samples studied are polycrystalline with thorium concentrations of 0, 0.006, 0.015, and 0.033. Magnetization measurements show that all four samples exhibit superconducting transitions at temperatures of 0.90 K, 0.71 K, 0.52 K, and 0.57 K, respectively. The superconducting transitions have also been observed in specific heat measurements on the pure UBe_{13},(6) the x=0.015 sample, and 2 transitions for the x=0.033 sample. Specific heat measurements have not yet been made on the x=0.006 sample.

2.2 Procedure

Transverse magnetoresistance measurements were made on the four samples in the temperature range of 1.2 - 6 K, and for fields up to 9 T. Four gold wires of 0.002" diameter provided electrical contact to the samples, which were mounted on a copper block. The block was thermally connected to a ^4He pumped pot by a 0.25" diameter brass rod. The temperature was determined by a calibrated carbon-glass resistor, and was varied by applying fixed current to a (field-independent) heater on the copper block. The resistance was measured using the standard 4-terminal ac technique, with frequencies between 15 and 20 Hz and rms current levels of less than 1 mA. The current dependence of the measured voltage across the samples was measured to ensure that no self-heating of the samples occured. Magnetic fields were applied perpendicular to the current flow using a 9 T superconducting magnet, and were ramped at a rate of approximately 0.3 T/min. No hysteresis effects between up and down sweeps was observed. The temperature for each field sweep was determined by measuring the calibrated thermometer just before and after the field sweep, and the observed temperature drift was always less than 1 % .

3. RESULTS

The resistivities of the samples at 4.2 K are: $\rho(x{=}0) = 130\ \mu\Omega$-cm, $\rho(x = 0.006) = 170\ \mu\Omega$-cm, $\rho(x = 0.015) = 110\ \mu\Omega$-cm, and $\rho(x = 0.033) = 110\ \mu\Omega$-cm. The estimated error is approximately 10 % . We have measured the resistivities versus temperature between 1.2 K and 6 K and find the characteristic shoulder for pure UBe_{13} at 2.5 K. This structure is suppressed to 2 K for the x= 0.006 thoriated sample, and is not observed at all for the samples with higher thorium concentrations.

We now discuss the MR measurements. The results for x=0 and 0.033 agree well with previous measurements.(3,4) For both samples we observe a large negative MR that becomes more pronounced at lower temperatures (a reduction of approximately 50 % at

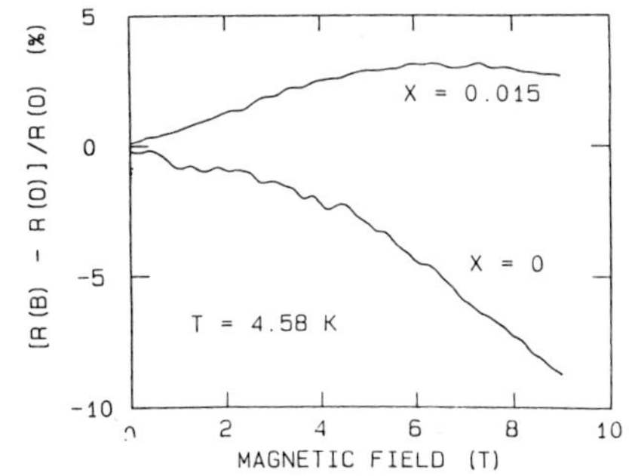

Figure 1
The fractional change in the resistance of UBe_{13} and of $U_{0.985}Th_{0.015}Be_{13}$ vs. magnetic field at T= 4.58 K.

1.2 K for UBe_{13}, and of about 20 % for x= 0.033 at 1.3 K). This MR is shown for UBe_{13} at 4.58 K in Figure 1. The behavior for the samples with *intermediate* thorium concentrations is more complicated. At low temperatures we observe the typical large (reductions of about 25-30 %), negative MR. As the temperature is raised, however, a low-field, positive MR becomes apparent, and results in a peak in the MR. This is also shown in Figure 1 for the sample with x= 0.015. In Figure 2 we show the temperature progression of the MR for the sample with x=0.015, and the development of a positive MR as the temperature increases is evident. In Figure 3 we have plotted the peak location in field, H_{max}, versus temperature for the x= 0.006 and 0.015 samples. As the temperature increases, this crossover field increases. The effect is evident at lower temperatures for the sample with higher thorium concentration, indicating that the effect is probably intrinsic; however we note that the positive MR is *not* observed at any temperature studied here for the x=0.033 sample.

4. DISCUSSION

The development of positive MR for samples with intermediate thorium impurities is difficult to understand. Clearly the nature of this effect is different than for $CeAl_3$, where the positive MR develops as the temperature is *lowered*, and is interpreted as signaling the onset of coherent Kondo scattering of electrons. One might conclude that the thorium impurities are simply

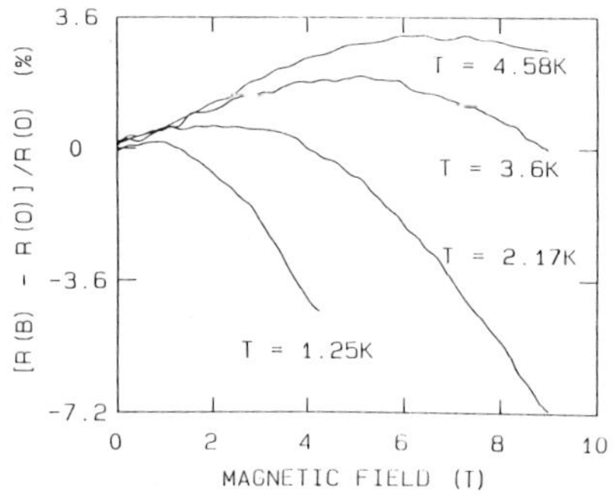

Figure 2
The fractional change in resistance of $U_{0.985}Th_{0.015}Be_{13}$ vs. magnetic field at several different temperatures.

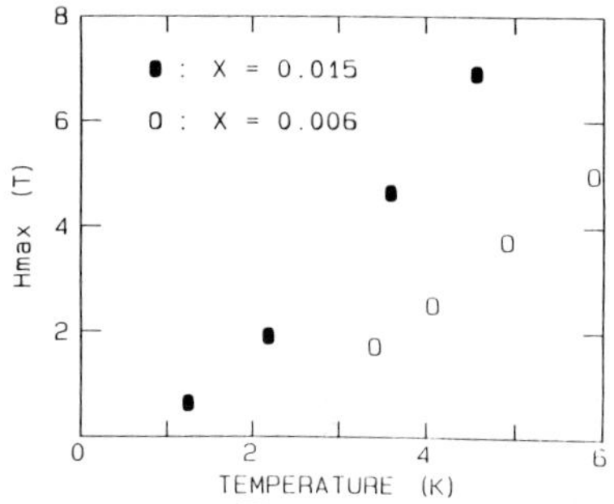

Figure 3
The variation of the field at which the maximum in magnetoresistance occurs with temperature for samples with thorium concentrations of x= 0.006 and x= 0.015.

weakening the Kondo scattering, and so at higher temperatures these samples are behaving more like normal metals. While the impurities could have an appreciable effect on coherent scattering by destroying the periodicity of magnetic scatterers, low levels of impurities should have little effect on incoherent Kondo scattering. Also, the large negative MR characteristic of incoherent Kondo scattering is observed once again for the sample with 3.3 % thorium. It is interesting to note that of the three thoriated samples studied here, the two which show positive MR have concentrations which, at low temperatures, have one superconducting transition, while the sample which shows negative MR has a concentration corresponding to two transitions at low temperatures.(7) Clearly more work is necessary to better understand this behavior.

5. CONCLUSION

We have measured the MR of $U_{1-x}Th_xBe_{13}$ for x= 0, 0.006, 0.015, and 0.033, and find that the samples with intermediate thorium concentrations have positive MR at temperatures above approximately 2 K. The pure and x=0.033 samples have the expected large negative MR at all temperatures studied. We are presently extending these studies.

REFERENCES

(1) U. Rauchschwalbe, F. Steglich, and H. Rietschel, Physica 148B (1987) 33, and references therein.
(2) C.D. Bredl, S. Horn, F. Steglich, B. Lü thi, and R.M. Martin, Phys. Rev. Lett. 52 (1984) 1982.
(3) B. Batlogg, D.J. Bishop, E. Bucher, B. Golding, Jr., A.P. Ramirez, Z. Fisk, and J.L. Smith, J. Magn. Magn. Mat. 63& 64 (1987) 441.
(4) H.M. Mayer, U. Rauchschwalbe, F. Steglich, G.R. Stewart, and A.L. Giorgi, Z. Phys. B 64 (1986) 299.
(5) M.C. Aronson, J.D. Thomson, J.L. Smith, Z. Fisk, and M.W. McElfresh, Phys. Rev. Lett. 63 (1989) 2311.
(6) M.J. Graf, N.A. Fortune, J.S. Brooks, J.L. Smith, and Z. Fisk, Phys. Rev. B40 (1989) 9358.
(7) H. R. Ott, H. Rudigier, Z. Fisk, and J.L. Smith, Phys. Rev. B31 (1985) 1651.

Physica B 165&166 (1990) 361–362
North-Holland

REENTRANCE OF THE SUPERCONDUCTING STATE IN A STRONG MAGNETIC FIELD

T.Maniv[a] *, R.S.Markiewicz[b], I.D.Vagner[c] and P.Wyder[c]

a) Department of Physics and Astronomy,The Johns Hopkins University, Baltimore,Maryland 21218; b) Physics Department, Northeastern University, Boston Mass. 02115, USA; c) Max-Planck-Institut für Festkörperforschung, Hochfeld-Magnetlabor, F-38042, 166X, Grenoble Cedex, France.

Strong quantum magneto-oscillations in the order parameter will result in the reentrance of the superconducting state at low temperatures in extreme type II highly two-dimensional superconductors. In systems with critical fields as high as in the high T_C cuprates, an additional oscillatory structure associated with the pairing correlation may be observable. The average order parameter over a single superconducting core region in the vortex phase in quantizing magnetic field near $H_{C_2}(T)$ is calculated and application is made to the experimental observation of quantum oscillations in the layered dichalcogenide compounds.

1. INTRODUCTION.

More than a decade ago, Graebner and Robbins (1) observed significant magnetothermal and dHvA oscillations well below H_{C_2} in investigating the quasi-two-dimensional dichalcogenide 2H-Nb Se_2. The new high T_C oxide superconductors, with their quasi-two-dimensional structure and their enormously high upper critical fields (2) are, in principle, very promising candidates for this type of investigation. The observation of quantum oscillations in the superconducting state would be of great importance as a Fermi surface probe and also as a way of sorting out the very nature of the pairing mechanism in this class of materials.

In (3) we have presented an analytical calculation of the linear kernel in the Gorkov equation, which provides the information on the quantum oscillations in H_{C_2}. Here we calculate the nonlinear term in the Gorkov equation and study, analytically and numerically, the quantum oscillations in the order parameter.

2. THEORY.

We consider a two-dimensional free electron gas model with a simple B.C.S. pairing interaction in the vicinity of $Hc_2(T)$ for an arbitrary temperature $0 < T < T_C$. We assume that many Landau levels are filled, i.e. we work with the semiclassical electronic wave functions. The extreme quantum case was considered recently by Tesanovic et al. (5). The Helfand-Werthamer ansatz, Ref. (4), which is an *exact* solution of the linearized Gorkov equation in quantizing magnetic field, is used as a variational solution, $\Delta_L(r) \equiv \Delta_0 \exp\{-1/2 (r/a_H)2\}$, for the nonlinear equation. The Gaussian form of $\Delta_L(r)$ is consistent with the semiclassical structure of the kernel R and the multiple integral associated with it can be reduced to a single integral by employing a combination of Gaussian (steepest descent) and contour integration methods. The only approximation made throughout the analysis is the semiclassical approximation. This provides a simple quadratic equation for Δ_0, which is solved to yield:

$$\Delta^2(H,T) = 2\,\frac{(2\pi k_B T_C)^2 \left(A - \frac{1}{\lambda}\right)}{B} \qquad [1]$$

with:

$$B \equiv \left(\frac{a_H}{\zeta}\right)\left(\frac{a_H}{\zeta_0}\right)^2 \sum_{\nu=0}^{\nu_D-1} (\mathrm{Re}\, q_\nu^2)\, \delta_\nu \qquad [2]$$

---

* Permanent address: Department of Chemistry and the Solid State Institute, Technion-IIT, Haifa 32000, Israel.

0921-4526/90/$03.50 © 1990 – Elsevier Science Publishers B.V. (North-Holland)

$$\delta_\nu = 4\pi \int_0^\infty e^{-2\alpha_\nu \rho - \rho^2} \left[\mathrm{erf}\left(\frac{\rho}{\sqrt{2}} \right) \right]^{-2} d\rho \quad [3]$$

The expressions for A and q_ν are presented in Ref. (3). Here: $\alpha_\nu = (2\nu + 1)\,\frac{a_H}{\zeta}$ and $\lambda \equiv N(0)V$ is the effective B.C.S coupling constant with N(0) the density of states at the Fermi energy, $T_c = 2T_D\,1.781\,e^{-1/\lambda}$ the transition temperature, $\zeta_0 \equiv \hbar v_F/\pi k_B T_c$, T_D - the cutoff temperature and $\nu_D \equiv (T_D/T - 1)/2$. $\hbar\omega_e = geH/m_0\,c$ is the Zeeman splitting of a Landau level. Note that our imaginary (Matzubara) frequency formalism is equivalent to the real frequency one only at discrete values of the temperature T for which ν_D is an exact integer. We therefore restrict the temperature in our calculation to these values.

3. DISCUSSION

Equating $\Delta^2(H,T)$ to zero in the entire temperature range $0 < T < T_c$, yields an equation for $H_{c_2}(T)$, which can be easily solved numerically. Strong QO will smear dramatically the transition to the superconducting state as a result of multiple reentrance transition. This phenomenon may account for the anomalously broad transition to the superconducting mixed state, reported in Ref. (1), for 2H-Nb Se_2 . For instance, by using $m_c = 0.26\ m_0$, which is a fairly good estimate of the experimentally observed effective mass, and fitting the upper edge of the reentrance regime to the experimental upper edge of the anomaly, we automatically reproduce the width of the transition. Another example is the deviation from the conventional theory of the measured $H_{c_2}(T)$ curves in the layered dichalcogenide compounds $Nb_{1-x}\,Ta_x\,Se_2$, reported in Ref. (6).

Another interesting feature of the present theory is revealed in model systems with much higher critical fields. Fig. (1) shows the calculated $\Delta^2(H,T)$ in the range of the relevant parameters which may characterize some of the high T_c oxides. Depending on the value of the (inplane) cyclotron mass, the dHvA-like peaks, associated with the field enhancement of the single electron density of states, may split at sufficiently low (~ 0.5 K) temperatures into doublets. This additional structure is due to a feedback effect associated with the enhanced pair-pair repulsion at the peaks of the density of states, which tends to reduce the superconducting order.

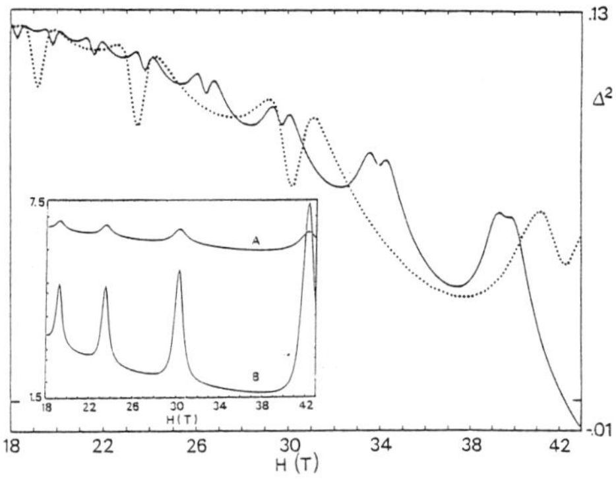

Fig. 1
Quantum oscillations in the order parameter. Here: T_D = 978.8K, T_c = 88.8K, ξ_0 = 12.86A, E_F = 0.039 eV. The temperature is T = 0.5K. The values of m_c are: (a) $m_c/m_0 = 2$ (dotted curve), (b) $m_c/m_0 = 3$ (solid curve).

We acknowledge valuable discussions with W. Joss, W. Biberacher, D. Shoenberg and Z. Tesanovic. This research was supported by a grant from the German-Israeli Foundation for Scientific Research and Development, No. G-112-279.7/88 .

REFERENCES

(1) J.E.Graebner and M.Robbins, Phys. Rev. Lett. **36** (1976) 422.
(2) T.K.Worthington, W.J.Gallagher, D.L.Kaiser, F.H.Holtzberg and T.R.Dinger, Physica C **153-155** (1988) 32.
(3) T.Maniv, R.S.Markiewicz, I.D.Vagner and P.Wyder, Physica C **153-155**, (1988) 1179 ; R.S.Markiewicz, I.D.Vagner, P.Wyder and T.Maniv, Solid State Comm **67** (1988) 43.
(4) E.Helfand and R.Werthamer, Phys. Rev. Lett. **13** (1964) 686; L.W.Gruenberg and L.Gunther, Phys. Rev. Lett. **16** (1966) 996; Phys. Rev. **176** (1968) 606 .
(5) Z. Tesanovic, M. Rasolt and L. Xing, Phys.Rev.Lett.**63** (1989) 2425.
(6) M.Ikebe, K.Katagiri, K.Noto and Y.Muto, Physica **99B** (1980) 209.

Physica B 165&166 (1990) 363–364
North-Holland

THE UPPER CRITICAL FIELD OF SUPERCONDUCTING UPt_3 ALLOYED WITH BORON.

T. VORENKAMP, A. DE VISSER, R. WESTER, A.A. MENOVSKY and J.J.M. FRANSE.

Natuurkundig Laboratorium der Universiteit van Amsterdam,
Valckenierstraat 65, 1018 XE, Amsterdam, The Netherlands.

Measurements of $H_{c2}(T)$, on two very pure samples of hexagonal $UPt_3B_{0.11}$, are presented for H//a, H//b and H//c. A "kink" in $H_{c2}(T)$ is observed for both basal plane directions. The values of dH_{c2}/dT are comparable with those of pure UPt_3.

1. INTRODUCTION

Heavy fermion superconductors attract a lot of interest because the superconducting order parameter in these compounds might be non-scalar, leading to the possible existence of more than one superconducting phase. In particular for UPt_3 there is growing evidence for a multi-component phase diagram. Specific heat measurements in zero field of UPt_3 (1) reveal two anomalies separated by 60 mK, suggesting that there are two nearly degenerate superconducting states. The upper critical field of UPt_3 shows large anisotropy when the field direction is rotated from the basal plane to the hexagonal axis (c-axis), and a change of slope is seen for a field direction in the basal plane (2-6) This change in slope is reported to be a sharp kink by some authors (3-5), and a positive curvature by others (2,6). Recently a theoretical approach is made (7-9), explaining the double peak in the specific heat with a coupling between the superconducting order parameter and the observed small antiferromagnetic moment M_s. In this model M_s is fixed to one basal plane direction and acts as a symmetry breaking field. Calculations predict a kink in $H_{c2}(T)$ for one direction in the basal plane, and anisotropy for field directions within the basal plane. No basal plane anisotropy has been observed by the authors of ref. (4), whereas in ref. (10) a small anisotropy has been reported. Previously it has been shown that alloying UPt_3 with small amounts of boron leads to an increase of T_c (11). Moreover it has been shown that the double peak in the zero field specific heat is not altered or depressed by the addition of small amounts of boron (12). In this paper we report on $H_{c2}(T)$ measurements of $UPt_3B_{0.11}$ (nominal concentration of boron) for field directions parallel to the a,b and c-axis. A comparison is made with the data for UPt_3. In our investigation special attention has been given to the question whether there is basal plane anisotropy, whether both H//a and H//b show a change in slope, and whether the change in slope is a sharp kink or a positive curvature.

2. EXPERIMENTAL

The two samples investigated are both cut from the same Czochralski-grown single crystalline rod. Both samples are annealed in the same way. Growing and annealing procedures were described previously (12). The homogeneity of the boron content was confirmed by microprobe analysis. The absolute value of the boron content, however, can not be checked in this way. Therefore we will refer to the 11% boron content as a nominal value. The samples are of high purity as is indicated by the low ρ_0-value of 0.29 $\mu\Omega$cm (for I//c-axis). The RRR-value as defined by R(294K)/R(0K) is 536, and the transition temperature in zero field is 559 mK. Sample 1 exhibits a small mosaik of 3°, and sample 2 of 5°, as was shown by Laue-pictures. Resistance was measured in a ^{3}He-cryostat with a four point AC-method, using a transformer coupled bridge and a current density of 0.2 A/cm^2 (sample 1) and 0.1 A/cm^2 (sample 2). The temperature was measured with a RuO_2 thermometer which was calibrated in field against a capacitance thermometer. The samples were cut into the shape of a rectangular bar and the demagnetization correction is estimated to be always less than 0.8 %.

3. RESULTS AND DISCUSSION

The results are presented in figs 1 and 2 for sample 1 (I//c) and sample 2 (I//b), respectively. The transition temperatures have been determined from the 50% value of the resistance in the normal state. The latter being defined as the point where the resistance deviates from the normal state T^2 behaviour. The width of the transition as determined from the 10% and 90% values of the normal state resistance is (16±2) mK and does not broaden with magnetic field up to 1.4 Tesla. Refering to the data in figs. 1 and 2, we stress 4 important facts.

0921-4526/90/$03.50 © 1990 – Elsevier Science Publishers B.V. (North-Holland)

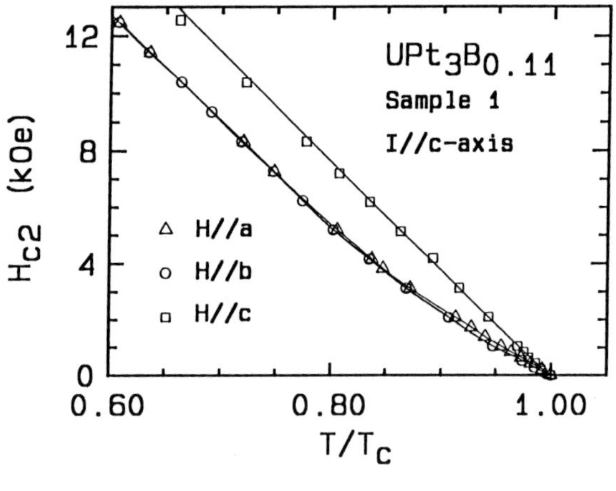

FIGURE 1
$H_{c2}(T)$ for $UPt_3B_{0.11}$ (sample 1, I//c), for H//a and H//b (solid lines are guides to the eye) and for H//c (solid line indicates the initial slope).

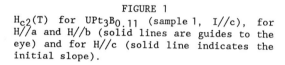

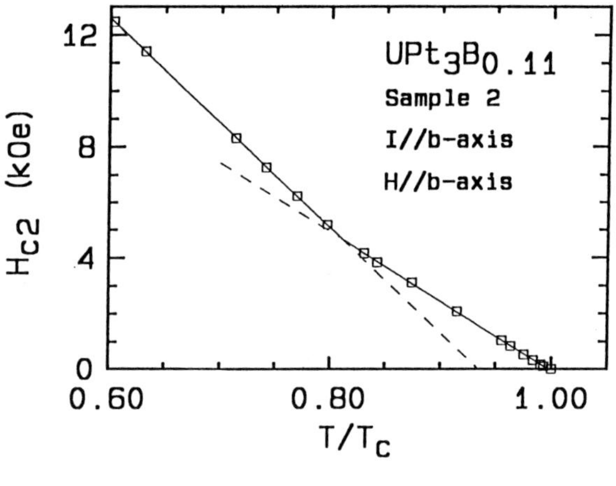

FIGURE 2
$H_{c2}(T)$ for $UPt_3B_{0.11}$ (sample 2, I//b) for H//b (solid lines indicate the two different slopes dH_{c2}/dT).

1) The addition of boron does not appreciably change the characteristics of $H_{c2}(T)$. For the basal plane field direction (sample 2) dH_{c2}/dT is -67 kOe/K below $T_H/T_c = 0.81$ and -45 kOe/K there above (T_H being the temperature where the change in slope occurs). For H//c dH_{c2}/dT is -69 kOe/K. These values are comparible to the values -64, -44 and -71 kOe/K, respectively, for pure UPt_3 (10).

2) For both H//a and H//b $H_{c2}(T)$ shows a change of slope which sets in at $T_H/T_c = 0.81$ (fig. 1). This implies that the overal behaviour of $H_{c2}(T)$ in the basal plane is isotropic, and that the earliest theoretical models, using a large in-plane anisotropy of the antiferromagnetic order, need some refinement.

3) Nevertheless, from the data in fig. 1 a small in-plane anisotropy remains visible. When we denote this as an increase of T_c by going from the H//b phase line to H//a phase line, the upper limit for the anisotropy is 5 mK for $T>T_H$ and 2.5 mK for $T<T_H$ This is in rough agreement with the findings of ref. (10).

4) The experimental results for sample 2 (H//b//I) indeed show a sharp kink at $T_H/T_c = 0.81$, whereas for sample 1 (H//b⊥I) the change of slope sets in above $T_H/T_c = 0.81$ and a small positive curvature is observed. The deviation between the curved phase line of sample 1 and the linear phase line of sample 2 is small, the maximum deviation being 5 mK. The difference between a sharp change of slope or a gradual change of slope (or less sharp kink) might be related to the difference in current densities (10), the current density being a factor two smaller in case of the sharp kink.

In conclusion we have presented data for $H_{c2}(T)$ of single-crystalline $UPt_3B_{0.11}$ and commented on the observed anisotropies.

ACKNOWLEDGEMENTS

The work of T.V. was part of the research program of the foundation FOM. The work of A.d.V. has been made possible by a fellowship of the Royal Netherlands Academy of Arts and Sciences.

REFERENCES

(1) R.A. Fisher et al., Phys. Rev. Lett. 62 (1989) 1411.
(2) J.W. Chen et al., Phys. Rev. B 30 (1984) 1583.
(3) U. Rauchschwalbe et al., Z. Phys. B Cond. Matt. 60 (1985) 379
(4) B.S. Shivaram et al., Phys. Rev. Lett. 57 (1986) 1259. B.S. Shivaram et al., Phys. Rev. Lett. 63 (1989) 1723.
(5) L. Taillefer et al., Physica C 153-155 (1988) 451
(6) W.K. KwoK, L.E. DeLong, G.W. Crabtree, D.G. Hinks and R. Joynt, (to be published).
(7) D.W. Hess et al., J. Phys. Condens. Matter 1 (1989) 8135.
(8) K. Machida et al., J. Phys. Soc. Jpn. 58 (1989) 4116.
(9) S.K. Sundaram and R. Joynt, Phys. Rev. B 40 (1989) 8780.
(10) L. Taillefer, Physica B 'in print'.
(11) T. Vorenkamp et al., J. Magn. Magn. Mat. 76&77 (1988) 531.
(12) T. Vorenkamp, et al., Physica B 'in print'.

Physica B 165&166 (1990) 365–366
North-Holland

VERY STRONG FLUX MOTION AT MILLIKELVIN TEMPERATURES IN THE HEAVY FERMION SUPERCONDUCTOR UPt_3

A. POLLINI*, A.C. MOTA*, P. VISANI*, G. JURI*, and J.J.M. FRANSE+

*Laboratorium für Festkörperphysik, ETH Hönggerberg, CH-8093 Zürich, Switzerland
+Natuurkundig Laboratorium der Universiteit van Amsterdam, 1018 XE Amsterdam, The Netherlands

Extremely strong decays of the remanent magnetization are observed in a single crystal of UPt_3 at temperatures as low as 7 mK. The decays deviate strongly from a logarithmic law. They are described well by stretched exponentials for $10^{-1} < t < 10^5$ sec with parameters that change only slightly with temperature. The total decay of M_{rem} from the beginning of the measured relaxation to equilibrium is 50% to 70% of its initial value with 30% occurring in the first 10000 sec. This novel behaviour points to strong motion of flux which is not thermally activated.

Strong flux motion at millikelvin temperatures was observed for the first time in the high-T_c superconducting oxides Sr-La-Cu-O and Ba-La-Cu-O [1]. At T = 20 mK, the low-field magnetization of a polycrystalline specimen of Ba-La-Cu-O decayed following a logarithmic law with a normalized decay rate $M^{-1}\, \partial M/\partial \log t$ = 1.3% per decade of time. A comparison with decay rates measured in classical type II superconductors at T = 4.2 K showed that the magnetization in HTSC materials relaxes faster by a factor of 40, even at a 200 times lower temperature. Strong flux motion has been observed now in all the HTSC. It has been explained[2] by using extensions of the Anderson-Kim and Bean models that include the fingerprints of HTSC, i.e. the high temperatures and the unusually small pinning energies due to short coherence lengths. However, such models do not account for the relaxation rates at temperatures below 1 K, where thermal activation becomes less important. It seems clear to us that a new and until now undetermined mechanism is present in the HTSC in addition to the thermally activated flux creep as observed in conventional type II superconductors. This mechanism might also be the cause of the very high decay rates of the magnetization observed in other unconventional superconductors, i.e. the heavy fermions[3] and the organic compounds[4].

Here we present relaxation data on the heavy fermion compound UPt_3. This superconductor shows extremely large, almost temperature-independent, non-logarithmic time decays of the remanent magnetization.

The specimen we investigated is a single crystal with a transition temperature T_c = 0.43 K. It has the dimensions 4 x 2.3 x 0.5 mm^3 with the c-axis perpendicular to the plane of the specimen. The magnetic field was applied perpendicular to the c-axis. The experimental arrangement and measuring methods have been described elsewhere[5].

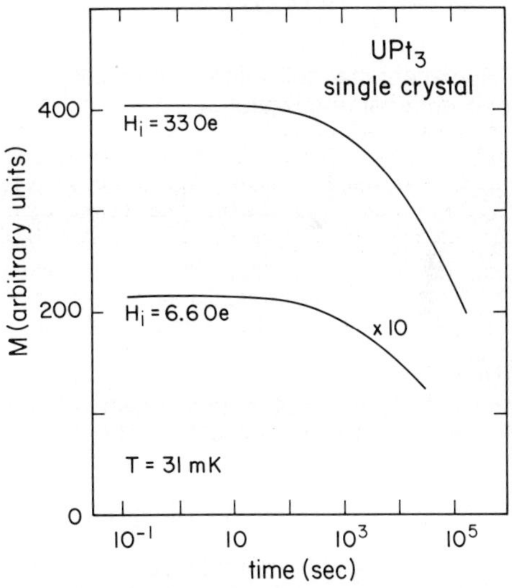

Fig. 1 - Remanent magnetization as a function of time for two different cycling fields. Each curve consists of typically 500 points which lie within the thickness of the drawn line.

In Fig. 1 we show decays of the remanent magnetization taken at T = 31 mK after cycling the zero field cooled specimen in a field H_i = 33 Oe (upper curve) and 6.6 Oe (lower curve). The same arbitrary units have been used for both decays with the lower curve expanded by a factor of 10. The initial values of the remanent magnetization $M_{rem}(0)$ were obtained independently from magnetization loops up to the given fields. We notice that after 10^4 sec $M_{rem}(0)$ is reduced by about 31% for H_i = 6.6 Oe

0921-4526/90/$03.50

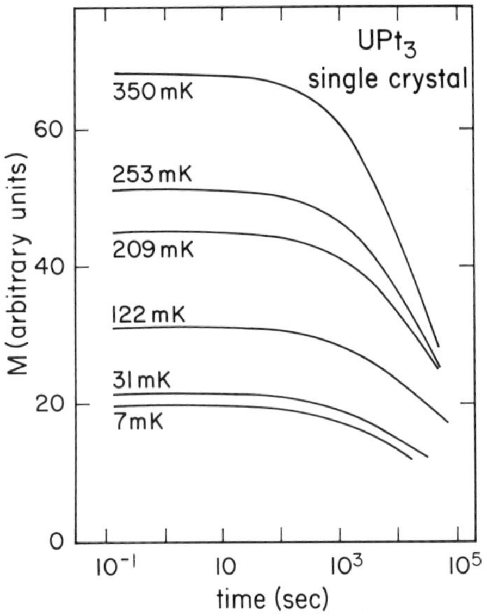

Fig. 2 - Remanent magnetization as a function of time for six different temperatures.

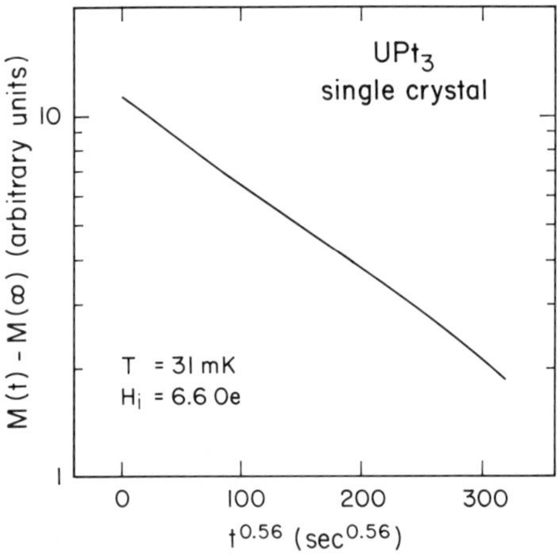

Fig. 3 - Relaxation data at T = 31 mK taken from Fig. 2 plotted as $M(t) - M(t \to \infty)$ vs t^{β}, where $M(t \to \infty)$ and ß are parameters obtained from the fit using the stretched exponential function given above.

and by about 20% for H_i = 33 Oe. In Fig. 2 we show similar decays at different temperatures after cycling the specimen to H_i = 6.6 Oe. It is interesting to notice that the percentage decay in the first 10^4 sec is about 30% ± 5% and practically independent of temperature for $7 < T < 350$ mK. Clearly, this extremely strong decay cannot be due to thermal activation.

A reasonably good fit to the data shown in Figs. 1 and 2 is given by stretched exponentials of the form:

$$M_{rem}(t)-M_{rem}(\infty) = [M_{rem}(0) - M_{rem}(\infty)]\exp[-(t/\tau_p)^{\beta}]$$

where $M_{rem}(\infty)$, τ_p and ß are taken as adjustable parameters. As an example, the relaxation curve $M_{rem}(t)$ at T = 31 mK, after cycling the specimen in a field H_i = 6.6 Oe, has been fitted with the following parameters: $M_{rem}(\infty)$ = 10.48 arbitrary units, τ_p = 10900 sec, ß = 0.56. The result of the fit is shown in Fig. 3, where we have plotted $M_{rem}(t)$ - $M_{rem}(\infty)$ in a logarithmic scale as a function of t^{β}. The quantity $b/M_{rem}(0) = [M_{rem}(0) - M_{rem}(\infty)]/M_{rem}(0)$ gives the relative decay of the remanent magnetization from t ≈ 0.1 sec to equilibrium at $t \to \infty$. For the decay in Fig. 3, $b/M_{rem}(0)$ amounts to 0.52.

Similar fits for the decays shown in Fig. 2 give values of ß between 0.5 and 0.6, values of τ_p between 8000 sec and 30000 sec and of $b/M_{rem}(0)$ from 0.5 to 0.7 as the temperature is increased from 7 mK to 350 mK.

In conclusion, we have found that UPt_3 shows much stronger decays of the remanent magnetization than the HTSC. For small magnetic inductions and temperatures as low as T = 7 mK, about 30% of M_{rem} decays in the first 10^4 sec. The decay law is very close to a stretched exponential in the observation time of this experiment: $10^{-1} < t < 10^5$ sec. This novel and extremely fast relaxation seems to be independent of temperature.

ACKNOWLEDGEMENTS

We are grateful to H.-U. Nissen and R. Wessicken for the determination of the UPt_3 single crystal orientation by electron channelling patterns. This work was partially supported by the Schweizerischer Nationalfonds zur Förderung der wissenschaftlichen Forschung.

REFERENCES

1. A.C. Mota, A. Pollini, P. Visani, K.A. Müller, and J.G. Bednorz, Phys. Scripta 37 (1988) 823
2. Y. Yeshurun and A.P. Malozemoff, Phys. Rev. Lett. 60 (1988) 2202
3. A.C. Mota, P. Visani, and A. Pollini, Physica C 153-155 (1988) 441, and Phys. Rev. B 37 (1988) 9830
4. A.C. Mota, P. Visani, A. Pollini, G. Juri, and D. Jérome, Physica C 153-155 (1988) 1153
5. A.C. Mota, G. Juri, P. Visani, and A. Pollini, Physica C 162-164 (1989) 1152

Physica B 165&166 (1990) 367–368
North-Holland

Superconducting phases in UPt_3

D. Maurer, V. Müller, E.-W. Scheidt and K. Lüders

Freie Universität Berlin, Fachbereich Physik, Arnimallee 14, D-1000 Berlin 33, FRG

A brief discussion is given concerning the superconducting properties of UPt_3 based on extensive ultrasonic investigations. Unsuccessful attempts in deriving a consistent picture for the complex behaviour observed, lead to the suggestion that not only superconductivity but also other types of ordering processes should be involved.

With the discovery of different superconducting phases in the heavy-fermion compound UPt_3 by means of ultrasonic experiments (1), the question immediately arose whether or not these phases should be a direct consequence of an unconventional pairing mechanism similar to the one found in superfluid ^3He. Especially the observation of a λ-shaped absorption peak in the vicinity of the superconducting transition for longitudinal sound waves only (2) supports this idea. Although the physical origin of the absorption peak is still an open question, our first attempt (1) to explain the frequency dependence of the absorption peak as a consequence of a phase transition rather than of the excitiation of collective modes like in superfluid ^3He, is now well established by means of other experimental methods e.g. by specific heat measurements (6).

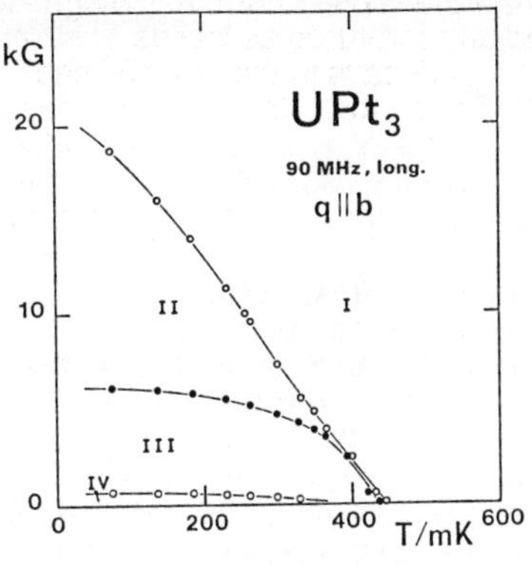

Fig. 1: Phase diagram of superconducting UPt_3 obtained from the longitudinal ultrasonic attenuation data. I refers to the normal-state phase and II, III and IV to phases in the superconducting state.

Fig.1 shows a typical example of an ultrasonically obtained superconducting phase diagram, for which the wavevector **q** of the sound wave as well as the external magnetic field were chosen to be parallel to the **b**-axis of the crystal, i.e. normal to the hexagonal symmetry axis of the UPt_3 unit cell. The similarity of the phase diagram to the one originally obtained for a magnetic field directed out of the hexagonal basal plane is in contradiction to theoretical concepts, which attribute the occurence of the different superconducting phases to flux lattice transformations caused by a strong influence of the hexagonal crystal structure (3).

The existence of at least three different phases in superconducting UPt_3 (II, III, IV in Fig.1) is now well established by several groups for different field orientations. While it becomes clear in the meantime, that the boundary between phase III and IV is most probably due to the transition from a highly damped viscomagnetic flux phase into the Meissner phase, the origin of the boundary between phase II and III identified by the above mentioned absorption peak remains obscure. In spite of several attempts (3,4) to explain this phenomenon by a transformation of the superconducting flux lattice in the whole or - in a more subtle way - by vortex-core transitions, the situation remains unsatisfactory with regard to the experimental fact that the observed viscomagnetic behaviour due to the flux lattice relaxation is not significantly altered when passing the phase boundary. It must be emphasized here that no anomalies at all could be detected by transverse sound.

The absorption peak occurs just below the superconducting transition temperature up to a critical field of about 4 - 7 kG depending on the magnetic field direction. Above this critical field a marked shift of the attenuation peak relative to the superconducting transition sets in, leading to the speculation of a multicritical behaviour in this region of the phase diagram. Nevertheless, the existence of a multicritical point is hard to confirm by our ultrasonic investigations, since with an increasing magnetic field the peak position shifts continuously to lower temperatures, resulting in a T^2-dependence of the phase boundary in the diagram for $T \leq 0.6\, T_c$, which suggests an intimate relationship to the penetration depth for which the same behaviour is found (5).

The polarization dependent temperature behaviour of the ultrasonic attenuation punctuates the complex nature of the superconducting state in UPt_3. Since the electronic part of the attenuation is essentially related to the number density of the normal conducting electrons, its temperature dependence below T_c reflects directly the condensation into the superconducting ground state and therefore yields valuable informations about the gap properties. The power law behaviour of the attenuation with respect to the magnetic field and the temperature as well as the orientation dependence reveal a high anisotropic gap structure, supporting the idea of nonconventional superconductivity in UPt_3 although no selfconsistent picture could be drawn until now.

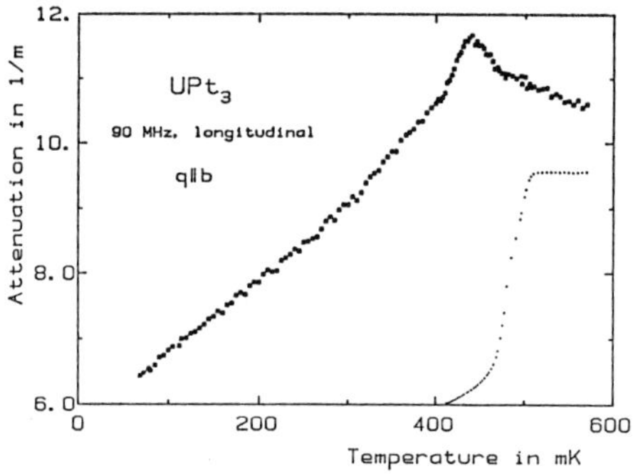

Fig. 2: Attenuation for longitudinal sound of 90 MHz propagating along the **b**-axis. The simultaneously measured ac-susceptibility is represented by the dotted line.

As an example Fig.2 shows for longitudinal ultrasound propagating along the **b**-axis the temperature dependent attenuation, the conspicuous linear behaviour of which differs markedly from the attenuation of other propagation directions and sound polarizations. Generally it is not mentioned that the longitudinal and transverse sound attenuation differs strong not only in the temperature dependence (Fig.2 and 3) but also in the absolute values. In contradiction to the behaviour of conventional superconductors where the electronic viscosity η differs only little for longitudinal and transverse sound waves (within the hydrodynamic limit, which is justified here), the ratio η_{long}/η_{trans} amounts to about 5...10 in UPt_3!

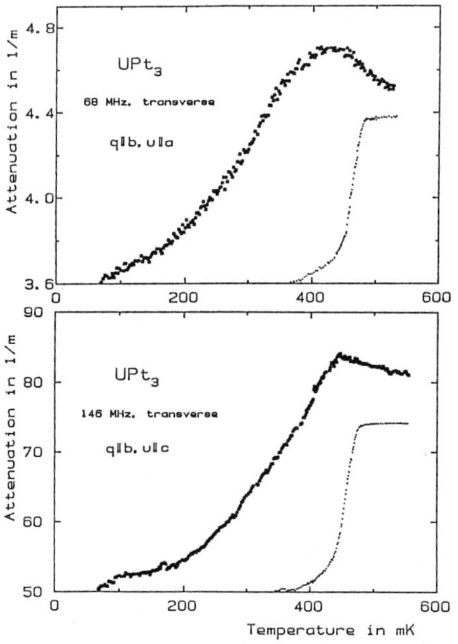

Fig. 3: Attenuation of transverse sound propagating along the **b**-axis for polarization normal (upper part) and parallel (lower part) to the hexagonal **c**-axis.

A simple explanation for these discrepancies could be the existence of additional temperature dependent contributions to the attenuation of longitudinal sound which are of other origin but the condensation of normal conducting electrons into Cooper pairs. Since the main difference between transverse and longitudinal sound waves is that the latter is not volume conserving, the assumed additional coupling mechanism should be of volumestrictive nature and therefore similar to the one responsible for the observed absorption peak.

With regard to neutron scattering results (7) it seems reasonable to attribute the unusual behaviour of the longitudinal ultrasound absorption to an interaction with the antiferromagnetic ordering observed for $T \leq 5$ K in UPt_3. In this context it is also noteworthy that the change in the transverse ultrasonic attenuation just below T_c depends sensitively on the polarization (Fig.3). This possibly indicates that the assumed interaction between the sound wave and the magnetic order is not mediated exclusively via volume coupling mechanisms.

Summarizing our findings, superconducting UPt_3 provides a number of exciting phenomena which were quite unexpected within traditional concepts. Although the origin of the different superconducting phases is far from being well understood, there are some indications that the complex behaviour observed may be due to a coexistence of antiferromagnetism and superconductivity.

This work was supported by the Deutsche Forschungsgemeinschaft.

References

(1) V. Müller, Ch. Roth, D. Maurer, E.-W. Scheidt, K. Lüders, E. Bucher, and H.E. Bömmel, Phys. Rev. Lett. **58**, 1224 (1987); Y.J. Qian, M.-F. Xu, A. Schenstrom, H.-P. Baum, J.B. Ketterson, D. Hinks, M. Levy, and Bimal K. Sarma, Solid State Comm. **63**, 599 (1987)

(2) V. Müller, D. Maurer, E.-W. Scheidt, Ch. Roth, K. Lüders, E. Bucher, and H.E. Bömmel, Solid State Comm. **57**, 319 (1986)

(3) G.E. Volovik, J. Phys. **C21**, L221 (1988)

(4) A. Schenstrom, M.-F. Xu, Y. Hong, D. Bein, M. Levy, B.K. Sarma, S. Adenwalla, Z. Zhao, T. Tokuyasu, D.W. Hess, J.B. Ketterson, J.A. Sauls, and D.G. Hinks, Phys. Rev. Lett. **62**, 332 (1989); G.E. Volovik, J. Phys. **C21**, L215 (1988)

(5) F. Gross, B.S. Chandrasekhar, K. Andres, U. Rauchschwalbe, E. Bucher, and B. Lüthi, Physica **C153-155**, 439 (1988)

(6) R.A. Fisher, S. Kim, B.F. Woodfield, N.E. Phillips, L. Taillefer, K. Hasselbach, J. Flouquet, A.L. Giorgi, and J.L. Smith, Phys. Rev. Lett. **62**, 1411 (1989); K. Hasselbach, L. Taillefer, and J. Floquet, Phys. Rev. Lett. **63**, 93 (1989)

(7) G. Aeppli, E. Bucher, C. Broholm, J.K. Kjems, J. Baumann, and J. Hufnagel, Phys. Rev. Lett. **60**, 615 (1988); G. Aeppli, D. Bishop, C. Broholm, E. Bucher, K. Siemensmeyer, M. Steiner, and N. Stüsser, Phys. Rev. Lett. **63**, 676 (1989)

Physica B 165&166 (1990) 369–370
North-Holland

ULTRASONIC STUDY OF THE SUPERCONDUCTING PHASES IN HEAVY FERMION COMPOUNDS

G.J.C.L.Bruls*, D.Weber*, B.Wolf*, B.Lüthi*,
A.A.Menovsky+, A.deVisser+, J.J.M.Franse+

* Physikalisches Institut, Universität Frankfurt, D-6000 Frankfurt
+ Natuurkundig Laboratorium, Universiteit Amsterdam, NL-1018 XE Amsterdam

We studied elastic constants, ultrasonic attenuation and ac-susceptibility in the superconducting states of the heavy fermion compounds UPt_3 and URu_2Si_2. Both materials show steps in some, but not all elastic constants at the superconducting transitions. For UPt_3 we present a B-T phase diagram for fields along a,b and c axis and compare it with existing theories.

1. UPt_3

The heavy fermion compound UPt_3 is already extensively investigated by ultrasonic experiments [1-4]. On a high quality single crystal we performed measurements of the elastic constants (c_{11}, c_{33}, c_{44}), the corresponding ultrasonic attenuation and the magnetic ac-susceptibility χ as function of temperature as well as magnetic field strength for various field directions. With these experiments we obtained a detailed phase diagram exhibiting a double superconducting transition. As a function of temperature, this splitting was seen in the specific heat data [5].

1.2 EXPERIMENTS

In the elastic constants $c_{11}$and c_{33} step like anomalies of the order of 10^{-5} were observed. Measurements as a function of temperature at constant magnetic field show a step going down at T_c^x and a kink at T_c^y. Versus magnetic field we observe two clear steps (see fig.1). The ultrasonic attenuation does not show much structure. It will be discussed elsewhere [6]. In the ac susceptibility we also observed the splitting in the phase transtition. This is surprising, since at a measuring frequency of 128 Hz one expects primarily shielding effects. Fig. 2 gives the phase diagram for B//a,b and B//c exhibiting three superconducting phases [6].

1.3 THEORETICAL CONSIDERATIONS

On the basis of group theoretical considerations together with a coupling of a spin density wave to the superconducting pairing function a splitting of T_c and a phase diagram similar to the observed one can be deduced [7].

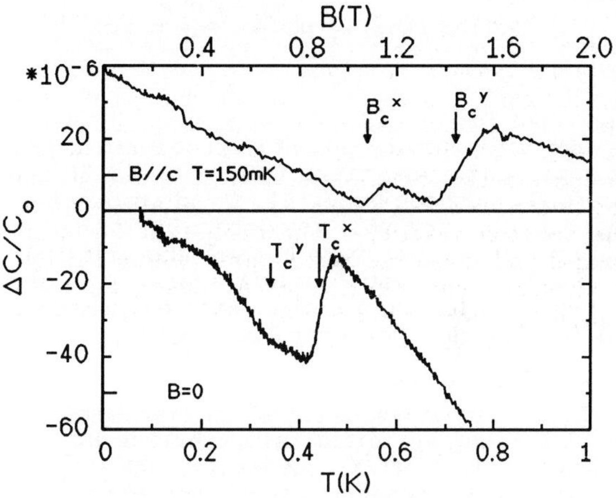

Fig.1. c_{11} mode in UPt_3 vs. B at T=150mK and vs. T at B=0T.

Theoretically, a coupling of the symmetrized strains to the square of the two component superconducting order parameter (η_x,η_y) lead for c_{11} and c_{33} to two steps of equal size as a function of temperature as well as of magnetic field. Experimentally, we do observe two steps as a function of field very clearly, whereas as a function of temperature we see one step and a kink (see Fig.1). The phase diagram in fig.2 is constructed from the T_c's and B_c's like the ones marked in Fig.1. It consists of phases ($\eta_x \neq 0$) bordered by the full lines and a phase ($\eta_y \neq 0$) marked by the dotted line. The ($\eta_y \neq 0$) phase is very similar for all three directions of the field, the ($\eta_x \neq 0$) phase is identical for B//a and B//b, but different for B//c. The phase at very low fields is probably due to B_{c1} [3]. A detailed analysis is following [6].

0921-4526/90/$03.50 © 1990 – Elsevier Science Publishers B.V. (North-Holland)

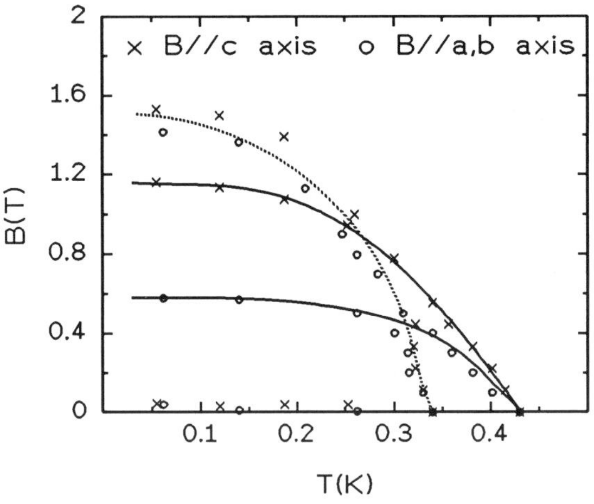

Fig.2. Superconducting phases in UPt_3. Full lines: η_x phases. Dotted line: η_y phase, see text.

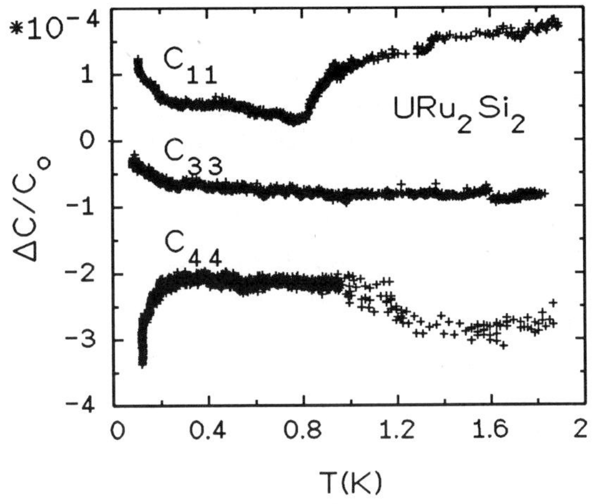

Fig.3. c_{11},c_{33} and c_{44} modes in URu_2Si_2 vs. T at B=0T (10-50MHz).

2. URu_2Si_2

URu_2Si_2 exhibits an antiferromagnetic and a superconducting transition at T_N=17.5K and T_c=1.2K, respectively [8]. To characterize our single crystal we determined the transition temperature T_c and the critical field B_{c2} via ac-susceptibility with B in the a,b-plane. The same ultrasonic measurements as in UPt_3 were performed in URu_2Si_2.

2.1 EXPERIMENTS

From our measurements we obtained T_c=1.13K and an initial slope of the B_{c2} - curve of 8.85 (T/K). Generally URu_2Si_2 exhibits a broader superconducting transition than UPt_3. For our crystal the width is nearly 200mK and similar to [9]. Fig.3 shows the temperature dependence of the elastic constants c_{11}, c_{33} and c_{44} in zero magnetic field. The superconducting transition is only visible in the c_{11} - mode and leads to a change of $5x10^{-5}$ after subtracting the background of the antiferromagnetic phase (as determined from a measurement in a 5 Tesla field). Below 0.25K all elastic constants show anomalies. In the longitudinal modes c_{11} and c_{33} an increase and in the transverse mode c_{44} a pronounced softening is observed. The softening seems to be independent of magnetic fields up to 2.5T. When measuring the magnetic field dependence at a fixed temperature we observed anomalies in all elastic constants at B_{c2}. We do not have yet the interpretation for these anomalies for T<0.25 K.

Part of this work was supported by SFB 252.

3. REFERENCES

(1) I.Kouroudis, D.Weber, M.Yoshizawa, B.Lüthi, P.Haen, J.Flouquet, G.Bruls, U.Welp, J.J.M.Franse, A.Menovsky, E.Bucher and J.Hufnagel, Phys.Rev.Lett. 58 (1987) 820.

(2) G.Bruls, D.Weber, G.Hampel, B.Wolf, I.Kouroudis, W.Sun and B.Lüthi, Physica B in press.

(3) V.Müller,D.Maurer, E.W.Scheidt, K.Lüders, E.Bucher and H.E.Bömmel, Phys.Rev.Lett. 58 (1987) 1224.

(4) A.Schenstrom, M.F.Xu, Y.Hong, D.Bein, M.Levy, B.K.Sarma, S.Adenwalla, Z.Zhao, T.Tokuyasu, D.W.Hess, J.Ketterson, J.Sauls and D.Hinks, Phys.Rev.Lett. 62 (1989) 332.

(5) R.A.Fisher, S.Kim, B.F.Woodfield, N.Phillips, L.Taillefer, K.Hasselbach, J.Flouquet, A.L.Giorgi and J.L.Smith, Phys.Rev.Lett. 62 (1989) 1411.

(6) G.Bruls, D.Weber, B.Wolf, P.Thalmeier and B.Lüthi, to be published

(7) K.Machida, M.Ozaki and T.Ohmi, J.Phys.Soc.Jap. 58 (1989) 4116. D.W.Hess,T.A.Tohuyasu and J.Sauls, J.Phys.Condens.Matter 1 (1989) 8135. R.Joynt, Superc.Sci.Technol.1 (1988) 210.

(8) T.T.M.Palstra, A.A.Menovsky, J.van den Berg, A.Dirkmaat, P.Kes, G.J.Nieuwenhuys and J.A.Mydosh, Phys.Rev.Lett. 55 (1985) 2727.

(9) K.J.Sun, A.Schenstrom, B.K.Sarma, M.Levy and D.Hinks, Phys.Rev.B 40 (1989) 11284.

Physica B 165&166 (1990) 371–372
North-Holland

ELECTROMAGNETIC RESPONSE IN THE E_{1g} MODEL STATE FOR THE HEAVY FERMION SUPERCONDUCTOR UPT_3

W.O. PUTIKKA
Theoretische Physik, E.T.H.-Hönggerberg, 8093 Zürich, Switzerland*

P.J. HIRSCHFELD
Department of Physics, University of Florida, Gainesville, FL. 32611 USA

P. WÖLFLE
Institut für Theorie der Kondensierten Materie, Universität Karlsruhe, D7500 Karlsruhe, Fed. Rep. Germany†

We calculate the real and imaginary parts of the dynamic electromagnetic response of a superconductor in the particular unconventional pairing state with E_{1g} symmetry thought to be realized as the zero magnetic field ground state of UPt_3. We discuss the temperature dependence and anisotropy of the electromagnetic penetration depth previously shown to depend sensitively on frequency ω even for $\omega \ll T_c$, and calculate the influence of resonantly scattering impurites.

1. INTRODUCTION

Thermodynamic and transport measurements in UPt_3 strongly suggest a highly anisotropic superconducting state.(1) Moreover, the discovery of additional transitions in the H-T phase diagram point to the possiblity that the order parameter in this material may be unconventional, i.e. transforms according to a nontrivial representation of the point group of the crystal.(2) Most existing experimental data may be understood on the basis of a model pair state belonging to the E_{1g} representation of the group D_{6h} of the form $\Delta_{\mathbf{k}} = 2\Delta_0 \hat{k}_z \left(\hat{k}_x + i\hat{k}_y\right)$.(3) However, a conclusive demonstration of the symmetry of the order parameter is still lacking. Finite frequency probes appear to be a more powerful tool for distinguishing among competing symmetries for the order parameter. Sound propagation experiments were employed early on to probe the structure of the order parameter more directly by varying the propagation direction and polarization. Comparison with theory is difficult because of the indirect interaction of sound with the electronic system. In addition, the experimentally accessible ultrasound frequencies are well below the T=0 gap frequency, which excludes observation of order parameter collective modes. In both respects electromagnetic probes are more promising.

2. CURRENT RESPONSE

As shown by the current authors (4), the electromagnetic response of any unconventional superconductor has substantial frequency dependence even for frequencies much less than T_c, due to virtual excitations of Bogoliubov quasiparticle-quasihole pairs. The wavevector and frequency-dependent current response function in the presence of impurities is given by

$$K_{ij}(\mathbf{q},\Omega) = -\frac{Ne^2}{mc}\left\{1 - \frac{3}{4}\int \frac{d\Omega}{4\pi}\hat{k}_i\hat{k}_j \times \right.$$
$$\left. \times \int \frac{d\omega\, d\omega'}{\pi} \frac{\left[th\left(\frac{\beta\omega}{2}\right) - th\left(\frac{\beta\omega'}{2}\right)\right]}{\omega - \omega' - \Omega - i0} F(\omega,\omega') \right\},$$

$$F(\omega,\omega') \equiv Im\left\{ \frac{\tilde{\omega}_+\tilde{\omega}'_+ - \eta\xi_{0+} + \tilde{\omega}_+^2}{\xi_{0+}\left[(\eta - \xi_{0+})^2 - \xi'^{\,2}_{0+}\right]} \right.$$
$$- \frac{\tilde{\omega}_+\tilde{\omega}'_- - \eta\xi_{0+} + \tilde{\omega}_+^2}{\xi_{0+}\left[(\eta - \xi_{0+})^2 - \xi'^{\,2}_{0-}\right]} + \frac{\tilde{\omega}_+\tilde{\omega}'_+ + \eta\xi'_{0+} + \tilde{\omega}_+^2}{\xi'_{0+}\left[(\eta + \xi'_{0+})^2 - \xi_{0+}^2\right]}$$
$$\left. + \frac{\tilde{\omega}_+\tilde{\omega}'_- - \eta\xi'_{0-} + \tilde{\omega}'^{\,2}_-}{\xi'_{0-}\left[(\eta - \xi'_{0-})^2 - \xi_{0+}^2\right]} \right\}$$

where $\tilde{\omega}_\pm = \omega \pm i0 - \Sigma_0(\omega)$, the self-energy Σ_0 is defined, e.g. in Ref. (5), and is valid for arbitrarily strong s-wave

*Partially supported by NATO fellowship 1988
†Work partially performed at Univ. Florida

0921-4526/90/$03.50 © 1990 – Elsevier Science Publishers B.V. (North-Holland)

scattering. We have defined $\xi_{0\pm} \equiv sgn(\omega)\sqrt{\tilde{\omega}_\pm^2 - \Delta_{\mathbf{k}}^2}$ and $\eta \equiv v_F\hat{\mathbf{k}} \cdot \mathbf{q}$.

3. RESULTS

We have evaluated the real part of this expression for the E_{1g} state in the clean limit, where it simplifies considerably, and within a simple phenomenological model simulating the effect of impurities in the resonant scattering limit. This is sufficient to obtain the frequency-dependent penetration depth $\Lambda(\omega)$ except in a temperature range very close to T_c, provided $\omega\tau$ <<1 (4). . The surprising result is that in an anisotropic superconductor the presence of a substantial number of thermally excited Bogoliubov quasiparticles even at low temperature enhances the screening currents. (See Figure 1 of Ref. (5)) In Figure 1 we plot penetration depth versus temperature for a value of the dimensionless parameter $\alpha \equiv \Omega\kappa/\Delta_0$, found in Ref (5) consistent with the low-frequency resonant oscillator experiments of Shivaram et al (6). Here κ is the Ginzburg-Landau parameter Λ_0/ξ_0. The curves shown are for the three experimental geometries of Ref (6), where the penetration depth was found to vary roughly as T^4 at low T, in contrast to the DC measurement of Gross et al (7), which found a T^2 dependence. The curves for $\mathbf{q} \parallel \mathbf{z}$, $\mathbf{A} \parallel \mathbf{x}$ and $\mathbf{q} \parallel \mathbf{y}$, $\mathbf{A} \parallel \mathbf{z}$ do in fact obey a rough T^4 dependence, and curve $\mathbf{q} \parallel \mathbf{x}$, $\mathbf{A} \parallel \mathbf{y}$ is considerably larger, again consistent with Shivaram et al. However the T-dependence of the last is linear, as expected for a state with lines of nodes at ω=0. We pointed out in Ref. (4) that resonant impurity scattering should depress the static penetration depth of this state to T^2, and speculated that impurities might depress the frequency-dependent skin depth to a power law closer to T^4. In Figure 1b, we show how the anisotropy changes upon addtion of impurites, based not on Equation 1 but on a phenomenological treatment described in Ref. (4). The impurity scattering parameter x_0 introduced there is roughly the width of the angular region in the neighborhood of the nodes where the angle-resolved density of quasiparticle states is enhanced due to pair-breaking, and scales with the impurity concentration.

The comparison with experiments on UPt_3 is thus encouraging but hardly conclusive. The impurity scattering model used above gives only a crude estimate of qualitative tendencies. A full numerical evaluation of Equation 1 is in progress.

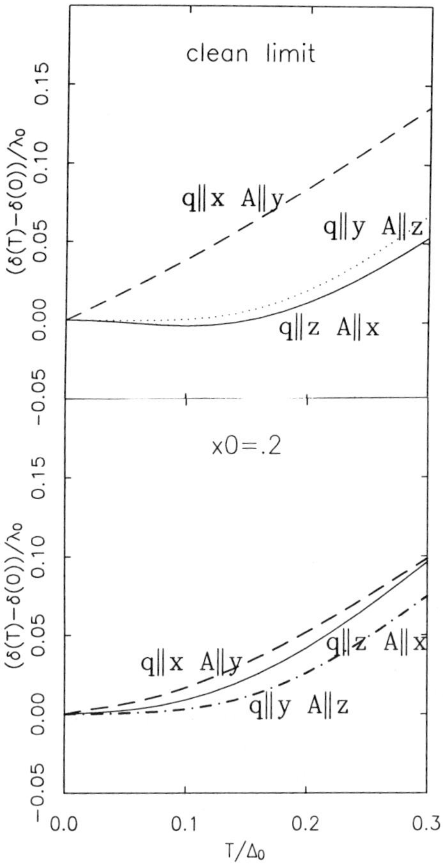

FIGURE 1

Normalized penetration depth variation in an E_{1g} state versus temperature. a) clean limit $\mathbf{q} \parallel \mathbf{z}$, $\mathbf{A} \parallel \mathbf{x}$, α=.3, $\mathbf{q} \parallel \mathbf{x}$, $\mathbf{A} \parallel \mathbf{y}$, α=.3, and $\mathbf{q} \parallel \mathbf{y}$, $\mathbf{A} \parallel \mathbf{z}$, α=.2; c) impurity parameter x_0=.2, $\mathbf{q} \parallel \mathbf{z}$, $\mathbf{A} \parallel \mathbf{x}$, α=.3, $\mathbf{q} \parallel \mathbf{x}$, $\mathbf{A} \parallel \mathbf{y}$, α=.3, and $\mathbf{q} \parallel \mathbf{y}$, $\mathbf{A} \parallel \mathbf{z}$, α=.2.

REFERENCES

1. H.R. Ott, in *Progress in Low Temperature Physics*, edited by D.F. Brewer (North-Holland, Amsterdam), 1987, Vol. **11**, p. 215.
2. for a review see L. Taillefer, Proc. Santa Fe Conf. on Heavy Fermion Materials (1989).
3. P. Hirschfeld, D. Vollhardt, and P. Wölfle, Sol. St. Comm. **59**, 111 (1986).
4. W.O. Putikka, P.J. Hirschfeld, and P. Wölfle, Phys. Rev. **B**, 1 April (1990)
5. P. J. Hirschfeld, P. Wölfle, J.A. Sauls, D. Einzel and W.O. Putikka, Phys. Rev. **B40**, 6695 (1989).
6. B.S. Shivaram, J.J. Gannon, Jr., and D.G. Hinks, unpublished.
7. F. Gross, B.S. Chandrasekhar, K. Andres, U. Rauchschwalbe, E.B. Bucher, B. Lüthi, Physica **C153–155**, 439 (1988).

Physica B 165&166 (1990) 373–374
North-Holland

$H_{c2}(T)$ FOR THE HEAVY–FERMION SUPERCONDUCTOR $U_{1-x}Th_xBe_{13}$ WITH $0.00 \leq x \leq 0.06$.

E.A. KNETSCH, J.A MYDOSH, R.H. HEFFNER* and J.L. SMITH*.

Kamerlingh Onnes Laboratorium der Rijksuniversiteit, P.O. Box 9506, 2300 RA, Leiden, The Netherlands and *Los Alamos National Laboratory, Los Alamos, New Mexico 87545, U.S.A.

We have measured the temperature dependence of the upper critical field, $H_{c2}(T)$, for the superconducting heavy–fermion system $U_{1-x}Th_xBe_{13}$ with $0.00 \leq x \leq 0.06$, using a DC–resistivity technique in magnetic fields up to 3 Tesla. The width of the resistive transitions remains independent of the magnetic field. There is, however, a strong temperature dependence of the resistivity just above the transition temperature, along with a large negative magnetoresistance which is independent of the thorium concentration. $H_{c2}(T)$ has a very high slope ($\approx$-40 T/K) near T_c for all thorium concentrations and starts to break away from this initial slope very close to T_c ($T/T_c \approx 0.95$). After a strong reduction in slope, the $H_{c2}(T)$ curves become linear again at lower temperatures. No anomalous behavior, such as was seen in our H_{c1} measurements, was observed in any of the samples; all exhibited a qualitatively similar temperature dependence of H_{c2}.

Thorium doping in the superconducting heavy–fermion system $U_{1-x}Th_xBe_{13}$ changes the superconducting transition temperature $T_c(x)$ in a very nonmonotonic way [1,2], creating a broad maximum around x=0.03 after an initial decrease and a final drop at high thorium concentrations. It is in the narrow concentration region $0.017 \leq x \leq 0.04$ that a second transition occurs below the superconducting one. The second transition, initially observed in specific heat measurements [3], was also found in ultrasonic attenuation [4], μSR [5], and H_{c1} [6,7] measurements. This second transition is now generally thought to be to a magnetic state with an extremely small value of the observed moment ($\approx 10^{-2}\mu_B$), as was determined from the μSR experiments [5]. From the theoretical point of view the formation of a SDW, coexisting with superconductivity, has been extensively studied [8]. The appearance of a SDW seems the most likely candidate for the second transition in $U_{1-x}Th_xBe_{13}$ since it is in very good agreement with the available experimental data.

In the present studies we investigated the upper critical field, $H_{c2}(T)$, on a wide range of thorium concentrations ($0.00 \leq x \leq 0.06$). The measurements were performed on the same set of samples used in $H_{c1}(T)$ [7] and μSR [9] experiments. This enables us to compare the properties of the complete $U_{1-x}Th_xBe_{13}$ system from both microscopic and macroscopic experiments.

To determine H_{c2} we used a standard 4–point DC–resistivity technique, with measurement currents of typically 1mA. The experiments were performed in a ^{3}He–cryostat, with magnetic fields up to 3 Tesla. H_{c2} was determined by recording the superconducting transitions as a function of temperature in a fixed magnetic field. The values of H_{c2} were taken to be the midpoints of the resistive transitions, which had a typical width (10–90%) of 30–40mK. Some typical transition curves are shown in Fig. 1. The typical values of the resistivities at T_c in all these materials are of order 40–80 μOhm cm in zero field.

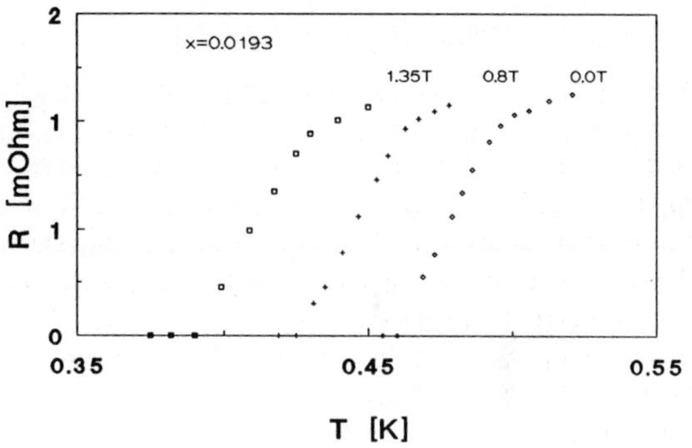

FIGURE 1

Superconducting transitions for $U_{0.9807}Th_{0.0193}Be_{13}$ in fields of 0.00, 0.80 *and* 1.35 *T. Such transition curves are typical for all measured samples.*

The salient features of these curves are the insensitivity of the width and shape of the transition to the magnetic field strength. This enables an accurate determination of H_{c2} for the entire field and temperature range. Note that the resistivity at T_c is not a constant, residual resistivity, for, just above T_c it still has a strong temperature and magnetic field dependence. The temperature variation is proportional to T^2 as predicted by Fermi–liquid theory. The magnetoresistance in these materials is large and negative near T_c. Furthermore, the size of the observed magnetoresistance is independent of the thorium concentration, as can be seen from Table I.

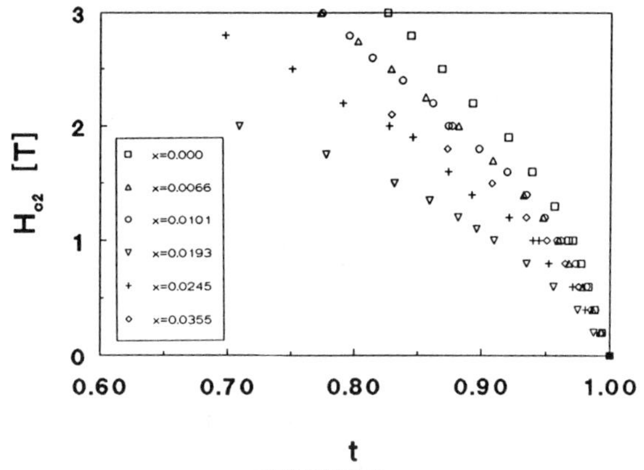

FIGURE 2

$H_{c2}(T)$ data plotted versus reduced temperature. Note the qualitatively similar behavior for all samples and the absence anomalies in the curve for x=0.0193 *at the second transition temperature in this sample.*

If we now consider the full set of $H_{c2}(T)$ curves given in Fig. 2, we observe a qualitatively similar behavior for all thorium concentrations, including the pure UBe_{13}. Our data are in good agreement with earlier results [6,10, 11,12,13] taken on various samples and concentrations. Ours, however is the first high precision set of H_{c2} data systemetically taken on a complete set of samples.

$H_{c2}(T)$ has a very high slope (≈-40 T/K) near T_c for all thorium concentrations (Table I). The slope values do not scale with T_c. Already at temperatures very close to T_c (T/T_c ≈0.95) $H_{c2}(T)$ starts to break away from the expected linear behavior predicted by Ginzburg–Landau theory. After a strong reduction in slope to ≈-10 T/K the curves become linear again. As might be expected, no sign of the second transition, or lower T_c is reflected in the features of $H_{c2}(T)$. Even in high fields, the second transition remains well below the upper, superconducting one, that is probed by these resistivity measurements.

The exotic shape of the $H_{c2}(T)$ curve in this material can be nicely fitted for the low–field region using the dirty limit approximation [6]. This reproduces the curvature near T_c, but fails at higher fields. A good fit for UBe_{13} has been obtained by assuming a field variation of the condensation energy [10]. A changeover in the type of superconductivity below T_{c1} (e.g *s*– to *d*–wave) is not very likely, taking into account the similarity in $H_{c2}(T)$ curves for the different thorium concentrations. However, a complete theory of the upper critical fields in this system is not yet available for comparison with our experiments.

This work was partially supported by the Nederlandse Stichting FOM.

Th. conc. [%]	T_c[K] (H=0T)	$H'_{c2}(T_c)$[T/K]	ratio
0.00	0.920	-35.5	1.5
			1.2*
0.66	0.743	-37.7	1.5
			1.2*
1.01	0.704	-47.0	1.5
			1.2*
1.93	0.476	-33.3	1.2*
2.45	0.649	-32.2	1.5
			1.2*
3.55	0.598	-44.7	1.2*

TABLE I

Collected transition temperatures and initial critical field slopes of the $U_{1-x}Th_xBe_{13}$ system. Ratio is defined as $R(T_c,H=0T)/R(T_c,H=3T)$.
**: Ratio = $R(T_c,H=0T)/R(T_c,H=2T)$*

References

[1] J.L. Smith et al., J. Appl. Phys., **55** (6), (1984) 1996.

[2] J.L. Smith et al., in: Theoretical and Experimental Aspects of Valence Fluctuations and Heavy–Fermions, eds. L.C. Gupta and S.K. Malik, (Plenum, New York, 1987), p11.

[3] H.R. Ott et al., Phys. Rev. B**31**, (1985) 1651.

[4] B. Batlogg et al., Phys. Rev. Lett., **55**, (1985) 1319.

[5] R.H. Heffner et al., Phys. Rev. B**40**, (1989) 806.

[6] U. Rauchschwalbe, Physica **147**B, (1987) 1, and references herein.

[7] E.A. Knetsch et al., in print.

[8] K. Machida and M. Kato, Phys. Rev. Lett., **58**, (1987) 1986. M. Kato and K. Machida, J. of the Phys. Soc. of Japan, **56**, (1987) 2136. K. Machida and M. Kato, Physica **147**B, (1987) 54.

[9] R.H. Heffner et al., to be published.

[10] J.P. Brison et al., J. of Low Temp. Physics, **76**, (1989) 453.

[11] M.B. Maple et al., Phys. Rev. Lett., **54**, (1985) 477.

[12] J.W. Chen et al., J. Appl. Phys., **57**, (1985) 3076.

[13] G.M. Schmiedeshoff et al., Phys. Rev. B**38**, (1988) 2934.

Physica B 165&166 (1990) 375–376
North-Holland

THERMAL EXPANSION OF HEAVY-FERMION SUPERCONDUCTORS

A. DE VISSER, F.E. KAYZEL, A.A. MENOVSKY, J.J.M. FRANSE, *K. HASSELBACH, *A. LACERDA, *L. TAILLEFER, *J. FLOUQUET and #J.L. SMITH

Natuurkundig Laboratorium der Universiteit van Amsterdam, Valckenierstraat 65, 1018 XE Amsterdam, The Netherlands
*CRTBT-CNRS, BP 166X, 38042 Grenoble Cédex, France
#Los Alamos National Laboratory, Los Alamos, New Mexico 87545, U.S.A.

We have measured the coefficients of linear thermal expansion (α) of single-crystalline samples of the heavy-fermion superconductors UPt_3, URu_2Si_2 and UBe_{13}. For the non-cubic systems α shows a remarkable anisotropy. On entering the superconducting state a strong reduction of the volume expansion is observed for all three systems.

1. INTRODUCTION

The occurrence of superconductivity in the heavy-fermion compounds is undoubtedly an intriquing phenomenon and has received a wide interest, because of the unusual behaviour in the superconducting state. The measured large anomalies in the specific heat at the superconducting transition temperature (T_c) have shown unambiguously that the heavy fermions take part in the superconducting condensate. It is then of interest to investigate the corresponding volume and shape effects. Our normal state volume thermal expansion (α_v) measurements (1-3) yield large linear electronic terms at low temperatures, α_v= aT, just as the specific heat, c= γT, and give rise to large electronic Grüneisen parameters, $\Gamma \propto a/\gamma \simeq$ 50 (4,5). From this we conclude that large volume changes can be expected at T_c. In this paper we present thermal-expansion measurements on the uranium based heavy-fermion compounds UPt_3, URu_2Si_2 and UBe_{13}. In order to study the anisotropy in the coefficients of linear thermal expansion, $\alpha = L^{-1}(dL/dT)$, the experiments have been performed on single-crystalline specimens.

2. EXPERIMENTAL

The samples were mounted in a dilatation cell (6) machined of OFHC copper. The thermal expansion was measured using a sensitive 3-terminal capacitance method with a detection limit of 0.1 Å. In the case of UPt_3, the cell was attached to the mixing chamber of a dilution refrigerator, whereas in the case of URu_2Si_2 and UBe_{13} it was attached to the cold plate of a ^{3}He cryostat. Data points were gathered stepwise (ΔT> 5 mK) in order to ensure thermal equilibrium of sample and cell. The data have been corrected for the cell effect, i.e. the signal of the cell with a copper sample.

3. RESULTS

The thermal expansion of hexagonal UPt_3 has been studied for two samples (7). In Fig.1 we present data taken on the specimen prepared by L.Taillefer (size 1x2x3 mm^3). The superconducting transition is most clearly observed in the expansion along the hexagonal axis ($\alpha_{//}$), where a negative jump occurs at a (midpoint) transition temperature of 490 mK.

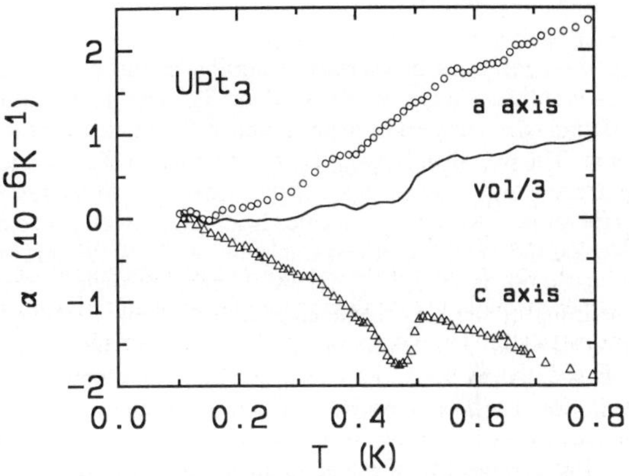

FIGURE 1
Coefficient of linear thermal expansion of UPt_3 for the a and c axis as indicated. The solid line represents $\alpha_v/3$.

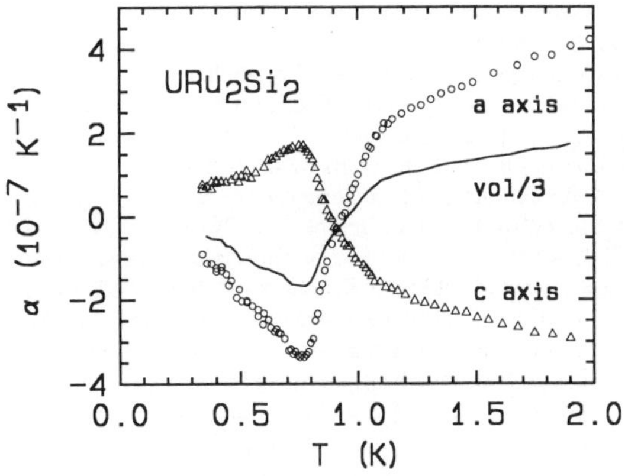

FIGURE 2
Coefficient of linear thermal expansion of URu_2Si_2 for the a and c axis as indicated. The solid line represents $\alpha_v/3$.

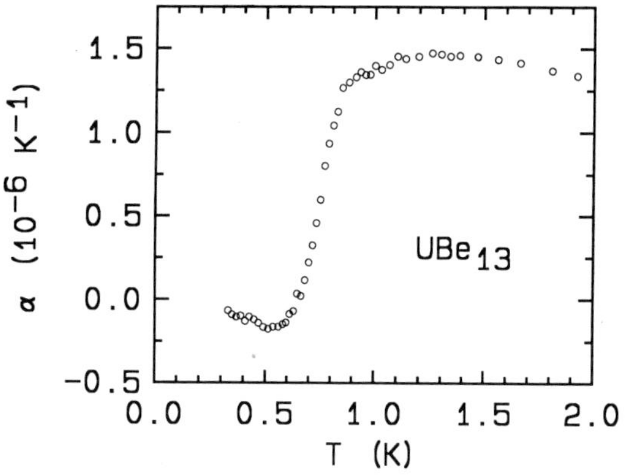

FIGURE 3
Coefficient of linear thermal expansion of UBe_{13}.

For the coefficient of thermal expansion in the basal plane ($\alpha_\perp$) no discontinuity is observed at T_c. The temperature variation of $\alpha_\perp$ changes from linear above T_c to quadratic there below. The overall anisotropy is preserved below T_c: the c-axis contracts and the basal plane expands with raising temperature. Recently, specific heat measurements have revealed the existence of two transitions at T_{c1}= 490 mK and T_{c2}= 430 mK in UPt_3 (8), that are thought to originate from a splitting of the superconducting transition by a symmetry breaking field. The transition at T_{c2} could not be detected from the data in Fig.1, but having increased the temperature resolution and the sensitivity of our dilatometer we have indeed resolved an anomaly at T_{c2} (9).

The coefficients of linear thermal expansion along ($\alpha_{//}$) and at right angles ($\alpha_\perp$) to the tetragonal axis of URu_2Si_2 have been measured in the temperature range 0.3-2 K (see Fig.2). The sample (size 5x5x5 mm^3) has been prepared by the Czochralski technique by A.A.Menovsky. Superconductivity sets in near 1.1 K, below which temperature broad anomalies are observed: $\alpha_\perp$ shows a rapid drop below T_c and attains a negative mimimum at 0.75 K, while $\alpha_{//}$ increases and has a positive maximum at 0.75 K. We would like to draw special attention to the reversal of anisotropy: well below T_c the tetragonal axis expands and the basal plane contracts, whereas in the normal state the contrary takes place.

The coefficient of linear thermal expansion of cubic UBe_{13} (prepared by J.L.Smith; sample size 2x4x6 mm^3) is shown in Fig.3, for $0.3<T<2$ K. The normal-state expansion shows a weak maximum near 1.3 K, that is attributed to the Kondo-lattice effect. Superconductivity sets in at 0.85 K, where α starts to fall rapidly. A negative mimimum is attained at 0.52 K. Similar data have been obtained by Ott (10).

For UPt_3 and URu_2Si_2 the volume expansion has been calculated from $\alpha_v = \alpha_{//} + 2\alpha_\perp$ (see Figs 1 and 2). For all three compounds we observe a strong reduction of α_v on entering the superconducting state, which indicates that a positive volume is associated with the formation of the ground state. The Grüneisen parameters drop from large positive values above T_c, to small negative values well below T_c.

4. DISCUSSION

In a first analysis the anomaly in α_v at T_c can be used to calculate the pressure dependence of T_c via the Ehrenfest relation: $dT_c/dP = V_m T_c \Delta\alpha_v/\Delta c_p$ (V_m is the molar volume, and $\Delta\alpha_v$ and Δc_p are the jumps in the volume expansion and specific heat at T_c, respectively). From the data in Figs 1-3 and published specific heat data we derive values for dT_c/dP of -16.3, -101 and -19.4 mK/kbar for UPt_3, URu_2Si_2 and UBe_{13}, respectively. The values for the depression of T_c can be compared with the ones obtained from resistivity measurements as function of pressure: -24 (11), -95 (12) and -16 mK/kbar (13), respectively.

The anisotropy in α near T_c can be used to determine the uniaxial stress effects on T_c. For instance, in the case of UPt_3 the anomaly at T_c for $\alpha_{//}$ indicates a decrease of T_c for uniaxial stress along the c-axis, whereas stress in the basal plane will hardly effect T_c (7). In the case of URu_2Si_2 stress along c will increase T_c, whereas stress applied in the basal plane will lead to a reduction of T_c. In order to study further the volume and shape effects at the superconducting transitions in the heavy-fermion superconductors thermal expansion measurements in an external field are in progress.

ACKNOWLEDGEMENTS

The work of A.d.V. has been made possible by a fellowship of the Royal Netherlands Academy of Arts and Sciences. A.L. is supported by CNPq (Brazil).

REFERENCES

(1) A. de Visser, J.J.M. Franse and A. Menovsky, J. Phys. F15 (1985) L53.
(2) A. de Visser, F.E. Kayzel, A.A. Menovsky, J.J.M. Franse, J. van den Berg and G.J. Nieuwenhuys, Phys. Rev. B34 (1986) 8168.
(3) A. de Visser et al., to be published.
(4) A. de Visser, J.J.M. Franse and J. Flouquet, Physica B161 (1989) 324.
(5) A. de Visser, J.J.M. Franse, A. Lacerda, P. Haen and J. Flouquet, Physica B, in print.
(6) A. de Visser, Ph.D. Thesis, University of Amsterdam (1986), unpublished.
(7) A. de Visser, A.A. Menovsky, J.J.M. Franse, K. Hasselbach, A. Lacerda, L. Taillefer, P. Haen and J. Flouquet, Phys. Rev. B41 (1990), in print.
(8) R.A. Fisher, S. Kim, B.F. Woodfield, N.E. Phillips, L. Taillefer, K. Hasselbach, J. Flouquet, A.L. Giorgi and J.L. Smith, Phys. Rev. Lett. 62 (1989) 1411.
(9) K. Hasselbach et al., to be published.
(10) H.R. Ott, Physica 126 B (1984) 100.
(11) K. Behnia, L. Taillefer and J. Flouquet, J. Appl. Phys., in print.
(12) M.W. McElfresh, J.D. Thompson, J.O. Willis, M.B. Maple, T. Kohara and M.S. Torikachvili, Phys. Rev. B35 (1987) 43.
(13) J.W. Chen et al., in: Proceedings of LT17, edited by U. Ekern, A. Schmid, W. Weber and H. Wuhl (North-Holland, Amsterdam, 1984) p. 325.

Physica B 165&166 (1990) 377–378
North-Holland

RF PENETRATION DEPTH IN SUPERCONDUCTING UPt_3 WITH DOUBLE HEAT CAPACITY ANOMALY

S.E. BROWN, Hauli LI, M.W. MEISEL
Department of Physics, University of Florida, Gainesville, FL 32611, USA.

J.L. SMITH, A.L. GIORGI, J.D. THOMPSON
Los Alamos National Laboratory, Los Alamos, NM 87545, USA.

The radiofrequency penetration depth has been measured in a polycrystalline sample of UPt_3 that has been observed to have a double–peak structure in the specific heat. At low temperatures, the penetration depth varies as T^4, consistent with similar measurements on single crystals. No change in the penetration depth, or its variation with temperature, is detected at the onset of the lower transition.

Several recent low temperature studies of UPt_3 have revealed an unusual phase diagram in the magnetic field vs. temperature plane for the superconducting state in this material.[1-4] While some features of the phase diagram are not always present, the measurements suggest the existence of a low temperature state, below T_c^-, which is distinct from the one present at the onset of superconductivity at T_c^+. Although a bit of confusion exists because some details of a particular measurement (such as the specific heat C_p) may vary from sample to sample,[3 5] the low temperature anomaly at T_c^- is thought to be associated with a change in symmetry of the superconducting state.[6]

If UPt_3 is an example of a superconductor with different symmetries, it is important to establish whether the transition between the states is continuous, and also how the transition manifests itself in different measurements. Herein, we report measurements of the penetration depth $\lambda(T)$ that were made at radio–frequencies (rf) in a polycrystalline sample that has shown the characteristic "double–bump" structure in the specific heat.[3] No evidence of a discontinuity at T_c^- is observed for $\lambda(T)$, or in its variation with temperature $d\lambda/dT$.

Using a low–power tunnel diode oscillating close to 14 MHz, $\lambda(T)$ was inferred by measuring the rf susceptibility of the sample. The resonant frequency is primarily determined by the inductance (L) and capacitance (C) of the tank circuit. We placed the sample in a 6 μH inductor, and the circuit had a stray capacitance, associated with the tunnel diode and lead wires, of ~ 25 pF. Since the experiments were the beginning of an investigation of the pressure dependent phase diagram, the inductor was loaded into a pressure bomb, set to exert ~ 1 bar at low temperatures. In the limit of high Q, changes in the inductance lead to resonant frequency changes of

$$\frac{\Delta\omega}{\omega} = -\frac{1}{2}\frac{\Delta L}{L}\left[1-\frac{2}{Q^2}\right] - \frac{2}{Q^2}\frac{\Delta R}{R}. \quad (1)$$

From the composition of the circuit elements, we estimate our Q to be on the order of 150 at 300 K with the sample inside; thus, the change in frequency is dominated by the change in inductance.

Because we are crossing from low temperatures into the normal state, it is necessary to verify that changes in the inductance are a result of changes in λ. In the absence of a complete description for the superconducting state of UPt_3, we will consider the simple picture provided by the London theory,[8] where the decay of an electromagnetic field into a sample is given by $\mathrm{Im}(1/K)$, with

$$-K^2 = 1/\lambda^2 + 2i/\delta^2, \quad (2)$$

where δ is the classical skin depth, $\delta^2 = c^2/2\pi\sigma\omega$. In the low frequency limit when $\delta >> \lambda$, the field penetration is determined by λ. However, in the regime just below T_c when λ is large, the analysis becomes somewhat more complicated because of eddy–current response from uncondensed carriers. Using Eq. 2, it is possible to estimate the importance of the quasiparticle response near T_c^-. From typical resistivity values for UPt_3,[7] the normal state δ is estimated to be several μm at 14 MHz. Since we presently do not have enough information to independently evaluate $\lambda(0)$, we will use the result of Ref. 9 of $\lambda(0) = 0.5$ μm. With these values, we conclude that near T_c^-, ~ 10% of the change in L that results from insertion of the sample is from the quasiparticles. Below T_c^-, the contribution falls off extremely fast. Therefore, we remain sensitive to the penetration depth up to temperatures exceeding T_c^-.

Our sample has been described previously in the literature.[10] It is polycrystalline and cut from an ingot that was annealed at 950°C for 90 hours. From Ref. 3, T_c^+ occurs at 470 mK, with the corresponding peak in C_p at ~ 400 mK. A shallow minimum in C_p was observed at ~ 365 mK with a second peak at 350 mK. A 50–60 mK split is typical for the difference $T_c^+ - T_c^-$.

Measurements of the resonant frequency were made from 60–700 mK with an overall change of 13.5 kHz out of 13.6 MHz. The entire range is shown in Fig. 1. As expected from the resistivity of UPt_3,[7] there is a variation of the normal state δ just above T_c^+. The onset of superconductivity is identified by the first break in slope at $T_c^+ = 470$ mK, precisely the same temperature as the C_p measurements.[3] Our data shows features similar to those reported by Shivaram *et al.*[4] on single crystals. In particular, a somewhat broad peak near the normal–superconducting phase transition is

0921-4526/90/$03.50

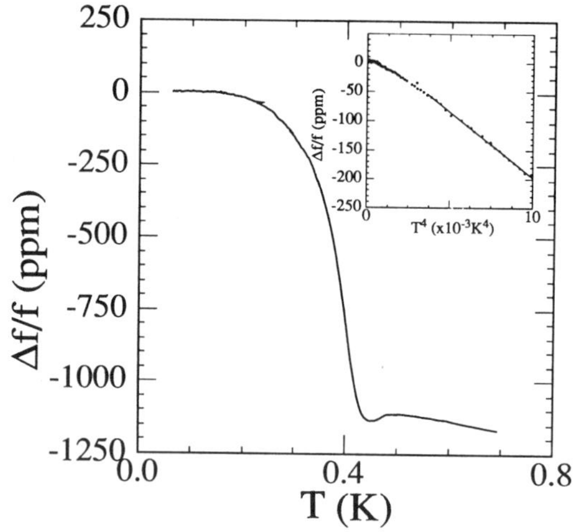

FIGURE 1. Change in resonant frequency vs. temperature over the range of the experiment. The inset shows the T^4 dependence at low temperatures.

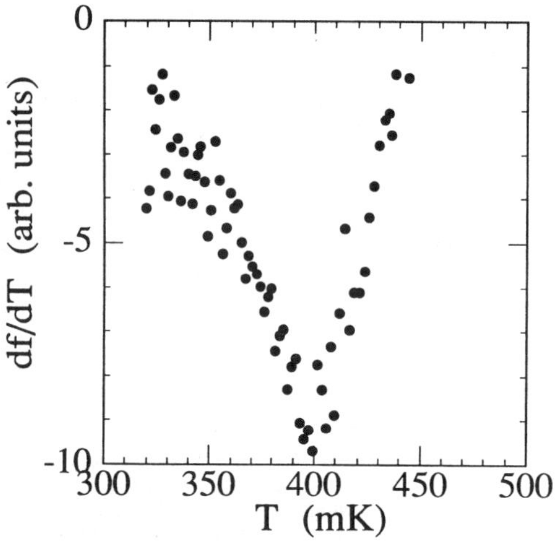

FIGURE 2. df/dT vs. T in the temperature range around $T_c^- = 360$ mK.

observed. Below this temperature, the frequency increases roughly as expected for a typical superconductor, with no distinguishable features arising from the lower transition which occurs at 365 mK.

A peak just below the superconducting temperature is also observed in conventional materials, and can be crudely thought of as a signature of the competetion between λ and δ. The details of the peak have been found to be extremely sensitive to the quality of the surface and impurity states.[11] Due to the strong surface dependence, it is not yet clear whether this affect explains the peak in UPt_3.

The change in frequency at low temperatures is plotted against T^4 in the inset of Fig. 1. The straight line indicates that our results are consistent with other rf measurements,[4] which show a T^4 variation for several orientations of the field with respect to the crystal axes, and at variance with the T^2 dependence observed using a dc SQUID.[9] Deviations at the lowest temperatures are likely attributable to thermal grandients between the sample and the thermometer, or to background Curie contributions.

An additional temperature sweep concentrating on the region near T_c^- was also made. The data appear completely smooth through this region. The derivative curve, plotted in Fig. 2, is also featureless, except for the maximum in the negative slope at 400 mK. Perhaps it is noteworthy that the point of largest negative slope coincides with the maximum of the high–temperature peak in the heat capacity measured by Fisher, *et al.*[3] on the same sample.

Since λ depends on the superfluid density, one might expect that a discontinuous transition would lead to a jump at T_c^-, whereas a continuous transition would result in a change in slope. For the former, we conclude an experimental upper bound of ~0.1%, and an upper bound on the change in slope of ~10%.

In summary, we have measured the rf penetration depth of UPt_3 at 14 MHz in a sample possessing the characteristic double transition in the specific heat. An upper bound on the change in the penetration depth is found to be ~0.1% for $\Delta\lambda/\lambda$ and ~10% for $[\Delta(d\lambda/dT)]/(d\lambda/dT)$. The lack of any detectable change could be related to the fact that the specific heat measurements probe the bulk material, whereas the rf susceptibility experiment studies the sample only to a depth of λ within the surface. At low temperatures, we see a T^4 dependence for λ, consistent with other rf measurements and in contrast to the T^2 dependence using a dc SQUID, suggesting that there is an unconventional frequency dependence below 1 MHz. We are presently checking for these possibilities.

S.E.B. and M.W.M. gratefully acknowledge partial support from UF DSR and discussions with Z. Fisk, P. Hirschfeld, P. Wölfle, W.O. Putikka, B.S. Shivaram, and J.S. Brooks. Work at Los Alamos was performed under the auspices of the U.S.Department of Energy.

1. R.N. Kleiman *et al.*, Phys. Rev. Lett. 62 (1989) 328.
2. A. Schenstrom *et al.*, *ibid.* 332.
3. R.A. Fisher *et al.*, *ibid.* 1411.
4. B.S. Shivaram *et al.*, Phys. Rev. Lett. 63 (1989) 1723, and (preprint).
5. J.L. Smith *et al.* (preprint).
6. D. Hess, T.A. Tokoyasu, and J.A. Sauls (preprint).
7. H.R. Ott and Z. Fisk, in Handbook of the Physics and Chemistry of the Actinides, ed. A.J. Freeman and G.H. Lander (Elsevier, Amsterdam, 1987), 85.
8. F. London, Superfluids (Wiley, New York, 1950).
9. F. Gross *et al.*, Physica C 153–155 (1988) 439.
10. It is sample #2 of Ref. 3, described therein as the μSR sample of D. Cooke, *et al.*, Hyperfine Interactions 31 (1986) 425.
11. C. Varmazis *et al.*, Phys. Rev. B 11 (1975) 3354.

Physica B 165&166 (1990) 379–380
North-Holland

THERMODYNAMIC PROPERTIES OF SUPERCONDUCTING SINGLE CRYSTALLINE $CeCu_2Si_2$

W. ASSMUS, W. SUN, G. BRULS, D. WEBER, B. WOLF and B. LÜTHI
Physikalisches Institut, Universität Frankfurt, SFB 252, D-6000 Frankfurt, FRG

M. LANG, U. AHLHEIM, A. ZAHN and F. STEGLICH
Institut für Festkörperphysik, Technische Hochschule Darmstadt, SFB 252, D-6100 Darmstadt, FRG

$CeCu_2Si_2$ single crystals (T_c : 0.63-0.68 K) have been studied utilizing calorimetric, dilatometric and ultrasound experiments. The strain dependencies of T_c along the a- and c-axis have opposite sign and indicate that superconductivity is favoured upon compensation of the local 4f moments. According to the $c_{11}(T,B)$ behaviour, T_c appears limited by an additional, presumably antiferromagnetic, instability of the Fermi liquid.

A great variety of exotic superconducting and magnetic properties of heavy-fermion (HF) superconductors, including non-exponential temperature dependencies of the specific heat, multi-component phase diagrams and the occurrence of magnetically ordered phases with very small moments have highlighted strongly anisotropic superconducting order parameters, $\Delta(\underline{k})$, and the relevance of magnetic coupling mechanisms [1]. Many problems remain as yet unsolved, e.g., the kind of $\Delta(\underline{k})$ anisotropy (conventional vs unconventional [2]) and the role [3] of the distinct coupling of HF to the breathing mode ("Grüneisen-parameter coupling") in the Cooper-pairing process.

Below we communicate thermodynamic properties (specific heat, thermal expansion, and elastic constant c_{11}) of $CeCu_2Si_2$ single crystals which, compared to previously studied crystals [4], exhibit larger dimensions, i.e., 1.7 x 1 x 1.5 mm^3 (#1) and 1.8 x 1.7 x 1.1 mm^3 (#2). The crystals were grown using the cold-boat technique [5]. Fig. 1 shows the specific heat of crystal #1 at zero magnetic field as C/T vs T. These results are very similar to earlier ones [6] on polycrystalline $CeCu_2Si_2$ samples with correspondingly high T_c (= 0.63 ± 0.015K). A strongly non-exponential C(T) variation is found below T_c which cannot be fitted by a power law. The data at $T \lesssim 0.1$K suggest an asymptotic C(T) dependence substantially stronger than T^2. The superconducting phase transition is characterized by a slightly broadened mean-field type anomaly which can be replaced by an idealized jump $\Delta C = (1.25 \pm 0.05)C_n(T_c)$, where $\gamma(T_c) = C_n(T_c)/T_c = 0.83$ J/K^2mole. The coefficient of the linear thermal expansion, $\alpha(T)$, is shown in fig. 2 for the same crystal measured along both the a- and c-axis. This measurement yields sharp discontinuities in $\alpha(T)$ at $T_c = (0.625 \pm 0.01)$K. Idealized jumps, $\Delta\alpha = \alpha_s(T_c)-\alpha_n(T_c)$, are found from an α/T vs T plot in the same way as described above for C/T vs T (fig. 1) : $\Delta\alpha_a = +(1.74 \pm 0.1)10^{-6}K^{-1}$ and $\Delta\alpha_c = -(2.8 \pm 0.1)10^{-6}K^{-1}$.

Combining the $\alpha(T)$ and C(T) results, we can derive the relative pressure derivatives of T_c for external pressure ($p_i \to 0$) applied along the respective a- and c-axis by using Ehrenfest's relation : $(\partial T_c/\partial p_i)_0 = T_c \cdot V_{mol} \cdot \Delta\alpha_i/\Delta C$; i : a = b, c. We find $\partial T_c/\partial p_a = +(9 \pm 1)$mK/kbar and $\partial T_c/\partial p_c = -(14.4 \pm 1)$mK/kbar. For the pressure derivative $\partial T_c/\partial p_h$ under hydrostatic pressure conditions, $\partial T_c/\partial p_h = 2\partial T_c/\partial p_a + \partial T_c/\partial p_c$, we get $\partial T_c/\partial p_h = (3.6 \pm 1.5)$mK/kbar, in excellent agreement with results from specific-heat experiments on polycrystalline $CeCu_2Si_2$ samples [7]. Since pressure applied along the a-axis will also lead to changes in the c-axis and vice versa, we were interested in determining the more essential uniaxial strain (ϵ_i) dependencies via : $\partial T_c/\partial\epsilon_a = -(c_{11}+c_{12})\partial T_c/\partial p_a - c_{13}\partial T/\partial p_c$ and $\partial T_c/\partial\epsilon_c = -2c_{13}\partial T_c/\partial p_a - c_{33}\partial T/\partial p_c$, where the elastic constants c_{ij} were determined from the sound-propagation experiments described below and a bulk modulus c_b = 1100 kbar [8] was used. This yields $\partial T_c/\partial\epsilon_a = -(12 \pm 4)$K and $\partial T_c/\partial\epsilon_c = +(5 \pm 3)$K. Reduction of the Ce-Ce separation (lattice parameter a) will reinforce the 4f-5d hybridization within the basal Ce-planes and raise the "lattice Kondo temperature" $T^*(\simeq 15$ K). This favours superconductivity as is proven by an increased T_c, cf. [6]. We stress that in a hydrostatic pressure experiment this effect predominates over the T_c depression due to the reduction of the c-axis. The latter observation has a less obvious explanation and may involve a strong reduction in the Cu-derived 3d density of states at the Fermi level.

Sound propagation measurements for temperatures between 0.08 and 1.1K at magnetic fields $B \lesssim 8$T have been performed for crystal #2. The results are shown in fig. 3 for the c_{11} mode (B ∥ a-axis). The Δc_{33} vs T and Δc_{33} vs B(∥c) dependencies are very similar to c_{11}, but the values are smaller by a factor of 10. Clearly, two step-like anomalies are observed for the isothermal (T' = 0.09K) measurement, i.e., at $B_{c2}(T')$ = 1.4T and $B_m(T')$ = 6.7T. They are ascribed to the respective superconducting transition and the high-field transition presumed to be an antiferromagnetic one [9]. Note that the "step" in $\Delta c_{11}(B)$ associated with the latter is smaller by a factor of five and of

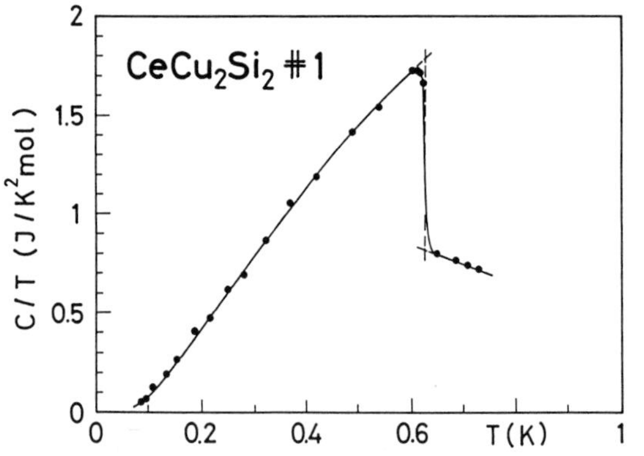

Fig. 1. Specific heat, C(T), of $CeCu_2Si_2$ single crystal #1 in a plot C/T vs T at B = 0T.

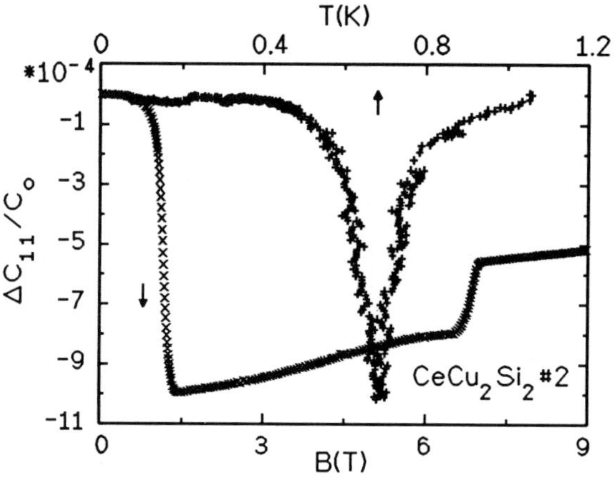

Fig. 3. Normalized change in the elastic constant, c_{11}, vs temperature at B = 0T (upper scale) and vs magnetic field B ∥ [100] (lower scale) at T' = 0.09K for $CeCu_2Si_2$ single crystal #2.

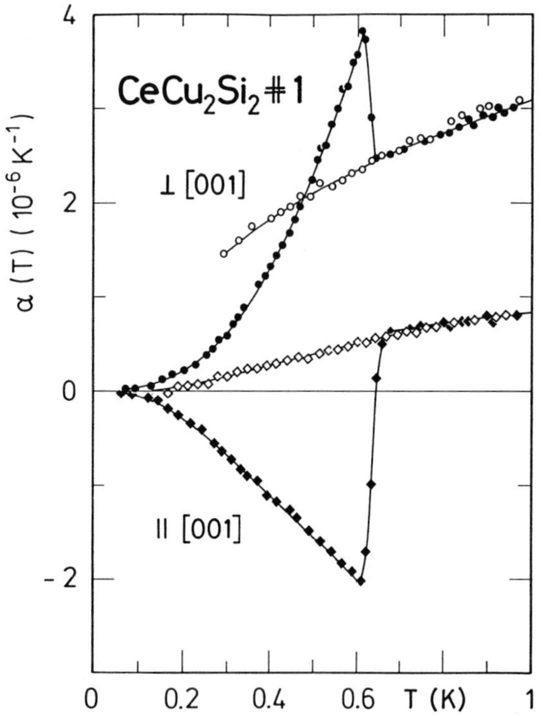

Fig. 2. Coefficient of thermal expansion, α, vs temperature for same crystal as in fig. 1, measured along [100] and [001] directions at B = 0T (full symbols) and B = 6T (open symbols).

opposite sign compared to the one at the superconducting transition. No magnetic-phase transition anomaly could be resolved either in C(T) at B = 0T or in $\alpha(T)$ measured at fixed B fields up to 6T. At B = 0T, for each of C(T), $\alpha(T)$ and $\Delta c_{11}(T)$ only one anomaly is found at T_c (= 0.68K, #2). Apparently, T_c coincides with the magnetic ordering temperature T_m. Though being of the same size as the change in $\Delta c_{11}(B)$ at the superconducting transition, this B = 0T anomaly is of a quite different shape. It indicates a fluctuation contribution to $\Delta c_{11}(T)$ when approaching $T_c \simeq T_m$. Utilizing this technique, we have determined the B-T phase diagrams for $CeCu_2Si_2$ single crystals with differing stoichiometry which will be discused in more detail in a forthcoming publication [10].

References

[1] Crystal Field Effects and Heavy Fermion Physics, W. Assmus, P. Fulde, B. Lüthi and F. Steglich (eds.), J. Magn. Magn. Mat. 76-77 (1988).
[2] P. Fulde et al., Solid State Phys. 41 (1988) 1.
[3] H. Razafimandimby et al., Z. Phys. B 54 (1984) 111. N. Grewe, Z. Phys. B 56 (1984) 111.
[4] H. Spille et al., Helv. Phys. Acta 56 (1983) 165. W. Assmus et al., Phys.Rev.Lett.52 (1984) 469.
[5] W. Sun, M. Brand, G. Bruls and W. Assmus, Z. Phys. B, in press.
[6] F. Steglich, Springer Series in Solid-State Sciences 62 (1985) 23.
[7] A. Bleckwedel and A. Eichler, Solid State Commun. 56 (1985) 693.
[8] I.L. Spain et al., Physica 139&140B (1986) 449.
[9] F. Steglich, J. Phys. Chem. Solids 50 (1989) 235, and references cited therein.
[10] G. Bruls et al., to be published.

Physica B 165&166 (1990) 381–382
North-Holland

^{195}Pt NMR STUDY IN UPt_3 AND $U(Pt\text{-}Pd)_3$

Yoh KOHORI, Masafumi KYOGAKU*, Takao KOHARA, Kunisuke ASAYAMA*, Hiroshi AMITSUKA**, and Yoshihito MIYAKO**

Basic Research Laboratory, Himeji Institute of Technology, Himeji, Hyogo, 671-22, Japan
*Department of Material Physics, Faculty of Engineering Science, Osaka University Toyonaka 560, Japan
**Department of Physics, Faculty of Science, Hokkaido University, Sapporo 060, Japan

Temperature dependence of $1/(T_1T)$ of ^{195}Pt in UPt_3 indicates that $\chi''(Q_{AF},\omega_0)$ increases gradually with decreasing temperature for 5K<T<20K, and becomes temperature independent below 5K. No anomaly is observed at 5K, where T_N is expected. $1/(T_1T)$ in $U(Pt\text{-}Pd)_3$ system indicates that U spin fluctuations in $U(Pt_{0.99}Pd_{0.01})_3$ are nearly the same as those in UPt_3, and increase near T_N in $U(Pt_{0.95}Pd_{0.05})_3$.

Heavy fermion systems discovered in rare earth or actinide intermetallic compound have been a subject of extensive research in recent years. Among them, $CeCu_2Si_2$, UBe_{13} and UPt_3 have occupied a unique position, whereby they exhibit transition to the superconducting state at low temperatures. Heavy fermion state appears nearby the magnetically ordered state. The long range ordering appears with the substitution of a small amount of impurities. In UPt_3 the antiferromagnetic ordering appears with Th or Pd doping [1-3]. In UPt_3, the antiferromagnetic order with the Neel temperature, 5K, is observed in the neutron diffraction experiment [4]. In this paper, we report the results of ^{195}Pt NMR study in UPt_3 and $U(Pt_{1-x}Pd_x)_3$.

A conventional pulsed NMR spectrometer and the superconducting magnet were used for the nuclear spin lattice relaxation time, T_1, measurements.

The spectrum arising from randomly oriented powders of UPt_3 has large negative shift and is quite anisotropic. The Knight shift analysis is published in the preceding paper [5]. NMR signal has been observed in wide temperature range by using the magnetically oriented powder samples, whose c-axis aligns perpendicular to external magnetic field [5]. The transverse component (to hexagonal c-axis of UPt_3) of the Knight shift, $K_\perp$, has a broad maximum around 20K, which is similar to the transverse component (to c-axis) of the susceptibility [6]. We measured T_1 in oriented samples(c-axis⊥H), $(T_1)_{c\perp H}$. The result of UPt_3 is shown in Fig.1. $1/(T_1T)_{c\perp H}$ is temperature independent for 0.5K<T<5K. Above 5K, $1/(T_1T)_{c\perp H}$ decreases gradually with increasing temperature. The longitudinal

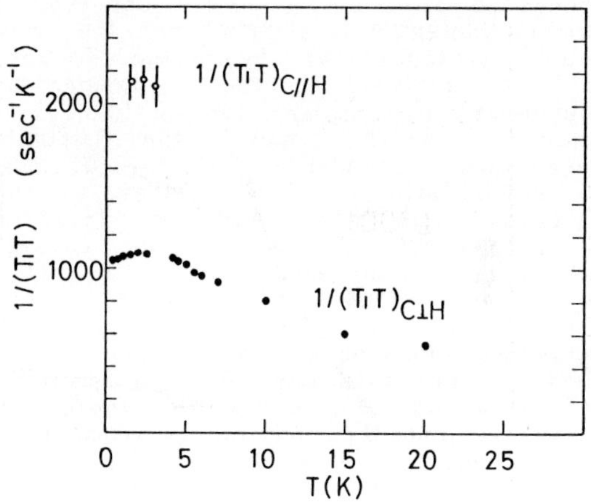

FIGURE 1
Temperature dependence of $1/(T_1T)_{c\perp H}$ and $1/(T_1T)_{c\parallel H}$ of ^{195}Pt in UPt_3.

component of $1/(T_1T)$ (c-axis∥H), $1/(T_1T)_{c\parallel H}$, is also measured at the edge of the spectrum obtained by the randomly oriented powders, which is also shown in the figure. This component is about 2 times larger than the transverse component, and is also temperature independent at low temperatures.
$1/T_1$ is expressed as,

$$1/T_1=2\gamma_N{}^2A^2k_BT\Sigma_q\chi_\perp''(q,\omega_0)/\omega_0,$$

where A is the hyperfine coupling constant, γ_N is the nuclear gyromagnetic ratio, ω_0 is the nuclear Larmor frequency, and $\chi_\perp''$ is the imaginary part and the perpendicular component (to the quantization axis of nuclear spin) of the dynamical susceptibility. The z-axis of nuclear moment (the quantization axis) is defined to be parallel to the external field.

UPt_3 is antiferromagnetic or at least nearly antiferromagnetic compound. Then $\chi_\perp''(q,\omega_0)$ will be peaked at the antiferromagnetic wave vector, Q_{AF}, and will provide the dominant contribution to $\Sigma_q \chi_\perp''(q,\omega_0)/\omega_0$. We consider that the temperature dependence of $1/(T_1T)_{c\perp H}$ indicates that $\chi_\perp''(Q_{AF},\omega_0)$ increases with decreasing temperature for 5K<T<20K, and becomes temperature independent below 5K. The temperatue dependence of $\chi_\perp''(Q_{AF},\omega)$ is different from that of the uniform susceptibility, which has a broad maximum around 20K, and decreases with decreasing temperature. The temperature dependence of $\chi_\perp''(Q_{AF},\omega)$ is observed in neutron inelastic scattering at the wave vector $Q_{AF}=(0,0,1)$ [4]. From the Kramers-Kronig analysis, the temperature dependence of $\chi(Q)$ for $Q=(0,0,1)$ is obtained, which has the similar temperature dependence of $\chi_\perp''(Q_{AF},\omega_0)$. The neutron data also show that $\chi''(Q_{AF},\omega)$ has maximum at q= (±1/2,0,2) in low energy ($\hbar\omega$=0.5meV). We consider that the dominant contribution to $1/(T_1T)$ is arising from this wave vector.

$1/(T_1T)_{c\parallel H}$ is determined by the transverse component (to c-axis) of the dynamical susceptibility. On the contrary, $1/(T_1T)_{c\perp H}$ is determined by both the transverse and the longitudinal components (to c-axis) of the dynamical susceptibility. In order to explain the anisotropy of $1/(T_1T)$ in Fig.1, we have to assume that the transverse component of the dynamical susceptibility is much larger than the longitudinal component.

$1/T_1$ of ^{195}Pt is measured in $U(Pt_{1-x}Pd_x)_3$ system for x=0.01 and 0.05. The results are shown in Fig.2 together with UPt_3. For $U(Pt_{0.99}Pd_{0.01})_3$, $1/T_1$ is nearly the same as that of UPt_3. Below 5K, $1/T_1$ is proportional to T, which indicates that the system sets in Fermi liquid state. Above 10K, $1/T_1$ is nearly temperature independent. The temperature dependence and the absolute value of $1/T_1$ are nearly the same as those of UPt_3. The charactor of U spin fluctuations is not affected with 1 at % Pd doping. With increasing Pd concentrations, the antiferromagnetic ordering appears [7]. In $U(Pt_{0.95}Pd_{0.05})_3$, T_N is 5K. Pt NMR is observed in the paramagnetic state above

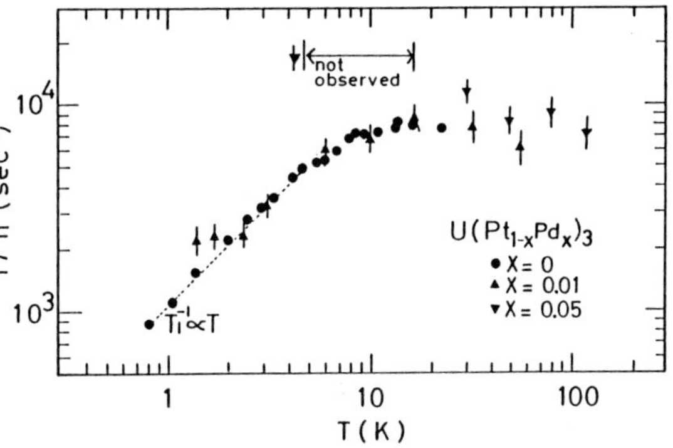

FUGURE 2
Temperature dependence of $1/T_1$ of ^{195}Pt in $U(Pt_{1-x}Pd_x)_3$. We could not observe NMR signal in $U(Pt_{0.95}Pd_{0.05})_3$ for 5K<T<20K.

30K, where $1/T_1$ is slightly larger than that of UPt_3. U spin fluctuations are nearly the same as those of UPt_3 at temperature higher than T_N. With decreasing temperature, the spin echo decay time becomes shorter and shorter associated with the critical slowing down, which makes the observation of NMR signal difficult for 5K<T<20K.

The signal is again observed in the ordered state. $1/T_1$ at 4.2K is shown in the figure, which is about 4 times larger than that of UPt_3. At 4.2K, which is near T_N, large U spin fluctuations still remain. This behavior is different from that of UPt_3, where the critical slowing down at T_N or the the rapid decrease of $1/T_1$ associated with the opening of the spin density wave below T_N is not observed.

References

[1] A.de.Visser, J.C.Klaase. M.Van.Sprang. J.J.M.Franse and A.A.Menovsky, J. Magn. Magn. Mat. 54-57(1986)375.
[2] B.Batlogg, D.J.Bucher, B.Golding Jr. A.P.Ramiez, Z.Fisk, J.L.Smith and H.R.Ott, J. Magn. Magn. Mat. 63-64 (1987)441
[3] G.R.Stewart, A.L.Giorgi, J.O.Willis and A.S.Cooper, Phys. Pev. Lett. 57(1986) 1072.
[4] G.Aepli, E.Bucher, A.I.Goldman, G.Shirane C.Broholm and J.K.Kjems, J. Magn. Magn. Mat. 76-77(1988)385.
[5] Y.Kohori, M.Kyogaku, T.Kohara, K.Asayama, H.Amitsuka and Y.Miyako Submitted to MPT90 (Yamada Conference)
[6] P.Frings, J.J.M.Franse, F.R.de Boer and A.Menovsky, J. Magn. Magn. Mat. 31-34 (1983)240
[7] J.J.M.Franse, K.Kadowaki, A.A.Menovsky, M.van Sprang, and A.de Visser, J. Appl. Phys. 61(1987)3380

Physica B 165&166 (1990) 383–384
North-Holland

CHARACTERISTIC TEMPERATURES IN THE HEAVY-FERMION SYSTEM $U(Pt,Pd)_3$

J.J.M.FRANSE, H.P.VAN DER MEULEN and A.DE VISSER

Natuurkundig Laboratorium der Universiteit van Amsterdam,
Valckenierstraat 65, 1018 XE Amsterdam, The Netherlands.

A summary is presented of the characteristic temperatures in the series of $U(Pt,Pd)_3$ heavy-fermion compounds as derived from thermodynamic and transport measurements.

1. INTRODUCTION

The heavy-fermion system $U(Pt,Pd)_3$ is an outstanding example in which many ingredients that are characteristic for heavy-fermion behaviour in cerium and uranium intermetallics are found to be present: a strongly correlated electron gas leading to large values for the electronic coefficient in the specific heat, spin-fluctuation and/or Kondo-type of behaviour in the resistivity, superconductivity, long-range antiferromagnetic order, short-range antiferromagnetic correlations etc. (1), see fig.1. Each of these phenomena can be represented by a characteristic temperature. These characteristic temperatures are well defined for the transition to an ordered state but are less established for the other phenomena. In this contribution values for these characteristic temperatures are collected as they can be deduced from mostly published data on the thermomagnetic and transport properties of the $U(Pt,Pd)_3$ system.

2. CHARACTERISTIC TEMPERATURES

Heavy-fermion systems are well known for their anomalous low-temperature behaviour in the specific heat that becomes prominent below a certain temperature. In general, the coefficient of the linear term in the specific heat is considered to be inverse proportional to the characteristic temperature, T^*, of the strongly interacting electron gas. Since no proper theory for describing the heavy-fermion part in the specific heat is available at present, the definition of this temperature on the basis of the specific heat results is rather ambiguous. One way to arrive at a value for this temperature is by adopting the result for a single-ion (spin-1/2) Kondo model: $\lim_{T\to 0} c/T = 0.68\ R/T_K$, where R is the gas constant and T_K the Kondo temperature (2). Identifying the characteristic temperature T^* with this Kondo temperature will lead in case of UPt_3 with $\lim_{T\to 0} c/T = \gamma^*$ equal to 420 mJ/mol K^2, to a value of 13.2 K. This identification, however, is not justified because of the coherence effects that dominate the low-temperature phenomena. Nevertheless, we shall use the specific-heat data that are available for the 1, 2, 5, 7, 10 and 15 at % Pd in order to derive in the above-given way the change in the characteristic temperature T^* with Pd content. There is an additional difficulty to study the variation of the heavy-fermion effect in the $U(Pt,Pd)_3$ system in this manner because of the long-range antiferromagnetic order below 6 K that sets in for Pd concentrations between 1 and 10 at %. The specific-heat anomaly associated with this long-range antiferromagnetic order is still present at temperatures down to 1.5 K and hampers the evaluation of the remaining heavy-fermion contribution to the specific heat.

Another method to follow in deriving a value for T^* is to separate the heavy-fermion contribution to the specific heat from the remaining contributions on the basis of an experimental approach. The method that has been followed for UPt_3 and $U(Pt_{0.95}Pd_{0.05})_3$ is the evaluation of the heavy-fermion part to the specific heat by applying a Grüneisen analysis (3). This analysis is most successful in compounds where different contributions with largely different values for the corresponding Grüneisen parameters are present. Extremely large values for the electronicGrüneisenparameter is a typical feature of most heavy-fermion compounds. In

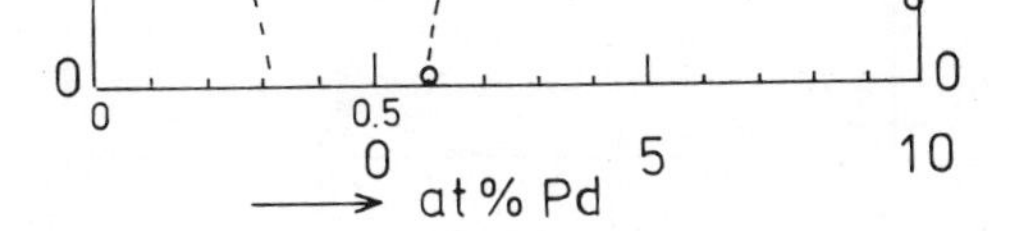

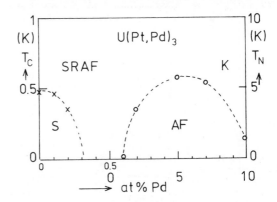

FIGURE 1
Phase diagram for the $U(Pt,Pd)_3$ system: S indicates the superconducting phase, AF the long-range antiferromagnetically ordered state, K the Kondo regime and SRAF the region where the short-range antiferromagnetic correlations are most pronounced.

UPt_3, this parameter takes a value of 71 to be compared with values of about two for the lattice and normal electronic contributions. In such a situation a combined analysis of specific-heat and thermal-expansion data can lead to a splitting of the heavy-fermion part and the lattice and normal electronic contributions to specific heat and thermal expansion. As a result of this analysis, a heavy-fermion contribution to the specific heat is obtained in a phenomenological way that shows some resemblance with the spin-1/2 single-ion Kondo specific heat. The entropy of this contribution turns out to saturate above 100 K and reaches a saturation value of almost exactly R ln 2. The heavy-fermion contribution has its maximal value around 12 K, not far from the temperature of 13.2 K deduced above. The Grüneisen analysis is more complicated in case of the $U(Pt_{0.95}Pd_{0.05})_3$ compound where long-range antiferromagnetic order sets in below 6 K. Analysing the data above 6 K, however, a clear shift of the heavy-fermion curve to lower temperatures is found, resulting in a three times smaller value for the temperature where the maximum in the heavy-fermion part of the specific heat is expected (4). This temperature (4.4 K) is given in fig.2 as well. Other

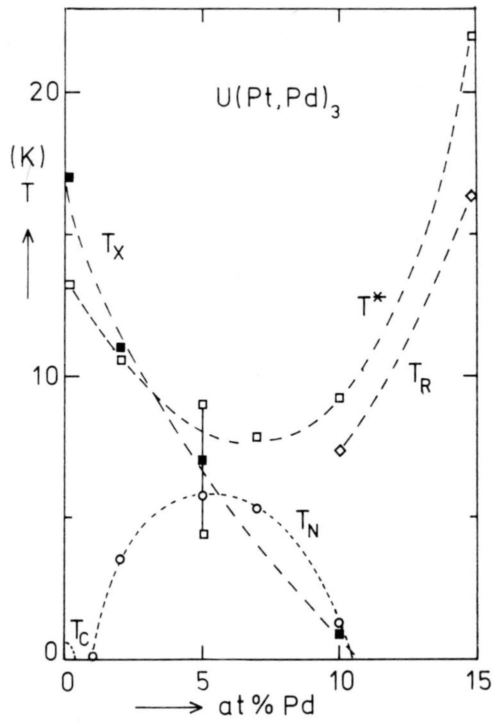

FIGURE 2
Characteristic temperatures for the $U(Pt,Pd)_3$ system: T^* from the specific heat, T_X from the susceptibility, T_R from the resistivity; T_C and T_N are the superconducting transition and the Néel temperature, respectively; data from ref.5.

characteristic temperatures are the superconducting transition temperature, T_C, and the Néel temperature, T_N, as deduced from the peak in the specific heat of the relevant compound. Two other characteristic temperatures remain to be discussed: the temperature T_X, reflecting the temperature where a maximum in the susceptibility is found in the hexagonal plane and the temperature T_R, a temperature characterising the temperature dependence of the resistivity. This latter temperature is rather badly defined because of the change-over from spin-fluctuation to Kondo-type of behaviour in the resistivity curves. An appropriate definition of T_R for UPt_3 is the temperature where the temperature derivative of the resistivity takes its maximum. Going from pure UPt_3 to the 10 at % Pd alloy, however, the quadratic temperature dependence at the lowest temperatures changes into a typical Kondo curve with increasing resistivity values for decreasing temperatures. For the 10 and 15 at % Pd alloys, T_R is best defined by the maximal (negative) slope in the ρ vs T curve. Only these results for T_R are shown in fig.2.

3. CONCLUDING REMARKS

At higher Pd concentrations, the characteristic temperatures T^* and T_R both increase with increasing Pd content, whereas T_N decreases, leading to a loss of antiferromagnetism above 10 at % Pd. This behaviour fits into a general picture in which the Kondo interactions become dominant over the RKKY interactions for larger values of the exchange coupling constant. It is tempting to deduce from this result an increase in the exchange constant with increasing Pd content.

At low Pd concentrations, the large values for T_X indicate strong short-range antiferromagnetic correlations. These correlations are weaker in the compounds where long-range antiferromagnetic order exists. These short-range correlations certainly contribute to the specific heat and consequently to the value of T^*. It must be concluded for that reason, that the T^* values within this range of Pd concentration are not directly comparable. This conclusion is supported by the change in sign of the initial field dependence of the coefficient γ^* between 7 and 10 at % Pd.

ACKNOWLEDGEMENTS

The work of A.d.V. has been made possible by a fellowship of the Royal Netherlands Academy of Arts and Sciences.

REFERENCES

(1) J.J.M.Franse, K.Kadowaki, A.Menovsky, M.van Sprang and A.de Visser, J.appl.Phys. 61 (1987) 3380.
(2) N.Andrei, K.Furuya and J.K.Loewenstein, Rev.Mod.Phys. 55 (1983) 331.
(3) J.J.M.Franse, M.van Sprang, A.de Visser and P.E.Brommer, Physica B 154 (1989) 379.
(4) P.E.Brommer, M.van Sprang and J.J.M.Franse, Physica B 161 (1989) 337.
(5) A.de Visser, A.A.Menovsky and J.J.M.Franse, Physica 147B (1987) 81.

Physica B 165&166 (1990) 385–386
North-Holland

LOW TEMPERATURE SPECIFIC HEAT OF URu_2Si_2 NEAR THE SUPERCONDUCTING TRANSITION

M.A. López de la Torre, S. Vieira, R. Villar, M.B. Maple(*) and M.S. Torikachvili(*)

Departamento de Física de la Materia Condensada, C-III, Universidad Autonoma de Madrid, Cantoblanco 28049-Madrid, Spain. (*) Department of Physics and Institute for Pure and Applied Physical Sciences, University of California, San Diego, La Jolla, California 92093, USA.

We report the observation of a new anomaly in the low temperature specific heat of the heavy-fermion superconductor URu_2Si_2 and the possible existence of spin fluctuations occurring well below the magnetic phase transition at T_N = 17.5 K. Electrical resistivity measurements confirm our heat capacity results.

1. INTRODUCTION

URu_2Si_2 is a heavy-fermion system in which superconductivity coexists with antiferromagnetic (SDW) ordering (1-3). This feature makes the study of this material very interesting. It has been speculated that superconductivity in heavy-fermion systems may be non-BCS in origin. Spin-one pairing as well as anisotropic energy gaps with nodes or lines of zeros on the Fermi surface have been suggested. In UPt_3 specific heat measurements seem to support some sort of unconventional superconductivity (4,5). Pairing could be induced by spin fluctuations which are present in the normal state. We have carried out this work with the aim of finding similar indications of unconventional superconductivity (a multicomponent superconducting transition, for example) in URu_2Si_2.

2. EXPERIMENTAL

Specific heat measurements were performed in a pumped He^3 cryostat using an adiabatic calorimeter and the standard heat pulse method. Electrical resistivity measurements were made in the same cryostat using the standard four-lead technique with a dc current adjusted to avoid self-heating. The temperature range of our measurements was between 0.5 K and 20 K. The temperature was measured using calibrated germanium resistance thermometers.

Our samples were cut from the same polycrystalline ingot and consisted of a rod (length ∿ 10 mm, diameter ∿ 3.1 mm) for electrical resistivity and a cylinder (6.3 mm heigth and a mass of 2.00 g) for specific heat. Microprobe analysis did not reveal the existence of impurities or a significant amount of any second phase.

3. RESULTS

Our results of specific heat measurements below 10 K are displayed in Fig. 1 as C/T vs. T. We observe a first peak at T_{c1} ∿ 1.15 K and a second "step" at T_{c2} ∿ 2.7 K. In Fig. 2 the

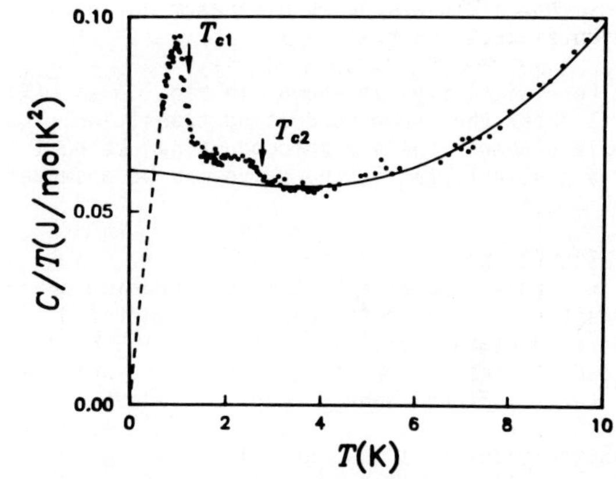

Figure 1.
Specific heat of URu_2Si_2 as C/T vs. T, with a fit to Eq. 2 (A = 60.4 mJ mol^{-1} K^{-2}, B = -1.19 mJ mol^{-1} K^{-4}, C = .685 mJ mol^{-1} K^{-4}) which includes a spin fluctuation contribution. The dashed line presents the entropy conservation construction cited in the text.

same results are shown as C/T vs. T^2. Above T_{c2} it is only possible to fit our experimental results to an expression of the form $C/T = a + bT^2$ (Eq. 1) with reasonable validity between 4 K and 8 K. However, a fit in the form $C/T = a + bT^2 + cT^2 \ln T$ (Eq. 2) is a fair approximation to our results in a less restricted range: from 3 K to 9 K (see Fig. 1). Such a fit represents the experimental results within 2% - 3%, which is an amount of the same order as the experimental uncertainty. Using this fit and the extrapolation of Fig. 1, we obtain an entropy at T_{c2} for the experimental results only 5% greater than that calculated from Eq. 2. Similar results were obtained by Fisher et al (4) for the double superconducting transition in UPt_3.

0921-4526/90/$03.50

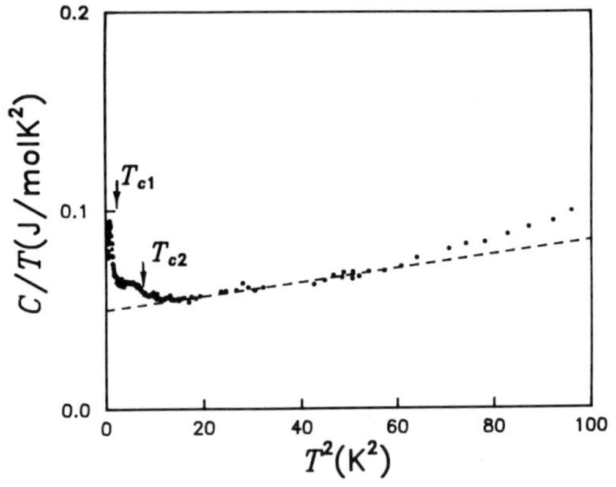

Figure 2.
Specific heat of URu_2Si_2 as C/T vs. T^2 which shows that a linear plus cubic term do not fit the data above 8 K.

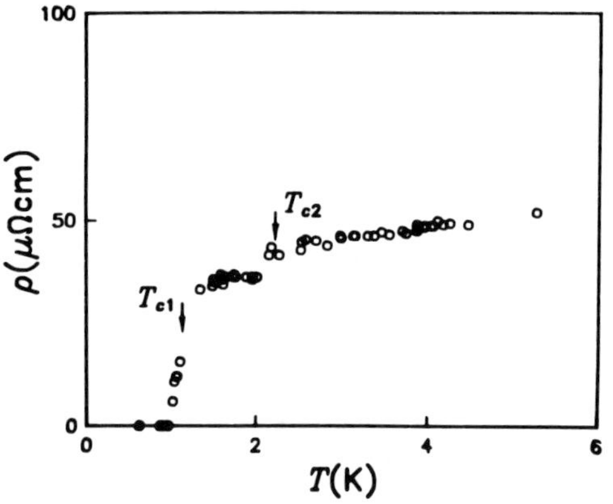

Figure 3.
Electrical resistivity of URu_2Si_2 near the superconducting transition. A second step is clearly visible above T_{c1}.

The resistivity is shown in Fig. 3 as ρ(T) vs. T near the superconducting transition. The sample becomes fully superconducting below 1 K and a distinct step in the curve may be observed at T ∿ 2.6 K.

4. DISCUSSION

In this material two electronic phase transitions have been reported: a superconducting transition at T_{c1} ∿ 1.5 K and a second BCS transition at T_N ∿ 17.5 K analyzed in terms of an energy gap of about 11 meV opening over part of the Fermi surface, presumably due to a SDW instability at T_N (Maple et al. (1)). Neutron scattering measurements revealed the development of an antiferromagnetic ordering with small ordered moment (U ∿ 0.03 μ_B) (6). This antiferromagnetic structure coexists with superconductivity. Recent results have suggested the possible existence of spin fluctuations and/or another magnetic transformation well below T_N (7,8). It is likely that the T^3lnT term in the specific heat can be attributed to these spin fluctuations.

A second-order magnetic transition may be the first possible explanation for the step at 2.7 K, because in considering URu_2Si_2 we can not discard some sort of rearrangement of the spin density wave below T_N. Such behaviour has been observed in other SDW systems (9). The order of the associated transition would depend on the form of this rearrangement. An alternative explanation could be a multicomponent superconducting transition as was observed in UPt_3 (4,5). In this form, the step in the ρ(T) vs. T curve is easily explained.

For a magnetic transition the shape of the electrical resistivity anomaly would depend on the type of rearrangement of the magnetic moments.

CONCLUSION

Our specific heat and electrical resistivity measurements in URu_2Si_2 show strong indications of the existence of a new phase transition in the vicinity of the superconducting one. The characterization of this transition requires further investigation.

ACKNOWLEDGEMENTS

We thank Dr. J. Ibáñez for microprobe analysis of our samples. Financial support from Comité Conjunto Hispano USA and DGICyT (Proyecto MAT88-0717) is also acknowledged.

REFERENCES

(1) M.B. Maple et al., Phys. Rev. Lett. 56 (1986) 185.
(2) W. Schlabitz et al. Z. Phys. B62 (1986) 171-177.
(3) T.T.M. Palstra et al., Phys. Rev. Lett. 55 (1985) 2727.
(4) R.A. Fisher et al., Phys. Rev. Lett. 62 (1989) 1411.
(5) K. Hasselbach et al., Phys. Rev. Lett. 63 (1989) 93.
(6) C. Broholm et al., Phys. Rev. Lett. 58 (1987) 1467.
(7) K.J. Sun et al., Phys. Rev. B40 (1989) 11284.
(8) T. Fukase et al. Japanese Journal of Applied Physics, Vol. 26 (1987) Supplement 26-3.
(9) "Antiferromagnetism in Metals", A. Arrott in "Magnetism", v. IIB (Edited by G.T. Rado and H. Suhl), Academic Press, 1966.

Physica B 165&166 (1990) 387–388
North-Holland

MAGNETISM AND PHASE - DIAGRAM OF HEAVY - FERMION COMPOUNDS

B. WELSLAU, N. GREWE

Institut für Festkörperphysik, Technische Hochschule Darmstadt, Hochschulstr. 8,
D - 6100 Darmstadt /FRG

Recent neutron - scattering experiments concerning the magnetic behaviour of Ce $(Cu_{1-x}Ni_x)_2$ Ge_2 with varying x are interpreted via the Kondo - lattice picture of Heavy - Fermion systems. The occurence of small ordered moments and small modulation wave - vectors finds a natural explanation in terms of itinerant antiferromagnetism of heavy quasi - particles. A study of the full susceptibility of the low - temperature Fermi - liquid allows for a discussion of possible deviations from a Lorentzian line shape.

The remarkable magnetic phase - diagram of $Ce(Cu_{1-x}Ni_x)_2Ge_2$ as studied by neutron - scattering [1] seems to furnish a good realization of the concept of band - magnetism of Heavy - Fermions (HF), as proposed in the last years on the basis of pertubational studies of the Anderson - model in the Kondo - lattice regime [2,3]. For sufficiently large x ($x \geq 0.5$) the characteristic low - temperature scale T* ("Kondo -lattice temperature ") distinctly exceeds the magnetic transition temperature T_N, indicating that moment - compensation and HF - formation is completed to a large extent before magnetic correlations develop. Whereas $T^*(x)$ rises with x up to $T^*(x = 1) \approx 30K$, T_N drops monotonously to below 2 K near $x \approx 0.7$. The magnetic order below T_N involves a large modulation - length of about 10 lattice - constants and the ordered moment seems to be smaller than $0.2\mu_B$. The good Lorentzian fit for the quasi - elastic scattering lines at temperatures well between T_N and T* points at a quasi - local magnetic relaxation mechanism, consistent with only weakly correlated compensation clouds for individual local moments [3].

In an effective site picture of the Kondo - lattice, which can be founded on the LNCA - approximation to the periodic Anderson - model [4], a quasi - local susceptibility $\chi_0(\nu, T)$ is derived with the following properties [4,7] :

$$\chi_0(0,T) \rightarrow \frac{1}{T^*}(T \rightarrow 0), \sim \frac{1}{T}(T \gg T^*) \qquad (1)$$

$$\chi_0(\nu,T) \approx \frac{1}{\Gamma + i\nu}, \Gamma \sim T^*(T \ll T^*). \qquad (2)$$

A ladder summation for the particle - hole propagator of the full interacting system, including local quasi-particle repulsion, leads to a Stoner - like expression for the magnetic susceptibility χ,

$$\chi^{-1}(\underline{q}, \nu) = \chi_0^{-1}(\underline{q}, \nu) - K(\underline{q}, \nu, T), \qquad (3)$$

in which the quasi - local term is corrected for by an interaction term. This contribution K can induce cooperative magnetic behaviour, either of RKKY - type or of a type which may best be described as exchange splitting of HF - bands, depending on whether the r.h.s. of equ.(3) becomes small at a temperature T> T* or T < T*.

Although quasi - particle damping is naturally taken into account in the LNCA - formalism, we chose for simplicity well defined quasi - particle bands $\tilde{\varepsilon}_{\underline{k},\alpha}$ for the present calculation, in which case the interaction contribution at all temperatures can be expressed in the form

$$K(\underline{q}, i\nu, T) = \frac{F(T)}{N} \sum_{\underline{k}} \sum_{\alpha,\alpha'} a_{\underline{k},\alpha} a_{\underline{k}+\underline{q},\alpha'} \cdot \frac{f(\tilde{\varepsilon}_{\underline{k}+\underline{q},\alpha'}) - f(\tilde{\varepsilon}_{\underline{k},\alpha})}{\tilde{\varepsilon}_{\underline{k},\alpha} - \tilde{\varepsilon}_{\underline{k}+\underline{q},\alpha'} + i\nu} \qquad (4)$$

$\tilde{\varepsilon}_{\underline{k},\alpha}$ approaches the unperturbed band - structure at high temperatures T≫T* and exhibits the typical hybridization structure leading to a narrow resonance with a pseudo - gap near the Fermi - level for T≪T*. The $a_{\underline{k},\alpha}$ are the weights of the quasi - particle resonances and the prefactor F(T) measures the temperature - dependent local interaction strength [2].

In our model - calculations we chose a tetragonal crystal symmetry for the unperturbed band - structure and modelled a variation of the composition - parameter x as

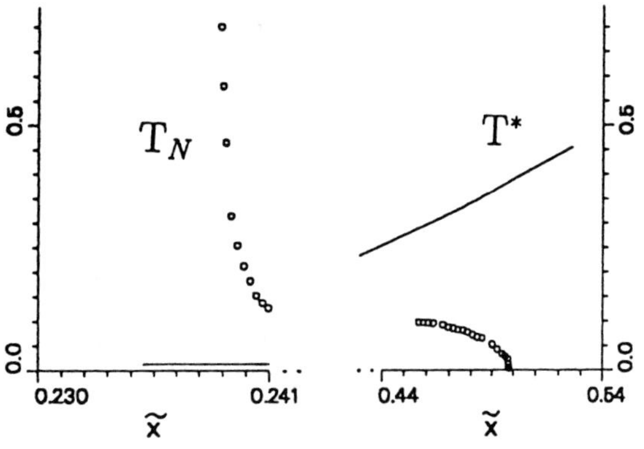

Figure 1: Magnetic transition-temperature T_N and characteristic temperature T^* as a function of $\tilde{x} = \mathcal{N}_F \cdot J$.

a variation of the effective exchange coupling $\tilde{x} = \mathcal{N}_F J$: $J \sim \frac{V^2}{|\Delta E|}$ is taken to increase with the hybridization matrix element V as a consequence of a decreasing lattice constant. The LNCA-theory then gives the functions $T^*(\tilde{x})$ and, via the vanishing r.h.s. of equ.(3), $T_N(\tilde{x})$. They are plotted in fig.(1). In a qualitative sense, Doniach's estimate for a competition of RKKY-interaction and moment compensation is reflected here [5] : on the l.h.s. (small $\tilde{x}$) $T_N \gg T^*$ is essentially determined by low order perturbational processes $\sim J^2$, which couple nearly unrenormalized local magnetic moments. On the r.h.s. (large $\tilde{x}$), however, the moments remaining at $T \approx T_N \ll T^*$ are very small and are stabilized via a long-wavelength modulation of the quasi-particle spin-density,involving a typical small wave-vector. The latter is connected with excitations over the hybridization-gap of the HF-bands [6] and competes with the well known $2k_F$-instability. In the experiments of [1] indeed a long-wavelength spin-density modulation is found below the Neel-temperature. Such critical wave-vectors and a possible competition between different types of magnetic order, in particular in the small-x regime with better localized moments, depend sensitively on details of the anisotropic bandstructure. This is born out by our model calculations, too. The competition between RKKY-correlations and local moment compensation involves roughly the balance between J^2 and the singlet binding energy $T^* \sim \exp(-1/\tilde{x})$. The HF-regime of the compound, where T_N and T^* are of comparable order, lies between the deep Kondo-range $\tilde{x} \ll 1$ and the intermediate-valence regime $\tilde{x} \approx 1$. This fact furnishes a particular meaning to the folklore that HF systems are always near to a magnetic instability. In the range where $T_N \leq T^*$ a close resemblance of fig.1 to the experimental phase-diagram of $Ce(Cu_{1-x}Ni_x)_2Ge_2$ [1] is obvious. Our present calculations do not intend to analyze the various magnetic interactions between the localized magnetic moments for small x. So the drop and following increase of T_N around $x \approx 0.1..0.2$ in [1] is not reproduced in fig.1.
The calculation of the quasi-elastic line shape in the HF-regime does not yield any significant deviations from a Lorentzian line shape, in accordance with experimental results [1]. In this regime, the main effect of the interaction K is to shift the quasi-elastic line width up to 10% from T^*. The calculations, however, indicate deviations from the Lorentzian line-shape to occur for small $\tilde{x}$ where the interaction of the localized magnetic moments is furnished by band electrons with full spectral weights. This is again in accordance with experimental findings.
In conclusion, we have demonstrated strong evidence for the picture of itinerant magnetism of HF's from the consistency between our model-calculation and neutron-scattering data on the system $Ce(Cu_{1-x}Ni_x)_2Ge_2$.

ACKNOWLEDGEMENT:
The authors would like to express their gratitude to Prof. F. Steglich for enlightening discussions.

REFERENCES:

[1] A. Loidl, A. Krimmel, K. Knorr, G. Sparn, M. Lang, S. Horn, F. Steglich, A.P. Murani, preprint; G. Sparn, C.Geibel, S.Horn, M.Lang, F.Steglich, A.Krimmel, A.Loidl, preprint, this conference

[2] N. Grewe, Sol. St. Comm. 66, 1053 (1988)

[3] N. Grewe, F. Steglich to be published in vol. 14 of the " Handbook of the Physics and Chemistry of Rare-Earth", ed. K.A. Gschneider jr., North-Holland-Publishing-Co., Amsterdam, 1990

[4] N. Grewe, T. Pruschke, H. Keiter, Z. Phys. B - Cond. Matter 71, 75 (1988)

[5] S. Doniach, Physica 91 B, 231, (1977)

[6] N. Grewe, B. Welslau, Sol. St. Comm. 65, 437 (1988)

[7] N.E. Bickers, D.L. Cox, J.W. Wilkins, Phys. Rev. B 36(4), 2036 (1987)

Physica B 165&166 (1990) 389–390
North-Holland

MAGNETIC FIELD DEPENDENCE OF THE SPECIFIC HEAT OF HEAVY-FERMION $YbCu_{4.5}$

A. AMATO, R.A. FISHER, N.E. PHILLIPS, D. JACCARD* and E. WALKER*

Materials and Chemical Sciences Division, Lawrence Berkeley Laboratory, Berkeley, CA 94720, USA
*Département de Physique de la Matière Condensée, Université de Genève, CH-1211 Genève, Switzerland

The specific heat of a polycrystalline sample of $YbCu_{4.5}$ has been measured between 0.3 and 20K in magnetic fields to 7T. At zero field a minimum in C/T is observed near 11K. Below that temperature C/T increases and below 0.5K exhibits an upturn ascribed to a hyperfine contribution. The increase in C/T below 11K is reduced by a factor 1.5 for H=7T, whereas the hyperfine term is enhanced due to the contribution of the ^{63}Cu and ^{65}Cu nuclei.

The hybridization of f electrons with the conduction band, in rare-earth and actinide based compounds, leads to various kind of phenomena. The most interesting one is probably the occurrence of an unusually large value of the coefficient (γ) of the linear term in the specific heat at low temperature for some compounds known as heavy-fermion compounds (HFC). To date, due to the difficulty of preparing good quality samples, the class of ytterbium-based HFC has not received as much attention as the cerium and uranium-based HFC. We report here on specific heat (C) measurements on a high purity polycrystalline sample of $YbCu_{4.5}$ between 0.3 and 20K in magnetic fields (H) to 7T.

Stoichiometric amounts of 3N ytterbium and 5N copper were melted by resistive heating in a tantalum tube with both ends sealed with an electron gun and squeezed in high-current clamps. Three passes through a high frequency vertical zone melting furnace were then made. Due to the thermal gradients in this process migration of Yb was observed throughout the sample. Nevertheless, we found that about half of the sample was a pure $YbCu_{4.5}$ phase as shown by metallography, X-ray diffraction and energy dispersive electron microprobe analysis.

Fig. 1 shows the temperature dependence of C/T in magnetic fields. The points below 0.7K are well represented by the sum of a hyperfine contribution, $A(H)T^{-2}$, and γT, and were analyzed on that basis. The dashed lines at low temperature represent C/T after subtraction of $A(H)T^{-2}$. Unlike previous data (1-3) limited to T>1.3K and measurements of non-zone refined samples, the present data do not show any anomaly at 2.2K, where the main parasitic phase, Yb_2O_3, orders antiferromagnetically, emphasizing the good quality of the sample. The behavior of C/T is similar to that reported for non-magnetic cerium-based HFC, such as $CeCu_6$, with different temperature dependencies in two regions of temperature: (i) above 12K, one observes the usual decrease with decreasing T; (ii) Below 10K, C/T exhibits an enormous increase ascribed to the formation of a resonance at the Fermi level due to the Kondo interaction between 4f and conduction electrons. The extrapolated value of $\gamma \simeq 635$ mJ/K^{2}mole is, to our knowledge, the largest among non-magnetic ytterbium-based HFC. As a first approximation, by applying the theory of dilute Kondo systems (4) our results lead to an estimation of the Kondo temperature $T_K \simeq 9$K. This value roughly corresponds to the temperature at which the electrical resistivity and the thermoelectric power reach their extrema (1).

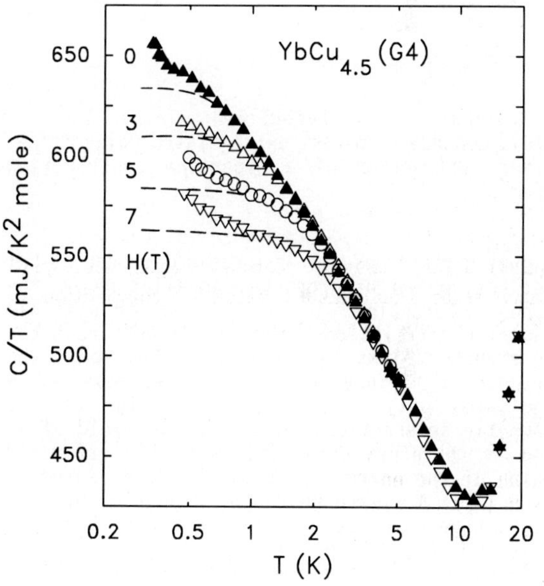

FIGURE 1
Field dependence of C/T. The dashed lines show C/T after subtraction of a hyperfine contribution.

The field dependence of the extrapolated values of γ are plotted in Fig. 2. Applying a

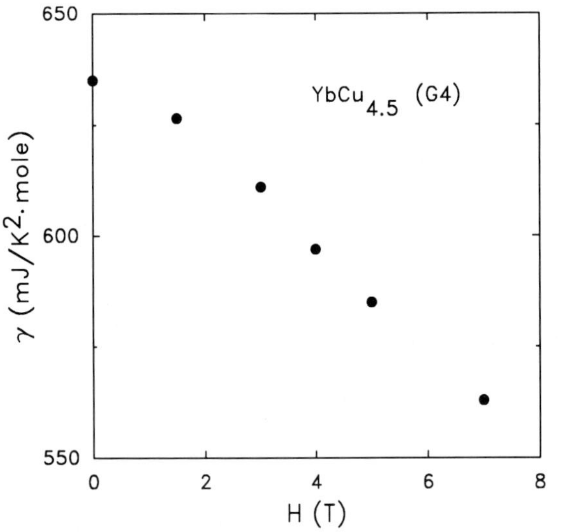

FIGURE 2
Field dependence of the extrapolated values of γ, after subtraction of a hyperfine contribution from C.

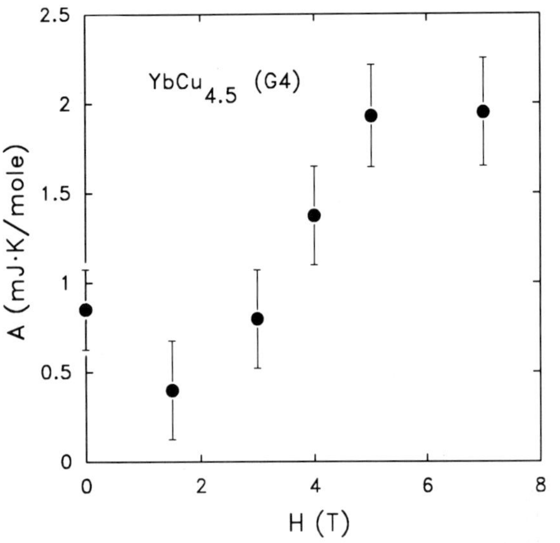

FIGURE 3
Field dependence of the coefficient A(H) of the hyperfine contribution $C_{hf}=A(H)/T^2$.

magnetic field results in a rapid decrease of C/T reflecting the progressive cancellation of the Kondo interaction (intrasite) and of possible magnetic correlations between two next neighbor rare-earth atoms (intersite). However, the absence of saturation in the decrease of γ indicates only a partial destruction of the magnetic interactions in a magnetic field of 7T and is therefore consistent with the estimated value of the energy scale of the Kondo interaction (T_K). A regime where the magnetic interactions have completely collapsed can be expected only at higher fields, at which the Zeeman energy becomes larger than the energy scales of the intrasite and intersite magnetic interactions.

The field dependence of the coefficient of the T^{-2} term is plotted in Fig. 3. The error bars represent the change in A(H) from the best-fit value that would increase the rms deviations by 20%. The uncertainty in the coefficient A(H) reflects the large value of the electronic term compared with the hyperfine contribution in the temperature range of the fit. At H=0, the coefficient A=850 μJK/mole probably arises mainly from pure quadrupolar splitting of ^{173}Yb. However, preliminary Mössbauer measurements (5), showed the presence of two different Yb sites with equal occupancy, leading to a quadrupolar contribution to the coefficient A of 320 μJK/mole. (The remaining part might possibly be ascribed to the quadrupolar splitting of ^{63}Cu and ^{65}Cu). For $0<H\leq 7T$, the contribution to A of ^{171}Yb and ^{173}Yb nuclei can be calculated to be negligible and the hyperfine contribution of ^{63}Cu and ^{65}Cu should be dominant. However, in this case also, the values of A are about 2.6 times greater than expected.

ACKNOWLEDGEMENTS
This work was supported by the Director, Office of Basic Energy Sciences, Division of Materials Sciences of the U.S. Department of Energy under Contract DE-AC03-76SF00098. Additional support for A.A. was provided by a grant from the Swiss National Science Foundation.

REFERENCES

(1) D. Jaccard, A. Junod and J. Sierro, Helv. Phys. Acta. 53 (1980) 583.
(2) N. Sato, H. Abe, M. Kontani, S. Yamagata, K. Adachi and T. Komatsubara, Physica B to be published.
(3) Z. Fisk, J.D. Thompson and H.R. Ott, J. Magn. Magn. Mat. 76&77 (1988) 637.
(4) V.T. Rajan, Phys. Rev. Lett. 51 (1983) 308.
(5) P. Bonville, private communication.

Physica B 165&166 (1990) 391–392
North-Holland

LOCAL HOLON AND SPINON IN THE ANDERSON MODEL

Norio KAWAKAMI and Ayao OKIJI *)

Research Institute for Fundamental Physics, Kyoto University, Kyoto 606, Japan
*) *Department of Applied Physics, Osaka University, Suita, Osaka 565, Japan*

Elementary excitations in the Anderson model are investigated by means of the exact solution. Two kinds of the dressed particles concerning the charge and spin excitations are introduced. Excitation spectra for these particles clearly describe the formation of the narrow Kondo resonance and the deep-lying charge-excitation hump. It is a crucial point that all kinds of elementary excitations can be described in terms of these dressed particles.

Kondo problem was solved exactly and its thermodynamic quantities were calculated precisely at arbitrary temperature and magnetic fields (1). Moreover, many transport and dynamical quantities have been investigated exactly with the aid of the local Fermi liquid theory (2). However, it is still hard to derive information about the density of states from the exact solution. In this brief report, we calculate the elementary excitation spectra in the Kondo problem by means of the exact solution of the Anderson model. The basic idea is very simple: we introduce two kinds of the dressed particles in the thermodynamic limit, *local holon* and *local spinon*, and calculate their one-particle excitation spectra exactly. It is to be noted that all kinds of the elementary excitations can be described in terms of these two kinds of the dressed particles, so far as the system is in the excited states close to the ground state.

We consider the Anderson model described by the Hamiltonian,

$$H = \sum_{k,\sigma}[\epsilon_k c^\dagger_{k\sigma} c_{k\sigma} + V(c^\dagger_{k\sigma} d_\sigma + h.c.)] + \epsilon_d \sum_\sigma n_\sigma + U n_\uparrow n_\downarrow, \qquad 1)$$

where each term has an ordinary meaning. For simplicity, we deal with the symmetric Anderson model. The Bethe Ansatz diagonalization of this model results in the set of the algebraic equations for two kinds of the rapidities k and Λ.

First, we investigate the excitation concerning the charge degree of freedom. As was shown in (3), the ground state of the Anderson model is constructed by the sea of the bound states of the antiparallel-spin electrons. We remove the one bound state from the sea of the ground state. The bare energy change associated with this procedure is denoted by $G(\Lambda_0)$, where $G(\Lambda) = -2Re[\Lambda + iU\Delta]^{1/2}$ (Δ:resonance width) and Λ_0 represents the quasi-momentum to be removed. Obviously, the change of the electron charge is $2e$. It is a crucial point that the many body effect is induced, by creating the $2e$ hole, so as to partially compensate the change of electron charge. As a result, the back-flow effect causes the partial cancellation of the electron charge to form the dressed hole with *charge* $e^* = e$ *with no spin*. This dressed hole is referred to as *local holon* in the present paper. The energy of the local holon is given by the superposition of the energy due to the bare and the back-flow ones,

$$\omega = \int_{-\infty}^{\infty} (k^2/\Delta)\mathrm{sech}[\pi(\Lambda_0 - k^2)/\Delta]\mathrm{d}k. \qquad 2)$$

It is noted that the quantum number I_0, which specifies Λ_0, can be chosen as a successive integer. This fact enables us to calculate the density of states for the local holon by the formula $D_c(\omega) = \delta I_0/\delta\omega$, which leads to the expression in terms of Λ_0,

$$D_c(\omega) = \frac{\int_{-\infty}^{\infty} F(k)\mathrm{sech}[\pi(\Lambda_0 - k^2)/\Delta]\mathrm{d}k}{\int_{-\infty}^{\infty} \mathrm{sech}[\pi(\Lambda_0 - k^2)/\Delta]\mathrm{d}k}, \qquad 3)$$

where $F(k) = (\Delta/\pi)[(k + U/2)^2 + \Delta^2]^{-1}$. Above formulas provide the density of states for the local holon. In Fig.1(a), the excitation spectrum for the local holon is shown for several values of the Coulomb interaction. In the absence of the Coulomb interaction, the shape of the spectrum is given by the Lorentzian, as should be. With the increase of the Coulomb interaction, it is seen that the hump structure is gradually developed around the impurity level $\epsilon_d = -U/2$. In the Kondo limit ($U >> \Delta$), this hump is described by the universal function, which is explicitly given below. It should be noted that $D_c(0)$ gives the exact value of the charge susceptibility.

Next, we consider the spin excitation. We first add an electron to the local singlet for the Kondo ground

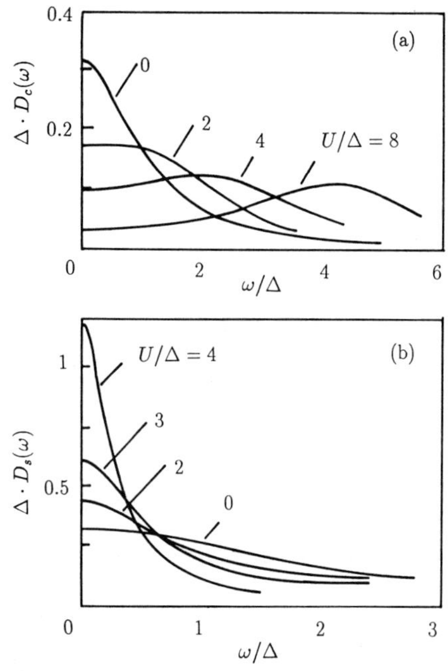

Fig.1 The excitation spectra for the local holon $D_c(\omega)$ and spinon $D_s(\omega)$ for the symmetric Anderson model.

state. In this case also, the back-flow effect is induced in the thermodynamic limit. In contrast to the formation of the local holon, the bare increase of the electron charge is exactly canceled by the back-flow effect. Consequently, the dressed particle with the effective *spin* $s^* = 1/2$ and *no charge* is formed in the thermodynamic limit. This dressed particle is referred to as *local spinon*. The renormalized energy for the local spinon is given in terms of the bare energy change k_0,

$$\omega = 2k\theta(k_0) + \frac{1}{\Delta}\int_{-\infty}^{\infty} x^2 \mathrm{cosech}\frac{\pi}{\Delta}(k_0^2 + x^2)\mathrm{d}x, \qquad 4)$$

where $\theta(x)$ is the step function. The density of states for the local spinon is obtained after a straightforward manipulation. We omit here the explicit expression for the density of states, because of rather complicated formula. In Fig.(b) , the excitation spectrum for the local spinon, $D_s(\omega)$, is shown for several choices of the Coulomb interaction. It is seen that the spinon spectrum becomes narrower with the increase of the Coulomb interaction. In the Kondo regime, this spectrum describes the Kondo resonance itself, which is discussed below. It is to be noted that the density of states $D_s(0)$ gives the exact value of the spin susceptibility.

In the Kondo limiting case ($U = -2\epsilon_d >> \Delta$), the charge and spin excitations are energetically separated from each other completely. In this regime, we obtain the density of states for the local holon as an explicit function of ω,

$$D_c(\omega) = \frac{1}{2\pi}\int_{-\infty}^{\infty} \mathrm{d}x \frac{\mathrm{sech}(\pi x/\nu\Delta)}{(\omega - \epsilon_d - x)^2 + \Delta^2}. \qquad 5)$$

As evident from this expression, the spectrum for the local holon takes the form of Lorentzian smeared by the hyperbolic-sec function. Similarly the universal function for the local spinon can be obtained in an expected form, $D_s(\omega) = (T_K/\pi)[\omega^2 + (T_K)^2]^{-1}$, which agrees with the one discussed by Andrei et al. in (1). Here the Kondo temperature T_K is defined by $T_K = \frac{1}{\pi}(2U\Delta)^{1/2}\exp[-\pi U/(8\Delta)]$. Note that the density of states for the local spinon has a constant weight irrespective of the strength of the Coulomb interaction. So, it is distinguished from the usual density of states obtained from the one-particle Green function which contains the wave-function renormalization factor of the order of T_K/Δ.

We have investigated the elementary excitation spectra for the Kondo effect with the use of the exact solutions of the Anderson model. Two kinds of the dressed particles, local holon and local spinon, are introduced to discuss the charge and spin excitations. It should be emphasized here that all kinds of the elementary excitations can be described completely in terms of these two kinds of the dressed particles, as far as the system is close to the ground state. This is the reason why we consider the excitation spectra for these particles to be important. We have already extended the present calculation to the orbital degenerate model including the crystalline field and also to the multi-channel Kondo model. The detailed results will be published elsewhere.

References

(1) see for reviews, *e.g.*,
N. Andrei, K. Furuya and J. H. Lowenstein, Rev. Mod. Phys. **55**(1983)331.
A. M. Tsvelick and P. B. Wiegmann, Adv. Phys. **32**(1983)453.
A. Okiji and N. Kawakami, Springer Series in Solid State Science **62**(1984)46.

(2) N. Kawakami and A. Okiji, Phys. Lett. **118A** (1986) 301.
N. Kawakami, A. Nakamura and A. Okiji, J. Phys. Soc. Jpn. **57**(1988)4359.
A. Okiji, Springer Series in Solid State Science **77**(1988)63.

(3) N. Kawakami and A. Okiji, Phys. Lett. **86A** (1981) 483.

Physica B 165&166 (1990) 393–394
North-Holland

Numerical Study of Slave Boson Mean Field Equations for the Hubbard Model

L. LILLY, A. MURAMATSU, W. HANKE

Physikalisches Institut der Universität Würzburg, 8700 Würzburg, FRG

We consider the one band Hubbard model in the Slave Boson formulation first introduced by Kotliar and Ruckenstein. Applying a modified Mean Field Approximation (MFA), which allows for broken symmetry states, we numerically determine the values of the involved Bose fields at the saddle point. Our data for local Fermion correlation functions agree surprisingly well with data obtained in recent Quantum Monte Carlo calculations.

1 Introduction

Slave Bosons were first introduced in solid state theory by Barnes[1], who considered a single site Anderson lattice. Since then there have been numerous applications to a wide class of models[2, 3]. In recent years, there has been an increasing interest in the Hubbard model[4] as a possible description of heavy Fermion systems and High Temperature Superconductors (HTS)[3, 5]. Here the Slave Boson technique seems to be one of the more promising approaches apart from Quantum Monte Carlo (QMC) calculations, since it is in principle capable of handling weak, intermediate and strong coupling situations on the same footing. The Slave Boson formulation is an exact rewriting of a purely Fermionic model to an effectively Bosonic model, which, as we will show, allows for approximations yielding good results.
We will use the notation introduced by Kotliar and Ruckenstein[5] and we will also include some of their results without rederiving them.

2 Slave Boson Formulation

Our starting point is the single band Hubbard model on a two dimensional (2-D) square lattice. The Hamiltonian is given by

$$H = -\sum_{i,j,\sigma} t_{ij} c^{\dagger}_{i,\sigma} c_{j,\sigma} + U \sum_{i} c^{\dagger}_{i\uparrow} c_{i\uparrow} c^{\dagger}_{i\downarrow} c_{j\downarrow} \ , \qquad (1)$$

where $c^{(\dagger)}_{i,\sigma}$ are annihilation (creation) operators for Fermions with spin projection σ at site i, U is the on-site interaction and $t_{ij} = t\delta_{i,j\pm1}$ is the nearest neighbor hopping matrix element. The Slave Boson formulation constitutes an exact rewriting of the Hubbard model in a functional integral form for the partition function Z in terms of four types of Bosons: e_i, $p_{i,\sigma}$ and d_i, which act as projectors on empty (e_i), singly occupied with a Fermion of spin σ ($p_{i,\sigma}$) and doubly occupied (d_i) sites i . The states in the enlarged (Fermionic *and* Bosonic) Hilbert space are ensured to be contained in the physical subspace by enforcing a set of constraints, which are introduced into the functional integral using Lagrange multipliers $\lambda^{(1)}_i$, $\lambda^{(2)}_{i,\sigma}$.

The partition function is rewritten as a functional integral over complex-valued and Grassmann fields. In the resulting equation the Fermions degrees of freedom are bilinear and hence can be integrated out exactly[6]. The operators $\tilde{z}_{i,\sigma}$ that renormalize the Fermion hopping amplitude[5] are chosen to ensure in the MFA the correct limiting behavior in the noninteracting case and also to reproduce the "Brinkman-Rice" transition of the Gutzwiller approximation (GA) to the Gutzwiller wavefunction (GW). For a more detailed discussion of this problem, we refer to the literature[3, 5].

3 Two Sublattice Mean Field Approach

Due to the bipartite structure of the 2-D square lattice, our model is susceptible to broken symmetry ground states that reflect this bipartite structure (i.e. two sublattices A and B). More specifically the Hubbard model on a square lattice displays strong Fermi surface nesting (with nesting vector $\mathbf{Q} = (\frac{\pi}{a}, \frac{\pi}{a})$, which occurs at (and close to) half filling[8]. For these reasons we chose a modified MFA.
In this two sublattice MFA, the time– and space–dependent fields are replaced by static and, on each of the two sublattices A and B, uniform variables.
From here we can easily calculate the free energy per lattice site to be

$$\begin{aligned} f &= -\beta^{-1} \log Z^{MF} = \\ &= -\beta^{-1} \frac{1}{2} \sum_{\substack{\mathbf{k},\sigma \\ j=1,2}} \log\left(1 + e^{-\beta\varepsilon^{(j)}_{\sigma}(\mathbf{k})}\right) + \\ &\quad + \frac{1}{2} \sum_{\alpha=A,B} \Big\{ U d^2_\alpha + \lambda^{(1)}_\alpha \big(\sum_\sigma p^2_{\alpha\sigma} + e^2_\alpha + d^2_\alpha - 1\big) - \\ &\qquad - \sum_\sigma \lambda^{(2)}_{\alpha\sigma} (p^2_{\alpha\sigma} + d^2_\alpha) \Big\} . \end{aligned} \qquad (2)$$

The single particle energy eigenvalues $\varepsilon^{(j)}_{\sigma}(\mathbf{k})$ are obtained by diagonalizing the Fermion determinant. The values of the boson fields and Lagrange multipliers are determined by the condition that the free energy f should be a saddle point in the enlarged configuration space of slave Boson fields and

0921-4526/90/$03.50 © 1990 – Elsevier Science Publishers B.V. (North-Holland)

Lagrange multipliers. We determine the values of the fields at this physical point in the 15–dimensional parameter space numerically by a constrained steepest descent algorithm. Depending on initial conditions we find more than one solution to the "saddle point" equations in certain areas of the parameter space. In these cases, the physically realized configuration is the one with the lowest free energy. The different states show various types of symmetry with respect to spin or sublattice index. From there we discern paramagnetic (PM), ferromagnetic (FM), antiferromagnetic (AFM) and charge density wave (CDW) states.

4 Numerical Results

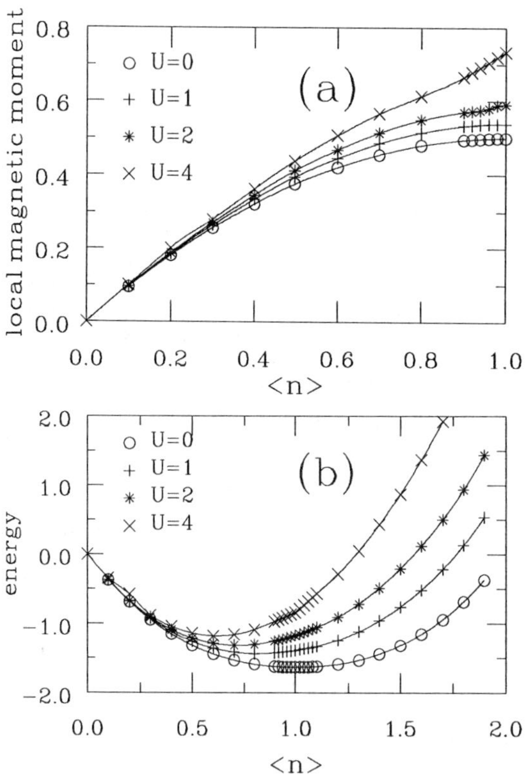

Figure 1: a)The local magnetic moment, a measure for the number of "spins" per site; b) The energy per site. (Results agree with ref.8 within less than 1%.)

The results of our numerical investigations compare astonishingly well with recent QMC results by Moreo et. al.[9]. To allow an easy comparison, we calculated all physical observables for finite temperatures ($\beta = 10$). We specifically looked at the local magnetic moment and the average energy per site for various values of the interaction parameter U (in units of t) and almost all band fillings $< n >$. Figure 1 a) and b) show these observables for values of $U = 0, 1, 2,$ and 4.

Figure 2 shows the magnetization and the staggered magnetization as a function of band filling $< n >$ for U=4. We see a strong AFM peak near half filling, and FM ordering for very dilute systems.

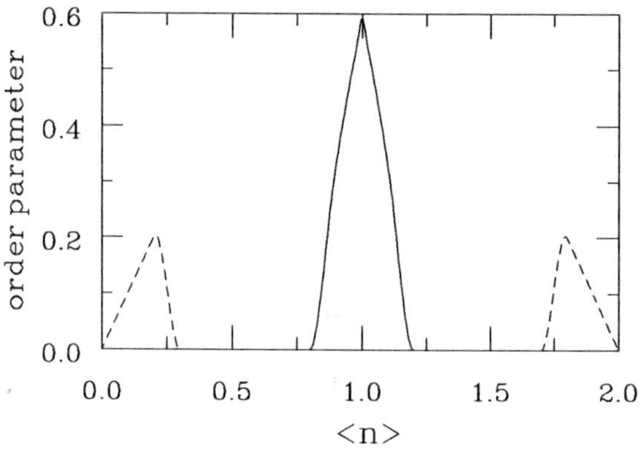

Figure 2: The magnetization, the order parameter for an FM state (dashed lines), and staggered magnetization, the order parameter for an AFM state (solid line) as a function of band filling $< n >$. Note the AFM peak near half filling and FM peaks for very dilute systems.

In general, we find an AFM phase for U > 1.8 near half filling and likewise a CDW phase for U < −1.8 again close to half filling. (These finite values of U are due to the finite temperature in our calculation.) For not too small U we also find a FM phase for the very dilute systems, as indicated in figure 2.

A more complete account of this work will be given elsewhere.

Acknowledgements

This work is in part supported by Bundesministerium für Forschung und Technologie project No. 13 N 5501-1. One of us (L.L.) gratefully acknowledges stimulating discussions with P. Wölfle.

References

[1] S. E. Barnes, J. Phys. F **6**, 1375 (1976), and **7**, 2637 (1977)

[2] N. Read, D. Newns, J. Phys. C **16**, 3273 (1983)

[3] T. Li, P. Wölfle and P. J. Hirschfeld, Phys. Rev. B **40**, 6817 (1989)

[4] J. Hubbard, Proc. Roy. Soc. London, Ser. A **296**, 100 (1966)

[5] Gabriel Kotliar and Andrei E. Ruckenstein, Phys. Rev. Lett. **57**,1362 (1986)

[6] For details of this calculation we refer to the excellent work by Kotliar and Ruckenstein, ref. 5.

[7] J. W. Rasul, Tiecheng Li, J. Phys. C **21**, 5119 (1988)

[8] D. J. Scalapino, E. Loh, J. E. Hirsch, Phys. Rev. B **34**, 8190 (1986)

[9] A. Moreo, D. J. Scalapino, R. L. Sugar, N. R. White and N. E. Bickers, preprint (1989)

Physica B 165&166 (1990) 395–396
North-Holland

AMPLITUDE RELATIONS NEAR A ZERO TEMPERATURE TRANSITION

Mucio A. CONTINENTINO*

Instituto de Física, Universidade Federal Fluminense, Outeiro de São João Batista s/nº, Niterói, 24020, Rio de Janeiro, Brasil

We obtain amplitude relations for thermodynamic quantities close to a zero temperature phase transition. The amplitude ratios along an isocritical line provide additional expressions between measurable exponents. These allow to check inequalities for the zero temperature exponents and to determine the dynamical exponent z which plays a central role in the theory of quantum critical phenomena.

Zero temperature phase transitions differ considerably from those occurring at finite temperatures where thermal fluctuations have a decisive role in desorganizing the system. At zero temperature the critical phenomena arise in general due to a competition between different parameters of the Hamiltonian (1). In these transitions static and dynamics are inextricably coupled due to the quantum character of the critical fluctuations (1). Dimensionality shift (2) near a zero temperature transition modifies the hyperscaling relation which is now expressed as $2 - \alpha = \nu(d + z)$ where α and ν are standard critical exponents, d the dimensionality of the system and z the dynamical critical exponent. This exponent plays a central role in the theory of quantum critical phenomena and it's determination is of fundamental importance.

The concepts of critical phenomena, including those at T=0, find a natural expression within the renormalization group theory (3). The zero temperature instability is associated in this approach with an unstable fixed point at a critical value of a number g_c (or a critical density) which is a ratio between competing parameters of the Hamiltonian.

When considering the effect of a finite temperature in the system, two typical situations arise. Since temperature, close to the zero temperature fixed point, scales (1,2) with the exponents z it may behave as a relevant ($z > 0$) or irrelevant field ($z < 0$) depending on the sign of z. The case where $z < 0$, is typical of random systems (2,4) and in principle the dynamic exponent, or the combination νz, can be determined experimentally from the anomalous critical slowing down close to a finite temperature critical line (5). We shall be concerned here with the case $z > 0$ in which temperature acts as a relevant field at g_c. Furthermore we shall consider the situation that stemming from the zero temperature fixed point there is a line of second order phase transitions signaling a finite temperature instability. Scaling theory implies (3) that the equation of this line, for $T \cong 0$, is given by $g_c(T) = g_c - aT + bT^{1/\phi}$, where $\phi = \nu z$, a and b are constants and a regular term α T has been included for completeness. The scaling form of the free energy close to the zero temperature fixed point is given by (2):

$$F \propto |\tilde{g}_c|^{2-\alpha} f[T/|g_c|^{\phi}] \qquad (1)$$

where $\tilde{g}_c = g - g_c$, α a standard critical exponent and $\phi = \nu z$. These exponents are associated with the fixed point at $T = 0$, $g = g_c$. Under the assumption of new critical behavior at finite temperature, as implied by the relevance of the temperature field, we can also write (3) ($T \geq 0$):

$$F \propto A_f(T)\ |\tilde{g}_T|^{2-\tilde{\alpha}} \qquad (2)$$

where $\tilde{g}_T = g - g_c(T)$ and $g_c(T) = g_c - aT - bT^{1/\phi}$ is the equation for the critical line. The exponent $\bar{\alpha}$ is associated with a new fixed point governing the critical behavior at finite temperatures. From Eqs.1 and 2 and the expression for the critical line, we can calculate the amplitude $A_f(T)$ which is given by $A_f(T) = T^{(\bar{\alpha}-\alpha)/\phi}$. Differentiating Eq.2 twice with respect to temperature, we get for the specific heat:

$$C \propto A_c(T)\ |\tilde{g}_T|^{-\bar{\alpha}} \qquad (3)$$

where $A_c(T) = T^{(2+\bar{\alpha}-\alpha-2\phi)/\phi}$. Using the modified hyperscaling relation, we can rewrite the amplitude exponent $y_1 = (2+\bar{\alpha}-\alpha-2\phi)/\phi = (\nu d+\bar{\alpha}-\phi)/\phi$.

---

*CNPq Research Fellow

Then varying g along an isocritical line, by for example applying pressure (since generally g depends on volume (6)),at small but finite temperature, we can obtain from the experimental ratio of the specific heat amplitudes the amplitude exponent y_1. Since the exponent $\bar{a}$ can be measured independently, we have found a numerical relation between the zero temperature critical exponents which can be used to check inequalities for these exponents obtained from physical arguments. In the case of the superfluid-insulator (7) where it is expected that z = d we get $y_1 = \bar{a}/\phi$ which together with the inequality (7) $\nu \geq 2/d$ implies $y_1 \leq \bar{a}/2$. For the superconductor-insulator (8) transition with $\nu \geq 2/d$ and z = 1 we find $y_1 \leq \bar{a}d/2$.

An interesting situation occurs when d+z $\geq d_c$ the upper critical dimensionality. In this case the exponents associated with the zero temperature fixed point assume classical (Landau) values (1,2) and consequently $y_1 = (2/z)(2+\bar{a}-z)$. So provided $\bar{a}$ is known the amplitude exponent allows for a direct determination of the dynamic exponent z. Note that if z > 2+$\bar{a}$, $y_1 < 0$ and the amplitude of the specific heat anomaly increases with decreasing critical temperature. In the classical case the amplitude relation for the diverging susceptibility is

$$\chi \propto T^{2(\tilde{\gamma}-1)/z} |\tilde{g}_T|$$. These results have direct relevance for magnetic heavy fermion systems (6).

REFERENCES

(1) J.A.Hertz, Phys.Rev.B 14, 1165 (1976).
(2) A.J.Bray and M.A.Moore, J.Phys. C 18, L927 (1985)
(3) M.E.Fisher, in *Critical Phenomena*, Vol.186 of *Lecture Notes in Physics*, edited by F.J. Habne (Springer, Berlin, 1983), p.1.
(4) D.S.Fisher, Phys.Rev.Lett. 56, 416 (1986).
(5) W.Kleemann, B.Igel and U.A.Leitão, to be published.
(6) M.A.Continentino, G.M.Japiassu and A.Troper, Phys.Rev. B 39, 9734 (1989).
(7) M.P.A.Fisher, P.B.Weichman, G.Grinstein and D.S.Fisher, Phys.Rev.B 40, 546 (1989).
(8) M.P.A.Fisher, G.Grinstein and S.M.Girvin, Phys.Rev.Lett. 64, 587 (1990).

Physica B 165&166 (1990) 397–398
North-Holland

GROUND STATE OF THE KONDO LATTICE

Kazuo Ueda and Kouki Yamamoto

Institute of Materials Science, University of Tsukuba, Tsukuba 305, Japan

The one-dimensional Kondo lattice is diagonalized numerically. It is shown that the ground state is ferromagnetic at low densities of conduction electrons. On the other hand at half-filling the ground state is a singlet with strong antiferromagnetic correlations which are the main mechanism to compensate the localized spins.

1. INTRODUCTION

The key question about the normal heavy Fermion state is why the nonmagnetic state instead of some magnetically ordered state is stabilized in spite of the strong Coulomb repulsion. Mean field type theories, the lattice version of 1/N expansion and the Gutzwiller type theory have succeeded in describing the normal heavy Fermion state with the agreement in the large degeneracy limit. However it is clear that a better theory than those mean field type theories is required to explain such magnetic properties as small antiferromagnetic moment in URu_2Si_2 (1) and UPt_3 (2) and metamagnetic-like behavior in $CeRu_2Si_2$ (3). In the present paper we report on the results of exact numerical diagonalization of the one-dimensional Kondo lattice.

2. MODEL

The one-dimensional Kondo lattice is written in the usual form

$$H=-\sum_{n=1}^{L}\sum_{\sigma}(a^+_{n+1\,\sigma}a_{n\sigma}+a^+_{n\,\sigma}a_{n+1\,\sigma})$$
$$-\frac{J}{2}\sum_{n=1}^{L}\{(a^+_{n\uparrow}a_{n\uparrow}-a^+_{n\downarrow}a_{n\downarrow})S_{nz} \tag{1}$$
$$+a^+_{n\uparrow}a_{n\downarrow}S_{n-}+a^+_{n\downarrow}a_{n\uparrow}S_{n+}\}$$

where L is the number of lattice sites. We impose the periodic boundary condition. As the exchange constant J= -0.2 is used in the following numerical results. We show the results of six and seven lattice sites. The exact numerical diagonalization was performed for all electron numbers N less than or equal to half filling, N≤L, changing magnetization of the system.

3. RESULTS AND DISCUSSIONS

Magnitude of total spin S for each electron number is summarized in Table I for the seven-sites Kondo lattice and in Table II for the six-sites Kondo lattice.

Table I Spin of the ground state of the seven-sites Kondo lattice. N is the conduction electron number.

N	1	2	3	4	5	6	7
S	3	1/2	0	5/2	0	1/2	0

Tabel II Spin of the ground state of the six-sites Kondo lattice.

N	1	2	3	4	5	6
S	5/2	0	1/2	2	1/2	0

The results for the six sites and the seven sites are very similar. In the following we discuss the case of L=7. To understand the ground state it is helpful to consider the energy levels of the conduction band at J=0

$$\varepsilon_k=-2\cos\frac{2\pi}{L}k \quad k=0,\pm1,\cdots \tag{2}$$

0921-4526/90/$03.50

Table III On-site correlation between the localized spin and conduciton electorn spins

N	1	2	3	4	5	6	7
$<\mathbf{s}_f(i)\cdot\mathbf{s}_c(i)>$	-0.054	-0.025	-0.088	-0.129	-0.089	-0.024	-0.100

Tab.IV Nearest neighbor correlation between the localized spins

N	1	2	3	4	5	6	7
$<\delta\mathbf{s}_f(i)\cdot\delta\mathbf{s}_f(i+1)>$	0.018	0.134	-0.069	0.026	-0.107	-0.384	-0.396

Let us start with the case of N=1. The ground state is a incomplete ferromagnetic state. The conduction electron occupies the lowest level k=0. The gain of exchange energy is maximized by aligning localized spin in one direction and at the same time fixing the spin of the conduction electron in opposite direction. Actually spin correlation between a localized spin and conduction electrons on a single site is -0.054. This magnitude is slightly bigger than $-1/4\cdot7$ =-0.036 which is a value expected when the spin alignment is static. The difference comes from quantum fluctuations due to the transverse components of the exchange term.

For N=2, the two conduction electrons are accommodated in the lowest level k=0 with antiparallel spins. Therefore the system is singlet as a whole. However the correlation between nearest neighbor localized spins are ferromagnetic, 0.134. This means that the double exchange mechanism which causes the ferromagnetic ground state when N=1 is still effective.

Now we turn to the opposite case of nearly half-filled conduction band (N=6, 7). The ground states have the smallest spin possible. In a single impurity Kondo problem it is well known that the localized spin is completely screened by conduction electron spins (Kondo cloud). As is pointed out by Nozieres (4), in a lattice Kondo cloud does not suffice to compensate the localized spins since only a part (T_K/W) of conduciton electrons take part in the Kondo cloud. The present results show that close to the half-filling the localized spins themselves are cancelling each other mainly via the nearest neighbor antiferromagnetic correlations.

In the vicinity of quater filling, antiferromagneic correlations and the ferromagnetic correleations are competing. For N=3 and 5, singlet states are realized but the nearest neighbor antiferromagnetic correlations are not strong. The origin of the ferromagnetic ground state at N=4 may be traced back to a special level scheme of the conduction electrons. The highest occupied level is four fold degenerate including spin degrees of freedom. Therefore without much loss of kinetic evergy two conduction electrons in the highest level can have the same spin opposite to the localized spins aligned in one direction.

REFERENCES

(1) C. Broholm et al. Phys. Rev. Lett. 58 (1987) 1467.

(2) G. Aeppli et al. Phys. Rev. Lett. 60 (1988) 615

(3) P. Haen et al. J. Low Temp. Phys. 67 (1987) 391.

(4) P. Nozieres, Ann. Phys.(Paris) 10 (1985) 19.

Physica B 165&166 (1990) 399–400
North-Holland

Numerical Calculations of Susceptibility for Two-Impurity Anderson Model

Takashi YANAGISAWA

Fundamental Physics Section, Electrotechnical Laboratory, 1-1-4 Umezono, Tsukuba, Ibaraki 305 Japan

Two-impurity Anderson model is investigated by the exact diagonalization method in small systems. In connection with the recent renormalization group and Monte Carlo calculations, we focus on the staggered susceptibility. It seems to have no tendency to diverge for various values of indirect exchange interaction (RKKY) J_{RKKY}. However, if we consider the direct exchange interaction between two impurities, $J_{dir}\mathbf{S}_1\cdot\mathbf{S}_2$, the staggered susceptibility will have a chance to diverge at the zero temperature.

1. INTRODUCTION

There are controversial questions regarding the interactions between two magnetic impurities. One such question concerns the interplay between the Kondo effect and the RKKY interaction among two impurities. Due to the latter effect two impurity spins will tend to form a singlet or triplet, while the former will be likely to break such a two-body state.

Recently, from the numerical renormalization group calculations[1] an anomalous fixed point was predicted at the special ratio J_{RKKY}/T_K of the RKKY coupling to the single-impurity Kondo temperature. Here the staggered susceptibility and the specific-heat coefficient were found to diverge. A simple model was proposed[2] to try to explain these anomalous behaviors. No evidence, however, for anomalous behavior of the staggered susceptibility was found in the Monte-Carlo calculations.[3]

In this paper we present exact diagonalization study of small systems. Staggered susceptibility was evaluated at low temperatures with varying the exchange coupling between two impurities. The model that we consider is the two-impurity Anderson model:

$$H=\sum_{k\sigma}\varepsilon_k c_{k\sigma}^{+}c_{k\sigma}+\frac{V}{N^{1/2}}\sum_{kj\sigma}(e^{ikR_j}c_{k\sigma}^{+}d_{j\sigma}+\mathrm{H.c.})+\varepsilon_d\sum_{j\sigma}n_{dj\sigma}+U\sum_{j}n_{dj\uparrow}n_{dj\downarrow}+J_{dir}\mathbf{S}_1\cdot\mathbf{S}_2. \quad (1)$$

Here $n_{dj\sigma}$ denotes an localized electron of spin σ at site R_j where j=1,2. The last term is the direct exchange interaction.

2. SUSCEPTIBILITY

Our model contains two impurity sites and a six-site conduction-electron chain, that is, the number of total site is eight. Calculations for rather large systems will be published in a separate paper. The Staggered susceptibility is defined by

$$\chi^{(s)}=\int_0^\beta d\tau \langle (S_1^z(\tau)-S_2^z(\tau))(S_1^z-S_2^z)\rangle. \quad (2)$$

The on-site (χ) and uniform ($\chi^{(u)}$) susceptibilities are defined similarly.

The difficulty in the two-impurity problem lies in that indirect exchange interaction between two spins appears, which is given by $J_{RKKY}\mathbf{S}_1\cdot\mathbf{S}_2$ where

$$J_{RKKY}\approx(\rho J_{sd})^2F(R) \text{ with } J_{sd}=2V^2\left(\frac{1}{\varepsilon_d}-\frac{1}{\varepsilon_d+U}\right).$$

$F(R)$ is a function of the impurity distance R. If RKKY interaction is dominant, the staggered susceptibility at low temperatures shows $\chi^{(s)}\approx 1/|J_{RKKY}|$. On the other hand in the momentum quenched regime the staggered and uniform susceptibilities are reduced considerably. Therefore $\chi^{(s)}$ can diverge only in the intermediate regime where $|J_{RKKY}|\approx T_K$. We propose one possible case in which we consider the direct exchange interaction $J_{dir}\mathbf{S}_1\cdot\mathbf{S}_2$. At the point where the singlet and triplet states are degenerate, $\chi^{(s)}$ may diverge as

$$\chi^{(s)}\approx\frac{2}{|J_{RKKY}+J_{dir}|}\quad (J_{RKKY}+J_{dir}\approx 0), \quad (3)$$

with an additional factor 1/3 for $J_{RKKY}+J_{dir}<0$. Because the Kondo effect will dominate over the exchange inter action if $J_{RKKY}+J_{dir}\approx 0$, this diverging behavior will appear in limited cases.

We can change the ratio J_{RKKY}/T_K varying ε_d or V. We consider two nearest-neighbor impurities. In

0921-4526/90/$03.50 © 1990 – Elsevier Science Publishers B.V. (North-Holland)

Fig.1 we show $T\chi^{(s)}$, $T\chi$ and $T\chi^{(u)}$ versus $\log_{10}T$ for half-filled conduction band (8 electrons).

Parameters are $U=\infty, \varepsilon_d=-2$, $V=0.5$ and $J_{dir}=0$. The indirect exchange interaction is antiferromagnetic here. As the temperature is lowered, RKKY interaction begins to be dominant so that $\chi^{(u)}$ decreases rapidly as $2\beta\exp(-\beta J_{RKKY})$ and $\chi^{(s)}$ is nearly constant $2/J_{RKKY}$. We show the spin-spin correlation function $\langle S_1^z S_2^z\rangle$ in Fig.2. Two nearest-neighbor impurities are well fitted by the RKKY prediction,

$$\langle S_1^z S_2^z\rangle = \frac{1}{4}\,\frac{e^{-\beta J_{RKKY}}-1}{1+3e^{-\beta J_{RKKY}}}\, n_d^2, \qquad (4)$$

where $J_{RKKY}=0.0017$ and n_d^2 is the normalization factor. For various ε_d, $\chi^{(s)}$ never diverges in our calculations. We can, however, found the behavior predicted in Eq.(3). We present in Fig.3 $\chi^{(s)}$ for several J_{dir}'s, as well as the right-hand side in Eq.(3) with $J_{RKKY}=0.0017$. One can see that $\chi^{(s)}$ shows diverging behavior for $J_{dir}\approx -J_{RKKY}$ and $\chi^{(s)}$ is well approximated by Eq.(3). One should think that T_K is very small in this case. Now we show another example where parameters are the same as above but the filling of conduction electrons is less than half (6 electrons). Since the indirect exchange interaction is slightly ferromagnetic here, $\chi^{(u)}$ is enhanced weakly and $\chi^{(s)}$ lies below (twice) the on-site susceptibility. These results suggest that T_K is larger than J_{RKKY}. In fact $\langle S_1^z S_2^z\rangle$ is considerably reduced, which is of order 0.004 at the zero temperature. Therefore the diverging behavior shown in Eq.(3) is not expected. In Fig.4 we show $\chi^{(s)}$ versus J_{dir}. No sign of the anomalous $\chi^{(s)}$ is found.

To summarize we have shown the example in which the diverging behavior of $\chi^{(s)}$ in Eq.(3) can appear. Now the question is whether this anomalous point will survive in the infinite systems.

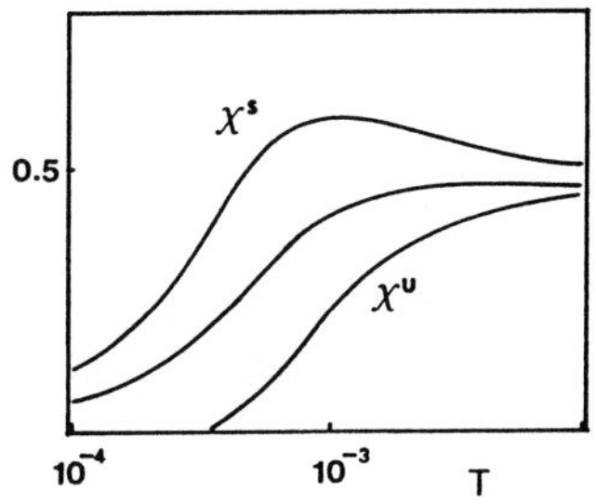

Fig.1. From the top $T\chi^{(s)}$, $T\chi$ (×2), and $T\chi^{(u)}$. $n_d=0.987$.

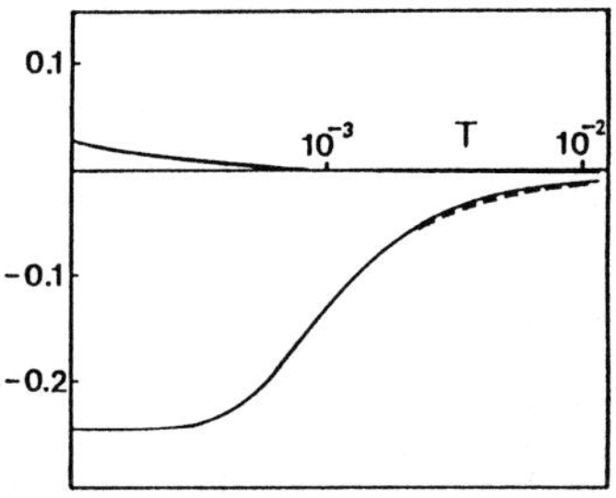

Fig.2. $\langle S_1^z S_2^z\rangle$ vs T. Dashed line shows function in Eq.(4). Upper line is for next nearest neighbor impurities (R=2).

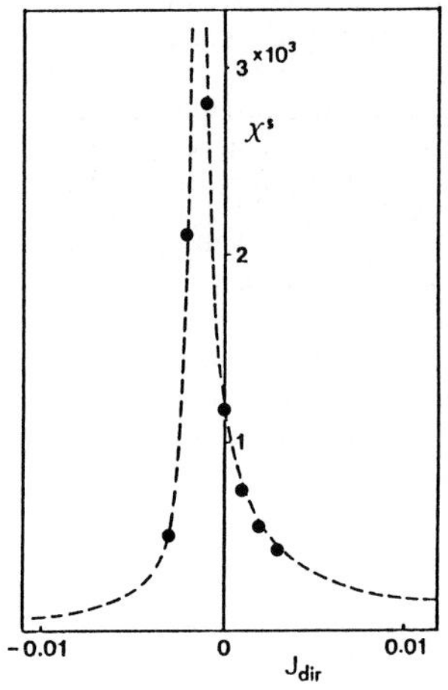

Fig.3. $\chi^{(s)}$ as a function of J_{dir}. Dashed line shows function in Eq.(3)

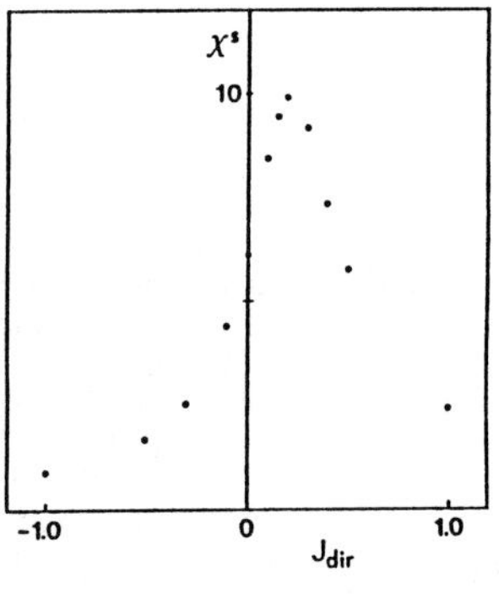

Fig.4. $\chi^{(s)}$ as a function of J_{dir} for 6 electrons.

REFERENCES

(1) B.A. Jones, C.M. Varma and J.W. Wilkins, Phys. Rev. Lett. 61 (1988) 125.
(2) J.W. Rasul and P. Schlottmann, Phys. Rev. Lett. 62 (1989) 1701.
(3) R.M. Fye and J.E. Hirsch, Phys. Rev. 40 (1989). 4780.

Physica B 165&166 (1990) 401–402
North-Holland

CORRELATED LATTICE ELECTRONS: EXACT MEAN FIELD DESCRIPTION IN INFINITE DIMENSIONS

Peter VAN DONGEN and Dieter VOLLHARDT

Institut für Theoretische Physik C, Technische Hochschule Aachen, D-5100 Aachen,
Federal Republic of Germany

It is shown that correlated Fermi systems in high dimensions can be *exactly* described in terms of an effective Hamiltonian with obvious mean field character. The method is applied to, and worked out for, a simplified Hubbard model. Thereby the exact solution in $d = \infty$ is recovered.

It has recently become clear [1] that the limit of high dimensions ($d \to \infty$) is an extremely useful starting point for studying correlated Fermi systems, since their theoretical description becomes simpler in this limit, while remaining non-trivial. Over the past two years, many interesting new results have been found [2-4], but one aspect remained obscure: whether some type of mean field theory becomes exact in $d \to \infty$ or, conversely, whether the exact solution of the model in $d = \infty$ has a mean field character. *That* mean field theory should become exact is, at least, what one would expect on the basis of classical [5] and quantum mechanical [6] spin systems. One result in this direction is easily obtained: the usual "mean field" description based on Hartree-Fock definitely does not become exact in $d = \infty$, [8] since, at weak coupling, the correlation energy for the Hubbard model obtained by Hartree-Fock is non-analytically small, in contrast to the exact result. In this paper it is shown how a mean field description, that does become exact, can be constructed.

Consider the Hubbard model [7], which is a standard model for itinerant magnetism and also for high-T_c-superconductivity. The Hubbard Hamiltonian is in its grand canonical form given by

$$H = H_t + H_U + H_\mu \tag{1a}$$

where

$$H_t = \sum_{\mathbf{ij},\sigma} t_{\mathbf{ij}\sigma} \hat{c}^+_{\mathbf{i}\sigma} \hat{c}_{\mathbf{j}\sigma} \tag{1b}$$

$$H_U = U \sum_{\mathbf{i}} \hat{n}_{\mathbf{i}\uparrow} \hat{n}_{\mathbf{i}\downarrow} \qquad (U > 0) \tag{1c}$$

$$H_\mu = -\sum_{\mathbf{i}\sigma} \mu_\sigma \hat{n}_{\mathbf{i}\sigma} \tag{1d}$$

Here $\hat{c}^+_{\mathbf{i}\sigma}(\hat{c}_{\mathbf{i}\sigma})$ is a creation (annihilation) operator for an electron with spin σ at site $\mathbf{i}$, and $\hat{n}_{\mathbf{i}\sigma} = \hat{c}^+_{\mathbf{i}\sigma}\hat{c}_{\mathbf{i}\sigma}$. In spite of the simplifications occurring in high dimensions [1,2], the Hubbard model has as yet resisted an exact solution. However, very recently the exact solution in $d = \infty$ has been found for a simplified version of the Hubbard model, the so-called Falicov-Kimball model [3], where only one type of electron is mobile. Apart from the deficiencies of Hartree-Fock for the Hubbard model as discussed above, this approximation is also poor for the Falicov-Kimball model, especially at large U, e. g. the predicted critical temperature ($T_c \sim U$) deviates drastically from the exact result [9]: $T_c \sim U^{-1}$.

To construct a Hamiltonian that becomes exact in high dimensions we consider the Falicov-Kimball model. Furthermore the lattice best suited for our analysis is the Bethe lattice. The limit of high dimensions then corresponds to a large coordination number (i. e. $Z \to \infty$). Note that both the Bethe lattice and the hypercubical lattice have AB-structure, so that the thermodynamical properties are essentially the same [9].

Therefore consider the Hamiltonian (1) with $t_{\mathbf{ij}\uparrow} = 0$ and $t_{\mathbf{ij}\downarrow} = -\bar{t}_{\mathbf{ij}}/Z^{1/2}$, where $\bar{t}_{\mathbf{ij}} = 1$ for nearest neighbors and $\bar{t}_{\mathbf{ij}} = 0$ otherwise. The Falicov-Kimball model describes immobile "nuclei" (spin $\uparrow$), interacting via mobile "electrons"(spin $\downarrow$). In view of the discrete symmetry (presence/absence of a nucleus) it is not surprising that the model contains an Ising-like phase transition at half filling [9].

To obtain a mean field description we note that a site $\mathbf{i}$ "sees" its surroundings through Green functions of the form $G(\tau - \tau') = \langle T_\tau \hat{c}_{\mathbf{i}\downarrow}(\tau)\hat{c}^+_{\mathbf{i}\downarrow}(\tau')\rangle$. The time developement of $G(\tau-\tau')$ is described by the equations of motion: the various derivatives $\partial G/\partial\tau$, $\partial^2 G/\partial\tau^2, \ldots$, involve the commutators $[H, c_{\mathbf{i}\downarrow}]$, $[U[H, c_{\mathbf{i}\downarrow}]]$, etc.. The basic point is that these commutators assume a very simple form in the limit $Z \to \infty$. For instance, the commutator $[H, c_{\mathbf{i}\downarrow}]$ can be described in terms of two new fermions, representing the entire first shell of neighbors around site $\mathbf{i}$:

$$\hat{c}_e \equiv [(1-\rho_1)Z]^{-1/2} \sum_{|\mathbf{i}-\mathbf{j}|=1} (1 - \hat{n}_{\mathbf{j}\uparrow})\hat{c}_{\mathbf{j}\downarrow} \tag{2a}$$

$$\hat{c}_f \equiv [\rho_1 Z]^{-1/2} \sum_{|\mathbf{i}-\mathbf{j}|=1} \hat{n}_{\mathbf{j}\uparrow}\hat{c}_{\mathbf{j}\downarrow} \tag{2b}$$

where ρ_1 is the density of nuclei in the first shell. Similarly the commutator $[H, [H, \hat{c}_{\mathbf{i}\downarrow}]]$ introduces four new fermions, and in general the n-th commutator yields 2^n new fermionic variables, corresponding to the 2^n possible ways of finding n sites either occupied or unoccupied by a nucleus. The new fermions are denoted by c_X, where $X = \{x_1, \ldots, x_n\}$ and $x_l = 0, 1$ indicates the absence/presence of a nucleus in the l-th shell around $\mathbf{i}$.

The same commutators can also be derived from an effective Hamiltonian H^{MF} having the same form (1a) as H, but with H_t^{MF}, H_U^{MF} and H_μ^{MF} depending only of c_X:

$$H_U^{MF} = U\hat{n}_{\mathbf{i}\uparrow}\hat{n}_{\mathbf{i}\downarrow} \tag{3a}$$

$$H_t^{MF} = \sum_{l(X)\geq 0} [t_{l+1}^0(\hat{c}_{X0}^+\hat{c}_X + \hat{c}_X^+\hat{c}_{X0}) + t_{l+1}^1(\hat{c}_{X1}^+\hat{c}_X + \hat{c}_X^+\hat{c}_{X1})] \quad (3b)$$

$$H_\mu^{MF} = -\Big\{\mu_\downarrow\hat{n}_{\mathbf{i}\downarrow} + \mu_\uparrow\hat{n}_{\mathbf{i}\uparrow} + \sum_{l(X)\geq 0} [\mu_\downarrow n_{X0} + (\mu_\downarrow - U)n_{X1}]\Big\} \quad (3c)$$

Here the hopping matrix elements are defined by $t_l^0 \equiv (1-\rho_l)^{1/2}$ and $t_l^1 \equiv \rho_l^{1/2}$, where ρ_l is the density of nuclei in the l-th shell. Furthermore $l(X)$ is the length (number of components) of X, and the notation $X0$ or $X1$ represents the $(l+1)$-dimensional vectors $(X,0)$ or $(X,1)$. Note that H^{MF} has the form of a tight-binding model on a Bethe lattice with coordination number $Z^{MF} = 3$. Also note that: (1) the Hubbard interaction remains only on site **i**; (2) a Hartree-Fock decoupling in H_U^{MF} does not occur, since the correlations between electrons and nuclei on-site **i** are in general not small; (3) H^{MF} depends on the nuclei surrounding site **i** only in an average manner, through the density ρ_l (order parameter). The last point makes clear that H^{MF} is truely a mean field Hamiltonian.

Obviously there are also great differences between the quantum Hamiltonian H^{MF} and classical mean field theory. First, H^{MF} contains many mean fields c_X, representing the various shells around **i**. This feature is due to the hopping. Secondly, the mean fields c_X in (3) are operators, not numbers, which emphasizes the quantum character of the model.

Having identified H^{MF} as a typical mean field reformulation of the original model one may wonder whether the phase transition described by H^{MF} also displays the usual mean field behavior of its order parameter. The order parameter in this case is the density difference Δ of the nuclei on the A- and B-sublattices.

To investigate this question we focus on the half-filled case ($\mu_\uparrow = \mu_\downarrow = U/2$). The Green functions $\langle T_\tau \hat{c}_{\mathbf{i}\downarrow}(\tau)\hat{c}_{\mathbf{j}\downarrow}^+(\tau')\rangle$, with $|\,\mathbf{i}-\mathbf{j}\,| = 0, 1$, can easily be calculated using the standard techniques developed for the Bethe lattice [10]. The assumption that the low temperature phase has AB-structure implies $\rho_{2l-1} = \rho_1$ and $\rho_{2l} = \rho_2 (l = 1, 2, \ldots)$, so that the order parameter is $\Delta = |\,\rho_1 - \rho_2\,|$. The Green functions depend explicitly on the parameter Δ. Knowing the Green functions, we can calculate the kinetic energy per site $e_t(\beta,\Delta)$, the interaction energy $e_U(\beta,\Delta)$, the internal energy $e \equiv e_t + e_U$ and, by integrating the relation $e = \partial\beta f/\partial\beta$, the free energy per site $f(\beta,\Delta)$. The physically realized value of Δ, denoted by $\Delta_0(T)$, is obtained by minimizing the free enegy with respect to Δ.

In order to localized the phase transition, we may expand $f(\beta,\Delta)$ in a power series around $\Delta = 0$. From this expansions the critical temperature $T_c(U)$ can readily be calculated for large and small values of U. One finds that

$$k_B T_c \sim (2U)^{-1} \qquad (U \to \infty) \quad (4a)$$

$$\sim \frac{U^2}{2\pi} ln(U^{-1}) \qquad (U \downarrow 0) \quad (4b)$$

so that T_c vanishes in both limits, as expected.

The temperature dependence of $\Delta_0(T)$ near T_c follows directly by minimizing $f(\beta,\Delta)$ with respect to Δ as

$$\Delta_0(T) \propto (T_c - T)^{1/2} \qquad (T \uparrow T_c) \quad (5)$$

Similarly one finds, by expanding $f(\beta,\Delta)$ in powers of $(1-\Delta)$, that the low temperature behavior of $\Delta_0(T)$ is given by

$$1 - \Delta_0(T) \propto \exp(-C/T) \qquad (T \downarrow 0) \quad (6)$$

where C is some positive constant. Equations (5) and (6) clearly show that the order parameter in this quantum mechanical model is indeed completely analogous to the magnetization of the Ising and Heisenberg models in $d = \infty$, thus stipulating its mean field character.

A more detailed discussion of the model discussed in this paper will be published elsewhere.

ACKNOWLEDGMENT

It is a pleasure to thank Drs. Florian Gebhard and Walter Metzner for many interesting discussions.

References

[1] W. Metzner and D. Vollhardt, Phys. Rev. Lett. **62**, 324 (1989).

[2] E. Müller-Hartmann, Z. Phys. **B74**, 507 (1989); H. Schweitzer and G. Czycholl, Solid State Comm. **69**, 171 (1989); P.G.J. van Dongen, F. Gebhard and D. Vollhardt, Z. Phys. B **76**, 199 (1989).

[3] U. Brandt and C. Mielsch, Z. Phys. **B75**, 365 (1989).

[4] For a short review, see D. Vollhardt, Int. J. Mod. Phys. **B3** 2189 (1989); E. Müller-Hartmann, Int. J. Mod. Phys. **B3** 2169 (1989).

[5] C.J. Thompson, Commun. Math. Phys. **36**, 255 (1974); R. Brout, Phys. Rev. **118**, 1009 (1960).

[6] P. A. Pierce and C.J. Thompson, Commun. Math. Phys. **58**, 131 (1978); T. Kennedy, E.H. Lieb and B.S. Shastry, J. Stat. Phys. **53**, 1019 (1988) and Phys. Rev. Lett. **61**, 2582 (1988).

[7] J. Hubbard, Proc. Roy. Soc. London **A276**, 238 (1963); M. C. Gutzwiller, Phys. Rev. Lett. **10**, 159 (1963); J. Kanamori, Prog. Theor. Phys. **30**, 275 (1963).

[8] P. Lizana, R. Strack and D. Vollhardt, to be published.

[9] T. Kennedy and E.H. Lieb, Physica **138A**, 320 (1986); E.H. Lieb, Physica **140A**, 240 (1986); U. Brandt and R. Schmidt, Z. Phys. **B63**, 43 (1987).

[10] E.N. Economou, "Green's Functions in Quantum Physics", Springer-Verlag, Berlin (1979).

Physica B 165&166 (1990) 403–404
North-Holland

HUBBARD MODEL IN HIGH DIMENSIONS: A PERTURBATION EXPANSION ABOUT THE ATOMIC LIMIT

Walter METZNER

Institut für Theoretische Physik C, Technische Hochschule Aachen, Sommerfeldstraße 26/28, D-5100 Aachen, Federal Republic of Germany

We discuss a perturbation expansion about the atomic limit of the Hubbard model, i.e. the inter-atomic hopping of electrons is treated as a perturbation. Diagrammatic rules for the calculation of the Green's function of the d-dimensional Hubbard model are described. In the limit of high lattice dimensions only the subclass of fully two-particle reducible diagrams contributes. In this case a proper renormalization of the perturbation expansion reduces the solution of the Hubbard model to a zero-dimensional problem. The renormalized expansion yields approximate solutions which interpolate between the atomic and the free fermion limit.

1. INTRODUCTION

Recently, the limit of infinite lattice dimensions $d \to \infty$ has been introduced as a new promising starting point towards a better understanding of correlated lattice fermions in finite dimensions. (1) In a short while a lot of perturbational and variational investigations of various infinite dimensional lattice models have been performed. (2) It turned out that all calculations are considerably simplified in $d \to \infty$ while many interesting properties of interacting fermions such as strong correlations, mass renormalization and magnetism are still retained in this limit. In this contribution a perturbation theory of the Hubbard model (3) about the atomic limit will be formulated and its application to the case $d \to \infty$ will be discussed.

2. PERTURBATION EXPANSION

The one-band Hubbard model on a d-dimensional lattice is given by the Hamiltonian $H = H_0 + H_1$, where

$$H_0 = U \sum_j n_{j\uparrow} n_{j\downarrow} \tag{1a}$$

is a local repulsion and

$$H_1 = \sum_\sigma \sum_{i,j} t_{ij}\, c^\dagger_{i\sigma} c_{j\sigma} \tag{1b}$$

is the kinetic energy of the model.

For $H_1 = 0$ the Hubbard Hamiltonian describes a collection of independent atoms (atomic limit). Taking the kinetic energy H_1 as a perturbation and using the interaction picture, one may express the thermal Green's function $G_{\sigma ij}(\tau,\tau')$ as a power series in the hopping matrix t_{ij}. The n-th order term involves the calculation of an $(n+1)$- particle Green's function G^0_{n+1} in the atomic limit; G^0_{n+1} can be expressed as a sum of products of cumulants (connected Green's functions) C^0_m, $m = 1,..,n+1$. (4) The cumulants are local (site-diagonal) because H_0 is a sum of local operators. Hence $C^0_m = C^0_{mj}(\tau_1\sigma_1, \tau_2\sigma_2, ..)$ is a function of 2m time and spin variables, but there is only one site variable. Thus a cumulant can be represented graphically as a local vertex with m entering and m leaving particles. In this way each term of the cumulant expansion can be represented as a diagram, where lines correspond to hopping matrix elements and vertices to cumulants. Disconnected diagrams cancel the denominator appearing in the interaction representation of G. Thus, the n-th order contribution to the exact Green's function G can be calculated by applying the following rules:

i) draw all topologically distinct connected diagrams containing n lines (see Fig. 1);

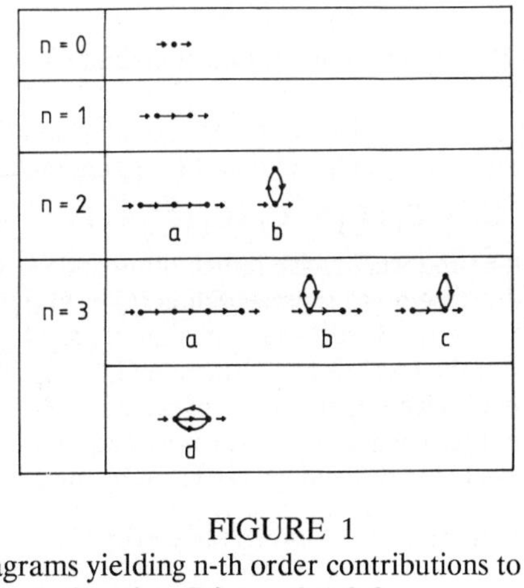

FIGURE 1
Diagrams yielding n-th order contributions to the Green's function G for n=0, 1, 2, 3.

ii) label each line with a spin and time index, label each vertex with a site index; a line from **i** to **j** yields a factor t_{ij}, a vertex at **j** connected to 2m lines yields a factor C^0_{mj} (...);
iii) determine the sign (−1 for an odd number of loops, +1 else) and the symmetry factor (4) of each diagram;
iv) integrate all internal time variables from 0 to β (inverse temperature), sum all internal spin- and site-indices and finally all diagrams.

These rules are an efficient tool for the evaluation of the leading hopping corrections to the atomic limit of the Hubbard model. A more detailed derivation and explicit evaluations will be published elsewhere.

3. LIMIT $d \to \infty$ AND RENORMALIZATION

To keep the average kinetic energy H_1 finite as $d \to \infty$ one must scale the hopping matrix, e.g. $t = t^*/(2d)^{1/2}$ (t^* fixed) for next-neighbor hopping. (1) As a consequence all diagrams containing two-particle irreducible parts (e.g. 3d in Fig. 1) can be neglected as $d \to \infty$.

We now renormalize the perturbation expansion derived above, i.e. we set up an expansion in terms of skeleton diagrams which are functionals of the exact Green's function G. We define a dressed one-particle cumulant C_1 as the sum of all possible local insertions as illustrated in Fig. 2. In $d \to \infty$, the exact Green's

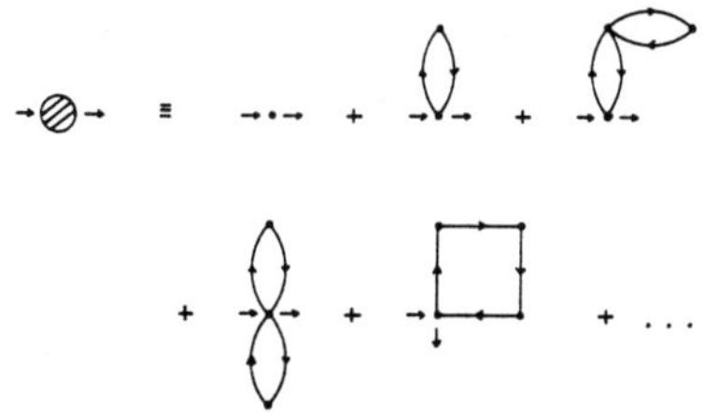

FIGURE 2
Definition of the dressed cumulant C_1.

function G is then given as the sum of all chain diagrams with vertices C_1 and lines t (hopping matrix), i.e.

$$G = C_1 + C_1 t C_1 + C_1 t C_1 t C_1 + \ldots \qquad (2)$$

where the products involve lattice sums and G, C_1 are inserted in frequency representation ($G = G_{\sigma ij}(\omega)$, $C_1 = C_{1\sigma i}(\omega)$). The dressed cumulant C_1 is a functional only of the local Green's function $G_{\sigma ii}(\omega)$. This functional is given by the series of skeleton diagrams shown in Fig. 3, where the vertices correspond to bare cumulants C^0_m and the loops are given by the quantity

$$\Phi_{\sigma i}(\omega) = [G_{\sigma ii}(\omega) - C_{1\sigma i}(\omega)]/[C_{1\sigma i}(\omega)]^2 \qquad (3)$$

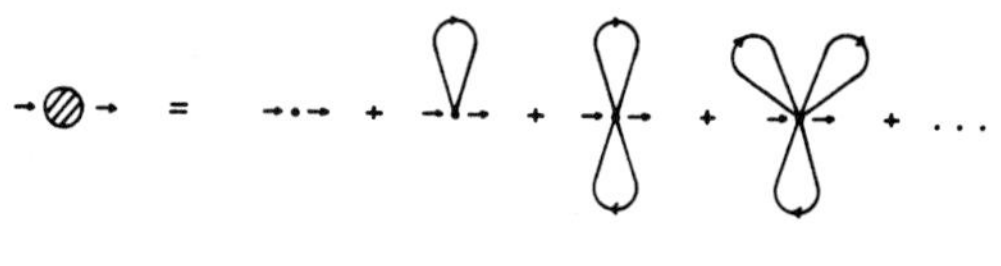

Figure 3
Sum of skeleton diagrams yielding C_1 in terms of G.

The chain-sum (2) and the skeleton-sum given in Fig.3 make up a complete system of equations for G in $d=\infty$. Equation (2), which describes unrenormalized hopping between dressed atoms is easy to handle. The skeleton-sum represents a single site in the presence of a fermion bath creating and annihilating particles on that site. This is a zero-dimensional (i.e. local) problem which, however, has not been solved so far.

Self-consistent approximations for G which can be obtained by truncating the skeleton-sum are presently being investigated. These approximations are automatically exact in both the atomic limit and the free fermion limit because $C^0_m = 0$ for all $m > 1$, if $U = 0$.

4. CONCLUSION

A diagrammatic perturbation expansion about the atomic limit may be a powerful tool for the investigation of Hubbard-type models. Such an expansion yields an intuitive real-space picture of the dynamics of lattice electrons and the simplifications arising in $d \to \infty$ become particularly clear. The renormalized theory is a source of self-consistent solutions which interpolate between the atomic and the free fermion limit. Generalizations to many-band models such as the periodic Anderson model are straightforward.

ACKNOWLEDGEMENTS
I would like to thank G.Czycholl, P. van Dongen, F. Gebhard and D. Vollhardt for interesting and useful discussions.

REFERENCES
(1) W. Metzner and D. Vollhardt, Phys. Rev. Lett. **62**, 324 (1989).
(2) For a brief review see: D. Vollhardt, Int. J. Mod. Phys. **B3**, 2189 (1989); E. Müller-Hartmann, Int. J. Mod. Phys. **B3**, 2169 (1989).
(3) J. Hubbard, Proc. Roy. Soc. London **A276**, 238 (1963); M.C. Gutzwiller, Phys. Rev. Lett. **10**, 159 (1963); J. Kanamori, Prog. Theor. Phys. **30**, 275 (1963).
(4) See, for example: J. W. Negele and H. Orland, Quantum Many-Particle Systems (Addison-Wesley, New York, 1988), chapter 2.

Physica B 165&166 (1990) 405–406
North-Holland

STUDY OF TWO IMPURITY KONDO PROBLEM BY SELF-CONSISTENT PERTURBATION APPROACH

Tetsuro Saso
Department of Physics, Tohoku University, Sendai, 980 Japan.

It is pointed out that two impurity Kondo problem resembles valence fluctuation of a Tm or Pr impurity, and that the theoretical method successfully used in the latter is applicable. Using the extended NCA approach and molecular-orbital representation, competition of the Kondo effect and magnetic interaction is discussed.

Study of the two-impurity Kondo problem[1, 2] is very important for understanding the physics of heavy fermions, oxide superconductors and other strongly correlated many-electron systems. Although recent renewed interest[3, 4] has activated the research, an essential physics of the problem still seems to remain to be fully understood.

We start with the Anderson Hamiltonian for two impurities placed at $\mathrm{R}_i = \pm\mathrm{R}/2$ $(i = 1, 2)$,

$$\begin{aligned} H &= \sum_{\mathbf{k}\sigma} \epsilon_{\mathbf{k}} c^{+}_{\mathbf{k}\sigma} c_{\mathbf{k}\sigma} + \sum_{i\sigma} E_d n_{di\sigma} \\ &+ V \sum_{i\mathbf{k}\sigma} (e^{i\mathbf{k}\cdot\mathbf{R}_i} d_{i\sigma} c^{+}_{\mathbf{k}\sigma} + h.c.) + U \sum_i n_{di\uparrow} n_{di\downarrow} \quad (1) \end{aligned}$$

with the ordinary notation. Similarly to [1] and following [4] we introduce even- and odd-parity d-orbitals

$$d_{p\sigma} = (d_{1\sigma} \pm d_{2\sigma})/\sqrt{2} \quad (p = e,\ o) \quad (2)$$

and conduction electron channels

$$c_{\epsilon p\sigma} \equiv N_{\epsilon p} \sum_{\mathbf{k}} \delta(\epsilon - \epsilon_{\mathbf{k}}) \times \begin{cases} \cos \frac{\mathbf{k}\cdot\mathbf{R}}{2} \frac{c_{\mathbf{k}\sigma} + c_{\mathbf{k}\sigma}}{\sqrt{2}} \ (p = e) \\ i \sin \frac{\mathbf{k}\cdot\mathbf{R}}{2} \frac{c_{\mathbf{k}\sigma} - c_{\mathbf{k}\sigma}}{\sqrt{2}} \ (p = o). \end{cases} \quad (3)$$

The normalization constant $N_{\epsilon p}$ is determined by the condition

$$[c_{\epsilon p\sigma}\ ,\ c^{+}_{\epsilon' p'\sigma}] = \delta(\epsilon - \epsilon')\delta_{\sigma\sigma'}\delta_{pp'} \quad (4)$$

yielding

$$N_{\epsilon p}^{-2} = \rho(\epsilon)(1 \pm \sin kR/kR), \quad (5)$$

where $k = |\mathbf{k}| = \sqrt{2m\epsilon}$, $R = |\mathrm{R}|$ and $\rho(\epsilon)$ is the density of states of conduction electrons.

The Hamiltonian (1) is transformed into

$$\begin{aligned} H &= \int d\epsilon\ \epsilon \sum_{p\sigma} \rho_p(\epsilon) c^{+}_{\epsilon p\sigma} c_{\epsilon p\sigma} + \sum_{p\sigma} E_d n_{dp\sigma} \\ &+ \int d\epsilon \sum_{p\sigma} V_p(\epsilon)(c^{+}_{\epsilon p\sigma} d_{p\sigma} + h.c.) + U\text{–term} \quad (6) \end{aligned}$$

where $\rho_p(\epsilon)$ is the density of states of the p-channel, which plays no role hereafter, and

$$V_p(\epsilon)^2 = \rho(\epsilon)V^2(1 \pm \sin kR/kR). \quad (7)$$

It is pointed out that the different results in [1] and [2] may originate from this energy-dependence of $V_p(\epsilon)$. We assume the limit $U \longrightarrow \infty$ in the following.

The Hamiltonian (6) is very similar to that for a valence fluctuating Tm impurity [5, 6, 7], or more precisely, a Pr impurity which fluctuates between $4f^2$ and $4f^1$ states. Hence the method used there is applicable also to the present problem.

In parallel with Pr, we may work with the Hilbert space consisting of d^1 and d^2. The latter is further confined to the states with $n_{d1} = n_{d2} = 1$, and in terms of the representation (3) it can be expressed as the singlet and the triplet states constructed from $d^{+}_{e\sigma} d^{+}_{o\sigma'}|0\rangle$.

In the following calculation, we add a direct spin-spin intereaction term $J\vec{S}^d_1 \cdot \vec{S}^d_2$ to (6), which operates only on the $n_{d1} = n_{d2} = 1$ states, for comparison with [1]. Then the difference from Pr is that the term $J\vec{S}^d_1 \cdot \vec{S}^d_2$ corresponds to the inclusion of the crystal field splitting in Pr.

Thermodynamics of the whole system is determined by the partition function $Z = Z_c \cdot Z_d$, where the d-part Z_d is expressed by use of the resolvents as [7]

$$Z_d = \int_{-D}^{D} d\epsilon\ e^{-\beta\epsilon}(-\frac{1}{\pi})\,\mathrm{Im}\,[2R_e(z) + 2R_o(z) + R_s(z) + 3R_t(z)] \quad (8)$$

where $2D$ is the conduction band width and the resolvents for d^1 (even and odd) and d^2 (singlet and triplet) sectors are written as

$$\begin{aligned} R_{e,o}(\epsilon) &= (z - E_d - \Sigma_{e,o}(z))^{-1}, \\ R_{s,t}(\epsilon) &= (z - E_{s,t} - \Sigma_{s,t}(z))^{-1}, \quad (9) \end{aligned}$$

with $E_s = 2E_d - (3/4)J$ and $E_t = 2E_d + J/4$. Self-energies for the resolvents are calculated in the NCA

0921-4526/90/$03.50

scheme as[7]

$$\Sigma_p(z) = \int_{-D}^{D} d\epsilon\ f(\epsilon)\Gamma_{\bar{p}}(\epsilon)[R_s(z+\epsilon) + 3R_t(z+\epsilon)], \quad (10)$$

$$\Sigma_s(z) = \Sigma_t(z)$$
$$= 2\int_{-D}^{D} d\epsilon\ (1-f(\epsilon))[\Gamma_o(\epsilon)R_e(z-\epsilon) + \Gamma_e(\epsilon)R_o(z-\epsilon)], \quad (11)$$

where $f(\epsilon)$ is the Fermi function, $\Gamma_p(\epsilon) \equiv V_p^2(\epsilon)/2$ and $\bar{p}$ denotes the counter-component of p.

If we neglect the difference between $\Gamma_e(\epsilon)$ and $\Gamma_o(\epsilon)$ and set equal to a constant Γ, then we obtain $\Sigma_e(z) = \Sigma_o(z) \equiv \Sigma_1(z)$, $R_e(z) = R_o(z) \equiv R_1(z)$ and $\Sigma_s(z) = \Sigma_t(z)$ but $R_s(z) \neq R_t(z)$. In this case, the "T-matrices"[7] for singlet and triplet channels are evaluated as

$$T^{s,t}(z) \simeq \bar{T}_0^{s,t}(z)/[1 - P(z)\bar{T}_0^{s,t}(z)] \quad (12)$$

where

$$\bar{T}_0^{\{s \atop t\}}(z) = \Gamma\langle[R_s(z+\epsilon+\epsilon') + \{{3 \atop 1}\}R_t(z+\epsilon+\epsilon')]\rangle_{\epsilon\epsilon'}, \quad (13)$$

$$P(z) = \int_{-D}^{D} d\epsilon\ (1-f(\epsilon))R_1(z+\epsilon). \quad (14)$$

and $\langle\cdots\rangle_{\epsilon\epsilon'}$ denotes an everage over $\epsilon\epsilon'$ [7]. In this case, $\Sigma_1(z)$ is calculated to the lowest order of Γ as

$$\Sigma_1(z) \simeq \Gamma \log\frac{|z-E_s|}{D} + 3\Gamma\log\frac{|z-E_t|}{D}. \quad (15)$$

For $J < De^{E_d/3\Gamma} \equiv T_0$, we find that the resolvent $R_1(z)$ behaves as

$$R_1(z) \simeq \frac{r}{z-\tilde{E}_d} + \frac{1-r}{z-E_d+4i\Gamma} \quad (16)$$

with $\tilde{E}_d \simeq E_t - De^{-(E_d-E_t)/3\Gamma}$ and the residue $r \propto (2-n_d)$ [7].

The T-matrix for the singlet bound state (Kondo singlet) is calculated as

$$\bar{T}^s(z) \propto \frac{1}{1 - r\Gamma\langle R_s + 3R_t\rangle\log(|z-\tilde{E}_d|/D)} \quad (17)$$

which has a pole at $z \sim \tilde{E}_d - De^{-1/r\Gamma\langle R_s+3R_t\rangle} \equiv E_0$. This pole quickly disappears for $J > T_0$, since the contribution from $\langle 3R_t\rangle$ rapidly decreases in comparison to $\langle R_s\rangle$, and the d^2 singlet state dominates the thermodynamics. So a continuous but rapid crossover takes place at around $J \sim T_0$, and the character of the dominant state changes from the singlet bound state(Kondo state) of d^2 triplet plus conduction electrons to the d^2 singlet. For $J < T_0$, the staggered susceptibility χ_s[1, 3] is determined by the transition from the d^2triplet state included in the Kondo singlet (E_0) to the d^2 singlet ($\tilde{E}_s$). For $J > T_0$, the transition from $\tilde{E}_s$ to $\tilde{E}_t$ contributes to χ_s. This picture precisely corresponds to the spin-model interpretation of [3]. In the present description, however, the internal degrees of freedom of conduction electrons are properly treated.

In the general cases of even-odd asymmetry $V_e(\epsilon) \neq V_o(\epsilon)$, we have to solve

$$T_{\epsilon\epsilon'}^{s,t}(z) = \Gamma_{eo}(\epsilon,\epsilon')R_2(z+\epsilon+\epsilon') +$$
$$\int d\epsilon'' \int d\epsilon''' f(\epsilon'')f(\epsilon''')\Gamma_{eo}(\epsilon,\epsilon'')R_2(z+\epsilon+\epsilon'')R_o(z+\epsilon'')$$
$$\times\Gamma_{oe}(\epsilon'',\epsilon''')R_2(z+\epsilon''+\epsilon''')R_e(z+\epsilon''')T_{\epsilon'''\epsilon'}^{s,t}(z)] \quad (18)$$

with $R_2 = R_s + 3R_t$ ($R_s + R_t$) for s (t) channel and $\Gamma_{eo}(\epsilon,\epsilon') = V_e(\epsilon)V_o(\epsilon')/2$. Because of the relative shift of the renormalized poles of R_e and R_o due to the aysmmetry of the mixing, R_s and R_t jointly contribute to the formation of the Kondo singlet ground state for $J < T_0$. Thus the rapid crossover of the ground state is smeared [2, 4].

The Author is much indebted to the useful discussion with Professor O. Sakai. Thanks are also due to Professor T. Kasuya for his continuous encouragement. The present work is supported by the Grant-in-Aids No. 01634014 and 01634008 from the Ministry of Education, Science and Culture.

References

[1] B.A. Jones and C.M. Varma, Phys. Rev. Lett. **58** (1987) 843.

[2] R.M. Fye and J.E. Hirsch, Phys. Rev. **B40** (1989) 4780.

[3] J.W. Rasul and P. Schlottmann, Phys. Rev. Lett. **62** (1989) 1701.

[4] O. Sakai, in preparation.

[5] Y. Yafet, C.M. Varma and B. Jones, Phys. Rev. **B32** (1985) 360.

[6] S.M.M. Evans and G.A. Gehring, J. Phys. Condensed Matter **1** (1989) 3095.

[7] T. Saso, J. Phys. Soc. Jpn. **58** (1989) 4064; to be published in Physica **B** (1990) [*Proc. Int. Conf. on the Physics of Highly Correlated Electron Systems*(Santa Fe, 1989 Sept.9-15)].

Physica B 165&166 (1990) 407–408
North-Holland

THE STATIC INTERACTING SUSCEPTIBILITY OF UPt_3 AND $CeAl_3$

H D LANGFORD† and W M TEMMERMAN‡

†Department of Theoretical Physics, Oxford University, Oxford OX1 3NP, UK

‡SERC Daresbury Laboratory, Warrington, WA4 4AD, UK

Within the local approximation to spin density functional theory (LSDA), we have calculated the fully - interacting static spin susceptibility of UPt_3 and $CeAl_3$ by performing self - consistent spin - polarized LMTO - ASA band structure calculations in an applied magnetic field. By investigating supercell configurations, we have evaluated the susceptibility as a function of wavevector along the c-axis of the hexagonal Brillouin zone. Although we do find considerable enhancement as compared with the single - particle susceptibility evaluated from the density of states for all values of q considered ($q = 0, \frac{\pi}{2c}, \frac{\pi}{c}$ and $\frac{2\pi}{c}$), we find that the experimental bulk susceptibility is still substantially larger that our calculations predict, indicating that both many - body effects beyond the LSDA and orbital effects are important in these systems.

1. Introduction

The so-called heavy fermion systems are characterized by a massively enhanced linear coefficient of specific heat at low temperatures compared with that of ordinary metals [1]. It is mainly among Ce and U containing alloys that the heavy fermion systems are found and it is believed that magnetic fluctuations are important in determining the low temperature properties so that a calculation of the spin-spin correlation functional $\chi(\mathbf{q}, \mathbf{q}'; \omega)$ is of great interest.

In this paper, we will use band theory within local spin-density functional theory [2] to evaluate the static interacting susceptibility as a function of wavevector for two heavy fermion systems, namely UPt_3 and $CeAl_3$. The method we employ is the non-perturbative approach of applying a magnetic field and self-consistently determining the response, *i.e.* the induced magnetic moment. The usefulness of this approach is that it includes both the effects of electrons close to the Fermi level flipping their spins and the changes in the wavefunctions of all states from the spin-polarization induced change in the potentials. This can be generalized to finite q by performing calculations for frozen spin-wave configurations commensurate with the lattice so that the susceptibility can be mapped out along a certain high-symmetry direction in the Brillouin zone [3].

2. Method

We have performed self-consistent linear muffin-tin orbital band calculations within the atomic sphere approximation [4] including induced frozen spin-waves for UPt_3 and $CeAl_3$. The structure of these two systems is hexagonal with two formula units per unit cell. Calculations were performed for four different q-values, namely $q = 0, \frac{\pi}{2c}, \frac{\pi}{c}$ and $\frac{2\pi}{c}$ having respectively 2, 8, 4 and 2 formula units in the resultant magnetic unit cell. Due to the large Hamiltonian dimensions thus incurred, the valence states were included in a semi-relativistic way.

3. Results

Table 1 shows the values of the Stoner enhancements for the four spin configurations and two compounds considered. The $q = 0$ state was analysed by applying a field of 100 Tesla in the same direction on each atom, corresponding to a ferromagnetic state. For the $q = \frac{2\pi}{c}$ configuration, opposite fields were applied on the two f-electron atoms in the unit cell to create an antiferromagnetic state. Similarly, the $q = \frac{\pi}{c}$ results were obtained by applying opposite fields in each of a pair of unit cells, while the $q = \frac{\pi}{2c}$ results needed a configuration containing four unit cells, the first two being given an external spin-splitting $+\Delta$ and the second two a splitting $-\Delta$. It was found necessary to use splittings corresponding to a field of 200 T for the $q = \frac{\pi}{c}$ and $q = \frac{\pi}{2c}$ states in order to stabilize the configuration and reduce the noise caused by having only a finite number of $\mathbf{k}$-points for the Brillouin zone integrations [5].

As can be seen from table 1, the Stoner enhancements for finite q are very similar to those for $q = 0$, implying that the interaction between different f-electron sites is small so that spin-fluctuation must be a local effect. Further, when we come to consider the actual value of $\chi(0)$, we find that $\chi_{Theory} = 3.34 \times 10^{-4}$ for UPt_3 and $\chi_{Theory} = 0.95 \times 10^{-4}$ for $CeAl_3$. Experimentally, $\chi_{Expt.} = 25.0 \times 10^{-4}$ for UPt_3 and $\chi_{Expt.} = 87.9 \times 10^{-4}$ for $CeAl_3$. This indicates that, although we have predicted some enhancement in $\chi(0)$, the effect is nowhere near big enough. It would be interesting to extend these calculations to the inclusion of spin-orbit coupling so that the effect of such relativistic effects could be analysed. This would also allow a quantitative estimate to be made of the effects of many-body corrections beyond the LSDA.

4. Conclusions

If we write the susceptibility as

$$\chi(\mathbf{q}) = \frac{\chi_s(\mathbf{q})}{1 - I(\mathbf{q})\chi_s(\mathbf{q})}, \tag{1}$$

in a generalized Stoner form, where χ_s is the single particle value of the susceptibility, we can see that, for $CeAl_3$ and UPt_3, there is not a particularly rapid variation in $I(\mathbf{q})$. This suggests single-ion behaviour and thus provides a possible reason for why the Wilson ratio for non-magnetic heavy fermion systems should be approximately unity. This is to be compared with the transition metal Pd, where it is the strong variation in $I(\mathbf{q})$ that is important and nearly produces a ferromagnetic instability [3], [5].

References

[1] Stewart G.R., Rev. Mod. Phys. **56** 755 (1984)

[2] von Barth U. and Hedin L., J. Phys. C **5** 1629 (1972)

[3] Jarlborg T., Solid State Commun. **57** 683 (1986)

[4] Andersen O.K., Phys. Rev. B **12** 3060 (1975)

[5] Langford H.D., D. Phil. Thesis, Oxford, unpublished

Table 1: Values of the Stoner enhancement S for UPt_3 and $CeAl_3$ for the values of q considered.

q	S(UPt_3)	S($CeAl_3$)
0	4.6	2.9
$\pi/2c$	1.9	5.1
π/c	3.1	6.3
$2\pi/c$	3.0	3.1

Physica B 165&166 (1990) 409–410
North-Holland

ON THE DYNAMIC SUSCEPTIBILITY OF HEAVY FERMION COMPOUNDS

J CHOI† and G A GEHRING‡

†Department of Theoretical Physics, Oxford University, Oxford OX1 3NP, UK

‡Department of Physics, Sheffield University, Sheffield S3 7RH, UK

Magnetic susceptibilities are calculated for the lattice Anderson model. From the Lindhard expresssion of the zero field magnetic susceptibility we calculate static and dynamic susceptibilities due to inter- and intra-band contributions. At zero temperature we find the interband term is much smaller than the intraband term. Relevance with experiments is briefly discussed.

In recent years much attention has been focused on mixed valent or heavy fermion compounds. In these materials highly correlated f electrons hybridize with conduction electrons. One of the sources of the most direct informations on these material is by neutron scattering experiments, where differential cross sections are directly related to the imaginary part of the dynamic susceptibility. It is thus natural to attempt to calculate the dynamic susceptibilty and compare with experimental results. So far the dynamic susceptibility has been calculated for the periodic Anderson model, which is regarded as a suitable model in describing many properties of mixed valent systems, using various methods.(1-3) According to recent neutron scattering data(4) for small momentum transfer there is a sharp peak near zero frequency on top of broad band, which can be fitted to two Lorenzian forms. It is attributed that these two Lorenzians come from intra- and interband contributions, respectively.

For the last few years much understanding has been obtained for the periodic Anderson model thanks to the so-called "slave boson technique".(5-7) One can calculate the dynamic susceptibility in the usual thermodynamic way. (7,8) In this approach the dynamic susceptibility is directly given from boson propagators and shown to be greatly enhanced. The dynamic susceptibility can also be derived using a retarded Green's function technique coupling magnetic fields to both of c- and f-bands.(9) The Lindhard expression for the dynamic susceptibility contains intra- and inter-band contributions. We refer these as Pauli and Van Vleck respectively.

$$\chi^P(q,\omega) = \sum_k \frac{f(E_+(k))-f(E_+(k+q))}{E_+(k+q)-E_+(k)-\omega}(g_c A_k^+ A_{k+q}^+ + g_f A_k^- A_{k+q}^-)^2 + \sum_k \frac{f(E_-(k))-f(E_-(k+q))}{E_-(k+q)-E_-(k)-\omega}(g_c A_k^- A_{k+q}^- + g_f A_k^+ A_{k+q}^+)^2 \quad (1)$$

$$\chi^{VV}(q,\omega) = \sum_k \frac{f(E_-(k))-f(E_+(k+q))}{E_+(k+q)-E_-(k)-\omega}(g_c A_k^- A_{k+q}^+ + g_f A_k^+ A_{k+q}^-)^2 + \sum_k \frac{f(E_+(k))-f(E_-(k+q))}{E_-(k+q)-E_+(k)-\omega}(g_c A_k^+ A_{k+q}^- + g_f A_k^- A_{k+q}^+)^2, \quad (2)$$

where g_c and g_f are the g factors for c- and f-band respectively and f is the usual fermi function. The band energy $E_\pm$ and $A^\pm$ are given as

$$E_\pm(k) = \tfrac{1}{2}(\epsilon_k + \epsilon_f \pm \sqrt{(\epsilon_k - \epsilon_f)^2 + 4\tilde{V}^2}) \quad (3)$$

and $A_k^\pm = (\partial E_\pm / \partial \epsilon_k)^{\frac{1}{2}}$. Here ϵ_f and $\tilde{V}^2$ are the renormalized f-level energy and hybridization energy from appropriate mean field solutions.(7) For the static susceptibilities we take the limit $q,\omega \to 0$. It is easy to evaluate these quantities at zero temperature since the contributions from the fermi functions of the upper level vanish. Assuming the constant density of states for the conduction band, we obtain

$$\chi^P(0,0) = \rho_0(1+x)(\tfrac{1}{1+x}g_c + \tfrac{x}{1+x}g_f)^2 \quad (4)$$

$$\chi^{VV}(0,0) = \rho_0(g_f - g_c)^2 \tfrac{x}{x+1}, \quad (5)$$

where ρ_0 denotes the density of states for the conduction band and $x = \tilde{V}^2/\epsilon_f^2$. It is noted here that x is related to the mass enhancement and becomes large in the Kondo limit. When $g_c \ll g_f$, as is often the case, the total susceptibility can be expressed as $\chi^T = \rho_0 g_f^2 x$ This is precisely same as that obtained in the usual thermodynamic way(7,8), as it should, *i.e.*, via taking the second derivative of the free energy with respect to the field, but it only gives us the total susceptibility not the Pauli- and Van Vleck term separately. We also note that this gives us the Wilson ratio R as 1 only in the Kondo limit since the linear coefficient of the specific heat is proportional to $x + 1$.(10) However, as can be seen in Eqs. (4) and (5), χ^P is much bigger than χ^{VV}, since x is very big in the Kondo limit. This implies that Van Vleck contribution can hardly be seen in the static susceptibility. We now turn our attention to the dynamic susceptibility. The imaginary part of the dynamic susceptibility can be evaluated in closed form as $T \to 0$. Setting $\omega \to \omega - i0^+$ in Eq. (1) and expanding fermi functions for small q we obtain

$$\frac{Im\chi^P(q,\omega)}{\omega} = \frac{\pi}{2}\rho_0(1+x)(\tfrac{1}{1+x}g_c+\tfrac{x}{1+x}g_f)^2\tfrac{1}{qv_F^*}\Theta(qv_F^*-\omega) \tag{6}$$

When g_c can be neglected and for large x this expression reduces to the form

$$\frac{Im\chi^P(q,\omega)}{\omega} = \frac{\pi}{2}\rho_0 g_f^2 x\frac{1}{qv_F^*}\Theta(qv_F^* - \omega), \tag{7}$$

which is same as that obtained by several authors(7,8) For the Van Vleck part we obtain as $q \to 0$ limit,

$$\frac{Im\chi^{VV}(0,\omega)}{\omega} = \pi\rho_0(g_c - g_f)^2\frac{\tilde{V}}{\omega^2}\Theta(\omega - \sqrt{\epsilon_f^2 + 4\tilde{V}^2}) \Theta(-\omega + \sqrt{(W + \epsilon_f)^2 + 4\tilde{V}^2}), \tag{8}$$

where W is the width of the conduction band. Again we note that the Pauli term is relatively bigger than the Van Vleck term. We plot these quantities as a function of ω in Fig. 1.

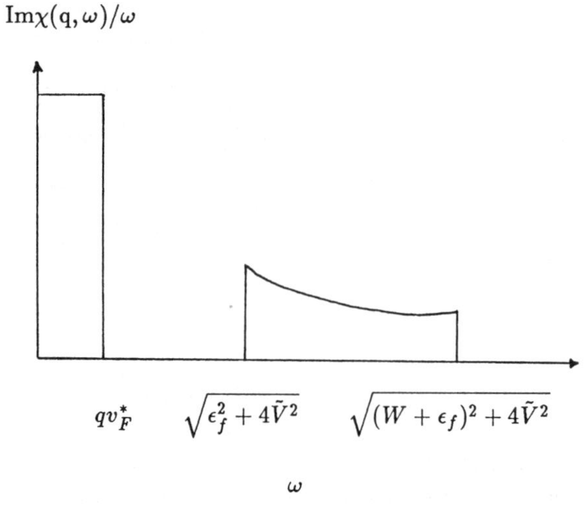

Fig.1 Imaginary part of Pauli and Van Vleck susceptibility as a function of ω in the limit $q \to 0$ at zero temperature.

As can be seen in Fig. 1, there is a gap between Pauli and Van Vleck term when $q \to 0$ limit and $\chi^{VV}(0,\omega)/\omega$ is a decreasing function of ω. We believe one should be able to see this for sufficiently small q at low temperatures. It is expected that the gap becomes smaller as we increase the temperature since $\tilde{V}$ is a decreasing function of temperature so that much contribution would come from interband transitions, which in turn gives the broad band down to near $\omega = 0$ while there is still a narrow peak due to the Pauli term as observed experimentally We expect that the behavior of χ^{VV} does not change much for finite q as long as q is not too big, while the narrow peak of $\chi^P(q,\omega)/\omega$ near $\omega = 0$ broadens as we increase q. Finally we note that the Fermi liquid relation still holds between static and dynamic susceptibility

$$\lim_{\omega\to 0}\frac{Im\chi^T(q,\omega)}{\omega} = \frac{\pi k_F}{2Nq}Re\chi^T(0,0)^2, \tag{9}$$

which is analogous to the Korringa relation for the impurity problems.

In summary we have calculated the static and dynamic susceptibility using the retarded Green' function technique to show that there are two contributions which we call Pauli and Van Vleck terms. We have further showed that Van Vleck term is very small compared to Pauli term in the Kondo limit but observable for very small q in the dynamic susceptibility.

We are grateful to G. G. Lonzarich for informative discussions. JC would like to thank SERC for financial support.

REFERENCES

(1) M. E. Foglio, J. Phys. C11 (1978), 4177
(2) Y. Kuramoto, Z. Physik B 37 (1980), 299
(3) P. S. Riseborough and D. L. Mills, Phys. Rev. B21 (1980), 5338
(4) G. G. Lonzarich, private communications
(5) N. Read and D. M. Newns, J. Phys. C16 (1983), 3273
(6) P. Coleman, Phys. Rev. B35 (1987) 5772
(7) A. Millis and P. A. Lee, Phys. Rev. B35 (1987), 3394
(8) S. M. M. Evans and G. A. Gehring, J. Phys. CM1 (1989) 15487
(9) J. Choi and G. A. Gehring, to be published
(10) J. Rasul and A. P. Harrington, J. Phys. C20 (1987), 4783

Physica B 165&166 (1990) 411–412
North-Holland

STRONGLY CORRELATED FERMIONS IN 2 D : VARIATIONAL MONTE CARLO STUDY OF MAGNETISM AND SUPRACONDUCTIVITY

Thierry GIAMARCHI† and Claire LHUILLIER
Laboratoire de Physique Théorique des Liquides‡, Université Pierre et Marie Curie,
4, Place Jussieu, 75252 Paris Cedex 05, France.

Variational Monte Carlo techniques have been used to investigate the nature of the fundamental state of the Hubbard's for repulsions in the range $4 < U/t < 10$ and for dopings up to 0.4. Away from half filling the antiferromagnetic phase is to be unstable against the formation of diagonal domain walls. This phase with diagonal domain walls is at our knowled stabler phase actually exhibited, but the antiferromagnetic order may still be stabilized by a superconducting pairing.

Heavy fermions and high Tc supraconductors have urged a renewal of interest in the Hubbard's and (t,J) models. The questions that have been addressed in this work concern two dimensionnal Hubbard model on a square lattice and are related to the existence and nature of the order in the fundamental (T=0) state.

At half filling there is a large consensus in favour of an antiferromagnetic order with a staggered magnetization ($m \sim 0.58$). Wether this magnetic order is rapidly superseded by a superconducting one when the doping is increased was an open question ? Variational calculations (1,2) tended to conclude that it was indeed the case in the (t,J) model, but these variational results were based on the assumption that the antiferromagnetic phase should remain commensurate, an hypothesis which was rapidly seriously questionned (3,4,5). It was then shown (3) that for small repulsions the holes localize in domain walls and the relative stability of vertical and diagonal walls was then studied in an Hartree Fock approximation. However the wave functions obtained with Hartree Fock are poor variational wavefunctions to describe these highly correlated fermions. In particular such a choice of one-body wave-functions enforces artificially the antiferromagnetic order, to reduce the double occupancies of sites. As it is well known, it is much better to use a Gutzwiller projector.

In this work we have used variational wave functions of the general form

$$| \phi > = P | \phi_0 > \qquad (1)$$

where P is a simple Gutzwiller projector of the form g^{N_d} (g being a variational parameter and N_d the number of doubly occupied sites) and $| \phi_0 >$ is a model wavefunction insuring the fermionic antisymmetry. Using Monte Carlo techniques for lattices from 6x6 up to 8x8 sites, we have computed the optimal energy of these Gutzwiller wavefunctions in the following variational subspaces:

C.A.F.	Commensurate antiferromagnetic w.f.
C.A.F. + S.U.	coexisting d.wave superconductivity and commensurate antiferromagnetism.

The results for a 8x8 lattice are in fig. 1. For the Hubbard's model a pure antiferromagnetic phase is stabler than a pure superconducting one (for doping up to .1). Enlargement of the variational space, allowing the coexistence of antiferromagnetism and superconductivity, slighly improves the variational energy. In as much as we could associate this effect with the superconducting pairing, this would lead to typical superconducting energy no larger than a few $10^{-3}t$, to be compared to an "antiferromagnetic energy" two orders of magnitude larger.

We then check the stability of the doped antiferromagnetic phase versus incommensurate modulation of the magnetization. This was done by introducing variationally adjusted domain walls in the antiferromagnetic order. In qualitative agreement with the Hartree Fock results (3,4) we found a large energy gain due to the

†Permanent address : Laboratoire de Physique des Solides, Université Paris Sud, 91405 Orsay, France.
‡Laboratoire associé au CNRS, URA n^{0}765

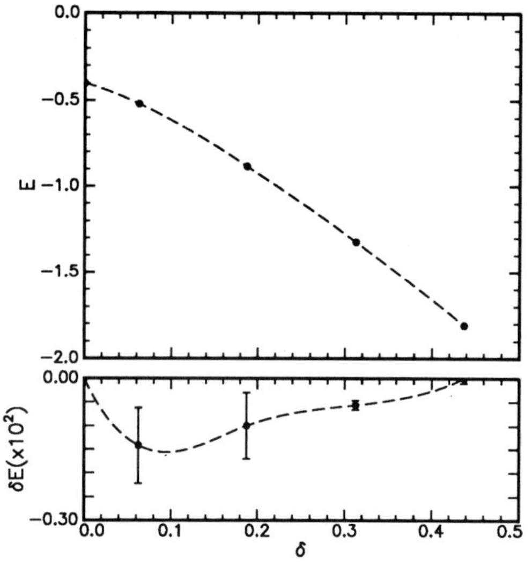

Fig 1. *Optimal variational energy per particle versus doping ($U/t = 10$) and the gain in energy due to the superconducting instability.*

incommensurate instablity, even for quite large doping ($\delta = 0.1$). Whereas at this doping, the antiferromagnetic order parameter is nearly destroyed ($m \sim 0.2$) in the commensurate phase; it keeps nearly its half-filling value ($m \sim 0.5$) in the incommensurate one, where the holes are seen to condense in the walls. The condensation energy of a hole near half filling is measured (for $U/t = 4, 7, 10$) to be larger for a diagonal wall than for a vertical one and respectively equals to ($.17t, .36t, .42t$). This type of phase (DW) is the stabler variational phase exhibited for the Hubbard's model in this range of parameters ; see the following table :

Optimal energy per particle for a 6×6 system with 4 holes ($U/t = 10, doping = .1$), for the various wavefunctions discussed in the text (t units).

$\mid \phi >$	Energy
SU	-0.634 (1)
CAF	-0.6428 (10)
CAF+SU	-0.6447 (15)
DW	-0.695

The question of coexistence at $T = 0$ of incommensurate antiferromagnetism and supraconductivity has been addressed : as in the commensurate case to allow for superconducting pairing lowers the energy by a few $10^{-3}t$ per particle.

More details are to be found in refs.(6a,b). The study of these questions for the (t,J) model is underway, as well as the analysis of the stability of this incommensurate antiferromagnetic phase versus the spin-liquid, one.

ACKNOWLEDGEMENTS. It is a pleasure to thank H.J. Schulz for having prompted key questions and useful remarks. We would also thank M. Gabay and E. Siggia for useful discussions.

REFERENCES [1] H. Yokoyama and H. Shiba, J. Phys. Soc. Jpn 57 (1988) 2482, 56 (1987) 3582, 56 (1987) 1490, 56(1987) 3570.

[2] C. Gros, Ann. Phys. 189 (1989) 53.

[3] H.J. Schulz, J. Physique and Phys. Rev. Lett. to be published.

[4] D. Poilblanc and T.M. Rice, Phys. Rev., B 39 (1989) 9749.

[5] B.I. Shraiman and E.D. Siggia, Phys. Rev. Lett. 61 (1988) 467, 62 (1989) 1564.

[6] a - T. Giamarchi and C. Lhuillier, in Proceedings of the XIII International Workshop on Condensed Matter Theory (to be published by Plenum). b - T. Giamarchi and C. Lhuillier submitted to Phys. Rev. B (1990).

Physica B 165&166 (1990) 413–414
North-Holland

A THEORETICAL STUDY OF THE ANISOTROPIC SUSCEPTIBILITY AND RESISTIVITY IN HEAVY FERMION COMPOUNDS.

S. M. M. Evans, A. K. Bhattacharjee and B. Coqblin.

Laboratoire de Physique des Solides, Bât. 510, Université Paris-Sud, 91405 Orsay Cédex, France.

The anisotropy of the susceptibility, χ, and resistivity, ρ, of the heavy fermion compounds is studied by including crystal field splitting in the slave boson approach to the periodic Anderson model. The low temperature results for ρ are compared with results from a high temperature expansion of the Coqblin-Schrieffer model. The agreement with experimental results is discussed.

1. INTRODUCTION.

The experimental properties of the normal state of the heavy fermion compounds depend in general on two features. Firstly there are the universal properties arising from a Kondo like effect. We observe a crossover from an enhanced Pauli like susceptibility (i.e. $\chi \sim T_K^{-1}$ where T_K is the 'Kondo' temperature) near $T = 0$ to a Curie-Weiss like susceptibility at high temperatures. The resistivity varies as AT^2 close to $T = 0$, where $A \sim T_K^{-2}$, goes through a maximum at temperatures somewhat lower than T_K and decreases as $\log T$ as T is further increased. Other features are nonuniversal and are due to the crystal field splitting, Δ, which is important in these compounds (1-5). The susceptibility only becomes truly Curie-Weiss like for relatively higher temperatures determined by Δ and the resistivity may have extra structure with a second peak at a temperature which again corresponds roughly to Δ. The magnetic susceptibility and transport properties are strongly anisotropic in the non-cubic Ce compounds.

The slave boson approach to the periodic Anderson model has had much success in modeling the low temperature properties of the heavy fermion compounds (6). We consider here the extension of this approach to the case where the f levels are split by crystal field effects. This gives us an anisotropic susceptibility which can be calculated as a function of temperature. The anisotropy of the T^2 term in the resistivity can also be calculated by including the anisotropy of the conduction electron relaxation time. We compare the results for this low temperature limit with the results from a high temperature expansion of the Coqblin-Schrieffer model (1).

2. CRYSTAL FIELD SPLITTING IN THE ANDERSON MODEL.

In the non-cubic cerium compounds which we study here the crystalline electric field splits the N fold degenerate f levels into doublets. As in the degenerate case, the slave boson method allows us to write an effective Hamiltonian which is equivalent to the periodic Anderson model in the limit $U \to \infty$ (6). In the 'mean field' limit we obtain

$$H = \sum_{km} \epsilon_k c^\dagger_{km} c_{km} + \sum_{i\mu} \epsilon_{f\mu} f^{i\dagger}_\mu f^i_\mu + \sum_{kim} (\tilde{V} e^{i\underline{R}_i \cdot \underline{k}} c^\dagger_{km} f^i_m + \mathrm{HC}) + i\lambda(\rho^2 - 1) \quad (1)$$

where $\tilde{V}^2 = \rho^2 V^2 = (1 - n_f)V^2$, with n_f denoting the mean f valence, and $\epsilon_{f\mu} = E_{0\mu} + i\lambda$ where $E_{0\mu}$ are the bare f level energies. ρ and $i\lambda$ are the mean field parameters which are determined by minimising the free energy. μ denotes the linear combination of m states which are eigenstates of the crystal field. We have taken the c electrons as having 'spin N' and V as independent of spin and $\underline{k}$. Within this approximation we will have 6 quasiparticle bands with strong peaks in the densities of states at the three values of $\epsilon_{f\mu}$.

By coupling a magnetic field to the f electrons, $\epsilon_{fm} \to \epsilon_{fm} - hm$, we can calculate the susceptibility both for a field along the c axis, χ_z, and perpendicular to it, χ_x, by taking the second derivative of the free energy with respect to h. In general we can write

$$\chi_\nu = \sum_{\underline{k},\alpha=\pm,\mu} \int d\omega f(\omega) \left(\delta'(\omega - E_{\alpha\mu}) \left(\frac{\partial E_{\alpha\mu}}{\partial h_\nu}\right)^2 + \delta(\omega - E_{\alpha\mu}) \frac{\partial^2 E_{\alpha\mu}}{\partial h_\nu^2} \right) \quad (2)$$

where $E_{\pm\mu} = 0.5(\epsilon_k + \epsilon_{f\mu} \pm \sqrt{(\epsilon_k - \epsilon_{f\mu})^2 + 4\tilde{V}^2})$ denotes the quasiparticle bands. $\partial E_{\alpha\mu}/\partial h_\nu = <\mu|J_\nu|\mu>$ and $\partial^2 E_{\alpha\mu}/\partial h_\nu^2 = 2\sum_{\mu' \neq \mu} | <\mu|J_\nu|\mu'> |^2/(E_{\alpha\mu} - E_{\alpha\mu'})$ where the matrix elements depend on the details of the crystal field levels.

The susceptibility can be calculated as a function of T. Within the mean field approximation this can be calculated up to reasonably high temperatures (7). The results depend strongly on the value of Δ and the configuration used. Results are given in the last section.

We want also to consider the effect the crystal field splitting has on the T^2 term in the resistivity, ρ. We assume the conduction electron scattering rate, τ_μ^{-1}, is given by $\tau_\mu^{-1} = 2Im\Sigma^c_\mu$ where Σ^c_μ is the conduction electron self energy. This is found by including the bose fluctuations around the mean field level (6). We find

$$Im\Sigma^c_\mu = \rho_0 \frac{\tilde{V}^4}{\epsilon^2_{f\mu}(\omega - \epsilon_{f\mu})^2}(\omega^2 + \pi^2 T^2). \quad (3)$$

If we use the analogy with the one impurity problem we can estimate the anisotopy of the resistivity by using

$$\frac{1}{\tau_\sigma} = \sum_\mu \frac{|<\underline{k}\sigma|k\mu>|^2}{\tau_\mu}. \quad (4)$$

$<\underline{k}\sigma|k\mu>$ is a linear combination of $<\underline{k}\sigma|km>$, and $<\underline{k}\sigma|km> = Y_{3m-\sigma}(k) < 3lm-\sigma, \frac{1}{2}\sigma|\frac{5}{2}m>$ where $Y_{3m-\sigma}(\underline{k})$ is the spherical harmonic for $l = 3$ and $< 3lm-\sigma, \frac{1}{2}\sigma|\frac{5}{2}m> = (4\pi)^{\frac{1}{2}}\left(\frac{7-4\sigma m}{14}\right)^{\frac{1}{2}}$. We note that the same result can be derived for the lattice by including the spin and $\underline{k}$ dependence of V, provided we neglect the spin dependence of the boson propagators. The electrical conductivity is found from

$$\sigma_\nu = \left(\frac{eh}{m_\nu}\right)^2 \frac{1}{v}\sum_{\sigma k} k^2_\nu \left(-\frac{\partial f(\epsilon_k)}{\partial \epsilon_k}\right)\tau_\sigma(\epsilon_k). \quad (5)$$

We note that for infinite crystal fields the integral diverges for ground states $|\pm 3/2>$ and $|\pm 5/2>$. It is therefore important to include all the levels (8). So far, however, we have neglected any impurity scattering which is not justified experimentally. In general we can write $\tau^{-1}_\sigma = \tau^{-1}_{i\sigma} + \tau^{-1}_{e\sigma}$ where $\tau_{i\sigma}$ is due to impurity scattering and $\tau_{e\sigma}$ due to electron - electron scattering. For small enough temperatures we can always expand $\tau_\sigma \approx \tau_{i\sigma} - \tau^2_{i\sigma}\tau^{-1}_{e\sigma}$. Using this we find very different results to those found for a pure system. In particular there is no tendency to divergence as $\Delta \to \infty$.

Calculations in the high temperature regime have already been performed and applied to $CePt_2Si_2$ (1). The calculation can be used for other compounds and the results are given in the following section.

3. COMPARISON BETWEEN THEORY AND EXPERIMENT.

In this section we present our theoretical results and compare with experiment. We start by looking at the compound $CePt_2Si_2$ which has crystal field configuration $|\pm 1/2>$, $|\pm 3/2>$, $|\pm 5/2>$ with $T_K \approx 50K$, $\Delta_1 \approx 1.6T_K$ and $\Delta_2 \approx 4.6T_K$ (1). For $T = 0$ we find $\chi_z/\chi_x = 0.2$. The agreement with the experimental value of 0.57 is not particularly good although in both cases $\chi_z < \chi_x$ (2). At finite temperatures the most pronounced feature is a minimum at $T \sim T_K$ in χ^{-1}_z which arises due to crystal field, rather than Kondo, effects. There is no such minimum for χ^{-1}_x. Experimentally both χ^{-1}_z and χ^{-1}_x exhibit this minimum. The discrepancy in the results for χ_x is somewhat puzzling.

For the low temperature resistivity we find $A_z/A_x = 1.5$ where A_z and A_x are respectively the coefficients of the term in T^2 for the current parallel to and perpendicular to the c axis. The high temperature expansion gives a peak in the resistivity for temperatures $\sim \Delta$ which agrees well with experiment. In order to obtain better agreement with experimental data which shows $\rho_z/\rho_x \sim 2$ it is convenient to treat the ratio of bare conduction electron effective masses along and perperdicular to the c axis (m_z and m_x respectively) as an adjustable parameter. Using the value $m_z/m_x \sim 1.6$ in the low temperature regime gives an anisotropy ~ 3.7. Experimental results for the T^2 term are not available.

Both $CeAl_3$ and $CeCu_2Si_2$ have crystal field levels given by $|0> = a|\pm 5/2> + b|\mp 3/2>$, $|1> = |\pm 1/2>$ and $|2> = a|\pm 3/2> - b|\mp 5/2>$. For $CeAl_3$ we have $a = 0.24$, $b = 0.97$, $T_K \approx 3K$, $\Delta_1 \approx 20T_K$ and $\Delta_2 \approx 30T_K$. Our theoretical value $\chi_z(0)/\chi_x(0) = 3.2$ compares reasonably well with the experimental value of 3.6 (3). At high temperatures both curves show a Curie-Weiss behaviour. As the temperature is reduced the effective moment crosses over from the full magnetic moment of the $j = 5/2$ state to that of the ground state doublet. The two curves cross for $T \sim 3T_K$. Similar behaviour is observed experimentally although the crossover is at rather higher temperatures, $T \sim 12T_K$.

For the low T resistivity we find $A_z/A_x = 1.1$. Again the high temperature calculation gives a good description of the peak in ρ. Adjusting m_z/m_x to give the correct experimental value of $\rho_z/\rho_x \approx 0.33$ at high T changes also the value of the low T anisotropy to 0.36 compared to the value 0.37 found experimentally (4).

Finally we look at $CeCu_2Si_2$ which has $a = 0.83$, $b = 0.56$, $T_K \approx 10K$, $\Delta_1 \approx 14T_K$ and $\Delta_2 \approx 36T_K$. We find $\chi_z(0)/\chi_x(0) = 1.5$ again in reasonable agreement with the experimental value of 1.2 (5). $\chi_z > \chi_x$ for all temperatures both experimentally and theoretically.

For the resistivity we find $A_z/A_x = 0.55$. Adjusting our parameters as before to get agreement with the high temperature anisotropy $\rho_z/\rho_x \approx 0.5$ reduces this to 0.27. No experimental data exists in this regime.

We conclude that the slave boson approach to the Anderson model with crystal field splitting gives a qualitative agreement with experiment which is rather better than we might expect given that we are using a $1/N$ technique with $N = 2$. The anisotropy of the resistivity as $T \to 0$ is much more significant than in the high temperature calculation. A more detailed paper is currently in preparation.

ACKNOWLEDGEMENTS.

One of us (S. M. M. E.) wishes to thank The Royal Society for financial support.

REFERENCES.

(1) A. K. Bhattacharjee, B. Coqblin, M. Raki, L. Forro, C. Ayache and D. Schmitt, J. Phys. France 50 (1989) 2781.
(2) D. Gignoux, D. Schmitt and M. Zerguine, Phys. Rev. B 37 (1988) 9882.
(3) D. Jaccard, R. Cibin, A. Bezinge, J. Sierro, K. Matho and J. Flouquet, J. Magn. Magn. Mater. 76&77 (1988) 255.
(4) D. Jaccard, R. Cibin, J. L. Jorda and J. Flouquet, Jap. J. App. Phys. 26 (1987) Suppl 26-3 517.
(5) Y. Ōnuki, Y. Furukawa and T. Komatsubara, J. Phys. Soc. Japan 53 (1984) 2197.
(6) A. J. Millis and P. Lee, Phys. Rev. B 35 (1987) 3394.
(7) S. M. M. Evans, T. Chung and G. A. Gehring, J. Phys. Condens. Mater. 1 (1989) 10473.
(8) S. Zhang and P. M. Levy, Phys. Rev. B 40 (1989) 7179.

Physica B 165&166 (1990) 415–416
North-Holland

WEAK-COUPLING TREATMENT OF CORRELATED ELECTRON SYSTEMS IN FINITE DIMENSIONS

Heinz SCHWEITZER and Gerd CZYCHOLL

Institut für Physik, Universität Dortmund, Postfach 500500, D-4600 Dortmund 50, Germany

The standard models for correlated electron systems, namely Hubbard model (HM) and periodic Anderson model (PAM) are investigated within the second order U-perturbation treatment around the non-magnetic Hartree-Fock solution. For a d-dimensional system the full k-dependent selfenergy can efficiently be calculated by means of a 1/d-expansion starting from the local approximation, which is correct in the limit $d\to\infty$.

Models of correlated electron systems as the Hubbard model (HM) and the periodic Anderson model (PAM) are presently of great interest, in particular for a theoretical description of high temperature superconductors and of heavy fermion systems. The HM is given by

$$H=\sum_{\mathbf{k}\sigma}\varepsilon_{\mathbf{k}}c^+_{\mathbf{k}\sigma}c_{\mathbf{k}\sigma} + U\sum_i c^+_{i\uparrow}c_{i\uparrow}c^+_{i\downarrow}c_{i\downarrow}$$

and the PAM is given by

$$H=\sum_{\mathbf{k}\sigma}\varepsilon_{\mathbf{k}}c^+_{\mathbf{k}\sigma}c_{\mathbf{k}\sigma}+\sum_{i\sigma}\left(E_f f^+_{i\sigma}f_{i\sigma}+V(c^+_{i\sigma}f_{i\sigma}+\text{h.c.})+\frac{U}{2}f^+_{i\sigma}f_{i\sigma}f^+_{i-\sigma}f_{i-\sigma}\right)$$

For the conduction band we assume a simple cubic tight binding band structure for dimension d, i.e

$$\varepsilon_{\mathbf{k}}=\frac{1}{d}\sum_{l=1}^{d}\cos(k_l)$$

Due to the prefactor 1/d energies are measured in units of the half conduction band width. - The simplest approximation for these models beyond the (trivial) Hartree-Fock treatment is provided by the second order perturbation theory (SOPT) with respect to the Coulomb correlation U. Within the SOPT relative to the non-magnetic Hartree-Fock solution the selfenergy is given by

$$\Sigma_{\mathbf{k}}(i\omega_n)=-\left(\frac{UT}{N}\right)^2\sum_{\substack{\omega_1\omega_2\\ \mathbf{pq}}}G_{\mathbf{k+q}}(i\omega_1)G_{\mathbf{p-q}}(i\omega_n+i\omega_2-i\omega_1)G_{\mathbf{p}}(i\omega_2)$$

Here N denotes the number of lattice sites, T the temperature, ω_i are Matsubara frequencies, and $G_{\mathbf{k}}(z)$ is the Hartree-Fock Green function, which for the HM is given by

$$G_{\mathbf{k}}(z)=\frac{1}{z-\langle c^+_{i-\sigma}c_{i-\sigma}\rangle U-\varepsilon_{\mathbf{k}}}$$

and for the PAM by

$$G_{\mathbf{k}}(z)=\frac{1}{z-E_f-U\langle f^+_{i-\sigma}f_{i-\sigma}\rangle-\frac{V^2}{z-\varepsilon_{\mathbf{k}}}}$$

But usually the explicit calculation of the **k**-dependent SOPT selfenergy $\Sigma_{\mathbf{k}}(z)$ is relatively complicated because of the double momentum (**p**,**q**-)summation. Therefore, in most of the existing SOPT selfenergy computations (1,2) the "local approximation" (LA) of a **k**-independent (site-diagonal) $\Sigma(z)$ has been used as an additional ad hoc approximation without having a criterion for the validity of the LA. But recently it has been pointed out (3,4) that the LA becomes correct only in the limit $d\to\infty$ (5). So the corrections to the LA for finite d can be classified as 1/d contributions. We found that in this manner, i.e. by a 1/d-expansion, which converges even for d=1, $\Sigma_{\mathbf{k}}(z)$ can much more efficiently be calculated than by the direct **p**,**q** double summation.

To be specific, we start with a Fourier transformation of the selfenergy

$$\Sigma_{\mathbf{k}}(z)=\Sigma_{\mathbf{R=0}}(z)+\sum_{\mathbf{R}\neq\mathbf{0}}\exp(i\mathbf{kR})\,\Sigma_{\mathbf{R}}(z)$$

The site matrix elements of the selfenergy are given by

$$\Sigma_{\mathbf{R}}(i\omega_n)=-U^2T^2\sum_{\omega_1\omega_2}G_{\mathbf{R}}(i\omega_n+i\omega_2-i\omega_1)G_{\mathbf{R}}(i\omega_1)G_{-\mathbf{R}}(i\omega_2)$$

where

$$G_{\mathbf{R}}(z)=\frac{1}{N}\sum_{\mathbf{k}}\exp(i\mathbf{kR})\,G_{\mathbf{k}}(z)=\int d\zeta\,\frac{\rho_{\mathbf{R}}(\zeta)}{z-\zeta}$$

denotes the site matrix elements of the Hartree Fock Green function, $\rho_{\mathbf{R}}(\zeta)=-\operatorname{Im}G_{\mathbf{R}}(\zeta+i0)/\pi$ its spectral function. Replacing the Matsubara frequency summations by Fermi function integrations and using a Laplace transformation as in (4) we finally obtain

$$\Sigma_{\mathbf{R}}(z)=U^2(-i)\int d\lambda\, e^{i\lambda z}\left(A^2_{\mathbf{R}}(\lambda)B_{-\mathbf{R}}(-\lambda)+B^2_{\mathbf{R}}(\lambda)A_{-\mathbf{R}}(-\lambda)\right)$$

with

$$A_{\mathbf{R}}(\lambda)=\int d\zeta\,\rho_{\mathbf{R}}(\zeta)\,f(\zeta)\,e^{-i\lambda\zeta}$$

$$B_{\mathbf{R}}(\lambda)=\int d\zeta\,\rho_{\mathbf{R}}(\zeta)\,(1-f(\zeta))\,e^{-i\lambda\zeta}$$

Thus the calculation of each site matrix element of the selfenergy $\Sigma_{\mathbf{R}}(z)$ can as efficiently be performed as the calculation of the LA or the $d=\infty$ selfenergy (4,5), and provided that the **R**-sum can be terminated after a finite number of neighbor shells, i.e. provided that the 1/d-expansion converges, we have a quick algorithm to determine the full **k**-dependent selfenergy $\Sigma_{\mathbf{k}}(z)$.

The convergence of the 1/d-expansion depends on the model under consideration, i.e. on the particular spectral function $\rho_{\mathbf{R}}(z)$, and it depends, of course, on the dimension d. Empirically we found out that for the HM one has to extend the **R**-sum up to the 3rd neighbor shell for d=3, to the 6th neighbor shell for d=2, and even up to the 50th neighbor shell for d=1. For the PAM the convergence is somewhat quicker: one has to go up to the 2nd neighbor shell for d=3, to the 5th neighbor shell for d=2 and to the 20th neighbor shell for d=1. So even for d=1 the 1/d-expansion converges, but the convergence is weak. But for d=2 and 3 the convergence is quick and the present **R**-summation method is certainly superior to the direct **p,q** double summation.

For an explicit example we present a few results obtained for the PAM in two dimensions (d=2). We choose the following parameters: $E_f=-0.5$, U=1., V=0.2 and a total number of 2 electrons per lattice site ("symmetric PAM"). Then the chemical potential must be at zero and the total f-electron occupation must be $n_f= \langle f^+_{i\uparrow} f_{i\uparrow}\rangle + \langle f^+_{i\downarrow} f_{i\downarrow}\rangle = 1.$, i.e. the effective Hartree-Fock f-level is at zero, too: $E_f + U\langle f^+_{i\sigma} f_{i\sigma}\rangle = 0$. Figures 1 a. and b. show the results for the momentum (**k**-) and frequency (z-) dependence of the SOPT selfenergy real and imaginary part for T=0. In the frequency dependence we observe, in particular, the vanishing of $\mathrm{Im}\Sigma_{\mathbf{k}}(z+i0)$ at the Fermi level, z=0., in accordance with the Luttinger theorem. The real part has a corresponding structure resulting in a strong negative slope at the Fermi level, which automatically provides for a strong mass enhancement. Furthermore we observe a weak (cosine like) momentum (**k**-)dependence. The resulting f-electron spectral function is shown in Fig.2. We observe, in particular, strong quasi-particle resonance peaks near to the chemical potential and a hybridisation gap around zero. The width of this hybridisation gap and the exact position of the resonance peaks is strongly influenced by the explicit **k**-dependence of the selfenergy.

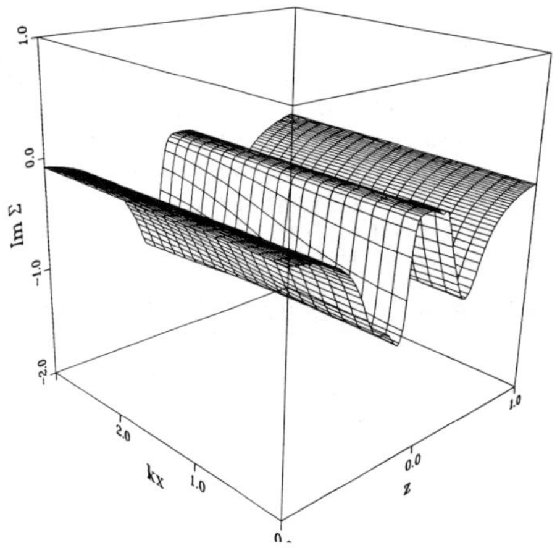

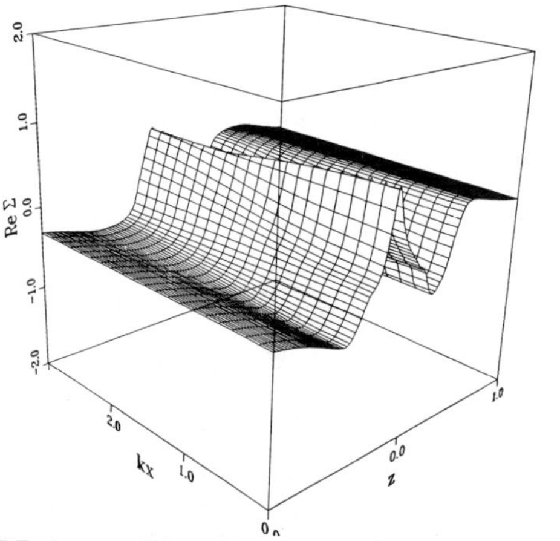

FIGURE 1 a and b: Momentum (**k**-) and frequency (z-) dependence of the real and imaginary part of the SOPT selfenergy $\Sigma_{\mathbf{k}}(z+i0)$ of the symmetric PAM in dimension d=2 for $E_f=-0.5$, U=1., V=0.2

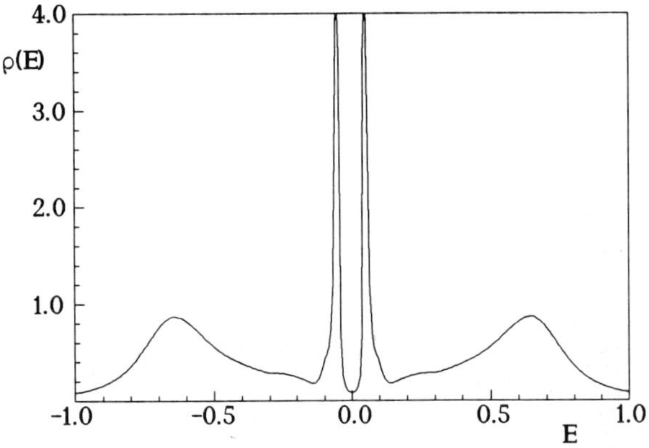

FIGURE 2: Corresponding frequency dependence of the SOPT f-electron spectral function of the symmetric PAM in dimension d=2

In summary we have demonstrated that within the SOPT the full **k**-dependent selfenergy can efficiently be determined by means of a Fourier transformation , i.e. a nearest neighbor shell summation, which can be interpreted as a 1/d-expansion. Of course this is not only valid for the SOPT, but it remains true also for higher order summations (e.g. T-matrix approximation and RPA). More details and results will be presented elsewhere (6).

REFERENCES

(1) G.Treglia, F. Ducastelle, D. Spanjaard, Phys. Rev. B 21, 3729 (1980); B 22, 6472 (1980); Journ. de Physique 41, 281 (1980)
(2) R.Taranko, E.Taranko, J.Malek, J.Phys. F 18, L87 (1988); J. Phys.: Condensed Matter 1, 2935 (1989)
(3) E. Müller-Hartmann, Z. Phys. B 76, 211 (1989)
(4) H. Schweitzer, G.Czycholl, Solid State Commun. 69, 171 (1989)
(5) W. Metzner, D. Vollhardt, Phys. Rev. Letters 62, 324 (1989)
(6) H.Schweitzer, G.Czycholl, Solid State Commun. (1990), in print, and to be published

Physica B 165&166 (1990) 417–418
North-Holland

SPIN DYNAMICS IN KONDO SYSTEMS AT LOW TEMPERATURES

N.G.FAZLEEV, G.I.MIRONOV

Department of Physics, Kazan State University, Kazan 420008, USSR

The theory of the electron paramagnetic resonance (EPR) of localized moments (LM) in Kondo systems at low temperatures ($kT<\omega_s$, ω, the resonance and external frequencies) is presented, having taken into account the dynamic nature of the exchange coupling between LM and conduction electrons (CE). The obtained EPR parameters of Kondo systems, which are valid at arbitrary temperatures, static magnetic fields and frequencies are studied at different conditions of the dynamic coupling of spin systems of LM and CE. The experimental data on EPR studies of the Kondo system Au:Yb are discussed.

1. INTRODUCTION

The investigation of static and dynamic characteristics of Kondo systems at low temperatures by means of EPR techniques has become an object of growing interest in recent years (1). This is due to the fact that unlike other experimental techniques, EPR investigations make it possible to study independently in the same experiment the Kondo anomalies in the static susceptibility and in the rate of transverse spin relaxation of LM. But increasing of the effectiveness of relaxation processes in spin subsystems of LM and CE due to s-d exchange with decreasing temperature to the Kondo temperature T_k leads to the emergence of a coupled motion of magnetic moments of impurities and CE in the Kondo system (2). Under these conditions, the spin subsystems of LM and CE are characterized by the summary dynamic response. The aim of the present work is the microscopic investigation of the spin dynamics of LM in Kondo systems at low temperatures taking into account the dynamic nature of the exchange interaction.

2. CALCULATION OF THE SUMMARY RESPONSE

The properties of LM and CE in a Kondo system in an external DC H_o and a weak AC $\mathbf{h}(t)$ magnetic fields are described by the Hamiltonian (2,3)

$$H(t) = H_o(t) + H_{es} + H_{eL} .$$

Here $H_o(t)$ is the Hamiltonian of LM and CE in the external magnetic fields; H_{es} and H_{eL} determine the interaction of CE with LM and nonmagnetic impurities, respectively.

Following the non-equilibrium statistical operator method (3) and confining ourselves to the linear approximation in the deviation of the spin subsystems from the equilibrium state and to first order terms in concentrations of impurities, after temporal Fourier transformation we obtain for the total response of LM and CE

$$\chi(\omega)=(c_sa_e-c_eb_e+c_ea_s-c_sb_s)(a_sa_e-b_sb_e)^{-1}.$$

The parameters a_i, b_i, c_i $(i=s,e)$ are defined by the expressions:

$$a_i= \omega_i' - \omega - \Theta_i\Xi_{ij}(\omega) - \Xi_{iL},$$

$$b_i= \lambda\chi_j(\Xi_{jL}- \omega_j) + g_i\Theta_i\Xi_{ij}(\omega) /g_j ,$$

$$c_i=\chi_i(\omega_i'-\Xi_{ij}(\omega)-\Xi_{iL})+g_i\chi_j\Xi_{ji}(\omega)/g_j,$$

where

$$\omega_i'=\omega_i(1+\lambda\chi_j) , \Theta_i=1+\lambda\chi_ig_j /g_i ,$$

$$\Xi_{ij}(\omega)=\mathrm{Re}\Xi_{ij}(\omega)+i\mathrm{Im}\Xi_{ij}(\omega),(ij=se,es,eL),$$

$$\chi_i=\langle M_i^z\rangle/H_i', H_i'=H_0+\lambda\langle M_j^z\rangle, \langle\ldots\rangle=\mathrm{Tr}(\ldots\rho),$$

ρ is the equilibrium statistical operator; λ is the molecular field constant; $\mathbf{M}_i$ $(i=s,e)$ is the operator of the spin magnetization.

The imaginary parts of the kinetic coefficients $\Xi_{ij}(\omega)$ define the transverse relaxation rates of spin magnetizations of LM and CE, caused by the s-d(-f) exchange and spin - orbit scattering of CE on nonmagnetic impurities. The real parts $\Xi_{ij}(\omega)$ determine shifts of the resonance frequencies of LM and CE.

Using the temperature Green's functions methods, the method of the dynamical renormalization groups (4), we get the following low temperature expressions for the kinetic coefficients

$$\mathrm{Re}\Xi_{ij}(\omega)=(g_i/g_j)\lambda\chi_j\omega_s\{[1/1+2dJL(\omega)]-1\},$$

$$\mathrm{Im}\Xi_{ij}(\omega)=\pi dJW_{ij}\{(\omega_s-\omega)\mathrm{cth}[(\omega_s-\omega)/2kT][1/1+2dJF(\omega)]+\omega\mathrm{cth}(\omega/2kT)/[1+2dJG(\omega)]\},$$

$$W_{se}=dJ,\ W_{es}=J\mathrm{th}(\omega_s/2kT)c/\omega_s,$$

$$L(\omega)=[\omega\ln|D/\omega|-(\omega_s-\omega)\ln|D/\omega_s-\omega|],$$

$$F(\omega)=[\omega_s\ln|D/\omega_s|-\omega\ln|D/\omega|]/(\omega_s-\omega),$$

$$G(\omega)=[\omega\ln|D/\omega_s-\omega|+\omega_s\ln|(\omega_s-\omega)/\omega_s|]/\omega,$$

$$\mathrm{Im}\Xi_{eL}=(2/3)d|V_{so}|^2c'\int d\Omega\sin\theta+D'q^2,$$

$$\mathrm{Re}\Xi_{eL}=0,\ D'=v_F^2\tau_p/3,\ \tau_p^{-1}=2\pi dc'|V_p|^2,$$

2D is the CE bandwidth; d is the density of states of CE on the Fermi level; J is the exchange constant; V_p and V_{so} are the spin-independent and spin-orbit potentials, respectively; c, c' are the concetrations of LM and nonmagnetic impurities, respectively. For high temperatures ($kT>>\omega$, ω_s), we obtain the following expressions for kinetic coefficients

$$\mathrm{Im}\Xi_{se}(\omega)=\pi kT/\ln^2(T/T_k),\ kT_k=D\exp(1/2dJ),$$

$$\mathrm{Im}\Xi_{es}(\omega)=\pi c/2d\ln^2(T/T_k),$$

$$\mathrm{Re}\Xi_{se}(\omega)=(g_s/g_e)\omega_s\lambda\chi_e\{[1/2dJ\ln(T_k/T)]-1\},$$

$$\mathrm{Re}\Xi_{es}(\omega)=\omega_e\lambda\chi_s\{[1/2dJ\ln(T_k/T)]-1\},$$

Poles of the total dynamic susceptibility ω_1 and ω_2 define the coupled spin modes of LM and CE. In the strong electron bottleneck (EBN) conditions, which are determined by the inequalities:

$$\mathrm{Im}[\Theta_s\Xi_{se}(\omega_s)+\Theta_e\Xi_{es}(\omega_s)]>>\mathrm{Im}\Xi_{eL},|\omega_s'-\omega_e'|,$$

the Kondo anomalies in EPR parameters of Kondo systems are fully suppressed. In the intermediate EBN regime, which is defined by the inequalities

$$|\omega_s'-\omega_e'|>\mathrm{Im}[\Theta_s\Xi_{se}(\omega_s)+\Theta_e\Xi_{es}(\omega_s)]>>\mathrm{Im}\Xi_{eL}$$

the poles of $\chi(\omega)$ contain Kondo anomalous terms and are equal to

$$\mathrm{Re}\omega_1=\omega_e'-\mathrm{Re}[\Theta_s\Xi_{se}(\omega_s)+\Theta_e\Xi_{es}(\omega_s)]+(1-Q)\mathrm{Re}\Xi_{se}(\omega_s),$$

$$\mathrm{Im}\omega_1=\mathrm{Im}\Xi_{eL}+\mathrm{Im}[\Theta_s\Xi_{se}(\omega_s)+\Theta_e\Xi_{es}(\omega_s)]-(1-Q)\mathrm{Re}\Xi_{se}(\omega_s),$$

$$\mathrm{Re}\omega_2=\omega_s'-Q\mathrm{Re}\Xi_{se}(\omega_s),\ \mathrm{Im}\omega_2=Q\mathrm{Im}\Xi_{se}(\omega_s),$$

where $Q=1-(g_e/g_s)\omega_s'\lambda\chi_s(\Theta_s+\Theta_e)/(\omega_e'-\omega_s')$.

In the absence of coupled motion of LM and magnetic moments of CE, when

$$\mathrm{Im}\Xi_{eL}>>\mathrm{Im}[\Theta_s\Xi_{se}(\omega_s)+\Theta_e\Xi_{es}(\omega_s)],|\omega_s'-\omega_e'|$$

The Kondo anomalies are manifested independently and completely both in EPR on LM and in EPR on CE in Kondo systems.

3. DISCUSSION OF THE RESULTS

To describe the temperature dependences of the observed shift in g-value of the impurity and the spin relaxation rate of LM in dilute Kondo system Au:Yb, neglecting dynamic coupling between the spin systems of LM and CE, the static and dynamic Kondo temperatures with considerably different values: $T_k^g=4\cdot10^{-8}$K and $T_k^r=2\cdot10^{-12}$K were introduced in (1). However, an analysis of experimental data in (1) shows that the EPR parameters depend on concentration of LM and an intermediate EBN appears in the system. Calculation of the actual Kondo temperature in Au:Yb, using expressions for ω_2, which define the effective parameters of EPR on LM in the intermediate EBN regime, and experimentally obtained T_k^g and T_k^r for $g_e=2$; $g_s=3,423$; $D=100$K; $\lambda\chi=-0,5$; $2dJ=-0,05$; gives $T_k^{ex}=2\cdot10^{-7}$ K, which coincides with its theoretical value (5), calculated for the same values of parameters. Consequently, there is no necessity to introduce two Kondo temperatures, specific to fit the results of EPR experiments in Kondo systems.

ACKNOWLEDGEMENTS

The authors express their gratitude to Kochelaev B.I. for a discussion of the results and for valuable comments.

REFERENCES

(1) Y.von Spalden et al. Phys.Rev. B28 (1983) 24.
(2) G.I.Mironov and N.G.Fazleev, Sov.J. Low Temp.Phys. 14 (1988) 522.
(3) N.G.Fazleev, Sov.J. Low Temp. Phys. 6 (1980) 693.
(4) M.Fowler and A.Zavadowski, Sol. St. Commun. 9 (1971) 471.
(5) A.M.Tsvelick and P.B.Wiegmann, Adv. Phys. 32 (1983) 453.

Physica B 165&166 (1990) 419–420
North-Holland

HIGH-FIELD HALL EFFECT AND CRYSTAL ORIENTATION DEPENDENCE OF MAGNETORESISTANCE IN THIN FILMS OF $CeAl_3$[+]

G.M. ROESLER, Jr. and P.M. TEDROW

Francis Bitter National Magnet Laboratory, MIT, Cambridge, MA 02139, USA[*]

Thin films of $CeAl_3$ on sapphire were prepared by metallic coevaporation and in vacuo annealing. Highly oriented growth with the hexagonal crystal basal plane parallel to the substrate plane was observed. The resistivity versus temperature curves of the best films were similar to bulk $CeAl_3$, including a T^2 dependence below 0.3 K. Transverse magnetoresistivity and Hall effect were measured in fields to 23 T and at temperatures down to 70 mK. Transverse magnetoresistivity was highly dependent on magnetic field orientation with respect to the basal plane. With the field perpendicular to the basal plane, the low-field magnetoresistivity became positive below 5 K, and the high-field magnetoresistivity exhibited a temperature-dependent maximum. With the field in the basal plane, the low-field magnetoresistivity remained negative down to 2 K, below which it exhibited a temperature-independent maximum at 1.8 T. The high-field Hall coefficient below 5 K decreased with field in a temperature-dependent manner, but was temperature-independent at all fields below 800 mK.

The heavy-fermion compound $CeAl_3$ displays a variety of transport properties which differ markedly from those of normal metals, including: a large negative magnetoresistance, with a change in the sign of the magnetoresistance as the temperature is lowered (1); a complex magnetic field dependence of the magnetoresistance (2); a large resistivity maximum at intermediate temperatures (3); and an anomalous large peak in the Hall coefficient (4). Polycrystalline samples have been used exclusively to obtain these data. The importance of the crystal field in determining the properties of hexagonal $CeAl_3$ has been recognized (5). The tendency of many hexagonal metallic thin films to grow with a preferred orientation suggests a way to study orientation-dependent transport properties of $CeAl_3$, without confronting the difficulties of single-crystal preparation resulting from the peritectic behavior of the Ce-Al system and the incongruent melting characteristics of $CeAl_3$. We have made such films and have measured their magnetoresistance and Hall coefficient for magnetic fields to 23 T and temperatures down to 70 mK.

Films of 110-250 nm thickness were prepared by electron beam coevaporation of Ce and Al with independent rate monitoring of both sources. Post-evaporation annealing for 1 to 25 minutes at 800-950°C was conducted *in situ*. The resistivity-temperatures curves of the best films obtained had negative slopes at room temperature, peaks at 30-40 K, and higher positive slopes below 5.5 K. One film was measured down to 70 mK and displayed a T^2 resistivity dependence below 0.4 K. The T^2 coefficient of the resistivity, less than 2 $\mu\Omega$ cm/K^2, was significantly smaller than reported values in bulk samples (6), presumably due to disorder in the films. Upper limits of the resistivities were 140-180 $\mu\Omega$-cm at room temperature and 160-220 $\mu\Omega$-cm at the peak. Upper limits of the residual resistivities of the films were approximately 10-30 $\mu\Omega$-cm.

X-ray diffraction analysis confirmed that the composition of the films was predominantly $CeAl_3$, occasionally with a $CeAl_2$ minority phase. An X-ray pole figure using the [002] $CeAl_3$ peak showed that most of the Bragg-reflected energy was concentrated in a narrow solid angle such that the [002] plane was in a highly preferred

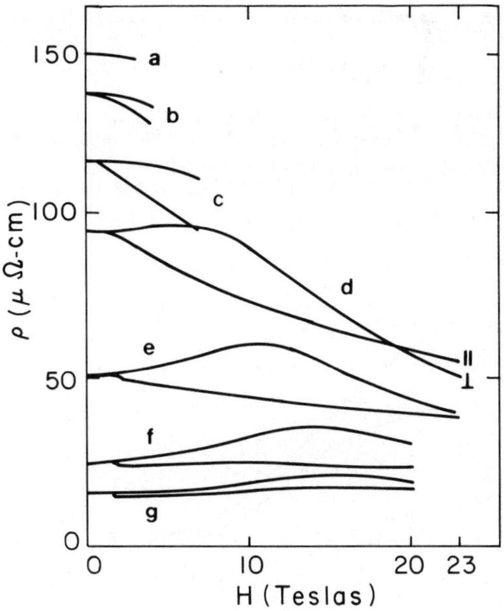

Fig. 1. Resistivity of $CeAl_3$ films in magnetic field: (a) 8.9 K, (b) 5.7 K, (c) 4.9 K, (d) 4.2 K, (e) 2.9 K, (f) 2.0 K, (g) 0.6 K. The upper curve is in perpendicular field except as noted.

[+]Supported by NSF Grant DMR-8618072. [*]The National Magnet Laboratory is supported at MIT by NSF.

orientation parallel to the substrate plane. This suggests that orientation dependence of magnetotransport properties in these films may be related to crystal direction.

The transverse resistivity in a magnetic field parallel and perpendicular to the basal plane are shown in Fig. 1. In a perpendicular field, the resistivity no longer has a negative slope below 5-6 K, and develops a temperature- dependent maximum with field. For higher fields, the resistivity parallels the negatively-sloped single-ion Kondo behavior displayed at higher temperatures. This magnetoresistance behavior is similar to that displayed by thin films of the heavy fermion compound $CePb_3$ below 1 K (7). One sample was measured at temperatures down to 70 mK. Its resistivity in a perpendicular field displayed additional structure below 0.7 K, in the form of plateaus around H=3 T and H=5 T, which were temperature-independent down to 70 mK.

The resistivity in a parallel field is markedly different from that in a perpendicular field. The slope of the resistivity is negative at temperatures down to 2 K. Below this temperature, a small, temperature-independent maximum is observed at H=1.8 T. At higher fields, another temperature-dependent maximum is observed.

A comparison can be made to magnetoresistance data on bulk samples (2). A higher onset of positive magnetoresistance is observed in the thin-film samples. The observed anisotropy of the magnetoresistance is qualitatively similar to that observed (8) in a single crystal of the heavy-fermion compound $CeCu_6$ below 2 K. Data is not available on $CeCu_6$ below 1.3 K to determine if a positive magnetoresistance ultimately occurs.

The Hall effect data are displayed in Fig. 2. The initial (low field) Hall coefficients are in agreement with those of polycrystalline samples (4). As field is increased, the Hall coefficient remains constant to about 3 T, then gradually decreases. Below a temperature of about 800 mK, the Hall coefficient is temperature-independent at all fields.

The resistivity maxima in a perpendicular field are plottted in Fig. 3. They form a curve in the H-T plane which is qualitatively similar to the phase boundary of a magnetic system. Further work is required to determine the underlying nature of the interactions which produce such widely different magnetotransport behavior for different field orientations, and to determine the relationship of these phenomena to the thermodynamic behavior of this heavy-electron material.

To summarize, we have made oriented thin films of $CeAl_3$ by electron beam codeposition and have measured their magnetoresistance and Hall coefficient for a wide range of H and T. Agreement with bulk measurements is good. A dependence of the magnetoresistance on crystalline direction was noted.

REFERENCES

(1) A.S. Edelstein, C.J. Tranchita, O.D. McMasters, and K.A. Gschneider, Jr., Solid State Commun. **15** (1974) 81.

(2) G. Remenyi, A. Briggs, J. Flouquet, O. Laborde, and F. Lapierre, J. Magn. Magn. Mater. **31-34** (1983) 407.

(3) K.H.J. Buschow, J.H. van Daal, F.E. Maranzana, and P.B. van Aken, Phys. Rev. B**3** (1971) 1662.

(4) F. Lapierre, P. Haen, R. Briggs, A. Hamzic, A. Fert, and J.P. Kappler, J. Magn. Magn. Mater. **63-64** (1987) 338.

(5) D.M. Newns and A.C. Hewson, J. Phys. F**10** (1980) 2429.

(6) K. Andres, J.E. Graebner, and H.R. Ott, Phys. Rev. Lett. **35** (1975) 1779.

(7) P.M. Tedrow (unpublished).

(8) Y. Onuki, Y. Shimizu, and T. Komatsubara, J. Phys. Soc. Jpn. **54** (1985) 304.

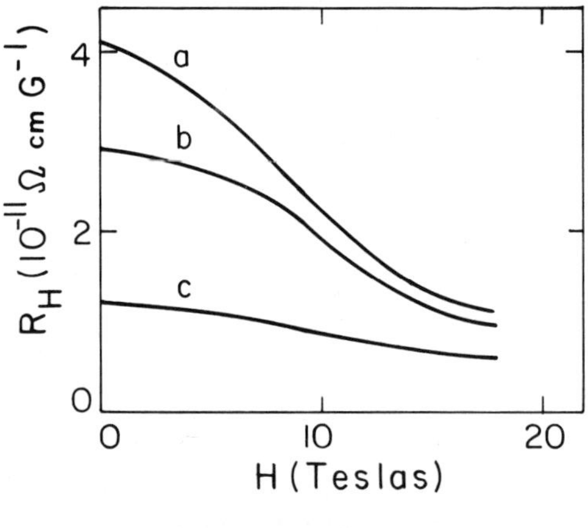

Fig. 2. High-field Hall coefficient of $CeAl_3$. (a) 4.2 K, (b) 3.3 K, (c) all temperatures 70 mK-1.0 K.

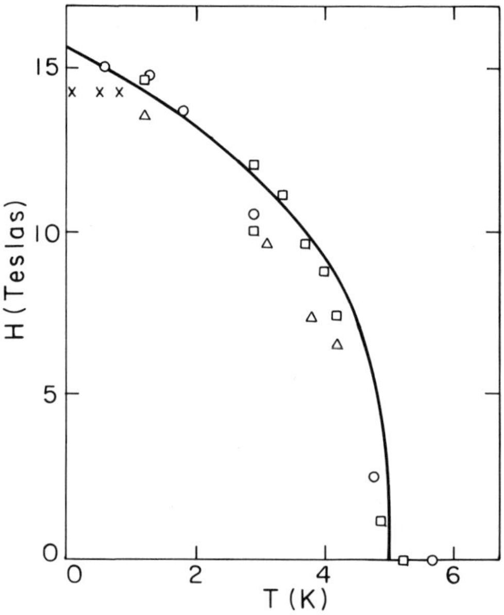

Fig. 3. Resistivity maxima of $CeAl_3$ with field perpendicular to basal plane. The symbols denote different samples used to obtain.

Physica B 165&166 (1990) 421–422
North-Holland

ELASTIC PROPERTIES OF HEAVY-ELECTRON COMPOUND CePdIn

Takashi SUZUKI, Minoru NOHARA, Toshizo FUJITA, Toshiro TAKABATAKE* and Hironobu FUJII*

Faculty of Science and *Faculty of Integrated Arts and Science, Hiroshima University, Hiroshima 730, Japan

The temperature dependence of the elastic constants C_{33}, $(C_{11} - C_{12})/2$ and C_{44} of CePdIn has been measured between 2 K and 150 K. All the elastic constants exhibit softening at low temperatures due to the crystalline electric field and the antiferromagnetic ordering.

1. INTRODUCTION

Rare earth compounds exhibit characteristic dependences of the elastic constants on temperature and magnetic field reflecting a variety of 4f-electron states (1). Particularly, elastic softening at relatively higher temperatures $T > T_K$, where T_K is the Kondo temperature and 4f-electrons are well localized, is explained by crystalline electric field (CEF) effect mediating strain-quadrupole interaction since the localized state of 4f-electrons has an electric quadrupole (2,3).

Recently, equiatomic ternary compound CePdIn has been discovered as a new antiferromagnetic heavy-electron compound with T_N = 1.8 K by our group (4-6). CePdIn crystallizes in the Fe_2P-type hexagonal (space group-D^3_{3h}) structure, in which the J = 5/2 sextet is expected to split into three Kramers-doublets. The magnetic susceptibility follows the Curie-Weiss law at high temperatures above 100 K with an effective moment of 2.56μ_B, which is very close to the value for Ce^{3+} free ion. Satoh et al. (6) estimated the Kondo temperature as T_K = 3.3 K and the CEF energy separation as Δ_1 = 100 K between the ground state and the first excited doublet from the magnetic part of the specific heat.

In this paper we report the temperature dependence of the elastic constants $(C_{11}-C_{12})/2$, C_{33} and C_{44} of CePdIn from 2 K to 150 K, and propose a preliminary but new CEF level scheme estimated by a theoretical fit of strain susceptibility which is the linear response for the strain-quadrupole interaction against strain.

2. EXPERIMENTS

Single crystals of CePdIn were grown by a Czochralski method using a triarc furnace. Crystal axes were determined by an X-ray Laue camera. Temperature dependence of sound velocity v(T) was measured by a hand-made phase-comparison-type ultrasonic apparatus with a liq.-He cryostat. The elastic constant C(T) was calculated from v(T) with the density ρ = 8.605 g/cm^3, for which we ignore the thermal expansion (7) because the contribution is less than one-tenth of that from the velocity-change.

3. RESULTS AND DISCUSSION

In Fig.1 (a), plotted with circles is the temperature dependence of longitudinal C_{33}-mode elastic stiffness with Γ_1 symmetry in the D_{3h} point group. As temperature is reduced, the characteristic softening due to the CEF effect is observed between 14 K and 150 K. The inset shows the lower temperature part of C_{33}, which exhibits a minimum around 14 K and renews softening below 6 K. Shown in Figs. 1(b) and (c) is the temperature dependence of transverse $(C_{11}-C_{12})/2$ with Γ_5 symmetry and C_{44} with Γ_6 symmetry, respectively. These elastic constants of the transverse modes reveal also the softening due to the CEF effect and the renewed softening without minimum.

The renewed softening is likely to be related to the antiferromagnetic ordering and not to the formation of dense Kondo state (1) because the elastic anomaly arising from the formation of dense Kondo state should be expected only in the longitudinal mode mediating a large electronic Grüneisen parameter.

Since the elastic constant is the second derivative of the free energy of the system with respect to the strain, the strain susceptibility χ includes both contributions of van Vleck-type and Curie-type on the analogy of the magnetic susceptibility. In case of CePdIn, the van Vleck term contributes only to the transverse modes of $(C_{11}-C_{12})/2$ and C_{44}, whereas the Curie term contributes only to the longitudinal mode of C_{33}. The strain susceptibility for the transverse modes exhibits a saturating behavior without minimum at low temperatures. In contrast with the transverse modes, the appreciable minimum is observed in the strain susceptibility for the longitudinal mode at low temperatures. The temperature dependences of elastic constants shown in Fig. 1 are in qualitative agreement with the theory. In order to estimate the CEF level scheme, we took account of only the strain-quadrupole interaction and tried a non-linear least squares fitting to $C(T) = C^0(T) - Ng^2\chi$ for the softening above 10 K, where C^0, N, g and χ are the elastic constant without the CEF effect, the number of Ce^{3+} ion per unit volume,

0921-4526/90/$03.50 © 1990 – Elsevier Science Publishers B.V. (North-Holland)

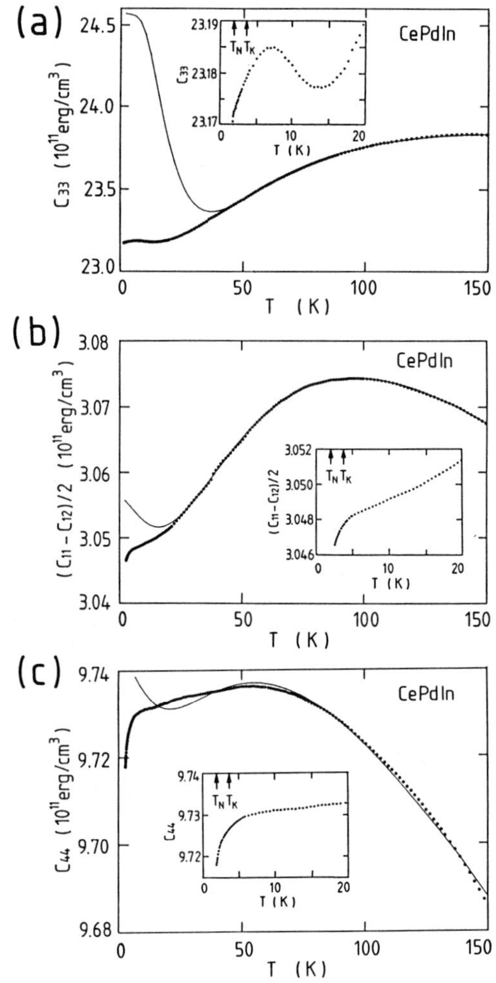

FIGURE 1
Temperature dependences of the elastic constants (a) for C_{33} mode, (b) for $(C_{11}-C_{12})/2$ mode and (c) for C_{44} mode. The inset is the low temperature part of each elastic constant.

coupling constant between the quadrupole and the strain, and the strain susceptibility, respectively. The quadrupole-quadrupole exchange interaction is not taken into consideration since quadrupolar ordering is hardly expected for Ce^{3+} ground state in hexagonal symmetry. In Figs. 1(a), (b) and (c), a theoretical fit is indicated by a solid curve for Δ_1 = 56 K and Δ_2 = 128 K, where Δ_1 and Δ_2 are the energy separations from the ground-state doublet $|\pm5/2\rangle$ to the first excited doublet $|\pm1/2\rangle$ and from $|\pm5/2\rangle$ to the second excited doublet $|\pm3/2\rangle$ by assuming a temperature-linear dependence of

	C_{33}	$(C_{11}-C_{12})/2$	C_{44}
$\|g\|$ (K)	625	126	152
a $(10^7$ erg/cm$^3\cdot$K)	-24.98	-4.371	-10.52
b $(10^{11}$ erg/cm$^3)$	24.583	3.1692	9.8881

TABLE I
The fitted values of $|g|$, a and b for C_{33}, $(C_{11}-C_{12})/2$ and C_{44} mode.

$C^0(T) = aT + b$. The values of $|g|$, a and b are tabulated in Table I. The calculated curves are in good agreement with the experimental results above 30 K. The discrepancy between the calculation and the experiment at low temperatures is caused by the appearance of the renewed softening due to the antiferromagnetic order and/or by the formation of dense Kondo state, even if it is unsuitable for the fitting to assume the linear temperature dependence of background elastic constant C^0 at low temperatures. The coupling constant $|g|$ for the C_{33} mode is extraordinarily large comparing with that for other modes. This suggests anisotropic c-f mixing in CePdIn which is also expected from the anisotropy of resistivity and magnetic susceptibility of single crystal (7).

In summary, we present the temperature dependence of the elastic constants for CePdIn, and propose a CEF level scheme for Ce^{3+} in this compound determined by means of the strain susceptibility.

ACKNOWLEDGMENTS
The authors would like to thank Dr.Y.Maeno for valuable discussion. This work was supported by a Grant-in-Aid for Scientific Research from the Ministry of Education, Science and Culture of Japan.

REFERENCES
(1) T.Goto, T.Suzuki, Y.Ohe, T.Fujimura and A.Tamaki, J. Magn. Magn. Mat. **76 & 77** (1988) 305.
(2) B.Lüthi, "Dynamical Properties of Solids", vol.3, eds. G.K.Horton and A.A.Maradudin (North-Holland, Amsterdam, 1980) p.245.
(3) P.Fulde, "Handbook on the Physics and Chemistry of Rare Earths", eds. K.A.Gshneider and L.Eyring (North-Holland, Amsterdam, 1978) chap.17.
(4) H.Fujii, Y.Uwatoko, M.Akayama, K.Satoh, Y.Maeno, T.Fujita, J.Sakurai, H.Kamimura and T.Okamoto, Jpn. J. Appl. Phys. **26-3** (1987) 549.
(5) Y.Maeno, M.Takahashi, T.Fujita, Y.Uwatoko, H.Fujii and T.Okamoto, Jpn. J. Appl. Phys. **26-3** (1987) 545.
(6) K.Satoh, T.Fujita, Y.Maeno, Y.Uwatoko and H.Fujii, J. Phys. Soc. Jpn. **2** (1990) 692.
(7) H.Fujii et al., this issue.

Physica B 165&166 (1990) 423–424
North-Holland

SOUND VELOCITY AND MAGNETIZATION MEASUREMENTS NEAR THE ANTIFERROMAGNETIC TRANSITION IN THE HEAVY FERMION SYSTEM URu_2Si_2

S. W. Lin[1], S. Adenwalla[2], Q. Z. Ran[1], K. J. Sun[3], D. G. Hinks[4], J. B. Ketterson[2], M. Levy[1] and Bimal K. Sarma[1],

[1] Physics Department, University of Wisconsin-Milwaukee, Milwaukee, WI 53201, USA,
[2] Physics Department, Northwestern University, Evanston, IL 60208, USA,
[3] NASA Langley Research Center, MS 231, Hampton, VA 23602, USA,
[4] Materials Science Division, Argonne National Laboratory, Argonne, IL 60439, USA.

The heavy fermion superconductor URu_2Si_2 has a superconducting transition at 1.4 K and an antiferromagnetic transition at 17.5K. We present measurements in the vicinity of the antiferromagnetic transition of both sound velocity and magnetization.

1. INTRODUCTION

URu_2Si_2 is one of several heavy fermion superconductors (HFS), in which magnetism and superconductivity coexist. Experimental evidence suggests that the superconductivity in at least some of the heavy fermion superconductors maybe unconventional, with antiferromagnetic (AFM) ordering and spin fluctuations playing an important role.

Neutron diffraction measurements (1) reveal an AFM ordering along the c-axis with a Neel temperature, T_N, of 17.5K (URu_2Si_2 is a body centered tetragonal crystal with c = 9.582 Å and a = 4.124 Å). The AFM moment is small, 0.03 μ_B, a very low value compared to the high temperature susceptibility along the tetragonal axis, with an equivalent moment of $\mu_{eff}= 3.51\mu_B$. Heat capacity and resistance measurements also reveal a signature at T_N. Heat capacity measurements (γ = 180 mJ/mole K^2) show a lambda anamoly at 17.5 K and along with the neutron measurements show the existence of finite-gap magnetic excitations. The spin-wave gap is measured to be 1.8 meV, and there is the possibility that the superconducting pairing is caused by the spin-wave bosons.

2. EXPERIMENTAL

We present magnetization and velocity measurements at the Neel transition on a single crystal of URu_2Si_2. Measurements were performed on a single crystal , 0.37 cm long and .173 g. Longitudinal ultrasonic waves up to 300 MHz were generated by means of an overtone polished X-cut quartz transducer. Shear waves up to 300 MHz were generated by means of a 10 MHz $LiNbO_3$ transducer. The attenuation was measured using a conventional pulsed spectrometer (MATEC 6600 series), and the velocity was measured by a computer controlled pulsed heterodyne phase sensitive spectrometer (MATEC MBS8000 unit). The dc magnetization measurements were performed on a commercial SQUID magnetometer (QUANTUM DESIGN). The velocity measurements were done at a frequency around 100 MHz, having a velocity resolution of ~1-2 ppm.

Magnetization measurements were performed along both the a-axis and the c-axis. Because of sample geometry and shape, longitudinal and transverse sound velocity and attenuation measurements were done only along the a-axis, the magnetic field applied was along both the a-axis, and the c-axis.

3. RESULTS AND DISCUSSION

Ultrasonic measurements have been performed by Sun et al. (4) and Fukase et al. (5), both measurements were carried out to below the superconducting state. Here we report the measurements only near the Neel temperature, where a sharp kink or anamoly is seen in both the ultrasonic and magnetization measurements. This is in clear contrast with that observed in UPt_3 (6) , where the AFM ordering moment is only a factor of two smaller.

In Fig. 1 we plot the longitudinal magnetization in a magnetic field of 100 gauss applied along the a-axis, and along the c-axis. In both cases there are sharp signatures at T_N. The magnetization drops abruptly at T_N. Measurements have been done in fields upto 5 Tesla, and the sharp transition is still visible. The shift in T_N with magnetic field is very small, and in agreement with the results reported by Fukase et al (3).

In Fig. 2 we plot the shear wave velocity, v_t, for sound propagating along the a-axis with polari-zation along the c-axis, and the longitudinal sound velocity, v_ℓ, along the a-axis. In both cases there is a sharp signature at T_N. Above T_N the velocity is increasing linearly with decreasing temperature. Below T_N there is a sharp increase in the velocity, showing a cusp like behavior.

0921-4526/90/$03.50

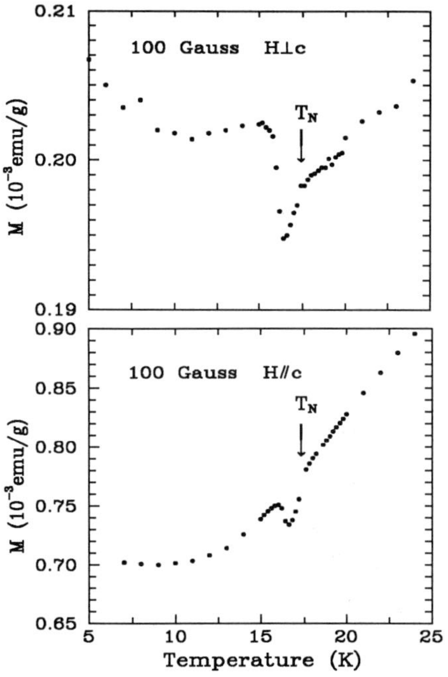

Fig. 1 dc magnetization for H⊥c and H//c

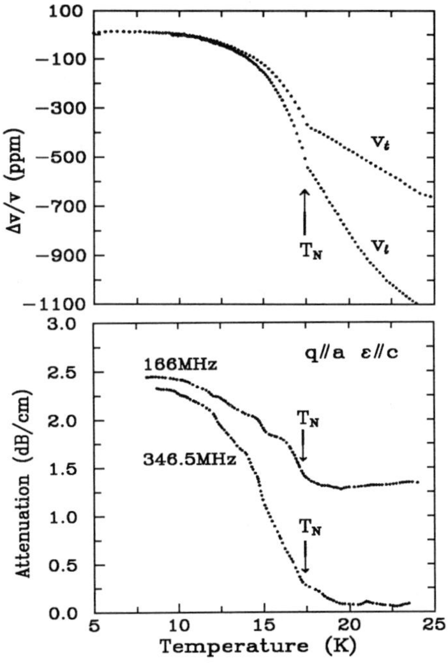

Fig. 2 velocity and attenuation along the c-axis

It is surprising that both the longitudinal and shear velocity show approximately the same behavior at T_N. Since any velocity measurement involves a time of travel measurement which is also affected by the dilation $\Delta\ell/\ell$, it is possible there is a dominating magnetostriction effect at the AFM transition. Similar observations have been made in the heavy fermion superconductor UBe_{13} at a temperature in the vicinity of 8.9 K (7). One should thus also measure the dilation to get a true measure of the velocity change at T_N.

In Fig. 2 we also plot the attenuation as a function of temperature, for the two frequencies 166 MHz and 346.5 MHz. Above T_N the attenuation is flat, and below T_N it starts rising. This is in contrast to a peak like behavior reported for longitudinal sound (4,5).

In conclusion we have performed ultrasonic and magnetization measurements in the heavy fermion superconductor URu_2Si_2 in the vicinity of the AFM transition, and find distinct signatures at T_N. Since there is the possibility that this system may also show unconventional pairing as in UPt_3, it would be interesting to see what role the AFM ordering plays over the subsequent superconducting ordering.

ACKNOWLEDGEMENTS

We thank Jia-Qi Zheng for help in the x-ray orientation of the samples, and M-F. Xu for help with the ultrasonics. This work was supported in parts by the Office of Naval Research (UWM), by the National Science Foundation under grant NO. DMR 89-07396 (Northwerstern) and by the Department of Energy, Office of BES-Materials Sciences Grant No. W-31-109-ENG-38 (ANL).

REFERENCES

1. C. Broholm, J. K. Kjems, W. J. L. Buyers, P. Mathews, T. T. M. Palstra, A. A. Menovsky and J. A. Mydosh, Phys. Rev. Lett. 58, 1467 (1987).
2. T. T. M. Palstra, A. A. Menovsky, J. van den Berg, A. J. Dirkmaat, P. H. Kes, G. J. Nieuwenhuys and J. A. Mydosh, Phys. Rev. Lett. 55, 2727 (1985).
3. M.B. Maple, J.W. Chen, Y. Dalichaouch, T. Kohara, C. Rossel, M.S. Torikachvili, M.W. McElfresh and J.D. Thompson, Phys. Rev. Lett. 56, 185 (1986).
4. K.J. Sun, A. Schenstrom, M. Levy, B.K. Sarma and D.G.Hinks, Phys. Rev. B59, 11284 (1989).
5. T. Fukase, Y. Koike, T. Nakanomyo, Y. Shiokawa, A. A. Menovsky, J. A. Mydosh and P. H. Kes, Proc. 18th Int. Conf. on Low Temperature Physics, Kyoto 1987, Jap. J. Appl. Phys., 26, Suppl. 26-3, 1249 (1987).
6. S. Adenwalla, et al., "Magnetization and Velocity Measurements in UPt_3 Near the Neel Transition", this volume.
7. J. Smith, private communication.

Physica B 165&166 (1990) 425–426
North-Holland

PRESSURE DEPENDENCE OF THE SPECIFIC HEAT OF HEAVY-FERMION $YbCu_{4.5}$

A. AMATO, R.A. FISHER, N.E. PHILLIPS, D. JACCARD* and E. WALKER*

Materials and Chemical Sciences Division, Lawrence Berkeley Laboratory, Berkeley, CA 94720, USA.
*Département de Physique de la Matière Condensée, Université de Genève, CH-1211 Genève, Switzerland.

The specific heat of a polycrystalline sample of $YbCu_{4.5}$ has been measured between 0.3 and 20K at pressures to 8.2 kbar. Unlike cerium-based heavy-fermion compounds, an increase of C/T is observed with increasing pressure, with the linear term enhanced by about 16% at 8.2 kbar. Above 7K, $(\partial C/\partial P)_T$ is negative. The nuclear contribution observed at P=0 is increased by roughly a factor of two at 8.2 kbar.

Due to the high sensitivity to pressure of the 4f and 5f electrons, the heavy-fermion compounds (HFC), in which the hybridization of the f electrons with the conduction band plays the key role in the phenomenon, represent an interesting field for measurements of the pressure dependence of the physical properties. We report here on specific heat (C) measurements on a high purity polycrystalline sample of $YbCu_{4.5}$ between 0.3 and 20K as a function of pressure (P) to 8.2 kbar.

$YbCu_{4.5}$ can be categorized as a HFC that does not undergo a transition to either a magnetically ordered or a superconducting state to the lowest temperatures at which it has been studied (T≈300mK). Magnetic measurements below room temperature (1,2) suggest that ytterbium is close to the trivalent state ($4f^{13}$) with an effective moment $\mu=4.22\mu_B$ which is near the value for a free trivalent ion $\mu=4.54\mu_B$.

Details of the preparation and characterization of the sample are given in Ref. 3. Fig. 1 shows the temperature dependence of C/T under pressure. The precision of the measurements is such that the uncertainty in C for the sample is about 0.4% at the lowest temperatures, increasing to about 6% at 20K. At P=0, C/T exhibits a minimum near 11K. Below 10K, one observes a dramatic increase with decreasing T reflecting the formation of a narrow resonance due to the Kondo interaction between 4f and conduction electrons. The points below 0.7K are well represented by the sum of a hyperfine contribution, AT^{-2}, and γT. They were analyzed on that basis to obtain $\gamma(P)$ and $A(P)$. $A(0)\approx 850$ μJK/mole, which is at least in part due to the quadrupole splitting of ^{173}Yb nuclei (3,4), and a possible contribution from ^{63}Cu and ^{65}Cu. There are two temperature regions that differ in the pressure dependence of C: (i) above about 7K $(\partial C/\partial P)_T$ is predominantly negative; (ii) Below 7K a rapid increase of C with increasing P is observed. The extrapolated values of the coefficient of the electronic term of the specific heat (γ), as a function of P, are plotted in Fig. 2. In the pressure range investigated γ is linear in P, $\gamma(P)=637+12.6P$ mJ/K^{2}mole. A similar behavior of $\gamma(P)$ has been found in the intermediate valence compound YbCuAl (5) and is in sharp contrast with the usual decrease in $\gamma(P)$ observed in non-magnetic uranium or cerium-based HFC, such as $CeCu_6$ -- see Fig. 2.

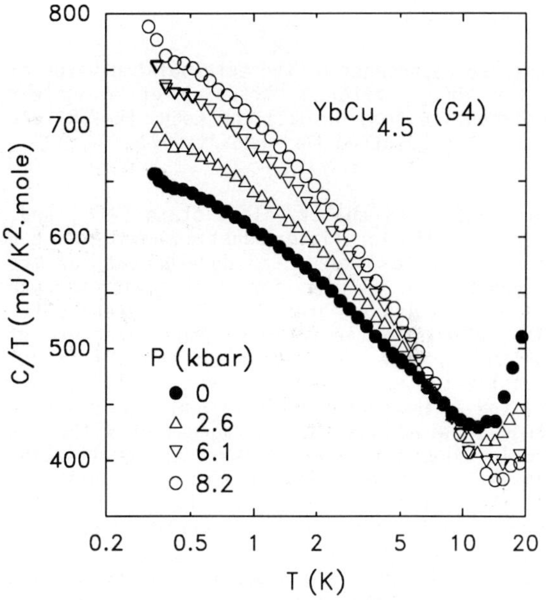

FIGURE 1
Pressure dependence of C/T for $YbCu_{4.5}$

The cerium (or uranium) compounds lose their localized magnetic moments under pressure because the single $4f^1$ (or $5f^1$) electron is squeezed out of the f-shell. Thus a gradual transition is observed from a heavy-fermion state to a valence fluctuating state and a decrease of γ occurs. On the other hand, for ytterbium compounds the pres-

0921-4526/90/$03.50

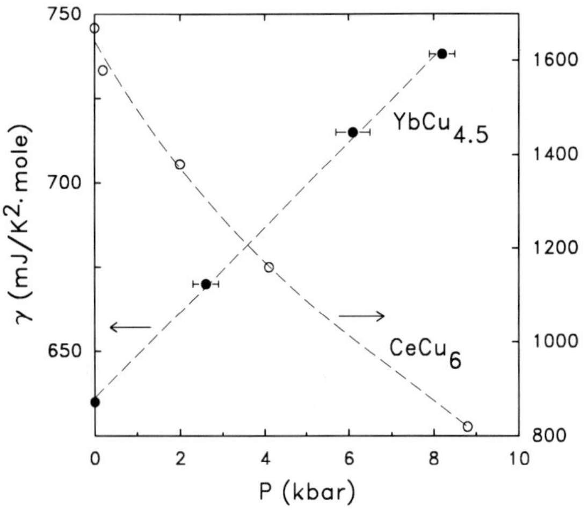

FIGURE 2
Pressure dependence of the extrapolated value of γ for $YbCu_{4.5}$ after subtraction of a nuclear contribution from C. Data for $CeCu_6$ (Ref.6) are plotted for comparison.

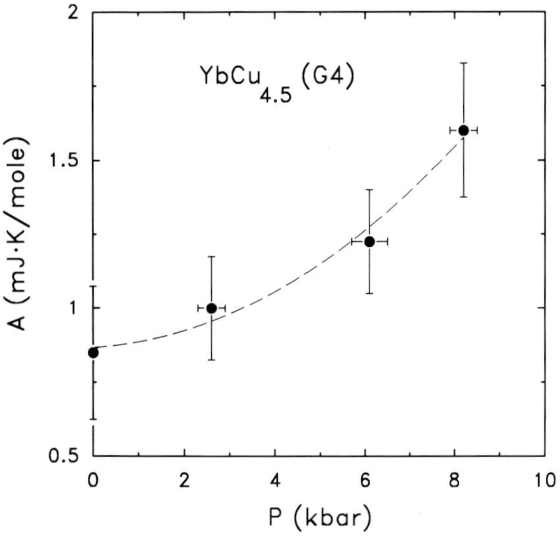

FIGURE 3
Pressure dependence of the coefficient A of the nuclear contribution.

sure stabilizes the trivalent state ($4f^{13}$) and, therefore, the localized magnetic moment. One can thus expect to observe the development, or the strengthening, of heavy-fermion behavior through the application of pressure. The present data, like the systematic decrease under pressure of the temperature (T_M) at which the electrical resistivity is maximum (7), with T_M presumably related to the energy scale (T_K) of the intrasite Kondo interactions, are in agreement with this simple viewpoint. Whereas there are qualitative similarities between the effects of the pressure and the magnetic field (H) on the specific heat for cerium-based HFC (namely a similar decrease), the responses of the ytterbium-based HFC to the action of P and H are opposite (3) emphasizing the different effects of the pressure (modifying the occupancy of the 4f-shell) and of the magnetic field (locking the spin-flip processes).

Fig. 3 is a plot of the pressure dependence of the coefficient of the T^{-2} term of the high temperature limit of the nuclear contribution. At P=8.2 kbar A is enhanced by roughly a factor of two, describing, presumably, an increase of the electrical field gradient at the ^{173}Yb, ^{63}Cu and ^{65}Cu sites.

ACKNOWLEDGEMENTS

This work was supported by the Director, Office of Basic Energy Sciences, Division of Materials Sciences of the U.S. Department of Energy under Contract DE-AC03-76SF00098. Additional support for A.A. was provided by a grant from the Swiss National Science Foundation.

REFERENCES

(1) A. Iandelli and A. Palenzona, J. Less Common Metals 25 (1971) 333.
(2) D. Jaccard, unpublished data.
(3) A. Amato et al., Magnetic field dependence of the specific heat of heavy-fermion $YbCu_{4.5}$, this volume.
(4) P. Bonville, private communication.
(5) A. Bleckwedel and A. Eichler, Specific heat measurements on intermediate valent YbCuAl under high pressure, in: Physics of Solids under High Pressure, eds. J.S. Schilling & R.N. Shelton (North-Holland, Amsterdam, 1981) pp.323-325.
(6) R.A. Fisher et al., Japn. J. Appl. Phys. 26, Suppl.3 (1987) 1257.
(7) J.D. Thompson et al., Response of Kondo lattice systems to pressure, in: Theoretical and Experimental Aspects of Valence Fluctuation and Heavy Fermions, eds. L.C. Gupta & S.K. Malik (Plenum Press, New York, 1987) pp. 151-158.

Physica B 165&166 (1990) 427–428
North-Holland

MAGNETIC ORDER IN THE HEAVY FERMION SYSTEM $Ce(Cu_{1-x}Ni_x)_2Ge_2$

G. SPARN, C. GEIBEL, S. HORN, M. LANG, F. STEGLICH, A. KRIMMEL[$], A. LOIDL[$]

Institut für Festkörperphysik, Technische Hochschule Darmstadt and SFB 252, D-6100 Darmstadt, F.R.G.
[$]Institut für Physik, Johannes Gutenberg-Universität Mainz and SFB 252, D-6500 Mainz, F.R.G.

The magnetic phase diagram of the heavy fermion (HF) systems $Ce(Cu_{1-x}Ni_x)_2Ge_2$ is discussed utilizing results of transport, thermodynamic and neutron-scattering measurements. While the Kondo temperature increases monotonically with x, a complex x-dependence is found for the Néel temperature, associated with a transition from local-moment to itinerant HF magnetism.

The tetragonal Ce-homologes of $ThCr_2Si_2$, while lacking direct 4f-wave function overlap, exhibit a distinct 4f-ligand hybridization. The latter, along with strong many-body renormalizations below the so called "lattice Kondo temperature" T^*, is effective in forming heavy-mass quasiparticles ("heavy fermions" (HF)). In the case of the compounds CeM_2X_2 (M : Cu, Ag, Au, Ru, Ni; X : Si, Ge), for large Ce-M distances r (≥ 3.3Å), magnetic ordering between local moments occurs at T_m , whereas in the limit of small r (≤ 3.2Å) a valence fluctuating state forms [1]. In the intermediate regime, the HF-compounds $CeNi_2Ge_2$, $CeRu_2Si_2$ and $CeCu_2Si_2$ are found, of which the latter is unique not only because it is a superconductor below 0.7K [2], but also it seems to exhibit an itinerant type of HF magnetism below approximately 0.8K (at zero magnetic field) and 7 Tesla (T→0) [3] .

In this communication, we report a brief summary of a systematic study on the quasibinary system $Ce(Cu_{1-x}Ni_x)_2Ge_2$, which reveals a transition from local-moment ordering to itinerant HF magnetism. These alloys are ideally suited for this purpose : The Kondo binding energy kT^* of $CeCu_2Ge_2$ ($T^*=8\pm2K$) is of the same order of magnitude as the magnetic interaction energy, kT_{RKKY} ($T_{RKKY}=7K$), at which short-range ordering sets in [4]. Long range magnetic order has been detected below T_m = T_{N1} = 4.1K [4,5]. On the other hand, $CeNi_2Ge_2$ ($T^*=30K$) exhibits the signature of a Kondo lattice or HF compound with no long-range magnetic order [6]. For the samples prepared as described in [1], no parasitic phases have been detected by X-ray diffractometry or electron microprobe analysis.

Fig.1 shows the concentration dependence of the magnetic ordering temperature (in reduced units) for $Ce(Cu_{1-x}Ni_x)_2Ge_2$ as determined from thermal expansion, specific heat and magnetic susceptibility measurements [7,8]. Even small Ni concentrations cause a dramatic decrease of $T_{N1}(x)$, which extrapolates to zero near x=0.2. However, for larger x a second branch, $T_{N2}(x)$, develops with a maximum near x=0.5. This new magnetically ordered phase can be monitored up to approximately x=0.75. Bulk measurements show clear evidence for a second transition below T_{N1} and T_{N2}, respectively in the range $0.02 \le x \le 0.3$ (hatched region) as discussed in [8].

The magnetic structure and the size of the ordered moments have been determined by neutron powder-diffraction experiments for the compounds with concentrations x=0, 0.1, 0.28 and 0.5. Incommensurate modulated spin arrangements with magnetic Bragg reflections at $|Q|=|\tau_{hkl}\pm q_0|$, where (hkl) is a vector of the reciprocal chemical lattice and q_0 the propagation vector, have been found. They are described by the following values. For x=0: $q_0=(0.28, 0.28, 0.54)$ and the effective moment $\mu_S=0.74\ \mu_B$; for x=0.1 : $q_0=(0.28, 0.28, 0.41)$, $\mu_S=0.5\mu_B$; for x=0.28 : $q_0=(0.11, 0.11, 0.25)$ and finally for x=0.5 : $q_0=(0, 0, 0.14)$, $\mu_S=0.3\mu_B$. For this latter sample the magnetic Bragg reflections are much broader than the experimental resolution of the spectrometer which presumably reflects a disturbance of long-range magnetic order by single-site (Kondo) interactions.

To gain further insight into this competition of single-site and inter-site interactions, the magnetic relaxation rate via quasielastic neutron scattering has been investigated in the temperature range 1.5K ≤ T ≤ 200K. The lattice Kondo temperature as a function of Ni concentration, as read off the residual quasielastic line widths at low temperatures ($T \ge T_N$), is also plotted in Fig.1 and compared to T^* as determined from thermal expansion and resistivity

0921-4526/90/$03.50

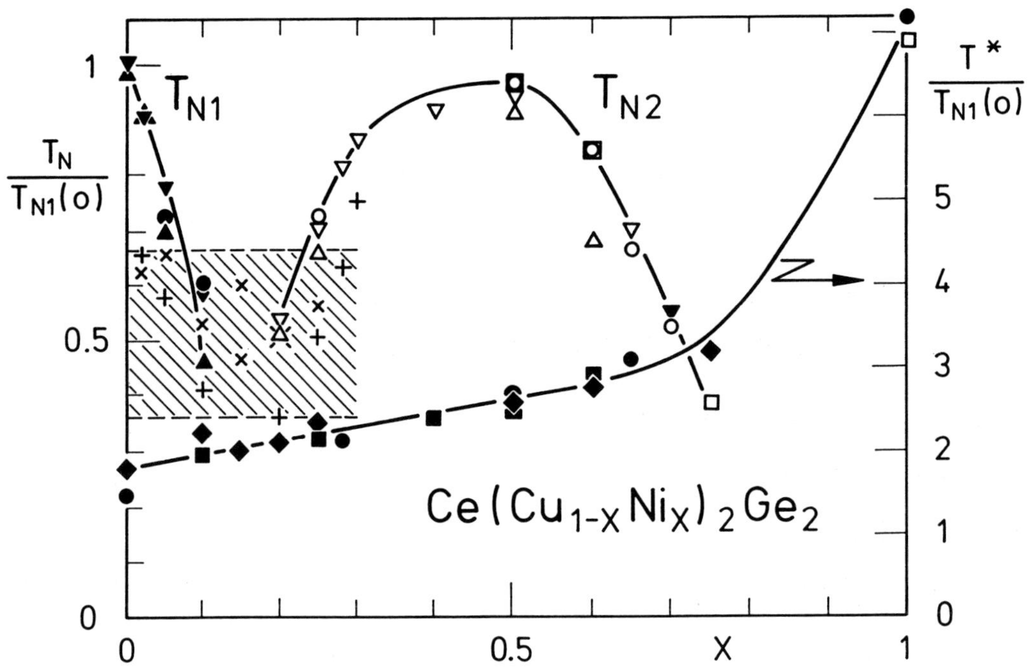

Fig.1. LEFT SCALE : x, T_N phase diagram for the quasi-binary compound $Ce(Cu_{1-x}Ni_x)_2Ge_2$ as determined from specific heat (▼, ▽, +), thermal expansion (▲, △, x) and dc susceptibility (●,O) measurements [9]. RIGHT SCALE : T^* as determined from the residual quasielastic neutron line width (●), thermal expansion (■) and resistivity (◆) peaks. Positions of the latter are scaled by factors 1.5 and 1.9, respectively (T^* and T_N are normalized to T_{N1} for x=0).

maxima. With increasing x, T^* increases continuously and, for x=0.65, the Kondo binding energy is found to be larger than the energy of the inter-site interactions by almost a factor of four.

The most exiting result of the present study is the occurence of an extremely short propagation vector for the x=0.5 sample which characterizes a static spin wave extending over almost ten lattice constants. Such short ordering wave vectors have been predicted by Grewe and Welslau [9], if HF band magnetism develops in a Kondo lattice. Recent calculations of Welslau and Grewe within this model [10] revealed the development of T_N and T^* as a function of the local exchange coupling constant $g=N_FJ$ (N_F : conduction band density of states at the Fermi level; $J<0$: exchange integral) in the full range between $T_N \gg T^*$ and $T_N \ll T^*$. These calculations are corroborated by our results near x=0 where $T_N \approx T^*$ and for $0.5< x \leq 0.75$ where $T_N \ll T^*$.

The intermediate composition range, in which

(i) two subsequent transitions are observed as a function of T for each sample ($0.02 \leq x \leq 0.3$) and

(ii) $T_{N2}(x)$ shows a monotonic increase ($0.2 \leq x \leq 0.5$), challenges further investigation.

REFERENCES

1. A. Loidl et al. to be published .
2. F. Steglich et al., Phys. Rev. Lett. 43 (1979) 1982.
3. Y. J. Uemura et al., Phys. Rev. B 39 (1989) 4726; H. Nakamura et al., J. Magn. Magn. Mat. 76&77 (1988) 517; U.Rauchschwalbe et al., J. Magn. Magn. Mat. 63&64 (1987) 447; F. Steglich J. Phys. Chem. Solids 50 (1989) 235.
4. G. Knopp et al., Z. Phys. B77 (1989) 95 .
5. F. R. de Boer et al., J. Magn. Magn. Mat. 63&64 (1987) 91 .
6. G. Knopp et al., J. Magn. Magn. Mat. 74 (1988) 341.
7. G. Sparn et al., J. Magn. Magn. Mat. 76&77 (1988) 153.
8. F. Steglich et al., Physica B, in press.
9. N. Grewe and B. Welslau, Solid State Commun.65 (1988) 437 .
10. B. Welslau and N. Grewe, this conference .

Physica B 165&166 (1990) 429–430
North-Holland

Electronic Transport Properties of $UCu_{4+x}Al_{8-x}$ at Low Temperatures

R. Köhler, C. Geibel, S. Horn, B. Strobel, S. Arnold, G. Sparn, A. Höhr, C. Kämmerer, and F. Steglich

Institut für Festkörperphysik, Technische Hochschule Darmstadt and SFB 252, D-6100 Darmstadt, FRG

Magneto-resistivity measurements show that in the compound $UCu_{4+x}Al_{8-x}$, which undergoes a transition from a magnetically ordered to a heavy fermion groundstate with increasing x, single ion Kondo processes dominate the low temperature properties. A comparison to Bethe Ansaty calculations indicates an angular momentum of the groundstate larger than two.

The compound UCu_4Al_8, which crystallizes in the $ThMn_{12}$-structure, orders antiferromagnetically below 30 K (1).In an earlier investigation it was shown (2) that $UCu_{4+x}Al_{8-x}$ forms in the same structure over a range $0.1 \leq x \leq 1.95$. Replacement of Al by Cu suppresses antiferromagnetic order for $x > 1.5$ and apparently results in a heavy fermion ground state with a linear specific heat coefficient $\gamma \approx 400$ mJ/mol K^2 at $x = 1.5$ increasing to $\gamma \approx 800$ mJ/mol K^2 for $x = 1.9$. The transition has been interpreted (3) in the spirit of Doniach s phase diagram (4), i.e. by an increasing exchange coupling constant $|J|$, so that the intra-site Kondo interaction dominates over the inter-site magnetic interaction and, therefore, suppresses long range magnetic order. The substitution of Al by Cu should generate substantial disorder in the material, since Al occupies two inequivalent sites in this structure, which are statistically occupied by the substituted Cu-ions. This disorder is reflected by the electrical resistivity, which exhibits high values over the whole temperature range and, for $x > 1.5$, shows only weak indication for an onset of coherent scattering at low temperatures. A clear observation of a transition from a magnetically ordered to a heavy fermion state is rare for U-based systems. It deemed, therefore, interesting to probe the effect of a magnetic field on the physical properties on both sides of this transition. Here we present results of the magneto-resistivity on samples with varying compositions x.

Fig. 1 shows the magneto-resistivity $\Delta\rho = \rho(B = 7T) - \rho(B = 0)$ as a function of temperature. For $x = 1.0$ $\Delta\rho$ remains negative for the whole temperature range measured. At T = 18 K a sharp minimum is observed, which coincides with the Néel temperature as determined from magnetic susceptibility measurements (4). A similar $\Delta\rho(T)$ dependence for T > 18 K is found for the x = 1.75 sample which, however, shows a monotonical decrease to higher negative values down to the lowest temperature measured. Finally, $\Delta\rho(T)$ of the x = 1.9 sample behaves correspondingly at low temperatures, but changes sign around $T \approx 30$ K and shows a broad maximum around $T \approx 150$ K. The negative values of $\Delta\rho(T)$ for the samples with x = 1.75 and x = 1.9 are smaller than for the x = 1.0 sample above 18 K.

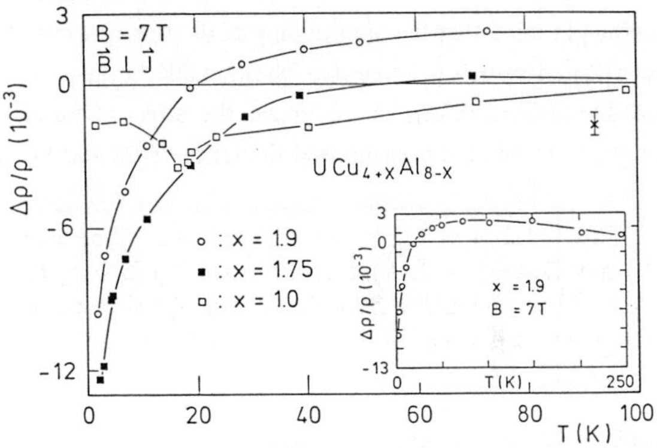

Fig. 1. Magneto-resistivity $\Delta\rho$ as a function of temperature.

The negative temperature characteristic of the zero-field resistivity (2) indicates the presence of the Kondo effect in this compound and, therefore, the most obvious explanation for the negative magneto-resistivity is the presence of independnet paramagnetic, i.e. Kondo, ions. Due to the disorder in the system coherency effects, as inferred from a change of sign for a temperature T_{coh} much smaller than the characteristic temperature T^* (5), are apparently absent above T = 2 K. If single ion Kondo processes dominate $\Delta\rho$, a comparison with Bethe Ansatz calculations of the magneto-resistivity is adequate. In fig.

0921-4526/90/$03.50

2 we have compared the magnetic field dependence of $\Delta\rho$ for the x = 1.9 sample at T = 3.27 K with theoretical results (6). This temperature was chosen to ensure $T_{coh} < T << T^*$. The experimental data are plotted as $\ln((\rho_o-\rho_B)/\rho_o)$ vs. $\ln(B/(B=1T))$, where $\rho_B = \rho(B)$ and $\rho_o = \rho(B = 0)$. In such a plot the theoretical curves can simply be shifted to fit the experimental data; a horizontal shift corresponds to a variation of the effective field B^* (T_H in ref. 6), which is a measure of the characteristic energy scale of the system. A vertical shift varies the ratio: resistivity contributions of magnetic origin ρ_{mag} to the total resistivity ρ_{tot}. Fig. 2 shows that the agreement between experimental data and theory improves with increasing value of the 5f angular momentum j, which determines the shape of the theoretical curve. This implies a degeneracy greater than four for the ground state of the uranium ions, keeping in mind that the applicability of the theory to the investigated system is somewhat questionable because of possible residual effects of coherence, the effect of finite temperature and the presupposed similarity of Ce and U-based systems.

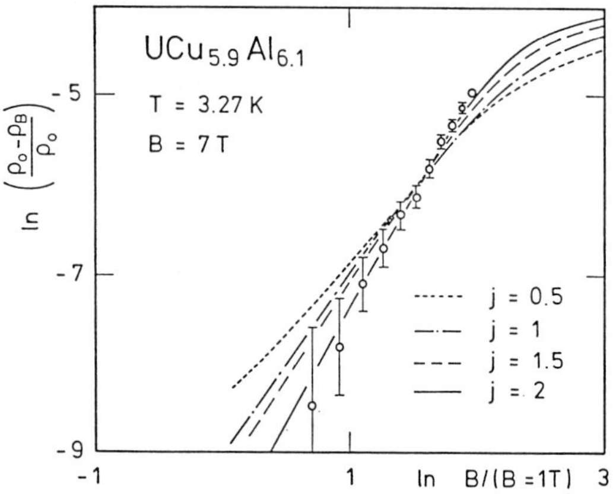

Fig. 2. Magneto-resistivity as a function of magnetic field, compared to Bethe Ansatz calculations (6). For details see text.

Since the data points lie on a straight line in fig.2 the ratio ρ_{mag}/ρ_{tot} and B^* cannot be determined independently. Taking the room temperature resistivity as an upper limit for the non-magnetic contributions to ρ_{tot}, we obtain a $B^* \approx 30$ T. This value is about the same as that for UBe_{13} at $T \approx 3$ K (7)

The change of sign $\Delta\rho(T)$ for the x = 1.9 sample is ascribed to a temperature-independent bandstructure derived positive contribution to the magneto-resistivity of all samples. The maximum appears to result from magnetic short range ordering effects on a characteristic temperature scale of about 100 K, which suppress the negative contribution to $\Delta\rho(T)$ around that temperature. This characteristic energy can be inferred from magnetic susceptibility results, which show a Curie-Weiss (CW) behavior for T > 60 K, with a CW temperature $\Theta \approx 100$ K and a pronounced upturn below 60 K.

In conclusion, the results of magneto-resistivity measurements are in accord with the picture of a single ion Kondo effect in the concentrated Kondo compound $UCu_{4+x}Al_{8-x}$. Comparison to theory suggests a ground state degeneracy of greater than four in this compound. A confirmation of this result by specific heat measurements would be desirable. Such experiments are in preparation.

Stimulating discussions with P. Schlottmann and W. Suski are gratefully acknowledged.

References

(1) A. Baran, W. Suski, O.J. Zogal, and T. Mydlay, J. Less Common Met. 110 (1985) 327.

(2) C. Geibel, U. Ahlheim, A.L. Giorgi, G. Sparn, H. Spille, F. Steglich, and W. Suski, Physica B, in press.

(3) F. Steglich, U. Ahlheim, C. Geibel, S. Horn, M. Lang, G. Sparn, A. Loidl, and A. Krimmel, J. Magn. Magn. Mat., in press.

(4) S. Doniach, Physica 91B (1977) 231.

(5) See, e.g., G. Remeny, A. Briggs, J. Flouquet, O. Laborde and F. Lapierre, J. Magn. Magn. Mat. 31 - 34 (1983) 407.

(6) P. Schlottmann, Z. Physik B 51 (1983) 223.

(7) U. Rauchschwalbe, Physica 147 B (1987) 1.

Physica B 165&166 (1990) 431–432
North-Holland

MAGNETISM AND LINEAR FIELD MAGNETORESISTIVITY IN UPt_3

K. Behnia, O. Laborde+, L. Taillefer and J. Flouquet

Centre de Recherches sur les Très Basses Températures, CNRS, BP 166X, 38042 Grenoble, France

+ also at Service National des Champs Intenses, CNRS, BP 166X, 38042 Grenoble, France

Abstract- The temperature dependence of the transverse magnetoresistivity (upto 7.6 tesla) for heavy fermion UPt_3 has been studied .The linear term in the field dependence of resistivity is found to decrease with increasing temperature and vanishes for T> 6 K. The possible connection of this new experimental result with the aniferromagnetic order is discussed.

The relation between magnetism and superconducting pairing is a key point in the understanding of heavy fermion compounds. UPt_3 is a particular interesting case as two successive superconducting transitions appear on cooling (1). One possible mechanism of the splitting of the superconducting transition is the coupling to antiferromagnetism (2). If neutron diffraction experiments performed on single crystal have shown the occurrence of antiferromagnetic Bragg reflection at $T_N \sim 5$ K corresponding to extremely small sublattice magnetization (~ 0.02 μ_B/U atoms) (3), there is up to now no convincing indication of the existence of magnetic ordering in other measurements : NMR (4), specific heat (5), resistivity and thermoelectric power (6). An unusual result is that at very low temperature (T < 1 K), the magnetoresistivity $\Delta\rho(H)$ for an applied magnetic field H in the easy magnetization (i.e. basal)plane follows a quasilinear field dependence ($\Delta\rho \sim a$ H) with a coefficient a independent of the residual resistivity in zero magnetic field (0.1 $\mu\Omega$ cm < ρ_0 < 1.5 $\mu\Omega$ cm) (7). In fact this can be understood by arguing that when $T \to 0$ K, as the Köhler law is obeyed , $\Delta\rho(H)/\rho$ depends only on H/ρ. Thus, a consequence of a linear magnetoresistivity would be an intrinsic (that is, ρ_o independent) linear dependence of $\Delta\rho$ on magnetic field. One may thus expect that the temperature dependence of the coefficient a of the linear H term of the magnetoresistivity is related to an intrinsic property of UPt_3.

In order to investigate any possible influence of magnetic ordering on resistivity, we performed systematic measurements of transverse magnetoresistivity (upto 7.6 T) in isothermal conditions for several temperatures . The sample was a single crystal described in reference 7 with a residual resistivity of 0.1 $\mu\Omega$.cm . The magnetic field was applied parallel to the b axis.

In our previous study on the same sample (7), focus was given on the field variation of the T^2 inelastic term and on the quantum regime $\omega_c \tau > 1$. In this paper we report on the observation of a possible signature of antiferromagnetic order in the magnetoresistivity of this compound. Our idea was to follow the temperature variation of the magnetoresistivity $\Delta\rho_T(H)$, which was found to be well decompositable in linear a(T) H plus quadratic b(T) H^2 terms :

$$\Delta\rho_T(H) = a(T)H + b(T)H^2$$

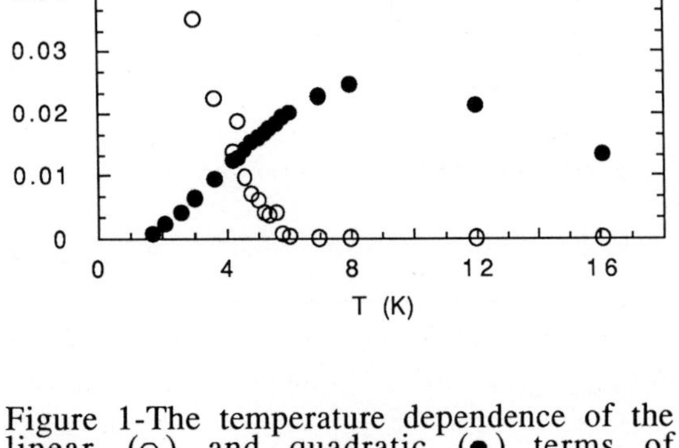

Figure 1-The temperature dependence of the linear (○) and quadratic (●) terms of magnetoresistivity. Note the appearance of the linear term below 6 K and its sysrematic increase with decreasing temperature. The quadratic term absent at low temperatures appears at 1.6 K and passes through a maximum at about 8 K.

Such a study has been done on the heavy fermion compound UBe_{13} in order to detect deviations from a single Kondo impurity behavior (8).

Figure 1 shows the temperature dependence of the linear and quadratic terms of absolute magnetoresistivity: The linear term, a(T) collapses for T ~ 6 K i.e. almost the same temperature where the sublattice magnetization vanishes while b(T) reaches a maximum at T ~ 8 K. The disappearance of the linear field term at $T \sim T_N$ suggests that it is related to the magnetic ordering and characteristic of its nature.

Similar effects where observed in antiferromagnetic Pt_3Fe (9). It was stressed that in the case of structures admitting the existence of weak ferromagnetism an odd component of the transverse magnetoresistivity can occur (10). A positive linear magnetoresistivity was also observed in the itinerant antiferromagnet ∝Mn (11). Calculations on itinerant antiferromagnet (12) with competing ferromagnetic interactions predict that a H^2 dependence may be obeyed only at very low magnetic field while a linear negative part emerges rapidly. With a strong uniaxial anisotropy, longitudinal positive magneto–resistivity is found. Up to now, there is no theoretical model which describes the situation of UPt_3.

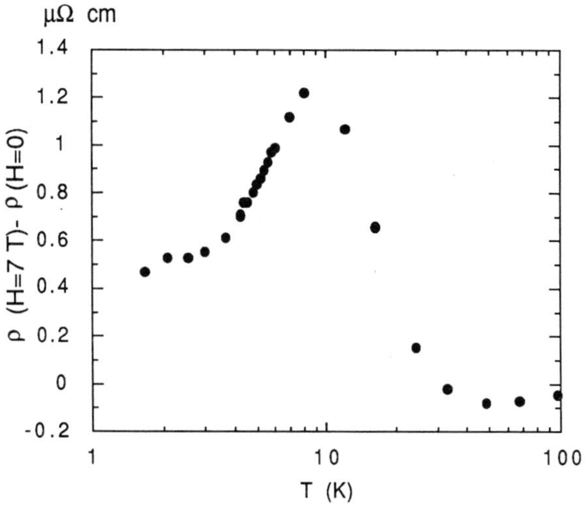

Figure 2 - The absolute change of resistivity in a field of 7 tesla versus temperature. The positive magnetoresistivity passes through a maximum at 8 K before changing sign at 30 K.

Figure 2 shows the variarion of magnetoresistivity with temperature between 1 K and 100 K. The positive magnetoresistivity at low temperatures passes through a maximum at about 8 K It changes sign at 30 K before passing through a minimum at T~ 50 K. At the moment, one could only speculate about a possible relation between this curve and the temperature dependence of the magnetic susceptiblity which passes through a maximum at 16 K (13). In the two cases, there seems to be a distinct transition from the high - temperature to low-temperature behaviour.

For the first time, a macroscopic manifestation of the onset of magnetic ordering has been observed in UPt_3. That will give the possibility to study the pressure or strain dependence of T_N over values difficult to achieve in microscopic experiments (notably neutron) and thus to study its link with the corresponding dependence of superconductivity.

REFERENCES

(1) R.A. Fisher, S. Kim, B.F. Woodfield, N.E. Phillips, L. Taillefer, K. Hasselbach, J. Flouquet, A.L. Giorgi and J.L. Smith, Phys. Rev. Lett., **62**, (1989), 1411
(2) See for example F. Machida and M. Ozaki, J. Phys. Soc. Jpn, **58**, 2244, (1989)
(3) G. Sepphi, E. Bücher, C. Broholm, J.K. Kjemsn J. Baumann and J. Hufnagl, Phys. Rev. Lett., **60**, (1988), 615
(4) Y. Kahari, H. Shibai, T. Kohana, Y. Oda, Y. Kitaoka and K. Asayama, J. Magn. Magn. Mat., **76-77**, (1988), 478
(5) J. Odin, E. Bucher, A.A. Menovsky, L. Taillefer and A. de Visser, J. Magn. Magn. Mat., **76-77**, (1988), 223 and N. Phillips, to be published
(6) D. Jaccard and C. Marcenat, to be published
(7) L. Taillefer, J. Flouquet and W. Joss, J. Magn. Magn. Mat., **76-77**, (1988), 218
(8) J.P. Brison, O. Laborde, D. Jaccard, J. Flouquet, P. Morin, Z. Fisk and J.L. Smith, J. Physique, **50**, 2795 (1989)
(9) N.I. Tourov, Yu N. Tisovkin and N.V. Volkenstein, Sov. Phys. State, **23**, 612 (1981)
(10)E.A. Turov and V.G. Shavrov, Izv. Akad. Nauk SSR, Ser. Fiz, **27** (1963), 1487
(11)S. Murayama and H. Nagasawa, J. Phys. Soc. Jpn, **43**, 1216, (1977)
(12)K. Usami, J. Phys. Soc. Jpn, **45**, 466 (1978)
(13)See J.J.M. France, A. de Visser, A. Menovsky and P.H. Frings, J. Magn. Magn. Mat., **52**, 61 (1985)

Physica B 165&166 (1990) 433–434
North-Holland

LOW TEMPERATURE PROPERTIES OF HEAVY-FERMION $CeRu_2Si_2$ STUDIED BY MAGNETIZATION EXPERIMENTS

C. Paulsen, A. Lacerda, J.L. Tholence and J. Flouquet

Centre de Recherches sur les Très Basses Températures, CNRS, BP 166X, 38042 Grenoble Cédex, France

We present magnetization experiments on single-crystalline heavy fermion $CeRu_2Si_2$ in the temperature range 1.3 K - 200 mK. This well known heavy fermion compound shows a metamagnetic transition for a field applied along the tetragonal axis ($B^* = 7.665$ T at T = 200 mK). The magnetization data M(T,B) are used to calculate the behaviour of the term linear in temperature (γ) of the specific heat as a function of magnetic field.

1. INTRODUCTION

The heavy fermion compound $CeRu_2Si_2$ (tetragonal structure) has a large coefficient of the linear temperature term in the electronic specific heat (1) ($\gamma \sim 350$ mJ/mol K^2). In an applied magnetic field a collapse of intersite fluctuations (2) occurs, resulting in an increase of the local character of the magnetization (M). The huge increase of the differential susceptibility, $\chi = dM/dH$, at B^* ($\chi(B^*)/\chi(B \to 0) \sim 10$ at 1.3 K (3-4)) leads to call B^* a metamagnetic field. However, $CeRu_2Si_2$ is considered to stay in a Pauli paramagnetic state whatever the applied field is ($B \gtrless B^*$).

2. EXPERIMENT

The single-crystalline sample of $CeRu_2Si_2$ was prepared by a tri-arc Czochralski technique. The magnetization measurements were made using a high-field low-temperature SQUID magnetometer developed at the CRTBT. At the core of the system is a miniature dilution refrigerator. A long copper tress holding the sample is attached to the cold finger of the mixing chamber. The tress is encased in a leak tight quartz tail section which fits into the highly uniform field of the 8 tesla magnet. The temperature dependence of the magnetization, at constant fields, M(T,B) has been measured for fields applied along the (c) axis. The purpose of this paper is to investigate the presence of a T^2 term in the magnetization at low temperatures: $M(T,B) = \beta(B)T^2$.

3. RESULTS

In Fig. 1 we show M as a function of T^2 for various magnetic fields. In the temperature range $0.200 \text{ K} < T < 0.800$ K, for fields $B \gtrless B^*$, a T^2 law is perfectly obeyed. However, in the vicinity of $B^* = 7.665$ T the T^2 law is a reasonable approximation only below 0.3 K. The difficulty in observing a T^2-term near B^* comes mainly from the weak temperature variation of B^* itself. At B^* β changes sign.

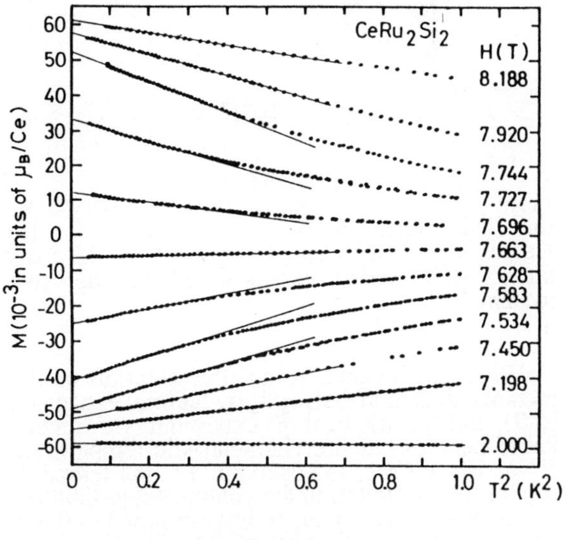

FIGURE 1
Magnetization (B//c) of $CeRu_2Si_2$ versus T^2 for magnetic fields as indicated. For clarity the curves have been vertically shifted by ΔM. For each curve ΔM is given by the intercept with the vertical axis.

Through the Maxwell relation: $(\partial M/\partial T)_{P,B}= (\partial S/\partial B)_P$, where we take in the low temperature limit, the magnetization proportional to T^2, the field dependence of the electronic term (γ) of the molar specific heat is obtained : $\partial\gamma/\partial B = 2\beta(B)$. Fig. 2 shows $\gamma(B)$ for pure $CeRu_2Si_2$ and for the compound in which 5 % of Ce has been replaced by La. In the latter compound B* has dropped to 5.2 T (at T ~ 150 mK (5)), and $\gamma(0)$ equals 500 mJ/molK^2 (1). $Ce_{0.95}La_{0.05}Ru_2Si_2$ has still a Pauli paramagnetic ground state. The relative enhancement of γ at B* is almost two times larger for the pure compound than for the doped compound. In the insert of Fig. 2 we show $\gamma(B)/\gamma(B^*)$ as a function of B/B*. We observe a difference between the widths of $\gamma(B)/\gamma(B^*)$ at B*, for the two compounds. This suggests that the itinerant character of the f electrons becomes less dominant when going to the doped system.

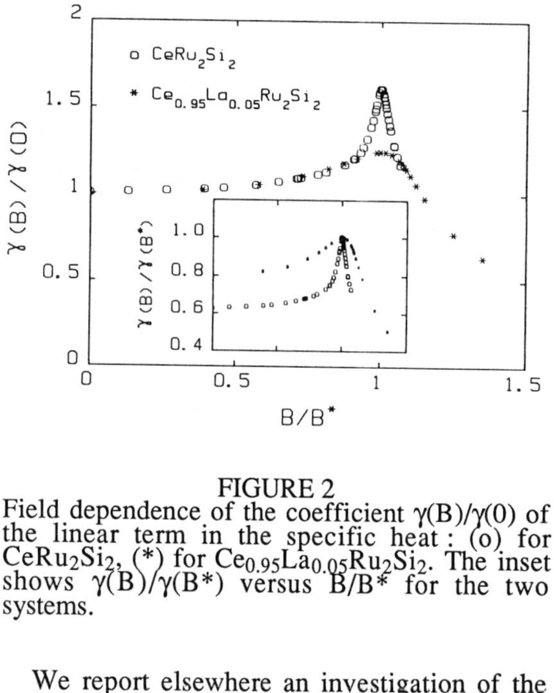

FIGURE 2
Field dependence of the coefficient $\gamma(B)/\gamma(0)$ of the linear term in the specific heat : (o) for $CeRu_2Si_2$, (*) for $Ce_{0.95}La_{0.05}Ru_2Si_2$. The inset shows $\gamma(B)/\gamma(B^*)$ versus B/B* for the two systems.

We report elsewhere an investigation of the metamagnetism in $CeRu_2Si_2$ by thermal expansion (6-7), and specific heat (8) experiments. A remarkable decrease of the characteristic temperature T* at B = B* ± ε, is observed. T*, the temperature of the extremum in the volume expansion, α_v, equals 9 K at zero field, and has dropped to 0.3 K for fields in the vicinity of B* (7). This large variation of T* can be coupled to the variation of the energy scale by means of thermodynamic relations. Assuming that at low temperature only one single scaling parameter prevails, we may define a heavy-fermion (HF) Grüneisen parameter for T → 0 (see for instance ref. 9). $\Gamma_{HF} = V_o a_v/\kappa\gamma$, where V_o is the molar volume, κ is the isothermal compressibility, and $a_v = \alpha_v/T$ is the linear term of the volume expansion. We can also define an experimental field dependent low temperature Grüneisen parameter: $\Gamma_{HF}(B)=V_o a_v(B)/\kappa(B)\gamma(B)$. Combining the thermal expansion data (6-7) with the field variation of γ (Fig. 2) we find Γ_{HF} (B*-ε,T→0) = +1474, and Γ_{HF} (B*+ε,T→0) = -1477 (ε = 0.07 tesla). Despite the difficulty to explain quantitatively the huge temperature variation of Γ_{HF} (B, T→0), this suggests that one main energy is extremely correlated with the magnetic field. In fact the key feature must be the coupling between the local fluctuations and the magnetic field (10). Experiments are actually in progress to study the role of ferromagnetic fluctuations for B→B*.

In summary, we have investigated the metamagnetism in heavy fermion $CeRu_2Si_2$ by means of low-temperature for magnetization experiments. These experiments allow for a quasicontinuous sweep of through the metamagnetic transition. The large variation of the magnetization in the vicinity of B*, and the huge variation of the Grüneisen parameter near B*, reveals that the one parameter scaling should be taken with same caution.

ACKNOWLEDGEMENTS

A.L. is supported by Conselho Nacional de Desenvolvimento Cientifico e Tecnologico, Brazil.

REFERENCES

(1) R.A. Fisher, N.E. Phillips, C.Marcenat, J. Flouquet, P. Haen, P. Lejay and J.M. Mignot, J. Physique (Paris), colloq. C8-49 (1988) 759 and references therein.
(2) J. Rossat-Mignod, L.P. Regnault, J.L. Jaccoud, C. Vettier, P. Lejay, J. Flouquet, E. Walker, D. Jaccard and A. Amato, J. Magn. Magn. Mater., 76 & 77 (1988), 376.
(3) P. Haen, J. Flouquet, F. Lapierre, P. Lejay, and G. Remenyi, J. Low Temp. Phys. 67 (1987), 39.
(4) A. Lacerda, A. de Visser, L. Puech, P. Lejay, P. Haen and J. Flouquet, J. Appl. Phys., in print.
(5) C. Paulsen, A. Lacerda, A. de Visser, K. Bakker, L. Puech and J.L. Tholence, submitted to the Yamada Conference XXV on Magnetic Phase Transition (Osaka, April 13-16, 1990).
(6) A. Lacerda, A. de Visser, L. Puech, P. Lejay, P. Haen, J. Flouquet, J. Voiron and F.J. Ohkawa, Phys. Rev., B40, (1989), 11429.
(7) A. Lacerda, C. Paulsen, L. Puech, A. de Visser, J.L. Tholence, P. Haen and J. Flouquet (to be published).
(8) R.A. Fisher, C. Marcenat, N.E. Phillips, J. Voiron, P. Haen, F. Lapierre, P. Lejay and J. Flouquet (to be published).
(9) A. de Visser, J.J.M. Franse, A. Lacerda, P. Haen and J. Flouquet , Physica B, in print.
(10) F. J. Ohkawa, Solid State Commun. 71(1989), 907.

Physica B 165&166 (1990) 435–436
North-Holland

ANISOTROPIC TRANSPORT AND MAGNETIC PROPERTIES OF Fe_2P-TYPE HEAVY FERMIONS CePdIn AND UPdIn

H. FUJII, M. NAGASAWA, H. KAWANAKA, T. INOUE AND T. TAKABATAKE

Faculty of Integrated Arts and Sciences,Hiroshima University, Hiroshima 730, Japan

Measurements of electrical resistivity and magnetic susceptibility have been performed on CePdIn and UPdIn single crystals with a hexagonal Fe_2P-type structure. For the two compounds, the transport and magnetic properties along the a and c-axes are significantly anisotropic and characteristic , suggesting the remarkable effects of anisotropic hybridizations of 4f or 5f-electrons with ligand s, p or d-electron states.

In the course of our research on new Kondo compounds, we have found that both CePdIn and UPdIn with a Fe_2P-type structure display some interesting properties reflecting the hybridization of f-electrons with ligand s, p or d-electrons. CePdIn is an antiferromagnetic heavy-fermion with T_N=1.8 K (1) and specific heat versus temperature ratio of C/T=700 mJ/mole·K^2 at 70 mK (2). On the other hand, UPdIn is a magnetic heavy-fermion with double magnetic transitions(3,4). Below T_N=21 K, a c-axis modulated structure with the propagation vector $\mathbb{k}$=(0,0,0.4) appears, which changes into a square-up structure below 8.5 K (5). The ratio C/T reaches 280 mJ/mole·K^2 at 1.5 K (3,6). In such a low symmetric crystal as a hexagonal Fe_2P-type structure, we can expect a remarkable anisotropy in transport and magnetic properties between the a and c-axes. In this paper, we present our results of electrical resistivity and magnetic susceptibility measurements of the CePdIn and UPdIn single crystals and discuss about the difference on their anisotropic.behavior.

Fig. 1 shows the temperature dependence of electrical resistivities along the a and c-axes for CePdIn and UPdIn, respectively. In CePdIn, both the a and c-axes resistivities, ρ_a and ρ_c, exhibit a double-peak structure characteristic of the Kondo lattice system. However, we notice that the a-axis resistivity ρ_a is about two times larger than ρ_c in magnitude and the double-peak structure is much sharper in ρ_a than in ρ_c. In a valence-fluctuating CeNiIn with the same Fe_2P-type structure (7), we observed further significantly different behavior between ρ_a(T) and ρ_c(T); ρ_a has a single-peak around 120 K and a lnT dependence in high temperature region, whereas ρ_c decreases with decreasing temperature like metal without showing any anomaly. These results indicate that the Kondo scattering is more significant in the plane perpendicular to the c-axis than along the c-axis, which may come from anisotropic hybridizations of 4f-electrons with ligand s, p or d-electron states.

For UPdIn, we notice that a much more remarkable anisotropy in the resistivity is observable along the a and c-axis: ρ_a rapidly decreases near T_N=21 K with decreasing temperature, exhibits another sharp drop around T_C=8.5 K and tends to decrease at 1.5 K followed by a T^2-law below 8 K. In contrast, ρ_c starts to increase near 50 K, rapidly turns up near T_N and reaches to saturation at 1.5 K (6). This suggests that a pseudogap over the Fermi surface along the c-axis is formed by appearance of the c-axis modurated structure, while the coherent Kondo-state develops in the c-plane below 4 K through anisotropic hybrizations.

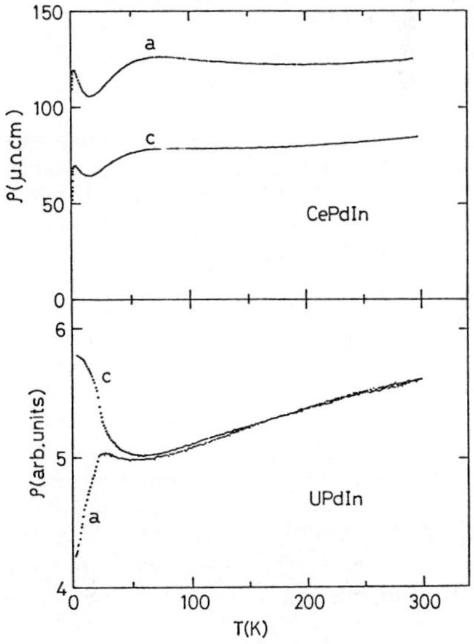

FIGURE 1
Temperature dependence of the electrical resistivity along the a and c-axes of the Fe_2P-type hexagonal structure for CePdIn and UPdIn.

0921-4526/90/$03.50

Such a significant difference in the transport properties between CePdIn and UPdIn might be ascribed to anisotropic hybridizations which come from the differences both of the spatial extension of the Ce-4f and U-5f wave functions and of the shape in their spatial extension around the quatization axis J in the ground state. In order to obtain information on the shape of their spatial extensions in the ground state, we measured magnetic susceptibilities on single crystals of CePdIn and UPdIn. The results are compared in Fig. 2. For CePdIn, only a small anisotropy is seen in χ^{-1} along the a and c-axes, in which χ^{-1} follows the CW law above 100 K. The effective number of Bohr's magneton is estimated as 2.61 μ_B per Ce atom for both the axes. The paramagnetic Curie temperatures along the a and c-axes are θ_a=-65 K and θ_c=-43 K, respectively. The difference between θ_a and θ_c, $\Delta\theta_p=\theta_c-\theta_a$=22 K, gives a measure of uniaxial anisotropy energy. As is shown in the inset of Fig. 2(a), we notice that χ_c has a small peak at T_N=1.8 K, but χ_a exhibits no anomaly around 1.8 K.

On the other hand, for UPdIn, a huge anisotropy appears in both M and χ^{-1} along both the axes. The ferromagnetic moment along the c-axis disappears at T_C=8.5 K, where the a-axis magnetization displays a small cusp. The χ^{-1} along both the axes follows the Curie Weiss relation above 100 K with $n^a_{eff}=n^c_{eff}$=2.87 μ_B, θ_a=-74 K and θ_c=34 K, respectively. The difference between θ_a and θ_c is obtained as $\Delta\theta_p$=108 K, which is very large compared to that for CePdIn.

The anisotropy in χ along the a and c-axes is considered to originate in both the CEF effects and the anisotropic hybridizations. In order to determine the crystalline electric field (CEF) parameters in this system, we therefore measured and analyzed the magnetization and susceptibility on a NdPdIn single crystal because NdPdIn has stable $4f^3$-electrons without any hybridization effects. Then, we used the effective single-ion Hamiltonian, including the CEF, molecular field and applied magnetic field, and determined the CEF parameters from a least-square fitting of the data of M and χ^{-1} in NdPdIn. The best fit was obtained for the following parameters: B_2^0=-0.46 K, $B_4^0=3.1\times10^{-2}$ K, $B_6^0=-2.8\times10^{-5}$ K, $B_6^6=-5.0\times10^{-3}$ K and λ=9.6 mole/emu, where λ is the molecular field coefficient. If the CEF in CePdIn or UPdIn is the same as in NdPdIn, then we can deduce $\Delta\theta_p^{CEF}$=47 K for CePdIn or $\Delta\theta_p^{CEF}$=38 K for UPdIn according to a scaling law on the basis of the above results. We notice that, for CePdIn, the observed value of $\Delta\theta_p^{ob}$=22 K is much smaller than $\Delta\theta_p^{CEF}$ expected from the CEF effects, whereas for UPdIn, $\Delta\theta_p^{ob}$=108 K is much larger than $\Delta\theta_p^{CEF}$. These results indicate that the anisotropic hybridizations act subtractively in CePdIn but additively in UPdIn on the CEF effects in the anisotropy of χ, reflecting the difference of the shape in spatial extension of Ce-4f and U-5f electrons in the ground state.

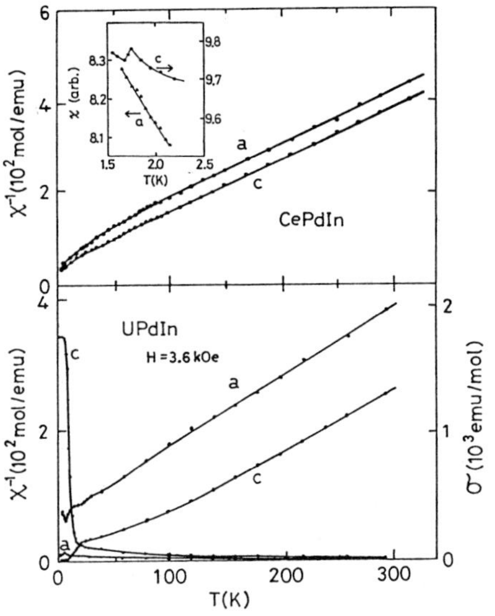

FIGURE 2
Temperature dependence of the magnetization, susceptibility and inverse susceptibility along the a and c-axes for CePdIn and UPdIn.

This work was supported by the project of priority area for the physics on actinide compounds, a Grant-in-Aid for Scientific Research from the Ministry of Education, Science and Culture of Japan.

REFERENCES

(1) H. Fujii, Y. Uwatoko, M. Akayama, K. Satoh, Y. Maeno, T. Fujita, J. Sakurai, H. Kamimura and T. Okamoto, Jpn. J. Appl. Phys. 26, Suppl. 26-3 (1987) 549.

(2) K. Satoh, Y. Maeno, T. Fujita, Y. Uwatoko and H. Fujii, J. de Phys. 49 (1988) C8-779, J. Phys. Soc. Jpn. 59 (1990) 692.

(3) E. Brück, F. R. De Boer, V. Sechovský and L. Havela, Europhys. Lett. 7 (1988) 177.

(4) H. Fujii, H. Kawanaka, T. Takabatake, E. Sugiura, K. Sugiyama and M. Date, to be published in J. Magn. Magn. Mat. (1990).

(5) T. Ekino, H. Fujii, M. Nagasawa, H. Kawanaka, M. Nishi, K. Motoya, T. Takabatake and Y. Ito, to be published in J. Phys. Soc. Jpn. (1990).

(6) H. Fujii, H. Kawanaka, M. Nagasawa, T. Takabatake, Y. Aoki, T. Suzuki, T. Fujita, E. Sugiura, K. Sugiyama and M. Date, to be accepted in J. Magn. Magn. Mat. (1990).

(7) H. Fujii, T. Inoue, Y. Andoh, T. Takabatake, K. Satoh, Y. Maeno, T. Fujita, J. Sakurai and Y. Yamaguchi, Phys. Rev. B39 (1989) 6840.

Physica B 165&166 (1990) 437-438
North-Holland

SEMICONDUCTING AND HEAVY-FERMION BEHAVIOR IN NEW CLASS OF MATERIALS OF $U_3T_3X_4$

T. TAKABATAKE, S. MIYATA, H. FUJII, Y. AOKI+, T. SUZUKI+ and T. FUJITA+

Faculty of Integrated Arts and Sciences, +Faculty of Science,
Hiroshima University, Hiroshima 730, Japan

Synthesis and results of transport, magnetic and calorimetric measurements are reported on $U_3T_3X_4$ (T= Ni,Cu,Pd,Pt,Au and X= Sn,Sb). The resistivities of $U_3T_3Sb_4$(T= Ni,Pd) are unusually high and show semiconducting behavior at temperatures above 200 K, whereas those of $U_3T_3Sn_4$ (T= Ni,Pt) are metallic and follow a T^2 law below 10 K. The low-temperature coefficients of specific heat of $U_3T_3Sn_4$ (T=Cu,Au) are much enhanced compared to that of $U_3Ni_3Sn_4$. The results indicate that the 5f-sp hybridization governs transport properties, whereas the degree of 5f-d hybridization is a crucial parameter for the develpoment of heavy-fermion state.

There has been considerable interest in ternary uranium compounds U-T-X (T= transition element and X= metalloid) because of their anomalous physical properties. The compounds of $U_3T_3Sb_4$ with T= Ni, Pd, Pt, Co, Rh and Ir are known to crystallize in the cubic $Y_3Au_3Sb_4$-type structure (1,2). Recently, we have reported that an isostructural compound $U_3Ni_3Sn_4$ is a moderately heavy-fermion system with γ = 92mJ/K^2molU (3). Through our investigation, we have found that $U_3T_3Sn_4$ with T= Cu, Pt and Au and $U_3Cu_3Sb_4$ crystallize with the same type structure. These isostructural compounds, therefore, provide an opportunity of a systematic study of the effect of hybridization of uranium 5f electrons with d- and sp-electrons on the physical properties. In this paper, our preliminary results are presented.

The samples of $U_3T_3Sn_4$ (T= Ni, Cu, Pt, Au) and $U_3T_3Sb_4$ (T= Ni, Cu, Pd, Pt) were prepared by arc melting the constituent metals under a purified flowing argon atmosphere. Since Sb has a high vapor pressure near the melting temperatures of the compounds, excess amount of Sb was added so as to compensate the weight loss. The samples were annealed at 800°C for 10 days. Powder X-ray diffraction analysis indicated all the samples to be almost single phased.

For three selected compounds of $U_3Ni_3Sn_4$ (a=9.380 Å), $U_3Cu_3Sn_4$ (9.522 Å) and $U_3Ni_3Sb_4$ (9.393 Å), the temperature dependences of electrical resistivity ρ and magnetic susceptibility χ are shown in Figs. 1 and 2, respectively. All the ρ(T) curves have negative slope near room temperature but exhibit considerablly different behavior at low temperatures. $U_3Ni_3Sn_4$ becomes metallic below 200 K, whereas ρ(T) of $U_3Cu_3Sn_4$ is almost independent of temperature down to 40 K and then increases with further decreasing temperature. For $U_3Ni_3Sb_4$, the absolute value of ρ at room temperature is 40 times larger than those of other two and it increases markedly like in a semiconductor. In fact, the temperature dependence above 200 K can be described by a simple activation law with an activation energy of 0.20 eV. However, at low temperatures below 30 K, the resistivity increases in proportion to lnT.

As reported previously (3), both χ(T) and the ratio of specific heat to temperature C/T of $U_3Ni_3Sn_4$ are strongly enhanced at low temperatures compared with those of usual uranium compounds. As is seen in Fig.2, χ(T) of $U_3Ni_3Sb_4$ and $U_3Cu_3Sn_4$ is still larger than that of $U_3Ni_3Sn_4$ and it follows the Curie-Weiss law at temperatures above 100 K with an effective

0921-4526/90/$03.50

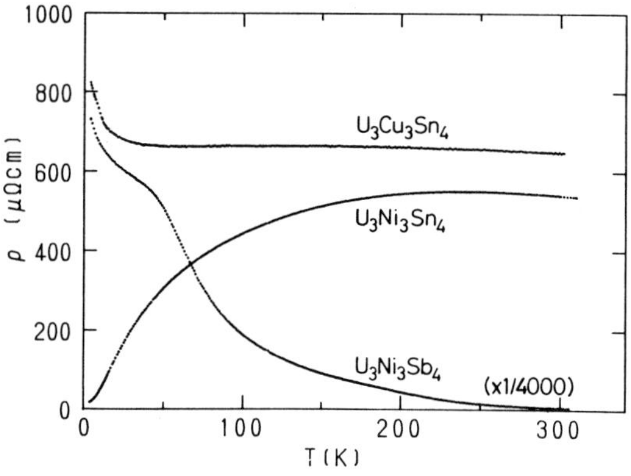

Fig.1 Electrical resistivity vs temperature of $U_3T_3Sn_4$(T= Ni,Cu) and $U_3Ni_3Sb_4$.

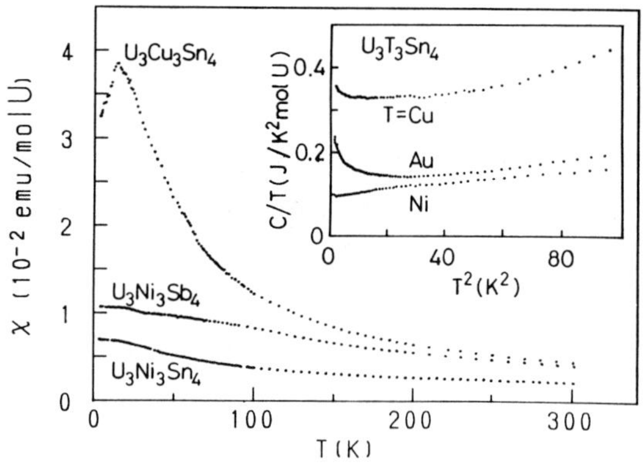

Fig.2 Magnetic susceptibility vs temperature of $U_3T_3Sn_4$(T= Ni,Cu) and $U_3Ni_3Sb_4$. Inset shows specific heat divided by temperature C/T vs T^2 of $U_3T_3Sn_4$(T= Ni,Cu,Au).

magnetic moment of 3.65 and 3.34 μ_B/Uatom, respectively. Furthermore, χ(T) of $U_3Cu_3Sn_4$ exhibits a peak at 14 K to mark an antiferromagnetic transition, which is confirmed by a λ-type anomaly in the specific heat. In the inset of Fig.2, low-temperature specific heats of $U_3T_3Sn_4$ for T= Ni, Cu and Au are displayed as C/T vs T^2. For the compounds with T= Cu and Au, the values of C/T are further enhanced than that of $U_3Ni_3Sn_4$. An upturn emerges in the C/T curves below 3 K and the extrapolation to 0 K yields 0.38 and 0.28 J/K^2molU for T= Cu and Au, respectively. It is worthy to note that C/T of $U_3Cu_3Sn_4$ is so largely enhanced even in the magnetically ordered state but ρ(T) sharply increases below T_N down to 4.2 K (see Fig.1). We note that similar increase in ρ(T) has been found for UCuSn below its T_N of 60 K (4).

The above experimental results demonstrate that the new class of compounds $U_3T_3X_4$ exhibit interesting properties such as semiconducting and heavy-fermion behavior depending on the combination of T and X atoms. For $U_3Ni_3Sn_4$, the ground state is a moderately heavy-fermion state, which changes into an antiferromagnetic heavy-fermion state by the replacement of Ni by Cu. The replacement gives rise to an increase both in the number of d-electrons and unit cell volume. This would lead to dehybridization of 5f states with d states as a result of lowering of the d-band with respect to the narrow 5f band. Thereby, an antiferromagnetically ordered heavy-fermion state as in $U_3Cu_3Sn_4$ may be stabilized. On the other hand, the number of sp-electrons in $U_3Ni_3Sn_4$ may be increased by the replacement of Sn by Sb, which has resulted in a semi-conducting behavior without a significant change in the unit cell volume. This fact implies that the degree of hybridization of 5f states with sp states is also an important parameter determining the physical properties of this class of compounds.

REFERENCES

1. A.E. Dwight, J. Nucl. Mat. 79(1979)417.
2. K.H.J. Bushow, D.B. de Mooij, T.T.M. Palstra, G.J. Nieuwenhuys and J.A. Mydosh, Philips J. Res. 40(1985)313.
3. T.Takabatake, H.Fujii, S.Miyata, H.Kawananaka Y.Aoki, T.Suzuki, T.Fujita, Y.Yamaguchi and J.Sakurai, J. Phys. Soc. Japan 59(1990)16.
4. H.Fujii, H.Kawanaka, T.Takabatake, E.Sugiura, K.Sugiyama and M.Date, to be published in J. Magn. Magn. Mater. 84(1990).

Physica B 165&166 (1990) 439–440
North-Holland

MAGNETIC SUSCEPTIBILITY OF UPt_3 IN HIGH MAGNETIC FIELDS

T. MÜLLER and W. JOSS

Max-Planck-Institut für Festkörperforschung, Hochfeld-Magnetlabor, B.P. 166 X,
F-38042 Grenoble-Cédex, France

L. TAILLEFER
Centre de Recherche sur les Très Basses Températures, CNRS,
B.P. 166 X, F-38042 Grenoble-Cédex, France

We measured the magnetic susceptibility on a high-quality single crystal of the heavy-fermion compound UPt_3 in the temperature range from 3.5 to 22 K in steady magnetic fields up to 28 T with the field direction along the *a* axis of the hexagonal structure. The maximum in the susceptibility at 17 K, due to spinfluctuations, becomes more pronounced with increasing magnetic field and shifts to 13 K at 10 T. The peak at the metamagnetic transition disappears smoothly with increasing temperature without changing the position at B_C=20 T. We propose a 'phase - diagram' for a crossover from the low temperature (T<17 K) low field (B<20 T) to the high temperature high field paramagnetic spinfluctuation state.

1. INTRODUCTION

The heavy fermion superconductor UPt_3 undergoes a field induced magnetic transition with a critical field of B_C=20 T (1). We investigated the correlation between this metamagnetic transition observed at low temperatures and a maximum in the susceptibility at 17 K (2) by ac-susceptibility measurements (f=117 Hz) in a continouus flow cryostat from 3.5 to 22 K. The magnetic field was always applied parallel to the *a* axis of the high-quality single crystal. The same sample was already investigated by de Haas-van Alphen measurements (3).

2. RESULTS

The temperature dependence of the magnetic susceptibility χ at several constant magnetic fields up to 10 T is shown in Figure 1. The curves are deplaced vertically by one arbitrary unit. The well-known maximum in χ (2) is due to spinfluctuations with a typical energy that is given by the characteristic temperature T^*=17 K of the maximum in a low magnetic field. Neutron scattering experiments (4) prooved, that the spinfluctuations are of the antiferromagnetic type. In higher external fields the maximum of χ shifts to T^*=13 K in 10 T and the maximum value of χ increases, which indicates that the characteristic temperature of the excitation spectrum is suppressed and the amplitudes of the spinfluctuations are enhanced. A measurement at 0.1 T, which was taken after having applied a field of 10 T, falls exactly on the curve shown here for 3.2 T. In previous publications (2) it was mentioned that the cusp in the susceptibility appears only under an applied minimal field of 1.2 T. We conclude that at low temperatures remanent magnetic contributions might be important.

The results of the high-field measurements

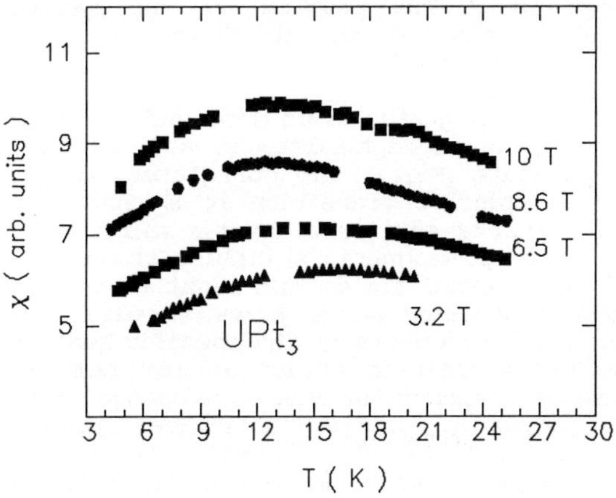

FIGURE 1
Temperature dependence of the ac-susceptibility of UPt_3, for magnetic fields of 3.2, 6.5, 8.6, and 10 T aligned parallel to the *a* axis of the single crystal. To separate the curves, they are displaced vertically by one arbitrary unit.

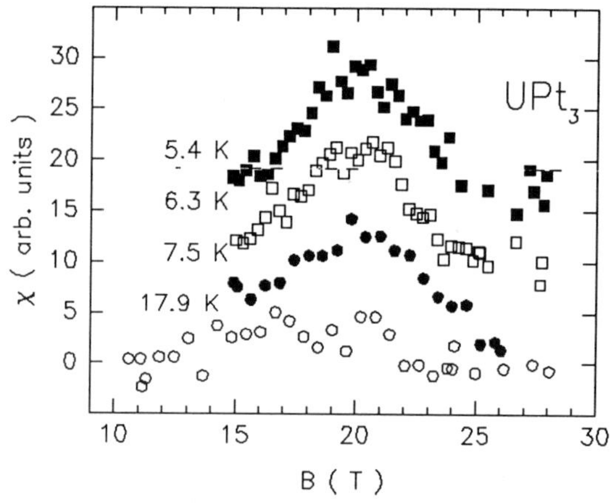

FIGURE 2
Magnetic field dependence of the ac-susceptibility of single cristalline UPt_3 at constant temperatures. With decreasing temperature, the maximum at the transition field B_C=20 T becomes more pronounced. The magnetic field, sweeped with a speed of 1 T/min, was aligned parallel to the *a* axis of the single crystal. The curves are displaced vertically by five units to separate them.

are shown in Figure 2. The curves are displaced by five arbitrary units to separate them. The maximum in χ at the metamagnetic transition becomes less pronounced with increasing temperature and has disappeard at about 18 K. The position of the maximum itself does not change with field.

3. DISCUSSION

The metamagnetic transition and the maximum in the susceptibility as a function of temperature (Fig. 1) are correlated. We propose a spinflop transition at B_C=20 T from the region of antiferromagnetic spinfluctuations to paramagnetic spinfluctuations in a strong external field. This might evolve to a phase transition at very low temperatures. Recent measurements of the specific heat and the magnetocaloric effect at low temperatures (5) support the idea of a change of the thermodynamic state at the critical field. At higher temperatures thermal fluctuations overwhelm the magnetic interactions and the transition becomes diffuse. Above a characteristic temperature of T^*=17 K in zero field, the system is in the paramagnetic state.

The importance of spinfluctuations for the heavy fermion behaviour of UPt_3 is reflected, for example, by the magnetic field dependence of the linear specific heat coefficient at low temperatures, which undergoes a pronounced maximum at the critical field of B_C=20 T (5).

Like other heavy fermion compounds, UPt_3 is very close to an antiferromagnetic instability. The dilution of the Pt by several percent of Pd leads to long range antiferromagnetism (6). Antiferromagnetic ordering with a critical temperature of T_N=5 K was observed in neutron scattering experiments (4). Here we find no sign for a phasetransition to long range order antiferromagnetism in the investigated temperature range.

In the heavy fermion system $CeRu_2Si_2$ a similar metamagnetic phase transition is observed at B_C=8 T (7). In this case neutron scattering measurements in magnetic fields up to 10 T clearly show the change of the spinfluctuation spectrum at B_C. In contrast to UPt_3 superconductivity has not been observed. The relation between the above described spinfluctuation phenomena and the superconductivity of UPt_3 at T_C=0.5 K (1) can only be explored by measurements at much lower temperatures.

In summary, we measured the susceptibility χ of UPt_3 in the vincinity of the metamagnetic phase transition at B_C=20 T. We believe that this transition and the cusp in the susceptibility in low fields at T^*=17 K are of the same origin, namely a transition of the antiferromagnetic spinfluctuations to paramagnetic spinfluctuations above 17 K and 20 T.

ACKNOWLEDGEMENTS

The authors acknowledge P. Wyder, J. Flouquet and A. de Visser for stimulating discussions. The work on the hybrid magnet was supported by P. Sala, J.-C. Vallier and C. Warth.

REFERENCES

(1) For a review see e. g. A. de Visser, A. Menovsky, and J.M.M. Franse, Physica 147B (1987) 81.
(2) P.H. Frings, J.J.M. Franse, F.R. de Boer, and A. Menovsky, J. Magn. Magn. Mat. 31 - 34 (1983) 240.
(3) L. Taillefer and G.G. Lonzarich, Phys. Rev. Lett. 60 (1988) 1570.
(4) G. Aeppli, E. Bucher, C. Broholm, K.J. Kjems, J. Baumann, J. Hufnagl, Phys. Rev. Lett. 60 (1988) 615.
(5) T. Müller, W. Joss, and L. Taillefer, Phys. Rev. B 40 (1989) 2614; H.P. van der Meulen, Z. Tarnawski, J. M. M. Franse, J.A.A.J. Perenboom, D. Althof and H. van Kempen, Physica B, to appear.
(6) A. de Visser, J.C.P. Klaasse, M. van Sprang, J.M.M. Franse, A. Menkovsky, and T.T.M. Palstra, J. Magn. Magn. Mat. 54 - 57 (1986) 375.
(7) P. Haen, J. Flouquet, F. Lapierre, P. Lejay, and G. Remenyi, J. Low Temp. Phys. 67 (1987) 391

Physica B 165&166 (1990) 441–442
North-Holland

HIGH FIELD SPECIFIC HEAT OF SINGLE-CRYSTALLINE $U(Pt_{0.95}Pd_{0.05})_3$

H.P.VAN DER MEULEN[a], J.J.M.FRANSE[a], A. DE VISSER[a], J.A.A.J.PERENBOOM[b] and H.VAN KEMPEN[b]

[a]Natuurkundig Laboratorium, Universiteit van Amsterdam, Valckenierstraat 65, 1018 XE Amsterdam, The Netherlands
[b]High Field Magnet Laboratory, University of Nijmegen, Toernooiveld, 6525 ED Nijmegen, The Netherlands

The hexagonal compound $U(Pt_{0.95}Pd_{0.05})_3$ shows both long-range antiferromagnetic order and high c/T-values at low temperature that are characteristic for heavy-fermion behaviour. Magnetic fields (of 12 T along the b-axis and 13 T along the a-axis) can suppress the ordering completely, but the large heavy-fermion contribution is still present at high fields.

1. INTRODUCTION

In the series $U(Pt_{1-x}Pd_x)_3$ superconductivity and long-range antiferromagnetic order occur besides spinfluctuation and Kondo phenomena (1). When substituting Pd for Pt in UPt_3, the superconductivity is suppressed and a state of long-range antiferromagnetic order (LRAF) is created (2). The antiferromagnetic order, of the spin-density wave type, is most pronounced in the 5 at% Pd-compound, that has a Néel temperature of 5.9 K. In the specific heat (c(T)) this ordering is reflected as a sharp peak superimposed on a large heavy-fermion background. Neutron scattering experiments (3) on single-crystalline $U(Pt_{0.95}Pd_{0.05})_3$ show that the ordered moments ($\sim 0.6\mu_B$/U-atom) are aligned along the b-axis. Specific heat measurements (4) in an applied field ($B \leq 5$ T) have shown a rather strong in-plane anisotropy: $T_N(B)$ decreases faster for B//b than for B//a.

As noted before (1,2) it is remarkable that the heavy-fermion behaviour is preserved as the LRAF order sets in. This suggests that only a part of the itinerant character of the f-electrons is lost and that in the ordered state strong fluctuations, probably of the Kondo type, persist (T_K decreases with increasing Pd contents for $x \leq 0.10$ (5)). Measurements of c(T) of pure UPt_3 in field show that c/T increases with B⊥c and has a maximum at 20 T where the metamagnetic transition occurs (6). However, in a field of 24.5 T, c/T is still larger than in 0 T. This shows that when the intersite correlations have collapsed at 20 T other fluctuations (possibly of the Kondo type) persist.

In this contribution we investigate the suppression of the long-range order in $U(Pt_{0.95}Pd_{0.05})_3$ in very large magnetic fields (B up to 20 T). We also investigate the heavy fermion part of c(T) as function of field.

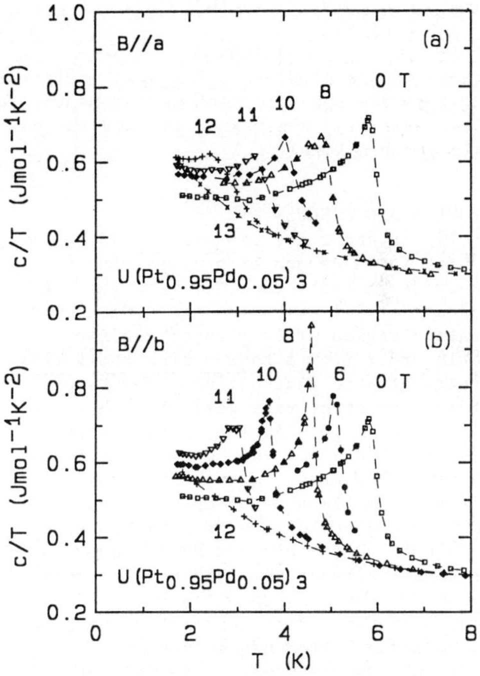

FIGURE 1
Specific heat of $U(Pt_{0.95}Pd_{0.05})_3$ with a) B//a and b) B//b, field values are indicated.

2. EXPERIMENTAL

The specific heat measurements were performed on a single-crystalline sample of a cubic shape (5x5x5 mm^3), cut out by spark erosion of a crystal grown by the Czochralski method in a tri-arc furnace. The high field specific heat measurements were carried out at the High Field Magnet Laboratory of the University of Nijmegen, using the 20 T Bitter-type of coil. The experiments were performed in an adiabatic way with a sapphire sample holder equipped with a ruthenium-oxide thermometer and a nickel-chromium film as a heater (6).

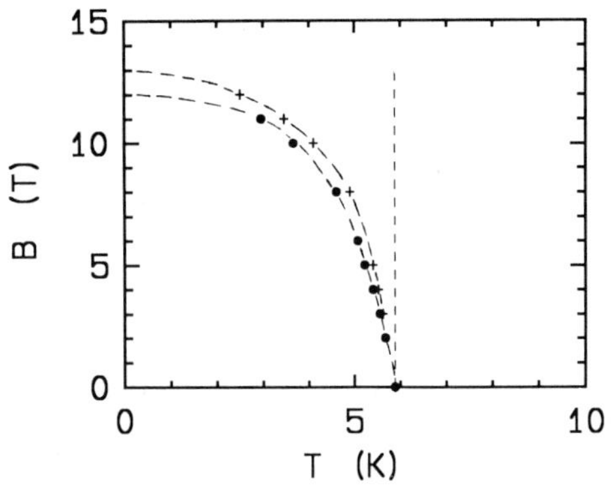

FIGURE 2
Phase diagram of $U(Pt_{0.95}Pd_{0.05})_3$, originating from the specific heat measurements, for the a-axis (+), the b-axis (•) and the c-axis (broken line).

3. RESULTS AND DISCUSSION

In Fig.1a and b the results ($B \leq 13$ T) are presented for B//a and B//b respectively. $T_N(B)$ is reported in Fig.2 together with the low field data of Ref.(4). The phase boundaries separate a region of long-range antiferromagnetic order from a region with short range antiferromagnetic fluctuations. The results obtained here agree very well with those obtained from magnetisation measurements. Fields directed along the hexagonal axis are known not to suppress T_N (4), this was confirmed by a field sweep up to 13 T at 5 K, in which no magnetocaloric effect was observed. Extrapolating $T_N(B)$ towards 0 K we obtain a total suppression of the LRAF ordered state at 13 T and 12 T for B//a and B//b respectively. A similar anisotropy for suppressing the LRAF could be deduced from magnetisation measurements where anomalies in M(B) at 1.3 K occur at 12.6 T and 11.8 T respectively (7).

A remarkable feature of the field measurements is seen in the shape of the ordering peaks. With the field along the b-axis the peak sharpens with increasing field up to 8 T, and looses its sharpness there above. This feature is absent for B//a. At present we cannot offer an explanation for this sharpening. On the other hand, the external field tends to break up the antiferromagnetic coupling thus causing the dominant overall suppression of T_N.

From the low-T part of our data it is clear that γ ($= \lim_{T\to 0} c/T$) remains large over the whole field range. Data at 20 T reveal a c/T of 450 $mJmol^{-1}K^{-2}$ at 2.25 K (8), which is still larger than the value for UPt_3 in 0 T. Thus the heavy-fermion behaviour is preserved in $U(Pt_{0.95}Pd_{0.05})_3$ at our maximum field.

Despite the difficulty to separate out the contribution of the LRAF order to c(T), we infer from our high-field data an initial increase of γ, as observed in pure UPt_3. This suggests that the scenario for UPt_3 (6) also applies to the remaining contribution of $U(Pt_{0.95}Pd_{0.05})_3$, i.e. the presence of competing magnetic interactions that can be separated into a Kondo on-site type of interaction and an inter-site interaction. It would be interesting to investigate at which field γ passes through a maximum by extending our temperature range towards lower temperatures. Presumably this occurs near ~12 T (7) compared to 20 T in UPt_3. The coexistence of three types of interactions demonstrates the complex nature of the electron-electron processes in $U(Pt_{0.95}Pd_{0.05})_3$.

ACKNOWLEDGEMENTS

This work was supported by the "Stichting Fundamenteel Onderzoek der Materie" (FOM) with financial support from the "Nederlandse Organisatie van Wetenschappelijk Onderzoek" (NWO). The work of A.d.V. has been made possible by a fellowship of the Royal Netherlands Academy of Arts and Sciences.

REFERENCES

(1) A.de Visser, A.Menovsky and J.J.M.Franse, Physica 147 B, 81 (1987).
(2) A.de Visser, J.C.P.Klaasse, M.van Sprang, J.J.M.Franse, A.A.Menovsky, T.T.M.Palstra and A.J.Dirkmaat, Phys.Lett. 113A (1986) 489.
(3) P.H.Frings, B.Renker and C.Vettier, J.Magn.Magn.Mat. 63&64 (1987) 202.
(4) M.van Sprang, R.A.Boer, A.A.Menovsky and J.J.M.Franse, Jpn.J.Appl. Phys. 26 (1987) 563; M.van Sprang, PhD thesis, University of Amsterdam (1989), unpublished.
(5) J.J.M.Franse, H.P.van der Meulen and A. de Visser, this conference.
(6) H.P.van der Meulen, Z.Tarnawski, J.J.M. Franse, A. de Visser, J.A.A.J.Perenboom, D.Althof and H.van Kempen, Phys.Rev.B, in print.
(7) A.de Visser, M.van Sprang, A.A.Menovsky, J.J.M.Franse, J.de Physique (Paris) 49-C8, (1988) 761.
(8) J.J.M.Franse, H.P.van der Meulen, A.A. Menovsky, A. de Visser, J.A.A.J. Perenboom and H.van Kempen, in Proc. Yamada Conf. on Magn. Phase Trans. (Osaka, April 13-16, 1990).

Physica B 165&166 (1990) 443–444
North-Holland

CRITICAL MAGNETIC BEHAVIORS IN $CeRuSn_x$ $(2.85 \leq x \leq 3.15)$

T. Fukuhara, I. Sakamoto and H. Sato

Department of Physics, Tokyo Metropolitan University, Setagaya-ku, Tokyo 158, Japan

$CeRuSn_3$ is reported to be a heavy fermion compound with an enormous coefficient of electronic specific heat $\gamma \sim 1.4$ J/mole·K^2. In order to gain a deeper insight into the heavy fermion state of $CeRuSn_3$, we have measured resistivity, magnetization, magnetic susceptibility, Hall effect and thermopower on $CeRuSn_x$ $(2.85 \geq x \geq 3.15)$. The low temperature resistivity shows a maximum at x=0 as a function of x. Sn deficient samples exhibit two phase transitions at 30K and 4K.

1. INTRODUCTION

The intermetallic ternary compound $CeRuSn_3$ crystallize in the cubic $Pr_3Rh_4Sn_{13}$ structure (1) (Pm3n space group). Our recent experiments (2) suggest that $CeRuSn_3$ is a new heavy fermion compound. According to Takayanagi et al.(3), the heat capacity exhibit only a small bump around 0.5K. The electronic specific heat coefficient reaches a maximum of about 1.4 J/mole·K^2 at 0.6K.

Transport properties of $CeRuSn_3$ show anomalous behaviors. As temperature decreases the electrical resistivity (ρ) increases continuously down to 0.7K without showing coherence effect. The Hall coefficient (R_H) also increases down to 1.7K as expected from the skew scattering by independent Ce impurities. We suspect that the coherence is destroyed by the random occupation of original Sn site (24k in Wykoff notation). According to our study, the $Pr_3Rh_4Sn_{13}$ structure is stable over a wide range of nonstoichiometric compounds. In order to get a better understanding about the heavy fermion state in $CeRuSn_3$, we have performed transport and magnetic measurements in the series $CeRuSn_x$ where $2.85 \leq x \leq 3.15$.

2. EXPERIMENT

The polycrystalline $CeRuSn_x$ samples were prepared by arc-melting appropriate amounts of the constituent elements (Ce 99.9%, Ru 99.98%, Sn 99.999%) on a water cooled cooper hearth. The ingots were wrapped in Ta and Zr foils and annealed at 950°C for 3 days in a evacuated quartz tube. The samples obtained were analysed by X-ray diffraction. No parasitic phase was detected in the sample of x=3, while X-ray diffraction patterns for $x \neq 3$ indicate very weak additional reflections. Single crystals of $CeRuSn_3$ have been grown using the Czochralski method in a SELEC triarc furnace. We have not observed any significant difference in bulk properties between single crystals and polycrystals.

3. RESULTS AND DISCUSSION

Figure 1 shows the temperature dependence of ρ for $CeRuSn_x$ $(2.85 \leq x \leq 3.15)$. Between 80K and 300K, $\rho(x=3.0)$ exhibits a logarithmic increase with decreasing temperature due to the incoherent Kondo scattering.

In the Sn excessive cases, the absolute value of ρ becomes smaller with increasing x, while the temperature dependence of ρ for x=3.03 and 3.09 is not so much different from that for x=3.0. The excess of 5 at% Sn (x=3.15) leads to drastic decrease of ρ at low temperature. In this sample the extraction of pure Sn was confirmed by X-ray diffraction, and the abrupt drop of $\rho(x=3.15)$ at 3.7K is due to the superconducting transition of Sn impurity. In the Sn deficient cases, ρ becomes smaller with decreasing x, and exhibits two step anomalies reflecting some kind of phase transitions near 33K and 4K. As to x dependence of the low temperature resistivity, a maximum was observed at x=3.0. This fact partly supports that the low temperature increase of ρ for $CeRuSn_3$ is not due to the random occupancy of Sn site as opposed to the above-mentioned suspicion.

Figure 2 shows the magnetic susceptibility χ of $CeRuSn_x$ (x=3.0, 2.91 and 2.85) down to 1.8K.

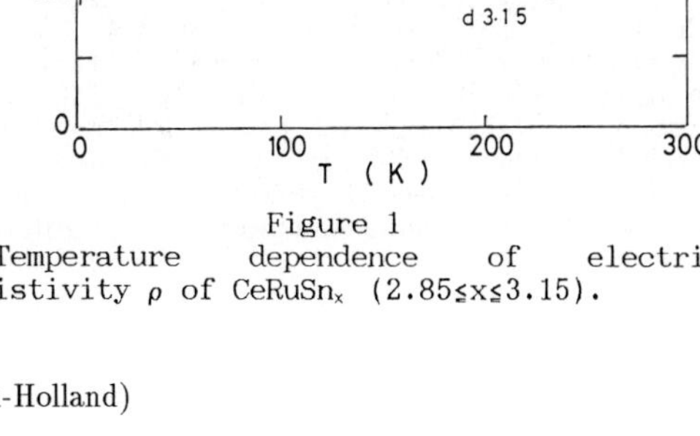

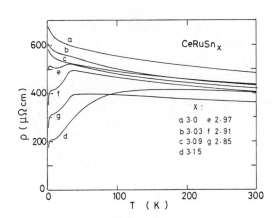

Figure 1
Temperature dependence of electrical resistivity ρ of $CeRuSn_x$ $(2.85 \leq x \leq 3.15)$.

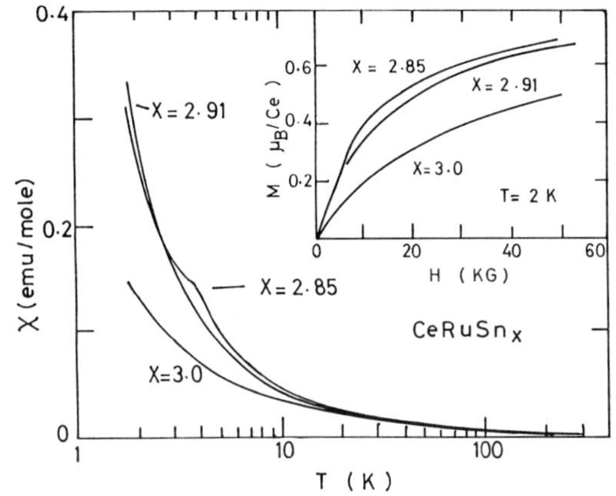

Figure 2
Temperature dependence of magnetic susceptivility χ of $CeRuSn_x$. The inset shows field dependence of magnetization at 2K.

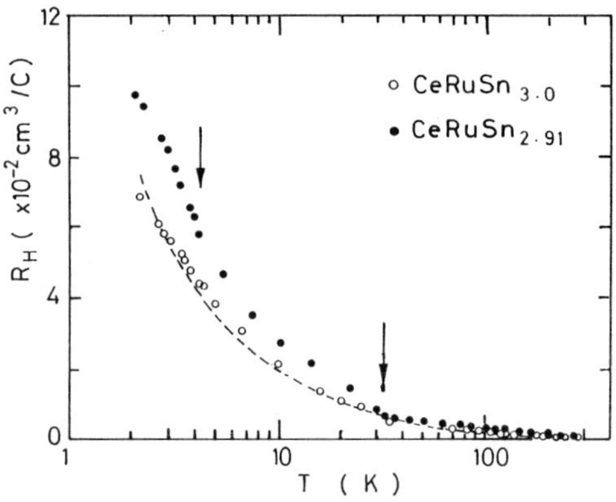

Figure 3
Temperature dependence of Hall coefficient R_H of $CeRuSn_{3.0}$ and $CeRuSn_{2.91}$ at 10KG. The broken curve is calculated from the skew scattering model.

absolute value of $\chi(x=3)$ at 1.8K is large in comparison with those of typical heavy fermion compounds, and is rather close to those of Kondo compounds with magnetic order (such as CeB_6 or $CeCu_2$). It suggests that the possibility of some magnetic order at lower temperature in $CeRuSn_3$. In the Sn deficient samples, χ is clearly enhanced compared to x=3.0 sample. The enhancement is especially evident below 30K, but no significant kink or bend can be seen around 30K. Corresponding to the lower temperature anomaly in ρ, the bump at 4K in χ-T curve gives a clear evidence of a magnetic transition for x=2.85. For x=2.91, χ-T curve only shows a small change in slope around 4K.

The inset in figure 2 shows the field dependence of magnetization (M) at 2K. M becomes larger with decreasing Sn content. M(x=2.85) also shows a metamagnetic behavior at about 6 KG. This fact together with the small bump in χ-T curve suggest that an antiferromagnetic long range order develops below 4K in $CeRuSn_{2.85}$.

Figure 3 shows the temperature dependence of the Hall coefficient of $CeRuSn_3$ and $CeRuSn_{2.91}$ measured under 10 KG. The broken curve is calculated from the recent theories based on the skew scattering by independent Ce impurities, and is proportional to $\rho_m\tilde{\chi}$ (4) or $\rho\tilde{\chi}(1-\tilde{\chi})$ (5). Here ρ_m and ρ is the magnetic and the bulk resistivity and $\tilde{\chi}$ is the normalized susceptibility. $R_H(x=3.0)$ can be fitted well with the calculated curve down to 2K. This suggests that the incoherent Kondo scattering is a dominant process in anomalous Hall effect even at such a low temperatures. As T decreases, $R_H(x=2.91)$ shows slight up turn around 4K reflecting the enhanced skew scattering due to the onset of antiferromagnetic ordering. The similar upturn is also seen at 30K, though it is not so clear as low temperature one.

Even at 300K, $R_H(x=2.91)$ is about 60% larger than $R_H(x=3.0)$. This fact implys some change of carrier concentration, since the estimated anomalous part $\rho_m\tilde{\chi}$ can not explain the difference. It is reasonable to expect that a deficiency of Sn will cause a change of the number of electrons in the conduction band.

To summarize, the competition between RKKY interaction and Kondo effect is very critical in $CeRuSn_3$. The deficiency of Sn induces two step phase transitions at 4K and 33K. The lower temperature one is an antiferromagnetic ordering, while the higher temperature one is not clear at present stage. The microscopic measurements, such as neutron scattering, are necessary to clean up its origin.

ACKNOWLEDGEMENTS
The authors are grateful to Professors S. Takayanagi and Y. Ōnuki and Dr I. Shiozaki for their helpful discussion.

REFERENCES

(1) P. Eisenmann and H. Schäfer, J. Les-Com. Met. 123 (1986) 2734.
(2) T. Fukuhara, I. Sakamoto, H. Sato, S. Takayanagi and N. Wada, J. Phys. Condens. Matter 1 (1989) 7487.
(3) S. Takayanagi, T. Fukuhara, H. Sato, N. Wada and Y. Yamada, this volume.
(4) A. Fert, A. Hamzic and P.M. Levy, J. Magn. & Magn. Matter 63&64 (1987) 535.
(5) P. Coleman, P.W. Anderson and T.V. Ramakrishnan, Phys. Rev. Lett. 55 (1985) 414.

Physica B 165&166 (1990) 445–446
North-Holland

MAGNETIC AND TRANSPORT PROPERTIES OF NEW URANIUM COMPOUNDS U_3Sn_5 AND $URuSn_x$ +

H. Sato, T. Fukuhara, I. Sakamoto and Y. Onuki*

Department of Physics, Tokyo Metropolitan University, Setagaya-ku, Tokyo 158, Japan
*Institute of Materials Science, University of Tsukuba, Tsukuba, Ibaraki 305, Japan

The electrical resistivity, the thermoelectric power, and the magnetization have been measured below 300K. U_3Sn_5 shows ferromagnetic transitions near 70K, while $URuSn_x$ shows no magnetic transition down to 1.8K. Above the transition temperature, the resistivity of U_3Sn_5 increases with decreasing temperature reflecting Kondo scattering. At low temperatures, U_3Sn_5 shows peculiar magnetic behaviors.

1. INTRODUCTION

Recently uranium intermetallic compounds have attracted much attention because of their wealth of different magnetic and electrical behaviors from the complete Pauli itinerant to the localized magnetic moment including heavy fermion state (1). It is believed that the direct 5f-5f overlap and 5f-ligand hybridization are the most important factors. We recently investigated a large 4f-4f distance Ce compound $CeRuSn_3$ (2), which was found to show heavy-fermion characters.

In this paper we report our trial to make $URuSn_x$ type compound. We also report the transport and the magnetic properties of U_3Sn_5, since we believe it important when we explore uranium ternary stannides. The physical properties of U_3Sn_5 has not been reported because of the pyrophoric nature of Sn-U alloys at intermediate compositions, though its existence is settled (3).

2. EXPERIMENT

The samples were prepared by arc melting appropriate amount of the constituent elements (99.9% U, 99.99% Ru and 99.999% Sn), and were turned over and remelted several times to ensure homogeneity. In case of $URuSn_x$, a single crystal was grown by SELEC triarc furnace. They were wrapped in Ta foil plus Zr foil and sealed in evacuated quartz tubes, and annealed at 950 C for three days. As to U_3Sn_5, a sample annealed for three weeks and an as cast one are also prepared. In a single crystal $URuSn_3$, only sharp X-ray diffraction peaks for the type III structure reported by Vandenberg (4) for RRh_xSn_y (R=rare earth) were observed except very weak Sn reflections, though small amount of Sn and URuSn was generally inevitable as impurity phase in polycrystals. We have not succeeded in obtaining reasonable X-ray diffraction pattern for U_3Sn_5 because of its pyrophoric nature. Electrical resistivity (ρ) was measured using a standard four probe dc technique and the magnetization (M) measurements were performed with a SQUID magnetometer.

+ supported by the project of priority area for the physics on actinide compounds, a Grant-in-Aid for Scientific Research from the Ministry of Education, Science and Culture of Japan.

3. RESULTS AND DISCUSSION

3.1 U_3Sn_5

Figure 1 shows the temperature dependences of ρ for U_3Sn_5. The temperature dependences of ρ on the samples annealed for three days and for three weeks are fundamentally same. It increases with decreasing temperatures down to near 70K (T_d) where it drops sharply. The higher temperature part of curve is reminiscent of the single ion Kondo effect. In the as cast sample, the temperature dependence is largely different. $d\rho/dT$ at higher temperature is positive, and the residual resistivity is an order of magnitude larger. T_d is, however, unaffected by the heat treatment. The smallness of the residual resistivity on the annealed samples (4.8 to 13 $\mu\Omega$cm) implys good crystalinity of these samples. Below about 20K

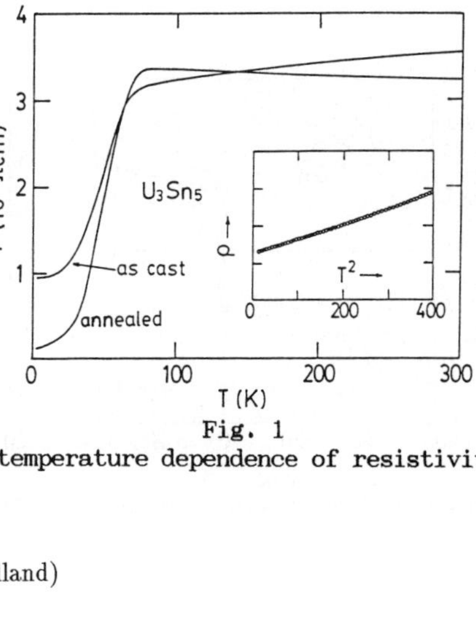

Fig. 1
The temperature dependence of resistivity for U_3Sn_5.

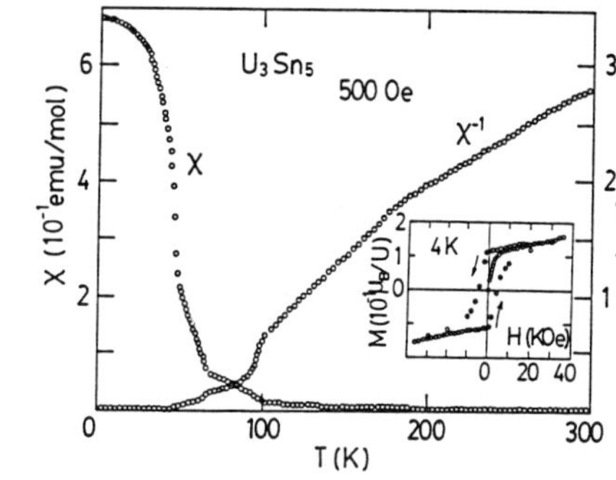

Fig. 2
The temperature dependence of susceptibility, and magetization at 4K.

(inset in Fig. 1), ρ approximately follows T^2 law which is characteristic of spin fluctuation or a Fermi liquid. Also in thermoelectric power, we observed a knick near T_d.

The magnetic susceptibility (χ) and χ^{-1} are shown in figure 2. Above 180K, χ follows Curie-Weiss law, and effective magnetic moment P_{eff} and Curie temperature Θ_p are estimated to be $3.06\mu_B$ and -30K respectively. At lower temperatures, χ-T curve has two shoulders near 100K and 70K, and the latter position agrees with T_d. Above 70K, M-H curve is linear up to 55KOe. χ-T curve have a bend also near 50K below which the remnant magnetization starts to grow. These facts suggest that U_3Sn_5 is ferromagnetic at least below 50K.

At the lower temperatures, M-H curve shows interesting behaviors. As is shown in figure 2, M shows a sharp knick in the hysteresis curve at 4K. At 1.8K, M shows more complex metamagnetic like dependence below 10KOe and near 40KOe (Fig. 3). One possible explanation at present stage is that U_3Sn_5 is a very anisotropic ferromagnet and all the moments in a grain flip spin in a group. A single crystal study is necessary in order to clear up its origin.

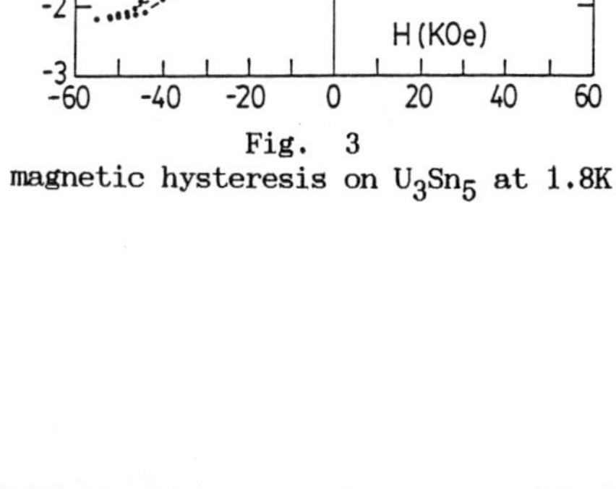

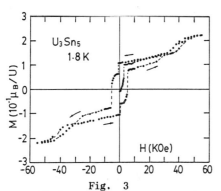

Fig. 3
The magnetic hysteresis on U_3Sn_5 at 1.8K.

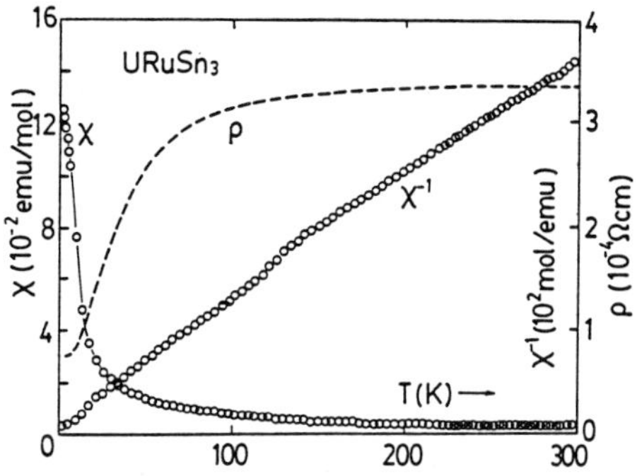

Fig. 4
The temperature dependence of susceptibility and resistivity for $URuSn_3$ single crystal.

3.2 $URuSn_x$

In polycrystalline samples, we always observed some sign of ferromagnetic ordering at 53K due to the impurity phase URuSn.

The temperature dependence of ρ for a single crystal (Fig. 4) shows a type of curve that is characteristic for many actinide compounds: a sharp rise at low temperature and tendency to saturate at large values near room temperature, which is ascribed to strong conduction electron scattering by the spin fluctuations. There is no sign of magnetic phase transition at least down to 3.7K where ρ shows superconductivity due to impurity Sn.

From the higher temperature slope of χ^{-1}-T curve (Fig. 4), $\chi_{eff} = 2.77\mu_B$ and Θ_p =-45K. There is also no clear sign of magnetic ordering down to 1.8K. One of the characteristics of this system is the large magnetic moment at low temperatures. This suggests the possibility of some magnetic order at the lower temperature. In order to clarify the low temperature ground state, the study on the lower temperature specific heat is necessary.

ACKNOWLEDGMENTS

We would like to thank Professor T. Komatsubara Dr. I. Shiozaki and Dr. A. Shinogi for useful discussions.

REFERENCES

(1) Z. Fisk, J.L. Smith, H.R. Ott and B. Batlogg, J. Mag. & Mag. Mat. 52 (1985) 79.
(2) T. Fukuhara, I. Sakamoto, H. Sato, S. Takayanagi and N. Wada, J. Phys.: Condens. Matter, 1 (1989) 7487.
(3) R.I. Sheldon, E.M. Foltyn and D.E. Peterson, Bull. of Alloy Phase Diagrams, 8 (1987) 347.
(4) J.M. Vandenberg, Mat. Res. Bull., 15 (1980) 835.

Physica B 165&166 (1990) 447–448
North-Holland

HEAVY FERMION BEHAVIOUR IN $CeRuSn_3$ COMPOUND

Shigeru TAKAYANAGI, [a]Tadasi FUKUHARA, [a]Hideyuki SATO, [b]Nobuo WADA, [c]Yuh YAMADA

Physics Department, Sapporo Campus, Hokkaido-Kyouiku University, Sapporo, 002, Japan.
[a]Department of Physics, Tokyo Metropolitan University, Setagaya-ku Tokyo 158, Japan.
[b]Department of Physics, College of Arts and Science, University of Tokyo, Komaba, Tokyo 153, Japan.
[c]National Research Institute for Metal, Nakameguro, Tokyo 153, Japan

The electrical resistivity under pressure (up to 23 kbar), magnetic susceptibility and specific heat of $CeRuSn_3$ with cubic Pm3n structure were measured between 30 mK and 300 K. The magnetic scattering of the resistivity behaves like a dilute Kondo type. Below 1 K, the ratio C/T remarkably increases and reaches a broad maximum around 0.5 K. The magnetic susceptibility also shows a maximum at the same temperature. These results indicate a antiferromagnetic-tpye ordering induced by the RKKY-interaction between 4f-electrons below 0.5 K. $CeRuSn_3$ is a new heavy fermion compound with specific heat coefficient of 1.67 $J/K^2\cdot mol$ at 0.6 K.

1. INTRODUCTION

The f-atom spacing in heavy fermion systems is an important parameter, presumably dominated the effective width of the hybridized f band. Heavy fermion compounds have f-f distance greater than 4 Å beyond which Hill point out that f-f overlap ceases and magnetism occurs. For example, $CeAl_3$ and $CeCu_6$ have relatively large Ce-Ce distance of 4.43 Å and 4.83 Å respectively. Almost all the cubic Kondo compounds, such as CeB_6 and $CeAl_2$, order magnetically at low temperatures except $CeInCu_2$ which have been recently reported to be a heavy fermion compound (1). $CeInCu_2$ also has large Ce spacing 4.80 Å which is close to that of $CeCu_6$.

The existence of compounds with the formula $RRuSn_3$ (R=La,Ce,Pr,Nd) was reported by Eisenmann and Schafer (2). $CeRuSn_3$ crystallize in cubic structure ($Pr_3Rh_4Sn_{13}$ structure: Pm3n), and Ce spacing is 4.86 Å. Recently, we have found this compound to be a new heavy fermion system with a large electronic specific heat coefficient (3) such as that of $CeCu_6$. It is of interest to know the ground state properties of $CeRuSn_3$ in detail. In this letter we report the extensive experimental studies of resistivity, magnetic susceptibility and specific heat.

2. EXPERIMENTAL

Polycrystals of $CeRuSn_3$ and $LaRuSn_3$ were obtained under argon atmosphere by arc melting stoichiometric amounts of 99.9%-Ce, -La and 99.98%-Ru. The lattice constants obtained by X-ray analysis agree with literature values $CeRuSn_3$: 9.73 Å, $LaRuSn_3$: 9.77 Å. Single crystal samples were grown by the Czochralski pulling method using a tri-arc furnace. Samples were annealed at 950 C for 3days. The electrical resistivity was measured under hydrostatic pressures up to 23 kbar using a standard four-lead method. The pressure was generated by means of a Cu-Be clamp type piston-cylinder apparatus. The specific heat measurements were carried out between 50 mK and 50 K using two adiabatic calorimeters; one was for 2-50 K measurements and the other was mounted in a dilution-refrigerator for low temperature measurements (50 mK-2 K).

3. RESULTS AND DISCUSSIONS

Figure 1 shows the temperature dependence of the electrical resistivity for $CeRuSn_3$ and

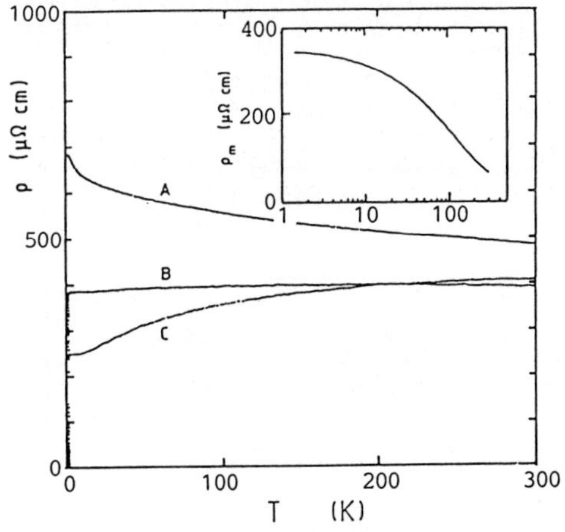

Figure 1. Temperature dependence of electrical resistivity of $CeRuSn_3$(A) and $LaRuSn_3$: as-grown(B), annealed(C). Inset shows temperature dependence of magnetic resistivity $\rho_m = \rho(CeRuSn_3) - \rho(LaRuSn_3)$

0921-4526/90/$03.50 © 1990 – Elsevier Science Publishers B.V. (North-Holland)

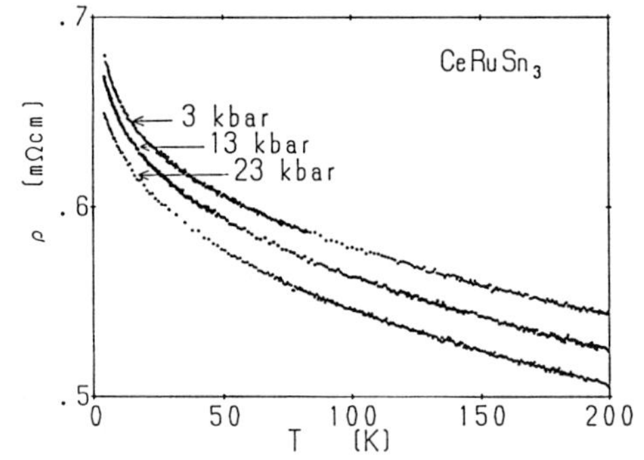

Figure 2. Temperature dependence of electrical resistivity under the hydrostatic pressure.

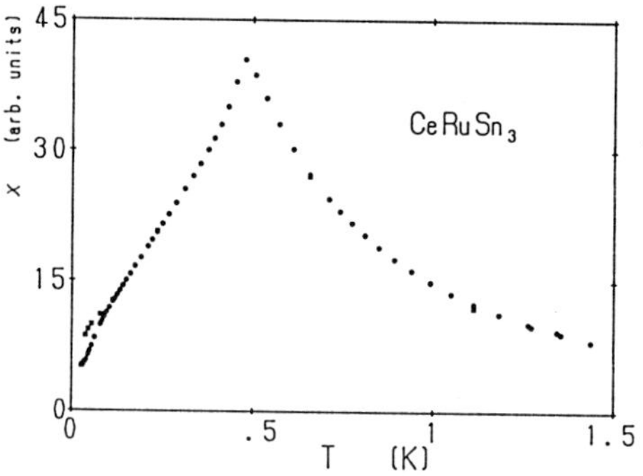

Figure 4. Temperature dependence of the ac-magnetic susceptibility below 1.5 K.

$LaRuSn_3$. We plotted magnetic resistivity, which looks like a dilute Kondo type, in the inset of figure 1. The ρ_m is proportional to -logT between 300 K and 70 K. $LaRuSn_3$ shows a transition into a superconducting state at T_c=1.5 K.

We discuss why the magnetic resistivity of $CeRuSn_3$ does not show a decrease due to the coherence effect at low temperatures. For heavy fermion systems, the coherence effect is destroyed easily by the existence of atomic disorder (1) and impurity. For example, in the case of $Ce_xLa_{1-x}Cu_6$ system, the Kondo lattice is formed above x=0.8. Figure 2 shows the pressure dependence of the resistivity of $CeRuSn_3$ between 1.7 K and 200 K. This behavior is similar to the result for amorphous system. It may indicates the existence of atomic disorder for $CeRuSn_3$ compound.

The C/T of $CeRuSn_3$ has a minimum around 7 K then increases at lower temperatures without any sign of phase transition down to 0.6 K (3). But, at 5 K, as shown in figure 3, the specific heat coefficient C/T reaches a maximum of 1.67 J/K^2mole which is comparable with those of $CeCu_6$, $CeAl_3$, and $CeInCu_2$. The C/T vs T curve has close resemblance to that of $CeInCu_2$ though C/T shows saturation below 1 K in $CeInCu_2$.

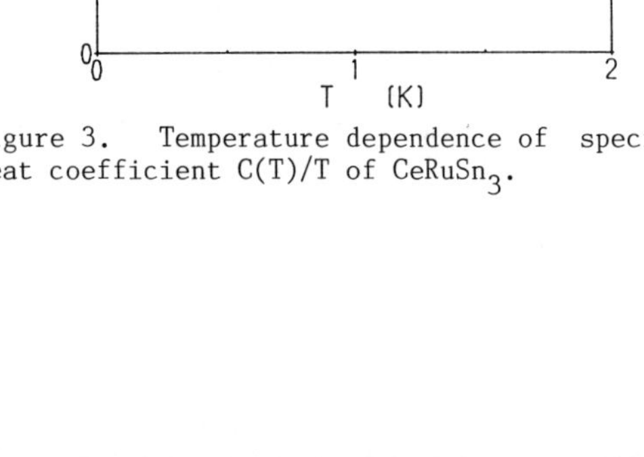

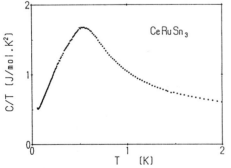

Figure 3. Temperature dependence of specific heat coefficient C(T)/T of $CeRuSn_3$.

The susceptibility follows the Curie-Weiss law above 100 K. The paramagnetic Curie temperature is -38 K and the effective Bohr magneton is 2.5μ_B/mole which is close to the value of Ce^{+3}. The susceptibility shows a large value of $1.5x10^{-1}$ emu/mole at 1.7 K, which is about five times as large as the values that have observed for $CeCu_6$ and $CeAl_3$ and $CeInCu_2$.

The ac-susceptibility measurement, as shown in fig. 4 , below 1.5 K reveals a sharp maximum around 0.5 K suggesting some antiferromagnetic ordering. But, the behaviour is very similar to the formation of a spin glass ordering. This is presumably caused by the presence of the atomic disorder of Ce-ions.

To summarize, the present experimental results suggest $CeRuSn_3$ is a new fermion compound which displays some (antiferromagnetic or spin glass) ordering below 0.6 K. Also it indicates that the atomic disorder plays a important role in the magnetic property of the heavy fermion $CeRuSn_3$ at lower temperature. For the purpose, measurements on samples with controlled stoichiometry and under the lower temperature are now in progress.

ACKNOWLEDGMENTS

The authors are grateful to Professors T. Komatsubara, Y. Onuki, K. Yonemitsu, and Dr. I. Shiozaki for their helpful discussion. This work was supported by Grant-In-Aid for the Scientific and Culture in Japan and by the Supply Center for low temperature liquids at Hokkaido University.

REFERENCES

(1) Y. Onuki, T. Yamazaki, A. Kobori, T. Omi, T. Komatsubara, S. Takayanagi, S. Kato and N. Wada, J. Phys. Soc. Jpn. 56 (1987) 4251.
(2) B. Eisenmann and H. Schafer, J. Less Common Met. 123 (1986) 89.
(3) T. Fukuhara, I. Sakamaoto, H. Sato, S. Takayanagi and N. Wada, J. Phys. C. 1 (1989) 7487.

Physica B 165&166 (1990) 449–450
North-Holland

TWO-DIMENSIONAL FERROMAGNETISM IN METALLIC FILMS

E.V. LINS DE MELLO and M.A. CONTINENTINO

Instituto de Física, Universidade Federal Fluminense, Outeiro de São João Batista s/n, Niterói, Cep 24020, RJ - Brazil

We suggest that uncoupled ferromagnetic films in the monolaver regime provide a physical realization of the two-dimensional anisotropic Heisenberg ferromagnet. Using a real space renormalisation group approach we calculate the spontaneous magnetisation of this model system. The calculations account for the observed slow decrease of the magnetisation at low temperatures, with reasonable values of the anisotropy, and also for it's sharp drop close to T_c.

1. INTRODUCTION

Recent work on ultrathin ferromagnetic films has revealed interesting features of the magnetic behaviour of these systems (1). Differently from the case the films are weakly coupled to a ferromagnetic bulk, the uncoupled films in the monolayer regime show a much slower decrease of the magnetisation with temperature which cannot be described by spin wave theory (2,1). In fact the interpretation of the experimental results seems to imply a quenching of the long wavelength magnons in the uncoupled films and probably the existence of a very large magnetocrystalline anisotropy (1). Another important feature of the magnetic behaviour of the uncoupled films is the sharp drop in the magnetisation as the critical temperature is approached.

The quenching of low energy magnons in ferromagnets of rare-earths or actinide elements, due to magneto-crystalline anisotropy, is a normal feature of these systems (3). Spin wave gaps a order 10^2K give rise to exponential terms in the magnetic contribution to the low temperature thermodynamic properties of these materials (3). On the other hand such large anisotropies on ferromagnets of transition metal elements are completely unusual although this is just the order of magnitude of the gaps required to describe the behaviour of the magnetisation in the uncoupled films (4).

In this article we suggest that the uncoupled films of transition metal elements in the monolayer regime provide a physical realization of the two-dimensional anisotropic Heisenberg ferromagnet. In order to substantiate our point of view, we have used the real space renormalisation group to obtain the spontaneous magnetisation of this model as a function of temperature.

2. FORMALISM

The Hamiltonian describing the system is:

$$H = \sum_{i,j} [W(\sigma_i^x\sigma_j^x + \sigma_i^y\sigma_j^y) + K\,\sigma_i^z\sigma_j^z] + H\sum_i \sigma_i^z + NC \qquad (1)$$

where the σ's are Pauli matrices for neighbour sites in a two dimensional hierarchical cell. $K = J/k_BT$, $W = J(1-\Delta)/k_BT$ and $H = \mu B/k_BT$. B is the external magnetic field, μ the magnetic moment per site, Δ the anisotropy factor, C a constant and N the total number of sites. The renormalization group method used to study the Hamiltonian given by Eq.1 is fully described elsewhere (5). The anisotropic case ($\Delta \neq 0$) presents long range magnetic order below a critical temperature T_c which is given, for small anisotropies, by a solution of the implicit equation:

$$K_BT_c/J = A/[1 - B\mathrm{Ln}(4J\Delta/3k_BT_c)]$$

where $A = 1.39$, $B = .52$ and $J/k_BT_c < 3/4$.

The spontaneous magnetisation is calculated for any value of $\Delta \in [0,1]$ and $K (> K_c = J/k_BT_c)$ from the attractor at $\Delta = 1$, $K \to \infty$. In the neighbourhood of this Ising-like attractor the spins are paralell and the magnetisation $M(T=0) = 1$. The magnetisation calculations are also described elsewhere (5).

3. RESULTS AND CONCLUSIONS

As a test of the accuracy of our method, we first compare our curve, for the case $\Delta = 1$, with the exact result of the Onsager solution. This is shown in Ref.5 where a discrepancy of order 10% occurs for a small interval below T_c. In Fig.1 we display the magnetisation calculated for several values of the anisotropy factor Δ. At low temperatures as Δ increases the magnetisation decreases slower with temperature due to a quenching of the magnetic excitations. For $\Delta = .3$ the magnetisation is almost constant up to approximately 1/3 of the critical temperature as observed in the uncoupled films studied in Ref.1. In Fig.2 we show our results for $\Delta = .3$ and for three different values of the exchange coupling J. The isotropic bulk value J_B is cal-

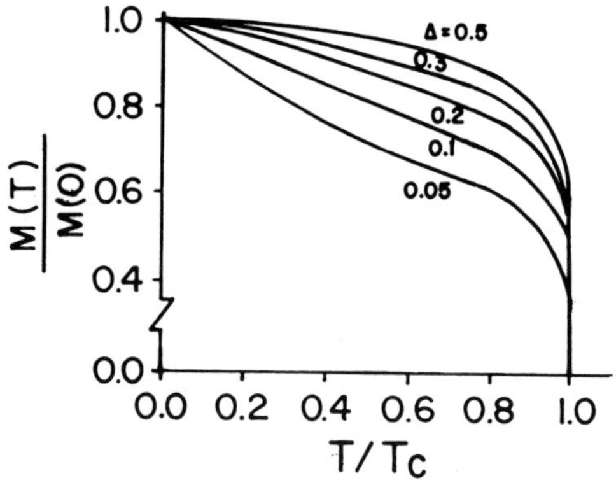

Fig. 1. The magnetisation as a function of temperature for different values of the anisotropy parameter Δ.

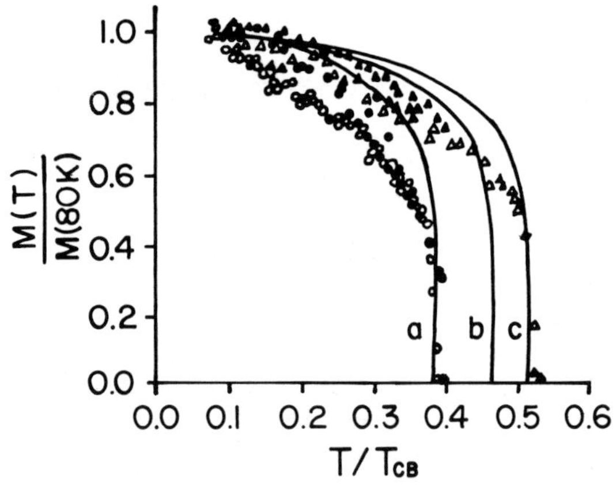

Fig. 2. Magnetisation curves as a function of temperature for $\Delta = .3$ and different values of the exchange parameter. Curve a, $J = .82J_B$, curve b, $J = J_B$ and curve c, $J = 1.11\ J_B$ where $J_B = 292K$. The experimental points are taken from Fig.4 of Mauri et al. (Phys.Rev. Lett. 62, p. 1900, 1989 and references therein) with the same meaning for the symbols. The proportionality between P(T), the low energy cascade spin polarization, and the magnetisation has been assumed.

culated using our method in three dimensions which yields $K_c = 2.94$ (see also Refs.5 and 6). The drop in the magnetisation close to T_c is well described by the theory. This sharp drop reflects the Ising criticality of the films. This is consistent with the renormalisation group since in this approach the critical behaviour along the critical line $T_c(\Delta)$ is governed by the Ising fixed point (5,6). The fact that the experimental points, in a temperature range below T_c, fall consistently below the theoretical curves, may be a consequence of disorder. It is well know (7) from the study of random magnets that disorder produces a flattening of the magnetisation curve and could account for the discrepancies found here at least for the permalloys films. We have performed calculations with the diluted Heisenberg model. Within the approximations of Ref.6 we found that the effect of vacancies is only to weaken the exchange constant J. Furthermore the renormalisation group method with finite cells understimates the spin wave contribution (5) for $k_BT > J\Delta$.

The effect of anisotropy is much more dramatic in 2D than in 3D Heisenberg systems. Indeed in the former case anisotropy itself is responsible for the existence of a finite temperature phase transition. Consequently in two dimensions there is a much larger effect in quenching the low energy excitations so that it is not necessary to assume unphysically large anisotropies to account for the experimental data.

ACKNOWLEDGEMENTS

We would like to thank Dr. D. Scholl from IBM for sending us useful information on the magnetic properties of thin films and CNPq of Brazil for partial financial support.

REFERENCES

(1) D. Mauri, D. Scholl, H.C. Siegmann and E. Kay, Phys.Rev.Lett. 62, 1900 (1989) and Appl.Phys. A49, 439 (1989).
(2) J. Mathon and S.B. Ahmad, Phys.Rev. B37, 660 (1988); J. Mathon, Physica 149B, 31 (1988).
(3) J.C. Fernandes, M.A. Continentino and A.P. Guimarães, Solid State Comm. 55, 1011 (1985).
(4) D. Scholl, private communication.
(5) E.V. de Mello and M.A. Continentino, J.Phys.: Cond.Matter; in print.
(6) R.B. Stinchcombe, J.Phys.C: Solid State Physics 14, 397 (1981); A.M.Mariz, C.Tsallis and A.O. Caride, J.Phys.C: Solid State Physics 18, 4189 (1985).
(7) M.A. Continentino and N. Rivier, J.Phys.C: Solid State Physics 10, 3613 (1977).

Physica B 165&166 (1990) 451–452
North-Holland

SPIN WAVES AT THE FERROMAGNETIC-ANTIFERROMAGNETIC INTERFACE

E. PESTANA, G.J. MATA, and N. PANTOJA

Departamento de Física, Universidad Simón Bolívar, Apartado 89000, Caracas 1080A, Venezuela.

We investigate the magnon modes in the neighbourhood of the interface between a ferromagnet and an antiferromagnet. The system is one-dimensional and it is characterized by three parameters $-J_F$, J_A, and J_0 which describe the exchange couplings in the ferromagnet, the antiferromagnet, and across the interface itself. We study the nature of the interface magnons as a function of these parameters.

1. INTRODUCTION

The study of magnetic phenomena at interfaces has been made specially interesting by modern superlattice-synthesis techniques, which open the possibility of realizing near-ideal magnetic interfaces in the laboratory. Of particular interest is the nature of magnetic excitations at interfaces (and in low-dimensional systems in general). Theoretical studies of interface modes (1-5) have been carried out by several authors. In this paper we study a one-dimensional system which consists of a semi-infinite antiferromagnet coupled to a semi-infinite ferromagnet. Elsewhere (6) we have developed a formalism which allows the local description of antiferromagnetic excitations. Here we use that formalism to investigate the nature of the interface modes for a range of values of the exchange integrals that characterize the system.

2. THE INTERFACE

The hamiltonian of the system can be expressed as

$$H = H_F + V + H_A \qquad (1)$$

where H_F, H_A, and V describe the ferromagnetic side, the antiferromagnetic side, and the coupling across the interface. A given spin interacts only with its nearest neighbours, via a Heisenberg-type coupling. The absolute values of the exchange integrals in the ferromagnetic and antiferromagnetic side are J_F and J_A, respectively. H_A contains the effect of an anisotropy-field h that stabilizes the antiferromagnetic order (6). The two sides are coupled by a single bond with exchange integral $\pm J_0$. According to the sign, this bond can be ferromagnetic or antiferromagnetic. We discuss both cases here. The temperature $T=0$.

The net magnetic moment added to an antiferromagnet when an excitation is created can take the values $\pm g\mu_B$, where g is the gyromagnetic ratio and μ_B is the Bohr magneton (12). The excitations are thus classified as α ($-g\mu_B$) and β ($+g\mu_B$) excitations. In the isolated ferromagnet only α excitations exist, since the ground-state magnetic moment is the maximum possible. One then expects that at the interface the behaviour of α and β excitations must be quite different.

The bulk excitations form a band in the interval $(0, 4J_F)$ for the ferromagnet and $([4J_Ah+h^2]^{1/2}, 2J_A+h)$ for the antiferromagnet. The interface modes can be referred to as acoustical if their energy is below the antiferromagnetic band and as optical if the latter is above the antiferromagnetic band.

We use the formalism in (5) and (6) to find the interface modes.

3. RESULTS AND DISCUSSION

In figure 1 we show a phase diagram for the type and number of interface states. We choose our energy scale so that $J_A+2J_F=1$ and define the variable $x = J_A-2J_F$. We take $h = 0.1J_A$. The first

0921-4526/90/$03.50 © 1990 – Elsevier Science Publishers B.V. (North-Holland)

and second numbers inside the parentheses indicate the number of acoustical and optical interface states. The top frame corresponds to α modes when the interface bond is ferromagnetic. (There are no β modes in this case.) The middle and bottom frames correspond to α and β modes when the interface bond is antiferromagnetic.

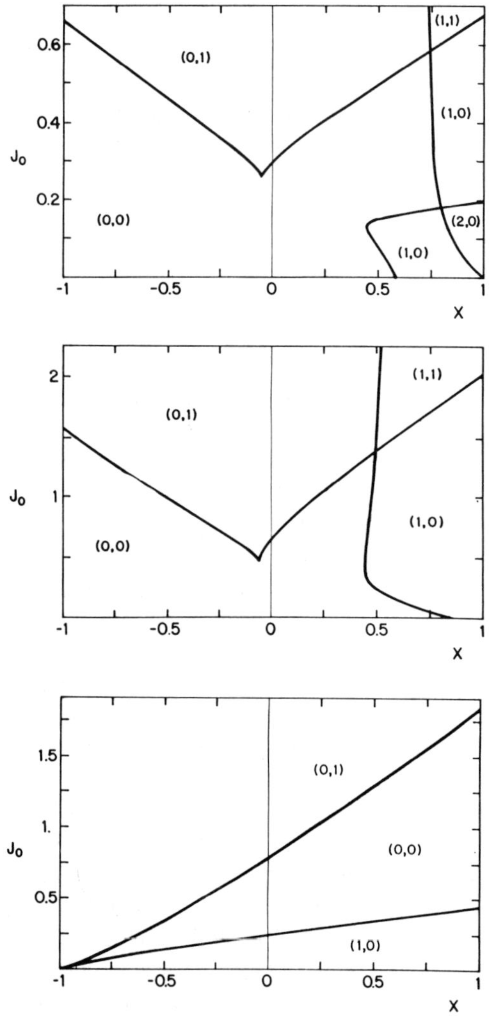

FIGURE 1
Phase diagram in parameter space for α modes and ferromagnetic interface bond (top frame); α (middle frame) and β (bottom frame) modes for antiferromagnetic interface bond

The difference in behaviour of α and β excitations is quite clear. The form of the phase diagrams can be qualitatively understood in terms of the local states for the uncoupled system ($J_0 = 0$) and their overlap with the continua. For example, in the lower right corner of the upper frame there is a region corresponding to a single acoustical α state. As one moves to the left in the diagram this state merges with the ferromagnetic continuum and becomes a resonance as it enters the (0,0) region. When the interface bond is antiferromagnetic this state corresponds to the acoustical β state in the lower right corner of the bottom frame. In this case as one moves to lower values of x the state remains localized despite the fact that it overlaps the ferromagnetic continuum. This indicates that β states cannot propagate into the ferromagnetic bulk.

In general one finds that the spectral weight of β excitations penetrates within the ferromagnetic side but decays quite rapidly. In the decay region there appears a proximity effect: the average spin of the ferromagnetic sites close to the interface diminishes by a small amount.

In conclusion we have found that the magnetic moment of the antiferromagnetic excitations plays a significant role on their behaviour at the interface, in particular in their localization. The antiferromagnet induces fluctuations in the ferromagnet and the magnetization profile is changed at the interface.

ACKNOWLEDGMENTS

We thank Gustavo Machado and José A. Vivas for technical help.

REFERENCES

(1) L.L. Hinchey and D.L. Mills, Phys. Rev. B 33 (1986) 3329.
(2) L. Dobrzynski, B. Djajari-Rouhani, and H. Puszkarski, Phys. Rev. B 33 (1986) 3251.
(3) B. Xing Xu, M. Mostoller, and A.K. Ragajopal, Phys. Rev. B 31 (1985) 7413.
(4) A. Yaniv, Phys. Rev. B 31 (1985) 402.
(5) G.J. Mata and E. Pestana, Phys. Rev. B, in print.
(6) G.J. Mata and E. Pestana, Phys. Rev. B 31 (1985) 7285.

Physica B 165&166 (1990) 453–454
North-Holland

PERPENDICULAR RESISTANCE OF THIN Co FILMS IN CONTACT WITH SUPERCONDUCTING Nb

C. FIERZ, S.-F. LEE, J. BASS, W.P. PRATT, JR. and P.A. SCHROEDER

Department of Physics and Astronomy, Michigan State University, East Lansing, MI 48824, USA

Perpendicular resistance measurements are reported for sputtered Nb/Co/Nb sandwiches where the thickness of the Co was varied from 6nm to 1200nm. These measurements yield a boundary resistance between the ferromagnetic Co and the superconducting Nb of (3.2 ± 0.2) $f\Omega m^2$.

Although a number of measurements have been made (see [1] and references therein) of the resistance parallel to the layers of layered metallic systems (LMS), few have been made of the more difficult quantity, the resistance perpendicular to the layers, R(perp). Of the latter, most have been made on superconducting/normal/superconducting (S/N/S) sandwiches.[2-5]

Superconducting strips (see Fig. 1) can be used as current and potential contacts to a LMS for R(perp) measurements, thereby ensuring that the contact surfaces will be equipotentials. However, using such contacts leads to a possible problem due to the proximity effect between the superconductor and the normal metals of the LMS. The proximity effect can be eliminated by placing a sufficiently thick layer of a ferromagnetic metal (F) such as Co between the superconducting strips and the LMS (about 2.5nm of Co appears to be sufficient to eliminate the proximity effect through 1-μm-thick film of Ag [6]). One must then determine the values of the two S/F boundary resistances (R_b^*) in order to subtract them from the total R(perp) measured for the LMS plus its contacts. (Up to now, our LMS samples have had the same F metal at the S/F boundary and in the LMS, so there is no additional boundary resistance associated with the F/LMS interface.) In this paper we limit our discussion to measurements of the temperature independent part of R_b^*, which is straightforward to determine since R(perp) is not strongly temperature dependent for 1.5K ≤ T ≤ 3K (i.e., ≤ 0.3 T_c).

As previously reported [6,7], our technique consists of placing 500-nm-thick Nb strips on opposite sides of a thin metallic film of interest that has a thickness t. We use an UHV-compatible sputtering apparatus in which the sapphire-substrate temperature is kept below 40 °C during fabrication. The 1-mm-wide strips act as both the current and voltage leads. As shown in Fig.1, the 1-mm^2 overlap region of the strips defines the effective cross sectional area (A) of the sample for R(perp). The Nb strips were formed by placing 0.1-mm-thick stainless steel masks as close as possible to the sample, resulting in strips whose edges were not sharply defined. The width of the strip was defined to be that part of the Nb film for which the thickness was ≥ 10nm, as measured with a Dektak IIA surface profile analyser. The uncertainty in A is estimated to be ± 5%. The profile analyser was also used to measure t, which was generally found to agree with the nominal thickness indicated by the quartz-crystal thickness monitors within our estimated random uncertainties for t. These uncertainties in t range from ±5% for t ≥ 200nm to ±20% for t ≤ 50nm. Typical values of R(perp) are 6nΩ to 100nΩ, and such R's can be measured to a precision of ±1pΩ. We used a small cryostat to measure R(perp) from 1.5K to 7.5K, utilizing our SQUID-based techniques for such measurements [8]. The standard resistor R_s, that acts as a reference for R(perp), was kept at 4.2 K; the systematic error in measuring R_s was less than 3%.

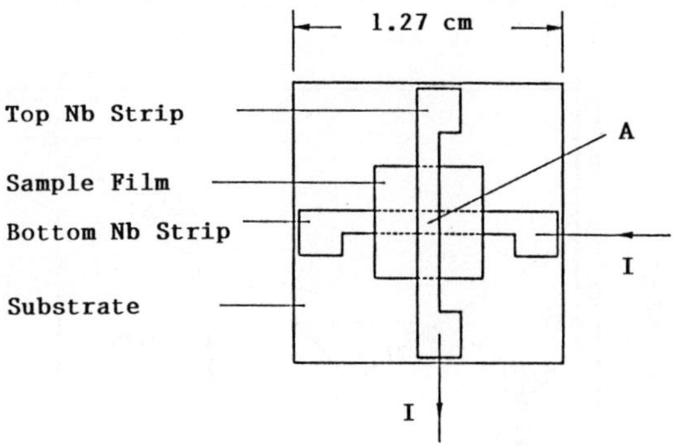

Fig. 1. Sample design.

To determine R_b^*, we made Nb/Co/Nb sandwiches where the thickness t of the Co was

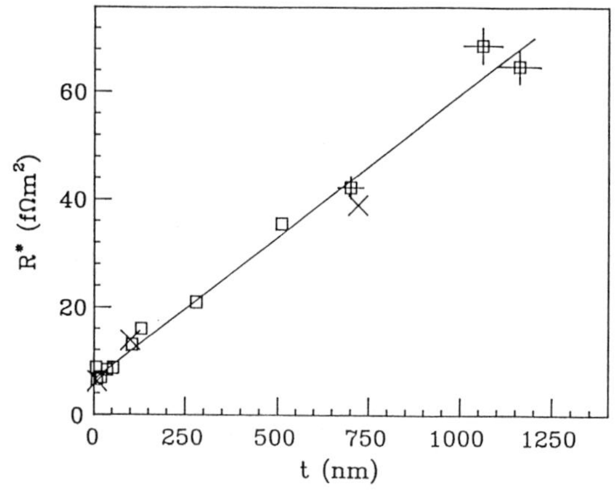

Fig. 2. R* vs t for Nb/Co/Nb sandwiches.

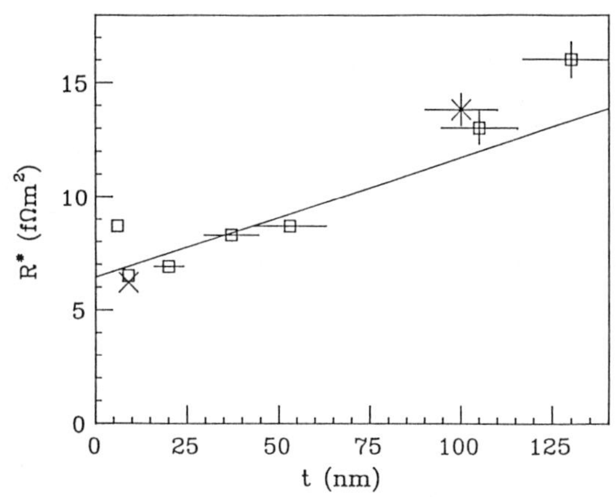

Fig. 3. A magnification of Fig. 2

systematically varied from 6nm to 1200nm. Our technique differs from previous studies [2-5] of S/N/S sandwiches in that we varied t to obtain R^*_b as well as ρ_{Co}, where ρ_{Co} is the resistivity of the Co film in the perpendicular direction.

We expect the following for R(perp) in our S/F/S sandwich:

$$R(perp) = \rho_{Co} t/A + 2R^*_b/A. \quad (1)$$

Fig. 2 shows a plot of R* vs t where R* = R(perp)A. The error bars indicate the above-mentioned uncertainties in A and t. The straight line in Fig. 2 is a least-squares fit of Eq. (1) to the data that is weighted by the vertical uncertainties of the different points. Fig. 3 shows a magnified plot for $t \leq 140$nm. The results are $R^*_b = (3.2 \pm 0.2)$ fΩm^2 and $\rho_{Co} = (53 \pm 3)$ nΩm. This value of ρ_{Co} agrees well with the value $\rho = (58 \pm 6)$ nΩm obtained from a measurement of ρ parallel to a 500-nm-thick Co film. The "X" symbols in Figs. 2 and 3 represent three samples where 10nm of Ag was placed between the Co and the Nb strips. Since these data points are not shifted systematically from the best-fit line, the introduction of this Ag layer does not appear to significantly alter the R^*_b of the Nb/Co boundary.

Our value of R^*_b is similar to those obtained in non-magnetic S/N/S sandwiches where S was dirty [2,3,5]. These earlier results were analysed in terms of models of the S/N interface [2-4] that predict $R^*_b \propto (\rho_S)^n$, where $0.5 \leq n \leq 1$ and ρ_S is the normal-state resistivity of the superconductor just above T_c. To permit a quantitative comparison with these models and prior results, we made parallel ρ_S measurements at 10 K on separately prepared 500-nm Nb films. We found a very large variability (30nΩm $\leq \rho_S \leq$ 200nΩm), part of which was attributable to our using two different sputtering targets. However, analysis of subsets of our R* vs t data showed no significant correlation between R^*_b and the apparent ρ_S. Thus it is not possible at this time to determine the extent to which the superconducting aspect of this Nb/Co boundary contributes to R^*_b. Direct measurements of ρ_S for each S/F/S sandwich may be required.

In summary, we have determined the Nb/Co boundary resistance and found that it appears not to be a significant function of ρ_S for pure Nb strips. With this information in hand, we will soon publish our analysis of R(perp) for Ag/Co LMS [9].

ACKNOWLEDGEMENTS

This work was supported in part by the US NSF under Low Temperature Physics Grant DMR-88-13287, the Michigan State University Center for Fundamental Materials Research, and the Swiss National Science Foundation.

REFERENCES

(1) M. Gurvitch, Phys. Rev. 34 (1986) 540.
(2) G.L. Harding, A.B. Pippard and J.R. Tomlinson, Proc. Roy. Soc A340 (1974) 1.
(3) T.Y. Hsiang and J. Clarke, Phys. Rev. B 21 (1980) 945.
(4) H.W. Lean and J.R. Waldram, J. Phys.: Condens. Matter 1 (1989) 1285, 1299.
(5) S.J. Battersby and J.R. Waldram, J. Phys. F:Met. Phys. 14 (1984) L109.
(6) J.M. Slaughter, W.P. Pratt, Jr. and P.A. Schroeder, Rev. Sci. Instrum. 60 (1989) 127.
(7) J. Slaughter *et al.*, Japanese J. of Appl. Phys. 26 Suppl. 26-3 (1987) 1451.
(8) D. Edmunds, W.P. Pratt, Jr. and J.A. Rowlands, Rev. Sci. Instrum. 51 (1980) 1516.
(9) J.M. Slaughter, Ph.D. Thesis, Michigan State University (1987) unpublished.

Physica B 165&166 (1990) 455–456
North-Holland

THICKNESS DEPENDENCE OF THE KONDO EFFECT IN AUFE FILMS

G. Chen and N. Giordano

Department of Physics, Purdue University, West Lafayette, IN 47907, USA

We have studied the Kondo effect in thin films of AuFe, through measurements of the temperature dependence of the resistivity, $\Delta\rho(T)$. We find that $\Delta\rho\sim -B\log(T)$, as expected for the Kondo effect. We also find that the coefficient B decreases as the film thickness is reduced. This result is discussed in terms of the effect of the film thickness on the conduction electron screening clouds which surround the magnetic impurities.

1. INTRODUCTION

The Kondo effect has been a subject of numerous investigations (1). The established picture is that at temperatures large compared to the Kondo temperature, T_K, the interaction of a localized magnetic moment with the conduction electrons has relatively little effect, but as the temperature is lowered the conduction electrons screen out the impurity moment with increasing efficiency. Eventually, for $T \lesssim T_K$ the impurity is fully screened. The size of the associated screening "cloud" is predicted to be of order $R_K \sim \hbar v_F / 2\pi k_B T_K$ (1). However, the direct observation of this screening cloud has proven to be extremely elusive, and there is essentially no direct experimental information concerning the value of R_K. In a sense, this is quite surprising, since for a system with a low Kondo temperature, R_K can be relatively large; for AuFe, which has $T_K \sim 0.3$ K, one finds $R_K \sim 3$ μm. However, the magnitude of the total magnetic moment of this cloud is equal to that of the impurity spin, so the large size of the cloud makes the average polarization quite small.

We have studied the Kondo effect in thin films of AuFe, a much studied and well characterized Kondo system (2). The aim was to examine films whose thickness, t, is much less than R_K, with the hope that restricting the volume of the screening cloud to quasi-two dimensions would modify the behavior. Our results suggest that this does indeed occur, and allow us to draw some conclusions concerning the value of R_K.

2. EXPERIMENTS

The AuFe films were prepared by flash evaporation from a single resistively heated source onto glass substrates. The thickness was varied by varying the distance from the source to the substrate, and the angular orientation of the substrate. In this way a single evaporation produced a series of films of different thicknesses, all with the same concentration of Fe. For the results reported here, this concentration was near 50 ppm. We have also obtained results for smaller concentrations, and they are similar to those shown below. The resistivity varied from 1.0 μΩ-cm in the thickest films, to 1.6 μΩ-cm in the thinnest samples.

Figure 1 shows some typical results. These films were prepared in a single evaporation, and hence all have the same concentration of Fe. In all cases $\Delta\rho\sim -\log(T)$ for $T \lesssim 3$ K. At higher temperatures the usual phonon contribution to the resistivity becomes increasingly important, as evidenced by a rapid increase of ρ above about 4 K. (This is clear from measurements well above 4 K which are not shown in Fig. 1.) This is why the logarithmic variation of ρ begins to breaks down above about 3 K. Nevertheless, there is a significant range over which the usual Kondo, i.e., logarithmic, variation of ρ is

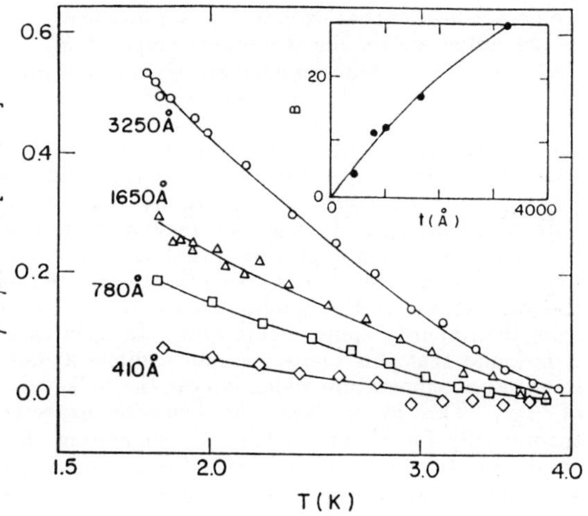

FIGURE 1
$[\rho(T)-\rho(4K)]$ as a function of $\log(T)$ for AuFe films with different thicknesses, as indicated in the figure. The inset shows B as a function of t. The units of B are 10^{-12} Ω-cm/(ppm-decade). The lines are guides to the eye.

seen in Fig. 1. We define the coefficient B by

$$\Delta\rho = -B \log(T) \quad , \qquad (1)$$

and obtain B from the data below 3 K in Fig. 1. Results for B as a function of film thickness are shown in the inset of Fig. 1. Measurements with bulk AuFe (2) yield a value of B which is 4 times higher than that found for the thickest sample in Fig. 1. This is consistent with the trend seen in our data. Namely, that the size of the Kondo effect decreases significantly as the thickness is reduced.

3. DISCUSSION

For these experiments, sample uniformity is a major concern. Au and Fe have very similar vapor pressures at $\sim 10^{-6}$ Torr (the pressure during the film deposition), and hence flash evaporation should produce uniform films (3). Another concern is oxidation of the Fe, which could make it nonmagnetic. One might imagine that oxidation could reduce the active Fe concentration, and since this effect would be largest in the thinnest films, it would produce a variation of B qualitatively similar to that seen in Fig. 1. To test for this possibility, we studied AuFe films from many different evaporations. In some cases, no special precautions were taken other than to store the samples in a vacuum dessicator. In some experiments the films were coated with photoresist immediately after removing them from the evaporator, so as to protect them from oxidation, while in other instances a thin layer of Au was evaporated on top of the AuFe films, before breaking vacuum, again to protect them from oxidation. The results were in all cases very similar to those seen in Fig. 1. We also remeasured several samples, and found that storage in vacuum at room temperature for a month or more had no effect on their behavior. For these reasons, we do not believe that oxidation was a problem.

As noted above, the theoretical value of R_K for AuFe is ~ 3 μm. Our samples are all much thinner than this, so one would not be surprised if the screening cloud were significantly distorted by the geometry of our samples. Intuitively we would expect that at a given temperature the screening would be less complete in quasi-two dimensions than in the bulk, and hence that B would decrease as t is made less than R_K. This is precisely what is observed in Fig. 1. However, previous workers (4,5) have studied the rate at which the conduction electrons scatter from the impurity spins in thin films. In most cases it is found that this scattering rate exhibits a maximum at a temperature which is near the bulk value of T_K. This is precisely the behavior expected theoretically for a bulk system (1), suggesting that the Kondo temperature, and hence the screening, are not appreciably affected by the quasi-two dimensionality of the samples. This is in contrast with our results for $\Delta\rho$, which imply that T_K is reduced as t is made smaller. Our results, when taken together with those for the conduction electron scattering rate, seem to make for an interesting puzzle.

The possibility that the finite thickness would affect the screening seems qualitatively to be quite reasonable. However, the simplest calculation of the Kondo divergence of $\Delta\rho$ in quasi-two dimensions does *not* appear to produce a suppression of B which is large enough to be compatible with our results (7). In addition, it seems likely that a number of other length scales might play an important role in this problem. First, both weak localization (WL) and electron-electron interactions (EEI) can make important contributions to the conductance (8). In two dimensions, these contributions both vary logarithmically with T, and hence have the potential to be confused with the Kondo effect, Eq. 1. However our samples have relatively low sheet resistances, and the calculated contributions of WL and EEI to $\Delta\rho$ are typically three orders of magnitude smaller that the variation seen in Fig. 1. Theoretical predictions concerning the interplay of WL and the Kondo effect (9) are also not consistent with our observations. It is possible that impurity-impurity interactions could be important. However, our results, and also those of other workers (2), suggest (but certainly do not prove) that for the concentration employed here impurity-impurity interactions do not have a significant effect.

In conclusion, we have presented what is, to the best of our knowledge, the first study of the Kondo contribution to $\Delta\rho$ in a quasi-two dimensional system. The results imply that the conduction electron screening is appreciably less efficient in this case as compared with the bulk, and that the Kondo screening cloud in AuFe is larger than $\sim$3000 Å. Further measurements of this kind should lead to a better picture of the screening cloud.

ACKNOWLEDGMENTS

We thank G. F. Guiliani, C. Jayaprakash, J. Liu, P. F. Muzikar, G. E. Santoro, and J. W. Wilkins for helpful discussions. This work was supported by the National Science Foundation through grants DMR-8614862 and DMR-8915574.

REFERENCES

(1) K. Fischer, in *Springer Tracts in Modern Physics,* Vol. 54, edited by G. Höhler, (Springer-Verlag, Berlin, 1970) p. 1.

(2) J. W. Loram, T. E. Whall, and P. J. Ford, Phys. Rev. B **2** (1970) 857.

(3) C. Van Haesendonck, H. Vloeberghs, Y. Bruynseraede, and R. Jonckheere, in *Nanostructure Physics and Fabrication,* edited by W. P. Kirk and M. A. Reed (Academic, San Diego, 1989), p. 467.

(4) R. P. Peters, G. Bergmann, and R. M. Mueller, Phys. Rev. Lett. **58** (1987) 1964.

(5) C. Van Haesendonck, J. Vranken, and Y. Bruynseraede, Phys. Rev. Lett. **58** (1987) 1968.

(7) G. E. Santoro, G. F. Guiliani, and P. F. Muzikar, (unpublished).

(8) P. A. Lee and T. V. Ramakrishnan, Rev. Mod. Phys, **57** (1985) 287.

(9) F. J. Ohkawa, H. Fukuyama, and K. Yosida, J. Phys. Soc. Jpn. **52** (1983) 1701. S. Suga, H. Kasai, and A. Okiji, *ibid* , **56** (1987) 863.

Physica B 165&166 (1990) 457–458
North-Holland

MAGNETIC PROPERTIES OF Nb/CuMn MULTILAYER SYSTEMS

M.L. WILSON, R. LOLOEE and J.A. COWEN

Department of Physics and Astronomy and Center for Fundamental Materials Research
Michigan State University, East Lansing, MI 48824 USA

The coupling between superconductivity and a metallic spin glass has been investigated in multilayer samples of Nb and $Cu_{1-x}Mn_x$. Initial results of the dependence of H_{c2} on temperature lead to the conclusion that thin films of the spin glass alloy can be forced into the superconducting state. Since both T_f and T_c can be tuned by appropriate dimensional adjustments, the system shows much flexibility for the study of magnetic-superconducting interactions.

We are reporting some preliminary results of a study of the interactions between superconductivity and spin glass behavior in multilayer films of $Cu_{1-x}Mn_x$ and Nb. Although our principle aim is to investigate the magnetic interactions between a superconductor, SC, and a spin glass, SG, we will discuss here only:
1) The characterization of the multilayers.
2) The finite size effects of decreasing CuMn layer thickness with thick Nb interlayers.
3) The effects of thin CuMn layers on the transition temperature, T_c, and the parallel critical field, $H_{c2||}$, of the Nb.

PREPARATION AND CHARACTERIZATION

We chose Nb and CuMn as our materials because of their immiscibility at room temperature, because they are easily sputtered and because of our previous observation of finite size effects in CuMn [1].

Our samples were fabricated by sputtering alternate layers of Nb and CuMn onto polished Si substrates.

From X-ray diffraction studies we found that the films grow in the Nb[110] and CuMn[111] directions. Crystallite sizes derived from the widths of these lines agree with single layer thicknesses obtained from the sputtering rates. Satellites surrounding these peaks have an angular separation related to the bilayer thickness L. These satellites, also, give thicknesses close to those determined from the sputtering rates.

FINITE SIZE EFFECTS

Our initial samples consisted of alternate layers of a few hundred Angstroms of Nb and 20Å of $Cu_{.96}Mn_{.04}$. At these thicknesses and concentration of Mn, the Nb remains in the normal state and there is a significant depression of the spin-glass transition temperature, T_f, due to the 2-D nature of the isolated films [1].

We examined a series of samples with 150Å Nb films interlayered with $Cu_{.96}Mn_{.04}$ films of varying thickness, λ_{CuMn}. We found a drop of nearly 75% in T_f as λ_{CuMn} is altered from 350Å down to 20Å, (Fig. 1). This drop in T_f is precisly analogous to finite size effects studied by Kenning et.al. in thin CuMn layers

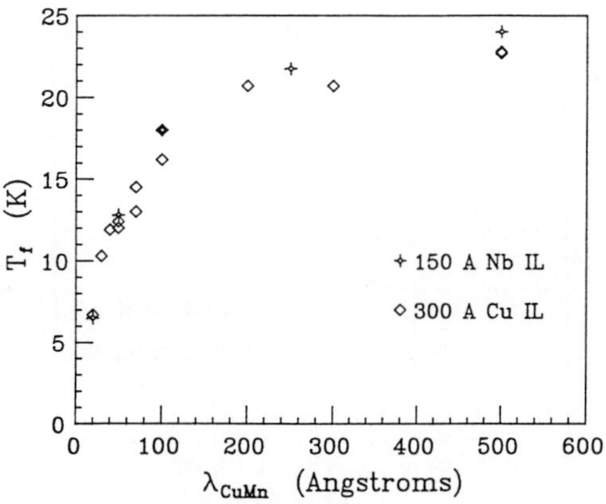

Figure 1

Spin glass freezing temperature, T_f, plotted as a function of CuMn thickness.

0921-4526/90/$03.50

with 300Å Cu interlayers [1]. They found that 300Å of Cu was able to magnetically decouple the CuMn layers and hence create a stack of isolated CuMn films . From the agreement of our data with those of Kenning et.al. we conclude that our CuMn layers, too, are magnetically decoupled.

T_c AND H_{c2} EFFECTS

In order to investigate the possibility of coexistence of SC and SG ordering, we fabricated two sets of samples with identical layers of Nb. One set was interlayered by 20Å of $Cu_{.99}Mn_{.01}$ the other by 20Å of pure Cu. We used a 1% Mn alloy to limit the pair breaking effect of the Mn ions and thus to examine the superconducting transition above our minimum measuring temperature, T = 2K.

In figure 2 we show the dependence of T_c on Nb interlayer thickness, λ_{Nb}, for these two interlayer materials. These data show little depression in T_c for the pure Cu interlayer as is expected from proximity effect coupling between Nb layers [2]. The inclusion of 1%Mn in Cu caused a depression of T_c which was so large that T_c was only measurable for $\lambda_{Nb} > 150Å$.

We measured $H_{c2||}$ by cooling each sample in zero field to a fixed temperature and then measuring the magnetization in increasing and decreasing fields. Above T_c we were unable to detect any effects due to spin glass freezing. However, as the number of Mn spins in these samples is near the noise limit of our equipment we cannot rule out such freezing.

The curves of $H_{c2||}$ vs. T are shown in figure 3. These show a linear response at high temperature with a roll off at low temperature. Banerjee et.al. [3] associated a linear dependence of $H_{c2||}(T)$ in similar Nb/Cu films with isotropic 3-D behavior. Hence we conclude that both of these samples are fully superconducting through the CuMn interlayer regions.

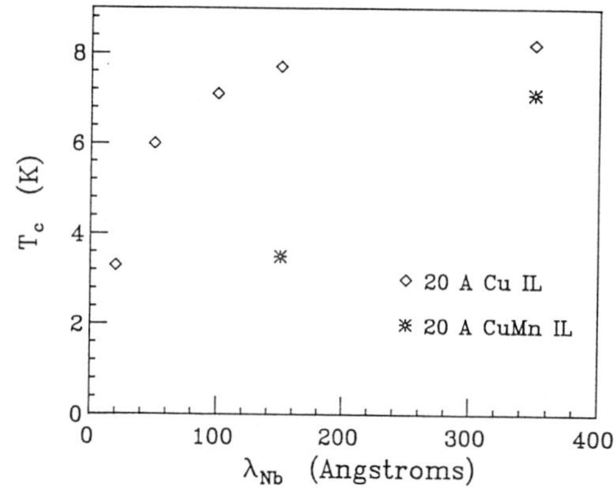

Figure 2

Superconducting transition temperature, T_c, plotted as a function of Nb thickness.

CONCLUSIONS

We have produced multilayer films of Nb and CuMn and observed that:

With thick Nb interlayers in the temperature regime where Nb remains in the normal state, the CuMn exhibits finite size effects identical to those observed with pure Cu interlayers. The T_c of Nb is very strongly depressed by very low concentrations of Mn in Cu. In addition, from the temperature dependence of $H_{c2||}$ we infer 3-D superconductivity with the result that the spin glass alloy has been forced into the superconducting state.

By adjusting the SC and SG layer thicknesses we can tune both T_c and T_f to regions where these transition temperatures lie in any arbitrary relationship with respect to one another. As a result, this system shows great flexibility for the study of the nature of magnetic-superconductive interactions.

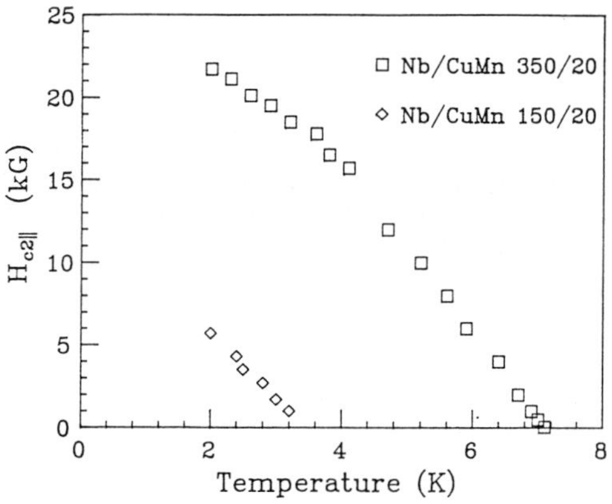

Figure 3

The parallel critical field $H_{c2||}$ plotted as a function of temperature.

ACKNOWLEDGEMENTS

We would like to acknowledge the support of the NSF under grant DMR-88-19429.

REFERENCES

(1) G.G. Kenning, Jon Slaughter, and J.A. Cowen, Phys Rev Letts. 59, (1987) 2596; G.G. Kenning, Jack Bass, W.P. Pratt, Jr., D. Leslie-Pelecky, Lilian Hoines, W. Leach, M. Wilson, R. Stubi, and J.A. Cowen, submitted to Phys. Rev. B.

(2) Indrajit Banerjee, Q.S. Yang, Charles M. Falco and Ivan K. Schuller, Solid State Commun. 41, (1982) 805.

(3) Indrajit Banerjee and Ivan K. Schuller, J. Low. Temp. Phys. 54, (1984) 501.

Physica B 165&166 (1990) 459–460
North-Holland

"UNIVERSALITY" OF FINITE SIZE EFFECTS IN CUMN AND AGMN SPIN-GLASSES

R. STUBI, J. A. COWEN, D. LESLIE-PELECKY, and J. BASS

Department of Physics and Astronomy and Center For Fundamental Materials Research
Michigan State University, East Lansing, MI 48824 USA

Recent measurements of the quasi-static spin-freezing temperature, T_f, of thin $Cu_{1-x}Mn_x$ layers (x = 0.04, 0.07, 0.14) in magnetron sputtered multilayer samples (MS) of CuMn alternated with Cu or Si revealed two different apparently "universal" curves for the ratio T_f/T_f^b (T_f^b is the bulk freezing temperature for a given x) for CuMn/Cu and CuMn/Si MS. We show that cooling the substrates leaves the CuMn/Cu MS data unchanged, but moves the CuMn/Si MS data closer to the "universal" curve for the CuMn/Cu MS. We show also that T_f/T_f^b for AgMn/Ag MS agrees with the CuMn/Cu MS "universal" curve.

We recently reported [1,2,3] that the quasistatic spin-freezing temperature, T_f, of thin $Cu_{1-x}Mn_x$ spin-glass (SG) layers with x = 0.04, 0.07, and 0.14 decreases dramatically with decreasing layer width W_{CuMn}, starting at values of W_{CuMn} as large as ≈ 1000Å. The values of T_f were obtained by measuring the locations of the maxima (peaks) in the zero-field-cooled (zfc) "dc" susceptibilities, χ, of multilayer samples (MS). The samples were produced by magnetron sputtering of the CuMn alternated with interlayers of a non-magnetic material. Two different interlayer materials were used in these MS: Cu and Si. 300Å of Cu or 70Å of Si was found to be sufficient to magnetically decouple the CuMn layers.

When the values of T_f for $Cu_{1-x}Mn_x$/Cu MS were each normalized to the bulk spin-freezing temperature, T_f^b, determined for shavings from the sputtering targets for the same value of x, the data were found to fall closely along a single "universal" curve, to within experimental uncertainties [3]. These data are shown as open symbols in Fig. 1.

When the values of T_f for $Cu_{1-x}Mn_x$/Si MS with the same values of x were similarly normalized, the data again [3] fell closely along a single curve, but this curve was lower than the one for the CuMn/Cu MS. The solid curve in Fig. 1 represents a smoothed fit to the data for $Cu_{.93}Mn_{.07}$/Si.

It was suggested [3] that the Si perturbed the CuMn layers so as to reduce their values of T_f, probably by Si atoms diffusing into the layers along grain boundaries. Evidence for this hypothesis was provided from both x-ray and resistivity measurements [3]. The most likely time for such diffusion to occur is while the multilayers are at the elevated temperatures (≈ 360K) reached during the sputtering process.

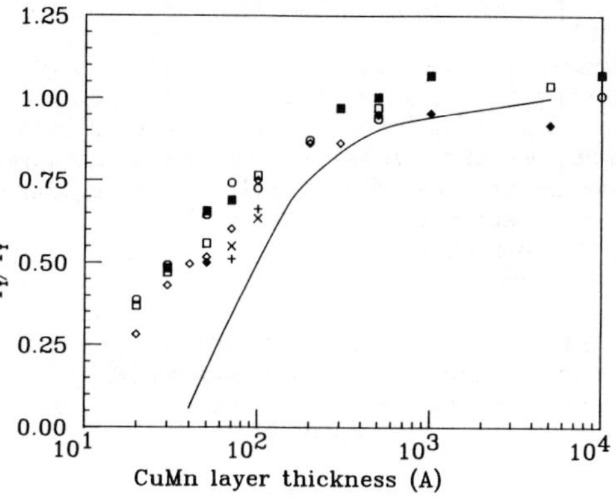

Figure 1
T_f/T_f^b for CuMn/Cu and CuMn/Si MS

	warm	cool	cold
$Cu_{.96}Mn_{.04}$/Cu	◇	◆	–
$Cu_{.93}Mn_{.07}$/Cu	□	■	–
$Cu_{.86}Mn_{.14}$/Cu	○	–	–
$Cu_{.93}Mn_{.07}$/Si	–	+	X

We have recently installed a substrate cooling system which permits us to investigate whether cooling the substrates during sputtering will bring the data for CuMn/Si MS closer to those for CuMn/Cu MS. We present here the first results for cooled CuMn/Si MS as well as for cooled AgMn/Ag MS.

Fig. 1 contains data for CuMn/Cu and CuMn/Si MS with both cooled and uncooled substrates. As already indicated, the open symbols for $Cu_{1-x}Mn_x$/Cu MS and the solid curve for $Cu_{.93}Mn_{.07}$/Si MS are for uncooled substrates. The filled symbols were obtained with $Cu_{.98}Mn_{.07}$/Cu and $Cu_{.93}Mn_{.07}$/Cu MS prepared keeping the substrates at ≈ 273K during sputtering. We see that the filled symbols are in excellent agreement with the open symbols.

The x symbols represent new data on $Cu_{.93}Mn_{.07}$/Si MS cooled to ≈ 273K during sputtering, and the + symbols represent new data on similar MS cooled to ≈ 180K. Each data point represents the average of measurements on two independent samples for which the bilayer thicknesses determined from x-rays bracketed the intended thicknesses. We see that cooling brings the CuMn/Si data considerably closer to the CuMn/Cu data. Such behavior is compatible with reduction of the penetration of Si into the CuMn, thereby yielding data closer to the "universal" behavior for the CuMn/Cu MS--at least for values of W_{CuMn} down to 70Å. Measurements of cooled CuMn/Si with still smaller values of W_{CuMn} are in progress.

AgMn is the spin-glass with characteristics closest to those of CuMn. Fig. 2 shows $T_f/T_f^{b'}$ versus spin-glass layer thickness for cooled $Ag_{.96}Mn_{.04}$/Ag MS compared to similar data for the cooled CuMn/Cu data of Fig. 1, except that here the data are normalized to $T_f^{b'}$, the T_f for 5000 - 10,000Å films, rather than to the T_f^b for target shavings. The reasons for using $T_f^{b'}$ for the AgMn/Ag MS are explained elsewhere [4].

We conclude that normalized values of T_f for CuMn and AgMn can be closely described by a single "universal" curve. Interpretations of this curve in terms of finite size scaling [5] and the cluster excitation model of Fisher and Huse [6] are given elsewhere [3].

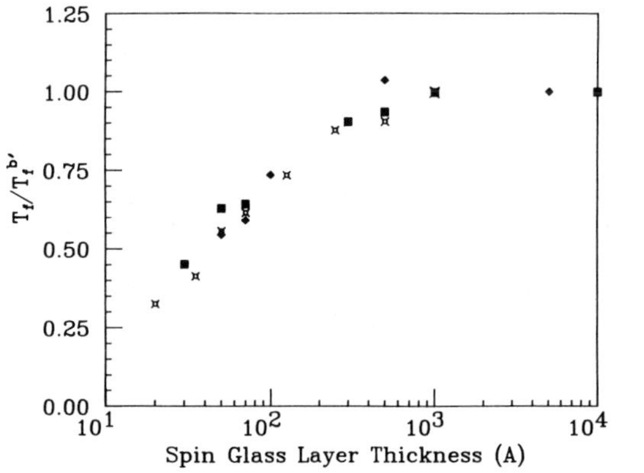

Figure 2

$T_f/T_f^{b'}$ for cooled CuMn/Cu and $Ag_{.96}Mn_{.04}$/Ag MS. The filled symbols are for the same samples as in Fig. 1. The open symbols are for the $Ag_{.96}Mn_{.04}$/Ag MS.

ACKNOWLEDGMENTS

The authors would like to acknowledge partial support for this research from the US NSF under grant DMR-88-19429 and from the Swiss National Science foundation for a postdoctoral fellowship to RS.

REFERENCES

(1) G.G. Kenning, J. Slaughter, and J.A. Cowen, Phys. Rev. Lett. 59 (1987) 2596.

(2) J.A. Cowen, G.G. Kenning, and J. Bass, J. Appl. Phys., 64 (1988) 5781.

(3) G.G. Kenning, J. Bass, W.P. Pratt Jr., D. Leslie-Pelecky, L. Hoines, W. Leach, M. Wilson, R. Stubi, and J.A. Cowen (Submitted for Publication).

(4) R. Stubi, J.A. Cowen, and J. Bass (Submitted for publication).

(5) M.N. Barber, in "Phase Transitions and Critical Phenomena", C. Domb and J.L. Levovitz, Eds., Academic Press, Vol. 8 (1983).

(6) D.S. Fisher and D.A. Huse, Phys. Rev. B36 (1987) 8937.

Physica B 165&166 (1990) 461–462
North-Holland

2D AND 3D SPIN GLASS DYNAMICS IN THIN Cu(Mn) FILMS

J. Mattsson+, P. Granberg+, P. Nordblad+, L. Lundgren+, R. Stubi*, D. Leslie-Pelecky*, J. Bass*, J. Cowen*

+ Uppsala University, Institute of Techology, Box 534, S-75121 Uppsala, Sweden
* Michigan State University, Department of Physics, East Lansing, Mich 48824, USA

Dynamic susceptibility measurements on multilayered Cu(Mn) spin glass films of thickness 40Å with interlayers of Cu of thicknesses in the range 10-1200Å are reported. The results yield new support for the applicability of domain growth theories to describe spin glass dynamics.

The lower critical dimension of spin glasses is today thought to be between two and three. Quite different magnetic relaxation behaviours are predicted for 2D and 3D spin glasses. Recently, J. Cowen et al have acquired the ability to sputter high quality multilayered Cu(Mn) spin glass films (1), which yield a practicable approach to comparative studies of the dynamics of 2D and 3D spin glasses of the same origin. Investigations (2) on films of composition Cu(13.5 at.%Mn) with different thicknesses separated by 300 Å thick interlayers of pure Cu have shown that a very thin (< 40 Å) film exhibits dynamics that resembles predictions for 2D spin glasses.

Here we report measurements of the dynamic susceptibility on samples consisting of 40 Å spin glass layers separated by interlayers of pure Cu of varying thickness (w) in the range 10-1200 Å. It is found that the sample with 10 Å Cu interlayers show typical bulk properties but a lower T_g than a bulk sample of the same composition, i e the interlayers do not decouple the spin glass layers but only act as additional dilutants. The sample with an interlayer of 1200 Å exhibits a typical two dimensional dynamic behaviour, i e the spin glass layers are fully decoupled. The samples with intermediate interlayer thicknesses indicate *crossovers* from 2D dynamics at short observation times to 3D dynamics at long observation times.

The dynamic susceptibility was investigated by zero-field-cooled (ZFC) measurements using a SQUID magnetometer. The ZFC measurements were performed by cooling the sample in zero field from a reference temperature, to the measurement temperature. After a wait time, t_w, a small magnetic field (10 G) was applied and the magnetization M(t) was recorded in the time interval $0.3<t<3\cdot10^4$ sec.

According to the droplet scaling theory (3) local equilibrium spin glass order occurs within domains with a characteristic length, L, growing in time as:

$$L\propto[T(\ln t)/\Delta T]^{1/\psi} \qquad (1)$$

where T is temperature, t is the time at constant temperature (age of the system) and ψ is a barrier exponent. In a ZFC experiment, the dynamics are probed on a certain length scale that grows with *observation time* according to eq. (1). As long as the probing length scale is much shorter than the size of the domains, i.e. $\ln(t) \ll \ln(t_w)$, processes within spin glass domains are probed, equilibrium dynamics is observed.

The earlier studies of Cu(Mn) (2) films have shown two significant features that distinguish the equilibrium magnetic relaxation of a 2D and a 3D system:

(i) The relaxation rate, dM/d(ln t), at a fixed temperature is smaller in 3D than in 2D.

(ii) In 2D the relaxation rate increases with increasing observation time but in 3D the rate decreases with time.

These observations are also in agreement with results from MC simulations on 2D (4) and 3D (5) systems. The droplet scaling theory (3) predicts that the magnetization (at spin glass equilibrium) should relax according to:

$$M(t)\propto(\ln t)^{1/\psi\upsilon} \qquad (2)$$

where υ is the correlation length exponent. The described behaviour of the relaxation rates then imply that $\psi\upsilon>1$ for a 3D and $\psi\upsilon<1$ for a 2D spin glass.

Using increasing and decreasing relaxation rate with time as the criteria for 2D and 3D dynamics, respectively, the samples with 300, 600 and 1200 Å Cu interlayers exhibit a 2D relaxation behavior at all temperatures measured. Fig 1 shows the ZFC susceptibility and the corresponding rate vs log(t) for the film 40 Å CuMn/150 Å Cu at different temperatures. The curves have been recorded after a wait time of 30 000 sec. We focus our interest to the region of 'equilibrium' relaxation, which is seen at t<30 sec (ln t <<ln t_W). At longer observation times a crossover to nonequilibrium dynamics starts and a drastic increase of the relaxation rate is seen. A decreasing 'equilibrium' relaxation rate is observed at temperatures slightly below the maximum of the field-cooled susceptibility curve. At lower temperatures an increasing rate is observed. As the interlayer thickness is reduced this crossover from 2D to 3D behaviour at a fixed observation time shifts towards lower temperatures.

In a pure 2D or 3D spin glass the equilibrium relaxation rate, at a specific observation time, continuously increases with temperature until the maximum relaxation time of the system equals the observation time, where the rate suddenly drops to zero. The equilibrium relaxation rates for samples with different Cu interlayer thicknesses are plotted in fig. 2 as functions of temperature. The relaxation

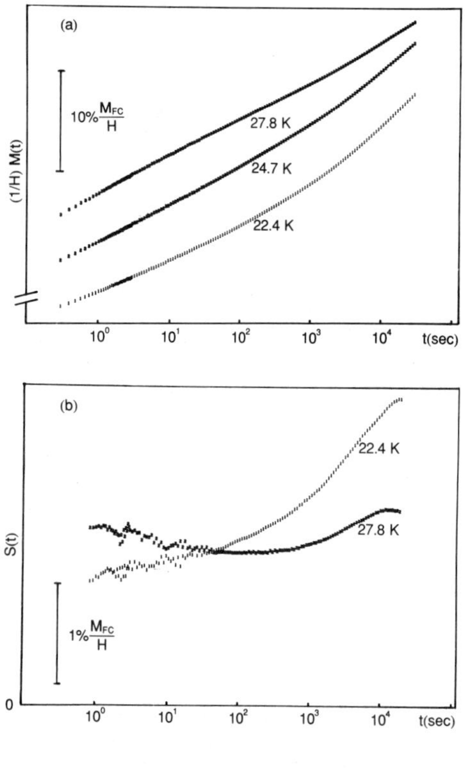

FIGURE 1
Zero-field-cooled susceptibility (a) and the corresponding relaxation rate (b) for 40Å CuMn/150Å Cu. The maximum in M_{FC} at T=31.7 K.

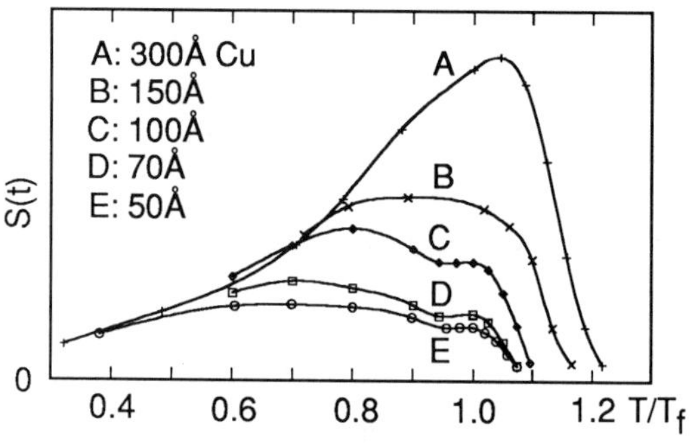

FIGURE 2
Relaxation rate at t=3 sec for samples with different interlayer thicknesses. t_W=30000 sec.

rate for the samples of intermediate Cu layer thickness starts to deviate from the rate of the 2D-system (300 Å Cu layers) and exhibits a maximum which is shifted towards lower temperatures with decreasing Cu layer thickness. Using criterium (i) above this deviation and maximum indicate a crossover from 2D to 3D dynamics. The temperatures for this crossover determined from this criterium or determined from criterium (ii) show good agreement for the different films

Interpretation: The RKKY interaction is mediated through the copper interlayers of thickness w to a strength $\Delta J(w)$. This interaction can be looked upon as a perturbation to the bonds in the spin glass layers. Within the droplet model in terms of Bray and Moore (6), the spin glass domains becomes unstable to such a bond perturbation at a length scale (L^*). When the 2D spin glass domains have grown to order L^*, the interlayer coupling $\Delta J(w)$ becomes relevant and adjacent spin glass layers become coupled. The continued domain growth then sees 3D interaction and the system attains 3D dynamical properties.

ACKNOWLEDGEMENT
Financial support from NFR is acknowledged.

REFERENCES

(1) G. Kenning, J. Slaughter and J Cowen; Phys. Rev. Lett. 59 (1987) 2596.
(2) P.Granberg et al, J. Appl. Phys. (in print).
(3) D. Fisher and D. Huse, Phys. Rev. B 38 (1988) 373, 38 (1988) 386, 36 (1987) 8937.
(4) K. Nemoto and H. Takayama, J Phys. C (GB) 16, (1983) 6835.
(5) A. Ogelski, Phys. Rev. B 32 (1985) 7384.
(6) A. Bray and M. Moore, Phys. Rev. Lett. 58 (1987) 57.

Physica B 165&166 (1990) 463–464
North-Holland

MAGNETIC REENTRANT BEHAVIOR OF PdFe-FILMS DUE TO DISORDER

Peter BRÜLL, Paul ZIEMANN

Fakultät Physik, Universität Konstanz, D-7750 Konstanz, FRG

$Pd_{1-x}Fe_x$-films (x=0.01) were prepared by vapor-quenching onto liquid-He cooled substrates. Insitu measurements of the ac-susceptibility performed in additionally applied dc-fields allow to check these films for a possible reentrant behavior. Reentrance is found for disordered films and substantiated by field-cooling experiments. Lowering the degree of disorder by stepwise annealing the films up to 330 K leads to a gradual disappearance of the reentrant behavior.

1. INTRODUCTION

Bulk $Pd_{1-x}Fe_x$ is a wellknown giant moment system exhibiting ferromagnetism (FM) above a percolation threshold of 0.1at% Fe (1). On the other hand recent measurements on corresponding films (2) provided some evidence for a reentrant transition from FM to a cluster glass (CG). To substantiate these observations, further experiments have been performed under improved preparation conditions (the magnetic behavior of PdFe is extremely sensitive to oxygen impurities built-in during the evaporation process (3)) and special emphasis has been put on the influence of superposed dc-magnetic fields applied after cooling the samples (ZFC) or during cooling (FC). The role of disorder on the reentrant behavior is tested by gradually annealing the films.

2. EXPERIMENTAL

The PdFe-films (1at% Fe, typical thickness 55 nm) were prepared by flash-evaporation of small pieces from a mother alloy onto liquid-He cooled sapphire substrates. The pressure during evaporation was less than $7\cdot10^{-9}$ mbar, a necessary prequisite to obtain reproducible results. The highly disordered films were stepwise annealed up to T_A and the ac-susceptibility χ was determined insitu between 5.5 K and T_A using a field-amplitude of 0.1 mT at 1049 Hz. In addition small ($\lesssim$3 mT) magnetic dc-fields were applied either after cooling to 5.5 K (ZFC) or at T_A (FC). More details on the susceptometer can be found in (4).

3. RESULTS AND DISCUSSION

Fig.1 shows the temperature dependence of the specific susceptibility χ for a film annealed up to T_A=240 K. Here different dc-fields as indicated in the figure were applied at 5.5 K. The results exhibit two main features: A steep increase of χ below 12 K indicating the transition into the FM-state (a Curie-temperature T_C=11.4 K is extrapolated from a tangent-construction) and a pronounced shoulder below T_f=7.2 K as indicated by the arrow in fig.1, which is interpreted as a transition into the CG-state. This interpretation is suggested by the observed effect of the dc-fields, which suppress the FM-part of χ above T_f much stronger than the CG-part below T_f (5). The reduced T_C-value as compared to bulk samples confirms earlier observations that defects are lowering the Curie-temperature (6). The above interpreta-

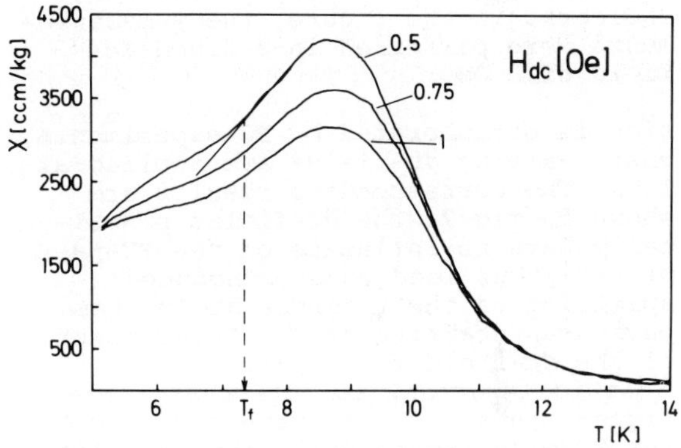

Fig. 1: Temperature dependence of the specific susceptibility χ for a quench condensed PdFe-film (1at% Fe) annealed at T_A=240 K. Different dc-fields as indicated were applied (ZFC).

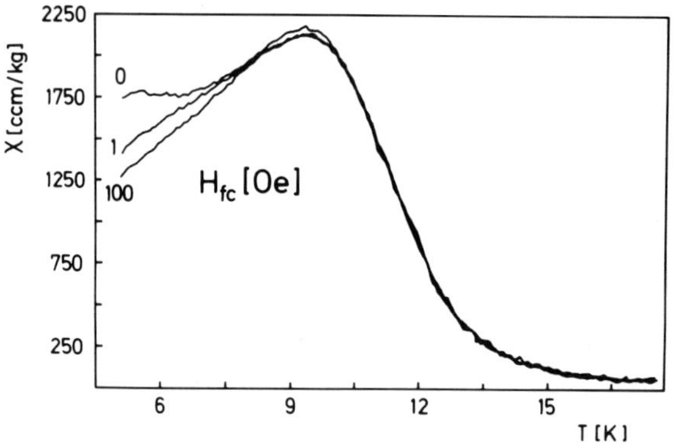

Fig. 2: Influence of dc-field-cooling on the temperature dependence of the susceptibility χ for a PdFe-film (1at% Fe). The different cooling fields are indicated in the figure, the χ-measurements were performed in a fixed dc-field of 2 Oe.

tion is corroborated by FC-experiments, where varying dc-fields are applied at $T>T_C$. The corresponding results are shown in fig.2: the dc-fields practically have no influence on the FM-part of $\chi(T)$, but lead to a pronounced splitting of the χ-curves at low temperatures starting at T_f, independent of the dc-field for $B_{dc}\lesssim 10$ mT.

It is important to note that for a higher annealing temperature T_A=330 K the FC-experiments still show a splitting of the $\chi(T)$-curves depending on the dc-fields, but starting not at a well defined temperature T_f. One rather observes a gradual onset starting at temperatures very close to T_C for fields $B_{dc}\gtrsim 0.1$ mT. This qualitatively different behavior is also found by the ZFC-experiments: For T_A=330 K $\chi(T)$ exhibits no longer a shoulder at low temperatures. This is interpreted as disappearance of the reentrant CG-state due to annealing.

Fig. 3 summarizes the magnetic behavior of the PdFe-films containing 1at% Fe. Here T_C and the reentrant temperature T_f are plotted versus T_A. While the T_A-dependence of T_C exhibits some structure reflecting the preferential annealing of a specific type of disorder (6), the corresponding T_f-behavior is structureless with a small positive slope within the observed range (at T_A=330 K, according to our criteria, no reentrance was found, at T_A=160 K the reentrance could not be observed above the lowest experimentally attained temperature of 5.5 K).

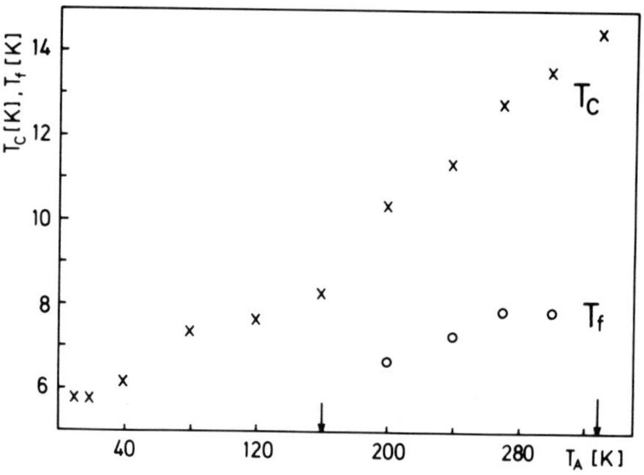

Fig. 3: Dependence of the Curie- and the reentrant temperatures (T_C, T_f) on the degree of disorder as indicated by the annealing temperature T_A. The meaning of the arrows is described in the text.

4. CONCLUSIONS

The experimental results on PdFe-films demonstrate that strong disorder as produced by vapor-quenching onto cold substrates can lead to a reentrant transition from FM to CG even in systems, which,as bulk samples, exhibit only FM. The reported effect is restricted to Fe-concentrations below 2at% suggesting that the coupling between moments via the Pd-host electrons is the most disorder-sensitive part.

ACKNOWLEDGEMENTS

We thank Ch. Neumann for his experimental help. This work was partly supported by the Deutsche Forschungsgemeinschaft, SFB 306.

REFERENCES

(1) G.J.Nieuwenhuys, Adv. Phys. 24, 515 (1975)
(2) M.Avirovic, P.Ziemann, Jap. J. Appl. Phys. 26, 785 (1987)
(3) H.Claus, N.C.Koon, Sol. State Commun. 60, 481 (1986)
(4) G.Ziebold, D.Korn, J. Phys. E: Sci. Instr. 12, 490 (1979)
(5) B.H.Verbeeck, thesis, Leiden (1979)
(6) M.Hitzfeld, P.Ziemann, W.Buckel, H.Claus, Phys. Rev. B 29, 5023 (1984)

Physica B 165&166 (1990) 465–466
North-Holland

CROSSOVER BEHAVIOUR IN MAGNETIC HEAT CAPACITY OF LAYERED MOLYBDATES BELOW T_c

Peter STEFÁNYI and Alexander FEHER

Department of Experimental Physics P.J.Šafárik University, nám. Febr.víť.9,
041 54 Košice, Czechoslovakia

The dimensional 2D-3D crossover below T_c in heat capacity of layered $KDy(MoO_4)_2$, $CsDy(MoO_4)_2$ and $CsDy_{0.95}Eu_{0.05}(MoO_4)_2$ crystals was observed which could by probably connected with the low-lying branches in energy spectra of magnetic excitations.

1. INTRODUCTION

Crystals of $MDy(MoO_4)_2$ (M=K,Cs,...) belong to the group of chain-layered crystals showing the structural low dimensionality. The members of this group undergo the cascade of structural transitions, some of which have the Jahn-Teller origin in which the electron-phonon interaction plays the dominant role (1). The phonon spectra of these layered molybdates shows unusual properties. The strong anisotropy of elastic interaction leads to the presence of the low frequency phonon branches in the phonon energy spectra (2). The magnetic ordering in these materials takes place at about 1K (3) because the spin-spin interaction is weaker than the electron-phonon interaction. The ground state $^6H_{15/2}$ of Dy^{3+} ion in low symmetry crystalline field is Kramers doublet with $J = \pm 15/2$ having strongly anisotropic behaviour. The magnetic system in these molybdates can be described by the Ising model (3), but the thermodynamic quantities calculated from the magnetic heat capacity didn´t fully correspond to those for the 2D Ising model.

This disagreement inspired us to study the critical behaviour of the heat capacity in the vicinity of the magnetic phase transition temperature T_c, which resulted in detecting the 2D-3D crossover behaviour in $KDy(MoO_4)_2$ below T_c (4). In this paper we investigated the critical heat capacity behaviour of two other molybdates $CsDy(MoO_4)_2$ and $CsDy_{0.95}Eu_{0.05}(MoO_4)_2$ and comparing the obtained results with those for $KDy(MoO_4)_2$ we tried to explain interaction magnitude on the crossover behaviour.

2. EXPERIMENTAL RESULTS

The heat capacity was measured by the quasi-adiabatic heat pulse method. The magnetic part to the heat capacity was obtained in an usual way by subtracting the lattice and Schottky heat capacity (if necessary) (3) from the total heat capacity. The precise evaluation of the phase transition temperatures was complicated by the "rounding off" effect. Using the characteristic behaviour of magnetic models and considering "rounding off" effect the most probable temperature interval for T_c was estimated. The value of T_c was established by plotting the critical heat capacity vs. the reduced temperature $\varepsilon = /1-T_c/T/$ with various T_c chosen from the estimated temperature interval. The choice of the correct value of T_c produces the longest linearity on this plot. The following temperatures were obtained: $T_c = (1.294\pm0.005)$K for $CsDy(MoO_4)_2$, $T_c = (1.000\pm0.005)$K for $KDy(MoO_4)_2$ and $T_c = (1.226\pm0.005)$K for $CsDyEu(MoO_4)_2$. The experimental data of heat capacity near T_c were compared directly with the theoretical prediction for the magnetic models (the analogical procedure was used for $KDy(MoO_4)_2$ in (4)). The behaviour of heat capacity of all three samples for $T > T_c$ strongly follows the logarithmic divergence form for 2D Ising model on a quadratic lattice (Figure 1). The most interesting feature of the investigated samples for $T < T_c$ is the crossover from the 2D behaviour to the 3D behaviour (Figure 2). The crossover temperature is as follows $\varepsilon_{cr} = 8.10^{-2}$ for $KDy(MoO_4)_2$, $\varepsilon_{cr} = 2.2\times10^{-1}$ for $CsDyEu(MoO_4)_2$ and $\varepsilon_{cr} = 3.1\times10^{-1}$ for $CsDy(MoO_4)_2$. The deviation from 3D behaviour for $CsDyEu(MoO_4)_2$ for $\varepsilon < 10^{-2}$ can be the result of the "rounding off" - effect due to impurities represented probably by Eu^{3+} ions.

3. DISCUSSION

It is shown that the ferromagnetic ordering is observed along the c-axis in these samples and antiferromagnetic ordering along the a-axis (5). In such a way the energy spectra must contain ferromagnetic and antiferromagnetic dispersion branches some of which must be low-lying, which might be responsible for the crossover observed in our samples. It may be deduced from

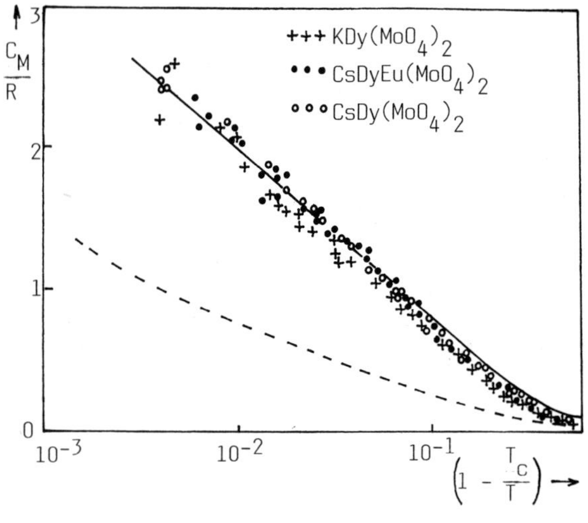

FIGURE 1
Critical behaviour of heat capacity for $T > T_c$: ——, theoretical prediction for a simple quadratic lattice; --- theoretical prediction for a simple cubic lattice.

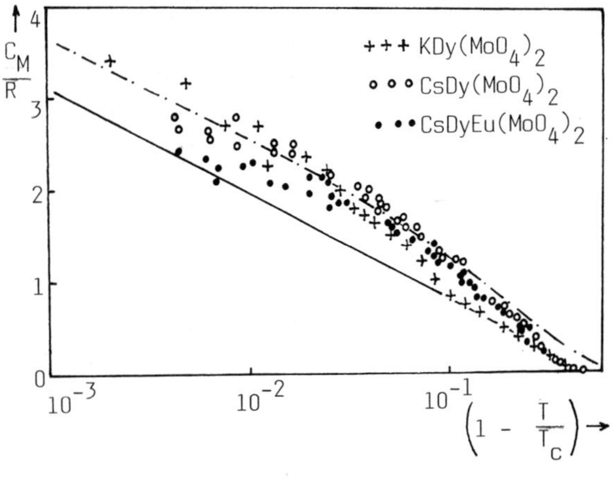

FIGURE 2
Critical behaviour of heat capacity for $T < T_c$: —— theoretical prediction for a simple quadratic lattice; -.-.- theoretical prediction for a tetrahedral lattice.

the comparison of ε_{cr} temperatures for investigated samples, that the behaviour of the low-lying magnetic dispersion branch approaches the behaviour typical of Ising model most apparently in $KDy(MoO_4)_2$, then in $CsDyEu(MoO_4)_2$ and in $CsDy(MoO_4)_2$. It follows from the measurement of heat capacity over the wide range of temperatures that substitution of larger Cs^- ion by the K^- ions didn't change the dimensionality of the magnetic system; similarly, dimensionality is not changed when 5% Dy^{3+} ions is substituted by Eu^{3+} ions. On the other hand, such a substitution of ions has an apparent influence on the structural dimensionality of the investigated samples (6). This, together with the above mentioned series of ε_{cr} values, leads to the conclusion, that the structural and magnetic planes of the easy ordering do not coincide, but are perpendicular to each other.

4. CONCLUSIONS

We observed the 2D-3D crossover in the heat capacity of the $KDy(MoO_4)_2$, $CsDy(MoO_4)_2$ and $CsDyEu(MoO_4)_2$ samples, which may be explained by the existence of weakly disperging curves in their energy spectra, responsible for the magnetic excitations along the weak coupling direction. But the connection between the dimensional crossover and the expected character of dispersion branches should yet to be proved by both ferro- and antiferromagnetic resonance techniques.

ACKNOWLEDGEMENTS

The authors would like to express their thanks to Professor A.I.Zvyagin for providing the samples and Dr.A.G.Anders for useful discussions.

REFERENCES

(1) A.I.Zvyagin et al. Sov.J.Low.Temp.Phys. 1 (1975) 79.
(2) S.D.Elchaninova, A.I.Zvyagin, Sov.J.Low. Temp.Phys. 9 (1983) 1200.
(3) P.Stefányi at al. J.Phys.: France 50 (1989) 1297.
(4) P.Stefányi, A.Feher and A.Orendáčová, J.Phys.: Condens.Matter 1 (1989) 7529.
(5) E.N.Khatsko, A.S.Cherny; Sov.J.Low.Temp. Phys. 11 (1985) 540.
(6) A.Feher, P.Stefányi and A.Orendáčová, to be published in phys.stat.sol.

Physica B 165&166 (1990) 467–468
North-Holland

SUPERCONDUCTIVE TUNNELING STUDY OF THE SPIN-FILTER AND PROXIMITY EFFECTS

Xin HAO, Jagadeesh S. MOODERA, and Robert MESERVEY

Francis Bitter National Magnet Laboratory, M. I. T, Cambridge, MA 02139, USA

Metal-EuS-Al junctions and EuS/Al bilayers have been studied by superconductive tunneling. In the former, the tunnel current showed a spin-polarization up to 85%, which was attributed to the splitting of the EuS conduction band. The EuS/Al bilayers displayed a large extra Zeeman splitting in the quasiparticle density of states in Al due to the exchange interaction between the conduction electrons and the localized and spontaneousely ordered Eu^{2+} moments. From the dependence of this internal (exchange) field on the Al film thickness, the exchange constant was found to be about 13 meV.

1. INTRODUCTION

The europium chalcogenides have been studied extensively for the past three decades. One study pertinent to the present work is tunneling through EuS barriers. It was shown that electrons field-emitted from EuS-coated tungsten tips were highly polarized; this was explained by the spin-filter effect (1). Early tunneling experiments also provided indirect evidence of this effect in the observation of junction resistance decrease below the ferromagnetic transition temperature (T_m) of the EuS barriers (2,3). By using a very thin EuS film and employing the spin-sensitive tunneling technique (4), we were able to measure the degree of spin-polarization in the tunnel current caused by the spin-filter effect (5).

The spin-filter effect arises from the conduction band splitting in the EuS below T_m, which, for bulk EuS, is 16.6K (6). This splitting of conduction band into spin-up, -down subbands lowers (raises) the tunnel barrier for spin-up (down) electrons: $\phi_{\uparrow,\downarrow} = \phi_0 \mp \Delta E_{ex}/2$ (where ϕ_0 is the average barrier height above T_m and ΔE_{ex} the amount of conduction band splitting); therefore, tunneling probability for spin-up electrons relative to that for spin-down electrons is greatly increased.

A separate consequence of the EuS tunnel barrier is the proximity effect at the ferromagnet-superconductor interface. The exchange interaction between conduction electrons in Al and the spontaneously ordered magnetic moments in EuS gives rise to a large effective magnetic field which acts only on the spins of the quasiparticles in Al. This effective field can cause Zeeman splitting in the density of states (DOS) of Al even in the absence of any applied field.

2. SAMPLES AND METHODS

For the study of the spin-filter effect, Au(or Ag, Al)/EuS/Al junctions were made, and for the proximity effect EuS/Al/Al_2O_3/Ag(Fe) junctions were made, using standard evaporation techniques. Very thin Al electrodes (4 to 12 nm) were deposited onto liquid-nitrogen-cooled substrates. The other metal electrode of Au, Ag, thick Al (50 nm) or Fe and the barrier EuS were deposited onto room temperature substrates. The barrier had a nominal thickness of 3.3 nm. A 5-nm EuS film was used to study the proximity effect. Tunneling conductance curves were taken at temperatures near 0.5K in magnetic fields parallel to the junction surface. R_J as a function of temperature was also measured for some of the junctions.

3. SPIN-FILTER EFFECT

Figure 1 shows the normalized tunnel conductance curves in various applied fields for a Au/EuS/Al junction. The high asymmetry indicates a large polarization in the tunnel current. Fitting the theory of high field superconductivity in thin films (7) to our data gave polarization P=80%.

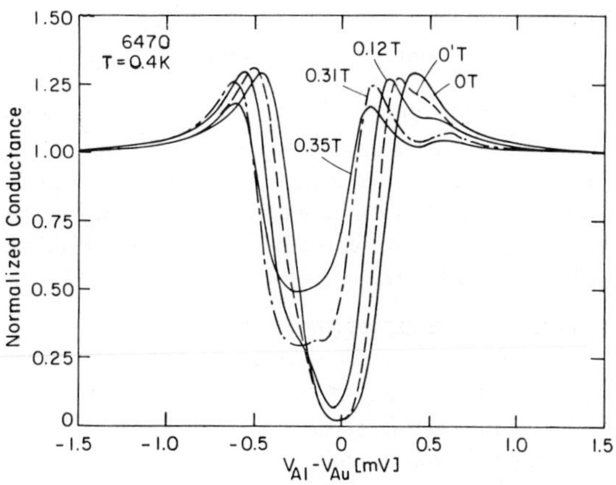

FIGURE 1

The amount of Zeeman splitting in Al quasiparticle DOS was much larger than that expected from the applied field H_a. For H_a=0.31 T, the Zeeman splitting corresponded to a total field of 4.3 T, with the extra field (B^*) being 4 T. A remanent Zeeman splitting persisted after the applied field was reduced to zero (see the curve labeled 0′ in Fig. 1). This junction also had a small initial zero-field Zeeman splitting corresponding to a field of B_i^*=0.5 T as determined by fitting the conductance curve. Tkaczyk *et al.* discovered similar, but less pronounced, extra Zeeman splitting in their study of rare earth (RE) or RE oxide/Al bilayers; they attributed this extra field to the exchange interaction (8).

Junction resistance was more or less constant from 77K down to 20K but showed a sharp decrease around 15 K. The resistance ratio, $R_J(4.2K)/R_J(25K)$, was 0.54. The decrease of junction resistance is another manifestation of the spin-filter effect, and the resistance change can be used to calculate the value of polarization by a simple tunneling theory (9).

Using the full expression for tunnel current density (J) of Ref.(9) for each spin direction and the optically measured $\Delta E_{ex}(T)$ (7), the resistance ratio as a function of temperature for $T < T_m$ can be calculated and fitted to the experimental data. We obtained, from the fits, the effective barrier thickness and height to be 18.2Å and 1.6 eV, respectively. From these values a theoretical $P = (J_\uparrow - J_\downarrow)/(J_\uparrow + J_\downarrow)$ was calculated to be 85%, somewhat larger than the value measured from the conductance asymmetry.

Electron spin-polarization in the tunnel current was observed in many other junctions studied, including superconductor-superconductor junctions. P ranged from 60% to 85%. On the whole, the value of polarization inferred from the change in R_J is always larger than that obtained from fitting the conductance curves, which suggests that additional factors responsible for the junction resistance decrease, besides the EuS conduction band splitting, may exist. Thompson *et al.* (3) observed a six-fold decrease in the resistance in a Schottky tunnel barrier between In and Gd-doped EuS. They attributed it to the suppression of magnetic scattering at temperatures below T_m. Their experiment with degenerate semiconducting EuS cannot be directly compared with ours. However, the possibility of such an effect in our junctions should be further investigated.

4. PROXIMITY EFFECT

As mentioned earlier the quasiparticle spins in Al experienced an internal magnetic field (B^*) due to the exchange interaction at the EuS/Al interface. The exchange field was usually so large that little or no external field was needed to obtain Zeeman splitting in the quasiparticle DOS. This exchange field does not affect electron orbital motion, which makes the EuS/Al bilayer an experimental system for studying superconductivity in magnetic fields which act only on quasiparticle spins, a case on which several theoretical studies have been carried out (10).

We measured junctions having EuS/Al as one of the tunnel electrodes with different Al film thicknesses (d_s=4, 8, or 12 nm). Preliminary results indicated that the saturation exchange field seen by the Al quasiparticles varied inversely with d_s, in agreement with de Gennes's prediction (11): $B^* = a\Gamma S/d_s$, where a is the lattice constant of Al, S the magnetic moment of Eu^{2+}, and Γ the exchange constant. This exchange constant was determined to be about 13 meV from the slope of the B^* vs. d_s plot.

5. CONCLUSIONS

We have measured directly the value of polarization in tunnel current caused by the spin-filter effect. The proximity effect between the EuS barrier and the Al electrode gives rise to a large exchange field which is inversely proportional to the supercoductor film thickness. The spin-filter effect may be used to provide low energy spin-polarized electrons. The large internal field seen by the superconductor in the EuS/Al bilayer makes spin-sensitive tunneling in zero (applied) magnetic field possible. Both may be useful in applications.

ACKNOWLEDGEMENTS

We thank Richard MacNabb for fabricating the junctions. This work is supported by the NSF under grants DMR-8619087 and DMR-8822744.

REFERENCES

(1) N. Müller, W. Echstein, W. Heiland, and W. Zinn, Phys. Rev. Lett. 29 (1972) 1651; E. Kisker, G. Baum, A. H. Mahan, W. Raith, and K. Schröder, Phys. Rev. Lett. 36 (1976) 982.
(2) L. Esaki, P. J. Stiles, and S. von Molnar, Phys. Rev. Lett. 19 (1967) 852.
(3) W. A. Thompson, F. Holtzberg, T. R. McGuire, and G. Petrich, AIP Conf. Proc. No. 5, eds. C.D. Graham Jr. and J. J. Rhyne, (1971), pp.827-36.
(4) R. Meservey, P. M. Tedrow, and J. S. Moodera, J. Mag. and Mag. Mat. 35 (1983) 1.
(5) J. S. Moodera, X. Hao, G. A. Gibson, and R. Meservey, Phys. Rev. Lett. 61 (1988) 637.
(6) P. Wachter, CRC Crit. Rev. of Solid State Sci. 3 (1972) 189.
(7) J. A. X. Alexander, T. P. Orlando, D. Rainer, and P. M. Tedrow, Phys. Rev. B31 (1985) 5811.
(8) J. E. Tkaczyk, Ph. D. thesis, MIT, (unpublished) 1988; P. M. Tedrow, J. E. Tkaczyk, and A. Kumar, Phys. Rev. Lett. 56 (1986) 1746.
(9) J. G. Simmons, J. Appl. Phys. 34 (1963) 1793.
(10) P. Fulde, Adv. Phys. 22 (1973) 667.
(11) P. G. de Gennes, Phys. Lett. 23 (1966) 10.

Physica B 165&166 (1990) 469–470
North-Holland

CRITICAL FIELD ANISOTROPY OF ARTIFICIAL $YBa_2Cu_3O_7/PrBa_2Cu_3O_7$ SUPERLATTICES.

O.BRUNNER, J.-M. TRISCONE, L. ANTOGNAZZA, M.G KARKUT*, and Ø. FISCHER.

University of Geneva, DPMC, 24 Quai E.-Ansermet, 1211 Geneva 4, Switzerland.

We have measured the resistive transitions of $YBa_2Cu_3O_7/PrBa_2Cu_3O_7$ (YBCO/PrBCO) superlattices in magnetic fields parallel and perpendicular to the CuO planes. This series was built with constant 24Å YBCO layers and increasing thickness of the PrBCO layers. We find that the anisotropy in field increases dramatically when the PrBCO separation of the YBCO layers increases from 12 to 24Å. For a 12Å PrBCO separation the anisotropy is about 7, whereas with 24Å PrBCO, the anisotropy is at least 180. For thicker PrBCO layers (60Å, 96Å, and 144Å) the resistive transitions become field independent for the parallel orientation and the effect of a 90kG magnetic field is below our experimental resolution.

1. INTRODUCTION

We have recently reported the growth of YBCO/DyBCO(1) and YBCO/PrBCO(2) superlattices. For both systems we have shown that artificially layered structures can be obtained with the High T_c Superconductors (HTS) on a unit cell scale. In the YBCO/PrBCO system we found striking modifications of the superconducting properties as a function of the PrBCO thickness (2). This multilayer structure is particularly interesting since PrBCO, while isostructural to YBCO, is an insulator. With this system the epitaxial growth of ultrathin (12Å or 24Å, i.e. one or two unit cells) YBCO layers separated by increasingly thicker PrBCO layers, allows us to study the behavior of isolated YBCO ultrathin layers. We believe this is important since numerous theories treat these materials as being essentially two dimensional systems, whereas a compound such as YBCO is a 3 dimensional anisotropic superconductor(3). With our multilayers we can study how the superconducting properties evolve as the coupling between the individual YBCO layers decreases.

In this paper we report measurements of the resistive transitions of YBCO/PrBCO superlattices in a magnetic field applied either parallel or perpendicular to the CuO planes. We will focus on a series of samples with a constant YBCO layer thickness of 24Å and various PrBCO thicknesses (24Å-YBCO/d-PrBCO). The samples had a total thickness between 1200 and 1500Å.

2. DEPOSITION TECHNIQUE

The samples were prepared on polished MgO substrates by an alternating deposition of YBCO and PrBCO using dc single target planar magnetron sputtering. The details of the preparation of thin films of the individual compounds and of multilayers can be found in refs. (4), and (1) respectively.

3. RESULTS

We find for this series of 24Å-YBCO/d-PrBCO superlattices that the superconducting transition temperature T_c, defined by the 10% of the normal state resistance, decreases with increasing thickness d, but then remains about constant with a value of ≈50K when d >≈60Å. The T_c behavior of this 24Å-YBCO/d-PrBCO series and of the 12Å-YBCO/d-PrBCO series is discussed in ref. 5. Fig.1 shows the resistive transitions in parallel and perpendicular fields (0, 10, 30, 60, 90kG) for a YBCO sample, a 24Å/12Å, 24Å/24Å, and 24Å/144Å YBCO/PrBCO multilayers. Starting with the perpendicular field measurements we note that the behavior of the multilayers resembles that of YBCO with the characteristic broadening of the transition instead of the shift observed in "classical" superconductors. We note that a decrease of T_c from 90 K to 50 K does not lead to a reduction of the broadening, but, on the contrary, to a slight increase for the 24Å/144Å (T_c≈50 K) sample.

The transitions in a parallel field are strikingly different. Starting with the 24Å/12Å sample we note a broadening similar to, but slightly smaller than the one observed in YBCO. The anisotropy for this system is about 7 as compared to 4.5 for YBCO. We define the anisotropy as $A=H_{c2}//(T)/H_{c2}\perp(T)$ where the temperature is taken at 10% of the normal state

* Present Address: Laboratoire de Chimie Minérale B, URA 254 CNRS, Université de Rennes I, avenue du Général Leclerc, 35042, Rennes CEDEX, France.

0921-4526/90/$03.50

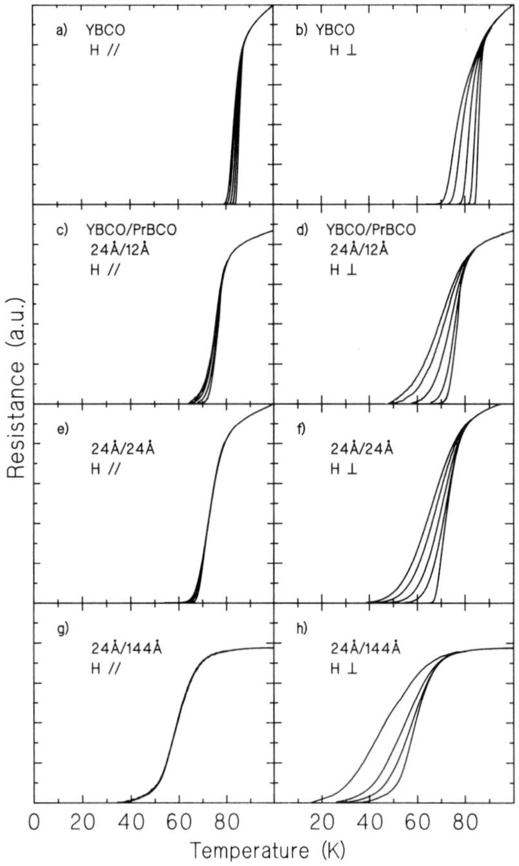

FIGURE 1
Resistive transitions in parallel and perpendicular fields of 0, 10, 30, 60 and 90kG. The 60kG perpendicular field curve for the 24Å/144Å sample is missing.

resistance in a parallel field of 90kG. (we use here the notation H_{c2}, with the understanding that H_{c2} determined resistively is probably not the real upper critical field).
The results for the 24Å/24Å and the 24Å/144Å multilayers (Fig. 1 e) and g)) are in sharp contrast to those found for YBCO and the 24Å/12Å YBCO/PrBCO sample. With a precise alignment of the field we find that parallel fields up to 90kG have only a small effect on the foot of the transition for the 24Å/24Å sample and no effect, within our experimental precision, on the 24Å/144Å multilayer. For the 24Å/24Å sample, $H_{c2}//=90$kG and $H_{c2}\perp=0.5$kG for T=68.9K giving an anisotropy of 180! For the 24Å/144Å sample the anisotropy is even larger and can simply not be defined by our criterion. These measurements show that our modulated structures are very different from single thin films of YBCO and reveal their highly two-dimensional behavior. Fig.2 shows the inverse anisotropy 1/A as a function of the PrBCO thickness. An abrupt change is clearly visible and occurs somewhere between 12 and 24Å of PrBCO separation. In ref. 2 we speculated that the linear decrease of T_c as a function of the PrBCO thickness in the series with 12Å YBCO could have its origin in the progressive decoupling of the YBCO layers. Following the same argument for the 24Å YBCO series leads to the conclusion that about 60Å of PrBCO are necessary to entirely decouple the YBCO layers. This PrBCO separation corresponds to the beginning of the T_c plateau discussed in ref. 5. The measurements presented here show that 24Å of PrBCO already produces a strong decoupling of the YBCO layers in the presence of a magnetic field. However, the complete decoupling seems to occur for a PrBCO layer thickness between 24Å and 60Å, not incompatible with the value suggested by the T_c behavior.

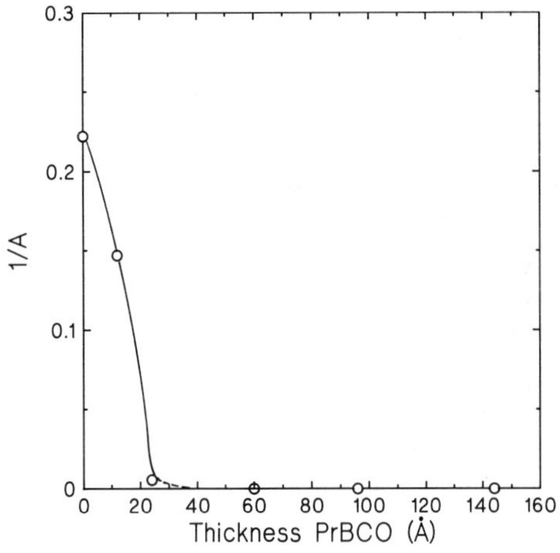

FIGURE 2
Inverse anisotropy as a function of d-PrBCO.

REFERENCES
(1) J.-M. Triscone, M.G. Karkut, L. Antognazza, O. Brunner, and Ø. Fischer, Phys. Rev. Lett. 63 (989) 1016.
(2) J.-M. Triscone, Ø. Fischer, O. Brunner, L. Antognazza, A.D. Kent, and M.G. Karkut, Phys. Rev. Lett. 64 (1990) 804.
(3) See for example, B. Oh, K. Char, A.D. Kent, M. Naito, M.R. Beasley, T.H. Geballe, R.H. Hammond, A. Kapitulnik, and J.M. Graybeal, Phys. Rev. B37 (1988) 7861.
(4) J.-M. Triscone, M.G. Karkut, O. Brunner, L. Antognazza, M. Decroux and Ø. Fischer, Physica C 158, 293 (1989).
(5) L. Antognazza, J.-M. Triscone, O. Brunner, M.G. Karkut, and Ø. Fischer, this volume.

Physica B 165&166 (1990) 471–472
North-Holland

SUPERCONDUCTING CRITICAL TEMPERATURES OF ARTIFICIAL $YBa_2Cu_3O_7/PrBa_2Cu_3O_7$ SUPERLATTICES.

L. ANTOGNAZZA, J.-M. TRISCONE, O. BRUNNER, M.G. KARKUT*, and Ø. FISCHER.

University of Geneva, DPMC, 24 Quai E.-Ansermet, 1211 Geneva 4, Switzerland.

We have prepared two series of $YBa_2Cu_3O_7/PrBa_2Cu_3O_7$ (YBCO/PrBCO) superlattices with constant, either 12Å or 24Å YBCO layers, and increasing thickness of the PrBCO layers. For both series the superconducting transition temperatures T_c first decreases linearly with increasing the PrBCO thickness, and then remains about constant. A plateau at ≈50K is clearly observed for the 24Å constant YBCO layers for d>60Å. For the 12Å constant YBCO layers a similar plateau seems to occur for d-PrBCO>70Å with a T_c value around 10K. The possible origins of this behavior are discussed.

1. INTRODUCTION

We have recently reported the growth of artificial YBCO/PrBCO superlattices(1) and shown that modulated structures can be made on a scale comparable to the unit cell size. Our interest in the YBCO/PrBCO system resides in the fact that PrBCO, while isostructural to YBCO, is not superconducting but insulating. In a previous publication(1) we have studied a series of samples with a constant 12Å YBCO layer thickness and various PrBCO thicknesses. In this paper we present the T_c behavior of a new series with YBCO layer thickness of 24Å (2 unit cells).

2. RESULTS AND DISCUSSION

The samples were prepared on polished MgO substrates by an alternating deposition of YBCO and PrBCO using dc single target planar magnetron sputtering. The preparation details can be found in Ref. (2).

We used the standard dc four point technique to determine T_c. We took the 10% point of the normal state resistance as the definition of T_c. Fig. 1 shows the resistively measured T_c as a function of the thickness of PrBCO (d-PrBCO) for the series with constant 24Å YBCO layer thickness (top) and constant 12Å YBCO layer thickness (bottom). For the 24Å-YBCO series, T_c decreases linearly with increasing d-PrBCO, and for d>60Å a clear plateau at about 50K is observed. In the 12Å-YBCO series the T_c reduction is more severe. T_c decreases linearly and seemed(1) to extrapolate to zero for a d-PrBCO around 90Å. For this paper we made a 12Å-YBCO/144Å-PrBCO multilayer and measured T_c to be 9.5K, leaving open the possibility that a plateau also occurs in the 12Å series. Clearly more samples are required to resolve this question.

The linear decrease of T_c as a function of increasing PrBCO layer spacing may have different origins: a) a proximity effect between YBCO and PrBCO. We consider this unlikely because the insulating nature of PrBCO makes a proximity effect over long distance improbable. b) strong interdiffusion between Y and Pr planes producing a modulated alloy -recall here that when Pr is substituted in YBCO T_c drops and is zero for about 60% Pr. c) the decoupling of the CuO bilayers by increasingly thicker insulating layers. This last point suggests that the interlayer coupling is important for high T_c superconductivity. We will now consider b): the possible effect of interdiffusion. We have shown in Ref.1 that the T_c behavior of the multilayers is different from that of an alloy. We compared a 12Å/12Å multilayer to the corresponding $(Y_{0.5}Pr_{0.5})BCO$ alloy and found the T_c of the

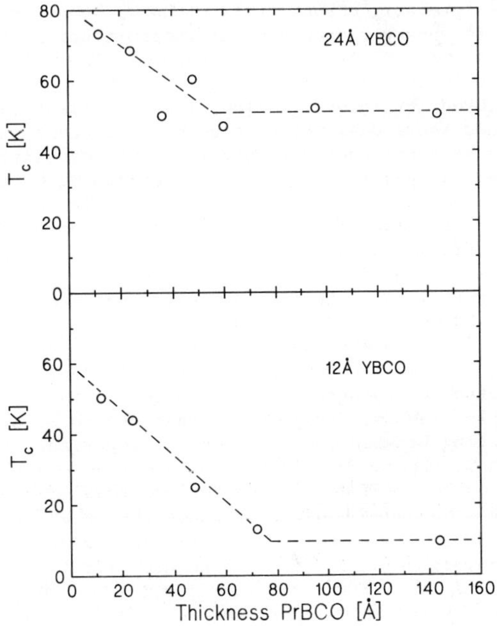

FIGURE 1
T_c as a function of the PrBCO thickness.

*Present Address: Laboratoire de Chimie Minérale B, URA 254 CNRS, Université de Rennes I, avenue du Général Leclerc, 35042, Rennes CEDEX, France.

multilayer to be twice the T_c of the alloy. This result reinforces the x-ray spectra which showed that modulated structures can be obtained on a unit cell scale and that interdiffusion is sufficiently weak so as not to destroy the modulation. However, there will be a certain amount of interdiffusion -one certainly does not expect 100% Y planes in these series because of the unavoidable "mechanical" mixing discussed in ref.1. Two questions to be addressed are first, what is the amplitude of the yttrium modulation? and second, does this amplitude depend on the thickness of the separation PrBCO layer? This second question is important because this effect could play a role in the T_c behavior as a function of d-PrBCO. If T_c(d-PrBCO) is the result of interdiffusion, it means that this interdiffusion occurs on a typical length on the order of 70Å or more. To test this point we made a 24Å-YBCO/144Å-PrBCO multilayer, 10 layers of each material. The growing time was 55 minutes and we let the sample at the growth temperature (650-700°C) for 3½ additional hours. After this treatment the sample was measured, the transition onset was 60K and the 10% point was at 22K (compare this with the T_c=50K sample in Fig.1). This result shows that some thermal interdiffusion is present but it also clearly demonstrates that it is mainly restricted to the adjacent planes since an interdiffusion over longer distances would destroy superconductivity (the Pr concentration in this particular multilayer is 86%). We conclude that thermal interdiffusion most probably cannot explain the T_c behavior in Fig.1. However, non-thermal interdiffusion can occur in strained multilayers(3). The driving force in this case are the strains which can be relieved by interdiffusion. The experiment mentioned above does not test such a process and thus we cannot definitively rule out interdiffusion as a possible explanation of the T_c behavior even though we think the strain-induced interdiffusion is probably slight considering the small mismatch between the YBCO and PrBCO lattices. In addition our x-ray results are consistent with weak interdiffusion between adjacent YBCO and PrBCO planes.

A more plausible explanation for the T_c decrease with increasing d-PrBCO is that it is related to a progressive decoupling of the YBCO single layers, suggesting that the interlayer coupling between CuO bilayers is important for high T_c superconductivity. The decrease observed for both series would then reflect this decoupling and the plateaus reflect the behavior of isolated 12Å or 24Å YBCO layers. Following these speculations one would conclude first, that ≈60-70Å of PrBCO are sufficient to entirely decouple the YBCO layers and second, that the T_c of a single 24Å, 12Å YBCO layers would be respectively 50K and 10K. This result means that two unit cells of YBCO are enough for superconductivity to occur. The recent result obtained here on the 12Å-YBCO/144Å-PrBCO multilayer still requires confirmation to understand whether one unit cell of YBCO is or is not superconducting.

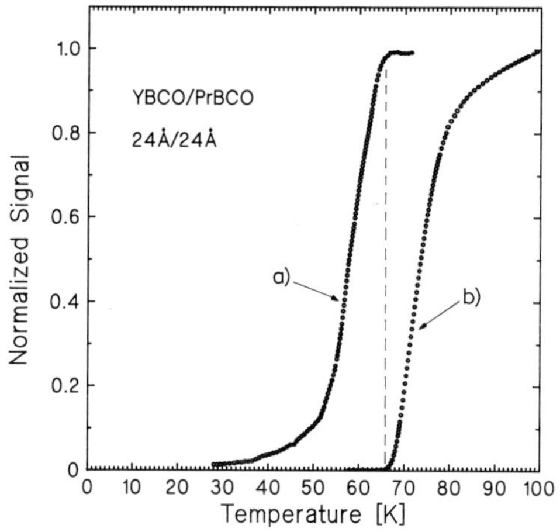

FIGURE 2
T_c measured a) inductively, b) resistively.

Fig. 2 shows the comparison between a resistive transition and an ac susceptibility transition for a 24Å-YBCO/24Å-PrBCO multilayer. The ac susceptibility measurement gives an onset which corresponds to the zero resistance of the resistive measurement. This behavior has been observed for all the multilayers measured by both methods and is also observed for YBCO single layers. The transition widths 10%-90% measured by ac susceptibility correlate very well with the results obtained by resistive measurements. This gives additional weight to the homogeneity of the samples.

3. CONCLUSIONS

We have made two series of superlattices: 12Å-YBCO/d-PrBCO and 24Å-YBCO/d-PrBCO. We find for both series that T_c decreases linearly with d-PrBCO. For the 24Å YBCO series a clear plateau with T_c=50K is observed for d>≈60Å. A plateau seems also to occur for the 12Å YBCO series with T_c=10K. We have argued that it is unlikely that these results can be explained by either interdiffusion or proximity effect. Instead, we believe that we are seeing the result of the decoupling of the ultrathin YBCO layers by increasingly larger insulating layers.

REFERENCES

(1) J.-M. Triscone, Ø. Fischer, O. Brunner, L. Antognazza, A.D. Kent, and M.G. Karkut, Phys. Rev. Lett. 64 (1990) 804.
(2) J.-M. Triscone, M.G. Karkut, L. Antognazza, O. Brunner, and Ø. Fischer, Phys. Rev. Lett. 63 (989) 1016.
(3) See, for example, A.L. Greer, and F. Spaepen, in Synthetic Modulated Structures, edited by L. Chang and B.C. Giessen (Academic, New York, 1984), p.419.

Physica B 165&166 (1990) 473–474
North-Holland

ANISOTROPY AND MAGNETIC FIELD DEPENDENCE OF CRITICAL CURRENTS IN Pb/Ge MULTILAYERS

D. NEERINCK, K. TEMST, M. DHALLE, C. VAN HAESENDONCK, Y. BRUYNSERAEDE
Laboratorium voor Vaste Stof-Fysika en Magnetisme, K.U. Leuven, B-3030 Leuven, Belgium
A. GILABERT
Laboratoire de Physique de la Matière Condensée, Université de Nice, F-06034 Nice Cedex, France
Ivan K. SCHULLER
Physics Department-B019, University of California-San Diego, California 92093, U.S.A.
T. KREKELS, G. VAN TENDELOO
Universiteit Antwerpen (RUCA), B-2020 Antwerpen, Belgium

We have measured critical currents in Pb/Ge multilayers as a function of direction and magnitude of a magnetic field perpendicular to the current. The higher J_c for a field parallel to the layers indicates flux pinning at the Pb/Ge interfaces. The critical current as a function of a perpendicular field displays a pronounced maximum at low fields. TEM reveals Pb grain sizes of about 1000 Å and indicates that the enhanced flux pinning may be explained by grain boundary pinning.

1. INTRODUCTION

Artificially prepared superconducting multilayers have received a renewed interest, since they offer the possibility to test phenomena related to the high-T_c superconductors. Up to now, most work in multilayers has been focussed on the dimensional effects in the critical fields (1), whereas the behaviour of the critical current has been much less studied. Critical current measurements in Nb/Ta (2) and NbN/AlN (3) multilayers both show an enhanced flux pinning for parallel magnetic fields due to the multilayering. Recently, Tachiki and Takahashi (4) presented a model to explain the critical current anisotropy in layered superconductors, which was then applied to single crystal high-T_c superconductors. In this paper we focus on the anisotropy and magnetic field dependence of artificial superconducting Pb/Ge multilayers.

2. EXPERIMENTAL

The Pb/Ge multilayer samples were prepared by electron beam evaporation in an UHV chamber. The base pressure of the system is 2×10^{-9} Torr, and the pressure increased during evaporation up to 10^{-8} Torr. The evaporation rates (5 Å/s for Pb, 2 Å/s for Ge) were kept constant using a quadrupole mass spectrometer (5). The layers were evaporated onto liquid nitrogen cooled oxidized silicon wafers. All multilayers consist of 10 bilayers. The top and bottom layers are always Ge, and the sample is covered with an extra 500 Å Ge protective layer. The four point pattern was defined photolithographically using a liftoff technique. The measurements were performed in a standard ^4He cryostat, equipped with a 7 Tesla superconducting coil. The critical current is defined as the current necessary to produce a 2 μV voltage across the sample at 4.2 K. Other voltage criteria yielded very similar results.
The structure of Pb/Ge (crystalline/amorphous) multilayers have been extensively studied and are described elsewhere (6).

3. RESULTS AND DISCUSSION

Fig. 1 shows the critical current density J_c for a Pb/Ge (220Å/50Å) multilayer as a function of the angle θ between an applied magnetic field $H = 10^3$ Gauss and

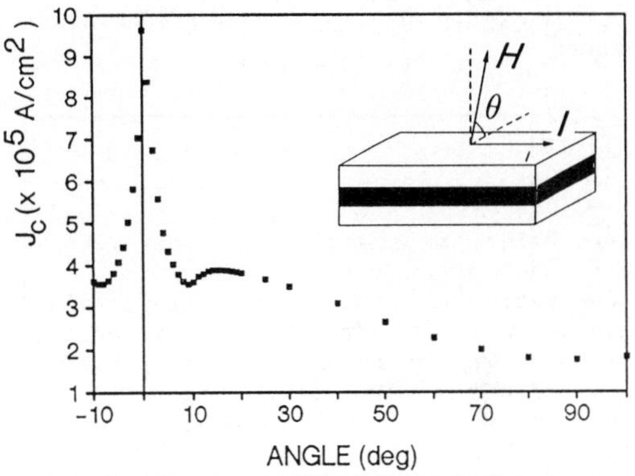

Fig. 1 : Critical current density at 4.2 K for a Pb/Ge (220Å/50Å) multilayer as a function of the angle between an applied magnetic field of 1000 Gauss and the layers. The inset shows the geometrical configuration.

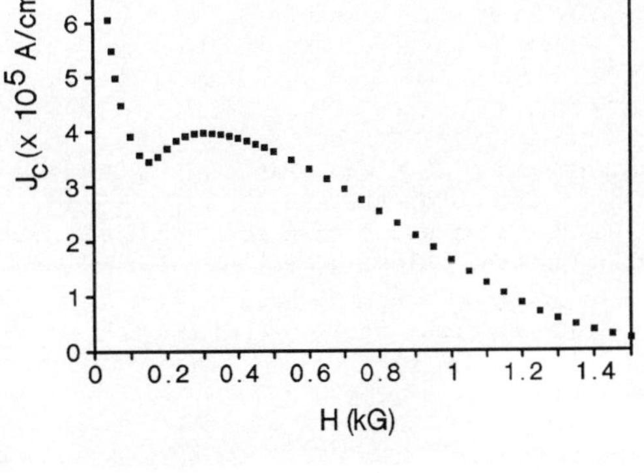

Fig. 2 : Critical current density at 4.2 K for a Pb/Ge (220Å/50Å) multilayer as a function of perpendicular magnetic field.

the film surface. The direction of the current was always perpendicular to the magnetic field. The much higher value of the depinning current for the parallel field shows the importance of flux pinning in the normal layers or at the superconductor/normal interfaces, as reported in Nb/Ta (2) and NbN/AlN (3) multilayers. However, the anisotropy of J_c in this case shows a remarkable new feature : the presence of a pronounced maximum at an angle $\theta_m \simeq 15$ degrees. Applying a larger magnetic field shifts this maximum to smaller angles, indicating that the maximum is caused by the perpendicular component $H_\perp = H sin\theta$. Indeed, the critical current density as a function of perpendicular magnetic field (θ=90 degrees) displays a maximum at $H_{\perp m} \simeq 260$ Gauss (Fig. 2). Qualitatively the same behaviour was found in a Pb/Ge (100Å/50Å) multilayer. The parallel field dependence of J_c will be the subject of future work. Assuming an equilateral triangular vortex lattice, the distance between the flux lines at which enhanced flux pinning occurs is $d = (\sqrt{3}\Phi_0)/2H_{\perp m})^{1/2} \simeq 2600$ Å, where Φ_0 is the flux quantum. The perpendicular critical field at 4.2 K is 2.2 kG, leading to a vortex normal core radius $\xi_{||} \simeq 390$ Å, much smaller than the flux line distance.

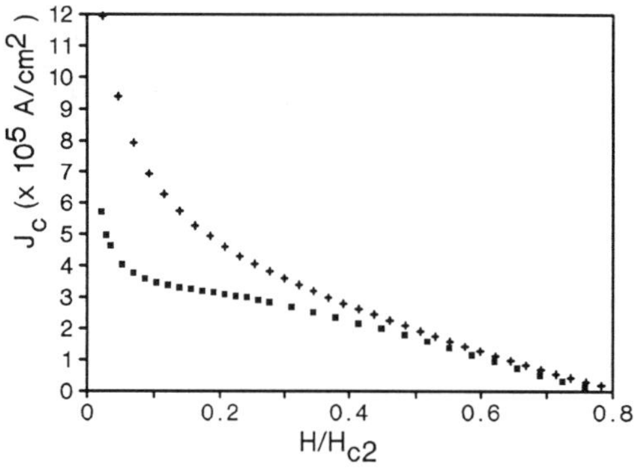

Fig. 3 : Critical current density for a 250 Å single Pb film as a function of perpendicular field (■) and parallel field (+), in reduced units ($H_{c2\perp}$=1.5 kG, $H_{c2||}$=21.7 kG).

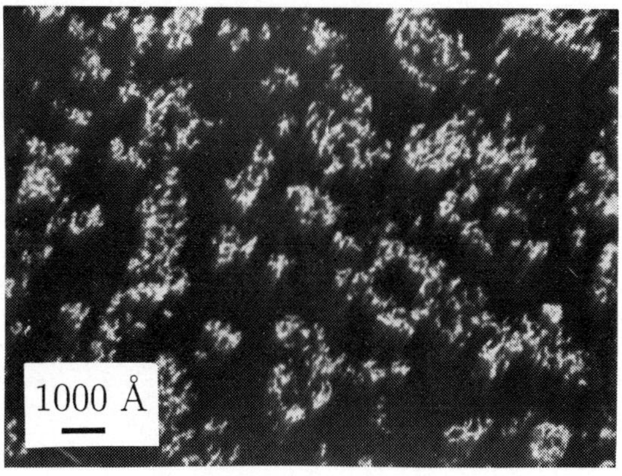

Fig. 4 : Dark Field TEM photograph of a 250 Å single Pb-film sandwiched between 50 Å Ge.

Since this enhanced flux pinning is caused by a perpendicular field, we expect the pinning mechanism to be the same in single Pb films. We prepared single Pb films of thicknesses 80 Å, 150 Å and 250 Å, sandwiched between 50 Å Ge, on locally thinned substrates, and in identical circumstances as the Pb/Ge multilayers. The critical current density as a function of perpendicular magnetic field (Fig. 3) for the 250 Å Pb film does not show a pronounced maximum, but does not decrease as monotonously as in the parallel field case (shown here for comparison). This behaviour is the same for the other film thicknesses. The microstructure of the thin films was investigated by dark field transmission electron microscopy and reveals Pb grains of dimensions varying between 500 Å and 1500 Å for all film thicknesses (Fig. 4). Since the in-plane grain dimensions are larger than the film thickness, the grain boundaries are almost perpendicular to the film surface. Therefore, grain boundaries can act as pinning centra for a perpendicular field. However, the grains have a large distribution in shape and dimensions, as seen in Fig. 4, so that an exact agreement between the vortex spacing and the grain size cannot be expected. The more pronounced maximum in the J_c versus $H_\perp$ for the Pb/Ge multilayer compared to the single film can be understood in this picture as resulting from an increase in surface density of grain boundaries.

In summary, we measured the critical current in Pb/Ge multilayers and single Pb films as a function of the perpendicular magnetic field and the angle between the layers and the field. The critical current in the parallel field case is larger than for a perpendicular field due to flux pinning in the Ge layers or at the Pb/Ge interfaces. The critical current as a function of perpendicular magnetic field displays a maximum in Pb/Ge multilayers, which may be interpreted in terms of grain boundary flux pinning.

ACKNOWLEDGEMENTS

This work was supported by the Belgian Inter-University Institute for Nuclear Sciences (I.I.K.W.), the Inter-University Attraction Poles (I.U.A.P.) and the Concerted Action (G.O.A.) Programmes (at K.U.L.), and by the U.S. Department of Energy under contract number DE-FG03-87ER45332 (at U.C.S.D.). International travel was provided by NATO. D.N. is a Research Assistant of the Belgian National Fund for Scientific Research (N.F.W.O.) K.T. is a Research Fellow of the I.I.K.W., M. D. is a Research Fellow of the I.W.O.N.L. and C.V.H. is a Research Associate of the N.F.W.O.

REFERENCES

(1) Ivan K. Schuller, J. Guimpel and Y. Bruynseraede, MRS Bulletin Vol. XV, 2 (1990) 29.
(2) P.R. Broussard and T.H. Geballe, Phys. Rev. B 37 (1988) 68.
(3) J.M. Murduck, D.W. Capone, II, I.K. Schuller, S. Foner and J.B. Ketterson, Appl. Phys. Lett. 52 (1988) 6.
(4) M. Tachiki and S. Takahashi, Solid State Comm., in print.
(5) W. Sevenhans, J.-P. Locquet and Y. Bruynseraede, Rev. Sci. Instr. 57 (1986) 937.
(6) D. Neerinck, H. Vanderstraeten, L. Stockman, J.-P. Locquet, Y. Bruynseraede and I.K. Schuller, Journal of Physics : Condensed Matter, in print.

Physica B 165&166 (1990) 475–476
North-Holland

PARALLEL CRITICAL FIELDS IN $Nb/Nb_{0.6}Zr_{0.4}$ MULTILAYERS

J. Aarts, K.-J. de Korver, W. Maj and P.H. Kes,

Kamerlingh Onnes Laboratory, University of Leiden,
P.O. Box 9506, 2300 RA Leiden, The Netherlands.

In parallel critical fields of multilayers of $Nb/Nb_{0.6}Zr_{0.4}$, both the usual 3D-2D transition and a second crossover from 2D to 3D behaviour may occur. By changing the Nb thickness d_{Nb} while keeping the NbZr thickness d_{Nz} constant, we show that the effective thickness of the layer causing 2D behaviour is the Nb thickness plus parts of the enclosing NbZr layers. As a consequence, a 3D-3D transition is found below a certain value of d_{Nz}.

1. INTRODUCTION

In the parallel critical field $B_{c2\|}$ of superconducting multilayers a dimensional crossover can occur from three-dimensional (3D) to two-dimensional (2D) behaviour. When both constituents have the same T_c but one layer is clean (large coherence length ξ) and the other is dirty (small ξ), Takahashi and Tachiki predicted (1) that a second crossover (2D to 3D) may be found. This occurs since the order parameter nucleates in the layer which gives rise to the highest critical field. In the 2D regime (if present) this is the clean layer, but if $\xi(T)$ becomes much less than the thickness of the dirty layer, nucleation in this layer is preferred, since this leads to a higher critical field. B_{c2} then increases sharply. Such crossovers were for the first time reported by Karkut et al. (2), and very recently by Kuwasawa et al. (3). However, a more quantitative understanding of the expected temperature (field) for the crossover points as function the layer thickness still seems to be lacking. We demonstrate here, using multilayers of $Nb/Nb_{0.6}Zr_{0.4}$, how such understanding may be gained by varying one layer thickness while keeping the other constant. Especially, by varying d_{Nb} we show that B_{c2} in the 2D regime is correlated with the Nb layer thickness but that parts of the enclosing layers also contribute. Also, by varying d_{Nz} we show that a minimal thickness of the NbZr layers is required to 'decouple' the Nb layers and so to observe the 2D regime. Part of this work has been reported elsewhere (4).

2. RESULTS AND DISCUSSION

Samples were prepared by sputtering. Details of preparation and analysis are given elsewhere (4). The sapphire substrates were kept near room temperature during deposition, minimizing diffusion. X-ray diffraction still showed crystalline (002) Bragg peaks of Nb and $Nb_{0.6}Zr_{0.4}$. Critical fields were determined by 4-point resistivity measurements and defined by 50 % of the transition.

Basic properties of the constituents were measured on films of Nb and $Nb_{0.6}Zr_{0.4}$. Nb layers showed T_c's of 9.2 K, transition widths around 0.02 K and a perpendicular critical field slope of about 0.25 T/K, corresponding to $\xi(0)$ = 120 Å. The layers therefore appear to be free of (oxygen) impurities; the small coherence length is due to a small electronic mean free path as a consequence of sputtering the films at room temperature. Films of $Nb_{0.6}Zr_{0.4}$ showed T_c's around 10.7 K, transition widths around 0.07 K and a perpendicular critical field slope of about 1.5 T/K, corresponding to $\xi(0)$ = 45 Å.

Multilayers, consisting of 21 single layers with NbZr as the first and the last layer, were prepared with a constant value d_{Nz} = 190 Å and values for d_{Nb} of 170 Å, 115 Å, 85 Å and 55 Å, as well as with a constant value d_{Nb} = 115 Å and values for d_{Nz} of 125 Å and 60 Å. T_c's are given in table I. Measured parallel critical fields as function of the reduced temperature $t = T/T_c$ are given in fig.1 (constant d_{Nz}) and fig.2 (constant d_{Nb}). For constant d_{Nb}, all samples show a (linear) 3D regime near T_c followed by a 3D -2D transition around t = 0.95. The 2D regime becomes less clearly visible with decreasing Nb thickness. This is due to the fact that the 3D slope near T_c steadily increases, since it is determined by the average Zr-concentration in one period of the multilayer (4). In these samples a second transition in the form of a steep upturn around t = 0.8 is present. Note that the field at which this transition takes place increases with decreasing d_{Nb}. The upturn asymptotically approaches bulk NbZr behaviour (4). The samples with constant d_{Nz} (fig.2) show 3D behaviour near T_c, and for d_{Nz} = 125 Å a 3D -2D transition, followed by a sharp upturn near t = 0.5. For d_{Nz} = 60 Å, $B_{c2\|}$ shows no 3D -2D transition. Instead, the slope near T_c of about 0.6 T/K changes to 1.14 T/K at low temperatures.

We try to describe the behaviour of $B_{c2\|}$ in the 2D regime by the formula for single thin films :

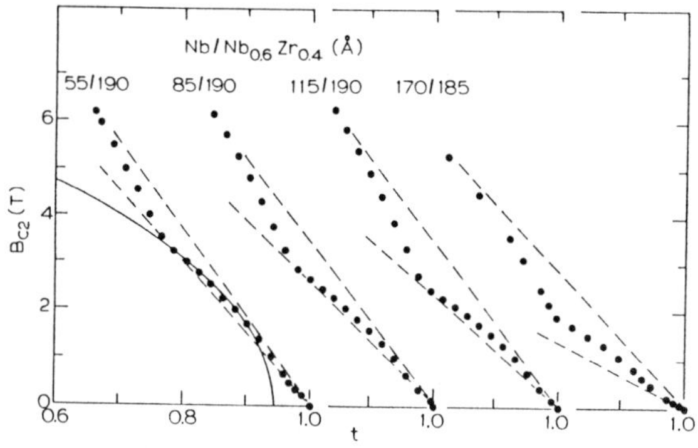

Fig. 1. $B_{c2\|}$ versus reduced temperature for samples with constant d_{Nz}.

$$B_{c2}(T) = \sqrt{12}\,\frac{\phi_o}{2\pi}\,\frac{1}{\xi(0)d}\,(1-T/T_{c2D})^{1/2} \qquad (1)$$

Here, d is the film thickness and T_{c2D} is the critical temperature of the film. The square root behaviour can be fitted well, as is shown for one sample in fig.1. The fit results in values for $\xi(0)d$ and T_{c2D}, given in table I. The values for $\xi(0)d$ are plotted in fig.3 against d_{Nb}. They fall on a straight line, which intercepts the thickness axis at minus 100 Å and has a slope $\xi(0)$ = 90 Å (lower than the value for Nb). We interpret this to mean that the entity responsible for the 2D behaviour consists of the Nb layer, "dressed" with about 50 Å NbZr on either side. The critical temperature T_{c2D} of this entity then should lie between the T_c's of Nb and $Nb_{0.6}Zr_{0.4}$, as is found. Note that the coherence length of the NbZr layer is about 50 Å, and that d_{Nz} is larger than 100 Å for all samples where 2D behaviour is found. It can then be expected that if d_{Nz} falls below 100 Å, decoupling of the entities is no longer possible and 2D behaviour cannot appear anymore. This is in agreement with the observed behaviour for d_{Nz} = 60 Å. The transition appearing in this sample indicates that there is still a difference between a regime near T_c where the critical field slope is due to averaging over many layers, and the low temperature regime, where nucleation still seems to be preferred in the NbZr layer. On the other hand, the value of 1.14 T/K shows that the behaviour is not simply bulk $Nb_{0.6}Zr_{0.4}$ anymore. A final remark concerns the low temperature (t = 0.6) of the anomalous upturn in the sample with d_{Nz} = 125 Å. We found before (4) that the temperature of the upturn t_{up} occurs when the coherence length which follows from the linear slope at T_c (ξ_{3D}) has decreased to about 0.4 Λ (Λ is the superlattice wavelength). The values for $\xi_{3D}(t_{up})/\Lambda$ are given in table I. The sample with d_{Nz} = 125 Å follows the above description and the low value for t_{up} is essentially due to the small value for Λ. It appears then that upon

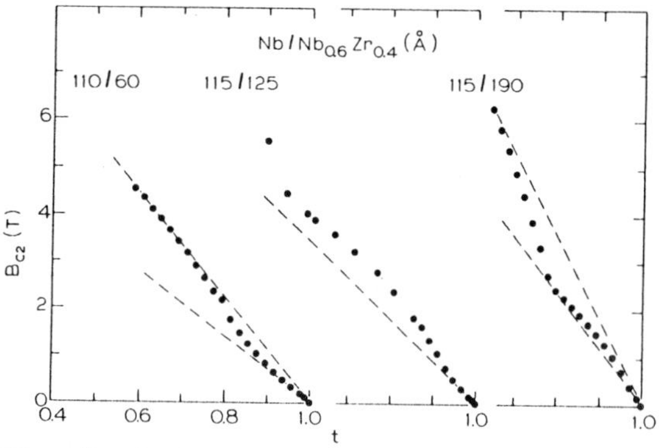

Fig. 2. $B_{c2\|}$ versus reduced temperature for samples with constant d_{Nb}.

decreasing d_{Nz} the 2D region first becomes larger, and may disappear abruptly around 100 Å.

This work is part of the research program of the Dutch Foundation 'F.O.M.'.

REFERENCES

(1) S. Takahashi and M. Tachiki, Phys.Rev. B34 (1986) 3162.
(2) M.G. Karkut, V. Matijasevic, L. Antognazza, J.-M. Triscone, N. Missert, M.R. Beasly, O. Fischer, Phys.Rev.Lett. 60 (1988) 1751.
(3) Y. Kuwasawa, U. Hayano, T. Tosaka, S. Nakano, S. Matuda, Physica C165 (1990) 173.
(4) J. Aarts, K.-J. de Korver and P.H. Kes, submitted to Europhys. Lett.

Table I. Multilayer parameters

d_{Nb}	d_{Nz}	T_c	$\xi(0)d$	T_{c2D}	$\xi_{3D}(t_{up})/\Lambda$
Å	Å	K	10^{-14} m^2	K	
170	185	10.3	237	9.8	0.44
115	190	10.5	197	9.9	0.37
85	190	10.4	165	9.9	0.38
55	190	10.4	143	9.9	0.40
115	125	10.4	177	9.7	0.38
110	60	9.8	-	-	-

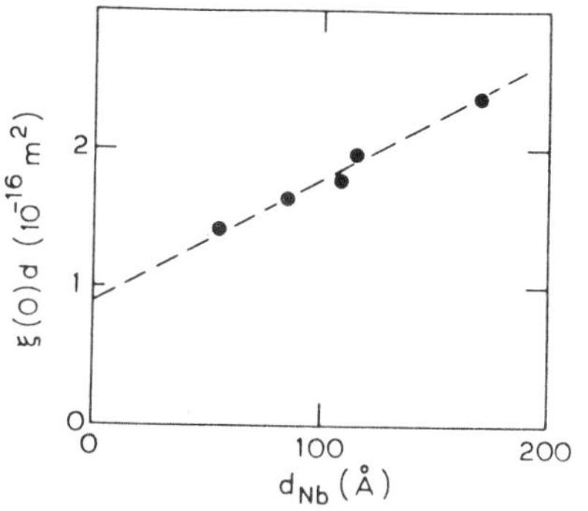

Fig. 3. Values of $\xi(0)d$ versus d_{Nb}. The dashed line is a guide to the eye.

Physica B 165&166 (1990) 477–478
North-Holland

ANOMALOUS UPPER CRITICAL FIELDS OF SUPERCONDUCTING $Nb/Nb_{0.5}Zr_{0.5}$ MULTILAYERS

Y. KUWASAWA, T. TOSAKA, A. UCHIYAMA, Y. HAYANO and S. NAKANO

Department of Physics, Faculty of Science, Chiba University, 1-33 Yayoicho, 280 Chiba, Japan

Upper critical fields have been studied for $Nb/Nb_{0.5}Zr_{0.5}$ multilayers grown onto the substrates of 300℃ by magnetron sputtering. The multilayers are constituted from two superconducting materials with different electronic diffusion constants and nearly equal critical temperatures. For the samples with period of 360-, 400- and 500-Å, the upturn features at lower temperature in addition to dimensional crossover were observed in parallel upper critical field $H_{c2\parallel}$. Noticeable changes were also found in the transition width $\Delta H_{c2\parallel}(T)$ and the angular dependence of H_{c2} in the region of temperature which the upturn took place. The positive curvatures of $H_{c2\perp}$ had a tendency to depend upon Nb layer thicknesses.

In artificially layered superconductors, the study of multilayer, stacked alternately by superconducting layers with different electronic diffusion constants, is of considerable interest. For the system constiuted from two materials with nealy equal critical temperature T_c, it has been predicted by Takahashi and Tachiki(1,2) that the parallel upper critical fields $H_{c2\parallel}(T)$ would exhibit anomalous upturn behavior at lower temperature in addition to the 3D to 2D crossover. Experimentally, Karkurt et al.(3) first verified such an effect in the multilayers of $Nb_{0.6}Ti_{0.4}/Nb$. Most recently, we observed the upturn feature in $Nb/Nb_{0.5}Zr_{0.5}$ multilayers and further we reported a variation in the characteristics of angular dependence of H_{c2} above and below temperature which the upturn occurred(4). In this paper, we present new data of upturn feature for sample with period $\Lambda = 360$ Å (d_{Nb}/d_{NbZr}=240 Å/120 Å) in comparision with our previous data.

Preparation of samples was made by using two magnetron sputtering guns. Multilayers were deposited onto the sapphire substrates heated at 300℃. The procedures for determining this temperature and sputtering conditions have been described in

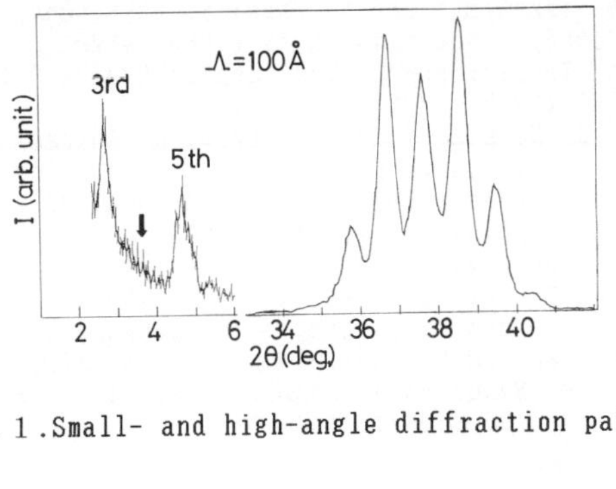

Fig. 1. Small- and high-angle diffraction patters.

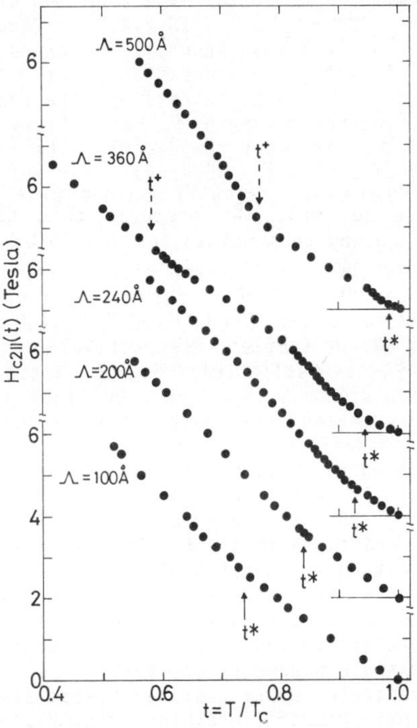

Fig. 2. $H_{c2\parallel}$ vs t. The t^* and t^+ indicate the 3D to 2D and upturn points, respectively.

Ref.4. The thicknesses were independently determined by the preparation condition and x-ray diffraction measurements. Fig. 1 shows diffractmeter scan for the low angle satellites and the satellites around main Bragg peak for a representative sample. The low-angle reflections have no even-order peaks. This implies that the regular structure was well developed. The thickness obtained

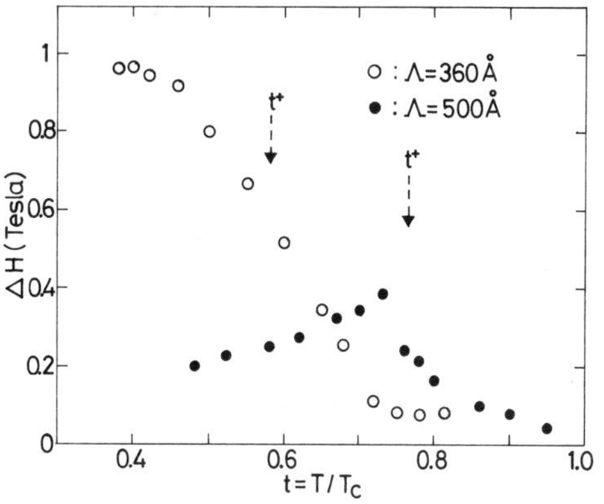

Fig. 3. Plots of the transition width ΔH for $H_{c2\parallel}$ as a function of t.

from these data was in agreement with that calculated from calibrated sputtering rates.

Parallel upper critical fields for various periodic samples are plotted as a function of reduced temperature $t = T/Tc$ in Fig. 2. Dimensional crossover(3D→2D) phenomenon is observed for all samples. The 3D to 2D transition point t^* is approaching to Tc with increasing the period Λ and the coherence length $\xi_\perp$ at t^* is nearly equal to Λ. At lower temperature the anomalous upturn features were observed. For multilayers with $\Lambda = 360$ Å(d_{NbZr}=120 Å) and 500 Å(d_{NbZr}=250 Å), the upturn transition temperatures t^+'s are defined here as the intersection of lineally extraporated lines in low and high temperature region on the $(H_{c2\parallel})^2$ vs. t curves. Thus we obtain $t^+ \fallingdotseq 0.58$, 0.76 and $\xi_\perp(t^+) \fallingdotseq 81$ Å, 114 Å for the 360- and 500-Å samples, respectively, in which $d_{NbZr} > \xi_\perp$ is satisfied. Fig. 3 exhibits the transition width $\Delta H_{c2\parallel}$ as a function of t in the upturn region for both samples, where the $\Delta H_{c2\parallel}$ is defined by the 10 - 90% transition points of normal resistivity for a magnetic field sweeping. As seen from this figure, the width is increasing abruptly with decreasing temperature near t^+ and there is a difference in the maximum values between two samples. The former may be interpreted as the enhanced fluctuation due to the situation of order parameter to nucleate, the latter may be attributed to the difference of d_{NbZr}. More works are needed at present.

Perpendicular upper critical fields are plotted against reduced temperature in Fig. 4. The positive curvature takes place in the 2D region of $H_{c2\parallel}(T)$ for any samples. The position $t^\perp$, which deviates from a straight line, is shifting toward higher temperature with increase of Λ, but there is no appreciable change between 360- and 500-Å samples. The upturn slopes in a lower temperature for all samples indicate same values which are very close to $(dH_{c2\perp}/dT)_{Tc}$ of our thick $Nb_{0.5}Zr_{0.5}$ film (Ref.4). The slope near Tc for samples with $d_{Nb} \geqq 200$ Å is close to that of a single Nb layer(Ref.4). From our data described

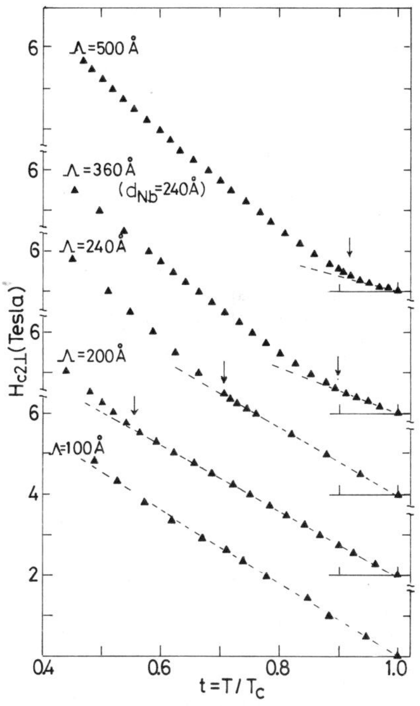

Fig. 4. $H_{c2\perp}$ vs t.

above, it is presumably suggested that the positive curvature strongly correlates with the cleanness of Nb layer, that is, its electronic diffusion constant D_{Nb}, which depends upon Nb-layer thickness. Similar behavior has been reported in Nb/Ta superlattice(5,6). This fact may be reflected to the characteristics of $H_{c2\parallel}$ through the ratio D_{Nb}/D_{NbZr}.

In summary, we have presented the upturn features in $H_{c2\parallel}(T)$ and the positive curvature in $H_{c2\perp}(T)$ in the system of multilayer composed of materials with equal transition temperatures and different electronic diffusion constants. These results are explained qualitatively by T.T.theory (1,2). Our results for angular dependence of H_{c2} will be presented at the LT19.

REFERENCES

(1) S. Takahashi and M. Tachiki, Phys. Rev. B 34 (1986) 3162; ibid. B 33 (1986) 4620.
(2) M. Tachiki and S. Takahashi, Physica C 153 - 155 (1988) 1702.
(3) M.G. Karkut, V. Matijasevic, L. Antognazza, J.-M.Triscone, N. Missert, M.R. Beasley and Ø. Fischer, Phys. Rev. Lett. 60(1988)1751; ibid. Physica C 153-155 (1988) 473.
(4) Y. Kuwasawa, U. Hayano, T. Tosaka, S. Nakano and S. Matuda, Physica C 165 (1990) 173.
(5) J. L. Cohn, J.J. Lin, F.J. Lamelas, H. He, R. Clarke and C. Uher, Phys. Rev. B 38(1988)2326
(6) T. R. Brousssard and T. H. Geballe, Phys. Rev. B 35 (1987) 1664.

Physica B 165&166 (1990) 479–480
North-Holland

VERTICAL RESISTIVE TRANSITION IN Si/Nb MULTILAYERS

S.N. Song and J.B. Ketterson

Department of Physics and Astronomy, Northwestern University, Evanston, Il. 60208, USA

The vertical and in-plane resistive transition in a series of Si/Nb multilayers has been studied. An anisotropic percolation model is adequate to explain the observed transition behavior. The measured temperature dependence of the supercurrent is compared with the Ambagaokar-Baratoff theory. The tail structure in $I_c(T)$ curves near T_c is interpreted as the short coherence length effect.

1. INTRODUCTION

High T_c oxide superconductors may be characterized as highly conductive Cu-O planes with "activated" conduction both between planes and across the grain boundaries. To understand their highly anisotropic transport and superconducting properties, a model system in which the important material parameters can be varied artificially is desirable. In this paper we report results on studies of the vertical resistive transition in a highly anisotropic layered structure using Si/Nb multilayer as a model system.

2. SAMPLE PREPARATION

Structures involving a thick Nb base layer-Si/Nb multilayer-thick Nb top layer were grown by e-beam evaporation in an UHV system on sapphire substrates. For in-plane transport measurements, identical Si/Nb multilayers without the base and top Nb layers were deposited in the same run with the same deposition parameters. Film growth was controlled by a computer in a constant thickness mode. Mesa structures were defined by standard photolithographic techniques followed by chemical etching.

3. RESULTS AND DISCUSSIONS

The vertical resistive transition in various multilayers was measured using the dc four probe method . Figure 1 shows the temperature dependence of the vertical resistance for a 60 period multilayer having a Si layer thickness d_{Si}=26Å and d_{Nb}=12Å(denoted as 26/12:60). The inset shows the superconducting transition of an identical multilayer prepared without the bottom and top thick-Nb layers. As can be seen, the true zero resistance state in the vertical direction is very sensitive to the current level, as expected for a weak link structure. In a 2D film having a thickness d, the fluctuation-enhanced conductivityit is given by[1]

$$\Delta\sigma = \frac{1}{16}\frac{e^2}{\hbar d}\frac{T_c}{T-T_c}. \quad (1)$$

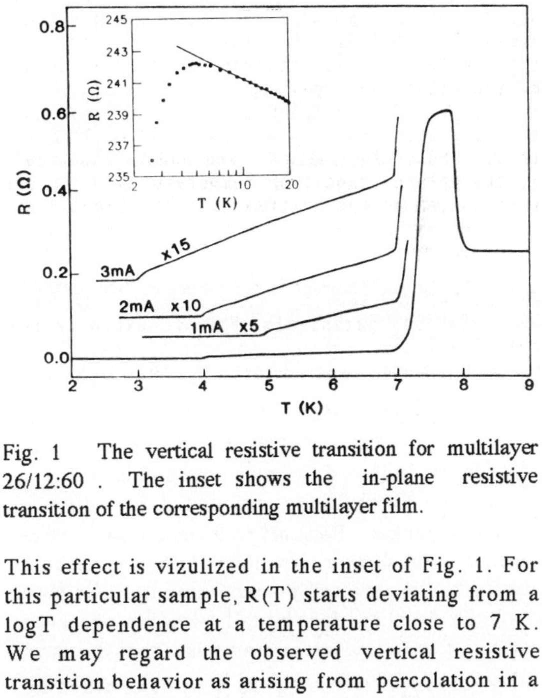

Fig. 1 The vertical resistive transition for multilayer 26/12:60 . The inset shows the in-plane resistive transition of the corresponding multilayer film.

This effect is vizulized in the inset of Fig. 1. For this particular sample, R(T) starts deviating from a logT dependence at a temperature close to 7 K. We may regard the observed vertical resistive transition behavior as arising from percolation in a highly anisotropic system ; $\Delta\sigma$ now serves as a measure of the superconducting volume fraction above T_c. With decreasing temperature a global phase coherence develops and the multilayer enters the zero resistance state. Because of the difference in the sample geometries involved in the vertical and in-plane measurements (the cross-sectional area of the mesa is about 0.5x0.5mm^2 and the total

0921-4526/90/$03.50 © 1990 – Elsevier Science Publishers B.V. (North-Holland)

thickness of the multilayer is about 3000 Å), the probability to form a percolation path in the vertical direction is much larger than that in the in-plane direction. Hence for multilayers having thinner Si layers, we may have $T_{c\perp} > T_{c//}$. For multilayers having thicker Si layers, because the interlayer coupling is weak ($R_\perp$ is large), both $T_{c\perp} \sim T_{c//}$ or $T_{c\perp} < T_{c//}$ have been observed. One might prefer to think that the observed effect simply arises from "pinholes" such defects are common in single barrier structure. However, a pinhole threading 60 layers is extremely unlikely; besides, cross-sectional TEM images show highly uniform metallic and semiconducting layers. It is worth noting that the vertical resistive transition is sensitive to a magnetic field.

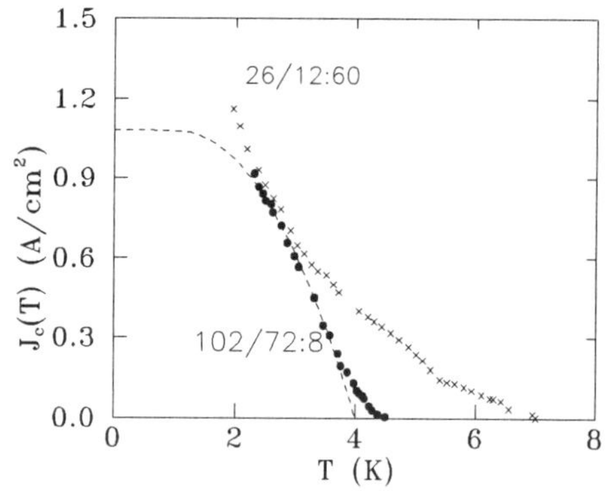

Fig.2 Critical supercurrent vs temperature . The dashed line is calculated from the AB equation.

The critical supercurrent ,I_c, through the multilayers has been measured using the appearance of a voltage of 5×10^{-8} V as a criterion. Data for two selected samples are shown in Fig. 2 in which the dashed line is calculated according to the Ambegaokar - Baratoff (AB) equation(2). For samples with thicker Nb layers, the agreement between the theory and the experimental results is satisfactory, except for the tail structure observed near T_c. The tail structure may arise from the short coherence length effect($\xi_z \sim 10$ Å as determined by upper critical field measurements). As suggested by Deutscher and Muller(3), in this case the critical current will have an anomalous temperature dependence near T_c such that $J_c \propto (T_c-T)^2$. This effect has also been observed in high T_c oxide superconductors(4). For the

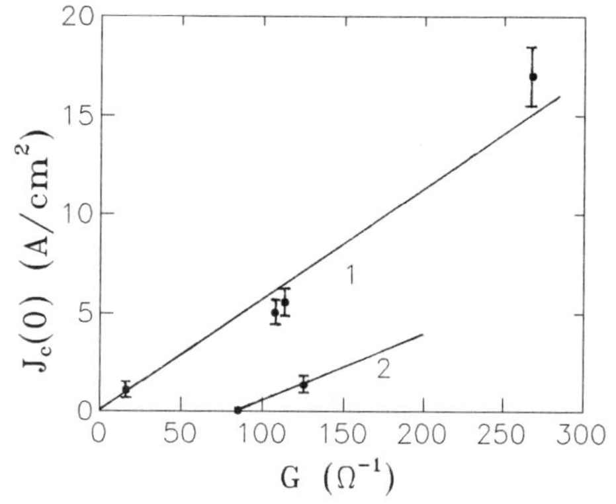

Fig. 3 $J_c(0)$ vs. the normalized junction conductance. Curves 1 and 2 correspond to samples with polycrystalline Nb layers ($d_{Nb} \sim 50$ Å-70 Å) and amorphous Nb layers ($d_{Nb} < 25$ Å) respectively.

26/12:60 multilayer, the tail persists into the lower temperature region due to structural disorder of the Nb layers. In Fig. 3 we plot $J_c(0)$ vs. the normal state conductance G for the multilayer junctions. The values of $J_c(0)$ were determined by fitting to the AB equation. In this figure, curve 1 corresponds to multilayers having polycrystalline Nb layers ($d_{Nb} \sim 50$ Å-70 Å) while curve 2 depicts samples with amorphous Nb layers ($d_{Nb} \leq 20$ Å). J_c is proportional to G as expected. Disorder seriously degrads J_c. In multilayers having $d_{Nb} <$ 5 Å, the supercurrent is absent. The measured I-V curves are highly nonlinear and show large sumgap values.

This work is supported by the NFS MRL program under grant DMR-8520280 and the NSF HTSC Science and Technology Center under grant DMR-89-117.

REFERENCES

(1) M. Tinkham, Introduction to Superconductivity, (Krieger, Melbourne, 1975).
(2) V. Ambagaokar and A. Baratoff, Phys. Rev. Lett. 10(1963) 486.
(3) G. Deutscher and K.A. Muller, Phys. Rev. Lett. 59(1987) 1745.
(4) J. Mannhart et al., Phys. Rev. Lett. 61(1988) 2476.

Physica B 165&166 (1990) 481–482
North-Holland

ENERGY GAP OF PROXIMITY-EFFECT-INDUCED SUPERCONDUCTIVITY IN Nb-Cu MULTILAYERS

G.-q. Zheng, Y. Kitaoka, Y. Oda, Y. Kohori, K. Asayama, Y. Obi+, H. Fujimori+, and R. Aoki#

Department of Material Physics, Faculty of Engineering Science, Osaka University, 560 Osaka, Japan.
+ Institute for Material Research, Tohoku University, Sendai 980, Japan.
# Department of of Electrical Engineering, Faculty of Engineering, Osaka University, Osaka 565.

The nuclear spin-lattice relaxation rate, T_1^{-1}, for ^{63}Cu in zero field from 4.2K down to 0.3K, has been measured in a series of Nb-Cu multilayers. A BCS-like temperature dependence of T_1^{-1} followed gradually by A Korringa-law-like one at low temperatures, has been found. It has been confirmed that such behavior of T_1^{-1} comes from the anisotropic energy gap in the Cu-layers.

1. INTRODUCTION

In recent years, artificially fabricated superlattices or multilayers have attracted much attention and been studied intensively. Several interesting phenomena due to the artificial structure, such as the dimension-crossover temperature dependence of the parallel upper critical field[1] and so on, have been observed in the superconductor(S)-normal metal(N) multilayers. The S-N multilayers also provide a good opportunity for studying the microscopic nature of superconductivity induced by proximity effect. NMR is powerful for studying superconductivity, and particularly suitable for studying a multilayered system since it can selectively measure the properties of one side without any influence from the other side.

In order to investigate the excitation spectrum in the Cu-layer, the nuclear spin-lattice relaxation rate for ^{63}Cu in zero field, has been measured down to 0.3K in a series of Nb-Cu multilayered thin films.

2. EXPERIMENTAL

The samples consisting of 30～50 layers of Nb and Cu were prepared by RF-sputtering. The initial vacuum of the chamber reached to better than 7×10^{-7} torr and the evaporation was carried out in 2×10^{-2} torr of Argon atmosphere. T_1^{-1} in zero field was measured by using the field cycling method, with a switching time of ～100 ms from ～10 kOe to zero.

3. RESULTS AND DISCUSSION

Figure 1 shows T_1^{-1} in zero field for the multilayers Nb200A-Cu200A, Nb200A-Cu400A and Nb200A-Cu600A. A common feature of the temperature dependence of T_1^{-1} is seen throughout the samples. T_1^{-1} initially followed the Korringa law as seen in a bulk Cu metal; at temperatures somewhat below T_c, it was once enhanced over the value in the normal state, then decreased sharply as temperature decreases; the behavior of T_1^{-1} so far was similar to the BCS one and the gap parameter of $\Delta=1.0\sim1.2\ k_BT_c$ was found to fit

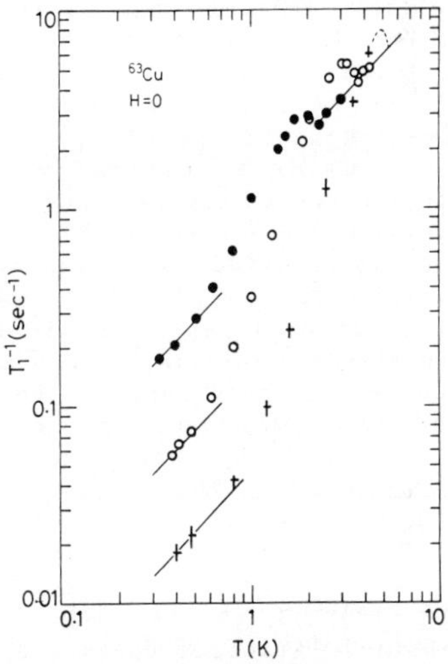

Fig. 1. Temperature dependence of T_1^{-1} in three samples. The closed, open circle and the cross mark indicate those for Nb200A-Cu600A, Nb200A-Cu400A and Nb200A-Cu200A, respectively.

the experimental data; as temperature was lowered further, however, the rapid decrease of T_1^{-1} was gradually replaced by a slow one as T_1^{-1} T.[2]

Kuboki and Fukuyama have recently tried to interpret our preliminary results theoretically.[3),4] They have calculated the relaxation rate both for Cu and Nb in a Nb-Cu multilayer, based on McMillan's tunneling model for a S-N bilayer,[5] and shown that T_1^{-1} should vary as $\exp(-1.75T_c/T)$ even at low temperatures.

It has been demonstrated experimentally that the relaxation at low temperatures is not caused by magnetic impurities or the spin diffusion to the trapped vortex cores in the sample, but in relation to the intrinsic nature of superconductivity in the Cu-layer.[2]

In McMillan's model, a simplification is made. that is, a unique τ_N, the time which an electron stays in the N layer, is assumed to

0921-4526/90/$03.50 © 1990 – Elsevier Science Publishers B.V. (North-Holland)

describe the electronic properties. This model leading to the existence of a uniform energy gap in the N layer, is expected to be applicable in a bilayer where the mean free path l_N in a bulk film is of the order of d_N.[5] In our samples, however, l_{Cu} in a bulk film was estimated to be as long as 6000 A, much larger than d_{Cu} (200∼600A). In such a sample, τ_N for the electrons propagated parallel to the Nb-Cu interface is much larger than that for the electrons moving normal to the interface. It is therefore believed that McMillan's model is not applicable to the present case. An anisotropic energy gap has been proposed to explain the experimental results.[2] That is, the electrons propagated perpendicularly to the interface, being frequently transferred to the Nb-layers, have a finite energy gap, while those traveling nearly parallel to the interface, experiencing no scattering by the attractive interaction and thus maintaining their directions, have no gap; these ungapped electrons are responsible for the gradual, Korringa-law-like decrease of T_1^{-1} at low temperatures. Assuming a critical value for θ beyond which the energy gap vanishes, where θ is the angle between the normal of the interface and the direction of electron momentum, one leads a linear relation between d_{Cu}^2 and the magnitude of $1/T_1T$ at low temperatures. This is consistent with the experimental results.[6]

This proposal has been shown to be acceptable by the results obtained in two modified samples.[6] In the first one a small amount of germanium was doped in the Cu-layer to shorten the electron path length parallel to the inter-face. In the other one the interfaces were waved so that the anisotropy of the electron path was diminished. Figure 2 is the comparison of the results obtained in the Ge-doped and almost undoped sample. For the latter one, the behavior of

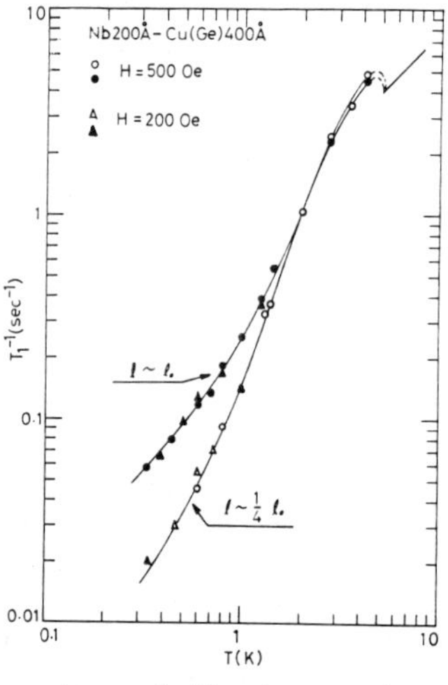

Fig. 2. Comparison of the temperature dependence of T_1^{-1} in Ge-doped Cu-layer (open mark) and in undoped one. (closed mark).

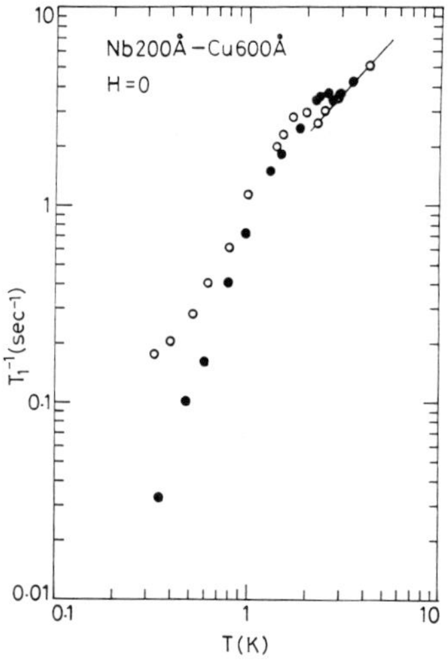

Fig. 3. Temperature dependence of T_1^{-1} for the sample with waved interfaces (closed circle), compared to that for the sample with flat ones (open circle).

T_1T=const. was observed again at low temperatures, while for the sample with substitution of Ge resulting in a 3/4 reduction of l_{Cu}, T_1^{-1} decreased by 3 times at low temperatures and deviated from the T_1T=const. relation. The sample prepared with waved interfaces had the up-down of 600 A with a transverse period of 3000∼4000 A in the surface, while the up-down of the surface for the samples prepared so far was less than 10 A. T_1^{-1} for this new sample is indicated by closed circle in Fig. 3 in comparison to that for the sample of the same dimension but with a flat surface (also shown in Fig. 1). No tendency to the T_1T=const. behavior was observed down to 0.3K. These results provide an evidence for the anisotropy of the energy gap in the original Nb-Cu multilayers.

4. CONCLUSION

The temperature dependence of T_1^{-1} for ^{63}Cu indicating the existence of the energy gap even in the Cu-layer, has been observed. It has also been shown that the energy gap has the anisotropic nature reflecting the geometrical feature of the thin layers.

REFERENCES

1) C. S. L. Chun, G. -G. Zheng, J. L. Vincent and I. K. Shuller: Phys. Rev. B29 (1984) 4915.
2) G. -q. Zheng, Y. Kohori, Y. Oda, K. Asayama, R. Aoki, Y. Obi and H. Fujimori: J. Phys. Soc. Jpn **58** (1989) 39.
3) R. Aoki, G. -q. Zheng, Y. Kohori, Y. Oda, K. Asayama and S. Wada: J. Phys. Soc. Jpn. **56** (1987) 4495.
4) K. Kuboki and H. Fukuyama: J. Phys. Soc. Jpn. **57** (1988) 3102.
5) W. L. McMillan: Phys. Rev. 175 (1968) 537.
6) G. -q. Zheng et al: to be submitted.

Physica B 165&166 (1990) 483–484
North-Holland

ALUMINUM THIN FILMS WITH IN-PLANE MODULATED SUPERCONDUCTING STRUCTURES

K. LIN, Y. K. KWONG, M. ISAACSON, and J. M. PARPIA

School of Applied and Engineering Physics and Department of Physics
Cornell University, Ithaca, NY 14853, USA

We have investigated superconducting properties of aluminum thin films with periodically modulated regions of different transition temperatures (Tc). These samples are made by microlithography and reactive ion etching techniques. They provide a simple and flexible alternative to metallic superlattices previously used in studying the effects of periodicity and anisotropy on superconductivity. Results indicate that these films show global superconducting behavior with properties determined by modulation length scales.

1. Preliminaries

Superconductors with modulated compositions have always been of great interest for studying the effects of periodicity (or controlled aperiodicity) and anisotropy. Well-known effects include enhanced Tc (1), critical field anisotropy, and various dimensional crossovers (2). All previous studies use artificial metallic superlattices (3), which are prepared by successive deposition of two (or more) different materials by sputtering or evaporation. Since the transport properties of these superlattices depend critically on the interfaces, experimenters must always carefully consider such factors as interlayer mixing and diffusion, sample purity, deposition vacuum conditions, interfacial mismatch, and layer uniformity (3). We present here a simple method for modifying the transition temperature, in selected regions of arbitrary shape or size, of a continuous aluminum thin film. We have used this method to study the resistive transitions of films with periodically spaced regions of different Tc's. The relationship between various transport and fabrication length scales are discussed.

2. Samples

In a previous investigation (4), we observed that reactive ion etching (RIE) using freon gas can alter the superconducting transition temperature of aluminum thin films without affecting other transport properties such as the diffusion constant, residual resistance, and inelastic scattering rates. Standard photo- or electron beam lithography followed by RIE can thus define any arbitrary configuration of regions of higher and lower Tc's in a single continuous film (Fig. 1). Aluminum films have the advantage of simple preparation and long coherence lengths (ξ). To be useful as alternatives to superlattices, these samples should have structure sizes comparable to or less than ξ. The elastic mean free path in our films is estimated to be ≥90Å, from which we estimate the zero-temperature coherence length, $\xi(0) \approx 0.1\mu m$. Since our data is taken close to Tc, $\xi(T)$ should be ≥1μm, much larger than minimum achievable lengths in microlithography. The interface between adjacent regions has a definition of ≤0.1μm. Thus, sensitivity to interface quality should be negligible. Since we have shown that other transport properties remain unaltered (4), the spatial variation of the superconducting potential can be studied as an isolated variable.

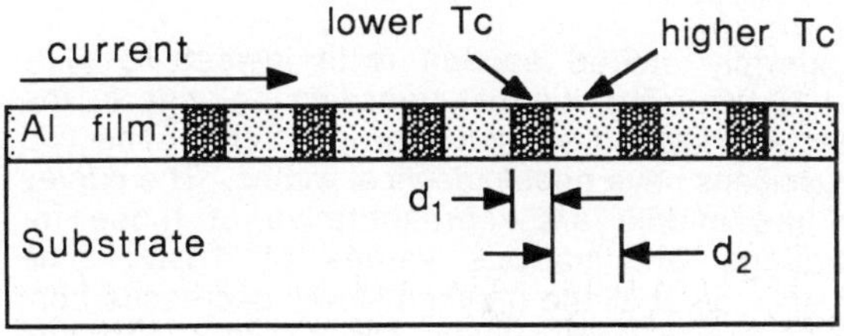

Figure 1: Schematic of aluminum film with modulated Tc regions.

3. Results

Our films consist of gratings of etched and unetched regions with width d_1 and d_2 respectively, for an overall periodicity of d_1+d_2 (Fig. 1). The resistive transitions are measured with currents perpendicular to the grating, using a four-point resistance bridge. Resistance thermometry is used to regulate sample temperatures to better than ±100μK in a He-3 cryostat. Figure 2 plots some observed transitions. The leftmost one is that of a

0921-4526/90/$03.50

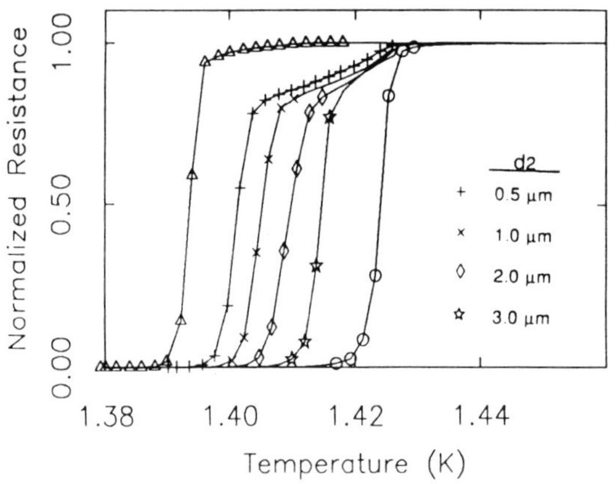

Figure 2: Resistive transitions of modulated superconducting films, (Δ) T_{c1}, (O)T_{c2}. All intermediate transitions are for d_1=1μm and d_2 as indicated.

uniformly etched section, with lowest Tc (T_{c1} =1.394K). The highest transition belongs to the totally unetched section (T_{c2} =1.424K). The two transitions have nearly identical widths. The curves in the middle are representative of those for gratings with various values of d_1/d_2. The resistance of all the gratings slowly decreases from their normal state value, Rn, to ≈0.8Rn in the temperature range between T_{c2} and their own "real" Tc, where a sharp transition to near-zero resistance occurs. We speculate that this gradual decrease is due to charge imbalance effects (5). Figure 3 plots the shift in Tc (T_c-T_{c1}) as a function of the ratio d_1/d_2. Only two values of d_1 (physical width of lower Tc region), 1.0μm and 5.0μm, are used. Tc appears to be determined by the ratio d_1/d_2 rather than the size of the region d_1, indicating that the characteristic length scale is ≥5μm. For example, the transitions for d_1/d_2= 1μm/0.5μm and 5μm/2.5μm coincide. This is contrary to our estimate of ξ(T). Using linearized Ginzburg-Landau equations and the assumption that ξ(T)>>d_1, d_2 everywhere, it can be shown that

$$T_c - T_{c1} \approx \frac{T_{c2} - T_{c1}}{1 + p\left(\frac{d_1}{d_2}\right)} \qquad (1)$$

where p is a factor of order unity which characterizes interface conditions (3). This function is shown in Figure 3, using p=1.3. It fits remarkably well, implying that the characteristic length may be considerably larger than our estimate.

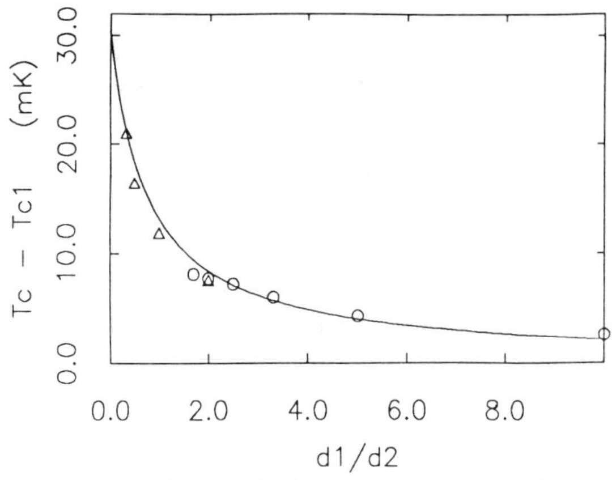

Figure 3: Change in Tc vs. the ratio (d_1/d_2). (Δ) d_1=1μm, (O) d_1=5μm. Solid curve is from Equation (1) using p=1.3.

4. Remarks

For the lengths d_1 and d_2 that we have used, the modulated films act as a single superconductor. As the lengths d_1 and d_2 are increased we expect that the film should act like different superconductors in series, with distinct transition temperatures. We plan to vary the lengths over a wider range which together with critical field measurements now underway, should provide more clues as to the origin and magnitude of the characteristic length scales. We will also investigate other pattern configurations in the future.

*This work is partially supported by the National Nanofabrication Facility under NSF grant ECS-8619094, the Cornell Materials Science Center under DMR85-16616, and AFOSR grant 90-0111.

1. M. Strongin, O. F. Kammerer, and J. E. Crow, Phys. Rev. Lett. **21** (1968) 1320.
2. I. Banerjee, Q. S. Yang, C. M. Falco, and I. K. Schuller, Phys. Rev. B **28** (1983) 5037 .
3. B. Y. Jin and J. B. Ketterson, Adv. in Phys.**38** (1989) 189 (Equation 6-66).
4. Y. K. Kwong, K. Lin, P. Hakonen, J. M. Parpia, and M. Isaacson, J. Vac. Sci. Tech. **B7** (1989) 2020.
5. J. Clarke, Experiments on charge imbalance in superconductors, in: Nonequilibrium Superconductivity, eds. D. N. Langenberg and A. I. Larkin (Elsevier, New York) pp. 1-63.

Physica B 165&166 (1990) 485–486
North-Holland

TEMPERATURE DEPENDENCE OF ENERGY AND LIFETIME OF ROTONS DETERMINED BY RAMAN SCATTERING

Kohji OHBAYASHI, Masayuki UDAGAWA, Hiroyuki YAMASHITA, Mitsuo WATABE and Norio OGITA

Faculty of Integrated Arts and Sciences, Hiroshima University, Hiroshima 730, JAPAN

Temperature dependence of the Raman spectrum of liquid ^4He at a pressure of 5 kg/cm^2 has been measured in the temperature region from 0.65 to 3.75 K. The temperature dependence has been found to be described by the one-component model with lorentzian temperature-broadening both above and below the lambda temperature T_λ. From the analysis, temperature dependence of the minimum energy and the lifetime of rotons have been determined. The roton energy shows weak temperature dependence, and the lifetime deviates from the Landau-Khalatnikov theory above 1.80 K.

1. INTRODUCTION

In spite of the extensive experimental and theoretical study, the temperature dependence of $S(k,\omega)$ determined by inelastic neutron scattering is not well understood. Woods and Svensson proposed a two-component model that the one-phonon peak has a weight proportional to macroscopic superfluid density $\rho_s(T)$ [1]. Talbot et al. claimed the simple multiphonon subtraction model [2]. At finite temperatures, the two models give different values of the roton parameters such as the roton energy Δ_0 and the roton width Γ. Theoretically, Griffin proposed that the Bose-Einstein condensation does not cause a drastic change to $S(k,\omega)$ and only abrupt narrowing is expected as we pass through T_λ from above.

The Raman spectrum of superfluid ^4He is related to the integral of $S(k,\omega)$ in k-space. Thus above mentioned difference of $S(k,\omega)$ may be studied by Raman scattering also. To explore the nature of the temperature dependence of it, we have carried out Raman scattering measurements of liquid ^4He at various temperatures.

2. EXPERIMENTAL PROCEDURE

The experimetal method is mainly the same with that used for other measurements [4]. In this experiment, temperature was kept within ± 0.015 K of desired value using a recirculation type ^3He cryostat and an electronic control circuit. The temperature was swept from 0.65 K to 3.75 K crossing T_λ. The pressure was kept at 5 kg/cm^2.

3. RESULTS AND DISCUSSION

The spectra measured at the three representative temperatures are shown in Fig.1. (a) The lowest temperature 0.65K at which the spectrum is broadened only by the instrumental profile. (b) The temperature 1.75 K below T_λ where temperature-broadening is not so small but the spectrum still keeps the feature apparently similar to that at the lowest temperature. (c) The temperature 2.45 K above T_λ at which considerable temperature-broadening makes the spectrum almost featureless. The sharp peak in Fig. 1 (a) is the 2-roton scattering.

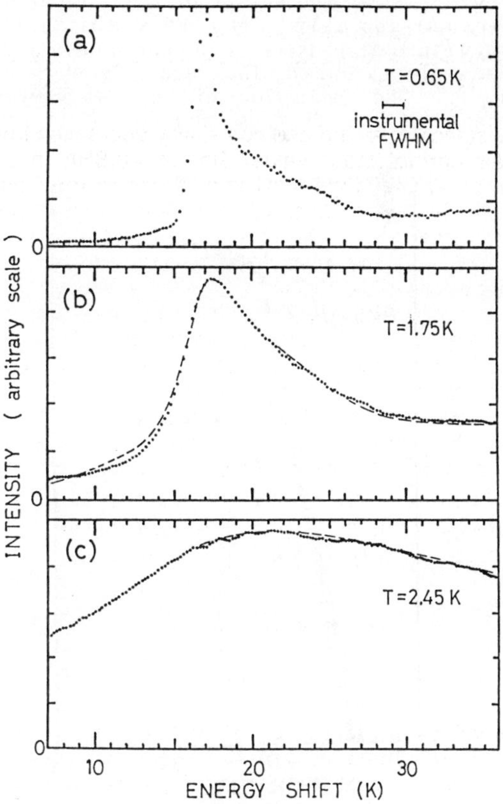

Figure 1
The Raman spectra of liquid ^4He under 5 kg/cm^2 measured at three different temperatures. The broken lines are the calculated curves.

After analyzing the spectra with a few models, the one-component model with lorentzian temperature-broadening was found to fit most excellently at all temperatures. Because the width of the sharp 2-roton peak in Fig. 1 (a) is equal to the instrumental width, the convolution integrals of it with lorentzian functions of various widths were calculated for comparing with the experimental spectra at higher temperatures. In the case of Fig. 1 (b) and (c), the best fit was observed for the widths of 2Γ = 1.34K and 2Γ = 11K, respectively. As shown in Fig. 1 (b) and (c) by broken curves, the both calculated curves fit excellently, demonstrating that there is no qualitative difference of the Raman spectrum above and below T_λ. The fact is consistent with Griffin's proposal. The inverse of Γ is the roton lifetimes. Γ is plotted in Fig. 2.

In Fig. 1, normalized spectra are plotted. The actual peak height decreases as the temperature increases. Present analysis can reproduce the temperature dependence of the height within the experimental accuracy.

In the spectral-shape-fitting process, we needed to shift the energy of the calculated spectrum slightly. From this shift, the temperature dependence of the roton energy was accurately estimated. The result is shown in Fig. 3. The estimation of the temperature dependence $\Delta_0(T)$ from neutron data was controversial [5]. Present result supports the neutron analysis which yielded a weak temperature dependence.

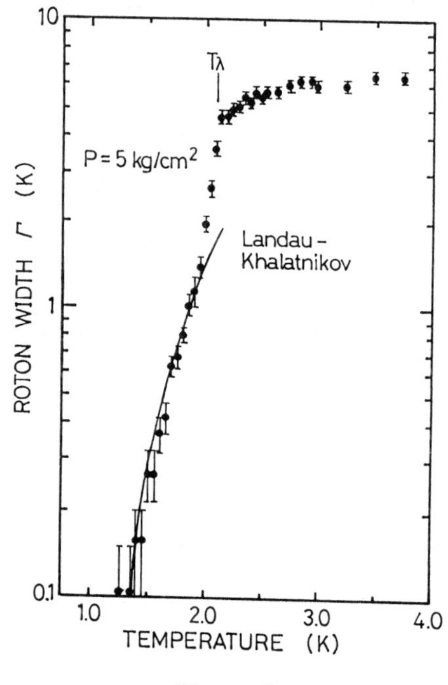

Figure 2
The temperature dependence of the roton width determined by Raman scattering. The solid curve is the Landau-Khalatnikov theory.

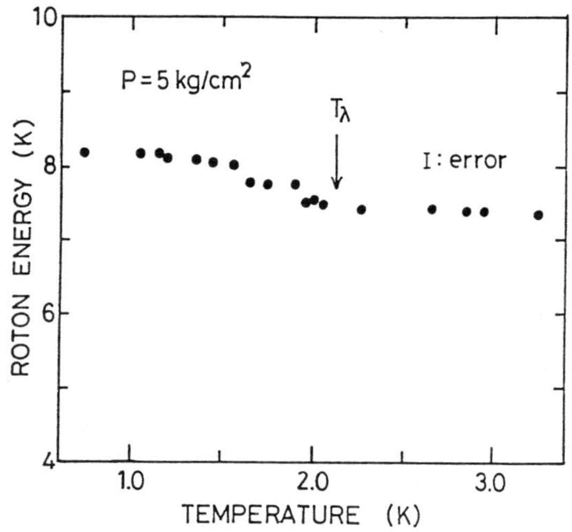

Figure 3
The temperature dependence of the roton energy determined by the Raman scattering.

Because the roton energy is given as the function of temperature with no ambiguity, we can make a comparison of the temperature dependence of the roton width with the well known Landau-Khalatnikov theory as shown by the solid curve in Fig. 2. Above about 1.80 K, the observed widths show significant upward deviation from the theory. This indicates that an additional broadening mechanism other than the roton-roton interactions appears near T_λ. A recent theory suggests that vortex-excitations exceed roton-excitations near T_λ [6]. And the deviation may be detecting the vortex-excitations through roton-vortex interactions.

The details of this work will be reported elsewhere soon.

REFERENCES

(1) A.D.B Woods and E. C. Svensson, Phys. Rev. Lett. **41** (1978) 974.

(2) E.F. Talbot, H.R. Glyde, W.G. Stirling and E.C. Svensson, Phys. Rev. B **38** (1988) 11229.

(3) A. Griffin, Elementary Excitations in Bose-Condensed Liquids and Gases at Finite Temperatures, in: **Elementary Excitations in Quantum Fluids**, eds. K. Ohbayashi and M. Watabe (Springer-Verlag, Berlin Heidelberg, 1989) pp. 23-30.

(4) K. Ohbayashi, Raman Scattering of Liquid ^{4}He, in: **ibid.**, pp. 32-52.

(5) E.C. Sensson, Temperature Dependence of S(Q,ω) for Liquid ^{4}He, in: **ibid.**, pp.59-95.

(6) G.A. Williams, Phys. Rev. Lett. B **33** (1987) 1926.

Physica B 165&166 (1990) 487–488
North-Holland

NEW INTERPRETATION OF THE QUASIPARTICLE WEIGHT Z(Q) FOR SUPERFLUID ^{4}HE

A. GRIFFIN

Department of Physics, University of Toronto, Toronto, Ontario, Canada M5S 1A7

E.C. SVENSSON

Atomic Energy of Canada Limited, Chalk River, Ontario, Canada K0J 1J0

Glyde and Griffin have recently given a new interpretation of the quasiparticle dispersion relation for superfluid ^{4}He. The low Q phonon peak in $S(\vec{Q},\omega)$ corresponds to a zero sound mode while the high Q maxon-roton peak is a strongly-renormalized single particle excitation. We propose that Z(Q) should be viewed as the superposition of the weights of zero sound and maxon-roton excitations. We also speculate that the ratio Z(maxon)/Z(roton) may reflect the value of the condensate fraction n_0.

Glyde and Griffin[1] have given a new interpretation of the well-known phonon-maxon-roton quasiparticle dispersion relation for superfluid ^{4}He. Their work is based on an analysis of the features exhibited by high resolution neutron scattering results[2–4] as a function of the temperature. GG argue that the phonon peak (which persists largely unchanged[4] through T_λ) is a collective zero sound mode, as suggested by Pines in 1965. In striking contrast, the weight of the sharp maxon-roton peak disappears completely as $T \to T_\lambda$, as noted by Woods and Svensson[2]. This mode is interpreted as a single-particle excitation.

That the sharp peak in $S(\vec{Q},\omega)$ has a different physical origin in the low and high Q regions is a natural consequence of microscopic theory.[1] Because of the Bose condensate, the single particle (SP) and collective zero sound (ZS) modes are coupled and hybridized, both appearing as poles in the single-particle Green's function $G_{\alpha\beta}(\vec{Q},\omega)$ and the density response function $\chi_{nn}(\vec{Q},\omega)$. The ZS mode dominates $S(\vec{Q},\omega)$ at low $Q(\lesssim 0.7\text{Å}^{-1})$, although it disappeared by $Q \gtrsim 1.5\text{Å}^{-1}$. At higher $Q(\gtrsim 0.8\text{Å}^{-1})$, there is a well-defined maxon-roton single particle excitation $\omega(Q)$ associated with $G_{\alpha\beta}(\vec{Q},\omega)$. Microscopic theory shows that[1]

$$\chi_{nn}(\vec{Q},\omega) = \sum_{\alpha,\beta} \Lambda_\alpha(\vec{Q},\omega) G_{\alpha\beta}(\vec{Q},\omega) \Lambda_\beta(\vec{Q},\omega) + \chi^N_{nn}(\vec{Q},\omega) \ , \quad (1)$$

where the Bose broken symmetry vertex function $\Lambda_\alpha(\vec{Q},\omega)$ determines the weight of the $\omega(Q)$ mode in $S(\vec{Q},\omega)$. Λ_α vanishes with $n_0(T)$.

This new picture has important implications for the interpretation of $S(\vec{Q},\omega)$ in the region $Q \lesssim 3\text{Å}^{-1}$. One now has to consider four physically distinct contributions from:

(a) Zero-sound excitation (phonons) at low Q
(b) Thermal scattering (or particle-hole) continuum
(c) Single-particle (maxon-roton) excitation at high Q
(d) High energy multiparticle continuum

The crossover region between the phonon and maxon excitation shown in Fig. 1 is not yet understood in detail. The shading is an attempt to show that the SP excitation disappears at low Q and the ZS excitation disappears at high Q. The broad multiparticle[5] contribution (d) can only be separated out cleanly at low T and is peaked at ~25 K.

With this new picture, one must think of $Z(\vec{Q})$ as being composed of the contributions from two quite distinct excitation branches,

$$Z(\vec{Q}) = Z_{ZS}(\vec{Q}) + Z_{SP}(\vec{Q}) \ . \quad (2)$$

At low Q, the zero sound mode (a) is dominant but, as Q increases, it will become broader and ultimately lose all intensity. In contrast, the maxon-roton SP excitation (c) starts to dominate $Z(Q)$ at $Q \gtrsim 1\text{Å}^{-1}$, with a weight which depends very much on Q and T (vanishing above T_λ). In Fig. 2, we have decomposed the svp results[5] for $Z(\vec{Q})$ at T=1.1 K in terms of this new picture. It gives a natural explanation of the shape of $Z(\vec{Q})$ which has been a puzzle for nearly 20 years. It would be valuable to also have plots of $Z(\vec{Q})$ at higher pressures.

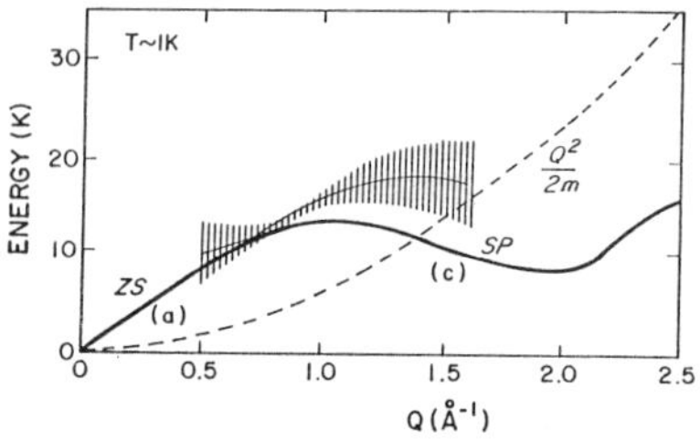

Figure 1
The quasiparticle dispersion relation, switching from a collective zero sound mode at low Q to a single-particle maxon-roton excitation at high Q.

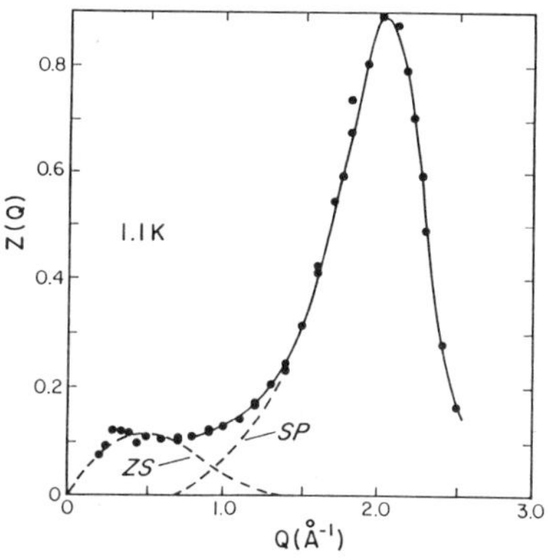

Figure 2
Quasiparticle contribution $Z(Q)$ to static structure factor $S(\vec{Q})$ (based on ref. 5). The dashed curves show a plausible decomposition into zero sound and single particle contributions.

We emphasize that $Z_{ZS}(\vec{Q})$ sketched in Fig. 2 is not the total integrated weight of the ZS mode but only its contribution to the sharp peak in $S(\vec{Q},\omega)$. As sketched in Fig. 1, a high energy broad ZS peak should also co-exist[1] with the sharp lower energy maxon SP peak up to $Q \sim 1.5$Å^{-1}. What is usually called the multiparticle (or multiphonon) component of $S(\vec{Q},\omega)$ is, in the GG picture, a combination of (a) and (d) in the region $0.8 \lesssim Q \lesssim 1.5$ Å^{-1}.

Above T_λ, only the (a) and (b) contributions remain. The peak position of this distribution at 4.2 K is plotted in ref. 6 and it shows the expected bendover at $Q \sim 1.4$ Å^{-1} (see Fig. 1). An experimental study of the peak position of $S(\vec{Q},\omega)$ in normal ^{4}He just above T_λ would be most valuable.

One of the long time puzzling features about the $Z(\vec{Q})$ data shown in Fig. 2 is why $Z(Q \sim 2$ Å$^{-1})$ is of order unity while $Z(Q \sim$Å$^{-1})$ is an order of magnitude smaller (~ 0.1). In the GG picture, this might be explained as follows. In Fig. 1, we note that at $Q_R \sim 2$Å^{-1}, $\omega(Q) \sim 8$ K while $\epsilon_Q \sim 25$ K; in contrast at $Q_M \sim 1$ Å^{-1}, $\omega(Q) \sim 13$ K while $\epsilon_Q \sim 7$ K. Taking $\epsilon_Q = Q^2/2m$ as a characteristic energy, this suggests that in the maxon region one can use the result[7] $\Lambda_\alpha(\vec{Q},\omega \to \infty) = \sqrt{n_0}$ and hence predict that $Z(Q_M) \sim n_0/n$ (which is 0.1 at low T). In contrast, in the roton region, the $\Lambda_\alpha(Q,\omega \to 0)$ limit would appear to be more relevant since $\omega(Q) \ll \epsilon_Q$. Microscopic calculations[7] indicate that at zero frequency, one can have significant renormalization effects on the bare high frequency value $\sqrt{n_0}$ with the possibility that $Z(Q_R)$ may be of order unity (at low T). In principal, one should be able to obtain information about $n_0(T)$ directly from the observed[2-4] strong temperature dependence of the intensity of the maxon-roton peak in $S(\vec{Q},\omega)$.

In the GG picture, the maxon-roton peak in $S(\vec{Q},\omega)$ is the low Q analogue of the long sought-after free particle condensate peak, expected[8] to exist at very large Q (the impulse approximation regime). We believe the SP excitation is easily observed in the region $1 \lesssim Q \lesssim 2$ Å^{-1} simply because of the sharpness of the excitation, in conjunction with high instrumental resolution. The instrumental width as well as the SP intrinsic width *both* increase with Q, so that at larger $Q \gtrsim 4$ Å^{-1}, the condensate peak does *not* stand out cleanly from the Doppler-broadened distribution (b).

ACKNOWLEDGEMENTS

This paper grew out of work done with H.R. Glyde and W.G. Stirling. A.G. was supported by NSERC of Canada.

REFERENCES

1. H.R. Glyde and A. Griffin, to be published
2. A.D.B. Woods and E.C. Svensson, *Phys. Rev. Letters* **41**, 974 (1978)
3. E.F. Talbot, H.R. Glyde, W.G. Stirling and E.C. Svensson, *Phys. Rev.* **B38**, 11229 (1988)
4. W.G. Stirling and H.R. Glyde, *Phys. Rev.* **B**.
5. R.A. Cowley and A.D.B. Woods, *Can. Journ. Phys.* **49**, 177 (1971)
6. A.D.B. Woods, E.C. Svensson and P. Martel, Proc. of LT14, ed. by M. Krusius and M.Vuorio (North Holland, 1975) vol. 1, p. 187
7. E. Talbot and A. Griffin, *Ann. Phys.* (N.Y.) **151**, 71 (1983)
8. P.C. Hohenberg and P.M. Platzman, *Phys. Rev.* **152**, 198 (1966)

Physica B 165&166 (1990) 489–490
North-Holland

ROTATING SUPERFLUID ^{4}He CLUSTERS

L. PITAEVSKII+ and S. STRINGARI*

+ *Institute for Physical Problems, 117334 Moscow, ul. Kosygina 2, USSR*
* *Dipartimento di Fisica, Università di Trento, 38050 Povo, Italy; INFN, Trento*

We study the response of ^{4}He clusters to an external field corresponding to a rotation with frequency ω. We discuss the dependence of the normal fraction of the system as a function of the temperature T and of the mass number N of the cluster. The critical behaviour at high rotational frequencies is also investigated.

Superfluid effects in ^{4}He clusters have been the object of a recent theoretical investigation via a path-integral Monte-Carlo approach [1]. In particular this calculation has revealed that in clusters the thermal excitation of the normal components of the system is, at low temperature, much more important than in the bulk. One of the aims of the present work is to understand such a behaviour and provide an explicit description of the temperature dependence of the superfluid fraction of the system at low temperature.

A first point to discuss is the range of temperatures which is relevant for ^{4}He clusters. At too high temperatures free helium clusters are expected to cool down very quickly due to evaporation. According to the estimates of ref.[2] it is reasonable that the temperatures relevant for the experiments of ref. [3-4] lie below $1K$. The idea of the present work is that, differently from bulk helium, at such temperatures the statistical properties of helium clusters are governed by the excitation of surface modes ("ripplons"). The same idea has been already used in ref.[2] to calculate the density of states as well as the rate of evaporation of ^{4}He clusters.

For a finite system a natural definition of superfluidity can be obtained in terms of the departure of the moment of inertia Θ from the rigid value Θ_{rig}. Other quantum finite systems (atomic nuclei) are known to exhibit [5] important deviations from the rigid value which are interpreted as a signature of their superfluid behavior. In the following we will consider clusters whith spherical ground state. Clearly the occurrence of permanent deformations in ^{4}He clusters, as suggested in ref.[6], would open new interesting perspectives in the study of their rotational motion.

It is well known that the hamiltonian H' of a body in a coordinate system rotating with the body is related to the hamiltonian H in the laboratory system by the equation [7]

$$H' = H - \omega L_z \qquad (1)$$

where ω is the angular frequency (which has been taken along the z-axis) and $L_z = \sum_i l_z^i$ is the z-component of the angular momentum of the system. Equation (1) shows that the angular momentum of the rotating body can be calculated as the response of the system to the conjugate "field" ω. Since the eigenstates of H' (as well as of H) can be classified in terms of the angular momentum (l, l_z), their eigenvalues are straightforwardly given by the relation $\epsilon'_{l,l_z} = \epsilon_l - \omega l_z$. The ratio

$$\frac{\Theta}{\Theta_{rig}} = \frac{1}{\Theta_{rig}}\frac{\partial}{\partial\omega} < \omega \mid L_z \mid \omega > \qquad (2)$$

between the moment of inertia $\frac{\partial}{\partial\omega} < \omega \mid L_z \mid \omega >$ of the system and the rigid value $\Theta_{rig} = \frac{2}{3}mN < r^2 >$ will be taken in the following as our definition of the normal (non superfluid) fraction of the system. In eq.(2) $\mid \omega >$ is the statistical state associated with the hamiltonian H' at a given temperature T. Equations (1-2) provide a natural generalization of the concept of superfluidity introduced by Landau [8] in bulk liquid and have been extensively employed in problems with cylindrical symmetry.

The angular momentum $< \omega \mid L_z \mid \omega >$ of the system can be written in the following way:

$$< \omega \mid L_z \mid \omega > = \sum_l \sum_{l_z=-l}^{+l} l_z \frac{1}{exp[\beta(\epsilon_l - \omega l_z)] - 1} \qquad (3)$$

where we have assumed statistical thermalization of the elementary excitations given by the hamiltonian H' and neglected possible interaction terms between phonons. Let us first consider the case of small angular frequencies. Eq.(3) then becomes

$$< \omega \mid L_z \mid \omega > = -\omega \sum_l \sum_{l_z=-l}^{+l} l_z^2 \frac{\partial}{\partial\epsilon_l}\frac{1}{exp[\beta\epsilon_l] - 1}. \qquad (4)$$

A compact expression for $< \omega \mid L_z \mid \omega >$ is easily obtained restricting the sum to the surface states

$$\epsilon_l = \sqrt{\frac{4\pi}{3} l(l-1)(l+2)\frac{\sigma}{m}N^{-\frac{1}{2}}} \qquad (5)$$

given by the liquid droplet model [5] and replacing the sum over angular momentum with a suitable integral.

The final result for the normal fraction of the system then takes the following form:

$$\frac{\Theta}{\Theta_{rig}} = \frac{50}{27}\frac{1}{mr_\circ^2}\left(\frac{3m}{4\pi\sigma}\right)^{\frac{4}{3}}\Gamma(\frac{5}{3})\zeta(\frac{5}{3})T^{5/3}N^{-\frac{1}{3}} \quad (6)$$

where Γ and ζ are the factorial and Reimann's functions, σ is the surface tension and we have used the expression $R = r_\circ N^{1/3}$ for the radius of the cluster ($r_\circ = (\frac{3}{4\pi\rho})^{1/3}$ is the unit radius). Result (6) is expected to provide a reasonable estimate of the normal fraction of clusters with $N = 10^3 - 10^5$ and $T = 0.3 - 0.8K$. In fact within the above ranges of N and T the thermal excitation of bulk modes is not expected to be crucial and the discretization of surface levels can be ignored in first approximation.

Equation (6) should be compared with the expression

$$\frac{\rho_n}{\rho} = \frac{2}{45}\frac{\pi^2}{\rho mc^5}T^4 \quad (7)$$

for the phonon contribution to the normal fraction holding in the bulk at low T ($\rho = \frac{N}{V}$ is the density of the system). The figure clearly shows that the thermal excitation of the normal component results to be more active in the clusters than in the bulk. This remains true (in the range of temperatures reported in the figure) also if one considers the roton contribution to the bulk normal density. The above behaviour agrees with the results found in the analysis of ref.[1] based on a path-integral Monte Carlo calculation in light clusters.

It is interesting to investigate what happens to the normal fraction of the system when we consider higher frequencies ω. At $T = 0$ the system remains spherical and superfluid until the critical frequency

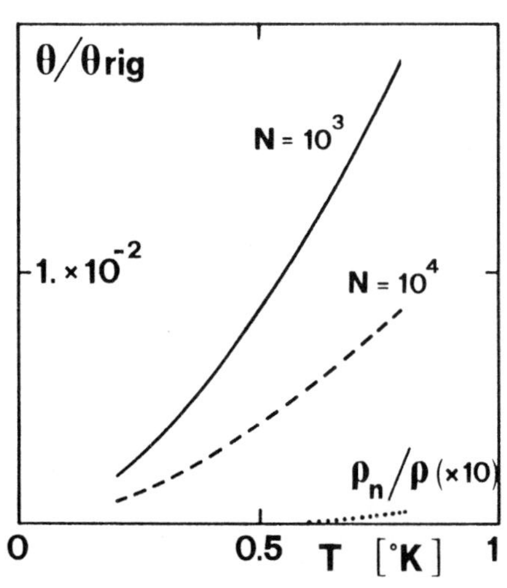

FIGURE 1
Normal fraction Θ/Θ_{rig} as a function of T for two different clusters. The bulk result (7) (multiplied by 10) is also reported for the comparison.

$$\omega_{cr} = min_l \frac{\omega_l}{l} \quad (8)$$

is reached. Result (8) is analogous to the condition $v_{cr} = min_q \frac{\epsilon(q)}{q)}$ for the critical velocity breaking superfluidity in bulk ^{4}He. If one takes the dispersion law (5) for the surface modes, the critical frequency is given by

$$\omega_{cr} = \frac{1}{2}\omega_2 = 5.1N^{-\frac{1}{2}}K\,. \quad (9)$$

For frequencies larger than ω_{cr} the energy ϵ_l of the $l = 2$ surface mode becomes negative and one has a condensation of surface phonons with the consequent occurrence of permanent deformations. The amount of such a condensation as well as the nature of the phase transition can be studied only including interaction terms between surface phonons, analogously to what is done for the condensation of rotons at v_c in the bulk [9].

Clearly at $T > 0$ one has condensation of surface phonons also for $\omega < \omega_c$. Such phonons contribute to the normal fraction of the system. We have calculated the ω-dependence of Θ/Θ_{rig} for a typical cluster with $N = 10^3$ atoms at $T = 0.5K$. The result reveals that the normal fraction of the system remains very small except for values of ω very close to ω_{cr}. This indicates that the "critical" behaviour associated with the onset of phonon condensation at ω_{cr} survives also at temperatures higher than ω_{cr}.

It is finally interesting to compare the "surface" critical frequency (9) with the critical frequency ω_{cr}^v associated with the creation of a quantized vortex line. A simple estimate yields

$$\omega_{cr}^v = \simeq N^{-\frac{2}{3}} lnNK \quad (10)$$

One can see that ω_{cr}^v at large N is smaller than the critical frequency given by eq.(9). However it is possible that vortex lines and ripplons are excited independently.

REFERENCES

[1] Ph. Sindzingre, M. L. Klein and D.M. Ceperley, Phys. Rev. Lett. **63**,1601 (1989);
[2] D.M. Brink and S. Stringari, Z. Phys. D , in press
[3] J. Gspann, in *Physics of Electrons and Atomic Collisions*, S. Datz ed. (North Holland, Amsterdam 1982);
[4] H. Buchenau, J. Northby, E.L. Knuth, J.P. Toennies and C. Winkler, J. Chem. Phys. , to be published;
[5] A.Bohr and B.R. Mottelson, *Nuclear Structure* (Benjamin, N.Y. 1975) vol.II;
[6] T. Regge, J. Low Temp. Phys. **9**, 123 (1972); M. Rasetti and T. Regge, in *Quantum Liquids*, eds. J. Ruvalds and T. Regge (North Holland, Amsterdam 1978);
[7] L.D. Landau and E.M. Lifshitz, *Statistical Mechanics*, (Pergamon Press, Oxford, 1980) Part 1, sect. 26;
[8] L.D. Landau, Zh. Eksp. Teor. Fiz. **11**,592 (1941);
[9] L.P. Pitaevskii, Pis'ma Zh. Eksp. Teor. Fiz. **39**,423 (1984) [JETP Lett. **39**,511 (1984)];

Physica B 165&166 (1990) 491–492
North-Holland

POSSIBILITIES FOR BOSE CONDENSATION OF POSITRONIUM (Ps)

P. M. PLATZMAN and A. MILLS, Jr.

AT&T Bell Laboratories, 600 Mountain Avenue, Murray Hill, NJ 07974, USA

We suggest that a dense ($n=10^{18}cm^{-3}$) gas of positronium (Ps) atoms will show ideal Bose condensation. Schemes for generation confinement and detection are discussed. Estimates of cooling and recombination rates demonstrate feasibility with existing sources.

Bose Einstein condensation is one of the most striking and most compelling phenomenon in statistical physics.(1) It is generally believed that the superfluid properties of He^4 below the λ point arise from a type of "non-ideal" Bose condensation. To date He^4 is unique. However, there have been, in the last decade, numerous attempts made to find other more "ideal" systems. In particular, spin polarized Hydrogen(2) and a gas of excitons in a semiconductor(3) have received the most attention. In this paper we would like to suggest that a dense gas of positronium (Ps) atoms is a good candidate for such a system. Recent investigations of the interactions of positrons e^+ and Ps with solids has led to extraordinary improvements in the kinds of low energy experiments we can do with the anti-electron the positron.(4)

Ps is comprised of an e^+–e^- tightly bound in a hydrogenic orbit. Its ground state is a spin singlet separated from an excited triplet by an energy of approximately $\cong 10°K$. Ps annihilates itself i.e., becomes high energy γ-rays. The singlet is short lived ($\tau_{sing} \cong 10^{-10}$sec). For a Ps atom at rest the decay occurs with the emission of two .5 MeV γ-rays which come out precisely in opposite directions (momentum conservation). If the Ps atom is moving with momentum p, the γ-rays come out with a small angle $\theta \cong p/mc$ relative to each other. The triplet on the other hand has a very different decay pattern. It decays by three γ-rays and has a lifetime of 10^{-7}sec. A magnetic field mixes triplet with singlet thus rapidly quenching the triplet.

In order to make the case for Bose condensation we need to understand at least qualitatively how a mildly energetic (≈ 10 keV) e^+ interacts with a semi-infinite slab of some simple solid. The e^+ enters the solid and begins to make electronic excitations in the material, plasmons, inner shell ionizations etc. Calculations and experiment clearly demonstrate that the e^+ reaches an energy of 10 eV or so in a short time $\tau < 10^{-10}$ sec. and does so in a distance of roughly 1000 Å for a 10 keV e^+. Subsequently emission of phonons and particle hole pairs insure that our e^+ will, thermalize and reach the surface before it annihilates ($\cong 10^{-9}$sec). Since the work function for e^+ at the surface of many materials is negative, low energy e^+ and Ps emission occurs with high probability.

The re-emitted positrons can be used to generate beams and to tailor the properties of existing beams. In particular it is possible to brightness enhance a beam by accelerating a large area beam, focusing it to a spot, cooling it off and then reaccelerating it. Each stage results in a brightness enhancement of a few hundred. Bunched five nanosecond beams of a few times 10^5 e^+ confined to a radius of ten microns have been generated.(5)

A detailed scenario for a possible Bose condensation experiment is sketched in Fig. 1. A bunched one

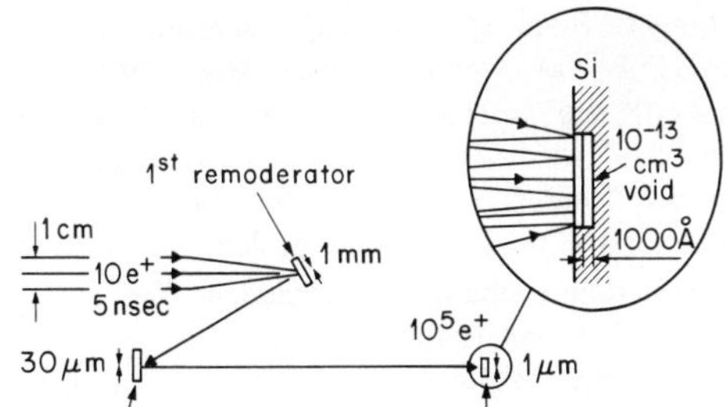

Fig. 1 - Schematic of a possible scenario for a Bose condensation of Ps experiment.

nanosecond brightness enhanced (diameter 1 μ) microbeam of 10^6 e^+ is incident at time t = 0 on the surface of Si which has at the point of entry of the beam a small cavity etched into it. The cyclinder is typically 1 μ in diameter and has a height of 10^3 Å. The volume of this cavity is $V \cong 10^{-13}$ cm^3. The e^+, as discussed, stop in a 1000Å cloud and a significant fraction are reemitted as energetic Ps into the cavity. The singlet Ps rapidly decays leaving us with a hot gas

0921-4526/90/$03.50

of Bosons, triplet Ps atoms, at a density $n = N/V$ 10^{18} cm^{-3} At this point the Ps atom begin to collide with the walls and with each other. Do we expect such a gas to cool to a Bose condensed state before it disappears? If such a state is reached then how would we observe it?

At some temperature T_c determined by the condition,(1) $n\lambda_D^3 = .25$, where the DeBroglie wavelength, $\lambda_D = 2\pi\ (2mT)^{-1/2}$, the chemical potential goes to zero and the number of particles in the zero momentum state $<n_o>$ develops a macroscopic value. The value increases as the temperature is lowered i.e. $<n_0>=(1-(T/Tc)^{3/2})$. For Ps atoms at $n = 10^{18}$ cm^{-3} $T_c \cong 800°K$!! Collisions with the wall will cool the Ps atoms. The elastic collisions occur at a rate, $\Gamma_W \cong v_{th}/d$. $d \cong 10^{-5}$ cm i.e., $\Gamma_w \cong 10^{12}$ sec^{-1} under our conditions. With each elastic collision there is some probability, which we estimated to be one percent, that there will be a surface phonon emitted. Since momentum parallel to the surface is conserved the energy of a typical phonon ω_t will be $\omega_t \cong k_D v_s$ where v_s is a surface phonon velocity Since $v_s \cong 10^5$cm/sec, we find for Ps that $\omega_t \cong 100$ °K. This means that every 10^{-10}sec, the energy drops by 100 °K i.e., our gas will cool to room temperature in some tens of nanoseconds. This is perfect assuming no other collisions quench the triplett Ps.

As far as we can determine the most dangerous collisions are the two body exchange collisions between Ps atoms. Collisions between triplett Ps atoms can definitely result in a high probability of spin flip (mixing in of the singlet). Since our estimates of the biggest (total spin singlet) scattering length are about 5Å we expect the two body rates $\Gamma_{2B} \cong 2\times10^{11} sec^{-1}$. However, if we start with spin polarized e^+ which is relatively easy to obtain from a radioactive source,(4) then we can show that only a mixture of the m=1 and m=0 triplett are present. In this case spin exchange is forbidden. We must use polarized e^+ in our beam.

In principle a collision with the wall can also flip the spin of either the bound e^+ or e^- in Ps. This process is weak because it is essentially a relativistic effect coming from the magnetic field of the electrons in the wall colliding with Ps when it is in contact with the walls. An estimate based on the known Bhaba scattering cross section for an e^+ by an e^- at tens of eV energy at solid densities (screening lengths of 1 Å) shows that the spin flip rate $\Gamma_{spin} \cong .1\ \Gamma_w \times (v/c)^2$. Since $(v/c)^2 \cong 10^{-6}$ for our Ps atoms this rate is too slow to be relevant. At a metal surface the exchange scattering would be very rapid and the Ps would quenched. Recombination of two Ps to form a molecule is also bad since it will lead to a mixing in of the singlet state. However this rate can be shown for our case to be very small.

Finally the Ps–Ps scattering rate in the dense gas must be fast enough to enable the Ps gas as it cools to come to equilibrium i.e., to develop a condensate. Since the Ps–Ps scattering length $\tilde{a}$ in the total spin one or two channel is roughly 1Å the rate for this process is, $\Gamma_{2B} \cong 2\pi n \tilde{a}^2 v_t \cong 10^{10}$ sec^{-1}. This is rapid enough to establish equilibrium.

These rough estimates show that it is possible to achieve Bose condensation of our 10^5 Ps atoms ($n \cong 10^{18}$ cm^{-3}). What is more for Ps because of its unique annihilation characteristics we will be able to observe the condensate directly. By putting the correct size magnetic field on the sample we can easily convert a small but significant fraction of triplets to singlets. As we have seen the singlet annihilates into two γ-rays. The angle between the two γ-rays directly mirrors the velocity of the Ps atom. Thus an angular correlation profile with enough resolution will exhibit a condensate. If the sample is below T_c it will resemble the angular correlation data shown in Fig. 2.

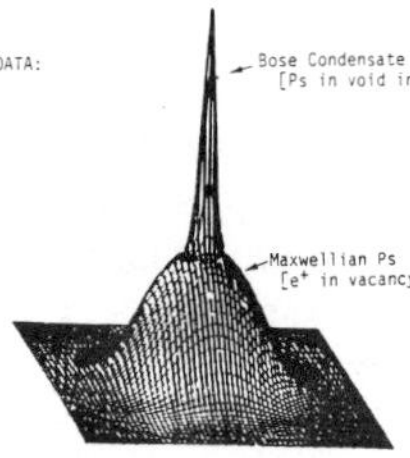

Fig. 2 - 2-D X-ray angular correlation data show a two component distribution similar to the Bose gas below Tc.

Under our conditions we will be able to "see" the condensate with good resolution, study the kinetics of the Bose condensation, understand in detail the size of then Ps atom collision cross section and just possibly see some deviations from ideal behavior.

REFERENCES

1. *Statistical Mechanics,* K. Huang, John Wiley & Sons; NY 1963.

2. H. Hess, et. al, *Phys. Rev. Letts.* 51 , 483 (1983), R. Sprik et.al, *Phys. Rev. Letts.* 51 , 479 (1983).

3. D. W. Snooke, J. P. Woltke and A. Mysyrowicz, *Phys. Rev. Letts.,* 59 , 827, (1988).

4. *Positron Studies of Solids, Surface and Atoms,* Ed. A. P. Mills, Jr., W. S. Crane and K. F. Canter, World Scientific, Singapore (1986).

5. T. N. Horsky, et. al. *Phys. Rev. Lett.* 62 , 1876 (1989).

Physica B 165&166 (1990) 493–494
North-Holland

TIME OF FLIGHT OF PHONON - ATOM QUANTUM EVAPORATION SIGNALS

M A H TUCKER and A F G WYATT

Dept. of Physics, University of Exeter, Exeter, EX4 4QL, UK

The propagation time of high energy phonons $(\hbar\omega / k_B > 10K)$ travelling through HeII has been measured using quantum evaporation. The observed minimum time of flight is slightly, but significantly faster than that expected from the group velocity obtained from neutron scattering measurements of the dispersion curve. Measurements at various liquid depths indicate that the effect is occurring in the bulk HeII.

1. INTRODUCTION

Quantum evaporation enables the study of high energy phonons in HeII because only phonons with energy greater than the binding energy (7.16K) can liberate a helium-4 atom from the liquid. The initial anomalous dispersion (fig 1) causes phonons with $\omega < \omega_c$ to rapidly decay to lower energy states. At higher energies, the dispersion becomes normal, resulting in a cut-off for decay at about $\hbar\omega_c/k_B = 10K$. Above this cut-off, it is impossible for a phonon to spontaneously decay into lower energy phonons, and conserve both energy and momentum. Consequently, only high energy phonons with $\omega > \omega_c$ have sufficiently long lifetimes to reach the liquid surface and quantum evaporate.

When a phonon liberates a helium atom from the liquid, energy and momentum parallel to the surface is conserved. Using these boundary conditions and the form of the dispersion curve $\omega(q)$, the time of flight of a phonon-atom quantum evaporation process can be calculated. The minimum arrival time of the signal occurs at different phonon wave-vectors for different liquid depths (1).

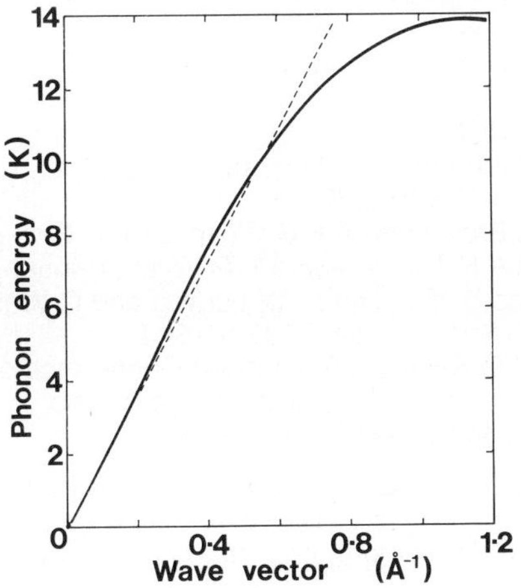

FIGURE 1 *Dispersion curve $\omega(q)$ for phonons in HeII. The dashed line is the ultrasonic velocity.*

In previous quantum evaporation experiments with equal liquid and vapour path lengths, it has been noted (2) that the minimum arrival times are too early by about 5%. Seeking an explanation for these faster signals leads us to question accepted numerical constants such as the binding energy of 7.16K and the exact form of the dispersion curve. Scattering processes can also lead to a faster signal. A fast phonon with energy $\omega \sim \omega_c$ can gain energy to produce a faster atom than would otherwise be evaporated. To get a significant time decrease, the interactions must take place near the liquid surface. The scattering could be with ambient thermal phonons, ripplons, or other excitations which are emitted by the heater.

We can distinguish between such scattering effects and the phonon velocity or energy being wrong by measuring the minimum signal times at various liquid depths. At small depths, the velocity of the atom dominates the time of flight, and so the arrival times are most sensitive to scattering near the surface. At greater liquid depths, the phonon time of flight is more important.

2. EXPERIMENT

A superconducting transition edge bolometer is at a fixed distance of 13.1±0.1 mm above a gold film heater immersed in isotopically pure ^{4}He (fig 2) The liquid depth is varied and measured to ±0.2 mm by a phonon sonar which detects the time of flight of low energy phonons reflected from the

0921-4526/90/$03.50 © 1990 – Elsevier Science Publishers B.V. (North-Holland)

surface.

The heater is pulsed (10mW,0.3μs), to create high energy phonons which travel ballistically at the ambient temperature of 60mK (3).The evaporation signal at the bolometer is acquired on a Biomation 8100, the sampling rate of which is controlled by a quartz oscillator. The signal is averaged to improve the signal to noise ratio.

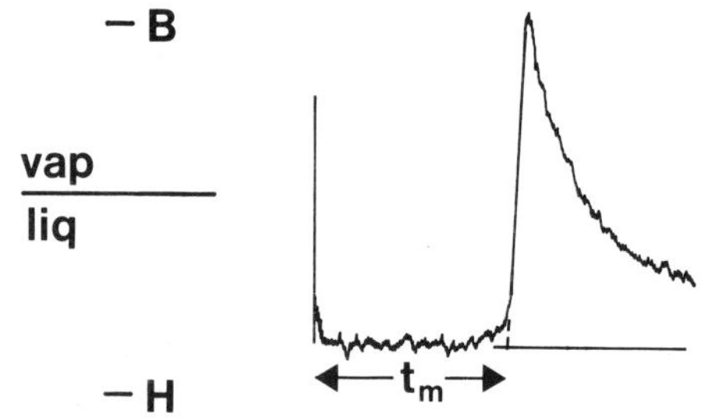

FIGURE 2 *On the left is a schematic diagram of the apparatus. The heater (H) is submerged in the liquid and the bolometer (B) is directly above in the vapour. On the right is an averaged signal. The minimum time* t_m *is the interval between the cross-talk from the heater pulse and the extrapolation of the leading edge at half maximum height through the baseline.*

3. RESULTS

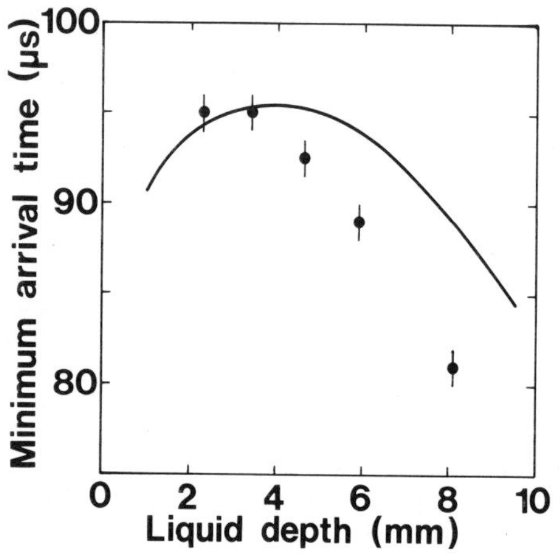

FIGURE 3 *Minimum arrival time plotted at different liquid depths. The solid line is the calculated minimum time as a function of depth.*

The minimum arrival time expected as a function of depth is compared in figure 3 with the measurements which are obtained by extrapolating the leading edge of the signal at half maximum height through the noise to the baseline (fig 2). The discrepancy between expected and observed minimum times increases with liquid depth, and is unaffected by increasing the ambient temperature up to ~150mK.

4. DISCUSSION

The results are consistent with the previously noted discrepancies of the order of 5% at a liquid depth of 6.5 mm. The purity of the helium eliminates the possibility that the fast part of the signal is due to ^3He atoms. Calculations show that changing the binding energy does not account for the observed variation of minimum time with depth. Also, varying the dispersion curve to fit our data results in values well outside the errors on the neutron scattering data points (4).

The lack of dependence on ambient temperature means that scattering with thermal phonons in the bulk liquid and ripplons on the liquid surface can be neglected as a cause for the early signal. Scattering near the free liquid surface can be dismissed since this would require the time discrepancy to be greater for shorter liquid depths, contrary to the data.

5. CONCLUSIONS

We have confirmed the discrepancy in minimum arrival times of phonon-atom quantum evaporation signals found in previous experiments, and found it to be depth dependent. The possible explanations for the effect have been discussed, but none account for the results within the presently accepted values of the binding energy and dispersion curve.

REFERENCES

(1) M J Baird, F R Hope and A F G Wyatt, Nature 304 (1983) 325

(2) M Brown and A F G Wyatt, J Phys.C, in print

(3) M A H Tucker and A F G Wyatt, Phonons 89, eds S Hunklinger, W Ludwig and G Weiss (World Scientific, 1990) 1044

(4) W G Stirling, 75th Jubilee Conference on Helium 4, ed J G M Armitage (World Scientific, 1983) 109

Physica B 165&166 (1990) 495–496
North-Holland

GENERATING BEAMS OF HIGH ENERGY PHONONS AND ROTONS IN LIQUID ^{4}HE.

A F G WYATT and M BROWN

Dept of Physics, University of Exeter, Stocker Road, Exeter, EX4 4QL, UK

High energy phonons ($\hbar\omega/k_B \geq 10$K) and R^+ rotons can be injected into liquid ^{4}He by a thin film heater. It is shown that these excitations may interact with the large number of low energy phonons which are also emitted. The conditions for avoiding this are analysed and compared to measurements.

1. INTRODUCTION

A very useful way of studying the high energy excitations in ^{4}He is to create ballistic beams and use time of flight measurements and quantum evaporation to distinguish between the different wave vectors. Excitations are most easily injected from a thin metal film which is heated by a current pulse. This creates a spectrum of excitations which is modified by decay processes and interactions within the liquid ^{4}He. It has been found that high energy phonons are best injected with a short and high power heater pulse. For rotons it is better to use longer and lower power pulses (1). This paper discusses the reasons for this.

A thin film heater at a temperature of 1-2K mainly injects phonons with $\omega < 10$K. Phonons with $\omega < \omega_c$ ($\omega_c = 10$K, the critical energy for spontaneous phonon decay) rapidly decay to produce a distribution characterised by a temperature of $\approx$1K. The high ω phonons injected with $\omega > \omega_c$ are stable against this decay and so can have long lifetimes. The only rotons injected by a heater are those with positive group velocity (R^+ rotons), and these are also stable against spontaneous decay. The energy fractions emitted into these three channels are 0.99, 5 10^{-3} and 5 10^{-3} respectively (2). These are order of magnitude estimates since the ratios have not yet been measured directly. Also it would be surprising if they did not depend on the heater temperature.

The high ω phonons and R^+ rotons are not affected by ambient excitations if the bulk temperature is less than ~50mK. However there can be scattering between the injected excitations. The most important interactions for the high ω phonons and R^+ rotons are with the low ω phonons and the other injected R^+ rotons. The R^+ rotons mainly have low group velocities and so concentrate near the heater. For this reason we consider the scattering from rotons but not from the high ω phonons, which have mainly high velocities.

2. EXPERIMENT AND RESULTS

To see how different pulse lengths and powers affect the injected densities, we measure the quantum evaporation signal as a function of time at the angles for phonon atom and R^+ roton - atom evaporation (1). The propagation distance is 6.5mm for the excitations and the atoms. The results for high ω phonons and fast rotons are shown in figure1. It can be seen that the high ω phonon signal initially increases with t_p and then at t_p~0.3μs (dependent on power) the signal saturates. On the other hand the fast R^+ roton signal increases with t_p, at least up to t_p = 10μs.

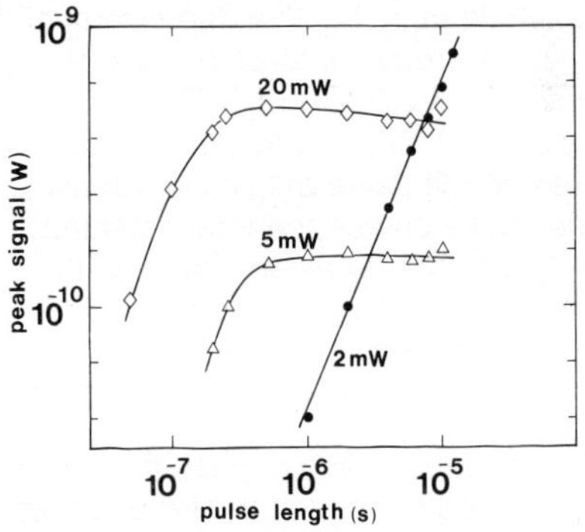

FIGURE 1 *The high energy phonon (open symbols) and fast roton signals (dots) are shown as a function of heater pulse length (t_p) for various heater powers per mm^2.*

3. DISCUSSION

The heater has a very short time constant of approximately 30ns, so it is at a constant temperature for pulses longer than this time. The low energy phonons are therefore created at a constant rate for essentially the whole duration of the pulse and propagate away from the heater ballistically, at velocity c = 238ms^{-1}, in all directions. As they are predominant we assume that they remain essentially unchanged although they scatter the high energy phonons and fast rotons. The number density of low energy phonons near the heater is given by $n_p = W/\hbar c\varpi$ where W is the heater power per unit area and $\hbar\varpi$ is the average phonon energy.

The high ω phonons travel with group velocities $v_g \leq 190\text{ms}^{-1}$ (the group velocity for $\omega = \omega_c$ phonons), and so low energy phonons emitted later in the heater pulse overtake the high energy ones. For a heater pulse of duration t_p, the low energy phonons become spatially separated from the high energy ones after an overlap time $t_{ov} = t_p c/(c-v_g)$.

Up to this time the high energy phonons can interact with the low energy ones via the four phonon process and so may decay to below ω_c after which they decay by the much stronger spontaneous process. The reduction of the high ω phonon flux is proportional to the product of the time of overlap of the two groups of phonons and the density n_p of the low energy phonons i.e. $n_p t_{ov}$. When the heater pulse is high powered and long, all but a few high ω phonons are energy down converted. The few that escape are those injected at the end of the pulse. After a short propagation distance these high ω phonons are left behind the low ω phonons and so then travel undisturbed.

If the critical value of $(n_p t_{ov})$ which produces a significant attenuation of the high ω phonons, e.g. 50% (3dB point), is $(n_p t_{ov})_{crit}$ then the time t_u of the input pulse which produces 'undisturbed' phonons is given by $W t_u c/(c-v_g)\hbar c\varpi = (n_p t_{ov})_{crit}$, i.e. $W t_u$ is a constant for this condition. At a given power, increasing the value of t_p much above t_u therefore does not increase the total detected signal. For $t_p \ll t_u$ the detected signal just increases as t_p since all the injected high ω phonons are undisturbed. This behaviour is qualitatively confirmed by the results shown in figure 1.

The high ω phonons have to pass through most of the rotons emitted during the heater pulse as the majority of the rotons have a very low group velocity. However, using the phonon-roton cross-section determined from neutron linewidth measurements (3), and an injected roton number density based on the fractions given earlier, the influence of scattering by rotons can be neglected for the values of t_p and W used here.

We now consider the injected rotons. We concentrate on fast rotons with $q \geq 2.15 Å^{-1}$ which can be selectively resolved using quantum-evaporation by time of flight. Measurements of the attenuation of a roton beam at ambient temperatures ~0.2K show that low energy phonons are as efficient at scattering these rotons as they are at scattering high energy phonons (4). This indicates that it is the low energy phonons which are the principle scatterers.

However, there is an important difference between the scattering of high ω phonons and fast rotons; the fast rotons travel at almost the same velocity as the low ω phonons so the overlap time is the whole propagation time. This means that the low ω phonon density must be much smaller for the fast rotons to produce a signal, and so the heater power needs to be much lower for rotons than for phonons. The pulse length does not affect the overlap time for fast rotons and so their signal increases linearly with pulse length as shown in figure 1.

The fast rotons have to pass through the slow rotons accumulated in front of the heater. Neutron line width measurements show that the fast roton - slow roton scattering probability is similar to the high ω phonon - slow roton one, and as the heater power used here for rotons is lower than that for high ω phonons we expect the slow rotons to affect the fast rotons even less than high ω phonons. The linear dependence of the fast roton signal on pulse length indicates that this is the case.

CONCLUSION

We have shown that injected low energy phonons are the predominant scatterers of both the high ω phonons and the fast rotons. However due to their different group velocities the dependence of their attenuation on pulse power and duration are dissimilar. The effect of the scattering is minimised for the high ω phonons by using short pulses to reduce the period of interaction. The signal can be increased up to the point where W t_p reaches a critical value. For the fast rotons the scattering can only be reduced by keeping the density of low ω phonons low, i.e. by using low powers. However, as there is no dependence of the scattering on pulse length, the fast roton signal can be increased by using a longer heater pulse, within the bound imposed by the required time resolution of the experiment.

REFERENCES

(1) M Brown and A F G Wyatt, J.Phys. : Condens. Matter, to be published.
(2) A F G Wyatt, N A Lockerbie and R A Sherlock, J .Phys. Condens. Matter 1 (1989) 3507-22.
(3) F Mezei and W G Stirling, Proceeding so the 75th Jubilee conference on liquid He. P111-2, ed. JGM Armitage, World Scientific Publishing Co, Singapore (1983) P111
(4) A F G Wyatt and M Brown, to be published

Physica B 165&166 (1990) 497–498
North-Holland

A DIRECT COMPARISON OF THE SCATTERING OF PHONONS AND ROTONS FROM ROTONS IN SUPERFLUID ^{4}HE

A C FORBES and A F G WYATT

Dept. of Physics, University of Exeter, Stocker Road, Exeter, EX4 4QL, UK

The technique of crossed ballistic beams of excitations is used to investigate the scattering of high energy phonons and rotons from rotons. A direct comparison of the cross-sections for these processes, using the same scattering rotons, is possible. The cross-section for phonon-roton scattering is found to be 0.6 times that of roton-roton scattering for rotons near the minimum. This is in excellent agreement with measurements made on thermal populations.

1. INTRODUCTION

We have recently published some details of an experiment which employs a new method for studying the scattering of excitations in superfluid ^{4}He(1). This involves producing two ballistic beams of excitations and crossing them. One beam is used as a probe and its attenuation is measured as a function of the density of excitations in the second beam. In this paper we compare the scattering of rotons and high energy phonons from the same group of scattering rotons, using this technique.

Other measurements of the scattering (2), (3), for example using inelastic neutron scattering, (4), (5), all involve thermal distributions of excitations, with consequent averages over energy and angle. The ballistic beam technique overcomes this limitation and allows the use of rotons away from the minimum to be used as the scattering excitations. There is still no satisfactory microscopic picture of the roton, and the scattering of excitations should provide a useful probe of their structure. Theoretical predictions of excitation scattering require this knowledge of the roton (6), (7).

2. EXPERIMENTAL

We pass a low density probe beam of ballistic excitations from h_1 through a second beam from h_2, as indicated in fig 1. The scattering angle is predominantly 90°. The beams are generated by thin gold film heaters. The phonon probe beam is generated by a 0.5μs 10mW heater pulse and a 10μs 3.15 mW heater pulse for the probe roton beam. These conditions maximise the concentration of desired excitations in the probe beam (8). The roton and phonon signals are separated at the free surface using quantum evaporation (9). In this process excitations impinging on the surface liberate single atoms, conserving total energy and momentum parallel to the surface. An angle of incidence of 10° separates the evaporated atoms from the beams of rotons and phonons into two distinct angles. These atoms are detected by superconducting zinc transition edge bolometers above the surface of the liquid.

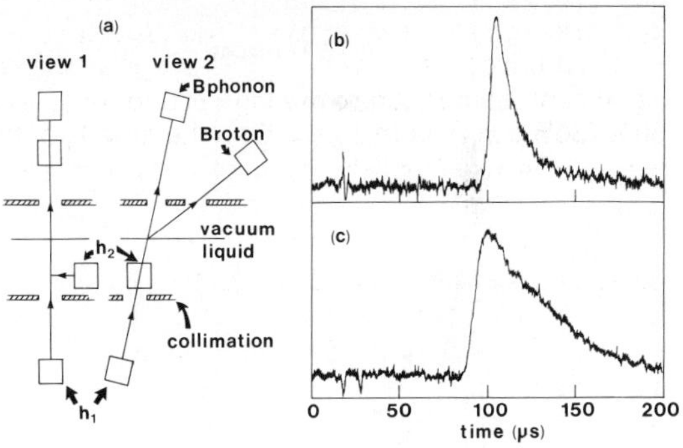

FIGURE 1(a) *Two views of the experiment are shown schematically h_1 and h_2 are the probe and scattering heaters respectively and B phonon and B roton are the bolometers detecting the probe phonons and rotons.*
(b) A typical phonon-atom signal shape.
(c) A typical roton atom signal shape

We can observe the effect of the same scattering (h_2) excitations on probe (h_1) excitations with different group velocities by allowing for the extra time taken by slower ones to reach the interaction volume. The number of h_2 excitations in the interaction volume is determined by the injected energy and the time elapsed between the h_2 pulse and the arrival of the probe excitations.

The scattering can be pictured as the probe propagating through a region which has density of scatterers n_2. If the cross-section for the process is σ then the change in the flux of the probe beam is given by $\phi = \phi_o \exp(-n_2\sigma L)$. Where L is the path length of the probe beam in the scattering region and σ and n_2 are the average values of cross-section and scattering excitation density respectively. Thus the quantity $- \ln (\phi/\phi_o)$ is proportional to σ and n_2. We have shown that $- \ln (\phi/\phi_o)$ is proportional to the h_2 heat pulse power and duration. These both affect n_2 linearly as the spectrum injected by h_2 does not change rapidly with power, and the pulse length is short compared with the dispersion in time of the roton pulse at the interaction volume.

The q dependence of the roton-roton (RR) scattering is accessible with this experiment but not the q dependence of phonon-roton (PR) scattering. The signal at different times for rotons can be associated with different wavevectors, however, this cannot be done for phonons, because there is the near cancellation of the phonon and atom dispersion, when the path lengths in liquid and vacuum are equal and for the spectrum of phonons injected. This is in contrast to the reinforcing dispersion of rotons and atoms. Typical phonon and roton signals are shown in fig 1(b) and (c).

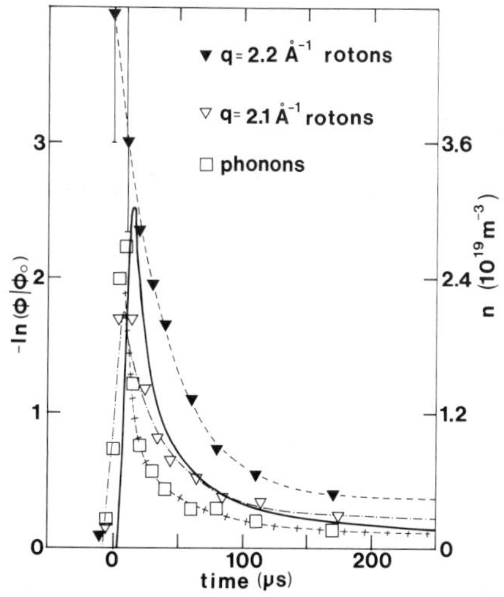

FIGURE 2 *The variation of the scattering of various excitations in the probe beam is shown as a function of the transit time of the scattering rotons from the heater h_2 to the scattering region. The solid line is the calculated scattering roton number density using the same assumptions outlined in (1).*

Before we make a direct comparison of the PR and RR cross sections we investigate the time development of the h_2 rotons using the probe phonons and rotons. Fig 2 shows the variation of $- \ln (\phi/\phi_o)$ with the delay of the h_1 pulse after the h_2 pulse for rotons with $q = 2.1Å^{-1}$ and $q = 2.2Å^{-1}$ and phonons. The difference in cross-sections for these processes is immediately apparent. The time axis has be adjusted in each case to compensate for the time taken for the probe excitations with different group velocities to reach the interaction volume. The solid line is the calculated n_2, which is proportional to $- \ln (\phi/\phi_o)$ for σ constant. The rotons are taken to have a 1.5K Bose-Einstien distribution. The general agreement shows the h_2 rotons propagate ballistically. The disagreement around the peak is due to the scattering cross section increasing for these wavevectors.

From a comparison of the reduction of the phonon and roton signal for delays which correspond to the same scattering rotons we find that $\sigma_{PR}/\sigma_{RR} = 0.6 \pm 0.1$ for $q = 0.55Å^{-1}$ phonons and $q = 2.1Å^{-1}$ rotons scattering from $q = 1.95Å^{-1}$ rotons. Previously we have found that the scattering does not change with q or $q<2.1Å^{-1}$. As the phonon and roton group velocities are the same for $q = 0.55$ and $2.1Å^{-1}$ respectively the ratio of cross sections is the same as the ratio of inverse lifetimes. So we can compare it with the ratio of neutron linewidths measured for similar phonons and rotons scattering from thermal rotons. This ratio is 0.65 ± 0.09 which is in agreement with the value above.

3. CONCLUSION

We have measured the relative scattering cross-sections for high energy phonons and rotons scattering from a beam of rotons and find that $\sigma_{PR}/\sigma_{RR} = 0.6 \pm 0.1$. This is in agreement with neutron scattering measurements.

REFERENCES

(1) A C Forbes and A F G Wyatt, Phys.Rev.Letts 'in print'
(2) D F Brewer and D O Edwards, Proc. Roy. Soc. A251 (1959) 247
(3) B Castaing and A Libchaber, J. Low. Temp. Phys. 31 No 5/6 (1978) 887
(4) F Mezei, Phys. Rev. Letts. 44 (1980) 1601
(5) F Mezei and W G Stirling, 75th Jubilee Conf. on ^{4}He. ed. JGM Armitage (World Scientific 1983) pp.111 - 112
(6) J Yau and M J Stephen, Phys. Rev. Letts 27 (1971) 481
(7) I M Khalatnikov, An introduction to the theory of superfluidity (Addison-Wesley 1989)
(8) A F G Wyatt and M Brown, This volume.
(9) F R Hope, M T Baird and A F G Wyatt, Phys. Rev. Letts. 52 (1984) 1528

Physica B 165&166 (1990) 499–500
North-Holland

DYNAMICAL STRUCTURE FACTOR S(k,ω) IN HeII AT FINITE TEMPERATURE

Koji ISHIKAWA

Department of Physics, Yokohama City University, 22-2, Seto, Kanazawa-ku, Yokohama 236, Japan

Kazuo YAMADA

Department of Physics, College of General Education, Nagoya University, Nagoya 464, Japan

The dynamical structure factors $S(k,\omega)$ at k=0.8A^{-1} from 1.4K to T$_\lambda$ are calculated on the basis of two-fluid dynamical approach,taking rotons as thermal excitations into account. The obtained $S(k,\omega)$ have spectra of Lorentzian type whose resonance energy has weak temperature dependence and width is narrower than inverse roton life time.

The dynamical structure factors $S(k,\omega)$ are obtained experimentally from the inelastic scattering of neutrons in HeII and one can know about the elementary excitation from $S(k,\omega)$. Woods and Svensson(1) proposed that $S(k,\omega)$ for superfluid He at finite temperatures is separated as

$$S(k,\omega) = \bar{\rho}_s S_s(k,\omega) + \bar{\rho}_n S_n(k,\omega),$$

where $\bar{\rho}_s = \rho_s/\rho$ and $\bar{\rho}_n = \rho_n/\rho$, and ρ, ρ_s and ρ_n denote total, superfluid and normal fluid density, respectively. They analyzed that $S_s(k,\omega)$ have Lorentzian type spectra in which a peak has the energy of the elementary excitation and its width gives inverse life time of roton in the region of temperature higher than 1.0K.

We previously derived the formula of $S(k,\omega)$ on the basis of the two-fluid dynamical approach in which we took phonons in ref.(2) into account and rotons in ref.(3) as thermal excitations. In this paper we calculated the $S(k,\omega)$ in the range of temperature from 1.4K to T$_\lambda$ by the similar approach to the preceding one(3), further taking the roton-roton interactions into account. From the calculated spectra of $S(k,\omega)$ at finite temperatures, we discuss the Wood and Svensson's proposal and what we know about the elementary excitations in HeII obtained from the observed $S(k,\omega)$.

We use three basic equations of two fluid dynamics(4); Conservation law of mass density ρ with mass current $\mathbf{j} = \rho\mathbf{v}_s + \sum \mathbf{p}N_\mathbf{p}$, equation of motion for the superfluid velocity $\mathbf{v}_s$ and the kinetic equation for the roton distribution function $N_\mathbf{p}$. To obtain the response function from the two fluid dynamical equations, they are regarded as operator equations for $\rho, \mathbf{v}_s$ and $N_\mathbf{p}$. Using the linearized and Fourier-transformed equations, the equations of motion for the response function defined by $\chi_{AB}(k,t) = -i\theta(t) < [A(t), B(0)] >$, are written as,

$$-i\omega\chi_{\rho\rho}(k,\omega) - \rho k^2\chi_{\phi\rho}(k,\omega) + i\sum_\mathbf{p}(\mathbf{kp})\chi_{N_\mathbf{p}\rho}(k,\omega) = 0,$$

$$-i\omega\chi_{\phi\rho}(k,\omega) + (\tilde{c}^2/\rho)\chi_{\rho\rho}(k,\omega) + (cp_0/\rho)\sum_\mathbf{p} U_p\chi_{N_\mathbf{p}\rho}(k,\omega) = 1,$$

$$-i(\omega + i\gamma_p - \mathbf{kv}_p)\chi_{N_\mathbf{p}\rho}(k,\omega)$$
$$-i(\mathbf{kv}_p)N'_\mathbf{p}[(cp_0/\rho)U_p\chi_{\rho\rho} + i(\mathbf{kp})\chi_{\phi\rho} + g\sum_{\mathbf{p}'}\chi_{N_{\mathbf{p}'}\rho}] = 0.$$

The ϕ_k is the velocity potential defined by $\mathbf{v}_s = i\mathbf{k}\phi_k$. $N'_\mathbf{p} = \partial N^0_\mathbf{p}/\partial E_p$ is the derivative of the roton distribution function at equilibrium $N^0_\mathbf{p}$ by the roton energy $E_p = \Delta + (p-p_0)^2/2\mu$. $\mathbf{v}_p = (\mathbf{p}-\mathbf{p}_0)/\mu$ is the roton group velocity. $(cp_0/\rho)U_p$ is phonon-roton coupling given by

$$(cp_0/\rho)U_p = \partial E_p/\partial\rho = \partial\Delta/\partial\rho - ((p-p_0)/\mu)\partial p_0/\partial\rho - ((p-p_0)^2/(2\mu^2))\partial\mu/\partial\rho.$$

$$\tilde{c}^2 = c^2 + \sum \rho N^0_\mathbf{p}\partial^2 E_p/\partial^2\rho.$$

The γ_p is the inverse roton life time. The $g = 2V_0$ where V_0 is roton-roton coupling constant. In order that these equations are valid, the wave vector k is restricted in the collision-less phonon region. Solving the equations of motion for the response function, we find

$$\chi_{\rho\rho}(k,\omega) = -Q_1(k,\omega)/D(k,\omega)$$

where

$$D(k,\omega) = P^2(k,\omega) - Q_1(k,\omega)[Q_2(k,\omega) + gR^2(k,\omega)/\Pi(k,\omega)],$$

$$Q_1(k,\omega) = \rho k^2 - \sum_p \frac{(\mathbf{kp})^2(\mathbf{kv}_p)N'_\mathbf{p}}{(\omega + i\gamma_p - \mathbf{kv}_p)} = \rho k^2[\bar{\rho}_s + 3\bar{\rho}_n(\omega/kv_T)^2\xi(k,\omega,v_T)],$$

$$Q_2(k,\omega) = (c^2/\rho)[(\tilde{c}/c)^2 + 3\bar{\rho}_n\{A^2 + (B^2 - 2AG)v_T^2 + G^2 v_T^4\}\xi(k,\omega,v_T)],$$

$$P(k,\omega) = \omega[1+3c\bar{\rho}_n B\xi(k,\omega,v_T)],$$

$$R(k,\omega) = -3cp_0^{-1}\bar{\rho}_n(A-Gv_T^2)\xi(k,\omega,v_T)$$

and

$$\Pi(k,\omega) = 1 - 3g\bar{\rho}_n\xi(k,\omega,v_T)/p_0^2.$$

with

$$\xi(k,\omega,v_T) = (\omega/2kv_T)[\log\frac{\omega+i\gamma+kv_T}{\omega+i\gamma-kv_T} - 1].$$

The coefficients A, B and G is defined by

$$A = (\rho/p_0c)\partial\Delta/\partial\rho,\ B = (\rho/p_0c)\partial p_0/\partial\rho$$

and $G = (\rho/2p_0c)\partial\mu/\partial\rho$.

To get above expression , we used the approximation to replace v_p by thermal velocity of roton $v_T = \sqrt{k_B T/\mu}$ and the relations $\rho_n = -\sum(\mathbf{kp})^2 N_\mathbf{p}'/k^2$ and $\rho - \rho_n = \rho_s$.

The dynamical structure factor $S(k,\omega)$ is obtained from the imaginary part of $\chi_{\rho\rho}$ and written as

$$S(k,\omega) = \frac{\rho_s k^2\omega\Gamma_k + 3\rho_n(\omega/v_T)^2[(\omega^2-\varepsilon_k^2)^2\xi_I + \omega\Gamma_k\xi_R]}{(1-e^{-\omega/k_BT})[(\omega^2-\varepsilon_k^2)^2 + (\omega\Gamma_k)^2]}$$

where

$$\varepsilon_k^2 = (ck)^2[\bar{\rho}_s(\tilde{c}/c)^2 + 3\bar{\rho}_n(\beta\xi_R + \eta\xi_I) + 9\bar{\rho}_n^2\alpha(\xi_R^2 - \xi_I^2) + 18\bar{\rho}_n^2\nu\xi_R\xi_I],$$

$$\omega\Gamma_k = 3\bar{\rho}_n[\beta\xi_I - \eta\xi_R + 6\bar{\rho}_n\alpha\xi_R\xi_I - 3\bar{\rho}_n\nu(\xi_R^2 - \xi_I^2)].$$

and ξ_I and ξ_R are the real and imaginary part of ξ.

The coefficients α, β, ν and η are the complex functions depending on k, ω and other parameters, not written here explicitly. To obtain the spectra of $S(k,\omega)$ numerically, we set the experimentally observed values for $\bar{\rho}_n$, $A, B, G, \tilde{c}^2$ with

$$\partial\Delta/\partial\rho = -0.94\Delta/\rho, \qquad \partial p_0/\partial\rho = 0.37p_0/\rho,$$
$$\partial\mu/\partial\rho = -1.1\mu/\rho, \qquad \partial^2\Delta/\partial\rho^2 = -5.2\Delta/\rho^2,$$

and $V_0 = -1.7\times10^{-38}$erg-cm^3 and $c = 2.39\times10^4$cm according to Khalatnikov and Chernikova(5). The roton width is calculated from the viscosity coefficient η_r of roton by $\gamma = \bar{\rho}_n k_B T/(5\mu\eta_r)$. The calculated $S(k,\omega)$ for k=0.8A^{-1} are shown in figure for the series of temperatures from 1.4K to T_λ.

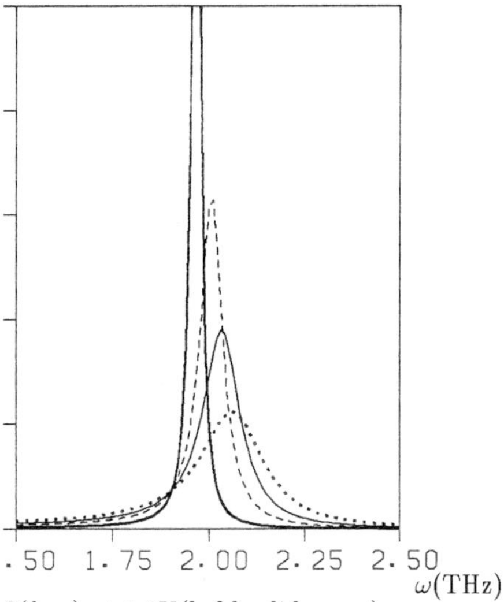

$S(k,\omega)$ at 1.8K(bold solid curve), 2.0K(dashed curve),2.1K(fine solid curve) and T_λ(dotted curve).

The obtained $S(k,\omega)$ have spectra of Lorentzian type, the peak frequencies increase slightly as temperature increases, in agreement with the observed one, but the width is narrower than the inverse roton life time. Although we examined the possibilities of separation of $S(k,\omega)$ obtained above into two components S_s and S_n, we have never obtained physically plausible results. Thus it was concluded that the validity of Woods and Svensson's proposal was not proved theoretically. Talbo and Griffin(6) calculated the $S(k,\omega)$ using the perturbation on the Green function at finite temperature and obtained similar results.

Finally it is shown that the other two response functions $\chi_{\phi\phi}$ and χ_{jj} are easily derived from our two-fluid dynamical equations, and these functions obey the following general relations(7) in the same way as Ref.(2).

$$\chi_{\phi\phi}(k,0) = (k^2\rho_s)^{-1},$$

$$\chi_{jj}^{\alpha\beta}(k,0) = \delta_{\alpha\beta} + k_\alpha k_\beta k^{-2}\rho_s.$$

where Greek letters denote the components of Cartesian coordinate.

(1) A.D.B.Woods and E.C.Svensson, Phys. Rev. Lett. 41(1978)974.
(2) K.Yamada, Prog. Theor. Phys. 63(1980)715.
(3) K.Ishikawa and K.Yamada, Two component character of $S(k,\omega)$ in superfluid He^4, in: Elementary Excitation in Quantum Fluids, eds. K.Ohbayashi and M.Watabe (Springer-Verlag,1989)pp.155-159.
(4) I.M.Khalatnikov, An Introduction to the Theory of Superfluid, (W.A.Benjamin Inc, New York,1965).
(5) I.M.Khalatnikov and D.M.Chernikova, Sov. Phys. JETP. 22(1966)1336.
(6) E.Talbot and A.Griffin, Phys. Rev. 29(1984)2531.
(7) P.C.Hohenberg and P.C.Martin, Ann. of Phys. 34 (1965)291.

Physica B 165&166 (1990) 501–502
North-Holland

MULTIPHONON SCATTERING FROM SUPERFLUID ^{4}HELIUM

W.G. STIRLING

Department of Physics, School of Physical Science and Engineering,
University of Keele, Staffs. ST5 5BG, U.K. and
S.E.R.C. Daresbury Laboratory, Warrington, WA4 4AD, U.K.

Neutron inelastic scattering has been used to investigate the form of the multi-phonon "continuum" scattering from superfluid ^{4}He. The results show clear evidence of structure arising from pairs of quasiparticles, with peaks at energies close to those of two rotons, two maxons and a roton-maxon pair.

1. INTRODUCTION

Although the principal feature of the dynamical structure factor S(Q,E) of superfluid ^{4}He as measured by neutron inelastic scattering is the one-phonon (or one-quasiparticle) peak, there is significant scattering at higher energies (1). This continuum arises from multiphonon (or multi-quasiparticle) excitations and has a long high energy tail, extending to at least 1.5 THz (72K). The shape of the multiphonon continuum is determined by the multiphonon density of states (in particular the two-phonon density of states) and by interactions between the excitations. Several studies have been made of the continuum scattering, notably by Svensson and collaborators (2) but instrumental resolution has been inadequate to distinguish separate contributions. Clear structure in the multiphonon scattering has been observed, however, in a recent high resolution experiment (3). This paper reports a continuation of the measurements reported in reference (3), under improved experimental conditions.

2. EXPERIMENTAL DETAILS

The current experiment was designed to extend the previous work described in reference (3). Neutron inelastic scattering measurements were made using the IN12 cold-neutron triple-axis spectrometer at the Institut Laue Langevin, Grenoble, France. The scattered neutron energy was held fixed at 54.1K, yielding an instrumental resolution width of 1.58K (0.033 THz; 1THz = 47.99K). Other experimental conditions were as in reference (3).

3. RESULTS

Figure 1 presents typical results for the roton (1.925Å^{-1}) and maxon (1.13Å^{-1}) wave vectors at 1.2K and saturated vapour pressure. The left-hand frames show that, as expected, the sharp one-phonon excitation dominates the spectrum; the multiphonon spectra are shown enlarged on the right-hand side. The arrows indicate the positions of twice the roton energy (2R: 0.36 THz), one roton plus one maxon (R+M: 0.47 THz), and twice the maxon energy (2M: 0.58 THz), respectively.

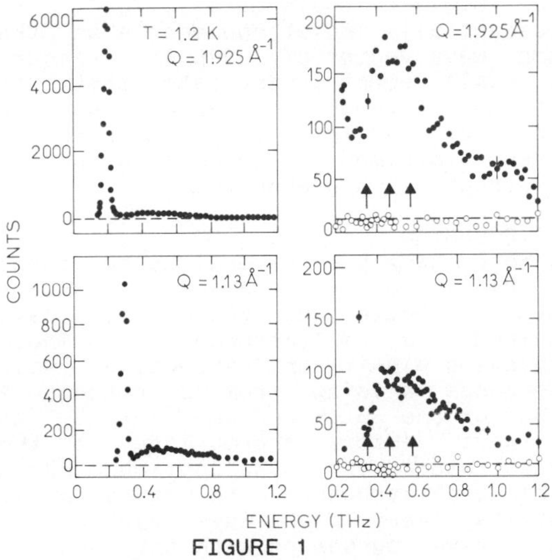

FIGURE 1
Energy spectra for the maxon and roton wave vectors; full circles are from the sample cell filled with ^{4}He, open circles are the empty-cell background. The arrows denote the energies 2R, R+M, and 2M, as explained in the text.

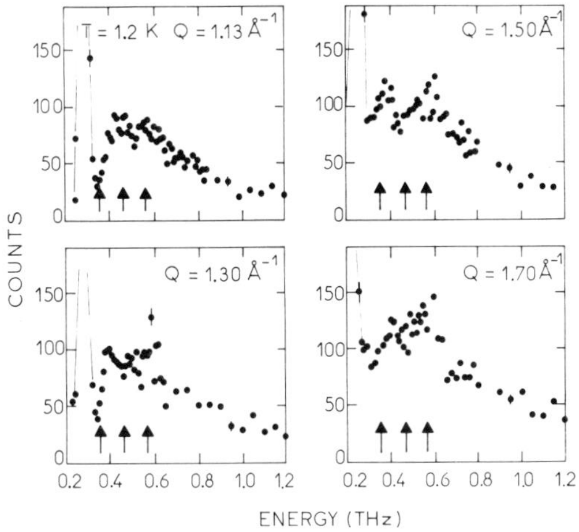

FIGURE 2
Net ^{4}He multiphonon scattering. The arrows are as in Figure 1. The single excitation peaks are indicated by the lines.

The wave vector variation of the multiphonon spectrum is shown in Figure 2. At 1.13 Å^{-1} the wide distribution has two peaks at close to R+M and 2M, the lower of which decreases in energy towards 2R at 1.30 Å^{-1}. By 1.50 Å^{-1} there are clearly 2 peaks (close to 2R and 2M) which begin to coalesce at 1.70 Å^{-1} to be essentially indistinguishable at the roton wave vector of 1.925 Å^{-1} (Figure 1). All spectra indicate that the continuum extends to frequencies in excess of 1.2 THz, the limit of the current experiments. On heating to 1.9K the structure is washed-out.

4. DISCUSSION

There have been several calculations of the liquid ^{4}He dynamical structure factor. Götze and Lücke (4) have obtained a multiphonon continuum displaying significant structure, with resonances arising from 2 rotons, 2 maxons or one roton and one maxon. At the wave vectors appropriate to the present experiment, the continuum has a double peak structure with maxima whose energies vary with wave vector in qualitative agreement with the present results. More recently, Manousakis and Pandharipande (5) have employed a perturbation technique, calculating explicitly the two-quasiparticle density of states. For wave vectors in the range 0.8 to 1.9 Å^{-1}, this function is found to display a strong peak at 23K (0.48 THz), arising from one roton plus one maxon. Weaker side peaks can be associated with 2 rotons and 2 maxons. The resulting S(Q,E) is dominated by a roton-maxon (R+M) peak at 1.125Å^{-1}, with 2 peaks of comparable intensity (2R, R+M) at 1.525 Å^{-1}. The 2M contribution is strongest at 1.925 Å^{-1} with a weaker 2R peak. Inspection of Figure 2 indicates that the R+M contributions are indeed dominant at 1.13 Å^{-1}, although a 2M component can also be distinguished, while at 1.50 Å^{-1} 2R and 2M contributions are apparent. It is difficult to separate individual contributions at 1.925 Å^{-1} (Figure 1). Overall, there is a satisfactory agreement between theory and experiment.

5. CONCLUSIONS

Structure in the multiphonon continuum scattering from superfluid ^{4}He has been observed which can be correlated with peaks in the calculated dynamical structure factor arising from pairs of excitations at the "flat" portions of the single excitation spectrum. New high resolution neutron time-of-flight results, currently being evaluated, should provide even more experimental detail for comparison with theoretical predictions of the form of the continuum scattering.

ACKNOWLEDGEMENTS
The author is grateful to the Institut Laue-Langevin for the provision of the experimental facilities used in this work. Valuable discussions with Prof. W. Götze and Prof. E. Manousakis are gratefully acknowledged.

REFERENCES
(1) R.A. Cowley and A.D.B. Woods, Can. J. Phys. 49 (1971) 177.
(2) E.C. Svensson, P. Martel, V.F. Sears, and A.D.B. Woods, Can. J. Phys. 54 (1976) 2178.
(3) W.G. Stirling, in: Proceedings of the Second International Conference on Phonon Physics, eds. J. Kollar, N. Kroo, N. Menyhard and T. Siklos (World Scientific,Singapore, 1985) pp.829-835.
(4) W. Götze and M. Lücke, Phys. Rev. B13 (1976) 3825.
(5) E. Manousakis and V.R. Pandharipande, Phys. Rev. B33 (1986) 150.

Physica B 165&166 (1990) 503–504
North-Holland

PHONON CONTRIBUTION AND 2-ROTON RAMAN SPECTRUM OF HeII

Norio OGITA, Mitsuo WATABE, Masayuki UDAGAWA and Kohji OHBAYASHI

Faculty of Integrated Arts and Sciences, Hiroshima University, Hiroshima, 730 JAPAN

2-roton Raman spectra of superfluid ^{4}He have been measured at T=0.65K under pressures from svp to P=20kg/cm^2. No intrinsic spectral width has been observed at all pressures. Contribution of phonons has been found in the lower energy region of the 2-roton peak. Considering it, the experimental spectra were compared with the theory by Iwamoto and Ruvalds-Zawadowski (IRZ model). A good fit of the experimental result to the theory has been observed for the svp spectrum. However, similar analysis of the spectrum at 20kg/cm^2 has not given a good fit.

1. INTRODUCTION

Raman scattering is a useful method to investigate elementary excitations in quantum fluids. Especially detailed analyses of interaction between them are possible. Greytak et al. [1,2] experimentally observed the 2-roton spectral shape at svp consistent with the IRZ model [3,4]. They showed that the roton parameters, such as roton energy (Δ_0), roton-roton interaction (E_i) and roton width (Γ), can be determined by parameter fitting to the model. However, Ohbayashi et al. [5] reported that the IRZ model is not good enough to describe experimental spectral shape at all pressures. Small but significant discrepancy existed between the experimental results of the two groups. The purpose of this paper is to find out the source of it and to analyze whether the spectra may be explained by the conventional theoretical model or not at various pressures.

2. EXPERIMENTAL

The sample was cooled down to 0.65K with a pumped ^{3}He cryostat. The light source was 514.5nm radiation from an argon ion laser. The spectrometer was a JASCO CT-1000D type double-grating monochromator. A great improvement in signal to noise ratio has been achieved by adopting a Surface Sciences Laboratory Model 2601A 2-dimensional Imasing Photon-Detector. To calibrate the energy shift scale, rotational Raman spectra of N_2 and O_2 gases were used as a standard. The absolute accuracy was ±0.04K.

In the former works by the above mentioned two groups, the tail of the central component (Brillouin scattering) made background to the 2-roton Raman scattering. The subtraction of it introduced some ambiguity to the final deduced data. This was the source of the discrepancy. In the present experiment, the tail of the central component was reduced to a negligible value at the 2-roton region by using the horizontally polarized incident beam, by reducing the stray light in the cell and by critically adjusting the widths of the slits in the monochromator.

3. RESULTS AND DISCUSSION

The 2-roton Raman spectra of superfluid ^{4}He measured at a temperature of 0.65K are shown in Fig.1. The two representative spectra at pressures of svp and 20kg/cm^2 are shown in this figure. Because the tail of the central component was very weak, we did not need to subtract it from the observed 2-roton spectrum. This eliminated ambiguity of the base. The

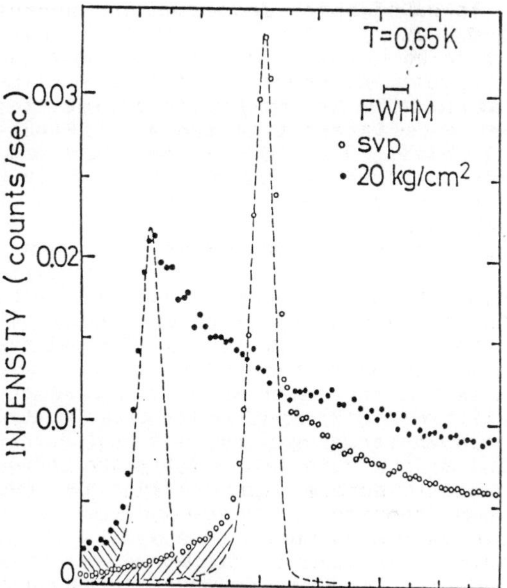

Fig.1 The Raman spectra of HeII at svp and at P=20kg/cm^2. The broken lines are instrumental profiles.

0921-4526/90/$03.50

instrumental profiles are shown by the broken curves. The instrumental FWHM was 0.45K. It is noticeable that the two spectral shapes are almost the same with the instrumental profile in the lower energy side of the peak. This means energy dependence of the spectra on this side is either step-like or delta-function-like without an observable intrinsic width ($\Gamma=0$). It is also noticeable that the weak response (hatched) remains and extends in the lower energy side of the peak outside the instrumental profile. Because only the phonon branch exists below the roton minimum energy of the dispersion curve, the weak response should be attributed to the phonon contribution. This has not been noticed in previuos works explicity.

The observed spectra are compared with the IRZ model taking into account the phonon contribution. In the model there are four parameters; the amplitude A of the spectrum, $2\Delta_0$, E_i and Γ. As mentioned above, the intrinsic width Γ is negligible. An additional adjustable constant C is included to take into account the phonon contribution. Weak energy-dependence of it is neglected. The convolution integral of the theoretical spectrum with the instrumental profile was calculated for various parameters to determine the best fit. As the profile, we adopted a lorentzian in our former work [5,6]. In practice, however, it is neither gaussian nor lorentzian. For precise comparison, we determined the profile function to fit the observed central component accurately.

As shown by the solid line in Fig.2, an excellent fit was obtained to the svp spectrum obtained with the following parameters; attractive interaction energy $E_i=0.2$K, $2\Delta_0=17.28$K and C/A=0.09. The 2-roton energy agrees with the neutron result $2\Delta_0=17.236$K ±0.018K within the experimental accuracy of ±0.04K. A change of E_i over ±0.05K makes the calculated curve deviate from the experimental data points appreciably. Thus the uncertainty of the interaction energy is estimate to be ±0.05K. The value of E_i agrees with the result obtained by Murray et al. [2]. Without the constant parameter C, we could not find a good fit. This is the reason why we could not get good fit of the IRZ model to the experimental data in our former work [5]. Although Murray et al. did not take into account the phonon contribution, it was implicitly subtracted included in the tail of the central component.

For 20kg/cm^2, data must be analyzed with the IRZ model with repulsive interaction because of positive shift of the peak from the 2-roton energy determined by neutron scattering experiments. A similar analysis to that of svp data, however, showed considerable deviation even at the best fit. To illustrate this, we show calculated curves for three repulsive interaction energies in Fig.2(b). The repulsive energy consistent with the Raman peak shift from the 2-roton energy determined by neutron scattering should be within 0.29K$\pm$0.16K. The constant base C was estimated at the lower energy side of the peak. As shown in Fig.2(b), the IRZ model cannot give a good description of the Raman spectrum for repulsive interaction case.

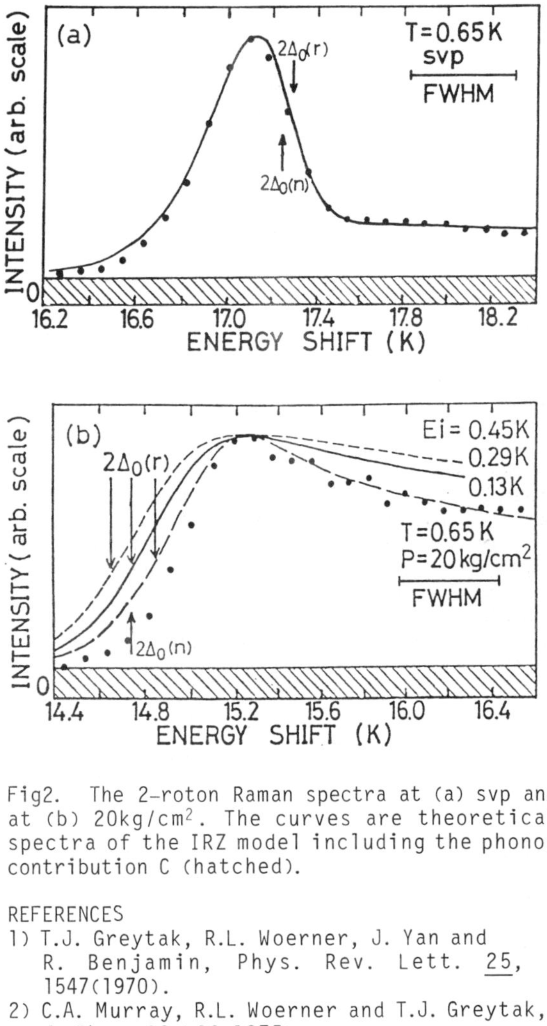

Fig2. The 2-roton Raman spectra at (a) svp and at (b) 20kg/cm^2. The curves are theoretical spectra of the IRZ model including the phonon contribution C (hatched).

REFERENCES

1) T.J. Greytak, R.L. Woerner, J. Yan and R. Benjamin, Phys. Rev. Lett. 25, 1547(1970).
2) C.A. Murray, R.L. Woerner and T.J. Greytak, J. Phys. C8 L90(1975).
3) F.Iwamoto, Prog. Theor. Phys. (Japan), 44, 1135(1970).
4) J. Ruvalds and A. Zawadowski, Phys. Rev. Lett. 25, 333(1970).
5) K. Ohbayashi, T. Akagi, N. Ogita, M. Watabe, and M. Udagawa, Jpn. J. Appl. Phys. Supplement 26-3, 5(1987).
6) K. Ohbayashi and M. Watabe (eds.), Elementary Excitations in Quantum Fluids. (Springer-Verlag, Berlin Heidelberg, 1989).

Physica B 165&166 (1990) 505–506
North-Holland

RIPPLON–LIMITED MOBILITIES OF IONS TRAPPED BELOW THE SURFACE OF SUPERFLUID ^{4}He

J. MEREDITH, C.M. MUIRHEAD, P.K.H. SOMMERFELD & W.F. VINEN

School of Physics and Space Research, University of Birmingham, Birmingham, B15 2TT, UK

Measurements of the damping of plasma waves in two-dimensional pools of positive ions trapped below the surface of superfluid ^{4}He have been used to obtain values of the mobility of the ions in a temperature range where the mobility is limited by ripplon scattering (10 mK to 120 mK). The mobility has been found to be inversely proportional to the temperature and to vary with the trapping depth (z_0) as $z_0^{3.5}$. These dependences are reproduced by a theory of the mobility based on the calculated scattering of capillary waves by an isolated submerged charged sphere, but the absolute magnitude of the mobility is predicted to be a factor of about 20 too high. Some experimental evidence has been found for a weak dependence of the mobility on the areal density of ions in the pool, which suggests that the theory ought to take account of the interactions of ripplons with the collective modes of the ion pool.

1. INTRODUCTION

Pools of ions can be trapped below the surface of ^{4}He by a combination of the image potential and an externally applied vertical electric field, the magnitude of which determines the depth, z_0, at which the ions are trapped. There has been extensive experimental study of two-dimensional plasma resonances in such pools (1–5): resonant frequencies have yielded information about ionic masses; and resonant linewidths have yielded information about ionic mobilities. At higher temperatures the pools exist as two-dimensional fluids; but at lower temperatures, typically below about 100 mK, the pools can form Wigner crystals, as has recently been demonstrated experimentally (6). At low temperatures the ionic mobilities are limited by ripplon scattering, as is clear from the fact that the mobilities become strongly dependent on z_0 (3,4), although reliable quantitative data have not so far been available. In this paper we aim to fill this gap in our knowledge by presenting the results of a detailed study of the ripplon-limited mobility for positive ions in the temperature range from 10mK to 120mK.

2. MEASUREMENT OF THE IONIC MOBILITY AT VERY LOW TEMPERATURES

Our cryostat and experimental cell are identical to those used by Barenghi *et al* (2). The damping of the plasma resonances, from which the mobility is obtained, has been measured in two ways: from the resonant linewidths associated with *forced* plasma oscillations, as in earlier work; and from the rate of decay of *free* plasma oscillations, occuring after the removal of a driving pulse. This latter method is of particular value at the lowest temperatures, where mobilities are so high that the linewidths are comparable with small fluctuations in resonant frequency caused by vibrationally excited gravity waves on the surface of the helium (1,2).

3. EXPERIMENTAL RESULTS

The dependence of the observed mobility on temperature and trapping depth is shown in figures 1 and 2. To a good approximation the mobility is seen to be inversely proportional to the temperature and to be proportional to $z_0^{3.5}$. The absolute magnitude of the mobility, for a given temperature and depth, is significantly larger than has been reported previously (3,4), suggesting that earlier measurements were affected by vibrational broadening of the resonance. We have observed no anomaly in the mobility at the temperature at which crystallization has been observed to take place, although our mobility measurements have not so far been carried out on a pool that has actually been observed by the technique of reference (6) to have undergone Wigner crystallization. This last proviso is necessary because experiments on the Wigner crystallization have shown that it can be inhibited by a small tilt of the

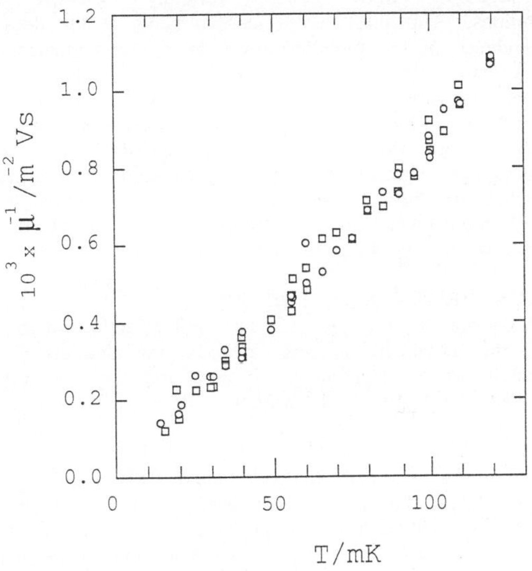

FIGURE 1
Temperature dependence of the mobility, μ, of the positive ions at very low temperatures; z_0 = 43.5 nm.

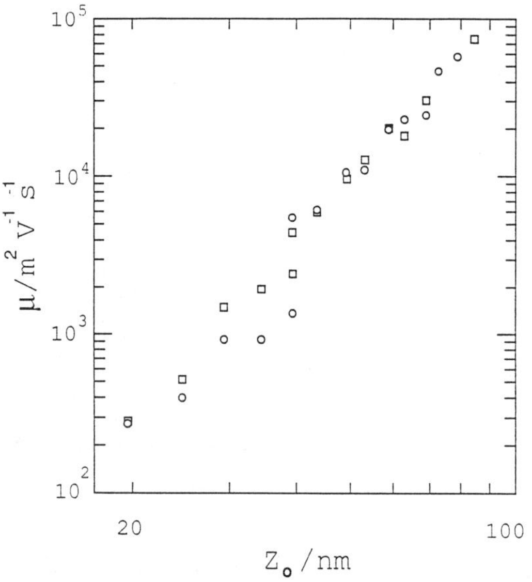

FIGURE 2
Dependence of the mobility, μ, of the positive ions on the depth, z_0, at which the ions are trapped below the liquid surface; T = 15 mK.

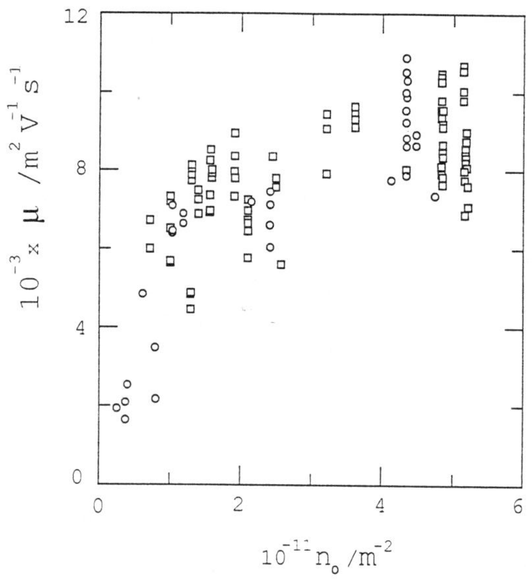

FIGURE 3
Dependence of the mobility, μ, of the positive ions on the areal density, n_0, of the pool; T = 15 mK; z_0 = 43.5 nm.

experimental cell.

As we explain in the next Section, it is of considerable interest to determine whether there is any dependence of the mobility on the areal density of the pool. A strong dependence was reported by Hannahs & Williams (3), although, as we have seen, their measurements were almost certainly affected by vibrations. Our most recent measurements of the density dependence of the mobility, taken at a low temperature, where any effect seems to be most pronounced, are shown in figure 3. There does seem to be a decrease in mobility with decreasing density, expecially at low densities, but the effect is much smaller than that reported by Hannahs & Williams. Measurements at low densities are difficult because of loss of sensitivity, and some improvement in our technique will be necessary before we can be sure that the effect is real.

4. COMPARISON WITH THEORY

According to a theory of the ripplon-limited mobility (4), the dominant process involves the scattering of thermally-excited ripplons by an ion, and the assumption is made that the ions scatter independently of one another. The scattering cross-section can be obtained from a classical model in which capillary waves are scattered by a small charged sphere submerged in a classical ideal fluid. The ions have a natural frequency of vertical oscillation in the potential well in which they are trapped, and most of the cross-section arises from resonant scattering at this frequency. The theory leads to a contribution to the inverse mobility that is proportional to the temperature and to $z_0^{3.5}$, in agreement with experiment. However, the predicted absolute magnitude of the mobility is too large by a factor of about 20. It is possible that this discrepancy arises from the assumption that the ions scatter independently of one another, a more correct theory treating the interaction of the ripplons with the whole assembly of ions and with their collective excitations. Confirmation that the mobility is density dependent would confirm the need for such a theory. However, if, as is probably the case, the mobility is unaffected by Wigner crystallization, the theory would need to account for the fact that neither the onset of long-range positional order in the ion pool nor the change in the collective modes on crystallization appears to affect the mobility.

ACKNOWLEDGEMENTS

We are grateful to Dr. C.J. Mellor for his invaluable help with the experiments and for many very useful discussions.

REFERENCES

(1) M.R. Ott-Rowland, V. Kotsubo, J. Theobald & G.A. Williams, Phys.Rev.Lett. **49** (1982) 1708.
(2) C.F. Barenghi, C.J. Mellor, C.M. Muirhead & W.F. Vinen, J.Phys.C: Solid State Phys. **19** (1986) 1135.
(3) S. Hannahs & G.A. Williams, Jap.J. of Applied Phys. **26**, Suppl. 26-3 (1987) 741.
(4) C.J. Mellor, C.M. Muirhead, J. Traverse and W.F. Vinen, Jap.J. of Applied Phys. **26**, Suppl. 26-3 (1987) 383.
(5) C.J. Mellor, C.M. Muirhead, J. Traverse and W.F. Vinen, J.Phys.C: Solid State Phys. **21** (1988) 325.
(6) C.J. Mellor and W.F. Vinen, Surface Science (1990), in the press.

Physica B 165&166 (1990) 507–508
North-Holland

SUPERFLUIDITY OF LIQUID HELIUM AS A CONSEQUENCE OF NON-LINEARITY OF ENERGY SPECTRUM

Shosuke SASAKI

Department of Physics, College of General Education, Osaka University, Toyonaka Osaka 560, Japan

In an interacting many boson system, the total energy is, generally, non-linear as a function of set $\{n_p\}$ where n_p denotes the number of bosons occupying the eigen-energy level p. Owing to the non-linearity, various quantum levels have a possibility that their level energies become minimum according to the changes of the number distribution. We show that the non-linear effect produces superfluidity of liquid helium. Based on the theory, we also predict several new phenomena in liquid helium, and propose possible experiments to detect them.

1. INTRODUCTION

We know that a few one-dimensional many-body problems can be solved exactly(see ref.(1)). Examining these exact solutions, we observe non-linear effects in their energy spectra. It is the non-linear effect that the excitation energies from one level to the other levels depend on the distribution of all the other particles in eigenergy levels. Namely, the particle distribution determines which energy level has the minimum level energy (that is energy increase by addition of a particle to the level). In the other words, a different distribution gives a different level as the one which has the minimum level energy (see Sec.2). The non-linear effect is not peculiar to one-dimensional systems, but must be a general character in any interacting many-particle system. We will prove the existence of two-fluid in superfluid helium as a consequence of the non-linear effect in the energy spectra (see Sec.3).

In this paper, we further investigate the non-linear effect for superfluid helium, and we will predict the following phenomena;

(1) existence of "elementary phonon" and temperature dependence of its velocity.

(2) softenning of the dispersion curve at λ-point.

(3) triple reflection and refraction of "dressed-boson" at a boundary of superfluid helium.

(4) existence of the Josephson oscillation in superfluid helium even when the relative phase of the initial state is completely indefinite.

2. NON-LINEAR EFFECT IN ENERGY SPECTRA

We take a many boson system with only two eigen-energy levels as an example to explain the non-linear effect. The total energy E of the system is assumed to be written as

$$E=E_1n_1+E_2n_2+f_{12}n_1n_2 \tag{1}$$

where n_1 and n_2 are the boson numbers occupying level 1 and 2, respectively, and we restrict ourselves to $E_1<E_2$. If the coefficient f_{12} is equal to zero, the total energy E is linear for n_1 and n_2. Then, level 1 is lower than level 2.

If f_{12} is not zero, an addition of a boson into level i leads to the following energy increase as,

$$\varepsilon_1=E_1+f_{12}n_2, \quad \varepsilon_2=E_2+f_{12}n_1 \tag{2}$$

which mean the level energies of the present system. In the case of $f_{12}>0$, ε_1 becomes higher than ε_2 when n_2 exceeds a certain value. Namely, level 2 can become lower than level 1 even if $E_1<E_2$ when the boson number occupying level 2 is large enough. It has been shown that a level inversion occurs by the non-linear effect. This mechanism gives a remarkable effect to liquid helium.

3. GENERAL FORM OF TOTAL ENERGY OF LIQUID HELIUM

The total Hamiltonian of liquid helium is given as

$$H=\sum_p (p^2/2m)a_p^*a_p+(1/V)\sum_{p,q,k} g(k)a_{p+k}^*a_{q-k}^*a_pa_q \tag{3}$$

where m is the mass of a helium atom, V is the volume of the system and $g(k)$ is the interatomic potential. The operators a_p^* and a_p create and annihilate a helium atom with momentum p. Since the Hamiltonian H is hermitian, it can be diagonalized by a unitary operator U. All the eigenstates of H can be obtained with the unitary transformation from all the free states of helium atoms, because there is no bound-state of plural helium atoms. Namely the eigen-equation of H is described as

$$HUa_{p_1}^*\cdots a_{p_N}^*|0\rangle=E(p_1,\cdots,p_N)\,Ua_{p_1}^*\cdots a_{p_N}^*|0\rangle \tag{4}$$

Let us introduce the inverse unitary transformation of the operators a_p^* and a_p as

$$A_p^*=Ua_p^*U^*, \quad A_p=Ua_pU^* \tag{5}$$

where A_p^* and A_p create and annihilate a "dressed-boson" with momentum p, which is a helium atom wearing the interaction cloud produced by the unitary transformation. As the vacuum state $|0\rangle$ is an eigenstate of H, we get $U|0\rangle=|0\rangle$ This property and the definitions (5) yield

$$\begin{aligned} Ua_{p_1}^*\cdots a_{p_N}^*|0\rangle &= Ua_{p_1}^*U^*U\cdots\cdots U^*Ua_{p_N}^*U^*U|0\rangle \\ &= A_{p_1}^*\cdots A_{p_N}^*|0\rangle \end{aligned} \tag{6}$$

which means that all the eigenstates of H can be generated by operating direct product of dressed-boson operators to the vacuum state.

From Eqs. (4) and (6), the total energy of liquid helium, E, depends upon the momentum distribution of dressed-bosons only. The general form of the total energy can be written down as

$$E=\sum_{p}(p^2/2m)n_p-\sum_{p}W_B n_p+\frac{1}{2V}\sum_{p,q}f_2(p-q;N/V)n_p n_q$$
$$+\frac{1}{3V^2}\sum_{p,q,r}f_3(p-q,p-r;N/V)n_p n_q n_r+\cdots \quad (7)$$

where n_p is the dressed-boson number with momentum p (see ref. (2-4)). The first term in the right-hand-side of Eq. (7) denotes the kinetic energy of dressed-bosons and the other terms are Galilean invariant. In this energy expression, the function $f_2(p-q;N/V)$ can be explicitly written in terms of the elementary excitation energy ε_p^0 at absolute zero temperature (see ref. (2-4));

$$f_2(p-q;N/V)=[\varepsilon^0_{p-q}-(p-q)^2/2m](V/N) \quad (8)$$

Since the Galilean invariant terms involve only relative momenta of dressed-bosons, the total energy neccessarily has a non-linear dependence on the number distribution of dressed-bosons. Therefore, an inversion of energy levels which is mentioned in Sec. 2 can occur in the system of liquid helium.

4. APPEARANCE OF TWO-FLUID BY NON-LINEAR EFFECTS

We can write down an energy increase by addition of a dressed-boson with mometum p as

$$\omega_p=\delta E/\delta n_p=p^2/(2m)-W_B-\frac{\partial W_B}{\partial N}N$$
$$+V^{-1}\sum_{q}f_2(p-q;N/V)n_q \quad (9)$$
$$+(2V)^{-1}\sum_{q,k}[\partial f_2(q-k;N/V)/\partial N]n_q n_k+\cdots\cdots$$

This quantity depends not only on the momentum p, but also on the number distribution because of the non-linear terms of Eq. (7). In order to clarify the non-linear effect, we discuss a simple case such that almost all dressed-bosons have momenta Q

Then, Eq. (9) becomes

$$\omega_p=p^2/(2m)-W_B-\frac{\partial W_B}{\partial N}N$$
$$+(n_Q/V)\quad f_2(p-Q;N/V) \quad (10)$$
$$+(\text{higher order of } (N-n_Q)/V\ )$$

As easily seen from Eq. (8), the function $f_2(p-Q;N/V)$ is proportional to $|p-Q|$ when momentum p is close to Q. Therefore, the dressed-boson energy ω_p takes a minimum value at $p=Q$ as long as $(N-n_Q)/V$ is not too large, also Q is not too large. Namely, there are infinitely many values of Q at which ω_p becomes minimum. Therefore, the momentum of a Bose-condensate of dressed-bosons can take on an arbitrary value within a certain region.

Namely, the velocity of the condensate (superfluid velocity) can take on any value in the region, while the velocity of the normal fluid (assembly of non-condensed dressed-bosons) is different from it(see ref. (2)). This establishes the existence of two-fluid in superfluid helium.

5. PREDICTIONS

We will predict a few properties of liquid helium caused by the non-linear effects.

5.1. Existence of elementary phonon

In our theory, the elementary excitation energy is given as

$$\varepsilon_p=[\delta E/\delta n_p-\delta E/\delta n_0] \quad (11)$$

where the velocity of superlfuid helium is assumed to be zero. The dispersion curve which is given by this function ε_p has a phonon-like behaviour, whose velocity is given by

$$c=\lim_{p\to 0}\frac{1}{p}(\delta E/\delta n_p-\delta E/\delta n_0) \quad (12)$$

This velocity c becomes zero as the condensate of dressed-bosons disappears. Moreover, the value of the velocity is equal to the first sound velocity at absolute zero temperature. Thus, our elementary excitation of a dressed-boson is differrent from either one of Landau's first sound and second sound. We name this small momentum excitation "elementary phonon" (see ref. (3)).

5.2. Triple reflection and refraction

A dressed-boson refelcts or refracts at a boundary of superfluid helium. The refraction phenomena were elegantly discussed by Wyatt and experimentally detected (see ref. (5)). After that, the present author has pointed out the existence of triple reflection and calculated the transmission rates in the reflections and refractions (see ref. (4)).

5.3. Josephson oscillation in superfluidhelium

The concept of dressed-boson is the analog in superfluid helium of the Cooper pair in superconductor. We consider a system composed of superfluid helium contained in two containers which are weakly coupled to each other through a small orifice. By solving the Heisenberg equation of Bose-condensed dressed-bosons, it is shown that there occurs Josephson oscillation of superfluid helium even when the relative phase between the two condensed dressed-bosons is initially completely indefinite (see ref. (6)). I would like to express my hearty thanks to Professor A. Ikeda for his valuable advice.

REFERENCES

(1) E.H. Lieb and W. Liniger, Phys. Rev. 130 (1963) 1605.
S. Sasaki and T. Kebukawa, Prog. Theor. Phys. 65 (1981) 1198, 1217, 1798; ibid 66 (1981) 831.
S. Sasaki, ibid 81 (1989) 1158.
(2) S. Sasaki, Jpn. J. Appl. Phys. Suppl. Proc. LT18 26-3 (1987) 23.
(3) S. Sasaki, Springer Series in Solid-State Sciences Vol. 79 (Springer, 1989) p. 160.
(4) S. Sasaki and T. Kunimasa, J. Phys. Soc. Jpn. 58 (1989) 3651.
(5) A.F.G. Wyatt, Physica 126B (1984) 392.
G.M. Wyborn and A.F.G. Wyatt, Jpn. J. Appl. Phys. Suppl. Proc. LT18 26-3 (1987) 2095.
(6) T. Kebukawa and S. Sasaki, J. Phys. Soc. Jpn. 59 (1990) 601.

Physica B 165&166 (1990) 509–510
North-Holland

BOSE-EINSTEIN CONDENSATION IN A RANDOM POTENTIAL

D K K LEE* and J M F GUNN+

*Cavendish Laboratory, Madingley Rd, Cambridge CB3 0HE, UK
+Rutherford Appleton Laboratory, Chilton, Didcot, Oxon OX11 0QX, UK

We present a suitable weak-repulsive limit of a dense Bose gas where an extended Bose condensate can be stable in a random potential, even though the non-interacting case is pathological. The condensate exists primarily because the interactions allow screening of the random potential, even when the chemical potential is in the Lifshitz tails of the single-particle case. Using the number-phase representation, we calculate the increase in the depletion of the condensate with increasing randomness. The physical picture discussed should be relevant to the understanding of helium thin films.

Helium absorbed on various substrates (1-3) loses superfluidity at low concentrations. One interpretation is that the ^{4}He atoms are localised randomly at locations of strong van der Waals attraction. We will explore the robustness of the condensed phase to disorder in the high-density limit. In particular, we will emphasise the rôle of the Bose condensate in the partial *screening* of the random potential. Although a noninteracting Bose gas is pathological in a random potential, we have argued previously (4) that there is a *form* of noninteracting limit that is useful. To recapitulate, consider the Anderson model with short-range repulsion:

$$H = -t \sum_{\langle \mathbf{nn}' \rangle} c^+_{\mathbf{n}} c_{\mathbf{n}'} + \sum_{\mathbf{n}} (\sigma V_{\mathbf{n}} - \mu) n_{\mathbf{n}} + \frac{1}{2} U \sum_{\mathbf{n}} n^2_{\mathbf{n}}$$

$V_{\mathbf{n}}$ represents a spatially uncorrelated random potential. In the zero-hopping case, each site "fills up" with an integral number of bosons until the energy to add one more is above the chemical potential, μ. At weak disorder ($\sigma \ll U\overline{n}$), the resulting "Hartree potential", $V_{\mathbf{n}} + U n_{\mathbf{n}}$, varies around the chemical potential from site to site by a fraction of the repulsive energy U. To obtain a completely smooth potential, we can take the limit of $U \to 0$. We also keep $U\overline{n}$ fixed (*i.e.* $\overline{n} \to \infty$) in order not to alter the Hartree potential. A particle added to the system now will see a totally flat landscape. We need to restore the hopping term so that we can have Bose-Einstein condensation into an extended state. The interacting Bose gas has a characteristic healing length, $\lambda = (2ta^2/U\overline{n})^{1/2}$, over which the condensate wavefunction may vary significantly so that fast variations in the random potential becomes imperfectly screened at finite t. (a is the lattice spacing.) Nevertheless, motivated by this argument, our strategy is to assume that a condensed phase exists and then to establish whether it is stable to zero-point fluctuations in our régime of a dense but weakly-interacting gas.

How good is the screening? In our dense limit, we can use the number-phase representation: $c^+_{\mathbf{n}} \leftrightarrow n^{1/2}_{\mathbf{n}} \exp i\phi_{\mathbf{n}}$. We get a lattice version of the nonlinear Schrödinger equation for the local condensate density: $-t \sum_{\mathbf{n}'} \widetilde{n}^{1/2}_{\mathbf{n}'} + W_{\mathbf{n}} \widetilde{n}^{1/2}_{\mathbf{n}} = 0$ where $W_{\mathbf{n}} = \sigma V_{\mathbf{n}} - \mu + U\widetilde{n}_{\mathbf{n}}$. We have summed over the nearest neighbours $\mathbf{n}'$ of site $\mathbf{n}$. W is the smoother "residual potential". At weak disorder, its Fourier transform, $W_{\mathbf{k}}$, has a power spectrum proportional to k^4 at length scales greater than the healing length. This means that there are few traps which are both wide and deep. We argue (4) that W does not trap any localised states at the low-energy end of the spectrum even if the chemical potential is in the Lifshitz tail of the unscreened potential V. Using this result, we expect our interacting system to have a spectrum with delocalised low-energy excitations, at least in a treatment which looks at quantum fluctuations around the "mean-field" condensate wavefunction.

We will now proceed to discuss these zero-point fluctuations which lead to a depletion of the condensate: $\Xi = \sum'_{\mathbf{k}} \langle c^+_{\mathbf{k}} c_{\mathbf{k}} \rangle$. From now on, $2U\overline{n}$ is used as the unit of energy. The most important quantum fluctuations should be the phase fluctuations (with their conjugate density fluctuations). To concentrate on these collective modes ("phonons"), we use the number-phase representation and look only at the case of $\omega_B \ll 1$ where the

0921-4526/90/$03.50

spectrum of the pure system is dominated by them. (ω_B is the bandwidth of the lattice model.) In the pure system, the fractional depletion $\Xi/\bar{n}$ is $1/2\omega_B^{1/2}\bar{n}$.

To examine the zero-point fluctuations, it is convenient to use the rescaled operators Δ and Φ defined by: $n_{\mathbf{n}} = \tilde{n}_{\mathbf{n}} + (2\bar{n})^{1/2}\Delta_{\mathbf{n}}$ and $\phi_{\mathbf{n}} = \Phi_{\mathbf{n}}/(2\bar{n})^{1/2}$. From this, we can now define "number-phase bosons" by the creation operator $a^+_{\mathbf{n}} = (\Delta_{\mathbf{n}} + \mathrm{i}\,\Phi_{\mathbf{n}})/\sqrt{2}$ satisfying $[a_{\mathbf{n}}, a^+_{\mathbf{n}'}] = \delta_{\mathbf{nn}'}$. To lowest order in $1/\bar{n}$, we need only treat the terms quadratic in quantum fluctuations. Truncated at this level, the pure Hamiltonian, H_0, is diagonalised by a Bogoliubov transformation $a^+_{\mathbf{k}} = \cosh\theta_{\mathbf{k}}\alpha^+_{\mathbf{k}} - \sinh\theta_{\mathbf{k}}\alpha_{-\mathbf{k}}$ giving us the spectrum $\Omega_{\mathbf{k}} = [\omega_{\mathbf{k}}(\omega_{\mathbf{k}}+1)]^{1/2}$ where $\omega_{\mathbf{k}}$ is the single-particle tight-binding spectrum and $\exp 2\theta_{\mathbf{k}} = \Omega_{\mathbf{k}}/\omega_{\mathbf{k}}$. In this representation, the residual disorder is off-diagonal and can be expressed in terms of the spatial variation $\sigma\nu$ of the condensate. The scattering processes, at the leading order in the randomness, are $H_1 =$

$$\frac{\sigma}{N}\sum_{\mathbf{k}\neq\mathbf{q}}[S_{\mathbf{kq}}\alpha^+_{\mathbf{k}}\alpha_{\mathbf{q}} + S^*_{\mathbf{kq}}\alpha_{\mathbf{k}}\alpha^+_{\mathbf{q}} + T_{\mathbf{kq}}\alpha^+_{\mathbf{k}}\alpha^+_{\mathbf{q}} + T^*_{\mathbf{kq}}\alpha_{\mathbf{k}}\alpha_{\mathbf{q}}]$$

$$S_{\mathbf{kq}} = [\sinh(\theta_{\mathbf{k}}+\theta_{\mathbf{q}})(\omega_{\mathbf{k}}+\omega_{\mathbf{q}}) - \cosh(\theta_{\mathbf{k}}+\theta_{\mathbf{q}})\omega_{\mathbf{k}-\mathbf{q}}]\nu_{\mathbf{k}-\mathbf{q}}$$

$$T_{\mathbf{kq}} = [\sinh(\theta_{\mathbf{k}}+\theta_{\mathbf{q}})\omega_{\mathbf{k}-\mathbf{q}} - \cosh(\theta_{\mathbf{k}}+\theta_{\mathbf{q}})(\omega_{\mathbf{k}}+\omega_{\mathbf{q}})]\nu_{\mathbf{k}+\mathbf{q}}$$

We perform a unitary transformation on the Hilbert space to eliminate the terms linear in the disorder $\sigma\nu$. Under this transformation, $H \to \mathrm{e}^{-A}H\mathrm{e}^{A}$. The appropriate choice for A satisfies $[H_0, A] = -H_1$. Then, we will average over the second-order randomness. Diagrammtically, we have included only the minimally crossed terms in the self-energy of the quasiparticles. This approximation breaks down as we approach the transition where condensation is destroyed. However, since we believe that the condensed phase exists, the averaged Hamiltonian still has physical interest in such a phase. It is again quadratic and can be diagonalised by a second Bogoliubov transformation whose "angles" $\theta'_{\mathbf{k}}$ are of the order of σ^2. (We can ignore the resonant scattering between degenerate states because the resulting decay at low k is slow by phase-space arguments.)

The modified $\mathbf{k}$-states have an altered $k \to 0$ group velocity — "phonon velocity". Compared with the pure case, it has changed by a factor of $(1-\sigma^2)^{1/2}$ for $\omega_B \ll 1$. The number of bosons not in the condensate is given by $\Xi = \sum'_{\mathbf{k}}\langle c^+_{\mathbf{k}}c_{\mathbf{k}}\rangle \simeq \sum'_{\mathbf{k}}\langle a^+_{\mathbf{k}}a_{\mathbf{k}}\rangle =$

$$\sum_{\mathbf{k}}{}' \sinh^2(\theta_{\mathbf{k}} + \theta'_{\mathbf{k}}) + \frac{4\sigma^2}{N}\sum_{\mathbf{kq}}{}' \frac{2\,|T_{\mathbf{kq}}|^2}{(\Omega_{\mathbf{k}} + \Omega_{\mathbf{q}})^2} + O(\sigma^4)$$

The first term shows the combined effect of the two Bogoliubov transformations. The second term comes from the unitary transformation to eliminate randomness at the lowest order. The depletion Ξ has increased — by a factor of $1 + C(\sigma/2U\bar{n})^2$ for $\omega_B \ll 1$ where $C = 3/2 - 2\ln 2$. We have shown that the RPA treatment of the dense condensed limit is well-behaved for the case of small bandwidth and good screening. The change in the contribution of the low-lying collective excitations to the depletion due to disorder can be small when the condensate adjusts well to the random potential, *i.e.* short healing length. While the depletion increases with a decreasing bandwidth, the condensate can adjust better to the bare potential. It can be seen that the depletion does not scale with the phonon velocity as in the pure case but an increase in depletion is still associated with a decrease in the phonon velocity.

Finally we speculate that there is a strong-disorder limit in which the question of Bose-Einstein condensation is of a percolative nature. At strong disorder (*i.e.* $\mu < 0$, $|\mu| > \sigma \gg t$) , there are particles in only a fraction of the lattice sites but the condensate is extended with exponential tails outside these dense "lakes". In our dense limit, there are many bosons even in the tails, offering hope of retaining phase coherence among the lakes. Consider the case when $\sigma \sim O(U\bar{n})$. This gives us a lattice coverage where two lakes are typically separated by a site with a large $\sigma V_{\mathbf{n}}$. The effective hopping t_{eff} between the two lakes is approximately $(t\bar{n}^{1/2})^2/\sigma \propto t^2\bar{n}$ since $\sigma \sim U\bar{n}$ is fixed. It can be seen that the fractional depletion $\Xi/\bar{n} \to 0$ as $\bar{n} \to 0$ provided that we do not reduce t faster than $1/\bar{n}^{3/2}$ as well. The lakes will be weakly coupled to each other. These Josephson-like links have random strengths so that the phase coherence across the entire system should be determined by a *percolative* criterion. We would expect the condensate to be destroyed when the lattice coverage, controlled by σ/μ, drops below the percolation threshold.

ACKNOWLEDGEMENTS

We would like to thank J.T. Chalker for useful discussions. One of us (DKKL) would like to thank Trinity College, Cambridge, for financial support and the Rutherford Appleton Laboratory for its hospitality.

REFERENCES

(1) J.D. Reppy, Physica **126B** (1984) 335
(2) D. Finotello, K.A. Gillis, A. Wong and M.H.W. Chan, Phys. Rev. Lett. **61** (1989) 1954
(3) D. McQuenney, G. Agnolet and J.D. Reppy, Phys. Rev. Lett. **52** (1984) 1325
(4) J.M.F. Gunn and D.K.K. Lee, Physica **162-164C** (1989) 1483

Physica B 165&166 (1990) 511–512
North-Holland

SUM RULES AND THE SPECTRAL FUNCTION OF SUPERFLUID ^{4}He

S. GIORGINI and S. STRINGARI*

Dipartimento di Fisica, Università di Trento, 38050 Povo,Italy
**INFN, gruppo collegato di Trento*

Sum rules for the density-particle spectral function of a Bose superfluid are derived and used to give predictions for the residue of the particle-particle Green's function in the phonon-maxon-roton branch of liquid ^{4}He at T=0.

The particle-particle and density-density Green's functions have been extensively investigated in quantum liquids. An interesting quantity to consider in Bose superfluids is also the density-particle Green's function. In fact the non vanishing of such a quantity is the direct consequence of a broken symmetry in the system. Such a peculiar property has been already used in ref.[1] to investigate the long-range order in one and two-dimensional Bose systems. The density-particle spectral function is defined by:

$$A_{\rho,a}(\mathbf{q},\omega) = \int dt e^{i\omega(t-t')} < [\rho_{\mathbf{q}}(t), a_{\mathbf{q}}(t')] >=$$

$$= \frac{1}{Z}\sum_{m,n}(e^{-\beta E_m} - e^{-\beta E_n})\langle m|\rho_{\mathbf{q}}|n\rangle\langle n|a_{\mathbf{q}}|m\rangle\delta(\omega - E_n + E_m) \qquad (1)$$

In eq.(1) $<>$ means statistical average in the restricted ensemble characterized by the property $< a_0 >=< a_0^\dagger >$ $= \sqrt{N_0}$ where N_0 is the number of atoms in the condensate, Z is the grand canonical partition function and E_m,E_n are eigenvalues of the grand canonical hamiltonian $H' = H - \mu N$. At $T = 0$ eq.(1) becomes

$$A^{T=0}_{\rho,a}(\mathbf{q},\omega) = \sum_n \langle 0|\rho_{\mathbf{q}}|n\rangle\langle n|a_{\mathbf{q}}|0\rangle\delta(\omega - \omega_n) -$$

$$-\langle 0|a_{\mathbf{q}}|n\rangle\langle n|\rho_{\mathbf{q}}|0\rangle\delta(\omega + \omega_n) \qquad (2)$$

with $\omega_n = E_n - E_0$.

The following sum rules for $A_{\rho,a}(\mathbf{q},\omega)$ can be derived ($\hbar = 1$):

$$\int_{-\infty}^{+\infty} A_{\rho,a}(\mathbf{q},\omega)d\omega =< [\rho_{\mathbf{q}}, a_{\mathbf{q}}] >= -\sqrt{N n_0(T)} \qquad (3)$$

$$\int_{-\infty}^{+\infty} A_{\rho,a}(\mathbf{q},\omega)\omega d\omega =< [[\rho_{\mathbf{q}}, H'], a_{\mathbf{q}}] >=$$

$$= \sqrt{N n_0(T)}\frac{q^2}{2m} \qquad (4)$$

where $n_0(T) = N_0(T)/N$ is the condensate fraction ($\simeq 9\%$ in liquid ^{4}He at s.v.p.).

Result (3) has been already discussed in ref.[1-2]. It follows from the Bose commutation rules and leads to a non zero value if there is a broken symmetry (long range order) in the system. Result (4) holds in general for systems interacting with velocity independent potentials ($[V, \rho_{\mathbf{q}}] = 0$). The sum rule (4) has important analogies with the f-sum rule [3] with respect to which it however explicitly depends on the long range order parameter n_0.

Results (2-4) can be usefully employed to give a prediction for the residue of the particle -particle Green's function at $T = 0$. In fact at $T = 0$ a reasonable hypothesis is that the density operator $\rho_{\mathbf{q}}$, when applied to the ground state $|0\rangle$, excites only one (collective) state $|\mathbf{q}\rangle = \rho_{\mathbf{q}}|0\rangle$ with frequency $\omega(q)$. This is the well known Feynman's ansatz [4] yielding the expression $\omega(q) = q^2/2mS(q)$ for the excitation energy ($S(q)$ is the static structure function). The above ansatz is rigorous at low q and provides a useful semi-quantitative description of the spectrum in the roton region. Applying the Feynman's ansatz to eq.(2) and using the sum rules (3-4) we find:

$$\langle 0|a_{\mathbf{q}}|\mathbf{q}\rangle = \frac{1}{2}\sqrt{\frac{n_0}{S(q)}}\left(\frac{q^2}{2m\omega(q)} + 1\right) \qquad (5)$$

$$\langle \mathbf{q}|a_{\mathbf{q}}|0\rangle = \frac{1}{2}\sqrt{\frac{n_0}{S(q)}}\left(\frac{q^2}{2m\omega(q)} - 1\right) \qquad (6)$$

from which one estimates the residue $Z_{a^\dagger,a}$ of the particle-particle Green's function at the phonon-maxon-roton pole:

$$Z_{a^\dagger,a}(q) = |\langle \mathbf{q}|a_{\mathbf{q}}|0\rangle|^2 = n_0\frac{1}{4S(q)}(S(q) - 1)^2 \qquad (7)$$

Equation (7) reproduces in the low q limit the rigorous result $Z_{a^\dagger,a}(q \to 0) = n_0\frac{mc}{2q}$ [5-6]. This indicates that the collective state (phonon) completely exhausts the sum rules (3) and (4) at small q. However, similarly to the Feynman's formula for $\omega(q)$, eq.(7) is expected to be only semiquantitatively accurate at higher values of q. This is the consequence of the excitation of multiphonon states (ignored in eqs(5-6)) which is particularly important in the energy weighted sum rule (4).

A better estimate of $Z_{a^\dagger,a}(q)$ can be obtained using the "fluctuation-dissipation" sum rule:

$$\int_{-\infty}^{+\infty} \frac{1}{1-e^{-\beta\omega}} A_{\rho,a}(\mathbf{q},\omega) d\omega =$$

$$=< \rho_{\mathbf{q}} a_{\mathbf{q}} >= (N n_0)^{-1/2} n(\mathbf{q},\mathbf{q}) \quad (8)$$

This non-energy weighted sum rule yields, at $T = 0$,

$$\sum_n \langle 0|\rho_{\mathbf{q}}|n\rangle\langle n|a_{\mathbf{q}}|0\rangle = (N n_0)^{-1/2} n_{T=0}(\mathbf{q},\mathbf{q}) \quad (9)$$

and is expected to be less influenced by the excitation of multiphonon states with respect to the energy weighted sum rule (4).

In eqs(8-9) $n(\mathbf{p},\mathbf{q}) =< \rho_{\mathbf{q}} a^\dagger_{\mathbf{p}-\mathbf{q}} a_{\mathbf{p}} >$ is the two-body momentum distribution recently investigated by Ristig and Clark [8]. Applying the Feynman's ansatz to the sum rule (9) we straightforwardly find the following result for the residue $Z_{a^\dagger,a}$

$$Z_{a^\dagger,a}(q) = n_0 \frac{F_1(q)^2}{S(q)} \quad (10)$$

where $F_1(q)$ is defined by the relation [8] $n(\mathbf{q},\mathbf{q}) = N n_0 F_1(q)$. Clearly result (10) requires an explicit microscopic calculation of the function $F_1(q)$ which is at present available only within the HNC/0 approximation. In the figure we report the predictions (7) and (10) employing the HNC/0 results of ref.[8] for n_0, S(q) and $F_1(q)$.

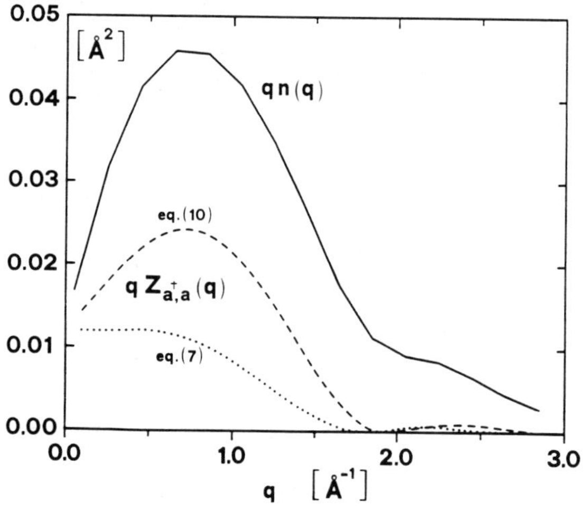

FIGURE 1
Residue of the particle-particle Green's function calculated with approximations (7) and (10). The momentum distribution calculated by Manousakis-Pandharipande [7] is also reported.

At T=0 both predictions correctly reproduce the low q behavior of $Z_{a^\dagger,a}(q)$ ($F_1(q=0) = 1/2$, see ref.[8]) and exhibit a characteristic bump in the roton region (see figure).

This bump is associated with the zero point motion of rotons and is very likely at the origin of the shoulder recently found in the calculation of the momentum distribution $n(q) = \sum_n |\langle n|a_{\mathbf{q}}|0\rangle|^2$ in liquid ^{4}He [7] (see full curve in the figure).

Finally we like to stress that in order to extract information on the residue of the particle-particle Green's function the sum rules derived in the present work for the density-particle spectral function are more advantageous than the corresponding ones for the particle-particle spectral function. The latter are in fact much more sensitive to the excitation of multi-phonon states which in some cases can wash out the information on the collective behavior. Consider for example the non energy-weighted sum rule for the particle-particle spectral function.

$$\int A_{a^\dagger,a}(\mathbf{q},\omega) d\omega =< [a^\dagger_{\mathbf{q}}, a_{\mathbf{q}}] >= -1 \quad (11)$$

The phonon-roton contribution to this sum rule is easily estimated starting from our results for $\langle 0|a_{\mathbf{q}}|\mathbf{q}\rangle$, $\langle \mathbf{q}|a_{\mathbf{q}}|0\rangle$.

Using for example eqs(5), (6) we find

$$|\langle 0|a^\dagger_{\mathbf{q}}|\mathbf{q}\rangle|^2 - |\langle 0|a_{\mathbf{q}}|\mathbf{q}\rangle|^2 = -n_0 \quad (12)$$

showing that the sum rule (11) is not in general exhausted by the collective pole, even in the low q limit. The situation becomes even more dramatic in the case of the energy-weighted sum rule for the particle-particle spectral function which diverges in the presence of hard core potentials [2].

ACKNOWLEDGEMENTS
We would like to thank J. Clark and G. Senger for providing us with their numerical results for S(q) and $F_1(q)$.

REFERENCES
[1] P.C.Hohenberg, Phys. Rev. **158**, 383(1967)
[2] H.Wagner, Z. Physik. **195**, 273(1966)
[3] D.Pines and P.Nozières, *Theory of Quantum Liquids*(W.A. Benjamin, N.Y. 1966) Vol.1, ch.5
[4] R.P.Feynman, Phys. Rev. **94**, 267(1954)
[5] J.Gavoret and P.Nozières, Ann. Phys. (N.Y.)**28**, 349 (1964)
[6] G.V.Chester and L.Reatto, Phys. Lett. **22**, 276 (1966)
[7] E.M.Manousakis, V.R.Pandharipande and Q.N. Usmani, Phys. Rev. B **31**, 7022(1985); P.Whitlock, R.M.Panoff, Can.J.Phys. **65**, 1409(1987)
[8] M.L.Ristig and J.W.Clark, Phys. Rev. B**40**, 4355 (1989)

Physica B 165&166 (1990) 513–514
North-Holland

POSSIBILITY OF *STABLE BUT INACCESSIBLE STATES* IN THEORY OF SUPERFLUIDITY

W. A. B. EVANS

Physics Laboratory, University of Kent, Canterbury, Kent CT2 7NR, U.K.

Model computer solutions within the formulation of the microscopic theory of superfluidity proposed by Dorre, Haug and Tran-Thoai, which embrace the *self-consistent* Bogoliubov-Beliaev solution, the normal solution, and the Pair solution as *alternative states* have been studied *over their entire range of existence.* Here the results for the free energies of the various solutions is presented. It is seen that the *lowest* free energy states can only be accessed via a first order transition. It is argued that the probability of this transition may be so small, that it is unlikely ever to be seen experimentally. This suggests that the observed behaviour of superfluid Helium may need to be understood in terms of *the metastable states of the theory* that do, in fact, give a better qualitative description.

1. INTRODUCTION AND THEORY.

Possibly the most rigorous and comprehensive presentation of the microscopic theory of the Bose superfluid is the work of Dörre, Haug and Tran-Thoai[1], herinafter referred to as DHT, who gave a microscopic development that was sufficiently general to embrace both the *c-number condensate theory*, the origins of which go back to Bogoliubov[2] and later workers[3,4] and the pair theory (i.e. the boson analogue of the BCS theory) proposed soon after the BCS theory came out by Valatin and Butler[5] and elaborated by others[6,7]. There had been a tendency to regard these as rival theories, but DHT's work showed how both the *self-consistent* c-number solution, as given by Beliaev[2] (but with a new closure equation [see **(14)** below] defining the chemical potential), and the pair theory, in the version given by Evans and Imry[6], naturally arise as *alternative states* for the bose system. DHT numerically investigated the free-energies and properties of the various solutions i.e. normal, pair *("P-)* solutions and their *new self-consistent* c-number condensate *("C-")* solutions. DHT found that, *for a wide class of potentials*, the C-solutions proved to have the lowest free-energy near T=0 K. However, their properties were wildly at variance with the observed behaviour of superfluid ^{4}Helium. This contrasts with the P-solutions, *especially of the condensate-less type* (hypothesised by Coniglio and Marinario[8] and found numerically by Evans and Harris[9]) which do quite plausibly mimic the observed behaviour of superfluid ^{4}Helium.

DHT base their development on the exact Gibbs-Bogoliubov[10,11] *"statistical variational principle"*

$$\Omega \le \Omega_0 + \langle(\mathcal{H} - \mathcal{H}_0)\rangle_0 \qquad (1)$$

where, $\Omega = -k_B T \ \ln\{\mathrm{Tr}\{\exp[-\beta\mathcal{H}]\}\}$, is the thermodynamic potential in the system, $\mathcal{H}$. $\mathcal{H}_0$ can be *any* (hermitian) Hamiltonian one cares to choose, but, of course, to apply the above, $\mathcal{H}_0$ must be sufficiently tractable that $\Omega_0 = -k_B T \ \ln\{\mathrm{Tr}\{\exp[-\beta\mathcal{H}_0]\}\}$ and also $\langle\mathcal{H}\rangle_0$ can be explicitly evaluated. DHT chose

$$\mathcal{H}_0 = \sum_k E_k \, \alpha_k^+ \alpha_k \qquad (2)$$

where the E_k's are variational patameters, and the α_k's are operators that are ***unitarily similar*** to the a_k's and so obey the same Bose commutation relations. It is worthwhile to verify that the unitary operator

$$U = \exp\left[\sum_p\left\{\tfrac{1}{2}\sigma_p a_p^+ a_{-p}^+ + \gamma_p a_p^+ - \tfrac{1}{2}\sigma_p^* a_{-p} a_p - \gamma_p^* a_p\right\}\right]$$

transforms a_k to

$$\alpha_k = U^+ a_k U = \cosh(|\sigma_k|)a_k + \frac{|\sigma_k|}{\sigma_k}\sinh(|\sigma_k|)a_{-k}^+ + \zeta_k$$

or

$$\alpha_k = u_k\, a_k + v_k\, a_{-k}^+ + \zeta_k \qquad (3)$$

where the c-number part is given by

$$\zeta_k = \gamma_k^* \sigma_k\left[\frac{\cosh(|\sigma_k|)-1}{|\sigma_k|^2}\right] + \gamma_k\left[\frac{\sinh(|\sigma_k|)}{|\sigma_k|}\right]$$

Note $u_k^2 - |v_k|^2 = 1$ (unitarity condition) **(4)**

Further, ζ_k may be easily shown to be a *completely independent parameter* from σ_k (and hence from u_k and v_k), in the sense that, for any σ_k, *any value* of ζ_k can be generated by a suitable choice of γ_k. This shows that the c-number term, ζ_k, arises *completely rigorously* as the result of a unitary transformation.

In principle, the technique is to evaluate the upper bound **(1)** on the thermodynamic potential and to minimise the latter with respect to all the variational parameters that have been introduced i.e. E_k, u_k, $|v_k|$, $|\zeta_k|$, $\arg(\zeta_k)$, and $\arg(v_k)$ subject to the unitarity constraint **(4)**. This leads to a coupled set of nonlinear integral equations, the self-consistent solution(s) of which are the possible thermodynamic state(s) of the system. It has not yet proved possible to carry out this program *fully* and allow for an arbitrary *complex* ζ_k for *all* **k**-values. Indeed this would seem a most promising path for the *further refinement of the microscopic theory that would be relevant for the normal phase description also.* In the spirit of Bogoliubov, DHT restrict the choice of ζ_k to

$$\zeta_k = \zeta\, \delta_{k,0} \qquad (5)$$

Then only the *pair* terms of the *full* Hamiltonian,

$$\mathcal{H} = \sum_k\left(\frac{k^2}{2m} - \mu\right)a_k^+ a_k + \frac{1}{2}\sum_{k,p,q} V_q\, a_{k+q}^+ a_{p-q}^+ a_p a_k \,,$$

contribute to $\Omega_0 + \langle(\mathcal{H} - \mathcal{H}_0)\rangle_0$. Minimising this, with respect to all variational parameters, one is led to the following set of nonlinear coupled equations

$$\tilde{\varepsilon}_k = \frac{k^2}{2m} - \mu + NV_0 + \sum_p V_{k-p}\, n_p + N_0 V_k \qquad (6)$$

$$\Delta_k = \sum_p V_{k-p}\chi_p + N_o V_k \tag{7}$$

$$\chi_k = -\frac{\Delta_k}{2E_k}\coth\left[\frac{1}{2}\beta E_k\right] \tag{8}$$

$$n_k = \frac{\tilde{\varepsilon}_k}{2E_k}\coth\left[\frac{1}{2}\beta E_k\right] - \frac{1}{2} \tag{10}$$

$$E_k = \sqrt{\tilde{\varepsilon}_k^{\,2} - |\Delta_k|^2} \tag{11}$$

$$N = \sum_k \langle a_k^+ a_k \rangle = N_o + \sum_k n_k \tag{12}$$

$$|\zeta|\left[-\mu + \sum_k (V_o + V_k)n_k + \Delta_o\right] = 0 \tag{13}$$

where $N_o = |\zeta|^2$. Focussing on (13), we appreciate that it allows two possibilities viz. i/ $\zeta=0$ that leads to the self-consistent equations of the boson pair theory[6] or ii/ $\zeta \neq 0$ provided the chemical potential is given by

$$\mu = \sum_k (V_o + V_k)n_k + \Delta_o \tag{14}$$

Eq. (14) is DHT's new closure condition which, with equations (6)→(12), constitute the equations of the *self-consistent* Bogoliubov theory (C-solutions). Also possible, of course, is the *normal* solution where $\zeta=0$ and *all* $\Delta_k=0$. This is the only solution known to exist at arbitrarily high temperatures.

2. RESULTS.

The present work investigates the Free Energies *(Helmholtz)* for all the various contending solutions, evaluated at constant density (ρ = 0.022 $Å^{-3}$) over their ***entire temperature range of existence.*** The model potential used had an attractive part and a strong repulsive core that is typical of the type needed for *condensate-less* P-solutions[9]. Here (lack of) space permits only the results to be discussed and these are illustrated in Figure 1.

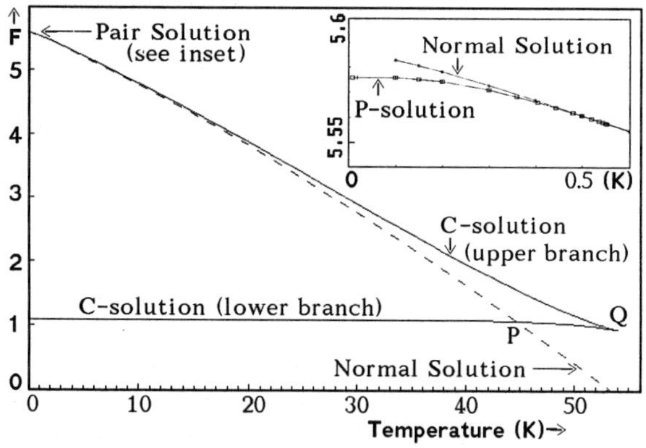

Fig. 1 Helmholtz Free Energies of the Various Solutions (inset shows enlargement of region below 0.6K)

It will be observed that the C-solution branch doubles back on itself. At low temperatures, in fact from 0 to 44K, the (lower) C-solution branch has the lowest F. Above 44K the normal solution takes over. Thus the *obvious* interpretation is that the normal solution would (at P in Fig. 1), undergo a *first order* transition to the C-solution state (which has a 93% condensate fraction!) and thereafter stay on the lower C-solution branch to 0K (where the condensate fraction is 96%). This agrees with DHT's conclusions.

We suggest an alternative possibility. Fig. 1 shows clearly that, to access the stable C-solution states from the cooled "normal" states, a *first order transition is essential.* Since the normal state and the "lower-branch" C-solution state at P are so dissimilar, it clearly demands a large fluctuation *"in Hilbert Space"* to realise this first order transition. We suggest that this, plausibly, would not happen (in *realizable* experimental durations). Instead, the system would remain in the (now metastable) normal state. Note the C-solution continues to 54.5K before *doubling back* along its upper branch. At its *turning point,* Q, we see a cusp in the Free-Energy plot (which results since $(\partial F/\partial T)_\rho = -S$ must be *negative* everywhere). Along its upper branch the C-solution becomes increasingly depleted and so gets *more like* the normal state, but is less stable (has a higher F) than the latter all the way down to 0.55K. At this point the C-solution becomes fully depleted (ζ vanishes *simultaneously* with all the Δ_k's) at which point it satisfies *exactly the same equations* as the normal solution and, moreover, those of the *nucleating condensate-less* P-solution which has its onset critical temperature at this point [for its Free-Energy plot see inset in Fig. 1]. The transition to this P-solution is *second order* and, *insofar as this model pertains to actual Helium,* it appears most likely that this state corresponds to the superfluid state. Plausibly, the analogue of the true (C-solution) stable state has never been observed i.e. we may here have a ***stable but inaccessible*** state. By analogy with inducing Graphite to form Diamond by subjecting it to *extreme high pressures,* one might hope to induce the transition if one had a similar *"experimental handle"* on the order parameter. Here, this possibility is probably denied by the subtle nature of the latter. *Indeed this may be common to many systems in Nature* - if all their possible states could be known *theoretically,* it may well be that the state that is found to describe the *observable* low temperature phase proves not to be the most stable.

ACKNOWLEDGEMENTS.

The author is grateful to Professor Sen-Gupta, Dr. A. M. Khan and the organisers of ***"The International Bose Symposium"*** (Dhaka University, December 1988) for inviting him to lecture on these topics.

REFERENCES.

1. P. Dörre, H. Haug and D. B. Tran-Thoai, *J. Low Temp. Phys.*, 35, 465.
2. N.N. Bogoliubov, *J. of Physics U.S.S.R.*, 11, 23.
3. S.T. Beliaev, *Soviet Physics J.E.T.P.*, 7, 289, 299.
4. P. Hohenberg and P.C. Martin, *Ann. Phys.*, 34, 291.
5. J. G. Valatin and P. Butler, *Nuovo Cimento*, 10, 37.
6. W. A. B. Evans and Y. Imry, *Nuovo Cimento*, 63B, 155.
7. M. Luban, Phys. Rev., 128, 965.
8. A. Coniglio and M. Marinaro, *Nuovo Cimento*, 48B, 249.
9. W. A. B. Evans and C. G. Harris, *J. de Physique*, Colloque C6, C6-237.
10. H. Falk, *Am. J. Phys.*, 38, 858.
11. N. N. Bogoliubov, *Nuovo Cimento*, 7, 794.

Physica B 165&166 (1990) 515–516
North-Holland

LIQUID ^{4}HE : A DENSITY FUNCTIONAL APPROACH WITH A FINITE RANGE INTERACTION.

J. Dupont-Roc and M. Himbert

Laboratoire de Spectroscopie Hertzienne†, Département de Physique de l'ENS, F-75231 Paris Cedex 05.

N. Pavloff and J. Treiner

Division de Physique Théorique*, Institut de Physique Nucléaire, F-91406 Orsay Cedex, France.

The properties of liquid helium-4 are investigated in a mean field approximation using a novel finite range effective interaction. The surface tension is found in agreement with the experimental value and the surface width equal to 5.7 Å.

1. INTRODUCTION

Among the various theoretical descriptions of liquid helium, density functional methods have the advantage of simplicity and appear attractive to treat inhomogeous situations such as interfaces, impurities in helium, clusters, *etc...* In contrast, their quantitative prediction capability appears to have been often spoiled by various shortcomings. For instance, on the free surface problem, Regge obtained an unrealistic density profile (1), Ebner and Saam a very low unrenormalized surface tension(2), Stringari and Treiner a wrong long-range behaviour in the mean field(3).

We propose to introduce for liquid helium a novel density functional with an effective interaction having a correct long-range behaviour and reasonable short range characteristics, and which matches actual properties of liquid helium, such as the pressure-density relation, the static density-density response function, and finally yields the correct surface tension without further adjustement.

2. THE FINITE RANGE EFFECTIVE INTERACTION.

The energy of a given volume V of the fluid is writen as:

$$E = E_{liq} + E_q = \int \mathcal{H}[\rho] d^3r \tag{1}$$

E_{liq} describes the internal energy of the fluid. For uniform density ρ, it is equal to :

$$E_{liq} = V\left(\frac{b}{2}\rho^2 + \frac{c}{2}\rho^{2+\gamma}\right) \tag{2}$$

The parameter b, c, γ are taken from Ref. (3) and ensure a correct pressure-density relation over a wide range of densities. For non-uniform situations, we take functional of the density from which we discard any term with a zero range.

$$E_{liq} = \frac{1}{2}\int\int d^3r d^3r' \rho(\mathbf{r})\rho(\mathbf{r}')V_l(|\mathbf{r}-\mathbf{r}'|) + \int d^3r \frac{c}{2}\bar{\rho}_{\mathbf{r}}^{2+\gamma} \tag{3}$$

It is indeed natural to take for V_l the standard Lennard-Jones potential describing the He-He interaction ($\epsilon = 10.22$ K, $\alpha = 2.556$ Å), screened in a simple way at distances shorter than a characteristic distance h, let $x = |\mathbf{r}-\mathbf{r}'|/\alpha$.

$$V_l(|\mathbf{r}-\mathbf{r}'|) = \begin{cases} 4\epsilon(x^{-12} - x^{-6}) & \text{for } |\mathbf{r}-\mathbf{r}'| \geq h \quad (4.a) \\ V_l(h)(\alpha x/h)^4 & \text{for } |\mathbf{r}-\mathbf{r}'| \leq h \quad (4.b) \end{cases}$$

The parameter h is adjusted so that the space integral of V_l is equal to b in order to preserve the correct equation of state. This gives $h = 2.378$ Å. The shape of the screened core potential (4.b) has been taken as a power law. The choice of the fourth power is not critical as will be seen below.

In the second term of Eq. 3, $\bar{\rho}_{\mathbf{r}}$ is the local density averaged over a sphere centered at $\mathbf{r}$ with a radius h: $\bar{\rho}_{\mathbf{r}} = \rho * \Pi_h$ with $\Pi_h(\mathbf{r}) = 3\theta(r-h)/(4\pi h^3)$, where θ is the Heaviside function. It is always positive, and accounts for the internal kinetic energy as well as for increasing contribution of the hard core when the density is increasing. When $\bar{\rho}_{\mathbf{r}}$ is expressed in term of $\rho(\mathbf{r})$, it appears clearly to describe many-body correlations with a range of order h.

The last term of Eq. 1 is the usual inhomogeneity correction to the quantum kinetic energy density.

$$E_q = \int d^3r \frac{\hbar^2}{2m}|\nabla\phi|^2 \tag{5}$$

where $\phi(\mathbf{r}) = \sqrt{\rho(\mathbf{r})}$. There is no adjustable parameter left. Although some choices may appear arbitrary, they meet several essential requirements. First, the equation of state of bulk helium is correctely described. Second, we will show now that the static response function of the model fluid is very close to that of real helium around the equilibrium density ρ_0.

† Unité de Recherche de l'Ecole Normale Supérieure et de l'Université Pierre et Marie Curie, associée au CNRS (UA28)

* Unité de Recherche des Universités Paris 11 et Paris 6 Associée au CNRS.

3. LINEAR RESPONSE THEORY

The state of the fluid in an external time dependent potential $\mathcal{V}_{ext}(\mathbf{r},t)$ is determined through the Time Dependent Hartree theory. Minimizing the action leads to a non-linear Schrödinger equation. Taking $\rho = \rho_0 + \delta\rho$ one obtains then, to first order in $\delta\rho$ and $\mathcal{V}_{ext}$, a R.P.A. equation for the helium density. A Fourier transform brings this equation to be algebraic, and yields the density-density response function. Its inverse reads as :

$$\chi^{-1}(\mathbf{q},\omega) = \frac{m\omega^2}{\rho_0 q^2} - \frac{\hbar^2 q^2}{4m\rho_0} - \left[V_l(q) - b\frac{2+\gamma}{2}\Pi_h^2(q)\right] \tag{6}$$

Fig. 1 presents the comparison of $\chi^{-1}(q,\omega=0)$ with the measurements of Cowley and Woods (4).

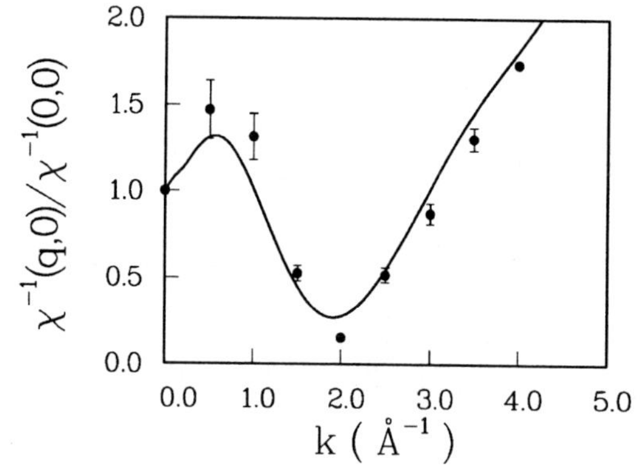

FIGURE 1

It is worth pointing out that the response $\chi^{-1}(q,0)$ depends only slightly of the power used to define V_l in the interval $r = 0$ to $r = h$. At this stage it appears that a simple density functional description of liquid ^{4}He can reproduce the static response function of the fluid even for large wave vectors. This is indeed essential to have this tool in order to study inhomogeous situations where the density may vary on a small scale.

4. THE FREE SURFACE

As an application to an inhomogeous situation, we study the free surface of liquid helium at zero temperature and pression. The interface is taken parallel to the x–y plane, the bulk liquid being in the negative z direction. One minimizes the grand potential $\Omega = E - \mu N$ with respect to $\phi(z)$ and one obtains again a non-linear self-consistent Schrödinger equation that we solve numerically. The resulting surface thickness is of 5.7 Å. The surface tension is

$$\sigma = \int dz(\mathcal{H}[\rho] - \mu\rho) \tag{7}$$

and is found to be 0.2741 K.Å^{-2}. Remarkably, this value is, within experimental error, equal to the measured surface tension of helium extrapolated at zero temperature. σ is not very sensitive to details of the model used for the fluid *provided* that it meets the criteria enumerated at the end of Sect. 2. For instance, changing the power law of the two body effective potential between 0 and h from 1 to 10 changes the value of σ by only $\pm 1\%$.

From this result one can get a hint about the origin of the surface energy of liquid helium in our approach. The quantum kinetic energy participates only for 6 % to the surface energy. The essential part of σ comes from the contribution of $E_{liq} - \mu N$. Thus, whereas the quantum energy is indeed located near the surface, the main contribution to σ is located inside the fluid and decay only slowly with the distance from the interface, as $1/z^3$. This is quite analogous to the situation in classical Van der Waals fluids : the surface energy results essentially from the lack of attraction due to the missing half-space of liquid above the surface.

5. CONCLUSIONS

It appears possible to describe liquid helium by a density functional involving a non-local interaction having a correct $1/r^6$ behaviour on the long range. A simple short range shape has been proposed. The effective interaction build on this prescription gives a correct response function in the bulk and the right surface tension. This modelisation of helium is likely to be a useful tool to extrapolate actual properties of helium over a wide range of configurations, even on nearly microscopic scale. It could be used to study several interesting problems such as surface states of ^{3}He atoms, scattering of ^{3}He or ^{4}He atoms by the surface, clusters, atomic impurities in liquid helium and thin ^{4}He films.

REFERENCES

(1) T. Regge, J. Low. Temp. Phys. 9 (1972) 123.
(2) C. Ebner and W.F. Saam, Phys. Rev. B 12, (1975) 923.
(3) S. Stringari and J. Treiner, Phys. Rev. B 36, (1987) 8369.
(4) R. Cowley and A.D.B. Woods, Can. J. Phys. 49 (1971) 177.

Physica B 165&166 (1990) 517–518
North-Holland

SOLUBILITY OF ^{4}HE IN LIQUID ^{3}HE AT VERY LOW TEMPERATURES

Masaki NAKAMURA, Gohzo SHIROTA, Toshinobu SHIGEMATSU, Keisuke NAGAO*, Yoshiko FUJII
Minoru YAMAGUCHI, Toyoichiro SHIGI

Department of Applied Physics, Okayama University of Science, Okayama 700, Japan
*Institute for Study of the Earth's Interior, Okayama University, Misasa, Tottori 682-02, Japan

As a continuation of our previous experiments (1), the phase separation temperatures of dilute solution of ^{4}He in liquid ^{3}He have been measured with increased surface area in the cell, in order to estimate the amount of ^{4}He adsorbed on the surface. The solubility of ^{4}He below 200 mK is larger than the value expected from the equation proposed by Saam et al.

1. INTRODUCTION

The solubility of ^{4}He in liquid ^{3}He was measured by Laheurte (2) down to 270 mK. Below this temperature, the solubility becomes very small, which makes measurements very difficult. Saam et al. (3) suggested a following formula for the solubility of ^{4}He in liquid ^{3}He.

$$X_4 = 0.85\ T^{3/2}\ e^{-0.56/T}\ , \qquad (1)$$

where X_4 is the ^{4}He molar concentration. Above 270 mK, this formula is consistent with the experimental results.

We measured the ^{4}He solubility below 300 mK using a sensitive capacitor and a film-flow-tight valve attached to the cell, and found that the solubility below 200 mK is larger than the value predicted by the eq. (1). The ^{4}He solubility curve, however, was not determined exactly, because the amount of adsorbed ^{4}He in the cell was not estimated experimentally.

In order to make this clear, we have continued this research with increased surface area in the cell.

2. EXPERIMENTAL

The experimental cell (Fig. 1) is thermally connected to the mixing chamber of a dilution refrigerator with copper brush, and is almost completely filled with the sample liquid ^{3}He containing small amount of ^{4}He, in which a capacitor is immersed. The surface area in the cell was increased to about 900 cm^2 from 350 cm^2 by putting Tetron mesh into the cell.

When the liquid ^{3}He is cooled down to the phase separation temperature, the superfluid ^{4}He-rich phase appears and covers all the surface in the cell. This is indicated by increase of capacitance due to difference between dielectric constants of liquid ^{4}He and ^{3}He. For sensitive detection, the capacitor was made to have very narrow gap (2μm) between electrodes.

In addition to this, a film-flow-tight valve is inserted near the cell. Otherwise, the

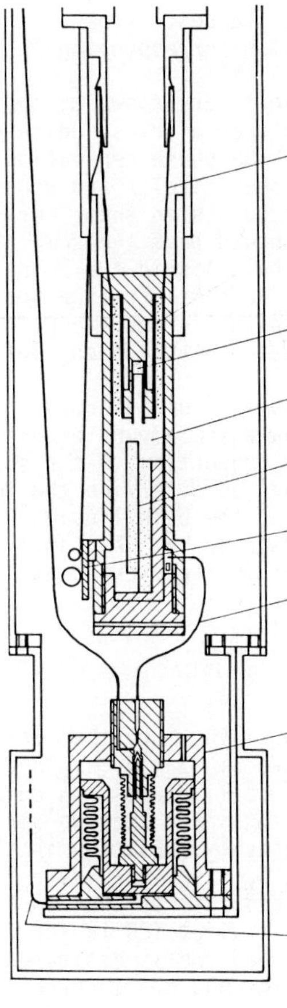

Figure 1
Experimental setup for measurements of the phase separation of dilute solution of ^{4}He in liquid ^{3}He.

separated superfluid ^4He film would move up to higher-temperature parts out of the cell along a filling tube.

The ^4He concentration of the sample gas was determined with a mass spectrometer specially designed for rare gas.

A CMN susceptibility thermometer using a SQUID as a null detector and a carbon resistor were employed for temperature measurements.

3. RESULTS AND DISCUSSION

In this work, the phase separation temperatures were measured for six mixtures of ^4He molar concentration, 80, 134, 180, 297, 600 and 1200 ppm. Below 80 ppm, phase separation was not observed obviously.

Owing to the stronger attractive force of the cell wall for ^4He, the surface phase separation commences at the temperature higher than that of the bulk phase separation. So, this temperature was obtained from extraporation of the tangential line with the steepest slope of the capacitance variation corresponding to the ^4He film formation.

The present results are shown by open squares in Fig. 2, whereas open circles indicate the previous results. Two phase separation curves are determined by open circles and squares respectively. The points on these curves at the same temperature should have the same ^4He concentration in the bulk liquid ^3He. Therefore, the difference of ^4He concentration between two curves is due to the different amount of ^4He which was not detected as the change of capacitance.

Assuming this comes from adsorbed ^4He in the cell, we can evaluate the amount of adsorbed ^4He which should be in proportion to the surface area in the cell and so determine the corrected ^4He concentration in the bulk liquid ^3He.

The amount of adsorbed ^4He (A_2) in a cell with the surface area S_2 is represented as

$$A_2 = \frac{S_2}{V_2} \frac{C_2 - C_1}{S_2/V_2 - S_1/V_1}, \qquad (2)$$

for the sample containing more than certain amount of ^4He, where S_1, S_2, V_1, V_2 and C_1, C_2 are surface areas, cell volumes and concentrations of the sample gas for the two series of experiments respectively. Putting in this formula S_1 = 350 cm^2, S_2 = 900 cm^2, V_1 = 3.6 cc, V_2 = 4.4 cc and the values of the two ^4He solubility curves, say, at 100 mK for C_1 and C_2, the amount of adsorbed ^4He is estimated to be about 140 ppm ^4He for 900 cm^2 surface area.

Subtracting this value from the ^4He concentration of all samples, the corrected concentrations ^4He are determined. The results are shown by broken curve in Fig. 2.

While, according to Chan et al.(4) the amount of ^4He adsorbed on the substrate was 31 μmole m^{-2}

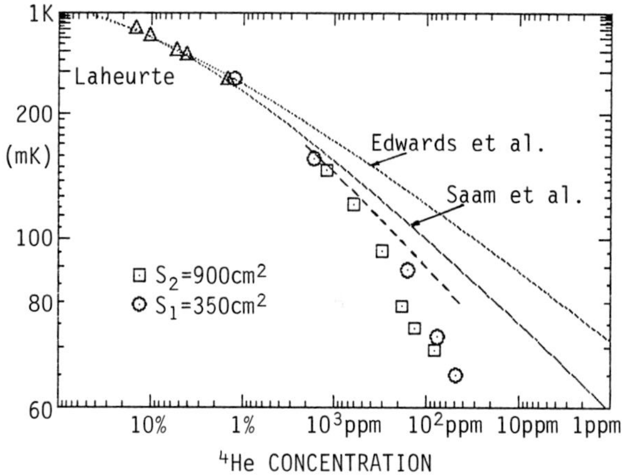

Figure 2
Phase separation temperatures of dilute solution of ^4He in liquid ^3He.

which did not contribute to the superflow at any temperature. In our cell, the adsorbed ^4He (140ppm) corresponds to 190 μmole m^{-2}. This large value is considered to mean that the ^4He film separated initially might be sucked in the texture of Tetron mesh. In fact, the capacitance variation for the sample with lower ^4He concentration was extremely small, compared with the case of 350 cm^2 surface area.

The previous cell, however, had not suffered from suction of ^4He film because of no Tetron mesh. Taking this into account, the solubility of ^4He might be larger than the value indicated by the broken curve.

4. CONCLUSION

The phase separation temperatures for ^3He-^4He liquid mixtures of the ^4He molar concentration, 80, 134, 180, 297, 600 and 1200 ppm were measured using a cell with the surface area of 900 cm^2. Comparing with the case of 350 cm^2 surface area, we estimated the solubility curve. The solubility of ^4He in liquid ^3He below 200 mK is larger than the value expected from eq. (1) proposed by Saam et al.(3)

REFERENCES

(1) M.Nakamura, Y.Fujii. T,Shigi et al. J. Phys. Soc. Jpn. 57 (1988) 1676
(2) J.P.Laheurte, J. Low Temp. Phys. 12 (1973) 127.
(3) W.F.Saam and J.P.Laheurte, Phys. Rev. A4 (1971) 1170.
(4) M.H.W.Chan, A.W.Yanof and J.D.Reppy: Phys. Rev. Lett. 32 (1974) 1347.

Physica B 165&166 (1990) 519–520
North-Holland

EXPERIMENTAL MEASUREMENTS OF THE ATTENUATION OF SECOND SOUND IN ULTRA DILUTE ^3He-^4He MIXTURES

J. R. Smith and A. Tyler

Department of Physics, The University, Manchester, M13 9PL, England.

We report measurements of the damping of second sound for concentrations of 0.1, 0.05, 0.01 and 0.001% ^3He in ^4He. These measurements were performed in the temperature range 0.1K to 1K and in the frequency range 1kHz to 6kHz.

1. INTRODUCTION

In 1979 Greywall (1) measured the damping of second sound in a 0.1% mixture using a c.w. resonance method for temperatures between 0.05K and 1K. Further experimental measurements for a 0.1% mixture have been made by Church et al (2) covering a wider frequency range (~700Hz to 12KHz).

For temperatures below approximately 0.5K a peak was observed to occur in the damping coefficient D_2. The temperature at which the peak occurred decreased with decreasing frequency. For temperatures above T~0.5K the damping was found to be frequency independent.

Hydrodynamic behaviour is defined by the conditions $\omega\tau \ll 1$ and $k\ell \ll 1$ where k is the phonon wave vector and ℓ the phonon mean free path. As the temperature is lowered the phonon mean free path increases and eventually becomes comparable to the second sound wavelength. The second inequality is broken and the system becomes non-hydrodynamic giving rise to a peak in the second sound damping.

Adamenko et al (3) and Bowley et al (4) have extended the theory of Baym, Saam and Ebner to include phonon-phonon interactions. As well as these there are also two types of phonon-^3He interaction to be considered: elastic scattering of phonons by the ^3He and phonon absorption by the ^3He. Husson et al (5) has shown that the phonon absorption rate has a concentration dependent enhancement, this has been confirmed by Guillon et al (6). For the high pressure measurements of Greywall, Bowley et al find good agreement between the theory and experiment. For the lower pressure measurements where 3-phonon processes are allowed the agreement between theory and experiment is not as good.

2. EXPERIMENTAL METHOD

The experimental cell was a copper cylinder resonator of length 1.30cm and diameter 0.6cm. Second sound was excited and detected by gold plated nuclepore filters. The temperature was measured using carbon resistance thermometers that were calibrated using measurements of the magnetic susceptibility of cerium magnesium nitrate as well as measurements of ^3He vapour pressure. One thermometer was a small sliver of carbon resistor with a very short time constant (~10^{-2}s) suspended in the fluid and thermally isolated from the wall.

Frequency sweeps between 1 and 6kHz were made at fixed temperatures in the range 0.1K to 1K. Measurements were made of the resonance frequency and half width for the longitudinal modes.

3. RESULTS

The half width of a second sound resonance results from damping from the following sources: losses in the bulk liquid, viscous dissipation at the surfaces and reflection losses at the ends of the cavity.

$$\Delta_{½} = \frac{\pi D_2 f^2}{U_2{}^2} + \frac{\rho_s}{2a\rho}\left[\frac{f\eta}{\pi\rho_n}\right]^{½} - \frac{f_1}{\pi}\,\ell_n(R) \qquad (1)$$

The surface loss and reflection loss contributions were subtracted from the measured half width and the second sound damping coefficient D_2 calculated as a function of temperature for a fixed frequency.

The results for the 0.1% mixture are in agreement with the measurements of Greywall and Church at 0.6K but are somewhat lower (~10%) than Greywall's for temperatures around the peak value. The results for the 0.05% and 0.01% mixture are given in graphs 1 and 2. As can be seen there is a large increase in the damping for these very dilute mixtures. At high temperatures, T>0.6K, the damping is observed to be frequency independent. The Q's of the resonances for the 0.01% mixture were extremely low (<10) and have not been included for temperatures 0.78>T>0.4K. Graph 3 shows our results at 3KHz for all the concentrations. Also shown are some high temperature results for a 0.001% solution for which we were unable to detect any signal below T~0.78K.

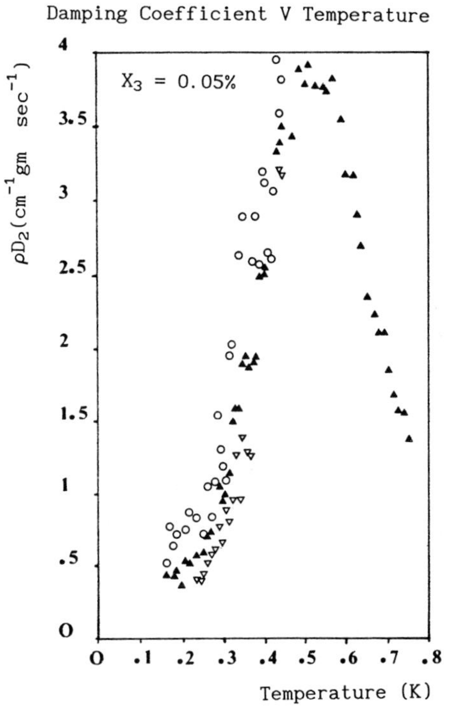

Fig. 1 Damping of Second Sound in a 0.05% solution. ○ 3KHz, ▲ 4KHz, ▽ 5KHz

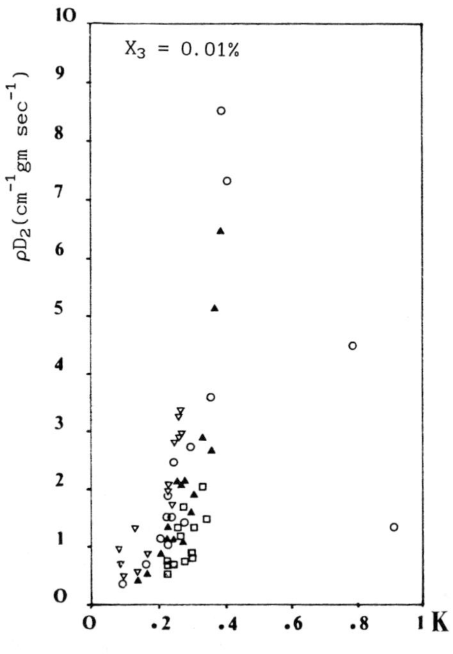

Fig. 2 Damping of Second Sound in a 0.01% solution. ▽ 2.5KHz, ○ 3KHz, ▲ 4KHz, □ 5KHz.

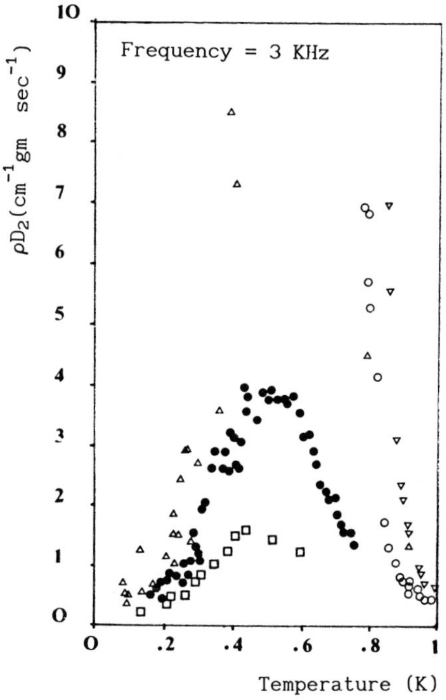

Fig. 3 Damping of Second Sound at 3KHz. □ 0.1%, ● 0.05%, △ 0.01%, ○ 0.001%, ▽ ^{4}He.

4. CONCLUSIONS

The results reported are the first c.w. resonance measurements for the damping in mixtures for concentrations of X_3<0.1%. The results show a peak in the damping coefficient. The temperature of the peak increases with decreasing concentration and decreases with decreasing frequency. The mixtures studied show a large increase in the attenuation as the concentration decreases.

ACKNOWLEDGEMENTS

This work was supported by SERC grant No 92096.

REFERENCES

(1) D. Greywall, Phys. Rev. Lett. 42 1758 (1979).

(2) R.J. Church, J.R. Owers-Bradley, P.C. Main, G. McHale and R.M. Bowley, Proceedings of LT 18 Kyoto (1987).

(3) I.N. Adamenko, E. Ya. Rudaviski, V.I. Tsyganok and V.K. Chagovets, J. Low Temp. Physics 71 No. 3/4 (1988).

(4) G. McHale, M.A. Jones and R.M. Bowley, Proceedings of LT 18 Kyoto (1987).

(5) L.P.J. Husson and R de Bruyn Ouboter, Physica 122B 201 (1983).

(6) F Guillon, J.P. Harrison, A. Sachrajda and D. Atkins, J. Low Temp. Phys. 57 95 (1984).

Physica B 165&166 (1990) 521–522
North-Holland

CODIMENSION-TWO AND HYDRODYNAMIC TRICRITICAL POINTS IN ^{3}HE-^{4}HE MIXTURES

T. ONIONS, M.R. ARDRON, P.G.J. LUCAS, M.D.J. TERRETT, and M.S. THURLOW

Department of Physics, The University, Manchester, M13 9PL, England.

Thermal convection measurements on a normal layer of mixture of 2.35% molar concentration of ^{3}He in ^{4}He heated from below in cylindrical geometry reveal a codimension-two point at 2.21K and a change from the bifurcation to stationary convection being hysteretic to non-hysteretic at 2.228K. There is weak evidence of tricritical behaviour at the latter temperature.

1. INTRODUCTION

When a horizontal layer of normal liquid ^{3}He-^{4}He mixture is heated from below the flow state of the fluid is determined by the magnitude of the Rayleigh number R, the value of the separation ratio S and the history of the route by which the state has been reached. ^{3}He-^{4}He mixtures have the useful property that S can be changed from large negative values to small positive values by increasing the mean fluid temperature from a few mK to over 100mK above the superfluid transition temperature T_λ. On theoretical grounds (1) a codimension-two point (R_{CT}, S_{CT}) is expected such that for increasing heat current and $S<S_{CT}$ the fluid should Hopf-bifurcate from conduction to a travelling-wave state while for $S>S_{CT}$ the fluid should bifurcate from conduction to a stationary convection state. Calculations of S_{CT} (1,2) are of order -10^{-3}, which correspond to about 80mK above T_λ for a mixture of 2.35% ^{3}He molar concentration. It is further expected (1) that there should be separate branches of backward bifurcations from (a) steady convection to conduction for $S<S_b$ with the dynamical behaviour at S_b near the convection threshold corresponding to a hydrodynamic tricritical point, and (b) travelling wave convection to conduction for $S<S_c$, where $S_c<S_{CT}<S_b$.

The bifurcation diagram of dilute ^{3}He-^{4}He mixtures for $|S|<10^{-2}$ has been investigated experimentally in circular geometry by Gao and Behringer (3) and in rectangular geometry by Ahlers and coworkers (4,5,6). The former observe evidence of a hydrodynamic tricritical point but not a codimension-two point, while the latter observe a cusp in the threshold Rayleigh number S-dependence which can be attributed to a codimension-two point and the expected bifurcation branches for $S<S_b$ and $S<S_c$ but without observing tricritical behaviour at $S=S_b$. Tricritical behaviour would correspond to $Nu-1 \propto \varepsilon^{1/2}$ where Nu is the Nusselt number and $\varepsilon=R/R_c-1$ where R_c is the critical Rayleigh number. We report here experiments on 2.35 molar percent mixture in circular geometry

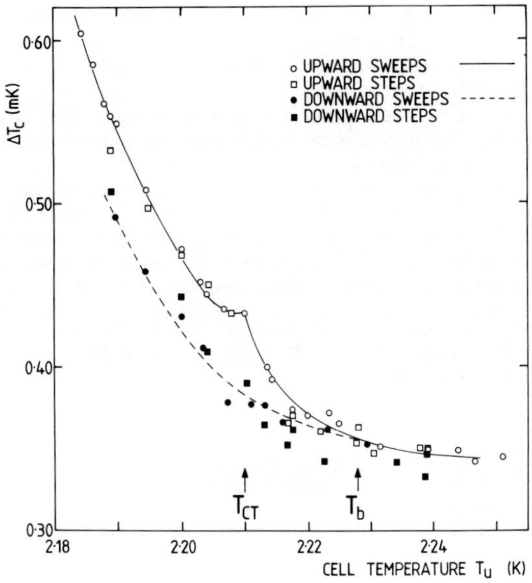

FIGURE 1
Critical temperature differences ΔT_c plotted against cell upper boundary temperature T_u. The curves are guides only.

radius 1.24cm, height 0.21cm undertaken to throw more light on the differences between the two previous sets of experiments.

2. EXPERIMENTAL

Our experimental technique was similar to that reported earlier by us (7,8). Briefly the upper boundary of the experimental cell was held fixed to within 10^{-6}K of some fixed temperature T_u above T_λ, while a heat current W applied to the lower boundary was varied in time as either a linear ramp or a series of steps. The resulting temperature difference ΔT between the upper and lower boundaries was measured both by differential resistance thermometry and a SQUID-

0921-4526/90/$03.50 © 1990 – Elsevier Science Publishers B.V. (North-Holland)

thermocouple sensor (9). Ramp rates were sufficiently slow that ΔT changed by less than $50\mu K\ hr^{-1}$. In the case of the steps ΔT changed by at most $20\mu K$ between each step, each period of constant heat current lasting 40 minutes or longer, data on ΔT being taken 20 minutes after each heat current adjustment. The vertical diffusion time was of order 8 minutes.

Throughout all the measurements our procedure was to measure critical temperature differences ΔT_c at different values of $T_u - T_\lambda$ between 40mK and 110mK using both the ramp and the step technique for both increasing and decreasing heat currents. T_λ was 2.139K for our mixture.

3. OBSERVATIONS

The measurements of ΔT_c are shown in Figure 1. There is a clear cusp at T_{CT}=2.21K which we identify with the codimension-two point observed by Ahlers and Rehberg (4). Time dependence was not observed for the convecting state reached with heat-increasing when $T<T_{CT}$, but was not expected since our sensor averaged over the cell boundaries and hence would not be sensitive to travelling waves.

The cusp is absent for heat-decreasing sequences and ΔT_c is clearly smaller than for heat-increasing sequences provided $T<T_b$. Above T_b the transition was not hysteretic and so we identify T_b with the point at which the bifurcation changes from backward ($T<T_b$) to forward ($T>T_b$). The data shown in Figure 2 give a better measure of T_b. Here we plot the temperature dependence of the jump ΔNu in the Nusselt number observed in Nu, ε plots. ΔNu is smaller for heat-decreasing than for heat-increasing as expected but vanishes above T_b=2.228K.

We looked for tricritical behaviour at T_b by checking the initial slopes of the Nusselt number plots in this region. This is difficult because of the scatter on our data, caused by the use of a rather large vertical spacing on our cell chosen for continuity with our earlier measurements (7,8) and the data are still being analysed. However there appears to be weak evidence that $Nu-1=\xi\varepsilon^{1/2}$ up to ε=0.2 with ξ=0.49. We note that Gao and Behringer (3) observe similar behaviour with ξ=0.46 while ξ=0.2 according to the calculations of Holton (10).

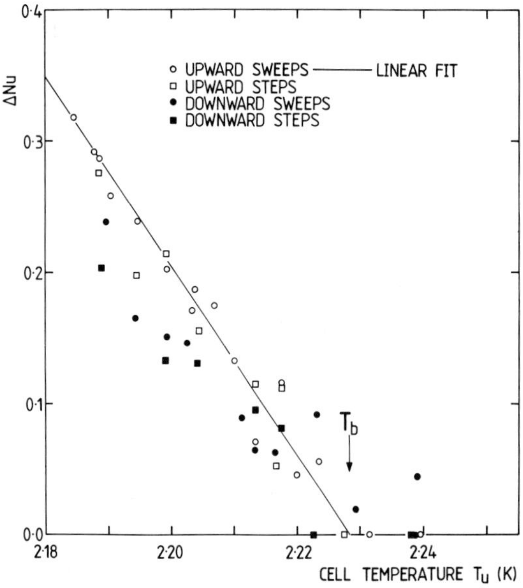

FIGURE 2
Dependence of Nusselt number discontinuities ΔNu, observed in plots of Nu against R/R_c-1, on cell upper temperature T_u. the straight line is a linear fit to the upward sweeps.

4. CONCLUSIONS

Our data show that the codimension-two point is observable in circular geometry. The change from the bifurcation to stationary convection being hysteretic to non-hysteretic is also observable in circular geometry. Thus to this extent our observations are similar to those of Ahlers and Rehberg (4) in rectangular geometry. However unlike these authors who do not observe tricritical behaviour at T_b our data support the measurements of Gao and Behringer (3) which demonstrate a concave slope of Nu-1 against ε up to $\varepsilon \simeq 0.15$. The evidence for this support is weak because of experimental scatter and is being analysed further.

REFERENCES

1) H. Brand, P.C. Hohenberg and V.Steinberg, Phys. Rev. A30 (1984) 2548.
2) S.J. Linz and M. Lücke, Phys. Rev. A35 (1987) 3997; W. Schöpf and W. Zimmermann, Europhys. Lett. 8 (1989) 41; D.R. Moore and E. Knobloch, in print.
3) H. Gao and R.P. Behringer, Phys. Rev A34 (1986) 697.
4) G. Ahlers and I. Rehberg, Phys. Rev. Lett. 56 (1986) 1373.
5) T.S. Sullivan and G. Ahlers, Phys. Rev. Lett. 61 (1988) 78.
6) T.S. Sullivan and G. Ahlers, Phys. Rev. A38 (1988) 3143.
7) G.W.T. Lee, P. Lucas and A. Tyler, J. Fluid Mech. 135 (1983) 235.
8) T.J. Bloodworth, M.R. Ardron, P.G.J. Lucas and J.K. Bhattacharjee, Jap. J. App. Phys. 26, Supp. 26-3 (1987) 59.
9) Y. Maeno, H.Haucke and J.C. Wheatley, Rev. Sci. Instrum. 54 (1983) 946.
10) D. Holton, Convective instabilities in binary fluids, Ph.D Thesis (1989), University of Warwick.

Physica B 165&166 (1990) 523–524
North-Holland

Magnetic Kapitza Resistance between Enhanced Nuclear Spin System and Liquid ^{3}He

Naoki MIZUTANI+, Haruhiko SUZUKI, and Masumi ONO

Department of Physics, Faculty of Science, Tohoku University, Sendai 980, Japan

We measured the Kapitza resistance between $TmVO_4$ and liquid ^{3}He in the temperature range 1-10mK. The Kapitza resistance was proportional to T^2 below about 3mK. We also observed the decreasing Kapitza resistance between $HoVO_4$ and liquid ^{3}He with decreasing temperature, but its value was two orders of magnitude smaller than that for $TmVO_4$. The experimental results can be qualitatively explained by the magnetic coupling between enhanced nuclear spins of Tm or Ho and nuclear spins of adsorbed solid ^{3}He on the surface of the crystal.

1. INTRODUCTION

The anomalous Kapitza resistance between CMN and liquid ^{3}He which decreases with decreasing temperature has been attributed to the magnetic coupling between the electron spins of Ce ions and the ^{3}He nuclear spins [1]. However, except for CMN, such anomalous temperature dependence of thermal resistance has not yet been observed. We have investigated the magnetic Kapitza resistance between other substances and liquid ^{3}He. We used the Van Vleck paramagnetic compound $TmVO_4$ as a sample, because its surface is stable and spin-lattice resistance is small [2]. Enhanced nuclear moment of Tm in $TmVO_4$ is about one tenth of Bohr magneton. Its magnetic ordering is not observed down to 80μK. In order to get the information on the anomalous resistance, we have studied the effect of ^{4}He adding into liquid ^{3}He, and measured the temperature and magnetic field dependences of the thermal resistance. Furthermore, we studied the Kapitza resistance by using another enhanced nuclear spin system, Ho in $HoVO_4$. Enhanced nuclear moment of Ho in $HoVO_4$ is about 3.8 times larger than that of Tm in $TmVO_4$ [3]. It orders antiferromagnetically at about 4.9mK.

2. EXPERIMENTAL PROCEDURE

Samples of $TmVO_4$ and $HoVO_4$ were powder whose sizes were less than 37μm. A thermal resistance between enhanced nuclear spins of Tm or Ho and liquid ^{3}He was measured by the thermal relaxation method. Measured relaxation times were much longer than spin-lattice relaxation times. Therefore, we can attribute the observed thermal relaxations to the boundary resistances.

3. RESULTS AND DISCUSSIONS

3.1. $TmVO_4$

As shown in Fig.1, for pure ^{3}He, the Kapitza resistance had a peak at about 5mK, and decreased with decreasing temperature below that. Adding ^{4}He of 3000ppm which corresponds to the amount of 8 layers, the resistance was proportional to T^{-3} down to the lowest temperature we could reach. Its value was almost same value as the calculated one by the acoustic mismatch theory. These results tell us that the magnetic Kapitza resistance which decreases with decreasing temperature was enlarged by ^{4}He.

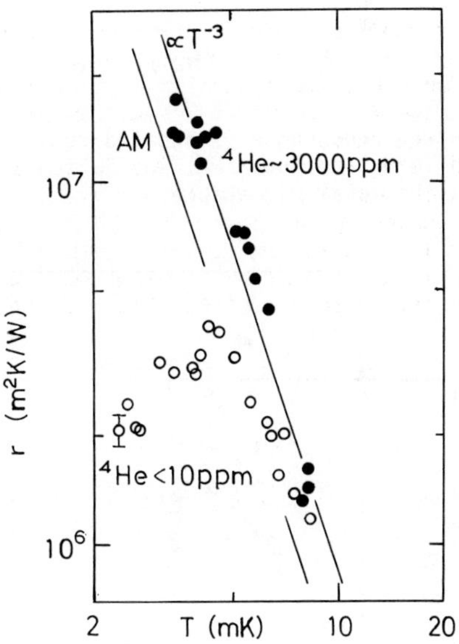

FIGURE 1
Effect of ^{4}He on Kapitza resistance between $TmVO_4$ and ^{3}He. The solid line denoted by AM, represents the value calculated by the acoustic mismatch theory.

Fig.2 shows the results extended to lower temperature for pure ^{3}He. The dependence on temperature is well explained by assuming the magnetic Kapitza resistance ($\propto T^2$) in parallel

+ Present address: Ulvac, Chigasaki 253, Japan

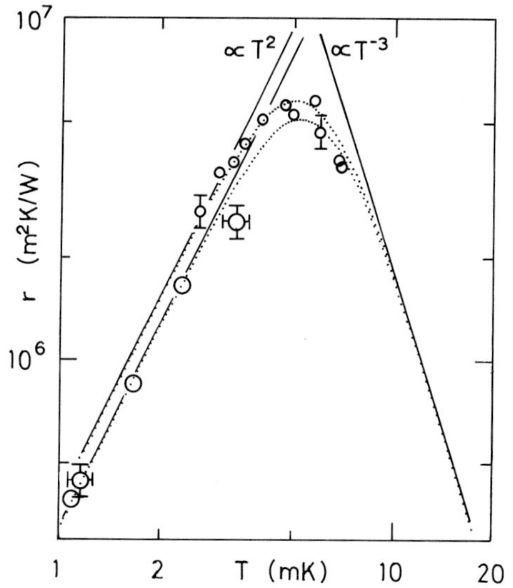

FIGURE 2
Temperature dependence of Kapitza resistance between $TmVO_4$ and pure ^3He. The dotted line represents the total resistance.

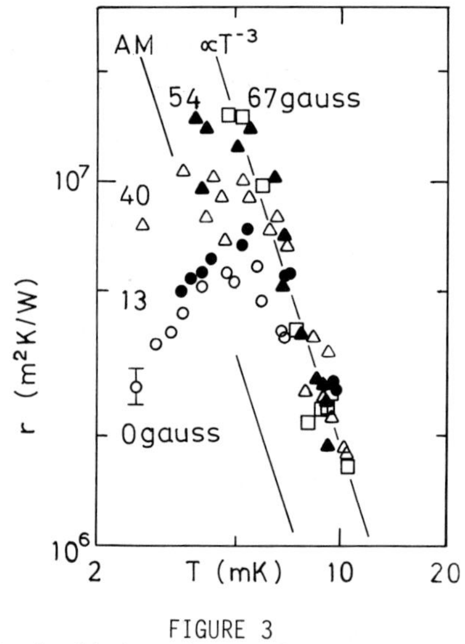

FIGURE 3
Magnetic field dependence of the resistance between $TmVO_4$ and pure ^3He.

with the resistance due to phonon mismatch ($\propto T^{-3}$). This T^2 dependence of the Kapitza resistance was observed for the first time.

Fig.3 shows the magnetic field dependence of the resistance. The result can be understood as follows. The magnetic Kapitza resistance increased with field, and in 67gauss, only the phonon mismatch resistance was observed.

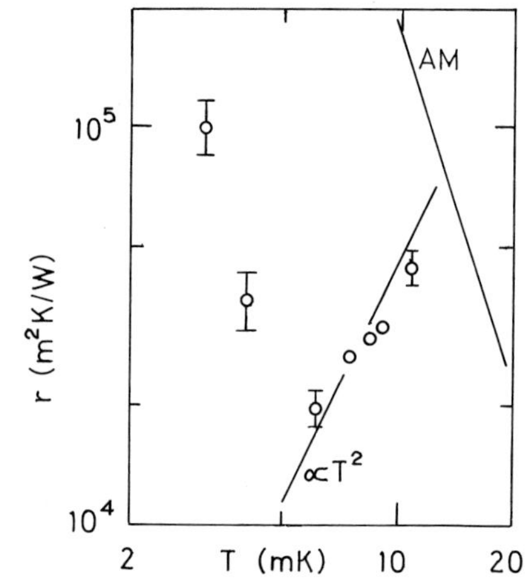

FIGURE 4
Kapitza resistance between $HoVO_4$ and pure ^3He.

3.2. $HoVO_4$

Fig.4 shows the Kapitza resistance between $HoVO_4$ and pure ^3He. In paramagnetic region of Ho, it decreased with decreasing temperature. However, its value was two orders of magnitude smaller than that for $TmVO_4$. This is attributed to the different values of the magnetic moment of Tm and Ho. In ordered state of Ho, the resistance increased with decreasing temperature. This difference of the dependence on temperature is attributed to the difference of the states of Ho spin system.

The Kapitza conductance due to the magnetic coupling with liquid ^3He was calculated by Leggett and Vuorio [4]. That theory was applied to various spin systems by many authors[5][6]. We have calculated the resistance between enhanced nuclear spins and nuclear spins of adsorbed solid ^3He on the surface of the crystal. We divided ^3He layers into 1st and 2nd layer. Experimental results were qualitatively explained by our model.

REFERENCES

(1) D. Marek, A.C. Mota, and J.C. Weber, J. Low Temp. Phys. 63 (1986) 401.
(2) H. Suzuki, T. Inoue, Y. Higashino, and T. Ohtsuka, Phys. Lett. 77A (1980) 185.
(3) B. Bleaney, Proc. R. Soc. Lond. A370 (1980) 313.
(4) A.J. Leggett and M. Vuorio, J. Low Temp. Phys. 3 (1970) 359.
(5) M.T. Beal-Monod and D.L. Mills, J. Low Temp. Phys. 30 (1978) 289.
(6) K. Hood, E. Zaremba, and T. McMullen, J. Low Temp. Phys. 68 (1987) 29.

Physica B 165&166 (1990) 525–526
North-Holland

KAPITZA RESISTANCE BETWEEN SILVER AND ^{3}He AT MILLIKELVIN TEMPERATURES*

G. J. STECHER, Y. HU,† T. J. GRAMILA,‡ and R. C. RICHARDSON

Laboratory of Atomic and Solid State Physics, Clark Hall, Cornell University, Ithaca, New York 14853-2501, U.S.A.

A glass capacitance thermometer was used to measure the Kapitza resistance R_k between a 0.5 μm silver film and liquid ^{3}He at temperatures from 3 to 40 mK. Temperature dependence of R_k was observed in three regimes: roughly proportional to T^{-1} below 15 mK, T^{-2} from 15 to 26 mK, and T^{-3} above 26 mK. R_k depended on pressure over the entire temperature range; at 22 mK, R_k was inversely proportional to $m^*(P)$, the effective mass of liquid ^{3}He. At 6.7 mK, R_k varied with applied magnetic field with a minimum at $H \approx 2.5$ mT. No magnetic field dependence was observed above 15 mK.

1. INTRODUCTION

Recent experiments on Kapitza resistance at milliKelvin temperatures have been conducted with powdered or sintered metals (1–3). Glass capacitance thermometry provides a simple method of measuring the Kapitza resistance of better characterized surfaces. We have measured the Kapitza resistance between ^{3}He and a silver surface, smooth on the scale of 0.1 μm, and its dependence on temperature, pressure, and magnetic field.

2. EXPERIMENT

The sample was made by evaporating 0.5 μm thick silver films onto both sides of a piece of Corning #2 coverslide (area 1.35 cm^2, thickness 0.024 cm). Superconducting leads were attached to each silver surface, forming a capacitor with the glass as the dielectric. Since the dielectric constant of the glass varies monotonically with temperature in the region of interest (4), the temperature of the sample can be determined from the capacitance reading. On one side of the sample, a small region of silver was removed, leaving the remaining silver in the shape of a 'U.' The resistance of this 'U' was 53.7 mΩ and remained constant to within 1% from 1 K to 50 mK. Power was applied to the sample by passing a known current through the 'U.' The sample was immersed in ^{3}He and the capacitance thermometer was calibrated using a melting curve thermometer. The 1986 Greywall temperature scale (5) was used. A small solenoid surrounding the sample was used to create magnetic fields of up to 8 mT. The Kapitza resistance was determined from the equation

$$R_k = \frac{T_{\rm Ag} - T_{\rm He}}{(\dot{Q}/A)}$$

where $T_{\rm Ag}$ and $T_{\rm He}$ are the temperatures of the sample and the ^{3}He, $\dot{Q}$ is the power applied to the sample through Joule heating, and A is the area of the silver. Further details of the apparatus and experiment can be found in Hu (4).

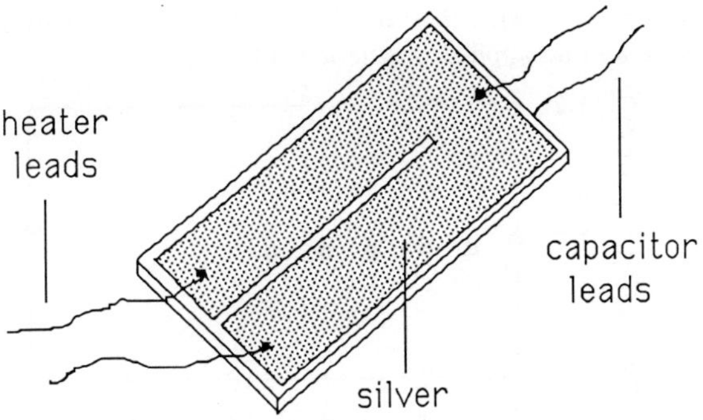

FIGURE 1
Diagram of sample.

3. RESULTS

R_k was observed to vary with T in three distinct regimes: roughly proportional to T^{-1} for T less than 15 mK, T^{-2} for T between 15 and 26 mK, and T^{-3} for T greater than 26 mK (fig. 2). R_k varied with ^{3}He pres-

*Supported by the National Science Foundation through grant number DMR-8418605.
†Present address: Physics Dept., Worcester Polytechnic Institute, Worcester, Mass. 01609, U.S.A.
‡Present address: AT&T Bell Laboratories, Murray Hill, New Jersey 07974, U.S.A.

sure between 0 and 10 bar over the entire temperature range. At 22 mK, R_k was found to be inversely proportional to $m^*(P)$, the effective mass of ^{3}He (fig. 3). Present theories predict $R_k \propto (m^*)^{-2}$ (6). Values for m^* were taken from Greywall (5). Temperature sweeps at pressures of 0, 5, and 10 bar indicate that the relation $R_k \propto 1/m^*(P)$ still holds for temperatures from 6 to 30 mK. Below 6 mK and above 30 mK, the relationship between R_k and $m^*(P)$ could not be determined reliably from the data. R_k was also observed to depend on applied magnetic field H for temperatures below 15 mK. At 6.7 mK, $R_k(H)$ drops with increasing field to a minimum near 2.5 mT (fig. 4). Above the minimum, from approximately 3 to 8 mT, R_k appears to be proportional to H.

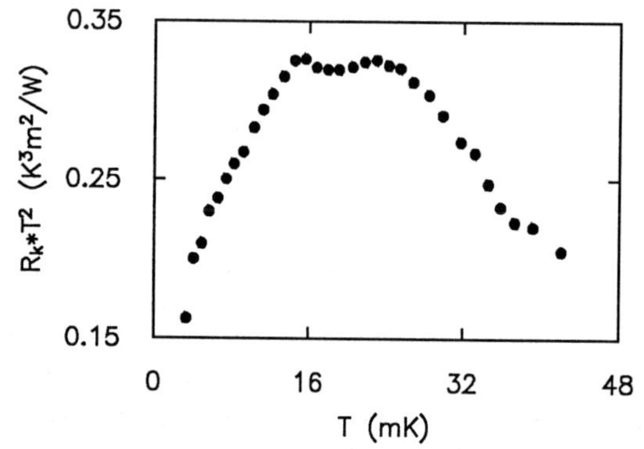

FIGURE 2
R_kT^2 vs. T with the ^{3}He under saturated vapor pressure and no applied magnetic field.

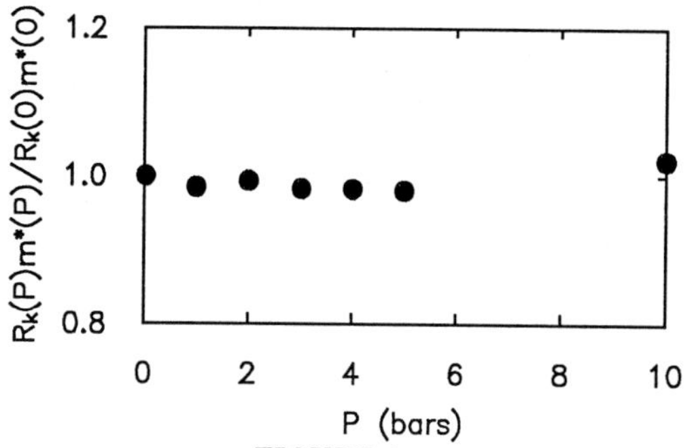

FIGURE 3
$R_k(P)m^*(P)/R_k(0)m^*(0)$ vs. P at 22 mK and no applied magnetic field.

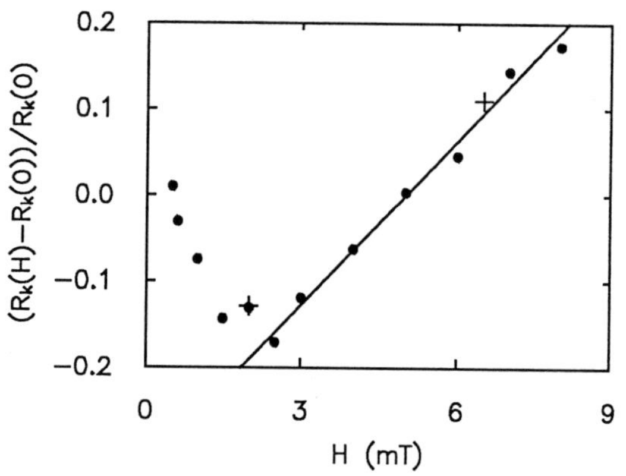

FIGURE 4
Fractional change of Kapitza resistance with applied magnetic field at 6.7 mK and saturated vapor pressure. The + signs represent points taken with the field in the opposite direction of the other points. The line illustrates $R_k \propto H$ above the minimum.

Preliminary measurements of the Kapitza resistance between ^{3}He and silver with ~4 monolayers of ^{4}He adsorbed were made. These measurements indicate an increase in R_k over that with no ^{4}He by a factor of 3 below approximately 10 mK. The difference diminishes at higher temperatures; at 38 mK the increase appears to be less than 20%.

REFERENCES

(1) T. Perry, Keith DeConde, J. A. Sauls, and D. L. Stein, Phys. Rev Lett. 48 (1982) 1831.
(2) D. D. Osheroff and R. C. Richardson, Phys. Rev. Lett. 54 (1985) 1178.
(3) S. N. Ytterboe, P. D. Saundry, L. J. Friedman, C. M. Gould and H. M. Bozler, Proc. LT-18 in Jap. J. Appl. Phys. 26 Suppl. 26-3 (1987) 379.
(4) Yue Hu, Ph.D. Thesis, Cornell Univ. (1990) unpublished.
(5) Dennis S. Greywall, Phys. Rev. B33 (1986) 7520.
(6) Tsuneyoshi Nakayama, Kapitza Thermal Boundary Resistance and Interactions of Helium Quasiparticles with Surfaces, in: Progress in Low Temperature Physics, Vol. 12, ed. D. F. Brewer (North-Holland, Amsterdam, 1990), pp.115–193.

Physica B 165&166 (1990) 527–528
North-Holland

KAPITZA RESISTANCE BETWEEN ^{3}HE–^{4}HE DILUTE SOLUTION AND METAL FINE POWDER BELOW 1 MILLIKELVIN

Gong–Hun OH, Masahito NAKAGAWA, Hikota AKIMOTO*, Osamu ISHIKAWA, Tohru HATA, and Takao KODAMA

Faculty of Science, Osaka City University, Osaka 558, Japan

We have cooled ^{3}He–^{4}He dilute solution down to 202 μK, and find that T^{-2} dependency of the Kapitza resistance between the solution and the metal powder is still valid at the lowest temperature. The estimated heat leak was 45 pW. We have, also, investigated the time evolution of the heat leak, and find that it is proportional to $t^{-0.32}$.

1. INTRODUCTION

Search for superfluidity of ^{3}He–^{4}He dilute solution is one of the great interests in ultra low temperature physics. There are several theoretical predictions of the transition temperature based on the attractive interaction between ^{3}He quasiparticles.(1) These predicted transition temperatures diverge from a few microkelvin to a millikelvin depending on the interaction amplitude. Today, it is not so difficult to get a few tens of microkelvin using a conventional copper nuclear demagnetization. However, the obtained lowest temperature of the dilute solution is about 210 μK, which is much higher than that of nuclear stage. This temperature difference comes from a large Kapitza resistance between the dilute solution and the metal fine powder as a heat exchanger. To find how to cool it below 210 μK, it is important to investigate the temperature dependence of the Kapitza resistance far below 1 mK.

2. EXPERIMENTAL

A double nuclear demagnetization was used to cool the dilute solution. The first stage consisted of 4.8 moles of $PrNi_5$ and 13.8 moles of copper. The coolant of the second stage was copper plates of 8.0 moles immersed in the dilute solution. We divided the sample cell into two parts so that the outer cell served as a thermal guard for the inner cell, which was a similar arrangement to Lancaster group.(2) The sample cell was made of stycast 1266 and only the side wall between the inner and outer cell was made of Kapton foil. These two cells were connected with a fine copper tube of 0.1 mm i.d. and 1 m in length. The dilute solution was cooled by two kinds of sintered metal powder on the copper plates, one was the composite of Ag and Pt for the inner cell, and the other was Ag only for the outer cell. The packing factor was about 40 % and the total surface area of the heat exchanger was 450 m² and 140 m² for the inner and outer cell, respectively.

Three Pt thermometers were used to measure the lattice temperature of the inner copper plate, the inner liquid and outer liquid. These were calibrated above 2 mK against the ^{3}He melting curve thermometer. Each thermometer to measure the liquid temperature had a sintered Pt+Ag pill as a heat exchanger whose surface area was 50 m². We used ^{3}He concentration of 5.5 % in this experiment.

Two nuclear stages were precooled down to 11 mK by the dilution refrigerator. The 2nd nuclear stage at 8 T and the dilute solution were precooled again to about 3.2 mK by the demagnetization of the 1st stage from 6 T. The final field of demagnetization of the 2nd stage was 27 mT which corresponded to a resonance frequency of the platinum thermometer.

3. RESULTS

The lattice temperature of the nuclear stage was 30 μK just after a demagnetization and it kept below 40 μK more than 2 weeks. The temperature of the inner liquid was 250 μK at the end of demagnetization and it reached to 202 μK

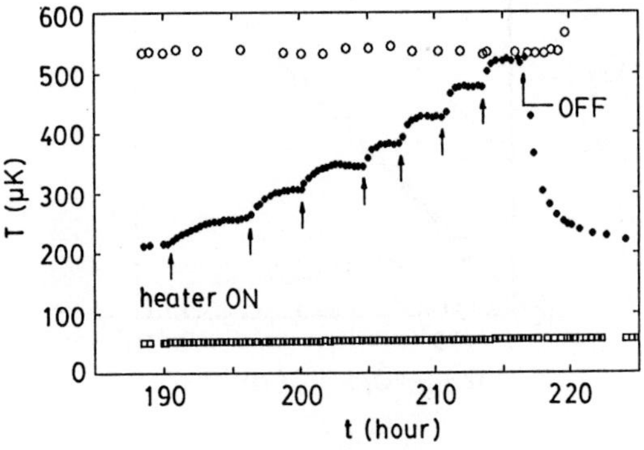

FIGURE 1
Time evolution of temperatures of the outer liquid (○), the inner lquid (●) and the copper plate (□). The origin of time is the end of the demagnetization. The arrows show the change of the heater power into the inner liquid.

* Present address: The Institute for Solid State Physics, The University of Tokyo, Tokyo 106, Japan

0921-4526/90/$03.50

after one week later. The lowest temperature of the outer liquid was 540 μK. We measured Kapitza resistance between the solution and the sintered metal powder by heating the solution. As shown in Fig.1, the lattice temperature of the copper plate and the outer liquid stayed constant while the heat was applied into the inner liquid. We used the equation as below,

$$\dot{Q}_{app} + \dot{Q}_{hl} = \frac{A}{R(n+1)}(T_l^{\,n+1} - T_w^{\,n+1}) \qquad (1)$$

to analyze our data, where we assumed that the Kapitza resistance was represented by the powers of temperature RT^{-n}. Here, $\dot{Q}_{app}$, $\dot{Q}_{hl}$, T_l, T_w, were an applied heat, an external heat leak, a liquid temperature and a heat exchanger temperature, respectively, and A is the surface area of the sintered powder. Assuming that $\dot{Q}_{hl}$ is constant during our measurement, we got n=2 as a best fit. The obtained heat leak $\dot{Q}_{hl}$ in this measurement was 45 pW and the data accurately follow Eq.(1) as shown in Fig.2.

The estimated Kapitza resistivity was RAT^2 = 24 m^2K^3/W for silver powder and 32 m^2K^3/W for platinum powder. This value of silver powder agrees well with the previous measurement.(3) In Fig.3, we showed our value of the Kapitza resistance comparing with that obtained in other measurements.(3,4)

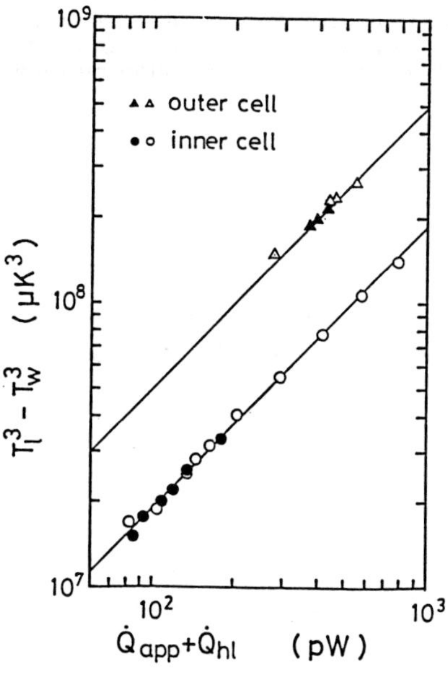

FIGURE 2
A plot of $T_l^3 - T_w^3$ versus $\dot{Q}_{app} + \dot{Q}_{hl}$ for each cell. The straight lines indicate the slope is 1. Closed circles and triangles are obtained in a different run of the demagnetization.

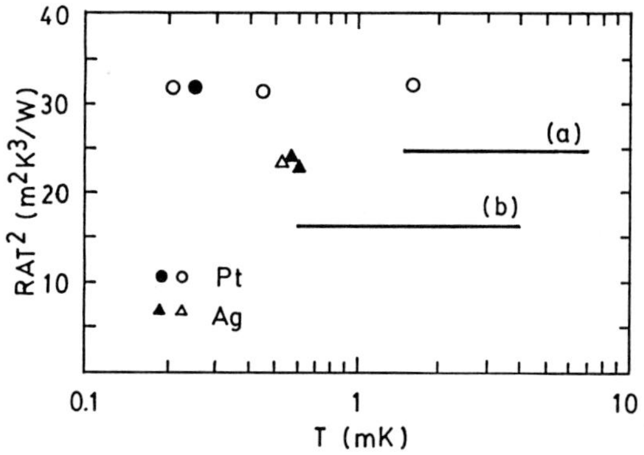

FIGURE 3
A plot of $R\dot{A}T^2$ as a function of temperature. This work is shown by (●○ ▲△), (a) Ref.(3) for 6.4 % ^{3}He concentration and (b) Ref.(4) for 8 %.

4. DISCUSSION

The Kapitza resistance between the dilute solution and the metal powder is proportional to T^{-2} far below 1 mK. This experimental result supports the recent theory of the magnetic coupling mechanism proposed by Nakayama.(5) According to Eq.(1), if we want to reach below 100 μK, we have to increase the ratio A/R 10 times larger than this experiment or decrease the heat leak to a few pW, because we have been able to get the low enough temperature of the nuclear stage.

Since the temperature of the outer liquid dose not affect the lowest temperature of the inner cell, the heat coming from the outer liquid to the inner seemed to be less than 10 pW. To investigate the origin of the heat leak, we calculated the time evolution of the heat leak into the inner and outer liquid using the obtained Kapitza resistance. The heat leak as a function of time t can be expressed by $t^{-0.32}$. This behaviour could be explained by the heat release from the plastic like stycast 1266 and we could reduce the heat leak by replacing the plastic by metal. However, the exponent disagrees with that obtained at higher temperature.(6)

REFERENCES

(1) J.Bardeen, G. Baym and D.Pines. Phys. Rev. 156 (1967) 207.
(2) D.I.Bradley et al. J. Low Temp. Phys. 57 (1984) 359.
(3) G.Frossati J. de Physique C6 (1987) 1578.
(4) D.D.Osheroff and L.R.Corruccini Phys. Lett. 82A (1981) 38.
(5) T.Nakayama. Prog. Low Temp. Phys. (North-Holland, Amsterdam, 1989)
(6) M.Schwark et al. J. Low Temp. Phys. 58 (1985) 171.

Physica B 165&166 (1990) 529–530
North-Holland

KAPITZA RESISTANCE AT THE SOLID LIQUID INTERFACE IN ^3He

J. AMRIT and J. BOSSY

Centre de Recherches sur les Très Basses Températures, C.N.R.S., BP 166 X, 38042 Grenoble-Cédex, France

We have measured the Kapitza resistance at the liquid-solid interface in ^3He for temperature between 50 mK and 200 mK. We found a T^3 behaviour ($R_K T^3$ = 0.03 cm^{2}K/W). The low value is in good agreement with theoretical predictions.

1. INTRODUCTION

The usual way of measuring the Kapitza resistance is by extrapolating the thermal gradients on each side of the interface. The resistance R_K is simply given by $R_K = (\Delta T/J_Q)$ where ΔT is the temperature difference across the interface, and J_Q is the heat flux through the interface. This method is ruled out in the case of ^3He, because of the low thermal conductivity of the liquid and of the low value of the resistance.

Instead of measuring the temperatures of the solid and the liquid near the interface, we measure the temperature of the solid and indirectly the temperature in the liquid side by measuring the pressure P in the liquid. The principle of the experiment is to fix the temperature of the crystal, and measure the pressure drop in the liquid when we change the heat flux across the interface. The chemical potentials in the crystal (c) and in the liquid (ℓ) are

$$d\mu_{c,\ell} = - S_{c,\ell}\, dT_{c,\ell} + V_{c,\ell}\, dP$$

where $S_{c,\ell}$ are the entropies of the crystal and the liquid, $T_{c,\ell}$ the temperatures and $V_{c,\ell}$ the molar volumes. The chemical potential difference $\Delta\mu$ between the crystal and the liquid for a fixed position of the interface given by the linear response theory is [1] :

$$\Delta\mu = \frac{\lambda \Delta T}{T}$$

where $\Delta T = T_\ell - T_c$ and λ is a coefficient. Assuming that $dT_c = 0$ and $\Delta T = R_K J_Q$ we find :

$$R_K = \frac{T(V_c - V_\ell)\delta P}{(\lambda - TS_\ell)J_Q}$$

Most of the theoreticians agree that $\lambda \simeq TS_c$ [2,3]. The Kapitza resistance takes the simple form

$$R_K = \frac{\delta P}{J_Q \left(\frac{dP}{dT}\right)_{mc}}$$

where $\left(\frac{dP}{dT}\right)_{mc}$ is the slope of the melting curve.

2. EXPERIMENTAL

The cell is shown in figure 1. The crystal is grown in a cylindrical capacitor which allows to measure the interface position, the latter can be moved by a bellows. The crystal grows at the hottest part of the cell when the temperature is below the minimum of the melting curve. Four 150 Ω Matsushita resistors act as thermometers. The pressure is measured with a capacitive gauge. During the experiment the interface is kept as near as possible of the highest thermometer. The temperature of the solid is regulated by means of heater 1. This situation implies a heat flux through the interface.

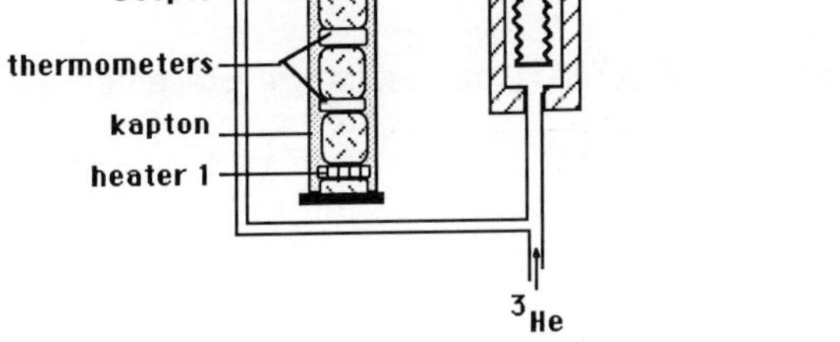

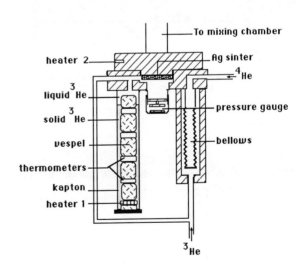

FIGURE 1
Schematic view of the experimental cell.

0921-4526/90/$03.50

We obtain a decrease of this heat flux when the liquid temperature is increased by using heater 2 without changing the temperature of the solid. During this process we observe a pressure drop proportional to the Kapitza resistance. The solid temperature should be regulated with a high precision ($\Delta T/T \sim 10^{-4}$) in order to prevent the masking of the effect due to the shift in the equilibrium pressure.

3. RESULTS AND COMMENTS

We have measured the Kapitza resistance between 50 mK and 200 mK on the same ^3He crystal containing less than 5 ppm ^4He. The results are presented in figure2 as R_KT^3 versus T. R_KT^3 is nearly constant in this temperature range.

$$R_KT^3 = 0{,}03\ cm^2K^4/W.$$

This result is in agreement with the experiment of Graner *et al.* [4] at the minimum of the melting curve. Our measurements are also in very good agreement with recent calculation based on the acoustic mismach theory [3]. The low value of the Kapitza resistance is explained by the good coupling between the transverse and longitudinal phonons of the crystal and the transverse and longitudinal zero sound mode in the liquid phase.

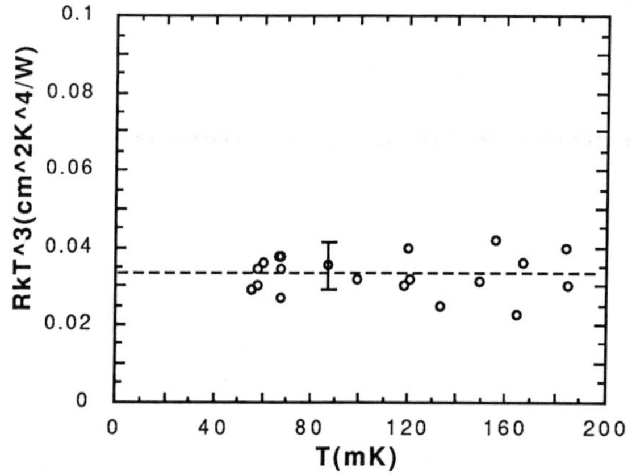

FIGURE 2
Kapitza resistance : The results are presented as R_KT^3 versus T.

4. CONCLUSION

Our method of measuring the Kapitza resistance by means of pressure measurements in the liquid has been very powerfull. This method gives actually the product $R_K(\lambda - TS_\ell)$. Experiments with a slow melting of the crystal should be tempted to check the assumption that $\lambda = TS_S$.

ACKNOWLEDGEMENTS

We are greatly indebted to B. Castaing for his interest in our work and his numerous advices. We thank also F. Graner for fruitful discussions.

REFERENCES

[1] B. Castaing and P. Nozières, J. Physique **41** (1980) 701.
[2] L. Puech, G. Bonfait, and B. Castaing, J. Low Temp. Phys. **62** (1986) 315.
[3] F. Graner, R.M. Bowley, P. Nozières, to be published in J. Low Temp. Phys.
[4] F. Graner, S. Balibar, and E. Rolley, J. Low Temp. Phys. **75** (1989) 69.

Physica B 165&166 (1990) 531–532
North-Holland

EXCITATION SPECTRUM AND DYNAMIC STRUCTURE FUNCTION OF ^{4}He CLUSTERS*

S. A. Chin and E. Krotscheck

Center for Theoretical Physics, Department of Physics, Texas A&M University, College Station, TX 77843, USA

We compute the collective excitations, transition densities, and the dynamic structure function of ^{4}He droplets at zero temperature using a generalized Feynman *ansatz* for the excitation function. The theory uses as inputs the exact ground state, one- and two-body densities sampled by a second order Diffusion Monte Carlo algorithm.

The ground-state structure and the excitation spectrum of clusters of ^{4}He is presently of considerable theoretical and experimental interest[1–3]. As a generalization of the Feynman theory of collective excitations[4], we have developed a generic method[3,5] of calculating the collective excitations spectrum and the dynamic structure function of finite quantum systems from exact simulations of ground-state distribution functions.

We start with a translationally invariant ground state wave function $\Psi_0(\mathbf{r}_1,\ldots\mathbf{r}_N) = \Psi_0(\mathbf{r}_1+\mathbf{r},\ldots\mathbf{r}_N+\mathbf{r})$. Following Feynman[4], we write a trial function of the excited state as

$$\Psi_F(\mathbf{r}_1,\ldots\mathbf{r}_N) = \sum_{i=1}^{N} f(\mathbf{r}_i - \mathbf{r}_{cm})\Psi_0(\mathbf{r}_1,\ldots\mathbf{r}_N), \quad (1)$$

where $\mathbf{r}_{cm}$ is the center-of-mass coordinate. For the exact ground-state wave function, the energy difference between the ground state and the trial excited state can be expressed as

$$\hbar\omega \equiv E_F - E_0 = \frac{\int d^3r d^3r' u(\mathbf{r}) H_1(\mathbf{r},\mathbf{r}') u(\mathbf{r}')}{\int d^3r d^3r' u(\mathbf{r}) S(\mathbf{r},\mathbf{r}') u(\mathbf{r}')}, \quad (2)$$

where $E_F = \langle\Psi_F \mid H \mid \Psi_F\rangle / \langle\Psi_F \mid \Psi_F\rangle$, E_0 is the ground state energy, and $u(\mathbf{r}) \equiv \sqrt{\rho_1(\mathbf{r})} f(\mathbf{r})$. The quantities $H_1(\mathbf{r},\mathbf{r}')$ and $S(\mathbf{r},\mathbf{r}')$ are defined in terms of the ground-state one- and two-body densities $\rho_1(\mathbf{r})$ and $\rho_2(\mathbf{r},\mathbf{r}')$:

$$S(\mathbf{r},\mathbf{r}') = \delta(\mathbf{r}-\mathbf{r}') + \frac{\rho_2(\mathbf{r},\mathbf{r}') - \rho_1(\mathbf{r})\rho_1(\mathbf{r}')}{\sqrt{\rho_1(\mathbf{r})\rho_1(\mathbf{r}')}}, \quad (3)$$

$$H_1(\mathbf{r},\mathbf{r}') = -\frac{\hbar^2}{2mN} \frac{\nabla_{\mathbf{r}} \cdot \nabla'_{\mathbf{r}} \rho_2(\mathbf{r},\mathbf{r}')}{\sqrt{\rho_1(\mathbf{r})}\sqrt{\rho_1(\mathbf{r}')}} - \left[1 - \frac{1}{N}\right] \delta(\mathbf{r}-\mathbf{r}') \frac{\hbar^2}{2m} \frac{1}{\sqrt{\rho_1(\mathbf{r})}} \nabla\rho_1(\mathbf{r}) \cdot \nabla \frac{1}{\sqrt{\rho_1(\mathbf{r})}}. \quad (4)$$

The lowest upper bound for the excitation energy $\hbar\omega$ is obtained by minimizing the energy difference (2) with respect to the excitation function $u(\mathbf{r})$. This leads to the desired generalization of the Feynman dispersion relation[4]

$$H_1 u(\mathbf{r}) = \hbar\omega \int d^3r' S(\mathbf{r},\mathbf{r}') u(\mathbf{r}'). \quad (5)$$

The transition densities can then be calculated as

$$\delta\rho_1(\mathbf{r}) = \sqrt{\rho_1(\mathbf{r})} \int d^3r' S(\mathbf{r},\mathbf{r}') u(\mathbf{r}'). \quad (6)$$

The eigenvalue problem (5) can be solved by partia wave expansion. The input to Eq. (5) are the one-bod) density and partial wave amplitudes of the two-body den- sity directly sampled by the algorithm of Ref. 6. The eigenvalue problem (5) has one discrete eigenvalue in both the monopole and the quadrupole state, and a continuum above the evaporation energy. Our results for the ground-state energy of the droplets, the *rms*-radii, and the discrete excitation energies of three different droplets are given in the following table.

N	E_0/N (K)	r_{rms} (Å)	$\hbar\omega_0$ (K)	$\hbar\omega_2$ (K)
40	-2.525	6.67	3.60	1.37
70	-3.188	7.79	3.94	1.50
112	-3.705	8.96	3.92	1.76

*Work supported in part by NSF grants PHY89-07986 (to S. A. C.), PHY-8806265 and the Texas Advanced Research Program under Grant No. 010366-012 (to E. K.).

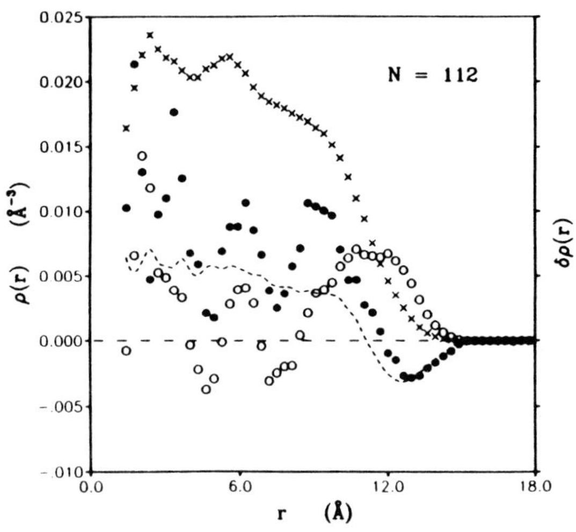

FIGURE 1
The one-body density (crosses) and transition densities for the monopole (dots) and the quadrupole (circles) collective mode of a $N = 112$ helium cluster. The dotted line is the monopole transition density as calculated from a variational wave function.

For a 112 particle cluster, our calculated ground state and transition densities are shown in Fig. 1. The quadrupole excitation has the expected form of a surface mode. The most remarkable feature of the monopole excitation is an oscillatory behavior of the transition density with a periodicity of approximatley the avarage particle separation. This behavior, which we found in all of our calculations independent on the size of the droplet and the sampling algorithm, appears to be a clear indication of a geometric shell structure. For comparison we also show the monopole transition density obtained from a *variational* Monte Carlo calculation with a simple MacMillan type Jastrow wave-function. None of the shell structure is observed in this case.

From the transition densities, one can readily construct the dynamic structure function

$$S_\ell(k,\omega) = |\delta\rho_{\ell,\omega}(k)|^2 , \qquad (7)$$

where $\delta\rho_{\ell,\omega}(k)$ is the Fourier-transform of the transition density $\delta\rho(\mathbf{r})$ defined in Eq. (6) corresponding to angular momentum ℓ and energy $\hbar\omega$. Typically, one has a discrete state and a continuum. Fig. 2 shows the $\ell = 0$ dynamic structure function for a 112 helium atom cluster. For comparison, we have drawn in the $k-\omega$ plane the graph of the Feynman spectrum[4] $\omega(k) = \hbar^2 k^2/2mS(k)$ for bulk ^{4}He. The dynamic structure function appears to be dominated by the zero sound excitation and its modulation by finite size effects.

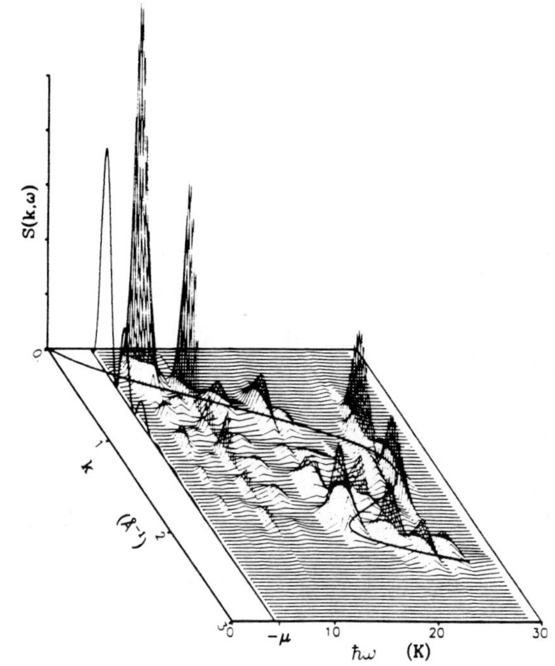

FIGURE 2
Monopole dynamic structure function for a 112 helium atom droplet. The single line at constant frequency slightly to the left of the continuum edge at $\hbar\omega = -\mu$ corresponds to the strength of the discrete collective mode.

REFERENCES

1. V. R. Pandharipande, J. G. Zabolitzky, S. C. Pieper, R. B. Wiringa, and U. Helmbrecht, Phys. Rev. Lett. **50**, 1676 (1983).
2. P. Sindzingre, M. J. Klein, and D. M. Ceperley, Phys. Rev. Lett. **63** 1601 (1989).
3. S. A. Chin and E. Krotscheck, TAMCTP preprint 1/90.
4. R. P. Feynman, Phys. Rev. **94**, 262 (1954).
5. S. A. Chin and E. Krotscheck, TAMCTP preprint 27/90..
6. S. A. Chin, Phys. Rev. A (in press).

Physica B 165&166 (1990) 533–534
North-Holland

MEASUREMENTS OF THE KAPITZA BOUNDARY RESISTANCE BETWEEN COPPER AND HELIUM ABOVE THE LAMBDA TRANSITION

Qiang LI, Talso C. P. CHUI and John A. LIPA

Department of Physics, Stanford University, Stanford, CA 94305 USA

Kapitza boundary resistance between copper and helium was measured above T_λ and found to be an order of magnitude smaller than the value previously proposed. Hence it is too small to explain the disagreement between thermal conductivity theory and some previous results.

Recent measurements (1) of the thermal conductivity of helium just above the lambda point have indicated some disagreement with the dynamic renormalization group predictions(2). In an attempt to reconcile the data with theory, Ahlers and Duncan (AD), (3), proposed that the disagreement may be explained by an anomalous boundary resistance term. Using theoretical arguments and results from thermal conductivity cells of various sizes, they deduced that a singular boundary resistance of the form $R_k = 1.55 \times 10^{-4}\, t^{-0.672}$ (cm^2 K/W) could fit the data, where $t = |1-T/T_\lambda|$.

In this paper we report preliminary results from an experiment specifically designed to measure R_k in the region just above the transition. We find that near $t = 10^{-6}$, R_k is about an order of magnitude smaller than that deduced by AD and is therefore too small to explain the original disagreement.

A cross-section of our apparatus is shown in fig. 1. The most significant feature is the ability to change the sample height without significantly affecting the Kapitza resistance of the end plates. The boundary resistance term is then obtained by extrapolation of the apparent thermal resistance, R, of the cell to zero gap. We have also eliminated all superfluid from the measurement area when the helium between the two plates is in the normal state. Since the shift of T_λ due to hydrostatic pressure always causes the superfluid to appear on top of the normal fluid, this goal was reached by moving the filling capillary to below the bottom plate, and by sealing the gap between the top plate and the side wall with epoxy. This design eliminated the changes in the temperature gradients inside the top copper block due to the variation of the superfluid level. A three terminal capacitance system was also added to allow plate separation measurements in situ, and the diameter of the cell was increased by a factor of three over previous work (1) to reduce wall effects.

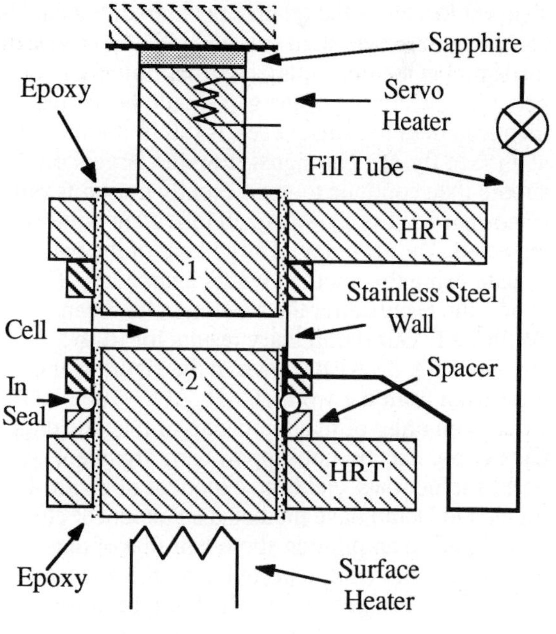

Fig. 1 Schematic view of the apparatus. Copper plates 1, 2 and the wall form a three terminal capacitance system. HRT stands for high resolution thermometer.

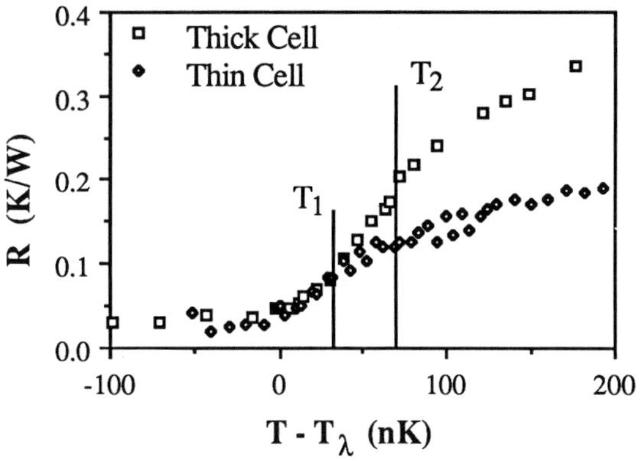

Fig. 2. Thermal resistance data near T_λ for two different cell spacings.

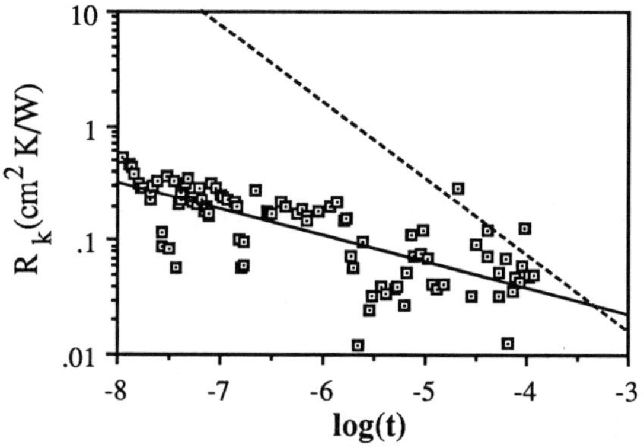

Fig. 3 Kapitza boundary resistance above T_λ. Solid line is a fit to data. Broken line is the curve used by Ahlers and Duncan.

Examples of cell resistance data near T_λ are shown in fig. 2, for sample heights of 0.25 and 0.5 mm. The maximum power density used in these measurements was 10^{-9} W/cm^2. After subtraction of the background cell resistance seen far below T_λ, the only remaining free parameter in the data comparison is the relative location of T_λ. The sharpest locater of the relative position of the data is based on the observation that the initial rapid rise of the signal in the two cells as normal fluid first occupies the bottom of the gap should be almost identical, until the smaller cell is full of normal fluid at T_1 (see fig.2). The signal from the larger cell should then continue to grow rapidly until it is full of normal fluid at T_2, where an inflection point is expected. The accuracy of alignment of the two sets of data using this method was $\pm$ 3 nanodeg. Fig. 2 shows the results after the alignment has been performed. Our preliminary results for R_k are shown in fig. 3. Also shown in the figure is the prediction from the work of AD. At $t = 10^{-6}$, AD's value is an order of magnitude larger than the data. Clearly our data do not support their argument. A best fit to our data yields an exponent of about -0.2. Frank and Dohm have predicted an exponent close to -0.3 and an amplitutde about half that of our data(4). For the work reported here, the effect of gravity has been neglected, which might cause some distortion of the results below $t \approx 5 \times 10^{-8}$. Additional experiments and analysis are in progress to improve the quality of the data.

Our results show that it is unlikely that the idea of an anomalous boundary layer can be used to explain the puzzling thermal conductivity results reported to date. It may be more fruitful to examine the experiments more closely to look for subtle effects that may perturb the data near T_λ. We believe our current apparatus is the first that is capable of giving reliable results to the resolution limit set by gravity in small cells. Hopefully further measurements will allow us to better understand the situation in very small cells, leading to a more comprehensive picture of the experimental situation. Of course, the results can also be used to obtain the bulk conductivity, but more extensive analysis is needed than has been completed so far.

This work was supported by JPL contract #957448 and by the NASA Ames Fund for Independent Research, grant #NAG 2-276.

REFERENCES

(1) J.A.Lipa and T.C.P.Chui, Phys. Rev. Lett. 58, 1340(1987). T.C.P.Chui, Q.Li and J.A.Lipa, Japan. J. Appl. Phys. 26, Suppl. 26-3, 371(1987).
(2) V. Dohm, Z.Phys. B61, 193(1985).
(3) G.Ahlers and R.V.Duncan, Phys. Rev. Lett. 61(7) 846(1988).
(4) D.Frank and V.Dohm, Phys. Rev. Lett. 62(16) 1864(1989).

Physica B 165&166 (1990) 535–536
North-Holland

SCALING THE SUPERFLUID DENSITY OF ^{4}He CONFINED IN PLANAR GEOMETRY *

Ilsu Rhee, David J. Bishop[a], Francis M. Gasparini

Department of Physics and Astronomy, State University of New York at Buffalo, Buffalo, N.Y. 14260, U.S.A.
[a]AT & T Bell Laboratory, Murray Hill, N.J. 07974, U.S.A.

We have measured the superfluid density of ^{4}He confined between two silicon surfaces. The reduction of ρ_s from the bulk value near the transition is observed for 0.106, 0.519, 2.8 and 3.9 μm confinement spacings,L. We find that the systematic reduction of ρ_s with confinement do not scale with the bulk correlation length, ξ. We find that the data scale with L, but not with the temperature dependence of ξ. We report calculations of ρ_s within ψ theory which include van der Waals effects. We also report analysis according to a suggesttion by Privman.

The physical properties of a finite system will in general differ from those of an infinite one. These finite-size effects may be negligibly small except in the region where the system is strongly correlated, that is, near a critical point. Even here these effects are likely to be beyond the experimental resolution or be at a level where they are mixed with other contributions such as those due to impurities, inhomogeneities or in some cases gravitational rounding. By comparison, size effects are relatively easily studied for ^{4}He at the superfluid transition.

In finite-size scaling one assumes that the only relevant length scale is the correlation length, ξ, therefore all quantities depend on $\frac{L}{\xi}$, where L is the smallest size of *homogeneous* confinement(1). The following scaling form may be assumed for the superfluid density,

$$\frac{\rho_s}{\rho} = \frac{\rho_{sb}}{\rho}(1 - f(L^\theta t)), \qquad (1)$$

where $\frac{\rho_s}{\rho}$ and $\frac{\rho_{sb}}{\rho}$ represent the superfluid fractions of the confined and the unconfined(bulk) systems respectively, and t is the reduced temperature $1 - \frac{T}{T_\lambda}$. The scaling function f must have properties such that $f(\infty) = 0$ and $f(L^\theta t_c) = 1$. The first condition guarantees the recovery of the bulk superfluid fraction for large L. The second says that the superfluid fraction vanishes at a temperature T_c which depends on the confinement spacing L. This condition implies a shift equation such that $L^\theta t_c =$ constant. To agree with correlation-length scaling one must have $\theta = \frac{1}{\nu}$, where ν is the critical exponent of the bulk correlation length, 0.672. This type of scaling seems to fail in a number of experiments in helium. This was observed for the specific heat(2) and superfluid density of helium confined in cylindrical pores(3). Experimentally, the difficulty of testing Eq. 1 is that the explicit form of the scaling function is not known.

* Work supported by NSF, Grant Numbers DMR 8601848 and DMR 8905771

One can obtain a function f from ψ theory(4) given a specific geometry and boundary conditions.

To test finite-size scaling accurately and systematically, the preparation of an experimental sample with definite confinement spacing is essential. In this experiment we use cells made of two silicon wafers. The construction and diagnostics of these cells have been described(5). The bonded silicon cell is attached to a Be-Cu high-Q torsional oscillator which is mounted on a temperature regulated plate. The helium sample is condensed into the cell through a hole in the torsional elements. The superfluid fraction of helium confined in 4 different cells is shown in Fig 1.

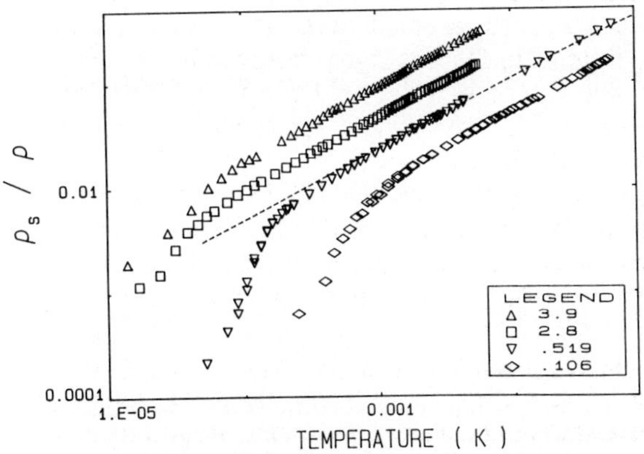

Fig 1. The superfluid fraction for helium confined in a planar geometry at the small dimensions indicated. Data are offset from one another by a factor of 2.

From the fractional period change of the oscillator, one can deduce the change in superfluid fraction(6). The bulk value of this, $\frac{\rho_{sb}}{\rho}$, follows the power law(7) kt^ν near the transition temperature, where the asymptotic value of k is 2.403(8). This bulk behavior for the 0.519μm cell is shown by a dashed line in this figure.

To avoid overlap in the power law region, the data are offset by factors of 2. In this figure we can see that $\frac{\rho_s}{\rho}$ deviates from the bulk behavior, a power law of $t^{0.672}$, below a certain temperature. This is due to the finite-size effects. Note that this deviation onsets earlier, the smaller the confinement spacing.

Using the power law behavior of $\frac{\rho_{sb}}{\rho}$, Eq. 1 can be rewritten as

$$\frac{\rho_s}{\rho}t^{-0.672} = k(1 - f(L^{\frac{1}{\nu}}t)) \equiv g(L^{\frac{1}{\nu}}t). \quad (2)$$

Therefore, if we plot $\frac{\rho_s}{\rho}t^{-0.672}$ vs $L^{\frac{1}{\nu}}t$, all data should collapse on one universal curve. We have done this in Fig. 2, and we can see that*the data do not collapse.* The finite-size effects, that is, the reductions of the superfluid density fraction, are amplified by plotting the data in this way. Here the bulk behavior for the $0.519\mu m$ cell is drawn as a dashed line.

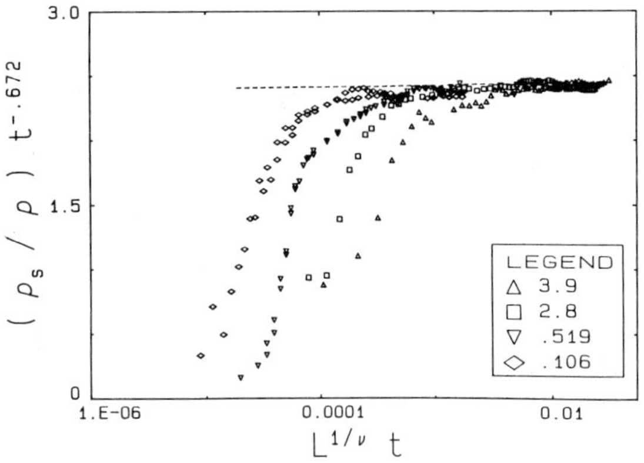

Fig 2. Scaling plot of $\frac{\rho_s}{\rho}$. If correlation-length scaling were valid all data would collapse on a universal curve.

On the assumption that $f(L^{\frac{1}{\nu}}t) \sim \frac{t^{-\nu}}{L}$ to leading order, we have $\frac{\rho_{sb}-\rho_s}{\rho_{sb}}L \sim t^{-0.672}$. Thus, we plot the left hand side of this equation vs. t in Fig 3. We can see that the data do collapse reasonably well on this plot. However, the data do not define a straight line of slope -0.672. We have also plotted on this figure calculations of $\frac{\rho_s}{\rho}$ for *each confinement* based on ψ theory with Dirichlet boundary conditions and *no adjustable parameters* and *including the van der Waals interaction* with the solid surfaces. This calculation does not agree with the data, but reveals some interesting features. A slight curvature of the calculated $\frac{\rho_s}{\rho}$ at large t is due to the van der Waals interaction[9].

A slight lack of collapse is also due to this. *The point is, however, that this is barely visible suggesting that van der Waals effects are not important in explaining the lack of correlation length scaling of the experimental data.*

We have also tried to scale the data as follows[10]

$$L\frac{\rho_s}{\rho} = F(L^{\frac{1}{\nu}}t) + u\ln\frac{L}{\Lambda} \quad (3)$$

where Λ and u are free parameters. We find that one can get some degree of collapse with $\Lambda = 5\mu m$ and $u = 8\times10^{-4}\mu m$. However, the collapse is not as good as that shown in Fig 3 and the unphysically large value of Λ (suggested to be of the order of the atomic distance) can not be explained.

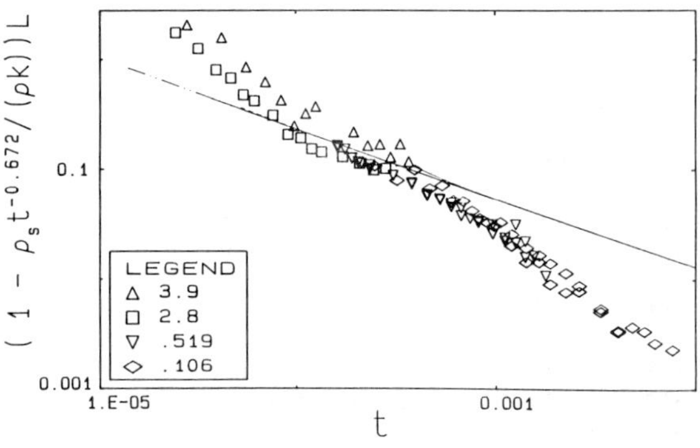

Fig 3. Scaled $\frac{\rho_s}{\rho}$ as function of t. The four, very nearly coincident lines are the calculated behavior from ψ theory *including effects due to the van der Waals field.* Note the slight lack of power-law behavior for the calculation.

(1) M.E. Fisher and M.N. Barber, Phys. Rev. Lett. **28**, 1516(1972)
(2) T.-P. Chen and F.M. Gasparini, Phys. Rev. Lett. **40**, 331(1978)
(3) F.M. Gasparini, G. Agnolet and J.D. Reppy, Phys. Rev. B**29**, 138(1984)
(4) V.L. Ginzburg and A.A. Sobyanin, Sov. Phys. Usp. **19**, 773(1976). [Usp. Fiz. Nauk **120**,153(1976)]
(5) I. Rhee, F.M. Gasparini, A. Petrou and D. Bishop, Rev. Sci. Inst. (1990). see also this volume.
(6) D. J. Bishop and J. Reppy, Phys. Rev. B**22**, 5171(1980)
(7) I. Rhee, F.M. Gasparini and D.J. Bishop, Phys. Rev. Lett. **63**, 410(1989)
(8) A. Singsaas and G. Ahlers, Phys. Rev. B**30**, 5103(1984)
(9) X.F. Wang, I. Rhee and F.M. Gasparini; this volume
10) V. Privman, private communication

Physica B 165&166 (1990) 537–538
North-Holland

SUPERFLUID DENSITY IN METASTABLE ^3He-^4He MIXTURES

J. BODENSOHN* and P. LEIDERER#

*Institut für Physik, Universität Mainz, D-6500 Mainz, FRG
#Fakultät für Physik, Universität Konstanz, D-7750 Konstanz, FRG

We have studied superfluid ^3He-^4He mixtures quenched into nonequilibrium states inside the miscibility gap by means of second sound . From the results for the second sound velocity we conclude that the superfluid density in the metastable state is well described by extrapolation from equilibrium values. The boundary of the metastable region, where nucleation processes set in rapidly, is reflected in a sharp increase of the second sound attenuation.

1. INTRODUCTION

The superfluid density ρ_s of ^4He and of ^3He-^4He mixtures, being one of the essential quantities of the superfluid state, has been studied in great detail. In particular, the behavior near the λ-line and along the coexistence curve near the tricritical point has been investigated with high accuracy (1). Little is known, however, about the superfluid density in the miscibility gap of ^3He-^4He, where the mixture is not thermodynamically stable, but where for some limited time homogeneous metastable states are accessible (2,3). In this work we address the question how ρ_s behaves in this non-equilibrium state as one approaches the so-called cloud line, where spontaneous nucleation of ^3He-rich droplets sets in.

2. EXPERIMENTAL

In order to prepare the mixture in a proper state inside the miscibility gap we used the pressure quench technique with an experimental set-up as described earlier (2). A schematic path in the phase diagram, plotted on a reduced temperature scale, is shown in the insert of Fig.1. The quench starts on the superfluid branch of the coexistence curve (A), and the system remains in a homogeneous, metastable state (B). Upon nucleation, which develops rapidly as the cloud point is reached, local decomposition into ^3He-rich droplets (C^+) in a superfluid ^4He-rich background phase (C^-) occurs within a few milliseconds. Subsequently the system "slides down" along the coexistence curve until the pressure quench is terminated at points D^+ and D^-. There the late stages of decomposition, coarsening and macroscopic phase separation by gravity, are also completed. The typical time scale for the whole decomposition process is a few seconds for ^3He-^4He.

The qualitative behavior of the second sound signals is illustrated in Fig.1, which shows examples of the various conditions (A-D) described above. The pulses were generated by heating a thin metal film and detected with a

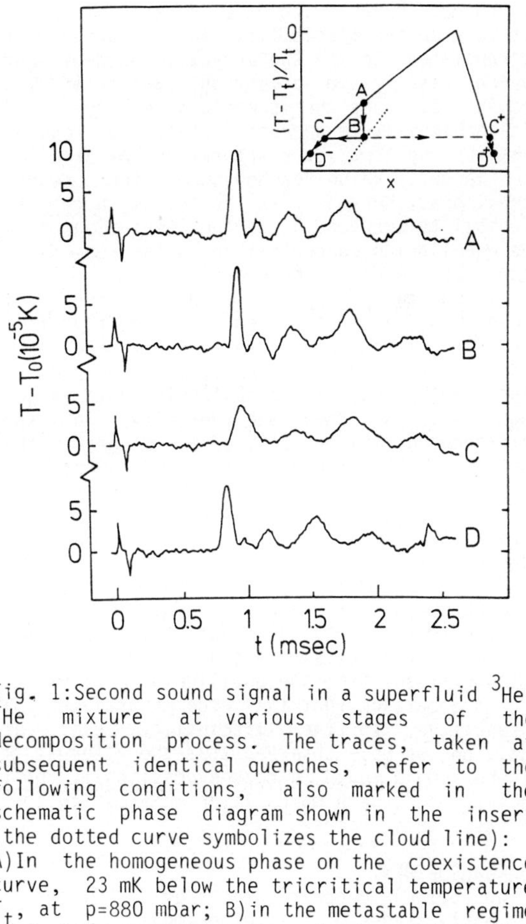

Fig. 1: Second sound signal in a superfluid ^3He-^4He mixture at various stages of the decomposition process. The traces, taken at subsequent identical quenches, refer to the following conditions, also marked in the schematic phase diagram shown in the insert (the dotted curve symbolizes the cloud line): A) In the homogeneous phase on the coexistence curve, 23 mK below the tricritical temperature T_t, at p=880 mbar; B) in the metastable regime just before reaching the cloud point; C) after nucleation, when the system has undergone decomposition on a local scale; D) at the end of the decomposition process, when the superfluid phase is homogeneous again (T_t-T=35mK, p=510mbar)

carbon bolometer at a distance of 0.7 mm. The traces display a sharp rise about 1 msec after the application of the heat pulse, corresponding to the transit time of second sound. The structure after the leading pulse is due to multiple reflections. It is seen that well-defined second sound signals are obtained not only in the superfluid phase on the coexistence curve, but also in the metastable state (B) and in the heterogeneous mixture (C).

3. RESULTS AND DISCUSSION

Here we concentrate on the metastable regime and the onset of nucleation. Data for the velocity and the attenuation of second sound in this region are shown in Fig 2a and 2b. The time axis is related to the quench depth, although in a nonlinear way (2). In the metastable mixture, which here corresponds to the time interval $0<t<0.12$sec, the velocity remains unchanged within our accuracy of about 1%. This is to be compared with data in the vicinity, but outside the miscibility gap (4), which when extrapolated to the thermodynamic path of our quench also yield a constant value of v_{II} within 1%. Thus the superfluid density in a metastable mixture appears to follow a regular behavior up to a supersaturation $\delta x/\Delta x=0.15$, the largest value reached under the present conditions.(Here $\Delta x=x^+-x^-$ is the width of the miscibility gap and δx is the deviation from the equilibrium concentration on the superfluid branch of the coexistence curve.)

The arrow in Fig.2 indicates the onset of nucleation, as determined independently from the optical transmissivity of the sample (2). Although some structure in v_{II} might be present near that point, it is obviously not larger than 1%, so that the velocity in the heterogeneous mixture ($t>0.12$sec) is nearly the same as in the homogeneous case. By contrast, the attenuation of the second sound is strongly affected by the normalfluid droplets developing during the nucleation process, as seen from Fig.2b.

4. CONCLUSIONS

The measurements show that the superfluid density in the metastable state agrees with an extrapolation from the equilibrium values of ρ_s, indicating that this order parameter is insensitive to the miscibility gap. In addition, we have observed that the attenuation of the second sound increases drastically upon nucleation, which makes it an interesting tool to study this process.

ACKNOWLEDGEMENT

We appreciate contributions by S. Klesy in the early stages of this experiment.

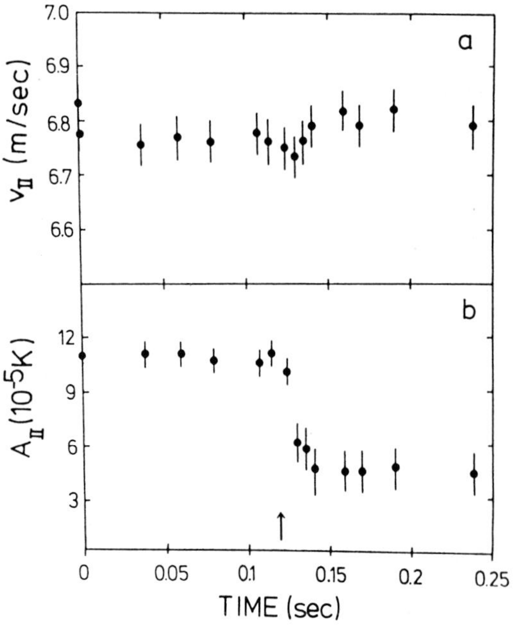

Fig.2 :a) velocity v_{II} and b)amplitude $A=T-T_0$ of a second sound pulse propagating through a ^{3}He-^{4}He mixture at various times after the start of the quench. T_0 is the temperature before the application of the heat pulse. The quench parameters were the same as in Fig. 1, except for the initial and final temperature on the coexistence curve, which was $T_t-T=17$ and 29 mK, respectively. The onset of nucleation is marked by an arrow.

REFERENCES

(1) G. Ahlers, in "The Physics of Liquid and Solid Helium" Part I, ed. by K.H.Bennemann and J.B. Ketterson (Wiley, New York 1976), p.85

(2) P. Alpern, Th. Benda and P. Leiderer, Phys. Rev. Lett. 49 (1982) 1267; J. Bodensohn, S. Klesy and P. Leiderer, Europhys. Lett. 8 (1989) 59

(3) J.K. Hoffer and N. Sinha, Phys. Rev. A33 (1986) 1918

(4) Ref. 1, eq. 2.2.52c

Physica B 165&166 (1990) 539–540
North-Holland

Quantitative Studies of Nonlinear Second Sound Near T_λ

Lori S. Goldner, Guenter Ahlers, and Ravi Mehrotra*

Department of Physics and Center for Nonlinear Science
University of California, Santa Barbara, CA 93106 U.S.A.

Experimental studies of the nonlinear evolution in time of planar second-sound pulses in superfluid ^{4}He have been carried out a few milliKelvin below T_λ. When thermal effects in the second-sound generator and detector are taken into consideration, the data can be quantitatively understood using Burgers' equation to describe sound propagation in the superfluid. A method for extracting the second-sound velocity and damping very near T_λ, where nonlinear effects are always important, is described.

1. INTRODUCTION

Studies of linear second sound in superfluid ^{4}He have often provided useful and straightforward measurements of thermo/hydrodynamic parameters that are of particular interest in the study of critical phenomena. Values of the superfluid fraction have been derived from second-sound velocity measurements, and values for second-sound damping have been used in the study of dynamic critical phenomena. However, for work very near T_λ, sound of even the smallest measurable amplitudes will behave in a nonlinear fashion. This complicates the usual resonance and time-of-flight methods for extracting second-sound parameters, requiring the inclusion of nonlinear terms in the two-fluid hydrodynamics. Measurements of nonlinear sound are therefore motivated by two factors; first a desire to check the predictions of the critical behavior near T_λ, and second to check the predictions of nonlinear two-fluid hydrodynamics. We report here on the results of a study of the weakly nonlinear evolution in time of planar second-sound pulses 2 to 25 mK below T_λ (1), and we discuss a method for extracting second-sound velocity and damping to within 1 μK of T_λ using a similar method.

2. THEORY

It has been shown theoretically (2 – 4) that the normal-fluid velocity v_n in a second-sound planar pulse is given (to lowest nonlinear order, ignoring coupling to first sound) by solutions of Burgers' equation with damping,

$$\frac{\partial v_n}{\partial t} = D_2 \frac{\partial^2 v_n}{\partial x^2} + (u_{20} + \alpha v_n)\frac{\partial v_n}{\partial x} \quad , \qquad (1)$$

where D_2 is the second-sound damping, u_{20} is the velocity of linear second sound, and α is the Khalatnikov nonlinear coefficient (5)

$$\alpha = \frac{\sigma T}{C_p}\frac{\partial}{\partial T}\ln\left(\frac{{u_{20}}^3 C_p}{T}\right). \qquad (2)$$

Here σ and C_p are the entropy and specific heat (at constant pressure) per unit mass, respectively. Near T_λ, $\alpha \sim t^{-1}$ (where $t \equiv (T_\lambda - T)/T_\lambda$), so that the nonlinearities will be apparent even for the smallest amplitudes of v_n if t is small enough. We show that one can account quantitatively for the measured time evolution of the shape of nonlinear planar second-sound pulses propagating in the fluid in terms of solutions of Eq. 1 (6). For this purpose we have undertaken a detailed study of pulses at large enough amplitudes and near enough to T_λ that nonlinear effects can be clearly seen, but far enough from T_λ so that the linear amplitude-range is also experimentally accessible. In the quantitative comparison between experiment and theory we found it necessary to include careful thermal modeling of the pulse generator (heater) and detector (bolometer). Our composite model thus includes heat diffusion in the heater and bolometer substrates, as well as propagation in the fluid as described by Burgers' equation.

3. EXPERIMENTAL TECHNIQUES

The apparatus consisted of two sealed sample cells of different lengths, filled through a capillary with ^{4}He and immersed in a bath of superfluid ^{4}He. The bath was regulated to a temperature stability better than 1 μK (7) using a five-wire germanium thermometer bridge (8) and a 0.3 kΩ heater. Each cell contained a second-sound generator (a large, planar chromium heater, 3.18×3.18 cm^2 and 200 Å thick on a $3.8 \times 3.8 \times 0.32$ cm^3 pyrex substrate) and a small, fast superconducting bolometer (1) (Pb-Au film on the same type of substrate) located on opposing walls. The cells were surrounded by a superconducting magnet, which was used to shift the transition temperatures of the bolometers to the desired range. A pulse launched at the heater travels back and forth in the cell and is detected each time it reflects from the cell wall containing the bolometer. The cells are quite broad and flat ($3.18 \times 3.18 \times 0.120$ cm^3 and $3.18 \times 3.18 \times .362$ cm^3) so that signals from the edges of the heater do not reach the bolometer until the conclusion of the second echo in the long cell or the

*Present address, National Physical Laboratory, Dr. K.S. Krishnan Road, New Delhi-110012, INDIA.

0921-4526/90/$03.50

sixth echo in the short cell. Until this time, planar (one-dimensional) pulses will be detected at the bolometer, greatly simplifying the data analysis.

The bolometers were biased at constant current, so that voltage fluctuations across them were proportional to their temperature fluctuations, $\delta V = i(dR/dT)\delta T$. With some signal averaging, sub-micro-Kelvin fluctuations could be resolved.

For a typical data set, a haversine voltage pulse launched at the heater is detected upon its successive arrivals (echoes) at the bolometer. The first few echoes are amplified and digitized, and after the cell has relaxed back to its equilibrium temperature, another pulse is launched. The procedure is repeated and the signals are averaged, typically 30 times. Accuracy of these measurements was somewhat better than 1% of the peak pulse amplitude.

4. RESULTS

A typical example of measured pulse shapes and fits of the model to them are shown in Figs. 1(a-d). Figure 1(a) shows the data at the bolometer upon the first arrival of the second-sound pulse in the short cell. Using this data, the temperature profile in the helium could be calculated and used as an initial condition for the solution of Burgers' equation. Figures 1(b-d) show the second, fourth and sixth arrivals of the pulse at the bolometer (open circles) and the corresponding result from the model (line). For clarity, not all data are shown. Second-sound damping is taken from Ref. 9. The nonlinear coefficient α is known from Eq. 2 and thermodynamic measurements. The linear second-sound velocity is determined by fitting the solutions of the model to the sixth echo. It is the only adjustable parameter in the model, and is used for our most accurate determination of temperature. There are no systematic differences between second sound velocities measured at different pulse amplitudes. Fits similar to or better than that shown in Fig. 1 are found throughout the region studied.

This study was performed at or near the minimum in the second-sound damping, and thus no fits to determine damping were possible. However closer to T_λ where the damping is larger, the play between the linear damping and nonlinear terms becomes quite noticeable; The diverging nonlinear term causes higher wavenumbers to evolve as a shock front starts to form, but larger wavenumbers are also more strongly attenuated. Thus closer to T_λ we can expect to extract both second-sound velocity and damping from fits to the data.

5. CONCLUSIONS

In summary, we find good agreement of our data with solutions of Burgers' equation in the region studied. The model takes into account the pulse distortion due to thermal effects in the detector, as well as nonlinear effects in the helium. We have demonstrated how this method can be used to measure linear second-sound velocities, and at only slightly smaller reduced temperatures we expect to be able to measure second-sound damping as well. Using this method, it should be possible to extend or improve upon such measurements (9 – 11) close to T_λ.

This work was supported by NSF grants DMR 84-14804 and DMR 89-18393.

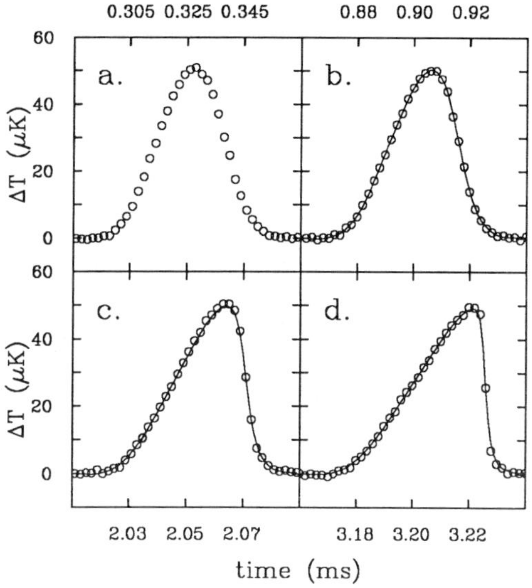

FIG. 1. Pulse evolution at $t = 2 \times 10^{-3}$. The open circles are data, the line is calculated from the model.

REFERENCES

(1) L.S. Goldner, G. Ahlers, and R. Mehrotra, submitted to *Phys. Rev. B*.

(2) S. Kitabatake and Y. Sawada, *J. Phys. Soc. Japan* **45** (1978) 345.

(3) V. Steinberg, private communication.

(4) S.K. Nemirovskii, as quoted in M.O. Lutset, S.K. Nemirovskii, and A.N. Tsoi, *Sov. Phys. JETP* **54** (1981) 127.

(5) I.M. Khalatnikov, *An Introduction to the Theory of Superfluidity* (Benjamin, N.Y., 1965).

(6) Very recently, solutions to this equation were compared with experimental results for the time evolution of the amplitude of cylindrical second sound by W. Fiszdon, Z. Peradzynski and G. Stamm, *Phys. Fluids A* **1**, 881 (1989).

(7) G. Ahlers, in *Phase Transitions*, edited by M. Levy, J-C. Le Guillou, and J. Zinn-Justin (Plenum, NY, 1982), p. 1.

(8) A. Singsaas and G. Ahlers, *Phys. Rev. B* **29** (1984) 4951.

(9) R. Mehrotra and G. Ahlers, *Phys. Rev. B* **30** (1984) 5116; Phys. Rev. Lett. **51** (1983) 2116.

(10) M.J. Crooks and B.J. Robinson, *Physica (Utrecht)* **107B** (1981) 339; *Phys. Rev. B* **27** (1983) 5433.

(11) D. Marek, J.A. Lipa, and D. Philips, *Phys. Rev. B* **38** (1988) 4465.

Physica B 165&166 (1990) 541–542
North-Holland

QUANTUM NUCLEATION AND DISSIPATION IN THE PHASE SEPARATION OF SUPERSATURATED LIQUID ^3He-^4He SOLUTION AT LOW TEMPERATURES

Serguei N. BURMISTROV and Leonid B. DUBOVSKII

I. V. Kurchatov Institute of Atomic Energy, Moscow 123182, USSR

The quantum stratification of supersaturated metastable liquid sol
is considered. The effects of the energy dissipation on the nucleat
rate are incorporated. For weak supersaturations diffusion determi
decay probability in normal solution. In superfluid solution heat
conduction must be also involved.

Kinetics of nucleation in a metastable phase excites constantly a great interest. At low enough temperatures the phase transitions in metastable states represent the possibility of observing quantum phenomena at a macroscopic level. The transition is initiated by the nucleation of the stable phase in the metastable one. At high temperatures the nucleation rate is governed by thermodynamical fluctuations. At low enough temperatures, however, transitions can only occur via a quantum tunneling through the barrier (1).

Much interest has recently been attracted to the kinetics of helium crystallization where the first-order transition exists up to the absolute zero (2,3). The supersaturated liquid or solid ^3He-^4He solution is also one of metastable systems where one can observe the macroscopic tunneling. Here, we consider the nucleation kinetics and stratification of liquid solution at low temperatures. At $T<T_t=0.87K$ the ^3He solubility in liquid ^4He is limited (5). The solutions with molar concentration of ^3He in the region $x_l<x<x_u$ are unstable and decay into the phases with the concentration $x_l(T,P)$ and $x_u(T,P)$. The lower phase is the ^3He solution in the superfluid ^4He. The upper phase is enriched by ^3He and represents the normal solution.

The approach developed for the analysis of the quantum kinetics of phase transitions is based on a transformation to an imaginary time and the derivation of solutions with a finite action (1). As it became clear the dissipative effects play an essential role in the decay probability (5). Thus, the effects of dissipation were incorporated into the calculation of the nucleation rate in a metastable liquid (6). Dissipation leads also to reduction of the decay rate and changes the behavior of decay probability at the temperature when the mean free path of excitations in metastable phase equals the critical nucleus size.

Quantum kinetics of nucleation and stratification of liquid solutions was first considered by Lifshits (7). The analysis was done in the assumption of the absence of any dissipation. As we include the influence of dissipative effects we have at once the different kinetics of stratification of the supersaturated lower and upper phases. The point is that the dissipative processes are drastically different in normal and superfluid liquids.

Two types of dissipative processes occur in the one-component phase as liquid helium. The first is heat conductivity due to the temperature gradients in the various parts of the system. The second is viscosity, which arises if there is the inhomogeneous velocity distribution in a medium. In solutions the presence of the second component leads to an additional mechanism of dissipation. The point is that the formation and growth of a nucleus in a solution is connected with the variation of the concentration distribution realized by two kinds of processes. The first is the pure mechanical mixing when each element together with liquid displaces as a whole one with the constant composition. If viscosity and heat conduction are ignored, such changing of the concentration is thermodynamically reversible. The second type is diffusion when the concentration of each element of solution is equalized by molecular transfer of the components. This transfer is dissipative.

Let the metastable phase be normal liquid solution with density ρ and with ^3He concentration c. The potential energy U of a spherical fluctuation of the stable phase with density ρ' and concentration c', of radius R, is given by $U(R)=4\pi\sigma R^2(1-R/R_c)$, where σ is the interphase surface tension and R_c is the critical size of a nucleus. The critical size is inversely proprtional to the supersaturation and characterizes the deviation from the equilibrium.

The kinetic energy and dissipation power are defined by the velocity and concentration distributions. Velocity and concentration can be obtained from the continuity equations and boundary conditions providing conservation of the mass current

$$div \rho v=0,\quad div(\rho c v+i)=0,$$

$$-\rho' c' \dot{R}=\rho c(v-\dot{R})+i\,\Big|_{r=R},\quad -\rho' \dot{R}=\rho(v-\dot{R})\Big|_{r=R}.$$

Here $i=-\rho D\nabla c$ is the density of diffusion current and D is the diffusion coefficient. The velocity and concentration fields are

$$v(r)=-[(\rho'-\rho)/\rho]R\dot{R}^2/r^2,\quad \text{if } r>R(t),$$

$$v(r)=0,\quad \text{if } r<R(t),$$

$$c(r)=c-(c'-c)\frac{\rho'}{\rho}\frac{R}{D}\frac{R}{r}\dot{R},\quad \text{if } r<R(t),$$

The total kinetic energy is given by the integral over the volume (1)

$$K=M(R)\dot{R}^2/2,\qquad M(R)=4\pi R^3(\rho-\rho')^2/\rho .$$

The dissipation power is also the integral over the volume from the expression for the density dissipation

$$\frac{dE}{dt}=-16\pi\left[\frac{\rho-\rho'}{\rho}\right]^2\eta R\dot{R}^2-4\pi(c'-c)^2\rho'^2\frac{\partial Z/\partial c}{\rho D}R^3\dot{R}^2=-\mu(R)\dot{R}^2 .$$

Here $\eta(T)$ is the viscosity coefficient and Z is one of the chemical potentials of the solution.

The nucleation rate of the metastable phase into the stable one is $\Gamma=exp(-S)$, where S is the effective action

$$S=\int_{-1/2T}^{1/2T}d\tau\left[\frac{M(R)\dot{R}^2}{2}+U(R)+\frac{1}{4\pi}\int_{-1/2T}^{1/2T}d\tau'[\gamma(R_\tau)-\gamma(R_{\tau'})]^2(\pi T)^2/\sin^2[\pi T(\tau-\tau')]\right]$$

evaluated along the extremal trajectory. $\gamma(R)$ is defined by $(\partial\gamma/\partial R)^2=\mu(R)$.

For small supersaturations of solution corresponding to large sizes of a nucleus $R_c \gg R_D(T)$ the diffusion is the main process which limits the nucleation rate. At low enough temperatures $T<T_0$ instead of thermal activation we obtain

$$\Gamma=exp[-4\pi(c'-c)^2\rho'^2\frac{\partial Z/\partial c}{\rho D}R_c^5] .$$

In Knudsen limit (8) when the mean free path $l(T)$ of excitations is greater then the critical size we have a crossover to another dependence on temperature and critical size.

Quantum nucleation in superfluid solution is more complicated. The mass transfer to a nucleus of the stable phase is realized by the normal and superfluid components of velocity $j=\rho_n v_n+\rho_s v_s$. ^{3}He atoms are assumed to participate only in normal motion. To find the spatial distributions of velocity and concentration we use the continuity equations. As a result, we obtain the kinetic energy in the same form but with another effective density

$$M(R)=4\pi\rho_{eff}R^3,\quad \rho_{eff}=\rho_n+(\rho'-\rho_s)^2/\rho_s .$$

Here ρ_n and ρ_s are the normal and superfluid densities, correspondingly.

In calculation of energy dissipation in superfluid liquid we must take into consideration heat conduction besides viscosity and diffusion. The point is that during the growth of a nucleus the nonuniform distribution of ^{3}He concentration causes the variation of temperature. Where the ^{3}He concentration is larger, there temperature is lower. Temperature gradients are propotional to the growth rate of a nucleus and for the dissipation power we have $dE/dt=-\mu(R)\dot{R}^2$. The friction coefficient $\mu(R)$ is defined by the similar expression

$$\mu(R)=16\pi\eta R+4\pi c'^2\rho'^2\frac{\partial Z/\partial c}{\rho D}R^3+$$

$$+4\pi c'^2\rho'^2\frac{\kappa}{\rho T D_{eff}^2}\left[\frac{\partial\mu_4/\partial c}{\partial\mu_4/\partial T}\right]^2R^3 .$$

Here $\mu_4(T,c)$ is the ^{4}He chemical potential in solution, κ is heat conductivity and

$$D_{eff}=D\left[1-\frac{K_T}{T}*\frac{\partial\mu_4/\partial c}{\partial\mu_4/\partial T}\right] .$$

The effective action has the above form. At low enough temperatures $T<T_0$ the decay rate of weak supersaturated superfluid solution is governed by the dissipative processes $\Gamma=exp[-\mu(R_c)R_c^2]$. Diffusion and heat conduction play the main role in decay.

(1) I.M.Lifshits and Yu.Kagan, Sov.Phys. JETP 35 (1972) 206.
(2) D.Lezak et al. Phys.Rev.B 37 (1988) 150
(3) V.L.Tsymbalenko, Pis'ma Zh.Eksp.Teor. Fiz. 50 (1989) 87.
(4) D.O.Edwards, D.F.Brewer et al., Phys. Rev.Lett. 15 (1965) 773.
(5) A.O.Caldeira and A.J.Leggett, Ann. Phys.149 (1983) 374.
(6) S.N.Burmistrov and L.B.Dubovskii, Sov. Phys. JETP 66 (1987) 414.
(7) I.M.Lifshits, V.N.Polesky and V.A.Khokhlov, Zh.Eksp.Teor.Fiz.74 (1978) 268.
(8) S.N.Burmistrov and L.B.Dubovskii, Phys. Lett.A 127 (1988) 79.

Physica B 165&166 (1990) 543–544
North-Holland

CRITICAL THERMAL RESISTANCE OF CONFINED ^{4}HE ABOVE T_λ

D. FRANK and V. DOHM

Institut für Theoretische Physik, Technische Hochschule Aachen, Germany

We present a theoretical prediction for the critical thermal boundary resistance of ^{4}He above the λ line and estimate the onset of finite-size effects. We find serious difficulties in interpreting the existing data for the thermal resistance in confined ^{4}He above T_λ, in contrast to the good agreement between our theory and experimental data below T_λ.

1. INTRODUCTION

In recent experiments /1/ on the thermal conductivity λ_T of ^{4}He above the λ line unexpected deviations from the theoretical (bulk) prediction /2/ have been found for $t=(T-T_\lambda)/T_\lambda < 10^{-6}$. It has been suspected /3/ that a reliable interpretation of these data requires a detailed theory of the thermal boundary resistance and of finite-size effects. Ahlers and Duncan /4/ have indeed found indications that the earlier /1/ are affected by the size of the ^{4}He cell. In the following we present the results of a more systematic theory /5/.

2. THEORY

We consider ^{4}He confined in a cell of thickness L in the presence of a heat current Q which induces a temperature difference ΔT across the cell. We assume that the total thermal resistance

$$R^{tot} = \lim_{Q\to 0} \Delta T/Q = R + 2\,R_K^o \quad (1)$$

can be decomposed into a contribution R, that results solely from the ^{4}He liquid, and a surface contribution R_K^o that depends also on the material of the cell. R_K^o may be different above and below T_λ but we assume that R_K^o does not have a significant critical t dependence. We expect that R can be sketched schematically as in Fig. 1. Not too close to T_λ, R essentially constists of the bulk contribution (dotted line)

$$R_b = \begin{cases} L/\lambda_T & , \; T > T_\lambda \; , \\ 0 & , \; T < T_\lambda \; . \end{cases} \quad (2)$$

In Eq. (5) below we replace the complicated t dependence of λ_T by $\lambda_T \sim t^{-x_\lambda}$ with an effective exponent x_λ. Closer to T_λ, but still for $\xi \ll L$ (shaded region), a surface contribution R_K should set in gradually,

$$R = R_b + 2\,R_K + O((\xi/L)^2) \quad . \quad (3)$$

On the basis of renormalization-group arguments, in the presence of Dirichlet boundary conditions (vanishing order parameter at the solid), R_K can be represented as

$$R_K = A_K^{\pm}\, \xi(t_\pm)/\lambda_T(t_\pm) \quad (4)$$

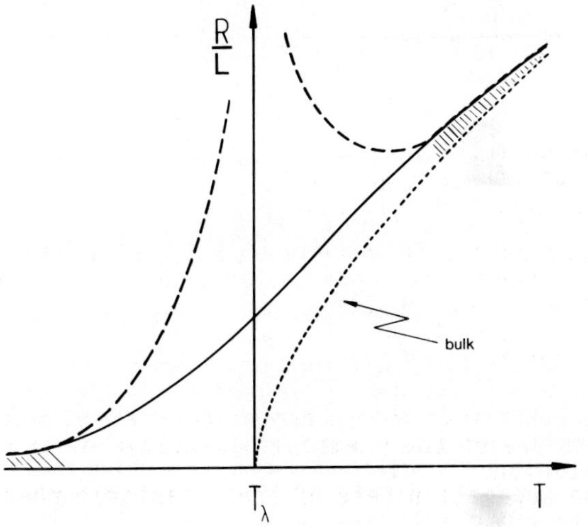

FIGURE 1
Schematic plot of R/L vs T at finite L (solid line). Dashed lines: $(R_b+2R_K)/L$. Dotted line: λ_T^{-1}

where $\xi(t) \sim \xi_o\, t^{-\nu}$ is the bulk correlation length above T_λ, with $t_+ = t$ and $t_- = -2t$ for $t>0$ and $t<0$, respectively. As ξ becomes comparable to L, finite-size effects start to suppress the initial weak divergence (dashed lines) due to R_K and lead to a smooth temperature dependence of R across T_λ (solid line). A rough estimate for the onset of noticable finite-size effects is $\xi \sim O(L/10)$ as indicated by the specific heat /7/. This is consistent with an estimate /5/ of the amplitude of the $(\xi/L)^2$ term in (3). From scaling arguments we obtain right at T_λ

$$R/L \sim L^{-x_\lambda/\nu} \quad (5)$$

Here we focus on the surface behavior (shaded area). Thus it suffices to consider halfspace geometry where $R = R_b + R_K(t)$. On the basis of model F with D.b.c. we have studied $R_K(t) = R_K(t_o) + R_K^s(t)$, $|t_o| \sim O(10^{-1})$. A one-loop calculation yields for the singular part above T_λ

$$R_K^s(t) = \int_{\ell(t)}^{\ell(t_o)} d\ell'\; G(\ell')/\,[\ell'\,\nu(\ell')] \quad , \quad (6)$$

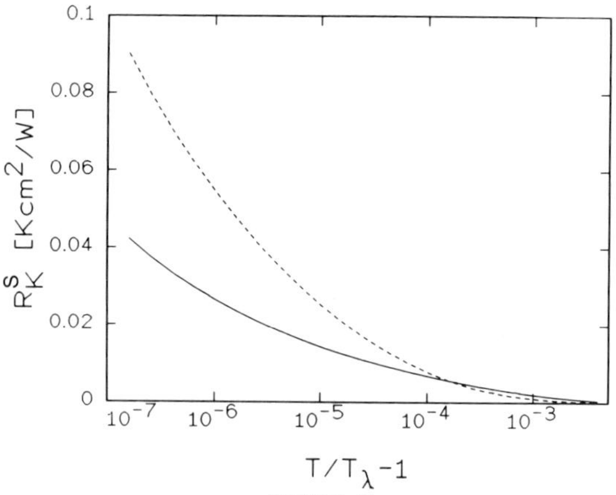

FIGURE 2
R_K^s vs t for P=0.05 bar (solid line) and 28 bar (dashed line) from (6)-(8).

$$G(\ell') = (2\ \lambda_T[\ell']\ell/\xi_o)^{-1}\ A^+(x(\ell'),y(\ell')), \quad (7)$$

$$A^+(x,y) = (6x)^{-1} + \pi x^2/4 - x/2 + (xy/2)\ln[(1+y)/x]. \quad (8)$$

Here $\lambda_T[\ell'(t')] = \lambda_T(t')$ with the flow parameter $\ell(t) = \xi_o/\xi(t)$. $x = |w'/w|$ and $y = |w''/w|$ are known from the model F parameter $w = w' + iw''$ for various pressures. $\nu(\ell')$ denotes the effective correlation length exponent. Fig. 2 shows the prediction of (6)-(8) for P= 0.05 and 28 bar. The major P dependence is due to λ_T. From (4) we obtain the universal amplitude ratio

$$A_K^+/A_K^- \simeq 0.28. \quad (9)$$

3. COMPARISON WITH EXPERIMENTS

Unlike below T_λ, a direct measurement of R_K above T_λ is not possible due to the finite bulk contribution R_b. Here we make a preliminary attempt to interpret the existing data /1,4,6/ in terms of bulk and surface contributions and to compare the resulting R_K^{exp} with our theory. Since in all experiments L is much larger than ξ we assume that (3) is applicable without the $(\xi/L)^2$ correction. Furthermore we assume that $(R_b/L)^{-1}$ is well described by the theoretical result /2/ λ_T^{th} which we distinguish from the measured /1,4,6/ conductivity λ_T^{exp} because the latter contains size effects. Thus we define

$$R_K^{exp} = [(\lambda_T^{exp})^{-1} - (\lambda_T^{th})^{-1}]\ L/2 \quad . \quad (10)$$

In order to test the effective exponent predicted by (4) we have plotted in Fig. 3 the dimensionless quantity $R_K^{exp}\lambda_T/\xi$ as obtained from the data of /1,4,6/. The crosses appear to approach a constant as $T \to T_\lambda$ which would support the exponent of (4). This is not confirmed by the data of Ref. 4. Most serious is the fact that the data sets for different cells in Fig. 3 differ considerably from each other. This is in conflict with the idea that R_K^{exp} should be a property solely of the ^{4}He liquid. The large discrepancy between our result $R_K^s\ \lambda_T/\xi \simeq 0.9$ and the data in Fig. 3 is surprising since our theory agrees well with experiments /8/ below T_λ. Also the ratio between R_K^{exp}, (10), and R_K below T_λ is unusually large. A reexamination of our analysis based on the quantity (10) is surely necessary as soon as better defined data on R_K^{exp} are available which do not suffer from uncertainties related to wall corrections.

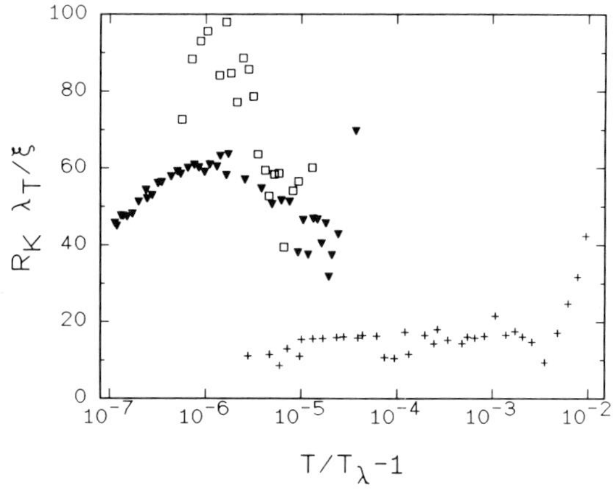

FIGURE 3
$R_K^{exp}\lambda_T/\xi$ vs t for the data of Ref.6, cell I (crosses), Ref.1 (triangles), Ref. 4 (squares).

Finally we comment on the model by Ahlers and Duncan /4/ who argue that the discrepancies between bulk theory /2/ and earlier data /1/ can be attributed to boundary effects. Their model yields $R_K \sim \xi$, in contrast to (4). Although they obtain good results for the onset of deviations from λ_T the departures at smaller t are unacceptably large (as anticipated in Ref. 4).

REFERENCES

/1/ J.A. Lipa and T.C.P. Chui, Phys. Rev. Lett. **58** (1987) 1340; T.C.P. Chui, Q.Li, J.A.Lipa, Jpn. J. Appl. Phys. Suppl. **26-3** (1987) 371; J.A. Lipa, Q. Li, T.C.P. Chui, and D. Marek, Nucl. Phys.**B**(Proc. Suppl.) **5A** (1988) 31.
/2/ V. Dohm, Z. Phys. **B 61** (1985) 193.
/3/ V. Dohm, J. Low Temp. Phys. **69** (1987) 51.
/4/ G. Ahlers and R. Duncan, Phys. Rev. Lett. **61** (1988) 846.
/5/ D. Frank and V. Dohm, Phys. Rev. Lett. **62** (1989) 1864; D. Frank, Ph. D. Dissertation, TH Aachen (1989).
/6/ W.Y. Tam and G. Ahlers, Phys. Rev. **B 32** (1985) 5932; Rev. **B 33** (1986) 183.
/7/ W. Huhn and V. Dohm, Phys. Rev. Lett. **61** (1988) 1368; R. Schmolke, A. Wacker, V. Dohm, D. Frank, paper at this conference.
/8/ R.V. Duncan, G. Ahlers, and V. Steinberg, Phys. Rev. Lett. **58** (1987) 377.

Physica B 165&166 (1990) 545–546
North-Holland

KOSTERLITZ-THOULESS SUPERFLUID TRANSITION IN ^{4}He FILMS ADSORBED ON POROUS GLASSES

K. SHIRAHAMA, M. KUBOTA, and S. OGAWA

Institute for Solid State Physics, University of Tokyo, Roppongi, Minato-ku, Tokyo 106, Japan

N. WADA (a) and T. WATANABE (b)

Department of Physics, Faculty of Science, Hokkaido University, Sapporo 060, Japan

We have systematically studied the size- or frequency- dependent superfluid transition in ^{4}He films adsorbed on porous glasses with pore diameters d's from 50 to 10000A, using two sets of torsional oscillators containing different pore size glasses with a common ^{4}He inlet. It is clarified that the transition temperature T_C is determined by the vortex pair unbinding on length scales comparable to the characteristic lengths, the half of the pore circumference $\pi d/2$ or the vortex diffusion length r_D.

1.INTRODUCTION

Superfluid transition of ^{4}He films confined in porous media is of considerable current interest. The role of the thermally activated vortex pairs on the transition is important on length scales comparable to the pore sizes, because the ^{4}He film is two dimensional (2D) on such length scales.

Study of the pore size dependence of the superfluid transition gives a possible clue for the understanding of the role of the vortices. We have systematically studied the pore size dependence of the superfluid transition in the ^{4}He films adsorbed on porous glasses with pore diameters d's from 50 to 10000A (1). In order to study quantitatively the pore size dependence of the transition, we use two sets of torsional oscillators containing different pore size glasses with a common ^{4}He gas inlet. This setup realizes an equivalent ^{4}He film thickness in both porous glasses. The pore size dependence of the transition temperature T_C reveals that the vortex-pair unbinding on length scales comparable to the half of the pore circumference $\pi d/2$ determines T_C (2).

In the torsional-oscillator experiment for the ^{4}He films in porous material, there is another characteristic length for the determination of T_C, namely the vortex diffusion length $r_D=(14D/\omega)^{1/2}$ (3), where D is the vortex diffusivity and ω is the angular frequency of oscillation. The transition is expected to occur where the 2D phase coherence length $\xi_2{}^+(T)$ becomes equal to the smaller one of $\pi d/2$ or r_D; i.e.

$$\xi_2{}^+(T_C)=\mathrm{Min}(\pi d/2,\ r_D). \qquad (1)$$

At $r_D=\pi d/2$, a crossover from a regime where T_C is determined by $\pi d/2$ to the one where T_C is determined by the oscillating frequency, i.e. r_D, is expected. Ultrasound experiment for the ^{4}He films on Vycor of about 60A pore diameter by Mulders and Beamish (4) suggested the crossover at MHz frequencies for this pore size. Our study of the pore size dependence in Ref.1 suggested the crossover in the region where $d\sim$2900A and the frequency $f\sim$300Hz. In order to confirm the crossover concept, we have studied the frequency dependence of the superfluid transition in ^{4}He films on the porous glass of 2900A pore diameter, using two oscillators of different frequencies with a common ^{4}He inlet. We discuss here the size dependence results together with our new frequency dependence one. For the detailed discussion of the pore size dependence, we refer to Ref.1.

2.EXPERIMENTAL SETUP

In order to study the frequency dependence, we set two 2900A-glass cells with different oscillating frequency, 361.6 and 1094.7Hz. The schematic illustration of the setup is shown in Fig.1 as an inset. We control the frequency by changing the diameter of the BeCu torsion rod and the mass of the sample cell.

3.RESULTS AND DISCUSSION

3-A.Pore Size Dependence

The results of the pore-size dependence study (1) show that T_C decreases as d increases up to 2900A. The d dependence of T_C has exactly the opposite trend to that of the full-pore ^{4}He, in which T_C decreases from the

(a): Present address: Inst. of Phys, College of Arts and Sciences, Univ. of Tokyo, Komaba, Meguro-ku, Tokyo 153, Japan
(b): Present address: Lab. of Phys, Fac. of Engineering, Yamagata Univ, Yonezawa 992, Japan

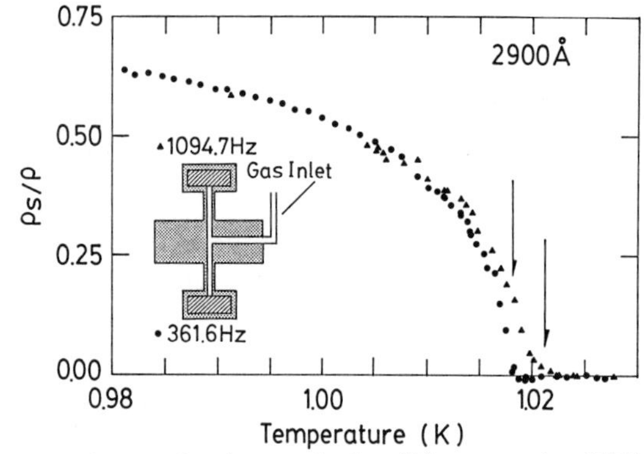

Fig.1 The ρ_s/ρ data of the films on the 2900A glasses for two different frequencies; circles: 361.6Hz; triangles: 1094.7Hz. Vertical arrows indicate T_c's. The inset is a schematic view of our experimental setup.

λ point with decreasing d. In addition, we find an dissipation peak at the transition. The dissipation peaks clearly show that the vortex pairs unbind at the superfluid transition. Minoguchi and Nagaoka (2) theoretically clarified that the vortex-pair unbinding occurs where $\xi_2{}^+(T)$ becomes equal to $\pi d/2$, which acts as a cutoff in the Kosterlitz-Thouless recursion relation (5). T_c is determined by the condition that

$$\pi d/2=\xi_2{}^+(T_c)=a_\theta \exp\left[\frac{2\pi}{b}\left(\frac{T_c}{T_{KT}}-1\right)^{-1/2}\right], \quad (2)$$

where a_θ is the diameter of vortex core, b is a constant, and T_{KT} is the transition temperature in the flat film case.

The pore size dependence of T_c observed for the 50~2900A films is consistent with Eq.(2). In Fig.2, we plot $(T_c/T_{KT}-1)^{-1/2}$ as a function of $\ln\pi d/2$ for T_{KT}/T_c(2900A)=0.975. A linear relation seen for 50~2900A shows a good agreement with the theoretical prediction.

From the fit, we obtain b=7.3±1.4 and the vortex core diameter a_θ=25±12A, which is much larger than that of the bulk ^{4}He.

3-B.Effect of Frequency

In the 2900~10000A range of d, T_c hardly depends on d, as shown in Fig.2. This fact suggests that T_c is determined by r_D in this range. From the fact that T_c(2900A,361.6Hz) is nearly equal to T_c(7000A,295Hz), we estimate r_D~2900π/2~4560A at 295Hz, and hence $D=2.8\times10^{-7}$cm^{2}sec^{-1}.

T_c(10000A, 475Hz) is slightly higher than T_c(7000A, 295Hz). We interpret this T_c difference as the effect of frequency. Assuming $D=2.8\times10^{-7}$cm^{2}sec^{-1}, we estimate r_D~3600A at 475Hz. We should use r_D instead of $\pi d/2$ for the 10000A data in the fit to Eq.(2). In Fig.2, we show $(T_c/T_{KT}-1)^{-1/2}$ for the 10000A data as a function of r_D. The data are consistent with the fit.

In Fig.1, we show the superfluid fraction ρ_s/ρ of the 2900A films with different frequencies. T_c increases as the frequency increases. The frequency effect is clearly seen. As for the discussion of the 10000A data, we estimate r_D(1094.7Hz)~2400A by assuming $D=2.8\times10^{-7}$cm^{2}sec^{-1}. In Fig.2, we plot $(T_c/T_{KT}-1)^{-1/2}$ from Fig.1, against $\ln r_D$. The data agree with the fit to Eq.(1) again.

The results of the above analyses support our picture of the crossover. Further experiment of the frequency dependence for the 2900A films is in progress.

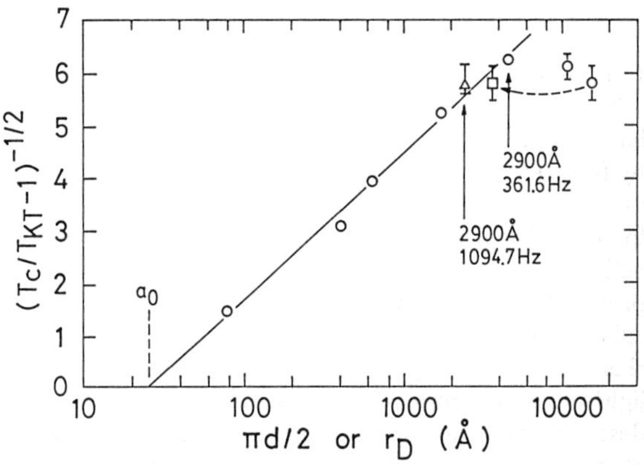

Fig.2 Circles: $(T_c/T_{KT}-1)^{-1/2}$ against $\ln\pi d/2$. The straight line for 50~2900A is a fit to Eq.(2) with T_{KT}/T_c(2900A)=0.975. Square: $(T_c/T_{KT}-1)^{-1/2}$ of the 10000A films at 475Hz as a function of r_D. Triangle: $(T_c/T_{KT}-1)^{-1/2}$ of the 2900A films at 1094.7Hz against $\ln r_D$.

4. SUMMARY

In the ^{4}He films on porous glasses, the characteristic length scale, $\pi d/2$ or r_D, determines T_c. The preliminary study of the frequency dependence for the 2900A films has assisted our suggestion of the crossover.

REFERENCES

(1)K. Shirahama et al, to be published in Phys. Rev. Lett. **64** (1989); K. Shirahama et al, in *Quantum Fluids and Solids-1989*, edited by G. G. Ihas and Y. Takano, AIP Conference Proceedings 194, p.195.
(2)T. Minoguchi and Y. Nagaoka, Prog. Theor. Phys. **80**, 397 (1988)
(3)V. Ambegaokar et al, Phys. Rev. B21, 1806 (1980); V. Ambegaokar and S. Teitel, Phys. Rev. B19, 1667 (1980)
(4)N. Mulders and J. R. Beamish, Phys. Rev. Lett. **62**, 438 (1989)
(5)V. Kotsubo and G. A. Williams, Phys. Rev. B33, 6106 (1986)

Physica B 165&166 (1990) 547–548
North-Holland

SIMULTANEOUS MEASUREMENT OF THE HEAT CAPACITY AND SUPERFLUID DENSITY OF ^{4}He FILMS IN VYCOR GLASS

S.Q. MURPHY AND J.D. REPPY

Laboratory of Atomic and Solid State Physics, and the Material Sciences Center, Clark Hall, Cornell University, Ithaca, New York, 14853-2501, U.S.A.

A simultaneous measurement of the heat capacity and superfluid density of thin ^{4}He films on Vycor glass has been made. We have conclusively established the 3D nature of the critical region by the observation of a heat capacity peak coincident with T_c of the film. In addition, we find reasonable agreement when two scale factor universality is used to compare the critical region of the film transition to that of the bulk lambda transition.

1. BACKGROUND

The nature of the superfluid transition in Vycor glass has been a subject of debate since the first experiments by Kiewiet *et al.* (1) on this system. Those fourth sound experiments showed that the superfluid exponent, ν, for the filled Vycor cell was the *same* as that for bulk helium, 2/3, with the transition temperature, T_c, suppressed by 217mK from that of T_λ. Further torsional oscillator studies by Berthold *et al.* (2), demonstrated that this bulk like exponent persisted even in thin films. In light of the Kosterlitz-Thouless theory (KT) (3) which describes an ordered state in 2D and its experimental observation by Bishop (4) and Rudnick (5) in 2D superfluidity on flat substrates, the demonstration of thin film superfluidity with bulk like exponents was surprising. It has been proposed that the combination of the Vycor surface with a 3D connectivity and the correlation length, ξ_0, which exceeds the diameter of the Vycor pore is responsible for the 3D nature of the transition (2), however, the observation of 3D behavior in other thermodynamic variables would be most encouraging; heat capacity is an obvious candidate.

Brewer *et al.* (6) and Joseph *et al.* (7) looked for and failed to see any singular behavior of helium in filled pore Vycor. (Theoretical predictions from two scale factor universality suggest that the singular part of C_v is extremely small in filled pore Vycor). Several years later, Tait *et al.* (8) examined the heat capacity of films of helium in Vycor. His data showed a change in slope for C_v/T *vs.* T where the kink was believed to lie at the T_c of the film. Most recently, high resolution measurements of thin film heat capacity were performed by Finotello *et al.* (9). For the first time peaks indicating singular behavior were seen in the heat capacity in the vicinity of the expected T_c. If the transition was truly 3D, one would expect the heat capacity peak and T_c to be coincident, however if the transition was of a 2D KT type any heat capacity peak should coincide with the development of short range order which occurs above T_c. The importance of determining this coincidence motivated our experiment.

2. RESULTS

The measurements were made in a combination pulsed heat capacity cell-torsional oscillator (10). In Fig.1 we present our results for a film with a T_c of 163.5mK; the addendum C_v due to the Vycor, heater and thermometers has been subtracted. The small peak in C_v located where ρ_s goes to zero is more easily seen in the expanded plot of C_v/T in Fig.2. We have also observed this coincidence for a film with a T_c of 49mK. This figure is evidence of the coincidence of the C_v peak and T_c to within 1mK. Measurements to lower temperatures proved impossible due to a large addendum C_v signal from the Evanohm heater (11). We intend to continue these experiments to lower temperatures by using Pt-W heaters (12) and thinner samples to reduce the background C_v and the internal time constant at low temperatures.

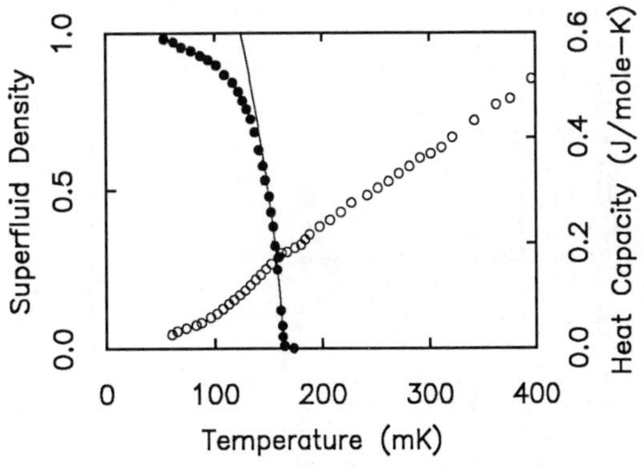

Figure 1: Superfluid density (filled circles) and heat capacity (open circles) for a film with T_c = 163.5mK.

Having established the 3D behavior of the critical region, we can now use two scale factor universality

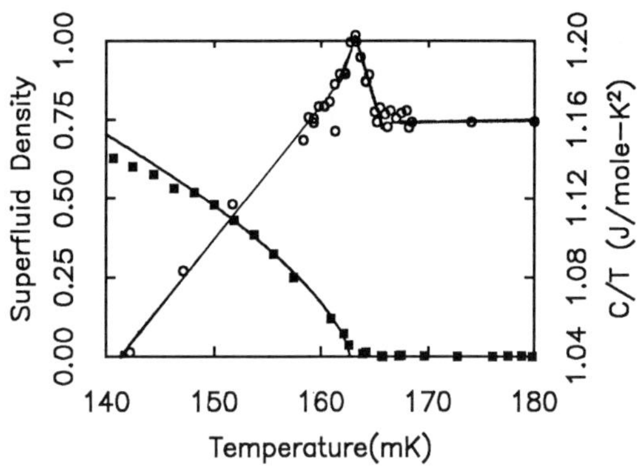

Figure 2: Superfluid density (filled squares) and heat capacity/temperature (open circles) for the film with $T_c = 163.5$mK.

(TSFU) (13) to compare the critical behavior of this system with that of the bulk lambda transition (14). TSFU states that the free energy per unit ξ_o^3 of volume is equal for systems in the same universality class. Thus the singular part of C_v for the film and bulk lambda transition are related via $C_{vfilm}/C_{vbulk} = [\xi_{obulk}/\xi_{ofilm}]^3$. Although we measure C_v, we cannot directly measure ξ_o and must extract it from ρ_s via $1/\xi_o = \hbar^2\rho_{so}/m^2k_BT_c$. For the film with T_c=163.5mK, ξ_{ofilm}=135Å compared to ξ_{obulk}, 3.6Å for the bulk; the factor to scale the bulk lambda peak is then $[3.6/135]^3 = 1.9\text{x}10^{-5}$. We compare the scaled singular part bulk lambda peak and the singular part of C_v for our 163.5mK transition in Fig.3. The bulk lambda peak has been smoothed by the appropriate factor to mimic the smoothing due to the nonzero temperature jump in the heat capacity measurement (15). The comparison is quite good considering the factor of 600 which separates the densities of these two systems.

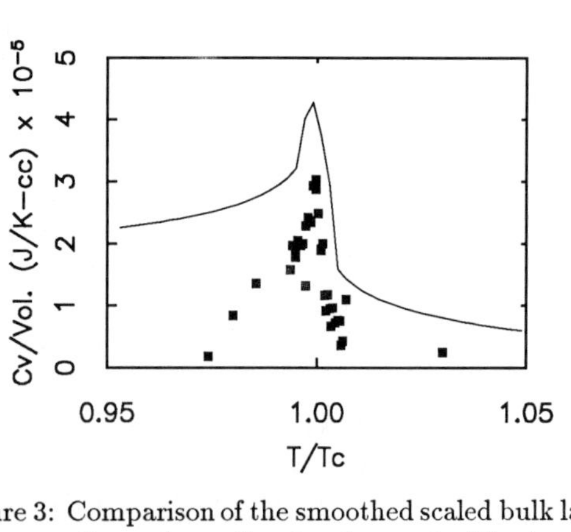

Figure 3: Comparison of the smoothed scaled bulk lambda heat capacity peak vs. the singular part of the heat capacity for a film with T_c=163.5mK.

3. CONCLUSIONS

In summary, we report the observation of heat capacity peaks coincident with T_c's of thin films of helium on Vycor glass. This observation conclusively shows that the nature of the critical behavior is 3D for films with T_c's in excess of 50mK. In addition, the comparison of the singular part of C_v for the film and the scaled singular part of C_v for the bulk transition are consistent with TSFU. None of these conclusions contradict the recent theoretical studies of vortices on complicated interconnected structures (16–18). Although, these theories invoke 2D mechanisms at low temperatures, at higher temperatures a 3D interaction of 'pore vortices' is responsible for 3D critical behavior, however details such as the size of the critical region and the effect of disorder have yet to be established.

ACKNOWLEDGMENTS

We wish to thank M.H.W. Chan, D. Finotello and K.A. Gillis for many informative discussions. This research is supported by the National Science Foundation through Grant No. DMR-8418605 and by the Materials Science Center through Grant No. DMR-8818558.

REFERENCES

(1) C.W. Kiewiet *et al.*, Phys. Rev. Lett. **35**, 1286 (1975).
(2) J.E. Berthold *et al.*, Phys. Rev. Lett. **39**, 348 (1977).
(3) J.M. Kosterlitz *et al.*, J. Phys. C**6**, 1181 (1973).
(4) D.J. Bishop *et al.*, Phys. Rev. B **22**, 5171 (1978).
(5) I. Rudnick, Phys. Rev. Lett. **40**, 1454 (1978).
(6) D.F. Brewer, J. Low Temp. Phys. **3**, 205 (1970).
(7) R.A. Joseph *et al.*, J. Phys. (Paris), Colloq. **39**, C6-310 (1978).
(8) R.H. Tait *et al.*, Phys. Rev. B **20**, 997 (1979).
(9) D. Finotello *et al.*, Phys. Rev. Lett. **61**, 1954 (1988).
(10) To be described at a later date.
(11) R.B. Stephens, Cryogenics **15**, 481 (1975).
(12) Available from Sigmund Cohn Corp., 121 South Columbus Ave., Mount Vernon, NY 10553; J.C. Ho *et al.*, Rev. Sci. Inst. **36**, 1382 (1965).
(13) P.C. Hohenberg *et al.*, Phys. Rev. B **13**, 2986 (1976).
(14) M.H.W. Chan, Superfluid Transition of ^{4}He Confined in Porous Glasses, in: Quantum Fluids and Solids-1989, AIP Conference Proceedings 194, eds. G.G. Ihas and Y. Takano (American Institute of Physics, New York, 1989), pp. 170-178.
(15) The nonzero temperature jump generated by the heating pulse in the C_v measurement effectively averages C_v over the width of the temperature jump. I have averaged C_v of the bulk over 1mK/163.5mK x 2.172K.
(16) T.Minoguchi *et al.*, Jpn. J. App. Phys., **26**, Suppl. 26-3, 327 (1987); Prog. Theor. Phys. **80**, 397 (1988).
(17) J. Machta *et al.*, Phys. Rev. Lett. **60**, 2054 (1988).
(18) F. Gallet *et al.*, Phys. Rev. B **39**, 4673 (1989).

Physica B 165&166 (1990) 549–550
North-Holland

MEASUREMENT OF THE SUPERFLUID DENSITY IN SILICA AEROGELS

P.A. CROWELL, G.K.S. WONG and J.D. REPPY

Laboratory of Atomic and Solid State Physics and the Material Science Center, Cornell University, Ithaca, New York 14853, U.S.A.

We have conducted torsional oscillator measurements of the superfluid density in ^{4}He-filled aerogel where the aerogel density is 0.20g/cc. The superfluid density ρ_s is observed to obey a power-law $\rho_s=\rho_{so}t^{\zeta}$ over a reduced temperature range $10^{-4.3}<t<10^{-1.8}$ with rms deviations of less than 2 percent. The exponent ζ is 0.812±0.004. This value is remarkably close to the exponent, 0.813±0.009, reported earlier for an aerogel sample of density 0.13g/cc.

There have been several recent experimental investigations of the effect of quenched disorder on the superfluid transition of ^{4}He[1,2]. In particular, the superfluid density in the presence of quenched disorder has been observed to obey a power law $\rho_s=\rho_{so}t^{\zeta}$ over some range of t, where t=1-T/Tc. The value of the exponent ζ depends on the nature of the disorder and is not necessarily equal to the bulk value of 0.674±0.001. Silica aerogels[3] comprise one of the classes of materials which have been used in these experiments. Small-angle neutron scattering measurements have shown that the structure of aerogels is self-similar over a range of length scales from several hundred angstroms down to molecular dimensions[4]. The effect of this structure on the critical behaviour of the ^{4}He-filled aerogel system is not understood; however, measurements of critical phenomena probe the system over a wide range of length scales. For example, the ^{4}He-filled aerogel discussed in Reference 1 exhibits power-law behaviour over a range of reduced temperature corresponding to correlation lengths 12nm<ξ<300nm. These lengths overlap the range over which a self-similar structure has been observed in aerogels[4].

We report here the results of torsional oscillator measurements of the superfluid density in ^{4}He-filled aerogel[5] where the aerogel density is 0.20g/cc. This is significantly larger than the density (0.14g/cc) of the sample used in the measurements reported in Reference 1. The sample used in the current experiment was sealed with epoxy into a cylindrical cell of inner diameter 5.64mm and length 5.12mm. The cell assembly was mounted on a heat-treated BeCu torsion rod. The resonant frequency of this torsional oscillator was 294 hz at a temperature of 2K.

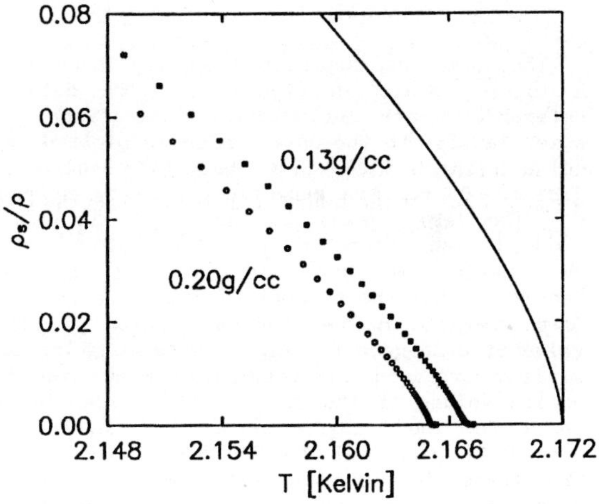

Figure 1

The measured superfluid density is shown in Figure 1 along with the results of Reference 1 for the less dense aerogel. The superfluid density of bulk helium[6] is indicated as a solid curve. The superfluid fraction ρ_s/ρ has been normalized to unity at T=0. In addition, a small contribution to the superfluid density from bulk liquid in the cell has been subtracted from the data. The presence of bulk liquid is believed to be due to gaps at the edge of the cell and/or large voids in the aerogel itself. The shift in the critical temperature, T-T$_c$, is 7.0 mk for the 0.20g/cc sample as opposed to 5.2mk for the 0.13g/cc sample.

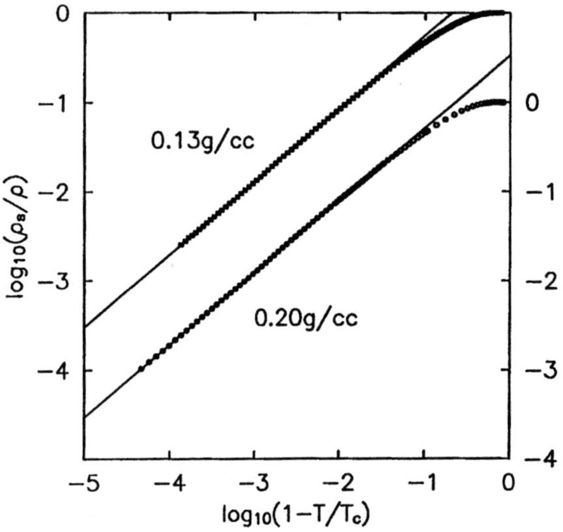

Figure 2

The normalized superfluid density is shown on a log-log scale in Figure 2. The data of Reference 1 are included for comparison. A power-law fit to the data, which is plotted as a solid line, yields a best-fit value of 0.812±0.004 for the superfluid density exponent ζ. The data conform to the power law over a reduced temperature range $10^{-4.3}<t<10^{-1.8}$ with rms deviations of less than 2 percent. Requiring that the superfluid density assume the form $\rho_s=\rho_{so}t^{\zeta}$ in the critical region yields a value of 0.369g/cc for ρ_{so}. In calculating ρ_{so}, we have averaged the superfluid mass over the entire volume of the cell[7]. The values of ρ_{so} and ζ for the 0.13g/cc aerogel of Reference 1 were 0.41 g/cc and 0.813±0.009, respectively. The transition in the more dense sample is somewhat sharper as the power-law behaviour extends to lower reduced temperatures.

The most striking feature of the results shown in Figure 2 is the close agreement between the critical exponents for the two different aerogels. This may be coincidental, but it suggests that measurements on aerogels over a wide range of densities should be made in order to investigate how the critical behaviour of ^{4}He-filled aerogel depends on the aerogel density. Such a study would be particularly relevant because the density-dependence of the structure of aerogels has been investigated experimentally[4].

The work reported here was supported by the National Science Foundation through grant **NSF-DMR84-18605** and by the Cornell Material Science Center through grant **NSF-DMR85-16616-A02**, MSC Report 6901.

P.A.C. acknowledges the support of an AT&T Bell Laboratories Ph.D Scholarship. G.K.S.W. acknowledges the support of an A.D. White Fellowship from Cornell University and a 1967 Postgraduates Fellowship from the Natural Sciences and Engineering Research Council of Canada.

References

1. M.H.W. Chan, K.I. Blum, S.Q. Murphy, G.K.S. Wong and J.D. Reppy, Phys. Rev. Lett. **61**, 1950 (1988); the scale for ρ_{so} in Table 2 should read 10^{-2}g/cc.
2. D. Finotello, K.A. Gillis, A. Wong, and M.H.W. Chan, Phys. Rev. Lett. **61**, 1954 (1988); G.K.S. Wong, P.A. Crowell, and J.D. Reppy, to be published.
3. G. Poelz and R. Riethmuller, Nucl. Instr. Methods **195**, 491 (1982); Aerogels, ed. by J. Fricke (Springer-Verlag, Berlin, 1986); J. Fricke, Sci. Am. **258**, 92 (May, 1988).
4. R. Vacher, T. Woignier, and J. Pelous, Phys. Rev. B **37**, 6500 (1988).
5. Airglass A.B. Attn: Sten Henning, S-24506 Stoffonstorp Sweden
6. D.S. Greywall and G. Ahlers, Phys. Rev. A **7**, 2145 (1973).
7. The motivation for taking such an average is the fact that the correlation length near the transition is larger than any characteristic pore size.

Physica B 165&166 (1990) 551–552
North-Holland

THE MELTING PROPERTIES OF HELIUM ON TWO DIFFERENT SURFACES

D F BREWER, J RAJENDRA, N SHARMA, A L THOMSON and JIN XIN*

Physics Division, University of Sussex, Brighton, Sussex BN1 9QH, Great Britain

Solid helium can apparently grow on certain surfaces while on others it is inhibited and solidification takes place away from the adsorbent walls. In this paper we present experimental data on helium inside grafoil, on which surface solid He-4 appears to be able to grow, and also inside the pores of Vycor glass in which evidence suggests that the solid phase forms in the centre of the pores. We present data on both the specific heat and the pressure-temperature relation for these two situations. Melting takes place at pressures below those of the bulk melting curve in the case of grafoil, and above in the case of Vycor. An analysis is made of the measured heat capacities below the melting anomalies, and their temperature dependence indicates that inside both grafoil and Vycor a possible explanation is that the main excitations present are rotons with an energy gap of about 6K to 7K. This would indicate the presence of a liquid phase down to our lowest temperatures, and this observation is backed up by the fact that the entropy under our latent heat anomalies is less than the value expected for bulk helium of the same mass.

The manner in which helium solidifies near an adsorbent wall appears to depend upon the nature of the wall. On most, there seems to be a preference for the liquid phase to be next to the adsorbed layers. On these surfaces solid helium forms 'off-the-wall' and hence inside porous media made of such materials the freezing (and melting) curves are shifted (1) (2) to lower temperatures. On the other hand on a few surfaces, such as graphite, solid helium grows 'at-the-wall'; and this has the consequence that the solid-liquid transition inside a porous material made of graphite tends to higher temperatures and lower pressures (3).

In this paper we present data on both the specific heat and the pressure, measured simultaneously, of helium confined at high pressures inside two materials which exhibit the contrasting behaviour just outlined. One of these is Vycor glass where the melting curve shifts to lower temperatures and higher pressures while the other is grafoil where the melting curve shifts to lower pressures. A principal aim of this work is to determine the nature of the confined helium and we present evidence to show that rotons may be the principal excitations which remain in these systems at the lower temperatures.

In the upper part of Figure 1, the pressure data taken during a specific heat run of our helium sample compressed inside grafoil is displayed. The plot shows the deviation of the pressure P from the pressure of the bulk melting curve P_m, plotted as a function of temperature. Near 2.5 K there is a small temperature region where the pressure deviations are zero and this is presumed to be where the melting of bulk helium surrounding the grafoil takes place. At temperatures immediately above this region the deviations are

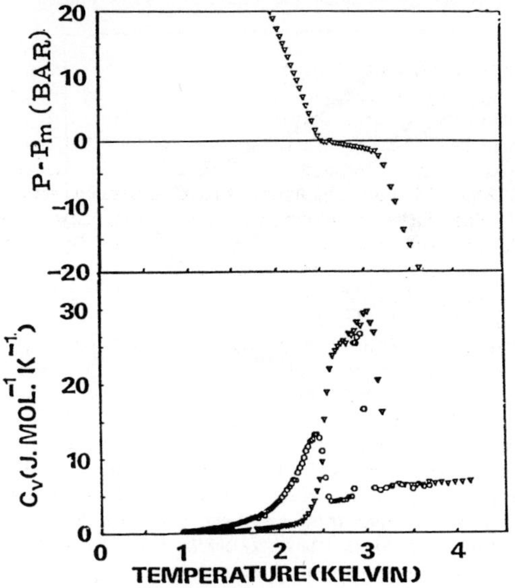

FIGURE 1

The upper part of this figure shows the deviations of pressure from the bulk melting curve recorded while the specific heat of helium compressed inside grafoil was being measured and this specific heat is displayed in the lower part of the figure by the triangular points. The open circle points in the lower part of the figure show the specific heat of helium compressed inside Vycor porous glass.

negative and increase in magnitude until the temperature 3.2 K is reached. These pressures, which are slightly below the bulk melting curve, are where the solidification 'at-the-wall' appears to take place. A prediction for these lowered pressures was made

---

* JIN XIN has now returned to Nanjing University, China.

0921-4526/90/$03.50

some time ago by Franchetti (4) in terms of the comparatively large van der Waals force exerted by an absorbent wall on adjacent helium and this force can be regarded as providing an equivalent extra pressure.

In the lower part of Figure 1 are plotted the specific heat of helium in grafoil, and in Vycor. The data on the grafoil were taken at the same time as the pressure variations discussed above, and show one large quite broad peak which is due to the latent heat of melting both outside and inside the grafoil, with no discernible drop in heat capacity separating the temperature regions where these take place. The Vycor data, on the other hand, show two very distinct peaks, with the broader lower temperature peak corresponding to the melting of helium inside the Vycor pores and the narrower peak at higher temperatures being caused by the melting of the helium surrounding the Vycor. Although not shown in this paper similar previous experiments (2) have shown that the melting inside the Vycor takes place at higher pressures than those of the bulk melting curve, which can be understood in terms of the curvature of the surface of solid helium in the centre of the pores, which alters the free energy of the system via the interfacial energy between the liquid and solid phases.

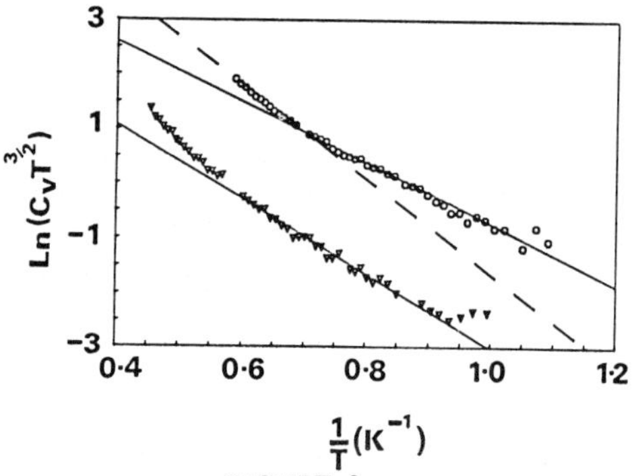

FIGURE 2

The open circle points refer to the measured specific heat of helium confined inside Vycor glass and the triangular points to the specific heat of helium inside grafoil. The straight lines represent best fits to these points, referred to in the text, while the dashed line indicates the specific heat of bulk helium.

Apart from the different character of the latent heat peaks in grafoil and in Vycor, another striking difference is the much larger specific heat in Vycor at low temperatures below the melting region. This may indicate a fundamental difference between the state of the helium in the two systems. To investigate this we have checked the fit of our data to a roton-type specific heat given by (5)

$$C_v = T^{3/2} e^{-\Delta/kT} \qquad (1)$$

in the region where $kT \ll \Delta$. We have therefore re-plotted the specific heat in the low temperature region as $\ln C_v T^{3/2}$ against T^{-1}, shown in Figure 2.

The upper set of points are for Vycor data while the lower ones are for those of grafoil. Both sets of points follow a reasonably good straight line indicating that the roton model may be an appropriate one. Furthermore, the slopes of these two lines are fairly similar with $\Delta \sim 5.6$ K for Vycor and $\Delta \sim 6.8$ K for the grafoil data. It would seem reasonable to infer that in both cases there remains at low temperatures a part of the sample in a superfluid state, as also implied by torsional oscillator measurements (6).

The difference in magnitude between the specific heats in the two different experiments is presumably due to the fact that in the much larger voids inside grafoil a larger proportion of the sample is solid at low temperatures. The solid phase has a comparatively small specific heat and therefore should contribute very little to the heat capacity. In fact the factor of five or so between the magnitudes of the specific heats appears to correspond approximately to the ratio of the void sizes in the two situations. Our Vycor was found from nitrogen adsorption measurements to have pores of 65Å diameter while the grafoil is reckoned to have slab shaped voids of about 400Å in width. Another consequence of the larger surface-to-volume ratio in the Vycor case is that the contribution of the adsorbed layers next to the walls should be more significant. It is interesting to note that if we substract an estimated amount for the T^2 specific heat (7) of the first two layers from the Vycor data the roton gap value is altered to become $\Delta \sim 6.4$ K and this is fairly close to values derived by a similar analysis on experiments (8), (9) carried out in Vycor at lower pressures.

ACKNOWLEDGEMENTS
The authors wish to gratefully acknowledge the support of SERC grant no. GR/F 51845.

REFERENCES
(1) S R Haynes, D F Brewer, N Sharma and A L Thomson, Jap. J. App. Phys. 26, (1987) 301.
(2) E D Adams, Y H Tang, K Uhlig and G E Haas, J. Low Temp. Phys. Vol 66, (1987) 85.
(3) J Landau and W F Saam, Phys. Rev. Lett. 38, (1977) 23.
(4) S Franchetti, NUOVO Cimento 4, (1956) 1504.
(5) J Wilks, 'The Properties of liquid and solid helium' pages 116–119.
(6) D F Brewer, Cao Liezhao, C Girit and J D Reppy, Physica 107B (1981) 583.
(7) D F Brewer, A Evenson and A L Thomson, J. Low Temp. Phys. 3, (1970) 603.
(8) D F Brewer, A J Symond and A L Thomson, Phys. Rev. Lett. 15 (1965) 182.
(9) H Tait amd K D Reppy, Phys. Rev. B 20, (1979) 997.

Physica B 165&166 (1990) 553–554
North-Holland

THE EFFECT OF ^{4}HE ON THE RELAXATION RATES OF THE TWO-LEVEL SYSTEMS IN POROUS VYCOR GLASS

Norbert Mulders* and John R. Beamish

Department of Physics and Astronomy, University of Delaware
Newark, Delaware, De 19716, USA

The sound speed in porous Vycor glass is strongly dependent on the amount of ^{4}He adsorbed in the pores. This is due to the interaction between the Two-Level Systems (TLS) in the glass and excitations in the helium film. As helium is added, the TLS relaxation rates gradually increase, to reach a maximum at approximately 90% of n_{crit}, the critical coverage for the onset of superfluidity. Very close to n_{crit} a local minimum in the relaxation rates is found. A further increase in the coverage causes again an increase in the relaxation rates. For thicker films and filled pores, the relaxation rates become essentially independent of the amount of helium adsorbed. We then find a new behavior in which the relaxation contribution to the sound speed scales with temperature and frequency as $T/\omega^{1/4}$. This indicates that an efficient relaxation channel becomes available for the TLS when there is superfluid in the pores. The rapid change in relaxation rates seen close to the completion of the "inert layer" indicates a change in excitation spectrum in the helium film at the onset of superfluidity.

1. INTRODUCTION

It has been known for some time that the presence of an adsorbed helium layer can have a noticeable effect on the low-temperature properties of amorphous solids (1,2). It is generally assumed that this is due to some interaction between the Two-Level Systems (TLS) in the amorphous material and excitations in the adsorbate. Because of the proximity of the majority of the TLS to the adsorbate, porous Vycor glass, with an average pore size of 6 nm and a surface area of approximately 120 m^2/g, is an ideal substrate to study this interaction.

Ultrasound is a particularly suitable tool to investigate the effect of an adsorbate on the TLS. The changes in sound speed are sufficiently large to be easily detected, and it is possible to do essentially simultaneous measurements over a wide frequency range. This makes it possible to study the frequency dependence of the relaxation contributions to the acoustic properties at fixed temperature and coverage.

In the tunneling model (3) often used to describe the low temperature properties of amorphous materials, the assumption is made that there are a large number of TLS dispersed throughout the solid with an essentially constant density of states and a very wide range of relaxation times. In Vycor, at temperatures below 5 K, there are two important relaxation mechanisms. At sufficiently low temperature only the direct process is of importance, in which a thermal phonon interacts resonantly with a TLS with appropriate energy splitting. At higher temperature a two phonon first order Raman process becomes dominant. Both relaxation processes cause a decrease in the sound speed. In Vycor they are easily distinguished because of their characteristic temperature/frequency dependence. The direct process gives a contribution that scales with T^3/ω while the Raman process gives one that scales with T^7/ω (4). Apart from the relaxation processes there is also a resonant interaction between the sound wave and the TLS. This causes a frequency independent change in the sound speed, $\Delta v/v_o = C \ln(T/T_o)$, with T_o some fiducial temperature and $C = P\gamma^2/\rho v^2$. Here P is the average density of states, γ is the coupling constant between the elastic strain and the TLS, ρ is the mass density and v is the sound speed. The combination of resonant and relaxation processes gives a maximum in the sound speed at a temperature T_{max}. An increase of the relaxation rates causes a shift of T_{max} to lower temperature and T_{max} can therefore be used as an indicator of the change in relaxation rates.

2. RESULTS AND CONCLUSIONS

We have measured the transverse sound speed in porous Vycor as a function of temperature, over a frequency range from 4 to 200 MHz and with helium coverages, n, ranging from zero to completely filled pores at a pressure of 25 atm. We observe that, at temperatures sufficiently below T_{max}, i.e., where only the resonant process is active, C is independent of coverage. This means that neither the interaction strength between the TLS and the sound wave nor the effective TLS density of states changes under the influence of the adsorbed helium, since it is unlikely that they individually change in such a way that their product remains constant. It also has the important

---

*Present address: Dept. of Physics, UCSB
Santa Barbara, California, CA 93106, USA

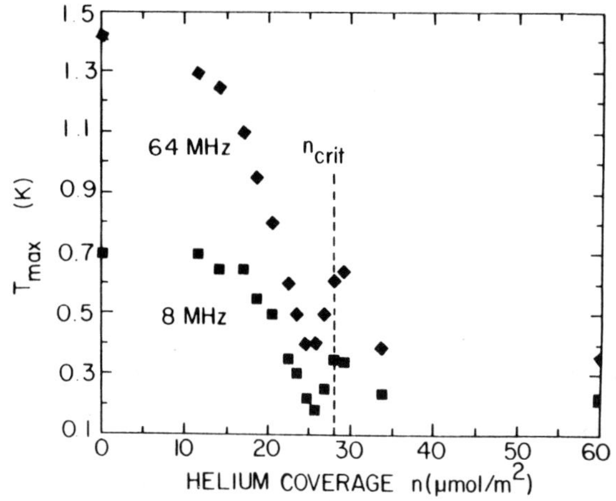

Figure 1 Temperature of the maximum of the transverse sound velocity in Vycor, T_{max}, as a function of the helium coverage, n. Data is shown for two frequencies, 8 MHz (■) and 64 MHz (♦). The isolated points on the right correspond to full pores at svp. The dashed line indicates the coverage for which the superfluid transition temperature extrapolates to zero.

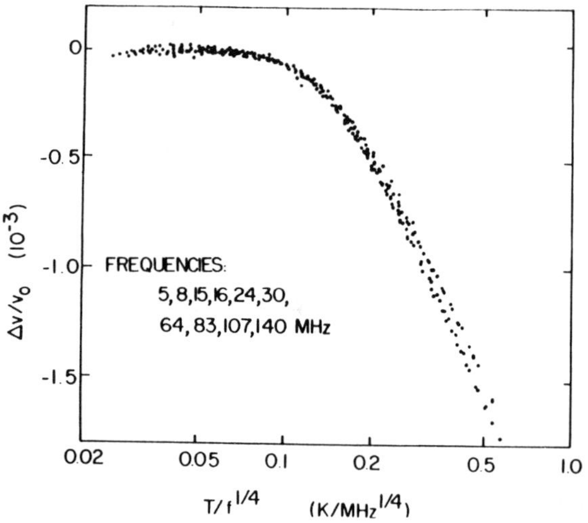

Figure 2 The relaxation contribution to the sound speed in Vycor (after subtracting the resonant contribution) when the pores are filled with helium at saturated vapor pressure. The temperature has been scaled by $f^{1/4}$, where f is the sound frequency, $\omega/2\pi$, in megahertz. The figure contains data at 10 frequencies from 4 to 140 MHz.

consequence that the resonant contribution to the sound speed can be subtracted systematically, which simplifies the study of the relaxation processes considerably.

In empty Vycor we see the expected scaling behavior characteristic of the direct and Raman processes. As helium is added, there is a gradual increase in the relaxation rates as judged from the change in T_{max} (Fig. 1) and this scaling behavior is lost. Approaching the critical coverage for onset of superfluidity, n_{crit}, (i.e., the "inert layer of about 28μmol/m^2, marked with the dashed line on Fig. 1), we find that the relaxation rates decrease again, to reach a local minimum very close to n_{crit}. A further increase in the coverage causes again a gradual increase in relaxation rates, which become essentially constant for $n > 35$ μmol/m^2. At the same time a new scaling behavior is observed. The whole data set, taken over a frequency range from 5 MHz to 140 MHz and from 75mK to 1.6 K collapses onto a single curve when plotted against $T/\omega^{1/4}$. Figure 2 shows this for full pores at saturated vapor pressure. Almost identical results were obtained for thick superfluid films and for helium under pressure.

The initial increase in the relaxation rates when helium is adsorbed may be due to the coupling of the TLS to the helium atoms via the van der Waals interaction (1). The relatively sharp change in relaxation rates close to the critical coverage indicates a change in excitation spectrum in the film, presumably connected with the zero-temperature superfluid onset transition as discussed by Fisher et al. (5). The $T/\omega^{1/4}$ scaling of the relaxation contribution to the sound speed shows that, when there is a significant amount of superfluid in the pores, a well defined and efficient relaxation channel becomes available for the TLS. Fourth sound phonons are a likely candidate for the TLS-adsorbate interaction.

ACKNOWLEDGEMENTS

This work was partially supported by a grant from the Research Corporation.

REFERENCES

(1) H. Schubert, P. Leiderer and H. Kinder, Phys. Rev.B**26** (1982) 2317.
(2) J. R. Beamish, A. Hikata, C. Elbaum, Phys. Rev. Lett. **52** (1984) 1790.
(3) For a review see for example: W. A. Phillips, Rep. Prog. Phys. **50** (1987) 1657.
(4) P. Doussineau, C. Frenois, R.G. Leisure, A. Levelut and J.-Y Prieur, J. Physique **41** (1980) 1193.
(5) M.P.A. Fisher, P.B. Weichmann, G. Grinstein and D.S. Fisher, Phys. Rev. **B40** (1989) 546.

Physica B 165&166 (1990) 555–556
North-Holland

FLUID SLIP AT WALLS WITH MESOSCOPIC SURFACE ROUGHNESS

P. Panzer+, D. Einzel* and M. Liu+

+ Institut für Theoretische Physik, Univ. Hannover, D-3000 Hannover, FRG
* Walther-Meißner-Institut für Tieftemperaturforschung, D-8046 Garching, FRG

The traditional picture of fluid slip at boundaries as a manifestation of a mean free path effect has to be revised whenever the boundary acquires a finite curvature on mesoscopic length scales. Our results explain discrepancies between experimentally observed fluid flow and conventional slip theory.

1. INTRODUCTION

In this Note we consider the problem of hydrodynamic flow of fluids past solid boundaries. Its purpose is to go beyond the traditional description of fluid slip at the container walls as a manifestation of mean free path effects. This is motivated by the results of two torsional oscillator experiments performed in normal fluid ^{3}He at low temperatures. The first of these [1] concludes from Poiseuille flux data, that the experimentally determined slip length ζ (defined as the fictive distance at which the macroscopic fluid velocity extrapolates to zero behind the wall) is (at low pressure) *smaller* than the theoretical result for *diffuse* scattering of quasiparticles. The second finds from surface impedance data [2], that in the presence of a ^{4}He coverage on the cell walls (i. e. enhanced specularity) the ratio of imaginary to real part of the surface impedance $Y/X \approx 0.6$ in the hydrodynamic limit, where one expects 1.

An assumption, common to almost all attempts to calculate slip lengths from a Maxwell- or Landau-Boltzmann equation, is, that in the case of *diffuse* scattering the fluid particles are reemitted with an azimuthally isotropic momentum distribution from a *flat* surface. However, the surfaces of measuring cells are rough on mesoscopic length scales [2] (typically $\leq 1\mu$m). A finite boundary curvature gives rise to additional quasiparticle backscattering events, which enhance the transfer of transverse momentum to the surface with respect to the case of purely diffuse scattering off a flat surface. As a consequence, the experimentally determined effective slip length ζ is reduced below its value ζ_0 for diffusely scattering flat surfaces. In what follows, we discuss this new effect on the level of phenomenological hydrodynamic theory using a simple model for mesoscopic surface roughness.

2. MODEL FOR THE SURFACE ROUGHNESS

We consider now 2-D fluid flow between parallel plates of distance L. The surface roughness is mimiced by the assumption of a symmetric periodic variation of the surface amplitude h(x) in the x-direction with wave number q and periodicity length $\Lambda = 2\pi/q$:

$$h(x) = \sum_{\mu,i} h^i_\mu \cos(\mu q x - i\pi/2) \; ; \quad i = 0,1 \tag{1}$$

Here h^i_μ denote the Fourier components of the amplitude variation. The fluid is then considered to be entrained between the upper wall at $y_U(x) = L + h(x)$ and a lower wall at $y_L(x) = -h(x)$.

A convenient description of 2-D flow between these boundaries is possible in terms of a streaming function $\Psi(x,y)$, which is related to the components $v_x(x,y)$ and $v_y(x,y)$ of the macroscopic velocity field in the usual way as $v_x = \partial\Psi/\partial y$, $v_y = -\partial\Psi/\partial x$.

The streaming function is obtained by solving the linearized Navier-Stokes equations for v_x and v_y, amended, by a generalized set of boundary conditions [3]. They are chosen such that (i) there is no velocity component *normal* to the surface, or equivalently:

$$\Psi(x,y_U(x)) \; ; \; \Psi(x,y_L(x)) = \text{const} \tag{2}$$

and (ii) there is slip, characterized by a (roughness independent) length ζ_0, of the *tangential* component of the velocity field:

$$v_i t_i = \mp \zeta_0 \, t_i \left[\frac{\partial v_i}{\partial x_j} + \frac{\partial v_j}{\partial x_i}\right] n_j \tag{3}$$

Here t_i, n_i, $i = x, y$ are the components of the surface tangent and normal vectors, respectively, and the plus (minus) sign refers to the lower (upper) surface. The slip length ζ_0 has been determined from microscopic theory [4] to be related to the transport mean free path λ through $\zeta_0 = [a_0(1+s)/(1-s)]\lambda$ with a_0 a slip coefficient for the case of diffuse quasiparticle scattering (s=0). The specularity enhancement factor in the brackets accounts for the presence of a fraction s $(0 \leq s \leq 1)$ of specularly reflected fluid particles.

In the case of Poiseuille flow, the streaming function Ψ is related to the total flux Q between the two plates:

0921-4526/90/$03.50

$$Q = \rho[\Psi(x,y_U(x)) - \Psi(x,y_L(x))]$$
$$\equiv -\frac{\rho L^3}{12\eta}\left(\frac{\partial P}{\partial x}\right)\left[1 + 6\,\frac{\zeta(\zeta_0)}{L}\right] \qquad (4)$$

where ρ denotes the fluid density, η the shear viscosity of the fluid, $\partial P/\partial x$ the external linear pressure gradient parallel to the plates. Eq. (4) defines an effective macroscopic slip length ζ which has to be determined as a function of ζ_0. Clearly, if $h(x) = 0$, then we recover the result of conventional slip theory, $\zeta = \zeta_0$.

It can be shown [4,5] that the same form of the macroscopic slip length ζ (to be specified below) also goes into the transverse surface impedance $Z_\perp$ which characterizes the dynamics of the plates, oscillating at a frequency ω in the limit where the viscous penetration depth $\delta = [2\eta/\rho\omega]^{1/2} \ll L$. The hydrodynamic result for $Z_\perp$ is found to be of the well known form:

$$Z_\perp(\omega) = X - iY = \frac{(1-i)\,\eta}{\delta + (1-i)\,\zeta(\zeta_0)} \qquad (5)$$

however, with the macroscopic slip length ζ replacing the microscopic one ζ_0.

We determine the slip length ζ by solving both the stationary and time dependent Navier-Stokes equation with the boundary conditions (2) and (3) and a wall profile of the form (1) by Fourier analysis. The details of this calculation will be published elsewhere [5].

3. DISCUSSION OF RESULTS

The effective slip length ζ is generally found to be characterized by two bounds $\zeta_w^0 = \zeta(\zeta_0 \to 0) < 0$ and $\zeta_w^\infty = \zeta(\zeta_0 \to \infty) < \infty$ which only depend on the wall shape. In the stick limit (small ζ_0) we find $\zeta = \zeta_0 + \zeta_w^0$, whereas in the specular limit (large ζ_0) the inverse lengths add according to $\zeta^{-1} = (\zeta_0)^{-1} + (\zeta_w^\infty)^{-1}$. In case of weak amplitude variation ($qh_n^i \ll 1$) we find for these lengths:

$$\zeta_w^0 = -\sum_{n,i} nq(h_n^i)^2$$
$$\zeta_w^\infty = \Big[\sum_{n,i} (nq)^3 (h_n^i)^2\Big]^{-1} \qquad (6)$$

Note the negativity of the slip length ζ in the stick limit, which reflects the effectively reduced cross sectional area.

At arbitrary ζ_0, ζ is a monotonous function of ζ_0, which smoothly connects the limiting values $\zeta_w^0 \leq \zeta \leq \zeta_w^\infty$. This behavior is illustrated in Fig. 1 for the special case of a cosine profile $h(x) = h_0 \cos(qx)$. We have plotted the exact numerical result for the dimensionless quantity $q\zeta$ in this case vs $q\zeta_0$. The parameter values $\kappa = qh_0$ range from weak to strong variation of the periodic roughness.

We believe that the almost temperature independent ratios $Y/X \approx 0.6$, $X/X_p \approx 0.7$ and $Y/Y_p \approx 0.45$ (where the index p refers to pure ^3He) of real and imaginary parts of the surface impedance observed in the presence of a ^4He coverage by Ritchie, Saunders and Brewer [2] are a consequence of the finiteness of the slip length in the specular limit. The ^4He film serves as to introduce specularity ($s \to 1$). The measured slip length in this specular limit does not diverge like $(1-s)^{-1}$, but is dominated by ζ_w^∞, which, assuming $q^{-1} = 1.8\ \mu m$ and $h_0 = 1\mu m$ is of the order of $6\mu m$. Inserting this into the expression (5) for the impedance, we obtain at 40 mK $Y/X = 0.56$, $X/X_p = 0.86$ and $Y/Y_p = 0.5$ in fair agreement with the experimental observation.

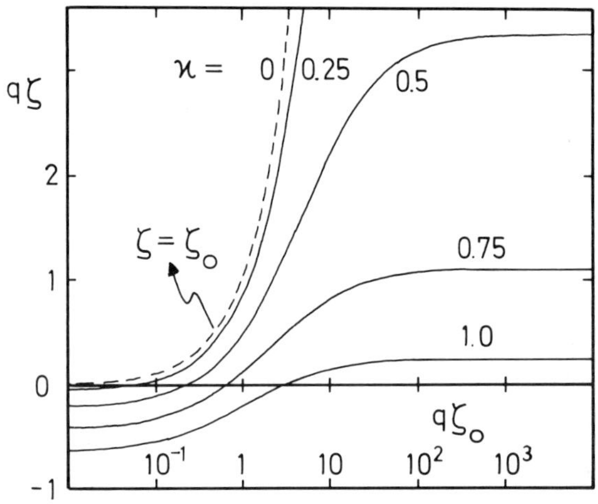

Fig. 1: Effective slip length ζ vs. ζ_0 for a cosine surface profile

In summary, we have demonstrated how a finite boundary curvature alters the flow properties (Poiseuille flow and shear impedance) and in particular the slip length of a fluid. For increasing values of the microscopic slip length ζ_0 the macroscopic slip length ζ varies from negative values in the stick limit to a finite asymptotic value in the specular limit, which is characterized by the details of the wall shape function. We conjecture that this effect is the cause for discrepancies between surface impedance experiments performed with specular walls and conventional slip theory.

REFERENCES

[1] D. Einzel and J. M. Parpia, Phys. Rev. Lett. **58**, 1937 (1987)
[2] D. A. Ritchie, J. Saunders and D. Brewer, Phys. Rev. Lett. **59**, 465 (1987)
[3] M. Grabinski and M. Liu Phys. Rev. Lett **58**, 800 (1987)
[4] H. Hojgaard Jensen, H. Smith, P. Wölfle, K. Nagai and T. Maak Bisgaard, J. Low Temp. Phys. **41**, 473 (1980)
[5] P. Panzer, M. Liu and D. Einzel, to be published

Physica B 165&166 (1990) 557–558
North-Holland

FLUCTUATION THEORY OF THE CRITICAL KAPITZA RESISTANCE NEAR THE LAMBDA POINT OF LIQUID HELIUM

Richard A. FERRELL

Center for Theoretical Physics of the Department of Physics and Astronomy, University of Maryland, College Park, Maryland 20742

The flow of heat in superfluid ^{4}He gives rise to a non-equilibrium situation at the boundaries. The normal component, arriving at the boundary where the heat is given up, piles up an excess density. A similar depletion of the normal component occurs at the other boundary where the heat is being carried away into the interior of the fluid. In steady state, the magnitude of the out-of-equilibrium deviation of the densities of the normal and superfluid components from their equilibrium values is determined by the normal-superfluid conversion rate. Because the fluctuations in ψ, the complex order parameter, near the λ point, are subject to critical slowing down, a relaxational bottleneck develops, resulting in a temperature gradient in the immediate vicinity of the boundary. Thanks to the recently developed techniques for high resolution thermometry (1), it is possible to detect the integrated strength of such temperature gradients, i.e., the critical Kapitza resistance. The present note reports a refinement on an earlier theoretical treatment of this problem (2).

The problem is most conveniently attacked by avoiding boundary effects and working entirely within the bulk fluid. This we accomplish by imagining that sources and sinks of heat are distributed inside the fluid, with a sinusoidal space dependence. If the wavelength of the impressed sinusoidal heat current is $2\pi k^{-1}$, a temperature modulation of the same wavelength will be generated, resulting from the finite relaxation rate $\tau_Q(k,\kappa)$ for the out-of-equilibrium disturbance of the quadratic "pair" field $Q \equiv |\psi|^2$. (The closeness to the λ point is specified by κ^{-1}, the correlation length.) This is the same relaxation that, in the long wavelength limit, $k \to 0$, determines the frequency dependent specific heat, and, hence, the ultrasonic attenuation (3). In steady state, for any given wave number, k, the amplitude of the temperature sine wave that is generated by the relaxational bottleneck is proportional to a kind of "non-local Kapitza resistance,"

$$\rho_K(k,\kappa) = \frac{k^2\tau_Q(k,\kappa)}{C(k,\kappa)} , \qquad (1)$$

where $C(k,\kappa)$ is the non-local constant pressure specific heat (i.e., the Fourier transform of the equilibrium energy-energy correlation function). This is a function that has been measured experimentally by light scattering (4) and has also been studied theoretically (5).

By superposing the sine waves for a continuous range of k values we can concentrate the heat source into a sheet, thereby simulating the boundary effect and yielding the critical Kapitza resistance,

$$\Delta R_K = \frac{2}{\pi}\int_0^\infty \frac{dk}{k^2}\,\frac{1}{\lambda_D + [\rho_K(k,\kappa)]^{-1}} , \qquad (2)$$

where λ_D is the diffusive thermal conductivity that is principally responsible for the damping of second sound. The implementation of Eq. (2) by the computation of Eq. (1) to one-loop order and its substitution into Eq. (2) yields results that are compared below with the experimental findings (1), (8). It should be noted that the range $0 < k < \kappa$ contributes heavily to the integral in Eq. (2), which would seem to render a hydrodynamic approximation (6) inappropriate. An alternative treatment (7) would seem to suffer from the similar shortcoming of using a local response function, in addition to neglecting the relaxational bottleneck.

In order to carry out the integration in Eq. (2) it is necessary to introduce some approximations. We note first that the effective wave number and temperature dependent relaxation time in Eq. (1) is mainly determined by the rate of relaxation of the longitudinal fluctuations of the order parameter, so

$$\tau_Q^{-1}(k,\kappa) \simeq D_\psi(k^2 + \kappa^2) , \qquad (3)$$

where D_ψ is the kinetic coefficient. In dealing with the non-local specific heat it is useful to first consider its $k = 0$ thermodynamic limit, C_P, which, at saturated vapor pressure, has the simple approximate dependence on t, the reduced temperature, of

$$C(0,\kappa) = C_P(t) \simeq 0.20\,\left(\ln\frac{2}{|t|} + 2\right)\,\frac{j}{cm^3K} . \qquad (4)$$

A portion of this, namely, $C_1 = 0.80\ j/cm^3K$, is attributable (5) to the order parameter, and has the non-local generalization

$$C_1(k,\kappa) = C_1\frac{\kappa^2}{k^2+\kappa^2} . \quad (5)$$

The remaining portion of $C(0,\kappa)$ acquires, for $k > 0$, a much weaker (logarithmic) dependence on k which, furthermore, sets in only for values of k of the order of 2κ and larger. Because this portion of $C(k,\kappa)$ occurs in the denominator of the integral in Eq. (2), multiplied by the more rapidly varying factor $k^2+\kappa^2$ (by virtue of Eq. (3)), we have,

$$C(k,\kappa) = C_1\frac{\kappa^2}{k^2+\kappa^2} + [C(k,\kappa) - C_1(k,\kappa)]$$
$$\simeq C_1\frac{\kappa^2}{k^2+\kappa^2} + [C(0,\kappa) - C_1]$$
$$= - C_1\frac{k^2}{k^2+\kappa^2} + C_P(t) . \quad (6)$$

Here we have separated out $C_1(k,\kappa)$ for special attention and have approximated the remaining bracketed portion by its thermodynamic limit.

Substituting Eqs. (3) and (6) into Eq. (1) determines the "non-local Kapitza resistance" by

$$\rho_K^{-1}(k,\kappa) \simeq -D_\psi C_1 + D_\psi k^{-2}(k^2+\kappa^2)C_P(t) . \quad (7)$$

Numerical evaluation of the first term on the right-hand side of Eq. (7) indicates that it is nearly cancelled by λ_D, so that the integral in Eq. (2) becomes

$$\Delta R_K(t) \simeq \frac{2/\pi}{D_\psi C_P(t)}\int_0^\infty \frac{dk}{k^2+\kappa^2} = \frac{\xi}{D_\psi C_P(t)} , \quad (8)$$

where $\xi = \kappa^{-1} \simeq 0.7\ \text{Å}\ |t|^{-2/3}$ is the correlation length. In the van Hove precritical range, $|t| > 10^{-5}$, we can approximate D_ψ by its non-critical background value, $B_\psi = 1.05 \times 10^{-4}\ cm^2/sec$. Substitution of this and Eq. (4) into Eq. (8) then yields

$$\Delta R_K(t) \simeq 3.2 \times 10^{-4}\frac{|t|^{-2/3}}{2+\ln\frac{2}{|t|}} \cdot \frac{cm^2K}{W} . \quad (9)$$

This is plotted below in Fig. 1 as $R_K(t) = R_K^{N.C.} + \Delta R_K(t)$ versus $|t|$ and compared with the data of Duncan et al. (1) and Zhong et al. (8). (The non-critical background value, $R_K^{N.C.}$, is chosen differently for the two sets of data.) The stronger temperature dependence exhibited by the Zhong data may be attributable to the non-linearity that sets in with increasing heat current. It is apparent that Eq. (9), valid only for a linear boundary resistance, would seem to give a satisfactory account of the Duncan data, taken at smaller heat currents. By a curious coincidence, the formula produced by Frank and Dohm (7), without apparently any allowance for the non-local relaxational bottleneck that is the basis for Eq. (9), agrees numerically almost exactly with Eq. (9) for $|t| \geq 5 \times 10^{-5}$. The Frank-Dohm formula falls below Eq. (9) for $|t| < 5 \times 10^{-5}$.

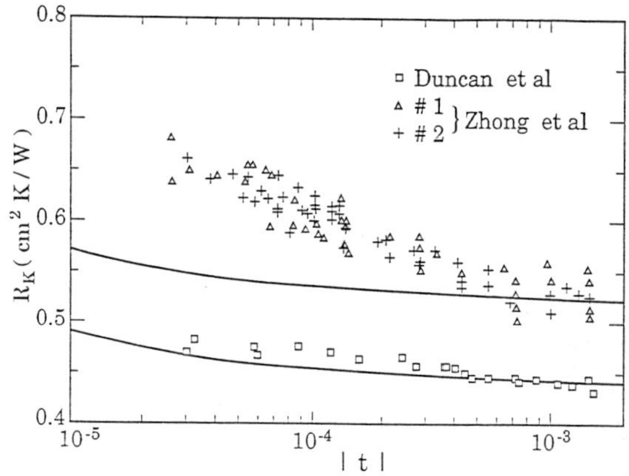

Fig. 1 Kapitza resistance versus reduced temperature: theory and three data sets.

ACKNOWLEDGEMENT

This work has been supported by the National Science Foundation under grants DMR-85-06009 and DMR-89-01723.

REFERENCES

(1) R.V. Duncan, G. Ahlers, and V. Steinberg, Phys. Rev. Lett. 58 (1987) 377.
(2) R.A. Ferrell, as quoted in Ref. 1.
(3) R.A. Ferrell and J.K. Bhattacharjee, Phys. Rev. Lett. 44 (1980) 403; Phys. Rev. B 23, 2434 (1981); R.A. Ferrell, B. Mirhashem, and J.K. Bhattacharjee, Phys. Rev. B 35 (1987) 4662.
(4) J. Tarvin, F. Vidal, and T.J. Greytak, Phys. Rev. B 15 (1977) 4193.
(5) R.A. Ferrell and J.K. Bhattacharjee, in: Recent Advances in Theoretical Physics, ed. R. Ramachandran, (World Scientific, Singapore, 1985) p. 257.
(6) M. Grabinski and M. Liu, Phys. Rev. B 40 (1989) 8720.
(7) D. Frank and V. Dohm, Phys. Rev. Lett. 62 (1989) 1864.
(8) F. Zhong, J. Tuttle, and H. Meyer, J. Low Temp. Phys. 79, 9 (1990).

Physica B 165&166 (1990) 559–560
North-Holland

CRITICAL DYNAMICS IN ^{3}HE-^{4}HE MIXTURES ABOVE T_λ

Günter MOSER and Reinhard FOLK

Instiut für Theoretische Physik, Universität Linz, A-4040 Linz, Austria

We calculate the hydrodynamic transport coefficients above the superfluid transition within the renormalized field theory of the complete model of Siggia and Nelson. The main improvement in the comparison with the experiment at higher concentrations of ^{3}He comes from the proper treatment of the temperature dependence of the static quantities.

1. INTRODUCTION

A quantitative comparison of the transport coefficients in ^{3}He ^{4}He mixtures above the superfluid transition with the results of a nonasymptotic critical theory was performed in [1]. The theory was based on the model of Siggia and Nelson [2] without taking into account the static couplings γ to the conserved quantities entropy density and concentration fluctuations (model E'). A good overall agreement was achieved, however some deviations remained, which were attributed to the neglection of the coupling γ (for the effects of γ in pure ^{4}He see [3]). (i) In the asymptotic region at higher concentrations (>5%) the thermal conductivity K did not reach a constant value but became smaller near T_λ; (ii) also in the background region $((T-T_\lambda)/T_\lambda > 10^{-2.5})$ no satisfactory agreement could be reached and (iii) the concentration dependence of the subleading behaviour (see the definition below) turned out to be qualitative different [4]. We therefore reconsider the theory taking into account the static couplings γ. These couplings although small in the asymptotics are quantitatively important in the experimental accessible region. Their inclusion allows a consistent treatment of the temperature dependence of the static susceptibilities and the transport coefficients and avoids ambiguities connected with the static parameter B of Ref. [1].

2. THE MODEL

We start with the complete model of Siggia and Nelson [2] (model F'), which is based on the hydrodynamic equations of Khalatnikov [5]. The Langevin equations for the fluctuations in the complex order parameter Ψ, the entropy density m_1 and the concentration fluctuation m_2 read

$$\frac{\partial\psi}{\partial t} = -2\Gamma \frac{\delta H}{\delta\psi^*} + i\psi \mathbf{g} \frac{\delta H}{\delta \mathbf{m}} + \theta_\psi$$

$$\frac{\partial \mathbf{m}}{\partial t} = \Lambda_{\mathbf{m}} \nabla^2 \frac{\delta H}{\delta \mathbf{m}} - 2\mathbf{g}\, \mathcal{Im}\left(\psi^* \frac{\delta H}{\delta\psi}\right) + \theta_{\mathbf{m}}$$

The mode couplings $\mathbf{g}$ are given by static quantities at the λ-line [6] and the static Hamiltonian H is the sum of

$$H_\psi\{\psi\} = \int d^dx \left\{ \frac{1}{2}\tau|\psi|^2 + \frac{1}{2}|\nabla\psi|^2 + \frac{u}{4!}|\psi|^4 \right\}$$

and

$$H_{\mathbf{m}}\{\psi,\mathbf{m}\} = \int d^dx \left\{ \frac{1}{2}\mathbf{m}^t \mathbf{A}\, \mathbf{m} + \frac{1}{2}\gamma^t \mathbf{m}|\psi|^2 \right\}$$

The matrix $\mathbf{A}$ is given by the background susceptibilities of the densities $\mathbf{m}$.

We study this model within renormalized field theory; for another approach see Ref. [6]. The theoretical treatment of this model in statics as well as in dynamics is similar to the pure ^{4}He model when first sound is taken into account [7]. All the renormalized static parameters needed, i.e. the couplings u, γ and $\mathbf{g}$, can be related to thermodynamic quantities. The dynamic parameters are obtained from the flow equations with the background values as initial parameters. These have in general to be determined by comparison with experiment.

3.RESULTS

We have calculated the transport coefficients from the two point vertex functions in one loop order. The static parts have been identified in the respective vertex functions and replaced by their experimental counterpart. A similar procedure in extracting the static parts has been performed in the flow equations [8]. The remaining dynamic part contains model F' terms in one loop order and model E' terms [1] in two loop order. Fig.1 shows the fit of our theoretical result to measurements of the Duke University group [9] at higher concentrations. In order to demonstrate point (i) the

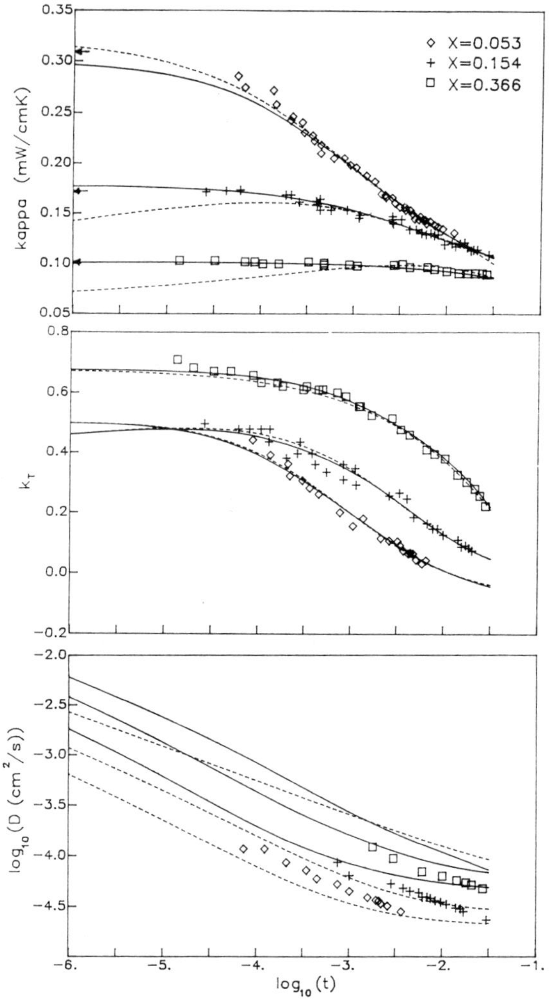

FIGURE 1
Comparison of a fit of κ and k_T in the region $10^{-3}<t<10^{-1.5}$ with measurements [8] of κ, k_T and D for different molar concentrations X. (solid lines complete model, dashed lines based on [1], arrows $\kappa(T_\lambda)$ [10]).

fit was restricted to the temperature region $10^{-3}<t<10^{-1.5}$ (no weight was given to the mass diffusion constant). However no substantial improvement could be reached by including the diffusion data, without destroying the excellent agreement in the other transport coefficients, in contrast to the simpler theory of [1]. We attribute this remaining deviation mainly to the neglection of two loop order terms in the flow equations.

Starting from the 5% fit we have extrapolated to the low concentration region [8]. The inputs

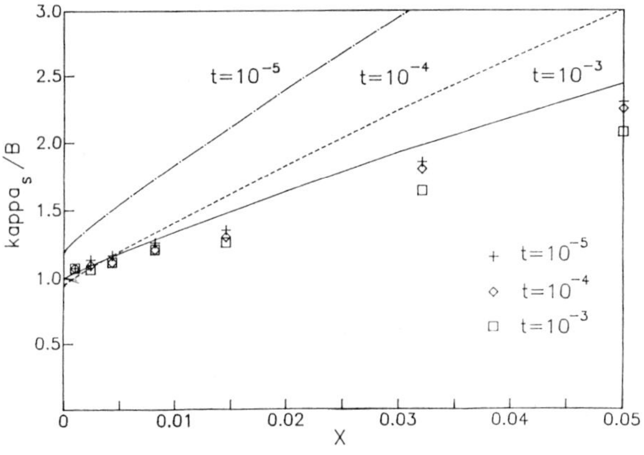

FIGURE 2
The subleading part of the thermal conductivity divided by $B = 1.19\, t^{0.44}$ as a function of concentration for different temperatures (upper lines complete model, lower lines based on ref.[1], data ref. [4]).

are the background values at $t=10^{-1.5}$ of the dynamic parameters, which we take from an appropriate fit, observing the low concentration behaviour of the experimental data [4]. We use this extrapolation to calculate the subleading temperature and concentration dependence in the thermal conductivity defined by [4,11] (for $\kappa(c,0)$ we take $\kappa(c,10^{-12})$)

$$\kappa_s(c,t) = [\kappa^{-1}(c,t) - \kappa^{-1}(c,0)]^{-1}$$

The quantitative improvement is shown in Fig.2. However we want to remark that the value of $\kappa(c,10^{-12})$ enters in a very sensitive way

REFERENCES

[1] V. Dohm and R. Folk Phys.Rev.B **28** (1983) 1332
[2] E.D. Siggia Phys.Rev.B **15** (1977) 2830; E.D. Siggia and D.R. Nelson Phys. Rev. **B15** (1977) 1427
[3] V. Dohm, Z. Phys. **61** (1985) 193
[4] F. Zhong, et al., J.Low Temp.Phys. **68** (1987) 55
[5] I.M. Khalatnikov (1965) "An Introduction to the Theory of Superfluidity" Benjamin
[6] A. Onuki, J.Low Temp.Phys.**53** (1983) 1,189
[7] J. Pankert and V. Dohm, Phys.Rev. **B40** (1989) 10842, 10856
[8] G. Moser, Thesis, Universität Linz (1989), G. Moser and R. Folk to be published
[9] D. Gestrich, R. Walsworth and H. Meyer J.Low Temp.Phys.**54** (1984) 37
[10] M. Dingus et al. J.Low Temp.Phys.**65** (1986) 213
[11] G. Ahlers, Phys.Rev.Lett. **24** (1970) 1333

Physica B 165&166 (1990) 561–562
North-Holland

BOUNDARY EFFECTS IN SECOND SOUND NEAR T_λ

D.R. Swanson, T.C.P. Chui, K.W. Rigby*, and J.A. Lipa

Department of Physics, Stanford University, Stanford, California 94305 U.S.A.

We investigate the effect of non-ideal boundary conditions applied to a second sound resonator near the lambda point. Appropriate boundary conditions for He-II are applied to a second sound resonator with a heater and detector at either end. We find the analytic boundary value solution for the homogeneous (zero gravity) differential equation. We show that the standing wave approximation is accurate to a few tenths of a percent at 10^{-8} K from the transition.

1. INTRODUCTION

Recent second sound experiments (1) near T_λ show large deviations from theory in the reduced temperature range $10^{-7} < \varepsilon \equiv 1\text{-}T/T_\lambda < 10^{-6}$. The theoretical model used in (1) made several simplifying assumptions. The gravitational effect was approximated with a time of flight method. In addition the boundaries of the resonator were assumed to be ideal insulators. For materials ordinarily used in the resonator, this is a good approximation. However we find that this is not necessarily accurate for the porous salt pill detector used in (1).

In this paper we derive the boundary conditions for the resonator and explore their effect for the simple homogeneous case (zero gravity), very close to the transition. We are currently investigating the more complete solution of numerically integrating the inhomogeneous differential equation for second sound and matching to the boundary conditions presented here.

2. BOUNDARY VALUE PROBLEM

In the absence of dissipation the linearized equation for second sound is usually written (2):

$$\partial^2 s(z,t)/\partial t^2 = u_{II}^2 \nabla^2 s(z,t), \qquad (1)$$

where $u_{II}^2 = \rho_s T S^2/\rho_n C_p$ is the velocity of second sound with the parameters defined in the usual way. Second sound is transmitted through He-II in the steady state with:

$$s(z,t) = \mathrm{Re}[\mathbf{s}(z)e^{i\omega t}], \qquad (2)$$

where the bold style indicates a complex number. To find **s** we apply appropriate boundary conditions at the heater and detector ends of the cell. The required boundary conditions for He-II at a solid wall are given by London (3):

$$(\rho S v_n - \kappa \nabla T/T)_\perp = q_\perp/T \qquad (3)$$

$$T = T' - Rq_\perp \qquad (4)$$

where ρ is the He-II density, v_n is the velocity of the normal fluid, κ is the thermal conductivity of the normal fluid, T' is the temperature of the wall and R is the Kapitza boundary resistance. Expressing v_n in terms of ∇T gives (2):

$$(-\rho S^2 T\rho_s/i\omega\rho_n - \kappa)_\perp \nabla \mathbf{T}_\perp = \mathbf{q}_\perp. \qquad (5)$$

With a harmonically driven heater, the boundary equation [5] at the heater end of the cell becomes:

$$\alpha_o \nabla \mathbf{T}_o = \mathbf{q}_o - \kappa'_o \nabla \mathbf{T}'_o \cong \mathbf{q}_o, \qquad (6)$$

where the term $\kappa'_o\nabla\mathbf{T}'_o$, denoting the heater backing material, is negligible compared to $\mathbf{q}_o$ and $\alpha \equiv \alpha_1 + i\alpha_2 = -\kappa + i\rho S^2 T\rho_s/\omega\rho_n$. The helium thermal conductivity is small compared to α_2 and we also neglect it. At the detector end of the cell the boundary condition [5] is similar to [6] but without

*Current address: General Electric R&D, P.O. Box 8, Schenectady, NY 12603.

0921-4526/90/$03.50

the driving term:

$$\alpha_L \nabla \mathbf{T}_L = -\kappa'_L \nabla \mathbf{T}'_L, \qquad (7)$$

where the primed symbols represent quantities in the wall and the subscript L indicates quantities to be evaluated at z=L, the interface between the fluid and the detector. Similarly [4] becomes:

$$\mathbf{T}_L = \mathbf{T}'_L + R\kappa'_L \nabla \mathbf{T}'_L. \qquad (8)$$

In a solid wall heat travels as a thermal diffusive wave which is given by (3):

$$T' = \mathrm{Re}[\mathbf{T}_b\, e^{-z'(i-1)/\delta} e^{i\omega t}], \qquad (9)$$

where $\mathbf{T}_b$ is the temperature at the boundary (z'=0), and the thermal penetration depth δ is given by:

$$\delta^2 = 2\,\kappa'_L/\rho' C_p'\omega, \qquad (10)$$

where κ'_L, ρ' and C_p' are properties of the detector wall material. Substituting [9] into [7] and [8] and simplifying we obtain:

$$s_L/\nabla s_L = -\alpha_L\,(\delta(1+i)/2\kappa'_L + R) \equiv \mathbf{C}_L \qquad (11)$$

3. EXACT SOLUTION IN ZERO GRAVITY

In a gravitational potential, ρ_s (thus u_{II} and α) vary with height in the resonator. The resulting non-linear problem requires numerical techniques which we will report on elsewhere. Here we estimate the frequency shift from the boundary effects in zero gravity. This may be useful in designing a future space based experiment.

The general solution to [1] is the sum of two waves traveling in opposite directions with wave vectors $\pm$k. :

$$\mathbf{s}(z) = \mathbf{A}e^{ikz} + \mathbf{B}e^{-ikz}. \qquad (12)$$

The complex coefficients **A** and **B** contain the amplitude and phase of each wave. Solving for **s**(L) using [6], [11] and [12] we obtain:

$$\mathbf{s}(L) = \mathbf{q}_o \mathbf{C}_L C_p / \alpha_o T(\cos kL + k\mathbf{C}\sin kL). \qquad (13)$$

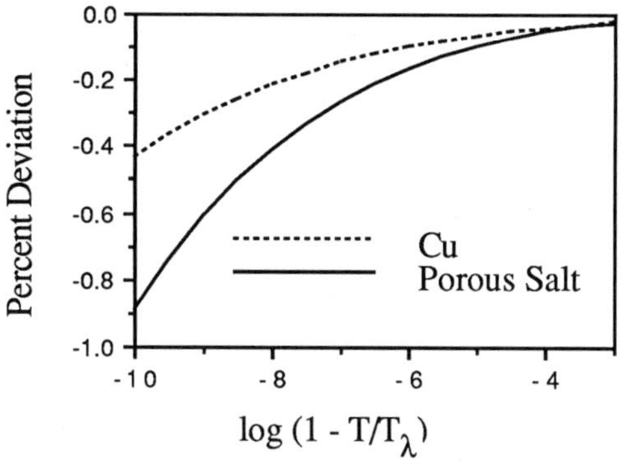

FIGURE 1

This result scales with $\mathbf{q}_o$, as expected for small excitations, and displays a resonance as a function of ω (=u_{II}k from [1], [2] and [12]).

Figure 1 shows the deviation of the resonant frequency obtained from [13] from the ideal standing wave approximation ($u_{II} = \omega L/\pi$) for two different boundary materials, Cu and the porous salt pill used in (1). For the salt pill we approximate the parameters in [10] as those of He-I. For both materials we use R = 0.5 cm^2K/W. The frequency shift for glass was several orders of magnitude smaller and is negligible on this scale.

4. CONCLUSION

It is clear from these results that the boundary effects alone are unable to explain the experimental discrepancies. The deviations shown in Fig. 1 are about on third are large as the experiment. Analysis of the more complete model mentioned above may produce a closer agreement with experiment.

ACKNOWLEDGMENTS

We thank U.S. National Aeronautics and Space administration for its support with contract JPL 957448.

REFERENCES

(1) D. Marek, J.A. Lipa, and D. Philips, Phys. Rev. B 38 (1988) 4465.
(2) I.M.Khalatnikov, An Introduction to the Theory Superfluidity (Benjamin, New York, 1965).
(3) F.London, Superfluids (John Wiley & Sons, London, 1954).
(4) H.S.Carslaw and J.C.Jaeger, Conduction of Heat in Solids, Second Edition (Clarendon, Oxford, 1959).

Physica B 165&166 (1990) 563–564
North-Holland

CRITICAL SHEAR VISCOSITY OF ^{4}HE ABOVE AND BELOW T_λ

R. SCHLOMS, J. PANKERT, V. DOHM

Institut für Theoretische Physik, Technische Hochschule Aachen, Germany

On the basis of a new stochastic dynamic model the critical behavior of the shear viscosity above and below the λ transition of ^{4}He is calculated. We show that two different mechanisms contribute to the shear viscosity. Both are described by couplings which are irrelevant in the sense of the renormalization-group theory. Good agreement with experiments is found.

1. INTRODUCTION

Detailed measurements have shown /1-4/ that the shear viscosity η has a weakly singular behavior near the λ transition of ^{4}He. No quantitative theory is available so far for this fundamental transport coefficient. In the following we present the results of a quantitative renormalization-group (RG) calculation for η both above and below T_λ. We define η via the correlation function of the components j_α of the transverse part $\mathbf{j}$ of the momentum density in the limit of small $\mathbf{k}$ and ω /5/,

$$< j_\alpha j_\beta > = \tau^k_{\alpha\beta}\, 2k_B T\, \eta\, k^2 (\omega^2 + \eta^2 k^4/\rho_n^2)^{-1}, \quad (1)$$

$$\tau^k_{\alpha\beta} = \delta_{\alpha\beta} - k_\alpha k_\beta / k^2 \quad , \quad (2)$$

where ρ_n is the normal fluid density ($\rho_n = \rho$ above T_λ). This correlation function will be calculated from the stochastic equations of a new model that includes reversible and dissipative couplings of $\mathbf{j}(x,t)$ to the order parameter $\psi(x,t)$. Althoug both types of couplings are irrelevant in the RG sense they are nonnegligible and lead to an observable singularity of η near T_λ in agreement with experiments.

2. MODEL

In constructing an appropriate model we have started from the equations of motion for ^{4}He as given in Ref. 6. We keep $\mathbf{j}(x,t)$, $\psi(x,t)$ and the entropy variable $m(x,t)$ and neglect the coupling to the first-sound mode. As a novel feature we include a ψ-dependent bare kinetic coefficient

$$\hat{\eta}(\psi)/k_B T = \eta_o + \eta_1 |\psi|^2 \quad (3)$$

in the dissipative part of the $\mathbf{j}$ equation, with two phenomenological constants η_o and η_1. Thus we propose the following model

$$\dot{\psi} = -2\Gamma \delta H/\delta\psi^* + ig\psi\delta H/\delta m - g_j \nabla\psi\cdot\delta H/\delta \mathbf{j} + \theta_\psi \quad ,(4)$$

$$\dot{m} = \lambda\nabla^2 \delta H/\delta m - 2g\mathrm{Im}\,(\psi^*\, \delta H/\delta\psi^*) + \theta_m \quad ,(5)$$

$$\dot{\mathbf{j}} = \tau[\hat{\eta}\nabla^2 \delta H/\delta \mathbf{j} + 2\, g_j\, \mathrm{Re}(\nabla\psi\; \delta H/\delta\psi) + \hat{\eta}^{1/2}\, \boldsymbol{\theta}_j] \quad , \quad (6)$$

$$H = H_F(\psi,m) + \int d^d x\, \chi_j^{-1}\, \mathbf{j}^2/2 \quad , \quad (7)$$

where $H_F(\psi,m)$ is the model F Hamiltonian /7/. The Gaussian Langevin forces $\theta_i(x,t)$ have correlations in accord with the equilibrium distribution $\sim \exp - H$. If the variable $\mathbf{j}$ is dropped we recover model F whose nonuniversal parameters are well known /7,8/. Since χ_j and g_j are also known /6/ we are left with two new parameters η_o and η_1 which can be determined from the measured η above T_λ. Then the calculation below T_λ does not contain any adjustable parameter. The Hamiltonian (7) could be extended to include also critical contributions to ρ_n and ρ which are neglected here.

3. CALCULATION AND RESULTS

Within our model we can define η in terms of the vertex tensor $\Gamma_{\tilde{j}\tilde{j}}(\mathbf{k},\omega)$ where $\tilde{\mathbf{j}}$ denotes the response field /9/ conjugate to $\mathbf{j}$,

$$\eta = \partial_{k^2} \mathrm{Tr}[\tau^k\, \Gamma_{\tilde{j}\tilde{j}}(\mathbf{k},0)]_{k=0}/[2(d-1)\, k_B T]. \quad (8)$$

To leading order in the couplings η_1 and $g_j = k_B T$ we find

$$\eta = \eta_o - \eta_1 <|\psi|^2> \quad (9)$$

$$+ g_j^2 \sum_{\alpha\beta\gamma\delta} \tau^k_{\alpha\delta}(k_\beta k_\gamma/k^2) <\pi_{\alpha\beta}\pi_{\gamma\delta}>^F / [2(d-1)\, k_B T]$$

with

$$\pi_{\alpha\beta}(x,t) = [\partial_\alpha\psi(x,t)\partial_\beta\psi(x,t)^* + \mathrm{c.c.}]/2 \quad ,(10)$$

where $<\ldots>^F$ denotes the Fourier transformed correlation function at $k = \omega = 0$, calculated within model F. The parameter η_o can be absorbed in the finite value η_λ of η at T_λ which we take from experiment. We have determined the critical exponent of the last term of (9) and have calculated its amplitude in one-loop order (for $T<T_\lambda$ we have also neglected effects due to 2. sound). The results in three dimensions above and below T_λ, respectively, are

$$\eta - \eta_\lambda = \eta_1(<|\psi|^2> - <|\psi|^2>_\lambda)$$

$$+ g_j^2\, A^\pm_\eta\, [\xi(t_\pm)\, \Gamma(t_\pm)\, k_B T]^{-1}, \quad (11)$$

0921-4526/90/$03.50

$$A_\eta^+ = -(16\pi)^{-1} \quad , \quad A_\eta^- = -(32\pi)^{-1} \quad , \tag{12}$$

where $t_+ = t$ and $t_- = -2t$, $t = (T-T_\lambda)/T_\lambda$. Here $\Gamma(t)$ denotes the known effective renormalized kinetic coefficient of the order parameter /7, 8/ and $\xi(t)$ is the correlation length above T_λ. Since the leading singular term of

$$\langle|\psi|^2\rangle - \langle|\psi|^2\rangle_\lambda = c_0 t|t|^{-\alpha} + c_1 t \tag{13}$$

is known from the specific heat, the result (11) contains only η_1 as an unknown parameter. We have adjusted η_1 so as to achieve agreement with the data /1-4/ at $t = 10^{-3}$ above T_λ. The second term $\sim g_1^2$ in (11) turns out to be one order of magnitude smaller than the first term and does not change sign at T_λ, in contrast to the term (13) and to the measured values of $\eta - \eta_\lambda$. The comparison between theory and the data is shown in Fig. 1. The temperature dependence above T_λ and both the magnitude and the temperature dependence below T_λ are well described by our theory without adjustments.

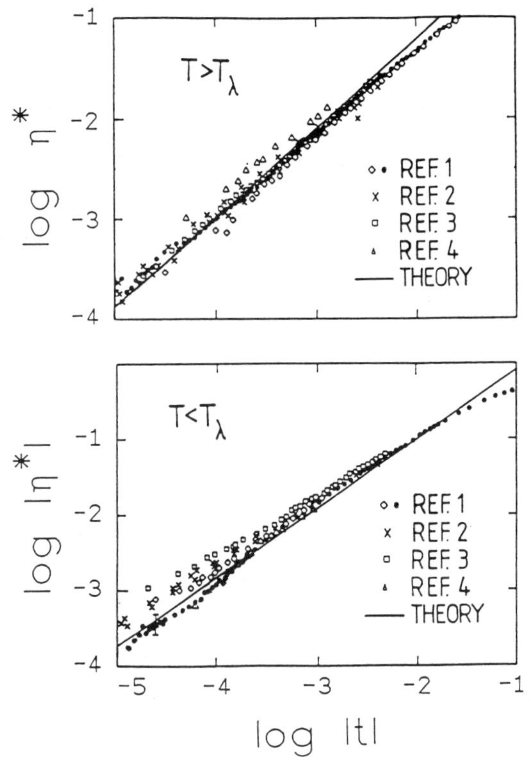

FIGURE 1
Critical shear viscosity $\eta^* = (\eta - \eta_\lambda)/\eta_\lambda$ vs t. Data from Refs. 1-4. The solid lines represent our expression (11) - (13).

4. DISCUSSION

We have treated the effects of two different contributions to the critical shear viscosity.

(i) The stress-tensor correlation[last term of (9)] is parallel to the known formula of fluctuating hydrodynamics of ordinary fluids /10/. According to the last term of (11) this effect alone would yield the wrong sign of $\eta - \eta_\lambda$ above T_λ and the wrong magnitude below T_λ.

(ii) The $|\psi|^2$ term in (3) models the leading dependence of the effective noncritical viscosity of superfluid ^{4}He on the superfluid fraction. Above T_λ this shows up as a local effect corresponding to fluctuating clusters of superfluid ^{4}He, as speculated earlier /11/. Below T_λ the $|\psi|^2$ term contributes significantly already at the hydrodynamic (mean field) level.

The semiquantitative considerations by Ferrell and collaborators /12,13/ do not take into account the effect (ii). Their results based on the effect (i) differ from ours both with respect to sign and magnitude. We doubt that their results provide a correct explanation of the effective exponent of $\eta - \eta_\lambda$ above T_λ.

REFERENCES

/1/ S. Wang, C. Howald, and H. Meyer, J. Low Temp. Phys., to be published.
/2/ L. Bruschi, G. Mazzi, M. Santini, and G. Torzo, J. Low Temp. Phys. **29** (1977) 63.
/3/ R. Biskeborn and R.W. Guernsey, Phys. Rev. Lett. **34** (1975) 455.
/4/ R.W.H. Webeler and G. Allen, Phys. Rev. **A 5** (1972) 1820.
/5/ P.C. Hohenberg and P.C. Martin, Ann. Phys. (N.Y.) **34** (1965) 291.
/6/ J. Pankert and V. Dohm. Phys. Rev. **B 40** (1989) 10842.
/7/ V. Dohm, Z. Phys. **B 61** (1985) 193.
/8/ W.Y. Tam and G. Ahlers, Phys. Rev. **B 33** (1986) 183; **B 37** (1988) 7898.
/9/ P.C. Martin, E.D. Siggia, and H.A. Rose, Phys. Rev. **A 8** (1973) 423.
/10/ L.D. Landau and E.M. Lifschitz, Sov. Phys. JEPT **5** (1957) 512.
/11/ K. Mendelssohn, in Encyclopedia of Physics, Vol. **XV**, ed. S. Flügge (Springer, Berlin, 1956).
/12/ R.A. Ferrell, J. Low Temp. Phys. **70** (1988) 435.
/13/ J.K. Bhattacharjee, R.A. Ferrell, and Z.Y. Chen, in Proceedings of 17th International Conference on Low Temperature Physics, eds. U. Eckern, A. Schmid, W. Weber, and H. Wühl (North-Holland, New York, 1984), p. 977.

Physica B 165&166 (1990) 565–566
North-Holland

THE EFFECTS OF GRAVITY IN SECOND-SOUND

T. C. P. CHUI and D. R. SWANSON

Physics Department, Stanford University, Stanford, California 94305 U.S.A.

We present a modified wave equation for second-sound propagation in an inhomogeneous medium. For extreme inhomogeneity we show that a new mode exists at low frequency, in which normal wave-like propagation is not possible. The nature of this mode is discussed.

1. INTRODUCTION

The gravitational inhomogeneity near T_λ can be easily understood. The pressure in a fluid varies with height due to gravity, while T_λ is depressed at higher pressure. Thus it can be shown (1) that $T_\lambda(z) = T_\lambda(0) + Az$, where $A = 1.27\times10^{-6}$ K/cm and z is the vertical height. Near T_λ the superfluid density, $\rho_s(z,T) = 2.4\,\rho\,\varepsilon(z,T)^{.674}$, where ρ is the density and $\varepsilon(z,T) = 1 - T/T_\lambda(z)$. Combining the expressions for ρ_s and ε gives:

$$\rho_s(z,T) = 2.4\,\rho\,[\,Az/T_\lambda(0) + \varepsilon(0,T)\,]^{.674}. \quad (1)$$

Recent developments of high resolution thermometers (2) and second-sound detectors (3) enable the measurements of second-sound (4) to $\varepsilon \sim 10^{-8}$. At these temperatures, ρ_s is highly inhomogeneous. Using two-fluid hydrodynamic theory, we rederived the wave equation for second-sound propagation taking into account the effect of inhomogeneity. For small amplitude excitations, we obtain:

$$\ddot{T} = U^2\{\,[\,\frac{\nabla(\rho_s/\rho_n)}{(\rho_s/\rho_n)} + \frac{\nabla\rho}{\rho}\,]\,\nabla T + \nabla^2 T\,\},$$

which can be approximated to within 10% by

$$\ddot{T} = (\nabla U^2)\,\nabla T + U^2\,\nabla^2 T, \quad (2)$$

where the local second-sound velocity is

$$U(z,T) = [S^2\,\rho_s(z,T)\,T\,/\,\rho_n\,C(z,T)]^{1/2}, \quad (3)$$

C and S are the heat capacity and the entropy.

2. THIN CELL APPROXIMATION

Equation 2 is highly non-linear, and can only be solved numerically. However, it is intuitive to make certain assumptions to simplify this equation to a point that an exact solution can be found. We assume that the experimental cell is thin, so that U does not vary much inside, and ∇U can be approximated by a constant. We seek a solution of the type $T(z,t) \sim e^{\,i(kz-\omega t)}$. The solution is given by:

$$k = -\,\alpha i \pm \beta \quad (4)$$

where $\alpha = \nabla U/U$ (5)

and $\beta = (\omega/U)\sqrt{1-(\nabla U/\omega)^2}$. (6)

Depending on $(\nabla U/\omega)$, β can be real or imaginary. Therefore, two distinct types of solutions exist. The usual propagative type corresponds to the case of real β. The case of imaginary β is a new mode in which wave-like propagation is not possible.

3. WAVE-LIKE MODE

For real β, the solution is a traveling wave with an amplitude that diverges in the -z direction and decays in the +z direction.

In a resonator, the resonant frequency can be found by solving the Eigenvalue problem with the boundary condition $\nabla T = 0$ (insulating walls). We obtain:

$$\omega = \sqrt{\omega_o^2 + (\nabla U)^2} \quad (7)$$

where ω_o is the resonant angular frequency in the

absence of the ∇U term. Thus the effect of the ∇U term is to increase the resonant frequency.

4. NON-WAVE-LIKE MODE

When β is imaginary, $T(z,t) \sim e^{-z/\delta - i\omega t}$ where,

$$\delta_{+,-} = \frac{U \nabla U}{\omega^2} \left[1 \pm \sqrt{1 - (\omega/\nabla U)^2} \right] \quad (8)$$

δ_+ and δ_- are decay lengths. In this case k is purely imaginary, and there is no phase shift between a sound generator and a detector at different locations. This situation would appear to violate causality, since a finite time is required for a signal to travel. To understand this, we consider the case of a sine-wave second-sound generator being activated at t = 0. The driving signal can be written as the product of a step function at t = 0 and a sine wave. The step function contains Fourier components of high ω, therefore the signal would propagate in the wave-like region until a steady state is reached in which no phase shift exists between the generator and the detector. Since this is only a steady state phenomenon, causality is not violated.

Figure 1 shows the expected steady state phase difference as a function of ω. Below the critical frequency of $\omega_c = \nabla U$, no phase shift exists. Also shown is the usual propagative mode in which the phase shift approaches zero as $\omega \to 0$.

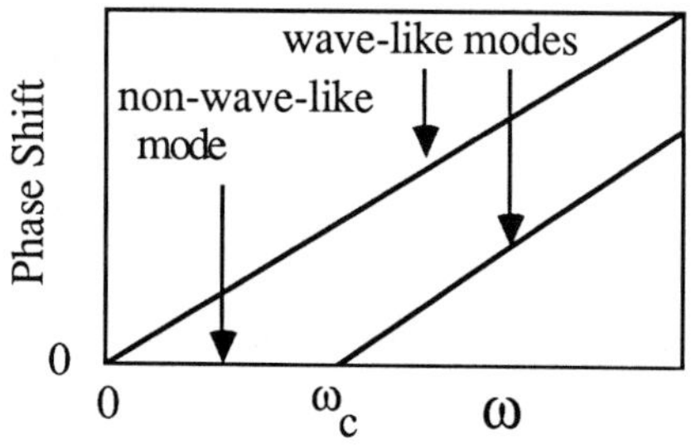

Figure 1. Phase shift between a second-sound generator and a detector.

5. ORDER OF MAGNITUDE ESTIMATION

We estimate the order of magnitude of the various quantities discussed above using $U \sim 46.28\ \varepsilon^{.387}$(m/sec) (5). Table 1 shows the quantities ∇U (or ω_c), and the frequency shift of a .3 cm resonator caused by the ∇U term (eq. 7), as a function of ε. Assuming an infinite column of helium, the quantity $\Lambda = T_\lambda \varepsilon / A$ is the distance from the normal-superfluid interface when a fluid element is at ε. This table, shows that the effects caused by ∇U should be observable. For example, if a generator is located at the bottom of a cell and a detector is placed 1.7 cm above, then $\omega_c \sim 5$ rad./sec., when the bottom is near T_λ.

Table 1 Estimation of the order of magnitude of several relevant quantities.

ε	$\nabla U = \omega_c$ (sec^{-1})	$(\omega-\omega_0)/\omega_0$ (.3cm cell)	Λ (cm)
1E-3	7.21E-2	2.32E-10	1.71E+03
1E-4	2.96E-1	2.32E-08	1.71E+02
1E-5	1.21E+0	2.32E-06	1.71E+01
1E-6	4.98E+0	2.32E-04	1.71E+00
1E-7	2.04E+1	2.30E-02	1.71E-01
1E-8	8.38E+1	1.38E+00	1.71E-02
1E-9	3.44E+2	2.06E+01	1.71E-03
1E-10	1.41E+3	2.15E+02	1.71E-04
1E-11	5.78E+3	2.16E+03	1.71E-05

ACKNOWLEDGMENTS

We thank Q. Li, J. A. Lipa and M. J. Adriaans for helpful comments, and NASA for its support with contract JPL 957448.

REFERENCES

(1) G. Ahlers, Phys. Rev. 171 (1968) 275 .
(2) T. C. P. Chui and J. A. Lipa, Proc. of LT-17, eds. U. Eckern, A. Schmid, W. Weber, and H. Wurhl, (North-Holland, Amsterdam, 1984) P.931.
(3) T. C. P. Chui and D. Marek, J. Low Temp. Phys. 73 (1988) 161.
(4) D. Marek, J. A. Lipa and D. Philips, Phys. Rev. B 38 (1988) 4465 .
(5) G. Ahlers, The Physics of Liquid and Solid

Physica B 165&166 (1990) 567–568
North-Holland

Immiscibililty of Bulk Liquid ^{3}He with ^{3}He Contained in a Porous Glass*

Gary G. Ihas and Gregory F. Spencer+

Department of Physics, University of Florida, Gainesville, FL 32611–2085, USA.

Magnetic susceptibility and pressure measurements on ^{3}He flow through porous Vycor glass show that bulk ^{3}He and the ^{3}He in the Vycor are immiscible fluids with an interfacial tension of $\sim 10^{-4}$ ergs/cm^2.

A number of excellent measurements[1] have recently begun to illuminate the nature of ^{3}He atoms at low temperatures on surfaces or in confined geometries. In particular, it seems clear that ^{3}He near surfaces at millikelvin temperatures has a ferromagnetic tendency. We present measurements here which indicate that ^{3}He contained in porous Vycor glass is fundamentally different from bulk liquid ^{3}He.

The study[2] of immiscible fluids in porous media at room temperature has shown: (a) when one fluid attempts to displace the other from the porous medium there is a range of applied pressures over which the meniscus is pinned to the walls, and (b) the random medium represents a set of metastable states among which transitions may be made at a steady rate. We have studied the flow of ^{3}He at millikelvin temperatures through Vycor glass using a low temperature pump and sensitive manometer. These measurements indicate that the bulk fluid which we attempt to push into the Vycor is immiscible with the ^{3}He in the pores. After the initial change in the shape of the meniscus between the two fluids, very slow flow occurs as one fluid transforms into the other fluid at a steady rate in the random medium. Magnetic susceptibility measurements show hydrostatic equilibrium is achieved relatively quickly throughout the porous rod.

The low temperature flow measurement device has been recently described elsewhere.[3] A pump drives fluid through a conductance while a manometer measures the pressure drop produced across the conductance. The entire apparatus is cooled to the operating temperature of the experiment. The flow conductance for the present experiment is shown in figure 1. A cylinder of Vycor glass is held in epoxy so that ^{3}He may enter a 1 mm length of its outer surface at one end and exit through a similar surface at the other end. End caps prevent the fluid from entering longitudinal cracks which might exist in the center of the rod. NMR pick–up coils are mounted directly on the rod, embedded in epoxy, allowing susceptibility measurements to be made on the ^{3}He contained in the rod.

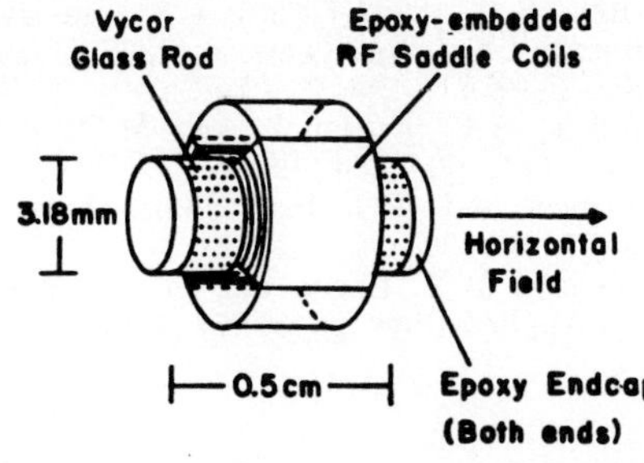

Figure 1. Flow conductance used in these experiments.

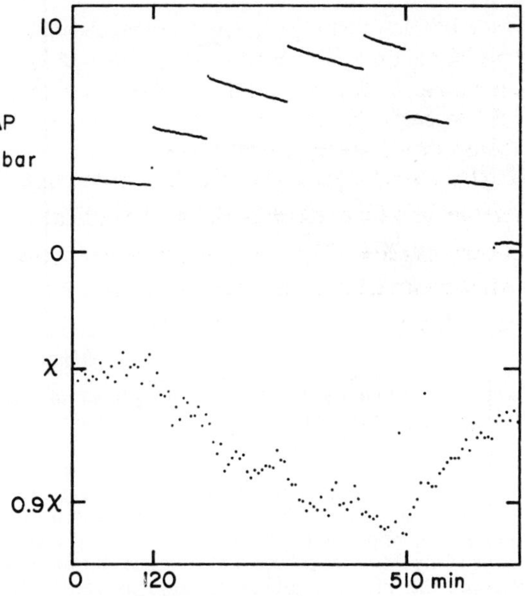

Figure 2. Relation between nuclear magnetic susceptibility and changes in the pressure gradient across the Vycor rod.

Figure 2 shows simultaneous flow and susceptibility data at 1.23 mK and a cell pressure of 5.45 bar. The upper part of the plot gives the pressure difference across the Vycor while a series of 4 forward strokes and 3 reverse strokes of the pump are executed. On the time scale of the plot these strokes are instantaneous. The lower part of the plot shows the relative susceptibility (χ) of the ^{3}He in the Vycor. The change in susceptibility is reflecting the work done on the fluid in a magnetic field H by the pump in an adiabatic process:

$$dW = -PdV = HdM.$$

For constant applied magnetic field and bulk compressibility β this becomes for a volume V:

$$V\beta P dP = H^2 d\chi.$$

For $\Delta P/P \ll 1$ we have

$$\Delta\chi = \frac{V\beta P}{H^2}\Delta P$$

which is verified in figure 2. The response of χ to a change of P shows that a uniform pressure gradient, which is nearly static, is set up across the Vycor rod in about 20 minutes. This is the key to understanding the type of flow data shown in figure 3.

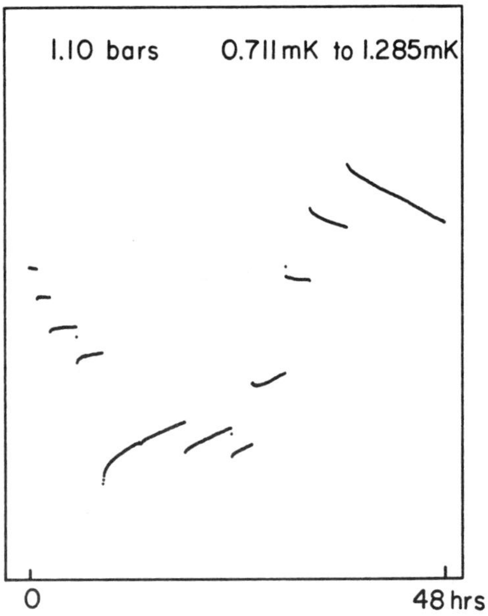

Figure 3. Manometer readings for a number of pump strokes in both directions.

The flow data in figure 3 show two important features: (a) a relatively rapid change in pressure across the Vycor occurs immediately after each pump stroke whose sign is opposite to the sign of the pressure charge produced by the stroke and (b) a much slower linear decay of the pressure difference towards zero. It is believed that feature (a) is due to the change in the shape of the meniscus between the two fluids (bulk and confined ^{3}He) at the surface of the Vycor rod. Feature (b) is due to a steady conversion of bulk (paramagnetic) ^{3}He to confined (near ferromagnetic) ^{3}He in the random porous surface structure of the Vycor.

The apparent rapid flow associated with feature (a) is due to the change in the contact angle of the meniscus with the walls of the Vycor pores. In the idealized case of one immiscible fluid being displaced by another in a circular tube of radius R, a changing pressure differential ΔP may be supported across the tube as the contact angle of the meniscus with the tube wall (θ) changes according to the the Young–Laplace equation:

$$\Delta P = \frac{2}{R}\gamma \text{ cor } \theta$$

where γ is the interfacial tension between the two fluids. A second type of measurement must be performed to separate γ from $\cos\theta$; however, it is reasonable to assume that $\cos\theta$ varies from 1 to 0 as the pressure differential is seen to change. With this assumption one obtains the following results from our data:

P =	1.1 bar	5.45 bar
γ =	3×10^{-4}erg/cm^2	1×10^{-4} erg/cm^2.

Temperature is not an important parameter as long as it is $\sim$ 1 mK. Normal and superfluid bulk liquids give the same results.

The reverse short–term flow seen after the seventh stroke in figure 3 shows that the meniscus responds faster than the pressure gradient can change in the Vycor. Then the pressure gradient reestablishes itself over the time of $\sim$20 minutes as measured by the susceptibility, allowing the meniscus to relax (ready to flex in which ever direction the next stroke occurs). The long term linear decay shown in figure 3 is caused by ^{3}He atoms moving into (or out of) the Vycor as they convert between paramagnetic and ferromagnetic states. This conversion occurs at $\sim10^{11}$ sites in the Vycor surface proceeding at $\sim10^{-10}$ mole/sec. Hence each pore has about 10^{-21} mole/sec of ^{3}He atoms converting their magnetic state (about 600 atoms/sec). Since a pore is about 30 Å in radius, taken as a hemisphere at the Vycor surface, its surface is $\sim$ 5400 Å^2 implying that each atom converts on $\sim$ 9 Å^2 of pore surface per second, a reasonable number.

Although this analysis is somewhat speculative, it explains better than previous attempts[4] the unusual flow characteristics of ^{3}He in a porous medium, and is consistent with other magnetic measurements on surface ^{3}He. This is the first demonstration of the dynamic conversion which occurs between these two states of ^{3}He matter.

The authors wish to thank E. Flint and M. Meisel for help in this work.

*Supported in part by U.S. NSF grant DMR 8519007.
+Present address: Department of Physics, Texas A.&M. University, College Stadium, Texas 77843 USA.

1. H.M. Bozler. D.M. Bates, and A.L. Thompson, Phys. Rev. B 27, 6992 (1983); D.S. Greywall and P.A. Busch, Phys. Rev. Lett. 6, 1868 (1989) and references therein.
2. J.P. Stokes, A.P. Kushnick, and M.O. Robbins, Phys. Rev. Lett. 60, 1386 (1988).
3. G.F. Spencer and G.G. Ihas, Journ. Low Temp. Physics. 77, 61 (1990).
4. G.F. Spencer, B.N. Engel, and G. G. Ihas, Jap. Journ. of Applied Phys. 26–3, 271 (1987).

Physica B 165&166 (1990) 569–570
North-Holland

LOW TEMPERATURE HYDROGEN WHICH HAS NOT SOLIDIFIED

D F BREWER, J RAJENDRA, N SHARMA and A L THOMSON

University of Sussex, Physics Division, Brighton, Sussex BN1 9QH, Great Britain

We have measured the specific heat of hydrogen inside the small pores of Vycor glass at temperatures extending down to 4K. The latent heat associated with freezing when cooling, and with melting when warming, is clearly seen. When this latent heat is compared with that of bulk hydrogen it appears that only a fraction of the hydrogen inside our pores takes part in the solid–liquid transition. Furthermore the specific heat measured at temperatures below the latent heat anomaly is clearly much larger than the specific heat of the bulk solid phase. It is also much larger than the T^2 specific heat expected for the monolayer of hydrogen next to the glass wall. If the large specific heat is due to phonon modes, they are unusually soft. It it interesting that the data also fit a roton-type specific heat, with a roton gap of about 23K.

Over the past few years interest in the properties of hydrogen in confined geometries has been steadily growing. It has already been established that the freezing and melting properties are considerably altered and the experimental situation (1) shows that, inside the porous media examined so far, the melting and freezing temperatures are different from each other and both are depressed to lower temperatures than those of the bulk phases. These lowered temperatures at which the solid– liquid transition takes place imply that the liquid phase exists at anomalously low temperatures inside small pores, and a principal reason for research on this system is that this liquid phase might exhibit superfluid properties. One of the first experiments (2) which sought such a phase was a flow experiment which produced a null result although similar experiments (3) on helium in several porous media have shown the flow expected of a superfluid. In this present paper we have measured the specific heat of hydrogen inside Vycor porous glass while both warming and cooling. The latent heat peaks associated with both melting and freezing were carefully measured and the entropy associated with the liquid–solid transition determined. In addition the temperature dependence and magnitude of the specific heat at temperatures below the latent heat peaks has been determined and indicates clearly that a considerable amount of the hydrogen sample does not exist in a simple solid phase.

The apparatus in which the heat capacity measurements were made has a berylium–copper sample chamber which was filled with Vycor glass. On top of the chamber a needle valve was situated and this ensured that the amount of sample hydrogen remained constant throughout each experiment. The hydrogen entirely filled the Vycor pores and the space around them inside the sample chamber and the molecular state of the hydrogen was believed to be 100% para. Shown in Figure 1 is a log–log plot of the specific heat of our confined hydrogen as a function of temperature measured while warming.
Also displayed in the graph is a solid line showing the specific heat of bulk hydrogen. The vertical line indicates the position of the triple point: there is good agreement between our data and the specific heat of liquid hydrogen at the highest temperatures.

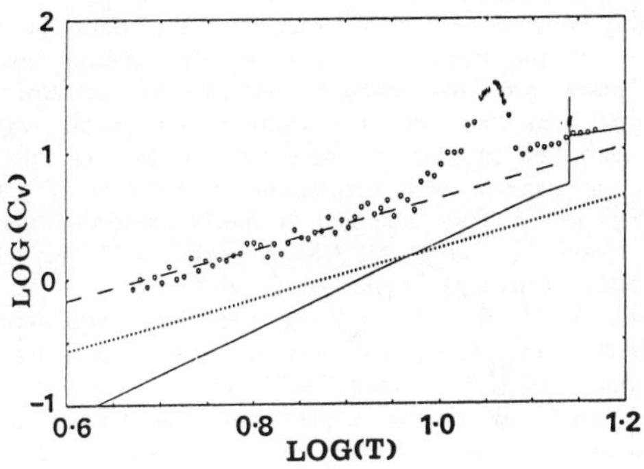

FIGURE 1
The logarithm of our specific heat plotted against the logarithm of the temperature. The dashed line represents a best fit for the lower temperature points. The dotted line represents the specific heat of the first adsorbed layer. The solid line represents the specific heat of hydrogen in its bulk phase.

A notable feature in Figure 1 is the large anomaly associated with part of the hydrogen melting which occurs substantially below the bulk triple point, around 11.2K. In a similar experiment when cooling, the peak of the anomaly was shifted downwards to about 10K, and this behaviour is similar to that already observed by another group (1). At temperatures well below the melting anomaly, the specific heat is much larger than that of bulk solid hydrogen, and the temperature dependence is close to T^2, unlike the solid dependence which is closer to T^3. A computer fit has been made to this low temperature specific heat and is represented by the dashed line which follows the equation

$$\log C_V = 2 \log T - 1.37 \quad (1)$$

which may be re-written as

$$C_V = 4.3 \times 10^{-2} T^2 \ \mathrm{J mole^{-1} K^{-1}} \quad (2)$$

It is therefore quite clear that a considerable part of our hydrogen does not exist in an ordinary bulk phase and it is interesting to speculate on what the character of this phase must be. A specific heat with a T^2 dependence has previously been observed for a monolayer of helium in Vycor (4) with a Debye θ_D of about 28K. If we analyse the hydrogen results in the same way and estimate θ_D = 120K, close to that of bulk solid, 122K (5), the result is the dotted line in figure 1. It is apparent that our data are higher than this line by a factor of about 3.

The data of Figure 1 were derived on the basis that all of the sample is in some homogeneous phase. Clearly this is not the case since the part of the sample that melts in the region of 11K is presumably in a solid phase at lower temperatures. In an attempt to ascertain the amount of hydrogen taking part in the melting process we have integrated the data in the region of the latent heat peak of the melting inside the pores and have hence derived the amount of entropy associated with the melting. The peak region was identified as starting where the specific heat points began to increase with temperature faster than T^2 and ending where they assumed a fairly constant value. The result is ΔS_{LS} = 2.5 J mole^{-1}K^{-1} for the warming (melting) experiment and ΔS_{LS} = 1.7 J mole^{-1}K^{-1} for the cooling (freezing) experiment. Both of these values are very far below the entropy difference of 8.5 J mole^{-1}K^{-1} which exists in the bulk phase (6) at the triple point and hence it is reasonable to assume that only a fraction of our sample, probably less than on half, takes part in the melting process. However, since this fraction should have a small heat capacity at low temperatures, the remaining part of the sample must have a correspondingly larger specific heat.

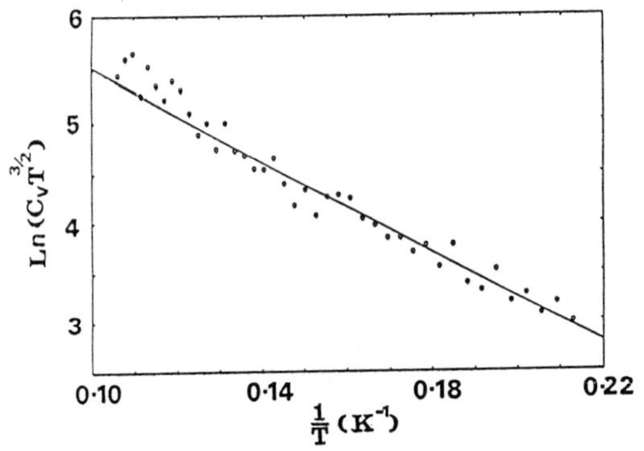

FIGURE 2
The natural logarithm of $(C_V T^{3/2})$ plotted against inverse temperature. The straight line represents a best fit to the experimental points and its slope indicates a roton gap energy of about 23K.

Although our data give quite a good fit on a T^2 plot, it is interesting to take a different approach and to examine the extent to which they might be explained by a model in which the excitations are rotons. Evidence has previously been obtained for the existence of rotons in helium inside Vycor glass at low pressures (7),(8) where plots of $\ln(C_V T^{3/2})$ versus T^{-1} showed a satisfactory fit to experimental data. Accordingly we present in Figure 2 a similar plot for our hydrogen data and the fit is reasonably good with the slope of the line indicating a gap energy for the rotons of approximately 23K. This is considerably larger than the values of about 6K to 7K of the confined helium but, bearing in mind the larger Van der Waals forces in hydrogen, is not unreasonable. Of course, rotons are normally regarded as excitations that exist in the superfluid phase of liquid helium.

CONCLUSIONS

The first result of this work is that the amount of entropy change associated with the melting of the hydrogen inside the porous glass indicates that a considerable fraction does not take part in this process. The second result is that at our lowest temperatures this fraction has a comparatively large specific heat and we can interpret this in different ways. The fact that it may be fitted to a T^2 dependence points to the possibility that it is produced by some interface layer(s) in which two dimensional phonons may prove to be the main excitations. However the data also appear to fit an analysis in which rotons are the principal excitations. On the relative quality of the fits to the two different ways of plotting our data it is, at this stage, not clear to regard one as being significantly better than the other. Clearly to resolve this it will be necessary to acquire more accurate data over a wider temperature range.

ACKNOWLEDGEMENTS

The authors wish to gratefully acknowledge the support of SERC grant number GR/F 51845.

REFERENCES.

(1) J.L. Tell and H.J. Maris, Phy.Rev. B Vol 28, (1983) 5122.
(2) M. Bretz and A.L. Thomson, Phy. Rev, B Vol 24, (1981) 467.
(3) D.F. Brewer, Cao Liezhao, C. Girit and J.D. Reppy, Physica 107B (1981) 583.
(4) D.F. Brewer, A. Evenson and A.L. Thomson, J. Low Temp. Phys. Vol 3, (1970) 603.
(5) R.J. Roberts and J.G. Daunt, J. Low Temp. Phys. Vol 6 (1972) 97.
(6) R.F. Dwyer, G.A. Cook, B.M. Shields and D.H. Stellrecht, J.Chem. Phys. Vol 42, (1965) 3809.
(7) D.F. Brewer, A.J. Symonds and A.L. Thomson Phys. Rev. Lett. Vol 15, (1965) 182.
(8) R.H. Tait and J.D. Reppy, Phys. Rev. B Vol 20, (1979) 997.

Physica B 165&166 (1990) 571–572
North-Holland

HEAT CAPACITIES OF ^{3}He AND ^{4}He IN ONE- AND THREE-DIMENSIONAL CHANNELS

Hiroyuki Deguchi, Kensuke Konishi, Kazuyoshi Takeda and Norikiyo Uryu

Department of Applied Science, Faculty of Engineering, Kyushu University, Fukuoka 812, Japan

Thermal properties of ^{3}He and ^{4}He adsorbed in high-silica zeolites have been studied in the temperature range 0.8-10 K for various helium concentration. In the case of ZSM-23 having one-dimensional(1-d) channels, helium behaves as a semi-quantum liquid. In ZSM-5 with three-dimensional(3-d) channel structure, the heat capacity of helium shows a characteristic shoulder around 2 K. This seems to be due to the motion of helium atoms trapped within the crossing space of the channels in ZSM-5.

1. INTRODUCTION

Recently, there have been several studies on quantum liquids and solids in restricted geometries such as zeolites.(1-3) In many zeolites, high-silica zeolites provide the very small channel and various structures, where the possibility of exchanging the position of helium atoms is limited. In the last conference(LT18), we reported the NMR study of ^{3}He adsorbed in high-silica zeolites, where we showed some new aspects of the Fermi degeneracy of ^{3}He(3): The nuclear susceptibility of ^{3}He in ZSM-5 fits well with the Fermi liquid theory, while that in ZSM-23 obeys the Curie's law below 1.5 K. The different experiments on ^{3}He in these two zeolites are very important. In this paper we report the heat capacity of adsorbed ^{3}He and ^{4}He to clarify the thermal property.

2. EXPERIMENTAL

ZSM-5 has the 3-d opening channels enclosed with the 10-membered atoms: Straight channels parallel to (010) have openings of size 5.4×5.6 A and the zigzag channels perpendicular to straight channels with openings of the size of 5.1×5.4 A intersect the straight channels at right angles.(4) ZSM-23 has 1-d channels parallel to (100) enclosed by the 10-membered atomic ring of size 5.6×5.3 A.(5) Each powdered specimen of which average size is about 1 μm is packed in a copper cell for the heat capacity measurement. After dehydration of the zeolites, sample gases of ^{3}He or ^{4}He is introduced into the cell and adsorbed at the liquid N_2 or ^{4}He temperature and cooled further down in a ^{3}He cryostat. After annealing up to above 5 K and cooling down to 0.8 K, we measured heat capacities by a conventional DC heat pulse method in the course of increasing temperature.

3. RESULTS AND DISCUSSION

Heat capacities are measured for various amount of adsorbed ^{3}He and ^{4}He gas in ZSM-23. The porosity ratio of ZSM-23 is small(0.21) and the ability of adsorbent is weak in channels. However, once He gas is adsorbed in channels, it is expected to behave without being trapped on inside walls. Small amount of sample gas is likely to be adsorbed at the surface of specimen and form a two-dimensional(2-d) solid of which heat capacity can be approximated to have a T^2 temperature dependence. Heat capacities of He atom in channels show T^1

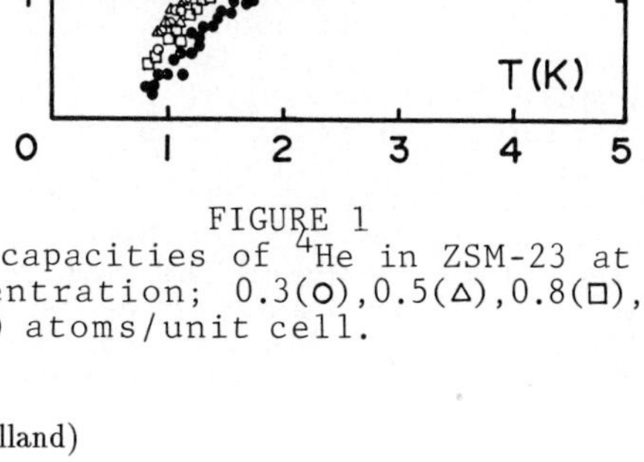

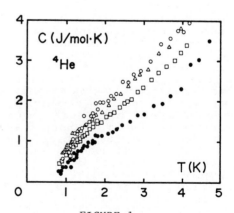

FIGURE 1
Heat capacities of ^{4}He in ZSM-23 at four concentration; 0.3(○),0.5(△),0.8(□), and 1.0(●) atoms/unit cell.

0921-4526/90/$03.50

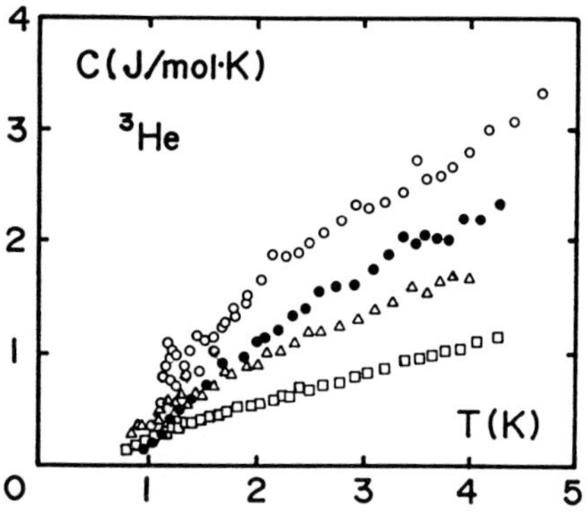

FIGURE 2
Heat capacities of ^3He in ZSM-23 at four concentration; 0.2(○),0.3(●),0.4(△) and 1.1(□) atoms/unit cell.

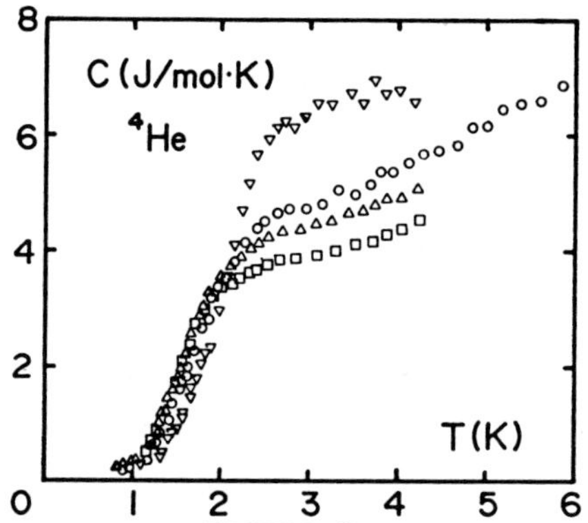

FIGURE 3
Heat capacities of ^4He in ZSM-5 at four concentration;1.5(▽),3.1(○),5.5(△) and 10.0(□) atoms/unit cell.

temperature dependence.(1,2) Actually, the both data for ^4He and ^3He fit well aT + bT^2 above 1.3 K. The value of b is smaller than a, so the second term contributes less than 20 % of the total heat capacity in this temperature range. Then we could subtracted the contribution of heat capacity of the 2-d solid with the Debye temperatures Θ_{2D}=38 K and 32 K for ^3He and ^4He, respectably. The results are shown in Figures 1 and 2 after the subtraction of the surface effect. Heat capacities of adsorbed ^3He and ^4He depend linearly on T as AT above 1.3 K, where the coefficient A is renormalized for He atoms in the 1-d channels. Here, the concentration of 1 atom/unit cell corresponds to 30 % of total pore volume. Below 1.3 K, the heat capacities becomes smaller than AT. In both case of ^3He and ^4He, A decreases monotonously with the concentration n of adsorbed gas ($0.57 \leq A \leq 0.97$(^4He), $0.27 \leq A \leq 0.97$(^3He)). Two possibilities are considered on the behavior of helium in the channel, of which heat capacity expressed by AT. One is in a 1-d solid state. However, it is not reasonable because the value of the Debye temperature varies largely with n if we fit the experimental results to the 1-d solid model. Theoretically, the Debye temperature should be independent of n or the length of cluster size. The other possibility is semi-quantum liquid state (6), which is observed in a bulky state of liquid helium (A=3.9(^4He),=2.3(^3He)) and helium in the restricted geometries such as zeolites ($1.0 \leq A \leq 2.2$).(1,2) The value of A in restricted geometries is proportional to the number of empty nearest sites to which the particle can tunnel(6), so it is expected to become small as the channel size is small or the density is large. The semi-quantum liquid state is probable to the present case.

Heat capacity of ^4He in ZSM-5 is shown in Figure 3. Here, 1 atom/unit cell corresponds to 2 % of total pore volume. The heat capacity increases abruptly with temperature increasing up to 2 K independently of the concentration n. Above 2 K, the heat capacity seems to depend linearly on temperature, though more data points are necessary to confirm so. Similar result is obtained for the case of ^3He. In ZSM-5, the space where channels are crossing is much larger than the space within a channel. Then, for the low concentration range as in the present case, He atom is supposed to be trapped in this larger space rather than in channel space. So we considered a model for a quantum particle in a small box. The energy gap of ^4He between ground and first excited states in a cubic box with a possible size of 5 A is estimated about 7 K and the heat capacity of ^4He in ZSM-5 below 2 K can be explained by this model. At high temperatures, helium seems to behave like the semi-quantum liquid state through 3-d channels.

REFERENCES

(1) H.Kato,N.Wada,T.Ito,S.Takayanagi and T.Watanabe: J.Phys.Soc.Jpn. 55 (1986) 246.
(2) H.Kato, K.Ishioh,N.Wada,T.Ito and T.Watanabe: J.LowTemp.Phys. 68 (1987) 321.
(3) H.Deguchi, Y.Moriyasu, K.Amaya, Y.Kitaoka,Y.Kohori, O.Terasakiand T.Haseda: Jpn.J.Appl.Phys. Suppl. 26-3 (1987) 311.
(4) G.T.Kokotailo,S.L.Lawton,D.H.Olsen and W.M.Meier:Nature 272 (1978) 437.
(5) L.M.Parker: Zeolite 3 (1983) 8.
(6) A.F.Andreev:Pis'ma Zh. Eksp.Teor. Fiz. 28 (1978) 603.

Physica B 165&166 (1990) 573–574
North-Holland

CRITICAL BEHAVIOR OF SUPERFLUID ^{4}He IN POROUS ALUMINA

J.R. BEAMISH and Kevin WARNER

Department of Physics, University of Delaware, Newark, Delaware 19716, U.S.A.

We have made ultrasonic measurements of the superfluid density and the dissipation in several porous alumina samples filled with ^{4}He. In one sample (formed by a sol-gel process) with high porosity (92%) and large pores (R=1600 Å), we found that the superfluid density disappeared at the transition with essentially the bulk exponent of 0.67, as expected. However, in a second material (slip-cast from fine powder) with lower porosity (52%) and much smaller pores (R=110 Å), the superfluid critical exponent was 0.83±0.02. The difference in critical behavior in the two media was also reflected in the dissipation. The sound attenuation in the first sample was due to viscous motion of the normal fluid and did not show any unusual behavior at the superfluid transition. In the small pore sample, on the other hand, there was a clear critical attenuation peak centered at the transition temperature.

1. INTRODUCTION

A number of recent experiments have shown that the critical behavior of helium contained in porous media is markedly different from that in bulk. Measurements using various porous silica glasses (1,2) have found a range of critical exponents, ζ, for the superfluid density. These include the bulk value (ζ=0.67) in Vycor glass, as well as new values in xerogel (ζ=0.89) and aerogels (ζ=0.81 in the lowest density glass, ζ=0.95 in a slightly higher density glass). The superfluid transition in aerogels, in particular, is remarkably sharp, although at a temperature slightly lower than in bulk. In aerogel, the specific heat has a prominent peak at the critical temperature, in contrast to Vycor where it varies smoothly throughout the transition region (3). These and surprisingly similar results for thin films adsorbed on Vycor or xerogel (1,4) lead one to look for structural properties of the porous medium which might determine the critical behavior.

We have recently used ultrasonic velocity and attenuation measurements to study the behavior of superfluid helium in porous media, including Vycor (5) and a silica aerogel (2). When a transverse sound wave is propagated through the porous material, any normal fluid is viscously locked to the substrate and contributes to the total density. Below the transition temperature, the superfluid fraction decouples from the substrate, reducing the effective density and increasing the sound speed. The superfluid density ρ_s is then related to the change $\Delta v/v_0$ in the sound velocity by

$$\rho_s \approx (2\alpha\rho_p/\phi)\Delta v/v_0 \qquad (1)$$

where α is the pore tortuousity, ϕ is the porosity and ρ_p is the density of the porous medium. The damping $1/Q$ can also be determined since it is proportional to the sound attenuation α_t.

We have used this technique to study the critical behavior of helium in two porous alumina samples. The first was formed by a sol-gel process similar to that used to make the silica aerogels, resulting in a low density material (ϕ=0.92) with a specific surface area of 35 m^2/g. This corresponds to an average pore radius of about 1600 Å, consistent with electron microscope images. With such large pores, deviations from bulk behavior are expected only within about 10^{-4} K of the transition where the correlation length exceeds the pore radius. The second sample was made by slip-casting alumina powder (nominal size 500 Å). After drying at 130 °C the porosity was 52% and the surface area 59 m^2/g, corresponding to an average pore radius of 110 Å. Since these samples are made from different material from the silica glasses and are expected to have different structure, it is interesting to compare their critical behavior.

2. RESULTS AND DISCUSSION

In Figure 1 we show the results for the first sample. The superfluid transition at 2.175 K is marked by a sharp increase in the superfluid density (as determined from the sound speed) and a decrease in the sound attenuation. The rather large attenuation in this sample is due to the viscous motion ("sloshing") of the normal fluid in the large pores. The transition temperature can be accurately determined from the sudden drop in attenuation when the normal fluid density begins to decrease.

Figure 2 shows the corresponding results for the second sample. The transition is somewhat less sharp, with the superfluid density showing rounding of about 20 mK above the transition (at 2.138 K). The attenuation is much smaller, since the smaller pores effectively lock the viscous fluid, and there is a critical peak centered at the transition.

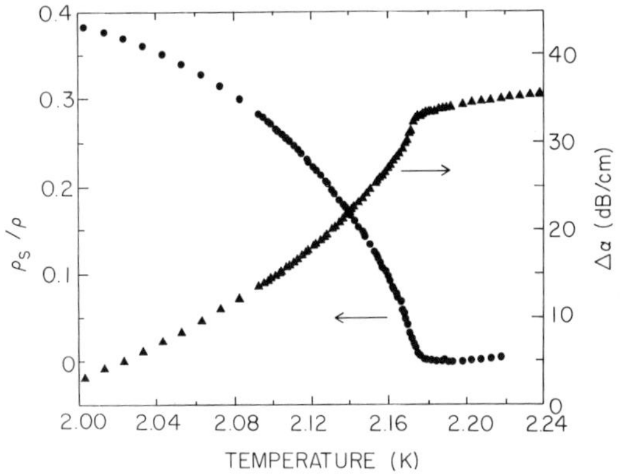

FIGURE 1
Superfluid fraction (circles, left axis) and attenuation of 4 MHz transverse sound (triangles, right axis) in a sol-gel formed alumina sample (porosity 92%, average pore radius 1600 Å).

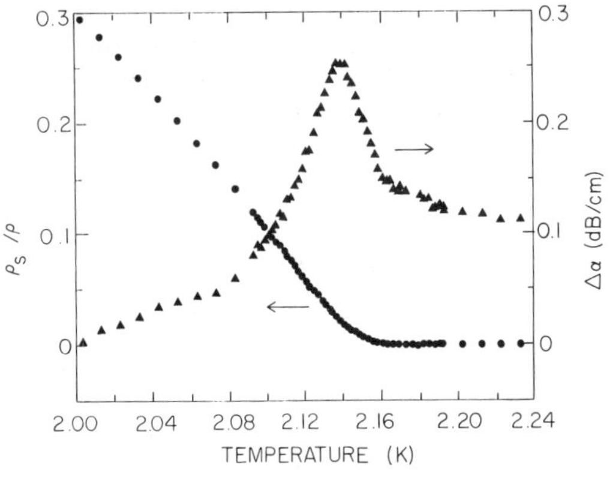

FIGURE 2
Superfluid fraction (circles, left axis) and attenuation of 8 MHz transverse sound (triangles, right axis) in a slip-cast alumina sample (porosity 52%, average pore radius 110 Å).

The superfluid density appears to disappear less rapidly in the second sample. This is confirmed in Figure 3 which shows the superfluid density versus the reduced temperature $t = (T_c - T)/T_c$ on a log-log scale. The solid lines are fits which give values for the critical exponents of ζ=0.67±0.02 for the first sample and ζ=0.83±0.02 for the second. The first is essentially the bulk exponent, as

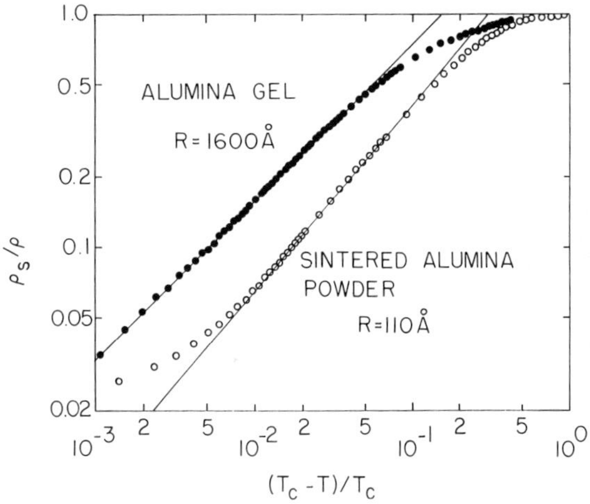

FIGURE 3
Superfluid density versus reduced temperature for the two samples of Figs. 1 and 2 above. The solid lines correspond to critical exponents ζ of 0.67 and 0.83 for the large and small pore samples, respectively.

expected over this range of reduced temperature given the large pore size. The second is distinctly different, giving yet another exponent for superfluid helium in porous media. The rounding of the transition for $t<10^{-2}$ for the second sample is comparable to that observed for Vycor and may well be associated with sample inhomogeneities.

ACKNOWLEDGEMENTS

We would like to thank G. Williams for providing the slip-cast sample. This research was supported in part by a grant from the Research Corporation.

REFERENCES

1. M.H.W. Chan, K.I. Blum, S.Q. Murphy, G.K.S. Wong and J.D. Reppy, Phys. Rev. Lett. 61 (1988) 1950.
2. J.R. Beamish and N. Mulders, in: Quantum Fluids and Solids-1989, eds. G.G. Ihas and Y. Takano (AIP, New York, 1989).
3. G.K.S. Wong and J.D. Reppy, M.H.W. Chan, private communications.
4. D. Finotello, K.A. Gillis, A. Wong and M.H.W. Chan, Phys. Rev. Lett. 61 (1988) 1954.
5. N. Mulders and J.R. Beamish, Phys. Rev. Lett. 62 (1989) 438.

Physica B 165&166 (1990) 575–576
North-Holland

SPECIFIC HEAT AND SUPERFLUID DENSITY OF CONFINED ^{4}HE NEAR T_λ

R. SCHMOLKE, A. WACKER, V. DOHM, D. FRANK

Institut für Theoretische Physik, Technische Hochschule Aachen, Germany

We study the effect of Dirichlet boundary conditions (D.b.c.) on the critical behavior of confined ^{4}He above and below T_λ by means of renormalized field theory. Scaling functions are presented for the specific heat and superfluid density in a geometry with D.b.c. in one direction and periodic b.c. in d-1 directions. The results are briefly discussed with regard to existing data.

1. INTRODUCTION

The field-theoretic approach to calculations of finite-size effects near critical points has been extended recently to the case of Dirichlet boundary conditions (D.b.c., vanishing order parameter at the boundaries) /1,2/. Application to the specific heat indicated that D.b.c. are fairly realistic in the case of the superfluid transition of ^{4}He. In the following we shall present further results on the specific heat and extend the theory to $T<T_\lambda$. Preliminary results for the superfluid density are also given. We shall consider a cube of length L and use the model for a two-component order parameter $\phi(x)$

$$H = \int d^dx\left(\frac{1}{2} r_o \phi^2 + \frac{1}{2} (\nabla\phi)^2 + u_o \phi^4\right) \qquad (1)$$

in the presence of D.b.c. in one direction. Furthermore we comment on the size-dependent reference temperature $T_c(L)$ discussed recently /1/.

2. THEORY FOR $T \gtrsim T_\lambda$

After integration over all standing-wave modes the partition function can be written as

$$Z = \int d^2 \phi_1 \; \exp - L^d (H_1 + \Gamma), \qquad (2)$$

$$H_1 = (1/2)\,(r_o + \pi^2/L^2)\,\phi_1^2 + (3/2)\,u_o\,\phi_1^4\,, \qquad (3)$$

where ϕ_1 is the lowest-mode amplitude./1/ An L-dependent shift $r_{oc}(u_o,L)$ of r_o can be defined by requiring that the quadratic term of $H_1 + \Gamma$ vanishes at $r_o=r_{oc}(u_o,L)=-\pi^2/L^2 + O(u_o)$. So far there exists no fully quantitative calculation of r_{oc}. It can be shown that in one- loop order H, (1), remains renormalizable, in the presence of D.b.c., for $d \leq 4$ after the *bulk* shift $r_o - r_{oc}(u_o,\infty)$ has been performed. Consider the hypothesis that this property remains valid to all orders and that no new Z factors are introduced by D.b.c.. We find that this implies

$$r_{oc}(u_o, \infty) - r_{oc}(u_o,L) \sim T_c(\infty) - T_c(L) \sim L^{-1/\nu} \qquad (4)$$

for $L \to\infty$ where ν is the bulk correlation-length exponent. For large L this would invalidate the applicability of the nonscaling expression for $r_{oc}(u_o,L)$ given in Ref. 1 (where the above hypothesis was not made). On the other hand Eq.(4) would be in conflict with the observed /3/ exponent ν .This problem, however, is not related to the two main effects of D.b.c. /1/: (i) the large reduction of the specific-heat maximum, (ii) the gradual onset of surface effects far from T_λ . These effects are substantiated by our one-loop results for the scaling functions of the specific heat C(t,L), with $t_o^\nu = \xi_o/L$,

$$C(t,L) - C(t_o,\infty) = L^{\alpha/\nu}\, f_1(tL^{1/\nu}), \qquad (5)$$

$$C(t,L) - C(t,\infty) = - L^{\alpha/\nu}\, f_2(tL^{1/\nu}). \qquad (6)$$

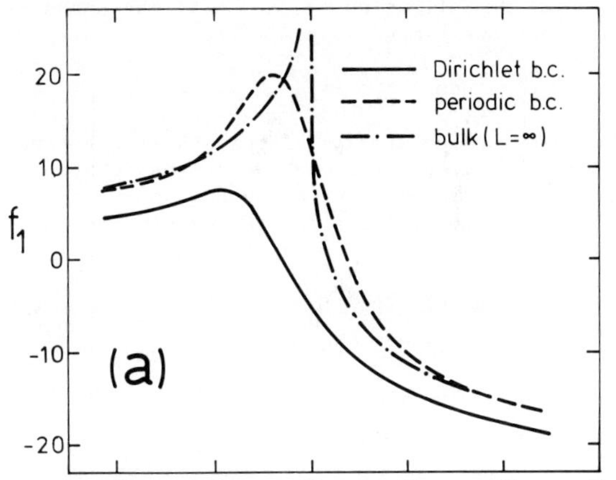

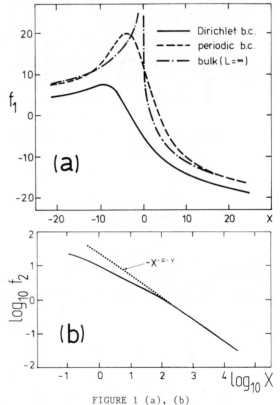

FIGURE 1 (a), (b)
Scaling functions of the specific heat C (in J mol^{-1} K^{-1}) vs $x=tL^{1/\nu}$,L in Å, $t=(T-T_\lambda)/T_\lambda$. The dotted line in (b) is $C_s \sim t^{-\alpha-\nu}/L$.

3. THEORY FOR $T < T_\lambda$

The sinusoidal mode functions used above T_λ are not appropriate below T_λ. Instead we use (i) a renormalized mean-field (MF) approach, (ii) a renormalized lowest-mode approximation. The MF order-parameter profile $\psi(z)$ is determined by

$$[-(d/dz)^2 + r_o + 4\,u_o\,\psi^2]\;\psi = 0 \qquad (7)$$

with $\psi(0) = \psi(L) = 0$. In the approach (i), we substitute $\phi=\psi$ into (1); in the approach (ii) we use $\phi(x) = \phi_o\, f_o(z)$ where $f_o(z)$ is a normalized mode function proportional to $\psi(z)$ and ϕ_o is a two-component amplitude. After integration over z the lowest-mode Hamiltonian becomes

$$H_o = \frac{1}{2}\,A(r_o,u_o,L)\;\phi_o^2 \;+\; B(r_o,u_o,L)\;\phi_o^4 \qquad (8)$$

which replaces H_1 in (3) and joins H_1 smoothly at $r_o = -\pi^2/L^2$. A and B have the appropriate bulk $(L\to\infty)$ limits r_o and u_o. In the approach (ii) we calculate the free energy

$$F = -\ln \int d^2\phi_o\;\exp - L^d\, H_o \quad . \qquad (9)$$

We have applied the theory to the specific heat C^- and the superfluid density ρ_s. For ρ_s we use (i) the MF definition /4/ and (ii) the unrenormalized lowest-mode approximation

$$\rho_s = \text{const}\;<|\phi_o|>_o^2 \quad . \qquad (10)$$

We have incorporated both approaches in a renormalization-group treatment which ensures the correct non-MF critical exponents for C^- and ρ_s. For the asymptotic scaling representations in Figs. 2 and 3 we use (5) and /5/

$$k - \rho_s|t|^{-\nu}/\rho = f_\rho(|t|\;L^{1/\nu}). \qquad (11)$$

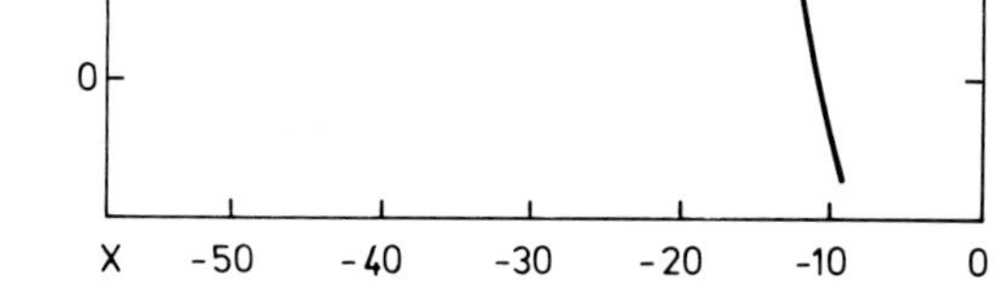

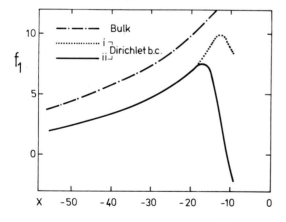

FIGURE 2
Scaling function of the specific heat below T_λ according to (5), as in Fig. 1(a).

Approach (i) is applicable only for $\xi_-/L < 1$ whereas approach (ii) is valid also for $\xi_-/L >> 1$ since it includes the fluctuations of the lowest mode. Approach (ii) can also be systematically extended to include the effects of higher modes.

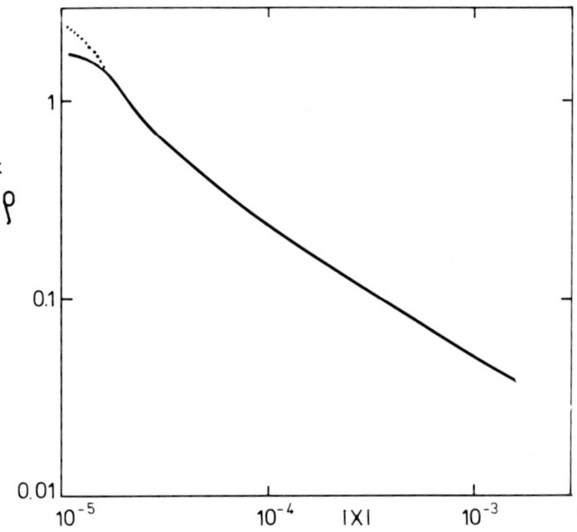

FIGURE 3
Scaling function of the superfluid density according to (11), with L in µm, k=2.35. Solid line: approach (ii). Dotted line: approach (i).

4. COMPARISON WITH EXPERIMENTS

The general features of our scaling functions f_1 and f_2 for D.b.c. agree with those of the data /3/. As far as the detailed L dependence of C is concerned, however, we do not have a clear-cut interpretation of the observed deviations from scaling /3/ because of both theoretical and experimental uncertainties. For $T_\lambda - T_m$ we estimate the error bars of the 300 Å and 2000 Å data to be larger than given in Ref. 6. Our scaling function for ρ_s can be compared with the data as presented in the top panel of Fig. 4 of Ref. 5. Our result clearly disagrees with these data which do not collapse on a single curve.

REFERENCES

/1/ W. Huhn and V. Dohm, Phys. Rev. Lett. **61** (1988) 1368.

/2/ V. Dohm, Z. Phys. **B 75** (1989) 109.

/3/ T.P. Chen and F.M. Gasparini, Phys. Rev. Lett. **40** (1978) 331.

/4/ V.L. Ginzburg and L.P. Pitaevski, Sov. Phys. JETP **7** (1958) 858.

/5/ I.Rhee, F.M. Gasparini, and D.J. Bishop, Phys. Rev. Lett. **63** (1989) 410.

/6/F.M. Gasparini, T.P. Chen, and B. Bhattacharyya, Phys. Rev. **B 23** (1980) 5797.

Physica B 165&166 (1990) 577–578
North-Holland

THE THERMAL PROPERTIES OF HELIUM CONFINED IN SMALL GEOMETRIES

D F BREWER, J RAJENDRA, N SHARMA, A L THOMSON and JIN XIN*

University of Sussex, Physics Division, Brighton, Sussex BN1 9QH, Great Britain

The phenomenon of the shifts in the freezing and melting temperatures of various elements inside media with small pores is well established. What is not yet well understood are the attendant changes in the properties of the liquid and solid phases at these anomalously low temperatures. In our present work we produce various analyses of our experimental data to produce estimates for the way in which the confinement of pressurized helium inside the pores of Vycor glass has changed its properties from those of the bulk phase. Simultaneous measurements of pressure and heat capacity, coupled with quantitative measurements on the amount of helium sample inside our porous glass, allow a significant number of parameters to be examined. In particular the entropy difference between the liquid and solid phases is presented, as well as the volume difference between these phases. At least the former of these two properties appears to have values substantially less than that observed in the bulk phase.

We present simultaneous measurements of the specific heat of helium confined in pores and the corresponding pressure changes. Coupled with accurate volumetric measurements we are able to calculate several properties which accompany the solid–liquid phase change including the entropy and volume differences. In addition, the interfacial energy between the two phases is determined and the compressibility of the liquid–solid mixture.

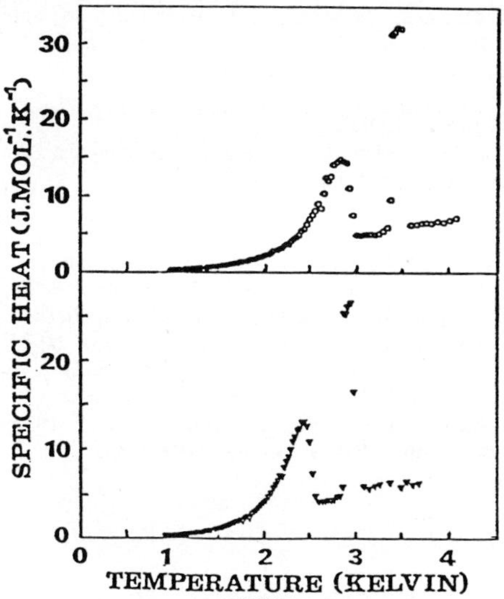

FIGURE 1

A plot against temperature of the heat capacity of our confined helium divided by the total number of moles in the sample chamber. The upper curve is for an experiment where the pressure at the lowest temperatures was around 87 bars while the lower curve is for an experiment whose lowest pressure was around 63 bars.

Displayed in Figure 1 are the results of two sets of measurements of the specific heat of helium in the pores of Vycor glass. Both sets contain two peaks, which are associated with melting from the solid to the liquid phase. The lower temperature and broader peak is due to melting of the helium inside the pores, while the narrow higher temperature peak is produced by melting of bulk helium present in the sample chamber. Both sets of data were taken at quite high pressures; at the lowest temperatures they were 63 and 87 bars. The peaks at the higher temperatures agree well with bulk values.

A most interesting quantity to measure is the entropy associated with the respective melting anomalies. Underneath our bulk melting curves we measure 3×10^{-2} JK^{-1} compared with approximately 6 $Jmole^{-1}K^{-1}$ for bulk (1) from which we can infer the fraction of our sample which is outside the pores. For the broader anomalies at lower temperatures we find that the difference in the entropies of our liquid and solid phase inside the Vycor is 5.1×10^{-2} JK^{-1} for the 63 bars run and 4.3×10^{-2} JK^{-1} for the 87 bars run. These values would imply that only a fraction of sample inside the pores participates in the melting process, if we were to assume that the latent heat had the bulk value; this is not necessarily true for very small samples.

We can also calculate from the specific heat the interfacial surface tension γ_{LS} between the liquid and solid phases for the helium inside the pores. This can be obtained approximately (2) from the equation

$$\gamma_{LS} \approx \frac{r}{2V_s} \int_{T_i}^{T_f} (T_m - T) \frac{\Delta C_v}{T} dT. \qquad (1)$$

---

*Jin Zin has now returned to Nanjing University, Nanjing, China

Here r is the radius of curvature of the liquid–solid interface inside the Vycor pores, V_s is the molar volume of the solid phase and ΔC_v is the part of our measured heat capacity adjudged to be caused by the latent heat of melting of helium inside the pores. The integral is carried out from the initial temperature T_i at which this melting anomaly begins up to the final temperature T_f at which it ends. This calculation yields the results:

$$\gamma_{LS} = \left[\frac{1}{x}\right] 0.130 \text{ erg/cm}^2 \quad \text{at } T \sim 2.4\text{K}$$

and $$\gamma_{LS} = \left[\frac{1}{x}\right] 0.142 \text{ erg/cm}^2 \quad \text{at } T \sim 2.8\text{K}$$

where x is the fraction of helium inside the pores which takes part in the melting/freezing process. Our value for γ_{LS} should be compared with that (3) derived from bulk measurements, γ_{LS} = 0.21 erg/cm^2 at T = 1.28K. Comparing these two results, and assuming that the interfacial energy is the same in small and large samples, we find x ~ 0.62 and 0.68. When these values are inserted back into the entropy difference calculation we obtain molar quantities which are

$$\Delta S_{LS} \sim 3.0 \text{ Jmole}^{-1}\text{K}^{-1} \quad \text{at } T \sim 2.4\text{K}$$

and $$\Delta S_{LS} \sim 2.3 \text{ Jmole}^{-1}\text{K}^{-1} \quad \text{at } T \sim 2.8\text{K}$$

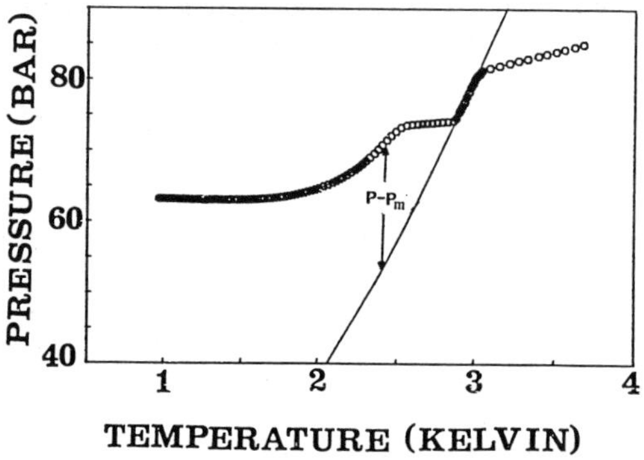

FIGURE 2

The pressure inside our sample chamber plotted as a function of temperature. The data points correspond to the specific heat run displayed in the lower half of Figure 1. The solid line represents the pressures P_m of the bulk melting curve.

In addition to the specific heat data we took simultaneous measurements of pressure inside the sample chamber, and Figure 2 contains part of these data corresponding to the lower curve of Figure 1. The higher melting pressure inside the pores compared with the bulk phase is given by $(P - P_m)$ where P_m is the pressure of the bulk melting curve. This pressure difference can be derived (4) in terms of the interfacial surface tension to be

$$P - P_m = \frac{V_s}{\Delta V_{LS}} \quad \frac{2\,\gamma_{LS} \cos\theta}{r} \qquad (2)$$

Here ΔV_{LS} is the molar volume difference between the liquid and solid phases and r is the radius of curvature of the liquid–solid interface, while θ is the angle made by the liquid–solid meniscus at the surrounding wall. We can use Equation (2) to calculate ΔV_{LS} by inserting experimental values (3) for γ_{LS} and θ and also a value of r = 25Å for the effective pore radius. The values for ΔV_{LS} calculated in this way are

$$\Delta V_{LS} = 1.4 \text{ cm}^3\text{mole}^{-1} \quad \text{at } T \sim 2.4\text{K}$$

and $$\Delta V_{LS} = 1.0 \text{ cm}^3\text{mole}^{-1} \quad \text{at } T \sim 2.8\text{K}$$

These values are similar to the values corresponding to the bulk phase which are 1.33 cm^{3}mole^{-1} and 1.25 cm^{3}mole^{-1}. They should be compared with a result from an experiment (5) in which much more bulk helium was present. This yielded the value ΔV_{LS} ~ 0.55 cm^{3}mole^{-1}.

CONCLUSIONS

The fact that our derived values for ΔS_{LS} are substantially different from the bulk values indicates that either the confined liquid or the confined solid (or even both) are physically altered in some way – perhaps by the presence (or absence) of some defects such as vacancies. However it is interesting to note that this different physical state does not appear to have affected ΔV_{LS} to any large extent.

ACKNOWLEDGEMENTS

The authors wish to gratefully acknowledge the support of SERC grant number GR/F 51845.

REFERENCES

(1) C A Swenson Phys. Rev. Vol 79, (1950) 626.
(2) N Sharma, Ph.D Thesis, Sussex University (1989).
(3) A V Babkin, K O Keshishef, O B Kopelioritch, Ya. Parshina, Zh. Pisma, Eksp. Teor. Fiz (1984) 579. (Sov. Phys. JETP Lett. 39, (1984) 633).
(4) L D Landau, E M Lifshitz, Statistical Physics.
(5) E D Adams, Y H Tang, K Uhlig and G E Haas, J. Low Temp. Physics Vol 66, (1987) 85.

Physica B 165&166 (1990) 579–580
North-Holland

UNIVERSAL EFFECT OF SURFACE ROUGHNESS ON THE CRITICAL CURRENT OF SUPERFLUID ^{4}He IN NARROW PORES

Mark W. Meisel and Pradeep Kumar

Department of Physics, University of Florida, 215 Williamson Hall, Gainesville, FL 32611, USA.

The nucleation of vortices in superfluid ^{4}He flowing through small orifices is discussed for the case where no primordial vortex lines exist. Vortex loops are generated in the vicinity of wall irregularities which have a characteristic length of approximately ten times the coherence length of the superfluid. Our analysis of the experiments seems to be consistent with this mechanism when the superfluid condensate density near the wall is taken to be temperature independent, due to surface correlations.

Questions concerning the nature of vortices and critical velocities in superfluid ^{4}He have been the focus of investigation for many years.[1] Presently, several characteristics appear to distinguish between various types of critical velocities. For example, *nonideal*[2] critical velocities, usually generated by normal fluid versus superfluid counter flow in large channels, typically >1000Å, appear to be well described by hydrodynamic interactions with a tangled mass of quantized vortex lines.[3] In addition, several theoretical descriptions,[4-5] involving the *ideal*[2] nucleation of vortex rings in the bulk superfluid, have been successful in describing some of the observed experimental features.[6]

Until recently, most experiments have studied *nonideal* critical velocities associated with the vortex tangle or with primordial vortices (created when the system is cooled into the superfluid state). However, more recent investigations of the critical current in flowing superfluid have attempted to avoid the existence of primordial vortices.[1] The results of these experimental investigations are not well described by any presently existing theory. Bowley and his coworkers[7] have noted that if vortices are created in geometries where the relevant length scale is much less than the distance between pre–existing vortices, as in the new experiments,[8-10] then it is reasonable to suppose that new vortices are nucleated. Schwarz[11] has analyzed the experiments in terms of the dynamics of primordial vortices that are pinned in the channel, and this picture has been debated in the literature.[12-13]

Before associating the initial occurrence of quantized dissipation in the flow of superfluid ^{4}He through a micro–orifice with the onset of an *ideal* critical velocity, two questions need to be addressed. Firstly, the measured[9-10] critical velocity has a linear temperature dependence that contains a scale temperature nearly equal to the superfluid transition temperature, T_c, and all other quantities in the expression are temperature independent. Secondly, when analyzed in any reasonable framework, there is also a length scale of approximately 50Å. In the following, we propose an explanation within the *ideal* picture.

We consider an orifice whose interior contains wall imperfections, with a characteristic dimension of several times the superfluid coherence length, $\xi \simeq 2$Å. The length scale of these defects has an upper bound due to the fabrication of the pores and a lower bound arising from the smoothing caused by the adsorbed ^{4}He layer. In the well known physical picture at the apex of this defect, the fluid velocity will be elevated compared to the velocity in the bulk fluid. As a consequence, a critical velocity will be achieved at this location, and a vortex will be nucleated. In the following discussion, the first layer of helium in contact with the surface will be assumed to be immobile and solid–like.[14] This layer will play no explicit role in the nucleation process. It will be assumed that the second layer of helium will be a superfluid and will have an effective pressure (corresponding to ≈ 20 bars) which arises from the surface attraction.[14] The nucleation process is envisioned as arising from the superfluid confined to this layer whose density is taken to be independent of the bulk fluid temperature and pressure, due to the quasi–two dimensional surface correlations. All additional layers are expected to exhibit bulk superfluid properties.

The nucleation of a vortex will require a rate which can be written

$$\nu = \nu_0 + A \exp(-E_a/kT) \ , \qquad (1)$$

where ν_0 represents a temperature independent nucleation rate (which includes the quantum mechanical limit[15] and parasitic vibrations of the walls arising from perturbations of the cryostat), E_a is an activation energy, and A is an attempt frequency. In the absence of external vibrations, Eq. 1 yields the Volovik picture[15] as $T \to 0$, and the Iordanski–Langer–Fisher description[4-5] as $T \to T_c$. The loops have a critical radius given by a stability condition for growth where the self–energy of the loop balances the energy of the loop provided by the flowing superfluid. Eventually, as T decreases, the critical radius of the loop will achieve length scales equal to the physical dimensions of the pore. As described above, it is natural for the wall imperfections to define this limiting critical radius for loop generation. In this picture,

$$E_a = E_{self} - p_0 v_s \ , \qquad (2)$$

with

$$E_{self} \approx \frac{1}{2} \rho_{sc} \kappa^2 R \left\{ \ln\left(\frac{8R}{\xi}\right) - \frac{7}{4} \right\} \qquad (3)$$

and

$$p_0 = \pi \rho_{sc} \kappa R^2 \ , \qquad (4)$$

0921-4526/90/$03.50 © 1990 – Elsevier Science Publishers B.V. (North-Holland)

where ρ_{sc} is the density of the superfluid condensate[16] (which, due to the influence of the surface, will be only weakly pressure and temperature dependent), R is the radius of the nucleated vortex loop and, in our model, should be the curvature of the hemispheric wall imperfection, κ is the quantum of circulation, $\kappa = h/m_4$, and p_0 is the impulse of the loop.

When the rate of nucleation reaches a critical level, the critical velocity of the superfluid is obtained, i.e. $\nu=\nu_c$ in Eq. 1 and $v_s=v_c$ in Eq. 2. Rewriting these equations leads to

$$v_c = v_{co} \left\{ 1 - \frac{T}{T_o} \right\} , \qquad (5)$$

where

$$v_{co} = \frac{E_{self}}{p_o} \quad \text{and} \quad T_o = \frac{E_{self}}{k\Gamma} , \qquad (6)$$

where $\Gamma=\ln\{A/(\nu_c-\nu_o)\}$. Equations 5–6 are the constitutive equations to be compared to the experimental observations.

In order to compare the above predictions to the experimental results, the values of the physical parameters must be set. The most critical choice will be for R, which is the radius of the nucleated vortex loop and is also assumed to be the approximate radius of the hemispheric wall defect, i.e. R $\simeq$ 50 Å. This choice for R has been discussed by Varoquaux *et al.*,[9] who have also estimated that, at least, $\nu_c \simeq 10^3$ s^{-1}. The surface influenced two dimensional superfluid will be assumed to have an effective pressure of approximately 20 bars.[14] From this assumption, $\rho_{sc} \simeq 1.7{\times}10^{-2}$ g cm^{-3} [Ref. 16,17]. With the idea that the nucleation arises from fluctuations of the atoms in the vicinity of the wall imperfection, then $A \simeq 10^{16}$ s^{-1} [Ref. 9]. Finally, in the absence of mechanical vibrations and outside the T→0 limit, $\nu_o << \nu_c$.

With the above values, Eq. 6 gives $v_{co} \simeq 11$ m/s and $T_o \simeq 3.4$ K. Before comparing these values with the experimental results, it is important to note that the experiments do not measure v_{co} but instead observe a global critical velocity, v_{gco}, of the flow averaged over the entire orifice, namely $v_{gco} \simeq (2/3)v_{co}$. Varoquaux *et al.*[9] reported $v_{gco} = 1.5$ m/s with an uncertainty of a factor of 3. Beecken and Zimmerman[10] observed v_{gco} values between 9 m/s and 20 m/s. As Varoquaux *et al.*[9] and Beecken and Zimmerman[10] have discussed, there are several experiments which indicate that $T_o \simeq 2.4$ K. Further comparisons of this model with experimental results are difficult because the majority of earlier experiments have studied *nonideal* rather than *ideal* critical velocities. In addition, external mechanical vibrations of the experimental cell introduce important temperature independent nucleation processes. If sufficiently strong, these vibrations would govern the nucleation mechanisms and would provide a temperature independent critical velocity.

In conclusion, this paper provides a possible description of the temperature dependence of *ideal* critical velocities studied by superfluid flow through micro–orifices. The universal nucleation length scale of ~50Å is associated with wall imperfections whose curvature is limited by fabrication techniques and by the adsorbed helium layer. The process involves thermal nucleation of vortex loops in the superfluid in the vicinity of the wall defects. The quasi–two dimensional surface correlations strongly influence this superfluid and cause the relevant density (that of the superfluid condensate) to be relatively independent of the bulk fluid pressure and temperature. This description and the experimental results for v_{cg} and T_o are in remarkably good agreement. Further refinements of the model would require a microscopic picture of Γ in Eq. 6. Despite the fact that the fabrication of channels with atomically smooth walls is not presently possible, a natural test of the picture would be to smooth the walls by adsorbing an inert gas layer prior to the introduction of the helium. Under these conditions, the *ideal* critical velocities would be associated with vortices nucleated in the bulk or at the lips of the orifice. This particular test is possible and worthy of experimental attention.

One of us (MWM) gratefully acknowledges fruitful conversations with Roger Bowley and John C. Waterman. This work was made possible in part by support received from the UF DSR (MWM) and NATO 88–703 and NSF INT–8815199 (PK).

1. E. Varoquaux, O. Avenel and M.W. Meisel, Can. J. Phys. 65, 1377 (1987); and references therein since space limitations do not permit a detailed listing.
2. W.F. Vinen, in *Liquid Helium*, Proc. Int. School of Physics, Enrico Fermi Course XXI, Academic Press, New York (1963), p. 336.
3. J.T. Tough, in *Progress in Low Temperature Physics*, Vol. VIII, D.F. Brewer ed., North–Holland, Amsterdam (1982), Ch. 3.
4. S.V. Iordanskii, Sov. Phys. JETP 21, 467 (1965).
5. J.S. Langer and M.E. Fisher, Phys. Rev. Lett. 19, 560 (1967).
6. J.S. Langer and J.D. Reppy, in *Progress in Low Temperature Physics*, Vol. VI, C.J. Gorter ed., North–Holland, Amsterdam (1970), p. 1.
7. P.C. Hendry, N.S. Lawson, P.V.E. McClintock, C.D.H. Willaims and R.M. Bowley, Phys. Rev. Lett. 60, 604 (1988). R.M. Bowley, AIP Conf. Proc. No. 194 (1989) 149.
8. O. Avenel and E. Varoquaux, Phys. Rev. Lett. 55, 2704 (1985).
9. E. Varoquaux, M.W. Meisel and O. Avenel, Phys. Rev. Lett. 57, 2291 (1986).
10. B.P. Beecken and W. Zimmermann, Jr., Phys. Rev. B 35, 1630 (1987).
11. K.W. Schwarz, Phys. Rev. Lett. 57, 1448 (1986).
12. K.W. Schwarz, Phys. Rev. Lett. 59, 1167 (1987).
13. O. Avenel and E. Varoquaux, Phys. Rev. Lett. 59, 1168 (1987).
14. D.F. Brewer, J. Low Temp. Phys. 3, 205 (1970).
15. G.E. Volovik, JETP Lett. 15, 81 (1972).
16. L.J. Campbell, Phys. Rev. B 27, 1913 (1983); AIP Conf. Proc. No. 103 (1983) 403. A.A. Sobyanin, A.A. Stratonnikov, JETP Lett. 45 , 613 (1987).
17. J. Wilks, *The Properties of Liquid and Solid Helium*, Clarendon Press, Oxford, 1967.

Physica B 165&166 (1990) 581–582
North-Holland

SECOND AND FOURTH SOUND MODES FOR SUPERFLUID HELIUM IN AEROGEL

M. J. McKENNA, TANIA M. SLAWECKI, and J. D. MAYNARD

Department of Physics, The Pennsylvania State University, University Park, PA 16802, USA†

Because of the interest in superfluid onset in porous media (high porosity aerogel in particular) one would like to measure the superfluid fraction as well as the heat capacity near T_c. However, standard techniques such as torsional oscillators and fourth sound have encountered difficulties because the compliant aerogel allows the viscous normal fluid to move. Using data we obtained for fourth sound propagation below 1.8 K, we develop a theory which fits the data with no adjustible parameters, and show that the theory predicts another second-sound-like mode which may not be damped near T_c.

Currently there is considerable interest in the onset of superfluidity for 4He in porous media, particularly in high porosity (~95%) aerogel. It would be desirable to complement the existing heat capacity measurements with measurements of the superfluid fraction ρ_s/ρ. However, standard techniques such as torsional oscillators and fourth sound have been unsuccessful because the compliant aerogel allows the viscous normal fluid to move; near T_c where the normal fluid density ρ_n/ρ is large, measurements of small ρ_s/ρ are limited by the large viscous damping. Recently we have made measurements of fourth sound in aerogel at temperatures up to 1.8 K, where ρ_n/ρ is not sufficiently large to damp the sound propagation. We have used this data to develop a theory which fits the data with no adjustible parameters. We find that the theory predicts a second-sound-like mode (originally proposed by J. D. Reppy) which may not be damped near T_c, and we obtain expressions for the relative normal fluid velocity, pressure amplitude, and temperature amplitude for the sound modes.

The apparatus which we used for measuring the fourth sound is shown in Fig. 1. The aerogel sample is a 2.0 mm diameter, 2.7 mm long cylinder inside teflon heat-shrink. The aerogel has a density ρ_a = .11 g/cm^3, and a sound speed C_a = 110 m/s. PVDF transducers are placed at each end of the sample, forming an acoustic resonator. The resonance frequency is converted to an effective ρ_s/ρ, normalized at a low temperature. Fig. 2 shows the data as triangles; the solid line is the theoretical prediction, and the dashed line is what would be obtained if the aerogel were perfectly rigid.

For the theory we begin with the usual two-fluid hydrodynamic equations, assuming a small normal fluid velocity v_n and small superfluid velocity v_s:

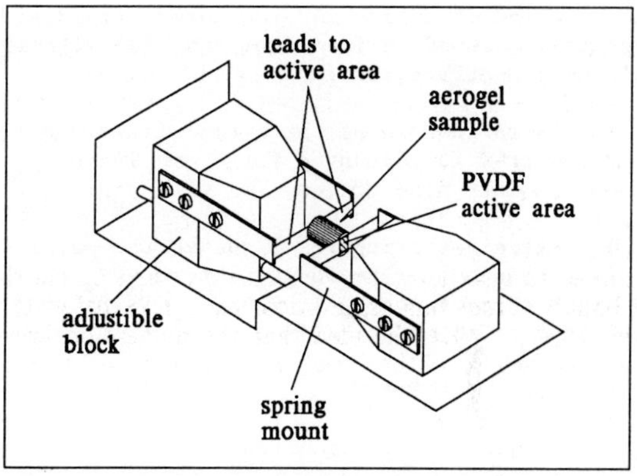

Figure 1. Experimental apparatus. The transducers cover the ends of the sample, forming a small acoustic resonator.

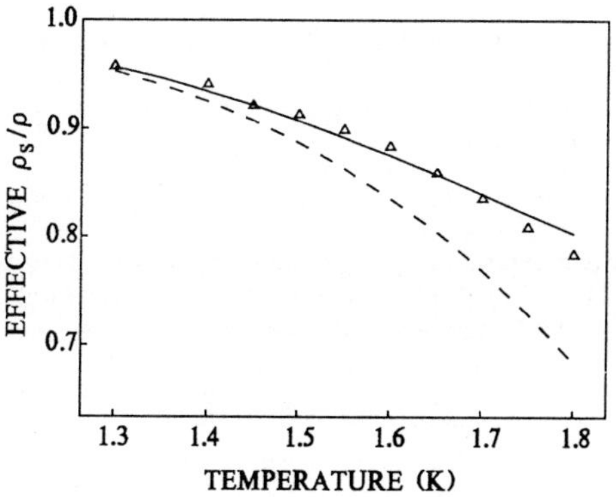

Figure 2. Aerogel fourth sound data (triangles) theory (———); bulk ρ_s/ρ (-----).

†Work supported by NSF DMR 8701682.

$$\frac{\partial\rho}{\partial t} + \vec{\nabla}\cdot(\rho_n\vec{v}_n + \rho_s\vec{v}_s) = 0 \qquad (1)$$

$$\frac{\partial\rho S}{\partial t} + \vec{\nabla}\cdot(\rho S\vec{v}_n) = 0 \qquad (2)$$

$$\frac{\partial\vec{v}_s}{\partial t} = -\frac{1}{\rho}\vec{\nabla}P + S\,\vec{\nabla}T \qquad (3)$$

$$\rho_n\frac{\partial\vec{v}_n}{\partial t} = -\frac{\rho_n}{\rho}\vec{\nabla}P - \rho_s S\,\vec{\nabla}T \qquad (4)$$

where T is temperature, P is pressure, S is entropy, and ρ is the helium mass density. In the usual derivation of fourth sound one assumes that the normal fluid is locked to a rigid porous media matrix, so that eqn. (4) is replaced with simply $v_n = 0$. Here we assume that the normal fluid is locked to the matrix, but that the matrix and the normal fluid move together with a velocity $v_n \neq 0$. We assume that the equation of motion for the normal fluid and the matrix can be derived from eqn. (4) with the following modifications:

The extra inertia of the normal fluid due to the matrix is accounted for by replacing ρ_n on the left side of eqn. (4) with $\rho_n + \rho_a$.

The extra restoring force due to the matrix is accounted for by adding a term $-\vec{\nabla}P_a$ on the right side of eqn. (4), where P_a is related to the speed of sound in the matrix C_a with

$$C_a^2 = (\partial P_a/\partial\rho_a) \qquad (5)$$

Eqn. (4) is thus replaced with

$$(\rho_n+\rho_a)\frac{\partial\vec{v}_n}{\partial t} = -\frac{\rho_n}{\rho}\vec{\nabla}P - \vec{\nabla}P_a - \rho_s S\,\vec{\nabla}T \qquad (6)$$

For conservation of mass for the matrix we have

$$\frac{\partial\rho_a}{\partial t} + \vec{\nabla}\cdot(\rho_a\vec{v}_n) = 0 \qquad (7)$$

For small displacements from equilibrium we write

$$P(x,t) = P + P'e^{ik(x-Ct)} \qquad (8)$$

$$T(x,t) = T + T'e^{ik(x-Ct)} \qquad (9)$$

Combining eqn. (1) - (3) and (5) - (9), to first order in P′ and T′, we obtain the following secular equation for the sound velocity C:

$$C^4 - C^2(C_1^2+\gamma C_2^2) + C_1^2C_2^2 + \frac{\rho_a}{\rho_n}(C^2-C_a^2)(C^2-C_4^2) = 0 \qquad (10)$$

where γ is the ratio of specific heats and, in the limit of small thermal expansion ($\gamma\simeq1$), C_1, C_2, and C_4 are the usual velocities of first, second, and fourth sound. In the limit $\rho_a\to0$, the solutions for C from eqn. (10) are C_1 and C_2, as they should be. In the limit $\rho_a\to\infty$, the solutions are C_4 and C_a, again as expected (the solution C_a is usually very large and ignored). For the values of ρ_a and C_a appropriate to our experiment, eqn. (10) yields solutions C_{4a} and C_{2a}, where C_{4a} is a velocity intermediate between C_4 and C_1, and C_{2a} is intermediate between C_2 and C_4. The temperature dependence of all of the velocities are shown in Fig. 3. Near T_c, C_{2a} is proportional to $\sqrt{\rho_s/\rho}$. The effective "fourth sound superfluid density", $(C_{4a}/C_1)^2$, is shown as the solid line in Fig. 2.

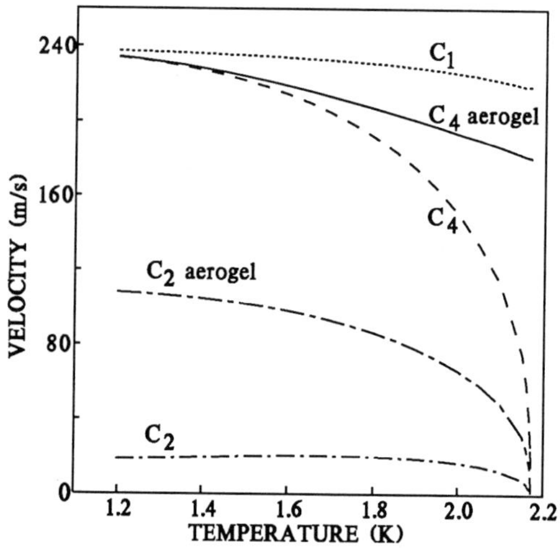

Figure 3. Sound velocities for superfluid helium in bulk and in aerogel.

As was found in the experimental discovery of second sound, the calculation of other quantities is important. Assuming small thermal expansion, the pressure and temperature amplitudes, relative to what they would be in ordinary fourth and second sound, are

$$P_a' \,/\, P_4' = A\,\frac{\rho_s}{\rho}\,C_1^2 \,/\, C^2 \qquad (11)$$

$$T_a' \,/\, T_2' = 1 - A\,\frac{\rho_s}{\rho}\,C_1^2 \,/\, C^2 \qquad (12)$$

where $A = (1 + \frac{\rho_n}{\rho_a}\,(C_1^2-C^2)/(C_a^2-C^2))^{-1}$.

The ratio of the normal fluid velocity to the superfluid velocity is given by

$$\frac{v_n}{v_s} = \frac{\rho_s}{\rho}\,B\,(C^2 + A\,(\frac{\rho_n}{\rho} - \frac{\rho_s}{\rho})\,C_1^2) \qquad (13)$$

where $B = (\frac{\rho_a}{\rho}\,(C_a^2 - C^2) - \frac{\rho_n}{\rho}\,C^2)^{-1}$.

For the C_{2a} mode near T_c, $v_n\to0$, so that this mode may not be too heavily damped. The temperature amplitude for this mode is ~4% that of ordinary second sound.

Physica B 165&166 (1990) 583–584
North-Holland

SUPERFLUID TRANSITION OF ^{4}He IN AEROGELS*

Raimundo R. dos Santos, N. S. Branco, and S. L. A. de Queiroz

Departamento de Física, Pontifícia Universidade Católica do Rio de Janeiro, C.P. 38071, 22452 Rio de Janeiro, Brasil

The superfluid transition of ^{4}He in some porous media is studied through a fractal model. We incorporate the fractal aspects of the medium in both the Harris criterion and the hyperscaling relations, so that the experimentally observed non-bulk exponents can be consistently explained as resulting from a crossover to new behavior. Available data for aerogel suggest that the superfluid transition in this case is a *fractal surface* (as opposed to *volume*) phenomenon, which can be modelled microscopically by means of an inhomogeneous effective mass.

The superfluid transition of ^{4}He in porous Vycor glass has been interpreted as that of bulk He in which the porous medium plays the role of quenched disorder (1,2). Since for the well known λ transition the specific heat exponent $\alpha < 0$, the critical behavior of the disordered system should be the same as that of the 'pure' (i.e., bulk) system, according to the Harris criterion (3). Indeed, the superfluid density exponent ς measured in these systems (1) is practically the same as that for bulk He. On the other hand, recent experiments (4) on the superfluid transition of ^{4}He in porous silica gels – aerogel and xerogel – revealed critical exponents different from the bulk ones, in apparent contradiction to the Harris criterion.

Aerogels are low density materials with open volume fractions typically of the order of 90% and their structure can be thought of as being composed of particles with diameters of the order of 1 nm that cluster together into chains, giving rise to a porous skeleton (5); thus, typical pore diameters span from a few nanometers to about 50 nm. Small-angle neutron scattering in aerogels revealed volume fractal behaviour for length scales between ~ 0.4 nm and ~ 40 nm; at shorter distances it is the fractal structure of the pore frames which is probed, thus giving rise to a crossover towards surface fractal behaviour (6). The respective volume and surface fractal dimensions are $d_f^v \simeq 2.4$ and $d_f^s \simeq 2.9$.

Here we consider a mechanism, which explicitly takes into account the fractal character of aerogels (6). We argue that the observed exponents differ from the bulk values due to the fractal nature of the supporting space, thus removing the inconsistency with the Harris criterion. This supporting space turns out to be the aerogel surface.

The Harris criterion (3) is derived by taking into account the fluctuations in the critical temperature induced by spatial inhomogeneities. When these inhomogeneities come about as a result of fractal behaviour, characterized by a dimensionality d_f, it can be shown (7) that the condition for having a sharp transition described by the 'pure' (*i.e.*, not disordered) exponents becomes

$$2 - \nu_0 d_f < 0 \tag{1}$$

where ν_0 is the correlation length exponent, the subscript 0 referring to the 'pure' case. Since $\nu_0 = 0.674 \pm 0.001$ for ^{4}He (Ref. (8)), this condition implies that a crossover to new critical behavior is possible for a fractal medium if the relevant dimension d_f is $\lesssim 2.96$. When the fractal aspect is not present (as in Vycor), the transition then takes place in the whole pore volume ($d_f = d$), and the usual condition $\alpha_0 < 0$ follows from hyperscaling. Conversely, when new exponents are observed, as in aerogel and in xerogel (4), one can immediately attribute fractal characteristics to these media. In a similar fashion, the euclidian dimension, d, is replaced by d_f in all relations between exponents (7). From measured values of $\varsigma = 0.813 \pm 0.009$ (4) and $\alpha = -0.69 \pm 0.15$ (9) in

* Work supported by FINEP, CAPES and CNPq.

aerogel we get $d_f = 2(2-\alpha)/(2-\alpha-\varsigma) = 2.86 \pm 0.07$. The fact that the resulting dimension is very close to the measured surface fractal dimension (6), ~ 2.9 is an evidence of dominant surface phenomena driving the superfluid transition, even when the pores are full.

Surface behavior dominating over bulk behavior has been observed experimentally in the context of magnetic phase transitions (10). It is therefore instructive to interpret this phenomenon for the superfluid transition within a pseudo-spin formulation. We first assume that the net effect due to the fractality of the medium is to introduce an inhomogeneous effective mass that distinguishes the motion of ^{4}He atoms near the pore surfaces from those deep within the pores. Then, a quantum lattice-gas model for hard-core bosons yields an XY-model in which the 'exchange' coupling (or hopping term) for sites on the fractal surface (J_s) is different from the coupling for sites in the bulk region (J_v); see Ref. (7). The couplings turn out to be inversely proportional to the respective effective masses m_s and m_v.

The competition between surface and bulk orderings in fractals has been extensively studied in recent years. In particular, spins with discrete symmetry (*i.e.*, Ising and Potts) on the sites of a Menger sponge (a fractal whose structure is similar to that of aerogels (11) have been investigated (12) by approximate real-space renormalization group (RG) methods; the trends observed in Ref. (12) can be carried over to the present case since all relevant dimensions are above the lower critical dimensionality for the XY-model. It was found that, regardless of the number of spin states and of geometrical details such as degree of lacunarity, there is a critical value of the ratio $R = J_s/J_v$ (usually > 1) above which surface fractal (as opposed to *volume* fractal) critical behavior dominates (12). In Fig. 1 we show a schematic phase diagram, based on the results of Ref. (12): as the temperature is lowered (in the constant-R path shown) the system first orders its fractal surface, followed by a bulk ordering at much lower temperatures. For R below the critical value, the system transitions directly between bulk order and paramagnetism.

The above analysis is consistent with the interpretation of the observed superfluid transition of ^{4}He in aerogels as arising from a surface ordering, provided $m_s^* < m_v$ (*i.e.*, $J_s > J_v$ in our pseudo-spin formulation). The experimental results for Vycor and for xerogels (4), including full pores and thin films can also be explained within this model; see Ref. (7).

Further experimental tests for this model would be to investigate both the possibility of another transition to bulk superfluidity at much lower temperatures,

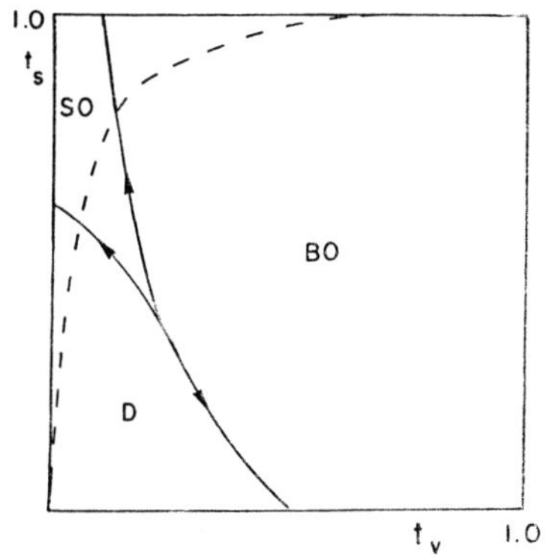

FIGURE 1
Schematic phase diagram (based on Ref. (12)) for Potts spins in the parameter space (t_s,t_v), where $t_r = (1-e^{-qJ_r})/(1+(q-1)e^{-qJ_r})$, with $r = s, v$ and q is the number of Potts states. The bold lines denote boundaries between the disordered (D), bulk ordered (BO) and surface ordered (SO); the dashed line represents a thermodynamic path in which the temperature is varied while the ratio J_s/J_v is kept constant. The arrows indicate RG flows.

as well as the low temperature spectrum of excitations in the superfluid phase; the latter should reveal a crossover to the fracton regime (13), thus confirming the dominant fractal behavior. We hope these predictions will stimulate further experimental work.

REFERENCES

(1) B. C. Crooker *et al.*, Phys. Rev. Lett. **51**, 666 (1983).
(2) P. B. Weichman *et al.*, Phys. Rev. **B 33**, 4632 (1986).
(3) A B Harris, J. Phys. C **7**, 1671 (1974).
(4) M. H. W. Chan *et al.*, Phys. Rev. Lett. **61**, 1950 (1988).
(5) J. Fricke, Sci. Am. **256**, 68 (1988).
(6) R. Vacher *et al.*, Phys. Rev. **B 37**, 6500 (1988).
(7) R. R. dos Santos, *et al.*, (unpublished).
(8) D. S. Greywall and G. Ahlers, Phys. Rev. **A 7**, 2145 (1973).
(9) see Phys. Today **42**, July 1989, page 24; S. Q. Murphy, (*private communication*).
(10) C. Rau *et al.*, Phys. Rev. Lett. **57**, 2311 (1986); **58**, 2714 (1987); J. Appl. Phys. **63**, 3667 (1988).
(11) A. Bourret, Europhys. Lett. **6**, 731 (1988).
(12) R. Riera and C. M. Chaves, Z. Phys. **B 62**, 387 (1986).
(13) S. Alexander and R. Orbach, J. Phys. Lett. (Paris) **43**, L625 (1982); Y. Tsujimi *et al.*, Phys. Rev. Lett. **60**, 2757 (1988).

Physica B 165&166 (1990) 585–586
North-Holland

^{4}He FILM GROWTH AND WETTING ON CURVED SURFACES

L. Wilen and E. POLTURAK

Department of Physics, Technion-Israel Institute of Technology, Haifa 32000, Israel.

We report adsorption experiments of ^{4}He film on graphite fibers having either a positive or a negative curvature. The curvature has a dramatic effect on the limiting thickness of the film at saturation (200 Å for positive curvature, over 10000Å for negative curvature). The data agrees very well with theoretical predictions. This agreement, however, is questionable in view of the convoluted surface morphology of the fibers.

The limiting thickness of an adsorbed film at saturated conditions indicates whether an adsorbate wets the substrate. The determination of the limiting thickness of the film becomes non-trivial when the surface is curved, since a positive curvature will inhibit film growth near saturation. We used the oscillating fiber technique[1] to investigate this effect in some detail[2]. We experimented with two kinds of graphite fibers, shown in Fig. 1, for which the curvatures are very different. In particular, the Fortafil fiber shown in Fig. 1b has a region of negative curvature. The fiber was mounted under tension above a vertical parallel capacitor. The capacitor was used to measure the density of the gas in the cell at pressures below saturation. At saturation, it measured the liquid level in the cell. Manipulating the liquid level in the cell is a very precise way to vary the chemical potential of the film very near the phase bounary. At 2K, with h=1cm (height of fiber above the liquid) the hydrostatic pressure head is less than 10^{-3} mmHg, i.e. $P/P_{sat}=3\times10^{-5}$. In our cell, 1cm<h<10cm. The decrease of the resonant frequency of the fiber is linearly proportional to the adsorbed mass[1]. For our system, we estimate a shift of ~ 1Hz/5 statistical layers. Under saturated conditions, the film thickness d is determined from the relation: $d=(\gamma(d)/(mgh+\sigma v/R))^{1/3}$, where $\gamma(d)$ is the thickness dependent Van der Waals constant[3], m is a mass of a ^{4}He atom, σ is the surface tension and R is the radius of curvature. Positive R leads to a finite film thickness at h=0, whereas for negative R the film thickness should diverge. This is precisely what is shown in Fig. 2. A detailed analysis of the data shows that the limiting film thickness for the Fortafil fiber is corresponds to the negative R region being totally filled by liquid, thus reducing the curvature of the surface and preventing the divergence of the film thickness[2]. The limiting thickness of the film is in excess of 10000Å, convincingly demonstrating complete wetting. The problem that we see with the theory is that

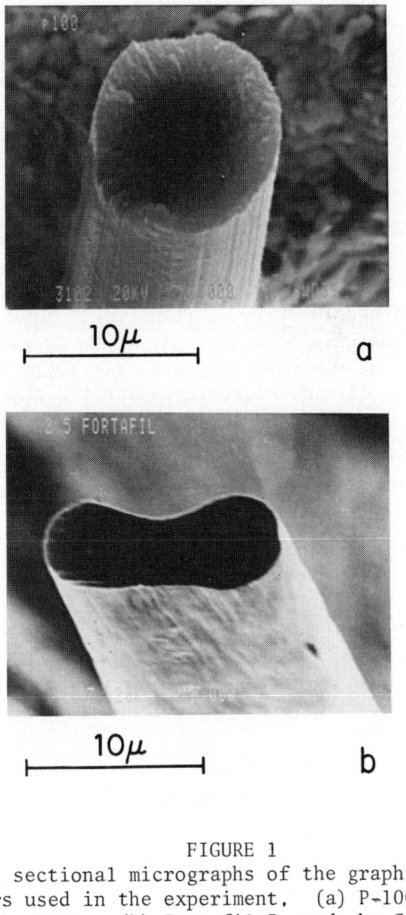

FIGURE 1
Cross sectional micrographs of the graphite fibers used in the experiment. (a) P-100 fiber made by UCAR. (b) Fortafil 5, made by Great Lakes Carbon.

0921-4526/90/$03.50

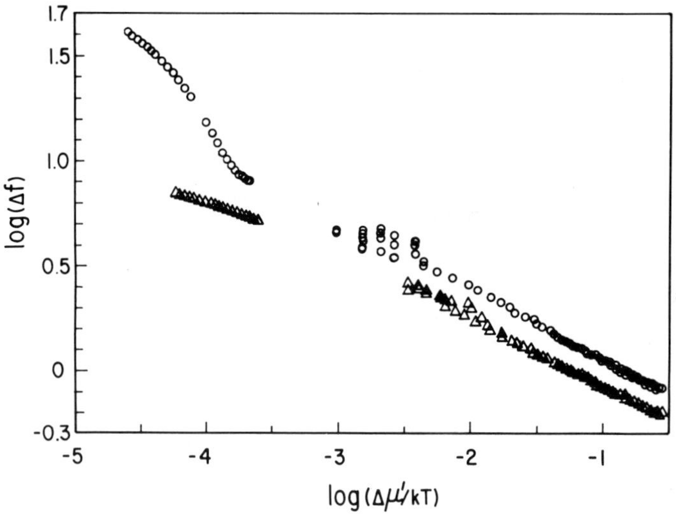

FIGURE 2
Frequency Shift vs. chemical potential difference $\Delta\mu = KT\ell n(\rho/\rho_0) + mgh$ for the Fortafil fiber(o) and for the P-100 (Δ).

we have to use an effective surface area in fitting the data[4]. This is acceptable, given the convoluted structure of the fibers' surface. However, as the film thickens and small pores fill with liquid, we would expect the effective surface area to decrease. What we find is that the effective area comes out independent of the film thickness. We regard this as a major puzzle with respect to the theoretical description of adsorption on realistic surfaces.

This work was supported by the Israeli Academy of Sciences and by the U.S.-Israel Binational Foundation.

REFERENCES

(1) C. Bartosch and S. Gregory, Phys. Rev. Lett. 54, 2513 (1985).
(2) L. Wilen and E. Polturak, to be published.
(3) E. Cheng and M. Cole, Phys. Rev. B38, 987 (1988).
(4) G. Zimmerli and M. Chan, Phys. Rev. B38, 8760 (1988).

Physica B 165&166 (1990) 587–588
North-Holland

REPRODUCIBLE FLOW RATE STRUCTURE IN ^{4}HE FILMS

R.R. TURKINGTON AND R.F. HARRIS-LOWE

Physics Department, Royal Military College, Kingston, Ontario, Canada*

Measurements of ^{4}He film flow rates taken with high resolution and repetition rates have revealed a reproducible structure highly correlated with the liquid level meniscus position. The details of this structure have been previously masked by a lack of experimental resolution.

1. INTRODUCTION

We have for some years been interested in attaining the ability to make accurate measurements of flow rates in superfluid helium films at sufficiently high repetition rates to allow the study of transient events in the flow. This work was stimulated by many, often conflicting, reports in the literature of sudden changes in flow rates and other flow parameters(1,2). Such changes have been associated with a presumed sudden restructuring of the dissipating vortex line configuration(3), but until now the time resolution and accuracy of any experimental observations has obscured the precise nature of such changes. By employing a capacitance level measuring technique(4) together with a high speed data acquisition system we have attained the ability to determine flow rates with a scatter of less than ±3% at a repetition rate of up to four measurements per second. The results taken over different glass substrates have revealed a number of new and startling features which we believe were previously masked by limited resolution and which help to explain many of the conflicting reports coming from earlier film flow work. In this and a following paper we will present some of the new features which have been observed to date.

2. RESULTS

Figure 1 presents a typical set of our measurements of the liquid level in a beaker (wrt an arbitrary fixed reference point) versus time, and this set resembles earlier results except that the data displayed consists of roughly 1000 level measurements which were taken over a span of less than 1 mm of level change. This data is sufficiently precise that sequential differences between readings can

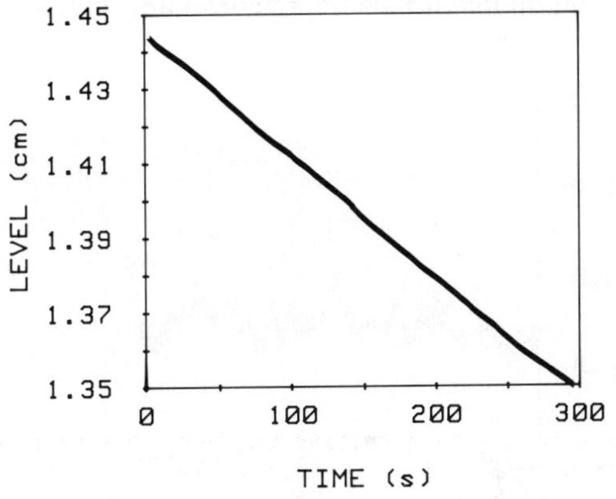

FIGURE 1

also provide precise determinations of the flow rate per unit limiting perimeter (the flow rate) and the rate results obtained from the data of Figure 1 are presented in Figure 2. This figure

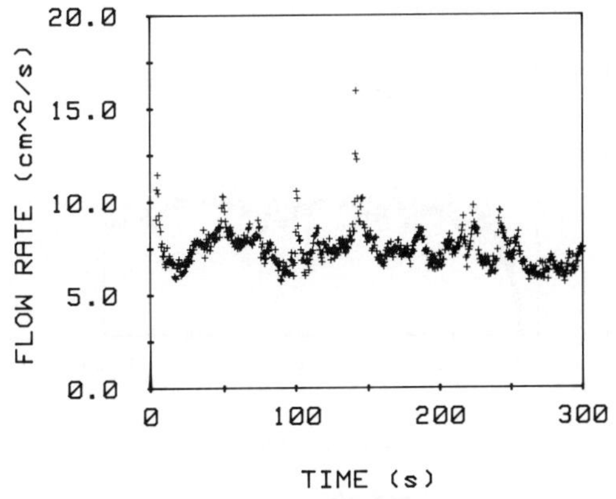

FIGURE 2

*Research supported by the Department of National Defence, Canada

0921-4526/90/$03.50 © 1990 – Elsevier Science Publishers B.V. (North-Holland)

and all of our other results reveal that even though the long term average of the flow rate has a value typical for glass substrates (~7.5×10^{-5} cm^2/s at T = 1.16 K), the flow rate is changing continuously, sometimes by dramatic amounts, over short time spans. Our first impression of these results was that the flow rate was changing randomly in time with preferential bursts to higher rates. However when we repeated the measurements with the liquid level in the same region of the beaker we were surprised to discover that the apparently random changes in rate were highly reproducible. Figure 3 is a replot of the data from Figure 2 shown as a function of liquid level in the beaker, and Figure 4 is a plot of flow rate data in the same region taken some 4.5 hours later. In the

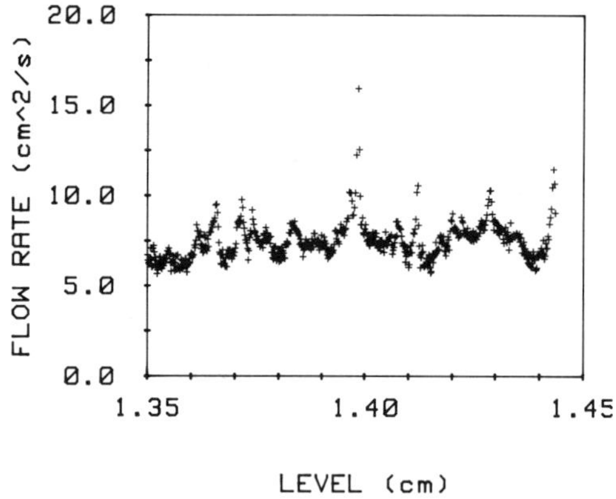

FIGURE 3

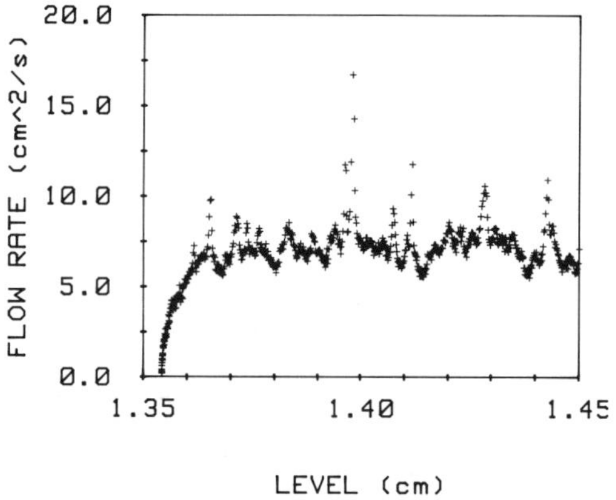

FIGURE 4

intervening period we had added helium to the system with a resulting increase in the equilibrium level position, and the start of film oscillations about this higher level is evident in the tail off of the data to the left of Figure 4. Even a cursory examination reveals that there is a high degree of correlation between the two sets of data. A more detailed examination of these results reveals that this correlation extends to features which well might be construed as noise when the data is plotted on the scale of Figure 3. In fact, replots of this data on a much enhanced scale show that there is an astonishing correspondence between positions and magnitudes of the flow rates, including many of the smaller noise-like excursions. Typically we can see clear correlations between excursions which are as closely spaced as 30 μm, and whose positions are easily reproducible to within ±5 μm. All of our data taken since we have been able to achieve this level of resolution has displayed similar behaviour. The flow rate varies in a dramatic but reproducible manner when considered as a function of the liquid level position in the beaker.

3. CONCLUSIONS

As stated earlier, these results can perhaps explain many of the apparently conflicting results which have been previously reported, conflicts which we believe were a consequence of the time interval required for rate determinations. The features reported here would not be revealed by results obtained from measurements requiring an interval of greater than 1 s for accurate rate determinations. Measurements taken at greater intervals would display a high level of scatter with varying statistics which would be crucially dependent upon the exact liquid level position. The degree of "scatter" would slowly diminish as the time interval increased.

Our current feeling upon considering these results is that they can be explained in terms of a sequence of highly localized but strong attractive interactions between specific sites on the substrates and the meniscus of the liquid level as it passes the sites, interactions which perhaps can be explained in terms of vortex line configurations and dynamics in these regions.

REFERENCES

(1) E.F. Hammel, W.E. Keller and R.H. Sherman, Phys. Rev. Lett. 24 (1970) 712.
(2) R.B. Hallock and E.B. Flint, Phys. Lett. 45A (1973) 245.
(3) R.F. Harris-Lowe, J. Low Temp. Phys. 28 (1977) 489.
(4) R.R. Turkington and R.F. Harris-Lowe, J. Low Temp. Phys. 10 (1973) 369.

Physica B 165&166 (1990) 589–590
North-Holland

^{4}HE FILM FLOW RATES IN THE OSCILLATORY REGIME

R.F. HARRIS-LOWE AND R.R. TURKINGTON

Physics Department, Royal Military College, Kingston, Ontario, Canada*

Measurements taken of film flow rates at high resolution and repetion rate in the oscillatory regime have revealed some startling new features previously masked by a lack of experimental resolution. The oscillations which have always appeared to be sinusoidal are clearly revealed as periodic but non-sinusoidal when the rates are observed with high resolution. The rate structure observed is shown to be a highly reproducible function of meniscus level position.

1. INTRODUCTION

In the previous paper(1) we presented results which established that helium film flow rates over a glass beaker substrate vary continuously but in a highly reproducible manner determined by the precise position of the meniscus in the beaker. The magnitude of the rate can exhibit positive excursions which can be greater than 100% but which are completed in time intervals of under 5 s during which time the level changes by less than 20 μm.

When we commenced our investigation of this flow rate "stucture", we had assumed that it was due to dissipation in the flow, and consequently that it might disappear when the flow was sub-critical. However, when we investigated the rates in the sub-critical oscillatory region, we discovered that the structure was still very much in evidence, and in fact was more amenable to investigation because of the ability to determine rates while sweeping the meniscus repeatedly through a limited range of levels, typically about 0.15 mm. The theory of film oscillations, see for example references 2 and 3, in a close to isothermal system, predicts that the oscillations can be represented by a lightly damped sinusoidal waveform. The damping is not a consequence of dissipation in the film but instead can be accounted for by effects associated with temperature differences between reservoirs which appear as a consequence of the flow.

2. RESULTS

Our results confirm that measurements of level versus time which have an effective time constant of greater than 5 s will likely appear to be damped sinusoidal as predicted. Figure 1 displays one set of our measurements, which in this case appears to be consistent with the theoretical prediction. The data shown is comprised of slightly under 1000 data points taken with an effective time constant of 0.5 s, and does show a slight damping.

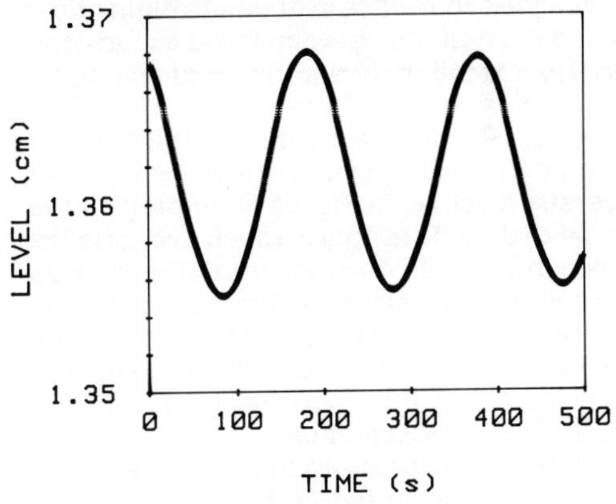

FIGURE 1

However, when sequential differences of this data are taken in order to determine the interval flow rates, a different picture emerges. Figure 2 displays the rate results obtained in this manner, and clearly the result is not sinusoidal. We have investigated the oscillatory region under many different circumstances, and have concluded that the flow rates, although periodic, cannot be described as a slightly-damped sinusoid, and the specific rate

*Research supported by the Department of National Defence, Canada

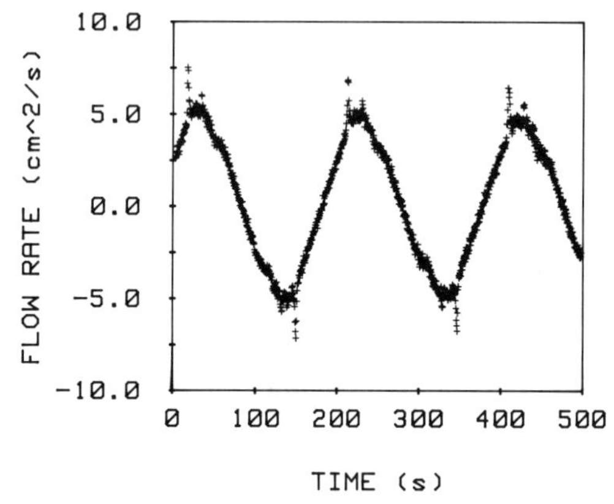

FIGURE 2

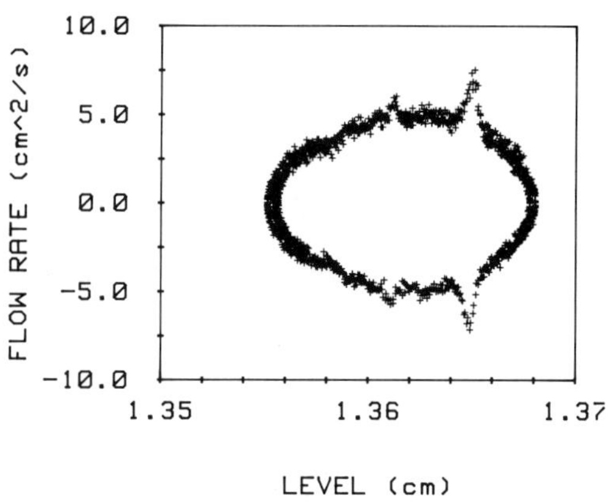

FIGURE 3

behaviour is once again determined by the location of the liquid level in the beaker. We have been able to draw some general conclusions about the rate behaviour in the oscillatory region. Once again we see significant "spikes" or excursions in the magnitude of the flow rate which occur at precise values of the helium level. In addition however, there appears to be a tendency for the excursions to be correlated with the occurrence of absolute higher flow rates and the regions between these excursions appear to exhibit a linear dependence upon time.

The very high degree of reproducibility of our flow rate data can be demonstrated by a plot of rate versus level, as in Figure 3, employing the same set of data. This figure, which overlaps the data from roughly 2½ oscillations, demonstrates the structured and reproducible nature of rate versus level, and demonstrates some of the points made above. The data appears to exhibit more scatter, but this increase is due to the attenuation of the oscillations.

Another conclusion is that the rate excursions are present during both dissipative and non-dissipative flow, and appear at the same liquid levels. We have been able to come to this conclusion by comparing the rates obtained during oscillations with those obtained during dissipative flow as the level passed the same positions during an experiment when the final equilibrium level was lower.

3. DISCUSSION

We have amassed a great deal of data which all exhibits the same general type of behaviour as presented here. At this stage we can attempt to make some speculative observations about these phenomena. The fact that the occurrence of very large rate spikes in the oscillatory region does not coincide with a large damping of the oscillations, and that the flow rate before and after the spikes is roughly the same, would appear to indicate that the mechanism responsible for the spikes is non-dissipative. The fact that the magnitude of the rate for these excursions always increases and then is restored to its previous value would suggest a mechanism whereby potential energy is converted into flow kinetic energy and then back into potential energy as the liquid level passes well-defined locations on the substrate, perhaps an indication of the existence of some type of local vortex line "attractor" centre. Finally, the generally linear behaviour of the rate versus time behaviour in the regions between rate spike positions seems to indicate that over significant regions there exists a potential energy storage mechanism which is linearly dependent upon the magnitudes of the level differences between the liquid level and the various attractor sites. As mentioned in our previous paper, we believe that the current observations can explain many of the discrepancies reported in earlier work on film flow, and that an explanation of the phenomena reported here will be found in terms of pinned vortex line dynamics.

REFERENCES

(1) R.R. Turkington and R.F. Harris-Lowe, Reproducible Flow Rate Structure in ^{4}He Films, this volume.
(2) J.F. Robinson, Phys. Rev. 82 (1951) 440.
(3) R.B. Hallock and E.B. Flint, Phys. Rev. A10 (1974) 1285.

Physica B 165&166 (1990) 591–592
North-Holland

Thickness Scaling of the Convective Conductance of ^{4}He Films*

D. FINOTELLO†, Y. Y. YU‡, X. F. WANG and F. M. GASPARINI

Department of physics, State University of New York at Buffalo, Buffalo, NY 14260, USA

We report results of analysis of the convective conductance, K_f, of pure ^{4}He films of thickness from 11.7 to 156 Å, and critical temperature, T_c, from 1.281 to 2.155 K. The data show a narrowing of the critical region with thickness as T_c approaches T_λ. In our analysis we have corrected for the small change of film thickness with temperature, and explored the consequences of using different values for the background conductance and a possible temperature dependence of the vortex diffusion constant, D. In all circumstances we find a strong thickness dependence of the parameters D/a^2 and b, with a the vortex core parameter and b a constant describing the divergence of the correlation length. The shift of T_c with thickness, which does not follow simple scaling predictions, can be partly understood in terms of ψ theory with the addition of the van der Waals interaction.

The thermal conduction in a film of ^{4}He near the superfluid transition is dominated by a convective process whereby liquid flows toward the source of heat and evaporating atoms return to a cold sink and release their latent heat. This process above the transition is limited by the presence of free vortices. Ambegaokar et al (1) pointed out that the thermal conduction should be inversely proportional to the number density of free vortices, hence the square of the two-dimensional ($2D$) correlation length. Specifically for a geometry of a film moving along parallel surfaces and the gas returning in the center, one obtains an expression for the film conductance given by Teitel (2),

$$K_f = F\frac{\mathcal{L}^2(W/L)k_B}{(h/m)^2(D/a^2)}\exp(\frac{4\pi}{b}t^{-1/2})$$

$$\equiv f(T)\frac{a^2}{D}\exp(\frac{4\pi}{b}t^{-1/2}) \qquad (1)$$

where $\mathcal{L}$ is the latent heat , and W and L are respectively the film perimeter and the distance the films flows. The unknown parameters in this equation are D/a^2, the ratio of diffusion constant to (core parameter)^2, the constant b, and T_c, in $t \equiv (T/T_c - 1)$. The constant F in a number of order unity which is not known and precludes obtaining an absolute value of D/a^2 from K_f.

We have reported measurements of K_f which initially covered unsaturated films in the range of 11 – 55 Å formed on a Mylar substrate (3). These were subsequently extended to saturated films up to 156 Å (4). Some of these data are shown in Fig. 1 . This is a plot of $K_f/f(T)$ versus $T-T_c$. We can see from this plot that all films show a strong singularity at T_c which extends over a progressively broader region as the film becomes thinner. These data were analyzed in a procedure, in which D/a^2, b and T_c are least squares adjusted to achieve an optimum fit to Eq. 1. In doing this we have tried a variety of procedures which principally involve the details of how one treats the regular, background conductance (5). As it turns out this background is larger than one would estimate and, more importantly, has a stronger temperature dependence than expected. In Fig. 1 we have plotted K_f obtained by using an *estimated* value for the background. We have also analyzed the data with a temperature dependence for D as suggested by Huber (6). In these different treatments one obtains slightly different values of parameters, but trends with thickness are retained (5). In particular a temperature dependent D has a very weak effect on our results because of the dominant exponential divergence of K_f. We can summarize the results of our analysis as follows.

1. The parameter D/a^2 *decreases* by about three order of magnitude as $T_\lambda - T_c$ changes from $1K$ to $0.01K$. It is about 10^{10} sec^{-1} for the thinnest films and about $10^7 sec^{-1}$ for the thickest. This can be understood to some extent if one assumes that a is proportional to the $3D$ correlation length. This leaves a much weaker temperature dependence for D which, if one assumes that $F \cong 2 - 3$, is consistent with direct measurements of D obtained for $T < T_c$ (7).
2. The parameter b *increases* by about an order of magnitude in the same range of $T_\lambda - T_c$. This parameter also describes the strength of the $t^{1/2}$ cusp in the superfluid density, σ_s, near T_c. Our results imply that it becomes progressively harder for thicker films to extract the universal jump in σ_s at T_c because of the increase of b and the expected more rapid temperature dependence of the background superfluid density. The increase of b can be understood using scaling arguments involving the vortex core energy and the shift in T_c with thickness (8).

* Work supported in part by the National Science Foundation DMR 8305742, DMR 8601848 and DMR 8905771.
† Permanent address: Department of Physics, Kent State University, Kent, OH 44242, USA.
‡ Present address: Department of Physics, Syracuse University, Syracuse, NY 13244, USA.

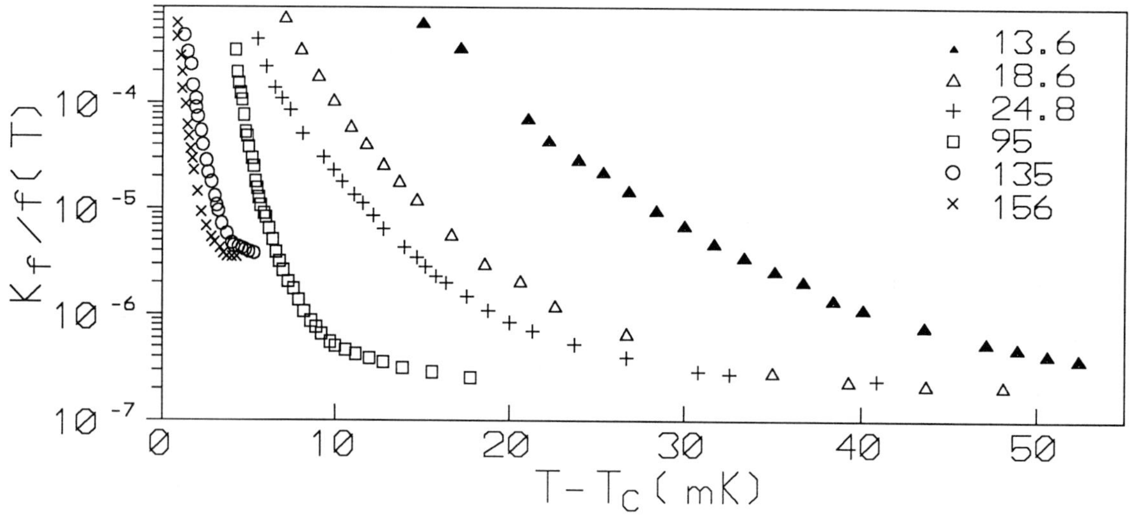

Fig.1 Scaled film conductance vs $T - T_c$ for films of various thickness. T_c for the 95 Å has been shifted by +3mK for clarity.

3. The critical temperature does not scale as expected on the basis of the $3D$ correlation length, $d \propto (T_\lambda - T_c)^{-\nu}$. One finds that for films thinner than about 20 Å one obtains an exponent of 0.52 ± 0.01 rather than 0.672 (4).

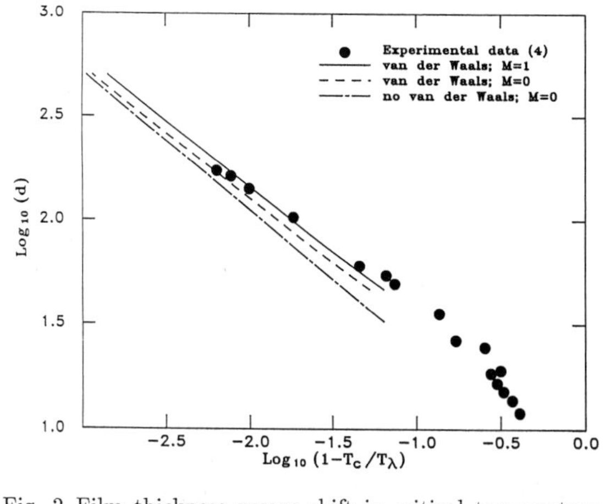

Fig. 2 Film thickness versus shift in critical temperature. The calculated lines are done with ψ theory under the conditions indicated. Note that presence of the van der Waals field destroys the correlation length scaling of the shift in T_c.

We have attempted to explain the lack of the correlation length scaling of T_c by calculating the effect of the van der Waals field on the transition. This was done in the context of ψ theory (9). With this theory one can obtain a differential equation for the order parameter which can be solved subject to specified boundary conditions and in the presence of a position-dependent external field. We have done this for a film formed on a Mylar substrate. Results are shown in Fig. 2. Here we have plotted on a log – log scale the thickness of the films versus t_c. The lines represent calculated T_c's with and without the van der Waals field. The line which agrees with the data best is for the parameter $M = 1$. This parameter indicates the strength of the ψ^6 term in the theory (8). It is clear from Fig. 2 that the effect of the van der Waals attraction causes the effective exponent in the region of the data to be less than ν. This is precisely the observed behavior. Further details of these calculations will be given elsewhere (10).

1. V. Ambegaokar, B. I. Halperin, D. R. Nelson and E. D. Siggia, Phys. Rev. B21, 1806(1980).
2. S. L. Teitel, J. Low Temp. Phys. 46, 77(1982).
3. D. Finotello and F. M. Gasparini, Phys. Rev. Lett. 55, 2156(1985).
4. Y. Y. Yu, D. Finotello and F. M. Gasparini, Phys. Rev. 39, 6519(1989).
5. D. Finotello, Y. Y. Yu and F. M. Gasparini, to appear in Phys. Rev. B.
6. D. L. Huber, Phys. Lett. 79A, 331(1980).
7. P. W. Adams, W. I. Glaberson, Phys. Rev. B35, 4633 (1987).
8. R. G. Petschek, Phys. Rev. Lett. 57, 501(1986).
9. V. L. Ginzburg, A. A. Sobyanin, Usp. Fiz. Nauk 120, 153(1976). [Sov. Phys. Usp. Vol.19, 773(1977)].
10. X. F. Wang, I. Rhee and F. M. Gasparini, this volume.

Physica B 165&166 (1990) 593–594
North-Holland

Calculation of van der Waals Effects on Correlation-length Scaling Near T_λ^*

X. F. WANG, I. RHEE and F. M. GASPARINI

Department of physics, State University of New York at Buffalo, Buffalo, NY 14260, USA

We report calculations of the superfluid density, ρ_s, of helium confined between Si surfaces, and of the onset temperature for helium films formed on a Mylar substrate. The calculations are done within ψ–theory and include the effect of helium–surface interaction. We find that this interaction precludes size scaling with the bulk correlation length. The calculated effect however is not enough to explain the experimental results. In the case of the onset temperature, the influence of the van der Waals field causes the effective shift exponent to be less than that of the correlation length.

In a number of experiments involving confined helium in well characterized geometry it is found that predictions of correlation-length scaling do not hold (1,2). One of the issues which arise in either complete confinement within rigid walls, or a situation of a film with a free surface is the role of the van der Waals interaction between helium and the confining surfaces. The central prediction of correlation-length scaling is the fact that thermodynamic properties should depend on the ratio of ξ/L, correlation length to smallest confining dimension (3). It is clear that the presence of the van der Waals interaction must break down this scaling because of the introduction of a new length scale associated with the range of this interaction. The issue which remains is a quantitative one regarding the magnitude of this effect. One can obtain an estimate of this in the context of ψ-theory.

In ψ-theory, one starts out with a mean field expansion of the free energy in terms of the order parameter, ψ, and introduces specific temperature dependence in the expansion coefficients to approximate the true critical behavior at the superfluid transition: a logarithmic divergence (rather than $\alpha \cong -0.01$) for the specific heat, and a 2/3 power (rather than 0.672) for the superfluid density ρ_{sb} (4,5).

How good these approximations are, and how far from T_c one can expect this procedure to hold has been discussed (6). For our purpose, the virtue of ψ–theory is the possibility of incorporating in the equation of confined helium the effect of the van der Waals interaction on ρ_s (6). For a planar geometry, with z as the coordinate measured from the liquid–solid interface, one has

$$\frac{\partial^2 y}{\partial z^2} = \frac{3}{3+M}\frac{1}{l^2(z)}\{-t^{4/3}y + (1-M)t^{2/3}\frac{y^3}{k(z)} + M\frac{y^5}{k^2(z)}\}, \qquad (1)$$

where $y = (\frac{\rho_s}{\rho})^{1/2}$, $t = 1 - \frac{T}{T_\lambda(z)}$, $l^2(z) = \frac{\hbar^2 k(z)}{2m^2 \Delta C_p(z) T_\lambda(z)}$, and M is the only unknown parameter. The quantities k, ΔC_p are respectively the prefactor in the power-law dependence of the bulk superfluid fraction, ρ_{sb}/ρ, and the jump in the specific heat. We calculated the z–dependence using bulk properties at the local pressure, $P(z)$, obtained from the van der Waals interaction $u(z)$, $P(z) = -\int v^{-1} du$. We write $u(z)$ as follows (8),

$$u(z) = \alpha(0)[1 + 1.64(\frac{z}{d_{1/2}})^{1.4}]^{-\frac{1}{1.4}}\frac{1}{z^3} \qquad (2)$$

For Si and Mylar surfaces we have respectively $\alpha(0) =$ 1950, 2174 K $\AA^3$, and $d_{1/2} =$ 230, 170Å(7,8). The form of Eq. 2 is useful since it includes retardation effects and applies to several hundred Å away from the surface (8). Eq. 1 is solved numerically to obtain $y(z,t)$. This is then integrated over the total volume to obtain $< \rho_s(t)/\rho >$. One can easily verify from Eq. 1 that in the absence of z–dependence in the coefficients of y, one obtains the scaling result $\rho_s/\rho = t^{2/3}f(t^{-2/3}/L)$.

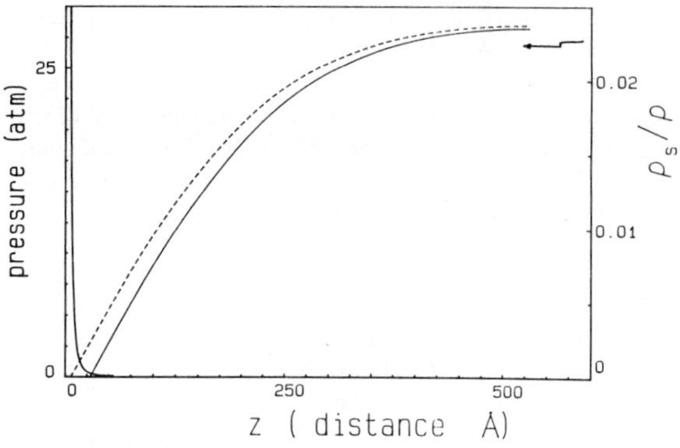

Fig.1 Spatial dependence of pressure and superfluid fraction ρ_s/ρ at $t = 10^{-3}$. Solid line: ρ_s vanishes at $T > T_\lambda(z)$; dashed line: ρ_s vanishes at $P > 30$ atm.

In Fig. 1 we show the spatial dependence of pressure due to van der Waals potential and y^2 for helium confined between two silicon surfaces a distance L apart. y is chosen to have $y'(L/2) = 0$, and, for the solid line, to vanish in the region where $T > T_\lambda(z)$. We also show, as a dashed line, y^2 when this condition is relaxed. Clearly the solid line incorporates the *largest effect* of the van der Waals force. We will use this solution and explore its consequence.

* Work supported by the National Science Foundation DMR 8905771.

0921-4526/90/$03.50

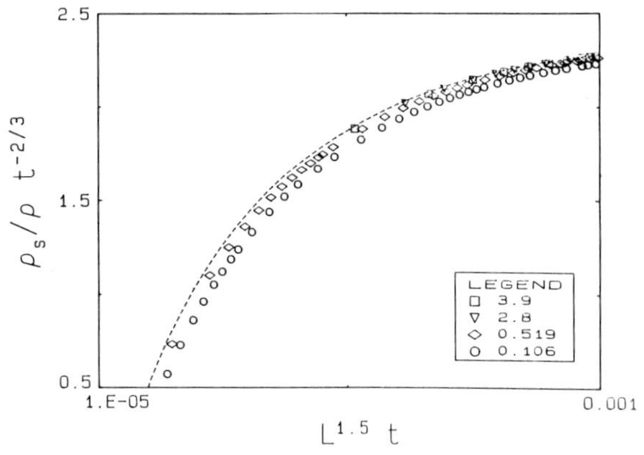

Fig.2 Scaling plot of ρ_s/ρ vs $L^{3/2}t$ for four confining sizes as indicated. Dashed line shows perfect scaling without van de Waals field. Four separate symbols show the extent van der Waals field destroys such scaling.

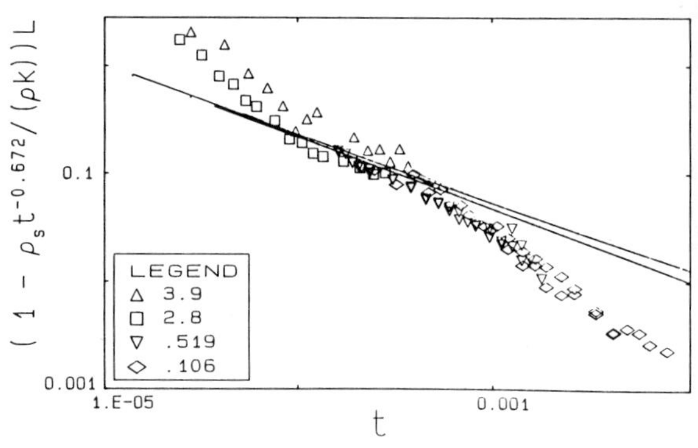

Fig.3 Scaling plot of $\frac{\Delta\rho_s}{\rho_{sb}}$ vs $\log t$ for the calculations with (upper line) and without (lower line) van der Waals field, along with experimental data (2).

In Fig. 2 we plot $(\rho_s/\rho)t^{-2/3}$ vs $L^{3/2}t$ for $L = 0.106$, 0.519, 2.8 and 3.9 μm. This, as discussed above, should yield a universal function. This is shown by the dashed line which is the solution of Eq. 1 without van der Waals interaction. With the interaction, the calculation for four separate confinement sizes shows by the lack of collapse the extent to which correlation-length scaling fails. To compare our calculation with the actual measurements we show in Fig. 3 the data of Rhee et. al. (2) along with the calculated value. This plot shows the scaled fractional difference between the bulk and confined superfluid density. This, *to leading order*, should be given by $(\rho_{sb} - \rho_s)L/\rho_{sb} \sim t^{-2/3}$. The data and the calculations do collapse to first order on this plot, but on separate lines. The data may be fit by a line of slope of $\sim$ -1.14 instead of -0.672 as expected. The calculation with no van der Waals interaction shows good collapse, a line of slope $-2/3$. The calculation even with van der Waals field show good collapse on this plot but on a *curve* with average slope of ~ -0.62. We would conclude from this numerical analysis that to the extent that van der Waals effects can be described within ψ-theory they do not explain the observed behavior of the data. The calculation does point out quite clearly that the presence of a surface field destroys correlation-length scaling.

The condition for y to vanish where $T > T_\lambda(z)$ seems more appropriate when there is a large gradient in the pressure and ρ_s goes to zero. We have calculated the temperature, T_c, at which $y = 0$, for films formed on Mylar and compared this with experimental results (7). We find that the van der Waals field causes the local exponent in the region where it can be compared with the data to be less than 2/3, $\sim$ 0.615 (9). This feature agrees qualitatively with the data which yield an exponent 0.52 rather than 0.672. Further the magnitude of the shift in T_c also agrees reasonably well with the data if $M \cong 1$. This latter, however, must be taken with a grain of salt since different boundary conditions, or identification of T_c with the universal jump in ρ_s, rather than $y = 0$, can move this curve around. We conclude that, at least to some degree, the disagreement with correlation length scaling in the *shift equation* can be attributed to the van der Waals interaction.

1. T. P. Chen and F. M. Gasparini, Phys. Rev. Lett., **40**, 331(1978); F. M. Gasparini, T. P. Chen and B. Bhattacharyya, Phys. Rev.B. **23**, 5797(1981).
2. I. Rhee, F. M. Gasparini and D. J. Bishop, Phys. Rev. Lett.**63**, 410(1989).
3. M. E. Fisher, in "Critical Phenomena", 51st Enrico Fermi Summer School, Varenna, Italy, M. S. Green ed. (Academic Press, New York, 1971).
4. V. L. Ginzburg and L. P. Pitaevskii, Zh. Eksp. Teor. Fiz. **34**, 1240(1959). [Sov. Phys. JETP **7**, 858(1958)].
5. Y. G. Mamaladze, Zh. Eksp. Theo. Fiz. **52**, 729(1967). [Sov. Phys. JETP **25**, 479(1967)].
6. V. L. Ginzburg, A. A. Sobyanin, Usp. Fiz. Nauk **120**, 153 (1976) [Sov. Phys. Usp. Vol.**19**, 773 (1977)]; J. Low Temp. Phys. **49**, 507(1982); Jap. J. of Appl. Phys. **26**, 1785(1987).
7. Y. Y. Yu, D. Finotello and F. M. Gasparini, Phys. Rev.B. **39**, 6519(1989).
8. E. Cheng and W. Cole, Phys. Rev.B. **38**, 987(1988).
9. D. Finotello, Y. Y. Yu, X. F. Wang and F. M. Gasparini, this volume.

Physica B 165&166 (1990) 595–596
North-Holland

BOSE-EINSTEIN CONDENSATION IN QUASI-TWO-DIMENSIONAL SYSTEMS

Humam B. GHASSIB and Yahya F. WAQQAD

Department of Physics, University of Jordan, Amman, Jordan

Bose-Einstein condensation in an ideal quasi-two-dimensional Bose gas is reconsidered with a view of establishing a simple criterion for crossover effects from two- to three-dimensional systems. The main attention is focused on the density of states involved. It is deduced that the critical value at which these effects occur is universal, apart from a geometrical factor of the order of unity. The implications for strongly interacting Bose systems are discussed very briefly.

1. INTRODUCTION

It is well known that Bose-Einstein condensation (BEC) does not occur in strictly two-dimensional ideal Bose systems (1-3), unlike the three-dimensional case. However, it remains an interesting problem to see how, in an ($\infty \times \infty \times d$) system, BEC reappears and, furthermore, how the critical value d_c can be determined at which crossover effects from two- to three-dimensional behaviour manifest themselves. This is attempted in Section 2, using the simplest possible techniques. The results obtained are then discussed in Section 3.

2. DEVELOPMENT

2.1 BEC in Strictly 2d-Systems

To establish the notation we first revisit the strictly two-dimensional ideal Bose system.

The number of excited states in this system is given by

$$N_e = \frac{S}{h^2}\int_0^\infty \frac{d\mathbf{p}}{z^{-1}\exp\{\beta\epsilon(\mathbf{p})\}-1}, \qquad (1)$$

where S is the area of the system, h is Planck's constant, β is the temperature parameter, $1/k_BT$, k_B being Boltzmann's constant and T the temperature, $\epsilon(\mathbf{p})$ is the energy of a Bose particle of momentum $\mathbf{p}$, and z is the fugacity $\equiv \exp(-\beta\mu)$, μ being the chemical potential.

Performing the angular integration in eq.(1) yields

$$\frac{N_e}{S} = \frac{2\pi m}{h^2\beta}\, g_1(z), \qquad (2)$$

where m is the mass of the boson and $g_n(z)$ is the familiar Bose function of order n and argument z (4,5). Clearly, N_e is a maximum only when $g_1(z)$ is a maximum – i.e., when $z \simeq 1$. However, $g_1(1)$ diverges; accordingly, all particles in the two-dimensional ideal Bose gas are excited. This means that no BEC occurs for systems whose density of states is constant, which is the case for infinite two-dimensional systems.

2.2 BEC in Quasi-2d-Systems

Suppose the ideal Bose gas is now confined to an infinite plane sheet whose thickness is d. BEC can now be probed in a simple manner by specifying those terms which depend on dimensionality and those which do not. While N_e is still given by

$$N_e = \int_0^\infty g(\epsilon;d)\langle n(\epsilon)\rangle d\epsilon, \qquad (3)$$

its spatial dependence now enters through the density of states $g\,(\epsilon;d)$. The problem is then reduced to evaluating this last function.

To calculate $g(\epsilon;d)$ the total number of states σ up to an energy ϵ should be determined first; the derivative $\partial\sigma/\partial\epsilon$ will then give $g(\epsilon;d)$. Specifically,

$$\sigma = S\sum_{n=-n_\circ}^{n_\circ}\int \frac{d\mathbf{p}}{h^2}\, f(n;d), \qquad (4)$$

where $f(n;d)$ is 1 if $n=0,\pm1,\pm2,\ldots$, and $n_\circ$ is the maximum value of n compatible with the total energy ϵ:

$$\epsilon(n) = \frac{h^2}{2m}\left[k^2 - (n\pi/d)^2\right]; \qquad (5)$$

so that

$$n_\circ \equiv n_{max} = [k_\circ]_{int}; \qquad k_\circ = 2\left(\frac{2m\epsilon}{h^2}\right)^{\frac{1}{2}} d, \qquad (6)$$

k being the wavenumber.

It follows that

$$\sigma(\epsilon(n);d) = \frac{2\pi S}{h^2}\sum_{n=-n_\circ}^{n_\circ}\left(m\epsilon - \frac{n^2\pi^2h^2}{2d^2}\right)$$
$$= \frac{8\pi V(2m\epsilon)^{3/2}}{h^3} + \frac{4\pi S}{h^2}\, m\epsilon$$

$$- \frac{S}{4d^2} \sum_{n=o}^{n} \int_{o}^{k_o} \int_{-\infty}^{\infty} k^2 e^{iR(k-n)} dRdk,$$

where $V \equiv Sd$ is the volume of the system.

On performing the simple contour integral involved, this gives

$$\sigma(\epsilon(n);d) = \frac{8\pi V(2m\epsilon)^{3/2}}{3h^3} + \frac{4\pi S\ m\epsilon}{h^2}$$

$$+ \frac{\pi S}{2d^2} \sum_{n=-\infty}^{\infty}{}' \left[k_o^2 \frac{\sin(2n\pi k_o)}{n\pi} + k_o \frac{\cos(2n\pi k_o)}{n^2\pi^2} - \frac{\sin(2n\pi k_o)}{2n^3\pi^3} \right], \quad (7)$$

where the prime on the summation sign indicates that the (n=o) term has been excluded; its value is incorporated into the first term on the right. This is, of course, the exact three-dimensional expression for the total number of energy states. The second term is the analogous expression for the exact two-dimensional case. The third terms is of especial interest: as $d \to \infty$, it reduces to zero, as indeed it must. Further, in the zero-d limit, the sum of all terms on the right vanishes except the surface term - again, as expected.

Finally, the sum of the last two terms on the right will be comparable in magnitude to the first term at $d=d_c$. To determine d_c, the summation is first evaluated readily with the aid of Bernoulli polynomials (6). The final result is the following quadratic equation for d_c:

$$\frac{2}{3h^3} (2m\epsilon)^{\frac{3}{2}} d_c^2 - \frac{m\epsilon}{h^2} d_c - \frac{1}{4} \frac{(2m\epsilon)^{\frac{1}{2}}}{3h} = 0; (8)$$

from which

$$d_c = \frac{3+\sqrt{17}}{8\sqrt{2}} \cdot \left(\frac{m\epsilon}{h^2}\right)^{-\frac{1}{2}} \sim 0(1) \frac{h}{(m\epsilon)^{\frac{1}{2}}} \quad . \quad (9)$$

3. DISCUSSION

The value of d_c given by eq. (9) represents that value of d above which bulk effects begin to dominate. It is simply a natural wavelength reflecting the "confinement effect". Apart from a geometrical factor, which is of the order of unity, d_c is clearly universal, in the sense that it is independent of the geometry involved.

In conclusion, it is observed that the confinement of the ideal Bose gas in space enhances the density fluctuations, thereby destroying the long-range order and, hence, BEC. The disruptive effects in the strongly-interacting system have a rather similar consequence. Accordingl the above conclusions are a fortiori applicable even to this system.

ACKNOWLEDGEMENTS

One of the authors (H.B.G.) wishes to thank Professor Abdus Salam, the International Atomic Energy Agency and UNESCO for hospitality at the International Centre for Theoretical Physics, Trieste, where this work was begun.

REFERENCES

1. M.F.M. Osborne, Phys. Rev. 76 (1949) 396.
2. J.M. Ziman, Philos. Mag. 44 (1953) 548.
3. D. Forster, Hydrodynamic Fluctuations, Broken Symmetry and Correlation Functions (Benjamin, Reading-Massachusetts 1975) pp. 214-251.
4. J.E. Robinson, Phys. Rev. 83 (1951)678.
5. F. London, Superfluids, Vol. II (Dover, New York, 1964) Appendix.
6. I.S. Gradshteyn and I.M. Ryzhik, Table of Integrals, Series, and Products (Academic Press, New York, 1980) p. 1077.

Physica B 165&166 (1990) 597–598
North-Holland

^{3}He-Layers on Graphite

H.J. LAUTER§, H. GODFRIN§, V.L.P. FRANK§* and H.P. SCHILDBERG+

§*Institute Laue-Langevin, BP 156X, F-38042 Grenoble, France*
**University of Konstanz, D-7750 Konstanz 1, F.R.G.*
+*BASF, D-7600 Ludwigshafen, F.R.G.*

Neutron diffraction studies show that in the monolayer of ^{3}He on graphite the commensurate phase is separated from the incommensurate phase by a domain wall phase. Under the pressure of the second adsorbed layer the first layer is driven into an 8*8 commensurability with the substrate. This coincides probably with events seen in NMR (1) and heat capacity measurements (2) in the same coverage region.

1. INTRODUCTION

The phase diagram of a quantum gas monolayer adsorbed on graphite is characterized by the pronounced appearance of the commensurate (C) phase (√3*√3)R30º, as shown for ^{3}He in fig.1. The phases between the C and the incommensurate (IC1) phase could be identified for D_2 (3,4) as a domain wall (D) phase and a domain wall liquid (DL) phase. We present here the identification of the D-phase of ^{3}He as a striped superheavy domain wall phase.

For the helium isotopes, the second adsorbed layer is less dense than the first one (5) in contrast to the hydrogen isotopes (6), where both layers have the same density. The weaker adsorption potential and the enhanced zero point motion of the helium atoms are clearly visible in the low density of the second layer. Under the pressure of the second layer the density of the first layer increases slightly and finally locks into a 8*8 and 9*9 overstructure for ^{3}He and ^{4}He (7), respectively. In the coverage range just below the lock-in two features are proposed: the solidification of the second layer (2) and a precursor of surface induced ferromagnetism (1). To further clarify this situation we made new measurements in this region .

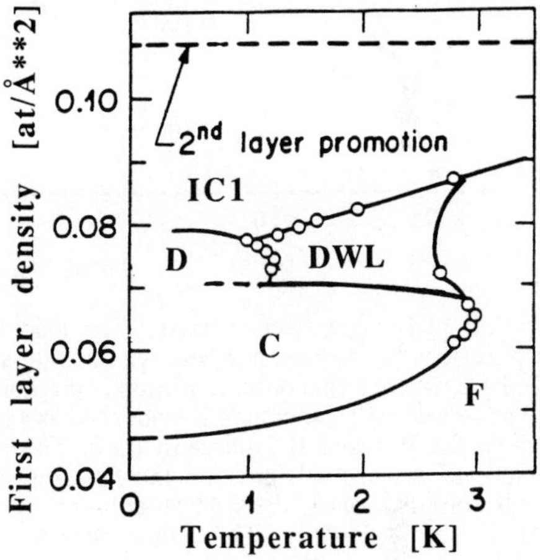

Figure 1: Phase diagram of the monolayer of ^{3}He on graphite based on Ref.12. Reattribution of phases with respect to Ref.2. C = commensurate phase, D = domain wall phase, DWL = domain wall liquid phase, IC1 = incommensurate phase and F = fluid phase.

2. EXPERIMENTAL RESULTS

Neutron diffraction data were obtained at a wavelength of 4.547 Å on the diffractometer D16 at the HFR of the ILL, Grenoble. The substrate used was ZYX-grade graphite (Union Carbide). The coverage range between 0.05 and 0.35 atom/Å^2 of ^{3}He on graphite was explored. The density of the adsorbed layer can be calculated from the diffraction peak position if a homogeneous triangular lattice is assumed. The first five points in fig.2 show all the same ordinate of 0.0636 at/Å^2, the density of the C-phase. The abscissa´s calibration (the total number of adsorbed atoms on the surface of the sample) is given in fig. 2 by the amount of ^{3}He that produces the highest Bragg-peak intensity within the C-phase. This defines the best C-phase: all adsorption sites of the C-phase on the graphite are occupied by an adsorbate atom. This calibration was first made with D_2 on graphite (3,7) and defined a value of 14.34 cc (STP) of gas. This agrees with ^{3}He, despite its low scattering power and high neutron absorption cross section.

The straight full line in fig.2 through the best C-phase point and the origin does not match the points in the IC1-phase. This shows clearly that the calibration of the C-phase and of the IC1-phase do not coincide. One possible explanation is that the effective surface area of the C-phase is smaller than the one of the IC1-phase. Another possibility may be that in the IC1-phase atoms are additionally adsorbed on other planes than the basal ones. The calibration of the IC1-phase is shown in fig.2 by the dashed line, which does not pass through the origin. The difference at monolayer completion amounts already to 6% .

The deduced densities in the D-phase (fig.2) show a behavior that is equivalent to the case of D_2 (3,8), which has been identified as a striped superheavy domain wall phase. The much less favorable scattering conditions of ^{3}He with respect to D_2 made it impossible to measure any satellites. However, the characteristic coverage dependence in fig.2 is a proof that this domain wall phase exists as proposed by heat capacity measurements for ^{4}He (9). The density deduced from the diffraction peak position does of course not represent any more the real density of the layer, but serves just for visualization. The densest stage is reached when the superheavy walls are separated by one row of atoms in commensurate position (7) as for D_2 (3,8).

The diffraction data show that in the discussed coverage

0921-4526/90/$03.50

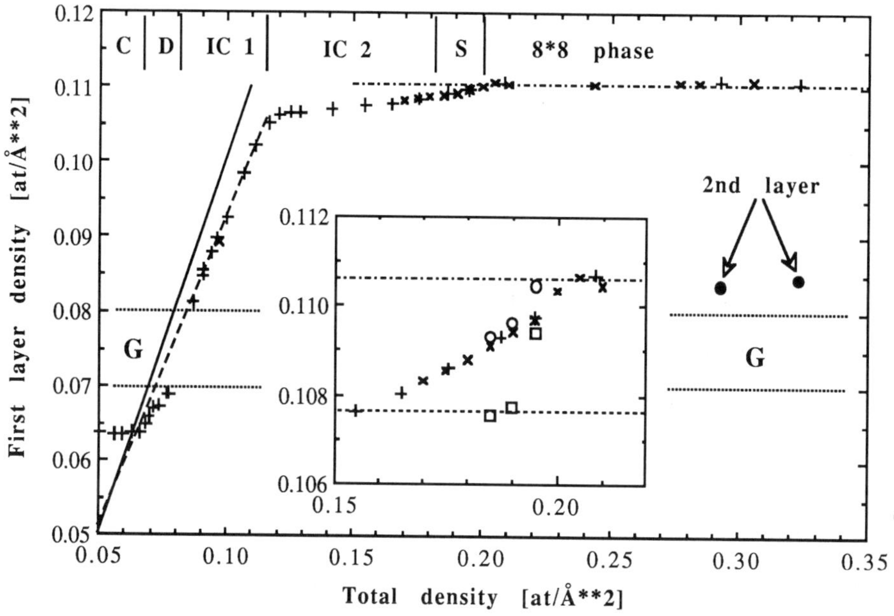

Figure 2: Measured density of the first ^{3}He layer vs. total coverage: at T=1.0K (+) and at T=0.06K (x). (□) and (o) indicate the splitting of the diffraction peak in the insert. (•) is the position of the second layer density. The identification of the different phases is described in the text. **G** marks a region disturbed by a graphite background reflection. The full and dashed line depict the monolayer behavior. The dashed-dotted line shows the density of the 8*8 phase. The dotted line (insert) indicates the 4*4 density.

range no phase coexistence exists (see also (10)). An exception may be the transition between the D-phase and the IC1-phase, where a first order transition is possible.

The second layer promotion is evidenced as a sharp knee between the IC1 and IC2-phase in fig.2. The increasing pressure of the second layer on the first one induces a further compression of the first layer, seen by the small slope in the IC2-phase. This slope increases slightly towards the S-phase, changing over to a zero slope at 0.2 at/Å^2. At this point the density of the first layer (0.1106 at/Å^2) corresponds to an 8*8 overstructure with respect to the graphite, having 37 atoms per unit cell. The approach to this clear lock-in is marked in the S-phase by some diffraction peaks, whose line shape is better described by a double peak (similar to the case of ^{4}He on Papyex at higher coverages (5)). This splitting is shown in the insert of fig.2. It can be interpreted as a slight distortion of the unit cell due to a one directional registry with the substrate. This partial registry agrees for two points with the 4*4 lattice spacing and for one point with the 8*8 one, before finally the pure 8*8 overstructure is reached.

The comparison to the NMR (1) and heat capacity measurements (2) is rendered difficult due to the different calibrations of the surface area, and probably by the different substrate qualities used. In these experiments the calibration was performed either on the incommensurate phase (^{3}He isotherm) or on the overfilled C-phase (N_2 isotherm). A recalibration of + 7 % in total density with respect to the C-phase is made (see discussion above). This shifts the proposed 2nd-layer coexistence region (2) right into the S-phase, where the first layer exhibits the first signs of registry.

3. CONCLUSIONS

We conclude that in the monolayer coverage a domain wall phase exists between the C and IC1-phase. The events in the bilayer seen by NMR (1) and heat capacity measurements (2) happen in a coverage range where we see substantial changes in the first layer. The strong variation of the density and the inhomogeneity of the first layer are not consistent with the simple picture (2) of a coexistence region in the second layer under a constant chemical potential. The evidence for a localized second layer (2) is certainly strong. At this small second layer density only a registered solid is stable. The proposed structure (2,11) and the details of the phase transformation, however, are inconsistent with the present data. Unfortunately, we could not yet see diffraction signals from the second layer in this coverage range. The second layer solid only becomes visible to neutron diffraction above 0.29 at/Å^2 (5,7).

This work has been partially supported by the Federal Ministry of Research and Technology (BMFT), F.R.G.

REFERENCES

(1) H.Franco, R.E.Rapp and H.Godfrin, Phys.Rev.Lett. 57, 1161 (1986)
(2) D.S.Greywall, Phys.Rev. B 41, 1842 (1990)
(3) H.P.Schildberg, H.J.Lauter, H.Freimuth, H.Wiechert and R.Haensel, Jpn.J.Appl.Phys. 26, 345 (1987)
(4) H.J.Lauter, in "Phonons 89", eds. S.Hunklinger, W.Ludwig, G.Weiss (World Scientific, 1990) p.871
(5) H.J.Lauter, H.P.Schildberg, H.Godfrin, H.Wiechert and R.Haensel, Can.J.Phys. 65, 1435 (1987)
(6) H.P.Schildberg, H.J.Lauter, H.Freimuth, H.Wiechert and R.Haensel, Jpn.J.Appl.Phys. 26, 343 (1987)
(7) H.P.Schildberg, Thesis (University of Kiel, 1988)
(8) J.Cui, S.C.Fain, H.Freimuth, H.Wiechert, H.P. Schildberg,H.J.Lauter,Phys.Rev.Lett.60,1848 (1987)
(9) F.C.Motteler, Thesis (Univ. of Washington, 1985)
(10) V.L.P.Frank, H.J.Lauter and P.Leiderer, Physica B 156 & 157, 276 (1989)
(11) V.Elser, Phys.Rev.Lett. 62, 2405 (1989)
(12) S.V.Hering, S.W.Van Sciver and O.E.Vilches, J.Low Temp.Phys. 25, 793 (1976)

Physica B 165&166 (1990) 599–600
North-Holland

^{3}He FILM FLOW: ^{4}He SUBSTRATE COATING EFFECT

Stephen C. STEEL*†, Peter ZAWADZKI*, John P. HARRISON* and Andy SACHRAJDA‡

*Dept. of Physics, Queen's University, Kingston, Ontario, Canada; ‡Division of Physics, National Research Council, Ottawa, Ontario, Canada

Superfluid ^{3}He film flow out of a copper beaker was measured without and then with a ^{4}He coating on the copper surface. The effect of the ^{4}He was to convert the substrate from a purely diffuse to a purely specular surface for the ^{3}He quasiparticles.

1. INTRODUCTION

Superfluid flow in thin films of ^{3}He adsorbed on a substrate offers a chance to observe the effects of a confined geometry on the superfluid properties, since typical film thicknesses are comparable to the coherence length ξ. This paper is a brief report of a larger study which was motivated by the need to improve the quality of the superfluid ^{3}He film flow experiment (1). During the course of this study Freeman *et al* showed that the superfluid properties of 350 nm thick layers of ^{3}He confined between mylar sheets were dramatically changed when the sheets were coated with two monolayers of ^{4}He (2). They concluded that this changed the boundary conditions for the ^{3}He quasiparticles from diffuse to specular. The focus of the results presented here is the similar influence of a ^{4}He coating on superfluid ^{3}He film flow.

2. EXPERIMENT

2.1 Film Flow

The substrate for the adsorbed film, a beaker made out of copper, had inner and outer diameters of 2.5 mm and 4.5 mm. The rim was machined to a semicircular cross-section with radius 0.5 mm in order to avoid film thinning at sharp edges. The beaker surface was mechanically and then electrolytically polished. Electron microscope pictures of a similarly prepared beaker showed a smooth surface with isolated round pits (~ 1 μm). Unfortunately, following these experiments, pictures of the actual beaker showed a larger density of such pits. Nevertheless, the results presented here have been verified with later beakers of glass, rough and smooth stainless steel, and rough and smooth copper.

The measurement techniques have been described elsewhere (1). In brief, the beaker is filled to the rim and held at at 100 mK until an equilibrium meniscus is established by normal flow. The system is slowly precooled to about 10 mK and then to a final temperature between 0.37 mK and 0.9 mK. The warm-up rate is low ($\leq$ 10 μK/day) so there is superflow along the adsorbed film at an almost constant temperature. After about 30 hours the level in the beaker has dropped by up to 1 mm and the experiment is recycled. The superfluid flow rate was determined by measurements of the ^{3}He levels inside and outside the beaker as a function of time.

For the experiments with ^{4}He monolayers, sufficient ^{4}He gas to provide one monolayer, and later 3 monolayers, coverage over the entire inside surface of the ^{3}He space was admitted to the cell at 20 K. The cryostat was then cooled slowly to about 1 K to ensure a uniform coverage of ^{4}He, and the ^{3}He sample was added.

2.2 Film Thickness

A separate Atkins' oscillations cell was built for the film thickness determination. A direct measurement on the film flow beaker was not made. However, a calibrated indirect measurement was made and will be described elsewhere. The van der Waals parameter (1) was deduced to be $\alpha_3 = 11.4 \pm 1.1 \times 10^{-9}\ m^{4/3}$.

3. RESULTS

Figures 1 and 2 show the rate at which the ^{3}He level inside the beaker drops as a function of the position of the level relative to the original meniscus for zero and one monolayer coverage of ^{4}He on the beaker surface respectively. For pure ^{3}He the flow rate decreases rapidly as the level drops, largely because the film thickness over the rim approaches the coherence length. This decrease is modified dramatically by the presence of one monolayer of ^{4}He. Compare, for instance, the 0.53 mK data in figure 1 with the 0.54 mK data in figure 2.

4. DISCUSSION OF RESULTS

The upper scale in figures 1 and 2 shows the calculated film thickness δ corresponding to the

†Present Address: Kamerlingh Onnes Lab., Postbus 9506, 2300 RA Leiden, The Netherlands

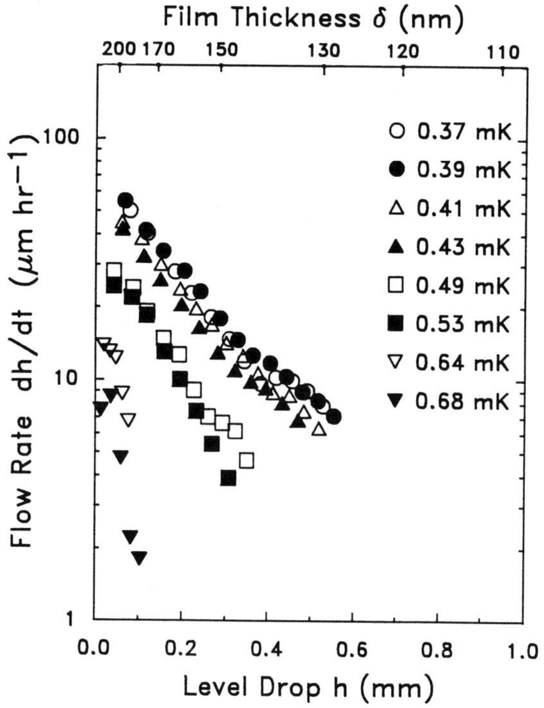

FIGURE 1
Flow rate versus level drop for pure ^{3}He.

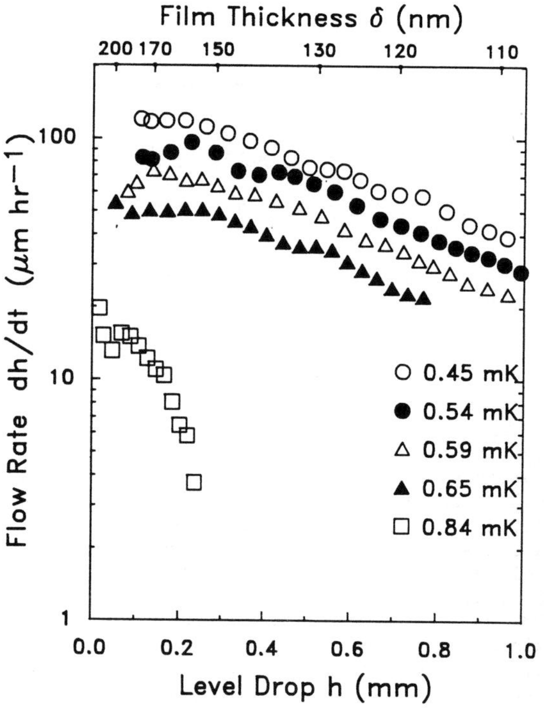

FIGURE 2
Flow rate versus level drop with ^{4}He monolayer.

^{3}He level given on the lower scale. Using this film thickness, the critical current density was calculated as a function of film thickness and temperature. For a given thickness, the superfluid transition temperature was then determined by extrapolating the current density to zero. These results are summarized in figure 3, which shows the transition temperature, normalized by that in bulk ^{3}He (T_c^B = 0.93 mK), as a function of δ^{-2}.

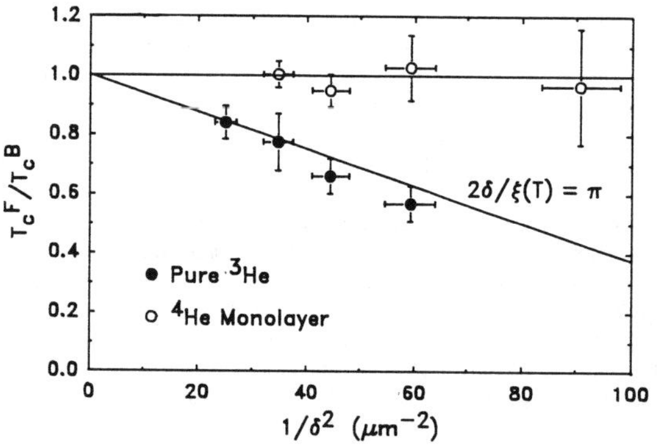

FIGURE 3
Reduced superfluid transition temperature in the ^{3}He film as a function of the film thickness.

5. CONCLUSIONS

In the case of pure ^{3}He, the transition temperatures in the adsorbed film were suppressed below that in the bulk. They fall along the line $2\delta/\xi(T) = \pi$ (shown in figure 3), where $\xi(T)$ is the temperature dependent coherence length. This is the Ginzburg-Landau theory prediction for the superfluid A-phase to normal transition in a layer of ^{3}He with one specular and one diffuse boundary (the free surface and beaker surface respectively) (3). With an adsorbed monolayer of ^{4}He, there was no suppression of the transition temperature, as predicted for purely specular boundaries.

ACKNOWLEDGEMENTS

This work was supported by NSERC.

REFERENCES

(1) J.G. Daunt, R.F. Harris-Lowe, J.P. Harrison, A. Sachrajda, S.C. Steel, R.R. Turkington and P. Zawadzki, *J. Low Temp. Phys.*, **70** (1988) 547.

(2) M.R. Freeman, R.S. Germain, E.V. Thuneberg and R.C. Richardson, *Phys. Rev. Lett.*, **60** (1988) 596.

(3) A.L. Fetter and S. Ullah, *J. Low Temp. Phys.*, **70** (1988) 515.

Physica B 165&166 (1990) 601–602
North-Holland

EVIDENCE FOR QUANTUM KINKS IN THE LAYER-BY-LAYER GROWTH OF SOLID 4HE ON GRAPHITE

T. P. BROSIUS*, M. J. McKENNA, and J. D. MAYNARD

Department of Physics, The Pennsylvania State University, University Park, PA 16802, USA†

We have made measurements of the layer-by-layer growth of solid 4He from the superfluid onto a graphite substrate, and at temperatures below 0.95 K we have observed remarkable and unexpected behavior which we believe is evidence of quantum kinks.

The 4He solid/superfluid interface is an interesting system to study because it demonstrates both classical and quantum behavior. For example, the surface of a 4He crystal undergoes a bulk roughening transition, a classical effect, but also propagates melting/freezing waves, which are possible because of the quantum nature of the system. Edwards, et. al. (1) have shown that quantum kinks can be used to explain several properties of the bulk solid/liquid interface. In this paper we report evidence for quantum kinks in the layer-by-layer growth of solid helium from the superfluid on a graphite substrate.

Our system consists of adsorbed solid layers of 4He on a graphite substrate (Grafoil) contained within an annular acoustic resonator pressurized with superfluid helium. By adjusting the cell pressure at a fixed temperature, we can move the system along a solid adsorption isotherm (2,3). To probe the interface, we propagate fourth sound through the superfluid and induce melting and freezing at the interface. The melting and freezing is sensitive to the condition of the interface and modifies the fourth sound speed. The observed sound speed, C_4, varies from the theoretical speed, C_{04}, according to the equation (4)

$$C_4 = C_{04}\ [1- \Gamma(\partial\Theta/\partial P)],$$

where Γ is a constant determined by parameters characterizing the graphite/helium system and $\partial\Theta/\partial P$ is the slope of the solid adsorption isotherm. (A more rigorous treatment by Uwaha (5) yields essentially the same acoustic equation.) The essence of this equation is demonstrated in Fig. 1 where (a) represents one step in a solid adsorption isotherm and (b) its negative derivative. Since the acoustic equation predicts the fourth sound speed should vary as $-\partial\Theta/\partial P$, we expect the fourth sound speed to follow Fig. 1(b) during a layering transition. The important point is that the nucleation of a solid layer is indicated by a local minimum in the fourth sound speed. By tracking a cell resonance, we make precise measurements of the fourth sound speed as a function of cell pressure to study solid layering transitions. The effect of several step-wise layering transitions on a fourth sound resonance at 1.10 K is illustrated by the top trace of Fig. 2.

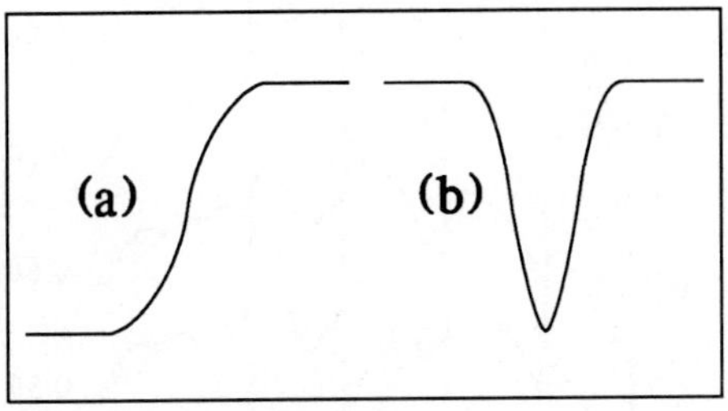

Figure 1. (a) Illustration of the step structure in a solid adsorption isotherm due to a layering transition. (b) Negative derivative of (a), predicted fourth sound feature.

In extending our measurements down to lower temperatures, we observed some remarkable and unexpected behavior. In addition to the normal minima (as typified by the 1.10 K trace of Fig. 2), new features became evident in the data below 0.95 K. These new features consisted of extra minima appearing on the low pressure side of each normal minimum. Each new feature became more prominent as the temperature was lowered. Fig. 2 presents a series of isotherms from 1.10 K to 0.50 K showing the development of the new minima (indicated by the vertical lines). Fig. 3 presents a detailed plot of some fully developed double minima.

*Present address: BDM International, Inc., 7915 Jones Branch Drive, McLean, VA 22102-3396.
†Work supported by NSF DMR 8701682.

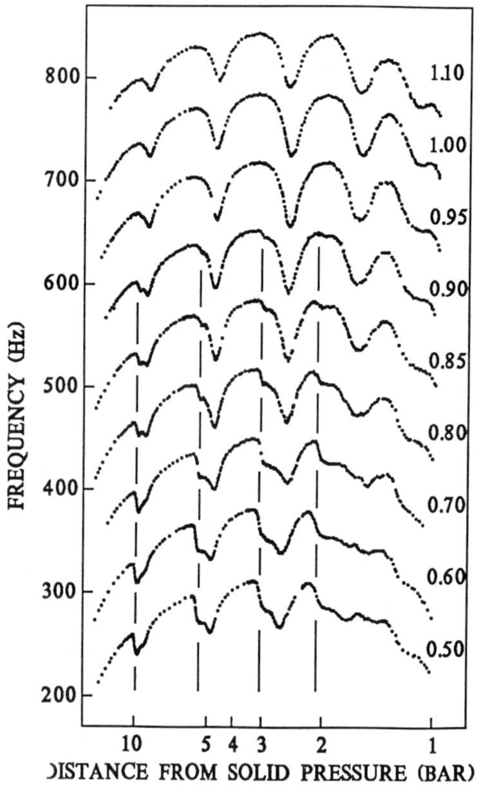

Figure 2. A series of isotherms showing the variations of a fourth sound resonance as a function of the distance from the bulk solidification pressure scaled so as to clarify the features (see Ref. 4). The vertical lines indicate the positions of features attributed to the running quantum kink growth mode. For plotting, all traces except 1.10 K have been displaced downward.

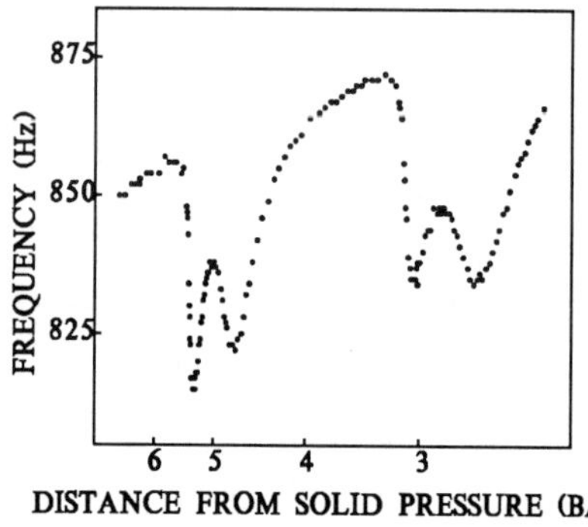

Figure 3. Expanded plot showing two sets of double minima caused by random classical accretion and running quantum kink growth modes.

When the new features were first observed, we thought they were some artifact, and we spent considerable time trying to eliminate them (e.g. cycling the cell through various temperature and pressure paths, baking the resonator under vacuum, and changing the Grafoil). In spite of our efforts, the new features persisted and behaved in a consistent manner. We thought the new features might be associated with the $1\bar{1}00$ bulk roughening transition since the onset of the new features appeared at approximately the same temperature as the $1\bar{1}00$ transition ($\simeq$0.95 K). However, we could imagine no clear connection between this bulk property and the formation of individual epitaxial layers. We believe the new features indicate a second growth mode for the adsorbed solid layer, and we can explain our data in the following way.

Solid layers grow epitaxially through the development of islands or clusters on the surface of a completed layer. The double minima in the data (Fig. 3) indicate that clusters grow in two distinct ways. We hypothesize that the two growth modes are random classical accretion and running quantum kinks. Random classical accretion growth involves atoms sticking to random points around the perimeter of a cluster. With no preference given to either kink or step sites, the clusters grow at a single (average) value of a driving force (pressure increase). This mode dominates at higher temperatures and accounts for the characteristic minima shown in the traces above 0.95 K in Fig. 2.

Running quantum kink growth involves atoms sticking to kink sites and propagating the kinks along step edges. The onset of this mode requires a smaller driving force and can be observed at cell pressures lower than the random classical accretion mode. As discussed by Edwards (1), the running kink mode is easily suppressed by scattering with the excitations in the liquid, so this growth mode only appears at lower temperatures. It should be noted that the dominant scattering excitations (rotons) rapidly decrease in number in the temperature range 0.95 K to 0.50 K, the same temperature range over which the new features in our data become observable. Furthermore, we find the addition of a small quantity of ^3He (2%) suppresses the new features, even at the lowest temperature of 0.50 K. We would expect this because the ^3He increases the number of liquid excitations and may also pin kink sites.

REFERENCES

(1) D.O. Edwards, S. Mukherjee, M.S. Pettersen, Phys. Rev. Lett. 64 (1990) 902.
(2) J. Weeks, Phys. Rev. B26 (1982) 3998.
(3) D.A. Huse, Phys. Rev. B30 (1984) 1371.
(4) S. Ramesh, Q. Zhang, G. Torzo and J.D. Maynard, Phys. Rev. Lett. 52 (1984) 2375.
(5) M. Uwaha, J. Physique 45 (1984) 1559.

Physica B 165&166 (1990) 603–604
North-Holland

OBSERVATION OF SOLITON-LIKE WAVES IN ADSORBED FILMS OF SUPERFLUID 4HE

M. J. McKENNA, R. J. STANLEY, ELAINE DiMASI, and J. D. MAYNARD

Department of Physics, The Pennsylvania State University, University Park, PA, 16802, USA

While there has been considerable theoretical attention given to nonlinear (soliton) effects for adsorbed films of superfluid 4He, there have been no systematic experimental studies. In this paper we report measurements of wave propagation in such films with drive levels varying over five orders of magnitude. In addition to the critical saturation of a third sound wave, we observe another wave whose velocity of propagation changes with drive amplitude (linearly over a limited range). This dependence agrees with theoretical predictions for soliton behavior.

Like shallow water waves, third sound waves in adsorbed films of superfluid 4He are intrinsically nonlinear. That is, the speed of sound depends on the depth of the fluid, and since a wave of finite amplitude modifies the depth, the equation of propagation becomes nonlinear. When soliton solutions were found to describe the behavior of shallow water waves, theoretical attention was directed toward understanding nonlinear effects in third sound propagation. Theoretical treatments (1,2,3) involved approaches from the Schrodinger equation as well as from the equations of two fluid hydrodynamics. The resulting full equations could not be solved, but if some terms were dropped, the Korteweg-deVries equation could be obtained. It was predicted (1,2) that a "train" of short wavelength ($\sim 10^{-7}$ cm) solitons could be observed in the tail of a third sound pulse.

Despite the extensive theoretical attention given to nonlinear effects in third sound propagation, there have been virtually no systematic experimental studies. Kono et. al (4) claimed to have verified a theory for the shape of the tail of a received pulse, but used only one relatively low drive level (0.1 nJ); thus, it is impossible to verify if the effect is nonlinear. Third sound data taken in other experiments in a linear regime show similar tails, and thermal effects of the transducer, substrate, and vapor cannot be discounted.

Recently we have undertaken a systematic study of the behavior of third sound as a function of drive level. Our experimental system consists of a glass substrate for the adsorbed helium film and two evaporated aluminum strip transducers, 3.2 cm. apart, one of which is driven as a resistive heater and the other is biased near its superconducting transition so as to act as a thermometer.(5) The thickness of the adsorbed helium film, determined from the unsaturated vapor pressure, was varied from 2.9 layers to 6.7 layers. The temperature of all our measurements was ~1.1 K. Pulses in the adsorbed film were launched by driving the heater with a square heat pulse of measured power and duration corresponding to input energies ranging from 20 pJ (20 μW peak power and 1 μs duration)to 1.2 μJ (40 mW peak power and 30 μs durationJ). Data subsequently recorded at the receive transducer was digitized and converted to a plot of the temperature change (ΔT) as a function of time.

At low drive levels, ordinary third sound is observed. As the drive level is increased, the third sound amplitude increases until the velocity of the superfluid component (v_s) in the third sound pulse approachs the critical velocity of the film (v_c), and the amplitude of the third sound pulse saturates and remains constant for larger drive levels. (For this film thickness, 3.9 layers, saturation occurs at a drive level of ~0.1 μJ.)

Data taken at a drive level of 0.2 μJ is shown in Fig. 1a. The feature having a

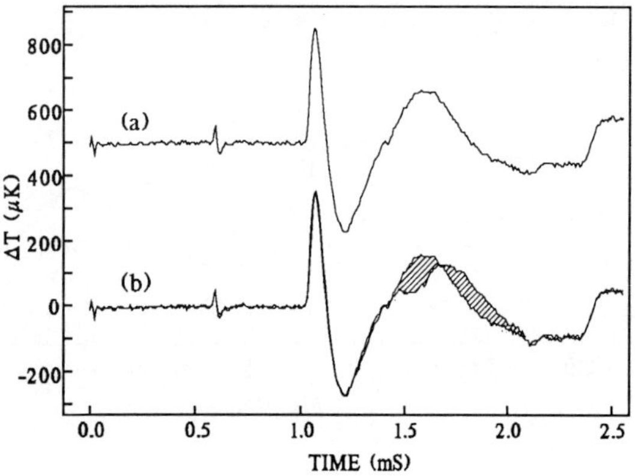

Figure 1. (a) Received signal with a 0.2 μJ drive level. (b) Superposition of (a) with the data from a 0.3 μJ drive level; the difference is emphasized with shading.

†Work supported by NSF DMR 8701682

time-of-flight of ~0.7 ms is due to ordinary sound in the vapor which is generated when the adsorbed helium evaporates at the drive. A third sound reflection from the end of the glass substrate appears at ~2.4 ms. The third sound pulse is received at ~1 ms.

For the drive level in Fig. 1a, the third sound pulse has saturated. If the drive level is increased further, a second pulse appears out of the cold tail of the third sound pulse. This effect is illustrated in Fig. 1b, which shows a duplicate of Fig. 1a (drive level of 0.2 μJ) superimposed on the data for a drive level of 0.3 μJ. Because the third sound pulse has saturated, there is little difference between the two traces for most of the data; however, the presence of the second pulse shows up as a difference in the two traces, as illustrated with shading in Fig. 1b. To clarify the shape of the second pulse, the two data sets were subtracted; the difference is shown in Fig. 2a. (The features in Fig. 2 at ~1 ms are due to noise in the third sound pulse which is amplified when data sets are subtracted.) Traces (b) through (h) (determined from the differences between successive data sets at higher drive levels) show the consequences of further increasing the the drive level. The time-of-flight of the second pulse clearly increases with drive level; such an amplitude dependent velocity (C_s) suggests soliton behavior. Both the third sound saturation, and the amplitude dependence of C_s have been observed for both

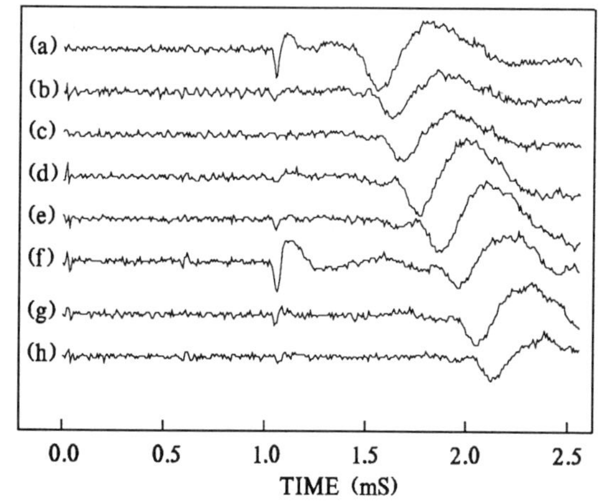

Figure 2. Change in the amplitude of the soliton-like wave for two different drive levels. (a): between drive levels of 0.3 and 0.2 μJ. (shown in trace (b) of Fig. 1), (b): 0.38 and 0.3 μJ, (c): 0.4 and 0.38 μJ, (d): 0.44 and 0.4 μJ, (e): 0.48 and 0.44 μJ, (f): 0.84 and 0.48 μJ, (g): 1.0 and 0.84 μJ, and (h): 1.2 and 1.0 μJ.

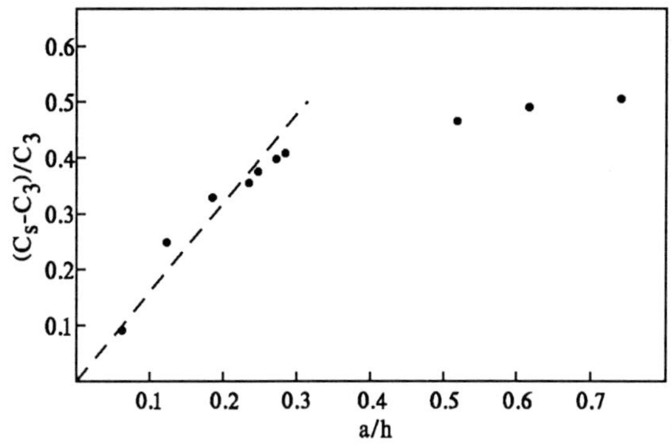

Figure 3. Difference in soliton velocity and third sound velocity for increasing soliton heights (increasing drive levels.)

thinner and thicker superfluid films.

Such an effect for the velocity C_s of a helium film soliton has been predicted by Condat and Guyer (1) and by Browne (2). According to theory, $|C_s-C_3|/C_3$ should scale with a/h, where C_3 is the velocity of third sound, a is the amplitude of the soliton and h is the film thickness. For low amplitudes, Bergman (3) finds that the amplitude of a third sound pulse is $\delta h \simeq -h(L/C_3^2)(\Delta T/T)$, where L is the latent heat of He II and ΔT is the temperature amplitude of the pulse. From our low amplitude data we can calibrate δh as a function of drive level. For the higher drive levels we assume that a/h scales with the drive level the same as δh/h.

Fig. 3 shows a plot of $|C_s-C_3|/C_3$ versus a/h. For the lower values of a/h the data agrees with the predictions of the soliton theory, but higher values deviate. It should be noted that for the higher values, our estimated amplitude of the pulse is close to the depth of the film. It should also be noted that neither the theory nor the experiment rigorously yields a Korteweg-deVries soliton. However the data clearly indicates some soliton-like behavior. At the present time we are pursuing experiments to study how two of our nonlinear pulses interact.

REFERENCES

(1) C.A. Condat and R.A. Guyer, Phys. Rev. B, 25, (1981), 3117.
(2) D.A. Browne, J. Low. Temp. Phys., 57, (1984), 207.
(3) D. Bergman, Phys. Rev. 188, (1969), 307.
(4) K. Kono, S. Kobayashi, and W. Sasaki, J. Phys. Soc. Japan, 50, (1981), 721.
(5) J.A. Roth, G.J. Jelatis, and J.D. Maynard, Phys. Rev. Lett., 44, (1980), 333.

Physica B 165&166 (1990) 605–606
North-Holland

HYBRIDIZATION OF ROTATIONAL STATES OF ADSORBED HD ON POROUS VYCOR GLASS

T.E. HUBER.* Lyman Laboratory. Harvard University. Cambridge, MA 02138 and C.A. HUBER.* Francis Bitter Nat.Magnet Laboratory. M.I.T.. Cambridge, MA 02139.

We present a study of the optical absorption due to vibrational and rotational transitions of hydrogen in porous Vycor glass. The absorption linewidths of HD are six times larger than those of the pure isotopes H_2 and D_2. Also we observe the transitions corresponding to $\delta J=3$ and $\delta J=5$, which are not observed for the pure isotopes, indicating strong mixing of the rotational states of HD on porous Vycor glass.

1. INTRODUCTION

We have previously reported on the fundamental and overtone infrared absorption of the hydrogens in porous Vycor glass (PVG) and silica gels.[1] In general, we find that the vibrational absorption by the pure isotopes, H_2 and D_2, is enhanced by the presence of the surface. The rotational state is little modified by the surface when compared to the free molecule and the corresponding absorption features are sharp (FWHM are typically 20 cm^{-1}). If quantum effects are considered HD stands as an intermediate between H_2 and D_2. However, we observe that the width of the various resonances of HD is 120 cm^{-1}, six times that of the pure isotopes. We also observe lines corresponding to $\delta J=3$ and 5, that are not allowed for the pure isotopes. This communication was prompted by recent, nuclear magnetic resonance (NMR) studies of HD and D_2 in amorphous silicon.[2] In these studies deuteron multiple spin echoes have been detected. Since the observations are made at low temperatures the rotational ground state, J=0, is believed to be mixed with higher J states similarly to our system.

2. EXPERIMENTAL METHODS

The Vycor sample was a disk 4 mm thick and was contained in a closely fitting copper cell with sapphire windows which was connected to a gas handling system through a capillary. The Vycor average pore diameter is 7.0 nm. The temperature of measurements is 20 K. From a BET fit to the low-coverage HD adsorption isotherms we obtain that the surface-layer coverage is $n_o=3.4\times10^{-3}$ mole/cm^3. Infrared-absorption spectra were measured with a single beam dispersive spectrometer with a resolution of 5 cm^{-1}. The high purity HD (99.5%) sample was obtained from a commercial supplier. Figure 1 shows the absorbance $A=\log_{10}(I_o/I)$ measured for various coverages. The reference intensity I_o is that which is transmitted through the bare Vycor at the same temperature. Three lines from adsorbed HD are easily noticeable for low coverages at 3850 cm^{-1}, 4040 cm^{-1}, and 4800 cm^{-1}. Their FWHM are 120 cm^{-1}. The large features at 3750 cm^{-1} and between 4550 and 4600 cm^{-1} arise from a shift upon adsorption of the hydroxyl peaks at those frequencies. For full pores a mixed signal is obtained which contains the surface lines and also the bulk features $Q_1+S_0(0)$ and $S_1(0)+S_0(0)$ arising from the HD in the pores. The absorption spectrum of bulk liquid HD, obtained as a control measurement with our set-up is also shown for comparison. It is also observed that the surface lines have no analog in the bulk. As anticipated, the surface spectrum is unusual in several respects. First, the HD roto-vibration absorption features, with a FWHM=120 cm^{-1} are broader than the corresponding H_2 and D_2 vibrational transitions (FWHM=20 cm^{-1}) and roto-vibrational transitions (50 cm^{-1}). Second, and more importantly, the frequency of the "surface" lines do not match the values of those for the bulk, suggesting that there is a relaxation of the selection rules. The selection rule $\delta J=2$ is applicable to the pure isotopes in the bulk and on PVG, and seems to work well for bulk HD and for the pure isotopes on Vycor yet this selection rule seems to be broken for HD on Vycor.

It has long been known that the rotation of HD is special. Unlike the pure isotopes, the center of mass and the center of rotation are not the same for HD. Also, phonon coupling could qualitatively explain the large bandwidths, the phonons of H_2 in Vycor have a wide energy distribution peaking at around 100 cm^{-1}. The coupling of

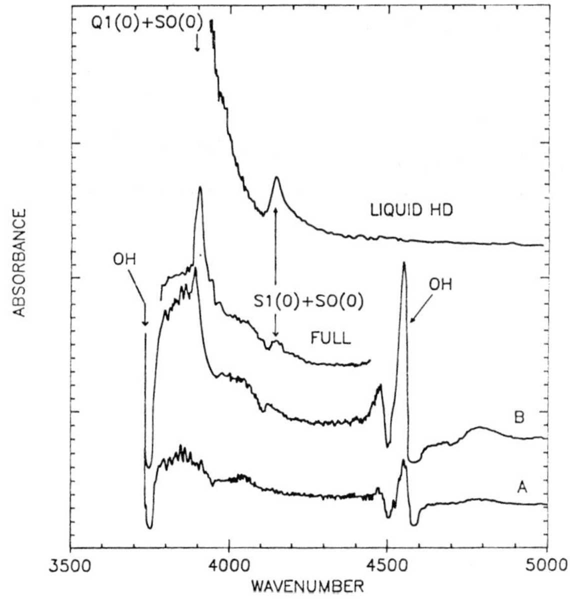

FIGURE 1

Infrared absorption spectra of HD on porous Vycor glass. Curve A and B correspond to surface coverages n=0.3 n_0 and 1.2 n_0, respectively. CurveThe spectra measured for full pores and of bulk liquid HD are also shown.

molecular rotations and phonons was treated by Babloyanz.[3] In the solid phase, where the effect is small, this phenomena was also studied by Bose and Poll.[4] The selection rules for roton-phonon processes have been given by Holleman and Ewing.[5] They find $\delta J=0, \pm1, \pm2$ etc, with δJ the difference between the final and initial rotational quantum numbers. In comparison, the solid and liquid hydrogens (H_2, D_2 and HD) $\delta J= 0, \pm2$. The absorption features, at 3850, 4040 and 4800 cm^{-1} are identified as the $\delta=2$, $\delta=3$ and $\delta J=5$ transitions. The observed frequencies correspond closely to the transition frequencies calculated from the well-known parameters of the molecule. The $\delta J=0$ and 1 transitions are outside the experimental frequency range. The $\delta J=4$ transition is superimposed to the hydroxyl feature and was not observed either.

3. CONCLUSIONS

In summary, we present optical measurements of infrared absorption of HD on PVG. These observations suggest that the rotational state of HD on PVG is strongly hybridized, resulting in anomalous bandwidths and breakdown of selection rules.

REFERENCES

1. T.E. Huber and C.A. Huber. Phys. Rev. B40 (1989) 12527 and references therein.
2. M.P. Volz, P. Santos-Fihlo, M.S. Conradi, P.A. Fedders, R.E. Norberg, W. Turner and W. Paul.- Phys. Rev. Lett. 63 (1989) 340.
3. A. Babloyanz. Mol. Phys. 2 (1958) 39
4. S.K. Bose and J.D. Poll. Can. J. Phys. 65 (1987) 1577.
5. G.W. Holleman and G.W.Ewing. J. Chem Phys. 44 (1966) 3121.
6. B.P. Stoicheff, Can. J. Phys. 35 (1957) 730.

Physica B 165&166 (1990) 607–608
North-Holland

WETTING OF p-H_2 ON GRAPHITE*

E. LERNER, A.L. PEREIRA and F.A.B. CHAVES

Instituto de Física, Universidade Federal do Rio de Janeiro
Rio de Janeiro, 21945, Brasil

Vapor pressure measurements of p-H_2 adsorbed on uncompressed exfoliated graphite foam were made, as a function of coverage, for the temperature range of 8.0 to 20.0 K. The isotherms for the temperatures of 13.64, 14.14, 14.53 and 15.11 K were measured up to about 20 layers and the others up to about 6 layers. These preliminary results seems to indicate a wetting transition below the bulk triple point.

1. INTRODUCTION

Several references (1-6) indicate that almost all wetting transitions on graphite and on chemically inert metal surfaces by noble gases and simple molecular adsorbates, occur at or near the bulk triple point.

A few systems have been studied which have their wetting transitions at temperatures lower than the respective bulk triple points. We could mention Ar(7), Xe(7), Kr(8), CF_4(9), CH_4(10) and possible Ne(11) adsorbed on graphite, as examples of these systems.

We measured isotherms of p-H_2 adsorbed on uncompressed exfoliated Graphite foam, for the temperature interval of 8 to 20 K. For the isotherms at 13.64, 14.14, 14.53 and 15.11 K about 20 layers were measured. The 8.10 K isotherm was measured up to about 5 layers, where it seems to reach its saturated vapor pressure. Analysis of the data seems to indicate that the wetting transition os p-H_2 adsorbed on Graphite is below the bulk triple point.

2. EXPERIMENTAL

The experimental arrangement was the same as the one used in reference (12), with the adsorbent being 5.60 g of uncompressed exfoliated Graphite foam in the shape of disks. The monolayer coverage obtained from the p-H_2 isotherms is 40.2 cm^3 (STP). The total area of the substrate was 126.9 m^2.

The temperatures were measured by a germanium thermometer, calibrated from the values of the saturated vapor pressures of some isotherms. The temperature uncertainty was of about $\pm$ 10mK. Pressures for P<10 Torr were measured using a MKS Baratron capacitance gauge and for P>10 Torr by a Texas Instruments precision gauge. The H_2 gas used, had a purity of 99.9999%, and the conversion from ortho to para, followed the conventional method.

3. RESULTS

The results plotted as $\mathrm{Log}(P_0/P)$ versus V_{ads}^{-3} in cm^{-9} (STP), for the 8.10 K, 13.64 K and 15.11 K isotherms, are shown in figures 1,2 and 3 respectively.

All three graphs, at the first view, seem to indicate that the Frankel-Hill-Halsey equation,

$$kT\,\mathrm{Ln}(P_0/P) = \alpha\, d^{-3}$$

where d is the film thickness and α the van der Waals coefficient, is not obeyed and therefore the p-H_2 doesn't wet the graphite for this temperature interval.

Analysing the data the p-H_2 doesn't wet the graphite at 8.10K, and the maximum number of layers is about five. For the 13.64 and 15.11K isotherms the intersections of the straight line fittings with the X axis at $P = P_0$, are 2.44×10^{-9} and 1.61×10^{-9} cm^{-9}, respectively. Calculating which would be the dispersion in temperature, corresponding to the ones in pressure for these isotherms to pass through the

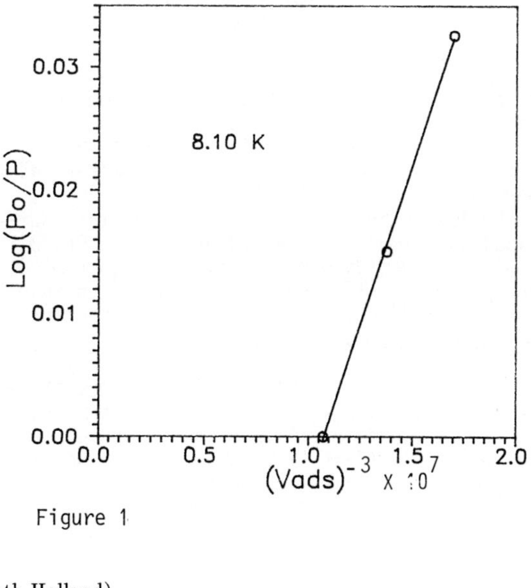

Figure 1

* Work supported by FINEP and CNPq.

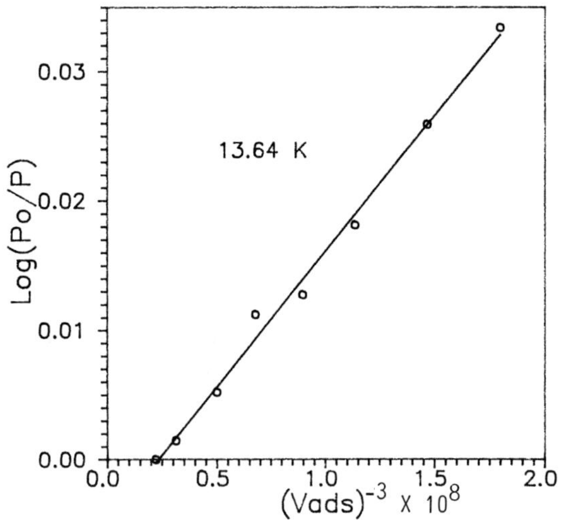

Figure 2

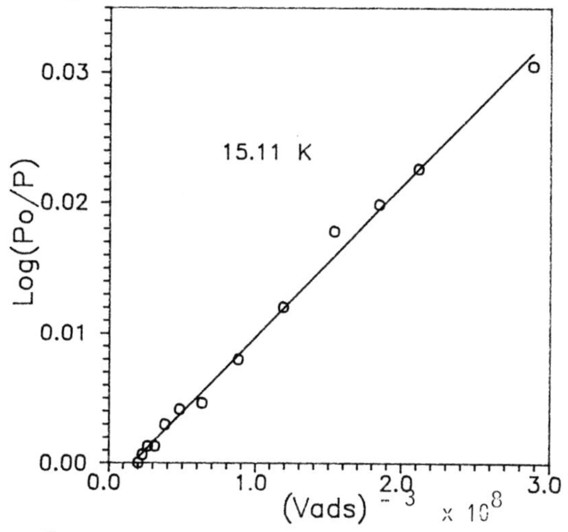

Figure 3

origin, we obtained 0.06 and 0.05%, which are within our experimental uncertainties. The other two isotherms for 14.14 and 14.53 K behave in the same manner, having similar dispersions.

Following the same procedure used in previous experiment of $p\text{-}H_2$ adsorbed on Grafoil (13), the results were plotted as V_{ads} (cm^{-3}) versus $(P/P_o)^{T/T*}$, where T* is any convenient temperature base. Figure 4 shows the adsorption isotherms, giving the amount adsorbed in cm^3 (STP) as a function of $(P/P_o)^{T/100}$, indicating that the ones in the interval of temperature between 13.64 and 15.11 K fall on a commun curve, as found in (13). In these curves we supressed the first layer and several points for better clearness of the figure.

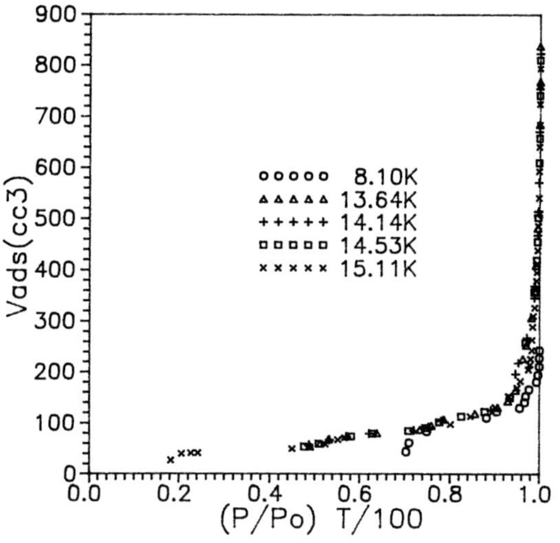

Figure 4

4. CONCLUSIONS

From the behavior of the isotherms shown in figures 1 to 3 and of the other two isotherms, these preliminary results seem to indicate that the $p\text{-}H_2$ wets graphite below the bulk triple point T = 13.80 K. This assumption seems to be confirmed by the "universal" curve followed by the isotherms from 13.64 to 15.11 K and moreover confirming the previous results (13) where there was evidence that even the 10 K isotherm fitted also the "universal" curve. From our results and the previous ones we suggest that the $p\text{-}H_2$ graphite system hes a wetting transition temperature around 10 K.

REFERENCES

1) M.Sutton et al,Phys.Rev.Lett 51 (1983) 407
2) M.Drir and G.B.Hess,Phys.Rev.B33 (1986)4758
3) M.J.Lysck et al,Phys.Lett. A-155(1986) 340
4) J.Krim et al,Phys.Rev.Lett. B6(1987) 584
5) L.Bruschi et al,Europ.Phys.Lett.58(1988)541
6) A.D.Migone et al,Phys.Rev. B37(1988) 5440 (and references within)
7) J.L.Seguin et al,Phys.Rev.Lett.51 (1983)122
8) M.Bienfait et al,Phys.Rev. B29 (1984) 983
9) J.Suzanne et al,Phys.Rev.Lett 52 (1984) 637
10) J.J.Hamilton and D.L.Goodstein,Phys.Rev.B28 (1983) 3838
11) E.Lerner et al, Surf Sci 160 (1985) L524
12) F.Hanono et al, J.Low Temp. Phys. 60 (1985) 73
13) J.G.Daunt et al,J.Low Temp.Phys.44(1981)207

Physica B 165&166 (1990) 609–610
North-Holland

LOW TEMPERATURE NMR STUDY OF TWO-DIMENSIONAL MOLECULAR MOTION

Richard F. Tennis, Feng Li, and Wiley P. Kirk

Physics Department, Texas A&M University, College Station, TX 77843-4242

The magnetic properties of surface layers of methane physisorbed on porous silicon have been studied in detail for several submonolayer coverages. At 0.6 monolayer, the spin-lattice relaxation time T_1 versus temperature showed two minima and two discontinuities between 1.3K and 73K, while at 0.3 monolayer the T_1 behavior was much more complex. At 0.8 monolayer, T_1 increased dramatically by several orders of magnitude. The restriction of the pore size is believed to be responsible for this unusual T_1 behavior. The relaxation times for the 0.3 and 0.8 monolayers were non-exponential, due to the presence of two or three groups of spins whose local environments were different from each other.

1. INTRODUCTION

It is expected intuitively that various types of motion of gas molecules physisorbed on a substrate will become greatly restricted by the geometry of the surface. Consequently, surface effects can dominate, especially at low temperatures, the motional correlation times of magnetic spin systems, which in turn is reflected by the appearance of striking spin relaxation time dependencies on substrate dimensions and morphology. Therefore, low temperature NMR studies of two-dimensional spin systems provide a powerful tool for probing the dynamics of motional effects in surface layers arising from a variety of adsorbate and substrate combinations. Past studies have shown that the magnetic features of two-dimensional surface layers typically include, T_1, spin-lattice relaxation times, uniformly smaller than those in the bulk system, a double minima in the T_1 verse temperature curve in the 4K to 100K temperature range, and spin-spin relaxation times T_2 always less than T_1 even in the motionally narrowed limit.[1,2,3] Also, some systems are found to relax non-exponentially.

In this study, we have made nuclear magnetic resonance measurements on methane physisorbed on porous silicon, a new substrate system that we recently developed.[4] Plates of porous silicon can be formed up to a thickness of 100 to 200 μm by anodic dissolution in hydrofluoric acid. Calculation of the distribution of pore sizes from the adsorbtion isotherm data showed that most of the pores were a uniform 18 Å in diameter, and 75% of the surface area was provided by pores of diameters less than 30 Å. Thus the porous silicon substrate provided a homogeneous environment for the adsorbed gas.

Spin-lattice relaxation times were usually measured by the standard $90°\text{-}\tau\text{-}90°\text{-}\tau_w$ pulse sequence. Spin-spin relaxation times were measured in different ways. For values of the natural linewidth much wider than the linewidth of the static field, T_2 was measured by fitting an exponential dacay to the tail of the FID. When the linewidth was closer to or smaller than the field width, the MGCP spin echo sequence[5,6] was used to measure T_2.

2. RESULTS

At a methane coverage of 0.3 of a statistical monolayer and temperatures below 20K, the spin-spin and spin-lattice relaxation were non-exponential. This was due to the presence of two or three groups of spins whose local environ-

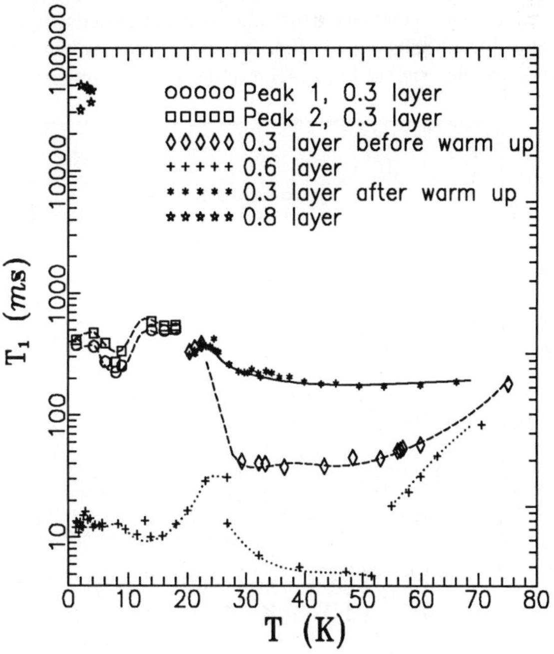

Figure 1. Spin-lattice relaxation times T_1 versus temperature T for different layers of coverage. Lines guide the eye.

0921-4526/90/$03.50

ments were different from each other. After the initial cooldown to 4.2K, the spectrum showed two peaks separated by 24.8 *k*Hz, which corresponds to a 5.8 Gauss difference in the local fields experienced by the two sets of protons. As the temperature of the sample increased, a third peak developed between the initial two peaks. This peak narrowed and grew in amplitude such that at a temperature of 30K it completely dominated the spectrum. This peak was assumed to be the result of motional narrowing. After measurements were taken up to 73K and the sample was quickly cooled back to 4.2K, the narrow middle peak was still present. This behavior suggested that some of the adsorbed molecules were thermally induced into second layer positions at high temperatures and were then constrained there during rapid cooling. These molecules would have to be hindered in some manner from returning to the substrate yet still be free enough to experience motional narrowing at lower temperatures than first layer molecules.

The spin-lattice relaxation times (Fig. 1) of methane for 0.6 monolayer were an order of magnitude smaller than those seen in bulk methane by DeWitt and Bloom[7] and those seen for one and four monolayers, and two orders of magnitude smaller than the 0.8 layer values. The position of the low temperature minimum was shifted to 15K with respect to the bulk where it occurs at ~8K. A 53K minimum was also seen as is the case for methane on graphite.[1,2] The sudden change in T_1 at 27K is in the opposite direction of the transition seen at 20K in the bulk data. For 0.3 monolayer, T_1 values were much larger and there were two additional jumps in the data, one at 20K, the position of a first order phase transition in the bulk, and the other at 11 to 14K near the T_1 minimum in the 0.6 layer data. The 0.3 layer T_1 values increased three to four times after warming up to 70K and rapid cooling back to 4.2K, except above 70K where the two sets of data merged. The data below 20K were split between the two broad "solid" peaks with equal T_1.

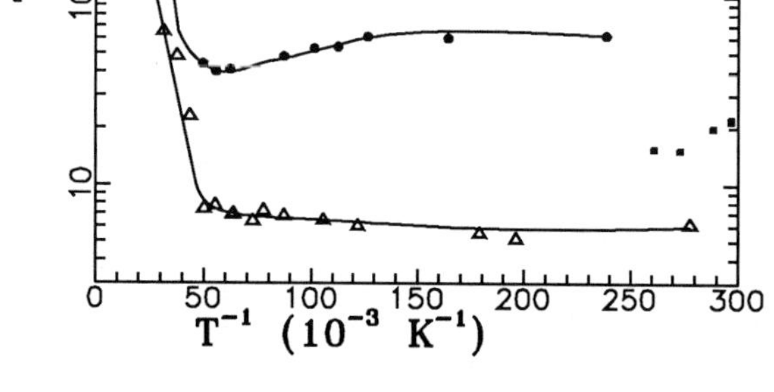

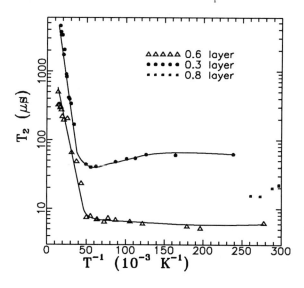

Figure 2. Spin-spin relaxation T_2 versus inverse temperature T^{-1}. T_2 is linear with T^{-1} above 22K, and becomes temperature independent below 22K. Lines guide the eye.

The temperature behavior of the T_2 spin-spin relaxation times for different coverages were very similar except for a difference in magnitude (Fig. 2). For temperatures below 22K, the dependence of T_2 on temperature was relatively small. The sudden break at 22K coincided with the onset of motional narrowing and a much stronger T dependence. Above the transition temperature, the correlation time for spin-spin relaxation was thermally activated and obeyed the Arrhenius' relation $\tau=\tau_0 e^{\varepsilon/kT}$, where ε is the activation energy. Below 22K, T_2 reached a limiting value where the Larmor period was much smaller than the molecular reorientation time.

3. CONCLUSIONS

Methane adsorbed on porous silicon has provided a new 2-D system with several interesting new details. The low coverage T_1 data is especially complex with multiple T_1 minima and discontinuities. The non-exponential relaxation and accompanying multiple peaks in the spectrum may with further study provide valuable insight into the orientation of the adsorbed surface layer at low temperatures. The jump in T_1 at 56K was much larger than the slight discontinuity seen by Quatemen and Bretz[2] for methane on grafoil. This indicates that the 2-D liquid state was more restricted rotationally by the narrow pores in the silicon than by the planes in grafoil. Translational motion may also be significantly hindered causing a rise in T_1. The different T_1 behavior showed that the 2-D solid that formed as a result of rapid cooling has less freedom of motion than the solid which was formed more slowly. This also suggests that the methane was "clumped" together in patches rather than being spread out over the surface when cooled too quickly.

ACKNOWLEDGEMENTS

We wish to thank G.F. Spencer and J.H. Ross for helpful suggestions. Work was supported in part by NSF DMR-8803559.

REFERENCES

(1) J.W. Riehl and K. Koch, J. Chem. Phys. **57** (1972) 2199.
(2) J.H. Quateman and M. Bretz, Phys. Rev. B. **29** (1984) 1159.
(3) J.B. Aubrey and A.A.V. Gibson, Phys. Rev. B. **39** (1989) 3959.
(4) R.F. Tennis and W.P. Kirk, Jpn. J. Appl. Phys. **26** (1987) 321.
(5) S. Meiboom and D. Gill, Rev. Sci. Instrum. **29** (1958) 688.
(6) H.Y. Carr and E.M. Purcell, Phys. Rev. **94** (1954) 630.
(7) G.A. DeWitt and M. Bloom, Can. J. Phys. **43** (1965) 986.

Physica B 165&166 (1990) 611–612
North-Holland

LATTICE DYNAMICS OF COMMENSURATE MONOLAYERS ADSORBED ON GRAPHITE

H. J. LAUTER[§], V. L. P. FRANK[§*], H. TAUB[+] and P. LEIDERER[*]

[§] *Institute Laue-Langevin, BP 156X, F-38042 Grenoble, France*
[*] *University of Konstanz, D-7750 Konstanz 1, West Germany*
[+] *University of Missouri-Columbia, Columbia, Missouri 65211, USA*

A review of the lattice dynamics of gases adsorbed on graphite in the commensurate phase is presented, taking into account their quantum character. The phonon dispersion curves are characterized by an energy gap in the acoustic branch at the zone center. Its magnitude is related to the corrugation of the in-plane adsorption potential. The energy of the zone boundary phonons, on the other hand, is determined by the interaction between adatoms. The measurement of these quantities allows the comparison with theoretical models.

1. INTRODUCTION

During the last decade, substantial progress in the knowledge of the properties of two-dimensional (2D) matter has been achieved thanks to studies done on monolayers of gases adsorbed on well characterized substrates like graphite. The substrate can impose its periodicity on the position of the adsorbed atoms, often leading to a commensurate structure. In this phase, the lack of translational invariance produces a gap (Δ) in the acoustic branch at the zone center of the phonon dispersion relation. The magnitude of Δ is related to the corrugation of the adsorption potential. Changes in temperature and/or coverage induce phase transitions that are the result of a delicate balance between the interaction of the adsorbed gas and the substrate, and between the adsorbed gas molecules themselves. Unfortunately, the knowledge of the details of these interaction potentials is sparse, since not many experimental techniques are available. The adsorption potential itself has been determined mainly with molecular beam scattering, but the magnitude of the in-plane corrugation of the adsorption potential is very difficult to obtain in this way. The temperature renormalization of the phonon spectrum gives insight into the anharmonic terms of the adsorption and intermolecular potentials. Recently, several measurements of the gap became available (1-8) and allowed a comparison with theoretical models of the adsorption potentials.

2. EXPERIMENTAL RESULTS

We present here a review of results of inelastic neutron scattering experiments that determined Δ and its dependence on temperature, for various substances. The studied adsorbates present all a registered ($\sqrt{3}*\sqrt{3}$)R 30º phase (a_{nn}=4.26Å) and can be grouped mainly into two classes according to their quantum character (see Table I). The hydrogen isotopes (H_2, HD, D_2) (1-3) and ^3He (4) are typical quantum gases: their interaction potential is weak, they exhibit a large zero point motion and a very large compressibility (the commensurate a_{nn} is much larger than the 3D-solid one). Nitrogen (N_2) (5,6), deuterated-methane (CD_4) (7) and Krypton (8), on the other hand, are much heavier molecules/atoms, with consequently a smaller zero point motion forming less compressible monolayers . In these cases the 3D lattice parameter matches within some percent the one of the 2D commensurate phase.

Table I: Parameters that characterize the adsorbate gas - graphite system in the commensurate phase, the adsorbates are ordered with increasing de Boer parameter.

[a] the de Boer parameter indicates the quantum character of a substance, $\Lambda = h / \sigma\sqrt{(m\varepsilon)}$, it is the ratio between the de Broglie wavelength of the relative motion of two atoms with energy ε and the minimum in the LJ potential.

[b] This column presents the width of the in-plane phonon density of states (DOS). These values are hard to determine with inelastic neutron scattering, since the intensity of the structure factor decreases with increasing energy.

[c] The phase diagrams of Kr, N_2 and CD_4 present a commensurate region that extends to higher temperatures when the coverage is slightly higher than the commensurate one. Details of the phase diagram can be found in refs. (9), (10) and (11), respectively.

Adsorbate	Mass [a.u.]	de Boer[a] parameter	Lennard-Jones parameter ε [K]	σ[Å]	Gap energy [K]	Gap ratio $\Delta_{meas}/\Delta_{calc}$	DOS width[b] [K]	Melting Temperature
Kr	83.8	0.10	165.3	3.63	8.7	1,0	-	~125[c]
CD_4	20.0	0.23	137.0	3.68	14.5	0,9	48	~55[c]
N_2	28.0	0.42	35.6	3.32	19.3	1,7	34	~72[c]
D_2	4.0	1.26	35.2	2.95	40.0	1,1	9.5	18.5
HD	3.0	1.43	35.9	2.95	43.2	1,1	14.7	19.4
H_2	2.0	1.74	36.7	2.95	47.3	1,0	27.5	20.5
^3He	3.0	3.10	10.2	2.56	10.9	0,3	40	3.05

One of the most striking differences between both types of adsorbates is the temperature dependence of the gap energy. On a flat substrate no gap exists, and the acoustic phonon branches go to zero at the zone center. In this case, a non-zero temperature produces thermal phonons . In the case of a commensurate phase, the acoustic phonon branches have a finite value at the zone center: the phonon gap. In order to thermally populate phonon modes a minimum temperature has to be reached (of the order of the gap energy). At lower temperatures mainly the zero point vibration remains. All the quantum gases melt at a temperature (T_m) that is between 30-50% of the energy of the gap: at T_m a negligible fraction of thermal phonons are excited. Thus, a small increase in $\sqrt{<u^2>}$ (12) due to these thermal phonons suffices to melt the commensurate monolayer. This increase occurs at a temperature that is very close to T_m. As a consequence, the gap energy does not change up to T_m and the phonon spectrum is practically not affected by temperature. On the other hand, the heavier gases melt at a temperature that is much higher (at least a factor 3) than the gap energy and a strong temperature renormalization of the phonon spectrum occurs (13,14). This can bee seen in figure 1, where the data for the phonon gap as a function of temperature are depicted for several gases.

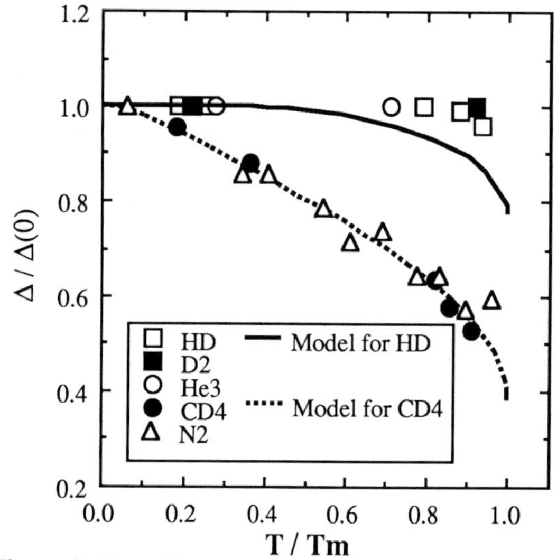

Figure 1: Normalized zone center phonon gap vs. reduced temperature for several gases adsorbed on graphite. The lines are the result of the model calculation described in ref. 13. (Δ stands for the zone center phonon gap)

The agreement between theoretical calculations of Δ (15, 16) and the measured values is indicated in the 7th column of Table I. Two elements present a large deviation from the expected magnitude: N_2 and ^3He. ^3He is particularly hard to model due to its high quantum character that makes it very difficult to take into account all the correlations and many body effects (16). A single ^3He atom does not stay on an adsorption site, but needs the presence of the other adatoms to localize. Already H_2 localizes at low coverages and the calculation (15) agree well with experiment. For N_2 no simple explanation fo this disagreement has been found and this suggests tha further work is required to improve our knowledge of th adatom-substrate potential (5).

ACKNOWLEDGEMENTS

This work has been partially supported by the Federal Ministry of Research and Technology (BMFT) of th Federal Republic of Germany.

REFERENCES

(1) Frank V.L.P., Lauter H.J. and Leiderer P., Jpn.J.Appl.Phys.Suppl. 26, 347 (1987)
(2) Frank V.L.P., Lauter H.J. and Leiderer P., Phys.Rev.Lett., 61, 436 (1988)
(3) Lauter H.J., Frank V.L.P., Leiderer P. and Wiechert H., Physica B, 156&157, 280 (1989)
(4) Frank V.L.P., Lauter H.J., Godfrin H. and Leiderer P., PHONONS 89, pp. 1001 (World Scientific, Singapore, 1990)
(5) Hansen F.Y., Frank V.L.P., Taub H., Bruch L.W., Lauter H.J. and Dennison J.R., Phys.Rev.Lett., 64, 764 (1990)
(6) Frank V.L.P., Lauter H.J., Hansen F.Y., Taub H., Bruch L.W. and Dennison J.R., PHONONS 89, pp.922 (World Scientific, Singapore, 1990)
(7) Moeller T., Lauter H.J., Frank V.L.P. and Leiderer P., PHONONS 89, pp.919 (World Scientific, Singapore, 1990)
(8) Frank V.L.P., Lauter H.J. and Taub H., unpublished data.
(9) Butler D.M., Litzinger J.A. and Stewart G.A., Phys.Rev.Lett. 44, 466 (1980)
(10) Chan M.H.W., Migone A.D., Miner K.D. and Li Z.R., Phys.Rev.B, 30, 2681 (1984)
(11) Kim H.K., Zhang Q.M., Chan M.H.W., Phys.Rev.B, 34, 4699 (1986)
(12) The rms vibrational amplitude ($\sqrt{<u^2>}$) at a given temperature can be expressed as a thermally weighted average over the phonon density of states (DOS).
(13) Frank V.L.P., Lauter H.J., Godfrin H. and Leiderer P., PHONONS 89, pp.913 (World Scientific, Singapore, 1990)
(14) Hakim T.M., Glyde H.R. and Chui S.T., Phys.Rev.B, 37, 974 (1988)
(15) Novaco A.D., Phys.Rev.Lett., 61, 436 (1988)
(16) Ni X.-Z. and Bruch L.W., Phys.Rev.B, 33, 4584 (1986)

Physica B 165&166 (1990) 613–614
North-Holland

ZERO SOUND EXPERIMENTS ON THE DISTORTED B PHASE AND IN THE BA TRANSITION REGION AT LOW PRESSURES

J.P. Pekola, J.M. Kyynäräinen, A.J. Manninen, and K. Torizuka

Low Temperature Laboratory, Helsinki University of Technology, 02150 Espoo, Finland

Zero sound at 8.9 and 26.8 MHz, propagating parallel with the magnetic field, shows an attenuation maximum along the T_{BA} line in a field of ~ 2 kG at low pressures. This is in accordance with the theoretical prediction that a critical magnetic field separates two types of B phases, with and without nodes in the energy gap at the transition line. The B→A phase change takes place via an intermediate state, with excess sound attenuation both at 8.9 and 26.8 MHz.

We have recently reported an ultrasonic experiment on distorted ^{3}He-B and on the nature of the BA transition (1). Here we review our main results and update our interpretation of them.

In zero magnetic field, ^{3}He-A is stabilized by the strong coupling effects only above 21 bar, and the BA transition is of first order. In a non-zero field, the A phase exists at all pressures as a consequence of a susceptibility difference between ^{3}He-A and ^{3}He-B. The A phase is known to be in an axial state, with two nodes in the energy gap. The isotropic energy gap of the B phase is distorted by the magnetic field; this makes the nature of the BA transition more complex.

Ashida and Nagai (2), AN hereafter, predict that in the weak coupling limit, ie. at the lowest pressures, the BA transition changes from second to first order at a critical magnetic field $H_C \sim 2$ kG. Below H_C, the energy gap Δ of the B phase is continuously deformed: Δ is almost isotropic at $T = 0$ but A phase like with two nodes at T_{BA}. Above H_C, the change is discontinuous, and the transition is of first order.

The dependence of the zero sound attenuation α in a magnetic field ($\vec{H} \parallel \hat{q}$, the wave vector of sound) is presented in Fig. 1, in the B phase along the BA transition line. Data taken both at $f = 8.9$ MHz and at $f = 26.8$ MHz show a clear anomaly in α at $H_C \sim 2$ kG. At 8.9 MHz the cusp is most prominent when the pressure $p = 0$.

A simple qualitative explanation for the data of Fig. 1 can be given by using the theory of AN (see the inset in Fig. 1). Below H_C, the energy gap minimum Δ_2, in the direction of $\vec{H}$ and $\hat{q}$, vanishes at the transition in the weak coupling limit, which leads to high attenuation owing to pair breaking at all temperatures. Above H_C, Δ_2 increases with the field ($\Delta_2 >> hf$), and α decreases. In other words, increasing the magnetic field in the vicinity of H_C increases Δ_2 rapidly so that it meets the requirement for pair breaking, $hf = 2\Delta_2$ and the conditions for other possible collective modes ($hf \sim \Delta_2$). The decrease of the maximum with increasing pressure occurs, at least partly, because the weak coupling result with $\Delta_2 = 0$ at $H < H_C$ is not strictly

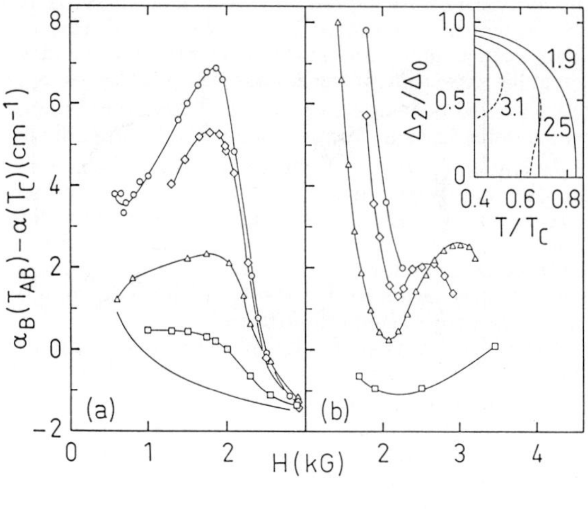

Fig. 1. Sound attenuation α vs. the magnetic field H in ^{3}He-B at T_{BA}. Circles correspond to $p = 0$, diamonds to $p = 0.5$ bar, triangles to $p = 2.3$ bar, and squares to $p = 6.6$ bar. The lines connecting the data points are just for guiding the eye. (a) $f = 8.9$ MHz. The lowest curve shows measurements at $H = 0$ as a function of T_{BA}(H). (b) $f = 26.8$ MHz. Inset: the minimum energy gap Δ_2, normalized by the $T = 0$ gap Δ_0, as a function of the reduced temperature for H =1.9, 2.5, and 3.1 kG, according to calculations by AN (2). The vertical lines show the positions of the first order BA transitions.

valid at non-zero pressures; the nodes in the gap develop only in the weak coupling limit (3).

The dependence of α on H at $f = 26.8$ MHz is in agreement with the model discussed above (see Fig. 1b). Attenuation has a minimum at ~ 2 kG after a steep decrease due to the proximity of the squashing mode. The maximum at 2.5 - 3 kG is again a manifestation of the collective modes of Δ_2; the condition for pair breaking is met at $\Delta_2/\Delta_0 \simeq 0.4$ (see the inset of Fig. 1), where Δ_0 is the energy gap of the B phase at $T = 0$. Further checks of this interpretation were provided by our experiments under rotation (see Refs. 1 and 4).

Schopohl (5) and Salomaa (6) have theoretically found several possibilities for the structure of the BA interface, with a continuous deformation of ^{3}He-B into ^{3}He-A via an axiplanar phase sheet, only a few coherence lengths thick. The possibility of the BA transition taking place via an intermediate phase has also been considered by AN (2). They propose the planar phase as an alternative to the axial phase at low pressures.

As the temperature in our experiment was allowed to drift slowly through T_{BA}, an intermediate regime with α higher than either in ^{3}He-A or in ^{3}He-B appeared (see Fig. 2). We interpret our data as evidence for an intermediate superfluid state between ^{3}He-B and ^{3}He-A, which is supported by the presence of relatively sharp corners and linear parts in the attenuation curve. Irregular nucleation of the A phase is not likely to cause the observed behaviour, because in ^{3}He-A the $\hat{\ell}$ vector, which determines α, always orients itself perpendicular to $\vec{H}$ (and to $\hat{q}$. Also, an interface less than 10 coherence lengths thick seems unable to give rise to the apparent attenuation increase of ~ 0.1 cm^{-1} in the intermediate state. The fact that a linear magnetic field gradient of ~ 10 G/mm is enough to suppress completely the increased attenuation during the transition is in support of an intermediate superfluid state.

The identification of this I phase as one of the axiplanar states is tempting because, eg., in a planar state the energy gap vanishes along $\hat{q}$, thus giving rise to strong pair breaking attenuation even at low temperatures of the experiment. There is, however, at least one serious argument against this interpretation: the temperature width of the I state (typically $\sim 2\mu$K) depends very weakly, if at all, on the pressure in the experiment. The transition temperature between an axiplanar state and the axial phase should depend entirely on strong coupling corrections (ie. pressure), whereas the transition between the B-phase and the axiplanar states is determined by the susceptibility difference in a magnetic field and by a difference in the condensation energies. Therefore, the two transition temperatures ought to have dissimilar pressure dependencies.

We wish to thank A. Babkin, J-P. Kneib, A. Balatsky, O.V. Lounasmaa, V. Mineev, M. Salomaa, E. Thuneberg and G. Volovik for their contributions. This work was supported by the Academy of Finland and the U.S.S.R. Academy of Sciences through project ROTA, and by the Körber Stiftung of Hamburg.

References:

1. J.M. Kyynäräinen, J.P. Pekola, A.J. Manninen, and K. Torizuka, Phys. Rev. Lett. 64 (1990) 1027
2. M. Ashida and K. Nagai, Prog. Theor. Phys. 74 (1985) 949
3. V.P. Mineev, private communication
4. A.J. Manninen, J.M. Kyynäräinen, K. Torizuka, and J.P. Pekola, this volume
5. N. Schopohl, Phys. Rev. Lett. 58 (1987) 1664
6. M.M. Salomaa, J. Phys. C: Solid State Phys. 21 (1988) 4425

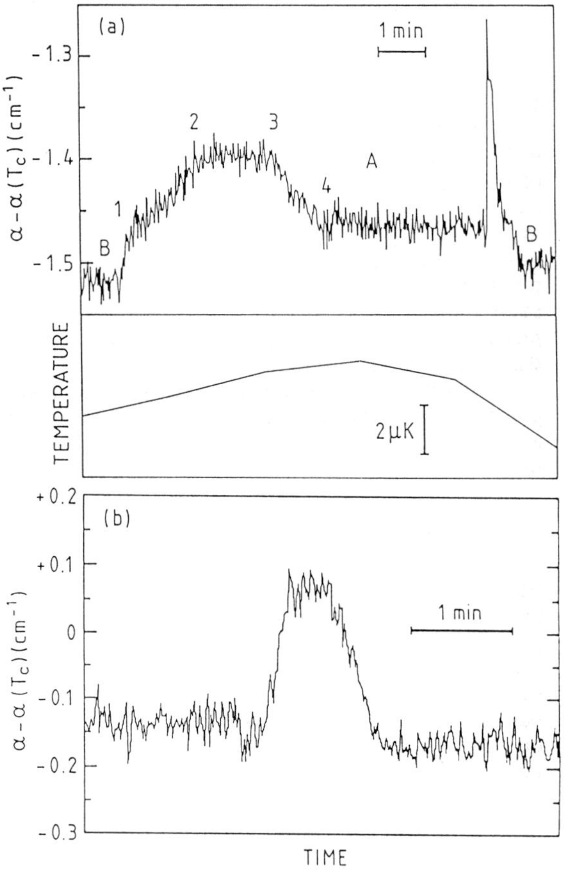

Fig. 2. Anomalous zero sound propagation at the BA transition.(a) α at the B$\rightarrow$A and A$\rightarrow$B transitions is plotted when $f = 8.9$ MHz, $p = 2.3$ bar, $T/T_c = 0.56$, and $H = 3.4$ kG. Approximate temperatures, measured by a Pt-NMR thermometer, are also shown. (b) α at the B$\rightarrow$A transition when $f = 26.8$ MHz, $p = 9.3$ bar, $T/T_c = 0.71$, and $H = 3.2$ kG.

Physica B 165&166 (1990) 615–616
North-Holland

SEARCH FOR FLOW INDUCED PRESSURE GRADIENTS IN SUPERFLUID ^{3}He-A*

P.D. SAUNDRY, L.J. FRIEDMAN**, C.M. GOULD, and H.M. BOZLER

Department of Physics, University of Southern California, Los Angeles, CA 90089-0484 USA

We have searched for the pressure drops which are a necessary consequence of the dissipation attendant textural motion in ^{3}He-A when uniform bulk superflow exceeds a critical value. We find no reliable pressure signature at a level significantly smaller than expected theoretically. We have modified our sensitive differential pressure gauge by replacing the plastic diaphragm with an extremely thin aluminum film, increasing the resolution by two orders of magnitude. Further investigations are underway.

1. INTRODUCTION

Studies into the nature of texture transitions and the stability of bulk superflow in superfluid ^{3}He-A have provided results that significantly disagree with theory. (1) These experiments used the attenuation of zero sound as a probe to detect changes in textures away from uniform superflow. Associated with the predicted texture transformations are pressure drops expected to be on the order of several thousand nbar. We have begun a search for these pressure drops by including in our apparatus a sensitive differential pressure gauge.

A qualitative analysis by Cross (2) indicates that the pressure drop due to the motion of a texture with inhomogeneities on the length scale λ along a flow channel of length L, for $v_s > v_c$, is given by:

$$\Delta P \sim 3\left(1-\frac{T}{T_c}\right)^{-1/2}\left(\frac{v_s}{1\ mm/s}\right)\left(\frac{L}{\lambda}\right)\ \text{nbar}$$

For the predicted helical textures $\lambda \sim 10\ \mu$m so that for typical values of L ~ 1 cm and $v_s \sim 1$ mm/s at T = 0.9 T_c a pressure drop of 10^4 nbar is expected. Paalanen and Osheroff (3) have observed pressure drops of several hundred nbar for $v_s \sim 1$ mm/s in a flow channel 1 cm long parallel to a 20 Oe or greater field. These results are quantitatively lower than might be expected from the above equation.

2. PRELIMINARY RESULTS

A sensitive capacitive strain gauge was constructed to operate with the earlier experimental arrangement used in our laboratory (1) where zero sound was the primary probe. The flow channel side of the new experiment is shown in Figure 1. The strain gauge had a compliance of 0.03 Å/nbar and a resolution of 30 nbar.

With this gauge we find that dissipation in flows with speeds up to v_s = 1.5 mm/s (in our flow channel which is 1.3 cm long) results in pressure drops significantly less than

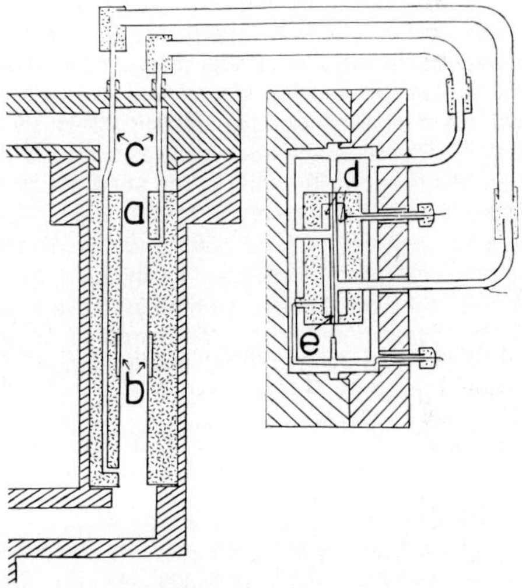

Figure 1 Experimental arrangement for detecting flow dissipation in ^{3}He-A: (a) flow channel; (b) zero sound crystals; (c) pressure lines; (d) fixed capacitor plates; and (e) diaphragm.

expected. We see no convincing and reproducible evidence for induced pressure drops substantially greater than the resolution of our gauge. This has led us to modify our strain gauge to obtain significantly higher resolution.

3. ORIGINAL STRAIN GAUGE

The gauge uses a flexible conducting diaphragm as a common electrode between two fixed plates approximately 200 μm distant on either side. The two resultant capacitors are used in an AC bridge which detects movement of the diaphragm as changes in the balance point. The strain gauge was not originally designed to obtain very high resolution but to be robust (in case of accidental

* Supported by NSF through grants DMR88-00291 (LJF, HMB) and DMR89-01701 (PDS, CMG).
** Present Address: Raytheon Co., 131 Spring St., Lexington, MA 02173

Table 1 Comparison of high resolution capacitive differential strain gauges.

Diaphragm Material	Ref.	Thickness (μm)	Area (cm^2)	Compliance (Å/nbar)	Spacing (μm)	Resolution (nbar)
Kapton	4	12.5	1.6	0.03	23.	1.3
Polycarbonate	5	10.	n/a	0.05	20.	1.6
Mylar	6	6.	1.3	0.05	19.	0.8-3.
Kapton	original	25.4	1.4	0.03	220.	30.
Aluminum	present	0.8	1.4	5.	208.	0.2

overpressuring) and to allow easy changes of the diaphragm.

The original diaphragm was a 25 μm thick kapton film coated with gold on one side. The gauge was calibrated in three independent ways at several different temperatures. It was calibrated *in situ* using normal fluid flows just above T_c where ^{3}He is relatively viscous. (This is obviously useful only near T_c. Room temperature calibrations used helium gas in an identical configuration.) The second calibration created pressure differences of up to 2000 nbar by applying electric fields between a fixed plate and the diaphragm. The third calibration process examined drumhead resonances of the diaphragm. Comparing several natural frequencies reveals much about the nature of stresses in the diaphragm and often allows the diaphragm to be described by a simple model. Recent calibrations by all three techniques agree with each other to within a factor of two.

4. DISCUSSION

Plastic films have been widely used as the moveable element of sensitive strain gauges (4) (5) (6) and the type of results obtainable are shown in Table 1. Plastics have the advantage of relatively high compliance at readily available thicknesses, but they have the disadvantages of requiring a metal coating before being usable in an electrical device, and of stiffening significantly when cooled. The metal overlayer may also induce thermal strains in the diaphragm.

From Table 1 we can see that the low temperature compliance of all of the surveyed plastic diaphragms are essentially the same. (In doing this comparison, note that the compliance also scales with the area of the diaphragm, but these are also essentially identical.) Among the plastic diaphragms the only quantity significantly affecting the resolution of the device is the physical spacing of the diaphragm from the fixed plates. Our gauge was intentionally built with a safety margin of a relatively large spacing.

Kapton films of thickness 12.5 μm and 8 μm were investigated but only moderate improvements found. Despite strenuous attempts to reduce the tension in these films, none of them approached the limiting resolution predicted for a thin plate with no applied tension. Films either behaved as a membrane under tension or were not describable by a simple model. This latter case was probably due to inhomogeneous stresses in the kapton.

5. IMPROVED STRAIN GAUGE

Rather than modifying the assembly of the strain gauge to obtain the desired increase in resolution, we attacked the problem of the diaphragm material. Although the bulk moduli of metals are generally higher than that of plastics, it is possible to obtain metal films which are significantly thinner than plastic films, and which may therefore exhibit much higher compliance.

We have now replaced our kapton diaphragm with a 0.8 μm Al film (7) which has provided the greatest sensitivity to date and a resolvable pressure difference of 0.2 nbar. There is a noticeable hysteresis in the diaphragm as has been seen in another strain gauge approaching this sensitivity. (6) As was noted there, hysteresis is reduced by cycling the gauge through pressure and temperature excursions. The origin of the hysteresis is thought to be inhomogeneous stresses in the Al film which also results in a complex set of natural resonances.

Using this high resolution strain gauge we are carrying out further investigations into superflow dissipation in ^{3}He-A.

REFERENCES

(1) D.M. Bates, S.N. Ytterboe, C.M. Gould, and H.M. Bozler, Phys. Rev. Lett. 53 (1984) 1574.
(2) M.C. Cross, Flow Dissipation in the ^{3}He Superfluids, in Quantum Fluids and Solids, eds. E.D. Adams and G.G. Ihas (American Institute of Physics, New York, 1983), pp. 325-335.
(3) M.A. Paalanen and D.D. Osheroff, Phys. Rev. Lett. 45 (1980) 362.
(4) A.P.M. Matthey, J.T.M. Walraven, and I.F. Silvera, Phys. Rev. Lett. 46 (1981) 668.
(5) J.P. Pekola, J.C. Davis, Zhu Yu-Qun, R.N.R. Spohr, P.B. Price, and R.E. Packard, J. Low Temp. Phys. 67 (1987) 47.
(6) B. Yurke, J.S. Denker, B.R. Johnson, N. Bigelow, L.P. Levy, D.M. Lee, and J.H. Freed, Phys. Rev. Lett. 50 (1983) 1137.
(7) Goodfellow Metals Cambridge Limited, Cambridge CB4 4DJ, England.

Physica B 165&166 (1990) 617–618
North-Holland

INTRINSIC MAGNUS EFFECT IN SUPERFLUID ^{3}He-A

R. H. SALMELIN* and M. M. SALOMAA

Low Temperature Laboratory, Helsinki University of Technology, SF-02150 Espoo, Finland

The perpetual coherent orbital circulation of the Cooper pairs in superfluid ^{3}He-A results in a novel, purely quantum-mechanical phenomenon, in which the orbital angular momentum of the condensate is being transferred into the linear momentum of a tiny object – such as a negative ion – moving in ^{3}He-A.

The classical Magnus force occurs when a rotating object moves in a viscous fluid, see Fig. 1.a. The object is at rest, while the liquid flows past it at the velocity $\vec{v}$; the circulation of the body is $\vec{\kappa}$. On the right-hand side, the circulation opposes the velocity, while on the left side the two flows reinforce one another.

According to the Bernoulli equation, the liquid exerts a higher pressure on the object at right, thus resulting in a net force which tends to deflect the body from its original trajectory. This is known as the Magnus force $\vec{F}_M \propto \vec{\kappa} \times \vec{v}$.

In the analogous quantum effect in ^{3}He-A, the roles of the object and the medium are reversed: here the superfluid itself has an internal angular momentum $\hat{\ell}$, a coherent property of all the Cooper pairs interacting with a negative ion, see Fig. 1.b. Thus, in the plane perpendicular to $\hat{\ell}$, the ^{3}He quasiparticles scatter differently when they encounter the ion with their circulation opposed to or coincident with its velocity.

This picture leads to a difference in the scattering "pressure" across the object. The ion moving in the plane perpendicular to $\hat{\ell}$ thus experiences a reactive deflecting force (1): the Cooper-pair angular momentum is transferred into the linear momentum of the ion.

The drift velocity $\vec{v}$ of a negative ion is found by equating the rate of momentum transfer from the ion to the quasiparticles, $d\vec{P}/dt$, with the driving force supplied by the electric field, $e\vec{\mathcal{E}}$. We perform the calculation for small fields, *i.e.*, for low $\vec{v}$. Using the definition $v_i \equiv \mu_{ij}\mathcal{E}_j$ for the ion-mobility tensor μ_{ij}, the components of the mobility may be expressed as

$$e\left(\mu^{-1}\right)_{ij} = n_3 p_F \int dE \left(-\frac{\partial f}{\partial E}\right) < \sigma_{ij}(E) > , \quad (1)$$

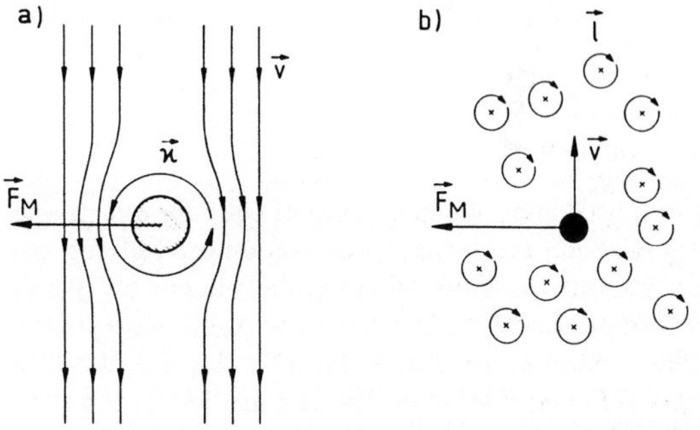

FIGURE 1
(a) Magnus force $\vec{F}_M$ illustrated in the classical hydrodynamical situation with liquid flow past a rotating object, and (b) in the quantum-mechanical case, where a non-rotating particle moves through a superfluid condensate possessing a spontaneous internal orbital angular momentum.

where $f(E)$ is the Fermi distribution function, n_3 the number density of ^{3}He, and p_F the Fermi momentum. Here $< \sigma_{ij}(E) >$ is the angular-averaged (ij)'th component of the scattering cross-section tensor at energy E; equal to the differential cross section $d\sigma/d\Omega$ weighed by momentum transfers in the directions i and j.

The differential cross section includes detailed features of the scattering process through the T matrix; the latter is found by solving the Lippmann-Schwinger equation $T = V + VGT$, where T is the scattering matrix, G is the propagator, and V the bare scattering potential; V is not affected by the superfluid pairing correlations. Eliminating V in favour of T_N results in $T_S = T_N + T_N(G_S - G_N)T_S$ which has been solved analytically (2) for the superfluid B phase.

* Present address: Service de Physique du Solide et de Résonance Magnetique, CEN-Saclay, 91191 Gif-sur-Yvette cedex, France.

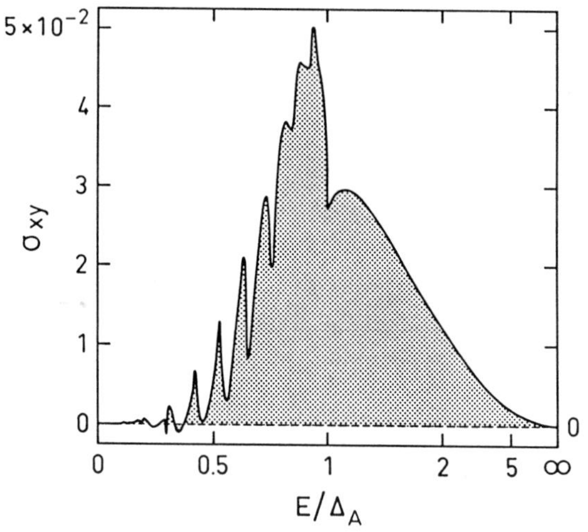

FIGURE 2
Computed energy dependence of the ^{3}He quasiparticle-negative ion scattering cross section σ_{xy}, giving rise to the intrinsic Magnus effect. Scale is linear for quasiparticle energies $E \leq \Delta_A$; for $E > \Delta_A$ it varies as E^{-1}. The scattering resonances for $E < \Delta_A$ are due to quasibound energy levels in the vicinity of the ion (4).

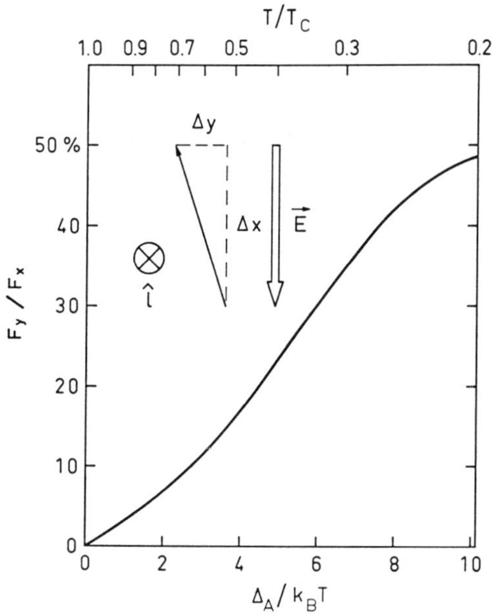

FIGURE 3
Magnitude of the intrinsic Magnus effect in the plane $\perp \hat{\ell}$ for negative ions in ^{3}He-A at $p = 29$ bar. An ion has a transverse component Δy, in addition to its motion (Δx) parallel to $\vec{\mathcal{E}}$. The calculated force ratio, F_y/F_x, is shown as a function of Δ_A/kT (lower scale) and T/T_c (upper scale, assuming weak-coupling).

For the anisotropic A phase, a numerical solution for the principal components $\mu_{\parallel}$ (along $\hat{\ell} \parallel \hat{z}$) and $\mu_{\perp}$ (in the plane perpendicular to $\hat{\ell}$) has first been found (3) by using an expansion in spherical harmonics.

We solve T by direct inversion (4) of complex matrices. There is found to exist, in addition to $\sigma_{\parallel}$ and $\sigma_{\perp}$, an additional contribution σ_{xy} in the plane perpendicular to $\hat{\ell}$, see Fig. 2. Thus the clockwise and counterclockwise directions are not equal, for scattering off a small object. The cross section peaks for quasiparticle energies E close to the maximum value of the energy gap, Δ_A, where the resonant effects between partial waves are strongest and vanishes towards $E/\Delta_A \to \infty$, where the excitations approach normal-state quasiparticles.

The drift velocity of the ion is now expressed as

$$\vec{v} = \left[\mu_{\parallel}(\hat{\ell}\cdot\hat{\mathcal{E}})\hat{\ell} + \mu_{\perp}\hat{\ell}\times(\hat{\ell}\times\hat{\mathcal{E}}) + \mu_{xy}(\hat{\ell}\times\hat{\mathcal{E}})\right]\mathcal{E} \ . \quad (2)$$

The largest deflection is thus obtained for the case $\hat{\ell} \perp \hat{\mathcal{E}}$. The calculated ratio of the transverse force F_y ($\equiv F_M$, see Fig. 1) and the component F_x driving the ion along $\hat{\mathcal{E}}$ is plotted into Fig. 3.

With $\hat{\ell}$ orientated at random in the plane perpendicular to $\hat{\mathcal{E}}$ (by applying a magnetic field $\vec{B} \parallel \vec{\mathcal{E}}$), it follows that the diameter of a cloud of ions moving the distance Δx widens by $2\Delta y$ when it reaches the collector. A detectable deflection is expected already at $T = 0.9\,T_c$, while at $T = 0.5\,T_c$ an expansion of as much as 26° is predicted.

We suggest further experiments to find this novel, purely quantum-mechanical, manifestation of the internal Cooper-pair orbital angular momentum in superfluid ^{3}He-A. Moreover, its effects on the motion of negative ions in the $\vec{\ell}$-vector vortex textures in ^{3}He-A are of interest; for low temperatures these become prominent.

ACKNOWLEDGEMENTS

We thank O. V. Lounasmaa, V. P. Mineev, C. J. Pethick, and G. E. Volovik for interest. Research supported through the Award for the Advancement of European Science by the Körber Foundation (Hamburg, FRG) and by the Academy of Finland.

REFERENCES

(1) R. H. Salmelin, M. M. Salomaa, and V. P. Mineev, *Phys. Rev. Lett.* **63** (1989) 868.
(2) G. Baym, C. J. Pethick, and M. M. Salomaa, *Phys. Rev. Lett.* **38** (1977) 845; *J. Low Temp. Phys.* **36** (1979) 431.
(3) M. M. Salomaa, C. J. Pethick, and G. Baym, *J. Low Temp. Phys.* **40** (1980) 297.
(4) R. H. Salmelin and M. M. Salomaa, *Phys. Rev. B* **41** (1990) 4142.

Physica B 165&166 (1990) 619–620
North-Holland

ULTRASONIC INVESTIGATIONS OF VORTEX TEXTURES IN ^{3}He-A

J.M. Kyynäräinen, J.P. Pekola, K. Torizuka, and A.J. Manninen

Low Temperature Laboratory, Helsinki University of Technology, SF-02150 Espoo, Finland

Zero sound attenuation has been used to probe vortex textures in rotating ^{3}He-A. The highly nonlinear rotation speed dependencies of the measured sound amplitudes can be explained partly as being due to sonic velocity gradients in the vortex cores. Calculations of sound propagation in the geometrical limit have been made to extract information on the types of vortex cores and their sizes using our experimental data.

1 INTRODUCTION

The question of the nature of vortices in ^{3}He-A is still unsettled (1). In a high magnetic field and an open geometry, Seppälä and Volovik (SV) have predicted singular and singly quantized vortices to exist at rotation speeds $\Omega \lesssim 1$ rad/s, whereas at higher values of Ω continuous, doubly quantized vortices should be preferred (2,3). The presence of continuous vortices have been verified in several experiments, while singular vortices seem to be much more difficult to create (4). In this paper we present our zero sound transmission data and discuss whether ultrasound can be used to distinguish between the vortex types.

2 SOUND PROPAGATION IN ^{3}He-A

Sound attenuation α in ^{3}He-A can be expressed by the well-known formula

$$\alpha = \alpha_{||}\cos^4\theta + 2\alpha_c\cos^2\theta\sin^2\theta + \alpha_\perp\sin^4\theta,$$

where the coefficients $\alpha_{||}$,α_c, and $\alpha_\perp$ depend on temperature, pressure, and the sound frequency, and θ is the angle between the $\hat{\ell}$-vector and the sound propagation direction $\hat{q}$. To calculate the received sound amplitude A in an $\hat{\ell}$-texture, the simplest approach is to average A over the experimental cell area S, $A = (1/S)\int_S A_0\exp(-\alpha(\vec{r})L)\,dS$, where L is the path length of sound. However, sound velocity c depends on θ as well:

$$c = c_0 - c_{||}\cos^4\theta - 2c_c\cos^2\theta\sin^2\theta - c_\perp\sin^4\theta,$$

where c_0 is the sound velocity at T_c. This leads to nontrivial sound propagation due to velocity inhomogeneity: Firstly, phase cancellation reduces A and secondly, using the geometrical picture, the wave front is curved. In Fig. 1 the profiles of c and α are shown in the core of the SV continuous vortex.

3 EXPERIMENTAL TECHNIQUES

Our experiments were carried out in a cylindrical volume with height L = 4 mm and 6 mm diameter. The sound propagated axially along the magnetic field $\vec{H}$ and the rotation axis $\hat{\Omega}$. Pressure was varied between 2.3 bar and 29.3 bar and the magnetic field between 1.6 kG and 3.5 kG. Sound frequencies of 8.9, 26.8, and 44.7 MHz and rotation speeds up to 5.5 rad/s were used.

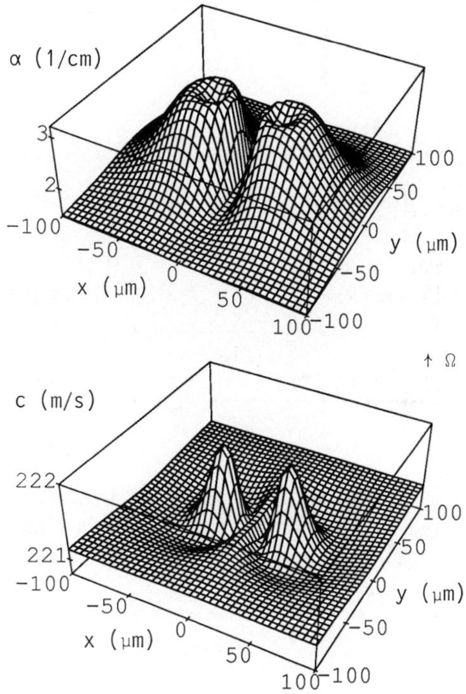

FIGURE 1: Calculated contours of α and c in a continuous SV vortex at p = 2.3 bar, f = 8.9 MHz, and T/T_c = 0.95, with $\hat{q}$ along the rotation axis.

4 EXPERIMENTAL RESULTS

In Fig. 2 we present the Ω-dependence of the relative change in A, measured at three pressures. The decrease in A is due to the fact that usually $\alpha_c > \alpha_\perp$; α_c is the relevant parameter in the vortex cores. The striking feature is the short range of the linear part in Ω. We expected a linear response up to rotation speeds at which vortex cores start to overlap; with a core radius on the order of the dipolar length $\xi_D \simeq 10\ \mu$m, this would happen only above $\simeq$ 20 rad/s. Note also that A disappears totally in the 6.6 bar data, already at $\sim$ 3 rad/s. These effects can be understood in terms of sound focusing (see Fig. 1): Sound is guided into regions of minimum velocity, corresponding at the same time to maximum attenuation, thus enhancing the

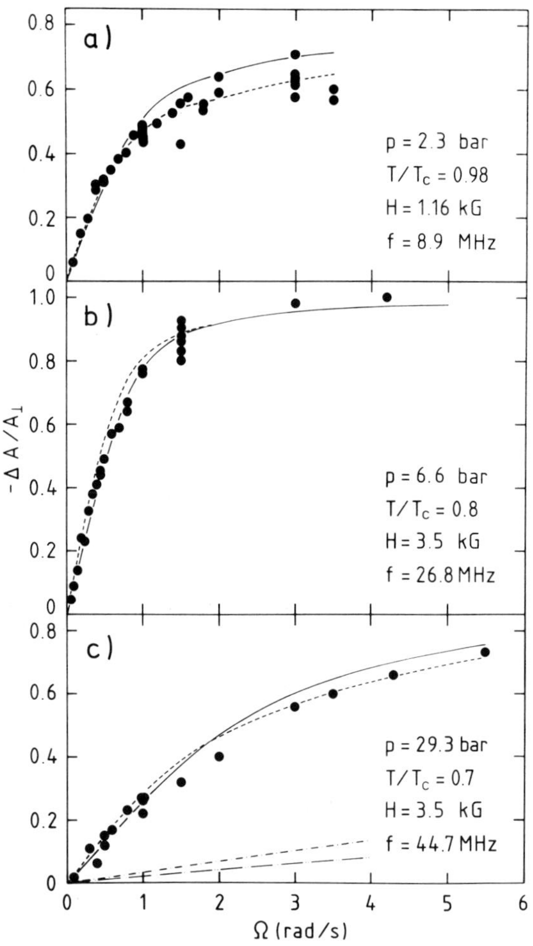

FIGURE 2: Measured effect of rotation on A at three pressures. The curves are calculated responses with vortex core radii a) 50 μm, b) 45 μm, and c) 30 μm. Solid and dashed lines correspond to continuous and singular SV vortices, respectively (2). In c), the dash-dotted and dash-double-dotted lines illustrate the theoretical prediction for continuous and singular SV vortices, with $\xi_D \simeq 10$ μm.

5 CALCULATIONS

We calculated sound responses using ray acoustics (5) with different vortex types and core sizes. This approach should be valid when the wavelength λ of sound is much smaller than the intervortex distance and the vortex cores. At 10 MHz, $\lambda \simeq 20$ μm, which is of the same order as core sizes, whereas vortex separation is much larger, about 100 μm at 1 rad/s. At higher frequencies, however, λ is smaller and the validity is improved. We expect the calculation to give an upper limit for effect due to vortices, i.e., a lower limit to their core size.

The accuracy of our calculations depends entirely on the parameters α_i and c_i; we have used the values provided by Wojtanowski (6). The difference between measured and theoretical values of $\alpha_\perp$ is 30% at worst; the other two components of α we cannot measure.

In Fig. 2 we have plotted the best fits of the core radius, corresponding to continuous and singular SV vortices. We have used an exponentially relaxing core model, with the angle between $\hat{\ell}$ and $\hat{q} \propto 1 - \exp(r/r_0)$, which approximates the solution obtained by Seppälä and Volovik in Ref. 2. They found that for the singular vortex $r_0 \simeq \xi_D$ and for the continuous vortex $r_0 \simeq 2\xi_D$. Vulovic et al. (7) have estimated core sizes using a linear model. Their result is that the radius of continuous SV vortices is twice that for singular SV vortices and about $3.5\xi_D$. At high pressures, where $\xi_D \simeq 10$ μm, our data suggest considerably larger vortex cores than expected, especially for the singular SV vortex (see the lowest two curves in Fig. 2c). Note how responses from different vortex types are almost equal at the same core size. The pressure dependence is weaker than expected from that of ξ_D. However, this may be due to the inaccuracy of our method at lower sound frequencies.

ACKNOWLEDGEMENTS

We thank W. Wojtanowski for providing us with computer programs for calculating α and c, E. Sonin for useful discussions, and A. Babkin for taking part in some of the experiments. This work was supported by the Academy of Finland and by the Körber-Stiftung (Hamburg).

REFERENCES

(1) P. Hakonen, O.V.Lounasmaa, and J. Simola, Physica B160 (1989) 1.
(2) H.K. Seppälä and G.E. Volovik, J. Low. Temp. Phys. 51 (1983) 279.
(3) A.L. Fetter, J. Low Temp. Phys. 67 (1987) 145.
(4) J.T. Simola, L. Skrbek, K.K. Nummila, and J.S. Korhonen, Phys. Rev. Lett. 58 (1987) 904.
(5) L.D. Landau and E.M. Lifshitz, Fluid Mechanics (Pergamon, 1986) 256.
(6) W. Wojtanowski, private communication (1989).
(7) V.Z. Vulovic, D.L. Stein, and A.L. Fetter, Phys. Rev. B29 (1984) 6090.

Physica B 165&166 (1990) 621–622
North-Holland

COMBINED ION AND NMR MEASUREMENTS ON ROTATING ^{3}He-A

K.K. NUMMILA, P.J. HAKONEN, J.S. KORHONEN

Low Temperature Laboratory, Helsinki University of Technology, 02150 Espoo, Finland

We have constructed an experimental chamber permitting simultaneous observation of both the transverse cw-NMR spectrum and the time-of-flight of negative ions from a rotating sample of vortices in ^{3}He. Our measurements in the A-phase confirm that the vortex state, causing the large delay and strong deformation of the ion pulses drifting parallel to the rotation axis, is also responsible for the satellite peak corresponding to continuos vortices in the NMR spectrum. Unfortunately, singular vortices, whose distinct ion response was detected earlier, could not be observed in the present setup.

1. INTRODUCTION

The complex structure of the order parameter in ^{3}He-A supports a variety of stable textures (1). Vortices have been mostly studied by using the transverse cw-NMR (2) and ion transmission techniques (3,4). Two basically different types of vortices, singular and continuous, have been observed so far in the A-phase. Recently also the feasibility of ultrasonic measurements to distinguish between the two types of vortices has been studied (5).

The soft cores of both types of vortices provide potential wells for localized spin wave modes. Based on the measured frequency shift and the intensity of the vortex satellite peak in NMR spectra, the vortex structure observed in earlier NMR experiments was interpreted as continuous (6). Owing to the smaller radial extent of their soft cores, singular vortices should support a spin wave mode, localized close to the continuum frequency and having about ten times larger intensity than the spin wave mode localized to continuous vortices. This behavior has never been seen in NMR measurements on ^{3}He-A (2).

Both types of vortices have, however, been detected in the ion transmission experiments where the time-of-flight of negative ions drifting through a rotating sample of ^{3}He-A was measured (4,7). The observed strong deformation and retardation of ion pulses, caused by continuos vortices, could be explained by assuming continual focusing of ions into vortex cores. Singular vortices caused no essential delay in the time-of-flight or changes in the pulse shape. Both types of vortices were focusing, which was seen as an increased collector current in a diverging electric field during rotation. These vortex type interpretations are based on the different schemes needed to nucleate singular and continuous vortices, on the nucleation times, on the focusing model, and on the behavior as a funtion of magnetic field (4,7).

2. EXPERIMENTAL

The experimental chamber, shown in Fig. 1, is a slight modification of our previous ion cell (4). The cylindrical wall was made of epoxy instead of copper to ensure penetration of the rf-excitation field into the sample volume. In order to reduce the build-up of static charge on the dielectric material, the ratio of the cell radius to its length was increased to withdraw the epoxy wall. The observed time-of-flight data was checked against a similar cell having a metallic casing (8).

Our transverse, continuous-wave NMR procedures and the ion transmission techniques have been described earlier (2,4). The measurements were performed at the pressure $P = 28$ bar in a static magnetic field of $H = 28.4$ mT.

3. RESULTS

The vortex state present in the cell depends on the rotation history of the sample. Continuous vortices can be created by a rapid acceleration (typically 0.04 rad/s^2) up to the final rotation speed, whereas the thermodynamic ground state, consisting of singular vortices can be reached by a continuous rotation at constant Ω through T_C or by a gradual stepwise acceleration at velocities $\Omega < \Omega_C \approx 1$ rad/s (4,7).

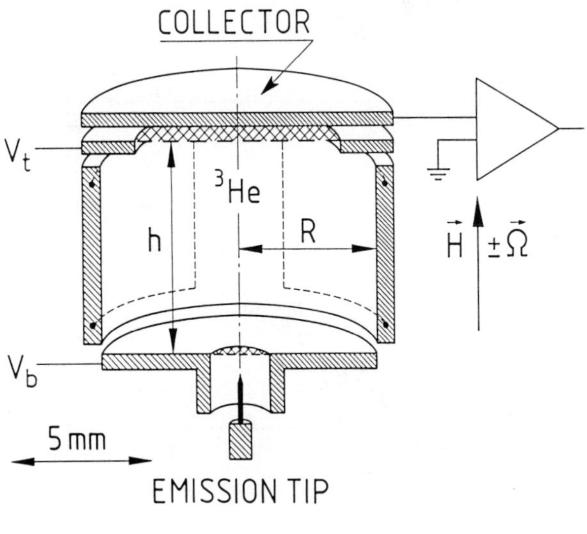

FIGURE 1
Our experimental cell. Ion pulses injected by the field emission tip are driven across the sample volume (h = 7 mm, R = 5 mm) and detected at the collector by a sensitive electrometer. The driving electric field was created by biasing the metallic top and bottom electrodes. The dashed lines depict the saddle shaped NMR coil.

0921-4526/90/$03.50

Figure 2 shows the simultaneously measured NMR and ion responses from vortex samples created by fast acceleration up to 2.5 rad/s and subsequent deceleration to the final rotation speed Ω. The data were recorded at $T = 0.82T_C$.

The lower frame displays the relative intensity of the vortex satellite peak as a function of Ω. The dashed line shows the result of our earlier measurements (2), $I_V/I_{tot} = 0.065\Omega$, corresponding to the equilibrium vortex density. Our data are in good agreement with Ref. 2, taking into account the rather high measurement temperature, at which the vortex satellite is not clearly separable from the main peak.

The upper frame shows the reduced time-of-flight as a function of Ω. $\langle t\rangle$ and $\langle t\rangle^*$ denote the flight times at $\Omega = 0$ and during rotation, respectively. The solid line is calculated from the focusing model (4) using $\alpha = 0.45$ and $\Omega_0 = 0.17$ rad/s for the two fitting parameters. Here α gives the effective mobility of the focused ions in the core and Ω_0 is proportional to the total area of focusing cores. These values agree with earlier results (4).

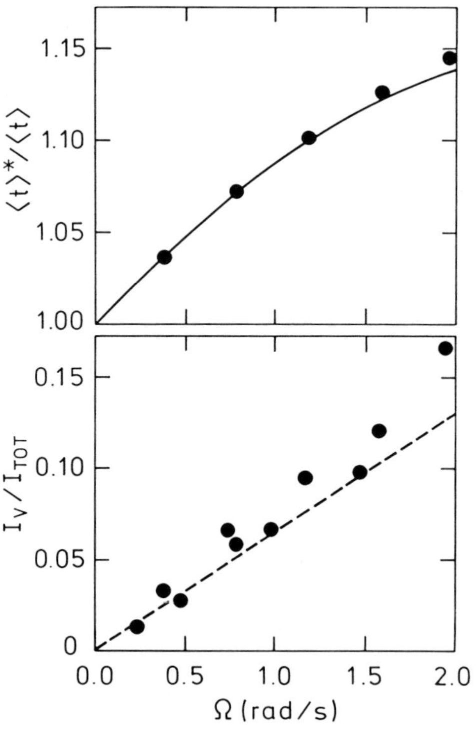

FIGURE 2
The relative time-of-flight of negative ions (upper frame) and the relative intensity of the vortex satellite peak (lower frame) as a function of rotation speed. The driving electric field corresponded to $V_b = -7.12$ V and $V_t = -2.37$ V. Pulses consisting of about 10^6 electrons were used.

The two preparation procedures for singular vortices were tried in various runs; another experimental cell, having the cylindrical wall made of gold-plated copper, was used as well. In spite of these efforts, no singular vortices were detected.

We have previously suggested two possible factors influencing the nucleation of singular vortices (7), namely the smoothness of the walls and the presence of ions in the sample. The present experiment has shown them inadequate. The smoothness of rotation seems to affect the nucleation process as well. We could not create pure samples of singular vortices in the old ion cell either when the air bearings of the cryostat became dirty (4).

The most important factor is probably the accidentally weak connection between the drift space and the sinter in the old cell. These two volumes were effectively separated so that vortices, first generated in the complicated geometry of the sinter, could not enter the geometrically well defined experimental volume in the drift space. The generation of vortices in the heat-exchanger volume has been detected earlier both in the A- and B-phases (9).

4. CONCLUSIONS

Our new experiments have verified the interpretations of the earlier separate NMR and ion experiments about continuous vortices. The nucleation of singular vortices seems problematic and further studies on the vortex nucleation process should be done.

ACKNOWLEDGEMENTS

We wish to thank O. Magradze for his contributions to the experiments and O.V. Lounasmaa and J.T. Simola for useful discussions.

REFERENCES

(1) P. Hakonen, O.V. Lounasmaa, and J. Simola, Physica B 160 (1989) 1.
(2) P.J. Hakonen, M. Krusius, and H.K. Seppälä, J. Low Temp. Phys. 60, (1985) 187.
(3) J.T. Simola, K.K. Nummila, A. Hirai, J.S. Korhonen, W. Schoepe, and L. Skrbek, Phys. Rev. Lett. 57, (1986) 1923.
(4) K.K. Nummila, J.T. Simola, and J.S. Korhonen, J. Low Temp. Phys. 75, (1989) 111.
(5) J.M. Kyynäräinen, J.P Pekola, K. Torizuka, and A.J. Manninen, Investigation of vortex textures by ultrasound in ^{3}He-A, these Proceedings.
(6) P.J. Hakonen, O.T. Ikkala, and S.T. Islander, Phys. Rev. Lett. 49, (1982) 1258; H.K. Seppälä, P.J. Hakonen, M. Krusius, T. Ohmi, M.M. Salomaa, J.T. Simola, and G.E. Volovik, Phys. Rev. Lett. 52, (1984) 1802.
(7) J.T. Simola, L. Skrbek, K.K. Nummila, and J.S. Korhonen, Phys. Rev. Lett. 58, (1987) 904 .
(8) K.K. Nummila, P.J. Hakonen, and J.S. Korhonen, to be published in Europhys. Lett. (1990).
(9) P.J. Hakonen and K.K. Nummila, Jap. J. Appl. Phys., Suppl. 26-3, (1987) 181; P.J. Hakonen and K.K. Nummila, Phys. Rev. Lett. 59, (1987) 1006.

Physica B 165&166 (1990) 623–624
North-Holland

ULTRASONIC EXPERIMENTS ON ROTATING SUPERFLUID ^3He-A_1

K. TORIZUKA, J.P. PEKOLA, J.M. KYYNÄRÄINEN, A.J. MANNINEN, and O.V. LOUNASMAA

Low Temperature Laboratory, Helsinki University of Technology, 02150 Espoo, Finland, and

G.K. TVALASHVILI

Institute of Physics of the Georgian Academy of Sciences, 380077 Tbilisi, USSR

The first experiments on rotating ^3He-A_1 have been done using pulsed zero sound transmission techniques at several pressures in magnetic fields up to 1 T. Preliminary analysis of our data by a geometrical acoustics method suggests that the vortex cores in ^3He-A_1 are larger than in ^3He-A_2. This result is based on the strong and highly non-linear rotation dependence of sound attenuation in ^3He-A_1. Conclusions about the type of vortices in ^3He-A_1 cannot be made at present owing to lack of theory.

The A_1 phase of superfluid ^3He is composed of Cooper pairs in one spin state ($\downarrow\downarrow$) only, unlike the A_2 phase which is a system of equal spin pairing ($\downarrow\downarrow$ and $\uparrow\uparrow$) in low magnetic fields. Transition from normal to the A_1 phase occurs at T_{A1} and the A_1 to the A_2 phase transition at T_{A2} : $T_{A1} > T_c > T_{A2}$ (T_c is the superfluid transition temperature in zero external magnetic field). The width of the A_1 phase, $T_{A1} - T_{A2} \simeq 27\ \mu K$ at 4.6 bar and in a 1 T field (1).

In this paper we report the first experimental results on the rotating A_1 phase and give a preliminary interpretation of the data.

In the A phase Seppälä-Volovik (SV) (2) vortices are known to exist in a magnetic field ($\geq$2.5mT). Kiviladze et. al. (3) have recently discussed the core size of nonsingular SV v-vortices in ^3He-A_2, near the A_1 phase. In their theory, an important parameter is

$$\beta = \Delta_\uparrow \Delta_\downarrow / \Delta^2,$$

where $\Delta_\uparrow$ ($\Delta_\downarrow$) is the energy gap of the up-spin (down-spin) component; β=0 in the A_1 phase and nonzero in the A_2 phase. They showed that the v-vortex cores are ellipsoidally distorted so that the ratio of the principal axes $\kappa \equiv r_x/r_y \simeq 0.8$ when $\beta \rightarrow 1$, and $\kappa \simeq 0.58$ when $\beta = 0$. Here r_x (r_y) is the component along x (y) axis of the orbital order parameter. With decreasing β, the cores grow: the overall size doubles along y axis and the increase is somewhat smaller along x axis.

To our knowledge, however, there is no theory which discusses the type of vortices in ^3He-A_1. We thus analyzed our experimental results in light of the v- vortex, paying attention to the size of the cores.

We have measured ultrasonic attenuation by a conventional pulse transmission technique in a cylindrical ^3He volume (the diameter is 6 mm, the distance between the two quartz crystals is L = 4 mm) using our rotating nuclear de-

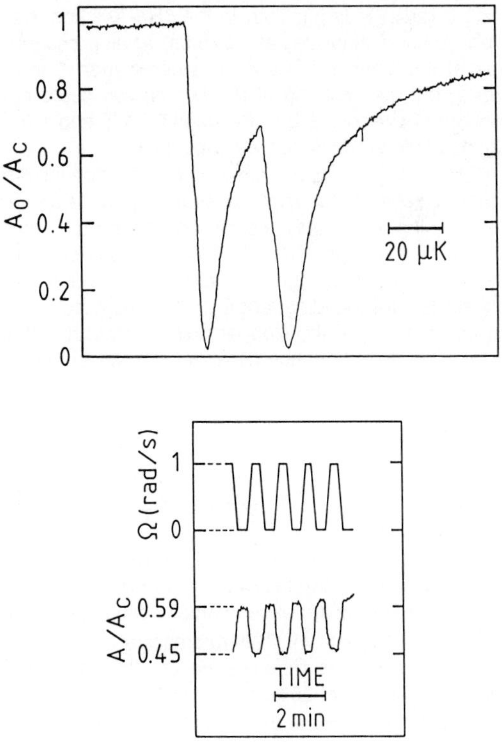

Fig.1
Top: Amplitude of the received sound signal in the vicinity of the superfluid transition in the stationary case, demonstrating the splitting of the clapping mode at p = 4.6 bar, H = 1.0 T.
Bottom: An example of start/stop experiments in ^3He-A_1.

magnetization cryostat. The sound, whose frequency is 8.9 MHz, propagates along the rotation axis $\hat{\Omega}$, and the magnetic field up to 1.0 T was also applied along $\hat{\Omega}$. We used a conventional Pt NMR thermometer and the Greywall temperature scale (4) to determine T_c and Sagan's results (1) to determine T_{A1} and T_{A2}. The temperature scale was calibrated against sound amplitude with the cryostat at rest. The splitting of the clapping mode (5) in a magnetic field H = 1.0 T is shown in Fig. 1. We first stabilized the temperature in the A_1 phase, then started and stopped the rotation repeatedly for short time intervals, and measured the received sound amplitude A, which is related to the attenuation α as A $\sim e^{-\alpha L}$. Figure 1 also illustrates an example of the start/stop experiments.

Figure 2 shows the relative change in the amplitude, $\Delta A/A_0$, against Ω at 4.6 bar, where A_0 denotes the amplitude at rest. Data were taken at $T/T_{A1} = T/T_{A2} \simeq 0.986$ in the A_1 and A_2 phases. In addition, we performed start/stop experiments at 6.5, 9.4, and 15.4 bar. $|\Delta A/A_0|$ decreases with increasing pressure because the attenuation itself becomes smaller. Important features of Fig. 2 are that the initial slope in ^{3}He-A_1 at slow rotation speeds ($\Omega \leq 1.0$ rad/s) is larger than in the ^{3}He-A_2, and that the A_1 phase curve saturates at ~2 rad/s. At all pressures the slope of the curve for ^{3}He-A_1 is steeper than ^{3}He-A_2 at slow Ω, but the crossing of the two curves could not be observed above p = 4.6 bar. Between 0.74 T and 1.0 T no magnetic field dependence was observed.

We have employed a geometrical acoustics (GA) method to calculate the received sound signal as a function of Ω by using theoretically determined attenuation and velocity parameters (See Ref.(6)). GA method is strictly valid only when the wavelength of the sound is much smaller than the textural bending lengths. Assuming that the two spin components of the Cooper pairs contribute independently to sound attenuation (See Ref.(7)), we obtain

$$\alpha_1(T) = \frac{1}{2}[\alpha(\frac{T}{T_{A1}}) + \alpha_{Tc}]$$

$$\alpha_2(T) = \frac{1}{2}[\alpha(\frac{T}{T_{A2}}) + \alpha(\frac{T}{T_{A1}})]$$

in ^{3}He-A_1 and ^{3}He-A_2, respectively. Here $\alpha(T/T_{A1,2})$ are values calculated by Wojtanowski in zero magnetic field for the A phase, α_{Tc} is the experimentally estimated attenuation at T_c ($\sim$ 1.5 cm^{-1}). This expression reproduces the observed temperature dependence of attenuation at rest within the accuracy of 20%.

Using these values of attenuation, we assumed that v-vortices are created even in the A_1 phase and searched for the size of a circular core that gives the best fit to the experimental data. We have used an exponentially relaxing core model: $\theta \sim 1\text{-}\exp(r/r_0)$ (r is the distance from the center of the core, r_0 the core size, θ the angle between $\vec{l}$ and $\vec{q}$, and $\vec{q}$ the direction of the sound). Note that the initial slope in the low Ω range, where the vortices cause attenuation independently of each other, is proportional to the core size.

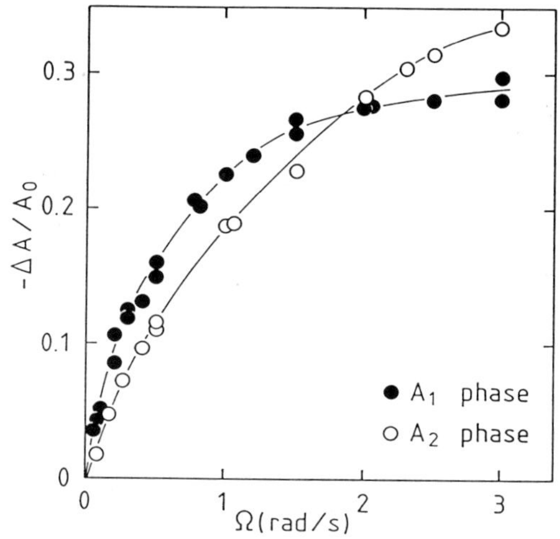

Fig.2
Sound absorption by the vortices in the A_1 and A_2 phases; p = 4.6 bar. Solid lines are guides for the eye.

By trying a circular core, the best fit was obtained with the core radius $r_c \simeq 70$ μm at slow Ω ($\leq$ 0.4 rad/s). But, the $r_c \simeq 70\,\mu$m curve is so steep that the fit rapidly becomes worse at higher Ω ($\geq$0.6 rad/s). If we try an ellipsoidal core, e.g., with the average size $\bar{r} \equiv \sqrt{r_x r_y} \simeq 65\,\mu$m, κ=0.2, the fit becomes somewhat better even at higher Ω. The GA approach is, however, doubtful at least at high Ω.

In conclusion:

1. The Ω dependence of the attenuation in ^{3}He-A_1 is stronger but saturates at lower Ω than in the A_2 phase.
2. It is difficult to fit the data on ^{3}He-A_1 over the whole Ω-range from zero up to 3 rad/s in the framework of the GA method.
3. The initial slope yields, however, an effective core size of $\sim$70 μm, which is clearly in excess of the core in He-A_2 ($\sim$ 45 μm).
4. Theoretical discussion on the types of vortices in ^{3}He-A_1 is needed.

Acknowledgements

This work was supported by the Academy of Finland and by the USSR Academy of Sciences and by the Körber Stiftung.

References

(1) D.C. Sagan, P.G.N. deVegvar, E. Polturak, L. Friedman, S-S, Yan, E.L. Ziercher, and D.M. Lee, Phys. Rev. Lett. 53 (1984) 1939
(2) H.K. Seppälä and G.E. Volovik, J. Low Temp. Phys. 51 (1983) 279
(3) B.N. Kiviladze, G.A. Kharadze, and J.M. Kyynärä inen, this volume
(4) D.S. Greywall, Phys. Rev. B33 (1986) 7520
(5) P. Wölfle, Physica 90B (1977) 96
(6) J.M. Kyynäräinen J.P. Pekola, K. Torizuka, and A.J. Manninen, this volume
(7) N. Schopohl, W. Marquardt, and L. Tewordt, J. Low Temp. Phys. 59 (1985) 469

Physica B 165&166 (1990) 625–626
North-Holland

THE INFLUENCE OF MAGNETIC FIELDS ON COLLECTIVE EXCITATIONS IN HELIUM-3-A

P.N. BRUSOV, M.Y. NASTEN'KA, T.V. FILATOVA-NOVOSELOVA, M.V. LOMAKOV, and V.N. POPOV

Physical Research Institute, Rostov-on-Don State University
Leningrad Branch of V.A. Steklov Mathematical Institute, USSR

A whole set of equations, which describe the collective excitation (CE) in He-3-A in arbitrary magnetic fields is firdst obtained with taking the CE damping into account by the path integration method, developed earlier by Brusov and Popov. These equations are solved for small fields and zero CE momenta and the linear Zeeman effect for the clapping and pair breaking modes is obtained: three-fold linear splitting of these modes, which could be observed in zero sound experiments. The influence ofm agnetic fields on CE damping is investigated too.

First theoretical investigations(1) of the collective excitations (CE) spectrum determined the energies of clapping (cl) $E=1.22\Delta_o$, flapping (fl)$E = 1.56\Delta_o$, and pairbreaking (pb) $E = 2\Delta_o$ modes (here Δ_o is the maximum value of gap $\Delta = \Delta_o \sin\theta$). These energy values were obtained without taking any CE damping into account. But it's clear that vanishing of gap along the orbital anisotrophy axis $\vec{l}$ leads to the damping of CE due to these decay into two fermions, because CE with nonzero energy and small momentum can always decay kinematically into two fermions whose momenta are almost opposite and close to axis $\vec{l}$. The whole CE spectrum with taking into account this damping was obtained first by Brusov and Popov(2). They obtained six cl- modes with energy $E = \Delta_o(1.17- i\ 0.13)$ and three pb-modes (superclapping modes in other terminology) with $E = \Delta_o(1.96- i\ 0.31)$. The difference between Re E in Ref. 1 and Ref. 2 is connected with the fact that taking CE damping into account (Im $E \neq 0$) leads via dispersion relations to renormalization of Re E. The recently precise measuring of the $c\ell$-mode frequency(3) is in a good agreement with Brusov-Popov path integration theory(2). The other interesting fact, obtained first in Ref. 2. is that the number of goldstone modes (gd) in the weak coupling approximation is equal to 9 rather than 5, which takes place in real He^3-A. The existence of 4 additional quasi-gd spin-orbit modes is the consequence of the latent symmetry of the system as Volovik(3) noted, these 4 modes are the analogy of the massless W-bosons in the theory of electroweak interaction. But in He^3-A the W- bosons have a mass because of the strong coupling corrections(2) (and gd-modes become fl-ones) in contrast with the Weinberg-Salam theory in which they acquire a mass because of the Higgs phenomenon.

Here by using the path integration method we obtain a whole set of equations, which describe the CE in He^3-A in a magnetic field for $T_c - T \sim T_c$, and solve them for small fields and zero momenta of CE.

The system is described by the hydrodynamic action functional

$$S_h = g^{-1} \sum_{p,i,a} c^+_{ia}(p)\, c_{ia}(p) + \frac{1}{2} \ln \det \hat{M}(c,c^+) / M(c^{(o)},c^{(o)+})$$

where $c_{ia}(p)$ are Bose fields, and $\hat{M}$ is the operator, dependent on quasifermion parameters and $c_{ia}(p)$. At a first approximation CE spectrum is determined by the quadratic part of functional S_h, which can be obtained by shift $c_{ia}(p) \rightarrow c^{(o)}_{ia}(p) + c_{ia}(p)$ in the Bose-fields $c_{ia}(p)$, where $c^{(o)}(p)$ is condensate wave function.

In accordance with Nasten'ka and Brusov's idea(5) for investigation CE spectrum in the presence of a magnetic field, we must take into account both the additional term in S_h and distortion of order parameter. The latter one in our case is equal

$$c_{ia}(p) = c\sqrt{\beta V}\, \delta_{po}\, (\delta_{a1}\alpha_+ + i\, \delta_{a2}\alpha_-)(\delta_{i1} + i\delta_{i2})$$

Here $\alpha_\pm = \dfrac{\Delta_\uparrow \pm \Delta_\downarrow}{2\Delta}$, $\Delta^2_{\uparrow\downarrow} = N(0)(\tau \pm \eta h)/2\beta_{245}$,

$\eta = (N(0)/N(0))T_c \ln (1.14\varepsilon_o/T_c)$,

$h = \frac{\mu_o H}{T_c}$, $\Delta = 2cZ$ -

is a single fermion spectrum gap, determined by next gap equation

$$\frac{1}{g} = - \frac{Z^2}{\beta V} \frac{1}{(\alpha_+^2 + \alpha_-^2)} \sum_p \left[\frac{(\alpha_+ + \alpha_-)^2 \sin^2\theta}{\omega^2 + (\xi - \mu H)^2 + \Delta^2 \sin^2\theta (\alpha_+ + \alpha_-)^2} + \frac{(\alpha_+ - \alpha_-)^2 \sin^2\theta}{\omega^2 + (\xi + \mu H)^2 + \Delta^2 \sin^2\theta (\alpha_+ - \alpha_-)^2} \right]$$

Equation det Q = 0, where Q is the matrix of quadratic form for S_h, gives us 18 equations which completely determine 18 collective modes in He^3-A in arbitrary magnetic field and arbitrary CE momenta.

After solving them for case of small $\vec{H}$ and $\vec{K} = 0$ we obtained for the energies of cl and pb-modes ($\gamma H = \alpha_-/\alpha_+$):

cl: $E_1 = (1.17 - i0.13)\Delta_o$

$E_{2,3} = (1.17 \pm 1.62\gamma H)\Delta_o - i(0.13 \pm 1.33\gamma H)\Delta_o$

pb: $E_1 = (1.96 - i0.31)\Delta_o$

$E_{2,3} = (1.96 \pm 1.18\gamma H)\Delta_o - i\,(0.31 \pm 0.69\,\gamma H)\Delta_o$

So for small $\vec{H}$ we have 3-fold splitting of the cl- and pb-modes; i.e. we got a linear Zeeman effect for these modes. Magnetic fields lift completely the degeneracy of pb-modes, and particularly lift the degeneracy of cl-modes, each branch of latter modes remain twice degenercity.

Note that magnetic field changes both the real parts of CE energies and the imaginary ones i.e. changes CE frequencies and damping, even in linear approximation. Magnetic fields decreases the number of gd-modes from 9 to 6 via the apperance of the gap $\sim\mu H$ in their spectrum.

REFERENCES

(1) P. Wölfle, Physica 90B (1977) 96.

(2) P.N. Brusov, V.N. Popov, Sov. Phys. JETP, 52, (1980) 943

(3) E.R. Dobbs et al., AIP Conference Proceeding, V. 194 (1989), 103.

(4) G.E. Volovik, JETP Lett, 43, (1986) 693.

(5) M.Y. Nasten'ka, P.N. Brusov, Phys. Lett. A, 136 (1989), 321.

Physica B 165&166 (1990) 627–628
North-Holland

SYMMETRY OF THE $\hat{d}$-VECTOR HEDGEHOGS IN SUPERFLUID ^{3}He-A

M. M. SALOMAA

Low Temperature Laboratory, Helsinki University of Technology, SF-02150 Espoo, Finland

Core structures of axisymmetric $\hat{d}$-vector monopole states in superfluid ^{3}He-A are classified in terms of discrete symmetries. The pointlike order-parameter singularity may be resolved via the formation of a topologically stable half-integer disclination ring; phase transitions may occur between these $\hat{d}$-hedgehogs.

1. INTRODUCTION

Linear singularities – such as quantized vortices – may in superfluid ^{3}He be in several different topological classes. For example, in ^{3}He-A, one may have smooth (continuous, analytic, *i.e.*: coreless) vortices, singular vortices (with integer quanta of circulation), and half-quantum vortices (HQV's).

Within a given topological class, there may yet exist many distinct vortices displaying different internal structures. Therefore, classifications of the symmetries of the quantized vortices have proven extremely useful for discussions of their physical properties (1).

Here we consider a corresponding symmetry classification for a special pointlike order-parameter singularity, the $\hat{d}$-vector hedgehog.

The $\hat{d}$-vector hedgehogs provide interesting physical models for the 't Hooft magnetic monopoles (2) within models of unified gauge field theories (3), see Fig. 1.

2. TOPOLOGICAL TRANSITION OF VORTICES

The $\hat{d}$-vector hedgehog may occur in actual experiments on rotating superfluid ^{3}He-A in low magnetic fields. For weak fields, the smooth (Mermin-Ho and Anderson-Toulouse) vortex textures are dipole locked; in this situation, the $\hat{d}$-field follows the $\hat{\ell}$-texture.

For larger magnetic fields ($H > H_d$, where $H_d \approx 25$ G is the dipole field), the $\hat{d}$-vector becomes uncoupled from the $\hat{\ell}$-field. This topological first-order transition in the vortex structures from the dipole locked texture in a low field to the dipole unlocked texture at high fields is accompanied with the motion of a monopole-like object, the $\hat{d}$-vector hedgehog (1).

3. SYMMETRY OF A-PHASE ASYMPTOTICS

In order to discuss the $\hat{d}$-field around a hedgehog, we may neglect the $\hat{\ell}$-vector, which we assume uniform: $\hat{\ell}(\vec{r}) \equiv \hat{\ell}_0 \equiv \hat{z}$. This is justified since the dipole length is much in excess of the hedgehog-core size.

Here we consider axisymmetric hedgehogs; it is convenient to expand the superfluid order parameter $A_{\alpha i}$ in terms of the pair amplitudes $a_{\mu\nu}$ in a cylindrical basis:

$$A_{\alpha i} = \sum_{\mu\nu} \lambda^{\mu}_{\alpha} \lambda^{\nu}_{i}\, a_{\mu\nu} \ , \qquad (1)$$

where $\lambda^0_{\alpha(i)} = \hat{z}_{\alpha(i)}$, and $\lambda^{\pm}_{\alpha(i)} = (\hat{x}_{\alpha(i)} \pm i\hat{y}_{\alpha(i)})/\sqrt{2}$.

For an axisymmetric solution, the amplitudes $a_{\mu\nu}$ are given by the Ansatz:

$$a_{\mu\nu} = e^{i(1-\mu-\nu)\varphi}\, C_{\mu\nu}(r,z) \ . \qquad (2)$$

The possible elements of discrete symmetry are:

$$P_1 = P \ , \quad P_2 = PTU_2 \ , \quad P_3 = TU_2 \ , \qquad (3)$$

where P is the space parity transformation, T denotes time inversion, and U_2 is a rotation by π around an axis perpendicular to $\hat{z}$.

Note that while, additionally, the axisymmetry of the $\hat{d}$-hedgehogs may be broken for energetical reasons, our present classification of the internal symmetries remains quite generally valid.

4. THE MOST SYMMETRIC $\hat{d}$-HEDGEHOG

Operating with the symmetry elements of Eq. (3) on the order parameter in Eq. (1) results in:

$$P_1\, C_{\mu\nu}(r,z) = (-1)^{\mu+\nu} C_{\mu\nu}(r,-z) \ , \qquad (4.a)$$

$$P_2\, C_{\mu\nu}(r,z) = C^{\star}_{\mu\nu}(r,z) \ , \qquad (4.b)$$

$$P_3\, C_{\mu\nu}(r,z) = (-1)^{\mu+\nu} C^{\star}_{\mu\nu}(r,-z) \ . \qquad (4.c)$$

The most symmetric o hedgehog obeys all these symmetries; therefore, on imposing

$$P_1 C_{\mu\nu} = P_2 C_{\mu\nu} = P_3 C_{\mu\nu} = C_{\mu\nu} \ , \qquad (5)$$

it follows from the symmetry P_2 that all the $C_{\mu\nu}$ are real for the maximally symmetric $\hat{d}$-hedgehog.

From the P_1 symmetry it then follows that $C_{\mu\nu}(z) = C_{\mu\nu}(-z)$ holds for even $\mu+\nu$, while $C_{\mu\nu}(z) = -C_{\mu\nu}(-z)$ for odd values of $\mu + \nu$.

Thus the o hedgehog has 9 real components $C_{\mu\nu}(r,z)$, of these 5 are even (with $\mu + \nu$ even) and 4 odd (whenever $\mu + \nu$ is odd).

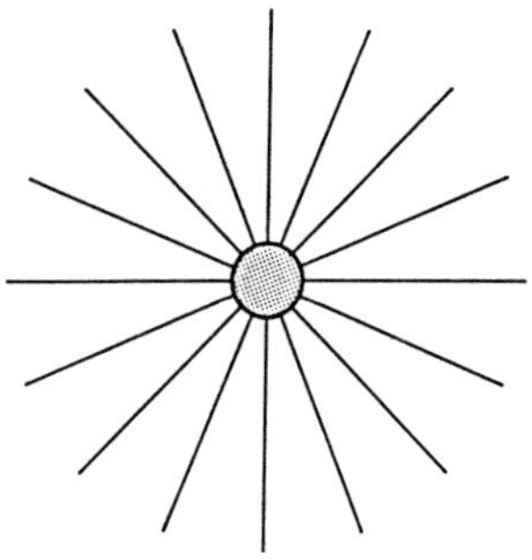

FIGURE 1
Asymptotic field configuration around a $\hat{d}$-vector hedgehog (lines) in ^{3}He-A; the $\hat{\ell}$-vector field is assumed constant ($\parallel \hat{z}$). Superfluidity may escape from the A phase into other p-wave pairing states in the hard-core region of the $\hat{d}$-monopole (dotted).

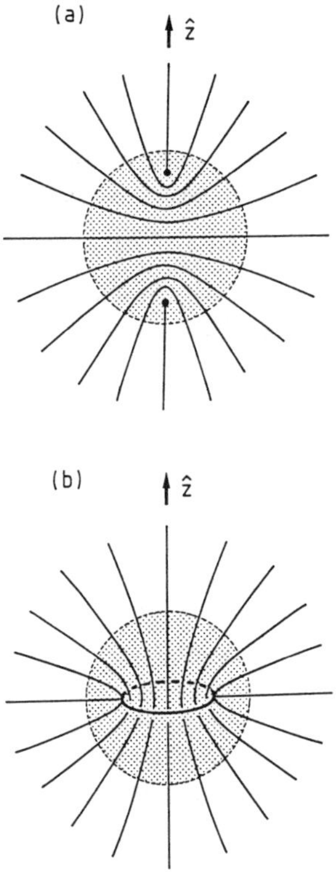

FIGURE 2
(a) Axisymmetric $\hat{d}$-monopole configuration displaying two-core structure. (b) A configuration forming a $\hat{d}$-vector disclination ring defect, with half-integer topological index. Outside of the dotted spheres, both configurations may be considered as point defects.

5. $\hat{d}$-HEDGEHOGS WITH BROKEN SYMMETRY

Following the symmetry classification of the vortices in superfluid ^{3}He, we call the P_1 symmetric $\hat{d}$ monopole the u hedgehog; it has 9 complex components $C_{\mu\nu}(r,z)$:

$$C_{\mu\nu}(r,z) = \begin{cases} C_{\mu\nu}(r,z), & \text{for } \mu+\nu \text{ even,} \\ -C_{\mu\nu}(r,-z), & \text{for } \mu+\nu \text{ odd.} \end{cases} \quad (6)$$

Correspondingly, the v hedgehog is P_2 symmetric, thus it has 9 real components without any parity constraint under the $z \leftrightarrow -z$ transformation.

The w hedgehog fulfills the discrete symmetry P_3; it has the following space-parity properties:

$$C_{\mu\nu}(r,z) = \begin{cases} C^{\star}_{\mu\nu}(r,-z), & \text{for } \mu+\nu \text{ even;} \\ -C^{\star}_{\mu\nu}(r,-z), & \text{for } \mu+\nu \text{ odd,} \end{cases} \quad (7)$$

i.e., for even $\mu+\nu$, $C_{\mu\nu}(r,z) \pm C_{\mu\nu}(r,-z)$ is real (imaginary), while for odd $\mu+\nu$, $C_{\mu\nu}(r,z) \pm C_{\mu\nu}(r,-z)$ is imaginary (real), correspondingly.

6. TRANSITION OF MOVING HEDGEHOGS

Consider describing the motion of a $\hat{d}$ hedgehog along $\hat{z}$ with application of a supercurrent, e^{iqz}. This violates parity P (and hence the symmetry element P_1), but conserves P_3 (*i.e.*, TU_2). Hence, under translational motion, the w hedgehog is maximally symmetric.

The above consideration implies that when in the stationary case ($q=0$) the v hedgehog is most favourable, then q would lift its degeneracy, and there appear two different (least symmetric) uvw hedgehogs for the case $q \neq 0$. This leads to the possibility to have phase transitions between $\hat{d}$-hedgehogs, driven by superflow.

In addition, it is possible that there occur spontaneous phase transitions between $\hat{d}$-hedgehogs, just like there are phase transitions for monopolelike structures in nematic liquid crystals (4), see Figs. 2.

ACKNOWLEDGEMENTS

The author thanks G. E. Volovik for a discussion. Research supported through an Award for the Advancement of European Science by the Körber Foundation (Hamburg, FRG) and by the Academy of Finland.

REFERENCES

(1) M. M. Salomaa and G. E. Volovik, *Rev. Mod. Phys.* **59** (1987) 533; M. M. Salomaa, *these Proceedings*.
(2) G. 't Hooft, *Nucl. Phys. B* **79** (1974) 276; J. Preskill, *Ann. Rev. Nucl. Sci.* **34** (1984) 461.
(3) M. Atiyah and N. Hitchin, **The Geometry and Dynamics of Magnetic Monopoles** (Princeton University Press, 1988).
(4) O. D. Lavrentovich and E. M. Terent'ev, *Zh. Eksp. Teor. Fiz.* **91** (1986) 2084 [*Sov. Phys. JETP* **64** (1986) 1237]; H. Mori and H. Nakanishi, *J. Phys. Soc. Japan* **57** (1988) 1281.

Physica B 165&166 (1990) 629–630
North-Holland

TOPOLOGY OF SUPERFLUID PHASE SLIPPAGE FOR ^{3}He FLOW IN A CHANNEL

N. B. KOPNIN* and M. M. SALOMAA

Low Temperature Laboratory, Helsinki University of Technology, SF-02150 Espoo, Finland

The momentum-space topology of phase slippage in superfluid ^{3}He flowing through a narrow channel is discussed, on the basis of numerical integration of the time-dependent Ginzburg-Landau equations. Here the phase-slip processes are associated with superfluid-core phase-slip centers – delocalized in space-time.

Recently, considerable interest has been devoted to phase slippage and the Josephson effect (1) and the flow of superfluid ^{3}He through narrow channels (2). Vortex formation is suppressed in pores with transverse dimensions on the order of $\xi(T)$, and one can expect the formation of phase-slip centers (PSC's) (3). This situation is nonstationary in time and nonuniform in space. Like quantized vortices, PSC structures in ^{3}He could prove more diverse than those in superconductors.

We integrated the dynamical equations for the order-parameter field $A_{\alpha i}$ numerically (see also (4)). We model the channel by a rectangular tube. Its width is a and height d, and we assume $a \gg d$; the length L is much larger than $\xi(T)$. The z-axis is along the tube and y upwards. The height $d \approx \xi_0$, so that the y components of the order parameter $A_{\alpha y}$ vanish owing to the boundary conditions at the horizontal surfaces. The width of the channel (along x) is assumed large: $a \gg \xi_0$, hence $A_{\alpha x}$ and $A_{\alpha z}$ are both finite. This is one of the simplest restricted geometries, yet preserving the multicomponent order-parameter structure.

We employ the time-dependent Ginzburg-Landau (TDGL) equations which can be derived microscopically for gapless superfluid ^{3}He (5). The TDGL equations can be used also more generally to model the order-parameter dynamics in superfluid ^{3}He. We consider all quantities as functions of only one spatial coordinate z and of time t. For further simplification, we separate the spin variable by setting $A_{\alpha i} = \hat{d}_\alpha a_i$, where $\hat{d}$ is a unit vector in spin space, and we assume constant $\hat{d}$. This factorization of $A_{\alpha i}$ is one of the possible solutions; it resembles the usual stationary A or polar phases.

Introducing the moduli and phases of the order-parameter components, $a_z = \rho_z \exp(i\phi_z)$ and $a_x = \rho_x \exp(i\phi_x)$, we impose boundary conditions at the ends of the computation interval ($z = 0$ and $z = L$):

$$\frac{\partial \rho_x}{\partial z} = \frac{\partial \rho_z}{\partial z} = 0 \ , \quad \frac{\partial \phi_x}{\partial t} + \mu = \frac{\partial \phi_z}{\partial t} + \mu = 0 \ ,$$

where μ is the chemical potential (pressure), measured with respect to the Fermi energy, and also one additional condition which can be of Class 1 or of Class 2,

$$\phi_z = \phi_x + \frac{\pi}{2} + \pi k \ , \qquad (1)$$

$$\phi_z = \phi_x + \pi k \ \text{(k is an integer)} \ , \qquad (2)$$

respectively. These conditions are compatible with the TDGL equations and comply with the periodicity of ρ_x and ρ_z as functions of z with periods L or $2L$. We specify a constant value for the mass current j_z and evaluate the functions $\mu(z,t)$, $a_x(z,t)$, and $a_z(z,t)$.

Phase-slip solutions are found for all classes of these boundary conditions. Their most interesting feature is that the jumps in ϕ_x and ϕ_z take place at different space-time points, such that the order-parameter components a_x and a_z never tend to zero simultaneously. Therefore, the phase slips in ^{3}He confined to a narrow channel possess superfluid cores. In Fig. 1, the time dependencies of both ρ_x and ρ_z are shown for the current $j_z = 0.23$ at $z = 0$.

We now discuss the topology of the phase-slip processes: for a phase slippage to occur, it is necessary that the order parameter $\Delta(\hat{p}) = a_i \hat{p}_i$ (here $\hat{p}_i$ is the unit vector in the momentum direction) would turn to zero somewhere for any direction of $\hat{p}$ on the Fermi sphere (6), see Fig. 2. The solutions that we found do satisfy this condition and have the further interesting feature that $\Delta(\hat{p})$ vanishes at different points and time instants for different directions of $\hat{p}$.

One can write $|\Delta(\hat{p})|^2 = a^2 + b^2 + 2ab \cdot \cos(\phi_x - \phi_z)$, where $a = \rho_x \hat{p}_x$ and $b = \rho_z \hat{p}_z$. Apart from the trivial case $a = b = 0$, $\Delta(\hat{p})$ vanishes (i) for $\phi_x - \phi_z = 2\pi n$

* Permanent address: L. D. Landau Institute for Theoretical Physics, USSR Academy of Sciences, 117334 Moscow, USSR

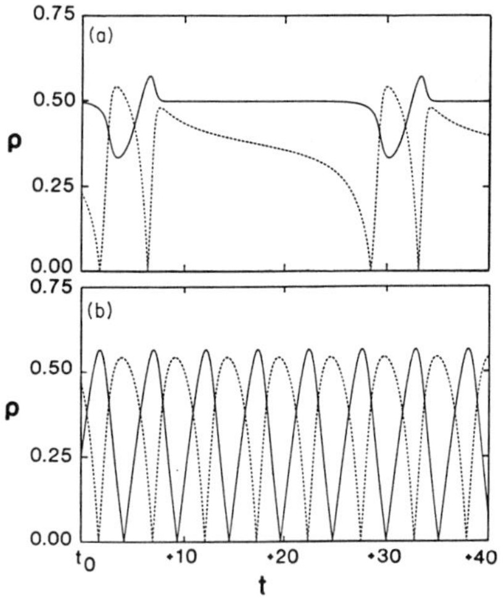

FIGURE 1
Time (t) dependencies of the superfluid ^{3}He amplitudes ρ_x (solid line) and ρ_z (dotted line) at $z = 0$. (a) Class 1 and (b) Class 2 solutions, respectively, for $j_z = 0.23$.

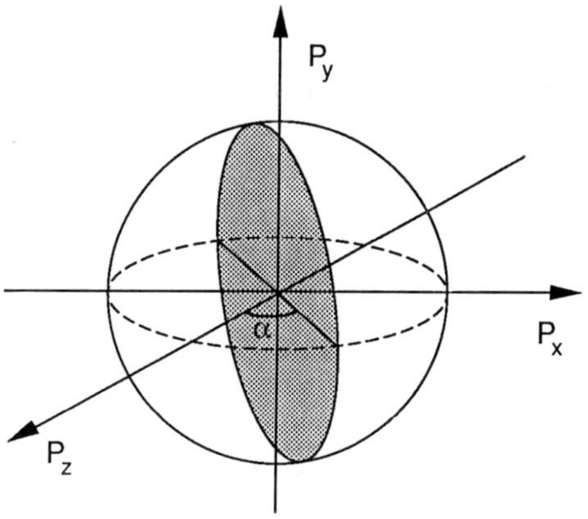

FIGURE 2
Unit Fermi sphere $|\hat{p}| = 1$ in the momentum space. The hatched surface, extending the angle α, represents the line of zeroes (nodes) in the energy gap, $\Delta(\hat{p})$.

and $a = -b$, or (ii) for $\phi_x - \phi_z = \pi(2n+1)$ and $a = b$. In Fig. 2, the unit sphere in the momentum space is shown. The angle α extended between the hatched plane and the z-axis is defined by $\tan\alpha = -\rho_z/\rho_x$ in case (i), and as $\tan\alpha = \rho_z/\rho_x$ in the second case (ii). The order parameter vanishes when the $\hat{p}$-vector is on the circular line of intersection of the unit sphere with the plane, *i.e.*, when $\hat{p}_x/\hat{p}_z = \tan\alpha$, provided that the phase relations (i) and (ii) are satisfied.

The solutions we found are such that, for any direction of $\hat{p}$, there exists a point (z,t) in space-time at which the angle α of the plane comprising $\hat{p}$ and the phase difference $\phi_x - \phi_z$ satisfy conditions (i) or (ii). In other words, the line of $\Delta(\hat{p})$-nodes maps a region of space-time (z,t) onto the unit sphere with a nonzero degree of mapping. This region we call the localization region for the PSC.

Consider first the Class 2 solutions. According to Eq. (2), the necessary phase relations needed for $\Delta(\hat{p})$ to be zero are fulfilled at the ends of the computation interval, *i.e.*, at fixed spatial points. In turn, the moduli ρ_x and ρ_z vanish at these points and thus induce variations of α in the intervals $(-\pi/2, 0)$ and $(0, \pi/2)$. The plane shown in Fig. 2 sweeps the sphere once during each period of oscillation. Hence, the Class 2 PCS is localized in space but delocalized in time.

The topology of the Class 1 solutions is more complicated. For a given point z in space, there exist two instants of time, $t_1(z)$ and $t_2(z)$, at which one of the phase relations (i) or (ii), respectively, is satisfied. At the time t_1 we can draw the hatched plane in Fig. 2 with $\alpha < 0$, and at t_2 one with $\alpha > 0$. The orientations of these planes vary with z. During one phase-slip event, both of these planes sweep together the unit sphere once. Therefore, the Class 1 PSC is delocalized both in space – and in time.

ACKNOWLEDGEMENTS

We thank O. V. Lounasmaa and G. E. Volovik for helpful comments, and O. Avenel and E. Varoquaux for useful discussions on their experiments. This work has been supported by the Körber Foundation (Hamburg, FRG) and by the Soviet Academy of Sciences and the Academy of Finland through the project ROTA.

REFERENCES

(1) O. Avenel and E. Varoquaux, *Phys. Rev. Lett.* **60** (1989) 416.
(2) J. P. Pekola, J. C. Davis, Zhu Yu-Qun, R. N. R. Spohr, P. B. Price, and R. E. Packard, *J. Low Temp. Phys.* **67** (1987) 47.
(3) B. I. Ivlev and N. B. Kopnin, *Adv. Phys.* **33** (1984) 47.
(4) N. B. Kopnin and M. M. Salomaa, *Phys. Rev.* **41** (1990) 2601.
(5) N. B. Kopnin, *J. Low Temp. Phys.* **65** (1986) 433.
(6) G. E. Volovik and V. P. Mineev, *Zh. Eksp. Teor. Fiz.* **83** (1982) 1025 [*Sov. Phys. JETP* **56** (1982) 579].

Physica B 165&166 (1990) 631–632
North-Holland

ROTATING ^{3}He-A_2 IN THE VICINITY OF THE TRANSITION TO THE A_1 PHASE

B.N. Kiviladze and G.A. Kharadze

Institute of Physics of the Georgian Academy of Sciences, 380077 Tbilisi, USSR, and

J.M. Kyynäräinen

Low Temperature Laboratory, Helsinki University of Technology, 02150 Espoo, Finland

We show that, owing to a decrease in the dipolar energy, the soft cores of the nonsingular, doubly quantized vortices in ^{3}He-A_2 expand and become ellipsoidal upon approaching the A_1 phase. Results of zero sound experiments in rotating ^{3}He-A_2, very near the A_2A_1 transition, support this conclusion. The experimental data was analyzed by a geometrical acoustics method, using theoretically calculated attenuation and velocity parameters.

In the past decade, vortex $\hat{\ell}$-textures of the type predicted by Seppälä and Volovik (SV) (1) were discovered and explored in rotating ^{3}He-A by means of NMR spectroscopy and ion focusing (2). Here, we consider these textures in the vicinity of the A_2A_1 transition and their observation in a zero sound experiment. The A_2 phase is a coherent mixture of Cooper condensates of up and down spin species, whereas in the A_1 phase only one spin state is present.

In the presence of a strong magnetic field and for a moderate angular velocity Ω, there appears a hierarchy of length scales, viz.,

$$\xi_H \ll \xi_D \ll b_\nu = (\nu h/4\pi m_3\Omega)^{1/2} \ , \qquad (1)$$

where ξ_H is the magnetic healing length, ξ_D denotes the dipolar healing length, and b_ν is the radius of the Wigner-Seitz cell (carrying ν quanta of circulation) of the vortex lattice. An $\hat{\ell}$-textural soft core, with dimensions on the order of $\xi_D \simeq 10\ \mu$m, builds up around each vortex. This happens in the presence of a quasi-homogeneous background in the spin part of the order parameter.

The structure of the soft vortex cores in ^{3}He-A is determined by the balance between the gradient and dipolar energies. In the vicinity of the continuous transition from the A_2-phase to the A_1 phase the character and strength of the dipole-dipole interaction is radically changing. This is easy to understand because some of the dipolar energy is responsible for the coupling between the two spin condensates.

The expression for the order parameter of ^{3}He-A, in the presence of a strong magnetic field, can be written as

$$A_{\mu i} = (\Delta/\sqrt{2})(\alpha_+\hat{d}_1 + i\alpha_-\hat{d}_2)_\mu(\hat{u}_1 + i\hat{u}_2)_i \ , \qquad (2)$$

where $\Delta^2 = (\Delta_\uparrow^2 + \Delta_\downarrow^2)/2$ and $\alpha_\pm = (\Delta_\uparrow \pm \Delta_\downarrow)/(2\Delta)$; $\Delta_\uparrow$ (respectively $\Delta_\downarrow$) is the gap in the quasiparticle energy spectrum of the superfluid condensate with spin configuration $\uparrow\uparrow$ ($\downarrow\downarrow$). In Eq. (2), pairs of orthogonal unit vectors, $\hat{d}_1$, $\hat{d}_2$ and $\hat{u}_1$, $\hat{u}_2$, define the quantization axes $\hat{s} = \hat{d}_1 \times \hat{d}_2$ and $\hat{\ell} = \hat{u}_1 \times \hat{u}_2$ of the spin and orbital momenta of the Cooper pairs, respectively.

For the superfluid state of Eq. (2), the dipolar energy density is expressed by

$$f_D = -\chi_N(\Omega_A/\gamma)^2[\alpha_+^2(\hat{d}_1\cdot\hat{\ell})^2 + \alpha_-^2(\hat{d}_2\cdot\hat{\ell})^2]/2 \ , \qquad (3)$$

where χ_N is the magnetic susceptibility of the normal phase of liquid ^{3}He, Ω_A denotes the longitudinal NMR frequency, and γ is the gyromagnetic ratio. We consider the configuration $\vec{H} \parallel \vec{\Omega} \parallel \hat{z}$. For $\hat{d}_1 = \hat{x}$ and $\hat{d}_2 = \hat{y}$ we have

$$f_D = -\chi_N(\Omega_A/\gamma)^2[(1+\beta)\ell_x^2 + (1-\beta)\ell_y^2]/4 \ , \qquad (4)$$

where $\beta = \Delta_\uparrow\Delta_\downarrow/\Delta^2$ gradually approaches zero when $A_2 \rightarrow A_1$ (for the A_1 phase $\beta = 0$).

We next turn our attention to the doubly quantized (ν = 2), nonsingular SV v-vortex (1), whose $\hat{\ell}$-texture is given by

$$\hat{\ell} = \sin\eta\hat{x} + \cos\eta(\sin\phi\hat{y} + \cos\phi\hat{z}) \ . \qquad (5)$$

The angle η is a smooth function of cylindrical coordinates (r,ϕ), with the asymptotic behavior $\eta \rightarrow \pi/2$, when $r \gg \xi_D$, and $\eta \rightarrow -\pi/2$, when $r \rightarrow 0$. It is easy to verify that for the vortex texture of Eq. (5), the dipolar energy density is

$$f_D = \chi_N(\Omega_A/\gamma)^2[(1+\beta)\cos^2\phi + 2\beta\sin^2\phi]\cos^2\eta(r,\phi)/4 + \text{const} \ . \qquad (6)$$

The dipolar rigidity along the x-axis ($\phi = 0, \pi$) is governed by the factor $1+\beta$. On the other hand, along the y-axis ($\phi = \pm\pi/2$) the rigidity is proportional to 2β which vanishes on reaching the A_1 phase. This shows that near the $A_2 \rightarrow A_1$ phase transition one has to expect a pronounced stretching of the vortex along the y-axis. Such an anisotropic structure of the soft core can be modelled as

$$\eta(r,\phi) = \begin{cases} \pi(r/r_D) - \pi/2, & 0 < r < r_D(\phi) \\ \pi/2, & r_D(\phi) < r < b_2 \ , \end{cases} \qquad (7)$$

where the soft core dimension $r_D(\phi) = r_\parallel\cos^2\phi + r_\perp\sin^2\phi$.

In order to obtain the temperature and pressure dependencies of $r_\parallel$ and $r_\perp$, the sum of the gradient and dipolar free energies was minimized. In the case of the fixed spin part of the order parameter, the results of this minimization

0921-4526/90/$03.50

are shown in Fig. 1 near T_{A_2} for our model of nonsingular v vortices (3).

The above calculation was made assuming a homogeneous background of the spin part. A variational calculation allows us to estimate that the maximal deviation of the $\hat{d}$-vectors from their asymptotic orientation is less than 5 degrees upon approaching the A_1 phase.

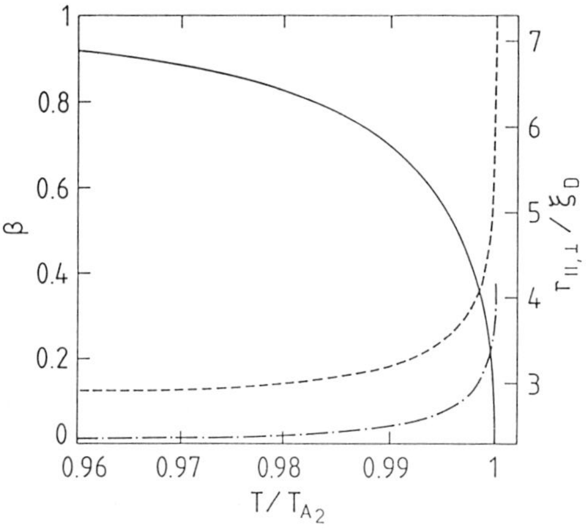

FIGURE 1: Calculated temperature dependencies of $\beta = \Delta_\uparrow\Delta_\downarrow/\Delta^2$ (solid line) and the vortex core radii $r_\parallel$ (dash-dotted line) and $r_\perp$ (dashed line) at p = 0 and H = 1 T.

The increase in the vortex core size was searched for in an ultrasonic transmission experiment. Sound attenuation in the core region is usually different from that in the bulk liquid, owing to different orientation of the $\hat{\ell}$-vector. The change in the received sound amplitude A is roughly proportional to the cross-sectional area of the vortex core. In Fig. 2 we show the relative change of A in the vicinity of the A_2A_1 transition, measured at a constant rotation speed.

To find the approximate core size, we have calculated the response, employing a geometrical acoustics method (4) and theoretical values for ultrasonic attenuation and velocity shifts. We evaluated these parameters by using zero field data provided by W. Wojtanowski (5), and by assuming that the components corresponding to different spin species are independent of each other (6). We have checked the validity of this assumption by comparing the measured and theoretical attenuations in the stationary liquid and found that they agree within 20% in this region.

The curves in top part of Fig. 2 correspond to different core radii, 45 μm and 60 μm, of the SV v-vortex. The predicted distortion in the shape of the core has been neglegted. At this pressure, the 45 μm radius agrees well with experimental data at all temperatures (also at temperatures not shown in Fig. 2), except very near the A_2A_1 transition, where the traces have opposite slopes. However, the calculated response with larger core fits to the measured points just below T_{A_2}. The lowermost curve in Fig. 2 shows the theoretical average increase of the core radius in the experimental temperature region. The observed change at T_{A_2}, about 30%, is less than the predicted value of about 80%. However, the accuracy of the measurement is only moderate because of the very narrow temperature interval. In conclusion, there is some, although preliminary, experimental evidence for the expansion of vortex cores near the A_2A_1 transition.

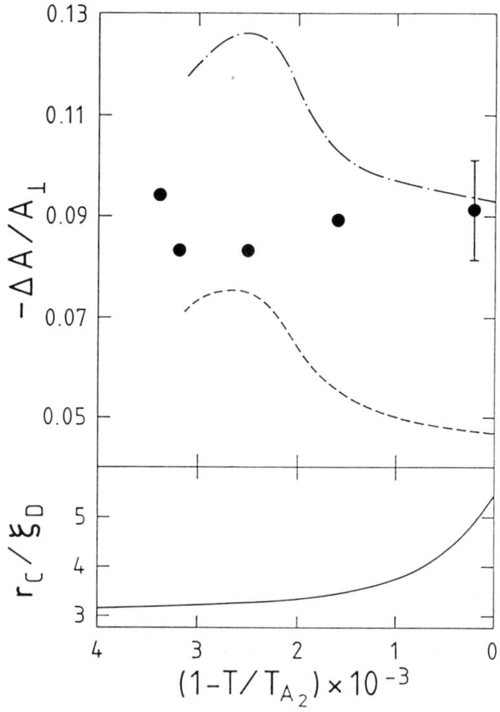

FIGURE 2. Top: Measured change in sound amplitude A at p = 4.6 bar and H = 0.72 T. Dashed and dash-dotted lines correspond to calculated responses with vortex core radii of 45 μm and 60 μm, respectively. Bottom: Average vortex core radius $\bar{r}_c$ at p = 0 and H = 1 T, according to theory.

ACKNOWLEDGEMENTS
We thank J. Pekola, K. Torizuka, and A. Manninen for providing their experimental data prior to publication. This work was supported by the Academy of Finland and by the USSR Academy of Sciences through project ROTA.

REFERENCES
(1) H.K. Seppälä and G.E. Volovik, J. Low Temp. Phys. 51 (1983) 279.
(2) P.J. Hakonen, O.V. Lounasmaa, and J.T. Simola, Physica B60 (1989) 1.
(3) B.N. Kiviladze, Ph.D. Thesis, Institute of Physics, Tbilisi (1988, unpublished).
(4) J.M. Kyynäräinen, J.P. Pekola, K. Torizuka, and A.J. Manninen, this volume.
(5) W. Wojtanowski, private communication (1989).
(6) N. Schopohl, W. Marquardt, and L. Tewordt, J. Low Temp. Phys. 59 (1985) 469.

Physica B 165&166 (1990) 633–634
North-Holland

MAGNETIC FIELD DEPENDENCE OF THE SUPERFLUID ^{3}He A-B TRANSITION*

Y.H. TANG, Inseob HAHN, M.R. THOMAN, H.M. BOZLER, and C.M. GOULD

Department of Physics, University of Southern California, Los Angeles, CA 90089-0484, USA

We are measuring the boundary between the superfluid A and B phases of ^{3}He as a function of magnetic field and pressure. The results yield new information on Landau-Ginzburg parameters, microscopic quasiparticle scattering amplitudes, superfluid specific heats, and potentially other normal liquid properties through comparison with models of the superfluid.

1. INTRODUCTION

Fundamental to our understanding of superfluid ^{3}He are the free energies of the various phases. Since the functional dependences of the free energies in each of these phases are similar, one of the most sensitive measurements of the free energies lies in the location of the phase boundaries between them. We are presently engaged in a precision measurement of the location of the boundary between the A and B phases as a function of magnetic field over the entire available phase diagram.

2. THEORETICAL CONSIDERATIONS

The qualitative fact that a magnetic field suppresses the B phase relative to the A phase has been known since the discovery of superfluid ^{3}He. (1) A survey of the field dependence of the A-B transition was undertaken by Feder et al. (2) at several pressures below the polycritical point. This survey confirmed predictions of the general trends in the limiting low temperature critical fields, and provided a convenient polynomial approximation to describe the location of the phase boundary, but the survey was not of sufficient density to allow a determination of Landau-Ginzburg parameters or other properties described below.

Since the A-B phase transition is of first order, it must obey a magnetic Clausius-Clapeyron equation:

$$\left(\frac{\partial T}{\partial H}\right)_{P,AB} = -\frac{M_A(T_{AB},H,P) - M_B(T_{AB},H,P)}{S_A(T_{AB},H,P) - S_B(T_{AB},H,P)}$$

and analogous relations including pressure dependence. In the limit as $H \to 0$ this equation shows that $T_{AB} \sim H^2$. At pressures above the polycritical point where T_{AB} is below T_c, the quadratic coefficient depends upon the latent heat and magnetic susceptibility discontinuity in zero field, both of which are presumed known from other measurements, but would be checked by this independent means.

At pressures below the polycritical points where T_{AB} is asymptotically coincident with T_c, the quadratic coefficient depends upon the limiting slopes of susceptibility and entropy differences, which have not been previously measured. Equivalently, the quadratic coefficient depends upon a new and previously unmeasured combination of Landau-Ginzburg parameters. (3) These parameters reflect the free energies directly, and depend themselves upon the microscopic quasiparticle scattering amplitudes in the normal liquid. (4)

The quadratic coefficients are thus valuable tools in a variety of settings, but the challenge to the experimentalist is to insure that the measurements are, in fact, being taken in the Landau-Ginzburg regime.

A sufficiently careful measurement at very low temperatures is expected to find that $T_{AB} \sim (H_c - H)^{1/4}$. If this behavior is confirmed, the prefactor would constitute the first measurement of the low temperature specific heat of the A phase. If the behavior differs from this form, however, it will have significant consequences for all models of the superfluid.

3. EXPERIMENTAL CONSIDERATIONS

With the exception of thermometry, all experimental details to acquire the requisite precision here are routine. The magnetic field is produced by a persistent shielded magnet of the style used earlier in our laboratory for a precision measurement of the A_1 phase diagram. (5) The transition itself is detected by the discontinuity in sound attenuations in the two phases. The sound propagation direction is along the magnetic field. In the rare cases where the attenuations in both phases are so large as to leave no signal, we simply shift to different sound frequencies. At low temperatures where attenuations get small, we examine successive sound echoes to locate the transition.

This experiment demands precise thermometry. There are two aspects to this demand. First, we achieve the necessary resolution in temperature with a standard LCMN thermometer (although it is driven by a unique digital bridge described elsewhere in these proceedings). The second aspect is that we must know the absolute temperature accurately. Above about 1 mK we have so far relied upon the Greywall temperature scale (6) though our measurements will allow us to test it further.

* Supported by NSF through grants DMR89-01701 (IH, CMG) and DMR88-00291 (YHT, MRT, HMB).

Below 1 mK there are no reliable guides, so we have chosen to develop a guaranteed linear Pt thermometer. This thermometer differs from the rectify-and-integrate style typical in ultralow temperature laboratories and available commercially. This type of detector suffers from several known problems, all traceable to the nonlinear rectification stage, leading to zero offsets (of several kinds arising from the synergistic interactions of system noise with diode effects and amplifier slew rate limits) and low- and high-amplitude curvatures which limit the dynamic range of useable temperatures without recalibration.

Instead, our system uses a purely linear detector, involving no nonlinear elements. The excitation of Pt spins is conventional, as is the first preamp stage. However, instead of presenting the signal at this point to a rectification stage, we digitize the free induction decay signal in its entirety, and fit it to a decaying sine wave. All issues of linearity are thus confined to the digitization stage which easily satisfies our demands. To control for long-term drifts in the overall system gain, we follow every Pt measurement with a calibration pulse (which is simply a long duration low level pulse) which is similarly digitized and analyzed. With this technique, our pulse-to-pulse reproducibility is limited only by the stability in our pulse transmitter.

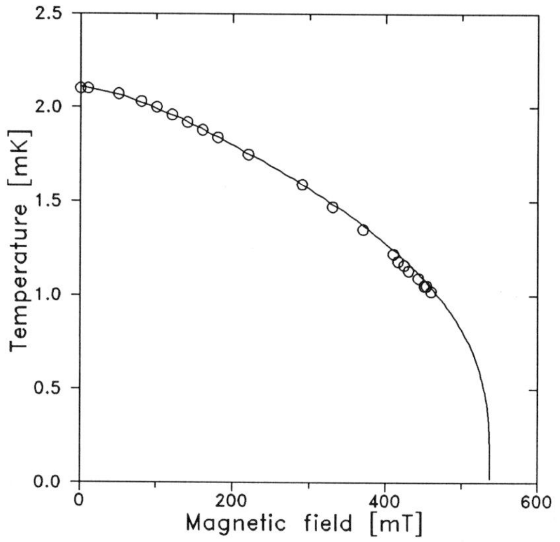

Figure 1. Field dependence of A-B transition temperature.

4. PROCEDURE

Once the A-B transition is located by the abrupt change of the sound level, we sweep the temperature back and forth several times with different temperature sweep rates, ranging from 0.5 to 2.0 μK/sec. The transition temperature shifts with sweep rate owing to effective time constants between the thermometer and cell of 30 sec. We use the zero sweep rate value as the true transition temperature. The resolution in determining the transition temperature is better than 1 μK. The first time ^{3}He goes through the A-B transition from high temperature we always see a clear supercooling effect which indicates the first-order nature of A-B transition. The temperature at which the B-phase first nucleates is about 200 μK lower than the true transition temperature. After forming the B phase in the ^{3}He cell the supercooling effect is substantially eliminated.

5. RESULTS

Figure 1 shows the current state of our data on the magnetic field dependence of the A-B phase boundary at a pressure of 29 bars. The smooth curve through the data points is the six parameter polynomial fit of Ref. 2, suitably modified for the differences in temperature scales. While the fit is reasonable over the region of current data, we would not claim fundamental significance to it, since the data is essentially featureless over its domain. It is clear that lower temperature data will be necessary to extract limiting values. Work in this direction and at other pressures is continuing.

REFERENCES

(1) D.D. Osheroff, R.C. Richardson, and D.M. Lee, Phys. Rev. Lett. 28 (1972) 885.
(2) J.D. Feder, D.O. Edwards, W.J. Gully, K.A. Muething, and H.N. Scholz, Phys. Rev. Lett. 47 (1981) 428.
(3) A.L. Fetter, Hydromagnetic Effects in Superfluid ^{3}He, in Quantum Statistics and the Many-Body Problem, eds. S.B. Trickey, W.P. Kirk, and J.W. Dufty (Plenum Press, New York, 1975) pp. 127-138.
(4) D. Rainer and J.W. Serene, Phys. Rev. B13 (1976) 4745.
(5) U.E. Israelsson, B.C. Crooker, H.M. Bozler, and C.M. Gould, Phys. Rev. Lett. 53 (1984) 1943, and 54 (1985) 254(E).
(6) D.S. Greywall, Phys. Rev. B33 (1986) 7520.

Physica B 165&166 (1990) 635–636
North-Holland

THE COLLECTIVE EXCITATIONS IN PLANAR 2D PHASE OF HELIUM-3.

P.N. BRUSOV, M.V. LOMAKOV

Physical Research Institute, Rostov-on-Don State University, USSR

For the first time whole spectrum of collective excitation in planar 2D phase is investigated by path integration method, developed by Brusov and Popov earlier. Among 18 collective modes there are 6 goldstone modes and 12 nonphonone ones. Spectrum of some of nonphonone modes depends on magnetic fields while of others does not. Dispersion lows of later ones are quite similar to these in helium-3-A.

1. INTRODUCTION

Among superfluid phases of He^3 there is so-called 2D phase with order parameter $c_{ia}^{(0)} = c\sqrt{\beta V}\,\delta_{po}(\delta_{i1}\delta_{a1}+\delta_{i2}\delta_{a2})$, where c = const, $\beta = T^{-1}$, V - volume, $p = (\vec{k},\omega)$ - 4 - momentum. This phase has not been observed yet experimentally, but it has been mentioned in a lot of theoretical papers. For example Alonso and Popov(1) have shown the stability of 2D-phase against small perturbations at $H > H_c$ and predicted the phase transition B → 2D at $H = H_c$. Fujita et al.(2) predicted that in case of B-phase in half limited volume one had 2D-phase closed to boundary. (Collective excitations for this case are calculated in paper by Brusov and Bukshpun, represented at LT-19(3)).

In this paper for the first time we have calculated the collective excitations in 2D-phase by path integration method.

2. METHOD OF CALCULATIONS.

All properties of the model of He^3 system within path integration technique are described by hydrodynamic action functional S_h, and its quadratic form which is obtained from S_h by shift $c_{ia}(p) \to c_{ia}^{(0)}(p) + c_{ia}(p)$ determines in first approximation the collective excitation spectrum (For more details see (4)). Gap equation has next form

$$g^{-1} + \frac{2z^2}{\beta V}\sum_p \frac{\sin^2\theta}{\omega^2+\xi^2+\Delta^2} = 0.$$

Here g is negative constant which is proportional to the amplitude of scattering of two fermions near Fermi surface, $\Delta = 2cZ$ is a gap in the single fermion spectrum, $\omega = (2n+1)\pi T$ are the Fermi frequencies, $\xi = c_F(k-k_F)$, c_F is the velocity at Fermi surface, $\Delta = \Delta_o \sin\theta$.

3. EQUATIONS FOR COLLECTIVE MODE SPECTRUM

After solving the equation det Q = 0, where Q is the matrix of mentioned above quadratic form, we have obtained next equations for collective modes $\left(v_{ij} = \mathrm{Im}\,c_{ia}^{(0)},\ u_{ia} = \mathrm{Re}\,c_{ia}^{(0)}\right)$.

$$\int_0^1 dx(1-x^2)(1+4c)I(c) = 0, \quad u_{11}-u_{22}\pm(v_{12}-v_{21})$$

$$\int_0^1 dx(1-x^2)(1+2c)I(c) = 0, \quad u_{11}+u_{22}\pm(v_{12}+v_{21}),\ v_{11}-v_{22}\pm(u_{12}-u_{21})$$

$$\int_0^1 dx(1-x^2)I(c) = 0, \quad v_{11}+v_{22}\pm(u_{12}+u_{21})$$

$$\int_0^1 dx\, x^2[(1+2c)I(c) - 1] = 0, \quad u_{31}\pm v_{32}, u_{32}\pm v_{31}$$

$$\int_0^1 dx(1-x^2)[(1+4c_+)I(c_+) + (1+4c_-)I(c_-)] = 0, \quad u_{13}, v_{23}$$

$$\int_0^1 dx(1-x^2)[I(c_+) + I(c_-)] = 0, \quad v_{13},\ v_{23}$$

$$\int_0^1 dx\, x^2[(1+4c_+)I(c_+)+(1+4c_-)-2] = 0, \quad u_{33}$$

$$\int_0^1 dx\, x^2[I(c_+) + I(c_-)-2] = 0, \quad v_{33}$$

where

$$I(c) = \frac{1}{\sqrt{1+4c}}\,\ell n\,\frac{\sqrt{1+4c}+1}{\sqrt{1+4c}-1}$$

$$c_{\pm} = \frac{\Delta_o^2(1-x^2)}{\omega^2+(c_F(\vec{k}\vec{n})\pm 2\mu H)^2},$$

$$c = \frac{\Delta_o^2(1-x)^2}{\omega^2+(c_F^2(\vec{k}\vec{n})^2}, \quad x = \cos\theta.$$

4. RESULTS

The analysis of obtained equation shows that spectrum consists of 6 goldstone (gd) modes E = 0 and 12 nonphonone ones, which include: 2 pairbreaking (pb) modes $E_{pb}=(1.96-i0,31)\Delta_o$, 4 clapping (cℓ) - $E_{c\ell} = (1.17-i0,13)\Delta_o$ (see (5)) which exists in He^3-A, and 6 modes, which spectrum depends on magnetic fields. It is interesting that for $\vec{k} = 0$ we obtain 2 quasi pb modes ($E_{pb}^2 \to E_{pb}^2 - 4\mu^2H^2$) and 2 quasi gd modes E = $2\mu H$. The collective modes could be excited in sound and NMR experiments similar to cases of A- and B-phase.

REFERENCES

(1) V. Alonso, V.N. Popov, Sov. Phys. JETP, V. 77 (1977), 1445.
(2) T. Fujita, M. Nakahara, T. Ohmi et al. Progr. Theor. Phys. 64 (1980), 396.
(3) P.N. Brusov, M.M. Bukshpun, this volume.
(4) P.N. Brusov, V.N. Popov, "Superfluity and collective properties of quantum fluids" (Nauka, Moskow, 1988).
(5) P.N. Brusov, V.N. Popov Sov. Phys. JETP 79 (1980), 1871.

Physica B 165&166 (1990) 637–638
North-Holland

EFFECTS OF ^{4}HE COATING ON THE SIZE EFFECTS IN SUPERFLUID ^{3}HE AS MEASURED BY FOURTH SOUND

Donglak KIM, Takashi MIKI, Osamu ISHIKAWA, Tohru HATA, Takao KODAMA, and Haruo KOJIMA*

Faculty of Science, Osaka City University, Sumiyoshi-ku, Osaka, 558 Japan
*Serin Physics Laboratory, Rutgers University, Piscataway, N. J. 08854

We studied the effects of ^{4}He coating of the surfaces of alumina powder grains on the superfluid fraction of ^{3}He. The superfluid fraction was measured by fourth sound techniques. The observed size effects in pure ^{3}He were similar to our previous results. When 1 layer of ^{4}He coverage was adsorbed onto the powder, no change in the superfluid fraction was observed. The attenuation of fourth sound was not affected either. When 3 layers of ^{4}He coverage was adsorbed, a substantial increase in the superfluid fraction was observed. An increase in the attenuation of fourth sound was also seen.

1. INTRODUCTION

Fourth sound propagation occurs when and only when a fluid makes a phase transition into a superfluid phase. The fourth sound has been a useful tool in superfluid physics. It served as a direct probe of superfluidity in ^{3}He and provided a precise direct measurement of superfluid fraction. It may even be used to study persistent currents. The condition for propagation of fourth sound is met by locking the viscous normal component in a suitable "superleak" structure through which the superfluid component carries pressure wave propagation. The most commonly used superleak in fourth sound propagation study is the porous medium provided by interstices of packed powder. The geometry is somewhat complicated, but there is advantage in the simplicity in fabrication and execution of measurement.

In our own studies (1,2), we have applied the fourth sound techniques to probe the size effects on the order parameter in superfluid ^{3}He. We were motivated to initiate the present experiments by the work of Freeman et al.(3) who discovered that the boundary condition (and the size effects) on the order parameter can be altered by introducing small amount of ^{4}He into their cell. The ^{4}He coating apparently increased the specularity of quasi-particles at the Mylar surface. Our goal here is to see how general this change in boundary condition is and how the propagation of fourth sound itself is affected by the ^{4}He coating.

2. EXPERIMENTAL TECHNIQUES

The basic scheme of excitation and detection of fourth sound by resonance was described in Ref.(1). The present experimental set up is very similar to Ref.(2). Two cylindrical resonant cavities were packed with alumina powders of nominanl sizes, 1 μm (resonator I) and 3 μm (resonator II). (Though a third resonator packed with 0.3 μm powder was included, it failed to operate owing to a transducer problem.) The average superfluid fraction is determined from the measured fourth sound resonance frequency using

$$\overline{\rho}_s/\rho = (2Lf/mnC_1)^{1/2}$$

where f is the resonant frequency of the m-th mode, n the index of refraction, L the length of the cavity and C_1 the speed of first sound. The resonators were both mounted into a cell attached to a copper nuclear demagnetization apparatus and were immersed in liquid and in good thermal contact with an LCMN thermometer tower. The thermometer indication was converted to Greywall temperature scale.

BET measurements showed that the surface areas present in the fourth sound cavities were 4.96 m^2 (I) and 2.73 m^2 (II) and 19.1 m^2 in the sintered silver powder heat exchanger. We estimated the thickness of the ^{4}He layer from the measured surface area and using 0.12 atoms/A^2 per layer for all surfaces.

3. RESULTS

We carried out measurements of superfluid fraction in the resonators first filled with high purity ^{3}He. This procedure enabled us to do diagnostic checks on the system and to compare the size effects in the present resonators with our earlier ones. The entire apparatus was then warmed up at 3.5 K to drive out ^{3}He and an appropriate amount of ^{4}He to produce 1 layer coverage was introduced into the cell. Subsequently the cell was cooled to 1 K and filled with ^{3}He. Then measurements of ^{3}He superfluid fraction were made with this ^{4}He coverage. The entire process was repeated again with 3 layer ^{4}He coverage. The measured average superfluid fractions in the two resonators at 10 bars are summarized in Fig. 1. The magnitude of suppres-

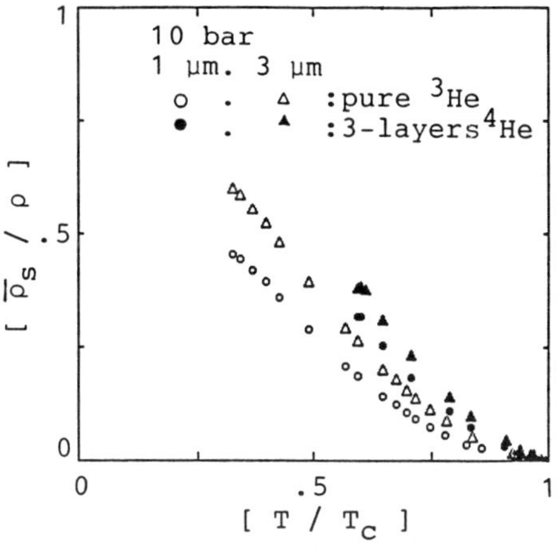

FIGURE 1
A substantial increase in the superfluid fraction was observed when 3 layers of ^{4}He coverage was adsorbed onto the alumina powders.

sion of superfluid fraction with pure ^{3}He (open circles (I), open triangles (II)) is consistent with our previous experiments (2). After one layer of ^{4}He was put into the cell, the measured superfluid fraction did not change within 1 % or so in both resonators from that with pure ^{3}He. So this data is not shown in Fig. 1. Thermal relaxation in the cell as observed during demagnetization was also not significantly affected. However, after the ^{4}He coverage was increased to 3 layers, we observed that the superfluid fraction increased dramatically in both resonators at all temperatures (closed circles (I), closed triangles (II)). The measurement became a very slow process as the thermal relaxation time in the cell and the minimum achievable temperature were significantly degraded. Even though we do observe a large enhancement of the superfluid fraction there still remains size effect depressions compared to the bulk values in both resonators. A similar set of data were taken at 5 bars. Aside from the greater depression in the pure ^{3}He owing to longer healing length, the enhancement at 5 bars with 3 layers of ^{4}He was quite similar to that at 10 bars.

In addition to superfluid fraction, we made observations on the qualitative effect of ^{4}He coating on the Q value of the resonances. At both pressures, the Q values with 1 layer ^{4}He coverage did not differ from that with pure ^{3}He. With 3 layers of ^{4}He coverage, we observed a decrease in Q value at a given temperature. The degradation in Q may be greater at lower pressures, but we can not be certain at this time.

4. DISCUSSION

The preliminary experiments presented here showed that the effective superfluid fraction of ^{3}He contained in the interstices of packed alumina powder can be greatly enhanced by coating the surfaces with a sufficient amount of ^{4}He. The surface of Mylar has already been shown to provide such enhancement (3). The role played by the surface condition on the ability of ^{4}He to modify the boundary condition on the order parameter is not yet clear. A superfluid fraction enhancement was searched for but not found in sintered packed silver powder nor in Aerogels (4).

Though the observed enhancement of superfluid fraction is very large in our experiment, it has not reached the bulk value as measured with torsion oscillator techniques. How thick a layer of ^{4}He is needed, if it is possible in our experiment, to achieve the bulk value is an open question at this time. It would be also interesting to measure pressure dependence on the enhancement as the ^{4}He coating layer changes from liquid to solid.

If incomplete locking of normal component within the pores (but the tangential velocity at the wall is still assumed zero) is the dominent mechanism of dissipation, we expect that the Q value should increase at a given temperature when the superfluid fraction increases. A slip boundary condition which allows for non-zero tangential velocity at the wall has been introduced in describing normal ^{3}He flow at low temperature (5). The observed decrease in Q value may be another evidence of increased specularity of quasi-particles which would be expected to suppress pair breaking near the wall.

ACKNOWLEDGEMENTS

One of us (H.K.) was supported in part by NSF, Low temperature Physics Program, Grant No. DMR-8815776. The other (D.K.) wants to thank Rotary Yoneyama Memorial Foundation for the financial aid.

REFERENCES

(1) T. Chainer, Y. Morii and H. Kojima, J. Low Temp. Phys. 55 (1984) 353.
(2) K. Ichikawa, et al. Phys. Rev. Lett. 58 (1987) 1949.
(3) M.R. Freeman et al. Phys. Rev. Lett. 60 (1988) 596.
(4) T.Hall and J.M. Parpia, Bull. Am. Phys. Soc. 34 (1989) 1197.
(5) H.H. Jensen et al. J. Low Temp. Phys. 41 (1980) 473.

Physica B 165&166 (1990) 639–640
North-Holland

VISCOSITY OF SUPERFLUID ^{3}HE IN HIGH MAGNETIC FIELD

L.P. ROOBOL, R. JOCHEMSEN, C.M.C.M. van WOERKENS, T. HATA, S.A.J. WIEGERS and G. FROSSATI

Kamerlingh Onnes Laboratorium der Rijksuniversiteit Leiden, P. O. Box 9506, 2300 RA Leiden, The Netherlands

We have studied the viscosity of ^{3}He in the A1 phase at various pressures with a vibrating wire viscometer in a magnetic field of 9.3 T. The data could be fitted surprisingly well using a simple model. The value of the minimum of the reduced viscosity is independent of pressure.

1. INTRODUCTION

The ^{3}He superfluid order parameter ψ can be described by a complex 3x3 matrix. Several superfluid phases are possible, characterized by different values of ψ. (For review articles about the superfluid phases see *e.g.* refs. (1,2)). In magnetic fields above 0.6 T only two phases are known to exist: the A1- (up-spin pairs) and the A2- (equal spin pairs) phase. They are anisotropic phases : all transport coefficients and the normal density are described by tensors rather than scalars.

2. EXPERIMENTAL SET UP

2.1 Experimental cell

The experiments were done in a compression cell which was also used for the experiments reported in refs. (3,4). Two NbTi vibrating wire viscometers were mounted in the cell, one at the top and one at the bottom. The upper [lower] viscometer was made of 80 [50] μm wire bent in a semicircle of 2 [2.7] mm diameter, giving a resonant frequency of 20 [10] kHz in a 9.3 T field. The cell was thermally anchored to a $PrNi_5$ demagnetization stage (5) which could cool the sample down to approximately 2 mK with the maximum field on.

2.2 Thermometry

A carbon resistor thermometer was mounted in the lower part of the cell, but below 10 mK carbon resistors saturate, and we had to use the melting curve and the viscometers to determine the temperature. The pressure was measured with a sapphire pressure transducer. At pressures below the melting curve only the viscometers could serve as thermometers. Given certain conditions on the wire and fluid properties the resonance amplitude of the viscometer is proportional to T in the normal phase (6). The A1 transition served as a fixed point. In the superfluid phases this temperature behaviour is no longer valid, so when off the melting curve we had no good means of determining the temperature.

3. DATA DERIVATION SCHEME

When we neglect the fact that we are dealing with an anisotropic fluid, but assume we measure some kind of average of the tensor components of the viscosity and normal density, the amplitude of the in-phase signal of the vibrating wire viscometer is given by:

$$V(f) = \frac{\gamma f\left[\gamma f + \sigma_n k' f^2\right]}{\left[f_0^2 - (1+\sigma_s+\sigma_n k)f^2\right]^2 + \left[\gamma f + \sigma_n k' f^2\right]^2} V_0$$

where f_0, V_0 and γ are respectively the resonance frequency, the amplitude and the linewidth in vacuum. σ_n and σ_s are respectively the normal and superfluid density divided by the density of the wire. k and k' are Stokes' dimensionless functions of the viscosity, wire diameter, normal density and frequency (7). The out-of phase signal vanishes at

$$f_{FB} = f_0(1+\sigma_s+\sigma_n k)^{-1/2} \qquad [1]$$

and at that frequency the amplitude will be

$$V(f_{FB}) = \frac{\gamma}{\gamma + \sigma_n k' f_{FB}} V_0 \qquad [2]$$

The constants f_0, V_0 and γ were determined by fitting the normal phase data. These constants were used to fit η (through k and k') from the superfluid data. Because of statistical errors it turned out to be impossible to fit the normal fraction independently, so we used the zero field values of ref. (9).

4. RESULTS

The lower viscometer data could be fitted well with f_0= 9990 Hz and γ = 30 Hz. The viscous damping was in the order of 1 kHz. Assuming the normal fraction decreases as in fig. 1, we derive the reduced viscosity shown in fig. 2. We have taken viscometer traces at various pressures between 17 bar and the melting pressure, and for all of them the minimum

0921-4526/90/$03.50

viscosity was about 65% of the viscosity at the A1 transition.

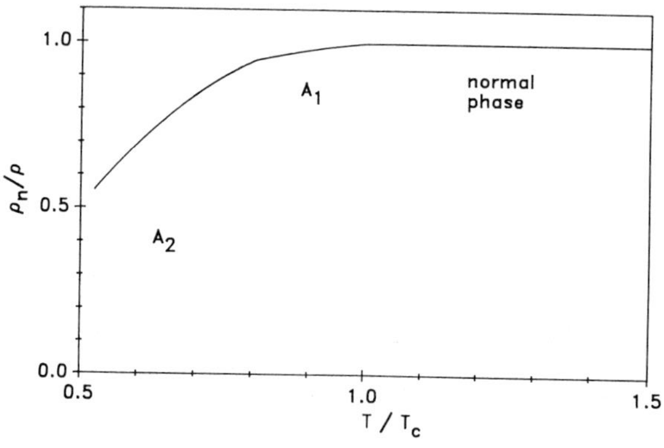

FIGURE 1

The normal fraction as function of T/T_{A1} (from Hook *et al.*)

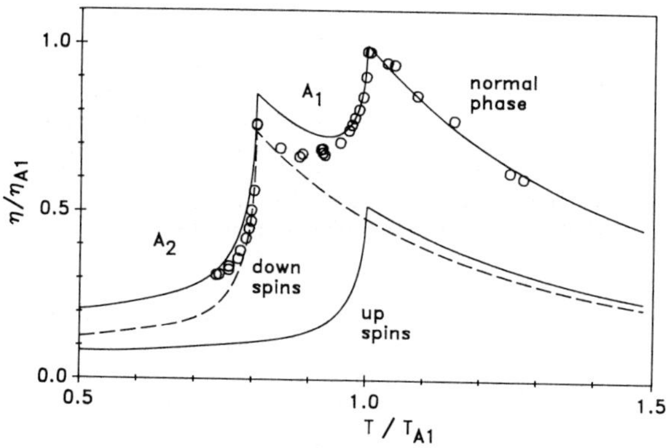

FIGURE 2

The reduced viscosity in 9.3 T as function of T/T_{A1}. circles: experimental points; upper solid line: model; dashed line : up-spin contribution, lower solid line: down-spin contribution

Above T_{A1} we see the viscosity increases like T^{-2}, as expected in a fermi liquid. Just below T_{A1} we see a sharp decrease proportional to the opening of the superfluid energy gap $\Delta \sim (1-T/T_{A1})^{-1/2}$ (8). Our assumption is that scattering processes involving down-spins are not affected by the pairing of the up-spins. This means that in the A1-phase the down-spins still behave as a fermi liquid. For the viscosity of the up-spins we have taken the zero-field data of Hook et al. (9). The liquid consists of 48% down- and 52% up-spins in 9.3 T, so we weight both contributions to the viscosity appropriately and sum them. The decrease in η because of pairing of up-spins competes with the increase because of the fermi-liquid viscosity of the down-spins, resulting in the minimum in η. The model fits surprisingly well.

Of course the assumptions of this model do not hold: the pairing of up-spins affects the rate at which up-spins collide with down-spins, which is the dominant collision type in the normal phase. The fit should get worse as the superfluid density increases, which indeed is the case. The best way to deal with this problem is to take into account all possible scattering processes and make a microscopic calculation of the viscosity in the A1 phase.

5. CONCLUSIONS

Like all transport phenomena the viscosity is a macroscopic manifestation of scattering processes at a microscopic level. Examining viscosity differences between the normal and the A1-phase therefore is a way to gain insight about these scattering processes. The observed value of the minimum is a useful test for theories of the viscosity in the A1 phase. Unfortunately, in order to draw any definitive conclusions from these experiments, it is necessary to make a full microscopic calculation of the viscosity in the A1 phase.

ACKNOWLEDGEMENTS

We like to thank Dr. D. Rainer for a useful discussion, and one of us (LPR) likes to thank P. Remeijer and T. van der Vorm for discussions on data fitting methods. This work is financially supported by the "Stichting voor Fundamenteel Onderzoek der Materie" F.O.M.

REFERENCES

(1) A.J. Leggett, Rev. Mod. Phys. 47 (1975) 331

(2) D.M. Lee and R.C. Richardson in: The Physics of Liquid and Solid Helium, eds. K.H. Benneman and J.B. Ketterson (Wiley-Interscience Publ., New York, 1978) p. 287

(3) T. Hata, S.A.J. Wiegers, R. Jochemsen and G. Frossati, Phys. Rev. Lett. 63 (1989) 2745

(4) S.A.J. Wiegers, C.C. Kranenburg, T. Hata, R. Jochemsen and G. Frossati, Europhys. Lett. 10 (1989) 477

(5) S.A.J. Wiegers, T.Hata, C.C. Kranenburg, R. Jochemsen and G. Frossati, Cryogenics, in print

(6) C.C. Kranenburg, L.P. Roobol, R. Jochemsen and G. Frossati, J. Low Temp. Phys. 77 (1989) 371

(7) G.G. Stokes, Mathematical and Physical Papers, 3 (1901) 38

(8) P. Bhattacharyya, C.J. Pethick and H. Smith, Phys. Rev. B 15 (1977) 3367

(9) J.R. Hook, E. Faraj, S.G. Gould and H.E. Hall, J. Low Temp. Phys. 74 (1989) 45

Physica B 165&166 (1990) 641–642
North-Holland

EFFECTS OF ^{4}HE COVERAGE AND INTERFACE CONDITION ON THE SPIN RELAXATION IN ^{3}HE-A_1

Qiang Jiang and Haruo Kojima

Serin Physics Laboratory, Rutgers University, Piscataway, NJ 08854

Effects of moving magnetic field relative to sensor and coating cell surfaces with ^{4}He on the spin density relaxation time in ^{3}He-A_1 were measured. In the upper temperature side of the phase, where A_1/N interface is present in the cell, the relaxation time is strongly modified by the the surface condition and magnet position changes. In the lower temperature side of the phase, where A_1/A_2 interface is present, the relaxation time is not affected by either changes.

1. INTRODUCTION

The spin polarized superfluid component in the ^{3}He-A_1 phase may be driven to flow by an applied magnetic field gradient. The mass superflow is one and same as the spin superflow in this phase. We have utilized this unique property to study spin dynamics phenomena with a novel mechanical device.(1,2) The details of measurements were described in our previous reports. Briefly, our device is a sensitive capacitive transducer which senses the fluid exchange between a small chamber and the surrounding large liquid reservoir. The small chamber and the liquid reservoir are connected by 25 μm height superleak channels through which superflow takes place. The field profiles (symmetrical about z=0) of static field(H_0) and gradient field(δH) are shown in Fig. 1. The device is cooled through a liquid column which extends in the positive z direction and connects to a silver powder heat exchanger region.

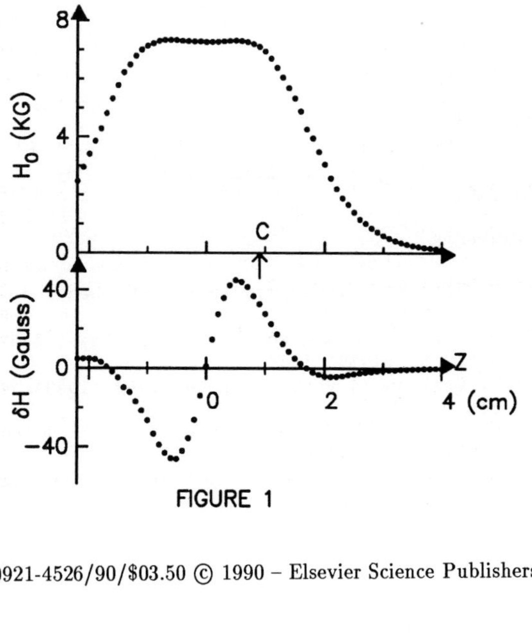

FIGURE 1

When a "step" of field gradient is applied across the superleak, a superflow into the chamber takes place. A is defined as the maximum deflection of the sensing element. Subsequently, the induced pressure decays with an exponential time dependence. A better understanding of the underlying decay mechanism has been the aim of our recent studies. To try to gain more insight we have investigated the effects on the relaxation time by (1)moving the magnetic field coils relative to the superleak and (2)coating the surfaces of the cell with magnetically inert ^{4}He. This paper is a preliminary report of these experiments.

2. RESULTS

(1)In all of our previous reports the magnets were placed such that their center (i.e. z=0 in Fig. 1) was located at the center of the superleak. Motivated in part by discussions with Grabinski we measured the effect of moving the magnetic field coils relative to the superleak center. In the measurements reported below the magnet center was moved such that the point C in Fig. 1 is coincident with the superleak center. The maximum amplitude A in a step experiment is shown as a function of reduced temperature in Fig. 2. T_{c1} and T_{c2} are the upper and lower transition temperature of A_1 phase, respectively. The open circles(H_0=7.5 KG) are from data with the magnets moved to new position and the dots(H_0=9.2 KG) from data with magnets at center. The directions of currents into the coils were identical in both measurements. The detailed temperature dependence depends on H_0, but what is peculiar is the sign of A. Since the sign of the field gradient at the superleak is reversed (see Fig. 1) when the magnets are moved, we expect the direction of superflow to reverse. That is indeed the case in the lower temperature range, but unexpectedly the direction does not reverse in the upper range.

The temperature dependences of relaxation time τ for the two positions of the magnets

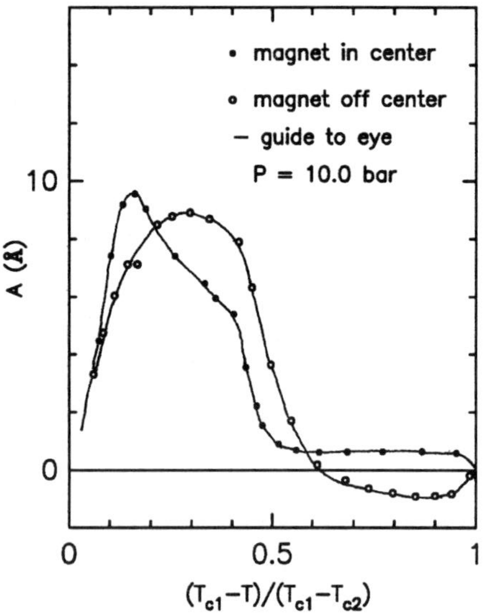

FIGURE 2

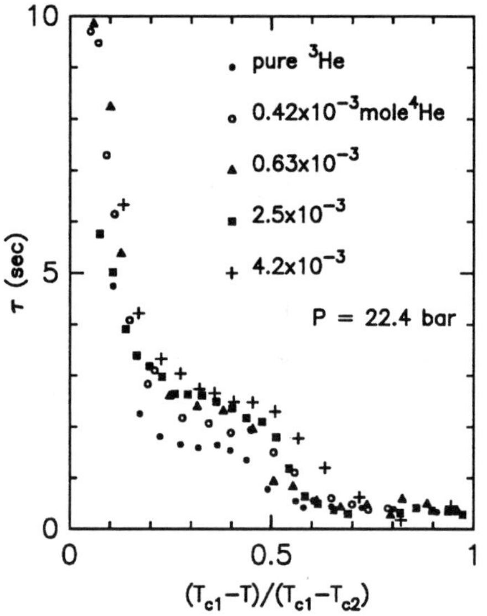

FIGURE 3

were already shown Ref. (3). The data are again qualitatively different depending on temperature. In the lower range τ does not depend on the position of the magnet, while in the upper range τ becomes longer when the magnets are shifted. There was little pressure dependence on τ when the magnets were shifted.

(2)Even though there is that unexpected sign reversal in the upper temperature range, the relaxation time itself was not altered greatly. Furthermore we expect that the change in τ due to position of magnet is independent of the possible changes due to ^{4}He coverage of surfaces. For these reasons we chose to do the ^{4}He coverage experiments with the magnets in the shifted position rather than the center location. All our previous measurements have been done with high purity ^{3}He with less than about 20 ppm of ^{4}He. To save the high purity ^{3}He aside we started with a 99.95 % grade obtained from Isotec. We carried out measurements with 6 different coverages of ^{4}He. In between the experiments with different coverages the whole dilution refrigerator system was warmed above 5 K before introducing a new ^{3}He-^{4}He mixture.

The results of the relaxation time for 4 of the ^{4}He coverages studied are shown in Fig. 3. We estimate 8.3×10^{-4} moles correspond to 1 layer of ^{4}He. The qualitative effects of ^{4}He coverage are again seen to divide into two distinct temperature ranges. The observed τ in the lower range is independent of ^{4}He coverage. In the higher range τ is affected by ^{4}He coverage. It appears that the relaxation time is saturating at the highest coverages. The ^{4}He coverage affected the maximum amplitude also. As the coverage was increased, the amplitude decreased in the upper range but did not change in the lower range.

3 SUMMARY

We carried out observations of the effects of shifting magnet position and ^{4}He coverage on the spin relaxation time in ^{3}He-A_1 phase. The effects are qualitatively different depending on temperature. In the lower range there exists an A_2/A_1 interface in the cell (near z=2~3 cm). In the upper range there exists an A_1/N interface. Our observations show that the interfacial condition is crucial. This is consistent with Grabinski(3) who has stressed the importance of the interface in understanding the discontinuity in τ at T_c. How the interfacial effects lead in detail to the observations presented here is yet to be clarified.

ACKNOWLEDGEMENTS

We thank Michael Grabinski for fruitful discussions. This research was supported in part by National Science Foundation, Low Temperature Physics Program, Grant No. DMR-8815776.

REFERENCES

(1) R. Ruel and H. Kojima, Phys. Rev. Lett. 54 (1985) 2238.
(2) S.T. Lu, Q. Jiang and H. Kojima, Phys. Rev. Lett. 62 (1989) 1639.
(3) M. Grabinski, Phys. Rev. Lett. 63 (1989) 814

Physica B 165&166 (1990) 643–644
North-Holland

THERMAL EXPANSION COEFFICIENT OF FERMI LIQUIDS AT LOW AND ELEVATED TEMPERATURES

Setsuo MISAWA and Hiroshi MORITA

Department of Physics, College of Science and Technology, Nihon University, Kanda-Surugadai, Tokyo 101, Japan

The thermal expansion coefficient of Fermi liquids such as liquid ^{3}He and some heavy fermion compounds is known to show a "v"-shape behaviour as a function of temperature. It is shown that this anomalous behaviour can be microscopically explained by the finite temperature version of the Fermi liquid model. We use an exact expression for the expansion coefficient in terms of the quasiparticle interaction function. The expansion coefficient is numerically evaluated to yield the result in good agreement with experiment. It is concluded that the v-shape behaviour of the coefficient is inherent in any strongly correlated Fermi liquid.

The observation of the thermal expansion coefficient α of Fermi liquids is rather a classic problem. Almost 30 years ago the expansion coefficient of liquid ^{3}He was measured as a function of temperature T and pressure P by several groups (1, 2), the result being shown in Figure 1. It shows an anomalous "v"-shape behaviour ; α decreases linearly with T at the lowest temperatures, then exhibits a broad minimum and turns to positive at elevated temperatures. Early theoretical treatments on this problem, however, were limited only to the lowest temperatures; the negative values of the coefficient were discussed by Brueckner and Atkins (3), and by Usui (4). Recently it has been found that, in some heavy fermion compounds such as $CeAl_3$ and UPt_3 (5), α shows the v-shape behaviour similar to liquid ^{3}He; interests for the thermal expansion coefficient are renewed.

A phenomenological explanation for the v-shape behaviour has been given by the present author (6) on the basis of the Fermi liquid model; both the negative values and the v-shape behaviour may be attributed to the Fermi liquid effect. Here we shall present a microscopic explanation for these features of the expansion coefficient. We shall use the finite temperature version of Landau's Fermi liquid theory due to Usui (4), where thermodynamic quantities are exactly expressed in terms of the f-function (or Fermi liquid function). On the basis of a suitable form of the f-function, one can numerically evaluate the thermal expansion coefficient as a function of temperature and obtain the result in good agreement with experiment.

In the Landau's theory of the Fermi liquid, a state of the whole system is specified by the distribution function of the quasiparticle, n_i, at the single particle state i. Thus the total number of particles (quasiparticles) N and the entropy S (exact at any temperature) are given by

$$N=\sum_i n_i\ , \quad S=-k\sum_i\{n_i\ln n_i+(1-n_i)\ln(1-n_i)\}$$

where k is the Boltzmann constant. The energy of the system, E, is a functional of the distribution function; its explicit form can be defined by its variation through the variation of n_i, δn_i,

$$\Delta E=\sum_i\varepsilon_i\delta n_i+\frac{1}{2}\sum_i\sum_j f_{ij}\delta n_i\delta n_j+\cdots\ ,$$

where ε_i is the energy of the quasiparticle and f_{ij} is the so-called Landau f-function which represents the interaction between quasiparticles.

We shall describe the system as a function of volume V, temperature T and chemical potential μ. At fixed values of V, T and μ, the equilibrium state is determined by the vanishing of the first-order variation of $\Delta E - T\Delta S-\mu\Delta N$, which yields $n_i=[e^{\beta(\varepsilon_i-\mu)}+1]^{-1}$ ($\beta\equiv 1/kT$). Thus the probability distribution of δn_i is given by

$$e^{-\beta(\Delta E-T\Delta S-\mu\Delta N)}$$
$$=\exp\left[-\tfrac{1}{2}\left\{\sum_i\frac{(\delta n_i)^2}{n_i(1-n_i)}+\beta\sum_i\sum_j f_{ij}\delta n_i\delta n_j\right\}\right]\ .$$

Thermodynamic quantities can be expressed in terms of fluctuations. In particular, the thermal expansion coefficient at constant pressure is given by

$$\alpha\ =\ V^{-1}(\partial V/\partial T)_P=\beta N^{-2}\overline{S\Delta N^2} \\ -k\beta^2N^{-1}\overline{\Delta N(\Delta E-\mu\Delta N)}\ . \qquad (1)$$

To evaluate this we only need an average of the form $\overline{\delta n_i\delta n_j}$. From a known result for the multi-Gaussian distribution, we obtain

$$\overline{\delta n_i\delta n_j}=(j|(1+\beta\tilde{f})^{-1}|i)[n_i(1-n_i)]^{1/2}[n_j(1-n_j)]^{1/2}\ ,$$

where $\widetilde{f_{ij}}=[n_i(1-n_i)]^{\frac{1}{2}}f_{ij}[n_j(1-n_j)]^{\frac{1}{2}}$ and $(j|(1+\beta\tilde{f})^{-1}|i)$ is the $(j,\ i)$ element of the inverse matrix $(1+\beta\tilde{f})^{-1}$. Inserting this into (1), and performing the sum over the spin states, we finally obtain

$$\alpha\ =\ -2k\beta^2N^{-1}\sum_i\sum_j(\varepsilon_i-\mu-\sigma)\left(j|(1+2\beta\varphi)^{-1}|i\right) \\ \times[n_i(1-n_i)]^{\frac{1}{2}}[n_j(1-n_j)]^{\frac{1}{2}}\ , \qquad (2)$$

0921-4526/90/$03.50

where $\sigma = ST/N$, $\varphi_{ij} = [n_i(1-n_i)]^{\frac{1}{2}} f^s_{ij} [n_j(1-n_j)]^{\frac{1}{2}}$ and f^s is the spin symmetric part of the f-function.

If the analytic forms of ε and f^s are explicitly known as functions of quasiparticle momentum $\mathbf{p}$, α can be numerically evaluated through (2). The f-function is related to the forward scattering amplitude for two quasiparticles in the many particle system. From the view of the many-body perturbation theory, however, the explicit form of the scattering amplitude for dense Fermi liquids is hardly known. Here we make an approach from the dilute Fermi liquid; the f-function is given by the sum of ladder type diagrams for particle-particle (or hole-hole) scatterings. This treatment is equivalent to Brueckner's t-matrix theory for nuclear matter and liquid ^{3}He. According to Galitskii (7), or Abrikosov and Khalatnikov (8), the f-function can be expressed in terms of the s-wave scattering amplitude a,

$$f^s(\mathbf{p},\mathbf{p}') = 4\lambda/3\pi - (8\lambda^2/3\pi^3) \int d\mathbf{p}_1 n_0(\mathbf{p}_1) \int d\mathbf{p}_2 \times \left[\frac{\delta(\mathbf{p}+\mathbf{p}'-\mathbf{p}_1-\mathbf{p}_2)}{p^2+p'^2-p_1^2-p_2^2} + (\mathbf{p}_1 \leftrightarrow \mathbf{p}) + (\mathbf{p}_1 \leftrightarrow \mathbf{p}') \right]$$

up to order a^2. In the same approximation, the quasiparticle energy is given by

$$\varepsilon(p) = p^2/2 + 2\lambda/3\pi + (\lambda^2/\pi^4) \int d\mathbf{p}_1 n_0(\mathbf{p}_1) \int d\mathbf{p}_2 n_0(\mathbf{p}_2) \times \int d\mathbf{p}_3 \left[\frac{\delta(\mathbf{p}+\mathbf{p}_3-\mathbf{p}_1-\mathbf{p}_2)}{p^2+p_3^2-p_1^2-p_2^2} - 2\frac{\delta(\mathbf{p}+\mathbf{p}_1-\mathbf{p}_2-\mathbf{p}_3)}{p^2+p_1^2-p_2^2-p_3^2} \right]$$

Here p, f and ε are measured in units of p_0 (Fermi momentum), $p_0^2/2m$ and p_0^2/m, respectively, m is the mass of a particle, $n_0(\mathbf{p})$ is the Fermi distribution function at $T = 0$, and $\lambda = k_F a$ (k_F = Fermi wave number) is a parameter which measures the interaction strength. Performing integrations in f and ε, and taking average over the angle between $\mathbf{p}$ and $\mathbf{p}'$ in f, we obtain the explicit forms for $f^s(p,p')$ and $\varepsilon(p)$.

On the basis of these functions we perform the numerical evaluation of α for various values of λ. For liquid ^{3}He near the saturated vapour pressure we take $\lambda = 1.8$ which corresponds to $m^*/m = 3.1$, where $m^* = p_0(\partial\varepsilon/\partial p)^{-1}_{p_0}$ is the effective mass of the quasiparticle; the result of calculation is shown is Figure 2. Agreement between theory and experiment is excellent in view of that the expansion coefficient changes rapidly with pressure (1, 2).

Because of the Pauli principle, the interaction necessarily excites particles within the Fermi sphere to the states above the Fermi momentum; the strong repulsion inhibits the particles to form momentum order near $T = 0$; to weaken the effect of the repulsion the liquid should expand as $T \to 0$ (3). Thus the negative thermal expansion is a direct consequence of the Fermi liquid effect. For strongly correlated Fermi liquids, where λ exceeds some critical value, α becomes always negative (6). The convex nature of α with T can also be attributed to the Fermi liquid effect. The Fermi liquid effect produces logarithmic terms such as $(p-p_0)^2 \ln|p-p_0|$ in f and hence the $T^3 \ln T$ variation of α. A sum of two terms, $\alpha_1 T + \alpha_2 T^3 \ln(T/T^*)$, $(\alpha_1, \alpha_2 < 0)$, gives a convex function of α with T (6).

In conclusion, the observed v-shape behaviour of the expansion coefficient can be thoroughly understood in terms of the Fermi liquid effect.

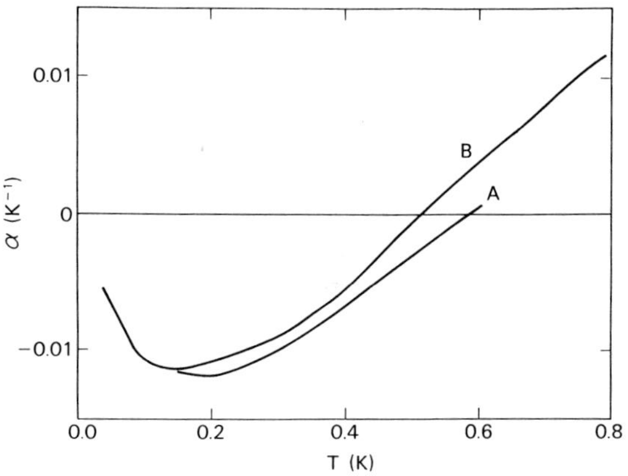

FIGURE 1
Earlier experimental works on the expansion coefficient of liquid ^{3}He plotted against temperature. Curves A and B are at 1.70 and 0.18 atm by Ref. (1) and (2), respectively.

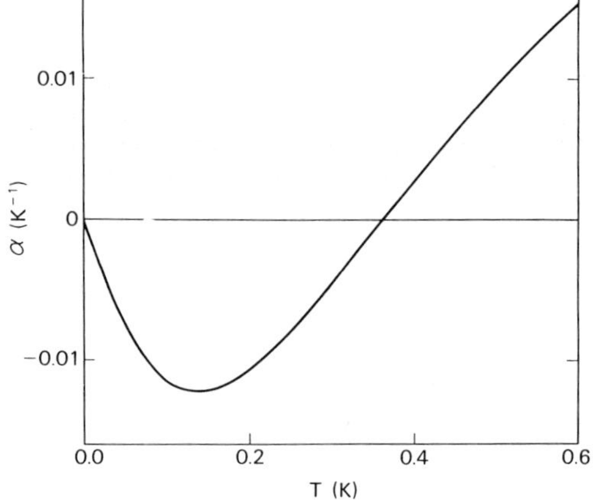

FIGURE 2
Theoretical curve of the expansion coefficient as a function of temperature for $\lambda = 1.8$ and $p_0^2/2mk = 5$K.

REFERENCES
(1) D.F. Brewer and J.G. Daunt, Phys. Rev. 115 (1959) 843.
(2) J.E. Rives and H. Meyer, Phys. Rev. Lett. 7 (1961) 217.
(3) K.A. Brueckner and K.R. Atkins, Phys. Rev. Lett. 1 (1958) 315.
(4) T. Usui, Phys. Rev. 114 (1959) 21.
(5) K. Andres, J.E. Graebner and H.R. Ott, Phys. Rev. Lett. 26 (1975) 1779.
(6) S. Misawa, Solid State Commun. 58 (1986) 63.
(7) V.M. Galitskii, Sov. Phys. JETP 7 (1958) 104.
(8) A.A. Abrikosov and I.M. Khalatnikov, Sov. Phys. JETP 6 (1958) 888.

Physica B 165&166 (1990) 645–646
North-Holland

SLIP OF NORMAL ^{3}HE ON SMOOTH SURFACES

S. M. THOLEN and J. M. PARPIA

Laboratory of Atomic and Solid State Physics, Cornell University, Ithaca, New York 14853-2501

Measurements of the slip length of normal ^{3}He confined by smooth silicon suggest that the boundary condition for ^{3}He scattering at surfaces is diffuse even for substrates which are smooth compared to the mean thermal wavelength of ^{3}He. Coating the surface with just two or three monolayers of ^{4}He greatly increases the specularity. In addition, the expected behavior of the bulk viscosity of ^{3}He appears to be modified by the reduced momentum transfer caused by the smooth surface and specular scattering condition.

1. INTRODUCTION

In classical hydrodynamics, one solves the Navier-Stokes equation for fluid flow with the boundary condition that the relative velocity of the fluid with respect to a surface go to zero. If the details of the scattering process at the surfaces are included, it is found that this velocity is not zero at a boundary; there is finite slip of the fluid. The amount of slip depends strongly on the type of scattering. If the surface scatters the liquid particles diffusively, there is significant momentum transfer between the solid and the liquid, and thus small slip. In the other extreme, if the scattering is specular, then little momentum is exchanged and the slip becomes very large.

The purpose of this experiment is to gain information on the scattering mechanism for normal ^{3}He particles at solid surfaces by measuring the slip at a surface of known characteristics. In particular, highly polished silicon with a roughness small compared to the mean thermal wavelength of ^{3}He particles was selected. We measured the microscopic roughness to be better than 50 Å. If the scattering process depends on geometrical roughness alone, then such smooth surfaces would be expected to scatter particles specularly.

2. THE EXPERIMENTAL APPARATUS

A flat cylindrical region for ^{3}He is produced by gluing together two flat silicon disks spaced by a thin glass washer of height 50μm. The height of the cell is made small to amplify the contribution of the fluid near the walls to the total dissipation of the oscillator. A small hole through the center of one disc acts as a fill line for the liquid. The oscillator is run at constant amplitude via capacitive electrodes; the drive voltage is thus directly proportional to the dissipation. The empty-cell dissipation is subtracted to yield the contribution from the fluid alone.

3. RESULTS

The coupling of the fluid to the solid cell controls the behavior of the oscillator. At low temperatures, the fluid is very viscous and its motion is well-locked to that of the cell. The moment of inertia (period) is large, while the dissipation is small. As temperature is increased, some fluid begins to unlock, so the period drops and the dissipation rises. At some point, the fluid is sufficiently decoupled from the cell and the dissipation falls while the period continues to decrease. The dissipation and period thus trace out a peaked curve.

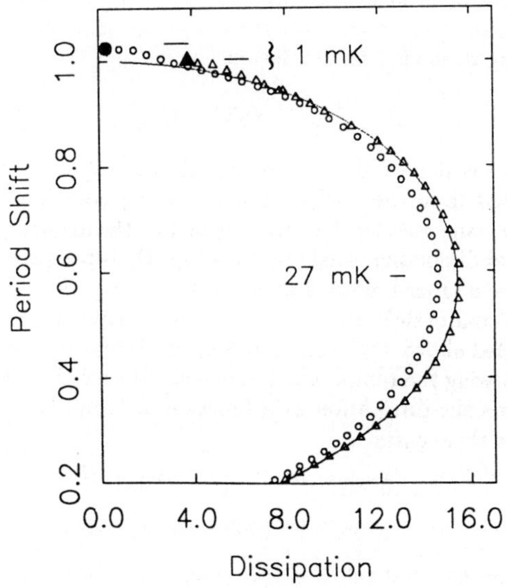

Figure 1

Period shift (relative to the full-cell period shift) versus dissipation (in arbitrary units). Note the dissipation offset between the two sets at low T.

In figure 1 we plot the dissipation versus period shift. Data were taken for three cases. First the cell was filled with pure ^{3}He. Next, a 50 μmole/m^2 layer of ^{4}He was admitted to coat the surfaces before the addition of the ^{3}He. In the third case, the ^{4}He coverage was increased to 75 μmole/m^2. The circles represent the pure ^{3}He and the triangles represent the data taken with surface ^{4}He added. The two ^{4}He-coated cases trace out very similar curves; for clarity, only the 50 μmole/m^2 data is shown here. The period endpoints were similar for all three cases. Thus, one theoretical curve applies to all the data sets. For the pure ^{3}He, the dissipation falls short of its expected value at the maximum, but for the cases with ^{4}He the data matches theory fairly well in this region. At low temperatures, the period overshoots theory for all data sets.

The presence of slip should not modify the hydrodynamic relation between dissipation and period shift. The temperature dependence of these variables is expected to change, however, when slip is present. In particular, theory predicts that, in the Poisieulle flow limit, slip simply adds a temperature-independent term to the inverse viscosity of the liquid; for normal ^{3}He, the viscosity is known to be proportional to $1/T^2$. Since the dissipation of a torsional oscillator is simply proportional to $1/\eta$ at low temperatures, the dissipation should be of the form:

$$D = D_s + D_b T^2,$$

where D_s is the extra dissipation caused by slip and $D_b T^2$ is the ordinary bulk dissipation. The size of this slip term is expected to increase with specularity according to the following relation:

$$D_s = D_o(1+s)/(1-s),$$

where s is the fraction of particles that are specularly reflected from the surface and D_o is the completely diffuse value for the slip term. Scaled to the arbitrary units of dissipation used for our setup, D_o is predicted to have a lower bound of about 0.20.

In none of three runs did the dissipation fit the form expected above. One can, however, produce a good fit by allowing the temperature exponent to vary. Figure 2 shows the dissipation as a function of temperature for the three cases.

$$D = 0.09 + .0786T^{1.70} \quad (\text{pure } ^3\text{He})$$

$$D = 3.70 + .0885T^{1.65} \quad (50\ \mu\text{mole/m}^2\ ^4\text{He})$$

$$D = 4.43 + 0.143T^{1.45} \quad (75\ \mu\text{mole/m}^2\ ^4\text{He})$$

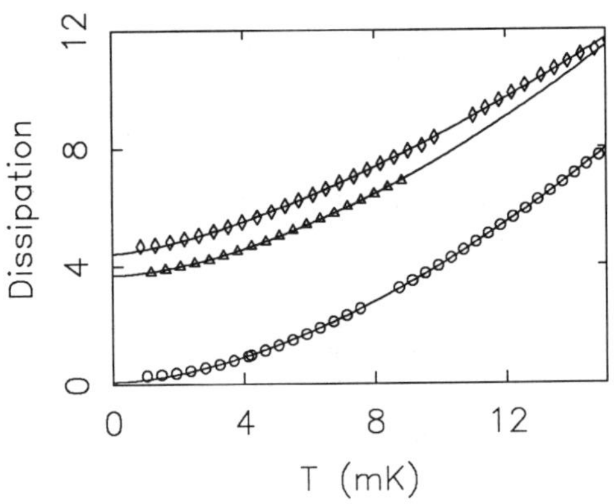

Figure 2
Dissipation as a function of temperature for the three cases described. The solid lines represent the fits to the above power laws. The circles represent pure ^{3}He; the smaller ^{4}He coverage is shown by triangles, and the greater with diamonds.

It is clear from this graph that there is an offset term in the dissipation which increases dramatically upon the addition of ^{4}He, as if the surface were becoming more specular. However, the remaining dissipation does not seem to behave as expected for bulk ^{3}He, but instead fits a power law with an exponent somewhat less than the expected 2.0. This exponent also seems to depend on the amount of ^{4}He coverage on the surfaces. An alternative fit for the ^{3}He data which does conform to a bulk $1/T^2$ behavior for the viscosity would require an additional scattering length on the order of 100 μm, independent of temperature.

4. CONCLUSIONS

The addition of ^{4}He affects the slip and the bulk behavior of the liquid ^{3}He. It is possible that the momentum transfer occurs only because of surface imperfections or the non-zero amount of ^{3}He dissolved in the ^{4}He layers. This effect may be dominating the dissipation and thus leading to the unusual temperature dependence for apparent viscosity.

ACKNOWLEDGEMENTS: Supported by the NSF through DMR-8820170

REFERENCES

(1) H. Jensen et al., JLTP **41**, 473 (1980)

(2) D. Ritchie et al., PRL **59**, 465 (1987)

(3) M. Freeman et al., LT-17 Proceedings (1988)

Physica B 165&166 (1990) 647–648
North-Holland

MICROSCOPIC LIQUID ^{3}He DROPLETS IN Ag: HEAT CAPACITY AND TEM

E.SYSKAKIS, Y. FUJII*, F.POBELL, and H. SCHROEDER+.

Phys. Inst., Universität Bayreuth, D-8580 Bayreuth, FRG.
+Inst. für Festkörperforschung, KFA Jülich, D-5170 Jülich, FRG.

We present heat capacity measurements of ^{3}He droplets of mean radii 25 Å < r < 77 Å obtained in 42 μm thick Ag foils after implantation with 0.1 at.% ^{3}He and annealing between 570 K and 1112 K. The heat capacity of liquid ^{3}He droplets at 15 mK ≤ T ≤ 1 K shows deviations from bulk behaviour whose origin will be discussed.

1. INTRODUCTION

Introduction of He in a metal matrix results in microscopic high-density He gas bubbles of radius <100 Å. Liquid or solid He droplets are obtained when such a He-doped metal is cooled to low temperatures.[1] For liquid ^{3}He droplets three effects might influence the properties of these microscopic Fermi systems. Firstly, the small number of Fermions/droplet may result in shell effects. Secondly, characteristic lengths of ^{3}He become larger than the size of the droplets in the mK temperature range. Finally, the effect of the metal wall on the properties of liquid ^{3}He has to be considered.

2. EXPERIMENTAL

We have implanted Ag foils of 9.5 mm ø and 42 μm thickness homogeneously with 0.1 at. % ^{3}He using a cyclotron. Twelve foils were annealed for 2h in 10^{-5} mbar at successively higher temperatures (see Table 1). We used eleven of them for heat capacity measurements performed after each annealing step and one for transmission electron microscopy, TEM.

The heat capacity was measured in a Ag microcalorimeter linked to the mixing chamber of a dilution refrigerator via Ag wires. We measured the thermal relaxation time τ of the foils and calculated the heat capacity $C = \tau/R$, with $R \propto T^{-1}$, the thermal resistance of the link.

Enlarged TEM micrographs of ^{3}He bubbles were evaluated to obtain size distributions, their half width and average radii, see Table 1. Using an equation of state for dense He gas[2] and assuming that the bubbles are at thermal equilibrium at the annealing temperature T_a, i.e. gas pressure $P(T_a) = 2\,\sigma_{Ag}/r$, with σ_{Ag}= 1.12 J/m^2, we calculate the density of ^{3}He and the number of ^{3}He atoms/bubble. According to these calculations we expect liquid ^{3}He droplets at low temperatures in bubbles of r≥ 40 Å while for r≥ 50 Å a liquid-gas equilibrium in the bubbles is expected.

*Permanent Address: Okayama University of Science, Okayama, Japan.

3. RESULTS AND DISCUSSION

After annealing the foils at 570 K, 750 K, and 900 K we did not observe any excess heat capacity and the data agree well to the calculated heat capacity of Ag. At these T_a's the bubble radii are smaller than 26 Å, containing solid ^{3}He at ≤ 1 K. After the foils were annealed at 940 K and higher T_a we observe a substantial increase of the heat capacity. This increase is due to the increase of the ratio of liquid ^{3}He/solid ^{3}He in the droplets due to the increase of their size with T_a. In Fig.1 we show the heat capacity of ^{3}He droplets in comparison to the heat capacity of bulk liquid at 0 and 30 bar for 3.2 μmol ^{3}He which is the calculated amount in our foils. Subtracting from the 3.2 μmol an amount of 0.6 μmol for the first solid layer of ^{3}He close to the Ag substrate, whose contribution is expected to be negligible, we

Table 1: Annealing temperatures, T_a; reduced annealing temperatures T_a/T_m (T_m= 1234 K, the melting temperature of Ag); mean bubble radii r, and half widths of the size distributions of ^{3}He bubbles in Ag. We calculate the gas pressure P, at T_a; density of ^{3}He ρ and numbers of ^{3}He atoms/bubble.

T_a (K)	T_a/T_m	r (Å)	δr (Å)	P (kbar)	ρ (g/cm^3)	number of at./bubble
570	0.462	---	---	---	---	---
750	0.608	---	---	---	---	---
900	0.729	26	15	8.7	0.177	$2.5\cdot10^3$
940	0.762	36	14	6.2	0.142	$5.6\cdot10^3$
960	0.778	---	---	---	---	---
976	0.791	40	23	5.6	0.129	$6.9\cdot10^3$
1003	0.813	---	---	---	---	---
1033	0.837	---	---	---	---	---
1069	0.866	67	48	3.4	0.085	$2.1\cdot10^4$
1112	0.901	77	58	2.9	0.073	$2.8\cdot10^4$

still have a heat capacity deficiency. Based on the TEM data we estimate the mass of ^{3}He in the droplets and conclude that the deficiency cannot be explained by a ^{3}He mass deficiency; it seems that the specific heat of liquid ^{3}He in the droplets is smaller than the specific heat of bulk liquid ^{3}He. Former investigations of ^{3}He in porous glasses [3,4] also showed smaller heat capacities which were attributed to a deficiency of the heat capacity of the second ^{3}He layer close to the substrate. However, we cannot account fully for the deficiency observed by us even if we assume a negligible contribution of the second layer.

The temperature dependence of the heat capacity of the droplets at 0.1 K $\leq$ T $\leq$1 K coincides with that of bulk ^{3}He if we shift the Fermi temperature of the droplets to T_F = 1.2 K < T_F^{bulk}(SVP) = 1.8 K, (see Fig.2). Fig. 2 also shows that we observe an almost T-independent contribution to the heat capacity of the droplets below about 0.1 K, (like for a classical gas or as in the plateau region of a Fermi liquid at T > $T_F/3$), similar to that observed for the second ^{3}He layer on Ag powder[5] or ^{3}He on Grafoil.[6]

We interpete our data in terms of a modified layer model by assuming that the first solid layer close to the Ag has a negligible heat

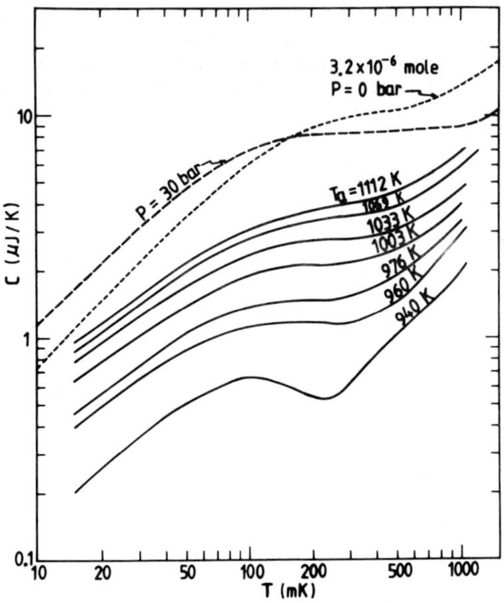

Fig.1: Heat capacity C of ^{3}He droplets in Ag foils. The heat capacity of 3.2 μmol of bulk liquid ^{3}He at P=0 bar and 30 bar is also given for comparison.

capacity, while the second liquid layer contributes $C_2 \simeq 0.15\ N_2\ k_B$ (as for ^{3}He on Ag[5]) also at 15 mK$\leq$ T $\leq$ 100 mK. For the rest of the liquid in the droplets we assume bulk behaviour, but with a Fermi temperature T_F =1.2 K. With

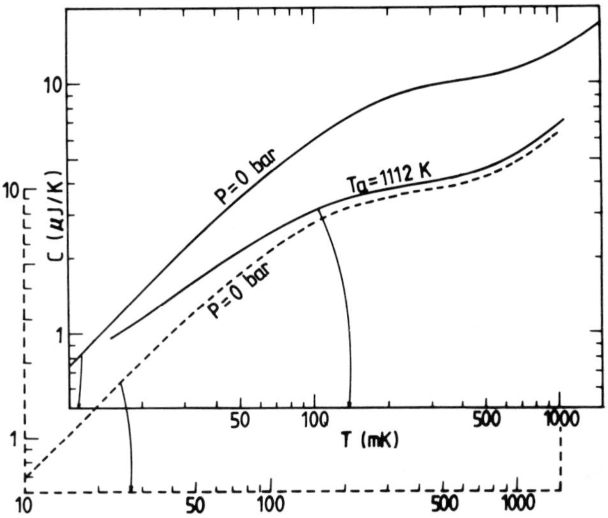

Fig.2: Heat capacity of ^{3}He in droplets after annealing at 1112 K in comparison to the heat capacity of bulk ^{3}He at P=0 bar.

these crude assumptions we obtain remarkably good agreement of the T-dependence of the calculated and measured heat capacity below 0.1K.

If the second layer of ^{3}He has a plateau specific heat in the investigated T range its Fermi temperature has to be T_F < 40 mK pointing to a strong enhancement of the magnetic interactions of the ^{3}He atoms close to the Ag substrate. Magnetization data of ^{3}He films on Ag and Grafoil also indicate enhancement of magnetic interactions.[7] But we have to distinguish between ^{3}He films and our isolated Fermi droplets whose dimensions are smaller than the mean free path of quasiparticles below 100 mK. In this temperature range we should have independent particle motion; but it is not obvious whether our data are influenced by shell effects recently observed for free clusters with N $\leq$ 10^3 atoms.[8]

4. REFERENCES.

(1) E. Syskakis, F.Pobell, and H.Ullmaier, Phys. Rev. Lett. 55,(1985) 2964.
(2) H. Trinkaus, Rad. Effects 78,(1983) 184.
(3) D.F. Brewer, The Physics of Liquid and Solid Helium, Vol.II (H. Wiley & Sons, New York,1978).
(4) M. Shimoda, T. Mizusaki, M. Hiroi, A. Hirai, and K.Egushi, J. Low Temp. Phys.64, (1986) 285.
(5) D.S. Greywall and P.A. Bush, Phys. Rev. Lett. 60, (1988) 1860.
(6) D.S. Greywall and P.A. Bush, Phys. Rev.Lett. 62, (1989) 1868; D. S. Greywall, preprint 1989.
(7) Y. Okuda, A.J.Ikushima and H.Kojima, Phys. Rev. Lett. 54, (1985) 130; H. Franco, R.E.Rapp, and H. Godfrin, Phys. Rev. Lett. 57, (1986) 1161.
(8) W. Miehle, O. Kandler, T. Leisner and O. Echt, J. Chem. Phys. (in print).

Physica B 165&166 (1990) 649–650
North-Holland

Observation of Vortex-like Spin Supercurrent in ^3He-B.

A.S.Borovik-Romanov, Yu.M.Bunkov, V.V.Dmitriev, Yu.M.Mukharskiĭ and D.A.Sergatskov

Institute for Physical Problems, Academy of Sciences of the USSR, Kosygin st.2, 117334, Moscow, USSR

By creating a homogeneously precessing domain in ^3He-B in a quadrupole RF field we have obtained a structure, which according to the theory must bear a circular spin supercurrent. This structure is analogous to a quantum vortex in ^4He or a magnetic flux line in superconductors.

1. Introduction

In previous experiments (1,2) we studied spin supercurrents in ^3He-B using so called *homogeneously precessing domain, HPD.* It was shown that the spin supercurrent inside HPD has the same properties as other types of supercurrents (mass supercurrents in superfluids or electric currents in superconductors). So, the analog of the potential of the superfluid velocity is the phase (α) of the precession of the magnetization and the value of the spin supercurrent is proprtional to $\nabla\alpha$ (3). The critical spin current was observed. The corresponding critical gradient of α defines the analog of Ginzburg-Landau coherence length for the spin system (3):

$$\xi_s = 1/(\nabla\alpha)_c = c_\perp/\sqrt{\omega\delta\omega} \qquad (1)$$

where $c_\perp$ is the spin wave velocity, $\delta\omega$ is the frequency shift from the local Larmor value and ω is the frequency of the precession. A nonhysteretic current-phase relationship for the spin current in an orifice (i.e. the analog of the Josephson phenomenon) was also observed. The fact that the spin supercurrent is the potential current results in a possibility of an existence of quantized spin vortices (4). The phase α is changing by $2\pi n$ along the closed path around the spin vortex. Recently I.A.Fomin has obtained the solution of equations of motion of the magnetization in ^3He-B which corresponds to the singly quantized spin vortex (5). In particular, he has shown that this vortex should have a core with the radius of the order of ξ_s and β is changing from 0 at the vortex axis to $\approx 104°$ far from the core (β is the angle of the deflection of the magnetization from the direction of a steady magnetic field $\vec{H}$). In this paper we describe experiments on creating the circular spin current (and consequentally the spin vortex) in the HPD.

Let us consider the HPD, which is maintained by the radiofrequency (RF) field $\vec{h}$. Inside the HPD the magnetization precesses in phase (even in an inhomogeneous magnetic field) and $\beta \geq 104°$. The frequency of the precession equals the Larmor value on the domain wall separating the HPD from the region where the magnetization is in equilibrium. The power, absorbed by the HPD from the RF field is proportional to $M_{xy}h\sin\phi$ (here M_{xy} is the projection of the magnetization on the plane perpendicular to $\vec{H}$; ϕ is the angle between $\vec{M}_{xy}$ and $\vec{h}$). This power compensates the magnetic relaxation in HPD. So, if we make experiments in the presence of a large RF field, then the phase of the magnetization is almost equal to the phase of the RF field. One can easily produce the RF field, whose phase is changing by 2π along the circular path. The insert to Fig.1 shows the distribution of the RF field of coils, connected in opposition. Numbers show the relative phase of the field along a circular path. The HPD created by such RF field should have the distribution of the precessing magnetization with a circular spin supercurrent. The spin vortex arising in the center should have a core with a diameter which depends only on the difference between the frequency of the precession and the local Larmor frequency in accordance with Eq.1. For our experimental conditions the core size should change along the vortex line and be equal to ~ 0.01 cm far from the HPD boundary.

One can observe the free induction decay signal (*FIDS*) of the HPD if CW RF pumping is switched off (6). In this case the magnetic relaxation decreases the size of the HPD and the frequency of the observed FIDS changes. The voltage induced in the NMR coil is proportional to an integral of $M_{xy}he^{i\phi}$ over the sample. Therefore, the signal from the quadrupole magnetization distribution will be zero in coils connected in series (dipole coils) and nonzero in coils connected in opposition (quadrupole coils). For the homogeneous distribution of the magnetization in the regular HPD we have the opposite situation: the signal is maximal in dipole coils and equals zero for quadrupole coils.

2. Experiment

Experiments were carried out in a magnetic field of 142 Oe ($\omega/2\pi$ = 460 kHz) and at the pressures of 11 and 29.3 bar. The field gradient ($\vec{\nabla}H$) of the order of 0.5 Oe/cm was applied to fix the position of the domain wall. The gradient was directed so that the HPD was situated in the upper part of the cell. The experimental cell was a cylinder with the vertical axis (along $\vec{H}$ and $\vec{\nabla}H$) with the length=diameter=6 mm. Two NMR coils were used. Two halfs of a main large coil, wound around the cell could be connected in series or in opposition. Two parts of a small saddle-shaped subsidiary coil were connected in series. This coil was situated at

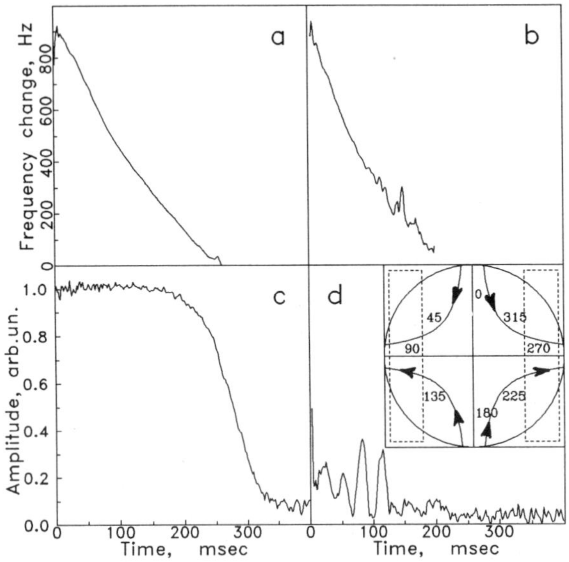

Figure 1. The FIDS in different situations. The temperature is about 0.5 T_c. P=29.3 bar. Left column corresponds to regular HPD (dipole connection of the main coil), right column is the HPD with the vortex (quadrupole connection of the main coil). The upper row shows the frequency of the FIDS in main coil, and the lower row shows the FIDS amplitude in subsidiary coil. Insert schematically shows top view on cell, quadrupole main coil (dashed rectangles) and the RF field distribution. The numbers show relative phase of the RF field.

the upper part of the cell. The HPD was formed by applying a sufficiently large CW RF field by the main coil while sweeping the steady magnetic field down, like in (1,2,6). The process was monitored by the conventional CW NMR spectrometer described in (6). When the HPD filled the cell, the RF field was switched off and the FIDS was observed.

If the two parts of the main coil were connected in series, the CW NMR signal and the FIDS indicated the formation of the regular HPD. The dependence of the frequency of the FIDS versus the time for this case is shown in Fig.1a. The change of the frequency is proportional to the magnetic dissipation and is due to the movement of the domain wall. The amplitude of the signal for this case from the subsidiary coil is shown in Fig.1b. It is seen that the amplitude is not changing during first 150 ms and then drops to zero. The point is that the subsidiary coil picks up the signal only from the upper part of the cell, so the amplitude starts to change only when the HPD boundary enters the sensitivity region of the coil. The frequency of the signal, however, at any moment coincides with the frequency of the signal from the main coil.

In order to excite the circular spin supercurrent we used the main coil connected in opposition. The CW NMR signal in this case looks like the signal from the regular HPD, which points out that we have inhomogeneous distribution of α over the sample. The FIDS obtained after the RF field had been switched off are shown in Fig.1c,d. The signal from the subsidiary coil (Fig.1d) is almost zero as it should be for the case of the vortex-like spin current. The frequency of the signal from the main coil (Fig.1c) changes slightly faster than in Fig.1a. This behaviour also corresponds to the expected one, since large gradients of the magnetization near the vortex core should cause an additional spin diffusion dissipation. This is also proved by the fact that the difference in the rate of the frequency change increases at 11 bars where the spin diffusion is larger. It is worth mentioning that the obtained state is stable: the FIDS does not indicate the transition to the regular HPD mode.

At 11 bars for the case of the quardrupol distribution of the RF field we sometimes obtained the state corresponding to the regular HPD. The FIDS recorded in this case from the subsidiary coil looks like the signal shown in Fig.1b. The reason for the creation of the regular HPD in this case is not yet clear. Possibly it is due to some assimmetry in positions of two halfs of the main coil.

Thus, we have shown that it is possible to create the state which does not induce the signal in the dipole NMR coil. The signal from the quadrupole coil, however, is similar to the signal of the regular HPD in the dipole coil. We may suggest only one explanation of this phenomenon: the quadrupole RF field creates the quadrupole magnetization distribution, i.e. the spin vortex.

Authors are grateful to I.A.Fomin for stimulating discussions.

References

(1) A.S.Borovik-Romanov, Yu.M.Bunkov, V.V.Dmitriev, Yu.M.Mukharsky and D.A.Sergatskov, *Phys.Rev.Lett.* **62** *(1989) 1631.*

(2) A.S.Borovik-Romanov, Yu.M.Bunkov, V.V.Dmitriev, Yu.M.Mukharsky and D.A.Sergatskov, *in Quantum Fluids and Solids 1989, AIP Conf.Proceedings* **194**, *N.Y. (1989) 27.*

(3) I.A.Fomin, *JETP Lett.* **45** *(1987) 135.*

(4) E.B.Sonin, *JETP Lett.* **45** *(1987)*

(5) I.A.Fomin, *Sov. Phys. JETP* **67** *(1988) 1148.*

(6) A.S.Borovik-Romanov et al, *Zh. Exp. Theor. Fiziky* **96** *(1989) 956.*

Physica B 165&166 (1990) 651–652
North-Holland

ANOMALOUS DYNAMIC BEHAVIOUR OF A GAS OF QUASIPARTICLES IN SUPERFLUID ^{3}He-B

S N FISHER, A M GUÉNAULT, C J KENNEDY and G R PICKETT

Department of Physics, Lancaster University, Lancaster, LA1 4YB, United Kingdom.

In ^{3}He-B it is possible to probe the unusual dynamics arising from the non-Newtonian dispersion relation of the excitations. Since the dispersion relation is tied to the rest frame of the condensate, it is distorted by condensate motion. A container of superfluid with a velocity gradient has a range of excitation dispersion curves, depending on velocity and thus the density of states available to the excitations varies with position. We show that, in consequence the behaviour of a moving object in the excitation gas shows unusual and often counter-intuitive behaviour. At low enough temperature, the force on a moving object becomes independent of velocity.

The dispersion relation of quasiparticles in ^{3}He-B is very unlike that of a Newtonian particle where $E = p^2/2m$. An unusual dispersion relation should lead to unusual dynamic properties. However, while unusual dispersion curves are common for excitations in condensed matter, they are rarely free of encumbering lattices.

The dispersion curve for quasiparticles in ^{3}He-B, is shown in figure 1. The group velocity is a minimum at $p = p_F$, not at $p = 0$, and there are two classes of excitation; quasiparticles with group velocity v_g and momentum p parallel, and quasiholes with v_g and p antiparallel. These two factors are the key to the unusual dynamical behaviour of a gas of these excitations. For example, near p_F v_g can change from positive to negative with only a small change in p. Thus a "tennis racket" can hit and reverse the direction of incoming particles with virtually no change of momentum. (This is Andreev [1] scattering.) Furthermore, when an excitation is "normally" scattered (i.e. right across the Fermi surface), an incoming quasiparticle experiences a momentum change of $2p_F$ whereas an incoming quasihole has a momentum change of $-2p_F$. This means that a scattered quasiparticle pushes but a scattered quasihole pulls.

Since the dispersion curve of figure 1 is tied to the rest frame of the liquid, it is sensitive to relative motion. If we move with velocity v relative to the liquid we

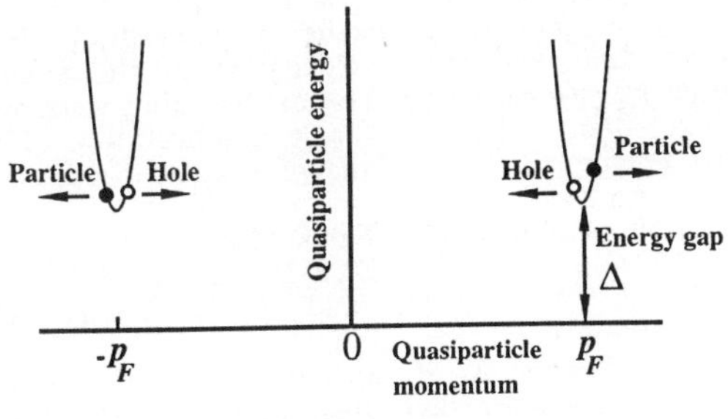

FIGURE 1
The dispersion curve for excitations in ^{3}He-B.

observe excitations with momenta p to have energies changed by $-v \cdot p$. This tips the dispersion curve of figure 1, which now looks as in figure 2. Scattering takes place at constant energy in the frame of the scatterer, and therefore figure 2 indicates the possible scattering processes for particles impinging on a scatterer moving at velocity v. Considering the oncoming excitations, we see that quasiparticles at A are free to undergo normal scattering processes since there are available states on the far side of the Fermi surface, whereas the quasiholes at C are not.

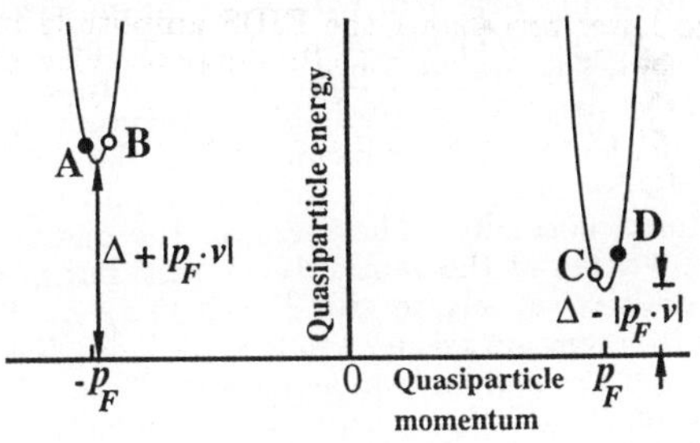

FIGURE 2
The dispersion curve for excitations in ^{3}He-B, viewed in a frame moving at velocity v.

We illustrate the unusual dynamics of the quasiparticle gas by looking at the drag force on a moving wire (a vibrating wire resonator) measured as a function of the velocity.

Were we to excite such a resonator in a gas of Newtonian particles, the damping would be linear in v and the resonance would be harmonic. In a 1-D picture we can represent the wire (of radius a) by a flat paddle of width $2a$, moving with constant velocity v in the (1-D) quasiparticle gas. The excitations have number density n, and v is small compared with the excitation group velocity v_g. The front side of the paddle intercepts $n\,a\,(v_g + v)$ excitations per unit time and the rear side $n\,a\,(v_g - v)$. For normal fermions, particles scattered by the leading side of the paddle exchange momentum $+2p_F$ and those scattered by the trailing side $-2p_F$. Hence the net force on the paddle

0921-4526/90/$03.50

is $F = 4a\,n\,p_F\,v$, i.e. linear in v.

In the superfluid the situation is quite different. In 3-D the motion of the wire sets up a pure potential backflow in the condensate in which the dispersion relation changes rapidly with position. We can approximate backflow in 1-D by assuming that near the paddle the condensate is moving with the paddle velocity, whereas at a distance the condensate is at rest with respect to the container walls. The dispersion relations in the paddle frame for near and far liquid are shown in figure 3.

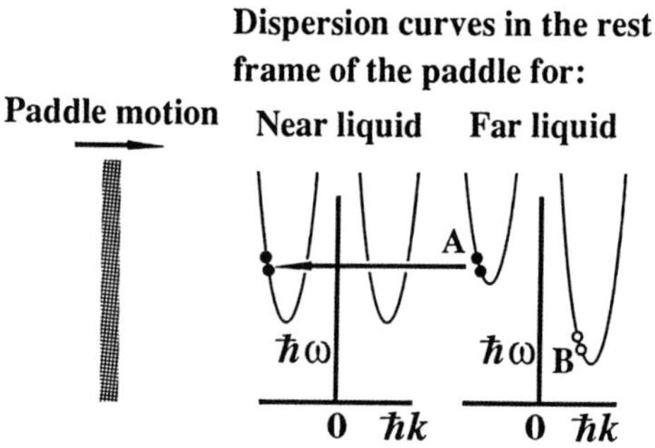

FIGURE 3
The dispersion curves far from and near to the paddle.

At the lowest temperatures only the extreme minima of the dispersion curve are populated. From the figure we see that quasiparticles approaching the paddle from the front (at A) can pass from far to near liquid and be normally reflected, with momentum change of $2p_F$, whereas quasiholes approaching from the front (at B) cannot reach the near liquid and are reflected by Andreev processes with negligible momentum change. Conversely, for excitations incident from the rear, only the quasiholes can penetrate the near liquid region to be scattered normally. The force on the paddle may again be written as the sum of two terms, but now the paddle velocity cancels, to give $F = 2\,n\,a\,p_F\,v_g$.

This is a very surprising result since it implies that the force on the paddle is independent of velocity, the reason being that only quasiparticles scatter normally from the front and only quasiholes from the rear and both retard the motion by exchanging momentum $2p_F$. If the paddle moves faster more quasiparticles are intercepted on the front side and fewer quasiholes on the rear side but the total momentum change remains the same.

This simple argument applies to low T ($kT \ll 2p_F v < \Delta$). At higher T we must remember that the incident ballistic excitations have a thermal distribution appropriate to the container walls. Now, not only can quasiparticles reach the front side of the paddle, but also those quasiholes with energy greater than $p_F v$ above the dispersion minimum. Similarly, the more energetic of the quasiparticles from the rear may also reach the paddle. The force is now calculated from an integral over the appropriate states in each branch.

We integrate the momentum transfer from $E = \Delta$ on the $-p$ branches with a contribution of opposite sign from $E = \Delta + p_F v$ on the $+p$ branches. The force F then becomes:-

$$F = -\int_{\Delta}^{\Delta + p_F v} 8\,p_F\,a\,g(E)\exp(-E/kT)v_g\,dE.$$

where $g(E)$ is the d.o.s. per branch. In 1-D $g(E) = (2/h)(dp/dE)$. Since the group velocity is (dE/dp), these terms cancel, and we find: $F = (16\,a\,p_F\,kT/h)\exp(-\Delta/kT)(1 - \exp(-p_F\,v/kT))$. At low velocities ($p_F v \ll kT$) the force is proportional to v, i.e. $F = (16\,a\,p_F^2/h)\exp(-\Delta/kT)\,v$, but it becomes independent of v at high v, with value $F = (16\,a\,p_F\,kT/h)\exp(-\Delta/kT)$.

We may now compare this result with the observed quasiparticle damping force on a vibrating wire wire in ^{3}He-B. In 3-D we may expect the damping to have the same general form, $F = F_0\exp(-\Delta/kT)(1 - \exp(-p_F v/kT))$. (A recent full 3-D calculation by Watts-Tobin and Fisher [2] confirms this assumption.)

The data comprise a series of quasiparticle damping measurements taken with resonator of 4.5 μm diameter NbTi filament and extend from 124 μK to 204 μK and from zero velocity to beyond the onset of pair breaking [3],[4]. The force derived from the data is plotted in reduced form against reduced velocity in figure 4. The agreement with the simple picture is excellent. The high degree of scatter at high v is a due to the large damping at the onset of pair-breaking.

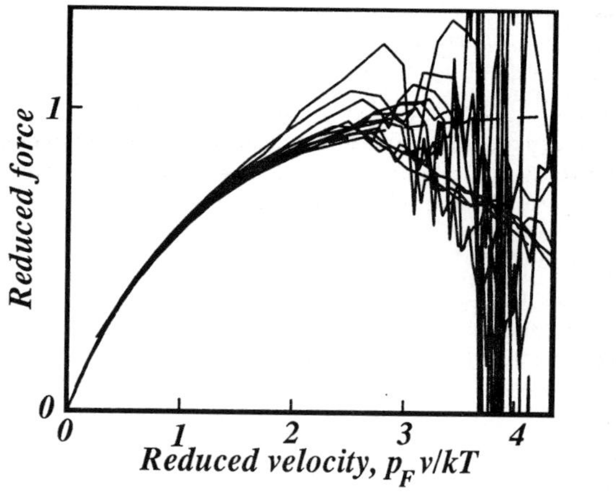

FIGURE 4
The measured reduced force plotted as a function of reduced velocity. The dashed line represents the calculated force given above.

The conclusion is that at low temperatures the force on a moving projectile in superfluid ^{3}He-B becomes independent of velocity when the velocity exceeds kT/p_F provided the pair-breaking velocity is not exceeded. This is a direct manifestation of the unusual dispersion curve of the quasiparticle excitations in the superfluid.

(1) A. F. Andreev, Zh. Eksp. Teor. Fiz. 46, 1823 (1964), [Sov. Phys. JETP 19, 1228 (1964)]
(2) R. J. Watts-Tobin and S. N. Fisher, this conference.
(3) J. P. Carney, A. M. Guénault, G. R. Pickett and G. F. Spencer, Phys. Rev. Lett. 62, 3042 (1989).
(4) S. N. Fisher, A. M. Guénault, C. J. Kennedy and G. R. Pickett, Phys. Rev. Lett. 63, 2566 (1989).

Physica B 165&166 (1990) 653–654
North-Holland

ON THE APPROACH TO CRITICALITY OF A VIBRATING, MACROSCOPIC OBJECT IN SUPERFLUID ^{3}HE-B

C.J. LAMBERT

Department of Physics, University of Lancaster, Lancaster LA1 4YB, United Kingdom

An intuitive picture of pair breaking by a vibrating wire is developed, which makes clear how quasi-particles can escape from the wire at critical velocities below the Landau critical velocity v_L. Assuming strong gap suppression at the wire surface, a critical velocity of $v^* = v_L/5$ is obtained, for ideal dipole flow around a wire of circular cross-section. In addition, a further transition to enhanced quasi-particle emission is predicted to occur at a velocity $v_1^* = v_L/3$. More generally, for non-dipole flow, $v^* = v_L/(1+2\alpha_l)$ and $v_1^* = v_L/(1+\alpha_l)$, where α_l characterizes the local superflow near the surface of the wire.

1. Introduction

An intuitive picture of pair breaking by a vibrating wire is developed, which makes clear how quasi-particles can escape from the wire at critical velocities below the Landau critical velocity v_L. Assuming strong gap suppression at the wire surface, a critical velocity of $v^* = v_L/5$ is obtained, for ideal dipole flow around a wire of circular cross-section. In addition, a further transition to enhanced quasi-particle emission is predicted to occur at a velocity $v_1^* = v_L/3$. More generally, for non-dipole flow, $v^* = v_L/(1+2\alpha_l)$ and $v_1^* = v_L/(1+\alpha_l)$, where α_l characterizes the local superflow near the surface of the wire.

Recent experiments (1) aimed at probing quantum properties of superfluid ^{3}He-B, have demonstrated that the response of micron size vibrating wires, immersed in the fluid, is a sensitive measure of quasi-particle (q.p.) properties and of the state of the superconducting order parameter. For example, measurements of the sub-critical, temperature dependent damping of these macroscopic objects (2) have recently provided firm evidence for Andreev scattering by superflows. While the behaviour of a vibrating wire at low drives is now at least qualitatively understood, the same is not true at higher velocities, where in contrast with theoretical predictions (3,4), pair breaking occurs at wire velocities of between 0.2 and 0.3 of the Landau critical velocity v_L. A full self-consistent treatment of the problem should take into account not only gap suppression at the surface of the wire but should also incorporate a condition for the escape of q.p.'s from the vicinity of the wire. The aim of this paper is to show that for a wire with strong gap suppression at its surface, such a treatment is easily obtained and yields critical velocities which are rather close to the measured values.

2 Analysis

Consider the superflow in the rest frame of the wire at zero temperature, described by a velocity field $\underline{v}(\underline{r})$. To avoid clutter, in what follows, all energies are measured in units of the Fermi energy $\epsilon_F = \frac{\hbar^2 k_F^2}{2M^*}$, momenta in units of $\hbar k_F$, lengths in units of k_F^{-1} and velocities in units of $v_F/2 = \hbar k_F/2M^*$. If the spatial change in the order parameter is slow on the scale of the coherence length, then the allowed q.p. energies can be obtained by solving the Bogoliubov-de Gennes equation locally, with a local superfluid velocity $\underline{v}_l$ and self-consistent local "gap" $\Delta_l(\underline{v}_l)$. To lowest order in the small quantities v_l, Δ_l, (k^2-1) this leads to locally allowed energies given by the positive values of

$$E_{\underline{k}} = \hat{\underline{k}} \cdot \underline{v}_l \pm \sqrt{(k^2-1)^2 + |\Delta_l(\underline{v}_l)|^2}$$

where $\hat{\underline{k}}$ is a unit vector parallel to $\underline{k}$. This expression does not of course imply that $\underline{k}$ is a good quantum number. Indeed locally, an eigenstate of energy E has the form of a superposition of plane waves, with $\underline{k}$ vectors satisfying $E_{\underline{k}} = E$. For this reason $\underline{k}$ is to be interpreted as the expectation value of the momentum of a q.p. In what follows, attention will be restricted to values of $\underline{k}$ parallel to the local velocity, since these are the q.p. states which determine the critical velocity v^*.

In order to compute v^*, we note that if $\underline{v}$ is the asymptotic superfluid velocity at large distances from the wire, then the local flow near the wire surface is of magnitude $v_l = \alpha_l\, v$. For dipole flow around a wire of circular cross-section, the maximum local velocity at the wire surface is $v_l = 2v$, so in the region where pair breaking occurs, $\alpha_l = 2$. Assuming (5) that in the vicinity of the wire surface, each component of the local gap satisfies $\Delta_l(0) << \Delta_\infty$, where Δ_∞ is the asymptotic value at large distances from the wire, leads to the asymptotic and local dispersion relations sketched in figure 1, for which a value $\alpha_l = 2$ was used. In drawing these curves, it was noted that for all $v < \Delta_\infty$ the asymptotic gap remains isotropic and velocity independent, while for $v_l >> \Delta_l(0)$, the self-consistent local gap $\Delta_l(v_l)$ vanishes.

For $\Delta_l(0) << \Delta_\infty$, pair breaking occurs almost immediately (on the scale of the Landau velocity $v_L = \Delta_\infty$), at a velocity v_0^* given by $\alpha_l v_0^* = \Delta_l(\alpha_l v_0^*)$. However the q.p.'s cannot escape, since the states at energies less than $E_\infty = \Delta_\infty - v$ are localised near the wire surface. To analyze the escape of q.p.'s, we restrict attention to slow changes in v_l, so that q.p. states transform adiabatically at constant k. Consider for example, the empty local q.p. states in figure 1(b) between energies $E_\infty(v)$ and $E_l(v) = v_l - \Delta_l(v_l)$. These are bound states with a momentum expectation value of $k \approx -1$, comprising a superposition of a particle and a hole with momenta ≈ -1 and a decaying contribution from evanescent waves arising from complex k vectors on the opposite side of the Fermi surface. Thus, for positive v, these states adiabatically decrease in energy as v increases, in accordance with equation (1). The transformation properties of bound states with energies less than $E_l(v)$ are more subtle, since contributions from both sides of the Fermi surface are present and in principle expectation values of the momentum which are very different from ± 1 can arise. In practice this is unlikely because a bound state can be regarded as a dynamical process in which a particle converting into a hole and vice versa at opposite sides of a velocity "potential well". During this conversion, the group velocity reverses, but the momentum remains almost

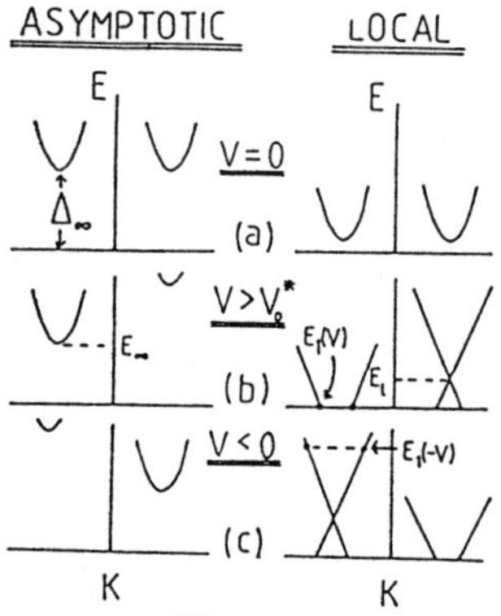

Figure 1

Asymptotic (left column) and local (right column) dispersion curves for a variety of velocities v.

unchanged. For this reason, we assume that for all q.p's $|k| \approx 1$ and therefore the bound states below E_l can be divided into those with $k \approx -1$, which in accordance with equation (1), decrease in energy as v increases and those with $k \approx +1$ which increase in energy. If this assumption is not correct, the final result for v^* is unaffected, provided at least some of the q.p.'s have $k \approx \pm 1$.

In accordance with equation (1), as v is slowly increased beyond v_0^*, states with positive k, filled by pair breaking at $E = 0$, are adiabatically lifted in energy. The critical velocity v_1^*, at which q.p.'s can escape from the wire, occurs when the energy of the highest filled state $E_l(v) = v_l - \Delta_l(v_l) \approx v_l$ equals the energy of the lowest outgoing channel $E_\infty(v)$. This yields a value $v_1^* = \Delta_\infty/(1+\alpha_l)$, where since $v_1^* < \Delta_\infty(0)$, the zero velocity asymptotic gap has been used.

While v_1^* is the velocity at which q.p.'s are first emitted in the presence of a monotonically increasing flow, it is not the critical velocity v^* of a vibrating macroscopic object. To see this, consider the effect of slowly reversing the superflow from v to $-v$. States filled by pair breaking on the $k \approx -1$ branch of figure 1(b) are adiabatically raised from

$$E_1(v) = -v_l + \sqrt{(k^2-1)^2 + \Delta_l^2(v_l)} = 0$$

to a maximum energy of

$$E_1(-v) = v_l + \sqrt{(k^2-1)^2 + \Delta_l{}^2(-v_l)}$$

Setting $\Delta_l(v_l) = \Delta_l(-v_l)$ and eliminating the square root from this equation yields $E_1(-v) = 2v_l = 2\alpha_l v$. The critical velocity v^* occurs when $E_1(-v^*) = E_\infty(v^*)$ which yields $v^* = \Delta_\infty/(1+2\alpha_l)$.

ACKNOWLEDGEMENTS

SERC support is acknowledged. I should also like to thank G.R. Pickett for several lively discussions relating to this work.

REFERENCES

(1) J.P. Carney, A.M. Guenault, G.R. Pickett and G.F. Spencer, Phys. Rev. Lett. 62 3042 (1989).

(2) R.W. Watts-Tobin (Preprint: these conference proceedings).

(3) D. Vollhardt and K. Maki, J. Low Temp. Phys. 31 457 (1978).

(4) D. Vollhardt, K. Maki and N. Schopohl, J. Low Temp. Phys. 39 79 (1980).

(5) L.J. Buchholtz, Phys. Rev. B33 1579 (1986).

Physica B 165&166 (1990) 655–656
North-Holland

VORTICES AND HOMOGENEOUSLY PRECESSING SPINS IN ROTATING ³He-B

V.V. DMITRIEV[+], Y. KONDO, J.S. KORHONEN, M. KRUSIUS, Yu.M. MUKHARSKIY[+], E.B. SONIN[*], AND
G.E. VOLOVIK[#]
Low Temperature Laboratory, Helsinki University of Technology, 02150 Espoo, Finland

NMR absorption by vortices in the homogeneously precessing spin domain is discussed in terms of a torque, which arises from spin-orbit coupling between the coherently precessing spins and the nonuniformity in the orbital order parameter in the vortex core. A new soft mode of the nonaxisymmetric vortex core has been observed to contribute to spin relaxation.

1. INTRODUCTION

In superfluid ³He-B local coherence of spin precession produces large NMR frequency shifts which are governed by the order parameter texture. The presence of an array of rectilinear vortices in the rotating superfluid gives rise to changes in the frequency shifts; one may thus study the structure of the vortices by their influence on the texture.

In contrast to these measurements at low rf exitation levels global coherence in spin precession can be created by the use of high rf excitation (1). In this resonance mode the spins are Homogeneously Precessing within a Domain (HPD) at a frequency which corresponds to the Larmor frequency ω_0 at the domain boundary separating the HPD mode from the region unperturbed by the resonance. In the rotating state an array of vortices produces additional enhancement of the rf absorption, found to be directly proportional to the total length of rectilinear vortices within the precessing domain (2). The additional absorption P_v depends critically on the vortex core structure: at the core transition temperature T_v the absorption increases discontinuously by a factor of 3 when the axisymmetric core transforms into its elliptically distorted form at $T < T_v$. In contrast, in conventional low level NMR the relative discontinuity in the frequency shift at T_v is only ~25%. In this case the magnetic anisotropy acquires from the superflow circulating in the out-of-core region a large contribution which is independent of the core structure and does not contribute to the HPD absorption (3).

Therefore the HPD absorption provides an exceptionally interesting and informative new tool to probe the vortex core more directly than before. Here we shall offer a first tentative explanation of some aspects of the intricate phenomena connected with vortex absorption in the HPD mode.

2. MEASUREMENTS

Some experimental details have been reported in Refs. (2), (3), and (4). The central observations are summarized in Figs. 1 and 2 for the case when the HPD completely fills the NMR cell. In Fig. 1 the measured vortex absorption is shown as a function of temperature for two fields: Within our experimental uncertainty of ±15%, determined by the procedure of comparing rf absorptions measured with the NMR spectrometer tuned to different frequencies (4), P_v appears to be independent of H. Its dependences on the two other externally controlable variables are such that P_v decreases with increasing temperature and decreasing pressure (2). In Fig. 1 the core transition is displayed as a prominent

discontinuity, which for still unknown reasons is slightly shifted below its usual value T_v=0.60 T_c at 29.3 bar. In Fig. 2 the dependence of P_v on the inclination angle η of **H** with respect to the rotation axis Ω is given: $P_v \propto (1+ A \cos^2\eta)$ where the constant A depends on the core structure.

A most remarkable phenomenon is illustrated in the insert of Fig. 2. In the region close to the axial field orientation, when $|\eta| \leq 20°$, P_v assumes a time dependent value below T_v: after a change in the rotation velocity Ω, which eg would be sufficient to cause the vortex lines to become unpinned, P_v is initially up to 15% larger. It then relaxes exponentially, with a time constant of ~2 min, to a lower level which remains stable as long as Ω is kept constant. A similar relaxation in the absorption is observed if the HPD mode is suddenly switched on. No time dependent signal, within our measuring resolution (4), has been observed above T_v.

3. VORTEX ABSORPTION

While a detailed understanding of the HPD vortex absorption has not yet been worked out we shall, nevertheless, argue that two broken symmetry properties of the vortex core and at least two different dissipation mechanisms are crucial. Both types of B-phase vortices are characterized by broken parity in the core and the low-temperature vortex, in addition, by broken axisymmetry of the core (5). The low-temperature vortex is bi-axial with a double-core; the unit vector **b** is used here to note the orientation of the long axis of the core in the transverse plane. Thus in addition to the orbital anisotropy of the superfluid properties along the vortex line, anisotropy perpendicular to the vortex axis also appears and couples to the uniform spin precession via the dipolar interaction.

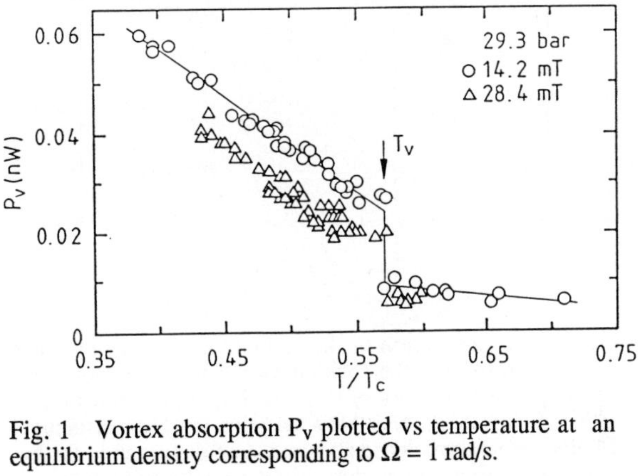

Fig. 1 Vortex absorption P_v plotted vs temperature at an equilibrium density corresponding to $\Omega = 1$ rad/s.

+ Institute for Physical Problems, Moscow 117334
* Ioffe Physical Technical Institute, Leningrad 194021
# Landau Insitute for Theoretical Physics, Moscow 117334

The dominant dissipation we attribute to the Leggett-Takagi mechanism, although it appears that other contributions, such as spin diffusion, may also be effective. A common ingredient in all cases is the fact that the influence from the core with broken parity extends well outside the region of the hard core order parameter nonuniformity and substantially enhances spin relaxation. This soft tail decays, as noted in Ref. (6) for the axisymmetric v-vortex with broken parity, as 1/r outside the core. The tail is not present in connection with the nonuniform order parameter distribution at the chamber wall and thus the contribution from surface relaxation to HPD absorption is substantially less than from vortices (4).

The orbital structure of the vortex core is highly rigid since it is governed by the large condensation energy of the core. In the presence of the homogeneous spin precession this gives rise to a local torque on the order parameter $\mathbf{R}(\mathbf{n},\Theta)$. In the vicinity of the core the order parameter deviates by $\delta\mathbf{R}=\Theta\times\mathbf{R}$ from its precessing value $\mathbf{R}(\mathbf{n}_o(t),\Theta_o)$ in bulk liquid where $\Theta_o=104^o$ and the unit vector $\mathbf{n}_o(t)$ precesses at ω_o in the plane $\perp\mathbf{H}$. $\Theta(\mathbf{r},t)$ can be written in terms of two parameters, A_1 and A_2, which are on the order of the core radius R_c, with $A_2=A_1$ for the axisymmetric vortex:

$$\Theta(\mathbf{r},t) = (A_1\, \mathbf{R}_y(t) \cos\phi - A_2\, \mathbf{R}_x(t) \sin\phi)/r\,. \quad (1)$$

Here ϕ is the azimuthal angle around the vortex axis and $\mathbf{x}$ is chosen to lie along $\mathbf{b}$. The resulting change in spin precession $\delta S = \chi d\Theta/dt$ leads to a Leggett-Takagi dissipation $P_v \sim \int(d\mathbf{S}/dt - \mathbf{S}\times\mathbf{H})^2 dV$, which has the measured dependence $P_v \propto [(A_1)^2+(A_2)^2 \cos^2\eta]$ on the orientation of $\mathbf{H}$. From Fig. 2 it can be inferred that below T_v the measured asymmetry of the core is rather large, $(A_2)^2/(A_1)^2 \sim 5$, but less than estimated from the calculations for the Ginzburg-Landau regime in Ref. (7). Above T_v the dependence on the orientation of the field is $\propto (1+\cos^2\eta)$ in agreement with Fig. 2. The Leggett-Takagi relaxation is estimated to give for P_v a value of order 0.1 nW, which is also consistent with experiment.

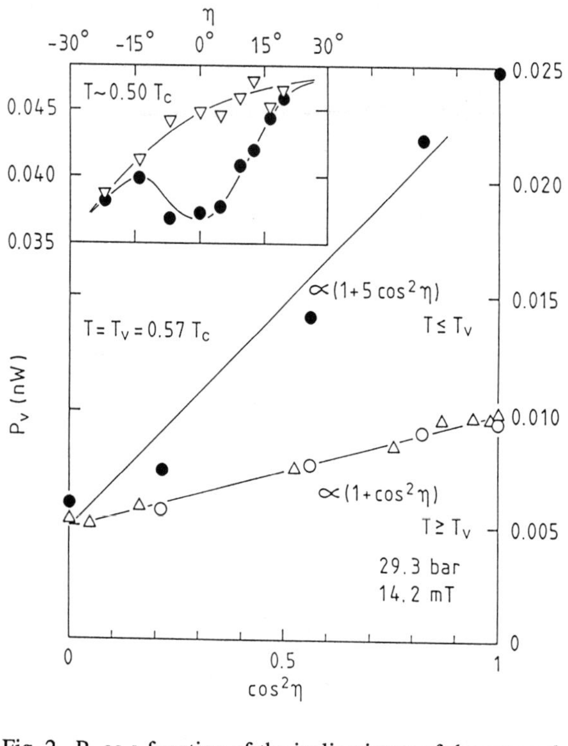

Fig. 2. P_v as a function of the inclination η of the external field with respect to the rotation axis. The insert shows the time dependent behavior of P_v at small η: ∇, P_v immediately after a change in Ω; ●, stable absorption level at constant Ω. The lop-sided asymmetric appearance of the graph in the insert is due to slow temperature drift during the measurement.

Another source of vortex absorption appears to involve a new type of relaxation mechanism, namely a soft mode associated with the orientation of $\mathbf{b}$ and with hydrodynamic dissipation via torsional motion of the nonaxisymmetric core. The B-phase anisotropy axis $\mathbf{n}$ precesses together with the magnetization and exerts via the dipolar interaction a torque on the rigid orbital core structure which gives rise to an energy term of the form $F_o = T_o(\mathbf{nb})^2 = T_o\cos^2(\omega_o t-\beta)$. Here β is the orientation of $\mathbf{b}$ in the transverse plane and T_o is of order of the dipole energy integrated over the core area πR_c^2. The torque $T = \partial F_o/\partial\beta = T_o \sin 2(\omega_o t-\beta)$ causes the asymmetric core to start a rocking oscillation about its equilibrium orientation, given by the equation of motion

$$f\,(\partial\beta/\partial t) + k\beta = T_o \sin\, 2(\omega_o t - \beta)\,. \quad (2)$$

Here $f=\alpha(\rho_s\rho_n/\rho)\kappa R_c^2$, α is a frictional parameter similar to the mutual friction parameter B which determines the losses in the translational motion of vortices, and k is a parameter characterizing the moment attempting to restore the equilibrium position of $\mathbf{b}$, defined by pinning, magnetic anisotropy in a tilted field, etc. Solving Eq. (2) to first order in $T_o/(\omega_o f)$ and calculating the average $\langle T\rangle = T_o^2\omega_o f/(2(\omega_o f)^2+k^2/2)$ to second order, one obtains $P_v = N\ell\,\omega_o\langle T\rangle$ where $N\ell$ is the total length of vortices in a domain with length ℓ. Note that this result for P_v does not depend on H if k is small.

In addition to the rocking oscillation of $\mathbf{b}$, a second order effect will be a slow rotational drift of $\mathbf{b}$ in the direction of the spin precession. If the ends of the vortex are pinned $\mathbf{b}$ will become multiply wound along the vortex axis. With increasing spiraling k increases in Eq. (2) suppressing the absorption. This may be the origin of the time dependent absorption signal. Suppose $P_v(t=0) - P_v(\infty) \approx 10^{-2}$ nW from Fig. 2, $R_c \approx 10^{-5}$ cm and $\alpha \approx 1$ (order of magnitude of the experimental value of B) one obtains $T_o \approx 3\cdot10^{-11}$ erg/cm from Eq. (2) while the value calculated from Ref. (7) is $3\cdot10^{-12}$ erg/cm in reasonable agreement. In a tilted field the magnetic anisotropy of the core gives rise to a torque with an orientational energy $F_H = T_H(\mathbf{h}\cdot\mathbf{b})^2$, where $\mathbf{h} = \mathbf{H}/H$, $T_H = a\lambda_2 H^2/(5n)$, and n is the vortex density. At about $\eta \sim 20^o$, $F_H = T_o^2/(2f\omega_o)$ and thus at larger inclinations $\mathbf{b}$ is fixed by the external field.

ACKNOWLEDGEMENTS

We wish to thank O.V. Lounasmaa and E.V. Thuneberg. This work has been supported by the Körber Stiftung, the Finnish Academy and the USSR Academy of Sciences.

REFERENCES

(1) A.S. Borovik-Romanov, Yu.M. Bunkov, V. Dmitriev, and Yu.M Mukharskiy, in Quantum Fluids and Solids 1989, AIP Conf. Proc. Vol. 194, N.Y. 1989, p. 15.
(2) J.S. Korhonen, Z. Janu, Y. Kondo et al, ibid. p. 147.
(3) J.S. Korhonen, V.V. Dmitriev et al, this volume.
(4) Y. Kondo, J.S. Korhonen et al, this volume.
(5) M.M. Salomaa and G.E. Volovik, Rev. Mod. Phys. 59 (1987) 533 .
(6) Y. Hasegawa, Progr. Theor. Phys. 73 (1985) 1258.
(7) E.V. Thuneberg, Phys. Rev. B 36 (1987) 3583.

Physica B 165&166 (1990) 657–658
North-Holland

Superfluid ^{3}He - B in a Realistic Thin Film: Gap Function

Louis J. Buchholtz

*Departmet of Physics, California State University , Chico
Chico, California 95926*

This paper reports a calculation of superfluid in a planar film with one diffusely reflecting surface and a facing specular surface. The intent of this work is to demonstrate the use of the nonlinear Quasiclassical boundary conditions in a fully self-consistent calculation of the superfluid gap function for a realistic anduseful geometry. The diffusely scattering boundary surface was represented by the versatile "RandomlyRippledWall" model. We display results for various surface qualities and film thicknesses.

1. Introduction

We have performed a fully self consistent calculation of the gap function for superfluid ^{3}He-B in a "realistic " thin film. We employ the quasiclassical equations throughout along with the full non-linear boundary conditions of reference [1]. The "Randomly Rippled Wall" model is used to describe a diffusely reflecting surface on which the film rests and the facing surface is taken as specular. The primary significance of this effort is technical in nature. We demonstrate that the body of descriptive equations, inclusive of the highly non-linear boundary conditions, lends itself to very straightforward and stable numerical solution. Moreover, these very nonlinearities scrupulously preserve those properties and the analytic structure necessary for a physical solution. The specific results displayed below are interesting in detail but unremarkable in form. We begin with a brief comment on the equations and the solution algorithm. This is followed by two sequences of graphs. These have been arrayed so as to better illustrate the separate effects of surface roughness and finite film width .

2. Equations and Algorithm

The basic equations are taken from reference [1] and include a differential equation and normalization condition for the quasiclassical Green's function along with a boundary condition. In turn, the boundary condition is defined in terms of a surface t-matrix which is modeled for the problem under discussion. For the diffusely reflecting boundary of the present problem we employ the versatile "Randomly Rippled Wall" formulation and a two parameter (Gaussian) model of the surface "bump" distribution. The two parameters are designated "a" and "b" and represent, so to speak, the average height and width of surface irregularites in units of the inverse Fermi wave vector. Finally, the gap function is evaluated with the p-wave gap equation which closes the circle of equations.

The numerical solution is effected iteratively and posed no stability problems. Indeed, the system seems to admit a very high degree of over-relaxation. The principle feature of note is that describing a diffuse surface requires solution of the differential equation for a large number of different quasiparticle directions simultaneously. The actual algorithm is thus two iterative procedures one inside the other. Complete details are to be published elsewhere.

3. Results

We display first the effects of surface roughness. Figures 1. - 5. show B-phase superfluid in a semi-infinite geometry bounded by a rough wall at a reduced temperature of T/Tc = .733 . Once again, the parameters "a" and "b" represent surface "bump height" , and "width". The length scale is given in units of the temperature dependent coherence length and the gap is scaled by its bulk value. We point out, especially, that all gap components are suppressed and the more so as bump height increases relative to width.

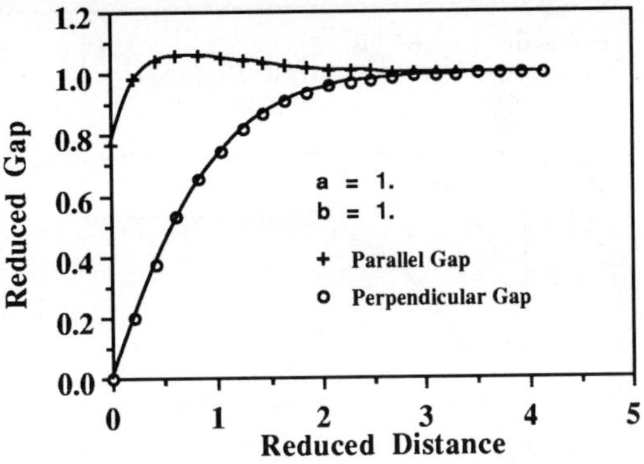

Figure 1.

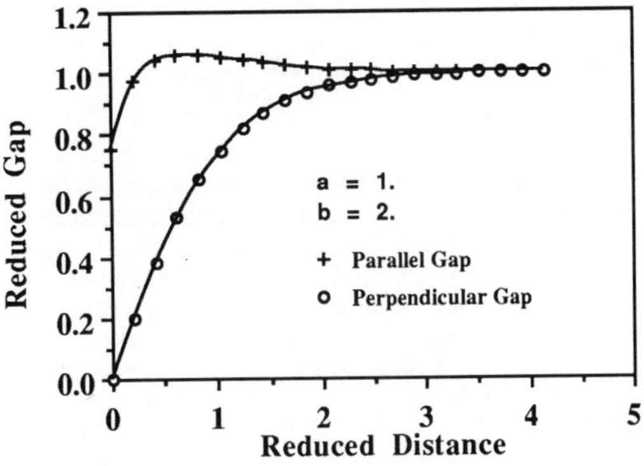

Figure 2.

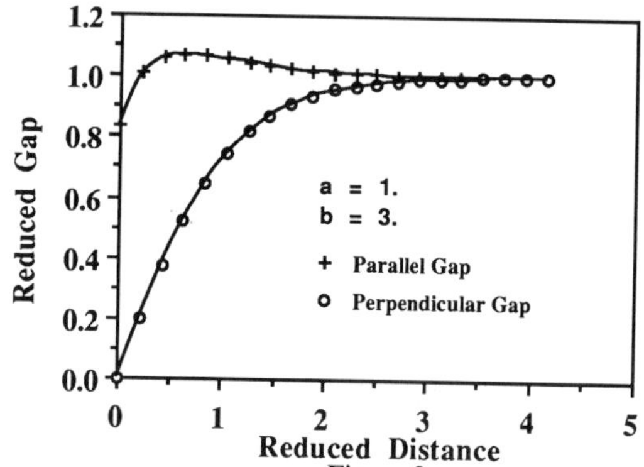

Figure 3.

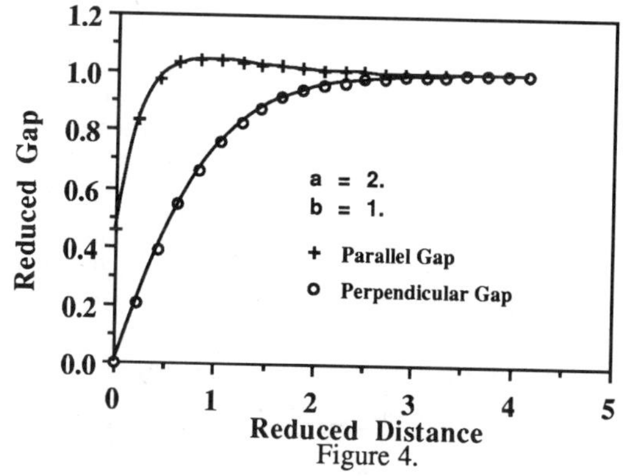

Figure 4.

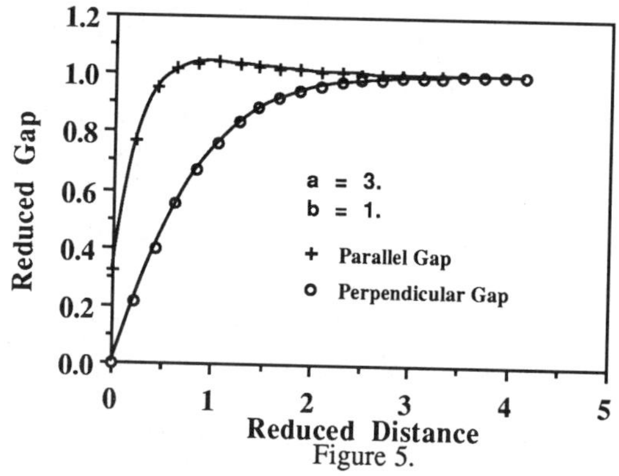

Figure 5.

Second, we display the gap function as it appears in a slab. The rough boundary is on the left and the specular boundary on the right. The asymmetry engendered is clearly evident. The unit of length is taken as the temperature independent coherence length and the gap is once again scaled by its bulk value. For this sequence we take a = 1 , b = 1, and T/Tc = .733 , and the graphs depict three different widths.

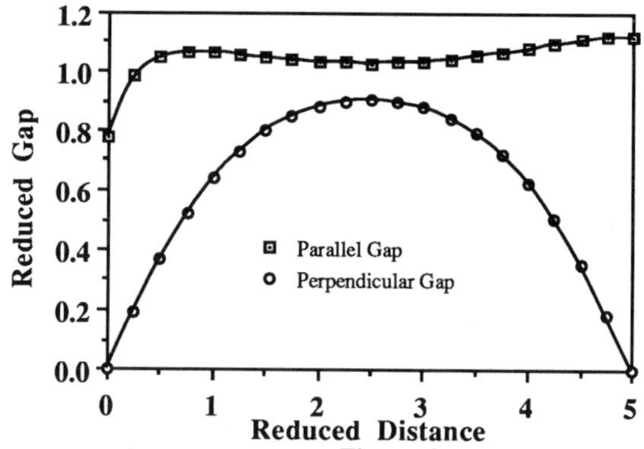

Figure 6.

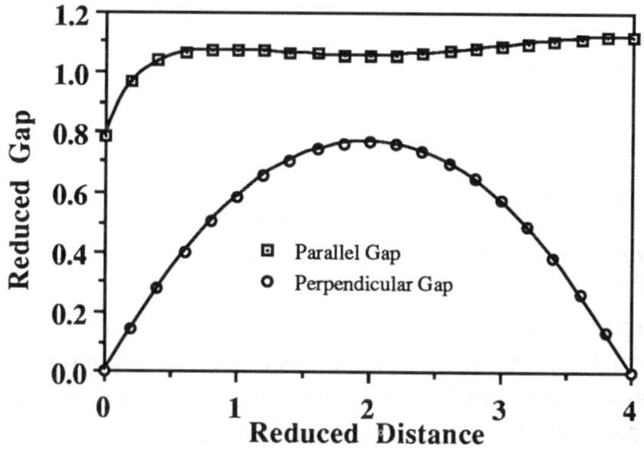

Figure 7.

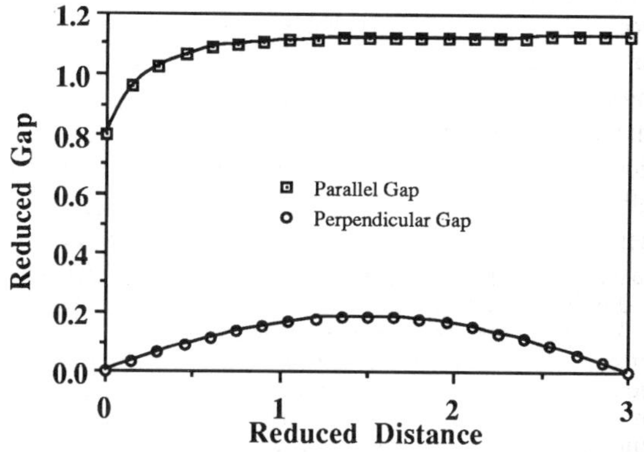

Figure 8.

4. Summary

Diffusely reflecting walls suppress all gap components of the p-wave gap-function in the vicinity of the surface. In conjunction with the geometry of a narrow slab, diffuse boundaries exercise a profound effect on the gap and corresponding superfluid quantities. The boundary conditions of reference [1] , though forbidding in appearence, provide a well behaved system of equations amenable to numerical evaluation.

References

[1] L. J. Buchholtz and D. Rainer, *Z. Phys. B* **35,** 151 (1979)

Physica B 165&166 (1990) 659–660
North-Holland

Superfluid ^{3}He - B at a Rough Surface: Density of States

Louis J. Buchholtz

Department of Physics, California State University Chico, Chico, California 95926

The density of states for superfluid ^{3}He-B at a rough wall is calculated from the quasiclassical equations using the full nonlinear boundary conditions and the "Randomly Rippled Wall " model. Increased surface roughness leads to a substantially enhanced total density of states at sub gap energies as well as to a more complicated spectrum structure. The calculation provides a sensitive test of boundary condition validity and demonstrates the ready applicability of the nonlinear boundary conditions.

1. Introduction

This paper reports the results of a fully self consistent numerical computation of the density of states for superfluid ^{3}He - B at a diffusely reflecting surface. The calculation employs quasiclassical Green's functions throughout and models the surface with the "Randomly Rippled Wall" model. The central conclusions of this work are as follows. First, we provide a strong affirmation of the original nonlinear boundary conditions as proposed a decade ago [1].These boundary conditions were unique among all those tried in that they preserved delicate analytic qualities required for a physical solution. Second, we demonstrate that, however forbidding these conditions may appear at the outset, in practice they manifest exceptional "user friendliness". Their application in this project is outlined below. Third, we display the density of states for typical energies to show just how strongly this quantity depends on the detailed spatial dependence of the gap.

2. Basic Equations

We collect here the basic equations of the theory. These consist of a differential equation (the E.L.O.E. equation) and normalization condition along with boundary conditions for the quasiclassical Green's function. As usual, we express the Green's function in 4x4 form:

$$\hat{g} = \begin{pmatrix} g & f \\ \tilde{f} & \tilde{g} \end{pmatrix}$$

The basic equations now appear:

$$i\hbar v_F \hat{\mathbf{k}}\cdot\vec{\nabla}\hat{g} + i\varepsilon_n(\hat{\tau}_3\hat{g} - \hat{g}\hat{\tau}_3) - (\hat{\tau}_3\hat{\sigma}\hat{g} - \hat{g}\hat{\sigma}\hat{\tau}_3) = 0 \qquad (1)$$

$$\left(\hat{\tau}_3\hat{g}(\hat{\mathbf{k}};\vec{\mathbf{R}}:\varepsilon_n)\right)^2 = -\pi^2\hbar^2\,\mathbb{1} \qquad (2)$$

$$\Delta\hat{g}\,\hat{\tau}_3\,\Delta\hat{g} = 0 \qquad (3)$$

where we define

$$\Delta\hat{g} = \frac{-i}{\hbar v_F \hat{\mathbf{k}}\cdot\hat{\mathbf{n}}}\left(\hat{\tau}_3\hat{t}\,\hat{g} - \hat{g}\,\hat{t}\,\hat{\tau}_3\right)$$

$$\hat{\sigma} = \begin{pmatrix} 0 & \vec{\Delta}\cdot\vec{\sigma}i\sigma_2 \\ -i\sigma_2\vec{\Delta}^{\dagger}\cdot\vec{\sigma} & 0 \end{pmatrix}$$

The gap function $\vec{\Delta}\cdot\vec{\sigma}i\sigma_2$ is taken as B - phase in the bulk which then generalizes near the wall. The quantity t is the surface t-matrix which is taken from the "Randomly Rippled Wall" model [1]

The density of states is extracted from the Green's function by analytically continuing it from the Matsubara energies to the real energy axis according to the prescription:

$$n(\hat{\mathbf{k}}:\varepsilon) = -\frac{1}{\pi}\,\mathrm{Im}\,\frac{1}{2}\mathrm{tr}_2\, g \qquad i\varepsilon_n \rightarrow \varepsilon + i\eta \quad .$$

3. Algorithm

The numerical evaluation of the system proceeds iteratively. We assume the gap function to have been previously determined. A central difficulty stems from the fact that both the normalization condition and the surface t-matrix involve the Green's function nonlinearly. This implies that a variety of solutions is allowed - from among which we must discern the true physical solution. Our method here is to begin the process far enough from the real energy axis that the solutions are well separated. Having selected the correct solution, one gently "descends" to the real axis by gradually decreasing the imaginary part and following the solution. Since the t-matrix depends on the amplitudes for all angles, one must discretize the angle intervals and simultaneously solve the E.L.O.E. equations for each set of angle values. In practice, the procedure proved very stable. However, the large number of simultaneous equations being solved meant that an enormous quantity of stored numbers had to be manipulated. This proved to be the limiting factor in the resolution achieved.

4. Results

The suppression of the p-wave gap function near a surface introduces a significant enhancement of the local density of excitation states at energies less than the bulk gap. This is true at specular surfaces and very much more pronounced at diffusely reflecting walls. In the specular case the density of states spectrum contains angle dependent delta-functions at sub gap energies. In the diffuse case these have broadened into bands which still contain a singular nonsymmetric peak. Regrettably, the structure in the immediate vicinity of the bulk gap was so rich that the discretization required to resolve it could not be handled by the available computing facilities. We display below the results for T/Tc = .733 and with the Randomly Rippled Wall roughness parameters a = 1. and b = 1.. All energies are in units of the bulk gap

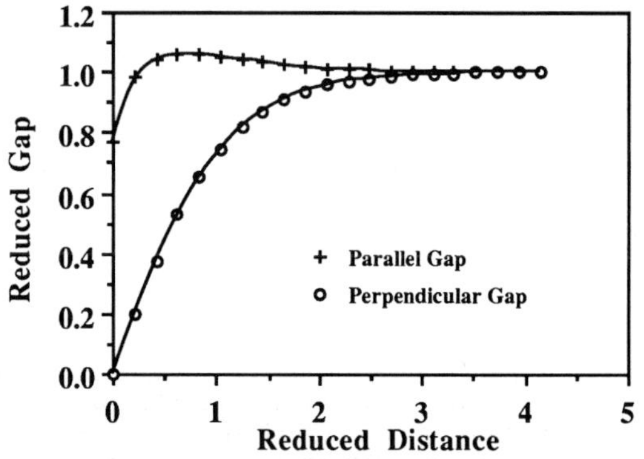

Figure 1.

Superfluid gap components near a diffusely reflecting wall. Length given in units of the temperature dependent coherence length.

Next, we display the local density of states at the wall for various energies. The graphs depict the density of states (in units of N(0)) versus the angle of quasiparticle momentum as measured from the surface normal. The designation D.O.S. signifies the total (integrated) density of states.

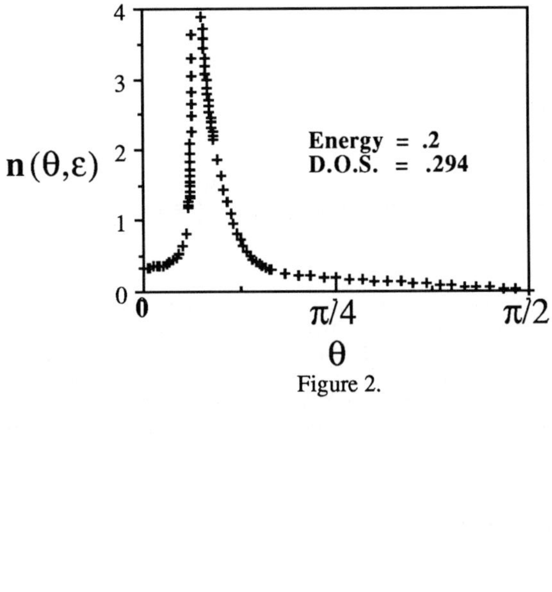

Figure 2.

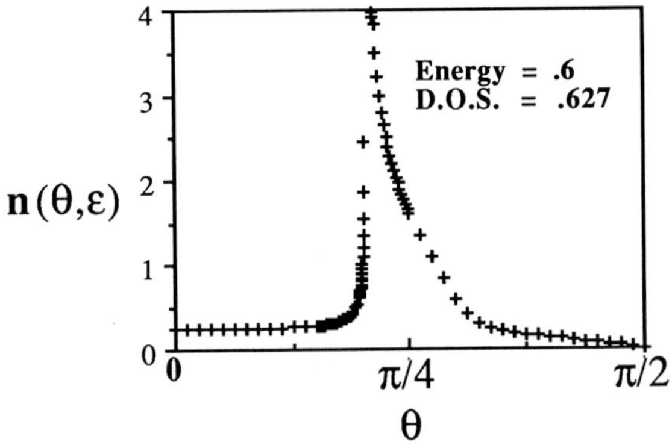

Figure 3.

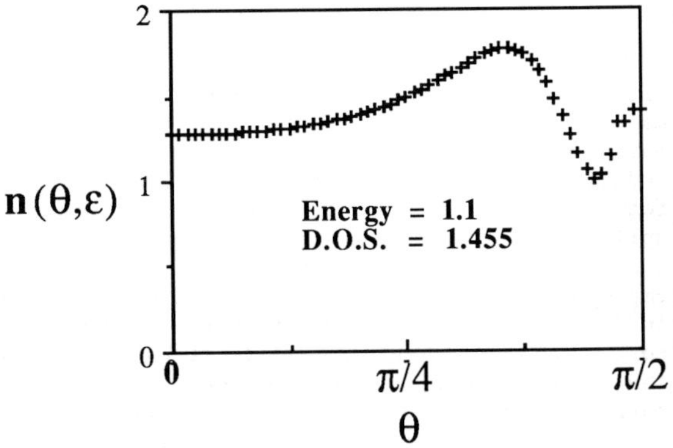

Figure 4.

5. Summary

The original nonlinear boundary conditions of reference [1] provide precisely the correct number of conditions for the quasiclassical differential equations. Moreover, they preserve the delicate analytic structure required of a physical solution. Their unusual character would appear forbidding but proved unusually "user friendly" and stable in practice. Surface roughness does indeed engender greatly increased spectral density and spectrum structure at sub-gap energies.

References

[1] L. J. Buchholtz and D. Rainer, *Z. Phys. B* **35,** 151 (1979)

Physica B 165&166 (1990) 661–662
North-Holland

Observation of New Structures in the Collective Mode Spectrum of ^{3}He-B

Zuyu Zhao*, S. Adenwalla*, M.C. Shih*, J.B. Ketterson* and Bimal K. Sarma†

*Department of Physics and Astronomy, Northwestern University, Evanston, IL 60208 USA
†Physics Department, University of Wisconsin at Milwaukee, Milwaukee, WI 53201 USA

The single-ended, cw, impedance technique has been employed to study the squashing mode of superfluid ^{3}He-B in a sound cell with a path length of 381 μm with ultrasound frequencies of 115.8 MHz and 141.6 MHz. A doublet splitting of the squashing mode has been observed and the magnitude of the splitting depends on the pressure. In addition, a "bump-like" structure, preceding the pair-breaking edge by about 3.5 MHz, has been observed with a sound frequency of 64.3 MHz in an applied magnetic field at pressures ranging from 4 to 7 bar.

We have observed two new features associated with the collective mode structure of ^{3}He-B using techniques similar to those of Ref. 1, but with the path length of the sound cell reduced by a factor of twenty. The first new feature to be discussed here is a doublet splitting of the squashing mode peak. Sound frequencies of 115.8 MHz and 141.6 MHz were employed. A typical temperature trace is shown in Fig. 1. The step-like feature in the trace corresponds to the superfluid transition and the onset of oscillations occurs at a temperature where acoustic pair breaking disappears.(2) These oscillations arise from phase velocity changes of the sound associated with the approach to the $J = 2^-$ collective mode. The oscillations disappear in the vicinity of the collective mode resonance due to the high attenuation; the peak itself, which is here observed to be split, actually arises from a shift in the acoustic impedance.

The splitting has been studied for pressures in the range from 19.2 bar to 27.7 bar in zero magnetic field. Measurements with the magnetic field perpendicular to $\vec{q}$ (the sound propagation direction) were also performed up to 1.36 kG at a single pressure of 27.7 bar. The doublet splitting is observed in both zero (Fig. 2) and finite magentic field (Fig. 1); i.e., there is no threshold field for the onset of the splitting. A Lorentzian fit of the form

$$y = \frac{\beta\cos\phi}{(T-T_0)^2+(w)^2} + \frac{(\beta/w)(T-T_0)\sin\phi}{(T-T_0)^2+(w)^2} \qquad (1)$$

yields a good representation of the data; here T_0, w and ϕ are, respectively, the peak position, the half-width at half-maximum (HWHM) and the phase of the detected signal (the appearance of an unknown phase angle results from peculiarities of the c.w. acoustic impedance method which we will not discuss). The area under a Lorentzian is given by $\frac{\pi\beta}{w}$.

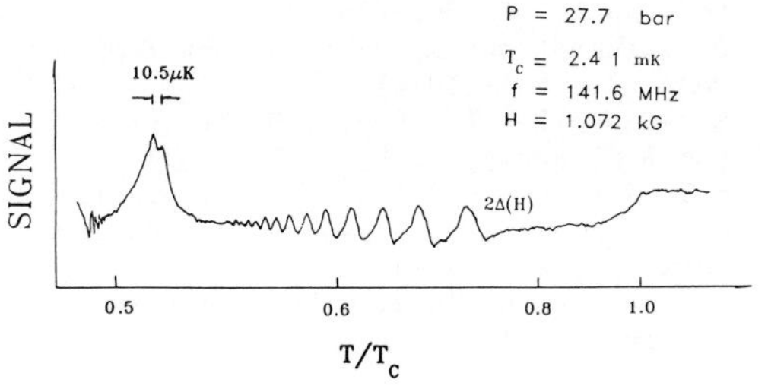

FIGURE 1
Typical demagnetization traces of the acoustic impedance signal. The traces are in real time, with the approximate temperatures as shown.

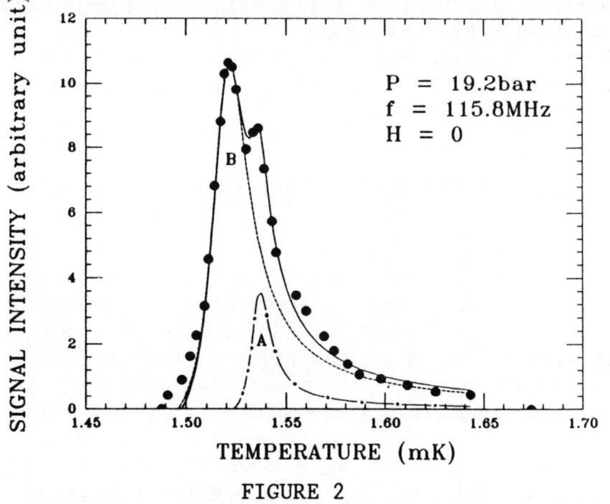

FIGURE 2
A Lorentzian fit to the doublet splitting of the imaginary squashing mode. The fits to the two components of the collective mode, referred to as A and B, are shown as the dot-dash and dashed lines respectively. The solid line is the resultant curve. The experimental data are shown as cirlces.

0921-4526/90/$03.50

Fig. 2 shows an example of a conventional non-linear least square fit to a temperature sweep performed at a pressure of 19.2 bar with a sound frequency of 115.8 MHz in zero field. In order to best represent the splitting feature, the fit was performed in the range 1.51 mK to 1.55 mK, although the resulting curve is plotted over a wider temperature range. Two Lorentzian lines were assumed which are depicted by the dot-dash line (A) and dashed line (B). The resultant trace, shown as the solid line, is in reasonably good agreement with the experimental data, which are depicted by the filled circles. The fitting yields the following parameters: $\phi = 25.5°$

$$\beta_A = 1.3\text{x}10^{-4},\quad T_{0A} = 1.533\text{K},\quad w_A^{-1} = 1.73\text{x}10^2\text{K}^{-1},$$

$$\beta_B = 13.7\text{x}10^{-4},\quad T_{0B} = 1.517\text{ K},\quad w_B^{-1} = 0.91\text{x}10^2\text{K}^{-1}.$$

The coupling strength, λ, of the collective mode components will be assumed to be proportional to the area under the corresponding Lorentzian; by this criteria the ratio of the coupling strength of these two components is $\lambda_A/\lambda_B \sim 0.18$. Avenel et al.(3) have generated a model to interpret the acoustic impedance signal when only one of the squashing mode components couples to zero sound; this model could easily be generalized. We adopted a Lorentzian form for simplicity. The splitting was observed to increase as the pressure was increased.

One mechanism for the splitting is superflow, which is always present(4) at some level due to the existence of temperature gradients within the acoustic cell.(5) A calculation of the superflow induced splitting of the squashing mode(6) predicts a <u>three fold</u> splitting with the frequencies given by

$$\omega_2^2 = \omega_{00}^2 + \frac{3(\alpha_0+3)}{5}\Omega_0^2,\qquad (J_z = \pm 2)\qquad (2\text{-}1)$$

$$\omega_1^2 = \omega_{00}^2 + \frac{6\alpha_0+11}{10}\Omega_0^2,\qquad (J_z = \pm 1)\qquad (2\text{-}2)$$

$$\omega_0^2 = \omega_{00}^2 + \frac{2\alpha_0+3}{2}\Omega_0^2,\qquad (J_z = 0)\qquad (2\text{-}3)$$

with the ordering $\omega_2 > \omega_1 > \omega_0$, where $\omega_{00} = \sqrt{12/5}\Delta(T)$; α_0 and Ω_0 are parameters that measure the gap distortion transverse, $\Delta_\perp^2 = \Delta^2 + \Omega_0^2$, and parallel, $\Delta_\parallel^2 = \Delta^2 + \alpha_0\Omega_0^2$, to the direction of superflow. Both α_0 and Ω_0 are functions of the superflow velocity, v_s, and the normalized temperature T/T_c; however it is extremely difficult to estimate v_s. Reasons for our only observing a doublet splitting may be that two of the components are nearly degenerate or that the coupling to the third component is much weaker. (A second factor

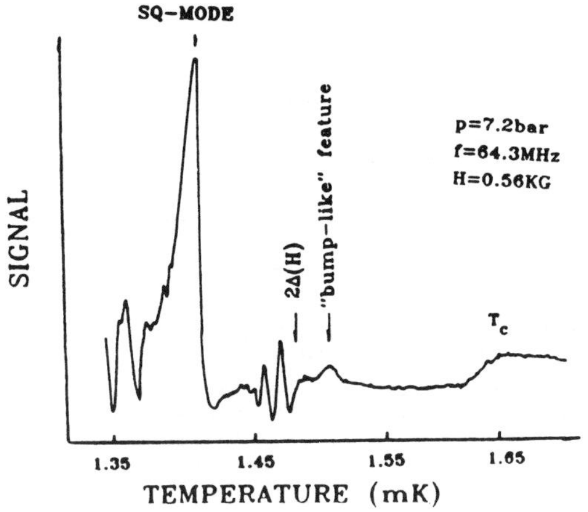

FIGURE 3
A "bump-like" structure in the vicinity of $2\Delta(H)$.

leading to a three fold mode splitting is dispersion and to lowest order the dispersion and superflow effects are additive.)

A second feature we observed is a bump-like structure (Fig. 3) preceding the pair-breaking edge by 20μK with an external magnetic field perpendicular to the sound propagation direction. This structure was observed at 64.3 MHz for pressures between 4 and 7; it was absent in zero field. The data cluster around the position of the $J_z = -1$, $J = 1^-$ mode predicted by Schopohl et al.(7)

A more lengthy discussion of the above phenomena can be found in Ref. 8. This work was supported by the National Science Foundation under grant DMR 89-07396.

REFERENCES

(1) B.S. Shivaram, M.W. Meisel, B.K. Sarma, W.P. Halperin and J.B. Ketterson, J. Low Temp, Phys. 63, 1/2 (1986).
(2) S. Adenwalla, Z. Zhao, J.B. Ketterson and B.K. Sarma, Phys. Rev. Lett. 63, 1811 (1989).
(3) O. Avenel, L. Piche, W.M. Saslow, E. Varoquaux and R. Combescot, Phys. Rev. Lett. 47, 803 (1981).
(4) D. Volhardt and K. Maki, J. Low Temp. Phys. 31, 457 (1978).
(5) S. Adenwalla, Z. Zhao, J.B. Ketterson and B.K. Sarma, JLTP 76, 1 (1989).
(6) M.Y. Nastenka and P.N. Brusov, preprint.
(7) N. Schopohl and L. Tewordt, J. Low Temp. Phys. 45, 67 (1981).
(8) Z. Zhao, Ph.D. Thesis, Northwestern University (1990).

Physica B 165&166 (1990) 663–664
North-Holland

MODES OF SUPERFLUID ^{3}He IN THE ENTIRE HYDRODYNAMIC REGION

Michael GRABINSKI* and Mario Liu#

* California Institute of Technology, Pasadena, CA 91125, USA

\# Universitat Hannover, 3000 Hannover 1, FRG

Dispersion relations for hydrodynamic modes are normally derived in the limit of small damping. This limit covers the entire hydrodynamic region as long as the region conincides with the small damping region (e.g., in ^{4}He-II). This is not the case in superfluid ^{3}He. In this paper the dispersion relations of second sound and sq-mode are generalized throughout the hydrodynamic region. It might have experimental implications for any ^{3}He experiment dealing with frequencies of Hz and higher. They are discussed in detail.

Second sound may be taken as a fingerprint for superfluidity (1). Its dispersion relation

$$q_2 = \pm \frac{\omega}{c_2}(1 + i\omega\tau_2) \tag{1}$$

is only valid in the limit $\omega\tau_2 \ll 1$. (τ_2 is a combination of various transport coefficients and c_2 is the second sound velocity (1).) In ^{4}He-II, τ_2 is of the same order as the quasi particle collision time $\tau_{micr} \approx \xi/c_2$, ξ being the mean free path. Eq. (1) is therefore valid throughout the hydrodynamic region, $\omega\tau_{micr} \ll 1$. In superfluid ^{3}He (2) $\tau_{micr} \approx \xi/\upsilon_p$ (3), where υ_p is the averaged velocity of the Bogoliobov quasi-particles. τ_2 of ^{3}He is still of order ξ/c_2, but $\upsilon_p \sim \upsilon_F \gg c_2$. $\omega\tau_2 \ll 1$ may be violated, although $\omega\tau_{micr} \ll 1$ still holds. It is therefore worthwhile to generalize Eq. (1) to arbitrary $\omega\tau_2$ (but, of course, $\omega\tau_{micr} \ll 1$). In order to get the general result one must solve the characteristic polynomial (1) without expanding in $\omega\tau$. In the hydrodynamic region, $\omega\tau_{micr} \ll 1$, first sound damping $\propto\omega\tau_1$ is small ($\tau_1 \ll \tau_{micr}$) and may be neglected. In doing so one finds the simple generalization of Eq. (1):

$$q_{2,3}^2 = -\frac{1 - i2\omega\tau_2}{2\lambda^2} \mp \sqrt{\left(\frac{1 - i2\omega\tau_2}{2\lambda^2}\right)^2 + \frac{\omega^2}{c_2{}^2\lambda^2}} \tag{2}$$

Together with the first sound dispersion relation $q_1 = \pm\omega/c_1$, Eq. (2) gives the complete mode structure in the entire hydrodynamic region. The newly introduced shock length $\lambda \approx 2c_2\tau_2$ (for a precise definition see Refs. (4),(5)). Its meaning becomes clear by expanding Eq. (2) in $\omega\tau_2$. It reproduces Eq. (1) and in addition

$$q_3 = \pm\frac{i}{\lambda}(1 - i\omega\tau_2), \tag{3}$$

the dispersion relation of the sq-mode (4). The opposite limit, $\omega\tau_2 \gg 1$, of Eq. (2) is also intriguing. Both, q_2^2 and q_3^2, become proportional $i\omega$. These two purely diffusive modes were derived earlier (6). They are phase and temperature diffusion, respectively. Eq. (2) gives the complete crossover from second sound and sq-mode to phase and temperature diffusion. Both modes cause a variation in temperature and $j^s = \rho^s(\upsilon^s - \upsilon^n)$. It is expressed by $j^s_\alpha = -\lambda_\alpha \delta T_\alpha$, $\alpha = 2$ or 3. (See Ref. (5) for the function $\lambda_\alpha = \lambda_\alpha(\omega, q_\alpha)$.)

Before discussing experimental implications, a few words of caution are necessary. Eq. (2) is of interest to describe superfluid ^{3}He. Its hydrodynamics (2) is in general much more complicated than that of ^{4}He-II (1). The latter was assumed here. It is a proper description for ^{3}He-B and ^{3}He-A, if their preferred directions in spin and orbital space are well clamped (for more detail see Ref. (5)). It is definitely incorrect for ^{3}He-A$_1$. By an analogous treatment, Ref. (5), one finds that spin-temperature diffusion and sq-mode cross over to a new diffusive and pure q-mode.

Because τ_2 is of order seconds in ^{3}He-A,B, Eq. (2) is important for almost every dynamic experiment dealing with superfluid ^{3}He. The most direct verification of

0921-4526/90/$03.50

Eq. (2) can be done by a periodically heated solid wall immersed in superfluid ^{3}He. It will emit the modes denoted by $q_{2,3}$ (second sound and sq-mode in the limit $\omega \dashrightarrow 0$). Measuring the temperature profile $\delta T(x,t)$ by, e.g. NMR can precisely verify Eq. (2). It will determine τ_2, λ, and c_2, which are not accurately known in ^{3}He-A,B. (Note that λ is only approximately given by $2c_2\lambda$. Therefore, one can measure two independent transport coefficient rather than one.)

Another common tool in superfluid low temperature physics is the porous membrane. For $\omega \to 0$ an oscillating membrane will emit second sound and sq-mode; or more general modes denoted by $q_{2,3}$, Eq. (2). Together with these modes energy is emitted. This gives rise to an extra damping force f_{mode}. Naturally, f_{mode} depends crucially on the dispersion relation. A straight forward calculation (5) shows

$$f_{mode} = 2\rho\sigma\lambda^2 \frac{q_2^2 - q_3^2}{\lambda_2 - \lambda_3} A \, \delta\dot{L} \, , \qquad (4)$$

where ρ is the density, σ the entropy per mass, A the membrane's area, and δL its displacement. Measuring the damping force f_{mode} could again verify Eq. (2) and determine τ_2, λ, and c_2.

The general dispersion relation for the collective modes of temperature and phase in superfluid ^{3}He was derived. It showed what happens when second sound becomes overdamped, at a few Hz, and how it crosses over to eventually become phase and temperature diffusion, at 10^2 Hz. Experimental implications were also discussed.

Acknowledgement

Financial support of the Deutsche Forschungsgemeinschaft is gratefully acknowledged.

References

(1) I.M. Khalatnikov, Introduction to the theory of superfluidity (Benjamin, New York, 1965).

(2) R. Graham, Phys. Rev. Lett. 33 (1974) 1431; M. Liu, Physics 109&110B (1982) 1615.

(3) P. Wölfe, Rep. Prog. Phys. 42 (1979) 269.

(4) M. Grabinski and M. Liu, Phys. Rev. Lett. 58 (1987) 800.

(5) M. Grabinski and M. Liu, J. of Low Temp. Phys. 78 (1990) 247.

(6) D.C. Carless, H.E. Hall, and J.R. Hook, J. of Low Temp. Phys. 50 (1983) 605; H.E. Hall, Physica 108B (1981) 1147.

Physica B 165&166 (1990) 665–666
North-Holland

SYMMETRY OF **n** SOLITON IN SUPERFLUID ^{3}He-B IN SLAB GEOMETRY

Takeo TAKAGI and Makoto TSUBOTA*

Department of Applied Physics, Fukui University, Fukui 910, Japan
Institute of Fluid Science, Tohoku University, Sendai 980, Japan*

The structures of **n** soliton in superfluid ^{3}He-B phase in slab geometry are studied. In the geometry, there are four degenerate equilibrium directions of the **n** vector, which allows the **n** soliton (domain wall) separating two domains with different directions of **n**. We get the structure of the soliton by minimizing its total free energy and find that the bending energy of **n** changes symmetry of the soliton. The result is compared with the NMR data obtained by Ishikawa et al.

1. INTRODUCTION

Recently Ishikawa et al. made cw NMR experiments on superfluid ^{3}He-B confined between two parallel plates with a small gap in between, and found in the signals some satellite peaks shifted from the main peak (1). In this paper we study new possible solitons (domain walls) of the anisotropic vector **n** in the slab geometry and compare our results with the experimental data (2).

2. EQUILIBRIUM DIRECTIONS OF **n** VECTOR

The condensed state of ^{3}He-B is obtained by rotating the spin space of the diagonal Balian-Werthamer(B-W) state with respect to the orbital space around an anisotropic axis **n** by the Leggett angle $\theta_L=\cos^{-1}(-1/4)=104°$. The direction of **n** is determined by minimizing the free energy expression (3)

$$F_{tot} = F_B{}^B + F_H{}^B + F_H{}^S, \tag{1}$$

where the three terms refer to the bulk bending energy, the bulk field anisotropy energy and the surface field anisotropy energy, respectively.

In the experiments of Ishikawa et al., the gap L between two plates is much smaller than the textural healing length, so that the texture of **n** vector is expected to be uniform perpendicular to the slabs. If the **n** vector is denoted as $\mathbf{n}=(\sin\beta\cos\alpha,\ \sin\beta\sin\alpha,\ \cos b)$, the minimization of the sum of $F_H{}^B$ and $F_H{}^S$ gives four degenerate equilibrium directions of n:

$$\cos\beta = \cos\beta_0 \equiv \sqrt{\frac{1}{5} + \frac{4aL}{25d}}$$

$$\tan\alpha = \tan\alpha_0 \equiv \sqrt{(5/3)\cos\beta_0}, \tag{2a}$$

with α in the first or third quadrant, and

$$\cos\beta = -\cos\beta_0\ ,\ \tan\alpha = -\tan\alpha_0 \tag{2b}$$

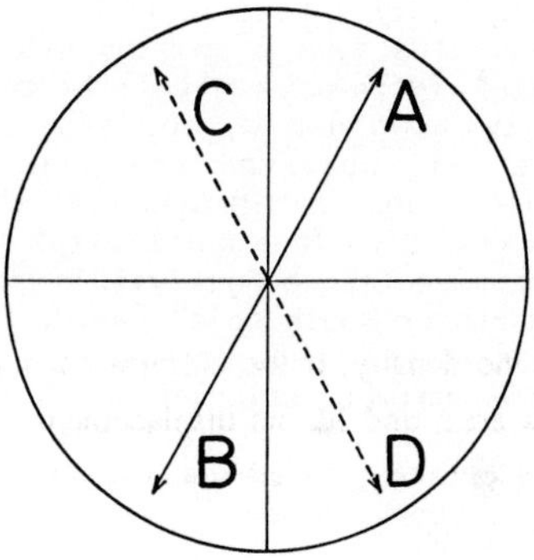

Fig. 1 Four equilibrium **n** directions projected on the n_x-n_y plane.

with α in the second or fourth quadrant. Here a and d are the coefficients of $F_H{}^B$ and $F_H{}^S$, respectively, and the magnetic field is applied parallel to the slabs (along the z axis). We call them A, B, C and D, respectively(Fig. 1).

This allows the possibility of **n** soliton that is a domain wall across which the **n** vector changes from an equilibrium direction in one half-space to another in the other. The soliton extends one-dimensionally in the x-z plane, where the y axis is perpendicular to the slabs. We assume that the soliton extends along the ζ axis, so that $\alpha = \alpha(\zeta)$ and $\beta = \beta(\zeta)$ with $\zeta = x\sin\theta + z\cos\theta$.

The **n** textures can be studied by NMR techniques, because the transverse resonance frequency is shifted from the Larmor frequency whenever the **n** vector makes a finite angle with the applied field. In the strong field limit $\omega_L >> \Omega_L$, the transverse resonance frequency is given by

$$\omega_t = \omega_L + (\Omega_L{}^2/2\omega_L)\sin^2\beta \tag{3}$$

0921-4526/90/$03.50 © 1990 – Elsevier Science Publishers B.V. (North-Holland)

where ω_L is the Larmor frequency and Ω_L is the longitudinal resonance frequency in the B phase. As discussed in reference (2), in the geometry studied by Ishikawa et al., the local oscillator approximation should be correct (3); the spectrum should be

$$P(\omega) \propto \int d\zeta\ \delta[\ \omega - \omega_L - (\Omega_L{}^2/2\omega_L)\sin^2\beta(\zeta)]. \quad (4)$$

The observed main peaks result from the regions with the equilibrium direction $\beta=\beta_0$.

Noting that $\sin\beta(\zeta)$ determines $P(\omega)$, we can classify the above solitons to two types. Type 1 solitons connect a direction in the upper half **n** space (A or B) and one in the lower (C or D). Type 2 solitons connect both directions in the upper half space (A and B) or both in the lower (C and D). Since the type 1 soliton is studied in reference (2), we will discuss the type 2 soliton in the present paper.

3. STRUCTURE OF THE TYPE 2 SOLITON

If we have the type 2 soliton whose minimum value of $\beta(\zeta)$ is zero, a satellite peak should appear at ω_L as shown by eq. (4). However, Ishikawa et al. observed the peak a little shifted from ω_L and determined that the minimum value of β was 18°. Thus our present problem is whether the **n** vector changes within the soliton via "the north or south pole" (**n**=(0, 0, ±1)) in the order parameter space or not.

Since the analytic calculation is very hard, we look for the soliton path minimizing the total free energy by means of the numerical variation method, which is described in the following. First, set the straight path C_0 connecting A and B in the order parameter space. Next, change the path C_0 by adding a sine-wave component C_1 and calculate the free energy for the new path(Fig. 2). Then we can determine the favorable amplitude of the component C_1 which minimizes the free energy. By continuing this operation for higher harmonic sine-waves C_2, C_3, •••, we can obtain the last soliton path. The paths for some values of θ are shown in Fig. 3.

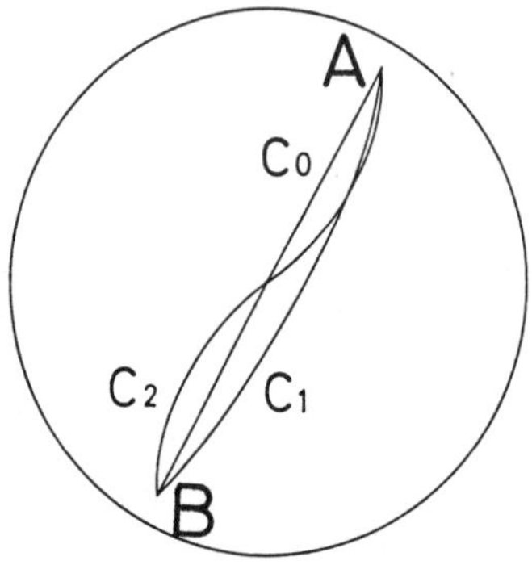

Fig. 2 Numerical variation method. See text.

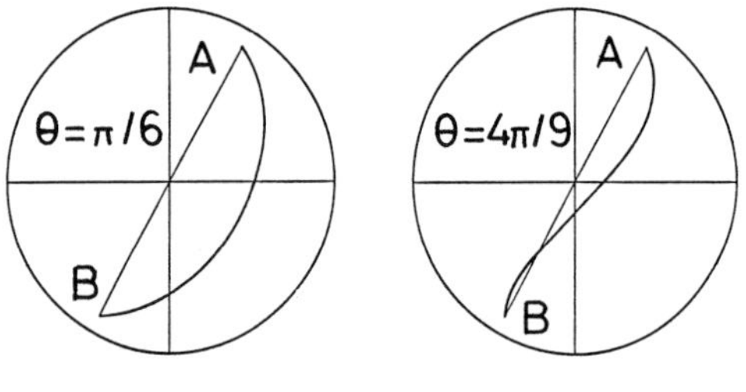

Fig. 3 Soliton paths on the n_x-n_y plane.

The obtained paths dodge the north pole except for θ=0 or π/2. It may seem strange because neither $F_H{}^B$ nor $F_H{}^S$ breaks the rotational symmetry around the north pole. The symmetry breaking is caused by the bending energy. We consider the soliton paths $\mathbf{n}(\zeta)$ and $\mathbf{n}'(\zeta)$ which have the rotational symmetry; those two paths are transformed to each other by the operation

$$(n_x,\ n_y,\ n_z) \rightarrow (-n_x,\ -n_y,\ n_z). \quad (5)$$

The bending energy density of the soliton extending along the direction θ is given by

$$F_B{}^B \propto [\ \partial_\zeta R_{\mu j}\partial_\zeta R_{\mu j} + 2\cos^2\theta\partial_\zeta R_{\mu z}\partial_\zeta R_{\mu z} + 2\sin^2\theta\partial_\zeta R_{\mu x}\partial_\zeta R_{\mu x} + 2\sin 2\theta\partial_\zeta R_{\mu x}\partial_\zeta R_{\mu z}]. \quad (6)$$

Here $R_{\mu j}(\mathbf{n},\theta_L)$ is the rotation matrix such that

$$R_{\mu j}(\mathbf{n},\theta_L) = \delta_{\mu j}\cos\theta_L + n_\mu n_j(1-\cos\theta_L) + \varepsilon_{\mu jk}n_k\sin\theta_L,$$

where $\varepsilon_{\mu jk}$ is the Levi-Civita symbol. The sign of the last term in eq. (6) is changed by the operation of eq. (5). Thus those two solitons $\mathbf{n}(\zeta)$ and $\mathbf{n}'(\zeta)$ have different bending energy except for θ=0 or π/2.

The soliton path and its minimum polar angle β_{min} depend on θ. The free energy of the soliton decreases with increasing θ, and becomes minimum at θ=π/2. On the other hand, the data of Ishikawa et al. gives β_{min}=18°, which corresponds to θ=3° or 60° in our analyses. The contradiction may be attributed to the following. Although our ideal system has a translational symmetry, the real soliton is expected to be trapped at defects on the wall. Then the observed parameter θ need not be fixed at π/2. Studies on the creation of the soliton are being carried out.

REFERENCES

(1) O. Ishikawa, Y. Sasaki, T. Mizusaki, A. Hirai and M. Tsubota, Jour. Low Tem. Phys. 75 (1989) 35.
(2) M. Tsubota and T. Takagi, Prog. Theor. Phys. 83 (1990) 207.
(3) L. Smith, W.F. Brinkman and S. Engelsberg, Phys. Rev. B15 (1977) 199.

Physica B 165&166 (1990) 667–668
North-Holland

INVESTIGATION OF THE RSQ MODE OF HELIUM-3-B USING NON-RESONANT TRANSDUCERS

Peter N. FRAENKEL, Robert KEOLIAN and John D. REPPY

Laboratory of Atomic and Solid State Physics, and the Material Science Center, Clark Hall, Cornell University, Ithaca, New York, 14853-2501

We present systematic measurements of the frequency of the real squashing mode (RSQ) in superfluid ^{3}He-B. The use of ultrasonic transducers with broadband response enabled us to track RSQ frequency down to its zero temperature limit with the sample held at constant pressure.

1. INTRODUCTION

Ultrasonic collective mode spectroscopy (1) invites comparison to other, more routine, forms of spectroscopy. Yet ultrasound studies of superfluid ^{3}He are far from routine, requiring extensive cold times but generally identifying modes at only a few points in the phase space defined by pressure, frequency, temperature and magnetic field. A major difficulty is that the resonant quartz transducers used to generate and detect sound in liquid helium allow phase space to be explored only on surfaces of constant frequency. Since other variables can be changed either slowly or not at all, the spectroscopic investigation is tedious.

In the present work (2–4), transducers were fashioned from piezoelectric polyvinylidene fluoride (PVDF) film (5–7) fastened to a solid substrate. The smooth broadband response up to ~100MHz of these composite transducers allowed a systematic investigation of the real squashing (RSQ, J=2$^+$) mode (8–10) in superfluid ^{3}He-B. The zero field mode frequency was identified at seven pressures between 0 and 29 bar, over a continuum of temperature. In particular, the zero temperature limit of RSQ frequency at constant pressure—a number impossible to observe with resonant transducers—was easily extrapolated from repeated frequency sweeps at constant temperatures.

2. BROADBAND TRANSDUCERS

Resonant quartz slabs (11) are often chosen for low temperature work because of their high mechanical Q's, but there are other quantities that determine the sensitivity of a transducer. Assuming matched transducers, the detected electrical signal for a given drive voltage is proportional to the square of the piezoelectric constant d_{33} (the ratio of induced charge per unit area to the applied mechanical stress). There are additional factors expressing the degree of acoustical impedance match between the sample helium and the piezoelectric and between the piezoelectric and whatever else it might contact. At the resonant frequency ω_0 of the transducers, signal is increased by a factor of Q^2; far from resonance, the Q is irrelevant, and signal decreases as $(\omega_0^2 - \omega^2)^{-2}$. For broadband, pulse use, a high Q is detrimental, since the parasitic signal at resonance can easily swamp the signal at the frequency under investigation. Cold PVDF film (5) has $Q \sim 10$ and $d_{33} = 8.4 \times 10^{-12}$m/V, as against $Q \sim 2000$ and $d_{33} = 2 \times 10^{-12}$m/V for quartz (11).

Figure 1 shows an exploded view of a transducer made with PVDF. To assure parallelism, the ground side of the film was laid against an optical flat while the epoxy cured. To produce signal at a useable level, PVDF transducers required approximately 50dB more drive than did quartz, yet heating was tolerable:

$$\dot{Q} \simeq 1\mu\mathrm{W}\left(\frac{f}{10\mathrm{MHz}}\right)^2\left(\frac{V}{1V_{\mathrm{pp}}}\right)^2,$$

or less than half a nanoJoule for a 2μs, 2.5V$_{pp}$ pulse at 50MHz. Note that, in contrast to the case for a ringing quartz crystal, a 2μs electrical pulse produces a sound pulse exactly 2μs long. Since the electrical pulse cannot be made much longer without causing feedthrough problems, the frequency distribution of the excitation is inevitably quite broad.

3. RSQ DATA

Details of the cryogenics have been described elsewhere (2). We note here only that the ^{3}He sample was cooled by the adiabatic demagnetization of $PrNi_5$ and that its temperature was determined with a torsional oscillator measurement of the superfluid density. Figure 2 shows transmitted pulse amplitude (arbitrary scale) as a function of frequency and normal fluid fraction for repeated sweeps at 28.88 bar. The RSQ mode is clearly visible as a temperature dependent furrow between 75 and 104MHz. The mode can be fitted by parametrizing a curve $f(\rho_n/\rho)$ and minimizing the amplitude in a Gaussian weighted region around it. Features at con-

0921-4526/90/$03.50

stant frequency are resonances of the transducers or other cell elements.

Data at six other pressures down to 0.06 bar, extrapolations to zero temperature, and a discussion of theoretical implications have been presented elsewhere (2,3). Table 1 summarizes some of the results. Note that, while the $T = 0$ frequencies cannot be compared to theory without assuming values of $T_c(p)$, they are insensitive to thermometric uncertainties of the experiment; for this reason they are more precise than quartz measurements, despite the larger uncertainty in the frequencies themselves. Precise constraints are also placed on the values of x_3^{-1} (a measure of the strength of f-wave coupling) and F_2^a (a Fermi interaction parameter). Comparison of these determinations with those from other experiments makes clear that "non-trivial" strong-coupling corrections to collective mode theory are needed in order to explain data at high pressures.

ACKNOWLEDGEMENTS

The work reported here was carried out at Cornell University in the Laboratory of Atomic and Solid State Physics with support from the National Science Foundation through grant **NSF-DMR-8418605** and with support from the Cornell Material Science Center through grant **NSF-DMR-82-17227A**, MSC Report 6903.

REFERENCES

(1) P. Wölfle, *Progress in Low Temperature Physics*, edited by D.R. Brewer (North Holland, Amsterdam, 1978), Vol. 7A, pp 191-281.
(2) P. N. Fraenkel, R. Keolian and J. D. Reppy, Phys. Rev. Lett **62**, 1126 (1989).
(3) P. N. Fraenkel, A.I.P. Conference Proceedings **194**, 96 (1989).
(4) R. Keolian, P. N. Fraenkel and J. D. Reppy (to be published).
(5) Hiroji Ohigashi, J. App. Phys. **47**, 949 (1976).
(6) Penwalt Corporation, Kynar Piezo Film Department, P.O. Box 799, Valley Forge, PA, 19482.
(7) Solvay Technologies, Inc., 500 Fifth Avenue, Suite 3000, New York, NY 10110.
(8) J. A. Sauls and J. W. Serene, Phys. Rev. B. **23**, 4798 (1981).
(9) D. B. Mast, Bimal K. Sarma, J. R. Owers-Bradley, I. D. Calder, J. B. Ketterson, and W. P. Halperin, Phys. Rev. Lett. **45**, 266 (1980).
(10) R. W. Gianetta, A. Ahonen, E. Polturak, J. Saunders, E. K. Zeise, R. C. Richardson, and D. M. Lee, Phys. Rev. Lett. **45**, 262 (1980).
(11) *Methods of Experimental Physics: vol. 19, Ultrasonics*, ed. Peter D. Edmonds (Academic Press, New York, 1981), p. 54.

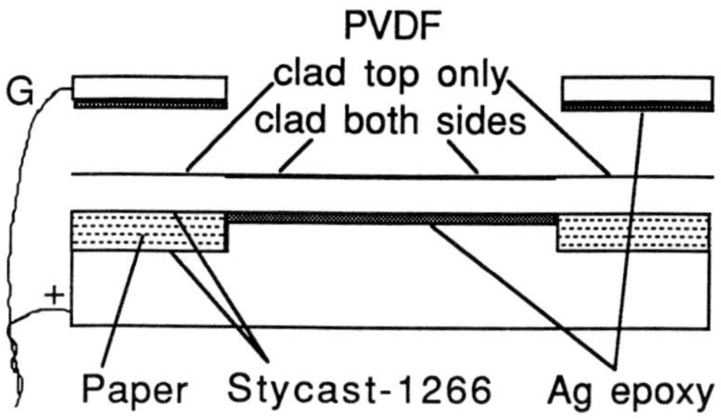

Figure 1. Exploded view of PVDF transducer.

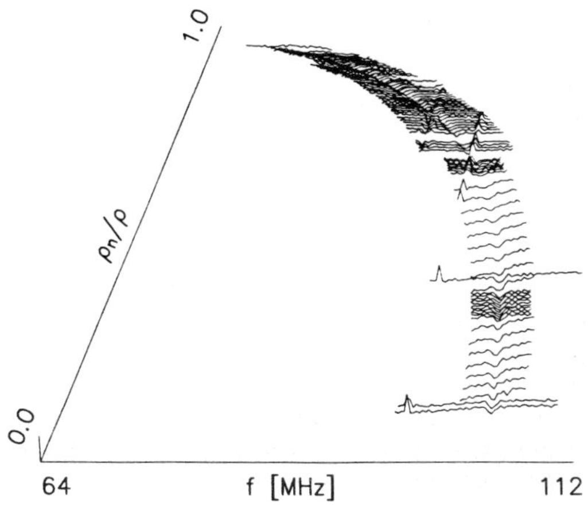

Figure 2. Amplitude of transmitted sound as a function of normal fluid fraction and frequency.

p	f_{RSQ}	f_{RSQ}	x_3^{-1}	F_2^a
	(T=0)	(T=q=0)	$(F_2^a = 0)$	$(x_3^{-1} = 0)$
bar	MHz	MHz		
0.06	42.42	41.64	-0.137	-0.614
2.12	53.09	52.45	-0.172	-0.750
5.06	65.58	65.09	-0.202	-0.860
10.07	80.66	80.30	-0.232	-0.964
15.42	90.92	90.66	-0.270	-1.094
23.00	100.32	100.13	-0.310	-1.218
28.88	103.86	103.71	-0.353	-1.345
±0.005	±0.10	±0.10	±0.010	±0.040

Table 1 Parameters from fits to RSQ data

Physica B 165&166 (1990) 669–670
North-Holland

ULTRASONIC INVESTIGATION OF ROTATING ^{3}He-B IN STRONG MAGNETIC FIELDS

A.J. MANNINEN, J.M. KYYNÄRÄINEN, K. TORIZUKA, and J.P. PEKOLA

Low Temperature Laboratory, Helsinki University of Technology, 02150 Espoo, Finland

We have studied ultrasonic attenuation in rotating ^{3}He-B at low pressures (< 10 bar) in magnetic fields up to 300 mT. In a strong field, the energy gap Δ in ^{3}He-B is anisotropic, and the attenuation α of zero sound depends on the orientation of Δ. Because vortices and the counterflow between the normal and superfluid components of the liquid turn the anisotropy axis of Δ, a clear dependence of α on the rotation velocity Ω can be observed. The magnitude of this response is not a linear, sometimes not even a monotonic function of Ω. Its dependence on the magnetic field is anomalous around H = 200 mT. We have also observed a metastable state in rotating ^{3}He-B, which can be created by rotating the sample while cooling through the AB-transition. The lifetime of this state is on the order of 10 minutes.

Even moderate magnetic fields ($H \gtrsim 100$ mT) distort the energy gap of ^{3}He-B: the "longitudinal" gap Δ_2 in the direction of $\hat{h} = \overleftrightarrow{R}(\hat{n},\theta)\,\vec{H}/H$ becomes smaller than the "transverse" gap Δ_1 perpendicular to $\hat{h}$ (1); here $\overleftrightarrow{R}$ is a rotation matrix and $\vec{H}$ the external magnetic field. Therefore, ultrasonic attenuation α is anisotropic, too, and depends on the angle γ between the wave vector $\hat{q}$ and the anisotropy axis $\hat{h}$ according to the general formula

$$\alpha = \alpha_{||}\cos^4\gamma + 2\alpha_c\cos^2\gamma\sin^2\gamma + \alpha_{\perp}\sin^4\gamma . \quad [1]$$

Coefficients α_i depend on temperature, pressure, frequency, and magnetic field, but their values in ^{3}He-B have not been calculated nor measured yet. The experimental determination of α_i's in ^{3}He-B, using tilted magnetic fields, would be more direct than in ^{3}He-A: In the B-phase the anisotropy axis is along $\vec{H}$, whereas in the A-phase it may choose any direction perpendicular to $\vec{H}$. On the other hand, in ^{3}He-B, α_i's depend very strongly on H, unlike in ^{3}He-A.

In rotating ^{3}He-B, vortices and the counterflow between the normal and superfluid components turn vector $\hat{h}$ towards the plane perpendicular to the rotation axis, and γ changes. When the external field is stronger than $\approx$100 mT, the magnetic healing length ξ_H is smaller than the intervortex distance, and $\hat{h}$ can vary locally. The coefficients α_i in Eq. [1] are unequal in strong fields, and the apparent attenuation depends on Ω, the angular velocity of rotation; in low fields (< 50 mT), rotation response can be observed only near the rsq-mode (2) or in the immediate vicinity of the AB-transition.

We have studied ultrasonic attenuation in ^{3}He-B at low pressures (< 10 bar) up to Ω = 5 rad/s using 8.9 MHz and 26.8 MHz pulses. The magnetic field, up to 300 mT, was parallel to $\vec{\Omega}$ and $\hat{q}$. Our experi-

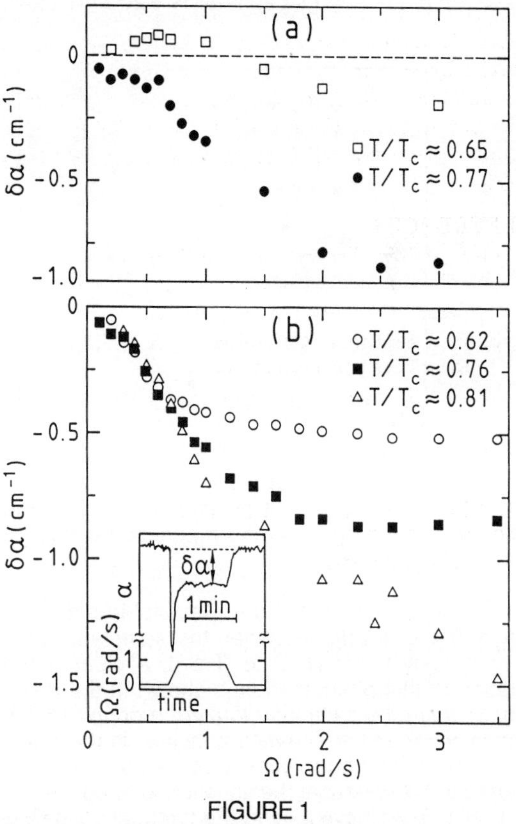

FIGURE 1
The rotation-induced shift of ultrasound attenuation, $\delta\alpha \equiv \alpha(\Omega) - \alpha(0)$, at different temperatures as a function of Ω; p = 2.3 bar, f = 8.9 MHz, and H = 220 mT (a) and H = 174 mT (b). Inset: A typical attenuation curve in our start-stop experiments; the lower part illustrates the changes of Ω with time.

0921-4526/90/$03.50

mental cell, which has been described elsewhere (2), is a 4 mm high cylinder whose diameter is 6 mm.

Immediately after rotation is started, a sharp minimum (or maximum) is observed in sound attenuation, caused by the counterflow state. After this, vortices start forming, and attenuation relaxes to a new level with a time constant $\tau \approx 5$ - 60 s, as shown in the inset of Fig. 1. The Ω-dependence of this equilibrium shift of attenuation in the vortex state, $\delta\alpha \equiv \alpha(\Omega) - \alpha(0)$, at different temperatures and magnetic fields is illustrated in the main parts of Fig. 1. The shift is far from being linear in Ω, and in some cases it is not even monotonic.

In Fig. 2 we have presented the magnetic field dependence of $\delta\alpha$, with Ω as a parameter. In these measurements, which were made at the approximately constant temperature $0.62 < T/T_c < 0.68$, one can see a similar change at $H \approx 200$ mT as was observed in measurements made near the AB-transition temperature in different magnetic fields (3). This change is due to the strong H-dependence in the distortion of the energy gap near H = 200 mT (1,4).

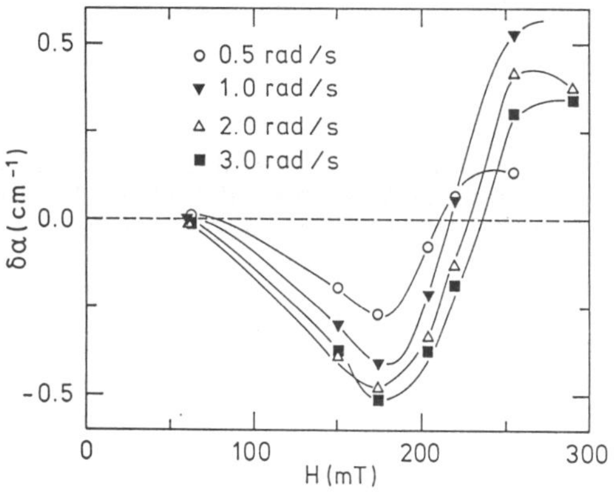

FIGURE 2
The rotation-induced shift of attenuation at different velocities as a function of the magnetic field; p = 2.3 bar and f = 8.9 MHz.

We have also observed a metastable state in rotating ^{3}He-B. It appears when the sample is rotated continuously from ^{3}He-A to ^{3}He-B. A clear demonstration of this phenomenon is shown in the inset of Fig. 3: when the sample is rotated through the A→B-transition, sound attenuation is higher in the rotating than in the stationary case, but when rotation is started in the B-phase, the situation is opposite.

In Fig. 3, we have plotted the attenuation in ^{3}He-B just below the AB-transition temperature, when rotation was started either in the A- or in the B-phase; the data on the metastable state were taken immediately after the A→B-transition, and those of the equilibrium state immediately before the B→A-transition.

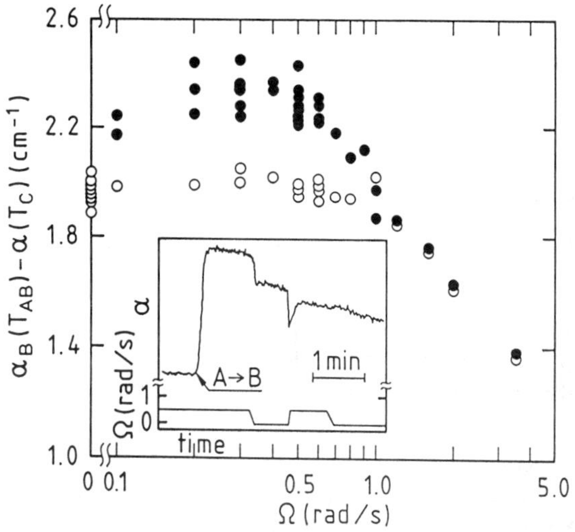

FIGURE 3
Metastable state in rotating ^{3}He-B at p = 2.3 bar, H = 116 mT: Attenuation of 8.9 MHz ultrasound immediately below the AB-transition, $\alpha_B(T_{AB})$, is higher when the sample is rotated through the A→B-transition (filled circles) than when rotation is started in the B-phase (open circles). Inset: Sound attenuation as a function of time, when the sample is first cooled through the AB-transition while rotating, rotation is then stopped and started again in the B-phase; the lower part illustrates the changes in the speed of rotation.

In this case, the metastable state below $\Omega = 1$ rad/s is characterized by higher and more varying attenuation than the equilibrium state. When rotation is continued in ^{3}He-B, after rotating through the AB-transition, the sound attenuation relaxes towards the value observed for the regular vortex state typically in about 10 minutes.

ACKNOWLEDGEMENTS

This work was financially supported by the Academy of Finland and the U.S.S.R. Academy of Sciences through project ROTA, and by the Körber-Stiftung of Hamburg.

REFERENCES

(1) N. Schopohl, J. Low Temp. Phys. 49 (1982) 347.
(2) R.H. Salmelin, J.P. Pekola, A.J. Manninen, K. Torizuka, M.P. Berglund, J.M. Kyynäräinen, O.V. Lounasmaa, G.K. Tvalashvili, O.V. Magradze, E. Varoquaux, O. Avenel, V.P. Mineev, Phys. Rev. Lett. 63 (1989) 620.
(3) J.M. Kyynäräinen, J.P. Pekola, A.J. Manninen, K. Torizuka, Phys. Rev. Lett. 64 (1990) 1027.
(4) M. Ashida and K. Nagai, Prog. Theor. Phys. 74 (1985) 949.

Physica B 165&166 (1990) 671–672
North-Holland

HOMOGENEOUSLY PRECESSING SPIN DOMAIN IN ROTATING VORTEX-FREE ^{3}He-B

J.S. KORHONEN, V.V. DMITRIEV+, Z. JANU*, Y. KONDO, M. KRUSIUS, AND Yu.M. MUKHARSKIY+

Low Temperature Laboratory, Helsinki University of Technology, 02150 Espoo, Finland

Superfluid counterflow in rotating vortex-free ^{3}He-B induces distinct changes in the homogeneously precessing resonance domain: its volume is reduced and the NMR absorption increased. Both effects disappear if the domain fills the whole experimental cell.

1. INTRODUCTION

Unlike in He-II and ^{3}He-A, a vortex-free state can be maintained in ^{3}He-B up to appreciable rotation velocities. In conventional cw NMR spectra the superfluid counterflow $\mathbf{v}_n$ - $\mathbf{v}_s$ of the vortex-free state is manifested as a rearrangement of the resonance absorption with frequency, implicating substantial changes in the counterflow dominated **n**-texture (1). At high rf excitation levels a linear field gradient ∇H of the static magnetic field stabilizes a resonance mode unique to ^{3}He-B (2). In this mode two domains with uniform spin behavior exist: in the low field homogeneously precessing domain (HPD) the magnetization **M** makes an angle $\beta > 104°$ with **H**, while in the high field domain **M** is parallel to **H**. Vortices and counterflow both affect the two domain structure. While vortices interact with the precessing spins in the HPD (3), changes induced by counterflow occur only if the narrow (< 1 mm) boundary region between the two domains is present.

2. EXPERIMENTAL SET-UP

Our measurements are performed in a rotating nuclear demagnetization cryostat at 29.3 bar pressure. The experimental chamber consists of two parts. The cylindrical NMR cell (diameter ϕ = 7 mm, length L = 7 mm is separated from the large heat-exchanger volume by a tube (ϕ = 1.5 mm, L = 5.5 mm) with an orifice (ϕ = 1 mm, L = 0.5 mm). This construction prevents vortices present in the main ^{3}He volume from entering the NMR cell. The rotation velocity for vortex creation depends on temperature and is ≈ 1 rad/s at $T = 0.5T_c$ and >2.8 rad/s at $T > 0.6T_c$. The magnetic field, H = 14.2 mT, is parallel with the rotation axis Ω and the symmetry axis of the cell. The field gradient points to the cell bottom, so that the HPD forms first in the upper part of the cell. The NMR spectrometer (4) is operated at the constant frequency of 460 kHz.

3. EXPERIMENTAL RESULTS

When the static magnetic field is swept downwards, the HPD is formed in the low field region. At the domain boundary the Larmor frequency becomes equal to that of the rf excitation; therefore by changing the magnitude of H, the boundary can be moved inside the cell. Fig. 1 shows the components of **M** perpendicular to **H**. The lowermost curve, measured in the stationary state, gives the trace of **M** while the domain boundary travels from the top of the cell (origin) to the bottom (A). The total magnetization $M = (M_\perp^2 + M_\parallel^2)^{1/2}$ grows proportional to the volume of the HPD. The phase of M is defined by the energy dissipation P: at constant M, ΔP is proportional to $\Delta M_\perp$ (4). Various relaxation processes in the HPD mode are discussed in Ref. 4; here we note only that energy dissipation connected with spin diffusion through the domain boundary is seen in Fig. 1 as the rapid increase in $M_\perp$, when the domain wall first enters the cell.

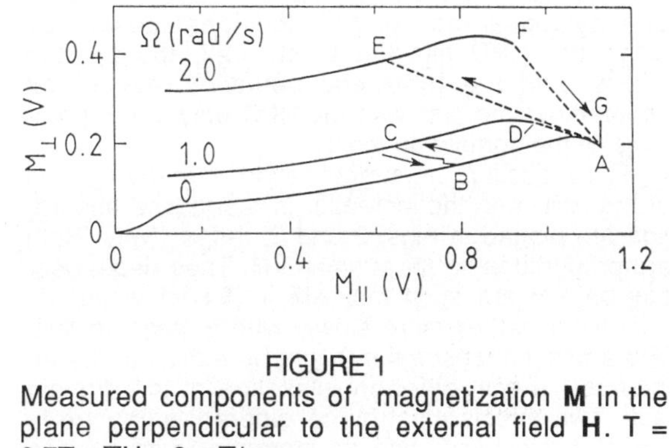

FIGURE 1
Measured components of magnetization **M** in the plane perpendicular to the external field **H**. $T = 0.7T_c$, ∇H = 3 mT/m.

+ Institute for Physical Problems, Moscow, 117 334 USSR.
* Institute of Physics, Rez, CS-25068 Czechoslovakia.

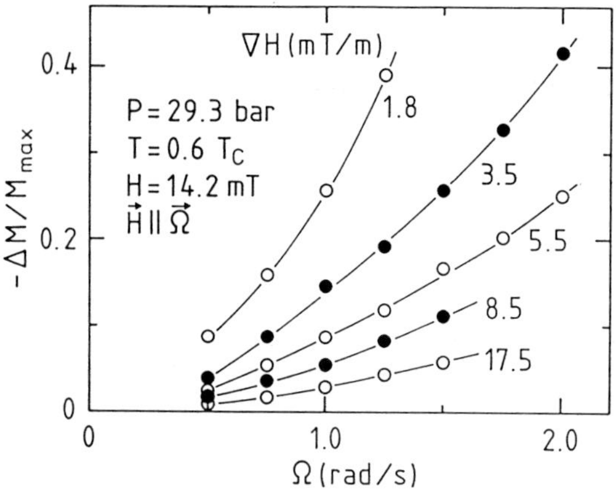

FIGURE 2
Reduction ΔM of the HPD magnetization as a funtion of rotation velocity at different ∇H. ΔM is normalized by the maximum value M_{max}, measured when the HPD fills the whole cell.

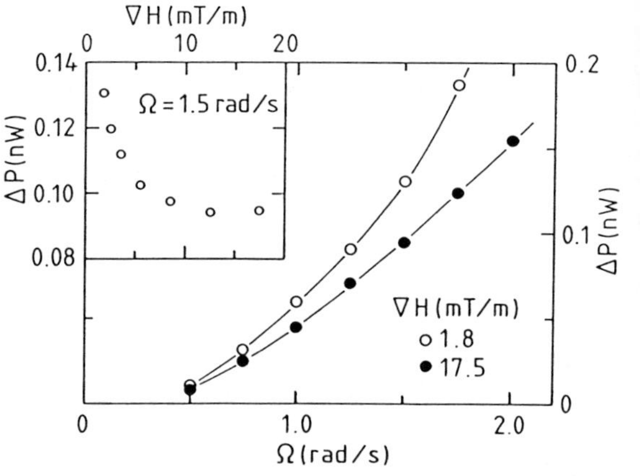

FIGURE 3
Additional absorption ΔP as a function of rotation velocity at two magnetic field gradients. The data was measured at constant magnetization $M = 0.55M_{max}$. The insert shows ΔP as a function of the field gradient at $\Omega = 1.5$ rad/s. P, T and H are as in Fig. 2.

If the domain boundary is swept to point B (see Fig. 1) and the cryostat is accelerated to $\Omega = 1$ rad/s, curve B-C is traversed and a clear reduction in the total magnetization, or in the volume of the HPD, is seen. Sweeping the magnetic field at Ω = const. gives the curve with points C, D and A. A similar curve with points E and F is recorded after acceleration to 2 rad/s. Both curves, measured in the rotating state, show an additional absorption independent of total magnetization.

The dotted lines depict sudden jumps from D and F to A (when H is reduced) and from A to E (when H is increased). If the magnetic field is further reduced after a jump to A, the trace A-G is independent of Ω. The disappearance of the additional absorption when the HPD fills the whole cell, the sudden jumps, and the hysteretic behavior reveal that counterflow interacts with the HPD only in the presence of the domain boundary.

The reduction in magnetization or in the volume of the HPD and the increase in resonance absorption are plotted in Figs. 2 and 3, respectively. Both are proportional to Ω^2 at small ∇H. Their dependences on ∇H are such that $\Delta M \propto (\nabla H)^{-1}$ while ΔP decreases rather more slowly with increasing ∇H. The additional absorption might be explained by an increase in spin diffusion relaxation at the domain wall. Two effects must be considered: a change in the wall thickness and an increase in its surface area due to bending of the domain wall in the inhomogeneous velocity distribution $\mathbf{v} = -\Omega \times \mathbf{r}$.

Experiments in a magnetic field tilted by 90° with respect to the rotation axis show a similar behavior of the HPD as in the axial field. A qualitative difference is that in the transverse field the additional absorption due to counterflow depends on the total magnetization: no increase in absorption is seen, if the HPD fills approximately half of the cell.

ACKNOWLEDGEMENTS
We wish to thank I.A. Fomin, A.D. Gongadze and G.E. Volovik for discussions. This work has been supported by the Körber Stiftung (Hamburg) and the Academy of Finland and the USSR Academy of Sciences through project ROTA.

REFERENCES

(1) P.J. Hakonen and K.K. Nummila, Phys. Rev. Lett. 59 (1987) 1006, and A.D. Gongadze, Z. Janu, Y. Kondo, J.S. Korhonen, M. Krusius, Yu. M. Mukharskiy, and E.V. Thuneberg, to be published.
(2) A.S. Borovik-Romanov, Yu. M. Bunkov, V.V. Dmitriev, Yu. M. Mukharskiy, and K. Flachbart, Sov.Phys. JETP 61 (1985) 1199, and I.A. Fomin, Sov. Phys. JETP 61 (1985) 1207.
(3) V.V. Dmitriev, Y. Kondo, J.S. Korhonen, M. Krusius, Yu.M. Mukharskiy, E.B. Sonin, and G.E. Volovik, this volume.
(4) Y. Kondo, J.S. Korhonen, and M. Krusius, this volume.

Physica B 165&166 (1990) 673–674
North-Holland

NMR ABSORPTION IN A HOMOGENEOUSLY PRECESSING DOMAIN OF ^{3}He-B

Y. KONDO, J.S. KORHONEN, and M. KRUSIUS

Low Temperature Laboratory, Helsinki University of Technology, 02150 Espoo, Finland

The measurement of cw NMR absorption in a homogeneously precessing spin domain of ^{3}He-B is discussed. The absorption is divided into several contributions, based upon their different dependences on the domain length. Within our experimental accuracy, we found no contribution linear with domain size which would be caused by surface relaxation on the cell wall.

1. INTRODUCTION

In superfluid ^{3}He-B the homogeneously precessing spin domain (HPD) is created at high rf excitation levels in the presence of a linear field gradient ∇H superimposed on the static polarization field H. In this domain the spins are uniformly precessing at ω, which corresponds to the Larmor frequency at the domain boundary. This has been demonstrated by the Moscow group both theoretically (1) and in a number of beautiful experiments (2).

We have used the HPD resonance mode for studying the B-phase vortices (3) and counterflow (4) in the rotating liquid. Vortices produce an additional NMR absorption which is proportional to their total length within the domain. When analyzing the mechanism for this additional spin relaxation it became necessary to compare the NMR absorption in different external fields.

In this report we discuss the method for extracting the absolute value of the NMR absorption, and divide it into different contributions on the basis of their dependences on the domain size in the stationary liquid. This procedure has been introduced by the Moscow group (2), but here we consider more extensively the term linear with domain length which would be caused by surface relaxation on the cell wall. Our experiments show that this linear contribution is less than 0.01 nW/cm^2 at 14.2 mT.

2. TANK CIRCUIT

An analysis of the cw NMR tank circuit has been carried out in Fig. 1 to determine the NMR absorption P_M. Here v_M is the voltage which is induced by the precessing total magnetization M in the pick-up coil, ie., $|v_M| \propto \omega |M| = \omega \chi_B H$ in the HPD mode rather than being proportional to the rf excitation field H_1 as in conventional low level NMR. The voltage v_o across the tank circuit, in the absence of the HPD ($v_M = 0$), is cancelled in the actual measurement with a compensation tank circuit connected to the second input of the differential preamplifier.

The remaining signal voltage, $-jQv_M$, is measured with a phase sensitive detector where v_M' and v_M'' are the dispersion and absorption signals, respectively. The phase setting of the detector is adjusted by enforcing the requirement that P_M is independent of v_c (see Fig. 1), ie., $v_M''|v_o|$ = const. Even if we add a

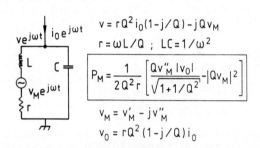

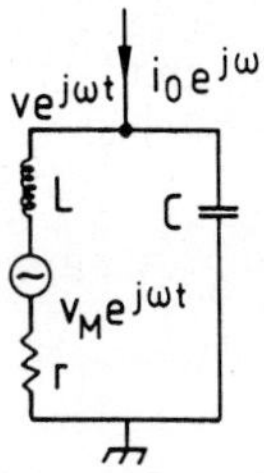

$$v = rQ^2 i_0(1-j/Q) - jQv_M$$
$$r = \omega L/Q \; ; \; LC = 1/\omega^2$$
$$P_M = \frac{1}{2Q^2 r}\left[\frac{Qv_M''|v_0|}{\sqrt{1+1/Q^2}} - |Qv_M|^2\right]$$
$$v_M = v_M' - jv_M''$$
$$v_0 = rQ^2(1-j/Q)i_0$$

FIGURE 1
Equivalent circuit of our cw NMR tank circuit.

resistance to damp the Q value of the tank circuit, our method of calculating P_M and adjusting the phase is still valid save for correction terms of order $1/Q^2$. In our experiments, the error in the phase adjustment is less than 0.3°. The limiting features are the accuracies of our synthesizer oscillator (Hewlett-Packard #3325A) and lock-in amplifier (Princeton Applied Research #5202).

3. MEASUREMENTS

Our NMR cell is a cylinder with diameter = height = 7 mm, with H and ∇H oriented along the cylinder axis (3 , 4). When H is swept downward, the HPD first starts forming at that end of the cylinder where the Larmor frequency begins to drop below ω. The domain boundary forms at the location where the Larmor frequency equals ω. Thus, by sweeping H with ∇H = const., one can control the size of the domain in such a way that, at each location of the domain boundary, $M \propto$ length of the domain ℓ and the tipping angle of the precessing spins is $\approx 104°$ (1).

In Fig. 2 the tank circuit voltage which is calculated from the lock-in amplifier output, while sweeping H, is shown as a plot with Qv_M'' vs Qv_M' (top): at the origin there is no HPD while at point A the domain fills the entire cell. The lower graph in Fig. 2 illustrates the corresponding P_M vs $Q|v_M| \propto$ M plot. Note the large correction due to $-|v_M|^2/r$.

The NMR absorption was fitted to a polynomial in M $= |M|$ with

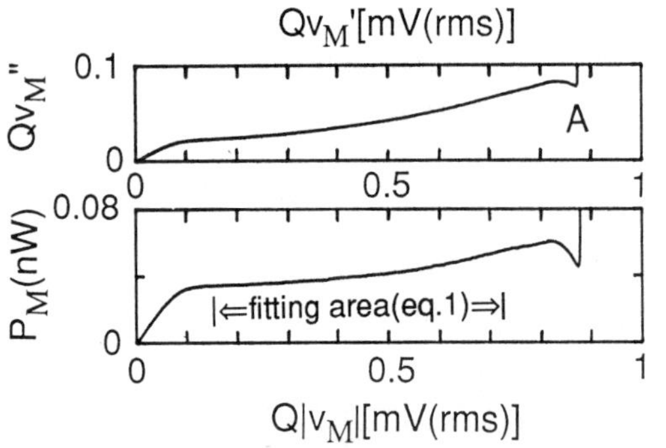

FIGURE 2
Plot of the in-phase Qv_M" vs the out-of-phase Qv_M' components of the HPD signal (top) and the corresponding NMR absorption P_M vs $Q|v_M| \propto M$ (bottom) at 14.2 mT, 11 mT/m, 0.48T_c, and 29.3 bar. Circuit constants are L = 60 μH, Q = 58, $|v_0|$ = 15.7 mV(rms).

$$P_M = a_1 + a_2 M + a_3 M^3 \qquad (1)$$

where the first constant term corresponds to relaxation in the domain boundary, $\propto H^{7/3} \nabla H^{1/3}$, the second to relaxation at the cylindrical cell wall with an area $\propto \ell$, and the third to Leggett-Takagi relaxation in the bulk domain, $\propto \ell^3 H^2 \nabla H^2$. In Fig. 3 the division into these 3 terms is shown as a function of ∇H. We found no significant linear term. The worst case phase error, 0.3°, makes 0.008 nW error in the linear term in this case at 14.2 mT. The current setting of the magnet producing the gradient for optimum field homogeneity was found by adjusting for minimum line width of the conventional NMR absorption signal in the normal Fermi liquid phase.

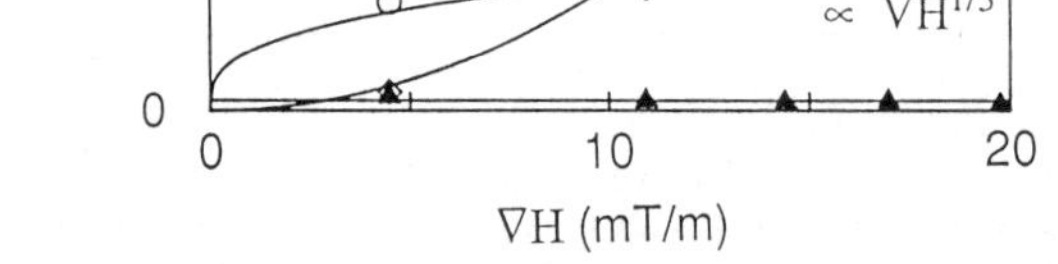

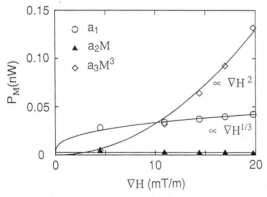

FIGURE 3
Division of the NMR absorption into three components according to Eq. (1), shown here as a function of ∇H at 14.2 mT, 0.48T_c and 29.3 bar. The terms $\propto M$ and M^2 are shown with their values when the HPD fills the entire cell.

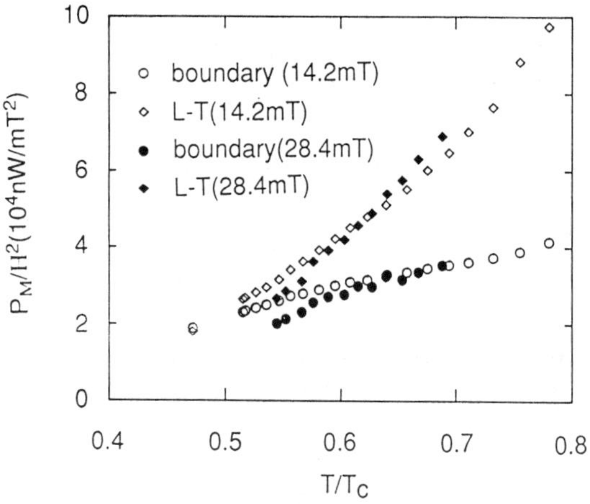

FIGURE 4
The domain boundary and Leggett-Takagi (L-T) components to P_M as a function of temperature measured at 10 mT/m and 29.3 bar, in two different fields 14.2 and 28.4 mT.

From these results we can determine the spin diffusion constant and the Leggett-Takagi relaxation time (2), and we find no significant surface relaxation. When the measurements of Fig. 3 were repeated at different temperatures and the results were plotted in normalized form, P_M/H^2, then Fig. 4 was obtained. Even in the higher field of 28.4 mT we could not find a meaningful contribution from surface relaxation which has been predicted to have the strong field dependence of H^4 (5).

ACKNOWLEDGEMENTS
We wish to thank V.V. Dmitriev and Yu.M. Mukharskiy for useful discussions. This work has been supported through the Award for the Advancement of European Science by the Körber Stiftung.

REFERENCES
(1) I.A. Fomin, Sov. Phys. JETP 61 (1985) 1207.
(2) A.S. Borovik-Romanov, Yu.M. Bunkov, V.V. Dmitriev, and Yu.M. Mukharskiy, Quantum Fluids and Solids 1989, AIP Conf. Proc. 194 (1989) 15.
(3) V.V. Dmitriev, Y. Kondo, J.S. Korhonen, M. Krusius, E.B. Sonin, and G.E. Volovik, this volume.
(4) J.S.Korhonen, V.V.Dmitriev, Z.Janu, Y.Kondo, M.Krusius, and Yu.M.Mukharskiy, this volume.
(5) T. Ohmi, M. Tsubota, and T. Tsuneto, Proc. 18th Internat. Conf. on Low Temp. Phys., Kyoto, Jap. J. Appl. Phys.26 (1987) 169.

Physica B 165&166 (1990) 675–676
North-Holland

Instability of the homogeneous precession in ^{3}He – B (catastrophic relaxation).

YU.M.BUNKOV, V.V.DMITRIEV, J.NYEKI[†], YU.M.MUKHARSKIĬ, D.A.SERGATSKOV, I.A.FOMIN[‡]

Institute for Physical Problems Academy of Sciences of the USSR, Kosygin st.2, 117334 Moscow, USSR
[†]*Institute for Experimental Physics, Košice, Czechoslovakia*
[‡]*L.D.Landau Institute for Theoretical Physics Academy of Sciences of the USSR*

At temperatures below $0.5T_c$ we observed a sharp change in a shape and a duration of a free induction decay signal of ^{3}He – B. Below this temperature signal bears clear signs of instability. The onset temperature of this phenomenon does not depend significantly on the pressure, the magnetic field or the coverage of a container walls with ^{4}He. Possible explanation is discussed.

1. INTRODUCTION

We have recently reported (1) on the very fast increase of the magnetic relaxation rate in ^{3}He – B in the so called *homogeneously precessing domain, HPD* (2) at temperatures below $\sim 0.45\,T_c$. This phenomenon was named as *catastrophic relaxation* because relaxation rate increases by two orders of magnitude in temperature interval of 0.05 T_c. At the same time numerious *cw* NMR experiments show no singularities below 0.5 T_c. To check if this phenomenon is the property of the HPD mode or can be generally attributed to the spin dynamics of ^{3}He – B we have performed pulsed NMR measurements without HPD.

A time of the formation of the HPD is determined by the inhomogeneity of the external magnetic field (H_0) and by the angle of the initial deflection of the magnetization (β_0). For $\beta_0 < 90°$ in the homogeneous field ($\delta H_0/H_0 \sim 10^{-4}$) this time becomes longer than the free induction decay signal (*FIDS*) duration and the HPD is not formed. The FIDS duration is determined by inhomogeneity of the external field and is $\sim$ 5 ms.

2. EXPERIMENT

Our cylindrical experimental cell has the height (6 mm) which is equal to the diameter and has an axis oriented vertically along $\vec{H}_0$. There is enough ^{4}He (15%) in the cell for the walls to be covered with thick ^{4}He film. Experiments are performed at pressures of 10.6, 20 and 27 bars, and in magnetic fields of 142 and 284 Oe. We use conventional pulsed NMR spectrometer with a single transmitting/receiving coil. Magnetization is tilted with a short ($\sim$ 5 periods) radiofrequency (*RF*) pulse. FIDS is recorded by a transient recorder and analyzed by a computer. The temperature is measured by platinum NMR thermometer calibrated using the superfluid transition of ^{3}He.

The FIDS amplitude is decreased smoothly with temperature down to $T \sim 0.5\,T_c$ because of the change in the magnetic susceptibility of ^{3}He – B. Change in the dependence of the FIDS amplitude versus time for lower temperatures is shown on Fig.1. At the temperature of

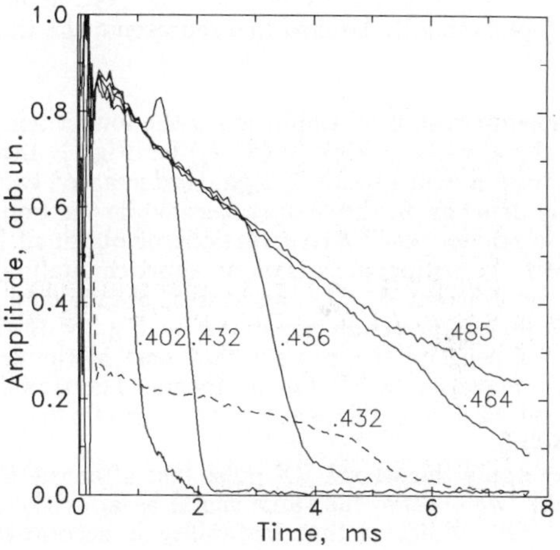

Figure 1: The amplitude of the FIDS versus time at different temperatures. $P = 20$ bar, $H = 142$ Oe, solid lines — $\beta_0 \approx 60°$, dashed line — $\beta_0 \approx 30°$.

0.485 T_c the shape of the FIDS is determined by the inhomogeneity of the magnetic field. As ^{3}He is cooled down, a kink develops on the signal. A time delay between RF pulse and the kink decreases with decreasing the temperature, part of the signal before the kink remaining essentially unchanged. Signal corresponding to lower $\beta_0 \approx 30°$ (dashed line) persists longer at the same temperature.

The existence of the unchanged part of the signal, followed by a sharp vanishing of the signal looks rather like a signature of an instability of the homogeneous precession. A time of the development of the instability is decreased with the decrease of the temperature and increase of the tipping angle, the latter explains why this phenomenon was not seen in *cw* NMR experiments. To determine the temperature of the onset of the instability we have divided FIDS amplitude at par-

0921-4526/90/$03.50

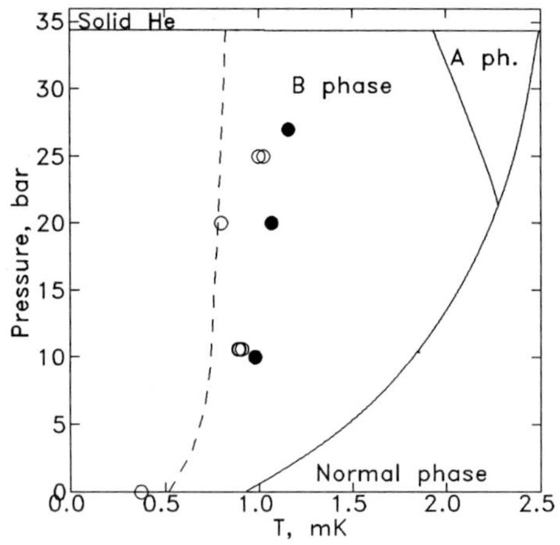

Figure 2: Phase diagram of the onset of the instability (filled circles) and the catastrophic relaxation in the HPD (open circles). Dashed line represents the theory.

ticular temperature by amplitude of the lowest temperature signal without kink ($0.485\,T_c$) for Fig.1). Inverse of the time at which resulting signal is decreased twofold linearly depends on the temperature while the temperature is not too low. Extrapolations of obtained lines intersect the temperature axis at approximately same point for different β_0. Temperatures, obtained in this way at various pressures are shown on Fig.2 with filled circles. Open circles represent the onset of the catastrophic relaxation in HPD experiments. The transition temperature does not depend on the magnetic field (142 or 284 Oe).

If we apply the second RF pulse just after the FIDS vanishes, we observe the same signal as after the first pulse. This indicates that instability is accompanied by the enhanced relaxation. In the HPD experiments instability may be suppressed by the spin supercurrents and manifested only as large dissipation.

3. Discussion

In temperature region of interest the effective collision time τ is not small, i.e. $\omega_L\tau \geq 1$, where ω_L is the Larmor frequency. Correct spin dynamics in that region must be based on the kinetic equation. Recently Markelov (3) proposed an explanation of the catastrophic relaxation, based on such type of a theory, but the physical criterion for the onset of the relaxation used in that paper is not quite convincing. We will discuss here an explanation of the observed anomaly within the framework of the theory of Leggett and Takagi and will obtain a simple criterion for the onset of the anomaly, which is different from that of the paper (3). It is not clear at the moment, whether this difference comes from the approximate character of the Leggett and Takagi equations or it is more essential.

The effective field, acting on the spins in a liquid ^{3}He is a sum of an external magnetic field and a "molecular field" $(\gamma Z_0)/(4\chi_{n0})\,\mathbf{S}$ in the notations of Leggett. The total spin $\mathbf{S}$ in the Leggett and Takagi theory is split in two parts — a spin of the condensate $\mathbf{S}_p$ and a spin of the quasiparticles $\mathbf{S}_q$. Each of this two vectors is treated as an independent dynamical variable. The $\mathbf{S}_p$ and $\mathbf{S}_q$ need not be parallel to $\mathbf{S}$. As the result they can precess around $\mathbf{S}$ under the influence of the torque $(\gamma^2 Z_0)/(4\chi_{n0})\,\mathbf{S}\times\mathbf{S}_p$. The ratio of the frequency of that precession ω_s to the Larmor frequency under the conditions of the experiments is $(Z_0\chi(T))/(4\chi_{n0})$, where $\chi(T)$ is the temperature dependent magnetic susceptibility of ^{3}He $- B$. The Landau parameter Z_0 for ^{3}He is close to -3 for all pressures. For such Z_0 the condition $(Z_0\chi(T))/(4\chi_{n0}) = -1$ can be met at a certain temperature T_r. At that temperature the precession of $\mathbf{S}_p$ and $\mathbf{S}_q$ around $\mathbf{S}$ comes to a resonance with the Larmor precession and the former precession can be excited at the expense of the latter one. This process gives rise to a leak of energy from the Larmor precession, which appears as a relaxation of this precession. The previous criterion with the use of the known expression for $\chi(T)$ can be expressed through the Yosida function $Y(T)$:

$$Y(T) = -2(3/Z_0 + 1) \tag{1}$$

For the values of Z_0 from (4) one arrives at the T_r varying from $0.55\,T_c$ at zero pressure to $0.33\,T_c$ at pressures above 20 bar.

Two interesting features of the criterion (1) are worth mentioning: *1)* T_r is very sensitive to the parameter $3 + Z_0$, which is small in ^{3}He. With a more refined theory the onset of the catastrophic relaxation can be used for the accurate measurement of Z_0. *2)* For low temperatures the two modes of precession in principle can come out of resonance again, but the dependence of χ on T at low temperatures is very flat, so it is not clear, whether such decoupling can really be achieved.

We have shown that the catastrophic relaxation in the HPD is just a particular example of more general phenomenon. With our interpretation this phenomenon is a direct manifestation of the two-fluid nature of the spin dynamics of ^{3}He at sufficiently low temperatures.

The authors are grateful to A.V.Markelov for useful discussions.

References

(1) Yu.M.Bunkov, V.V.Dmitriev, Yu.M.Mukharskiĭ, J.Nyeki, D.A.Sergatskov, *Europhysics Lett. v.***8** *(1989) 645.*

(2) A.S.Borovik-Romanov, Yu.M.Bunkov, V.V.Dmitriev, Yu.M.Mukharskiĭ and K.Flachbart, *Sov. Phys. JETP* **61** *(1985) 1199.*

(3) A.V.Markelov, *To be published in Europhysics letters.*

(4) J.C.Wheatley, *Rev.Mod.Phys.* **47** *(1975) 415.*

Physica B 165&166 (1990) 677–678
North-Holland

ULTRASONIC SPECTROSCOPY OF THE GAP MODE RESONANCE IN SUPERFLUID 3He–B *

E.R. DOBBS, R. LING+ and J. SAUNDERS

Department of Physics, Royal Holloway & Bedford New College, University of London, Egham, Surrey, TW20 0EX, U.K.

The gap (J = 1-) mode in superfluid 3He–B has been studied spectroscopically by measuring the ultrasonic attenuation at pressures less than 4 bar and at frequencies 44.2 and 54.0 MHz in magnetic fields up to 80 mT. The field dependence of two Zeeman components (Jz = ± 1) is compared with the theory of Schopohl and Tewordt.

The gap ($J=1^-$) mode has been shown (1) to split into two components ($J_z=\pm1$), whose total frequency splitting ($\omega_+ - \omega_-$) is in good agreement with the theory of Schopohl and Tewordt (2). The frequencies of these components are given by:

$$\omega_\pm = 2\Delta \pm g\,\Omega ,$$

where the calculated Landé g factor is 0.392 and Ω is the renormalised Larmor frequency:

$$\Omega = \gamma B \{ 1 + (F_0^a/3)(2+Y) \}^{-1}.$$

Here γ is the gyromagnetic ratio, F_0^a is a Landau parameter and Y is the Yosida function. The frequency ω_- produces a peak in the ultrasonic attenuation, but since the frequency ω_+ exceeds 2Δ, it produces a negative contribution to the total attenuation, an antipeak, while as predicted, the $J_z=0$ component is not seen.

If the observed reduced temperatures of the peak and antipeak are denoted by t_- and t_+ respectively, then the reduced frequency splitting of the two components is:

$$\delta(\Delta/\Delta_0) = \{\Delta(t_-) - \Delta(t_+)\}/\Delta_0 ,$$

where Δ_0 is the BCS gap at T=0 and the reduced gap $\Delta(t)/\Delta_0$ is taken at low pressures to be of the BCS form, as tabulated by Mühlschlegel (3). The g factor is then given by $g = (\Delta_0/\Omega)\delta(\Delta/\Delta_0)$. Although the total splitting is in good agreement with theory, when in figure 1 we plot the evolution of the frequency of each component in reduced energy gap units, Δ/Δ_0, as a function of applied field, it is clear that the shift in ω_+ is much larger than in ω_- for the same field. It is possible that this arises from shifts in the frequency of the mode due to gap distortion, but it may also be due to a systematic shift in our measured reduced temperatures with magnetic field.

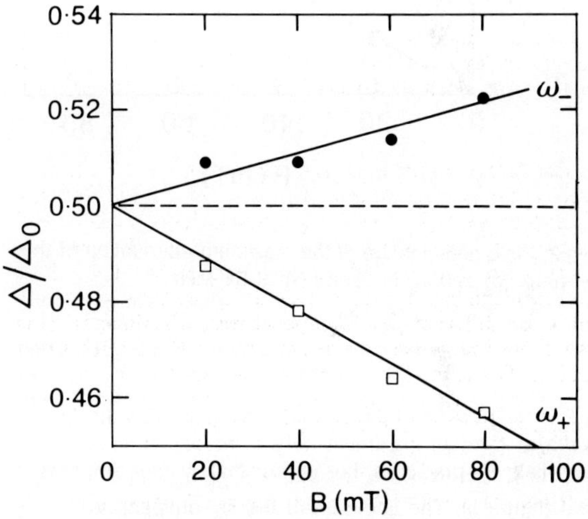

Fig.1. Field dependence of the reduced frequency shift for each component of the $J=1^-$ resonance at 54 MHz, showing their asymmetry.

There are a number of other features of our results which are not in agreement with the theory of Schopohl and Tewordt (2):

1. Theoretically the maximum attenuation, α_{peak}, is proportional to $(\Delta_1 - \Delta_2)^2\Gamma^{-1}$ and hence to Ω^3 or B^3 for both the peak and antipeak. Experimentally the field dependencies of each component are quite different, as shown in figure 2. On the one hand α_{peak} for the ω_- peak increases approximately linearly with B, while on the other, that for the ω_+ antipeak saturates at quite a low level.

* Supported by the U.K. Science and Engineering Research Council with grants GR/D/03628, GR/D/50356 and GR/E/84297.

+ Present address: Oxford Instruments plc, Eynsham, Oxford, OX8 1TL, U.K.

0921-4526/90/$03.50

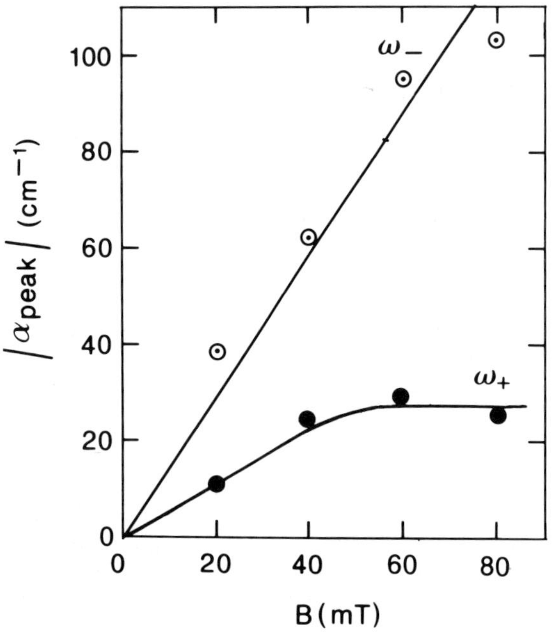

Fig.2. Field dependence of the maximum attenuation of the ω_- peak (o) and ω_+ antipeak (•) at 54 MHz.

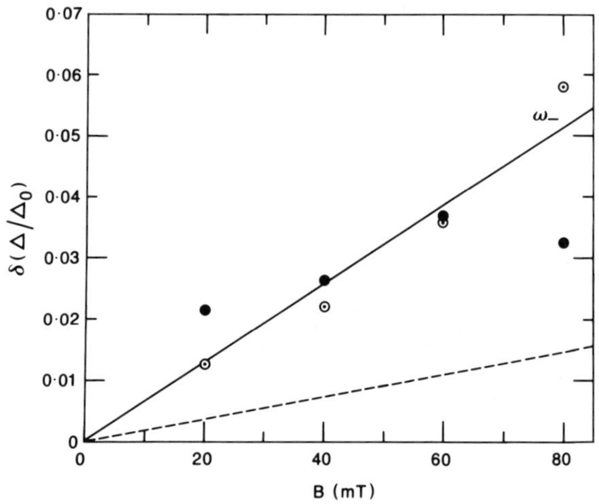

Fig.3. Field dependence at 54 MHz of the line-widths of the ω_- peak (o) and ω_+ antipeak (•) in reduced units compared with theory, $\Gamma/2\Delta_0$(---).

2. The linewidth of the ω_- peak, measured as the full width at the half maximum height, increases linearly with field as predicted, but approximately four times as fast (figure 3). The linewidth of the ω_+ antipeak is unfortunately much more subject to experimental error, but it does show weaker field dependence.

3. The shape of the ω_+ component of the resonance is different from that predicted. In the theory there is an antipeak together with a smaller, broader peak at a higher temperature, but experimentally only the antipeak has been seen. A physical interpretation of the theoretical lineshape might be quantum interference between the discrete state $J=1^-$, $J_z=+1$ and the continuum of excited pair states, by analogy with Fano resonance in atomic physics. In the case of ^{3}He-B mixing would occur between macroscopic wavefunctions, but the effects would only be observable at sufficiently long quasiparticle lifetimes, which may account for their absence from our data.

Schopohl and Tewordt (2) made a number of assumptions in their theory, including neglect of the $J=0^+$ mode, which couples to zero sound through particle hole asymmetry, and the influence of the $J_z=0$ component of the $J=1^-$ mode. In addition, in the presence of a finite $l=3$ pairing interaction, a $J=4^-$ mode is possible. The existence of so many modes in the vicinity of 2Δ, together with possible effects of mode-mode coupling may be responsible for the discrepancies between theory and experiment, when taken together with collision effects. Further details of our data are being published elsewhere (4).

REFERENCES

(1) R.Ling, J.Saunders and E.R.Dobbs, Phys.Rev. Lett. 59 (1987) 461.
(2) N.Schopohl and L.Tewordt, J.Low Temp.Phys. 57 (1984) 601.
(3) B.Mühlschlegel, Z.Phys. 155 (1959) 313.
(4) J.Saunders, R.Ling, W.Wojtanowski and E.R. Dobbs, J.Low Temp.Phys. (in print).

Physica B 165&166 (1990) 679–680
North-Holland

THE INFLUENCE OF VAN DER WAALS FORCES ON T_c IN THIN FILM OF He^3

G.Haran, L.Borkowski*, L.Jacak

Institute of Physics,Technical University of Wroclaw, Wroclaw, Poland,
*Physics Dept.,University of Florida, Gainesville, Fl 32611, USA

Recently it has been reported that the transition temperature to superfluid state of He^3 in thin film is highly affected by various types of substratum. In this note we try to elucidate this effect by the inclusion of the Van der Waals forces into consideration . Though the model used is very simplified the proper behaviour is found.

Recently Harrison et al (1) have reported that the transition temperature to superfluid state of He^3 in thin film is higly reduced from its bulk value. This suppression decreases , however, when the substrate is coated with a monolayer of He^4. Some authors suggested (2) that the absorbed He^4 changes the nature of scattering on substrate from diffuse to specular. In the present short note we intend to check up the influence on transition temperature in thin He^3 film of the Van der Waals forces in the range of substratum . This forces could be also used as the convenient model of various types of substrata (with inclusion of He^4 atoms coating too). Many authors describe this forces via the Lenhard- Jones potential (instead of the infinite potential wall which models the surface of the thin film). In order to check the tendency only we simplify highly this model and consider the potential well in the z direction (perpendicular to the infinite film) according to the formula:

$$V(z) = \{ V>0 \text{ for } 0<z<=a;\ 0 \text{ for } a<z<d \text{ and } \infty \text{ for } z<=0 \text{ or } z>=d \}.$$

The one-particle energy attains then the form: $e_{p,\nu}=f_\nu+p^2/2m$, with $p=(p_x,p_y)$. The spectrum f_ν was found numerically by the use of the quasiclassical methods. It is clear that instead of the homogeneous set of Fermi circles (as for simple well, cf. (3)) we deal now with similar crcle set but inhomogeneous one. Note also that, rougly speaking, in the vicinity of substratum we deal with some mumber (which is dependent of well parameters) of "bound states" . This states will influence the chemical potential of He^3 thin film system. There is easy also to calculate the transition temperature to superfluid A- - phase in such described thin film . The gap equation has the form:

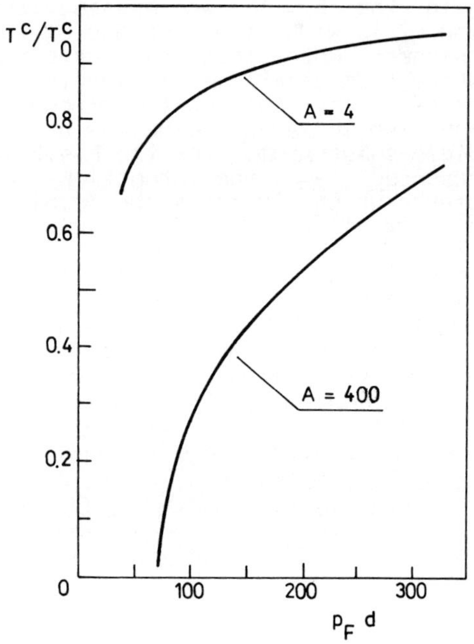

Figure 1
T_c versus the thickness of the film for two types of Van der Waals forces ($A=2mVa^2/\pi^2$)

0921-4526/90/$03.50 © 1990 – Elsevier Science Publishers B.V. (North-Holland)

$$1 = 3gmS/2\pi \sum_{v} \int_0^c de \frac{\sin^2\theta_v}{E_v} \tanh \frac{E_v}{2T}$$

with

$$E_v = (e^2 + \Delta^2 \sin^2\theta_v)^{1/2} ;$$
$$\sin\theta_v = (1 - f_v/\mu)^{1/2},$$

where μ is the chemical potential in thin film for the same density of He^3 as in bulk; S is the area of the film; g is the pairing constant; c is energetical cut-off parameter.
For the transition temperature the folowing formula holds:

$$T_c/T^c_o = \exp((1-1/(3\pi/(2p_F d)\sum \sin^2\theta_v))* \ln(2\, \mathbf{c}\, e^{\gamma} /\pi T^c_o)),$$

where T^c_o is the transition temperature to superfluid state in the bulk and p_F is the Fermi radius in the bulk.
The results for the transition temperature T_c in the thin film of He^3 are presented in the Figs 1 and 2, for several values of parameters for the model potential. In calculations we have assumed $c/T_o^c = 100$, taking in mind that the cut-off parameter in the thin film can differ from its value in the bulk. In the Fig.1 we present the decrease of T_c with the decrease of the film thicness. Note that the curve for A=4 (where $A=2mVa^2/\pi^2$) is similar to the T_c dependence for the substratum coated with He^4, while A=400 coincides better with clean substratum. In the Fig.2 T_c is presented as the function of parameter a which imitates the "radius" of Van der Waals forces.

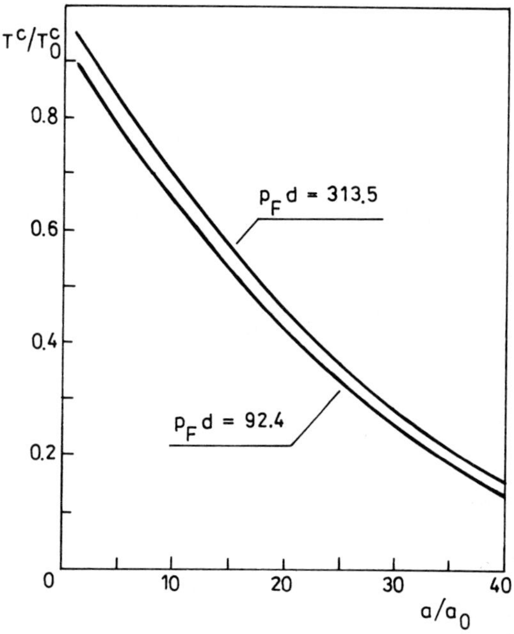

Figure 2
T_c versus parameters of the Van der Waals forces ($2mVa_o^2=4\pi^2$).

REFERENCES
(1) J.P.Harrison, A.Sachrajda, S.C.Steel, P.Zawadzki ,preprint
(2) M.R.Freeman et al, Phys.Rev.Lett. 60 (1988) 596
(3) G.Haran, L.Borkowski,L.Jacak, Physica B 159 (1989) 223

Physica B 165&166 (1990) 681–682
North-Holland

MAGNETIC RELAXATION IN SUPERFLUID ^{3}He-B.

A.S.BOROVIK-ROMANOV, Yu.M.BUNKOV, A.deWAARD*, V.V.DMITRIEV, A.V.MARKELOV, Yu.M.MUKHARSKIY

Institute for Physical Problems, 117334, Ul. Kosygina 2, Moscow, USSR.

D.EINZEL

Walther-Meissner-Institute für Tieftemperaturforschung, Garching, FRG.

We present new experimental and theoretical studies of magnetic relaxation in ^{3}He-B. Methods are developed for the measurement of relaxation using either CW or Pulsed NMR. Both methods make use of the properties of the Homogeneously Precessing Domain. Terms corresponding to intrinsic (Leggett-Takagi), spin diffusion and surface relaxation are distinguished. Theoretical results for the spin diffusion coefficient and the Leggett-Takagi relaxation time, which display nonhydrodynamic anomalies, are compared with experiment.

1. INTRODUCTION

The puzzling properties of magnetic relaxation in superfluid ^{3}He are connected to the transport of magnetization by spin supercurrents (see invited talks at LT-18 (1) and at this conference (2)). In pulsed NMR, the spin supercurrent after an excitation pulse leads to a nontrivial spatial structure of precessing magnetization. In CW NMR, the structure can be excited in the nonlinear NMR regime (3). This spatial structure arises in an inhomogeneous magnetic field and consist of two domains. One domain with stationary magnetization is located in the higher field region. The magnetization in the second domain located in the low field region, precesses uniformly throughout the domain in spite of the inhomogeneity of the magnetic field. We call this the *Homogeneous Precessing Domain, HPD.* The magnetic relaxation of the structure occurs by the spin transport of quasiparticles (*Bogolon gas*), which causes the HPD to shrink and an increase in the size of stationary domain.

Intrinsic magnetic relaxation processes in superfluid ^{3}He are intimately connected with the nonconservation property of the Bogolon magnetization. As a consequence, a mutual disequilibrium between contributions of quasiparticles and condensate to the total magnetization relaxes on the time scale of the intrinsic (Leggett-Takagi) relaxation time (4). A second mechanism of relaxation is connected with the spin diffusion of quasiparticles and takes place in spatially inhomogeneous structures. A third (surface) relaxation mechanism, proposed for ^{3}He-B in Ref.(5), is connected with the distortion of the bulk liquid precession mode near the walls. It turns out that these three relaxation mechanisms have different influences on the decay of the HPD, and can therefore be distinguished experimentally.

2. EXPERIMENTAL RESULTS

The measurement of spin diffusion relaxation from the properties of HPD in pulsed NMR experiments was discussed in Ref.(1). CW NMR studies of the HPD have revealed possibilities for the investigation of different relaxation mechanisms. The HPD is formed by CW NMR in sufficiently strong RF field when the magnetic field is swept down. Details of recent investigations of CW NMR excitation of the HPD and its use for magnetic relaxation measurements can be found in Ref.(3). They are based on the fact that the energy absorbed from the RF field is equal to the energy dissipated in the HPD.

Let us consider the three relaxation processes mentioned above, which lead to an energy dissipation W of the HPD. For a cylindrical chamber of radius R and length L_o with the axis oriented along the field and its gradient one finds:

$$W = \chi H^2 \pi R^2 \left(\frac{4D_\perp \sigma}{5d} + \frac{5}{16}\kappa_{LT}(\nabla\omega)^2 F(L)\right) + W_s 2\pi RL$$

The two terms in brackets refer to spin diffusion through the domain wall and intrinsic relaxation. Here d is the thickness of the domain wall, $D_\perp$ is the transverse component of spin diffusion tensor, σ is a domain wall shape-dependent constant of the order of unity, κ_{LT} is the effective Leggett-Takagi relaxation time and L is the length of HPD. $F(L) = L^3$ if the domain boundary is in the chamber, and $F(L) = (L_o + l)^3 - l^3$ if the HPD fills all the chamber. ($l = \Delta H/\nabla H$, ΔH is the difference between the highest field in the chamber and $-\omega_{rf}/\gamma$). The third term represents the surface relaxation with W_s the surface relaxation rate.

All these terms have different dependences on the dimension L of the HPD. The first two terms are

* Present address North-Holland P.C., Amsterdam, The Netherlands.

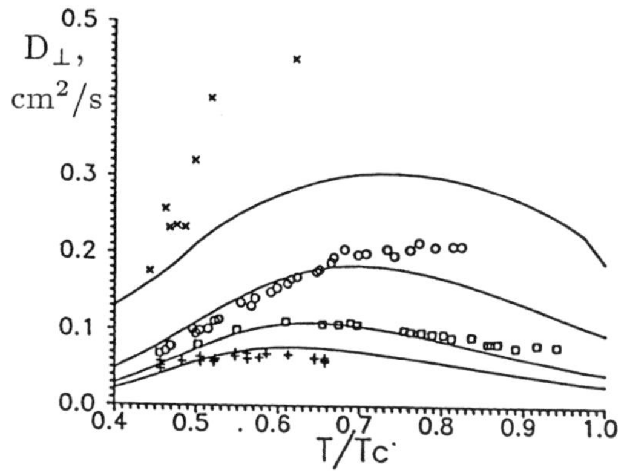

FIGURE 1

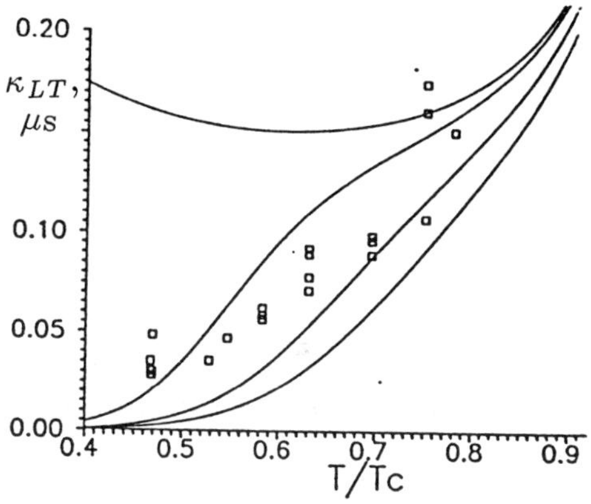

FIGURE 2

determined from NMR data with good accuracy. The third term can be mixed with an additional large dispersion signal. The parameters $D_\perp$ and κ_{LT} can be extracted from first two terms. The results of recent systematic measurements of these parameters are presented in Figs.1 and 2. In Fig.1 points are ploted for the values of $D_\perp$ as a function of temperature in a magnetic field 144 Oe and at 0 bar (+), 11 bar ($\diamond$), 20 bar ($\circ$) and 29 bar ($\times$). From these data one can clearly see the depression of the effective diffusion at low temperatures. In Fig. 2 points are plotted for the value of κ_{LT} at 920 kHz and 29.3 bar. The depression of the Leggett-Takagi relaxation time is also very clear.

In order to estimate the value for the surface relaxation we have performed additional measurements in a cell with increased surface area using mylar foils. The surface area of this cell was 3 times larger, but otherwise the dimensions were the same as the first. The value of W_s for pressure 11 bar and $T = 0.6T_c$ was estimated as $0.01 nW/cm^2$ with 50% accuracy. It is in agreement a with theoretical estimate of $0.02 nW/cm^2$ (5).

3. COMPARISON WITH THEORY

In the collisionless limit $\omega\tau \gg 1$ the quasiparticle dynamics of a (superfluid) Fermi liquid is known to be governed by the reactive forces originating from the Fermi liquid interaction. One example is the blocking of the spin diffusion in a normal Fermi liquid, known as the Leggett-Rice effect. In short our theoretical results for the quasiparticle spin diffusion in ^{3}He-B can be summarized by saying that the Leggett-Rice effect carries over into the superfluid, where it remains as a property of the Bogolon gas. The major difference is that in the superfluid phase the decay time τ_D of the quasiparticle magnetization current increases exponentially at low temperature $\propto \exp(\Delta/kT)$ instead of $\propto T^{-2}$ in the normal Fermi liquid. This means that the blocking of diffusion happens much more rapidly in the superfluid. The details of a calculations of the transverse component of the quasiparticle spin diffusion tensor will be published elsewhere (6). The full lines in Fig.1 are our theoretical result for transverse spin diffusion. Except for the lowest preassure, where the applicability of the theory fails, there is quantitative agreement between theory and experiment indicating that the observed maximum structure is indeed a signature of a "*superfluid Leggett-Rice effect*".

It can be shown that the nonhydrodynamic effect also modifies the intrinsic relaxation. The effective time constant of Leggett-Takagi relaxation at low temperatures should be depressed as (7) $\kappa_{LT}(\omega) = \kappa_{LT}(0)/(1 + A(\omega\tau_{LT}(0))^2)$ where A is a parameter below 1. Its temperature dependence is recently under theoretical investigation. The theoretical results for the effective Leggett-Takagi relaxation time are plotted as a full lines in Fig.2 for the value of the parameter A = 0, 0.1, 0.5 and 1. These results clearly demonstrate the nonhydrodynamic depression of this relaxation parameter.

CONCLUSION

In summary we have demonstrated theoretically and verified experimentally that the spin transport properties of superfluid ^{3}He-B, namely both the transverse spin diffusion coefficient and the Leggett-Takagi time, are strongly reduced in the low temperature limit. In conclusion, there are no remaining mysteries in the relaxation processes in superfluid ^{3}He-B at temperatures down to $0.4T_c$. At lower temperatures, however, a new very fast process of relaxation, named catastrophic, takes place[8], and this awaits an explanation.

REFERENCES

(1) Yu.M.Bunkov Jap. J. Appl. Phys. **26**(3), 1809, (1987).
(2) I.A.Fomin This conference.
(3) A.S.Borovik-Romanov, Yu.M.Bunkov, V.V.Dmitriev, Yu.M.Mukharskiy, K.Poddyakova, O.D.Timofeevskaya, Zh. Exp. Teor. Fiz. (Sov. Phys. JETP), **96**, 956, (1989).
(4) A.J.Leggett, S.Takagi. Ann. Phys. **106**, 79, (1977).
(5) T.Ohmi, M.Tsubota, T.Tsuneto. Jap. J. Appl. Phys. **26**(1), 169 (1987).
(6) Yu.M.Bunkov, V.V.Dmitriev, A.V.Markelov, Yu.M.Mukharskiy, D.Einzel, to be published.
(7) A.V.Markelov. Sov.Phys JETP **64**, 1218 (1986).
(8) Yu.M.Bunkov, V.V.Dmitriev, Yu.M.Mukharskiy, J.Nyeki, D.A.Sergatskov. Europhys.Lett., **8**, 645 (1989).

Physica B 165&166 (1990) 683–684
North-Holland

THE FORCE ON A WIRE MOVING THROUGH SUPERFLUID ^{3}He-B

R. J. WATTS-TOBIN and S. N. FISHER

Department of Physics, University of Lancaster, Lancaster LA14YB, England

When a wire moves through superfluid ^{3}He it experiences a force due to quasiparticle scattering. We develop a three-dimensional microscopic model to calculate the force. At large (subcritical) wire velocity v the force tends to an asymptotic value as $1/v^2$, rather than exponentially as in a one- dimensional calculation. At low v the force is proportional to v. In this range our calculated form of the variation of the force with v is in good agreement with experiment.

Fisher *et al* [1] have measured the force on a wire moving through the B phase of ^{3}He. They find the force on the wire tends to zero linearly as $v \to 0$, and it tends to an asymptotic value for $v \gg kT/\hbar k_F$. We have calculated the force by means of an approximate solution of the Bogoliubov equations.

The wire is modelled by a smooth cylinder. Owing to its macroscopic size, the superfluid flow round it is purely dipolar (for $v < \Delta/\hbar k_F$). The quasiparticle spectrum can be calculated by the method of Cyrot [2]. The scale on which the superfluid velocity varies is very long compared with the superfluid coherence length. The helium is imagined to be divided into pieces, large enough to treat each piece as a bulk sample, and small enough so that the superfluid velocity is constant in it.

In the rest frame of the wire a quasiparticle with excitation energy ξ above the Fermi level and momentum $\hbar k_F \hat{\mathbf{p}}$ has energy

$$E = v_F \hat{\mathbf{p}} \cdot \mathbf{q} + \sqrt{\xi^2 + \Delta^2}.$$

Here $\mathbf{q}$ is the local superfluid momentum per atom. The quasiparticle has amplitude $u = \sqrt{\frac{1}{2}[1 + \xi/(E - v_F \hat{\mathbf{p}} \cdot \mathbf{q})]}$ to be a particle and amplitude $v = \sqrt{1 - u^2}$ to be a hole.

Specular reflection at an element $\hat{\mathbf{n}}\, dS$ of the wire surface gives a normal force

$$\mathbf{F} = -2\hbar k_F \hat{\mathbf{n}}\, dS \sum (\hat{\mathbf{p}} \cdot \hat{\mathbf{n}})^2 (u^2 - v^2).$$

The sum is over quasiparticles which are moving towards the wire, i.e. $\hat{\mathbf{p}} \cdot \hat{\mathbf{n}} (u^2 - v^2) < 0$. The terms in the sum need to be weighted by a thermal occupation factor. We assume that this is an equilibrium distribution in the frame of the container, and that only quasiparticle states which extend to the container are populated. In the wire frame the superfluid momentum near the container is $\mathbf{Q} = -m\mathbf{v}$, where $\mathbf{v}$ is the wire velocity in the container frame. The probability of occupation of a quasiparticle state with energy E is $\exp(-E + v_F \hat{\mathbf{p}} \cdot \mathbf{Q})/kT$ for $E > \Delta + v_F \hat{\mathbf{p}} \cdot \mathbf{Q}$. This shift of the occupation probability with $\mathbf{Q}$ and the the asymmetry of the excitation spectrum due to the local value of $\mathbf{q}$ enhances the force by several orders of magnitude by comparison with that predicted for a classical gas of particles with momenta p_F.

At a point on the wire with polar angle θ the superfluid momentum is tangential with $q = 2Q \sin\theta$. For simplicity we have assumed that in a quasiparticle state with energy E the excitation energy ξ varies with the local value of $\mathbf{q}$, but not the direction $\hat{\mathbf{p}}$. The variation of $\hat{\mathbf{p}}$ with $\mathbf{q}$ is being taken into account in a calculation which extends the theory to triplet pairing with $\hat{\mathbf{p}}$-dependent Δ.

The force in the direction of $\mathbf{Q}$ per unit length on the wire becomes

$$F = \frac{32}{3\pi} \hbar k_F v_F a N(0) kT \exp\left(-\frac{\Delta}{kT}\right) \int_0^{\frac{\pi}{2}} \sin^3 x\, dx \int_0^{\frac{\pi}{2}} \sin y\, dy \left(1 - \exp \frac{v_F Q \sin x \sin y}{kT}\right).$$

a is the radius of the wire, and x and y are derived from the polar angles of $\hat{\mathbf{p}}$ and the position θ on the wire. For $v_F Q \ll kT$,

$$F \sim \frac{\pi}{2} \hbar k_F v_F{}^2 a N(0) \exp\left(-\frac{\Delta}{kT}\right) Q;$$

and for $kT \ll v_F Q < \Delta$,

$$F \sim \frac{64}{9\pi} \hbar k_F v_F a N(0) kT \exp\left(-\frac{\Delta}{kT}\right) \left[1 - \frac{3}{2}\left(\frac{kT}{v_F Q}\right)^2\right].$$

0921-4526/90/\$03.50

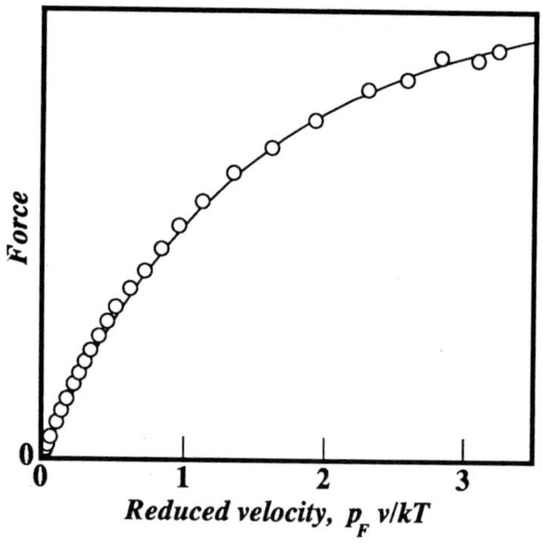

The Figure shows the experimental data of Fisher *et al* for a vibrating wire below its critical velocity, at a temperature below $0.2\,T_C$. The experimental points are represented by circles. Also shown is our theoretical curve adjusted (in magnitude only) by a factor of order unity. The agreement is fair. However the comparison does not take full account of the fact that in the experiment different parts of the wire are not all moving with the same velocity.

REFERENCES

1. Fisher, S. N., Guénault, A. M., Kennedy, C. J., and Pickett, G. R., Phys. Rev. Lett. 63 (1989) 2566.
2. Cyrot, M., Phys. kondens. Materie 3 (1965) 374.

Physica B 165&166 (1990) 685–686
North-Holland

THE INFLUENCE OF TEXTURES ON COLLECTIVE MODES IN SUPERFLUID HELIUM-3

P.N. BRUSOV, M.M. BUKSHPUN

Physical Research Institute, Rostov-on-Don State University, USSR

By path integration method developed by Brusov and Popov earlier we investigated the influence of textures on collective excitations in superfluid He-3. For arbitrary textures we have obtained the quadratic form of the hydrodynamic action functional, which determined the collective mode spectrum. Using this quadratic form we have calculated the whole collective mode spectrum for B-phase of helium-3 closed to boundary. The texture splits the real squashing, squashing and pair breaking-modes and leads to appearance of gaps in goldstone-mode spectrum.

1. INTRODUCTION

The textures in superfluid phases of He^3 are studied by ultrasound and NMR experiments. Magnetic fields and restricted geometry create different kinds of textures. In A-phase we have $\vec{\ell}$- and $\vec{d}$-textures, in B-phase-$\vec{n}$- textures. The theoretical base for studying the textures is the theory of collective excitations. Unfortunately we have not a microscopic theory of influence of textures on collective mode spectrum up to now. We could mention only the phenomenological theory of six-fold splitting (ten-fold really) of real squashing (rsq) mode in He^3-B by Volovik(1) and its quasiclassical development by Sauls and Fishman.(2)

In this paper we tried to develop general approach to studying the influence of textures on collective mode spectrum in superfluid phases of He^3.

2. MODEL

We use the path integration technique, which was developed for investigation of superfluid He^3 by Brusov and Popov(3). They used the model of He^3 which was described by the hydrodynamic action function S_h, which was obtained after integration over the "fast" and "slow" fermi-fields

$$S_h = g^{-1}\sum_p c^+_{ia}(p)c_{ia}(p) + \frac{1}{2}\,\ell n \det \frac{M(c^+, c)}{M(c^{+(0)}, c^{(0)})}$$

Here $c_{ia}(p)$ are bose-fields, which describe the Cooper pairs of quasifermions near the Fermi-surface, p - is the 4-momentum. In first approximation the collective spectrum is determined by quadratic form of S_h, which is obtained after shift $c_{ia}(p) \to c^{(0)}_{ia}(p) + c_{ia}(p)$ in S_h, where $c^{(0)}_{ia}(p)$ is the condensate wave function. In homogeneous case $c^{(0)}_{ia}(\vec{x},\tau) \sim \delta_{\omega o} \sum_q e^{i\vec{q}\vec{x}}\, \tilde{c}_{ia}(\vec{k})$. Using this order parameter we obtain the quadratic form of S_h in nonhomogenous case (with some simplifications).

3. COLLECTIVE MODE SPECTRUM IN He^3-B WITH BOUNDARY

We apply our quadratic form of S_h to investigate the influence of boundary on collective mode spectrum in He^3-B. As Fujita et al.(4) have shown in this case we have B-phase far from boundary and 2D phase close to it. In intermediate region we have the order parameter(4)

$$c^{(0)}_{ia} = f(z)\begin{pmatrix}1&0&0\\0&1&0\\0&0&1\end{pmatrix} + \big(1-f(z)\big)\lambda'_0\begin{pmatrix}1&0&0\\0&1&0\\0&0&0\end{pmatrix}$$

or in $\vec{k}$ - representation

$$c^{(0)}_{ia} \sim \begin{pmatrix}\Delta_1&0&0\\0&\Delta_1&0\\0&0&\Delta_2\end{pmatrix}$$

Here $\hat{z}$ is perpendicular to the boundary. Using this order parameter, we have calculated the whole bose-spectrum in He^3-B in presence of texture in terms of Δ_1 and Δ_2.

4. RESULTS

Below we represent the results for frequencies of collective modes

a) gd-modes

$$\omega^2 = \frac{2}{3}\Delta_1(\Delta_2-\Delta_1), \qquad u_{21}-u_{12}$$

$$\omega^2 = \frac{16}{5}\Delta_1(\Delta_2-\Delta_1), \qquad u_{13}-u_{31},\ u_{23}-u_{32}$$

$$\omega^2 = A(\gamma), \qquad v_{11}+v_{22}+v_{33},$$

here $A(\gamma) \sim 0(\gamma)$, where $\gamma = \dfrac{\Delta_2-\Delta_1}{\Delta_1}$.

b) pb-modes

$$\omega^2 = \left(4 + \frac{32}{15}\gamma\right)\Delta_1^2, \qquad v_{13}-v_{31},\ v_{23}-v_{32}$$

$$\omega^2 = \left(4 + \frac{4}{3}\gamma\right)\Delta_1^2, \qquad v_{21}-v_{12}$$

$$\omega^2 = B(\gamma), \qquad u_{11}+u_{22}+u_{33},$$

here $B(\gamma) \to 4\Delta^2$, if $\gamma \to 0$.

c) rsq-modes

$$\omega^2 = \frac{4\Delta_1}{15}(\Delta_1 + 5\Delta_2), \qquad u_{21}-u_{12}$$

$$\omega^2 = \frac{8}{15}(4\Delta_2-\Delta_1), \qquad u_{13}-u_{31},\ u_{32}-u_{23}$$

$$\left.\begin{aligned}\omega^2 &= c_1(\gamma)\\ \omega^2 &= c_2(\gamma)\end{aligned}\right\} \begin{aligned}&u_{11}+u_{22}-2u_{33}\\ &u_{11}-u_{22}\end{aligned}$$

here $c_1(\gamma)$, $c_2(\gamma) \to \left(\frac{8}{5}\Delta\right)^2$, if $\gamma \to 0$.

d) sq-modes

$$\omega^2 = \frac{4}{5}\left(3-\frac{2}{3}\gamma\right)\Delta_1^2, \qquad u_{13}-u_{31},\ u_{23}-u_{32}$$

$$\omega^2 = \frac{4}{5}\left(3+\frac{5}{3}\gamma\right)\Delta_1^2, \qquad v_{21}-v_{12}$$

$$\left.\begin{aligned}\omega^2 &= D_1(\gamma)\\ \omega^2 &= D_2(\gamma)\end{aligned}\right\} \qquad \begin{aligned}&v_{11}-v_{22},\\ &v_{11}+v_{22}-2v_{33}\end{aligned}$$

here $D_1(\gamma)$; $D_2(\gamma) \to \left(\frac{12}{5}\Delta\right)^2$, if $\gamma \to 0$.

5. CONCLUSIONS

We could see that there is 3-fold splitting of gd-modes and pb-modes and 4-fold splitting of rsq- and sq-modes. Twice degeneracy of one mode of each type is connected with the symmetry of x and y axes.

REFERENCES

(1) G.E. Volovik, Soviet Phys. JETP Letters, 39 (1984), 304.

(2) J.A. Sauls and R.S. Fishman, Phys. Rev. Letts. 61 (1988) 2871.

(3) P.N. Brusov, V.N. Popov "The superfluity and collective properties of quantum fields" Nauka, Moscow, 1988.

(4) T. Fujita, M. Nakahara, T. Ohmi et al. Progr. Theor. Phys. 64 (1980), 396.

Physica B 165&166 (1990) 687–688
North-Holland

THE INFLUENCE OF ELECTRIC FIELDS ON COLLECTIVE EXCITATIONS IN HELIUM-3-B

P.N. BRUSOV, M.V. LOMAKOV

Physical Research Institute, Rostov-on-Don State University, USSR

The collective excitation spectrum of He-B is investigated in the presence of electric fields (EF). We have taken the gap distortion caused by EF into account and considered the case without dipole interaction as well as with it. It ha s been shown that a 3-fold splitting of the real squashing, squashing and pair breaking modes took place. To observe this splitting experimentally one needs to have the fields of order from $5*10^5$ up to $5*10^6$ V/cm (in dependence on pressure). Thus this splitting could be observed at part of phase diagram (at not too high pressure). EF does not change the goldstone mode spectrum (it changes only sound velocity).

1. INTRODUCTION

In the absence of electric field (FE) He^3 atoms have not electric dipole moments and thus electric dipole interaction is equal to zero. External electric field polarizes the He^3 -atoms and interaction appearances between their induced moments. This interaction as well as magnetic dipoles one leads to set of interesting effects. Delrie(1) and Maki(2) have shown that in anisotropic A-phase EF causes the orientational effect-orbital moments $\vec{\ell}$ of Cooper pairs turn perpendicular to EF $\vec{E}$ in fields of order 10^4 V/cm. As Fomin et al. have shown(3) Fermi-liquid corrections increase this value up to $10^5 \div 10^6$ V/cm (in dependence on pressure).Another interesting effect is the change of collective mode spectrum in EF.

2. INFLUENCE OF EF ON COLLECTIVE MODES.

Brusov and Popov(4) have investigated the influence of collective modes by path integration technique and shown that EF changed only one goldstone (gd) mode spectrum the sound one, which velocity decreased in perpendicular to EF direction and did not change it along EF. Nonphonone modes are not changed. But as Brusov has shown later (5) this conclusion was connected with the fact that in (4) authors did not take the gap distortion caused by EF into account. The order parameter in B-phase, for example, changes under turning on the EF from

$$c^{(0)}_{ia}(p) \sim \Delta \begin{pmatrix} \cos\theta & -\sin\theta & 0 \\ \sin\theta & \cos\theta & 0 \\ 0 & 0 & 1 \end{pmatrix} \quad \text{to}$$

$$c^{(0)}_{ia}(p) \sim \begin{pmatrix} \Delta_1 & 0 & 0 \\ 0 & \Delta_1\cos\theta & -\Delta_1\sin\theta \\ 0 & \Delta_2\sin\theta & \Delta_2\cos\theta \end{pmatrix}$$

Where Δ, Δ_1, Δ_2 are gaps in Fermi-spectrum without EF, perpendicular and parallel to EF consequently, θ is the rotation angle of spin system of coordinate in relative of orbital one over $\vec{n}$-axis. Last form of order parameter takes two effects of EF into account: orientational ($\vec{n} \perp \vec{E}$) and distortion ($\Delta_1 \neq \Delta_2$) one. In absence of the dipole interaction (DI) angle θ remains arbitrary while in presence of it $\theta \approx \arccos(-1/4)$. For our calculation, we used Brusov and Popov model of He^3, which was described by hydrodynamic action functional

$$S_h = g^{-1} \sum_{p,i,a} c^+_{ia}(p)c_{ia}(p) + \frac{1}{2}\ell n\det [\hat{M}(c,c^+)/\hat{M}(c^{(0)},c^{(0)+})]$$

In first approximation the bose-spectrum is determined by the quadratic form of S_h, obtaining by the shift $c_{ia}(p) \to c^{(0)}_{ia}(p) + c_{ia}(p)$ in S_h

$$\sum_p A_{ijab}(p)c^+_{ia}(p)c_{jb}(p) + \frac{1}{2}B_{ijab}(p)[c_{ia}(p)c_{jb}(-p) + c^+_{ia}(p)c^+_{jb}(-p)]$$

The equation for bose-spectrum is det Q = 0, where Q is the matrix of the quadratic from which tensor coefficients are proportional to the integrals (sums) of the products of Green functions of quasifermions and equal

$$A_{ijab} = \delta_{ab}\left\{\frac{\delta_{ij}}{\xi} + \frac{4z^2}{\beta V}\sum_{p_1+p_2=p} n_{1i}n_{1j}\right.$$

$$\left.\frac{(i\omega_1+\xi_i)(i\omega_2+\xi_2)}{[\omega_1^2+\xi_1^2+\Delta^2(\theta')][\omega_2^2+\xi_2^2+\Delta^2(\theta')]}\right\}$$

$$B_{ijab} = -\frac{4z^2}{\beta V}\sum_{p_1+p_2=p} n_{1i}n_{1j}$$

$$\frac{\left(2f_af_b-f_i^2\delta_{ab}\right)}{[\omega_1^2+\xi_1^2+\Delta^2(\theta')][\omega_2^2+\xi_2^2+\Delta^2(\theta')]}$$

Here

$$\Delta^2(\theta') = \Delta_1^2\left(n_1^2+n_2^2\right)+\Delta_2^2n_3^2 =$$

$$\Delta^2+\Omega_0^2\left[\alpha_o+\left(n_1^2+n_2^2\right)\left(1-\alpha_o\right)\right]$$

$$n_1^2+n_2^2=\sin^2\theta',\quad \vec{f}=\left\{n_1\Delta_1, n_2\Delta_1\cos\theta' +n_3\Delta_2\sin\theta'; -n_2\Delta_1\sin\theta'+n_3\Delta_2\cos\theta'\right\}.$$

3. RESULTS

We have calculated whole collective mode spectrum in B-phase in presence of EF by method described above for both cases with DI and without it (in this case we put $\theta = 0$ for simplicity) and came to next conclusions. In accordance with result of Brusov and Popov(4) we obtained that all gd-modes remained gapless in the EF. All kinds of nonphonone modes (squashing (sq) with $\omega = \sqrt{12/5}\Delta$, real squashing (rsq) with $\omega = \sqrt{8/5}\Delta$ and pair breaking (pb) with $\omega = 2\Delta$ obtained the corrections of order E^2. EF leads to 3-fold splitting of all of these modes (sq, rsq, pb). We estimated the EF value, which is needed to observe the splitting of sq-, rsq- and pb-modes experimentally. We obtained the value (with taking the Fermi liquid correction into account) $E \sim 5\cdot10^5 \div 5\cdot10^6$ V/cm. Because the critical value of EF in He^3 is equal $E_c = 2.7\cdot10^5$ V/cm, it's possible to observe 3-form splitting of nonphonone modes at part of phase diagram (at not too high pressure). The experiments over investigation of influence of EF on collective modes in superfluid He^3 is under preparation at Northwestern University now.

We would like to thank V. Popov, I. Fomin, G. Volovik, M.Y. Nasten'ka, T.V. Filatova-Novoselova, and J.B. Ketterson for discussions.

REFERENCES

(1) T.M. Delrieu, J. de Physique Lett., 35 (1974), 189.
(2) K. Maki, J. Low Temp. Phys., 24 (1976), 755.
(3) I.A. Fomin, C.I. Pethick, I.W. Serene, Phys. Rev.Lett, 40 (1978) 1144.
(4) P.N. Brusov, V.N. Popov, Soviet J. of Theor. and Math. Phys. 57 (1983), 249.
(5) P.N. Brusov, Soviet Phys. JETP, 88 (1985), 1197.

Physica B 165&166 (1990) 689–690
North-Holland

NMR MEASUREMENTS OF ^{3}HE IN THIN ^{3}HE-^{4}HE MIXTURE FILMS

R. H. Higley, D. T. Sprague, and R. B. Hallock

Laboratory for Low Temperature Physics, Department of Physics and Astronomy
University of Massachusetts, Amherst, Massachusetts 01003

Evolution of ^{3}He from a 2-d Fermi liquid to bulk-like liquid on thin ^{3}He-^{4}He mixture films shows stepped structure in magnetization and NMR relaxation times as a function of ^{3}He coverage d_3. Spin-echo measurements in a weakly polarizing field over the temperature range $30 < T < 250$ mK for $0.01 < d_3 < 4$ layers are compared at three ^{4}He coverages $d_4 = 1.77$, 2.14, 2.91 atomic layers.

^{3}He-^{4}He mixture films in which the ^{4}He is thick enough so that adsorbed ^{3}He is fluid rather than solid provide an excellent system for studies of the evolution of the ^{3}He from a dilute 2-d gas to bulk-like fluid. They may also allow studies of the nature of the ^{4}He surface. NMR spin-echo measurements of the magnetization (1) and relaxation times of the ^{3}He in such mixture films adsorbed on Nuclepore show interesting structure as a function of ^{3}He coverage.

The measurements reported here were taken in three experimental runs, at fixed ^{4}He coverages $d_4 = 40, 44, 54\,\mu$mol/m^2 (thicknesses 1.77, 2.14, 2.91 atomic layers). In each run successive amounts of ^{3}He were added to span ^{3}He coverages $0.01 < d_3 < 4$ layers, where $d_3 = 1$ layer $\equiv 1\text{atom}/(3.93\text{Å})^2$, and the magnetization and T_2 were measured by the $90 - \tau - 180$ spin-echo technique over the temperature range (30 or 40) $< T \leq 250$mK in an $H_0 = 2$T field. The runs were in the order $d_4 = 2.14$, 2.91, 1.77 layers; systematic T_1 measurements were made only in the final run.

The lowest temperature data of each run is shown in Fig. 1. For each ^{4}He coverage d_4, the magnetization of the sample as a function of d_3 shows a plateau, a step near $d_3 = 0.75$ layers, a second plateau, and, for $d_3 > 2$ layers, a linear increase in magnetization with slope roughly that expected for bulk ^{3}He liquid. Below 0.05 layers, the Curie law behavior at higher temperatures provides an absolute calibration of the spectrometer with an uncertainty of 6%. Using this calibration, the data are normalized by $M_0 = N\mu_m^2 H_0/k_B T_F(m_3)$, the magnetization at $T = 0$ of a weakly polarized 2-d ideal Fermi gas of N particles each with the mass m_3 of a bare ^{3}He atom. Note that in two dimensions quasiparticles occupy a Fermi disk, and the Fermi temperature T_F is proportional to N/A. ($A = 1.77m^2 \pm 10\%$ is the area of the sample cell.) Thus M_0 is independent of coverage.

The first plateau shows the degenerate magnetization of a 2-d Fermi liquid. The magnetization is greater than $M/M_0 = 1$ and increases with coverage, due to interactions among the quasiparticles. The step-like approximate doubling of the magnetization near $d_3 = 0.75$ layers is interpreted as occupation of a second Fermi disk, corresponding to a ^{3}He state of higher discrete energy level in the mixture film. ^{3}He cover-

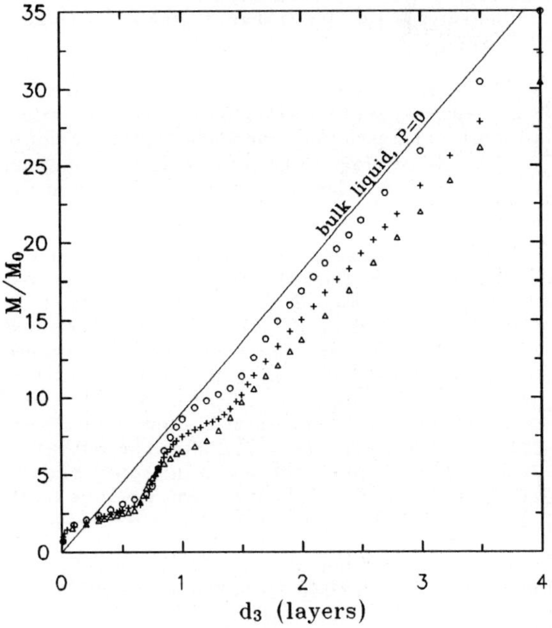

FIG. 1. Magnetization of the sample vs ^{3}He coverage d_3 for the ^{4}He coverages 1.77 (circles, 30 mK, uppermost data), 2.14 (plusses, 40 mK), and 2.91 atomic layers (triangles, 30 mK).

ages just less than that of the step show magnetization increasing with temperature, indicating thermal excitation into the second level. The temperature and ^{3}He coverage dependence of M/M_0 seen in the region of the two plateaus and step has been analyzed in terms of a 2-level 2-d Fermi liquid (2). Near $d_3 = 1.4$ layers is a hint of a second step, but the following linear increase of magnetization as for bulk liquid suggests that the model of the ^{3}He liquid as multiple levels of a 2-d Fermi liquid breaks down beyond two levels. Heat capacity measurements on mixture films of greater d_4 showed a merging of multi-level 2-d, and 3-d models beyond a 2-level 2-d model (3); no step structure was seen in surface tension measurements of ^{3}He on a bulk mixture (4). A model of a free standing ^{3}He film of variable thickness yields an unlimited sequence of levels (5).

Each of the three ^{4}He coverages shows common structure. Differences are noticeable, though, perhaps due to the increased pressure of the ^{3}He atop thinner coverages of ^{4}He, or to the detailed nature of the ^{4}He surface (6,7). Linear fits to the magnetization vs d_3 for each ^{4}He coverage in Fig. 1 for $0.1 < d_3 < 0.6$ layers, extrapolated to $d_3 = 0$ give $m_{hydro}/m_3 = 1.45$, 1.41, 1.34, and normalized slopes $(m_3/m_{hydro})d(M/M_0)/dd_3 = 2.26$, 1.87, 1.65. The increase in hydrodynamic mass m_{hydro}, and the increase in slope of the first plateau (8), with decreasing ^{4}He film thickness is as expected for quasiparticles on a denser film (9,10). The factor by which the magnetization increases at the step is slightly greater for smaller ^{4}He coverage.

The slope of magnetization vs d_3 in the range $2 < d_3 < 4$ layers is seen to be close to the bulk liquid value, and constant for each run. One might expect the slopes for each ^{4}He coverage to be converging to a common value before $d_3 = 4$ layers, but this is not seen in the data.

Relaxation measurements in this system show a step-like increase at the same coverage as the first step in magnetization. Fig. 2 shows measurements of T_1 at $T = 40\,\mathrm{mK}$ and $d_4 = 1.77$ layers, as determined by $90 - \tau_{long} - 90 - \tau - 90$ spin echos. The pulse energy caused negligible heating during these measurements. T_2 measurements, affected somewhat by pulse heating at the lowest temperatures, show qualitatively similar behavior, with $T_2 \approx T_1/200$. The relaxation measurements are more difficult to interpret, but the step feature near $d_3 = 0.75$ layers must be due to an overall change in transport of the ^{3}He.

This work was supported by the National Science Foundation through DMR 85-17939 and 88-20517.

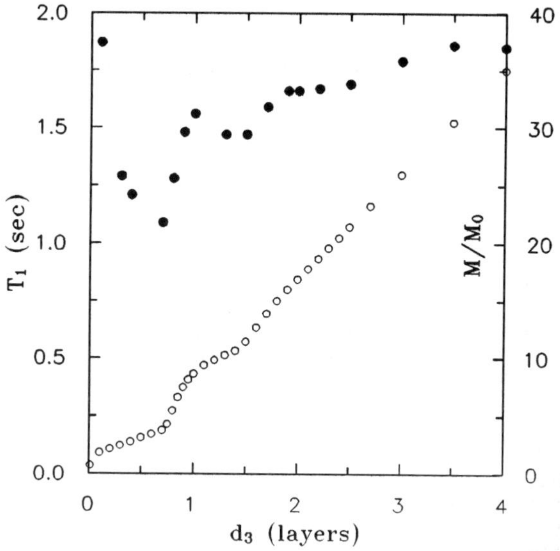

FIG. 2. Relaxation time T_1 measured at $d_4 = 1.77$ layers and $T = 40\,\mathrm{mK}$ (solid circles), shown with magnetization at $T = 30\,\mathrm{mK}$ (open circles).

REFERENCES

(1) R. H. Higley, D. T. Sprague, and R. B. Hallock, in *QuantumFluidsandSolids – 1989*, ed. G. G. Ihas and Y. Takano, AIP Conference Proc. No. 194 (AIP, New York, 1989), p. 225.
(2) R. H. Higley, D. T. Sprague, and R. B. Hallock, Phys. Rev. Lett. **63**, 2570 (1989).
(3) M. J. Dipirro and F. M. Gasparini, Phys. Rev. Lett. **44**, 269 (1980).
(4) H. M. Guo, D. O. Edwards, R. E. Sarwinski, and J. T. Tough, Phys. Rev. Lett. **27**, 1259 (1971).
(5) R. A. Guyer, K. R. McCall and D. T. Sprague, Phys. Rev. B **40**, 7417 (1989).
(6) N. Pavloff and J. Treiner (preprint); F. Dalfovo and S. Stringari, Physica Scripta **38**, 204 (1988).
(7) E. Krotscheck, Phys. Rev. B **32**, 5713 (1985); E. Krotscheck, M. Saarela and J. L. Epstein, Phys. Rev. B, **38**, 111 (1988).
(8) E. Krotscheck, M. Saarela and J. L. Epstein, Phys. Rev. Lett. **61**, 1728 (1988).
(9) X. W. Wang and F. M. Gasparini, Phys. Rev. B, **38**, 11245 (1988).
(10) J. L. Epstein, E. Krotscheck and M. Saarela, Phys. Rev. Lett. **64**, 427 (1990).

Physica B 165&166 (1990) 691–692
North-Holland

FERMI LIQUID PARAMETERS OF A 2D ^3He FILM

C.P.LUSHER, J.SAUNDERS and B.P.COWAN

Millikelvin Laboratory, Royal Holloway and Bedford New College, University of London, Egham, Surrey, TW20 0EX, U.K.

A temperature independent magnetic susceptibility has been observed for the second layer of ^3He on graphite for second layer surface densities less than 0.055 $Å^{-2}$, consistent with 2D Fermi liquid behaviour. The Landau parameter F_0^a is determined using known values of m^*/m. The relative dependence of these two parameters is in good agreement with almost localised Fermion theory, as is the case in bulk liquid ^3He.

The first layer of ^3He on graphite forms a highly compressed incommensurate solid . The second layer is fluid up to a surface density in the second layer of 0.055 $Å^{-2}$ whereupon a solid-fluid coexistence region is entered due to the formation of a 2D solid with antiferromagnetic exchange (1,2). It is of interest to establish whether the second layer fluid is a 2D Fermi liquid and , if so, what is the density dependence of the interactions. Since there is no evidence for condensation of this system it can be studied from the weakly interacting gas through to the strongly interacting fluid. Furthermore this system should be more strongly two dimensional than the related system of ^3He bound at the surface of liquid ^4He films due to the greater binding energy and the incompressibility of the "substrate".

The heat capacity of ^3He films adsorbed on graphite at total surface densities in the range $0.11<\rho<0.16$ has recently been measured (2) to be linear in temperature below approximately 50mK. This result was interpreted as evidence for 2D Fermi liquid behaviour and the effective mass inferred. A small intercept at T=0 was attributed to weak substrate heterogeneity. At these temperatures the heat capacity of the first highly compressed solid layer is negligible.

Another thermodynamic quantity of interest is the nuclear magnetic susceptibility $\chi(T)$. For an ideal 2D Fermi gas $\chi(0)$ is independent of surface density, while for an interacting Landau Fermi liquid it is enhanced by the density dependent factor $(m^*/m)(1+F_0^a)^{-1}$. For ^3He adsorbed on graphite the large Curie susceptibility of the solid first layer obscures the fluid signal at lower temperatures. We have overcome this problem by plating the grafoil with a single layer of ^4He , which is preferentially bound to the surface and also forms a compressed 2D solid.

The graphite substrate used in this work was Grafoil, a commercial form of exfoliated graphite. Grafoil sheets, 125μm thick, were bonded to thin copper foils for thermal contact and oriented in the plane of the static and radio frequency magnetic fields. The nuclear magnetic susceptibility measurements were made by integrating the area under the absorption line obtained using continuous wave NMR at 1.08 MHz. The temperature was determined by a ^3He melting curve thermometer.

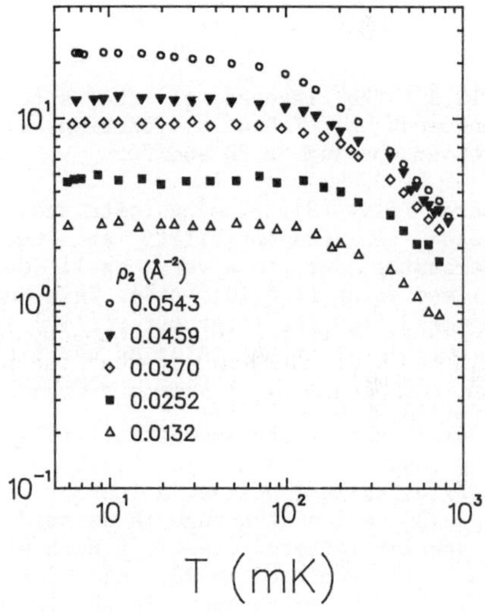

FIG.1. Susceptibility of the second layer fluid as a function of temperature.

The fluid signal normalised by the T=0 ideal gas susceptibility $\chi_0(0)$ is shown in fig.1 where the large susceptibility enhancements above the ideal value are apparent. Values of the Curie constant determined from the limiting value of χT at high temperatures are found to be proportional to surface density, as expected, and are in good agreement with the more precise determination in the second layer solid. The observed susceptibility enhancements are significantly greater than those observed in

0921-4526/90/$03.50

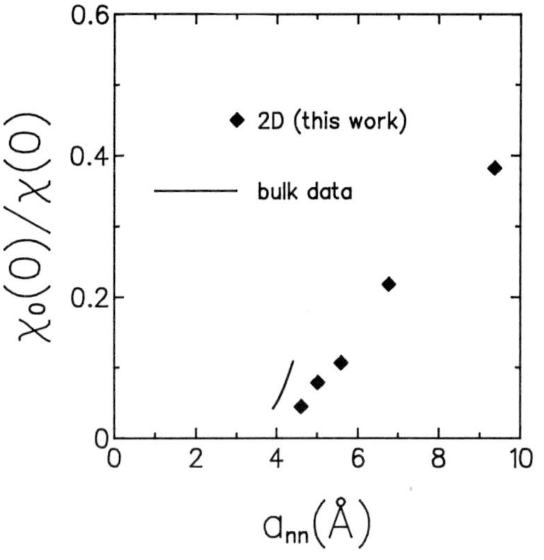

FIG.2. The inverse of the susceptibility enhancement at T=0 as a function of nearest neighbour spacing in 2D and 3D.

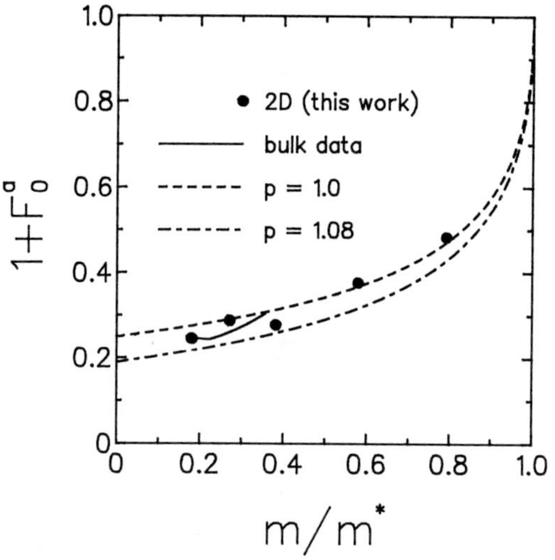

FIG.3. The relative dependence of the Landau parameters $1+F_0^a$ and m^*/m. The theoretical curves are the prediction of the almost localised Fermion model (5).

mixture films (3). The raw data show a slight increase in susceptibility at the lowest temperatures due to a very small density of localised spins ($< 5.10^{-4}$ $Å^{-2}$). This density was determined from the intercept $\chi T(T\to 0)$ and may be attributable to atoms entering the first layer solid as it is slightly compressed with increasing total density.

A comparison of the susceptibility enhancement in 2D and 3D is made in Fig.2 by plotting $\chi_0(0)/\chi(0)$ against nearest neighbour separation, a_{nn}. In 3D an fcc structure is assumed and in 2D a triangular lattice. Clearly a much wider range of a_{nn} is explored in 2D, and for a given separation the enhancement in 2D is stronger. This is consistent with calculations of Landau parameters in the low density limit including only binary collisions.

The almost localised fermion model (4,5) treats ^3He in terms of a Hubbard model with a half-filled band. This model has been solved in the Gutzwiller approximation which, it is suggested (5), is insensitive to details such as the absence of a lattice in liquid ^3He. Expressions for the Landau parameters are then obtained in terms of the microscopic parameters of the model. In terms of the normalised on site interaction potential $I=U/U_c$ it is found that $I=[1-m/m^*]^{1/2}$ and $F_0^a=-p[1-1/(1+I)^2]$, where p depends on the precise form of the density of states ($p\approx 1$ for a symmetric band). Within this model, therefore, the relation between the two Landau parameters m^* and F_0^a is determined by the single parameter p.

We have used the effective mass data of ref(2) to infer F_0^a from the measured susceptibility enhancement, assuming that the first solid layer is inert and that its isotopic composition does not influence the interactions in the fluid second layer. In Fig.3 we show the relative dependence of $1+F_0^a$ and m/m^* in 2D, together with the bulk results. The theoretical curves are given for two values of the parameter p: $p\approx 1.08$ for a half-elliptical density of states, while p=1 for nearest neighbour hopping on a cubic lattice. The agreement is good both in 2D and 3D; these new 2D results provide independent support for this microscopic model and test it over a wider range of interaction parameter. It would be of interest to compare the full temperature dependence of the susceptibility with the extension to finite temperatures of the almost localised Fermion theory (6).

Further support for the hypothesis that the second layer fluid is a 2D Landau Fermi liquid would be provided by the measurement of a transport property such as the spin diffusion coefficient.

References
(1)J.Saunders,C.P.Lusher,B.P.Cowan these proceedings.
(2)D.S.Greywall and P.A.Busch, Phys.Rev.Lett. 62 (1989) 1868. D.S.Greywall, Phys.Rev. B41 (1990) 1842
(3)J.M.Valles,R.H.Higley,B.R.Johnson and R.B.Hallock, Phys.Rev.Lett. 60 (1988) 428
(4)P.W.Anderson and W.F.Brinkman, The Physics of Liquid and Solid Helium, eds. K.H.Benneman and J.B.Ketterson (Wiley,NY) pp177-286.
(5)D.Vollhardt, Rev.Mod.Phys. 56 (1984) 99.
(6)K.Seiler,C.Gros,T.M.Rice,K.Ueda, and D.Vollhardt, J.Low Temp. Phys. 64 (1986) 195

Physica B 165&166 (1990) 693–694
North-Holland

NUCLEAR MAGNETIC SUSCEPTIBILITY OF SUBMONOLAYER ^{3}HE FILMS ADSORBED ON GRAPHITE

J.SAUNDERS, C.P.LUSHER and B.P.COWAN

Millikelvin Laboratory, Royal Holloway and Bedford New College, University of London
Egham, Surrey, TW20 0EX, U.K.

The nuclear magnetic susceptibility of submonolayer films of ^{3}He adsorbed on a graphite substrate has been measured for temperatures between 6 and 800mK. Three distinct surface density regimes are identified for fractional monolayer coverages below perfect registry. These appear to be ; a fluid phase with some atoms localised by substrate heterogeneity, a solid-fluid coexistence region, and a registered solid phase with defects.

The submonolayer film of helium on graphite is known to exhibit a variety of phases (1) . The helium atoms are strongly bound to the surface and at low densities tunnel rapidly under the influence of the periodic substrate potential so that above 1K a two dimensional fluid results. At a surface density of 0.064 $Å^{-2}$ the interplay of interatomic and substrate interactions leads to the formation of a registered solid; with further increases in surface density an incommensurate solid is formed followed by eventual promotion of atoms to the second layer (at 0.108 $Å^{-2}$ for ^{3}He films). The surface density may be expressed in coverage units x as a fraction of this monolayer density.

We have measured the nuclear magnetic susceptibility of the submonolayer film adsorbed on grafoil to low millikelvin temperatures, with the aim of establishing the nature of the "fluid" phase. Such a measurement was expected to give clear information on the degree of localisation or quantum degeneracy of the film. For coverages below registry we found no direct evidence of a constant low temperature susceptibility as expected for a degenerate Fermi fluid, although substantial reductions in susceptibility below the Curie value were observed. The results for the susceptibility, normalised by the Curie constant for x=1 (C_0), are shown as a function of temperature in Fig.1. C_0 was determined from the measured Curie law susceptibility in the incommensurate solid at coverages 0.742,0.785, 0.827,0.870. For $x \leq 0.521$ a good fit to the data is given by

$$\chi T = x_s C_0 + x_f C_0 (T/T_F)(1- \exp(-T_F/T)) \quad [1]$$

where $C=C_0 x$, x_s/x_f and T_F are adjustable temperature independent parameters. As can be seen from the insets the coverage dependence of χT shows a distinct break at $x \approx 0.3$ and this onset coverage appears constant up to a temperature of 300mK, above which the transition is not discernible in the susceptibility data.

Physically eq.[1] describes the total susceptibility of a film consisting of solid and fluid, with $x=x_s+x_f$. The fluid term is the form of the susceptibility of an ideal 2D gas with Fermi temperature T_F. The coverage dependence of these parameters is shown in Fig.2. In these terms the break in χT at $x \approx 0.3$ corresponds to an increase in x_s.

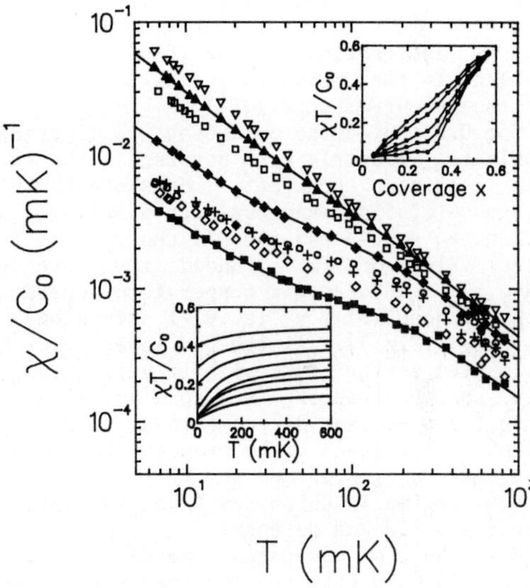

FIG.1. Normalised susceptibility as a function of temperature at coverages 0.144(■), 0.195(◇), 0.244(+), 0.291(○), 0.337(◆), 0.385(□), 0.430(▲), 0.475(▽).Curves through data illustrate fit to eq.(1). Inset(lower left); fit to normalised χT for these coverages. Inset (upper right); coverage dependence of χT at 10, 50, 100, 300, 1000 mK (reading upwards) with linear interpolations between measured coverages as guide to eye.

We identify three surface density regimes below registry.

(a) for x<0.3 the parameter x_s is approximately constant at 0.02. This is attributable to the monolayer fraction of spins localised by sub-

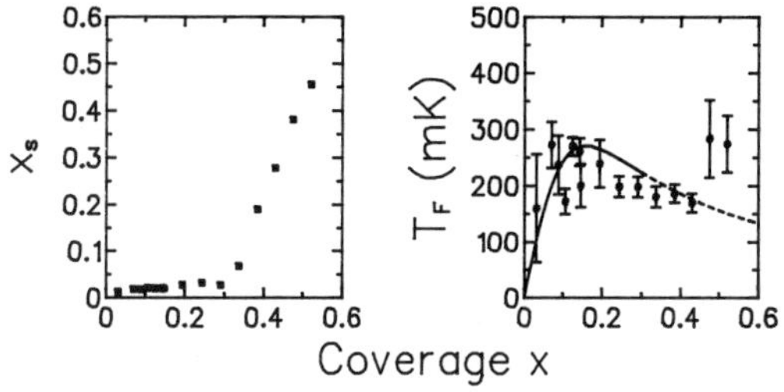

FIG.2. Coverage dependence of susceptibility fit parameters, see eq.(1) and text.

strate heterogeneity. Addition of a comparable monolayer fraction of ^{4}He and annealing the sample results in a substantial reduction of x_s due to preferential adsorption of the heavier isotope at the deep sites. The coverage dependence of T_F is consistent with a density dependent interaction in the fluid component equivalent to that measured in the second layer fluid (solid curve Fig.2.) (2).

(b) for $0.3<x<0.45$ the parameter x_s increases approximately linearly with coverage, the value of T_F remaining constant, consistent with coexistence of fluid and registered solid.

(c) Between x=0.43 and x=0.475 there is a jump in the value of T_F. Indeed for coverages 0.475,0.521,0.565 the low temperature dependence of the susceptibility is well described by Curie's law (Fig.3). At the two lower coverages the low temperature susceptibility is significantly smaller than the free spin value as can be seen fom the clear decrease in χT between 1K and 50mK. This leads us to propose the density range between x~0.45 and registry as a third distinct regime in which the film consists of registered solid and defects.

(d) For the two coverages immediately above registry (0.61,0.655) the low temperature susceptibility follows Curie's law, but exhibits a small but significant enhancement above the free spin value observed at higher temperatures. For coverages in the range $0.74<x<0.87$, in the incommensurate solid phase, there are no significant deviations from Curie's law at these temperatures.

The instability of the uniform fluid phase in region (b) revealed by these experiments had not been inferred from earlier heat capacity data (3). For ^{4}He films at coverages below registry distinct heat capacity peaks are found (3,4) around 1K; these have most recently been interpreted (4) as the signature of a registered solid-gas coexistence region with $x_{gas}\to 0$ as $T\to 0$. By contrast,in the ^{3}He commensurate solid-fluid coexistence region found in our experiment the fluid component persists at constant density to T=0, consistent with the finite and temperature independent onset coverage x~0.3. It is interesting to note that since on each grafoil platelet the film consists of a coexistence of solid and fluid, then if the solid is in the

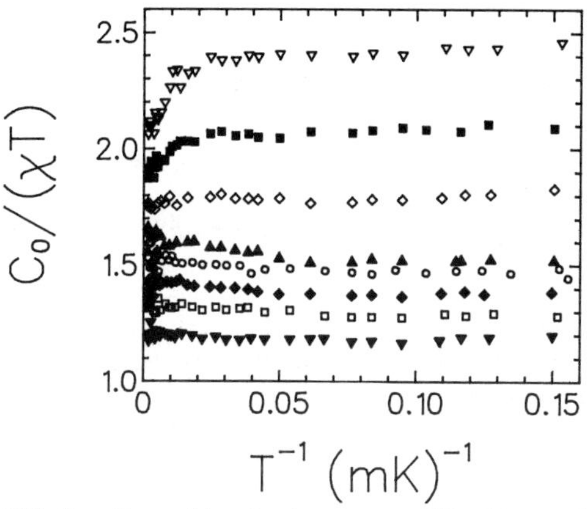

FIG.3. Normalised inverse effective Curie constant as a function of inverse temperature, for coverages 0.475(▽),0.521(■),0.565(◇),0.611 (▲),0.655(○),0.698(◆),0.742(□),0.827(▼).

form of clusters the high quantum mobility will result in fluctuations in the configuration due to exchange within and between clusters and interchange between fluid and solid. Theories of ^{3}He as a "nearly solid" liquid (5) may be useful in understanding the instability of the homogeneous fluid at these densities.

In region (c) the most appropriate description of the film may be as a quantum glass with a distribution of exchange energies, those spins in the vicinity of defects having a high quantum mobility and hence exhibiting the effects of degeneracy. The spin contribution to the heat capacity inferred from the susceptibility in such a model reproduces the weak temperature dependence and small absolute value of the direct measurement (3). The stability of the registered solid with respect to a homogeneous disordered phase below perfect registry is predicted (6), and there is some experimental evidence in neutron scattering data (7).

In region (d) the small susceptibility enhancements may arise from rapid three particle, and hence ferromagnetic, exchange involving two registered atoms and an interstitial.

References
(1) M.Schick, Phase Transitions in Surface Films, Plenum NY(1980)
(2) C.P.Lusher,J.Saunders,B.P.Cowan, these proceedings
(3) M.Bretz,J.G.Dash,D.C.Hickernell,E.O.McLean, O.E.Vilches Phys.Rev. A8 (1973) 1589
(4) R.E.Ecke,Q-S.Shu,T.S.Sullivan,O.E.Vilches, Phys.Rev. B31 (1985) 448
(5) J.P.Bouchard,C.Lhuillier, Europhys. Lett. 3 (1987) 1273
(6) W.Kinzel and M.Schick, Phys.Rev. B23 (1981) 3435
(7) M.Nielsen,J.P.McTague,W.Ellenson, J.de Phys. 38 (1977) C4-10

Physica B 165&166 (1990) 695–696
North-Holland

CHANGES IN THE FIELD DEPENDENCE OF THE ^{3}He RELAXATION PROCESS WITH ^{3}He COVERAGE*

T. J. Gramila, F. W. Van Keuls, and R. C. Richardson

Laboratory of Atomic and Solid State Physics, Cornell University, Ithaca, New York 14853-2501, U.S.A.

Measurements of the magnetic relaxation rates of both adsorbed ^{3}He and the F^{19} spins in a powdered CaF_2 substrate have been made for various ^{3}He coverages as a function of applied field. These measurements show a gradual transition to the anomalous linear frequency dependence of the relaxation, beginning at submonolayer coverages. The implications of our NMR measurements to the process responsible for the relaxation are discussed.

1. INTRODUCTION

Measurements of the temperature dependence of relaxation rates for ^{3}He and substrate spin species on a number of substrates (1–3) confirm that the anomalous linear increase in the ^{3}He relaxation rate with decreasing temperature is well described by the model of Hammel and Richardson (1). The physical process responsible for the relaxation, however, has not yet been identified.

Our previous measurements for ^{3}He on a substrate of CaF_2 also confirm this model of the temperature dependence (4), and indicate that the relaxation can be assigned to two distinct channels (5). One of these is due to the ^{3}He in the very first adsorbed layer and accounts for the bulk of the relaxation when only a few layers are present. The development of this process, as monitored by the substrate relaxation rate, is found to be completed by a ^{3}He coverage of roughly 1.5 atomic layers. Substantial relaxation rates are observed for both the ^{3}He and substrate spins for ^{3}He coverages less than second layer promotion.

The relaxation rate of the second channel is found to increase slowly as the amount of liquid in the cell increases. This relaxation cannot be accounted for by changes in the first layer relaxation, and is assigned to the ^{3}He in the second or subsequent layers. At .32 Tesla, this channel accounts for more than three quarters of the full cell relaxation rate. Extrapolation of the linear dependence on coverage observed in the multilayer data predicts a full cell relaxation rate in good agreement with the measured rate.

The measurements presented in this paper involve changes in the field dependence of the relaxation rates as the ^{3}He coverage is varied. The roughly linear dependence of the ^{3}He T_1 on applied field, which is characteristic of the surface dominated relaxation of ^{3}He, develops quite gradually as the ^{3}He coverage is increased.

2. DISCUSSION

The substrate used in these measurements is a powder of high purity CaF_2 . The average particle size was roughly 0.5 microns. NMR measurements were made using a pulsed spectrometer with a very short recovery time, enabling us to easily observe the short free induction decay of the F^{19}. The surface area of the substrate was measured using an Argon BET; the resulting ^{3}He coverages for layer promotion correspond closely to distinct features observed in the variation of the ^{3}He linewidth. Second layer promotion occurs at .95 STP-CC per gram.

Figures 1a and b show the changes in the field dependence of the relaxation times at 100 mK for the ^{3}He and the F^{19} spins, respectively. The data is plotted as T_1 divided by the field to emphasize deviations from linear dependence, which would appear here as a flat line. A line connects the fields measured for each coverage, and the data for each coverage is offset from its neighbor by an arbitrary amount to prevent overlap. This allows better comparison in trends for the different coverages, but removes information about the magnitudes of T_1 from coverage to coverage. The bar representing a single decade is consistent with the scale for each coverage. The coverage in STP-CC per gram is adjacent to each data set.

We have characterized the changes in the frequency dependence of the relaxation times with ^{3}He coverage by fitting to the high field region for each coverage. The region of fit and resulting best line fits are represented in the Figures 1a and b by the dark solid lines. The region of fit for the lowest ^{3}He coverages is reduced to account for the significant curvature of these data sets at high fields. The exponents describing the dependence of T_1 on field as determined by these fits are shown in Figure 2 for both the ^{3}He (circles) and the F^{19} (triangles). Full cell exponents are shown as two dashed horizontal lines.

---

*Supported by the Cornell MSC through NSF grant number DMR-8818558.
‡Present address: AT&T Bell Laboratories, Murray Hill, New Jersey 07974, U.S.A.

0921-4526/90/$03.50

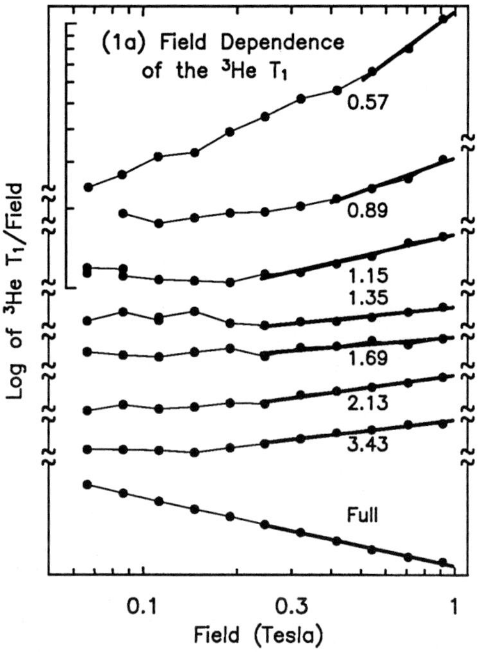

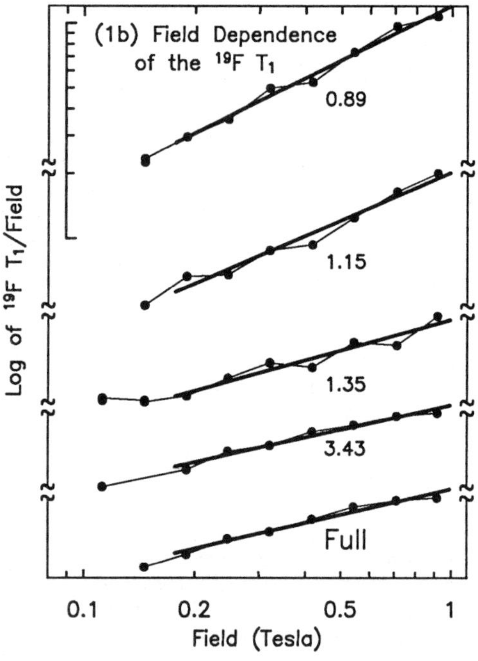

FIGURES 1a and 1b

Plots of the field dependence of the relaxation times for both ^{3}He and F^{19}. Data is plotted as the log of T_1 divided by field. See text for explanation of axes.

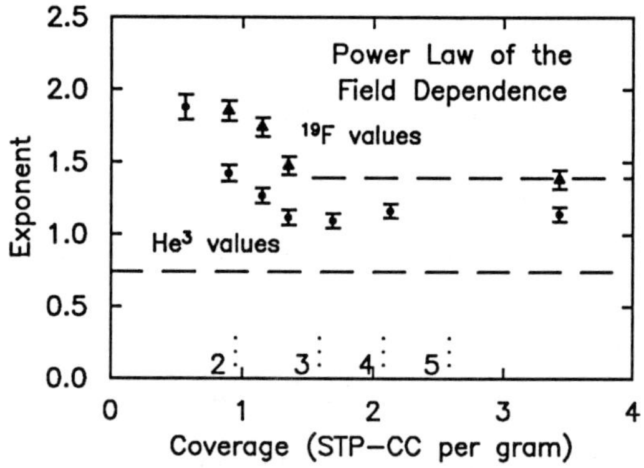

FIGURE 2

The dashed vertical lines in the figure indicate the coverages for various layer promotions.

The most striking feature of these data is the gradual nature of the change in the field dependence. In the range of coverages plotted, we have determined that the relaxation is dominated by a process in the first layer of ^{3}He (5). There is a substantial tendency toward linear field dependence even below second layer promotion, with no abrupt changes once atoms are in the second layer. The frequency dependence for atomic exchange has been characterized (6), and it is known that the ^{3}He exchange frequency approaches zero near first layer completion (7). The exponent for exchange would increase as layer completion is approached. It is therefore unlikely that atomic exchange is responsible for this behavior.

The development of this process is completed before third layer promotion. It is possible that the changes seen above second layer promotion are simply related to changes in the first layer density as the second layer is filled.

The difference between the multilayer ^{3}He exponent and the full cell value is due to the growth of a second relaxation process which we have previously identified as being due to the ^{3}He in the second or subsequent layers.

REFERENCES

(1) P. C. Hammel and R. C. Richardson, Phys. Rev. Lett. **52**, 1441 (1984).
(2) A. Schuhl, S. Maegawa, M. W. Meisel, and M. Chapellier, Phys. Rev. B **36**, 6811 (1987).
(3) H. Franco, H. Godfrin, H. Lauter, and D. Thouleze LT-17, Karlsruhe, Germany, 1984, (North-Holland, Amsterdam, 1984) Vol. 2, p. 723.
(4) T. J. Gramila, Ph.D. dissertation, Cornell University, Ithaca, N.Y., 1990 (unpublished).
(5) T. J. Gramila, F. W. Van Keuls, and R. C. Richardson, to be published.
(6) B Cowan, *et al.*, Phys. Rev. Lett. **58**, 2308 (1987).
(7) D. S. Greywall, Phys. Rev. B **41**, 1842 (1990).

Physica B 165&166 (1990) 697–698
North-Holland

TWO DIMENSIONAL SOLID ^{3}He: A FRUSTRATED MAGNETIC SYSTEM?.

Michel ROGER

S.P.S.R.M., Orme des Merisiers, C.E.A C.E.N.S., 91191 Gif sur Yvette Cedex, France

Like in the bcc three-dimensional solid, competing cyclic ring exchanges govern the magnetic properties of ^{3}He films. Three-particle exchange dominates at high and moderate areal density, leading to the ferromagnetic properties previously observed. At the low densities corresponding to the observation of commensurate phases, higher order (four and six-spin) antiferromagnetic exchanges are more important and lead to a frustrated antiferromagnetic system. A variety of anomalies recently observed are interpreted.

Recent experimental investigations (1) on nuclear spin ordering in a second layer of ^{3}He atoms, adsorbed on grafoil on the top of a dense solid first layer, have revealed an extremely rich behaviour.

1) For coverages near third layer promotion, a striking phase transition with entropy change of only $0.5k_B\ln 2$ per second-layer spin is signalled by a sharp specific-heat peak around T=2.5 mK (2). For similar coverages, the susceptibility shows a slightly negative deviation with respect to the Curie law (3). This transition might correspond to some kind of antiferromagnetic order in a *low density* two-dimensional (2D) solid, probably commensurate with the first layer (4).

2) At higher coverages (more than 2.5 layers), an incommensurate 2D solid *with a 20% higher density* is observed and the susceptibility shows a quasi-2D ferromagnetic behaviour (1,3).

Like in the three-dimensional bcc solid, we expect that competing cyclic ring exchanges govern the nuclear magnetic properties of two-dimensional ^{3}He films (5-8). Delrieu early predicted that three-particle ring exchanges (J_T), leading to ferromagnetism should dominate in hcp and 2D-triangular lattices (5). This prediction was confirmed a few years later for the hcp solid (9) and accounts qualitatively for the ferromagnetism observed in the incommensurate second layer on Grafoil (10). However, we expect higher order antiferromagnetic ring exchanges such as four (K) and six-particle (S) cyclic exchanges to have more weight in the much looser incommensurate structures. Rough estimates from multidimensional WKB approximations (7) and preliminary Monte-Carlo simulations (11) predict that four and six-spin ring exchanges should be of order one third of three-particle exchange for the densities corresponding to the region near third layer promotion, where the intriguing anomaly in the heat capacity is observed (see Ref. (12) for more quantitative explanations)

Hence 2D solid ^{3}He might appear as a very original magnetic system in which the ratio between ferromagnetic and antiferromagnetic exchange frequencies can be tuned as a function of density, -and also under the influence of the substrate potential-, to evolve from a ferromagnet at the higher densities corresponding to the incommensurate phases, to an intricate frustrated antiferromagnet with competing interactions for the much looser commensurate phases. We must emphasize that RKKY interactions between the second solid layer and a third liquid layer might also enhance, for the higher densities, the ferromagnetic tendency of the second layer (13). Hence we take the following model Hamiltonian:

$$H_{ex} = J\sum^{(1)} P_2 + K\sum(P_4 + P_4^{-1}) + S\sum(P_6 + P_6^{-1})$$

where the P_n's denote cyclic permutations of n particles; the first sum runs over distinct first neighbor pairs, the others over the most compact four- and six-spin cycles. Using the property that three-spin exchange operators can be expressed in terms of pair transpositions (6), three-particle exchange (J_T) and pair exchange between first neighbors (J_{NN}) have been included in an effective-pair-exchange frequency: $J = J_{NN} - 2J_T$. Note that J might also include the first-neighbor part of exchange mediated through RKKY interactions (13).

The high-temperature coefficients corresponding to this model are $\theta = -3(J+3K+5S/8)$ for the Curie-Weiss constant and:

$$e_2 = 9(J^2+33K^2/4+5JK+SJ/2+3SK/2+193S^2/192)$$

for the leading term in the heat capacity.

We mainly study this model through exact diagonalization for a 4x4 particle cluster with periodic boundary conditions. As a first approach, we take $S = K$, -a reasonable assumption, taking into account rough predictions (7,12)-, and vary the ratio $r = K/J$.

The heat capacity and entropy per second layer spin is shown in Fig. 1, as a function of reduced temperature $T/\sqrt{e_2}$, for a few values of r. When r is varied from $r = 0$ (2D-ferromagnetic Heisenberg model) to $r = -0.25$ (multiple exchange Hamiltonian with 'maximum frustration': θ is the sum of two opposite terms which almost cancel) the specific heat evolves from a single-peak to a double-peak structure with about half height, and the low temperature entropy increases substantially. This excess entropy and peculiar shape of the heat capacity is directly related to an anomalously large density of quasidegenerate low-lying energy states resulting from the frustration. These theoretical curves are compared to the data of Greywall (1) at second layer densities $\rho_2 = 0.0694$ atoms/Å^2 (second-layer completion) and $\rho_2 = 0.0813$ atoms/Å^2 (for total coverage around 2.5 layers) with temperature scaled respectively to $\sqrt{e_2} = 7.8mK$ and $\sqrt{e_2} = 5.3mK$.

The inset shows the susceptibility. From $r = 0$ to $r = -0.25$, it evolves from a ferromagnetic behaviour to an almost Curie law with a small antiferromagnetic deviation at low temperatures. The theoretical results are compared to the data of Godfrin and coworkers at second layer completion (full circles) and around 2.5 layers (open squares); the temperature is scaled respectively to $\sqrt{e_2} = 7.8mK$ and $\sqrt{e_2} = 7.5mK$. A Curie contribution of the first layer is added and the result is normalized to the number of atoms contained in the two first layers.

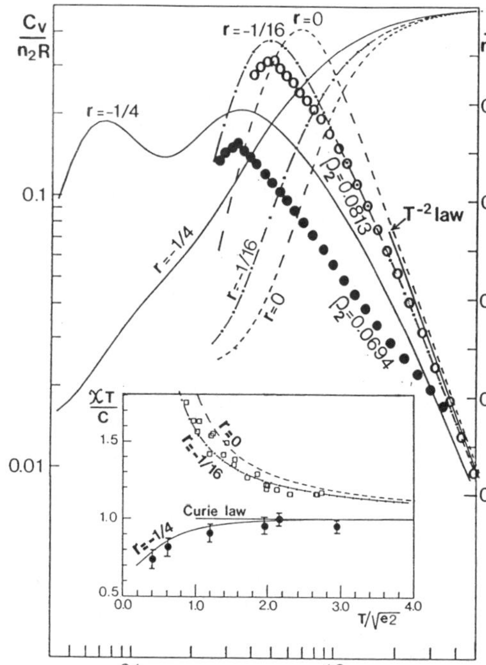

FIGURE 1

The multiple exchange model offers a coherent interpretation of a variety of anomaly observed. As shown in Fig.1, in the incommensurate phase, the negative deviation of the specific heat with respect to the T^{-2} law at high temperature might be due to the presence of small but significant four and six-spin exchanges (the susceptibility is less sensitive and can be fit as well with a pure Heisenberg model (1,3)). In the low-density phase, the large remaining entropy observed by Greywall below the transition is compatible with the model with large frustration ($r \approx -0.25$). We do not know whether the double peak shape is intrinsic or is an artifact of the finite-size calculation. However, we expect for the infinite system a second anomaly at lower temperatures (at least a shoulder or flat region) as inferred by Greywall from his measurements. For $r \approx -0.25$, the susceptibility is also in good agreement with Godfrin's measurements at second layer completion.

In the presence of a magnetic field, with $r \approx -0.25$, the stable state should be a canted three-sublattice antiferromagnetic phase, analogous to the 'pf' phase observed in bcc ^{3}He (6). From mean fied calculations and exact diagonalization, we expect at moderate field (10 T) a high magnetization of order 75% of the saturation magnetization (see Fig. 2 in Ref. (12)). Its observation would be a crucial test of the multiple exchange Hamiltonian, with respect to other theoretical models (4,14).

I am indebted to J. M. Delrieu for years of fruitful collaboration on nuclear magnetism in solid ^{3}He.

REFERENCES

(1) H. Godfrin, this conference (invited talk)
(2) D. S. Greywall and P. A. Busch, Phys. Rev. Lett. 62 (1989) 1868; D. S. Greywall, Phys. Rev. B 41 (1990) 1842.
(3) H. Franco, E Rapp, and H. Godfrin, Phys. Rev. Lett. 57 (1986) 1161; H. Franco, thesis, University of Grenoble, France (1985); H. Godfrin, R. R. Ruel, and D. D. Osheroff, Phys. Rev. Lett. 60 (1988) 305.
(4) V. Elser, Phys. Rev. Lett. 62 (1989) 2405.
(5) J. M. Delrieu, M. Roger, and J. H. Hetherington, J. Low. Temp. Phys. 40 (1980) 71.
(6) M. Roger, J. H. Hetherington, and J. M. Delrieu, Rev. Mod. Phys. 55 (1983) 1.
(7) M. Roger, Phys. Rev. B 30 (1984) 6432.
(8) D. M. Ceperley and G. Jacucci, Phys. Rev. Lett. 58 (1987) 1648.
(9) Y. Takano et al., Phys. Rev. Lett. 55 (1985) 1490.
(10) M. Roger and J. M. Delrieu, Jpn. J. Appl. Phys. 26, Suppl. 26-3 (1987) 267.
(11) B. Bernu, C. Lhuillier, and D. M. Ceperley (to be published)
(12) M. Roger, Phys. Rev. Lett. 64 (1990) 297.
(13) H. Jichu and Y. Kuroda, Prog. Theor. Phys. 67 (1982) 715.
(14) K. Machida and M. Fujita, preprint.

Physica B 165&166 (1990) 699–700
North-Holland

SPIN HEAT CAPACITY OF THE SECOND SOLID LAYER OF ^{3}He ADSORBED ON GRAPHITE

Dennis S. GREYWALL

AT&T Bell Laboratories, 600 Mountain Avenue, Murray Hill, NJ 07974

Precise heat capacity data have been obtained for multilayers of ^{3}He adsorbed on graphite in the temperature range between 2 and 200 mK. As a function of total coverage, these data show an accurate correspondence with special features in recent magnetization measurements but suggest models very different from those proposed to explain the magnetization data alone. The heat capacity results indicate that the ferromagnetic anomaly observed in the magnetization measurements, occurs in the second layer when this layer exists in a state intermediate between a registered and an incommensurate solid.

At coverages corresponding to somewhat more than two adsorbed atomic layers, the ^{3}He-on-graphite system exhibits a ferromagnetic anomaly, with the magnetization rising well above the free spin value (1). The peak magnetization at 0.24 atoms/Å^2 has been interpreted as arising from the first-order freezing of the second adsorbed layer into a two dimensional solid with an anomalously large exchange energy. Magnetization measurements (2) at $\rho = 0.24$ extending to below 1 mK have also been used to claim that at this coverage the solid layer represents an ideal 2D nuclear Heisenberg ferromagnet with an exchange J of 2.1 mK. In this paper we discuss zero-field heat capacity measurements (3,4) on this system which span a range of coverages centered around 0.24 atoms/Å^2 and which show, as a function of coverage, a detailed correspondence with the special features observed in the magnetization measurements. However, the new data are not consistent with the models proposed to explain the magnetization data alone. The heat capacity data (3) indicate, in fact, that the second layer freezes into a registered phase at a much lower coverage and that the ferromagnetic anomaly occurs when the second layer is in a state intermediate between registered and incommensurate structures. Furthermore, and contrary to the magnetization data (2), the heat capacity data (4) at 0.24 are not consistent with the nearest-neighbor Heisenberg Hamiltonian.

The heat capacity measurements were made in the temperature range 2 to 200 mK using the standard heat pulse technique and a fast-thermal-response CMN thermometer (5). The graphite substrate consisted of many discs of 0.1 mm Grafoil (6) with a total surface area of 203 m^2.

Four heat capacity isotherms are plotted in Figure 1. Above 50 mK the heat capacity is dominated by the Fermi fluid contributions. The abrupt increases in the 50 and 200 mK isotherms, shown in the upper portion of the figure, locate the coverages at which layer promotions occur. Prior to third layer promotion, the sudden *decrease* corresponds to the first-order freezing of the second layer into a solid which is registered relative to the underlying incommensurate solid ^{3}He first layer. The freezing of the second layer is also evident in the 2.5 and 5 mK isotherms shown in the bottom portion of Figure 1. Here the heat capacity, now dominated by the nuclear spin contribution from the second layer solid, suddenly increases.

At higher coverages atoms are promoted into the third and eventually into higher layers, while the spin heat capacity evolves through the ferromagnetic anomaly. Changes occur in the second layer spin

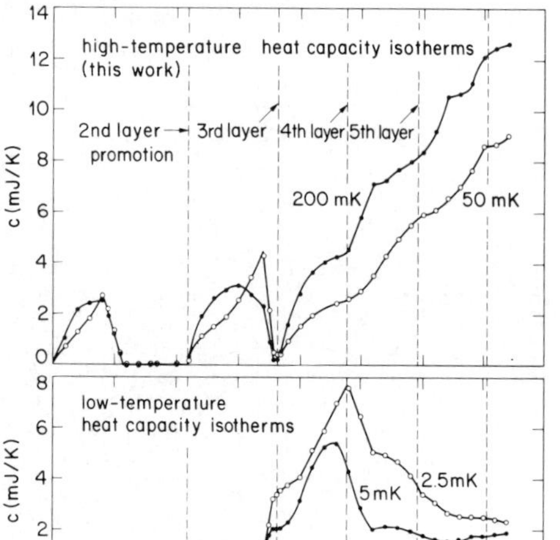

FIGURE 1. Heat capacity isotherms.

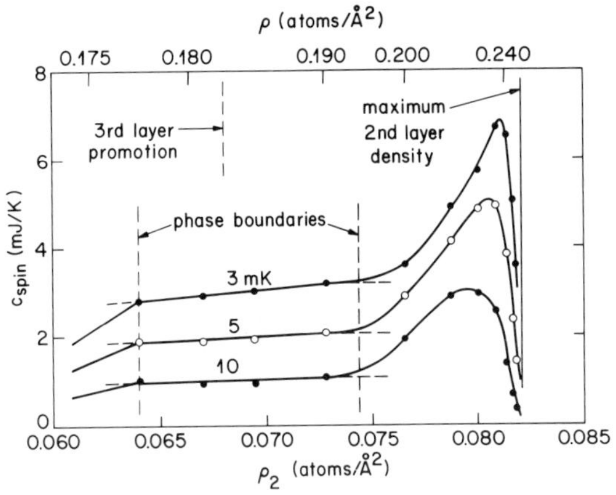

FIGURE 2. Spin heat capacity isotherms as a function of the second layer density.

system as a consequence of the compression of this layer due to the overlying fluid layers. At a coverage of 0.297 atoms/Å^2 neutron scattering measurements (7) indicate that the second layer exists as an incommensurate solid. Consequently the ferromagnetic anomaly must occur when the second layer exists as an incommensurate solid or as the second layer transforms from the registered phase to the incommensurate solid phase.

Because the second layer atoms are believed to be entirely responsible for the spin heat capacity, it is more appropriate to plot the isotherms as a function of the second layer density ρ_2 rather than as a function of the total coverage ρ. Such a plot is given in Figure 2 where the ρ-to-ρ_2 conversion is based on an analysis (4) of the second layer melting peak data of Ref. 8. The break in the curves at $\rho_2 = 0.064$ locates the higher-coverage boundary of the fluid-registered phase coexistence referred to above. The region between 0.064 and 0.074 is proposed to correspond to a coexistence of two registered phases. The transition between the higher-density registered phase and the incommensurate solid phase takes place between 0.074 and 0.082. Note that because the peaks in the isotherms are not positioned at the same coverage, the anomaly can not correspond to the crossing of a phase boundary.

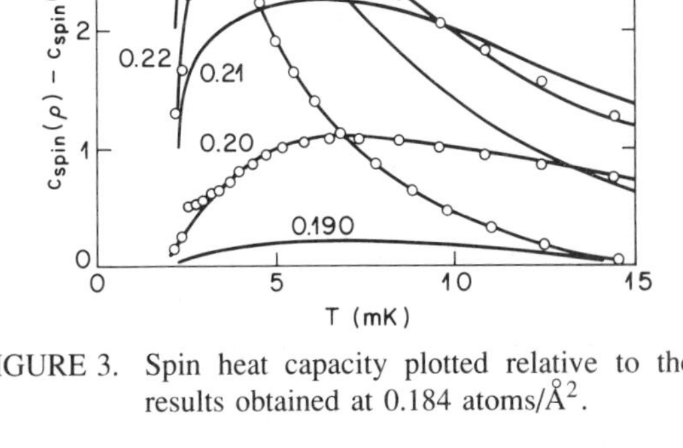

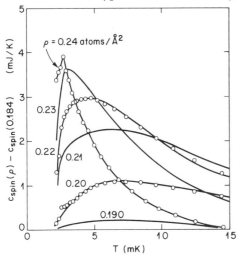

FIGURE 3. Spin heat capacity plotted relative to the results obtained at 0.184 atoms/Å^2.

Figure 3 shows the spin heat capacity plotted relative to the results at 0.184 atoms/Å^2. It is intriguing that these difference plots do not exhibit peaks at 2.5 mK since this is the temperature location of the spin heat capacity peaks. Instead there is a rounded maximum at a higher temperature which shifts toward lower temperatures and sharpens as the density increases toward $\rho = 0.24$. The temperature and density dependence of these difference curves is reminiscent of the signatures for the 2D fluid, which suggests the possible existence of a high density fluid separating the coexisting registered and incommensurate solid phases. This fluid would then be responsible for the observed ferromagnetism.

When compared with the 8- and 10-term series expressions for the ideal 2D nearest neighbor Heisenberg ferromagnet, the measured spin heat capacity shows large deviations at 0.24 and at 0.25, but good agreement at 0.26 atoms/Å^2. This is explained by our claim that at this highest coverage the second layer has just entered into the fully incommensurate solid regime. A multiphase coexistence at 0.24 would also explain the measured (2) saturation magnetization being 20% smaller than expected.

REFERENCES

(1) H. Franco, R. E. Rapp, and H. Godfrin, Phys. Rev. Lett. *57* (1986) 1161.
(2) H. Godfrin, R. R. Ruel, and D. D. Osheroff, Phys. Rev. Lett. *60* (1988) 305.
(3) D. S. Greywall and P. A. Busch, Phys. Rev. Lett. *62* (1989) 1868.
(4) D. S. Greywall, Phys. Rev. B *41* (1990) 1842.
(5) D. S. Greywall and P. A. Busch, Rev. Sci. Instrum. *60* (1989) 471.
(6) Grafoil is an exfoliated graphite manufactured by Union Carbide.
(7) H. J. Lauter, H. P. Schildberg, H. Godfrin, H. Wiechert, and R. Haensel, Can. J. Phys. *65* (1987) 1435.
(8) S. W. Van Sciver and O. E. Vilches, Phys. Rev. B *18* (1978) 285; S. W. Van Sciver, Phys. Rev. B *18* (1978) 277.
(9) G. A. Baker, H. E. Gilbert, J. E. Rushbrooke, and G. S. Rushbrooke, Phys. Lett. *25A* (1967) 207.

Physica B 165&166 (1990) 701–702
North-Holland

^{3}He FILMS AND THE RKKY INTERACTION

R.A. GUYER

Laboratory for Low Temperature Physics, Department of Physics, University of Massachusetts, Amherst, MA 01003

^{3}He films on Grafoil at areal density $n \gtrsim 0.25$ atoms/Å^2 have unusual magnetic properties. This system is modeled as a two dimensional solid in interaction with a two dimensional fermi liquid. A calculation of the magnetic behavior of this model leads to an RKKY exchange interaction that has several features in reasonable accord with experiment.

Several recent thermodynamic measurements(1-3) have focused attention on the neutral, two dimensional fermi systems that reside in thin ^{3}He-^{4}He films. Guyer, McCall, and Sprague(4) have explained the qualitative behavior of some of these data, i.e. the magnetization steps(3), in terms of evolution of the states available to fermions in an adjustable box, the box size being driven by n.

In this paper we argue that the ^{3}He liquid layer atop the second solid layer studied by Greywall(1), is much the same as the ^{3}He liquid layer studied by Higley et al(3). Using this model we find that the magnetic behavior of the second solid layer, at $n \gtrsim 0.25$, may be understood in terms of an RKKY exchange interaction between particles in this layer that is mediated by the liquid layer. This exchange interaction is very sensitive to the behavior of particles in the liquid layer.

We model the system with a two dimensional hexagonal solid layer of area A on which ^{3}He particles, having wave function

$$\psi_R(\vec{x}) = \psi_R(\vec{r})\,\psi_\perp(z) \qquad (1)$$

are localized near lattice sites $\vec{R}_1, \ldots \vec{R}_N$. Adjacent to this layer is a liquid layer of thickness d in which ^{3}He particles move parallel to the solid layer in plane wave states, $\psi_q(\vec{r}) = \frac{1}{\sqrt{A}} e^{i\vec{q}\cdot\vec{r}}$. These particles have perpendicular energy states, the m states, corresponding to

$$\psi_m(z) = \sqrt{\frac{2}{d}} \sin\frac{m\pi z}{d}\,, \qquad (2)$$

where d is a functional of n_L; $n_L = n - n_1 - n_2$, $n_1 = 0.114$, $n_2 = 0.082$. The energies of the m states, e_m, and their occupation numbers, N_m, are functions of the liquid layer thickness(4). For small n the states occupied by the particles in the liquid layer lie on a sequence of disks, each labeled by m, of radii $q_m = \sqrt{q_F^2 - pm^2/d^2}$, where q_F is the fermi wave vector of disk m (see Fig. 2 of reference (4)). When there are few particles in the liquid layer all of them are in the m = 1 disk, $N_1 = n_L$. As n_L increases, the energies of the m-states, $e_m \propto m^2/d^2$, decrease and at $n_L \approx 0.07$ the fermi energy crosses from below to above e_2, the m = 2 disk is occupied for the first time, the magnetization doubles. This event is seen as a step of a factor of two in the specific heat. We believe that the "layer promotion" events seen by Greywall at $n \gtrsim 0.240$ are these "disk occupation" events.

This model of the ^{3}He film admits a number of processes that lead to a magnetic interaction among the solid layer particles. An RKKY interaction results from a particle at site R of spin s in the solid layer, (R, s) making a transition to the empty state (m′, q′, s) in the liquid layer with a particle from an occupied state in the liquid layer (m, q, s′) going to (R, s′). The intermediate state, a particle-hole pair, has energy $e_{m'q'} - e_{mq} > 0$. The further transition, (S, s″) → (m, q, s″), (m′, q′, s) → (S,s), leads to exchange of the spins (s, s″) among the solid layer sites (R, S) since s′ = s″. This is a 3 particle cyclic permutation. We have

$$H_x = \sum_{RS} K(RS)\, \vec{\sigma}_R \cdot \vec{\sigma}_S \qquad (3)$$

with the strength of the exchange interaction given by

$$K(RS) = \sum_{mm'qq'} \langle mq\sigma'', S\sigma|V|S\sigma'', m'q'\sigma\rangle$$

$$\frac{1}{-\varepsilon_{m'q'} + \varepsilon_{mq}} \langle R\sigma', m'q'\sigma|V|mq\sigma', R\sigma\rangle \; N_{mq}(1 - N_{mq'}), \tag{4}$$

where N_{mq} is the Fermi-Dirac probability distribution and V the two particle interaction. Because K(RS) < 0 the interaction in Eq. (3) is ferromagnetic.

We focus on how K(RS) depends on the behavior of the liquid layer. Using Eqs. (1) and (2) and a contact interaction, $V \propto d(\vec{x} - \vec{x}')$, we can write K(RS) in the form

$$K(RS) \propto \sum_m N_m \cdot \frac{1}{d} \left(\frac{m^2}{m^2 + \left(\frac{d}{2\pi \ell}\right)^2} \right)^2 \tag{5}$$

This equation is an approximation to K(RS) as a sum of terms, one for each disk. Each term involves exchange of a particle between two sites in the solid layer through "tunneling" out into the liquid, into the states $(m, \vec{q})$, and then back. The out and back process is controlled by B_{mm}. The motion of the particles parallel to the surface, going from $\vec{R}$ to $\vec{S}$ in the liquid, is controlled by N_m, a measure of the number of particles in the disk m.

In Fig. 1 we show $|K|^2$ vs. d from Eq. (5). In constructing this figure we have used the ^{3}He film model of Guyer, McCall, and Sprague which provides a relationship between n_L, d and the N_m. The values of $|K|^2$ in Fig. 1 have been normed to 1 at large d, i.e. normed to the value of $|K|^2$ appropriate to the solid layer interacting with bulk liquid. On this figure we also show the square of the coupling constant found by Greywall from an analysis of specific heat data, his Fig. 27. We have normed this data in the same way as above. Our result in Fig. 1 shows that oscillatory features in $|K|^2$ as d increases are due to the participation of additional disks in the exchange process. As d in-creases the contribution of each disk diminishes in a way that is compensated by the increase in the number of participating disks.

Thanks to D. T. Sprague and K. R. McCall for valuable discussions.

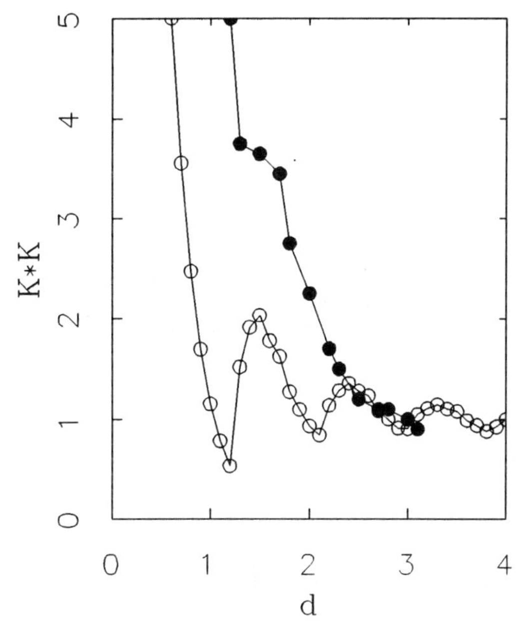

FIGURE 1
Coupling Constant. The value of the coupling constant from Eq. (5), squared, is plotted as a function of n_L (smooth curve). n_L is measured in layers; 1 layer = 0.7 atoms/Å^2. Open circles are an experimental measure of the coupling constant squared from Fig. 27 of Greywall. The horizontal axis on this figure is the liquid layer thickness in units of layers. For the data of Greywall use $d_3 = 2 + (n - 0.182)/0.70$ and $d = d_3 - 2$, so that fourth layer promotion corresponds to $d \approx 1$. Both curves have their magnitude scaled by the asymptotic value ($d \to +\infty$) to facilitate comparison and to remove factors from the theory that do not depend on n, e.g. the second solid layer wavefunction.

REFERENCES

(1) D.S. Greywall, Phys. Rev. B 41 (1990) 1842.
(2) J. Saunders, C. Lusher and B. Cowan, private communication.
(3) R.H. Higley, D.T. Sprague and R.B. Hallock, Phys. Rev. Lett. 63 (1989) 2570.
(4) R.A. Guyer, K.R. McCall and D.T. Sprague, Phys. Rev. B 40 (1989) 7414.

Physica B 165&166 (1990) 703–704
North-Holland

ANISOTROPIC MAGNETIC ORDER IN ADSORBED ^{3}He FILMS

Shuichi TASAKI

Service de Chimie Physique II, Université Libre de Bruxelles, Campus Plaine, C.P.231, Bd. du Triomphe, 1050 Brussels, Belgium*

Based on a theoretical model proposed by the author, new NMR modes observed in adsorbed ^{3}He films by Friedman et al. will be explained in terms of an anisotropic magnetic phase (V_2 phase: the ground-state phase of a 2 dimensional antiferromagnet on a triangular lattice).

1. INTRODUCTION

Magnetic properties of adsorbed ^{3}He films are now extensively studied. Recently Friedman et al. (1) have found that the bulk-^{3}He-graphite interface shows new NMR modes under low external fields;

a) When the external field is vertical to the interface, negative frequency shift always appears and positive frequency shift will appear at very low fields ($\leq$ 4 G).
b) When the field is parallel to the interface, two modes will appear; one is linearly shifted from the Larmor frequency as a function of the field strength ("shifted mode") and the other is constantly displaced from the Larmor frequency ("displaced mode").

Although the negative frequency shift for the vertical field and "displaced mode" for the parallel field can be explained in terms of demagnetizing field and ferromagnetic order (1), the origin of the positive frequency shift and "shifted mode" is not known. To reveal it is our aim.

2. V_2 PHASE

In order to explain the coverage dependence of the Curie-Weiss temperature, the author has proposed a three-layer lattice model (one solid layer + two liquid layers)(2), where the magnetic properties of the interface are assumed to be dominated by the indirect spin exchange via liquid. At very high coverage, this model predicts that the nearest neighbor spin exchange is antiferromagnetic, that the next nearest neighbor one is ferromagnetic and that the latter is much larger than the former (2,3). Because of the triangular symmetry of the system, it implies that the system can admit the so-called V_2 phase (Fig.1) as its ground state. [2] As is easily shown, the V_2 phase is antiferromagnet-like and has an anisotropic direction. Therefore it is expected to behave like a uniaxial antiferromagnet.

3. EQUATION OF MOTION AND NMR MODES

We will investigate the V_2 phase by a cluster approximation where the unit is a triangle shown in Fig.1. This is because single site is not an appropriate unit for the V_2 phase and the ordinary mean field approximation gives qualitatively incorrect results. Since, in the indirect exchange model, exchange constants up to 5th or 6th nearest neighbors can not be neglected, our approximation contain the following parameters; $\bar{J}$ exchange constant between nearest neighbor spins, J_0 mean

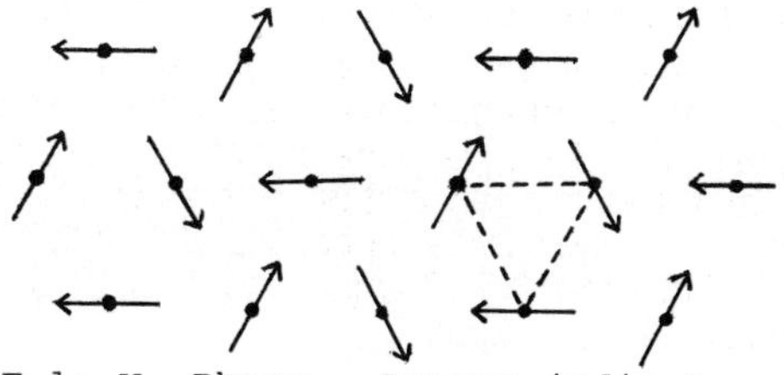

FIGURE 1; V_2 Phase. Arrows indicate the directions of spins and a triangle is a unit of our cluster approximation.

* This work was carried out when the author was in the Department of Physics, Kyoto University, Kyoto 606, Japan.

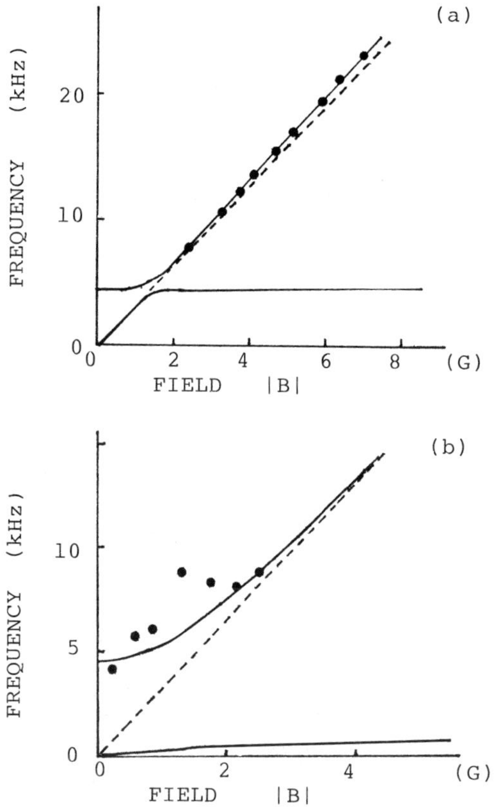

FIGURE 2; NMR modes. (a) Parallel field and (b) vertical field. Theoretical results are shown by solid curves and experimental ones (Ref.(1)) by solid circles. Broken line is Larmor frequency.

exchange constant between parallel spins and J_1 mean exchange constant between unparallel spins. And the magnetic part of the free energy F_m is given by

$$F_m = \frac{1}{2} X^{-1} S_t^2 + \bar{K}(d\cdot S_t)^2 + \frac{\lambda}{2}(d\cdot l)^2 - g^* B\cdot S_t \quad (1)$$

where S_t is the total spin, d the unit normal of a plane spanned by three sublattice spins, l the unit normal of the interface plane, B the external field, X and $\bar{K}$ the isotropic and anisotropic parts of the susceptibility, λ the strength of the dipole interaction and g^* the effective gyromagnetic ratio of ^{3}He atom (see Ref.(2)). This gives

$$\dot{S}_t = g\, S_t \times B - (g/g^*)\, \lambda\, (d\cdot l)\, d \times l \qquad (2a)$$

$$\dot{d} = d \times (g\, B - (g/g^*)\, X^{-1}\, S_t)\,, \qquad (2b)$$

where g is the bare gyromagnetic ratio of ^{3}He. These are the same as the equations proposed by Osheroff, Cross and Fisher (4) for the uudd phase of solid ^{3}He (OCF equation).

Equations (2) have the following modes as small oscillations;

$$\omega^2 = \frac{1}{2}\left(\omega_L^2 + \omega_0^2 \pm \sqrt{(\omega_L^2+\omega_0^2)^2 - 4\,\omega_L^2\omega_0^2\sin^2\theta}\right) \quad (3$$

where $\omega_L = g\,|B|$ is the Larmor frequency, θ the angle between B and l at equilibrium and ω_0 the zero-field resonant frequency;

$$\omega_0 = \frac{g}{g^*}\sqrt{X^{-1}\lambda} = 2.29\,\frac{g^2}{g^*}\bar{s}\sqrt{\frac{-J_1}{a^3}}\,, \quad (4)$$

where $\bar{s}$ is the length of a sublattice spin divided by 1/2 of the Planck constant and a the lattice constant. The demagnetization effects are taken into account by replacing ω_L to

$$\omega_L^v = g\,(1-f)\,|B| \qquad \text{(vertical field)}$$

$$\omega_L^p = g\sqrt{1+f}\,|B| \qquad \text{(parallel field)}$$

where $f = 4.84\,\mu^2/(-J_1 a^3)$ is a form factor with μ the bare magnetic moment of ^{3}He. The parameter values are set so as to be consistent with the previous calculations of Ref.(2); $\bar{J} = -0.02$ mK, $J_0 = 0.37$ mK, $J_1 = -0.002$ mK, $\bar{s} = 0.25$, a=3.7 A and $g^*/g = 0.96$. They give $\omega_0/2\pi = 4.52$ kHz and $f = 0.04$.

In Fig.2, theoretical predictions are compared with the experiments of Ref.(1). The agreement is good. Notice that the Néel temperature is of order of J_0 (not J_1) and is also consistent with the experiments.

ACKNOWLEDGEMENT
The author would like to thank Prof. T.Tsuneto for continuous encouragement during this work and is grateful to Prof.H.Godfrin, Prof.A.Hirai and Dr. T.Ohmi for fruitful discussions and useful comments.

REFERENCES

(1) L.J.Friedman et al.,Phys. Rev. Lett. 57 (1986),2943; ibid 62 (1989),1635.
(2) S.Tasaki, Prog. Theor. Phys. 79 (1988),1311;80 (1988),922E; 81 (1989),946.
(3) S.Tasaki, Prog. Theor. Phys. 82 (1989), 1032.
(4) D.D.Osheroff, M.C.Cross and D.S. Fisher, Phys. Rev. Lett. 44 (1980), 792.

Physica B 165&166 (1990) 705–706
North-Holland

VERY LOW FIELD SUSCEPTIBILITY OF THE BOUNDARY LAYERS OF ^{3}HE ON GRAFOIL*

L.J. FRIEDMAN[(a)], A.L. THOMSON[(b)], C.M. GOULD, H.M. BOZLER

Department of Physics, University of Southern California, Los Angeles, CA 90089-0484 USA

P.B. WEICHMAN and M.C. CROSS

Department of Physics, California Institute of Technology, Pasadena, CA 91125 USA

The low-field NMR spectra of ^{3}He boundary layers on graphite show collective modes for $T < 1$mK One of these modes has a nonzero frequency and amplitude at zero field. We show the frequency and amplitude of these modes as a function of the applied magnetic field and the total amplitude as a function of temperature for a number of fields.

The boundary layer of ^{3}He on graphite provides an example for two dimensional magnetism. This layer has a ferromagnetic tendency that is most likely due to physical interchanges between the adsorbed atoms. Although a two-dimensional Heisenberg system has a phase transition only at $T = 0$, the presence of a weak dipole energy or finite size effects can give the appearance of order at finite temperatures.(1) Our recent experiments(2) on the boundary layer of ^{3}He indicate that this layer exhibits ordered behavior in the submillidegree regime.

The most striking feature of these experiments is the splitting in the NMR line of the surface spins at low temperatures and low fields. This result is summarized in Figures 1 and 2. The NMR line is split into a shifted mode whose frequency is proportional to the applied field, and a displaced mode whose frequency is roughly frequency independent as shown in Figure 1. (There is also a smaller residual unshifted signal, presumably due to localized spins that do not participate in the ordering.) Figure 2 shows the amplitude of these modes as the applied field approaches zero. We see that both the frequency and amplitude of one of the modes remains finite in zero field.

Our experiments at low applied fields were initially designed to study the growth of magnetization of the boundary layer of ^{3}He in the limit of low fields. The reason we used such low fields was motivated by the idea that one cannot distinguish the existence of order when a magnetic system is substantially polarized by the applied field. In the case of ^{3}He spins on

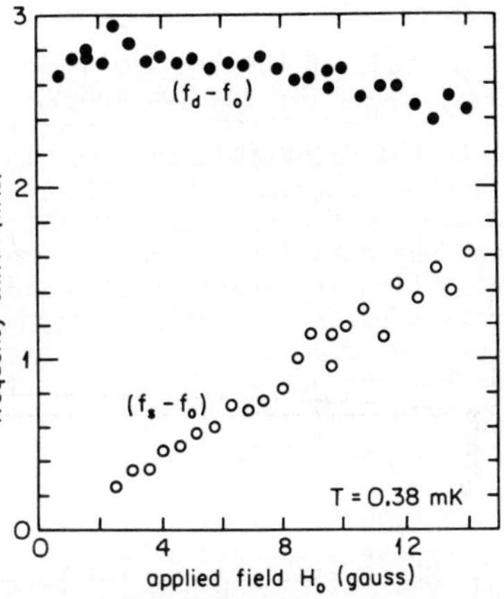

Figure 1. Field dependence for the frequency shifts of the two modes relative to the Larmor frequency.

graphite, this requires applied fields below ~ 3 gauss. The presence of the new modes was unexpected, and in terms of interpretation of the total magnetization, complicates this study. At much higher applied fields, and at a selected coverage (where the effective exchange interaction J is a maximum) Godfrin et al.(3) have been able to make precise measurements of the susceptibility and compare it with a pure Heisenberg exchange model. In such a comparison. the effective exchange constant is determined from the high temperature behavior ($T \gtrsim J$).

In our experiment, a SQUID with a flux transformer(4) is used to sense the precessing magnetization. The SQUID signal is directly

* This work has been supported by the National Science Foundation through grants DMR88-00291 (LJF,HMB), DMR89-01701 (CMG), DMR87-15474 (P.B.W.,M.C.C) and by the Science and Engineering Research Council (U.K.) (ALT).

(a) Permanent address: Raytheon Company, Research Division, 131 Spring St. Lexington, MA 0217
(b) Permanent address: Department of Physics, University of Sussex, Falmer, Brighton BN1 9QH, England.

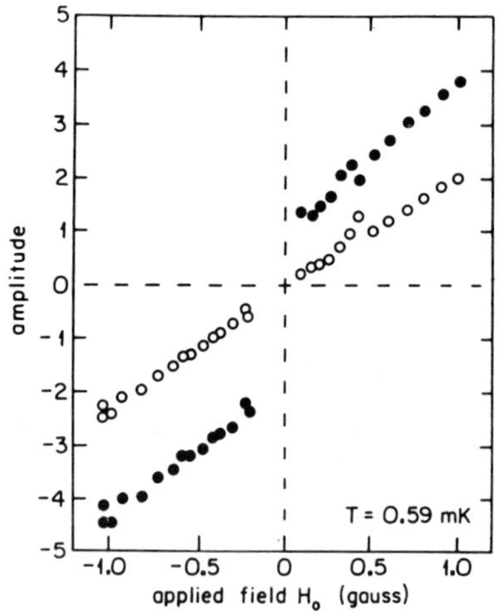

Figure 2. Amplitude of the NMR signal at T = 0.59 mK vs. field for the two modes.

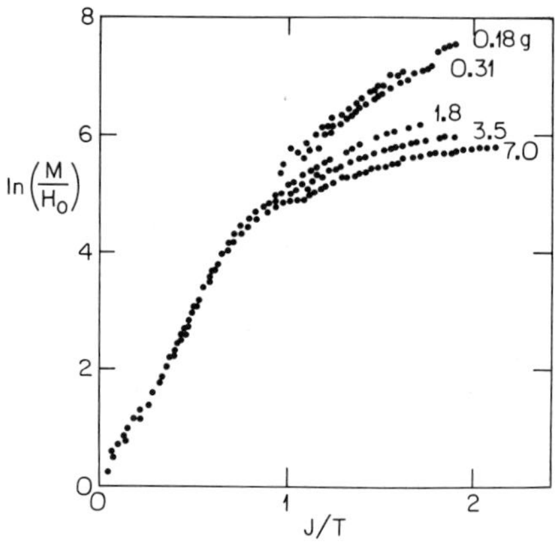

Figure 3. The temperature dependence of the total amplitude of the NMR signal at a variety of applied magnetic fields (in gauss).

proportional to the magnetization. For this reason, no renormalization of the data is necessary to compare the various magnetic fields. Figure 3 shows our data for the entire amplitude of the magnetization at several applied fields. Taking the entire surface signal, we find a value for J = 0.8 mK. We emphasize that the data in Figure 3 does not include a subtraction of a monolayer signal as was done by Godfrin et al. Such a subtraction will tend to increase the estimated value of J.

We see from Figure 3, that there is a departure in the low temperature behavior of M/H_0 from the high temperature exponential behavior expected for a Heisenberg model even in the lowest applied fields. This picture is qualitatively similar to the data of Godfrin et al.(3), however progressive saturation appears at much lower fields and values of J/T than one might expect. Despite the rather suggestive exponential behavior observed in Figure 3, the evidence for collective behavior as demonstrated in Figures 2 and 3 indicate that the system is ultimately not Heisenberg.

The maximum polarizations observed at these low fields by measuring M are much lower (less than 1%) than would be inferred by the frequency shifts (~ 30%). This may indicate that the large frequency shifts involve only a fraction of those spins which reside in the boundary layer. However, these polarizations are far in excess of the free spin values. Earlier experiments using filled pores(5) at higher applied magnetic fields ($H_0 \geq 84$ gauss) also showed relatively low polarizations.

If we treat our data by subtracting the amplitude corresponding to the spins in the first layer, we infer significantly higher polarizations. However, these are still well below the polarization reported by Godfrin et al. at a similar J/T. Of course, we also have a much smaller applied field. It is not yet possible to entirely reconcile the magnetic behavior of the boundary layer with filled pores at very low fields with the higher field film experiments. It is possible that differences exist beyond simply rescaling J. The recent studies of heat capacity by Greywall(6) indicate that at the 2.5 layer coverage where the exchange appears to be a maximum, there is promotion to the fourth layer.

REFERENCES

(1) Y. Yafet, J. Kwo, E.M. Gyorgy, Phys. Rev. B 33, (1986) 6519.
(2) L.J. Friedman, A.L. Thomson, C.M. Gould, H.M. Bozler, P.B. Weichman, and M.C. Cross, Phys. Rev. Lett. 62, (1989) 1635; L.J. Friedman, A.L. Thomson, C.M. Gould, H.M. Bozler, P.B. Weichman, and M.C. Cross, Quantum Fluid and Solids-1989, ed. by G.G. Ihas, Y. Takano, AIP Conference Proceedings 194, (Gainsville, 1989) p. 201.
(3) H. Godfrin, R.R. Ruel and D.D. Osheroff, Phys. Rev. Lett. 60, (1988) 305.
(4) L.J. Friedman, A.K.M. Wennberg, S.N. Ytterboe, H.M. Bozler, Rev. Sci. Instrum. 57, (1986) 410.
(5) H.M. Bozler, D.M. Bates, A.L. Thomson, Phys. Rev. B 27, (1983) 6992.
(6) D.S. Greywall, Phys. Rev. B. 41, (1990) 1842; D.S. Greywall, Quantum Fluid and Solids-1989, ed. by G.G. Ihas, Y. Takano, AIP Conference Proceedings 194, (Gainsville, 1989) p. 213.

Physica B 165&166 (1990) 707–708
North-Holland

SPIN DIFFUSION AND EXCHANGE IN 2D SOLID HELIUM-3 FILMS

Brian COWAN, Michael FARDIS, Tom CRANE and Laila ABOU-EL-NASR

Millikelvin Laboratory, Royal Holloway and Bedford New College, University of London, Egham Hill, Egham, Surrey, TW20 0EX, U.K.

Measurements of the exchange frequency and the spin diffusion coefficient in solid helium-3 monolayers on graphite are reported. It is shown that the diffusion coefficient may be found directly from the low frequency spin relaxation times, as a purely hydrodynamic consequence. Values of D are consisent with those calculated from the exchange frequency.

We report NMR measurements of the spin diffusion coefficient and the exchange frequency in solid submonolayer films of ^3He on a graphite surface. Experiments have been performed at a temperature of 1K, using a substrate of grafoil. Previously (1) we have obtained such exchange frequencies from spin-lattice relaxation time measurements and these have now been extended. Diffusion measurements for this system using the conventional NMR technique, involving a magnetic field gradient have not been successful, mainly because of the anisotropic demagnetising fields within the substrate. We are currently overcomming these technical difficulties, but here we discuss measurement of diffusion in a completely different way.

For a planar system the NMR relaxation times depend on the orientation of the surface with respect to the applied static magnetic field (2) because of the anisotropy of the internuclear dipolar interaction. Furthermore, it is a consequence of the dimensionality of the system that at long times the dipolar autocorrelation function decays as t^{-1}, leading to a logarithmic divergence of the spectral density, the function which determines the relaxation times. This leads to anomalous behaviour of the spin relaxation of adsorbed systems (3).

The general representation for the spectral density in 2D is as the sum of two independent functions, with coefficients which depend on the orientation of the surface (2). One of the functions displays the logarithmic divergence; the other does not. And when the normal to the surface points along the applied magnetic field the coefficient of the anomalous term becomes zero. Thus for such an orientation (β=0) the system should exhibit "normal" relaxation behaviour. We are concerned here with measurements made at this orientation. In figure 1 we show spin lattice relaxation measurements. The data are plotted in reduced form, according to the procedure we have outlined in (1). The tendency to flatten off at

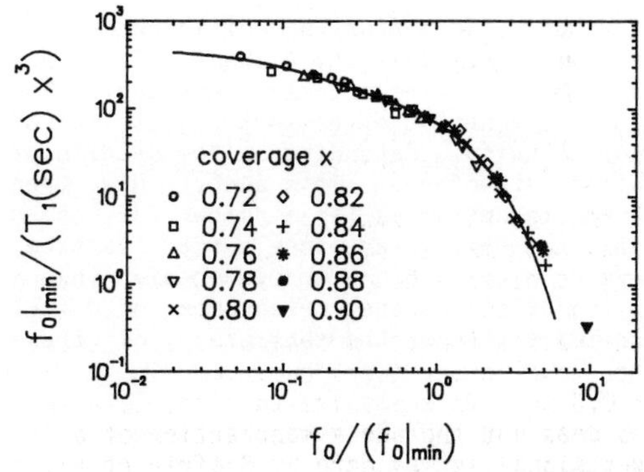

Figure 1

low frequencies is indicative of "normal" behaviour.

Although the exchange process really involves interchange cycles with different numbers of spins moving at different rates, spin relaxation measurements are sensitive to some average exchange frequency. This may be regarded as an effective pair exchange frequency. Observations of minima in T_1 give, to within a multiplicative constant (1), the magnitude of this exchange rate. In figure 2 we have plotted the Larmor frequency of such minima against fractional monolayer completion x. The solid line is a least squares fit through the points which is found to have precisely the same form as that calculated, for J, by Roger (4):

$$\omega_{min} = ax^{19/4}\exp(-19x^{5/2}).$$

We have been interested in spin diffusion in these systems for a number of reasons. At long times the dipolar correlation function is determined by the diffusion. Thus knowledge of the diffusion coefficient is needed in approximating the form of the spectral density functions, a necessary step in fully interpreting the relaxation time data. On very

0921-4526/90/$03.50

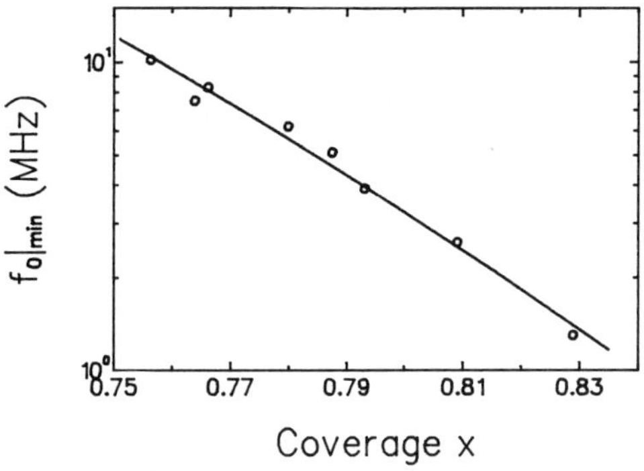

Figure 2

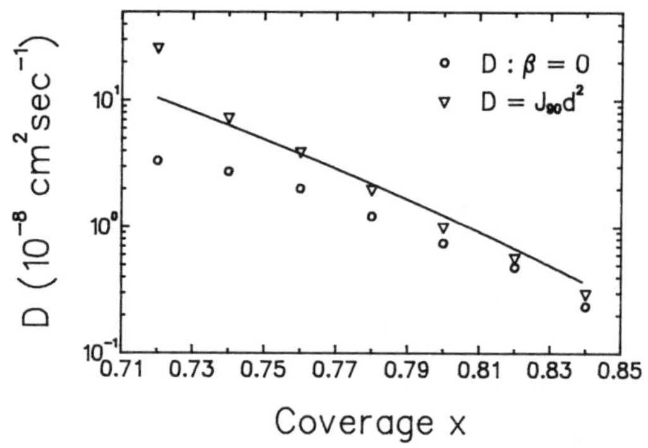

Figure 3

general grounds the diffusion coefficient D must vary as Ja^2, where J is the exchange frequency and a is the interparticle spacing. The dimensionless constant of proportionality is of the order of unity, depending on the details of the lattice structure. This number has been calculated recently (5) for a number of cases including the triangular close packed lattice, of relevance here. Since we can obtain values (1) for J, it follows that measurement of D will then provide a test of the consistency of these ideas.

At orientation β=0 the remaining term of the correlation function has the long-time behaviour

$$g_2(t) \sim \frac{\pi}{36}\frac{\alpha}{D^2}t^{-2}.$$

This is independent of the microscopic details of the atomic motion, the coefficient of t^{-2} relying only on the areal density α and the diffusion coefficient. By twice integrating by parts the Fourier transform integral we find that the coefficient of t^{-2} gives a term linear in ω in the corresponding spectral density function:

$$j_2(\omega) = j_2(0) - \frac{\pi^2}{36}\frac{\alpha}{D^2}\omega + \ldots\ldots$$

leading to

$$\frac{1}{T_1(\omega)} = \frac{1}{T_1(0)} - \frac{\pi^2}{16}\hbar^2\gamma^4\alpha\frac{\omega}{D^2} + \ldots$$

Analysis of the low-frequency T_1 data according to this formula thus gives directly the spin diffusion coefficient. In figure 3 the circles show values for D we have obtained in this way. For comparison the triangles indicate valued derived from previously established values for J (1) and using the relation between J and D as obtained from (5). Similarly the solid curve uses this relation in conjunction with the fitted line of figure 2. The agreement is reasonable, considering the errors in the experimental data.

Finally we consider the zero point phonon renormalisation of the dipolar interaction. For bulk systems it is known that this has the effect of reducing slightly the moments of the resonance line (6), (7). This may be investigated through certain sum rules which the relaxation times must obey (6). Thus the frequency integral of $1/T_1$ is equal to π times the second moment. Plotting the data in reduced form (as in figure 1, but with linear axes) we may perform a numerical integration over all points which can be compared with the calculated M_2. At β=0 one finds a renormalisation constant of ~0.73, which compares with ~1.0 for the β=90° orientation.

There are two possible explanations for this difference. It follows from the anisotropy of the dipolar interaction that at β=90° motion is relatively inefficient at averaging its effects away (2): thus little influence of the zero point phonons. Conversely, at β=0 normal behaviour prevails. Another possibility is the effect of platelet misorientation within the grafoil. We have established (8) that a distribution of surface orientations tends to decrease the β=90° moments while increasing the β=0 values. When taken with the "natural" renormalisation of the interaction, this is a competing explanation.

REFERENCES

(1) B.Cowan, L.A.El-Nasr, M.Fardis and A.Hussain, Phys.Rev.Lett.58(1987)2308.
(2) B.Cowan, J.Phys.C, 13 (1980) 4575.
(3) B.Cowan, J.Low Temp.Phys. 50 (1983) 135.
(4) M.Roger, Phys. Rev. B, 30 (1984) 6432.
(5) B.Cowan, W.J.Mullin and E.Nelson, J.Low Temp.Phys. 77 (1989) 181.
(6) A.Landesman, Ann. de Phys. 8 (1973) 53.
(7) B.Cowan and M.Fardis, elsewhere, this vol.
(8) B.Cowan and M.Fardis, to be published.

Physica B 165&166 (1990) 709–710
North-Holland

STATIC MAGNETIZATION OF MULTILAYERS OF ^{3}HE ADSORBED ON SINTERED SILVER

Akio FUKUSHIMA, Yuichi OKUDA† and Shinji OGAWA

Institute for Solid State Physics, University of Tokyo, Roppongi, Minato-ku, Tokyo 106, JAPAN
†Department of Applied Physics, Tokyo Institute of Technology, O-okayama, Meguro-ku, Tokyo 152, JAPAN

The precise static magnetization of multilayers of ^{3}He adsorbed on sintered silver was measured with SQUID magnetometer as a function of ^{3}He coverages. The temperature range of each measurement is from 10 mK down to 0.2 mK, and the coverage is varied from monolayer up to about nominal 65 layers. The ferromagnetic exchange is rather small ($J/k_B \approx 80$ μK) compared to that adsorbed on graphite. Around 0.2 mK the unique structure was observed in the coverage dependence of the static magnetization. The structure has a few broad peaks around 0.24 atoms/Å^2, 0.29 atoms/Å^2 and 0.40 atoms/Å^2. It is similar to the structure of the specific heat result of ^{3}He multilayers adsorbed on sintered silver.

1.Introduction

The magnetic measurement of ^{3}He multilayers adsorbed on sintered silver without bulk liquid revealed the coverage dependence of the surface ferromagnetism of ^{3}He. The ferromagnetic interaction was found to have its origin in the first few adsorbed layers [1]. The specific heat of this system was measured by Greywall in the temperature range from 0.5 mK up to 7 mK [2]. The double-peaked structur(in the specific heat vs coverage was observed. On the other hand, the coverage dependence of the magnetization haven't been measured with high accuracy at the temperature below 1 mK.

From a geometrical point of view, graphite is cleaner and more ideal for a substrate of ^{3}He multilayers. So recent studies had a great attention to the system, and the coverage dependences of the magnetization [3] and the specific heat [4] were measured extensively. Especially, the specific heat work revealed detailed structures of ^{3}He multilayers. Another important point of the graphite system is that the ferromagnetic interaction is much larger than the cases of other substrates, which is favorable to study the system experimentally.

We have measured the static magnetization on sintered silver more precisely to clarify what is the essential difference between graphite and sintered silver for the magnetism of ^{3}He multilayers.

2.Experiments

Experimental situation and method are almost the same as that of previous papers [1,5]. The magnetization was measured by a SQUID magnetometer under the field of 48 Oe. The only different point of this experiment is the surface area of the substrate. We performed experiments with two different experimental cells. The one has 8.94 m^2 of surface area with accuracy of 2% and the other has 117.2 m^2 with the same accuracy. These surface areas were determined using the vapor pressure isotherms of adsorbed nitrogen at 77.25K with standard surface area of 16.2 Å^2 per molecule. The measurements were done for 34 different coverages from 0.104 atoms/Å^2 up to 4.5 atoms/Å^2. Corresponding number of layers was calculated from first layer's areal density 0.108 atoms/Å^2, second layer's 0.092 atoms/Å^2, and 0.070 atoms/Å^2 for a layer above the second layer. 4.5 atoms/Å^2 corresponds to 65 layers if the layer-by-layer picture holds up to this thickness.

3.Results and Discussions

Data at 0.2 mK and 0.5 mK are shown in Fig.1. Vertical axis denotes SQUID output in Volt, which can be converted to the absolute value of the magnetization. The dotted lines are calculated free spin values. By fitting data from 0.2 mK to 2 mK to the high temperature series expansion of 2D Heisenberg ferromagnet, Curie constant for layers above the second layer completion was found to be constant with coverage. So the line above the second layer is drawn constant.

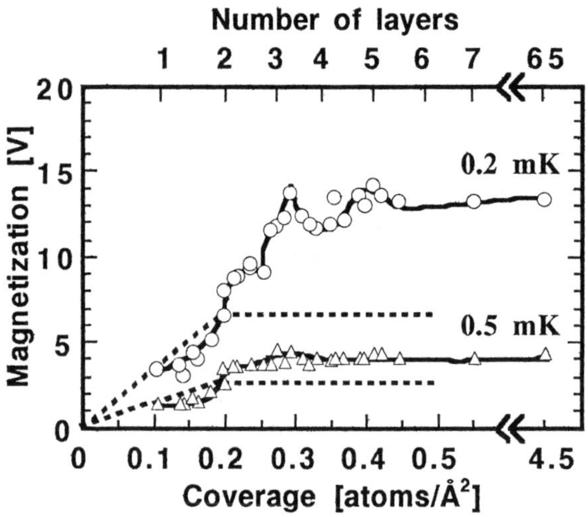

Fig.1 The magnetization as a function of ^3He coverage. Solid curves are guides for eyes. See text for the dotted lines.

In the low coverage regime from 0.10 atoms/Å^2 to 0.20 atoms/Å^2, the magnetization is smaller than the free spin value, and each temperature dependence is well described by Curie law. It means that ^3He atoms promoted to the second layer behave like a degenerate Fermi liquid. On the graphite system at the coverage of 0.184 atoms/Å^2, ^3He atoms solidify into the phase commensurate with the solid first layer below 1 K, and its magnetization has a possibility to be antiferromagnetic [3,4]. In our case, however, Curie constant does not increase with the coverage as the free spins (dotted lines in Fig.1). It is smaller than the free spin value. Around the second layer completion coverage the magnetization grows rapidly, which implies the solidification of the second layer.

In the high coverage region above 0.20 atoms/Å^2, no structure can be seen at 0.5 mK. It is almost independent of the coverage. But at 0.2 mK we observed a unique structure, which has one shoulder at 0.24 atoms/Å^2 and two clear peaks at 0.29 atoms/Å^2 and 0.40 atoms/Å^2. The coverage of the first shoulder corresponds to 2.5 atomic layer. Two peaks around 0.29 atoms/Å^2 and 0.40 atoms/Å^2 were not seen on the graphite system. But the similar peaks in the specific heat data on sintered silver were observed. The origin of the structure is not known at present. The ferromagnetic interaction appears for the coverage above 0.20 atoms/Å^2, and the maximum exchange interaction J/k_B is found to be about 80 μK.

We also observed a strange temperature dependence of the magnetization for coverages more than 0.4 atoms/Å^2 from 10 mK to 3 mK. The typical result is shown in Fig.2 at the coverage of nominal 4.5 atoms/Å^2. Though the magnetization below 2 mK obeys the 2D Heisenberg ferromagnet, it has a minimum around 3 mK and in creases as the temperature is raised. This funny behavior was not observed for the coverage lower than 0.4 atoms/Å^2. It is very mysterious, since the liquid layers of ^3He must be degenerated and the magnetization should be independent of the temperature at this low temperature.

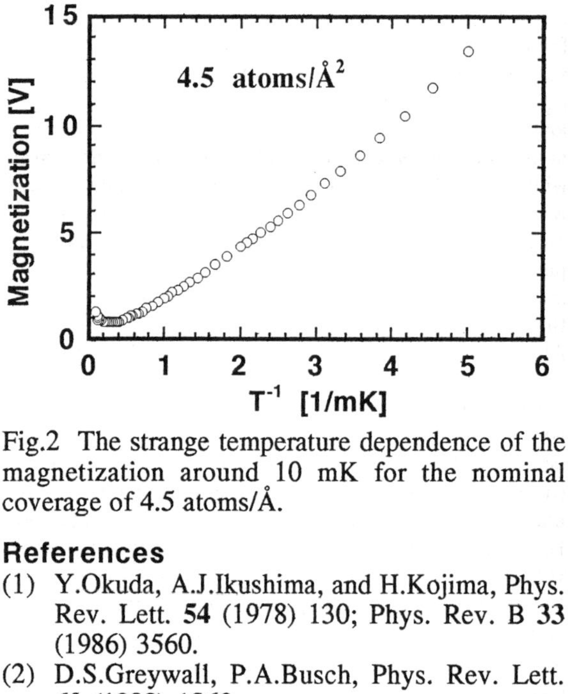

Fig.2 The strange temperature dependence of the magnetization around 10 mK for the nominal coverage of 4.5 atoms/Å.

References

(1) Y.Okuda, A.J.Ikushima, and H.Kojima, Phys. Rev. Lett. **54** (1978) 130; Phys. Rev. B **33** (1986) 3560.
(2) D.S.Greywall, P.A.Busch, Phys. Rev. Lett. **60** (1988) 1860.
(3) H.Franco, R.Rapp, and H.Godfrin, Phys. Rev. Lett. **57** (1986) 1161.
(4) D.S.Greywall, P.A.Busch, Phys. Rev. Lett. **62** (1989) 1868; preprint.
(5) Y.Okuda, A.Fukushima, and A.J.Ikushima, Jpn. J. Appl. Phys. **26** suppl. **26**-3 (1987) 269.

Physica B 165&166 (1990) 711–712
North-Holland

Magnetic Structure of Vortex Lattice State in the Adsorbed Second Layer of ^{3}He Films

Mitsutaka FUJITA and Kazushige MACHIDA[A]

Institute of Materials Science, University of Tsukuba, Tsukuba 305, Japan.
[A]Department of Physics, Kyoto University, Kyoto 606, Japan.

Stimulated by a recent discovery of a sharp specific heat anomaly at 2.5mK in a second layer of adsorbed ^3He on grafoil by Greywall et al., we study a lattice gas model or Hubbard model on a triangular lattice within the mean field approximation. In the case of nearly half fillings, we find a novel type of the magnetic structure forming vortex lattice states, which is non-collinear and characterized by triple Q vectors keeping the three-fold symmetry.

A mysterious phase transition signaled by a sharp specific heat peak around T=2.5mK in ^{3}He films adsorbed on grafoil is found by Greywall and Busch[1] and arouses much attention[2]. The experiment was done at surface coverages $\rho = 0.160, 0.175, 0.184$atoms/$\mathring{A}^2$ above the first layer is completed ($\rho=\rho_1 \sim 0.114$), but before the third layer promotion ($\rho \sim 0.24$). Since Franco *et al*[3] already found a discernible negative deviation of the magnetization expected from the free spin value in this coverage regime, this phenomenon is certainly related to some kind of nuclear magnetic ordering, possibly antiferromagnetic, of the second layer ^{3}He atoms sat on the first layer which itself forms a triangular lattice incommensurate with the underlying graphite. As for the solidification of second layer, Van Sciver and Vilches[4] showed a melting peak to the $\rho=0.18$ coverage in their specific heat experiment. Abraham et al.[5] found a registry solid phase in the low density by a numerical Monte Carlo simulation. Although the neutron experiment[6] on higher coverage $\rho=0.203$ failed to detect solidification of second layer, it sounds that depending upon the density the solid phases, which have different type of registry with the triangular lattice of the first layer, exist for very narrow ranges of coverage. These lower commensurate solid phases should play a role as a reference system for nearby densities. For example, say $\rho_2=\rho-\rho_1=0.064$ the system finds the $\sqrt{7}\times\sqrt{7}$ registry phase because ρ_2/ρ_1 almost equals to 4/7 [7]. Since the density ρ_2 is rather randomly chosen in an actual experimental situation, all ^{3}He atoms would never be precisely accommodated in an appropriately selected registry solid phase. There generally exist excess or deficit atoms in this registry phase.

From the view point that in the ^{3}He bulk the higher-order cyclic exchanges make significant contributions[8] to the magnetism, it is worth examining this two dimensional magnetic phenomenon from the itinerant picture rather than the localized one[7,9]. We study here a lattice gas model or Hubbard model at T=0 on a triangular lattice; $H = -t\sum_{<ij>\sigma} c^\dagger_{i\sigma}c_{j\sigma} + U\sum_i n_{i\uparrow}n_{i\downarrow}$ where t(=1) is the hopping integral between the nearest neighbor sites and U is the on-site hard core repulsion. We confine our discussion to coverages well before the second layer is completed and ignore the first layer, thus consider only the second layer. If the number of ^{3}He are equal to the lattice sites of a registered triangular lattice, it corresponds to the "half-filling". A deviation from that atom number inevitably introduces a few excess or deficit atoms, namely defects. This corresponds to the "nearly half-filling case" of the Hubbard model. We take here the simplest mean field approach under the assumption that there is a long-range order in the ground state. In fact, as stressed by Friedman et al.[10] the dipole interaction, although its coupling energy is extremely weak ($\sim 0.1\mu K$), guarantees its existence and makes the nuclear spins confine in the plane of the substrate, therefore we take account of only two spin components S^z and S^x. Then the mean field Hamiltonian is given by

$$H = -t\sum_{<ij>\sigma} c^\dagger_{i\sigma}c_{j\sigma} + \frac{U}{2}\sum_i\{(n_i - S^z_i)n_{i\uparrow} + (n_i + S^z_i)n_{i\downarrow} - S^x_i(c^\dagger_{i\uparrow}c_{i\downarrow} + h.c.)\} - \frac{U}{4}\sum_i\{n_i^2 - (S^z_i)^2 - (S^x_i)^2\} \quad (1)$$

where the order parameters; the moments S^z_i, S^x_i and atom density n_i depend on the ith site, are determined self-consistently by $S^z_i = <n_{i\uparrow}> - <n_{i\downarrow}>$, $S^x_i = <c^\dagger_{i\uparrow}c_{i\downarrow}> + <c^\dagger_{i\downarrow}c_{i\uparrow}>$ and $n_i = <n_{i\uparrow}> + <n_{i\downarrow}>$ $(\mu_B = 1)$.

To seek a stable spin and density configuration by increasing the unit cell size of a periodic pattern under a given filling, we numerically solve (1) and self-consistent equations iteratively, starting from various kinds of initial configurations $\{S^z_i, S^x_i, n_i\}(i = 1, \cdots N)$ under the periodic boundary condition. As for the initial condition both cases of the collinear and non-collinear spin patterns are considered. For larger unit cell, we restrict ourselves to the case that the system holds three-fold symmetry, namely the pattern is characterized by the triple Q wave vectors. We note that the $\sqrt{3}\times\sqrt{3}$ sublattice is characterized by the single Q vector.

0921-4526/90/$03.50

Let us first examine the half-filling case where the atom number per site is $n=1$. There are several possible magnetic phases: (1) antiferromagnetic phase (AF) with the 2×1 sublattice, (2) ferromagnetic phase (FM), (3) ferrimagnetic phase (FR) with the $\sqrt{3}\times\sqrt{3}$ sublattice , (4) the 120° structure (V_2) in which the three spins on a unit triangle take three distinctive directions differed by 120° each other and other collinear and non-collinear cases with a larger unit cell. We examine the relative stability among these states and find that V_2 is the lowest among them for a wide range of U ($\geq U_c = 5.5$)[11]. For example, when $U = 10.0$ the relative energies per site are $-2.9231(V_2)$, $-2.8925(\mathrm{FR})$, -2.8817 (AF) and -2.4997 (FM). In the V_2 phase, the six bands, which come from up and down spins and three sublattice structure, group together into two separate bands and open a gap only in the middle of them, making V_2 energetically favorable for $n=1$ but not for $n\neq 1$. Below U_c the indirect gap between them vanishes and V_2 is destabilized.

Having seen that the non-collinear V_2 is stable over the other collinear states for $n=1$, we seek a generic magnetic structure with larger unit cell sizes for nearly half-filling cases where defects should be neatly accommodated. We display a typical example of the stable magnetic ordering pattern in the self-consistent solution in Fig.1 for the case that$N = 9\times 9$ lattice with 93 atoms, i.e. $n=31/27$ when $U=8.0$. It forms $\sqrt{27}\times\sqrt{27}$ sublattice structure. We see that (1) there are three vortices contained in a unit cell and they form a vortex lattice structure[11]. There are two kinds of vorticity or winding number; one is +1 situated at the center and three corners of the hexagon while the other is -2. The total vorticity vanishes. (2) The gross feature of this vortex lattice structure (VLS) can be constructed by the triple Q wave vectors whose directions are 120° apart and whose magnitudes $|Q| = 8\pi/9$ in this case. (3) All the moments on each site almost equal except for the vortex sites where the moment vanishes and excess (or deficit) density exclusively accumulated, forming a localized object or a triple Q soliton lattice. This implies that higher harmonics play an important role. (4) The sum of all the moments vanishes, thus there is no net ferromagnetic component in the system. (5) The spin configuration far from the vortex sites tends to refer the V_2 configuration which is the most stable state for the half-filling case.

In other fillings near $n=1$ we can always find a self-consistent solution of VLS with the differernt sizes of the unit cell, provided that the combination between the triple Q wave vectors and the atomic filling is properly chosen. As the deviation from $n=1$ becomes small, the unit cell of the VLS becomes large by changing the fundamental wave vectors $|Q|$ to accommodate defects.

In conclusion, the second ^{3}He layer is found to be an interesting test ground for the study of two-dimensional

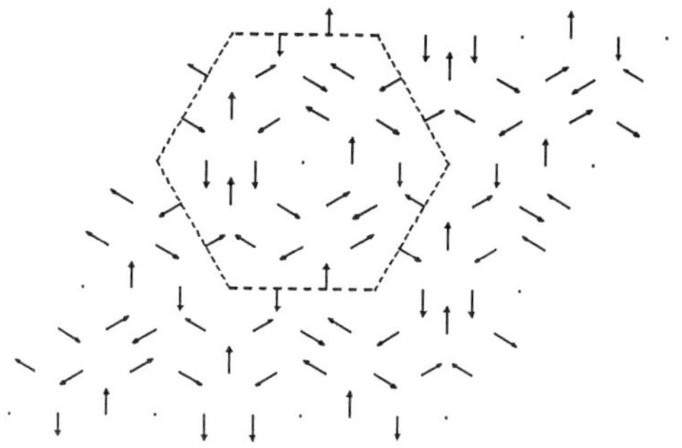

Fig.1 Magnetic structure of Vortex lattice state with the $\sqrt{27}\times\sqrt{27}$ sublattice when $U=8.0$ and $n=31/27$. The broken line indicates the unit cell.

spin $\frac{1}{2}$ system on a frustrated triangular lattice with defects whose number and interaction strength U/t can be controlled by varying coverages. We have found a novel type of the periodic ordering pattern of the vortex lattice structure for accommodating defects which normally exist in an actual experimental situation of the absorbed ^{3}He films on grafoil. The present study might help to understand the magnetism in itinerant electron systems such as fcc 3d-transition metal alloys[12] where because of non-bipartite crystalline structure there appear various non-collinear types of magnetic patterns with triple Q vectors. The physical origin for this largely remains unexplored.

References

1) D.S. Greywall and P.A. Busch, Phys.Rev.Lett. **62**, 1868 (1989).
2) P.V.E. McClintock, Nature **340**, 98 (1989).
3) H. Franco, R.E. Rapp and H. Godfrin, Phys.Rev.Lett. **57**, 1161 (1986).
4) S.W. Sciver and O.E. Vilches, Phys.Rev. B**18**, 285 (1978).
5) F.F. Abraham, J.Q. Broughton, P.W. Leung and V. Elser, preprint.
6) C. Tiby, H. Wiechert, H.J. Lauter and H. Godfrin, Physica **107B**, 209 (1981).
7) V. Elser, Phys.Rev.Lett. **62**, 2405 (1989).
8) D.M. Ceperley and G. Jacucci, Phys.Rev.Lett. **58**, 1648 (1987).
9) M. Roger, Phys.Rev.Lett. **64**, 297 (1990).
10) K. Machida and M. Fujita, preprint.
11) L.J. Friedman *et al.*, Phys.Rev.Lett. **62**, 1635 (1989).
12) See for example, T. Jo and K. Hirai, J.Phys.Soc.Jpn. **53**, 3183(1984). S. Kawarazaki *et al.*, Phys.Rev.Lett. **61**, 471 (1988). C.L. Henley, *ibid* **62**, 2056 (1989).

Physica B 165&166 (1990) 713–714
North-Holland

MEASUREMENTS OF THE EXCHANGE INTERACTION IN 2D SOLID ^{3}HE FILMS ADSORBED ON GRAPHITE

J.SAUNDERS, C.P.LUSHER and B.P.COWAN

Millikelvin Laboratory, Royal Holloway and Bedford New College, University of London, Egham, Surrey, TW20 0EX, U.K.

Measurements of the nuclear magnetic susceptibility of multilayer films of ^{3}He adsorbed on graphite plated with a monolayer of ^{4}He are reported for ^{3}He surface densities up to 0.27 Å^{-2} and temperatures in the range 6 to 600mK. The existence of a second layer solid with antiferromagnetic exchange has been confirmed directly. Following promotion of atoms to the third fluid layer the exchange becomes ferromagnetic. The dependence on total surface density of the exchange interaction and the density of the second layer solid have been determined and support the proposed RKKY mechanism. The second layer solid density evolves in a stepwise fashion.

The exchange interaction in two dimensional solid ^{3}He layers adsorbed on graphite has been investigated previously by studies of the magnetisation (1) and heat capacity (2). In this work we have measured the nuclear magnetic susceptibility over a wide temperature range, using Grafoil plated with a single monolayer of solid ^{4}He as a substrate, in order to determine precisely the density dependence of the exchange interaction of the second layer solid. Further experimental details are provided in ref.(3).

A clear signature of the solidification of the second layer, which we observe to commence at a second layer areal density of ρ_3=0.055 Å^{-2} in agreement with ref.(2), is provided by a sharp increase in the low temperature susceptibility. We describe our surface densities in terms of $\rho_T=\rho_3+0.114$ Å^{-2} to facilitate comparison with earlier work. At densities 0.180,0.182 Å^{-2}, following the completion of solidification, the data provide direct evidence for a solid with antiferromagnetic exchange.

The temperature dependence of the magnetic susceptibility is shown in fig.1 where we have plotted $(\chi T)^{-1}$ against T^{-1} to emphasise departures from Curie's law. In the antiferromagnetic solid at these two densities a fit to a Curie Weiss law, $\chi \propto (T-\theta)^{-1}$, gives θ= -1.70, -1.36 mK respectively. The heat capacity data at these densities have been interpreted (4) assuming that the second layer consists of a 2D triangular lattice with multiple spin exchange. Antiferromagnetic four, six and two particle exchange cycles compete with ferromagnetic three particle ring exchange. The parameters inferred from comparing the heat capacity data at 0.184 Å^{-2} to a calculation for a sixteen spin cluster imply a positive θ=1.35 mK, contrary to our result. However the calculation of the susceptibility of such a cluster shows antiferromagnetic deviations from Curie's law below ≈ 10 mK (4).

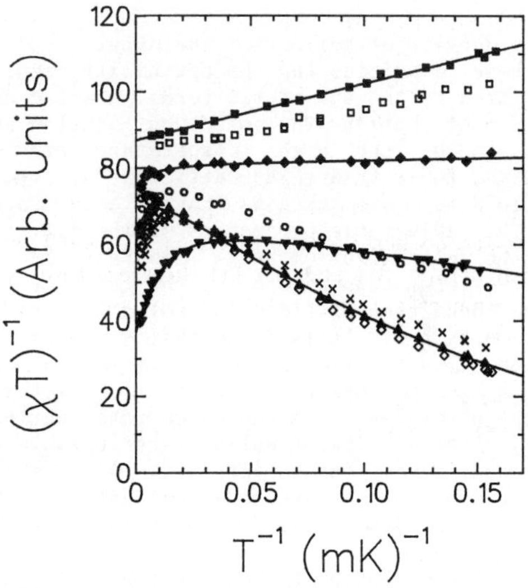

FIG.1 Temperature dependence of the susceptibility; $(\chi T)^{-1}$ vs. T^{-1}, at surface densities ρ_T (Å^{-2}), 0.180 (■), 0.182 (□), 0.190 (◆), 0.198 (○), 0.210 (▲), 0.220 (◇), 0.230 (×), 0.270 (▼). Solid curves are fits to data, see text.

We note that the observed density dependence of the exchange is significantly weaker than expected ($J\sim\rho^{-26}$) for a 2D solid (5), presumably due to motion perpendicular to the substrate during exchange. Pair exchange mediated by a virtual third layer liquid state (6) is another possibility.

With further increasing density a crossover to ferromagnetic exchange occurs. A fluid signal develops (for ρ_T>0.183 Å^{-2}), as shown by the increase in χT at the highest temperatures, due to promotion to the third layer. It is possible to infer both the exchange interaction and the

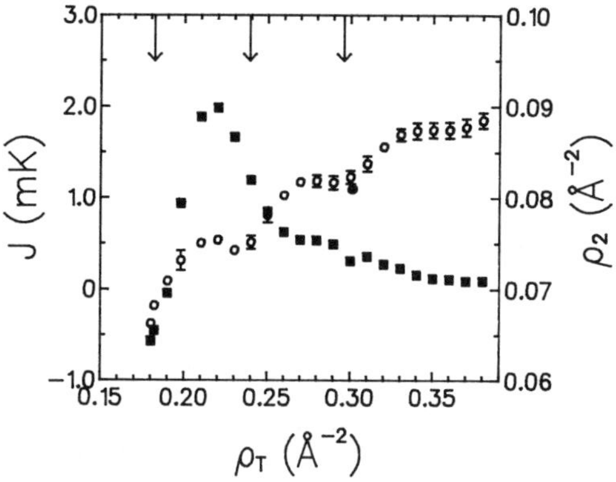

FIG.2. Exchange constant J (■) and second layer surface density ρ_2 (○) as a function of total surface density. (●) ρ_2 determination of ref.(8). Arrows indicate "layer promotion" as identified in ref.(2).

surface density of solid from the data.

At these densities the susceptibility results were fitted to the sum of two terms, a solid and a fluid contribution. The solid contribution was taken to be the high temperature series expansion for a triangular lattice (7), while the fluid term was taken as that of a 2D Fermi gas. The adjustable parameters of the fit are thus the effective exchange constant J, an effective Fermi temperature T_F, the ratio of the surface densities of fluid and solid α, and a Curie constant C. It is found that C scales as expected with total surface density and so can be constrained leaving three fit parameters. From the parameter α and the known total surface density, the density of solid, attributable to the second layer, may be inferred. Fig.2 shows the dependence of the exchange constant J, and second layer density ρ_2 on the total surface density.

The observed "stepwise compression " of the second layer is correlated with the "layer promotions" identified from heat capacity data (2) and arrowed in the figure. The measurement of the second layer density by neutron scattering (8) agrees well with our data. Furthermore if, as suggested in ref.(2), the heat capacity peaks observed in ref.(9) are due to melting of the second layer solid, the density dependence of ρ_2 observed here accounts naturally for the density dependence of the melting temperature. It *may* not be necessary to invoke any registered structure for the second layer solid, as advocated in (2). This could form as an incommensurate triangular lattice and be subsequently compressed as found here.

This measurement of the second layer density is valuable in helping to understand the importance of various exchange mechanisms. In-plane exchange should be strongly density dependent, and influenced by any structural changes occurring in the second layer solid. On the other hand indirect exchange of the RKKY type, involving the fluid film above the second solid layer, should be sensitive to the state of the fluid overlayer (10) and may show oscillatory features. In our data there are a number of indications of such behaviour. J dips around 0.30 $Å^{-2}$, where a "layer promotion" also occurs. Further the large variations in J between 0.20 and 0.24 $Å^{-2}$ occur with the density of the second layer solid remaining constant, while a strong peak in the fluid parameter T_F approximately coincides with the maximum in J. In addition, in common with the heat capacity results, there is a plateau in J centered around 0.27 $Å^{-2}$. We discuss the oscillatory properties of the fluid overlayer and their origin elsewhere (11).

The peak in the exchange interaction we observe occurs at a slightly lower surface density than that of the maximum magnetisation in the 3mK isotherm of ref.(1). The later precise determination of J from the magnetisation at the single density 0.233 $Å^{-2}$ (1) found J=2.1mK compared to our result J=1.5mK. The discrepancy may be due in part to the different temperature scales. A comparison of the exchange energies with those inferred from the heat capacity data is a more complex issue. Firstly at $\rho<0.24$ $Å^{-2}$ the (zero field) spin heat capacities do not fit the Heisenberg series expansion in contrast to the susceptibility results. Secondly the susceptibility is known to be more sensitive to competition between antiferromagnetic and ferromagnetic exchange processes (4) a result also seen experimentally in submonolayer films.

We have established that the strong ferromagnetic tendency in ^3He films is also present when the first layer is replaced by ^4He. Any exchange between first and second layer ^3He solid is therefore a weak effect and the first solid layer is essentially inert.

References
(1) H.Franco, R.Rapp, H.Godfrin, Phys.Rev.Lett. 57 (1986) 1161 ; H.Godfrin, R.Ruel, D.Osheroff, Phys.Rev.Lett. 60 (1988) 305
(2) D.S.Greywall and P.A.Busch, Phys.Rev.Lett. 62 (1989) 1868; D.S.Greywall Phys.Rev. B41 (1990) 1842
(3) C.P.Lusher,J.Saunders,B.P.Cowan, these proceedings
(4) M.Roger, Phys.Rev.Lett. 64 (1990) 297
(5) B.P.Cowan et.al., Phys.Rev.Lett. 58 (1987) 2308
(6) M.Heritier, J. de Phys.Lett. 40 (1979) 451
(7) G.A.Baker et.al. Phys.Lett. 25A (1967) 207
(8) H.Lauter et.al. Can.J.Phys. 65 (1987) 1435
(9) S.van Sciver, Phys.Rev. B18 (1978) 277
(10) S.Tasaki, Prog.Theor.Phys. 79 (1988) 1311 R.Guyer, preprint
(11) J.Saunders,C.P.Lusher,B.P.Cowan, these proceedings

Physica B 165&166 (1990) 715–716
North-Holland

OBSERVATION OF QUANTUM SIZE EFFECTS IN THIN FILMS OF LIQUID ^3He ADSORBED ON GRAPHITE

C.P.LUSHER, J.SAUNDERS and B.P.COWAN

Millikelvin Laboratory, Royal Holloway and Bedford New College, University of London, Egham, Surrey, TW20 OEX, U.K.

The nuclear magnetic susceptibility of multilayer liquid ^3He films adsorbed on the surface of graphite has been measured. The surface density dependence shows evidence of distinct step structure that has been studied up to five liquid layers. This is attributed to quantum size effects arising from the finite film thickness, leading to a description of the growth of the film in terms of the population of a set of two dimensional Fermi liquids. The Fermi liquid interactions in this system are discussed.

Calculations of the density profile of helium films adsorbed on a solid surface show variations in density suggestive of a layered structure (1). As the film thickens the influence of the substrate potential becomes less pronounced. However even when the substrate potential is neglected these films may show finite size effects due to size quantisation arising from the finite thickness of the film. Thus while the atoms are free to move parallel to the surface, the energy levels associated with motion perpendicular to the surface are those of particle in a box states (2). In this case the size of the box is determined by the thickness of the self bound film. This should be contrasted with multiple sub-band occupancy in quasi-two-dimensional charged systems where the sub-band energies are fixed by an external potential.

In this work we have studied multilayers of ^3He adsorbed on exfoliated graphite (grafoil). The surface density dependence of the fluid component of the nuclear magnetic susceptibility shows a distinct step structure which we have explored out to five liquid "layers" and which we believe reflects size quantisation due to the finite film thickness. The first observation of magnetisation steps was made in mixture films adsorbed on Nuclepore (3); in this case one pronounced step was observed and there were indications of a further less distinct step, which is presumably smeared by the effects of substrate heterogeneity. In our experiments with a grafoil substrate the first two layers are solid, the first layer being ^4He. The nuclear magnetic susceptibility has been measured by NMR at 1.08 MHz. The fluid susceptibility was determined from the total signal after subtracting a contribution from the second solid layer. This was done by fitting the signal to a solid term plus a fluid term of the form $\chi=(C/T_F)(1-\exp(-T_F/T))$, which is the form of the susceptibility of an ideal 2D gas; here T_F is treated as an adjustable parameter. More details

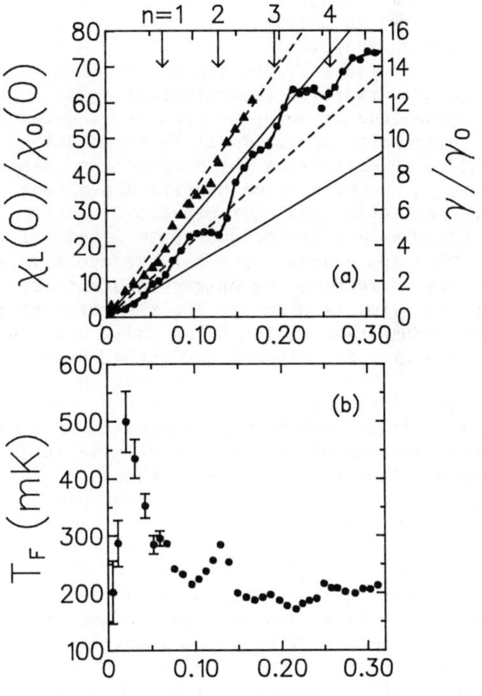

FIG.1. Dependence on surface density of a) low temperature limiting value of fluid susceptibility (this work) (●) and temperature coefficient of linear heat capacity ref.(5) (▲). b) characteristic degeneracy temperature T_F. In a) the straight lines indicate the bulk liquid values for γ/γ_0 (dashed) and χ/χ_0 (solid). In each case the upper (lower) line uses melting pressure (zero pressure) parameters. For comparison with ref.(3) arrows indicate the completion of liquid "layers" at intervals of surface density 0.065 Å^{-2}.

are provided elsewhere in these proceedings (4). This procedure allows a determination of the surface densities of the solid and fluid components.

In figure 1. the inferred low temperature limit of the fluid susceptibility is plotted against the surface density of fluid. The data have been normalised by the susceptibility of an ideal 2D gas, χ_0, which is independent of surface density. The plot shows clear evidence of step structure, where each step is followed by a plateau-like feature. We also show the dependence on fluid density of the parameter T_F, where $T_F=C/\chi(0)$, which shows up more clearly the structure at the lowest surface densities. The peaks in T_F are associated with the start of each riser in the step structure. Isotherms of the fluid susceptibility show this structure becoming far less distinct above 100mK. The structure we observe is well correlated with that observed in the heat capacity of multilayer ^{3}He films on grafoil (5). The figure also shows values of $\gamma=c_{fluid}/T$ determined from fits to the total heat capacity. The values of γ have here been normalised by γ_0, the density independent value appropriate to an ideal 2D Fermi gas.

A model of such films which predicts similar structure in the surface density dependence of the magnetisation and heat capacity is that of ref.(2). Neglecting the substrate potential the ^{3}He states are taken as $\psi \sim sin(\pi mz/d)exp(i\underline{k}.\underline{r})$, where the z part of the wavefunctions are the "particle in a box" states of energy $\hbar^2\pi^2m^2/2m_3d^2$. The size of the box, the film thickness d, is determined by the number of atoms and the atomic interactions, most importantly the hard core. Thus the available states form a series of 2D Fermi fluids with quantum numbers m=1,2,3,....where the highest occupied m state increases steplike as a function of surface density. However the thickening of the film decreases the energies of the m states and the Fermi level is found to be approximately constant (2). Since the total susceptibility is the sum of that of each of the m occupied 2D Fermi systems and the magnitude of each of these at $T=0$ is independent of number density for an ideal gas, the susceptibility of an ideal film evolves with a sequence of steps of equal magnitude.

The question then arises , how does the crossover from 2D to 3D behaviour occur? The general trend of both the susceptibility and heat capacity data is to increase in proportion to surface density, as might be expected if the thickening of the film were described as the growth of bulk liquid at constant density. Pressure dependent interactions are known to have a strong influence on bulk properties. In the figure we indicate bulk liquid behaviour both at the melting curve and at zero pressure and it is striking to note that the trend of both sets of data is more in accord with the melting curve parameters. However calculations of the film densities under the influence of the substrate potential find densities appropriate for low pressure liquid (6).

Rather we propose that the Fermi liquid parameters of each "filled" m state are those appropriate for a 2D Fermi fluid. As we have shown elsewhere (7), the Landau parameters of a 2D Fermi liquid at a surface density ~0.055 Å^{-2} are approximately the same as for bulk liquid at the melting pressure. These would be the right parameters to use for multilayer films if the interactions between quasiparticles in different m states were small. The orthogonality of the particle in a box wavefunctions and the relatively large energy difference between different m states may result in such a weak interaction. It is also likely that the mass that determines the energy of the m states is enhanced above the bare mass. The disappearance of the structure is expected when the separation of the m states (which decreases as d^{-2}) becomes comparable with k_BT.

The relative values of the effective m_3^*/m_3 and F_0^a which characterise the trend in the heat capacity and susceptibility are consistent with the "universal behaviour" expected from the almost localised Fermion model (7). Thus the proposal, in the framework of paramagnon theories of interactions in a Fermi liquid, that the presence of a solid boundary would enhance the Stoner factor further towards a ferromagnetic instability (8) does not appear to receive support from these data.

Developments of the model of ref.(2) which took into account the limitations imposed by the ^{3}He binding energy on the depth of the potential well, as well as the influence of the substrate interaction and quasiparticle interactions would be valuable.

The surface density interval between magnetisation steps we observe is greater than seen in mixture films. The structure seems to wash out more rapidly with increasing temperature consistent with the higher effective mass in films of grafoil.

We gratefully acknowledge a communication from R.A.Guyer which stimulated these experiments.

References

(1) E.Krotscheck Phys.Rev. B32 (1985) 5713
(2) R.Guyer,K.McCall,D.Sprague Phys.Rev. B40 (1989) 7417
(3) R.Higley,D.Sprague,R.Hallock Phys.Rev.Lett. 63 (1989) 2570
(4) J.Saunders,C.Lusher,B.Cowan these proceedings
(5) D.Greywall Phys.Rev. B41 (1990) 1842
(6) D.F.Brewer, Physics of Liquid and Solid Helium, ed. (K.Benneman and J.Ketterson) (Wiley, N.Y.)
(7) C.Lusher,J.Saunders,B.Cowan these proceedings
(8) D.Spanjaard,D.Mills,M.Beal-Monod J.Low Temp.Phys. 34 (1979) 307

Physica B 165&166 (1990) 717–718
North-Holland

ENHANCED MAGNETIC RELAXATION OF ^{3}He AT SUBSTRATE QUADRUPOLE FREQUENCY*

F.W. VAN KEULS, T.J. GRAMILA†, L.J. FRIEDMAN‡, and R.C. RICHARDSON

Laboratory of Atomic and Solid State Physics, Cornell University, Ithaca New York 14853-2501, USA

Nuclear magnetic resonance measurements have been performed on thin films of ^{3}He on a CaF_2 substrate. The CaF_2 was preplated with a monolayer of N_2 which has a quadrupolar resonance. The ^{3}He spin-lattice relaxation time was measured from 2 to 15 MHz at a temperature of 96 mK. Similar measurements were made on a system of liquid ^{3}He above two layers of N_2 on a Carbolac substrate at .4 K. Sharp minima in the ^{3}He T_1 are observed at the N_2 quadrupole frequency and at twice the quadrupole frequency in both systems.

1. INTRODUCTION

The spontaneous cross-polarization of ^{3}He and substrate spins has been observed in ^{3}He -^{19}F and other systems (1,7). In order to study the transfer of energy across the ^{3}He -substrate boundary, we have provided a substrate with an approximately discrete energy spectrum. Nitrogen 14 is a spin 1 nucleus with a gyromagnetic ratio .095 that of ^{3}He. Solid nitrogen has a quadrupole resonance at 3.49 MHz (2). A magnetic field splits the degenerate m = +/-1 states by $\Delta E = 2\gamma\hbar H_0 cos(\theta)$ where θ is the angle between the interatomic axis and the magnetic field. In our randomly oriented sample, the magnetic field should broaden the quadrupole resonance over a $2\gamma\hbar H_0$ frequency width.

2. EXPERIMENTAL DETAILS

The measurements that used a CaF_2 substrate were made at 96.2 mK. The ^{3}He data was collected using a crossed-coil configuration. This geometry excluded the N_2 quadrupole signal and allowed a fast recovery time. The ^{3}He T_1 was determined by a standard 90-τ-90 pulse sequence.

The underlying substrate is a powder of CaF_2 crystallites ground and filtered to an average particle size of 0.5 microns. The sample surface area was calibrated by nitrogen and argon BET's performed at liquid nitrogen temperatures. The resulting surface area was 2.34 m^2/gm for our 1.85 gram sample.

The amount of N_2 that was introduced to the cell was equal to a monolayer assuming an area of 15.75 $Å^2$/molecule, as has been measured for nitrogen on graphite (6). The nitrogen layer was annealed at room temperature to allow it to distribute more evenly on the surface. After cooling the sample, a layer of ^{3}He was then adsorbed on this surface and annealed at 8 K. The amount of ^{3}He required to form a monolayer on the bare CaF_2 substrate has been determined from many measurements of T_2 for coverages between .26 to 3.4 layers (4). The features of this curve were compared to measured ratios of layer densities of ^{3}He on Grafoil to decide where n^{th} layer promotions occurred (5). The value obtained for the density at monolayer completion is within a few percent of that found on Grafoil.

The T_1 minima at the quadrupolar frequency was first observed but not reproduced by Friedman (3) several years ago. The substrate used was 90 Å diameter Carbolac powder. The cell was filled with enough N_2 to form two monolayers and contained enough ^{3}He to fill the interstitial regions with liquid. These measurements were taken at .4 K.

3. RESULTS

The presence of the nitrogen monolayer did not significantly alter the basic magnetic properties of the system. Judging from a frequency sweep done on bare CaF_2 with 3.5% less ^{3}He and from coverage measurements done at a single frequency of 10.5 MHz, the N_2 caused the T_1 of the same amount of ^{3}He to increase by approximately 20% over its bare CaF_2 value.

The ^{3}He T_2 is 39% higher on the nitrogen coating than the same amount of ^{3}He on the bare CaF_2. This

*Research supported by NSF/MSC through grant number DMR-8818558.

†Present address: AT&T Bell Laboratories, Murray Hill, New Jersey 07974, USA
‡Present address: Raytheon Research, Lexington, Massachusetts 02173, USA

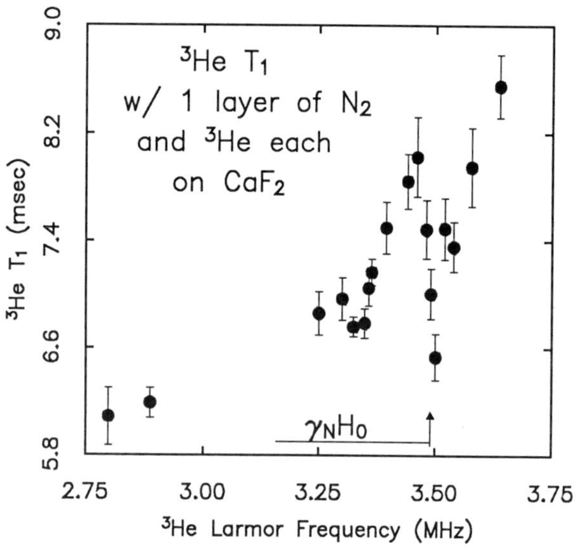

FIGURE 1

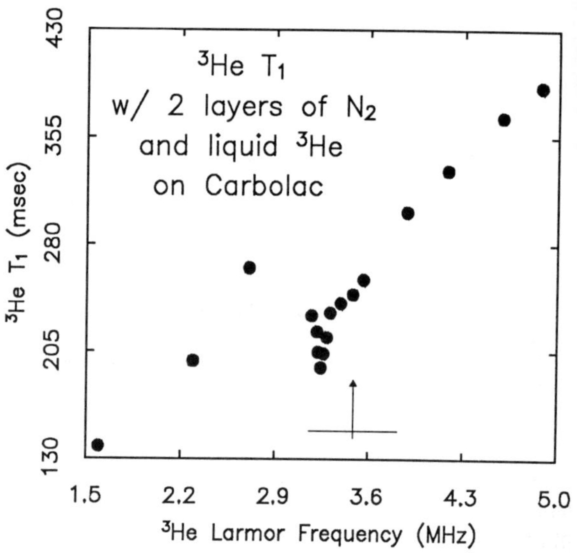

FIGURE 2

The horizontal bar shows the expected $\gamma_N H_0$ width of the N_2 quadrupole line.

effect may be due to an increase in the mobility of the ^{3}He atoms on the nitrogen surface, or the N_2 may mask large electronic magnetic moments on the CaF_2 surface. The ^{3}He magnetization is unchanged by addition of the N_2 layer within the limits of this measurement.

The T_1 minima occurs very close to the nitrogen quadrupole frequency on the CaF_2 substrate as shown in Fig. 1. The frequency dependence of the ^{3}He T_1 is basically linear far from the minima. The magnitude of the minima is 20% of T_1 at the quadrupole frequency and 9% of T_1 at twice that frequency. The dip at the nitrogen quadrupole frequency is quite sharp, sharper than one would expect given the smearing of the resonance by the nitrogen dipole splitting. Perhaps only N_2 molecules with their axes aligned in specific directions with respect to the magnetic field exchange energy with the ^{3}He spins. There appears to be a second dip at the lower frequency end of the smearing region. An anisotropic electric field could create a small splitting of the nitrogen quadrupole resonance resulting in two dips.

The measurements taken on Carbolac show larger dips at both the quadrupole frequency and twice that frequency, approximately 30% of T_1 and 44% of T_1 respectively. The minima on this substrate occurs well below the quadrupole resonance frequency as shown in Fig. 2. Since a precise determination of the surface area is not available, the two monolayer figure is only an estimate. N_2 quadrupole echos at 3.48 MHz were observed to have a T_1 of about 1 second on the Carbolac surface but have not yet been observed on the CaF_2 surface.

As an additional probe of this spin coupling, the spin-lattice relaxation times of the ^{19}F in the CaF_2 substrate were also measured. The addition of the N_2 monolayer caused the ^{19}F T_1 to increase by a factor of approximately 2.5. However, the ^{19}F T_1 without any ^{3}He on the surface was additional 33 times larger than this value. The ^{19}F showed no sign of a minima in T_1 where it crosses the N_2 quadrupole resonance. Thus, the ^{19}F experiences considerable relaxation, but is not subject to the process responsible for the resonant coupling.

REFERENCES

(1) P. C. Hammel and R. C. Richardson, Phys. Rev. Lett. 52 (1984) 1441.
(2) J. R. Brookeman, M. M. McEnnan, and T. A. Scott, Phys. Rev. B 10 (1971) 3661.
(3) L. J. Friedman, Ph.D. Thesis, (Cornell University, 1982) unpublished.
(4) T. J. Gramila, Ph.D. Thesis, (Cornell University, 1989) unpublished
(5) D. S. Greywall, Phys. Rev. B 41 (1990) 1842.
(6) J. K. Kjems et al. Phys. Rev. B 13 (1976) 1446.
(7) O. Gonen, P. L. Kuhns, and J. S. Waugh, in print

Physica B 165&166 (1990) 719–720
North-Holland

FERMIONS OCEANS

R.A. GUYER, K.R. MCCALL*, and D.T. SPRAGUE**

Laboratory for Low Temperature Physics, Department of Physics and Astronomy, University of Massachusetts, Amherst, MA 01003

Thin films of self bound fermions, fermion oceans, show striking evidence of the evolution of their single particle states. The thermodynamic properties of these films are an oscillatory function of thickness. A density functional theory that treats the single particle states self-consistently is developed. The structure of the film, its thermodynamic properties and its transport properties follow.

Two dimensional systems of fermions, fermion films, fall into two broad categories: the "atmospheres" that are not self bound (gas) and held in two dimensional conformation at a density determined by an external field and the "oceans" that are self-bound (liquid) and achieve approximately uniform density as a consequence of self interaction. The typical fermion atmosphere is a charged system, e.g., heterostructures or electrons on ^4He(1). The typical fermion ocean is uncharged, e.g., ^3He.

It is the latter system that we are interested in. ^3He films have been the subject of a wide range of investigations. Three recent works illustrate this(2). In this paper we sketch a density functional theory of self bound films that provides a good qualitative understanding of much experimental data and lets us predict the behavior of a wide variety of thermodynamic and transport properties(3).

We consider a ^3He film made up of N ^3He atoms on the surface of a ^4He film of area A. For N of order or greater than a "monolayer" ($N \gtrsim A/a_3^2$, where a_3 is the interparticle spacing in low pressure bulk ^3He), the ^3He film is self bound. Take the single particle states available to the ^3He atoms to be those of a box <u>of the atoms own making</u>. That is, we take the N atoms of the film to be confined to a region of space of size Ad and assign them, in accordance with the Pauli principle, to the single particle states

$$\varphi_{\vec{K},m}(\vec{\rho},z) \propto \sin\frac{m\pi z}{d}\exp(i\vec{K}\cdot\vec{\rho})$$

having energy

$$\varepsilon_{n,m} = \varepsilon_0\left(\frac{n^2}{N_0} + \frac{m^2}{(d^*)^2}\right) \qquad (1)$$

where

$$\varepsilon_0 = \frac{\hbar^2\pi^2}{2m_3a_3^2},\quad N_0 = \frac{A}{a_3^2},\quad \vec{K} = \frac{\pi\vec{n}}{\sqrt{A}},\quad d^* = \frac{d}{a_3}.$$

In addition to the kinetic energy associated with occupying the single particle states, the atoms have an energy of interaction that causes a suitably large number of them to attempt to achieve bulk ^3He density. For this interaction energy, we take the density functional approximation(4)

$$V(d,N) = \int v_0\left\{-2\left(\frac{n(z)^2}{N_0 d}\right) + \left(\frac{n(z)^3}{N_0^2 d^2}\right)\right\} dz \qquad (2)$$

where $n(z) = \sum_m N_m \varphi_m(z)^2$, $N_m = \sum_m n(\varepsilon_{n,m}, \varepsilon_F)$, $n(\varepsilon_{n,m}, \varepsilon_F)$ is the occupation probability of the state $(\vec{n},m)$ and ε_F is the fermi energy. Thus, the free energy of the system at finite temperature is taken to be

$$F(T,N;d) = \sum_\alpha \varepsilon_\alpha n_\alpha - TS + V(N,d), \qquad (3)$$

where S is the entropy,

$$S = -k_B\left[n_\alpha \ln(n_\alpha) + (1-n_\alpha)\ln(1-n_\alpha)\right] \qquad (4)$$

$\alpha = (\vec{n},m)$, ε_α is given by Eq. (1) and $n_\alpha = n(\varepsilon_\alpha, \varepsilon_F)$. At fixed (N, d), $n_\alpha = [\exp\beta(\varepsilon_\alpha - \varepsilon_F) + 1]^{-1}$ with the requirement

$$N = \sum_\alpha n_\alpha \qquad (5)$$

* Supported by Schlumberger - Doll Research, Ridgefield, CT.
** Supported by NSF DMR 85-17939 and 88-20517.

fixing ε_F. To place the film in a substrate potential, V(z), we add $U = \int dz\, V(z)\, n(z)$ to Eq. (3). At fixed N we minimize E(N,d) with respect to d; i.e. we find the best thickness for a film of N particles. Employing the apparatus described above we find the single particle energies and magnetization shown in Figure 1.

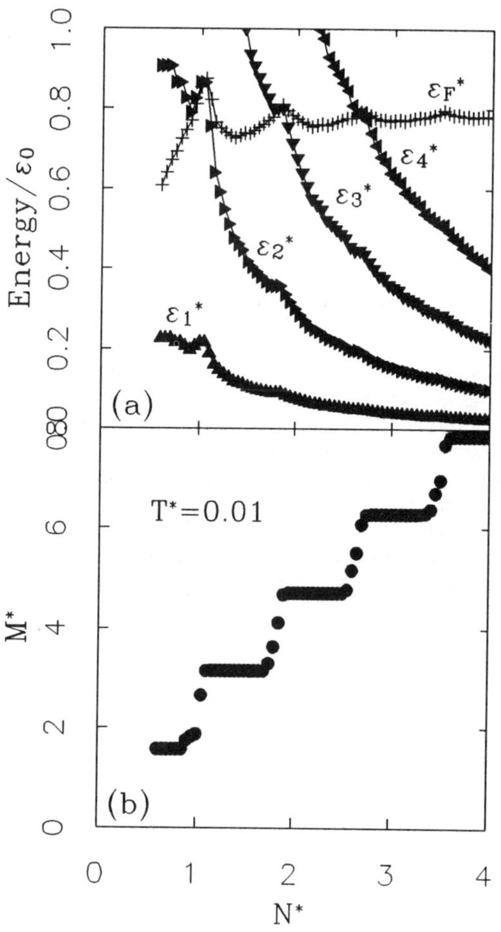

FIGURE 1
Behavior of a Fermion Film. (a) Single particle energies vs N/N_0. (b) Magnetization at $T^* = .01$ vs N/N_0.

Notable features are:

1) Structure of the film profile due to evolution of the states available to the particles (see Figure 2).

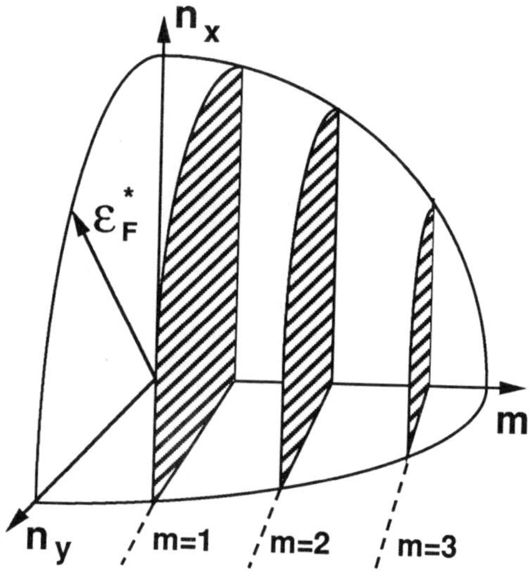

FIGURE 2
$\vec{k}$ - Space. The states available to the particles are closely spaced in n since $N_0 \approx 10^{15}$ and widely spaced in m since $d^* \approx 1$.

2) Quantized steps in the behavior of the magnetization. These steps and qualitatively similar features in all thermodynamic properties are understood in terms of the evolution of the film with n.

We have introduced a density functional model of thin fermion films that focuses attention on the self consistent energy levels occupied by the particles in the film. This model attributes the important features in the thermodynamic properties of the film to excitations of a small number of two dimensional systems of particles.

REFERENCES

(1) T. Ando, A.B. Fowler, and F. Stern, Rev. Mod. Phys. 54 (1982) 437.
(2) D.S. Greywall, Phys. Rev. B 41 (1990) 1842; Saunders, Lusher and Cowan (private communication); R.H. Higley, D.T. Sprague, and R.B. Hallock, Phys. Rev. Lett. 63 (1989) 2570.
(3) For a preliminary report see R.A. Guyer, K.R. McCall and D.T. Sprague, Phys. Rev. B 40 (1989) 7417.
(4) R.A. Guyer, Jnl. Low Temp. Phys. 64 (1986) 49.

Physica B 165&166 (1990) 721–722
North-Holland

MEASUREMENT OF LONGITUDINAL SPIN DIFFUSION IN DILUTE ^{3}He-^{4}He SOLUTIONS.

G. NUNES, Jr., C. JIN, A.M. PUTNAM and D.M. LEE

Laboratory of Atomic and Solid State Physics, Cornell University, Ithaca, NY 14853-2501, USA.

We have used pulsed NMR to investigate longitudinal spin diffusion in a dilute (350 ppm) solution of ^{3}He in ^{4}He at temperatures between 100 and 4 mK. The design of our experiment allows us to measure $D_\parallel$ independently of $D_\perp$. Our preliminary results are in good agreement with theoretical calculations, both in the classical ($T \gg T_F$) and intermediate ($T \sim T_F$) regimes.

1. BACKGROUND

Several years ago, Gully and Mullin reported the first measurement of the coefficient of longitudinal spin diffusion, $D_\parallel$, in a dilute spin polarized ^{3}He-^{4}He solution (1). Their experiment was performed against a theoretical background in which $D_\parallel$ had been calculated by Lhuillier and Laloë (2) for the case of a spin polarized system in which Boltzman statistics still applied, and by Meyerovich (3) and Mullin and Miyake (4) for the case of a fully degenerate Fermi system. While at the time there was no theory applicable to the intermediate region between these extremes, it was expected that $D_\parallel$ would vary in some smooth way between an approximately $\sqrt{T}$ dependence in the Boltzmann regime and a $1/T^2$ dependence in the degenerate regime. Instead, Gully and Mullin found that at a temperature where $D_\parallel$ might have been expected to level off or even rise with decreasing temperature, it began to drop more steeply, and showed no sign of turning around at the lowest temperatures obtained in the experiment.

A possible explanation for this discrepancy between expectation and experiment was first given by Meyerovich (5), who pointed out that in a degenerate and spin polarized system, the transverse component of the magnetization (perpendicular to the applied field) should relax on a time scale very different from the longitudinal component. Gully and Mullin determined $D_\parallel$ in their experiment with a spin-echo technique, which actually probes $D_\perp$. More recently, Jeon and Mullin (6) have calculated $D_\parallel$ as a function of temperature and polarization, and find that it should, in fact, vary smoothly between the Boltzmann and degenerate regimes.

2. EXPERIMENT

In order to investigate spin diffusion in the intermediate and degenerate regimes, we have designed an experimental cell (Fig. 1) that allows us to measure $D_\parallel$ independently of $D_\perp$. The main body of the cell is made from epoxy and is divided into two cylindrical chambers separated by a small channel. The lower chamber is enclosed by a 300 MHz resonant cavity (7) which is thermally anchored to an intermediate temperature stage of our dilution refrigerator. Thermal contact to the ~12 cm^3 of liquid sample is provided by a sintered silver heat exchanger with an area of ~25 m^2. The sample was prepared by careful titration at room temperature and loaded into the cell at about 1.5 K. In order to prevent heat flush effects from distorting the ^{3}He concentration, we use a level detecting capacitor immediately above the sample cell to ensure that the liquid does not extend up the fill line to warmer regions of the cryostat. The entire cell assembly is mounted below the mixing chamber on a thermal link that extends about 30 cm into the bore of our NMR magnet. We determine the temperature from the zero magnetic field melting curve of ^{3}He (Greywall scale).

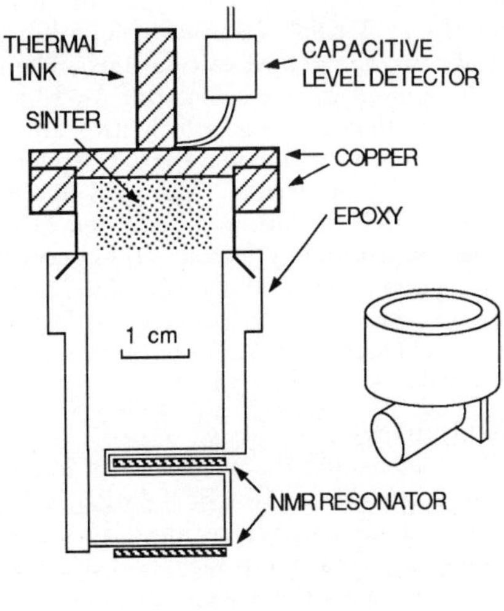

Figure 1
Schematic view of the sample cell. The smaller figure on the right indicates the relative orientation of the upper and lower cylindrical chambers. Note that the copper resonator does not touch the cell.

0921-4526/90/$03.50 © 1990 – Elsevier Science Publishers B.V. (North-Holland)

The actual measurement of $D_{\parallel}$ is quite straightforward. We apply a π pulse to invert the magnetization in the small chamber and use a series of small (~2°) tip angle pulses to probe M_z as the spins diffuse through the channel. We fit the digitized free induction decay signals from these small pulses to obtain an amplitude proportional to M_z, and then fit the entire series of amplitudes to an exponential recovery, which gives us a characteristic relaxation time τ_0 for the diffusion through the channel.

3. ANALYSIS

Following Johnson (8), we then obtain $D_{\parallel}$ from τ_0 and the geometry of our sample cell by analogy with a solution of the problem of current flow through a cylindrical wire connecting two large electrodes (9). This solution should be an adequate approximation, provided that the time it takes a spin to diffuse across the cell is much less than τ_0 , which must in turn be much less than T_1, the time scale for the magnetization throughout both chambers to return to equilibrium. The first of these conditions amounts to the assumption that the magnetization in each chamber is uniform except for a small region near the channel. The second allows the magnetization to be treated as a conserved quantity during the recovery. For our sample cell, these quantities are in the approximate ratio 1:3:9. We expect that a future, more sophisticated analysis might change our values for $D_{\parallel}$ by some over-all scale factor, but it should not affect the temperature dependence of our result.

4. RESULTS

Figure 2 shows our measured values of $D_{\parallel}$ as a function of temperature for a 350 ppm mixture (T_F = 12.8 mK) in a 9.2 T field. Also shown superimposed on the data are two theoretical calculations. The dashed line is a Boltzmann regime calculation by Ebner (10), scaled for our concentration, which uses an effective ^{3}He -^{3}He interaction with some spatial dependence to give the scattering amplitude. The solid line is the s-wave approximate calculation of Jeon and Mullin (6), again scaled appropriately for our ^{3}He concentration. In evaluating Jeon and Mullin's value for $D_{\parallel}$, we have used the s-wave scattering length given by Bashkin and Meyerovich (11) $|a_s| = 1.5$ Å. For both calculations, we used $m^* = 2.34 m_3$.

5. CONCLUSIONS

We have measured the coefficient of longitudinal spin diffusion in a dilute ^{3}He-^{4}He solution in both the intermediate and Boltzmann regimes. In the intermediate regime, we find our measured values to be in good agreement with the s-wave calculation of Jeon and Mullin. Above about 30 mK, however, we find that the s-wave approximation no longer adequately describes the scattering, and the data instead follow the result of Ebner. Our plans for future work include extending the temperature range of these experiments, as well as investigating solutions with different Fermi temperatures.

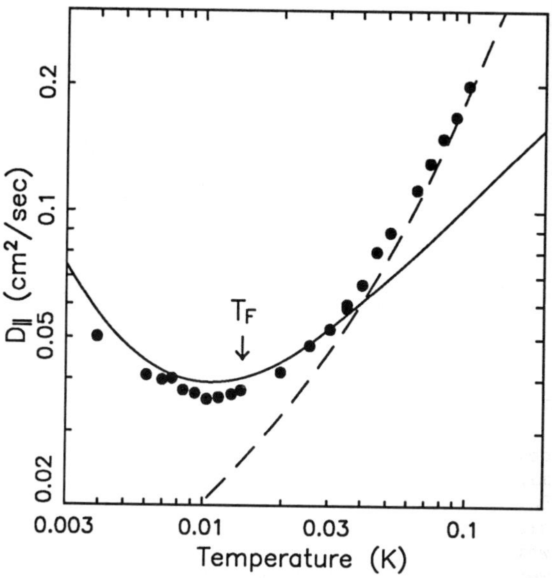

Figure 2
Measured $D_{\parallel}$ compared with the calculations of Ebner (dashed line) and Jeon and Mullin (solid line).

ACKNOWLEDGEMENTS

The authors are grateful for the support and assistance of both past and present members of the Cornell Low Temperature Group, especially Ben Tigner, Nick Bigelow, Eric Smith and Dean Hawthorne. We have also benefitted greatly from conversations with J.H. Freed. G.N. acknowledges the support of an IBM pre-doctoral fellowship. A.M.P. acknowledges the support of an NSERC 1967 Fellowship. This work was supported by the NSF under grants DMR-8616727 and DMR-8516616.

REFERENCES

(1) W.J. Gully and W.J. Mullin, Phys. Rev. Lett. **52**, 1810 (1984).
(2) C. Lhuillier and F. Laloë, J. Phys. (Paris) **43**, 197,255 (1982).
(3) A.E. Meyerovich, J. Low Temp. Phys. **47**, 271 (1982).
(4) W.J. Mullin and K. Miyake, J. Low Temp. Phys. **53**, 313 (1983).
(5) A.E. Meyerovich, Phys. Lett. **A 107**, 177 (1985).
(6) J.W. Jeon and W.J. Mullin, J. Low Temp. Phys. **67**, 421 (1987).
(7) G. Nunes, jr., to be published.
(8) B.R. Johnson, Ph.D. thesis, Cornell University, unpublished, 1984.
(9) R. Holm, "Electric Contacts", Springer-Verlag, Berlin, 1967.
(10) C. Ebner, Phys. Rev. **156**, 222 (1967).
(11) E.P. Bashkin and A.E. Meyerovich, Adv. Phys. **30**, 1 (1981).

Physica B 165&166 (1990) 723–724
North-Holland

NON-LINEAR EFFECT ON SPIN WAVES IN ^{3}HE-^{4}HE DILUTE SOLUTION

Hikota AKIMOTO*, Osamu ISHIKAWA, Gong-Hun OH, Masahito NAKAGAWA, Tohru HATA, and Takao KODAMA

Faculty of Science, Osaka City University, Osaka 558, Japan

Spin dynamics of ^{3}He-^{4}He dilute solution have been studied in the collisionless regime by using pulsed NMR methods. We observed non-linear effect on the large amplitude spin waves, the frequency of each spin wave mode shifted with changing a tipping angle. These behaviour are compared with the numerical calculation based on Leggett-Rice theory.

1. INTRODUCTION

The motion of the local magnetization of the degenerate Fermi liquid (normal ^{3}He or ^{3}He-^{4}He mixture) is described by the equation derived by Leggett-Rice(L-R).(1) Especially, in the collisionless regime, the spin waves of such liquids, which were observed in the cw NMR experiments, were well explained by using the linearized L-R equation.(2) But, there has been no study concerning with the non-linear effect on such liquids in the collisionless regime.

Here we report the spin dynamics of ^{3}He spin system of ^{3}He-^{4}He dilute solution in the collisionless regime by using the pulsed NMR method. After an rf pulse, we could observe a few signals in the Free Induction Decay (FID) whose frequencies were shifted with changing a tipping angle in the field gradient. These signals came from the large amplitude spin waves which were attributed to the non-linear effect.

2. EXPERIMENTAL

We used a double nuclear stage refrigerator for cooling the ^{3}He-^{4}He dilute solution. The ^{3}He concentration was 6.4 % and the pressure was 0 bar. The details of cooling procedure is shown in elsewhere.(3)

The ^{3}He NMR cell had dimensions of 2.85 mm x 2.85 mm x 4.0 mm. One end of the cell was closed and the other was connected to the bulk liquid through a narrow channel. The homogeneity of the rf field was better than 90 % over the NMR cell. The inhomogeneity of the static magnetic field was less than 1×10^{-4} over the NMR cell. The NMR measurements were performed at the frequencies of 920 kHz and 1.84 MHz with the field gradient between 6 mT/m and 50 mT/m which direction was the same as the uniform field.

The FID signal after an rf pulse was mixed with the local frequency which was usually a few kHz lower than the signal frequency. After a low pass filter the heterodyne signal was digitized to get a Fourier transformed spectrum.

*Present address: The Institute for Solid State Physics, The University of Tokyo, Tokyo 106, Japan

3. RESULTS AND DISCUSSIONS

The power spectra of FID signals during the 10 msec after an rf pulse are shown in Fig.1(a) for tipping angles from 20° to 180° at the Larmor frequency of 920 kHz in the field gradient of 32.0 mT/m at 280 μK. The peak frequencies in power spectra are plotted by solid symbols as a function of the tipping angle in Fig.2.

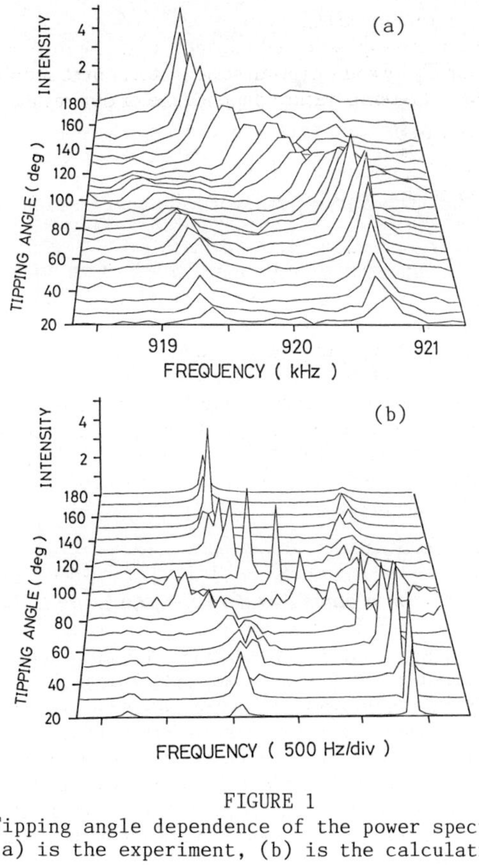

FIGURE 1
Tipping angle dependence of the power spectra, (a) is the experiment, (b) is the calculation.

The characteristic aspects of Fig.1(a) and Fig.2 are as follows: [1] for the tipping angle β_p below 40^0, peak frequencies are constant and correspond with 1st and 2nd spin wave frequencies in the cw NMR experiment. [2] $40^0 \leq \beta_p \leq 140^0$, on increasing a tipping angle each peak shifts to lower frequency side and the signal which corresponds to the 2nd spin wave mode disappears above 90^0 tipping angle and there appears a new small peak in higher frequency side at the same time. [3] $140^0 \leq \beta_p \leq 180^0$, the largest peak is always in lower frequency side than small peaks and each peak frequency is constant.

To compare these features with the theory, we made the numerical calculation of the L-R equation with the zero spin current at the wall. We divided the NMR cell into 31 sections along the field gradient, and calculated M(i,t) (i=1,2,,,,31) in each cell with the phenomenological parameter determined in the cw NMR and spin echo experiments. The M(t) obtained by summing up M(i,t) was Fourier transformed to get the power spectrum.

The calculated power spectra are shown in Fig.1(b) and peak frequencies are plotted by open symbols in Fig.2. The qualitative behavior agrees well with experimental results, but quantitatively there are some discrepancies near 90^0 and 270^0 tipping angles. The feature [1] and [3] are attributed to the motion of the magnetization near the magnetic field direction and the reverse direction, respectively. It is thought that with these transverse magnetization, the spin waves can exist in the same way as the small amplitude oscillation, which is characterized by no motion of the longitudinal magnetization. But the large transverse magnetization as

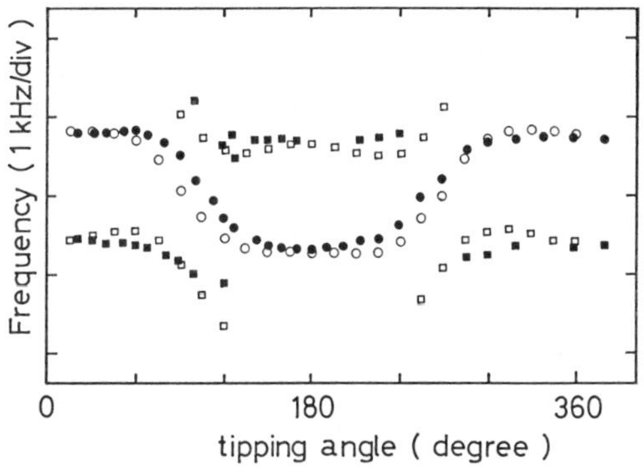

FIGURE 2
Peak frequency as a function of a tipping angle. Solid symbols are the experimental results, and open symbols are the calculated results. Circles correspond to the 1st spin waves and squares correspond to the 2nd.

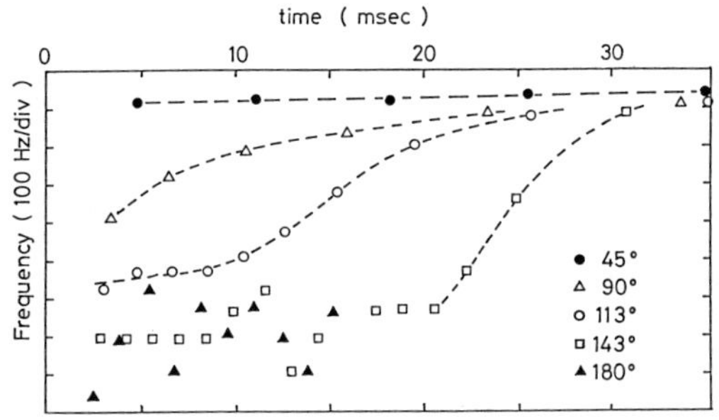

FIGURE 3
Time evolution of the 1st peak frequency with changing the tipping angle at the Larmor frequency of 1.84 MHz in the field gradient of 13.7 mT/m at 300 μK.

the case of the feature [2] causes the non-linear effect on the longitudinal motion, so this modifies the transverse motion and the feature [2] appears.

The precession frequencies of the 1st mode in the FID signal are plotted in Fig.3 as a function of time after the tipping pulse at the Larmor frequency of 1.84 MHz. It is clear that the relaxation process has the tipping angle dependence, too. But we could not get these features from the numerical calculation and the cause of these relaxation process is unclear.

Since the frequencies of Fig.2 are determined by a brute Fourier transformation of these time evolving signals, it might be some discrepancies between the peak frequency of the power spectrum and the actual precession frequency just after an rf pulse for the tipping angle between 60^0 and 130^0.

4. SUMMARY

After an rf pulse, we observed the tipping angle dependent frequency shifts of the large amplitude spin wave mode of ^3He-^4He dilute solution in the collisionless regime in the field gradient. The numerical calculation of Leggett-Rice equation are in good agreement with the experimental results except the relaxation process.

REFERENCES

(1) A.J.Leggett and M.Rice, Phys. Rev. Lett. 20 (1968) 586, and A.J.Leggett, J. Phys. C3 (1970) 448.
(2) D.Candela et al. J. Low Temp. Phys. 63 (1986) 369, and J.R.Owers-Bradley et al. Phys. Rev. Lett. 61 (1983) 2120, and H.Ishimoto et al. Phys. Rev B38 (1988) 6422.
(3) H.Akimoto et al. to be published in JLTP.

Physica B 165&166 (1990) 725–726
North-Holland

NMR Relaxation Times for a Dilute Polarized Fermi Gas

W. J. MULLIN[(a)], F. LALOË[(b)], and M. G. RICHARDS[(c)]

(a) Department of Physics, Hasbrouck Lab, University of Massachusetts, Amherst, MA 01003, USA
(b) Laboratoire de Physique, Ecole Normale Superieure, 24, rue Lhomond, 75005 Paris, France
(c) MAPS, University of Sussex, Falmer. Brighton, Sussex BN1 9QH, UK

We calculate the longitudinal relaxation time T_1 for a polarized Boltzmann spin-½ Fermi gas. We show that T_1 is independent of polarization of the gas. At high T, where the thermal wavelength λ is small compared to the scattering length a, T_1 is proportional $T^{1/2}$, while at low T, such that λ is greater than a, T_1 is proportional to $T^{-1/2}$. T_1 thus has a minimum at some intermediate temperature confirming the numerical results of Shizgal. The existence of the minimum does not depend on the presence of an attractive part of the potential. As an example of the expected temperature dependence we calculate T_1 numerically for a hard core gas.

1. Introduction

In 1973 Shizgal [1] predicted that T_1 for ^3He gas would have a minimum at about 1 K. His result was from numerical results based on an analytic derivation via a kinetic equation. In the highest temperature region studied his numerical results gave $T_1 n \sim T^{0.5}$, where n is the density. Experimental behavior [2] very close to this dependence has been observed in gaseous ^3He, but sufficiently low temperatures have not yet been reached to allow the observation of the predicted minimum. The polarization dependence of T_1 has not been studied experimentally.

In a separate development there has been considerable interest in the last few years, in the equilibrium and transport properties of polarized systems, including ^3He gas. In a highly polarized system, collisions between opposite spins become rare so that almost only triplet interactions may take place. These collisions, which, in a Fermi system at very at low T, are p-wave in lowest order, are much weaker than the s-wave interactions that can occur for a singlet interaction. Thus polarization can cause much longer mean-free paths and enhanced transport coefficients. By analogy one might therefore expect a strong magnetization dependence of T_1 for Fermions and the occurrence of an interesting nonexponential decay of the magnetization due to particle indistinguishability.

The physics behind the minimum predicted by Shizgal can be understood by simple heuristic arguments. We use an argument originally given by Chapman and Richards [3], who show that $T_1^{-1} \sim (\tau\omega_z)^2\nu_{coll}$. Here τ is the time for a collision of two atoms; ω_z is the local field during the collision, i. e., the dipolar field; γ is the gyromagnetic ratio; and ν_{coll} is the frequency of collisions. At high temperature T, with a the distance of closest approach, we have $\nu_{coll} \sim n\, v\, a^2$ where v is the average velocity, of order $\sqrt{k_BT/m}$. This gives

$$\frac{1}{T_1} \sim \frac{\hbar^2\gamma^4}{a^6}\left(\frac{a}{v}\right)^2 nva^2 = \frac{\hbar^2\gamma^4}{a^2}\frac{n\sqrt{m}}{\sqrt{k_BT}} \qquad (1)$$

Thus we predict that bulk relaxation will have $T_1 n \sim T^{1/2}$.

The argument given in the previous paragraph breaks down if the temperature is sufficiently low that the thermal wavelength λ, where $\lambda = \hbar/\sqrt{mk_BT}$, becomes larger than a. In this case we must replace a by λ in Eq.(1). Then we find

$$\frac{1}{T_1} \sim \frac{\hbar^2\gamma^4}{\lambda^6}\left(\frac{\lambda}{v}\right)^2 n v\lambda^2 = \gamma^4 m^{3/2} n\sqrt{k_BT} \qquad (2)$$

Here we have $T_1 n \sim T^{-1/2}$. Note that in this low temperature range T_1 is independent of the distance of closest approach (or scattering length) a and takes the form characteristic of a noninteracting gas. For even lower temperatures the system would become degenerate and $T_1 \sim T^{-2}$.

By interpolating between Eqs.(1) and (2) we see that there must be a minimum in T_1 at a temperature such that $\lambda \sim a$. This argument is independent of the form of the potential; in particular, it does not depend on there being an attractive part. Indeed, by carrying out a rigorous computation for a model system, we will see that the minimum occurs even for atoms interacting via a hard core.

The minimum and rapid increase of T_1 at low temperatures is reminiscent of the behavior of transport coefficients in fully polarized systems at decreasing temperature.[4] However, the polarization of the system never enters into the arguments leading to Eqs. (1) and (2). Indeed, our rigorous results for T_1 are *independent* of polarization, and no nonexponential relaxation is expected. The reason for this is that the dipolar interaction connects only triplet spin states.

2. Derivation

We use the Boltzmann equation to derive an expression for T_1 valid for all polarizations. The magnetization is $\mathbf{M} = \frac{1}{\Omega}\sum_{p\sigma}\sigma n_{p\sigma}\hat{\mathbf{z}}$ where Ω is the system volume, $n_{p\sigma}$ is the distribution function for particles of momentum $\mathbf{p}$ and spin $\sigma = \pm 1$. The magnetization is taken to be in the direction of the unit vector $\hat{\mathbf{z}}$. By writing a Boltzmann equation for $n_{p\sigma}$, we can examine $d\mathbf{M}/dt$ and look for the form $\frac{d\mathbf{M}}{dt} = -\frac{1}{T_1}\mathbf{M}(t)$ which identifies T_1. (We work with zero external field.)

We find the result

$$\frac{1}{T_1} = \frac{48}{5}\sqrt{2}\, n\lambda^3\hbar\gamma^4 m \times \int_0^\infty dk\, k^3 e^{-\beta\hbar^2k^2/m^*} \sum_{\ell\ell'(\mathrm{odd})} |r_{\ell,\ell'}(k)|^2 P_{\ell,\ell'} \quad (3)$$

with

$$P_{\ell,\ell+2} = \frac{15}{8\pi}\frac{(\ell+1)(\ell+2)}{(2\ell+3)}, \quad P_{\ell,\ell} = \frac{5}{4\pi}\frac{\ell(\ell+1)(2\ell+1)}{(2\ell+3)(2\ell-1)},$$
$$P_{\ell,\ell-2} = \frac{15}{8\pi}\frac{\ell(\ell-1)}{(2\ell-1)} \quad (4)$$

All other P's vanish. Our formula (3) is the same as the result derived in Ref. 1. In our derivation we make *no* assumption of small polarization, so the result is valid for all M.

We evaluate T_1 for free particles and for a hard-core potential. The results are shown in Fig. 1 where we plot $T_1 n$ versus T. The result for free particles is $T_1 n \sim T^{-1/2}$ for all temperatures. For a hard core potential T_1 can be evaluated numerically and we see that there is indeed a minimum in T_1 at an intermediate temperature even though the potential is completely repulsive. The existence of the minimum does require that there be an interaction potential and occurs when the thermal wavelength is approximately equal to the hard-core parameter a. The behavior for $\lambda > a$ and for $\lambda < a$ are as discussed heuristically in the introduction.

3. Conclusions

We have shown that T_1 for a dilute spin-$\frac{1}{2}$ Fermi gas has a minimum at a temperature for which $\lambda \sim a$, where a is a measure of the range of the potential. We expect, from the heuristic arguments given in Sec. 1, that the existence of this minimum does not depend crucially on the form of the potential. We have shown by our calculation with a hard-core potential that, in particular, the minimum does not depend on the existence of an attractive region of the potential. Calculations with a more realistic potential have already been reported in Refs. 1 and 2. The existence of a minimum is a quantum mechanical effect involving the transition from $\lambda < a$ to $\lambda > a$. The changeover causes an inversion of the T dependence because the dipolar Hamiltonian connects only triplet states. There is a similarity in this behavior to transport coefficients of highly polarized Fermi systems. In that case both singlet and triplet interactions are allowed by the interaction, but an increase in M gradually switches off the $\ell = 0$ (singlet) channel, causing the coefficient to diverge at T=0. In the case of relaxation, the dipolar Hamiltonian already eliminates the singlet states, only odd ℓ values contribute, and the rate diverges at T=0 for any M and indeed T_1 is *independent* of M.

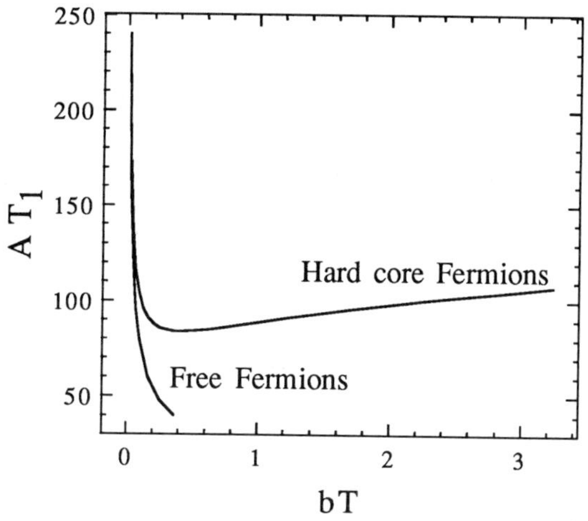

Fig. 1: Plot of bulk dipolar relaxation time T_1 versus T for a Fermi gas interacting via a hard-core potential; and for free-particle Fermions. The constants are $A = 48\sqrt{\pi}\frac{\hbar\gamma^4 nm}{a}$ and $b = \frac{a^2 m k_B}{\hbar^2}$.

Since it is difficult to observe bulk T_1 in ^3He gas at temperatures low enough for the minimum to appear, perhaps dilute solutions of ^3He in liquid ^4He would be a more suitable system. For this system there is little loss of ^3He density to condensation or sticking on the walls so that bulk relaxation should be observable at lower temperatures than in ^3He gas. While our hard-core calculation might be a reasonable first approximation to describe this system, it would be better to have a more realistic calculation that takes into account the interactions of ^3He atoms more accurately. We leave that case for future work.

Acknowledgement
This work was supported in part by a travel grant from the NSF/CNRS international exchange program.

References

1. B. Shizgal, *Jour. of Chem. Phys.* **58**, 3424 (1973).
2. R. Chapman, *Phys. Rev.* **A12**, 2333 (1975).
3. R. Chapman and M. G. Richards, *Phys. Rev. Lett.* **33**, 18 (1974).
4. C. Lhuillier, *J. Physique* **44**, 1 (1983).

Physica B 165&166 (1990) 727–728
North-Holland

MULTIPLE SPIN ECHOES IN FERMI LIQUIDS

A S Bedford, R M Bowley, D Wightman and J R Owers-Bradley

Department of Physics, University of Nottingham, Nottingham NG7 2RD, UK.

We have observed multiple spin echoes in a 2.6% ^3He-^4He mixture, similar to those seen by Einzel et al in pure ^3He. The theory of these echoes is in good agreement with experiment provided μM is less than 1.

Non linear terms can enter the Bloch equations in two ways; either through the demagnetising field caused by the dipolar interaction or through the spin current becoming a non-linear function of the magnetisation. Both of these non-linear terms can lead to multiple spin echoes MSE, that is echoes that appear at a time $n\tau$ where τ is the time between the tipping pulses. MSE have been seen in solid (1) and liquid ^3He (2) and more recently in water (3) at room temperature in a magnetic field of 11 tesla.

In figure 1 we show MSE obtained by us for a 2.6% mixture in a field of 10.6 tesla at 10.5 mK. The first echo saturates the recorder, the second and third echoes are much smaller but subsequent echoes increase in amplitude and then decay. There is also evidence that the echoes get wider for longer times. We have also measured the in-phase and quadrature response and hence the phases of these echoes.

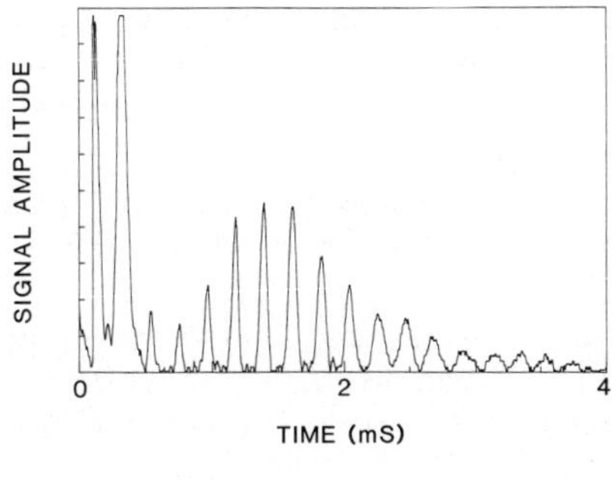

Fig 1 MSE seen at 10.5 mK in a 2.6% ^3He-^4He mixture. The first peak is the second tipping pulse, the next peak is the first spin echo.

Can we predict the amplitude and phases of the MSE? As a first attempt we have developed the numerical technique, proposed by Einzel et al(2), to calculate the echoes and have used our results to fit the data of Einzel et al and more recent data taken by us for a ^3He-^4He mixture. Both sets of data involve degenerate Fermi systems in which the spin current is given by the equation, due to Leggett and Rice (4).

$$J_z = \frac{-D}{1+\mu^2 M^2}\left(\frac{\partial M}{\partial z} + \mu M \times \frac{\partial M}{\partial z} + \mu^2 M \left(M . \frac{\partial M}{\partial z}\right)\right) \quad (1)$$

The transverse diffusion constant D and the parameter μ, both of which are proportional to the transverse relaxation time, vary with temperature as T^{-2}. Equation (1) simplifies if the quantity M^2 is spatially uniform for then the last term in the equation above is zero.

The Bloch equations can be written as

$$\frac{\partial M}{\partial t} = \gamma M \times [H_o + Gz)k - H_1 + H_D] - \frac{\partial J_z}{\partial z} \quad (2)$$

where H_o is the constant magnetic field along the z direction, with G the gradient field, H_D the dipolar field and H_1 the rf field used to tip the spins. We solved these equations by expanding $M^+ = M_x + iM_y$ and M_z in Fourier series as described by Einzel et al, and then solved the resulting differential equation numerically. There are three parameters, $\alpha = \mu M \propto T^{-2}$, $\beta = 4\pi\omega_o\chi_n\tau D^{*-1/3} \propto T^{2/3}$ and $D^* = D\gamma^2 G^2 \tau^3$ which can be adjusted to fit the data.

At a pressure of 28 bar and 60 mK MSE are due to the demagnetising field as shown by Einzel et al (2). We fitted their data at 60 mK to obtain β and then assumed it varied as $T^{2/3}$. Then we fitted their data at 8.2 mK with an extrapolated value of β and found that we could choose two different values of α both of which gave fairly good agreement with the data as shown in figure 2. We can now predict the variation of amplitude as a function of D^* for all other temperatures for these different values of α. The peak amplitudes can easily be calculated and these are shown in figure 3. If α is positive then there is a single peak as shown by the dashed line whereas if α is negative then there are two peaks as shown by

0921-4526/90/$03.50 © 1990 – Elsevier Science Publishers B.V. (North-Holland)

the continuous line. Clearly the fit is much better when α is negative. The largest value of $|\alpha|$ is less than 0.6 for the data taken at 28 bar.

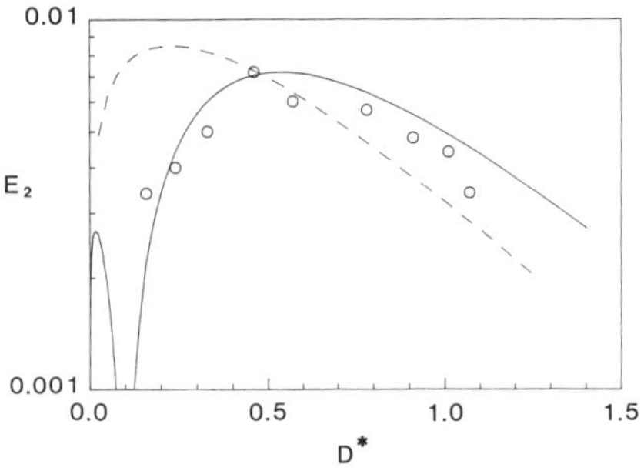

Fig 2 The height of the second echo as a function of D^* for ^{3}He at 28 bar and 8.2 mK. The dashed line is for α = .055, the solid line for α = -.16.

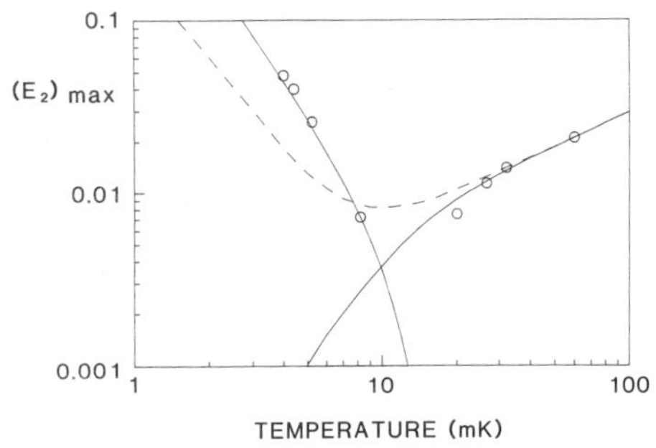

Fig 3 The maximum height as a function of temperature. The dashed line corresponds to α positive, the solid line to α negative.

We have also fitted data taken by the same group at 0 bar (kindly provided by G Eska) and the fits are good for $|\alpha|$ below 1 but get progressively worse as $|\alpha|$ increases above 1. At this pressure it is not possible to determine the sign of α.

Now we turn to our data for the ^{3}He-^{4}He mixture. The amplitude of the second echoes as a function of D^* is shown in figure 4 for temperatures of 11.5 and 16 mK together with the best theoretical fits. The agreement is fair except at longer times where the amplitudes are much larger than expected.

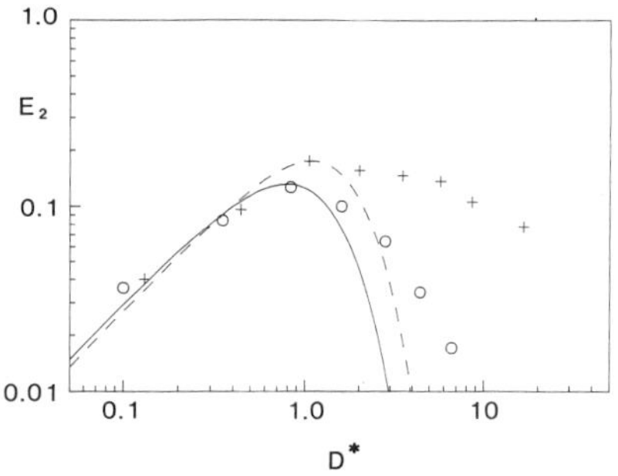

Fig 4 The height of the second echo as a function of D^* for the 2.6% mixture. The circles are data at 16 mK, the crosses at 11.5 mK. The solid line is for α = 1.3, the dashed line for α = 1.7.

We conclude that discrepancies appear for values of $|\alpha|$ larger than 1. One possible cause is that M^2 is not spatially uniform so the term which we neglected in equation (1) is significant particularly for large $|\alpha|$.

Other effects (5) have been observed which indicate that Gd is comparable to H_1, d being the length of the cell parallel to the gradient. In particular MSE are seen when the second tipping pulse is nominally 180°, for which, according to the theory, no MSE should then be visible. If Gd is comparable or larger than H_1 then the spins away from the centre of the cell are tipped by more than 180°. These spins will then form MSE.

When Gd is comparable to H_1 the longitudinal magnetisation, M_z, is not uniform spatially, so as the transverse magnetisation decays, M^2 starts to vary along the z direction. This gives rise to extra terms in the spin current which come into effect at long times and complicate the analysis. It appears that this may be the explanation for the discrepancy between theory and experiment.

Further experiments are planned in which the dimensions of the cell are reduced and the H_1 field increased so that this problem is ameliorated.

REFERENCES

(1) G Deville, M Bernier and J H Delrieux, Phys Rev B19 (1979) 5666.
(2) D Einzel, G Eska, Y Hirayoshi, T Kopp and P Wölfle, Phys Rev Lett 53 (1984) 2312.
(3) R Bowtell, R M Bowley and P Glover, to be published in Journal of Magnetic Resonance.
(4) A J Leggett and M Rice, Phys Rev Lett 20 (1986) 586 and 21 (1968) 506, A J Leggett, J Phys C12 (1970) 448.
(5) J R Owers-Bradley, D Wightman, R M Bowley and A Bedford: this volume.

Physica B 165&166 (1990) 729–730
North-Holland

HIGH FIELD NMR EXPERIMENTS ON A ^{3}HE-^{4}HE SOLUTION

J R Owers-Bradley, D Wightman, R M Bowley and A Bedford.

Department of Physics, University of Nottingham, Nottingham NG7 2RD.

We report pulsed nmr experiments on a 2.6% ^{3}He-^{4}He solution at 347 MHz and at temperatures between 10 mK and 1 K. By analysing the phase of the spin echoes we have determined the spin rotation parameter. The decay of the transverse magnetisation is not consistent with the Leggett-Rice theory.

The nmr behaviour of a dilute solution of ^{3}He in ^{4}He is dramatically altered when it is cooled to a temperature where $\omega\lambda\tau = 1$. Here ω is the Larmor frequency, τ the diffusion time constant and $\lambda = (1+F^a)^{-1} - (1+F^a/3)^{-1}$ a combination of Fermi liquid parameters. In this region the molecular field arising from the interactions between quasiparticles leads to the generation of spin waves in continuous wave nmr experiments (1) and the spin rotation effects (2) and multiple spin echoes (3) that are observed with pulsed nmr.

We are concerned here with the echo produced following a θ-t-180-t-echo pulse sequence where θ is the initial tipping pulse and 180° the refocussing pulse. The echo may be characterised by its height h, width Δt and phase ϕ. At high temperature the heights decay as expected as a function of t: T_1 and T_2 are long and the transverse magnetisation is damped by spin diffusion in the magnetic field gradient with the usual dependence, $\ln h = -(\gamma G)^2 Dt^3/12$. D is the transverse diffusion coefficient. Both Δt and ϕ remain constant.

At low temperatures when $\omega\lambda\tau \sim 1$ the behaviour is markedly different. In our experiments we have chosen a ^{3}He concentration of 2.6% with a Fermi temperature of 230 mK so that at the temperatures of interest ($\omega\lambda\tau=1$ at around 30 mK) the system is degenerate. Leggett (4) has treated this case and we write down his equations describing the evolution of the echo heights and echo phases.

$$(1+\mu^2\cos^2\theta)\ln h(t) + \tfrac{1}{2}\mu M_o^2\sin^2\theta[h^2(t) - 1]$$

$$= -\frac{1}{12}(\gamma G)^2 Dt^3 \qquad (1)$$

$$\phi = -\mu M_o \cos\theta \ln h(t) \qquad (2)$$

Here μ is the spin rotation parameter and M_o the initial magnetisation.

The experiments were performed at 10.6 tesla corresponding to a Larmor frequency of 347 MHz. The magnetic field gradient was approximately 0.05 Tm^{-1} and the rf field had an amplitude of 0.5mT. The ^{3}He-^{4}He solution was confined to a 5 mm diameter right cylinder with its axis perpendicular to the static magnetic field. The maximum polarisation was 5%.

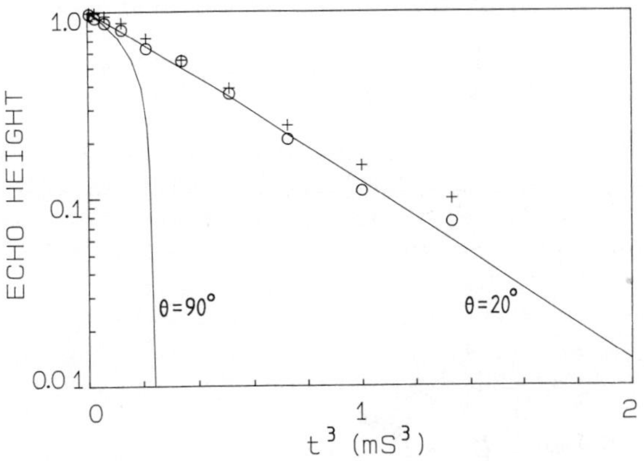

Fig 1 Echo heights vs t at 10 mK. The data are for $\theta = 20°$ (0) and $\Theta = 90°$ (X). The lines are the fit to $\Theta = 20$ and prediction for $\Theta = 90°$.

The longitudinal relaxation was non-exponential at all temperatures with a shorter component of around 15 seconds and a longer component of around 5 minutes. Therefore, to allow recovery, single shot measurements were taken at intervals of 30 minutes.

We consider first the echo heights at low temperature. Equation 1 predicts a dependence on θ. For small θ the decay should have approximately the same time dependence as at high temperature while for large θ e.g. 90° the decay should be much faster particularly at longer times. In fact we find quite a different behaviour. A comparison of theory and experiment is made in fig 1 where we plot ln h vs t^3 for $\theta = 20°$ and $\theta = 90°$. It is not possible to fit consistently both sets of data so in fig. 1 we have simply fitted the $\theta = 20°$ data and predicted the $\theta = 90°$ curve using eq 1 Another possible influence on the data is the presence of multiple spin echoes (3) which are present when $\theta = 90°$ but cannot be seen when $\theta = 20°$. It is difficult for us to determine the transverse diffusion coefficient since we have no satisfactory theory at present.

0921-4526/90/$03.50

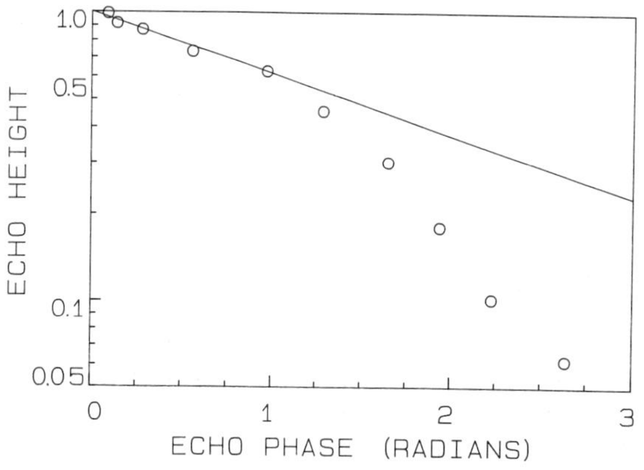

Fig 2 Echo height vs echo phase at 13 mK. The line is the fit to eq. 2 for short times.

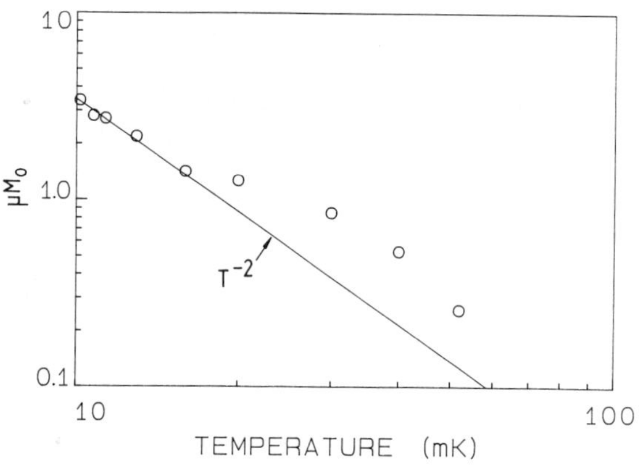

Fig 3 The parameter μM_0 vs temperature. The solid line shows a variation of T^{-2}.

The phase shift of the signal is analogous to the frequency shift associated with spin waves. It can be determined quite easily in our experiments by fitting to the in-phase and quadrature echo signals. Equation 2 suggests that a plot of ln h vs ϕ should yield a straight line whose slope is a measure of μM_o. In fig. 2 we show results and the fit to eq. 2 for small t. Obviously at longer times there are departures from the prediction of eq. 2. We obtain the spin rotation parameter from our data by examining the phase shift for short times. This is plotted on a log-log plot in fig. 3 where we see that the temperature dependence of μM_o is not quite the expected T^{-2}.

The echo width is a measure of the magnetic field gradients and should be constant in a uniform linear field gradient provided the spin density is homogeneous. In fig. 4 we see that

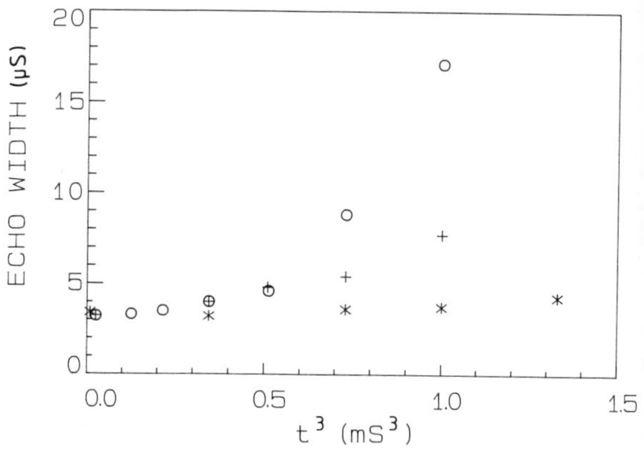

Fig 4 The echo width vs t^3 at temperatures of 10 mK (0), 20 mK (+) and 100 mK (*).

at 100 mK, this is approximately true, though there is a small increase at longer times. However, at the lowest temperatures, there is a large increase in echo width as a function of t. We argue that this is not caused by a non-linear gradient since we follow the decay over the same range of echo height at all temperatures. Rather, we believe there is longitudinal spin diffusion to a small central region of the cell hence reducing the gradient experienced by the spins and increasing the echo width. The equations 1 and 2 are derived assuming that M^2 is constant over the cell at all times. In practice this is not true. Away from the centre of the cell the field gradients cause the spins to be tipped through a range of different angles by the Θ pulse. At later times the transverse components of the magnetisation decay away, due to dephasing of the spins leaving an inhomogeneous M^2 which drives the longitudinal spin diffusion.

Gully and Mullin studied a very dilute system (0.037% ^3He) at temperatures of order the Fermi temperature. They determined μM_o and D by fitting to the echo heights and found that μM_o decreased rather than increased as the temperature was lowered. This discrepancy has not yet been explained but the present work indicates that care is required in interpreting measurements on this system.

REFERENCES

(1) J R Owers-Bradley, H Chocholacs, R M Mueller, Ch Buchal, M Kubota and F Pobell, Phys Rev Lett 51 (1983) 2120.
(2) W J Gully and W J Mullin, Phys Rev Lett 52 (1984) 1810.
(3) A Bedford, R M Bowley, D Wightman and J R Owers-Bradley, this volume.
(4) A J Leggett, J Phys C: Solid St Physics 3 (1970) 448.

Physica B 165&166 (1990) 731–732
North-Holland

MEASUREMENT OF SPIN TRANSPORT IN VERY DILUTE, POLARIZED ^{3}He-^{4}He MIXTURES*

D. CANDELA, D.R. McALLASTER and L-J. WEI

Physics and Astronomy Department, University of Massachusetts, Amherst, Massachusetts 01003, USA

Transverse spin transport is measured in a 0.1% ^{3}He-^{4}He mixture polarized by an 8T magnetic field. Spin diffusion as well as spin rotation effects are observed by exciting spin-wave modes in a small spherical cavity. A preliminary analysis of the data shows that the spin-rotation parameter $\Omega_{int}\tau_\perp$ increases continuously as the temperature is lowered through the degeneracy temperature T_F. This appears to resolve a long-standing disagreement between theory and earlier experiments.

1. INTRODUCTION

At present, a very dilute ^{3}He-^{4}He mixture provides the only example of a neutral fermion gas that may be cooled below its degeneracy temperature T_F without condensing into a dense phase (1). If T_F and the temperature T are both sufficiently low, the gas may be spin-polarized by available laboratory fields. This leads to a variety of interesting transport effects, including spin-wave propagation. Under these same conditions the quasiparticle interaction should be accurately described by s-wave scattering, making a quantitative confrontation between theory and experiment possible (2).

The only previous spin transport measurements (3) on this system that covered a wide range of T/T_F gave anomalous results for the spin rotation parameter $\Omega_{int}\tau_\perp$ (also called $-\lambda\omega_0\tau_\perp$ and μM in the literature). Agreement with theory was good at high temperatures $T>>T_F$, where the spin rotation is described by the "identical spin rotation" effect. At lower temperatures $T\approx T_F$, $\Omega_{int}\tau_\perp$ was much lower than expected, and in fact appeared to decline with decreasing T. It has been suggested that this is due to anisotropic spin relaxation $\tau_\perp<\tau_\parallel$, which must occur for $T<<T_F$ when the polarization P is near one (4). A recent calculation by Jeon and Mullin (5), which is valid for arbitrary T/T_F and P, makes this explanation appear unlikely. Thus a serious disagreement persists between theory and experiment.

We have undertaken a new series of experiments measuring spin transport in very dilute, polarized solutions, using a different experimental technique than Ref.(3). The experiment reported here studies a 0.1% ^{3}He solution at zero pressure (T_F=26mK), polarized by an 8T static field. At 10mK, the spin polarization is P=0.3. Preliminary results indicate that $\Omega_{int}\tau_\perp$ increases monotonically as the temperature is lowered through T_F, reaching approximately 10 at T=8mK.

*Work supported by NSF grant DMR-8720746.

2. EXPERIMENTAL METHOD

Rather than employ spin echoes as in earlier experiments (3), we observe the free induction decay following a single, small tipping angle pulse for a sample confined to a spherical cavity. The lower end of the cavity is open to a sintered silver heat exchanger linked to the mixing chamber of a dilution refrigerator (Fig.1). Thermometry is provided by a ^{195}Pt NMR thermometer (which has its own sintered heat exchanger), calibrated above 30mK against a melting-curve thermometer on the mixing chamber. A vibrating-wire viscometer (not shown in Fig.1) serves as a secondary thermometer and a sensitive indicator of the ^{3}He concentration.

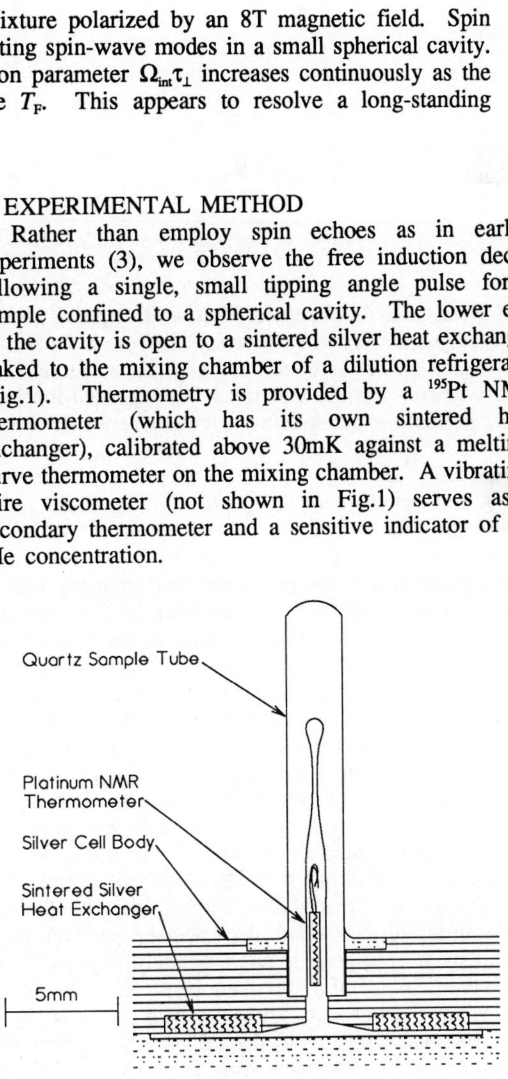

FIGURE 1
The sample cell. Electrical resonators for ^{3}He and ^{195}Pt NMR (not shown) surround but do not touch the sample tube.

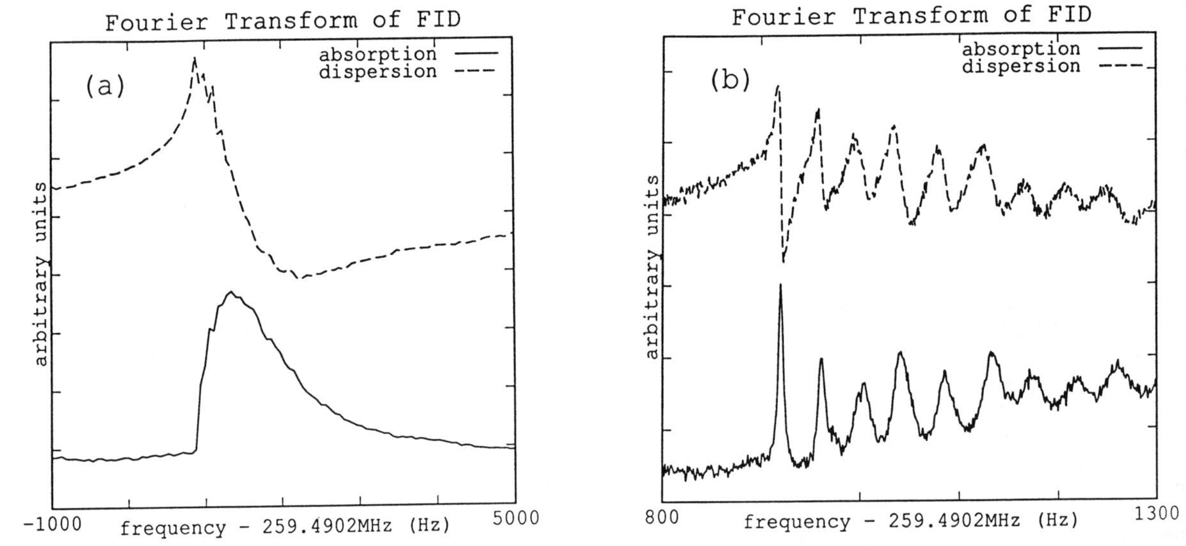

FIGURE 2
NMR signals observed after a 9° tipping pulse, for a nominal ^{3}He concentration of 0.1% and a sample temperature T=8mK. For (a), the free induction decay was digitized for 17msec after the pulse; the absorption signal has an inverted parabolic shape due to the spherical shape of the sample bulb and the 5G/cm axial field gradient. For (b) the signal was digitized for 1sec under the same conditions as (a). Now the frequency scale is greatly expanded and a regular series of spin-wave resonances is visible.

A field gradient of about 5G/cm is applied, so that the total field is weakest at the spherical tip of the cavity. In this configuration standing spin wave modes are trapped against the tip by the field gradient, and the open bottom of the tube should have little effect upon the modes (6). As the temperature is decreased below about 50mK we observe a series of sharp peaks at the low-frequency edge of the NMR line (Fig.2), similar to spin wave spectra observed previously in other quantum fluids (6). A quantitative analysis of this data requires numerical solution of the wave equation in a spherical geometry; this will be described in a future publication. Nevertheless, it is possible to make a rough estimate of $\Omega_{int}\tau_\perp$ by comparing the widths of the peaks to their frequency separation. A WKB treatment of the wave equation in one dimension (7) suggests the ratio between the splitting of the (n+1)th peak and the nth peak to the width of the nth peak is approximately $\Omega_{int}\tau_\perp/n^{1/3}$. We find that the splitting between peaks is only weakly temperature dependent, but the widths decrease rapidly and monotonically with decreasing temperature (we reach $T/T_F\approx0.3$). It is clear from the spectra (Fig.2) that $\Omega_{int}\tau_\perp$ is of order 10 at the lowest temperature.

Fitting the peak frequencies and widths to solutions of the wave equation should provide a rigorous test of the spin transport theory for the crossover temperature range $T\sim T_F$ (5). In addition, new information may be gained about the s-wave scattering length between quasiparticles; this is important for all of the transport properties as well as the BCS transition which is currently being sought in ultra-low temperature experiments (2). Finally, it is hoped that definitive evidence can be obtained for the predicted anisotropic spin relaxation (5) at $T<T_F$ for significantly polarized systems.

ACKNOWLEDGEMENT
We thank Gerard Vermeulen for participating in the design and construction of portions of the apparatus.

REFERENCES
(1) G. Baym and C. Pethick in: The Physics of Liquid and Solid Helium Part II, eds. K.H. Bennemann and J.B. Ketterson (Wiley, New York, 1978) p123.
(2) A.E. Meyerovitch in: Progress in Low Temp. Phys. Vol. XI, ed. D.F. Brewer (North Holland, Amsterdam, 1987) p1.
(3) W.J. Gully and W.J. Mullin, Phys. Rev. Lett. 52, 1810 (1984).
(4) A.E. Meyerovitch, Phys. Lett. 107A, 177 (1985).
(5) J.W. Jeon and W.J. Mullin, Phys. Rev. Lett. 62, 2691 (1989).
(6) D.Candela, N.Masuhara, D.S.Sherrill and D.O.Edwards, J. Low Temp. Phys. 63, 369 (1986).
(7) L.P. Lévy, Phys. Rev. B 31, 7077 (1985).

Physica B 165&166 (1990) 733–734
North-Holland

ENHANCEMENT OF THE SUPERFLUID TRANSITION TEMPERATURE IN LIQUID ^{3}HE DUE TO TRANSIENT POLARIZATION

S.A.J. WIEGERS*, T. HATA+, P.G. VAN DE HAAR, L.P. ROOBOL, C.M.C.M. VAN WOERKENS, R. JOCHEMSEN and G. FROSSATI

Kamerlingh Onnes Laboratorium, Postbus 9506, 2300 RA Leiden, The Netherlands

We have produced polarized liquid ^{3}He by rapid melting of polarized solid ^{3}He grown from the superfluid phase. The superfluid transition temperature of transiently polarized ^{3}He is shown to be above the equilibrium transition tempearture in the external field of 9.2 T. The quadratic term in the increase of the transition temperature with polarization seems lower than is predicted recently.

An exciting new branch of physics has originated from the possibility of polarizing liquid ^{3}He in a transient way by means of the Castaing-Nozieres effect (1). This method of polarizing the liquid by rapidly melting the polarized solid provides experimental access to the high magnetization properties of liquid ^{3}He. The study of the superfluid transition temperature of polarized ^{3}He can provide information about the normal state quasi-particle interactions in this system.

For low polarizations we know that T_{A1} increases linearly with the polarization Δ, about 100 μK per percent of polarization (2). This is supposed to be the result of the change in the density of states at the up-spin Fermi surface, which is induced by the polarization. At higher polarization, also the character of the interaction between the quasi particles should change from s-wave to p-wave and one can expect deviations from linearity in T_{A1} and T_{A2} caused by the polarization dependence of the interaction. A calculation of this has been performed within the 'nearly metamagnetic' model (3,4). A large increase in T_{A1} and T_{A2} as a function of polarization is predicted. Starting at 2.49 mK at zero polarization a maximum in the transition temperature of 24 mK at 0.35 polarization and 34.36 bar is calculated.

We report rapid melting experiments, in which we have observed for the first time polarized superfluid ^{3}He to persist at temperatures above the equilibrium superfluid transition temperature. The transition from the superfluid to the normal liquid and the temperature at which this occurs were both measured with a vibrating wire viscometer. In the normal phase of liquid ^{3}He, the amplitude of a vibrating wire viscometer is proportional to the temperature at low enough temperatures (about 30 mK in our case). When cooling through T_{A1}, the amplitude increases sharply. This feature plus the value of the amplitude allows us to determine the temperature at which the liquid becomes normal or superfluid.

In Fig. 1 we show typical recorder traces of the viscometer signals during several decompressions from an initial state with a few percent of solid. The initial temperature was close to or slightly below the A_2 transition and one can clearly observe the liquid warming through this transition, going over the maximum in the viscometer amplitude in the A_1 phase, and finally crossing the A_1 transition. For comparison, the horizontal dotted and drawn lines indicate where respectively the amplitude maximum and the transition point are located in equilibrium when the liquid is slowly cooled. One can see that both the maximum and the transition in the slow decompression are at significantly higher values than in the equilibrium case. We interpret this as evidence for an increase in T_{A1}, caused by the non-equilibrium polarization enhancement due to the melting of highly polarized solid. Since the liquid is still superfluid before the transition point, spin diffusion is fast and the liquid will have a rather homogeneous magnetization and temperature as can be deduced from the fact that both viscometers show the same feature.

To verify our interpretation, we have performed a fast decompression from the all liquid state, starting from the A_2 phase, shown in Fig. 2. In this decompression the liquid warmed through both transitions until about 6.5 mK. Afterwards the liquid cooled back into the

*Present address: CRTBT, 25 avenue des Martyrs, 166 X-Centre de tri, 38042 Grenoble Cedex, France
+Permanent address: Osaka City University, Sugimoto 3-3-138, Sumiyoshi-ku, Osaka 558, Japan

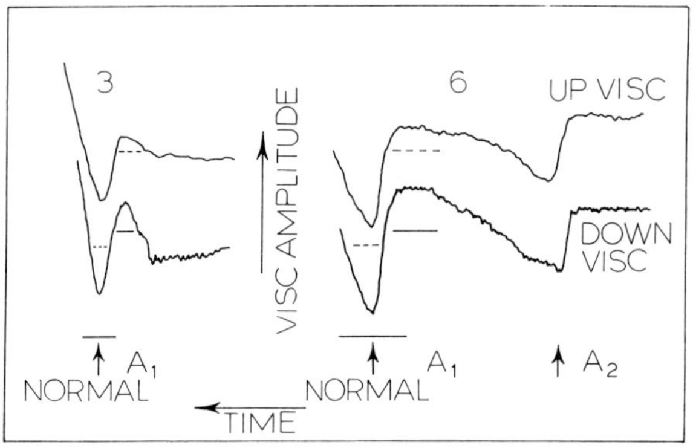

Fig. 1. Typical recorder traces of the viscometer signals in a decompression, showing an enhanced superfluid transition temperature. Time runs from right to left. All vertical scales are identical. The horizontal lines indicate where the A_1 transition and the amplitude maximum in the A_1 phase are located in equilibrium. The time elapsed in the traces is about 10 s and 50 s respectively.

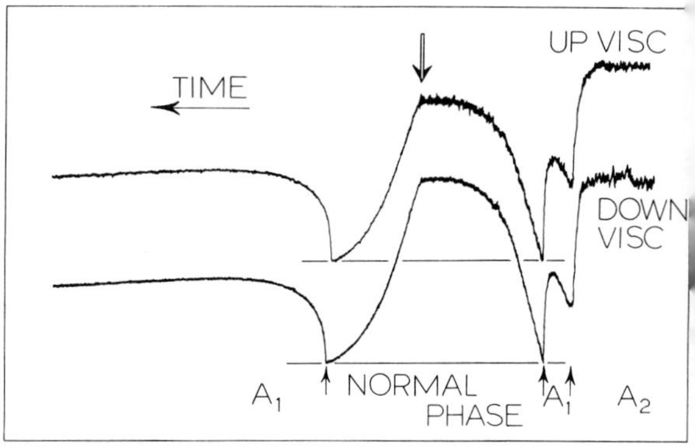

Fig. 2. A recorder trace of the viscometer amplitudes in a decompression from the all liquid state initiated in the A_2 phase. After the decompression the liquid cools again to the superfluid A_1 phase. At some time, marked by a double arrow, the recorder speed has been decreased. The viscometer amplitude at the A_1 transition in both the decompression and the slow cooldown are identical. The horizontal lines mark this amplitude.

A_1 phase. We observed no departure from the equilibrium transition temperatures in this experiment, thereby excluding any possible influence of the decompression itself on the transition temperatures. Another check we have performed is a decompression in low magnetic field, when the polarization of the solid can be neglected. In this case also no enhancement of T_{A1} was observed.

The magnitude of the observed enhancement in T_{A1} can be obtained from the increase in viscometer amplitude at the transition. We have summarized in Fig. 3 the enhancements of T_{A1} in the decompressions together with the estimated polarization of the liquid. The polarization was estimated by noting how much solid had melted at the transition point and neglecting any relaxation. Additionally, we have plotted an extrapolation of the measurements of Sagan et al. (2) to 34 bar, with (solid line) and without (dashed line) their estimated non-linear term. Our data cannot distinguish between the two curves. Also shown in Fig. 3 is the prediction of Ref.4 (dotted line). This curve does not seem to be confirmed by our measurements, although it should be noted that our estimate for the polarization could be too high.

This work was financially supported by the Stichting "FOM".

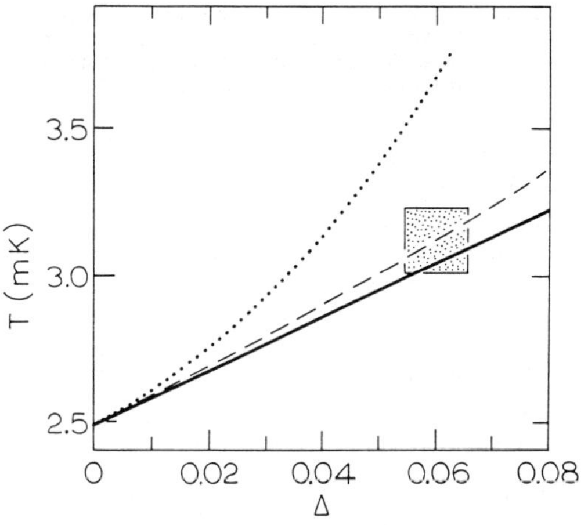

Fig. 3. Phase diagram of superfluid polarized ^3He. See text for an explanation.

REFERENCES

1. B. Castaing and P. Nozieres, J. Phys. (Paris) **40**, 257 (1979).
2. D.C. Sagan et al., Phys. Rev. Lett. **53**, 1939 (1984).
3. G. Frossati, K.S. Bedell, S.A.J. Wiegers and G.A. Vermeulen, Phys. Rev. Lett. **57**, 1032 (1986).
4. C. Sanchez-Castro, K.S. Bedell and S.A.J. Wiegers, Phys. Rev. B. **40**, 437 (1989).

Physica B 165&166 (1990) 735-736
North-Holland

HEAT CONDUCTION IN SPIN POLARIZED GASEOUS HELIUM-3 : RECENT THEORETICAL AND EXPERIMENTAL RESULTS

Brigitte BRAMI, Françoise JOLY, Claire LHUILLIER

Laboratoire de Physique Théorique des Liquides, Université Pierre et Marie Curie, 4 place Jussieu, 75252 Paris Cedex 05
and
Christian LARAT, Michèle LEDUC, Pierre-Jean NACHER, Geneviève TASTEVIN and Gérard VERMEULEN
Laboratoire de Spectroscopie Hertzienne de l'ENS (associé au CNRS et à l'Université Pierre et Marie Curie), 24 rue Lhomond, 75231 Paris Cedex 05

The transport properties of dilute spin polarized gases exhibit interesting dependences on the polarization M at low temperature, due to quantum indistinguishability between colliding particles. For the heat conduction κ of polarized gaseous helium 3, recent calculations have checked the convergence of the moment expansion of the out of equilibrium density matrix. Four to five moments are necessary to achieve a 1% accuracy. At the same time the changes of κ with M have been reliably measured in the 1.3-4.2 K range, with great care to control all the heat fluxes reaching the cell. The temperature dependence of the changes of κ with M is now found in satisfactory agreement with theory.

For dilute spin polarized gases, changes of the transport properties occur if the temperature is low enough, namely if the de Broglie wavelength is comparable to the range of the interactions between particles [1]. This derives from various interference effects induced by quantum indistinguishability between colliding particles. Proper symmetrization of the wave functions leads to scattering cross-sections which depend on the spin polarization. The Boltzmann theory has to be reformulated to incorporate the Pauli principle [2]. This leads to quantitative results for all the transport coefficients and their low temperature variations as a function of the nuclear polarization M. For the heat conduction coefficient κ of gaseous ^3He, such variations were observed in 1987 [3]. They exhibit a strong and non intuitive dependence on temperature T. However the preliminary experimental results of [3] did not show a good quantitative agreement with theory [2]. This fact motivated us both to derive new considerations on the problem of quantum exchange in Boltzmann theory [4] and to push the previous experiment to a more advanced stage [5].

Improving the calculation of transport coefficients of polarized gases, we first examined the influence of the choice of the interatomic potentials, using a few of the more recently proposed ones. We then checked the convergence of the Sonine expansion commonly used for classical gases, but which may be questionable in the case of spin polarized gases, because all the elements of the single particle density matrix do not relax towards equilibrium with the same collision cross-section. For the helium 3 gas at low temperature, the convergence of the expansion is rather rapid in the case of the coefficient κ for the non polarized gas. On the other hand for the totally polarized gas the result is much more sensitive to the approximative form of the trial density distribution which is chosen. The reason is that the new effect is due to quantum interferences and its magnitude is the result of a summation of large terms with alternate signs; this cancellation is total at 300K, explaining why interference effects do not show in thermodynamical measurements at room temperature; it is accidental around 0.3 and 0.9K; in the temperature range 1K to 4K, κ happens to increase with M. Even for slightly polarized samples in the range 1K to 4K, it is necessary to use as much as 3 or 4 Sonine polynomials to achieve 1% accuracy. For M = 100% the first order approximation is off from the "exact" result by 20 to 25% and one has to use 4 to 5 terms in the Sonine expansion to reach the desired convergence. These results are shown in figure 2, where the dotted line corresponds to the predictions with the first approximation, the full line with the fifth one. These curves were calculated using the potential of reference [6]. Little differences were found with other potentials.

0921-4526/90/$03.50

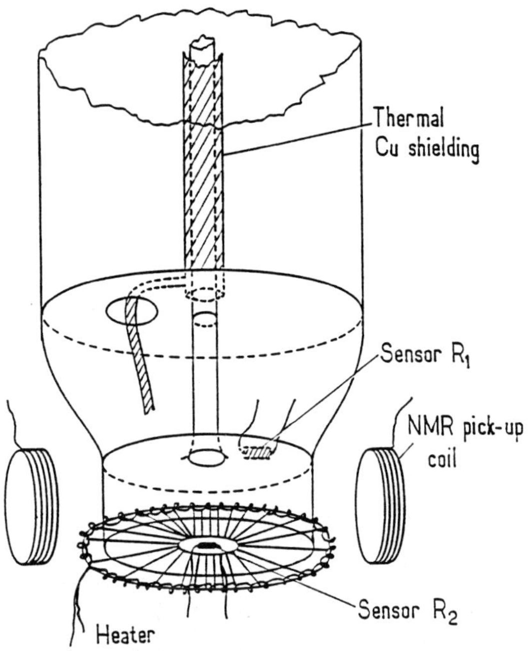

FIGURE 1
Sketch of the experiment (low part of the cell) measuring the heat conductivity of gaseous ^3He as a function of spin polarization.

As for the experiment here reported, its principle is similar to that described earlier [3], but the temperature was extended down to 1.3K by several improvements to the cryostat and to the ^3He cell. Also greater attention was payed to the control of heat fluxes reaching the measurement cells, in order to avoid systematic errors on the derived value of the coefficient κ. The polarization of the ^3He gas is achieved by optical pumping with an arc lamp pumped LNA laser. The low temperature part of the cell is shown in figure 1. The polarization of the measurement cylinder is monitored by NMR coils. Carbon resistors R_1 and R_2 measure the temperatures of the end plates and electrical heating can be applied to the bottom plate by means of a copper grid. The cell is shielded against radiation and spurious degasing by an aluminized coil foil screen. Additional experiments were performed to directly measure the thermal impedance of the pyrex cell walls and substract their contribution to the measure of κ. For the non polarized gas the measured values of κ were in good agreement with those of reference [7]. For the polarized gas, κ was found to vary linearly with M^2 at various temperatures, M ranging between 0 and 30%. The experimental values of $[\kappa(M)-\kappa(0)]/M^2$ are plotted in figure 2. The uncertainty on the absolute value of M^2 being large (but not on its relative value), a scaling factor for M was used to fit the theoretical curve at the arbitrarily chosen tempera ture of 2K.

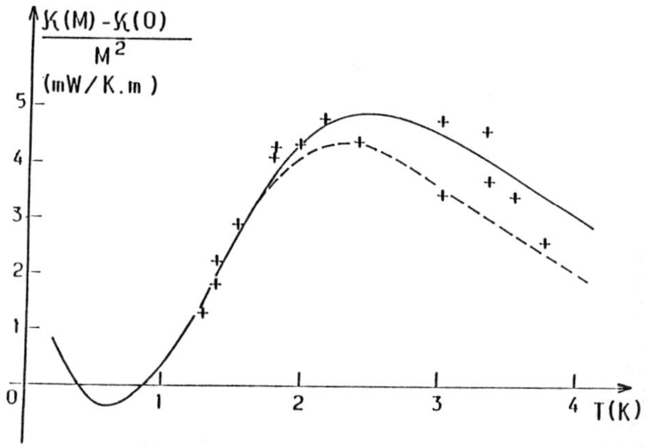

FIGURE 2
Changes of the heat conduction coefficient of gaseous ^3He with spin polarization M, plotted as a function of temperature T.

The comparison of the improved and now reliable experiments results with the new calculations is shown in figure 2. The agreement is satisfactory and even remarkably good below 2K. The problem of heat conductivity changes in ^3He with spin polarization can now be considered as well understood.

[1] "Spin Polarized Quantum Systems", Proceedings of the I.S.I. Conference (Torino), Stringari Editor, World Scientific (1988).
[2] C. Lhuillier and F. Laloë, J. Physique 43 (1982) 197 and 225; C. Lhuillier, J. Physique 43 (1983) 1.
[3] M. Leduc et al. Europhys. Letters 4 (1987) 59.
[4] B. Brami et al. Physica (1990) to be published.
[5] C. Larat et al., to be published in J. of Low Temp. Phys.
[6] R.A. Aziz et al. J. Chem. Phys. 70 (1979) 4330.
[7] D.S. Betts and J. Marshall, J. of Low Temp. Phys. 1 (1969) 595.

Physica B 165&166 (1990) 737–738
North-Holland

VISCOSITY CHANGE FOR LIQUID ^{3}HE IN A 10 TESLA FIELD

G.A. VERMEULEN, A. SCHUHL, F.B. RASMUSSEN*, J. JOFFRIN, and M. CHAPELLIER

Laboratoire de la Physique des Solides, Universite Paris-Sud, Orsay, France

By a combination of NMR and Viscosity measurements, the difference in viscosity between weakly polarized and unpolarized liquid ^{3}He has been observed. The relative difference was found to be + 0.3 ± 0.15% at the following conditions: temperature 50 mK, pressure 3 MPa, magnetic field 10 Tesla, corresponding to 3.9% equilibrium polarization.

1. INTRODUCTION

Spin polarization is expected to cause large changes in the transport coefficients of quantum liquids and gases (1). For liquid ^{3}He, a strongly interacting system, predictions based on reliable microscopic calculations do not exist. It is generally thought, however, that the probability of the dominant scattering mechanism (s-wave scattering) will be reduced with increasing polarization, one concequence being an enhancement of the viscosity.

A series expansion of the viscosity change $\delta\eta$ in terms of the polarization yields the expression

$$\frac{\delta\eta}{\eta} = a\, m^2 \qquad \text{(A)}$$

to third order in m. The qualitative expectation mentioned above amounts to the prediction that a is positive and of the order unity. The dependence on m^2 means that a substantial viscosity change requires a high polarization.

Liquid ^{3}He polarized 50% or more can be produced by the rapid melting (Castaing-Nozieres) method. Unfortunately, this procedure leads to a polarization which is spatially inhomogeneous, barring nearly any precision measurement. As polarization relaxes, the temperature of the liquid rises, with a speed that depends on the local polarization, i.e. the temperature is not constant, neither in time nor in space. The viscosity depends strongly on temperature and probably, this problem lies behind the discrepancy between the results of Kopietz et al (2) and of Kranenburg et al (3).

Our experiment uses liquid which is polarized under equilibrium conditions ("brute force method"). So, polarization is homogeneous over the sample volume, but with the parameters mentioned in the abstract, m is only about 4%. Consequently, the expected viscosity change is tiny, of order 0.2%.

2. EXPERIMENTAL PRINCIPLE

The precision needed to measure the estimated viscosity change is endangered by two obstacles: a) The quality factor and resonance frequency of our vibrating wire (superconducting) viscometer varies with magnetic field. b) viscosity is strongly dependent on temperature. Therefore, a simple comparison between viscometer signals in, say, 10 Tesla and 0 Tesla, could never detect the tiny changes expected here. Instead, the external field is kept at a constant value B_0.

The polarization is reduced from its equilibrium value m_0 to a smaller value m by RF irradiation, and by analyzing the subsequent viscometer response we can distinguish viscosity changes due to the inevitable temperature variations from those due to variations in polarization.

Using the model depicted in fig. 1, we have calculated the (almost instantaneous) viscometer response δQ_{sat} due to the saturating RF pulse, and the integrated change δQ_{rel} during the subsequent relaxation. Their ratio is:

$$\frac{\delta Q_{sat}}{\delta Q_{rel}} = \frac{1 + \gamma a T^2/T^{**}}{(m_0-m)/(m_0+m) - \gamma a T^2/T^{**}} \qquad \text{(B)}$$

Here, γ is defined from the molar specific heat C_V by $C_V = \gamma RT$, T^{**} means the "magnetic temperature" (characterizing the magnetic susceptibility), and T is the temperature of the experiment. All temperature changes are assumed small compared with T, which again is considered small compared with the Fermi temperature. From measurements of the ratio (B), temperature T, and m/m_0, we can determine a.

---

* Permanently at H.C. Ørsted Institute, Physics Lab., Universitetsparken 5, DK-2100 Copenhagen.

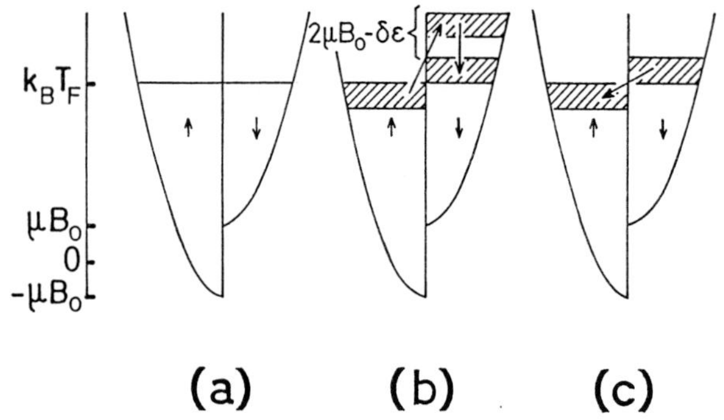

FIGURE 1

Coupling between the reservoirs of nuclear spin magnetic energy and quasiparticle kinetic energy in liquid ^{3}He.
(a) equilibrium in field B_0 (μ is the nuclear moment).
(b) spins in a band $\delta\epsilon = \mu B_0 (m_0-m)/m_0$ are excited by RF irradiation. The quasiparticles rapidly release their surplus kinetic energy as heat.
(c) the spin system relaxes back to (a) and more heat is generated from quasiparticle kinetic energy; this proces, proceeding with the spin-lattice relaxation time, is slow compared with (b). In the case of a complete saturation ($m = 0$), the amounts of heat released in (b) and (c) are equal.

3. RESULTS

Viscometer data from a typical saturation and relaxation are shown in fig.2. In addition to these data, the polarization was measured as a function of time, in order to find the polarization m right after saturation and the spin-lattice relaxation time. From a number of measurements like these we derive $a = +2\pm1$. Simulations of the behaviour in time support the analysis. This value is in direct conflict with a prediction based on the nearly metamagnetic model ($a = -30$) and does not support paramagnon calculations, either. The experiment is described in more detail in ref. (4).

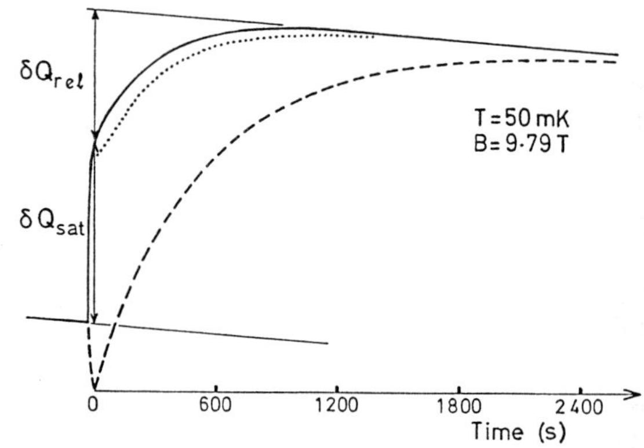

FIGURE 2

Viscometer quality factor (amplitude) as observed during saturation and subsequent relaxation.
Full line: recorder trace; the ratio $\delta Q_{sat}/\delta Q_{rel} = 1.36$, from which we derive $a = 2$.

Dotted line: curve calculated with $a = 0$.
Dashed line: curve calculated with $a = -30$.

REFERENCES
(1) A.E. Meyerovich, in: Anomalous Phases of ^{3}He eds., W.P. Halperin and L.P. Pitaevski (North Holland, Amsterdam 1989).
(2) P. Kopietz, A. Dutta, and C.N. Archie, Phys.Rev.Lett., 57 (1986) 1231.
(3) C.C. Kranenburg, S.A. Wiegers, L.P. Roobol, P.G. van de Haar, R. Jochemsen, and G. Frossati, Phys.Rev.Lett., 61 (1988) 1372.
(4) G.A. Vermeulen, A. Schuhl, F.B. Rasmussen, J. Joffrin, and M. Chapellier, J.Low.Temp. Physics, 76 (1989) 43.

Physica B 165&166 (1990) 739–740
North-Holland

Spin Wave Spectroscopy of Spin Polarized Atomic Hydrogen: Measurement of the Longitudinal Spin Diffusion Coefficient

N. P. Bigelow*, J. H. Freed and D. M. Lee

Laboratory of Atomic and Solid State Physics, and Baker Laboratory of Chemistry, Cornell University, Ithaca, NY 14853 USA

We describe the measurement of the longitudinal spin diffusion coefficient D_o in spin polarized atomic hydrogen through a detailed spectroscopic study of the small tipping angle, pulsed NMR frequency spectrum. Measurements are discussed for temperatures between 160 and 537 mK and for densities between 7×10^{15} and $5\times10^{16}\text{cm}^{-3}$ and are in good agreement with the theory. At low densities ($< 10^{16}\text{cm}^{-3}$), results are consistent with expected inertial corrections.

The transport properties of spin polarized atomic hydrogen have been shown to differ radically from those of a classical hard sphere gas as a result of identical particle quantum exchange (1,2,3). Perhaps one of the most striking consequences of such exchange effects has been the observation of spin waves in the NMR spectrum of the gas (4,5). In this paper we present the result of detailed spectroscopic studies of these spin wave modes. In particular, we present results on the measurements of the longitudinal diffusion coefficient, D_o as a function of sample temperature, T, and sample density, n. The experiments described here differ from earlier measurements of D_o in that here we have extracted the spin diffusion coefficient by studying the **transverse** spin dynamics.

For the case of small tipping angle pulsed NMR experiments (i.e. $M_+ \ll M_z \approx |M|$) the transport equations for the magnetization M can be linearized, and in a rotating frame,

$$i\frac{\partial M_+}{\partial t} = \{\gamma\delta H_o + iD_o\Big[\frac{(1+i\mu P)}{(1+\mu^2P^2)}\Big]\nabla^2\}M_+ \qquad (1)$$

where $M_+ = M_+(\vec{\mathbf{r}},t) = M_x + iM_y$ is the complex transverse magnetization density and γ is the effective nuclear gyromagnetic moment. D_o is the self diffusion coefficient which is independent of polarization $P=(n_a - n_b)/(n_a + n_b)$ where n_a and n_b are the populations of the lowest two hyperfine states. δH_o is the residual static magnetic field gradient in the rotating frame, where the static field $\vec{H}_o(\vec{\mathbf{r}}) = H_o(\vec{\mathbf{r}})\hat{\mathbf{z}}$. μ is a molecular field or quality factor for the spin wave modes (1,2,3). A detailed study of the temperature dependence of μ has also been carried out as part of the spectroscopic analysis described here and has been discussed in detail elsewhere (5).

The diffusion data was obtained by loading the sample cell with atomic hydrogen to a high density ($n\approx 10^{17}\text{cm}^{-3}$) and allowing the sample to become polarized by preferential recombination (6). The temperature of the sample cell was then held constant and small tipping angle NMR pulses were applied. Measurements are presented for temperatures between 160 and 537 mK and densities between $7\times10^{15}\text{cm}^{-3}$ and $5\times10^{16}\text{cm}^{-3}$. The experiments were performed using a linear static field gradient of 1.7 or 2.1 G/cm applied along the long axis of our cylindrical loop gap NMR resonator. This experimental configuration has been described in greater detail elsewhere (4,5).

Under these experimental conditions, the pulsed NMR frequency spectrum is characterized by a broad background lineshape (the inhomogeneously broadened lineshape) on which are superimposed a series of sharp resonant peaks (3,4,5). The sharp peaks in the NMR spectrum are attributed to the spin wave modes as confined by one wall of the NMR resonator and by the linear static magnetic field gradient δH_o (3,4). We determine D_o through a detailed analysis of these pulsed NMR spectra.

The spectra were analyzed in a multi-step, self consistent process (5). Preliminary values for μ and D_o were determined by fitting the positions and widths of the sharp resonances (the spin wave modes) to a simplified (one-dimensional) solution of eq.1. The positions and widths of the individual spin wave modes used in this step were determined by fitting the NMR free induction decays using a linear prediction technique (7). The values of μ and D_o found in this manner were then used as seed values for a least squares minimization fit of the full NMR spectrum to a numerical solution of eq. 1. The numerical solution was a fully three dimensional analysis using the known boundary conditions of the NMR sample cell (4,5).

Results for nD_o as a function of sample temperature, T, and of sample density , n, are summarized

*Present Address : A.T.&T.Bell Laboratories, Holmdel, N. J. 07733, USA

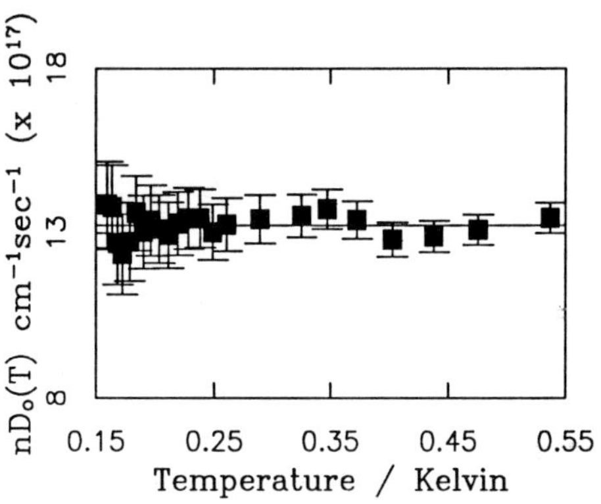

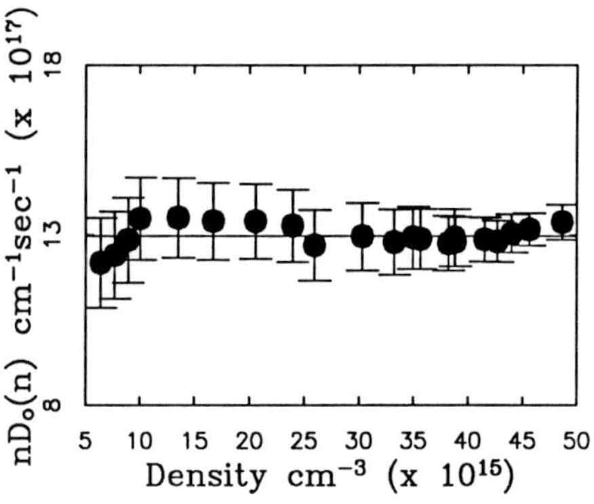

in the figures. As expected in the hydrodynamic limit (mean free path $\ll$ sample cell size), nD_o is essentially temperature and density independent over the experimental range (8). We find n$D_o = 1.3 \times 10^{18}$cm^{-1}sec^{-1} to be compared to the theoretical value of 1.5×10^{18}cm^{-1} sec^{-1}. Over this temperature range, we are unable to discern the weak ($\approx$ 2%) temperature variation predicted by the theory (8).

These results may be compared to other experimental work on the measurement of longitudinal spin diffusion. The value of nD_o from the present work is a factor (9) of 1.9 or a factor (4) of 1.2 smaller than our earlier experiments, and in exact agreement with the results of Berkhout et. al. (10). In our earlier experiments the diffusion coefficient was measured by inverting the magnetization inside the resonator with a π pulse, and measuring the rate of recovery of magnetization with a series of small probe pulses. The recovery was attributed to the diffusion of magnetization into the resonator through small hole in the bottom. In extracting the diffusion coefficient it was necessary to correct for the flow dynamics of the hydrogen atoms through the flow constriction. The experiments by the Berkhout et.al. have found that there can be substantial deviations from expected behavior in the dynamics of flow of atomic hydrogen through a capillary channel. We suggest that such deviations may be a major source of the discrepancy between our recent measurements and our earlier work. In particular, the larger value of nD_o found in our earlier experiments may be due to an overestimate of the effective flow impedance presented by our flow constriction. For this reason, we prefer the results presented here because of their insensitivity to such flow dynamics of the gas.

At the lowest densities (n$< 10^{16}$cm^{-3}), the values of nD_o as derived from the fit of our data to eq. 1 appear to deviate consistently toward smaller D_o (see FIG. 2). At lower densities, the hydrodynamic limit which is used to derive eq. 1 is no longer completely accurate (8). We have estimated the low density inertial corrections needed to the Chapman-Enskog expansion used in the theoretical work of ref. 2 and 8 to derive eq. 1 (11). We find that the neglect of these inertial corrections at the lower densities can be partially compensated for in the fitting process by the use of too small a value of D_o, in agreement with the observed results of the fits.

ACKNOWLEDGEMENTS

The authors would like to thank B. W. Statt for his many contributions. This work was supported by NSF Grant No. DMR-8616727.

REFERENCES

(1)E. P. Bashkin, Pis'ma Zh. Eksp. Teor. Fiz. 33 (1981) 11 [JETP Lett. 33 (1981) 8].

(2) C. Lhuillier and F. Laloë, J. Phys. (Paris) 43 (1982) 197; 43 (1982) 225.

(3) L. P Lévy and A. E. Ruckenstein, Phys. Rev. Lett. 52 (1984) 1512.

(4) B. R. Johnson, J. S. Denker, N. Bigelow, L. P. Lévy, J. H. Freed and D. M. Lee, Phys. Rev. Lett. 52 (1984) 1508.

(5) N. P. Bigelow, J. H. Freed and D. M. Lee, Phys. Rev. Lett. 63 (1989) 1609.

(6) T. J. Greytak and D. Kleppner, in: New Trends in Atomic Physics, Proceedings of Les Houches Summer School, Session XXXVIII, Eds. G. Grynberg and R. Stora (North-Holland, Amsterdam, 1984), and refs. therein.

(7) N. P. Bigelow, J. H. Freed, D. M. Lee and B. W. Statt, in: Spin Polarized Quantum Systems, ed. S. Stringari (World Scientific, Teaneck, NJ 1981)p. 231 and refs. therein.

(8) C. Lhuillier, J. Phys. (Paris) 44 (1983) 1, and J. P. Bouchaud and C. Lhiullier, J. Phys. (Paris) 46 (1985) 1135.

(9) N. P. Bigelow, J. S. Denker, B. W. Statt, D. M. Lee and J. H. Freed, Jap. J. Appl. Phys. 26(1987) 233 (Proc. LT18).

(10) J. J. Berkhout, E. J. Wolters, R. van Roijen and J. T. M. Walraven, Phys. Rev. Lett. 57 (1986) 2387.

(11) to be published.

Physica B 165&166 (1990) 741–742
North-Holland

Atomic Deuterium at 1K

M. W. Reynolds*, M. E. Hayden, and W. N. Hardy

Department of Physics, University of British Columbia, Vancouver V6T 2A6, Canada

Hyperfine magnetic resonance has been used to study atomic deuterium (D) at temperatures just above 1 K. The D atoms were observed to dissolve into the liquid helium film coating the walls, resulting in a fast exponential decay of the sample. The temperature dependence of the lifetime yields a preliminary value for the solvation energy E_s of 13.6(6) K.

1. INTRODUCTION

We have studied atomic deuterium (D) at temperatures just above 1 K. At such low temperatures, it is necessary to coat the walls of the sample container with a film of liquid helium (l^4He) in order to reduce wall adsorption and subsequent recombination.

We discovered that D atoms can penetrate the l^4He wall coating. In our experiments, the D dissolved into the film and irreversibly adsorbed onto the substrate. This semi-permeability of the film resulted in a fast exponential decay of the sample and rendered it extremely susceptible to contamination by atomic hydrogen (H), which does not decay by entering into the film.

In this paper we briefly describe our experiment and its interpretation. Details may be found in a PhD thesis (1).

2. EXPERIMENT

Our low-field magnetic resonance experiment was conceptually very simple, and similar to earlier experiments on H (2). A sealed pyrex cell containing D_2 and ^{4}He was immersed in a bath of l^4He, which was pumped to 1 K. The ^{4}He in the cell then formed a saturated film; its function was to suppress adsorption of D on the walls. One end of the cell was in a coil which generated an RF discharge to dissociate the frozen D_2. The other end was in a resonator which was used to detect the D through pulsed magnetic resonance on the β-δ hyperfine transition. This is a longitudinal transition, driven by an oscillating magnetic field along the bias magnetic field. The resonance frequency has a minimum of about 309 MHz at a bias field B_0 = 39 G. We operated at this minimum to reduce the effect of field inhomogeneities.

Our measurements span a rather narrow range of temperatures, 1.08 K to 1.16 K, limited below by the speed of our pumping system, and above by short sample lifetime (and resulting low density). The D density n_D was quite low, typically of order $5\times10^{10}\,cm^{-3}$. We also measured the relaxation times T_1 and T_2, both of which were very short (about 10 ms) and nearly independent of n_D. Spin-exchange in D-D collisions, which we had expected to be the dominant relaxation mechanism, is too slow to explain the observed relaxation rates and, in any case, should give $T_1, T_2 \propto 1/n_D$. The most plausible relaxation mechanism was

* Present Address: Lyman Laboratory of Physics, Harvard University, Cambridge MA 02138, USA.

spin-exchange in collisions between the D atoms and impurity H atoms. This would imply a H density of about $4\times10^{11}\,cm^{-3}$, far larger than n_D (3). This H impurity could have been produced from HD in the D_2. Fractionation during cooldown would greatly enhance the concentration of HD at the surface.

Our most important observation was that n_D decayed exponentially in time, thus ruling out D-D recombination as the rate-limiting step. The sample lifetime was short, decreasing rapidly with increasing temperature from 27 s at 1.08 K to 10 s at 1.16 K. Given the apparently high density of the H impurity, we initially suspected that the decay of n_D was due to scavenging ($H+D+{}^4He\rightarrow HD+{}^4He$), which would increase rapidly with temperature along with the ^4He vapour density. To check this, we contaminated a D_2 cell with H_2. A measurement of T_1 implied that the H density had been doubled. However, the D lifetime was unaffected.

3. FILM PENETRATION

We propose that the fast exponential decay of the D density is due to incomplete confinement of the atoms by the l^4He film lining the cell. The picture is that energetic atoms (from the high energy tail of the Boltzmann distribution) dissolve into the l^4He and are subsequently adsorbed onto the D_2 substrate. An important quantity characterizing this process is E_s, the energy required to force a D atom into l^4He. Theory estimates that E_s/k_B is about 11 K (4); the uncertainty is probably several K.

Assuming that the rate limiting step in sample decay is the solution of the D, and treating the dissolved D as an ideal gas of quasi-particles, we obtain a sample lifetime

$$\tau = (4V/A\ v\,\mu\alpha_{lv})\exp(E_s/k_BT) \qquad (1)$$

for a cell of volume V and surface area A. Here, $v=(8k_BT/\pi m)^{1/2}$ is the mean thermal velocity, $\mu=m^*/m$ is the ratio of the mass m^* of a D quasi-particle to the bare mass m, and α_{lv} is the thermally averaged probability that a D incident on the surface from below will pass into the vapour. Assuming that $\mu\alpha_{lv}$ is independent of temperature, our measurements of the temperature dependence of the lifetime yield $E_s/k_B = 13.6(6)\,K$; $\mu\alpha_{lv}$ is not well determined due to the exponential dependence of τ on E_s, and falls between 0.6 and 11. Due to possible complications resulting from the high H density, temperature dependence of $\mu\alpha_{lv}$, and scattering of the D within the film by rotons, this result must be considered tentative. Nevertheless, the consistency with theory supports our contention that the fast sample decay was due to film penetration.

This work was supported by the Natural Sciences and Engineering Research Council of Canada. We thank Dr. Richard Cline for useful discussions.

REFERENCES

(1) M. W. Reynolds. PhD thesis, Univerisity of British Columbia, 1989.

(2) W. N. Hardy, M. Morrow, R. Jochemsen, and A. J. Berlinsky. Physica 109 & 110B, 1964 (1982).

(3) The presence of H impurities was verified in a different apparatus using the H zero-field transition at 1420 MHz; a sizeable signal was observed.

(4) K. E. Kürten and M. L. Ristig. Phys. Rev. B, 31, 1346 (1985).

Physica B 165&166 (1990) 743–744
North-Holland

PROTON POLARIZATION OF ATOMIC HYDROGEN AT 0.5 K BY ESR PUMPING

A.Ya. KATUNIN, S.A. VASILYEV, I.I. LUKASHEVICH and A.I. SAFONOV

I.V. Kurchatov Institute of Atomic Energy, 123182 Moscow, USSR

E. TJUKANOV and S. JAAKKOLA

Wihuri Physical Laboratory and Department of Physical Sciences, University of Turku, SF-20500 Turku, Finland

Nuclear polarizations up to about 95 % have been achieved in atomic hydrogen gas at temperatures around 0.5 K by powerful excitation of the a → c ESR transition at 140 GHz.

1. INTRODUCTION

The hyperfine energy levels and expressions for the corresponding spin states of a hydrogen atom in magnetic field B are depicted in the upper part of figure 1. For our experimental conditions, B = 5 T and T = 0.5 K, a system of H atoms (1) can be considered well electron spin-polarized (H↓), because of the small values of the population ratio $(n_d+n_c)/n = \exp(-2\mu_B B/k_B T) = 1.5 \cdot 10^{-6}$ and the mixing parameter $\varepsilon = 5 \cdot 10^{-3}$. Here n_d and n_c are partial densities and n the total density. The ratios $P = (n_b-n_a)/(n_b+n_a)$ and $P' = n_b/n_a$ describe the degree of proton spin-polarization (⇕) which is an important factor in controlling the stability of H gas. In the mutual collisions of doubly polarized b atoms (H↓⇕) exchange interactions are inoperative to cause transitions to the singlet molecular state. The rate of density decay then depends on that of the b → a nuclear spin relaxation and P (and P') grows due to the more rapid a - a and a - b recombinations. Spontaneous nuclear polarizations as high as P = 99.8 % have been reported (1). Since the exchange recombination occurs primarily in surface-adsorbed H↓ layer, the evolution of P becomes, however, sluggish at higher temperatures, say, 0.5 K.

Mattheij et al (2) attempted to enhance P by exciting the ESR transition a → d (and b → c). The negative result was attributed to the very fast spin-exchange event (d,b) → (c,a) that occurs soon after the d state has been created. We report here on ESR pumping experiments where we have achieved significant P values by driving the longitudinal resonance transition a → c with mm-wave fields. In this case the excited H atom undergoes the spin-exchange relaxation (c,a) → (d,b) which increases n_b, ie. P. The a → c transition is forbidden to first order, its probability being by the factor $4\varepsilon^2$ smaller than the a → d probability. Therefore a

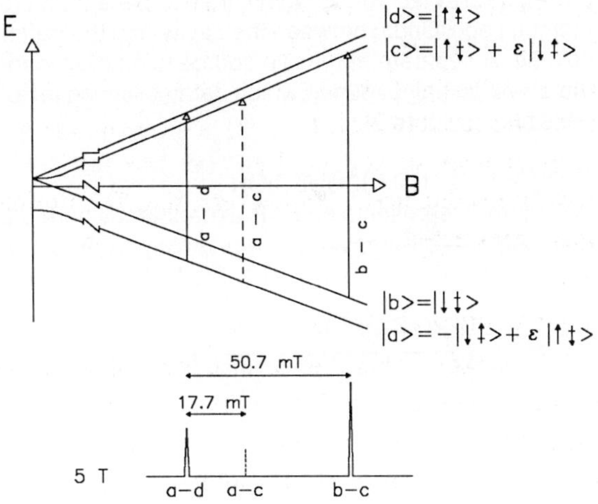

FIGURE 1
The hyperfine energy levels with corresponding wave functions of an isolated H atom as functions of external magnetic field. ESR absorption lines in a field sweep near 5 T across the a → d and b → c transitions are also shown together with the position of the forbidden a → c transition.

high ESR pumping power is needed. In this work power levels from 0.3 mW to 0.7 mW have been used.

2. EXPERIMENTAL

We have measured the decay of n_a and n_b respectively from the intensities of the a → d and b → c ESR absorption (cp figure 1) and dispersion signals with a homodyne 2-mm ESR spectrometer of the Kurchatov Institute. An InSb crystal detects changes, δ, in the voltage reflected from a TE_{011}-mode microwave cavity which is connected to a 15 times larger buffer volume (sample cell). The H↓ inlet to the latter can be closed off with liquid helium. The cavity-cell system is cooled by a ^{3}He refrigerator, the lowest temperature being in this work 0.42 K. Densities are extracted from the ESR absorption lineshape δ employing a simplified version of the procedure presented by Statt et al (3). The data thus obtained have been checked calorimetrically by increasing the incident ESR power for a few seconds to such a high level that the heating effect of the ESR-induced first-order recombination can be well monitored by cavity thermometer and the temperature excursion then simulated with a heater wound on the cavity. For $n \leq 6 \cdot 10^{15}$ cm^{-3}, ie relatively weak ESR signals, agreement is very good.

The sensitivity of the spectrometer is about $5 \cdot 10^{12}$ cm^{-3} when the ESR-induced recombination loss of atoms is in the range from 0.2 % to 0.5 % per sweep of all atoms in the cell corresponding to the incident powers of 1 nW to 10 nW which are still low enough to maintain equilibrium between the cavity and the buffer volume. A separate mm-wave source is employed for the a → c pumping, during which data collecting is not possible (cp figure 2).

3. RESULTS AND DISCUSSION

Figure 2 represents an ESR experiment where a 6 min long a → c pumping at the relatively slow rate of $W_{ac} \approx 2.5 \cdot 10^{-3}$ s^{-1} leads to a modest and rapidly

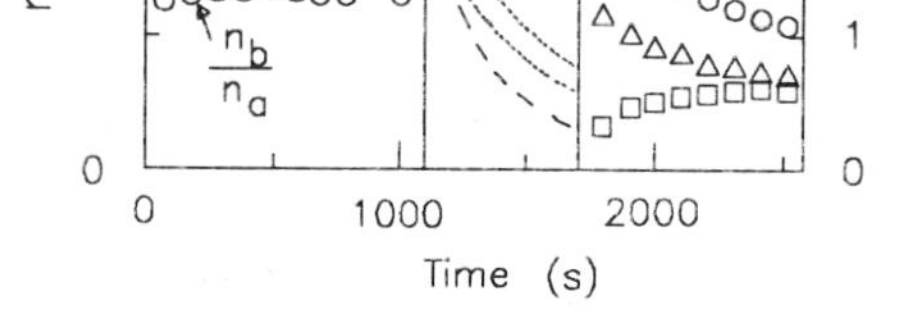

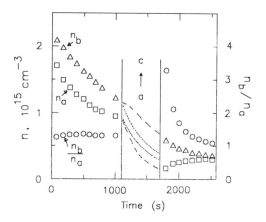

FIGURE 2
Densities of a and b atoms and their ratio n_a/n_b before and after a → c microwave pumping of H↓ at 0.5 K using pure ^{4}He coverage in the sample cell. The dashed lines show the result of a computer simulation during pumping while dots represent a simulated a → d pumping.

decaying nuclear polarization of P ≈ 55 %. By fitting the density decay curves to the rate equations for n_a and n_b the rate constant $G_i \approx 7 \cdot 10^{-5}$ cm/s is extracted for the first- order nuclear a → b relaxation. This process is attributed to magnetic impurities embedded in the sample cell walls whose depolarizing influence decreases when the distance between the H↓ gas and the wall is increased by a ^{4}He superfluid film and a solid H_2 layer (3). The highest a → c excited P we have observed is about 96 % when the H_2 layer has been of order 1 μm thick, 1 % ^{3}He added to the ^{4}He film (in order to decrease the adsorption energy), $W_{ac} \approx 9 \cdot 10^{-3}$ s^{-1}, $G_i = 1.8 \cdot 10^{-5}$ cm/s at T = 0.45 K and the pumping time has been 5 min. Likely P has been even larger, but the a → d signal could be resolved only after P' ≤ 50. Also shown in figure 2 is a computer simulated effect of a → d pumping on n_a and n_b at the same effective rate as for a → c pumping. An increased loss of a and b atoms is found but no gain in P' in agreement with the previous experiments (2).

Also the so-called resonance recombination of a-state atoms has been observed to enhance the nuclear polarization of H↓. For example when increasing temperature from 0.48 K to 0.68 K a 10-fold increase in the rate constant K_{aa} has resulted.

The dynamic proton polarization method discussed here may turn out to be of importance in proposed strong compression of H↓ (4) or generally in experiments where high polarization P is required at T ≥ 0.35 K and $n \leq 10^{15}$ cm^{-3} or in accelerator experiments where H↓‡ would be used as a polarized target and filter or as a high-intensity polarized proton source.

ACKNOWLEDGEMENTS

We would like to thank N.A. Chernoplekov for his support and Yu. Kagan and G.V. Shlyapnikov for valuable discussions.

REFERENCES

(1) I.F. Silvera and J.T.M. Walraven, Spin-polarized atomic hydrogen, in: Progress in Low Temperature Physics, Vol. X, ed. D.F. Brewer (North-Holland, Amsterdam 1986) pp. 139 - 370.
(2) A.P.M. Mattheij, J. van Zwol, J.T.M. Walraven and I.F. Silvera, Phys. Rev. B37 (1985), 4831.
(3) B.W. Statt, W.N. Hardy, A.J. Berlinsky and E. Klein, J. Low Temp. Phys. 61 (1985) 471.
(4) Yu. Kagan and G.V. Shlyapnikov, Phys. Lett. 130A (1988), 483.

Physica B 165&166 (1990) 745–746
North-Holland

THE RECOMBINATION OF ATOMIC HYDROGEN BELOW 1K

Saps Buchman*, Yueming Xiao*, Lois Pollak#, Daniel Kleppner#, and Thomas J. Greytak#

*GP-B Program, Hansen Laboratories of Physics, Stanford University, Stanford, CA 94305 U.S.A.
#Department of Physics, Massachusetts Institute of Technology Cambridge, MA 02139 U.S.A.

We have conducted an NMR study of the hydrogen recombination process in the temperature range 0.2K - 0.6K. Starting with a gas of doubly polarized atomic hydrogen we produce molecular ortho hydrogen. The nuclear spin polarization is quickly lost after the recombination process. The time constant for the de-polarization is less than 0.1s, and the initial nuclear spin temperature of the solid is larger than 4K.

The recombination of atomic hydrogen is one of the few chemical reactions simple enough to allow calculations from first principles. It is therefore highly desirable to study this fundamental process under well controlled conditions, so that the experimental results can be compared in detail with the theoretical predictions.

The experimental method used for this study is as follows. We first stabilize a gas of atomic hydrogen by polarizing the electron spins at a temperature of 0.2K to 0.6K in a magnetic field of 6.7T. Stable nuclear polarization follows automatically, with polarization times of the order of 1000 seconds (1,2). Figure 1 shows the hyperfine energy levels of a hydrogen atom in a magnetic field. It is the *b* hyperfine state which is stable and referred to as doubly spin polarized hydrogen. The recombination is then induced by flipping the electron spin. The techniques we use to produce and confine the doubly spin polarized hydrogen are described in our earlier work (2).

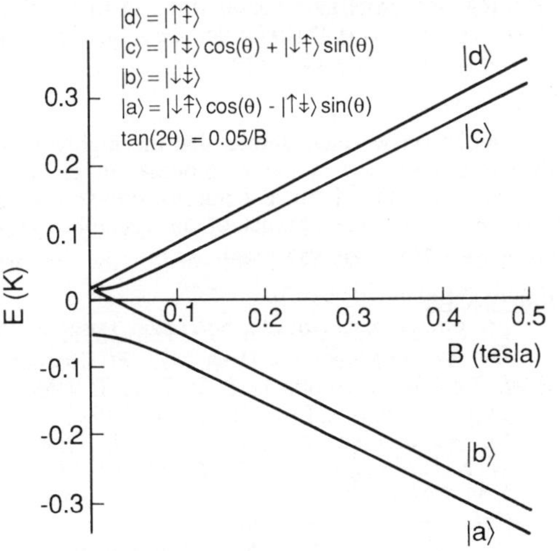

FIGURE 1
The hyperfine diagram of atomic hydrogen in a magnetic field. The plain and crossed arrows indicate the polarization of the electron and the proton spin.

The typical polarized atomic gas densities achieved in this experiment are of the order of 10^{16}atoms/cc at 0.3K. At these densities the dominant recombination process is the two-body surface recombination (3), with molecules forming close to the highest energy bound state; vibrational quantum number $V = 14$ and rotational quantum number $J = 1$ or 3 (4,5). We choose the *b+c* recombination channel since it has the following advantages: 1) The recombination rate is easily controlled by the *b* to *c* conversion rate. 2) The para hydrogen production is negligible. 3) Molecules form in a state with nuclear spin quantum numbers $I = 1$, $M_I = -1$. Note that the nuclear polarization has the opposite sign from the thermal nuclear polarization, a characteristic easily detectable by NMR. The *c* atoms are produced by ESR from *b* atoms at a frequency of 187GHz, for the 6.7T field.

Figure 2 is a diagram of our experimental cell. The cell contains a polarization chamber for the atomic gas, and an NMR chamber, where the stimulated recombination into molecular hydrogen takes place. The two chambers are connected by a 10mm length of 0.2mm diameter tube. By changing the helium level in the cell the chambers can be isolated from each other. We refer to this system as the helium level valve. A bolometer is used to ascertain the open/closed state of the valve. The pressure in the polarization chamber is monitored with a capacitive pressure transducer, while two carbon resistance thermometers measure the cell temperature. The NMR chamber contains a 285MHz resonator with Q=250. The sensitivity of the NMR spectrometer is 10^{14} fully polarized protons. A millimeter waveguide ends in the NMR chamber making it a Q=1 ESR cavity. The microwave power input into the cavity is adjustable from 1μW to 100μW over the 183GHz to 189GHz frequency range, allowing *b* to *c* conversion rates (and therefore recombination rates) from 10^{15}atoms/s to 10^{17}atoms/s. The *a* to *d* resonance can also be driven, and is used for calibration purposes.

The experiments have been carried out over a temperature range from 0.2K to 0.6K, for both a pure ^{4}He substrate and a ^{3}He-^{4}He mixture substrate. Since the results are similar, we choose to discuss the 0.3K experiments with a ^{4}He substrate. The experiments start with the helium level valve closed. Atomic hydrogen gas

0921-4526/90/$03.50

is admitted into the polarization chamber and reaches close to 100% nuclear polarization at about 10^{16}atoms/cc in 30 minutes. The helium level valve is then opened, allowing the doubly polarized atomic hydrogen gas to flow into the NMR chamber. The *b* to *c* transition is driven by the 187GHz ESR, with the *b+c* recombination taking place in the NMR chamber. Pressure and temperature measurements allow monitoring of the recombination process. Pulsed proton NMR (285MHz) measures the polarization of the molecular hydrogen sample during and following the recombination process.

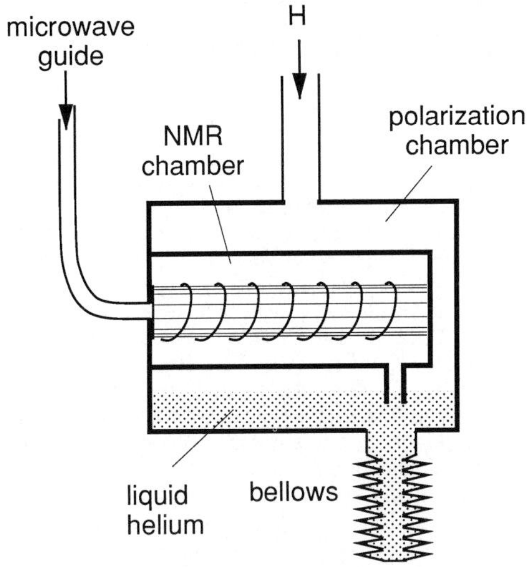

FIGURE 2
Schematic diagram of the experimental cell.

Figure 3 shows a 40minute time sequence of the NMR integrated amplitude. Molecular production rates are 10^{17}molecules/s for this fill. The solid line is a fit to the data by the standard form for the recovery of a saturated spin system to its equilibrium value, using a single relaxation time constant:

$$A = A_0\{1 - \exp[(t_0-t)/T_1]\} \quad (1)$$

For the 0.3K cell temperature, and the 6.7T magnetic field, the thermal polarization is about 3.5% of the full polarization of the protons. The initial nuclear polarization of the sample is 0.0±0.3%. This corresponds to a lower limit of 4K for the initial nuclear spin temperature. The first data point of figure 3 is taken 0.5s after the recombination event which lasts about 1s. Given that the molecules form in a fully polarized state, the low polarization at t=0.5s places an upper limit $T_1<0.1$s on the initial nuclear spin relaxation time. The fast nuclear spin relaxation and the high spin temperature indicate that the highly vibrational and rotational excited molecules do not dissolve in or adsorb on the liquid helium.

The nuclear spin relaxation constant from the fit in figure 3 is: T_1 = 11±1min, in agreement with measurements by other workers (6), when scaled to the ortho concentration and the temperature of the present experiment. The NMR line-width is 150kHz, consistent with the intramolecular dipolar broadening for the fcc phase of solid hydrogen. From the ortho to para conversion rate of the solid we find the initial ortho concentration to be X_0 = 99+1/-2%, therefore confirming the recombination model. An absolute calibration of the signal amplitude shows that the recombination is completely contained in the NMR chamber. The nuclear spin relaxation rate, the ortho to para conversion rate, and the line-width indicate that the solid hydrogen formed by the stimulated recombination of the polarized atomic gas is a standard compact fcc solid.

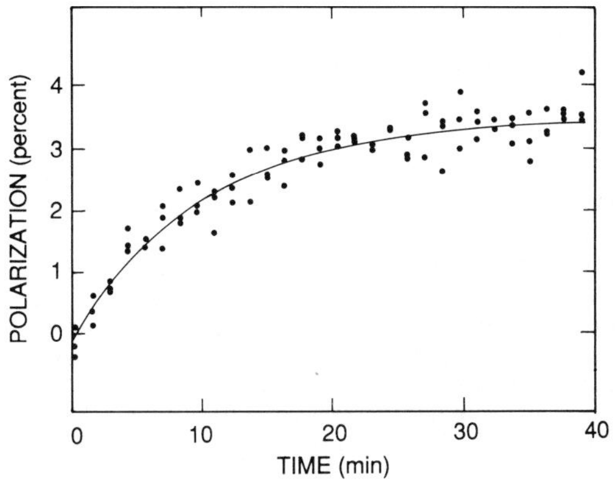

FIGURE 3
NMR integrated amplitude versus time. Sample size in the NMR chamber is $0.7*10^{17}$molecules. Solid line is fit to data using eq 1. T_1 = 11±1min.

In conclusion we have shown that the highly excited molecular gas formed in the recombination process of atomic hydrogen looses nuclear polarization in less than one second. The final product of the recombination is compact fcc 100% ortho hydrogen.

REFERENCES

(1) I.F. Silvera, Phys. Rev. B 29 (1984) 3899.
(2) D. A. Bell, H. F. Hess, G. P. Kochanski, S. Buchman, L. Pollack, Y. M. Xiao, D. Kleppner, and T. J. Greytak, Phys. Rev. B 34 (1986) 7670.
(3) M. W. Reynolds, I. Shinkoda, R. W. Cline, and W. N. Hardy, Phys. Rev B 34 (1986) 4912.
(4) J. M. Greben, A. W. Thomas, and A. J. Berlinsky, Can. J. Phys. 59 945 (1981) 945.
(5) Y. Kagan, I. A. Vartanyantz, and G. V. Shlyapnikov, Zh. Eksp. Teor. Fiz. 81 (1981) 1113. [Sov. Phys. JETP 54 (1981) 590].
(6) N. S. Sullivan and R. V. Pound, Phys. Rev. A 6 (1972) 1102.

Physica B 165&166 (1990) 747–748
North-Holland

BOUNDARY CONDITIONS AND SPIN WAVES IN SPIN-POLARIZED QUANTUM GASES

Alexander E. MEYEROVICH

Department of Physics, University of Rhode Island, Kingston, RI 02881 USA

Macroscopic boundary conditions are studied for spin dynamics in spin-polarized quantum gases. The constants in the boundary conditions are related to all possible microscopic processes (scattering, depolarization, sticking, etc.). Important changes start with the formation of an absorbed boundary layer at low temperatures.

1. INTRODUCTION

Spin-polarized quantum gases are usually studied experimentally at low temperatures and densities when the mean free paths are relatively long and boundary effects may become noticeable (see, e.g. review (1) and references therein). The interaction of gas particles with the walls is usually studied in terms of individual scattering probabilities or in the frames of kinetic equation (see, e.g., (2)). However, some of the major phenomena in quantum gases, including spin waves, are often described macroscopically in the frames of "hydrodynamic" approach. Therefore, it seems important to bridge the gap between kinetic and hydrodynamic boundary conditions, and to understand a microscopic meaning and an order of magnitude of macroscopic boundary constants.

The simplest macroscopic (hydrodynamic) boundary condition for spin dynamics has the form $0 = \mathbf{M} + \Lambda n_i \nabla_i \mathbf{M}$, where $\mathbf{M}$ is the density of magnetic moment, n_i is the unit vector normal to the boundary and directed into the gas. The parameter Λ contains all the information about the boundary scattering. The situations occurring most often correspond either to $\Lambda \to \infty$ (zero magnetic flow through the boundary; in quantum gases it occurs, e.g., for H↑ at relatively high temperatures) or to $\Lambda \to 0$ (complete magnetic relaxation on the walls; such case takes place for spin waves in ^{3}He systems when there is a localized absorbed helium layer on the walls; for details see Ref.(1)). The question is to formulate conditions when it is possible to express all relevant boundary information through the single constant Λ, and to relate Λ to microscopic boundary parameters. Another interesting question is what happens with H↑ when the temperature is lowered, mean free paths increase, collisions with the walls become more and more important, and, eventually a noticeable surface absorption of hydrogen atoms begins. Some experimental results have recently been reported, (3). With intermediate values of Λ, the positions and widths of spin-wave resonances, e.g., in Ref.(3), are described by the equation $\phi(x = 0) + \Lambda\phi'_x(x = 0) = 0$, where ϕ is the Airy function describing the density of the magnetic moment,

$$\mathbf{M}(x) = -\mathbf{M}_0\,\phi([-x+(\omega\tau T/Gm)(1+\Omega^2\tau^2)/(1+i\Omega\tau)][-iG]^3),$$

$\mathbf{M}_0$ is the scaling factor, τ is the bulk exchange relaxation time, Ω is the (bulk) molecular field frequency, and G is the magnetic field gradient.

2. SURFACE SCATTERING

At low temperatures (but higher than the beginning of intensive absorption) the main mechanism of interaction with non-magnetic walls is scattering giving rise to some additional spin relaxation and spin diffusion along and perpendicular to the wall. If the walls are located at x = 0 and x = L, then the integration of kinetic equation (1) with an additional collision integral corresponding to the boundary scattering leads to the following general macroscopic equations in $\mathbf{M}$ and spin current $\mathbf{J}_i$:

$$(\partial/\partial t)\mathbf{M} + (\partial/\partial r_i)\mathbf{J} = -\mathbf{M}_i(\mathbf{r})\ell[\delta(x) + \delta(x - L)]/\tau^*,$$

$$(\Omega + i/\tau)\mathbf{J} + i(T/m)(\partial/\partial r_i)\mathbf{M} = -i\ell(T/m)[(1/D -$$

$$- 1/D)n_i(n_k\mathbf{J}_k) + \mathbf{J}_i/D][\delta(x) + \delta(x - L)],$$

where τ^* is the surface spin non-conserving (e.g., dipole) relaxation time, D and D are the (surface) spin diffusion coefficients along and perpendicular to the surface, and $\ell \geq a$ is the dimensionality factor. Then the coefficient Λ in the macroscopic boundary condition is equal to

$$\Lambda = (i/\ell)[(T/m)q^2/(\Omega + i/\tau)D + (m/T)(\Omega + i/\tau)/\tau^*]$$

where $\mathbf{q}$ is the component of the wave vector along the boundary, $\mathbf{k} = \mathbf{q} + \mathbf{k}_b$.

The above description is valid only if τ^* and mD/T are much smaller than the bulk relaxation time τ, the effective "thickness" of the boundary layer is small, $k_b\ell \ll 1$, and the boundary condition on the "outer" side of the boundary (i.e., deep inside the wall) corresponds to the absence of spin current. The last condition is violated in case of magnetic and, to some extent, metal walls, and the first - if the density of magnetic "imperfections" of the wall is high.

If there are very few magnetic impurities or centers on the wall, the term with τ^* is less significant than the first one where D characterizes general imperfections of the wall and the difference of the reflection from the specular one [in simplest cases, $D \sim \ell T/m(1 - p)$, and p is the coefficient of specularity of reflection]. This term is responsible for the temperature dependence (or for the lack of it - while D is still large) of spin wave resonances. Certainly, this does not depend on whether the system is in hydrodynamic or Knudsen regime. From this point of view, the surface scattering is somewhat analogous to the elastic scattering by non-magnetic impurities in the bulk, which results only in some renormalization of the mass- (and, therefore, spin-) diffusion coefficient while the molecular field frequency, Ω, responsible for the existence of spin waves and the position of resonances, remains unchanged. With lowering temperatures Λ usually remains imaginary, and practically does not change. The change in density (and a possible transition to the Knudsen regime) does not affect $\Omega\tau$, but changes the surface contribution determined by the dimensionless parameter $\Omega Dm/T$.

3. INTERACTION WITH THE ABSORBED LAYERS

The formation of the relatively dense boundary layers by absorbed hydrogen particles at low temperatures leads to two effects: the set of boundary constants becomes different, and the number of equations doubles reflecting the appearance of additional macroscopic variables: the surface density of the magnetic moment **m** and the surface spin current **j**.

If one neglects the coupling of bulk and surface modes which may be weak because of possible large difference in densities and characteristic frequencies) then, effectively, the boundary conditions remain the same as above, but with considerable changes in the parameters involved. The most important change is the appearance of a large imaginary part in the coefficient $D = D' + iD''$ (the real part in Λ). This results in some effective increase in a spin waves attenuation, and was observed experimentally (3).

The origin of the coefficient D" is, in fact, not a dissipative, but a dephasing one. Since the boundary layer consists of the same particles as the bulk of the gas, the exchange with the boundary layer leads to some renormalization - near the surface - of the molecular field Ω. The simplest way to incorporate the "surface" molecular field $\Omega_s \sim \alpha hn/m$ (n is the density of these 2D layers with spin polarization α) into the above equations is to substitute D by a renormalized value with a large imaginary part, $D \rightarrow D/(1 - im\Omega_s/TD)$. The large dephasing associated with this imaginary part is more noticeable experimentally than all other changes: it leads to a broadening of the lines which were very narrow before (τ and τ^* are very large), while all other effects lead only to (small) changes of the order $n/\ell N$.

At even lower temperatures, one should also take into account the coupling of the surface and bulk modes. This process is much more complicated, and will be discussed separately. However, it is worth mentioning that here one deals with two different frequency regimes corresponding to different degrees of equilibration between surface and bulk variables during the oscillation. As a result, one will observe not only changes in values of boundary parameters, but also some frequency dispersion of all characteristics.

4. SUMMARY

The macroscopic boundary conditions for the spin waves in spin-polarized quantum gases are discussed. The relations are established between the coefficients in macro- and microscopic boundary conditions, thus allowing the explanation of the observed temperature and density dependencies of spin-wave resonances. The most interesting effects occur when the absorbed boundary layer starts to form leading to an appearance of the surface molecular field and to broadening of resonances because of additional surface dephasing. All possible surface processes are taken into account, including the surface scattering, magnetic dipole and exchange interactions, etc.

(1) A.E. Meyerovich, Spin-polarized phases of ^{3}He, in: Helium Three, eds. W.P. Halperin and L.P. Pitaevski (North - Holland, Amsterdam, 1990).
(2) J.J. Berkhout and J.T.M. Walraven, Sticking and thermal accommodation of atomic hydrogen on liquid helium surfaces, in: Spin-Polarized Quantum Systems, ed. S. Stringari (World Scientific, Singapore, 1988, pp. 201 - 208.
(3) N.P. Bigelow, J.M. Freed and D.M. Lee, Phys. Rev. Lett. **63** (1989), 1609.

Physica B 165&166 (1990) 749–750
North-Holland

CRITICAL FLOW OF SUPERFLUID HELIUM-4 THROUGH A SUBMICRON APERTURE: WIDTH OF THE CRITICAL TRANSITION

W. Zimmermann, Jr.‡

Service de Physique du Solide et de Résonance Magnétique, CEN–Saclay, 91191 – Gif–sur–Yvette Cedex, Laboratoire de Physique des Solides, Université Paris–Sud, 91405 – Orsay, France

O. Avenel

Service de Physique du Solide et de Résonance Magnétique, CEN–Saclay, 91191 – Gif–sur–Yvette Cedex, France.

E. Varoquaux

Laboratoire de Physique des Solides, Université Paris–Sud, 91405 – Orsay, France

We have observed the critical onset of dissipation in the flow of superfluid ^{4}He through a submicron aperture as a function of velocity to have a finite width that increases with temperature. The results suggest a thermal activation barrier of the form $E_0(1-v/v_{c0})$ with E_0 of the order of 1.5×10^{-21} J.

1. EXPERIMENT

We have observed that the critical velocity of the flow of superfluid ^{4}He through a submicron aperture exhibits fluctuations which increase with temperature. We have analyzed these fluctuations in order to gain information about the velocity dependence of the probability of occurrence of dissipative phase–slip events. These measurements were made with the same apparatus and during the same run as the study of the pressure dependence of v_c [1]. Data were taken at 0 bar, at four temperatures 16.3, 165, 329 and 447 mK. At each temperature, sets of about 1000 data values were recorded at each of ~ 6 different drive levels just above critical. These values were proportional to the maximum (absolute) deflections of the double-hole resonator diaphragm for successive half–cycles. The inset in fig. 1 shows a small portion of such a set. Sudden decreases in value represented phase slips and fell into discrete groups of from 1 to 4 units of change, with occasional larger changes. Single slips were by far the most frequent. After a small correction for resonator drive and losses, each value was proportional to the subsequent maximum flow rate through both holes, aperture and main passage together, in the absence of phase slip.

2. TREATMENT OF THE DATA

In order to determine a quantity y proportional to the actual maximum flow rate through the aperture itself, it was necessary to take into account the hydrodynamic circulation trapped in the loop threading the aperture and main passage. Using the association of phase slips with changes of circulation, it turned out to be possible by close inspection of the data to assign almost unambiguously a consistent quantum circulation to every point recorded.

The probability $p(y)$ of occurence of phase–slip at amplitude y could then be constructed by event counting. Separate distributions for each flow direction, *in* and *out* of the chamber, such as the one shown for outward flow in fig. 1, were in fact constructed to allow for different critical velocities in the two directions as well as for possible bias currents [2]. It can be seen that the critical velocity threshold has an appreciable width in velocity. This width was found to increase with temperature.

3. THE ACTIVATION ENERGY

These results can be interpreted in the framework of the thermal activation model put forward in [3] to account for the linear temperature dependence of v_c observed over a broad range of T [1,3,4]. In this model the rate of occurrence of phase–slips is given by

$$\Gamma = \Gamma_0 \exp\left\{ -E_0(1-v/v_{c0})/k_b T \right\}, \tag{1}$$

where the attempt rate Γ_0, the energy barrier at zero velocity E_0, and the zero–temperature critical velocity v_{c0} are phenomenological parameters.

From eq.(1), we derived an analytical expression for $p(y)$ which can be characterized by two parameters y_c and y_w such that $p(y_c) = 0.5$ and $y_w = (dp/dy)^{-1}$ at y_c. The central value y_c of the critical transition decreases nearly linearly with T and the width y_w is proportional to T:

$$y_w(T) = \frac{2k_b T}{\ln 2}\frac{y_c(0)}{E_0}. \tag{2}$$

‡ Permanent address: Tate Laboratory of Physics, University of Minnesota, Minneapolis, MN – 55455 – USA

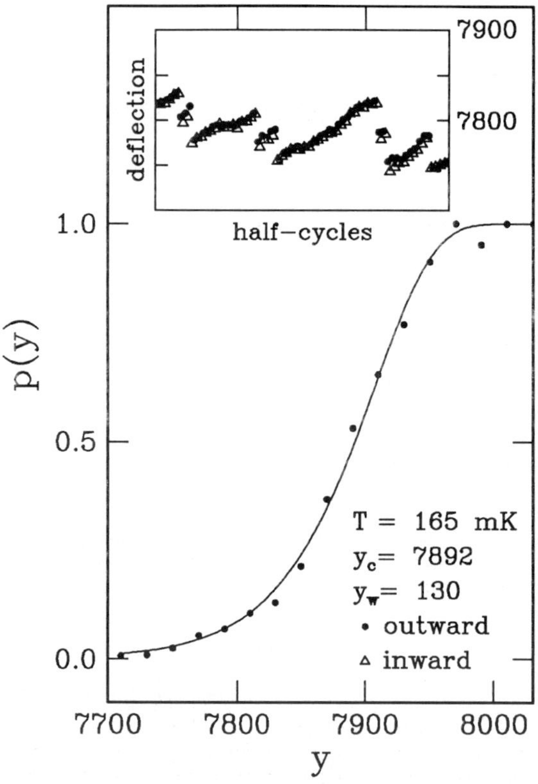

Figure 1. Probability distribution for phase slip.

Fits of $p(y)$ to the experimental distributions, such as shown by the curve in fig.1, yield the values of parameters y_c and y_w that are shown in fig.2. The values of y_c reflect the form of $v_c(T)$ observed previously [1,3,4], and, for the upper three temperatures, are consistent with the model above. For those same temperatures, y_w is approximately linear in T. If we assume there that y_w contains a constant additive contribution, due at least in part to residual vibration, and fit the slope of eq.(2) to that of the data, we obtain $E_0 = 106$ K $= 1.5\times10^{-21}$ J. The data at 16.3 mK presumably correspond to a different (^{3}He–related) type of behavior [3].

4. DISCUSSION

There is much uncertainty associated with this interpretation and estimate for E_0. It is not clear that the vibrational and temperature-dependent widths are simply additive nor is it clear that ^{3}He impurities have a negligible effect on the fluctuations of v_c in the high temperature régime. Nevertheless, the value of E_0 found by this statistical analysis is consistent with the upper bound of ~ 110 K estimated in [3].

The value of E_0 is 4 orders of magnitude less than the energy dissipated per 2π phase slip, estimated in earlier work with the same aperture to be 1.2×10^{-17} J. The fluctuation energy necessary to trigger the slip, be it by vortex nucleation or by vortex depinning as

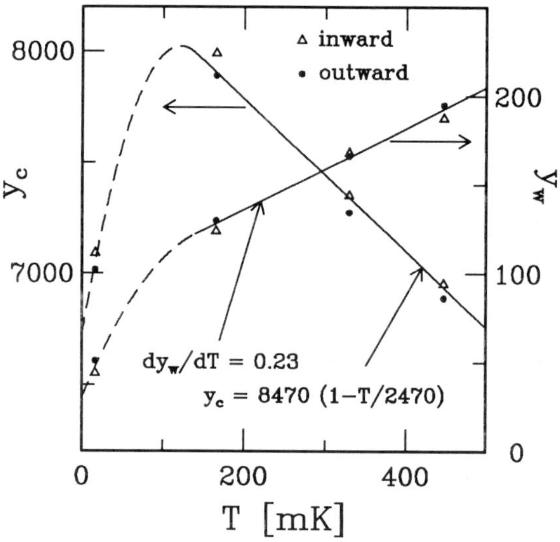

Figure 2. y_c and y_w versus temperature.

proposed by Schwarz [5], is very small indeed. But it is still much larger than the energy barrier of 3 K calculated by Muirhead, Vinen and Donnelly [6] for the case of negative ion propagation and which accounts for ion critical velocities [7] that are an order of magnitude larger than those found in the present work [1]. Our values for the critical velocities seem to be smaller than necessary for half vortex rings nucleated at the wall to grow spontaneously [5]. Sharp microscopic surface defects can however enhance the local superflow velocity somewhat and may assist vortex growth.

It is noteworthy that our value for E_0 is of the same magnitude as the energy barrier estimated by Schwarz for the depinning of a vortex from a typical 50 Å asperity [5]. The depinning process may offer an alternative to vortex growth at the wall although it suffers from known objections [8] as well as from the additional difficulty raised by the pressure dependence of v_c [1]. Further work is needed to elucidate the mechanism for phase-slips in small orifices.

REFERENCES

[1] O. Avenel, E. Varoquaux, W. Zimmermann, Jr., these proceedings.
[2] O. Avenel, E. Varoquaux, Jpn. J. Appl. Phys. **26–3** (1987) 1798.
[3] E. Varoquaux, M.W. Meisel, O. Avenel, Phys. Rev. Lett. **57** (1986) 2291.
[4] B.P. Beecken, W. Zimmermann, Jr., Phys. Rev. **B35** (1987) 1630.
[5] K.W. Schwarz, Phys. Rev. Lett. **59** (1987) 1167.
[6] C.M. Muirhead, W.F. Vinen, R.J. Donnelly, Phil. Trans. R. Soc. Lond. A **311** (1984) 433.
[7] R.M. Bowley, Quantum Fluids and Solids–1989, eds. G.G. Ihas, Y. Takano (AIP, New-York, 1989) p.149.
[8] O. Avenel, E. Varoquaux, Phys. Rev. Lett. **59** (1987) 1168.

Physica B 165&166 (1990) 751–752
North-Holland

CRITICAL FLOW OF SUPERFLUID HELIUM-4 THROUGH A SUBMICRON APERTURE: DEPENDENCE ON PRESSURE

O. Avenel

Service de Physique du Solide et de Résonance Magnétique, CEN–Saclay, 91191 – Gif–sur–Yvette Cedex, France.

E. Varoquaux

Laboratoire de Physique des Solides, Université Paris–Sud, 91405 – Orsay, France

W. Zimmermann, Jr.‡

Service de Physique du Solide et de Résonance Magnétique, CEN–Saclay, 91191 – Gif–sur–Yvette Cedex, Laboratoire de Physique des Solides, Université Paris–Sud, 91405 – Orsay, France

We report the existence of a pressure dependence of the critical velocity of flow of superfluid ^{4}He through a submicron aperture. Combined with the already-known temperature dependence of the critical velocity at very low temperatures, this observation puts further experimental constraints on the mechanism responsible for the nucleation of phase slips in ^{4}He.

1. INTRODUCTION

We report detailed measurements of the critical threshold for the onset of dissipative phase slips in the flow of superfluid ^{4}He through a submicron aperture as a function of pressure and temperature from 0 to 24 bar and from 16 to 600 mK. Previous experiments of this type have found at 0 bar a temperature-dependent critical velocity obeying the relationship

$$v_c(T) = v_{c0}(1-T/T_0) \,, \tag{1}$$

from ~150 mK to ~1.9 K. Here v_{c0} is of the order of 0.5 – 10 m/s and T_0 is ~2.5 K [1,2]. In ^{4}He purified of residual ^{3}He, this temperature dependence has been seen to extend down to at least 5 mK [1]. Little is known about the dependence of v_c on the shape of the aperture [2]. No consistent theory for v_c and its temperature dependence has been given.

This work confirms the temperature dependence of eq.(1) and the value of T_0 and also reports the existence of a dependence on pressure of both v_{c0} and T_0.

2. EXPERIMENTAL RESULTS

The present experiments were performed with the same cell and aperture as used previously [1], modified by the addition of a relatively long and open main passage in parallel with the aperture [3]. Attempts to purify the ^{4}He that we used here of residual ^{3}He failed, in part because of ^{3}He contamination of the cell. The ^{3}He concentration measured subsequently by mass spectrometry was 2.1×10^{-6}.

In our apparatus oscillatory flow at $\nu \sim 10$ Hz took place between a small resonator chamber and an outer bath through the two parallel passages. The critical velocity was reached in the small aperture, nominally 0.3 μm by 5 μm in a Ni film 0.2 μm thick, while the flow in the main passage remained subcritical. The motion was driven resonantly by applying an electrostatic force to a thin elastic diaphragm forming a wall of the small chamber. The same diaphragm together with a SQUID sensor was used to monitor the resulting fluid motion.

At each pressure and temperature, the amplitude of the drive was slowly swept from subcritical to supercritical levels. The critical level of the average diaphragm deflection amplitude, a quantity proportional to v_c/ν, was identified by an abrupt decrease in the slope of this amplitude versus drive amplitude. This critical threshold was caused by the appearance of discrete dissipative phase slips, which were resolved on the recorded traces, and constituted the edge of the first plateau of the staircase pattern [3].

To ensure high consistency in the data, measurements at a given pressure were made according to a fixed protocol in which the temperature was increased in steps between points. The cell was isolated at each pressure plateau by a low temperature valve. Between plateaus, which were separated by amounts of roughly equal magnitude in density, the pressure was reduced very gradually and smoothly. No noticeable change of cell operation or of aperture geometry by variation of the ^{3}He concentration or other contaminants such as air or dust particles occurred during the whole run.

In this work, staircase patterns were less distinctly resolved and multiple slips, sometimes causing a near-collapse of the resonator motion between 100 and 200 mK, were far more numerous than in earlier work.

‡ Permanent address: Tate Laboratory of Physics, University of Minnesota, Minneapolis, MN – 55455 – USA

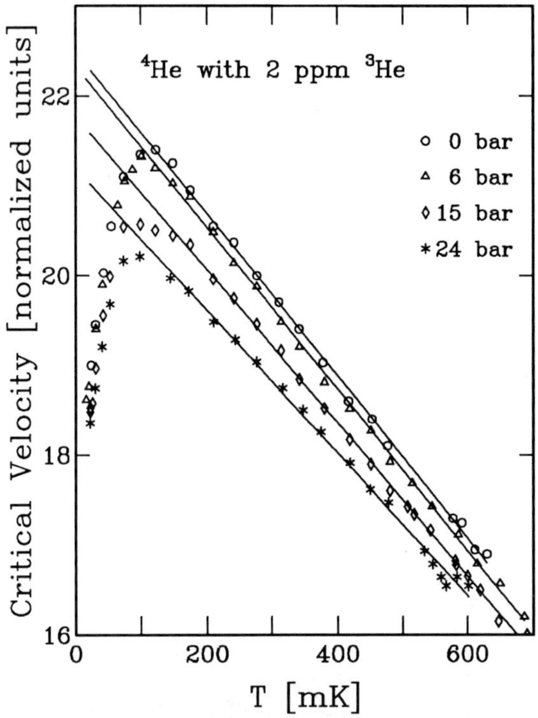

Figure 1. Critical velocity versus temperature.

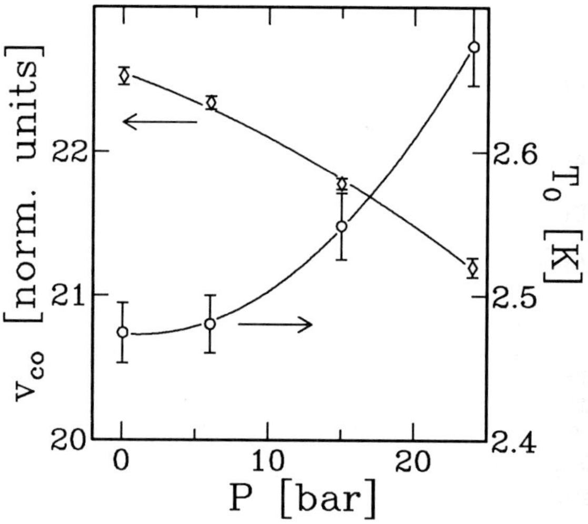

Figure 2. v_{c0} (◇) and T_0 (○) versus pressure.

Our data are shown in fig. 1, in which a clear dependence of v_c on pressure is exhibited. Absolute values of velocity are known only to within a factor of 3 and range around 1.5 m/s. The data above ~150 mK confirm the linear dependence of v_c on T seen earlier. Previous evidence suggests that the non-linear behavior observed below 150 mK is due to the presence of residual ^{3}He [1]. The straight lines are least-square fits to all data above 140 mK. The values of v_{c0} and T_0 for each pressure, as obtained from both least-square and least-absolute deviation minimization with negligible differences, are shown in fig. 2, together with the larger of the statistical deviations given by the two fitting methods. The curves in fig. 2, intended to guide the eye, are quadratic fits to the data. The magnitude of this effect of pressure is consistent with our previous, less accurate observation [1].

The overall changes of v_{c0} and T_0 with the density ρ comparable to those of $\rho^{-1/3}$ and $\rho^{1/3}$ respectively. But in both cases the signs of the curvatures of the fitted curves in fig. 2 are opposite to those of these simple dependences on ρ.

3. IMPLICATION FOR THEORY

The appearance at the critical velocity of discrete dissipative events involving energy losses of $(2\pi\hbar/m_4)J$, where J is the rate of mass flow through the aperture, provides evidence for quantized vortex motion being the agent of dissipation. Vortices have to cut all the streamlines threading the orifice to generate an overall 2π phase change. These moving vortices must find their origin somehow, in ample number, by nucleation [1] or depinning processes [4,5]. The temperature dependence in eq.(1) strongly suggests that thermal activation plays a leading rôle. The homogeneous nucleation theory of Iordanskii, Langer and Fisher [6] fails at very low temperature, and, in all likelihood, we are dealing with a wall-dominated mechanism.

However, the precise nature of the instability responsible for $v_c(T)$ here is unknown. The models proposed so far, based on vortex depinning [4] or growth at the walls [7], involve a density-independent v_{c0} and a T_0 proportional to ρ. A consequence of the dependence of v_{c0} and T_0 on ρ observed here is that these models, which are based on purely hydrodynamical arguments, do not apply as such.

Furthermore, the detailed shapes of the pressure dependence do not appear to be well described by simple functions of ρ. If confirmed by new experiments on different orifices, this result could indicate that the density in the bulk, albeit playing a rôle, is not a directly relevant parameter in this problem.

REFERENCES

[1] E. Varoquaux, O. Avenel, M. Meisel, Can. J. Phys. **65** (1987) 1377.
[2] B.P. Beecken, W. Zimmermann, Jr., Phys. Rev. **B35** (1987) 1630.
[3] O. Avenel, E. Varoquaux, Jpn. J. Appl. Phys. **26-3** (1987) 1798.
[4] K.W. Schwarz, Phys. Rev. Lett. **59** (1987) 1167.
[5] O. Avenel, E. Varoquaux, Phys. Rev. Lett. **59** (1987) 1168.
[6] see e.g. J.S. Langer, J.D. Reppy, Prog. Low Temp. Phys., Vol. 6, ed. C.J. Gorter (North-Holland, Amsterdam, 1970) Ch.1.
[7] W. Zimmermann, Jr., unpublished.

Physica B 165&166 (1990) 753–754
North-Holland

PHASE-SLIPS IN THE FLOW OF SUPERFLUID ^{4}HE THROUGH A SUBMICRON ORIFICE

Ajay AMAR, J.C.DAVIS, R.E.PACKARD and R.L.LOZES*

Department of Physics, University of California, Berkeley, CA 94720, U.S.A.

* Perkin-Elmer, Electron Beam Technology, Hayward, CA 94545, U.S.A.

We have constructed a 6 Hz Helmholtz oscillator to study phase-slip phenomena in superfluid ^{4}He.This method was pioneered by Zimmermann (1-2) and by Avenel and Varoquaux (3-6). A superconducting displacement transducer employing a D.C. SQUID and capable of resolving a displacement of 2×10^{-13} m in one second is used. We report preliminary flow studies at 0.3K using a slit with cross-sectional dimensions of $0.4\mu \times 4.6\mu$ and 0.08μ length. We detect sudden dissipation events, the smallest being consistent with phase-slips of 2π .

1. INTRODUCTION

An understanding of phase coherence properties and the microscopic processes leading to dissipation and onset of critical flow in a superfluid has been of fundamental interest to physicists for nearly four decades.It is believed that when the critical flow rate is reached in a channel the phase of the complex order parameter should slip resulting in a well defined loss of flow energy (7).

If a singly quantized vortex line traverses the entire flow channel, the phase will slip by 2π. However, phase changes of any arbitrary size can, in principle, be produced by either partial or full traverses by several lines.

If phase slips are to be used as a tool to study macroscopic phase coherence, one presumably needs to create a situation in which reproducible and empirically predictable phase slips occur. Zimmermann (1) has suggested that the most favorable geometry for such events would be a small channel with effective length and width well below one micron (8).

In an apparatus meeting the above criterion, Avenel and Varoquaux (AV) have observed highly reproducible,discrete transitions corresponding to phase-slips of 2π (3-4). The reproducibility of these slips, combined with quantization of circulation and a high degree of phase coherence has led them to produce staircase patterns analogous to those in an rf SQUID (5-6).

This paper reports the first attempts in Berkeley to reproduce the features seen by AV with the long-term goal of understanding the phase slip mechanisms and exploiting the quantum phase coherence to create a sensitive detector of rotation (9).

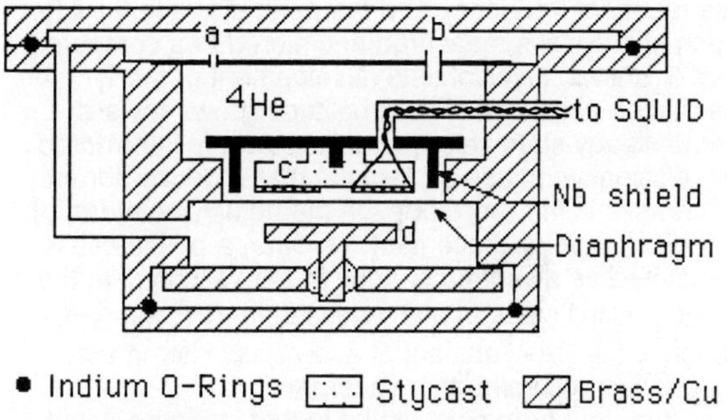

Figure 1. Experimental Cell

2. APPARATUS

Our Helmholtz resonator, shown in Fig. 1, is generically similar to that of AV. The small flow channel (a) is a slit of dimensions $0.43\mu \times 4.63\mu$ made with e-beam lithography in an 80nm thick membrane of silicon nitride. The parallel flow channel (b) was chosen to have an area to length ratio 3.4 times that of the main channel.The restoring force is provided by a 12μ thick mylar diaphragm with 150nm of sputtered Nb on each side, whose displacement is read by a planar superconducting coil (c) with the help of a D.C.SQUID (10-11). Displacement calibration comes from the capacitor formed by one side of the diaphragm and a copper plate (d) placed about 100μ from it. In the absence of liquid in the cell, the detector can resolve 1.7×10^{-13}m(Hz)$^{-1/2}$ for a current of 300 mA in the planar coil. With superfluid He^4 in the cell at temperatures where ρ_s/ρ is essentially 1.0, the resonator is seen to oscillate at 6.225 Hz .This frequency is within 5% of 5.98 Hz calculated from the measured spring constant of the diaphragm and nominal dimensions of the flow channels.

0921-4526/90/$03.50

The experimental cell is connected to the bottom of a single-shot charcoal-pumped He^3 fridge which, in the absence of the cell, can run for several days at T=225 mK (12). A needle valve on the fill line is used to isolate the cell from fill-line perturbations.The fridge is held rigidly at the top and bottom with respect to the dewar. During each run, the entire dewar is suspended from a vibration isolation stage consisting of a 1000 kg platform resting on soft air-springs. The laboratory is located at bedrock level in the lower basement of a building. An enclosure with sound-proofing foam is placed around the dewar to reduce acoustical input into the experiment. There is no liquid nitrogen in the dewar and the ^4He bath is pumped down to 1 torr and slowly warms toward 4K during the data collection. We believe that the ^3He in the fridge is not boiling owing to its low vapour pressure at 0.3 K. Thus every effort is made to eliminate external sources of vibration. The actual data are taken in the middle of the night when the disturbances in the building are minimal.

3. OPERATION

In a given run a small constant drive is applied at the resonant frequency and the peak displacement for each half cycle is measured and stored in a computer. Fig. 2 shows a typical time development of the oscillation amplitude. The amplitude grows toward some steady-state value but this growth is interrupted by dissipative phase-slip events that produce abrupt decreases in the amplitude.On the actual waveform of the diaphragm displacement vs. time, a phase-slip is registered as a sudden change in the velocity (i.e. the slope), taking place in a time less than 10^{-3} second, which is the time constant of a low pass filter in the electronics recording the waveform.

It is clear from data similar to that in figure 2 that with this particular apparatus the phase slips occur at random times and have varying sizes. From the waveform of actual diaphragm displacement one observes that the phase slips do not always occur at the instantaneous maximum of the velocity. Both these observations suggest that the dynamical processes causing these events are not intrinsic to the liquid. Rather, the slips may be triggered, perhaps well below the intrinsic critical velocity in the small channel, by some external mechanical disturbance.

A spectrum of the diaphragm's displacement indicates that the Helmholtz resonance is randomly driven at amplitudes of about 0.01 nm. The scatter in the data of figure 2 is about 0.004 nm. These values can be compared to the typical amplitude where phase slips begin to occur (0.8 nm) and the actual size of the smallest slips which appear (0.023 to 0.026 nm). It is interesting to note that these smallest slips are within 10% of the estimated size for 2π phase-slips.

Based on these preliminary observations we plan to change the size and shape of the small channel and improve the vibration isolation. We hope these changes will shed light on the mechanisms that give rise to reproducible 2π phase-slips.

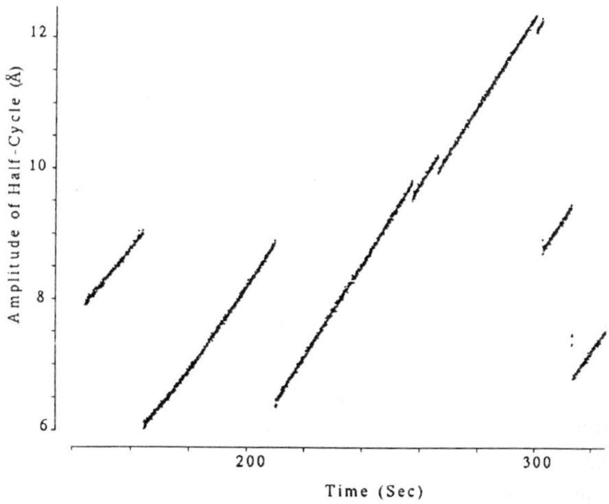

Figure 2. Phase-slips in the plot of amplitude vs. time for a constant drive. Each dot represents a half-cycle. The smallest jumps are close to 0.024 nm, the expected size for a 2π slip.

ACKNOWLEDGEMENTS

It is a pleasure to thank P.Moyland and W.Moeur for help during the construction of the experiment, and Kerwin Ng for data analysis. We are very grateful to J.Clarke for advice and sharing of his expertise on superconducting devices. Thanks are also due to F.Wellstood and N. Fan for help and valuable suggestions.This preliminary research was supported by NSF grant No. DMR-88-19110.

REFERENCES

(1) B.P.Beecken and W.Zimmermann,Jr., Phys. Rev.B 35,74 (1987)
(2) B.P.Beecken and W.Zimmermann,Jr.,Phys. Rev.B 35, 1630 (1987)
(3) O.Avenel and E.Varoquaux, Phys. Rev. Lett. 55, 2704 (1985)
(4) E.Varoquaux, O.Avenel and M.W.Meisel, Can. Jour. of Phys. 65, 1377 (1987)
(5) O.Avenel and E.Varoquaux, Japanese Jour. of App. Phys. 26, 1798 (1987)
(6) O.Avenel and E.Varoquaux, Phys. Rev. Lett. 60, 416 (1988)
(7) See for instance P.W.Anderson, Rev. of Mod. Phys. 38, 298 (1966)
(8) See also H.Xia, D.Jing, W.Y. Guan, J.L.T.P. 74, 417 (1989)
(9) R.E.Packard (to be published)
(10) H.J.Paik, J. Appl. Phys 47,1168 (1976)
(11) O.Avenel and E.Varoquaux, Proc. ICEC 11, 587 (1986)
(12) Based on a design by B.Neuhauser and B.Cabrera. Valuable suggestions by L.Duband are gratefully acknowledged.

Physica B 165&166 (1990) 755–756
North-Holland

QUASICLASSICAL THEORY OF THE JOSEPHSON EFFECT IN SUPERFLUID ^{3}He*

E.V THUNEBERG [1], J. KURKIJÄRVI [2], and J.A. SAULS [3]

[1]TFT, Helsinki University, Siltavuorenpenger 20C, 00170 Helsinki, Finland
[2]Institutionen för fysik, Åbo Akademi, Porthansgatan 3, 20500 Åbo, Finland
[3]Department of Physics, Northwestern University, 2145 Sheridan Road, Evanston, IL 60208, USA

* Supported by NSF (INT-8813867) and the Academy of Finland.

In an ideal Josephson junction the current is purely sinusoidal in the phase difference φ across the junction; $J = J_c \sin\varphi$, where J_c is the critical current. In narrow channels or orifices, the current-phase relation is not a simple sinusoidal function of φ, and may be a multi-valued, but periodic function of φ. The importance of the precise form of $J(\varphi)$ has been recently discussed in the context of weak links in superfluid ^{3}He. It is argued that dissipative events are associated with phase-slip events and are evidence for multi-valued current-phase relation [1]. We are interested in the theoretical constraints of geometry and surface scattering on the operation of a Josephson weak link in superfluid ^{3}He, and more generally in the current-phase relation and the onset of dissipation. Although superfluid ^{3}He is well described by the BCS pairing theory there are important features of ^{3}He that differentiate it from conventional superconductors and are important to the study of critical currents in narrow channels and the Josephson effect in weak links. The order parameter $A_{i\alpha}$, describing the spin (i) and orbital (α) degrees of freedom of the Cooper pairs, is particularly sensitive to boundaries. Even a specular boundary suppresses the components $A_{i\alpha}n_\alpha$ normal to the surface; it is this boundary condition that locks the l vector normal to the wall in ^{3}He-A and leads to the vanishing of the supercurrent in A-phase when the l vectors are antiparallel on the two sides of the weak link [2]. Surface roughness and magnetism introduce additional pairbreaking [3]. There is evidence from measurements of the critical currents in channels and thin films that surface roughness plays an important role in modifying the structure of the order parameter near a surface [4]. Realistic calculations of the Josephson effect must take account of these effects. The problem is particularly delicate also because the components of the order parameter that survive near a wall are typically distinct from those that survive inside a channel. It has been recently shown that a qualitatively new form of the current-phase relation arises in finite length channels, suggesting that the Josephson effect may be useful in identifying a many-component order parameter [5].

A simple weak link is a small orifice in a wall between two vessels of ^{3}He. The Josephson effect in such a geometry has been recently observed by Avenel and Varoquaux [1]. For a theoretical investigation of weak links the quasiclassical theory [6] is particularly suitable. It allows the calculation of the order parameter and current-phase relation in weak links of various shapes and sizes, circular or rectangular, short or long. It can handle the spatial dependence of the order parameter near surfaces as well as the dependence on external variables such as temperature, superflow parallel to the wall and magnetic fields. All results presented here are restricted to thin walls (*i.e.* thickness $<< \xi_0 = \hbar v_F/2\pi k_B T_c$). Results for more general weak links than pinholes are moreover restricted to specular scattering. We study the effects of a finite size radius aperture on the order parameter and Josephson current, and compare with the pinhole in the infinitely thin wall (*i.e.* radius $<< \xi_0$), which can be calculated straight-forwardly for both specular and diffuse surfaces.

The quasiclassical transport equation is a first-order differential equation for a propagator defined along classical trajectories of quasiparticles. It can be solved efficiently with the explosion method. It must be augmented by the self-consistency equation for the order parameter and Landau molecular fields, and boundary conditions describing surface scattering; a brief review of different boundary conditions is given in [3]. For a specular surface the boundary condition is simply the continuity of the propagator along the specularly reflected trajectory; rough surfaces are more difficult to treat. We use the 'thin dirty layer' model (original Refs. in [3]), but we have recently tested an alternative linear boundary condition for fully diffuse scattering which is more efficient in calculations and has so far reproduced the results of the 'thin dirty layer model.' The calculation of the Josephson current is relatively easy for a pinhole ($a \ll \xi_0$). Since the hole does not to distort the order parameter beyond what is already present due to the wall, the current is obtained by a straight-forward integration of the quasiclassical transport equation with the order parameter for a wall without a hole as input. In Fig. 1 we show the maximum Josephson current as a function of the temperature for a pinhole in a specular wall and a pinhole in a diffusely scattering wall.

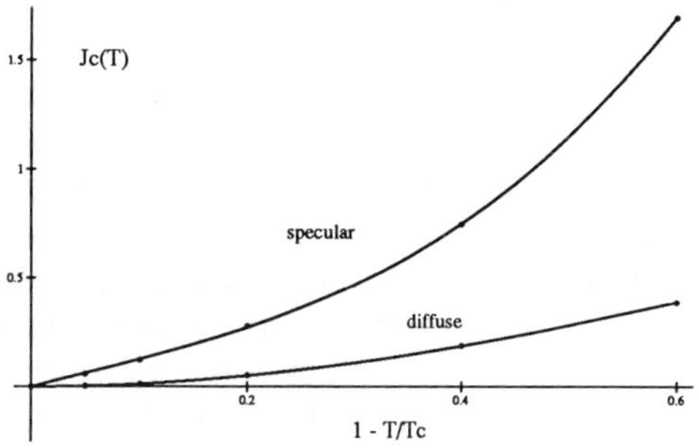

Figure 1. Josephson critical current for a pinhole.

The critical current for a diffuse-wall is quadratic in $(T_c - T)$ as predicted by Kopnin [7]. Also $J(\varphi)$ is asymmetric at lower temperatures, but remains single-valued. The asymmetry is slightly more pronounced in the specular case than in the diffuse case. At $T = 0.4T_c$ the maximum in the current is at $\varphi = 1.96$ and $\varphi = 1.77$, respectively.

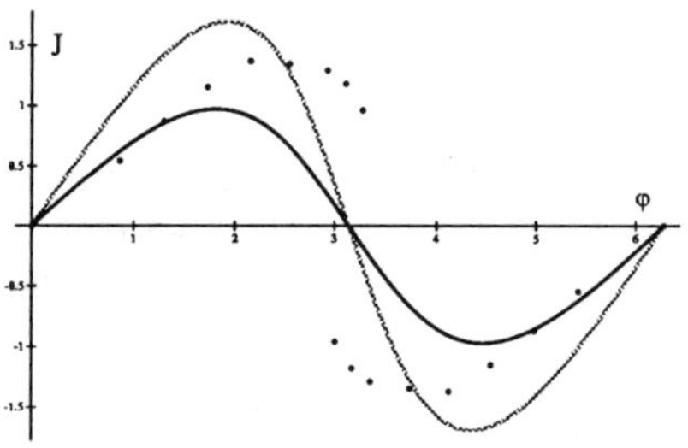

Figure 2. Josephson current: (gray) specular pinhole, (black) diffuse pinhole, (dots) specular aperture scaled by 3.

For finite size apertures the current and order parameter must be calculated self-consistently for each value of the phase difference. For small apertures the current-phase relation is qualitatively similar to that of the pinhole. However, for a hole of diameter $a = 3\xi_0$, $J(\varphi)$ develops a second branch at low temperatures signaling the onset of dissipation via phase slip (Fig. 2). The current is very large near the edge of the orifice, which leads to a very deformed order parameter. For phase differences near $\varphi = \pi$ the order parameter develops a structure which appears similar to the core of a B-phase vortex [8], including an induced magnetization density shown in Fig. 3. The vortex-like order parameter and magnetization move toward the center of the channel and their magnitudes increase as the phase difference increases, suggesting that this static structure is the source of phase slip in a dissipative Josephson weak link for ^{3}He-B.

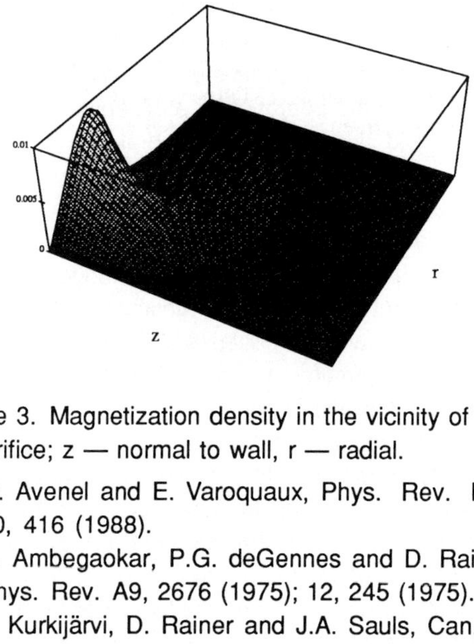

Figure 3. Magnetization density in the vicinity of the orifice; z — normal to wall, r — radial.

[1] O. Avenel and E. Varoquaux, Phys. Rev. Lett. 60, 416 (1988).

[2] V. Ambegaokar, P.G. deGennes and D. Rainer, Phys. Rev. A9, 2676 (1975); 12, 245 (1975).

[3] J. Kurkijärvi, D. Rainer and J.A. Sauls, Can. J. Phys. 65, 1440 (1987).

[4] M.R. Freeman, R.S. Germain, E.V. Thuneberg, R.C. Richardson, Phys. Rev. Lett. 60, 596(1988); S. Steel, Ph.D thesis, Queens Univ. (1989).

[5] E.V. Thuneberg, Europhys. Lett. 7, 441 (1988).

[6] J. Serene, D. Rainer, Phys. Rep. 101, 221 (1983).

[7] N.B. Kopnin, JETP Lett. 43, 700 (1986).

[8] E.V. Thuneberg. Phys. Rev. Lett. 56, 359 (1986).

Physica B 165&166 (1990) 757–758
North-Holland

PRESSURE DEPENDENCE OF VORTEX TUNNELING IN HE II

P C HENDRY, N S LAWSON, P V E McCLINTOCK, C D H WILLIAMS[1]

Department of Physics, Lancaster University, Lancaster, LA1 4YB.

R M BOWLEY

Department of Physics, The University, Nottingham, NG7 2RD, UK.

Measurements of the rate ν at which negative ions nucleate charged vortex rings in isotopically pure superfluid ^{4}He for electric fields E, temperatures T and pressures P within the ranges $10^3 \leq E \leq 10^6$ Vm^{-1}, $75 \leq T \leq 500$ mK, $12 \leq P \leq 23$ bar show that the energy barrier impeding the tunneling process is strongly dependent on pressure. Minima observed in $\nu(T)$ are ascribed to phonon damping.

1. INTRODUCTION

Experimental evidence for the energy barrier predicted (1) to impede the creation of quantized vortices in He II has been obtained (2) through studies of the motion of negative ions. Observations (3) made at relatively low pressures ($P \simeq 12$ bar) are consistent with the macroscopic quantum tunneling (MQT) model proposed (4) by Muirhead, Vinen and Donnelly (MVD). Measurements of the vortex nucleation rate ν as a function of temperature T at fixed electric field E show a constant value of ν below ~ 0.2 K and an exponential rise at higher T, interpreted as tunneling through the barrier, and thermal activation over it, respectively. At higher pressures, this simple picture fails because of the appearance (2) of pronounced minima in the $\nu(T^{-1})$ curves. These minima do not appear (5) to be caused by a reduction of the drift velocity $\bar{v}$ of the ion due to phonon scattering.

In this paper we report new and detailed measurements of $\nu(E,T,P)$ in isotopically pure ^{4}He. We propose that the minima in $\nu(T^{-1})$ are due to phonon damping of the tunneling mechanism and, on this basis, analyse the data to extract the pressure dependence of the barrier height.

2. EXPERIMENTAL DETAILS

The nucleation rate was measured by an electric induction technique that has been described in detail elsewhere (6). The sample of ^{4}He was isotopically purified (7) to eliminate complications (8) caused by the 0.2 ppm of ^{3}He present in gas-well helium.

3. RESULTS

Some typical experimental measurements, plotting ν as a function of reciprocal temperature, are shown in Fig 1. For the lowest pressure (top curve) there is no minimum in $\nu(T^{-1})$ but, as the pressure is increased, a clearly defined minimum appears and moves towards higher temperatures with increasing pressure.

4. DISCUSSION

At $P = 12$ bar, the data are well-described (2, 3) by a relation of the form

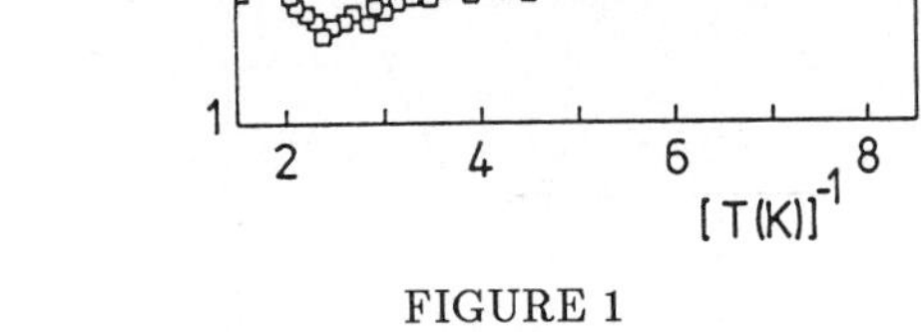

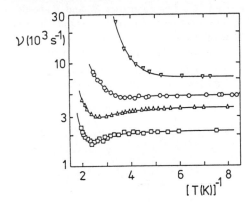

FIGURE 1

Measurements (data points) of the nucleation rate ν compared with fits (curves) of Eq (2), for various pressures and fields: 12 bar, 1.77×10^4 Vm^{-1} ($\bigtriangledown$); 15 bar, 3.10×10^4 Vm^{-1} ($\circ$); 17 bar, 6.20×10^4 Vm^{-1} ($\triangle$); 19 bar, 9.70×10^4 Vm^{-1} ($\square$).

[1]Present address: Department of Physics, University of Exeter, Exeter, EX4 4QL, UK.

$$\nu(T) = \nu_0 + AT\exp(-E_b/k_B T) \quad (1)$$

where the term ν_0 describes tunneling, and the second term involves thermal activation. The prefactor AT arises from the assumption that the latter process involves interaction with a single phonon of sufficient energy to kick the system over the energy barrier; other temperature dependences of the prefactor (including temperature independence) would also be consistent with the measurements. Using AT, we obtain $E_b/k_B = 2.4$K for all the electric fields used at 12 bar.

For higher pressures, this simple approach cannot be used because of the appearance of the dip in $\nu(T)$. We suggest that the effect may be due to phonons scattering off the vortex loop *as it tunnels.* Such a process would reduce the tunneling rate by a factor (9) of $\exp(-\phi\eta D^2)$ where η is a dissipation coefficient, D is the distance the vortex must tunnel through the barrier and ϕ is a numerical coefficient. This idea is closely related to a suggestion by MVD that roton emission at a rate τ_r^{-1} could decrease the tunneling rate by a factor of $\exp(-\phi m d D^2/\hbar\tau_r)$, where md is the mass of a vortex loop of length d; here, we are proposing that the phonon scattering rate should be included as well. Unlike the τ_r, which for our conditions depends only on E, τ_{ph} will depend not only on the scattering cross-section but also on the thermal phonon density: it will consequently be strongly temperature dependent. Assuming for simplicity that η varies as T^3, like the phonon number density, we would then expect

$$\nu(T) = \nu_0\exp(-BT^3) + AT\exp(-E_b/k_B T) \quad (2)$$

in place of Eq (1). The quantities E_b and B should be almost independent of E, but may be expected to vary smoothly with P.

It is found that Eq (2) gives a good fit to the data

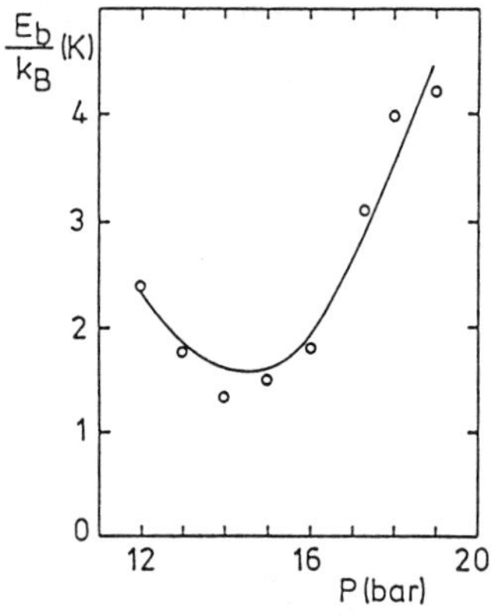

FIGURE 2
The energy barrier E_b deduced by fitting Eq (2) to the data, plotted as a function of pressure P.

at all pressures. At 12 bar, the constant A is relatively large, so that the thermally activated region occurs at a low enough temperature to mask the damping factor $\exp(-BT^3)$, with the consequences that no dip is seen. As the pressure increases, A decreases, the effect of the damping factor becomes relatively more pronounced, and the dip then appears. Full details of the fitting procedure will be given elsewhere (10).

The values of E_b derived in this way are plotted as a function of P in Fig 2. There appears to be a shallow minimum in $E_b(P)$ near 15 bar. The existence of the minimum depends on the values found at 12 and 13 bar, both of which are quite reliable. The minimum therefore appears to be a real, albeit somewhat puzzling observation; it was not predicted by MVD.

5. CONCLUSION

Our main conclusion is that the thermally activated process may be interpreted in terms of a phonon-assisted nucleation mechanism, exactly analogous to the roton-assisted process found previously (6) at higher pressures. The deduced barrier heights are consistent with MVD and confirm the conclusion that the initial vortex is a tiny loop rather than an encircling ring.

REFERENCES

(1) W. F. Vinen, in: Liquid Helium: Proceedings of the International School of Physics "Enrico Fermi", course XXI, ed. G. Careri (Academic Press, New York, 1963), p 336.
(2) P. C. Hendry, N. S. Lawson, C. D. H. Williams, P. V. E. McClintock and R. M. Bowley, Jpn. J. Appl. Phys. Suppl. 26, No 3 (1987) 73.
(3) P. C. Hendry, N. S. Lawson, P. V. E. McClintock, C. D. H. Williams and R. M. Bowley, Phys. Rev. Lett. 60 (1988) 604.
(4) C. M. Muirhead, W. F. Vinen and R. J. Donnelly, Phil. Trans. R. Soc. Lond. A 311 (1984) 433.
(5) P. C. Hendry, N. S. Lawson, C. D. H. Williams, P. V. E. McClintock and R. M. Bowley, in Elementary Excitations in Quantum Fluids, ed. K. Ohbayashi and M. Watabe (Springer-Verlag, Heidleberg, 1989) pp 184-188.
(6) R. M. Bowley, P. V. E. McClintock, F. E. Moss, G. G. Nancolas and P. C. E. Stamp, Phil. Trans. R. Soc. Lond. A 307 (1982) 201.
(7) P. C. Hendry and P. V. E. McClintock, Cryogenics 27 (1987) 131.
(8) R. M. Bowley, G. G. Nancolas and P. V. E. McClintock, Phys. Rev. Lett. 52 (1984) 659.
(9) A. O. Caldeira and A. J. Leggett, Ann. of Phys. 149 (1983) 374.
(10) P. C. Hendry, N. S. Lawson, P. V. E. McClintock, C. D. H. Williams and R. M. Bowley, to be published.

'hysica B 165&166 (1990) 759–760
'orth-Holland

FLOW STATE OF ^{3}HE IN ADIABATIC FLOW OF HEII

Takeo SATOH and Mineo OKUYAMA

Department of Physics, Faculty of Science, Tohoku University, Sendai 980, Japan

In adiabatic flow of HeII, the normal fluid components are forced to flow by the action of the superfluid component through vortices. The flow state of ^{3}He, added into the system, in the flow path was studied with the pulsed NMR method in the presence of field gradient. The results show that the flow is almost Poiseuille type in contrast with the highly turbulent state of the superfluid component. The method also directly gives the ^{3}He flow velocity in the flow path. The results show the existence of the characteristic velocity v_{c2} predicted by the new-type mutual friction force.

1. INTRODUCTION

It has long been noticed that the pulsed NMR technique applied at the position of the flow path, where field gradient presents, gives us various information about the flow state (1) such as the average flow velocity and the velocity profile of the flowing liquids. Furthermore, the method is used to distinguish whether the flow is laminar or turbulent. For liquid helium, the technique was applied to study the convective flow in ^{3}He-^{4}He mixtures (2), but has never been used to study the flow state of HeII. The main reason may be that it is not easy to maintain steady flow of ^{3}He in HeII. The dilute phase path in a ^{3}He-circulating dilution refrigerator may have steady ^{3}He flow in it. However, in that case, it may be just the flow of ^{3}He itself in HeII (3) and it is not certain whether it gives any information about the flow state of HeII.

In adiabatic flow of HeII, the normal fluid components flow by the action of the flow of the superfluid component via the mutual friction force transferred by vortices. Therefore, the study of the motion of ^{3}He may give information about the motion of vortices. In this sense, the ^{3}He-probe is very much like the ion-probe which has long been used to study the flow of HeII. The big advantageous aspect of the ^{3}He-probe with NMR technique is that it does not need any measuring assembly inside of the flow path, which may disturb the flow to study. The disadvantageous point of ^{3}He in adiabatic flow is that the concentration of ^{3}He in the flow path is not constant in time but decreases as a function of time. Fortunately, however, the time dependence is approximately expressed with a single exponential function as we have shown recently (4). So the attenuation of the spin-echo amplitude due to the concentration change in time can be taken into account to deduce the change of the spin-echo amplitude caused only by the flow.

2. EXPERIMENT

Experimental situation is schematically shown in Fig.1. We applied the crossed coil type pulsed NMR method. The flow path, which connects the chamber and the outlet connected to the inlet and outlet superleaks, respectively, consists of a Cu-Ni tube(1 mm i.d. and 100 mm length), a Stycast tube(1 mm i.d. and 300 mm length) and a Cu-Ni tube(1 mm i.d. and 100 mm length). The average ^{3}He concentration is about 6%.

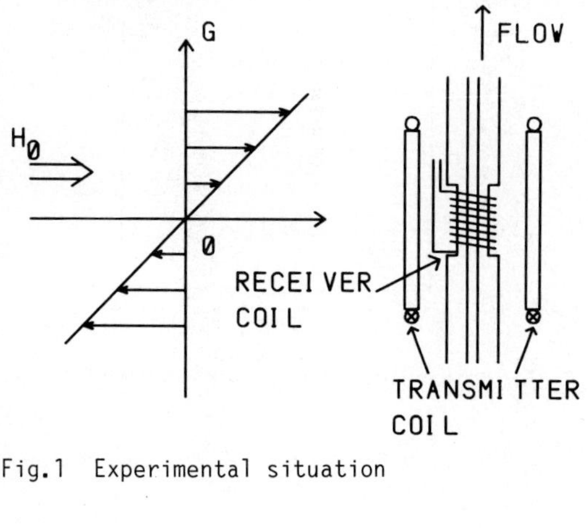

Fig.1 Experimental situation

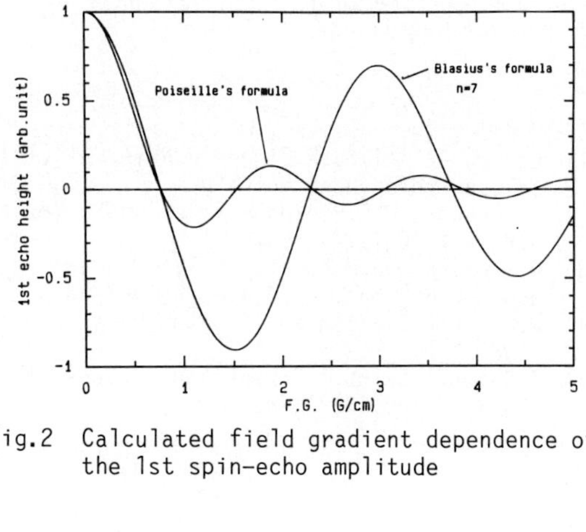

Fig.2 Calculated field gradient dependence of the 1st spin-echo amplitude

0921-4526/90/$03.50

3. RESULTS AND DISCUSSIONS

In Fig.2,we show the calculated field gradient dependence of the 1st spin-echo amplitude for the Poiseuille flow and the Blassius flow,in which we assume the same average velocity.Two points should be noticed.The first is that the first zero-crossing point is almost same for the both type flow.Therefore,we may obtain the average flow velocity of ^{3}He irrelevant to the flow state.The second is that the field gradient dependence shows remarkable difference between the two types of flow.So we may assign which type of flow is realized for ^{3}He.

Examples of the 1st spin-echo amplitude as a function of field gradient are shown in Fig.3. Here, the corrections due to (a) the spin diffusion in the presence of the field gradient, (b) the deviation from the resonance condition because of the presence of field gradient and (c) the ^{3}He concentration change due to flow are already made. In order to make the correction (c),the average flow velocity should be known, which is obtained from the first zero-crossing point as mentioned above.It is noted that the position of the zero-crossing point is irrelevant to these corrections.

In Fig.4, thus obtained average velocity of ^{3}He,v_3 is shown as a function of the average flow velocity of ^{4}He,v_4.It is clearly seen that the characteristic velocity v_{c2},at which v_3 starts increasing very rapidly,exists and its value is about 4 cm/s. It is noted that the methods of determining v_3 are completely different in the present experiment and in the previous experiment. Therefore,we are now fully convinced of the behavior of the normal fluid component velocity predicted with the new-type mutual friction force (5).

The solid curve in Fig.3 is a calculated one by assuming the Poiseuille type flow as in Fig.2. It is seen that the experimental points roughly lie on the curve.This means that the flow state of ^{3}He in adiabatic flow is almost Poiseuille type. This is rather surprising result,if we consider that the ^{3}He flow is produced by the highly turbulent superfluid component flow. Discussion will be given with the data of the 2nd spin-echo amplitude.

REFERENCES

(1) Review Article:J.R.Singer,J.Phys.E 11(1978) 281 and references therein.

(2) P.G.Luucas,D.A.Penman,A.Tyler and E.Vavasour J.Phys.E 10(1977)1150.

(3) J.Zeegers,J.G.M.Kuerten,A.T.A.M.de Waele and H.M.Gijsman,Jap.J.A.P.26(1987)63

(4) M.Okuyama,T.Satoh and T.Satoh,Physica B 154 (1988)116.

(5) T.Satoh,T.Satoh,T.Ohtsuka and M.Okuyama, Physica 146B(1987)379.

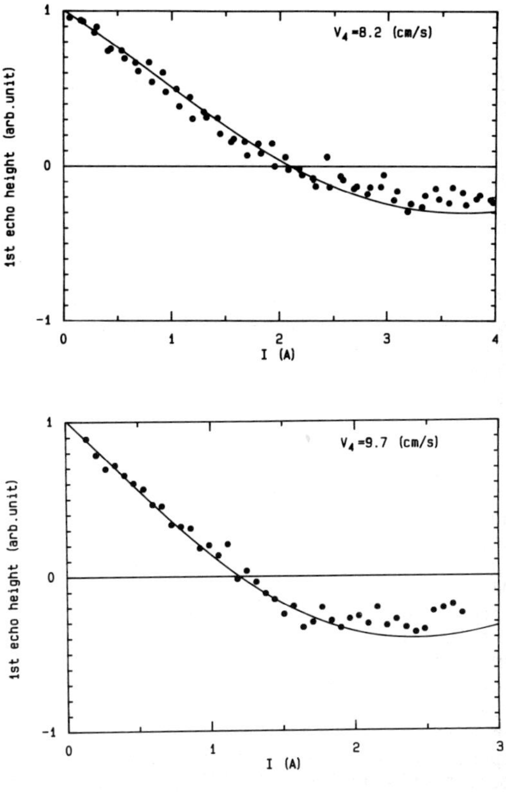

Fig.3 The 1st spin-echo amplitude as a function of field gradient at v_4=8.2 cm/s and 9.7 cm/s

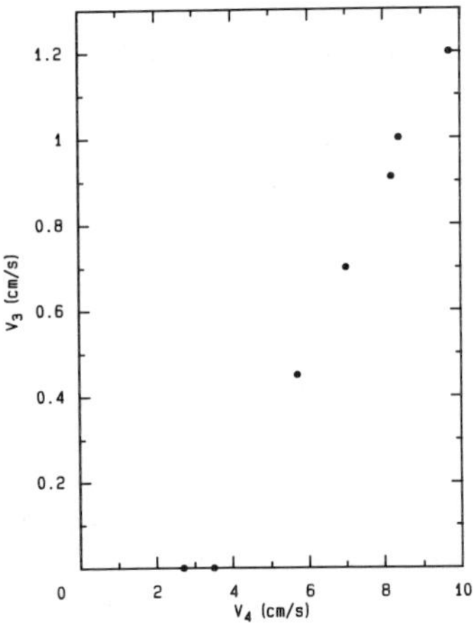

Fig.4 Average velocity of ^{3}He as a function of v_4

Physica B 165&166 (1990) 761–762
North-Holland

EVIDENCE FOR A LOW DIMENSION STRANGE ATTRACTOR FOR TEMPERATURE VARIATIONS INSIDE HELIUM-II COUNTERFLOW JETS*

Paul J. DOLAN, JR.[+] and Charles W. SMITH

Department of Physics and Astronomy, University of Maine, Orono, ME 04469 U.S.A.

Time series measurements of the temperature of liquid helium inside the chamber of thermal counterflow jets driven by a constant heat source show four distinct regions of fluid flow as a function of heater power. Detailed analysis suggests that the temperature variations for the lowest power oscillatory region may be characterized by a strange attractor of fractal dimension, ∿4.

1. INTRODUCTION

Thermal counterflow of helium II has been the subject of numerous experimental and theoretical studies (1-4) beginning with Kapitza's pioneer work on heat transfer in 1941 (5). This famous paper reports several observations including the displacement of a vane to monitor the flow of the normal fluid component at the nozzle of a thermal counter-flow jet and a transition to "erratic" flow above a fairly well defined critical heat flux. Our work was motivated in part by this historic paper, together with an interesting series of measurements in the mid-1970's on steady state thermal counterflow jets (6). The latter studies were able to conclude, among other things, that the jet plume into the isothermal bath, above a critical heat flux, was characterized by some type of turbulent velocity fluctuations. We report here an extension of this type of study to higher heater power while monitoring both the temperature inside the thermal counterflow jet chamber and the flow into the bath.

2. EXPERIMENTAL TECHNIQUES AND OBSERVATIONS

We employed thermal counterflow jets of cast acrylic: nominal chamber size, 2.5 cm in length by 1.0 cm in diameter and nozzle length 1.0 cm by 1.0 mm diameter (see insert, Figure 2). Each contained a 52.3 ohm metal film resistor, as a heater and a 3300 ohm carbon composite resistor as a thermometer. Time series measurements of the temperature of the liquid helium inside the jet chamber were digitally recorded by logging the current through the thermometer resistor at constant voltage as a function of time. Mass flow at the jet nozzle was indicated by the displacement of a sensitive microbalance vane located in the temperature controlled bath and viewed directly through a window in the dewar.

Four different regions of jet behavior were observed. In the lowest power range, a constant efflux from the jet was indicated as a constant deflection of the microbalance vane and a constant temperature inside the jet chamber. Kapitza's pioneer work on heat transfer in helium was carried out in this range. Above a critical power (nominally ∿25mW), the efflux from the nozzle and the temperature inside the jet chamber become oscillatory. Each decrease in temperature is in phase with a strong efflux from the nozzle, corresponding to out-going flow. This behavior is expected based upon conventional phenomenology for thermal counterflow (7). For a range of input power (up to about 50 mW) oscillations persist with the temperature inside the jet chamber never exceeding the lambda point. We therefore term this power range the superfluid region, S. As the power is increased further, again oscillations are observed, however the temperature in the jet chamber is above the lambda point but below the boiling temperature. The oscillations of the microbalance vane and variation of the temperature are not well correlated and show little variation in period with power. We term this power range the normal fluid region, N. If power is increased still further, the liquid helium around the heater is driven into the vapor phase.

3. RESULTS AND DISCUSSION

The remainder of this paper deals exclusively with the variation of the temperature inside the jet chamber in the superfluid region S, just above onset for oscillation. Figure 1 shows a typical plot of the temperature as a function of time. Each oscillation shows a slow warm-up interval, at the end of which flow from the nozzle appears restricted (the microbalance

*Supported by the National Science Foundation (USA), DMR-8312492, Low Temperature Physics Program.
+Present Address: Department of Physics, Northeastern Illinois University, Chicago, IL 60625 U.S.A.

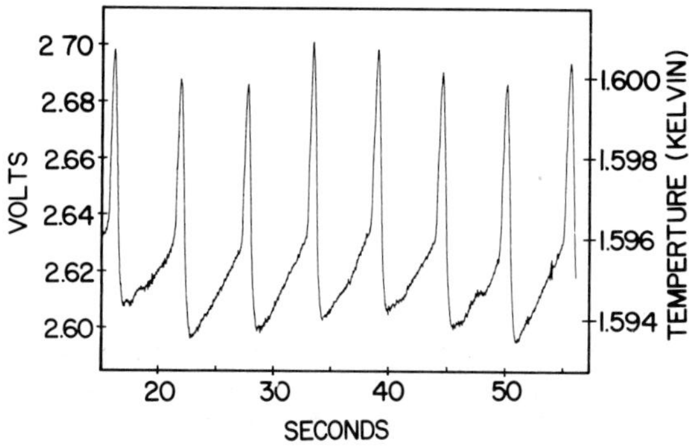

FIGURE 1
Temperature inside the jet chamber as a function of time. Bath temperature 1.77 K.

vane shows little displacement) and the temperature steeply rises. Flow then restarts as observed by a sharp temperature decrease caused by the inflow of superfluid component from the bath and a microbalance vane deflection away from the nozzle indicating outflow of normal fluid component. The cycle then repeats. As the input power is increased, the warm-up interval decreases, the magnitude of the temperature excursion increases and we observe a power range that will support continuous oscillations. The warm-up interval for fixed power monatonically increases with chamber volume, as expected.

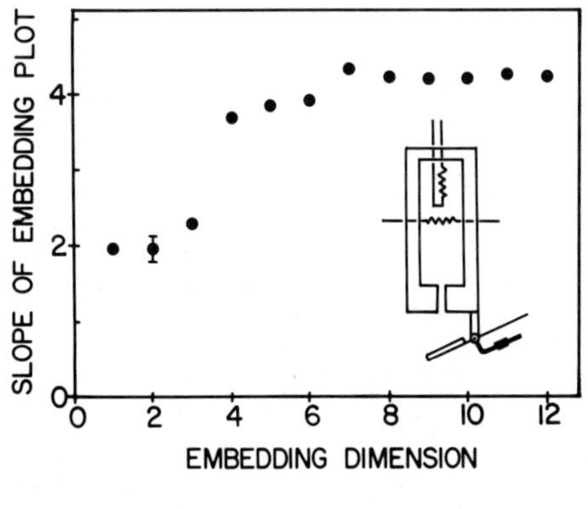

FIGURE 2
Slope from embedding plots versus embedding dimension. Insert shows thermal counterflow jet.

We have applied the techniques of nonlinear dynamics to attempt to characterize the temperature variations in the S region. Using the Grassberger-Procaccia algorithm (8), we have measured the fractal dimension of many time series records. The slopes for embedding plots (log of the correlation integral versus log of the correlation length) for embedding dimensions 1 thru 12 are shown in Figure 2. While the results for the N region exhibit substantial variation, the dimensionality for the S region consistently suggest behavior characterized by a strange attractor of low dimension.

4. CONCLUSION

We have investigated the oscillatory behavior of thermal counterflow jets operating under a range of conditions but primarily just above the constant flow region. Analysis of time series records of the temperature of the liquid helium inside the jet chamber suggests that the complicated oscillations observed can be characterized by a strange attractor of low fractal dimension, $\sim$4.

REFERENCES

(1) W.F. Vinen, Proc. Roy. Soc. A240 (1957) 114, 128; A242 (1957) 493; A243 (1957) 400.

(2) R.J. Donnelly, Proc. Roy. Soc. A281 (1964) 130; with P. H. Roberts, A312 (1969) 519; with C.E. Swanson, J. Low Temp. Phys. 61 (1985) 363; J. Fluid Mech. 173 (1986) 387.

(3) K.W. Schwarz, Phys. Rev. B18 (1978) 245; Phy. Rev. Lett. 49 (1982) 283; Phys. Rev. 1331 (1985) 5782.

(4) J.T. Tough, Prog. in Low Temp. Phys., ed. D.F. Brewer (North-Holland, Amsterdam) Vol. VIII (1982) 133.

(5) P.L. Kapitza, J. of Phys. USSR 4 (1941) 181.

(6) P.E. Dimotakis and J.E. Broadwell, Phys. Fluids 16 (1973) 1787; G.A. Laguna, Phys. Rev. B12 (1975) 4874; P.E. Dimotakis and G.A. Laguna, Phys. Rev. 1315 (1977) 5240.

(7) R.J. Donnelly and C.E. Swanson, J. Fluid Mech. 173 (1986) 387.

(8) P. Grassberger and I. Procaccia, Phys. Rev. Lett. 50 (1983) 346.

Physica B 165&166 (1990) 763–764
North-Holland

Josephson Oscillation in Superfluid Helium

Takeji KEBUKAWA*) and Shosuke SASAKI

Department of Physics, College of General Education, Osaka University, Toyonaka 560, Japan

A microscopic theory is presented which interprets the Josephson oscillation in superfluid helium in the light of the Heisenberg equations for dressed-bosons. The equations are exactly solved to yield the solutions in the form of a linear combination of two eigen-oscillations. Consequently it is shown that the interference of these eigen-oscillations produces the Josephson oscillation in superfluid helium, and that the Josephson oscillation occurs even when the quantum fluctuation of the relative phase is infinitely large. These results lead us to propose an experiment to detect the tunneling effect of the dressed-bosons. The formulation in this paper can be applied to the Josephson oscillation in superconductors.

1. INTRODUCTION

The research for the Josephson [1] effect has had a lot of progress to produce the wide application [2] to the technology. For superfluid helium, Richards and Anderson [3] discussed the ac Josephson effect in superfluid helium by using the order parameter in Beliave's paper [4]. Many later works [5] have been carried out on the basis of their work. The order parameter is defined by the approximate expectation value of an annihilation operator a_o of a helium atom. However the expectation value by the eigenstate with a definite number of helium atoms should become zero. This means that it is necessary to improve their theory of the ac Josephson effect by finding out the analogue as the Cooper-pair.

We can find this analogue in the theory [6,7] (hereafter referred to as I and II) of liquid helium recently developed by Sasaki. There have been introduced the dressed-bosons by the unitary transformation which diagonalizes the total Hamiltonian of liquid helium. The dressed-boson is the analogue of the Cooper-pair. In this paper we will develop a theory of the ac Josephson effect on the basis of the concept of dressed-bosons.

2. JOSEPHSON OSCILLATION

We consider such a system as illustrated in Figure, where two superfluid-heliums in each glass tubes are weakly coupled to each other by a small orifice. Initially the orifice is closed so that both superfluid-heliums are in equilibrium of temperature T_1 (T_2) and pressure P_1 (P_2). At a time the orifice is opened to weakly couple both superfluid-heliums to each other. Then our problem is to know what phenomena occur after this setting.

According to I and II the liquid helium is described by the concept of dressed-bosons. The number of a dressed-boson with momentum **p** is defined by $n_p \equiv A^*_p A_p$ in terms of its creation and annihilation operators. These operators are connected to the creation (a^*_p) and annihilation (a_p) operators of a helium atom as $A^*_p = U a^*_p U^*$ and $A_p = U a_p U^*$ by the unitary transformation U, which diagonalizes the total Hamiltonian of liquid helium. When the orifice is opened, there occurs an interaction between two superfluid-heliums. Representing the interaction in terms of dressed-bosons, we have

$$H_{int} = \sum_{p,q} g_{p,q} (A^*_{1p}A_{2q} + A^*_{2q}A_{1p}) + \text{(multi-operator terms)},$$

where we have taken the number conservation of dressed-bosons into account, and have used the suffix 1 or 2 according to whether each quantities are in the first superfluid-helium or the second one. The quantity $g_{p,q}$ is a coupling constant and its magnitude is proportional to the small cross section of the orifice. The coupling constants in the multi-operator terms have the magnitude of the higher power to be neglected as compared with the first term.

When the orifice is closed, both systems are in their equilibrium with no flow of superfluid. In this case, as has been shown in I and II, there occurs the condensation of dressed-bosons at zero momentum, that is, the number $\bar{n}_{10}$ and $\bar{n}_{20}$ in equilibrium take macroscopic values. This means that the matrix element of $A^*_{10}A_{20} + A^*_{20}A_{10}$ takes a large value. Furthermore the dressed-bosons with zero momentum are able to pass through the small orifice without friction, but the other dressed-bosons are hard to pass through it due to their viscosity. Therefore the dominant interaction proves to be $g_{00}(A^*_{10}A_{20} + A^*_{20}A_{10})$.

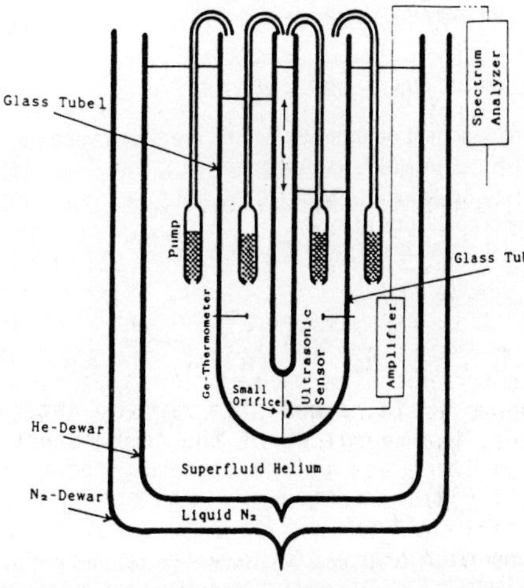

When the orifice is opened, we have small deviations $\delta\bar{n}_{1o}$ and $\delta\bar{n}_{2o}$ from their equilibrium values $\bar{n}_{1o}$ and $\bar{n}_{2o}$. The dynamical properties of these deviations can be discussed on the basis of an effective Hamiltonian in the following way. These deviations are supposed to be so small that we can neglect the second and higher order terms in the expansion of the total energy $E_j(\{\bar{n}_{j,q}\})$ in terms of these deviations. Then the excess energy due to these deviations is given by $(\delta E_j/\delta\bar{n}_{j,o})\,\delta\bar{n}_{j,o}$ (j=1,2).

Replacing $\delta\bar{n}_{jo}$ by the operator form $A^*_{jo}A_{jo} - \bar{n}_{jo}$, it is represented as const $+ (\delta E_j/\delta\bar{n}_{j,o})\,A^*_{jo}A_{jo}$. We can use this as an effective Hamiltonian under the restriction that the expectation value of the number operator $A^*_{jo}A_{jo}$ should be close to the average value n_{jo}. Then the effective Hamiltonian which describes the dynamical properties of those deviations becomes

$$H = \mu_1 A^*_{10}A_{10} + \mu_2 A^*_{20}A_{20} + g_{00}\,(A^*_{10}A_{20} + A^*_{20}A_{10}), \quad (1)$$

where μ_j denote chemical potentials and we have made use of the results $\delta E_j/\delta\bar{n}_{j,o} = \mu_j$ from I and II.

*) a temporary present address; Department of Chemistry, Baker Laboratory, Cornell University, Ithaca, New York 14853, USA

The Heisenberg equations for $A_{10}(t)$ and $A_{20}(t)$ become

$$i\hbar \frac{dA_{10}(t)}{dt} = [A_{10}(t), H] = \mu_1 A_{10}(t) + g_{00}A_{20}(t), \quad (2)$$

$$i\hbar \frac{dA_{20}(t)}{dt} = [A_{20}(t), H] = \mu_2 A_{20}(t) + g_{00}A_{10}(t), \quad (3)$$

from the effective Hamiltonian (1), where $\hbar$ is (Planck's constant)/(2π). The coupled equations (2) and (3) can be solved to yield the solutions

$$A_{10}(t) = e^{-i\nu_+ t/\hbar} B_+ + e^{-i\nu_- t/\hbar} B_-, \quad (4)$$

$$A_{20}(t) = \frac{\nu_+-\mu_1}{g_{00}} e^{-i\frac{\nu_+ t}{\hbar}} B_+ + \frac{\nu_--\mu_1}{g_{00}} e^{-i\frac{\nu_- t}{\hbar}} B_-, \quad (5)$$

where the eigen-energies ν_+ and ν_- are given by

$$\nu_+ = (\mu_1+\mu_2)/2 + \sqrt{((\mu_1-\mu_2)/2)^2 + g_{00}^2}, \quad (6)$$

$$\nu_- = (\mu_1+\mu_2)/2 - \sqrt{((\mu_1-\mu_2)/2)^2 + g_{00}^2}. \quad (7)$$

The time-independent operator B_+ and B_- are determined as

$$B_+ = [-(\nu_- - \mu_1)A_{10}(0) + g_{00}A_{20}(0)]/(\nu_+ - \nu_-), \quad (8)$$

$$B_- = [+(\nu_+ - \mu_1)A_{10}(0) - g_{00}A_{20}(0)]/(\nu_+ - \nu_-), \quad (9)$$

from the initial conditions at $t = 0$.

From the equations (4) and (5), we have

$$n_{10}(t) \equiv A_{10}^\dagger(t)A_{10}(t)$$

$$= B_+^\dagger B_+ + B_-^\dagger B_- + \left(e^{i\frac{\nu_+-\nu_-}{\hbar}t} B_+^\dagger B_- + e^{-i\frac{\nu_+-\nu_-}{\hbar}t} B_-^\dagger B_+\right), \quad (10)$$

$$n_{20}(t) \equiv A_{20}^\dagger(t)A_{20}(t) = ((\nu_+-\nu_-)/g_{00})^2 B_+^\dagger B_+ + ((\nu_--\mu_1)/g_{00})^2 B_-^\dagger B_-$$

$$+ (\nu_+-\mu_1)(\nu_--\mu_1)/g_{00}^2 \left(e^{i\frac{\nu_+-\nu_-}{\hbar}t} B_+^\dagger B_- + e^{-i\frac{\nu_+-\nu_-}{\hbar}t} B_-^\dagger B_+\right). \quad (11)$$

By noting that $(\nu_+-\mu_1)(\nu_--\mu_1)/g_{00} = -1$ from Eqs. (6) and (7), the time dependent terms in Eqs. (10) and (11) cancel out to yield the number conservation

$$A_{10}^\dagger(t)A_{10}(t) + A_{20}^\dagger(t)A_{20}(t) = A_{10}^\dagger(0)A_{10}(0) + A_{20}^\dagger(0)A_{20}(0).$$

These results (10) and (11) show that the oscillating terms in $n_{10}(t)$ and $n_{20}(t)$ come from the interference between the two eigen-oscillations.

We will calculate the superfluid current in the case of the following two initial states, and will show that the interference causes its oscillation.

2.1. Case of a definite number distribution

The initial state with a definite number distribution $\{n_{10}, n_{20}, n_{1p}, n_{2q}\}$ of dressed-bosons is represented by

$$|\psi_H\rangle = C_H (A_{10}^\dagger(0))^{n_{10}} (A_{20}^\dagger(0))^{n_{20}} \prod_{p\neq 0} (A_{1p}^\dagger(0))^{n_{1p}} \prod_{q\neq 0} (A_{2q}^\dagger(0))^{n_{2q}} |0\rangle,$$

where C_H is the normalization constant and $|0\rangle$ is the vacuum state. The current J of the dressed-bosons, then, becomes

$$J \equiv -\frac{d}{dt}\langle\psi_H|n_{10}(t)|\psi_H\rangle = 2\frac{g_{00}^2}{\hbar(\nu_+-\nu_-)}(n_{10}-n_{20}) \sin\left(\frac{\nu_+-\nu_-}{\hbar}t\right), \quad (12)$$

where J indicates the current from the tube 1 to the tube 2.

2.2. Case of a coherent state

Adopting the coherent state as our initial state, we have

$$|\psi_c\rangle = C\, e^{\alpha_{10}A_{10}^\dagger(0)} e^{\alpha_{20}A_{20}^\dagger(0)} \prod_{p\neq 0} e^{\alpha_{1p}A_{1p}^\dagger(0)} \prod_{q\neq 0} e^{\alpha_{2q}A_{2q}^\dagger(0)} |0\rangle$$

where α_{jp} is a complex amplitude of A_{jp} and C is a normalization constant.

The current of dressed-bosons from tube 1 to tube 2 is

$$J = -\frac{d}{dt}\langle\psi_c|n_{10}(t)|\psi_c\rangle = \frac{\nu_+-\nu_-}{\hbar}\left[\eta_1 \sin\left(\frac{\nu_+-\nu_-}{\hbar}t\right) - \zeta_1 \cos\left(\frac{\nu_+-\nu_-}{\hbar}t\right)\right]. \quad (13)$$

$$\zeta_1 = -\frac{2g_{00}\sqrt{(\mu_1-\mu_2)^2 + 4g_{00}^2}}{(\nu_+-\nu_-)^2}\sqrt{\bar{n}_{10}\bar{n}_{20}}\, \sin(\varphi_{10}-\varphi_{20}),$$

$$\eta_1 = \frac{-2g_{00}(\mu_1-\mu_2)}{(\nu_+-\nu_-)^2}\sqrt{\bar{n}_{10}\bar{n}_{20}}\, \cos(\varphi_{10}-\varphi_{20}) + \frac{2g_{00}^2}{(\nu_+-\nu_-)^2}(\bar{n}_{10}-\bar{n}_{20}),$$

where we have represented α_{jp} as $\alpha_{jp} = \sqrt{\bar{n}_{jp}}\, e^{i\varphi_{jp}}$.

If we neglect the second and higher order terms of g_{00}, we get approximately

$$J = -\frac{2g_{00}}{\hbar}\sqrt{\bar{n}_{10}\bar{n}_{20}}\, \sin\left[\left(\varphi_{10} - \frac{\mu_1}{\hbar}t\right) - \left(\varphi_{20} - \frac{\mu_2}{\hbar}t\right)\right],$$

which has the same form as the Josephson current in superconductor, and oscillates with the frequency $(\mu_1-\mu_2)/(2\pi\hbar)$. However it should be noted here that the exact result for the current J is given by eq. (13), which has a frequency $(\nu_+-\nu_-)/(2\pi\hbar)$.

From the eqs. (12) and (13), we can see that the oscillation of the superfluid current originates in the interference of the two eigen-oscillations and that the Josephson frequency has no dependence on the relative phase. In our theory, thus, the Josephson oscillation occurs even when the quantum fluctuation of the relative phase is infinitely large as in the case of 2.1..

3. PROPOSAL OF EXPERIMENT

In the previous sections we have discussed that the tunnelling effect of the dressed-bosons yields the oscillating current of superfluid. Here we propose an experiment to directly detect the tunnelling effect. The experimental apparatus is illustrated in Figure. By using pumps which work due to the fountain effect, the level heights of superfluid helium in each glass tubes are controlled to have a difference l. Then the difference of chemical potentials at a small orifice becomes mgl. Thus we get the oscillation with a frequency $f = (mgl)/(2\pi\hbar)$, because the coupling constant g_{00} is negligibly small. This oscillation is transformed into electric signals by an ultrasonic sensor. The electric signals are amplified by an amplifier which works in superfluid to avoid heat noises. Analyzing the output signals by a spectrum analyzer, we will be able to observe and confirm the frequency $f = (mgl)/(2\pi\hbar)$ of the oscillation.

The confirmation tells us that the carriers of the superfluid current have the same mass as helium atoms. Moreover an eigenstate of superfluid helium must be represented by the carriers in the limit of an infinitely small orifice. These carriers are nothing but the condensed dressed-bosons, which have the same mass as helium atoms and represent an eigenstate of superfluid helium.

ACKNOWLEDGMENTS

One of the authors (T.K.) would like to express his sincere thanks to Professor B. Widom and all members of his group in Baker Laboratory for his encouragement and their help.

REFERENCES

1. B.D. Josephson, Phys. Letters 1 (1962) 251.
2. R.D. Parks: Superconductivity (Dekker, 1969) Vols. I & II.
A. Barone and C. Paterno: Physics and Applications of the Josephson Effect (Wiley, 1982).
H. Ohta, SQID and Appl. (1977) 35.
3. P.L. Richards and P.W. Anderson, Phys. Rev. Lett., 14 (1965) 540.
4. S.T. Beliave, JETP 34 (1958) 289.
5. B.P. Beecken and W. Zimmermann, Jr., Phys. Rev. B35 (1987) 74. This paper has enough references to see many other works.
6. S. Sasaki, Jpn. J. Appl. Phys. Suppl. Proc. LT18, 26 (1987) 23.
7. S. Sasaki, Sci. Rep. Col. Gen. Educ. Osaka Univ. 35-2 (1986) 1.

Physica B 165&166 (1990) 765–766
North-Holland

PROGRESS TOWARDS A SUPERFLUID ^{4}He ANALOG OF THE SUPERCONDUCTING rf SQUID

Michele Bonaldi, Massimo Cerdonio and Stefano Vitale

Dipartimento di Fisica, Universita' di Trento and Istituto Nazionale di Fisica Nucleare, sezione di Trento, I-38050 Povo, Trento, Italy

We are working to realize a superfluid ^{4}He analog of the superconducting rf SQUID to be used as a gyroscope. As a first step we developed an experimental method to induce an ac superflow across a 5 μm radius orifice by using a rotation. The orifice was supported by a septum placed inside a hollow torus filled with liquid helium and the flow was induced by rotating the whole torus, that was the inertial member of a torsional oscillator. We report here measurements of the critical velocity for the orifice, obtained by detecting the occurrence of a dissipation in the oscillator's motion. The energy and the number of circulation quanta involved in the dissipative process are evaluated.

1. INTRODUCTION

The behaviour of a ^{4}He superfluid (HeII) analog of the superconducting rf SQUID was previously analyzed [1]. The system, a hollow torus filled with HeII and interrupted by a Josephson weak link , was predicted to be highly sensible to the rotation of the laboratory in respect to the local inertial frame. The Josephson junction had the only function to ensure a reproducible critical angular velocity for quantized energy dissipation. As a Josephson junction is probably impossible to realize in HeII, we proposed in substitution to use in the torus a septum with a submicrometric orifice as a weak link [2].

In an experiment we now demonstrate the existence of a reproducible critical angular velocity Ω_c for the HeII contained in a hollow torus of radius R=7 mm, interrupted by a septum with an orifice of radius $r = 5\ \mu m$. We discuss how much smaller should be the orifice to allow us to observe the SQUID effect.

2. THEORY OF OPERATION

A rotation of the torus induces in the orifice a superflow proportional to the angular velocity Ω_t. The flow due to the circulation quantum n in the torus is symply superposed to this rotationally induced flow. The total flow velocity in the orifice is:

$$v_c = 2(nk - 2\pi R^2 \Omega_t)/(\pi r) \qquad (1)$$

where as usual $k = h/m_{He}$. The HeII flows in the orifice without friction up to a critical velocity v_c characterizing the orifice, at which dissipation sets in. For each value of n the states described by Eq. 1 are metastable provided that:

$$-v_c < v_h < v_c. \qquad (2)$$

When $v_h = v_c$ the state becomes unstable and vortex rings appear in the hole. During their evolution the vortex core interacts with the normal fluid, that is at rest with the boundary, allowing the exchange of energy and angular momentum with the torus. When the vortex finally annihilate, the superfluid is left in a different circulation state [3]. The state n will be a metastable state of the system only for the angular velocities Ω_t satisfying the relation:

$$(n - \frac{v_c r \pi}{2k}) < \Omega_t(\frac{2\pi R^2}{k}) < (n + \frac{v_c r \pi}{2k}). \qquad (3)$$

At the two limiting values $\pm\Omega_c(n)$ the system undergoes a transition to another circulation state with a value of n for which Eq. 4 still holds. The number of different circulation states where the system can go at $\pm\Omega_c$ is $N_c = v_c r\pi/k$. If ω_t is swept between different values a hysteresys may occur with a related energy dissipation. The amount of dissipated energy depends on the jump Δn that the circulation quantum undergoes in a transition. N_c sets obviously an upper limit for Δn, as $\Delta n \leq N_c$.

3. EXPERIMENTAL APPARATUS

The experimental set-up will be described in more details elsewhere [3]. In brief the aluminum torus used in the experiment was the inertial member of a torsional pendulum with quality factor Q: by driving the pendulum at its resonance $\omega_0 \approx 30$ Hz we made the torus to oscillate along its axis and we could modulate Ω_t. To stop the preexisting superfluid vorticity the filling line outlet toward the torus was guarded by a porous filter and an open cell foam filled the torus except for a 5 mm zone on both side of the septum. The momentum of inertia I_t of the empty torus and of the helium inside were respectively 1.5 $g - cm^2$ and 0.09 $g - cm^2$.

An oscillating forcing torque was appied to the oscillator by an electrostatic transducer. The angular displacement of the torus was measured by detecting with a superconducting SQUID the position of two superconducting disks attached to the torus and moving in a uniform magnetic field. The resulting angular resolution was calculated to be 5×10^{-7} rad/(s $\sqrt{Hz}$).

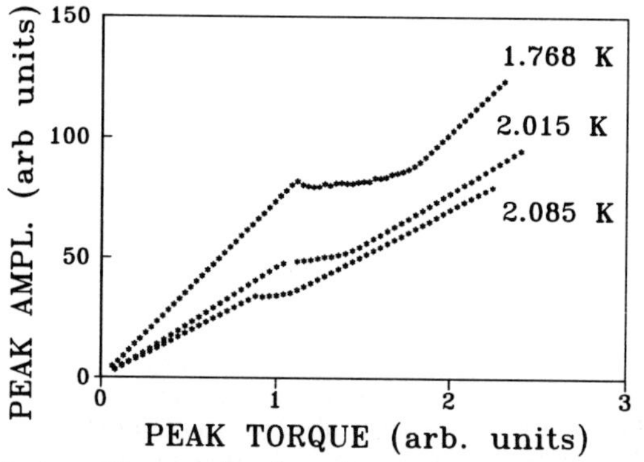

Fig.1. The amplitude unit corresponds to 1 μrad with a 30% accuracy, the torque unit corresponds to 10^{-9} $N-m$ with a 50% accuracy.

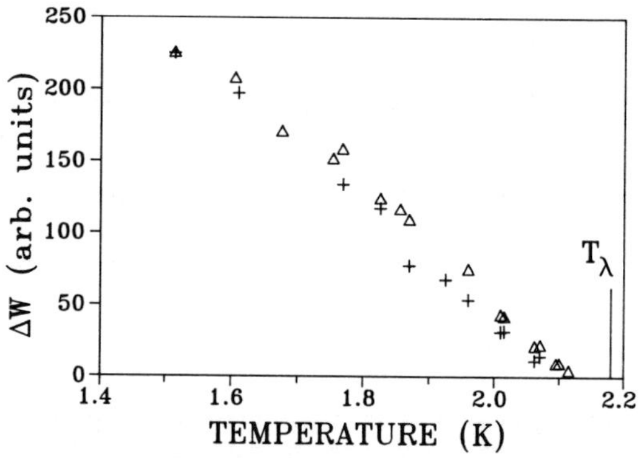

Fig.2. Dissipated energy ΔW vs. T (K); circles were calculated from the plateau amplitude, triangles from the decay time variation. The energy unit corresponds to $10^{-15}J$ with an estimated calibration error of 50%.

4. RESULTS

Typical experimental results of the oscillation peak amplitude θ_A as a function of the oscillating peak torque T_A at different temperatures are reported in Fig. 1. When the torus oscillates at low amplitude, θ_A is a linear function of T_A in agreement with the standard theory of the forced resonator: $\theta_A = T_A Q/(I_t\omega_0^2)$. At each temperature a plateau is then clearly recognizable: in this plateau region the torus acts no more as a linear passive device and θ_A is almost independent on T_A. The starting point of the plateau θ_c was observed to be highly reproducible, within the experimental errors, both within the same run and after cycling the system at 4.2 K. Then comes a supracritical region, where the oscillation amplitude increases again monotonically as a function of the drive.

For each value of θ_c the corresponding value of the flow velocity in the orifice can be calculated from Eq. 3 using $\Omega_t = \omega_0\theta_c$ in the hypothesys $n = 0$ [3]:

$$v_c\theta_c = (4\omega_0 R^2)/r = (7 \pm 3) \times 10^3\ m(s\ rad) \quad (4)$$

The temperature dependence of the critical velocities up to 2.11 K were measured. We found the typical values (from 0.15 to 0.9 m/s) and temperature behaviour of the intrinsic critical velocities for a micrometric orifice. Data taken for T≤1.9 K display a linear dependence on the temperature and extrapolate to zero at 2.65 K, confirming the results obtained with the Helmholtz resonator technique [4]. The high temperature data, 1.9≤T≤2.115 K, tend to zero at T_λ, in agreement with the result of gravitationally induced dc flow experiment.

The dissipative behaviour reveals itself in the plateau region as the occurrence of sudden collapses of the oscillation amplitude, while in the supracritical region only a steady eccess dissipation is recognizable. At each temperature value we calculated the excess dissipation ΔW per cycle in the supracritical region (Fig. 2) as $\Delta W = \pi\theta_c\Delta T$, where ΔT is the width of the plateau along the torque axis. We also calculated ΔW by an independent method based on the difference of the free oscillation decay times in the subcritical and supracritical regions. These values of ΔW are consistent with an energy dissipation corresponding to multiple circulation transitions in the torus, $30 > \Delta n > 500$ depending on temperature.

5.CONCLUSIONS

Jumps $\Delta n \approx 1$ in the circulation quantum are needed at Ω_c in order to observe in the θ_A *vs.* T_A plot the analogous of the staircase pattern of the superconducting rf SQUID. Any slow superimposed angular velocity would then modulate periodically such a pattern again in analogy with the SQUID. Though this second phenomenon has never been observed a staircase pattern has been reported in an experiment using a submicrometric orifice in a Helmholtz resonator [5].

In order to detect the SQUID behaviour we are now working to include a submicron orifice in our torus. In fact $N_c = v_c r\pi/k$ sets an upper limit for Δn and to obtain $N_c \approx 1$ at 2 K we should reduce r to 0.2 μm. We could eventually change v_c and N_c by operating at different temperatures. Note also that the evolution time τ of the vortices reduces with the temperature, and the occurrence of multiple transitions $\Delta n \gg 1$ should be inhibited when τ is shorter than the flow oscillation period. In fact the staircase pattern has been obtained only in the lowest frequency ($\nu_0 \approx 2\ Hz$) Helmholtz resonator though also in this case N_c was as large as $N_c \approx 26$.

REFERENCES

[1] M.Cerdonio, S.Vitale, Phys.Rev. B 29 1, 481 (1984)
[2] M.Bonaldi et al., Proc of the IV Marcel Grossman Meeting on General Relativity, R.Ruffini ed. (Elsevier, 1986) p. 1306
[3] M.Bonaldi, Ph.D. Thesis, 1990; M.Bonaldi et al., to be published
[4] B.P.Beecken, W.Zimmerman, Jr, Phys.Rev. B 35, 1630 (1987); E.Varoquaux, M.W.Meisel, O.Avenel, Phys.Rev.Lett. 57, 2291 (1986)
[5] O.Avenel, E.Varoquaux, Phys.Rev.Lett. 60, 416 (1988)

Physica B 165&166 (1990) 767–768
North-Holland

^{3}He-INDUCED FREE VORTICES IN THIN SUPERFLUID FILMS

D. Finotello[a,b], Y. Y. Yu[b,c], F. M. Gasparini[a]

a) Department of Physics, SUNY at Buffalo, Buffalo, NY, U.S.A.
b) Department of Physics, Kent State University, Kent, OH, U.S.A.
c) Department of Physics, Syracuse University, Syracuse, NY, U.S.A.

We report analysis of measurements of the thermal conductance of ^{3}He-^{4}He mixture films near the Kosterlitz-Thouless transition. It is found that on the superfluid side of the transition our data can be understood by the existence of free vortices induced by the addition of ^{3}He to ^{4}He films.

In 2D helium films near the superfluid transition, heat is transported via a convective mechanism that involves mass transport across the opposite ends of the experimental cell. The superfluid film, driven towards the hot end, evaporates by absorbing latent heat. A pressure gradient drives the gas to the cold end where it recondenses liberating its latent heat. By solving [1,2] the appropriate hydrodynamic equations it has been shown that

$$K_f = Fl^2(W/L)(h/m)^{-2}(a^2k_B/D)\exp(4\pi t^{-1/2}/b) \quad (1)$$

and

$$K_g = l^2(W/L)\rho_g^2 d^3/(24\eta T) \quad (2)$$

with W/L the perimeter for film flow, l=TS_g the latent heat, S_g being the entropy, D the vortex diffusion constant, a the vortex core diameter, b a non-universal constant, $t \equiv T/T_c-1$ the reduced temperature, ρ_g and η the density and viscosity of the gas, d the refluxing channel width, K_f and K_g the superfluid and the gas conductances, and n_f the free vortex density. The constant F reflects the proportionality between the vortex density and the correlation length,

$$n_f \sim \xi^{-2} = a^{-2}\exp(-4\pi t^{-1/2}/b) \quad (3)$$

Measurements of the thermal conductance of helium films provide information on the correlation length as well as on the number of free vortices present in the film. Since for $T<T_c$ vortices are paired, $K_f^{-1}=0$. In a measurement, to get heat in and out of the helium, one has to overcome a boundary resistance K_k^{-1} in series with the convective film mechanism. One must also allow for the parallel heat conduction through the structural materials of the cell, the diffusive conduction of the gas and that of the film. If K_b represents this small background conductance, the measured conductance is then given by [3,4]

$$K_m = K_b+(1/K_f+1/K_g+1/K_k)^{-1} \quad (4)$$

which for $T<T_c$ reduces to

$$K_m = K_u = (1/K_g+1/K_k)^{-1} \quad (5)$$

with K_u the largest measurable conductance dependent on the experimental cell geometry.

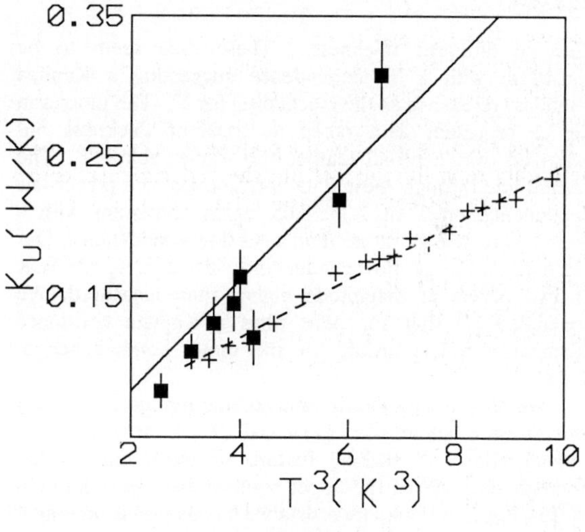

FIGURE 1
Full symbols, the highest value of measured conductance, K_u, for ^{4}He films of different thickness. Plusses, the temperature dependence of K_u for a thick film. The solid line is drawn to guide the eye. The dashed line is a power law fit.

We performed thermal conductance measurements on superfluid helium films whose thickness ranged from 11.7 to 36.5 A. Thicker films up to 156A were studied on a different cell and are discussed elsewhere [4]. Values for the upper conductance, K_u, as function of temperature are shown in Fig. 1. The symbols with the error bars represent K_u for ^{4}He

* Supported by NSF Grants DMR-8601848 and DMR-8905771

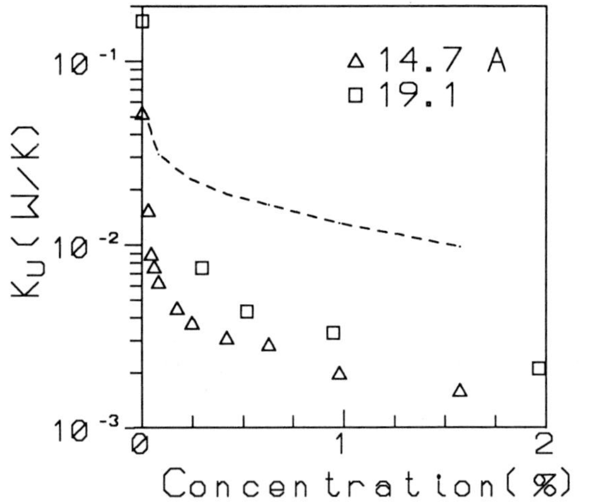

FIGURE 2
Dependence of the measured upper conductance on the percent ^{3}He concentration. Dashed line is a calculation for the 14.7 A mixture films including the scattering mechanism. See text.

films of different thickness. These data seem to be consistent with a T^3 dependence suggesting a Kapitza boundary resistance as the mechanism for K_u. The plusses in Fig. 1 represent data taken at constant thickness but extended over a broad temperature range below T_c. The dashed line through these data corresponds to a power-law dependence on T of 2.35±0.05, again consistent with a boundary resistance rather than a gas flow conductance. Our estimate for this gas flow conductance (Eq. 2) is K_g=28 W/K or two orders of magnitude higher than measured. We concluded [3] that for pure films a Kapitza resistance mechanism is responsible for the upper conductance at $T<T_c$.

We also studied the thermal conduction mechanism for two different helium films (14.7 and 19.1 A) at several ^{3}He concentrations. A striking feature of these data is the dramatic decrease of the upper conductance upon addition of ^{3}He, Fig. 2. A boundary resistance mechanism is no longer consistent with our data. Measurements with bulk mixtures [5] have shown that the Kapitza conductance is not affected by small amounts of ^{3}He. Changes in a Kapitza conductance mechanism could be expected if the added ^{3}He replaced the inert layer of ^{4}He strongly bound to the substrate. This is energetically unfavorable.

A possible mechanism for the decrease of the upper bound is the behavior of ^{4}He in the vapor. Given that most of the added ^{3}He resides in the vapor [4], it will impede the free refluxing of the ^{4}He gas, scattering the ^{4}He atoms. When this ^{3}He-^{4}He scattering mechanism is included we find

$$K_g(x) = K_{g0}[12/(d\alpha)^2 - 24\tanh(d\alpha/2)/(d\alpha)^3] \qquad (6)$$

Thus, K_g, through α, is concentration dependent. Assuming that no free vortices are present for $T<T_c$ we write

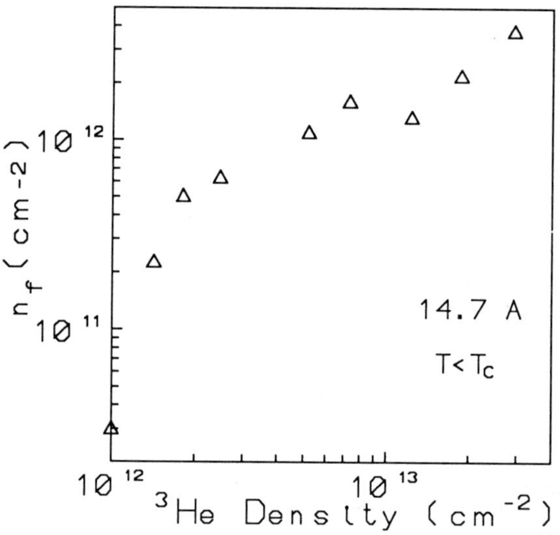

FIGURE 3
Number of induced free vortices for $T<T_c$ as a function of the ^{3}He number density.

$$K_u(x)/K_u(0) = (K_k/K_g(x)+1)^{-1} \qquad (7)$$

with $K_u(0)=K_k$ the upper conductance in the pure film case. The calculated conductance for the 14.7A mixtures is indicated by the dashed line in Fig. 2. The measured conductance values are clearly lower than what is expected from the scattering mechanism. This difference is most likely due to the assumption of zero free vortices. We thus suggest that besides the scattering mechanism, the addition of ^{3}He induces free vortices on the superfluid side of the transition. From the difference between the calculated and measured data in Fig. 2 we calculate the density of free vortices n_f in the film. This is plotted as a function of the number density of ^{3}He in Fig. 3. Note that for the mixture films studied, n_f is about one order of magnitude smaller than the ^{3}He number density. We conclude that the decrease in the largest measured conductance is due to the gas scattering mechanism together with the generation of free vortices in the film.

REFERENCES
(1) V. Ambegaokar, B. I. Halperin, D. R. Nelson, and E. D. Siggia Phys. Rev. B21, 1806 (1980)
(2) S. L. Teitel, J. Low Temp. Phys. 46, 77 (1982)
(3) D. Finotello and F. M. Gasparini, Phys. Rev Lett. 55, 2156 (1985)
(4) D. Finotello, Y. Y. Yu, and F. M. Gasparini, Phys. Rev. Lett. 57, 843 (1986); to appear in Phys. Rev. B.
(5) M. Dingus, F. Zhong, and H. Meyer, J. Low Temp. Phys. 65, 185 (1986).

Physica B 165&166 (1990) 769–770
North-Holland

VORTICES AND FINITE-SIZE SCALING OF SUPERFLUID PHASE TRANSITIONS*

Gary A. WILLIAMS

Physics Department, University of California, Los Angeles, CA 90024, USA

A vortex-ring theory of the three-dimensional XY model is compared with recent Monte Carlo simulations. Good agreement with the helicity modulus (superfluid density) is obtained when the recursion relations are truncated at a maximum ring diameter of about one-half the size of the Monte Carlo lattice. Near T_c the vortex theory is shown to satisfy the predictions of finite-size scaling.

Vortex theories of the superfluid transition in two and three dimensions (2D and 3D) are able to provide an intuitive and physical picture of the transition[1,2]. The real space renormalization method used makes finite-size effects particularly simple to treat[1], where the recursion relations are terminated at the scale of the largest vortex excitation that can fit into the size of the system. In 2D, where the excitations are vortex pairs, this procedure[3] has been verified by direct numerical simulations of helium films[4]. In the present paper we show that a similar procedure for the 3D vortex ring theory gives good agreement with recent Monte Carlo simulations of the 3D XY model[5], and that near T_c the theory satisfies finite-size scaling in detail.

The ring model of Ref. 1 is utilized, but incorporating the modifications proposed by Shenoy[2] to take into account non-circular rings. The recursion relations in this case are

$$\frac{\partial(1/K)}{\partial \ell} = -\frac{1}{K} + A_o y \tag{1}$$

$$\frac{\partial y}{\partial \ell} = \left[6 - \pi^2 K\left(\ell n(K^{-\theta}) + 1\right)\right] y \tag{2}$$

$$K_r = K(\ell)e^{-\ell}\ . \tag{3}$$

Here K and y are the XY coupling constant and fugacity as in Ref. 2, and K_r is the (normalized) superfluid density at scale $\ell = \ell n(a/a_o)$, where a is the average diameter of a ring and a_o the lattice constant. The recursion starts at $a = a_o$ where $K = K_o$ and the fugacity is $y_o = \exp(-4\pi^2 K_o/3)$, as found in Ref. 6. The value of $A_o = 1265$ is determined by requiring that the critical coupling constant be the known value $K_{oc} = 0.454$. The exponent θ is taken to be the Flory value for the self-avoiding walk in 3D, $\theta = 3/(D+2) = 0.6$. Iterating the above equations to large ℓ (cutting off at ℓ greater than the correlation length $\xi = a_o/K_r$) gives a power-law phase transition $K_r \sim (K_o - K_{oc})^{\nu}$ with an exponent closely matching the Shenoy prediction $\nu = 0.672$.

To compare with Monte Carlo calculations of the helicity modulus[5], the quantity $\rho_s/\rho = (K_r/K_o)(\rho_s^o/\rho)$ is computed where

$$\frac{\rho_s^o}{\rho} = 1 - \frac{1}{6}\left(\frac{1}{K_o}\right) - a\left(\frac{1}{K_o}\right)^2 + 0\left(\frac{1}{K_o}\right)^3 \ldots \tag{4}$$

is the "bare" superfluid density from spin-wave excitations. Only the coefficient 1/6 of the term linear in T is known for the 3D XY model; we have determined the second coefficient to be a = 0.05 by fitting to the four lowest-temperature Monte Carlo data points of Fig. 1. Because the Monte Carlo calculations are performed on a finite lattice size L, it is necessary to

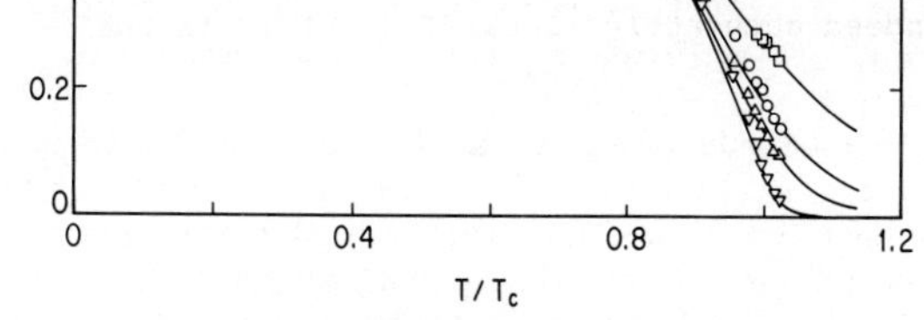

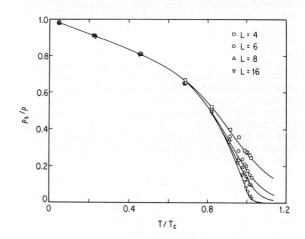

FIGURE 1
Vortex theory compared to Monte Carlo helicity modulus data of Ref. 5.

* Work supported in part by the US National Science Foundation, Contract DMR-8912825.

0921-4526/90/$03.50

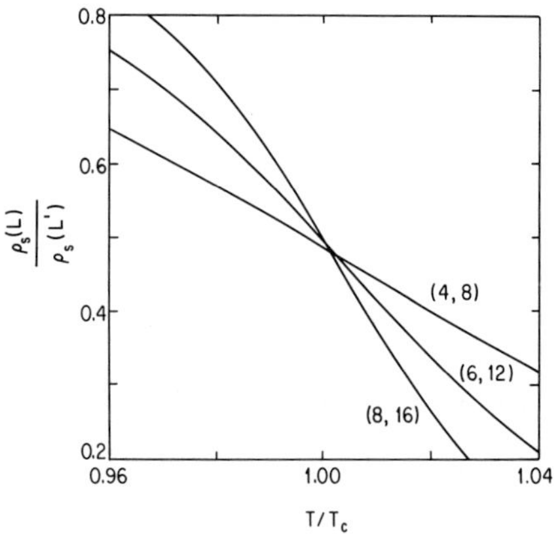

FIGURE 2
Superfluid density ratios of two lattice sizes (L, L').

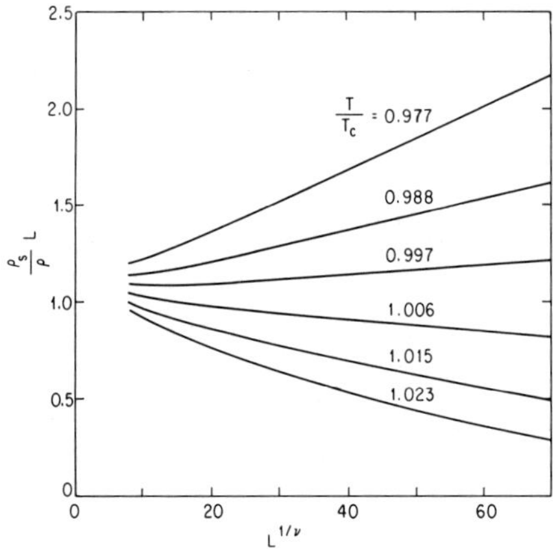

FIGURE 3
Scaling dependence of the superfluid density for temperatures near T_c.

terminate the recursion at a scale of order L, i.e. at a ring diameter $\exp(\ell) = a/a_o = \beta L$, where β is a parameter expected to be less than one, and which will depend on the particular boundary conditions employed (periodic in Ref. 5). In Fig. 1 β is the only parameter that is varied in order to match the theory to the data points; the reasonable result $\beta = 0.43$ is seen to give a good description of four different lattice sizes.

The finite-size scaling properties of Eqns. 1-4 can be checked following Ref. 5. In Fig. 2 ratios $\rho_s(L)/\rho_s(L')$ are shown for pairs of lattice sizes (L, L'). The larger lattices (6,12) and (8,16) intersect very close to a ratio of 0.5 at T_c, in agreement with finite-size scaling[5], since the theory is already known to satisfy the Josephson scaling relation[1], due to the form of Eq. 3. Fig. 3 shows the quantity $\rho_s L/\rho$ versus $L^{1/\nu}$ for various temperatures around T_c, which from finite-size scaling predictions is expected to be linear at large enough lattice size. As seen in Fig. 3 the curves for $T < T_c$ are indeed accurately linear for L>10. In the region $T>T_c$ the quadratic terms in the scaling expansion become appreciable and it can be seen in Fig. 3 there is really no linear regime, a trend visible in the data of Fig. 5 of Ref. 5.

It is perhaps not surprising that the theory satisfies finite-size scaling, since the coherence length of the theory is known to be correct[1]. The good agreement with the XY model for various lattice sizes (at least to within the uncertainties of the Monte Carlo data and the spin-wave expansion) is strong evidence that Eqns. 1-4 do correctly describe the scale dependence of the superfluid density.

REFERENCES

(1) G.A. Williams, Phys. Rev. Lett. 59 (1987) 1926.
(2) S.R. Shenoy, Phys. Rev. B40 (1989) 5056.
(3) V. Kotsubo and G.A. Williams, Phys. Rev. B33 (1986) 6106.
(4) D. Ceperly and E. Pollock, Phys. Rev. B39 (1989) 2084.
(5) Y. Li and S. Teitel, Phys. Rev. B40 (1989) 9122.
(6) G. Kohring, R. Schrock, and P. Wills, Phys. Rev. Lett. 57 (1986) 1358.

Physica B 165&166 (1990) 771–772
North-Holland

SUPERFLUID FRICTIONAL FORCE IN ADIABATIC FLOW OF HEII

Takeo SATOH and Mineo OKUYAMA

Department of Physics, Faculty of Science, Tohoku University, Sendai 980, Japan

The superfluid frictional force and/or the pressure drop across the flow path produced by the superfluid component flow were obtained up to about 70 cm/s of its velocity, v_s, in the Cu-Ni capillary of 0.2 mm i.d. and 480 mm length. The results show that the pressure drop is simply proportional to v_s^2 in the region of v_s>15 cm/s, although the vortex line density, L_0, shows rather complex v_s dependence. Discussions are given in the light of the new-type mutual friction force.

1. INTRODUCTION

In the physics of the flow of superfluid HeII, the superfluid frictional force, F_s, may be the most unclear concept. There seems two ideas for the physical content of F_s, one is the eddy viscosity(1) and the other is the vortex-wall interaction(2). Experimentally, it has been discussed with the thermal counterflow(1) and the pure superfluid flow(3). The velocity concerned with these experiments is rather small.

Our recent study of adiabatic flow(4), in which the velocity up to 100 cm/s is easily realized, showed that the rather complex temperature distribution along the flow path can be reproduced by the calculation within the frame work of the two fluid model assuming the new-type mutual friction force. This is written as

$$F_{sn} = \frac{2}{3}\,\frac{B\rho_n\rho_s\kappa}{2\rho}\,L_0(v_s)\cdot(v_s - v_n) \qquad (1)$$

The point is that the steady state vortex line density, L_0, is a function of only v_s. This means the importance of the interaction between vortices and the capillary wall. Therefore, it is interesting to see how the superfluid frictional force behaves in adiabatic flow.

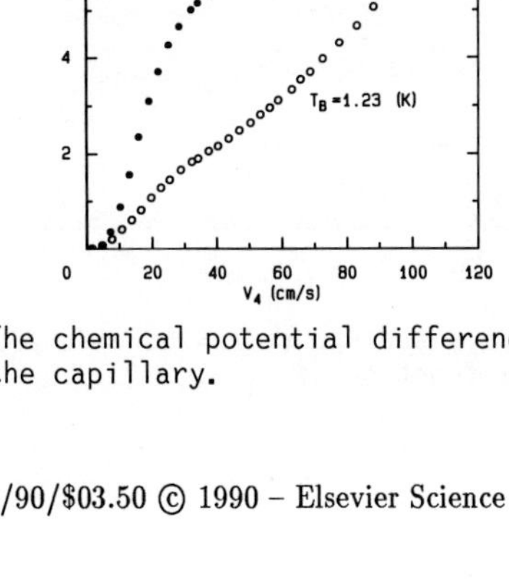

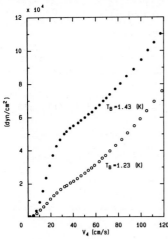

Fig.1 The chemical potential difference across the capillary.

2. EXPERIMENT

Apparatus used is almost same as previously described except the connections to the diaphragm pressure gauge both from the chamber and the outlet. In the previous experiment, they were done with Cu-Ni capillaries, which caused some ambiguity about the measured quantity. In the present experiment, they were made with superleaks. Therefore, the gauge measures the chemical potential difference across the flow capillary irrespective with the temperature difference between the chamber and outlet.

3. RESULTS AND DISCUSSIONS

The measured chemical potential difference, $\Delta\mu$, across the flow path is shown in Fig.1 as a function of the net mass flow velocity, v_4, which is almost equal to v_s. The pressure difference is obtained with the relation

$$\rho\ \Delta\mu = \Delta p + \int_{T_A}^{T_B} \rho S\, dT \qquad (2)$$

where T_A and T_B are the temperature of the chamber and the outlet, respectively. The two-fluid thermohydrodynamic description of adiabatic flow is made(4) with the equations of motion

$$\frac{dT}{dx} = \frac{1}{\rho S}\,\frac{dp}{dx} + \frac{1}{\rho_s S}(F_s + F_{sn}) \qquad (3)$$

$$\frac{dT}{dx} = -\frac{\rho_n}{\rho\rho_s S}\,\frac{dp}{dx} + \frac{1}{\rho_s S}(F_{sn} - F_n) \qquad (4)$$

and the energy conservation law

$$\rho S T v_n = (p_A - p)v + \rho v \int_{T_A}^{T} S\,dT + K\frac{dT}{dx} \qquad (5)$$

Solving eqs.(4) and (5) simultaneously for dT/dx and v_n, we obtain the temperature distribution along the flow path to be compared with experiments. L_0 in F_{sn} is a fitting parameter. For the details of the procedure, we refer the previous work(4). Then with eq.(3) or with the relation

$$\ell(F_s + F_n) = \Delta p \quad (6)$$

we obtain F_s. One thing to be mentioned is about dp/dx. In the previous work, we made the assumption of $d\mu/dx$ = const. along the capillary to obtain dp/dx. With this assumption, however, we found that F_s became negative near the outlet, which may contradict to the concept of the frictional force. So in the present study, we take the way to assume dp/dx = const. along the capillary.

The steady state vortex line density determined from the best fitting of T(x) is shown in Fig.2 as a function of v_4. The characteristic feature of its v_4 dependence, the existence of five stages, is same as in the previous study.

In Fig.3, F_s is plotted as a function of v_4. The three frictional forces F_s, F_n and F_{sn} are plotted together in Fig.4. These are the values at the position x = 30 cm. The overall features are almost same along the flow path. The interesting thing is that, in spite of the complex v_s dependence of L_0, F_s shows a very simple v_s dependence

$$F_s \propto v_s^{\,2} \quad (7)$$

in the velocity region above about 15 cm/s. Unfortunately, the present diaphragm gauge has no accuracy for the small pressure difference. The coefficient of the proportionarity is about 5 g/cm^3. This simple relation may be understood with the vortex-pinning model, if we consider that the pinning centers already fully interact with the flowing vortices in this velocity region(5) and the force between them is proportional to v_s. From Fig.4, it is seen that the pressure drop across the flow path is mainly due to F_s. Discussions will also be given to explain several characteristic aspects of the behavior of the vortex-cooler, which remain unsolved.

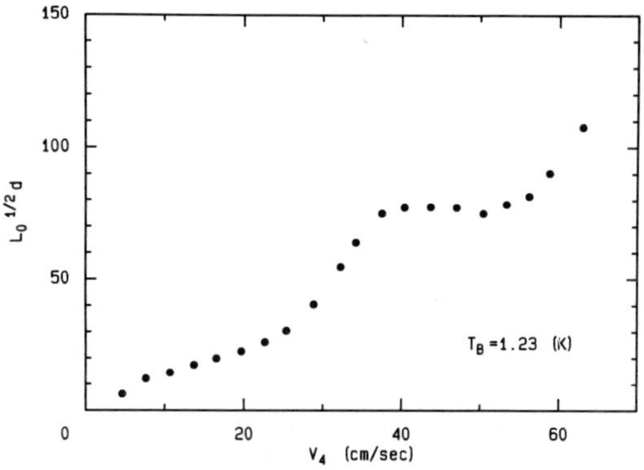

Fig.2 The steady state vortex line density determined from the best fitting of T(x) plotted as a function of v_4.

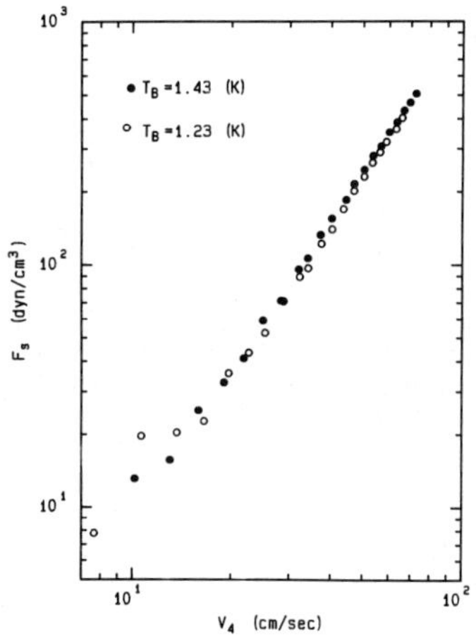

Fig.3 The superfluid frictional force as a function of v_4.

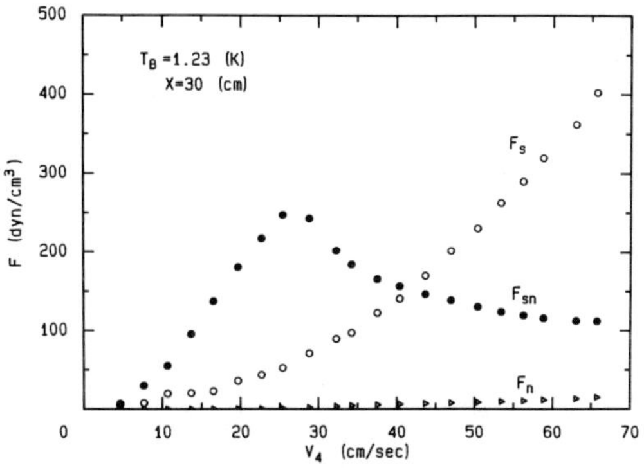

Fig.4 F_s, F_n, and F_{sn} at x = 30 cm/s plotted as a function of v_4.

REFERENCES

(1) D.F.Brewer and D.O.Edwards, Phyl.Mag.6 (1961) 1173.
R.K.Childers and J.T.Tough, Phys.Rev.B13 (1976) 1040.
(2) J.Yamauchi and K.Yamada, Physica 128B (1985) 45.
(3) G.van der Heijden, J.J.Giezen and H.C. Kramers, Physica 61(1972) 566.
W.de Haas, A.Hartoog, H.van Beelen, R.de Bruyn Ouboter and K.W.Taconis, Physica 75(1974)311.
(4) T.Satoh, T.Satoh, T.Ohtsuka and M.Okuyama, Physica 146B(1987) 379.
(5) M.Okuyama, T.Satoh and T.Satoh, Physica B154 (1988) 116.

Physica B 165&166 (1990) 773–774
North-Holland

KINETIC THEORY OF MUTUAL FRICTION FORCE IN SUPERFLUID ^{4}He WITH VORTEX LINES

Kazuo YAMADA, Kazumasa MIYAKE and Shohei KASHIWAMURA

Department of Physics, College of General Education, Nagoya University, Nagoya 464-01, Japan

Derivation of the mutual friction force between normal fluid and superfluid flows mediated by vortex lines is formulated on the basis of the kinetic theory for quasiparticle gas (phonons and rotons), in the case where two-fluid hydrodynamics is valid. Correspondence with the phenomenological theory is shown unambiguously, with respect to the issue about the Iordanskii force, the origin of the force on the vortex line due to quasiparticles, and the force balance relation between the Magnus force and the force from the normal fluid.

1. INTRODUCTION

Phenomenological theory [1,2] on the mutual friction involves two fundamental forces on the unit length of vortex line, the Magnus force $\mathbf{f}_M$ and the force from normal fluid $\mathbf{f}_N$, which are written as

$$\mathbf{f}_M = \rho_s \kappa \hat{\boldsymbol{\omega}} \times (\mathbf{v}_L - \mathbf{v}_s)$$
$$\mathbf{f}_N = -D(\mathbf{v}_L - \mathbf{v}_n) - (D' - \rho_n \kappa)\hat{\boldsymbol{\omega}} \times (\mathbf{v}_L - \mathbf{v}_n)$$

where notation in Ref.[2] was used. For simplicity, we here neglect the difference of normal fluid velocities near and far from the vortex line. Further the theory put on the balance relation of two forces, $\mathbf{f}_M + \mathbf{f}_N = 0$, from the fact that the inertia mass of vortex line is zero.

Although above arguments are physically plausible and have offered the basis for the analysis of experiments, it seems to us that some problems have not yet been completely settled [2,3,4] ; the microscopic derivation and meaning of coefficients D, D', the Iordanskii force $\mathbf{f}_I = \rho_n \kappa \hat{\boldsymbol{\omega}} \times (\mathbf{v}_L - \mathbf{v}_n)$, and of $\mathbf{f}_N$ proportional to the relative velocity $\mathbf{v}_L - \mathbf{v}_n$, and of the force balance relation.

The purpose of this note is to examine those problems on the basis of the kinetic theory for quasiparticles and to show in a simple form that the results so obtained are consistent in essence with those in the phenomenological theory.

2. HYDRODYNAMICAL EQUATIONS

2.1 Formulation

Let us start with the conservation equations for the mass and momentum density, ρ and $\mathbf{j} = \rho \mathbf{v}_s + \mathbf{j}_0$, and the kinetic equation for quasiparticle gas (we hereafter think of roton)[5],

$$\frac{\partial \rho}{\partial t} + \nabla \cdot \mathbf{j} = 0, \tag{1}$$

$$\frac{\partial j_i}{\partial t} + \frac{\partial}{\partial r_j}(\rho v_{si} v_{sj} + v_{si} j_{0j} + v_{sj} j_{0i} + \pi_{0ij}) = 0, \tag{2}$$

$$\frac{\partial n_\mathbf{p}}{\partial t} + \nabla_\mathbf{r} n_\mathbf{p} \cdot \nabla_\mathbf{p} E_\mathbf{p} - \nabla_\mathbf{p} n_\mathbf{p} \cdot \nabla_\mathbf{r} E_\mathbf{p} = I[n_\mathbf{p}], \tag{3}$$

where $n_\mathbf{p}$ and $\mathbf{j}_0 = \sum_\mathbf{p} \mathbf{p} n_\mathbf{p}$ are the distribution function of roton and the momentum density in the frame moving with superfluid velocity $\mathbf{v}_s$, and π_{0ij} is the pressure tensor in the same frame. In (3) $E_\mathbf{p}$ is the roton energy in the rest frame, $E_\mathbf{p} = \epsilon_\mathbf{p} + \mathbf{p} \cdot \mathbf{v}_s$, and $I[n_\mathbf{p}]$ is the *collision* terms due to roton-roton and roton-vortex interactions. Using (3) the equation of motion for $\mathbf{j}_0$ is obtained as

$$\frac{\partial j_{0i}}{\partial t} + \frac{\partial}{\partial r_j}[j_{0i} v_{sj} + \sum_\mathbf{p} n_\mathbf{p} p_i \frac{\partial \epsilon_\mathbf{p}}{\partial p_j}] + \sum_\mathbf{p} n_\mathbf{p} \frac{\partial E_\mathbf{p}}{\partial r_i} = \sum_\mathbf{p} p_i I[n_\mathbf{p}] \equiv -F_{exi}, \tag{4}$$

where $\mathbf{F}_{ex}$ is the force density on the vortex line by the excitation gas (no contribution from roton-roton collision arises because of the momentum conservation). Then from (1),(2) and (4) the acceleration equation for $\mathbf{v}_s$ is derived in the following form

$$\rho[\frac{\partial \mathbf{v}_s}{\partial t} + \nabla(\mu + \frac{\mathbf{v}_s^2}{2})] = (\rho \mathbf{v}_s + \mathbf{j}_0) \times (\nabla \times \mathbf{v}_s) + \mathbf{F}_{ex}, \tag{5}$$

where we used the definitions of μ and π_{0ij} given in Ref.[5]. In the two-fluid hydrodynamical region, $n_\mathbf{p}$ is the local equilibrium distribution, $n_\mathbf{p} = n^0_\mathbf{p}(\epsilon_\mathbf{p} + \mathbf{p} \cdot (\mathbf{v}_s - \mathbf{v}_n))$ which leads to $\mathbf{j}_0 = \rho_n(\mathbf{v}_n - \mathbf{v}_s)$. The equation (5), except $\mathbf{F}_{ex}$, was derived by a different approach [6].

For simplicity, consider a single vortex line moving with velocity $\mathbf{v}_L$, then $\mathbf{v}_s$ consists of $\bar{\mathbf{v}}_s$, the mean velocity, and $\mathbf{v}_{s\ell}$, one induced by the vortex line. Further assume the case $\nabla \times \bar{\mathbf{v}}_s = 0$ so that the vorticity is carried only by the vortex line, $\boldsymbol{\omega} = \nabla \times \mathbf{v}_{s\ell}$. Then (5) can be written, with use of the relation $\rho = \rho_s + \rho_n$, as

$$\rho[\frac{\partial \mathbf{v}_s}{\partial t} + \nabla(\mu + \frac{\mathbf{v}_s^2}{2})] = \rho \mathbf{v}_L \times \boldsymbol{\omega} + \rho_s \boldsymbol{\omega} \times (\mathbf{v}_L - \mathbf{v}_s) + \rho_n \boldsymbol{\omega} \times (\mathbf{v}_L - \mathbf{v}_n) + \mathbf{F}_{ex}. \tag{6}$$

2.2 Force balance

On the other hand, taking account of the vorticity conservation, the acceleration equation for $\mathbf{v}_s$ is given with ambiguity of an irrotational term in the following form [1,7,8]

0921-4526/90/$03.50

$$\frac{\partial \mathbf{v}_s}{\partial t} + \nabla(\mu + \frac{\mathbf{v}_s^2}{2}) = \mathbf{v}_L \times \boldsymbol{\omega}. \qquad (7)$$

In comparison of (6) with (7), we have a balance relation for the force densities:

$$\rho_s \boldsymbol{\omega} \times (\mathbf{v}_L - \mathbf{v}_s) + \rho_n \boldsymbol{\omega} \times (\mathbf{v}_L - \mathbf{v}_n) + \mathbf{F}_{ex} = 0. \qquad (8)$$

Although we will discuss on $\mathbf{F}_{ex}$ in section 3, here assume the following form

$$\mathbf{F}_{ex} = L^{-2}[-D\,\mathbf{v}_{Ln} - D'\hat{\boldsymbol{\omega}} \times \mathbf{v}_{Ln}],$$

where L^2 means the two dimensional area of system perpendicular to the vortex line and the abbreviation $\mathbf{v}_{Ln} \equiv \mathbf{v}_L - \mathbf{v}_n$ was used. To get the force per unit length of the line, $\mathbf{f}$, we integrate (8) over two dimensional area L^2 around the line, assuming the spatially slow variation of hydrodynamical velocities. Taking account of the familiar relation $\int d\mathbf{S} \cdot \boldsymbol{\omega} = \kappa$, we see that each term in (8) reduces respectively, to $\mathbf{f}_M$, $\mathbf{f}_I$, and $\mathbf{f}_{ex}$, and then (8) leads to the force balance relation $\mathbf{f}_M + \mathbf{f}_N = 0$. These results correspond completely to those given in Ref.[2], which discussed clearly the physical meanings of $\mathbf{f}_I$ and of the force balance relation. An alternative and transparent discussion was given for the force balance relation, considering the small but finite inertia mass of vortex line [8].

3. DISCUSSIONS ON $\mathbf{F}_{ex}$

By considering the collision term in (4) due to the elastic scattering and the absorption-emission processes, the expression of $\mathbf{F}_{ex}$ are examined paying attention to the dependence of the relative velocities, $\mathbf{v}_{sL}$, $\mathbf{v}_{ns}$ and $\mathbf{v}_{nL}$ and to the interrelation of the coefficients D and D' with the symmetry properties of the transition probability.

In the elastic scattering process, from the initial roton momentum $\mathbf{p}$ to the final one $\mathbf{p}'$, the transition probability $w(\mathbf{p},\mathbf{p}')$ consists generally of symmetric and antisymmetric parts with respect to the interchange $\mathbf{p} \leftrightarrow \mathbf{p}'$; $w = w_+ + w_-$. Further it is to be noted that if we use as usual the w calculated in the frame fixed on the vortex moving with $\mathbf{v}_L$, then the unperturbed energy of roton should be expressed as $\tilde{E}_\mathbf{p} = \epsilon_\mathbf{p} + \mathbf{p} \cdot \mathbf{v}_{sL}$. Then the elastic contribution of $\mathbf{F}_{ex}$ is given by

$$\mathbf{F}_{ex}^{el} = \frac{1}{2}\sum_{\mathbf{p},\mathbf{p}'}(\mathbf{p}-\mathbf{p}')\Big[w_+(\mathbf{p},\mathbf{p}')(n_\mathbf{p}^0 - n_{\mathbf{p}'}^0) + w_-(\mathbf{p},\mathbf{p}')(n_\mathbf{p}^0 + n_{\mathbf{p}'}^0)\Big]\,\delta(\tilde{E}_\mathbf{p} - \tilde{E}_{\mathbf{p}'})$$

$$\simeq \frac{1}{2}\sum_{\mathbf{p},\mathbf{p}'}(\mathbf{p}-\mathbf{p}')\Big[w_+(\mathbf{p},\mathbf{p}')((\mathbf{p}-\mathbf{p}')\cdot\mathbf{v}_{Ln}) + w_-(\mathbf{p},\mathbf{p}')((\mathbf{p}+\mathbf{p}')\cdot\mathbf{v}_{Ln})\Big] \times \frac{\partial n^0}{\partial \epsilon_\mathbf{p}}\delta(\epsilon_\mathbf{p} - \epsilon_{\mathbf{p}'}).$$

To get the last expression, energy conservation and the lowest approximation in small $\mathbf{v}$'s were taken into account. After the momentum's integrations we can expect a form

$$\mathbf{F}_{ex}^{el} = L^{-2}[-D_{el}\mathbf{v}_{Ln} - D'_{el}\,\hat{\boldsymbol{\omega}} \times \mathbf{v}_{Ln}]\ ,$$

where the longitudinal and transverse coefficients, D_{el} and D'_{el}, depend on the details of w_+ and w_- respectively. This is the desired form proportional to $\mathbf{v}_{Ln}$.

Next consider the absorption-emission process of rotons associated with the transition between eigen states of vortex mode with the energy, E_n and E_0. We note that the transition probability in this process satisfys the detailed balance condition, $w_a(\mathbf{p},0,n) = w_e(\mathbf{p},n,0)$, and assume the occupation probability of vortex state E_n, $P_n \propto \exp(-E_n/T)$. Then we can write,

$$\mathbf{F}_{ex}^{a-e} = \sum_{\mathbf{p},n}\mathbf{p}[w_e\,P_n(1+n_\mathbf{p}^0) - w_a\,P_0 n_\mathbf{p}^0] \times \delta(E_n - E_0 + \tilde{E}_\mathbf{p})$$

$$\simeq \sum_{\mathbf{p},n}\mathbf{p}w_e P_0[(\mathbf{p}\cdot\mathbf{v}_{sL}) - (\mathbf{p}\cdot\mathbf{v}_{sn})] \times \frac{\partial n^0}{\partial \epsilon_\mathbf{p}}\delta(E_n - E_0 + \epsilon_\mathbf{p})\ ,$$

where the energy conservation and the expansion in small $\mathbf{v}$'s were used. We can expect that above formula contributes to the longitudinal force proportional to $\mathbf{v}_{nL}$ with coefficient D_{a-e}. In Ref.[3] the estimation of the force equivalent to this $\mathbf{F}_{ex}^{a-e}$ was done by the classical approximation.

Finally we note that the Iordanskii force appeared independently from $\mathbf{F}_{ex}$ in (6), which originated in the drift term with the dynamical character in the left hand side of (3) or (4), so we should avoid the double counting of it in the evaluation of the collision term.

REFERENCES

[1] H.E. Hall and W.F. Vinen, Proc. R. Soc. 238A (1956) 204, 215; H.E. Hall, Adv. Phys. 9 (1960) 89.
[2] C.F. Barenghi, R.J. Donnelly and W.F. Vinen, J. Low Temp. Phys. 52 (1983) 189.
[3] A.J. Hillel, J. Phys. 14C (1981) 4027; A.J. Hillel and W.F. Vinen, J. Phys. 16C (1983) 3267.
[4] E.B. Sonin, Zh. Eksp. Teor. Fiz. 69 (1975) 921.[Sov. Phys.-JETP 42 (1976) 1027]; Rev. Mod. Phys. 59 (1987) 87.
[5] I.M. Khalatnikov, Introduction to the Theory of Superfluidity (Benjamin, New York, 1965), Sec.18.
[6] J.A. Geurst, Physica A 152 (1988) 1.
[7] A.J. Hillel, H.E. Hall and P. Lucas, J. Phys. 7C (1974) 3341.
[8] G. Baym and E. Chandler, J. Low Temp. Phys. 50 (1983) 57; E. Chandler and G. Baym, J. Low Temp. Phys. 62 (1986) 119.

Physica B 165&166 (1990) 775-776
North-Holland

SUPERFLUID TURBULENCE IN NON-UNIFORM THERMAL COUNTERFLOW

J.F.KAFKALIDIS and J.T.TOUGH

Department of Physics, The Ohio State University, 174 W. 18th. Ave., Columbus, Ohio, 43210, USA

We report preliminary results from a study of superfluid turbulence in a non-uniform thermal counterflow. These results are important for understanding how a vortex tangle is distributed in a spacially varying velocity field, and for describing the radial transport of heat in He II. Using a specially designed flow tube and a very high sensitivity thermocouple we are able to measure the temperature difference associated with thermal counterflow in a known non-uniform flow. We observe laminar flow, a critical heat current, and a single state of turbulent flow. The results indicate that the vortex line density in the supefluid turbulence is determined simply by the local velocity.

1. INTRODUCTION

Superfluid turbulence in He II has been studied extensively in a wide variety of flows but almost exclusively in tubes of uniform cross section. In the case of rectangular tubes, both pure superflow and thermal counterflow reveal a single turbulent state which can be identified with a homogeneous distribution of quantized vortex lines. This homogeneous state is quantitatively described by the theory of Schwarz (1). Little is known about an inhomgeneous disbribution of lines that would result from a non-uniform flow.

In this paper we give a preliminary account of an experiment designed to explore superfluid turbulence in a controlled non-uniform thermal counterflow. We have constructed a wedge-shaped flow tube having a rectangular cross section of constant height d whose width expands linearly with distance x in the direction of heat flow. In this flow tube the normal and superfluid velocities decrease as 1/x since the flow is incompressible. This particular non-uniform thermal counterflow also has a special practical significance since it characterizes the radial heat flow from a heated cylinder in He II.

2. APPARATUS

The apparatus is shown very schematically in figure 1. The flow tube, to be discussed below, connects a small lower cell containing a heater to a small upper cell which in turn is connected to a large reservoir of He II regulated at a temperature T_0. When a steady heat current $\dot{Q}$ is established through the flow tube the temperature in the lower cell increases to T_1 where

$$T_1 = T_0 + \Delta T$$

The temperature difference ΔT is measured with an AuFe thermocouple (2) connected between the upper and lower cells. (In future experiments the upper end of the thermocouple will be moved to a tab which makes thermal contact with the helium in the flow tube at some intermediate position.) The thermocouple is connected to a SQUID used as an ammeter giving an overall sensitivity of better than 1 μK in ΔT.

The flow tube is fabricated from Stycast 1266 epoxy. All internal surfaces are polished smooth. The tube cross section is a rectangle of height d = 0.025 cm., and width varying from w_1 = 0.2 cm. to w_2 = 1.0 cm. The tube length ℓ is 10.0 cm., and the angular opening of the tube α is 0.08 radians. A 0.063 cm. diameter wire is sealed into the flow tube wall and machined flush with the inner surface. This wire is soldered to a copper tab which can be be thermally connected to the upper thermocouple junction.

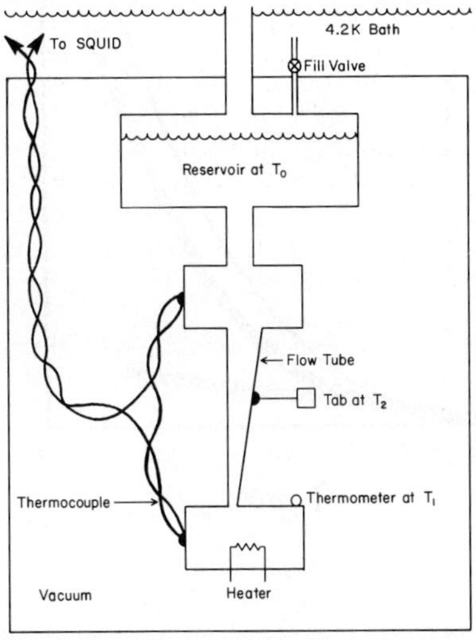

Fig. 1. Schematic diagram of the apparatus

3. RESULTS

The temperature difference ΔT produced by the steady heat flow $\dot{Q}$ through the tube is shown by the data in figure 2 for a base temperature T_0 of 1.4 K. The data appear qualitatively like those for thermal counterflow in a uniform rectangular tube (3), and reveal a laminar region, a critical heat current $\dot{Q}_c$ and a single non-linear region characteristic of superfluid turbulence. The hysteresis associated with the extension of the laminar region above the critical heat current is also typical of thermal counterflow in uniform retangular tubes.

A detailed examination of the laminar region shows that the temperature difference is approximately linear in $\dot{Q}$. This dependence is predicted by the laminar solution of the normal fluid equations of motion for similar geometries, but an exact solution for this specific geometry has not yet been obtained. An approximate solution gives a result within ten percent of our laminar flow data.

The critical heat current $\dot{Q}_c$ we define to be the lowest value of the heat current at which a turbulent flow can be observed. For example, the data at 1.4 K shown in figure 2 indicate a value of the critical heat current of about 0.4 mW. Results obtained at 1.3, 1.4, 1.5, 1.6, and 1.7 K show a temperature dependence for $\dot{Q}_c$ that is essentially identical

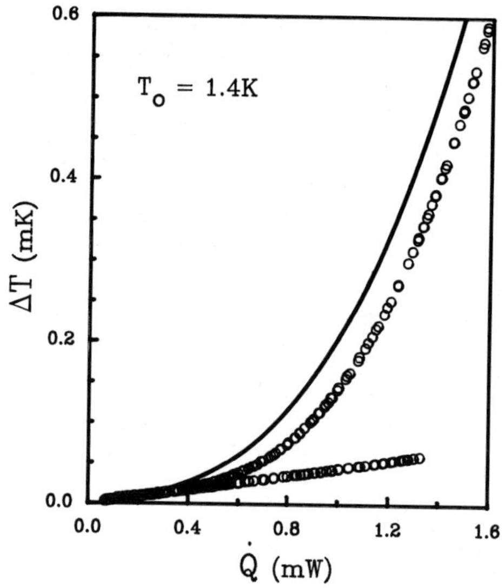

Fig. 2. Temperature difference data at 1.4K.

to that observed in uniform rectangular tubes (3), but is about a factor of two too large. That is, if the dimensional scaling established in uniform rectangular tubes ($\dot{Q}_c \propto d$) is applied to our flow tube the critical heat current at 1.4 K would be about 0.2 mW rather than the observed value of 0.4 mW. Because of the non-uniform flow it is not obvious that this is a valid comparison however. There may be a range of critical heat currents varying from 1.0 mW at the wide end to 0.2 mW at the narrow end of our tube. The data show no evidence of two critical heat currents, and it is not clear at present exactly what the critical condition represents.

The principal motivation for these experiments is to understand the influence of a non-uniform flow on the vortex line density in the superfluid turbulence. How do the vortex lines respond to a driving velocity that varies along the direction of flow? The most simple possibility is that the line density at some position x is determined by the local velocity at that point. In this way the $1/x$ velocity field would produce a $1/x^2$ distribution of vortex line density. The solid line in figure 2 is the result of a calculation using the Schwarz theory (1) based upon this assumption. The small discrepancy between the theory and data suggests that this simple local approximation is correct. In future experiments we will move the upper thermocouple junction to the tab (see figure 1) and in this way we will be able to verify more directly how the vortex line density is distributed along the flow tube.

4. CONCLUSIONS

Our experiments have revealed the characteristic features of superfluid turbulence in a non-uniform thermal counterflow. They reveal for the first time the laminar flow and the critical heat current in such a geometry. Further, they suggest that the vortex line density in the turbulence simply follows the local velocity field. Experiments are in progress to determine the distribution of vortex lines in the non-uniform flow. The effect of the non-uniform flow on the critical heat current is not presently understood.

ACKNOWLEDGMENTS

This work has been supported by the National Science Foundation under grant number CBT-8613459.

REFERENCES

(1) K.W.Schwarz, Phys. Rev. B 38 (1988) 2398.
(2) Y.Maeno, H.Hauke, and John Wheatley, Rev. Sci. Instr. 54 (1983) 946.
(3) D.R.Ladner and J.T.Tough, Phys. Rev. B 20 (1979) 2690.

Physica B 165&166 (1990) 777–778
North-Holland

EXPERIMENTAL EXAMINATION OF THE APPEARANCE OF VORTICES IN HE II TAYLOR COUETTE FLOW

Chris J. Swanson, Charles E. Swanson, and Russell J. Donnelly

University of Oregon, Eugene, OR 97403

We are investigating the appearance of quantized vorticity in rotating Couette flow of He II. Free energy arguments of Fetter precisely predict the rotational velocity for first appearance of vorticies in solid body rotation. (1) We have found, in agreement with other experiments, a value slightly lower than the theoretical value. The calculations of Fetter have also been generalized to non equilibrium shear flow in infinite cylinders by Swanson and Donnelly. (2) As with the solid body case, our results are somewhat different from these predictions which may be due to our finite geometry.

A recent review of Taylor Couette flow in He II by Donnelly and LaMar (3) has motivated further experimental study of the subject. It is well known that the vorticity in rotating He II in equilibrium can be successfully calculated using free energy minimization techniques. Specifically, in a rotating annulus Fetter has predicted that at a specified critical rotation rate a row of equally spaced vortices should appear around the annulus with an inter line spacing on the order of the gap size. (1) Recently Swanson and Donnelly have generalized these techniques to the non equilibrium case and predict a critical rotation rate for any flow. (2) Some data has been taken for various special flows like solid body rotation, potential flow, rotation of the outer cylinder only, and of the inner cylinder only. However, few of these experiments have had the sensitivity to look carefully at the onset and no data is available for general Couette flow.

One expects that in rotating He II there will be regions near the walls which are free of vortex lines. The size of these regions depends on the rotational velocity, Ω, of the cavity. For small rotation rates, when the vortex free strips become smaller than the gap size, vortices will appear. An exact critical rotational velocity has been calculated for the equilibrium solid body case as

$$\Omega_0 = \kappa/\pi d^2 \ln(2d/\pi a).$$

This prediction has been roughly confirmed by Bendt and Donnelly (4) and Shenk and Mehl. (5) Shenk and Mehl however report an experimental onset value consistently low by 15%.

We are able to detect the absolute magnitude of the vorticity by measuring the attenuation of a second sound resonance in our cavity in accordance with the method of Hall and Vinen.(6) Our resonant modes have both azimuthal and axial components and are excited by an AC voltage applied to long thin silver strips painted parallel to the axis of rotation. The second sound waves are detected by a DC biased, carbon strip similarly painted on the face of the inner cylinder. The signal is analyzed with a two channel phase sensitive lock-in from which we directly obtain the phase. The frequency dependence of the phase near the resonant frequency gives us the width of the resonance. We are interested in the length of line per unit volume (line density) which is proportional to $(\Delta - \Delta_0)$ where Δ is the width of the resonant curve with rotation, Δ_0 the width without rotation. We have used this method to obtain a best resolution of about 20 lines per cm^2. Out apparatus consists of two concentric cylinders of radii close to 2 cm and separated by a gap of .047 cm. By selecting a small gap, the line density at $\Omega = 1.05\Omega_0$ should be around 400 lines per cm^2 (well within our sensitivity range). Thus we are able to examine in detail the prediction of a sharp rise in line density immediately after onset.

In Fig. 1, the top curve represents solid body rotation and it can be seen that the onset is gradual and not sharp as predicted. Furthermore, in agreement with Shenk and Mehl, we find the onset to be low by 15% to 30%. We have defined the onset to be the intersection of curves fitted to the upper and lower portion of the data since there is no well defined break.

The other curves in Fig. 1 represent other types of Couette flow. Swanson and Donnelly have conjectured that since the laminar normal fluid flow is the sum of a solid body part and a potential flow part, the super fluid can mimic the potential flow part by virtual vorticity in the center. Thus the only part which will create actual vorticity in the gap is the solid body portion which can be treated with equilibrium techniques. Hence, they predict the onset of vortices when the dimensionless rotation rate A has the

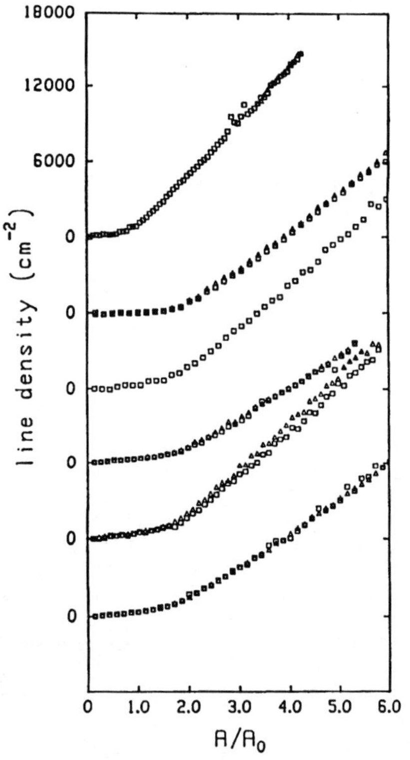

FIGURE 1
Plot of line density versus dimensionless rotation rate for various Ω_{in}/Ω_{out}'s ranging from 0(bottom curve) to 1 (top curve). Vortices are predicted to appear at $A/A_0 = 1$.

following critical value,

$$A = A_0 = \kappa/\pi d^2 \ln(2d/\pi a)$$

where $$A = (\Omega_2 R_2^2 - \Omega_1 R_1^2)/(R_2^2 - R_1^2).$$

It should be noted, however, that this theory is for infinite cylinders and cannot be exactly right since the ends do not rotate with the laminar Couette profile. The bottom curve in Fig. 1 represents a ratio $\Omega_{in}/\Omega_{out} = 0$. The other curves are for other relative rotations, ranging from .2 to .8 by .2's. It can be seen that for all values of relative rotation other than 1 (solid body rotation) the onset of vortices appears to occur at 1.5 times the predicted rotation rate of 1. We believe that this discrepancy is best explained by end effects for two reasons. First, the break has no gradual drift toward the solid body break as Ω_{in}/Ω_{out} increases from 0. Another run at $\Omega_{in}/\Omega_{out} = .9$ not shown here similarly has a break around 1.5. One would expect that very near solid body rotation the results should asymptotically approach the solid body results since the velocity of the end plates approaches the velocity of the laminar normal fluid. However with the present geometry, relative rotation can cause some turbulence *under* the end plates which would be detected as increased vorticity for almost any relative rotation. A second reason for suspecting end effects has to do with the variable values of the mutual friction coefficient B. B in equilibrium should be related to the slope of these plots above the break. We have found in solid body rotation and at many different temperatures that B is higher than the currently accepted values by 50%. Furthermore, our measured B for relative rotation can be seen to vary widely and apparently unsystematically. At this point our best explanation of these phenomena is the poorly defined end conditions and associated turbulence. We note also that the data taken in Fig. 1 was taken at 2.1 K. Other runs at other temperatures yield the same results in all aspects except the value of the break in relative rotation which decreases with increasing temperatures.

It seems that the equilibrium case is poorly understood. Both Shenk and Mehl and we find an onset of vorticity at rotation rates consistently lower than expected. Also the theory of Swanson and Donnelly can only be approximately tested since it assumes infinite cylinders and it is not clear how important the ends are in Helium. We plan to improve and adjust our geometry to make more specific measurements of end effects. It is our hope that these steps might partially or even completely eliminate their influence.

REFERENCES

(1) A.L. Fetter, Phys. Rev. 153 (1967) 285
(2) C.E. Swanson and R.J. Donnelly, J. Low Temp. Phys. 67 (1987) 185
(3) R.J. Donnelly and M.M. LaMar, J. Fluid Mech. 186 (1988) 163
(4) P.J. Bendt and R.J. Donnelly, Phys. Rev. Lett. 19 (1967) 214
(5) D.S. Shenk and J.B. Mehl, Phys. Rev. Lett. 27 (1971) 1703
(6) H.E. Hall and W.F. Vinen, Proc. Roy Soc., Ser. A 238 (1956) 204,215

Physica B 165&166 (1990) 779–780
North-Holland

NUCLEAR ORDER IN COPPER: NEW TYPE OF ANTIFERROMAGNETISM IN AN IDEAL FCC SYSTEM

A.J. Annila[(1)], K.N. Clausen[(2)], P.-A. Lindgård[(2)], O.V. Lounasmaa[(1)], A.S. Oja[(1)], K. Siemensmeyer[(3)], M. Steiner[(3)], J.T. Tuoriniemi[(1)], and H. Weinfurter[(4)]

[(1)]Low Temperature Laboratory, Helsinki University of Technology, SF-02150 Espoo, Finland,
[(2)]Physics Department, Risø National Laboratory, DK-4000 Roskilde, Denmark,
[(3)]Johann Gutenberg Universität, D-6500 Mainz, Federal Republic of Germany,
[(4)]Hahn-Meitner Institut, D-1000 Berlin 39, Federal Republic of Germany

The new antiferromagnetic reflection ($0\,{}^2/_3\,{}^2/_3$) has been found by neutron diffraction experiments at nanokelvin temperatures in the nuclear spin system of a ^{65}Cu single crystal. The corresponding three-sublattice structure has not been observed previously in any fcc antiferromagnet.

A new type of fcc antiferromagnetic order has been found. By neutron diffraction experiments, four equivalent nuclear magnetic Bragg reflections $\pm(0\,{}^2/_3\,{}^2/_3)$, $(1\,{}^1/_3\,{}^1/_3)$ and $(1\,{-}^1/_3\,{-}^1/_3)$ have been observed in the spin system of a ^{65}Cu single crystal. It was unexpected that the order proved to be simply commensurate with a three-sublattice structure. The discovery was made when the reciprocal space was searched along the high symmetry directions. This is the first time that conventional scanning has been employed at nanokelvin temperatures.

In the first neutron diffraction experiments (1,2) on copper, the nuclear antiferromagnetic (100) Bragg peak, X (see Fig. 1), was observed in two regions of the external magnetic field, applied along the $[01\bar{1}]$ crystalline direction: near zero field and around 0.15 mT. It was puzzling that no neutron intensity was found at intermediate fields around B = 0.10 mT, even though the simultaneously measured longitudinal susceptibility indicated order. The interpretation was that the spin structure at intermediate fields is characterized by a yet unknown propagation vector.

On the basis of first principles calculations (3,4) it has been shown that the fundamental ordering vector is expected at X corresponding to the observed (100) reflection. Another vector along ΓK, $(0\eta\eta)$ (see Fig. 1), is very close in energy (4), in particular if the strength of the Ruderman-Kittel interaction relative to the dipolar force is slightly reduced from the experimental value. Fluctuations could then stabilize a ΓK structure at the boundary between the two (100) phases (5). An ordering vector along ΓX, $(\zeta 00)$, resulting from a helical spin structure, has also been proposed (3,6,7).

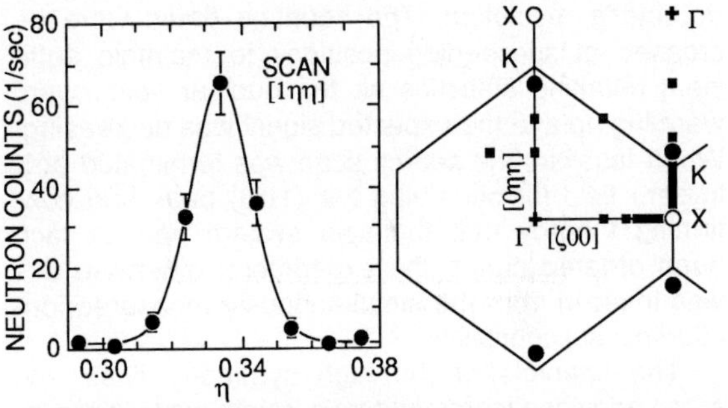

FIGURE 1
Left: Neutron intensity vs. position along ΓK, [1ηη], showing the discovery of the ($1\,{}^1/_3\,{}^1/_3$) Bragg reflection. Right: The scattering plane in the Brillouin zone of an fcc lattice. The search scans are marked with thick lines, the observed $\pm(0\,{}^2/_3\,{}^2/_3)$, ($1\,{}^1/_3\,{}^1/_3$) and ($1\,{-}^1/_3\,{-}^1/_3$) reflections by ● and the (100)-type reflections by ○. No intensity was observed at the commensurate positions indicated by ▪.

We attempted to find this possible new reflection. The extremely low nuclear ordering temperature in copper imposes severe experimental restrictions which make it difficult to use conventional ω–2θ scans, traditionally employed in similar investigations of electronic magnetism at higher temperatures.

The measuring time is limited because the spins can be cooled below the Néel temperature, $T_c = 58 \pm 10$ nK, only once every 36 hours by means of cascade nuclear demagnetization and because the spin-lattice relaxation process warms up the nuclei from the ordered state to the paramagnetic region in about five minutes. Mechanical movements of the cryostat and the detector during the scans contribute additional vibrational heat, thus further reducing the measuring time. The rigid sample mounting, a necessity for good thermal contact between the sample and the precooling first stage, restricts the scans with our present sample orientation to the plane spanned by the [ζ00] and [0ηη] axes (see Fig. 1).

The search scans were performed as follows: Prior to nuclear demagnetization, the diffractometer was positioned at the starting point for the scan. The neutron wavelenght $\lambda = 4.7$ Å was found to give optimal signal to noise ratio and access to the symmetry lines of interest. Second and third order reflections from the monochromator were used to align the sample so that the structural fcc peaks were seen. This assured that the scan was along the desired direction. During experiments higher order reflections were removed by a cooled BeO filter. The scan was programmed to proceed to the next reciprocal lattice position in steps equal to the half-width of the spectrometer's resolution. The counting times were increased at succeeding positions to maintain sufficient counting statistics as the nuclear spins were warming up and the expected signal was decreasing. When feasible, the search scan was terminated at X in zero field for observing the (100) peak and confirming thereby that the spin system had, in fact, been ordered during the experiment; otherwise this was inferred from the simultaneously measured longitudinal susceptibility.

The intervals of the high symmetry lines and some additional commensurate points which were investigated are shown in Fig. 1; the search scans were performed at $B = 0$ and $B \approx 0.8$ mT. The scans at zero field were made because during the first minute the (100) intensity was lower than expected.

Our search for a new reflection was rewarded (8) when a very clear Bragg peak was found at (1ηη), with $\eta = 0.33 \pm 0.01$, in the external field $B = 0.07$ mT (see Fig. 1). The order is, within the experimental accuracy, commensurate with the lattice structure. Later the equivalent ±(0 2/3 2/3) and (1 -1/3 -1/3) reflections were observed as well.

To confirm that $\pm(2\pi/a)$(0 2/3 2/3) are the fundamental propagation vectors of the structure, the (0 1/3 1/3) position was carefully searched for over the whole region below the critical field $B_c = 0.25$ mT, but no neutrons above the background (1–4 counts/sec) were detected. Also several other commensurate points were investigated (see Fig. 1) without observing any neutron signal.

There are 12 symmetry-related **k**-vectors for the (0 2/3 2/3) reflection. Comparing the intensity of the nuclear magnetic signal to the intensity of the nuclear peak observed at another wavelength and correcting for extinction, flux and for geometric effects, we can estimate the nuclear magnetic structure factor. This comparison is consistent with a spin structure characterized by the two ordering vectors $\pm(2\pi/a)$(0 2/3 2/3) alone, which implies a three-sublattice structure (see Fig. 2).

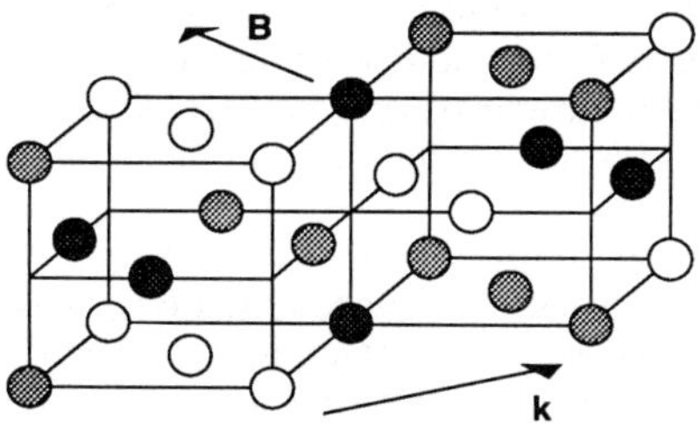

FIGURE 2
Planar three-sublattice structure (9) *consistent with the observed (0 2/3 2/3) reflection.*

In conclusion, we have shown that scanning can be used, with some restrictions, for investigations of a spin system at nanokelvin temperatures. The commensurate ordering vector $(2\pi/a)$(0 2/3 2/3), not observed in any electronic fcc antiferromagnet, was found in the ideal fcc system of nuclear spins in copper.

REFERENCES

(1) T.A. Jyrkkiö, M.T. Huiku, O.V. Lounasmaa, K. Siemensmeyer, K. Kakurai, M. Steiner, K.N. Clausen, and J. K. Kjems, Phys. Rev. Lett. 60 (1988) 2418.
(2) T.A. Jyrkkiö, M.T. Huiku, K. Siemensmeyer, K.N. Clausen, J. Low Temp. Phys. 74 (1989) 435.
(3) L.H. Kjäldman and J. Kurkijärvi, Phys. Lett. 71A (1979) 454.
(4) P.-A. Lindgård, X.-W. Wang, and B.N. Harmon, J. Magn. & Magn. Mater. 54-57 (1986) 1052.
(5) P.-A. Lindgård, Phys. Rev. Lett. 61 (1988) 629 and P.-A. Lindgård, J. Physique C8 (1988) 2051.
(6) A.S. Oja and P. Kumar, Physica 126B (1984) 451.
(7) A.S. Oja, X.-W. Wang, and B. N. Harmon, Phys. Rev. B 39 (1989) 4009.
(8) A.J. Annila, K.N. Clausen, P.-A. Lindgård, O.V. Lounasmaa, A.S. Oja, K. Siemensmeyer, M. Steiner, J.T. Tuoriniemi, and H. Weinfurter, Phys. Rev. Lett. (in press)
(9) H.E. Viertiö and A.S. Oja, Phys. Rev. Lett. (submitted)

Physica B 165&166 (1990) 781–782
North-Holland

NUCLEAR ANTIFERROMAGNETIC PHASES OF COPPER IN A FIELD

A.J. Annila[(1)], K.N. Clausen[(2)], P.-A. Lindgård[(2)], O.V Lounasmaa[(1)], A.S. Oja[(1)], K. Siemensmeyer[(3)], J.T. Tuoriniemi[(1)], and H. Weinfurter[(4)]

[(1)]Low Temperature Laboratory, Helsinki University of Technology, SF-02150 Espoo, Finland,
[(2)]Physics Department, Risø National Laboratory, DK-4000 Roskilde, Denmark,
[(3)]Johann Gutenberg Universität, D-6500 Mainz, Federal Republic of Germany,
[(4)]Hahn-Meitner Institut, D-1000 Berlin 39, Federal Republic of Germany

The phase diagram of the nuclear spin system in copper has been studied by neutron diffraction. Three antiferromagnetic phases appear as the external magnetic field aligned along the $[01\bar{1}]$ crystal axis is varied between zero and $B_c = 0.25$ mT. The results are compared with simultaneous susceptibility measurements.

The nuclear magnetic phase diagram of copper has been investigated by neutron diffraction techniques at nanokelvin temperatures. The intensities of the nuclear magnetic Bragg reflections for the (1 0 0) and (0 2/3 2/3) spin structures depend on the strength of the external magnetic field applied along the $[01\bar{1}]$ crystal axis. This reveals the presence of three different antiferromagnetic phases.

The first studies of the phase diagram by susceptibility measurements (1) showed the existence of three different antiferromagnetic regions as a function of the external magnetic field, aligned along the [0 0 1] direction. In order to determine the different spin structures, neutron diffraction experiments were initiated. (2)

The ordered state spin structures have been investigated recently by spin-wave theory (3), by second-order perturbation theory for a cluster of spins (4,5), and by Monte Carlo simulations (6,7). These calculations have yielded phase diagrams exhibiting different ordered structures as a function of the strength and direction of the external magnetic field. In order to map the phase diagram, the intensities of the (1 1/3 1/3), equivalent to (0 2/3 2/3) under cubic symmetry, and the (100) reflections were measured in different external magnetic fields.

A two-stage adiabatic nuclear demagnetization cryostat (8) was employed in these experiments. For the measurements the spins were first polarized at $B \approx 4.6$ T and $T \approx 100\,\mu$K and then rapidly demagnetized to low fields, where the long-range antiferromagnetic order appears, resulting in a magnetic nuclear Bragg reflection. The neutron signal was monitored in a constant field as a function of time while the spin system warmed to the paramagnetic region owing to spin-lattice relaxation (see inset of Fig. 1). In Fig. 1 we give three examples of the warmup curves in intermediate fields. Temperature, which could be determined in the paramagnetic phase only, is monotonically increasing with time.

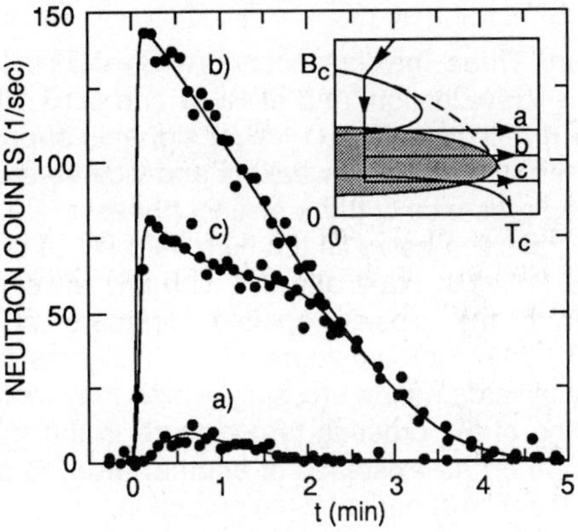

FIGURE 1
Time dependence of the (1 1/3 1/3) reflection at a) B = 0.13 mT, b) 0.10 mT, and c) 0.05 mT. A schematic phase diagram in the B-T plane is shown in the inset; the arrows indicate entrance, along an isentrop (S = 0.15·Rln4), into the ordered phase and subsequent measurements in a constant field; a, b and c correspond to curves a), b) and c), respectively.

From the neutron count vs. time curves we have constructed a neutron intensity contour diagram shown in Fig. 2. Our new data have been supplemented with earlier measurements (8) on the (1 0 0)

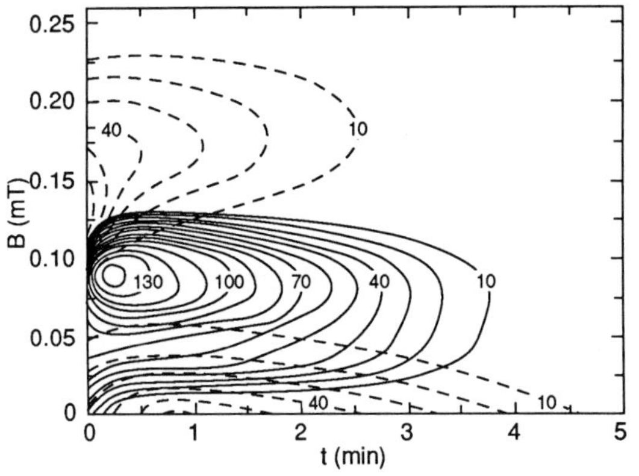

FIGURE 2
The neutron intensity contour diagram of the (1 1/3 1/3) (solid) and the (1 0 0) (dashed) Bragg peaks as a function of time and the external magnetic field. The outermost contours, 10 counts/sec, show approximately when the long range order disappears.

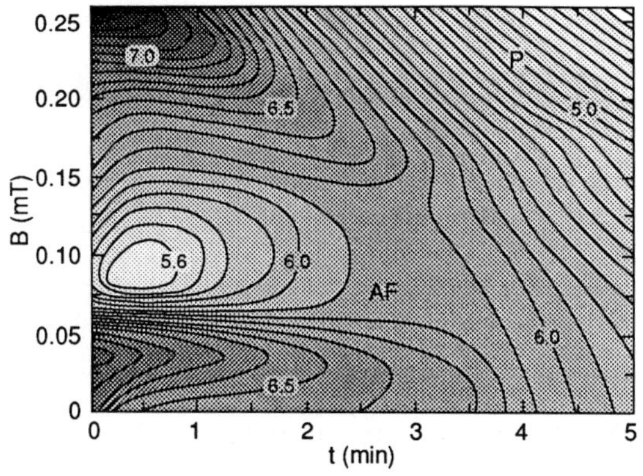

FIGURE 3
Contour diagram of the longitudinal susceptibility as a function of time and the external magnetic field. The antiferromagnetic phase (AF) is seen as a region bounded by the paramagnetic phase (P) where the contours are diagonal and uniformly spaced.

reflection. Three maxima occur: At B = 0.09 mT for the (1 1/3 1/3) reflection, and at B = 0 and 0.15 mT for the (1 0 0) reflection. The (1 1/3 1/3) signal is strongest when the (1 0 0) signal is weakest and vice versa, implying the presence of three distinct phases.

The transition between the high-field (1 0 0) phase and the (1 1/3 1/3) state appears abruptly as can be seen from the closely spaced contours at B ≈ 0.12 mT. The outer contours of both signals bend strongly inwards before crossing, which may indicate a lowering of the ordering temperature in the transition region or the existence of another ordered state with neither (1 0 0) nor (1 1/3 1/3) reflection.

The lower transition from the (1 1/3 1/3) peak back to the (1 0 0) reflection takes place over a wide field region. Between B = 0.02 mT and 0.07 mT both peaks are clearly visible, but with different time evolutions. This might be taken as evidence for the coexistence of two structures, although a 2**k**-phase with two simultaneous ordering vectors (7) cannot be excluded. The rapid disappearance of the (1 1/3 1/3) neutron signal at very low fields shows that the spin structure is stabilized by the field.

The longitudinal, low frequency (f = 20 Hz) susceptibility, χ'_L, was measured simultaneously with the neutron signal. In the ordered state, χ'_L is constant or increases with time, while it decreases exponentially due to spin-lattice relaxation in the paramagnetic phase. A contour diagram (Fig. 3) similar to the intensity diagram (Fig. 2) has been compiled from measurements at constant fields.

The clear minimum in χ'_L at intermediate fields coincides in field and time with the maximum of the (1 1/3 1/3) neutron intensity. The phase transitions, however, seem to manifest themselves in a rather different way in these graphs; the sharp upper transition is accompanied with a smooth decrease of the susceptibility, whereas the wide lower transition shows a rapid increase in the amplitude between B = 0.05 mT and 0.08 mT.

In conclusion, the neutron intensities of the (1 1/3 1/3) and the (1 0 0) nuclear magnetic peaks establish the presence of three antiferromagnetic phases as a function of the external magnetic field along the [0 1 $\bar{1}$] direction.

REFERENCES

(1) M.T. Huiku, T.A. Jyrkkiö, J.M. Kyynäräinen, A.S. Oja, and O.V. Lounasmaa, Phys. Rev. Lett. 53 (1984) 1692.
(2) T.A. Jyrkkiö, M. T. Huiku, O.V. Lounasmaa, K. Siemensmeyer, K. Kakurai, M. Steiner, K.N. Clausen, and J.K. Kjems, Phys. Rev. Lett. 60 (1988) 2418.
(3) H.E. Viertiö and A.S. Oja, Phys. Rev. B 36 (1987) 3805
(4) P.-A. Lindgård, Phys. Rev. Lett. 61 (1988) 629.
(5) P.-A. Lindgård, J. Physique C8 (1988) 2051.
(6) S.J. Frisken and D.J. Miller, Phys. Rev. Lett. 61 (1988) 1017.
(7) H.E. Viertiö and A.S. Oja, Phys. Rev. Lett. (submitted).
(8) T.A. Jyrkkiö, M.T. Huiku, K. Siemensmeyer, K.N. Clausen, J. Low Temp. Phys. 74 (1989) 435.

Physica B 165&166 (1990) 783–784
North-Holland

KINETICS, HYSTERESIS AND NONADIABATICITY OF THE PHASE TRANSITIONS IN THE NUCLEAR SPIN SYSTEM OF COPPER

A.J. Annila(1), K.N. Clausen(2), A.S. Oja(1), K. Siemensmeyer(3), M. Steiner(3), J.T. Tuoriniemi(1), and H. Weinfurter(4)

(1)Low Temperature Laboratory, Helsinki University of Technology, SF-02150 Espoo, Finland,
(2)Physics Department, Risø National Laboratory, DK-4000 Roskilde, Denmark,
(3) Johann Gutenberg Universität, D-6500 Mainz, Federal Republic of Germany,
(4)Hahn-Meitner Institut, D-1000 Berlin 39, Federal Republic of Germany

Antiferromagnetic phase transitions in the nuclear spin system of copper have been studied in varying external magnetic fields by neutron diffraction. Hysteresis and nonadiabaticity were observed at the boundary between the high-field and intermediate-field phases. The growth and decay of the respective antiferromagnetic Bragg reflections take place during a time span of tens of seconds.

The extremely low energy scale associated with nuclear magnetic systems(1) makes it possible to investigate the kinetics of phase transitions, taking place below 60 nK in copper. The observed (2,3) time span of up to one minute for the growth or decay of antiferromagnetic order in certain external magnetic fields is orders of magnitude longer than that for analogous phenomena in electronic magnets at much higher temperatures. Time-dependent effects were found also in the early susceptibility measurements (4) on copper. A large entropy increase, 0.12·Rln4 per spin, occurred in a few seconds when the boundary between the low- and intermediate-field phases was crossed.

Recently, we found a new type of antiferromagnetic long-range order in copper by neutron diffraction when the external magnetic field was aligned along the $[01\bar{1}]$ crystal axis (5). At high fields, between 0.12 and 0.25 mT, an antiferromagnetic (1 0 0) Bragg peak had been observed (2) while the order at intermediate fields around 0.09 mT gave (0 2/3 2/3)-type reflections (5). At low fields, from 0.02 to 0.07 mT, both reflections existed simultaneously but close to zero field only the (1 0 0) peak remained.

The data in Fig. 1 show the time evolution of the two antiferromagnetic reflections at the lower and upper transition regions. At the starting time t = 0 the spins have been demagnetized to the final field and into the ordered state. The reflected neutron intensity at λ = 4.7 Å was monitored when the spin-lattice relaxation process warmed up the nuclear spin system to the paramagnetic phase.

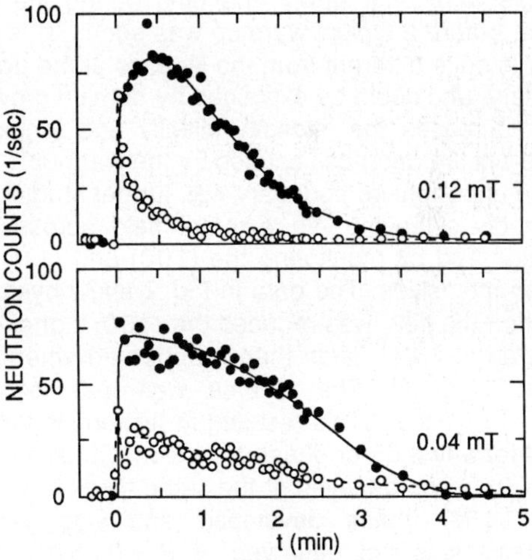

FIGURE 1
Time dependencies of the (1 1/3 1/3) (•) and (1 0 0) (○) neutron intensities at the phase boundary between the upper and middle phases (top) and between the middle and lower phases (below).

At B = 0.12 mT, the (1 0 0) signal disappeared while the (1 1/3 1/3) reflection, equivalent to the (0 2/3 2/3) peak under the fcc symmetry, appeared in a few tens of seconds. This can be explained by a formation of the (1 0 0) structure during the field sweep through the upper phase, which then decayed again as the middle phase was formed. After the transients, both signals disappeared in a few minutes as

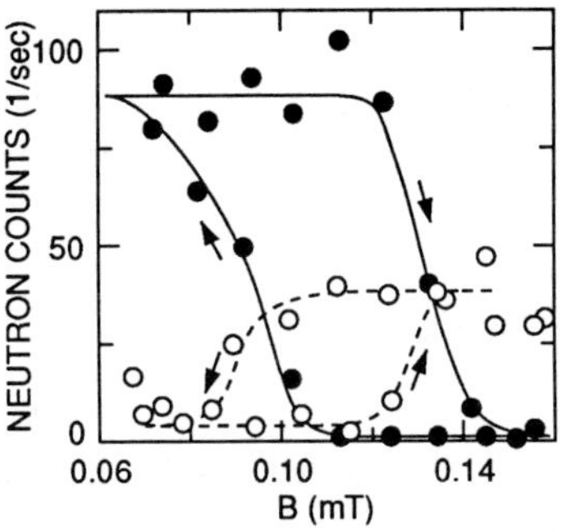

FIGURE 2
The neutron intensities of the (1 1/3 1/3) (●) and the (1 0 0) (○) Bragg peaks as a function of the external magnetic field, swept up and down across the phase boundary between the high- and intermediate-field phases at a constant rate, 4 μT/s.

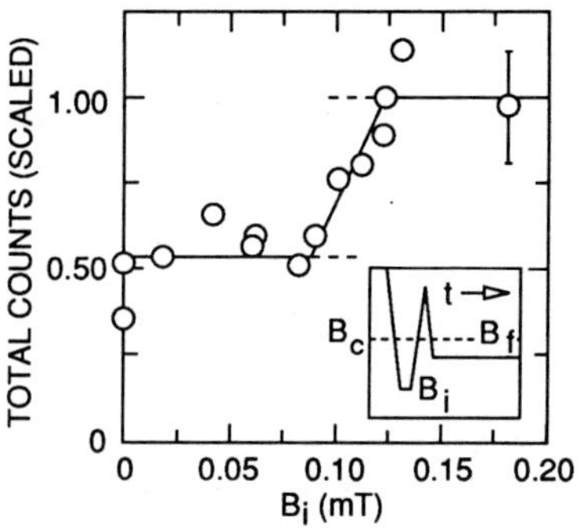

FIGURE 3
Total (1 0 0) neutron counts collected at $B_f = 0.18$ mT after a sweep to a chosen field B_i below, at, or above the phase boundary near $B = 0.12$ mT. The field sweeps are schematically illustrated in the inset.

the antiferromagnetic sublattice polarizations decreased due to the warmup.

At $B = 0.04$ mT the (1 0 0) and (1 1/3 1/3) neutron intensities were only slowly changing during the first minute, before a typical warmup was seen. This behavior is quite different from the kinetics at the upper boundary, and could be explained by domain growth, which increases the intensity initially and tends to counteract the decrease caused by the warmup.

The upper phase boundary was further studied by repeatedly sweeping the external field across the transition and by monitoring the (1 0 0) and (1 1/3 1/3) neutron intensities. The data in Fig. 2 show hysteresis: when the field was reduced the (1 0 0) signal disappeared in lower fields than it appeared when the field was raised. The reverse was true for the (1 1/3 1/3) reflection. This resulted in hysteresis loops, typical for a first order phase transition. Consistently, the (1 0 0) phase decayed at the same fields at which the (1 1/3 1/3) phase developed and vice versa. Hysteresis was not observed at $B_c = 0.25$ mT, the phase boundary against the paramagnetic region, which suggests that this transition is of second order, in agreement with previous measurements (4).

In the hysteresis experiments it became apparent that every time the boundary was crossed some neutron intensity was lost. To demonstrate this, a series of measurements was performed during which the external field was first demagnetized to a chosen field B_i below, at, or above the transition. The system was then brought, via the paramagnetic region, to a constant field $B_f = 0.18$ mT where the (1 0 0) intensity was recorded. This assured that the measuring field B_f was reached in the same way, independently of B_i. The counting statistics was improved by summing all neutrons collected during the warmup at 0.18 mT.

The total neutron counts, normalized to a reference measurement where $B_i = B_f$, is shown in Fig. 3 as a function of B_i. The drop in the neutron counts, when the field is reduced, extending from 0.09 to 0.12 mT, shows the transition region between the upper and middle phases. We correlate the loss of intensity to a gain of entropy at the first order phase transition. We cannot conclude, within the experimental accuracy, wether or not any entropy increase takes place below 0.09 mT.

In conclusion, neutron diffraction experiments in the $[01\bar{1}]$ external magnetic field direction have shown that the phase boundary at $B = 0.12$ mT, between the high- and intermediate-field phases of the nuclear spin system of copper, is associated with hysteresis and an entropy increase, indicating a first order phase transition. The time scale for kinetics was found to be as long as tens of seconds.

REFERENCES

(1) O.V. Lounasmaa, Phys. Today Oct. (1989) 26
(2) T.A. Jyrkkiö, M.T. Huiku, O.V. Lounasmaa, K. Siemensmeyer, K. Kakurai, M. Steiner, K.N. Clausen, and J. K. Kjems, Phys. Rev. Lett. 60 (1988) 2418.
(3) K. Siemensmeyer, K. Kakurai, M. Steiner, T.A. Jyrkkiö, M.T. Huiku, and K.N. Clausen, J. Appl. Phys. C (in press)
(4) M.T. Huiku, T.A. Jyrkkiö, J.M. Kyynäräinen, M.T. Loponen, O.V. Lounasmaa, and A.S. Oja, J. Low Temp. Phys. 62 (1986) 433.
(5) A.J. Annila, K.N. Clausen, P.-A. Lindgård, O.V. Lounasmaa, A.S. Oja, K. Siemensmeyer, M. Steiner, J.T. Tuoriniemi, and H. Weinfurter, Phys. Rev. Lett. (in press)

Physica B 165&166 (1990) 785–786
North-Holland

STUDIES OF NUCLEAR MAGNETISM IN SILVER AT POSITIVE AND NEGATIVE NANOKELVIN TEMPERATURES

Pertti HAKONEN and Shi YIN

Low Temperature Laboratory, Helsinki University of Technology, SF-02150 Espoo, Finland

We have measured the entropy and susceptibility of silver nuclei down to T = 0.8 nK and T = - 4.3 nK. A ferromagnetic Curie-Weiss law was found at negative absolute spin temperatures, in contrast to the antiferromagnetic law at T > 0. For the Curie-Weiss θ we obtain -4.4 ± 1.0 nK. These results provide the first direct demonstration that the antiferromagnetic alignment at positive temperatures transforms into a ferromagnetic preference at T < 0 in a nuclear system where exchange interaction dominates.

1. INTRODUCTION

Nuclear magnetism in pure metals has been investigated in copper (1,2), thallium (3), and silver (4,5). In all cases, spin ordering is expected at low enough temperatures, when the interaction energy is on the order of the thermal energy. We have studied silver which is an exchange dominated nuclear system that resembles closely the ordinary electronic magnets.

Natural silver has two isotopes: 51.8% of ^{107}Ag and 48.2% of ^{109}Ag, both with spin S = 1/2. These isotopes also have different gyromagnetic moments, -0.113μ_N and -0.130μ_N, respectively; here μ_N is the nuclear magneton. The Hamiltonian of our spin system $\{S_i\}$ is of the form

$$H = H_D + H_Z - (1/2)\Sigma_{ij} J_{ij} \mathbf{S}_i \cdot \mathbf{S}_j , \qquad (1)$$

where H_D and H_Z are the dipolar and Zeeman interactions, respectively, and J_{ij} are the exchange coefficients due to the indirect Ruderman-Kittel interaction (6). According to the NMR experiments (7), $|J_{ij}|/h \approx 26.5$ Hz for the nearest neighbour interaction, and the sign of J is expected to be negative on the basis of the spherical Fermi-surface approximation (6).

Monte-Carlo calculations predict nuclear ordering in silver into a simple type I antiferromagnetic state at 0.5 nK (8). However, the antiferromagnetic tendency may distinctly be observed already above T_c by looking for a Curie-Weiss type of behavior $\chi = C/(T - \theta)$, where C = 2.0 nK is the Curie constant and $\theta = S(S+1)\, \Sigma_j J_{ij}/3k_B \approx -4$ nK for a spherical specimen (9).

2. EXPERIMENTAL

The sample (4) consists of 78 silver foils with dimensions 25 μm x 4.5 mm x 40 mm. The nominal purity of the material is 99.99+%. The foils were selectively oxidized at 750 C for 20 h in $1 \cdot 10^{-4}$ Torr of dry air to eliminate the magnetic impurities which might shorten the spin-lattice relaxation time τ_1 in low magnetic fields. The foils were coated by a 7 μm layer of SiC powder to reduce eddy current heating during demagnetization.

The experiments were performed in our cascade nuclear demagnetization cryostat (1). The sample itself forms the second cooling stage in the initial field of 7.35 T, which was reduced to zero in 20 - 30 min. Two coaxial μ-metal cylinders were used as magnetic shields around the specimen. They were allowed to relax for 4 min in a field of 0.2 - 0.8 mT before starting the NMR measurements. The remanent field at the sample was on the order of 10 μT but it could be compensated to zero within ± 2 μT.

The highest initial polarizations in these experiments were 72 % at T > 0 and 40 % at T < 0. The population inversion for obtaining absolute negative temperatures was done by inverting the applied magnetic field rapidly.

NMR spectra were measured using an rf-SQUID amplifier located at 4 K. Its gain and phase were calibrated after each experiment to allow for a proper separation of the absorption and the dispersion signals.

One of the problems in studying an isolated spin system is thermometry. The nuclear spin temperature can be found by using the second law of thermodynamics $T = \Delta Q/\Delta S$ directly. Here ΔQ is the heat applied to the system and ΔS is the corresponding entropy change. At negative temperatures, ΔQ < 0 when the entropy increases, which shows that heat is being removed from the system. This is also evident when the system is heated by NMR excitation, $\Delta Q = \pi f \chi'' B_{rf}^2 \Delta t/\mu_o$ where the absorptive part of the susceptibility $\chi'' < 0$ when T < 0; Δt is duration of the rf-excitation, f is the frequency, and μ_o is the permeability.

In high magnetic fields, $B >> B_{loc} = 35$ μT, the polarization is given by $p = A \int \chi'' df$ (10), where A is a calibration constant. The entropy can be obtained from p, using the equation for the paramagnetic state. The polarization was determined around 0.2 mT because the identity of the silver spins with different gyromagnetic ratios is lost in low fields, resulting in a single exchange-narrowed NMR line which can be integrated with good accuracy.

0921-4526/90/$03.50

3. RESULTS AND DISCUSSION

The inverse static susceptibility $\chi'(0)^{-1}$ is displayed in Fig. 1. The susceptibility $\chi'(0)$ was obtained by integrating χ''/f from 30 to 180 Hz usually ; extrapolation to zero frequency was carried out by using the sum of two Lorentzian lines, one positive at $f > 0$ and one negative at $f < 0$. The figure shows an antiferromagnetic Curie-Weiss law $\chi = C/(T - \theta_A)$, with $\theta_A = -4.8$ nK when $T > 0$. At $T < 0$, we obtained a ferro-magnetic law $\chi = C/(|T| - \theta_F)$, with $\theta_F = 2.8$ nK.

Both sets of data in Fig. 1 follow the Curie-Weiss law to the lowest temperatures to an amazing accuracy. In fact, the susceptibility displays very little saturation, usually expected to be present close to an antiferromagnetic transition when $T > 0$ (1).

To obtain the true susceptibility χ_S (= susceptibility for a spherical specimen) from the measured data, $\chi'(0)$ must be corrected for the dipolar interaction and for the shape of the sample: $\chi_S'(0)^{-1} = \chi'(0)^{-1} + L - D$, where $L = 1/3$ is the Lorentz factor and D is the demagnetization coefficient. We estimate that $D \approx 0$, which is the demagnetization factor along the excitation field. Then we obtain for the Curie-Weiss $\theta = -[(|\theta_A| + |\theta_F|)/2 + LC] = -4.4 \pm 1.0$ nK

Combining the value of $\Sigma_j J_{ij}^2/h^2 = 8430$ Hz^2, obtained from NMR experiments (7), with our measured $\Sigma_j J_{ij}/h = 3k_B\theta/(hS(S+1)) = -370 \pm 80$ Hz, we may conclude that nearest neighbour interactions strongly dominate in silver. For an fcc lattice the molecular field theory then yields $\theta/T_c = -3$ for antiferromagnetic ordering (9). For silver this ratio is ≤ -5 which indicates that fluctuations may be suppressing T_c in this metal.

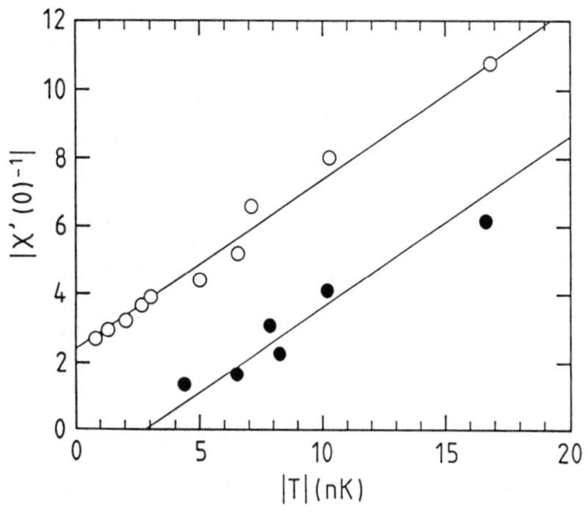

FIGURE 1
The inverse of the static susceptibility $\chi'(0)$ vs. the absolute value of the temperature measured at $T > 0$ (o) and at $T < 0$ (o). Solid lines are least square fits to $\chi'(0)^{-1} = T/C + \Delta$ keeping the Curie-constant $C = 2.0$ nK fixed.

4. CONCLUSIONS

We have studied nuclear magnetism in silver concentrating on the differences between negative and positive absolute temperatures. We have measured the susceptibility and entropy down to a record-low 0.8-nK temperature and, at $T < 0$ up to -4.3 nK. A Curie-Weiss law was observed for susceptibility both at $T > 0$ and $T < 0$. The nearest neighbour antiferromagnetic exchange interaction was seen to dominate in silver. For the Curie-Weiss θ we obtain -4.4 ± 1.0 nK.

5. ACKNOWLEDGEMENTS

We are grateful to A. Annila, Y. Takano and, especially, to A. Oja for discussions and for their contributions to the experimental setup. Continuous support and interest by O. V. Lounasmaa is gratefully acknowledged. We have also benefited from discussions with M. Goldman, M. Huiku, J. Jacquinot, J. Kurkijärvi, and H. Viertiö. This work was financially supported by the Academy of Finland.

REFERENCES

(1) M.T. Huiku, T.A. Jyrkkiö, J.M. Kyynäräinen, M.T. Loponen, O.V. Lounasmaa, and A. Oja, J. Low Temp. Phys. 62 (1986) 433.
(2) T.A. Jyrkkiö, M.T. Huiku, K. Siemensmayer, and K.N. Clausen J. Low Temp. Phys. 74 (1989) 435; A.J. Annila, K.N. Clausen, P.-A. Lindgård, O.V. Lounasmaa, A.S. Oja, K. Siemensmayer, M. Steiner, J.T. Tuoriniemi, and H. Weinfurter, Phys. Rev. Lett., (March 1990).
(3) G. Eska and E. Schuberth, Jap. J. Appl. Phys. Suppl. 26-3 (1987) 435; G. Eska, in Quantum Fluids and Solids - 1989, G.G. Ihas and Y. Takano eds., AIP Conf. Proc. 194 (New York, 1989), p. 316.
(4) A. Oja, A. Annila, and Y. Takano, to be published.
(5) P. J. Hakonen and S. Yin, to be published.
(6) M. Ruderman and C. Kittel, Phys. Rev. 96 (1954) 99.
(7) J. Pointrenaud and J.M. Winter, J. Phys. Chem. Solids 25 (1964) 123 .
(8) H.E. Viertiö and A.S. Oja, in Quantum Fluids and Solids - 1989, G.G. Ihas and Y. Takano eds., AIP Conf. Proc. 194 (New York, 1989), p. 305; H.E. Viertiö, Physica Scripta, in print.
(9) See e.g. J.S. Smart, Effective Field Theories of Magnetism (Saunders, Philadelphia, 1966).
(10) See e.g. A. Abragam and M. Goldman, Nuclear Magnetism: Order and Disorder (Clarendon Press, Oxford, 1982).

Physica B 165&166 (1990) 787–788
North-Holland

ANTIFERROMAGNETIC RESONANCE OF HYPERFINE ENHANCED NUCLEAR SPINS OF Ho IN $HoVO_4$

Haruhiko SUZUKI and Masumi ONO

Department of Physics, Faculty of Science, Tohoku University, Sendai 980, Japan

Hyperfine-enhanced nuclear antiferromagnetic resonance of $^{165}Ho(I=7/2)$ in $HoVO_4$ was observed below T_N=5 mK following demagnetization cooling. The temperature dependences of the resonance field and of the spectrum width were measured. The resonance frequency vs. applied field diagram was obtained.

1. INTRODUCTION

A Van Vleck paramagnetic compound $HoVO_4$ has a tetragonal zircon structure, space group D_{4h}^{19}. An antiferromagnetic ordering of the hyperfine enhanced nuclear spins in $HoVO_4$ takes place at about 5 mK (1). The nuclear spin ordering, produced by adiabatic demagnetized cooling of the specimen has been studied in some detail by the magnetic susceptibility (1),the nuclear orientation (2) and the neutron scattering (3) experiments. The ground state of the 4f electrons of $Ho^{3+}(4f^{10}, {}^5I_8)$ ion in HoVO4 is a singlet with a doublet at about 21cm^{-1}.When the nuclear spin of $^{165}Ho(I=7/2)$ in $HoVO_4$ couples with the induced electronic moment through the magnetic hyperfine interaction, the nuclear spin has an enhanced magnetic moment. The gyromagnetic ratio of the enhanced nuclear spin was found to be $\gamma_a/2\pi$=1526 MHz/T and $\gamma_c/2\pi \leq 20$ MHz/T. The coefficient of quadrupole interaction $P/\hbar$ was 25.9 MHz. The lowest levels of the enhanced nuclear spin are $I_z=\pm 1/2$. Here c axis is taken as the z axis. Since the gyromagnetic ratio of the enhanced nuclear spin is so highly anisotropic,the nuclear spin aligns in the a-a' plane of the tetragonal structure. From a calculation of the dipole interactions between enhanced nuclear moments, the enhanced nuclear spin of the antiferromagnetic state $HoVO_4$ can align along any direction in the a-a' plane (4). All of experiments described above, revealed the spin flop transition. Applying a magnetic field parallel to one of the equivalent a and a' axes, the spin flop transition takes place in a field of 60 or 80 Oe. This spin flop transition suggests the anisotropy existing in the a-a' plane. To get the information of this anisotropy and also of the dynamic properties of the antiferromagnetic state of the hyperfine enhanced nuclear spins, the antiferromagnetic resonance experiment was carried out below T_N=5 mK following demagnetization self cooling.

2. EXPERIMENT AND RESULTS

The single crystal used in our investigation was the same crystal used in our neutron scattering experiments (3) and roughly rectangular in shape, 4x4x13(mm), with the long dimension along the c-axis. The crystal was attached to the mixing chamber of a ^3He-^4He dilution refrigerator with thin copper wires and GE 7031 varnish. The crystal was mounted with one of a-axis parallel to the direction of the applied field. The cw NMR measurement was made by sweeping the magnetic field through resonance at some fixed rf frequencies. A NMR coil wound around the crystal produced a rf field parallel to another a-axis of the crystal. This NMR coil and a long coaxial cable formed the LC circuit. The modulation coil of the magnetic field was wound outside the vacuum jacket of the dilution refrigerator. The spin temperature of the specimen was determined from the measured magnetic susceptibility of the specimen itself. By this method we succeeded in observing the antiferromagnetic resonance of $HoVO_4$. Starting from an initial field of 1.5 T the crystal was self cooled by adiabatic demagnetization to 2.5 mK. Typical resonance spectra are shown in Figure 1. The resonance in Fig.1a was at the field of 320 Oe and the line width was about 120 Oe. The resonance in Fig.1b at the temperature just below T_N=5 mK was at the field of 215 Oe and the line width was about 220 Oe. The temperature dependences of the width of this spectrum are shown in Figure 2. Another typical resonance spectrum is shown in Figure 3. This spectrum consists of two resonance lines, one is at 8 Oe and the other at 260 Oe. The first one must be the resonance of the sublattice magnetic moment parallel to an applied field, while the other must be the resonance of the magnetic moment in the spin flop state. We can see the change in the spectrum around 100 Oe,corresponding to the spin flop transition. The noticeable feature of this spectrum is the line width. The resonance at lower field is rather sharp;The line width is 110 Oe. While the resonance at higher field, corresponding to the spin flop state show rather broad line with the width 230 Oe. In Figure 5 the resonance frequency versus the applied field diagram at the lowest

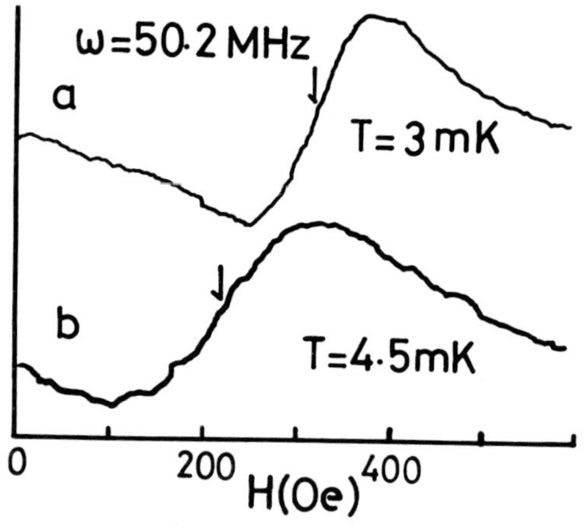

FIGURE 1
Typical antiferromagnetic resonance spectra.

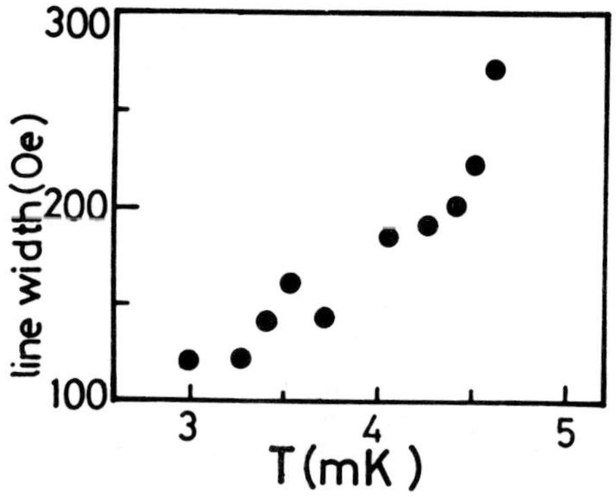

FIGURE 3
Temperature dependence of the line width. Measuring frequency was 50.2 MHz.

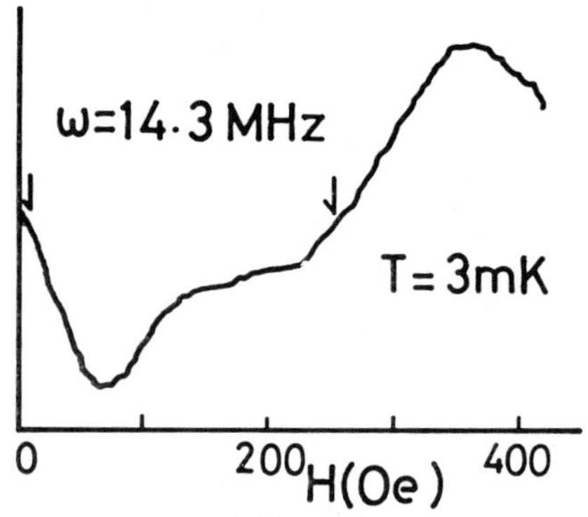

FIGURE 2.
Another resonance spectrum of $HoVO_4$ at ω=14.3MHz

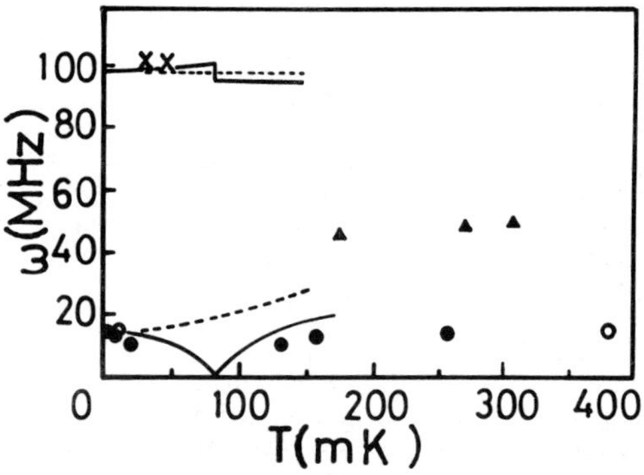

FIGURE 4
Resonance frequency vs. magnetic field diagram. Different marks represent different branches.

temperatures in our measurement was shown. In the figure we also plotted the calculated field dependences of the antiferromagnetic resonance of $HoVO_4$. Since two magnetic domains with sublattice moments parallel or perpendicular to the applied magnetic field exist with almost equal probability, as described in INTRODUCTION chapter, we can observe four branches in the resonance frequency vs. magnetic field diagram. Solid curves in the figure correspond to the calculated resonance with magnetic field parallel to the sublattice moment and dotted lines, perpendicular magnetic field. In this calculation we used the following values; The internal field H_E=111 gauss, that is, the calculated dipole-dipole interaction. The anisotropy field in a-a' plane H_A^a=31 gauss and the anisotropy field for the c axis H_A^c=543 gauss which is mainly due to the quadrupole interaction of the nuclear spin, were obtained to fit the experimental value at zero field to the calculated value. We also took the anisotropic gyromagnetic ratio into account for the calculation. At present we cannot fit the calculated lines to whole experimental points. More detailed analysis is in progress.

ACKNOWLEDGMENT
The support of Yamada Science Foundation are greatly acknowledged.

REFERENCES
(1) H.Suzuki, N.Nambudripad, B.Bleaney, A.L. Allsop, G.J.Bowden, I.A.Campbell and N.J. Stone, J.de Physique c6 39 (1978) 800.
(2) A.L.Allsop, B.Bleaney, G.J.Bowden, N. Nambudripad, N.J.Stone and H.Suzuki, Proc. R. Soc. A372 (1980) 19.
(3) H.Suzuki, T.Ohtsuka, S.Kawarazaki, N. Kunitomi, R.M.Moon and R.M.Nicklow, Solid State Commun. 49 (1984) 1157.
(4) B.Bleaney, Proc. R. Soc. A370 (1980) 313.

Physica B 165&166 (1990) 789–790
North-Holland

ANOMALOUSLY WEAK NUCLEAR SPIN-LATTICE COUPLING OF Cu BELOW 1 mK

K. GLOOS, P. SMEIBIDL, and F. POBELL

Phys. Inst., Universität Bayreuth, D-8580 Bayreuth, FRG

The nuclear spin-electron coupling of Cu has been determined in $2\ mT \le B \le 0.44\ T$ and at $15\ \mu K \le T \le 1\ mK$ using a constant heatflow method. At low temperatures the measured Korringa constant increases by at least an order of magnitude when compared to the high temperature value. Possible explanations for this result are discussed.

1. INTRODUCTION

The last decade has seen considerable progress in refrigerating samples to low temperatures using adiabatic demagnetization of Cu (1, 2). Whereas it is relatively easy to produce nuclear spin temperatures of 1 μK and lower (3), it is very difficult to transfer this low nuclear spin temperature to the electrons or lattice; today the lowest measured electronic temperatures are about 10 μK (1). Here we describe experiments to determine the nuclear spin-lattice coupling of Cu, which seems to be a limiting part in reducing the electronic temperatures.

2. EXPERIMENT

The sample, the second stage of a nuclear refrigerator (1), is a n = 1.0 mole, 5x20x70 mm^3 Cu plate of 5N purity and RRR ≃ 700. It is connected to a 104 mole Cu nuclear stage via a Ag rod interrupted by an Al s.c. heatswitch. Two Pt wire NMR thermometers in the field compensated region between both stages and on the sample itself monitor the electronic temperature of the sample. A 20 cm long, 25 μm ϕ WPt wire glued onto one side of the sample serves as resistive heater.

3. RESULTS

Precooling the sample to 2 to 6 mK in 9 Tesla and demagnetization to the final field B is conventional. Then we heat the sample using constant powers P ranging from 0.1 nW to 200 nW depending on the applied field. After each change of P we wait for a stationary reading of the NMR thermometers to get the electronic temperature T_e. From the measured energy Q necessary to warm up the Cu nuclear spin system we obtain the nuclear spin temperature T_n.

As a typical result we show in Fig. 1 the measured T_eQ versus $1/T_e^2$ for various powers P in a field of 55 mT. T_eQ depends linearly on $1/T_e^2$ and on P. In the high temperature approximation one would expect

$$T_eQ = n\varepsilon B^2 + \kappa P, \qquad (1)$$

where $\varepsilon = 3.19 \cdot 10^{-6}$ JK/T^2 mole. Our results can be interpreted as if Korringa's constant

$$\kappa = \kappa_o + \Gamma/T_e^2, \qquad (2)$$

with κ_o = 1.2 Ks, the high temperature value of Cu, and Γ a field dependent parameter.

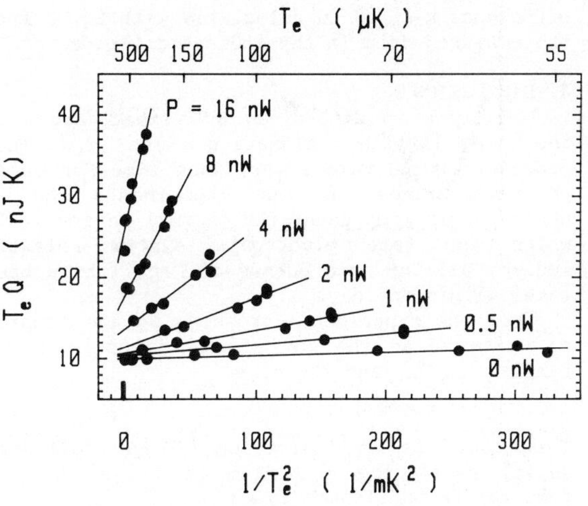

FIGURE 1
Measured data in 55 mT. The measured heatleak of order 50 pW has been taken into account.

Using the general formulae for: a) the heatflow between nuclei and electrons (4),

$$P = \frac{n\varepsilon B^2}{\kappa}\ \frac{e^{-Xe} - e^{-Xn}}{1 - e^{-Xe}}\ \frac{3}{2}\ \frac{I(I+1)-\langle m^2\rangle+|\langle m\rangle|}{I(I+1)} \qquad (3)$$

where $X_{e,n} = |g\mu_n|B/k_B T_{e,n}$, and b) the heat content of the nuclear spins, results in only small corrections to κ at the lowest T_e. Fig. 2 shows the difference $\kappa - \kappa_o$ calculated using the general formulae as a function of T_e for various fields. One clearly sees the T_e dependence of κ.

0921-4526/90/$03.50

For further support of these results we re-interpret earlier data of the minimum T_e reached by our first Cu nuclear stage in 2 mT to 16 mT (1). These data agree with the above results. The measured coefficient Γ depends linearly on field with $\Gamma = 0.87\ (mK^3s/mT)\ B$.

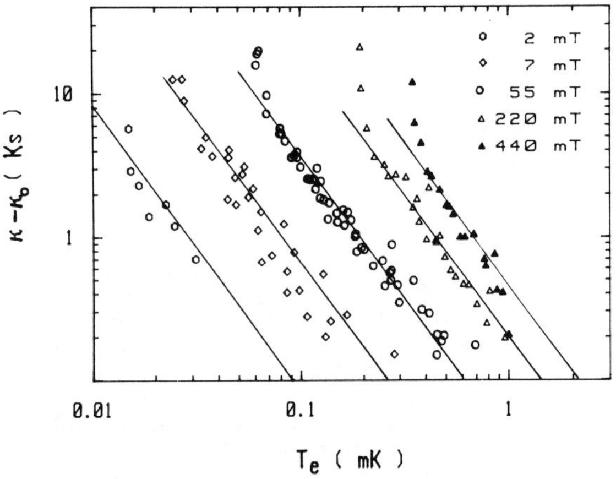

FIGURE 2
Difference κ-κ_o of Cu calculated with Eq. 3 from the measured data in the indicated fields.

4. DISCUSSION

Usually κ is determined by a measurement of the spin-lattice relaxation time τ_1. This procedure would take a very long time for Cu at μK temperatures. In our experiments the Cu nuclear spins represent the thermal reservoir to which the fast electronic system relaxes. Indeed, all the data points of Fig. 1 have been taken within one day.

From the experimental results we are tempted to write τ_1 as the sum of two times, one the familiar κ_o/T_e, and the other

$$\tau = \Gamma/T_e^3. \qquad (4)$$

This could be an indication of an additional serial scattering mechanism to cause the low temperature bottleneck in κ.

One possibility to explain an anomalous κ would be an anomalous behaviour of the nuclear spins in a magnetic field. We may expect some analogy with the high field phase of solid ^{3}He (5) or the appearance of nuclear spin waves (6). Since $\kappa = \kappa\ (T_e)$, and because we obtain the correct nuclear heat capacity, this possibility is unlikely. Another explanation would be an anomaly in the electronic system independent of the nuclear spins, e. g. a decreasing electronic density of states at E_F. Up to now there are no other measurements in this T and B range, to indicate any Fermi surface effects in Cu.

In the spin-lattice relaxation process the electrons are not only scattered to an energetically different state but they have to change their momentum because spin up and spin down bands are separated by the large electronic Zeeman energy. One may assume that the crystal as a whole takes the momentum. But it could be that phonons are involved in this momentum transfer. At high T many phonons having the required momentum are excited, so this process would not be the limiting one. The situation may change at lower temperatures when phonons freeze out much faster than electronic excitations. Anomalies in κ may then be expected at $T \simeq 2$ mK $\cdot$ (B/Tesla) in Cu, which would agree with our results.

According to Eq. (3) in the low-T limit the minimum electronic temperature achieveable using nuclear refrigeration of Cu, is

$$T_{min} \simeq (\Gamma_o \frac{9}{I(I+1)^2} \frac{1}{|g\mu_n|} \cdot \frac{P}{N_A})^{1/3}. \qquad (5)$$

Thus a heatleak of 10 pW/mole would yield results in agreement with todays measured minimum temperatures (1). To reduce T_{min} by an order of magnitude one would have to reduce P by three orders of magnitude which seems to be extremely difficult.

REFERENCES

(1) K. Gloos, P. Smeibidl, C. Kennedy, A. Singsaas, P. Sekowski, R.M. Mueller, F. Pobell J. Low Temp. Phys. 73, 101 (1988).

(2) F. Pobell, Physica 109&110 B, 1485 (1982); H. Ishimoto, N. Nishida, T. Furubayashi, M. Shinohara, Y. Takano, Y. Miura, K. Ono J. Low Temp. Phys. 55, 17 (1984).

(3) A. Abragam and M. Goldman, Nuclear Magnetism: Order and Disorder (Clarendon Press, Oxford, 1982); M.T. Huiku and M.T. Loponen, Phys. Rev. Lett. 49, 1288 (1984).

(4) F. Bacon, J.A. Barclay, W.D. Brewer, D.A. Shirley, J.E. Templeton, Phys. Rev. B 5, 2397 (1972); E. Klein in: Low Temperature Nuclear Orientation, eds. N.J. Stone and H. Postma (North-Holland, Amsterdam, 1986) D. Rainer, Int. Report, Univ. Bayreuth 1989.

(5) E.D. Adams, E.A. Schubert, G.E. Haas, and D.M. Bakalyar, Phys. Rev. Lett. 44, 789 (1980).

(6) G. Eska, AIP Conf. Proc. 194, Quantum Fluids and Solids, Gainesville, Fl. 1989, ed. G.G. Ihas, Y. Takano, p. 316.

hysica B 165&166 (1990) 791–792
orth-Holland

ANOMALOUS NUCLEAR MAGNETIC RESONANCE BEHAVIOUR OF ^{115}In IN $AuIn_2$

K. GLOOS, R. KÖNIG, P. SMEIBIDL, and F. POBELL

Phys. Institut, Universität Bayreuth, D-8580 Bayreuth, FRG

We have performed NMR experiments on ^{115}In in $AuIn_2$ refrigerated indirectly to 30 μK. Below a line in a temperature/field diagram of T/B = 4.7 (mK/T) and in fields of 6 mT to 53 mT we have observed strong deviations from a simple nuclear paramagnetic behaviour.

1. INTRODUCTION

At present the lowest electronic or lattice temperatures of almost 10 μK have been achieved by adiabatic nuclear demagnetization of Cu (1). These temperatures seem to be limited by heat leaks of order 10 pW/mole to the refrigerant, by its electronic thermal conductivity, and by a weakening of the nuclear spin-electron coupling of Cu in the microkelvin temperature range (2). Among the metals suitable for nuclear refrigeration to very low temperatures the intermetallic compound $AuIn_2$, in which both atoms are located in cubic lattice symmetry, seems to be promising because of its low superconducting critical field, its small Korringa constant, and the large nuclear moment of In (3-5).

Recently, we have performed nuclear demagnetization experiments with $AuIn_2$ which have led to a rather high minimum temperature of 86 μK (6). Consequently we did nuclear magnetic resonance investigations on ^{115}In in $AuIn_2$ to find the reason for the mentioned high final demagnetization temperature.

2. EXPERIMENT

The $AuIn_2$ sample (6 mm long, 5 mm diam.) has been produced by melting the stochiometric amounts of Au (5 N) and In (6 N) in a graphite crucible. After annealing for 4 h at 410°C, its RRR is about 70. Using less than 20 mg of Ga the sample was connected to a $(2 \times 7)\ mm^2$ Ag cold finger on the Cu nuclear stage of our nuclear refrigerator (1). A 4 mm long signal coil of three layers of 25 μm diam. Cu wire on a 6 μm thick mylar coil former was glued with a minute amount of diluted GE 7031 onto the $AuIn_2$ sample. The static field coil on a Cu coil former was sitting inside of a Nb cylinder for field shielding. For the NMR experiments typical pulse lengths were between 10 μsec and 100 μsec.

3. RESULTS

In Fig. 1 we show the FFT spectra in 53 mT and 26 mT at various temperatures. Above about 0.3 (0.2) mK at 53 (26) mT, the signal amplitude increases proportional to T^{-1} and the free induction decays exponentially with a constant $\tau_2^* \simeq 0.1$ ms. At these temperatures a single line with a temperature independent width is seen. At lower temperatures the NMR intensity decreases, the NMR line broadens, and the free induction decay is no longer exponential. We also observe a small shift of the resonance frequency to higher values, and at the lowest temperatures a doublet instead of a single line. Our data are not accurate enough to decide whether the original line splits into two lines or whether it stays and a new line appears at higher frequencies. The separation of the two lines of about 13 kHz seems to be independent of temperature and of magnetic field; it would correspond to an internal field of 1.4 mT. Both lines shift proportional to the applied magnetic field if this is slightly varied, without changing signal form or size.

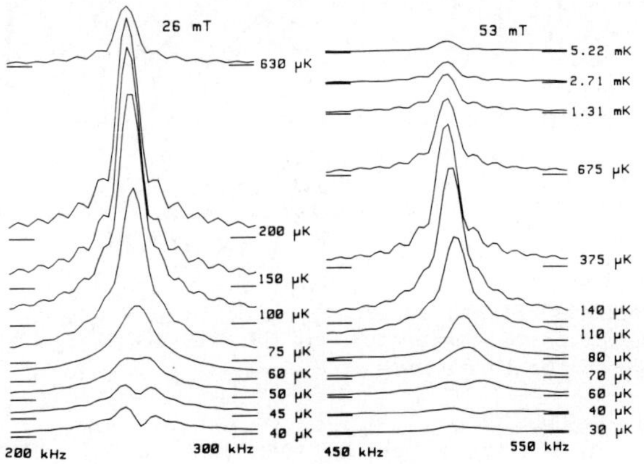

FIGURE 1
FFT NMR spectra of ^{115}In in $AuIn_2$ in fields of 26 mT and 53 mT.

Except for the line splitting we have observed the same effects in 13 mT and 6 mT; for these fields the line splitting should occur at lower temperatures.

In Fig. 2 we show the maximum FFT amplitudes A at constant field as a function of temperature. At high temperatures the dependence is $A \propto T^{-1}$, with a constant line width. This means that the nuclear susceptibility follows a Curie law. At a characteristic field-dependent temperature T_{ch} the amplitudes reach a maximum and then decrease drastically. T_{ch} increases linearly with magnetic field, according to $T_{ch} \cong 4.7$ (mK/T) B, corresponding to a 17 % polarization of the In nuclei. Plotting the reciprocal of the high temperature NMR amplitude A vs T, yields intercepts at 128 ± 70 (70 ±50; -12 ± 46) μK for 53 (26; 13) mT, which could indicate a ferromagnetic tendency.

Demagnetizing the $AuIn_2$ NMR sample from 53 mT in steps by a factor of two to 6 mT, and starting T between 0.1 mK to 3 mK, results again in too high final temperatures in agreement with the former results (6). Warming up the demagnetized sample we obtain T_{ch} in 6 mT which is given in Fig. 2 as well.

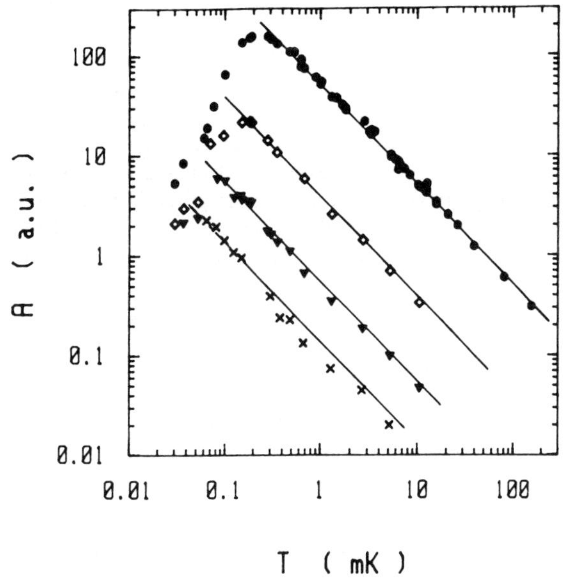

FIGURE 2
Maximum FFT NMR amplitudes of ^{115}In in $AuIn_2$ as a function of temperature at 53, 26, 13, and 6 mT (from top to bottom).

4. DISCUSSION

In (4) it was shown, that in bulk $AuIn_2$ the susceptibility results from nuclear magnetic interactions and quadrupolar effects are negligible. But even though $AuIn_2$ is cubic, there is the possibility that quadrupolar effects exist due to lattice distortions in our sample which has a lower RRR. Since we observe the resonance line at the correct Larmor frequency being proportional to the applied field, we conclude that this line is of magnetic origin. If the magnetic interaction is much smaller than quadrupolar interactions, we have either the ± 9/2 or the ± 1/2 doublet as the lowest levels. In the former case transitions within this doublet are not allowed and in the latter case the magnetization does not go through a maximum, opposite to what we observe. From these arguments and from the observation that $T_{ch} \propto B$ and the possibility to demagnetize $AuIn_2$ to below T_{ch}, we conclude that quadrupolar effects are not influencing our observations in a noticeable way.

The anomalous behaviour of ^{115}In nuclear spins in $AuIn_2$ occurs at temperatures which are at least two orders of magnitude larger than the nuclear spin temperatures where spontaneous long-range nuclear magnetic ordering is expected (5) and which increase with increasing magnetic field. A possible explanation of our observations would be an electronic polarization induced by the nuclear polarization due to the rather strong hyperfine coupling in $AuIn_2$. One may then observe an ordering of the hyperfine induced electronic moments similar to those seen in hyperfine enhanced Van Vleck paramagnets (7). But the high-temperature properties of $AuIn_2$ give no indication for such an electronic effect (3-5).

Qualitatively similar NMR anomalies have been observed for Tl in a comparable T-B range (8); but whereas we observe two NMR lines for a spin 9/2 system, the spin 1/2 system Tl shows even more NMR lines at low temperatures. In addition, in Tl the amplitude of the NMR line is larger than expected for a paramagnetic susceptibility whereas for $AuIn_2$ we observe a decrease of the signal intensity below the characteristic temperature.

REFERENCES:

(1) K. Gloos, P. Smeibidl, C. Kennedy, A. Singsaas, P. Sekowski, R.M. Mueller, and F. Pobell, J. Low Temp. Phys. 73, 101 (1988).
(2) K. Gloos, P. Smeibidl, and F. Pobell, to be publ. 1990.
(3) W.W. Warren Jr., R.W. Shaw Jr., A. Menth, F.J. Di Salvo, A.R. Storm, and J.H. Wernick, Phys. Rev. B 7, 1247 (1973).
(4) K. Andres and J.H. Wernick, Rev. Sci. Instr. 44, 1186 (1973).
(5) K. Andres and B. Millimill, J. de Physique 39, C 6-796 (1978).
(6) K. Gloos, R. König, P. Smeibidl, and F. Pobell, to be publ.
(7) J. Babcock, J. Kiely, T. Manley and W. Weyhmann, Phys. Rev. Lett. 43, 380 (1979); M. Kubota, H.R. Folle, C. Buchal, R.M. Mueller and F. Pobell, Phys. Rev. Lett. 45, 1812 (1980).
(8) G. Eska and E. Schuberth, Jap. J. Appl. Phys. 26-3, 435 (1987).

Physica B 165&166 (1990) 793–794
North-Holland

NQR STUDIES OF SCANDIUM METAL AT LOW TEMPERATURES

L. POLLACK, E.N. SMITH, R.E. MIHAILOVICH, J.H. ROSS Jr.[§], P. HAKONEN[†], E. VAROQUAUX[‡], J.M. PARPIA, R.C. RICHARDSON

Materials Science Center, Cornell University, Ithaca, NY 14853, USA

We have performed susceptibility and relaxation time measurements on scandium metal using nuclear quadrupole resonance (NQR) techniques at temperatures down to 100μK.

1. INTRODUCTION

In systems which possess nuclear spins greater than I = 1/2 the nuclei have electric quadrupole moments. When these atoms are in a crystal which does not have cubic symmetry, the coupling of the quadrupole moments of the nucleus to the electric field gradient causes a splitting of the nuclear levels. Since this interaction depends solely on the orientation of the nucleus, in the absence of an applied magnetic field the $\pm\ m_I$ levels are degenerate. Typically these splittings are on the order of 10's of MHz, but for some systems where the asymmetry of the crystalline axes is slight the splittings can be significantly smaller. One such system is scandium metal (Sc). The nuclear spin of Sc is 7/2, thus the effect of the electric field gradient on the nuclei is to split the nuclear levels into four doubly-degenerate levels. The fundamental quadrupolar splitting of scandium is 130kHz and the sign of the quadrupole interaction is such that the ground state is the $\pm$ 7/2 state. Thus the splitting between the lowest two states, the ($\pm$ 7/2 $\leftrightarrow$ $\pm$ 5/2) is 390kHz; the spacing between the next two levels ($\pm$ 5/2 $\leftrightarrow$ $\pm$ 3/2) is 260kHz; and the ($\pm$ 3/2 $\leftrightarrow$ $\pm$ 1/2) level spacing is 130kHz. In temperature these splittings are 18 μK, 12 μK and 6 μK respectively. Since this temperature is close to the expected operating range of our system, scandium metal should be interesting to observe as a function of temperature.

2. APPARATUS

In our first attempt at studying this system, ten 0.13 mm thick foils of polycrystalline Sc were clamped to a piece of copper which was firmly attached to the nuclear stage of a 65 mole copper nuclear demagnetization bundle. A single 500 turn pickup coil serves as the excitation and receiver coil. The coil surrounds the foils, but does not come into thermal contact with them. The circuit can be tuned with the addition of a capacitor external to the cryostat, allowing us to look for any of the three transitions. A superconducting shield encloses the experiment to protect the sample from fringe fields. The population differences between various states can be measured using pulsed-NQR techniques (1). Due to the variations in the gradient in the local electric field, the T_2^* of the Sc is on the order of 40 μsec. Since the recovery time of the receiving amplifier is comparable to the T_2^* it is only possible to observe spin-echoes at the "high" temperature end of the experiment (T> 1mK). At lower temperatures, as the magnetization increases, it is possible to observe tails of the free induction decay signals (FIDs) from the Sc nuclei. Other possible sources of broadening of the signal come from the inhomogeneities between different foils, from the possible reduction of the H_1 field at the innermost foils due to shielding from the outermost foils, or from the variation of the tipping pulse within a skin depth for a single foil. To minimize these problems a single Sc foil is currently being used. The thermal contact between this foil and the copper stage is made by melting potassium and "soldering" the scandium to the copper holder.

A second coil which contains several of the Sc foils is mounted near the center of the bundle, inside the superconducting magnet. This coil is not shielded from the demagnetizing field (H_o). It can be used to measure the changes in susceptibility at fixed field, and also to explore the field dependence of the quadrupole transitions. The T_2^* of the signal changes as a function of H_o, and at reasonably low values of H_o, such as the chosen final field present at the end of a demagnetization, it is possible to observe FID signals from this coil which persist into the "high temperature" regime (T $\approx$ 10mK) thus observation of this signal is relatively

Present addresses: [§] Physics Dept., Texas A&M Univ., College Station, TX [†] Low Temp. Lab Otakaari 3A, SF-02150 Espoo, Finland [‡] Universite de Paris Sud, Orsay, France easy. All values of the temperature are derived from measurements of Pt nuclear susceptibility.

3. RESULTS

Preliminary measurements of the susceptibility of the Sc in the second coil in a small field ($H_o = 300$ Gauss) show that Curie's law is satisfied down to temperatures on the order of 0.1 mK. A typical FID signal, taken at 1mK, is shown in figure 1. We have observed changes in the susceptibility in both the 260kHz and the 390 kHz transitions. We have not been able to observe the 130kHz transition on either coil, which is surprising since its expected signal size is greater than that from either of the other transitions (2).

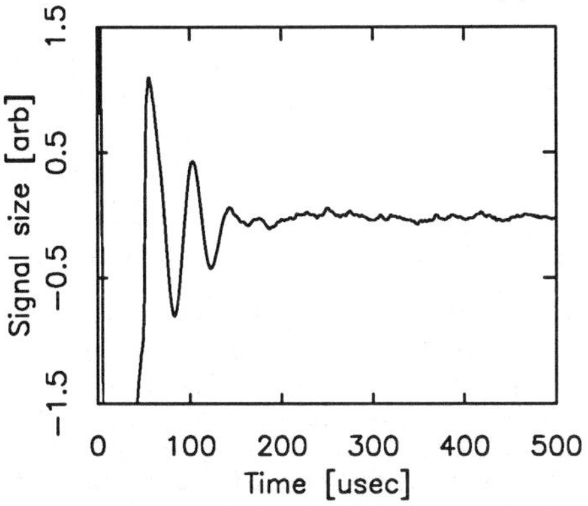

Figure 1 - A typical free induction decay signal

It is also interesting to measure the spin-lattice relaxation times of these systems. NQR systems are expected to show multi-exponential relaxation rates, since the relaxation of each set of levels to equilibrium occurs at an independent rate. For Sc, since there are three sets of levels, a three exponential decay time is expected. The time constants for these rates range from 10 to 60 seconds at 10 mK (3). Measurements made of the return to equilibrium after "saturation" indicate that there are several contributions to the total relaxation, but the shortest time constant we observe is about 10 msec which is significantly shorter than any of the expected spin-lattice relaxation times. If a "fast" process exists that equilizes the spin temperatures of the different levels, the time constant that dominates the equilibration of the system is the shortest spin-lattice relaxation time. A short recovery time would be advantageous in very low temperature experiments. This fast relaxation rate seems to be more or less independent of temperature. It is possibly due to incomplete saturation of the spins since the dimensions of the sample are large compared to the skin depth of the RF. The time constants were measured at the high temperatures by attempting to saturate the spins with the two-pulse sequence that generates the largest echo, then by the examination of a second echo at a variable later time. At lower temperatures smaller pulses were used to minimize heating. The results are displayed in figure 2.

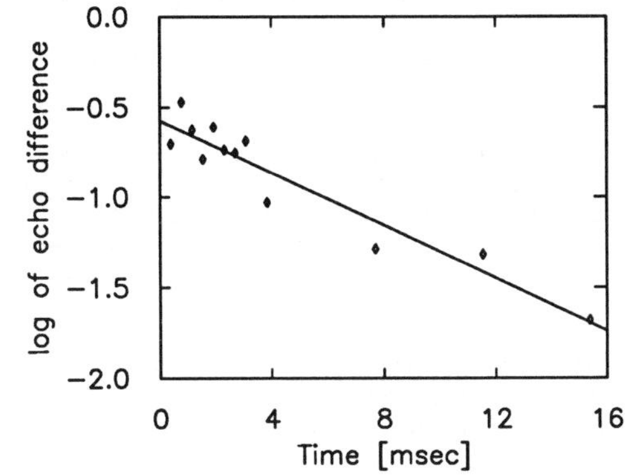

Figure 2 - A relaxation time of 14 msec results from fitting the difference in height of an echo and the "infinite time" echo as a function of the time delay between the saturating sequence and the echo-generating pulses.

4. CONCLUSIONS

Our initial measurements on Sc have yielded some unexpected results. We are currently modifying the system to allow for the detection of FIDs on the magnetically shielded coil. Study of these signals will eliminate questions about heating caused by the 2-pulse echo sequences and should simplify the analysis.

ACKNOWLEDGEMENTS

This work has been supported by the Materials Research Division of the NSF through grant number DMR-8818558. We would like to acknowledge the assistance of: P. Fraenkel, T. Gramila, L. Gunderson and G. Wong. We would also like to thank Professor J. Burlitch for his assistance in preparing the potassium solder joint.

REFERENCES

(1) M. Bloom and R.E. Norberg, Phys. Rev. 93, (1954) 638.
(2) E.I. Fukushima and S.B.W. Roeder, Experimental Pulse NMR (Addison-Wesley, Reading, MA 1981).
(3) A. Narath and T. Fromhold Jr., Phys. Lett. 25A (1967) 49 and M.H. Cohen and F. Reif, Quadrupole Effects in NMR Studies in Solids, in: Solid State Physics, Vol. 5, eds. F. Seitz and D. Turnbull, (Academic Press, NY 1957) pp. 322-448.

Physica B 165&166 (1990) 795–796
North-Holland

EXPERIMENT ON NUCLEAR SPIN ORDERING IN Sc METAL

Haruhiko SUZUKI ,Takuo SAKON and Naoki MIZUTANI

Department of Physics, Faculty of Science, Tohoku University, Sendai 980, Japasn.

In order to achieve the nuclear spin ordering in Sc metal, We constructed two stage demagnetization cryostat. Before doing the two stage demagnetized cooling experiment, we measured a temperature dependence of the magnetic susceptibility of Sc metal above 0.25 mK by the one stage demagnetized cooling. Result can be understood by the spin glass phenomenon of iron impurity in Sc metal.

1. INTRODUCTION

Among a large number of pure metals, only in Cu the nuclear spins have been cooled down below their spontaneous ordering temperature (1). Though great efforts have been devoted to achieve the nuclear spin order, the clear evidence of ordering has not yet been observed in other metals (2),(3) so far. Even in a simple metal as Cu, the observed nuclear spin structures are not so simple. In the case of Cu, the indirect exchange interaction, mediated by conduction electrons is comparable in magnitude to the nuclear dipole-dipole interaction.

We are interested in the nuclear spin order in Pd and Sc metals which are known as the highly exchange enhanced Pauli paramagnetic metals. Paradium, however, consists of a large number of isotopes, of which some have no nuclear magnetic moment. This makes it rather hard to cool nuclear spins in Pd below the ordering temperature by the demagnetized self cooling. In the case of Sc metal the natural abundance of ^{45}Sc (I=5/2) is 100 %. We chose Scandium metal as a candidate substance for the investigation of the nuclear spin order.

We will describe some features of Scandium metal in comparison with Cu.
Nuclear spin: I=7/2 (Cu: I=3/2), Magnetic moment: 4.756 μ_N (2.275 μ_N), Crystal structure: hcp (fcc), Lattice constant: a=3.29 A, c=5.25 A (a=3.615 A), Spin-lattice relaxation time: T_1T=1.6 Sec.K (1.2 Sec.K).

Assuming that only the dipole-dipole interaction between nuclear spins exist,the ordered state of nuclear spins in Scandium can be expected to be an antiferromagnetic state. The calculated dipole-dipole interaction of nuclear spins in hcp Sc is 7.31 Oe (4.56 Oe).

2. EXPERIMENTAL PROCEDURE AND RESULT

We constructed a two stage nuclear demagnetization cryostat. The second stage is our sample, Scandium rod, the purity of the sample is 99.9 %. The main magnetic impurity concentration in our sample is 50 at.ppm Fe. Since it is very hard to weld Sc metal, the Scandium rod was tightly fitted to the OFHC copper thermal link which is conducted to the first nuclear stage. The first stage is composed of 33 mol effective copper which is cut from a bulk OFHC copper. The magnetic susceptibility was measured by a SQUID magnetometer. To reduce the residual field at the specimen, a high metal shield was set just outside the radiation shield from the still. By the first stage demagnetization from 7.5 T to 0.1 T, the sample is pre-cooled in a magnetic field of 7.5 T to sub-milikelvin temperature. Before doing the two stage demagnetizated cooling experiment, we studied the properties of Sc metal at low temperatures by demagnetizing the first nuclear stage. The temperature dependence of the magnetic susceptibility of Sc metal was shown against the temperature in Figure 1. In our present experiment the temperature was measured by a LCMN thermometer which was immersed in the liquid ^{3}He. The liquid ^{3}He cell was installed on the experimental space of the first nuclear stage. Above about 3mK the LCMN thermometer followed the nuclear stage temperature rather quickly when the magnetic field was stopped at some fixed values. The temperature measured by the LCMN thermometer versus the magnetic field diagram showed rather good straight line which can be expected from the entropy relation of adiabatic demagnetized cooling. Below about 3 mK the LCMN temperature started to deviate from the straight line. On the other hand Sc metal showed rather quick response down to the lowest temperature of our experiment. The relaxation time of the spin in Sc measured by the magnetic susceptibility was about 3 hours at 250 μK. From the Korringa relation the nuclear spin-lattice relaxation time is expected to be 1.8 hours, same magnitude as the value in our experiment. However, the magnetic susceptibility shown in Figure 1 must be contributed mainly from the electronic one of the iron impurity. The spin-lattice relaxation time of the iron impurity should be much faster. Therefore the observed relaxation can be attributed to the boundary resistance between the Sc rod and the Cu thermal link. Since the heat leak into our nuclear stage is less than a few nW, we could keep the temperature of Sc metal below about 250 μK more than two days. Therefore, even though the thermal relaxation time is slightly longer than the intrinsic spin-lattice relaxation time, it will be possible to cool the nuclear spins in Sc metal in the field of 7.5 T low enough by waiting a few days.

Though Sc metal is one of the typical 3d metals, it has not yet been investigated so much. The major obstacle for the investigation of Sc metal has been a lack of ability of high purity Scandium metal, especially with respect to the iron content. The presence of a small amount of the iron impurity greatly affects the values of physical properties of Sc metal (4). The magnetic susceptibility of Sc metal at higher temperatures was measured by several authors (5,6). But there have been no measurement below 0.7 K. The measurement by R.J.Stierman (6) et al. was made for the high purity specimen which contained the iron impurity less than 1 at.ppm. Their result of the magnetic susceptibility showed the maximum at about 30 K and the minimum at around 6 K.

As shown in the Figure, the temperature dependence of Sc at low temperatures showed a maximum just below 1 mK. The magnetic moment of electrons in Sc metal has a possibility of ordering at low temperatures. However, the maximum of the magnetic susceptibility in the Figure might be correspond to the freezing temperature of spin glass of Fe impurity. In Pd metal spin glass phenomena of Fe impurity were observed at the same temperature region as our present experiment by Julich group (7). They also investigated the variation of the maximum of magnetic susceptibility with changing the concentration of Fe impurity from 2.2 ppm to 100 ppm. In the case of Sc metal the second phase Sc_3Fe can exist even in low Fe concentration Sc metal of about 85 at.ppm (4). This makes it rather hard to estimate the concentration of Fe effective for spin glass phenomena.

Anyhow, the fact that the iron impurity freezes at such high temperatures like mk or sub mK will be favourable for experiments on nuclear spin ordering. The iron impurity will not disturb the phenomena of the nuclear spin order and also will not disturb the observation of the nuclear spin order by the magnetic susceptibility measurement. The two stage nuclear demagnetization experiment is in progress.

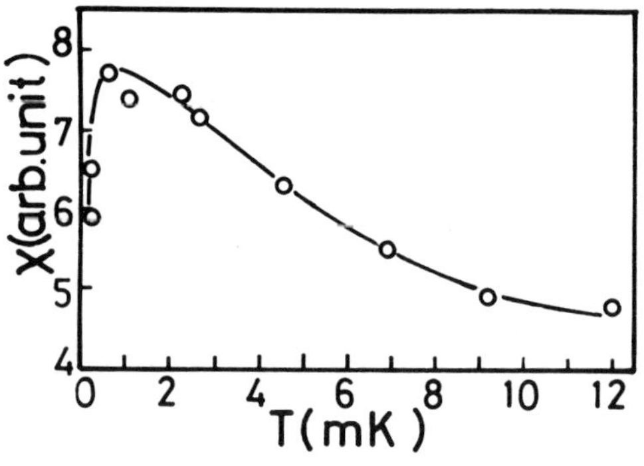

Figure 1
Temperature dependence of the magnetic susceptibility of Sc metal.

ACKNOWLEDGMENT

The support of Yamada Science Foundation are greatly acknowledged.

REFERENCES

(1) M.T.Huiku, T.A.Jyrkkio and M.T.Leponen, Phys. Rev. Letters 50 (1983) 1516.
(2) A.S.Oja, A.J.Annila and Y.Takano, unpublished.
(3) G.Eska and E.Schuberth, Proc. LT-18; Jap. J. Appl. Phys. Suppl.26-3 (1987) 435.
(4) T.W.E.Tsang and K.A.Gschneider,Jr, J. Less-Common Metals 80 (1981) 257.
(5) F.H.Spedding and J.J.Croat, J. Chem. Phys. 58 (1973) 5514.
(6) R.J.Stierman, K.A.Gschneider,Jr., T.W.E. Tsang, F.A.Schmidt, P.Klavins, R.N.Shelton, J.Queen and S.Legvald, J. Magn. Magn. Mat. 36 (1983) 249.
(7) R.P.Peters, Ch.Buchal, M.Kubota, R.M. Mueller, and F.Pobell, Phys. Rev. Letters 53 (1984) 1108.

Physica B 165&166 (1990) 797–798
North-Holland

COMPETITION OF ANTIFERROMAGNETIC "UP-DOWN" AND "UP-UP-DOWN" ORDER IN THE NUCLEAR SPIN SYSTEM OF COPPER

Hanna VIERTIÖ[1] and Aarne OJA[2]

[1]Research Institute for Theoretical Physics, University of Helsinki, SF-00170 Helsinki, Finland
[2]Low Temperature Laboratory, Helsinki University of Technology, SF-02150 Espoo, Finland

The magnetic phase diagram of nuclear spins in copper, an ideal fcc system, has been investigated using mean-field theory. At nanokelvin temperatures, the strongly anisotropic spin interactions stabilize field-induced antiferromagnetic "up-up-down" order, while the spin structure at zero field is of "up-down" type. At low fields, a single magnetic domain displays both up-up-down and up-down order. Our theory explains the recent neutron-diffraction data.

Recently, Annila et al. (1) reported observation of the $(0,^2/_3,^2/_3)$ neutron Bragg reflection in the nuclear spin system of copper at nanokelvin temperatures. Three antiferromagnetic phases were found as a function of the external magnetic field below 0.25 mT, aligned in the $[01\bar{1}]$ crystalline direction. In low fields, the previously observed (1,0,0) signal (2) and the novel $(0,^2/_3,^2/_3)$ reflection are present simultaneously. Around 0.08 mT, only the $(0,^2/_3,^2/_3)$ reflection was seen, while the order in still higher fields is of the (1,0,0) type. These observations were quite surprising: the $(0,^2/_3,^2/_3)$ reflection is not consistent with any previous experimental or theoretical spin structures in an fcc lattice.

We have recently developed a theory (3) which explains the neutron results. The observed $(0,^2/_3,^2/_3)$ antiferromagnetic reflection is found to correspond to an "up-up-down" order, *uud* for short, with spins along the external field. The theoretical spin arrangement at low fields is most remarkable: the *uud* modulation coexists with up-down (*ud*) ordering perpendicular to the field, which gives rise to simultaneous $(0,^2/_3,^2/_3)$ and (1,0,0) Bragg reflections.

Nuclear spins in copper interact via the dipolar force and the conduction-electron-mediated indirect interaction. Since the dipolar force dominates, the coupling $\mathbf{A}(\vec{r}_{ij})$ between spins $\vec{S}_i$ and $\vec{S}_j$ is strongly anisotropic. In the mean field (MF) theory, the characteristic energy of an antiferromagnetic modulation with ordering vector $\vec{k}$ can be found from the eigenvalue equation

$$\mathbf{A}(\vec{k}) \cdot \vec{e}_n(\vec{k}) = \lambda_n(\vec{k})\vec{e}_n(\vec{k}), \quad n = 1,2,3\,, \tag{1}$$

where $\lambda_n(\vec{k})$ are the eigenvalues and $\vec{e}_n(\vec{k})$ the eigenvectors of the Fourier sum $\mathbf{A}(\vec{k}) = \Sigma'_j \mathbf{A}(\vec{r}_{ij})\exp(-i\vec{k}\cdot\vec{r}_{ij})$ of the interactions.

First-principles calculations (4) have revealed a double-well structure of $\lambda_n(\vec{k})$ as a function of $\vec{k}$: there is a minimum at the Brillouin-zone boundary at $\vec{k} = (1,0,0)$ and another in the [011] direction close to $\vec{k} = (0,^2/_3,^2/_3)$, as illustrated schematically in Fig. 1. The calculated (5) $\lambda_n(\vec{k})$ for the *uud* order is only 8 % larger than that for a *ud* structure. As the reported uncertainty was 10 %, it is not clear which one is the global minimum.

There is a unique eigenvector corresponding to the $(0,^2/_3,^2/_3)$ order, which gives rise to an easy-axis Ising-type anisotropy, while for (1,0,0) there are two degenerate eigenvectors, which span an easy plane of anisotropy. Taking into account the cubic symmetry, there are altogether 12 possible ordering vectors - and Bragg reflections - for the *uud* order and 3 ordering vectors for the *ud* order. The easy-plane anisotropy makes it possi-

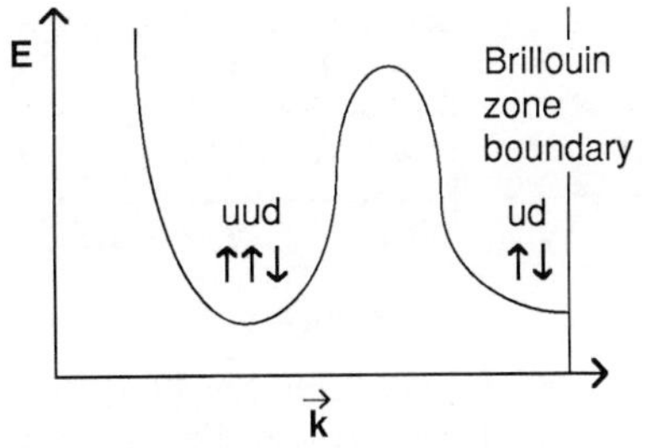

FIGURE 1
Schematic illustration of the double-well structure of energy E for the ordered states. The *ud* order corresponds to a local minimum of E, whereas the global minimum in our theory corresponds to a *uud* state. (6)

ble to find stable *ud* states at arbitrary magnetic fields, whereas the Ising feature limits the *uud* structures to discrete values of the field. An interesting situation occurs if λ_{uud} is slightly lower than λ_{ud}. The system then chooses the *uud* state at these discrete fields for which it is stable; at other fields the system must compromise.

In Ref. 3 we presented the stability conditions for the ground state rigorously by using the concept of permanent spin structures. The analysis can be performed analytically in the limit of degenerate *uud* and *ud* order, more precisely, for $\lambda_{ud} - \lambda_{uud} \rightarrow 0_+$.

Here we illustrate the general idea by using the double-well picture of Fig. 1. At $T \ll T_c$, the sublattice polarization is saturated and hence a pure *uud* phase has the magnetization (1+1+(-1))/3 = 1/3. Exactly at $B = B_c/3$, this is the ground state because the intrinsic ferromagnetic component then matches the induced magnetization. In lower fields, the too large intrinsic magnetization makes a pure *uud* structure unstable. Therefore, the system makes a linear combination of the best and the second-best solutions: when the field is lowered below $B_c/3$, the *uud* order is decreased so that the intrinsic ferromagnetic component matches the induced magnetization in that field, and the second-best antiferromagnetic order *ud* is increased.

The ground-state spin structures are shown in Fig. 2 for the $[01\bar{1}]$ field direction. The MF equation for the ground state is

$$\vec{S}_i = (0, d_0, -d_0) + (0, d_1, -d_1)\cos((0,{}^2/_3,{}^2/_3) \cdot \vec{r}_i) + (0, d_2, d_2)\cos((1,0,0) \cdot \vec{r}_i) + (d_3, 0, 0)\cos((0,1,0) \cdot \vec{r}_i) \; , \quad (2)$$

where $-d_1/4 = d_0 = B/(\sqrt{2}B_c)$ and $2d_2^2 + d_3^2 = p^2 - 9d_0^2$.

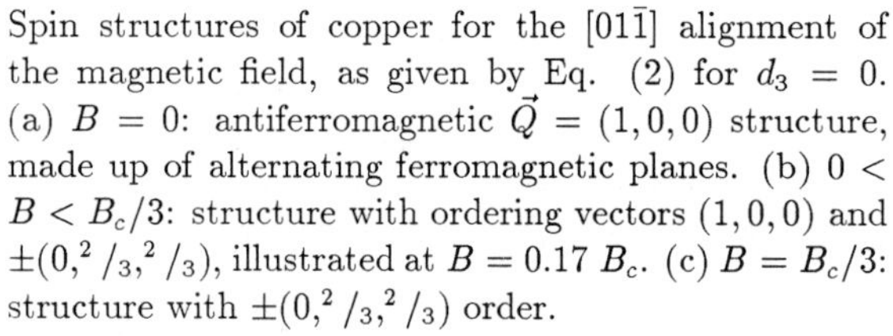

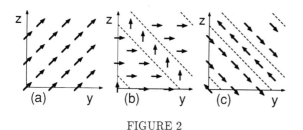

FIGURE 2

Spin structures of copper for the $[01\bar{1}]$ alignment of the magnetic field, as given by Eq. (2) for $d_3 = 0$. (a) $B = 0$: antiferromagnetic $\vec{Q} = (1,0,0)$ structure, made up of alternating ferromagnetic planes. (b) $0 < B < B_c/3$: structure with ordering vectors (1,0,0) and $\pm(0,{}^2/_3,{}^2/_3)$, illustrated at $B = 0.17\ B_c$. (c) $B = B_c/3$: structure with $\pm(0,{}^2/_3,{}^2/_3)$ order.

Here p is the sublattice polarization, which depends on temperature. When the field is reduced, the *ud* order grows continuously while the *uud* component goes to zero at $B = 0$. At fields $B/B_c > p/3$, the above structure does not exist. Therefore, a first-order transition to the *ud* phase takes place. Note the selection rule for the 12 symmetry-related *uud* reflections: only two, $\pm(0,{}^2/_3,{}^2/_3)$, are present and with equal intensity. All these findings are in excellent agreement with the neutron diffraction results. (1)

The temperature dependence of the *uud* and *ud* order is nontrivial. The *ud* order decreases continuously towards zero when $T \rightarrow T_c$, whereas the *uud* order in Eq. (2) does not depend on temperature. This explains the observed different time behaviors of the antiferromagnetic Bragg reflections during warmup. (1)

In general, the *uud* order can coexist either with *ud* or with *uudd* order in widely different systems having Hamiltonians of suitable anisotropy.

ACKNOWLEDGEMENTS

We thank O.V. Lounasmaa for valuable comments. This work has been supported by the Academy of Finland. H.V. acknowledges support by the Magnus Ehrnrooth Foundation.

REFERENCES

1. A.J. Annila, K.N. Clausen, P.-A. Lindgård, O.V. Lounasmaa, A.S. Oja, K. Siemensmeyer, M. Steiner, J.T. Tuoriniemi, and H. Weinfurter, to appear in Phys. Rev. Lett. (March, 1990); see also three contributions by Annila et al. in this volume.
2. T.A. Jyrkkiö, M.T. Huiku, O.V. Lounasmaa, K. Siemensmeyer, K. Kakurai, M. Steiner, K.N. Clausen, and J.K. Kjems, Phys. Rev. Lett. 60 (1988) 2418.
3. H.E. Viertiö and A.S. Oja, Phys. Rev. Lett. (submitted); H.E. Viertiö and A.S. Oja, Nuclear antiferromagnetism in copper: predictions for spin structures in a magnetic field, in this volume.
4. P.-A. Lindgård, X.-W. Wang, and B.N. Harmon, J. Magn. & Magn. Mater. 54-57 (1986) 1052.
5. A.S. Oja, X.-W. Wang, and B.N. Harmon, Phys. Rev. B39 (1989) 4009.
6. Close to T_c the global minimum of free energy in the MF theory corresponds to an ordering vector $(0, 2/3+\delta, 2/3+\delta)$, rather than the commensurate *uud* order. At low temperatures, 6th order terms in the appropriate Ginzburg-Landau expansion lock the ordering vector into $(0,{}^2/_3,{}^2/_3)$.

Physica B 165&166 (1990) 799–800
North-Holland

NUCLEAR ANTIFERROMAGNETISM IN COPPER: PREDICTIONS FOR SPIN STRUCTURES IN A MAGNETIC FIELD

Hanna VIERTIÖ[1] and Aarne OJA[2]

[1]Research Institute for Theoretical Physics, University of Helsinki, SF-00170 Helsinki, Finland
[2]Low Temperature Laboratory, Helsinki University of Technology, SF-02150 Espoo, Finland

The stable structures of nuclear spins in copper in various directions of the external magnetic field are predicted. At low fields the "up-up-down" order coexists with the "up-down" order in a single magnetic domain. A scheme for experimentally manipulating the domain populations is suggested.

The antiferromagnetic $(0,{}^2/_3,{}^2/_3)$ neutron Bragg reflection was recently observed in the nuclear spin system of copper at nanokelvin temperatures. (1) The new reflection, together with the previously observed (1,0,0) peak (2), indicated the presence of three antiferromagnetic phases as a function of the external magnetic field, aligned in the $[01\bar{1}]$ crystalline direction. We have recently developed a theory (3) which explains the observed interplay of these Bragg reflections corresponding to "up-up-down" (*uud*) and "up-down" (*ud*) order. Here we extend our analysis to the other high-symmetry directions of the field and predict the ground-state spin structures. Following the lines of our previous work (3), we perform a mean-field analysis, assuming almost degenerate *ud* and *uud* order, the latter being slightly preferred.

In the [100] field direction, which was investigated in the early susceptibility measurements (4), we find two types of stable configurations with coexisting *ud* and *uud* order. In the low-field phase, denoted by AF1, the $\pm({}^2/_3,{}^2/_3,0)$ and $\pm({}^2/_3,-{}^2/_3,0)$ modulations are singled out with equal amplitudes, see Fig. 1(a). AF1 is stable at $B < B_c/3$; it is given by

$$\begin{aligned}\vec{S}_i &= (d_0,0,0)\\ &+(d_1,-d_1,0)\cos(({}^2/_3,{}^2/_3,0)\cdot\vec{r}_i)\\ &+(d_1,d_1,0)\cos(({}^2/_3,-{}^2/_3,0)\cdot\vec{r}_i)\\ &+(0,0,d_2)\cos((1,0,0)\cdot\vec{r}_i)\ ,\end{aligned} \tag{1}$$

where $-d_1/2 = d_0 = B/B_c$ and $d_2^2 = 1-(3B/B_c)^2$. A structure in which the (1,0,0) modulation is replaced by (0,1,0) has the same energy if fluctuations are neglected. The y and z indeces can be interchanged. Hence, an equal number of domains with either $\pm({}^2/_3,{}^2/_3,0)$ and $\pm({}^2/_3,-{}^2/_3,0)$ or $\pm({}^2/_3,0,{}^2/_3)$ and $\pm({}^2/_3,0,-{}^2/_3)$ order will appear in a diffraction experiment if the field is oriented perfectly along [1,0,0]. Importantly, there should be no neutron intensity at $\pm(0,{}^2/_3,{}^2/_3)$ or at $\pm(0,{}^2/_3,-{}^2/_3)$ reflections.

The structure AF2, at fields $B_c/3 < B < B_c/\sqrt{5}$, is modulated either by the vectors $\pm({}^2/_3,{}^2/_3,0)$ or $\pm({}^2/_3,-{}^2/_3,0)$. The AF2 configuration, illustrated in Fig. 1(b), is given by

$$\begin{aligned}\vec{S}_i &= (d_0,0,0)\\ &+(d_1,-d_1,0)\cos(({}^2/_3,{}^2/_3,0)\cdot\vec{r}_i)\\ &+(0,0,d_2)\cos((1,0,0)\cdot\vec{r}_i)\\ &+(0,0,d_3)\cos((0,1,0)\cdot\vec{r}_i)\\ &+(d_4,d_4,0)\cos((0,0,1)\cdot\vec{r}_i)\ ,\end{aligned} \tag{2}$$

where $-d_1/2 = d_0 = B/B_c$. The continuously degenerate *ud* order is determined from $d_0d_4 + d_2d_3 = 0$ and $5d_0^4 + 2d_2^2d_3^2 + d_0^2(d_2^2+d_3^2) = d_0^2$. Other domains are ob-

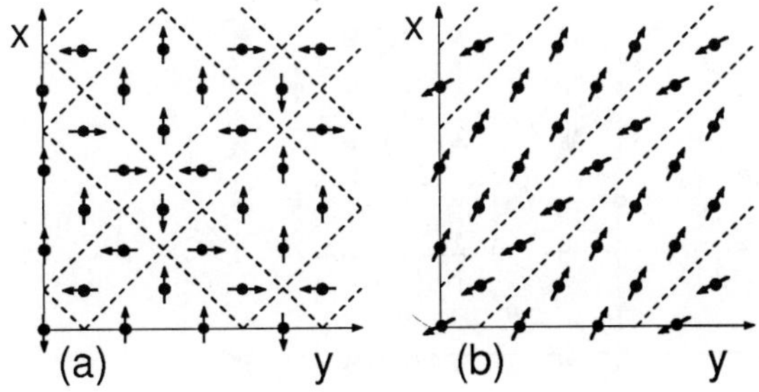

FIGURE 1
Spin structures in the [100] alignment of the magnetic field. The $\pm({}^2/_3,{}^2/_3,0)$ and $\pm({}^2/_3,-{}^2/_3,0)$ order is in the xy plane while the (1,0,0) order points out of the plane. (a) $B < B_c/3$. (b) $B_c/3 < B < B_c/\sqrt{5}$.

tained by interchanging indeces y and z. The high-field structure AF3, at $B > B_c/\sqrt{5}$, is a pure *ud* configuration. Preliminary results of our Monte Carlo simulations suggest that it is a 2-$\vec{k}$ state with $(0,1,0)$ and $(0,0,1)$ order, as was found in our earlier spin-wave analysis. (5)

The critical fields $B_c/3 \approx 0.13$ mT and $B_c/\sqrt{5} \approx 0.18$ mT coincide closely with the measured values, 0.12 mT between AF1 and AF2, and 0.17 mT between AF2 and AF3, respectively. The transitions are of first order, as observed. (4)

We have also found a metastable 12-$\vec{k}$ structure with all the *uud* modulation vectors present, see Fig. 2. Ther- mal fluctuations or a difference in the characteristic energies of the *uud* and *ud* modulations may stabilize this structure.

In low fields in the $[11\bar{1}]$ direction, we again find a stable structure which combines the *ud* and *uud* order. At $B < \sqrt{3/19}\, B_c$, the configuration is

$$\vec{S}_i = (d_0, d_0, -d_0) + (d_1, d_1, 0)\cos((^2/_3, -^2/_3, 0)\cdot\vec{r}_i) + (d_2, -d_2, 0)\cos(0,0,1)\cdot\vec{r}_i) \ , \qquad (3)$$

where $-d_1/4 = d_0 = 1/\sqrt{3}\ B/B_c$ and $2d_2^2 = 1 - (19/3)(B/B_c)^2$. At fields above $\sqrt{3/19}B_c$, the system is in a pure *ud* state; according to our earlier spin-wave results (5) this is a 3-$\vec{k}$ configuration.

It is possible to find at least one low-field structure in which the modulations *uud* and *ud* coexist for an arbitrary direction of the magnetic field in the plane $B_y = -B_z$, which includes all the investigated symmetry directions $[01\bar{1}]$, $[11\bar{1}]$, and $[100]$. The equation for the configuration is

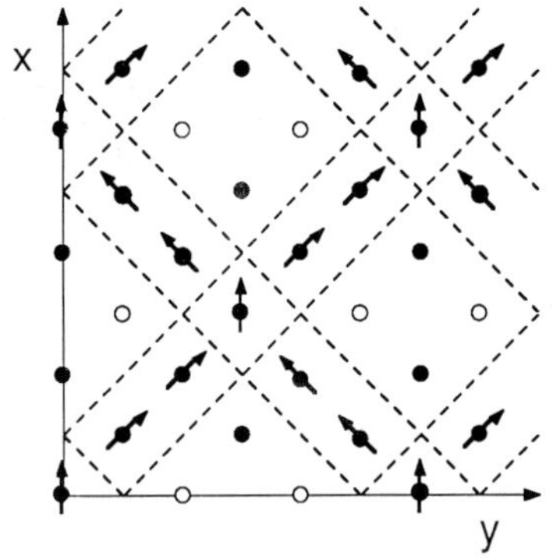

FIGURE 2
Metastable 12-$\vec{k}$ structure of pure *uud* order in a magnetic field $B \approx B_c/3$, applied in the $[100]$ direction.

$$\vec{S}_i = (d_0^x, d_0^{yz}, -d_0^{yz}) + (0, d_1, -d_1)\cos((0,^2/_3,^2/_3)\cdot\vec{r}_i) + (0, d_2, d_2)\cos((1,0,0)\cdot\vec{r}_i) \ , \qquad (4)$$

where $d_1/4 = -d_0^{yz} = B/B_c$ and $2d_2^2 = 1 - (d_0^x)^2 - 18(d_0^{yz})^2$. The structure coincides with particular solutions of Eq. (2) in Ref. (6) and the preceding Eq. (3) in the $[01\bar{1}]$ and $[11\bar{1}]$ field directions, respectively. When the field is turned towards the $[100]$ axis, the *uud* modulation decreases to zero.

The relative populations of different domains depend on the way in which the system enters a particular field. For example, if the ordered state is reached by lowering the field along $[11\bar{1}]$ below $\sqrt{3/19}B_c$, equal reflections at $\pm(^2/_3, -^2/_3, 0)$, $\pm(^2/_3, 0,^2/_3)$, and $\pm(0,^2/_3,^2/$ will be observed. However, if the field is decreased along $[01\bar{1}]$ below $B_c/3$, only $\pm(0,^2/_3,^2/_3)$ order is present. If the field is then tilted to the $[11\bar{1}]$ direction, domains with $\pm(^2/_3, -^2/_3, 0)$ and $\pm(^2/_3, 0,^2/_3)$ are not likely to be nucleated at all and only $\pm(0,^2/_3,^2/_3)$ reflections are observed.

ACKNOWLEDGEMENTS
We thank O.V. Lounasmaa for useful comments. This work has been supported by the Academy of Finland. H.V acknowledges a scholarship by the Magnus Ehrnrooth Foundation.

REFERENCES
1. A.J. Annila, K.N. Clausen, P.-A. Lindgård, O.V. Lounasmaa, A.S. Oja, K. Siemensmeyer, M. Steiner, J.T. Tuoriniemi, and H. Weinfurter, to appear in Phys. Rev. Lett. (March, 1990); see also three contributions by Annila et al. in this volume.
2. T.A. Jyrkkiö, M.T. Huiku, O.V. Lounasmaa, K. Siemensmeyer, K. Kakurai, M. Steiner, K.N. Clausen, and J.K. Kjems, Phys. Rev. Lett. 60 (1988) 2418.
3. H.E. Viertiö and A.S. Oja, Phys. Rev. Lett. (submitted).
4. M.T. Huiku, T.A. Jyrkkiö, J.M. Kyynäräinen, A.S. Oja, and O.V. Lounasmaa, Phys. Rev. Lett. 53 (1984) 1692.
5. H.E. Viertiö and A.S. Oja, Phys. Rev. B36 (1987) 3805.
6. H.E. Viertiö and A.S. Oja, Competition of the antiferromagnetic "up-down" and "up-up-down" order in the nuclear spin system of copper, this volume.

Physica B 165&166 (1990) 801–802
North-Holland

SPIN RELAXATION OF NUCLEAR ORDERED BCC SOLID ^{3}He IN THE LOW FIELD PHASE

Yutaka SASAKI, Yoshihiro HARA, Takao MIZUSAKI and Akira HIRAI

Department of Physics, Kyoto University, Kyoto 606, Japan

The transverse NMR linewidth of nuclear ordered bcc solid ^{3}He is measured in the low field phase as a function of applied magnetic field and, temperature and crystal orientation. While analyzing the measured linewidth by using the Tsubota-Tsuneto relaxation term (1), we could separate three kinds of relaxation mechanisms. We also find that the variation of linewidth with the crystal size has a strange dependence on the magnetic field.

1. INTRODUCTION

Nuclear ordered solid ^{3}He in the low field phase offers a unique opportunity to study nuclear spin dynamics in the nuclear ordered antiferromagnet with a uniaxial anisotropy. Osheroff et al. proposed the U2D2 spin structure and the spin dynamic equation (OCF eqs.) through their analysis of cw-NMR spectrum (2). Kusumoto et al. (3) did a pulsed NMR experiment and found an anomalous free induction decay signal after a strong RF pulse. We reported a measurement of the cw-NMR linewidth to investigate intrinsic relaxation process (4). Several relaxation mechanisms was proposed by Ohmi, Tsubota and Tsuneto (1,6,7,8). Tsubota and Tsuneto (T-T) introduced the phenomenological relaxation term into OCF eqs. These equations were solved for small amplitude oscillations around the equilibrium configuration (4). The two resonance frequencies $\omega_\pm$, which are denoted by the upper mode and the lower mode, and the linewidth $\Gamma_{T\text{-}T}$ of each mode are given by

$$\omega^2_\pm = \frac{1}{2}[\omega_L{}^2 + \Omega_0{}^2 \pm \{(\omega_L{}^2 - \Omega_0{}^2)^2 + 4\omega_L{}^2\Omega_0{}^2\cos^2\theta\}^{1/2}] \quad [1]$$

$$\Gamma_{T\text{-}T} = \frac{\omega^2_\pm - \omega_L{}^2\sin^2\theta}{2\omega^2_\pm - \omega_L{}^2 - \Omega_0{}^2}\,\mu \quad [2]$$

where $\omega_L = \gamma H$ is the larmor frequency of the external magnetic field. Ω_0 is the resonance frequency for zero magnetic field. θ is an angle between $\hat{l}$ and **H**, where $\hat{l}$ is the anisotropy axis which corresponds to (100) axis of bcc lattice or two other equivalent axes. μ is a relaxation parameter. This mechanism may be a candidate in the case of cw-NMR.

This experiment is an extension of our previous linewidth measurements as a function of a wider range of external magnetic field, temperature, angle between the anisotropy axis and magnetic field and the volume of the crystal.

2. EXPERIMENTAL SETUP

Our experiment was done for a single crystal along the melting curve. The crystal was created from overpressurized ^{3}He-B by applying a heat pulse. The crystal was carefully melted to a small seed and re-growned to remove the possible defects inside the crystal. By using NMR imaging we found that the single crystal was made of several large magnetic domains, typically three domains. We also proposed the possibility of the lattice distortion due to magnetostriction (5).

The inhomogeneity of the applied magnetic field is 50ppm in the RF coil whose diameter is 3mm and length is 4mm. The method of detecting a cw-NMR signal is essentially a Q-meter method. We sweep either the static magnetic field or the RF frequency whichever is more appropriate. When a linewidth is measured by sweeping the magnetic field, it is converted to the true linewidth by normalizing with the field derivative of the resonance frequency (4).

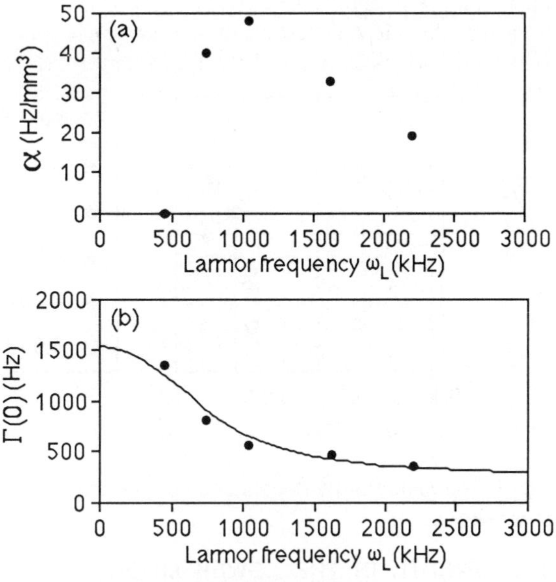

Fig. 1(a) The field dependence of the coefficient α. (b) The field dependence of the linewidth at zero volume $\Gamma(0)$. T/T_N=0.57, $\cos^2\theta$=0.62.

3. LINEWIDTH IN THE UPPER MODE

Here we show the results of our linewidth measurement in the upper mode. First, we studied the volume dependence of the linewidth at various magnetic fields. The linewidth increased linearly with the volume V of a crystal.

$$\Gamma = \Gamma(0) + \alpha V \quad [3]$$

As shown in Fig. 1(a), α is always positive, but it has a strange field dependence with a maximum around ω_L = 1 MHz. We have considered several possibilities to explain this. For example, a two magnon process due to surface scattering, spin diffusion, field inhomogeneity, the distribution of the direction of anisotropy axis and the temperature variation in the crystal. But none of them could explain the field dependence of α.

As shown in Fig. 1(b) $\Gamma(0)$ has the feature predicted by $\Gamma_{T\text{-}T}$. To obtain a good fit of our data to the theory, we introduce Γ_B such that $\Gamma(0) = \Gamma_{T\text{-}T}+\Gamma_B$ where Γ_B is the phenomenological linewidth which is assumed to be independent of the field. The solid line is the fitted curve with choosing μ and Γ_B as parameters. This good fit indicates the validity of the T-T relaxation term in U2D2 ^3He. In order to investigate the temperature dependence of the linewidth, we took the data for a crystal of 3mm^3 volume. The effect of this finite volume is not significant so that we can analyze our data by the same scheme as for $\Gamma(0)$. From Fig. 2, one can conclude that μ has a very strong temperature dependence, going as T^6 roughly. The Γ_B does not depend much on temperature and is in the order of several hundred Hz.

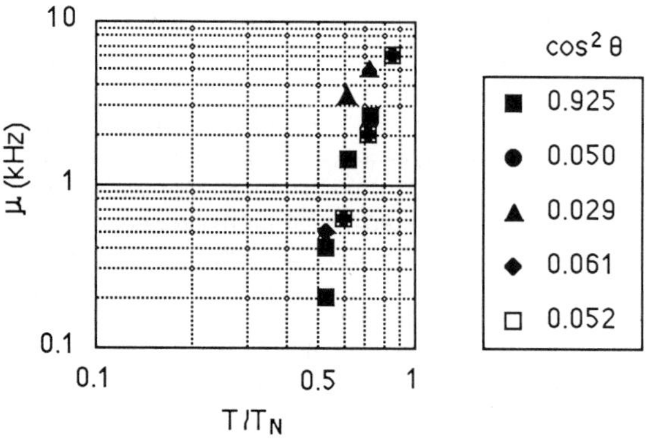

Fig. 2 The temperature dependence of the relaxation parameter μ.

4. LINEWIDTH IN THE LOWER MODE

Fig. 3 shows the measured linewidth in the upper mode (solid circles) and in the lower mode (open circles) for a crystal with 3mm^3 volume as a function of magnetic field. The linewidth in the lower mode agrees well with $\Gamma_{T\text{-}T}+\Gamma_B$ (solid line) in the region around $\omega_L = \Omega_0$ with values of μ and Γ_B chosen to fit the measured linewidth in the upper mode of the same domain. This result also indicates the validity of the T-T relaxation term in the lower mode.

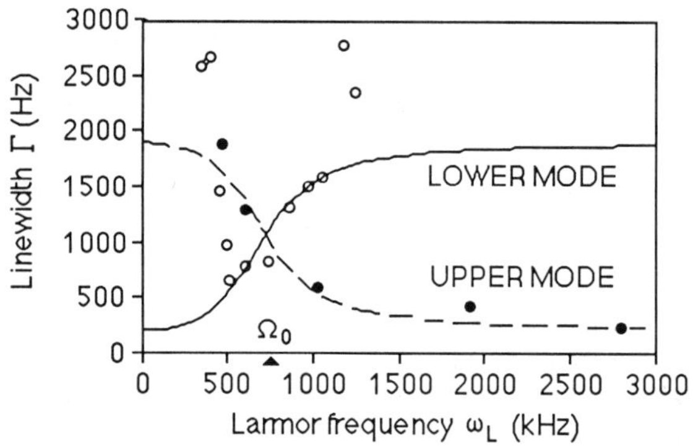

Fig. 3 The field dependence of the linewidth in the upper mode and the lower mode. T/T_N=0.68, $\cos^2\theta$=0.22, V=3mm^3.

Rapid departure from the theoretical curve is observed on both sides of Ω_0 at two critical fields, one above and one below Ω_0. The value of these two critical fields agree well with the fields where the resonance frequency of the upper mode is equal to twice the resonance frequency of the lower mode. In the field region between these two critical fields, the three magnon process is prohibited due to energy and momentum conservation. Outside this region the three magnon process is active and produces the increase in linewidth. This agrees well with the measured field dependence of the linewidth. Thus the three magnon process coexists with the T-T relaxation process in the lower mode.

It should be noted that this kind of sudden change of the linewidth was absent in the upper mode.

REFERENCES

(1) M. Tsubota and T. Tsuneto, Proceedings of LT-17 (1984) 241.
(2) D.D. Osheroff, M.C. Cross and D.S. Fisher, Phys. Rev. Lett. **24** (1980) 792.
(3) T.Kusumoto, O. Ishikawa, T. Mizusaki and A. Hirai, J.L.T.P.**59** (1985) 269.
(4) Y. Sasaki, K. Sasayama, T. Mizusaki and A. Hirai, Proceedings of LT-18 (1987) 417.
(5) Y. Sasaki, Y. Hara, T. Mizusaki and A. Hirai, AIP Proc. No.194 (1989) 286.
(6) T. Ohmi, M. Tsubota and T. Tsuneto, Prog. Theor. Phys.**73** (1985) 1075.
(7) M. Tsubota, T. Ohmi and T. Tsuneto, Prog. Theor. Phys.**76** (1986) 1222.
(8) M. Tsubota, Prog. Theor. Phys.**79** (1988) 47.

Physica B 165&166 (1990) 803–804
North-Holland

NMR STUDIES OF VACANCIES IN SOLID HYDROGEN*

D. ZHOU, M. RALL, J.P. BRISON and N.S. SULLIVAN

Department of Physics, University of Florida, Gainesville, FL 32611, USA.

We report measurements of the vacancy-induced nuclear spin-lattice relaxation in solid hydrogen. The results are used to deduce (1) the characteristic motional frequency of the vacancies and (2) the formation energy of the vacancies at low densities.

1. INTRODUCTION

Studies of the motion of defects in quantum crystals allow one to explore qualitatively new effects (1),(2) such as quantum diffusion which is one of the clearest manifestations of the quantum nature of these solids. As a result of the relatively large zero-point kinetic motion (33% of the atomic spacing in solid ^{3}He and 18% in solid H_2) there is an appreciable overlap of the wave functions of particles on neighboring sites and this allows defects such as vacancies or isotopic impurities to exchange positions with nearest neighbor atoms or molecules. Whereas in classical solids, the defects can be regarded as localized objects which move only occasionally by thermal excitation over intervening potential energy barriers, the defects will become delocalized in the quantum solids, and Andreev (1) has predicted that they will move through the crystal in a coherent manner by quantum tunneling. These excitations are called vacancy waves or impurity waves.(2),(3)

The possibility of observing quantum tunneling of vacancies in solid H_2 is especially interesting, because there are two "bound" states for the molecule-vacancy interaction potential (4), and one expects to observe three distinct temperature regimes for vacancy-induced diffusion (5): (i) thermal activation over the full potential barrier at high temperatures, (ii) quantum tunneling combined with thermal activation from the ground state on one side of the well to the first excited (but bound) state on the other side at intermediate temperatures, and (iii) purely quantum tunneling through the potential barrier from ground state to ground state at low temperatures. Each regime is expected to be characterized by a different activation energy. Only the first regime (the purely classical regime) was observed in earlier studies (6)-(9), and we report here high sensitivity tests for vacancy quantum tunneling in the low temperature regime.

2. NMR STUDIES OF VACANCIES

NMR experiments are sensitive to vacancies because of their motion and the consequent time dependence of the position coordinates of the nuclear spins. Measurements of the nuclear spin-lattice relaxation rates provide information about the spectral densities $J(\omega_L)$ of the microscopic motion at the Larmor frequency ω_L. X-ray studies measure vacancy concentrations $X_v(T)$ = exp (Φ/k_BT) where Φ is the formation energy, while NMR studies measure the product $X_v(T)\omega_v(T)$ where $\omega_v(T)$ is the characteristic motional frequency for the vacancies. For classical activation $X_v(T)\omega_v(T)$ varies as $\exp[-(\Phi + V)/k_BT]$ where V is the potential energy barrier, and for purely quantum tunneling the temperature dependence is $\exp[-\Phi/k_BT]$. A comparison of the x-ray studies and NMR measurements for bcc ^{3}He (10) has led to the conclusion that vacancies in solid ^{3}He move by quantum tunneling at low densities.

In order to test for quantum tunneling in solid H_2, we needed to observe vacancy-induced relaxation at temperatures significantly lower than those previously studied. We therefore chose to study samples containing low ortho-H_2 concentrations (0.5-2.5%) and prepared with small concentrations of HD impurities (typically 1%). The ortho-H_2 molecules (total nuclear spin I = 1, orbital angular momentum J = 1) have molecular electric quadrupole moments and their electrostatic interaction with other ortho molecules lead to a very rapid nuclear spin-lattice relaxation time, $T_{1Q} \approx$ msec for these concentrations, and this masks the weaker vacancy-induced relaxation except at very high temperatures. (The para-H_2 molecules, I = 0, J = 0, are not observed by NMR and do not participate in the relaxation.) The direct spin-lattice relaxation of the ^{1}H nuclei of HD molecules is very slow, and one observes a cross-relaxation, mediated by the spin-spin coupling between the HD molecules and the ortho-H_2 molecules.

*Work supported by NSF-low temperature physics—DMR-8913999.

Studies of the cross-relaxation are especially useful for investigating vacancy motion because the cross-relaxation rate T^{-1}_X depends on the spectral density J(0) at zero frequency. The number of vacancies is very small at low temperatures, and the characteristic frequency of the vacancy motion $X_v(T)\omega_v(T) \ll \omega_L$, and as a result $J(0) \gg J(\omega_L)$. Cross-relaxation studies can therefore provide information about the vacancy motion which cannot be deduced from studies of the direct relaxation.

The cross-relaxation rate between the ortho-H_2 molecules (a spins) and the HD molecules (b spins) is given by

$$\frac{1}{T_x^{ba}} = \frac{2}{9} f M_2^{ab}(x) J_{20}(0)$$

$N_a I_a(I_a + 1)/N_b I_b(I_b + 1)$ where $N_{a(b)}$ are the number densities for the two spin species $I_{a(b)}$. M_2^{ab} is the intermolecular rigid lattice second moment and $J_{20}(0)$ is the spectral weight of the fluctuations of the component of the spin-spin interactions that transforms as the spherical harmonic Y_{20}.

3. EXPERIMENTAL RESULTS

The relaxation was observed to consist of a short-time contribution (<2 msec.) due to the ortho-H_2 relaxation, and a long-time component attributed to the cross-coupling relaxation. The observed temperature dependence of the long-time relaxation is shown in Figure 1.

Three distinct temperature regimes are observed: (i) At low temperatures $1.5 < T < 7$K, T_x has only a weak (approximately linear) temperature dependence; (ii) at intermediate temperatures, $7 < T < 11$K, T_x passes through a minimum, and (iii) above 11K, T_x increases exponentially until the sample melts.

The experimental data cannot be described in terms of a single activation energy for the vacancy motion. We therefore included a quantum tunneling term analogous to that proposed by Ebner and Sung (4) to obtain an accurate fit. The best fit was obtained using

$$X(T)\omega_v(T) = 4.2\times10^8 e^{-91/T} \times [1 + 5.1\times10^4 e^{-107/T}] \mathrm{sec}^{-1}.$$

The formation energy $\Phi = 91$K and the potential barrier height $V = 107$K.

The low temperature relaxation is attributed to particle-particle quantum tunneling of the HD impurities with host molecules (para-H_2) with exchange frequency $J_t/2\pi = 9.5 \times 10^3$ Hz. The minimum in T_1 occurs because on warming above 7K, the coherent motion of the impurities is interrupted by scattering with vacancies. A minimum was also observed in studies of T_2, the transverse relaxation time, but the effect is much more significant for T_1.

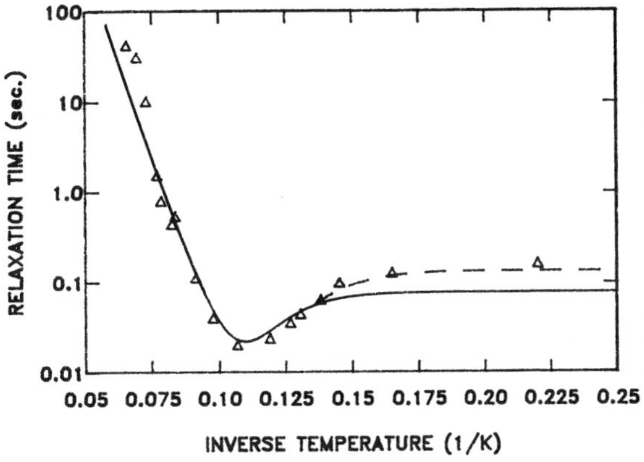

FIGURE 1.
Comparison of the observed temperature dependence of the nuclear spin-lattice relaxation of HD impurities (1.1%) in solid H_2 (1.8% ortho) with the calculated dependence (solid line). The broken line represents the correction due to the broadening of the ortho-H_2 line shape at low temperatures.

4. CONCLUSION

The strong temperature dependence of the nuclear spin-lattice relaxation of HD impurities in solid H_2 can be understood over the full temperature range explored in terms of the motion of vacancies if one includes a quantum tunneling contribution for the vacancy motion. The dependence at high temperatures is in excellent agreement with earlier studies.

REFERENCES

(1) A.F. Andreev and J.M. Lifshitz, Sov. Phys. JETP 29 (1969) 1107.
(2) R.A, Guyer, J. Low Temp. Phys. 8 (1972) 427.
(3) A. Landesman, Ann. Phys. 9 (1975) 69.
(4) C. Ebner and C.C. Sung, Phys. Rev. A5 (1972) 2625.
(5) D. Zhou, M. Rall, N.S. Sullivan, C.M. Edwards and J.P. Brison, Solid State Commun. 72 (1989) 657.
(6) M. Bloom, Physica 23 (1957) 767.
(7) F. Weinhaus and H. Meyer, Phys. Rev. B7 (1973) 2974.
(8) R.F. Buzerak, M. Chan and H. Meyer, J. Low Temp. Phys. 28 (1977) 415.
(9) W.P.A. Haas, N.J. Poulis and J.J.W. Borleffs, Physica 27 (1961) 1037.
(10) S.M. Heald, D.R. Baer and R.O. Simmons, Phys. Rev. B30 (1984) 2531.

Physica B 165&166 (1990) 805–806
North-Holland

EXPERIMENTAL MEASUREMENTS OF ^{4}He SOLID LIQUID INTERFACE INERTIA.

Jacqueline POITRENAUD and Pierre LEGROS

Université Pierre et Marie Curie, Laboratoire d'Ultrasons, URA CNRS n° 800, Tour 13, 4 place Jussieu, 75252 Paris Cedex 05, France.

We have measured the transmission of acoustic waves through a rough ^{4}He superfluid interface at 15, 45 and 75 MHz between 0.4 and 1.1 K. The phonon transmission is enhanced at low tempertures for the higher frequencies, in place of rapidly decreasing with T as explained by the high mobility interface theory. These experimental results are very well interpreted with the help of the surface inertia introduced by Puech and Castaing.

1. INTRODUCTION

^{4}He crystal draws considerable attention because of its extremely fast growth at low temperature. Among all the resultant fascinating properties, the sound transmission through the L/S rough interface [1, 2, 3] was found to decrease as the temperature is lowered although the acoustic mismatch theory predicts temperature independence. Until now heat and mass transport accross the interface as well as their coupling were well understood through the knowledge of the kinetic coefficients ; R_K, the Kapitza resistance, K the growth coefficient and λ the thermal sharing coefficient. The fast melting theory including the effect of the work of tension forces at the interface [4] predicts a T^{-5} variation of R_K. Puech and Castaing (PC) [5] have pointed out that the resulting order of magnitude is too high and were led to take into account the mass σ associated to the growth of the surface. This notion was first introduced by Gibbs [6] on the thermodynamical grounds, later on Kosevitch et al [7] in the case of a moving interface. PC have noticed that at the melting a strong structural rearrangement occurs. Thus when there is growing or melting of the crystal a kinetic energy has to be added, which they write $\sigma J^2/\rho_C\rho_L$ (ρ_C and ρ_L being the specific masses of the liquid and the solid, J the mass current). This new concept reconcilies the measurement and the theory of the Kapitza resistance. But the calculation incorporates the effect of a wide band of phonons. And the interest of the study of the transmission of a high frequency monochromatic ultrasound wave through a rough ^{4}He interface becomes evident.

2. THEORY

Surface movement can be written :

$$\Delta\mu = K^{-1} J + \frac{\sigma}{\rho_C\rho_L}\dot{J}$$

From which the energy transmission coefficient with the help of the usual mass conservation relation and pressure equality on each side of the surface [8] is easily deduced :

$$\tau = \frac{4 z_L z_C}{(z_L + z_C + (z_L z_C/\zeta))^2} \quad (1)$$

z_L and z_C are the acoustical impedances and ζ can be interpreted as a surface impedance :

$$\zeta = \frac{\rho_C\rho_L}{(\rho_C - \rho_L)^2}\left(K^{-1} + \frac{i\omega\sigma}{\rho_C\rho_L}\right) \quad (2)$$

The mobility of a rough interface is limited by the bulk thermal excitations, that is, by the rotons and the phonons in the liquid, by the phonons in the solid and,

$$\rho_C K^{-1} = A + BT^4 + C\exp(-\Delta/T)$$

A is a residual damping, Δ the minimum roton energy. From (1) and (2), it can be seen that at sufficiently high frequency, τ instead of continuously decreasing when T is lowered, tends to a constant value frequency dependence. From the estimated value of σ calculated by PC, the phenomenon is expected to be observable for a frequency ν higher than 30 MHz.

3. EXPERIENCE AND RESULTS

The experiments have been performed at three frequencies 15, 45 and 75 MHz and between .4 and 1.1 K. The thickness of the crystal and the velocity of sound C_C, from which we deduced the crystallographic orientation θ of the sample, were determined from the analysis of successive transmitted pulses. Figure 1 shows an example of the temperature dependence of τ for one sample at three frequencies after correction for attenuation of the bulk liquid and solid. One remarks the different behaviours of τ with T at different ν : for the same sample at high temperatures (< 0.8 K) the variation is identical whatever the frequency ; on the contrary at low temperatures we distinguish between the frequencies : at 15 MHz τ continuously decreases with T until becoming unmeasurable ; for the higher two frequencies τ tends towards a constant value. Taking Δ = 7.2 K, A, B, C and σ were deduced from (1) (2) in two steps. For high temperatures T > .8 K, the rotons

0921-4526/90/$03.50

TABLE 1
Values of A, B, C and σ, calculated from the data for three different orientated samples.

sample	C_C (m/s)	frequency (MHz)	A (cm/s)	B (cm/s/K^4)	C x 10^{-5}	σ x 10^{10} (g/cm^2)
a	538	15	.04 ± .04	5.4 ± .6	.72 ± .13	3.1 ± .2
(θ = 8°)		45	.65 ± .2	.02 ± .01	.72 ± .13	3.1 ± .2
b	520	45	.6 ± .2	8.5 ± .8	2.9 ± .3	7.1 ± .4
(θ = 18°)		75	1.34 ± .2	18 ± 2	2.5 ± .3	7.1 ± .4
c	540	15	.04 ± .2	1.95 ± .2	.29 ± .10	1.6 ± .2
(θ < 3°)		45	0.0 ± .01	3.3 ± .1	.29 ± .10	1.5 ± .1
		75	.06 ± .02	3.1 ± .2	.29 ± .10	1.7 ± .1

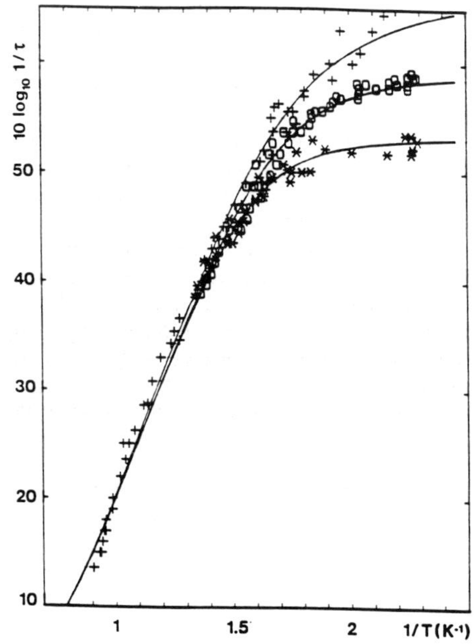

FIGURE 1
Comparison between the temperature dependence of the sound transmission data for sample c and the fit which is described in the text. Δ = 7.2 K .

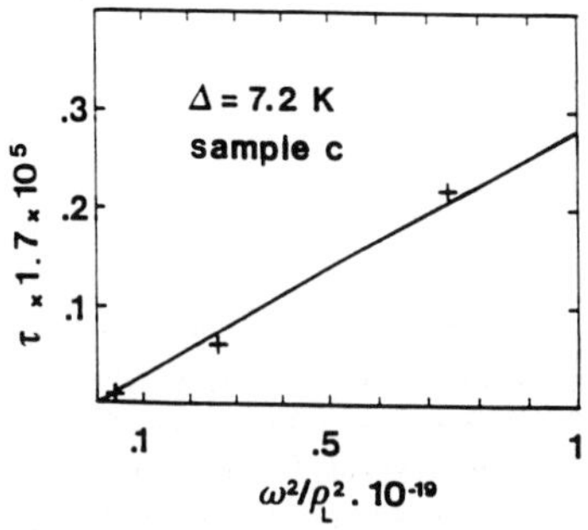

FIGURE 2
T = 0 extrapolated values of τ *vs.* ω^2/ρ_L^2 for sample c. The slope of the obtained straight line is σ^2.

monitor the interface mobility and C is deduced while, for low temperatures, T < .8 K, the acoustic impedances z_L and z_C become negligible with regard to ζ, then A, B and σ can be determined. The subsequent fit for sample c between experimental points and calculated curves is illustrated by figure 1. Table I gives A B, C and σ values for three different orientated samples. σ values are to be compared with PC estimation of 2.4 x 10^{-10} g/cm^2 deduced from R_K experimental values analysis. θ is the angle between the c axis of the crystal and the ultrasound wave vector.

We wish to emphasize that for the same sample the same σ value permits to take into account the variation of τ at all the frequencies. Figure 2 shows that the extrapolated τ values at T = 0 , at 15, 45 and 75 MHz vary quadratically with ν as is expected. Furthermore the anisotropy of σ has been put in evidence.

4. CONCLUSION

We have presented an extension of previous measurements of ultrasound transmission through a rough ^{4}He interface at frequencies which allow to get a clear evidence of liquid-solid interface mass inertia σ and its anisotropy. Additionally the values of the mobility coefficient K determined from our data agree with those of previous studies.

[1] CASTAING, B., BALIBAR, S. and LAROCHE, C., *J. Physique* **41** (1980) 897.
[2] CASTAING, B., *J. Physique Lett.* **45** (1984) L-233.
[3] MOELTER, M.J., MANNING, M.B., ELBAUM, C., *Phys. Rev.* ***B*** **34** (1986) 4924.
[4] MARCHENKO, V.I., PARSHIN, A., *J.E.T.P. Lett.* **31** (1980) 724.
[5] PUECH, L., CASTAING, B., *J. Physique Lett.* **43** (1982) L-601.
[6] GIBBS, J. W., *Scientific Papers* (Dover, N.-Y., 1961).
[7] KOSEVITCH, A. M., and KOSEVITCH, Yu., *Sov. J. Low Temp. Phys.* **7** (1981) 394.
[8] UWAHA, M. and NOZIERES, P., *J. Physique* **46** (1985) 109.

Physica B 165&166 (1990) 807–808
North-Holland

ORIENTATION DEPENDENCE OF THE KAPITZA RESISTANCE BETWEEN LIQUID AND HCP ^{4}He

M. S. PETTERSEN and D. O. EDWARDS

Department of Physics, The Ohio State University, 174 W. 18th Ave., Columbus, OH 43210 USA

We present a calculation of the Kapitza resistance between liquid and hcp ^{4}He, assuming that the interface has an inertia due to the motion of quantum kinks on the solid surface, and taking into account the orientation dependence of the surface stiffness and the anisotropy of the crystal.

1. INTRODUCTION

Marchenko and Parshin (1) first pointed out that the high mobility at low temperature of the rough (non-facetted) liquid-solid interface in ^{4}He (2), which allows for the existence of melting-freezing waves (3), makes it difficult to transmit energy across the interface: pressure fluctuations in the liquid are relieved by motion of the interface with no stress propagating into the crystal. They suggested that the only coupling mechanism is provided by surface tension, resulting in a Kapitza resistance R_K that varies at low temperature as T^{-5} rather than T^{-3}, as appears in the usual acoustic mismatch theory. This dependence was later observed experimentally (4-6). However, the magnitude of the resistance was lower than expected from the theory.

Kosevich and Kosevich (7) suggested that that the surface has an inertia which affects high frequency melting-freezing waves. They associated the inertia with the motion of steps on the crystal surface. Puech and Castaing (8) formulated the appropriate boundary conditions at the interface, and showed that the surface inertia enhances the transmission of sound across the interface, explaining the Kapitza resistance data quantitatively.

Microscopically, the growth of the crystal proceeds by motion of kinks along steps on the surface. Thus the interface inertia can be related to m_k, the effective mass of the kink. According to (7), m_k is due to the motion of the superfluid towards the moving kink and is estimated to be $\sim 0.002 m_4$. Recently we proposed a theory of the scattering of rotons in the liquid by kinks, which contributes to the growth resistance of the crystal, and also calculated the Kapitza resistance using an isotropic model for the crystal (9). In this paper we present a more elaborate calculation of the Kapitza resistance including the effects of the orientation dependence of the surface tension and the anisotropic elasticity.

2. RESULTS AND CONCLUSIONS

We begin by calculating the transmission coefficient for a sound wave, using the boundary conditions given by Puech et al. (8) to take into account the surface inertia ν_S and surface stiffness $\tilde{\alpha}$. We neglect the effects of the surface strain (10) and the Gibbs adsorption (11).

The surface stiffness tensor $\tilde{\alpha}$ depends on the orientation of the surface, specified by the polar coordinates θ and ϕ of the surface normal in the coordinate system of the crystal lattice. The data (12) indicate a negligible dependence on ϕ; for want of further experimental information, we approximate $\tilde{\alpha}$ by a scalar which depends only on θ. For the surface inertia, we take the results of ref. (9). The kink mass increases with temperature because when the kink velocity approaches the speed of sound in the liquid, the compressibility of the liquid makes a large increase in the energy. At low temperatures the surface inertia has the value 2.1×10^{-10}g/cm^2, assumed, for the moment, to be independent of orientation.

With these boundary conditions, the transmission coefficient τ for a phonon incident from the liquid is computed, using the full anisotropic elastic tensor for the He crystal (13). Fig. 1 shows how τ (plotted as a function of the angle of incidence θ_i) varies with the orientation θ of the surface.

The Kapitza resistance is then computed by integrating $\tau(\theta_i, \phi_i, \omega)$ over the angles of incidence θ_i and ϕ_i and the frequency ω. We express our results in terms of an average transmission coefficient $\bar{\tau}$ defined by Maris et al. (13):

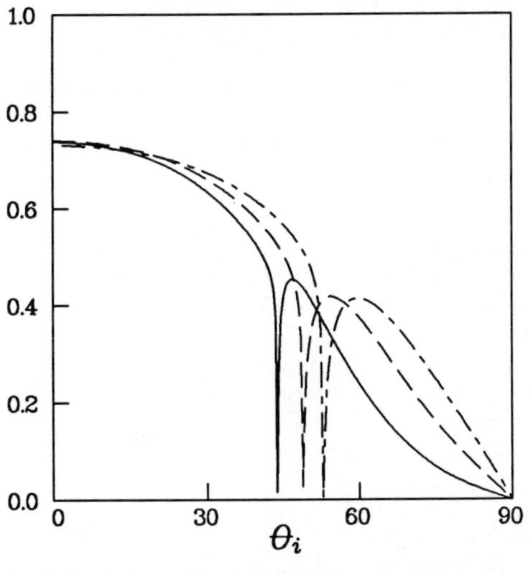

Fig. 1. Dependence of the transmission coefficient on angle of incidence for phonons in the liquid at 0.3 THz. The angle θ between the surface normal and the c-axis is: 80°, solid line; 60°, dashed line; 40°, dot-dash line. The incident wave vector lies in the plane of the c-axis and the normal.

$$\bar{\tau}(T) = \frac{\int_0^1 d\cos\theta_i \int_0^{2\pi} d\phi_i \int_0^\infty d\omega\, \omega^4 \cos\theta_i\, \bar{n}(\bar{n}+1)\, \tau(\theta_i, \phi_i, \omega)}{\int_0^1 d\cos\theta_i \int_0^{2\pi} d\phi_i \int_0^\infty d\omega\, \omega^4 \cos\theta_i\, \bar{n}(\bar{n}+1)}$$

where $\bar{n} = [\exp(\hbar\omega/k_B T) - 1]^{-1}$. The Kapitza conductance is related to this quantity as $R_K^{-1} = (\pi^2 k_B^4/30\hbar^3 c_\ell^2)T^3\bar{\tau}$, where c_ℓ is the speed of sound in the liquid. In fig. 2 we show the dependence of $\bar{\tau}$ on temperature for a few angles. The variation with surface orientation is comparable to the scatter in the data (plotted in ref. (9)).

At low temperatures, $\bar{\tau} \propto T^2$. We show the low T limit of $\bar{\tau}/T^2$ in fig. 3 for a variety of models. A comparison of the two upper curves shows that the orientation dependence of $\tilde{\alpha}$ has more effect on the Kapitza resistance than the elastic anisotropy. The graph also shows data from refs. (5) and (6). (Puech et al. (5) remark that the curvature of the surface in their experiment means that $\bar{\tau}$ was averaged over a range of angles, so the dependence on θ may be larger than their data indicate.) While the trend of the data generally agrees with the model, the remaining discrepancy may be due to the orientation dependence of ν_S. This could be due to the dependence of the kink mass on the surface lattice constants, and also to the dependence of the density of kinks and steps on the surface. Further experiments will be helpful in determining the variation of ν_S with surface orientation.

This work was supported by NSF grant DMR 8403441 and DMR 8905385. We thank S. Balibar for many useful discussions.

REFERENCES

(1) V.I. Marchenko and A.Ya. Parshin, Pis'ma Zh. Eksp. Teor. Fiz. **31** (1980) 767 [JETP Lett. **31** (1980) 724].

(2) A.F. Andreev and A.Ya. Parshin, Zh. Eksp. Teor. Fiz. **75** (1978) 1511 [Sov. Phys. JETP **48** (1978) 763].

(3) K.O. Keshishev, A.Ya. Parshin and A.B. Babkin, Zh. Eksp. Teor. Fiz. **80** (1981) 716 [Sov. Phys. JETP **53** (1981) 362].

(4) T.E. Huber and H.J. Maris, J. Low Temp. Phys. **48** (1982) 463.

(5) L. Puech, B. Hebral, D. Thoulouze and B. Castaing, J. Physique-Lett. **43** (1982) L809.

(6) P.E. Wolf, D.O. Edwards and S. Balibar, J. Low Temp. Phys. **51** (1983) 489.

(7) A.M. Kosevich and Yu.A. Kosevich, Fiz. Nizk. Temp. **7** (1981) 809. [Sov. J. Low Temp. Phys. **7** (1981) 394].

(8) L. Puech and B. Castaing, J. Physique-Lett. **43** (1982) L601.

(9) D.O. Edwards, S. Mukherjee and M.S. Pettersen, Phys. Rev. Lett. **64** (1990) 902.

(10) D.E. Wolf and P. Nozières, Z. Phys. **B70** (1988) 507.

(11) D.O. Edwards, J.R. Eckardt and F.M. Gasparini, Phys. Rev. **A9** (1974) 2070.

(12) O.A. Andreeva and O. Keshishev, Pis'ma Zh. Eksp. Teor. Fiz. **46** (1987) 160 [JETP Lett. **46** (1987) 200]; A.V. Babkin, D.B. Kopeliovich and A.Ya. Parshin, Zh. Eksp. Teor. Fiz. **89** (1985) 2288 [Sov. Phys. JETP **62** (1985) 1322]; P.E. Wolf, F. Gallet, S. Balibar and E. Rolley, J. Physique **46** (1985) 1987; A.V. Babkin, K.O. Keshishev, D.B. Kopeliovich and A.Ya. Parshin, Pis'ma Zh. Eksp. Teor. Fiz. **39** (1984) 519 [JETP Lett. **39** (1984) 633].

(13) H.J. Maris and T.E. Huber, J. Low Temp. Phys. **48** (1982) 99.

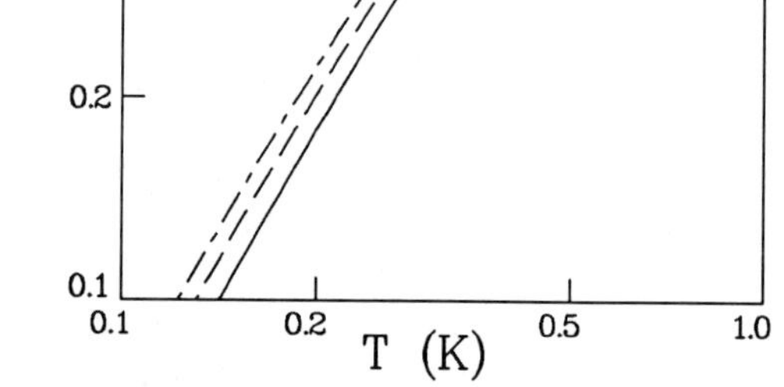

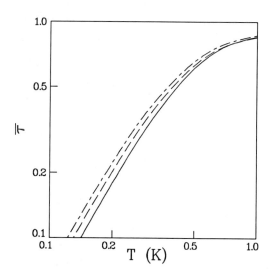

Fig. 2. Dependence of the average transmission coefficient on temperature for the same angles as in fig. 1.

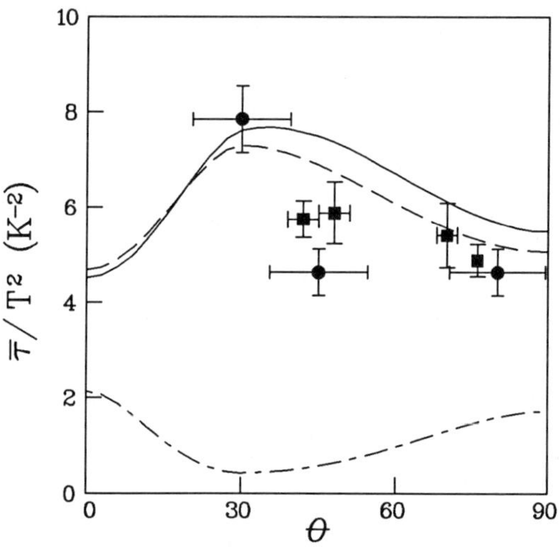

Fig. 3. Dependence of $\bar{\tau}/T^2$, where $\bar{\tau}$ is the mean thermal phonon transmission coefficient from the liquid, on the surface orientation angle θ. The solid curve shows the fully anisotropic theory; the dashed curve has orientation-dependent $\tilde{\alpha}$ but isotropically averaged elastic constants (13); the dot-dash curve is like the dashed curve, except that ν_S is set to 0. The circles are from ref. (5), squares from ref. (6).

Physica B 165&166 (1990) 809–810
North-Holland

LOW-TEMPERATURE CLUSTERING OF o-H_2 IMPURITIES IN p-H_2 CRYSTALS

Alexander E. MEYEROVICH

Department of Physics, University of Rhode Island, Kingston, RI 02881 USA

The presented explanation of anomalies in temperature dependence of pairing time for o-H_2 impurities in p-H_2 quantum crystals is based on an accurate analysis of differences in diffusion trajectories for different mechanisms of quantum diffusion. The transition between different regimes is accompanied by changes in characteristic lengths of pairing trajectories. Later stages of clustering may be influenced by coherent diffusion of pairs and triads.

1. INTRODUCTION

Several years ago puzzling experimental results (1) were reported on quantum diffusion of o-H_2 impurities in solid p-H_2. NMR data revealed a sharp drop in a pairing time for o-H_2 molecules below 0.3K, Fig. 1. Since all reasonable mechanisms of quantum diffusion predict either an increasing pairing time or, at least, a saturation at very low temperatures, the data (1) suggest an existence of some new mechanism of quantum diffusion. However, the existence of a new mechanism at such low temperatures does not seem very plausible.

Nevertheless, the data (1) can be interpreted on the basis of some geometric considerations in the frames of known mechanisms of quantum diffusion: the transition between diffusion regimes is accompanied by changes in patterns of diffusion trajectories with different characteristic lengths. Sometimes this leads to a decreasing pairing time even if the individual hopping rates are temperature independent or increase with the lowering temperature.

2. QUANTUM DIFFUSION AND PAIRING OF o-H_2 MOLECULES

The quantum diffusion, as well as a classical one, is, by definition, a random walk. However, in the process of pairing, two o-H_2 molecules spend most of the time close to each other when the dominant shifts in their energy levels (energy mismatches) on the different lattice sites are due to their electric quadrupole-quadrupole interaction with each other. Since the hopping rates for quantum diffusion strongly depend on energy mismatches (see,e.g.,(2)-(5)), and the mismatches change in a regular manner when the particles approach each other, the quantum diffusion may lose its random character at small separations. And the diffusion (pairing) trajectories do not resemble random walk anymore.

Different hopping mechanisms result in different diffusion trajectories, and the pairing trajectories change with a transition between diffusion mechanisms. There are three basic mechanisms of quantum diffusion, and not all of them favor random walks. The effective hopping rates are determined by the relation between the bare tunneling frequency J_0 and energy mismatches δE between the adjacent sites, and by the means to overcome these mismatches.

In the simplest case of negligible mismatches $J_0 >> \delta E$, one observes a really random diffusion with the diffusion coefficient $D \sim J_0 al$ (a is the lattice constant, l is the mean free path, and h=1). Other mechanisms (with $\delta E >> J_0$) have effective rates quadratic in J_0.

One of the most important mechanisms of quantum diffusion is associated with two-phonon processes when the mismatches are overcome because of phonons taking care of an energy balance. Then the effective hopping rate is (4),(5) $J = J_0^2 T_0/(T_0^2 + \delta E^2)$, where $T_0 = \theta(T/\xi\theta)^\beta$, θ is the Debye temperature, index β is 7 or 9, and ξ is unknown, $0.1 < \xi < 1$. If $T_0 >> \delta E$, then the effective rate J does not depend on δE, and the diffusion trajectory is random.

As soon as $\delta E << T_0$, J strongly depends on mismatches, and the most probable direction of motion is perpendicular to the energy gradient. If this mechanism dominates, the effective pairing trajectories are very long (much longer than in the case of random walk), especially because the interaction energy - and mismatches - very rapidly increase with decrease in impurity separation. It is possible to show that the effective pairing time for this mechanism is $N^{3\alpha}\delta E_{max}^2/J_0^2 T_0$, where the index α is not universal and lies between 1 and 3, δE_{max} is the maximal value of mismatch (e.g., the energy of EQQ interaction at the smallest pair radius), and the effective path, N, is larger than an initial particles separation, N_0, (in h.c.p. lattice $N > 2N_0$).

At larger mismatches there is a possibility of one-phonon processes when the energy balance is ensured by emission of phonons. Then the motion is irreversible, with the hopping rate $J_0^2(\delta E/\zeta\theta)^3/\theta$, $0.1 < \zeta < 1$. The trajectory is directed almost along the energy gradient, is very short, and has practically no randomness.

As a result, both the hopping rate and the length of the pairing trajectory depend on mismatches. On the final stages of pairing, the situation is clear: the potential relief and the mismatches δE depend only on EQQ interaction of the pairing particles. Then the hopping rate and the trajectory are determined only by the current pair radius, r: $\delta E = \delta E(r)$, $J = J(r)$.

At larger separations, the interactions with all other impurities are equally important, and the values of δE are random filling some energy interval E_0 with the constant density of states. Then it is possible to show that both the unassisted tunneling through the "windows" of the bare width J_0 and two-phonon processes averaged over the whole interval E_0 give the same hopping rate $J \sim J_0^2/E_0$, and lead to a random diffusion while the one-phonon processes are negligible.

The problem is to understand the temperature dependence of the pairing times at $0.05K < T < 2K$ and o-H_2 concentration $0.1\% < x < 1\%$ when the average distance between impurities is about 4 ÷ 5 lattice constants. At the distances less than 2 ÷ 3 the mismatches are due to the interaction within the given pair, and at larger separations the mismatches may be considered as random. [Except for high concentrations, x > 1%, none of the above regimes can alone account for the whole pairing process: the pairing corresponds to a combination of different regimes on different stages of pairing. At higher concentrations when the whole pairing process takes place with the mismatching caused by all other, randomly distributed, impurities]. Definitely, most of the pairing time is spent in overcoming the large mismatches at small distances. The only thing which can happen at $T < 0.3K$ is the transition from the two-phonon regime with $J \sim J_0^2 T_0/\delta E^2$ to the one-phonon mechanism, $J \sim J_0^2(\delta E/\zeta\theta)^3/\theta$. If one does not consider the pairing trajectory carefully enough, this transition results only in a disappearance of the temperature dependence of the pairing time. The account of the diffusion trajectories makes the results different.

All hopping rates discussed above contain one or more unknown constants. The calculation of the pairing time by means of simple numerical simulations of the diffusion problem with all these constants as fitting parameters provides an easy fit to the experimental data because of the large total number of parameters. The situation is even worse: the h.c.p. structure of the lattice dictates the doubling of the number of parameters.Therefore one may prefer to reduce a number of unknown parameters by simplifying the diffusion problem. Two such approaches are described below.

The first one involves only the transition from the two-phonon to the one-phonon regime with the drastic shortening of pairing trajectories. The set of unknown parameters is reduced to only two: one describes the product of J with the length, N, of the trajectory for the two-phonon regime, and the second - the ratio of coefficients ξ and ζ (curve 1 in Fig. 1). The first of these parameters is not very important being responsible for the overall scale. The second one was used to fit the position of maximum. The form of the curve

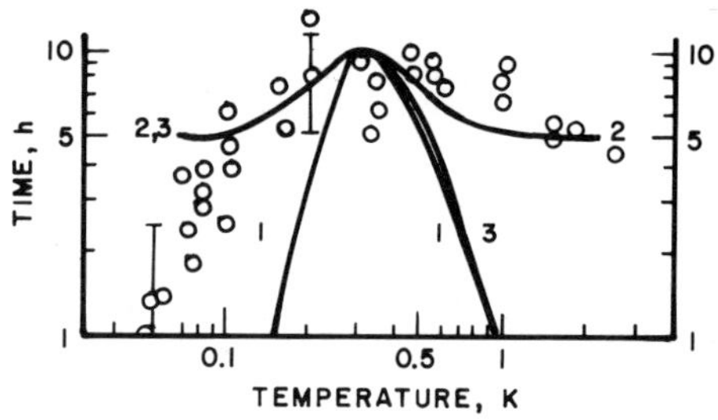

was independent of parameters. This mean-field, or continual, limit is not supposed to provide a very good agreement with the experiment: it does not use neither the exact lattice structure important at small pairs' radii (low-temperature wing), nor the randomness of mismatches responsible for the high-temperature wing of the curve.

The second approach is more accurate describing the diffusion in hexagonal lattice at small pairs' radii in combination with an approximate description of behavior at larger radii. The fitting parameters J and ξ/ζ gave the location of the maximum. An additional parameter described the ratio of jumps made at small and large separations (curve 2 in Fig. 1 corresponds to the case when nearly all time is spent at small separations, curve 3 - to the opposite case); the low-temperature side of the curve is independent of this parameter. One may improve the agreement with experiment, but the price would be the introduction of additional parameters: on the low-temperature side one should use the exact description of jumps including the second sublattice, and to increase the range of exact description (some admixture of the low-temperature wing of curve 1); on the high-temperature side, the improvement demands a better description of transition from random towards regular mismatches.

The lack of concentration dependence also gets a simple explanation.

3. SUMMARY

The large drop in the pairing time for ortho-impurities is due not only to changes in diffusion mechanisms, but also to a large difference in lengths and forms of diffusion trajectories associated with these mechanisms. The most important diffusion problem corresponds to the two-phonon mechanism of quantum diffusion. The fit to the experimental data may be achieved by an accurate analysis of transitions between the regimes with random and determinate mismatches.

REFERENCES

(1) H. Meyer, J. Phys. (Canada) 65 (1987) 1453.
(2) J. Van Kranendonk, Solid Hydrogen (Plenum Press, NY, 1983), Chapter 9.
(3) A. F. Andreev, Defects and surface phenomena in quantum crystals, in: Quantum Theory of Solids, ed.: I. M. Lifshitz (MIR, Moscow, 1982) p.p. 11 - 69.
(4) A. F. Andreev and A. E. Meyerovich, Sov.Phys. - JETP 40 776 1974.
(5) Yu.Kagan and L.A.Maksimov, Sov.Phys.- JETP 57 (1983) 459.

Physica B 165&166 (1990) 811–812
North-Holland

PINNING AND UNPINNING OF CHARGED GRAIN BOUNDARIES IN BCC ^{4}He WITH 5% ^{3}He

Andrew I. GOLOV, and Leonid P. MEZHOV-DEGLIN

Institute of Solid State Physics, Ac. Sci. USSR, CHERNOGOLOVKA, Moscow distr., 142432, USSR.

In bcc ^{4}He with 5% ^{3}He were investigated the quasiperiodic bursts of negative charges, corresponding to the motion of charged grain boundaries in an electric field. The lower threshold voltages and the upper threshold temperatures were observed for the bursts to occur, they are due to pinning and unpinning of the dislocations by ^{3}He atoms.

1. INTRODUCTION

Dahm and co-workers (1,2) observed periodic escapes of large portions of space charge superimposed on a steady-state diode current in strained hcp ^{4}He and bcc ^{3}He crystals. A model was proposed, in which low-angle grain boundary traps charges and moves under an external force to collector, and relaxes to its equilibrium after discharge (2).

We have studied similar bursts of negative charges in samples of bcc ^{4}He containing x = 5% ^{3}He.

2. EXPERIMENT

Crystals were grown in a planar diode of dimensions 6 x 15 mm^2 and separation L = 0.3 mm (see another letter of the same authors in this volume). Bursts were observed only in those samples that start to transform into hcp phase on cooling (molar volume V_m = 21.2 cm^3/mole). The difference between bcc and hcp molar volumes V_m is 1%, that causes a plastic flow and strain.

3. RESULTS

The bursts (Fig.1) were observed in some interval of voltages U, dependent on T (Fig.2). When the regular bursts took place, our results are in qualitative agreement with those of (2): burst duration, period t and time of flight of distance L by free charges were rising exponentially with reciprocal temperature (activation energies were 8 - 10 K). The deposited charge Q was 10^6 - 10^7 e,

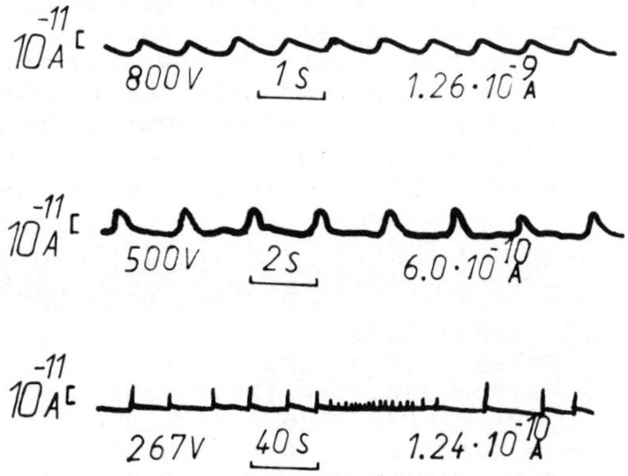

FIGURE 1
The records of bursts of negative current for different voltages U (indicated on the left). T = 0.9K. Time scales are different. The values of direct current are shown on the right.

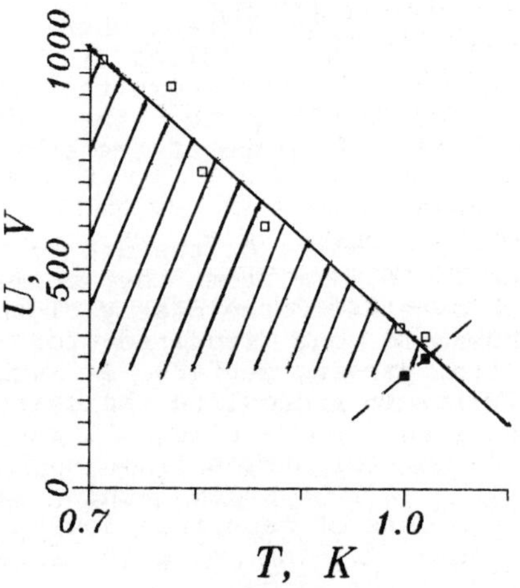

FIGURE 2
The range of voltages U and temperatures T, where regular bursts were observed.

and the integral bursts current Q/t was 0.4% - 2% of steady-state value in different samples. When crossing the upper boundary (□ at Fig.2), charge Q is vanishing gradually; it is due to thermal escape of charges under external force.

When the lower threshold U(T) is crossed (■ in Fig.2), bursts either become irregular (like in the lower record in Fig.1) or disappeared suddenly with finite charge Q. When increasing U once again they appear with a hysteresis.

In Fig.3 the record of collector current for U = -267V is shown on cooling and heating near the threshold temperature. When the sample is cooled, the bursts appear at T = 0.967K and the first is 1.6 times as large as the following ones (in this case its charge $Q = 6\cdot10^{-13}$ C). On heating they disappear also suddenly with finite Q, but at T = 0.990K.

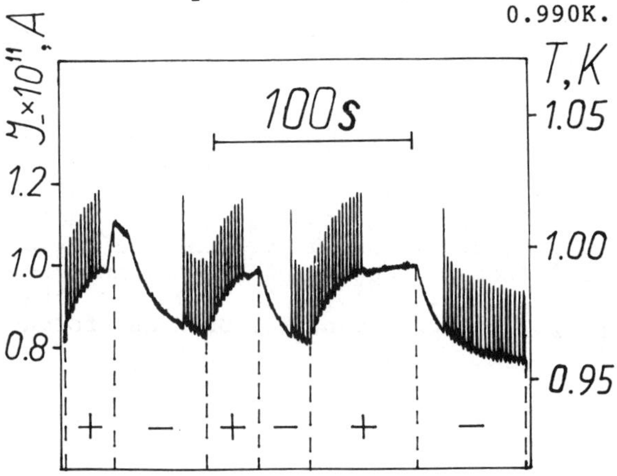

FIGURE 3
The record of diode current when sample was heating (+) and cooling (-). U = -267V. On the right is shown the temperature corresponding to the direct current (lower envelope of the curve).

4. DISCUSSION

Take a 2-d defect of the area S at a distance $l \sim 0.1$ mm from the collector under an external force nearly close to the threshold one required for its escape from pinning centers. At voltage U = 267V it can accumulate the charge $Q = \varepsilon US/L$, thus for $Q = 6\cdot10^{-13}$ C we have $S = 3\cdot10^{-2}$ mm. The stress is $\sigma_c = QE/S = \varepsilon U^2/2l^2 \sim 3\cdot10^2$ dyne/cm^2. This stress is three orders of magnitude lower than flow stresses for the plastic flow in bcc solid helium (3-5).

However, the dislocations forming a low-angle grain boundary of sufficiently small mismatch angle θ could begin to move at such stress. For ordinary impure bcc crystals the shear stress τ_c for dislocation motion is $\tau_c = (10^{-2} - 10^{-3})\mu$, where μ is the shear modulus. In bcc ^{4}He it is $\mu \sim 2\cdot10^{8}$ dyne/cm^2 (6). For gliding dislocations $\tau_c = \sigma_c/\theta$, and we obtain $\theta \sim 10^{-3}$, that confirms the low-angle grain boundary model of the authors (2).

The pinning centers for dislocations in this case are ^{3}He atoms. The hysteresis phenomena near the lower U-T threshold could be related with the fact that oscillating dislocation moves in a concentration x = 5% ^{3}He, but the localized dislocation accumulate $x_o = x\cdot\exp(W_o/T)$ (W_o is the binding energy of ^{3}He atom and dislocation). From the ratio 1.6 between the first and the other bursts in Fig.3 we can estimate W_o. $1.6 > x_o/x = \exp(W_o/T)$, thus $W_o < 0.97K\cdot\ln1.6 = 0.46K$. . This value is in agreement with that one obtained from ultrasonic measurements (7).

The existence of upper temperature threshold and the decrease of the lower voltage threshold with temperature (■ in Fig.2) in such a way means that the shear stress for dislocation motion τ_c decreases with temperature at T < 1K in our samples. Note that in a number of experiments (4,5,7,8) dislocations began more mobile as temperature was lowered down below 1K.

5. CONCLUSION

In terms of the model used for bursts to occur it is necessary to fulfill three conditions for U and T: charges do not escape from dislocations, dislocations are mobile enough and charge mobility is not very small. Possibly we didn't observe the bursts of positive current in bcc ^{4}He and the authors (1) didn't observe bursts of negative current in hcp ^{4}He because of their small mobilities at low enough temperatures.

REFERENCES

(1) S.C.Lau, A.J.Dahm, and W.A.Jeffers, J.Phys.(Paris)Suppl.,39(1978) C6-86.
(2) B.M.Guenin and A.J.Dahm, Phys.Rev.B, 23 (1981) 1139.
(3) A.Sakai, Y.Nishioka and H.Suzuki, J. Phys.Soc.Jap., 46 (1979) 881.
(4) D.J.Sanders,H.Kwun,A.Hikata,C.Elbaum, Phys.Rev.Lett., 40 (1978) 458.
(5) M.B.Manning,M.J.Moelter,C.Elbaum, Phys.Rev.B, 33 (1986) 1634.
(6) R.Wanner, Phys.Rev.A, 3 (1971) 448.
(7) I.Iwasa and H.Suzuki, J. Phys. Soc. Jap., 49 (1980) 1722.
(8) J.R.Beamish and J.P.Franck, Phys. Rev.Lett., 47 (1981) 1736.

Physica B 165&166 (1990) 813–814
North-Holland

MOTION OF INJECTED CHARGES IN SOLID ^{4}He CONTAINING UP TO 5% ^{3}He

Andrew I. GOLOV, and Leonid P. MEZHOV-DEGLIN

Institute of Solid State Physics, Ac. Sci. USSR, CHERNOGOLOVKA,
Moscow distr., 142432, USSR.

The temperature and field dependences of positive and negative charges velocities are measured in hcp and bcc ^{4}He with 0.16% and 5% ^{3}He. Small concentration 0.16% ^{3}He didn't change the charge transport markedly, and 5% ^{3}He suppress the low field mobility of positive charges in hcp ^{4}He at low temperature. The charges mobilities in bcc ^{4}He are closed to ones from bcc ^{3}He crystals of nearly the same molar volumes, and differ principally from those ones from hcp ^{4}He.

1. INTRODUCTION

The investigation of motion of injected charges in solid helium is now the only convenient way to study the microscopic processes of diffusion in quantum crystals under the influence of an external force. Many experimental results are obtained in this field (1,2), theoretical approaches are proposed, mainly suggesting the charge transport by means of vacancy motion (1,3). However, some experimental results are not explained in the frames of current theories.

What is the main origin of difference between charge behaviors in hcp ^{4}He and bcc ^{3}He: different crystal structure, different isotope mass and statistics or simply different impurity concentration (note that usually ^{4}He can be purified better than ^{3}He)? It is well known that ^{3}He impurities pin dislocations and localize each other, what is their influence on the charge motion?

To answer these questions we have measured charge velocities in hcp and bcc samples of ^{4}He with ^{3}He concentration 0.16% and 5% and compared them with those of pure hcp ^{4}He, bcc ^{4}He and bcc ^{3}He samples of the same molar volumes.

2. EXPERIMENT

The method for sample preparation and time-of-flight technique for charge velocity measurement are the same as in (2). Crystals were growing at constant pressure in the chamber containing a planar diode with separation L = 0.3 mm. One of electrodes made of tritium β-source produced charges in thin layer near it. The charge's time of flight τ of path L defined as the location of maximum at transients of collector current. Thus, the velocity is V = L/S.

3. RESULTS

3.1. Hcp ^{4}He. In the range of scattering of the data observed in samples of anisotropic hcp ^{4}He (2), introduction of 0.16% ^{3}He did not change charge velocities markedly.

Adding of 5% ^{3}He results in rising of activation energy ε_+ of positive charge motion in low fields E at T > 1K (Fig.1). An analogous increase of activation energy is observed for ^{3}He-atoms diffusion in hcp ^{4}He, when concentration x_3 exceeds 2% (4). It is possible that properties of vacancies, such as formation energy or mobility depend on x_3.

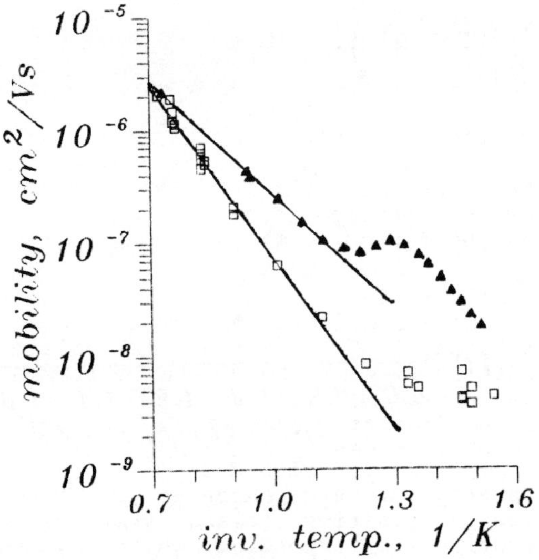

FIGURE 1
The temperature dependences of positive charge mobility (U=300V) in hcp samples:
^{4}He: ▲ – V_m = 20.55 cm^3/mole
^{4}He + 5% ^{3}He: □ – V_m = 20.5 cm^3/mole

0921-4526/90/$03.50

At T < 1K in samples with 5% ^{3}He non-monotonic dependences $V_+(E)$ and $V_+(T)$ are also observed at nearly the same E and T as in pure ^{4}He ,but they look smoother (Fig.1).

3.2 Bcc ^{4}He. The mobilities μ of charges were investigated earlier in narrow temperature range 1.46K - 1.77K of existence of pure bcc ^{4}He (5). By solving 5% ^{3}He we first realized the opportunity to investigate charge motion in this phase down to 0.5K.

The mobilities of both species depend on temperature nearly exponentially (Fig.2). It's interesting, that as the pressure is raised the mobility of positive ones μ_+ increases, and the μ_- decreases (Fig.2). Similar behavior was observed for negative charges in hcp ^{4}He and bcc ^{3}He in some ranges of pressure (7,8). On our opinion, it could be explained by the hypothesis of existence of stable states of electronic bubble in solid helium at some pressures (6), when "pinning" of bubble by crystal lattice create an additional potential barrier for motion. If it is true, the activation energy of negative charge motion $\varepsilon_-(p)$ has a local minimum at pressures p, when nonstable states take place and the additional barrier is small or zero.

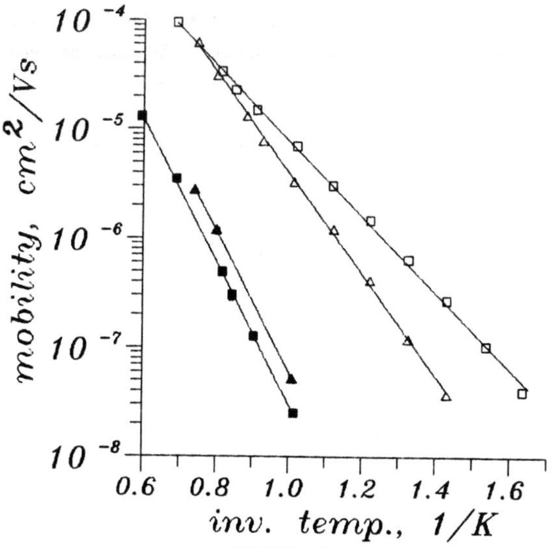

FIGURE 2
The temperature dependences of mobilities of positive (black symbols) and negative (corresponding white symbols) charges in bcc ^{4}He + 5% ^{3}He samples. Molar volumes are: □ - 21.2 cm^3/mole, and Δ - 21.3 cm^3/mole.

The control measurements of mobilities $\mu_+(T)$ and $\mu_-(T)$ in samples of bcc ^{3}He of V_m near 22 cm^3/mole, grown in the same diode, have shown that at all temperatures under study they are close to values from bcc ^{4}He with 5% ^{3}He of V_m = 21.3 cm^3/mole. At the same time they differ from data of hcp ^{4}He by many orders of magnitude.

At strong fields $E > 2\ 10^4$ V/cm, T > 1K, for positive charges sublinear dependences $V_+(E)$ are obtained , similar to those, observed in some hcp ^{4}He samples at near the same V_m = 21 cm^3/mole (2,7). The occurrence of such dependences is proposed by theory (3), at eEu ~ T, where u is the distance between two positions of charge of equal energy. Thus, in the frame of this theory we conclude u ~ 0.5 Å for positive charges in hcp and bcc ^{4}He.

4. CONCLUSIONS

Even though there are some qualitative similarities between dependences V(E,T) in hcp and bcc ^{4}He, the quantitative coincidence is observed only between bcc ^{4}He and ^{3}He. The most surprising fact is that though in bcc helium the mobility of vacancies and impurities are higher than in hcp, positive charges move in bcc lattice more slowly. Therefore, the type of crystal structure around the charge, that was usually ignored , needs to be used, when the theory of charge motion is designed.

REFERENCES

(1) A.J. Dahm, Charge motion in solid helium, in: Prog.Low.Temp.Phys., Vol. 9, ed. D.F.Brewer, North-Holl, 1985.
(2) A.I. Golov, V.B. Efimov, L.P. Mezhov-Deglin, Zh.Eksp.Teor.Fiz., 94 (1988) 198. (Sov. Phys. JETP, 67 (1988) 325).
(3) A.F Andreev and A.D. Savishchev, Zh. Eksp. Teor. Fiz., 96 (1989) 1109.
(4) V.A.Mikheev, V.A.Maidanov and N.P.Mikhin, Fiz.Nizk.Temp., 8 (1982) 1000.
(5) A.I. Golov, V.B. Efimov and L.P. Mezhov-Deglin, Piz'ma Zh. Eksp. Teor. Fiz., 40 (1984) 293.
(6) A.I. Golov, Piz'ma Zh. Eksp. Teor. Fiz., 49 (1989) 346.
(7) K.O. Keshishev, Zh. Eksp.Teor.Fiz., 72 (1977) 521.
(8) D. Marty and F.I.B. Williams, J. de Physique, 34 (1973) 989.

Physica B 165&166 (1990) 815–816
North-Holland

FLUCTUATION EFFECTS IN PHASE SEGREGATION

P. Kumar
Department of Physics, University of Florida-Gainesville, Fl.32611, U.S.A
M. Bernier
Service de Physique du Solide et de Résonance Magnétique
CEA-CEN Saclay, 91191 Gif sur Yvette, Cedex, France

The excess specific heat observed above the phase separation temperature in solid mixtures of ^{3}He and ^{4}He is attributed to the fluctuations that represent the symmetry of the condensing state. In a ^{3}He rich solid the fluctuations of the ^{4}He rich phase are 2-dimensional while in an even mixture they are 3-dimensional, the characteristic length scale is approximately the interparticle spacing.

I. INTRODUCTION

The segregation of a ^{3}He $-^4$ He mixture at high pressures, on cooling, has been studied by Edwards, Mc Williams and Daunt (EMD) [1]. The temperature dependent specific heat of mixing measured by EMD adds to the usual Debye specific heat of the solid mixture and is in excellent agreement with the regular solution theory. There is however some excess specific heat above the phase separation temperature $T_0(x)$, where x is the molar fraction of ^{3}He, that increases as the temperature T approaches $T_0(x)$. This excess specific heat can be understood as the contribution of the fluctuations, which depends on the dimensionality of the short range order.

In a ^{3}He rich mixture, the ^{4}He has been known to condense at the surface while the lattice for the ^{4}He rich component is hcp and thus, one might expect planar domains for the short range order. Furthermore the condensation anywhere else should consist of spherical domains i.e. 3-d fluctuations for the short range order. This is essentially what we find when analyzing the data of EMD [1] along these lines: the excess specific heat can be fitted with the Gaussian fluctuations with the dimension $d = 3$ for x in the neighborhood of 0.5 and $d = 2$ for x close to one.

II. DESCRIPTION OF THE FLUCTUATIONS

Let us consider a fictitious spin model proposed by Bernier and Landesman [2]: a lattice gas where the occupancy of the ^{4}He is characterized by the spin component $-1/2$ while the occupancy of ^{3}He is described by the spin component $+1/2$ on a lattice; a more elaborate version of the lattice gas model using a spin 1 is given in Ref.[3]. The Hamiltonian of these models, in the mean field theory, reduces to the regular solution theory used by Edwards and Balibar [4]. We can ignore the non Ising terms describing atomic exchange between the impurity and host atoms because they are significant only at much lower temperatures. We can then derive the fluctuation Hamiltonian H_f by defining the fluctuation variable as $\Delta = x - x_0$ where x_0 is the given molar concentration of ^{3}He in the mixture, and expanding the Hamiltonian for small Δ [5] . The fluctuation Hamiltonian H_f can also be simply written down following the symmetry requirements. Note that the parameter Δ is indeed expected to be small and the Gaussian fluctuations need only the first, quadratic, term. Since it must also represent the instability for phase separation, the coefficient must be proportional to $T - T_0(x)$. Thus the Gaussian fluctuations are fully characterized by

$$H_f/T_0(x) = [(T/T_0(x)) - 1]\,\Delta^2 + \sum_{\ell} \xi_\ell^2\, |(\nabla + ik_\ell)\Delta|^2 \quad (1)$$

where $T_0(x)$ is the phase separation temperature defined by EMD [1] and the k_ℓ's are the reciprocal lattice vectors of the condensing state. The last term denotes the contribution of the inhomogeneous fluctuations. The above form of the energy ensures that only the fluctuations away from the proper lattice symmetry increase the energy. Each of these fluctuations is characterized by a length scale ξ_ℓ measured in units of the lattice constant.

The fluctuation contribution to the thermodynamic free energy can then be written as [5]

$$F_f = -k_B T\ \ell n \left[\prod_q \int \mathrm{d}\ \Delta_q\ \exp\{-\beta\ H_f\,[\Delta_q]\}\right] \quad (2)$$

Where Δ_q is the Fourier transform of $\Delta(r)$. Eq.(2) leads to a specific heat [5]

$$\frac{C_f}{R} = \frac{1}{2}\frac{A_d}{(2\pi)^d}\frac{x(1-x)}{\xi_0^d}\left[\frac{T}{T_0(x)} - 1\right]^{d\nu-2} \quad (3)$$

Here A_d is the surface area of a unit sphere in d dimensions, ξ_0 is a characteristic average of ξ_ℓ and ν is the coherence length exponent. In Eq.(3) the fluctuation specific heat has been expressed as a function of d, the dimensionality of the dominant fluctuations. In critical phenomena [5] this result is better known as the exponent $\alpha = 2 - d\nu$. In the Gaussian approximation used here $\nu = 0.5$ and therefore the power law of the diverging excess specific heat becomes a clear measure of d. In the region of the phase diagram x near one, the condensing fluctuation goes into an hcp lattice which has a planar symmetry and the dominant fluctuations are in the basal plane, characterized by a length $\xi_{||}$. The effective dimension in Eq.(3) is then $d = 2$. For x near 0.5 the condensation takes place in the more isotropic bcc phase for which all the length scales are supposed to be similar; the effective $d = 3$.

Fig.1 shows the log-log plot of the excess specific heat for $x = 0.5$ at two different pressures, 2.7 and 3.6 MPa as given in Ref.[1]. The transition temperature and the Debye specific heat are from Ref.[1]. The exponent is clearly close to 0.5 giving the dimension $d = 3$. The data from Ref.[1] for $x = 0.997$ and 0.999 at $P = 3.6$ MPa are shown in Fig.2. This time the exponent is closer to one indicating $d = 2$. The coefficient $\xi_{||}$ can be determined from the experimental data as well. The length scale in units of the interparticle distance for $x = 0.5$ is found to be 1.2 while for smaller x it is 1.3 and 0.67 respectively for $x = 0.997$ and 0.999. In all cases it is the characteristic length scale for the density variation namely the interparticle distance.

These results support the phase diagram proposed by Edwards and Balibar [4] for the high pressure solid mixtures, based on an analysis of the excess specific heat in the experiments by EMD [1]. We find that the dimensionality of the fluctuations should be 3 for $x < 0.7$ and 2 for larger x. In as much as most of the data analyzed here are for large x, there is a need for measurements in the small x range. Panczyk et al. [6] have reported results for the phase diagram in that region, however the data cannot be analyzed in the same way for the excess specific heat. A more detailed discussion can be found in Ref.[7].

REFERENCES

[1] D.O. Edwards, A.S. McWilliams and J.G. Daunt, Phys. Rev. Lett. **9**, 195 (1962)

[2] M. Bernier and A. Landesman, Jour. de Physique **32**, C5a-213 (1971)

[3] J. Lajzerowicz and J. Sivardiere, Phys. Rev. A**11**, 2079 (1975) and references therein.

[4] D.O. Edwards and S. Balibar, Phys. Rev. **39**, 4083 (1989)

[5] S. K. Ma, "Modern Theory of Critical Phenomena", Benjamin, New York (1974)

[6] M.F. Panczyk, R.A. Scribner, J.R. Gonano and E.D. Adams, Phys. Rev. Lett. **21**, 594 (1968)

[7] P. Kumar and M. Bernier, to be published in J. Low Temp. Phys. **79** (1990)

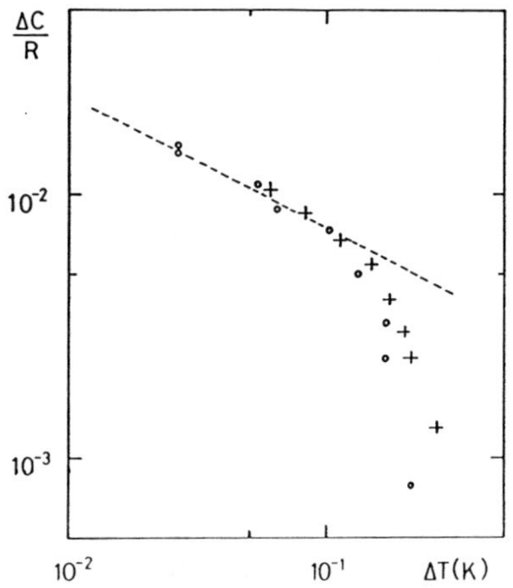

Fig.1 : The excess specific heat as a function of $T - T_0(x)$ for $x = 1/2$. The crosses denote the data of EMD (Ref.[1]) at a pressure of 3.58 MPa and the circles represent the pressure of 2.7 MPa. The dashed line represents an exponent $\alpha = 0.5$.

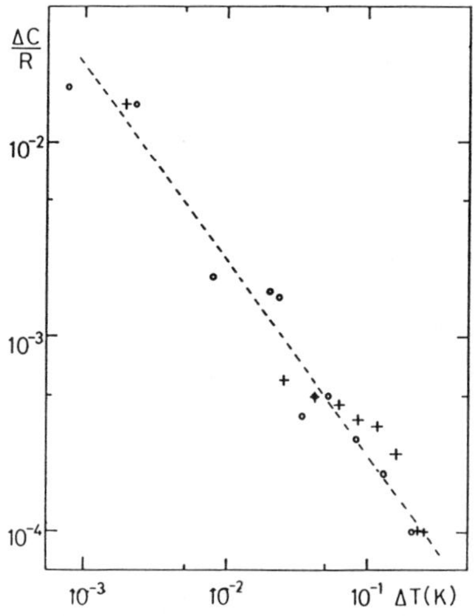

Fig.2 : The excess specific heat at small x. The pressure is 3.58 MPa. The circles represent the excess specific heat for $x = 0.997$ and the crosses are for $x = 0.999$. The dashed line in this figure represents an exponent $\alpha = 1$.

Physica B 165&166 (1990) 817–818
North-Holland

HEAT CAPACITY AND PRESSURE AT PHASE SEPARATION AND SOLIDIFICATION OF ^{3}HE IN HCP ^{4}HE

R. SCHRENK, O. FRIZ, Y. FUJII, E. SYSKAKIS, AND F. POBELL

Phys. Inst., Universität Bayreuth, D-8580 Bayreuth, FRG

We have measured heat capacity and pressure of 0.45 % and 0.9 % ^{3}He-^{4}He mixtures at pressures between 27 bar and 33 bar and temperatures between 18 mK and 300 mK. The data show the latent heats and the pressure changes associated with the solid phase separation (or remixing) and with the liquification (or solidification) of the resulting ^{3}He droplets in the hcp ^{4}He matrix. These data as well as the heat capacity of the liquid ^{3}He droplets show that all ^{3}He phase-separates, but only part of it liquifies. The amount of liquid depends on the history of the experiment.

1. INTRODUCTION

Liquid as well as solid ^{3}He-^{4}He mixtures phase-separate when cooled to millikelvin temperatures. There exists a pressure range between about 25 bar and about 34 bar where almost pure solid hcp ^{4}He coexists with solid bcc ^{4}He-rich or ^{3}He-rich phases, with a ^{3}He-rich liquid or with a ^{4}He-rich liquid depending on pressure and temperature. Many studies of the complicated phase diagram in this pressure range have been published with some disagreement on details (1). In recent studies it was shown that one may succeed in creating liquid droplets in almost pure hcp ^{4}He (2, 3). Here we report on an extension of our former preliminary study (3) to a wider pressure range and with two mixtures of 0.45 % and 0.9 % ^{3}He.

2. EXPERIMENTAL

The measurements were performed in a Cu calorimeter of 1.9 cm^3 inner volume. Carbon resistance thermometers and a PtW wire heater were glued to the outside of its wall. The bottom of the calorimeter contained a capacitive manometer. It could be thermally linked to the mixing chamber of a dilution refrigerator by closing a superconducting heat switch. During cooldown a solid plug was formed in the fill capillary to the calorimeter; therefore all our experiments are at constant volume.

3. RESULTS AND DISCUSSION

We describe as an example the behaviour of a 0.9 % ^{3}He mixture in two consecutive runs starting at a pressure of 31.94 bar at 0.4 K. The second run was started immediately after the first one and after keeping the mixture at 0.4 K for 1 h. In the first (second) cooldown the phase separation of the solid mixture started at 99 (128) mK and lasted for 20 (33) h until no heating could be observed anymore. The total pressure change after cooling to 18 mK was 151 (112) mbar. After the final pressure of 32.060 (32.018) bar was reached, we started the heat capacity measurements, monitoring simultaneously the pressure at constant volume. The results are shown in Fig. 1.

We first observe a linear increase of the heat capacity with T at constant P from 18 mK to 40 mK. The slope of this increase (after subtracting the background from the calorimeter) is $\gamma = 15.4$ (8.2) mJ/K^2 in the two runs, corresponding to a liquid amount of 454 (230) μmole if compared to the bulk heat capacity at this pressure. This would mean that only 54 (28) % of the ^{3}He is contributing. In addition we see a temperature independent contribution of about 0.2 mJ/K, similar to what has been observed recently for liquid ^{3}He on Ag or grafoil substrats (4) or for ^{3}He droplets in Ag (5).

At about 40 mK the heat capacity starts to increase more rapidly, and at about 50 mK the pressure starts to decrease slightly; we associate these changes with the start of remixing. In both runs we then observe a very sharp latent heat peak at about 68 mK which is superimposed on the mentioned slight increase of C, and a rather sudden drop of the pressure by 89 (45) mbar. At a pressure of 32 bar the melting line of bulk ^{3}He is hit at 71 mK. We therefore associate these sharp features with the solidification of the formerly liquid ^{3}He droplets in the hcp ^{4}He matrix. The latent heat occurs due to the difference in spin entropy of liquid and solid ^{3}He. Taking the difference in molar volume of liquid and solid ^{3}He and the compressibility of the solid hcp ^{4}He matrix, we find that the observed pressure changes correspond to the solidification of 432 (225) μmole or 51 (27) % of the ^{3}He in our calorimeter. These numbers are in excellent agreement with the amount of liquid deduced from the specific heat data at $T < 40$ mK. We therefore conclude that in the cooldown process only part of the ^{3}He in the calorimeter was liquified.

0921-4526/90/$03.50

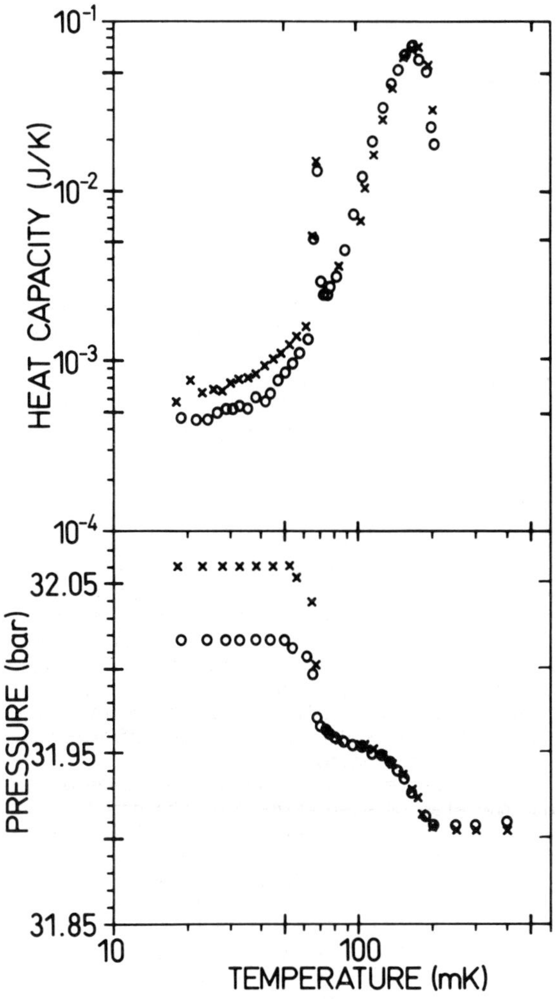

FIGURE 1
Heat capacity and pressure of a 0.9 % mixture from two consecutive runs.

During the further temperature increase we see a broad latent heat peak with a maximum at 170 mK. The associated pressure change is 65 mbar; the pressure stays constant at $0.2\ K \leq T \leq 0.4\ K$. These features are due to a total remixing of the two isotopes. The latent heat is the heat of mixing and the pressure decreases because of the difference in molar volumes of the ^{3}He-rich bcc and the ^{4}He-rich hcp phases. The features are identical in both runs; we conclude therefore that the same amount (very probably all) of the ^{3}He is phase-separated.

The pressure changes during warm-up, 155 (109) mbar, agree well with the pressure changes during cool-down, 151 (112) mbar.

The same observations with correspondingly different numbers were made in several runs with the 0.9 % mixture starting at 32.9, 30.6, 30.3, 28.3, and 26.9 bar, and with the 0.45 % mixture starting at 32.8 bar, 31.3, and 28.4 bar. The only difference being that a clear separation of phase separation and liquid/solid transition was only possible at $P > 31$ bar.

From all our observations we conclude that all ^{3}He phase-separates during cooldown and remixes during warm-up. This process - except for its duration for up to two days - was reproducible in each run. But only part of the phase-separated solid ^{3}He - which is probably in the form of small droplets in the hcp ^{4}He matrix - liquifies. This latter amount varied between 20 % and 70 %, and depends on the history of the experiment, in particular on the starting conditions. We may speculate that the geometry of the phase-separated ^{3}He varies from run to run, and that some portion of it is in a shape which does not allow liquification under the investigated experimental conditions, even though we are in the liquid part of the bulk phase diagram. - The temperature independent contribution to the heat capacity, too, varied from run to run.

The observed latent heats, $Q = T\,\Delta S = \int C\,dT$ result from the entropy of mixing, $\Delta S_{mix} = n\,R\,[x \ln x + (1-x) \ln (1-x)]$ and from the difference of the spin entropy of liquid and solid ^{3}He, $\Delta S_{spin} = n\,R \ln 2\,(T/T_F-1)$. From the results of the above discussed runs with the 0.9 % mixture at about 32 bar, we find $Q_{mix} = 7.1$ mJ, and $Q_{spin} \sim 0.1\,Q_{mix}$. These values agree to the corresponding numbers calculated with the mentioned equations using $T_{melt} = 68$ mK, $T_{PS} = 170$ mK, and the above mentioned amount of liquified ^{3}He. Similar agreement is found for the other run, supporting our interpretation of the observations.

REFERENCES

(1) P.M. Tedrow and D.M. Lee, Phys. Rev. 181 (399) 1969; V.L. Vvedenskii, JETP Lett. 24, 133 (1976); B.v.d. Brandt, W. Griffioen, G. Frossati, H.v. Beelen, and R. de Bruyn-Ouboter, Physica 114 B, 295 (1982); V.N. Lopatnik, Sov. Phys. JETP 59, 284 (1984); D.O. Edwards and S. Balibar, Phys. Rev. B 39, 4083 (1989).

(2) A.S. Greenberg, W.C. Thomlinson, and R.C. Richardson, J. Low Temp. Phys. 8, 3 (1972); B. Hébral, A.S. Greenberg, M.T. Béal-Monod, M. Papoular, G. Frossati, H. Godfrin, and D. Thoulouze, Phys. Rev. Lett. 46, 42 (1981).

(3) E. Syskakis, M. Gebhardt, and F. Pobell, in Spin Polarized Quantum Systems, ed. S. Stringari (World Scientific, Singapore, 1988) p. 365.

(4) D.S. Greywall and P.A. Busch, Phys. Rev. Lett. 54, 130 (1985); 60, 1860 (1988); and preprint (1989)

(5) E. Syskakis, Y. Fujii, F. Pobell, and H. Schroeder, to be publ. in J. Low Temp. Phys. (1990).

Physica B 165&166 (1990) 819–820
North-Holland

KINETICS OF THE PHASE SEPARATION OF THE DILUTE ^{3}He-^{4}He HCP SOLUTIONS AND MAGNETIC PROPERTIES OF ^{3}He DISPERSED IN SOLID ^{4}He MATRIX.

A.A.GOLUB, V.A.GONCHAROV, V.A.MIKHEEV, V.P.RUSNAK, V.A.SHVARTS

Institute for Low Temperture Physics and Engineering, Lenin's Avenue, Kharkov, USSR

Pulsed NMR technique is employed to observe the kinetics of the phase separation and to measure the magnetic properties of the decomposed ^{3}He-^{4}He dilute solutions. Observations of the kinetics made on the HCP solutions in the pressure range 3.5-3.8 MPa at 0.5 ± 0.1% and 1.2 ± 0.2% concentrations of ^{3}He before separation display different behaviour. Magnetic susceptibility of the samples at 0.5% ^{3}He concentration obeys Curie-Weiss' law at temperatures above 10 mK. The Weiss' constants of two samples from three investigated are 2 - 3 times more than the appropriate BCC ^{3}He values. One of the samples demonstrates a flat peak in the magnetic susceptibility near 3 mK. The magnetic phase ordering of the decomposed solutions was never found above 0.7 mK.

1. INTRODUCTION

The phase separation of dilute ^{3}He-^{4}He HCP solutions occuring at superlow temperatures displayed an unusual kinetics of this process in the samples of 0.5% ^{3}He concentration under the pressure P=3.2 MPa (1). Moreover, the NMR study of such an object undertaken at T=100 mK gave an unexpectedly high diffusion coefficient which was close to the value for liquid ^{3}He.

These results inspired a work reported in two directions. The first one was to get some additional information concerning the phase separation kinetics of dilute solid solutions, the other one was to investigate magnetic properties of ^{3}He dispersed in HCP ^{4}He down to 0.7 mK.

2. THE PHASE SEPARATION KINETICS

The studies of kinetics of phase separation were performed at 77.7 Gs magnetic field. It turned out that the spin-lattice relaxation times in both novel phases are short enough to be measured simultaneously. The relaxation time for the ^{3}He rich phase $T_{1.1}$=0.2-0.3 s doesn't practically change during the separation process, while for dilute phase $T_{1.2}$ >2 s are growing due to decreasing of ^{3}He concentration.

To monitor the process we measured the longitudinal magnetization of the samples by supplying the long series of "90^0" declining pulses which were following with time intervals either 1 s (U_1) or 31 s and more (U_∞). In first case the magnetization of dilute phase couldn't return to its equilibrium value because of large $T_{1.2}$. So it was convenient to monitor the process of phase separation by means determining the factor Ks =U_1/U_∞. Its value was increasing from 0.4 before separation to 0.98 after separation was completed.

2.1. Solutions of 1.2% ^{3}He concentration

The method of observations was following. The temperature of the samples was rapidly lowered (in 10 min intervals) step by step of 10-20 mK magnitude and stabilizing for a long time ensuring the factor Ks reached its constant values after rather smooth rise. Then Ks values could stay unchanged for the time intervals which weren't available experimentally (>16 h) It indicated the conservation of ^{3}He distribution between the coexisting phases, however a small temperature drop (10 mK) was sufficient to revive the process.

A hysteresis of Ks values was observed during the step by step warm-up of the samples. The process of phase separation was 2-3 times faster at the repeated cool-downs.

To determine the relative amounts of ^{3}He atoms in the coexisting phases we measured the time dependencies of the restoring longitudinal magnetization when Ks was unchanged. These measurements showed that the coexistance of two phases with different distribution of ^{3}He atoms between them takes place at the same temperature due to different thermal prehistory of the samples. Hence the phase separation line of the T - X diagram of solid ^{3}He-^{4}He solutions can be restored experimentally with great uncertainty.

The T_1 measurements revealed also that the annealing of the samples started at temperatures near 0,3 K.

2.2. Solutions of 0.5% ^{3}He concentration

The samples of the same density as before but smaller ^{3}He concentration diplayed very different behaviour. Despite the significant overcooling of the samples the separation process was stopping itself for time intervals which weren't available experimentally (>24h). To revive the process it was successfully tested the thermocycling of the samples at temperatures below 100 mK to prevent the annealing effects.

Thus the phase separation of the dilute HCP

0921-4526/90/$03.50 © 1990 – Elsevier Science Publishers B.V. (North-Holland)

solutions ^{3}He in ^{4}He are obstructed with decreasing of ^{3}He initial concentration in spite of growth of the quantum diffusion coefficient. It indicates the important role of phenomena near the interface boundaries and possibly is connected with the decreasing of the amount of defects involved into the samples during growth.

3. MAGNETIC PROPERTIES OF ^{3}He DISPERSED PHASE

Values of spin-lattice T_1 and spin-spin T_2 relaxation times also as spin-diffusion coefficient values of the samples of 0.5% initial concentration measured at T=80 mK were very close to the BCC ^{3}He ones. Magnetic properties of the samples were investigated down to temperatures as low as 0.7 mK.

Magnetic susceptibility of the samples studied obeys Curie-Weiss´ law at temperatures above 10 mK. The Weiss´ constants of the samples were determined graphically from the plots of U^{-1} vs T, where U is the magnitude of the free induction signal in arbitrary units (fig.1).

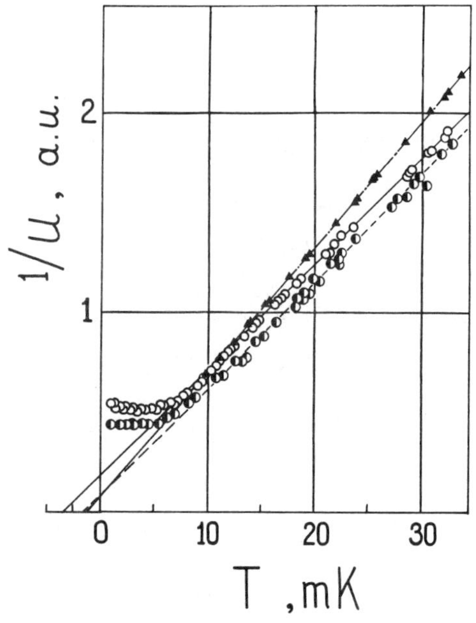

FIGURE 1

Simbols (▲ , ○ , ◐) denote our data on BCC ^{3}He sample under the pressure 3.8 MPa, the samples with 0.5% ^{3}He initial concentration under the pressures 3.71 MPa and 3.53 MPa correspondently. The Weiss´ constants displayed the following values: for BCC ^{3}He under P=3.8 MPa - θ =-1.2 mK, for disrersed ^{3}He under P=3.78 MPa - θ =-4.5 mK, for dispersed ^{3}He under P=3.71 MPa - θ =-3.2 mK, for dispersed ^{3}He under P=3.53 MPa - θ =-1.7 mK. Hence for two samples of phase separated solutions from three investigated the Weiss´ constants are 2-3 times more than appropriate BCC ^{3}He values.

One can see in fig.1 there exists deviations of magnetic susceptibility from Curie-Weiss´ law below 10 mK. One of the samples (P= 3.71 MPa) demonstrates a flat peak of the magnetic susceptibility near 3 mK. For other samples peaks weren´t found. One can suppose the existance of heat generation inside the latter samples due to phase separation process wasn´t completed entirely. In such a case one can estimate the values of the heat input and thermal resistance between sample and coolant using the observed deviations of magnetic susceptibility from Curie-Weiss´ law. Such an estimate gives heat input value of $2*10^{-10}$ W. Using estimates for the thermal resistance and heat capacity of the samples one can obtain the values of the time constant which are the same order of magnitude as experimental ones.

The spin-lattice T_1 and spin-spin T_2 relaxation times are rather short - 0.26-0.28 s and close to each other and to published data for BCC ^{3}He of appropriate density within 40-80 mK interval of temperature. Such a situation corresponds to the case of low frequency NMR at low temperatures when the bottle-neck of relaxation of the magnetic energy exists between Zeeman and exchange systems.

When the temperature of the samples was decreasing below 30 mK the abrupt rise of spin-lattice times starts. The restoration of the longitudinal magnetization occures in two stages. At first the temperature of the low capacity Zeeman system turns back to the exchange system temperature with time constant Tze. After that more slower relaxation process takes place in the course of which Zeeman system serves as thermometer for exchange system.

The process of the total thermal relaxation has exponential character and can be discribed by time costant T_1 in spite of the BCC ^{3}He case at temperatures below 100 mK. The absence of the well-known law $M(t) \approx t^{1/2}$ is a manifestation of disparsted state of ^{3}He phase. The temperature dependence of the T_1 time is close to T^{-5}.

At the temperatures below 7-9 mK the T_1 time reaches the plateau with T_1=4.5 ± 1.5 min while the magnetic susceptibility is significantly changing yet.

Measurements of the spin-spin relaxation time shows the trend of T_2 values to drop with decreasing the temperature. The absence of the peculiarities of the T_2 time behaviour down to 1 mK indicates the absence of transition to the state of magnetic phase ordering.

REFERENCIES

(1) V.A.Mikheev, V.A.Maidanov, V.P.Mikhin. Sov. J. Low Temp. Phys. 12(6) (1986) 375

Physica B 165&166 (1990) 821–822
North-Holland

UNIVERSAL BEHAVIOUR OF SPIN LATTICE RELAXATION IN SOLID HELIUM-3

Brian COWAN and Michael FARDIS

Millikelvin Laboratory, Royal Holloway and Bedford New College, University of London,
Egham Hill, Egham, Surrey, TW20 0EX, U.K.

We have studied the exchange-mediated spin lattice relaxation in solid helium-3. The frequency dependence of the relaxation times follows a universal curve when the data are scaled by observed T_1 minima. Such behaviour follows from the assumption that all significant exchange processes vary with density in a similar way. Some consequences of this are explored.

Quantum exchange in solid ^{3}He is a complex process with interchange cycles involving different numbers of particles (1). Conventional spin relaxation time experiments cannot distinguish these different processes; they measure a single characteristic frequency, some average of the various exchange frequencies. From the calculations of Matsumoto et.al (2) one can deduce this "effective" exchange frequency for the bcc and the hcp lattice.

There is evidence, both experimental and theoretical (3) that the Gruneisen constants for the different exchange processes are similar. Within this approximation a measured quantity is a unique function of the (effective) exchange frequency J for all densities. We have investigated the consequence of this for the case of NMR of ^{3}He films (4), and here we consider the implications for bulk solid ^{3}He.

The arguments follow closely those of (4). The conclusion is that the spectral density function $J(\omega)$ must have the form $(M_2/3J)K(\omega/J)$ for some function K, where M_2 is the second moment of the resonance line. The consequences, as they relate to the present case, are:

(a) Both ω/T_1 and J/T_1 depend on ω and J only through the combination ω/J. Thus they are "universal" functions of ω/J.

(b) A minimum in T_1 at Larmor frequency ω is observed when $J=\omega/C$, where C is a dimensionless constant of order unity, depending on the shape of the function $K(\omega/J)$.

(c) The value of T_1 at the minimum is given by $\omega/T_1=M_2CK(C)/3$.

We see that T_1 is proportional to the Larmor frequency at the minimum. Clearly this linearity holds only insofar as the different exchange frequencies scale similarly with density. In figure 1 we have plotted available T_1 minimum values against Larmor frequency. (We divide the relaxation times by the square of the molar volume to allow for the variation of M_2 with density.) Apart from one point, all minima observations were made at frequencies below 10MHz. Allowing for experimental errors these points are consistent with a straight line

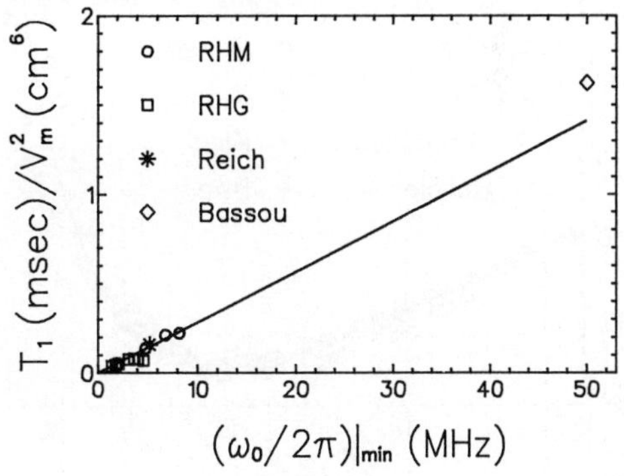

Figure 1

through the origin. And this is reinforced by the relatively new observation (5) of a minimum at 50MHz.

From (b) we see that to within a multiplicative constant, observation of a minimum in T_1 gives the magnitude of the motion frequency. The exchange frequency is a rapidly varying function of density. In figure 2 we plot available values of T_1 minima against molar volume. The solid line is a power law function which we have fitted through the points. Two functions have been tried:

$$\omega_{min}/2\pi = 34.065(V_m/24)^{18.13}\ \text{MHz}$$

and

$$\omega_{min}/2\pi = 35.139(V_m/24)^{18.40}\ \text{MHz}$$

In the upper expression the exponent was fixed to that adopted by Panczyk and Adams (6) while in the lower expression it was free. The difference is not discernable on the plot.

There is a further reason for the importance of the observation of minima in T_1. Since J is proportional to ω at the minimum and ω/T_1 is a function of ω/J, it follows that ω/T_1 is a function of ω/ω_{min}. When plotted in this "reduced" way all experimental data should fall on a single curve. We have collected together published measurements of T_1 in bcc ^{3}He in the exchange regime, which we have plotted in

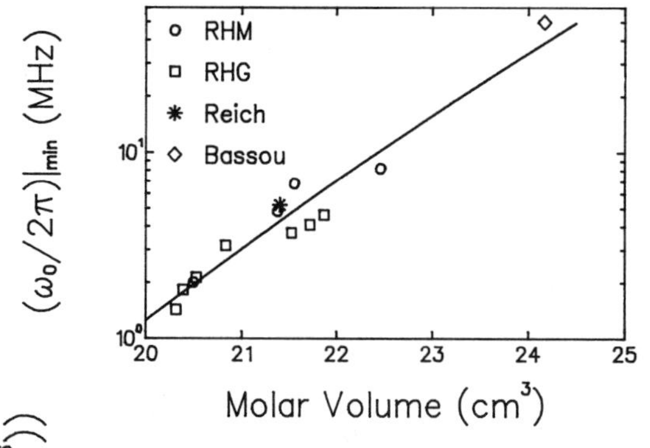

Figure 2

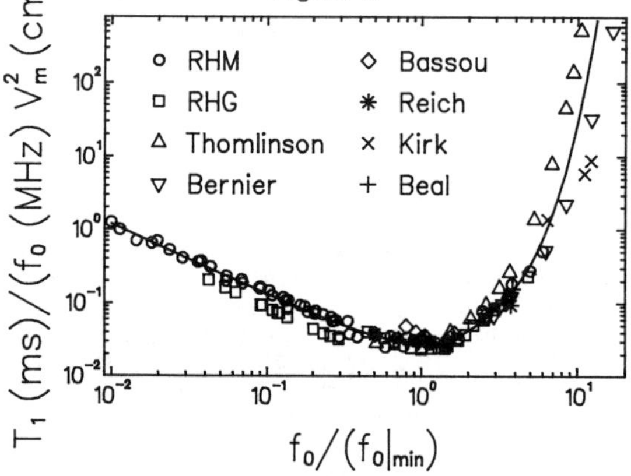

Figure 3

figure 3. Most data fall very close to a single curve although at low frequencies one set of points (7) falls consistently below the others while at high frequencies the data appear to split into two lines (8),(9). Notwithstanding these deviations, the disposition of the points (10) is encouraging, supporting the hypothesis of the common Gruneisen constant.

We mention briefly a number of other things which follow from this universal behaviour of T_1 data. There are a number of sum rules which must be satisfied (11). The area under $1/T_1(\omega)$ is equal to πM_2. However the second moment must be renormalised to account for the crystal lattice vibrations. Harris (12) has argued that the renormalisation factor should be approximately 0.8. Landesman found an experimental value of 0.89 for measurements at a molar volume of 20.15 cm^3. Unfortunately there are not many points for any one molar volume. But by representing data in reduced form, points for all densities may be exploited. Figure 4 shows such a plot. The area gives M_2, and a numerical estimation yields a renormalisation factor of 0.787, a more realistic figure.

Another use for the reduced representation of the T_1 data is in finding a mathematical expression which is a good approximation to the spectral density function. From this the

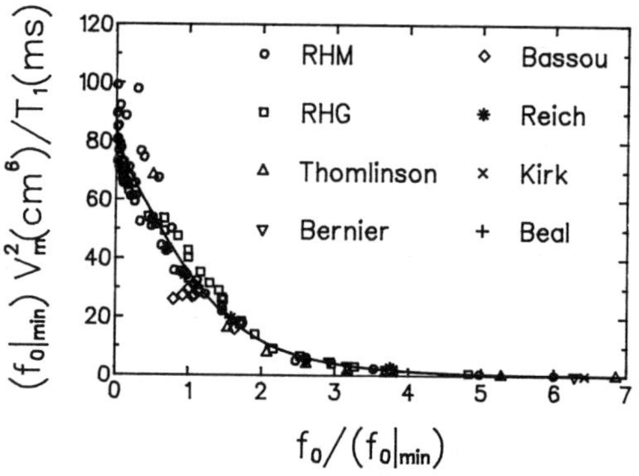

Figure 4

constant C can be obtained so that numerical values for the J may be found. In constructing such a function it is best to start in the time domain with the autocorrelation function G(t). The short-time behaviour may be calculated directly by a moment expansion and the long-time behaviour from hydrodynamic arguments. And the J(ω) must give the correct slope for the T_1 minimum against frequency curve.

From this function the solid curves in the above figures are generated. The constant C is found to be 2.42 so that $J=0.413\omega_{min}$. Also, the relation between the zero frequency T_1 and the moments M_2 and M_4 can be given a reliable numerical constant. We find $T_2=0.207M_4^{1/2}/M_2^{3/2}$. This compares with the constant $3/5\sqrt{(2\pi)}=0.239$ for a Gaussian function, as used by Matsumoto et.al.(2). Although only a 15% difference, this could be important in establishing bounds on the consistency of measured data with various exchange models (2).

REFERENCES

(1) M.Roger, J.H.Hetherington and J.M.Delrieu, Rev. Mod. Phys. 55 (1983) 1.
(2) K.Matsumoto, A.Yakayuki and T.Izuyama, J. Phys. Soc. Jap. 58 (1989) 1149.
(3) D.M.Ceperley and G.Jacucci, Phys. Rev. Lett. 58 (1987) 1648.
(4) B.Cowan, L.El-Nasr,M.Fardis and A.Hussain, Phys. Rev. Lett. 58 (1987) 2308.
(5) M.Chapellier et.al. J. Low Temp. Phys. 59 (1985) 45.
(6) M.F.Panczyk and E.D.Adams, Phys. Rev. A1 (1970) 1356.
(7) M.G.Richards, J.Hatton and R.P.Giffard, Phys. Rev. 139 (1965) A91.
(8) W.C.Thomlinson, Phys.Lett. 38A (1972) 531.
(9) M.Bernier and G.Guerrier, Physica 121B (1983) 202, and W.P.Kirk and E.D.Adams, Proc. LT13 (1972) 149.
(10) H.A.Reich, Phys. Rev. 129 (1963) 630. B.Beal et.al. Phys.Rev.Lett. 12 (1964) 394 R.Richardson, E.Hunt and H.Meyer, Phys. Rev. 138A (1965) 1326.
(11) A.Landesman, Ann. de Phys. 8 (1973) 53.
(12) A.B.Harris, Sol. St. Comm. 9 (1971) 2255

hysica B 165&166 (1990) 823–824
Jorth-Holland

MEASUREMENT OF SPIN DIFFUSION IN SOLID HELIUM-3

Brian COWAN and Michael FARDIS

Millikelvin Laboratory, Royal Holloway and Bedford New College, University of London, Egham Hill, Egham, Surrey, TW20 0EX, U.K.

It is shown that spin diffusion may be measured directly from the behaviour of spin relaxation times at low frequencies. The analysis is based on purely hydrodynamic arguments; the result is thus independent of any microscopic details of the motion. This technique is particularly suited to the measurement of very slow diffusion, where conventional methods are inapplicable.

Regardless of the precise details of the quantum exchange Hamiltonian, on the long time scale it will result in the phenomenon of spin diffusion. That is, a spin inhomogeneity of wave vector Q will relax exponentially with a time constant DQ^2. Here D is identified as the spin diffusion coefficient. The relation between D and the exchange parameter(s) has been evaluated in a number of cases (1),(2),(3). Thus measurement of D gives information about the magnitude of the exchange interaction.

The NMR technique of spin-echoes (5) is frequently utilised for the measurement of self-diffusion. Central to the method is the application of a magnetic field gradient over the specimen by which means positional information is encoded as phase variations. Using such methods Reich and Garwin (6),(7) and Thompson, Hunt and Meyer (8) have measured the spin diffusion coefficient in solid ^{3}He over a range of densities.

In this paper we show that values for the spin diffusion coefficient may be obtained directly from measurements of spin relaxation times. We shall demonstrate, using general hydrodynamic arguments together with a mathematical theorem of Watson (9), that the low frequency behaviour of both T_1 and T_2 is related directly to the diffusion coefficient.

Starting from the dipolar autocorrelation function expressed in its usual form (10),

$$G_m(t) = \frac{6\pi\hbar^2\gamma^4}{5N}\sum_{i<j}\frac{Y_2^{-m}(\Omega_{ij}(0))Y_2^{m}(\Omega_{ij}(t))}{r_{ij}^3(0)r_{ij}^3(t)},$$

the summation may be replaced by integration over a probability function which in the diffusive limit becomes $(1/8\pi Dt)^{3/2}\exp(-r^2/8Dt)$. Successive integrations by parts then gives an asymptotic series in inverse powers of time. In the diffusive limit one thus obtains

$$G(t) \sim \frac{\sqrt{\pi}}{60\sqrt{2}}\hbar^2\gamma^4\alpha\,(Dt)^{-3/2} + \ldots\ldots.$$

The $t^{-3/2}$ behaviour is characteristic of a three dimensional system; in n dimensions it is a general result that at long times the correlation function goes as $t^{-d/2}$. This leading term in the series depends only on the hydrodynamic property of the system: the diffusion coefficient, together with well-defined quantities such as the spin density α and the magnetogyric ratio γ. This is not the case for the higher order terms, which are dependent on details of the microscopic nature of the atomic motion.

The spin relaxation times are given in terms of the spectral functions, the Fourier transform of the dipolar autocorrelation functions. A feature of one function will manifest itself in the behaviour of its Fourier transform. Thus derivatives at the origin of the time function are related to moment integrals of the frequency function. There are also relations involving coefficients of asymptotic expansions. Kubo (11) has given some examples of these relating to integral power series.

We are involved with relations involving *fractional* powers, where the usual techniques are of no use. The mathematical result we require is embodied in what is known as Watson's Lemma (9). This states that if a function F(x) has an expansion

$$F(x) = \sum_{s=0}^{\infty} a_s\, x^{(s+\lambda-\mu)/\mu}$$

then its Laplace transform (variable t) may be

expressed by the asymptotic series

$$\sum_{s=0}^{\infty} \Gamma\left(\frac{s+\lambda}{\mu}\right) a_s\, t^{-(s+\lambda)/\mu} \quad \text{as} \quad t \to \infty$$

where s, λ and μ are positive integers. There are convergence conditions which we do not need to consider here. Fundamentally the lemma tells us that there is a relation between the coefficients of the small argument expansion and those of the large argument expansion of the dual function. However some terms are lost in going from the Laplace to the Fourier transform; for expansions inolving even integer powers only all are lost for the cosine transform pair!

The consequence of Watson's lemma for the problem in hand is that at low frequencies the spectral density function has the form:

$$J(\omega) = J(0) - (1/30)\hbar^2\gamma^4\pi\alpha\omega^{1/2}/D^{3/2} + \ldots\ldots$$

Although the value of J(0) depends somewhat on the microscopic details of the motion, the coefficient of $\omega^{1/2}$ does not. This term alone is determined solely by macroscopic/hydrodynamic quantities. A further consequence of the $\omega^{1/2}$ law is that the traditional functions adopted to represent the exchange-modulated spectral density in solid ^{3}He are quite wrong. Discussion of this is pursued elsewhere (12).

In the low-frequency limit the expressions for T_1 and T_2 are then

$$\frac{1}{T_1} = \text{const} - \frac{(1+4\sqrt{2})}{30}\pi\hbar^2\gamma^4\alpha\omega^{1/2}/D^{3/2} + \ldots$$

$$\frac{1}{T_2} = \text{const} - \frac{(5+2\sqrt{2})}{60}\pi\hbar^2\gamma^4\alpha\omega^{1/2}/D^{3/2} + \ldots$$

We have analysed spin relaxation measurements in solid ^{3}He made by a number of research groups (13),(14). By analysing the frequency dependence of the relaxation times according to the above formulae we have extracted the spin diffusion coefficients. These are plotted in the figure together with spin diffusion measurements (6),(7),(8) performed in the conventional manner. The agreement is quite satisfactory. Also on the figure we have shown the solid line (15) indicating the diffusion coefficient obtained from the accepted variation of the exchange frequency with molar volume.

It is clear that this method for measuring spin diffusion is a reliable one. It has the advantage that calibration of the field gradient is not required and it is capable of measuring small diffusion coefficients, wher conventionally the method of pulsed gradient and/or stimulated echoes would be used. We se on the figure that values of D have been foun for the high density solid, where th traditional method was not viable.

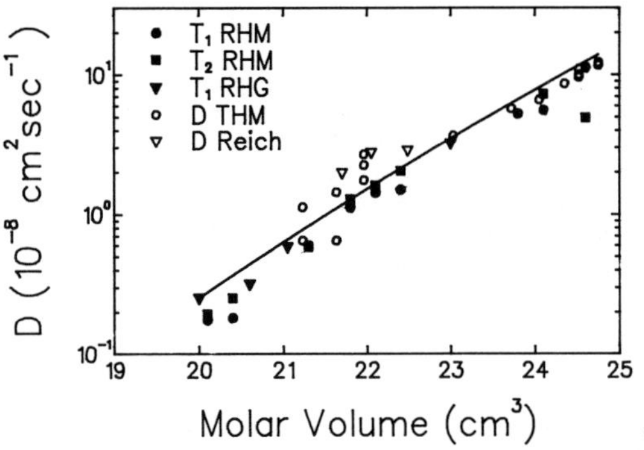

Spin Diffusion coefficient for bcc solid ^{3}He plotted against molar volume. Open points are conventional diffusion measurements. Filled points are obtained from relaxation time measurements using this analysis.

The obvious extension of the method is to use rotating frame measurements of $T_{1\rho}$. Such a technique is capable of very accurate determination of very small diffusion coefficients.

REFERENCES

(1) A.G.Redfield, Phys. Rev. 116 (1959) 315.
(2) P.G.deGennes, J. Phys. Chem. Solids 4 (1959) 223.
(3) B.P.Cowan, W.J.Mullin and E.Nelson, J. Low Temp. Phys. 77 (1989) 181.
(4) H.C.Torrey, Phys. Rev. 104 (1956) 563.
(5) J.Karger, H.Pfeifer and W.Heink, Adv. Mag. Res. 12 (1988) 1.
(6) R.L.Garwin and H.A.Reich, Phys. Rev. 115 (1959) 1478.
(7) H.A.Reich, Phys. Rev. 129 (1963) 630.
(8) J.R.Thompson, E.R.Hunt and H.Meyer, Phys. Lett. 25A (1967) 313.
(9) G.N.Watson, Proc. Lond. Math. Soc. 17 (1918) 116.
(10) B.P.Cowan, J. Phys. C, 13 (1980) 181.
(11) R.Kubo, Summer School in Theor. Phys.1958, Colorado, 120 (1959) Interscience, NY.
(12) B.P.Cowan and M.Fardis, to be published.
(13) R.C.Richardson, E.Hunt and H.Meyer, Phys. Rev. 138 (1965) A1326.
(14) M.G.Richards, J.Hatton and R.P.Giffard, Phys. Rev. 139 (1965) A91.
(15) M.F.Panczyk and E.D.Adams Phys. Rev. A1 (1970) 1356.

Physica B 165&166 (1990) 825–826
North-Holland

MELTING PRESSURE OF SOLID ^{3}HE NEAR THE MAGNETIC TRICRITICAL POINT

Tohru Okamoto, Hiroshi Fukuyama, Tsuneo Fukuda, Hidehiko Ishimoto, and Shinji Ogawa

Institute for Solid State Physics, University of Tokyo, Roppongi, Minato-ku, Tokyo 106, Japan

We have performed precise measurements of the melting pressure of solid ^{3}He near the magnetic tricritical point. The line which surrounds the low-field ordered phase on a melting pressure vs magnetic field phase diagram has a sharp turn at 3.92kG over a field interval no more than 0.03kG. This means that the entropy and the magnetization of the solid rapidly change across the transition from the paramagnetic phase to the high-field ordered phase in a narrow temperature range.

1. INTRODUCTION

Nuclear magnetism of bcc solid ^{3}He is understood to arise from several kinds of multiple exchange interactions among ^{3}He atoms. Competition between ferro and antiferromagnetic interactions causes three different magnetic phases; the paramagnetic disordered phase (PP), the low-field ordered phase (LFP), and the high-field ordered phase (HFP). While the transitions to LFP from either HFP or PP have been established as first order, there still remains a controversy on the order of the transition from PP to HFP. The NMR measurements by Osheroff (1) and the specific heat measurements by Sawada et al. (2) revealed the transition to be most likely first order. On the other hand, Tang, Adams, and Uhlig (3) claimed, from their measurements on the time evolution of the melting pressure, that there is no latent heat associated with the transition over the range of 4 to 5kG. More recently, Greywall and Busch (4) made specific heat measurements at 6 and 10kG. Their results also show no latent heat and a rather broad λ-type transition. In this paper we report precise measurements of the melting pressure at the magnetic transitions as a function of magnetic field near the tricritical point ($T\sim 0.8$mK, $B\sim 3.9$kG) to obtain further information on the nature of the PP-HFP transition.

2. EXPERIMENTAL

Design of the sample cell is similar to that of our previous work in zero magnetic field (5) except that all materials used in the present work are pure silver or silver based alloys. The cell was surrounded by a small active-shielded superconducting solenoid (6) which could produce magnetic fields up to 7kG with homogeneity of 5×10^{-4} over the total volume (0.55cm^3) of the sample. The field strength was determined by NMR of the ^{3}He sample itself. Field dependence of a background signal from a pressure gauge has not been detected above 0.5kG within an experimental resolution. To monitor the temperature, a platinum-wire NMR thermometer surrounded by another shielded solenoid was thermally linked to the sample cell.

The magnetic transitions of solid ^{3}He could be easily identified as a sharp kink on the time evolution of the melting pressure. Most of

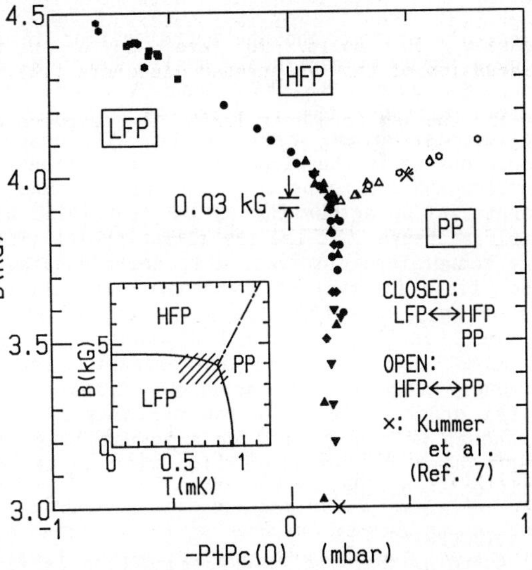

Figure 1

Melting pressures at the magnetic transitions vs magnetic field near the tricritical point. Data from five separate runs are plotted by different symbols. A volume fraction of solid in these measurements were from 2 to 4% at zero temperature. A small offset correction for the measured melting pressure has been made on each run. Inset: Schematic B-T phase diagram is shown for comparison. The present measurements were made for the hatched region.

the data were taken on cooling-back after applying a heat pulse into the sample keeping the temperature of a nuclear refrigerator below the transitions. The cooling rate was typically from 0.02 to 0.5mK/hour. The transition pressure in a constant magnetic field was reproducible within 60μbar on cooling with different cooling rates and also on warming. This indicates that a thermal equilibrium between the solid surface and the liquid was well established owing to a low boundary resistance and a high thermal conductance within the superfluid.

3. RESULTS

Fig. 1 shows the melting pressure P_c at the transitions as a function of magnetic field. The slope of the first order transition line surrounding LFP rapidly changes at 3.92kG over a field interval no more than 0.03kG. Along this line, a phase equilibrium among three phases, PP-LFP-liquid or HFP-LFP-liquid, is established. The slope of the transition line between the two solid phases (a:LFP, b:PP or HFP) can be expressed as

$$\frac{dP_c}{dB} = \frac{M_aS_b-M_bS_a}{(V_l-V_s)(S_a-S_b)} , \qquad [1]$$

since the entropy and the magnetization in the liquid phase is negligibly small. Here M_a and M_b, S_a and S_b, V_l and V_s are the magnetizations, the entropies, liquid and solid molar volume, respectively. The observed rapid change in the slope of the transition line surrounding LFP means that the entropy and the magnetization across the PP-HFP transition also rapidly change at the tricritical point (TCP). The temperature interval, ΔT, corresponding to the field interval of ΔB=0.03kG, is estimated to be only 3μK from the slope of the PP-HFP transition line in the T-B plane, $dT_c/dB \sim$ 0.1mK/kG (1,3). Therefore the present results strongly support that the PP-HFP transition is first order at least in the vicinity of TCP, although we cannot exclude a possibility of a very sharp λ-type transition with an unusual critical behavior.

4. DISCUSSION

Assuming that the PP-HFP transition is first order, we can deduce the jumps in entropy and magnetization (ΔS, ΔM) at the transition by applying Eq. [1] for the three experimental values of dP_c/dB at TCP, 0.03, 1.09, and -3.23 mbar/kG for the PP-LFP, the HFP-LFP, and the PP-HFP transition, respectively. The magnetization in HFP, M_{HFP}, was deduced to be (0.42$\pm$0.02)M_{sat} from the magnetic Clausius-Clapeyron equation and our own data of $(\partial P/\partial B)_T$ at 0.82mK. Also the entropy, S_{LFP}, and the magnetization, M_{LFP}, in LFP at TCP was calculated to be 0.15Rln2 and 0.078M_{sat} based on the previous results, on the assumption that the melting pressure P in this phase can be written as

$$P(T,B)=P(T,0)+P(0,B)-P(0,0) . \qquad [2]$$

Here P(T,0) is taken from Ref. 8 and P(0,B) is from Ref. 9. By use of these three values as well as the known volume difference V_l-V_s (10), the remaining, S_{PP}, S_{HFP}, and M_{PP}, are obtained as 0.51Rln2, 0.32Rln2, and 0.28M_{sat}. Finally, we have ΔS=0.19Rln2 and ΔM=-0.14M_{sat}.

For the PP-HFP transition, Tang, Adams, and Uhlig (3) excluded the possibility of a first order transition with an entropy jump larger than 0.05Rln2 in the vicinity of TCP. Their point is based on the absence of the sloping plateau on their time evolution of the melting pressure, which appeared at the transitions from LFP to either PP or HFP. Generally the appearance of the plateau, however, should be affected by many factors such as a size of the solid ^{3}He, the cooling or warming rate, and the thermal conductivity within the solid in each magnetic phase. Thus the absence of the plateau does not necessarily contradict to the first order transition.

Recent specific heat data across the PP-HFP transition at 6kG and 10kG by Greywall and Busch (4) indicate that the solid entropy gradually changes by ΔS=0.19Rln2 in a temperature width of $\Delta T/T_c \sim$0.1. On the other hand, our data near TCP show much sharper transition with $\Delta T/T_c \leqq$0.003 for the same entropy change. The fact might suggest that the PP-HFP transition changes its nature over the range from 4 to 6kG.

ACKNOWLEDGMENTS

We thank T. Tazaki and K. Sakayori for their technical assistance, and H. Akimoto for his experimental contributions.

REFERENCES

(1) D.D. Osheroff,Physica **109&110B**, 1461 (1982).
(2) A. Sawada, H. Yano, M.Kato, K. Iwahashi, and Y. Masuda, Phys. Rev. Lett. **56**, 1587 (1986).
(3) Y.H. Tang, E.D. Adams, and K. Uhlig, Phys. Rev. Lett. **57**, 222 (1986).
(4) D.S. Greywall, and P.A. Busch, Phys. Rev. B **36**, 6853 (1987).
(5) H. Fukuyama, H. Ishimoto, T. Tazaki, S. Ogawa, Phys. Rev. B **36**, 8921 (1987).
(6) U.E. Israelsson and C.M. Gould, Rev. Sci. Instrum. **55**, 1143 (1984).
(7) R.B. Kummer, R.M. Mueller, and E.D. Adams, J. Low Temp. Phys. **27**, 319 (1977).
(8) D.D. Osheroff, C. Yu, Phys. Lett. 77A, 458 (1980).
(9) D.D. Osheroff, H. Godfrin, and R. Ruel, Phys. Rev. Lett. **58**, 2458 (1987).
(10) W.P. Halperin, F.B. Rasmussen, C.N. Archie, and R.C. Richardson, J. Low Temp. Phys. **31**, 617 (1978).

Physica B 165&166 (1990) 827-828
North-Holland

MELTING PRESSURE OF SOLID ^{3}He IN MAGNETIC FIELD NEAR THE NUCLEAR ORDERING TEMPERATURE

A. SAWADA, A. SHINOZAKI, W. ITOH, K. TORIZUKA*, M. SUGA+, T. SATOH, and T. KOMATSUBARA

Department of Physics, Faculty of Science, Tohoku University, Sendai 980, Japan.

The melting pressure of ^{3}He was measured in magnetic field. The phase transitions of solid ^{3}He were determined by the anomaly of the melting curves during the temperature and the magnetic field sweep. With thus determined B-T phase diagram, the slope of the transition line near the triple point suggests the first order transition of the paramagnetic to the high field phase transition. However, any trace of the latent heat was not observed within our experimental accuracy. The spin wave velocity in magnetic field was also obtained.

1. INTRODUCTION

Solid ^{3}He shows interesting nuclear magnetic properties which are particular to the quantum solid. Many experimental results have established that three magnetic phases exist in bcc solid ^{3}He:a paramagnetic phase (PP) for temperatures higher than 1mK, a low field antiferromagnetic phase (LFP) for temperatures lower than 1mK, and a high field magnetically ordered phase (HFP) for field greater than 0.4T. The transitions to the LFP from the PP or the HFP have been clarified as first order. However, the order of the HFP-PP transition is still controversial[1-4].

Measurement of the melting pressure, P, provides the solid entropy through the Clausius-Clapeyron equation:

$$dP/dT=(S_s-S_l)/(V_s-V_l). \qquad (1)$$

Here S_s, S_l and V_s, V_l are the molar entropies and the molar volumes of solid and liquid, respectively. Using the facts $S_s>>S_l$, and $V_s-V_l=-1.314cm^3$/mole near 1mK, we obtain S_s from the measurement of dP/dT.

2. EXPERIMENTAL METHOD

The blocked-capillary method was used to confine the ^{3}He sample (1ppm ^{4}He) to constant average density. Pressure was measured by a Straty-Adams type strain gauge[5]. The cell was cooled by the demagnetization of Cu, 30 moles. The minimum temperature and the heat leak of nuclear stage were 81µK and 480pW, respectively. This nuclear stage maintains temperature less than 1mK for about two months. Superconducting magnets with Nb shield for a Pt pulsed NMR thermometer and for the ^{3}He sample were prepared independently. The NMR thermometer was calibrated by comparison with the melting curve thermometer, for which the magnetic transition point, 0.93mK was taken as he standard point. All of the data were digitally recorded by a automatic measuring system.

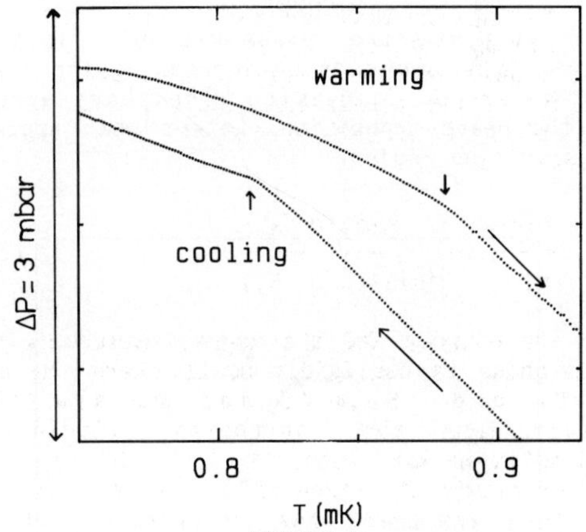

FIGURE 1
The melting Pressure versus cell temperature for 0.462T.
The short arrows show the phase transition. Transition temperature in warming differs from one in cooling, because of temperature difference between the cell and solid ^{3}He.

3. MEASUREMENTS

The behavior of the melting pressure at warming up and cooling down process is useful for indicating the order of the transition. The warming and the cooling curves in magnetic field less than 0.39T show the plateau due to the latent heat of first order LFP-PP transition. As shown in fig.1 the curves at 0.462T do not have such a plateau, and show only the

*Present address: Low Temperature Laboratory, Helsinki University of Technology, Otaniemi 02150 Espoo 15, Finland.
+Present address: Central Research Laboratory, Hitachi, Ltd. Kokubunji, Tokyo 185, Japan.

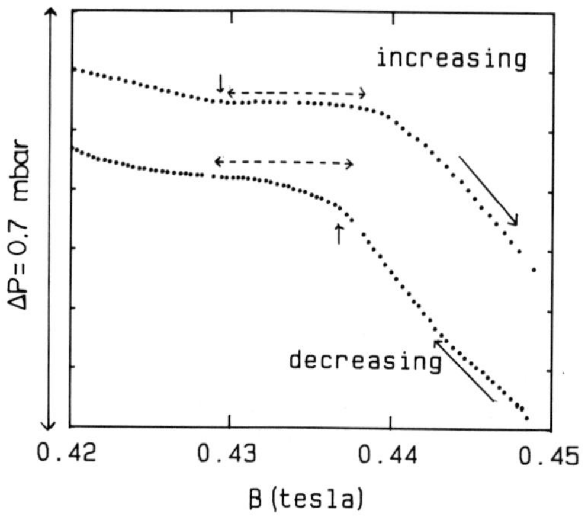

FIGURE 2
The melting pressure versus magnetic field at 0.842mK. Upper curve is on increasing the field and lower is on decreasing. The short arrows show the phase transition. Latent heat appears at dashed line region.

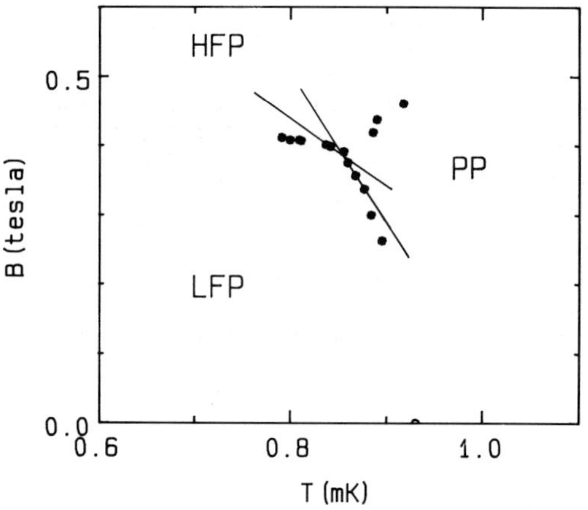

FIGURE 3
The B-T phase diagram at melting pressure. The error of temperature in each point is less than 5μK.

sharp change of the slope. The curves of the melting pressure versus magnetic field are shown in Fig.2. They clearly indicate the plateau due to the first order LFP-HFP transition. However those obtained for the HFP-PP transition at 1.1mK do not show such behavior. From these observation, we may conclude that the HFP-PP transition is not first order. This agrees with Tang et al[3].

The field sweep method is suitable for determine the LFP-HFP phase transition line, because its magnetic field dependence is very week. On the other hand, the pressure points on LFP-PP line and HFP-PP line were detected by sweeping the temperature, and the precise temperature, of the transition was determined by extrapolating from the thermal equilibrium points near the transition. Thus determined B-T phase diagram of solid ^{3}He on the melting pressure is shown in Fig.3. Along the LFP boundary, dB/dT is discontinuous at triple point. The entropy jump of the HFP-PP transition calculated from the discontinuity is about 0.03Rln2. This small jump of entropy could not be detected by our apparatus. We may say that the HFP-PP transition near the triple point is first order accompanied with a very small entropy discontinuity. Our thermometers, Pt NMR and the melting pressure, assigned the triple point at 0.86mK, which is slightly higher than the widely accepted value, 0.82mK. This difference is not within the accuracy of our temperature measurment.

The melting pressure P(T) of the ordered state in magnetic field is described well with T^4 term, which provides information of the field dependence of the spin wave[3].

$$P(T)=P_0-\alpha T^4. \qquad (2)$$

where α is related to the spin wave velocity c. Results for various fields are given in Table 1.

TABLE 1
Fitting coefficients and average velocity of spin wave.

B(T)	α(mbar/mK^4)	c(cm/sec)	Phase
0	3.03	7.5	LFP
0.36	3.61	7.0	LFP
0.47	3.86	6.9	HFP

ACKNOWLEDGMENT
We wish to thank T. Ohtsuka, T. Izuyama and Y. Masuda for useful discussions and supports. This work was supported by the Grant-in-Aid for Scientific Research (B) of the MInistry of Education, Science and Culture Japan, and by the Kurata Research of the Kurata Foundation.

REFERENCES
(1) D.D. Osheroff, Physica B+C 109&110(1982) 1461.
(2) A. Sawada, H. Yano, M. Kato, K. Iwahashi and Y. Masuda, Phys. Rev. Lett. 56(1986) 1587.
(3) Y.H. Tang, E.D. Adams, and K. Uhlig, Phys. Rev. Lett. 57(1986) 222.
(4) D.S. Greywall and F.A. Busch, Phys. Rev. B36 (1987) 6853.
(5) G.C. Straty and E.D. Adams, Rev. Sci. Instrum. 10(1969) 1393.

Physica B 165&166 (1990) 829–830
North-Holland

MONTE CARLO STUDY OF THE MULTIPLE EXCHANGE MODEL IN SOLID ^{3}He

Kazuyuki MATSUMOTO

Physics Laboratory, Nippon Dental University, Niigata Hamauracho 1-8, Niigata, 951*

By the mean field approximation, the Stipdonk and Hetherington's (SH) model which explains the magnetism of solid ^{3}He exhibits two successive phase transitions at H=0. Experimentally, there is only one phase transition, however. Then more detailed analysis is needed to remove this discrepancy. In this study, the Monte Carlo simulation have done for the SH model. From the result, we can see that this mismatch is due to inaccuracy of the mean field approximation. We have also determined the phase boundary (U2D2-PARA and U2D2-HFP) in the H-T phase diagram.

1. INTRODUCTION

Solid ^{3}He is called as an "ideal magnet" in which the magnetic behavior at 1mK is dominated by the exchange tunnelings. The dipole anisotoropy and hyperfine mediated interactions can be neglected in the same temperature range. Since solid ^{3}He is regard as a collection of spin 1/2 fermions, and its structure is quite simple, its magnetic behavior is expected to be understood from a microscopic principle.[1,2] One may suspect the nearest neighbor antiferromagnetic Heisenberg model well describe the magnetism of solid ^{3}He.

Through the powerful investigations in recent years, it has been revealed that solid ^{3}He has more exotic natures, however. The low field ordered phase (U2D2) has a tetragonal symmetry which is impossible based on the nearest neighbor antiferromagnetic Heisenberg model.[3] Now, this simplest view, i.e. the nearest neighbor antiferromagnetic Heisenberg model, is altered by the multiple exchange model. This model includes more complex exchange processes such as the three and the four particle cyclic exchanges.

There are some multiple exchange models according to the set of the exchange processes and their parameters. Among them Stipdonk and Hetherington's model (SH model) is attracted much attention, because it is specified by the familiar parameters, J_{NN}, J_t, and K_P.[4] Though it has three parameters, it is still tractable. The SH model is described by the Hamiltonian

$$H = - \sum J_{NN} (P_{ij}) + \sum J_t(P_{ijk} + P_{ijk}^{-1}) - \sum K_P(P_{ijkl} + P_{ijkl}^{-1}) \qquad (1)$$

*Present Address: Muroran Institute of Technology, Muroran 050

with

$$J_{NN} = - 0.377 \text{ mK} ,$$
$$J_t = - 0.155 \text{ mK} , \qquad (2)$$
$$K_P = - 0.327 \text{ mK} .$$

Stipdonk and Hetherington analyzed this Hamiltonian using the mean field approximation. The magnetic phase diagram that they obtained resembles the observed one as a whole. However, at at H=0 the high field phase (HFP) remain stable still above U2D2 phase. This seems contradict to the experimental observations. One can think that this contradiction is due to inaccuracy of the mean field approximation. If we can go beyond the mean field approximation, this contradiction may be removed. In this study, we carry out the Monte Carlo simulation to discuss this point.

2. METHODOLOGY

For a spin system, the Metropolis dynamics is familiar in the Monte Carlo simulation. However, this dynamics is not so efficient in the continuous spin systems such as the classical Heisenberg model. It needs much computing time, because in this dynamics the target direction of the spin is picked up randomly before trying a update. In low temperatures, it is rare that a transition to this direction is accepted. Then, the system may miss thermal equilibrium.

Recently, "heat bath" dynamics is developed to avoid this difficulty.[5] This dynamics is the following. Pick up a spin and calculate the effective field H based on the Hamiltonian (1). Taking H as the z-direction, the local spin coordinate is expressed by the polar angle θ and the azimuth angle Φ. For each Monte Carlo step (MCS), θ is determined by the weight function,

0921-4526/90/$03.50

$$\theta = \cos^{-1}\{(T/H) \times \log[(1-r)\exp(H/T)+r\exp(-H/T)]\} \quad (3)$$

where T denotes the temperature, and r is a random number within 0<r<1. The azimuth angle Φ is updated randomly in each step. By this dynamics we can accelerate the computation so much, and a micro-computer can be used in this simulation.[6]

We have four sub-lattice in U2D2 phase, then the system size of $4xn^3$ is useful. That is, samples used in this study are long in x-direction. The ferromagnetic sheets in the U2D2 structure are arranged to this direction. The periodic boundary condition is taken. In the present simulation, two samples are prepared.

For the U2D2-PARA transition, the sample is composed of $4x3^3$=108 atoms. Most of the initial condition is taken as U2D2, and the simulations are heat up process. In these simulations, 200 MCS are discarded for thermal equilibrium, and 2000 MCS are taken for the measurement of the susceptibility.

For the HFP-U2D2 transition, the system size is $4x2^3$=32 atoms. The initial condition is ferromagnetic, and the magnetic field is swept from HFP range to U2D2 one. In this calculation, 500 MCS are discarded to thermal equilibrium and 5000 MCS are taken for measurement. In both calculation, the stabilizing field which forces a spin to z-direction with 0.1 mK is applied.

In the U2D2-PARA case, the cooling process is also examined.

3. RESULTS

The internal energy of the SH model was obtained by Hetherington.[7] Then we focus our attention to the magnetic susceptibility. In Fig. 1, several runs are presented. In this figure, the cooling process is also presented. By using "heat bath" dynamics, the system may be in equilibrium all the time. Actually, we have no hysteresis through the U2D2-PARA transition.(cf. ref.[7]) Then the transition temperature can be determined with enough precision.

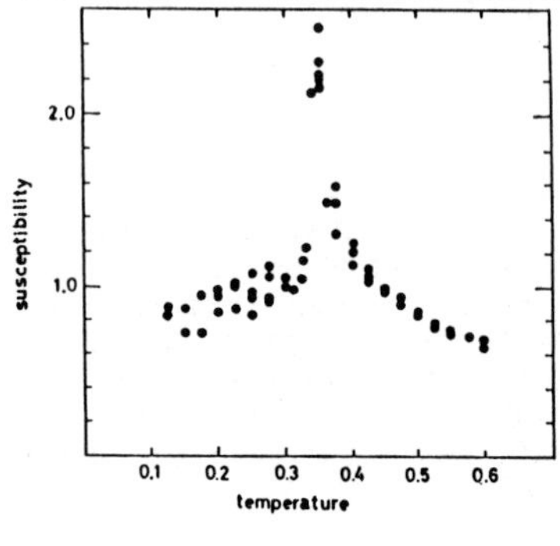

FIG.1: The magnetic susceptibility

As in the results of the mean field approximation, if HFP are exist just above the U2D2 phase in H=0, two anomaly should be observed in the magnetic susceptibility. We can observe only one anomaly in Fig. 1, however. This fact means that the second order phase transition from PARA to HFP at H=0 is fictitious of the mean field approximation.

We can extend our calculation under the magnetic field Hex. The transition temperature is determined in the same manner at each Hex. The magnetic phase diagram obtained is shown in Fig. 2. Open circles are the critical field of the HFP-U2D2 transition. These are determined by the field sweep process.

In conclusion, one may think that the SH model explain the important aspects of the magnetism of solid ^{3}He.

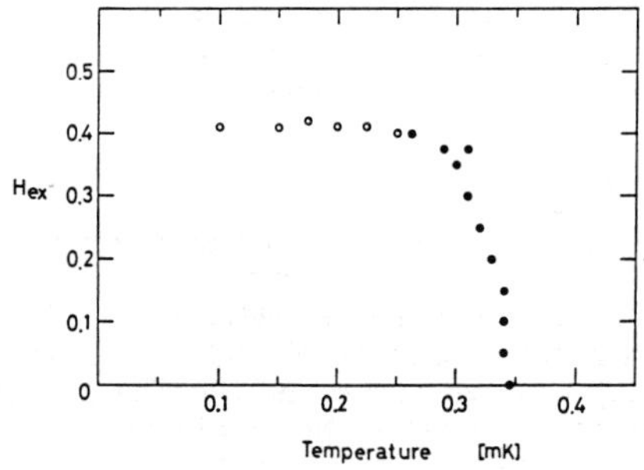

FIG.2: H-T phase diagram

Closed circles are transition temperatures determined from the heat up process for n=3 sample(108 atoms). Open circles are by the field sweep process for n=2 sample(32 atoms).

REFERENCES

(1) M. C. Cross and D. S. Fisher: Rev. Mod. Phys. 57 (1985) 881.
(2) M. Roger, J. M. Delrieu and J. H. Hetherington: Rev. Mod. Phys. 55 (1983) 1.
(3) D. D. Osheroff, M. C. Cross and D. S. Fisher: Phys. Rev. Lett. 44 (1980) 792.
(4) H. L. Stipdonk and J. H. Hetherington: Phys. Rev. B31 (1985) 4684.
(5) Y. Miyake, M. Yamamoto, J. J. Kim, M. Toyonaga, and O. Nagai: J. Phys. C19(1986)2539.
(6) P. Bak: PHYSICS TODAY, December(1983) 25.
(7) J. H. Hetherington: J. Low Temp. Phys. 66 (1987) 145.

Physica B 165&166 (1990) 831-832
North-Holland

MAGNETIC PROPERTIES OF THE HIGH FIELD PHASE IN BCC SOLID ^3He

K. IWAHASHI

Department of Physics, Nagoya University, Chikusa-ku, Nagoya 464-01, Japan

Y. MASUDA

Department of Liberal Arts, Aichi Gakuin University, Aichi-ken 470-01, Japan

Using the Green-function method with Tyablikov decoupling and taking into account all exchange processes up to 6 spins, we have considered theoretically the magnetic properties of the high-field magnetically ordered phase of bcc solid ^3He at low temperatures. The magnetic phase diagram was calculated. The result shows that the upper critical field $h_{c2}(T)$, which separates the high field phase and the paramagnetic phase, decreases as $T^{3/2}$ from $h_{c2}(0)$ as temperature T rises. Namely, in the low temperature region near 0 K, the $T^{3/2}$-law should definitely hold. This result differs completely from the result by the molecular field approximation, in which $h_{c2}(T)$ decreases exponentially. The zero temperature magnetization and the temperature dependence of the sublattice magnetization will be discussed.

1. INTRODUCTION

In the magnetic ordering of nuclear spins of bcc solid ^3He at low temperatures three phases exist: a paramagnetic phase (PP), a low-field antiferromagnetic phase (LFP) and a high-field magnetically ordered phase (HFP). The structure of the LFP is believed to be of the uudd type. On the other hand, the structure of the HFP is uncertain yet and we assumed to be the canted normal antiferromagnetic type.

The transition to the LFP from either the PP or the HFP is definitely to be first order. However, the experimental results on the PP to HFP transition remain controversial. In order to make sure these behaviors we developed the Green-function theory with Tyablikov decoupling, in which all exchange processes up to 6 spins are included. (1,2) We discuss the behaviors at finite temperatures of an upper critical field, the temperature and magnetic field dependences of the sublattice magnetization and magnetization.

2. CALCULATION

The exchange Hamiltonian, in which all exchange processes up to 6 spins were taken into account, was used. (2,3,4) According to Godfrin and Osheroff (4), the 18 types of exchange cycles were considered in this Hamiltonian. In the presence of the magnetic field B, the spin axes (x,y,z) at the corner and body-center sites of the bcc structure make angles of $\pm\theta$ with respect to the quantized axis z' of the original spin direction (x',y',z') respectively. The exchange parameter J_n, (3,4), are obviously influenced by the presence of the magnetic field. The detailed expressions for these effective parameters denoted as $\tilde{J}_n$, $\tilde{J}_{nx}$ and $\tilde{J}_{ny}$, were given in ref. 2. Using the nuclear magnetic moment of the ^3He nucleus μ, we put $h=\mu B$.

In the first place, we calculate the magnetization of sublattice spin, $p=2\langle S_z\rangle$. An effective field $\tilde{h}$ is obtained as

$$\tilde{h} = h\cos\theta + p\sum_{n=1}^{7} z_n \tilde{J}_{nx} ,$$

where z_n is number of the n-th neighbors.

The self-consistent condition for the sublattice magnetization p can be obtained in the following:

$$p^{-1} = N^{-1}\sum_k[\sqrt{F_k/f_k}\coth(\tfrac{1}{2}\beta p\sqrt{F_k f_k})]$$

where

$$F_k = p^{-1} h\cos\theta + \sum_{n=1}^{7} z_n \tilde{J}_{nx}(1-\gamma_{nk}),$$

$$f_k = \sum_{n=1}^{7} z_n \tilde{J}_{ny}(\mp 1-\gamma_{nk}), \quad \text{for } \theta\neq 0$$

$$= F_k, \quad \text{for } \theta=0$$

$$\gamma_{nk} = z_n^{-1}\sum_j \exp[ik\cdot(R_i-R_j)]$$

and $\beta=(k_BT)^{-1}$ and the upper sign in a parenthesis corresponds to the case of n=1, 4 and 7 and the lower sign to n=2, 3, 5 and 6 and R_j is the position vector of the n-th neighbors of the atom at R_i. k is a wave vector restricted in the first Brillouin zone of bcc structure.

The self-consistent condition for the angle θ can be obtained as,

0921-4526/90/$03.50

$$h \sin\theta + p \tan\theta \sum_{n=1}^{7} z_n(\tilde{J}_{nx} \pm \tilde{J}_{ny}) = 0,$$

where the (±) sign operates in the same way as in the equation of f_k.

3. MAGNETIC PROPERTIES

3.1 Upper critical field at finite temperatures, $h_{c2}(T)$

In the case of $\theta=0$, the relations $\tilde{J}_{nx}=\tilde{J}_{ny}=\tilde{J}_n$ and $F_k=f_k$ would hold and therefore

$$p^{-1} = 1 + (2/N)[\sum_k \exp(\beta p F_k)-1]^{-1},$$

$$h + 16p(\tilde{J}_1+3\tilde{J}_4+3\tilde{J}_7) = 0.$$

As example of the calculation of the $h_{c2}(T)$, which separates the HFP and the PP, is shown in Fig. 1.

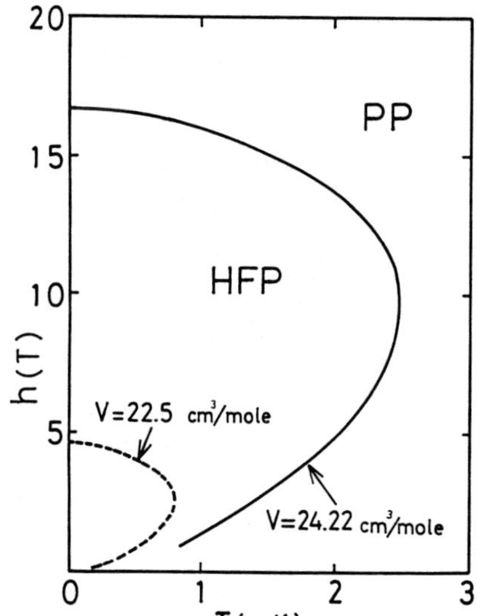

FIGURE 1

The temperature dependence of $h_{c2}(T)$ is shown in the following:

$$h_{c2}(T) \simeq h_{c2}(0)-[-2A-6(2B+C)-30D]$$

$$\times 2(4\pi J_e)^{-3/2}\zeta(3/2)(k_BT)^{3/2},$$

$$h_{c2}(0)=-16(\tilde{J}_1+3\tilde{J}_4+3\tilde{J}_7),$$

$$J_e=[-\tilde{J}_1-11\tilde{J}_4-19\tilde{J}_7+\tilde{J}_2+4\tilde{J}_3+4\tilde{J}_5]_{p=p_c},$$

where $A=8(J_1+3J_4+3J_7)$, $B=12(K_F+K_P-4F+S_1+3S_2)$, $C=6(4K_A-4F+S_1+3S_2)$ and $D=3(S_1+3S_2)$. p_c is the sublattice magnetization at $h_{c2}(T)$ and $\zeta(n)$ is the Riemann ζ-function. The calculated result shows that $h_{c2}(T)$ decreases as $T^{3/2}$ from $h_{c2}(0)$ as temperature T rises, which differs completely from the result by the molecular field approximation, in which $h_{c2}(T)$ decreases exponentially as T increases. In the low temperature region near 0 K, the $T^{3/2}$-law should definitely hold.

3.2 Zero temperature magnetization

The sublattice magnetization at 0 K, p_0, was calculated numerically. The results obtained are as follows: The magnetic field dependence of p_0 is quite interesting. As the magnetic field increases, p_0 begins to decrease first and then reaches minimum at h=4.0 T, where the sublattice magnetization fluctuates and becomes unstable somewhat. Then p_0 grows up to the saturated value. Contrary to the case of p_0, the magnetization at 0 K, M_0, is shown to be a monotonous increasing function of h.

3.3 Temperature dependence of the sublattice magnetization

The results of the numerical calculations of p are shown in Fig. 2.

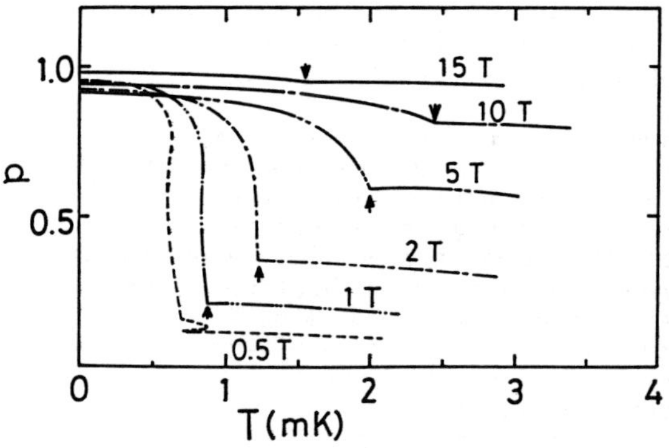

FIGURE 2

The calculated result shows that the system has two transitions of first- and second-order just above the LFP. In the case of the lowest field examined (0.5 T), another transition of first order exists. This extra transition seems to be due to an inaccuracy of the approximation used, in a low field and low temperature ranges. In either case, however, it also has a second-order transition at higher magnetic fields.

REFERENCES

(1) Y. Masuda, K. Iwahashi, H. Yano, M. Kato and A. Sawada, Can. J. Phys. 65 (1987) 1346.

(2) K. Iwahashi and Y. Masuda, J. Phys. Soc. Jpn. 59 (1990) No. 4 (in press).

(3) D. M. Ceperley and G. Jacucci, Phys. Rev. Lett. 58 (1987) 1648.

(4) H. Godfrin and D. D. Osheroff, Phys. Rev. B38 (1988) 4492.

Physica B 165&166 (1990) 833-834
North-Holland

MAGNETIZATION OF BCC SOLID ^{3}He IN A HIGH MAGNETIC FIELD

Tsuneo FUKUDA, Hiroshi FUKUYAMA, Tohru OKAMOTO, Hidehiko ISHIMOTO and Shinji OGAWA

The Institute for Solid State Physics, the University of Tokyo, Roppongi, Minato-ku, Tokyo 106, Japan.

A bulk bcc solid ^{3}He with a molar volume of V_m=23.06 cm^3/mol was cooled down to 0.5 mK in a magnetic field of 5.729 T. Magnetization was measured across the nuclear ordering transition to the high field phase by use of NMR method. The magnetization as large as 90 % of the saturation magnetization was observed at the lowest temperature.

1. INTRODUCTION

Competitive ferro- and antiferro-magnetic exchange interactions arising from many body exchange among ^{3}He atoms cause a peculiar magnetic phase diagram with two distinct ordered phases, the low field phase (u2d2 phase) and the high field phase (HFP)[1]. For microscopic identification of the ordered structures, NMR played a crucial role in low magnetic fields [2]. In contrast, NMR measurement of the HFP in a low temperature and high field region, where the nuclear magnetization is close to the saturated value, has not been made because of several experimental difficulties, such as poor thermal conductivity and large Kapitza boundary resistance of solid ^{3}He, and large heat dissipation associated with an rf power. Our recipe for cooling a bulk solid ^{3}He is to make it thin enough to transport the Zeeman energy to the heat exchanger through a spin diffusion process within an allowable time. The resultant low filling factor of the sample has been overcome by developing a novel high-sensitivity and low-power NMR system. In this paper, we report the first NMR measurements of the HFP with almost saturated magnetization in complete thermal equilibrium.

2. EXPERIMENTS

A bulk sample of tubular shape, 4 mm in diameter, 10 mm long and 25 micron thick, was confined between a plastic tube and a silver post. A silver sponge heat exchanger was sintered around the post with a thickness of 0.5 mm (Fig. 1). The post acted as an inductive part of an NMR resonator as well as a thermal path to a copper nuclear refrigerator. A capacitive part of the resonator consisted of two electrodes; one was attached to the end of the silver post, and the other was connected to the outer conductor of the resonator via silver strips and was hydraulically actuated by liquid ^{4}He to change the gap between the electrodes. Owing to this unique tuning mechanism, the resonant frequency could be varied between

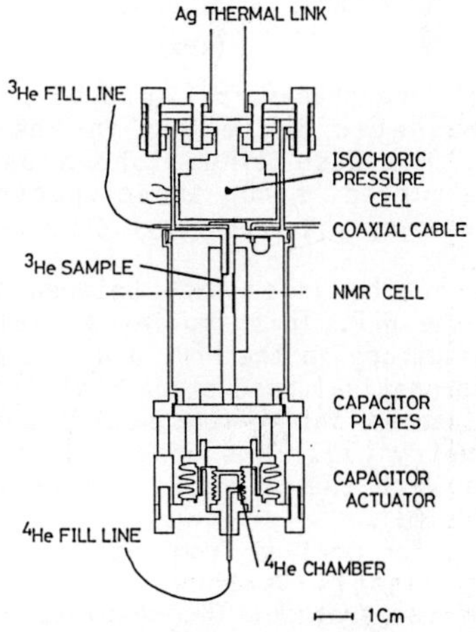

Fig. 1 NMR cell. The center of the magnetic field is shown as a dash dot line.

120 MHz and 360 MHz, although the maximum frequency was practically limited to 230 MHz by a maximum field of our magnet. The resonator was inductively coupled to a superconducting coaxial cable and a low loss (<0.2dB/m) coaxial cable to room temperature. A return loss of the resonator including the cable gave us an NMR signal as a small change of the SWR (Standing Wave Ratio) at any frequencies over the tuning range. Although the filling factor was only about 0.1 %, heat dissipation necessary for a sufficient signal to noise ratio was less than 5 pW.

3. RESULTS AND DISCUSSIONS

Typical NMR absorption signals at B=5.729 T are demonstrated in Fig. 2. The line width was partly caused by inhomogeneity of the static magnetic field (6×10^{-5} over the sample region), and another part of broadening was associated with inhomogeneity of the local field. It should be noted that we could not see

0921-4526/90/$03.50

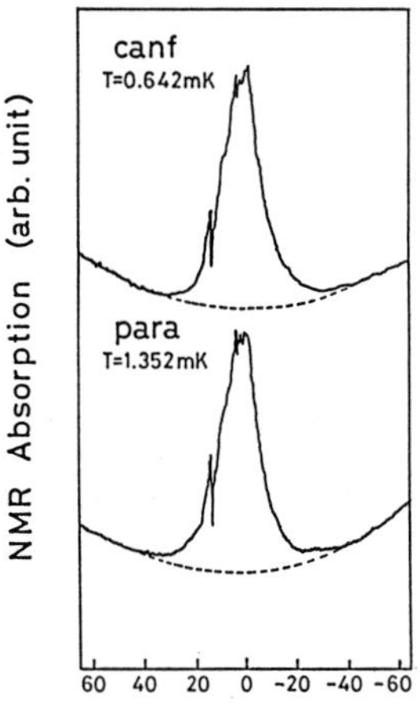

Fig. 2 NMR line shapes for V_m=23.06 cm^3/mol in the paramagnetic phase and in the HFP at B=5.729 T. The base lines (shown as broken lines) are curved, since these spectrum were taken by sweeping the frequency.

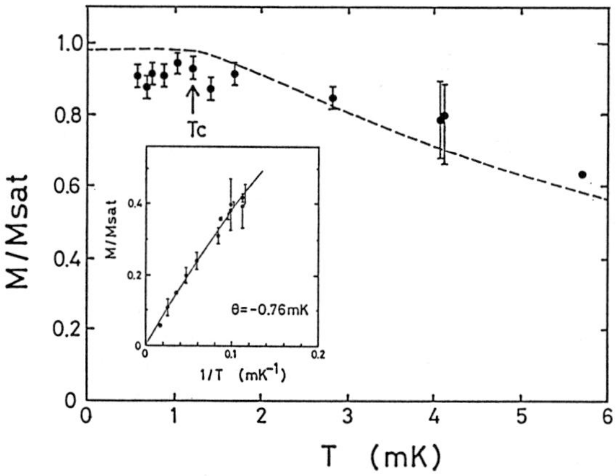

Fig. 3 Magnetization for V_m=23.06cm3/mol as a function of temperature at B=5.729 T. Molecular field calculation (Ref. 5) is represented as a broken line on the assumption of $V_m{}^{18}$ dependence of the exchange parameters. The magnetic ordering temperature (T_c) to the HFP is indicated as an arrow. Inset : Fitting the high temperature data of the magnetization to the molecular field approximation with a Weiss temperature, θ=−0.76 mK (Ref. 7).

any change of the line shape between the para phase and the HFP. This implies a small or no dipolar anisotropy in the HFP, and is consistent with the theoretical prediction that the HFP is the canf (canted antiferromagnetic) phase with cubic symmetry [1]. This observation is also consistent with the previous measurements at lower fields [3].

Magnetization derived from integration of the absorption signals is shown in Fig. 3 as a function of temperature. The absolute value was determined by fitting the magnetization to the mean field theory between 10 and 60 mK (see inset of Fig. 3). Its maximum value was about 90 % of the saturation magnetization (M_{sat}) at the lowest temperature. No abrupt change was seen across the phase transition between the para phase and the canf phase at 1.2 mK, which was determined by the isochoric pressure measurements in the same experiment [4]. This is not surprising, because, as shown in the mean field calculation [5], the magnetization of the present sample is large enough even in the para phase and its change should be small across the transition compared with an experimental accuracy.

In the course of NMR measurements, we performed a selective saturation in the absorption line by applying a large rf pulse for a few seconds. We observed an asymmetric valley and peak structure in the line shape, which was attributed to a change of the local field associated with reduction of the magnetization in a narrow region of the sample [6]. No spectral broadening was observed within our spatial resolution, because the estimated saturated region, 0.5 mm, along the static field was much larger than a spin diffusion length of an order of 10 micron within our measured time of about 1 hour. However, the structure disappeared with a time constant between 1000 and 2000 sec. The relaxation is thought to be determined by energy diffusion from the bulk portion of the sample to the region in the sintered heat exchanger. The relaxation had no temperature dependence in the para phase between 20 mK and 1.2 mK and its order of magnitude is consistent with the spin diffusion time across the sample thickness. Since all of the magnetization data in Fig. 3 was taken after waiting for much longer than the relaxation time, it is considered to be in thermal equilibrium.

REFERENCES

[1] M. Roger J.H. Hetherington and J.M. Delrieu, Rev. Mod. Phys. **55** (1983) 1.
[2] M. Cross, Jpn J. Appl. Phys. Supplement **26-3** (1987) 1855, and references therein.
[3] E.D. Adams et al., Phys. Rev. Lett. **44** (1980) 789; D.D. Osheroff, Physica **109&110B** (1982) 1461.
[4] H. Fukuyama et al., this conference.
[5] H.L. Stipdonk G. Frossati and J.H. Hetherington, J. Low Temp. Phys. **61** (1985) 185.
[6] R.T. Johnson et al., J. Low. Temp. Phys. **10** (1973) 35.
[7] C.T. Van Degrift, Quantum Fluids and Solids-1983, AIP Conference Proceedings, No. 103, eds. E.D. Adams and G.G. Ihas (American Institute of Physics, New York, 1983) pp. 16-31.

Physica B 165&166 (1990) 835-836
North-Holland

EXCHANGE CONTRIBUTION TO THE PRESSURE IN hcp SOLID ^{3}He

Y. MIURA, S. ABE, N. MATSUSHIMA, S. SUGIYAMA, and T. MAMIYA

Department of Physics, Nagoya University, Chikusa-ku, Nagoya 464, Japan

The thermodynamic pressure has been measured in hcp solid ^{3}He at molar volumes from 18.63 cm^3/mole to 19.37 cm^3/mole and for temperatures from 8 mK to 0.93 mK. The exchange contribution to the pressure is observed and compared to the expression of the high-temperature series expansion of the spin Hamiltonian including two- and three-particle exchange.

1. INTRODUCTION

The nuclear magnetic properties of solid ^{3}He are believed to arise from exchange processes, including at least two-particle exchange, three-, and four-particle cyclic exchange(1). In hcp solid ^{3}He, three-particle exchange which favors ferromagnetism is expected to be dominant because of the lattice symmetry. The magnetization measurements have revealed the ferromagnetic behaviour of hcp ^{3}He(2). In previous work, an increase in pressure due to magnetic fields have been demonstrated in hcp solid ^{3}He, indicating the ferromagnetic behavior(3). In this paper we report the exchange energy obtained from measurements of the thermodynamic pressure of the hcp ^{3}He at several molar volumes in zero magnetic field.

2. EXPERIMENT

The hcp solid was prepared by the blocked capillary method and annealed at temperatures just below the melting point and the bcc-hcp phase boundary. The pressure of hcp solid ^{3}He was measured with a beryllium copper capacitive pressure transducer(3,4). The sample cell design has previously been described in detail (3). Briefly, the cell chamber was packed with fine silver powder to provide a surface area of 2.3 m^2 for thermal contact with solid ^{3}He. In order to improve the resolution and the stability of the capacitance bridge, a large diaphragm, 16 mm in diameter, was adopted and cable capacitances in the bridge was minimized, resulting in high pressure resolution 1 μbar at 130 bar. The solid ^{3}He was cooled by adiabatic demagnetization of copper nuclei. The temperature of the sample cell was determined using the ^{3}He melting curve thermometer, which was calibrated against the superfluid ^{3}He-A, ^{3}He-B transition, and the magnetic transition of bcc solid.

3. RESULTS AND DISCUSSION

Fig.1 shows the typical change of the raw data of the pressure and the cell temperature with time at 19.14 cm^3/mole. The temperature of the cell was changed stepwise in order to attain a thermal equilibrium between the solid ^{3}He and the thermometer. The relaxation time was found to be less than 30 mim. down to 0.93 mK. The temperature of the cell was kept constant until the pressure reached the equilibrium value at each step. In Fig.1, a transient shift of the capacitance bridge corresponding to the pressure shift of 3 μbar is shown at 12 hour. This can not be considered a real shift of the pressure of the solid but a shift of a balance point of the bridge arising from some electronic disturbance such as a sudden change in the ground potential. The raw data shown in Fig.1 was corrected to eliminate such a clear transient shift of the capacitance bridge. In Fig.1, a large disturbance is also shown at 32 hour corresponding to a liquid helium transfer into the cryostat. This shift of the bridge is found to relax to a new stable

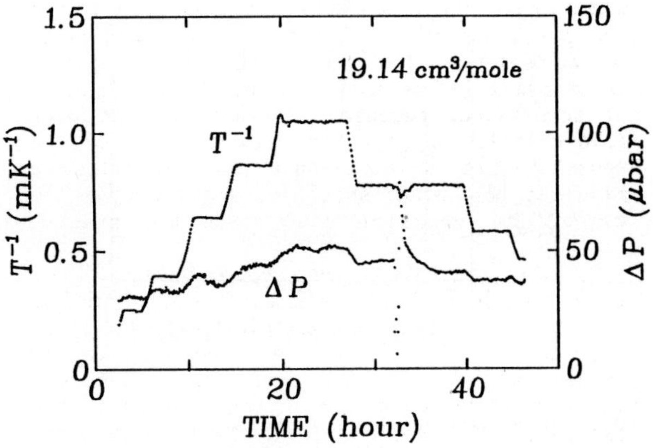

FIGURE 1
The typical change of the pressure and the cell temperature with time at 19.14 cm^3/mole.

value after 5 hours contrary to the former shift. The temperature of the cell was kept constant for a period of 8 hours after the transfer in order to find a new balance point of the bridge. Then the change in the balance point before and after the transfer was eliminated from the raw data of the pressure. The thermodynamic pressure $P(T)-P_0$ of the hcp solid ^3He was found to be inversely proportional to temperature below 8 mK, indicating a contribution from the exchange interactions. The pressure was then compared to the following high-temperature series expansion with the first term only, because higher order terms are negligible above 0.9 mK in hcp solid ^3He.

$$P(T) - P_0 = (R/8)(\partial\tilde{e}_2/\partial v)(1/T)$$

where P_0 is a pressure at high temperature limit, R is the gas constant, $\tilde{e}_2$ is leading coefficient in the high-temperature series expansion, and v is the molar volume.

In order to convert the raw pressure data to equivalent constant-volume values, we should determined a correction factor which depends on the elastic constant of the diaphragm of the cell, the open volume of bulk ^3He, and the compressibility of the solid ^3He. The factor was determined by comparing our pressure data of bcc solid ^3He at 24.08 cm^3/mole to the constant-volume pressure data (5). This volume correction factor varies from 1.47 at a molar volume of 19.37 cm^3 to 1.66 at 18.63 cm^3.

Fig.2 shows the molar volume dependence of $\partial\tilde{e}_2/\partial v$ which was corrected for constant-volume. The open symbols represent data on warming; closed ones, while cooling. The scatter is

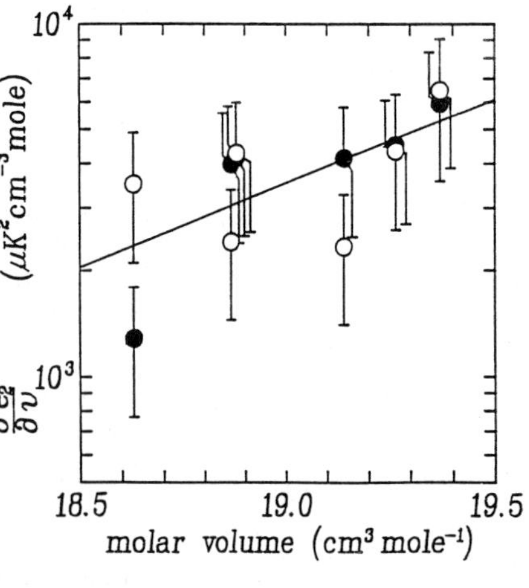

FIGURE 2
The molar volume dependence of $\partial\tilde{e}_2/\partial v$. The straight line represents the least-squares fit to the data as dicussed in the text. The open symbols represent data on warming; closed ones, while cooling.

indicative of the long-term stability of the capacitance bridge which can not be corrected except for the transient shift of the balance point of the bridge. Since our data cover only a limited range in molar volume, it can be approximated the volume dependence of the exchange parameter by a Grüneisen relation $\tilde{e}_2 \propto v^\gamma$. The solid line in Fig.2 is a power-law fit with exponent 20.8±9.9 leading that $\tilde{e}_2 = (3.1\ ^{+3.4}_{-1.3})\times10^3(v/19)^{21.8\pm9.9}\ \mu K^2$. Taking the spin Hamiltonian including two- and three-particle exchange interaction, the relation $\tilde{e}_2=18J_{eff}^2$ is obtained, where J_{eff} is an effective exchange energy. It is determined from the value of $\tilde{e}_2$ that $J_{eff}=(13.1\ ^{+5.8}_{-3.1})(v/19)^{10.9\pm5.0}\ \mu K$. The prefactor and the exponent of J_{eff} determined simultaneously in the fit are strongly correlated, for example an overestimate of the prefactor leads to an underestimate of the exponent of J_{eff}. The value of the exponent of J_{eff} in the present work is consistent with the result of the magnetization measurements (2), however, is smaller than that in nmr measurements(6). The prefactor of J_{eff} in this work is larger than those in the magnetization and nmr measurements.

4. CONCLUSION

The exchange contribution to the pressure is observed as a function of molar volume in hcp solid ^3He. Comparing the experimental results to the high-temperature series expansion of the spin Hamiltonian including two- and three-particle exchange interaction, the effective exchange energy J_{eff} is for the first time determined from pressure measurements.

ACKNOWLEDGEMENTS
We wish to thank S. Inoue, H. Shibayama, W. Itoh, and T. Kurokawa for help with the experiment.

REFERENCES

(1) See a review by M. Roger, J. H. Hetherington, and J. M. Delrieu, Rev. Mod. Phys. 55(1983)1.
(2) Y. Takano, N. Nishida, Y. Miura, H. Fukuyama, H. Ishimoto,and S. Ogawa, Phys. Rev. Lett. 55 (1985) 1490.
(3) Y. Miura, S. Abe, S. Sugiyama, T. Mamiya, and R. C. Richardson, in; Quantum Fluids and Solids, AIP Conference Proceedings vol. 194, eds. G. G. Ihas and Y. Takano (New York, 1989) pp 267.
S. Abe, Y. Miura, N. Matsushima, S. Sugiyama, T. Mamiya, and R. C. Richardson, preprint.
(4) G. C. Straty and E. D. Adams, Rev. Sci. Instrum. 40 (1969) 1393.
(5) D. G. Wildes, M. R. Freeman, J. Saunders, and R. C. Richardson, J. Low Temp. Phys. 62 (1986) 67.
(6) See a review by R. A. Guyer, R. C. Richardson, and L. I. Zane, Rev. Mod. Phys. 43 (1971) 532.

Physica B 165&166 (1990) 837–838
North-Holland

NUCLEAR MAGNETISM OF hcp SOLID ^{3}He ABOVE THE ORDERING TEMPERATURE

T.MAMIYA, H.YANO, H.KONDO, T.SUZUKI, T.KATO, Y.MINAMIDE, Y.MIURA and S.INOUE

Department of Physics, Nagoya University, Chikusa-ku, Nagoya, 464-01 Japan

We have precisely measured the magnetization and ac susceptibility of hcp solid ^{3}He as a function of molar volume in the temperature range down to 38 μK. We have obtained the Weiss temperature and the next higher order term in the high temperature expansion in magnetization at lower temperatures. The obtained molar volume dependence of the Weiss temperature is in good agreement with that of the bcc phase, though the sign of the Weiss temperature is opposite. The susceptibility data are discussed associated with the exchange-Zeeman relaxation time.

We report precise measurements of the magnetization and susceptibility of hcp solid ^{3}He, which have first been achieved in the temperature range down to 38 μK. The details of experimental procedures were described elsewhere(1).

In brief, the ^{3}He sample is cooled by means of two stage demagnetization. In order to make it possible to measure ac susceptibility at ultralow temperatures, we made a hybrid cell which had an inner part of pure silver and an outer part of alloy of silver and copper. A large penetration depth of 8.2 mm for ac magnetic field of 19 Hz resulted from this configuration. The sample cell was packed with silver powder and had a surface area of 1.1 m^2 and volume of 0.045 cm^3. The sample was in a constant magnetic field of 1.9 mT trapped by a concentric cylinder of superconducting titanium. The samples were made by the blocked capillary method. The molar volume of the samples was known to an accuracy of ±0.01 cm^3 by use of a pressure sensor of the Straty-Adams type which was installed on the second nuclear stage.

A SQUID was used for magnetization measurements and for ac susceptibility measurements, as well. For the latter purpose it was used as a null detector in combination with an ac mutual inductance bridge. We installed the SQUID sensor at the bottom of the cryostat, so that a superconducting loop connected to the SQUID was not exposed to the strong magnetic field. NMR measurement of platinum provided thermometry. Calibration of platinum thermometers was achieved against a ^{3}He melting curve thermometer. Temperatures were usually measured once every three hours at 125 kHz at the lowest temperatures and more frequently at higher temperatures. Overall inaccuracy in thermometry was less than 2 %.

We measured the susceptibility and magnetization with increasing temperature from the lowest temperature attainable, intermittently applying heat pulses of the order of 100 to 20 μJ to the second stage by using the heater wound around the stage. The thermal time constant toward magnetization equilibrium for the sample with 19.29cm^3/mol was 3100 and 580 sec at 59 and 109 μK, respectively. The temperature dependence of the time constant was expressed to be 6.2×10^8 T^{-3}sec, where temperarure, T, is denoted in μK. Our measurements were done during very slow warmup under the residual heat leak in much long time scale compared to the time constant, so that the data taken were considered to be in thermal equilibrium except for the period just after applying heat pulses.

First, we concentrate on the results of magnetization measurements at lower temperature range investigated from 38 to 200 μK. In this temperature range magnetization varied approximately 950 ϕ_o, where ϕ_o is the flux quantum unit. The temperature variation of the backgound magnetization in this range is approximately 10 ϕ_o. Correction for this small variation of background has not been done. Inacccuracy due to neglecting correction was less than 1 % in deducing the Weiss Θ. The magnetization data for the sample 19.29cm^3/mol was least-square fitted with the equation; $M= C/(T-\Theta+B/T)+M_o$, where C, Θ, B and M_o are fitting parameters. One obtains Θ=22.1±0.5 μK and B=203 ± 20 μK^2 for this molar volume. The value of Θ is to be compared to 40 μK by Takano et al.(2) and 60 μK, a value corrected for theoretical fit including four-particle exchange integral by Roger et al.(3). Our Θ is in good agreement with Ceperly's calculation(4), 25.9 μK for solid with 19.4 cm^3/mole. The ordering temperature T_c would be in the 10 μK range by considering $T_c \leq 2/3\Theta$. The value of B is first observed in hcp solid, thanks to the precise measurements made down to 38 μK. The existence of B is clearly evidenced from an upward curvature deviating from a straight line in a plot of $1/(M-M_o)T$ versus $1/T$.

From combination of Θ and B, we can deduce the exchange interactions involved in hcp solids. According to Roger et al.(3), the following expressions are valid for hcp solid,

$\Theta=3J_1+3J_2+9K$, $B=3J_1^2+3J_2^2+3/2J_1K+3/2J_2K+27/4K^2$, where J_1 and J_2 are three-particle exchange integrals in and out of the hexagonal plane, respectively. K is square four-particle exchange integral. On the assumption that $J_1=J_2$, we obtain $J_1=5.98$ μK and K=-1.53 μK; that is K=-0.26 J_1. The values of J_1 and K are in good agreement with Ceperly's calculation 6.75 μK and -1.4 μK, respectively, for the sample with 19.4cm^3/mol(4).

The values of Θ for all molar volumes investigated are shown in Fig.1 as a function of molar volume. The molar volume dependence of Θ which is expressed as $V^{18\pm1}$ is clearly different from the previous work, $V^{11\pm1}$(2). The dependence $V^{18\pm1}$ is very close to that in the bcc phase. This fact reflects the profound nature in the direct exchange interaction common to both the hcp and bcc phases.

We will discuss susceptibility measurements. As temperature was lower, the change of mutual inductance was somewhat smaller than that of magnetization. This is because the exchange-Zeeman relaxation time becomes larger as temperature decreases. Inductance m of the sample coils is decomposed into the real part m' and the imaginary part m'', as follows, m = m'+im''. The mutual inductance is related to the susceptibility, χ, of solid ^{3}He which is encapsuled in the metallic cell, by the following equations; $m' + im'' = Ae^{i\delta}(\chi'+i\chi'')$. A and δ indicate the amplitude and the phase shift of the applied fields at the site of solid ^{3}He due to the partial shielding of the fields in the metallic cell. Calculation, taken into account the electrical conductivity of the metals used for the hybrid cell, yields A =0.985±0.003 and δ =6.4±1.6° for the frequency of 19 Hz used in this work. With use of the measured m and calculated A and δ, we can obtain χ.

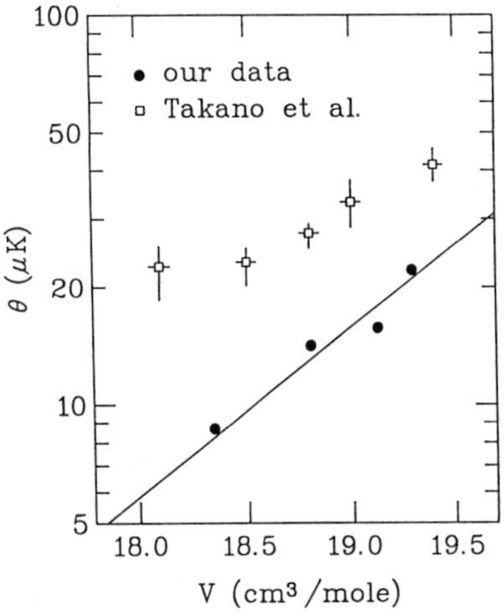

Fig.1 Molar volume dependence of the Weiss temperature. Error bar of our data is too small to display it. The line shows V^{18}.

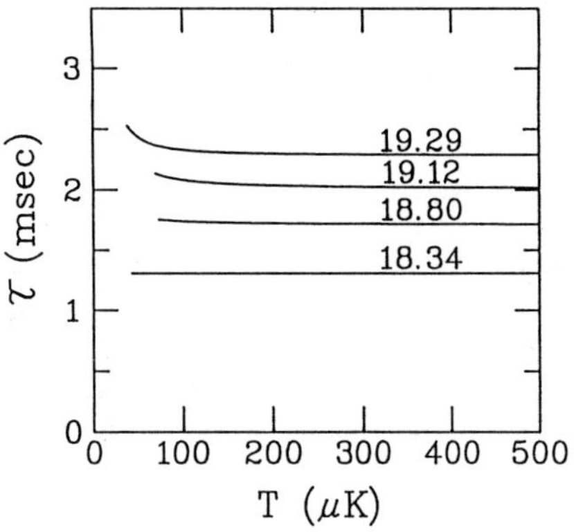

Fig.2 Nuclear relaxation time against temperature. The figures by the lines show molar volume in cm^3.

Next, we will derive the relaxation time from χ. The expression of Casimir-du Pre, which states the measured susceptibility in terms of isothermal and adiabatic susceptibilities in presence of the relaxation in responding to the applied field, resulted in $\chi'=\chi_0/(1+\omega^2\tau^2)$ and $\chi''=\omega\tau\chi_0/(1+\omega^2\tau^2)$. This expression was derived from the fact that the adiabatic susceptibility is null because there is no interaction which is rapid compared to the relaxation time involved in this experiment. In the above expression χ_0 is isothermal susceptibility and τ, spin-spin relaxation time. τ is to be taken approximately the same as the exchange-Zeeman relaxation time in case that the frequency of the applied field is much smaller than the exchange frequency. τ is obtained from $\chi''/\omega\chi'$ and plotted in Fig.2. At lower molar volumes τ is nearly constant and for larger molar volumes τ increses appreciably with decreasing temperature. This suggests that τ increases as one approaches the ordering temperature for larger molar volumes, which indication is first observed for hcp solid ^{3}He. The molar volume dependence of the relaxation time at the high temperature limit is in accordance with that of the exchange interaction, which is also supported by theoretical treatment done for NMR results at high temperatures(5).

We thank S.Abe, H.Shibayama, W.Ito and K.Kurokawa for experimental support.

REFERENCES

(1) H.Yano et al., J.Low Temp. Phys. 78(1990)165.
(2) Y.Takno et al., Phys.Rev.Lett. 55(1985) 1490.
(3) M.Roger et al., Phys.Rev. B35(1987) 2091.
(4) D.Ceperly private communication.
(5) R.A.Guyer et al., Rev.Mod.Phys. 43(1971) 532.

LIST OF CONTRIBUTORS